ESSENTIAL CELL BIOLOGY

third edition

Alberts Bray Hopkin Johnson Lewis Raff Roberts Walter

Garland Science
Taylor & Francis Group
NEW YORK AND LONDON

Garland Science
Vice President: Denise Schanck
Senior Editor: Michael Morales
Production Editor and Layout: Emma Jeffcock
Project Editor: Sigrid Masson
Assistant Editor: Katherine Ghezzi
Editorial Assistant: Monica Toledo
Text Editors: Eleanor Lawrence, Sherry Granum, Elizabeth Zayatz
Copyeditor: Bruce Goatly
Book and Cover Design: Matthew McClements, Blink Studio
Illustrator: Nigel Orme
Cover Illustration: Jose Ortega
Indexer: Liza Furnival
Permission Coordinator: Mary Dispenza

Essential Cell Biology Interactive
Artistic and Scientific Direction: Peter Walter
Narrated by: Julie Theriot
Producer: Michael Morales
Interface Design: Matthew McClements, Blink Studio
Programming: Tom McElderry

Bruce Alberts received his Ph.D. from Harvard University and is Professor of Biochemistry and Biophysics at the University of California, San Francisco. He is the editor-in-chief of *Science* magazine. For 12 years he served as President of the U.S. National Academy of Sciences (1993-2005).
Dennis Bray received his Ph.D. from Massachusetts Institute of Technology and is currently an active emeritus professor at University of Cambridge. In 2006 he was awarded the Microsoft European Science Award.
Karen Hopkin received her Ph.D. in biochemistry from the Albert Einstein College of Medicine and is a science writer in Somerville, Massachusetts.
Alexander Johnson received his Ph.D. from Harvard University and is Professor of Microbiology and Immunology and Director of the Biochemistry, Cell Biology, Genetics, and Developmental Biology Graduate Program at the University of California, San Francisco.
Julian Lewis received his D.Phil. from the University of Oxford and is a Principal Scientist at the London Research Institute of Cancer Research UK.
Martin Raff received his M.D. from McGill University and is at the Medical Research Council Laboratory for Molecular Cell Biology and Cell Biology Unit and in the Biology Department at University College London.
Keith Roberts received his Ph.D. from the University of Cambridge and is Emeritus Fellow at the John Innes Centre, Norwich.
Peter Walter received his Ph.D. from The Rockefeller University in New York and is Professor and Chairman of the Department of Biochemistry and Biophysics at the University of California, San Francisco, and an Investigator of the Howard Hughes Medical Institute.

Published by Garland Science, Taylor & Francis Group, LLC, an informa business, 270 Madison Avenue, New York, NY 10016, USA, and 2 Park Square, Milton Park, Abingdon, OX14 4RN, UK.

Printed in the United States of America
15 14 13 12 11 10 9 8 7 6 5 4 3 2 1

Library of Congress Cataloguing-in-Publication Data

Essential cell biology / Bruce Alberts ... [et al.]. -- 3rd ed.
p. cm.
ISBN 978-0-8153-4129-1 (hardcover) -- ISBN 978-0-8153-4130-7 (pbk.)
1. Cytology. 2. Molecular biology. 3. Biochemistry. I. Alberts, Bruce.
QH581.2.E78 2009
571.6--dc22

2009001400

Visit our website at http://www.garlandscience.com

ESSE[illegible] edition
CELL BIOLOGY

Preface

In our world there is no form of matter more astonishing than the living cell: tiny, fragile, marvelously intricate, continually made afresh, yet preserving in its DNA a record of information dating back more than three billion years, to a time when our planet had barely cooled from the hot materials of the nascent solar system. Ceaselessly re-engineered and diversified by evolution, extraordinarily versatile and adaptable, the cell still retains a core of complex self-replicating chemical machinery that is shared and endlessly repeated by every living organism on the face of the Earth, in every animal, every leaf, every bacterium in a piece of cheese, every yeast in a vat of wine.

Curiosity, if nothing else, should drive us to study cell biology; but there are practical reasons, too, why cell biology should be a part of everyone's education. We are made of cells, we feed on cells, and our world is made habitable by cells. We need to understand cell biology to understand ourselves; to look after our health; to take care of our food supplies; and to protect our endangered ecosystems. The challenge for scientists is to deepen knowledge and find new ways to apply it. But all of us, as citizens, need to know something of the subject to grapple with the modern world, from our own health affairs to the great public issues of environmental change, biomedical technologies, agriculture, and epidemic disease.

Cell biology is a big subject, and it has links with almost every other branch of science. The study of cell biology therefore provides a great scientific education. However, it is easy to become lost in the detail and distracted by an overload of information and technical terminology. In this book we therefore focus on providing a digestible, straightforward, and engaging account of only the essential principles. We seek to explain, in a way that can be understood even by a reader approaching modern biology for the first time, how the living cell works: to show how the molecules of the cell—especially the protein, DNA, and RNA molecules—cooperate to create this remarkable system that feeds, responds to stimuli, moves, grows, divides, and duplicates itself.

The need for a clear account of the essentials of cell biology became apparent to us while we were writing *Molecular Biology of the Cell* (*MBoC*), now in its fifth edition. *MBoC* is a large book aimed at advanced undergraduates and graduate students specializing in the life sciences or medicine. Many students and educated lay people who require an introductory account of cell biology would find *MBoC* too detailed for their needs. *Essential Cell Biology* (*ECB*), in contrast, is designed to provide the fundamentals of cell biology that are required by anyone to understand both the biomedical and the broader biological issues that affect our lives.

In this third edition, we have brought every part of the book up to date, with new material on chromosome structure and epigenetics, microRNAs and RNAi, protein quality control, cell-cell recognition, genetic variation, stem cells and their medical potential, rational cancer treatments, genome evolution, and many other topics. We have improved our discussion of energetics and thermodynamics, integrated the cell cycle and cell division into a single chapter, and updated the "How We Know" sections, describing experiments that illustrate how biologists tackle important questions and how their experimental results shape future ideas.

As before, the diagrams in *ECB* emphasize central concepts and are stripped of unnecessary details. The key terms introduced in each chapter are highlighted when they first appear and are collected together at the end of the book in a large, illustrated glossary. We have not listed references for further reading: those wishing to explore a subject in greater depth are encouraged to consult the reading lists in *MBoC5* or look for recent reviews in the current literature through one of the powerful search engines, such as Pubmed (http://www.ncbi.nlm.nih.gov) or Google Scholar (http://scholar.google.com).

A central feature of the book is the many questions that are presented in the text margins and at the end of each chapter. These are designed to provoke students to think about what they have read and to encourage them to pause and test their understanding. Many questions challenge the student to place the newly acquired information in a broader biological context, and some have more than one valid answer. Others invite speculation. Answers to all the questions are given at the end of the book; in many cases these provide a commentary or an alternative perspective on material presented in the main text.

For those who want to develop their active grasp of cell biology further and to get a deeper understanding of how cell biologists extract conclusions from experiments, we recommend *Molecular Biology of the Cell, Fifth Edition: A Problems Approach*, by John Wilson and Tim Hunt. Though written as a companion to *MBoC*, this contains questions at all levels of difficulty and is a goldmine of thought-provoking problems for teachers and students. We have drawn upon it for some of the questions in *ECB*, and we are very grateful to its authors.

The explosion of new imaging and computer technologies continues to give fresh and spectacular views of the inner workings of living cells. We have tried to capture some of the excitement of these advances in a revised and enlarged version of the *Essential Cell Biology Interactive* media player on the DVD-ROM included with each copy of the book. It contains over 130 video clips, animations, molecular structures, and high-resolution micrographs—all designed to complement the material in individual book chapters. One cannot watch cells crawling, dividing, segregating their chromosomes, or rearranging their surface without a sense of wonder at the molecular mechanisms that underlie these processes. For a vivid sense of the marvel that science reveals beneath the surface of everyday things, it is hard to match the movie of DNA replication included on the DVD. We hope that *ECB Interactive* will motivate and intrigue students while reinforcing basic concepts covered in the text, and thereby will make the learning of cell biology both easier and more rewarding.

As with *MBoC*, each chapter of *ECB* is the product of communal effort, with individual drafts circulating from one author to another. In addition, many people have helped us, and these are credited in the Acknowledgments that follow. Despite our best efforts, it is inevitable that there will be errors in the book. We encourage readers who find them to let us know at science@garland.com, so that we can correct these errors in the next printing.

Acknowledgments

The authors acknowledge the many contributions of professors and students from around the world in the creation of this Third Edition. In particular, we are grateful to the students who participated in the focus groups; they provided invaluable feedback about their experiences using the book and multimedia, and many of their suggestions were implemented in this edition.

We would also like to thank the professors who helped organize the focus groups at their schools: Chris Brandl at University of Western Ontario, David L. Gard at University of Utah, Juliet Spencer at University of San Francisco, and Keren Witkin and Linda Huang at University of Massachusetts, Boston. We greatly appreciate their hospitality.

We also received detailed reviews from instructors using the second edition, and we would like to thank them for their work: Margarida D. Amaral, University of Lisbon; Lynne Arneson, American University; Karl Aufderheide, Texas A&M University; David K. Banfield, The Hong Kong University of Science and Technology; Stephen F. Baron, Bridgewater College; Deborah Bielser, University of Illinois at Urbana-Champaign; Barbara D. Boyan, Georgia Institute of Technology; Chris Brandl, University of Western Ontario; Keith Brown, University of Bristol; Jane Bruner, California State University Stanislaus; Patrick Bryan, Middlesex Community College; Sharon K. Bullock, Virginia Commonwealth University; Mike Clemens, St. George's Hospital Medical School, University of London; Anne Cordon, University of Toronto at Mississauga; Andrew Dalby, University of Exeter; Dan Eshel, Brooklyn College; Nicolas Forraz, Kingston University; David L. Gard, University of Utah; Mark Grimes, University of Montana; Hollie Hale-Donze, Louisiana State University; Lynn Hannum, Colby College; Na'il M. Hasan, Birzeit University; Jeannette M. Loutsch, Arkansas State University; Charles Mallery, University of Miami; Kathy Martin-Troy, Central Connecticut State University; Gordon T.A. McEwan, Institute of Medical Sciences, University of Aberdeen; Colin McGuckin, Kingston University; Gerard McNeil, York College, The City University of New York; Roger W. Melvold, University of North Dakota, School of Medicine & Health Sciences; Cristina Murga, Universidad Autónoma de Madrid; T. Page Owen, Jr., Connecticut College; Martin Rumsby, University of York; Esther Siegfried, University of Pittsburgh at Johnstown; Roger D. Sloboda, Dartmouth College; Julio Soto, San Jose State University; Juliet Spencer, University of San Francisco; Paul H. Tomasek, California State University Northridge; Gary Wessel, Brown University; Esther F. Wheeler, Texas Tech University; Keren Witkin, University of Massachusetts, Boston.

Special thanks go to David Morgan for his help in formulating and reviewing the reorganized chapter on cell division.

We are very grateful, too, to the readers who alerted us to errors they had found in the previous edition.

Many Garland staff contributed to the creation of this book and made our work on it a pleasure. First of all, we owe a special debt to Michael Morales, our editor, who coordinated the whole enterprise. He organized the initial reviewing and the focus groups, worked closely with the authors on their chapters, urged us on when we fell behind, and played a major part in the design, assembly, and production of *Essential Cell Biology Interactive*. Sigrid Masson managed the flow of chapters through the production process, proof-read the entire book, and oversaw the writing of the accompanying question bank. Kate Ghezzi and Monica Toledo gave editorial assistance. Nigel Orme took original drawings created by author Keith Roberts and redrew them on a computer, or occasionally by hand, with skill and flair. To Matt McClements goes the credit for the graphic design of the book and of the DVD. Emma Jeffcock did a brilliant job in laying out the whole book and meticulously incorporating our endless corrections. Eleanor Lawrence and Sherry Granum did the developmental editing of individual chapters, repairing many rough edges, and Eleanor not only read the book from beginning to end for clarity and consistency, but also revised and extended the Glossary. Adam Sendroff and Lucy Brodie gathered user feedback and launched the book into the wide world. Denise Schanck, the Vice President of Garland Science, orchestrated all of this with great taste and diplomacy. We give our thanks to everyone in this long list.

Last but not least, we are grateful, yet again, to our families, our colleagues, and our childminders for their support and tolerance.

Instructor and Student Resources

Art of Essential Cell Biology, Third Edition

The images from the book are available in two convenient formats: Powerpoint® and JPEG. They are located in folders on the Media DVD-ROM that accompanies the book or can be downloaded on the Web from Classwire™. On Classwire the individual JPEGs are searchable by figure number, figure name, or by keywords used in the figure legend from the book.

Media DVD-ROM

Every copy of the book includes a DVD-ROM with the following student and instructor resources:

Essential Cell Biology Interactive
This multimedia player contains over 130 movies (animations, videos, and molecular tutorials), a self-test quiz for every chapter, and a cell explorer program that facilitates investigation of high-resolution micrographs. The movies are referenced directly in the textbook through movie "call outs" highlighted in red, for example (**Movie 1.1**). If you enter the movie number from the book into the movie-locator window in the media player, the relevant movie will automatically appear. This feature, which was requested by students, should make it easier to integrate movies into an active learning process.

Student Self-Quizzes
The quizzing feature, which is new to this edition, allows students to test themselves in basic reading comprehension of each chapter. It is accessed through the *Essential Cell Biology Interactive* media player.

The Art of Essential Cell Biology, Third Edition
This folder archive contains the figures from the book in JPEG and PowerPoint format as described above.

Movie Vault
This archive contains all of the movies from the *Essential Cell Biology Interactive* media player in three handy formats: WMV, QuickTime®, and iPod®. The WMV versions are suitable for importing movies into PowerPoint for Windows®. The QuickTime versions are suitable for importing the movies into PowerPoint for the Macintosh®. And the iPod versions have been formatted specifically for iPod and iTunes® use.

Media Guide
This PDF overviews the contents of the DVD and contains the text of the voice-over narration for all of the movies.

Figure-Integrated Lecture Outlines

The section headings, concept headings, and figures from the text have been integrated into PowerPoint presentations. These will be useful for instructors who would like a head start creating lectures for their course. Like all of our PowerPoint presentations, the lecture outlines can be customized. For example, the content of these presentations can be combined with videos from the DVD-ROM and questions from the book or "Question Bank," in order to create unique lectures that facilitate interactive learning in the classroom. This resource is available on Classwire.

Question Bank

Written by Linda Huang, University of Massachusetts, Boston, and Cheryl D. Vaughan, Harvard University Division of Continuing Education, the revised and expanded question bank includes a variety of question formats: multiple-choice, fill-in-the-blank, true-false, matching, essay, and challenging "thought" questions. There are approximately 50-60 questions per chapter, and a large number of the multiple-choice questions will be suitable for use with personal response systems (i.e., clickers). The *Question Bank* was created with the philosophy that a good exam should do much more than simply test students' ability to memorize information; it should require them to reflect upon and integrate information as a part of a sound understanding. It provides a comprehensive sampling of questions that can be used either directly or as inspiration for instructors to write their own test questions. Instructors can obtain the *Question Bank* by emailing: science@garland.com.

Exploring the Living Cell DVD-Video

Created by Christian Sardet, Centre National de la Recherche Scientifique (CNRS), and directed by Véronique Kleiner, this unique DVD takes us on a journey through the basic unit of life: the cell. Using the earliest drawings and exciting imagery taken with today's microscopes, renowned biologists and young scientists explain their research and share their discoveries. Learn how cells were discovered, how they function, how they relate to health and disease, and what the future holds.

Classwire™

The Classwire course management system, available at www.classwire.com/garlandscience, allows instructors to build Websites for their courses easily. It also serves as an online archive for instructor's resources. After registering for Classwire, you will be able to download all the figures from the book, as well as the movies from the DVD. Additional instructor's resources for Garland Science textbooks are also available on Classwire. Please contact science@garland.com for information on accessing the Classwire system. (Classwire™ is a trademark of Chalkfree, Inc.)

Contents and Special Features

Chapter 1	**Introduction to Cells**	**1**
Panel 1–1	Microscopy	8–9
Panel 1–2	Cell architecture	25
How We Know: Life's common mechanisms		30–31
Chapter 2	**Chemical Components of Cells**	**39**
How We Know: What are macromolecules?		60–61
Panel 2–1	Chemical bonds and groups	64–65
Panel 2–2	The chemical properties of water	66–67
Panel 2–3	An outline of some of the types of sugar	68–69
Panel 2–4	Fatty acids and other lipids	70–71
Panel 2–5	The 20 amino acids found in proteins	72–73
Panel 2–6	A survey of the nucleotides	74–75
Panel 2–7	The principal types of weak noncovalent bonds	76–77
Chapter 3	**Energy, Catalysis, and Biosynthesis**	**81**
Panel 3–1	Free energy and biological reactions	94–95
How We Know: Using kinetics to model and manipulate metabolic pathways		101–103
Chapter 4	**Protein Structure and Function**	**119**
Panel 4–1	A few examples of some general proteins	120
Panel 4–2	Four different ways of depicting a small protein	128–129
Panel 4–3	Making and using antibodies	144–145
How We Know: Probing protein structure		158–160
Panel 4–4	Cell breakage and initial fractionation of cell extracts	164–165
Panel 4–5	Protein separation by chromatography	166
Panel 4–6	Protein separation by electrophoresis	167
Chapter 5	**DNA and Chromosomes**	**171**
How We Know: Genes are made of DNA		174–176
Chapter 6	**DNA Replication, Repair, and Recombination**	**197**
How We Know: The nature of replication		200–202
Chapter 7	**From DNA to Protein: How Cells Read the Genome**	**231**
How We Know: Cracking the genetic code		248–249
Chapter 8	**Control of Gene Expression**	**269**
How We Know: Gene regulation—the story of *Eve*		282–284

Chapter 9 How Genes and Genomes Evolve **297**
How We Know: Counting genes 318–319

Chapter 10 Analyzing Genes and Genomes **327**
How We Know: Sequencing the human genome 348–349

Chapter 11 Membrane Structure **363**
How We Know: Measuring membrane flow 382–383

Chapter 12 Membrane Transport **387**
How We Know: Squid reveal secrets of membrane excitability 412–413

Chapter 13 How Cells Obtain Energy from Food **425**
Panel 13–1 Details of the 10 steps of glycolysis 430–431
How We Know: Unraveling the citric acid cycle 440–441
Panel 13–2 The complete citric acid cycle 442–443

Chapter 14 Energy Generation in Mitochondria and Chloroplasts **453**
How We Know: How chemiosmotic coupling drives ATP synthesis 468–469
Panel 14–1 Redox potentials 471

Chapter 15 Intracellular Compartments and Transport **495**
How We Know: Tracking protein and vesicle transport 520–521

Chapter 16 Cell Communication **531**
How We Know: Untangling cell signaling pathways 560–562

Chapter 17 Cytoskeleton **571**
Panel 17–1 The three major types of protein filaments 573
How We Know: Pursuing motor proteins 586–588

Chapter 18 The Cell Division Cycle **609**
How We Know: Discovery of cyclins and Cdks 615–616
Panel 18–1 The principal stages of M phase in an animal cell 626–627

Chapter 19 Sex and Genetics **651**
Panel 19–1 Some essentials of classical genetics 674
How We Know: Reading genetic linkage maps 680–681

Chapter 20 Cellular Communities: Tissues, Stem Cells, and Cancer **689**
How We Know: Making sense of the genes that are critical for cancer 725–726

Answers to Questions **A:1**

Glossary **G:1**

Index **I:1**

Detailed Contents

Chapter 1 Introduction to Cells 1

UNITY AND DIVERSITY OF CELLS 2

Cells Vary Enormously in Appearance and Function 2

Living Cells All Have a Similar Basic Chemistry 3

All Present-Day Cells Have Apparently Evolved from the Same Ancestor 5

Genes Provide the Instructions for Cellular Form, Function, and Complex Behavior 5

CELLS UNDER THE MICROSCOPE 6

The Invention of the Light Microscope Led to the Discovery of Cells 6

Cells, Organelles, and Even Molecules Can Be Seen Under the Microscope 7

THE PROCARYOTIC CELL 11

Procaryotes Are the Most Diverse of Cells 14

The World of Procaryotes Is Divided into Two Domains: Bacteria and Archaea 15

THE EUCARYOTIC CELL 16

The Nucleus Is the Information Store of the Cell 16

Mitochondria Generate Usable Energy from Food to Power the Cell 17

Chloroplasts Capture Energy from Sunlight 18

Internal Membranes Create Intracellular Compartments with Different Functions 19

The Cytosol Is a Concentrated Aqueous Gel of Large and Small Molecules 21

The Cytoskeleton Is Responsible for Directed Cell Movements 22

The Cytoplasm Is Far from Static 23

Eucaryotic Cells May Have Originated as Predators 23

MODEL ORGANISMS 26

Molecular Biologists Have Focused on *E. coli* 27

Brewer's Yeast Is a Simple Eucaryotic Cell 28

Arabidopsis Has Been Chosen Out of 300,000 Species as a Model Plant 28

The World of Animals Is Represented by a Fly, a Worm, a Fish, a Mouse, and the Human Species 29

Comparing Genome Sequences Reveals Life's Common Heritage 33

Essential Concepts 35

End-of-Chapter Questions 36

Chapter 2 Chemical Components of Cells **39**

CHEMICAL BONDS 40

Cells Are Made of Relatively Few Types of Atoms 40

The Outermost Electrons Determine How Atoms Interact 41

Ionic Bonds Form by the Gain and Loss of Electrons 44

Covalent Bonds Form by the Sharing of Electrons 45

Covalent Bonds Vary in Strength 46

There Are Different Types of Covalent Bonds 47

Electrostatic Attractions Help Bring Molecules Together in Cells 47

Water Is Held Together by Hydrogen Bonds 48

Some Polar Molecules Form Acids and Bases in Water 49

MOLECULES IN CELLS 50

A Cell Is Formed from Carbon Compounds 50

Cells Contain Four Major Families of Small Organic Molecules 51

Sugars Are Energy Sources for Cells and Subunits of Polysaccharides 52

Fatty Acids Are Components of Cell Membranes 54

Amino Acids Are the Subunits of Proteins 55

Nucleotides Are the Subunits of DNA and RNA 56

MACROMOLECULES IN CELLS 58
Macromolecules Contain a Specific Sequence of Subunits 59
Noncovalent Bonds Specify the Precise Shape of a Macromolecule 59
Noncovalent Bonds Allow a Macromolecule to Bind Other Selected Molecules 63
Essential Concepts 78
End-of-Chapter Questions 79

Chapter 3 Energy, Catalysis, and Biosynthesis 81

THE USE OF ENERGY BY CELLS 82
Biological Order Is Made Possible by the Release of Heat Energy from Cells 82
Photosynthetic Organisms Use Sunlight to Synthesize Organic Molecules 84
Cells Obtain Energy by the Oxidation of Organic Molecules 86
Oxidation and Reduction Involve Electron Transfers 87
FREE ENERGY AND CATALYSIS 88
Enzymes Lower the Energy Barriers That Prevent Chemical Reactions from Occurring 89
The Free-Energy Change for a Reaction Determines Whether It Can Occur 91
The Concentration of Reactants Influences the Free-Energy Change and a Reaction's Direction 92
The Standard Free-Energy Change Makes it Possible to Compare the Energetics of Different Reactions 92
Cells Exist in a State of Chemical Disequilibrium 92
The Equilibrium Constant is Directly Proportional to $\Delta G°$ 93
In Complex Reactions, the Equilibrium Constant Depends on the Concentrations of All Reactants and Products 96
The Equilibrium Constant Indicates the Strength of Molecular Interactions 96
For Sequential Reactions, the Changes in Free Energy are Additive 97
Rapid Diffusion Allows Enzymes to Find Their Substrates 98
V_{max} and K_M Measure Enzyme Performance 99
ACTIVATED CARRIER MOLECULES AND BIOSYNTHESIS 104
The Formation of an Activated Carrier Is Coupled to an Energetically Favorable Reaction 104
ATP is the Most Widely Used Activated Carrier Molecule 105
Energy Stored in ATP is Often Harnessed to Join Two Molecules Together 106
NADH and NADPH Are Important Electron Carriers 107
Cells Make Use of Many Other Activated Carrier Molecules 109
The Synthesis of Biological Polymers Requires an Energy Input 110
Essential Concepts 114
End-of-Chapter Questions 115

Chapter 4 Protein Structure and Function 119

THE SHAPE AND STRUCTURE OF PROTEINS 121
The Shape of a Protein Is Specified by Its Amino Acid Sequence 121
Proteins Fold into a Conformation of Lowest Energy 124
Proteins Come in a Wide Variety of Complicated Shapes 125
The α Helix and the β Sheet Are Common Folding Patterns 127
Helices Form Readily in Biological Structures 131
β Sheets Form Rigid Structures at the Core of Many Proteins 132
Proteins Have Several Levels of Organization 133
Few of the Many Possible Polypeptide Chains Will Be Useful 134
Proteins Can Be Classified into Families 135
Large Protein Molecules Often Contain More Than One Polypeptide Chain 135
Proteins Can Assemble into Filaments, Sheets, or Spheres 136
Some Types of Proteins Have Elongated Fibrous Shapes 138
Extracellular Proteins Are Often Stabilized by Covalent Cross-Linkages 138
HOW PROTEINS WORK 140
All Proteins Bind to Other Molecules 140
The Binding Sites of Antibodies Are Especially Versatile 142
Enzymes Are Powerful and Highly Specific Catalysts 143
Lysozyme Illustrates How an Enzyme Works 143
Most Drugs Inhibit Enzymes 148
Tightly Bound Small Molecules Add Extra Functions to Proteins 148
HOW PROTEINS ARE CONTROLLED 149
The Catalytic Activities of Enzymes Are Often Regulated by Other Molecules 150
Allosteric Enzymes Have Binding Sites That Influence One Another 150
Phosphorylation Can Control Protein Activity by Triggering a Conformational Change 152

GTP-Binding Proteins Are Also Regulated by the Cyclic Gain and Loss of a Phosphate Group 153
Nucleotide Hydrolysis Allows Motor Proteins to Produce Large Movements in Cells 154
Proteins Often Form Large Complexes That Function as Protein Machines 155
Covalent Modification Controls the Location and Assembly of Protein Machines 156

HOW PROTEINS ARE STUDIED 157
Cells Can Be Grown in a Culture Dish 157
Purification Techniques Allow Homogeneous Protein Preparations to Be Obtained from Cell Homogenates 161
Large Amounts of Almost Any Protein Can be Produced by Genetic Engineering Techniques 163
Automated Studies of Protein Structure and Function Are Increasing the Pace of Discovery 163

Essential Concepts 168
End-of-Chapter Questions 169

Chapter 5 DNA and Chromosomes 171

THE STRUCTURE AND FUNCTION OF DNA 172
A DNA Molecule Consists of Two Complementary Chains of Nucleotides 173
The Structure of DNA Provides a Mechanism for Heredity 178

THE STRUCTURE OF EUCARYOTIC CHROMOSOMES 179
Eucaryotic DNA Is Packaged into Multiple Chromosomes 179
Chromosomes Contain Long Strings of Genes 181
Chromosomes Exist in Different States Throughout the Life of a Cell 182
Interphase Chromosomes Are Organized Within the Nucleus 184
The DNA in Chromosomes Is Highly Condensed 184
Nucleosomes Are the Basic Units of Eucaryotic Chromosome Structure 185
Chromosome Packing Occurs on Multiple Levels 187

THE REGULATION OF CHROMOSOME STRUCTURE 188
Changes in Nucleosome Structure Allow Access to DNA 188
Interphase Chromosomes Contain Both Condensed and More Extended Forms of Chromatin 190
Changes in Chromatin Structure Can Be Inherited 191

Essential Concepts 192
End-of-Chapter Questions 193

Chapter 6 DNA Replication, Repair, and Recombination 197

DNA REPLICATION 198
Base-Pairing Enables DNA Replication 198
DNA Synthesis Begins at Replication Origins 199
New DNA Synthesis Occurs at Replication Forks 203
The Replication Fork Is Asymmetrical 204
DNA Polymerase Is Self-correcting 205
Short Lengths of RNA Act as Primers for DNA Synthesis 206
Proteins at a Replication Fork Cooperate to Form a Replication Machine 208
Telomerase Replicates the Ends of Eucaryotic Chromosomes 210

DNA REPAIR 211
Mutations Can Have Severe Consequences for a Cell or Organism 211
A DNA Mismatch Repair System Removes Replication Errors That Escape the Replication Machine 212
DNA Is Continually Suffering Damage in Cells 213
The Stability of Genes Depends on DNA Repair 215
Double-Strand Breaks Can be Repaired Rapidly But Imperfectly 216
A Record of the Fidelity of DNA Replication and Repair Is Preserved in Genome Sequences 217

HOMOLOGOUS RECOMBINATION 218
Homologous Recombination Requires Extensive Regions of Sequence Similarity 218
Homologous Recombination Can Flawlessly Repair DNA Double-strand Breaks 218
Homologous Recombination Exchanges Genetic Information During Meiosis 220

MOBILE GENETIC ELEMENTS AND VIRUSES 221
Mobile Genetic Elements Encode the Components They Need for Movement 222
The Human Genome Contains Two Major Families of Transposable Sequences 222
Viruses Are Fully Mobile Genetic Elements That Can Escape from Cells 223
Retroviruses Reverse the Normal Flow of Genetic Information 225

Essential Concepts 227
End-of-Chapter Questions 228

Chapter 7 From DNA to Protein: How Cells Read the Genome 231

FROM DNA TO RNA 232
Portions of DNA Sequence Are Transcribed into RNA 233
Transcription Produces RNA Complementary to One Strand of DNA 234

Several Types of RNA Are Produced in Cells 235
Signals in DNA Tell RNA Polymerase Where to Start and Finish 236
Initiation of Eucaryotic Gene Transcription Is a Complex Process 238
Eucaryotic RNA Polymerase Requires General Transcription Factors 239
Eucaryotic RNAs Are Transcribed and Processed Simultaneously in the Nucleus 240
Eucaryotic Genes Are Interrupted by Noncoding Sequences 241
Introns Are Removed by RNA Splicing 242
Mature Eucaryotic mRNAs Are Selectively Exported from the Nucleus 243
mRNA Molecules Are Eventually Degraded by the Cell 244
The Earliest Cells May Have Had Introns in Their Genes 245

FROM RNA TO PROTEIN 246
An mRNA Sequence Is Decoded in Sets of Three Nucleotides 246
tRNA Molecules Match Amino Acids to Codons in mRNA 247
Specific Enzymes Couple tRNAs to the Correct Amino Acid 251
The RNA Message Is Decoded on Ribosomes 251
The Ribosome Is a Ribozyme 253
Codons in mRNA Signal Where to Start and to Stop Protein Synthesis 254
Proteins Are Made on Polyribosomes 257
Inhibitors of Procaryotic Protein Synthesis Are Used as Antibiotics 257
Carefully Controlled Protein Breakdown Helps Regulate the Amount of Each Protein in a Cell 258
There Are Many Steps Between DNA and Protein 259

RNA AND THE ORIGINS OF LIFE 261
Life Requires Autocatalysis 261
RNA Can Both Store Information and Catalyze Chemical Reactions 261
RNA Is Thought to Predate DNA in Evolution 263

Essential Concepts 264
End-of-Chapter Questions 266

Chapter 8 Control of Gene Expression 269

AN OVERVIEW OF GENE EXPRESSION 270
The Different Cell Types of a Multicellular Organism Contain the Same DNA 270
Different Cell Types Produce Different Sets of Proteins 270
A Cell Can Change the Expression of Its Genes in Response to External Signals 272
Gene Expression Can Be Regulated at Many of the Steps in the Pathway from DNA to RNA to Protein 272

HOW TRANSCRIPTIONAL SWITCHES WORK 273
Transcription Is Controlled by Proteins Binding to Regulatory DNA Sequences 273
Transcription Switches Allow Cells to Respond to Changes in the Environment 275
Repressors Turn Genes Off, Activators Turn Them On 276
An Activator and a Repressor Control the *Lac* Operon 277
Eucaryotic Transcription Regulators Control Gene Expression from a Distance 278
Packing of Promoter DNA into Nucleosomes Affects Initiation of Transcription 279

THE MOLECULAR MECHANISMS THAT CREATE SPECIALIZED CELL TYPES 280
Eucaryotic Genes Are Regulated by Combinations of Proteins 280
The Expression of Different Genes Can Be Coordinated by a Single Protein 281
Combinatorial Control Can Create Different Cell Types 285
Stable Patterns of Gene Expression Can Be Transmitted to Daughter Cells 287
The Formation of an Entire Organ Can Be Triggered by a Single Transcription Regulator 288

POST-TRANSCRIPTIONAL CONTROLS 289
Riboswitches Provide An Economical Solution to Gene Regulation 289
The Untranslated Regions of mRNAs Can Control Their Translation 290
Small Regulatory RNAs Control the Expression of Thousands of Animal and Plant Genes 290
RNA Interference Destroys Double-Stranded Foreign RNAs 291
Scientists Can Use RNA Interference to Turn Off Genes 292

Essential Concepts 293
End-of-Chapter Questions 294

Chapter 9 How Genes and Genomes Evolve 297

GENERATING GENETIC VARIATION 298
In Sexually Reproducing Organisms, Only Changes to the Germ Line Are Passed Along To Progeny 299
Point Mutations Are Caused by Failures of the Normal Mechanisms for Copying and Maintaining DNA 300
Point Mutations Can Change the Regulation of a Gene 301

DNA Duplications Give Rise to Families of Related Genes 302
The Evolution of the Globin Gene Family Shows How Gene Duplication and Divergence Can Give Rise to Proteins Tailored to an Organism and Its Development 304
Whole Genome Duplications Have Shaped the Evolutionary History of Many Species 305
New Genes Can Be Generated by Repeating the Same Exon 306
Novel Genes Can Also Be Created by Exon Shuffling 306
The Evolution of Genomes Has Been Accelerated by the Movement of Mobile Genetic Elements 307
Genes Can Be Exchanged Between Organisms by Horizontal Gene Transfer 308

RECONSTRUCTING LIFE'S FAMILY TREE 309
Genetic Changes That Provide a Selective Advantage Are Likely to Be Preserved 309
Human and Chimpanzee Genomes Are Similar in Organization As Well As in Detailed Sequence 310
Functionally Important Regions Show Up As Islands of Conserved DNA Sequence 310
Genome Comparisons Show That Vertebrate Genomes Gain and Lose DNA Rapidly 312
Sequence Conservation Allows Us to Trace Even the Most Distant Evolutionary Relationships 313

EXAMINING THE HUMAN GENOME 315
The Nucleotide Sequence of the Human Genome Shows How Our Genes Are Arranged 316
Accelerated Changes in Conserved Genome Sequences Help Reveal What Makes Us Human 320
Genetic Variation Within the Human Genome Contributes to Our Individuality 320
The Human Genome Contains Copious Information Yet to Be Deciphered 321

Essential Concepts 323
End-of-Chapter Questions 324

Chapter 10 Analyzing Genes and Genomes 327

MANIPULATING AND ANALYZING DNA MOLECULES 329
Restriction Nucleases Cut DNA Molecules at Specific Sites 329
Gel Electrophoresis Separates DNA Fragments of Different Sizes 330
Hybridization Provides a Sensitive Way to Detect Specific Nucleotide Sequences 332
Hybridization Is Carried Out Using DNA Probes Designed to Recognize a Desired Nucleotide Sequence 332

DNA CLONING 333
DNA Ligase Joins DNA Fragments Together to Produce a Recombinant DNA Molecule 334
Recombinant DNA Can Be Copied Inside Bacterial Cells 334
Specialized Plasmid Vectors Are Used to Clone DNA 335
Genes Can Be Isolated from a DNA Library 336
cDNA Libraries Represent the mRNA Produced by a Particular Tissue 338
The Polymerase Chain Reaction Amplifies Selected DNA Sequences 340

DECIPHERING AND EXPLOITING GENETIC INFORMATION 343
DNA Can Be Rapidly Sequenced 345
Completely Novel DNA Molecules Can Be Constructed 347
Rare Proteins Can Be Made in Large Amounts Using Cloned DNA 347
Reporter Genes and *In Situ* Hybridization Can Reveal When and Where a Gene Is Expressed 350
Hybridization on DNA Microarrays Monitors the Expression of Thousands of Genes at Once 352
Genetic Approaches Can Reveal the Function of a Gene 354
Animals Can be Genetically Altered 354
RNA Interference Provides a Simple Way to Test Gene Function 356
Transgenic Plants Are Important for Both Cell Biology and Agriculture 357
Essential Concepts 358
End-of-Chapter Questions 360

Chapter 11 Membrane Structure 363

THE LIPID BILAYER 364
Membrane Lipids Form Bilayers in Water 365
The Lipid Bilayer Is a Two-dimensional Fluid 368
The Fluidity of a Lipid Bilayer Depends on Its Composition 369
The Lipid Bilayer Is Asymmetrical 370
Lipid Asymmetry Is Preserved During Membrane Transport 371

MEMBRANE PROTEINS 372
Membrane Proteins Associate with the Lipid Bilayer in Various Ways 373
A Polypeptide Chain Usually Crosses the Bilayer as an α Helix 374
Membrane Proteins Can Be Solubilized in Detergents and Purified 375
The Complete Structure Is Known for Relatively Few Membrane Proteins 376
The Plasma Membrane Is Reinforced by the Cell Cortex 377

Cells Can Restrict the Movement of Membrane Proteins 379
The Cell Surface Is Coated with Carbohydrate 380
Essential Concepts 384
End-of-Chapter Questions 385

Chapter 12 Membrane Transport 387

PRINCIPLES OF MEMBRANE TRANSPORT 388
The Ion Concentrations Inside a Cell Are Very Different from Those Outside 388
Lipid Bilayers Are Impermeable to Solutes and Ions 389
Membrane Transport Proteins Fall into Two Classes: Transporters and Channels 389
Solutes Cross Membranes by Passive or Active Transport 390

TRANSPORTERS AND THEIR FUNCTIONS 391
Concentration Gradients and Electrical Forces Drive Passive Transport 392
Active Transport Moves Solutes Against Their Electrochemical Gradients 393
Animal Cells Use the Energy of ATP Hydrolysis to Pump Out Na^+ 394
The Na^+-K^+ Pump Is Driven by the Transient Addition of a Phosphate Group 394
The Na^+-K^+ Pump Helps Maintain the Osmotic Balance of Animal Cells 396
Intracellular Ca^{2+} Concentrations Are Kept Low by Ca^{2+} Pumps 397
Coupled Transporters Exploit Gradients to Take Up Nutrients Actively 398
H^+ Gradients Are Used to Drive Membrane Transport in Plants, Fungi, and Bacteria 400

ION CHANNELS AND THE MEMBRANE POTENTIAL 400
Ion Channels Are Ion-selective and Gated 401
Ion Channels Randomly Snap Between Open and Closed States 403
Different Types of Stimuli Influence the Opening and Closing of Ion Channels 405
Voltage-gated Ion Channels Respond to the Membrane Potential 405
Membrane Potential Is Governed by Membrane Permeability to Specific Ions 407

ION CHANNELS AND SIGNALING IN NERVE CELLS 409
Action Potentials Provide for Rapid Long-Distance Communication 409
Action Potentials Are Usually Mediated by Voltage-gated Na^+ Channels 410
Voltage-gated Ca^{2+} Channels Convert Electrical Signals into Chemical Signals at Nerve Terminals 415
Transmitter-gated Channels in Target Cells Convert Chemical Signals Back into Electrical Signals 415
Neurons Receive Both Excitatory and Inhibitory Inputs 417
Transmitter-gated Ion Channels Are Major Targets for Psychoactive Drugs 418
Synaptic Connections Enable You to Think, Act, and Remember 419
Essential Concepts 420
End-of-Chapter Questions 421

Chapter 13 How Cells Obtain Energy from Food 425

THE BREAKDOWN AND UTILIZATION OF SUGARS AND FATS 426
Food Molecules Are Broken Down in Three Stages 426
Glycolysis Is a Central ATP-producing Pathway 427
Fermentations Allow ATP to Be Produced in the Absence of Oxygen 432
Glycolysis Illustrates How Enzymes Couple Oxidation to Energy Storage 433
Sugars and Fats Are Both Degraded to Acetyl CoA in Mitochondria 436
The Citric Acid Cycle Generates NADH by Oxidizing Acetyl Groups to CO_2 436
Many Biosynthetic Pathways Begin with Glycolysis or the Citric Acid Cycle 439
Electron Transport Drives the Synthesis of the Majority of the ATP in Most Cells 444

REGULATION OF METABOLISM 445
Catabolic and Anabolic Reactions Are Organized and Regulated 445
Feedback Regulation Allows Cells to Switch from Glucose Degradation to Glucose Biosynthesis 447
Cells Store Food Molecules in Special Reservoirs to Prepare for Periods of Need 448
Essential Concepts 450
End-of-Chapter Questions 451

Chapter 14 Energy Generation in Mitochondria and Chloroplasts 453

Cells Obtain Most of Their Energy by a Membrane-based Mechanism 454

MITOCHONDRIA AND OXIDATIVE PHOSPHORYLATION 456
A Mitochondrion Contains an Outer Membrane, an Inner Membrane, and Two Internal Compartments 456
The Citric Acid Cycle Generates High-Energy Electrons 458
A Chemiosmotic Process Converts the Energy From Activated Carrier Molecules into ATP 458

The Electron-Transport Chain Pumps Protons Across the Inner Mitochondrial Membrane 460
Proton Pumping Creates a Steep Electrochemical Proton Gradient Across the Inner Mitochondrial Membrane 460
The Electrochemical Proton Gradient Drives ATP Synthesis 461
Coupled Transport Across the Inner Mitochondrial Membrane Is Also Driven by the Electrochemical Proton Gradient 463
Oxidative Phosphorylation Produces Most of the Cell's ATP 464
The Rapid Conversion of ADP to ATP in Mitochondria Maintains a High ATP/ADP Ratio in Cells 465

MOLECULAR MECHANISMS OF ELECTRON TRANSPORT AND PROTON PUMPING 466
Protons Are Readily Moved by the Transfer of Electrons 466
The Redox Potential Is a Measure of Electron Affinities 467
Electron Transfers Release Large Amounts of Energy 470
Metals Tightly Bound to Proteins Form Versatile Electron Carriers 470
Cytochrome Oxidase Catalyzes the Reduction of Molecular Oxygen 473
The Mechanism of H^+ Pumping Can Be Studied in Atomic Detail 474
Respiration Is Amazingly Efficient 475

CHLOROPLASTS AND PHOTOSYNTHESIS 476
Chloroplasts Resemble Mitochondria but Have an Extra Compartment 477
Chloroplasts Capture Energy from Sunlight and Use It to Fix Carbon 478
Sunlight is Absorbed by Chlorophyll Molecules 479
Excited Chlorophyll Molecules Funnel Energy into a Reaction Center 480
Light Energy Drives the Synthesis of Both ATP and NADPH 481
Chloroplasts Can Adjust their ATP Production 483
Carbon Fixation Uses ATP and NADPH to Convert CO_2 into Sugars 484
Sugars Generated by Carbon Fixation Can Be Stored As Starch or Consumed to Produce ATP 486

THE ORIGINS OF CHLOROPLASTS AND MITOCHONDRIA 486
Oxidative Phosphorylation Might Have Given Ancient Bacteria an Evolutionary Advantage 487
Photosynthetic Bacteria Made Even Fewer Demands on Their Environment 488
The Lifestyle of *Methanococcus* Suggests That Chemiosmotic Coupling Is an Ancient Process 490
Essential Concepts 491
End-of-Chapter Questions 492

Chapter 15 Intracellular Compartments and Transport 495

MEMBRANE-ENCLOSED ORGANELLES 496
Eucaryotic Cells Contain a Basic Set of Membrane-enclosed Organelles 496
Membrane-enclosed Organelles Evolved in Different Ways 498

PROTEIN SORTING 500
Proteins Are Imported into Organelles by Three Mechanisms 500
Signal Sequences Direct Proteins to the Correct Compartment 501
Proteins Enter the Nucleus Through Nuclear Pores 502
Proteins Unfold to Enter Mitochondria and Chloroplasts 505
Proteins Enter the Endoplasmic Reticulum While Being Synthesized 505
Soluble Proteins Are Released into the ER Lumen 507
Start and Stop Signals Determine the Arrangement of a Transmembrane Protein in the Lipid Bilayer 508

VESICULAR TRANSPORT 510
Transport Vesicles Carry Soluble Proteins and Membrane Between Compartments 510
Vesicle Budding Is Driven by the Assembly of a Protein Coat 511
Vesicle Docking Depends on Tethers and SNAREs 512

SECRETORY PATHWAYS 514
Most Proteins Are Covalently Modified in the ER 514
Exit from the ER Is Controlled to Ensure Protein Quality 516
The Size of the ER Is Controlled by the Amount of Protein that Flows Through It 516
Proteins Are Further Modified and Sorted in the Golgi Apparatus 517
Secretory Proteins Are Released from the Cell by Exocytosis 518

ENDOCYTIC PATHWAYS 522
Specialized Phagocytic Cells Ingest Large Particles 522
Fluid and Macromolecules Are Taken Up by Pinocytosis 523
Receptor-mediated Endocytosis Provides a Specific Route into Animal Cells 524
Endocytosed Macromolecules Are Sorted in Endosomes 525
Lysosomes Are the Principal Sites of Intracellular Digestion 526
Essential Concepts 527
End-of-Chapter Questions 529

Chapter 16 Cell Communication 531

GENERAL PRINCIPLES OF CELL SIGNALING 532

Signals Can Act over a Long or Short Range 532

Each Cell Responds to a Limited Set of Signals, Depending on Its History and Its Current State 534

A Cell's Response to a Signal Can Be Fast or Slow 536

Some Hormones Cross the Plasma Membrane and Bind to Intracellular Receptors 537

Some Dissolved Gases Cross the Plasma Membrane and Activate Intracellular Enzymes Directly 538

Cell-Surface Receptors Relay Extracellular Signals via Intracellular Signaling Pathways 539

Some Intracellular Signaling Proteins Act as Molecular Switches 541

Cell-Surface Receptors Fall into Three Main Classes 542

Ion-channel–coupled Receptors Convert Chemical Signals into Electrical Ones 544

G-PROTEIN–COUPLED RECEPTORS 544

Stimulation of GPCRs Activates G-Protein Subunits 545

Some G Proteins Directly Regulate Ion Channels 547

Some G Proteins Activate Membrane-bound Enzymes 547

The Cyclic AMP Pathway Can Activate Enzymes and Turn On Genes 548

The Inositol Phospholipid Pathway Triggers a Rise in Intracellular Ca^{2+} 551

A Ca^{2+} Signal Triggers Many Biological Processes 552

Intracellular Signaling Cascades Can Achieve Astonishing Speed, Sensitivity, and Adaptability 554

ENZYME-COUPLED RECEPTORS 555

Activated RTKs Recruit a Complex of Intracellular Signaling Proteins 555

Most RTKs Activate the Monomeric GTPase Ras 556

RTKs Activate PI 3-Kinase to Produce Lipid Docking Sites in the Plasma Membrane 558

Some Receptors Activate a Fast Track to the Nucleus 559

Multicellularity and Cell Communication Evolved Independently in Plants and Animals 564

Protein Kinase Networks Integrate Information to Control Complex Cell Behaviors 564

Essential Concepts 567

End-of-Chapter Questions 569

Chapter 17 Cytoskeleton 571

INTERMEDIATE FILAMENTS 572

Intermediate Filaments Are Strong and Ropelike 574

Intermediate Filaments Strengthen Cells Against Mechanical Stress 575

The Nuclear Envelope Is Supported by a Meshwork of Intermediate Filaments 576

MICROTUBULES 577

Microtubules Are Hollow Tubes with Structurally Distinct Ends 578

The Centrosome Is the Major Microtubule-organizing Center in Animal Cells 579

Growing Microtubules Show Dynamic Instability 580

Microtubules Are Maintained by a Balance of Assembly and Disassembly 581

Microtubules Organize the Interior of the Cell 582

Motor Proteins Drive Intracellular Transport 583

Organelles Move Along Microtubules 584

Cilia and Flagella Contain Stable Microtubules Moved by Dynein 585

ACTIN FILAMENTS 590

Actin Filaments Are Thin and Flexible 591

Actin and Tubulin Polymerize by Similar Mechanisms 591

Many Proteins Bind to Actin and Modify Its Properties 592

An Actin-rich Cortex Underlies the Plasma Membrane of Most Eucaryotic Cells 594

Cell Crawling Depends on Actin 594

Actin Associates with Myosin to Form Contractile Structures 597

Extracellular Signals Control the Arrangement of Actin Filaments 597

MUSCLE CONTRACTION 599

Muscle Contraction Depends on Bundles of Actin and Myosin 599

During Muscle Contraction Actin Filaments Slide Against Myosin Filaments 600

Muscle Contraction Is Triggered by a Sudden Rise in Ca^{2+} 602

Muscle Cells Perform Highly Specialized Functions in the Body 604

Essential Concepts 605

End-of-Chapter Questions 606

Chapter 18 The Cell Division Cycle 609

OVERVIEW OF THE CELL CYCLE 610

The Eucaryotic Cell Cycle Is Divided into Four Phases 611

A Cell-Cycle Control System Triggers the Major Processes of the Cell Cycle 612

Cell-Cycle Control is Similar in All Eucaryotes 613

THE CELL-CYCLE CONTROL SYSTEM 613

The Cell-Cycle Control System Depends on Cyclically Activated Protein Kinases called Cdks 614

The Activity of Cdks Is Also Regulated by Phosphorylation and Dephosphorylation 614
Different Cyclin–Cdk Complexes Trigger Different Steps in the Cell Cycle 617
The Cell-Cycle Control System Also Depends on Cyclical Proteolysis 618
Proteins that Inhibit Cdks Can Arrest the Cell Cycle at Specific Checkpoints 618

S PHASE 620
S-Cdk Initiates DNA Replication and Helps Block Re-Replication 620
Cohesins Help Hold the Sister Chromatids of Each Replicated Chromosome Together 621
DNA Damage Checkpoints Help Prevent the Replication of Damaged DNA 621

M PHASE 622
M-Cdk Drives Entry Into M Phase and Mitosis 622
Condensins Help Configure Duplicated Chromosomes for Separation 623
The Cytoskeleton Carries Out Both Mitosis and Cytokinesis 624
M Phase Is Conventionally Divided into Six Stages 624

MITOSIS 625
Centrosomes Duplicate To Help Form the Two Poles of the Mitotic Spindle 625
The Mitotic Spindle Starts to Assemble in Prophase 628
Chromosomes Attach to the Mitotic Spindle at Prometaphase 628
Chromosomes Aid in the Assembly of the Mitotic Spindle 630
Chromosomes Line Up at the Spindle Equator at Metaphase 630
Proteolysis Triggers Sister-Chromatid Separation and the Completion of Mitosis 631
Chromosomes Segregate During Anaphase 631
Unattached Chromosomes Block Sister-Chromatid Separation 633
The Nuclear Envelope Re-forms at Telophase 634

CYTOKINESIS 634
The Mitotic Spindle Determines the Plane of Cytoplasmic Cleavage 634
The Contractile Ring of Animal Cells Is Made of Actin and Myosin 635
Cytokinesis in Plant Cells Involves the Formation of a New Cell Wall 636
Membrane-Enclosed Organelles Must Be Distributed to Daughter Cells When a Cell Divides 638

CONTROL OF CELL NUMBER AND CELL SIZE 638
Apoptosis Helps Regulate Animal Cell Numbers 638
Apoptosis Is Mediated by an Intracellular Proteolytic Cascade 639
The Death Program Is Regulated by the Bcl2 Family of Intracellular Proteins 641
Animal Cells Require Extracellular Signals to Survive, Grow, and Divide 642
Animal Cells Require Survival Factors to Avoid Apoptosis 643
Mitogens Stimulate Cell Division 644
Growth Factors Stimulate Cells to Grow 645
Some Extracellular Signal Proteins Inhibit Cell Survival, Division, or Growth 645
Essential Concepts 647
End-of-Chapter Questions 649

Chapter 19 Sex and Genetics 651

THE BENEFITS OF SEX 652
Sexual Reproduction Involves Both Diploid and Haploid Cells 652
Sexual Reproduction Gives Organisms a Competitive Advantage 654

MEIOSIS AND FERTILIZATION 655
Haploid Germ Cells Are Produced From Diploid Cells Through Meiosis 655
Meiosis Involves a Special Process of Chromosome Pairing 656
Crossing-Over Can Occur Between Maternal and Paternal Chromosomes 657
Chromosome Pairing and Recombination Ensure the Proper Segregation of Homologs 658
The Second Meiotic Division Produces Haploid Daughter Cells 659
Haploid Cells Contain Reassorted Genetic Information 661
Meiosis Is Not Flawless 662
Fertilization Reconstitutes a Complete Diploid Genome 663

MENDEL AND THE LAWS OF INHERITANCE 664
Mendel Chose to Study Traits That Are Inherited in a Discrete Fashion 665
Mendel Could Disprove the Alternative Theories of Inheritance 665
Mendel's Experiments Were the First to Reveal the Discrete Nature of Heredity 666
Each Gamete Carries a Single Allele for Each Character 667
Mendel's Law of Segregation Applies to All Sexually Reproducing Organisms 668
Alleles for Different Traits Segregate Independently 669
The Behavior of Chromosomes During Meiosis Underlies Mendel's Laws of Inheritance 671

Chromosome Crossovers Can Be Used to Determine the Order of Genes 671
Mutations in Genes Can Cause a Loss of Function Or a Gain of Function 673
Each of Us Carries Many Potentially Harmful Recessive Mutant Alleles 673

GENETICS AS AN EXPERIMENTAL TOOL 675
The Classical Approach Begins with Random Mutagenesis 675
Genetic Screens Identify Mutants Deficient in Specific Cellular Processes 676
A Complementation Test Reveals Whether Two Mutations Are in the Same Gene 677
Single-Nucleotide Polymorphisms (SNPs) Serve as Landmarks for Genetic Mapping 678
Linked Groups of SNPs Define Haplotype Blocks 682
Haplotype Blocks Give Clues to our Evolutionary History 683
Essential Concepts 684
End-of-Chapter Questions 685

Chapter 20 Cellular Communities: Tissues, Stem Cells, and Cancer 689

EXTRACELLULAR MATRIX AND CONNECTIVE TISSUES 690
Plant Cells Have Tough External Walls 691
Cellulose Microfibrils Give the Plant Cell Wall Its Tensile Strength 692
Animal Connective Tissues Consist Largely of Extracellular Matrix 693
Collagen Provides Tensile Strength in Animal Connective Tissues 694
Cells Organize the Collagen That They Secrete 696
Integrins Couple the Matrix Outside a Cell to the Cytoskeleton Inside It 696
Gels of Polysaccharide and Protein Fill Spaces and Resist Compression 698

EPITHELIAL SHEETS AND CELL JUNCTIONS 700
Epithelial Sheets Are Polarized and Rest on a Basal Lamina 700
Tight Junctions Make an Epithelium Leak-proof and Separate Its Apical and Basal Surfaces 701
Cytoskeleton-linked Junctions Bind Epithelial Cells Robustly to One Another and to the Basal Lamina 703
Gap Junctions Allow Ions and Small Molecules to Pass from Cell to Cell 705

TISSUE MAINTENANCE AND RENEWAL 707
Tissues Are Organized Mixtures of Many Cell Types 709
Different Tissues Are Renewed at Different Rates 710
Stem Cells Generate a Continuous Supply of Terminally Differentiated Cells 711
Specific Signals Maintain the Stem-Cell Populations 713
Stem Cells Can Be Used to Repair Damaged Tissues 714
Therapeutic Cloning Could Provide a Way to Generate Personalized ES Cells 715

CANCER 717
Cancer Cells Proliferate, Invade, and Metastasize 718
Epidemiology Identifies Preventable Causes of Cancer 718
Cancers Develop by an Accumulation of Mutations 719
Cancer Cells Evolve Properties that Give Them a Competitive Advantage 721
Many Diverse Types of Genes Are Critical for Cancer 722
Colorectal Cancer Illustrates How Loss of a Gene Can Lead to Growth of a Tumor 723
An Understanding of Cancer Cell Biology Opens the Way to New Treatments 727
Essential Concepts 729
End-of-Chapter Questions 731

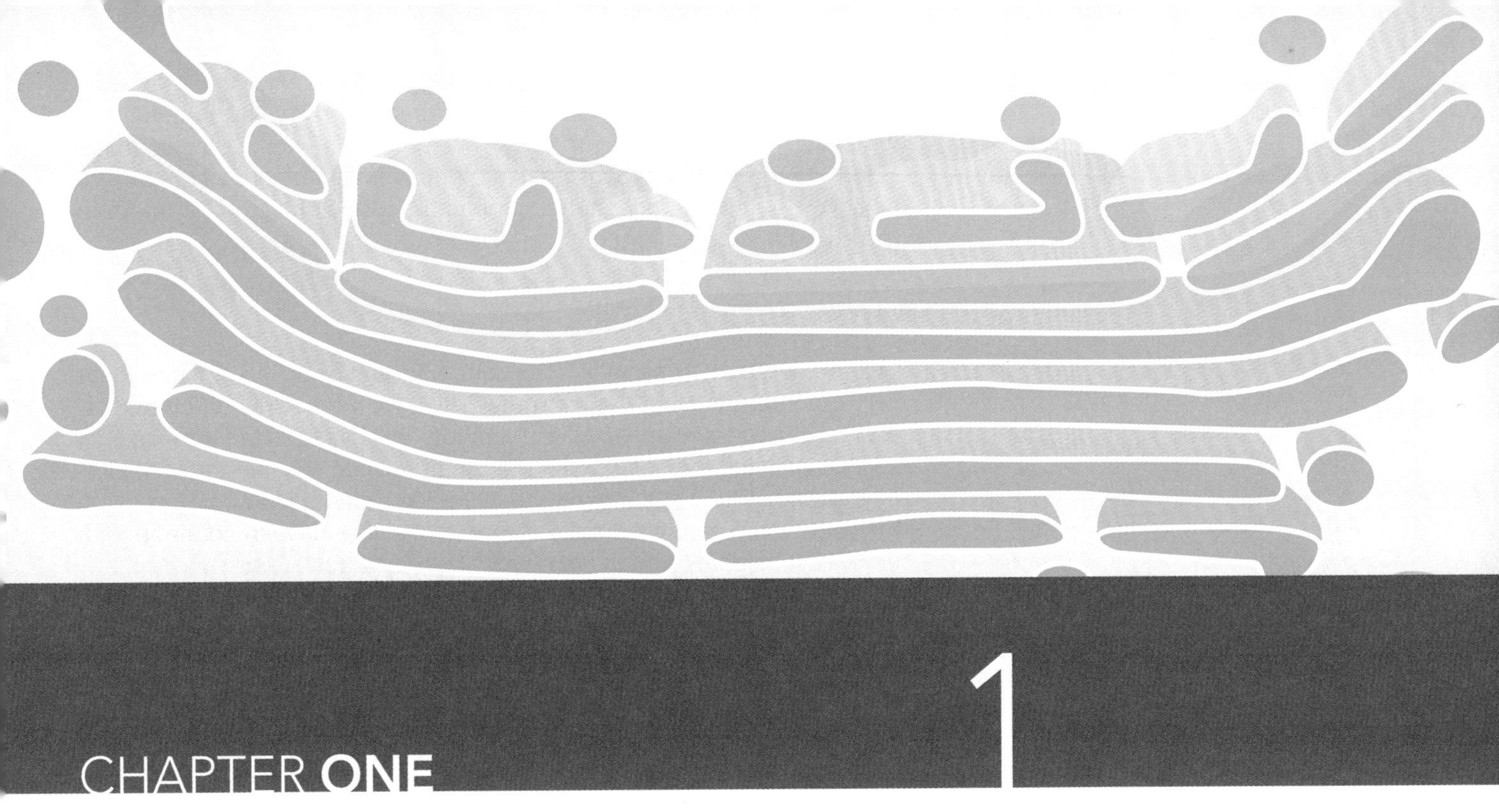

CHAPTER **ONE** 1

Introduction to Cells

UNITY AND DIVERSITY OF CELLS

CELLS UNDER THE MICROSCOPE

THE PROCARYOTIC CELL

THE EUCARYOTIC CELL

MODEL ORGANISMS

What does it mean to be living? People, petunias, and pond scum are all alive; stones, sand, and summer breezes are not. But what are the fundamental properties that characterize living things and distinguish them from nonliving matter?

The answer begins with a basic fact that is taken for granted by biologists now, but marked a revolution in thinking when first established 170 years ago. All living things are made of **cells**: small, membrane-enclosed units filled with a concentrated aqueous solution of chemicals and endowed with the extraordinary ability to create copies of themselves by growing and dividing in two. The simplest forms of life are solitary cells. Higher organisms, including ourselves, are communities of cells derived by growth and division from a single founder cell: each animal, plant, or fungus is a vast colony of individual cells that perform specialized functions coordinated by intricate systems of communication.

Cells, therefore, are the fundamental units of life, and it is to *cell biology* that we must look for an answer to the question of what life is and how it works. With a deeper understanding of the structure, function, behavior, and evolution of cells, we can begin to tackle the grand historical problems of life on Earth: its mysterious origins, its stunning diversity, and its invasion of every conceivable habitat. At the same time, cell biology can provide us with answers to the questions we have about ourselves: Where did we come from? How do we develop from a single fertilized egg cell? How is each of us different from every other person on Earth? Why do we get sick, grow old, and die?

In this chapter we begin by looking at the great variety of forms that cells can show and we take a preliminary glimpse at the chemical machinery that all cells have in common. We then consider how cells are made visible under the microscope and what we see when we peer inside them. Finally, we discuss how we can exploit the similarities of living things to achieve a coherent understanding of all the forms of life on Earth—from the tiniest bacterium to the mightiest oak.

UNITY AND DIVERSITY OF CELLS

Cell biologists often speak of "the cell" without specifying any particular cell. But cells are not all alike; in fact, they can be wildly different. It is estimated that there are at least 10 million—perhaps 100 million—distinct species of living things in the world. Before delving deeper into cell biology, we must take stock: what does a bacterium have in common with the cells of a butterfly; what do the cells of a rose have in common with those of a dolphin? And in what ways do they differ?

Cells Vary Enormously in Appearance and Function

Let us begin with size. A bacterial cell—say a *Lactobacillus* in a piece of cheese—is a few **micrometers**, or μm, in length. That's about 25 times smaller than the width of a human hair. A frog's egg—which is also a single cell—has a diameter of about 1 millimeter. If we scaled them up to make the *Lactobacillus* the size of a person, the frog's egg would be half a mile high.

Cells vary no less widely in their shapes and functions. Consider the gallery of cells displayed in Figure 1–1. A typical nerve cell in your brain is enormously extended; it sends out its electrical signals along a fine protrusion that is 10,000 times longer than it is thick, and it receives signals from other cells through a mass of shorter processes that sprout from its body like the branches of a tree. A *Paramecium* in a drop of pond water is shaped like a submarine and is covered with thousands of *cilia*—hairlike extensions whose sinuous beating sweeps the cell forward, rotating as it goes. A cell in the surface layer of a plant is a squat, immobile prism that surrounds itself in a rigid box of cellulose, with an outer waterproof coating of wax. A *Bdellovibrio* bacterium is a sausage-shaped torpedo driven forward by a rotating corkscrew-like *flagellum* that is attached to its stern, where it acts as a propeller. A neutrophil or a macrophage in the body of an animal crawls through tissues, constantly pouring itself into new shapes and engulfing debris, foreign microorganisms, and dead or dying cells.

Some cells are clad only in a flimsy membrane; others augment this delicate cover by cloaking themselves in an outer layer of slime, building themselves rigid *cell walls*, or surrounding themselves with a hard, mineralized material, such as that found in bone.

Cells are also enormously diverse in their chemical requirements and activities. Some require oxygen to live; for others it is deadly. Some consume little more than air, sunlight, and water as their raw materials; others need a complex mixture of molecules produced by other cells. Some appear to be specialized factories for the production of particular substances, such as hormones, starch, fat, latex, or pigments. Some are engines, like muscle, burning fuel to do mechanical work; others are electricity generators, like the modified muscle cells in the electric eel.

Some modifications specialize a cell so much that they spoil its chances of leaving any descendants. Such specialization would be senseless for a cell that lived a solitary life. In a multicellular organism, however, there is a division of labor among cells, allowing some cells to become specialized to an extreme degree for particular tasks and leaving them dependent on

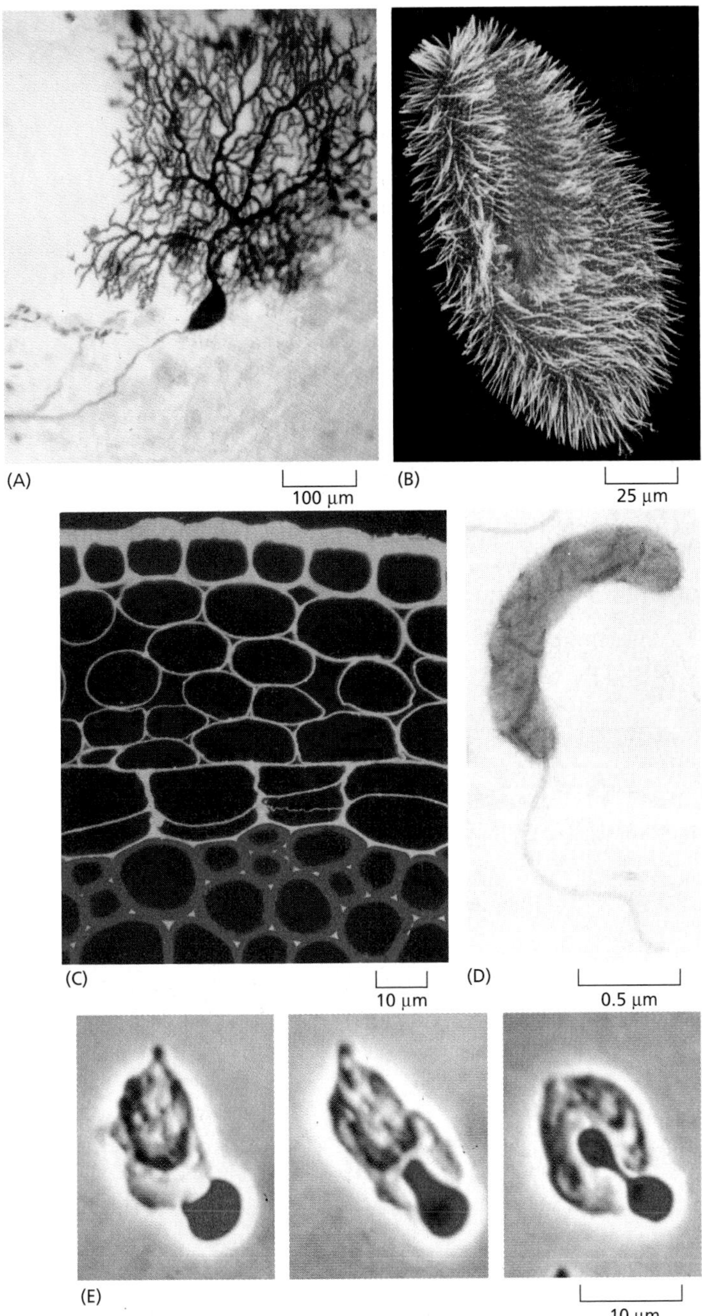

Figure 1–1 **Cells come in a variety of shapes and sizes.** (A) A nerve cell from the cerebellum (a part of the brain that controls movement). This cell has a huge branching tree of processes, through which it receives signals from as many as 100,000 other nerve cells. (B) *Paramecium*. This protozoan—a single giant cell—swims by means of the beating cilia that cover its surface. (C) A section of a young plant stem in which cellulose is stained *red* and another cell wall component, pectin, is stained *orange*. The outermost layer of cells is at the top of the photo. (D) A small bacterium, *Bdellovibrio bacteriovorus*, that uses a single terminal flagellum to propel itself. This bacterium attacks, kills, and feeds on other, larger bacteria. (E) A human white blood cell (a neutrophil) approaching and engulfing a red blood cell. (A, courtesy of Constantino Sotelo; B, courtesy of Anne Fleury, Michel Laurent, and André Adoutte; D, courtesy of Murry Stein; E, courtesy of Stephen E. Malawista and Anne de Boisfleury Chevance.)

their fellow cells for many basic requirements. Even the most basic need of all, that of passing on the genetic instructions to the next generation, is delegated to specialists—the egg and the sperm.

Living Cells All Have a Similar Basic Chemistry

Despite the extraordinary diversity of plants and animals, people have recognized from time immemorial that these organisms have something in common, something that entitles them all to be called living things. With the invention of the microscope, it became clear that plants and animals are assemblies of cells, that cells can also exist as independent organisms, and that individual cells are living in the sense that they can grow, reproduce, convert energy from one form into another, control their internal workings, respond to their environment, and so on. But while it seemed easy enough to recognize life, it was remarkably difficult to say in what sense all living things were alike. Textbooks had to settle for defining life in abstract general terms related to growth and reproduction.

QUESTION 1–1

"Life" is easy to recognize but difficult to define. The dictionary defines life as "The state or quality that distinguishes living beings or organisms from dead ones and from inorganic matter, characterized chiefly by metabolism, growth, and the ability to reproduce and respond to stimuli." Biology textbooks usually elaborate slightly; for example, according to a popular text, living things:

1. Are highly organized compared to natural inanimate objects.
2. Display homeostasis, maintaining a relatively constant internal environment.
3. Reproduce themselves.
4. Grow and develop from simple beginnings.
5. Take energy and matter from the environment and transform it.
6. Respond to stimuli.
7. Show adaptation to their environment.

Score yourself, a vacuum cleaner, and a potato with respect to these characteristics.

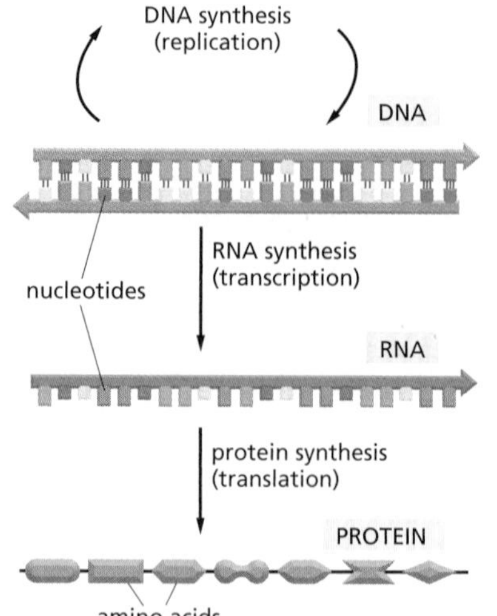

Figure 1–2 **In all living cells, genetic information flows from DNA to RNA (transcription) and from RNA to protein (translation).** Together these processes are known as gene expression.

The discoveries of biochemistry and molecular biology have made this problem disappear in a most spectacular way. Although they are infinitely varied when viewed from the outside, all living things are fundamentally similar inside. We now know that cells resemble one another to an astonishing degree in the details of their chemistry, sharing the same machinery for the most basic functions. All cells are composed of the same sorts of molecules that participate in the same types of chemical reactions (discussed in Chapter 2). In all living things, genetic instructions—*genes*—are stored in DNA molecules, written in the same chemical code, constructed out of the same chemical building blocks, interpreted by essentially the same chemical machinery, and duplicated in the same way to allow the organism to reproduce. Thus, in every cell, the long **DNA** polymer chains are made from the same set of four monomers, called *nucleotides,* strung together in different sequences like the letters of an alphabet to convey different information. In every cell, the instructions in the DNA are read out, or *transcribed,* into a chemically related set of polymers called **RNA** (Figure 1–2). RNA molecules have a variety of functions, but the major class serve as *messenger RNA*: the messages carried by these molecules are in turn *translated* into yet another type of polymer called a *protein*.

Protein molecules dominate the behavior of the cell, serving as structural supports, chemical catalysts, molecular motors, and so on. Proteins are built from *amino acids*, and every living thing uses the same set of 20 amino acids to make its proteins. But the amino acids are linked in different sequences, giving each type of protein molecule a different three-dimensional shape, or *conformation*, just as different sequences of letters spell different words. In this way, the same basic biochemical machinery has served to generate the whole gamut of living things (Figure 1–3). A more detailed discussion of the structure and function of proteins, RNA, and DNA is presented in Chapters 4 through 8.

Figure 1–3 **All living organisms are constructed from cells.** A colony of bacteria, a butterfly, a rose, and a dolphin are all made of cells that have a fundamentally similar chemistry and operate according to the same basic principles. (A, courtesy of Tony Brain and Science Photo Library; C, courtesy of the John Innes Foundation; D, courtesy of Jonathan Gordon, IFAW.)

If cells are the fundamental unit of living matter, then nothing less than a cell can truly be called living. Viruses, for example, are compact packages of genetic information—in the form of DNA or RNA—usually encased in protein, but they have no ability to reproduce themselves by their own efforts. Instead, they get themselves copied by parasitizing the reproductive machinery of the cells that they invade. Thus, viruses are chemical zombies: they are inert and inactive outside their host cells but exert a malign control once they gain entry.

All Present-Day Cells Have Apparently Evolved from the Same Ancestor

A cell reproduces by duplicating its DNA and then dividing in two, passing a copy of the genetic instructions encoded in its DNA to each of its daughter cells. That is why daughter cells resemble the parent cell. However, the copying is not always perfect, and the instructions are occasionally corrupted by *mutations* that change the DNA. That is why daughter cells do not always match the parent cell exactly.

Mutations can create offspring that are changed for the worse (in that they are less able to survive and reproduce), changed for the better (in that they are better able to survive and reproduce), or changed neutrally (in that they are genetically different but equally viable). The struggle for survival eliminates the first, favors the second, and tolerates the third. The genes of the next generation will be the genes of the survivors. Intermittently, the pattern of descent may be complicated by sexual reproduction, in which two cells of the same species fuse, pooling their DNA; the genetic cards are then shuffled, re-dealt, and distributed in new combinations to the next generation, to be tested again for their survival value.

These simple principles of genetic change and selection, applied repeatedly over billions of cell generations, are the basis of **evolution**—the process by which living species become gradually modified and adapted to their environment in more and more sophisticated ways. Evolution offers a startling but compelling explanation of why present-day cells are so similar in their fundamentals: they have all inherited their genetic instructions from the same common ancestor. It is estimated that this ancestral cell existed between 3.5 billion and 3.8 billion years ago, and we must suppose that it contained a prototype of the universal machinery of all life on Earth today. Through mutation, its descendants have gradually diverged to fill every habitat on Earth with living things, exploiting the potential of the machinery in an endless variety of ways.

QUESTION 1–2

Mutations are mistakes in the DNA that change the genetic plan from the previous generation. Imagine a shoe factory. Would you expect mistakes (i.e., unintentional changes) in copying the shoe design to lead to improvements in the shoes produced? Explain your answer.

Genes Provide the Instructions for Cellular Form, Function, and Complex Behavior

A cell's **genome**—that is, the entire library of genetic information in its DNA—provides a genetic program that instructs the cell how to function, and, for plant and animal cells, how to grow into an organism with hundreds of different cell types. Within an individual plant or animal, these cells can be extraordinarily varied, as we discuss in Chapter 20. Fat cells, skin cells, bone cells, and nerve cells seem as dissimilar as any cells could be. Yet all these *differentiated cell types* are generated during embryonic development from a single fertilized egg cell, and all contain identical copies of the DNA of the species. Their varied characters stem from the way that individual cells use their genetic instructions. Different cells *express* different genes—that is, they use their genes to produce some proteins and not others, depending on the cues that they and their ancestor cells have received from their surroundings.

The DNA, therefore, is not just a shopping list specifying the molecules that every cell must have, and a cell is not just an assembly of all the items on the list. Each cell is capable of carrying out a variety of biological tasks, depending on its environment and its history, using the information encoded in its DNA to guide its activities. Later in this book, we shall see in detail how DNA defines both the parts list of the cell and the rules that decide when and where these parts are to be made.

CELLS UNDER THE MICROSCOPE

Today we have the technology to decipher the underlying principles that govern the structure and activity of the cell. But cell biology started without these tools. The earliest cell biologists began by simply looking at tissues and cells, then breaking them open or slicing them up and attempting to view their contents. What they saw was to them profoundly baffling—a collection of tiny and scarcely visible objects whose relationship to the properties of living matter seemed an impenetrable mystery. Nevertheless, this type of visual investigation was the first step toward understanding, and it remains essential in the study of cell biology.

Cells, in general, are very small—too small to be seen with the naked eye. They were not made visible until the seventeenth century, when the **microscope** was invented. For hundreds of years afterward, all that was known about cells was discovered using this instrument. *Light microscopes*, which use visible light to illuminate specimens, are still vital pieces of equipment in the cell biology laboratory.

Although these instruments now incorporate many sophisticated improvements, the properties of light itself set a limit to the fineness of detail they can reveal. *Electron microscopes*, invented in the 1930s, go beyond this limit by using beams of electrons instead of beams of light as the source of illumination, greatly extending our ability to see the fine details of cells and even making some of the larger molecules visible individually. A survey of the principal types of microscopy used to examine cells is given in Panel 1–1 (pp. 8–9).

QUESTION 1–3

You have embarked on an ambitious research project: to create life in a test tube. You boil up a rich mixture of yeast extract and amino acids in a flask along with a sprinkling of the inorganic salts known to be essential for life. You seal the flask and allow it to cool. After several months, the liquid is as clear as ever, and there are no signs of life. A friend suggests that excluding the air was a mistake, since most life as we know it requires oxygen. You repeat the experiment, but this time you leave the flask open to the atmosphere. To your great delight, the liquid becomes cloudy after a few days and under the microscope you see beautiful small cells that are clearly growing and dividing. Does this experiment prove that you managed to generate a novel life form? How might you redesign your experiment to allow air into the flask, yet eliminate the possibility that contamination is the explanation for the results? (For a ready-made answer, look up the experiments of Louis Pasteur.)

The Invention of the Light Microscope Led to the Discovery of Cells

The development of the light microscope depended on advances in the production of glass lenses. By the seventeenth century, lenses were refined to the point that they could be used to make simple microscopes. Using such an instrument, Robert Hooke examined a piece of cork and in 1665 reported to the Royal Society of London that the cork was composed of a mass of minute chambers, which he called "cells." The name "cell" stuck, even though the structures Hooke described were only the cell walls that remained after the living plant cells inside them had died. Later, Hooke and his Dutch contemporary Antoni van Leeuwenhoek were able to observe living cells, revealing a previously unseen world teeming with motile microscopic organisms.

For almost 200 years, the light microscope remained an exotic instrument, available only to a few wealthy individuals. It was not until the nineteenth century that it began to be widely used to look at cells. The emergence of cell biology as a distinct science was a gradual process to which many individuals contributed, but its official birth is generally said to have been signaled by two publications: one by the botanist Matthias Schleiden in 1838 and the other by the zoologist Theodor Schwann in 1839. In these papers, Schleiden and Schwann documented the results of a systematic investigation of plant and animal tissues with the light microscope, showing that cells were the universal building blocks of all

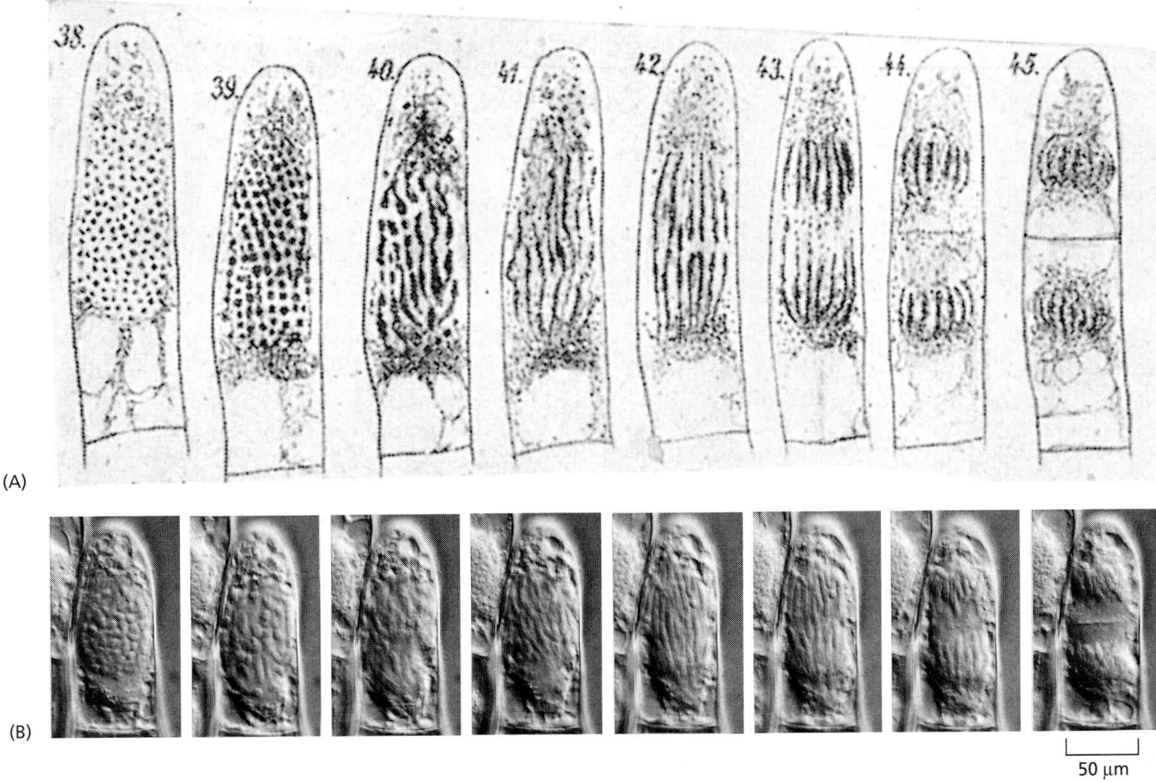

Figure 1–4 **New cells form by division of existing cells.** (A) In 1880, Eduard Strasburger drew a living plant cell (a hair cell from a *Tradescantia* flower), which he observed dividing into two daughter cells over a period of 2.5 hours. (B) A comparable living cell photographed recently through a modern light microscope. (B, courtesy of Peter Hepler.)

living tissues. Their work, and that of other nineteenth-century microscopists, slowly led to the realization that all living cells are formed by the division of existing cells—a principle sometimes referred to as the *cell theory* (Figure 1–4). The implication that living organisms do not arise spontaneously but can be generated only from existing organisms was hotly contested, but it was finally confirmed by experiments performed in the 1860s by Louis Pasteur.

The principle that cells are generated only from preexisting cells and inherit their characteristics from them underlies all of biology and gives the subject a unique flavor: in biology, questions about the present are inescapably linked to questions about the past. To understand why present-day cells and organisms behave as they do, we need to understand their history, all the way back to the misty origins of the first cells on Earth. Darwin's theory of evolution, published in 1859, provided the key insight that makes this history comprehensible, by showing how random variation and natural selection can drive the production of organisms with novel features, adapted to new ways of life. The theory of evolution explains how diversity has arisen among organisms that share a common ancestry. When combined with the cell theory, it leads us to a view of all life, from its beginnings to the present day, as one vast family tree of individual cells. Although this book is primarily about how cells work today, we shall encounter the theme of evolution again and again.

Cells, Organelles, and Even Molecules Can Be Seen Under the Microscope

If you cut a very thin slice of a suitable plant or animal tissue and place it under a light microscope, you will see that the tissue is divided into thousands of small cells. These may be either closely packed or separated from one another by an *extracellular matrix*, a dense material often made of protein fibers embedded in a polysaccharide gel (Figure 1–5). Each cell is typically about 5–20 μm in diameter (Figure 1–6). If you have taken care to keep your specimen under the right conditions, you will see that the

PANEL 1–1 Microscopy

THE LIGHT MICROSCOPE

The light microscope allows us to magnify cells up to 1000 times and to resolve details as small as 0.2 μm (a limitation imposed by the wavelike nature of light, not by the quality of the lenses). Three things are required for viewing cells in a light microscope. First, a bright light must be focused onto the specimen by lenses in the condenser. Second, the specimen must be carefully prepared to allow light to pass through it. Third, an appropriate set of lenses (objective and eyepiece) must be arranged to focus an image of the specimen in the eye.

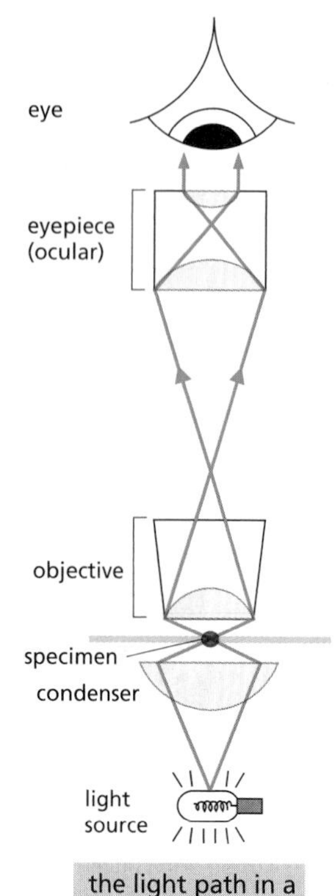

the light path in a light microscope

FLUORESCENCE MICROSCOPY

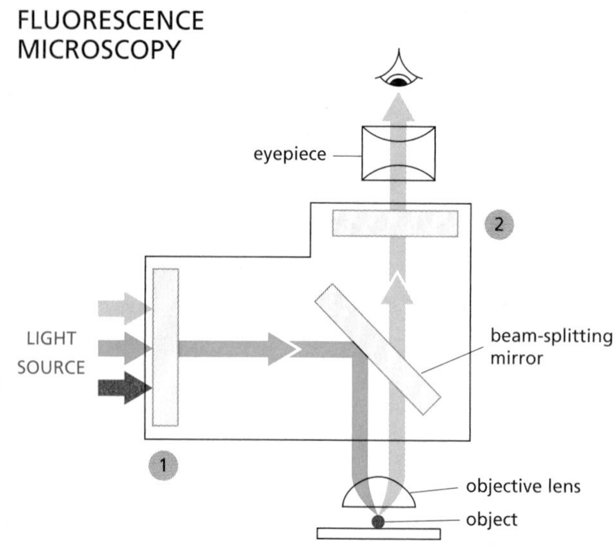

Fluorescent dyes used for staining cells are detected with the aid of a *fluorescence microscope*. This is similar to an ordinary light microscope except that the illuminating light is passed through two sets of filters. The first (1) filters the light before it reaches the specimen, passing only those wavelengths that excite the particular fluorescent dye. The second (2) blocks out this light and passes only those wavelengths emitted when the dye fluoresces. Dyed objects show up in bright color on a dark background.

LOOKING AT LIVING CELLS

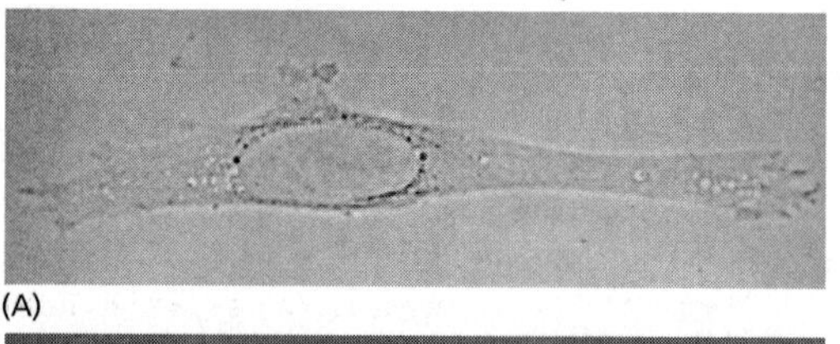

(A)

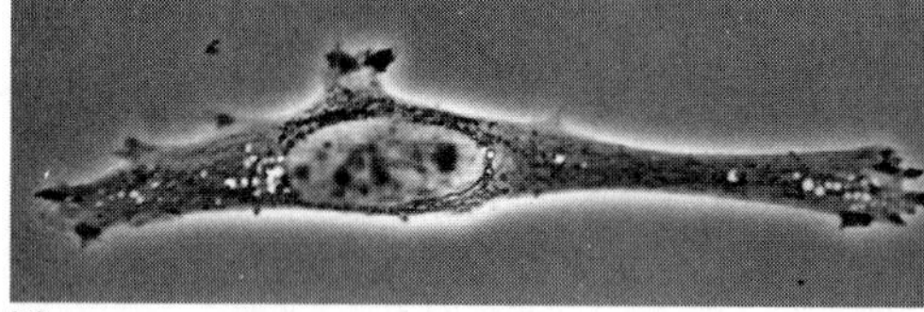

(B)

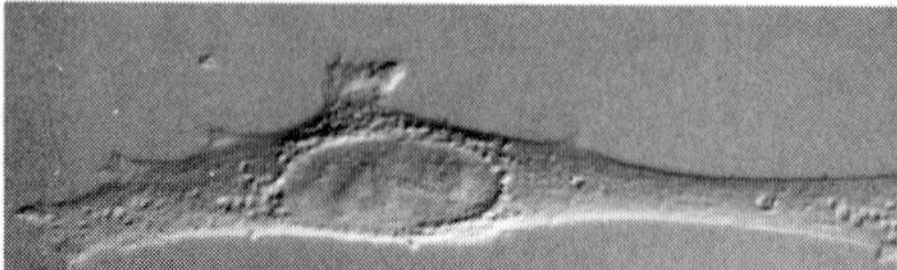

(C)

50 μm

The same unstained, living animal cell (fibroblast) in culture viewed with (A) straightforward (bright-field) optics; (B) phase-contrast optics; (C) interference-contrast optics. The two latter systems exploit differences in the way light travels through regions of the cell with differing refractive indexes. All three images can be obtained on the same microscope simply by interchanging optical components.

FIXED SAMPLES

Most tissues are neither small enough nor transparent enough to examine directly in the microscope. Typically, therefore, they are chemically fixed and cut into very thin slices, or *sections*, that can be mounted on a glass microscope slide and subsequently stained to reveal different components of the cells. A stained section of a plant root tip is shown here (D). (Courtesy of Catherine Kidner.)

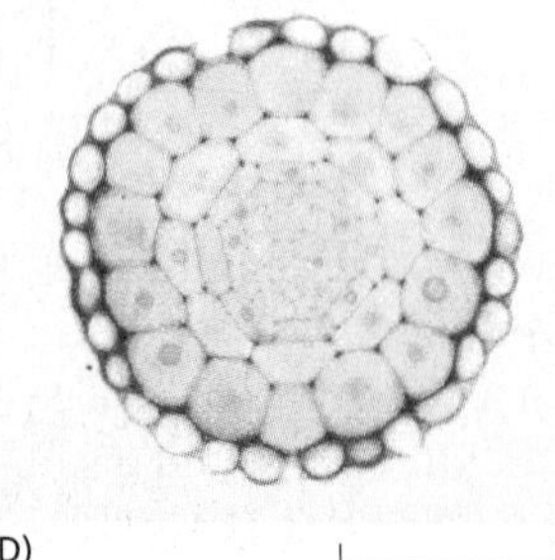

(D)

50 μm

FLUORESCENT PROBES

Dividing nuclei in a fly embryo seen with a fluorescence microscope after staining with specific fluorescent dyes.

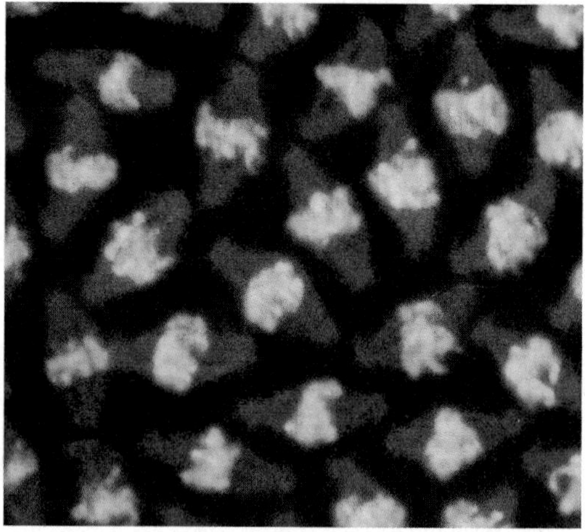

Fluorescent dyes absorb light at one wavelength and emit it at another, longer wavelength. Some such dyes bind specifically to particular molecules in cells and can reveal their location when examined with a fluorescence microscope. An example is the stain for DNA shown here (*green*). Other dyes can be coupled to antibody molecules, which then serve as highly specific and versatile staining reagents that bind selectively to particular macromolecules, allowing us to see their distribution in the cell. In the example shown, a microtubule protein in the mitotic spindle is stained *red* with a fluorescent antibody. (Courtesy of William Sullivan.)

CONFOCAL MICROSCOPY

A confocal microscope is a specialized type of fluorescence microscope that builds up an image by scanning the specimen with a laser beam. The beam is focused onto a single point at a specific depth in the specimen, and a pinhole aperture in the detector allows only fluorescence emitted from this same point to be included in the image. Scanning the beam across the specimen generates a sharp image of the plane of focus—an *optical* section. A series of optical sections at different depths allows a three-dimensional image to be constructed. An intact insect embryo is shown here stained with a fluorescent probe for actin (a filamentous protein). (A) Conventional fluorescence microscopy gives a blurry image due to the presence of fluorescent structures above and below the plane of focus. (B) Confocal microscopy provides an optical section showing the individual cells clearly. (Courtesy of Richard Warn and Peter Shaw.)

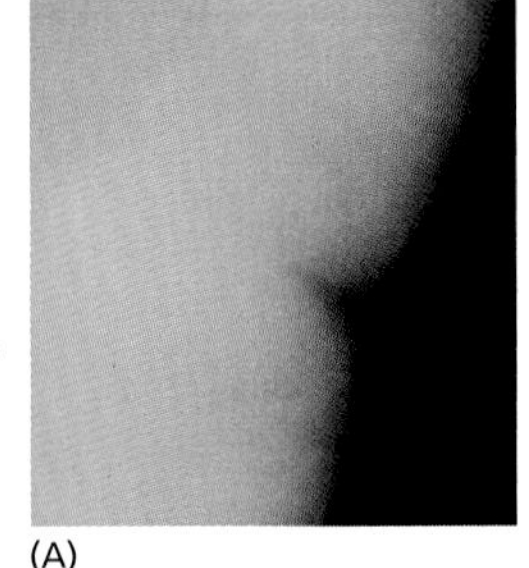

(A)

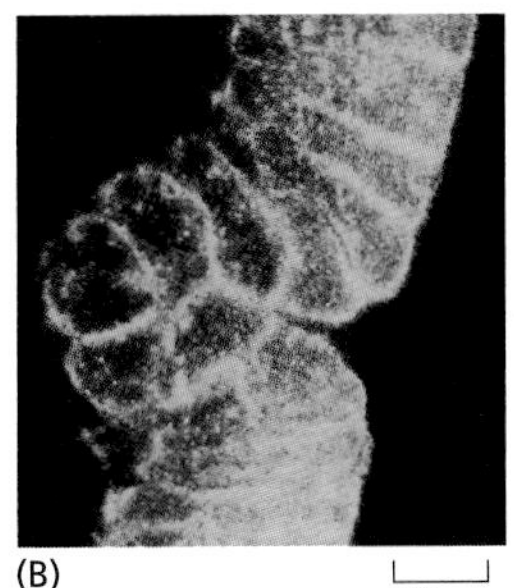

(B)

10 μm

TRANSMISSION ELECTRON MICROSCOPY

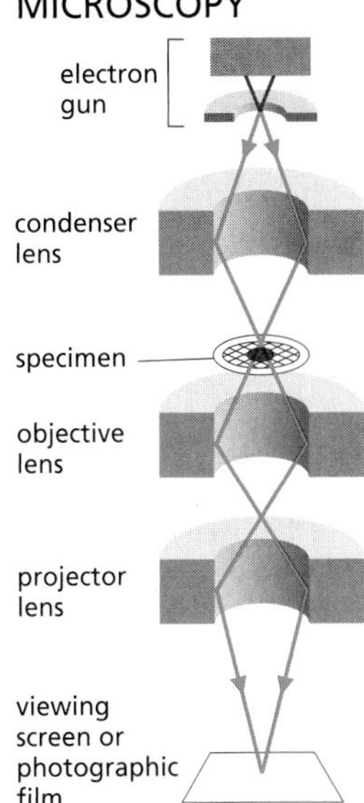

Courtesy of Philips Electron Optics, with permission from FEI Co.

The electron micrograph below shows a small region of a cell in a piece of testis. The tissue has been chemically fixed, embedded in plastic, and cut into very thin sections that have then been stained with salts of uranium and lead. (Courtesy of Daniel S. Friend.)

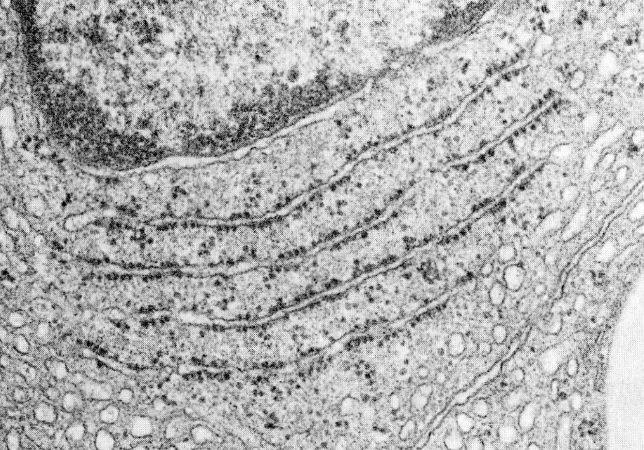

0.5 μm

The transmission electron microscope (TEM) is in principle similar to a light microscope, but it uses a beam of electrons instead of a beam of light, and magnetic coils to focus the beam instead of glass lenses. The specimen, which is placed in a vacuum, must be very thin. Contrast is usually introduced by staining the specimen with electron-dense heavy-metals that locally absorb or scatter electrons, removing them from the beam as it passes through the specimen. The TEM has a useful magnification of up to a million-fold and with biological specimens can resolve details as small as about 2 nm.

SCANNING ELECTRON MICROSCOPY

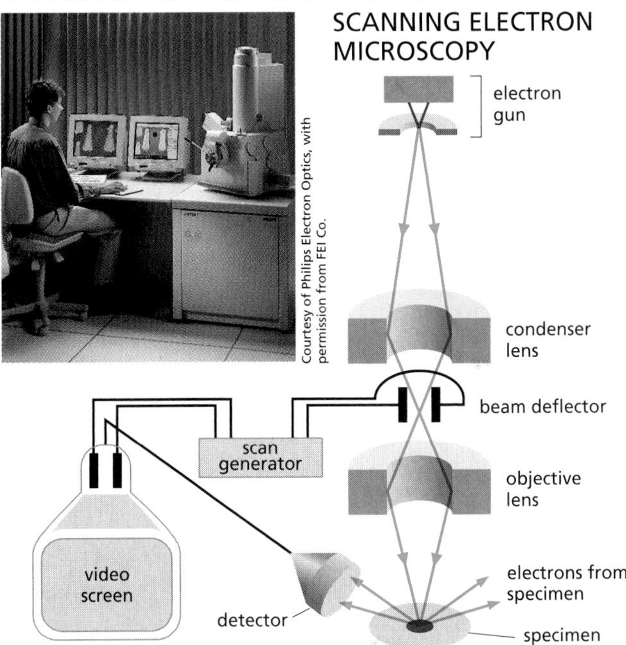

Courtesy of Philips Electron Optics, with permission from FEI Co.

In the scanning electron microscope (SEM) the specimen, which has been coated with a very thin film of a heavy metal, is scanned by a beam of electrons brought to a focus on the specimen by the electromagnetic coils that, in electron microscopes, act as lenses. The quantity of electrons scattered or emitted as the beam bombards each successive point on the surface of the specimen is measured by the detector, and is used to control the intensity of successive points in an image built up on a video screen. The microscope creates striking images of three-dimensional objects with great depth of focus and can resolve details down to somewhere between 3 nm and 20 nm, depending on the instrument.

1 μm

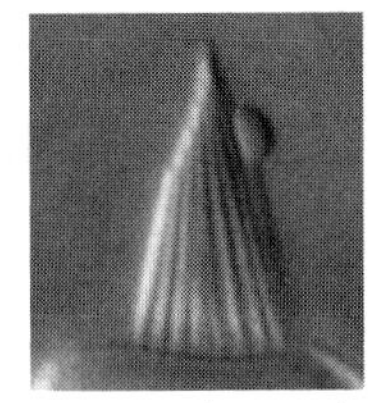

5 μm

Scanning electron micrograph of the stereocilia projecting from a hair cell in the inner ear (*left*). For comparison, the same structure is shown by light microscopy, at the limit of its resolution (*above*). (Courtesy of Richard Jacobs and James Hudspeth.)

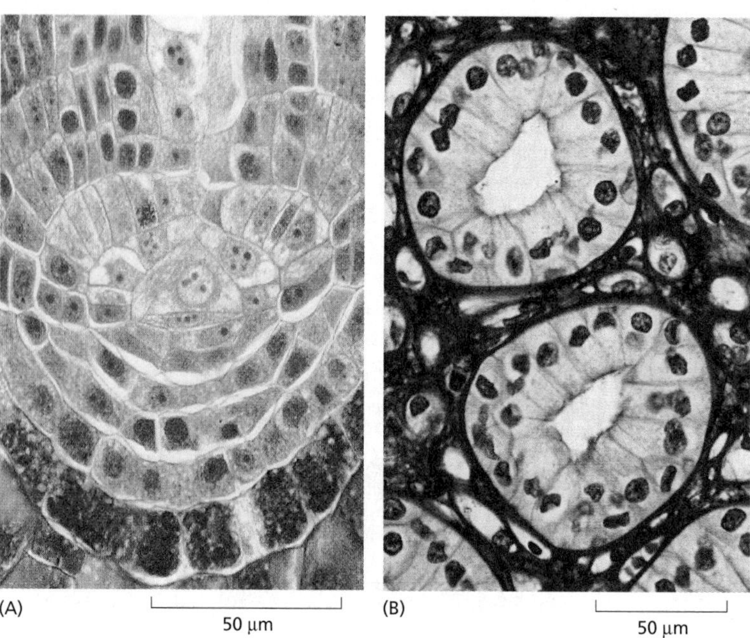

Figure 1–5 **Cells form tissues in plants and animals.** (A) Cells in the root tip of a fern. The nuclei are stained *red* and each cell is surrounded by a thin cell wall *(blue)*. (B) Cells in the urine-collecting ducts of the kidney. Each duct appears in this cross section as a ring of closely packed cells (with nuclei stained *red*). The ring is surrounded by extracellular matrix, stained *purple*. (A, courtesy of James Mauseth; B, from P.R. Wheater et al., Functional Histology, 2nd ed. Edinburgh: Churchill Livingstone, 1987. With permission from Elsevier.)

cells show signs of life: particles move around inside them, and if you watch patiently you may see a cell slowly change shape and divide into two (see Figure 1–4 and the speeded-up video of cell division in a frog embryo in Movie 1.1).

To see the internal structure of a cell is difficult, not only because the parts are small but also because they are transparent and mostly colorless. One approach is to stain cells with dyes that color particular components differently (see Figure 1–5). Alternatively, one can exploit the fact that cell components differ slightly from one another in refractive index, just as glass differs in refractive index from water, causing light rays to be deflected as they pass from the one medium into the other. The small differences in refractive index can be made visible by specialized optical techniques, and the resulting images can be enhanced further by electronic processing (see Panel 1–1, pp. 8–9).

The cell thus revealed has a distinct anatomy (Figure 1–7). It has a sharply defined boundary, indicating the presence of an enclosing membrane. A large, round body, the *nucleus*, is prominent in the middle of the cell. Around the nucleus and filling the cell's interior lies the **cytoplasm**, a transparent substance crammed with what seems at first to be a jumble of miscellaneous objects. With a good light microscope, one can begin to distinguish and classify specific components in the cytoplasm (Figure 1–7B). However, structures smaller than about 0.2 μm—about half the wavelength of visible light—cannot be resolved with a conventional light microscope (points closer than this are not distinguishable and appear as a single blur).

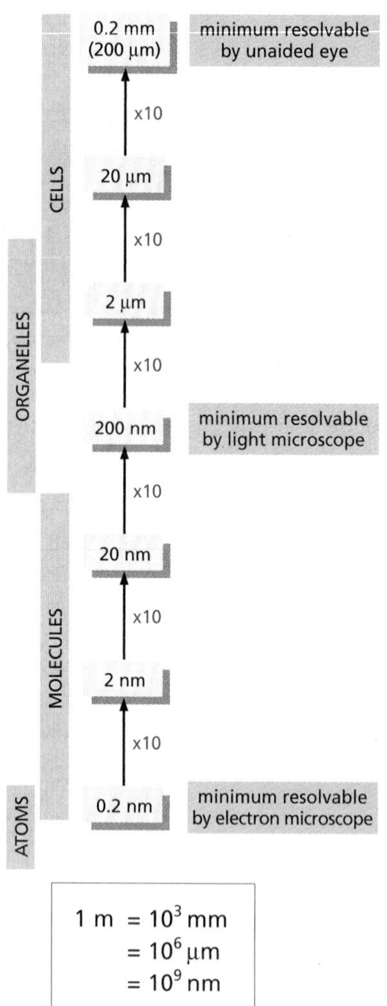

Figure 1–6 **What can we see?** This schematic diagram shows the sizes of cells and of their component parts, as well as the units in which they are measured.

In the past few years, new types of fluorescence microscopes have been developed, using sophisticated methods of illumination and image analysis to see detail several times finer than this. However, for the highest magnification and the best resolution, one must turn to an electron microscope, which can reveal details down to a few **nanometers**, or nm (see Figure 1–6). Cell samples for the electron microscope require painstaking preparation. Even for light microscopy, a tissue usually has to be *fixed* (that is, preserved by pickling in a reactive chemical solution), supported by *embedding* in a solid wax or resin, cut or *sectioned* into thin slices, and *stained* before it is viewed. For electron microscopy, similar procedures are required, but the sections have to be much thinner and there is no possibility of looking at living, wet cells.

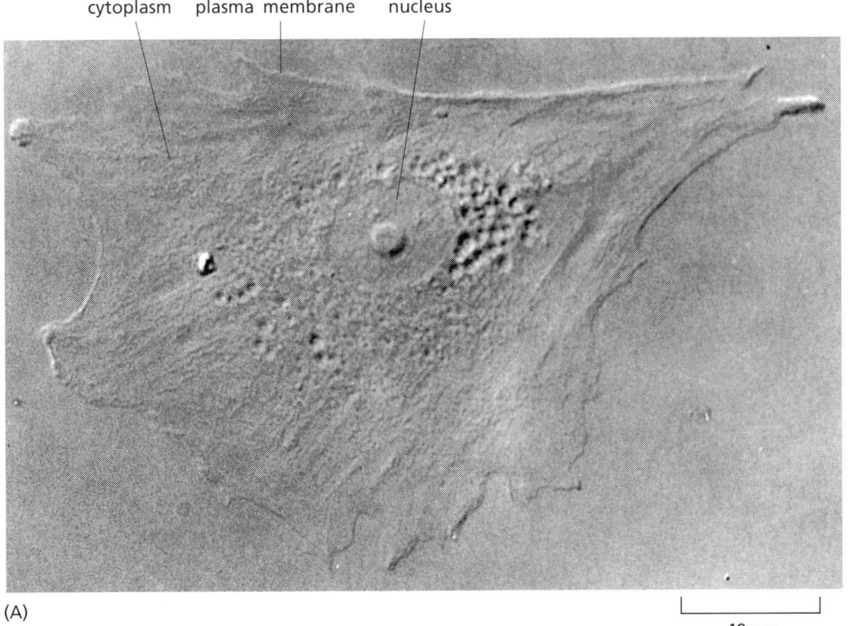

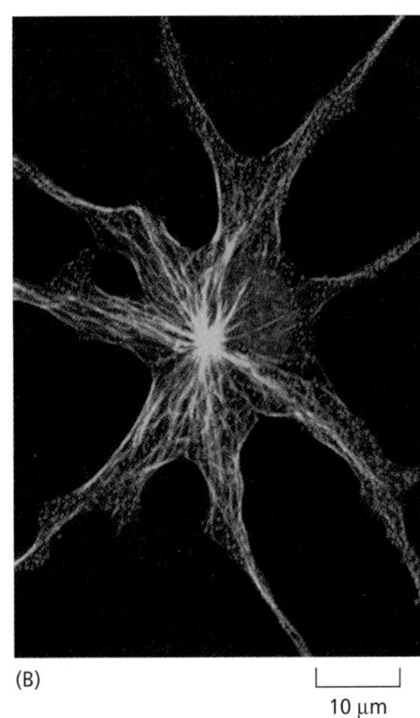

Figure 1–7 **The internal structures of a living cell can be seen with a light microscope.** (A) A cell taken from human skin and grown in tissue culture was photographed through a light microscope using interference-contrast optics (see Panel 1–1, pp. 8–9). The nucleus is especially prominent. (B) A pigment cell from a frog, stained with fluorescent dyes and viewed with a confocal microscope (see Panel 1–1). The nucleus is shown in *blue*, the pigment granules in *red*, and the microtubules—a class of filaments built from protein molecules in the cytoplasm—in *green*. (A, courtesy of Casey Cunningham; B, courtesy of Steve Rogers and the Imaging Technology Group.)

When thin sections are cut, stained, and placed in the electron microscope, much of the jumble of cell components becomes sharply resolved into distinct *organelles*—separate, recognizable substructures that are only hazily defined under the light microscope. A delicate membrane, about 5 nm thick, is visible enclosing the cell, and similar membranes form the boundary of many of the organelles inside (Figure 1–8A, B). The external membrane is called the *plasma membrane*, while the membranes surrounding organelles are called *internal membranes*. With an electron microscope, even some of the large molecules in a cell can be seen individually (Figure 1–8C).

The type of electron microscope used to look at thin sections of tissue is known as a *transmission electron microscope*. This is, in principle, similar to a light microscope, except that it transmits a beam of electrons rather than a beam of light through the sample. Another type of electron microscope—the *scanning electron microscope*—scatters electrons off the surface of the sample and so is used to look at the surface detail of cells and other structures (see Panel 1–1, pp. 8–9). Electron microscopy enables biologists to see the structure of biological membranes, which are only two molecules thick (discussed in detail in Chapter 11). Even the most powerful electron microscopes, however, cannot reveal the individual atoms that make up molecules (Figure 1–9).

The microscope is not the only tool that modern biologists use to study the details of cell components. Techniques such as X-ray crystallography, for example, can be used to determine the three-dimensional structure of protein molecules (discussed in Chapter 4). We shall describe other methods for probing the inner workings of cells as they arise throughout the book.

THE PROCARYOTIC CELL

Of all the types of cells revealed by the microscope, *bacteria* have the simplest structure and come closest to showing us life stripped down to its essentials. Indeed, a bacterium contains essentially no organelles—not even a nucleus to hold its DNA. This property—the presence or absence of a nucleus—is used as the basis for a simple but fundamental classification

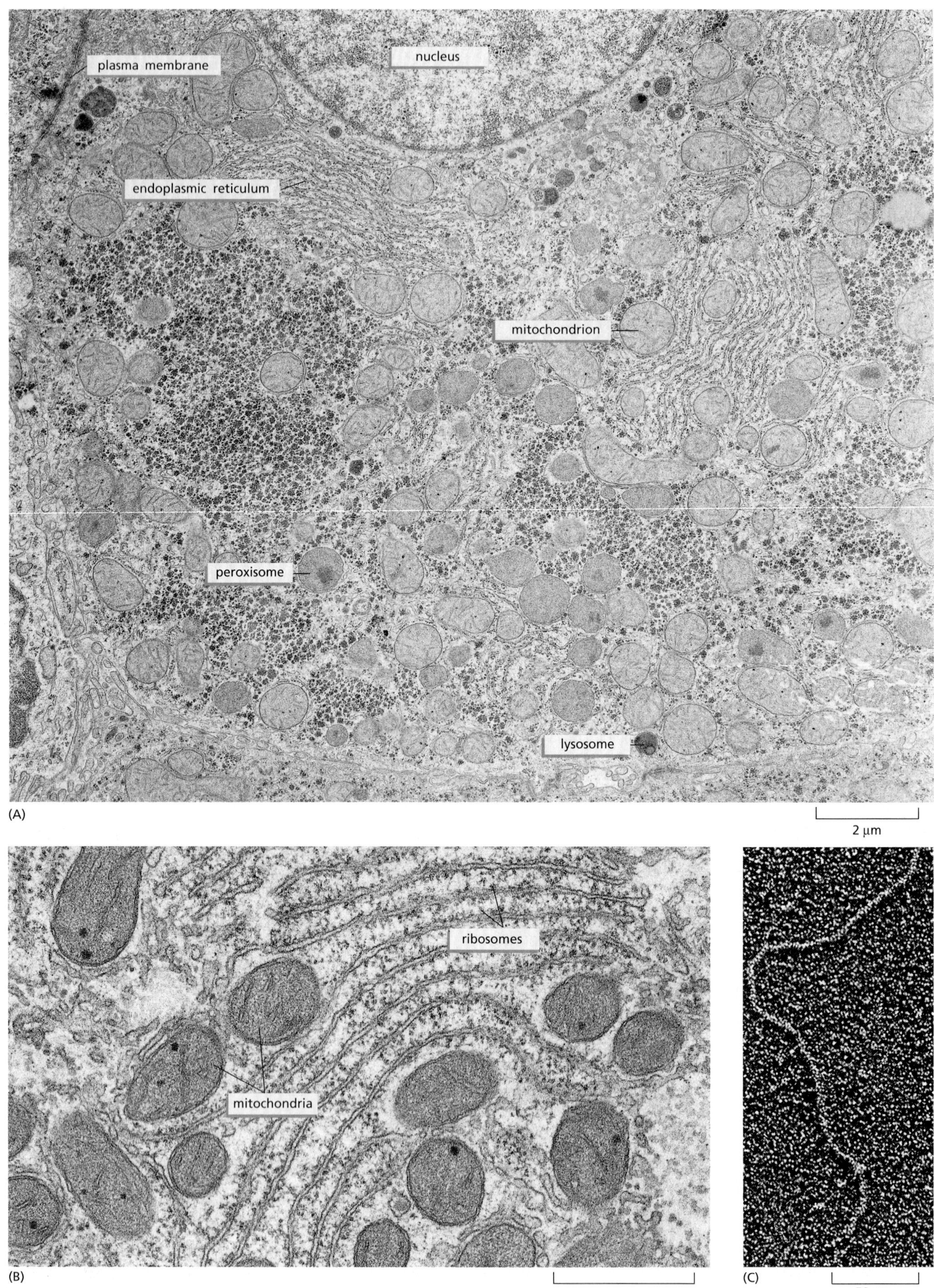
plasma membrane
nucleus
endoplasmic reticulum
mitochondrion
peroxisome
lysosome
(A)
2 μm
ribosomes
mitochondria
(B)
2 μm
(C)
50 nm

Figure 1–8 ***(opposite page)*** **The fine structure of a cell can be seen in a transmission electron microscope.** (A) Thin section of a liver cell showing the enormous amount of detail that is visible. Some of the components to be discussed later in the chapter are labeled; they are identifiable by their size and shape. (B) A small region of the cytoplasm at higher magnification. The smallest structures that are clearly visible are the ribosomes, each of which is made of 80–90 or so individual large molecules. (C) Portion of a long, threadlike DNA molecule isolated from a cell and viewed by electron microscopy. (A and B, courtesy of Daniel S. Friend; C, courtesy of Mei Lie Wong.)

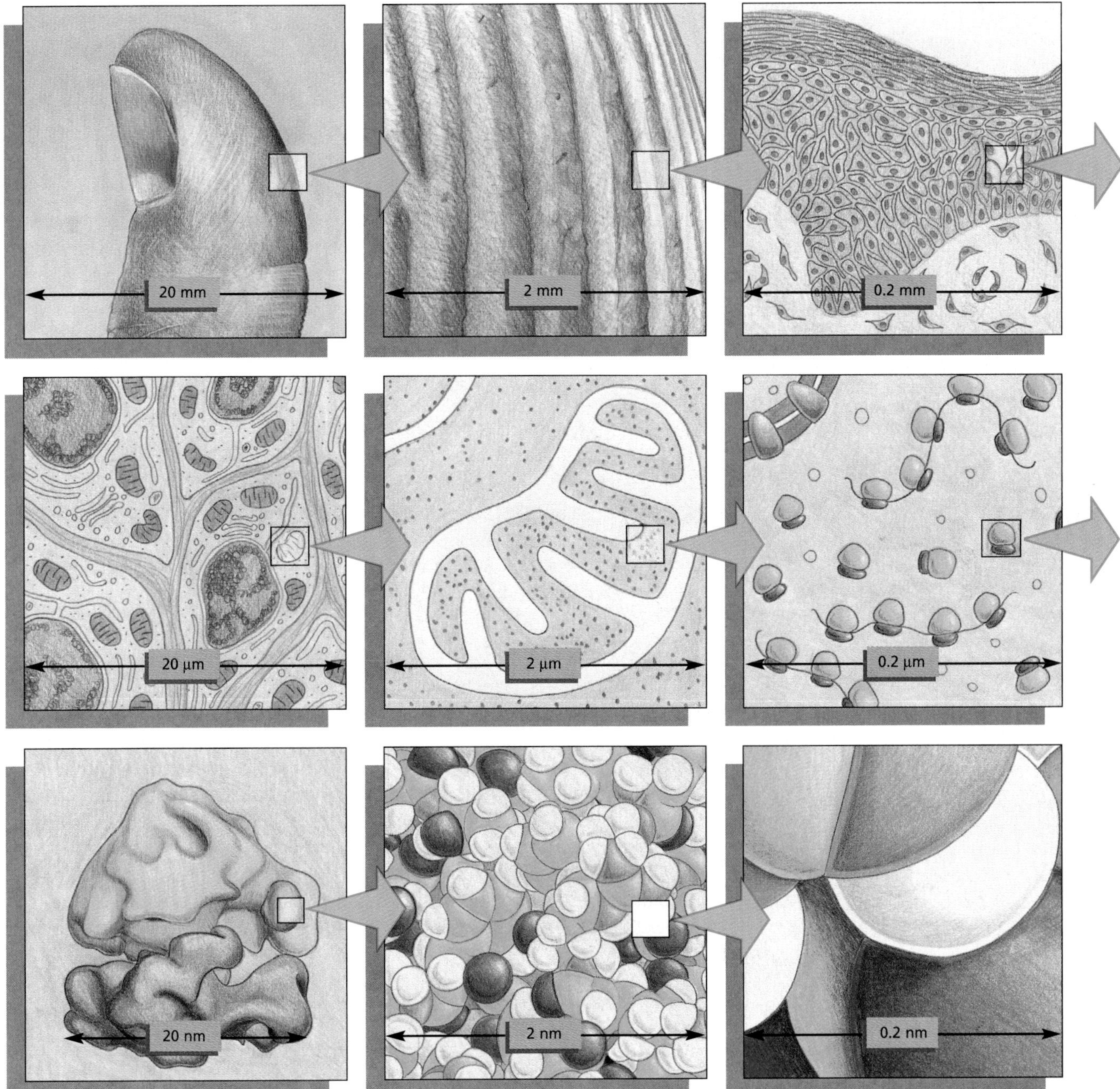

Figure 1–9 **How big is a cell and how big are its parts?** These drawings convey a sense of scale between living cells and atoms. Each panel shows an image that is then magnified by a factor of 10 in an imaginary progression from a thumb, to skin, to skin cells, to a mitochondrion, to a ribosome, and ultimately to a cluster of atoms forming part of one of the many protein molecules in our bodies. Details of molecular structure, as shown in the last two panels, are beyond the power of the electron microscope.

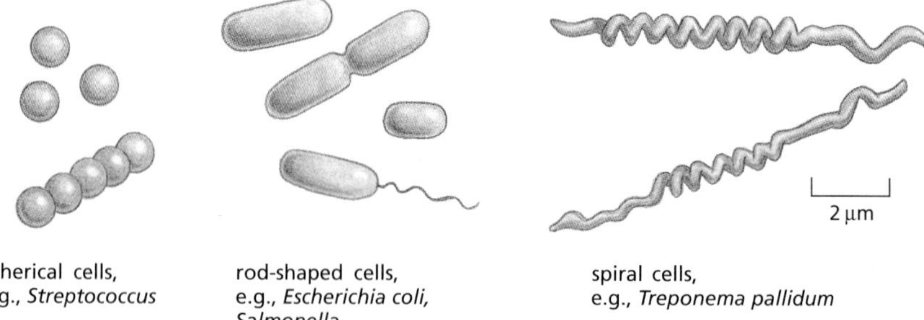

Figure 1–10 **Bacteria come in different shapes and sizes.** Typical spherical, rodlike, and spiral-shaped bacteria are drawn to scale. The spiral cells shown are the organisms that cause syphilis.

of all living things. Organisms whose cells have a nucleus are called **eucaryotes** (from the Greek words *eu*, meaning "well" or "truly," and *karyon*, a "kernel" or "nucleus"). Organisms whose cells do not have a nucleus are called **procaryotes** (from *pro*, meaning "before"). The terms "bacterium" and "procaryote" are often used interchangeably, although we shall see that the category of procaryotes also includes another class of cells, the *archaea* (singular archaeon), which are so remotely related to bacteria that they are given a separate name.

Procaryotes are typically spherical, rodlike, or corkscrew-shaped, and small—just a few micrometers long (Figure 1–10), although there are some giant species as much as 100 times longer than this. They often have a tough protective coat, called a cell wall, surrounding the plasma membrane, which encloses a single compartment containing the cytoplasm and the DNA. In the electron microscope, the cell interior typically appears as a matrix of varying texture without any obvious organized internal structure (Figure 1–11). The cells reproduce quickly by dividing in two. Under optimum conditions, when food is plentiful, a procaryotic cell can duplicate itself in as little as 20 minutes. In 11 hours, by repeated divisions, a single procaryote can give rise to more than 8 billion progeny (which exceeds the total number of humans presently on Earth). Thanks to their large numbers, rapid growth rates, and ability to exchange bits of genetic material by a process akin to sex, populations of procaryotic cells can evolve fast, rapidly acquiring the ability to use a new food source or to resist being killed by a new antibiotic.

QUESTION 1–4

A bacterium weighs about 10^{-12} g and can divide every 20 minutes. If a single bacterial cell carried on dividing at this rate, how long would it take before the mass of bacteria would equal that of the Earth (6×10^{24} kg)? Contrast your result with the fact that bacteria originated at least 3.5 billion years ago and have been dividing ever since. Explain the apparent paradox. (The number of cells N in a culture at time t is described by the equation $N = N_0 \times 2^{t/G}$, where N_0 is the number of cells at zero time and G is the population doubling time.)

Procaryotes Are the Most Diverse of Cells

Most procaryotes live as single-celled organisms, although some join together to form chains, clusters, or other organized multicellular structures. In shape and structure, procaryotes may seem simple and limited, but in terms of chemistry, they are the most diverse and inventive class of cells. These creatures exploit an enormous range of habitats, from hot puddles of volcanic mud to the interiors of other living cells, and they vastly outnumber other living organisms on Earth. Some are aerobic, using oxygen to oxidize food molecules; some are strictly anaerobic and are killed by the slightest exposure to oxygen. As we discuss later in this chapter, *mitochondria*—the organelles that generate energy for the

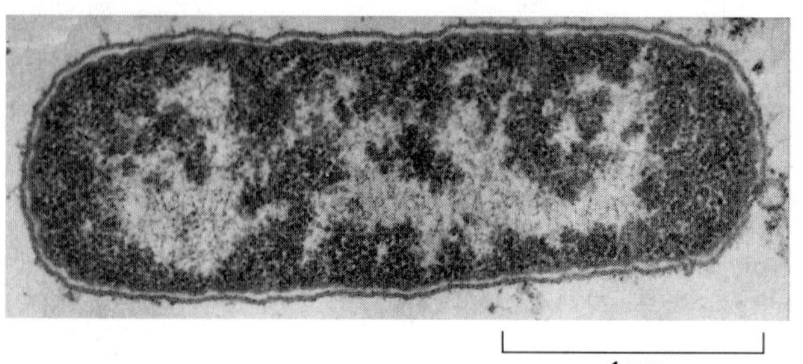

Figure 1–11 **The bacterium *Escherichia coli (E. coli)* is understood more thoroughly than any other living organism.** An electron micrograph of a longitudinal section is shown here; the cell's DNA is concentrated in the lightly stained region. (Courtesy of E. Kellenberger.)

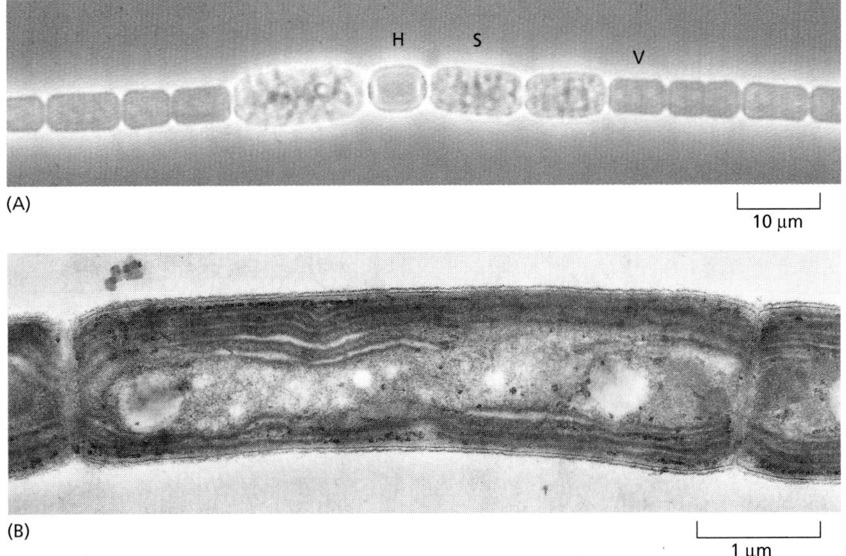

Figure 1–12 **Some bacteria are photosynthetic.** (A) *Anabaena cylindrica* forms long, multicellular filaments. This light micrograph shows specialized cells that either fix nitrogen (that is, capture N_2 from the atmosphere and incorporate it into organic compounds; labeled *H*), fix CO_2 (through photosynthesis; *V*), or become resistant spores (*S*). (B) An electron micrograph of a related species, *Phormidium laminosum*, shows the intracellular membranes where photosynthesis occurs. These micrographs illustrate that even some procaryotes can form simple multicellular organisms. (A, courtesy of David Adams; B, courtesy of D.P. Hill and C.J. Howe.)

eucaryotic cell—are thought to have evolved from aerobic bacteria that took to living inside the anaerobic ancestors of today's eucaryotic cells. Thus our own oxygen-based metabolism can be regarded as a product of the activities of bacterial cells.

Virtually any organic material, from wood to petroleum, can be used as food by one sort of bacterium or another. Even more remarkably, some procaryotes can live entirely on inorganic substances: they can get their carbon from CO_2 in the atmosphere, their nitrogen from atmospheric N_2, and their oxygen, hydrogen, sulfur, and phosphorus from air, water, and inorganic minerals. Some of these procaryotic cells, like plant cells, perform *photosynthesis*, getting energy from sunlight (Figure 1–12); others derive energy from the chemical reactivity of inorganic substances in the environment (Figure 1–13). In either case, such procaryotes play a unique and fundamental part in the economy of life on Earth: other living things depend on the organic compounds that these cells generate from inorganic materials.

Plants, too, can capture energy from sunlight and carbon from atmospheric CO_2. But plants unaided by bacteria cannot capture N_2 from the atmosphere, and in a sense even plants depend on bacteria for photosynthesis. It is almost certain that the organelles in the plant cell that perform photosynthesis—the *chloroplasts*—have evolved from photosynthetic bacteria that found a home inside the plant cell's cytoplasm.

The World of Procaryotes Is Divided into Two Domains: Bacteria and Archaea

Traditionally, all procaryotes have been classified together in one large group. But molecular studies reveal that there is a gulf within the class of procaryotes, dividing it into two distinct *domains* called the **bacteria** (or sometimes *eubacteria*) and the **archaea**. Remarkably, at a molecular level, the members of these two domains differ as much from one another as either does from the eucaryotes. Most of the procaryotes familiar from everyday life—the species that live in the soil or make us ill—are bacteria. Archaea are found not only in these habitats, but also in environments hostile to most other cells: there are species that live in concentrated brine, in hot acid volcanic springs, in the airless depths of marine sediments, in the sludge of sewage treatment plants, in pools beneath the frozen surface of Antarctica, and in the acidic, oxygen-free

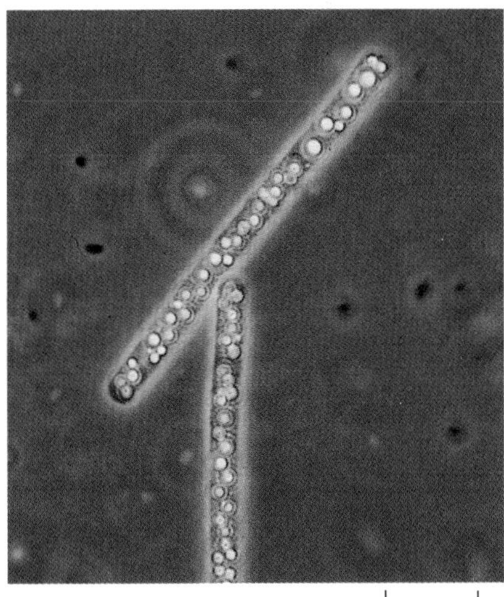

Figure 1–13 **A sulfur bacterium gets its energy from H_2S.** *Beggiatoa*, a procaryote that lives in sulfurous environments, oxidizes H_2S to produce sulfur and can fix carbon even in the dark. In this light micrograph, yellow deposits of sulfur can be seen inside the cells. (Courtesy of Ralph W. Wolfe.)

environment of a cow's stomach, where they break down cellulose and generate methane gas. Many of these environments resemble the harsh conditions that must have existed on the primitive Earth, where living things first evolved, before the atmosphere became rich in oxygen.

THE EUCARYOTIC CELL

Eucaryotic cells, in general, are bigger and more elaborate than bacteria and archaea. Some live independent lives as single-celled organisms, such as amoebae and yeasts (Figure 1–14); others live in multicellular assemblies. All of the more complex multicellular organisms—including plants, animals, and fungi—are formed from eucaryotic cells.

By definition, all eucaryotic cells have a nucleus. But possession of a nucleus goes hand-in-hand with possession of a variety of other organelles, subcellular structures that perform specialized functions. Most of these are likewise common to all eucaryotic organisms. We will now take a look at the main organelles found in eucaryotic cells from the point of view of their functions.

The Nucleus Is the Information Store of the Cell

The **nucleus** is usually the most prominent organelle in a eucaryotic cell (Figure 1–15). It is enclosed within two concentric membranes that form

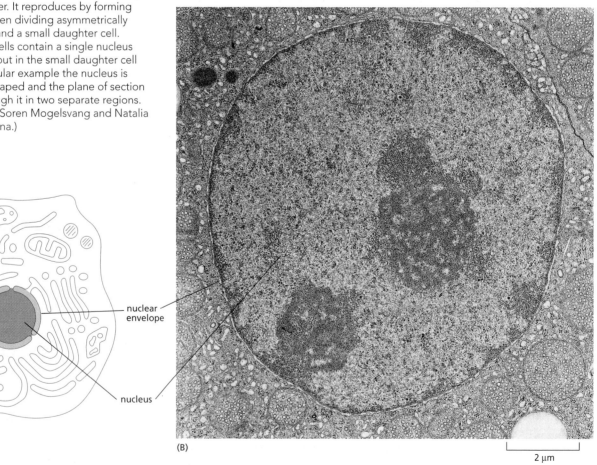

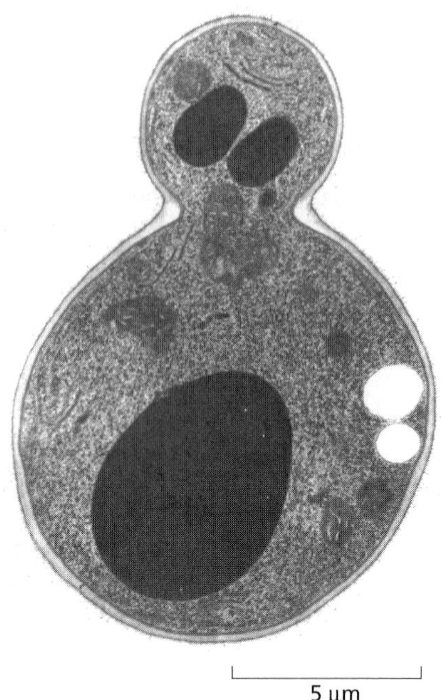

Figure 1–14 **Yeasts are simple free-living eucaryotes.** The cell shown in this electron micrograph belongs to the same species, *Saccharomyces cerevisiae*, that makes dough rise and turns malted barley juice into beer. It reproduces by forming a bud and then dividing asymmetrically into a large and a small daughter cell. Both these cells contain a single nucleus (dark stain); but in the small daughter cell in this particular example the nucleus is irregularly shaped and the plane of section has cut through it in two separate regions. (Courtesy of Soren Mogelsvang and Natalia Gomez-Ospina.)

(A) nuclear envelope, nucleus (B) 2 μm

Figure 1–15 **The nucleus contains most of the DNA in a eucaryotic cell.** (A) In this drawing of a typical animal cell—complete with its extensive system of membrane-enclosed organelles—the nucleus is colored *brown*, the nuclear envelope is *green*, and the cytoplasm (the interior of the cell outside the nucleus) is *white*. (B) An electron micrograph of a nucleus in a mammalian cell. Individual chromosomes are not visible because the DNA is dispersed as fine threads throughout the nucleus at this stage of the cell's growth. (B, courtesy of Daniel S. Friend.)

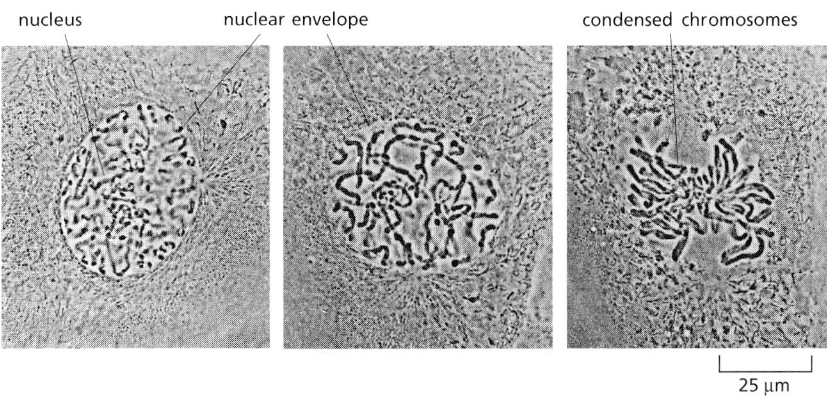

Figure 1–16 **Chromosomes become visible when a cell is about to divide.** As a eucaryotic cell prepares to divide, its DNA becomes compacted or condensed into threadlike chromosomes that can be distinguished in the light microscope. The photographs show three successive steps in this process in a cultured cell from a newt's lung. (Courtesy of Conly L. Rieder.)

the *nuclear envelope*, and it contains molecules of DNA—extremely long polymers that encode the genetic information of the organism. In the light microscope, these giant DNA molecules become visible as individual **chromosomes** when they become more compact as a cell prepares to divide into two daughter cells (Figure 1–16). DNA also stores the genetic information in procaryotic cells; these cells lack a distinct nucleus not because they lack DNA, but because they do not keep their DNA inside a nuclear envelope, segregated from the rest of the cell contents.

Mitochondria Generate Usable Energy from Food to Power the Cell

Mitochondria are present in essentially all eucaryotic cells, and they are among the most conspicuous organelles in the cytoplasm (Figure 1–17). These organelles have a very distinctive structure when seen with an electron microscope: each mitochondrion appears sausage- or worm-shaped, from one to many micrometers long; and each is enclosed in two separate membranes. The inner membrane is formed into folds that project into the interior of the mitochondrion (Figure 1–18). Mitochondria contain their own DNA and reproduce by dividing in two. Because mitochondria resemble bacteria in so many ways, they are thought to derive from bacteria that were engulfed by some ancestor of present-day eucaryotic cells (Figure 1–19). This evidently created a *symbiotic* relationship in which the host eucaryote and the engulfed bacterium helped one another to survive and reproduce.

Observation under the microscope by itself gives little indication of what mitochondria do. Their function was discovered by breaking open cells and then spinning the soup of cell fragments in a centrifuge; this separates the organelles according to their size, shape, and density. Purified mitochondria were then tested to see what chemical processes they could perform. This revealed that mitochondria are generators of chemical energy for the cell. They harness the energy from the oxidation of food molecules, such as sugars, to produce *adenosine triphosphate*, or ATP—the basic chemical fuel that powers most of the cell's activities. Because the mitochondrion consumes oxygen and releases carbon dioxide in the course of this activity, the entire process is called *cellular respiration*—essentially, breathing on a cellular level. The process of cellular respiration is considered in detail in Chapter 14.

Without mitochondria, animals, fungi, and plants would be unable to use oxygen to extract the maximum amount of energy from the food molecules that nourish them. Oxygen would be a poison for them, rather than an essential requirement—that is, they would be *anaerobic*. Many procaryotes are anaerobic, and there are even a few anaerobic eucaryotes, such as the intestinal parasite *Giardia*, that lack mitochondria and live only in environments that are low in oxygen.

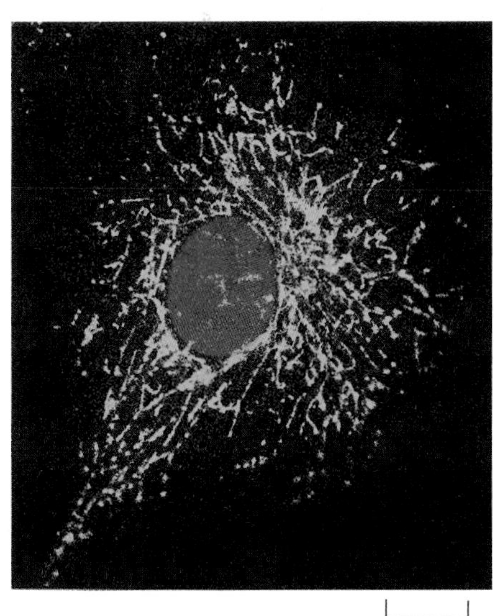

Figure 1–17 **Mitochondria can be variable in shape.** In this light micrograph of a cultured mammalian cell, the mitochondria are stained *green* with a fluorescent dye and appear wormlike. The nucleus is stained *blue*. Mitochondria are power generators that oxidize food molecules to produce useful chemical energy in almost all eucaryotic cells. (Courtesy of Lan Bo Chen.)

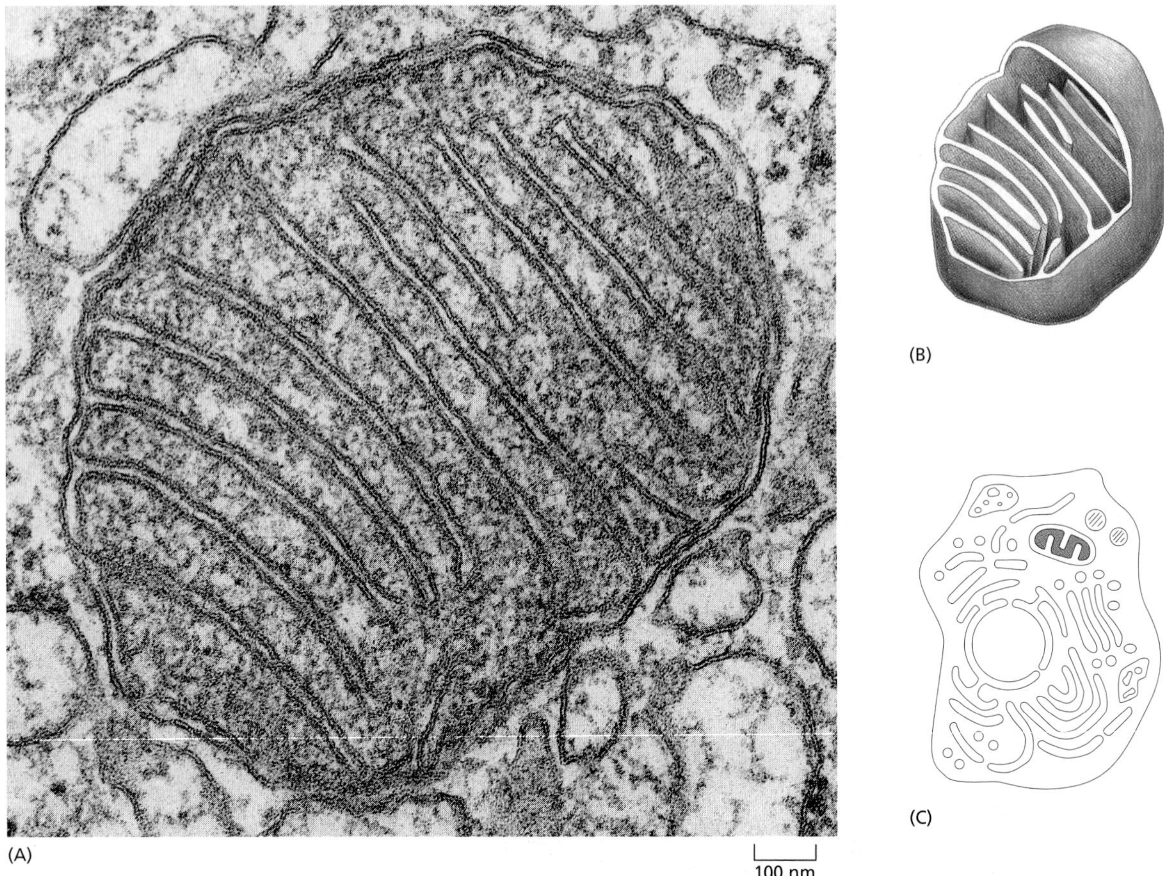

Figure 1–18 **Mitochondria have a distinctive structure.** (A) An electron micrograph of a cross-section of a mitochondrion reveals the extensive folding of the inner membrane. (B) This three-dimensional representation of the arrangement of the mitochondrial membranes shows the smooth outer membrane and the highly convoluted inner membrane. The inner membrane contains most of the proteins responsible for cellular respiration, and it is highly folded to provide a large surface area for this activity. (C) In this schematic cell, the interior space of the mitochondrion is colored. (A, courtesy of Daniel S. Friend.)

Chloroplasts Capture Energy from Sunlight

Chloroplasts are large green organelles that are found only in the cells of plants and algae, not in the cells of animals or fungi. These organelles have an even more complex structure than mitochondria: in addition to their two surrounding membranes, chloroplasts possess internal stacks of membranes containing the green pigment *chlorophyll* (Figure 1–20). When a plant is kept in the dark, its greenness fades; when put back in the light, its greenness returns. This suggests that the chlorophyll—and the chloroplasts that contain it—are crucial to the special relationship that plants and algae have with light. But what is that relationship?

Animals and plants all need energy to live, grow, and reproduce. Animals can use only the chemical energy they obtain by feeding on the products of other living things. But plants can get their energy directly from

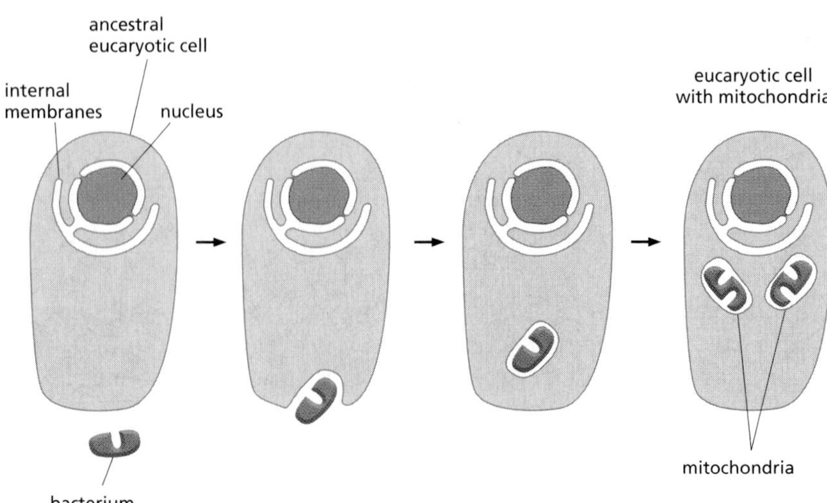

Figure 1–19 **Mitochondria most likely evolved from engulfed bacteria.** It is virtually certain that mitochondria originate from bacteria that were engulfed by an ancestral eucaryotic cell and survived inside it, living in symbiosis with their host.

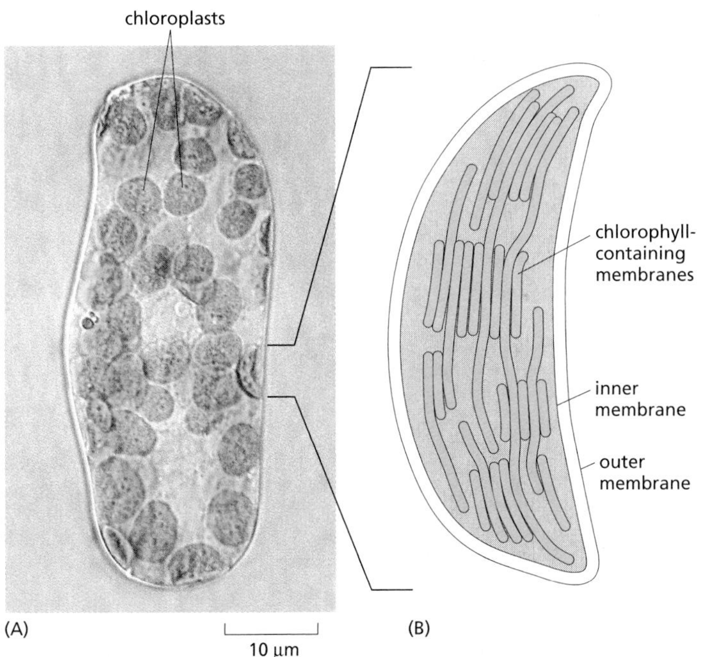

Figure 1–20 **Chloroplasts in plant cells capture the energy of sunlight.** (A) A single cell isolated from a leaf of a flowering plant, seen in the light microscope, showing many green chloroplasts. (B) A drawing of one of the chloroplasts, showing the highly folded system of internal membranes containing the green chlorophyll molecules that absorb light energy. (A, courtesy of Preeti Dahiya.)

sunlight, and chloroplasts are the organelles that enable them to do so. From the standpoint of life on Earth, chloroplasts carry out an even more essential task than mitochondria: they perform photosynthesis—that is, they trap the energy of sunlight in chlorophyll molecules and use this energy to drive the manufacture of energy-rich sugar molecules. In the process they release oxygen as a molecular by-product. Plant cells can then extract this stored chemical energy when they need it, by oxidizing these sugars in their mitochondria, just as animal cells do. Chloroplasts thus generate both the food molecules and the oxygen that all mitochondria use. How they do so is explained in Chapter 14.

Like mitochondria, chloroplasts contain their own DNA, reproduce by dividing in two, and are thought to have evolved from bacteria—in this case from photosynthetic bacteria that were somehow engulfed by an early eucaryotic cell (Figure 1–21).

Internal Membranes Create Intracellular Compartments with Different Functions

Nuclei, mitochondria, and chloroplasts are not the only membrane-enclosed organelles inside eucaryotic cells. The cytoplasm contains a profusion of other organelles—most of them surrounded by single membranes—that perform many distinct functions. Most of these structures

QUESTION 1–5

According to Figure 1–19, why does the mitochondrion have both an outer and an inner membrane? Which of the two mitochondrial membranes should be—in evolutionary terms—derived from the cell membrane of the ancestral eucaryotic cell? In the electron micrograph of a mitochondrion in Figure 1–18A, identify the space that contains the mitochondrial DNA, i.e., the space that corresponds to the cytosol of the bacterium that was internalized by the ancestral eucaryotic cell shown in Figure 1–19.

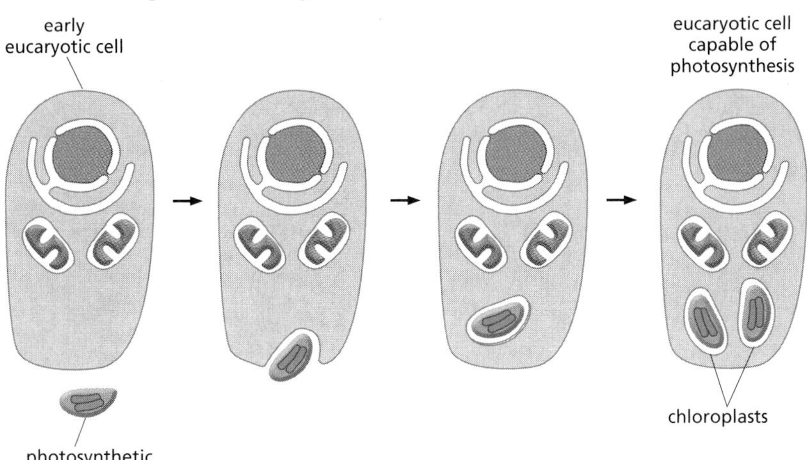

Figure 1–21 **Chloroplasts almost certainly evolved from engulfed bacteria.** Chloroplasts are thought to have originated from symbiotic photosynthetic bacteria, which were taken up by early eucaryotic cells that already contained mitochondria.

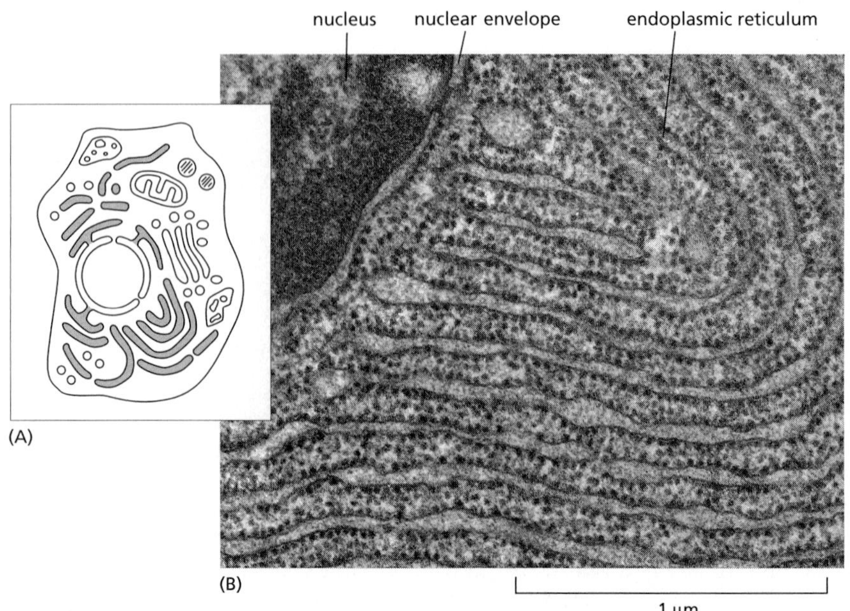

Figure 1–22 **Many cellular components are produced in the endoplasmic reticulum.** (A) Schematic diagram of an animal cell shows the endoplasmic reticulum in *green*. (B) Electron micrograph of a thin section of a mammalian pancreatic cell shows a small part of the endoplasmic reticulum (ER), of which there are vast tracts in this cell type, which is specialized for protein secretion. Note that the ER is continuous with the membrane of the nuclear envelope. The black particles studding the particular region of the ER shown here are ribosomes—the molecular assemblies that perform protein synthesis. Because of its appearance, ribosome-coated ER is often called "rough ER." (B, courtesy of Lelio Orci.)

are involved with the cell's ability to import raw materials and to export manufactured substances and waste products. Some of these membrane-enclosed organelles are enormously enlarged in cells that are specialized for the secretion of proteins; others are particularly plentiful in cells specialized for the digestion of foreign bodies.

The *endoplasmic reticulum (ER)*—an irregular maze of interconnected spaces enclosed by a membrane (Figure 1–22)—is the site where most cell membrane components, as well as materials destined for export from the cell, are made. Stacks of flattened membrane-enclosed sacs constitute the *Golgi apparatus* (Figure 1–23), which receives and often chemically modifies the molecules made in the endoplasmic reticulum

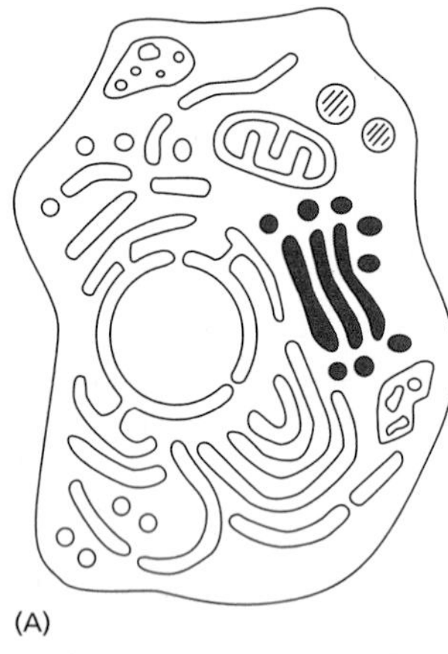

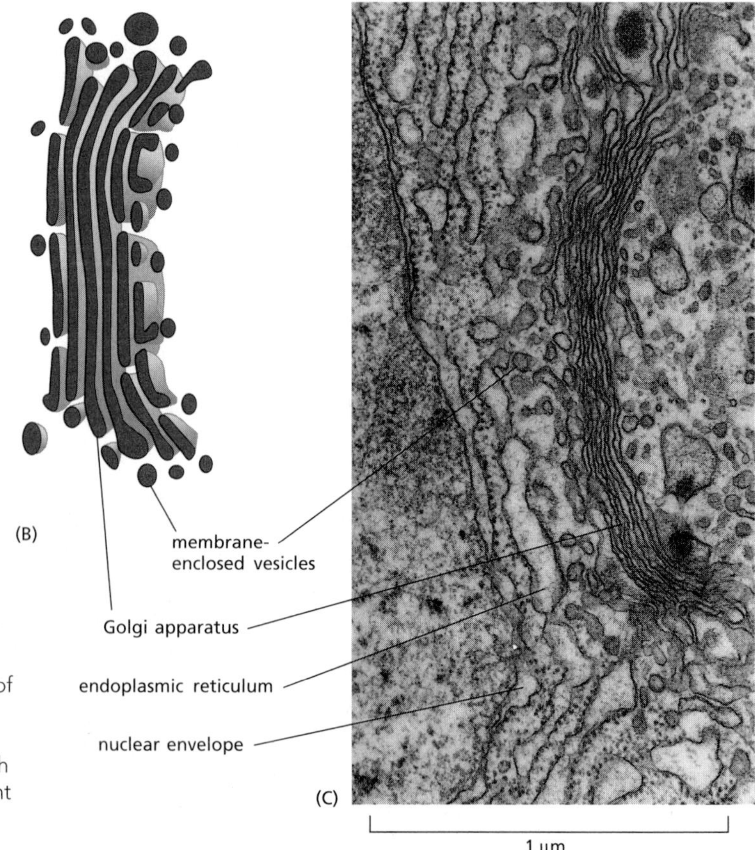

Figure 1–23 **The Golgi apparatus resembles a stack of flattened discs.** This organelle is just visible under the light microscope but is often inconspicuous. The Golgi apparatus is involved in the synthesis and packaging of molecules destined to be secreted from the cell, as well as in the routing of newly synthesized proteins to the correct cellular compartment. (A) Schematic diagram of an animal cell with the Golgi apparatus colored *red*. (B) Drawing of the Golgi apparatus reconstructed from electron microscope images. The organelle is composed of flattened sacs of membrane stacked in layers. Many small vesicles are seen nearby; some of these have pinched off from the Golgi stack, while others are destined to fuse with it. Only one stack is shown here, but several can be present in a cell. (C) Electron micrograph of the Golgi apparatus from a typical animal cell. (C, courtesy of Brij J. Gupta.)

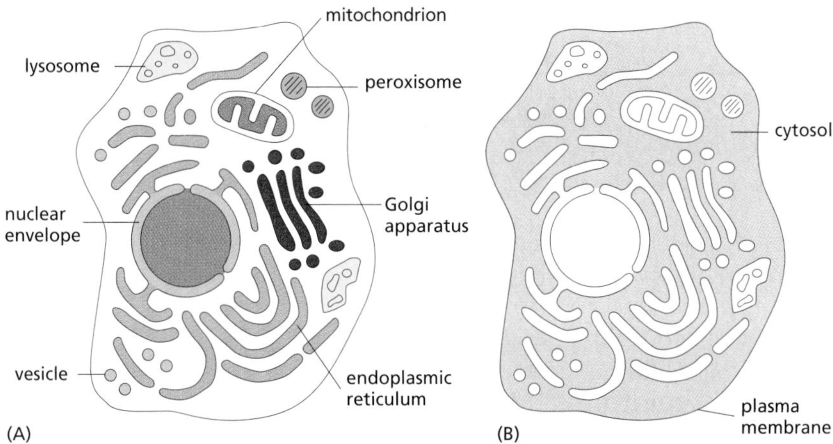

Figure 1–24 **Membrane-enclosed organelles are distributed throughout the cytoplasm.** (A) A variety of membrane-enclosed compartments exist within eucaryotic cells, each specialized to perform a different function. (B) The rest of the cell, excluding all these organelles, is called the cytosol (colored *blue*).

and then directs them to the exterior of the cell or to various locations inside the cell. *Lysosomes* are small, irregularly shaped organelles in which intracellular digestion occurs, releasing nutrients from food particles and breaking down unwanted molecules for recycling or excretion. And *peroxisomes* are small, membrane-enclosed vesicles that provide a contained environment for reactions in which hydrogen peroxide, a dangerously reactive chemical, is generated and degraded. Membranes also form many different types of small *vesicles* involved in the transport of materials between one membrane-enclosed organelle and another. This whole system of related organelles is sketched in Figure 1–24A.

A continual exchange of materials takes place between the endoplasmic reticulum, the Golgi apparatus, the lysosomes, and the outside of the cell. The exchange is mediated by small vesicles that pinch off from the membrane of one organelle and fuse with another, like tiny soap bubbles budding from and rejoining larger bubbles. At the surface of the cell, for example, portions of the plasma membrane tuck inward and pinch off to form vesicles that carry material captured from the external medium into the cell (Figure 1–25). These vesicles fuse with membrane-enclosed *endosomes*, which mature into lysosomes, where the imported material is digested. Animal cells can engulf very large particles, or even entire foreign cells, by this process of *endocytosis*. The reverse process, *exocytosis*, whereby vesicles from inside the cell fuse with the plasma membrane and release their contents into the external medium, is also a common cellular activity (see Figure 1–25). Most hormones, neurotransmitters, and other signaling molecules are secreted from cells by exocytosis. How the membrane-enclosed organelles transport proteins and other molecules from place to place inside the cell is discussed in more detail in Chapter 15.

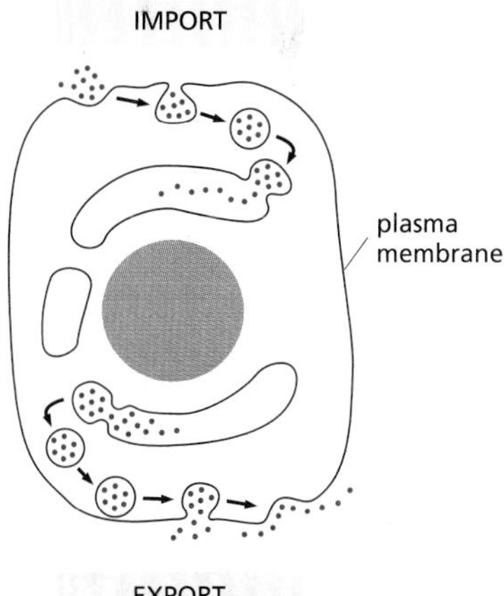

Figure 1–25 **Cells engage in endocytosis and exocytosis.** Cells can import materials from the external medium by capturing them in vesicles that pinch off from the plasma membrane. The vesicles ultimately fuse with lysosomes, where intracellular digestion occurs. By a converse process, cells export materials they have synthesized in the endoplasmic reticulum and Golgi apparatus: the materials are stored in the intracellular vesicles and released to the exterior when these vesicles fuse with the plasma membrane.

The Cytosol Is a Concentrated Aqueous Gel of Large and Small Molecules

If we were to strip the plasma membrane from a eucaryotic cell and then remove all of its membrane-enclosed organelles, including nucleus, endoplasmic reticulum, Golgi apparatus, mitochondria, chloroplasts, and so on, we would be left with the **cytosol** (see Figure 1–24B). In other words, the cytosol is the part of the cytoplasm that is not partitioned off within intracellular membranes. In most cells, the cytosol is the largest single compartment. It contains a host of large and small molecules, crowded together so closely that it behaves more like a water-based gel than a liquid solution (Figure 1–26). The cytosol is the site of many chemical reactions that are fundamental to the cell's existence. The early steps in the breakdown of nutrient molecules take place in the cytosol, for example, and it is here that the cell performs one of its key synthetic proc-

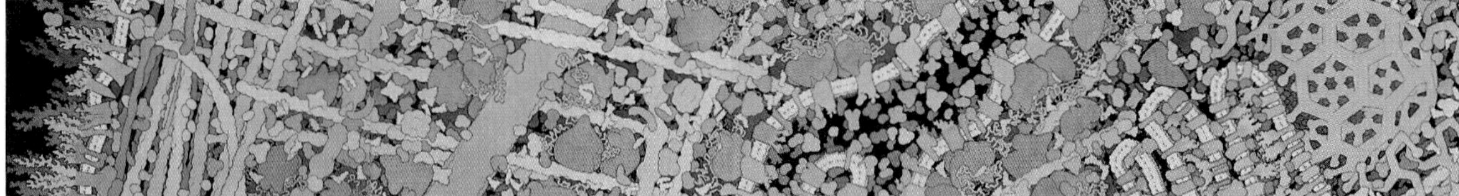

Figure 1–26 **The cytoplasm is stuffed with organelles and a host of large and small molecules.** This schematic drawing, based on the known sizes and concentrations of molecules in the cytosol, shows how crowded the cytoplasm is. The panorama begins on the far left at the cell surface; moves through the endoplasmic reticulum, Golgi apparatus, and a mitochondrion; and ends on the far right in the nucleus. Note that some ribosomes *(large pink objects)* are free in the cytosol, while others are attached to the ER. (Courtesy of D. Goodsell.)

esses—the manufacture of proteins. **Ribosomes**, the molecular machines that make the protein molecules, are visible with the electron microscope as small particles in the cytosol, often attached to the cytosolic face of the endoplasmic reticulum (see Figures 1–8B and 1–22B).

The Cytoskeleton Is Responsible for Directed Cell Movements

The cytoplasm is not just a structureless soup of chemicals and organelles. Using an electron microscope, one can see that in eucaryotic cells the cytosol is criss-crossed by long, fine filaments of protein. Frequently the filaments are seen to be anchored at one end to the plasma membrane or to radiate out from a central site adjacent to the nucleus. This system of filaments is called the **cytoskeleton** (Figure 1–27). The thinnest of the filaments are *actin filaments*, which are present in all eucaryotic cells but occur in especially large numbers inside muscle cells, where they serve as part of the machinery that generates contractile forces. The thickest filaments are called *microtubules*, because they have the form of minute hollow tubes. In dividing cells they become reorganized into a spectacular array that helps pull the duplicated chromosomes in opposite directions and distribute them equally to the two daughter cells (Figure 1–28). Intermediate in thickness between actin filaments and microtubules are the *intermediate filaments*, which serve to strengthen the cell mechanically. These three types of filaments, together with other proteins that attach to them, form a system of girders, ropes, and motors that gives the cell its mechanical strength, controls its shape, and drives and guides its movements (see Movie 1.2 and Movie 1.3).

Because the cytoskeleton governs the internal organization of the cell as well as its external features, it is as necessary to a plant cell—boxed in by a tough wall of extracellular matrix—as it is to an animal cell that freely bends, stretches, swims, or crawls. In a plant cell, for example, organelles such as mitochondria are driven in a constant stream around the cell interior along cytoskeletal tracks. And animal cells and plant cells

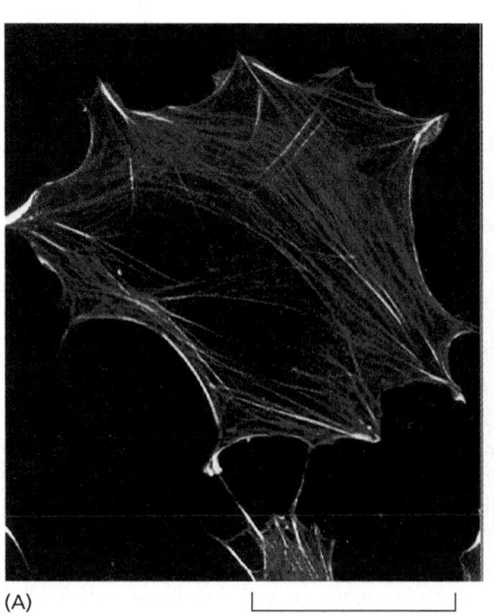

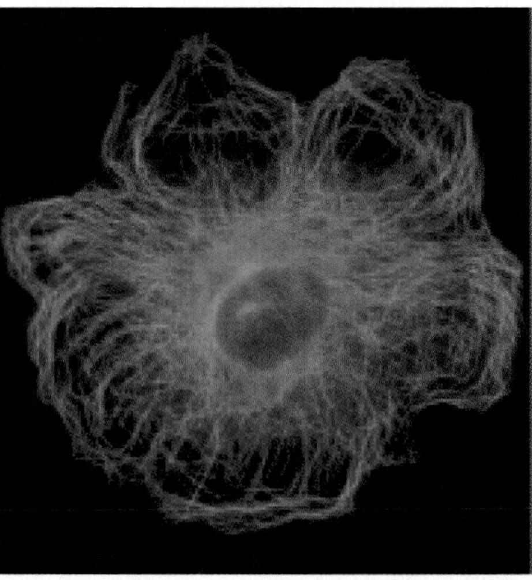

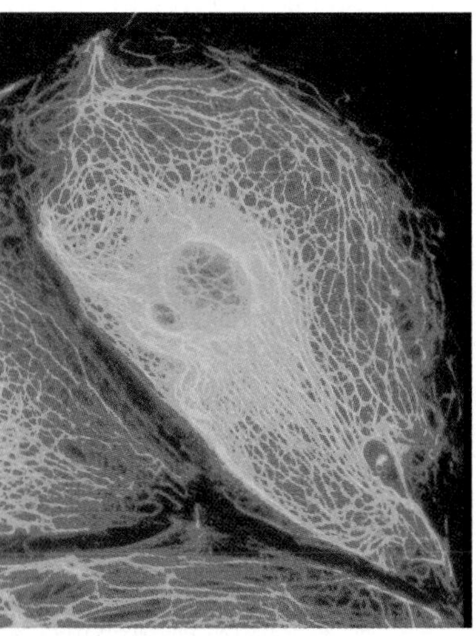

Figure 1–27 **The cytoskeleton is a network of filaments criss-crossing the cytoplasm of the eucaryotic cell.** Filaments made of protein provide all eucaryotic cells with an internal framework that helps organize the internal activities of the cell and underlies its movements and changes of shape. Different types of filaments can be detected using different fluorescent stains. Shown here are (A) actin filaments, (B) microtubules, and (C) intermediate filaments. (A, courtesy of Simon Barry and Chris D'Lacey; B, courtesy of Nancy Kedersha; C, courtesy of Clive Lloyd.)

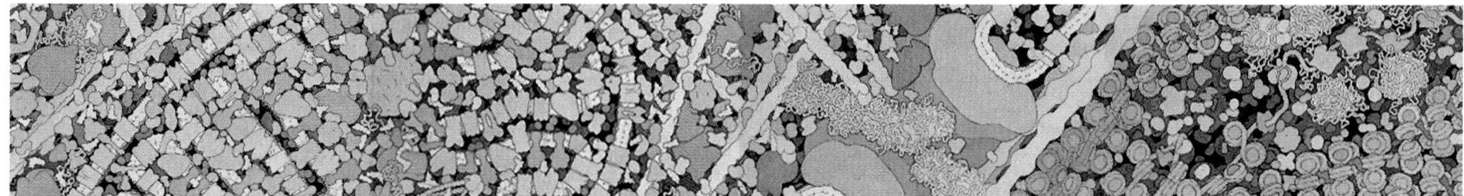

alike depend on the cytoskeleton to separate their internal components into two daughter sets during cell division. Its role in cell division may be the most ancient function of the cytoskeleton; even bacteria contain proteins that are distantly related to those of eucaryotic actin filaments and microtubules and that form filaments that play a part in procaryotic cell division. We examine the cytoskeleton in detail in Chapter 17 and discuss its role in cell division in Chapter 18, and its responses to signals from the environment in Chapter 16.

The Cytoplasm Is Far from Static

The cell interior is in constant motion. The cytoskeleton is a dynamic jungle of ropes and rods that are continually being strung together and taken apart; its filaments can assemble and then disappear in a matter of minutes. Along these tracks and cables, organelles and vesicles hurry to and fro, racing across the width of the cell in a second or so. The endoplasmic reticulum and the molecules that fill every free space are in frantic thermal commotion—with unattached proteins buzzing around so fast that, even though they move at random, they visit every corner of the cell within a few seconds, constantly colliding with an even more tumultuous dust storm of smaller organic molecules.

Of course, neither the bustling nature of the cell's interior nor the details of cell structure were appreciated when scientists first peered at cells in a microscope; our knowledge of cell structure accumulated slowly. A few of the key discoveries are listed in Table 1–1. Panel 1–2 summarizes the differences between animal, plant, and bacterial cells.

QUESTION 1–6

Suggest a reason why it would be advantageous for eucaryotic cells to evolve elaborate internal membrane systems that allow them to import substances from the outside, as shown in Figure 1–25.

QUESTION 1–7

Discuss the relative advantages and disadvantages of light and electron microscopy. How could you best visualize (a) a living skin cell, (b) a yeast mitochondrion, (c) a bacterium, and (d) a microtubule?

Eucaryotic Cells May Have Originated as Predators

Eucaryotic cells are typically 10 times the length and 1000 times the volume of procaryotic cells (although there is huge size variation within each category). In addition, they possess a whole collection of features—a cytoskeleton, mitochondria, and other organelles—that set them apart from bacteria and archaea.

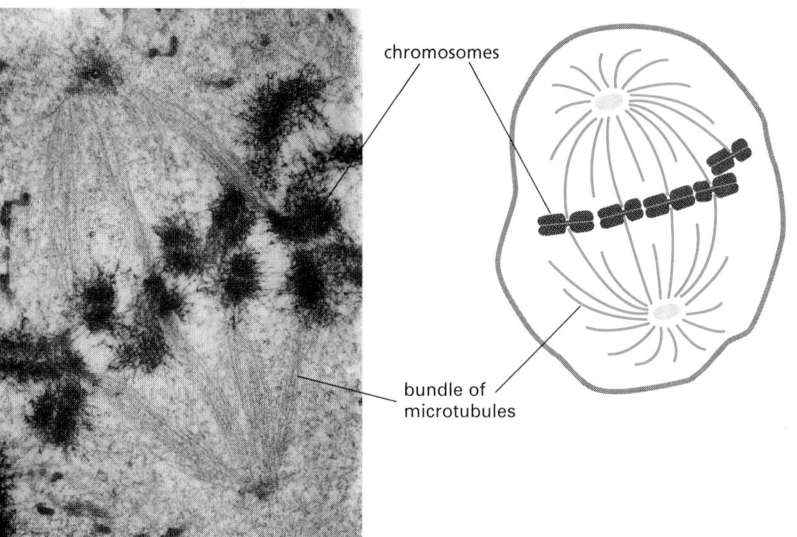

Figure 1–28 **Microtubules help distribute the chromosomes in a dividing cell.** When a cell divides, its nuclear envelope breaks down and its DNA condenses into pairs of visible chromosomes, which are pulled apart into separate cells by microtubules. In this transmission electron micrograph, the microtubules are seen to radiate from foci at opposite ends of the dividing cell. (Photomicrograph courtesy of Conly L. Rieder.)

TABLE 1–1 HISTORICAL LANDMARKS IN DETERMINING CELL STRUCTURE

1665	Hooke uses a primitive microscope to describe small pores in sections of cork that he calls "**cells**."
1674	Leeuwenhoek reports his discovery of **protozoa**. Nine years later, he sees **bacteria** for the first time.
1833	Brown publishes his microscopic observations of orchids, clearly describing the cell **nucleus**.
1838	Schleiden and Schwann propose the **cell theory**, stating that the nucleated cell is the universal building block of plant and animal tissues.
1857	Kölliker describes **mitochondria** in muscle cells.
1879	Flemming describes with great clarity **chromosome behavior during mitosis** in animal cells.
1881	Cajal and other histologists develop staining methods that reveal the structure of **nerve cells** and the organization of neural tissue.
1898	Golgi first sees and describes the **Golgi apparatus** by staining cells with silver nitrate.
1902	Boveri links **chromosomes and heredity** by observing chromosome behavior during sexual reproduction.
1952	Palade, Porter, and Sjöstrand develop methods of **electron microscopy** that enable many intracellular structures to be seen for the first time. In one of the first applications of these techniques, Huxley shows that muscle contains arrays of protein filaments—the first evidence of a **cytoskeleton**.
1957	Robertson describes the bilayer structure of the **cell membrane**, seen for the first time in the electron microscope.
1960	Kendrew describes the first detailed **protein structure** (sperm whale **myoglobin**) to a resolution of 0.2 nm using **X-ray crystallography**. Perutz proposes a lower-resolution structure for **hemoglobin**.
1965	Christian de Duve and his colleagues use a **cell fractionation** technique to separate **peroxisomes, mitochondria**, and **lysosomes** from a preparation of rat liver.
1968	Petran and collaborators make the first **confocal microscope**.
1974	Lazarides and Weber use **fluorescent antibodies** to stain the cytoskeleton.
1994	Chalfie and collaborators introduce **green fluorescent protein** (**GFP**) as a marker to follow the behavior of proteins in living cells.

When and how eucaryotes evolved these systems remains something of a mystery. Although eucaryotes, bacteria, and archaea must have diverged from one another very early in the history of life on Earth (discussed in Chapter 14), the eucaryotes did not acquire all of their distinctive features at the same time (Figure 1–29). According to one theory, the ancestral eucaryotic cell was a predator that fed by capturing other cells. Such a way of life requires a large size, a flexible membrane, and a cytoskeleton to help the cell move and eat. The nuclear compartment may have evolved to keep the DNA segregated from this physical and chemical hurly-burly, so as to allow more delicate and complex control of the way the cell reads out its genetic information.

Such a primitive eucaryote, with a nucleus and cytoskeleton, was most likely the sort of cell that engulfed the free-living, oxygen-consuming

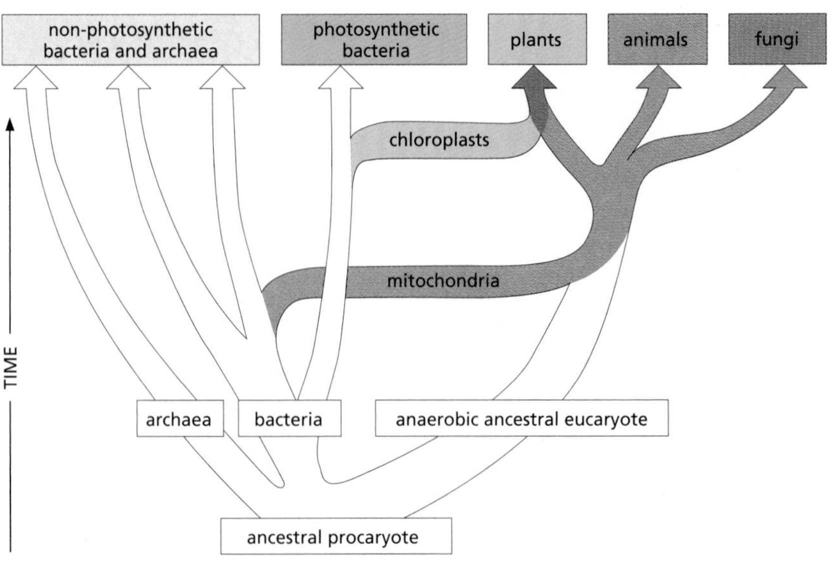

Figure 1–29 **Where did eucaryotes come from?** The eucaryotic, bacterial, and archaean lineages diverged from one another very early in the evolution of life on Earth. Some time later, eucaryotes are thought to have acquired mitochondria; later still, a subset of eucaryotes acquired chloroplasts. Mitochondria are essentially the same in plants, animals, and fungi, and therefore were presumably acquired before these lines diverged.

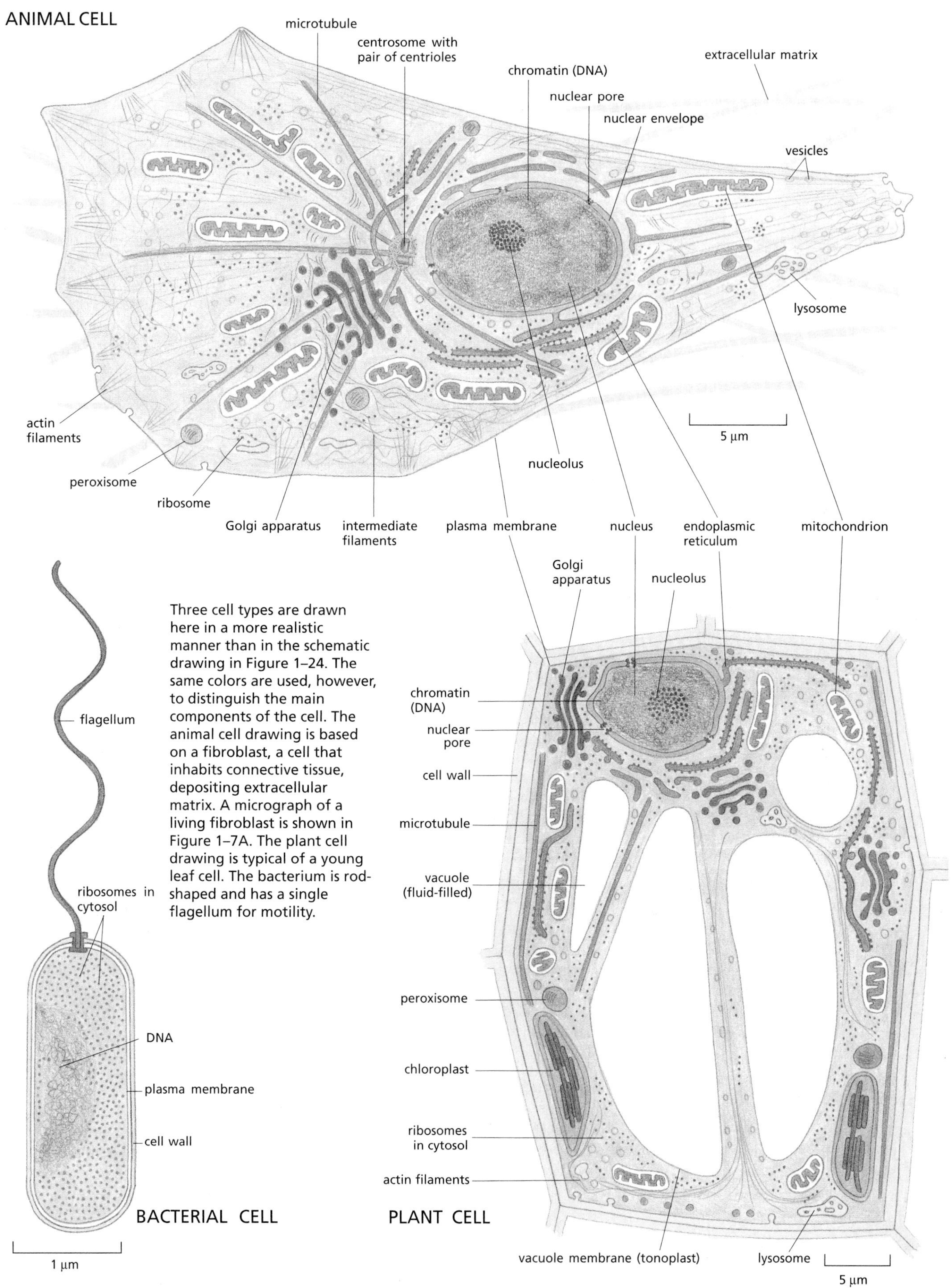

Three cell types are drawn here in a more realistic manner than in the schematic drawing in Figure 1–24. The same colors are used, however, to distinguish the main components of the cell. The animal cell drawing is based on a fibroblast, a cell that inhabits connective tissue, depositing extracellular matrix. A micrograph of a living fibroblast is shown in Figure 1–7A. The plant cell drawing is typical of a young leaf cell. The bacterium is rod-shaped and has a single flagellum for motility.

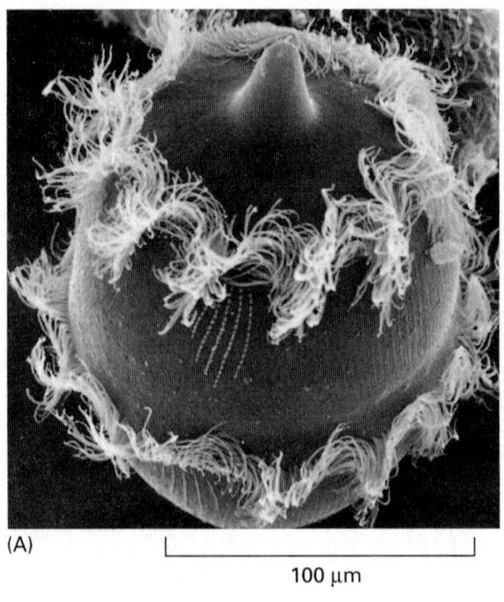

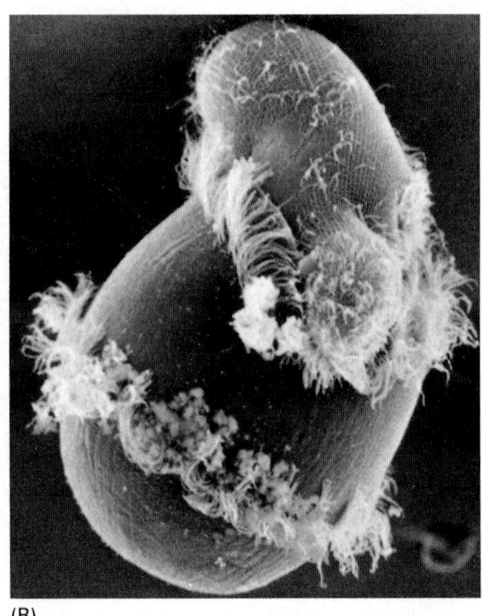

Figure 1–30 **One protozoan eats another.** (A) The scanning electron micrograph shows *Didinium* on its own, with its circumferential rings of beating cilia and its "snout" at the top. (B) *Didinium* is seen ingesting another ciliated protozoan, *Paramecium*. (Courtesy of D. Barlow.)

bacteria that were the ancestors of the mitochondria (see Figure 1–19). This partnership is thought to have been established 1.5 billion years ago, when the Earth's atmosphere first became rich in oxygen. A subset of these cells later acquired chloroplasts by engulfing photosynthetic bacteria (see Figures 1–21 and 1–29).

That single-celled eucaryotes can prey upon and swallow other cells is borne out by the behavior of many of the free-living actively motile microorganisms called **protozoans**. *Didinium*, for example, is a large, carnivorous protozoan with a diameter of about 150 μm—roughly 10 times that of the average human cell. It has a globular body encircled by two fringes of cilia, and its front end is flattened except for a single protrusion rather like a snout (Figure 1–30). *Didinium* swims at high speed by means of its beating cilia. When it encounters a suitable prey, usually another type of protozoan, it releases numerous small, paralyzing darts from its snout region. *Didinium* then attaches to and devours the other cell, inverting like a hollow ball to engulf its victim, which is almost as large as itself.

Protozoans include some of the most complex cells known. Figure 1–31 conveys something of the variety of forms of protozoans, and their behavior is just as diverse: they can be photosynthetic or carnivorous, motile or sedentary. Their anatomy is often elaborate and includes such structures as sensory bristles, photoreceptors, beating cilia, stalklike appendages, mouthparts, stinging darts, and musclelike contractile bundles. Although they are single cells, protozoans can be as intricate and versatile as many multicellular organisms.

MODEL ORGANISMS

All cells are thought to be descended from a common ancestor, whose fundamental properties have been conserved through evolution. Thus, knowledge gained from the study of one organism contributes to our understanding of others, including ourselves. But certain organisms are easier than others to study in the laboratory. Some reproduce rapidly and are convenient for genetic manipulations; others are multicellular but transparent, so that one can directly watch the development of all their internal tissues and organs. For reasons such as these, large communities of biologists have become dedicated to studying different aspects of the biology of a few chosen species, pooling their knowledge to gain a deeper understanding than could be achieved if their efforts were spread

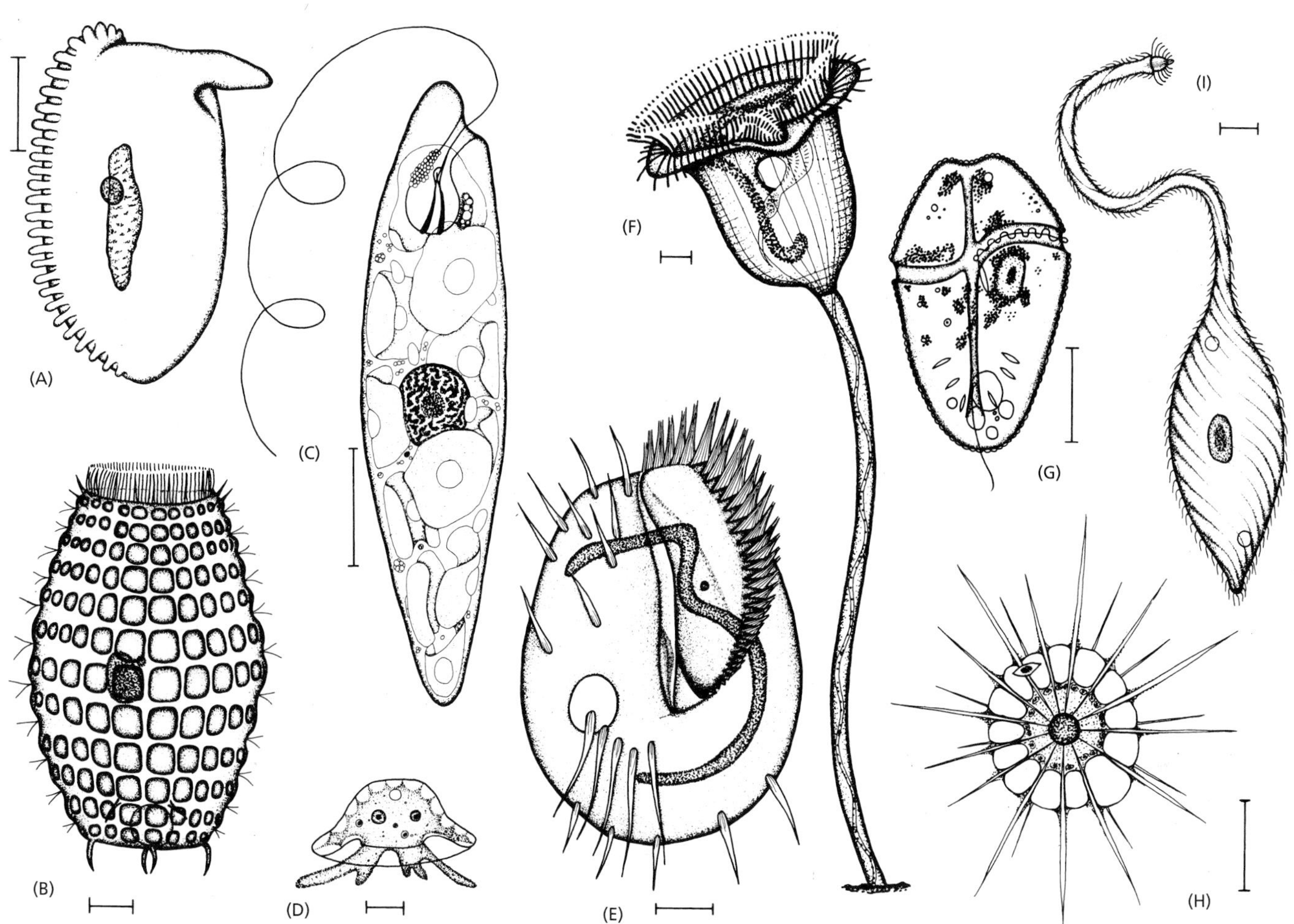

over many different species. Information obtained from these studies contributes to our understanding of how all cells work. Although the roster of these representative organisms is continually expanding, a few stand out in terms of the breadth and depth of information that has been accumulated about them over the years. In the following sections, we examine some of these **model organisms** and review the benefits each offers to the study of cell biology and, in many cases, to the promotion of human health.

Figure 1–31 **An assortment of protozoans illustrates the enormous variety within this class of single-celled microorganisms.** These drawings are done to different scales, but in each case the scale bar represents 10 μm. The organisms in (A), (B), (E), (F), and (I) are ciliates; (C) is a euglenoid; (D) is an amoeba; (G) is a dinoflagellate; and (H) is a heliozoan. To see a euglenoid in action, watch Movie 1.4. (From M.A. Sleigh, The Biology of Protozoa. London: Edward Arnold, 1973. With permission from Edward Arnold.)

Molecular Biologists Have Focused on *E. coli*

In the world of bacteria, the spotlight of molecular biology has fallen chiefly on just one species: *Escherichia coli*, or *E. coli* for short (see Figure 1–11). This small, rod-shaped bacterial cell normally lives in the gut of humans and other vertebrates, but it can be grown easily in a simple nutrient broth in a culture bottle. *E. coli* copes well with variable chemical conditions in its environment, and it reproduces rapidly. Its genetic instructions are contained in a single, circular, double-stranded molecule of DNA, approximately 4.6 million nucleotide pairs long, and it makes 4300 different kinds of proteins.

In molecular terms, we understand the workings of *E. coli* more thoroughly than those of any other living organism. Most of our knowledge of the fundamental mechanisms of life—including how cells replicate their DNA and how they decode these genetic instructions to make proteins—has come from studies of *E. coli*. Subsequent research has confirmed that these basic processes occur in essentially the same way in our own cells as they do in *E. coli*.

QUESTION 1–8

Your next-door neighbor has donated $100 in support of cancer research and is horrified to learn that her money is being spent on studying brewer's yeast. How could you put her mind at ease?

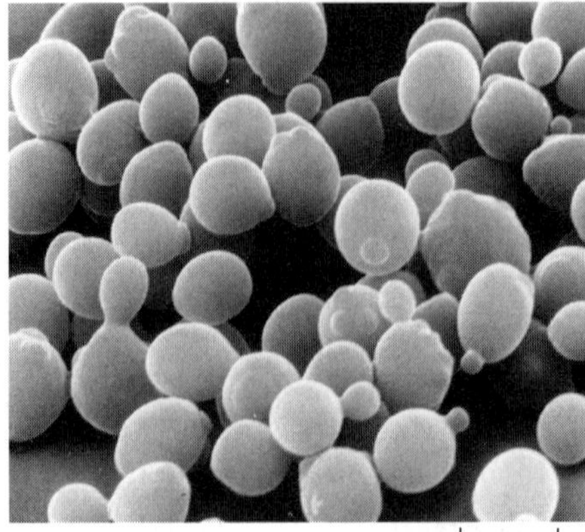

Figure 1–32 **The yeast *Saccharomyces cerevisiae* is a model eucaryote.** In this scanning electron micrograph, a few yeast cells are seen in the process of dividing, which they do by budding. Another micrograph of the same species is shown in Figure 1–14. (Courtesy of Ira Herskowitz and Eric Schabatach.)

Brewer's Yeast Is a Simple Eucaryotic Cell

We tend to be preoccupied with eucaryotes because we are eucaryotes ourselves. But human cells are complicated and difficult to work with, and if we want to understand the fundamentals of eucaryotic cell biology, it is often more effective to concentrate on a species that, like *E. coli* among the bacteria, is simple and robust and reproduces rapidly. The popular choice for this role of minimal model eucaryote has been the budding yeast *Saccharomyces cerevisiae* (Figure 1–32)—the same microorganism that is used for brewing beer and baking bread.

S. cerevisiae is a small, single-celled fungus and thus, according to modern views, is at least as closely related to animals as it is to plants. Like other fungi, it has a rigid cell wall, is relatively immobile, and possesses mitochondria but not chloroplasts. When nutrients are plentiful, it reproduces almost as rapidly as a bacterium. As its nucleus contains only about 2.5 times as much DNA as *E. coli*, this yeast is also a good subject for genetic analysis. Even though its genome is small (by eucaryotic standards), *S. cerevisiae* carries out all the basic tasks every eucaryotic cell must perform. Genetic and biochemical studies in yeast have been crucial to understanding many basic mechanisms in eucaryotic cells, including the cell-division cycle—the chain of events by which the nucleus and all the other components of a cell are duplicated and parceled out to create two daughter cells. In fact, the machinery that governs cell division has been so well conserved over the course of evolution that many of its components can function interchangeably in yeast and human cells. If a mutant yeast lacks a gene essential for cell division, providing it with a copy of the corresponding gene from a human will cure the yeast's defect and enable it to divide normally (see How We Know, pp. 30–31).

Arabidopsis Has Been Chosen Out of 300,000 Species as a Model Plant

The large multicellular organisms that we see around us—both plants and animals—seem fantastically varied, but they are much closer to one another in their evolutionary origins, and more similar in their basic cell biology, than the great host of microscopic single-cell organisms. Whereas bacteria, archaea, and eucaryotes separated from each other more than 3 billion years ago, plants, animals, and fungi are separated by only about 1.5 billion years, fish and mammals by only about 450 million years, and the different species of flowering plants by less than 200 million years.

The close evolutionary relationship among all flowering plants means that we can get insight into their cell and molecular biology by focusing on just a few convenient species for detailed analysis. Out of the several hundred thousand species of flowering plants on Earth today, molecular biologists have recently focused their efforts on a small weed, the common wall cress *Arabidopsis thaliana* (Figure 1–33), which can be grown indoors in large numbers and produces thousands of offspring per plant within 8–10 weeks. *Arabidopsis* has a genome of approximately 110 million nucleotide pairs, about eight times as many as yeast, and its complete sequence is known. By examining the genetic instructions that *Arabidopsis* carries, we are beginning to learn more about the genetics, molecular biology, and evolution of flowering plants, which dominate nearly every ecosystem on land. Because genes found in *Arabidopsis* have

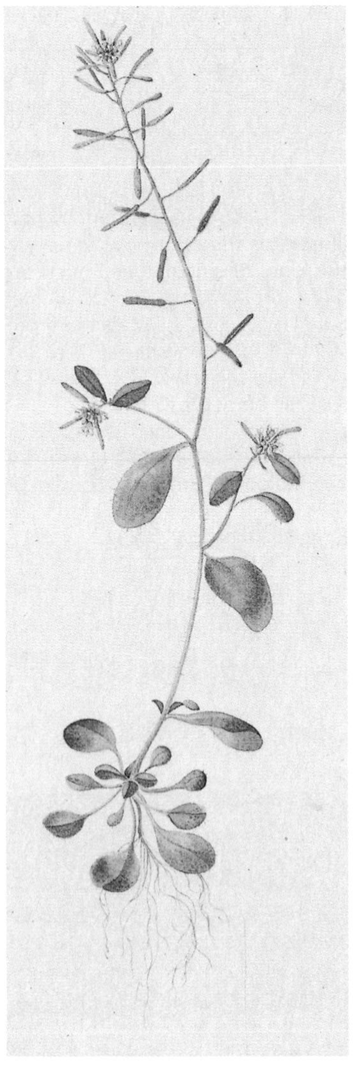

Figure 1–33 ***Arabidopsis thaliana*, the common wall cress, is a model plant.** This small weed has become the favorite organism of plant molecular and developmental biologists. (Courtesy of Toni Hayden and the John Innes Centre.)

counterparts in agricultural species, studying this simple weed provides insights into the development and physiology of the crop plants upon which our lives depend, as well as all the other plant species that are our companions on Earth.

The World of Animals Is Represented by a Fly, a Worm, a Fish, a Mouse, and the Human Species

Multicellular animals account for the majority of all named species of living organisms, and the majority of animal species are insects. It is fitting, therefore, that an insect, the small fruit fly *Drosophila melanogaster* (Figure 1–34), should occupy a central place in biological research. In fact, the foundations of classical genetics were built to a large extent on studies of this insect. More than 80 years ago, studies of the fruit fly provided definitive proof that genes—the units of heredity—are carried on chromosomes. In more recent times, a concentrated systematic effort has been made to elucidate the genetics of *Drosophila*, and especially the genetic mechanisms underlying its embryonic and larval development. Through this work, we are at last beginning to understand in detail how living cells achieve their most spectacular feat: how a single fertilized egg cell (or *zygote*) develops into a multicellular organism comprising vast numbers of cells of differing types, organized in an exactly predictable way. *Drosophila* mutants with body parts strangely misplaced or oddly patterned have provided the key to identifying and characterizing the genes that are needed to make a properly structured adult body, with gut, wings, legs, eyes, and all the other bits and pieces in their correct places. These genes—which are copied and passed on to every cell in the body—define how each cell will behave in its social interactions with its sisters and cousins; in this way, they control the structures that the cells create. *Drosophila*, more than any other organism, has shown us how to trace the chain of cause and effect from the genetic instructions encoded in the DNA to the structure of the adult multicellular organism. Moreover, the genes of *Drosophila* have turned out to be amazingly similar to those of humans—far more similar than one would suspect from outward appearances. Thus the fly serves as a model for studying human development and disease. The fly genome—185 million nucleotide pairs comprising just over 13,000 genes—contains counterparts of most human genes including most of those known to be critical in human diseases.

Figure 1–34 ***Drosophila melanogaster*** **is a favorite among developmental biologists and geneticists.** Molecular genetic studies on this small fly have provided a key to the understanding of how all animals develop. (Courtesy of E.B. Lewis.)

HOW WE KNOW: LIFE'S COMMON MECHANISMS

All living things are made of cells, and all cells—as we have discussed in this chapter—are fundamentally similar inside: they store their genetic instructions in DNA molecules, which ultimately direct the production of proteins, and proteins in turn carry out the cell's chemical reactions, give it its shape, and control its behavior. But how deep do these similarities really run? Are parts from one type of cell interchangeable with parts from another? Would an enzyme that digests glucose in a bacterium be able to break down the same sugar if it were asked to function inside a yeast, a lobster, or a human? What about the molecular machines that copy and interpret genetic information? Are they functionally equivalent from one organism to another? Are their component molecules interchangeable? Answers have come from many sources, but most strikingly from experiments on one of the most fundamental processes of life: cell division.

Divide or die

All cells come from other cells, and the only way to make a new cell is through division of a preexisting cell. To reproduce, a parent cell must execute an orderly sequence of reactions through which it duplicates its contents and divides in two. This critical process of duplication and division, known as the *cell cycle*, is complex and carefully controlled. Defects in any of the proteins involved in the cell cycle can be fatal.

Unfortunately, the lethal effects of cell-cycle mutations present a problem if one wants to discover the components of the cell-cycle control machinery and find out how they work. Scientists depend on mutants to identify genes and proteins on the basis of their functions: if a gene is essential for a given process, a mutation that disrupts the gene will show up as a disturbance of that process. By analyzing the misbehavior of the mutant organism, one can pinpoint the function for which the gene is needed, and by analyzing the DNA of the mutant one can track down the gene itself.

For such an analysis, however, a single mutant cell is not enough: one needs a large colony of cells carrying the mutation. And this is the problem. If the mutation disrupts a process critical to life, such as cell division, how can one ever obtain such a colony? Geneticists have found an ingenious solution. Mutants defective in cell-cycle genes can be maintained and studied if their defect is *conditional*—that is, if the gene product fails to function only under certain specific conditions. In particular, one can often find mutations that are temperature-sensitive: the mutant protein functions correctly when the organism is kept cool, allowing the cells to reproduce; but when the temperature is raised, the heat disrupts the protein's structure and destroys its performance, allowing the cells to display their interesting defect (Figure 1–35). The study of such conditional mutants in yeast has allowed the discovery of the genes that control the cell-division cycle—the *Cdc* genes—and has led to an understanding of how they work.

The same temperature-sensitive mutants, it turns out, offer an opportunity to see whether proteins from one organism can function interchangeably in another. Can a protein from a different organism cure a cell-cycle defect in a mutant yeast and enable it to reproduce normally? The first experiment was performed using two different species of yeast.

Next of kin

Yeasts—unicellular fungi—are popular organisms for studies of cell division. They are eucaryotes, like us, yet they are small, simple, rapidly reproducing, and easy to manipulate experimentally. *Saccharomyces cerevisiae*, the most widely studied yeast, divides by forming a small bud that grows steadily until it separates from the mother cell (see Figures 1–14 and 1–32). A second species of yeast, *Schizosaccharomyces pombe*, is also popular for studies of cell growth and division. Named after the African beer from which it was first isolated, *S. pombe* is a rod-shaped yeast that grows by elonga-

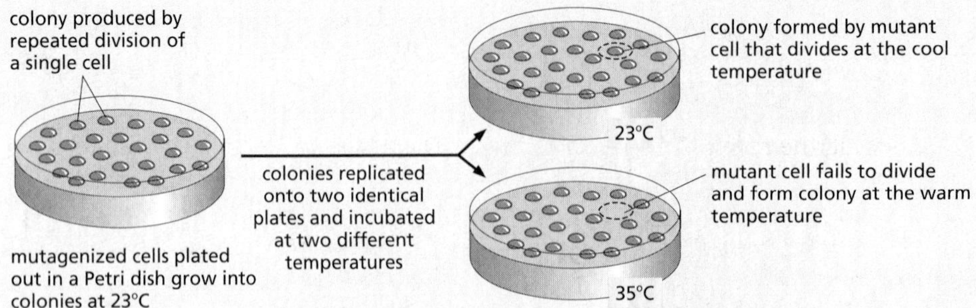

colony produced by repeated division of a single cell
mutagenized cells plated out in a Petri dish grow into colonies at 23°C
colonies replicated onto two identical plates and incubated at two different temperatures
colony formed by mutant cell that divides at the cool temperature
23°C
mutant cell fails to divide and form colony at the warm temperature
35°C

Figure 1–35 **Yeast cells that contain a temperature-sensitive mutation can be generated in the laboratory.** Yeast cells are incubated with a chemical that causes mutations in their DNA. They are then spread onto a plate and allowed to proliferate at a cool temperature. Under those conditions, cells containing a temperature-sensitive mutation, or no mutation at all, divide normally—each producing a visible colony. The colonies are transferred to two identical Petri dishes using a technique called replica plating. One of these plates is incubated at the cool temperature, the other at a warm temperature at which the mutant protein cannot function but the normal protein can. Cells containing a temperature-sensitive mutation in a gene essential for proliferation can divide at the cool temperature but fail to divide at the warm temperature.

Figure 1–36 **Temperature-sensitive *S. pombe* mutants defective in a cell-cycle gene can be rescued by the equivalent gene from *S. cerevisiae*.** DNA is collected from *S. cerevisiae* and broken into large fragments, which are introduced into a culture of temperature-sensitive *S. pombe* mutants. We discuss how DNA can be manipulated and transferred into different cell types in Chapter 10. The yeast cells that received foreign DNA are then spread onto a plate and incubated at the warm temperature. The rare cells that survive and proliferate on these plates have been rescued by incorporation of a foreign gene that allows normal division even at the warm temperature.

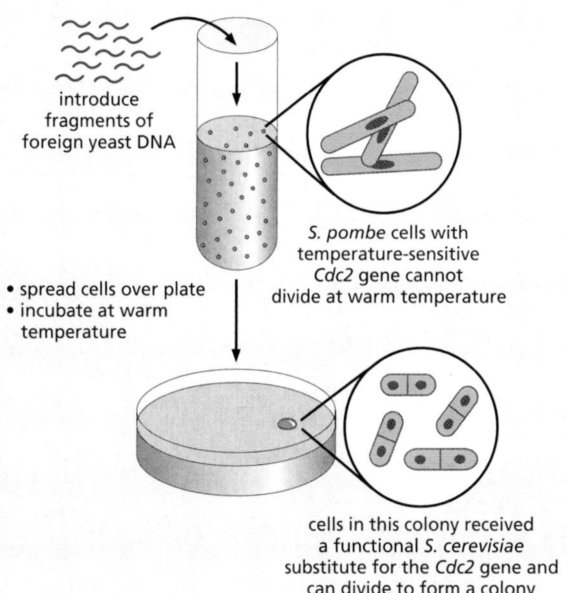
introduce fragments of foreign yeast DNA
S. pombe cells with temperature-sensitive *Cdc2* gene cannot divide at warm temperature
• spread cells over plate
• incubate at warm temperature
cells in this colony received a functional *S. cerevisiae* substitute for the *Cdc2* gene and can divide to form a colony

tion at its ends and divides by fission into two, through the formation of a partition in the center of the rod.

Although they differ in their style of cell division, both the budding yeast and the fission yeast must copy their DNA and pass on this material to their progeny. To establish whether the proteins controlling the whole process in *S. cerevisiae* and *S. pombe* are functionally equivalent, Paul Nurse and his colleagues set out to determine whether *S. pombe* cell-cycle mutants could be rescued by a gene from *S. cerevisiae.* The starting point was a colony of temperature-sensitive *S. pombe* mutants that were incapable of proceeding through the cell cycle when grown at a warm 35°C. These mutant cells had a defect in a gene called *Cdc2*, which is required to trigger several key events in the cell-division cycle. The researchers then introduced into these defective cells a collection of DNA fragments prepared from *S. cerevisiae* (Figure 1–36).

When these cultures were incubated at 35°C, the researchers found that some of the cells had regained the ability to reproduce: spread onto a plate of medium, they could divide again and again, forming visible colonies each containing millions of yeast cells (see Figure 1–35). These "rescued" yeast cells, the researchers discovered, had received a fragment of DNA containing the equivalent gene from *S. cerevisiae*—a gene that was already familiar from pioneering cell-division-cycle studies (by Lee Hartwell and colleagues) in the budding yeast.

Perhaps the result is not all that surprising. How different can one yeast be from another? What about more distant relatives? To find out, the researchers performed the same experiment, this time using human DNA to rescue the yeast cell-cycle mutants. The results were the same. The equivalent human gene rescued the yeast mutants, enabling the mutant cells to divide normally.

Reading genes

Not only are the human and yeast proteins functionally equivalent, they are almost exactly the same size and consist of amino acids strung together in a very similar order. When Nurse's team examined the amino acid sequences of the proteins, they found that human Cdc2 protein is identical to the *S. pombe* Cdc2 protein in 63% of its amino acids and 58% identical to the equivalent protein from *S. cerevisiae* (Figure 1–37).

These experiments show that proteins from different organisms can be functionally interchangeable. In fact, the molecules that orchestrate cell division in eucaryotes are so fundamentally important that they have been conserved almost unchanged over more than a billion years of eucaryotic evolution.

The same experiment highlights another, even more basic, point. The mutant yeast cells were rescued, not by direct injection of the human protein, but by introduction of a piece of human DNA. The yeast cells could read and use this information correctly, because the molecular machinery for reading the information encoded in DNA is also similar from cell to cell and from organism to organism. A yeast cell has all the equipment it needs to interpret the instructions encoded in a human gene and to use that information to direct the production of a fully functional human protein.

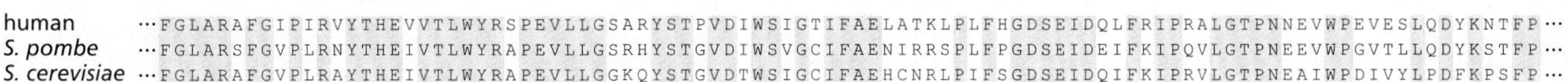
human ···FGLARAFGIPIRVYTHEVVTLWYRSPEVLLGSARYSTPVDIWSIGTIFAELATKLPLFHGDSEIDQLFRIPRALGTPNNEVWPEVESLQDYKNTFP···
S. pombe ···FGLARSFGVPLRNYTHEIVTLWYRAPEVLLGSRHYSTGVDIWSVGCIFAENIRRSPLFPGDSEIDEIFKIPQVLGTPNEEVWPGVTLLQDYKSTFP···
S. cerevisiae ···FGLARAFGVPLRAYTHEIVTLWYRAPEVLLGGKQYSTGVDTWSIGCIFAEHCNRLPIFSGDSEIDQIFKIPRVLGTPNEAIWPDIVYLPDFKPSFP···

Figure 1–37 **The cell-division-cycle proteins from yeasts and human are very similar in their amino acid sequences.** Identities between the amino acid sequences of a region of the human Cdc2 protein and a similar region of the equivalent proteins in *S. pombe* and *S. cerevisiae* are indicated by *green* shading. Each amino acid is represented by a single letter.

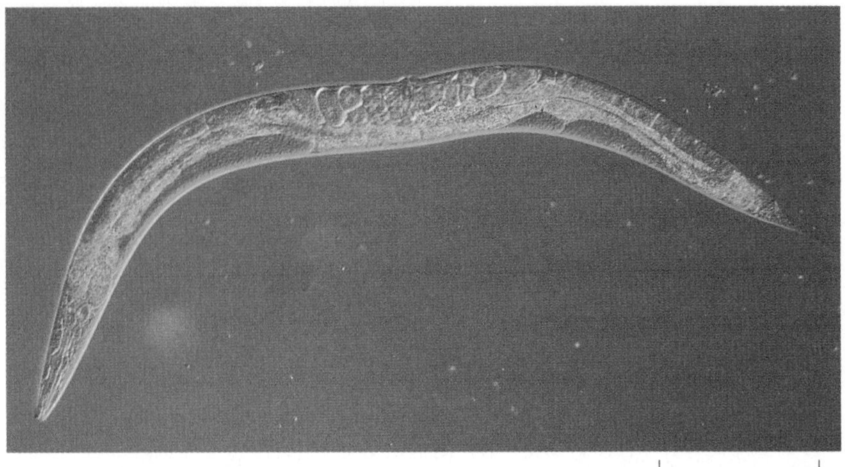

Figure 1–38 ***Caenorhabditis elegans* was the first multicellular organism to have its complete genome sequenced.** This small nematode worm lives in the soil. Its development from the fertilized egg cell to the 959 body cells of the adult has been traced in extraordinary detail, and a great deal is known about the underlying genetic mechanisms. Most individuals are hermaphrodites, producing both eggs and sperm. The coloring in this photograph is due to a special form of illumination used to enhance the contrast of the image; the worm itself is transparent and colorless. (Courtesy of Ian Hope.)

Another widely studied organism, smaller and simpler than *Drosophila*, is the nematode worm *Caenorhabditis elegans* (Figure 1–38), a harmless relative of the eelworms that attack the roots of crops. This creature develops with clockwork precision from a fertilized egg cell into an adult with exactly 959 body cells (plus a variable number of egg and sperm cells)—an unusual degree of regularity for an animal. We now have a minutely detailed description of the sequence of events by which this occurs—as the cells divide, move, and become specialized, according to strict and predictable rules. Its genome—some 97 million nucleotide pairs containing about 19,000 genes—has been sequenced, and a wealth of mutants are available for testing how these genes function. It appears that 70% of human proteins have some counterpart in the worm, and *C. elegans*, like *Drosophila*, has proved to be a valuable model for many of the processes that occur in our own bodies. Studies of nematode development, for example, have led to a detailed molecular understanding of *programmed cell death*, a process by which surplus cells are disposed of in all animals—a topic of great importance for cancer research (discussed in Chapters 18 and 20).

Another organism that has provided insights into developmental processes, particularly in vertebrates, is the *zebrafish* (Figure 1–39). Because this creature is transparent for the first 2 weeks of its life, it provides an ideal system in which to observe how cells behave during development in a living animal.

Figure 1–39 **Zebrafish are popular models for study of vertebrate development.** These small hardy tropical fish are convenient for genetics and have transparent embryos, so that one can observe cells moving and changing their characters in the living organism as it develops. (With permission from Steve Baskauf.)

At the other extreme, mammals are among the most complex of animals, with twice as many genes as *Drosophila*, 25 times as much DNA per cell, and millions of times more cells in their adult bodies. The mouse has long been used as the model organism in which to study mammalian genetics, development, immunology, and cell biology. New techniques have given it even greater importance. It is now possible to breed mice with deliberately engineered mutations in any specific gene, or with artificially constructed genes introduced into them. In this way, one can test what a given gene is required for and how it functions. And almost every human gene has a counterpart in the mouse, with a similar DNA sequence and function.

But humans are not mice—or fish or worms or flies or yeast—and so we also study human beings themselves. Research in many areas of cell biology has been largely driven by medical interests, and a great deal of what we know has come from studies of human cells. The medical database on human cells is enormous, and although naturally occurring mutations in any given gene are rare, the consequences of mutations in thousands of different genes are known without resort to genetic engineering. This

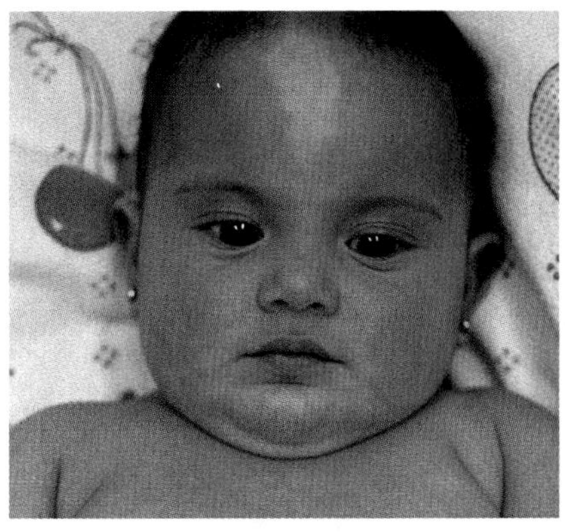

Figure 1–40 **Different living species share similar genes.** The human baby and the mouse shown here have similar white patches on their foreheads because they both have defects in the same gene (called *Kit*), which is required for the development and maintenance of pigment cells. (Courtesy of R.A. Fleischman, from *Proc. Natl. Acad. Sci. USA* 88:10885–10889, 1991. With permission from the National Academy of Sciences.)

is because humans demonstrate the unique behavior of reporting on and recording their own genetic defects; in no other species are billions of individuals so intensively examined, described, and investigated.

Nevertheless, the extent of our ignorance is still daunting. The mammalian body is enormously complex, and one might despair of ever understanding how the DNA in a fertilized mouse egg cell makes it generate a mouse, or how the DNA in a human egg cell directs the development of a human. Yet the revelations of molecular biology have made the task seem possible. As much as anything, this new optimism has come from the realization that the genes of one type of animal have close counterparts in most other types of animals, apparently serving similar functions (Figure 1–40). We all have a common evolutionary origin, and under the surface it seems that we share the same molecular mechanisms. Flies, worms, fish, mice, and humans thus provide a key to understanding how animals in general are made and how their cells operate.

Comparing Genome Sequences Reveals Life's Common Heritage

At a molecular level, evolutionary change has been remarkably slow. We can see in present-day organisms many features that have been preserved through more than 3 billion years of life on Earth, or about one-fifth of the age of the universe. This evolutionary conservatism provides the foundation on which the study of molecular biology is built. To set the scene for the chapters that follow, therefore, we end this chapter by considering a little more closely the family relationships and basic similarities among all living things. This topic has been dramatically clarified in the past few years by analysis of genome sequences—the sequences in which the four universal nucleotides are strung together to form the DNA of a given species (as discussed in more detail in Chapter 9).

DNA sequencing has made it easy to detect family resemblances between genes: if two genes from different organisms have closely similar DNA sequences, it is highly probable that both genes descended from a common ancestral gene. Genes (and gene products) related in this way are said to be **homologous**. Given the complete genome sequences of representative organisms from all three domains of life—archaea, bacteria, and eucaryotes—one can search systematically for homologies that span this enormous evolutionary divide. In this way, we can begin to take stock of the common inheritance of all living things and to trace life's origins back to the earliest ancestral cells. There are difficulties in this enterprise: some ancestral genes are lost, and some have changed so much that they are not readily recognizable as relatives. Despite these uncertainties, comparing genome sequences from the most widely separated branches of the tree of life can give us a sense of which genes are fundamental necessities for living cells.

A comparison of the complete genomes of five bacteria, one archaeon, and one eucaryote (a yeast) revealed a core set of 239 families of protein-coding genes that have representatives in all three domains. Most of these genes can be assigned a function, with the largest number of shared gene families being involved in amino acid metabolism and transport, and in the production and function of ribosomes. Thus the minimum number of genes needed for a cell to be viable in today's environments is probably not much less than 200–300. The most streamlined genome recorded to date is that of a bacterium called *Carsonella ruddii*, which lives inside specialized cells in plant lice and has 182 genes. This organism, however, depends on the genes of its insect host to carry out many of its essential functions.

Most organisms possess significantly more than the estimated minimum couple of hundred genes. Even procaryotes—frugal cells that carry very little superfluous genetic baggage—typically have genomes that contain at least 1 million nucleotide pairs and encode 1000–8000 genes. With these few thousand genes, bacteria are able to thrive in even the most hostile environments on Earth.

The compact genomes of typical bacteria are dwarfed by the genomes of typical eucaryotes. The human genome, for example, contains about 700 times more DNA than the *E. coli* genome, and the genome of a fern contains about 100 times more than that of a human (Figure 1–41). In terms of gene numbers, however, the differences are not so great. We have only about seven times as many genes as *E. coli*, if we count a gene as the stretch of DNA that contains the specifications for a protein molecule. Moreover, many of our 24,000 protein-coding genes and corresponding proteins themselves fall into closely related family groups, such as the family of hemoglobins, which has nine closely related members in humans. The number of fundamentally different proteins in a human is thus not very many times more than in a bacterium, and the number of human genes that have identifiable counterparts in the bacterium is a significant fraction of the total.

The rest of our human DNA—the vast bulk that does not code for protein or for functional RNA molecules—is a mixture of sequences that help regulate the expression of the genes, and sequences that seem to be dispensable junk, retained like a mass of old papers because, if there is no pressure to keep an archive small, it is easier to save everything than to

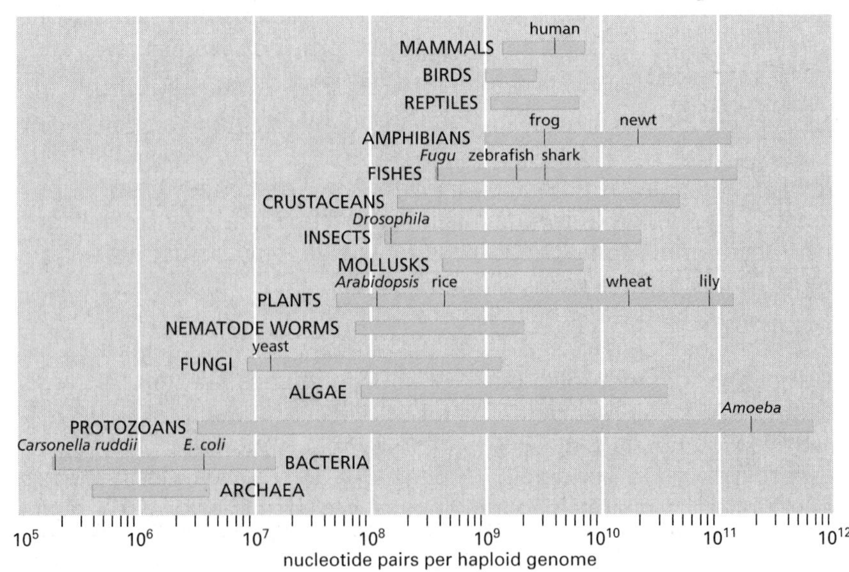

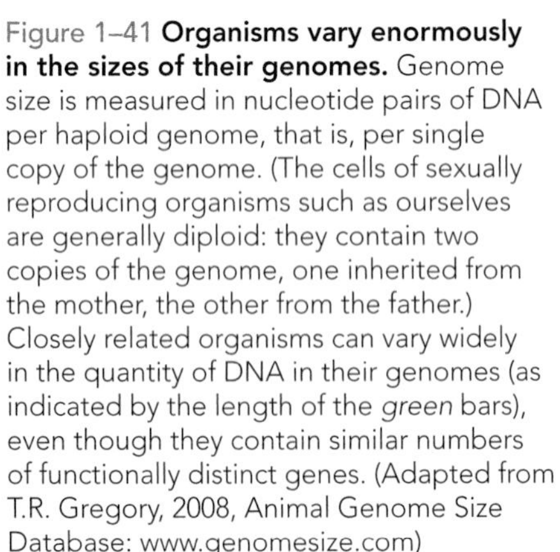

Figure 1–41 **Organisms vary enormously in the sizes of their genomes.** Genome size is measured in nucleotide pairs of DNA per haploid genome, that is, per single copy of the genome. (The cells of sexually reproducing organisms such as ourselves are generally diploid: they contain two copies of the genome, one inherited from the mother, the other from the father.) Closely related organisms can vary widely in the quantity of DNA in their genomes (as indicated by the length of the *green* bars), even though they contain similar numbers of functionally distinct genes. (Adapted from T.R. Gregory, 2008, Animal Genome Size Database: www.genomesize.com)

sort out the valuable information and discard the rest. The large quantity of regulatory DNA allows for enormous complexity and sophistication in the way different genes in a eucaryotic multicellular organism are brought into action at different times and places. But the basic list of parts—the set of proteins that our cells can make, as specified by the DNA—is not much longer than the parts list of an automobile, and many of those parts are common not only to all animals, but also to the entire living world.

That a length of DNA can program the growth, development, and reproduction of living cells and complex organisms is truly an amazing phenomenon. In the rest of this book, we will try to explain how cells work—by examining their component parts and seeing how these parts work together, and by investigating how the genome of each cell directs the manufacture of the components so as to reproduce and run each living thing.

ESSENTIAL CONCEPTS

- **Cells are the fundamental units of life. All present-day cells are believed to have evolved from an ancestral cell that existed more than 3 billion years ago.**
- **All cells, and hence all living things, grow, convert energy from one form to another, sense and respond to their environment, and reproduce themselves.**
- **All cells are enclosed by a plasma membrane that separates the inside of the cell from the environment.**
- **All cells contain DNA as a store of genetic information and use it to guide the synthesis of RNA molecules and of proteins.**
- **Cells in a multicellular organism, though they all contain the same DNA, can be very different. They turn on different sets of genes according to their developmental history and to cues they receive from their environment.**
- **Cells of animal and plant tissues are typically 5–20 μm in diameter and can be seen with a light microscope, which also reveals some of their internal components, or organelles.**
- **The electron microscope permits the smaller organelles and even individual large molecules to be seen, but specimens require elaborate preparation and cannot be viewed alive.**
- **The simplest of present-day living cells are procaryotes: although they contain DNA, they lack a nucleus and other organelles and probably resemble most closely the ancestral cell.**
- **Different species of procaryotes are diverse in their chemical capabilities and inhabit an amazingly wide range of habitats. Two fundamental evolutionary subdivisions are recognized: bacteria and archaea.**
- **Eucaryotic cells possess a nucleus and other organelles not found in procaryotes. They probably evolved in a series of stages. An important step appears to have been the acquisition of mitochondria, which are thought to have originated from bacteria engulfed by an ancestral eucaryotic cell.**
- **The nucleus is the most prominent organelle in most plant and animal cells. It contains the genetic information of the organism, stored in DNA molecules. The rest of the cell's contents, apart from the nucleus, constitute the cytoplasm.**
- **The cytoplasm includes all of the cell's contents outside the nucleus. It contains a variety of membrane-enclosed organelles with specialized chemical functions. Mitochondria carry out the oxidation of food**

molecules. In plant cells, chloroplasts perform photosynthesis. The endoplasmic reticulum, the Golgi apparatus, and lysosomes permit cells to synthesize complex molecules for export from the cell and for insertion in cell membranes, and to import and digest large molecules.

- Outside the membrane-enclosed organelles in the cytoplasm is the cytosol, a concentrated mixture of large and small molecules that carry out many essential biochemical processes.
- The cytoskeleton extends throughout the cytoplasm. This system of protein filaments is responsible for cell shape and movement and for the transport of organelles and molecules from one location to another in the cytoplasm.
- Free-living, single-celled eucaryotic microorganisms include some of the most complex eucaryotic cells known, and they are able to swim, mate, hunt, and devour food.
- An animal, plant, or fungus consists of diverse eucaryotic cell types all derived from a single fertilized egg cell; the number of such cells cooperating to form a large multicellular organism such as a human runs into thousands of billions.
- Biologists have chosen a small number of model organisms to study closely. These include the bacterium *E. coli*, brewer's yeast, a nematode worm, a fly, a small plant, a fish, a mouse, and the human species itself.
- Although the minimum number of genes needed for a viable cell is less than 400, most cells contain significantly more. Yet even such a complex organism as a human has only about 24,000 protein-coding genes—twice as many as a fly and seven times as many as *E. coli*.

KEY TERMS

archaeon	eucaryote	nanometer
bacterium	evolution	nucleus
cell	genome	organelle
chloroplast	homologous	procaryote
chromosome	micrometer	protein
cytoplasm	microscope	protozoan
cytoskeleton	mitochondrion	ribosome
cytosol	model organism	RNA
DNA		

QUESTIONS

QUESTION 1–9

By now you should be familiar with the following cellular components. Briefly define what they are and what function they provide for cells.

A. cytosol

B. cytoplasm

C. mitochondria

D. nucleus

E. chloroplasts

F. lysosomes

G. chromosomes

H. Golgi apparatus

I. peroxisomes

J. plasma membrane

K. endoplasmic reticulum

L. cytoskeleton

QUESTION 1–10

Which of the following statements are correct? Explain your answers.

A. The hereditary information of a cell is passed on by its proteins.

B. Bacterial DNA is found in the cytosol.

C. Plants are composed of procaryotic cells.

D. All cells of the same organism have the same number of chromosomes (with the exception of egg and sperm cells).

E. The cytosol contains membrane-enclosed organelles, such as lysosomes.

F. Nuclei and mitochondria are surrounded by a double membrane.

G. Protozoans are complex organisms with a set of specialized cells that form tissues, such as flagella, mouthparts, stinging darts, and leglike appendages.

H. Lysosomes and peroxisomes are the site of degradation of unwanted materials.

QUESTION 1–11

To get a feeling for the size of cells (and to practice the use of the metric system) consider the following: the human brain weighs about 1 kg and contains about 10^{12} cells. Calculate the average size of a brain cell (although we know that their sizes vary widely), assuming that each cell is entirely filled with water (1 cm^3 of water weighs 1 g). What would be the length of one side of this average-sized brain cell if it were a simple cube? If the cells were spread out as a thin layer that is only a single cell thick, how many pages of this book would this layer cover?

QUESTION 1–12

Identify the different organelles indicated with letters in the electron micrograph shown below. Estimate the length of the scale bar in the figure.

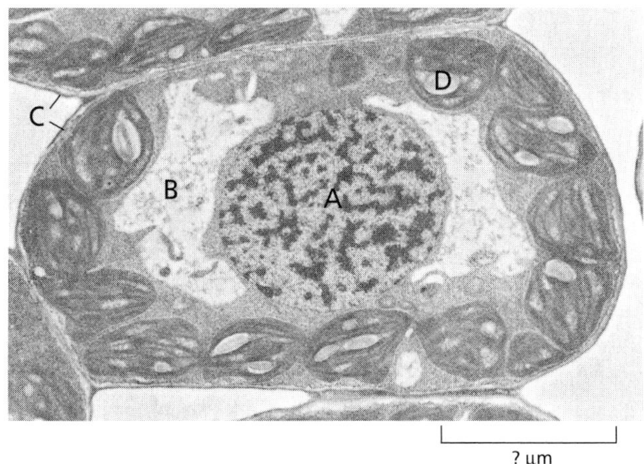

QUESTION 1–13

There are three major classes of filaments that make up the cytoskeleton. What are they and what are the differences in their functions? Which cytoskeletal filaments would be most plentiful in a muscle cell or in an epidermal cell making up the outer layer of the skin? Explain your answers.

QUESTION 1–14

Natural selection is such a powerful force in evolution because cells with even a small proliferation advantage quickly outgrow their competitors. To illustrate this process, consider a cell culture that contains 1 million bacterial cells that double every 20 minutes. A single cell in this culture acquires a mutation that allows it to divide faster, with a generation time of only 15 minutes. Assuming that there is an unlimited food supply and no cell death, how long would it take before the progeny of the mutated cell became predominant in the culture? (Before you go through the calculation, make a guess: do you think it would take about a day, a week, a month, or a year?) How many cells of either type are present in the culture at this time? (The number of cells N in the culture at time t is described by the equation $N = N_0 \times 2^{t/G}$, where N_0 is the number of cells at zero time and G is the generation time.)

QUESTION 1–15

When bacteria are grown under adverse conditions, i.e., in the presence of a poison such as an antibiotic, most cells grow and proliferate slowly. But it is not uncommon that the growth rate of a bacterial culture kept in the presence of the poison is restored after a few days to that observed in its absence. Suggest why this may be the case.

QUESTION 1–16

Apply the principle of exponential growth of a culture as described in Question 1–14 to the cells in a multicellular organism, such as yourself. There are about 10^{13} cells in your body. Assume that one cell acquires a mutation that allows it to divide in an uncontrolled manner (i.e., it becomes a cancer cell). Some cancer cells can proliferate with a generation time of about 24 hours. If none of the cancer cells died, how long would it take before 10^{13} cells in your body would be cancer cells? (Use the equation $N = N_0 \times 2^{t/G}$, with t, the time, and G, the length of each generation. Hint: $10^{13} \cong 2^{43}$.)

QUESTION 1–17

Discuss the following statement: "The structure and function of a living cell are dictated by the laws of physics and chemistry."

QUESTION 1–18

What, if any, are the advantages in being multicellular?

QUESTION 1–19

Draw to scale the outline of two spherical cells, one a bacterium with a diameter of 1 μm, the other an animal cell with a diameter of 15 μm. Calculate the volume, surface area, and surface-to-volume ratio for each cell. How would the latter ratio change if you included the internal membranes of the cell in the calculation of surface area (assume internal membranes have 15 times the area of the plasma membrane)? (The volume of a sphere is given by $4\pi R^3/3$ and its surface by $4\pi R^2$, where R is its radius.) Discuss the following hypothesis: "Internal membranes allowed bigger cells to evolve."

QUESTION 1–20

What are the arguments that all living cells evolved from a common ancestor cell? Imagine the very early days of evolution of life on Earth. Would you assume that the primordial ancestor cell was the first and only cell to form?

QUESTION 1–21

In Figure 1–26, proteins are blue, nucleic acids are orange or red, lipids are yellow, and polysaccharides are green. Identify major organelles and other important cellular structures shown in this slice through a eucaryotic cell.

QUESTION 1–22

Looking at some pond water under the microscope, you notice an unfamiliar rod-shaped cell about 200 μm long. Knowing that some exceptional bacteria can be as big as this or even bigger, you wonder whether your cell is a bacterium or a eucaryote. How will you decide? If it is not a eucaryote, how will you discover whether it is a bacterium or an archaeon?

CHAPTER **TWO**

2

Chemical Components of Cells

CHEMICAL BONDS

MOLECULES IN CELLS

MACROMOLECULES IN CELLS

It is at first sight difficult to accept that living creatures are merely chemical systems. Their incredible diversity of form, their seemingly purposeful behavior, and their ability to grow and reproduce all seem to set them apart from the world of solids, liquids, and gases that chemistry normally describes. Indeed, until the nineteenth century it was widely believed that animals contained a vital force—an "animus"—that was responsible for their distinctive properties.

We now know that there is nothing in living organisms that disobeys chemical or physical laws. However, the chemistry of life is indeed a special kind. First, it is based overwhelmingly on carbon compounds, the study of which is known as *organic chemistry*. Second, it depends almost exclusively on chemical reactions that take place in a watery, or *aqueous*, solution and in the relatively narrow range of temperatures experienced on Earth. Third, it is enormously complex: even the simplest cell is vastly more complicated in its chemistry than any other chemical system known. Fourth, it is dominated and coordinated by collections of enormous *polymeric molecules*—those formed from chains of chemical **subunits** linked end-to-end—whose unique properties enable cells and organisms to grow and reproduce and to do all the other things that are characteristic of life. Finally, it is tightly regulated: cells deploy a variety of mechanisms to make sure that all their chemical reactions occur at the proper place and time.

Chemistry, in a sense, dictates all of biology. In this chapter, therefore, we briefly survey the chemistry of the living cell. We will meet the molecules from which cells are made and examine their structures, their shapes, and their chemical properties. These molecules determine the size, structure,

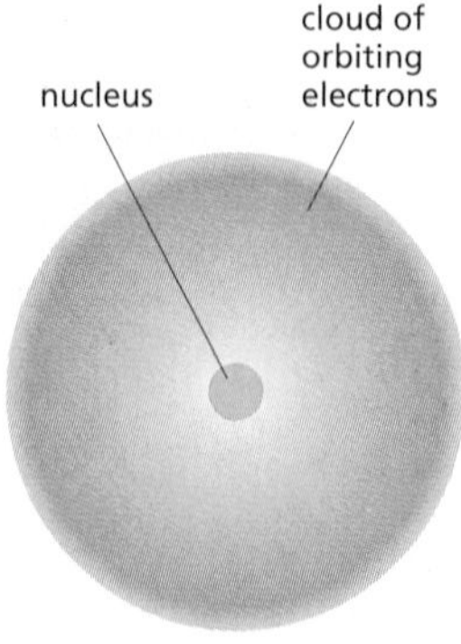

Figure 2–1 **An atom consists of a nucleus surrounded by an electron cloud.** The dense, positively charged nucleus contains most of the atom's mass. The much lighter and negatively charged electrons occupy space around the nucleus, as governed by the laws of quantum mechanics. The electrons are depicted as a continuous cloud, as there is no way of predicting exactly where an electron is at any given instant of time. The density of shading of the cloud is an indication of the probability that electrons will be found there. The diameter of the electron cloud ranges from about 0.1 nm (for hydrogen) to about 0.4 nm (for atoms of high atomic number). The nucleus is very much smaller: about 2×10^{-5} nm for carbon, for example.

and function of living cells. By understanding how these molecules interact, we can begin to see how cells exploit the laws of chemistry and physics to stay alive.

CHEMICAL BONDS

Matter is made of combinations of *elements*—substances such as hydrogen or carbon that cannot be broken down or converted into other substances by chemical means. The smallest particle of an element that still retains its distinctive chemical properties is an *atom*. The characteristics of substances other than pure elements—including the materials from which living cells are made—depend on which atoms they contain, and the way these atoms are linked together in groups to form *molecules*. In order to understand how living organisms are built from inanimate matter, therefore, it is crucial to know how the chemical bonds that hold atoms together in molecules are formed.

Cells Are Made of Relatively Few Types of Atoms

Each **atom** has at its center a dense, positively charged nucleus, which is surrounded at some distance by a cloud of negatively charged **electrons**, held in orbit by electrostatic attraction to the nucleus (Figure 2–1). The nucleus consists of two kinds of subatomic particles: **protons**, which are positively charged, and *neutrons*, which are electrically neutral. The number of protons present in an atomic nucleus determines its *atomic number*. An atom of hydrogen has a nucleus composed of a single proton; so hydrogen, with an atomic number of 1, is the lightest element. An atom of carbon has six protons in its nucleus and an atomic number of 6 (Figure 2–2). The electric charge carried by each proton is exactly equal and opposite to the charge carried by a single electron. Because the whole atom is electrically neutral, the number of negatively charged electrons surrounding the nucleus is equal to the number of positively charged protons that the nucleus contains; thus the number of electrons in an atom also equals the atomic number. All atoms of a given element have the same atomic number, and we shall shortly see that this number dictates the chemical behavior of the element.

Neutrons are uncharged subatomic particles with essentially the same mass as protons. They contribute to the structural stability of the nucleus—if there are too many or too few, the nucleus may disintegrate by radioactive decay—but they do not alter the chemical properties of the atom. Thus an element can exist in several physically distinguishable but chemically identical forms, called *isotopes*, each isotope having a different number of neutrons but the same number of protons. Multiple isotopes of

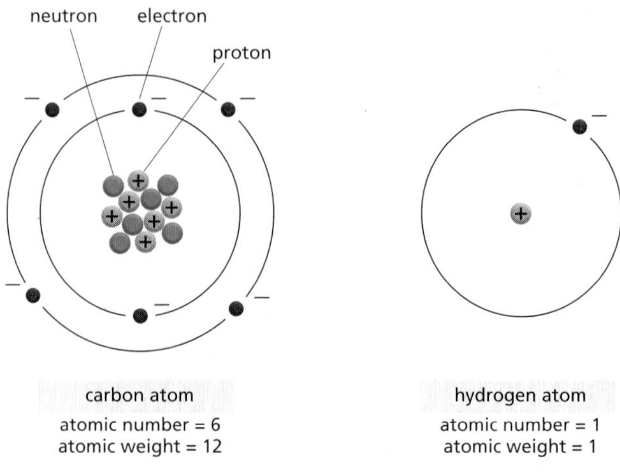

Figure 2–2 **The number of protons in an atom determines its atomic number.** Schematic representations of an atom of carbon and an atom of hydrogen. The nucleus of every atom except hydrogen consists of both positively charged protons and electrically neutral neutrons. The number of electrons in an atom is equal to the number of protons, so that the atom has no net charge. In contrast to Figure 2–1, the electrons are shown here as individual particles. The concentric *black* circles represent in a highly schematic form the "orbits" (that is, the different distributions) of the electrons. The neutrons, protons, and electrons are in reality minute in relation to the atom as a whole; their size is greatly exaggerated here.

almost all the elements occur naturally, including some that are unstable. For example, while most carbon on Earth exists as the stable isotope carbon-12, with six protons and six neutrons, there are also small amounts of an unstable isotope, the radioactive carbon-14, whose atoms have six protons and eight neutrons. Carbon-14 undergoes radioactive decay at a slow but steady rate, which is the basis for the technique of carbon-14 dating of organic material in archaeology.

The **atomic weight** of an atom, or the **molecular weight** of a molecule, is its mass relative to that of a hydrogen atom. This is essentially equal to the number of protons plus neutrons that the atom or molecule contains, because the electrons are so light that they contribute almost nothing to the total mass. Thus the major isotope of carbon has an atomic weight of 12 and is symbolized as ^{12}C. The unstable carbon isotope just mentioned has an atomic weight of 14 and is written as ^{14}C. The mass of an atom or a molecule is generally specified in *daltons*, one dalton being an atomic mass unit approximately equal to the mass of a hydrogen atom.

Atoms are so small that it is hard to imagine their size. An individual carbon atom is roughly 0.2 nm in diameter, so that it would take about 5 million of them, laid out in a straight line, to span a millimeter. One proton or neutron weighs approximately $1/(6 \times 10^{23})$ gram. Hydrogen has only one proton, with an atomic weight of one, so 1 gram of hydrogen contains 6×10^{23} atoms. For carbon, with an atomic weight of twelve, 12 grams of carbon contain 6×10^{23} atoms. This huge number (6×10^{23}, called **Avogadro's number**) is the key scale factor describing the relationship between everyday quantities and numbers of individual atoms or molecules. If a substance has a molecular weight of M, a mass of M grams of the substance will contain 6×10^{23} molecules. This quantity is called one *mole* of the substance (Figure 2–3). The concept of mole is used widely in chemistry as a way to represent the number of molecules that are available to participate in chemical reactions.

A mole is X grams of a substance, where X is its relative molecular mass (molecular weight). A mole will contain 6×10^{23} molecules of the substance.

1 mole of carbon weighs 12 g
1 mole of glucose weighs 180 g
1 mole of sodium chloride weighs 58 g

Molar solutions have a concentration of 1 mole of the substance in 1 liter of solution. A molar solution (denoted as 1 M) of glucose, for example, has 180 g/l, and a millimolar solution (1 mM) has 180 mg/l.

The standard abbreviation for gram is g; the abbreviation for liter is l.

Figure 2–3 **What's a mole?** Some sample calculations of moles and molar solutions.

There are 92 naturally occurring elements, each differing from the others in the number of protons and electrons in its atoms. Living organisms, however, are made of only a small selection of these elements, four of which—carbon (C), hydrogen (H), nitrogen (N), and oxygen (O)—make up 96.5% of an organism's weight. This composition differs markedly from that of the nonliving inorganic environment (Figure 2–4) and is evidence of a distinctive type of chemistry.

The Outermost Electrons Determine How Atoms Interact

To understand how atoms come together to form the molecules that make up living organisms, we have to pay special attention to their electrons. Protons and neutrons are welded tightly to one another in the nucleus and change partners only under extreme conditions—during radioactive decay, for example, or in the interior of the sun or of a nuclear reactor. In living tissues, only the electrons of an atom undergo rearrangements. They form the accessible part of the atom and specify the rules of chemistry by which atoms combine to form molecules.

Electrons are in continuous motion around the nucleus, but motions on this submicroscopic scale obey different laws from those we are familiar with in everyday life. These laws dictate that electrons in an atom can exist only in certain discrete regions of movement—roughly speaking, discrete orbits—and that there is a strict limit to the number of electrons that can be accommodated in an orbit of a given type, a so-called *electron shell*. The electrons closest on average to the positive nucleus are attracted most strongly to it and occupy the inner, most tightly bound shell. This innermost shell can hold a maximum of two electrons. The second shell

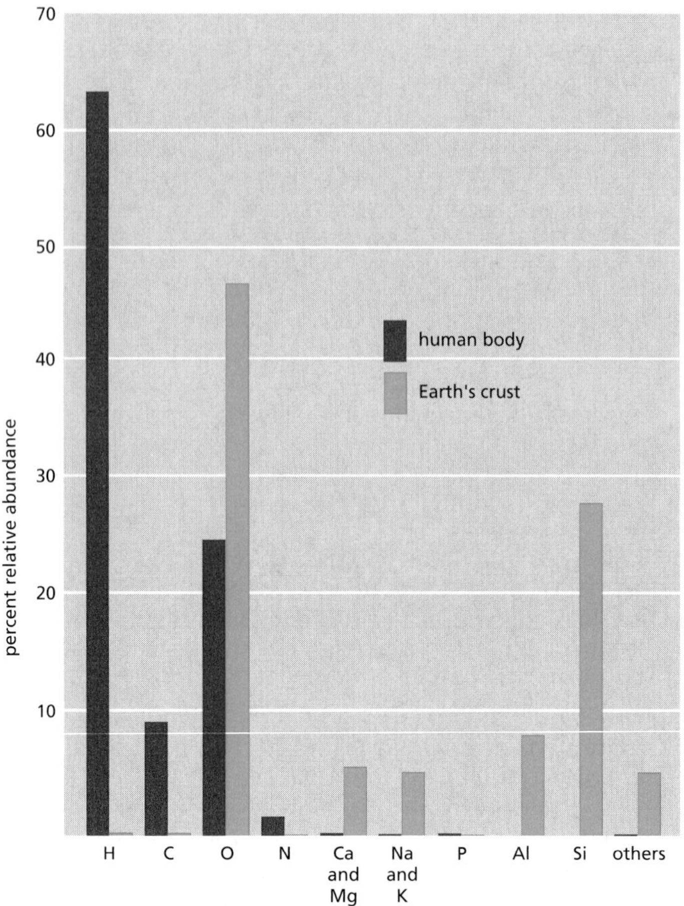

Figure 2–4 **The distribution of elements in Earth's crust differs radically from that in the tissues of an animal.** The abundance of each element is expressed here as a percentage of the total number of atoms present in a biological or geological sample, including water. Thus, for example, more than 60% of the atoms in a living organism are hydrogen atoms. The relative abundance of elements is similar in all living things.

QUESTION 2–1

A cup of water, containing exactly 18 g, or 1 mole, of water, was emptied into the Aegean Sea 3000 years ago. What are the chances that the same quantity of water, scooped today from the Pacific Ocean, would include at least one of these "Greek" water molecules? Assume perfect mixing and an approximate volume for the world's oceans of 1.5 billion cubic kilometers (1.5×10^9 km^3).

is farther away from the nucleus, and its electrons are less tightly bound. This second shell can hold up to eight electrons. The third shell contains electrons that are even less tightly bound; it can also hold up to eight electrons. The fourth and fifth shells can hold 18 electrons each. Atoms with more than four shells are very rare in biological molecules.

The arrangement of electrons in an atom is most stable when all the electrons are in the most tightly bound states that are possible for them—that is, when they occupy the innermost shells, closest to the positively charged nucleus. Therefore, with certain exceptions in the larger atoms, the electrons of an atom fill the shells in order—the first before the second, the second before the third, and so on. An atom whose outermost shell is entirely filled with electrons is especially stable and therefore chemically unreactive. Examples are helium with 2 electrons (and an atomic number of 2), neon with 2 + 8 (atomic number 10), and argon with 2 + 8 + 8 (atomic number 18); these are all inert gases. Hydrogen, by contrast, has only one electron, which leaves its outermost shell half-filled, so it is highly reactive. The atoms found in living tissues all have incomplete outer electron shells and are therefore able to react with one another to form molecules (Figure 2–5).

Because an unfilled electron shell is less stable than a filled one, atoms with incomplete outer shells have a strong tendency to interact with other atoms so as to either gain or lose enough electrons to achieve a completed outermost shell. This electron exchange can be achieved either by transferring electrons from one atom to another or by sharing electrons between two atoms. These two strategies generate the two types of **chemical bonds** that bind atoms to one another: an *ionic bond* is formed when electrons are donated by one atom to another, whereas a *covalent bond* is formed when two atoms share a pair of electrons (Figure 2–6). In the case of the covalent bond, the pair of electrons is often shared

atomic number

	element	I	II	III	IV
		electron shell			
1	Hydrogen	●			
2	Helium	●●			
6	Carbon	●●	●●●●		
7	Nitrogen	●●	●●●●●		
8	Oxygen	●●	●●●●●●		
10	Neon	●●	●●●●●●●●		
11	Sodium	●●	●●●●●●●●	●	
12	Magnesium	●●	●●●●●●●●	●●	
15	Phosphorus	●●	●●●●●●●●	●●●●●	
16	Sulfur	●●	●●●●●●●●	●●●●●●	
17	Chlorine	●●	●●●●●●●●	●●●●●●●	
18	Argon	●●	●●●●●●●●	●●●●●●●●	
19	Potassium	●●	●●●●●●●●	●●●●●●●●	●
20	Calcium	●●	●●●●●●●●	●●●●●●●●	●●

Figure 2–5 **An element's chemical reactivity is based on how its outermost electron shell is filled.** All of the elements commonly found in living organisms have unfilled outermost shells *(red)* and can thus participate in chemical reactions with other atoms. Inert gases *(yellow)*, in contrast, have only filled shells and are chemically unreactive.

unequally, with one atom attracting the shared electrons more than the other; this results in a *polar covalent bond*, as we will discuss later.

An H atom, which needs only one more electron to fill its shell, generally acquires it by sharing—forming one covalent bond with another atom; in many cases this bond is polar. The other most common elements in living cells—C, N, and O, which have an incomplete second shell, and P and S, which have an incomplete third shell (see Figure 2–5)—generally share electrons and achieve a filled outer shell of eight electrons by forming several covalent bonds. The number of electrons an atom must acquire or lose (either by sharing or by transfer) to attain a filled outer shell determines the number of bonds the atom can make.

Because the state of the outer electron shell determines the chemical properties of an element, when the elements are listed in order of their atomic number we see a periodic recurrence of elements with similar properties: an element with, say, an incomplete second shell containing one electron will behave in much the same way as an element that has filled its second shell and has an incomplete third shell containing one electron. The metals, for example, all have incomplete outer shells with just one or a few electrons, whereas, as we have just seen, the inert gases have full outer shells. This arrangement gives rise to the famous *periodic table* of the elements, which is outlined in Figure 2–7. Elements found in living organisms are highlighted.

QUESTION 2–2

A carbon atom contains six protons and six neutrons.

A. What are its atomic number and atomic weight?

B. How many electrons does it have?

C. How many additional electrons must it add to fill its outermost shell? How does this affect carbon's chemical behavior?

D. Carbon with an atomic weight of 14 is radioactive. How does it differ in structure from nonradioactive carbon? How does this difference affect its chemical behavior?

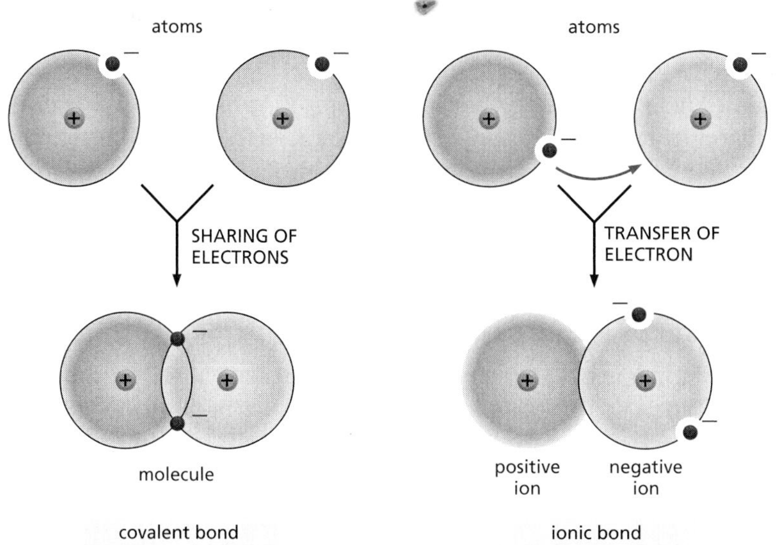

Figure 2–6 **Atoms can attain a more stable arrangement of electrons in their outermost shell by interacting with one another.** A covalent bond is formed when electrons are shared between atoms. An ionic bond is formed when electrons are transferred from one atom to the other. The two cases shown represent extremes; often, covalent bonds form with a partial transfer (unequal sharing of electrons), resulting in a polar covalent bond (see, for example, Figure 3–11).

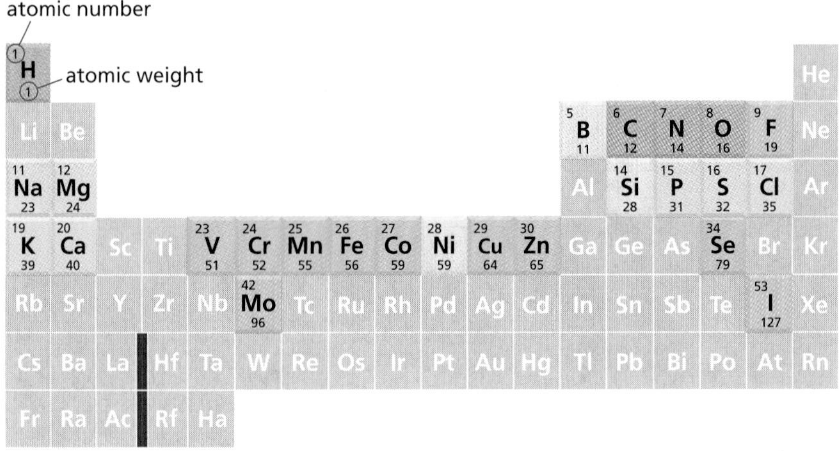

Figure 2–7 **Elements ordered by their atomic number form the periodic table.** Elements fall into groups that show similar properties based on the number of electrons each element possesses in its outer shell. Atoms in the same vertical column must gain (or lose) the same number of electrons to attain a filled outer shell, and they thus behave similarly in bond or ion formation. Thus, Mg and Ca tend to give away the two electrons in their outer shells.

The four elements highlighted in *red* constitute 99% of the total number of atoms present in the human body. An additional seven elements, highlighted in *blue*, together represent about 0.9% of the total. Other elements, shown in *green*, are required in trace amounts by humans. It remains unclear whether those elements shown in *yellow* are essential in humans or not. The chemistry of life, it seems, is therefore predominantly the chemistry of lighter elements.

Atomic weights, given by the sum of the protons and neutrons in the atomic nucleus, can vary with the particular isotope of the element. The atomic weights shown here are those of the most common isotope of each element.

Ionic Bonds Form by the Gain and Loss of Electrons

Ionic bonds are most likely to be formed by atoms that have just one or two electrons in their unfilled outer shell or are just one or two electrons short of acquiring a filled outer shell. These atoms can generally attain a completely filled outer electron shell most easily by giving electrons to—or accepting electrons from—another atom, rather than by sharing them. For example, returning to Figure 2–5, we see that a sodium (Na) atom, with atomic number 11, can strip itself down to a filled shell by giving up the single electron external to its second shell. By contrast, a chlorine (Cl) atom, with atomic number 17, can complete its outer shell by gaining just one electron. Consequently, if a Na atom encounters a Cl atom, an electron can jump from the Na to the Cl, leaving both atoms with filled outer shells. The offspring of this marriage between sodium, a soft and intensely reactive metal, and chlorine, a toxic green gas, is table salt (NaCl).

When an electron jumps from Na to Cl, both atoms become electrically charged **ions**. The Na atom that lost an electron now has one less electron than it has protons in its nucleus; it therefore has a net single positive charge (Na^+). The Cl atom that gained an electron now has one more electron than it has protons and has a single negative charge (Cl^-). Positive ions are called *cations*, and negative ions *anions*. Ions can be further classified according to how many electrons are lost or gained. Na and potassium (K) have one electron to lose, so they form cations with a single positive charge (Na^+ and K^+); magnesium (Mg) and calcium (Ca) have two electrons to lose and form cations with two positive charges (Mg^{2+} and Ca^{2+}).

Because of their opposite charges, Na^+ and Cl^- are attracted to each other and are thereby held together in an **ionic bond**. A salt crystal contains astronomical numbers of Na^+ and Cl^- packed together in a precise three-dimensional array with their opposite charges exactly balanced—a crystal only 1 mm across contains about 2×10^{19} ions of each type (Figure 2–8). Substances such as NaCl, which are held together solely by ionic bonds, are generally called *salts* rather than molecules.

Ionic bonds are a type of **electrostatic attraction**—an attractive force that occurs between oppositely charged atoms (see Panel 2–7, pp. 76–77). Because of the favorable interaction between ions and water molecules (which are polar), many salts (including NaCl) are highly soluble in water. They dissociate into individual ions (such as Na^+ and Cl^-), each surrounded by a group of water molecules. We discuss electrostatic attractions—and the other *noncovalent bonds* that can exist between atoms—later in the chapter.

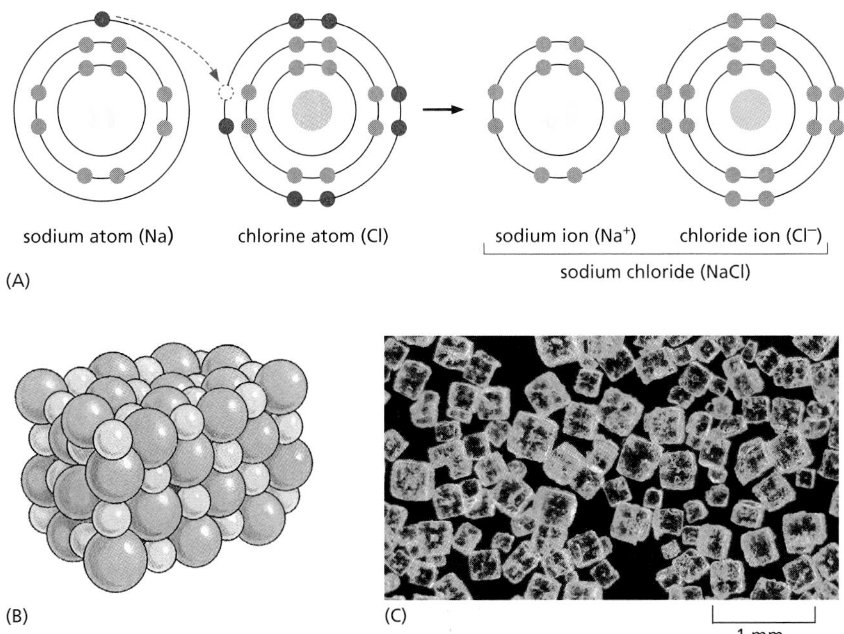

Figure 2–8 **Sodium chloride is held together by ionic bond formation.** (A) An atom of sodium (Na) reacts with an atom of chlorine (Cl). Electrons of each atom are shown in their different energy levels; electrons in the chemically reactive (incompletely filled) shells are shown in *red*. The reaction takes place with transfer of a single electron from sodium to chlorine, forming two electrically charged atoms, or ions, each with complete sets of electrons in their outermost levels. The two ions with opposite charge are held together by electrostatic attraction. (B) The product of the reaction between sodium and chlorine, crystalline sodium chloride, contains sodium and chloride ions packed closely together in a regular array in which the charges are exactly balanced. (C) Color photograph of crystals of sodium chloride.

Covalent Bonds Form by the Sharing of Electrons

All of the characteristics of a cell depend on the molecules it contains. A **molecule** is a cluster of atoms held together by **covalent bonds**, in which electrons are shared rather than transferred between atoms. The shared electrons complete the outer shells of both atoms. In the simplest possible molecule—a molecule of hydrogen (H_2)—two H atoms, each with a single electron, share their two electrons, thus filling their outermost shells. The shared electrons form a cloud of negative charge that is densest between the two positively charged nuclei. This electron density helps to hold the nuclei together by opposing the mutual repulsion between the like charges that would otherwise force them apart. The attractive and repulsive forces are in balance when the nuclei are separated by a characteristic distance, called the *bond length* (Figure 2–9).

Whereas an H atom can form only a single covalent bond, the other common atoms that form covalent bonds in cells—O, N, S, and P, as well as the all-important C—can form more than one. The outermost shells of these atoms, as we have seen, can accommodate up to eight electrons, and they form covalent bonds with as many other atoms as necessary to reach this number. Oxygen, with six electrons in its outer shell, is most stable when it acquires two extra electrons by sharing with other atoms, and it therefore forms up to two covalent bonds. Nitrogen, with five outer electrons, forms a maximum of three covalent bonds, while carbon, with four outer electrons, forms up to four covalent bonds—thus sharing four pairs of electrons (see Figure 2–5).

When one atom forms covalent bonds with several others, these multiple bonds have definite orientations in space relative to one another, reflect-

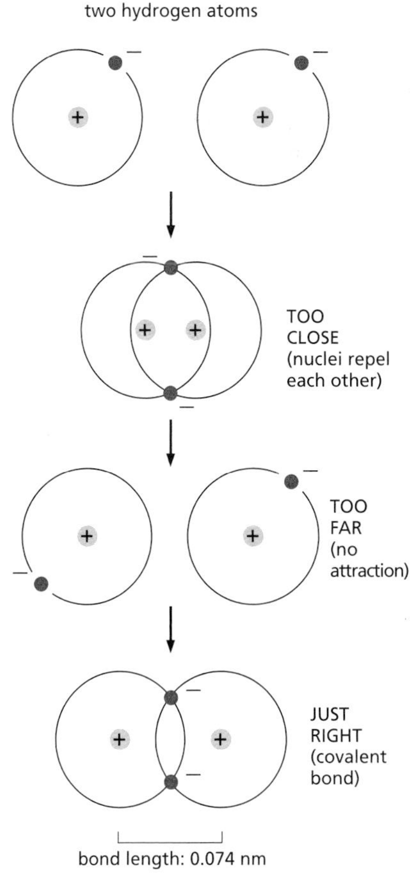

Figure 2–9 **The hydrogen molecule is held together by a covalent bond.** Each hydrogen atom in isolation has a single electron, which means that its first (and only) electron shell is incompletely filled. By coming together, the two atoms are able to share the two electrons, and each obtains a completely filled first shell, with the shared electrons adopting modified orbits around the two nuclei. The covalent bond between the two atoms has a definite length (0.074 nm). If the atoms were closer together, the positive nuclei would repel each other; if they were farther apart than this distance, they would not be able to share electrons as effectively.

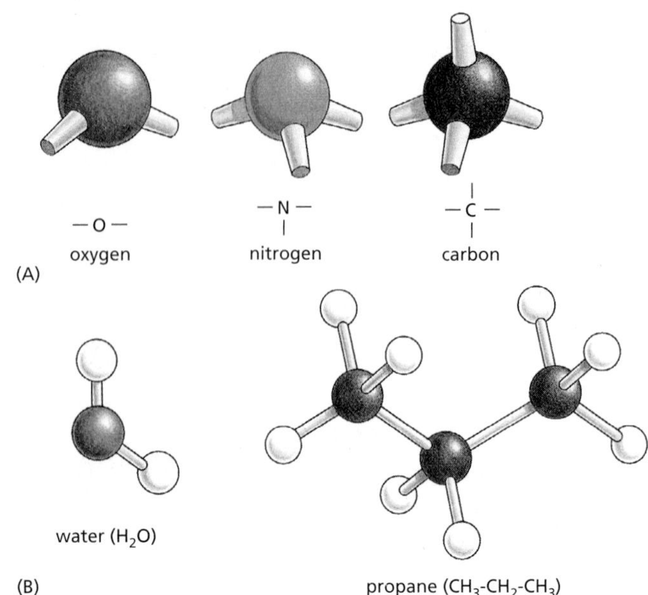

Figure 2–10 **Covalent bonds are characterized by particular geometries.** (A) The spatial arrangement of the covalent bonds that can be formed by oxygen, nitrogen, and carbon. (B) Molecules formed from these atoms therefore have a precise three-dimensional structure defined by the bond angles and bond lengths for each covalent linkage. A water molecule, for example, forms a "V" shape with an angle close to 109°. In these ball-and-stick models, the different colored balls represent the atoms, and the sticks represent the covalent bonds. The colors traditionally used to represent the different atoms—*black* for carbon, *white* for hydrogen, *blue* for nitrogen, and *red* for oxygen—were established by the chemist August Wilhelm Hofmann in 1865 when he used a set of colored croquet balls to build molecular models for a public lecture on the "combining power" of atoms.

ing the orientations of the orbits of the shared electrons. Covalent bonds between multiple atoms are therefore characterized by specific bond angles as well as bond lengths and bond energies (Figure 2–10). The four covalent bonds that can form around a carbon atom, for example, are arranged as if pointing to the four corners of a regular tetrahedron. The precise orientation of the covalent bonds around carbon is the basis for the three-dimensional geometry of organic molecules.

Covalent Bonds Vary in Strength

We have already seen that the covalent bond between two atoms has a characteristic length that depends on the atoms involved. A further crucial property of any bond—covalent or noncovalent—is its strength. *Bond strength* is measured by the amount of energy that must be supplied to break a bond, usually expressed in units of either kilocalories per mole (kcal/mole) or kilojoules per mole (kJ/mole). A kilocalorie is the amount of energy needed to raise the temperature of 1 liter of water by 1°C. Thus if 1 kilocalorie of energy must be supplied to break 6×10^{23} bonds of a specific type (that is, 1 mole of these bonds), then the strength of that bond is 1 kcal/mole. The other unit, kJ/mole, derived from the SI units (Système Internationale d'Unités) universally employed by physical scientists, is increasingly used by cell biologists. One kilocalorie is equal to about 4.2 kJ. Typical strengths and lengths of the main classes of chemical bonds are given in Table 2–1.

To get an idea of what bond strengths mean, it is helpful to compare them with the average energies of the impacts that molecules continually undergo owing to collisions with other molecules in their environment—their thermal, or heat, energy. Typical covalent bonds are stronger than these thermal energies by a factor of 100, so they are resistant to being pulled apart by thermal motions—heating—and are normally broken only during specific chemical reactions with other atoms and molecules. The making and breaking of covalent bonds are violent events, and in living cells these events are carefully controlled by highly specific catalysts, called *enzymes*. Noncovalent bonds as a rule are much weaker; we shall see later that they are critically important in the cell in the many situations where molecules have to associate and dissociate readily to carry out their functions.

TABLE 2–1 LENGTH AND STRENGTH OF CHEMICAL BONDS

BOND TYPE		LENGTH (nm)	STRENGTH IN kcal/mole	
			IN VACUUM	IN WATER
Covalent		0.15	90 (377)**	90 (377)
Noncovalent:	ionic bond*	0.25	80 (335)	3 (12.6)
	hydrogen bond	0.30	4 (16.7)	1 (4.2)
	van der Waals attraction (per atom)	0.35	0.1 (0.4)	0.1 (0.4)

*An ionic bond is an electrostatic attraction between two fully charged atoms.
**Values in parentheses are kJ/mole. 1 calorie = 4.184 joules.

Table 2–1 **Covalent and noncovalent chemical bonds have different lengths and strengths.** Bond strengths are measured by the energy required to break them, in kilocalories or kilojoules per mole (see Glossary for definitions of these units). The length of a hydrogen bond X–H–X is defined as the distance between the two nonhydrogen atoms (X). The bond strengths and lengths listed are approximate, because the exact values will depend on the atoms involved. The different types of noncovalent bonds are described later in the chapter (see Panel 2–7, pp. 76–77).

There Are Different Types of Covalent Bonds

Most covalent bonds involve the sharing of two electrons, one donated by each participating atom; these are called *single bonds*. Some covalent bonds, however, involve the sharing of more than one pair of electrons. Four electrons can be shared, for example, two coming from each participating atom; such a bond is called a *double bond*. Double bonds are shorter and stronger than single bonds and have a characteristic effect on the three-dimensional geometry of molecules containing them. A single covalent bond between two atoms generally allows the rotation of one part of a molecule relative to the other around the bond axis. A double bond prevents such rotation, producing a more rigid and less flexible arrangement of atoms (Figure 2–11). This restriction has a major influence on the three-dimensional shape of many macromolecules. Panel 2–1 (pp. 64–65) reviews the chemical bonds commonly encountered in biological molecules.

Some molecules contain atoms that share electrons in a way that produces bonds that are intermediate in character between single and double bonds. The highly stable benzene molecule, for example, is made up of a ring of six carbon atoms in which the bonding electrons are evenly distributed (although the arrangement is sometimes depicted as an alternating sequence of single and double bonds, as shown in Panel 2–1).

When the atoms joined by a single covalent bond belong to different elements, the two atoms usually attract the shared electrons to different degrees. Compared with a C atom, for example, O and N atoms attract electrons relatively strongly, whereas an H atom attracts electrons relatively weakly (because of the relative differences in the positive charges of the nuclei of C, O, N, and H). By definition, a **polar** structure (in the electrical sense) is one in which the positive charge is concentrated toward one end of the molecule (the positive pole) and the negative charge is concentrated toward the other end (the negative pole). Covalent bonds in which the electrons are shared unequally in this way are therefore known as polar covalent bonds. For example, the covalent bond between oxygen and hydrogen, –O–H, or between nitrogen and hydrogen, –N–H, is polar (Figure 2–12). The bond between carbon and hydrogen, –C–H, by contrast, has the electrons attracted much more equally by both atoms and is relatively nonpolar.

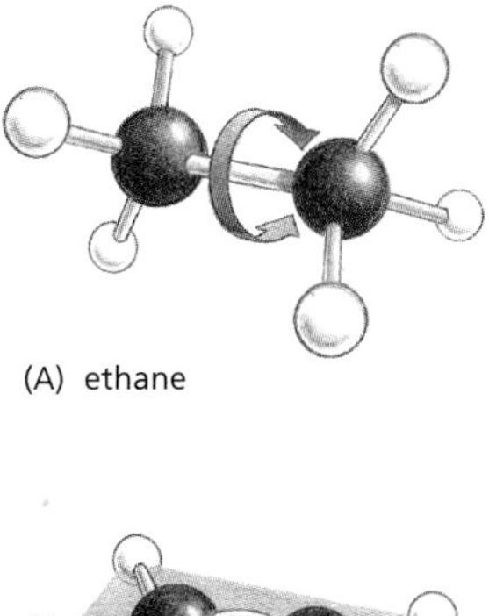

Figure 2–11 **Carbon–carbon double bonds are shorter and more rigid than carbon–carbon single bonds.** (A) The ethane molecule, with a single covalent bond between the two carbon atoms, shows the tetrahedral arrangement of single covalent bonds formed by carbon. One of the CH_3 groups joined by the covalent bond can rotate relative to the other around the bond axis. (B) The double bond between the two carbon atoms in a molecule of ethene (ethylene) alters the bond geometry of the carbon atoms and brings all the atoms into the same plane; the double bond prevents the rotation of one CH_2 group relative to the other.

Electrostatic Attractions Help Bring Molecules Together in Cells

In aqueous solutions, covalent bonds are 10–100 times stronger than the other attractive forces between atoms, allowing their connections to define the boundaries of one molecule from another. But much of biology depends on the specific binding of different molecules to each other. This binding is mediated by a group of noncovalent attractions that are

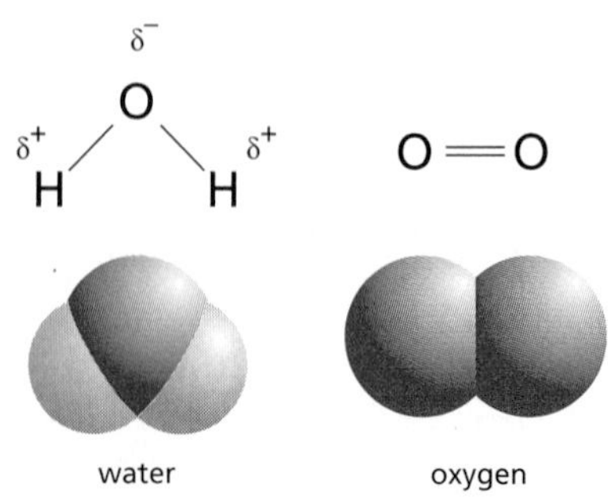

Figure 2–12 **In polar covalent bonds, the electrons are shared unequally.** Comparison of electron distributions in polar molecules such as water (H_2O) and nonpolar molecules such as oxygen (O_2). δ^+ indicates partial positive charge; δ^- indicates partial negative charge.

QUESTION 2–3

Discuss whether the following statement is correct: "An ionic bond can, in principle, be thought of as a very polar covalent bond. Polar covalent bonds, then, fall somewhere between ionic bonds at one end of the spectrum and nonpolar covalent bonds at the other end."

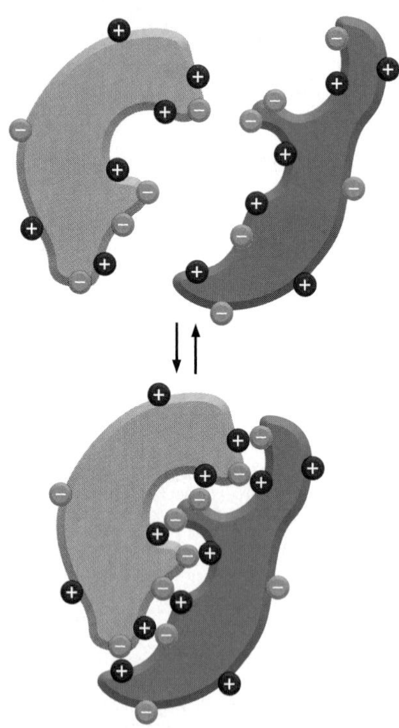

Figure 2–13 **Large molecules, such as proteins, can bind to one another through complementary charges on their surfaces.**

individually quite weak, but whose energies can sum to create an effective force between two separate molecules. We have already described the ionic bonds that hold together the Na^+ and Cl^- ions in a salt crystal. Electrostatic attractions are strongest when the atoms involved are fully charged—as Na^+ and Cl^- are. But a weaker electrostatic attraction also occurs between molecules that contain polar covalent bonds.

Polar covalent bonds are thus extremely important in biology because they allow molecules to interact through electrical forces. Any large molecule with many polar groups will have a pattern of partial positive and negative charges on its surface. When such a molecule encounters a second molecule with a complementary set of charges, the two will be attracted to each other by an electrostatic attraction that resembles (but is weaker than) the ionic bonds that hold together salts such as NaCl. When enough of these weak noncovalent bonds form between two large molecules, their surfaces will stick specifically to each other, as illustrated in Figure 2–13. However, water greatly reduces the attractiveness of these charges for each other in most biological settings.

Water Is Held Together by Hydrogen Bonds

Water accounts for about 70% of a cell's weight, and most intracellular reactions occur in an aqueous environment. Life on Earth is thought to have begun in the ocean, and the conditions in that primeval environment put a permanent stamp on the chemistry of living things. Thus the chemistry of life has been shaped by the properties of water.

In each molecule of water (H_2O) the two H atoms are linked to the O atom by covalent bonds. The two bonds are highly polar because the O is strongly attractive for electrons, whereas the H is only weakly attractive. Consequently, there is an unequal distribution of electrons in a water molecule, with a preponderance of positive charge on the two H atoms and negative charge on the O (see Figure 2–12). When a positively charged region of one water molecule (that is, one of its H atoms) comes close to a negatively charged region (that is, the O) of a second water molecule, the electrical attraction between them can establish a weak bond called a **hydrogen bond**. These bonds are much weaker than covalent bonds and are easily broken by the random thermal motions due to the heat energy of the molecules, so each bond lasts only an exceedingly short time. But the combined effect of many weak bonds is far from trivial. Each water molecule can form hydrogen bonds through its two H atoms to two other water molecules, producing a network in which hydrogen bonds are being continually broken and formed. It is because of these interlocking hydrogen bonds that water at room temperature is a liquid—with a high boiling point and high surface tension—and not a gas. Without hydrogen bonds, life as we know it could not exist. The biologically significant properties of water are reviewed in Panel 2–2 (pp. 66–67).

Not all hydrogen atoms form hydrogen bonds. In general, a hydrogen bond can form whenever a positively charged H held in one molecule by a polar covalent linkage comes close to a negatively charged atom—typically an oxygen or a nitrogen—belonging to another molecule. Hydrogen bonds can also occur between different parts of a single large molecule, where they often help create special shapes. But the hydrogen bond is just one member of a family of weak noncovalent bonds that play a crucial role in allowing large molecules to fold up in unique ways and to bind selectively to other molecules, as we will discuss later in this chapter.

Molecules, such as alcohols, that contain polar bonds and that can form hydrogen bonds mix well with water. As mentioned previously, molecules carrying positive or negative charges (ions) likewise dissolve readily in water. Such molecules are termed **hydrophilic**, meaning that they are 'water-loving.' A large proportion of the molecules in the aqueous environment of a cell necessarily fall into this category, including sugars, DNA, RNA, and a majority of proteins. **Hydrophobic** 'water-fearing' molecules, by contrast, are uncharged and form few or no hydrogen bonds, and so do not dissolve in water. Hydrocarbons are an important example of hydrophobic cellular constituents (see Panel 2–1, pp. 64–65). In these molecules the H atoms are covalently linked to C atoms by a largely nonpolar bond. Because the H atoms have almost no net positive charge, they cannot form effective hydrogen bonds to other molecules. This makes the hydrocarbon as a whole hydrophobic—a property that is exploited by cells, whose membranes are constructed from molecules that have long hydrocarbon tails, as we shall see in Chapter 11. Because they do not dissolve in water, the hydrophobic hydrocarbons can form the thin membrane barriers that keep the aqueous interior of the cell separate from the surrounding, also aqueous, environment.

Some Polar Molecules Form Acids and Bases in Water

One of the simplest kinds of chemical reaction, and one that has profound significance in cells, takes place when a molecule possessing a highly polar covalent bond between a hydrogen and another atom dissolves in water. The hydrogen atom in such a molecule has given up its electron almost entirely to the companion atom, and so exists as an almost naked positively charged hydrogen nucleus—in other words, a *proton* (H^+). When the polar molecule becomes surrounded by water molecules, the proton will be attracted to the partial negative charge on the O atom of an adjacent water molecule; this proton can dissociate from its original partner and associate instead with the oxygen atom of the water molecule, generating a **hydronium ion** (H_3O^+) (Figure 2–14A). The reverse reaction also takes place very readily, so one has to imagine an equilibrium state in which billions of protons are constantly flitting to and fro between one molecule in the aqueous solution and another.

Substances that release protons when they dissolve in water, thus forming H_3O^+, are termed **acids**. The higher the concentration of H_3O^+, the more acidic the solution. H_3O^+ is present even in pure water, at a concentration of 10^{-7} M, as a result of the movement of protons from one water molecule to another (Figure 2–14B). By tradition, the H_3O^+ concentration is usually referred to as the H^+ concentration, even though most protons in an aqueous solution are present as H_3O^+. To avoid the use of unwieldy numbers, the concentration of H_3O^+ is expressed using a logarithmic scale called the **pH scale**, as illustrated in Panel 2–2. Pure water has a pH of 7.0 and is thus neutral—that is, neither acidic (pH < 7) nor basic (pH > 7).

QUESTION 2–4

What, if anything, is wrong with the following statement: "When NaCl is dissolved in water, the water molecules closest to the ions will tend to preferentially orient themselves so that their oxygen atoms face the sodium ions and face away from the chloride ions"? Explain your answer.

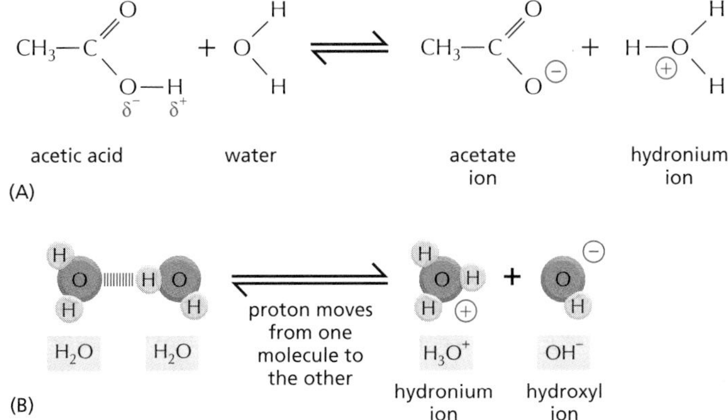

Figure 2–14 **Protons move continuously in aqueous solutions.** (A) The reaction that takes place when a molecule of acetic acid dissolves in water. At pH 7, nearly all of the acetic acid is present as acetate ion. (B) Water molecules are continually exchanging protons with each other to form hydronium and hydroxyl ions. These ions in turn rapidly recombine to form water molecules.

Acids are characterized as being strong or weak, depending on how readily they give up their protons to water. Strong acids, such as hydrochloric acid (HCl), lose their protons quickly. Acetic acid, on the other hand, is a weak acid because it holds onto its proton more tightly when dissolved in water. Many of the acids important in the cell—such as molecules containing a carboxyl (COOH) group—are weak acids (see Panel 2–2, pp. 66–67). Their tendency to dissociate with some reluctance is a useful characteristic; it renders the surfaces of large molecules sensitive to conditions in the cellular environment.

Because the proton of a hydronium ion can be passed readily to many types of molecules in cells, altering their character, the concentration of H_3O^+ inside a cell (the acidity) must be closely regulated. Acids—especially weak acids—will give up their protons more readily if the concentration of H_3O^+ in solution is low and will tend to receive them back if the concentration in solution is high.

The opposite of an acid is a **base**. Any molecule capable of accepting a proton is called a base. Just as the defining property of an acid is that it raises the concentration of H_3O^+ ions by donating a proton to a water molecule, so the defining property of a base is that it raises the concentration of hydroxyl (OH^-) ions by removing a proton from a water molecule. Thus sodium hydroxide (NaOH) is basic (the term *alkaline* is also used) because it dissociates in aqueous solution to form Na^+ ions and OH^- ions. Because NaOH dissociates readily in water, it is called a strong base. More important in living cells, however, are the weak bases—those that have a weak tendency to reversibly accept a proton from water. Many biologically important weak bases contain an amino (NH_2) group. This group can generate OH^- by taking a proton from water: $-NH_2 + H_2O \rightarrow -NH_3^+ + OH^-$ (see Panel 2–2, pp. 66–67).

QUESTION 2–5

A. Are there any H_3O^+ ions present in pure water at neutral pH (i.e., at pH = 7.0)? If so, how are they formed?

B. If they exist, what is the ratio of H_3O^+ ions to H_2O molecules at neutral pH? (Hint: the molecular weight of water is 18, and 1 liter of water weighs 1 kg.)

Because an OH^- ion combines with a H_3O^+ ion to form two water molecules, an increase in the OH^- concentration forces a decrease in the concentration of H_3O^+, and vice versa. A pure solution of water thus contains an equal concentration (10^{-7} M) of both ions, rendering it neutral. The interior of a cell is also kept close to neutrality by the presence of **buffers**: weak acids and bases that can release or take up protons near pH 7, keeping the environment of the cell relatively constant under a variety of conditions.

MOLECULES IN CELLS

Having looked at the ways atoms combine into small molecules, and how these molecules behave in an aqueous environment, we now examine the main classes of small molecules found in cells and their biological roles. Amazingly, we will see that a few basic categories of molecules, formed from a handful of different elements, give rise to all the extraordinary richness of form and behavior shown by living things.

A Cell Is Formed from Carbon Compounds

If we disregard water, nearly all of the molecules in a cell are based on carbon. Carbon is outstanding among all the elements in its ability to form large molecules; silicon—an element with the same electron configuration in its outer shell—is a poor second. Because carbon is small and has four electrons and four vacancies in its outermost shell, a carbon atom can form four covalent bonds with other atoms. Most importantly, one carbon atom can join to other carbon atoms through highly stable covalent C–C bonds to form chains and rings and hence generate large and complex molecules with no obvious upper limit to their size (see

Panel 2–1, pp. 64–65). The small and large carbon compounds made by cells are called *organic molecules*. All other molecules, including water, are said to be *inorganic* by contrast.

Certain combinations of atoms, such as the methyl ($-CH_3$), hydroxyl ($-OH$), carboxyl ($-COOH$), carbonyl ($-C{=}O$), phosphoryl ($-PO_3^{2-}$), and amino ($-NH_2$) groups, occur repeatedly in organic molecules. Each such **chemical group** has distinct chemical and physical properties that influence the behavior of the molecule in which the group occurs: whether they tend to gain or lose protons and which molecules they interact with, for example. Becoming familiar with these groups and their chemical properties greatly simplifies one's view of the chemistry of life. The most common chemical groups and some of their properties are summarized in Panel 2–1.

Cells Contain Four Major Families of Small Organic Molecules

The small organic molecules of the cell are carbon compounds with molecular weights in the range 100–1000 that contain up to 30 or so carbon atoms. They are usually found free in solution in the cytoplasm and have many different fates. Some are used as *monomer* subunits to construct the giant polymeric *macromolecules*—the proteins, nucleic acids, and large polysaccharides—of the cell. Others act as energy sources and are broken down and transformed into other small molecules in a maze of intracellular metabolic pathways. Many small molecules have more than one role in the cell—acting, for example, as both a potential subunit for a macromolecule and as an energy source. It is critical to recognize that small organic molecules are much less abundant than the organic macromolecules in living organisms, accounting for only about one-tenth of the total mass of organic matter in a cell (Table 2–2). As a rough guess, there may be a thousand different kinds of these small molecules in a typical cell.

All organic molecules are synthesized from—and are broken down into—the same set of simple compounds. Both their synthesis and their breakdown occur through sequences of simple chemical changes that are limited in variety and follow definite step-by-step rules. As a consequence, the compounds in a cell are chemically related and most can be classified into a small number of distinct families. Broadly speaking, cells contain four major families of small organic molecules: the *sugars*, the *fatty acids*, the *amino acids*, and the *nucleotides* (Figure 2–15). Although many compounds present in cells do not fit into these categories, these

TABLE 2–2 THE APPROXIMATE CHEMICAL COMPOSITION OF A BACTERIAL CELL

	PERCENTAGE OF TOTAL CELL WEIGHT	NUMBER OF TYPES OF EACH MOLECULE
Water	70	1
Inorganic ions	1	20
Sugars and precursors	1	250
Amino acids and precursors	0.4	100
Nucleotides and precursors	0.4	100
Fatty acids and precursors	1	50
Other small molecules	0.2	~300
Macromolecules (proteins, nucleic acids, polysaccharides, and phospholipids)	26	~3000

Figure 2–15 **Sugars, fatty acids, amino acids, and nucleotides constitute the four main families of small organic molecules in cells.** They form the monomeric building blocks, or subunits, for most of the macromolecules and other assemblies of the cell. Some, like the sugars and the fatty acids, are also energy sources.

building blocks of the cell		larger units of the cell
SUGARS	→	POLYSACCHARIDES
FATTY ACIDS	→	FATS, LIPIDS, MEMBRANES
AMINO ACIDS	→	PROTEINS
NUCLEOTIDES	→	NUCLEIC ACIDS

four families of small organic molecules, together with the macromolecules made by linking them into long chains, account for a large fraction of a cell's mass (see Table 2–2).

Sugars Are Energy Sources for Cells and Subunits of Polysaccharides

The simplest **sugars**—the *monosaccharides*—are compounds with the general formula $(CH_2O)_n$, where n is usually 3, 4, 5, or 6. Sugars, and the molecules made from them, are also called *carbohydrates* because of this simple formula. Glucose, for example, has the formula $C_6H_{12}O_6$ (Figure 2–16). The formula, however, does not fully define the molecule: the same set of carbons, hydrogens, and oxygens can be joined together by covalent bonds in a variety of ways, creating structures with different shapes. Glucose, for example, can be converted into a different sugar—mannose or galactose—simply by switching the orientations of specific –OH groups relative to the rest of the molecule (Panel 2–3, pp. 68–69). Each of these sugars, moreover, can exist in either of two forms, called the D-form and the L-form, which are mirror images of each other. Sets of molecules with the same chemical formula but different structures are called *isomers*, and mirror-image pairs of molecules are called *optical isomers*. Isomers are widespread among organic molecules in general, and they play a major part in generating the enormous variety of sugars. A more complete outline of sugar structures and chemistry is presented in Panel 2–3.

Monosaccharides can be linked by covalent bonds—called glycosidic bonds—to form larger carbohydrates. Two monosaccharides linked together make a disaccharide, such as sucrose, which is composed of a glucose and a fructose unit. Larger sugar polymers range from the *oligosaccharides* (trisaccharides, tetrasaccharides, and so on) up to giant *polysaccharides,* which can contain thousands of monosaccharide units. In most cases, the prefix "oligo-" is used to refer to macromolecules made

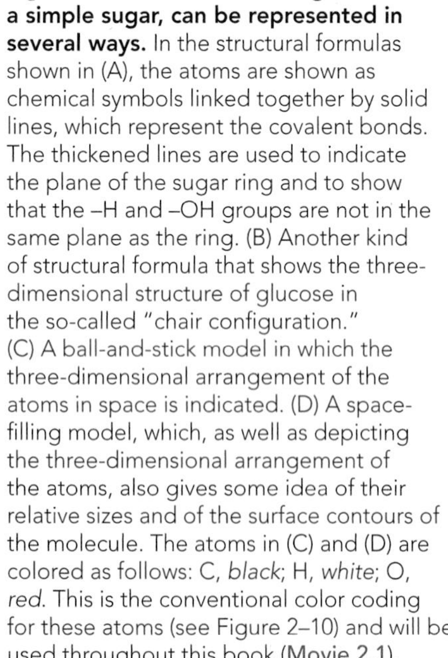

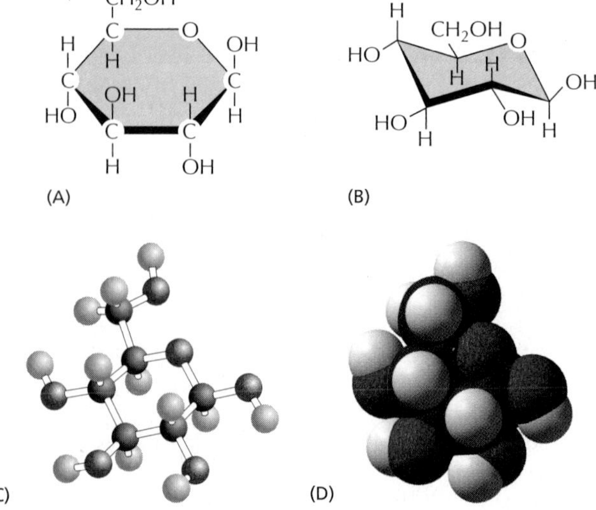

Figure 2–16 **The structure of glucose, a simple sugar, can be represented in several ways.** In the structural formulas shown in (A), the atoms are shown as chemical symbols linked together by solid lines, which represent the covalent bonds. The thickened lines are used to indicate the plane of the sugar ring and to show that the –H and –OH groups are not in the same plane as the ring. (B) Another kind of structural formula that shows the three-dimensional structure of glucose in the so-called "chair configuration." (C) A ball-and-stick model in which the three-dimensional arrangement of the atoms in space is indicated. (D) A space-filling model, which, as well as depicting the three-dimensional arrangement of the atoms, also gives some idea of their relative sizes and of the surface contours of the molecule. The atoms in (C) and (D) are colored as follows: C, *black*; H, *white*; O, *red*. This is the conventional color coding for these atoms (see Figure 2–10) and will be used throughout this book (Movie 2.1).

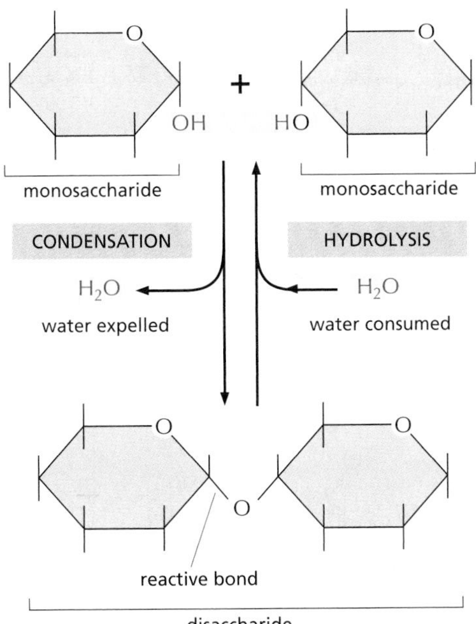

Figure 2–17 **Two monosaccharides can be linked to form a disaccharide.** This reaction belongs to a general category of reactions termed *condensation reactions*, in which two molecules join together as a result of the loss of a water molecule. The reverse reaction (in which water is added) is termed *hydrolysis*.

of a small number of monomers, between 3 and 50 or so. Polymers, in contrast, can contain hundreds or thousands of subunits.

The way sugars are linked together illustrates some common features of biochemical bond formation. A bond is formed between an –OH group on one sugar and an –OH group on another by a **condensation reaction**, in which a molecule of water is expelled as the bond is formed (Figure 2–17). Subunits in other biological polymers, such as nucleic acids and proteins, are also linked by condensation reactions in which water is expelled. The bonds created by all of these condensation reactions can be broken by the reverse process of **hydrolysis**, in which a molecule of water is consumed (see Figure 2–17).

Because each monosaccharide has several free hydroxyl groups that can form a link to another monosaccharide (or to some other compound), sugar polymers can be branched, and the number of possible polysaccharide structures is extremely large. For this reason it is much more difficult to determine the arrangement of sugars in a polysaccharide than to determine the nucleotide sequence of a DNA molecule, where each unit is joined to the next in exactly the same way.

The monosaccharide *glucose* has a central role as an energy source for cells. It is broken down to smaller molecules in a series of reactions, releasing energy that the cell can harness to do useful work, as we will explain in Chapter 13. Cells use simple polysaccharides composed only of glucose units—principally *glycogen* in animals and *starch* in plants—as long-term stores of glucose, held in reserve for energy production.

Sugars do not function exclusively in the production and storage of energy. They are also used, for example, to make mechanical supports. The most abundant organic molecule on Earth—the *cellulose* that forms plant cell walls—is a polysaccharide of glucose. Another extraordinarily abundant organic substance, the *chitin* of insect exoskeletons and fungal cell walls, is also a polysaccharide—in this case a linear polymer of a sugar derivative called *N*-acetylglucosamine (see Panel 2–3, pp. 68–69). Other polysaccharides, with their tendency to be slippery when wet, are the main components of slime, mucus, and gristle.

Smaller oligosaccharides can be covalently linked to proteins to form *glycoproteins*, or to lipids to form *glycolipids* (Panel 2–4, pp. 70–71), which are both found in cell membranes. The surfaces of most cells are

QUESTION 2–6

Have a close look at the space-filling representations of the glucose molecule shown in Figure 2–16D. Note that there are hydrogen atoms of two different sizes. Do we need to apologize because the artist made a mistake? Explain your answer.

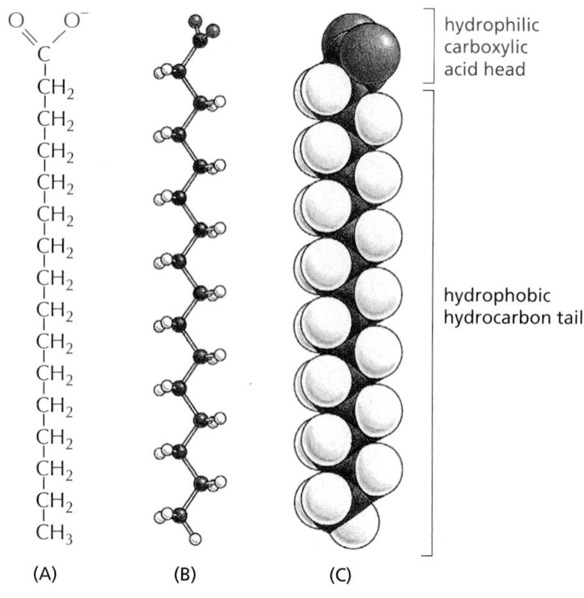

Figure 2–18 **Fatty acids have both hydrophobic and hydrophilic components.** The hydrophobic hydrocarbon chain is attached to a hydrophilic carboxylic acid group. Palmitic acid is shown here. Different fatty acids have different hydrocarbon tails. (A) Structural formula. The carboxylic acid head group is shown in its ionized form, as it exists in water at pH 7. (B) Ball-and-stick model. (C) Space-filling model (Movie 2.2).

decorated with sugar polymers that belong to glycoproteins and glycolipids in the plasma membrane. These sugar side chains are often recognized selectively by other cells. Differences in the types of cell-surface sugars form the molecular basis for different human blood groups.

Fatty Acids Are Components of Cell Membranes

A **fatty acid** molecule, such as *palmitic acid* (Figure 2–18), has two chemically distinct regions. One is a long hydrocarbon chain, which is hydrophobic and not very reactive chemically. The other is a carboxyl (–COOH) group, which behaves as an acid (carboxylic acid): it is ionized in solution ($-COO^-$), extremely hydrophilic, and chemically reactive. Almost all the fatty acid molecules in a cell are covalently linked to other molecules by their carboxylic acid group (see Panel 2–4, pp. 70–71). Molecules such as fatty acids, which possess both hydrophobic and hydrophilic regions, are termed *amphipathic*.

The hydrocarbon tail of palmitic acid is *saturated*: it has no double bonds between its carbon atoms and contains the maximum possible number of hydrogens. Stearic acid, another one of the common fatty acids in animal fat, is also saturated. Some other fatty acids, such as oleic acid, have *unsaturated* tails, with one or more double bonds along their length. The double bonds create kinks in the molecules, interfering with their ability to pack together in a solid mass, and it is the absence or presence of these double bonds that accounts for the difference between hard (saturated) and soft (polyunsaturated) margarine. Fatty acids are also found in cell membranes, where the tightness of their packing affects the fluidity of the membrane. The many different fatty acids found in cells differ only in the length of their hydrocarbon chains and in the number and position of the carbon–carbon double bonds (see Panel 2–4).

Fatty acids serve as a concentrated food reserve in cells: they can be broken down to produce about six times as much usable energy, weight for weight, as glucose. Fatty acids are stored in the cytoplasm of many cells in the form of droplets of *triacylglycerol* molecules—compounds made of three fatty acid chains joined to a glycerol molecule (see Panel 2–4). These molecules are the animal fats found in meat, butter, and cream, and the plant oils such as corn oil and olive oil (Figure 2–19). When a cell needs energy, the fatty acid chains can be released from triacylglycerols and broken down into two-carbon units. These two-carbon units are identical

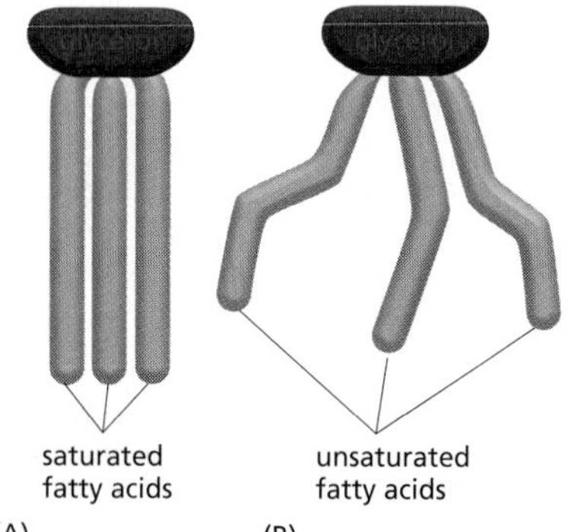

Figure 2–19 **The properties of fats depend on the fatty acid side chains they carry.** Fatty acids are stored in the cytoplasm of many cells in the form of droplets of *triacylglycerol* compounds made of three fatty acid chains joined to a glycerol molecule. (A) Saturated fats, such as tristearate, are found in meat and dairy products. The lack of double bonds in the fatty acid chains allows these molecules to pack together tightly, which is why butter and lard are solid at room temperature. (B) Plant oils, such as corn oil, contain unsaturated fatty acids, which may be monounsaturated (containing one double bond) or polyunsaturated (containing multiple double bonds). The double bonds produce kinks in the fatty acid chains that prevent the fats from packing close together; for this reason, plant oils are liquid at room temperature. Although fats are essential in the diet, saturated fats raise the concentrations of cholesterol in the blood and can cause arteries to become clogged with fat, a condition that can lead to heart disease. For this reason, manufacturers have been eliminating saturated fats from packaged foods.

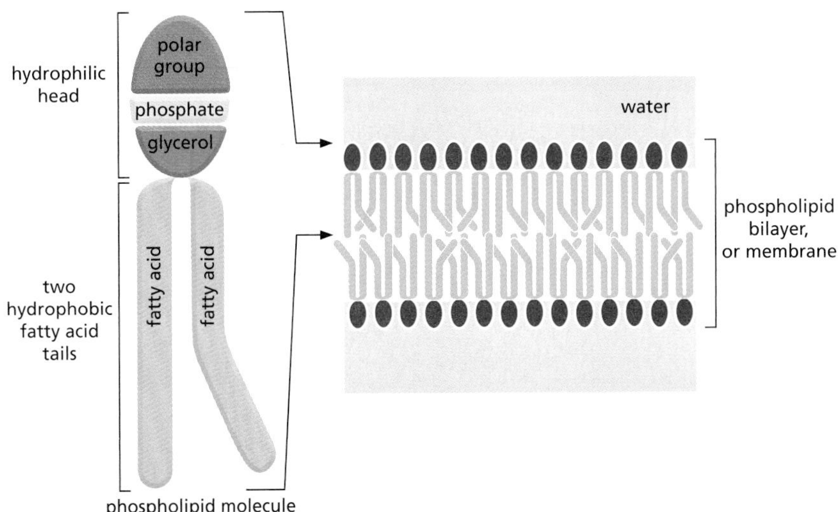

Figure 2–20 **Phospholipids aggregate to form cell membranes.** Phospholipids are composed of two hydrophobic fatty acid tails joined to a hydrophilic head. In an aqueous environment, the hydrophobic tails pack together to exclude water, forming a lipid bilayer with the hydrophilic heads of the phospholipid molecules facing the aqueous environment.

to those derived from the breakdown of glucose, and they enter the same energy-yielding reaction pathways, as will be described in Chapter 13.

Fatty acids and their derivatives, including triacylglycerols, are examples of *lipids*. This class of biological molecules is loosely defined, with the common feature that the molecules in the class are insoluble in water and soluble in fat and organic solvents such as benzene. Lipids typically contain long hydrocarbon chains, as in the fatty acids and *isoprenes*—or multiple linked aromatic rings, as in the *steroids* (see Panel 2–4, pp. 70–71).

The most important function of fatty acids in cells is in the formation of membranes. These thin sheets enclose all cells and surround their internal organelles. They are composed largely of *phospholipids*, which are small molecules that, like triacylglycerols, are constructed mainly from fatty acids and glycerol. In phospholipids, however, the glycerol is joined to two fatty acid chains, rather than to three as in triacylglycerols. The "third" site on the glycerol is linked to a hydrophilic phosphate group, which in turn is attached to a small hydrophilic compound such as choline (see Panel 2–4). Phospholipids are strongly amphipathic: each phospholipid molecule has a hydrophobic tail, composed of the two fatty acid chains, and a hydrophilic head, where the phosphate is located. This gives them different physical and chemical properties from triacylglycerols, which are predominantly hydrophobic. Other lipids present in the cell membrane contain one or more sugars instead of a phosphate group. Several of these *glycolipids* play an important role in intracellular cell signaling, as we will see in Chapter 16.

The membrane-forming property of phospholipids results from their amphipathic nature. Phospholipids will spread over the surface of water to form a monolayer of phospholipid molecules, with the hydrophobic tails facing the air and the hydrophilic heads in contact with the water. Two such molecular layers can readily combine tail-to-tail in water to make a phospholipid sandwich, or *lipid bilayer*, which forms the structural basis of all cell membranes (Figure 2–20; discussed further in Chapter 11).

Amino Acids Are the Subunits of Proteins

Amino acids are a varied class of molecules with one defining property: they all possess a carboxylic acid group and an amino group, both linked to the same carbon atom called the α-carbon (Figure 2–21). Their chemical variety comes from the side chain that is also attached to the α-carbon. Cells use amino acids to build **proteins**, which are polymers of amino acids joined head-to-tail in a long chain that is then folded into a three-dimensional structure unique to each type of protein.

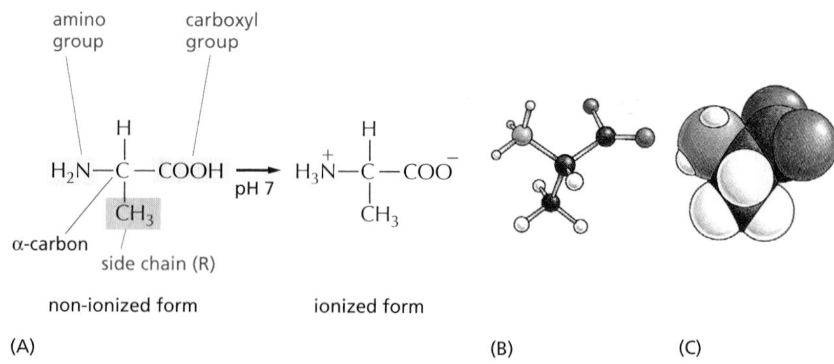

Figure 2–21 **Alanine is one of the simplest amino acids.** (A) In the cell, where the pH is close to 7, the free amino acid exists in its ionized form; but when it is incorporated into a polypeptide chain, the charges on the amino and carboxyl groups disappear. (B) A ball-and-stick model and (C) a space-filling model of alanine (C, *black*; H, *white*; O, *red*; N, *blue*).

QUESTION 2–7

Why do you suppose only L-amino acids and not a random mixture of the L- and D-forms of each amino acid are used to make proteins?

The covalent linkage between two adjacent amino acids in a protein chain is called a *peptide bond*; the chain of amino acids is also known as a polypeptide (Figure 2–22). Peptide bonds are formed by condensation reactions that link one amino acid to the next. Regardless of the specific amino acids from which it is made, the polypeptide always has an amino (NH_2) group at one end (its *N-terminus*) and a carboxyl (COOH) group at its other end (its *C-terminus*). This gives a protein or polypeptide a definite directionality—a structural (as opposed to electrical) polarity.

Twenty types of amino acids are commonly found in proteins, each with a different side chain attached to the α-carbon atom (Panel 2–5, pp. 72–73). The same 20 amino acids occur over and over again in all proteins, whether they hail from bacteria, plants, or animals. How this precise set of 20 amino acids came to be chosen is one of the mysteries surrounding the evolution of life; there is no obvious chemical reason why other amino acids could not have served just as well. But once the selection had been locked into place, it could not be changed; too much chemistry had evolved to exploit it. Switching the types of amino acids used by cells would require every living creature to retool its entire metabolism to cope with the new building blocks.

Like sugars, all amino acids (except glycine) exist as optical isomers in D- and L-forms (see Panel 2–5). But only L-forms are ever found in proteins (although D-amino acids occur as part of bacterial cell walls and in some antibiotics). The origin of this exclusive use of L-amino acids to make proteins is another evolutionary mystery.

N-terminus of polypeptide chain

Phe: N–H, H–C–CH_2–(phenyl), O=C
Ser: N–H, H–C–CH_2–OH, O=C
Glu: N–H, H–C–CH_2–CH_2–C(=O)–O^-, O=C
Lys: N–H, H–C–CH_2–CH_2–CH_2–CH_2–N(H)(H)–H^+, O=C

C-terminus of polypeptide chain

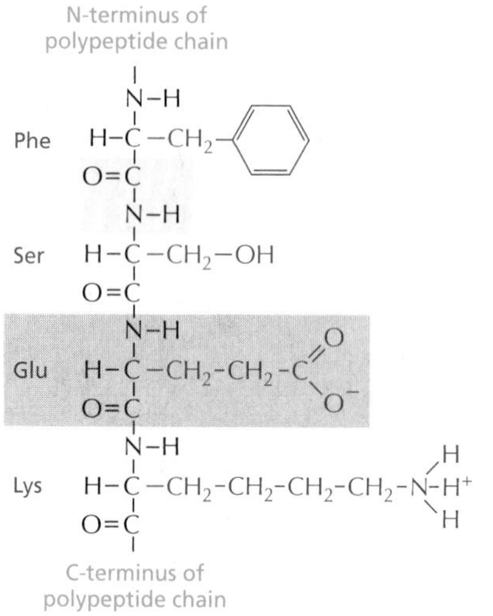

Figure 2–22 **Proteins are held together by peptide bonds.** The four amino acids shown are linked together by three peptide bonds, one of which is highlighted in *yellow*. One of the amino acids is shaded in *gray*. The amino acid side chains are shown in *red*. The two ends of a polypeptide chain are chemically distinct. One end, the N-terminus, is capped by an amino group, and the other, the C-terminus, ends in a carboxyl group. The sequence of amino acids in a protein or polypeptide is abbreviated using either a three-letter or a one-letter code, and the sequence is always read from the N-terminus (see Panel 2–5, pp. 72–73). In the example given, the sequence is Phe-Ser-Glu-Lys (or FSEK).

The chemical versatility that the 20 standard amino acids provide is vitally important to the function of proteins. Five of the 20 amino acids have side chains that can form ions in solution and can therefore carry a charge (lysine and glutamic acid, for example, shown in Figure 2–22). The others are uncharged. Some amino acids are polar and hydrophilic, and some are nonpolar and hydrophobic (see Panel 2–5). As we will discuss in Chapter 4, the collective properties of the amino acid side chains underlie all the diverse and sophisticated functions of proteins.

Nucleotides Are the Subunits of DNA and RNA

A *nucleoside* is a molecule made of a nitrogen-containing ring compound linked to a five-carbon sugar, which can be either ribose or deoxyribose (Panel 2–6, pp. 74–75). A nucleoside sporting one or more phosphate groups attached to its sugar is called a **nucleotide**. Nucleotides containing ribose are known as *ribonucleotides*, and those containing deoxyribose as *deoxyribonucleotides*.

The nitrogen-containing rings are generally referred to as *bases* for historical reasons: under acidic conditions they can each bind a H^+ (proton) and thereby increase the concentration of OH^- ions in aqueous solution.

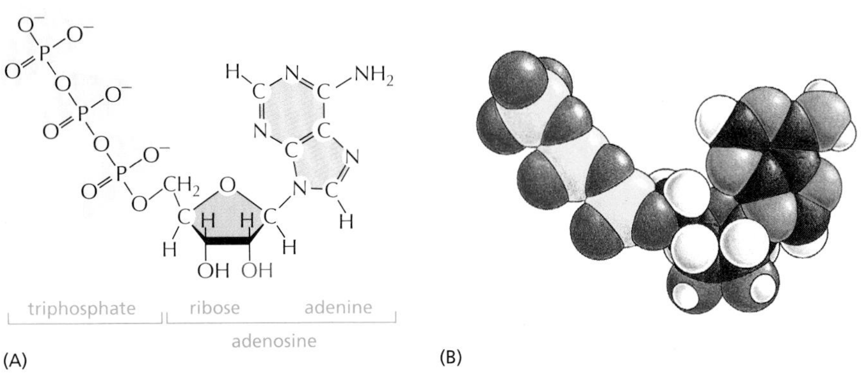

Figure 2–23 **Adenosine triphosphate (ATP) is a nucleotide whose reactivity resides in its terminal phosphate groups.** (A) Structural formula. The three phosphate groups are shaded in *yellow*. (B) Space-filling model (Movie 2.3). In (B) the colors of the atoms are C, *black*; H, *white*; N, *blue*; O, *red*; and P, *yellow*. The deoxyribonucleotide version of adenosine triphosphate (dATP) differs only in that a hydrogen atom replaces the hydroxyl group *(red)* in (A).

There is a strong family resemblance between the different nucleotide bases. *Cytosine* (C), *thymine* (T), and *uracil* (U) are called *pyrimidines* because they all derive from a six-membered pyrimidine ring; *guanine* (G) and *adenine* (A) are *purine compounds*, which bear a second, five-membered ring fused to the six-membered ring. Each nucleotide is named after the base it contains (see Panel 2–6, pp. 74–75).

Nucleotides can act as short-term carriers of chemical energy. Above all others, the ribonucleotide **adenosine triphosphate**, or **ATP** (Figure 2–23), participates in the transfer of energy in hundreds of cellular reactions. ATP is formed through reactions that are driven by the energy released by the breakdown of foodstuffs. Its three phosphates are linked in series by two *phosphoanhydride bonds* (see Panel 2–6). Rupture of these phosphate bonds releases large amounts of useful energy. The terminal phosphate group in particular is frequently split off by hydrolysis (Figure 2–24). In many situations, transfer of this phosphate to other molecules releases energy that drives energy-requiring biosynthetic reactions. Other nucleotide derivatives serve as carriers for the transfer of other chemical groups. All of this will be described in Chapter 3.

The most fundamental role of nucleotides in the cell is in the storage and retrieval of biological information. Nucleotides serve as building blocks for the construction of *nucleic acids*—long polymers in which nucleotide subunits are covalently linked by the formation of a *phosphodiester bond*

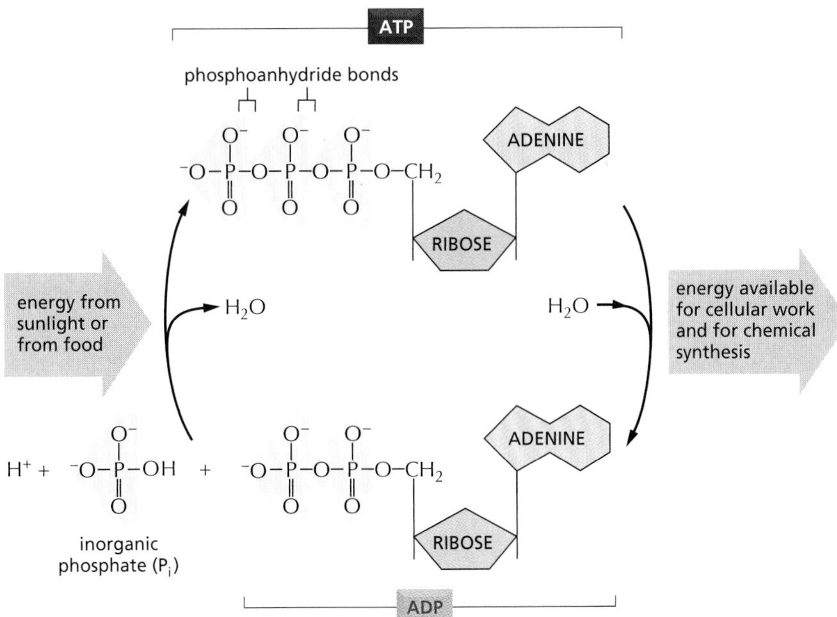

Figure 2–24 **ATP serves as an energy carrier in cells.** The energy-requiring formation of ATP from ADP and inorganic phosphate is coupled to the energy-yielding oxidation of foodstuffs (in animal cells, fungi, and some bacteria) or to the capture of light (in plant cells and some bacteria). The hydrolysis of this ATP back to ADP and inorganic phosphate in turn provides the energy to drive many cellular reactions. Together these reactions form the ATP cycle.

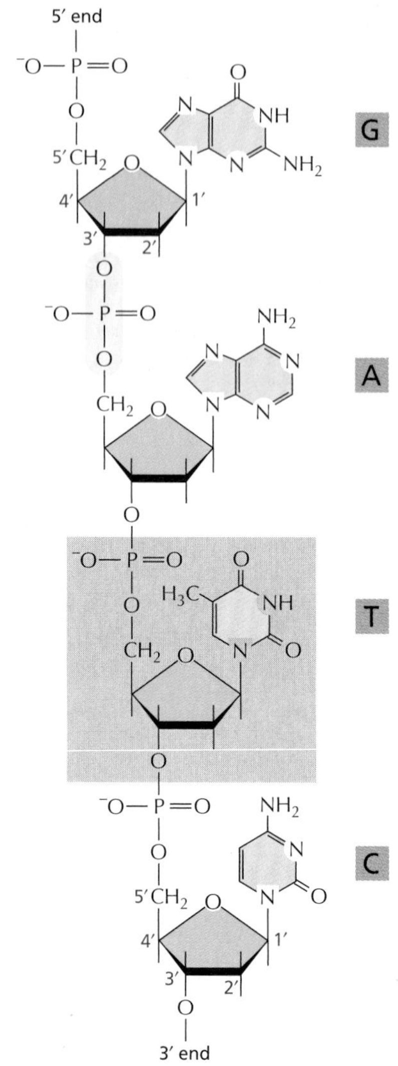

between the phosphate group attached to the sugar of one nucleotide and a hydroxyl group on the sugar of the next nucleotide (Figure 2–25). Nucleic acid chains are synthesized from energy-rich nucleoside triphosphates by a condensation reaction that releases inorganic pyrophosphate during phosphodiester bond formation (see Panel 2–6, pp. 74–75).

There are two main types of nucleic acids, which differ in the type of sugar they use in their sugar–phosphate backbone. Those based on the sugar ribose are known as **ribonucleic acids**, or **RNA**, and contain the bases A, G, C, and U. Those based on deoxyribose (in which the hydroxyl at the 2′ position of the ribose carbon ring is replaced by a hydrogen; see Panel 2–6) are known as **deoxyribonucleic acids**, or **DNA**, and contain the bases A, G, C, and T (T is chemically similar to the U in RNA) (see Figure 2–25). RNA usually occurs in cells in the form of a single-stranded polynucleotide chain, but DNA is virtually always in the form of a double-stranded molecule: the DNA double helix is composed of two polynucleotide chains running antiparallel to each other, being held together by hydrogen-bonding between the bases of the two chains (Panel 2–7, pp. 76–77).

The linear sequence of nucleotides in a DNA or an RNA molecule encodes genetic information. The two nucleic acids, however, have somewhat different roles in the cell. DNA, with its more stable, hydrogen-bonded helices, acts as a long-term repository for hereditary information, while single-stranded RNA is usually a more transient carrier of molecular instructions. The ability of the bases in different nucleic acid molecules to recognize and pair with each other by hydrogen-bonding (called *base-pairing*)—G with C, and A with either T or U—underlies all of heredity and evolution, as explained in Chapter 5.

Figure 2–25 **A short length of one chain of a deoxyribonucleic acid (DNA) molecule shows the bonds linking four consecutive nucleotides.** Nucleotides are joined by a phosphodiester linkage between specific carbon atoms of the sugar ring, known as the 5′ and 3′ atoms. For this reason, one end of a polynucleotide chain, the 5′ end, will have a free phosphate group and the other, the 3′ end, a free hydroxyl group. One of the nucleotides is enclosed in a *gray* box; a phosphodiester bond formed by a phosphate group of one nucleotide linking to the adjacent nucleotide is highlighted in *yellow*. The linear sequence of nucleotides in a polynucleotide chain is commonly abbreviated by a one-letter code, and the sequence is always read from the 5′ end. In the example illustrated the sequence is G-A-T-C.

MACROMOLECULES IN CELLS

On the basis of weight, macromolecules are by far the most abundant of the carbon-containing molecules in a living cell (Figure 2–26). They are the principal building blocks from which a cell is constructed and also the components that confer the most distinctive properties on living things. Intermediate in size and complexity between small molecules and cell organelles, **macromolecules** are polymers that are constructed simply by covalently linking small organic molecules (called **monomers**, or *subunits*) into long chains, or **polymers** (Figure 2–27 and How We Know, pp. 60–61). Yet they have many unexpected properties that could not have been predicted from their simple constituents. As one example, DNA and RNA molecules (the nucleic acids) store and transmit hereditary information.

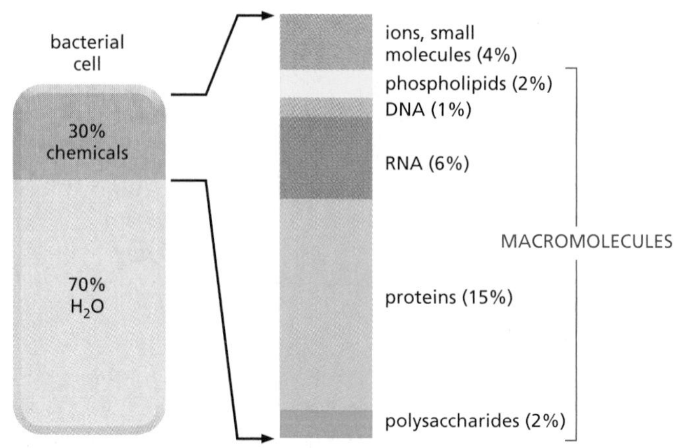

Figure 2–26 **Macromolecules are abundant in cells.** The approximate composition of a bacterial cell is shown. The composition of an animal cell is similar.

Proteins are especially versatile and perform thousands of distinct functions in cells. Many proteins act as enzymes that catalyze the chemical reactions that take place in the cell, including all of the reactions whereby cells extract energy from food molecules. Enzymes are also required to synthesize the many different molecules that a cell needs. For example, an enzyme called ribulose bisphosphate carboxylase, found in chloroplasts, converts CO_2 to sugars in plants; this protein thereby creates most of the organic matter used by the rest of the living world. Other proteins are used to build structural components: tubulin self-assembles to make the cell's long, stiff microtubules (see Figure 1–27B). Histone proteins pack the cell's DNA in chromosomes. Yet other proteins act as molecular motors to produce force and movement, as in the case of myosin in muscle. Proteins also have a wide variety of other functions. We shall examine the molecular basis for many specific functions later in this book. Here we consider only the general principles of macromolecular chemistry that make all of these functions possible.

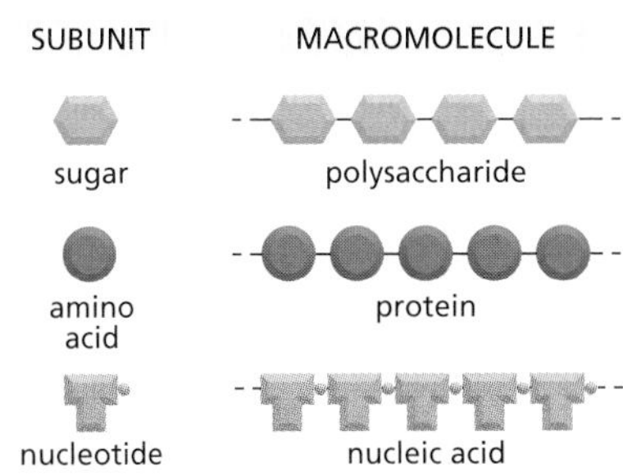

Figure 2–27 **Polysaccharides, proteins, and nucleic acids are made from monomeric subunits.** Each macromolecule is a polymer formed from small molecules (called monomers or subunits) linked together by covalent bonds.

Macromolecules Contain a Specific Sequence of Subunits

Although the chemical reactions for adding subunits to each polymer are different in detail for proteins, nucleic acids, and polysaccharides, they share important features. Each polymer grows by the addition of a monomer onto one end of the polymer chain via a condensation reaction, in which a molecule of water is lost with each subunit added (Figure 2–28; see also Figure 2–17). In all cases the reactions are catalyzed by specific enzymes, which ensure that only monomers of the appropriate type are incorporated.

The stepwise polymerization of monomers into a long chain is a simple way to manufacture a large, complex molecule, because the subunits are added by the same reaction performed over and over again by the same set of enzymes. In a sense, the process resembles the repetitive operation of a machine in a factory—with some important differences. First, apart from some of the polysaccharides, most macromolecules are made from a set of monomers that are slightly different from one another, for example, the 20 different amino acids from which proteins are made (see Panel 2–5, pp. 72–73). Second, and most important, the polymer chain is not assembled at random from these subunits; instead the subunits are added in a particular order, or **sequence**.

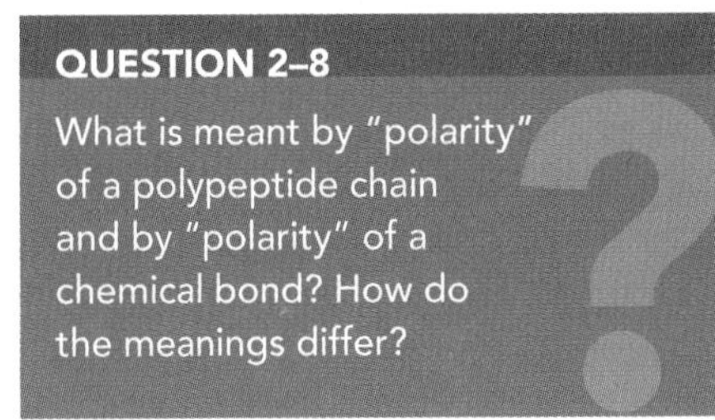

QUESTION 2–8
What is meant by "polarity" of a polypeptide chain and by "polarity" of a chemical bond? How do the meanings differ?

The mechanisms that specify polymer sequence in the cell are discussed in Chapters 6 and 7. These mechanisms are central to biology because the biological functions of proteins, nucleic acids, and many polysaccharides are absolutely dependent on the particular sequence of subunits in the linear chains. The possibility of varying the sequence of subunits creates enormous diversity in the polymeric molecules that can be produced. Thus, for a protein chain 200 amino acids long, there are 20^{200} possible combinations (20 × 20 × 20 × 20... multiplied 200 times), while for a DNA molecule 10,000 nucleotides long (small by DNA standards), with its four different nucleotides there are $4^{10,000}$ different possibilities, an unimaginably large number. Thus the machinery of polymerization must be subject to a sensitive control that allows it to specify exactly which subunit should be added next to the growing polymer end.

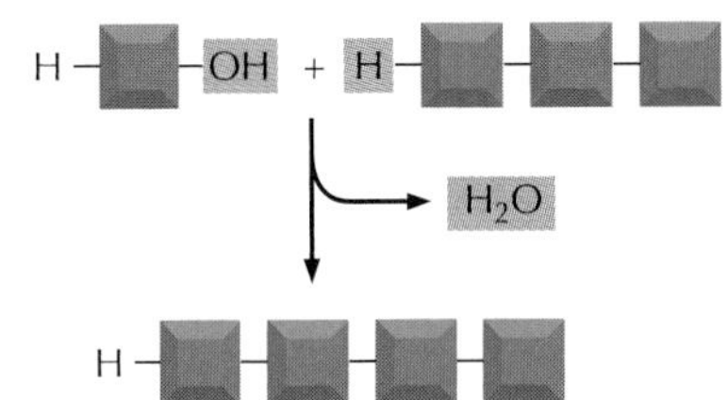

Figure 2–28 **Macromolecules are formed by adding subunits to one end.** In a condensation reaction, a molecule of water is lost with the addition of each monomer to one end of the growing chain. The reverse reaction—the breakdown of the polymer—occurs by the addition of water (hydrolysis).

Noncovalent Bonds Specify the Precise Shape of a Macromolecule

Most of the single covalent bonds in a macromolecule allow rotation of the atoms they join, so that the polymer chain has great flexibility. In principle, this allows a macromolecule to adopt an almost unlimited number

HOW WE KNOW:
WHAT ARE MACROMOLECULES?

The idea that proteins, polysaccharides, and nucleic acids are large molecules that are constructed from smaller subunits, linked one after another into long molecular chains, may seem fairly obvious today. But this was not always the case. In the early part of the twentieth century, few scientists believed in the existence of such macromolecules—polymers built from repeating units held together by covalent bonds. The notion that such "frighteningly large" compounds could be assembled from simple building blocks was considered "downright shocking" by chemists of the day. Instead, they thought that proteins and other seemingly large molecules were simply heterogeneous aggregates of small molecules held together by weak "association forces" (Figure 2–29).

The first hint that proteins and other polymers are so large came from observing their behavior in solution. At the time, scientists were working with various proteins and carbohydrates derived from foodstuffs and natural materials—albumin from egg whites, casein from milk, collagen from gelatin, and cellulose from wood. Their chemical compositions seemed simple enough—like other organic molecules they contained carbon, hydrogen, oxygen, and, in the case of proteins, nitrogen. But they behaved oddly in solution, showing, for example, an inability to pass through a fine filter.

Why these molecules misbehaved in solution was a puzzle. Were they really giant molecules, composed of an unusual number of covalently linked atoms? Or were they more like a colloidal suspension of particles—a big, sticky hodgepodge of simpler molecules that associate only loosely?

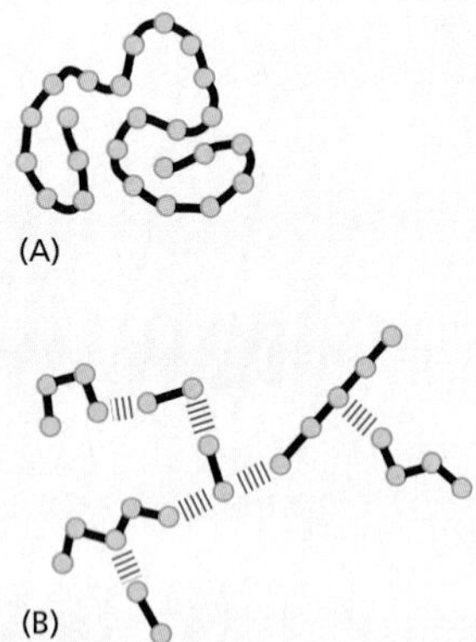

Figure 2–29 **What might a macromolecule look like?** Chemists in the early part of the twentieth century debated whether proteins, polysaccharides, and other apparently large molecules were (A) discrete particles made of an unusually large number of covalently linked atoms or (B) a loose aggregation of heterogeneous small molecules held together by weak forces.

One way to distinguish between the two possibilities was to determine the actual size of one of these molecules. If a substance such as serum albumin were made of molecules all identical in size, that would support the existence of true macromolecules. Conversely, if albumin were instead a conglomeration of miscellaneous peptides, these should show a whole range of molecular sizes in solution.

Unfortunately, the techniques available to scientists in the early 1900s were not ideal for measuring the sizes of such large molecules. Some chemists estimated a protein's size by determining how much it would lower a solution's freezing point; others measured the osmotic pressure of protein solutions. These methods were susceptible to experimental error and gave variable results. Different techniques, for example, suggested that cellulose was anywhere from 6000 to 103,000 daltons in mass (where 1 dalton is approximately equal to the mass of a hydrogen atom). Such variation helped to fuel the hypothesis that proteins and carbohydrates were loose aggregates rather than macromolecules.

Many scientists simply had trouble believing that molecules heavier than about 4000 daltons—the largest compound that had been synthesized by organic chemists—could exist at all. Take hemoglobin, the oxygen-carrying protein in red blood cells. Researchers tried to estimate its size by breaking it down into its chemical components. In addition to carbon, hydrogen, nitrogen, and oxygen, hemoglobin contains a small amount of iron. Working out the percentages, it appeared that hemoglobin had one atom of iron for every 712 atoms of carbon—and a minimum weight of 16,700 daltons. Could a molecule with hundreds of carbon atoms in one long chain remain intact in a cell and perform specific functions? Emil Fischer, the organic chemist who determined that the amino acids in proteins are linked by peptide bonds, thought that a polypeptide chain could grow no longer than about 30 or 40 amino acids. As for hemoglobin with its purported 700 carbon atoms, the existence of molecular chains of such "truly fantastic lengths" was deemed "very improbable" by leading chemists.

Definitive resolution of the debate had to await the development of new techniques. Convincing evidence that proteins are macromolecules came from studies using the ultracentrifuge—a device that uses centrifugal force to separate molecules according to their size (Figure 2–30; see also Panel 4–4, pp. 164–165). Theodor Svedberg, who designed the machine in 1925, performed the first studies. If a protein were really an aggregate of smaller molecules, he reasoned, it would appear as a smear of molecules of different sizes when sedimented in an ultracentrifuge. Using hemoglobin as his test protein, Svedberg found that the centrifuged

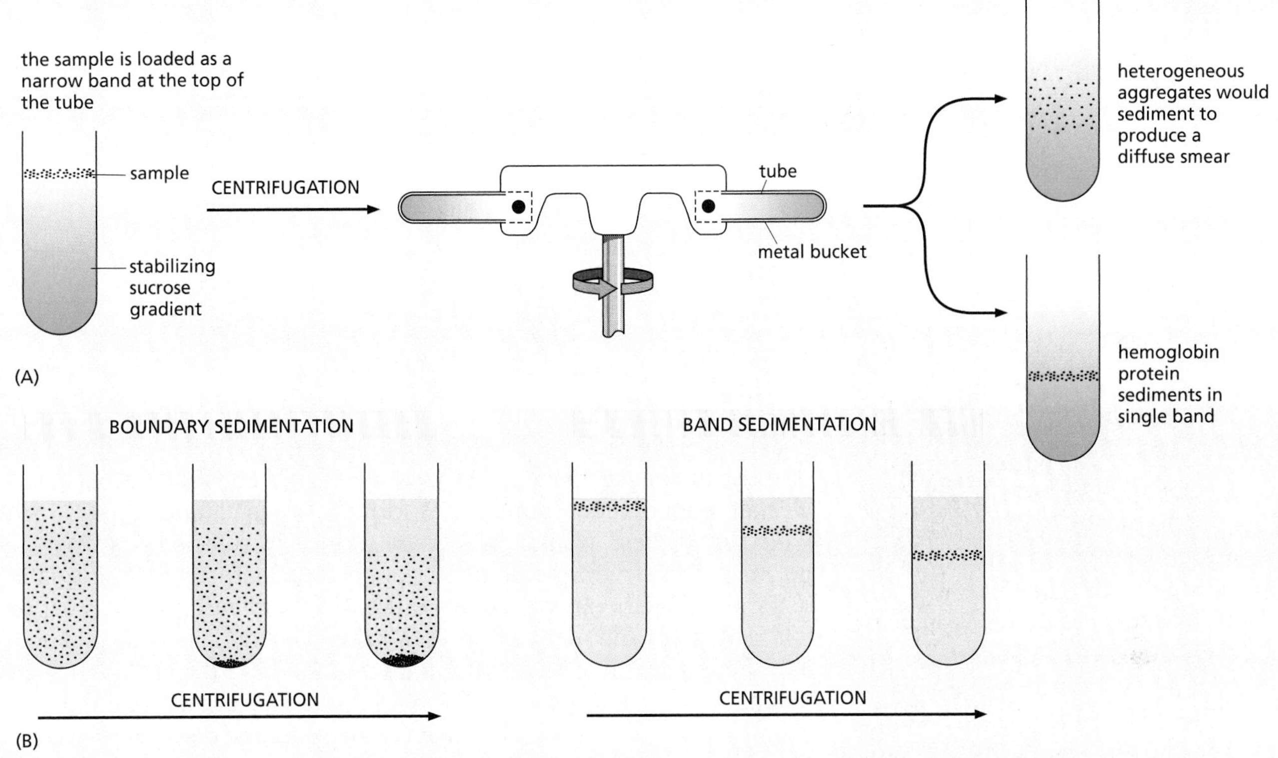

Figure 2–30 **The ultracentrifuge helped to settle the macromolecular debate.** In the ultracentrifuge, centrifugal forces exceeding 500,000 times the force of gravity can be used to separate proteins or other large molecules. (A) In a modern ultracentrifuge, samples are loaded in a thin layer on top of a gradient of sucrose solution formed in a tube. The tube is placed in a metal rotor that is rotated at high speed in an ultracentrifuge. Molecules of different sizes sediment at different rates, and these molecules will therefore move as distinct bands in the sample tube. If hemoglobin were a loose aggregate of heterogeneous peptides, it would show a broad smear of sizes after centrifugation (*top* tube). Instead it appears as a sharp band with a molecular weight of 68,000 daltons (*bottom* tube). Although the ultracentrifuge is now a standard, almost mundane, fixture in most biochemistry labs, its construction was a huge technological challenge. The centrifuge rotor must be capable of spinning at high speeds for many hours at constant temperature and with high stability; otherwise convection occurs in the sedimenting solution and ruins the experiment. In 1926 Svedberg won the Nobel Prize in Chemistry for his ultracentrifuge design and its application to chemistry. (B) In his actual experiment, Svedberg filled a special tube in the centrifuge with a homogeneous solution of hemoglobin; by shining light through the tube, he then carefully monitored the moving boundary between the sedimenting protein molecules and the clear aqueous solution left behind (so-called boundary sedimentation). The more recently developed method shown in (A) is a form of "band sedimentation."

sample revealed a single, sharp band with a molecular weight of 68,000 daltons. His results strongly supported the theory that proteins are true macromolecules.

Additional evidence continued to accumulate throughout the 1930s, as other researchers began to prepare crystals of pure protein that could be studied by X-ray diffraction. Only molecules with a uniform size and shape can form highly ordered crystals and diffract X-rays in such a way that their three-dimensional structure can be determined, as we shall see in Chapter 4. A heterogeneous suspension could not be studied in this way.

We now take it for granted that large macromolecules carry out many of the most important activities in living cells. But chemists once viewed the existence of such polymers with the same sort of skepticism that a zoologist might show on being told that somewhere in Africa live elephants that are 100 meters long and 20 metres tall. It took decades for researchers to master the techniques required to convince everyone that molecules ten times larger than anything they had ever encountered were a cornerstone of biology. As we shall see throughout this book, such a labored pathway to discovery is not unusual, and progress in science is often driven by advances in technology.

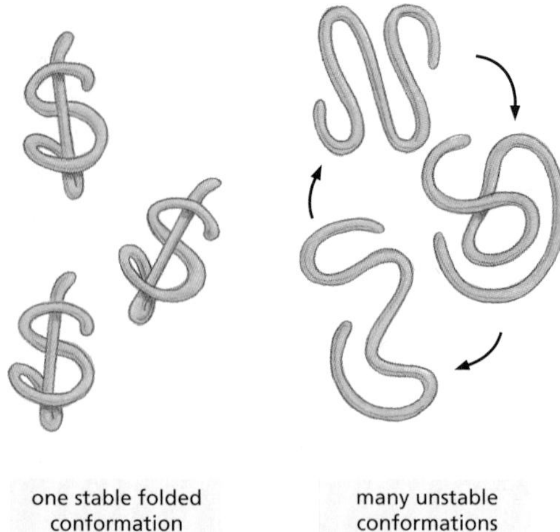

Figure 2–31 **Most proteins and many RNA molecules fold into a particularly stable three-dimensional shape, or conformation.** If the weak bonds maintaining this stable conformation are disrupted, the molecule becomes a flexible chain that loses its biological activity.

of shapes, or **conformations**, as the polymer chain writhes and rotates under the influence of random thermal energy. However, the shapes of most biological macromolecules are highly constrained because of weaker **noncovalent bonds** that form between different parts of the molecule. If these weaker bonds are formed in sufficient numbers, the polymer chain may preferentially adopt one particular conformation, determined by the linear sequence of monomers in its chain. Most protein molecules and many of the RNA molecules found in cells fold tightly into one highly preferred conformation in this way (Figure 2–31). These unique conformations—shaped by natural selection—determine the chemistry and activity of these macromolecules and dictate their interactions with other biological molecules.

QUESTION 2–9

In principle, there are many different, chemically diverse ways in which small molecules can be linked to form polymers. For example, the small molecule ethene ($CH_2{=}CH_2$) is used commercially to make the plastic polyethylene (...–CH_2–CH_2–CH_2–CH_2–CH_2–...). The individual subunits of the three major classes of biological macromolecules, however, are all linked by similar reaction mechanisms, i.e., by condensation reactions that eliminate water. Can you think of any benefits that this chemistry offers and why it might have been selected in evolution?

The noncovalent bonds important in biological molecules include two types described earlier in this chapter—electrostatic attractions and hydrogen bonds (see Panel 2–7, pp. 76–77). Electrostatic attractions, although strong on their own, are quite weak in water. This is because charged or partially charged (polar) groups are shielded by their interactions with water molecules or with other salts present in the aqueous solution. Electrostatic attractions, however, are very important in biological systems. An enzyme that binds a positively charged substrate will often use a negatively charged amino acid side chain to guide its substrate into the proper position. And we have already mentioned the importance of hydrogen bonds in establishing the unique properties of water. Hydrogen bonds also hold the two strands of the DNA double helix together. Because individual hydrogen bonds are weak, enzymes can easily unzip the helix—for example, when a cell needs to copy its genetic material.

A third type of weak bond results from *van der Waals attractions*, which are a form of electrical attraction caused by fluctuating electric charges that arise whenever two atoms come within a very short distance of each other. Although van der Waals interactions are weaker than hydrogen bonds, in large numbers they play an important role in the attraction between large molecules with complementary shapes. All of these noncovalent forces are reviewed in Panel 2–7.

Another important noncovalent force is created by the three-dimensional structure of water, which forces hydrophobic groups together in order to minimize their disruptive effect on the hydrogen-bonded network of water molecules (see Panel 2–7 and Panel 2–2, pp. 66–67). This expulsion from the aqueous solution generates what is sometimes thought of as a fourth kind of weak noncovalent bond, called a *hydrophobic interaction*. This interaction forces phospholipid molecules together in cell membranes, and it also gives most protein molecules a compact, globular shape.

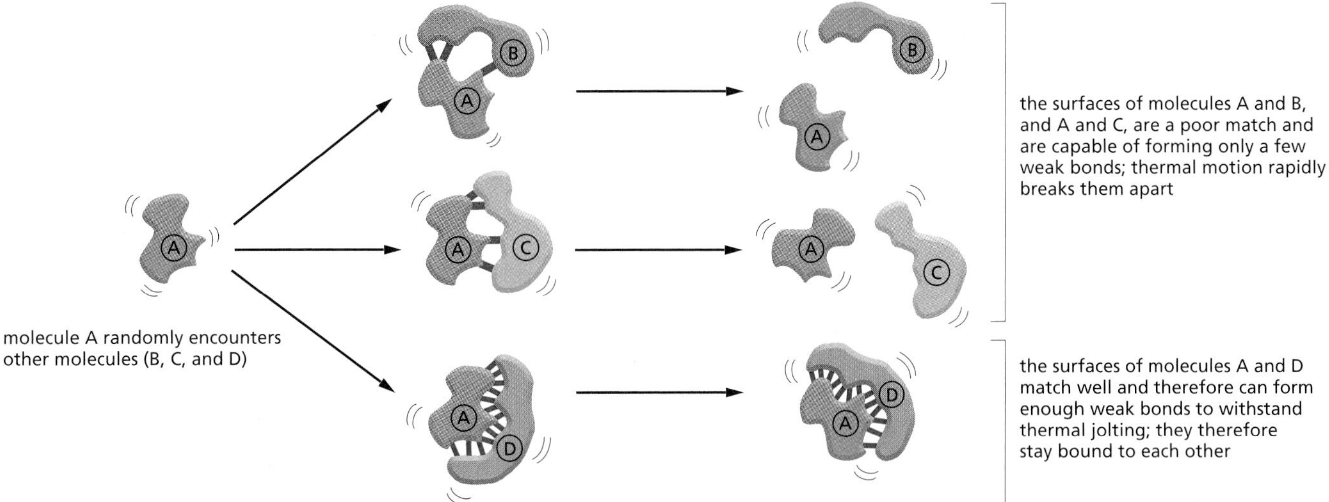
molecule A randomly encounters other molecules (B, C, and D)
A
B
C
D
the surfaces of molecules A and B, and A and C, are a poor match and are capable of forming only a few weak bonds; thermal motion rapidly breaks them apart
the surfaces of molecules A and D match well and therefore can form enough weak bonds to withstand thermal jolting; they therefore stay bound to each other

Figure 2–32 **Noncovalent bonds mediate interactions between macromolecules.**

Noncovalent Bonds Allow a Macromolecule to Bind Other Selected Molecules

Although noncovalent bonds are individually very weak, they can add up to create a strong attraction between two molecules when these molecules fit together very closely, like a hand in a glove, with many noncovalent bonds between them (see Panel 2–7, pp. 76–77). This form of molecular interaction provides for great specificity in the binding of macromolecules to other small and large molecules, because the multipoint contacts required for strong binding make it possible for a macromolecule to select—through binding interactions—just one of the many thousands of different molecules present inside a cell. Moreover, because the strength of the binding depends on the number of noncovalent bonds that are formed, interactions of almost any strength are possible.

QUESTION 2–10

Why could covalent bonds not be used in place of noncovalent bonds to mediate most of the interactions of macromolecules?

Binding of this type underlies all biological catalysis, making it possible for proteins to function as enzymes. Noncovalent bonds can also stabilize associations between two different macromolecules if their surfaces match closely (Figure 2–32 and Movie 2.4). These bonds thereby allow macromolecules to be used as building blocks for the formation of much larger structures. For example, proteins often bind together into multiprotein complexes, thereby forming intricate machines with multiple moving parts that perform such complex tasks as DNA replication and protein synthesis (Figure 2–33). Thus noncovalent bonds account for much of the specificity that we associate with living cells.

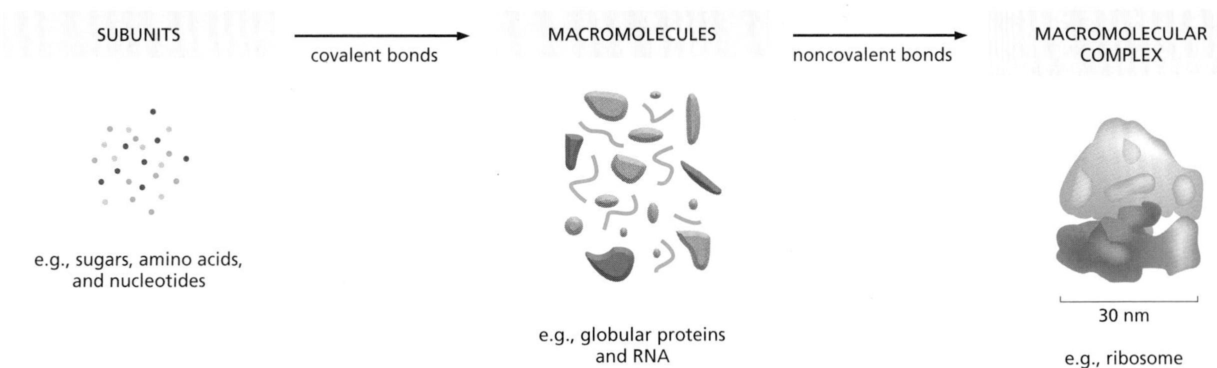
SUBUNITS
covalent bonds
MACROMOLECULES
noncovalent bonds
MACROMOLECULAR COMPLEX
e.g., sugars, amino acids, and nucleotides
e.g., globular proteins and RNA
30 nm
e.g., ribosome

Figure 2–33 **Small molecules join together to form macromolecules, which can assemble into large macromolecular complexes.** Subunits, proteins, and a ribosome are drawn to scale. Ribosomes are part of the machinery the cell uses to make proteins. Each ribosome is composed of about 90 macromolecules (proteins and RNA molecules), and it is large enough to see in the electron microscope (see Figure 7–30).

PANEL 2–1 Chemical bonds and groups

CARBON SKELETONS

Carbon has a unique role in the cell because of its ability to form strong covalent bonds with other carbon atoms. Thus carbon atoms can join to form chains

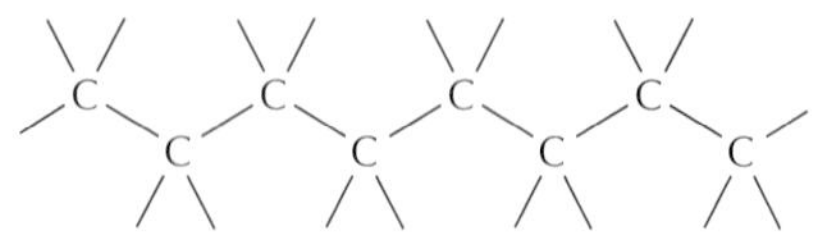

also written as

or branched trees

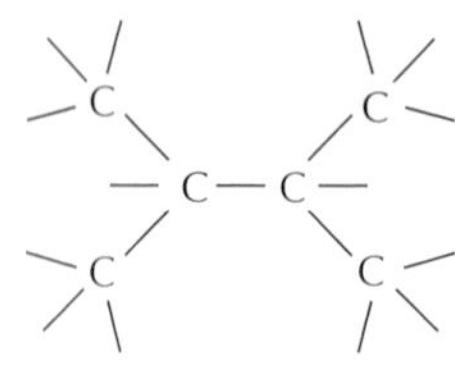

also written as

or rings.

also written as

COVALENT BONDS

A covalent bond forms when two atoms come very close together and share one or more of their electrons.

Each atom forms a fixed number of covalent bonds in a defined spatial arrangement.

SINGLE BONDS: two electrons shared per bond

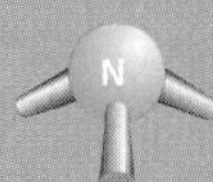

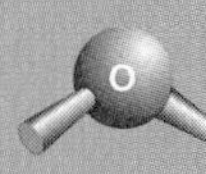

DOUBLE BONDS: four electrons shared per bond

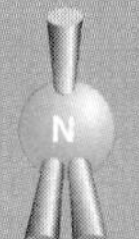

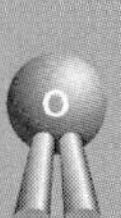

The precise spatial arrangement of covalent bonds influence the three-dimentional structure—and chemistry—of molecules. In this review panel, we see how covalent bonds are used in a variety of biological molecules.

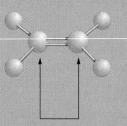

Atoms joined by two or more covalent bonds cannot rotate freely around the bond axis. This restriction has a major influence on the three-dimensional shape of many macromolecules.

C–H COMPOUNDS

Carbon and hydrogen together make stable compounds (or groups) called hydrocarbons. These are nonpolar, do not form hydrogen bonds, and are generally insoluble in water.

methane

methyl group

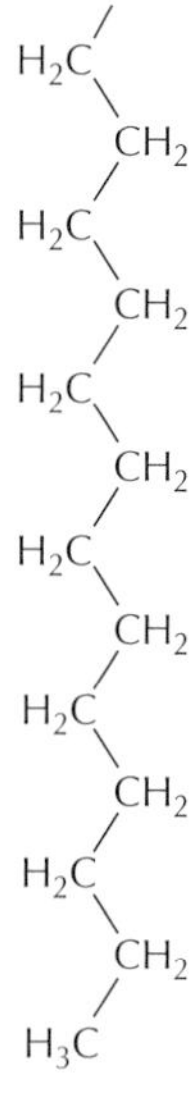

part of the hydrocarbon "tail" of a fatty acid molecule

ALTERNATING DOUBLE BONDS

A carbon chain can include double bonds. If these are on alternate carbon atoms, the bonding electrons move within the molecule, stabilizing the structure by a phenomenon called resonance.

the truth is somewhere between these two structures

Alternating double bonds in a ring can generate a very stable structure.

benzene

often written as

C–O COMPOUNDS

Many biological compounds contain a carbon bonded to an oxygen. For example,

alcohol

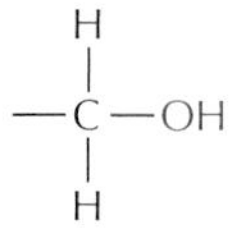

The –OH is called a hydroxyl group.

aldehyde

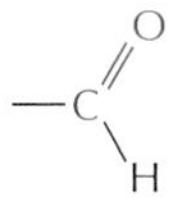

ketone

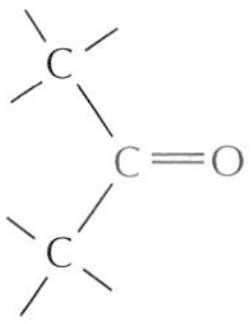

The C=O is called a carbonyl group.

carboxylic acid

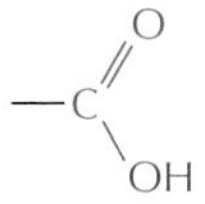

The –COOH is called a carboxyl group. In water this loses an H^+ ion to become $-COO^-$.

esters

Esters are formed by combining an acid and an alcohol.

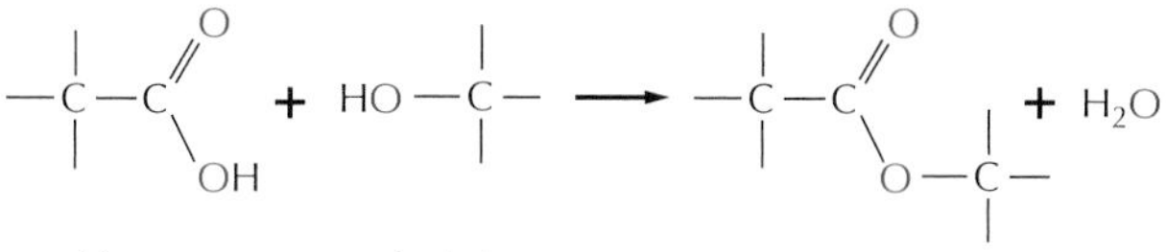

acid alcohol ester

C–N COMPOUNDS

Amines and amides are two important examples of compounds containing a carbon linked to a nitrogen.

Amines in water combine with an H^+ ion to become positively charged.

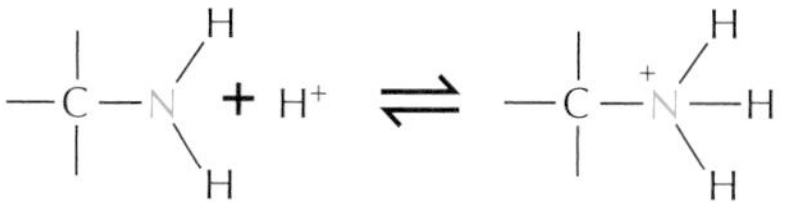

Amides are formed by combining an acid and an amine. Unlike amines, amides are uncharged in water. An example is the peptide bond that joins amino acids in a protein.

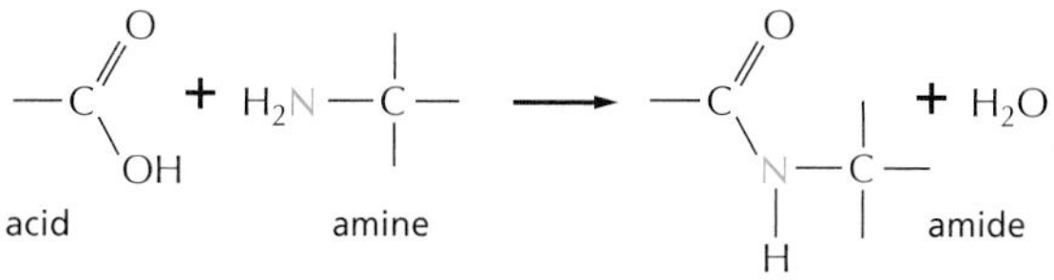

acid amine amide

Nitrogen also occurs in several ring compounds, including important constituents of nucleic acids: purines and pyrimidines.

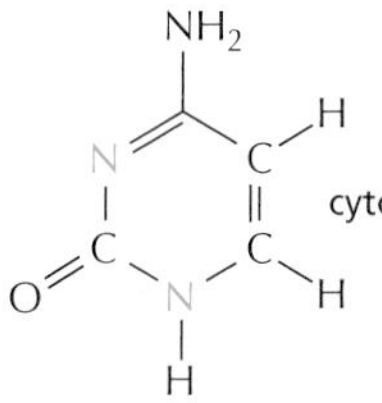

cytosine (a pyrimidine)

PHOSPHATES

Inorganic phosphate is a stable ion formed from phosphoric acid, H_3PO_4. It is often written as P_i.

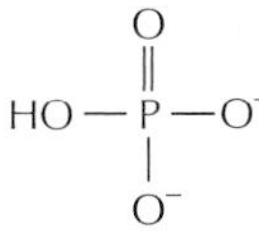

Phosphate esters can form between a phosphate and a free hydroxyl group. Phosphoryl groups are often attached to proteins in this way.

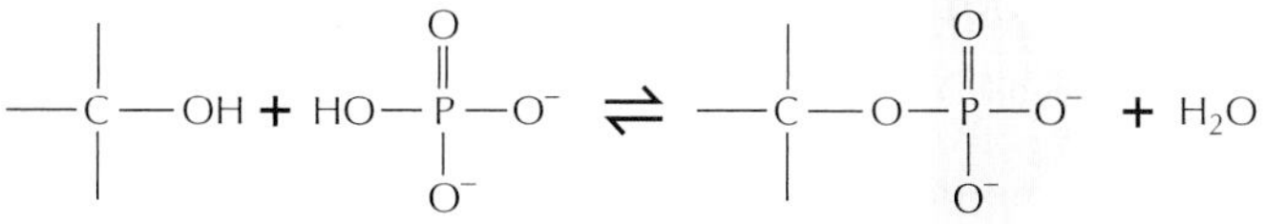

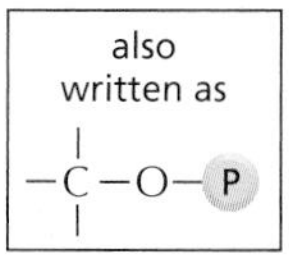

also written as

The combination of a phosphate and a carboxyl group, or two or more phosphate groups, gives an acid anhydride.

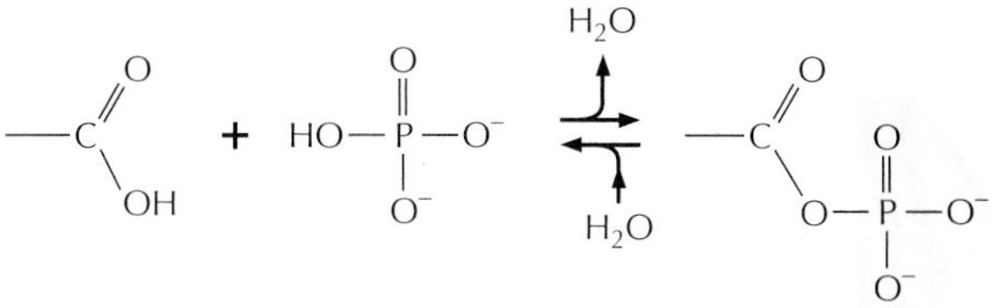

high-energy acyl phosphate bond (carboxylic–phosphoric acid anhydride) found in some metabolites

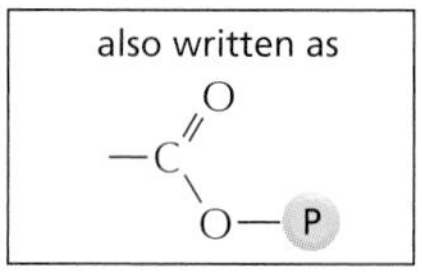

also written as

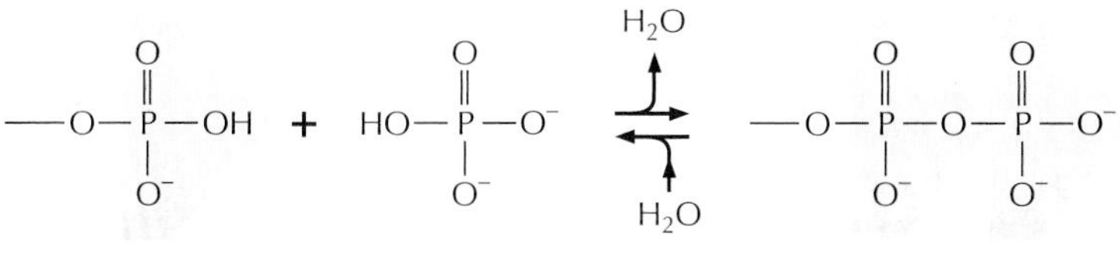

phosphoanhydride—a high-energy bond found in molecules such as ATP

also written as

PANEL 2–2 The chemical properties of water

HYDROGEN BONDS

Because they are polarized, two adjacent H_2O molecules can form a linkage known as a hydrogen bond. Hydrogen bonds have only about 1/20 the strength of a covalent bond.

Hydrogen bonds are strongest when the three atoms lie in a straight line.

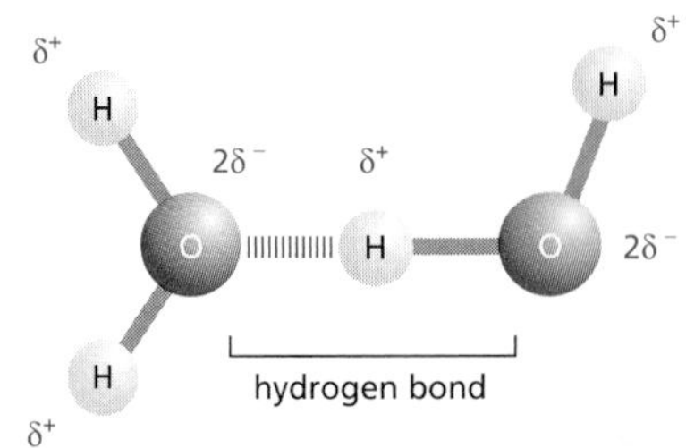

bond lengths

hydrogen bond 0.27 nm

0.10 nm covalent bond

WATER

Two atoms connected by a covalent bond may exert different attractions for the electrons of the bond. In such cases the bond is polar, with one end slightly negatively charged (δ−) and the other slightly positively charged (δ+).

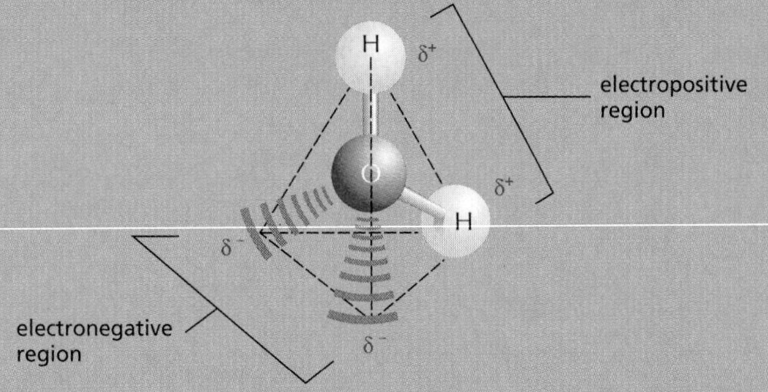

Although a water molecule has an overall neutral charge (having the same number of electrons and protons), the electrons are asymmetrically distributed, making the molecule polar. The oxygen nucleus draws electrons away from the hydrogen nuclei, leaving these nuclei with a small net positive charge. The excess of electron density on the oxygen atom creates weakly negative regions at the other two corners of an imaginary tetrahedron. On these pages we review the chemical properties of water and see how water influences the behavior of biological molecules.

WATER STRUCTURE

Molecules of water join together transiently in a hydrogen-bonded lattice.

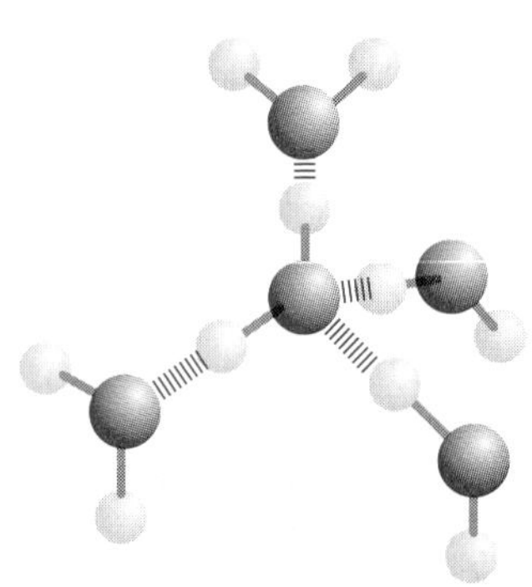

The cohesive nature of water is responsible for many of its unusual properties, such as high surface tension, specific heat, and heat of vaporization.

HYDROPHILIC MOLECULES

Substances that dissolve readily in water are termed hydrophilic. They are composed of ions or polar molecules that attract water molecules through electrical charge effects. Water molecules surround each ion or polar molecule on the surface of such a solid and carry it into solution.

Ionic substances such as sodium chloride dissolve because water molecules are attracted to the positive (Na^+) or negative (Cl^-) charge of each ion.

Polar substances such as urea dissolve because their molecules form hydrogen bonds with the surrounding water molecules

HYDROPHOBIC MOLECULES

Substances that contain a preponderance of nonpolar bonds are usually insoluble in water and are termed hydrophobic. Water molecules are not attracted to their molecules and so have little tendency to surround them and carry them into solution.

Hydrocarbons, which contain many C–H bonds, are especially hydrophobic.

WATER AS A SOLVENT

Many substances, such as household sugar, dissolve in water. That is, their molecules separate from each other, each becoming surrounded by water molecules.

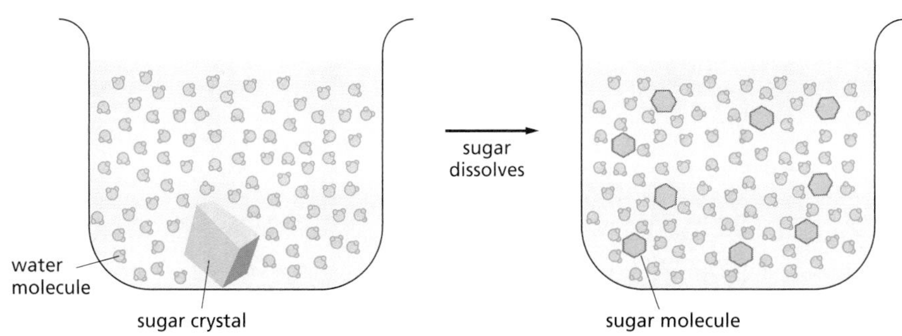

When a substance dissolves in a liquid, the mixture is termed a solution. The dissolved substance (in this case sugar) is the solute, and the liquid that does the dissolving (in this case water) is the solvent. Water is an excellent solvent for many substances because of its polar bonds.

ACIDS

Substances that release hydrogen ions into solution are called acids.

$$HCl \longrightarrow H^+ + Cl^-$$

HCl: hydrochloric acid (strong acid); H^+: hydrogen ion; Cl^-: chloride ion

Many of the acids important in the cell are not completely dissociated, and they are therefore weak acids—for example, the carboxyl group (–COOH), which dissociates to give a hydrogen ion in solution.

$$-C(=O)OH \rightleftharpoons H^+ + -C(=O)O^-$$

(weak acid)

Note that this is a reversible reaction.

HYDROGEN ION EXCHANGE

Positively charged hydrogen ions (H^+) can spontaneously move from one water molecule to another, thereby creating two ionic species.

$$H_2O \cdots H{-}OH \rightleftharpoons H_3O^+ + OH^-$$

hydronium ion (water acting as a weak base); hydroxyl ion (water acting as a weak acid)

often written as: $H_2O \rightleftharpoons H^+ + OH^-$

hydrogen ion; hydroxyl ion

Because the process is rapidly reversible, hydrogen ions are continually shuttling between water molecules. Pure water contains a steady-state concentration of hydronium ions and hydroxyl ions (both 10^{-7} M).

pH

The acidity of a solution is defined by the concentration of hydronium ions it possesses, generally abbreviated as H^+. For convenience we use the pH scale, where

$$pH = -\log_{10}[H^+]$$

For pure water

$[H^+] = 10^{-7}$ moles/liter

pH = 7.0

	H^+ conc. moles/liter	pH
ACIDIC	10^{-1}	1
	10^{-2}	2
	10^{-3}	3
	10^{-4}	4
	10^{-5}	5
	10^{-6}	6
	10^{-7}	7
ALKALINE	10^{-8}	8
	10^{-9}	9
	10^{-10}	10
	10^{-11}	11
	10^{-12}	12
	10^{-13}	13
	10^{-14}	14

BASES

Substances that reduce the number of hydrogen ions in solution are called bases. Some bases, such as ammonia, combine directly with hydrogen ions.

$$NH_3 + H^+ \longrightarrow NH_4^+$$

ammonia; hydrogen ion; ammonium ion

Other bases, such as sodium hydroxide, reduce the number of H^+ ions indirectly, by making OH^- ions that then combine directly with H^+ ions to make H_2O.

$$NaOH \longrightarrow Na^+ + OH^-$$

sodium hydroxide (strong base); sodium ion; hydroxyl ion

Many bases found in cells are partially associated with H^+ ions and are termed weak bases. This is true of compounds that contain an amino group ($-NH_2$), which has a weak tendency to reversibly accept an H^+ ion from water, increasing the quantity of free OH^- ions.

$$-NH_2 + H^+ \rightleftharpoons -NH_3^+$$

MONOSACCHARIDES

Monosaccharides usually have the general formula $(CH_2O)_n$, where *n* can be 3, 4, 5, or 6, and have two or more hydroxyl groups. They either contain an aldehyde group ($-C\lessgtr^{O}_{H}$) and are called aldoses, or a ketone group ($>C=O$) and are called ketoses.

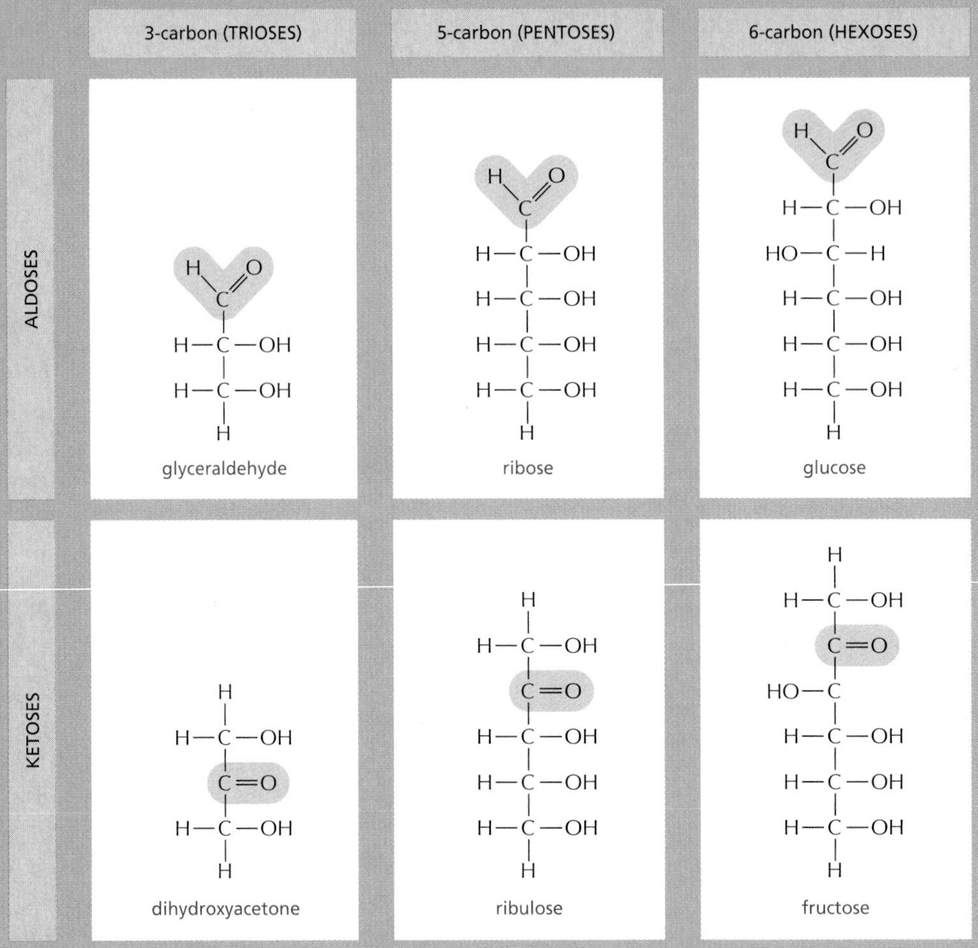

	3-carbon (TRIOSES)	5-carbon (PENTOSES)	6-carbon (HEXOSES)
ALDOSES	glyceraldehyde	ribose	glucose
KETOSES	dihydroxyacetone	ribulose	fructose

RING FORMATION

In aqueous solution, the aldehyde or ketone group of a sugar molecule tends to react with a hydroxyl group of the same molecule, thereby closing the molecule into a ring.

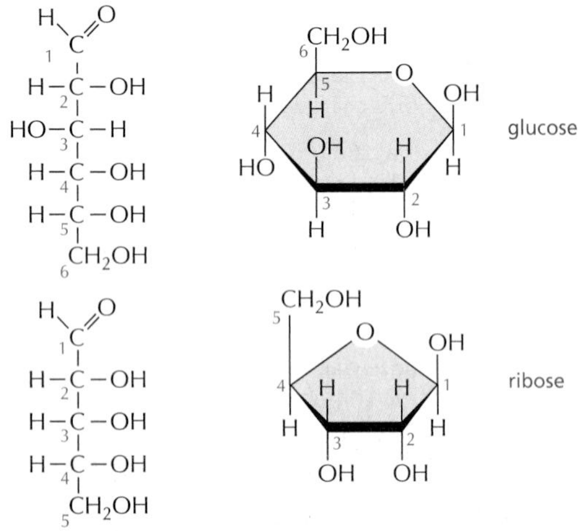

glucose

ribose

Note that each carbon atom has a number.

ISOMERS

Many monosaccharides differ only in the spatial arrangement of atoms—that is, they are isomers. For example, glucose, galactose, and mannose have the same formula ($C_6H_{12}O_6$) but differ in the arrangement of groups around one or two carbon atoms.

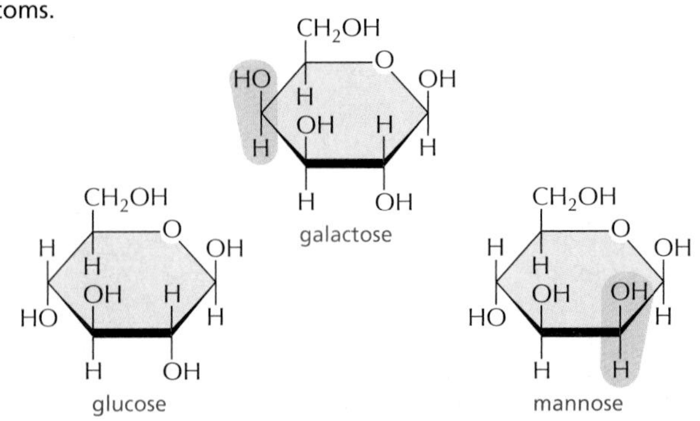

galactose

glucose

mannose

These small differences make only minor changes in the chemical properties of the sugars. But they are recognized by enzymes and other proteins and therefore can have important biological effects.

α AND β LINKS

The hydroxyl group on the carbon that carries the aldehyde or ketone can rapidly change from one position to the other. These two positions are called α and β.

β hydroxyl ⇌ α hydroxyl

As soon as one sugar is linked to another, the α or β form is frozen.

SUGAR DERIVATIVES

The hydroxyl groups of a simple monosaccharide can be replaced by other groups. For example,

glucosamine

glucuronic acid

N-acetylglucosamine

DISACCHARIDES

The carbon that carries the aldehyde or the ketone can react with any hydroxyl group on a second sugar molecule to form a disaccharide. Three common disaccharides are

- maltose (glucose + glucose)
- lactose (galactose + glucose)
- sucrose (glucose + fructose)

The reaction forming sucrose is shown here.

α glucose + β fructose → sucrose + H_2O

OLIGOSACCHARIDES AND POLYSACCHARIDES

Large linear and branched molecules can be made from simple repeating units. Short chains are called oligosaccharides, and long chains are called polysaccharides. Glycogen, for example, is a polysaccharide made entirely of glucose units joined together.

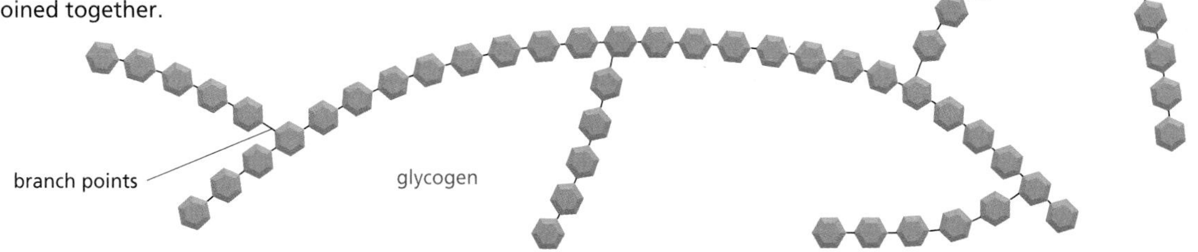

COMPLEX OLIGOSACCHARIDES

In many cases a sugar sequence is nonrepetitive. Many different molecules are possible. Such complex oligosaccharides are usually linked to proteins or to lipids, as is this oligosaccharide, which is part of a cell-surface molecule, that defines a particular blood group.

PANEL 2–4 Fatty acids and other lipids

FATTY ACIDS

All fatty acids have carboxyl groups with long hydrocarbon tails.

stearic acid (C_{18})	palmitic acid (C_{16})	oleic acid (C_{18})
COOH	COOH	COOH
CH_2	CH_2	CH_2
CH_2	CH_2	CH_2
CH_2	CH_2	CH_2
CH_2	CH_2	CH_2
CH_2	CH_2	CH_2
CH_2	CH_2	CH_2
CH_2	CH_2	CH_2
CH_2	CH_2	CH
CH_2	CH_2	CH
CH_2	CH_2	CH_2
CH_2	CH_2	CH_2
CH_2	CH_2	CH_2
CH_2	CH_2	CH_2
CH_2	CH_2	CH_2
CH_2	CH_3	CH_2
CH_2		CH_2
CH_3		CH_3

Hundreds of different kinds of fatty acids exist. Some have one or more double bonds in their hydrocarbon tail and are said to be unsaturated. Fatty acids with no double bonds are saturated.

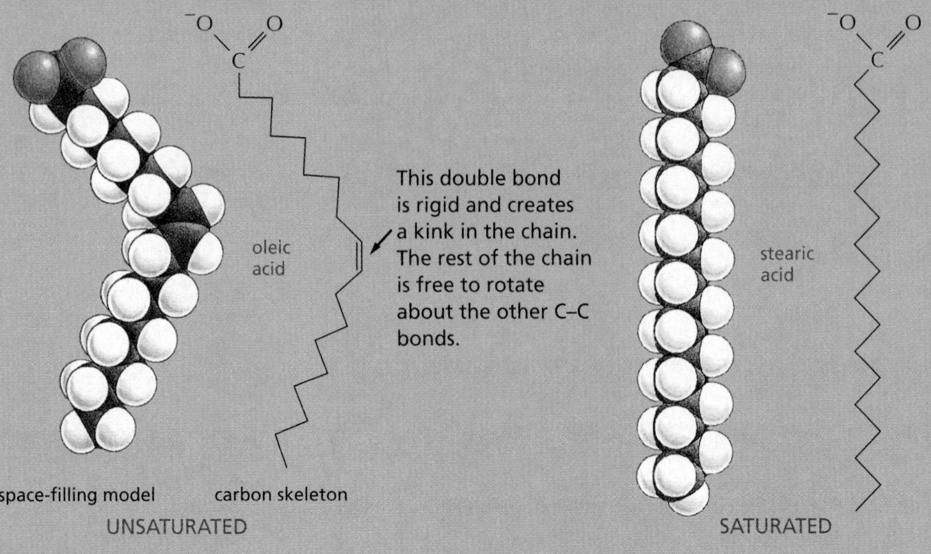

TRIACYLGLYCEROLS

Fatty acids are stored as an energy reserve (fats and oils) through an ester linkage to glycerol to form triacylglycerols.

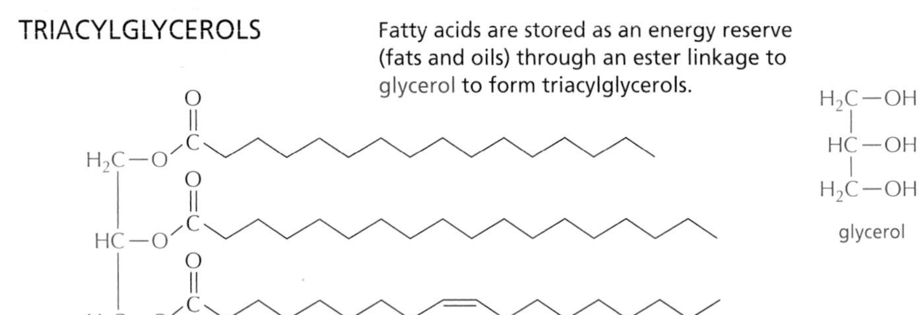

CARBOXYL GROUP

If free, the carboxyl group of a fatty acid will be ionized.

But more often it is linked to other groups to form either esters

or amides.

PHOSPHOLIPIDS

Phospholipids are the major constituents of cell membranes.

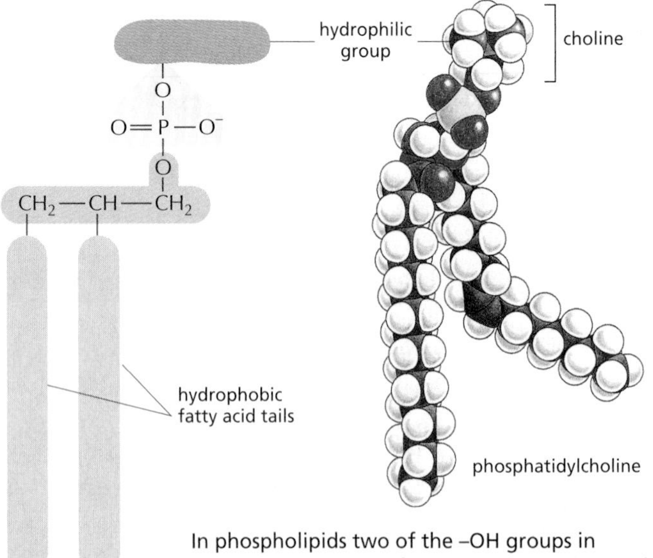

In phospholipids two of the –OH groups in glycerol are linked to fatty acids, while the third –OH group is linked to phosphoric acid. The phosphate is further linked to one of a variety of small polar groups (alcohols).

LIPID AGGREGATES

Fatty acids have a hydrophilic head and a hydrophobic tail.

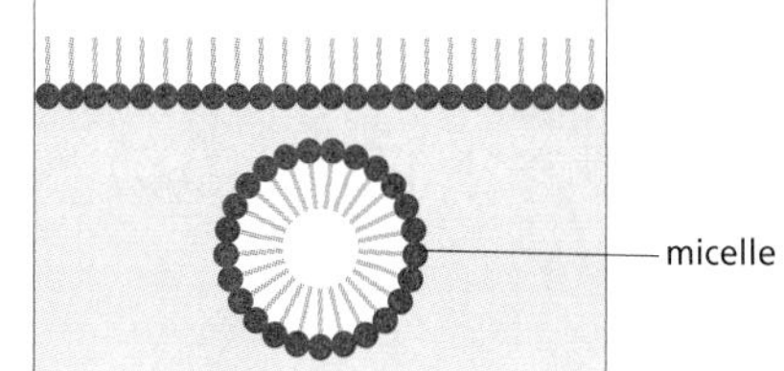

In water they can form a surface film or form small micelles.

Their derivatives can form larger aggregates held together by hydrophobic forces:

Triacylglycerols form large spherical fat droplets in the cell cytoplasm.

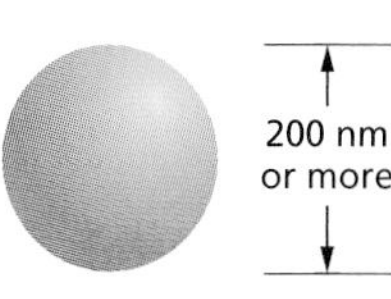

Phospholipids and glycolipids form self-sealing lipid bilayers that are the basis for all cellular membranes.

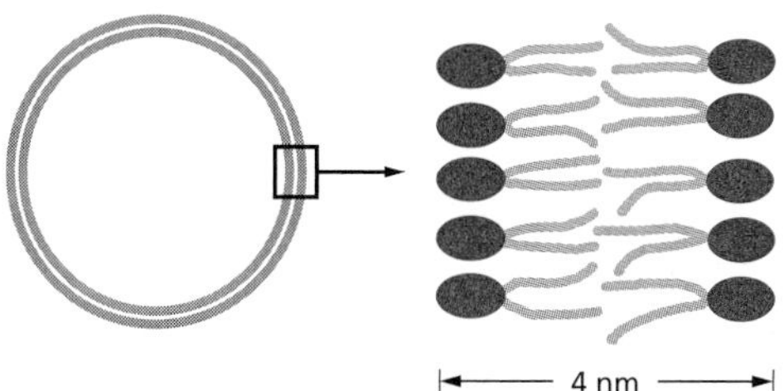

OTHER LIPIDS

Lipids are defined as the water-insoluble molecules in cells that are soluble in organic solvents. Two other common types of lipids are steroids and polyisoprenoids. Both are made from isoprene units.

CH_3 — C — CH=CH_2, =CH_2

isoprene

STEROIDS

Steroids have a common multiple-ring structure.

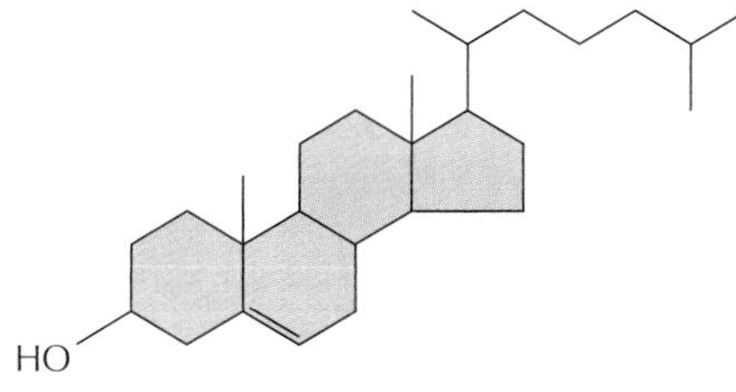

cholesterol—found in many membranes

OH

O

testosterone—male steroid hormone

GLYCOLIPIDS

Like phospholipids, these compounds are composed of a hydrophobic region, containing two long hydrocarbon tails, and a polar region, which, however, contains one or more sugars and no phosphate.

galactose

sugar

OH, H, O, CH_2, C, NH, O

hydrocarbon tails

a simple glycolipid

POLYISOPRENOIDS

long-chain polymers of isoprene

O^-, O=P—O^-, O

dolichol phosphate—used to carry activated sugars in the membrane-associated synthesis of glycoproteins and some polysaccharides

PANEL 2–5 The 20 amino acids found in proteins

FAMILIES OF AMINO ACIDS

The common amino acids are grouped according to whether their side chains are

- acidic
- basic
- uncharged polar
- nonpolar

These 20 amino acids are given both three-letter and one-letter abbreviations.

Thus: alanine = Ala = A

BASIC SIDE CHAINS

lysine (Lys, or K)

$-NH-CH(-C(=O)-)-CH_2-CH_2-CH_2-CH_2-NH_3^+$

arginine (Arg, or R)

$-NH-CH(-C(=O)-)-CH_2-CH_2-CH_2-NH-C(=NH_2^+)NH_2$ ($^+H_2N=C-NH_2$)

This group is very basic because its positive charge is stabilized by resonance.

histidine (His, or H)

$-NH-CH(-C(=O)-)-CH_2-$ imidazole ring (C, HN, CN, HC=NH$^+$)

These nitrogens have a relatively weak affinity for an H^+ and are only partly positive at neutral pH.

THE AMINO ACID

The general formula of an amino acid is

$$H_2N-C(H)(R)-COOH$$

amino group (H_2N); α-carbon atom (C); carboxyl group (COOH); side-chain group (R)

R is commonly one of 20 different side chains. At pH 7 both the amino and carboxyl groups are ionized.

$$^{\oplus}H_3N-C(H)(R)-COO^{\ominus}$$

These pages present the amino acids found in proteins and show how they are linked to form them.

OPTICAL ISOMERS

The α-carbon atom is asymmetric, allowing for two mirror-image (or stereo-) isomers, L and D.

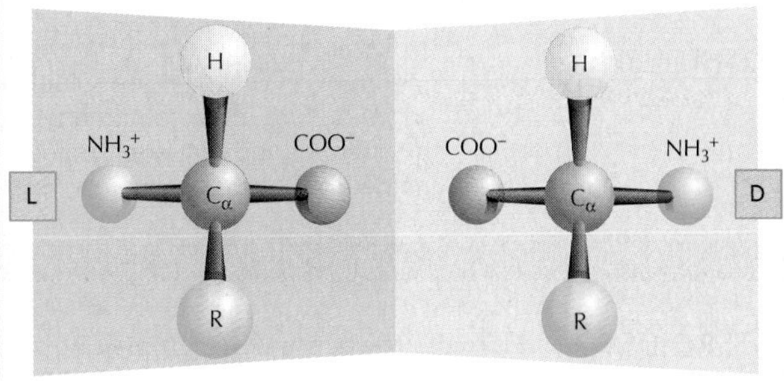

Proteins consist exclusively of L-amino acids.

PEPTIDE BONDS

Amino acids are commonly joined together by an amide linkage, called a peptide bond.

The four atoms in each peptide bond *(gray box)* form a rigid planar unit. There is no rotation around the C–N bond.

$H_2N-CH(R)-COOH \; + \; H_2N-CH(R)-COOH$

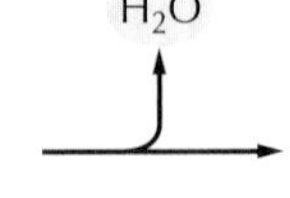

$H_2N-CH(R)-C(=O)-NH-CH(R)-COOH$

Proteins are long polymers of amino acids linked by peptide bonds, and they are always written with the N-terminus toward the left. The sequence of this tripeptide is histidine-cysteine-valine.

amino, or N-, terminus

$^+H_3N-CH(CH_2\text{-imidazole})-C(=O)-NH-CH(CH_2SH)-C(=O)-NH-CH(CH(CH_3)_2)-COO^-$

carboxyl, or C-, terminus

These two single bonds allow rotation, so that long chains of amino acids are very flexible.

ACIDIC SIDE CHAINS

aspartic acid
(Asp, or D)

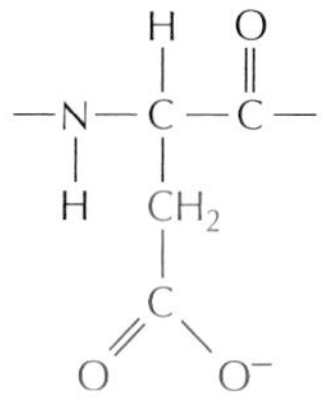

glutamic acid
(Glu, or E)

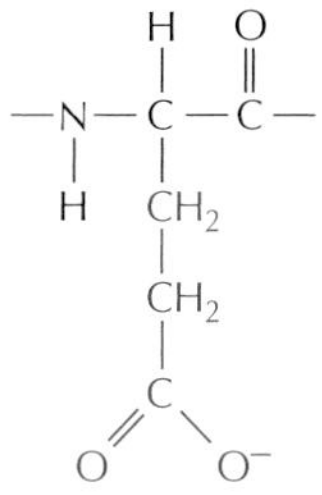

UNCHARGED POLAR SIDE CHAINS

asparagine
(Asn, or N)

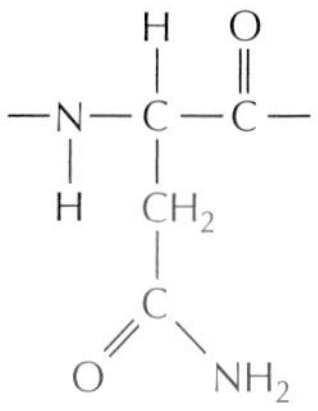

glutamine
(Gln, or Q)

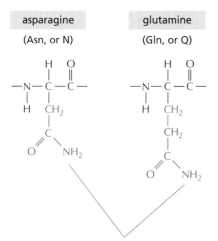

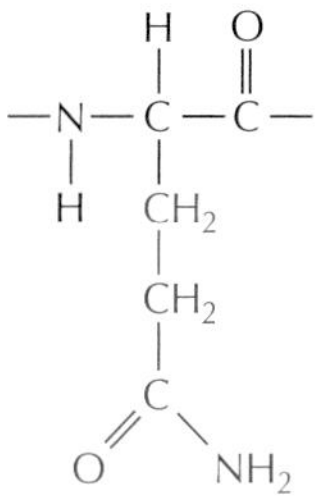

Although the amide N is not charged at neutral pH, it is polar.

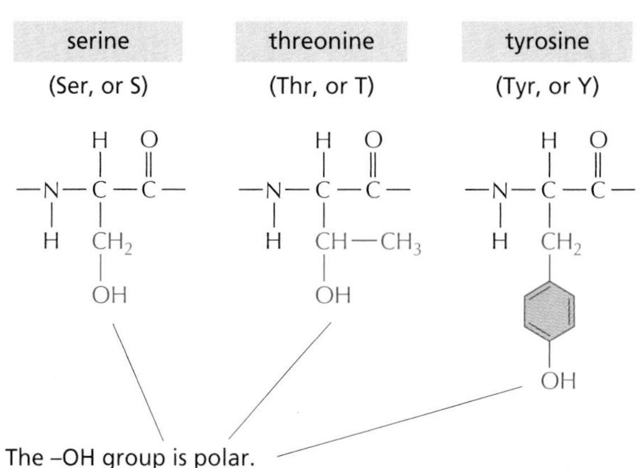

NONPOLAR SIDE CHAINS

alanine
(Ala, or A)

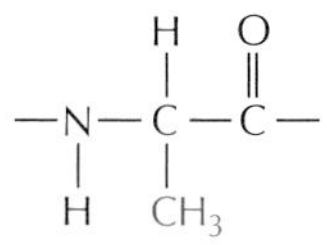

valine
(Val, or V)

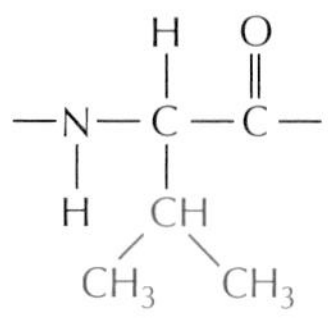

leucine
(Leu, or L)

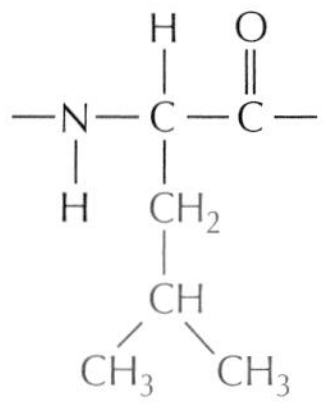

isoleucine
(Ile, or I)

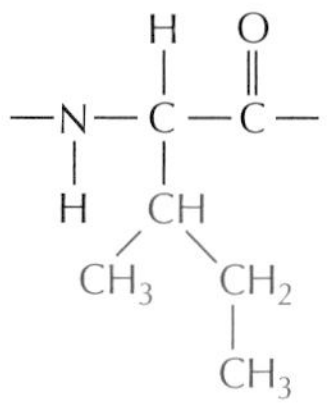

proline
(Pro, or P)

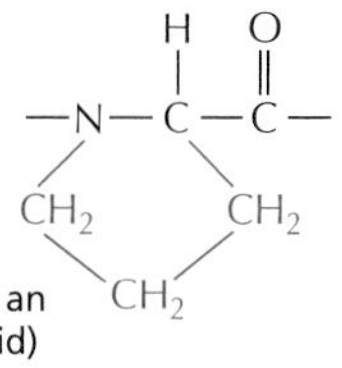

(actually an imino acid)

phenylalanine
(Phe, or F)

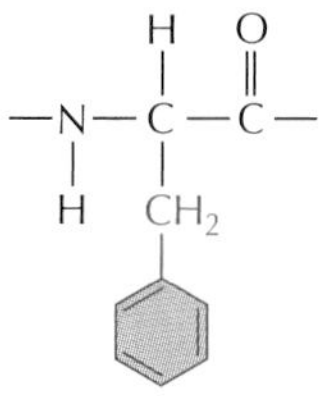

methionine
(Met, or M)

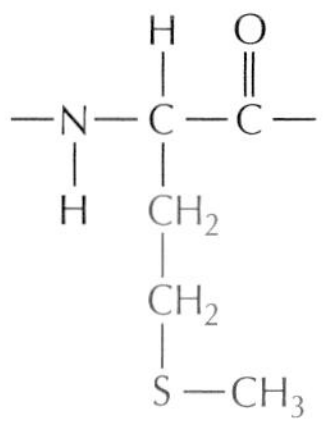

tryptophan
(Trp, or W)

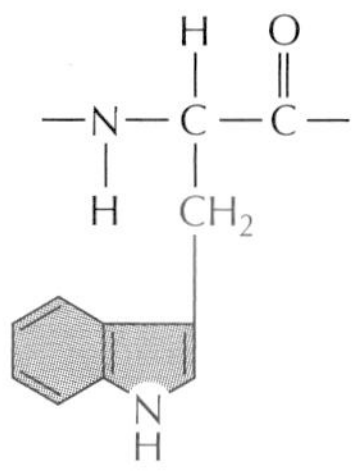

glycine
(Gly, or G)

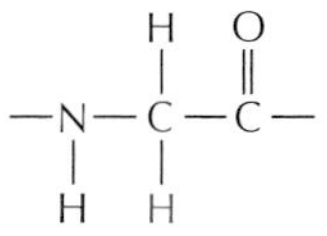

cysteine
(Cys, or C)

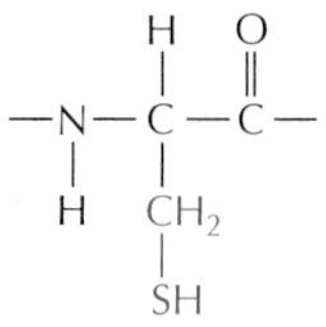

Disulfide bonds can form between two cysteine side chains in proteins.

$$- - CH_2 - S - S - CH_2 - -$$

PANEL 2–6 A survey of the nucleotides

BASES

The bases are nitrogen-containing ring compounds, either pyrimidines or purines.

PHOSPHATES

The phosphates are normally joined to the C5 hydroxyl of the ribose or deoxyribose sugar (designated 5'). Mono-, di-, and triphosphates are common.

as in AMP

as in ADP

as in ATP

The phosphate makes a nucleotide negatively charged.

NUCLEOTIDES

A nucleotide consists of a nitrogen-containing base, a five-carbon sugar, and one or more phosphate groups.

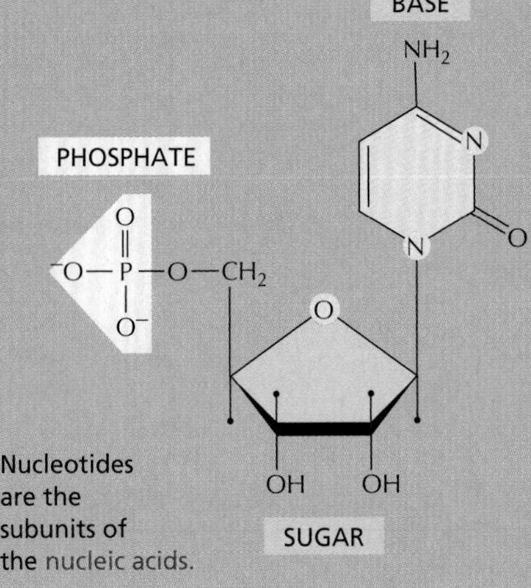

Nucleotides are the subunits of the nucleic acids.

BASIC SUGAR LINKAGE

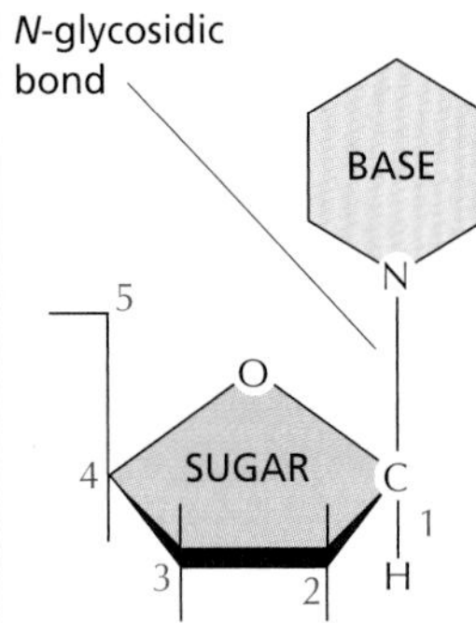

The base is linked to the same carbon (C1) used in sugar–sugar bonds.

SUGARS

PENTOSE

a five-carbon sugar

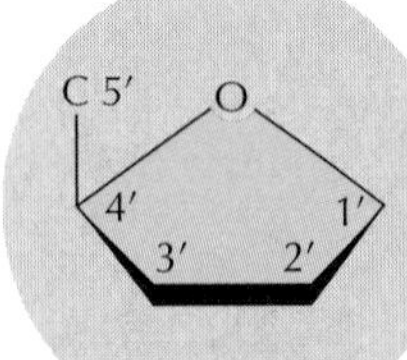

two kinds are used

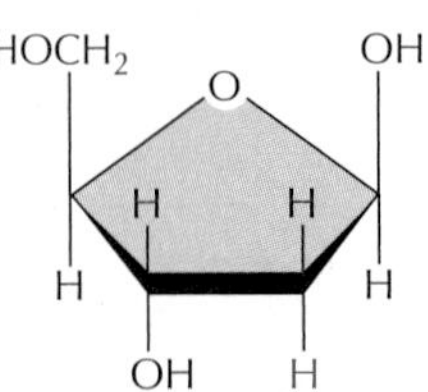

β-D-ribose
used in ribonucleic acid

β-D-2-deoxyribose
used in deoxyribonucleic acid

Each numbered carbon on the sugar of a nucleotide is followed by a prime mark; therefore, one speaks of the "5-prime carbon," etc.

NOMENCLATURE

The names can be confusing, but the abbreviations are clear.

BASE	NUCLEOSIDE	ABBR.
adenine	adenosine	A
guanine	guanosine	G
cytosine	cytidine	C
uracil	uridine	U
thymine	thymidine	T

Nucleotides are abbreviated by three capital letters. Some examples follow:

AMP = adenosine monophosphate
dAMP = deoxyadenosine monophosphate
UDP = uridine diphosphate
ATP = adenosine triphosphate

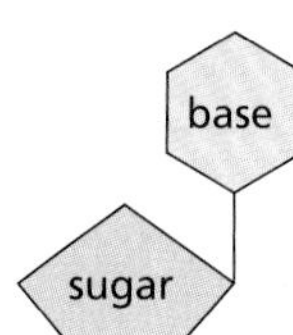

BASE + SUGAR = NUCLEOSIDE

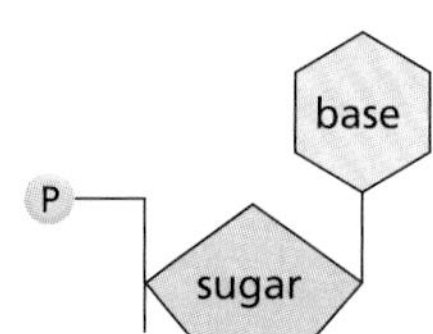

BASE + SUGAR + PHOSPHATE = NUCLEOTIDE

NUCLEIC ACIDS

Nucleotides are joined together by a phosphodiester linkage between 5′ and 3′ carbon atoms to form nucleic acids. The linear sequence of nucleotides in a nucleic acid chain is commonly abbreviated by a one-letter code, A—G—C—T—T—A—C—A, with the 5′ end of the chain at the left.

5′ end of chain

phosphodiester linkage

example: DNA

3′ end of chain

NUCLEOTIDES HAVE MANY OTHER FUNCTIONS

1 They carry chemical energy in their easily hydrolyzed phosphoanhydride bonds.

phosphoanhydride bonds

example: ATP (or **ATP**)

2 They combine with other groups to form coenzymes.

example: coenzyme A (CoA)

3 They are used as signaling molecules in the cell.

example: cyclic AMP

PANEL 2–7 The principal types of weak noncovalent bonds

WEAK CHEMICAL BONDS

Organic molecules can interact with other molecules through three types of short-range attractive forces known as *noncovalent bonds:* van der Waals attractions, electrostatic attractions, and hydrogen bonds. The repulsion of hydrophobic groups from water is also important for ordering biological macromolecules.

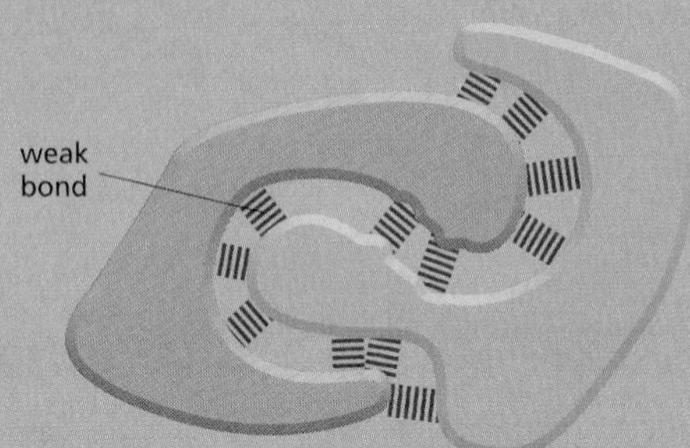

Weak chemical bonds have less than 1/20 the strength of a strong covalent bond. They are strong enough to provide tight binding only when many of them are formed simultaneously.

VAN DER WAALS ATTRACTIONS

If two atoms are too close together they repel each other very strongly. For this reason, an atom can often be treated as a sphere with a fixed radius. The characteristic "size" for each atom is specified by a unique van der Waals radius. The contact distance between any two non-covalently bonded atoms is the sum of their van der Waals radii.

C N

H	C	N	O
0.12 nm radius	0.2 nm radius	0.15 nm radius	0.14 nm radius

At very short distances any two atoms show a weak bonding interaction due to their fluctuating electrical charges. The two atoms will be attracted to each other in this way until the distance between their nuclei is approximately equal to the sum of their van der Waals radii. Although they are individually very weak, van der Waals attractions can become important when two macromolecular surfaces fit very close together, because many atoms are involved.

Note that when two atoms form a covalent bond, the centers of the two atoms (the two atomic nuclei) are much closer together than the sum of the two van der Waals radii. Thus,

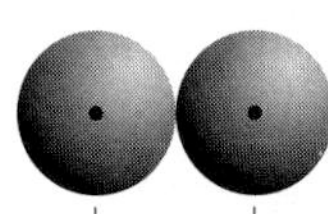

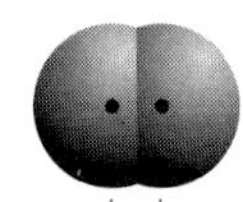

0.4 nm	0.15 nm	0.13 nm
two non-bonded carbon atoms	single-bonded carbons	double-bonded carbons

HYDROGEN BONDS

As already described for water (see Panel 2–2, pp. 66–67) hydrogen bonds form when a hydrogen atom is "sandwiched" between two electron-attracting atoms (usually oxygen or nitrogen).

Hydrogen bonds are strongest when the three atoms are in a straight line:

O—H ||||||||| O   N—H ||||||||| O

Examples in macromolecules:

Amino acids in polypeptide chains hydrogen-bonded together.

C=O ||||||||| H—N
R—C—H   R—C—H   H—C—R
C=O ||||||||| H—N

Two bases, G and C, hydrogen-bonded in DNA or RNA.

N—H||||||O
N ||||||||H—N
O||||||H—N

HYDROGEN BONDS IN WATER

Any molecules that can form hydrogen bonds to each other can alternatively form hydrogen bonds to water molecules. Because of this competition with water molecules, the hydrogen bonds formed between two molecules dissolved in water are relatively weak.

peptide bond

$2H_2O$

$2H_2O$

HYDROPHOBIC FORCES

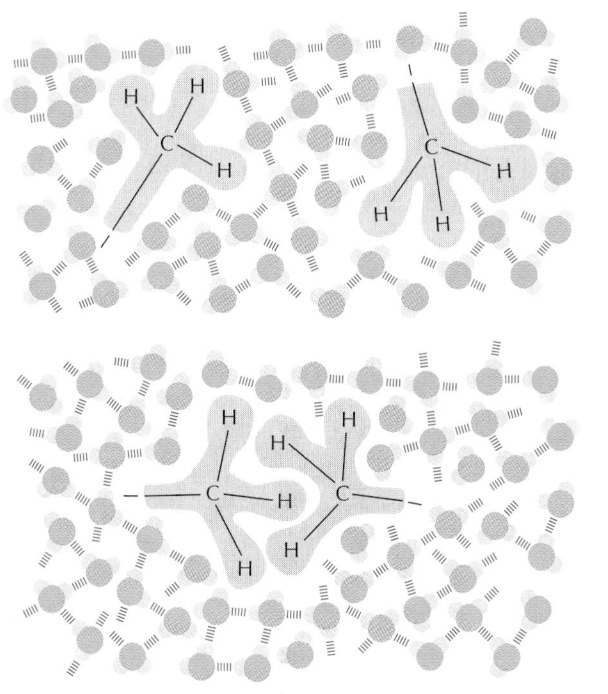

Water forces hydrophobic groups together in order to minimize their disruptive effects on the hydrogen-bonded water network. Hydrophobic groups held together in this way are sometimes said to be held together by "hydrophobic bonds," even though the attraction is actually caused by a repulsion from the water.

ELECTROSTATIC ATTRACTIONS IN AQUEOUS SOLUTIONS

Charged groups are shielded by their interactions with water molecules. electrostatic attractions are therefore quite weak in water.

Similarly, ions in solution can cluster around charged groups and further weaken these attractions.

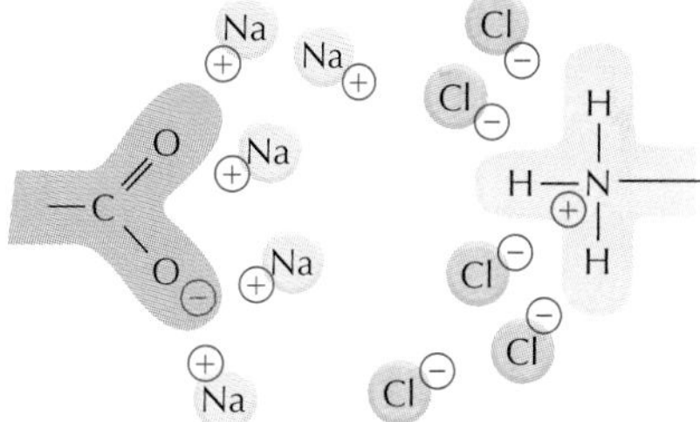

Despite being weakened by water and salt, electrostatic attractions are very important in biological systems. For example, an enzyme that binds a positively charged substrate will often have a negatively charged amino acid side chain at the appropriate place.

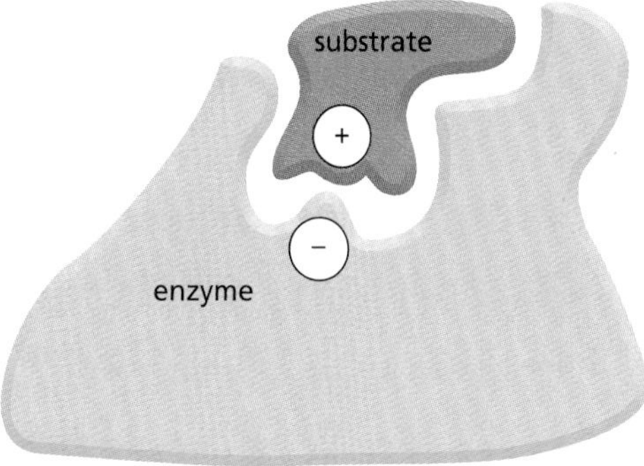

ELECTROSTATIC ATTRACTIONS

Attractive interactions occur both between fully charged groups (ionic bond) and between partially charged groups on polar molecules.

The force of attraction between the two charges, δ^+ and δ^-, falls off rapidly as the distance between the charges increases.

In the absence of water, electrostatic forces are very strong. They are responsible for the strength of such minerals as marble and agate, and for crystal formation in common table salt, NaCl.

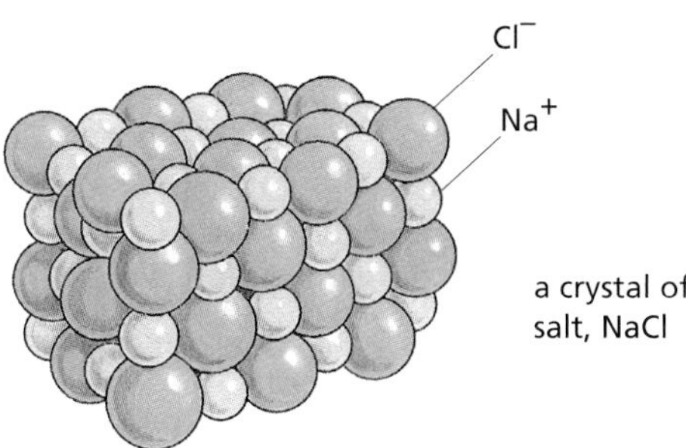

a crystal of salt, NaCl

ESSENTIAL CONCEPTS

- Living cells obey the same chemical and physical laws as nonliving things. Like all other forms of matter, they are composed of atoms, which are the smallest units of a chemical element that retains the distinctive chemical properties of that element.
- Atoms are made up of smaller particles. The nucleus of an atom contains protons, which are positively charged, and uncharged neutrons. The nucleus is surrounded by a cloud of negatively charged electrons.
- The number of electrons in an atom is equal to the number of protons in its nucleus. The nuclei of different isotopes of the same element contain the same number of protons but different numbers of neutrons.
- Living cells are made up of a limited number of elements, four of which—C, H, N, O—make up 96.5% of their mass.
- The chemical properties of an atom are determined by the number and arrangement of its electrons. An atom is most stable when all of its electrons are at their lowest possible energy level and when each electron shell is completely filled.
- Chemical bonds form between atoms as electrons move to reach a more stable arrangement. Clusters of two or more atoms held together by chemical bonds are known as molecules.
- When an electron jumps from one atom to another, two ions of opposite charge are generated; ionic bonds can then arise by the mutual attraction of these charged atoms.
- A covalent bond consists of a pair of electrons shared between two adjacent atoms. If two pairs of electrons are shared, a double bond is formed.
- Living organisms contain a distinctive and restricted set of small carbon-based molecules that are essentially the same for every living species. The main categories are sugars, fatty acids, amino acids, and nucleotides.
- Sugars are a primary source of chemical energy for cells and can be incorporated into polysaccharides for energy storage.
- Fatty acids are also important for energy storage, but their most essential function is in the formation of cell membranes.
- The vast majority of the dry mass of a cell consists of macromolecules, formed as polymers of sugars, amino acids, or nucleotides.
- Macromolecules are intermediate in both size and complexity between small molecules and cell organelles. They have many remarkable properties that are not easily deduced from the subunits from which they are made.
- The remarkably diverse and versatile class of macromolecules known as proteins are polymers formed from amino acids.
- Nucleotides play a central part in energy transfer and are the subunits from which the informational macromolecules, RNA and DNA, are made.
- Protein, RNA, and DNA molecules are synthesized from subunits by repetitive condensation reactions. Each of these biological macromolecules has a unique sequence of subunits.
- Weak noncovalent bonds form between different regions of a macromolecule. These can cause the macromolecule to fold into a unique three-dimensional shape (conformation) with a special chemistry, as seen most conspicuously in proteins.

KEY TERMS

acid	DNA	molecular weight
amino acid	electron	monomer
atom	electrostatic attraction	noncovalent bond
atomic weight	fatty acid	nucleotide
ATP	hydrogen bond	pH scale
Avogadro's number	hydrolysis	polar
base	hydronium ion	polymer
buffer	hydrophilic	protein
chemical bond	hydrophobic	proton
chemical group	ion	RNA
condensation reaction	ionic bond	sequence
conformation	macromolecule	subunit
covalent bond	molecule	sugar

QUESTIONS

QUESTION 2–11

Which of the following statements are correct? Explain your answers.

A. An atomic nucleus contains protons and neutrons.

B. An atom has more electrons than protons.

C. The nucleus is surrounded by a double membrane.

D. All atoms of the same element have the same number of neutrons.

E. The number of neutrons determines whether the nucleus of an atom is stable or radioactive.

F. Both fatty acids and polysaccharides can be important energy stores in the cell.

G. Hydrogen bonds are weak and can be broken by thermal energy, yet they contribute significantly to the specificity of interactions between macromolecules.

QUESTION 2–12

To gain a better feeling for atomic dimensions, assume that the page on which this question is printed is made entirely of the polysaccharide cellulose, whose molecules are described by the formula $(C_nH_{2n}O_n)$, where *n* can be a quite large number and is variable from one molecule to another. The atomic weights of carbon, hydrogen, and oxygen are 12, 1, and 16, respectively, and this page weighs 5 g.

A. How many carbon atoms are there in this page?

B. In cellulose, how many carbon atoms would be stacked on top of each other to span the thickness of this page (the size of the page is 21.2 cm × 27.6 cm, and it is 0.07 mm thick)?

C. Now consider the problem from a different angle. Assume that the page is composed only of carbon atoms. A carbon atom has a diameter of 2×10^{-10} m (0.2 nm); how many carbon atoms of 0.2 nm diameter would it take to span the thickness of the page?

D. Compare your answers from parts B and C and explain any differences.

QUESTION 2–13

A. How many electrons can be accommodated in the first, second, and third electron shells of an atom?

B. How many electrons would atoms of the elements listed below preferentially gain or lose in order to obtain completely filled sets of energy levels?

helium	gain __	lose __
oxygen	gain __	lose __
carbon	gain __	lose __
sodium	gain __	lose __
chlorine	gain __	lose __

C. What do the answers tell you about the bonds that can form between sodium and chlorine?

QUESTION 2–14

Oxygen and sulfur have similar chemical properties because both elements have six electrons in their outermost electron shells. Indeed, both elements form molecules with two hydrogen atoms, water (H_2O) and hydrogen sulfide (H_2S). Surprisingly, water is a liquid, yet H_2S is a gas, despite the fact that sulfur is much larger and heavier than oxygen. Explain why this might be the case.

QUESTION 2–15

Write the chemical formula for a condensation reaction of two amino acids to form a peptide bond. Write the formula for its hydrolysis.

QUESTION 2–16

Which of the following statements are correct? Explain your answers.

A. Proteins are so remarkably diverse because each is made from a unique mixture of amino acids that are linked in random order.

B. Lipid bilayers are macromolecules that are made up mostly of phospholipid subunits.

C. Nucleic acids contain sugar groups.

D. Many amino acids have hydrophobic side chains.

E. The hydrophobic tails of phospholipid molecules are repelled from water.

F. DNA contains the four different bases A, G, U, and C.

QUESTION 2–17

A. How many different molecules composed of (a) two, (b) three, and (c) four amino acids, linked together by peptide bonds, can be made from the set of 20 naturally occurring amino acids?

B. Assume you were given a mixture consisting of one molecule each of all possible sequences of a smallish protein of molecular weight 4800 daltons. If the average molecular weight of an amino acid is, say, 120 daltons, how much would the sample weigh? How big a container would you need to hold it?

C. What does this calculation tell you about the fraction of possible proteins that are currently in use by living organisms (the average molecular weight of proteins is about 30,000 daltons)?

QUESTION 2–18

This is a biology textbook. Explain why the chemical principles that are described in this chapter are important in the context of modern cell biology.

QUESTION 2–19

A. Describe the similarities and differences between van der Waals attractions and hydrogen bonds.

B. Which of the two bonds would form (a) between two hydrogens bound to carbon atoms, (b) between a nitrogen atom and a hydrogen bound to a carbon atom, and (c) between a nitrogen atom and a hydrogen bound to an oxygen atom?

QUESTION 2–20

What are the forces that determine the folding of a macromolecule into a unique shape?

QUESTION 2–21

Fatty acids are said to be "amphipathic." What is meant by this term, and how does an amphipathic molecule behave in water? Draw a diagram to illustrate your answer.

QUESTION 2–22

Are the formulas in Figure Q2–22 correct or incorrect? Explain your answer in each case.

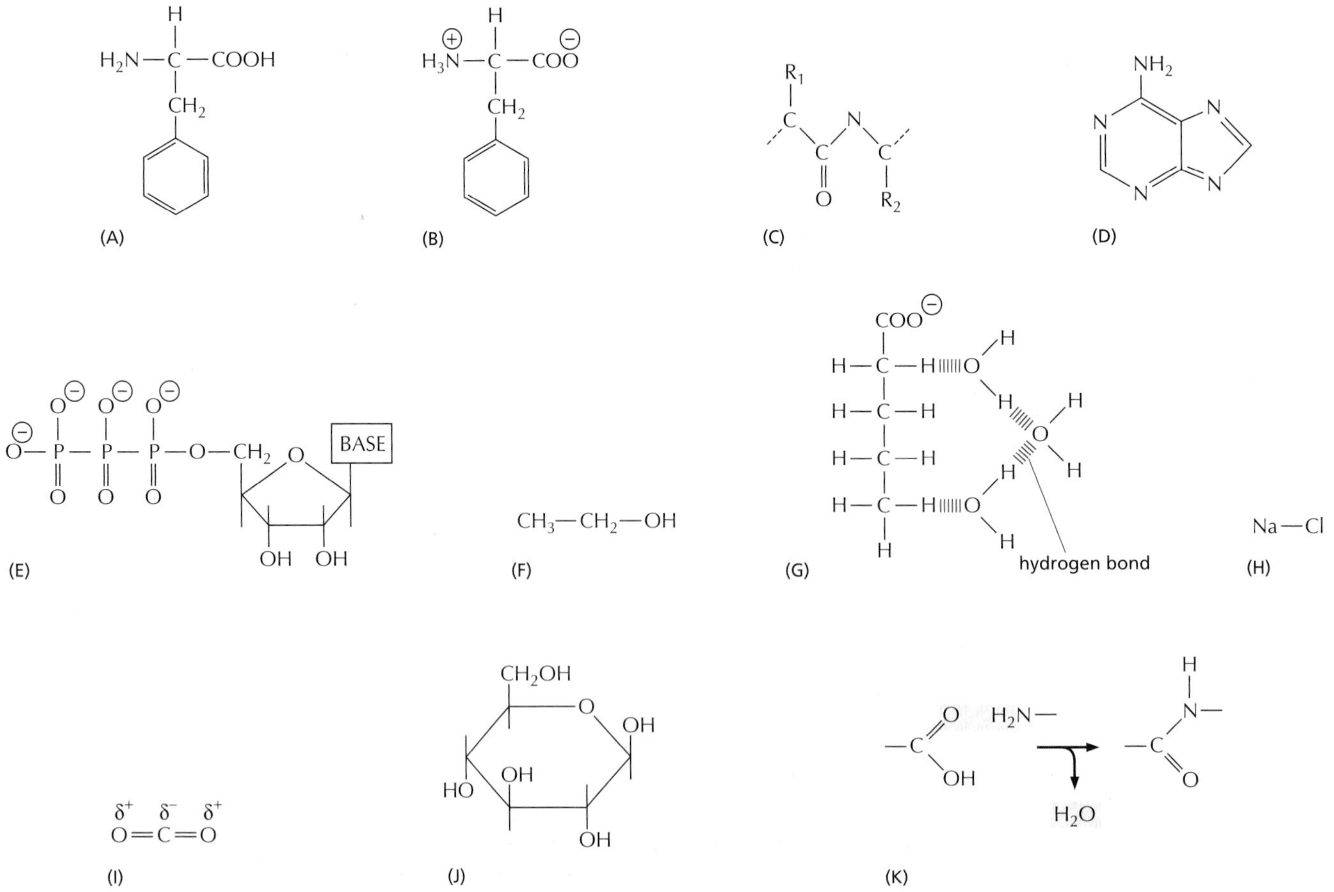

Figure Q2–22

CHAPTER **THREE** 3

Energy, Catalysis, and Biosynthesis

THE USE OF ENERGY BY CELLS

FREE ENERGY AND CATALYSIS

ACTIVATED CARRIER MOLECULES AND BIOSYNTHESIS

One property above all makes living things seem almost miraculously different from nonliving matter: they create and maintain order in a universe that is tending always toward greater disorder. To create this order, the cells in a living organism must carry out a never-ending stream of chemical reactions. In some of these reactions, small organic molecules—amino acids, sugars, nucleotides, and lipids—are taken apart or modified to supply the many other small molecules that the cell requires. In other reactions, these small molecules are used to construct an enormously diverse range of proteins, nucleic acids, and other macromolecules that endow living systems with all of their most distinctive properties. Each cell can be viewed as a tiny chemical factory, performing many millions of reactions every second.

To carry out the many chemical reactions needed to sustain it, a living organism requires both a source of atoms in the form of food molecules and a source of energy. The atoms and the energy must both come, ultimately, from the nonliving environment. In this chapter we discuss why cells require energy, and how they use energy and atoms from their environment to create the molecular order that makes life possible.

Most of the chemical reactions that cells perform would normally occur only at temperatures that are much higher than those inside a cell. Each reaction therefore requires a major boost in chemical reactivity to enable it to proceed rapidly within the cell. This boost is provided by specialized proteins called *enzymes*, each of which accelerates, or *catalyzes*, just one of the many possible kinds of reactions that a particular molecule might undergo. These enzyme-catalyzed reactions are usually connected in series, so that the product of one reaction becomes the starting material

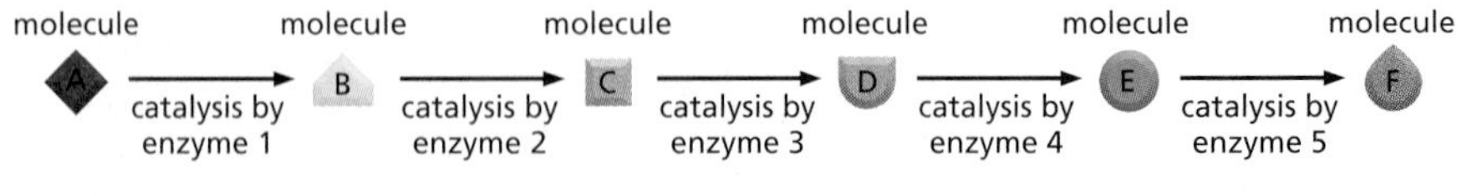

Figure 3–1 **A series of enzyme-catalyzed reactions forms a metabolic pathway.** Each enzyme catalyzes a chemical reaction involving a particular molecule. In this example, a set of enzymes acting in series converts molecule A to molecule F, forming a metabolic pathway.

for the next (Figure 3–1). The long linear reaction pathways, or *metabolic pathways*, that result are in turn linked to one another, forming a complex web of interconnected reactions.

Rather than being an inconvenience, the necessity for *catalysis* is a benefit, as it allows the cell to precisely control its **metabolism**—the sum total of all the chemical reactions it needs to carry out to survive, grow, and reproduce. This control is central to the chemistry of life.

Two opposing streams of chemical reactions occur in cells, the *catabolic* pathways and the *anabolic* pathways. The catabolic pathways (**catabolism**) break down foodstuffs into smaller molecules, thereby generating both a useful form of energy for the cell and some of the small molecules that the cell needs as building blocks. The anabolic, or *biosynthetic*, pathways (**anabolism**) use the energy harnessed by catabolism to drive the synthesis of the many molecules that form the cell. Together, these two sets of reactions constitute the metabolism of the cell (Figure 3–2).

The details regarding the individual reactions that comprise cell metabolism are part of the subject matter of *biochemistry*, and they need not concern us here. But the general principles by which cells obtain energy from their environment and use it to create order are central to cell biology. We begin this chapter with a discussion of why a constant input of energy is needed to sustain living organisms. We then discuss how enzymes catalyze the reactions that produce biological order. Finally, we describe the molecules that carry the energy that makes life possible.

THE USE OF ENERGY BY CELLS

Nonliving things left to themselves eventually become disordered: buildings crumble and dead organisms decay. Living cells, by contrast, not only maintain, but actually generate, order at every level, from the large-scale structure of a butterfly or a flower down to the organization of the molecules that make up these organisms (Figure 3–3). This property of life is made possible by elaborate cellular mechanisms that extract energy from the environment and convert it into the energy stored in chemical bonds. Biological structures are therefore able to maintain their form, even though the materials of which they are made are continually being replaced and recycled. Your body has the same basic structure it had 10 years ago, even though you now contain atoms that, for the most part, were not in your body then.

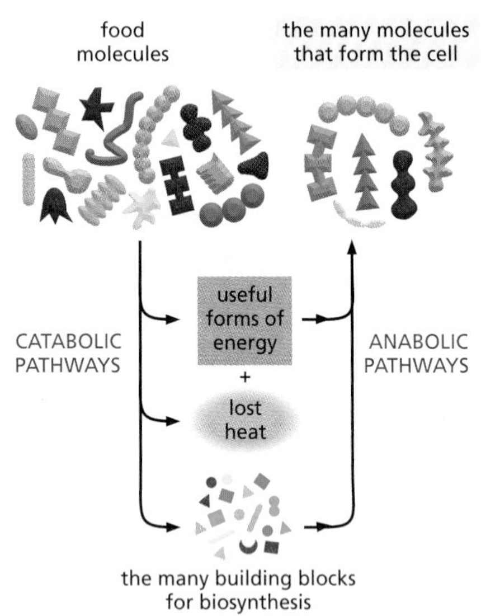

Figure 3–2 **Catabolic and anabolic pathways together constitute the cell's metabolism.** Note that a major portion of the energy stored in the chemical bonds of food molecules is dissipated as heat. Thus, only some of this energy can be converted to the useful forms of energy needed to drive the synthesis of new molecules.

Biological Order Is Made Possible by the Release of Heat Energy from Cells

The universal tendency of things to become disordered is expressed in a fundamental law of physics, the *second law of thermodynamics*. This law states that in the universe, or in any isolated system (a collection of matter that is completely isolated from the rest of the universe), the degree of disorder can only increase. The second law of thermodynamics has such profound implications for living things that it is worth restating in several ways.

We can express the second law in terms of probability by stating that *systems will change spontaneously toward those arrangements that have the greatest probability.* Consider a box of 100 coins all lying heads up. A

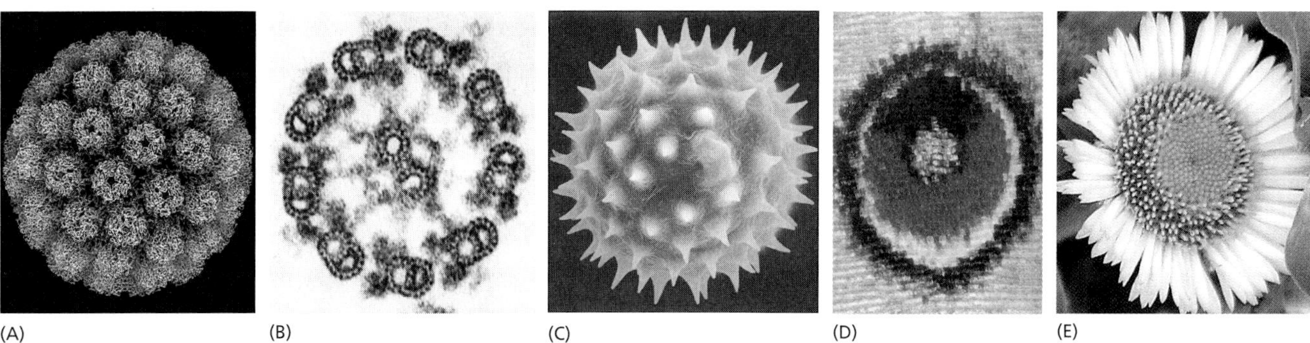

Figure 3–3 **Biological structures are highly ordered.** Well-defined, ornate, and beautiful spatial patterns can be found at every level of organization in living organisms. In order of increasing size: (A) protein molecules in the coat of a virus (a parasite that, although not technically alive, contains the same types of molecules as those found in living cells); (B) the regular array of microtubules seen in a cross section of a sperm tail; (C) surface contours of a pollen grain (a single cell); (D) close-up of the wing of a butterfly showing the pattern created by scales, each scale being the product of a single cell; and (E) flower with a spiral array of seeds, each made of millions of cells. (A, courtesy of Robert Grant, Stéphane Crainic, and James M. Hogle; B, courtesy of Lewis Tilney; C, courtesy of Colin MacFarlane and Chris Jeffree; D, courtesy of Kjell B. Sandved.)

series of events that disturbs the box will tend to move the arrangement toward a mixture of 50 heads and 50 tails. The reason is simple: there are a huge number of possible arrangements of the individual coins that can achieve the 50–50 result, but only one possible arrangement that keeps them all oriented heads up. Because the 50–50 mixture accommodates a greater number of possibilities and places fewer constraints on the orientation of each individual coin, we say that it is more "disordered." For the same reason, one's living space will become increasingly disordered without an intentional effort to keep it organized. Movement toward disorder is a spontaneous process, requiring a periodic effort to reverse it (Figure 3–4).

The measure of a system's disorder is called the **entropy** of the system, and the greater the disorder, the greater the entropy. Thus, another way to express the second law of thermodynamics is to say that systems will change spontaneously toward arrangements with greater entropy. Living cells—by surviving, growing, and forming complex communities and even whole organisms—are generating order and thus might appear to defy the second law of thermodynamics. This is not the case, however, because a cell is not an isolated system. Rather, it takes in energy from its environment—in the form of food, inorganic molecules, or photons of light from the sun—and it then uses this energy to generate order within itself, forging new chemical bonds or building large macromolecules. In the course of performing the chemical reactions that generate order, chemical bond energy is converted into heat. Heat is energy in its most disordered form—the random jostling of molecules (analogous to

Figure 3–4 **The spontaneous drive toward disorder is an everyday experience.** The restoration of order, however, does not occur spontaneously. Thus reversing the natural tendency toward disorder requires an intentional effort and an input of energy. In fact, from the second law of thermodynamics, we can be certain that the human intervention required will release enough heat to the environment to more than compensate for the reordering of the items in this room.

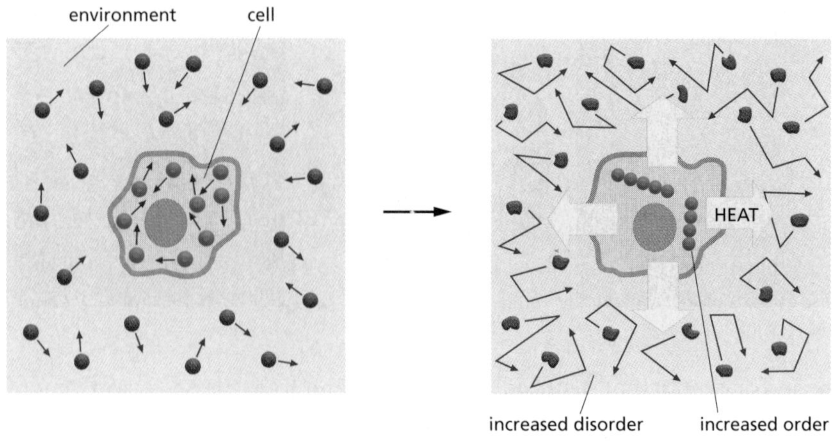

Figure 3–5 **Living cells do not defy the second law of thermodynamics.** In the diagram on the *left*, the molecules of both the cell and the rest of the universe (the environment) are depicted in a relatively disordered state. In the diagram on the *right*, the cell has taken in energy from food molecules and released heat by carrying out a reaction that orders the molecules that the cell contains. Because the heat increases the disorder in the environment around the cell (depicted by the *jagged* arrows, which represent increased thermal motion, and distorted molecules, which indicate enhanced molecular vibration and rotation), the second law of thermodynamics, which states that the amount of disorder in the universe must always increase, is satisfied even as the cell grows and divides.

the random jostling of the coins in the box). Because the cell is not an isolated system, the heat energy that its reactions generate is quickly dispersed into the cell's surroundings. There the heat increases the intensity of the thermal motions of the resident molecules, thereby increasing the entropy of the environment (Figure 3–5).

The amount of heat released by a cell must be great enough that the order generated inside the cell is more than compensated for by the decrease in order in the environment. Only in this case is the second law of thermodynamics satisfied, because the total entropy of the system—that of the cell plus its environment—will increase as a result of the chemical reactions inside the cell.

Where does the heat released by the cell come from? According to the *first law of thermodynamics*, energy can be converted from one form to another, but it cannot be created or destroyed. The interconversions of some forms of energy are illustrated in Figure 3–6. The amount of energy present in different forms will change as a result of the chemical reactions inside the cell, but the first law tells us that the total amount of energy in the universe must always be the same. For example, as an animal cell breaks down foodstuffs, some of the energy in the chemical bonds in the food molecules (chemical bond energy) is converted into the thermal motion of molecules (heat energy). This conversion of chemical energy into heat energy is essential if the reactions inside the cell are to cause the universe as a whole to become more disordered—as required by the second law.

The cell cannot derive any benefit from the heat energy it produces, however, unless the heat-generating reactions inside the cell are directly linked to processes that maintain molecular order. It is the tight coupling of heat production to an increase in order that distinguishes the metabolism of a cell from the wasteful burning of fuel in a fire. Later in this chapter we shall illustrate how this coupling occurs. For the moment, it is sufficient to recognize that—by directly linking the "burning" of food molecules to the generation of biological order—cells are able to create and maintain an island of order in a universe tending toward chaos.

Photosynthetic Organisms Use Sunlight to Synthesize Organic Molecules

All animals live on energy stored in the chemical bonds of organic molecules made by other organisms, which they take in as food. These food molecules also provide the atoms that animals need to construct new living matter. Some animals obtain their food by eating other animals. At the bottom of the animal food chain, however, are animals that consume

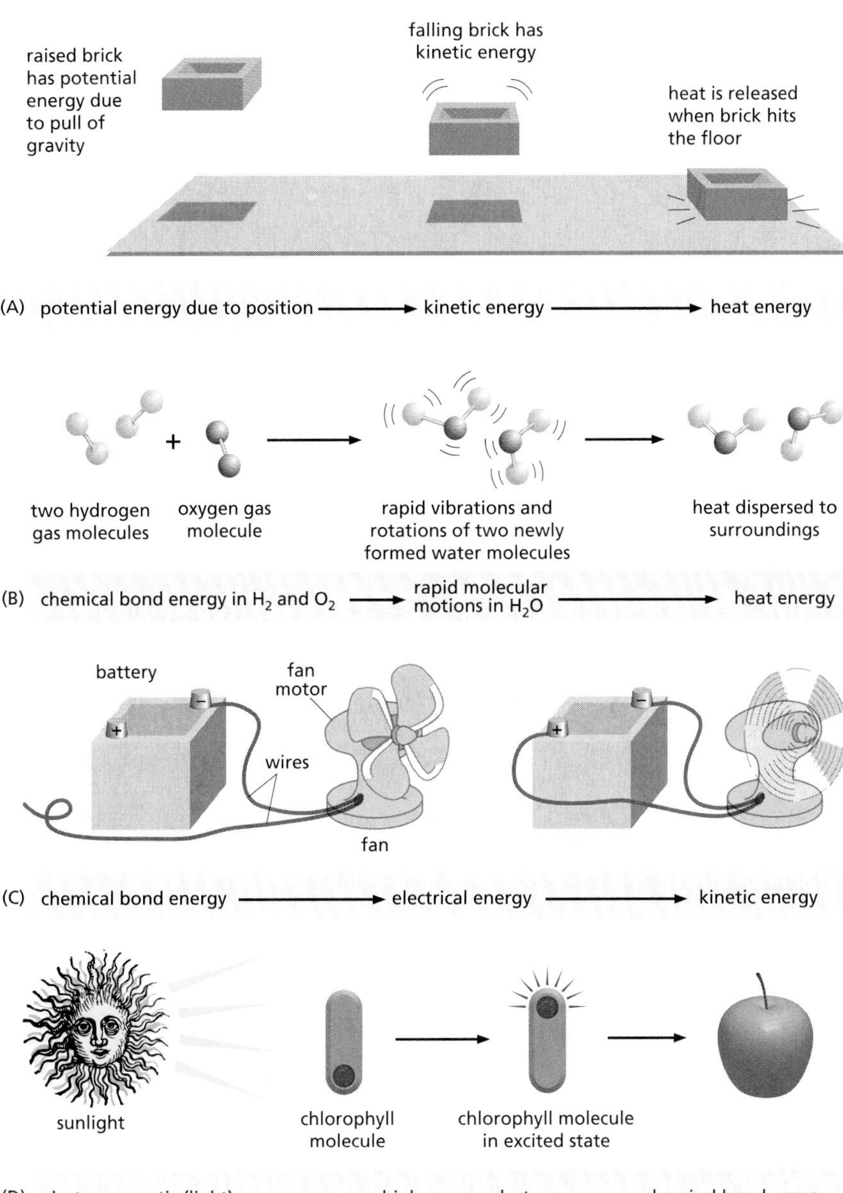

Figure 3–6 **Different forms of energy are interconvertible.** Energy can be converted from one form to another, but the total amount of energy must be conserved. In (A), we can use the height and weight of the brick to predict exactly how much heat will be released when it hits the floor. In (B), the large amount of chemical bond energy released when water is formed is initially converted to very rapid thermal motions in the two new water molecules; however, collisions with other molecules almost instantaneously spread this kinetic energy evenly throughout the surroundings (heat transfer), making the new molecules indistinguishable from all the rest. (C) Cells can convert chemical bond energy into kinetic energy to drive, for example, molecular motors, although without the intermediate conversion to electrical energy that an appliance such as a fan requires. (D) Cells also harvest the energy from sunlight to form chemical bonds via photosynthesis.

plants or other photosynthetic organisms. Before being eaten, these organisms trapped energy from sunlight. So, the energy animals obtain by eating plants came originally from the sun (Figure 3–7).

Solar energy enters the living world through **photosynthesis**, a process that converts the electromagnetic energy in sunlight into chemical bond energy in cells. Photosynthetic organisms—including plants, algae, and some bacteria—are able to obtain all of the atoms they need from inorganic sources. Plants, for example, use carbon from atmospheric carbon dioxide, hydrogen and oxygen from water, nitrogen from ammonia and nitrates in the soil, and other elements needed in smaller amounts from inorganic salts in the soil. They use the energy they derive from sunlight to form chemical bonds between these atoms, linking them into small chemical building blocks such as sugars, amino acids, nucleotides, and fatty acids. These small molecules in turn are converted into the macromolecules—the proteins, nucleic acids, polysaccharides, and lipids—that form the plant. All of these substances serve as nutrients for animals, fungi and nonphotosynthetic bacteria that may later feed on the plant.

Figure 3–7 **With few exceptions, the radiant energy of sunlight sustains all life.** Trapped by plants and some microorganisms through photosynthesis, light from the sun is the ultimate source of all energy for humans and other animals. (*Wheat Field Behind Saint-Paul Hospital with a Reaper* by Vincent van Gogh. Courtesy of Museum Folkwang, Essen.)

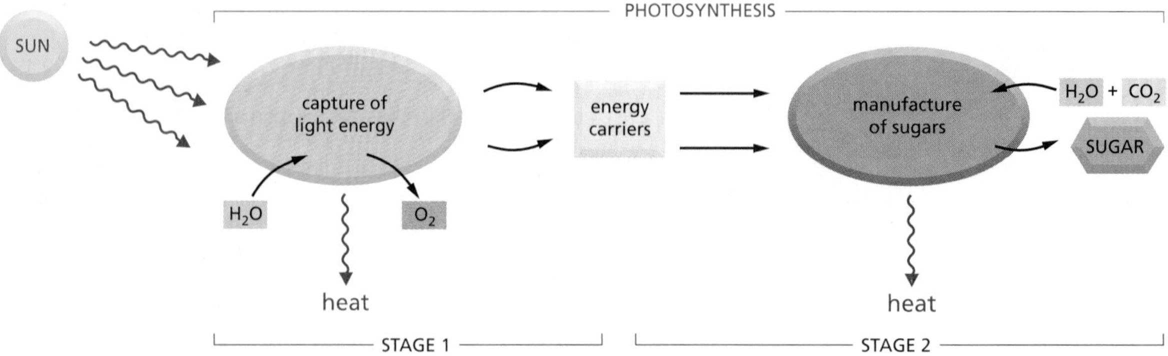

Figure 3–8 **Photosynthesis takes place in two stages, the first of which is light dependent, and the second of which is independent of light.** The energy carriers created in the first stage are two molecules that we will discuss shortly: ATP and NADPH.

The reactions of photosynthesis take place in two stages: one that depends on light and another that does not (Figure 3–8). In the first, light-dependent stage, energy from sunlight is captured and transiently stored as chemical bond energy in specialized small molecules that carry energy in their reactive chemical groups. (We discuss these activated carrier molecules in more detail later in the chapter.) Molecular oxygen (O_2 gas), derived from the splitting of water by light, is released as a by-product of this first stage.

In the second stage of photosynthesis, the molecules that serve as energy carriers are used to help drive a *carbon-fixation* process in which sugars are manufactured from carbon dioxide gas (CO_2) and water (H_2O). By producing sugars, these light-independent reactions generate an essential source of stored chemical bond energy and materials—both for the plant itself and for any animals that eat it. We describe the elegant mechanisms that underlie these two stages of photosynthesis in detail in Chapter 14.

The net result of both stages of photosynthesis, as far as the green plant is concerned, can be summarized simply in the equation

$$\text{light energy} + CO_2 + H_2O \rightarrow \text{sugars} + O_2 + \text{heat energy}$$

The sugars produced are then used both as a source of chemical bond energy and as a source of materials to make the many other small and large organic molecules that are essential to the plant cell.

QUESTION 3–1

Consider the equation

light energy + CO_2 + H_2O →
sugars + O_2 + heat energy

Would you expect this reaction to occur in a single step? Why must heat be generated in the reaction? Explain your answers.

Cells Obtain Energy by the Oxidation of Organic Molecules

All animal and plant cells require the chemical energy stored in the chemical bonds of organic molecules—either the sugars that a plant has photosynthesized as food for itself or the mixture of large and small molecules that an animal has eaten. To use this energy to live, grow, and reproduce, organisms must extract it in a usable form. In both plants and animals, energy is extracted from food molecules by a process of gradual *oxidation*, or controlled burning.

Earth's atmosphere is about 21% oxygen, and in the presence of oxygen the most energetically stable form of carbon is CO_2 and that of hydrogen is H_2O. A cell is therefore able to obtain energy from sugars or other organic molecules by allowing the carbon and hydrogen atoms in these molecules to combine with oxygen—that is, become *oxidized*—to produce CO_2 and H_2O, respectively—a process known as cellular **respiration**.

Photosynthesis and respiration are complementary processes (Figure 3–9). This means that the transactions between plants and animals are not all one way. Plants, animals, and microorganisms have existed together on this planet for so long that many of them have become an essential part of each other's environments. The oxygen released by photosynthesis is consumed by nearly all organisms for the combustion of

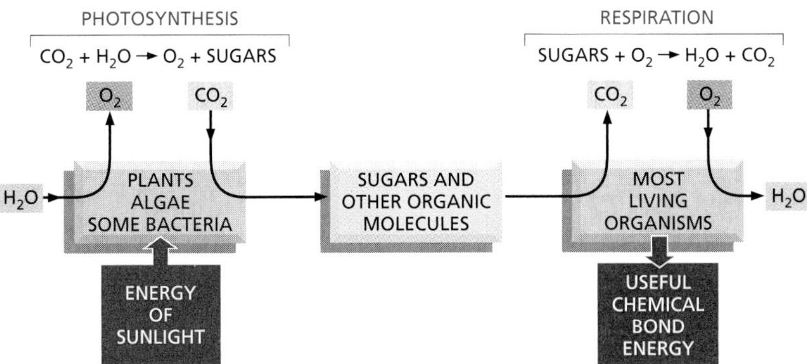

Figure 3–9 **Photosynthesis and cellular respiration are complementary processes in the living world.** The *left* side of the diagram shows how photosynthesis—carried out by plants and photosynthetic microorganisms—uses the energy of sunlight to produce sugars and other organic molecules from the carbon atoms in CO_2 in the atmosphere. In turn, these molecules serve as food for other organisms. The *right* side of the diagram shows how cellular respiration in these organisms uses O_2 to oxidize food molecules, releasing the same carbon atoms in the form of CO_2 back to the atmosphere. In the process, the organisms obtain the chemical bond energy that they need to survive. The first cells on Earth are thought to have been capable of neither photosynthesis nor respiration (discussed in Chapter 14). However, photosynthesis must have preceded respiration on the Earth, because there is strong evidence that billions of years of photosynthesis were required before O_2 had been released in sufficient quantity to create an atmosphere rich in this gas to support respiration.

organic molecules. And some of the CO_2 molecules that are fixed today into organic molecules by photosynthesis in a green leaf were released yesterday into the atmosphere by the respiration of an animal, fungus or bacterium decomposing dead organic matter—or by the respiration of the plant itself. Carbon utilization therefore forms a huge cycle that involves the *biosphere* (all of the living organisms on Earth) as a whole, crossing boundaries between individual organisms (Figure 3–10). Similarly, atoms of nitrogen, phosphorus, and sulfur move between the living and nonliving worlds in cycles that involve plants, animals, fungi, and bacteria. Procaryotes, in fact, are estimated to contain nearly half the carbon stored in living organisms. And they are the single largest reservoir of nitrogen and phosphorus on Earth, containing 10 times more of these nutrients than plants.

Oxidation and Reduction Involve Electron Transfers

The cell does not oxidize organic molecules in one step, as occurs when organic material is burned in a fire. Through the use of enzyme catalysts, metabolism carries the molecules through a large number of reactions that only rarely involve the direct addition of oxygen. Before we consider some of these reactions and the purpose behind them, we need to explain what is meant by oxidation.

The term **oxidation** literally means the addition of oxygen atoms to a molecule. More generally, though, oxidation is said to occur in any reaction in which electrons are transferred from one atom to another. Oxidation, in this sense, refers to the removal of electrons. The converse reaction, called **reduction**, involves the addition of electrons. Thus, Fe^{2+} is oxidized when it loses an electron to become Fe^{3+}, whereas a chlorine atom is reduced when it gains an electron to become Cl^-. Because the number of electrons is conserved in a chemical reaction (there is no net loss or gain), oxidation and reduction always occur simultaneously: that is, if one molecule gains an electron in a reaction (reduction), a second

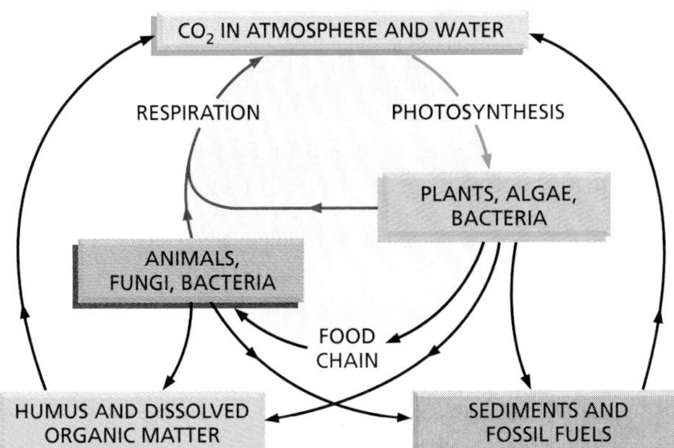

Figure 3–10 **Carbon atoms cycle continuously through the biosphere.** Individual carbon atoms are incorporated into organic molecules of the living world by the photosynthetic activity of plants, algae, and bacteria. They pass to animals, microorganisms, and into organic material in soil and oceans via cyclic pathways. CO_2 is restored to the atmosphere when organic molecules are oxidized by cells during respiration or burned by humans as fossil fuels.

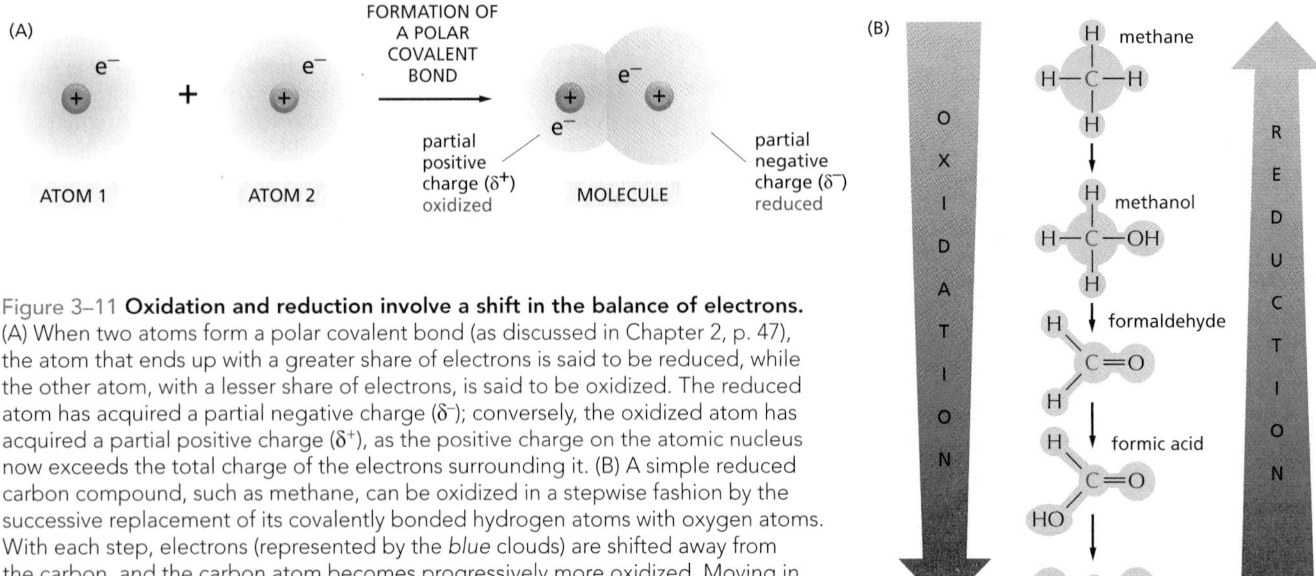

Figure 3–11 **Oxidation and reduction involve a shift in the balance of electrons.** (A) When two atoms form a polar covalent bond (as discussed in Chapter 2, p. 47), the atom that ends up with a greater share of electrons is said to be reduced, while the other atom, with a lesser share of electrons, is said to be oxidized. The reduced atom has acquired a partial negative charge (δ^-); conversely, the oxidized atom has acquired a partial positive charge (δ^+), as the positive charge on the atomic nucleus now exceeds the total charge of the electrons surrounding it. (B) A simple reduced carbon compound, such as methane, can be oxidized in a stepwise fashion by the successive replacement of its covalently bonded hydrogen atoms with oxygen atoms. With each step, electrons (represented by the *blue* clouds) are shifted away from the carbon, and the carbon atom becomes progressively more oxidized. Moving in the opposite direction, carbon dioxide becomes progressively more reduced as its oxygen atoms are replaced by hydrogens to yield methane.

molecule must lose the electron (oxidation). When a sugar molecule is oxidized to CO_2 and H_2O, for example, the O_2 molecules involved in forming H_2O gain electrons and thus are said to have been reduced.

The terms oxidation and reduction apply even when there is only a partial shift of electrons between atoms linked by a covalent bond. When a carbon atom becomes covalently bonded to an atom with a strong affinity for electrons—oxygen, chlorine, or sulfur, for example—it gives up more than its equal share of electrons and forms a polar covalent bond. The positive charge of the carbon nucleus now slightly exceeds the negative charge of its electrons, thus the carbon atom acquires a partial positive charge (δ^+) and is said to be oxidized. Conversely, the carbon atom in a C–H bond has somewhat more than its share of electrons; it acquires a partial negative charge (δ^-), and so is said to be reduced (Figure 3–11A).

When a molecule in a cell picks up an electron (e^-), it often picks up a proton (H^+) at the same time (protons being freely available in water). The net effect in this case is to add a hydrogen atom to the molecule:

$$A + e^- + H^+ \rightarrow AH$$

Even though a proton plus an electron is involved (instead of just an electron), such *hydrogenation* reactions are reductions, and the reverse, *dehydrogenation*, reactions are oxidations. An easy way to tell whether an organic molecule is being oxidized or reduced is to count its C–H bonds: reduction occurs when the number of C–H bonds increases, whereas oxidation occurs when the number of C–H bonds decreases (Figure 3–11B).

As we shall see later in this chapter, cells use enzymes to catalyze the oxidation of organic molecules in small steps, through a sequence of reactions that allows useful energy to be harvested.

FREE ENERGY AND CATALYSIS

Enzymes, like cells, obey the second law of thermodynamics. Although they can speed up energetically favorable reactions—those that produce disorder in the universe—enzymes cannot by themselves force energetically unfavorable reactions to occur. Cells, however, must do just that in order to grow and divide: they must build highly ordered and energy-rich molecules from small and simple ones—a process that requires an input of energy.

To understand how enzymes can catalyze such energetically unfavorable—but life-sustaining—reactions, we first need to examine the energetics involved. In this section, we consider how the free energy of molecules contributes to their chemistry, and we see how free-energy changes—which reflect how much disorder is generated by a reaction—influence whether and how that reaction will proceed. We also discuss how enzymes exploit differences in free-energy changes to produce biological order. But first we shall review how **catalysis** occurs rapidly and with great precision inside living cells.

QUESTION 3–2

In which of the following reactions does the *black* atom undergo an oxidation?

A. Na → Na^+ (Na atom → Na^+ ion)
B. Cl → Cl^- (Cl atom → Cl^- ion)
C. CH_3CH_2OH → CH_3CHO (ethanol → acetaldehyde)
D. CH_3CHO → CH_3COOH (acetaldehyde → acetic acid)
E. $CH_2{=}CH_2$ → CH_3CH_3 (ethene → ethane)

Enzymes Lower the Energy Barriers That Prevent Chemical Reactions from Occurring

Paper burns readily, releasing into the atmosphere water and carbon dioxide as gases and energy as heat:

$$\text{paper} + O_2 \rightarrow \text{smoke} + \text{ashes} + \text{heat} + CO_2 + H_2O$$

But this occurs in only one direction: smoke and ashes never spontaneously gather carbon dioxide and water from the heated atmosphere and reconstitute themselves into paper. When paper burns, its chemical energy is dissipated as heat—not lost from the universe, since energy can never be created or destroyed, but irretrievably dispersed in the chaotic random thermal motions of molecules. At the same time, the atoms and molecules of the paper become dispersed and disordered. In the language of thermodynamics, there has been a release of *free energy*—that is, of energy that can be harnessed to do work or drive chemical reactions. This release reflects a loss of orderliness in the way the energy and molecules had been stored in the paper. We will discuss free energy in more detail shortly, but the general principle can be summarized as follows: chemical reactions proceed only in the direction that leads to a loss of free energy. In other words, the spontaneous direction for any reaction is the direction that goes 'downhill.' A 'downhill' reaction in this sense is said to be energetically favorable.

Although the most energetically favorable form of carbon under ordinary conditions is CO_2, and that of hydrogen is H_2O, a living organism will not disappear in a puff of smoke, and the book in your hands will not burst spontaneously into flames. This is because the molecules in both the living organism and the book are in a relatively stable state, and they cannot be changed to lower-energy states without an initial input of energy. In other words, a molecule requires a boost over an energy barrier before it can undergo a chemical reaction that moves it to a lower-energy (more stable) state (Figure 3–12A). This boost is known as the **activation**

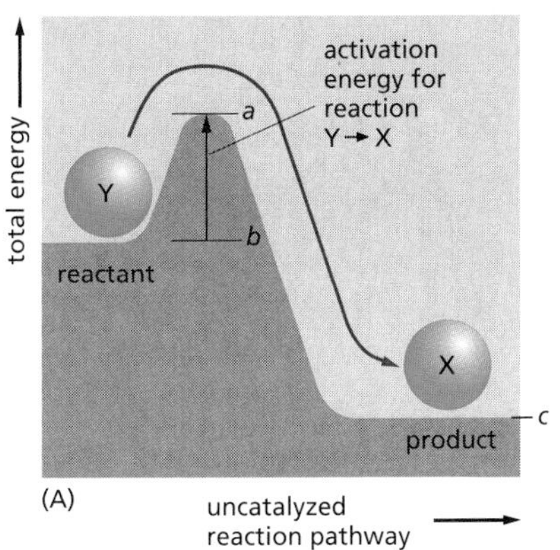

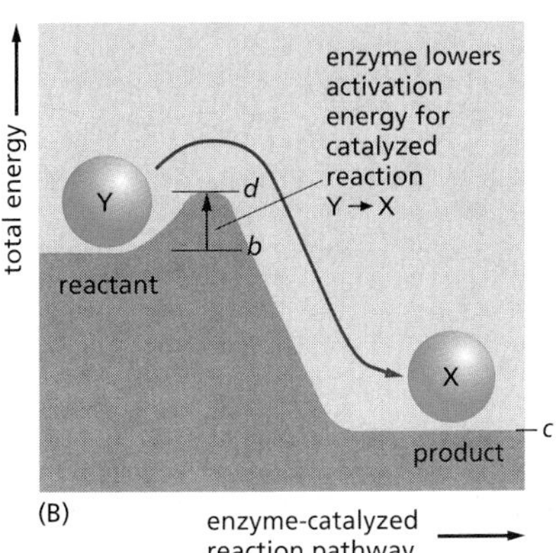

Figure 3–12 **Even energetically favorable reactions require activation energy to get them started.** (A) Compound Y (a reactant) is in a relatively stable state; thus energy is required to convert it to compound X (a product), even though X is at a lower overall energy level than Y. This conversion will not take place, therefore, unless compound Y can acquire enough activation energy (*energy* **a** minus *energy* **b**) from its surroundings to undergo the reaction that converts it into compound X. This energy may be provided by means of an unusually energetic collision with other molecules. For the reverse reaction, X → Y, the activation energy will be much larger (*energy* **a** minus *energy* **c**); this reaction will therefore occur much more rarely. Activation energies are always positive. The total energy change for the energetically favorable reaction Y → X, however, is *energy* **c** minus *energy* **b**, a negative number. (B) Energy barriers for specific reactions can be lowered by catalysts, as indicated by the line marked ***d***. Enzymes are particularly effective catalysts because they greatly reduce the activation energy for the reactions they catalyze.

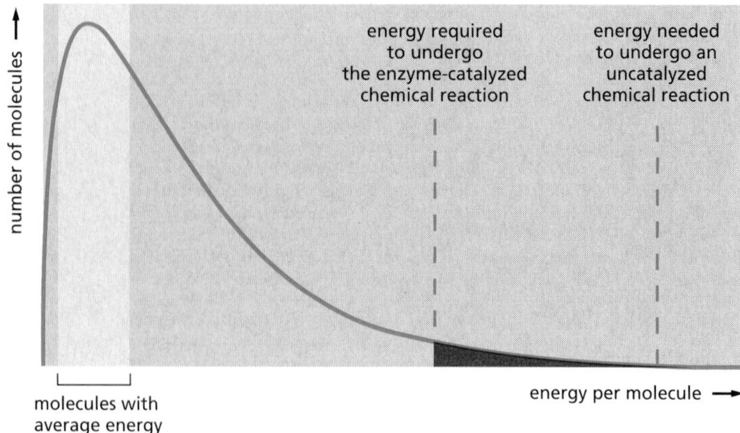

Figure 3–13 **Lowering the activation energy greatly increases the probability that a reaction will occur.** At any given instant, a population of identical substrate molecules will have a range of energies, distributed as shown on the graph. The varying energies come from collisions with surrounding molecules, which make the substrate molecules jiggle, vibrate, and spin. For a molecule to undergo a chemical reaction, the energy of the molecule must exceed the activation energy barrier for that reaction (*dashed lines*); for most biological reactions, this almost never happens without enzyme catalysis. Even with enzyme catalysis, only a small fraction of substrate molecules reach an energy state that is high enough for them to undergo a reaction (*red shaded* area).

energy. In the case of a burning book, the activation energy is provided by the heat of a lighted match. For the molecules in the watery solution inside a cell, the boost is delivered by an unusually energetic random collision with surrounding molecules—collisions that become more violent as the temperature is raised.

At the temperatures in living cells, the push over the energy barrier is greatly aided by specialized proteins called **enzymes**. Each enzyme binds tightly to one or two molecules, called **substrates**, and holds them in a way that greatly reduces the activation energy needed to facilitate a specific chemical interaction between them (Figure 3–12B). A substance that can lower the activation energy of a reaction is termed a **catalyst**; catalysts increase the rate of chemical reactions because they allow a much larger proportion of the random collisions with surrounding molecules to kick the substrates over the energy barrier, as illustrated in Figure 3–13 and Figure 3–14A. Enzymes are among the most effective catalysts known. They can speed up reactions by a factor of as much as 10^{14} (that is, trillions of times faster than the same reactions would proceed without an enzyme catalyst). Enzymes therefore allow reactions that would not otherwise occur to proceed rapidly at normal temperatures. Without enzymes, life could not exist.

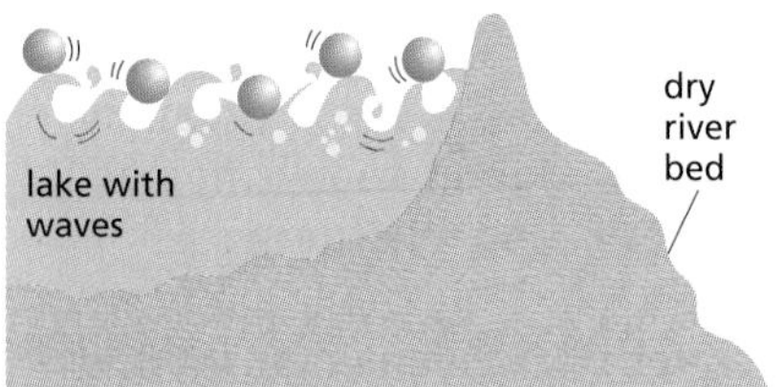

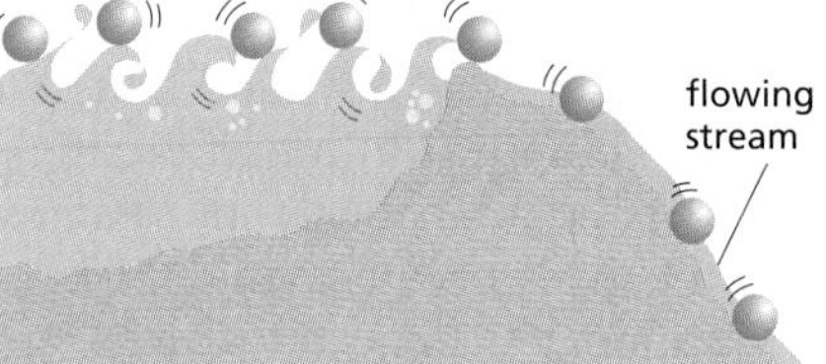

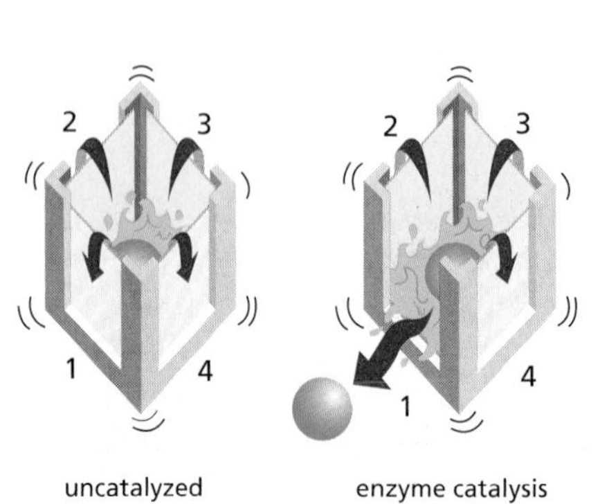

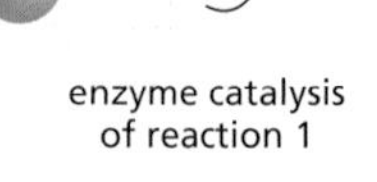

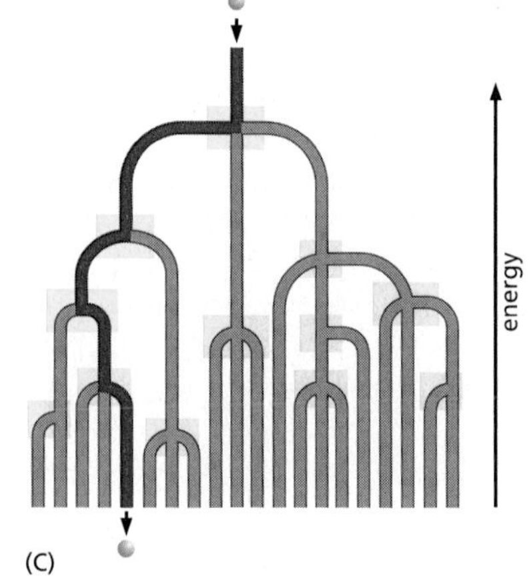

Figure 3–14 **Enzymes catalyze reactions by lowering the activation energy barrier.** (A) The dam represents the activation energy, which is lowered by enzyme catalysis. The *green ball* represents a potential substrate that is bouncing up and down in energy level owing to constant encounters with waves, an analogy for the thermal bombardment of the substrate with the surrounding water molecules. When the barrier—the activation energy—is lowered significantly, the balls (substrates) with sufficient energy can roll downhill, an energetically favorable movement. (B) The four walls of the box represent the activation energy barriers for four different chemical reactions that are all energetically favorable because the products are at lower energy levels than the substrates. In the left-hand box, none of these reactions occurs because even the largest waves are not large enough to surmount any of the energy barriers. In the right-hand box, enzyme catalysis lowers the activation energy for reaction number 1 only; now the jostling of the waves allows the molecule to pass over this energy barrier, inducing reaction 1 (Movie 3.1). (C) A branching river with a set of barrier dams (*yellow* boxes) serves to illustrate how a series of enzyme-catalyzed reactions determines the exact reaction pathway followed by each molecule inside the cell by controlling specifically which reaction will be allowed at each junction.

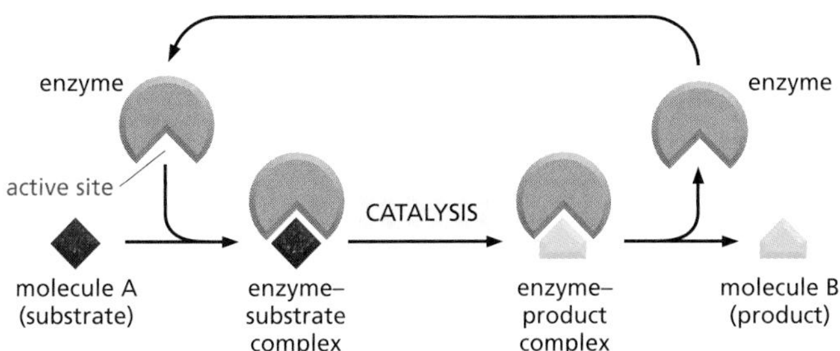

Figure 3–15 **Enzymes convert substrates to products while remaining unchanged themselves.** Each enzyme has an active site to which one or two substrate molecules bind, forming an enzyme–substrate complex. A reaction occurs at the active site, generating an enzyme–product complex. The product is then released, allowing the enzyme to bind additional substrate molecules and repeat the reaction.

Unlike temperature, enzymes are highly selective. Each enzyme usually speeds up only one particular reaction out of the several possible reactions that its substrate molecules could undergo. In this way, enzymes direct each of the many different molecules in a cell along specific reaction pathways (Figure 3–14B and C), thereby producing the compounds that the cell actually needs.

The success of living organisms can be attributed to the cell's ability to make enzymes of many types, each with precisely specified properties. Each enzyme has a unique shape containing an *active site*, a pocket or groove in the enzyme into which only particular substrates will fit (Figure 3–15). Like all catalysts, enzyme molecules themselves remain unchanged after participating in a reaction and therefore can function over and over again. In Chapter 4, we will discuss further how enzymes work, after we have looked in detail at the molecular structure of proteins.

The Free-Energy Change for a Reaction Determines Whether It Can Occur

According to the second law of thermodynamics, a chemical reaction can proceed only if it results in a net (or overall) increase in the disorder of the universe (see Figure 3–5). Disorder increases when useful energy that could be harnessed to do work is dissipated as heat. The criterion for an increase of disorder can be expressed most conveniently in terms of the **free energy**, ***G***, of a system. The value of G is of most interest when a system undergoes a change, so the **free-energy change**, denoted **Δ*G*** ("Delta G"), is the term we most often see. Suppose that the system being considered is a collection of molecules. Because of the way free energy is defined, ΔG measures the amount of disorder created in the universe when a reaction takes place that involves these molecules. *Energetically favorable* reactions, by definition, are those that create disorder by decreasing the free energy of the system to which they belong; in other words, they have a *negative* ΔG (Figure 3–16).

A reaction can occur spontaneously only if ΔG is negative. On a macroscopic scale, a familiar example of an energetically favorable reaction with a negative ΔG is the "reaction" by which a compressed spring relaxes to an expanded state, releasing its stored elastic energy as heat to its surroundings; an example on a microscopic scale is the dissolving of salt in water. Conversely, *energetically unfavorable* reactions, with a *positive* ΔG—such as those in which two amino acids are joined together to form a peptide bond—by themselves create order in the universe. These reactions cannot occur spontaneously. Energetically unfavorable reactions can take place only if they are coupled to a second reaction with a negative ΔG so large that the net ΔG of the entire process is negative (Figure 3–17). Life is possible because enzymes can create biological order by coupling energetically unfavorable reactions with energetically favorable ones. These critical concepts are summarized, with examples, in Panel 3–1 (pp. 94–95).

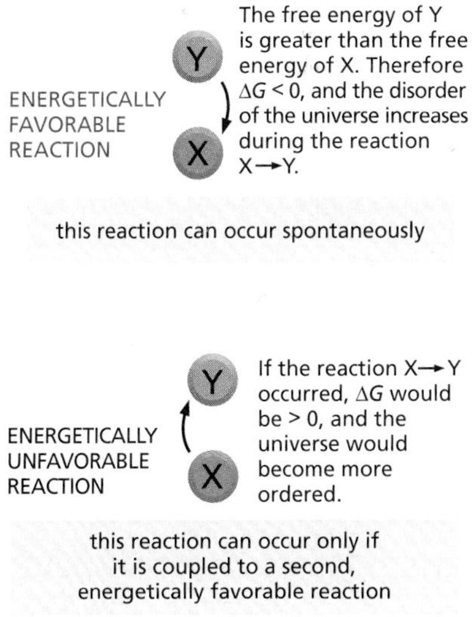

ENERGETICALLY FAVORABLE REACTION

Y
X

The free energy of Y is greater than the free energy of X. Therefore $\Delta G < 0$, and the disorder of the universe increases during the reaction X→Y.

this reaction can occur spontaneously

ENERGETICALLY UNFAVORABLE REACTION

Y
X

If the reaction X→Y occurred, ΔG would be > 0, and the universe would become more ordered.

this reaction can occur only if it is coupled to a second, energetically favorable reaction

Figure 3–16 **Energetically favorable reactions have a negative Δ*G*, whereas energetically unfavorable reactions have a positive Δ*G*.** Energetically favorable reactions are also called exergonic reactions because they cause a net release of free energy. Energetically unfavorable reactions are also called endergonic reactions because they must harness free energy from their surroundings to occur.

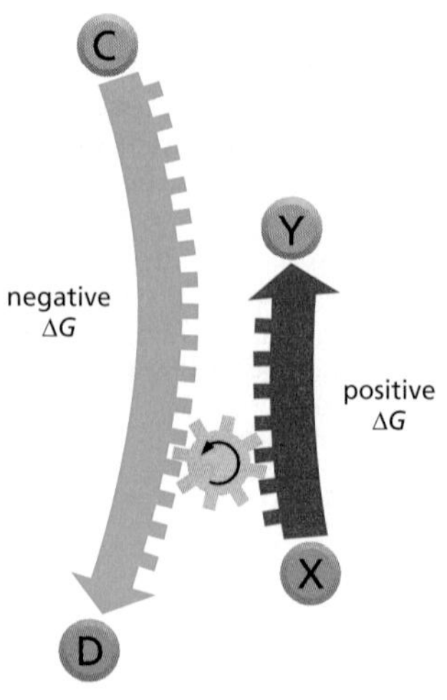

the energetically unfavorable reaction X→Y is driven by the energetically favorable reaction C→D, because the net free-energy change for the pair of coupled reactions is less than zero

Figure 3–17 **Reaction coupling can drive an energetically unfavorable reaction.** The energetically unfavorable ($\Delta G > 0$) reaction $X \rightarrow Y$ cannot occur unless it is coupled to an energetically favorable ($\Delta G < 0$) reaction $C \rightarrow D$, such that the net free-energy change for the coupled reactions is negative (less than 0).

The Concentration of Reactants Influences the Free-Energy Change and a Reaction's Direction

As we have just described, a reaction $Y \rightleftharpoons X$ will go in the direction $Y \rightarrow X$ when the associated free-energy change, ΔG, is negative, just as a tensed spring left to itself will relax and lose its stored energy to its surroundings as heat. For a chemical reaction, however, ΔG depends not only on the energy stored in each individual molecule, but also on the concentrations of the molecules in the reaction mixture. Remember that ΔG reflects the degree to which a reaction creates a more disordered—in other words, a more probable—state of the universe. Recalling our coin analogy, more coins in a jiggling box will flip from a head to a tail orientation when the box contains 90 heads and 10 tails, than when the box contains 10 heads and 90 tails.

The same is true for a chemical reaction. For the reversible reaction $Y \rightleftharpoons X$, a large excess of Y over X will tend to drive the reaction in the direction $Y \rightarrow X$; that is, there will be a tendency for there to be more molecules making the transition $Y \rightarrow X$ than there are molecules making the transition $X \rightarrow Y$. Thus, as the ratio of Y to X increases, the ΔG becomes more negative for the transition $Y \rightarrow X$ (and more positive for the transition $X \rightarrow Y$).

The Standard Free-Energy Change Makes it Possible to Compare the Energetics of Different Reactions

Because ΔG depends on the concentrations of the molecules in the reaction mixture at any given time, it is not a particularly useful value for comparing the relative energies of different types of reactions. Such comparisons are necessary, for example, to predict whether an energetically favorable reaction is likely to have a ΔG negative enough to drive an energetically unfavorable reaction. To level the playing field and place reactions on a comparable basis, we need to turn to the *standard free-energy change* of a reaction, $\boldsymbol{\Delta G°}$. The $\Delta G°$ is independent of concentration; it depends only on the intrinsic characters of the reacting molecules, based on their behavior under ideal conditions where the concentrations of all the reactants are set to the same fixed value of 1 mole/liter.

For the simple reaction $Y \rightarrow X$ at 37°C, $\Delta G°$ is related to ΔG as follows:

$$\Delta G = \Delta G° + RT \ln \frac{[X]}{[Y]}$$

where ΔG is in kilocalories per mole, [Y] and [X] denote the concentrations of Y and X in moles/liter, ln is the natural logarithm, and RT is the product of the gas constant, R, and the absolute temperature, T. At 37°C, $RT = 0.616$. (A mole is 6×10^{23} molecules of a substance.)

A large body of thermodynamic data has been collected from which $\Delta G°$ can be calculated for most metabolic reactions. Some common reactions are compared in terms of their $\Delta G°$ in Panel 3–1 (pp. 94–95).

QUESTION 3–3

Consider the analogy of the jiggling box containing coins that was described on page 83. The reaction, the flipping of coins that either face heads up (H) or tails up (T), is described by the equation $H \rightleftharpoons T$, where the rate of the forward reaction equals the rate of the reverse reaction.

A. What are ΔG and $\Delta G°$ in this analogy?

B. What corresponds to the temperature at which the reaction proceeds? What corresponds to the activation energy of the reaction? Assume you have an "enzyme," called jigglase, which catalyzes this reaction. What would the effect of jigglase be and what, mechanically, might jigglase do in this analogy?

Cells Exist in a State of Chemical Disequilibrium

From the equation above, we can see that when the molar concentrations of Y and X are equal, that is $[X]/[Y] = 1$, the value of ΔG equals the value of $\Delta G°$, because $\ln 1 = 0$. But as the favorable reaction $Y \rightarrow X$ proceeds, the concentration of the product X increases and the concentration of the substrate Y decreases. This change in relative concentrations will cause $[X]/[Y]$ to become larger, making the initially favorable ΔG less and less negative (the natural logarithm is positive for a number more than 1 and negative for a number less than 1).

Chemical reactions will generally proceed until they reach a state of **equilibrium**. At that point, the rates of the forward and reverse reactions

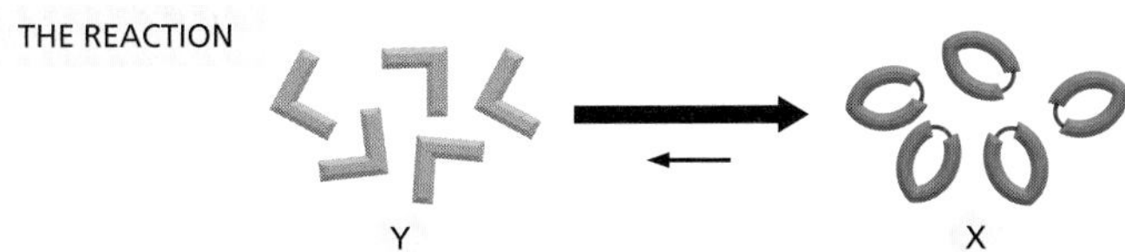
THE REACTION
Y
X

The formation of X is energetically favored in this example. In other words, the ΔG of Y → X is negative and the ΔG of X → Y is positive. But because of thermal bombardments, there will always be some X converting to Y and vice versa.

SUPPOSE WE START WITH AN EQUAL NUMBER OF Y AND X MOLECULES

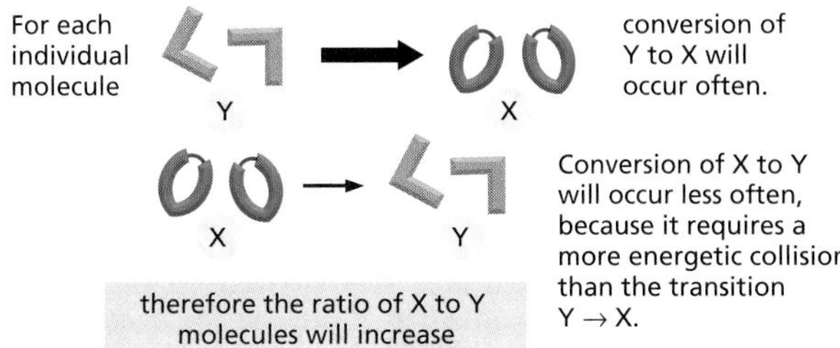
For each individual molecule
Y
X
conversion of Y to X will occur often.
X
Y
Conversion of X to Y will occur less often, because it requires a more energetic collision than the transition Y → X.

therefore the ratio of X to Y molecules will increase

EVENTUALLY there will be a large enough excess of X over Y to just compensate for the slow rate of X → Y. Equilibrium will then have been attained.

Y X

AT EQUILIBRIUM the number of Y molecules being converted to X molecules each second is exactly equal to the number of X molecules being converted to Y molecules each second, so that there is no net change in the ratio of X to Y. At equilibrium the ΔG is zero.

Figure 3–18 **Reactions will eventually reach a chemical equilibrium.** At that point, the forward and the backward fluxes of reacting molecules are equal and opposite.

are equal, and there is no further net change in the concentrations of substrate or product (Figure 3–18). For reactions at chemical equilibrium, $\Delta G = 0$, so the reaction will not proceed forward or backward, and no work can be done. Because the maintenance of order within the cell requires a continuous input of energy, any cell whose reactions have all reached chemical equilibrium is dead.

Living cells avoid reaching a state of equilibrium because they are constantly exchanging materials with their environment: taking in nutrients and eliminating waste products. Many of the individual reactions in the cell's complex metabolic network are kept in disequilibrium because the products of one reaction are continually being siphoned off to become the substrates in a subsequent reaction. Rarely do products and substrates reach concentrations at which the forward and reverse reaction rates are equal.

The Equilibrium Constant is Directly Proportional to $\Delta G°$

As we have seen, at chemical equilibrium, when the forward and reverse reaction rates are equal, the ratio of substrate to product will remain constant. This state makes it possible to calculate a reaction's **equilibrium constant**, *K*:

$$K = \frac{[X]}{[Y]}$$

where [X] is the concentration of the product and [Y] is the concentration of the reactant at equilibrium. This expression describes the situation at the point at which the concentration effect just balances the push given to the reaction by $\Delta G°$, so that $\Delta G = 0$ and there is no net change of free energy to drive the reaction in either direction (see Panel 3–1, p. 95).

FREE ENERGY

This panel reviews the concept of free energy and offers examples showing how changes in free energy determine whether—and how—biological reactions occur.

The molecules of a living cell possess energy because of their vibrations, rotations, and movement through space, and because of the energy that is stored in the bonds between individual atoms.

The free energy, G (in kcal/mole), measures the energy of a molecule that could in principle be used to do useful work at constant temperature, as in a living cell. Energy can also be expressed in joules (1 cal = 4.184 joules).

REACTIONS CAUSE DISORDER

Think of a chemical reaction occurring in an isolated cell with constant temperature and volume. This reaction can produce disorder in two ways.

1 Changes of bond energy of the reacting molecules can cause heat to be released, which disorders the environment.

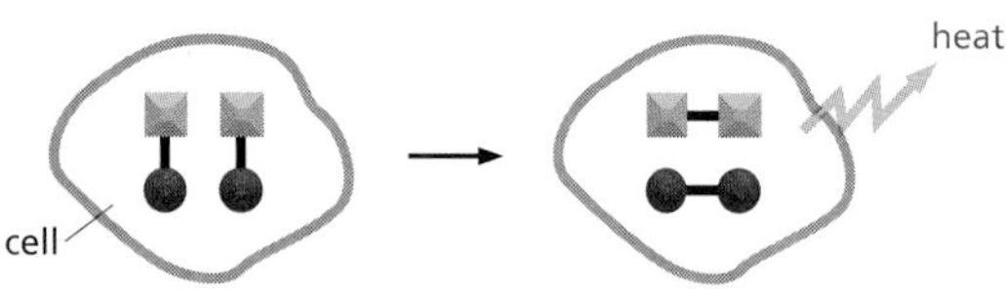

2 The reaction can decrease the amount of order in the reacting molecules—for example, by breaking apart a long chain of molecules, or by disrupting an interaction that prevents bond rotations.

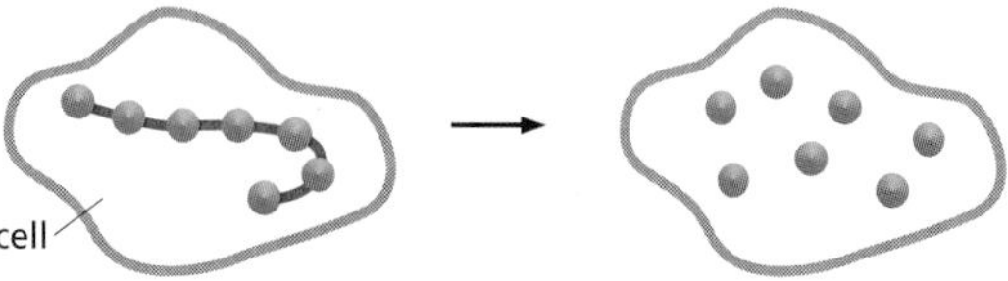

REACTION RATES

A spontaneous reaction is not necessarily an instantaneous reaction: a reaction with a negative free-energy change (ΔG) will not necessarily occur rapidly by itself. For the combustion of glucose in oxygen:

$$\text{glucose} + 6O_2 \longrightarrow 6CO_2 + 6H_2O$$

$$\Delta G^\circ = -686 \text{ kcal/mole}$$

But even this highly favorable reaction may not occur for centuries unless there are enzymes to speed up the process. Enzymes are able to catalyze reactions and speed up their rate, but they do not change the ΔG° of the reaction.

ΔG ("DELTA G")

Changes in free energy occurring in a reaction are denoted by ΔG, where "Δ" indicates a difference. Thus for the reaction:

$$A + B \longrightarrow C + D$$

$$\Delta G = \text{free energy } (C + D) \text{ minus free energy } (A + B)$$

ΔG measures the amount of disorder caused by a reaction: the change in order inside the cell, plus the change in order of the surroundings caused by the heat released.

ΔG is useful because it measures how far away from equilibrium a reaction is. Thus the reaction

$$ATP \longrightarrow ADP + P_i$$

has a large negative ΔG because cells keep it a long way from equilibrium by continually making fresh ATP. However, if the cell dies, then most of its ATP becomes hydrolyzed, until equilibrium is reached (forward and backward reactions occur at equal rates) and $\Delta G = 0$.

SPONTANEOUS REACTIONS

From the second law of thermodynamics, we know that the disorder of the universe can only increase. ΔG is *negative* if the disorder of the universe (reaction plus surroundings) *increases*.

In other words, a chemical reaction that occurs spontaneously must have a negative ΔG:

$$G_{products} - G_{reactants} = \Delta G < 0$$

EXAMPLE: The difference in free energy of 100 ml of 10 mM sucrose (common sugar) and 100 ml of 10 mM glucose plus 10 mM fructose is about –5.5 calories. Therefore, the hydrolysis reaction (sucrose → glucose + fructose) can proceed spontaneously.

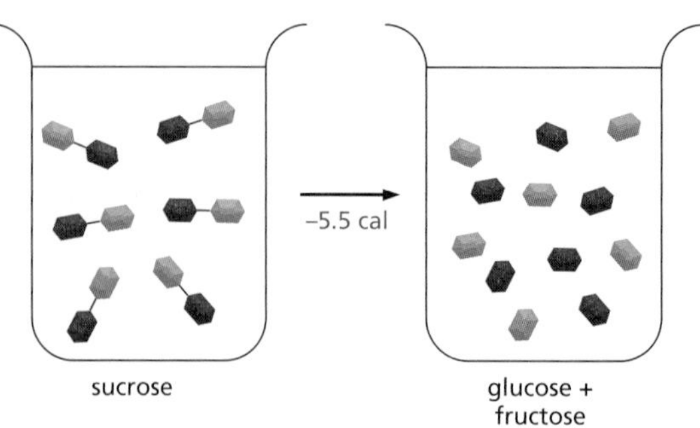

In contrast, the reverse reaction (glucose + fructose → sucrose), which has a ΔG of +5.5 calories, could not occur without an input of energy from a coupled reaction.

PREDICTING REACTIONS

To predict the outcome of a reaction (Will it proceed to the right or to the left? At what point will it stop?), we must measure its standard free-energy change ($\Delta G°$). This quantity represents the gain or loss of free energy as one mole of reactant is converted to one mole of product under "standard conditions" (all molecules present at a concentration of 1 M and pH 7.0).

driving force

$\Delta G°$ for some reactions	
glucose-1-P → glucose-6-P	–1.7 kcal/mole
sucrose → glucose + fructose	–5.5 kcal/mole
ATP → ADP + P_i	–7.3 kcal/mole
glucose + $6O_2$ → $6CO_2$ + $6H_2O$	–686 kcal/mole

CHEMICAL EQUILIBRIA

A fixed relationship exists between the standard free-energy change of a reaction, $\Delta G°$, and its equilibrium constant K. For example, the reversible reaction

$$Y \rightarrow X$$

will proceed until the ratio of concentrations [X]/[Y] is equal to K (note: square brackets [ ] indicate concentration). At this point the free energy of the system will have its lowest value.

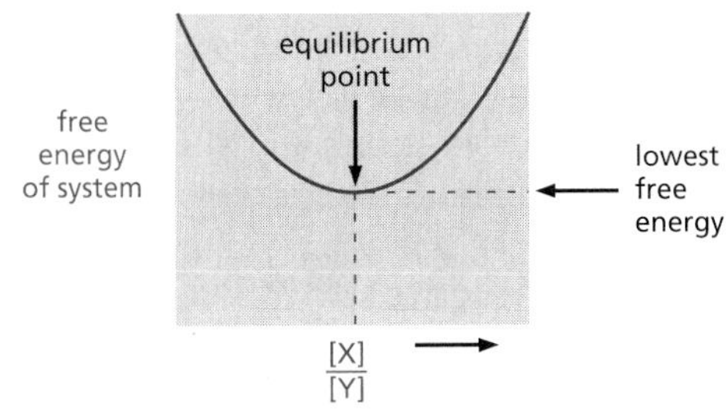

At 37°C, $\Delta G° = -1.42 \log_{10} K$ (see text, p. 96)

$$K = 10^{-\Delta G°/1.42}$$

For example, the reaction

CH_2OH glucose-1-P (O–P) → CH_2O–P glucose-6-P (OH)

has $\Delta G° = -1.74$ kcal/mole. Therefore, its equilibrium constant

$$K = 10^{(1.74/1.42)} = 10^{(1.23)} = 17$$

So the reaction will reach steady state when

[glucose-6-P]/[glucose-1-P] = 17

COUPLED REACTIONS

Reactions can be "coupled" together if they share one or more intermediates. In this case, the overall free-energy change is simply the sum of the individual $\Delta G°$ values. A reaction that is unfavorable (has a positive $\Delta G°$) can for this reason be driven by a second, highly favorable reaction.

SINGLE REACTION

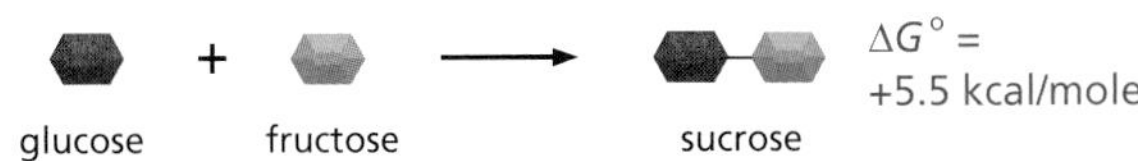

NET RESULT: will not occur!

COUPLED REACTION

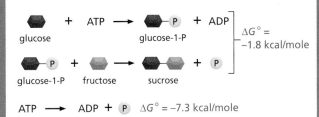

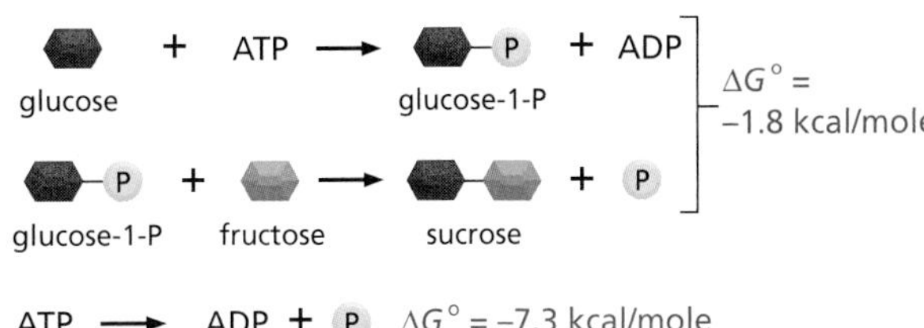

ATP → ADP + P $\Delta G° = -7.3$ kcal/mole

NET RESULT: Sucrose is made in a reaction driven by the hydrolysis of ATP.

HIGH-ENERGY BONDS

One of the most common reactions in the cell is hydrolysis, in which a covalent bond is split by adding water.

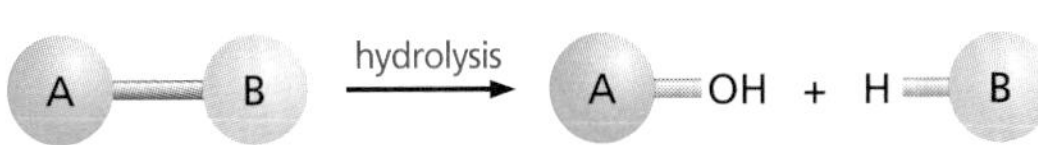

The $\Delta G°$ for this reaction is sometimes loosely termed the "bond energy." Compounds such as acetyl phosphate and ATP that have a large negative $\Delta G°$ of hydrolysis are said to have "high-energy" bonds.

		$\Delta G°$ (kcal/mole)
acetyl P →	acetate + P_i	–10.3
ATP →	ADP + P_i	–7.3
glucose-6-P →	glucose + P_i	–3.3

(Note that, for simplicity, water is omitted from the above equations.)

TABLE 3–1 RELATIONSHIP BETWEEN THE STANDARD FREE-ENERGY CHANGE, $\Delta G°$, AND THE EQUILIBRIUM CONSTANT

EQUILIBRIUM CONSTANT $\frac{[X]}{[Y]}$	FREE ENERGY OF X MINUS FREE ENERGY OF Y IN kcal/mole
10^5	–7.1
10^4	–5.7
10^3	–4.3
10^2	–2.8
10	–1.4
1	0
10^{-1}	1.4
10^{-2}	2.8
10^{-3}	4.3
10^{-4}	5.7
10^{-5}	7.1

Values of the equilibrium constant were calculated for the simple chemical reaction $Y \rightleftharpoons X$ using the equation given in the text.

The $\Delta G°$ values given here are in kilocalories per mole at 37°C. As explained in the text, $\Delta G°$ represents the free-energy difference under standard conditions (where all components are present at a concentration of 1 mole/liter).

From this table, we see that if there is a favorable free-energy change of –4.3 kcal/mole for the transition $Y \rightarrow X$, there will be 1000 times more molecules in state X than in state Y.

So, because

$$\Delta G = \Delta G° + RT \ln \frac{[X]}{[Y]},$$

(as stated on p. 92), then at equilibrium at 37°C, where $\Delta G = 0$ and $RT = 0.616$, the equation becomes:

$$\Delta G° = -0.616 \ln \frac{[X]}{[Y]}$$

or,

$$\Delta G° = -0.616 \ln K.$$

Converting this equation from natural log (ln) to the more commonly used base-10 logarithm (log), we get

$$\Delta G° = -1.42 \log K$$

This equation reveals how the equilibrium ratio of Y to X (expressed as the equilibrium constant, K) depends on the intrinsic character of the molecules, as expressed in the value of $\Delta G°$ (Table 3–1). Note that for every 1.42 kcal/mole difference in free energy at 37°C, the equilibrium constant changes by a factor of 10. Thus, the more energetically favorable the reaction, the more product will accumulate if the reaction proceeds to equilibrium.

In Complex Reactions, the Equilibrium Constant Depends on the Concentrations of All Reactants and Products

We have so far discussed the simplest of reactions, $Y \rightarrow X$, in which a single substrate is converted into a single product. But what happens in the more common situation where two reactants combine to form a single product, $A + B \rightleftharpoons AB$?

The same principles apply, except that now the equilibrium constant K depends on the concentrations of both of the reactants and that of the product:

$$K = [AB]/[A][B]$$

The concentrations of both substrates are multiplied because the formation of product AB depends on the collision of A and B, and these encounters occur at a rate that is proportional to $[A] \times [B]$. As with single-substrate reactions, $\Delta G° = -1.42 \log K$ at 37°C.

The Equilibrium Constant Indicates the Strength of Molecular Interactions

The concept of free-energy change does not only apply to chemical reactions where covalent bonds are being broken and formed, but also to interactions where one molecule binds to another by means of noncovalent bonds (see Chapter 2, pp. 62–63). These types of interactions are immensely important to cells. They include the binding of substrates to enzymes, the binding of gene regulatory proteins to DNA, and the binding of one protein to another to make the myriads of different structural and functional protein complexes that constitute a living cell.

Two molecules will bind to each other if the $\Delta G°$ of the interaction is negative; that is, the free energy of the resulting complex is lower than the sum of the free energies of the two partners when unbound. Because the equilibrium constant of a reaction is related directly to $\Delta G°$, K is commonly employed as a measure of the binding strength of a noncovalent interaction between two molecules. The binding strength is a very useful quantity to know because it also indicates how specific the interaction is between the two molecules.

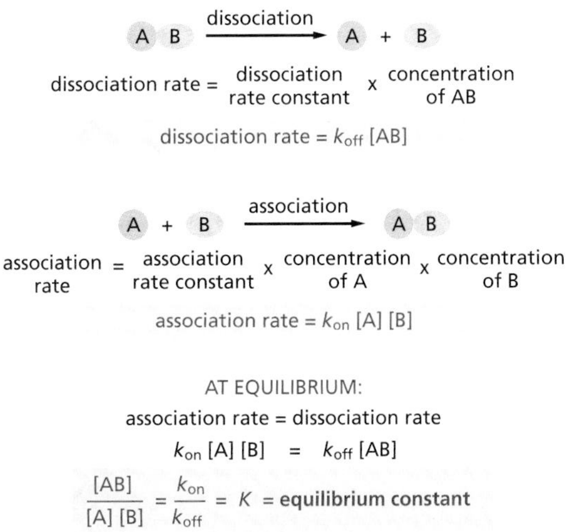

AB dissociation → A + B
dissociation rate = dissociation rate constant x concentration of AB
dissociation rate = k_{off} [AB]
A + B association → AB
association rate = association rate constant x concentration of A x concentration of B
association rate = k_{on} [A] [B]
AT EQUILIBRIUM:
association rate = dissociation rate
k_{on} [A] [B] = k_{off} [AB]
$\frac{[AB]}{[A][B]} = \frac{k_{on}}{k_{off}} = K$ = equilibrium constant

Figure 3–19 **The energy of binding interactions is reflected in the equilibrium constant.** The equilibrium between molecules A and B and the complex AB is maintained by a balance between the two opposing reactions shown. Molecules A and B must collide in order to interact, and the association rate is therefore proportional to the product of their individual concentrations [A] × [B]. As shown, the ratio of the rate constants k_{on} and k_{off} for the association and the dissociation reactions, respectively, is equal to the equilibrium constant (*K*) for the interaction. For two interacting components, *K* involves the concentrations of both substrates in addition to that of the product. However, the relationship between *K* and $\Delta G°$ is the same as that displayed in Table 3–1. The larger the value of *K*, the stronger is the binding between A and B.

Consider the reaction shown in Figure 3–19, where molecule A interacts with molecule B to form the complex AB. The reaction proceeds until it reaches equilibrium, at which point the number of association events precisely equals the number of dissociation events; at this point, the concentrations of reactants and of the complex AB can be used to determine the equilibrium constant.

K becomes larger as the *binding energy*—that is, the energy released in the binding interaction—increases. In other words, the larger *K* is, the greater is the drop in free energy between the dissociated and associated states, and the more tightly the two molecules will bind. Even a change of a few noncovalent bonds can have a striking effect on a binding interaction, as illustrated in Figure 3–20. In this example, eliminating a few hydrogen bonds from a binding interaction can be seen to cause a dramatic decrease in the amount of complex that exists at equilibrium.

For Sequential Reactions, the Changes in Free Energy are Additive

Now we return to our original concern: how can enzymes catalyze reactions that are energetically unfavorable? One way they do so is by directly coupling energetically unfavorable reactions with energetically favorable ones. Consider, for example, two reactions,

$$X \rightarrow Y \text{ and } Y \rightarrow Z$$

where the $\Delta G°$ values are +5 and –13 kcal/mole, respectively. (Recall that a mole is 6×10^{23} molecules of a substance.) The unfavorable reaction, X → Y, will not occur spontaneously. However, it can be driven by the favorable reaction Y → Z, provided that the second reaction follows the first. That's because the overall free-energy change for the coupled reaction is equal to the sum of the free-energy changes for each individual step. In this case, the $\Delta G°$ for the coupled reaction will be –8 kcal/mole, making the overall pathway energetically favorable.

Cells can therefore cause the energetically unfavorable transition, X → Y, to occur if an enzyme catalyzing the X → Y reaction is supplemented by a second enzyme that catalyzes the energetically favorable reaction, Y → Z. In effect, the reaction Y → Z acts as a "siphon," pulling the conversion of all of molecule X to molecule Y, and thence to molecule Z (Figure 3–21). For example, several of the reactions in the long pathway that converts sugars into CO_2 and H_2O are energetically unfavorable. The pathway nevertheless proceeds rapidly to completion, however, because the total $\Delta G°$ for the series of sequential reactions has a large negative value.

Consider 1000 molecules of A and 1000 molecules of B in a eucaryotic cell. The concentration of both will be about 10^{-9} M.
If the equilibrium constant (*K*) for A + B ⇌ AB is 10^{10}, then at equilibrium there will be
270 A molecules | 270 B molecules | 730 AB complex
If the equilibrium constant is a little weaker at 10^8, which represents a loss of 2.8 kcal/mole of binding energy from the example above, or 2–3 fewer hydrogen bonds, then there will be
915 A molecules | 915 B molecules | 85 AB complex

Figure 3–20 **Small changes in the number of weak bonds can have drastic effects on a binding interaction.** This example illustrates the dramatic effect of the presence or absence of a few weak noncovalent bonds in a biological context.

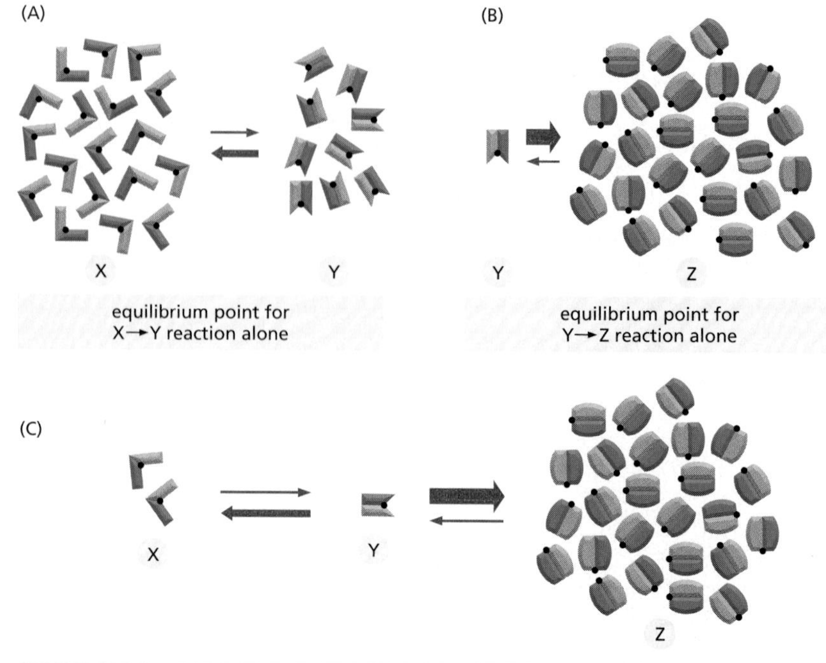

Figure 3–21 **An energetically unfavorable reaction can be driven by a second reaction that acts as a chemical siphon.** (A) At equilibrium, there are twice as many X molecules as Y molecules, because X is of lower energy than Y. (B) At equilibrium, there are 25 times more Z molecules than Y molecules, because Z is of much lower energy than Y. (C) If the reactions in (A) and (B) are coupled, nearly all of the X molecules will be converted to Z molecules, as shown. In terms of energetics, the $\Delta G°$ of the Y → Z reaction is so negative that when coupled to the X → Y reaction, it lowers the ΔG of X → Y, because the ΔG of X → Y decreases as the ratio of Y to X declines.

Forming a sequential pathway, however, is not the answer for all metabolic needs. Often the desired reaction is simply X → Y, without further conversion of Y to some other product. Fortunately, there are other, more general ways of using enzymes to couple reactions together, involving the production of activated carrier molecules that can shuttle energy from one reaction site to another. We will see shortly how these systems work, but before we do, we shall look at how enzymes encounter their substrates and how enzyme performance is measured

QUESTION 3–4

For the reactions shown in Figure 3–21, sketch an energy diagram similar to that in Figure 3–12 for the two reactions alone and for the combined reactions. Indicate the standard free-energy changes for the reactions X → Y, Y → Z, and X → Z in the graph. Indicate how enzymes that catalyze these reactions would change the energy diagram.

Rapid Diffusion Allows Enzymes to Find Their Substrates

Enzymes and their substrates are both present in relatively small amounts in a cell, yet a typical enzyme can capture and process about a thousand substrate molecules every second. This means that an enzyme must be able to release its product and bind a new substrate in a fraction of a millisecond. How can these molecules find each other so quickly inside the cell?

Rapid binding is possible because motions are enormously fast at the molecular level. Because of heat energy, molecules are in constant motion and consequently will explore the space inside the cell very efficiently by wandering randomly through it—a process called **diffusion**. In this way, every molecule in a cell collides with a huge number of other molecules each second. As the molecules in a liquid collide and bounce off one another, an individual molecule moves first one way and then another, its path constituting a *random walk* (Figure 3–22). In such a walk, the average distance that each molecule travels (as the crow flies) from its starting point is proportional to the square root of the time it takes; that is, if it takes a molecule 1 second on average to travel 1 μm, it will take 4 seconds to travel 2 μm, 100 seconds to travel 10 μm, and so on. Thus, diffusion only works well for very short distances. To move molecules quickly over larger distances, cells need to rely on more active and directed methods of transport—processes that inevitably require an expenditure of cellular energy.

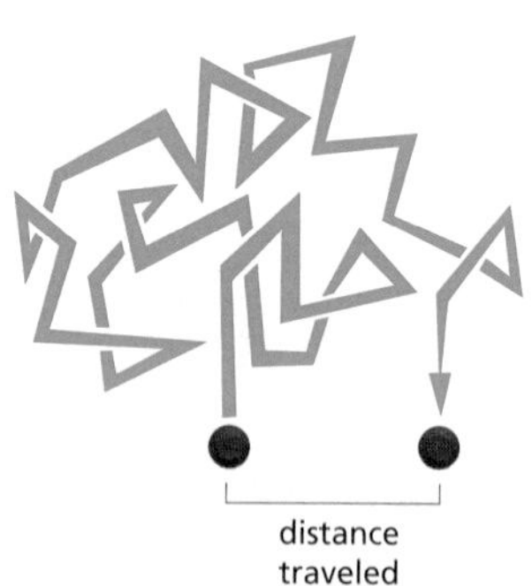

Figure 3–22 **A molecule traverses the cell by taking a random walk.** Molecules in solution move in a random fashion due to the continual buffeting they receive in collisions with other molecules. This movement allows small molecules to diffuse rapidly from one part of the cell to another (Movie 3.2).

The inside of a cell is very crowded (Figure 3–23). Nevertheless, experiments in which fluorescent dyes and other labeled molecules are injected

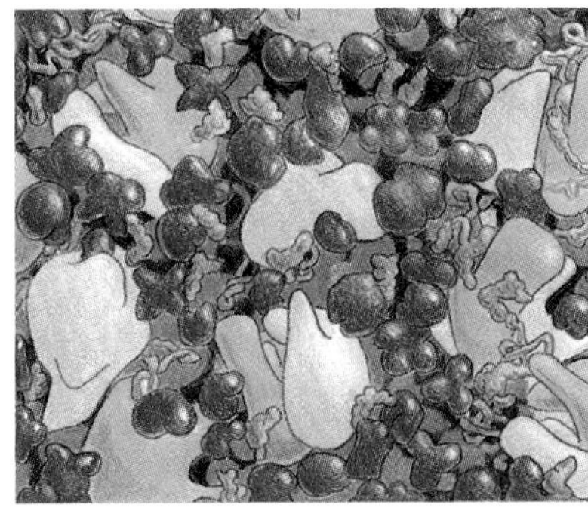

Figure 3–23 **The cell cytoplasm is crowded with various molecules.** The drawing is approximately to scale. Only the macromolecules are shown: RNAs are shown in *blue*, ribosomes in *green*, and proteins in *red*. Enzymes and other macromolecules diffuse relatively slowly in the cytoplasm, in part because they interact with many other macromolecules; small molecules, by contrast, diffuse nearly as rapidly as they do in water. (Adapted from D.S. Goodsell, *Trends Biochem. Sci.* 16:203–206, 1991. With permission from Elsevier.)

into cells show that small organic molecules diffuse through the dense aqueous gel of the cytosol nearly as rapidly as they do through water. A small organic molecule, such as a substrate, takes only about one-fifth of a second on average to diffuse a distance of 10 μm. Diffusion is therefore an efficient way for small molecules to move limited distances in the cell.

Enzymes and other macromolecules, however, diffuse through the cytoplasm much more slowly than do small molecules. In some cases, enzymes that interact with other proteins are actually held in close proximity to their partners by scaffold proteins that draw together sets of interacting proteins at specific locations in the cell. But even if they're not physically sequestered in one place, enzymes move much more slowly than small molecules. So we can think of them as standing still. Thus, the rate of encounter of each enzyme molecule with a small-molecule substrate depends on the concentration of the substrate molecule. For example, some abundant substrates are present in the cell at a concentration of 0.5 mM. Because pure water is 55 M, there is only about one such substrate molecule in the cell for every 10^5 water molecules. Nevertheless, the active site on an enzyme molecule that binds this substrate will be bombarded by about 500,000 random collisions with the substrate molecule per second. For a substrate concentration tenfold lower (0.05 mM), the number of collisions drops to 50,000 per second, and so on. A random encounter between the surface of an enzyme and the matching surface of its substrate molecule often leads immediately to the formation of an enzyme–substrate complex that is ready to react. A reaction in which a covalent bond is broken or formed can now occur extremely rapidly. Once one appreciates how quickly molecules move and react, the observed rates of enzymatic catalysis do not seem so amazing.

When an enzyme and substrate have collided and snuggled together properly at the active site, they form multiple weak bonds with each other that persist until random thermal motion causes the molecules to dissociate again. These weak interactions can include hydrogen bonds, van der Waals attractions, and electrostatic attractions (as discussed in Chapter 2). In general, the stronger the binding of enzyme and substrate, the slower their rate of dissociation, as discussed earlier in this chapter. When two colliding molecules have poorly matching surfaces, few noncovalent bonds are formed and their total energy is negligible compared with that of thermal motion. In this case the two molecules dissociate as rapidly as they come together (see Figure 2–32). This is what prevents incorrect and unwanted associations from forming between mismatched molecules, such as an enzyme and the wrong substrate.

QUESTION 3–5

The enzyme carbonic anhydrase is one of the speediest enzymes known. It catalyzes the conversion of CO_2 to HCO_3^-:
$CO_2 + H_2O \rightleftharpoons HCO_3^- + H^+$.
The rapid conversion of CO_2 gas into the much more soluble bicarbonate ion (HCO_3^-) is very important for the efficient transport of CO_2 from tissue, where CO_2 is produced by respiration, to the lungs, where it is exhaled. Carbonic anhydrase accelerates the reaction 10^7-fold, hydrating 10^5 CO_2 molecules per second at its maximal speed. What do you suppose limits the speed of the enzyme? Sketch a diagram analogous to the one shown in Figure 3–13 and indicate which portion of your diagram has been designed to display the 10^7-fold acceleration.

V_{max} and K_M Measure Enzyme Performance

To catalyze a reaction, an enzyme must first bind its substrate. The substrate then undergoes a reaction to form the product, which initially remains bound to the enzyme. Finally, the product is released and diffuses away, leaving the enzyme free to bind another substrate molecule and catalyze another reaction (see Figure 3–15). The rates of the different

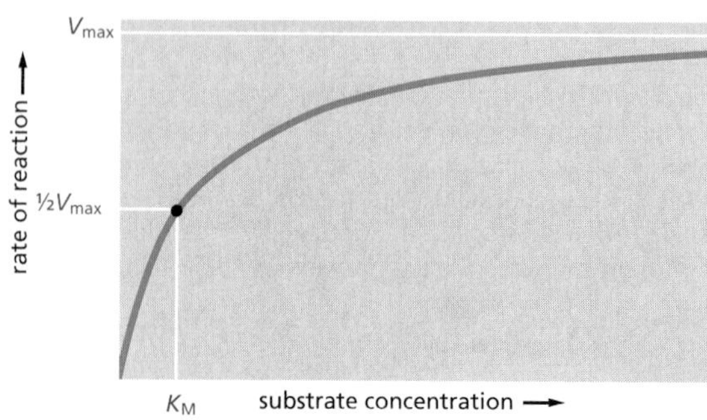

Figure 3–24 **An enzyme's performance depends on how rapidly it can process its substrate.** The rate of an enzyme reaction (V) increases as the substrate concentration increases until a maximum value (V_{max}) is reached. At this point all substrate-binding sites on the enzyme molecules are fully occupied, and the rate of reaction is limited by the rate of the catalytic process on the enzyme surface. For most enzymes the concentration of substrate at which the reaction rate is half-maximal (K_M) is a direct measure of how tightly the substrate is bound, with a large value of K_M (a large amount of substrate needed) corresponding to weak binding.

steps vary widely from one enzyme to another, and they can be measured by mixing purified enzymes and substrates together under carefully defined conditions.

In such experiments, if the concentration of the substrate is increased progressively from a very low value, the concentration of the enzyme–substrate complex—and therefore the rate at which product is formed—initially increases in a linear fashion in direct proportion to substrate concentration. However, as more and more enzyme molecules become occupied by substrate, this rate increase tapers off, until at a very high concentration of substrate it reaches a maximum value, termed $\mathbf{V_{max}}$. At this point, the active sites of all enzyme molecules in the sample are fully occupied by substrate, and the rate of product formation depends only on how rapidly the substrate molecule can be processed. For many enzymes, this **turnover number** is of the order of 1000 substrate molecules per second, although turnover numbers between 1 and 10,000 have been measured.

QUESTION 3–6

In cells, an enzyme catalyzes the reaction AB → A + B. It was isolated, however, as an enzyme that carries out the opposite reaction A + B → AB. Explain the paradox.

The concentration of substrate needed to make the enzyme work efficiently is often measured by a different parameter, the Michaelis' constant, $\mathbf{K_M}$, named after one of the biochemists who worked out this relationship. An enzyme's K_M is the concentration of substrate at which the enzyme works at half its maximum speed ($0.5V_{max}$; Figure 3–24). In general, a low value of K_M indicates that a substrate binds very tightly to the enzyme, and a large value corresponds to weak binding. For a discussion of how we measure these parameters, and how we can use them to model biochemical pathways—and potentially to design better catalysts—see How We Know, pp. 101–103.

It is important to recognize that when an enzyme (or any catalyst) lowers the activation energy for the reaction Y → X, it also lowers the activation energy for the reverse reaction X → Y by exactly the same amount (see Figure 3–12). The forward and backward reactions will therefore be accelerated by the same factor by an enzyme, and the equilibrium point for the reaction (and thus $\Delta G°$) remains unchanged (Figure 3–25).

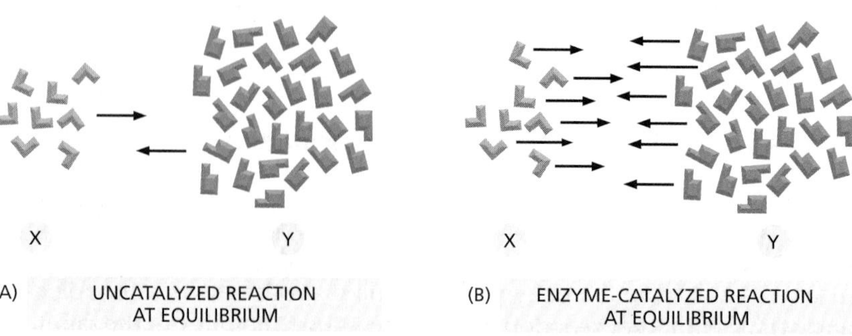

Figure 3–25 **Enzymes cannot change the equilibrium point for reactions.** Enzymes, like all catalysts, speed up the forward and reverse rates of a reaction by the same factor. Therefore, for both the (A) uncatalyzed and (B) catalyzed reactions shown here, the number of molecules undergoing the transition X → Y is equal to the number of molecules undergoing the transition Y → X when the ratio of Y molecules to X molecules is 3.5 to 1, as illustrated. In other words, both the catalyzed and uncatalyzed reactions will eventually reach the same equilibrium point.

USING KINETICS TO MODEL AND MANIPULATE METABOLIC PATHWAYS

At first glance, it seems that a cell's metabolic pathways have been pretty well mapped out, with each reaction proceeding predictably to the next—substrate X is converted to product Y, which is passed along to enzyme Z. So why would anyone need to know exactly how tightly a particular enzyme clutches its substrate? Or whether it can process 100 or 1000 substrate molecules every second?

In reality, such elaborate metabolic maps merely suggest which pathways a cell *might* follow as it converts nutrients into small molecules, chemical energy, and the larger building blocks of life. They do not reveal precisely how a cell will behave under a particular set of conditions: which pathways it will take when it is starving, when it is well fed, when oxygen is scarce, when it is stressed, or when it decides to divide. The study of an enzyme's *kinetics*—how fast it operates, how it handles its substrate, how its activity is controlled—makes it possible to predict how an individual catalyst will perform, and how it will interact with other enzymes in a network. Such knowledge leads to a deeper understanding of cell biology, and it opens the door to learning how to harness enzymes to perform desired reactions.

Speed

The first step to understanding how an enzyme performs involves determining the maximal velocity, V_{max}, for the reaction it catalyzes. This is accomplished by measuring, in a test tube, how rapidly the reaction proceeds in the presence of different concentrations of substrate: the rate should increase as the amount of substrate rises until the reaction reaches its V_{max}. The velocity of the reaction is measured by monitoring either how quickly the substrate is consumed or how rapidly the product accumulates. In many cases, the appearance of product or the disappearance of substrate can be observed directly with a spectrophotometer. This instrument detects the presence of molecules that absorb light at a selected wavelength; NADH, for example, absorbs light at 340 nm, while its oxidized counterpart, NAD^+, does not. So, a reaction that generates NADH (by reducing NAD^+) can be monitored by following the formation of NADH at 340 nm spectrophotometrically.

To determine the V_{max} of a reaction, you would set up a series of test tubes, where each tube contains a different concentration of substrate. For each tube, add the same amount of enzyme and then measure the velocity of the reaction: the number of micromoles of substrate consumed or product generated per minute. Because these numbers will tend to decrease over time, the rate used is the velocity measured early in the reaction. These initial velocity values (v) are then plotted against the substrate concentration, yielding a curve like the one shown in Figure 3–26.

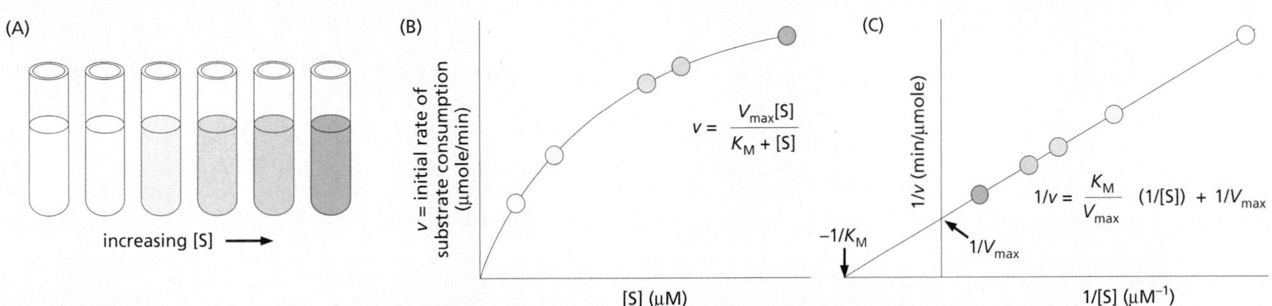

Figure 3–26 **Reaction rate data are plotted to determine V_{max} and K_M of an enzyme-catalyzed reaction.** (A) A series of increasing substrate concentrations is prepared, enzyme is added, and initial velocities are determined. (B) The initial velocities (v) plotted against the substrate concentrations [S] give a curve described by the general equation $y = ax/(b + x)$. Substituting our kinetic terms, the equation becomes $v = V_{max}[S]/(K_M + [S])$, where V_{max} is the asymptote of the curve (the value of y at an infinite value of x), and K_M is equal to the substrate concentration where v is one-half V_{max}. This is called the *Michaelis–Menten equation*, named for the biochemists who provided evidence for this enzymatic relationship. (C) In a double-reciprocal plot, $1/v$ is plotted against 1/[S]. The equation describing this straight line is $1/v = (K_M/V_{max})(1/[S]) + 1/V_{max}$. When $1/[S] = 0$, the y intercept ($1/v$) is $1/V_{max}$. When $1/v = 0$, the x intercept (1/[S]) is $-1/K_M$. Plotting the data this way allows V_{max} and K_M to be calculated more precisely. By convention, lowercase letters are used for variables (hence v for velocity) and uppercase letters are used for constants (hence V_{max}).

Looking at this plot, however, it is difficult to determine the exact value of V_{max}, as it is not clear where the reaction rate will reach its plateau. To get around this problem, the data are converted to their reciprocals and graphed in a "double-reciprocal plot," where the inverse of the velocity ($1/v$) appears on the y-axis and the inverse of the substrate concentration ($1/[S]$) on the x-axis (see Figure 3–26C). This graph yields a straight line whose y intercept (the point where the line crosses the y-axis) represents $1/V_{max}$ and whose x intercept corresponds to $-1/K_M$. These values are then converted to values for V_{max} and K_M.

Enzymologists use this technique to determine the kinetic parameters of many enzyme-catalyzed reactions (although these days computer programs automatically plot the data and spit out the sought-after values). Some reactions, however, happen too fast to be monitored in this way; the action is essentially complete—the substrate entirely consumed—within thousandths of a second. For these reactions, a special piece of equipment must be used to follow what happens during the first few milliseconds after enzyme and substrate meet (Figure 3–27).

Control

Substrates are not the only molecules that influence how well or how quickly an enzyme works. Many enzymes can also be controlled by products, substrate lookalikes, inhibitors, toxins, and other small molecules that either increase or decrease their activity. This regulation allows cells to control when and how rapidly various reactions occur, a process we will consider in more detail in Chapter 4.

Determining how an inhibitor decreases an enzyme's activity can reveal how a metabolic pathway is regulated—and can suggest how those control points can be circumvented by carefully designed mutations in specific genes.

The effect of an inhibitor on an enzyme's activity is monitored in the same way that we measured the enzyme's kinetics. A curve is generated showing the velocity of the uninhibited reaction between enzyme and substrate, as described previously. Additional curves are also produced for reactions in which the inhibitor molecule has been included in the mix.

Comparing these curves, with and without inhibitor, can also reveal how a particular inhibitor impedes enzyme activity. For example, some inhibitors bind to the same site on an enzyme as its substrate. These *competitive inhibitors* block enzyme activity by competing directly with the substrate for the enzyme's attention. They resemble the substrate enough to tie up the enzyme, but they differ enough in structure to avoid getting converted to product. This blockage can be overcome by adding enough substrate so that enzymes are more likely to encounter a substrate molecule than an inhibitor. From the kinetic data, we can see that competitive inhibitors do not change the V_{max} of a reaction; in other words, add enough substrate and the enzyme will encounter mostly substrate molecules and will reach its maximum velocity (Figure 3–28).

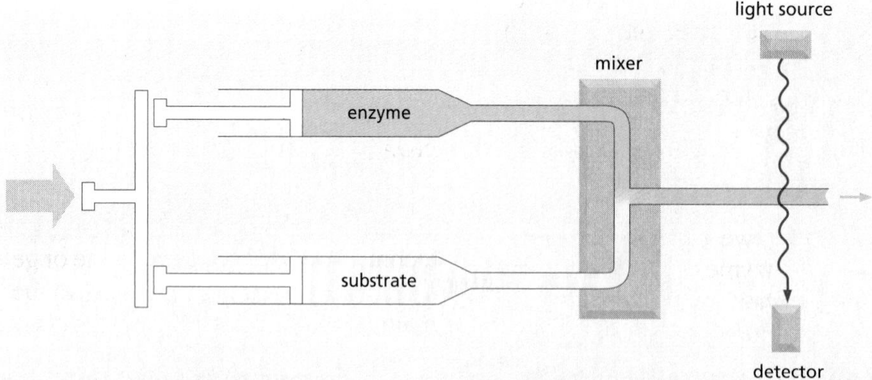

Figure 3–27 **A stopped-flow apparatus is used to observe reactions during the first few milliseconds.** In this piece of equipment, the enzyme and substrate are rapidly injected into a mixing chamber through two syringes. The enzyme and its substrate meet as they shoot through the mixing tube at flow rates that can easily reach 1000 cm/sec. They then enter another tube and zoom past a detector that monitors, say, the appearance of product. If the detector is located within a centimeter of where the enzyme and substrate meet, it is possible to observe reactions when they are only a few milliseconds old.

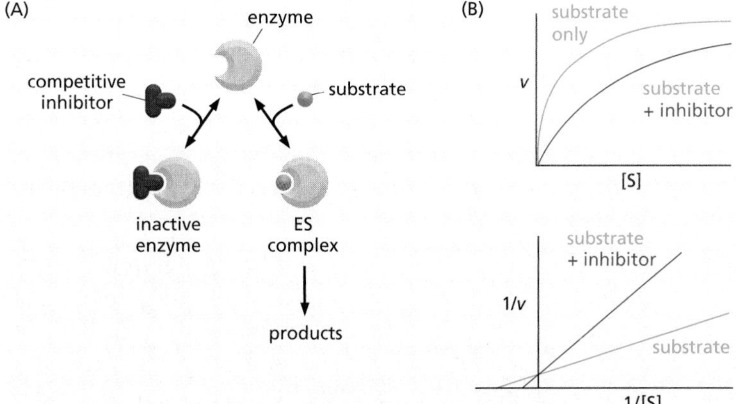

Figure 3–28 **A competitive inhibitor directly blocks substrate binding.** (A) The active site of an enzyme can bind either the competitive inhibitor or the substrate, but not both together. (B) The upper plot shows that inhibition by a competitive inhibitor can be overcome by increasing the substrate concentration. The double-reciprocal plot below shows that the V_{max} of the reaction is not changed in the presence of the competitive inhibitor: the *y* intercept is identical for both the curves.

Competitive inhibition can be used to treat patients who have been poisoned by ethylene glycol, an ingredient in commercially available antifreeze. Although ethylene glycol is itself not fatally toxic, a by-product of its metabolism—oxalic acid—can be lethal. To prevent oxalic acid from forming, the patient is given a large (though not quite intoxicating) dose of ethanol. Ethanol competes with the ethylene glycol for binding to alcohol dehydrogenase, the first enzyme in the pathway to oxalic acid formation. As a result, the ethylene glycol goes mostly unmetabolized and is safely eliminated from the body.

Other types of inhibitors may interact with sites on the enzyme distant from where the substrate binds. For example, chelating agents that bind reversibly to ions such as Mg^{2+} will inhibit enzymes that require such metals for their activity. In this case, the substrate can bind, but the enzyme–substrate complex may not form as quickly as it would in the absence of inhibitor. Such inhibition is not overcome by the addition of more substrate.

Design

With the kinetic data in hand, we can use modeling programs to predict how an enzyme will perform, and even how a cell will respond when exposed to different conditions—such as the addition of a particular sugar or amino acid to the culture medium, or the addition of a poison or a pollutant. Seeing how a cell manages its resources—which pathways it favors for dealing with particular biochemical challenges—can also suggest strategies we can follow to design better catalysts for reactions of medical or commercial importance (e.g., producing drugs or detoxifying industrial waste). Using such tactics, bacteria have even been engineered to produce large amounts of indigo—the dye, originally extracted from plants, that makes your blue jeans blue.

Several computer programs have been developed to facilitate the dissection of complex reaction pathways. One such program requires information about individual reactions, such as velocities and the concentrations of enzymes, substrates, products, inhibitors, and other regulatory molecules. The program then predicts how molecules will flow through the pathway, which products will be generated, and where any bottlenecks might be. The process is not unlike balancing an algebraic equation in which every atom of carbon, nitrogen, oxygen, and so on, must be properly accounted for. Such careful accounting makes it possible to rationally design ways to manipulate the pathway, such as re-routing it around a bottleneck, eliminating an important inhibitor, redirecting the reactions to favor the generation of predominantly one product, or extending the pathway to produce a novel molecule. Of course such models must be tested and validated in cells, which may not always behave as predicted.

Producing designer cells that spew out commercial products generally requires using genetic engineering techniques to introduce the gene or genes of choice into a cell, usually a bacterium, that can be manipulated and maintained in the laboratory. We discuss these methods at greater length in Chapter 10. Harnessing the power of cell biology for commercial purposes—even to produce something as simple as the amino acid tryptophan—is currently a billion-dollar industry. And, as more genome data come in, presenting us with more enzymes to exploit, it may not be long before vats of custom-made bacteria are churning out drugs and chemicals that represent the biological equivalent of pure gold.

ACTIVATED CARRIER MOLECULES AND BIOSYNTHESIS

The energy released by the oxidation of food molecules must be stored temporarily before it can be channeled into the construction of either other small organic molecules or the larger and more complex molecules needed by the cell. In most cases, the energy is stored as chemical bond energy in a small set of activated "carrier molecules," which contain one or more energy-rich covalent bonds. These molecules diffuse rapidly throughout the cell and thereby carry their bond energy from the sites of energy generation to the sites where energy is used for **biosynthesis** and other needed cell activities (Figure 3–29).

Activated carriers store energy in an easily exchangeable form, either as a readily transferable chemical group or as high-energy electrons, and they can serve a dual role as a source of both energy and chemical groups for biosynthetic reactions. The most important of the activated carrier molecules are ATP and two molecules that are closely related to each other, NADH and NADPH. Cells use activated carrier molecules like money to pay for reactions that otherwise could not take place.

The Formation of an Activated Carrier Is Coupled to an Energetically Favorable Reaction

When a fuel molecule such as glucose is oxidized in a cell, enzyme-catalyzed reactions ensure that a large part of the free energy that is released by oxidation is captured in a chemically useful form, rather than being released wastefully as heat. (Burning sugar in a cell allows you to power metabolic reactions, whereas burning a chocolate bar in the street will get you nowhere, energetically speaking: it produces no metabolically useful energy.) In living systems, this energy capture is achieved by means of a **coupled reaction**, in which an energetically favorable reaction is used to drive an energetically unfavorable one that produces an activated carrier molecule or some other useful molecule. Coupling mechanisms require enzymes, and they are fundamental to all of the energy transactions of the cell.

The nature of a coupled reaction is illustrated by a mechanical analogy in Figure 3–30, in which an energetically favorable chemical reaction is represented by rocks falling from a cliff. The energy of falling rocks would normally be entirely wasted in the form of heat generated by friction when the rocks hit the ground (Figure 3–30A). By careful design, however, part of this energy could be used to drive a paddle wheel that lifts a bucket of water (Figure 3–30B). Because the rocks can now reach the ground only after moving the paddle wheel, we say that the energetically favorable

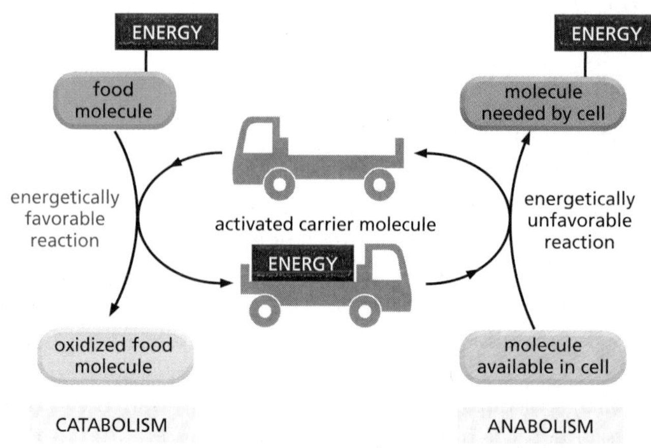

Figure 3–29 **Activated carriers can store and transfer energy needed for metabolism.** By serving as energy shuttles, activated carrier molecules perform their function as go-betweens that link the breakdown of food molecules and the release of energy (catabolism) to the energy-requiring biosynthesis of small and large organic molecules (anabolism).

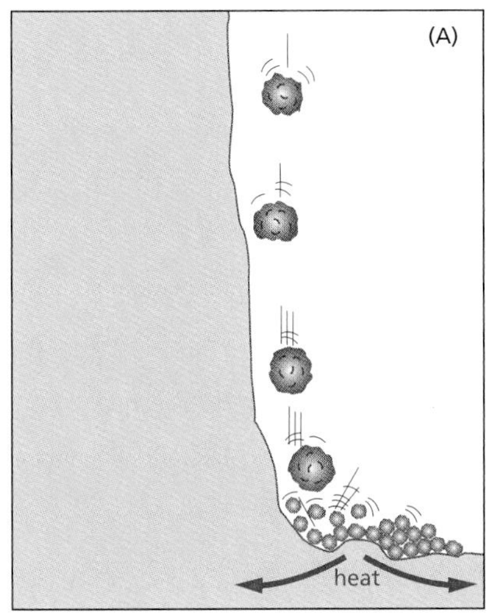

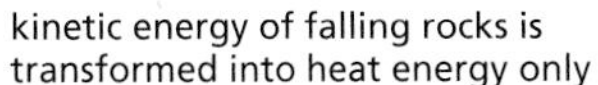

kinetic energy of falling rocks is transformed into heat energy only

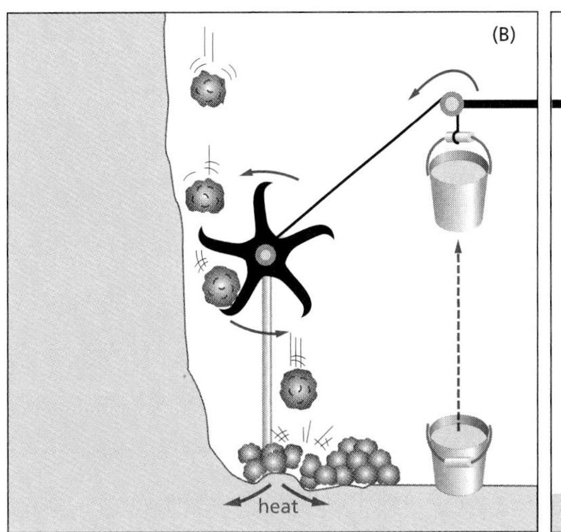

part of the kinetic energy is used to lift a bucket of water, and a correspondingly smaller amount is transformed into heat

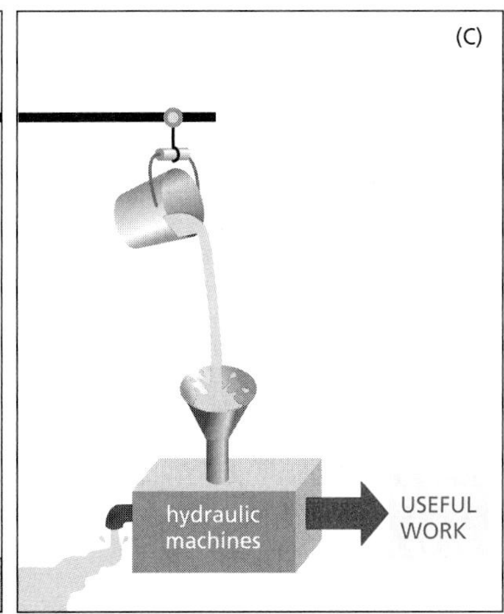

the potential kinetic energy stored in the raised bucket of water can be used to drive hydraulic machines that carry out a variety of useful tasks

Figure 3–30 **A mechanical model illustrates the principle of coupled chemical reactions.** The spontaneous reaction shown in (A) could serve as an analogy for the direct oxidation of glucose to CO_2 and H_2O, which produces heat only. In (B) the same reaction is coupled to a second reaction; this second reaction could serve as an analogy for the synthesis of activated carrier molecules. The energy produced in (B) is in a more useful form than in (A) and can be used to drive a variety of otherwise energetically unfavorable reactions (C).

reaction of rocks falling has been directly coupled to the energetically unfavorable reaction of lifting the bucket of water. Because part of the energy is used to do work in (B), the rocks hit the ground with less velocity than in (A), and correspondingly less energy is wasted as heat. The energy saved can be used to do useful work (Figure 3–30C).

Analogous processes occur in cells, where enzymes play the role of the paddle wheel in Figure 3–30B. By mechanisms that will be discussed in Chapter 13, enzymes couple an energetically favorable reaction, such as the oxidation of foodstuffs, to an energetically unfavorable reaction, such as the generation of an activated carrier molecule. As a result, the amount of heat released by the oxidation reaction is reduced by exactly the amount of energy that is stored in the energy-rich covalent bonds of the activated carrier molecule. The activated carrier molecule in turn picks up a packet of energy that is large enough to power a chemical reaction elsewhere in the cell.

ATP is the Most Widely Used Activated Carrier Molecule

The most important and versatile of the activated carriers in cells is **ATP** (adenosine 5′-triphosphate). Just as the energy stored in the raised bucket of water in Figure 3–30B can be used to drive a wide variety of hydraulic machines, ATP serves as a convenient and versatile store, or currency, of energy to drive a variety of chemical reactions in cells. As shown in Figure 3–31, ATP is synthesized in an energetically unfavorable *phosphorylation* reaction in which a phosphate group is added to **ADP** (adenosine 5′-diphosphate). When required, ATP gives up this energy packet in an energetically favorable hydrolysis to ADP and inorganic phosphate (P_i). The regenerated ADP is then available to be used for another round of the phosphorylation reaction that forms ATP, creating an ATP cycle in the cell.

The energetically favorable reaction of ATP hydrolysis is coupled to many otherwise unfavorable reactions through which other molecules are synthesized. We shall encounter several of these reactions later in this chapter and see exactly how this is done. Such hydrolysis reactions often involve the transfer of the terminal phosphate in ATP to another molecule, as illustrated in Figure 3–32. Any reaction that involves the transfer of a phosphate group to a molecule is termed a phosphorylation reaction. Phosphorylation reactions are examples of condensation reactions (see

QUESTION 3–7

Use Figure 3–30B to illustrate the following reaction driven by the hydrolysis of ATP:

$$X + ATP \rightarrow Y + ADP + P_i$$

A. In this case, which molecule or molecules would be analogous to (i) rocks at top of cliff, (ii) broken debris at bottom of cliff, (iii) bucket at its highest point, and (iv) bucket on the ground?

B. What would be analogous to (i) the rocks hitting the ground in the absence of the paddle wheel in Figure 3–30A and (ii) the hydraulic machine in Figure 3–30C?

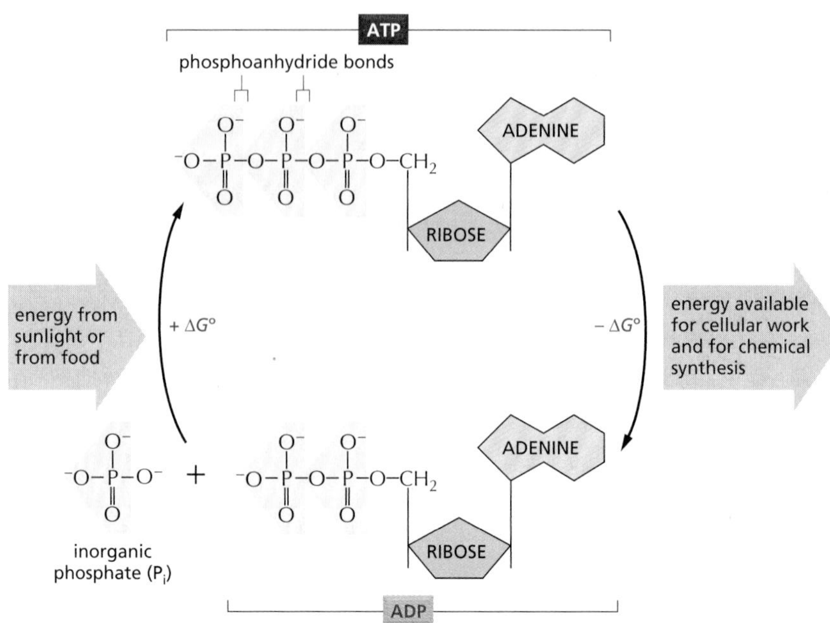

Figure 3–31 **The interconversion of ATP and ADP occurs in a cycle.** The two outermost phosphates in ATP are held to the rest of the molecule by high-energy phosphoanhydride bonds and are readily transferred. Water can be added to ATP to form ADP and inorganic phosphate (P_i). This hydrolysis of the terminal phosphate of ATP yields between 11 and 13 kcal/mole of usable energy. Although the $\Delta G°$ of this reaction is –7.3 kcal/mole, the ΔG is much more negative because the ratio of ATP to the products ADP and P_i is so high inside the cell.

The large negative $\Delta G°$ of the reaction arises from a number of factors. Release of the terminal phosphate group removes an unfavorable repulsion between adjacent negative charges; in addition, the inorganic phosphate ion (P_i) released is stabilized by resonance and by favorable hydrogen-bond formation with water. The formation of ATP from ADP and P_i reverses the hydrolysis reaction. Because this condensation reaction is energetically unfavorable, it must be coupled to an energetically more favorable reaction to occur.

Figure 2–25) and are involved in many important cellular functions: they activate substrates, they facilitate the exchange of chemical energy, and they help to control cell-signaling processes.

ATP is the most abundant energy carrier in cells. It is used, for example, to supply energy for many of the pumps that transport substances into and out of the cell (discussed in Chapter 12); it also powers the molecular motors that enable muscle cells to contract and nerve cells to transport materials from one end of their long axons to another (discussed in Chapter 17). Why evolution selected this particular nucleotide over the others as the major carrier of energy, however, remains a mystery.

Energy Stored in ATP is Often Harnessed to Join Two Molecules Together

We have previously discussed one way in which an energetically favorable reaction, Y → Z, can be coupled to an energetically unfavorable reaction, X → Y, so as to enable it to occur. In that scheme a second enzyme catalyzes the energetically favorable reaction Y → Z, pulling all

QUESTION 3–8

The phosphoanhydride bond that links two phosphate groups in ATP in a high-energy linkage has a $\Delta G°$ of –7.3 kcal/mole (–30.5 kJ/mole). Hydrolysis of this bond liberates from 11 to 13 kcal/mole of usable energy. How can this be? Why do you think a range of energies is given, rather than a precise number as for $\Delta G°$?

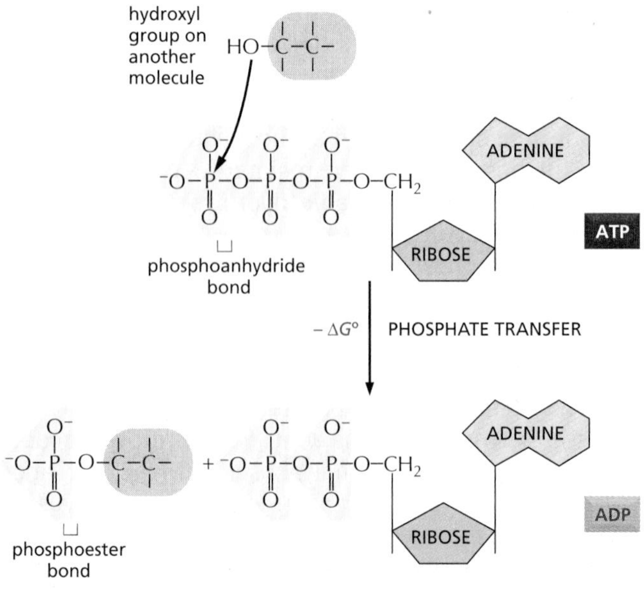

hydroxyl group on another molecule

phosphoanhydride bond

ATP

– ΔG° PHOSPHATE TRANSFER

phosphoester bond

ADP

Figure 3–32 **The terminal phosphate of ATP can be readily transferred to other molecules.** Because an energy-rich phosphoanhydride bond in ATP is converted to a less energy-rich phosphoester bond in the phosphate-accepting molecule, this reaction is energetically favorable, having a large negative $\Delta G°$. Phosphorylation reactions of this type are involved in the synthesis of phospholipids and in the initial steps of the reactions that catabolize sugars, as well as in many other metabolic events.

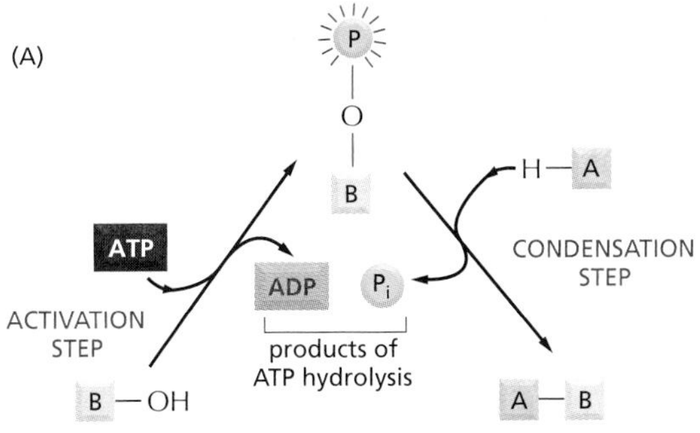

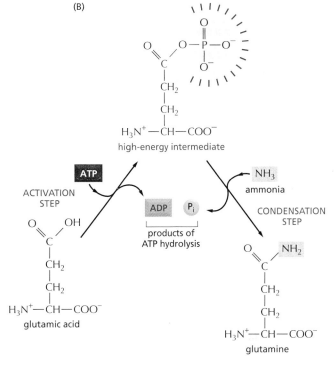

Figure 3–33 **An energetically unfavorable biosynthetic reaction can be driven by ATP hydrolysis.** (A) Schematic illustration of the formation of A–B in the condensation reaction described in the text. (B) The biosynthesis of the amino acid glutamine. Glutamic acid is first converted to a high-energy phosphorylated intermediate (corresponding to the compound B–O–PO_3 described in the text), which then reacts with ammonia (corresponding to A–H) to form glutamine. In this example both steps occur on the surface of the same enzyme, glutamine synthase. For clarity, the glutamic acid side chain is shown in its uncharged form. ATP hydrolysis can drive this energetically unfavorable reaction because it yields more energy ($\Delta G°$ of –7.3 kcal/mole) than the synthesis of glutamine from glutamic acid plus NH_3 consumes ($\Delta G°$ of +3.4 kcal/mole).

of the X to Y in the process (see Figure 3–21). This mechanism is not useful, however, when the required product is Y and not Z.

A common type of reaction that is needed for biosynthesis is one in which two molecules, A and B, are joined together to produce A–B in the energetically unfavorable condensation reaction:

$$\text{A–H} + \text{B–OH} \rightarrow \text{A–B} + H_2O$$

ATP hydrolysis can be coupled indirectly to this reaction to make it go forward. That is, energy from ATP hydrolysis is first used to convert B–OH to a higher-energy intermediate compound, which then reacts directly with A–H to give A–B. The simplest possible mechanism involves the transfer of a phosphate from ATP to B–OH to make B–O–PO_3, in which case the reaction pathway contains only two steps:

1. $\text{B–OH} + \text{ATP} \rightarrow \text{B–O–}PO_3 + \text{ADP}$
2. $\text{A–H} + \text{B–O–}PO_3 \rightarrow \text{A–B} + P_i$

Net result: $\text{B–OH} + \text{ATP} + \text{A–H} \rightarrow \text{A–B} + \text{ADP} + P_i$

The condensation reaction, which by itself is energetically unfavorable, has been forced to occur by being coupled to ATP hydrolysis in an enzyme-catalyzed reaction pathway (Figure 3–33A).

A biosynthetic reaction of exactly this type is employed to synthesize the amino acid glutamine, as illustrated in Figure 3–33B. We will see later in the chapter that very similar (but more complex) mechanisms are also used to produce nearly all of the large molecules of the cell.

NADH and NADPH Are Important Electron Carriers

Other important activated carrier molecules participate in oxidation–reduction reactions and are commonly part of coupled reactions in cells. These activated carriers are specialized to carry both high-energy electrons and hydrogen atoms. The most important of these electron carriers are **NAD^+** (nicotinamide adenine dinucleotide) and the closely related molecule **$NADP^+$** (nicotinamide adenine dinucleotide phosphate). NAD^+

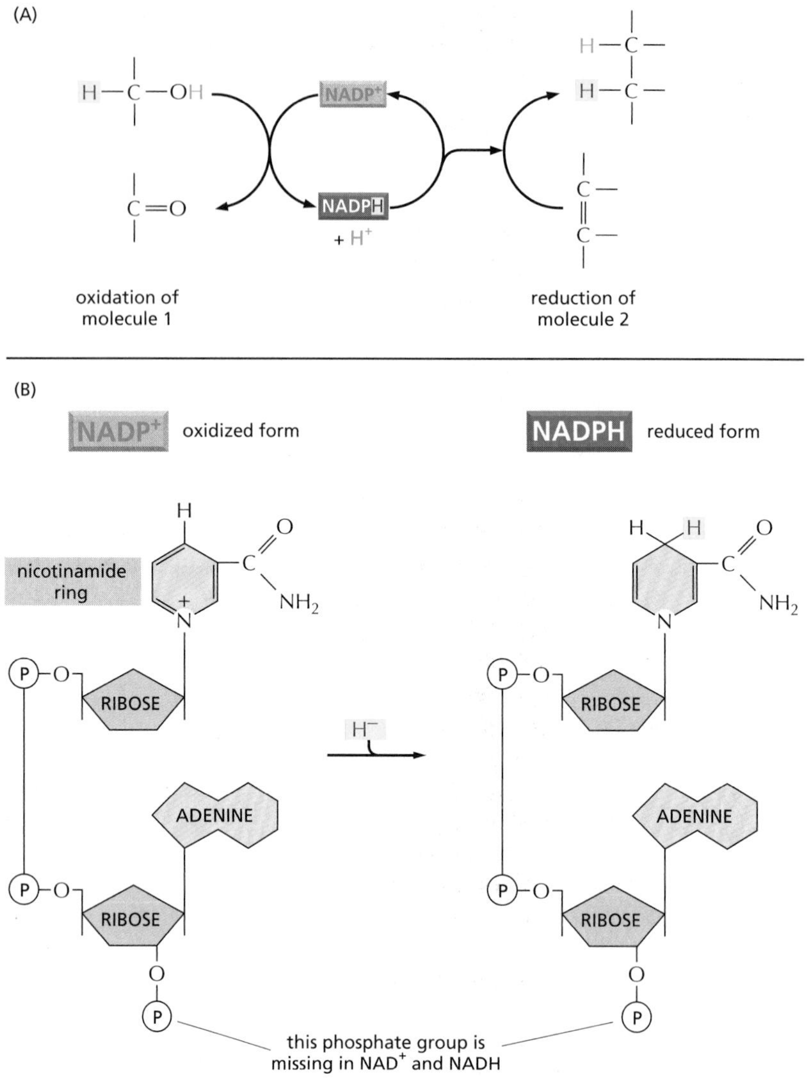

Figure 3–34 **NADPH is an important carrier of electrons.** (A) NADPH is produced in reactions of the general type shown on the left, in which two hydrogen atoms are removed from a substrate. The oxidized form of the carrier molecule, $NADP^+$, receives one hydrogen atom plus an electron (a hydride ion), and the proton (H^+) from the other H atom is released into solution. Because NADPH holds its hydride ion in a high-energy linkage, the added hydride ion can easily be transferred to other molecules, as shown on the right. (B) The structure of $NADP^+$ and NADPH. The part of the $NADP^+$ molecule known as the nicotinamide ring accepts two electrons together with a proton (the equivalent of a hydride ion, H^-), forming NADPH. NAD^+ and NADH are identical in structure to $NADP^+$ and NADPH, respectively, except that they lack the indicated phosphate group.

and $NADP^+$ each pick up a "packet of energy" in the form of two high-energy electrons plus a proton (H^+), becoming **NADH** (reduced nicotinamide adenine dinucleotide) and **NADPH** (reduced nicotinamide adenine dinucleotide phosphate), respectively. These molecules can therefore also be regarded as carriers of hydride ions (the H^+ plus two electrons, or H^-).

Like ATP, NADPH is an activated carrier that participates in many important biosynthetic reactions that would otherwise be energetically unfavorable. NADPH is produced according to the general scheme shown in Figure 3–34. During a special set of energy-yielding catabolic reactions, a hydrogen atom and two electrons are removed from the substrate molecule and added to the nicotinamide ring of $NADP^+$ to form NADPH. This is a typical oxidation–reduction reaction; the substrate is oxidized and $NADP^+$ is reduced.

The hydride ion carried by NADPH is given up readily in a subsequent oxidation–reduction reaction, because the ring can achieve a more stable arrangement of electrons without it. In this subsequent reaction, which regenerates $NADP^+$, the NADPH becomes oxidized and the substrate becomes reduced—thus completing the NADPH cycle. NADPH is efficient at donating its hydride ion to other molecules for the same reason that ATP readily transfers a phosphate: in both cases the transfer is accompanied by a large negative free-energy change. One example of the use of NADPH in biosynthesis is shown in Figure 3–35.

Figure 3–35 **NADPH participates in the final stage in one of the biosynthetic routes leading to cholesterol.** As in many other biosynthetic reactions, the reduction of the C=C bond is achieved by the transfer of a hydride ion from the carrier molecule NADPH, plus a proton (H^+) from the solution.

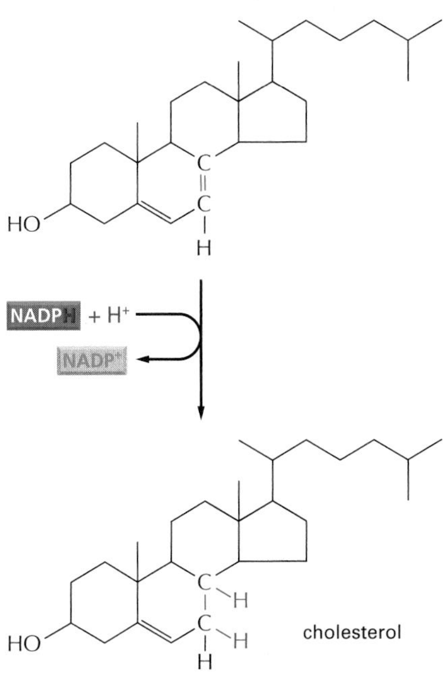

The difference of a single phosphate group has no effect on the electron-transfer properties of NADPH compared with NADH, but it is crucial for their distinctive roles. The extra phosphate group on NADPH is located far from the region involved in electron transfer (see Figure 3–34B). However, it serves to give a molecule of NADPH a slightly different shape from that of NADH, making it possible for NADPH and NADH to bind as substrates to different sets of enzymes. These two types of carriers are thereby used to deliver electrons (or hydride ions) to different destinations.

Why should there be this division of labor? The answer lies in the need to regulate two sets of electron-transfer reactions independently. NADPH operates chiefly with enzymes that catalyze anabolic reactions, supplying the high-energy electrons needed to synthesize energy-rich biological molecules. NADH, by contrast, has a special role as an intermediate in the catabolic system of reactions that generate ATP through the oxidation of food molecules, as we discuss in Chapter 13. The genesis of NADH from NAD^+ and that of NADPH from $NADP^+$ occur by different pathways and are independently regulated, so that the cell can adjust the supply of electrons for these two contrasting purposes. Inside the cell, the ratio of NAD^+ to NADH is kept high, whereas the ratio of $NADP^+$ to NADPH is kept low. This arrangement provides plenty of NAD^+ to act as an oxidizing agent and plenty of NADPH to act as a reducing agent—as required for their special roles in catabolism and anabolism, respectively.

Cells Make Use of Many Other Activated Carrier Molecules

Other activated carriers also pick up and carry a chemical group in an easily transferred, high-energy linkage (Table 3–2). For example, *$FADH_2$*, like NADH and NADPH, also carries hydrogen and high-energy electrons (see Figure 13–12). Coenzyme A, on the other hand, can carry an acetyl group in a readily transferable linkage. This activated molecule, called **acetyl CoA** (acetyl coenzyme A), is shown in Figure 3–36. It is used, for example, to add two-carbon units in the biosynthesis of the hydrocarbon tails of fatty acids.

In acetyl CoA and the other carrier molecules in Table 3–2, the transferable group makes up only a small part of the molecule. The rest consists of a large organic portion that serves as a convenient "handle," facilitating the recognition of the carrier molecule by specific enzymes. As with

TABLE 3–2 SOME ACTIVATED CARRIER MOLECULES WIDELY USED IN METABOLISM

ACTIVATED CARRIER	GROUP CARRIED IN HIGH-ENERGY LINKAGE
ATP	phosphate
NADH, NADPH, $FADH_2$	electrons and hydrogens
Acetyl CoA	acetyl group
Carboxylated biotin	carboxyl group
S-adenosylmethionine	methyl group
Uridine diphosphate glucose	glucose

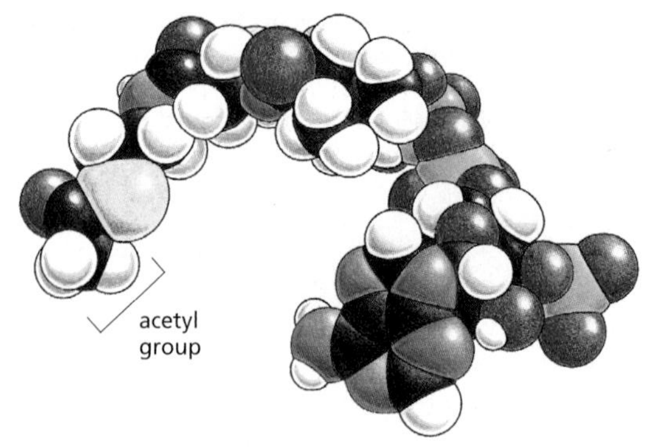

Figure 3–36 **Acetyl coenzyme A (CoA) is another important activated carrier molecule.** A space-filling model is shown above the structure of acetyl CoA. The sulfur atom *(yellow)* forms a thioester bond to acetate. Because the thioester bond is a high-energy linkage, it releases a large amount of free energy when it is hydrolyzed; thus, the acetyl group carried by CoA can be readily transferred to other molecules.

acetyl CoA, this handle portion very often contains a nucleotide. This curious fact may be a relic from an early stage of cell evolution. It is thought that the main catalysts for early life forms on Earth were RNA molecules (or their close relatives) and that proteins were a later evolutionary addition (discussed in Chapter 7). It is therefore tempting to speculate that many of the carrier molecules that we find today originated in an earlier RNA world, where their nucleotide portions would have been useful for binding these carriers to RNA enzymes.

In addition to the transfer reactions catalyzed by the activated carrier molecules ATP (transfer of phosphate) and NADPH (transfer of electrons and hydrogen), other important reactions involve the transfers of methyl, carboxyl, and glucose groups from activated carrier molecules for the purpose of biosynthesis. The activated carriers are usually generated in reactions coupled to ATP hydrolysis (Figure 3–37). Therefore, the energy that enables their groups to be used for biosynthesis ultimately comes from the catabolic reactions that generate ATP. Similar processes occur in the synthesis of the very large molecules of the cell—the nucleic acids, proteins, and polysaccharides—which we discuss next.

The Synthesis of Biological Polymers Requires an Energy Input

The macromolecules of the cell constitute the vast majority of its dry mass—that is, the mass not due to water. These molecules are made from *subunits* (or monomers) that are linked together by bonds formed during an enzyme-catalyzed condensation reaction. The reverse reaction—the breakdown of polymers—occurs through enzyme-catalyzed hydrolysis. This hydrolysis reaction is energetically favorable, whereas the biosynthetic reactions require an energy input and are more complex (Figure 3–38).

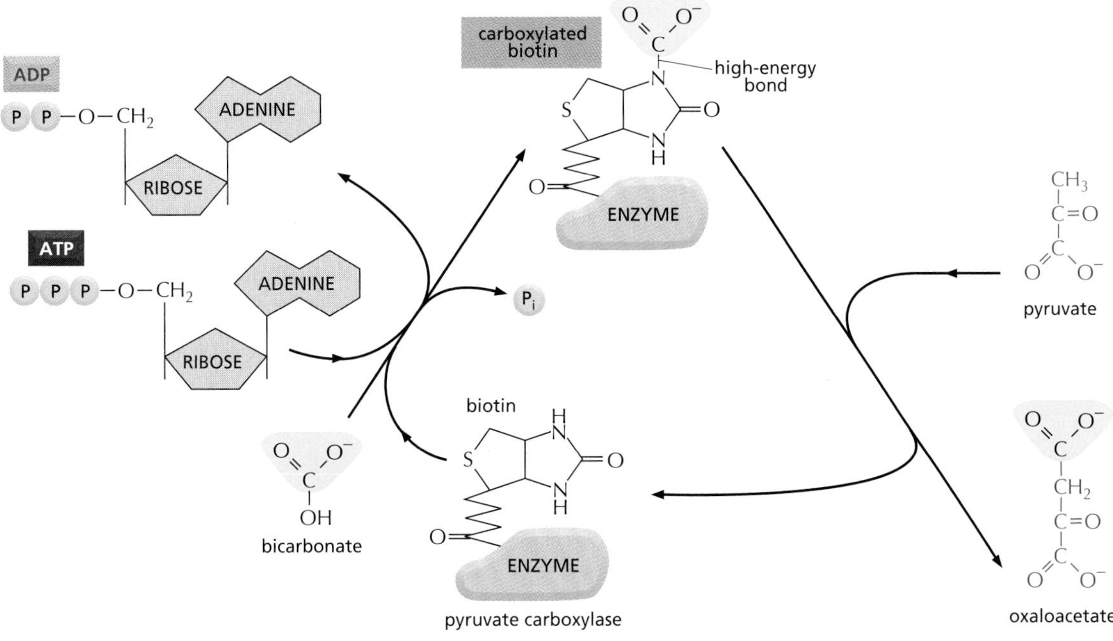

Figure 3–37 **An activated carrier molecule transfers a carboxyl group to a substrate molecule.** Carboxylated biotin is used by the enzyme *pyruvate carboxylase* to transfer a carboxyl group for the production of oxaloacetate, a molecule needed in the citric acid cycle. The acceptor molecule for this group-transfer reaction is pyruvate. Other enzymes use biotin to transfer carboxyl groups to other acceptor molecules. Note that the synthesis of carboxylated biotin requires energy derived from ATP—a general feature of many activated carriers.

The nucleic acids (DNA and RNA), proteins, and polysaccharides are all polymers that are produced by the repeated addition of a subunit onto one end of a growing chain. The mode of synthesis of each of these macromolecules is outlined in Figure 3–39. As indicated, the condensation step in each case depends on energy from the hydrolysis of a nucleoside triphosphate. And yet, except for the nucleic acids, there are no phosphate groups left in the final product molecules. How, then, are the reactions that release the energy of ATP hydrolysis coupled to polymer synthesis?

For each type of macromolecule, an enzyme-catalyzed pathway exists which resembles that discussed previously for the synthesis of the amino acid glutamine (see Figure 3–33). The principle is exactly the same, in that the OH group that will be removed in the condensation reaction is first activated by forming a high-energy linkage to a second molecule. The mechanisms used to link ATP hydrolysis to the synthesis of proteins and polysaccharides, however, are more complex than that used for

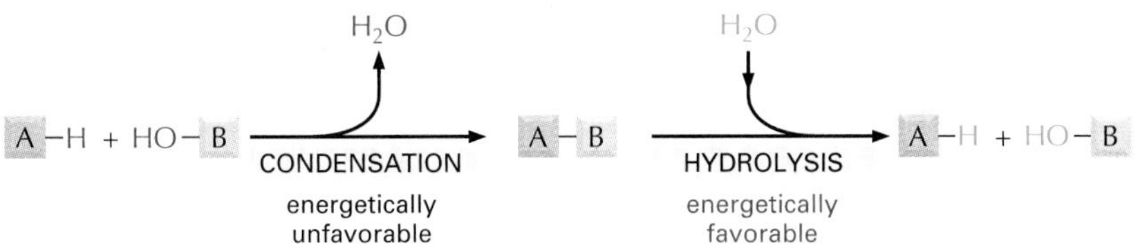

$$A{-}H + HO{-}B \xrightarrow[\text{CONDENSATION}]{-H_2O} A{-}B \xrightarrow[\text{HYDROLYSIS}]{+H_2O} A{-}H + HO{-}B$$

CONDENSATION: energetically unfavorable; HYDROLYSIS: energetically favorable

Figure 3–38 **Condensation and hydrolysis are opposite reactions.** The macromolecules of the cell are polymers that are formed from subunits (or monomers) by a condensation reaction and broken down by hydrolysis. Condensation reactions are all energetically unfavorable.

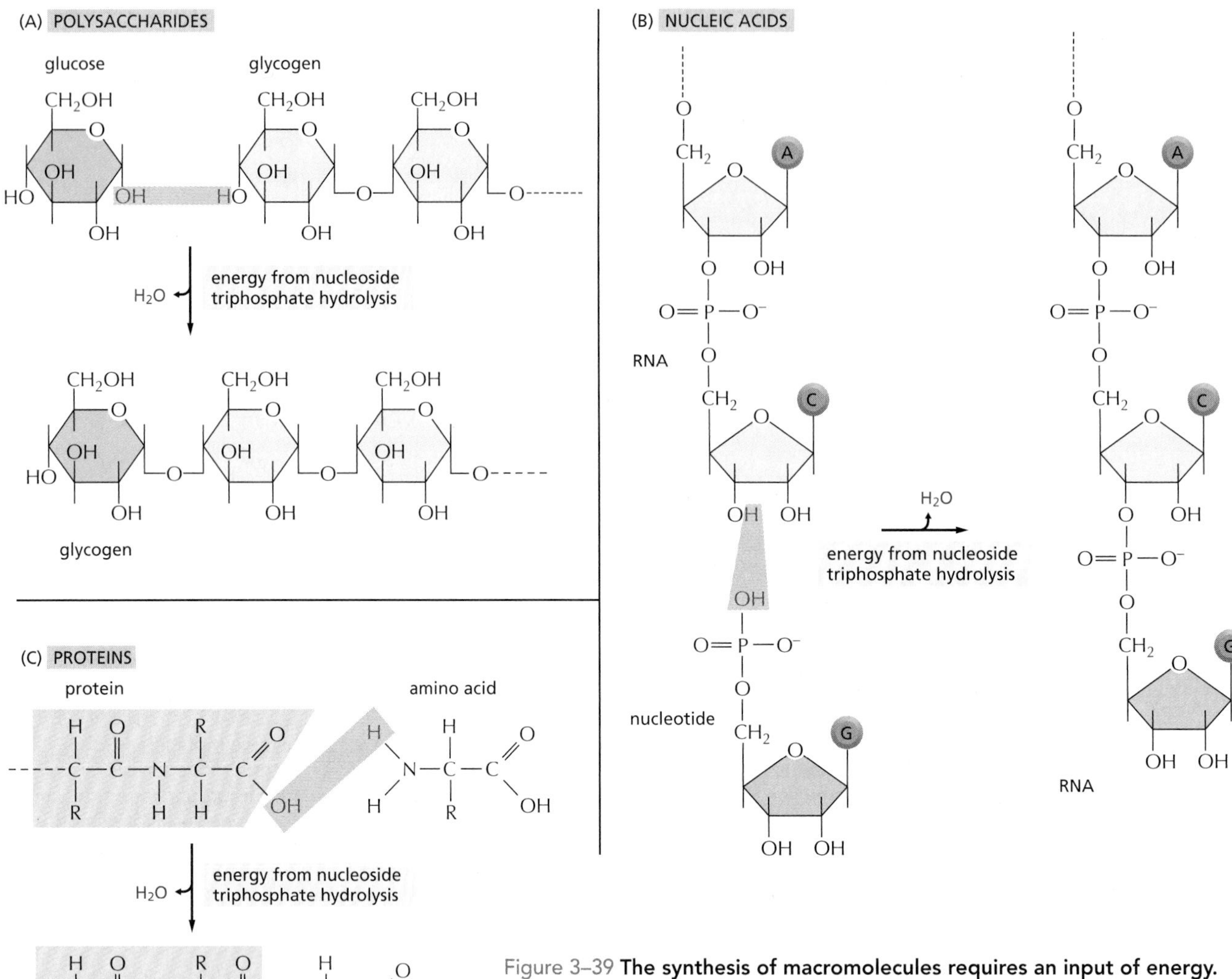

Figure 3–39 **The synthesis of macromolecules requires an input of energy.** Synthesis of a portion of (A) a polysaccharide, (B) a nucleic acid, and (C) a protein is shown here. In each case, synthesis involves a condensation reaction; the atoms involved are shaded in pink. Not shown is the consumption of high-energy nucleoside triphosphates that is required to activate each subunit prior to its addition. In contrast, the reverse reaction—the breakdown of all three types of polymers—occurs through the simple addition of water (hydrolysis).

QUESTION 3–9

Which of the following reactions will occur only if coupled to a second, energetically favorable reaction?

A. glucose + O_2 → CO_2 + H_2O
B. CO_2 + H_2O → glucose + O_2
C. nucleoside triphosphates → DNA
D. nucleotide bases → nucleoside triphosphates
E. ADP + P_i → ATP

glutamine synthesis. In the biosynthetic pathways leading to these macromolecules, a series of high-energy intermediates generates the final high-energy bond that is broken during the condensation step (as discussed in Chapter 7 for protein synthesis).

There are limits to what each activated carrier can do in driving biosynthesis. For example, the ΔG for the hydrolysis of ATP to ADP and inorganic phosphate (P_i) depends on the concentrations of all of the reactants, and under the usual conditions in a cell it is between –11 and –13 kcal/mole. In principle, this hydrolysis reaction can be used to drive an unfavorable reaction with a ΔG of, perhaps, +10 kcal/mole, provided that a suitable reaction path is available. For some biosynthetic reactions, however, even –13 kcal/mole may be insufficient. In these cases, the path of ATP hydrolysis can be altered so that it initially produces AMP and pyrophosphate (PP_i), which is itself then hydrolyzed in solution in a subsequent step (Figure 3–40). The whole process makes available a total ΔG of about –26 kcal/mole. An important biosynthetic reaction that is driven in this way, nucleic acid (polynucleotide) synthesis, is illustrated in Figure 3–41.

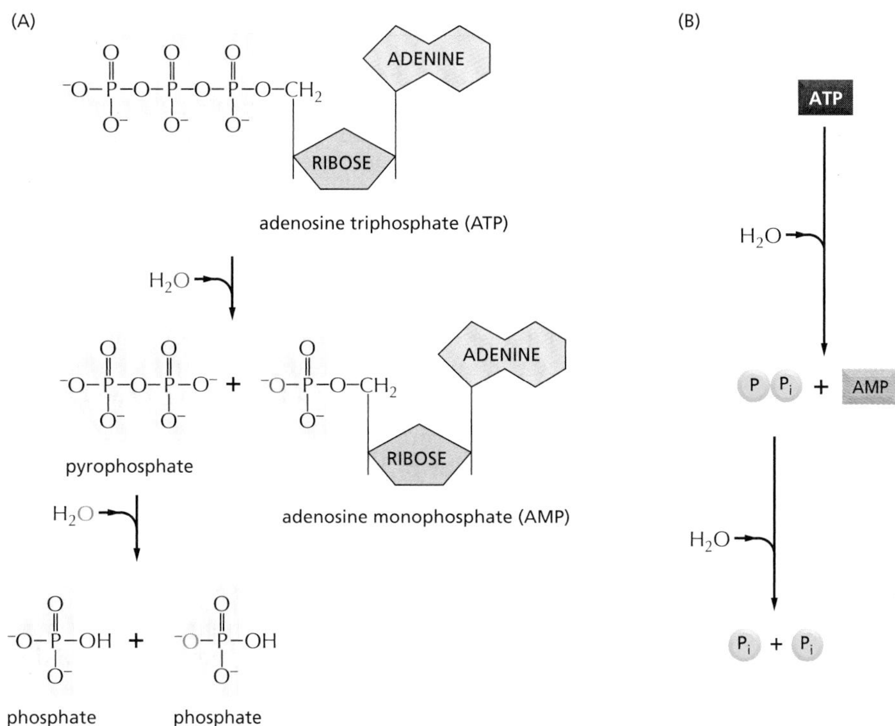

Figure 3–40 **In an alternative route for the hydrolysis of ATP, pyrophosphate is first formed and then hydrolyzed in solution.** This route releases about twice as much free energy as the reaction shown earlier in Figure 3–31. (A) In each of the two successive hydrolysis reactions, an oxygen atom from the participating water molecule is retained in the products, whereas the hydrogen atoms from water form free hydrogen ions, H^+. (B) Overall reaction shown in summary form.

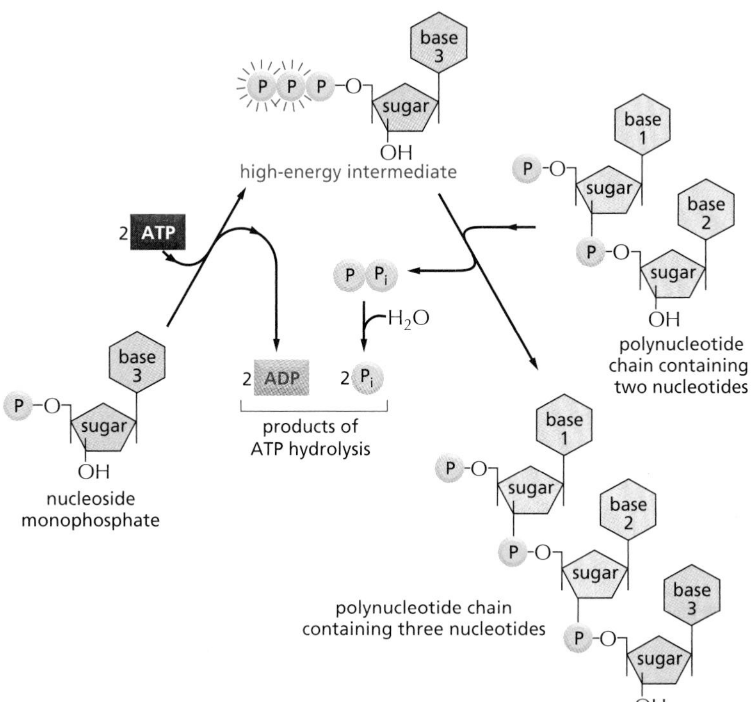

Figure 3–41 **Synthesis of a polynucleotide, RNA or DNA, is a multistep process driven by ATP hydrolysis.** In the first step, a nucleoside monophosphate is activated by the sequential transfer of the terminal phosphate groups from two ATP molecules. The high-energy intermediate formed—a nucleoside triphosphate—exists free in solution until it reacts with the growing end of an RNA or a DNA chain with release of pyrophosphate. Hydrolysis of the latter to inorganic phosphate is highly favorable and helps to drive the overall reaction in the direction of polynucleotide synthesis.

ESSENTIAL CONCEPTS

- Living organisms are able to exist because of a continual input of energy. Part of this energy is used to carry out essential functions—reactions that support cellular metabolism, growth, and reproduction—and the remainder is lost in the form of heat.
- The primary source of energy for most living organisms is the sun. Plants, algae, and photosynthetic bacteria use solar energy to produce organic molecules from carbon dioxide. Animals obtain food by eating plants or by eating animals that feed on plants.
- Each of the many hundreds of chemical reactions that occur in a cell is specifically catalyzed by an enzyme. Large numbers of different enzymes work in sequence to form chains of reactions, called metabolic pathways, each performing a different function in the cell.
- Catabolic reactions break down food molecules through oxidative pathways and release energy. Anabolic reactions generate the many complex molecules needed by the cell, and they require an energy input. In animal cells, both the building blocks and the energy required for the anabolic reactions are obtained by catabolism.
- Enzymes catalyze reactions by binding to particular substrate molecules in a way that lowers the activation energy required for making and breaking specific covalent bonds.
- The rate at which an enzyme catalyzes a reaction depends on how rapidly it finds its substrates and how quickly the product forms and then diffuses away. These rates vary widely from one enzyme to another, and they can be measured after mixing purified enzymes and substrates together under a set of defined conditions.
- The only chemical reactions possible are those that increase the total amount of disorder in the universe. The free-energy change for a reaction, ΔG, measures this disorder, and it must be less than zero for a reaction to proceed spontaneously.
- The free-energy change for a chemical reaction, ΔG, depends on the concentrations of the reacting molecules, and it may be calculated from these concentrations if the equilibrium constant (K) of the reaction (or the standard free-energy change, $\Delta G°$, for the reactants) is known.
- Equilibrium constants govern all of the associations (and dissociations) that occur between macromolecules and small molecules in the cell. The larger the binding energy between two molecules, the larger the equilibrium constant and the more likely that these molecules will be found bound to each other.
- By creating a reaction pathway that couples an energetically favorable reaction to an energetically unfavorable one, enzymes can make otherwise impossible chemical transformations occur.
- A small set of activated carrier molecules, particularly ATP, NADH, and NADPH, plays a central part in these coupling events. ATP carries high-energy phosphate groups, whereas NADH and NADPH carry high-energy electrons.
- Food molecules provide the carbon skeletons for the formation of larger molecules. The covalent bonds of these larger molecules are typically produced in reactions that are coupled to energetically favorable bond changes in activated carrier molecules such as ATP and NADPH.

KEY TERMS

acetyl CoA	free energy, G
activated carrier	free-energy change, ΔG
activation energy	K_M
ADP, ATP	metabolism
anabolism	NAD^+, NADH
biosynthesis	$NADP^+$, NADPH
catabolism	oxidation
catalysis	photosynthesis
catalyst	reduction
coupled reaction	respiration
diffusion	standard free-energy change, $\Delta G°$
entropy	substrate
enzyme	turnover number
equilibrium	V_{max}
equilibrium constant, K	

QUESTIONS

QUESTION 3–10

Which of the following statements are correct? Explain your answers.

A. Some enzyme-catalyzed reactions cease completely if their enzyme is absent.

B. High-energy electrons (such as those found in the activated carriers NADH and NADPH) move faster around the atomic nucleus.

C. Hydrolysis of ATP to AMP can provide about twice as much energy as hydrolysis of ATP to ADP.

D. A partially oxidized carbon atom has a somewhat smaller diameter than a more reduced one.

E. Some activated carrier molecules can transfer both energy and a chemical group to a second molecule.

F. The rule that oxidations release energy, whereas reductions require energy input, applies to all chemical reactions, not just those that occur in living cells.

G. Cold-blooded animals have an energetic disadvantage because they release less heat to the environment than warm-blooded animals do. This slows their ability to make ordered macromolecules.

H. Linking the reaction $X \rightarrow Y$ to a second, energetically favorable reaction $Y \rightarrow Z$ will shift the equilibrium constant of the first reaction.

QUESTION 3–11

Consider the transition of $X \rightarrow Y$ in Figure 3–18. Assume that the only difference between X and Y is the presence of three hydrogen bonds in Y that are absent in X. What is the ratio of X to Y when the reaction is in equilibrium? Approximate your answer by using Table 3–1 (p. 96), with 1 kcal/mole as the energy of each hydrogen bond. If Y instead has six hydrogen bonds that distinguish it from X, how would that change the ratio?

QUESTION 3–12

Protein A binds to protein B to form a complex, AB. A cell contains an equilibrium mixture of protein A at a concentration of 1 μM, protein B at a concentration of 1 μM, and protein AB (produced when A binds to B) also at 1 μM.

A. Referring to Figure 3–19, calculate the equilibrium constant for the reaction $A + B \rightleftharpoons AB$.

B. What would the equilibrium constant be if A, B, and AB were each present in equilibrium at the much lower concentrations of 1 nM each?

C. How many extra hydrogen bonds would be needed to hold A to B at this lower concentration so that a similar portion of the molecules are found in the AB complex? (Remember that each hydrogen bond contributes about 1 kcal/mole.)

QUESTION 3–13

Discuss the following statement: "Whether the ΔG for a reaction is larger, smaller, or the same as $\Delta G°$ depends on the concentration of the compounds that participate in the reaction."

QUESTION 3–14

A. How many ATP molecules could maximally be generated from one molecule of glucose, if the complete oxidation of 1 mole of glucose to CO_2 and H_2O yields 686 kcal of free energy and the useful chemical energy available in the high-energy phosphate bond of 1 mole of ATP is 12 kcal?

B. As we shall see in Chapter 14 (Table 14–1), respiration produces 30 moles of ATP from 1 mole of glucose. Compare this number with your answer in part (A). What is the overall efficiency of ATP production from glucose?

C. If the cells of your body oxidize 1 mole of glucose, by how much would the temperature of your body (assume that your body consists of 75 kg of water) increase if the heat were not dissipated into the environment? [Recall that a kilocalorie (kcal) is defined as that amount of energy that heats 1 kg of water by 1°C.]

D. What would the consequences be if the cells of your body could convert the energy in food substances with only 20% efficiency? Would your body—as it is presently constructed—work just fine, overheat, or freeze?

E. A resting human hydrolyzes about 40 kg of ATP every 24 hours. The oxidation of how much glucose would produce this amount of energy? (Hint: Look up the structure of ATP in Figure 2–23 to calculate its molecular weight; the atomic weights of H, C, N, O, and P are 1, 12, 14, 16, and 31, respectively.)

QUESTION 3–15

A prominent scientist claims to have isolated mutant cells that can convert 1 molecule of glucose into 57 molecules of ATP. Should this discovery be celebrated, or do you suppose that something might be wrong with it? Explain your answer.

QUESTION 3–16

In a simple reaction $A \rightleftharpoons A^*$, a molecule is interconvertible between two forms that differ in standard free energy $G°$ by 4.3 kcal/mole, with A* having the higher $G°$.

A. Use Table 3–1 (p. 96) to find how many more molecules will be in state A* compared with state A at equilibrium.

B. If an enzyme lowered the activation energy of the reaction by 2.8 kcal/mole, how would the ratio of A to A* change?

QUESTION 3–17

A reaction in a single-step biosynthetic pathway that converts a metabolite into a particularly vicious poison in a mushroom is energetically highly unfavorable (metabolite $\rightleftharpoons$ poison). The reaction is normally driven by ATP hydrolysis. Assume that a mutation in the enzyme that catalyzes the reaction prevents it from utilizing ATP, but still allows it to catalyze the reaction.

A. Do you suppose it might be safe for you to eat the mutant mushroom? Base your answer on an estimation of how much less poison the organism would produce, assuming the reaction is in equilibrium and most of the energy stored in ATP is used to drive the unfavorable reaction.

B. Would your answer be different for another mutant mushroom whose enzyme couples the reaction to ATP hydrolysis but works 100 times more slowly?

QUESTION 3–18

Consider the effects of two enzymes, A and B. Enzyme A catalyzes the reaction

$$ATP + GDP \rightleftharpoons ADP + GTP$$

whereas enzyme B catalyzes the reaction

$$NADH + NADP^+ \rightleftharpoons NAD^+ + NADPH$$

Discuss whether the enzymes would be beneficial or detrimental to cells.

QUESTION 3–19

Discuss the following statement: "Enzymes and heat are alike in that both can speed up reactions that—although thermodynamically feasible—do not occur at an appreciable rate because they require a high activation energy. Diseases that can be treated by the careful application of heat, such as by the ingestion of hot chicken soup, are therefore likely to be due to the insufficient function of an enzyme."

QUESTION 3–20

The curve shown in Figure E3–25/3–24 is described by the Michaelis–Menten equation:

$$\text{rate} = V_{max}\,[S]/([S] + K_M)$$

Can you convince yourself that the features qualitatively described in the text are accurately represented by this equation? In particular, how can the equation be simplified when the substrate concentration is in one of the following ranges: (A) the substrate concentration [S] is much smaller than the K_M, (B) the substrate concentration [S] equals the K_M, and (C) the substrate concentration [S] is much larger than the K_M?

QUESTION 3–21

The rate of a simple enzyme reaction is given by the standard Michaelis–Menten equation:

$$\text{rate} = V_{max}\,[S]/([S] + K_M)$$

If the V_{max} of an enzyme is 100 µmole/sec and the K_M is 1 mM, at what substrate concentration is the rate 50 µmole/sec? Plot a graph of rate versus substrate (S) concentration for [S] = 0 to 10 mM. Convert this to a plot of 1/rate versus 1/[S]. Why is the latter plot a straight line?

QUESTION 3–22

Select the correct options in the following and explain your choices. If [S] is much smaller than K_M, the active site of the enzyme is mostly occupied/unoccupied. If [S] is very much greater than K_M, the reaction rate is limited by the enzyme/substrate concentration.

QUESTION 3–23

A. The reaction rates of the reaction $S \rightarrow P$ catalyzed by enzyme E were determined under conditions such that only very little product was formed. The following data were measured:

Substrate concentration (µM)	Reaction rate (µmole/min)
0.08	0.15
0.12	0.21
0.54	0.7
1.23	1.1
1.82	1.3
2.72	1.5
4.94	1.7
10.00	1.8

Plot the above data as a graph. Use this graph to estimate the K_M and the V_{max} for this enzyme.

B. Recall from the How We Know essay (pp. 101–103) that to determine these values more precisely, a trick is generally used in which the Michaelis–Menten equation is transformed so that it is possible to plot the data as a straight line. A simple rearrangement yields

$$1/\text{rate} = (K_M/V_{max})\,(1/[S]) + 1/V_{max}$$

which is an equation of the form $y = ax + b$. Calculate 1/rate and 1/[S] for the data given in part (A) and then plot 1/rate versus 1/[S] as a new graph. Determine K_M and V_{max} from the intercept of the line with the axis, where 1/[S] = 0, combined with the slope of the line. Do your results agree with the estimates made from the first graph of the raw data?

C. It is stated in part (A) that only very little product was formed under the reaction conditions. Why is this important?

D. Assume the enzyme is regulated such that upon phosphorylation its K_M increases by a factor of 3 without changing its V_{max}. Is this an activation or inhibition? Plot the data you would expect for the phosphorylated enzyme in both the graph for (A) and the graph for (B).

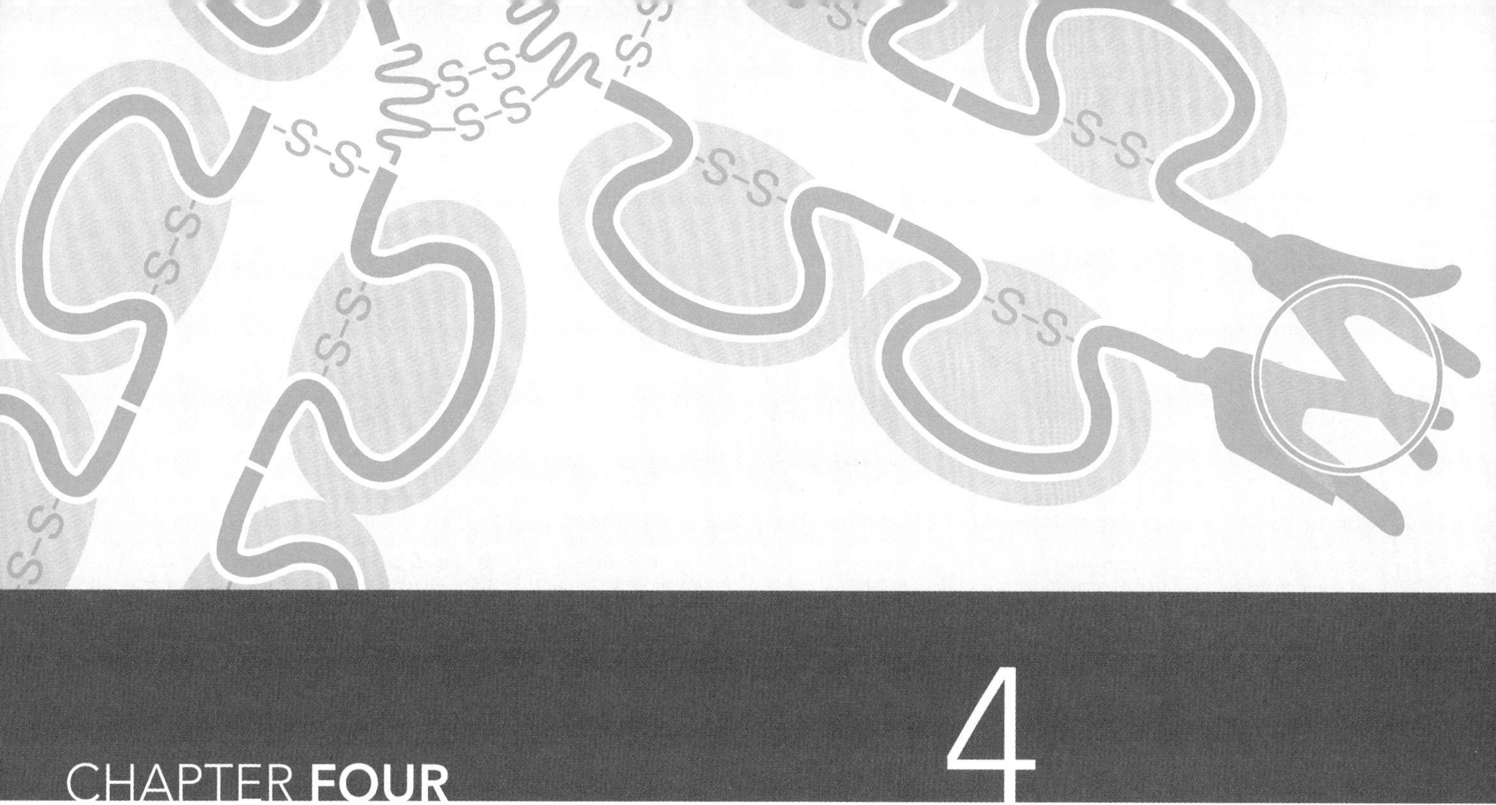

CHAPTER **FOUR** 4

Protein Structure and Function

When we look at a cell through a microscope or analyze its electrical or biochemical activity, we are, in essence, observing the handiwork of proteins. Proteins are the building blocks from which cells are assembled, and they constitute most of the cell's dry mass. But in addition to providing the cell with shape and structure, proteins also execute nearly all its myriad functions. **Enzymes** promote intracellular chemical reactions by providing intricate molecular surfaces, contoured with particular bumps and crevices, that can cradle or exclude specific molecules. Proteins embedded in the plasma membrane form the channels and pumps that control the passage of nutrients and other small molecules into and out of the cell. Other proteins carry messages from one cell to another, or act as signal integrators that relay information from the plasma membrane to the nucleus of individual cells. Yet others serve as tiny molecular machines with moving parts: some proteins, such as kinesin, propel organelles through the cytoplasm; others, such as helicases, pry open double-stranded DNA molecules. Specialized proteins also act as antibodies, toxins, hormones, antifreeze molecules, elastic fibers, or luminescence generators. Before we can hope to understand how genes work, how muscles contract, how nerves conduct electricity, how embryos develop, or how our bodies function, we must understand proteins.

The multiplicity of functions carried out by proteins (Panel 4–1, p. 120) arises from the huge number of different shapes they adopt. So we begin our description of these remarkable macromolecules by discussing their three-dimensional structures and the properties that these structures confer. We next look at how they work: how enzymes catalyze chemical

THE SHAPE AND STRUCTURE OF PROTEINS

HOW PROTEINS WORK

HOW PROTEINS ARE CONTROLLED

HOW PROTEINS ARE STUDIED

PANEL 4–1 A few examples of some general protein functions

ENZYME

function: Catalyzes covalent bond breakage or formation.

examples: Living cells contain thousands of different enzymes, each of which catalyzes (speeds up) one particular reaction. Examples include: *tryptophan synthetase*—makes the amino acid tryptophan; *pepsin*—degrades dietary proteins in the stomach; *ribulose bisphosphate carboxylase*—helps convert carbon dioxide into sugars in plants; *DNA polymerase*—copies DNA; *protein kinase*—adds a phosphate group to a protein molecule.

STRUCTURAL PROTEIN

function: Provides mechanical support to cells and tissues.

examples: Outside cells, *collagen* and *elastin* are common constituents of extracellular matrix and form fibers in tendons and ligaments. Inside cells, *tubulin* forms long, stiff microtubules and *actin* forms filaments that underlie and support the plasma membrane; *α-keratin* forms fibers that reinforce epithelial cells and is the major protein in hair and horn.

TRANSPORT PROTEIN

function: Carries small molecules or ions.

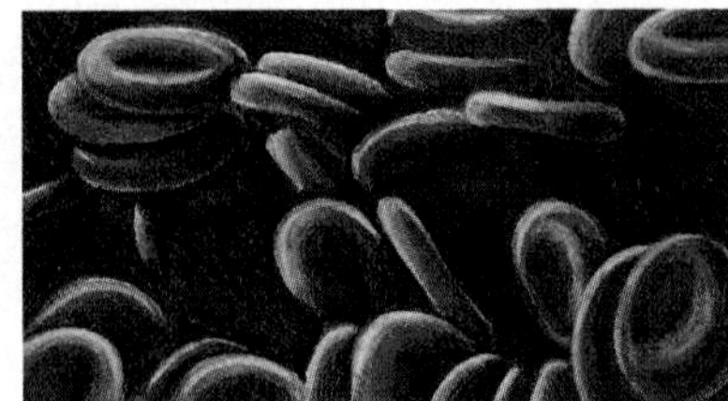

examples: In the bloodstream, *serum albumin* carries lipids, *hemoglobin* carries oxygen, and *transferrin* carries iron. Many proteins embedded in cell membranes transport ions or small molecules across the membrane. For example, the bacterial protein *bacteriorhodopsin* is a light-activated proton pump that transports H^+ ions out of the cell; the *glucose carrier* shuttles glucose into and out of liver cells; and a *Ca^{2+} pump* in muscle cells pumps the calcium ions needed to trigger muscle contraction into the endoplasmic reticulum, where they are stored.

MOTOR PROTEIN

function: Generates movement in cells and tissues.

examples: *Myosin* in skeletal muscle cells provides the motive force for humans to move; *kinesin* interacts with microtubules to move organelles around the cell; *dynein* enables eucaryotic cilia and flagella to beat.

STORAGE PROTEIN

function: Stores small molecules or ions.

examples: Iron is stored in the liver by binding to the small protein *ferritin*; *ovalbumin* in egg white is used as a source of amino acids for the developing bird embryo; *casein* in milk is a source of amino acids for baby mammals.

SIGNAL PROTEIN

function: Carries signals from cell to cell.

examples: Many of the hormones and growth factors that coordinate physiological function in animals are proteins; *insulin*, for example, is a small protein that controls glucose levels in the blood; *netrin* attracts growing nerve cells in a specific direction in a developing embryo; *nerve growth factor (NGF)* stimulates some types of nerve cells to grow axons; *epidermal growth factor (EGF)* stimulates the growth and division of epithelial cells.

RECEPTOR PROTEIN

function: Detects signals and transmits them to the cell's response machinery.

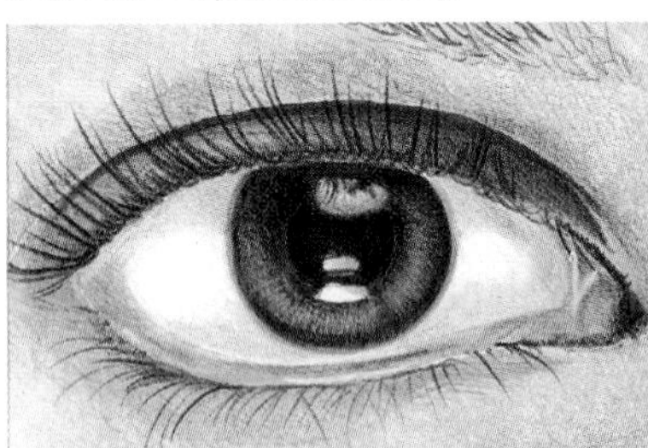

examples: *Rhodopsin* in the retina detects light; the *acetylcholine receptor* in the membrane of a muscle cell receives chemical signals released from a nerve ending; the *insulin receptor* allows a liver cell to respond to the hormone insulin by taking up glucose; the *adrenergic receptor* on heart muscle increases the rate of heartbeat when it binds to adrenaline.

GENE REGULATORY PROTEIN

function: Binds to DNA to switch genes on or off.

examples: The *lactose repressor* in bacteria silences the genes for the enzymes that degrade the sugar lactose; many different *homeodomain proteins* act as genetic switches to control development in multicellular organisms, including humans.

SPECIAL-PURPOSE PROTEIN

function: Highly variable.

examples: Organisms make many proteins with highly specialized properties. These molecules illustrate the amazing range of functions that proteins can perform. The *antifreeze proteins* of Arctic and Antarctic fishes protect their blood against freezing; *green fluorescent protein* from jellyfish emits a green light; *monellin*, a protein found in an African plant, has an intensely sweet taste; mussels and other marine organisms secrete *glue proteins* that attach them firmly to rocks, even when immersed in seawater.

reactions, how some proteins act as molecular switches, and how others generate coherent movement. We then examine how cells control the activity and location of the proteins they contain. Finally, we present a brief description of the techniques that biologists use to work with proteins, including methods for purifying them—from tissues or cultured cells—and determining their structures.

THE SHAPE AND STRUCTURE OF PROTEINS

From a chemical point of view, proteins are by far the most structurally complex and functionally sophisticated molecules known. This is perhaps not surprising, considering that the structure and activity of each protein has been developed and fine-tuned over billions of years of evolutionary history. We start by considering how the position of each amino acid in the long string of amino acids that forms a protein determines its three-dimensional shape, a structure stabilized by noncovalent interactions between different parts of the molecule. Understanding protein structure at the atomic level will allow us to describe how the precise shape of each protein determines its function in the cell.

The Shape of a Protein Is Specified by Its Amino Acid Sequence

Proteins, as you may recall from Chapter 2, are assembled from a set of 20 different amino acids, each with different chemical properties. A **protein** molecule is made from a long chain of these amino acids, each linked to its neighbor through a covalent peptide bond (Figure 4–1). Proteins are therefore referred to as **polypeptides** or **polypeptide chains**. In each type of protein, the amino acids are present in a unique order, called the **amino acid sequence**, which is exactly the same from one molecule of that protein to the next. One molecule of insulin, for example, has the same amino acid sequence as every other molecule of insulin. Many thousands of different proteins have been identified; each has its own distinct amino acid sequence.

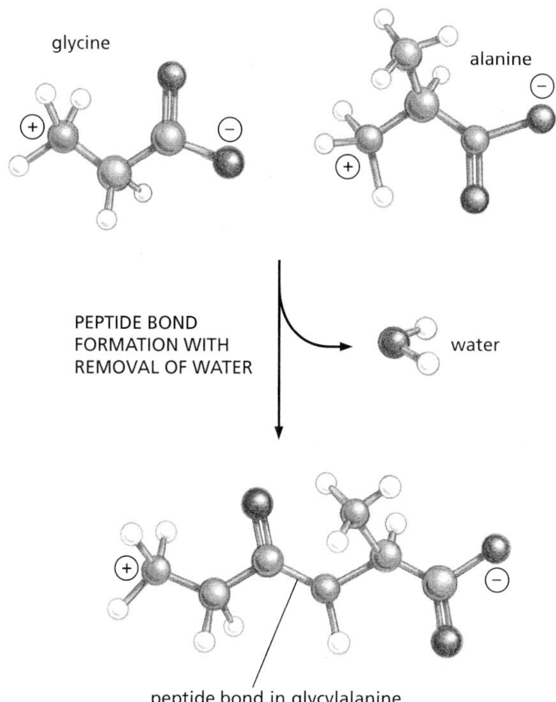

Figure 4–1 **Amino acids are linked together by peptide bonds.** A covalent peptide bond forms when the carbon atom of the carboxyl group of one amino acid shares electrons with the nitrogen atom (*blue*) from the amino group of a second amino acid. A molecule of water is generated during this condensation reaction. In this diagram, carbon atoms are *gray*, nitrogen *blue*, oxygen *red*, and hydrogen *white*.

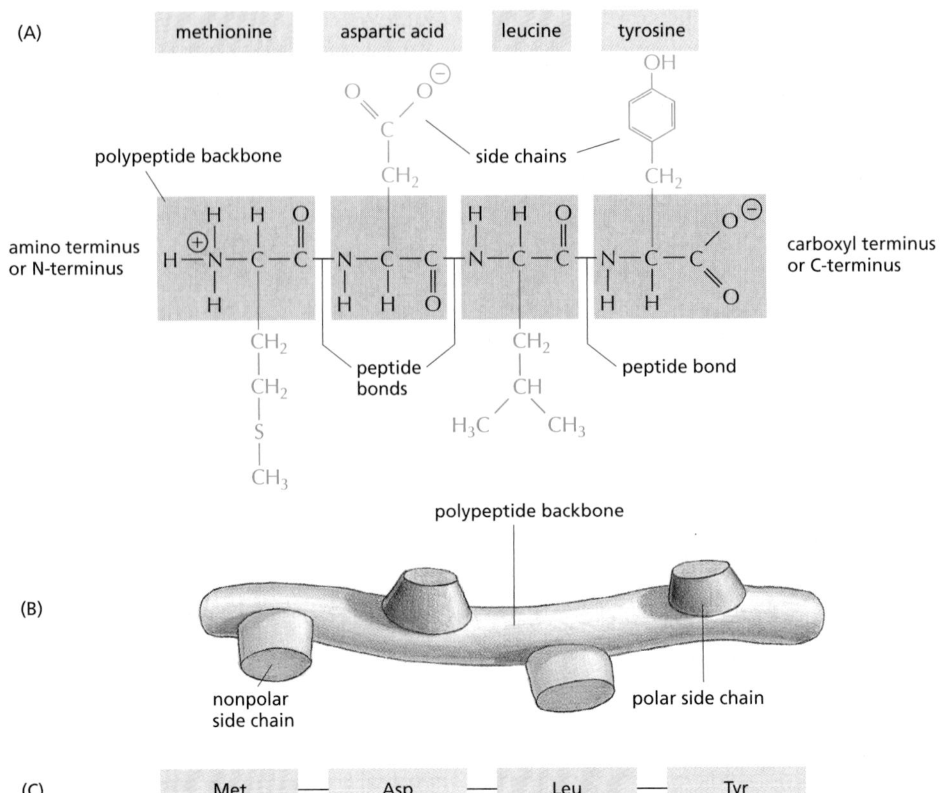

Figure 4–2 **A protein is made of amino acids linked together into a polypeptide chain.** (A) Amino acids linked together by the reaction shown in Figure 4–1 form a polypeptide backbone of repeating structure (*gray boxes*), from which the side chains of the amino acids project. A small polypeptide of just four amino acids is shown here. Proteins are typically made up of chains of several hundred amino acids. The two ends of a polypeptide chain are chemically different: the end that carries the free amino group (NH_3^+, also written NH_2) is called the amino, or N-, terminus; the end carrying the free carboxyl group (COO^-, also written COOH) is the carboxyl, or C-, terminus. (B) The positions of the different chemically distinct side chains, for example, polar and nonpolar side chains, give each protein its individual properties. (C) The amino acid sequence of a protein is always presented in the N to C direction, reading from left to right.

Each polypeptide chain consists of a backbone that supports the different amino acid side chains. The **polypeptide backbone** is made from the repeating sequence of the core atoms of the amino acids that form the chain. Projecting from this repetitive backbone are any of the 20 different amino acid **side chains**—the parts of the amino acids that are not involved in forming the peptide bond (Figure 4–2). These side chains give each amino acid its unique properties. Some are nonpolar and hydrophobic ("water-fearing"), some are negatively or positively charged, some are chemically reactive, and so on. The atomic structures of the 20 amino acids in proteins are presented in Panel 2–5 (pp. 72–73), and a brief list of amino acids, with their abbreviations, is provided in Figure 4–3.

Long polypeptide chains are very flexible: many of the covalent bonds that link carbon atoms in an extended chain of amino acids allow free rotation of the atoms they join. Thus proteins can in principle fold in an

AMINO ACID			SIDE CHAIN
Aspartic acid	Asp	D	negative
Glutamic acid	Glu	E	negative
Arginine	Arg	R	positive
Lysine	Lys	K	positive
Histidine	His	H	positive
Asparagine	Asn	N	uncharged polar
Glutamine	Gln	Q	uncharged polar
Serine	Ser	S	uncharged polar
Threonine	Thr	T	uncharged polar
Tyrosine	Tyr	Y	uncharged polar

POLAR AMINO ACIDS
(hydrophilic)

AMINO ACID			SIDE CHAIN
Alanine	Ala	A	nonpolar
Glycine	Gly	G	nonpolar
Valine	Val	V	nonpolar
Leucine	Leu	L	nonpolar
Isoleucine	Ile	I	nonpolar
Proline	Pro	P	nonpolar
Phenylalanine	Phe	F	nonpolar
Methionine	Met	M	nonpolar
Tryptophan	Trp	W	nonpolar
Cysteine	Cys	C	nonpolar

NONPOLAR AMINO ACIDS
(hydrophobic)

Figure 4–3 **Twenty different amino acids are found in proteins.** Both three-letter and one-letter abbreviations are given. There are equal numbers of polar and nonpolar side chains. For the atomic structures of the amino acids see Panel 2–5 (pp. 72–73).

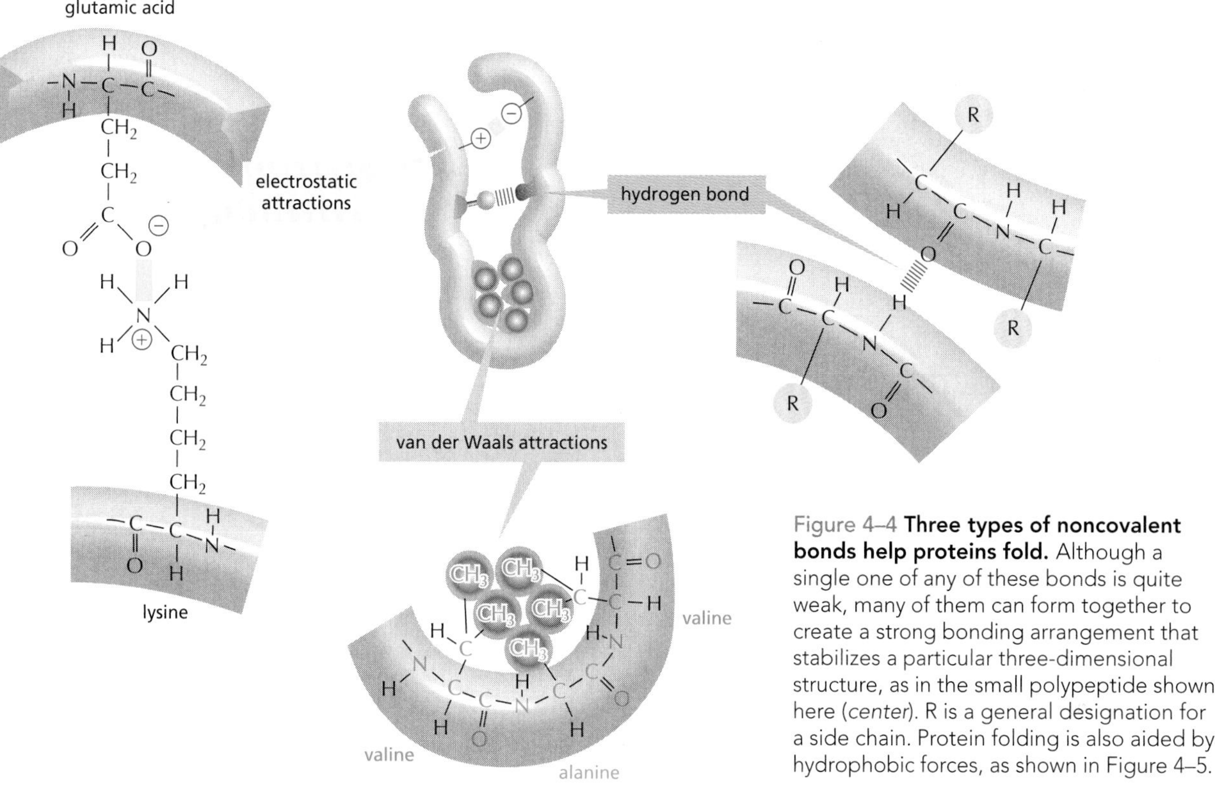

Figure 4–4 **Three types of noncovalent bonds help proteins fold.** Although a single one of any of these bonds is quite weak, many of them can form together to create a strong bonding arrangement that stabilizes a particular three-dimensional structure, as in the small polypeptide shown here (*center*). R is a general designation for a side chain. Protein folding is also aided by hydrophobic forces, as shown in Figure 4–5.

enormous number of ways. Each folded chain is constrained by many different sets of weak noncovalent bonds that form within proteins. These bonds involve atoms in the polypeptide backbone as well as atoms in the amino acid side chains. The noncovalent bonds that help proteins maintain their shape include hydrogen bonds, electrostatic attractions, and van der Waals attractions, which are described in Chapter 2 (see Panel 2–7, pp. 76–77). Because individual noncovalent bonds are much weaker than covalent bonds, it takes many noncovalent bonds to hold two regions of a polypeptide chain together tightly. The stability of each folded shape will therefore be affected by the combined strength of large numbers of noncovalent bonds (Figure 4–4).

A fourth weak force, hydrophobic interaction, also plays a central role in determining the shape of a protein. In an aqueous environment, hydrophobic molecules, including the nonpolar side chains of particular amino acids, tend to be forced together to minimize their disruptive effect on the hydrogen-bonded network of the surrounding water molecules (see Panel 2–2, pp. 66–67). Therefore, an important factor governing the folding of any protein is the distribution of its polar and nonpolar amino acids. The nonpolar (hydrophobic) side chains—which belong to amino acids such as phenylalanine, leucine, valine, and tryptophan—tend to cluster in the interior of the folded protein (just as hydrophobic oil droplets coalesce to form one large droplet). Tucked away inside the folded protein, hydrophobic side chains can avoid contact with the aqueous cytosol that surrounds them inside a cell. In contrast, polar side chains—such as those belonging to arginine, glutamine, and histidine—tend to arrange themselves near the outside of the folded protein, where they can form hydrogen bonds with water and with other polar molecules (Figure 4–5). When polar amino acids are buried within the protein, they are usually hydrogen-bonded to other polar amino acids or to the polypeptide backbone (Figure 4–6).

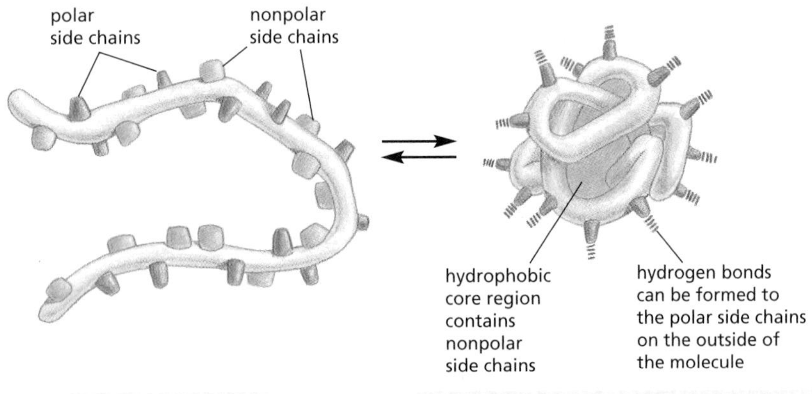

Figure 4–5 **Hydrophobic forces help proteins fold into compact conformations.** The polar amino acid side chains tend to fall on the outside of the folded protein, where they can interact with water; the nonpolar amino acid side chains are buried on the inside to form a highly packed hydrophobic core of atoms that are hidden from water. In this very schematic drawing, the protein contains only about 30 amino acids.

Proteins Fold into a Conformation of Lowest Energy

Each type of protein has a particular three-dimensional structure, which is determined by the order of the amino acids in its chain. The final folded structure, or **conformation**, adopted by any polypeptide chain is determined by energetic considerations: a protein generally folds into the shape in which the free energy (*G*) is minimized (see p. 91). Protein folding has been studied in the laboratory using highly purified proteins. A protein can be unfolded, or *denatured*, by treatment with solvents that disrupt the noncovalent interactions holding the folded chain together. This treatment converts the protein into a flexible polypeptide chain that has lost its natural shape. When the denaturing solvent is removed, the protein often refolds spontaneously, or *renatures*, into its original conformation (Figure 4–7). The fact that a renatured protein can, on its own, regain the correct conformation indicates that all the information necessary to specify the three-dimensional shape of a protein is contained in its amino acid sequence.

Each protein normally folds into a single stable conformation. This conformation, however, often changes slightly when the protein interacts with other molecules in the cell. This change in shape is crucial to the function of the protein, as we shall see later in this chapter.

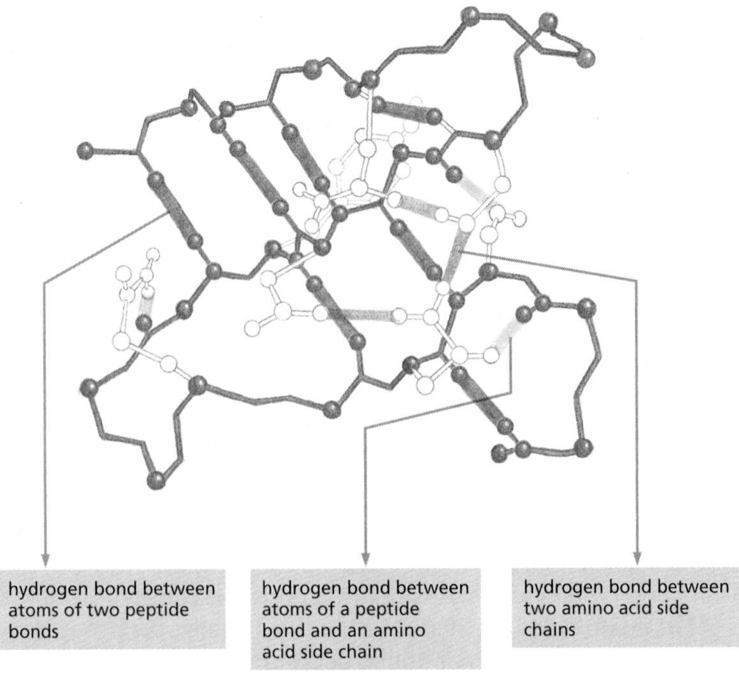

Figure 4–6 **Hydrogen bonds within a protein molecule help stabilize its folded shape.** Large numbers of hydrogen bonds form between adjacent regions of the folded polypeptide chain. The structure depicted is a portion of the enzyme lysozyme. The polypeptide backbone is colored *green*. Hydrogen bonds between backbone atoms are shown in *red*; those between atoms of a peptide bond and a side chain in *yellow*; and those between atoms of two side chains in *blue*. Note that the same amino acid side chain can make several hydrogen bonds. (After C.K. Mathews, K.E. van Holde, and K.G. Ahern, Biochemistry, 3rd ed. San Francisco: Benjamin Cummings, 2000.)

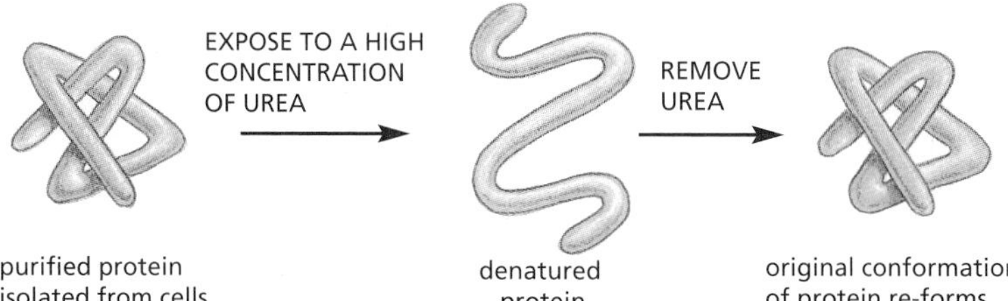

Figure 4–7 **Denatured proteins can recover their natural shapes.** This type of experiment demonstrates that the conformation of a protein is determined solely by its amino acid sequence. Renaturation works best for small proteins.

When proteins fold incorrectly, they sometimes form aggregates that can damage cells and even whole tissues. Aggregated proteins underlie a number of neurodegenerative disorders, including Alzheimer's disease and Huntington's disease. Prion diseases—such as scrapie in sheep, bovine spongiform encephalopathy (BSE, or "mad cow" disease) in cattle, and Creutzfeldt–Jacob disease (CJD) in humans—are also caused by protein aggregates. The prion protein, PrP, can adopt a misfolded form that is considered "infectious" because it can convert properly folded PrP proteins in the infected brain into the abnormal conformation (Figure 4–8). This allows the misfolded form of PrP to spread rapidly from cell to cell, causing the death of the infected animal or human.

Although a protein chain can fold into its correct conformation without outside help, protein folding in a living cell is generally assisted by special proteins called *molecular chaperones*. These proteins bind to partly folded chains and help them to fold along the most energetically favorable pathway, as we shall discuss in Chapter 7. Chaperones are vital in the crowded conditions of the cytoplasm, because they prevent newly synthesized protein chains from associating with the wrong partners. Nevertheless, the final three-dimensional shape of the protein is still specified by its amino acid sequence; chaperones merely make the folding process more efficient and reliable.

QUESTION 4–1

Urea used in the experiment shown in Figure 4–7 is a molecule that disrupts the hydrogen-bonded network of water molecules. Why might high concentrations of urea unfold proteins? The structure of urea is shown here.

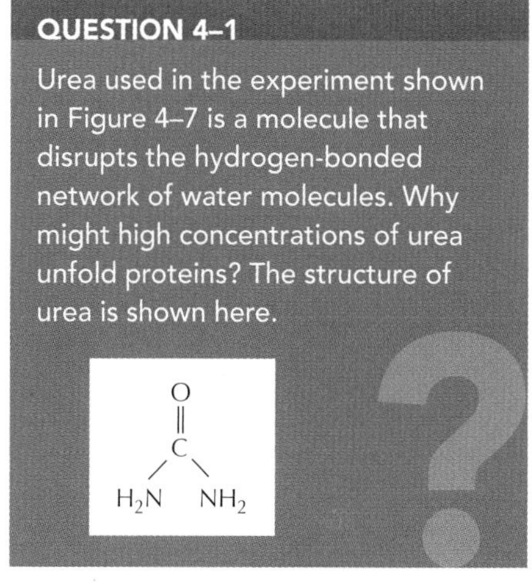

Proteins Come in a Wide Variety of Complicated Shapes

Proteins are the most structurally diverse macromolecules in the cell. Although they range in size from about 30 amino acids to more than 10,000, the vast majority of proteins are between 50 and 2000 amino acids long. Proteins can be globular or fibrous; they can form filaments, sheets, rings, or spheres. Figure 4–9 presents a sampling of proteins whose exact structures are known. We will encounter many of these proteins later in this chapter and throughout the book.

Resolving a protein's structure often begins with determining its amino acid sequence, a task that can be accomplished in several ways. For many years, protein sequencing was accomplished by directly analyzing the amino acids in the purified protein; the first protein sequenced was the hormone *insulin*, in 1955. Today we can determine the order of amino acids in a protein much more easily by sequencing the gene that encodes it (discussed in Chapter 10). Once the order of the nucleotides in the DNA that encodes a protein is known, this information can be converted into an amino acid sequence by applying the genetic code (discussed in Chapter 7). The amino acid sequences of millions of proteins have already been determined in this way, and they have been collected into vast electronic databases that allow users to obtain the amino acid sequence of any protein almost instantaneously.

(A) prion protein can adopt an abnormal, misfolded form

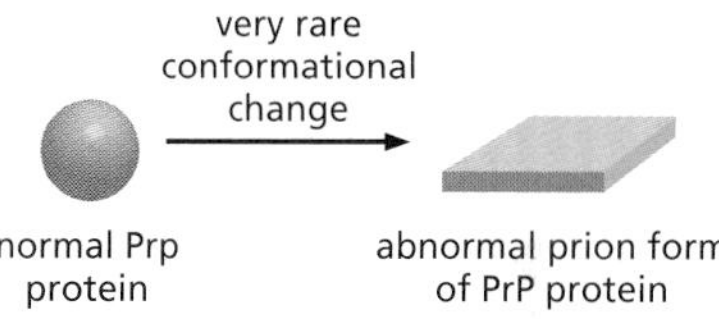

(B) misfolded protein can induce formation of protein aggregates

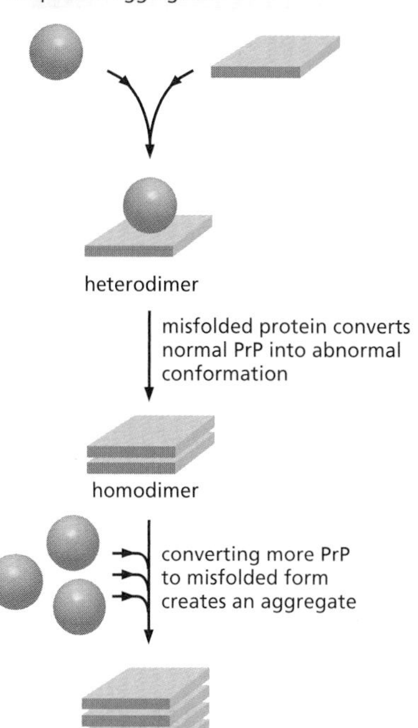

Figure 4–8 **Prion diseases are caused by rare proteins whose misfolding is infectious.** The mammalian protein PrP is the best known of these proteins, but other examples are known. (A) The protein undergoes a rare conformational change to give an abnormally folded prion form. (B) The abnormal form causes the conversion of normal PrP proteins in the host's brain into the misfolded form, which forms protein aggregates that disrupt brain function and cause disease.

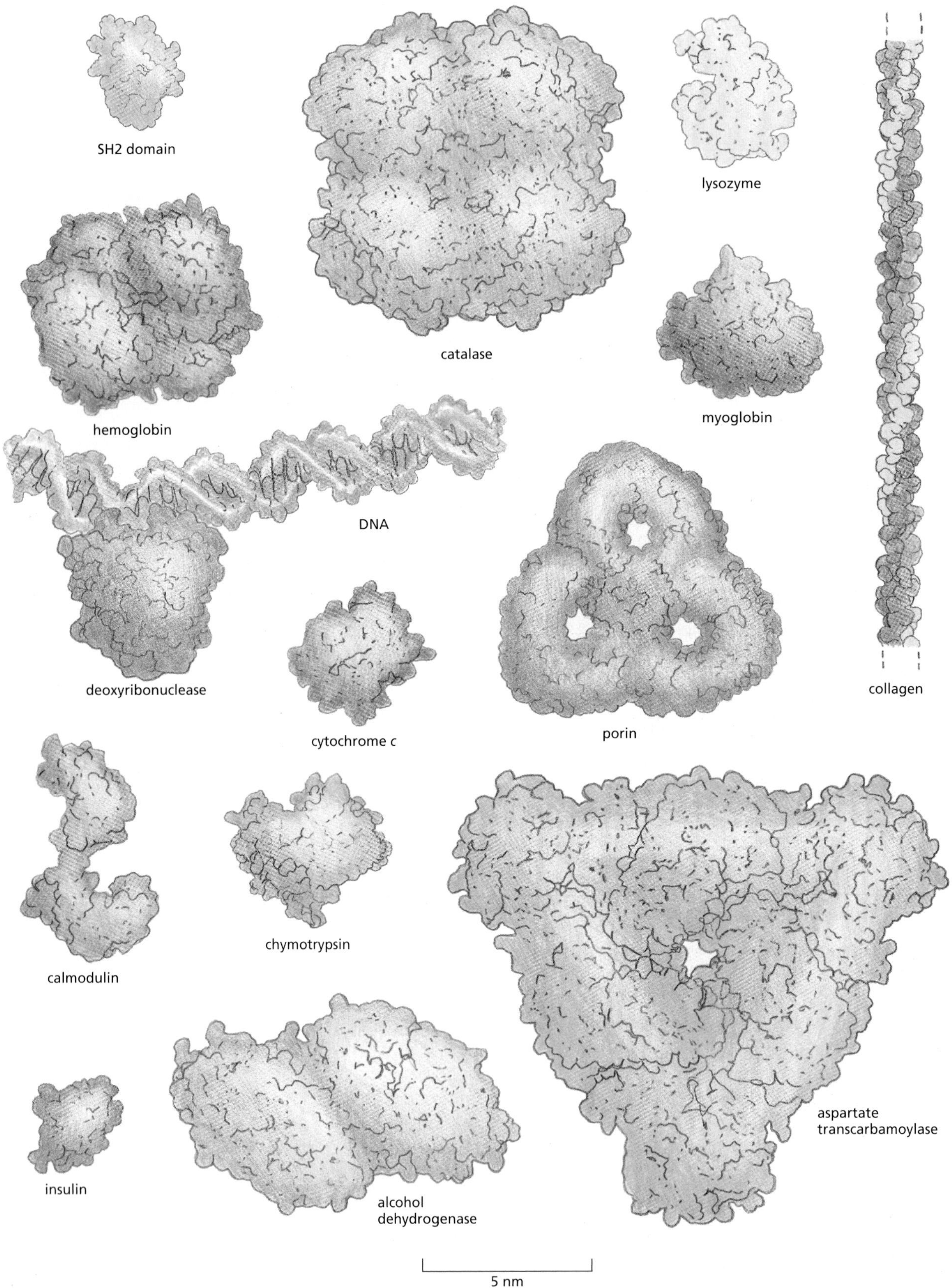

Figure 4–9 **Proteins come in a variety of shapes and sizes.** Each folded polypeptide is shown as a space-filling model, represented at the same scale. In the *top left corner* is the SH2 domain, which is featured in greater detail in Panel 4–2 (pp. 128–129). For comparison, part of a DNA molecule (*gray*) bound to a protein is illustrated. (After David S. Goodsell, Our Molecular Nature. New York: Springer-Verlag, 1996. With permission from Springer Science and Business Media.)

Although all the information required for a polypeptide chain to fold is contained in its amino acid sequence, we have not yet learned how to reliably predict a protein's detailed three-dimensional conformation—the spatial arrangement of its atoms—from its sequence alone. At present, the only way to discover the precise folding pattern of any protein is by experiment, using either X-ray crystallography or nuclear magnetic resonance (NMR) spectroscopy, as we discuss later in the chapter. So far, the structures of about 20,000 different proteins have been completely analyzed by these techniques. Most have a three-dimensional conformation so intricate and irregular that their structure would require an entire chapter to describe in detail.

Because the structure of a large protein can be overwhelming to look at, we will illustrate the intricacies of protein conformation by examining the structure of a smaller *protein domain*. As we discuss shortly, most proteins are formed from multiple domains, each folding into a compact three-dimensional structure. In Panel 4–2 (pp. 128–129), we present four different depictions of SH2, a protein domain that—as a part of proteins involved in cell signaling—has important functions in eucaryotic cells. Built from a string of 100 amino acids, the structure is displayed as (A) a polypeptide backbone model, (B) a ribbon model, (C) a wire model that includes the amino acid side chains, and (D) a space-filling model. As indicated in the panel, each model emphasizes different features of the polypeptide. The three horizontal rows show the SH2 domain in different orientations, and the images are colored to distinguish the path of the polypeptide chain, from its N-terminus (*purple*) to its C-terminus (*red*). We will describe the different structural elements in this protein domain shortly.

From Panel 4–2, we can clearly see how amazingly complex protein conformation is, even for a small domain like SH2. To visualize such complicated structures, scientists have developed various graphical and computer-based tools that generate a variety of images of a protein, some of which are depicted in Panel 4–2. These images can be displayed on a screen and rotated to view all aspects of the structure (Movie 4.1). In addition, describing and presenting such complex protein structures is made easier by recognizing that several common folding patterns underlie these conformations, as we discuss next.

The α Helix and the β Sheet Are Common Folding Patterns

When the three-dimensional structures of many different protein molecules are compared, it becomes clear that, although the overall conformation of each protein is unique, two regular folding patterns are often present. Both were discovered more than 50 years ago from studies of hair and silk. The first folding pattern to be discovered, called the **α helix**, was found in the protein *α-keratin*, which is abundant in skin and its derivatives—such as hair, nails, and horns. Within a year of the discovery of the α helix, a second folded structure, called a **β sheet**, was found in the protein *fibroin*, the major constituent of silk. (Biologists often use Greek letters to name their discoveries, with the first example receiving the designation α, the second β, and so on.)

These two folding patterns are particularly common because they result from hydrogen bonds that form between the N–H and C=O groups in the polypeptide backbone. Because the amino acid side chains are not involved in forming these hydrogen bonds, α helices and β sheets can be generated by many different amino acid sequences. In each case, the protein chain adopts a regular, repeating form or motif. These structural features, and the shorthand cartoon symbols that are often used to represent them in models of protein structures, are presented in Figure 4–10.

PANEL 4–2 Four different ways of depicting the small SH2 protein domain

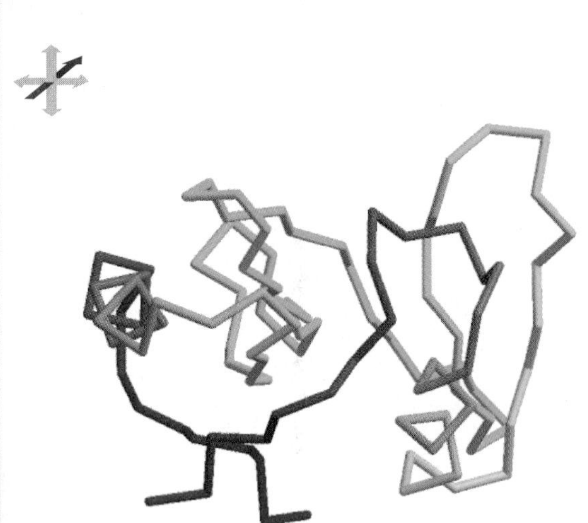

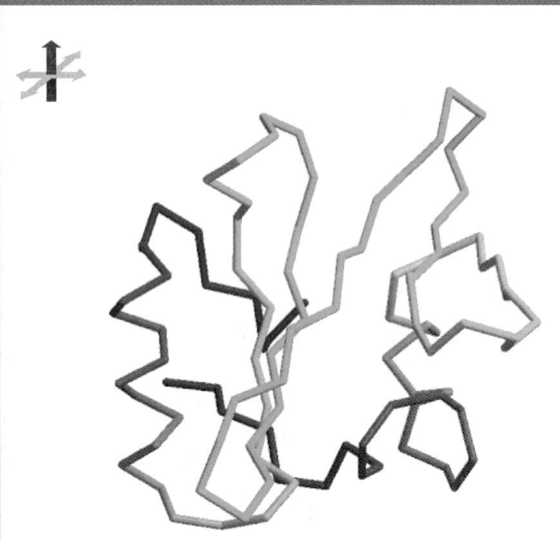

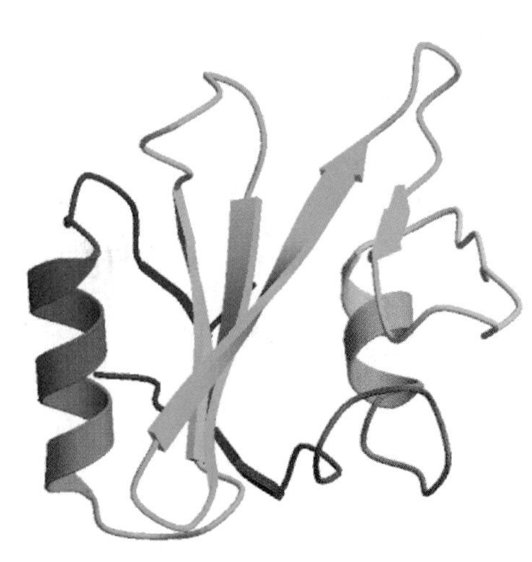

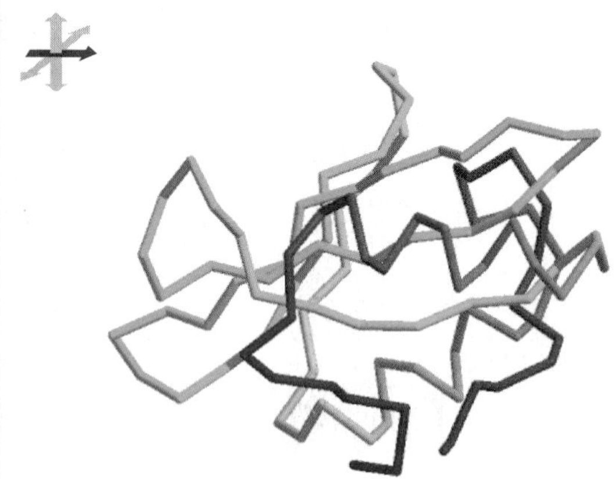

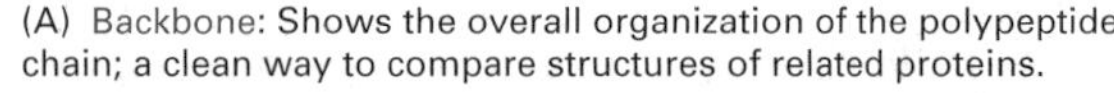

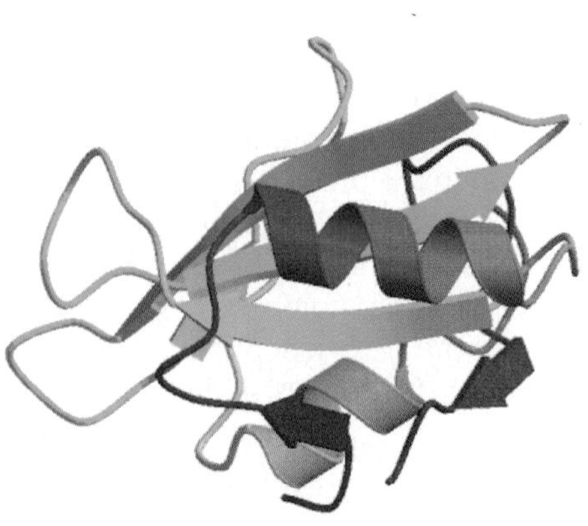

(A) Backbone: Shows the overall organization of the polypeptide chain; a clean way to compare structures of related proteins.

(B) Ribbon: Easy way to visualize secondary structures, such as α helices and β sheets.

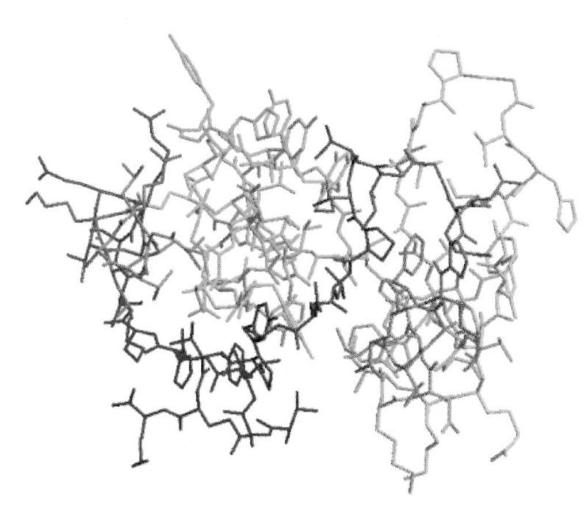

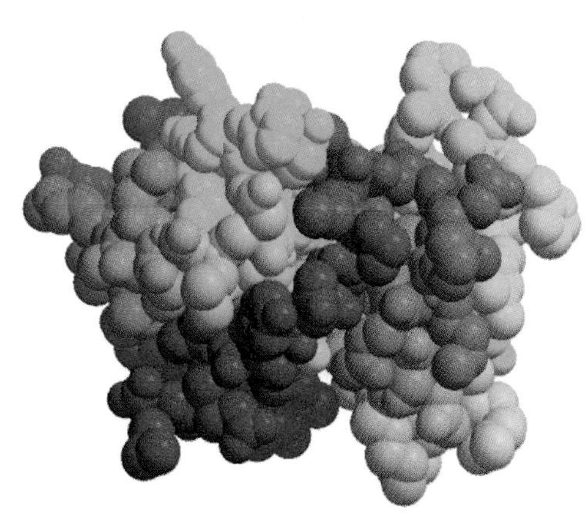

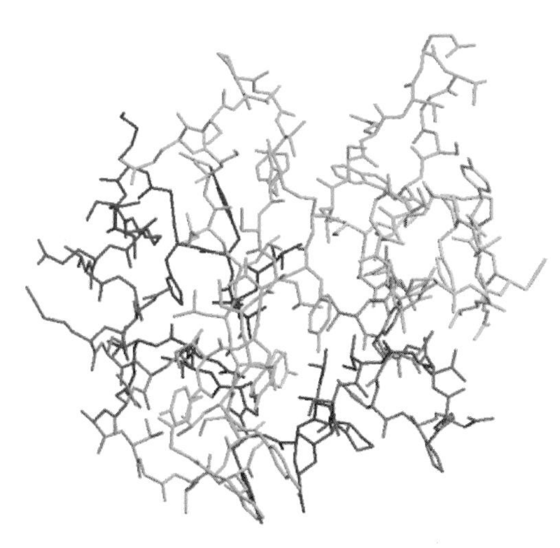

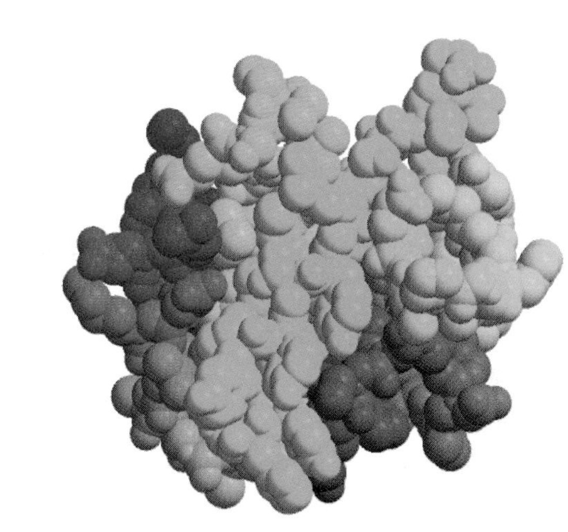

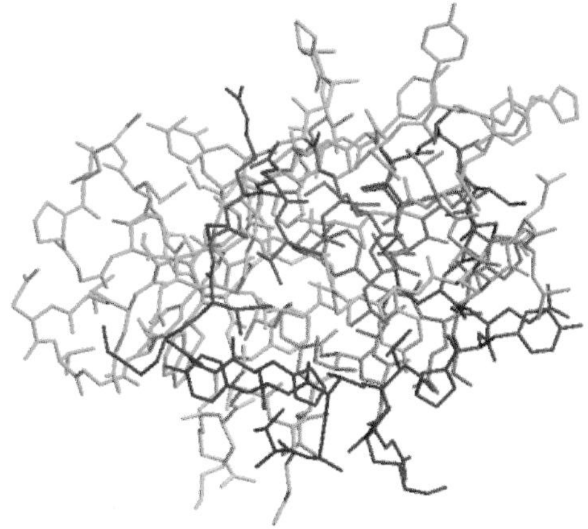

(C) Wire: Highlights side chains and their relative proximities; useful for predicting which amino acids might be involved in a protein's activity, particularly if the protein is an enzyme.

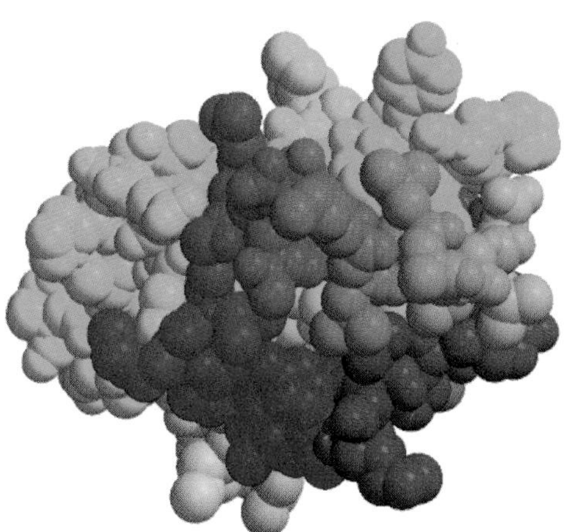

(Courtesy of David Lawson.)

(D) Space-filling: Provides contour map of the protein; gives a feel for the shape of the protein and shows which amino acid side chains are exposed on its surface. Shows how the protein might look to a small molecule, such as water, or to another protein.

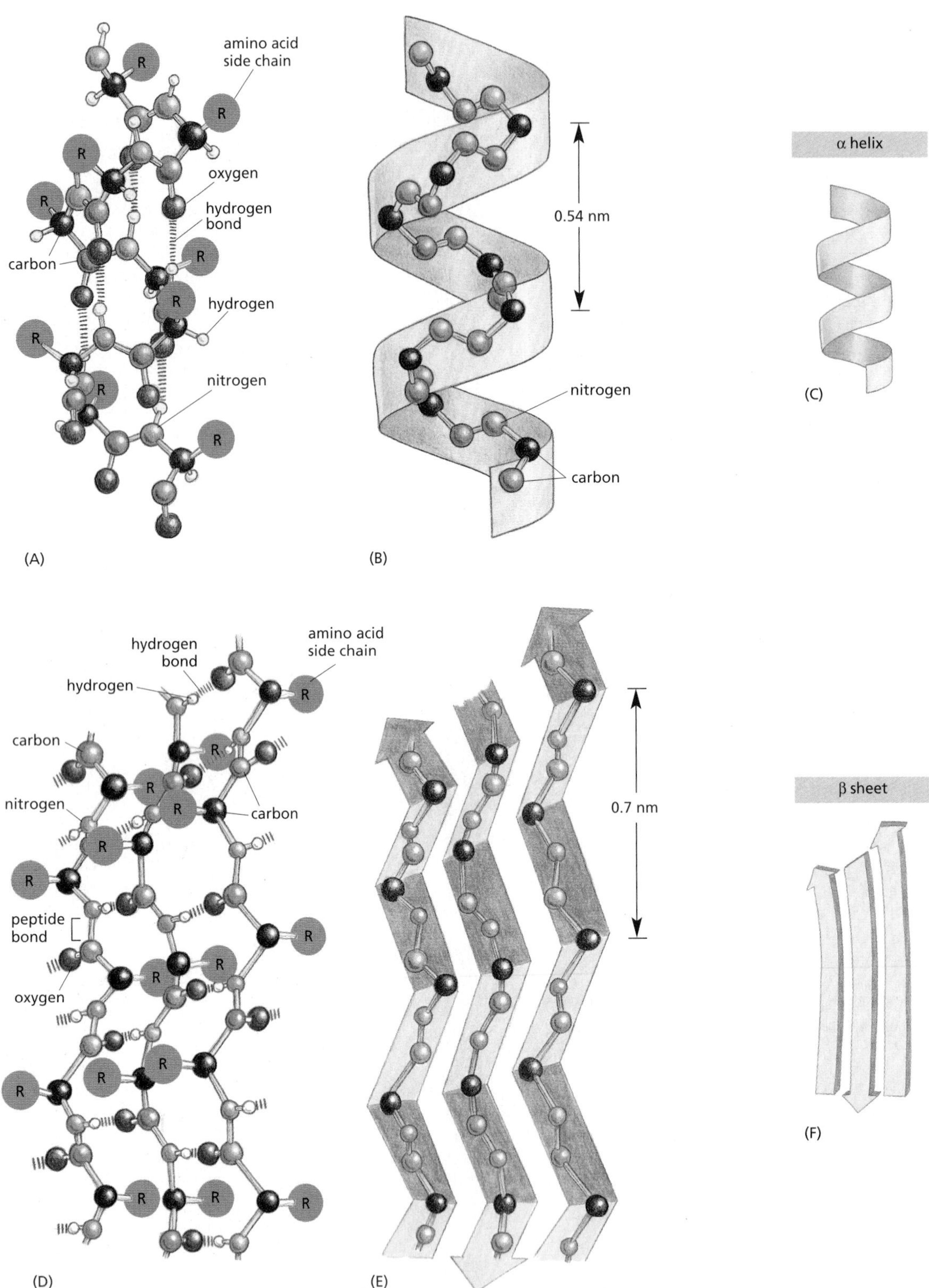

Figure 4–10 **Polypeptide chains often fold into one of two orderly repeating forms known as the α helix and the β sheet.** (A–C) In an α helix, the N–H of every peptide bond is hydrogen-bonded to the C=O of a neighboring peptide bond located four amino acids away in the same chain. (D–F) In the case of the β sheet, the individual polypeptide chains (strands) in the sheet are held together by hydrogen-bonding between peptide bonds in different strands, and the amino acid side chains in each strand project alternately above and below the plane of the sheet. In the example shown, the adjacent peptide chains run in opposite directions, forming an antiparallel β sheet. (A) and (D) show all of the atoms in the polypeptide backbone, but the amino acid side chains are denoted by R. (B) and (E) show the carbon and nitrogen backbone atoms only, while (C) and (F) display the cartoon symbols that are used to represent the α helix and the β sheet in ribbon models of proteins (see Panel 4–2B, p. 129).

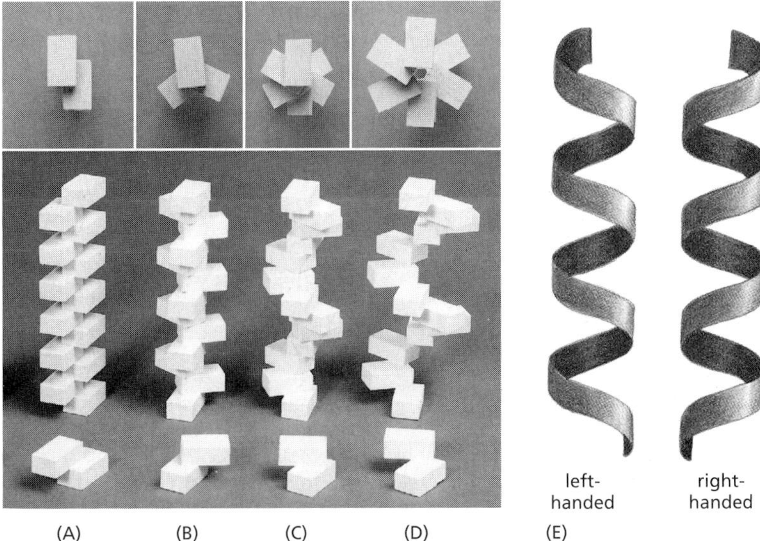

Figure 4–11 **The helix is a regular biological structure.** A helix will form when a series of subunits bind to each other in a regular way (A–D). At the bottom, the interaction between two subunits is shown; behind them are the helices that result. These helices have two (A), three (B), or six (C and D) subunits per helical turn. At the top, the arrangement of subunits has been photographed from directly above the helix. Note that the helix in (D) has a wider path than that in (C), but the same number of subunits per turn. (E) A helix can be either right-handed or left-handed. As a reference, it is useful to remember that standard metal screws, which insert when turned clockwise, are right-handed. Note that a helix preserves the same handedness when it is turned upside down.

QUESTION 4–2

Look at the models of the protein domain in Panel 4–2, pp. 128–129. Are the α helices right- or left-handed? Are the three chains that form the largest region of the β sheet parallel or antiparallel? Starting at the N-terminus (the *purple* end), trace your finger along the peptide backbone. Are there any knots? Why, or why not?

Helices Form Readily in Biological Structures

The abundance of helices in proteins is, in a way, not surprising. A **helix** is a regular structure that resembles a spiral staircase, as illustrated in Figure 4–11. It is generated simply by placing many similar subunits next to each other, each in the same strictly repeated relationship to the one before. Because it is very rare for subunits to join up in a straight line, this arrangement will generally result in a helix. Depending on the twist of the staircase, a helix is said to be either right-handed or left-handed (Figure 4–11E). Handedness is not affected by turning the helix upside down, but it is reversed if the helix is reflected in the mirror.

An α helix is generated when a single polypeptide chain turns around itself to form a structurally rigid cylinder. A hydrogen bond is made between every fourth amino acid, linking the C=O of one peptide bond to the N–H of another (see Figure 4–10A). This gives rise to a regular helix with a complete turn every 3.6 amino acids (Movie 4.2).

Short regions of α helix are especially abundant in the proteins located in cell membranes, such as transport proteins and receptors. We will see in Chapter 11 that those portions of a transmembrane protein that cross the lipid bilayer usually form an α helix that is composed largely of amino acids with nonpolar side chains. The polypeptide backbone, which is hydrophilic, is hydrogen-bonded to itself in the α helix, and it is shielded from the hydrophobic lipid environment of the membrane by its protruding nonpolar side chains (Figure 4–12).

Sometimes two (or three) α helices will wrap around one another to form a particularly stable structure known as a **coiled-coil**. This structure forms when the α helices have most of their nonpolar (hydrophobic) side chains on one side, so that they can twist around each other with these side chains facing inward—minimizing their contact with the aqueous cytosol (Figure 4–13). Long, rodlike coiled-coils form the structural framework for many elongated proteins. Examples include α-keratin, which forms the intracellular fibers that reinforce the outer layer of the skin, and myosin, the protein responsible for muscle contraction (discussed in Chapter 17).

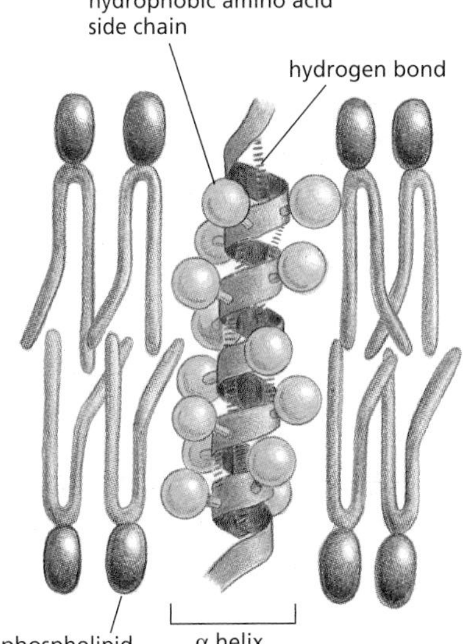

Figure 4–12 **A segment of α helix can cross a lipid bilayer.** The hydrophobic side chains of the amino acids forming the α helix contact the hydrophobic hydrocarbon tails of the phospholipid molecules, while the hydrophilic parts of the polypeptide backbone form hydrogen bonds with one another in the interior of the helix. About 20 amino acids are required to span a membrane in this way.

Figure 4–13 **Intertwined α helices can form a coiled-coil.** In (A) a single α helix is shown, with successive amino acid side chains labeled in a sevenfold sequence "abcdefg." Amino acids "a" and "d" in such a sequence lie close together on the cylinder surface, forming a stripe (*shaded in red*) that winds slowly around the α helix. Proteins that form coiled-coils typically have nonpolar amino acids at positions "a" and "d." Consequently, as shown in (B), the two α helices can wrap around each other with the nonpolar side chains of one α helix interacting with the nonpolar side chains of the other, while the more hydrophilic amino acid side chains are left exposed to the aqueous environment. (C) The atomic structure of a coiled-coil made of two α helices, as determined by X-ray crystallography. The *red* side chains are nonpolar. Coiled-coils can also form from three α helices (Movie 4.3).

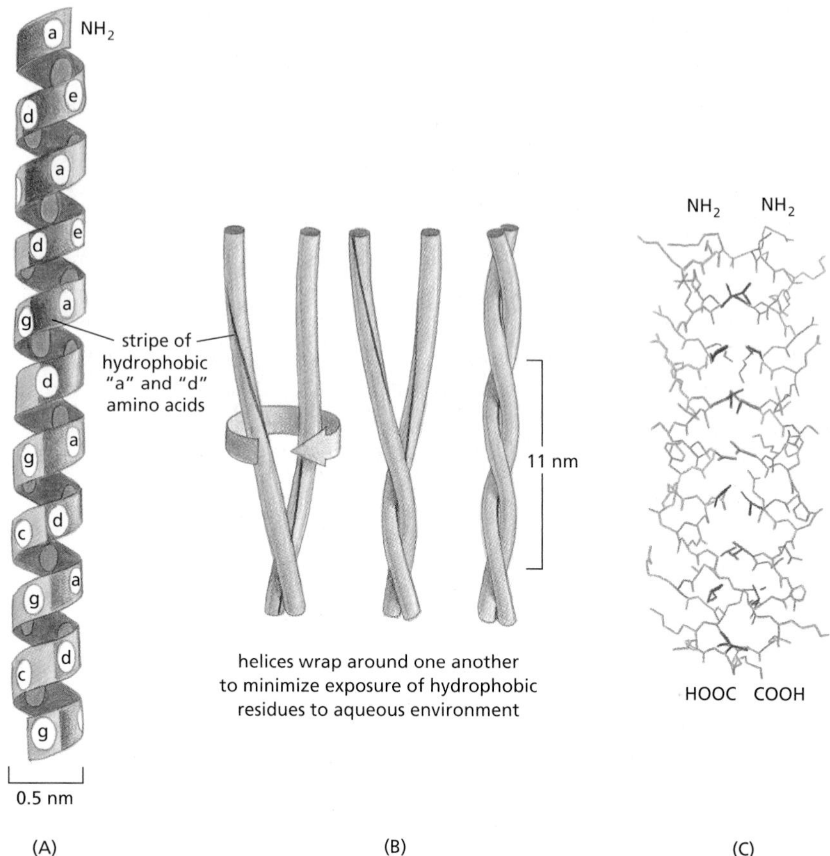

QUESTION 4–3

Remembering that the side chains projecting from each polypeptide backbone in a β sheet point alternately above and below the plane of the sheet (see Figure 4–10D), consider the following protein sequence: Leu-Lys-Val-Asp-Ile-Ser-Leu-Arg-Leu-Lys-Ile-Arg-Phe-Glu. Do you find anything remarkable about the arrangement of the amino acids in this sequence when incorporated into a β sheet? Can you make any predictions as to how the β sheet might be arranged in a protein? (Hint: consult the properties of the amino acids listed in Figure 4–3.)

β Sheets Form Rigid Structures at the Core of Many Proteins

SH2, the small protein structure we examined in Panel 4–2, contains both α helices and a β sheet. β sheets are made when hydrogen bonds form between segments of polypeptide chains lying side by side (see Figure 4–10D). When the structure consists of neighboring polypeptide chains that run in the same orientation (say, from the N-terminus to the C-terminus), it is considered a *parallel β sheet*; when it forms from a polypeptide chain that folds back and forth upon itself—with each section of the chain running in the direction opposite to that of its immediate neighbors—the structure is an *antiparallel β sheet* (Figure 4–14). Both types of β sheet produce a very rigid, pleated structure, and they form the core of many proteins. SH2, for example, has an antiparallel β-sheet core.

β sheets have remarkable properties. They give silk fibers their extraordinary tensile strength. And they can help to keep insects from freezing in the cold. In an antifreeze protein isolated from a beetle that lives in cold climates, a series of parallel β sheets forms a beautifully flat surface along one side of the protein molecule (Figure 4–15). This array appears to offer a perfect platform for binding to the regularly spaced water molecules that are present in an ice lattice. By adhering to the ice crystals that form when water is cooled below its freezing point, the antifreeze protein prevents the ice crystals from growing—thereby keeping the insects' cells from freezing solid.

(A)

(B)

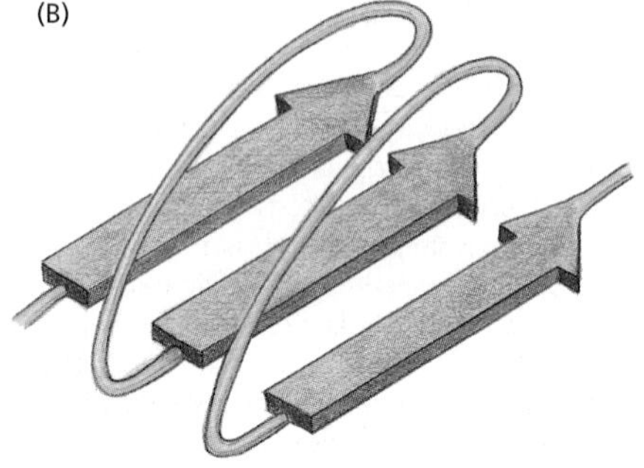

Figure 4–14 **β sheets come in two varieties.** (A) Antiparallel β sheet (see also Figure 4–10D). (B) Parallel β sheet. Both of these structures are common in proteins. By convention, the arrows point toward the C-terminus of the polypeptide chain (Movie 4.4).

Proteins Have Several Levels of Organization

A protein's structure does not end with α helices and β sheets; there are additional levels of organization. These levels are not independent but are built one upon the next until the three-dimensional structure of the entire protein has been fully defined. A protein's structure begins with its amino acid sequence, which is thus considered its *primary structure*. The next level of organization includes the α helices and β sheets that form within certain segments of a polypeptide chain; these folds are elements of the protein's **secondary structure**. The full, three-dimensional conformation formed by an entire polypeptide chain—including the α helices, β sheets, random coils, and any other loops and folds that form between the N- and C-termini—is sometimes referred to as the *tertiary structure* (see the structures shown in Panel 4–2, for example). Finally, if a particular protein molecule is formed as a complex of more than one polypeptide chain, then the complete structure is designated its *quaternary structure*.

Studies of the conformation, function, and evolution of proteins have also revealed the importance of a level of organization distinct from those just described. This organizational unit is the **protein domain**, which is defined as any segment of a polypeptide chain that can fold independently into a compact, stable structure, as we have seen in the case of the SH2 domain (see Panel 4–2, pp. 128–129). A domain usually consists of 100–250 amino acids (folded into α helices and β sheets and other elements of secondary structure), and it is the modular unit from which many larger proteins are constructed (Figure 4–16). The different domains of a protein are often associated with different functions. For example, the bacterial *catabolite activator protein (CAP)*, illustrated in Figure 4–16, has two domains: the small domain binds to DNA, while the large domain binds cyclic AMP, an intracellular signaling molecule. When the large domain binds cyclic AMP, it causes a conformational change in the protein that enables the small domain to bind to a specific DNA sequence and promote the expression of adjacent genes. The SH2 domain (see Panel 4–2) is found in many different polypeptides, where it serves as a binding domain to promote protein–protein interactions, as we shall see in Chapter 16.

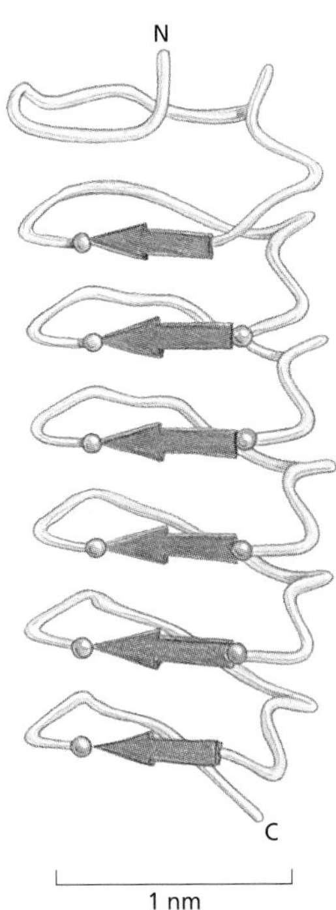

Figure 4–15 **β sheets provide an ideal ice-binding surface in an antifreeze protein.** The six parallel β strands, shown here in *red*, form a flat surface with 10 hydroxyl groups *(blue)* arranged at distances that correspond to water molecules in an ice lattice. The protein can therefore bind to ice crystals, preventing their growth. (After Y.C. Liou et al., *Nature* 406:322–324, 2000. With permission from Macmillan Publishers Ltd.)

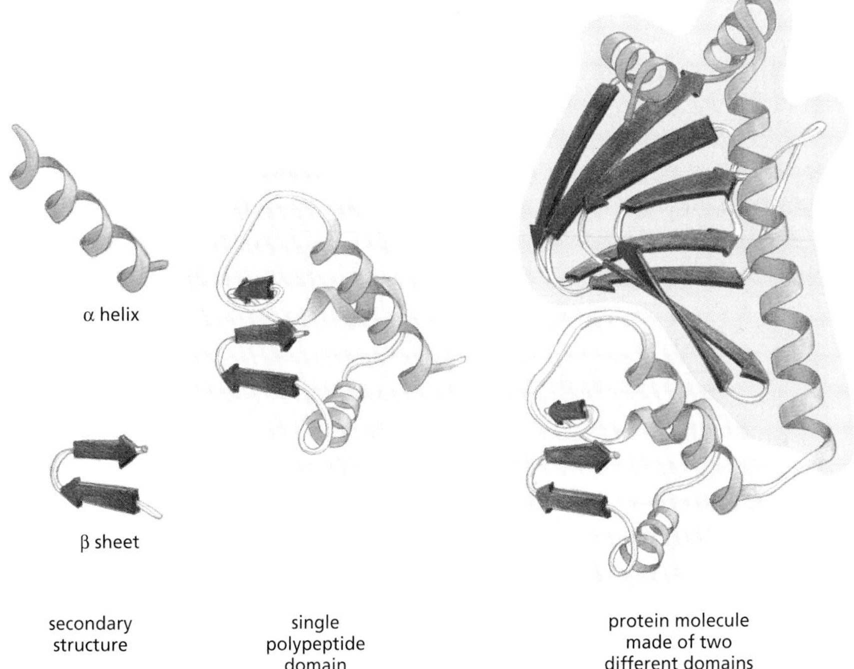

Figure 4–16 **Many proteins are composed of separate functional domains.** Elements of secondary structure such as α helices and β sheets pack together into stable, independently folding globular elements called domains. A typical protein molecule is built from one or more domains, linked by a region of polypeptide chain that is often relatively unstructured. The ribbon diagram on the right is of the bacterial gene regulatory protein CAP, with one large domain (outlined in *blue*) and one small domain (outlined in *yellow*).

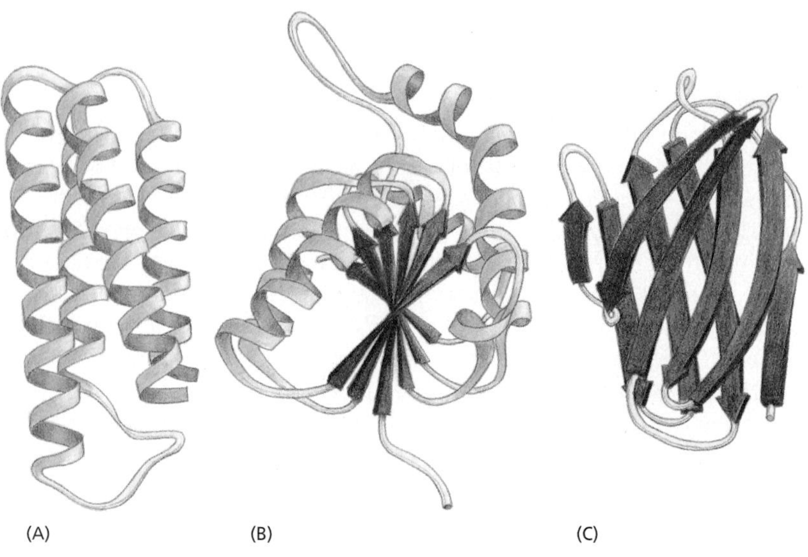

Figure 4–17 **Ribbon models show three different protein domains.** (A) Cytochrome b_{562}, a single-domain protein involved in electron transfer in *E. coli*, is composed almost entirely of α helices. (B) The NAD-binding domain of the enzyme lactic dehydrogenase is composed of a mixture of α helices and β sheets. (C) The variable domain of an immunoglobulin (antibody) light chain is composed of a sandwich of two β sheets. In these examples, the α helices are shown in *green*, while strands organized as β sheets are denoted by *red arrows*. The protruding loop regions (*yellow*) often form the binding sites for other molecules. (Redrawn from originals courtesy of Jane Richardson.)

Small protein molecules, such as the oxygen-carrying muscle protein myoglobin, contain only a single domain (see Figure 4–9). Larger proteins can contain as many as several dozen domains, which are usually connected by relatively unstructured lengths of polypeptide chain. Ribbon models of three differently organized domains are presented in Figure 4–17.

Few of the Many Possible Polypeptide Chains Will Be Useful

In theory, a vast number of different polypeptide chains could be made. Because each of the 20 amino acids is chemically distinct and each can, in principle, occur at any position in a protein chain, a polypeptide chain four amino acids long has $20 \times 20 \times 20 \times 20 = 160{,}000$ different possible sequences. In other words, for a polypeptide that is n amino acids long, 20^n different chains are possible. For a typical protein length of 300 amino acids, more than 20^{300} (that's 10^{390}) structurally different polypeptide chains could theoretically be made.

However, only a very small fraction of this unimaginably large number of polypeptide chains would fold into a stable, well-defined three-dimensional conformation. The vast majority of individual protein molecules would have many different conformations of roughly equal stability, each conformation having different chemical properties. So why do most proteins present in cells adopt unique and stable conformations? The answer is that a protein with many different conformations and variable properties would not be biologically useful, for it would be like a tool that unexpectedly changes its function. Such proteins would therefore have been eliminated by natural selection through the long trial-and-error process that underlies evolution (discussed in Chapter 9).

Because of natural selection, the amino acid sequences of most present-day proteins have evolved to guarantee that the polypeptide will adopt an extremely stable conformation—a structure that bestows upon the protein the exact chemical properties that will enable it to perform a particular catalytic or structural function in the cell. Proteins are so precisely built that the change of even a few atoms in one amino acid can sometimes disrupt the structure of a protein and thereby eliminate its function. In fact, many protein structures are so stable and effective that they have been conserved throughout evolution among many diverse organisms. The three-dimensional structures of the DNA-binding domains from the

QUESTION 4–4

Random mutations only very rarely result in changes in a protein that improve its usefulness for the cell, yet useful mutations are selected in evolution. Because these changes are so rare, for each useful mutation there are innumerable mutations that lead to either no improvement or inactive proteins. Why, then, do cells not contain millions of different proteins that are of no use?

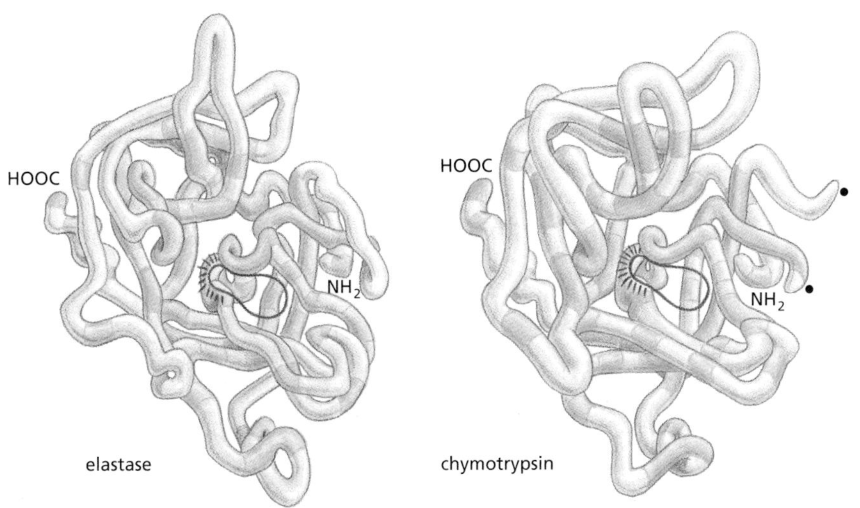

Figure 4–18 **Serine proteases comprise a family of proteolytic enzymes.** The backbone conformations of two serine proteases, elastase and chymotrypsin, are illustrated. Although only those amino acids in the polypeptide chain shaded in *green* are the same in the two proteins, the two conformations are very similar nearly everywhere. The active site of each enzyme is circled in *red*. This is where the peptide bonds of the proteins that are the substrates of these enzymes are bound and cleaved by hydrolysis. The serine proteases derive their name from the amino acid serine, whose side chain is part of the active site of each enzyme and directly participates in the cleavage reaction. The two dots on the *right side* of the chymotrypsin molecule mark the two ends created where the enzyme has cleaved its own backbone.

yeast α2 protein and the *Drosophila* Engrailed protein, for example, are almost completely superimposable, even though these organisms are separated by more than a billion years of evolution.

Proteins Can Be Classified into Families

Once a protein had evolved a stable conformation with useful properties, its structure could be modified over time to enable it to perform new functions. We know that this occurred quite often during evolution, because many present-day proteins can be grouped into **protein families**, in which each family member has an amino acid sequence and a three-dimensional conformation that closely resembles that of the other family members.

Consider, for example, the *serine proteases*, a family of protein-cleaving (proteolytic) enzymes that includes the digestive enzymes chymotrypsin, trypsin, and elastase, as well as several proteases involved in blood clotting. When any two of these enzymes are compared, portions of their amino acid sequences are found to be nearly the same. The similarity of their three-dimensional conformations is even more striking: most of the detailed twists and turns in their polypeptide chains, which are several hundred amino acids long, are virtually identical (Figure 4–18). The various serine proteases nevertheless have distinct enzymatic activities, each cleaving different proteins or the peptide bonds between different types of amino acids. Slight differences in structure allow each of these proteases to prefer different substrates; thus each carries out a distinct function in an organism.

Large Protein Molecules Often Contain More Than One Polypeptide Chain

The same weak noncovalent bonds that enable a polypeptide chain to fold into a specific conformation also allow proteins to bind to each other to produce larger structures in the cell. Any region on a protein's surface that interacts with another molecule through sets of noncovalent bonds is termed a *binding site*. A protein can contain binding sites for a variety of molecules, large and small. If a binding site recognizes the surface of a second protein, the tight binding of two folded polypeptide chains at this site will create a larger protein whose quaternary structure has a precisely defined geometry. Each polypeptide chain in such a protein is called a **subunit**. Each of these protein subunits may contain more than one domain, a portion of the polypeptide chain that folds up separately.

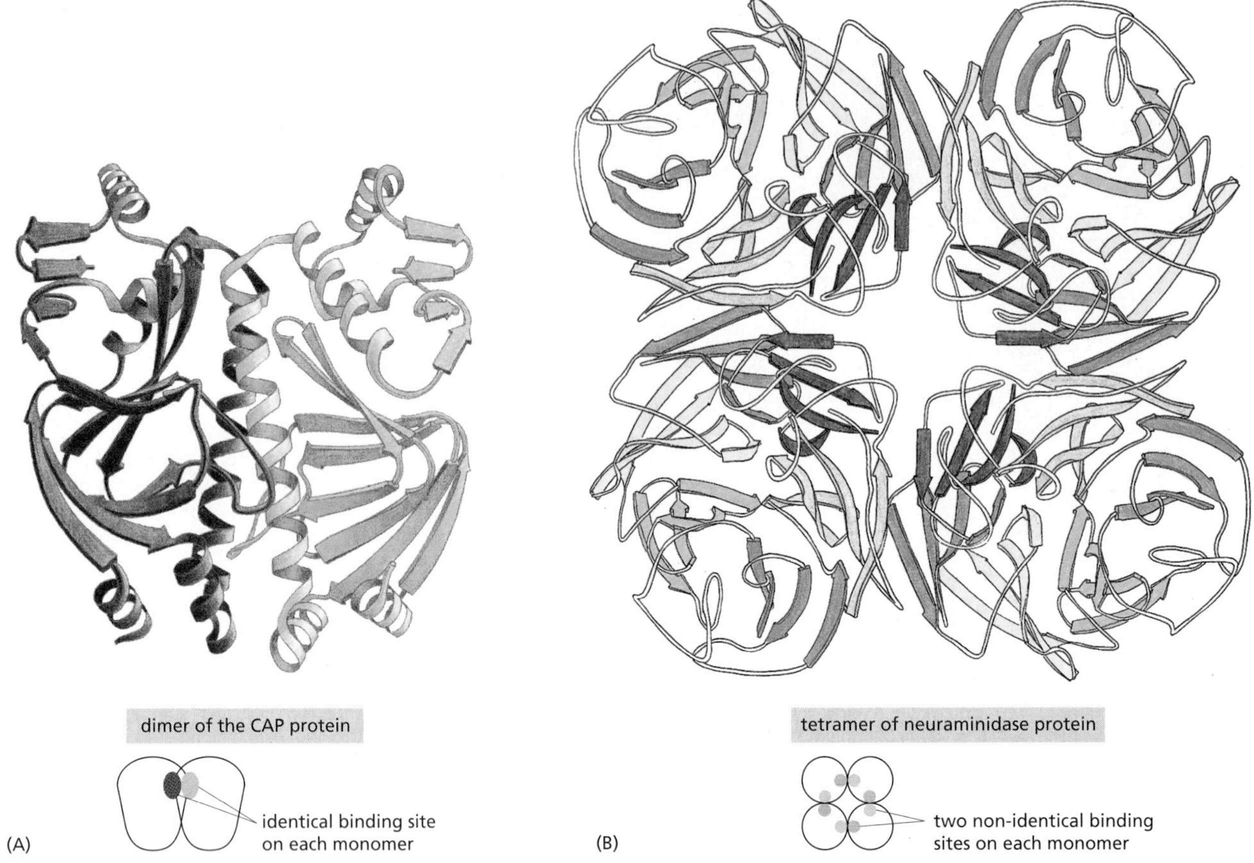

Figure 4–19 **Many protein molecules contain multiple copies of a single protein subunit.** (A) A symmetrical dimer. The CAP protein exists as a complex of two identical polypeptide chains (see also Figure 4–16). (B) A symmetrical tetramer. The enzyme neuraminidase exists as a ring of four identical polypeptide chains. For both (A) and (B), a small schematic below the structure emphasizes how the repeated use of the same binding interaction forms the structure. In (A), the use of the same binding site on each monomer (represented by *brown* and *green* ovals) causes the formation of a symmetrical dimer. In (B), a pair of nonidentical binding sites (represented by *orange* and *blue* circles), causes the formation of a symmetrical tetramer.

In the simplest case, two identical folded polypeptide chains form a symmetrical complex of two protein subunits (called a *dimer*) that is held together by interactions between two identical binding sites. The CAP protein in bacterial cells is a dimer (Figure 4–19A) that is formed from two identical copies of the protein subunit shown previously in Figure 4–16. Many other symmetrical protein complexes, formed from multiple copies of a single polypeptide chain, are commonly found in cells. The enzyme *neuraminidase*, for example, consists of a ring of four identical protein subunits (Figure 4–19B).

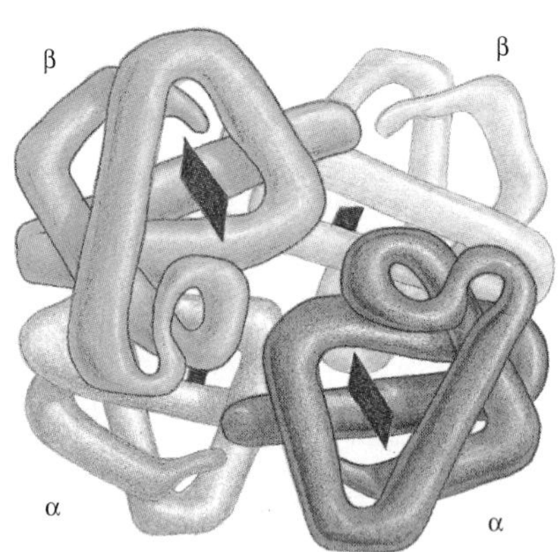

Figure 4–20 **Some proteins are formed as a symmetrical assembly of two different subunits.** Hemoglobin, a protein abundant in red blood cells, contains two copies of α-globin and two copies of β-globin. Each of these four polypeptide chains contains a heme molecule (*red* rectangle), where oxygen (O_2) is bound. Thus, each molecule of hemoglobin in the blood carries four molecules of oxygen.

Other proteins contain two or more different types of polypeptide chains. *Hemoglobin*, the protein that carries oxygen in red blood cells, is a particularly well-studied example (Figure 4–20). The protein contains two identical α-globin subunits and two identical β-globin subunits, symmetrically arranged. Many proteins contain multiple subunits, and they can be very large (Movie 4.5).

Proteins Can Assemble into Filaments, Sheets, or Spheres

Proteins can form even larger assemblies than those discussed so far. Most simply, a chain of identical protein molecules can be formed if the binding site on one protein molecule is complementary to another region on the surface of another protein molecule of the same type. Because

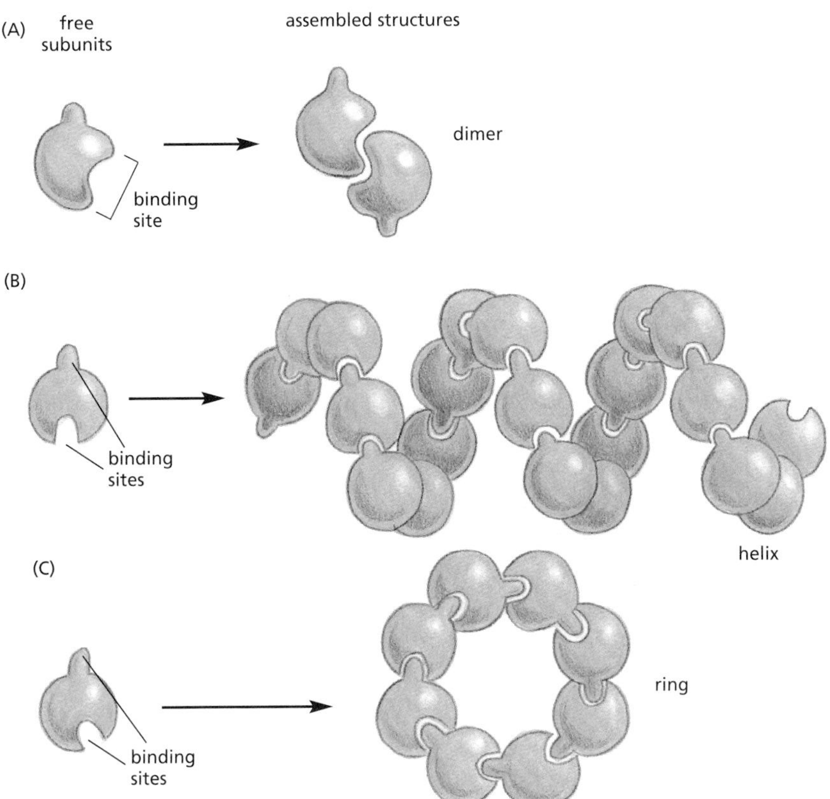

Figure 4–21 **Proteins can assemble into complex structures.** (A) A protein with just one binding site can form a dimer with another identical protein. (B) Identical proteins with two different binding sites will often form a long helical filament. (C) If the two binding sites are disposed appropriately in relation to each other, the protein subunits will form a closed ring instead of a helix (see also Figure 4–19B).

each protein molecule is bound to its neighbor in an identical way, the molecules will often be arranged in a helix that can be extended indefinitely (Figure 4–21). This type of arrangement can produce an extended protein filament. An actin filament, for example, is a long helical structure formed from many molecules of the protein actin (Figure 4–22). Actin is extremely abundant in eucaryotic cells, where it forms one of the major filament systems of the cytoskeleton (discussed in Chapter 17). Other sets of proteins associate to form extended sheets or tubes, as in the microtubules of the cell cytoskeleton (Figure 4–23); or cagelike spherical shells, as in the protein coats of virus particles (Figure 4–24).

Many large structures, such as viruses and ribosomes, are built from a mixture of one or more types of protein plus RNA or DNA molecules. These structures can be isolated in pure form and dissociated into their constituent macromolecules. It is often possible to mix the isolated components back together and watch them reassemble spontaneously into the original structure. This demonstrates that all the information needed for assembly of the complicated structure is contained in the macromolecules themselves. Experiments of this type show that much of the structure of a cell is self-organizing: if the required proteins are produced in the right amounts, the appropriate structures will form.

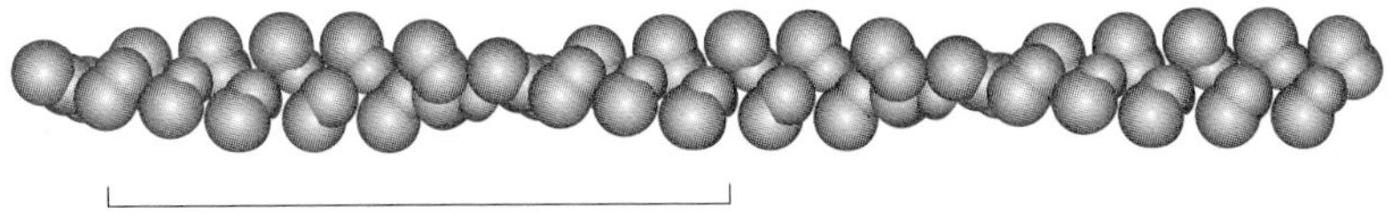

Figure 4–22 **An actin filament is composed of identical protein subunits.** The helical array of actin molecules often contains thousands of molecules and extends for micrometers in the cell.

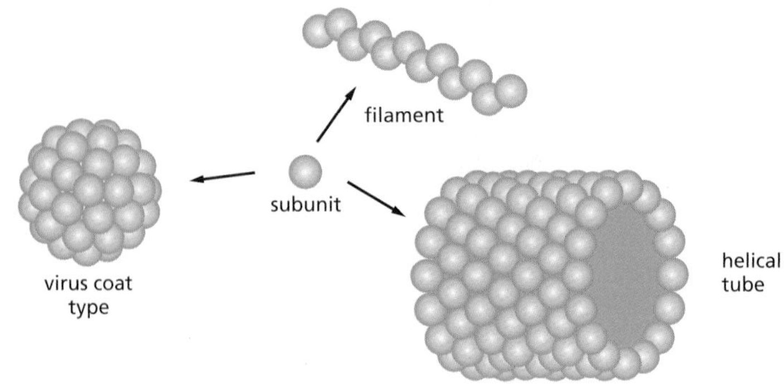

Figure 4–23 **Single protein subunits can pack to form a filament, a tube, or a spherical shell.**

Some Types of Proteins Have Elongated Fibrous Shapes

Most of the proteins we have discussed so far are **globular proteins**, in which the polypeptide chain folds up into a compact shape like a ball with an irregular surface. Enzymes tend to be globular proteins: even though many are large and complicated, with multiple subunits, most have a quaternary structure with an overall rounded shape (see Figure 4–9). In contrast, other proteins have roles in the cell that require them to span a large distance. These proteins generally have a relatively simple, elongated three-dimensional structure and are commonly referred to as **fibrous proteins**.

One large class of intracellular fibrous proteins resembles α-keratin, which we met earlier. Keratin filaments are extremely stable: long-lived structures such as hair, horn, and nails are composed mainly of this protein. An α-keratin molecule is a dimer of two identical subunits, with the long α helices of each subunit forming a coiled-coil (see Figure 4–13). These coiled-coil regions are capped at either end by globular domains containing binding sites that allow them to assemble into ropelike *intermediate filaments*—a component of the cell cytoskeleton that creates a structural scaffold for the cell's interior (discussed in Chapter 17).

Fibrous proteins are especially abundant outside the cell, where they form the gel-like *extracellular matrix* that helps cells bind together to form tissues. These proteins are secreted by the cells into their surroundings, where they often assemble into sheets or long fibrils. *Collagen* is the most abundant of these fibrous proteins in animal tissues. The collagen molecule consists of three long polypeptide chains, each containing the nonpolar amino acid glycine at every third position. This regular structure allows the chains to wind around one another to generate a long regular triple helix with glycine at its core (Figure 4–25A). Many collagen molecules bind to one another side-by-side and end-to-end to create long overlapping arrays—thereby generating the extremely strong collagen fibrils that hold tissues together, as described in Chapter 19.

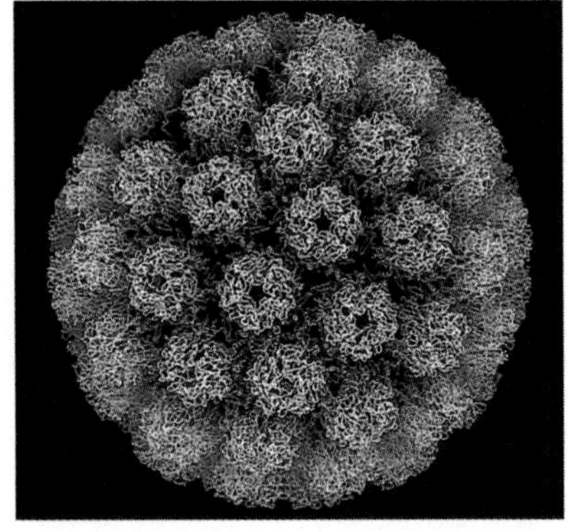

Figure 4–24 **Many viral capsids are more or less spherical protein assemblies.** These capsids are polyhedrons formed from many copies of a small set of protein subunits. The nucleic acid of the virus (DNA or RNA) is packaged inside. The structure of the simian virus SV40, shown here, was determined by X-ray crystallography and is known in atomic detail. (Courtesy of Robert Grant, Stephan Crainic, and James M. Hogle.)

In complete contrast to collagen is another protein in the extracellular matrix, *elastin*. Elastin molecules are formed from relatively loose and unstructured polypeptide chains that are covalently cross-linked into a rubberlike elastic meshwork. The resulting elastic fibers enable skin and other tissues, such as arteries and lungs, to stretch and recoil without tearing. As illustrated in Figure 4–25B, the elasticity is due to the ability of the individual protein molecules to uncoil reversibly whenever they are stretched.

Extracellular Proteins Are Often Stabilized by Covalent Cross-Linkages

Many protein molecules are either attached to the outside of a cell's plasma membrane or secreted as part of the extracellular matrix. All such

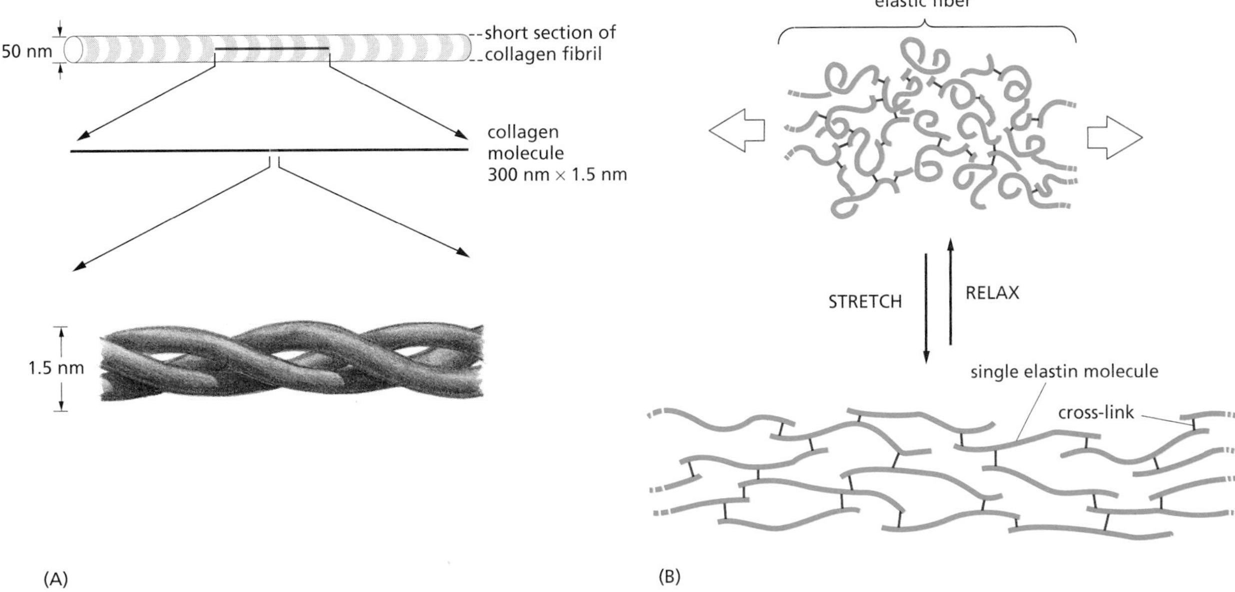

Figure 4–25 **Collagen and elastin are abundant fibrous proteins.** (A) Collagen is a triple helix formed by three extended protein chains that wrap around one another. Many rodlike collagen molecules are cross-linked together in the extracellular space to form collagen fibrils (*top*) that have the tensile strength of steel. The striping on the collagen fibril is caused by the regular repeating arrangement of the collagen molecules within the fibril. (B) Elastin polypeptide chains are cross-linked together to form rubberlike, elastic fibers. Each elastin molecule uncoils into a more extended conformation when the fiber is stretched and will recoil spontaneously as soon as the stretching force is relaxed.

proteins are directly exposed to extracellular conditions. To help maintain their structures, the polypeptide chains in such proteins are often stabilized by covalent cross-linkages. These linkages can tie together two amino acids in the same protein, or connect different polypeptide chains in a multisubunit protein. The most common covalent cross-links in proteins are sulfur–sulfur bonds. These **disulfide bonds** (also called *S–S bonds*) form as proteins are being exported from cells. Their formation is catalyzed in the endoplasmic reticulum by a special enzyme that links together two –SH groups from cysteine side chains that are adjacent in the folded protein (Figure 4–26). Disulfide bonds do not change the conformation of a protein, but instead act as a sort of "atomic staple" to reinforce its most favored conformation. For example, lysozyme—an enzyme in tears, saliva, and other secretions that can dissolve bacterial cell walls—retains its antibacterial activity for a long time because it is stabilized by such cross-linkages.

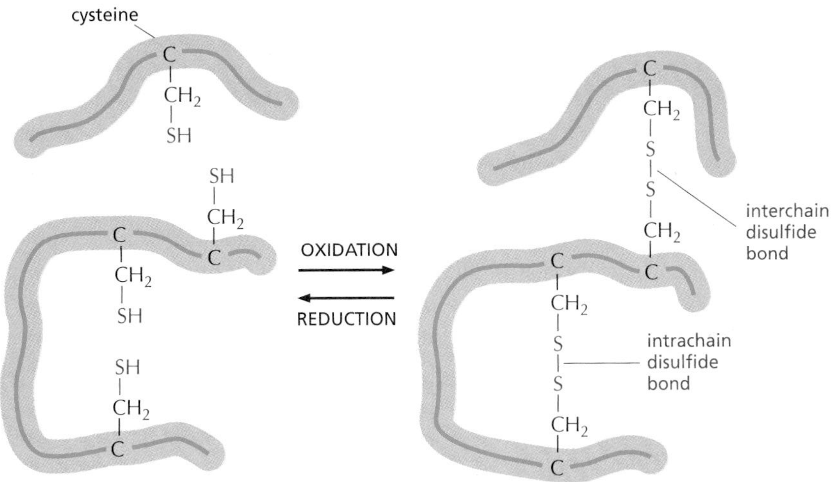

Figure 4–26 **Disulfide bonds help stabilize a favored protein conformation.** This diagram illustrates how covalent disulfide bonds form between adjacent cysteine side chains. As indicated, these cross-linkages can join either two parts of the same polypeptide chain or two different polypeptide chains. Because the energy required to break one covalent bond is much larger than the energy required to break even a whole set of noncovalent bonds (see Table 2–1, p. 47), a disulfide bond can have a major stabilizing effect on a protein (**Movie 4.6**).

Disulfide bonds generally do not form in the cell cytosol, where a high concentration of reducing agents converts such bonds back to cysteine –SH groups. Apparently, proteins do not require this type of structural reinforcement in the relatively mild environment inside the cell.

HOW PROTEINS WORK

QUESTION 4–5

Hair is composed largely of fibers of the protein keratin. Individual keratin fibers are covalently attached to one another (cross-linked) by many disulfide bonds. If curly hair is treated with mild reducing agents that break a few of the cross-links, pulled straight, and then oxidized again, it remains straight. Draw a diagram that illustrates the different stages of this chemical and mechanical process at the molecular level, focusing on the disulfide bonds. What do you think would happen if hair were treated with strong reducing agents that break all disulfide bonds?

Proteins are not inert lumps of material. Their different amino acid sequences give rise to an enormous variety of different protein shapes, each with a unique surface topography of chemical groups. And each protein's conformation endows it with a unique function based on its chemical properties. This union of structure, chemistry, and activity gives proteins the extraordinary ability to orchestrate the dynamic processes that occur in living cells.

Thus, for proteins, form and function are inextricably linked. But the fundamental question remains: how do proteins accomplish their function? In this part of the chapter, we will see that the activity of proteins depends on their ability to bind specifically to other molecules, allowing them to act as catalysts, structural supports, signal receptors, and tiny motors. The examples we review here by no means exhaust the vast functional repertoire of proteins. However, the specialized functions of many of the proteins you will encounter elsewhere in this book are based on the same principles.

All Proteins Bind to Other Molecules

The biological properties of a protein molecule depend on its physical interaction with other molecules. Antibodies attach to viruses or bacteria as part of the body's defenses, the enzyme hexokinase binds glucose and ATP to catalyze a reaction between them, actin molecules bind to each other to assemble into long filaments, and so on. Indeed, all proteins stick, or bind, to other molecules. In some cases this binding is very tight; in others it is weak and short-lived. As we saw in Chapter 3, the affinity of an enzyme for its substrate is reflected by its K_M: the lower the K_M, the tighter the binding. Regardless of its strength, the binding of proteins to other biological molecules always shows great *specificity*: each protein molecule can bind to just one or a few molecules out of the many thousands of different molecules it encounters. Any substance that is bound by a protein—whether it is an ion, a small molecule, or a macromolecule—is referred to as a **ligand** for that protein (from the Latin *ligare*, "to bind").

The ability of a protein to bind selectively and with high affinity to a ligand is due to the formation of a set of weak, noncovalent bonds—hydrogen bonds, electrostatic attractions, and van der Waals attractions—plus favorable hydrophobic interactions (see Panel 2–7, pp. 76–77). Each individual bond is weak, so that an effective interaction requires many weak bonds to be formed simultaneously. This is possible only if the surface contours of the ligand molecule fit very closely to the protein, matching it like a hand in a glove (Figure 4–27).

When molecules have poorly matching surfaces, few noncovalent bonds are formed and the two molecules dissociate as rapidly as they come together. This is what prevents incorrect and unwanted associations from forming between mismatched molecules. At the other extreme, when many noncovalent bonds are formed, the association can persist for a very long time. Strong interactions occur in cells whenever a biological function requires molecules to remain tightly associated for a long time—for example, when a group of macromolecules come together to form a subcellular structure such as a ribosome.

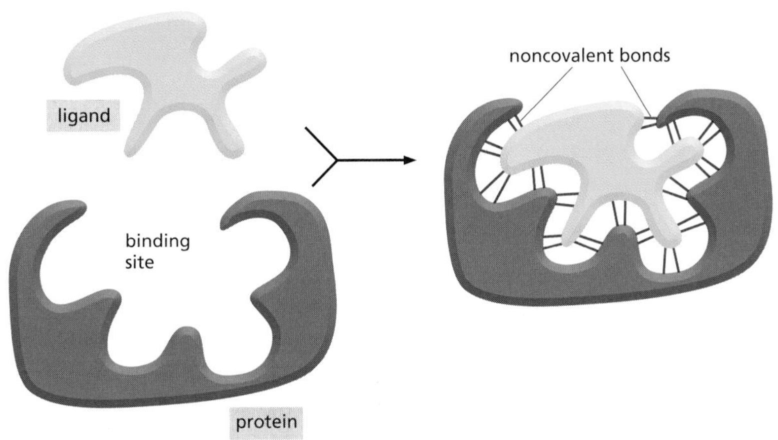

Figure 4–27 **The binding of a protein to another molecule is highly selective.** Many weak bonds are needed to enable a protein to bind tightly to a second molecule (a ligand). The ligand must therefore fit precisely into the protein's binding site, like a hand into a glove, so that a large number of noncovalent bonds can be formed between the protein and the ligand.

The region of a protein that associates with a ligand, known as its **binding site**, usually consists of a cavity in the protein surface formed by a particular arrangement of amino acids. These amino acids can belong to widely separated regions of the polypeptide chain that are brought together when the protein folds (Figure 4–28). Other regions on the surface often provide binding sites for different ligands, allowing the protein's activity to be regulated, as we shall see later. Yet other parts of the protein may be required to attract or attach the protein to a particular location in the cell—for example, the hydrophobic α helix of a membrane-spanning protein allows it to be inserted into the lipid bilayer of a cell membrane (discussed in Chapter 11).

Although the atoms buried in the interior of the protein have no direct contact with the ligand, they provide an essential scaffold that gives the surface its contours and chemical properties. Even small changes to the amino acids in the interior of a protein molecule can change its three-dimensional shape and destroy the protein's ability to function.

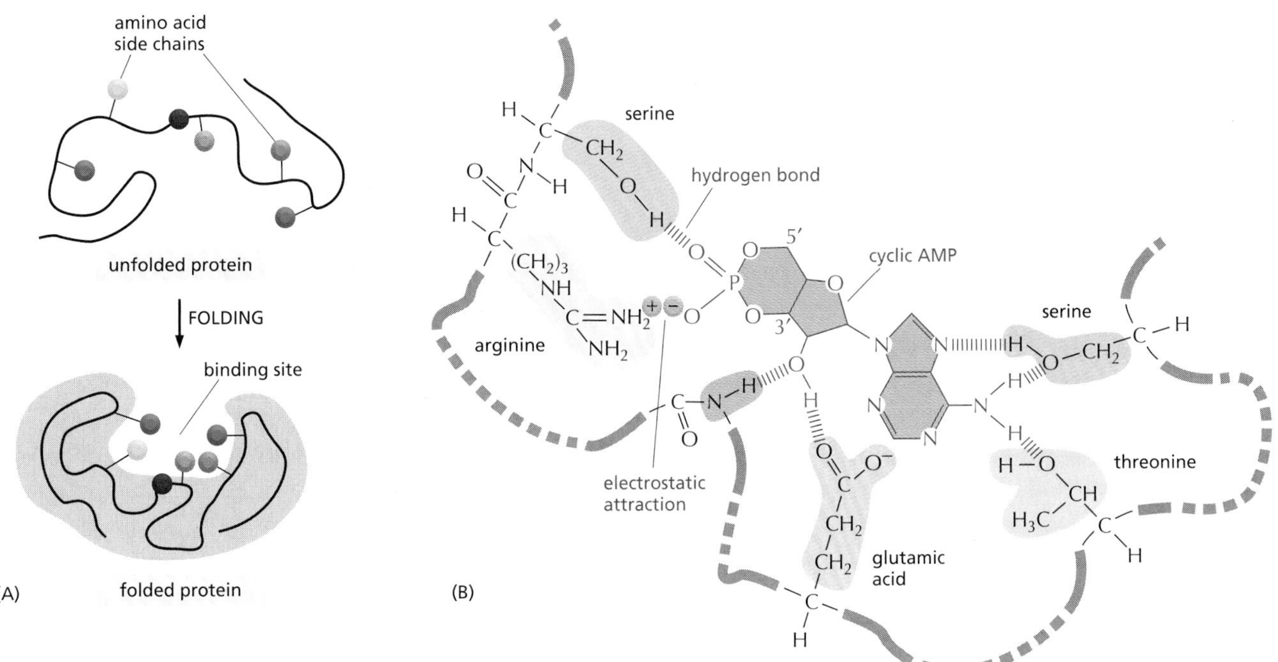

Figure 4–28 **Binding sites allow a protein to interact with specific ligands.** (A) The folding of the polypeptide chain typically creates a crevice or cavity on the protein surface. This crevice contains a set of amino acid side chains disposed in such a way that they can form a set of noncovalent bonds only with certain ligands. (B) Close-up view of an actual binding site showing the hydrogen bonds and ionic interactions formed between a protein and its ligand (in this example, the bound ligand is cyclic AMP, shown in *pink*).

The Binding Sites of Antibodies Are Especially Versatile

All proteins must bind to particular ligands to carry out their various functions. But this binding capacity seems to have been most highly developed for proteins in the antibody family: our bodies have the capacity to produce a unique antibody that is capable of recognizing and binding tightly to the structure of any molecule imaginable.

Antibodies, or immunoglobulins, are proteins produced by the immune system in response to foreign molecules, such as those on the surface of an invading microorganism. Each antibody binds to a particular target molecule extremely tightly, either inactivating the target directly or marking it for destruction. An antibody recognizes its target (called an **antigen**) with remarkable specificity, and because there are potentially billions of different antigens that a person might encounter, we have to be able to produce billions of different antibodies.

Antibodies are Y-shaped molecules with two identical binding sites that are each complementary to a small portion of the surface of the antigen molecule. A detailed examination of the antigen-binding sites of antibodies reveals that they are formed from several loops of polypeptide chain that protrude from the ends of a pair of closely juxtaposed protein domains (Figure 4–29). The amino acid sequence in these loops can be changed by mutation without altering the basic structure of the antibody. An enormous diversity of antigen-binding sites can be generated by changing only the length and amino acid sequence of the loops, which is how the wide variety of different antibodies is formed (Movie 4.7).

With their unique combination of specificity and diversity, not only are antibodies indispensable for fighting off infections, but they are also valuable in the laboratory, where they can be used to purify and study other molecules (Panel 4–3, pp. 144–145).

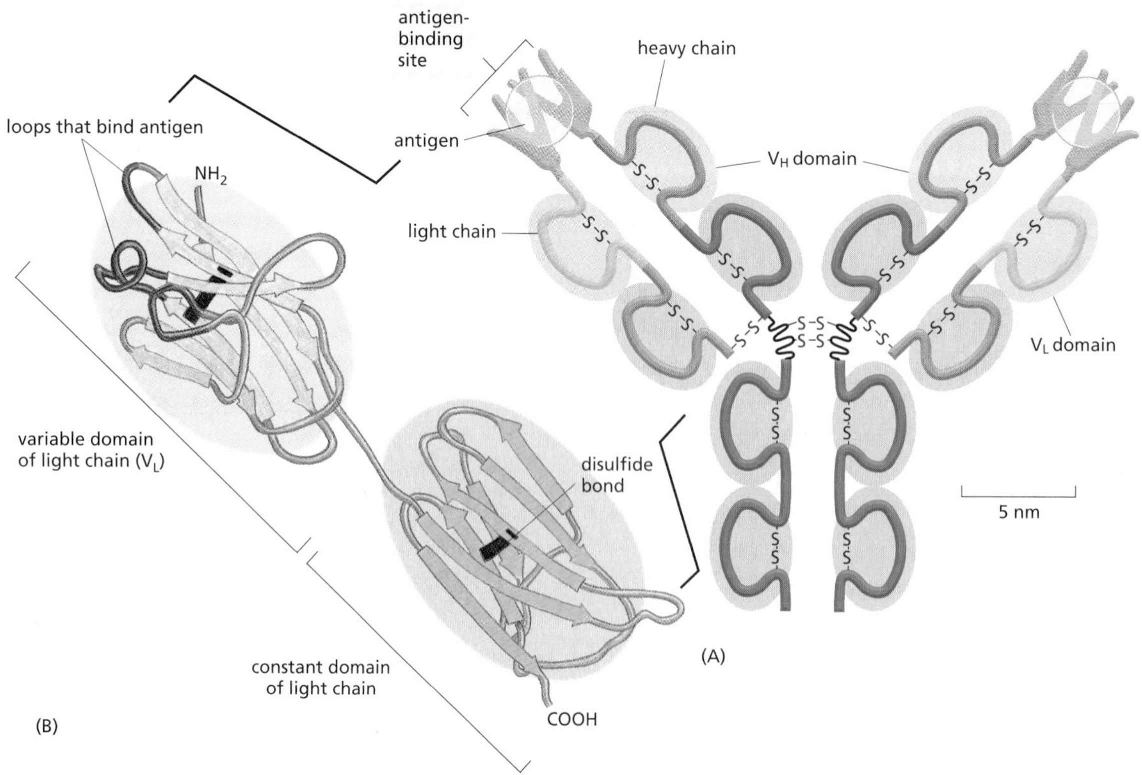

Figure 4–29 **An antibody is Y-shaped and has two identical binding sites for its antigen, one on each arm of the Y.** (A) Schematic drawing of a typical antibody molecule. The protein is composed of four polypeptide chains (two identical heavy chains and two identical and smaller light chains) held together by disulfide bonds. Each chain is made up of several different domains, here shaded either *blue* or *gray*. The antigen-binding site is formed where a heavy-chain variable domain (V_H) and a light-chain variable domain (V_L) come close together. These are the domains that differ most in their sequence and structure in different antibodies. (B) Ribbon drawing of a single light chain showing the parts of the V_L domain most closely involved in binding to the antigen in *red*; these contribute half of the fingerlike loops that fold around each of the antigen molecules in (A).

Enzymes Are Powerful and Highly Specific Catalysts

For many proteins, binding to another molecule is their only function. An antibody molecule need only bind to its target molecule on the surface of a bacterium or a virus and its job is done; an actin molecule need only associate with other actin molecules to form a filament. There are proteins, however, for which ligand binding is simply a necessary first step in their function. This is the case for the large and very important class of proteins called enzymes. These remarkable molecules perform nearly all of the chemical transformations that occur in cells. Enzymes bind to one or more ligands, called substrates, and convert them into chemically modified products, doing this over and over again with amazing rapidity. As we saw in Chapter 3, they speed up reactions, often by a factor of a million or more, without themselves being changed—that is, enzymes act as catalysts that permit cells to make or break covalent bonds at will. This catalysis of organized sets of chemical reactions by enzymes creates and maintains the cell, making life possible.

Enzymes can be grouped into functional classes based on the chemical reactions they catalyze (Table 4–1). Each type of enzyme is highly specific, catalyzing only a single type of reaction. Thus, *hexokinase* adds a phosphate group to D-glucose but will ignore its optical isomer L-glucose; the blood-clotting enzyme *thrombin* cuts one type of blood protein between a particular arginine and its adjacent glycine and nowhere else. As discussed in detail in Chapter 3, enzymes often work in tandem, with the product of one enzyme becoming the substrate for the next. The result is an elaborate network of *metabolic pathways* that provides the cell with energy and generates the many large and small molecules that the cell needs.

QUESTION 4–6

Explain how an enzyme (such as hexokinase, mentioned in the text) can distinguish substrates (here D-glucose) from their optical isomers (here L-glucose). (Hint: remembering that a carbon atom forms four single bonds that are tetrahedrally arranged and that the optical isomers are mirror images of each other around such a bond, draw the substrate as a simple tetrahedron with four different corners and then draw its mirror image. Using this drawing, indicate why only one compound might bind to a schematic active site of an enzyme.)

Lysozyme Illustrates How an Enzyme Works

To explain how enzymes catalyze chemical reactions, we will use the example of lysozyme—an enzyme that acts as a natural antibiotic in egg white, saliva, tears, and other secretions. Lysozyme severs the polysaccharide chains that form the cell walls of bacteria. Because the bacterial

TABLE 4–1 SOME COMMON FUNCTIONAL CLASSES OF ENZYMES

ENZYME CLASS	BIOCHEMICAL FUNCTION
Hydrolase	General term for enzymes that catalyze a hydrolytic cleavage reaction.
Nuclease	Breaks down nucleic acids by hydrolyzing bonds between nucleotides.
Protease	Breaks down proteins by hydrolyzing peptide bonds between amino acids.
Synthase	General name used for enzymes that synthesize molecules in anabolic reactions by condensing two molecules together.
Isomerase	Catalyzes the rearrangement of bonds within a single molecule.
Polymerase	Catalyzes polymerization reactions such as the synthesis of DNA and RNA.
Kinase	Catalyzes the addition of phosphate groups to molecules. Protein kinases are an important group of kinases that attach phosphate groups to proteins.
Phosphatase	Catalyzes the hydrolytic removal of a phosphate group from a molecule.
Oxido-reductase	General name for enzymes that catalyze reactions in which one molecule is oxidized while the other is reduced. Enzymes of this type are often called oxidases, reductases, or dehydrogenases.
ATPase	Hydrolyzes ATP. Many proteins with a wide range of roles have an energy-harnessing ATPase activity as part of their function, including motor proteins such as myosin and membrane transport proteins such as the sodium–potassium pump.

Enzyme names typically end in "-ase," with the exception of some enzymes, such as pepsin, trypsin, thrombin, lysozyme, and so on, which were discovered and named before the convention became generally accepted at the end of the nineteenth century. The name of an enzyme usually indicates the substrate and the nature of the reaction catalyzed. For example, citrate synthase catalyzes the synthesis of citrate by a reaction between acetyl CoA and oxaloacetate.

PANEL 4–3 Making and using antibodies

THE ANTIBODY MOLECULE

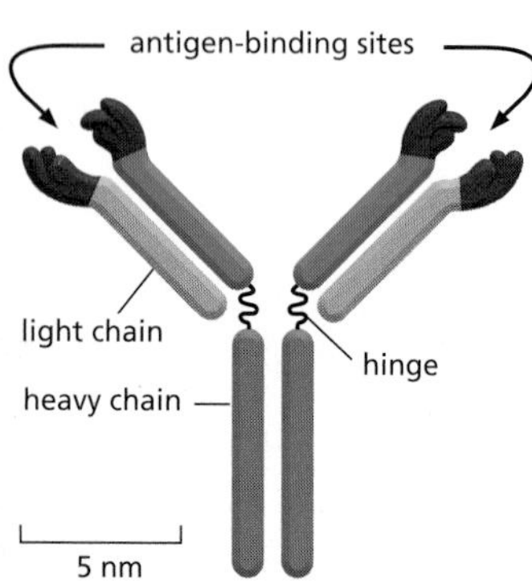

Antibodies are proteins that bind very tightly to their targets (antigens). They are produced in vertebrates as a defense against infection. Each antibody molecule is made of two identical light chains and two identical heavy chains, so the two antigen-binding sites are identical.

ANTIBODY SPECIFICITY

heavy chain

antigen

light chain

An individual animal can make billions of different antibody molecules, each with a distinct antigen-binding site. Each antibody recognizes its antigen with great specificity.

ANTIBODIES DEFEND US AGAINST INFECTION

foreign molecules

viruses

bacteria

ANTIBODIES (Y) FORM AGGREGATES

Antibody and antigen aggregates are ingested by phagocytic cells.

Special proteins in blood kill antibody-coated bacteria or viruses.

B CELLS PRODUCE ANTIBODIES

Antibodies are made by a class of white blood cells called B lymphocytes, or B cells. Each resting B cell carries a different membrane-bound antibody molecule on its surface that serves as a receptor for recognizing a specific antigen. When antigen binds to this receptor, the B cell is stimulated to divide and to secrete large amounts of the same antibody in a soluble form.

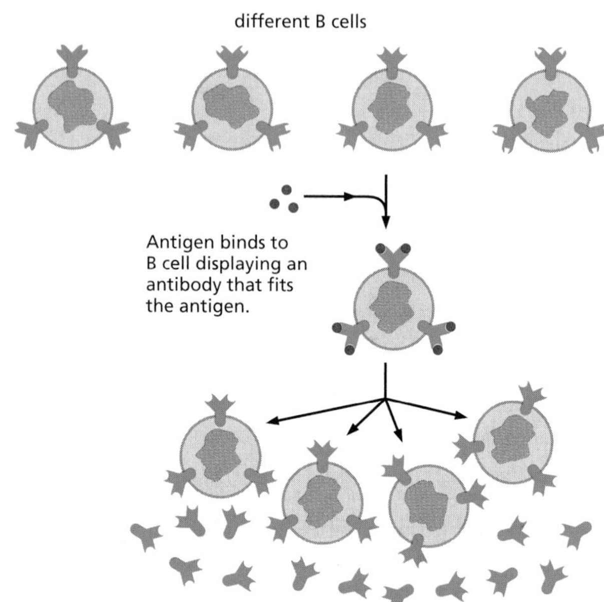

B cell is stimulated to make and secrete more of same antibody.

RAISING ANTIBODIES IN ANIMALS

Antibodies can be made in the laboratory by injecting an animal (usually a mouse, rabbit, sheep, or goat) with antigen A.

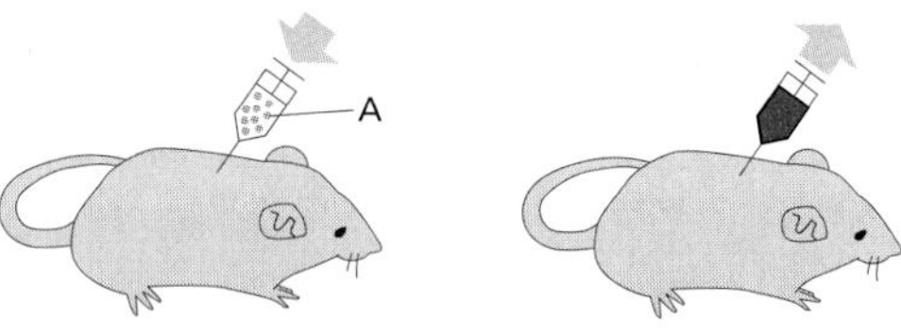

Repeated injections of the same antigen at intervals of several weeks stimulate specific B cells to secrete large amounts of anti-A antibodies into the bloodstream.

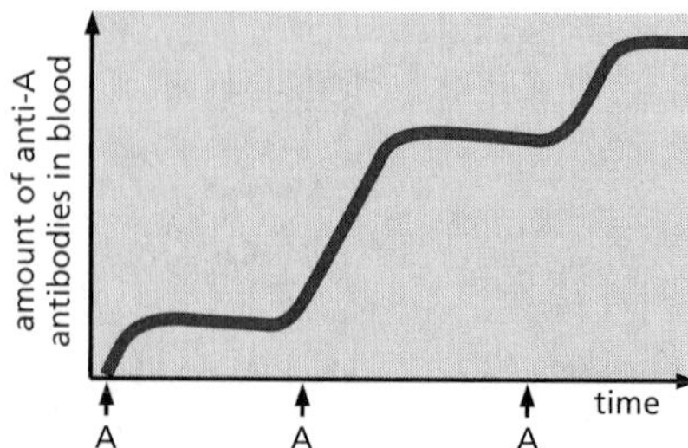

Because many different B cells are stimulated by antigen A, the blood will contain a variety of anti-A antibodies, each of which binds A in a slightly different way.

USING ANTIBODIES TO PURIFY MOLECULES

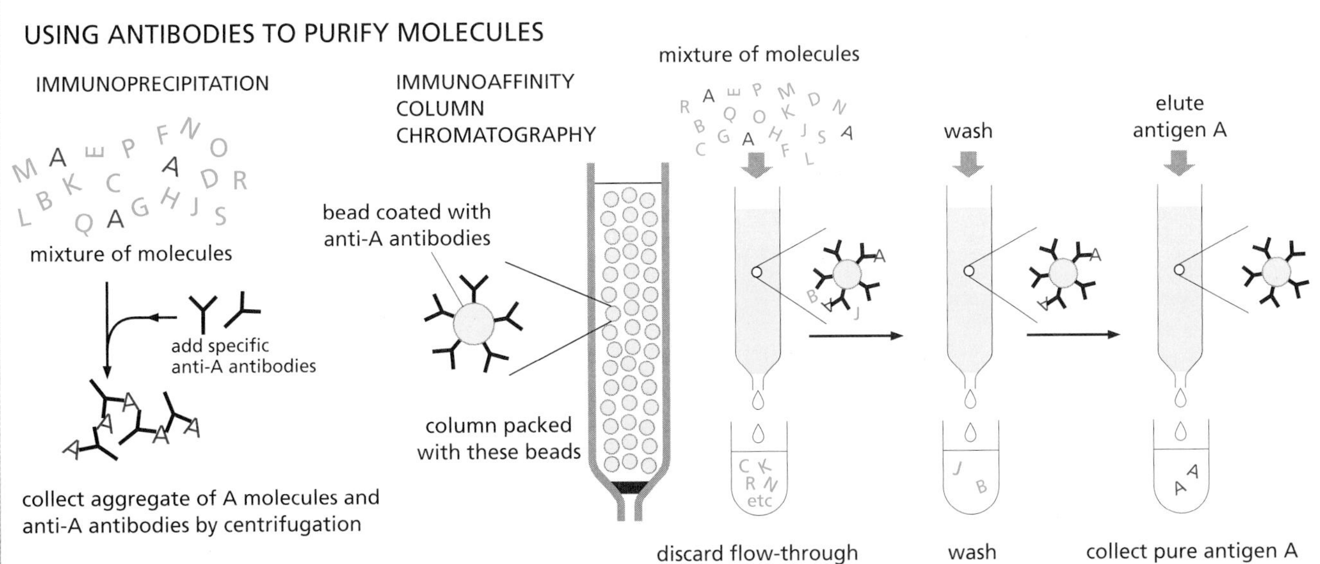

MONOCLONAL ANTIBODIES

Large quantities of a single type of antibody molecule can be obtained by fusing a B cell (taken from an animal injected with antigen A) with a tumor cell. The resulting hybrid cell divides indefinitely and secretes anti-A antibodies of a single (monoclonal) type.

B cell from animal injected with antigen A makes anti-A antibody but does not divide forever.

Tumor cell from cell culture divides indefinitely but does not make antibody.

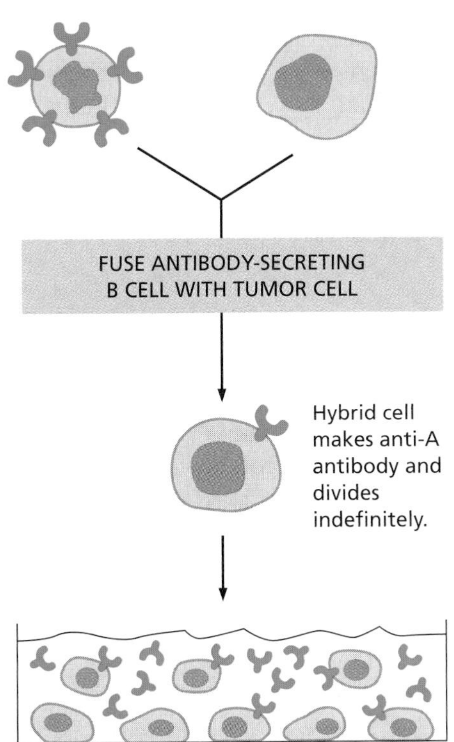

USING ANTIBODIES AS MOLECULAR TAGS

couple to fluorescent dye, colloidal gold particle, or other special tag

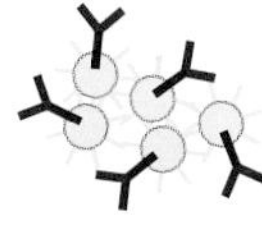

specific antibodies against antigen A

labeled antibodies

MICROSCOPIC DETECTION

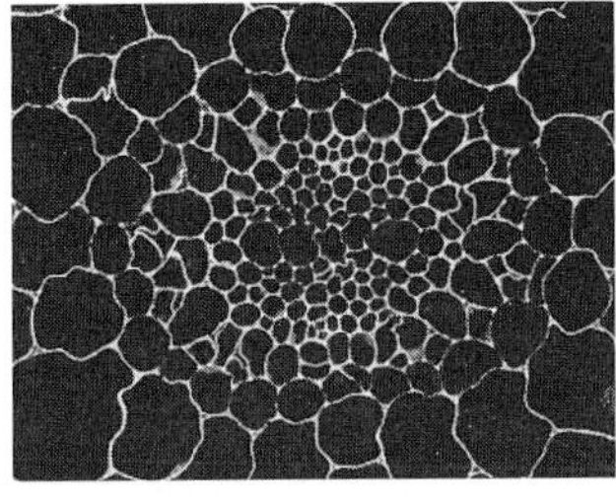

Fluorescent antibody binds to antigen A in tissue and is detected by fluorescence in a light microscope. The antigen here is pectin in the cell walls of a slice of plant tissue.

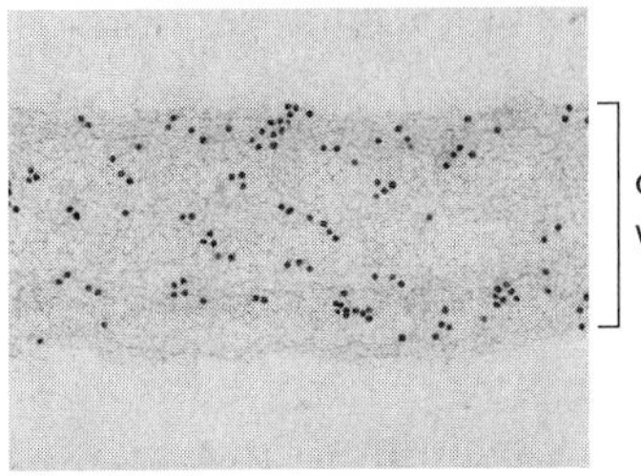

Gold-labeled antibody binds to antigen A in tissue and is detected in an electron microscope. The antigen is pectin in the cell wall of a single plant cell.

BIOCHEMICAL DETECTION

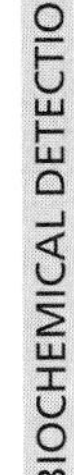

Antigen A is separated from other molecules by electrophoresis.

Incubation with the labeled antibodies that bind to antigen A allows the position of the antigen to be determined.

Note: In all cases, the sensitivity can be greatly increased by using multiple layers of antibodies. This "sandwich" method enables smaller numbers of antigen molecules to be detected.

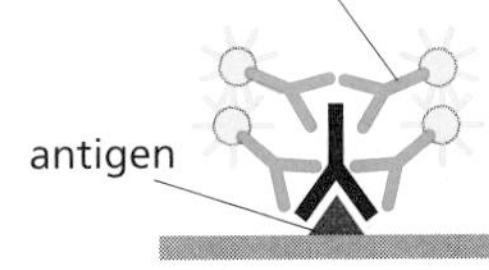

cell is under pressure due to osmotic forces, cutting even a small number of polysaccharide chains causes the cell wall to rupture and the bacterium to burst. Lysozyme is a relatively small and stable protein that can be isolated easily in large quantities. For these reasons, it has been intensively studied, and it was the first enzyme whose structure was worked out in atomic detail by X-ray crystallography.

The reaction catalyzed by lysozyme is a hydrolysis: the enzyme adds a molecule of water to a single bond between two adjacent sugar groups in the polysaccharide chain, thereby causing the bond to break. The reaction is energetically favorable because the free energy of the severed polysaccharide chain is lower than the free energy of the intact chain. However, the pure polysaccharide can sit for years in water without being hydrolyzed to any detectable degree. This is because there is an energy barrier to such reactions, called the *activation energy* (discussed in Chapter 3). For a colliding water molecule to break a bond linking two sugars, the polysaccharide molecule has to be distorted into a particular shape—the **transition state**—in which the atoms around the bond have an altered geometry and electron distribution. To distort the molecule in this way requires a large input of energy from random molecular collisions. In aqueous solution at room temperature, the energy of such collisions almost never exceeds the activation energy; therefore, hydrolysis occurs extremely slowly, if at all.

This is where the enzyme comes in. Like all enzymes, lysozyme has a special binding site on its surface, termed an **active site**, that cradles the contours of its substrate molecule. Here the catalysis of the chemical reaction occurs. Because its substrate is a polymer, lysozyme's active site is a long groove that holds six linked sugars at the same time. As soon as the polysaccharide binds to form an enzyme–substrate complex, the enzyme cuts the polysaccharide by catalyzing the addition of a water molecule to one of its sugar–sugar bonds. The severed chain is then quickly released, freeing the enzyme for further cycles of reaction (Figure 4–30).

The chemistry that underlies the binding of lysozyme to its substrate is the same as that for antibody binding—the formation of multiple noncovalent bonds. However, lysozyme holds its polysaccharide substrate in such a way that one of the two sugars involved in the bond to be broken is distorted from its normal, most stable conformation. The bond to be

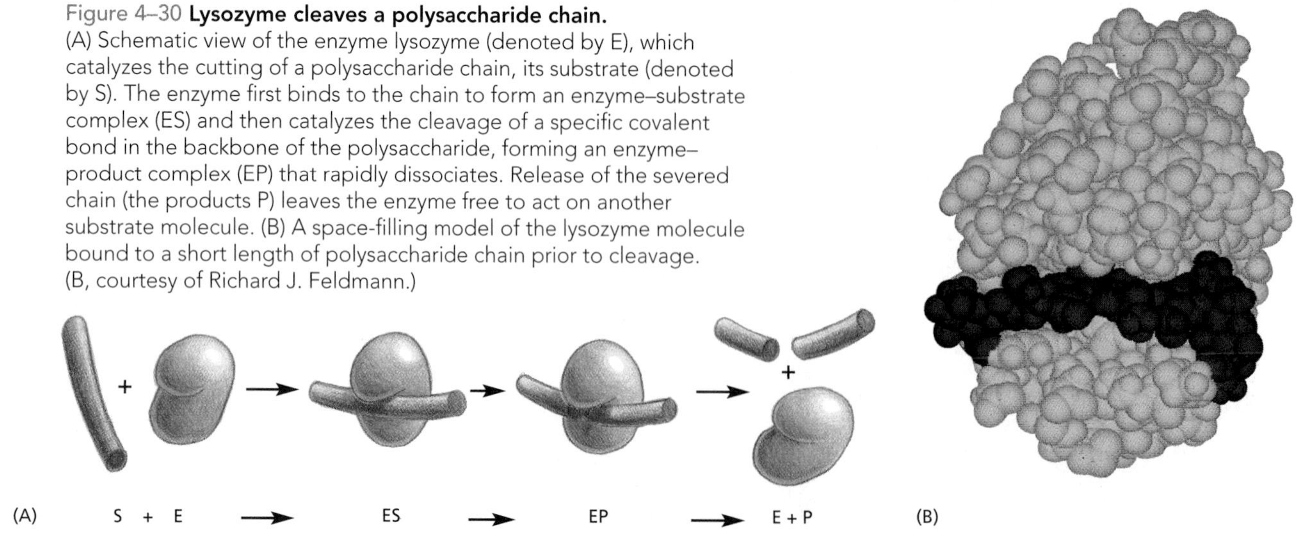

Figure 4–30 **Lysozyme cleaves a polysaccharide chain.** (A) Schematic view of the enzyme lysozyme (denoted by E), which catalyzes the cutting of a polysaccharide chain, its substrate (denoted by S). The enzyme first binds to the chain to form an enzyme–substrate complex (ES) and then catalyzes the cleavage of a specific covalent bond in the backbone of the polysaccharide, forming an enzyme–product complex (EP) that rapidly dissociates. Release of the severed chain (the products P) leaves the enzyme free to act on another substrate molecule. (B) A space-filling model of the lysozyme molecule bound to a short length of polysaccharide chain prior to cleavage. (B, courtesy of Richard J. Feldmann.)

(A) S + E → ES → EP → E + P (B)

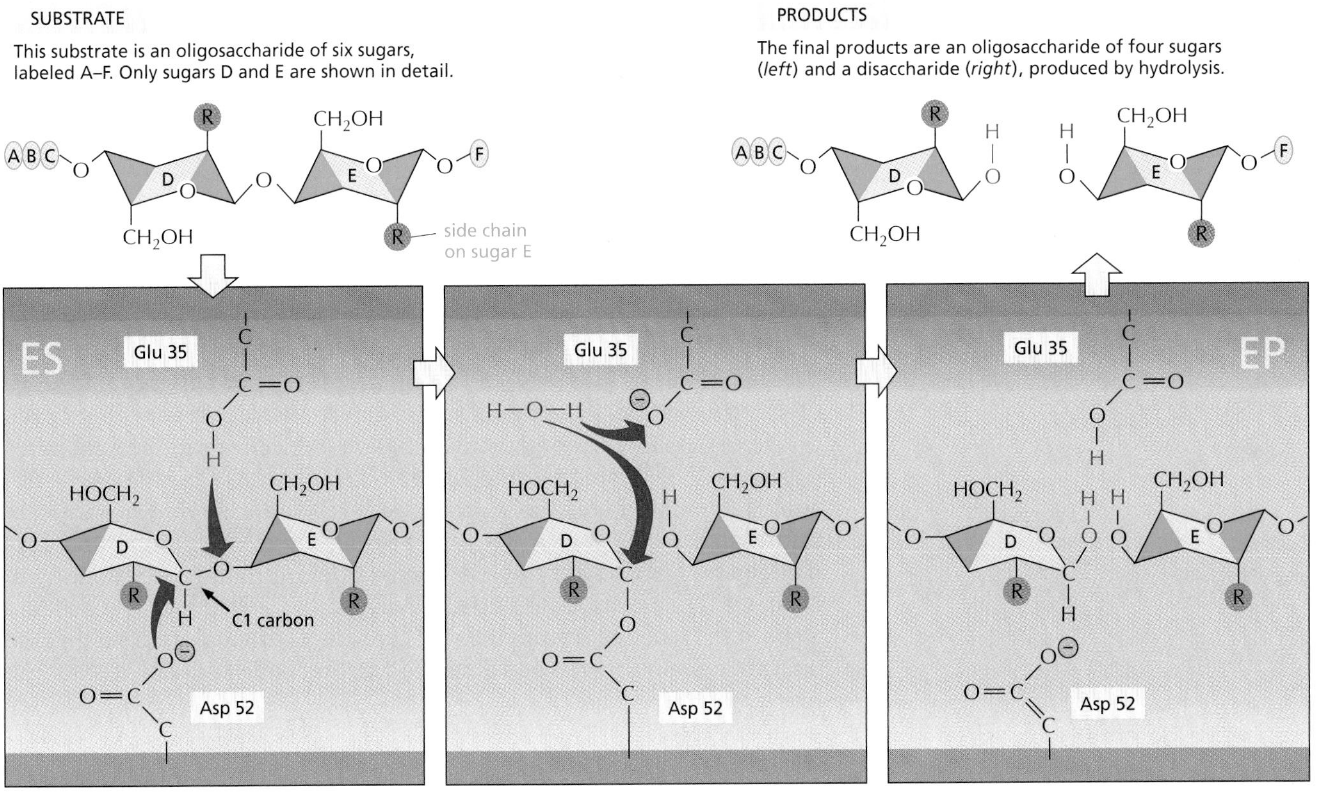

SUBSTRATE

This substrate is an oligosaccharide of six sugars, labeled A–F. Only sugars D and E are shown in detail.

PRODUCTS

The final products are an oligosaccharide of four sugars (*left*) and a disaccharide (*right*), produced by hydrolysis.

In the enzyme–substrate complex (ES), the enzyme forces sugar D into a strained conformation, with Glu 35 positioned to serve as an acid that attacks the adjacent sugar–sugar bond by donating a proton (H^+) to sugar E, and Asp 52 poised to attack the C1 carbon atom.

The Asp 52 has formed a covalent bond between the enzyme and the C1 carbon atom of sugar D. The Glu 35 then polarizes a water molecule (*red*), so that its oxygen can readily attack the C1 carbon atom and displace Asp 52.

The reaction of the water molecule (*red*) completes the hydrolysis and returns the Asp 52 to its initial state, forming the final enzyme–product complex (EP).

Figure 4–31 **In the active site of lysozyme, bonds are bent and broken.** The top row shows the free substrate and the free products. The three lower panels depict sequential events at the enzyme active site during which the sugar–sugar bond is broken. Note the change in the conformation of sugar D in the enzyme–substrate complex compared with the free substrate; this conformation favors the formation of the transition state. The reaction, and the structure of lysozyme bound to its product, are shown in Movie 4.8 and Movie 4.9. (Based on D.J. Vocadlo et al., *Nature* 412:835–838, 2001. With permission from Macmillan Publishers Ltd.)

broken is also held close to two amino acids with acidic side chains: a glutamic acid and an aspartic acid within the active site.

Conditions are thereby created in the microenvironment of the lysozyme active site that greatly reduce the activation energy necessary for the hydrolysis to take place. Figure 4–31 shows the main intermediates in this enzymatically catalyzed reaction.

1. The enzyme stresses its bound substrate (here an oligosaccharide of six sugars) by bending some critical chemical bonds in one sugar, so that the shape of this sugar more closely resembles the shape of high-energy transition states formed during the reaction.
2. The negatively charged aspartic acid reacts with the C1 carbon atom on the distorted sugar, breaking this sugar–sugar bond and leaving the aspartic acid covalently linked to the site of bond cleavage.
3. Aided by the negatively charged glutamic acid, a water molecule reacts with the C1 carbon atom, displacing the aspartic acid and completing the process of hydrolysis.

The overall chemical reaction, from the initial binding of the polysaccharide on the surface of the enzyme to the final release of the severed chains, occurs many millions of times faster than it would in the absence of enzyme.

Other enzymes use similar mechanisms to lower activation energies and speed up the reactions they catalyze. In reactions involving two or more substrates, the active site also acts like a template or mold that brings the reactants together in the proper orientation for chemistry to occur

Figure 4–32 **Enzymes can encourage catalysis in several ways.** (A) Holding substrates together in a precise alignment. (B) Charge stabilization of reaction intermediates. (C) Altering bond angles in the substrate to increase the rate of a particular reaction.

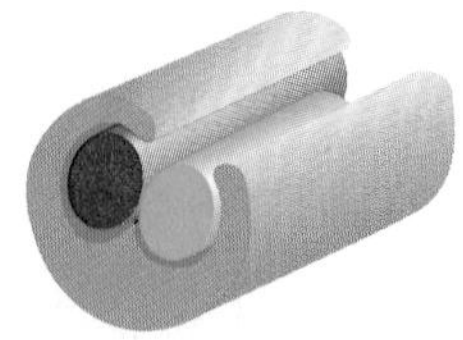

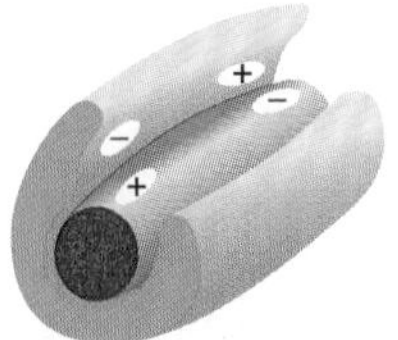

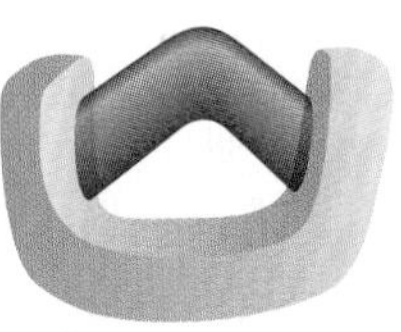

(A) enzyme binds to two substrate molecules and orients them precisely to encourage a reaction to occur between them

(B) binding of substrate to enzyme rearranges electrons in the substrate, creating partial negative and positive charges that favor a reaction

(C) enzyme strains the bound substrate molecule, forcing it toward a transition state to favor a reaction

between them (Figure 4–32A). As we saw for lysozyme, the active site of an enzyme contains precisely positioned chemical groups that speed up the reaction by altering the distribution of electrons in the substrates (Figure 4–32B). Binding to the enzyme also changes the shapes of substrates, bending bonds so as to drive a substrate toward a particular transition state (Figure 4–32C). Finally, like lysozyme, many enzymes participate intimately in the reaction by briefly forming a covalent bond between the substrate and a side chain in the active site. Subsequent steps in the reaction restore the side chain to its original state, so that the enzyme remains unchanged after the reaction and can go on to catalyze many more reactions.

Most Drugs Inhibit Enzymes

Many of the drugs we take to treat or prevent illness work by blocking the activity of a particular enzyme. Cholesterol-lowering statins inhibit HMG-CoA reductase, an enzyme involved in the synthesis of cholesterol by the liver. Methotrexate kills some types of cancer cells by shutting down dihydrofolate reductase, an enzyme that produces a compound required for DNA synthesis. Because cancer cells have lost important intracellular control systems, some of them are unusually sensitive to treatments that interrupt chromosome replication, making them susceptible to methotrexate.

Pharmaceutical companies often develop drugs by first using automated methods to screen massive libraries of compounds to find chemicals that are able to inhibit the activity of an enzyme of interest. They can then chemically modify the most promising to make them even more effective, enhancing their binding affinity and specificity for the target enzyme. As we shall see in Chapter 20, the anticancer drug Gleevec was designed to specifically inhibit an enzyme whose aberrant behavior is required for the growth of a type of cancer called chronic myeloid leukemia. The drug binds tightly in the substrate-binding pocket of the enzyme, blocking its activity (see Figure 20–53).

Tightly Bound Small Molecules Add Extra Functions to Proteins

Although the order of amino acids in proteins gives these molecules their shape and the versatility to perform different functions, sometimes the amino acids by themselves are not enough. Just as we use tools to enhance and extend the capabilities of our hands, so proteins often employ small nonprotein molecules to perform functions that would be difficult or impossible using amino acids alone. Thus the photoreceptor protein *rhodopsin*, which is the purple, light-sensitive pigment made by the rod cells in the retina, detects light by means of a small molecule, *retinal*, which is attached to the protein by a covalent bond to a lysine side chain (Figure 4–33A). Retinal changes its shape when it absorbs a photon

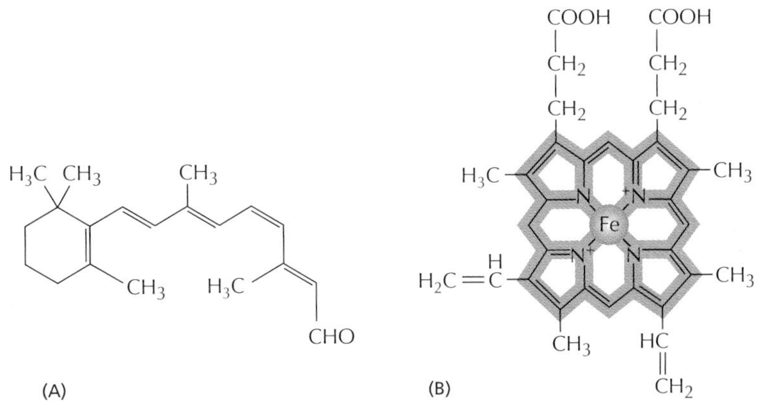

Figure 4–33 **Retinal and heme enhance the function of certain proteins.** (A) The structure of retinal, the light-sensitive molecule attached to rhodopsin in our eyes. (B) The structure of a heme group, shown with the carbon-containing heme ring colored *red* and the iron atom at its center in *orange*. A heme group is tightly bound to each of the four polypeptide chains in hemoglobin, the oxygen-carrying protein whose structure was shown in Figure 4–20.

of light, and this change is amplified by rhodopsin to trigger a cascade of enzymatic reactions that eventually leads to an electrical signal being carried to the brain.

Another example of a protein that contains a nonprotein portion essential for its function is *hemoglobin* (see Figure 4–20). A molecule of hemoglobin carries four noncovalently bound *heme* groups, ring-shaped molecules each with a single central iron atom (Figure 4–33B). Heme gives hemoglobin (and blood) its red color. By binding reversibly to dissolved oxygen gas through its iron atom, heme enables hemoglobin to pick up oxygen in the lungs and release it in the tissues.

When these small molecules are attached covalently and permanently to their protein, they become an integral part of the protein molecule itself. We will see in Chapter 11 that proteins are often anchored to cell membranes through covalently attached lipid molecules. And membrane proteins exposed on the surface of the cell, as well as proteins secreted outside the cell, are often modified by the covalent addition of sugars and oligosaccharides.

Enzymes, too, make use of nonprotein molecules: they frequently have a small molecule or metal atom associated with their active site that assists with their catalytic function. *Carboxypeptidase*, an enzyme that cuts polypeptide chains, carries a tightly bound zinc ion in its active site. During the cleavage of a peptide bond by carboxypeptidase, the zinc ion forms a transient bond with one of the substrate atoms, thereby assisting the hydrolysis reaction. In other enzymes, a small organic molecule serves a similar purpose. *Biotin*, for example, is found in enzymes that transfer a carboxylate group ($-COO^-$) from one molecule to another (see Figure 3–37). Biotin participates in these reactions by forming a transient covalent bond to the $-COO^-$ group to be transferred, thereby forming an activated carrier molecule (see Table 3–2, p. 109). This small molecule is better suited for this function than any of the amino acids used to make proteins. Because biotin cannot be synthesized by humans, it must be provided by the diet; thus biotin is classified as a *vitamin*. Other vitamins are similarly needed to make small molecules that are essential components of our proteins; vitamin A, for example, is needed in the diet to make retinal, the light-sensitive part of rhodopsin.

HOW PROTEINS ARE CONTROLLED

So far we have examined how proteins do their jobs: how binding to other proteins or small molecules allows them to perform their specific functions. But inside the cell, most proteins and enzymes do not work continuously, or at full speed. Instead, their activities are regulated in a coordinated fashion so that the cell can maintain itself in an optimal

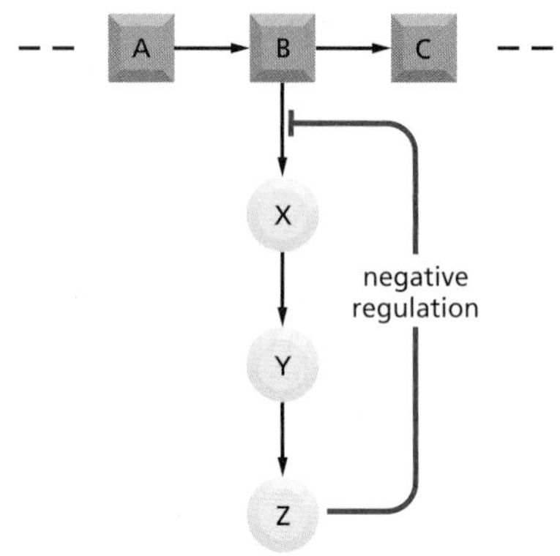

Figure 4–34 **Feedback inhibition regulates the flow through biosynthetic pathways.** B is the first metabolite in a pathway that gives the end product Z. Z inhibits the first enzyme that is unique to its synthesis and thereby controls its own concentration in the cell. This is an example of negative regulation.

state, generating only those molecules it requires to thrive under the current conditions. By coordinating when—and how vigorously—proteins function, the cell ensures that it does not deplete its energy reserves by accumulating molecules it does not need or waste its stockpiles of critical substrates. We now consider how cells control the activity of proteins and enzymes.

The regulation of protein activity occurs at many levels. At one level, the cell controls how many molecules of each enzyme it makes by regulating the expression of the gene that encodes that protein (discussed in Chapter 8). At another level, the cell controls enzymatic activities by confining sets of enzymes to particular subcellular compartments, often—but not always—enclosed by distinct membranes (discussed in Chapters 14 and 15). But the most rapid and general process used to adjust reaction rates operates at the level of the enzyme itself. Although proteins can be switched on—or switched off—by a variety of mechanisms, as we see next, all in some way cause the protein to alter its shape, and therefore its function.

The Catalytic Activities of Enzymes Are Often Regulated by Other Molecules

A living cell contains thousands of enzymes, many of which are operating at the same time and in the same small volume of the cytosol. By their catalytic action, enzymes generate a complex web of metabolic pathways composed of chains of chemical reactions in which the product of one enzyme becomes the substrate of the next. In this maze of pathways there are many branch points where different enzymes compete for the same substrate. The system is so complex that elaborate controls are required to regulate when and how rapidly each reaction occurs.

The most common type of control occurs when a molecule other than a substrate specifically binds to an enzyme at a special regulatory site outside of the active site, altering the rate at which the enzyme converts its substrates to products. In **feedback inhibition**, an enzyme acting early in a reaction pathway is inhibited by a late product of that pathway. Thus, whenever large quantities of the final product begin to accumulate, the product binds to an earlier enzyme and slows down its catalytic action, limiting further entry of substrates into that reaction pathway (Figure 4–34). Where pathways branch or intersect, there are usually multiple points of control by different final products, each of which works to regulate its own synthesis (Figure 4–35). Feedback inhibition can work almost instantaneously, and is rapidly reversed when product levels fall.

Feedback inhibition is a *negative regulation*: it prevents an enzyme from acting. Enzymes can also be subject to *positive regulation*, in which the enzyme's activity is stimulated by a regulatory molecule rather than being shut down. Positive regulation occurs when a product in one branch of the metabolic maze stimulates the activity of an enzyme in another pathway. As one example, the accumulation of ADP activates several enzymes involved in the oxidation of sugar molecules, thereby stimulating the cell to convert more ADP to ATP.

QUESTION 4–7

Consider the drawing in Figure 4–34. What will happen if, instead of the indicated feedback,

A. Feedback inhibition from Z affects the step B → C only?
B. Feedback inhibition from Z affects the step Y → Z only?
C. Z is a positive regulator of the step B → X?
D. Z is a positive regulator of the step B → C?

For each case, discuss how useful these regulatory schemes would be for a cell.

Allosteric Enzymes Have Binding Sites That Influence One Another

One feature of feedback inhibition was initially puzzling to those who discovered it: the regulatory molecule often has a shape that is totally different from the shape of the enzyme's preferred substrate. Indeed, when this form of regulation was discovered in the 1960s, it was termed

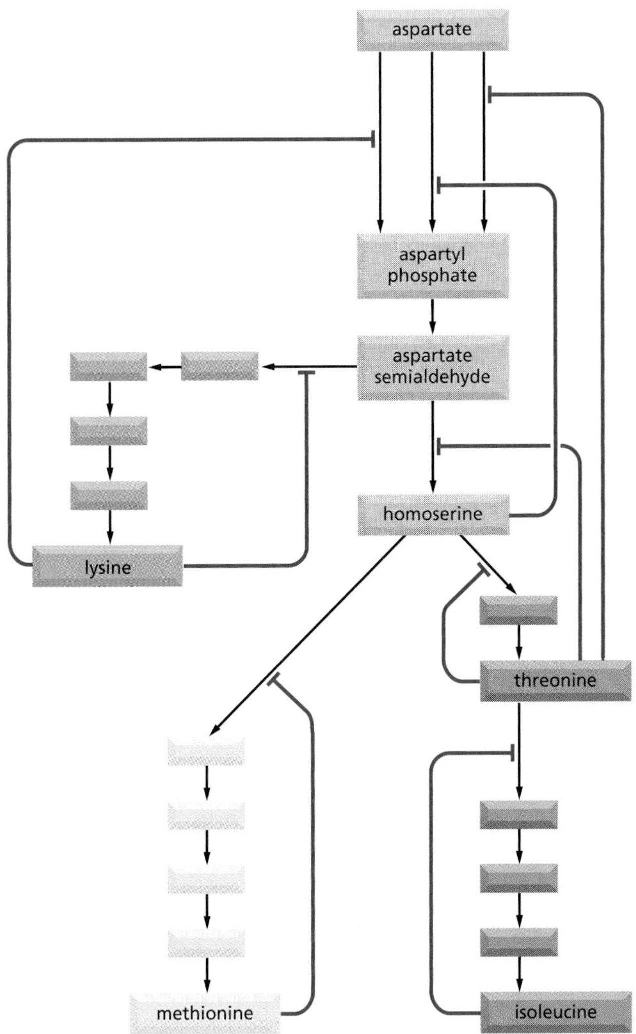

Figure 4–35 **Feedback inhibition at multiple sites regulates connected metabolic reactions.** Illustrated are the biosynthetic pathways for four different amino acids in bacteria starting from the amino acid aspartate. The *red* lines indicate positions at which products feed back to inhibit enzymes. In this example, each amino acid controls the first enzyme specific to its own synthesis, thereby controlling its own levels and avoiding a wasteful buildup of intermediates. The products can also separately inhibit the initial set of reactions common to all the syntheses because three different enzymes catalyze the initial reaction from aspartate to aspartyl phosphate; each of these enzymes is inhibited by a different product.

allostery (from the Greek *allo,* "other," and *stere,* "solid" or "shape"). As more was learned about feedback inhibition, researchers realized that many enzymes must have at least two different binding sites on their surface: the active site that recognizes the substrates and one or more sites that recognize regulatory molecules. Furthermore, the substrate and regulatory sites must somehow "communicate" in a way that allows the catalytic events at the active site to be influenced by the binding of the regulatory molecule at its separate site.

The interaction between sites that are located on separate regions of a protein molecule is now known to depend on *conformational changes* in the protein: binding at one of the sites causes a shift in the protein's structure from one folded shape to a slightly different folded shape. Many enzymes have two conformations that differ in activity, each stabilized by the binding of different ligands. During feedback inhibition, for example, the binding of an inhibitor at one site on the protein causes the protein to shift to a conformation in which its active site—located elsewhere in the protein—becomes less accommodating to the substrate molecule (Figure 4–36).

Many—if not most—protein molecules are **allosteric**: they can adopt two or more slightly different conformations, and by a shift from one to another, their activity can be regulated. This is true not only for enzymes but also for many other proteins—including receptors, structural proteins, and motor proteins. The chemistry involved here is extremely simple in

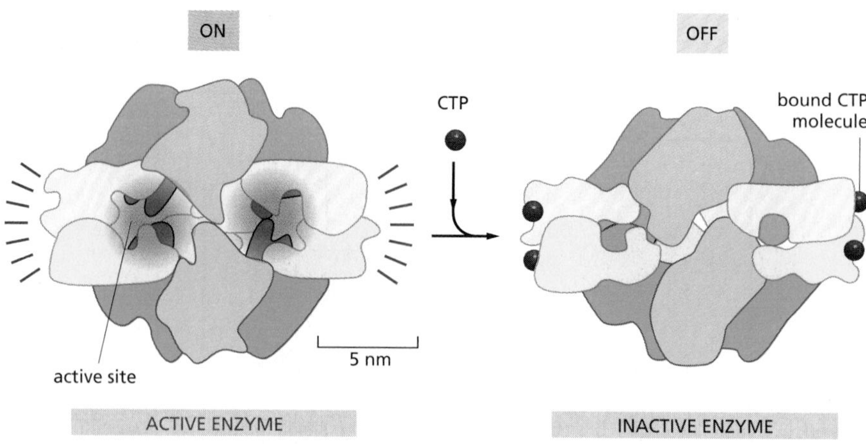

Figure 4–36 **Feedback inhibition triggers a conformational change.** An enzyme used in early studies of allosteric regulation was aspartate transcarbamoylase from *E. coli*. This large multisubunit enzyme (see Figure 4–9) catalyzes an important reaction that begins the synthesis of the pyrimidine ring of C, U, and T nucleotides. One of the final products of this pathway, cytosine triphosphate (CTP), binds to the enzyme to turn it off whenever CTP is plentiful. This diagram shows the conformational change that occurs when the enzyme is turned off by CTP binding. Note that the aspartate transcarbamoylase shown in Figure 4–9 is seen from the top. This figure depicts the enzyme as seen from the side.

concept: because each protein conformation will have somewhat different contours on its surface, the protein's binding sites for ligands will be altered when the protein changes shape. Each ligand will stabilize the conformation that it binds to most strongly, and at high enough concentrations the ligand will tend to "switch" the population of proteins to the conformation that it favors (Figure 4–37).

Phosphorylation Can Control Protein Activity by Triggering a Conformational Change

Enzymes are not only regulated by the binding of small molecules. Another method commonly used by eucaryotic cells to regulate protein activity involves attaching a phosphate group covalently to one of its amino acid side chains. Because each phosphate group carries two negative charges, the enzyme-catalyzed addition of a phosphate group to a protein can cause a major conformational change by, for example, attracting a cluster of positively charged amino acid side chains. This conformational change can, in turn, affect the binding of ligands elsewhere on the protein surface—thus altering the protein's activity. Removal of the phosphate group by a second enzyme returns the protein to its original conformation and restores its initial activity.

This reversible **protein phosphorylation** controls the activity of many different types of proteins in eucaryotic cells; in fact, this method is used so extensively that more than one-third of the 10,000 or so proteins in a typical mammalian cell appear to be phosphorylated at any one time. The addition and removal of phosphate groups from specific proteins often occurs in response to signals that specify some change in a cell's state. For example, the complicated series of events that takes place as a

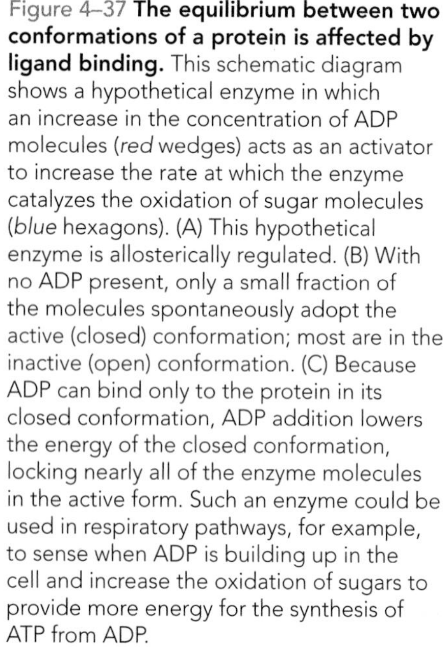

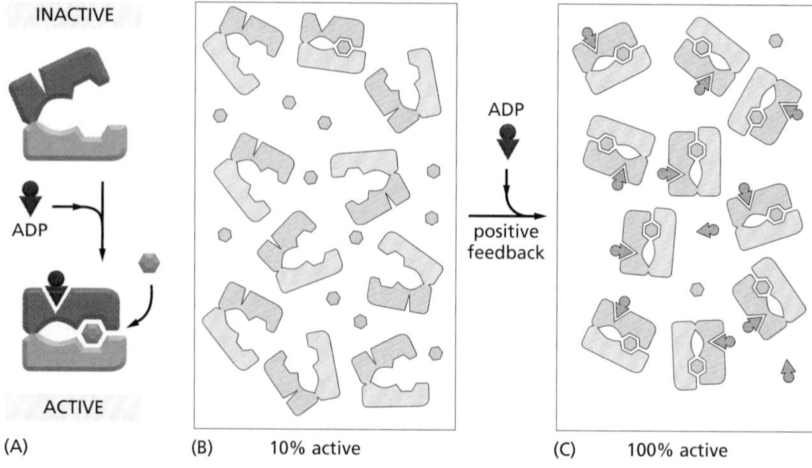

Figure 4–37 **The equilibrium between two conformations of a protein is affected by ligand binding.** This schematic diagram shows a hypothetical enzyme in which an increase in the concentration of ADP molecules (*red* wedges) acts as an activator to increase the rate at which the enzyme catalyzes the oxidation of sugar molecules (*blue* hexagons). (A) This hypothetical enzyme is allosterically regulated. (B) With no ADP present, only a small fraction of the molecules spontaneously adopt the active (closed) conformation; most are in the inactive (open) conformation. (C) Because ADP can bind only to the protein in its closed conformation, ADP addition lowers the energy of the closed conformation, locking nearly all of the enzyme molecules in the active form. Such an enzyme could be used in respiratory pathways, for example, to sense when ADP is building up in the cell and increase the oxidation of sugars to provide more energy for the synthesis of ATP from ADP.

eucaryotic cell divides is timed in this way (discussed in Chapter 18). And many of the signals generated by hormones and neurotransmitters are carried from the plasma membrane to the nucleus by a cascade of protein phosphorylation events (discussed in Chapter 16).

Protein phosphorylation involves the enzyme-catalyzed transfer of the terminal phosphate group of ATP to the hydroxyl group on a serine, threonine, or tyrosine side chain of the protein. This reaction is catalyzed by a **protein kinase**. The reverse reaction—removal of the phosphate group, or *dephosphorylation*—is catalyzed by a **protein phosphatase** (Figure 4–38A). Phosphorylation can either stimulate protein activity or inhibit it, depending on the protein involved and the site at which it is being phosphorylated (Figure 4–38B). Cells contain hundreds of different protein kinases, each responsible for phosphorylating a different protein or set of proteins. Cells also contain many different protein phosphatases; some of these are highly specific and remove phosphate groups from only one or a few proteins, whereas others act on a broad range of proteins. The state of phosphorylation of a protein at any moment in time, and thus its activity, will depend on the relative activities of the protein kinases and phosphatases that act on it.

For many proteins, a phosphate group is added to a particular side chain and then removed in a continuous cycle. Phosphorylation cycles of this kind allow proteins to switch rapidly from one state to another. The more rapidly the cycle is "turning," the faster the concentration of a phosphorylated protein can change in response to a sudden stimulus that increases its rate of phosphorylation. However, keeping the cycle turning costs energy because one molecule of ATP is hydrolyzed with each turn of the cycle.

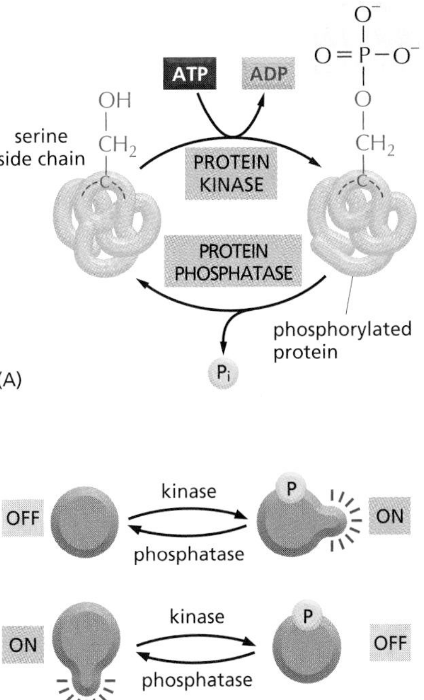

Figure 4–38 **Protein phosphorylation is a very common means of regulating protein activity.** Many thousands of proteins in a typical eucaryotic cell are modified by the covalent addition of one or more phosphate groups. (A) The general reaction, shown here, entails transfer of a phosphate group from ATP to an amino acid side chain of the target protein by a protein kinase. Removal of the phosphate group is catalyzed by a second enzyme, a protein phosphatase. In this example, the phosphate is added to a serine side chain; in other cases, the phosphate is instead linked to the –OH group of a threonine or a tyrosine in the protein. (B) The phosphorylation of a protein by a protein kinase can either increase or decrease the protein's activity, depending on the site of phosphorylation and the structure of the protein.

GTP-Binding Proteins Are Also Regulated by the Cyclic Gain and Loss of a Phosphate Group

Eucaryotic cells have a second way to regulate protein activity by phosphate addition and removal. In this case, instead of being enzymatically transferred from ATP to the protein, the phosphate is part of a guanine nucleotide—either guanosine triphosphate (GTP) or guanosine diphosphate (GDP)—that is bound tightly to the protein. Such **GTP-binding proteins** are in their active conformations with GTP bound; the protein itself then hydrolyzes this GTP to GDP—releasing a phosphate—and flips to an inactive conformation. As with protein phosphorylation, this process is reversible. The active conformation is regained by dissociation of the GDP, followed by the binding of a fresh molecule of GTP (Figure 4–39).

A large number of related GTP-binding proteins function as molecular switches in cells. The dissociation of GDP and its replacement by GTP, which turns the switch on, is often stimulated in response to a signal received by the cell. The GTP-binding proteins bind to other proteins to control their activities, and their crucial role in intracellular signaling pathways will be discussed in detail in Chapter 16. Here we shall look at their general mechanism of action by examining the bacterial elongation

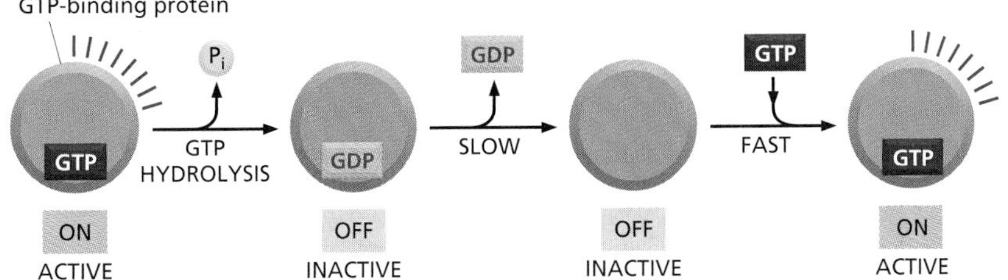

Figure 4–39 **GTP-binding proteins form molecular switches.** The activity of a GTP-binding protein (*green*) generally requires the presence of a tightly bound GTP molecule (switch on). Hydrolysis of this GTP molecule produces GDP and inorganic phosphate (P_i), and it causes the protein to convert to a different, usually inactive, conformation (switch off). As shown here, resetting the switch requires that the tightly bound GDP dissociate, a slow step that is greatly accelerated by specific signals; once the GDP dissociates, a molecule of GTP is quickly re-bound.

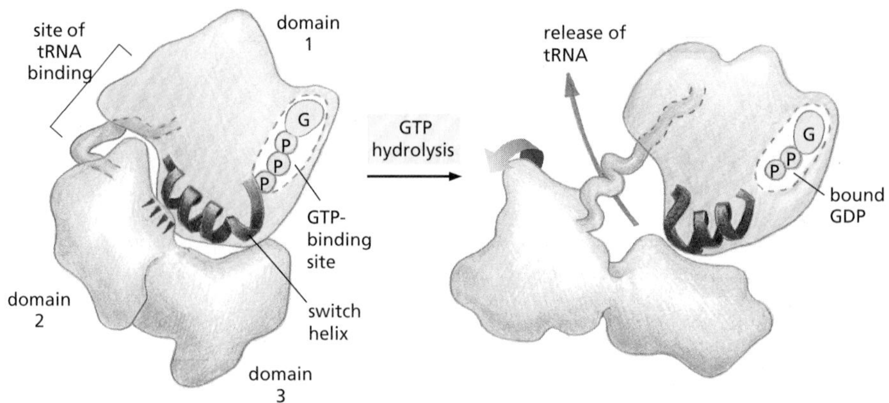

Figure 4–40 **A large conformational change is produced in EF-Tu in response to nucleotide hydrolysis.** The GTP-binding protein EF-Tu binds tRNA on the ribosome and plays a role in the elongation of a polypeptide chain during protein synthesis. In its GTP-bound form, EF-Tu binds a tRNA molecule tightly. The hydrolysis of bound GTP in EF-Tu causes only a minute change in the position of amino acids at the nucleotide-binding site (equivalent to a few times the diameter of a hydrogen atom). But this small change is magnified by conformational changes within the protein to produce a much larger movement. The hydrolysis of GTP releases an intramolecular bond, like a "latch" (*red* dashes in the structure on the *left*), which allows domains 2 and 3 to twist free and rotate by about 90° toward the viewer. This creates a major change in shape that releases the tRNA, as required to allow protein synthesis to proceed (**Movie 4.10**). These structures were determined by X-ray crystallography.

factor EF-Tu, a small GTP-binding protein that helps to load tRNA molecules onto ribosomes during protein synthesis (discussed in Chapter 7).

Analysis of the three-dimensional structure of EF-Tu has revealed how an allosteric transition triggered by the gain or loss of a phosphate on the bound guanine nucleotide can cause a major shape change in a GTP-binding protein. Figure 4–40 shows how the loss of a single phosphate group, which initially causes only a tiny movement of 0.1 nm or so at the binding site, is magnified by the protein to create a movement 50 times larger. Dramatic shape changes of this type also underlie the very large movements created by the types of proteins that we consider next.

Nucleotide Hydrolysis Allows Motor Proteins to Produce Large Movements in Cells

We have seen how conformational changes in proteins play a central part in enzyme regulation and cell signaling. But conformational changes also play another important role in the operation of the cell: they enable proteins whose major function is to move other molecules, the **motor proteins**, to generate the forces responsible for muscle contraction and many of the dramatic movements of cells. Motor proteins also power smaller-scale intracellular movements: they help move chromosomes to opposite ends of the cell during mitosis (discussed in Chapter 18), move organelles along molecular tracks within the cell (discussed in Chapter 17), and move enzymes along a DNA strand during the synthesis of a new DNA molecule (discussed in Chapter 6). An understanding of how proteins can operate as molecules with moving parts is therefore essential for understanding the molecular basis of cell behavior.

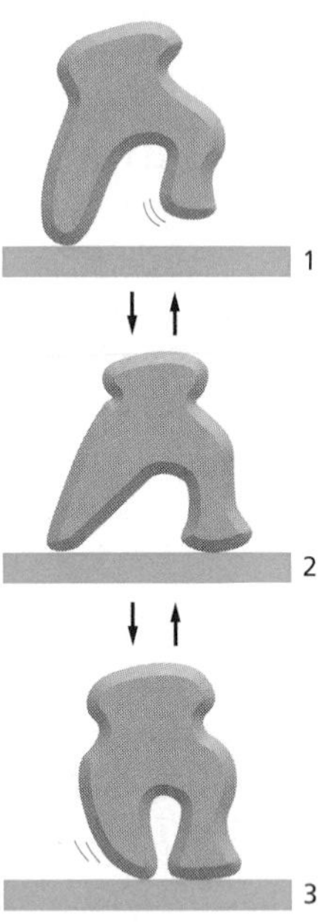

Figure 4–41 **Changes in conformation allow a protein to "walk" along a filament or thread.** This protein's three different conformations allow it to wander randomly back and forth while bound to a thread or a filament. But without an input of energy to drive its movement in a single direction, the protein will only shuffle aimlessly, getting nowhere.

How are shape changes in proteins used to generate orderly movements in cells? If, for example, a protein is required to walk along a narrow thread such as a DNA molecule, it can do this by undergoing a series of conformational changes—as illustrated in Figure 4–41. However, with nothing to drive these changes in an orderly sequence—in one direction only—they will be perfectly reversible and the protein will wander randomly back and forth.

To make the series of conformational changes unidirectional—and force the entire cycle to proceed in one direction—it is enough to make any one of the steps irreversible. For most proteins that are able to walk in a single direction for long distances, this irreversibility is achieved by coupling one of the conformational changes to the hydrolysis of an ATP molecule bound to the protein—as illustrated in the schematic model in Figure 4–42. The mechanism is similar to the one that drives allosteric shape changes by GTP hydrolysis. Because a great deal of free energy is released when ATP (or GTP) is hydrolyzed, it is very unlikely that the nucleotide-binding protein will undergo a reverse shape change—as

Figure 4–42 **An allosteric motor protein, driven by ATP hydrolysis, moves in one direction.** An orderly transition among three conformations is driven by the hydrolysis of a bound ATP molecule. Because one of these transitions—a change from conformation 2 to conformation 3—is coupled to the hydrolysis of ATP, the entire cycle is essentially irreversible. By repeated cycles the protein moves continuously to the right along the thread.

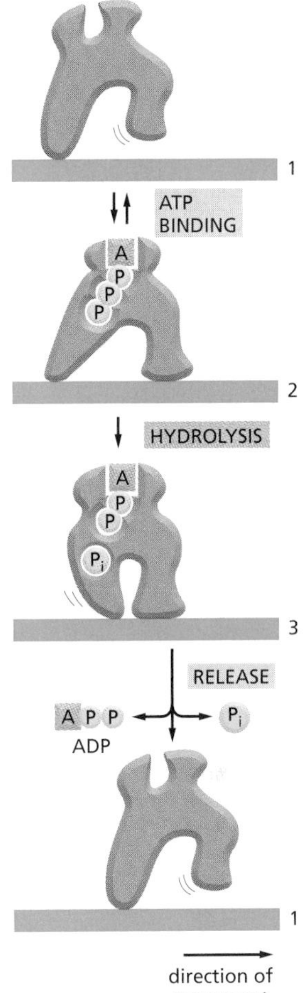

required for moving backward—since this would require that it also reverse the ATP hydrolysis by adding a phosphate molecule to ADP to form ATP. Such behavior is thermodynamically unfavorable and therefore highly unlikely. As a consequence, the protein moves rapidly forward.

Many motor proteins generate directional movement in this general way, including the muscle motor protein *myosin*—which "runs" along actin filaments to generate muscle contraction (discussed in Chapter 17)—and the *kinesin* protein involved in chromosome movements at mitosis (discussed in Chapter 18). Such movements can be rapid: some of the motor proteins involved in DNA replication propel themselves along a DNA strand at rates as high as 1000 nucleotides per second.

Proteins Often Form Large Complexes That Function as Protein Machines

As one progresses from small, single-domain proteins to large proteins formed from many domains, the functions that the proteins can perform become more elaborate. The most impressive tasks, however, are carried out by large protein assemblies formed from many protein molecules. Now that it is possible to reconstruct biological processes in cell-free systems in the laboratory, it is clear that each central process in a cell—such as DNA replication, protein synthesis, vesicle budding, and transmembrane signaling—is catalyzed by a highly coordinated, linked set of 10 or more proteins. In most such **protein machines** the hydrolysis of bound nucleoside triphosphates (ATP or GTP) drives an ordered series of conformational changes in some of the individual protein subunits, enabling the ensemble of proteins to move coordinately. In this way, the appropriate enzymes can be moved directly into the positions where they are needed to carry out successive reactions in a series as, for example, in protein synthesis on a ribosome (discussed in Chapter 7), or in DNA replication—where a large multiprotein complex moves rapidly along the DNA. A simple mechanical analogy is illustrated in Figure 4–43.

QUESTION 4–8

Explain how phosphorylation and the binding of a nucleotide can both be used to regulate protein activity. What do you suppose are advantages of either form of regulation?

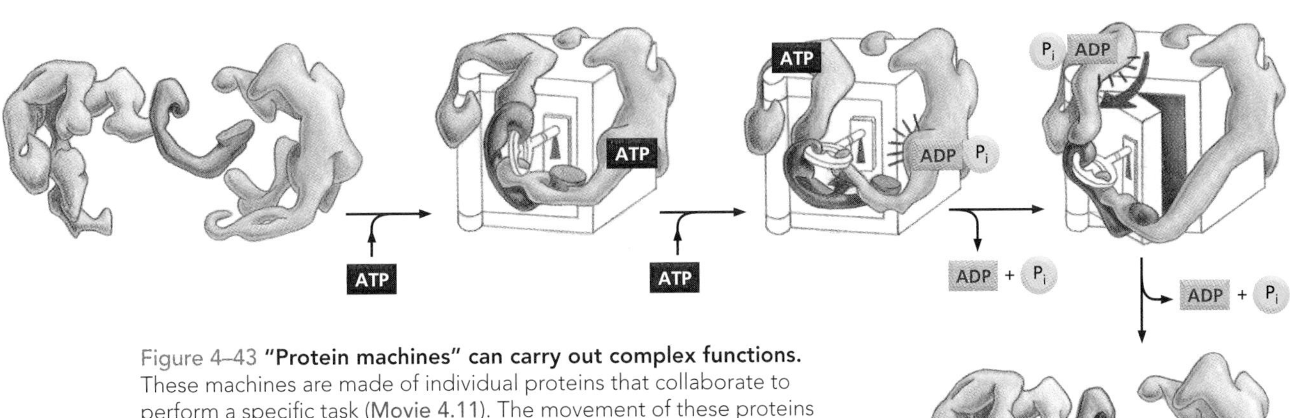

Figure 4–43 **"Protein machines" can carry out complex functions.** These machines are made of individual proteins that collaborate to perform a specific task (Movie 4.11). The movement of these proteins is often coordinated by the hydrolysis of a bound nucleotide (see Figure 4–40). Conformational changes of this type are especially useful to the cell if they occur in a large protein assembly in which the activities of several different protein molecules can be coordinated by the movements within the complex.

QUESTION 4–9

Explain why the enzymes in Figure 4–43 have a great advantage in opening the vault if they work as a protein complex, as opposed to working in an unlinked, sequential manner.

Through evolution, cells have built protein machines that are capable of carrying out most biological reactions. Cells employ protein machines for the same reason that humans have invented mechanical and electronic machines: for almost any task, manipulations that are spatially and temporally coordinated through linked processes are much more efficient than is the sequential use of individual tools.

Covalent Modification Controls the Location and Assembly of Protein Machines

Protein machines and other protein complexes play an important role in the life of the cell. But these complexes do not sit around in the cell interior, preassembled and poised for action. It has recently become clear that most protein machines form at specific sites in the cell and are activated only when and where they are needed. This mobilization is generally accomplished by the covalent addition of a modifying group to one or more specific amino acid side chains on the participating proteins.

More than 200 types of covalent modifications can occur in the cell, each helping to regulate protein function. We have so far focused on the addition and removal of phosphate groups. As we saw earlier, phosphorylation can increase or decrease a protein's activity (see Figure 4–38). But it can also promote the assembly of proteins into larger complexes. For example, extracellular signals can cause the phosphorylation of a class of transmembrane proteins called *receptor tyrosine kinases*; this phosphorylation often triggers the assembly and activation of elaborate intracellular complexes of signaling proteins, which pass along the message to grow or divide (see Figure 16–30). Other covalent modifications have different regulatory roles. Addition of the fatty acid palmitate to a cysteine residue drives a protein to associate with cell membranes. Attachment of ubiquitin, a 76-amino-acid polypeptide, can target a protein for degradation, as we shall see in Chapter 7. Each of these groups is enzymatically added or removed depending on the needs of the cell.

A large number of proteins are now known to be modified on more than one amino acid side chain. The p53 protein, which plays a central part in controlling how a cell responds to DNA damage, can be modified at 20 different sites (Figure 4–44A). Because an enormous number of combina-

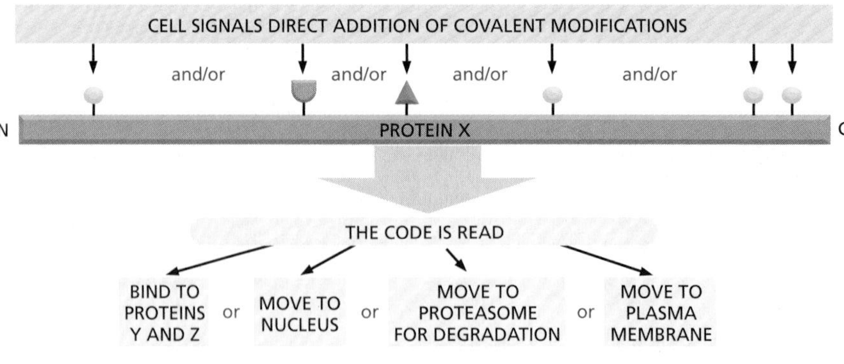

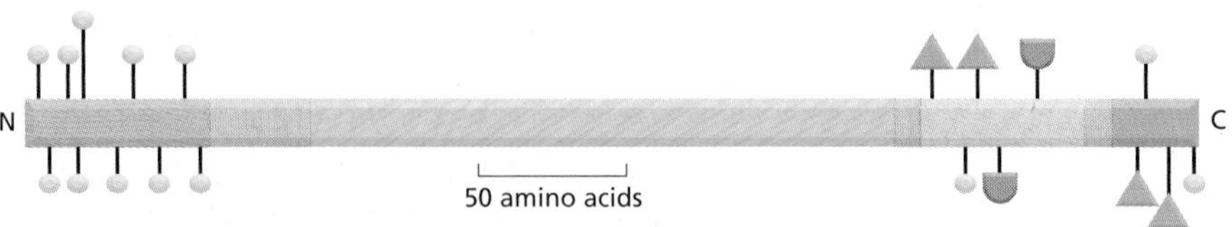

Figure 4–44 **The modification of a protein at multiple sites produces a regulatory code that controls the protein's behavior.** (A) The particular combination of modifications is dictated by different signals from inside or outside the cell (see Chapter 16) and is used by the cell to direct a protein's activity and/or location. For example, phosphorylation (*yellow* symbol) usually influences protein location and activity, whereas the presence of ubiquitin (*green* symbol) generally directs the protein to be destroyed (see Chapter 7). (B) Shown are some of the covalent modifications that control the activity and degradation of the protein p53, an important gene regulatory protein that regulates a cell's response to damage (see Chapter 18). Note that not all of these modifications—phosphorylation, ubiquitylation, and acetylation (*blue* symbols)—will be present at the same time. Colors along the body of the protein represent distinct protein domains.

tions of these 20 modifications is possible, the protein's behavior can in principle be altered in a huge number of ways.

The set of covalent modifications that a protein contains at any moment constitutes an important combinatorial **regulatory protein code**. The attachment or removal of these modifying groups controls the behavior of a protein, changing its activity or stability, its binding partners, or its location inside the cell (see Figure 4–44). This regulatory code enables the cell to make optimal use of its proteins, and it allows the cell to respond rapidly to changes in its condition or environment.

HOW PROTEINS ARE STUDIED

Understanding how a particular protein functions requires detailed structural and biochemical analyses—both of which require large amounts of pure protein. But isolating a single type of protein from the thousands of other proteins present in a cell is a formidable task. For many years, proteins had to be purified directly from the source: the tissues in which they are most plentiful. That approach was inconvenient, entailing, for example, early-morning trips to the slaughterhouse. More importantly, the complexity of intact tissues and organs is a major disadvantage when trying to purify particular molecules, because a long series of chromatography steps is generally required. These procedures not only take weeks to perform but they also yield only a few milligrams of pure protein.

Nowadays, proteins are more often isolated from cells that are grown in a laboratory. Often these cells have even been "tricked" into making large quantities of a given protein using the genetic engineering techniques that we will describe in Chapter 10. Such cells generally allow large amounts of pure protein to be obtained in only a few days.

In this section—and in Panels 4–4 to 4–6 (pp. 164–167)—we outline how cells can be grown in culture and how proteins are purified from these and other cells. In the How We Know section (pp. 158–160), we describe how these proteins are analyzed to determine their amino acid sequence and their three-dimensional structure. Finally, we shall discuss how efforts to probe protein structure and function are being conducted on a large scale, with the hope of obtaining a deeper understanding of how sets of proteins cooperate to make life possible.

Cells Can Be Grown in a Culture Dish

Given the appropriate surroundings, most plant and animal cells will live, proliferate, and even express specialized properties in a tissue-culture dish. Experiments performed using cultured cells are sometimes said to be carried out *in vitro* (literally, "in glass") to contrast them with experiments on intact organisms, which are said to be carried out *in vivo* (literally, "in the living"). These terms can be confusing, however, because they are often used in a very different sense by biochemists. In the biochemistry laboratory, *in vitro* refers to reactions carried out in a test tube in the absence of cells, whereas *in vivo* refers to any reaction taking place inside a living cell, even cells that are growing in culture.

Although not true for all types of cells, most cells grown in culture display the differentiated properties appropriate to their origin: fibroblasts, the precursor cells that give rise to connective tissue, continue to secrete collagen; cells derived from embryonic skeletal muscle fuse to form muscle fibers that contract spontaneously in the culture dish; nerve cells extend axons that are electrically excitable and make synapses with other nerve cells; and epithelial cells form extensive sheets with many of the

HOW WE KNOW: PROBING PROTEIN STRUCTURE

As you've no doubt already concluded in reading this chapter, structure dictates function. For proteins, the three-dimensional shape of the molecule determines what that protein does. So to learn more about how a protein works, it helps to know what it looks like.

The problem is that most proteins are too small to be seen in any detail, even with a powerful electron microscope. To follow the path of an amino acid chain that is folded into a protein molecule, you need to be able to "see" its individual atoms. Scientists use two main methods to map the location of atoms in a protein. The first involves the use of X-rays. Like light, X-rays are a form of electromagnetic radiation. But they have a wavelength that's much shorter: 0.1 nanometer (nm) as opposed to the 400–700 nm wavelength of visible light. That tiny wavelength—which is the approximate diameter of a hydrogen atom—allows scientists to probe the structure of very small and detailed objects at the atomic level.

A second method, called nuclear magnetic resonance (NMR) spectroscopy, takes advantage of the fact that the nuclei of many atoms are intrinsically magnetic. When exposed to a larger magnet, these nuclei act like tiny bar magnets and align themselves with the magnetic field. If they are then excited with a blast of radio waves, the nuclei will wobble around their magnetic axes, and, as they relax back into position, they will give off a signal that can be used to reveal their relative positions in a protein.

Using these techniques, investigators have painstakingly pieced together thousands of protein structures. With the help of computer graphics programs, they have been able to climb inside these proteins, exploring the nooks where ATP likes to nestle or examining the loops and helices that proteins use to grab hold of a partner or wrap around a segment of DNA. If the protein happens to belong to a virus or to a cancer cell, seeing its structure can provide clues to designing drugs that might thwart an infection or eliminate a tumor.

But before you can bombard your protein with X-rays or load it into a giant magnet, large amounts of the protein have to be isolated in a pure form (see Panels 4–4, 4–5, and 4–6 on pp. 164–167). You must also know its amino acid sequence, so that the data you collect about the way its atoms are arranged can be interpreted. So we will begin our discussion of solving protein structure with a brief review of the approaches that are used to determine a protein's amino acid sequence.

Fingerprints

Before a protein is sequenced (i.e., the order of its amino acids is determined), it is generally broken into smaller pieces with a selective protease. The enzyme trypsin, for example, cleaves polypeptide chains on the carboxyl side of a lysine or an arginine. So if a protein has nine lysines and seven arginines, digestion with trypsin should cut it into 17 peptide fragments. The identities of the amino acids in each fragment can then be determined chemically. Sometimes these peptide fragments are partially sequenced using chemical methods, and the partial sequences are then used to search a database to find the complete protein from which they came.

Another, faster way to determine amino acid sequences—especially when large numbers of proteins are being separated and identified at the same time—is to use a method called mass spectrometry. This technique determines the exact mass of every peptide fragment—information that will allow you to identify your protein from the list of all the proteins produced by the organism. The process works like this. The peptides from a tryptic digest are dried onto a metal plate. The sample is then blasted with a laser, which heats the peptides, causing them to become electrically charged (ionized) and ejected from the plate in the form of a gas. Accelerated by a powerful electric field, the peptide ions then fly toward a detector, and the time it takes them to get there is related to their mass and their charge. (The larger the peptide is, the more slowly it moves, while the more highly charged it is, the faster it moves.) The set of exact masses of the protein fragments produced by trypsin cleavage serves as a "fingerprint" that allows the identification of the protein—and its corresponding gene—from the sequence databases (Figure 4–45).

X-rays

Once you have the amino acid sequence of a protein and have produced enough of that protein to work with, the real challenge begins. To determine a protein's structure using X-ray crystallography, you first need to coax the protein into forming crystals: large, highly ordered arrays of the pure protein in which every molecule has the same conformation and is perfectly aligned with its neighbors. Growing high-quality protein crystals is still something of an art and is largely a matter of trial and error—selecting the right solvent conditions, and so on. It can take years to find the right conditions, and some proteins resist crystallization altogether.

If you're lucky enough to get good crystals, you are ready for the X-ray analysis. When a narrow beam of X-rays is directed at a protein crystal, the atoms in the protein molecules scatter the incoming X-rays. These scattered waves either reinforce or cancel one another, producing a complex diffraction pattern that is collected by electronic detectors. The position and intensity of each spot in the diffraction pattern contains information about the position of the atoms in the protein crystal (Figure 4–46).

Because these patterns are so complex—even a small protein can generate 25,000 discrete spots—computers are used to interpret them and transform them by complex mathematical calculations into maps of the relative spatial positions of atoms. By combining information obtained from such maps with the amino acid sequence of the protein, you can eventually generate an atomic model of the protein's structure. To determine whether the protein undergoes conformational changes in its structure when it binds a ligand that boosts its activity, you might subsequently try crystallizing it in the presence of its ligand. With crystals of sufficient quality, even small atomic movements can be detected by comparing the structures obtained in the presence and absence of stimulatory or inhibitory ligands.

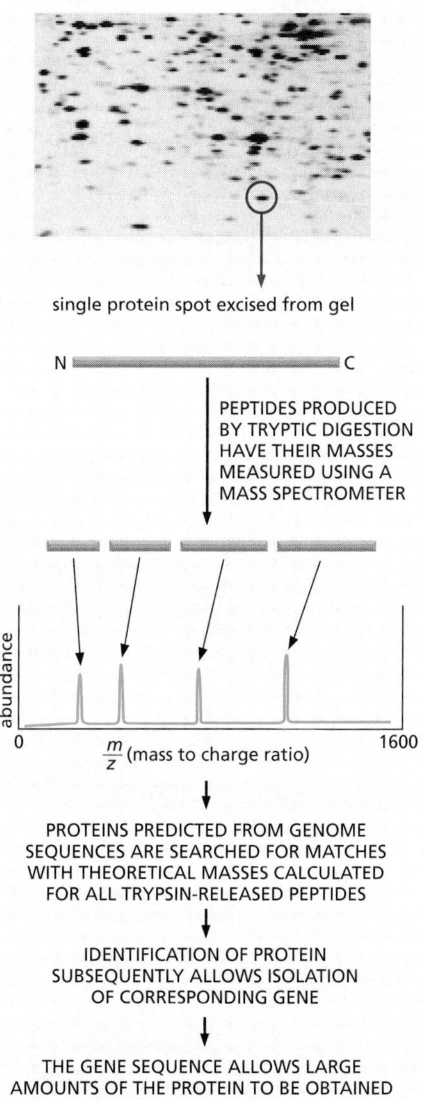
single protein spot excised from gel
N
C
PEPTIDES PRODUCED BY TRYPTIC DIGESTION HAVE THEIR MASSES MEASURED USING A MASS SPECTROMETER
abundance
0
1600
m/z (mass to charge ratio)
PROTEINS PREDICTED FROM GENOME SEQUENCES ARE SEARCHED FOR MATCHES WITH THEORETICAL MASSES CALCULATED FOR ALL TRYPSIN-RELEASED PEPTIDES
IDENTIFICATION OF PROTEIN SUBSEQUENTLY ALLOWS ISOLATION OF CORRESPONDING GENE
THE GENE SEQUENCE ALLOWS LARGE AMOUNTS OF THE PROTEIN TO BE OBTAINED BY GENETIC ENGINEERING TECHNIQUES

Figure 4–45 **Mass spectrometry can be used to identify proteins by determining the precise masses of peptides derived from them.** As indicated, this in turn allows the proteins to be produced in the large amounts needed for determining their three-dimensional structure. In this example, the protein of interest is excised from a polyacrylamide gel after two-dimensional electrophoresis (see Panel 4–6, p. 167) and then digested with trypsin. The peptide fragments are loaded into the mass spectrometer and their exact masses are measured. Sequence databases are then searched to find the protein whose calculated tryptic digest profile matches these values. Mixtures of proteins can also be analyzed in this way. (Image courtesy of Patrick O'Farrell.)

Magnets

The trouble with X-ray crystallography is that you need crystals. And not all proteins like to form such orderly assemblies. Some might have floppy sections that wiggle around too much to stack neatly into a crystalline array. Others might not crystallize in the absence of the membranes in which they normally reside.

There is another way to solve the structure of a protein, one that does not require protein crystals. If the protein is small—say, 50,000 daltons or less—you can determine its structure by NMR spectroscopy. In this technique, a concentrated solution of pure protein is placed in a strong magnetic field and then bombarded with radio waves of different frequencies. Its hydrogen nuclei, in particular, will generate an NMR signal that can be used to determine the distances between the atoms in different parts of the protein. This information is then used to build a model of how these atoms are arranged in space. Again, combined with the known amino acid sequence, an NMR spectrum can allow you to compute the three-dimensional structure of the protein (Figure 4–47). If the protein is larger than 50,000 daltons, you can try to break it up into its constituent functional domains and analyze each domain by NMR.

Because determining the precise conformation of a protein is so time-consuming and costly—and the resulting insights so valuable—scientists routinely make their structures freely available by submitting the information to a publicly accessible database. Thanks to such databases, anyone interested in viewing the structure of, say, the ribosome—a complex cellular machine made of several RNAs and more than 50 proteins—can easily do so. In the future, improvements in X-ray crystallography and NMR spectroscopy should permit rapid analysis of many more proteins and protein machines. And once enough structures have been determined, it might become possible to generate algorithms for accurately predicting structure solely on the basis of a protein's amino acid sequence. After all, it is the sequence of the amino acids alone that determines how each protein folds up into its three-dimensional shape.

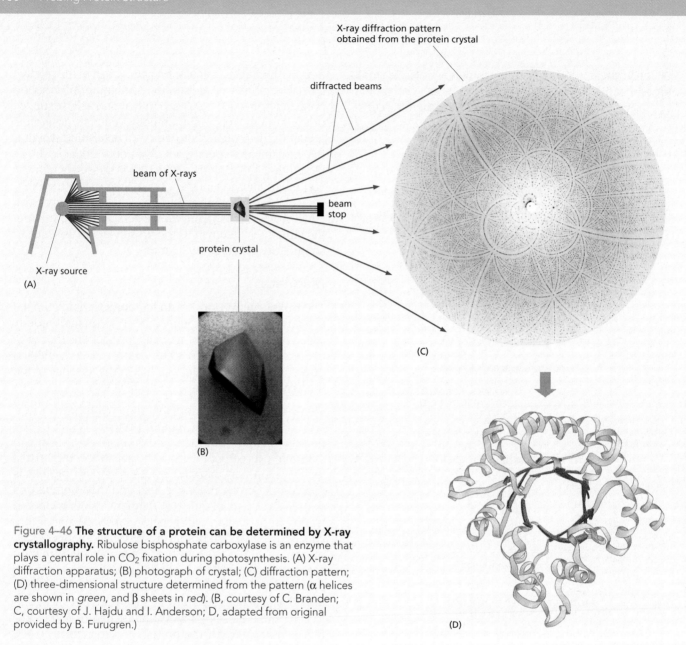

Figure 4–46 **The structure of a protein can be determined by X-ray crystallography.** Ribulose bisphosphate carboxylase is an enzyme that plays a central role in CO_2 fixation during photosynthesis. (A) X-ray diffraction apparatus; (B) photograph of crystal; (C) diffraction pattern; (D) three-dimensional structure determined from the pattern (α helices are shown in *green*, and β sheets in *red*). (B, courtesy of C. Branden; C, courtesy of J. Hajdu and I. Anderson; D, adapted from original provided by B. Furugren.)

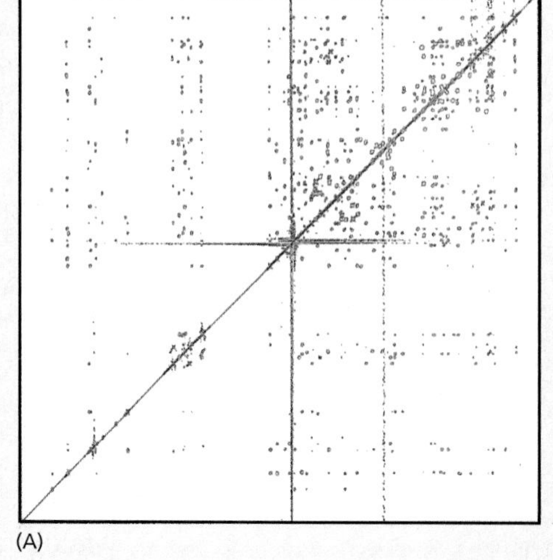

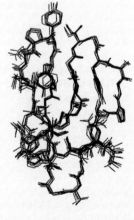

Figure 4–47 **NMR spectroscopy can be used to determine the structure of small proteins or protein domains.** (A) Two-dimensional NMR spectrum derived from the C-terminal domain of the enzyme cellulase. The spots represent interactions between neighboring H atoms. (B) The set of overlapping structures shown all satisfy the distance constraints equally well. (Courtesy of P. Kraulis.)

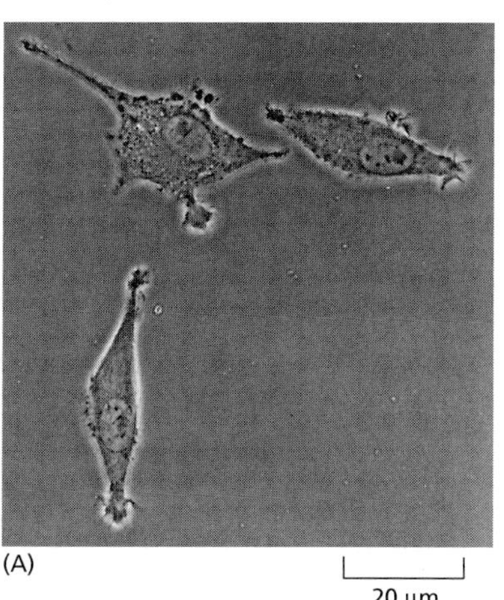

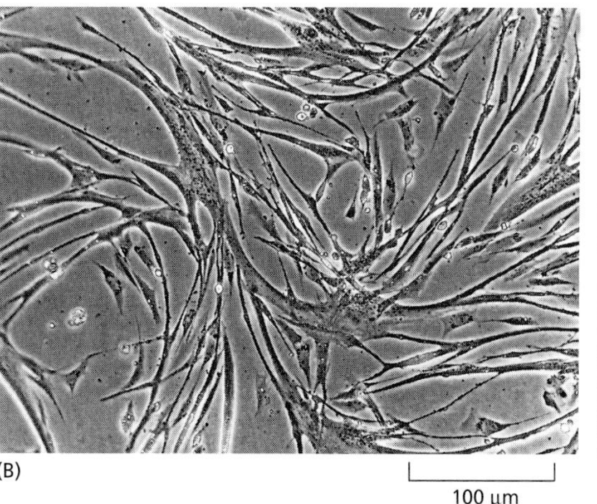

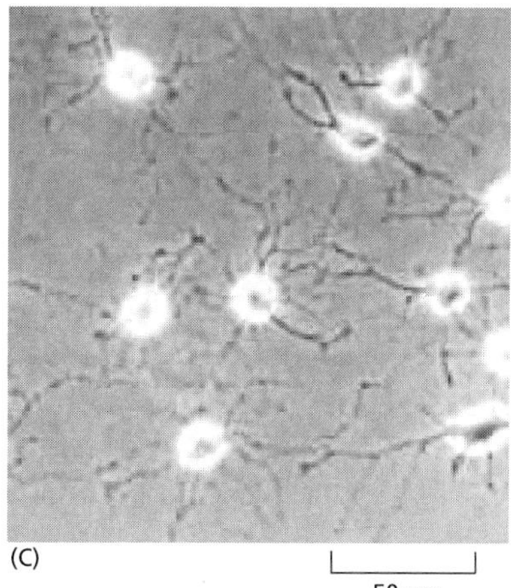

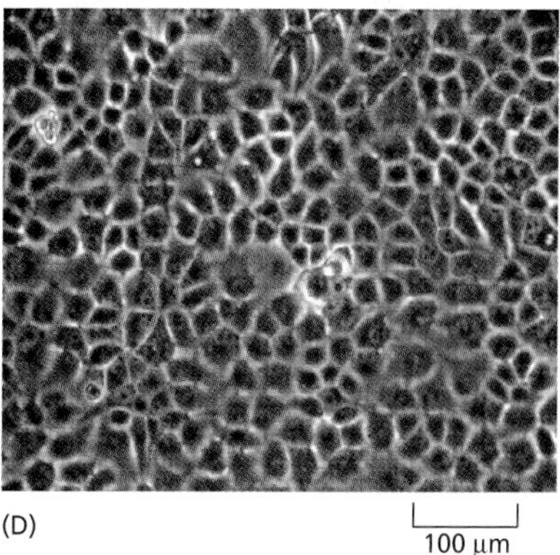

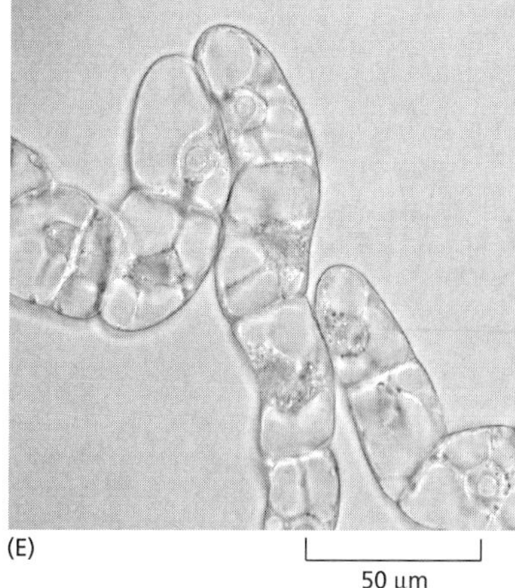

Figure 4–48 **Cells in culture often display properties that reflect their origin.** (A) Phase-contrast micrograph of fibroblasts in culture. (B) Micrograph of cultured myoblasts, the precursor cells that give rise to muscle, shows cells fusing to form multinucleate muscle cells. (C) Cultured precursor cells that give rise to oligodendrocytes, the glial cells that support and nurture neurons in the brain. (D) Cultured epithelial cells can form cell sheets. (E) Tobacco cells, from an immortal cell line, grown in liquid culture. (A, courtesy of Daniel Zicha; B, courtesy of Rosalind Zalin; C, from Tang et al., *J. Cell Biol.* 148:971–984, 2000, with permission from The Rockefeller University Press; D, from K.B. Chua et al., *Proc. Natl. Acad. Sci. USA* 104:11424–11429, 2007, with permission from the National Academy of Sciences; E, courtesy of Gethin Roberts.)

properties of an intact epithelium (Figure 4–48). Because these phenomena occur in culture, in a controlled environment, they are accessible to study in ways that are often not possible in intact tissues. For example, cultured cells can be exposed to hormones or growth factors, and the effects that these molecules have on the shape or behavior of the cells—or on the proteins they produce in response—can be easily explored.

Cultured cells can also provide a ready source of raw materials for biologists interested in purifying and studying a particular protein or protein machine, as we see next.

Purification Techniques Allow Homogeneous Protein Preparations to Be Obtained from Cell Homogenates

Whether starting with a fibroblast culture, a piece of liver, or a vat of cells that have been engineered to produce the protein of interest, the first step in any purification procedure is to break open the cells to release their contents; the resulting slurry is called a *cell homogenate.* This physical disruption is followed by an initial fractionation procedure to separate out the class of molecules of interest—for example, all the soluble proteins in the cell (Panel 4–4, pp. 164–165).

With this collection of proteins in hand, the job is then to isolate the desired protein. The standard approach involves purifying the protein through a series of **chromatography** steps, which separate the individual components of a complex mixture into different portions, or fractions. After each such step, one uses some sort of assay—for example, a test for the protein's activity—to determine which fractions contain the protein of interest. Such fractions are then subjected to additional chromatography steps until the desired protein is obtained in pure form. The most popular forms of protein chromatography separate polypeptides on the basis of their size, their charge, or their ability to bind to a particular chemical group (Panel 4–5, p. 166). If antibodies that recognize a particular protein are available, they can be used to help extract that protein from a mixture (see Panel 4–3, pp. 144–145).

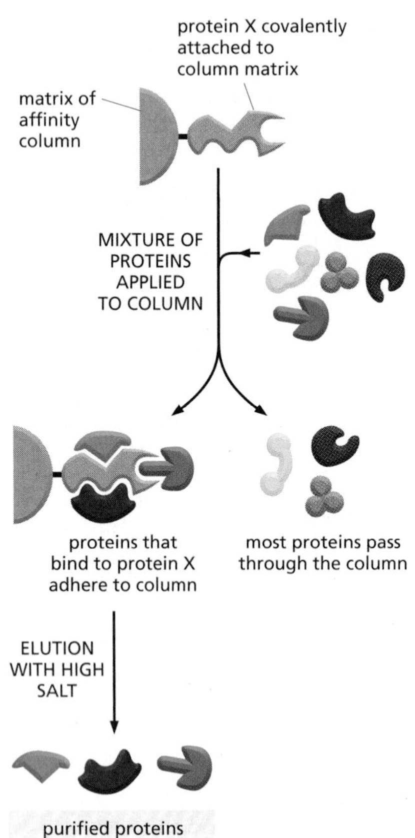

Figure 4–49 **Affinity chromatography can be used to isolate the binding partners of a protein of interest.** The purified protein of interest, here protein X, is covalently attached to the matrix of a chromatography column. An extract containing a mixture of proteins is then loaded onto the column. Those proteins that associate with protein X inside the cell will bind to it on the column. Proteins not bound to the column pass right through, and the proteins that are bound tightly to protein X can then be released by changing the pH or ionic composition of the washing solution.

A similar approach can be used to isolate those proteins that interact physically with the protein being studied. In this case, the purified protein of interest is attached to the matrix of the chromatography column; the proteins that bind to this protein will collect in the column and can then be eluted by changing the composition of the washing solution (Figure 4–49).

Proteins can also be separated by **electrophoresis**. In this technique, a mixture of proteins is loaded onto a polymer gel and subjected to an electric field; the polypeptides will then migrate through the gel at different speeds depending on their size and net charge (Panel 4–6, p. 167). If too many proteins are present in the sample, or if the proteins are very similar in their migration rate, they can be resolved further using two-dimensional gel electrophoresis (see Panel 4–6). These electrophoretic approaches yield a number of bands or spots that can be visualized by staining, each one containing a different protein. Electrophoresis and chromatography—each developed more than 50 years ago—have been instrumental in building an understanding of what proteins look like and how they behave (Table 4–2). Both techniques are still used very frequently in laboratories.

Once a protein has been obtained in pure form, it can be used in biochemical assays to study the details of its activity, and it can be subjected to techniques that reveal its amino acid sequence and precise three-dimensional structure (see How We Know, pp. 158–160).

TABLE 4–2 HISTORICAL LANDMARKS IN OUR UNDERSTANDING OF PROTEINS	
1838	The name "**protein**" (from the Greek proteios, "primary") was suggested by Berzelius for the complex nitrogen-rich substance found in the cells of all animals and plants.
1819–1904	Most of the 20 common **amino acids** found in proteins were discovered.
1864	Hoppe-Seyler crystallized, and named, the protein **hemoglobin**.
1894	Fischer proposed a **lock-and-key** analogy for enzyme–substrate interactions.
1897	Buchner and Buchner showed that cell-free extracts of yeast can ferment sucrose to form carbon dioxide and ethanol, thereby laying the foundations of **enzymology**.
1926	Sumner crystallized urease in pure form, demonstrating that proteins could possess the **catalytic activity** of enzymes; Svedberg developed the first **analytical ultracentrifuge** and used it to estimate the correct molecular weight of hemoglobin.
1933	Tiselius introduced **electrophoresis** for separating proteins in solution.
1934	Bernal and Crowfoot presented the first detailed **X-ray diffraction** patterns of a protein, obtained from crystals of the enzyme pepsin.
1942	Martin and Synge developed **chromatography**, a technique now widely used to separate proteins.
1951	Pauling and Corey proposed the structure of a helical conformation of a chain of amino acids—the **α helix**—and the structure of the **β sheet**, both of which were later found in many proteins.
1955	Sanger obtained the **amino acid sequence of insulin**, the first protein whose amino acid sequence was determined.
1956	Ingram produced the first **protein fingerprints**, showing that the difference between sickle-cell hemoglobin and normal hemoglobin is due to a change in a single amino acid (Movie 4.12).
1960	Kendrew described the first detailed **three-dimensional structure** of a protein (sperm whale myoglobin) to a resolution of 0.2 nm, and Perutz proposed a lower-resolution structure for hemoglobin.
1963	Monod, Jacob, and Changeux recognized that many enzymes are regulated through **allosteric changes** in their conformation.

Large Amounts of Almost Any Protein Can be Produced by Genetic Engineering Techniques

Advances in genetic engineering techniques now permit the production of large quantities of almost any desired protein. In addition to making life much easier for biochemists interested in purifying specific proteins, this ability to churn out huge quantities of protein has given rise to an entire biotechnology industry (Figure 4–50). Companies now use bacteria, yeast, or cultured mammalian cells to mass produce all sorts of proteins that are used therapeutically, such as insulin, human growth hormone, and even the fertility-enhancing drugs used to boost egg production in women undergoing *in vitro* fertilization. Preparing these proteins previously required the collection and processing of vast amounts of tissue and other biological products—including, in the case of the fertility drugs, the urine of postmenopausal nuns.

Using the same sorts of genetic engineering techniques, scientists can also design proteins that perform novel tasks: metabolizing toxic wastes, synthesizing life-saving drugs, or operating under conditions that would destroy most biological catalysts. We shall discuss these methods in great detail in Chapter 10.

Figure 4–50 **Biotechnology companies produce mass quantities of useful proteins.** Shown is a photograph of the fermenters used to grow the cells needed for such large-scale protein production. (Courtesy of Bioengineering AG, Switzerland.)

Automated Studies of Protein Structure and Function Are Increasing the Pace of Discovery

Biochemists have made enormous progress in understanding the structure and function of proteins over the past 150 years (see Table 4–2, p. 162). These advances are the fruits of decades of painstaking research on isolated proteins, performed by individual scientists working tirelessly on single proteins or protein families, one by one, sometimes for their entire careers. But many future advances may come from **proteomics**, the large-scale study of cellular proteins in which the activities or structures of hundreds—even thousands—of proteins are analyzed by highly sensitive, automated techniques. If scientists can perfect such methods, they might some day be able to monitor all of the proteins that are present in a cell: assessing whether they are switched on (or off) and seeing which proteins they are partnered with—all in a single experiment.

Large-scale analyses of protein structures are already under way. Techniques that have been miniaturized and automated allow researchers to rapidly clone genes, produce proteins, grow crystals, and collect X-ray diffraction data for hundreds of proteins at a time. Through **X-ray crystallography** and **nuclear magnetic resonance** (**NMR**) **spectroscopy** (see How We Know, p. 160), we now know the three-dimensional shapes of more than 20,000 proteins. These structures are archived in large, publicly available databases (Movie 4.13). By analyzing the conformations of these proteins, biologists have come to the conclusion that the vast majority of protein domains fold up into a limited number of patterns—perhaps as few as 2000. The structures of about 800 of these protein folds have been determined so far. By studying how these patterns form, scientists hope to develop computational methods that will be able to take any amino acid sequence and predict both the structure and the function of the protein.

Even with such information, it will still be an enormous challenge to decipher exactly how all these proteins—about 400 in the smallest bacterium and 10,000 in a typical human cell—work together to form a living cell. Figuring out exactly how proteins collaborate to create and maintain order in a universe that is always tending toward disorder will require both the continual development of new techniques and a great deal of human ingenuity. But the closer we get to answering this question, the closer we will be to understanding the fundamental basis for life.

BREAKING CELLS AND TISSUES

The first step in the purification of most proteins is to disrupt tissues and cells in a controlled fashion.

Using gentle mechanical procedures, called homogenization, the plasma membranes of cells can be ruptured so that the cell contents are released. Four commonly used procedures are shown here.

The resulting thick soup (called a homogenate or an extract) contains large and small molecules from the cytosol, such as enzymes, ribosomes, and metabolites, as well as all of the membrane-enclosed organelles.

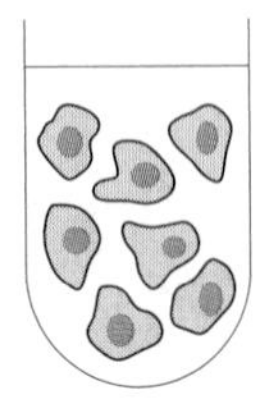

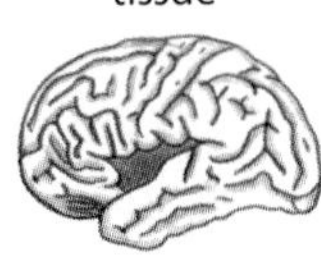

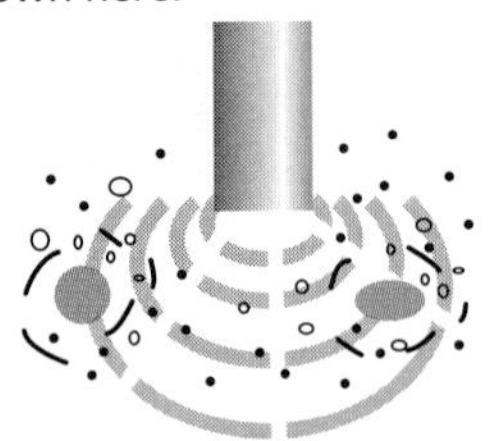

1 Break cells with high-frequency sound.

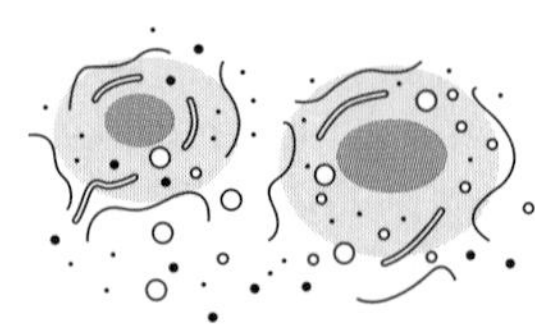

2 Use a mild detergent to make holes in the plasma membrane.

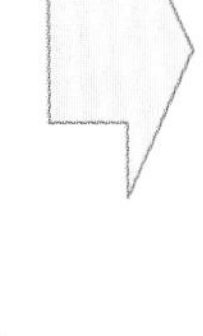

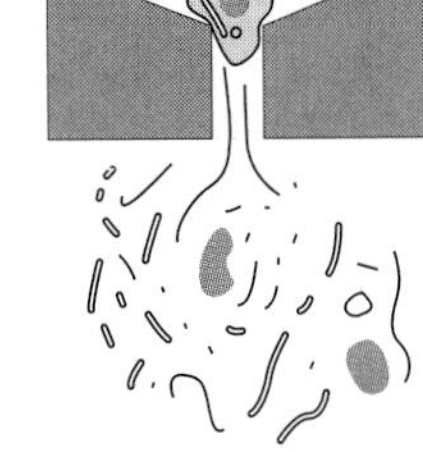

3 Force cells through a small hole using high pressure.

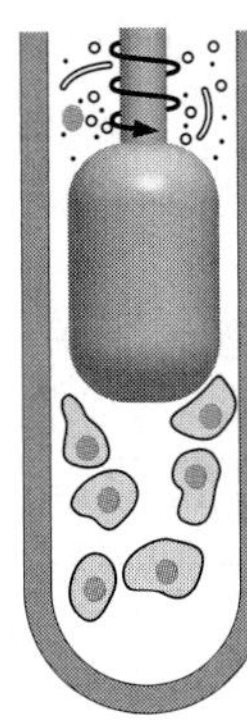

4 Shear cells between a close-fitting rotating plunger and the thick walls of a glass vessel.

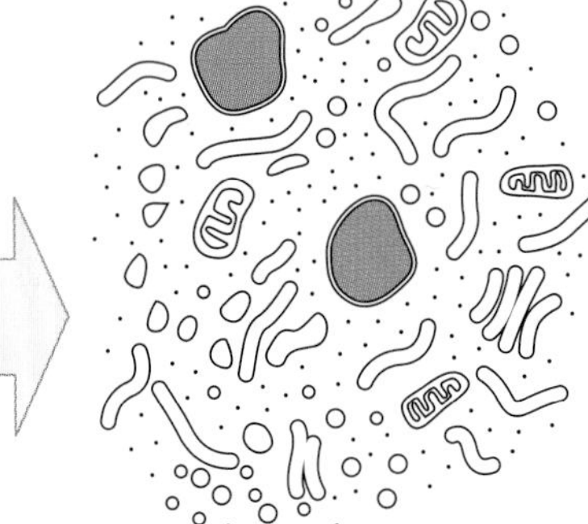

When carefully conducted, homogenization leaves most of the membrane-enclosed organelles intact.

THE CENTRIFUGE

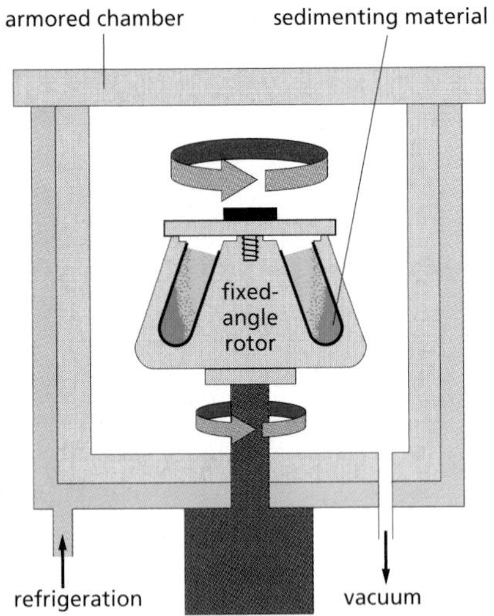

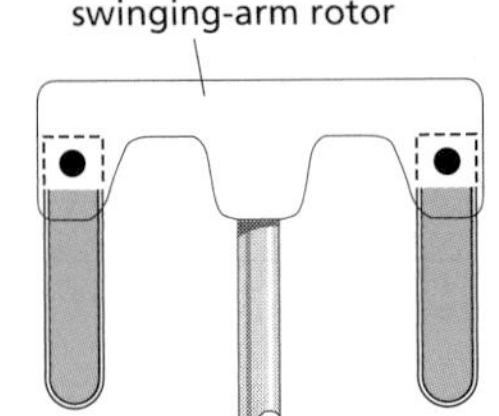

Many cell fractionations are done in a second type of rotor, a swinging-arm rotor.

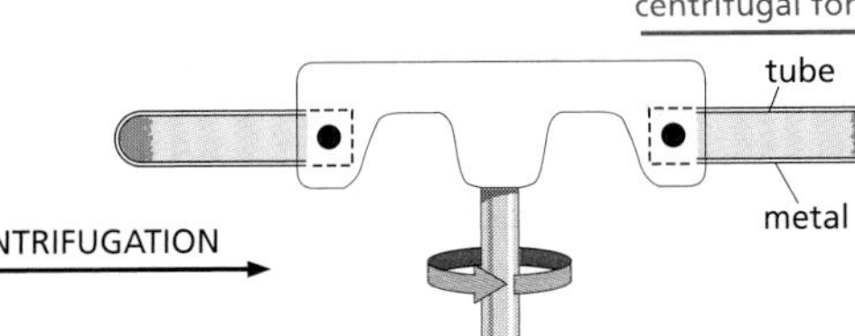

The metal buckets that hold the tubes are free to swing outward as the rotor turns.

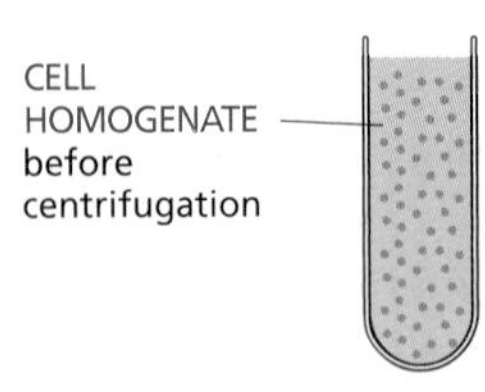

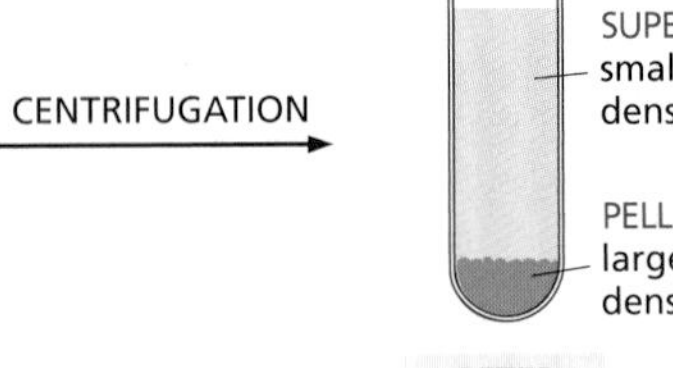

Centrifugation is the most widely used procedure to separate a homogenate into different parts, or fractions. The homogenate is placed in test tubes and rotated at high speed in a centrifuge or ultracentrifuge. Present-day ultracentrifuges rotate at speeds up to 100,000 revolutions per minute and produce enormous forces, as high as 600,000 times gravity.

Such speeds require centrifuge chambers to be refrigerated and evacuated so that friction does not heat up the homogenate. The centrifuge is surrounded by thick armor plating, because an unbalanced rotor can shatter with an explosive release of energy. A fixed-angle rotor can hold larger volumes than a swinging-arm rotor, but the pellet forms less evenly.

DIFFERENTIAL CENTRIFUGATION

Repeated centrifugation at progressively higher speeds will fractionate cell homogenates into their components.

Centrifugation separates cell components on the basis of size and density. The larger and denser components experience the greatest centrifugal force and move most rapidly. They sediment to form a pellet at the bottom of the tube, while smaller, less dense components remain in suspension above, a portion called the supernatant.

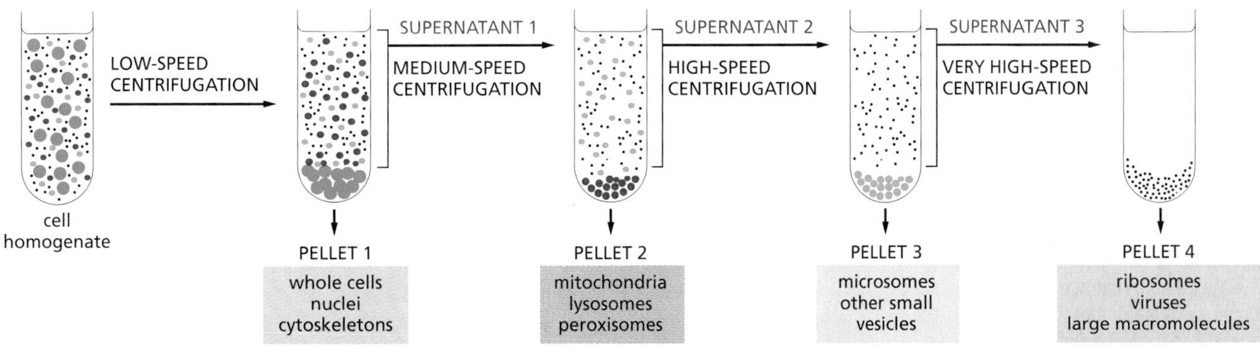

VELOCITY SEDIMENTATION

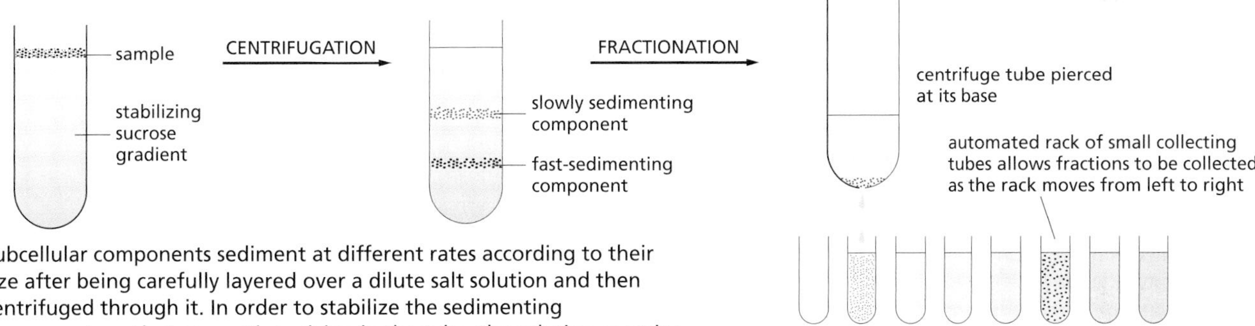

Subcellular components sediment at different rates according to their size after being carefully layered over a dilute salt solution and then centrifuged through it. In order to stabilize the sedimenting components against convective mixing in the tube, the solution contains a continuous shallow gradient of sucrose that increases in concentration toward the bottom of the tube. This is typically 5–20% sucrose. When sedimented through such a dilute sucrose gradient, different cell components separate into distinct bands that can be collected individually.

After an appropriate centrifugation time the bands may be collected, most simply by puncturing the plastic centrifuge tube and collecting drops from the bottom, as shown here.

EQUILIBRIUM SEDIMENTATION

The ultracentrifuge can also be used to separate cell components on the basis of their buoyant density, independently of their size or shape. The sample is usually either layered on top of, or dispersed within, a steep density gradient that contains a very high concentration of sucrose or cesium chloride. Each subcellular component will move up or down when centrifuged until it reaches a position where its density matches its surroundings and then will move no further. A series of distinct bands will eventually be produced, with those nearest the bottom of the tube containing the components of highest buoyant density. The method is also called density gradient centrifugation.

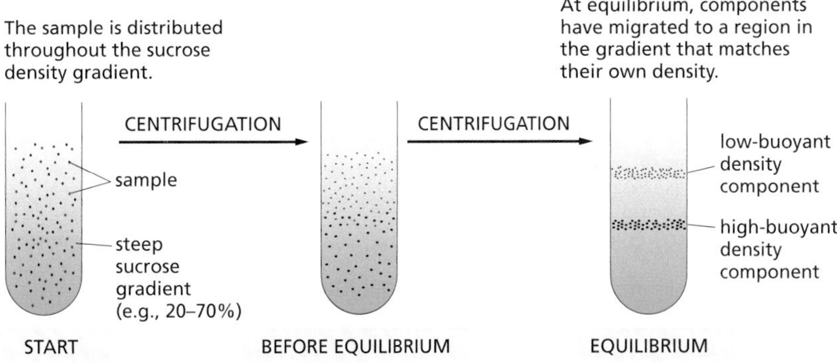

A sucrose gradient is shown here, but denser gradients can be formed with cesium chloride that are particularly useful for separating the nucleic acids (DNA and RNA).

The final bands can be collected from the base of the tube, as shown above.

PROTEIN SEPARATION

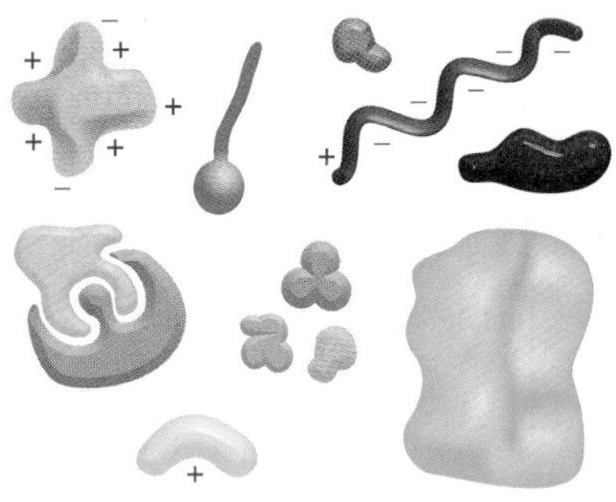

Proteins are very diverse. They differ in size, shape, charge, hydrophobicity, and their affinity for other molecules. All of these properties can be exploited to separate them from one another so that they can be studied individually.

COLUMN CHROMATOGRAPHY

Proteins are often fractionated by column chromatography. A mixture of proteins in solution is applied to the top of a cylindrical column filled with a permeable solid matrix immersed in solvent. A large amount of solvent is then pumped through the column. Because different proteins are retarded to different extents by their interaction with the matrix, they can be collected separately as they flow out from the bottom. According to the choice of matrix, proteins can be separated according to their charge, hydrophobicity, size, or ability to bind to particular chemical groups (see *below*).

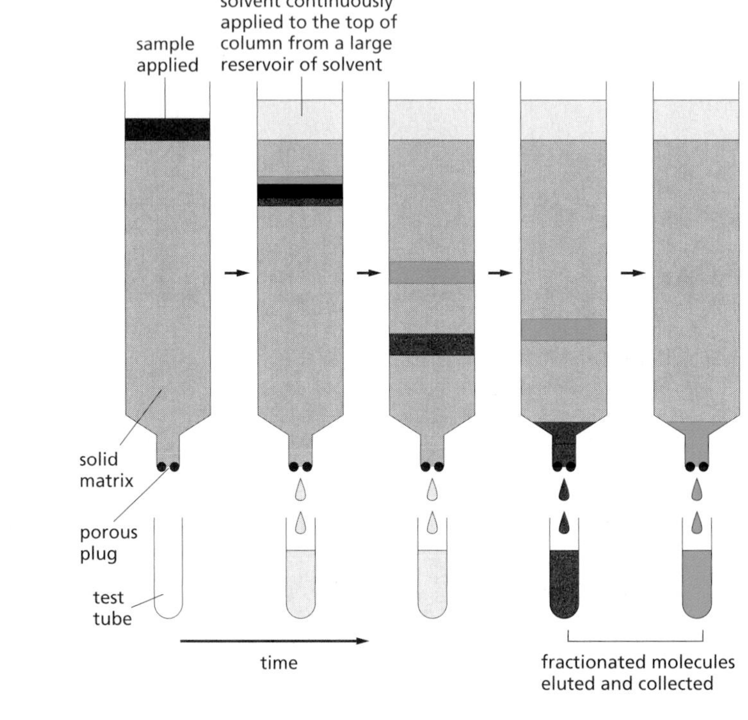

THREE KINDS OF CHROMATOGRAPHY

Although the material used to form the matrix for column chromatography varies, it is usually packed in the column in the form of small beads. A typical protein purification strategy might employ in turn each of the three kinds of matrix described below, with a final protein purification of up to 10,000-fold.

Purity can easily be assessed by gel electrophoresis (Panel 4–6).

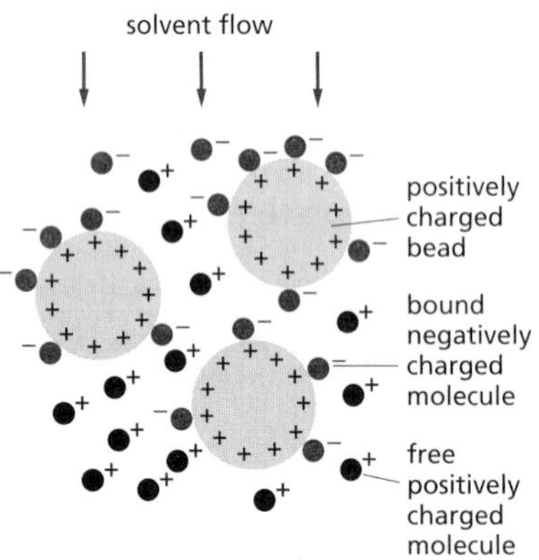

(A) ION-EXCHANGE CHROMATOGRAPHY

Ion-exchange columns are packed with small beads carrying either positive or negative charges that retard proteins of the opposite charge. The association between a protein and the matrix depends on the pH and ionic strength of the solution passing down the column. These can be varied in a controlled way to achieve an effective separation.

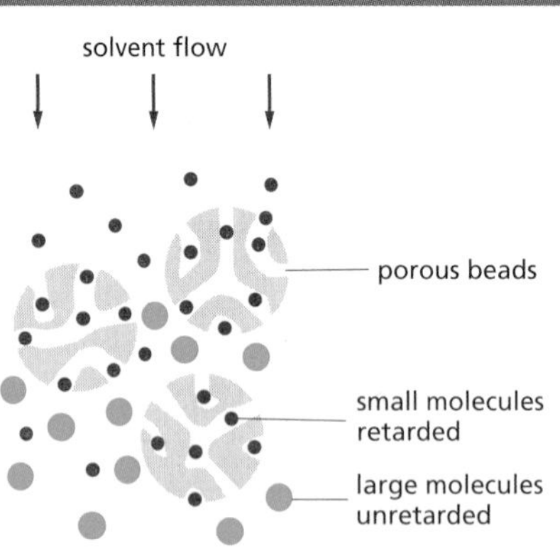

(B) GEL-FILTRATION CHROMATOGRAPHY

Gel-filtration columns separate proteins according to their size. The matrix consists of tiny porous beads. Protein molecules that are small enough to enter the holes in the beads are delayed and travel more slowly through the column. Proteins that cannot enter the beads are washed out of the column first. Such columns also allow an estimate of protein size.

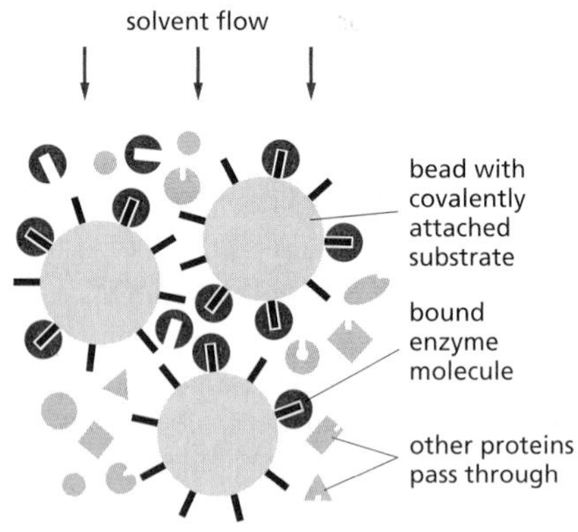

(C) AFFINITY CHROMATOGRAPHY

Affinity columns contain a matrix covalently coupled to a molecule that interacts specifically with the protein of interest (e.g., an antibody, or an enzyme substrate). Proteins that bind specifically to such a column can subsequently be released by a pH change or by concentrated salt solutions, and they emerge highly purified (see also Figure 4–49).

PANEL 4–6 Protein separation by electrophoresis

GEL ELECTROPHORESIS

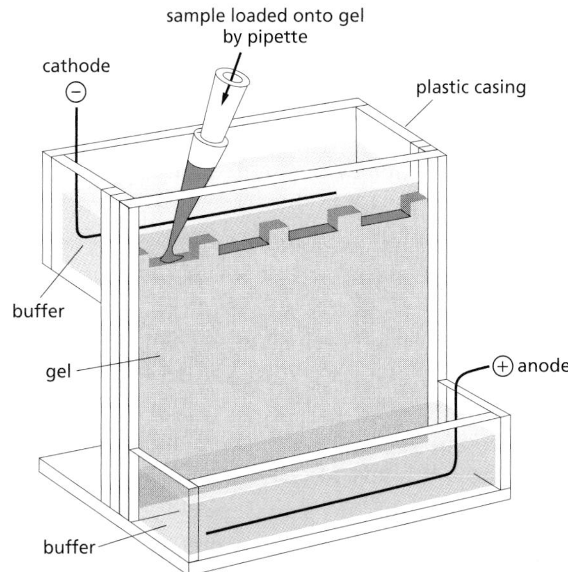

When an electric field is applied to a solution containing protein molecules, the molecules will migrate in a direction and at a speed that reflects their size and net charge. This forms the basis of the technique called electrophoresis.

The detergent sodium dodecyl sulfate (SDS) is used to solubilize proteins for SDS polyacrylamide-gel electrophoresis.

$$CH_3-(CH_2)_{11}-O-SO_2-O^{\ominus}\ Na^{\oplus}$$

SDS

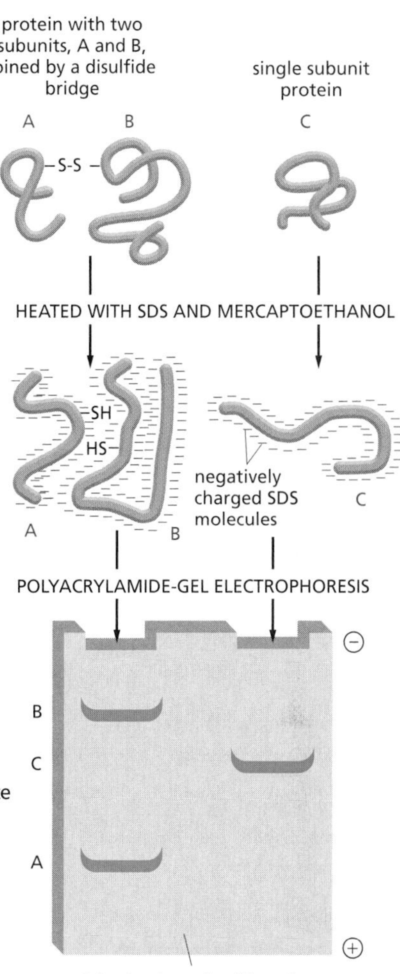

SDS polyacrylamide-gel electrophoresis (SDS-PAGE) Individual polypeptide chains form a complex with negatively charged molecules of sodium dodecyl sulfate (SDS) and therefore migrate as a negatively charged SDS–protein complex through a slab of porous polyacrylamide gel. The apparatus used for this electrophoresis technique is shown above (*left*). A reducing agent (mercaptoethanol) is usually added to break any –S–S– linkages in or between proteins. Under these conditions, proteins migrate at a rate that reflects their molecular weight.

ISOELECTRIC FOCUSING

For any protein there is a characteristic pH, called the isoelectric point, at which the protein has no net charge and therefore will not move in an electric field. In isoelectric focusing, proteins are electrophoresed in a narrow tube of polyacrylamide gel in which a pH gradient is established by a mixture of special buffers. Each protein moves to a point in the gradient that corresponds to its isoelectric point and stays there.

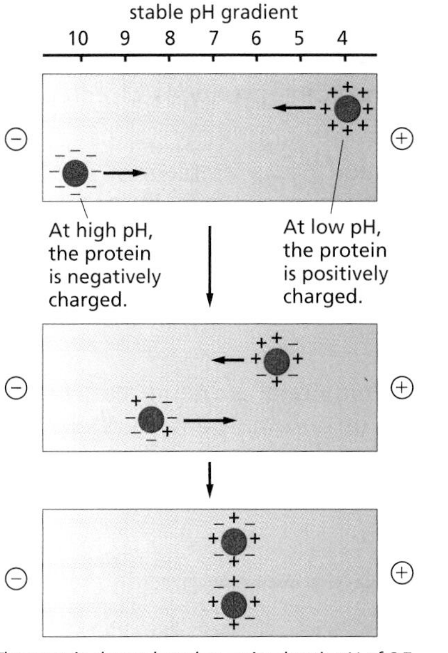

The protein shown here has an isoelectric pH of 6.5.

TWO-DIMENSIONAL POLYACRYLAMIDE-GEL ELECTROPHORESIS

Complex mixtures of proteins cannot be resolved well on one-dimensional gels, but two-dimensional gel electrophoresis, combining two different separation methods, can be used to resolve more than 1000 proteins in a two-dimensional protein map. In the first step, native proteins are separated in a narrow gel on the basis of their intrinsic charge using isoelectric focusing (see *left*). In the second step, this gel is placed on top of a gel slab, and the proteins are subjected to SDS-PAGE (see *above*) in a direction perpendicular to that used in the first step. Each protein migrates to form a discrete spot.

All the proteins in an *E. coli* bacterial cell are separated in this 2-D gel, in which each spot corresponds to a different polypeptide chain. They are separated according to their isoelectric point from left to right and to their molecular weight from top to bottom. (Courtesy of Patrick O'Farrell.)

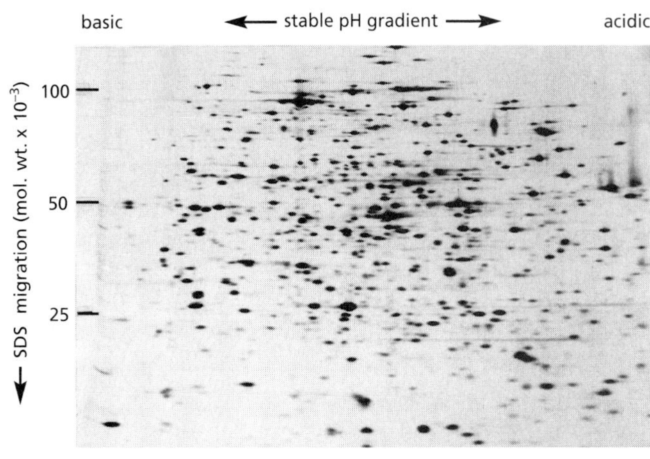

ESSENTIAL CONCEPTS

- Living cells contain an enormously diverse set of protein molecules, each made as a linear chain of amino acids covalently linked together.
- Each type of protein has a unique amino acid sequence that determines both its three-dimensional shape and its biological activity.
- The folded structure of a protein is stabilized by noncovalent interactions between different parts of the polypeptide chain.
- Hydrogen bonds between neighboring regions of the polypeptide backbone often give rise to regular folding patterns, known as α helices and β sheets.
- The structure of many proteins can be subdivided into smaller globular regions of compact three-dimensional structure, known as protein domains.
- The biological function of a protein depends on the detailed chemical properties of its surface and how it binds to other molecules, called ligands.
- When a protein catalyzes the formation or breakage of covalent bonds in a ligand, the protein is called an enzyme and the ligand is called a substrate.
- At the active site of an enzyme, the amino acid side chains of the folded protein are precisely positioned so that they favor the formation of the high-energy transition states that the substrates must pass through to be converted to product.
- The three-dimensional structure of many proteins has evolved so that the binding of a small ligand can induce a significant change in protein shape.
- Most enzymes are allosteric proteins that can exist in two conformations that differ in catalytic activity, and the enzyme can be turned on or off by ligands that bind to a distinct regulatory site to stabilize either the active or the inactive conformation.
- The activities of most enzymes within the cell are strictly regulated. One of the most common forms of regulation is feedback inhibition, in which an enzyme early in a metabolic pathway is inhibited by its binding to one of the pathway's end products.
- Many thousands of proteins in a typical eucaryotic cell are regulated either by cycles of phosphorylation and dephosphorylation, or by the binding and hydrolysis of GTP by a GTP-binding protein.
- The hydrolysis of ATP to ADP by motor proteins produces directed movements in the cell.
- Highly efficient protein machines are formed by assemblies of allosteric proteins in which conformational changes are coordinated to perform complex cellular functions.
- A regulatory protein code based on the covalent modification of multiple amino acid side chains allows each cell to control the location and assembly of its protein complexes.
- Starting from crude cell homogenates, individual proteins can be obtained in pure form by using a series of chromatography steps. Purification allows the detailed properties of a protein to be revealed by biochemical techniques and its exact three-dimensional structure to be determined.

KEY TERMS

active site
allosteric
α helix
amino acid sequence
antibody
antigen
β sheet
binding site
chromatography
coiled-coil
conformation
disulfide bond
electrophoresis
enzyme
feedback inhibition
fibrous protein
globular protein
GTP-binding protein
helix
ligand
motor protein
nuclear magnetic resonance (NMR) spectroscopy
polypeptide, polypeptide chain
polypeptide backbone
protein
protein domain
protein family
protein kinase
protein machine
protein phosphatase
protein phosphorylation
proteomics
regulatory protein code
secondary structure
side chain
subunit
transition state
X-ray crystallography

QUESTIONS

QUESTION 4–10

Which of the following statements are correct? Explain your answers.

A. The active site of an enzyme usually occupies only a small fraction of its surface.

B. Catalysis by some enzymes involves the formation of a covalent bond between an amino acid side chain and a substrate molecule.

C. A β sheet can contain up to five strands, but no more.

D. The specificity of an antibody molecule is contained exclusively in loops on the surface of the folded light-chain domain.

E. The possible linear arrangements of amino acids are so vast that new proteins almost never evolve by alteration of old ones.

F. Allosteric enzymes have two or more binding sites.

G. Noncovalent bonds are too weak to influence the three-dimensional structure of macromolecules.

H. Affinity chromatography separates molecules according to their intrinsic charge.

I. Upon centrifugation of a cell homogenate, smaller organelles experience less friction and thereby sediment faster than larger ones.

QUESTION 4–11

What common feature of α helices and β sheets makes them universal building blocks for proteins?

QUESTION 4–12

Protein structure is determined solely by a protein's amino acid sequence. Should a genetically engineered protein in which the order of all amino acids is reversed therefore have the same structure as the original protein?

QUESTION 4–13

Consider the following protein sequence as an α helix: Leu-Lys-Arg-Ile-Val-Asp-Ile-Leu-Ser-Arg-Leu-Phe-Lys-Val. How many turns does this helix make? Do you find anything remarkable about the arrangement of the amino acids in this sequence when folded into an α helix? (Hint: consult the properties of the amino acids in Figure 4–3.)

QUESTION 4–14

Simple enzyme reactions often conform to the equation

$$E + S \rightleftharpoons ES \rightarrow EP \rightleftharpoons E + P$$

where E, S, and P are enzyme, substrate, and product, respectively.

A. What does ES represent in this equation?

B. Why is the first step shown with bidirectional arrows and the second step as a unidirectional arrow?

C. Why does E appear at both ends of the equation?

D. One often finds that high concentrations of P inhibit the enzyme. Suggest why this might occur.

E. Compound X resembles S and binds to the active site of the enzyme but cannot undergo the reaction catalyzed by it. What effects would you expect the addition of X to the reaction to have? Compare the effects of X and of accumulation of P.

QUESTION 4–15

Which of the following amino acids would you expect to find more often near the center of a folded globular protein? Which ones would you expect to find more often exposed to the outside? Explain your answers. Ser, Ser-P (a Ser residue that is phosphorylated), Leu, Lys, Gln, His, Phe, Val, Ile, Met, Cys–S–S–Cys (two Cys residues that are disulfide-bonded), and Glu. Where would you expect to find the most N-terminal amino acid and the most C-terminal amino acid?

QUESTION 4–16

Assume you want to make and study fragments of a protein. Would you expect that any fragment of the polypeptide chain would fold the same way as the corresponding sequence folds in the intact protein? Consider the protein shown in Figure 4–16. Which fragments do you suppose are most likely to fold correctly?

QUESTION 4–17

An enzyme isolated from a mutant bacterium grown at 20°C works in a test tube at 20°C but not at 37°C (37°C is the temperature of the gut, where this bacterium normally lives). Furthermore, once the enzyme has been exposed to the higher temperature, it no longer works at the lower one. The same enzyme isolated from the normal bacterium works at both temperatures. Can you suggest what happens at the molecular level to the mutant enzyme as the temperature increases?

QUESTION 4–18

A motor protein moves along filaments in the cell. Why are the elements shown in the illustration not sufficient to provide unidirectionality to the movement (Figure Q4–18)? With reference to Figure 4–42, modify the illustration shown here to include other elements that are required to create a unidirectional motor, and justify each modification you make to the illustration.

Figure Q4–18

QUESTION 4–19

Gel-filtration chromatography separates molecules according to size (see Panel 4–5, p. 166). Smaller molecules diffuse faster in solution than larger ones, yet smaller molecules migrate more slowly through a gel-filtration column than larger ones. Explain this paradox. What should happen at very rapid flow rates?

QUESTION 4–20

Both an α helix and the coiled-coil that forms from it are helical structures, but do they have the same handedness (refer to Figure 4–11)?

QUESTION 4–21

How is it possible for a change in a single amino acid in a protein of 1000 amino acids to destroy its function, even when that amino acid is far away from any binding site?

CHAPTER **FIVE** 5

DNA and Chromosomes

THE STRUCTURE AND FUNCTION OF DNA

THE STRUCTURE OF EUCARYOTIC CHROMOSOMES

THE REGULATION OF CHROMOSOME STRUCTURE

Life depends on the ability of cells to store, retrieve, and translate the genetic instructions required to make and maintain a living organism. This hereditary information is passed on from a cell to its daughter cells at cell division, and from generation to generation in multicellular organisms through the reproductive cells. These instructions are stored within every living cell in its *genes*—the information-containing elements that determine the characteristics of a species as a whole and of the individuals within it.

At the beginning of the twentieth century, when genetics emerged as a science, scientists became intrigued by the chemical nature of genes. The information in genes is copied and transmitted from cell to daughter cells millions of times during the life of a multicellular organism, and it survives the process essentially unchanged. What kind of molecule could be capable of such accurate and almost unlimited replication, and also be able to direct the development of an organism and the daily life of a cell? What kind of instructions does the genetic information contain? How are these instructions physically organized so that the enormous amount of information required for the development and maintenance of even the simplest organism can be contained within the tiny space of a cell?

The answers to some of these questions began to emerge in the 1940s, when it was discovered from studies in simple fungi that genetic information consists primarily of instructions for making proteins. Proteins are the macromolecules that perform most of the cell's functions: they serve as building blocks for cell structures, they form the enzymes that catalyze the cell's chemical reactions, they regulate gene expression, and they enable cells to move and to communicate with one another. With hindsight, it is hard to imagine what other type of instructions the genetic information could have contained.

The other crucial advance made in the 1940s was the recognition that deoxyribonucleic acid (DNA) was the likely carrier of this genetic information. But the mechanism whereby the hereditary information is copied for transmission from cell to cell, and how proteins are specified by the instructions in the DNA, remained completely mysterious until 1953, when the structure of DNA was determined by James Watson and Francis Crick. The structure immediately revealed how DNA might be copied, or replicated, and it provided the first clues about how a molecule of DNA might encode the instructions for making proteins. Today, the fact that DNA is the genetic material is so fundamental to biological thought that it is difficult to appreciate what an enormous intellectual gap this discovery filled.

In this chapter, we begin by describing the structure of DNA. We see how, despite its chemical simplicity, the structure and chemical properties of DNA make it an ideal raw material for making genes. The genes of every cell on Earth are made of DNA, and insights into the relationship between DNA and genes have come from experiments in a wide variety of organisms. We then consider how genes and other important segments of DNA are arranged in the long molecules of DNA that are present in the chromosomes of cells. Finally, we discuss how eucaryotic cells fold these long DNA molecules into compact chromosomes that are contained inside the nucleus. This packing has to be done in an orderly fashion so that the chromosomes can be replicated and apportioned correctly between the two daughter cells at each cell division. It must also allow access of chromosomal DNA to enzymes that repair it when it is damaged and to the specialized proteins that direct the expression of its many genes.

This is the first of five chapters that deal with basic genetic mechanisms—the ways in which the cell replicates, repairs, expresses, and occasionally improves the genetic information carried in its DNA. The totality of this information in each cell is called its *genome*. In Chapter 6, we discuss the mechanisms by which the cell accurately replicates and repairs DNA; we also describe how DNA sequences can be rearranged through the process of genetic recombination. Gene expression—the process by which the information encoded in DNA is interpreted by the cell to guide the synthesis of proteins—is the main topic of Chapter 7. In Chapter 8, we describe how gene expression is controlled by the cell to ensure that each of the many thousands of proteins encrypted in its DNA is manufactured at the proper time and place in the life of the cell. We turn in Chapter 9 to a discussion of how present-day genes and genomes evolved from distant ancestors. An account of the experimental techniques used to study both DNA and its role in fundamental cellular processes will be presented in Chapter 10.

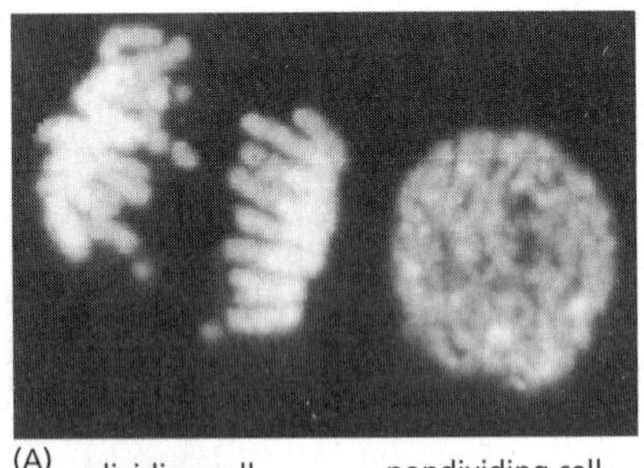

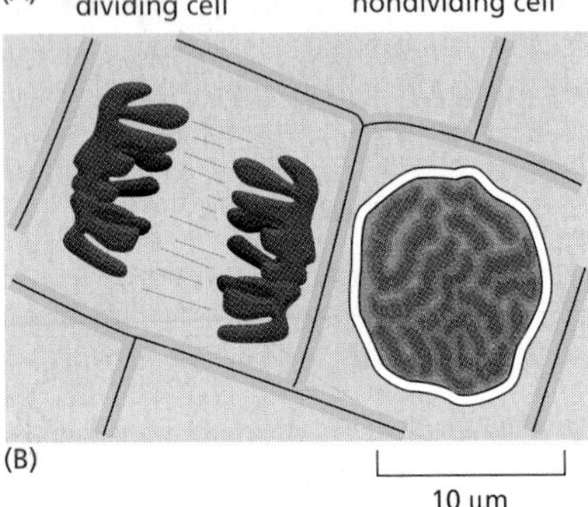

Figure 5–1 **Chromosomes become visible as cells prepare to divide.** (A) Two adjacent plant cells photographed through a light microscope. DNA is fluorescing brightly with a dye (DAPI) that binds to it. The DNA is present in chromosomes, which become visible as distinct structures in the light microscope only when they condense in preparation for cell division, as shown on the *left*. The cell on the *right*, which is not dividing, contains the identical chromosomes; they cannot be distinguished as individual chromosomes in the light microscope at this phase in the cell's life cycle because the DNA is in a much more extended conformation. (B) Schematic diagram of the outlines of the two cells along with their chromosomes. (A, courtesy of Peter Shaw.)

THE STRUCTURE AND FUNCTION OF DNA

Well before biologists understood the structure of DNA, they had recognized that inherited traits and the genes that determine them were associated with the chromosomes. Chromosomes (named from the Greek *chroma*, "color," because of their staining properties) were discovered in the nineteenth century as threadlike structures in the nucleus of the eucaryotic cell that become visible as the cell begins to divide (Figure 5–1). As biochemical analysis became possible, researchers learned that chromosomes contain both DNA and protein. But which of these components encoded the organism's genetic information was not at all clear.

We now know that the DNA carries the hereditary information of the cell, and that the protein components of chromosomes function largely to package and control the enormously long DNA molecules. But biologists in the 1940s had difficulty accepting DNA as the genetic material because of the apparent simplicity of its chemistry (see How We Know, pp. 174–176). DNA was thought of as simply a long polymer composed of only four types of subunits, which are chemically very similar to each other.

Then, early in the 1950s, DNA was examined by X-ray diffraction analysis, a technique for determining the three-dimensional atomic structure of a molecule (discussed in Chapter 4, pp. 158–160). The early results indicated that DNA is composed of two strands wound into a helix. The observation that DNA is double-stranded was of crucial significance. It provided one of the major clues that led, in 1953, to a correct model for the structure of DNA.

As soon as the Watson–Crick model of DNA structure was proposed, DNA's potential for replication and information encoding became apparent. In this section, we examine the structure of the DNA molecule and explain in general terms how it is able to store hereditary information.

A DNA Molecule Consists of Two Complementary Chains of Nucleotides

A molecule of **deoxyribonucleic acid (DNA)** consists of two long polynucleotide chains. Each of these *DNA chains*, or *DNA strands*, is composed of four types of nucleotide subunits, and the two chains are held together by hydrogen bonds between the base portions of the nucleotides (Figure 5–2, and see Panel 2–7, pp. 76–77, for a description of hydrogen bonds).

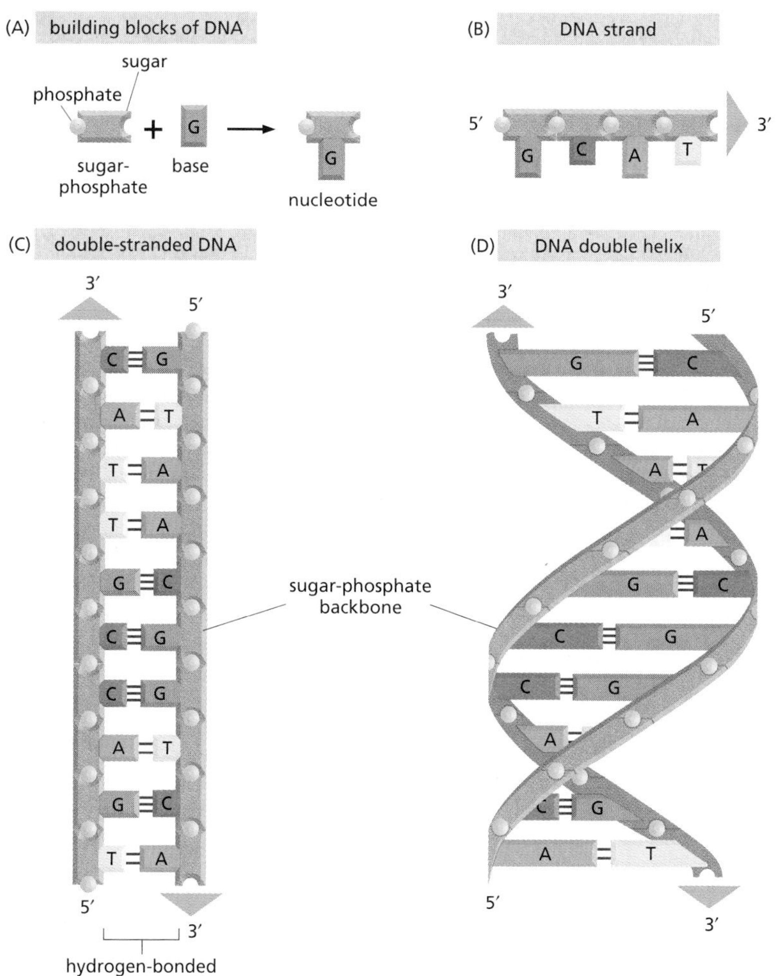

Figure 5–2 **DNA is made of four nucleotide building blocks.** (A) Each nucleotide is composed of a sugar–phosphate covalently linked to a base. (B) The nucleotides are covalently linked together into polynucleotide chains, with a sugar–phosphate backbone from which the bases (A, C, G, and T) extend. (C) A DNA molecule is composed of two polynucleotide chains (DNA strands) held together by hydrogen bonds between the paired bases. The *arrows* on the DNA strands indicate the polarities of the two strands, which run antiparallel to each other in the DNA molecule. (D) Although the DNA is shown straightened out in (C), in reality, it is wound into a double helix, as shown here.

GENES ARE MADE OF DNA

By the 1920s, scientists generally agreed that genes reside on chromosomes, and they knew that chromosomes are composed of both DNA and proteins. But because DNA is so chemically simple, researchers naturally assumed that genes had to be made of proteins, which are much more chemically diverse. Even when the experimental evidence suggested otherwise, this assumption proved hard to shake.

Messages from the dead

The case for DNA began to take shape in the late 1920s, when a British medical officer named Fred Griffith made an astonishing discovery. He was studying *Streptococcus pneumoniae* (pneumococcus), a bacterium that causes pneumonia. As antibiotics had not yet been discovered, infection with this organism was usually fatal. When

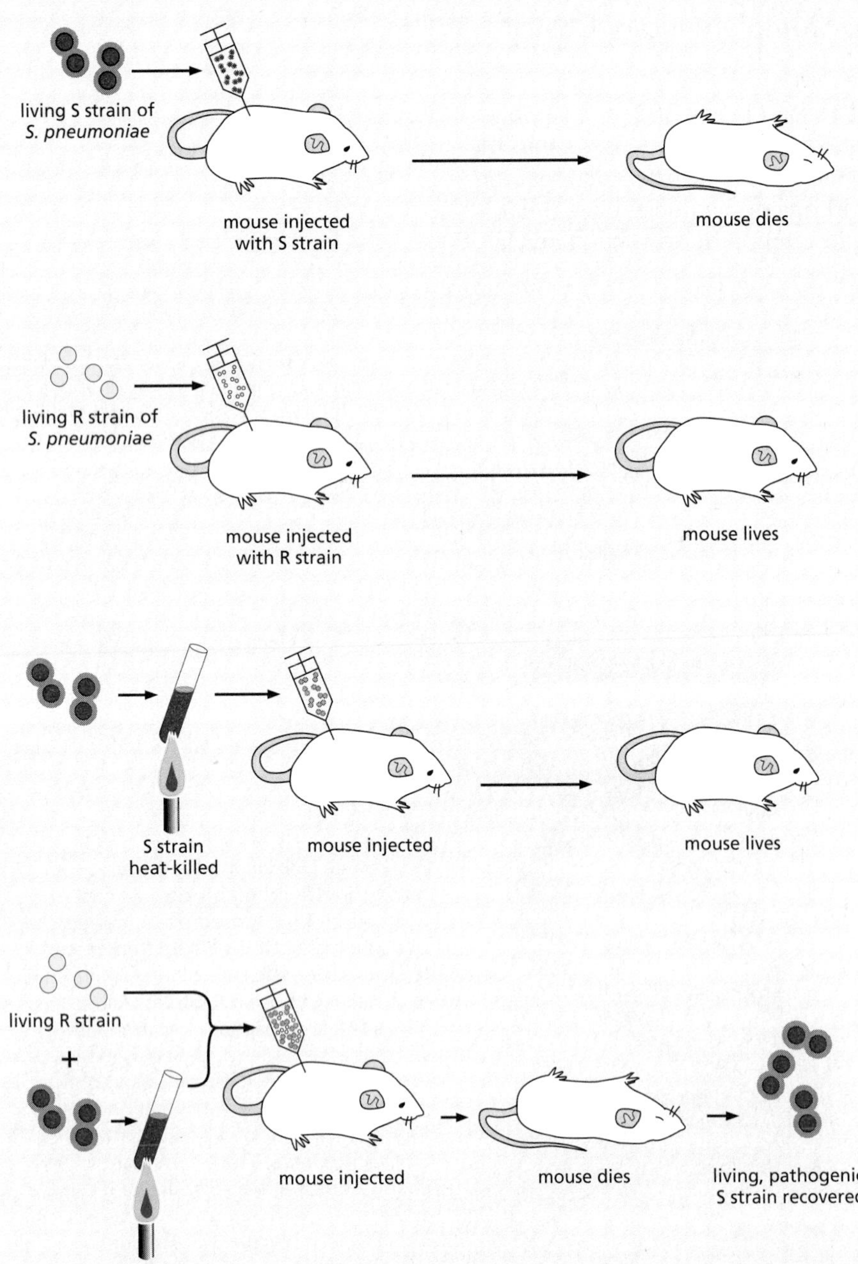

Figure 5–3 **Griffith showed that heat-killed bacteria can transform living cells.** The bacterium *Streptococcus pneumoniae* comes in two forms that differ from one another in their microscopic appearance and in their ability to cause disease. Cells of the pathogenic strain, which are lethal when injected into mice, are encased in a slimy, glistening polysaccharide capsule. When grown on a plate of nutrients in the laboratory, this disease-causing bacterium forms colonies that look dome-shaped and smooth; hence it is designated the S form. The harmless strain of the pneumococcus, on the other hand, lacks this protective coat; it forms colonies that appear flat and rough—hence, it is referred to as the R form. As illustrated, Griffith found that a substance present in the pathogenic S strain could permanently change, or transform, the nonlethal R strain into the deadly S strain.

grown in the laboratory, pneumococci come in two forms: a *pathogenic* form that causes a lethal infection when injected into animals, and a harmless form that is easily conquered by the animal's immune system and produces no infection.

In the course of his investigations, Griffith injected various preparations of these bacteria into mice. He showed that pathogenic pneumococci that had been killed by heating were no longer able to cause infection. The surprise came when Griffith injected both heat-killed pathogenic bacteria and live harmless bacteria into the same mouse. This combination proved lethal: not only did the animal die of pneumonia, but Griffith found that its blood was teeming with live bacteria of the pathogenic form (Figure 5–3). The heat-killed pneumococci had somehow converted the harmless bacteria into the lethal form. What's more, Griffith found that the change was permanent: he could grow these "transformed" bacteria in culture, and they remained pathogenic. But what was this mysterious material that turned harmless bacteria into killers? And how was this change passed on to progeny bacteria?

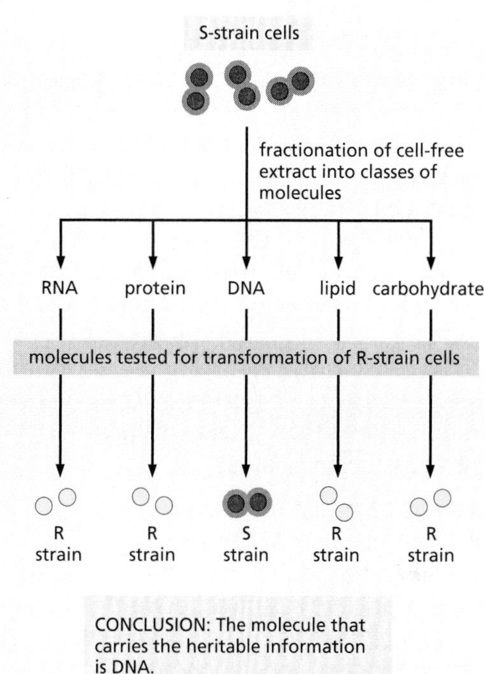

Figure 5–4 **Avery, MacLeod, and McCarty demonstrated that DNA is the genetic material.** The researchers prepared an extract from the disease-causing S strain and showed that the "transforming principle" that would permanently change the harmless R-strain pneumococci into the pathogenic S strain is DNA. This was the first evidence that DNA could serve as the genetic material.

Blowing bubbles

Griffith's remarkable finding set the stage for the experiments that would provide the first strong evidence that genes are made of DNA. The American bacteriologist Oswald Avery, following up on Griffith's work, discovered that the harmless pneumococcus could be transformed into a pathogenic strain in a culture tube by exposing it to an extract prepared from the pathogenic strain. It would take another 15 years, however, for Avery and his colleagues Colin MacLeod and Maclyn McCarty to successfully purify the "transforming principle" from this soluble extract and to demonstrate that the active ingredient was DNA. Because the transforming principle caused a heritable change in the bacteria that received it, DNA must be the very stuff of which genes are made.

The 15-year delay was in part a reflection of the academic climate—and the widespread supposition that the genetic material was likely to be made of protein. Because of the potential ramifications of their work, the researchers wanted to be absolutely certain that the transforming principle was DNA before they announced their findings. As Avery noted in a letter to his brother, also a bacteriologist, "It's lots of fun to blow bubbles, but it's wiser to prick them yourself before someone else tries to." So the researchers subjected the transforming material to a battery of chemical tests (Figure 5–4). They found that it exhibited all the chemical properties characteristic of DNA; furthermore, they showed that enzymes that destroy proteins and RNA did not affect its ability to transform bacteria, while enzymes that destroy DNA inactivated it. And like Griffith before them, the investigators found that their purified preparation changed the bacteria permanently: DNA from the pathogenic species was taken up by the harmless species, and this change was faithfully passed on to subsequent generations of bacteria.

This landmark study offered rigorous proof that purified DNA can act as genetic material. But the resulting paper, published in 1944, drew remarkably little attention. Despite the meticulous care with which these experiments were performed, geneticists were not immediately convinced that DNA is the hereditary material. Many argued that the transformation might have been caused by some trace protein contaminant in the preparations. Or that the extract might contain a mutagen that alters the genetic material of the harmless bacteria—converting the bugs to the pathogenic form—rather than containing the genetic material itself.

Virus cocktails

The debate was not settled definitively until 1952, when Alfred Hershey and Martha Chase fired up their laboratory blender and demonstrated, once and for all, that genes are made of DNA. The researchers were studying T2—a virus that infects and eventually destroys the bacterium *E. coli*. These bacteria-killing viruses behave like little molecular syringes: they inject their genetic material into the host cell, while the empty virus heads remain outside the infected bacterium (Figure 5–5A). Once inside the cell, the viral genes direct the formation of new virus particles. In less than an hour, the infected cells explode, spewing thousands of new viruses into the medium. These then infect neighboring bacteria, and the process begins again.

The beauty of T2 is that these viruses contain only two kinds of molecules: DNA and protein. So the genetic material had to be one or the other. But which? The experiment was fairly straightforward. Because the viral DNA enters the bacterial cell, while the rest of the virus particle remains outside, the researchers decided to radioactively label the protein in one batch of virus and the DNA in another. Then, all they had to do was follow the radioactivity to see whether viral DNA or viral protein wound up inside the bacteria. To do this, Hershey and Chase incubated their radiolabeled viruses with *E. coli*; after allowing a few minutes for infection to take place, they poured the mix into a Waring blender and hit "puree." The blender's spinning blades sheared the empty virus heads from the surfaces of the bacterial cells. The researchers then centrifuged the sample to separate the heavier, infected bacteria, which formed a pellet at the bottom of the centrifuge tube, from the empty viral coats, which remained in suspension (Figure 5–5B).

As you have probably guessed, Hershey and Chase found that the radioactive DNA entered the bacterial cells, while the radioactive proteins remained with the empty virus heads. They found that the radioactive DNA was also incorporated into the next generation of virus particles.

This experiment demonstrated conclusively that viral DNA enters bacterial host cells, whereas viral protein does not. Thus, the genetic material in this virus had to be made of DNA. Together with the studies done by Avery, MacLeod, and McCarty, this evidence clinched the case for DNA as the agent of heredity.

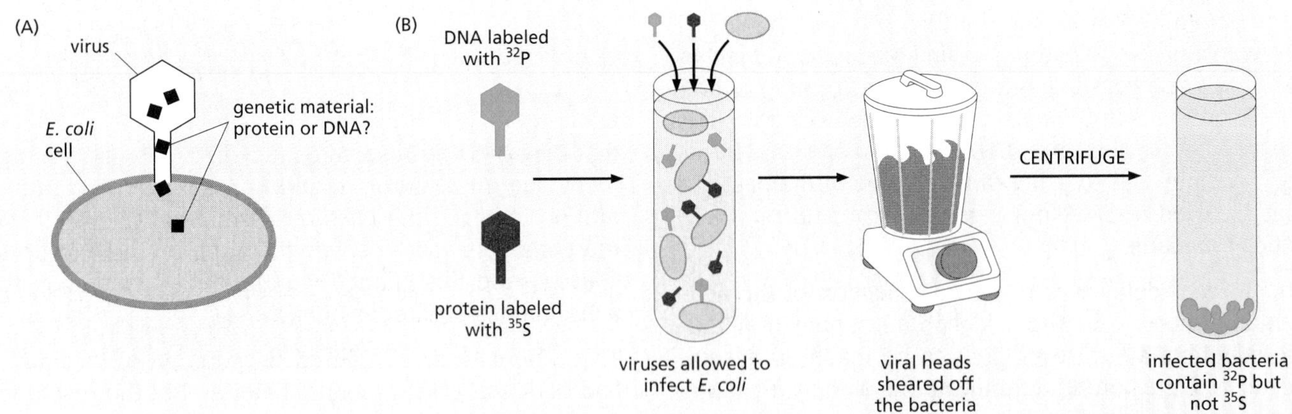

Figure 5–5 **Hershey and Chase showed definitively that genes are made of DNA.** (A) The researchers worked with T2 viruses, which are made entirely of protein and DNA. Each virus acts as a molecular syringe, injecting its genetic material into a bacterium; the empty viral capsule remains attached to the outside of the cell. (B) To determine whether the genetic material of the virus is protein or DNA, the researchers radioactively labeled the DNA in one batch of viruses with ^{32}P and the proteins in a second batch of viruses with ^{35}S. Because DNA lacks sulfur and the proteins lack phosphorus, these radioactive isotopes provided a handy way for the researchers to distinguish these two types of molecules. These labeled viruses were then allowed to infect *E. coli*, and the mixture was disrupted by brief pulsing in a Waring blender to separate the infected bacteria from the empty viral heads. When the researchers measured the radioactivity, they found that most of the ^{32}P-labeled DNA had entered the bacterial cells, while the vast majority of the ^{35}S-labeled proteins remained in solution with the spent viral particles.

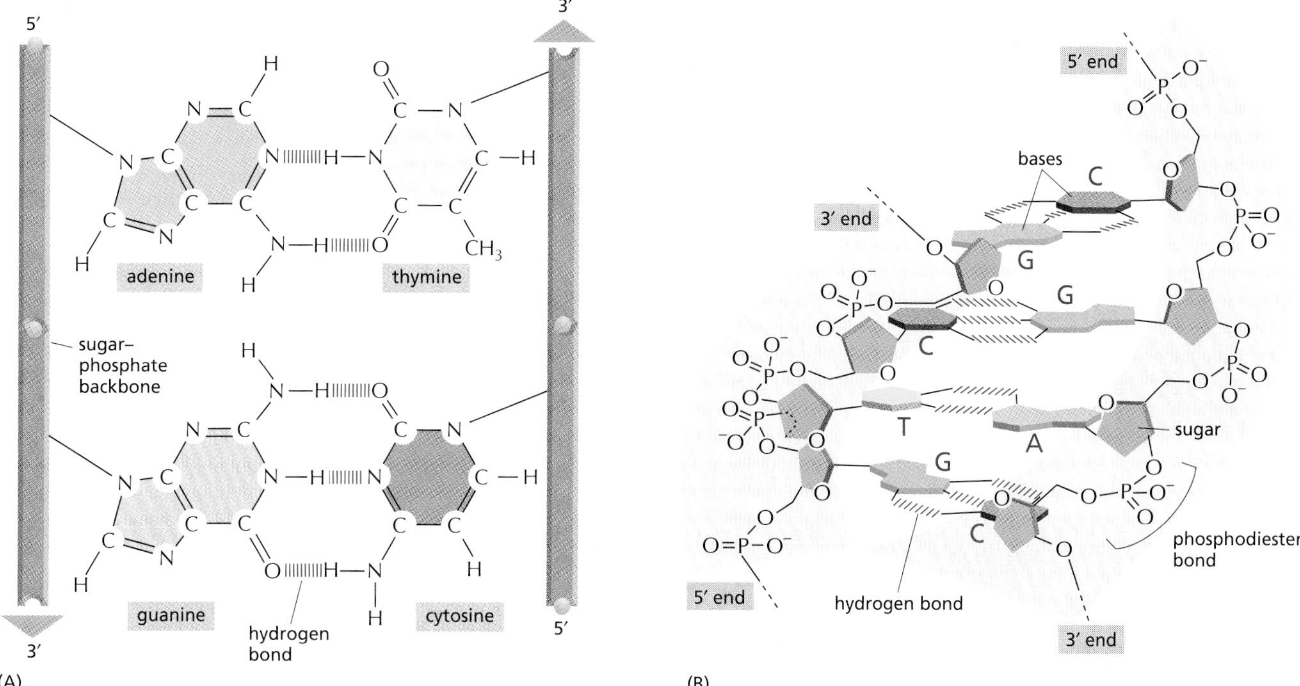

Figure 5–6 **The two strands of the DNA double helix are held together by hydrogen bonds between complementary base pairs.** (A) The shapes and chemical structure of the bases allow hydrogen bonds to form efficiently only between A and T and between G and C, where atoms that are able to form hydrogen bonds (see Panel 2–2, pp. 66–67) can be brought close together without perturbing the double helix. Two hydrogen bonds form between A and T, whereas three form between G and C. The bases can pair in this way only if the two polynucleotide chains that contain them are antiparallel—that is, oriented in opposite polarities. (B) A short section of the double helix viewed from its side. Four base pairs are shown. The nucleotides are linked together covalently by phosphodiester bonds through the 3′-hydroxyl (–OH) group of one sugar and the 5′-phosphate ($-PO_4$) of the next. This linkage gives each polynucleotide strand a chemical polarity; that is, its two ends are chemically different. The 3′ end carries an unlinked –OH group attached to the 3′ position on the sugar ring; the 5′ end carries a free phosphate group attached to the 5′ position on the sugar ring.

As we saw in Chapter 2 (Panel 2–6, pp. 74–75), nucleotides are composed of a five-carbon sugar to which are attached one or more phosphate groups and a nitrogen-containing base. For the nucleotides in DNA, the sugar is deoxyribose attached to a single phosphate group (hence the name deoxyribonucleic acid); the base may be either *adenine (A)*, *cytosine (C)*, *guanine (G)*, or *thymine (T)*. The nucleotides are covalently linked together in a chain through the sugars and phosphates, which thus form a "backbone" of alternating sugar–phosphate–sugar–phosphate (see Figure 5–2B). Because it is only the base that differs in each of the four types of subunits, each polynucleotide chain in DNA can be thought of as a necklace (the backbone) strung with four types of beads (the four bases A, C, G, and T). These same symbols (A, C, G, and T) are also commonly used to denote the four different nucleotides—that is, the bases with their attached sugar and phosphate groups.

The way in which the nucleotide subunits are linked together gives a DNA strand a chemical polarity. If we imagine that each nucleotide has a knob (the phosphate) and a hole (see Figure 5–2A), each chain, formed by interlocking knobs with holes, will have all of its subunits lined up in the same orientation. Moreover, the two ends of the chain can be easily distinguished, as one will have a hole (the 3′ hydroxyl) and the other a knob (the 5′ phosphate). This polarity in a DNA chain is indicated by referring to one end as the *3′ end* and the other as the *5′ end*. This convention is based on the details of the chemical linkage between the nucleotide subunits.

The two polynucleotide chains in the DNA **double helix** are held together by hydrogen-bonding between the bases on the different strands. All the bases are therefore on the inside of the helix, with the sugar–phosphate backbones on the outside (see Figure 5–2D). The bases do not pair at random, however: A always pairs with T, and G always pairs with C (Figure 5–6). In each case, a bulkier two-ring base (a purine, see Panel 2–6, pp. 74–75) is paired with a single-ring base (a pyrimidine). Each purine–pyrimidine pair is called a **base pair**, and this *complementary base-pairing* enables the base pairs to be packed in the energetically most favorable arrangement in the interior of the double helix. In this arrangement, each base pair is of similar width, thus holding the sugar–phosphate back-

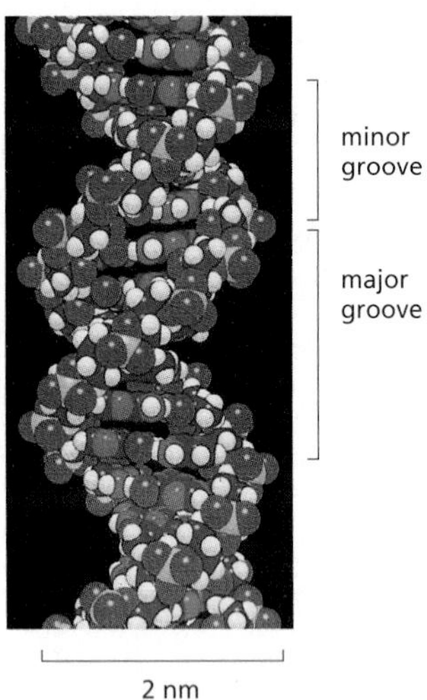

Figure 5–7 **A space-filling model shows the conformation of the DNA double helix.** The two strands of DNA wind around each other to form a right-handed helix with 10 bases per turn. Shown here are 1.5 turns of the DNA double helix. The coiling of the two strands around each other creates two grooves in the double helix. The wider groove is called the major groove and the smaller one the minor groove. The colors of the atoms are: N, *blue*; O, *red*; P, *yellow*; and H, *white*.

bones an equal distance apart along the DNA molecule. The members of each base pair can fit together within the double helix because the two strands of the helix run antiparallel to each other—that is, they are oriented with opposite polarities (see Figure 5–2C and D). The antiparallel sugar-phosphate strands then twist around each other to form a double helix containing 10 base pairs per helical turn (Figure 5–7). This winding also contributes to the energetically favorable conformation of the DNA double helix.

A consequence of the double helix base-pairing requirements is that each strand of a DNA molecule contains a sequence of nucleotides that is exactly **complementary** to the nucleotide sequence of its partner strand—an A always matches a T on the opposite strand, and a C always matches a G. This complementarity is of crucial importance when it comes to both copying and repairing the DNA, as we discuss in Chapter 6. An animated version of DNA structure can be seen in Movie 5.1.

QUESTION 5–1

Which of the following statements are correct? Explain your answers.
A. A DNA strand has a polarity because one end of the strand is more highly charged than the other.
B. G–C base pairs are more stable than A-T base pairs.

The Structure of DNA Provides a Mechanism for Heredity

Genes carry biological information that must be copied and transmitted accurately when a cell divides to form two daughter cells. This situation poses two central biological problems: how can the information for specifying an organism be carried in chemical form, and how is it accurately copied? The discovery of the structure of the DNA double helix was a landmark in twentieth-century biology because it immediately suggested answers to these two questions and thereby resolved the problem of heredity at the molecular level. In this chapter, we outline the answer to the first question; in the next chapter, we address in detail the answer to the second.

DNA encodes information in the order, or sequence, of the nucleotides along each strand. Each base—A, C, T, or G—can be considered as a letter in a four-letter alphabet that is used to spell out biological messages in the chemical structure of the DNA (Figure 5–8). Organisms differ from one another because their respective DNA molecules have different *nucleotide sequences* and, consequently, carry different biological messages. But how is the nucleotide alphabet used to make up messages, and what do they spell out?

It had already been established some time before the structure of DNA was determined that genes contain the instructions for producing proteins. The DNA messages, therefore, must somehow encode proteins. Consideration of the chemical character of proteins makes the problem easier to define. As discussed in Chapter 4, the function of a protein is determined by its three-dimensional structure, and its structure in turn is determined by the sequence of the amino acids in its polypeptide chain. The linear sequence of nucleotides in a gene must therefore somehow spell out the linear sequence of amino acids in a protein.

The exact correspondence between the 4-letter nucleotide alphabet of DNA and the 20-letter amino acid alphabet of proteins—the *genetic code*—is not obvious from the structure of the DNA molecule, and it took more than a decade after the discovery of the double helix to work it out. In Chapter 7, we describe this code in detail in the course of explaining the process, known as *gene expression*, through which a cell *transcribes* the nucleotide sequence of a gene into the nucleotide sequence of an RNA molecule, and then *translates* that information into the amino acid sequence of a protein (Figure 5–9).

The complete set of information in an organism's DNA is called its **genome** (the term is also used to refer to the DNA that carries this information). The total amount of this information is staggering: written out in the four-letter nucleotide alphabet, the nucleotide sequence of a very small gene from humans occupies a quarter of a page of text, while the complete sequence of the human genome would fill more than 1000 books the size of this one. Herein lies a problem that affects the architecture of all eucaryotic chromosomes: how can all this information be packed neatly into every cell nucleus? In the remainder of this chapter, we discuss the answer to this question.

(A) molecular biology is...

(B) 

(C) •— •—•• •• —•—• •

(D) 细胞生物学乐趣无穷

(E) TTCGAGCGACCTAACCTATAG

Figure 5–8 **Linear messages come in many forms.** The languages shown are, (A) English, (B) a musical score, (C) Morse code, (D) Chinese, and (E) DNA.

THE STRUCTURE OF EUCARYOTIC CHROMOSOMES

Large amounts of DNA are required to encode all the information needed to make even a single-celled bacterium, and far more DNA is needed to encode the instructions for the development of multicellular organisms like ourselves. Each human cell contains about 2 m of DNA; yet the cell nucleus is only 5–8 μm in diameter. Tucking all this material into such a small space is the equivalent of trying to fold 40 km (24 miles) of extremely fine thread into a tennis ball.

In eucaryotic cells, very long double-stranded DNA molecules are packaged into structures called **chromosomes**, which not only fit readily inside the nucleus but can be easily apportioned between the two daughter cells at each cell division. The complex task of packaging DNA is accomplished by specialized proteins that bind to and fold the DNA, generating a series of coils and loops that provide increasingly higher levels of organization and prevent the DNA from becoming an unmanageable tangle. Amazingly, the DNA is compacted in a way that allows it to remain accessible to all of the enzymes and other proteins that replicate it, repair it, and direct the expression of its genes.

Bacteria typically carry their genes on a single, circular DNA molecule. This molecule is also associated with proteins that condense DNA, but these proteins differ from the ones that package eucaryotic DNA. Although this procaryotic DNA is called a bacterial "chromosome," it does not have the same structure as eucaryotic chromosomes, and less is known about how it is packaged. Our discussion of chromosome structure in this chapter will therefore focus entirely on eucaryotic chromosomes.

Eucaryotic DNA Is Packaged into Multiple Chromosomes

In eucaryotes, such as ourselves, the DNA in the nucleus is distributed among a set of different chromosomes. The human genome, for example, contains approximately 3.2×10^9 nucleotides parceled out into 24 chromosomes. Each chromosome consists of a single, enormously long, linear DNA molecule associated with proteins that fold and pack the fine thread of DNA into a more compact structure. The complex of DNA and protein is called *chromatin*. In addition to the proteins involved in packaging the DNA, chromosomes are also associated with many other proteins involved in gene expression, DNA replication, and DNA repair.

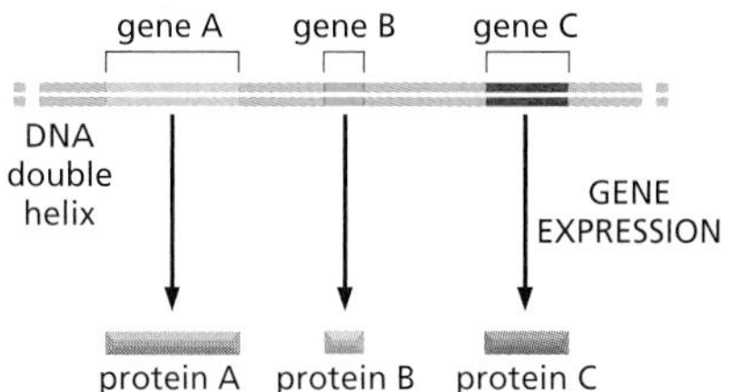

Figure 5–9 **Genes contain information to make proteins.** As we discuss in Chapter 7, each protein-coding gene is used to produce RNA molecules, which then direct the production of the specific protein molecules. In this diagram, the RNA intermediates are not shown.

With the exception of the germ cells (sperm and eggs) and highly specialized cells that lack DNA entirely (such as the mature red blood cell), human cells each contain two copies of each chromosome, one inherited from the mother and one from the father. The maternal and paternal chromosomes of a pair are called *homologous chromosomes* (*homologs*). The only nonhomologous chromosome pairs are the sex chromosomes in males, where a *Y chromosome* is inherited from the father and an *X chromosome* from the mother.

In addition to being different sizes, human chromosomes can be distinguished from one another by a variety of techniques. Each chromosome

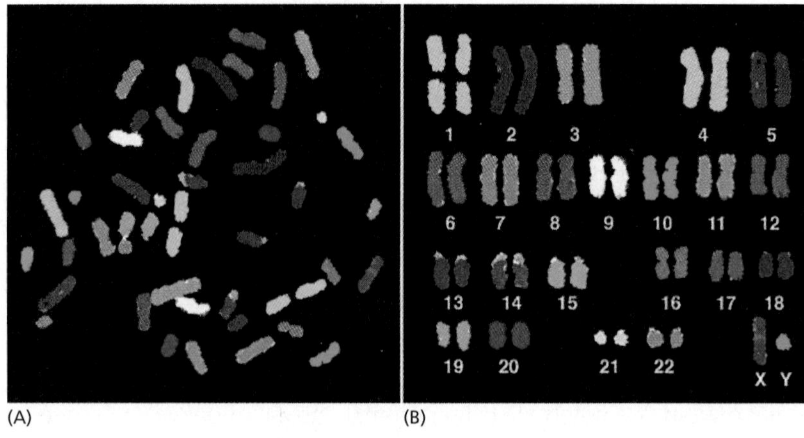

Figure 5–10 **Each human chromosome can be "painted" a different color to allow its unambiguous identification under the light microscope.** The chromosomes shown here were isolated from a cell undergoing nuclear division (mitosis) and are therefore in a highly compact state. Chromosome painting is carried out by exposing the chromosomes to a collection of human DNA molecules that have been coupled to a combination of fluorescent dyes. For example, DNA molecules derived from Chromosome 1 are labeled with one specific dye combination, those from Chromosome 2 with another, and so on. Because the labeled DNA can form base pairs, or hybridize, only to its chromosome of origin (discussed in Chapter 10), each chromosome is differently labeled. For such experiments, the chromosomes are treated so that the double-helical DNA separates into individual strands, to enable base-pairing with the labeled, single-stranded DNA while keeping the chromosome structure relatively intact. (A) The chromosomes as visualized as they originally spilled from the lysed cell. (B) The same chromosomes have been artificially lined up in order. In this so-called *karyotype*, the homologous chromosomes are numbered and arranged in pairs; the presence of a Y chromosome indicates that the DNA was isolated from a male. (From E. Schröck et al., *Science* 273:494–497, 1996. With permission from AAAS.)

can be "painted" a different color using sets of chromosome-specific DNA molecules coupled to different fluorescent dyes (Figure 5–10). This involves the technique of *DNA hybridization*, which will be described in detail in Chapter 10. A more traditional way of distinguishing one chromosome from another is to stain the chromosomes with dyes that bind to certain types of DNA sequences. These dyes mainly distinguish between DNA that is rich in A-T nucleotide pairs and DNA that is G-C rich, and they produce a striking and reliable pattern of bands along each chromosome (Figure 5–11). The pattern of bands on each type of chromosome is unique, allowing each chromosome to be identified and numbered.

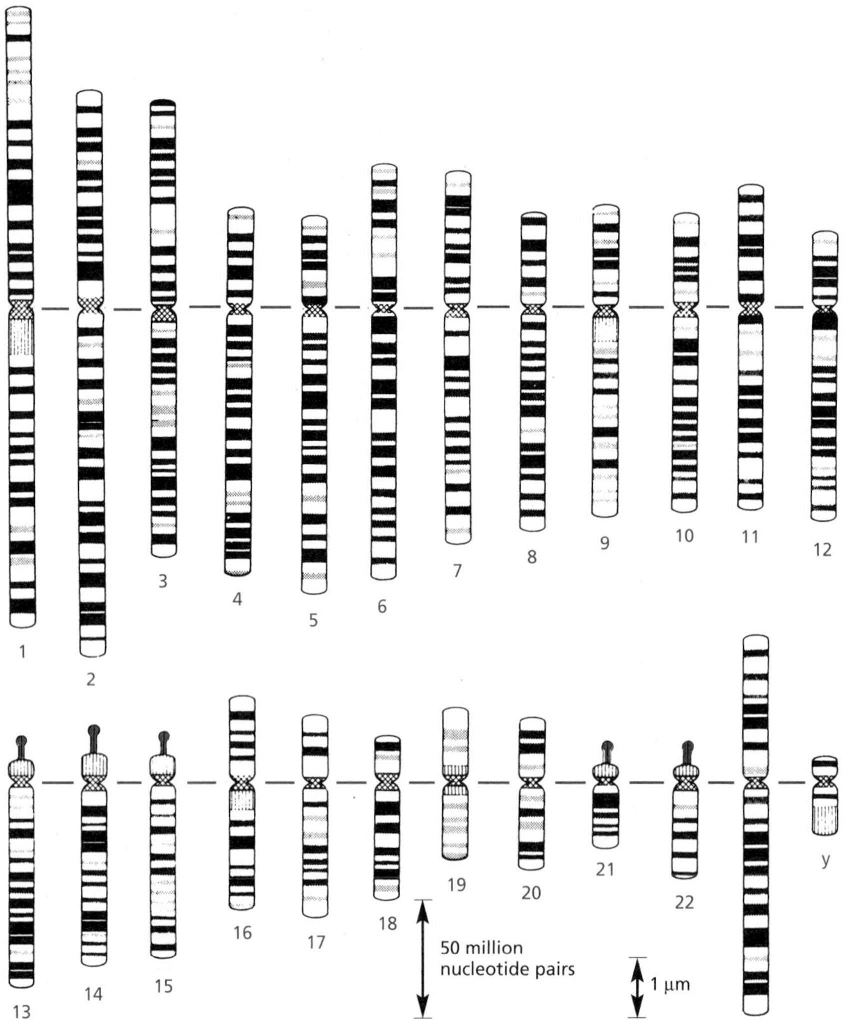

Figure 5–11 **Unique banding patterns allow the identification of each human chromosome.** Chromosomes 1 through 22 are numbered in approximate order of size. A typical human somatic (that is, non-germ) cell contains two of each of these chromosomes plus two sex chromosomes—two X chromosomes in a female, one X and one Y chromosome in a male. The chromosomes used to make these maps were stained at an early stage in mitosis, when the DNA is compacted, but not so heavily compacted as it will be later in mitosis. The horizontal *red* line represents the position of the centromere, which appears as a constriction on mitotic chromosomes; the knobs (*red*) on Chromosomes 13, 14, 15, 21, and 22 indicate the positions of genes that code for the large ribosomal RNAs (discussed in Chapter 7). These patterns are obtained by staining chromosomes with Giemsa stain, which produces dark bands in regions rich in A-T nucleotide pairs. (Adapted from U. Franke, *Cytogenet. Cell Genet.* 31:24–32, 1981. With permission from S. Karger AG.)

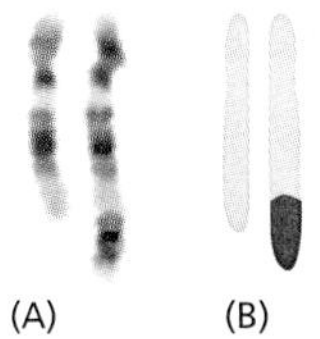

Figure 5–12 **Abnormal chromosomes are associated with some inherited genetic defects.** (A) A pair of Chromosomes 12 from a patient with inherited ataxia, a disease characterized by progressive deterioration of motor skills. The patient has one normal Chromosome 12 (*left*) and one abnormal, longer Chromosome 12. The additional material contained on the abnormal Chromosome 12 was determined from its banding pattern to be a piece of Chromosome 4 that had become inappropriately attached to Chromosome 12. (B) In this diagram, the chromosomes in (A) have the segment corresponding to Chromosome 4 DNA "painted" *red* and the parts corresponding to Chromosome 12 DNA "painted" *blue*. (From E. Schröck et al., *Science* 273:494–497, 1996. With permission from AAAS.)

A display of the full set of 46 human chromosomes is called the human **karyotype**. If parts of a chromosome are lost, or switched between chromosomes, these changes can be detected by changes in the banding patterns. Cytogeneticists use alterations in banding patterns to detect chromosomal abnormalities that are associated with some inherited defects (Figure 5–12) and with certain types of cancer.

QUESTION 5–2

In a DNA double helix, there is one nucleotide pair per 0.34 nm. Use Figure 5–11 to estimate the length of the DNA in human Chromosome 1 if it were unraveled and stretched out. If the actual length of Chromosome 1 at this stage of mitosis is approximately 10 μm, what is the degree of compaction of the DNA in this state?

Chromosomes Contain Long Strings of Genes

The most important function of chromosomes is to carry the genes—the functional units of heredity (Figure 5–13). A **gene** is usually defined as a segment of DNA that contains the instructions for making a particular protein (or, in some cases, a set of closely related proteins). Although this definition fits the majority of genes, some genes direct the production of an RNA molecule, instead of a protein, as their final product. Like proteins, these RNA molecules perform a diverse set of structural and catalytic functions in the cell, as we will see in later chapters.

As might be expected, some correlation exists between the complexity of an organism and the number of genes in its genome. For example, the total number of genes ranges from less than 500 for a simple bacterium to about 25,000 for humans. Bacteria and some single-celled eucaryotes have especially compact genomes: the DNA molecules that make up their chromosomes are little more than strings of closely packed genes. However, chromosomes from many eucaryotes (including humans) contain, in addition to genes, a large excess of interspersed DNA, the majority of which does not seem to carry critical information. This DNA is sometimes called "junk DNA," because its usefulness to the cell has not yet been clearly demonstrated. Although the particular nucleotide sequence of most of this DNA might not be important, the DNA itself—acting as spacer material—may be crucial for the long-term evolution of the species and for the proper activity of genes. In addition, comparisons of the genome sequences from many different species reveals that a portion of this extra DNA is highly conserved among related species, indicating that it serves an important function—although we don't yet know what that is. We take up this question in greater detail in Chapter 9.

In general, the more complex an organism, the larger is its genome. But this relationship does not always hold true. The human genome, for example, is 200 times larger than that of the yeast *S. cerevisiae*, but 30

Figure 5–13 **Genes are arranged along the chromosomes.** This figure shows a small region of one chromosome from the budding yeast *S. cerevisiae*. The *S. cerevisiae* genome contains about 6300 genes—more than 12 million nucleotide pairs—spread across 16 chromosomes. Note that, in each gene, only one of the two DNA strands actually encodes the information to make a protein or RNA molecule, and this can be either strand, as indicated by the *light red* bars. However, a gene is generally denoted to contain both this strand and its complement, as in Figure 5–9. The high density of genes is characteristic of the species.

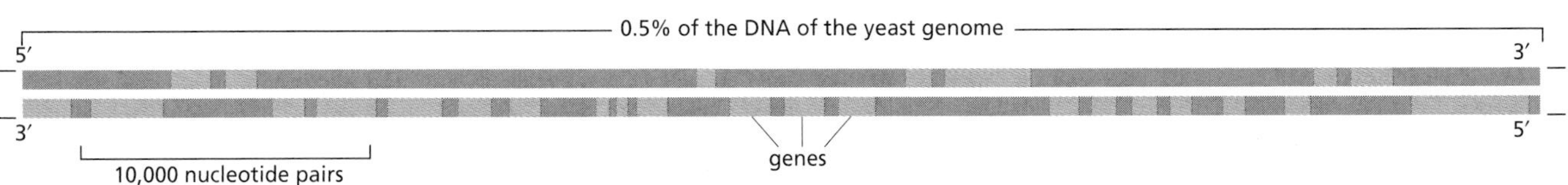

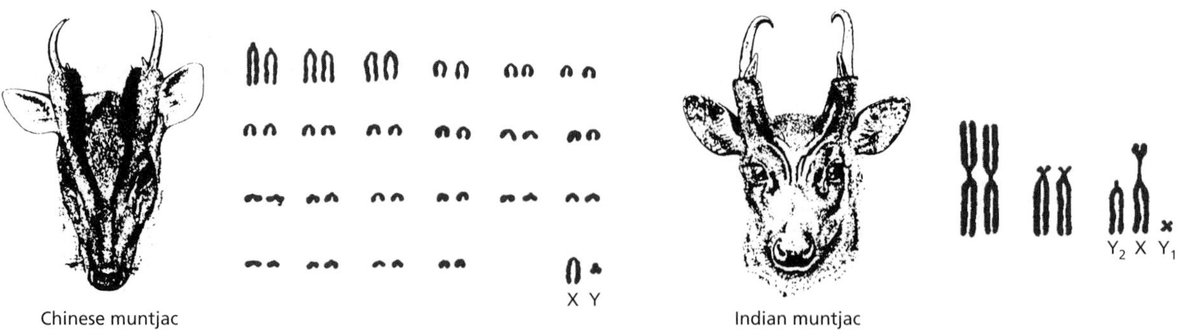

Figure 5–14 **Closely related species can have very different chromosome numbers.** In the evolution of the Indian muntjac deer, chromosomes that were initially separate, and that remain separate in the Chinese species, fused without having a major effect on the animal. The two species shown have roughly the same number of genes. (From B.K. Hall and B. Hallgrimsson, Strickberger's Evolution, 4th ed. Sudbury, MA, 2008. With permission from Jones & Bartlett Publishers.)

times smaller than that of some plants and at least 60 times smaller than some species of amoeba. Furthermore, how the DNA is apportioned over chromosomes also differs from one species to another. Humans have 46 chromosomes, but a species of small deer has only 6, while some carp species have more than 100. Even closely related species with similar genome sizes can have very different numbers and sizes of chromosomes (Figure 5–14). Thus, although gene number is roughly correlated with species complexity, there is no simple relationship between gene number, chromosome number, and total genome size. The genomes and chromosomes of modern species have each been shaped by a unique history of seemingly random genetic events, acted on by specific selection pressures.

Chromosomes Exist in Different States Throughout the Life of a Cell

To form a functional chromosome, a DNA molecule must do more than simply carry genes: it must be able to be replicated, and the replicated copies must be separated and partitioned reliably into daughter cells at each cell division. These processes occur through an ordered series of events, known collectively as the **cell cycle**. This cycle of cell growth and division is briefly summarized in Figure 5–15 and will be discussed in detail in Chapter 18. Only two broad stages of the cell cycle need concern us in this chapter: *interphase*, when chromosomes are duplicated; and *mitosis*, when they are distributed to the two daughter nuclei.

During interphase, the chromosomes are extended as long, thin, tangled threads of DNA in the nucleus and cannot be easily distinguished in the light microscope. We refer to chromosomes in this extended state as *interphase chromosomes*. Specialized DNA sequences found in all eucaryotes ensure that the interphase chromosomes replicate efficiently

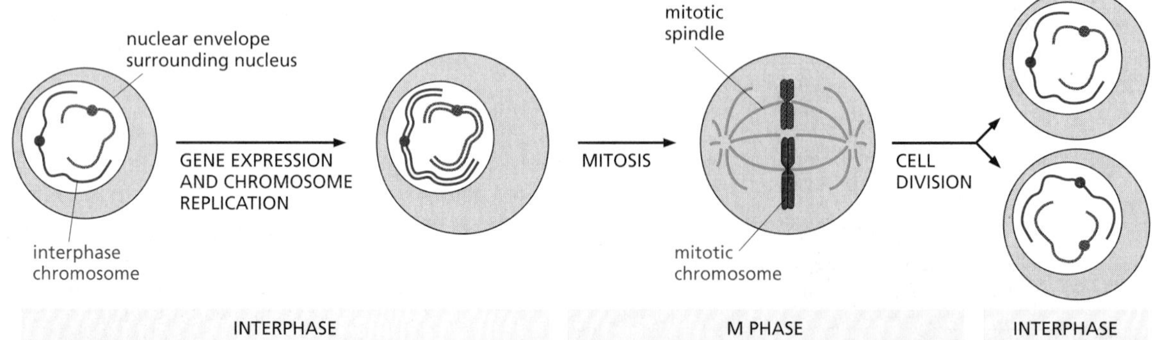

Figure 5–15 **The replication and segregation of chromosomes occurs through an ordered cell cycle in proliferating cells.** During interphase, the cell is actively expressing its genes. For part of this time, the DNA is being replicated and the chromosomes are duplicated. Once DNA replication is complete, the cell can enter *M phase*, when mitosis occurs. Mitosis is the division of the nucleus. During this stage, the chromosomes condense, gene expression largely ceases, the nuclear envelope breaks down, and the mitotic spindle forms from microtubules and other proteins. The condensed chromosomes are then captured by the mitotic spindle, and one complete set of chromosomes is pulled to each end of the cell. A nuclear envelope forms around each chromosome set, and in the final step of M phase, the cell divides to produce two daughter cells. Only two different chromosomes are shown here for simplicity.

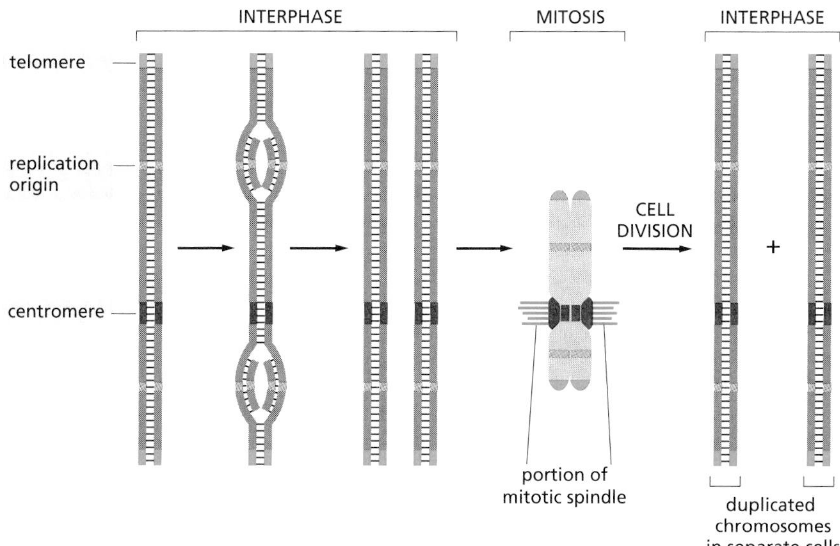

Figure 5–16 **Three DNA sequence elements are needed to produce a eucaryotic chromosome that can be replicated and then segregated at mitosis.** Each chromosome has multiple origins of replication, one centromere, and two telomeres. The sequence of events that a typical chromosome follows during the cell cycle is shown schematically. The DNA replicates in interphase, beginning at the origins of replication and proceeding bidirectionally from the origins across the chromosome. In M phase, the centromere attaches the duplicated chromosomes to the mitotic spindle so that one copy is distributed to each daughter cell when the cell divides. The centromere also helps to hold the duplicated chromosomes together until they are ready to be pulled apart. The telomeres form special caps at each chromosome end.

(Figure 5–16). One type of nucleotide sequence acts as a **replication origin**, where duplication of the DNA begins, as we discuss in Chapter 6. Eucaryotic chromosomes contain many replication origins to ensure that the entire chromosome can be replicated rapidly. Another DNA sequence forms the **telomeres** found at each of the two ends of a chromosome. Telomeres contain repeated nucleotide sequences that enable the ends of chromosomes to be replicated, as we discuss in Chapter 6. They also cap the end of the chromosome, preventing it from being mistaken by the cell as a broken DNA molecule in need of repair.

When the cell cycle reaches M phase, the DNA coils up, adopting a more and more compact structure, ultimately forming highly compacted, or condensed, *mitotic chromosomes*. This is the state in which chromosomes are most easily visualized. In this highly condensed state, duplicated chromosomes can be readily separated when the cell divides (see Figure 5–16). Once the chromosomes have condensed, it is the presence of the third specialized DNA sequence, the **centromere**, that allows one copy of each duplicated chromosome to be apportioned to each daughter cell (Figure 5–17). We describe the central role that centromeres play in cell division in Chapter 18.

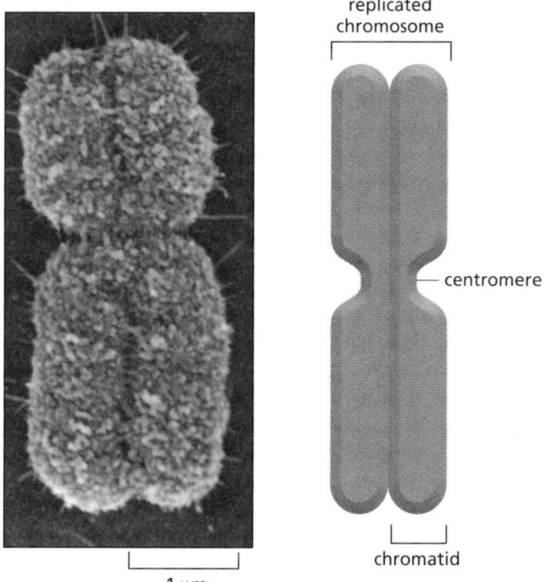

Figure 5–17 **A typical mitotic chromosome is highly compact.** Because of chromosome replication during interphase, each mitotic chromosome contains two identical daughter DNA molecules (see Figure 5–16). Each of these very long DNA molecules, with its associated proteins, is called a *chromatid.* (A) A scanning electron micrograph of a mitotic chromosome. The two chromatids are tightly joined together. The constricted region reveals the position of the centromere. (B) A cartoon representation of a mitotic chromosome. (A, courtesy of Terry D. Allen.)

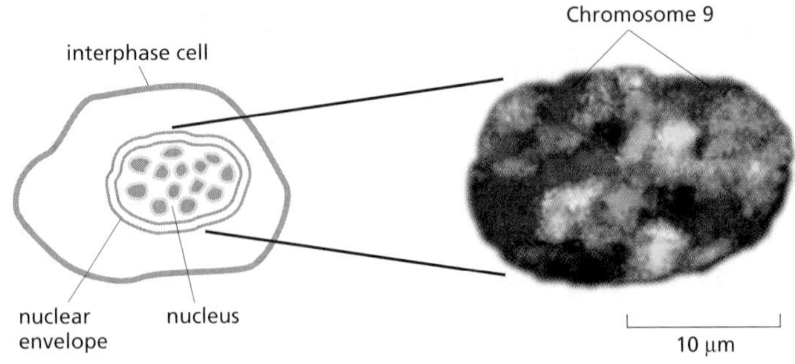

Figure 5–18 **Interphase chromosomes occupy different territories within the nucleus.** DNA probes coupled with different fluorescent markers are used to paint individual interphase chromosomes in a human cell. Viewed under a fluorescence microscope, each interphrase chromosome is seen to occupy its own discrete territory within the nucleus, rather than being mixed with the other chromosomes like spaghetti in a bowl. Note that the pairs of homologous chromosomes present in a diploid cell (e.g., the two copies of Chromosome 9 indicated) are not generally located in the same position. (From M.R. Speicher and N.P. Carter, *Nat. Rev. Genet.* 6:782–792, 2005. With permission from Macmillan Publishers Ltd.)

Interphase Chromosomes Are Organized Within the Nucleus

The nucleus is surrounded by a *nuclear envelope* formed by two concentric membranes. The nuclear envelope is punctured at intervals by nuclear pores, which actively transport selected molecules to and from the cytosol (described in detail in Chapter 15), and is supported by the *nuclear lamina*, a network of protein filaments that forms a thin layer underlying and the inner nuclear membrane (discussed in Chapter 17).

Inside the nucleus, the interphase chromosomes—although longer and finer than mitotic chromosomes—are nonetheless organized in various ways. First, each interphase chromosome tends to occupy a particular region of the nucleus, and so different chromosomes do not become extensively entangled with one another (Figure 5–18). In addition, specific regions of chromosomes are attached to sites on the nuclear envelope or the nuclear lamina.

The most obvious example of chromosome organization in the interphase nucleus is the **nucleolus** (Figure 5–19). The nucleolus is where the parts of the different chromosomes carrying genes for ribosomal RNA (the red knobs in Figure 5–11) cluster together. Here, ribosomal RNAs are synthesized and combined with proteins to form ribosomes, the cell's protein-synthesizing machines (discussed in Chapter 7).

The DNA in Chromosomes Is Highly Condensed

As we have seen, all eucaryotic cells, whether in interphase or mitosis, package their DNA tightly into chromosomes. Human Chromosome 22, for example, contains about 48 million nucleotide pairs; stretched out end-to-end, its DNA would extend about 1.5 cm. Yet, during mitosis,

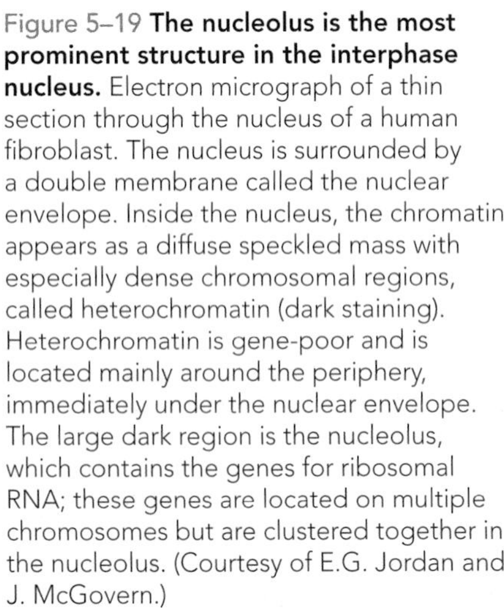

Figure 5–19 **The nucleolus is the most prominent structure in the interphase nucleus.** Electron micrograph of a thin section through the nucleus of a human fibroblast. The nucleus is surrounded by a double membrane called the nuclear envelope. Inside the nucleus, the chromatin appears as a diffuse speckled mass with especially dense chromosomal regions, called heterochromatin (dark staining). Heterochromatin is gene-poor and is located mainly around the periphery, immediately under the nuclear envelope. The large dark region is the nucleolus, which contains the genes for ribosomal RNA; these genes are located on multiple chromosomes but are clustered together in the nucleolus. (Courtesy of E.G. Jordan and J. McGovern.)

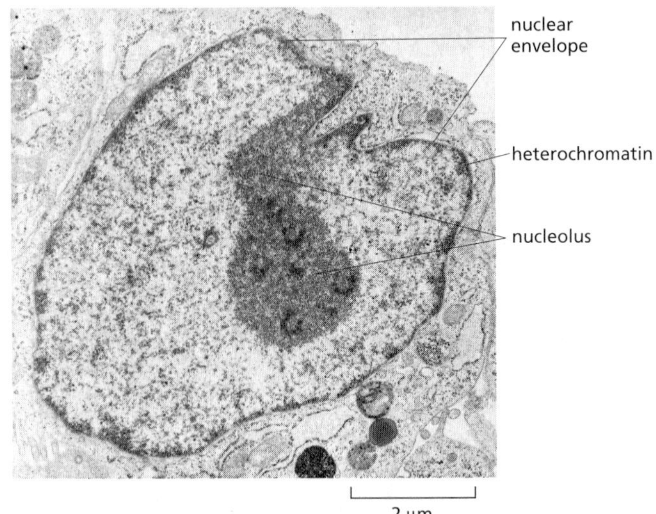

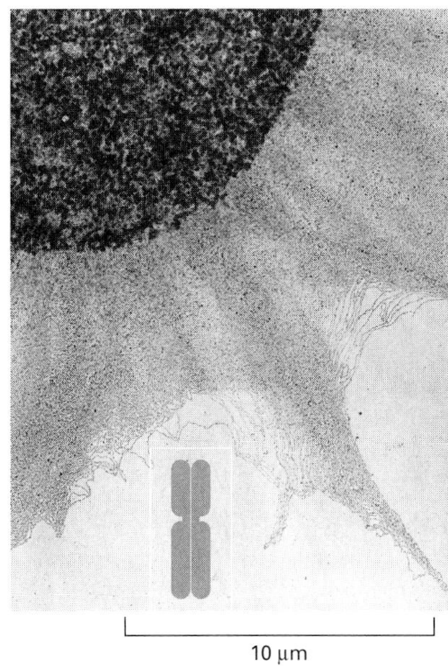

Figure 5–20 **DNA in interphase chromosomes is less compact than that in mitotic chromosomes.** An electron micrograph showing an enormous tangle of chromosomal DNA (with its associated proteins) spilling out of a lysed interphase nucleus. For comparison, an outline of a condensed, human mitotic chromosome is shown at the same scale. (Courtesy of Victoria Foe.)

Chromosome 22 measures only about 2 μm in length—that is, nearly 10,000 times more compact than the DNA in its extended form. This remarkable feat of compression is performed by proteins that coil and fold the DNA into higher and higher levels of organization. The DNA of interphase chromosomes, although less condensed than that of mitotic chromosomes (Figure 5–20), is still packed tightly, with a compaction ratio of about 500-fold.

In the next sections, we introduce the specialized proteins that make this compression possible. Bear in mind, though, that chromosome structure is dynamic. Not only do chromosomes condense and relax in concert with the cell cycle, but different regions of the interphase chromosome must unpack to allow cells to access specific DNA sequences for replication, repair, or gene expression. Chromosome packaging must be flexible enough to allow rapid, localized, on-demand access to the DNA.

Nucleosomes Are the Basic Units of Eucaryotic Chromosome Structure

The proteins that bind to the DNA to form eucaryotic chromosomes are traditionally divided into two general classes: the **histones** and the nonhistone chromosomal proteins. Histones are present in enormous quantities (more than 60 million molecules of several different types in each cell), and their total mass in chromosomes is about equal to that of the DNA itself. The complex of both classes of protein with nuclear DNA is called **chromatin**.

Histones are responsible for the first and most fundamental level of chromatin packing, the **nucleosome**, which was discovered in 1974. When interphase nuclei are broken open very gently and their contents examined under the electron microscope, most of the chromatin is in the form of fibers, each with a diameter of about 30 nm (Figure 5–21A). If this chromatin is subjected to treatments that cause it to unfold partially, it can then be seen under the electron microscope as a series of "beads on a string" (Figure 5–21B). The string is DNA, and each bead is a *nucleosome core particle* that consists of DNA wound around a core of proteins formed from histones.

QUESTION 5–3

Assuming that the histone octamer forms a cylinder 9 nm in diameter and 5 nm in height and that the human genome forms 32 million nucleosomes, what volume of the nucleus (6 μm in diameter) is occupied by histone octamers? (Volume of a cylinder is $\pi r^2 h$; volume of a sphere is $4/3\ \pi r^3$.) What fraction of the total volume of the nucleus do the histone octamers occupy? How does this compare with the volume of the nucleus occupied by human DNA?

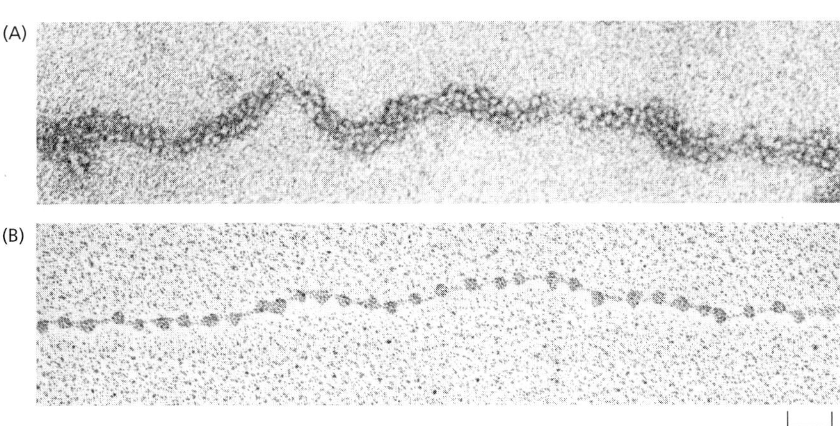

Figure 5–21 **Nucleosomes can be seen in the electron microscope.** (A) Chromatin isolated directly from an interphase nucleus appears in the electron microscope as threads 30-nm thick; a part of one such fiber is shown here. (B) This electron micrograph shows a length of a chromatin fiber that has been experimentally unpacked, or decondensed, after isolation to show the nucleosomes. (A, courtesy of Barbara Hamkalo; B, courtesy of Victoria Foe.)

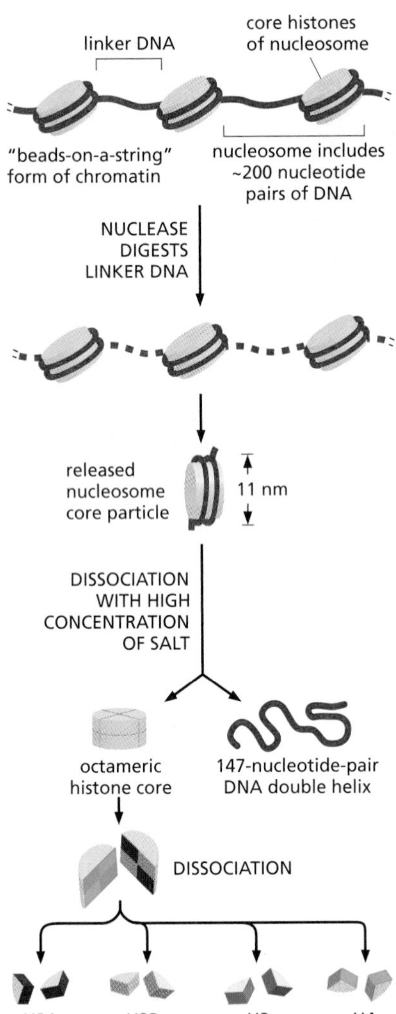

Figure 5–22 **Nucleosomes contain DNA wrapped around a protein core of eight histone molecules.** In a test tube, the nucleosome core particle can be released from chromatin by digestion of the linker DNA with a nuclease, an enzyme that breaks down DNA. (The nuclease can degrade the exposed DNA but cannot attack the DNA wound tightly around the nucleosome core.) After dissociation of the isolated nucleosome into its protein core and DNA, the length of the DNA that was wound around the core can be determined. Its length of 147 nucleotide pairs is sufficient to wrap almost twice around the histone core.

The structure of the nucleosome core particle was determined after first isolating nucleosomes from the unfolded chromatin by digestion with particular enzymes (called nucleases) that break down DNA by cutting between the nucleotides. After digestion for a short period, only the exposed DNA between the core particles, the *linker DNA*, is degraded. An individual nucleosome core particle consists of a complex of eight histone proteins—two molecules each of histones H2A, H2B, H3, and H4—and the double-stranded DNA, 147 nucleotide pairs long, that winds around this *histone octamer* (Figure 5–22). The high-resolution structure of the nucleosome core particle was solved in 1997, revealing in atomic detail the disc-shaped histone complex around which the DNA is tightly wrapped, making 1.7 turns in a left-handed coil (Figure 5–23).

The linker DNA between each nucleosome core particle can vary in length from a few nucleotide pairs up to about 80. (The term nucleosome technically refers to a nucleosome core particle plus one of its adjacent DNA linkers (see Figure 5–22), but it is often used to mean just the nucleosome core particle.) The formation of nucleosomes converts a DNA molecule into a chromatin thread approximately one-third of its initial length, and it provides the first level of DNA packing.

All four of the histones that make up the nucleosome core are relatively small proteins with a high proportion of positively charged amino acids (lysine and arginine). The positive charges help the histones bind tightly to the negatively charged sugar–phosphate backbone of DNA. These numerous interactions explain in part why DNA of virtually any sequence can bind to a histone core. Each of the core histones also has a long N-terminal amino acid "tail," which extends out from the nucleosome core particle (see Figure 5–23). These histone tails are subject to several types of covalent chemical modifications that control many aspects of chromatin structure.

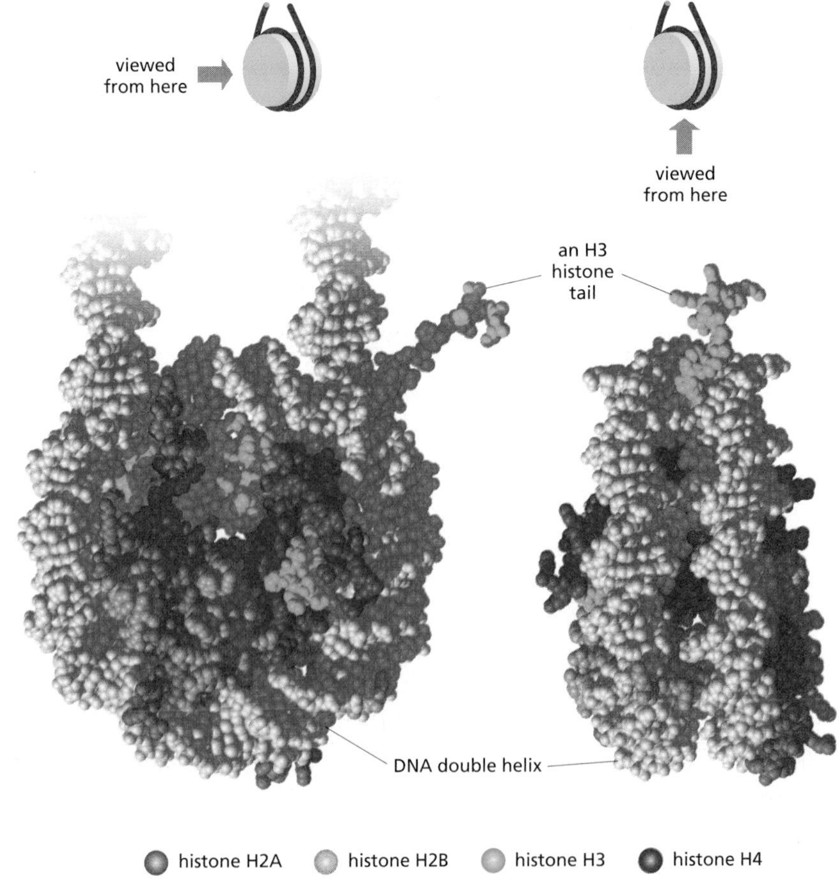

Figure 5–23 **The structure of the nucleosome core particle, as determined by X-ray diffraction analysis, reveals how DNA is tightly wrapped around a disc-shaped histone core.** Two views of a nucleosome structure are shown here. The DNA helix is *gray*. A portion of an H3 histone tail (*green*) can be seen extending from the nucleosome, but the tails of the other histones have been truncated. (Reprinted by permission from K. Luger et al., *Nature* 389:251–260, 1997. With permission from Macmillan Publishers Ltd.)

The histones that form the nucleosome core are among the most highly conserved of all known eucaryotic proteins: there are only two differences between the amino acid sequences of histone H4 from peas and cows, for example. This extreme evolutionary conservation reflects the vital role of histones in controlling eucaryotic chromosome structure. Recently, histones have also been found in archaea—procaryotes that form a phylogenetic kingdom distinct from eucaryotes and bacteria (discussed in Chapter 1).

Chromosome Packing Occurs on Multiple Levels

Although long strings of nucleosomes form on most chromosomal DNA, chromatin in the living cell rarely adopts the extended beads-on-a-string form seen in Figure 5–21B. Instead, the nucleosomes are further packed upon one another to generate a more compact structure, the *30-nm fiber* (see Figure 5–21A). Packing of nucleosomes into the 30-nm fiber depends on a fifth histone called histone H1, which is thought to pull the nucleosomes together into a regular repeating array. This "linker" histone changes the path the DNA takes as it exits the nucleosome core, allowing it to form a more compact structure (Figure 5–24). The 30-nm fiber that results is illustrated in Movie 5.2 and in Figure 5–25, which provides an overview of the various levels of chromosome packing.

We know that the 30-nm chromatin fiber can be compacted still further. We saw earlier in this chapter that during mitosis chromatin becomes so highly condensed that the chromosomes can be seen under the light microscope. How is the 30-nm fiber folded to produce mitotic chromo-

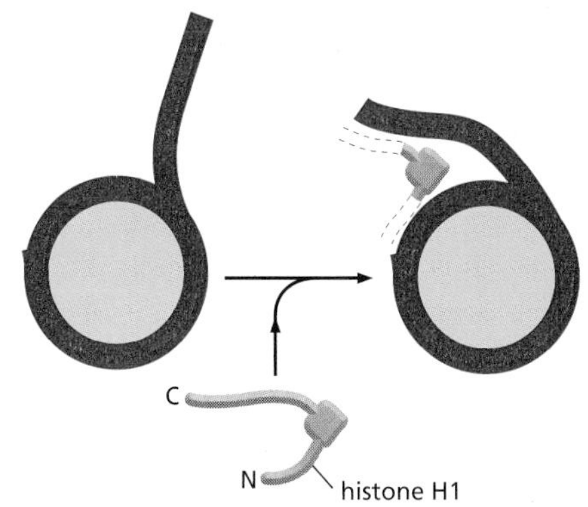

Figure 5–24 **A linker histone helps to pull nucleosomes together into the 30-nm fiber.** Histone H1 consists of a globular region plus a pair of long tails at its C-terminal and N-terminal ends. The globular region constrains an additional 20 base pairs of the DNA where it exits from the nucleosome core, an activity that is thought to be important for the formation of the 30-nm fiber. The long C-terminal tail is required for H1 to bind to chromatin, but its position—and the position of the N-terminal tail—are not known.

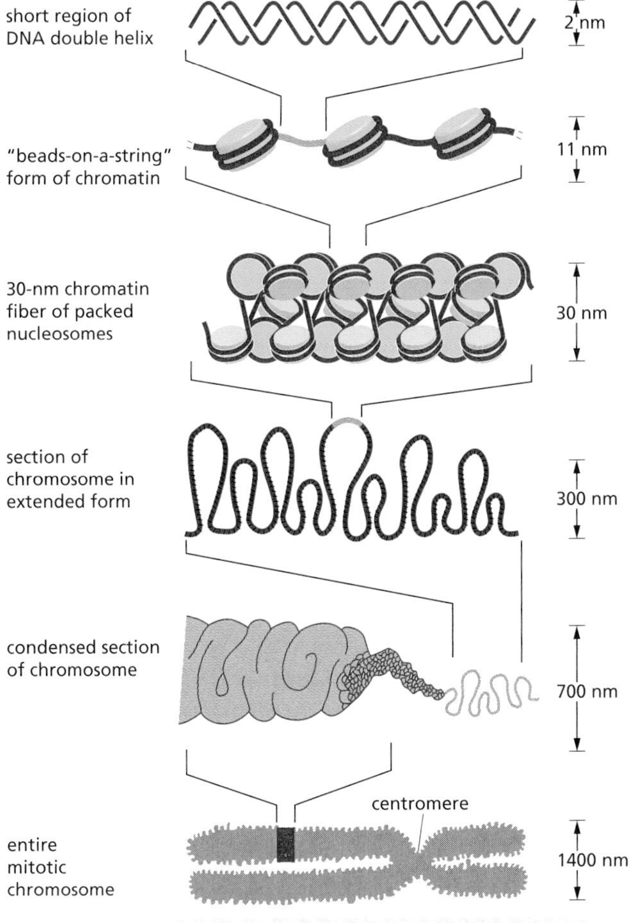

Figure 5–25 **DNA packing occurs on several levels in chromosomes.** This schematic drawing shows some of the levels thought to give rise to the highly condensed mitotic chromosome.

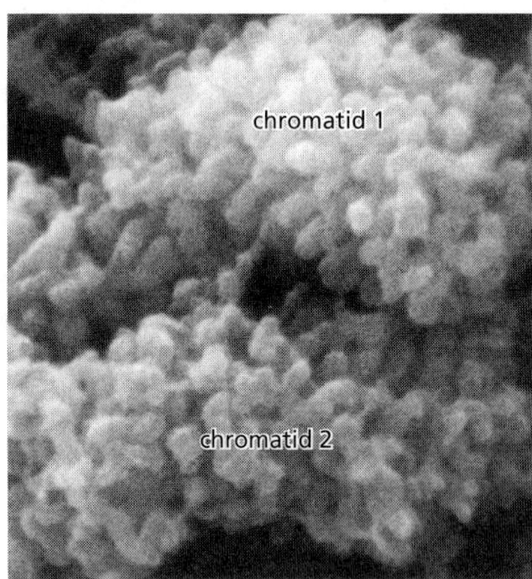

Figure 5–26 **The mitotic chromosome is formed from tightly packed chromatin.** This scanning electron micrograph shows a region near one end of a typical mitotic chromosome. Each knoblike projection is believed to represent the tip of a separate loop of chromatin. The chromosome has duplicated, and the two new chromosomes (called chromatids) are still held together (see Figure 5–17). The ends of the two chromosomes can be distinguished on the right of the photo. (From M.P. Marsden and U.K. Laemmli, *Cell* 17:849–858, 1989. With permission from Elsevier.)

somes? The answer is not yet known in detail, but it is known that the 30-nm fiber is folded into a series of loops, and that these loops are further condensed to produce the interphase chromosome. Finally, this compact string of loops is thought to undergo at least one more level of packing to form the mitotic chromosome (Figure 5–26, and see Figure 5–25).

THE REGULATION OF CHROMOSOME STRUCTURE

So far, we have discussed how DNA is packed carefully and tightly into chromatin. We now turn to the question of how this packaging can be dynamic, allowing rapid access to the underlying DNA. The DNA in cells carries enormous amounts of coded information, and cells must be able to get to this information as needed.

In this section, we discuss how a cell can alter its chromatin structure to expose localized regions of DNA and allow access to specific proteins, particularly those involved in gene expression and in DNA replication and repair. We then discuss how chromatin structure is established and maintained—and how a cell can pass on some forms of this structure to its descendants. The regulation and inheritance of chromatin structure play an important part in the development and growth of eucaryotic organisms.

QUESTION 5–4

Histone proteins are among the most highly conserved proteins in eucaryotes. Histone H4 proteins from a pea and a cow, for example, differ in only 2 of 102 amino acids. Comparison of the gene sequences shows many more differences, but only two that change encoded amino acids. These observations indicate that mutations that change amino acids must be selected against. Why do you suppose that amino acid-altering mutations in histone genes are deleterious?

Changes in Nucleosome Structure Allow Access to DNA

Eucaryotic cells have several ways to adjust the local structure of their chromatin rapidly. One way takes advantage of **chromatin-remodeling complexes**, protein machines that use the energy of ATP hydrolysis to change the position of the DNA wrapped around nucleosomes. By pushing on the tightly bound DNA as they move along, these complexes can loosen (decondense) the underlying DNA, making it more accessible to other proteins in the cell (Figure 5–27). During mitosis, at least some of the chromatin-remodeling complexes are inactivated, which may help mitotic chromosomes maintain their tightly packed structure.

Another way of altering chromatin structure relies on the reversible chemical modification of the histones. The tails of all four of the core histones are particularly subject to these covalent modifications. For example, acetyl, phosphate, or methyl groups can be added to and removed from the assembled nucleosome by enzymes that reside in the nucleus. These modifications of the histone tails have little direct effect on the stability of an individual nucleosome. But some seem to directly affect the stability of the 30-nm chromatin fiber and the higher-order structures discussed earlier.

Most importantly, these modifications affect the ability of the histone tails to bind specific proteins and thereby recruit them to particular stretches of chromatin. Different patterns of histone tail modifications attract different proteins, some of which cause further condensation of the chromatin, whereas others facilitate access to the DNA by decondensing chromatin. Specific combinations of tail modifications and the proteins that bind to

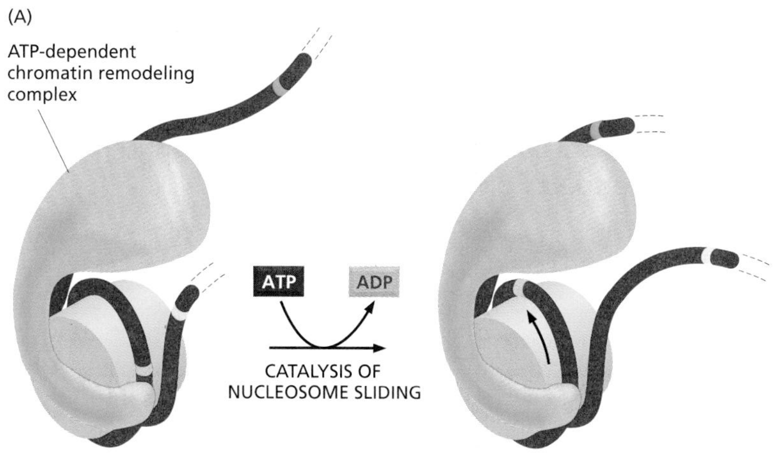

Figure 5–27 **Chromatin-remodeling complexes reposition the DNA wrapped around nucleosomes.** (A) Repeated cycles of ATP hydrolysis allow the chromatin-remodeling complex to loosen the nucleosomal DNA by pushing it along the histone core. This repositions (slides) the nucleosome, exposing the DNA to other DNA-binding proteins. Here, the nucleosome has been moved slightly to the left along the DNA. Many cycles of ATP hydrolysis are required to produce such a shift. (B) Multiple rounds of nucleosome sliding can decondense chromatin, making the underlying DNA accessible to other proteins in the cell. Conversely, other types of sliding can condense the chromatin in a particular chromosomal region.

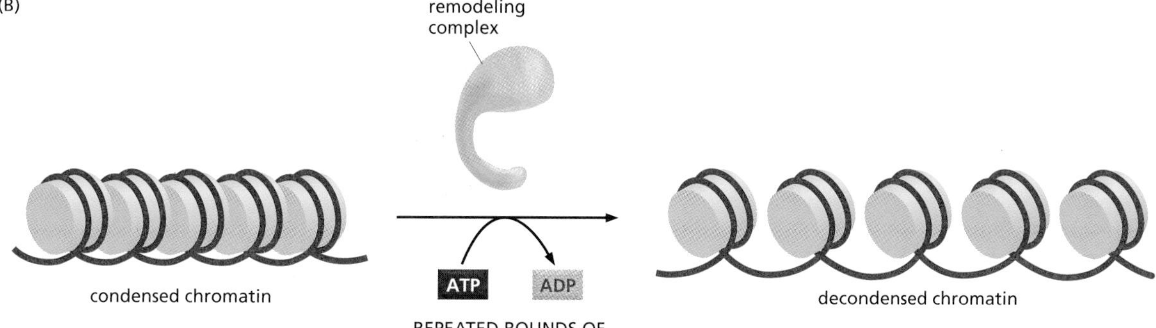

them have different meanings for the cell: for example, one pattern might indicate that a particular stretch of chromatin has been newly replicated, whereas another indicates that the genes in that chromatin should be expressed (Figure 5–28).

Like the chromatin-remodeling complexes, the enzymes that modify histone tails are tightly regulated. They are brought to a particular region of chromatin by other cues, particularly by interactions with proteins that bind to specific sequences in DNA (we discuss these proteins in Chapter 8). The histone-modifying enzymes work in concert with the chroma-

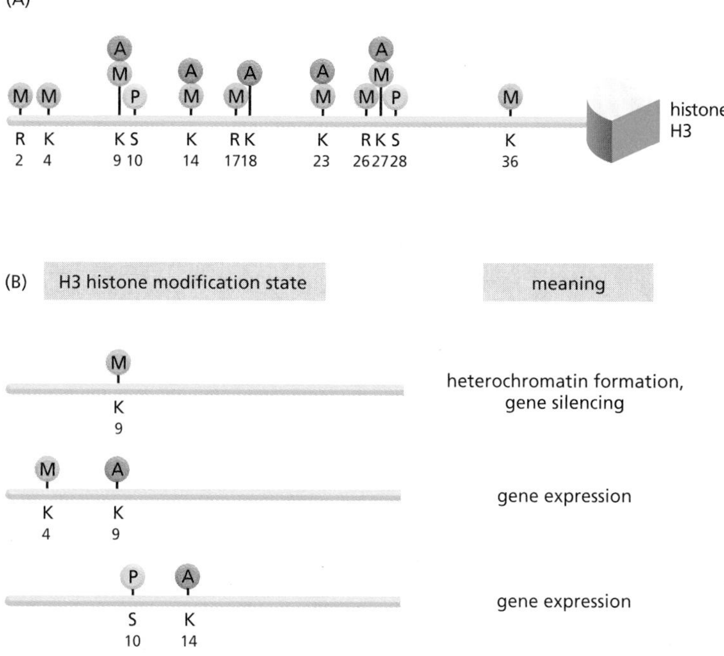

Figure 5–28 **The pattern of modification of histone tails can dictate how a stretch of chromatin is treated by the cell.** (A) Each histone can be modified by the covalent attachment of a number of different chemical groups. Histone H3, for example, can receive an acetyl group (A), a methyl group (M), or a phosphate (P). The numbers denote the positions of the modified amino acid in the protein chain. Note that some positions (e.g., lysines 9, 14, 23, and 27) can be modified in more than one way. Moreover, lysines can be modified with either one, two, or three methyl groups (not shown). Note that histone H3 contains 135 amino acids, most of which are in its globular portion (wedge), and that most modifications are on its N-terminal tail. (B) Different combinations of histone tail modifications can confer a specific meaning on the stretch of chromatin on which they occur, as indicated. Only a few of these "meanings" are known.

tin-remodeling complexes to condense or decondense stretches of chromatin, allowing local chromatin structure to change rapidly according to the needs of the cell.

Interphase Chromosomes Contain Both Condensed and More Extended Forms of Chromatin

The localized alteration of chromatin packing by remodeling complexes and histone modification has important effects on the large-scale structure of interphase chromosomes. The chromatin in these chromosomes is not uniformly packed. Instead, regions of the chromosome that contain genes that are being expressed are generally more extended, while those that contain quiescent genes are more compact. Thus, the detailed structure of an interphase chromosome can differ from one cell type to the next, helping to determine which genes are expressed.

The most highly condensed form of interphase chromatin is called **heterochromatin** (from the Greek *heteros*, "different," plus chromatin). It was first observed under the light microscope in the 1930s as discrete, strongly staining regions within the mass of chromatin. Heterochromatin typically makes up about 10% of an interphase chromosome, and in mammalian chromosomes, it is concentrated around the centromere region and in the telomeres at the ends of the chromosomes. The formation of the most common form of heterochromatin is induced by a particular set of histone tail modifications, including the methylation of lysine residue 9 in histone H3 (see Figure 5–28). These modifications attract a set of heterochromatin-specific proteins, which then induce the same histone tail modifications in adjacent nucleosomes. The new tail modifications in turn recruit the same set of heterochromatin-specific proteins, causing a spreading wave of condensed chromatin to propagate along the chromosome. In this manner, an extended region of heterochromatin is established along the DNA.

Most DNA that is permanently folded into heterochromatin in the cell does not contain genes. Because heterochromatin is so compact, genes that accidentally become packaged into heterochromatin usually fail to be expressed (Figure 5–29). Such inappropriate packaging of genes in heterochromatin can cause disease: in humans, the gene that encodes β-globin—which forms part of the oxygen-carrying hemoglobin molecule—is situated next to a region of condensed chromatin. If, because of an inherited DNA deletion, that region of heterochromatin spreads, the β-globin gene is poorly expressed and the person develops a severe form of anemia.

Perhaps the most striking example of the use of heterochromatin to keep genes shut down, or *silenced*, is found in the interphase X chromosomes of female mammals. Female cells contain two X chromosomes, where male cells contain one X and one Y. Because a double dose of X-chromosome

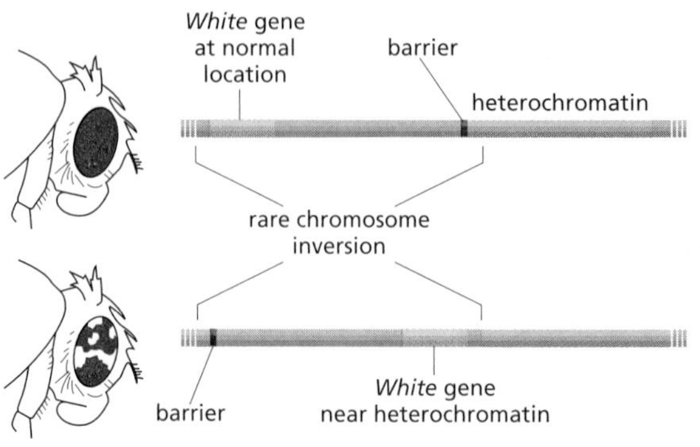

Figure 5–29 **Expression of a gene can be altered by moving it to another location in the genome.** This figure shows an example of a *position effect*, in which the activity of a gene depends on its position along a chromosome. The *White* gene in the fruit fly *Drosophila* controls eye pigment production and is named after the mutation that first identified it. Wild-type flies with a normal *White* gene (*White*$^{+}$) have normal pigment production, which gives them red eyes, but if the *White* gene is mutated and inactivated, the mutant flies (*White*$^{-}$) make no pigment and have white eyes. In wild-type flies, the *White* gene (here shown in *yellow*) is located some distance from the nearest region of heterochromatin (*green*). A special *barrier sequence* (*red*) keeps the heterochromatin from spreading to neighboring areas on the chromosome. In flies in which an inversion of a region of the chromosome moves the normal *White*$^{+}$ gene closer to the region of heterochromatin and displaces the barrier sequence, the eyes are mottled, with both red and white patches. The white patches represent cells where the *White*$^{+}$ gene has been silenced by the effects of the heterochromatin; the red patches represent cells that express the *White*$^{+}$ gene because the heterochromatin had not spread across this gene at the time in early development when the founder cell of these patches was formed. The presence of large patches of red and white cells indicates that the particular state of the gene (either active or silenced) is established early in development and is inherited by progeny cells thereafter.

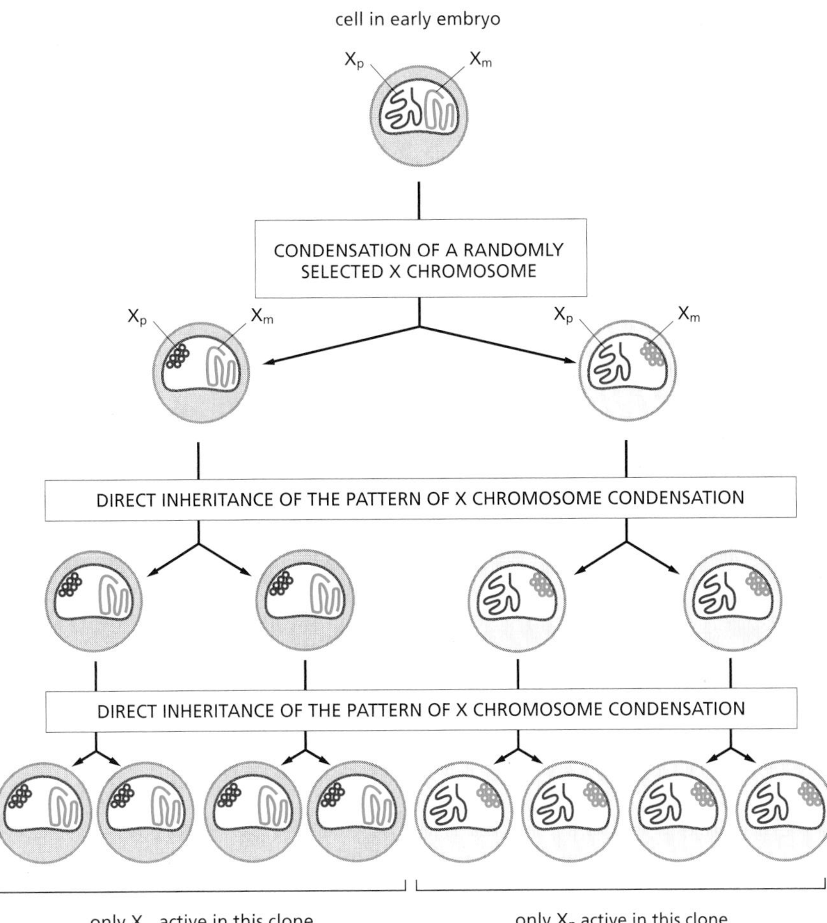

Figure 5–30 **An X chromosome can be inactivated by heterochromatin formation.** Cells in the early female mammalian embryo each contain two X chromosomes, one from the mother (X_m) and the other from the father (X_p). At an early stage of development, one of these two chromosomes becomes condensed into heterochromatin in each cell, apparently at random. At each cell division after this stage, the same chromosome becomes condensed in all the descendants of that original cell. In mice, X-chromosome inactivation occurs between the third and sixth days of embryonic development. In humans, too, X-inactivation occurs very early in embryonic development, before cells have been allocated to any particular developmental pathway. Thus, all mammalian females end up as mosaics of cells bearing maternal or paternal inactivated X chromosomes. In most of their tissues and organs, about half the cells will be of one type and the other half will be of the other.

products would be lethal, female mammals have evolved a means of permanently inactivating one of the two X chromosomes in each cell. At random, one or other of the two X chromosomes in each cell becomes highly condensed into heterochromatin early in embryonic development. Thereafter, the condensed and inactive state of that X chromosome is inherited in all of the many descendants of those cells (Figure 5–30).

The rest of the interphase chromatin is called **euchromatin** (from the Greek *eu*, "true" or "normal," plus chromatin). Although we use the term euchromatin to refer to chromatin that exists in a more extended state than heterochromatin, it is now clear that both euchromatin and heterochromatin are composed of mixtures of different chromatin structures, each established and maintained by different sets of histone tail modifications that attract distinct sets of non-histone proteins (Figure 5–31).

QUESTION 5–5

Mutations in a particular gene on the X chromosome result in color blindness. All men carrying a mutant gene are color-blind. Most women carrying a mutant gene have proper color vision but see color images with reduced resolution, as though functional cone cells (the cells that contain the color photoreceptors) are spaced farther apart than normal in the retina. Can you give a plausible explanation for this observation? If a woman is color-blind, what could you say about her father? About her mother? Explain your answers.

Changes in Chromatin Structure Can Be Inherited

As we have just seen, certain types of chromatin structure can be passed from a cell to its descendants: the progeny of a cell in which the maternal copy of the X chromosome is condensed and inactivated, for example, will also condense and inactivate their maternal X chromosome. How is such inheritance of chromatin structure possible? When a cell replicates

Figure 5–31 **The structure of chromatin varies along a single interphase chromosome.** As schematically indicated by different colors, heterochromatin and euchromatin each represent a set of different chromatin structures with different degrees of extension or condensation. Overall, heterochromatin is more condensed than euchromatin.

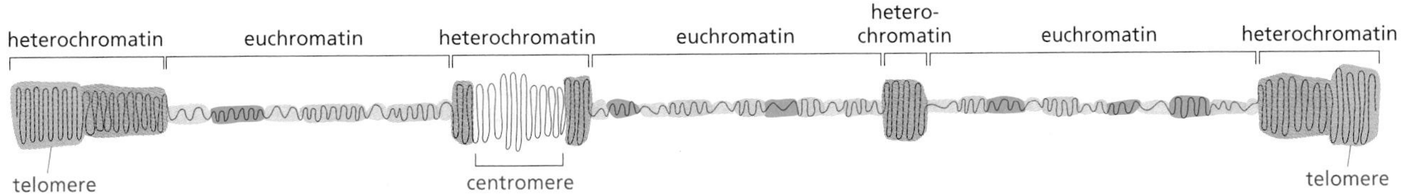

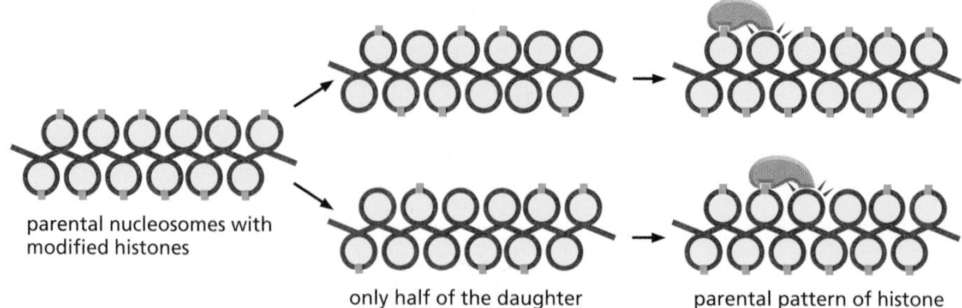
parental nucleosomes with modified histones
only half of the daughter nucleosomes have modified histones
parental pattern of histone modification re-established by proteins that recognize the same modifications they catalyze

Figure 5–32 **How histone modifications may be inherited by daughter chromosomes.** When a chromosome is replicated, its resident histones are distributed more or less randomly to each of the two daughter DNA helices. Thus, each daughter chromosome will inherit about half of its parent's collection of modified histones. The remaining stretches of DNA receive newly synthesized, not-yet-modified histones. At this point, proteins that recognize a particular modification can bind to the chromatin and catalyze the formation of the same modification on the new histones. This can restore the parental modification pattern and, ultimately, allow the inheritance of the parental chromatin structure. This mechanism appears to apply to some but certainly not all types of histone modifications.

its genome, each daughter DNA helix receives half of its parent's histone proteins. With those histone proteins come the covalent modifications associated with the type of chromatin structure that was present in each particular region of that parental chromosome. Thus, each daughter chromosome will initially contain an intermixed set of two types of nucleosomes: those that contain the modified histones inherited from its parent chromosome, and those that contain newly synthesized histones, which have not yet been modified. At this point, proteins that recognize the modified histones may bind to the parental histones and deposit the same type of modification on the nearby virgin histones, re-establishing the pattern of chromatin structure found in the parent (Figure 5–32).

The ability to inherit localized chromatin structure helps eucaryotic cells to "remember" whether a gene was active in its parental cell, a process that appears to be critical for the establishment and maintenance of different cell types, tissues, and organs during the development and growth of a complex multicellular organism. This type of inheritance does not involve passing along specific DNA sequences from one cell generation to the next, but instead depends on passing along specifically modified histone proteins. It is an example of **epigenetic inheritance** (from the Greek *epi-*, "on"), because it is superimposed on genetic inheritance based on DNA. Other forms of epigenetic inheritance are discussed in Chapter 8.

ESSENTIAL CONCEPTS

- **Life depends on the stable and compact storage of genetic information.**
- **Genetic information is carried by very long DNA molecules and encoded in the linear sequence of nucleotides A, T, G, and C.**
- **Each molecule of DNA is a double helix composed of a pair of complementary strands of nucleotides held together by hydrogen bonds between G-C and A-T base pairs.**
- **A strand of DNA has a chemical polarity due to the linkage of alternating sugars and phosphates in its backbone. The two strands of the DNA double helix run antiparallel—that is, in opposite orientations.**
- **The genetic material of a eucaryotic cell is contained in a set of chromosomes, each formed from a single, enormously long DNA molecule that contains many genes.**
- **When a protein-coding gene is expressed, part of its nucleotide sequence is copied into RNA, which then directs the synthesis of a specific protein.**
- **The DNA that forms each eucaryotic chromosome contains, in addition to genes, many replication origins, one centromere, and two telomeres. These sequences ensure that the chromosome can be replicated efficiently and passed on to daughter cells.**

- Chromosomes in eucaryotic cells consist of DNA tightly bound to a mass of specialized proteins. These proteins fold the DNA into a compact form. The complex of DNA and protein in chromosomes is called chromatin.
- The most abundant chromosomal proteins are the histones, which pack DNA into a repeating array of DNA–protein particles called nucleosomes.
- Nucleosomes pack together, with the aid of histone H1 molecules, to form a 30-nm fiber. This fiber is generally coiled and folded, producing more compact chromatin structures.
- Chromatin structure is dynamic: by temporarily decondensing its structure—using chromatin remodeling complexes and enzymes that covalently modify histone tails—the cell can ensure that proteins involved in gene expression, replication, and repair have rapid, localized access to the necessary DNA sequences.
- Some forms of chromatin have a pattern of histone tail modification that causes the DNA to become so highly compacted that the packaged genes cannot be expressed to produce RNA and protein.
- Chromatin structure can be transmitted from one cell generation to the next, producing a form of epigenetic inheritance that helps a cell to remember the state of gene expression in its parent cell.

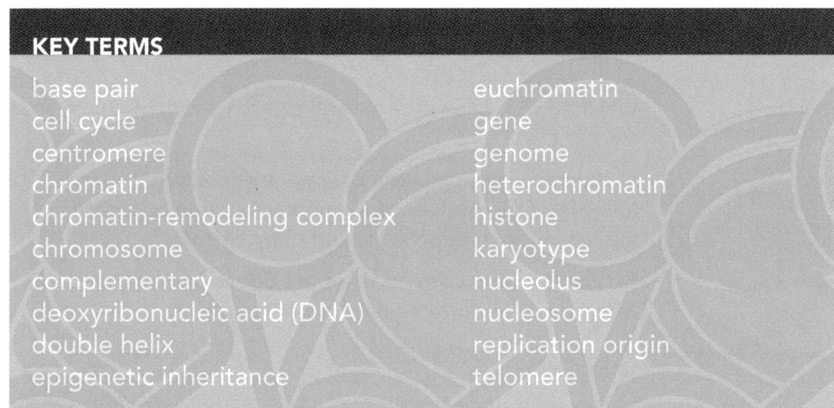

KEY TERMS

base pair
cell cycle
centromere
chromatin
chromatin-remodeling complex
chromosome
complementary
deoxyribonucleic acid (DNA)
double helix
epigenetic inheritance
euchromatin
gene
genome
heterochromatin
histone
karyotype
nucleolus
nucleosome
replication origin
telomere

QUESTIONS

QUESTION 5–6

A. The nucleotide sequence of one DNA strand of a DNA double helix is

5′-GGATTTTTGTCCACAATCA-3′.

What is the sequence of the complementary strand?

B. In the DNA of certain bacterial cells, 13% of the nucleotides are adenine. What are the percentages of the other nucleotides?

C. How many possible nucleotide sequences are there for a stretch of DNA that is *N* nucleotides long, if it is (a) single-stranded or (b) double-stranded?

D. Suppose you had a method of cutting DNA at specific sequences of nucleotides. How many nucleotides long (on average) would such a sequence have to be in order to make just one cut in a bacterial genome of 3×10^6 nucleotide pairs? How would the answer differ for the genome of an animal cell that contains 3×10^9 nucleotide pairs?

QUESTION 5–7

An A–T base pair is stabilized by only two hydrogen bonds. Hydrogen-bonding schemes of very similar strengths can also be drawn between other base combinations, such as the A–C and the A–G pairs shown in Figure Q5–7. What would happen if these pairs formed during DNA replication and the inappropriate bases were incorporated? Discuss why this does not happen often. (Hint: see Figure 5–6.)

QUESTION 5–8

A. A macromolecule isolated from an extraterrestrial source superficially resembles DNA but upon closer analysis reveals quite different base structures (Figure Q5–8). Bases V, W, X, and Y have replaced bases A, T, G, and C. Look at these structures closely. Could these DNA-like molecules have been derived from a living organism that uses principles of genetic inheritance similar to those used by cells on Earth? If so, what can you say about its properties?

B. Simply judged by their potential for hydrogen-bonding, could any of these extraterrestrial bases replace terrestrial A, T, G, or C in terrestrial DNA? Explain your answers.

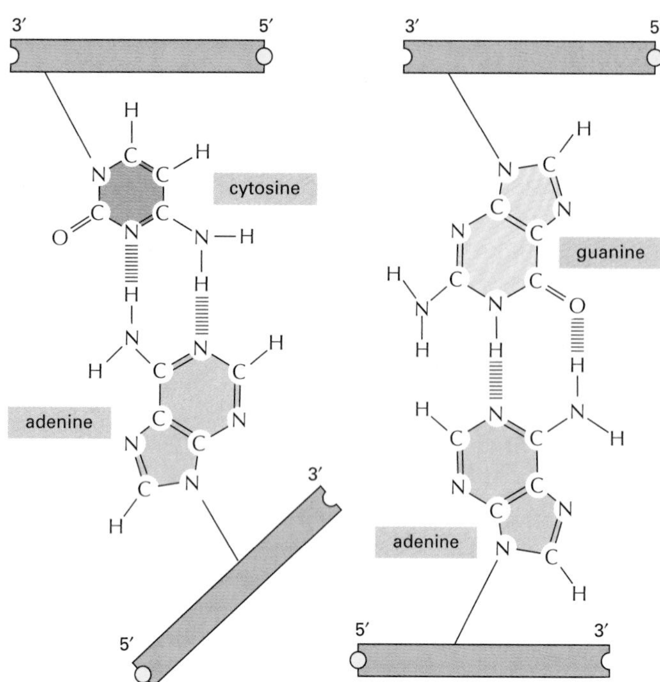

Figure Q5–7

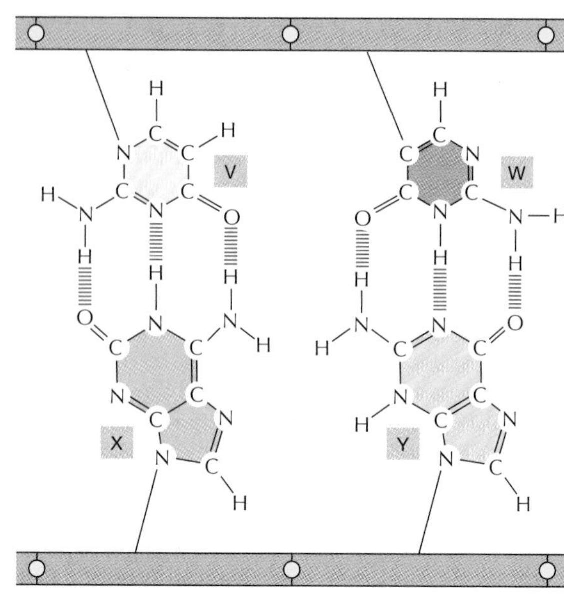

Figure Q5–8

QUESTION 5–9

The two strands of DNA double helix can be separated by heating. If you raised the temperature of a solution containing the following three DNA molecules, in what order do you suppose they would "melt"? Explain your answer.

A. 5′-GCGGGCCAGCCCGAGTGGGTAGCCCAGG-3′

3′-CGCCCGGTCGGGCTCACCCATCGGGTCC-5′

B. 5′-ATTATAAAATATTTAGATACTATATTTACAA-3′

3′-TAATATTTTATAAATCTATGATATAAATGTT-5′

C. 5′-AGAGCTAGATCGAT-3′

3′-TCTCGATCTAGCTA-5′

QUESTION 5–10

The total length of DNA in the human genome is about 1 m, and the diameter of the double helix is about 2 nm. Nucleotides in a DNA double helix are stacked at an interval of 0.34 nm. If the DNA were enlarged so that its diameter equaled that of an electrical extension cord (5 mm), how long would the extension cord be from one end to the other (assuming that it is completely stretched out)? How close would the bases be to each other? How long would a gene of 1000 nucleotide pairs be?

QUESTION 5–11

A compact disc (CD) stores about 4.8×10^9 bits of information in a 96 cm^2 area. This information is stored as a binary code—that is, every bit is either a 0 or a 1.

A. How many bits would it take to specify each nucleotide pair in a DNA sequence?

B. How many CDs would it take to store the information contained in the human genome?

QUESTION 5–12

Which of the following statements are correct? Explain your answers.

A. Each eucaryotic chromosome must contain the following DNA sequence elements: multiple origins of replication, two telomeres, and one centromere.

B. Nucleosome core particles are 30 nm in diameter and, when lined up, form 30-nm filaments.

QUESTION 5–13

Define the following terms and their relationships to one another:

A. Interphase chromosome

B. Mitotic chromosome

C. Chromatin

D. Heterochromatin

E. Histones

F. Nucleosome

QUESTION 5–14

Carefully consider the result shown in Figure Q5–14. Each of the two colonies shown is a clump of approximately 100,000 yeast cells that has grown up from a single cell that is now somewhere in the middle of the colony. The yeast *Ade2* encodes one of the enzymes of adenine biosynthesis, and the absence of the *Ade2* gene product leads to the accumulation of a red pigment. At its normal chromosome location, *Ade2* is expressed in all cells. When it is positioned near the telomere, which is highly condensed, *Ade2* is no longer expressed. Explain why the white sectors have formed near the rim of the colony. Based on the existence of these sectors, what can you conclude about

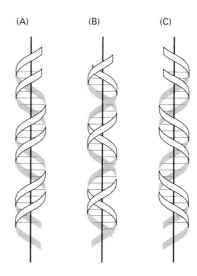

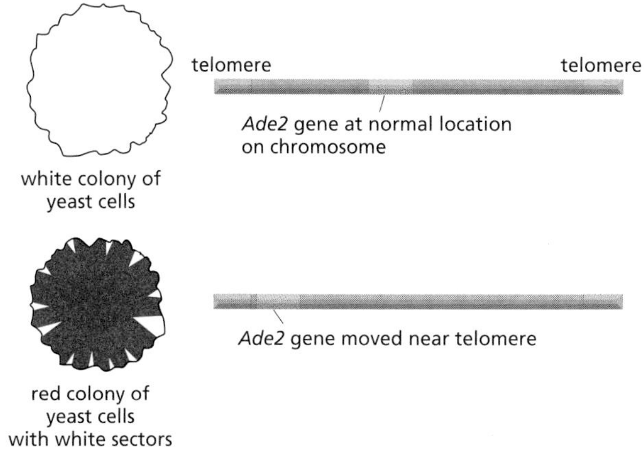

Figure Q5–14

the propagation of the transcriptional state of the *Ade2* gene from mother to daughter cells?

QUESTION 5–15

The two electron micrographs in Figure Q5–15 show nuclei of two different cell types. Can you tell from these pictures which of the two cells is transcribing more of its genes? Explain how you arrived at your answer. (Micrographs courtesy of Don W. Fawcett.)

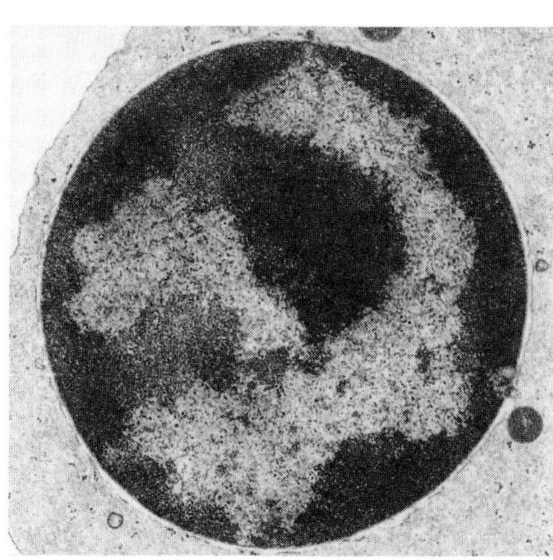

(A)

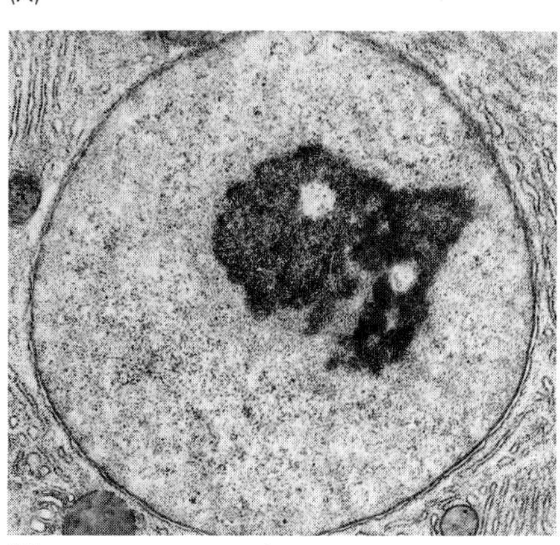

(B)

Figure Q5–15

(A) (B) (C)

Figure Q5–16

QUESTION 5–16

DNA forms a right-handed helix. Pick out the right-handed helix from those shown in Figure Q5–16.

QUESTION 5–17

A single nucleosome is 11 nm in diameter and contains 147 bp of DNA (the DNA double helix measures 0.34 nm/bp). What packing ratio (ratio of DNA length to nucleosome diameter) has been achieved by wrapping DNA around the histone octamer? Assuming that there are an additional 53 bp of extended DNA in the linker between nucleosomes, how condensed is "beads-on-a-string" DNA relative to fully extended DNA? What fraction of the 10,000-fold condensation that occurs at mitosis does this first level of packing represent?

QUESTION 5–18

The epigenetic inheritance of chromatin structure is thought to play an important role in specifying the different cell types in vertebrate organisms. Why might this mechanism of cell-to-cell inheritance be preferable to a hypothetical mechanism that alters the DNA sequence at specific DNA sites in selected cells during embryonic development?

CHAPTER **SIX** 6

DNA Replication, Repair, and Recombination

DNA REPLICATION

DNA REPAIR

HOMOLOGOUS RECOMBINATION

MOBILE GENETIC ELEMENTS AND VIRUSES

The ability of a cell to maintain order in a chaotic environment depends on the accurate duplication of the vast quantity of genetic information carried in its DNA. This duplication process, called *DNA replication*, must occur before a cell can produce two genetically identical daughter cells. Maintaining order in a cell also requires the continual surveillance and repair of its genetic information, as DNA is subject to damage by chemicals and radiation from the environment, and by reactive molecules that are generated inside the cell. In this chapter, we describe the protein machines that replicate and repair the cell's DNA. These machines catalyze some of the most rapid and accurate processes that take place within cells, and their actions reflect the elegance and efficiency of cell chemistry.

Despite these systems for protecting the genetic instructions from copying errors and accidental damage, permanent changes, or *mutations*, sometimes do occur. Although many mutations do not affect the organism in any noticeable way, some have profound consequences. Occasionally, these changes can benefit the organism: for example, mutations can make bacteria resistant to antibiotics that are used to kill them. Indeed, the accumulation of changes in DNA over millions of years provides the variety in genetic material that makes one species distinct from another, as we discuss in Chapter 9. Mutations also produce the smaller variations that underlie many of the differences between individuals of the same species that we can easily see in humans and other animals (Figure 6–1).

However, mutations can also be detrimental: in humans, they are responsible for thousands of inherited diseases, and for many types of cancer.

Figure 6–1 **Hereditary information is passed faithfully from one generation to the next.** Differences in the DNA, however, can produce the variations that underlie the differences between individuals of the same species—or, over time, the differences between one species and another. In this family photo, the children resemble one another and their parents more closely than they resemble other people because they inherit their particular genes from their parents. The cat shares many features with humans, but during the millions of years of evolution that have separated humans and cats, both have accumulated many hereditary changes that now make us quite different species. The chicken is an even more distant relative.

Thus, the survival of a cell or organism depends on keeping changes in its DNA to a minimum. Without the cellular systems that are continually monitoring and repairing damage to DNA, it is questionable whether life could exist at all.

We begin this chapter by reviewing the mechanisms that are responsible for copying and maintaining DNA with minimal changes. We next consider some of the intriguing ways in which genetic information can be altered, including *homologous recombination* and the movement of the special DNA sequences in our chromosomes called *mobile genetic elements*. Finally, we consider viruses—little more than genes protected by a protein coat—which can move from cell to cell.

DNA REPLICATION

At each cell division, a cell must copy its genome with extraordinary accuracy. In this section, we explore how the cell achieves this feat, while duplicating DNA at rates as high as 1000 nucleotides per second.

Base-Pairing Enables DNA Replication

In the preceding chapter, we saw that each strand of the DNA double helix contains a sequence of nucleotides that is exactly complementary to the nucleotide sequence of its partner strand. Each strand can therefore act as a **template**, or mold, for the synthesis of a new complementary strand (Figure 6–2). In other words, if we designate the two DNA strands as S and S′, strand S can serve as a template for making a new strand S′, while strand S′ can serve as a template for making a new strand S (Figure 6–3). Thus, the genetic information in DNA can be accurately copied by the beautifully simple process in which strand S separates from strand

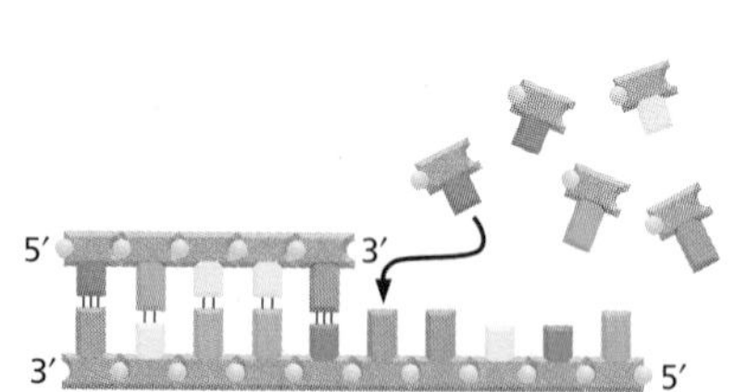

Figure 6–2 **A DNA strand can serve as a template.** Preferential binding occurs between pairs of nucleotides (A with T, and G with C) that can form base pairs. This enables each strand to act as a template for forming its complementary strand.

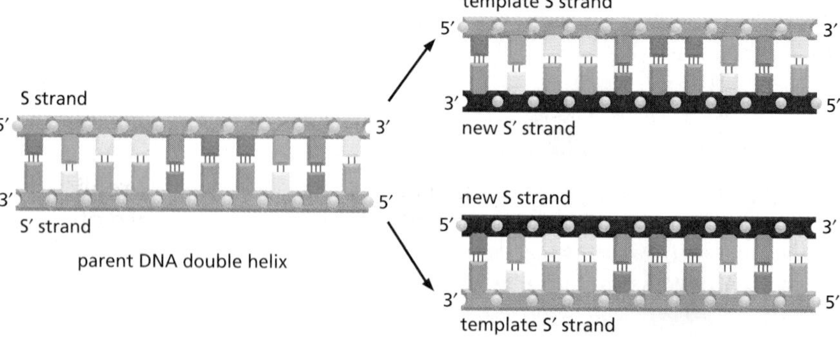

Figure 6–3 **DNA acts as a template for its own duplication.** Because the nucleotide A will successfully pair only with T, and G with C, each strand of DNA in the double helix—labeled here as the S strand and its complementary S′ strand—can serve as a template to specify the sequence of nucleotides in its complementary strand. In this way, double-helical DNA can be copied precisely. Keep in mind that although they are colored differently here, the template strands (*orange*) and the new strands (*red*) are chemically identical.

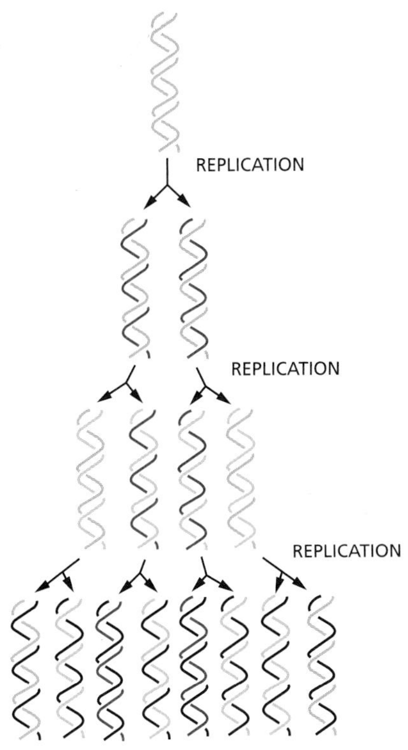

Figure 6–4 **In each round of replication, each of the two strands of DNA is used as a template for the formation of a complementary DNA strand.** The original strands therefore remain intact through many cell generations. DNA replication is 'semiconservative' because each daughter DNA double helix is composed of one conserved strand and one newly synthesized strand.

5′, and each separated strand then serves as a template for the production of a new complementary partner strand that is identical to its former partner.

The ability of each strand of a DNA molecule to act as a template for producing a complementary strand enables a cell to copy, or *replicate*, its genes before passing them on to its descendants. But the task is awe-inspiring, as it can involve copying billions of nucleotide pairs every time a cell divides. The copying must be carried out with speed and accuracy: in about 8 hours, a dividing animal cell will copy the equivalent of 1000 books like this one and, on average, get no more than a letter or two wrong. This feat is performed by a cluster of proteins that together form a *replication machine*.

DNA replication produces two complete double helices from the original DNA molecule, each new DNA helix identical (except for rare copying errors) in nucleotide sequence to the parental DNA double helix (see Figure 6–3). Because each parental strand serves as the template for one new strand, each of the daughter DNA double helices ends up with one of the original (old) strands plus one strand that is completely new; this style of replication is said to be *semiconservative* (Figure 6–4). In How We Know, pp. 200–202, we discuss the experiments that first demonstrated that DNA is replicated in this way.

DNA Synthesis Begins at Replication Origins

The DNA double helix is normally very stable: the two DNA strands are locked together firmly by the large numbers of hydrogen bonds between the bases on both strands (see Figure 5–2). As a result, only temperatures approaching those of boiling water provide enough thermal energy to separate these strands. To be used as a template, however, the double helix must first be opened up and the two strands separated to expose unpaired bases. How does this occur at the temperatures found in living cells?

The process of DNA replication is begun by initiator proteins that bind to the DNA and pry the two strands apart, breaking the hydrogen bonds between the bases (Figure 6–5). Although the hydrogen bonds collectively make the DNA helix very stable, individually each hydrogen bond is weak (as discussed in Chapter 2). Separating a short length of DNA a few base pairs at a time therefore does not require a large energy input, and it can occur with the assistance of these proteins at normal temperatures.

The positions at which the DNA is first opened are called **replication origins**, and they are usually marked by a particular sequence of nucleotides. In simple cells such as those of bacteria or yeast, replication origins span approximately 100 nucleotide pairs; they are composed of DNA sequences that attract the initiator proteins, as well as stretches of DNA that are especially easy to open. We saw in Chapter 5 that an A-T base pair is held together by fewer hydrogen bonds than is a G-C base pair. Therefore, DNA rich in A-T base pairs is relatively easy to pull apart, and A-T-rich stretches of DNA are typically found at replication origins.

A bacterial genome, which is typically contained in a circular DNA molecule of several million nucleotide pairs, has a single replication origin.

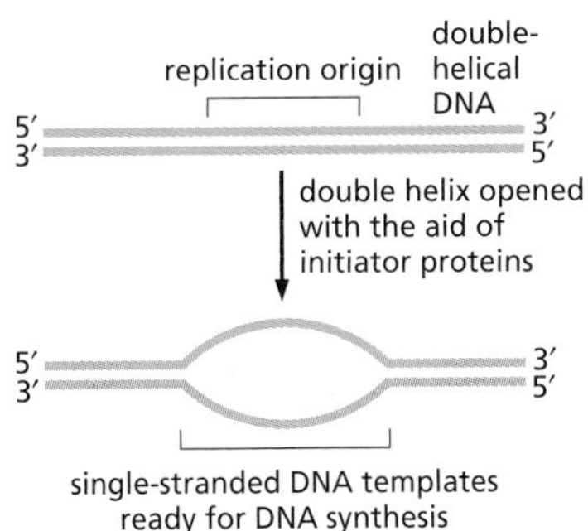

Figure 6–5 **A DNA double helix is opened at its replication origin.** Replication initiator proteins recognize sequences of DNA at replication origins and locally pry apart the two strands of the double helix. The exposed single strands can then serve as templates for copying the DNA.

HOW WE KNOW: THE NATURE OF REPLICATION

In 1953, James Watson and Francis Crick published their famous two-page paper describing a model for the structure of DNA (see Figure 5–2). In it, they proposed that complementary bases—adenine and thymine, guanine and cytosine—pair with one another along the center of the double helix, holding together the two strands of DNA. At the very end of this succinct scientific blockbuster, the researchers comment, almost as an aside, "It has not escaped our notice that the specific pairing we have postulated immediately suggests a possible copying mechanism for the genetic material."

Indeed, one month after the classic paper appeared in print in the journal *Nature*, Watson and Crick published a second article, suggesting how DNA might be duplicated. In this paper, they proposed that the two strands of the double helix unwind, and that each serves as a template for the synthesis of a complementary daughter strand. In their model, dubbed *semiconservative* replication, each new DNA molecule consists of one strand derived from the original parent molecule and one newly synthesized strand (Figure 6–6A).

We now know that Watson and Crick's model for DNA replication was correct—but it was not universally accepted at first. Respected physicist-turned-geneticist Max Delbrück, for one, got hung up on what he termed "the untwiddling problem;" that is: how could the two strands of a double helix, twisted around one another so many times all along their great length, possibly be unwound without making a big tangled mess? Watson and Crick's conception of the DNA helix opening up in a zipper-like fashion seemed, to Delbrück, physically unlikely and simply "too inelegant to be efficient."

Instead, Delbrück proposed that DNA replication proceeds through a series of breaks and reunions, in which the DNA backbone is broken and the strands copied in short segments—perhaps only 10 nucleotides at a time—before being rejoined. In this model, which was later dubbed *dispersive*, the resulting copies would be patchwork collections of old and new DNA, each strand containing a mixture of both (Figure 6–6B). No unwinding was necessary.

Yet a third camp promoted the idea that DNA replication might be *conservative*: that the parent helix would somehow remain entirely intact after copying, and the daughter molecule would contain two entirely new DNA strands (Figure 6–6C). To determine which of these models was correct, an experiment was needed—one that would reveal the composition of the newly synthesized DNA strands. That's where Matt Meselson and Frank Stahl came in.

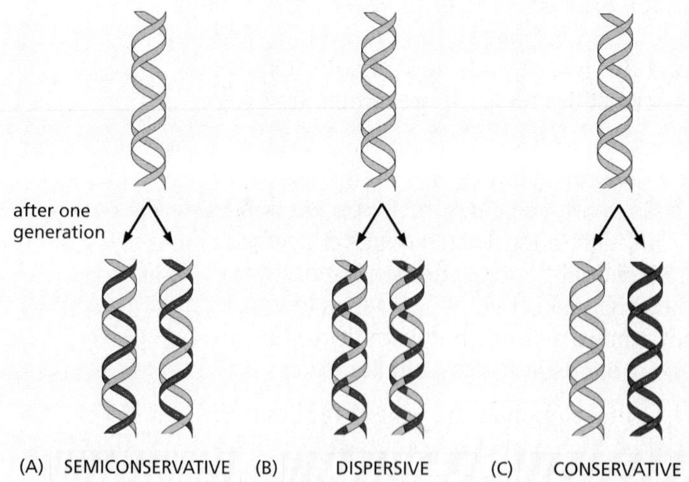

Figure 6–6 **Three models for DNA replication make different predictions.** (A) In the semiconservative model, each parent strand serves as a template for the synthesis of a new daughter strand. The first round of replication would produce two hybrid molecules, each containing one strand from the original parent in addition to one newly synthesized strand. A subsequent round of replication would yield two hybrid molecules and two molecules that contain none of the original parent DNA. (B) In the dispersive model, each generation of daughter DNA will contain a mixture of fragments from the parent strands along with newly synthesized material. (C) In the conservative model, the parent molecule remains intact after being copied. In this case, the first round of replication would yield the original parent helix along with an entirely new daughter molecule. For each model, parent DNA molecules are shown in *orange*; newly replicated DNA is *red*. Note that only a very small segment of DNA is shown for each model.

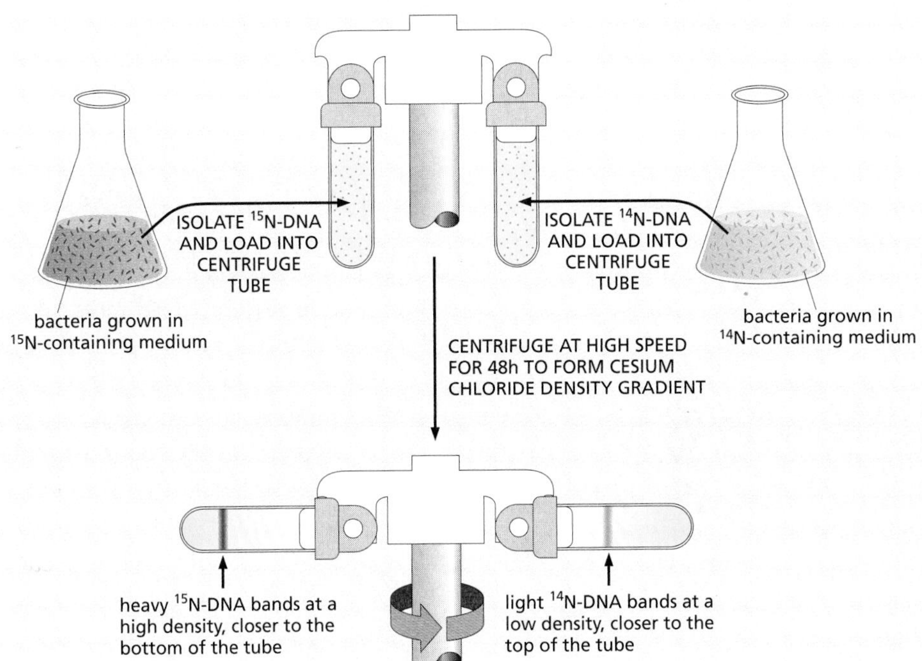

Figure 6–7 **Centrifugation in a cesium chloride gradient allows the separation of heavy and light DNA.** Bacteria are grown for several generations in a medium containing either ^{15}N (the heavy isotope) or ^{14}N (the light isotope) to label their DNA. The cells are then broken open and the DNA is loaded into an ultracentrifuge tube containing a cesium chloride salt solution. These tubes are centrifuged at high speed for two days to allow the DNA to collect in a region where its density matches that of the salt surrounding it. The heavy and light DNA molecules collect in different positions in the tube.

As a graduate student working with Linus Pauling, Meselson was toying with a method for telling the difference between old and new proteins. After chatting with Delbrück about Watson and Crick's replication model, it occurred to Meselson that the approach he'd envisaged for exploring protein synthesis might also work for studying DNA. In the summer of 1954, Meselson met Stahl, who was then a graduate student in Rochester, NY, and they agreed to collaborate. It took a few years to get everything working, but the two eventually performed what has come to be known as "the most beautiful experiment in biology."

Their approach, in retrospect, was stunningly straightforward. They started by growing two batches of *E. coli* bacteria, one in a medium containing a heavy isotope of nitrogen, ^{15}N, the other in a medium containing the normal, lighter ^{14}N. The nitrogen in the nutrient medium gets incorporated into the nucleotide bases and, from there, makes its way into the DNA of the organism. After growing cell cultures for many generations in either the ^{15}N- or ^{14}N-containing medium, the researchers had two flasks of bacteria, one whose DNA was *heavy*, the other whose DNA was *light*. Meselson and Stahl then broke open the bacterial cells and loaded the DNA into tubes containing a high concentration of the salt cesium chloride. When these tubes are centrifuged, the cesium chloride forms a density gradient, and the DNA molecules float or sink within the solution until they reach the point at which their density equals that of the surrounding salt (see Panel 4–4, pp. 164–165). Using this method, called equilibrium density centrifugation, Meselson and Stahl found that they could distinguish between heavy (^{15}N-containing) DNA and light (^{14}N-containing) DNA by observing the positions of the DNA within the cesium-chloride gradient. Because the heavy DNA was more dense than the light DNA, it collected at a position nearer to the bottom of the centrifuge tube (Figure 6–7).

Once they had established this method for differentiating between light and heavy DNA, Meselson and Stahl set out to test the various hypotheses proposed for DNA replication. To do this, they took a flask of bacteria that had been grown in heavy nitrogen and transferred the culture into a medium containing the light isotope. At the start of the experiment, all the DNA would be heavy, but as the bacteria divided, the newly synthesized DNA would be light. They could then monitor the accumulation of light DNA and see which model, if any, best fit the data. After one generation of growth, the researchers found that the parental, heavy DNA molecules—those made of two strands containing ^{15}N—had disappeared and were replaced by a new species of DNA that banded at a density halfway between those of ^{15}N-DNA and ^{14}N-DNA (Figure 6–8). These newly synthesized daughter helices, Meselson and Stahl reasoned, must be a hybrid—containing both heavy and light isotopes.

Right away, this observation ruled out the conservative model of DNA replication, which predicted that the parental DNA would remain entirely heavy, while

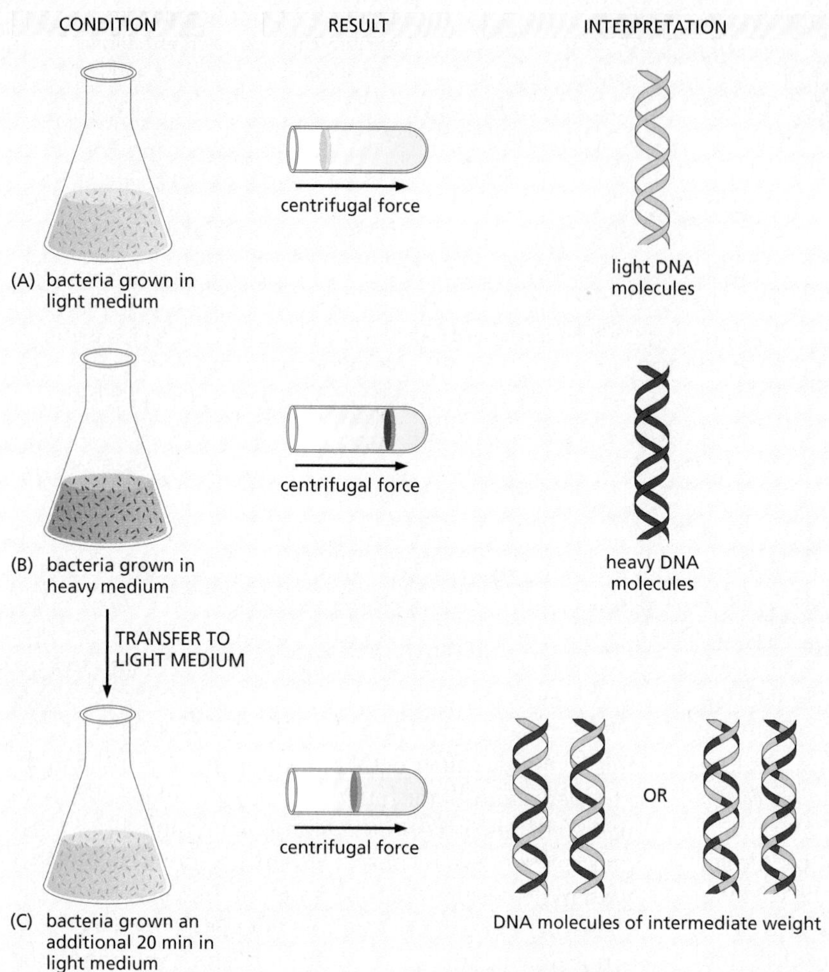

Figure 6–8 **The first part of the Meselson–Stahl experiment ruled out the conservative model of DNA replication.** (A) Bacteria grown in light medium (containing ^{14}N) yield DNA that bands high up in the centrifuge tube, whereas bacteria grown in ^{15}N-containing heavy medium (B) produce DNA that migrates further toward the bottom of the tube. When bacteria grown in a heavy medium are transferred to a light medium and allowed to continue dividing, they produce a band whose position falls somewhere between that of the parent bands (C). These results rule out the conservative model of replication but do not distinguish between the semiconservative and dispersive models, both of which predict the formation of hybrid daughter DNA molecules.

The fact that the results came out looking so clean—with discrete bands forming at the expected positions for newly replicated hybrid DNA molecules—was a happy accident of the experimental protocol. The researchers used a hypodermic syringe to load their DNA samples into the ultracentrifuge tubes (see Figure 6–7). In the process, they unwittingly sheared the large bacterial chromosome into smaller fragments. Had the chromosomes remained whole, the researchers might have isolated DNA molecules that were only partially replicated, because many cells would have been caught in the middle of copying their DNA. Molecules in such an intermediate stage of replication would not have separated into such discrete bands. Because the researchers were instead working with smaller pieces, the likelihood that any given fragment had been fully replicated—and contained a complete parent and daughter strand —was high, thus yielding nice, clean results.

the daughter DNA would be 100% light (see Figure 6–6C). The data matched with the semiconservative model, which predicted the formation of hybrid molecules containing one strand of heavy DNA and one strand of light (see Figure 6–6A). The results, however, were also consistent with the dispersive model, in which hybrid molecules would contain a mixture of heavy and light DNA (see Figure 6–6B).

To distinguish between those latter two models, Meselson and Stahl turned up the heat. When DNA is subjected to high temperature, the hydrogen bonds holding the two strands together break and the helix comes apart, leaving a collection of single-stranded DNAs. When the researchers heated their hybrid molecules before centrifuging, they discovered that one strand of the DNA was heavy, whereas the other was light. This observation further supported the semiconservative model; if the dispersive model were correct, the resulting strands, each containing a mottled assembly of heavy and light fragments, would have all banded together at an intermediate density.

According to historian Frederic Lawrence Holmes, the experiment was so elegant and the results so clean that Stahl—when being interviewed for a position at Yale University—was unable to fill the 50 minutes allotted for his talk. "I was finished in 25 minutes," said Stahl, "because that is all it takes to tell that experiment. It's so totally simple and contained." Stahl did not get the job at Yale, but the experiment convinced biologists that Watson and Crick had been correct. In fact, the results were accepted so widely and rapidly that the experiment was described in a textbook before Meselson and Stahl had even published the data.

The human genome, which is very much larger, has approximately 10,000 such origins. In humans, beginning DNA replication at many places at once greatly shortens the time a cell needs to copy its entire genome.

Once an initiator protein binds to DNA at the replication origin and locally opens up the double helix, it attracts a group of proteins that carry out DNA replication. These proteins form a protein machine, with each member of the group carrying out a specific function. We will introduce each of them shortly, after we present an overview of the process of DNA replication.

New DNA Synthesis Occurs at Replication Forks

DNA molecules in the process of being replicated contain Y-shaped junctions called **replication forks** (Figure 6–9). At these forks, the replication machine moves along the DNA, opening up the two strands of the double helix and using each strand as a template to make a new daughter strand. Two replication forks are formed at each replication origin, and they move away from the origin in opposite directions, unzipping the DNA as they go. DNA replication in bacterial and eucaryotic chromosomes is therefore termed *bidirectional*. The forks move very rapidly—at about 1000 nucleotide pairs per second in bacteria and 100 nucleotide pairs per second in humans. The slower rate of fork movement in humans (indeed, in all eucaryotes) may be due to the difficulties in replicating DNA through the more complex chromatin structure found in these organisms.

At the heart of the replication machine is an enzyme called **DNA polymerase**, which synthesizes new DNA using one of the old strands as a template. This enzyme catalyzes the addition of nucleotides to the 3′ end of a growing DNA strand by forming a phosphodiester bond between this end and the 5′-phosphate group of the incoming nucleotide (Figure 6–10). Nucleotides enter the reaction initially as nucleoside triphosphates, which provide the energy for polymerization. The hydrolysis of one high-energy bond in the nucleoside triphosphate fuels the reaction that links the nucleotide monomer to the chain and releases pyrophosphate (PP_i).

QUESTION 6–1

Look carefully at the micrograph in Figure 6–9.

A. Using the scale bar, estimate the lengths of the DNA strands between the replication forks. Numbering the replication forks sequentially from the left, how long will it take until forks 4 and 5, and forks 6 and 7, respectively, collide with each other? (Recall that the distance between the bases in DNA is 0.34 nm, and eucaryotic replication forks move at about 100 nucleotides per second.) For this question disregard the nucleosomes seen in the micrograph and assume that the DNA is fully extended.

B. The fly genome is about 1.8×10^8 nucleotide pairs in size. What fraction of the genome is shown in the micrograph?

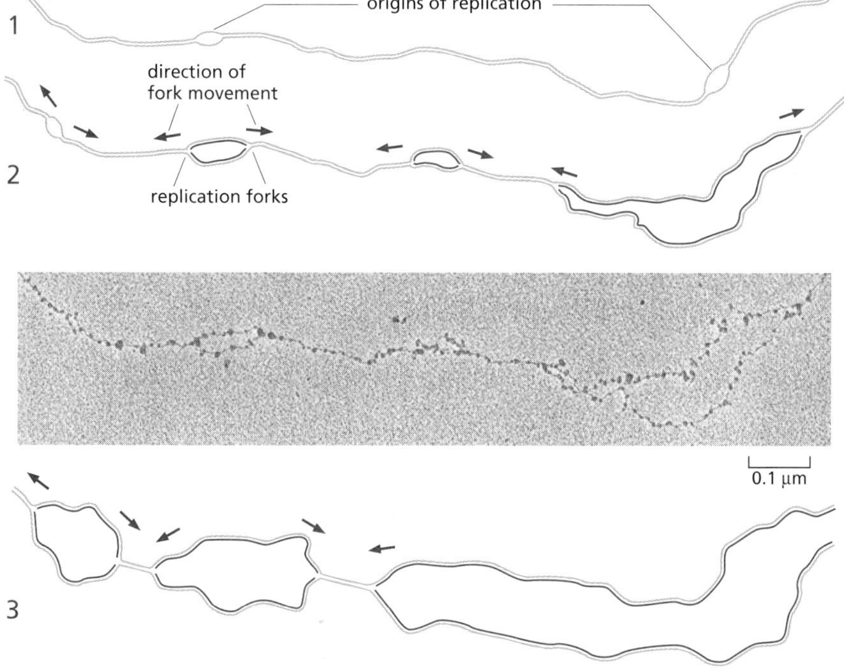

Figure 6–9 **Replication forks move away in opposite directions from multiple replication origins in a eucaryotic chromosome.** The electron micrograph shows DNA replicating in the early embryo of a fly. The particles visible along the DNA are nucleosomes, structures made of DNA and the protein complexes around which it is wrapped (see Chapter 5). (1), (2), and (3) are drawings of the same portion of a DNA molecule as it might appear at successive stages of replication, drawn from electron micrographs. (2) is drawn from the electron micrograph shown here. The *orange* lines represent the parental DNA strands; the *red* lines represent the newly synthesized DNA. (Electron micrograph courtesy of Victoria Foe.)

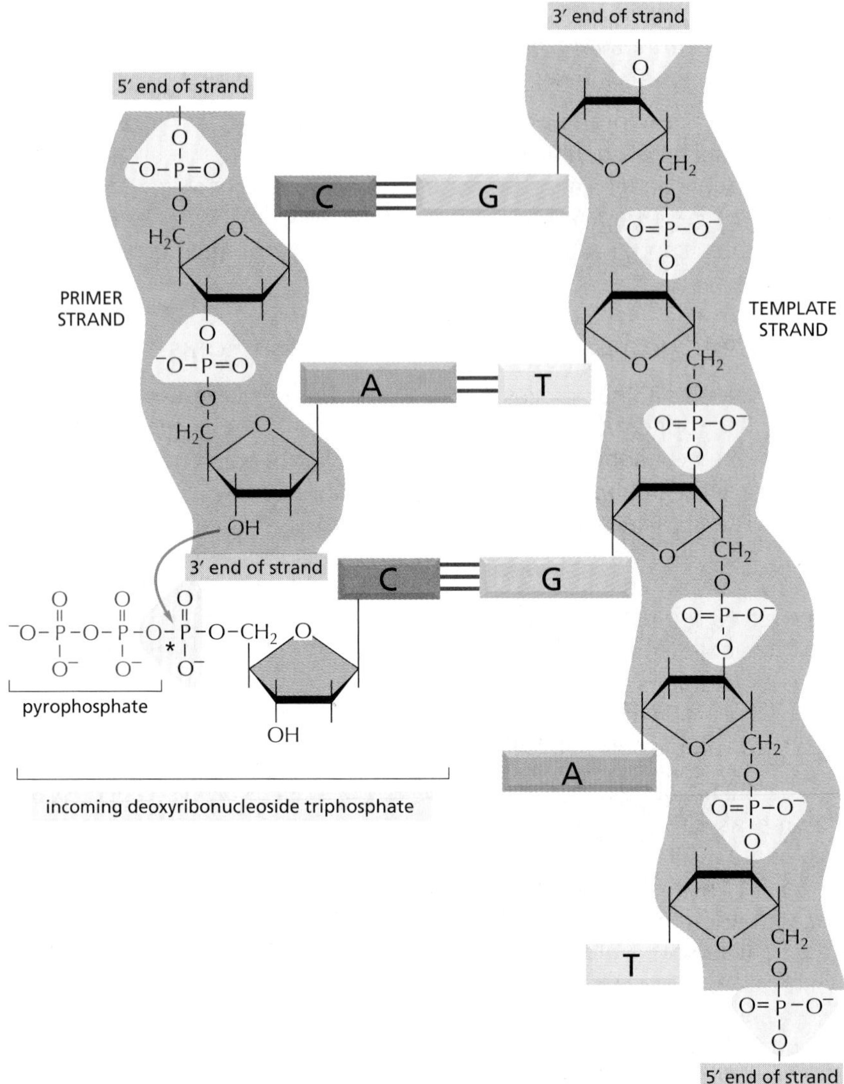

Figure 6–10 **DNA is synthesized in the 5′-to-3′ direction.** Addition of a deoxyribonucleotide to the 3′-hydroxyl end of a polynucleotide chain is the fundamental reaction by which DNA is synthesized; the new DNA chain is therefore synthesized in the 5′-to-3′ direction. The nucleotides enter the reaction as nucleoside triphosphates. Base-pairing between the incoming deoxyribonucleotide and the template strand guides the formation of a new strand of DNA that is complementary in nucleotide sequence to the template chain (see Figure 6–2). The enzyme DNA polymerase catalyzes the addition of nucleotides to the free 3′ hydroxyl on the growing DNA strand. Breakage of a phosphoanhydride bond (indicated by the *asterisk*) in the incoming nucleoside triphosphate releases a large amount of free energy and thus provides the energy for the polymerization reaction.

The DNA polymerase couples the release of this energy to the polymerization reaction. Pyrophosphate is further hydrolyzed to inorganic phosphate (P_i), which makes the polymerization reaction effectively irreversible (see Figure 3–41).

DNA polymerase does not dissociate from the DNA each time it adds a new nucleotide to the growing chain; rather, it stays associated with the DNA and moves along the template strand stepwise for many cycles of the polymerization reaction. Movie 6.1 shows a DNA polymerase molecule in action. We will see later in this chapter that a special protein keeps the polymerase attached to the DNA, as it repeatedly adds new nucleotides to the growing chain.

The Replication Fork Is Asymmetrical

The 5′-to-3′ direction of the DNA polymerization mechanism poses a problem at the replication fork. We saw in Figure 5–2 that the sugar–phosphate backbone of each strand of a DNA double helix has a unique chemical direction, or polarity, determined by the way each sugar residue is linked to the next, and that the two strands in the double helix run in opposite orientations. As a consequence, at the replication fork, one new DNA strand is being made on a template that runs in one direction (3′ to 5′), whereas the other new strand is being made on a template that runs in the opposite direction (5′ to 3′) (Figure 6–11). The replication fork

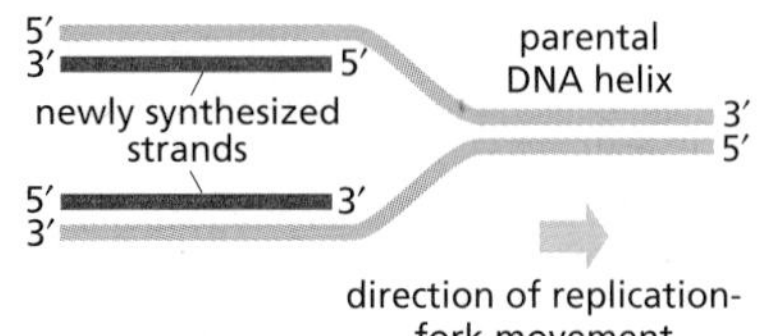

Figure 6–11 **At a replication fork, the two newly synthesized DNA strands are of opposite polarities.**

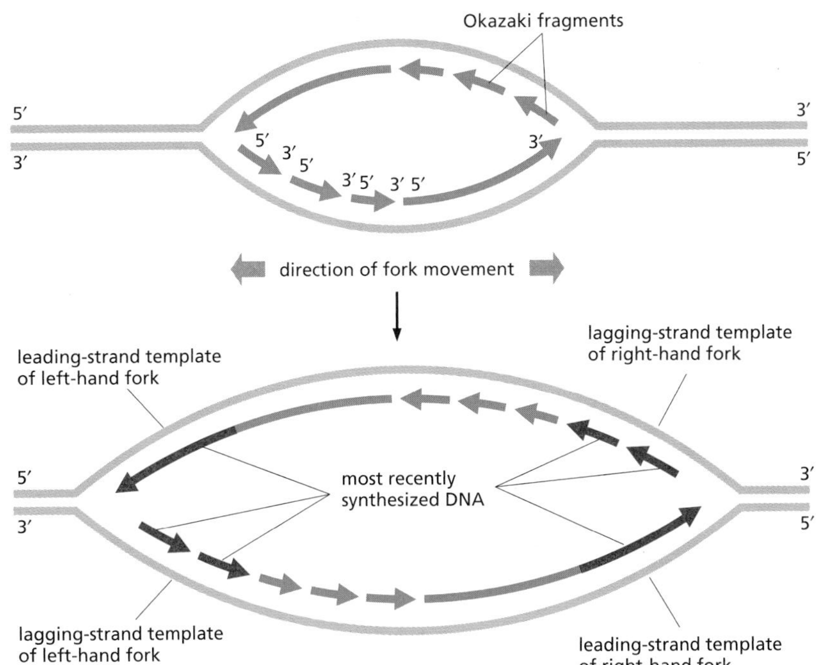

Figure 6–12 **DNA replication forks are asymmetrical.** Because both of the new strands are synthesized in the 5′-to-3′ direction, the lagging strand of DNA must be made initially as a series of short DNA strands that are later joined together. The upper diagram shows two replication forks moving in opposite directions; the lower diagram shows the same forks a short time later. To synthesize the lagging strand, DNA polymerase must 'backstitch': it synthesizes short pieces of DNA (called Okazaki fragments) in the 5′-to-3′ direction, and then must move in the opposite direction along the template strand (toward the fork) before synthesizing the next fragment.

is therefore asymmetrical. At first glance, both of the new DNA strands appear to be growing in the same direction, that is, the direction in which the replication fork is moving. On the face of it, this suggests that one strand is being synthesized in the 3′-to-5′ direction and one is being synthesized in the 5′-to-3′ direction.

DNA polymerase, however, can catalyze the growth of the DNA chain in only one direction: it can add new subunits only to the 3′ end of the chain (see Figure 6–10). As a result, a new DNA chain can be synthesized only in a 5′-to-3′ direction. This can easily account for the synthesis of one of the two strands of DNA at the replication fork, but not the other. One might have expected a second type of DNA polymerase to synthesize the other DNA strand—one that works by adding subunits to the 5′ end of a DNA chain. However, no such enzyme exists. Instead, the problem is solved by the use of a 'backstitching' maneuver. The DNA strand whose 5′ end must grow is made *discontinuously*, in successive separate small pieces, with the DNA polymerase moving backward with respect to the direction of the replication fork, as each new piece is made in the 5′-to-3′ direction.

The resulting small DNA pieces—called **Okazaki fragments** after the biochemist who discovered them—are later joined together to form a continuous new strand (Figure 6–12). The DNA strand that is synthesized discontinuously in this way is called the **lagging strand**; the other strand, which is synthesized continuously, is called the **leading strand**.

Although they differ in subtle details, the replication forks of all cells, procaryotic and eucaryotic, have leading and lagging strands. This common feature arises from the fact that all DNA polymerases work in the 5′-to-3′ direction only. An important advantage of this seemingly complicated molecular maneuver is discussed next.

DNA Polymerase Is Self-correcting

DNA polymerase is so accurate that it makes only about one error in every 10^7 nucleotide pairs it copies. This error rate is much lower than can be explained simply by the accuracy of complementary base-pairing. Although A-T and C-G are by far the most stable base pairs, other, less stable base pairs—for example, G-T and C-A—can also be formed. Such

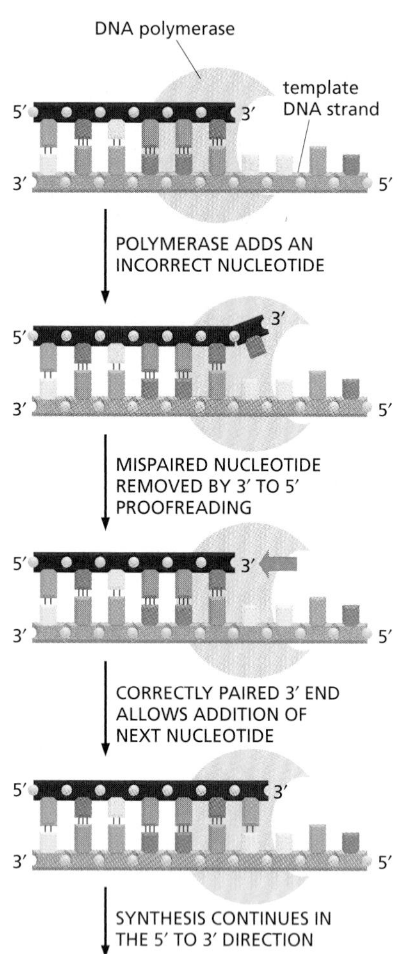
DNA polymerase
template DNA strand
5′
3′
3′
5′
POLYMERASE ADDS AN INCORRECT NUCLEOTIDE
MISPAIRED NUCLEOTIDE REMOVED BY 3′ TO 5′ PROOFREADING
CORRECTLY PAIRED 3′ END ALLOWS ADDITION OF NEXT NUCLEOTIDE
SYNTHESIS CONTINUES IN THE 5′ TO 3′ DIRECTION

Figure 6–13 **During DNA synthesis, DNA polymerase proofreads its own work.** If an incorrect nucleotide is added to a growing strand, the DNA polymerase will cleave it from the strand and replace it with the correct nucleotide before continuing.

incorrect base pairs are formed much less frequently than correct ones, but if allowed to remain, they would kill the cell through an accumulation of mutations. This catastrophe is avoided because DNA polymerase has two special qualities that greatly increase the accuracy of DNA replication. First, the enzyme carefully monitors the base-pairing between each incoming nucleotide and the template strand. Only when the match is correct does DNA polymerase catalyze the nucleotide addition reaction. Second, when DNA polymerase makes a rare mistake and adds the wrong nucleotide, it can correct the error through an activity called **proofreading**.

Proofreading takes place at the same time as DNA synthesis. Before the enzyme adds the next nucleotide to a growing DNA chain, it checks whether the previous nucleotide is correctly base-paired to the template strand. If so, the polymerase adds the next nucleotide; if not, the polymerase clips off the mispaired nucleotide and tries again (Figure 6–13). Thus, DNA polymerase possesses a highly accurate 5′-to-3′ polymerization activity, as well as a 3′-to-5′ proofreading activity. This proofreading is carried out by a nuclease that cleaves the phosphodiester backbone. Polymerization and proofreading are tightly coordinated, and the two reactions are carried out by different domains within the polymerase molecule (Figure 6–14).

This proofreading mechanism explains why DNA polymerases synthesize DNA only in the 5′-to-3′ direction, despite the need that this imposes for a cumbersome backstitching mechanism at the replication fork. As shown in Figure 6–15A, a hypothetical DNA polymerase that synthesized in the 3′-to-5′ direction (and would thereby circumvent the need for backstitching) would be unable to proofread: if it removed an incorrectly paired nucleotide, the polymerase would create a chain end that is chemically dead, in the sense that it could no longer be elongated. Thus, for a DNA polymerase to function as a self-correcting enzyme that removes its own polymerization errors as it moves along the DNA, it must proceed only in the 5′-to-3′ direction (Figure 6–15B).

Short Lengths of RNA Act as Primers for DNA Synthesis

We have seen that the accuracy of DNA replication depends on the requirement of the DNA polymerase for a correctly base-paired end before it can add more nucleotides. But because the polymerase can join a nucleotide only to a base-paired nucleotide in a DNA double helix, it cannot start a completely new DNA strand. A different enzyme is needed—one that

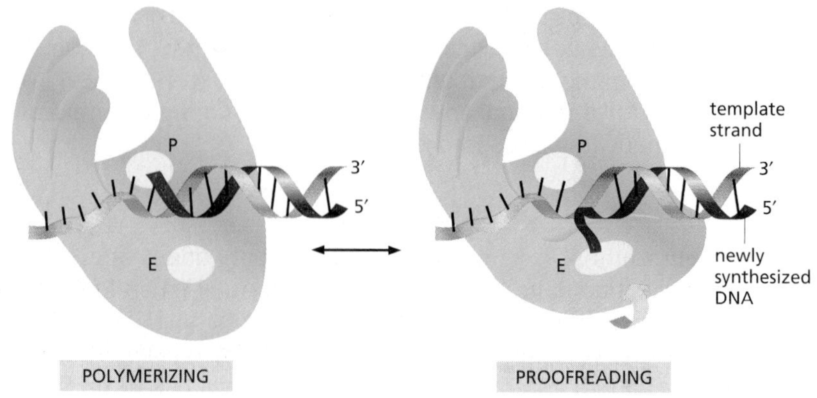
P
E
3′
5′
template strand
newly synthesized DNA
POLYMERIZING
PROOFREADING

Figure 6–14 **DNA polymerase contains separate sites for DNA synthesis and proofreading.** The diagrams are based on the structure of an *E. coli* DNA polymerase molecule, as determined by X-ray crystallography. Roughly speaking, the enzyme resembles a right hand in which the palm, fingers, and thumb cradle the DNA. DNA polymerase is shown with the DNA template in the polymerizing mode *(left)* and in the proofreading mode *(right)*. The catalytic sites for the polymerization activity (P) and error-correcting proofreading activity (E) are indicated. When an incorrect nucleotide is added, the newly synthesized DNA *(red)* transiently unpairs from the template *(orange)*, and the polymerase undergoes a conformational change (indicated by the *gray* arrow) that moves the error-correcting catalytic site into a position where it can remove the most recently added nucleotide.

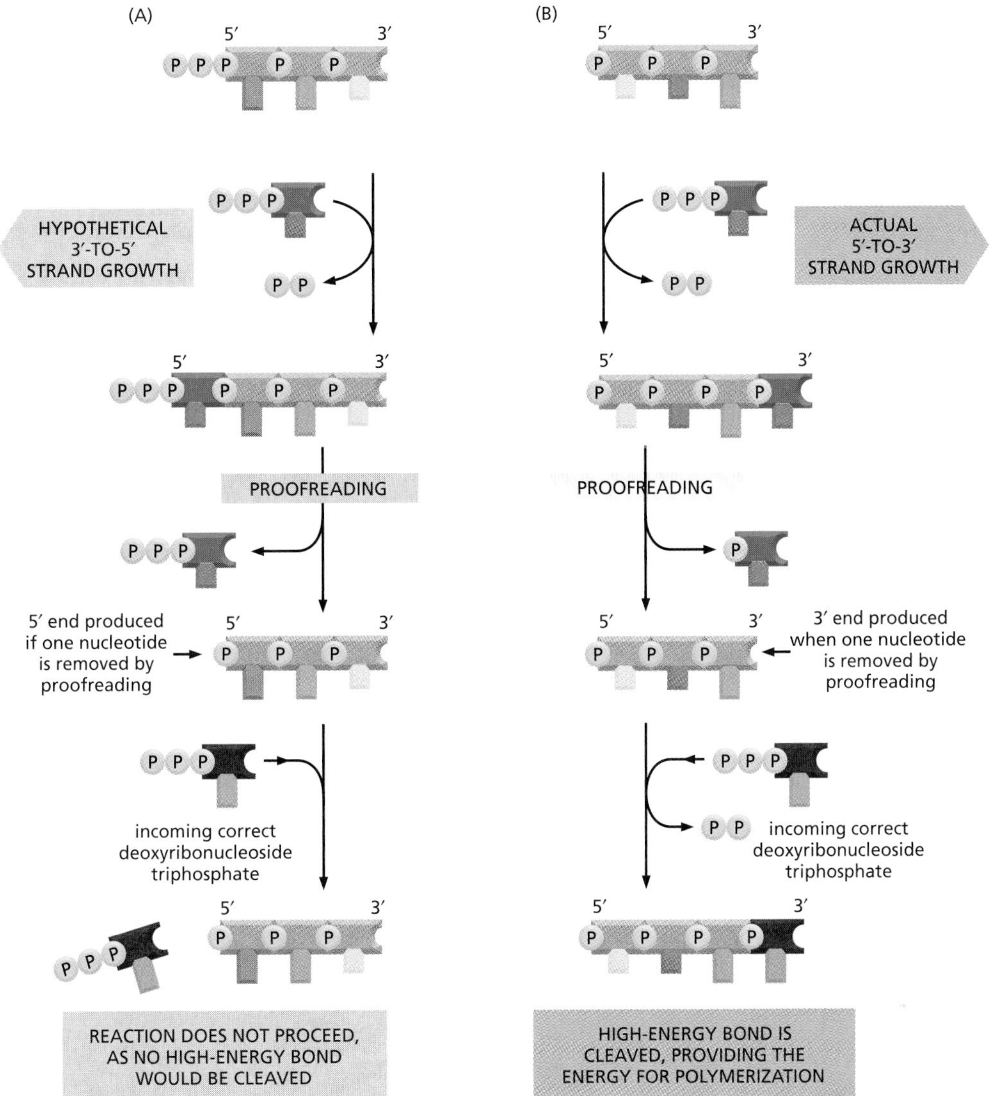
(A)
5′
3′
HYPOTHETICAL 3′-TO-5′ STRAND GROWTH
PROOFREADING
5′ end produced if one nucleotide is removed by proofreading
incoming correct deoxyribonucleoside triphosphate
REACTION DOES NOT PROCEED, AS NO HIGH-ENERGY BOND WOULD BE CLEAVED
(B)
ACTUAL 5′-TO-3′ STRAND GROWTH
PROOFREADING
3′ end produced when one nucleotide is removed by proofreading
incoming correct deoxyribonucleoside triphosphate
HIGH-ENERGY BOND IS CLEAVED, PROVIDING THE ENERGY FOR POLYMERIZATION

Figure 6–15 **A need for proofreading explains why DNA chains are synthesized only in the 5′ to 3′ direction.** (A) In the hypothetical 3′-to-5′ polymerization scheme, proofreading would remove an incorrect nucleotide (*dark green*), which would then block addition of the correct nucleotide (*red*) and thereby prevent further chain elongation. (B) Growth in the 5′-to-3′ direction allows the chain to continue to be elongated when an incorrect nucleotide has been added and then removed by proofreading (see Figure 6–14).

can begin a new polynucleotide chain simply by joining two nucleotides together without the need for a base-paired end. This enzyme does not, however, synthesize DNA. It makes a short length of a closely related type of nucleic acid—**RNA (ribonucleic acid)**—using the DNA strand as a template. This short length of RNA, about 10 nucleotides long, is base-paired to the template strand and provides a base-paired 3′ end as a starting point for DNA polymerase. It thus serves as a *primer* for DNA synthesis, and the enzyme that synthesizes the RNA primer is known as *primase*.

Primase is an example of an *RNA polymerase*, an enzyme that synthesizes RNA using DNA as a template. A strand of RNA is very similar chemically to a single strand of DNA except that it is made of ribonucleotide subunits, in which the sugar is ribose, not deoxyribose; RNA also differs from DNA in that it contains the base uracil (U) instead of thymine (T) (see Panel 2–6, pp. 74–75). However, because U can form a base pair with A, the RNA primer is synthesized on the DNA strand by complementary base-pairing in exactly the same way as is DNA.

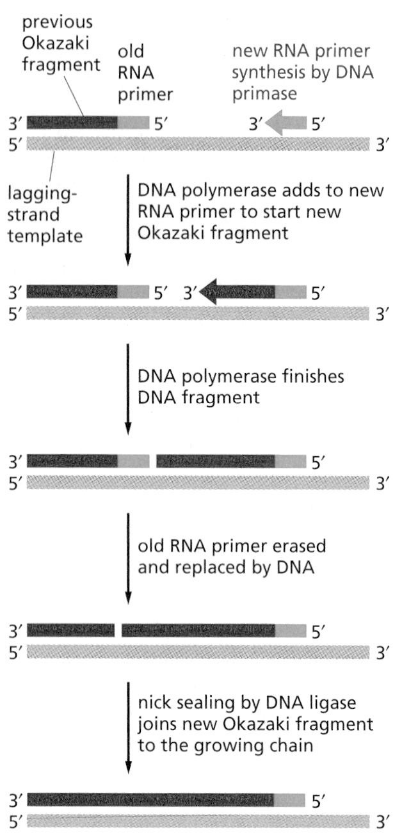
previous Okazaki fragment
old RNA primer
new RNA primer synthesis by DNA primase
3′
5′
lagging-strand template
DNA polymerase adds to new RNA primer to start new Okazaki fragment
DNA polymerase finishes DNA fragment
old RNA primer erased and replaced by DNA
nick sealing by DNA ligase joins new Okazaki fragment to the growing chain

Figure 6–16 **On the lagging strand, DNA is synthesized in fragments.** In eucaryotes, RNA primers are made at intervals of about 200 nucleotides on the lagging strand, and each RNA primer is approximately 10 nucleotides long. In the bacterium *E. coli*, the primers and Okazaki fragments are about 5 and 1000 nucleotides long, respectively. Primers are removed by nucleases that recognize an RNA strand in an RNA/DNA helix and degrade it; this leaves gaps that are filled in by a DNA repair polymerase that can proofread as it fills in the gaps. The completed fragments are finally joined together by an enzyme called DNA ligase, which catalyzes the formation of a phosphodiester bond between the 3′-OH end of one fragment and the 5′-phosphate end of the next, thus linking up the sugar–phosphate backbones. This nick-sealing reaction requires an input of energy in the form of ATP or NADH.

For the leading strand, an RNA primer is needed only to start replication at a replication origin; once a replication fork has been established, the DNA polymerase is continuously presented with a base-paired 3′ end as it tracks along the template strand. But on the lagging strand, where DNA synthesis is discontinuous, new primers are needed continually (see Figure 6–12). As the movement of the replication fork exposes a new stretch of unpaired bases, a new RNA primer is made at intervals along the lagging strand. DNA polymerase adds a deoxyribonucleotide to the 3′ end of this primer to start a DNA strand, and it will continue to elongate this strand until it runs into the next RNA primer (Figure 6–16).

To produce a continuous new DNA strand from the many separate pieces of nucleic acid made on the lagging strand, three additional enzymes are needed. These act quickly to remove the RNA primer, replace it with DNA, and join the DNA fragments together. Thus, a nuclease breaks apart the RNA primer, a DNA polymerase called a *repair polymerase* then replaces this RNA with DNA (using the end of the adjacent Okazaki fragment as a primer), and the enzyme *DNA ligase* joins the 5′-phosphate end of one new DNA fragment to the adjacent 3′-hydroxyl end of the next (see Figure 6–16).

Primase can begin new polynucleotide chains, but this activity is possible because the enzyme does not proofread its work. As a result, primers contain a high frequency of mistakes. But because primers are made of RNA instead of DNA, they stand out as 'suspect copy' to be automatically removed and replaced by DNA. This DNA is put in by DNA repair polymerases, which, like the replicative polymerases, proofread as they synthesize. In this way, the cell's replication machinery is able to begin new DNA chains and, at the same time, ensure that all of the DNA is copied faithfully.

QUESTION 6–2

Discuss the following statement: "Primase is a sloppy enzyme that makes many mistakes. Eventually, the RNA primers it makes are disposed of and replaced with DNA synthesized by a polymerase with higher fidelity. This is wasteful. It would be more energy-efficient if a DNA polymerase made an accurate copy in the first place."

Proteins at a Replication Fork Cooperate to Form a Replication Machine

As mentioned earlier, DNA replication requires a variety of proteins acting in concert. Here, we discuss the proteins that, together with DNA polymerase and primase, form the protein machine that powers the replication fork forward and synthesizes new DNA behind it.

For DNA synthesis to proceed, the double helix must be unzipped ahead of the replication fork so that the incoming deoxyribonucleoside triphosphates can form base pairs with the template strand. Two types of replication proteins—DNA helicases and single-strand binding proteins—cooperate to carry out this task. At the very front of the replication machine is the *helicase*, a protein that uses the energy of ATP hydrolysis to pry apart the double helix as it speeds along the DNA (Figure 6–17A and Movie 6.2). The *single-strand binding protein* clings to the single-stranded DNA exposed by the helicase, transiently preventing it from re-forming

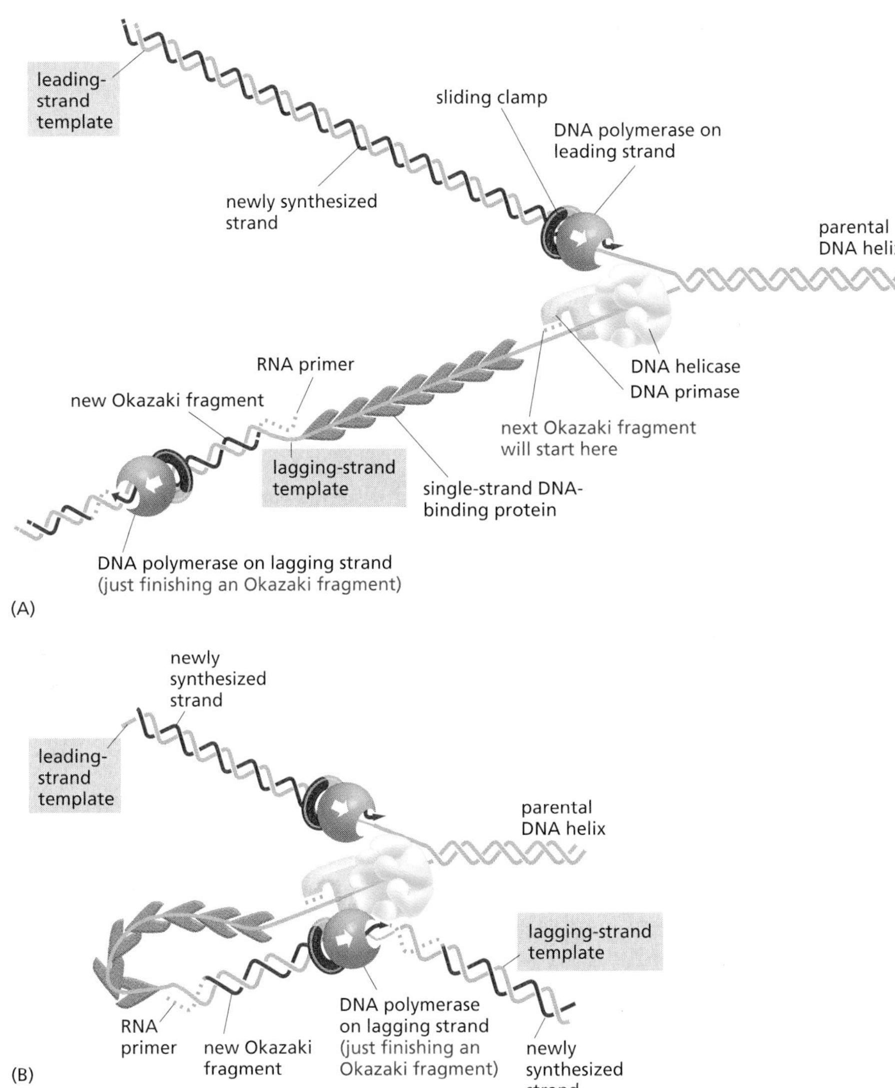

Figure 6–17 **DNA synthesis is carried out by a group of proteins that act together as a replication machine.** (A) Two molecules of DNA polymerase are shown, one on the leading strand and one on the lagging strand. Both are held on to the DNA by a circular protein clamp that allows the polymerase to slide. On the lagging-strand template, a clamp loader (not shown) is required to re-attach the sliding clamp each time a new Okazaki fragment is begun. At the head of the fork, a DNA helicase uses the energy of ATP hydrolysis to propel itself forward and thereby separate the strands of the parental DNA double helix. Single-strand DNA-binding proteins maintain these separated strands as single-stranded DNA to provide access for the primase and polymerase. For simplicity, this figure shows the proteins working independently; in the cell they are held together in a large replication machine, as shown in (B). (B) This diagram shows a current view of how the replication proteins are arranged at the replication fork when the fork is moving. The structure in (A) has been altered by folding the DNA on the lagging strand to bring the lagging-strand DNA polymerase molecule in contact with the leading-strand DNA polymerase molecule. This folding process also brings the 3′ end of each completed Okazaki fragment close to the start site for the next Okazaki fragment. Because the lagging-strand DNA polymerase is held to the rest of the replication proteins, it can be reused to synthesize successive Okazaki fragments; in this diagram, it is about to let go of its completed DNA fragment and move to the RNA primer that will be synthesized nearby, as required to start the next DNA fragment on the lagging strand. To watch the replication complex in action, see Movie 6.5.

base pairs and keeping it in an elongated form so that it can readily serve as a template for DNA polymerase.

An additional replication protein, called a *sliding clamp*, keeps the polymerase firmly attached to the template while it is synthesizing new strands of DNA. Left on their own, most DNA polymerase molecules will synthesize only a short string of nucleotides before falling off the DNA template. The sliding clamp forms a ring around the DNA helix and, by tightly gripping the polymerase, allows it to move along the template strand without falling off as it synthesizes new DNA (see Figure 6–17A and Movie 6.3).

Assembly of the clamp around DNA requires the activity of another replication protein, the *clamp loader*, which hydrolyzes ATP each time it locks a clamp around the DNA. This loading needs to occur only once per replication cycle on the leading strand; on the lagging strand, however, the clamp is removed and then reattached each time a new Okazaki fragment is made.

Most of the proteins involved in DNA replication are held together in a large multienzyme complex that moves as a unit along the DNA, enabling DNA to be synthesized on both strands in a coordinated manner. This complex can be likened to a miniature sewing machine composed of protein parts and powered by nucleoside triphosphate hydrolysis (Movie 6.4). Although the structures of the individual protein components of the replication machine have been determined, how these components fit together and work as a team is not entirely understood. Despite this lack of detailed knowledge, some ideas about the general appearance of the complex have been proposed (Figure 6–17B).

Telomerase Replicates the Ends of Eucaryotic Chromosomes

Having discussed how DNA replication begins at origins and how movement of the replication fork proceeds, we now turn to the special problem of replicating the very ends of chromosomes. As we discussed previously, the fact that DNA is synthesized only in the 5′-to-3′ direction means that the lagging strand of the replication fork is synthesized in the form of discontinuous DNA fragments, each of which is primed with an RNA primer laid down by a separate enzyme (see Figure 6–16). When the replication fork approaches the end of a chromosome, however, the replication machinery encounters a serious problem: there is no place to lay down the RNA primer needed to start the Okazaki fragment at the very tip of the linear DNA molecule. Without a strategy to deal with this problem, some DNA will inevitably be lost from the ends of a DNA molecule each time it is replicated.

QUESTION 6–3

A gene encoding one of the proteins involved in DNA replication has been inactivated by a mutation in a cell. In the absence of this protein the cell attempts to replicate its DNA for the very last time. What DNA products would be generated if each of the following proteins were missing?

A. DNA polymerase
B. DNA ligase
C. Sliding clamp for DNA polymerase
D. Nuclease that removes RNA primers
E. DNA helicase
F. Primase

Bacteria solve this 'end-replication' problem by having circular DNA molecules as chromosomes. Eucaryotes solve it by having special nucleotide sequences at the ends of their chromosomes which are incorporated into *telomeres*. These telomeric DNA sequences attract an enzyme called **telomerase** to the chromosome. Using an RNA template that is part of the enzyme itself, telomerase replenishes the nucleotides that are lost each time a eucaryotic chromosome is duplicated by adding multiple copies of the same short DNA sequence to the chromosome ends. This extended, repetitive DNA sequence then acts as a template that allows replication of the lagging strand to be completed by conventional DNA replication (Figure 6–18).

In addition to allowing replication of chromosome ends, telomeres serve additional functions: for example, the repeated telomere DNA sequences, together with the regions adjoining them, form structures that are recognized by the cell as the true end of a chromosome. This allows each chromosome end to be distinguished from the double-strand breaks that sometimes occur accidentally in the middle of chromosomes. These breaks must be immediately repaired, as we see in the next section.

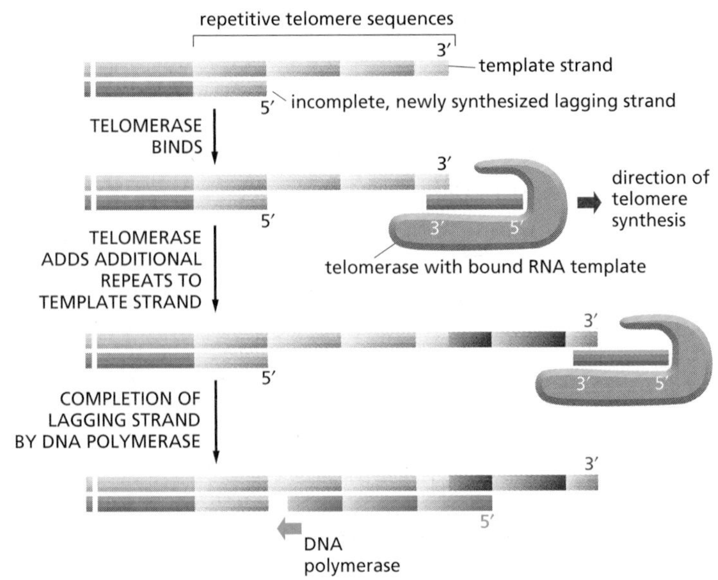

Figure 6–18 **Telomeres allow the completion of DNA synthesis at the ends of eucaryotic chromosomes.** To synthesize the lagging strand at the very end of a eucaryotic chromosome, the machinery of DNA replication requires a length of template DNA extending beyond the DNA that is to be copied. In a linear DNA molecule, synthesis of the lagging strand thus stops short just before the end of the template. The enzyme telomerase adds a series of repeats of a DNA sequence to the 3′ end of the template strand, which then allows the lagging strand to be completed by DNA polymerase, as shown. In humans, the nucleotide sequence of the repeat is GGGGTTA. The telomerase enzyme carries a short piece of RNA *(blue)* whose sequence is complementary to the DNA repeat sequence; this RNA acts as the template for the telomerase DNA synthesis. To see telomerase in action, view Movie 6.6.

DNA REPAIR

The diversity of living organisms and their success in colonizing almost every part of the Earth's surface depend on genetic changes accumulated gradually over millions of years. These changes allow organisms to adapt to changing conditions and to thrive in new habitats. However, in the short term, and from the perspective of an individual organism, alterations to gene sequences can be detrimental—especially in multicellular organisms, where they can upset an organism's extremely complex and finely tuned development and physiology. To survive and reproduce, individuals must be genetically stable. This stability is achieved not only through the extremely accurate mechanism for replicating DNA that we have just discussed, but also through the work of a variety of protein machines that continually scan the genome for damage and fix it. Indeed, most DNA damage is only temporary because it is immediately corrected by processes collectively called **DNA repair**.

Mutations Can Have Severe Consequences for a Cell or Organism

Only rarely do the cell's DNA replication and repair processes fail and allow a permanent change in the DNA. Such permanent changes are called **mutations**, and they can have profound consequences. A mutation that affects just a single nucleotide pair can severely compromise an organism's fitness if the change occurs in a vital position in the DNA sequence. Because the structure and activity of each protein depend on its amino acid sequence, a protein with an altered sequence may function poorly or not at all. For example, humans use the protein hemoglobin to transport oxygen in the blood (see Figure 4–20). A permanent change in a single nucleotide can cause cells to make a hemoglobin with an incorrect sequence of amino acids. One such mutation causes the disease *sickle-cell anemia* (Figure 6–19). The sickle-cell hemoglobin is less soluble than normal hemoglobin and forms fibrous precipitates, which produce the characteristic sickle shape of affected red blood cells. Because these cells are more fragile and frequently break in the bloodstream, patients with this potentially life-threatening disease have fewer red blood cells than usual (Figure 6–19C), a deficiency that can cause weakness, dizziness, headaches, pain, and organ failure.

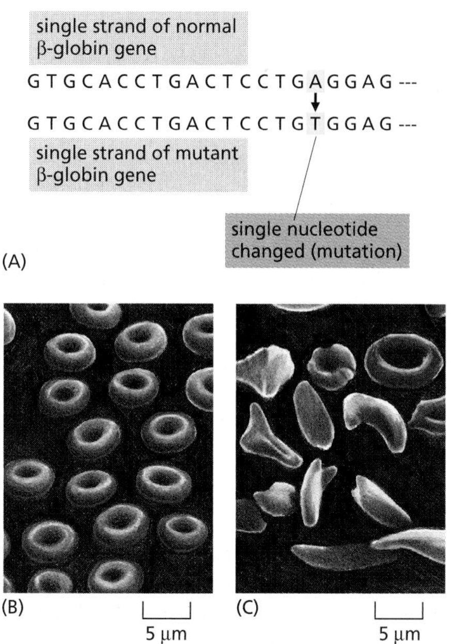

Figure 6–19 **A single nucleotide change causes the disease sickle-cell anemia.** (A) β-globin is one of the two types of subunit that form hemoglobin (see Figure 4–20). A single nucleotide change (mutation) in the β-globin gene produces a β-globin subunit that differs from normal β-globin only by a change from glutamic acid to valine at the sixth amino acid position. (Only a small portion of the gene is shown here; the β-globin subunit contains a total of 146 amino acids.) Humans carry two copies of each gene (one inherited from each parent); a sickle-cell mutation in one of the two β-globin genes generally causes no harm to the individual, as it is compensated for by the normal gene. However, an individual who inherits two copies of the mutant β-globin gene displays the symptoms of sickle-cell anemia. Normal red blood cells are shown in (B), and those from an individual suffering from sickle-cell anemia in (C). Although sickle-cell anemia can be a life-threatening disease, the mutation responsible can also be beneficial. People with the disease, or those who carry one normal gene and one sickle-cell gene, are more resistant to malaria than unaffected individuals, because the parasite that causes malaria grows poorly in red blood cells that contain the sickle-cell form of hemoglobin.

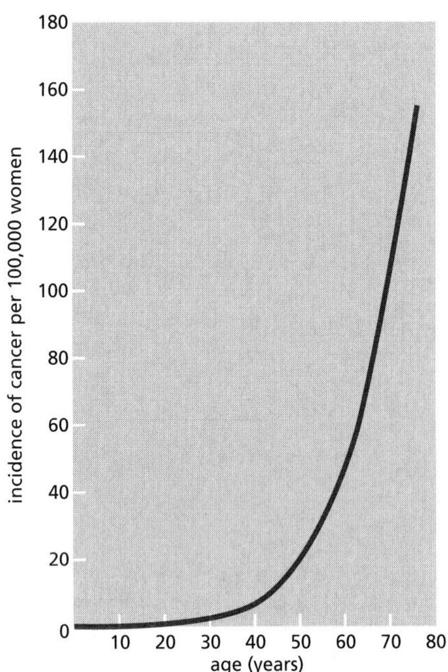

Figure 6–20 **Cancer incidence increases dramatically as a function of age.** The number of newly diagnosed cases of cancer of the colon in women in England and Wales in one year is plotted as a function of age at diagnosis. Colon cancer is caused by the accumulation of multiple mutations. Because cells are continually experiencing accidental changes to their DNA that accumulate and are passed on to progeny cells, the chance that a cell will become cancerous increases greatly with age. (Data from C. Muir et al., Cancer Incidence in Five Continents, Vol. V. Lyon: International Agency for Research on Cancer, 1987.)

The example of sickle-cell anemia, which is an inherited disease, illustrates the importance of protecting reproductive cells (*germ cells*) against mutation. A mutation in a germ cell will be passed on to all the cells in the body of the multicellular organism that develops from it, including the germ cells for production of the next generation.

The many other cells in a multicellular organism (its *somatic cells*) must be protected from the genetic changes that arise during the life of an individual. Nucleotide changes that occur in somatic cells can give rise to variant cells, some of which grow and divide in an uncontrolled fashion at the expense of the other cells in the organism. In the extreme case, an unchecked cell proliferation known as cancer results. This disease, which is responsible for about 30% of the deaths that occur in Europe and North America, is due largely to a gradual accumulation of changes in the DNA sequences of somatic cells that is caused by random mutation (Figure 6–20). Increasing the mutation frequency even two- or threefold could cause a disastrous increase in the incidence of cancer by accelerating the rate at which somatic cell variants arise.

Thus the high fidelity with which DNA sequences are replicated and maintained is important both for the reproductive cells, which transmit the genes to the next generation, and for the somatic cells, which normally function as carefully regulated members of the complex community of cells in a multicellular organism. We should therefore not be surprised to find that all cells possess a sophisticated set of mechanisms to reduce the number of mutations that occur in DNA.

A DNA Mismatch Repair System Removes Replication Errors That Escape the Replication Machine

In the first part of this chapter, we saw that the high fidelity of the cell's replication machinery generally prevents mistakes in copying. Despite these safeguards, however, such errors do occur. Fortunately, the cell has a backup system—called *DNA mismatch repair*—which is dedicated to correcting these rare mistakes. The replication machine itself makes approximately one error per 10^7 nucleotides copied; DNA mismatch repair corrects 99% of these errors, increasing the overall accuracy to one mistake in 10^9 nucleotides copied. This level of accuracy is much higher than that generally encountered in our day-to-day lives (Table 6–1).

Whenever the replication machinery makes a copying mistake, it leaves a mispaired nucleotide (commonly called a *mismatch*) behind. If left uncorrected, the mismatch will result in a permanent mutation in the next round of DNA replication (Figure 6–21A). A complex of mismatch repair proteins recognizes these DNA mismatches, removes (excises) one of the two strands of DNA involved in the mismatch, and resynthesizes the missing strand (Figure 6–22). To be effective in correcting replication mistakes, this mismatch repair system must always excise only the newly

TABLE 6–1 ERROR RATES	
US Postal Service on-time delivery of local first-class mail	13 late deliveries per 100 parcels
Airline luggage system	1 lost bag per 200
A professional typist typing at 120 words per minute	1 mistake per 250 characters
Driving a car in the United States	1 death per 10^4 people per year
DNA replication (without mismatch repair)	1 mistake per 10^7 nucleotides copied
DNA replication (including mismatch repair)	1 mistake per 10^9 nucleotides copied

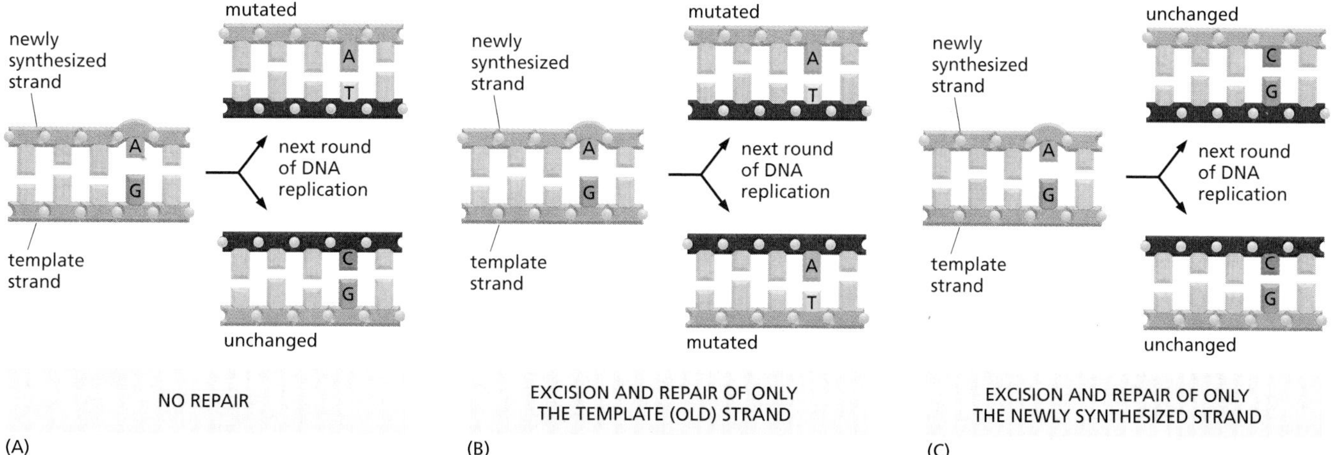

synthesized DNA strand: excising the other strand (the old strand) would double the mistake instead of correcting it (Figure 6–21B and C).

Figure 6–21 **Errors made during DNA replication must be corrected to avoid mutation.** (A) If uncorrected, the mismatch will lead to a permanent mutation in one of the two DNA molecules produced by the next round of DNA replication. (B) If the mismatch is 'repaired' using the newly synthesized DNA strand as the template, both DNA molecules produced by the next round of DNA replication will contain a mutation. (C) If the mismatch is corrected using the original template (old) strand as the template, the mutation is eliminated. The scheme shown in (C) is used by cells to repair mismatches, as shown in Figure 6–22.

In eucaryotes, it is not yet known for certain how the mismatch repair machinery distinguishes between the two DNA strands. However, there is evidence that newly replicated DNA strands—both leading and lagging—are preferentially nicked; it is these nicks (single-stranded breaks) that appear to provide the signal that directs the mismatch repair machinery to the appropriate strand (see Figure 6–22).

Mismatch repair plays a particularly important role in preventing cancer. An inherited predisposition to certain cancers (especially some types of colon cancer) is caused by mutations in genes that encode mismatch repair proteins. Humans inherit two copies of these genes (one from each parent), and individuals who inherit one damaged mismatch repair gene show no symptoms until the undamaged copy of the same gene is randomly mutated in a somatic cell. When this mutant cell divides, it will give rise to a cluster of somatic cells that, because they are deficient in mismatch repair, accumulate mutations more rapidly than do normal cells. Because most cancers arise from cells that have accumulated multiple mutations (see Figure 6–20), a cell deficient in mismatch repair has a greatly enhanced chance of becoming cancerous. Thus, inheriting a damaged mismatch repair gene predisposes an individual to cancer.

DNA Is Continually Suffering Damage in Cells

Rare mistakes in DNA replication, as we have seen, can be corrected by the mismatch repair mechanism. But DNA can be damaged in many other ways, and these require different mechanisms for their repair. Just

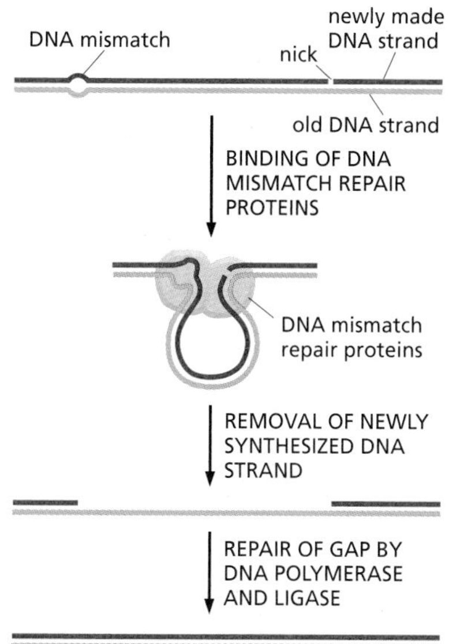

Figure 6–22 **DNA mismatch repair proteins correct errors that occur during DNA replication.** A DNA mismatch, formed when an incorrectly matched base is incorporated into a newly synthesized DNA chain, distorts the geometry of the double helix. This distortion is subsequently recognized by the DNA mismatch repair proteins, which then remove the newly synthesized DNA. The gap in the newly synthesized DNA is replaced by a DNA polymerase that proofreads as it synthesizes and is sealed by DNA ligase. As shown in the figure, a nick in the DNA has been proposed as the signal that allows the mismatch repair proteins to distinguish the newly synthesized DNA (which contains the mistake) from the old DNA. Such nicks are known to occur in the lagging strands (see Figure 6–12) and are observed also to occur, although less frequently, in the leading strands. These nicks remain for only a short period after a replication fork has passed (see Figure 6–16), so that mismatch repair must occur quickly.

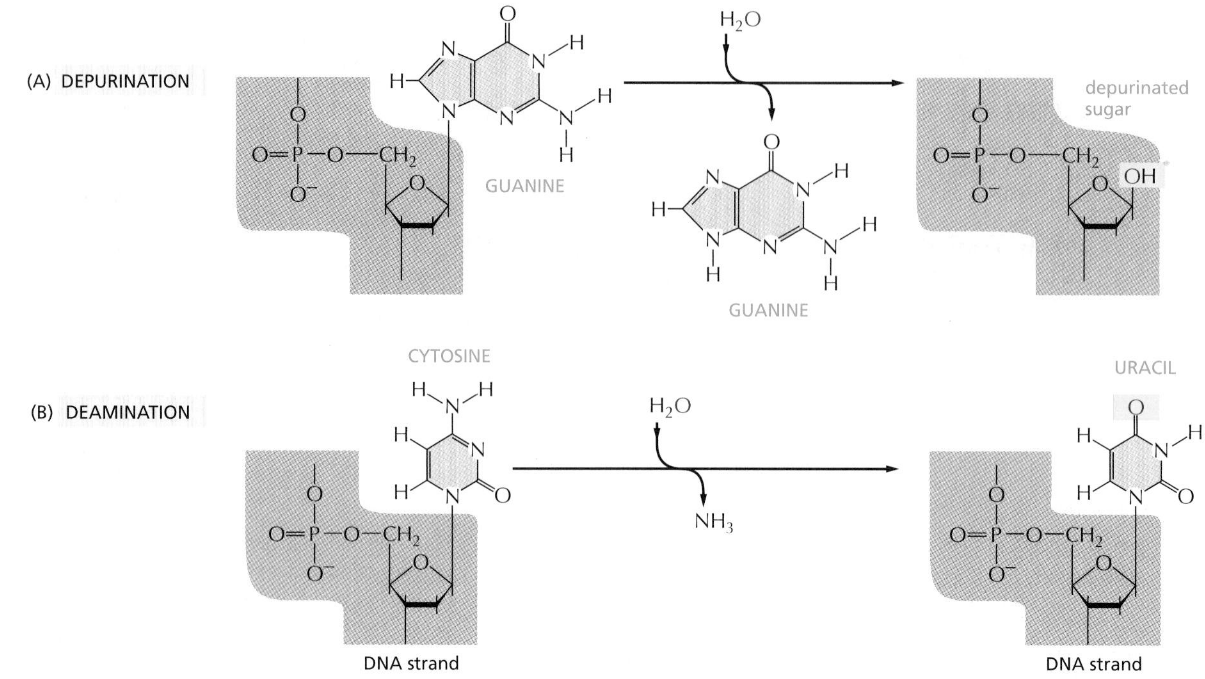

(A) DEPURINATION

GUANINE

H_2O

GUANINE

depurinated sugar

(B) DEAMINATION

CYTOSINE

H_2O

NH_3

URACIL

DNA strand

DNA strand

Figure 6–23 **Depurination and deamination are the most frequent chemical reactions known to create serious DNA damage in cells.** (A) Depurination can release guanine and also adenine, from DNA. (B) The major type of deamination reaction converts cytosine to an altered DNA base, uracil, but deamination can occur on other bases as well. Both of these reactions take place on double-helical DNA and neither break the phosphodiester backbone (highlighted in orange); for convenience, only one DNA strand is shown.

like any other molecule in the cell, DNA is continually undergoing thermal collisions with other molecules. These often result in major chemical changes in the DNA. For example, during the time it takes to read this sentence, a total of about a trillion (10^{12}) purine bases (A and G) will be lost from DNA in the cells of your body by a spontaneous reaction called *depurination* (Figure 6–23). Depurination does not break the phosphodiester backbone but, instead, gives rise to lesions that resemble missing teeth. Another common reaction is the spontaneous loss of an amino group (*deamination*) from cytosine in DNA to produce the base uracil (see Figure 6–23). Some chemically reactive by-products of cell metabolism also occasionally react with the bases in DNA, altering them in such a way that their base-pairing properties are changed. The ultraviolet radiation in sunlight is also damaging to DNA; it promotes covalent linkage between two adjacent pyrimidine bases, forming, for example, the *thymine dimer* shown in Figure 6–24.

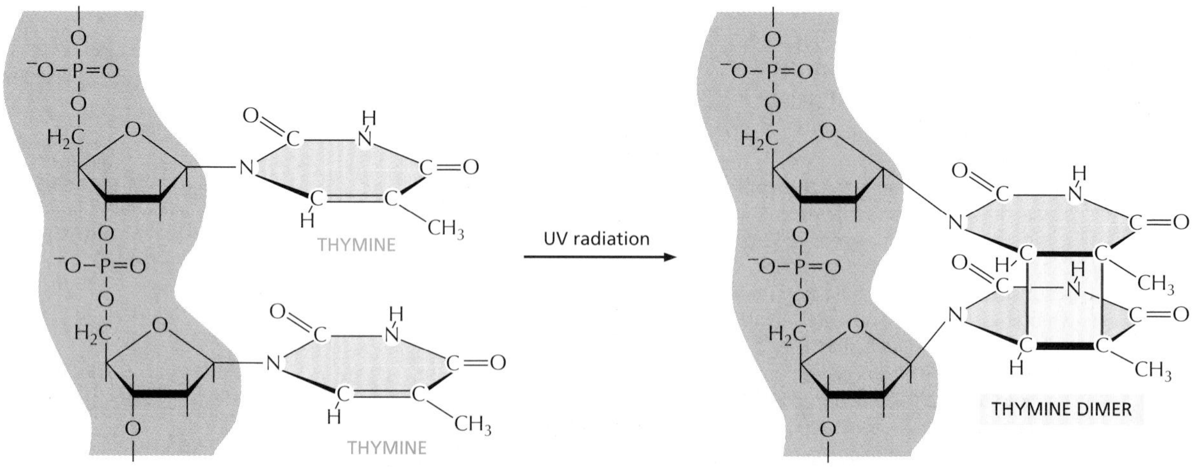

THYMINE

THYMINE

UV radiation

THYMINE DIMER

Figure 6–24 **The ultraviolet radiation in sunlight causes DNA damage.** Two adjacent thymine bases have become covalently attached to one another to form a thymine dimer. Skin cells that are exposed to sunlight are especially susceptible to this type of DNA damage.

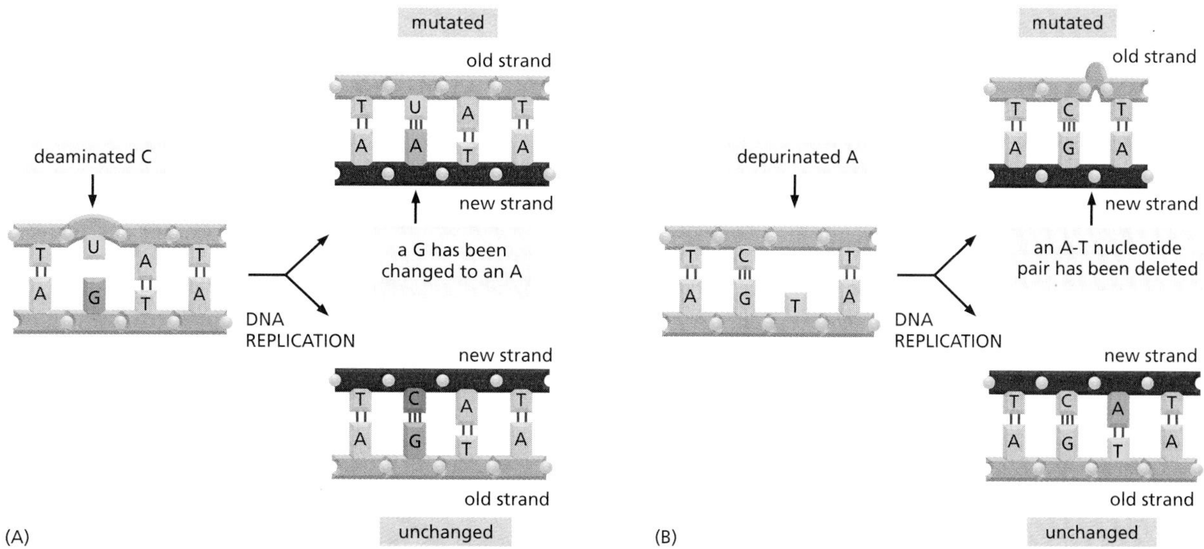

Figure 6–25 **Chemical modifications of nucleotides, if left unrepaired, produce mutations.** (A) Deamination of cytosine, if uncorrected, results in the substitution of one base for another when the DNA is replicated. As shown in Figure 6–23, deamination of cytosine produces uracil. Uracil differs from cytosine in its base-pairing properties and preferentially base-pairs with adenine. The DNA replication machinery therefore inserts an adenine when it encounters a uracil on the template strand. (B) Depurination, if uncorrected, can lead to the loss of a nucleotide pair. When the replication machinery encounters a missing purine on the template strand, it can skip to the next complete nucleotide, as shown, thus producing a nucleotide deletion in the newly synthesized strand. In other cases (not shown), the replication machinery places an incorrect nucleotide across from the missing base, again resulting in a mutation.

These are only a few of many chemical changes that can occur in our DNA. If left unrepaired, many of them would lead either to the substitution of one nucleotide pair for another as a result of incorrect base-pairing during replication or to deletion of one or more nucleotide pairs in the daughter DNA strand after DNA replication (Figure 6–25). Some types of DNA damage (thymine dimers, for example) often stall the DNA replication machinery at the site of the damage. All of these types of damage, if unrepaired, would have disastrous consequences for an organism.

QUESTION 6–4

Discuss the following statement: "The DNA repair enzymes that fix deamination and depurination damage must preferentially recognize such damage on newly synthesized DNA strands."

The Stability of Genes Depends on DNA Repair

The thousands of random chemical changes that occur every day in the DNA of a human cell, through metabolic accidents or exposure to DNA-damaging chemicals, are repaired by a variety of mechanisms, each catalyzed by a different set of enzymes. Nearly all these mechanisms depend on the existence of two copies of the genetic information, one in each strand of the DNA double helix: if the sequence in one strand is accidentally damaged, information is not lost irretrievably, because a backup version of the altered strand remains in the complementary sequence of nucleotides in the other strand. Most damage creates structures that are never encountered in an undamaged DNA strand; thus the good strand is easily distinguished from the bad.

The basic pathway for repairing damage to DNA is illustrated schematically in Figure 6–26. As indicated, it involves three steps:

1. The damaged DNA is recognized and removed by one of a variety of different mechanisms. These involve nucleases, which cleave the covalent bonds that join the damaged nucleotides to the rest of the DNA molecule, and they leave a small gap on one strand of the DNA double helix in the region.
2. A repair DNA polymerase binds to the 3′-hydroxyl end of the cut DNA strand. It then fills in the gap by making a complementary

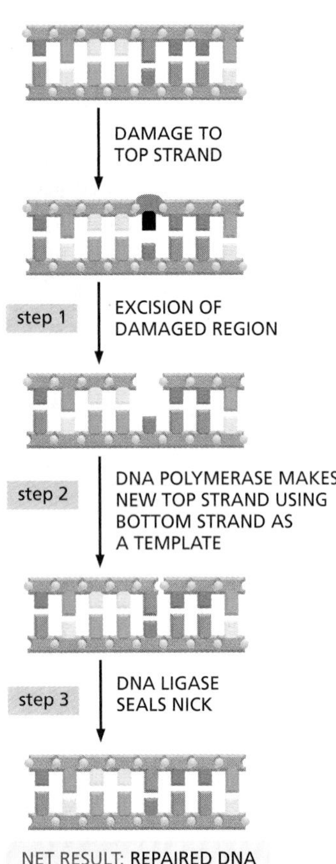
DAMAGE TO
TOP STRAND
step 1
EXCISION OF
DAMAGED REGION
step 2
DNA POLYMERASE MAKES
NEW TOP STRAND USING
BOTTOM STRAND AS
A TEMPLATE
step 3
DNA LIGASE
SEALS NICK
NET RESULT: REPAIRED DNA

Figure 6–26 **The basic mechanism of DNA repair involves three steps: excision, resynthesis, and ligation.** In step 1 (excision), the damage is cut out by one of a series of nucleases, each specialized for a type of DNA damage. In step 2 (resynthesis), the original DNA sequence is restored by a repair DNA polymerase, which fills in the gap created by the excision events. In step 3 (ligation), DNA ligase seals the nick left in the sugar–phosphate backbone of the repaired strand. Nick sealing, which requires energy from ATP hydrolysis, remakes the broken phosphodiester bond between the adjacent nucleotides. Some types of DNA damage (the deamination of cytosine [Figure 6–23], for example) involve the replacement of a single nucleotide, as shown in the figure. For the repair of other kinds of DNA damage, such as thymine dimers (see Figure 6–24), a longer stretch of 10–20 nucleotides is removed from the damaged strand.

copy of the information stored in the undamaged strand. Although different from the DNA polymerase that replicates DNA, repair DNA polymerases synthesize DNA strands in the same way. For example, they elongate chains in the 5′-to-3′ direction and have the same type of proofreading activity to ensure that the template strand is copied accurately. In many cells, this is the same enzyme that fills in the gap left after the RNA primers are removed during the normal DNA replication process (see Figure 6–16).

3. When the repair DNA polymerase has filled in the gap, a break remains in the sugar–phosphate backbone of the repaired strand. This nick in the helix is sealed by DNA ligase, the same enzyme that joins the lagging-strand DNA fragments during DNA replication.

Steps 2 and 3 are nearly the same for most types of DNA repair, including mismatch repair. However, step 1 uses a series of different enzymes, each specialized for removing different types of DNA damage.

The importance of these repair processes is indicated by the large investment that cells make in DNA repair enzymes. Single-celled organisms such as yeasts contain more than 50 different proteins that function in DNA repair, and DNA repair pathways are even more complex in humans. The importance of these DNA repair processes is also evident from the consequences of their malfunction. Humans with the genetic disease *xeroderma pigmentosum*, for example, cannot repair thymine dimers (see Figure 6–24) because they have inherited a defective gene for one of the proteins involved in this repair process. Such individuals develop severe skin lesions, including skin cancer, because of the accumulation of thymine dimers in cells that are exposed to sunlight and the consequent mutations that arise in the cells that contain them.

Double-Strand Breaks Can be Repaired Rapidly But Imperfectly

A particularly dangerous type of DNA damage occurs when both strands of the double helix are broken, leaving no intact template strand to guide proper repair. Ionizing radiation, mishaps at the replication fork, strong oxidizing agents, and metabolites produced in the cell can all cause breaks of this type. If these lesions are left unrepaired, they would lead quickly to the fragmentation of chromosomes and cause a loss of genes when the cell divides.

Several mechanisms have evolved to repair this potentially disastrous type of damage. In human somatic cells, the most common means of repairing double-strand breaks is by a mechanism called **nonhomologous end-joining**. In this process, the two broken ends are simply

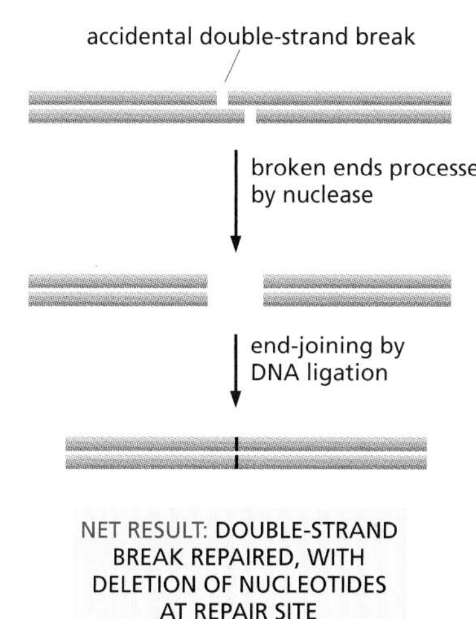

Figure 6–27 **Cells can use nonhomologous end-joining to repair double-strand breaks.** This 'quick and dirty' mechanism alters the original DNA sequence during the process of repair. The alterations are usually short deletions.

brought together by a specialized group of enzymes and rejoined by DNA ligation. Although this mechanism repairs the break, nucleotides are usually lost at the site of repair (Figure 6–27). Because such a small portion of mammalian genomes contains useful information, this 'quick and dirty' mechanism is apparently an acceptable solution to the problem of repairing broken chromosomes.

Cells also have an alternative, error-free strategy for repairing double-strand breaks, particularly those that occur in newly replicated DNA. This mechanism, called *homologous recombination*, is discussed in the next section of the chapter.

A Record of the Fidelity of DNA Replication and Repair Is Preserved in Genome Sequences

We have seen in this chapter how DNA sequences are replicated and maintained with remarkable fidelity. As a consequence, changes in DNA accumulate remarkably slowly in the course of evolution. Natural selection also plays a role: although the majority of mutations do neither harm or good to an organism, those that have harmful consequences are usually eliminated from the population through the death or reduced fertility of individuals carrying the altered DNA. But even where no selection operates—at the many sites in the DNA where a change of nucleotide has no effect on the fitness of the organism—the genetic message has been faithfully preserved over tens of millions of years. Thus humans and chimpanzees, after about 5 million years of divergent evolution, still have DNA sequences that are at least 98% identical. Even humans and whales, after 10 or 20 times this amount of time, have chromosomes that are unmistakably similar in their DNA sequence, and many proteins have amino acid sequences that are almost identical (Figure 6–28). Thus, in our genomes, we and our relatives contain a message from the distant past. Thanks to the faithfulness of DNA replication and repair, 100 million years have scarcely changed its essential content.

whale GTGTGGTCTCGTGATCAAAGGCGAAAGGTGGCTCTAGAGAATCCC

human GTGTGGTCTCGCGATCAGAGGCGCAAGATGGCTCTAGAGAATCCC

Figure 6–28 **The sex-determination genes from humans and whales are unmistakably similar.** Although their body plans are strikingly different, humans and whales are built from the same proteins. Despite the many millions of years that have passed since humans and whales diverged, the nucleotide sequences of many of their genes are closely similar. The sequences of a part of the gene that determines maleness in humans and in whales are shown, one above the other, with the positions where the two are identical *shaded.*

HOMOLOGOUS RECOMBINATION

Thus far we have discussed how DNA replication and repair mechanisms maintain the nucleotide sequences in cells from generation to generation with very little change. As we have emphasized, these mechanisms rely on the redundancy inherent in the DNA double helix, in which each strand is paired with a second strand that contains the complementary sequence. If nucleotides on one strand are damaged, they can be repaired using the information provided by the complementary strand.

But what happens to genetic information when both members of a nucleotide pair are damaged simultaneously—for example, when a double-strand break occurs? As we saw earlier, one approach to fixing such damage is to use nonhomologous end-joining to rapidly heal the wound. However, that mechanism usually sacrifices the information contained at the site of the injury. A more elegant solution is to use the genetic information provided by an entirely separate DNA duplex to repair the break accurately. This strategy is carried out by a set of reactions collectively known as **homologous recombination**. Its central feature is the exchange of genetic information between a pair of homologous DNA molecules: that is, DNA duplexes that are similar or identical in nucleotide sequence. In this process, the information present in an intact, undamaged DNA duplex is used as a template to accurately repair a broken DNA double helix.

In addition to its role in repair, homologous recombination is also responsible for generating genetic diversity during *meiosis*, the specialized form of cell division by which sexually reproducing organisms make germ cells. In this case, homologous recombination physically swaps genetic information between the homologous maternal and paternal chromosomes to produce chromosomes with novel DNA sequences. The potential evolutionary benefit of this type of gene mixing is that it generates an array of new, perhaps beneficial, combinations of genes that will be passed along to an organism's offspring.

The flawless repair of double-strand breaks and the exchange of genetic information in meiosis seem—on the surface—to be two unrelated processes. But in this section we will see that their underlying mechanisms are very similar and that they are based on a similar set of reactions and protein components.

Homologous Recombination Requires Extensive Regions of Sequence Similarity

Whether the end product is DNA repair or the exchange of nucleotide sequences during meiosis, the hallmark of homologous recombination is that it takes place only between DNA duplexes that contain extensive regions of sequence similarity (homology). A pair of DNA molecules can assess this homology by 'sampling' each other's nucleotide sequences when a single strand from one DNA duplex engages in an extensive bout of base-pairing with the complementary strand from the other duplex. The match need not be perfect for homologous recombination to succeed, but it must be very close.

Homologous Recombination Can Flawlessly Repair DNA Double-strand Breaks

Homologous recombination is often initiated when a double-strand break occurs shortly after a stretch of DNA has been replicated; at that time, the duplicated helices are still in close proximity to one another (Figure 6–29A). To begin the repair, a nuclease generates single-stranded ends at the break by chewing back one of the complementary DNA strands (Figure 6–29B). With the help of specialized enzymes, one of these single strands then 'invades' the homologous DNA duplex by forming base pairs with its complementary strand. If this sampling results in extensive base pairing, a *branch point* is created where the two DNA strands—one from each duplex—cross (Figure 6–29C). At this point, the invading strand is elongated by a repair DNA polymerase, using the complementary strand as a template (Figure 6–29D). The branch point then 'migrates' as the base pairs holding together the duplexes break, and new ones form (Figure 6–29E). Repair is completed by additional DNA synthesis, followed by DNA ligation (Figure 6–29F). The net result is two intact DNA helices, where the genetic information from one was used as a template to repair the other.

Homologous recombination can also be used to repair many other types of DNA damage, making it perhaps the most versatile DNA repair mech-

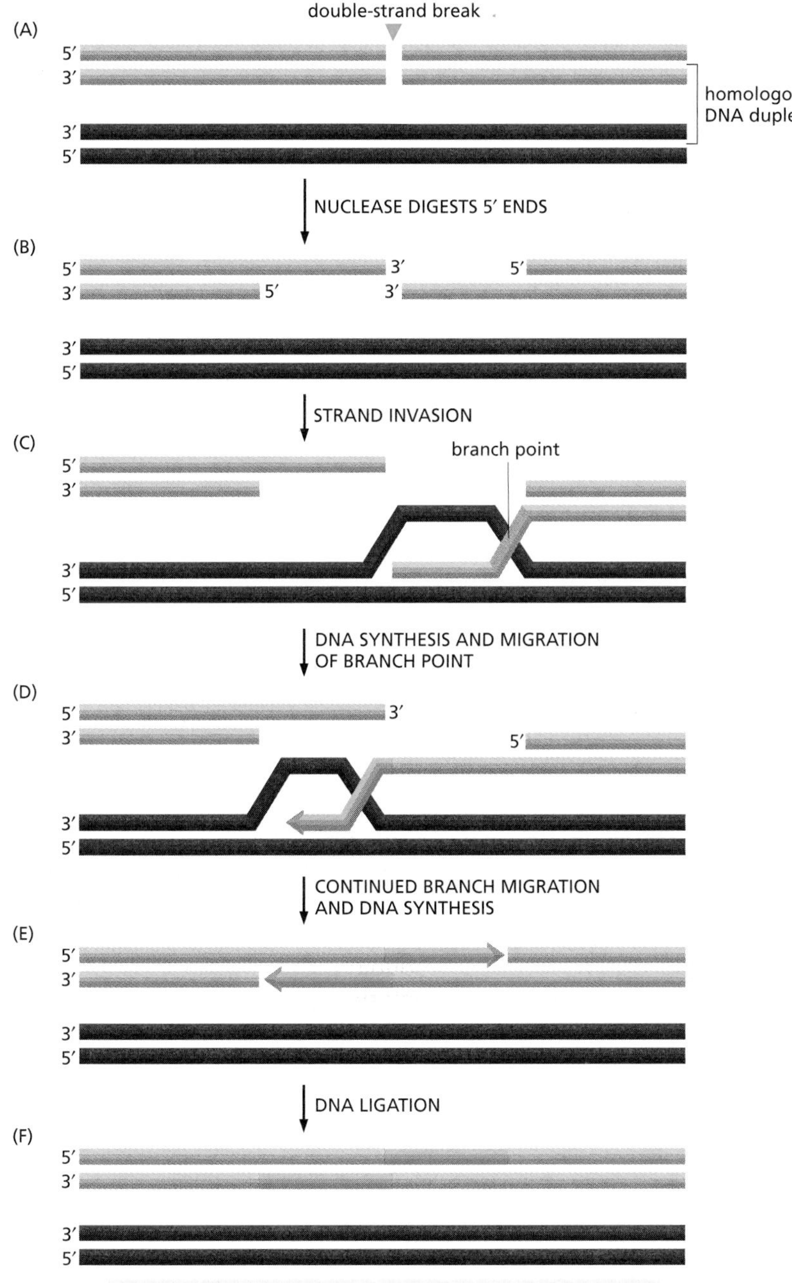

Figure 6–29 **Homologous recombination allows the flawless repair of DNA double-strand breaks.** This is the preferred method for repairing double-strand breaks that arise shortly after the DNA has been replicated but before the cell has divided. (Adapted from M. McVey et al., *Proc. Natl. Acad. Sci. USA* 101:15694–15699, 2004. With permission from the National Academy of Sciences.)

anism available to the cell: all that is needed is an intact homologous chromosome to use as a partner—and such paired identical chromosomes are produced transiently each time a chromosome is replicated. The 'all-purpose' nature of recombinational repair probably explains why this mechanism, and the proteins that carry it out, have been conserved in virtually all cells on Earth.

Homologous Recombination Exchanges Genetic Information During Meiosis

Sexually reproducing organisms depend on the process of meiosis to generate germ cells—sperm and eggs, in the case of mammals. Homologous recombination is essential for this process to proceed properly, as we discuss in more detail in Chapter 19. A particularly important consequence of homologous recombination during meiosis is the formation of chromosomal *crossovers*. Here, two homologous chromosomes—one from the father, one from the mother—come together and undergo a genetic exchange (Figure 6–30). The site of exchange can be anywhere in the homologous nucleotide sequences of the two participating DNA molecules. The cleavage and rejoining events that mediate the exchange occur so precisely that not a single nucleotide is lost or gained.

Because the paternal and maternal chromosomes typically differ slightly in their DNA sequences, this crossing-over generates new combinations of DNA sequences in each chromosome. The benefit of such gene mixing for the progeny organisms is apparently so great that the reassortment of genes by homologous recombination is not confined to sexually reproducing organisms; it is also widespread in asexually reproducing organisms, for example when a bacterium acquires a homologous chromosome from another bacterial cell through *horizontal gene transfer*, as we will see in Chapter 9.

Homologous recombination during meiosis begins with a bold stroke: a specialized enzyme deliberately slices through both strands of one of the recombining chromosomes, creating a double-strand break. At this point, some of the same proteins that function in recombinational double-strand break repair converge on the 'damage.' However, these recombination proteins are now directed by meiosis-specific proteins to perform their tasks differently, producing—via one or more crossovers—two molecules with novel DNA sequences (Figure 6–31). This outcome is possible because, in meiosis, recombination occurs preferentially between maternal and paternal chromosomes rather than between newly replicated, identical DNA strands, as happens when homologous recombination mediates double-strand break repair.

Crossing-over during meiosis ensures that each of our chromosomes contains a combination of DNA sequences from our two parents. As we discuss in detail in Chapter 19, this type of chromosomal rearrangement generates large amounts of genetic diversity in the offspring of sexually reproducing organisms, and it has thereby contributed greatly to the stunning variety of life forms present on the planet.

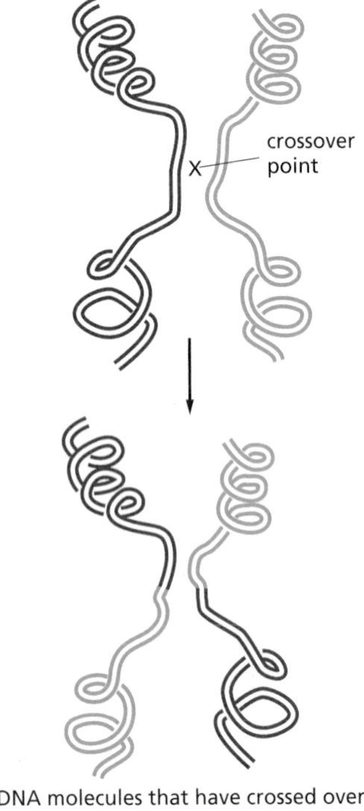

Figure 6–30 **Homologous recombination takes place between DNA molecules with similar nucleotide sequences.** The breaking and rejoining of two homologous DNA double helices creates two DNA molecules that have "crossed over." Although the two original DNA molecules must have similar nucleotide sequences in order to cross over, they do not have to be identical; thus, a crossover can create DNA molecules with novel nucleotide sequences.

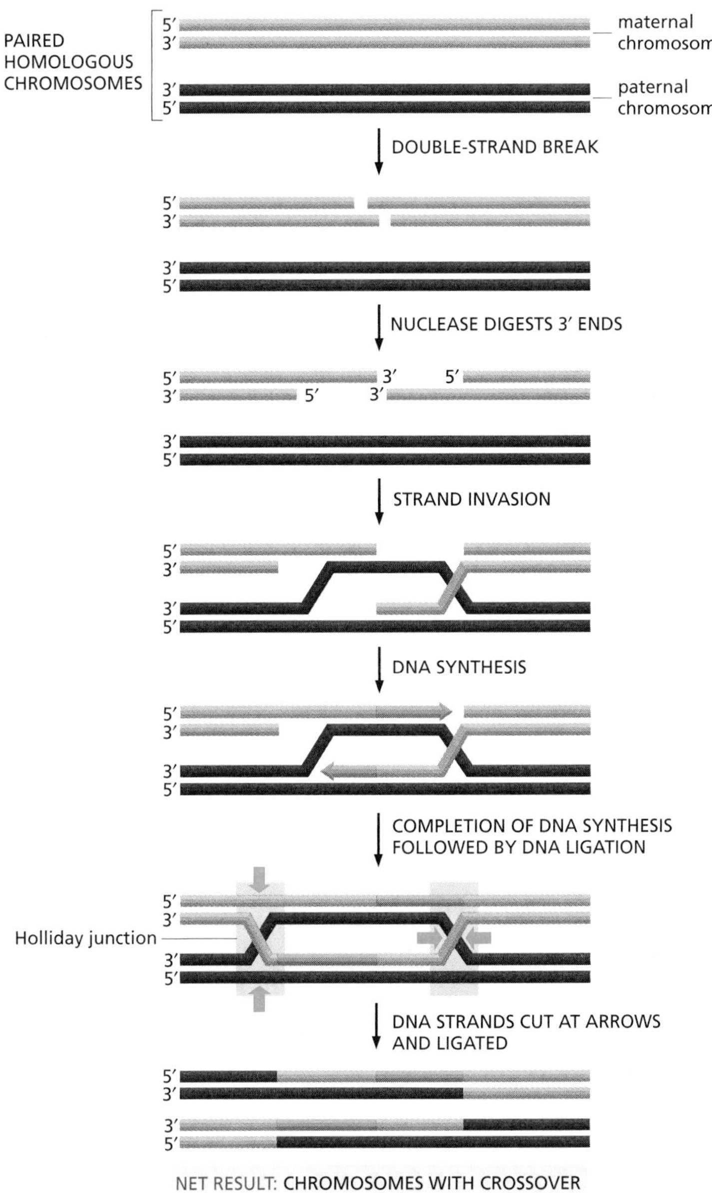

Figure 6–31 **Homologous recombination in meiosis generates crossovers.** Once meiosis-specific proteins have broken the DNA duplex and processed the ends, homologous recombination proceeds through the formation of two Holliday junctions (*blue* boxes)—the sites where the DNA duplexes cross, named after the scientist who discovered them. For a glimpse of these structures under the electron microscope see Movie 6.7. Many of the steps that produce chromosome crossovers during meiosis resemble those used to repair DNA double-strand breaks (compare with Figure 6–29).

MOBILE GENETIC ELEMENTS AND VIRUSES

We have seen that homologous recombination can promote the exchange of DNA sequences between chromosomes. But such exchanges are generally conservative: the order of the genes on the recombined chromosomes remains the same, because homologous recombination occurs only between chromosomes that are very similar in sequence.

But genomes are also subject to more dramatic forms of genetic variation—changes that alter the order of genes on a chromosome or even add new information. It is this more radical form of genetic resculpturing we turn to next. The agents underlying these drastic genetic changes are **mobile genetic elements**, sometimes known informally as jumping genes. Found in virtually all cells, these elements are short, specialized segments of DNA that can move from one position in the cell's genome to another. Although they can insert themselves into virtually any sequence within the genome, most mobile genetic elements lack the ability to leave the cell in which they reside. Their movement is therefore restricted to a single cell and its descendants.

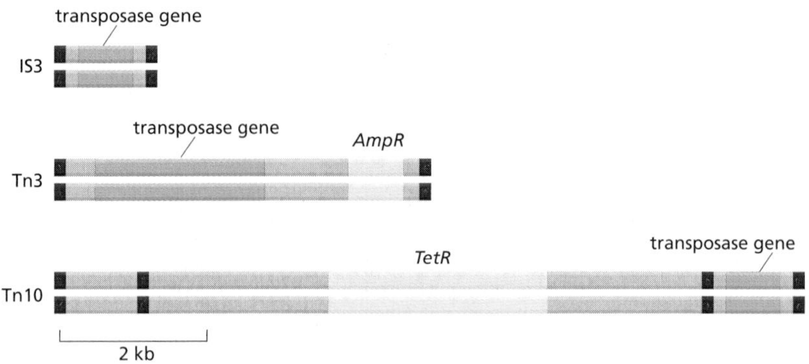

Figure 6–32 **Bacteria contain many DNA-only transposons, three of which are shown.** Each of these elements contains a gene that encodes a transposase (*blue*), an enzyme that catalyzes the movement of that element. Each also carries DNA sequences (indicated in *red*) that are recognized only by its own transposase and that are necessary for its movement.

Some mobile genetic elements carry, in addition, genes that encode enzymes that inactivate antibiotics such as ampicillin (*AmpR*) and tetracycline (*TetR*). Movement of these genes is a serious problem in medicine, as many disease-causing bacteria have become resistant to many of the antibiotics developed during the twentieth century.

This is not the case for *viruses*. Essentially just strings of genes wrapped in a protective coat, viruses are the ultimate mobile DNA, in that they can escape from one cell and infect another. We end this section with a brief discussion of viruses, which, although they are occasionally beneficial to cells, are also responsible for some of our most devastating diseases.

Mobile Genetic Elements Encode the Components They Need for Movement

Unlike homologous recombination, the movement of mobile genetic elements does not require DNA sequence similarity. Instead, each element typically encodes a specialized recombination enzyme that mediates its movement (Figure 6–32). These enzymes recognize and act on unique DNA sequences that are present on each mobile genetic element. Many mobile genetic elements also carry other genes. For example, mobile genetic elements that carry antibiotic resistance genes have contributed greatly to the widespread dissemination of antibiotic resistance in bacterial populations.

Mobile genetic elements are also called **transposons** and are typically classified according to the mechanism by which they move or *transpose*. In bacteria, the most common mobile genetic elements are the *DNA-only transposons*. The name is derived from the fact that, during their movement, the element remains as DNA, rather than being converted to RNA as for other elements that we discuss below. Bacteria contain many different DNA-only transposons. Some move to the target site using a simple cut-and-paste mechanism, whereby the element is simply excised from the genome and inserted into a different site; other DNA-only transposons replicate their DNA before inserting into the new chromosomal site, leaving the original copy intact at its previous location (Figure 6–33).

The Human Genome Contains Two Major Families of Transposable Sequences

Amazingly, nearly half of the human genome is made up of millions of copies of various mobile genetic elements, which form a very large part of our DNA. Some of these elements have moved from place to place within the human genome using the cut-and-paste mechanism discussed above for DNA-only transposons (see Figure 6–33A). However, most have moved not as DNA but via an RNA intermediate. These are called **retrotransposons** and are, as far as is known, unique to eucaryotes.

One abundant human retrotransposon, the *L1 element* (sometimes referred to as *LINE-1*), is transcribed into RNA by the host cell's RNA polymerases. A DNA copy of this RNA is then made using an enzyme called **reverse transcriptase**, an unusual DNA polymerase that can use RNA as a tem-

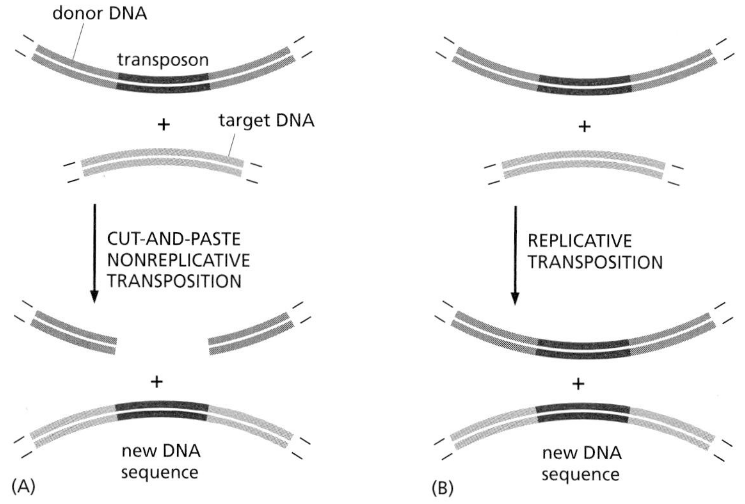

Figure 6–33 **DNA-only transposons move by two types of mechanisms.** (A) In cut-and-paste transposition, the element is cut out of the donor DNA and inserted into the target DNA, leaving behind a broken donor DNA molecule. (The donor DNA can be repaired in a variety of ways, but this sometimes results in deletions or rearrangements of the donor molecule.) (B) In the course of replicative transposition, the element is copied by DNA replication. The end products are a molecule that appears identical to the original donor and a target molecule that has a mobile genetic element inserted into it. In general, a particular type of transposon moves by only one of these mechanisms. However, the two mechanisms have many enzymatic similarities, and a few transposons can move by either mechanism. The donor and target DNAs can be part of the same DNA molecule or reside on different molecules.

plate. The reverse transcriptase is encoded by the *L1* element itself. The DNA copy of the element can then reintegrate into another site in the genome (Figure 6–34).

L1 elements constitute about 15% of the human genome. Although most copies are immobile due to the accumulation of deleterious mutations, a few still retain the ability to transpose. Their movement can sometimes result in human disease: for example, about 40 years ago, movement of an *L1* element into the gene that encodes Factor VIII—a protein essential for proper blood clotting—caused hemophilia in an individual with no family history of the disease.

Another type of retrotransposon, the *Alu sequence*, is present in about 1 million copies in our genome. *Alu* elements do not encode their own reverse transcriptase and thus depend on enzymes already present in the cell to help them move.

Comparisons of the sequence and locations of the *L1* and *Alu* elements in different mammals suggest that these sequences have proliferated in primates relatively recently in evolutionary time (Figure 6–35). These highly abundant elements, scattered throughout our genome, must have had major effects on the expression of many of our genes. It is humbling to contemplate how many of our uniquely human qualities we might owe to these genetic parasites.

Viruses Are Fully Mobile Genetic Elements That Can Escape from Cells

Viruses were first noticed as disease-causing agents that, by virtue of their tiny size, passed through ultrafine filters that can hold back even the smallest bacterial cell. We now know that viruses are essentially genes

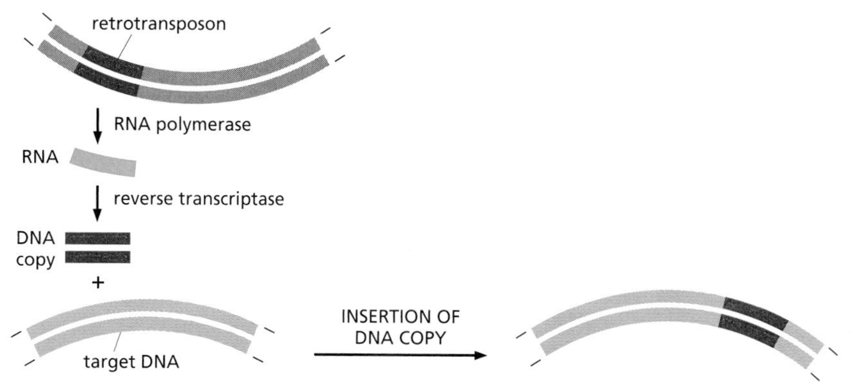

Figure 6–34 **Retrotransposons move via an RNA intermediate.** These transposable elements are first transcribed into an RNA intermediate. A DNA copy of this RNA is then synthesized by the enzyme reverse transcriptase. Next, the DNA copy of the transposon is inserted into the target location, which can be on the same or a different DNA molecule. The donor retrotransposon remains at its original location, so each time it transposes, it duplicates itself. These mobile genetic elements are called retrotransposons because at one stage in their transposition their genetic information flows backward, from RNA to DNA.

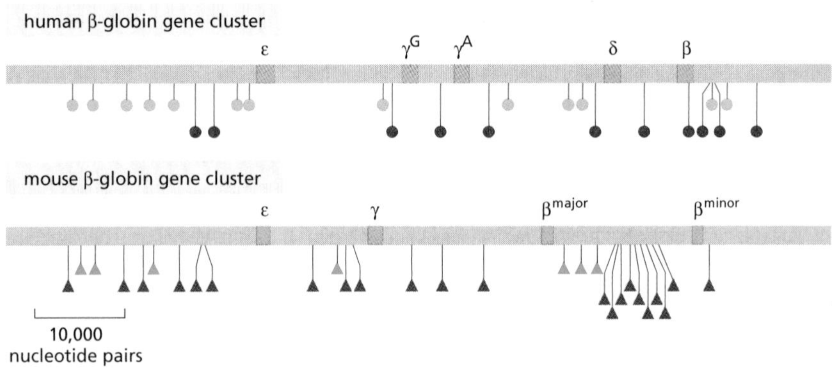

Figure 6–35 ***L1* and *Alu* elements have multiplied to high copy numbers relatively recently in evolutionary time.** The human genome contains a cluster of five globin genes (*top*). Each gene (shown in *orange* and designated by a Greek letter) encodes a protein that carries oxygen in the blood. The comparable region from the mouse genome (*bottom*) contains only four globin genes. The positions of the human *Alu* sequences are indicated by *green* circles, and the human *L1* elements by *red* circles. The mouse genome contains different transposable elements: the positions of *B1* elements (which are related to the human *Alu* sequences) are indicated by *blue* triangles, and the positions of the mouse *L1* elements (which are related to the human *L1*) are indicated by *brown* triangles. Because the DNA sequences of the mouse and human transposable elements are distinct and because the positions of these transposable elements in the β-globin gene cluster differ considerably between human and mouse, it is believed that they accumulated in mammalian genomes relatively recently in evolutionary time. (Courtesy of Ross Hardison and Webb Miller.)

enclosed by a protective protein coat. However, these genes must enter a cell and use the cell's machinery to express their genes to make proteins and to replicate their chromosomes. Virus reproduction is often lethal to the cells in which it occurs; in many cases the infected cell breaks open (lyses), releasing the progeny viruses, which can then attack neighboring cells. Many of the symptoms of viral infections reflect the lytic effect of the virus. The cold sores formed by herpes simplex virus and the blisters caused by the chickenpox virus, for example, both reflect the localized killing of skin cells. Although the first viruses that were discovered attack mammalian cells, it is now recognized that many types of viruses exist. Some of these infect plant cells, while others use bacterial cells as their hosts.

Viral genomes can be made of DNA or RNA and can be single-stranded or double-stranded (Table 6–2). The amount of DNA or RNA that can be packaged inside a protein shell is limited, and it is too small to encode the many different enzymes and other proteins that are required to replicate even the simplest virus. For this reason, viruses must hijack their host's biochemical machinery to reproduce themselves. Viral genomes typically encode the viral coat proteins as well as proteins that co-opt the host enzymes needed to replicate their genome (Figure 6–36).

TABLE 6–2 VIRUSES THAT CAUSE HUMAN DISEASE

VIRUS	GENOME TYPE	DISEASE
Herpes simplex virus	double-stranded DNA	recurrent cold sores
Epstein–Barr virus (EBV)	double-stranded DNA	infectious mononucleosis
Varicella-zoster virus	double-stranded DNA	chickenpox and shingles
Smallpox virus	double-stranded DNA	smallpox
Hepatitis B virus	part single-, part double-stranded DNA	serum hepatitis
Human immunodeficiency virus (HIV)	single-stranded RNA	acquired immune deficiency syndrome (AIDS)
Influenza virus type A	single-stranded RNA	respiratory disease (flu)
Poliovirus	single-stranded RNA	paralysis
Rhinovirus	single-stranded RNA	common cold
Hepatitis A virus	single-stranded RNA	infectious hepatitis
Hepatitis C virus	single-stranded RNA	non-A, non-B type hepatitis
Yellow fever virus	single-stranded RNA	yellow fever
Rabies virus	single-stranded RNA	rabies
Mumps virus	single-stranded RNA	mumps
Measles virus	single-stranded RNA	measles

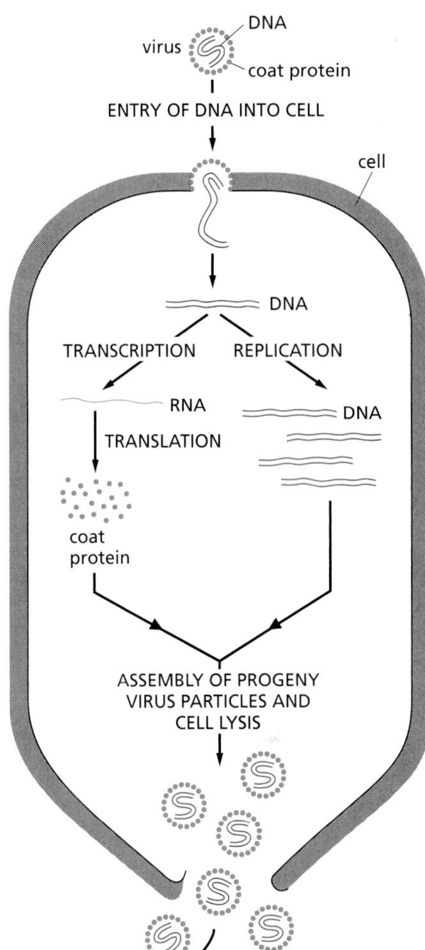

Figure 6–36 **Viruses commandeer the host cell's machinery to reproduce.** The hypothetical simple virus illustrated here consists of a small double-stranded DNA molecule that encodes just a single type of viral coat protein. To replicate, the viral genome must enter the cell. This is followed by replication of the viral DNA to form many copies. At the same time, the viral gene is expressed to make the coat protein through the sequential steps of *transcription* and *translation*, which are described in Chapter 7 (see Figure 7–2). The viral genomes assemble spontaneously with the coat protein to form new virus particles.

The simplest viruses consist of a protein coat made up primarily of many copies of a single polypeptide chain surrounding a small genome composed of as few as three genes. More complex viruses have larger genomes of up to several hundred genes, surrounded by an elaborate shell composed of many different proteins (Figure 6–37).

Retroviruses Reverse the Normal Flow of Genetic Information

Although there are many similarities between bacterial and eucaryotic viruses, one important type of virus—the **retrovirus**—is found only in eucaryotic cells. In many respects, retroviruses resemble the retrotransposons we discussed earlier. A key feature of the life cycle of both these genetic elements is a step in which DNA is synthesized using RNA as a template. The *retro* prefix refers to the reversal of the usual flow of DNA information to RNA (see Figure 7–2) and the enzyme that carries out this step is a *reverse transcriptase*. The retroviral genome (which is single-stranded RNA) encodes this enzyme, and a few molecules of the enzyme are packaged along with the RNA genome in each virus particle.

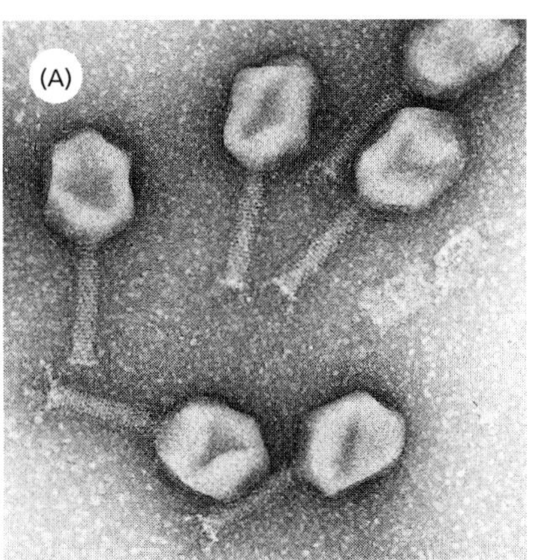

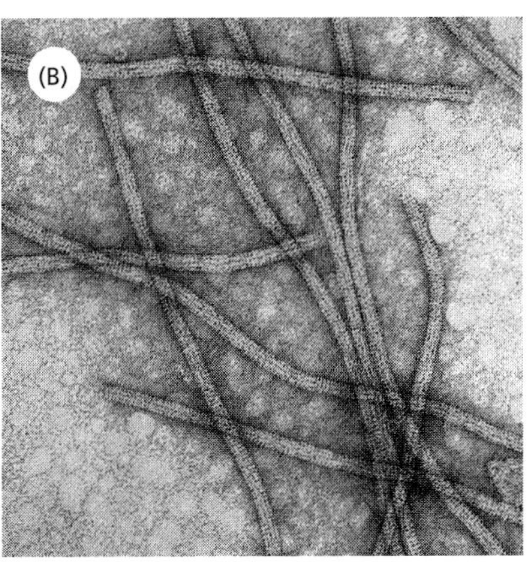

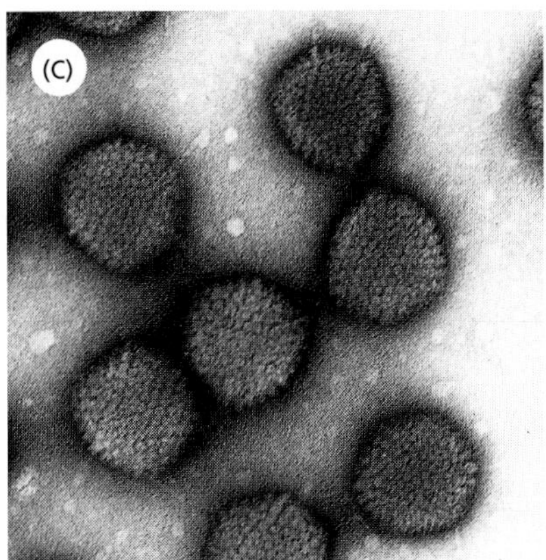

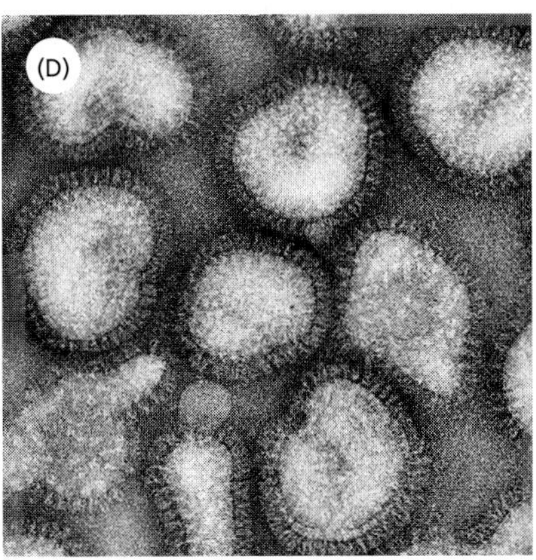

Figure 6–37 **Viruses come in different shapes and sizes.** These electron micrographs of virus particles are all shown at the same scale. (A) T4, a large DNA-containing virus that infects *E. coli* cells. The DNA is stored in the bacteriophage head and injected into the bacterium through the cylindrical tail. (B) Potato virus X, a plant virus that contains an RNA genome. (C) Adenovirus, a DNA-containing virus that can infect human cells. (D) Influenza virus, a large RNA-containing animal virus whose protein coat is further enclosed in a lipid-bilayer-based envelope. The spikes protruding from the envelope are viral proteins embedded in the membrane bilayer (see Figure 6–38). (A, courtesy of James R. Paulson; B, courtesy of Graham Hills; C, courtesy of Mei Lie Wong; D, courtesy of R.C. Williams and H.W. Fisher.)

QUESTION 6–5

Reverse transcriptases do not proofread as they synthesize DNA using an RNA template. What do you think the consequences of this are for the treatment of AIDS?

The life cycle of a retrovirus is shown in Figure 6–38. When the single-stranded RNA genome of the retrovirus enters a cell, the reverse transcriptase brought in with it makes a complementary DNA strand to form a DNA/RNA hybrid double helix. The RNA strand is removed, and the reverse transcriptase (which can use either DNA or RNA as a template) now synthesizes a complementary strand to produce a DNA double helix. This DNA is then inserted, or integrated, into a randomly selected site in the host genome by a virally encoded *integrase* enzyme. In this state, the virus is latent: each time the host cell divides, it passes on a copy of the integrated viral genome, which is known as a *provirus*, to its progeny cells.

The next step in the replication of a retrovirus—which can take place long after its integration into the host genome—is the copying of the integrated viral DNA into RNA by a host-cell RNA polymerase, which produces large numbers of single-stranded RNAs identical to the original infecting genome. The viral genes are then expressed by the host cell machinery to produce the protein shell, the envelope proteins, and reverse transcriptase—all of which are assembled with the RNA genome into new virus particles.

The human immunodeficiency virus (HIV), which is the cause of AIDS, is a retrovirus. As with other viruses of this type, the HIV genome can persist in the latent state as a provirus embedded in the chromosomes of an infected cell. This ability of the virus to hide within host cells complicates attempts to treat the infection with antiviral drugs. But because the HIV reverse transcriptase is not used by cells for any purpose of their own, this enzyme is one of the prime targets of drug development against AIDS.

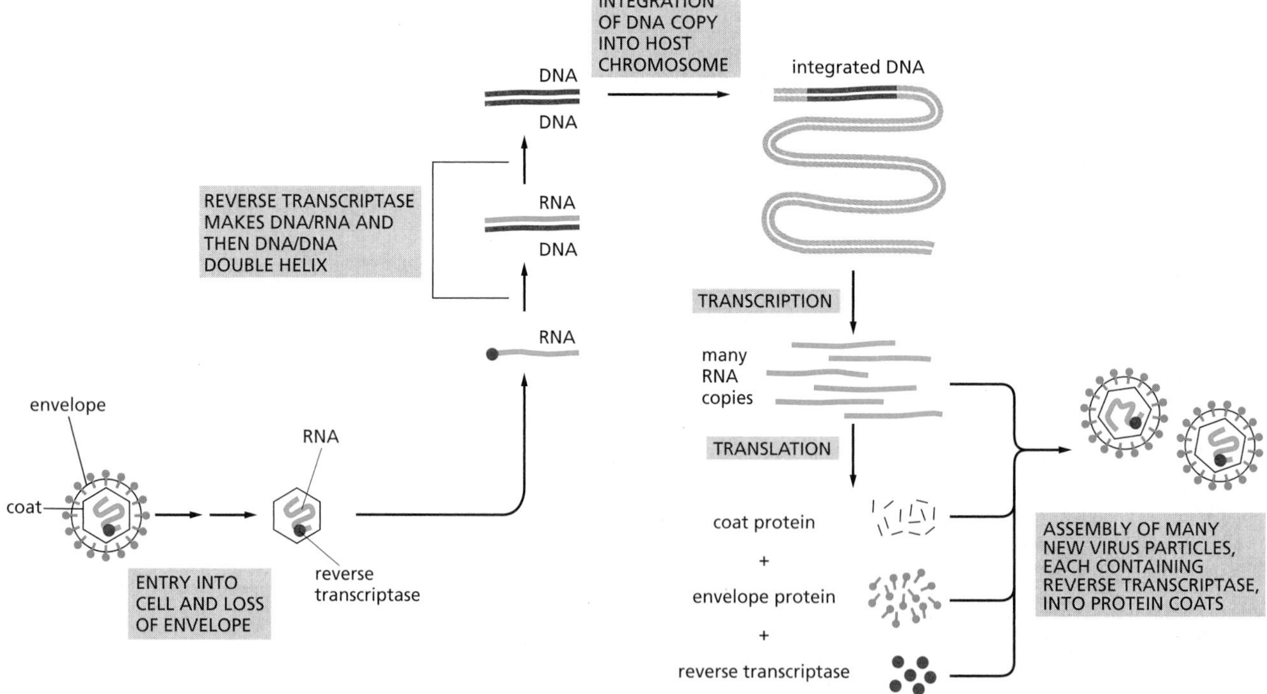

Figure 6–38 **The life cycle of a retrovirus includes reverse transcription and integration into the host genome.** The retrovirus genome consists of an RNA molecule (*blue*) of about 8500 nucleotides. It is packaged inside a protein coat surrounded by a lipid bilayer (called an envelope) which contains virus-encoded envelope proteins (*green*). The enzyme reverse transcriptase (*red* circle), encoded in the viral genome and packaged with its RNA, first makes a DNA copy of the viral RNA molecule and then a second DNA strand, generating a double-stranded DNA copy of the RNA genome. The integration of this DNA double helix into the host chromosome is then catalyzed by a virus-encoded *integrase* enzyme. This integration is required for the synthesis of new viral RNA molecules by a host-cell RNA polymerase.

ESSENTIAL CONCEPTS

- The ability of a cell to maintain order in a chaotic environment depends on the accurate duplication of the vast quantity of genetic information carried in its DNA.
- Each of the two DNA strands can act as a template for the synthesis of the other strand. A DNA double helix thus carries the same information in each of its strands.
- A DNA molecule is duplicated (replicated) by the polymerization of new complementary strands using each of the old strands of the DNA double helix as a template. Two identical DNA molecules are formed, enabling the genetic information to be copied and passed on from cell to daughter cell and from parent to offspring.
- As a DNA molecule replicates, its two strands are pulled apart to form one or more Y-shaped replication forks. DNA polymerase enzymes, situated at the fork, produce a new complementary DNA strand on each parental strand.
- DNA polymerase replicates a DNA template with remarkable fidelity, making less than one error in every 10^7 bases read. This accuracy is made possible, in part, by a proofreading process in which the enzyme removes its own polymerization errors as it moves along the DNA.
- Because DNA polymerase synthesizes new DNA in only one direction, only the leading strand at the replication fork can be synthesized in a continuous fashion. On the lagging strand, DNA is synthesized in a discontinuous backstitching process, producing short fragments of DNA that are later joined by the enzyme DNA ligase to complete that DNA strand.
- DNA polymerase is incapable of starting a new DNA chain from scratch. Instead, DNA synthesis is primed by an RNA polymerase called primase, which makes short lengths of RNA (primers) that are elongated by DNA polymerase. The primers are subsequently erased and replaced with DNA.
- DNA replication requires the cooperation of many proteins; these form a multienzyme replication machine that catalyzes DNA synthesis.
- In eucaryotes, a special enzyme called telomerase replicates the DNA at the ends of the chromosomes.
- The rare copying mistakes that slip through the DNA replication machinery are dealt with by the mismatch repair proteins. The overall accuracy of DNA replication, including mismatch repair, is one mistake per 10^9 nucleotides copied.
- DNA damage caused by unavoidable chemical reactions is repaired by a variety of enzymes that recognize damaged DNA and excise a short stretch of the DNA strand that contains it. The missing DNA is resynthesized by a repair DNA polymerase that uses the undamaged strand as a template.
- Nonhomologous end-joining allows the rapid repair of double-strand DNA breaks; the process often alters the DNA sequence at the site of the repair.
- Homologous recombination can faithfully repair double-strand DNA breaks using a homologous chromosome sequence as a guide. During meiosis, a related homologous recombination process causes a shuffling of genetic information that creates DNA molecules with novel sequences.
- Mobile genetic elements, or transposons, move from place to place in the genomes of their hosts, providing a source of genetic variation.

- Nearly half of the human genome consists of mobile genetic elements. Two classes of these elements have multiplied to especially high copy numbers.
- Viruses are little more than genes packaged in protective protein coats. They require host cells to reproduce themselves.
- Some viruses have RNA instead of DNA as their genomes. One group of RNA viruses—the retroviruses—must copy their RNA genomes into DNA to replicate.

KEY TERMS

DNA polymerase	replication fork
DNA repair	replication origin
DNA replication	retrotransposon
homologous recombination	retrovirus
lagging strand	reverse transcriptase
leading strand	RNA (ribonucleic acid)
mobile genetic element	telomerase
mutation	template
nonhomologous end-joining	transposon
Okazaki fragment	virus
proofreading	

QUESTIONS

QUESTION 6–6

DNA mismatch repair enzymes preferentially repair bases on the newly synthesized DNA strand, using the old DNA strand as a template. If mismatches were simply repaired without regard for which strand served as template, would this reduce replication errors? Explain your answer.

QUESTION 6–7

Suppose a mutation affects an enzyme that is required to repair the damage to DNA caused by the loss of purine bases. This mutation causes the accumulation of about 5000 mutations in the DNA of each of your cells per day. As the average difference in DNA sequence between humans and chimpanzees is about 1%, how long will it take you to turn into an ape? What is wrong with this argument?

QUESTION 6–8

Which of the following statements are correct? Explain your answers.

A. A bacterial replication fork is asymmetrical because it contains two DNA polymerase molecules that are structurally distinct.

B. Okazaki fragments are removed by a nuclease that degrades RNA.

C. The error rate of DNA replication is reduced both by proofreading by DNA polymerase and by DNA mismatch repair.

D. In the absence of DNA repair, genes are unstable.

E. None of the aberrant bases formed by deamination occur naturally in DNA.

F. Cancer can result from the accumulation of mutations in somatic cells.

QUESTION 6–9

The speed of DNA replication at a replication fork is about 100 nucleotides per second in human cells. What is the minimum number of origins of replication that a human cell must have if it is to replicate its DNA once every 24 hours? Recall that a human cell contains two copies of the human genome, one inherited from the mother, the other from the father, each consisting of 3×10^9 nucleotide pairs.

QUESTION 6–10

Look carefully at the structures of the compounds shown in Figure Q6–10. One or the other of the two compounds is added to a DNA replication reaction.

A. What would you expect if compound A were added in large excess over the concentration of the available deoxycytosine triphosphate (dCTP)?

B. What would happen if it were added at 10% of the concentration of the available dCTP?

C. What effects would you expect if compound B were added under the same conditions?

QUESTION 6–11

The genetic material of a hypothetical organism is structurally indistinguishable from conventional DNA. Surprisingly, analyses reveal that the DNA is synthesized

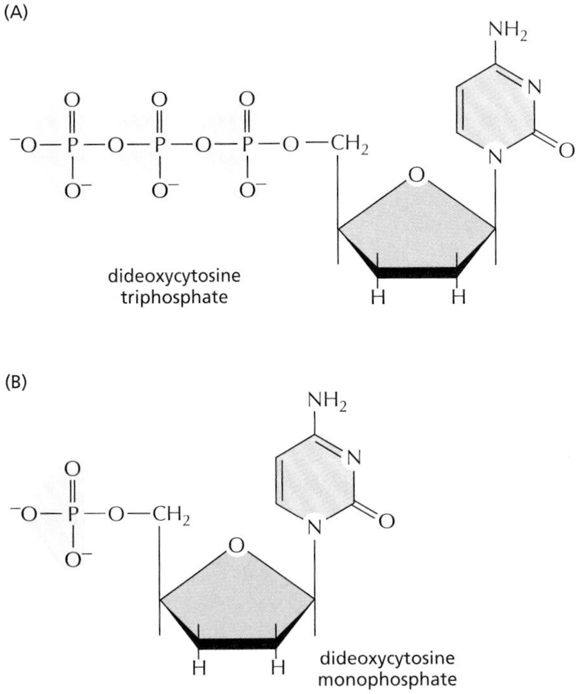

(A)

dideoxycytosine triphosphate

(B)

dideoxycytosine monophosphate

Figure Q6–10

from nucleoside triphosphates that contain free 5′-hydroxyl groups and triphosphate groups at the 3′ position. In what way must this organism's DNA polymerase differ from that of normal cells? Could it still proofread?

QUESTION 6–12

Figure Q6–12 shows a snapshot of a replication fork in which the RNA primer has just been added to the lagging strand. Using this diagram as a guide, sketch the path of the DNA as the next Okazaki fragment is synthesized. Indicate the sliding clamp and the single-strand binding protein as appropriate.

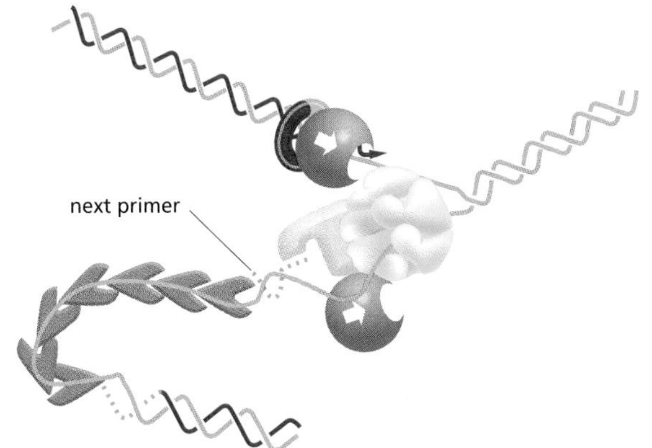

Figure Q6–12

QUESTION 6–13

Approximately how many high-energy bonds are used to replicate a bacterial chromosome? Compared with its own dry weight of 10^{-12} g, how much glucose does a single bacterium need to provide enough energy to copy its DNA once? The number of nucleotide pairs in the bacterial chromosome is 3×10^6. Oxidation of one glucose molecule yields about 30 high-energy phosphate bonds. The molecular weight of glucose is 180 g/mole. (Recall from Figure 2–3 that a mole consists of 6×10^{23} molecules.)

QUESTION 6–14

What, if anything, is wrong with the following statement: "DNA stability in both reproductive cells and somatic cells is essential for the survival of a species." Explain your answer.

QUESTION 6–15

A common type of error in DNA is produced by a spontaneous reaction termed *deamination* in which a nucleotide base loses an amino group (NH_2), which is replaced by a keto group (C=O) by the general reaction shown in Figure Q6–15. Write the structures of the bases A, G, C, T, and U and predict the products that will be produced by deamination. By looking at the products of this reaction—and remembering that, in the cell, these will need to be recognized and repaired—can you propose an explanation why DNA cannot contain uracil?

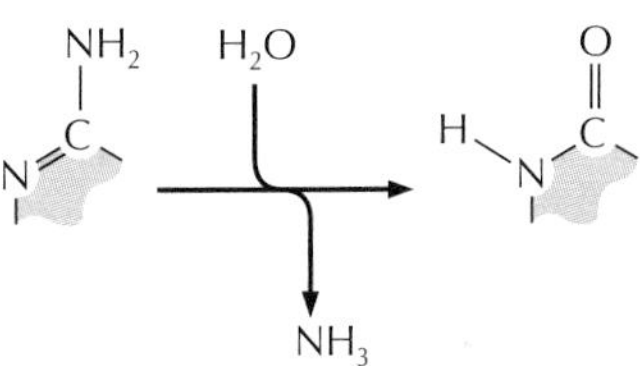

Figure Q6–15

QUESTION 6–16

A. Explain why telomeres and telomerase are needed for replication of eucaryotic chromosomes but not for replication of a circular bacterial chromosome. Draw a diagram to illustrate your explanation.

B. Would you still need telomeres and telomerase to complete eucaryotic chromosome replication if DNA primase always laid down the RNA primer at the very 3′ end of the template for the lagging strand?

QUESTION 6–17

Discuss the following statement: "Viruses exist in the twilight zone of life: outside cells they are simply dead assemblies of molecules; inside cells, however, they are alive."

QUESTION 6–18

Many transposons move within a genome by replicative mechanisms (such as those shown in Figures 6–33 and 6–34). They therefore increase in copy number each time they transpose. Although individual transposition events are rare, many transposons are found in multiple copies in genomes. What do you suppose keeps the transposons from completely overrunning their hosts' genomes?

QUESTION 6–19

Describe the consequences that would arise if a eucaryotic chromosome

A. Contained only one origin of replication:

(i) At the exact center of the chromosome

(ii) At one end of the chromosome

B. Lacked one or both telomeres

C. Had no centromere

Assume that the chromosome is 150 million nucleotide pairs in length, a typical size for an animal chromosome, and that DNA replication in animal cells proceeds at about 100 nucleotides per second.

CHAPTER **SEVEN**

7

From DNA to Protein: How Cells Read the Genome

FROM DNA TO RNA

FROM RNA TO PROTEIN

RNA AND THE ORIGINS OF LIFE

Once the structure of DNA (deoxyribonucleic acid) had been determined in the early 1950s, it became clear that the hereditary information in cells is encoded in DNA's sequence of nucleotides. We saw in Chapter 6 how this information can be passed on unchanged from a cell to its descendants through the process of DNA replication. But how does the cell decode and use the information? How do genetic instructions written in an alphabet of just four 'letters'—the four different nucleotides in DNA—direct the formation of a bacterium, a fruit fly, or a human? We still have a lot to learn about how the information stored in an organism's genes produces even the simplest unicellular bacterium, let alone how it directs the development of complex multicellular organisms like ourselves. But the DNA code itself has been deciphered, and the language of genes can be read.

Even before the DNA code had been broken, it was known that the information contained in genes somehow directed the synthesis of proteins. Proteins are the principal constituents of cells and determine not only their structure but also their functions. In previous chapters, we have encountered some of the thousands of different kinds of proteins that cells can make. We saw in Chapter 4 that the properties and function of a protein molecule are determined by the linear order—the *sequence*—of the different amino acid subunits in its polypeptide chain: each type of protein has its own unique amino acid sequence, and this sequence dictates how the chain will fold to give a molecule with a distinctive shape and chemistry. The genetic instructions carried by DNA must therefore specify the amino acid sequences of proteins. We shall see in this chapter exactly how this is done.

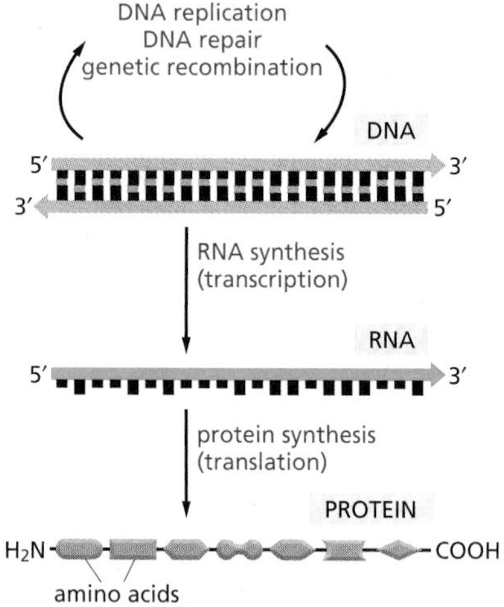

Figure 7–1 **Genetic information directs the synthesis of protein.** The flow of genetic information from DNA to RNA (transcription) and from RNA to protein (translation) occurs in all living cells.

DNA does not direct protein synthesis itself, but acts rather like a manager, delegating the various tasks to a team of workers. When a particular protein is needed by the cell, the nucleotide sequence of the appropriate section of an immensely long DNA molecule in a chromosome is first copied into another type of nucleic acid—*RNA* (*ribonucleic acid*). These RNA copies of short segments of the DNA are then used to direct the synthesis of the protein. Many thousands of these conversions from DNA to protein occur each second in every cell in our bodies. The flow of genetic information in cells is therefore from DNA to RNA to protein (Figure 7–1). All cells, from bacteria to humans, express their genetic information in this way—a principle so fundamental that it has been termed the *central dogma* of molecular biology.

In this chapter, we explain the mechanisms by which cells copy DNA into RNA (a process called *transcription*) and then use the information in RNA to make protein (a process called *translation*). We also introduce a few of the key variations on this basic scheme. Principal among these is *RNA splicing*, a process whereby RNA transcripts are cut and stitched back together before eucaryotic cells translate them into proteins. These alterations can change the message conveyed by RNA molecules and are therefore crucial for understanding how cells decode the genome. In the final section of this chapter, we consider how the present scheme of information storage, transcription, and translation might have arisen from simpler systems in the earliest stages of cellular evolution.

FROM DNA TO RNA

Transcription and translation are the means by which cells read out, or *express*, their genetic instructions—their *genes*. Many identical RNA copies can be made from the same gene, and each RNA molecule can direct the synthesis of many identical protein molecules. Because each cell contains only one or two copies of any particular gene, this successive amplification enables cells to rapidly synthesize large amounts of protein whenever necessary. At the same time, each gene can be transcribed and translated with a different efficiency, providing the cell with a way to make vast quantities of some proteins and tiny quantities of others (Figure 7–2). Moreover, as we shall see in Chapter 8, a cell can change (or regulate) the expression of each of its genes according to the needs of the moment. In this section, we discuss the production of RNA, the first step in *gene expression*.

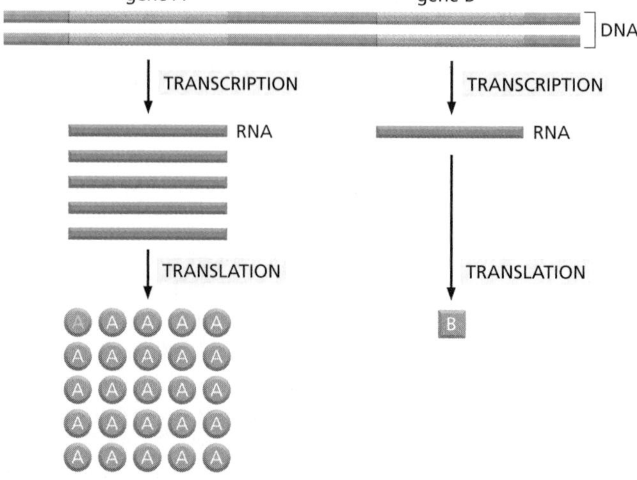

Figure 7–2 **Genes can be expressed with different efficiencies.** Gene A is transcribed and translated much more efficiently than gene B. This allows the amount of protein A in the cell to be much higher than that of protein B. In this and later figures, the untranscribed portions of the DNA are shown in *gray*.

Portions of DNA Sequence Are Transcribed into RNA

The first step a cell takes in reading out one of its many thousands of genes is to copy the nucleotide sequence of that gene into RNA. The process is called **transcription** because the information, though copied into another chemical form, is still written in essentially the same language—the language of nucleotides. Like DNA, **RNA** is a linear polymer made of four different types of nucleotide subunits linked together by phosphodiester bonds (Figure 7–3). It differs from DNA chemically in two respects: (1) the nucleotides in RNA are *ribonucleotides*—that is, they contain the sugar ribose (hence the name *ribo*nucleic acid) rather than deoxyribose; (2) although, like DNA, RNA contains the bases adenine (A), guanine (G), and cytosine (C), it contains uracil (U) instead of the thymine (T) found in DNA. Because U, like T, can base-pair by hydrogen-bonding with A (Figure 7–4), the complementary base-pairing properties described for DNA in Chapter 5 apply also to RNA.

Although their chemical differences are small, DNA and RNA differ quite dramatically in overall structure. Whereas DNA always occurs in cells as a double-stranded helix, RNA is single-stranded. This difference has important functional consequences. Because an RNA chain is single-stranded, it can fold up into a variety of shapes, just as a polypeptide chain folds up to form the final shape of a protein (Figure 7–5); double-stranded DNA cannot fold in this fashion. As we shall see later in this chapter, the ability to fold into a complex three-dimensional shape allows RNA to carry out functions in cells in addition to conveying information between DNA and protein. Whereas DNA functions solely as an information store, RNA comes in different varieties, some having structural and even catalytic functions.

QUESTION 7–1

Consider the expression "central dogma," which refers to the flow of genetic information from DNA to RNA to protein. Is the word "dogma" appropriate in this context?

(A) ribose — used in ribonucleic acid (RNA); deoxyribose — used in deoxyribonucleic acid (DNA)

(B) uracil — used in RNA; thymine — used in DNA

(C) 5′ end … 3′ end; bases C, A, U, G; ribose

Figure 7–3 **The chemical structure of RNA differs slightly from that of DNA.** (A) RNA contains the sugar ribose, which differs from deoxyribose, the sugar used in DNA, by the presence of an additional –OH group. (B) RNA contains the base uracil, which differs from thymine, the equivalent base in DNA, by the absence of a $-CH_3$ group. (C) A short length of RNA. The chemical linkage between nucleotides in RNA is the same as that in DNA.

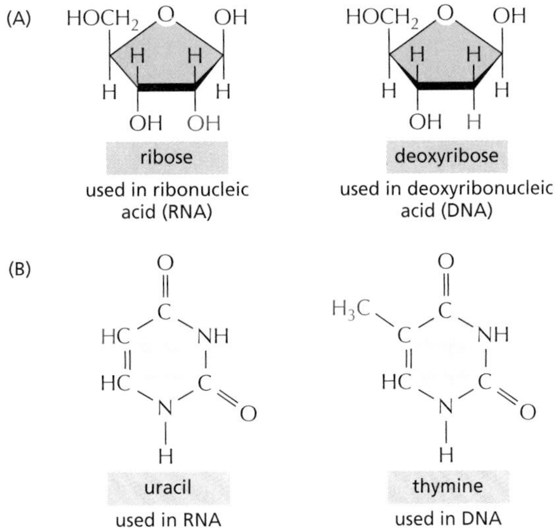

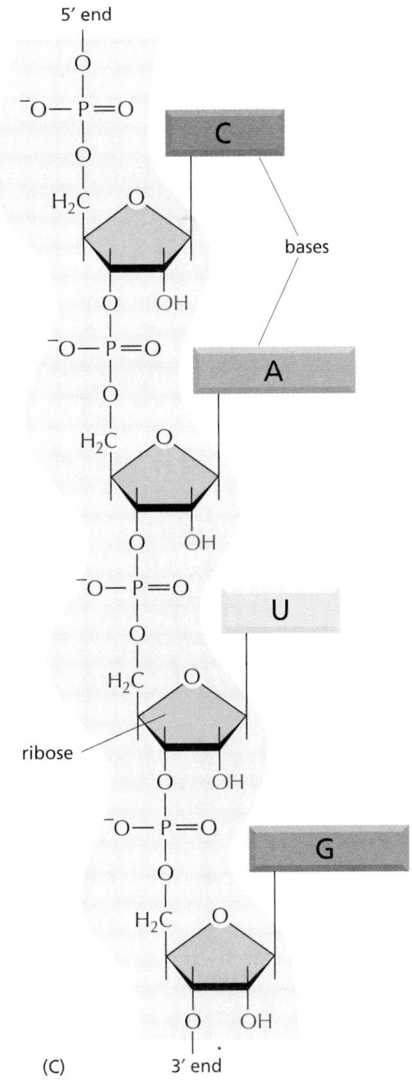

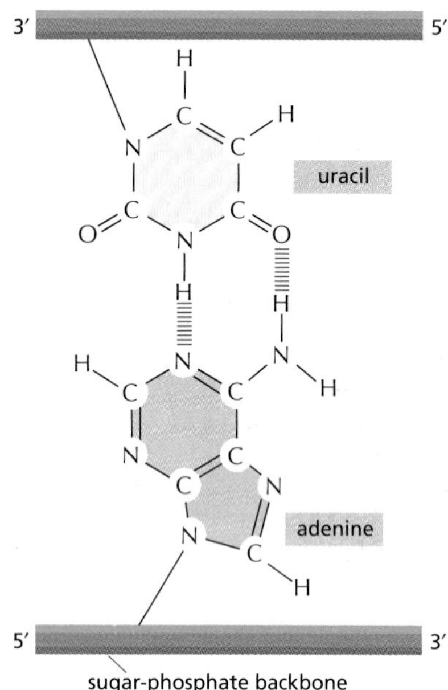

Figure 7–4 **Uracil forms a base pair with adenine.** Despite the absence of a methyl group, uracil has the same base-pairing properties as thymine. Thus, U-A base pairs closely resemble T-A base pairs (see Figure 5–6A).

Transcription Produces RNA Complementary to One Strand of DNA

All of the RNA in a cell is made by transcription, a process that has certain similarities to DNA replication (discussed in Chapter 6). Transcription begins with the opening and unwinding of a small portion of the DNA double helix to expose the bases on each DNA strand. One of the two strands of the DNA double helix then acts as a template for the synthesis of RNA. Ribonucleotides are added, one by one, to the growing RNA chain, and as in DNA replication, the nucleotide sequence of the RNA chain is determined by complementary base-pairing with the DNA template. When a good match is made, the incoming ribonucleotide is covalently linked to the growing RNA chain in an enzymatically catalyzed reaction. The RNA chain produced by transcription—the *transcript*—is therefore elongated one nucleotide at a time and has a nucleotide sequence exactly complementary to the strand of DNA used as the template (Figure 7–6).

Transcription, however, differs from DNA replication in several crucial features. Unlike a newly formed DNA strand, the RNA strand does not remain hydrogen-bonded to the DNA template strand. Instead, just behind the region where the ribonucleotides are being added, the RNA chain is displaced and the DNA helix re-forms. For this reason—and because only one strand of the DNA molecule is transcribed—RNA molecules are single-stranded. Further, as RNAs are copied from only a limited region of DNA, these molecules are much shorter than DNA molecules; DNA molecules in a human chromosome can be up to 250 million nucleotide pairs long, whereas most RNAs are no more than a few thousand nucleotides long, and many are much shorter than that.

The enzymes that carry out transcription are called **RNA polymerases**. Like the DNA polymerase that carries out DNA replication (discussed in Chapter 6), RNA polymerases catalyze the formation of the phosphodiester

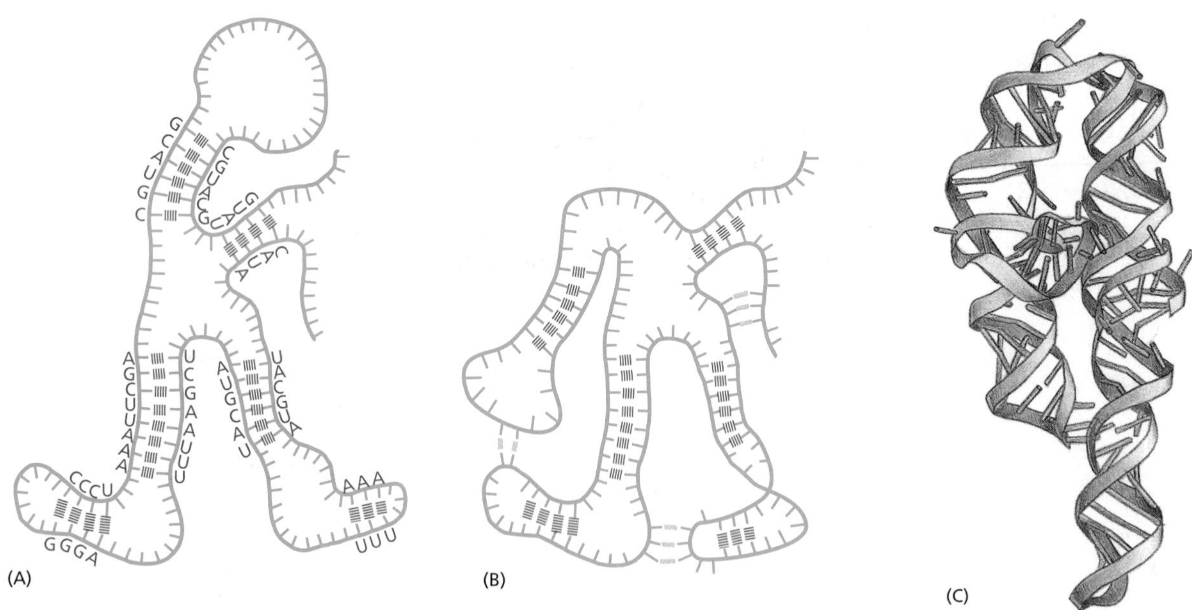

Figure 7–5 **RNA molecules can form intramolecular base pairs and fold into specific structures.** RNA is single-stranded, but it often contains short stretches of nucleotides that can base-pair with complementary sequences found elsewhere on the same molecule. These interactions, along with "nonconventional" base-pair interactions, allow an RNA molecule to fold into a three-dimensional structure that is determined by its sequence of nucleotides. (A) A diagram of a hypothetical, folded RNA structure showing only conventional (that is, Watson–Crick) base-pairing interactions; (B) the same hypothetical structure with both conventional (*red*) and nonconventional (e.g., A-G) base-pair interactions (*green*); (C) structure of an actual RNA molecule that is involved in RNA splicing. Each conventional base-pair interaction is indicated by a "rung" in the double helix. Bases in other configurations are indicated by broken rungs. For an additional view of RNA structure, see Movie 7.1.

bonds that link the nucleotides together and form the sugar–phosphate backbone of the RNA chain. The RNA polymerase moves stepwise along the DNA, unwinding the DNA helix just ahead to expose a new region of the template strand for complementary base-pairing. In this way, the growing RNA chain is extended by one nucleotide at a time in the 5′-to-3′ direction (Figure 7–7). The enzyme uses ribonucleoside triphosphates (ATP, CTP, UTP, and GTP), whose high-energy bonds provide the energy that drives the reaction forward (see Figure 6–10).

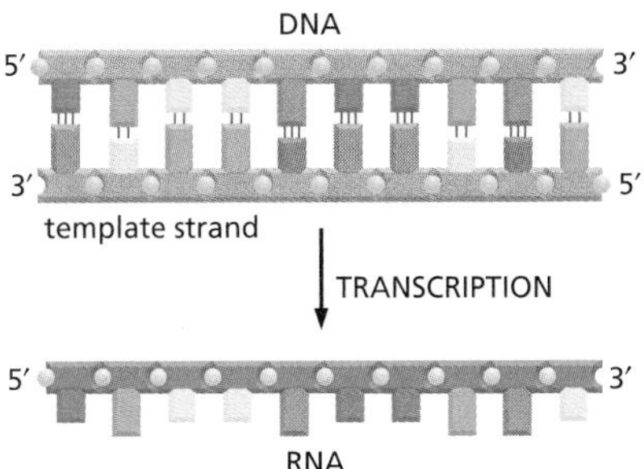

Figure 7–6 **Transcription produces an RNA complementary to one strand of DNA.** The nontemplate strand of the DNA (the *top* strand in this example) is sometimes called the *coding strand* because its sequence is equivalent to the RNA product.

The almost immediate release of the RNA strand from the DNA as it is synthesized means that many RNA copies can be made from the same gene in a relatively short time; the synthesis of the next RNA is usually started before the first RNA has been completed (Figure 7–8). A medium-sized gene (say, 1500 nucleotide pairs) requires approximately 50 seconds for a molecule of RNA polymerase to transcribe it (Movie 7.2). At any given time, there could be as many as 15 polymerases speeding along this single stretch of DNA, hard on one another's heels, allowing more than 1000 transcripts to be synthesized in an hour. For most genes, however, the amount of transcription is much less than this.

Although RNA polymerase catalyzes essentially the same chemical reaction as DNA polymerase, there are some important differences between the two enzymes. First, and most obviously, RNA polymerase catalyzes the linkage of ribonucleotides, not deoxyribonucleotides. Second, unlike the DNA polymerase involved in DNA replication, RNA polymerases can start an RNA chain without a primer. This difference may exist because transcription need not be as accurate as DNA replication; unlike DNA, RNA is not used as the permanent storage form of genetic information in cells, so mistakes in RNA transcripts have relatively minor consequences. RNA polymerases make about one mistake for every 10^4 nucleotides copied into RNA, compared with an error rate for DNA polymerase of about one in 10^7 nucleotides.

Several Types of RNA Are Produced in Cells

The vast majority of genes carried in a cell's DNA specify the amino acid sequence of proteins, and the RNA molecules that are copied from these genes (and that ultimately direct the synthesis of proteins) are collectively called **messenger RNA (mRNA)**. In eucaryotes, each mRNA typically carries information transcribed from just one gene, coding for a single protein; in bacteria, a set of adjacent genes is often transcribed as a single mRNA that therefore carries the information for several different proteins.

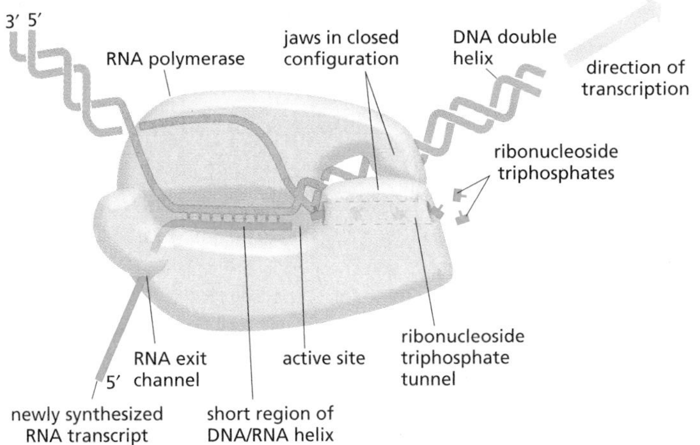

Figure 7–7 **DNA is transcribed by the enzyme RNA polymerase.** The RNA polymerase (*pale blue*) moves stepwise along the DNA, unwinding the DNA helix in front of it. As it progresses, the polymerase adds nucleotides (*small "T" shapes*) one by one to the RNA chain at the polymerization site using an exposed DNA strand as a template. The resulting RNA transcript is thus single-stranded and complementary to one of the two DNA strands. As it moves along the DNA template, the polymerase displaces the newly formed RNA, allowing the two strands of DNA behind the polymerase to rewind. A short region of hybrid DNA/RNA helix (approximately nine nucleotides in length) therefore forms only transiently, causing a 'window' of DNA/RNA helix to move along the DNA with the polymerase (Movie 7.2).

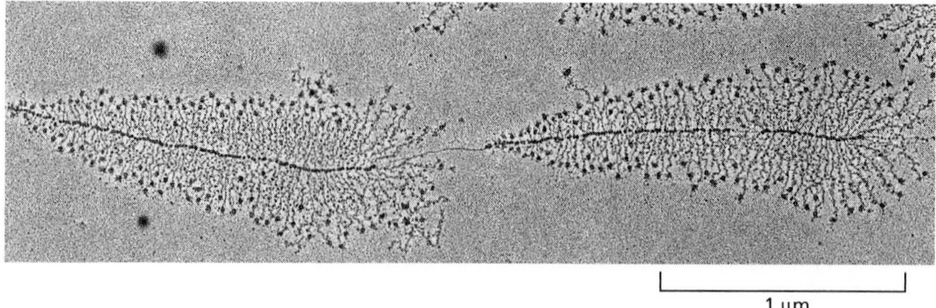

Figure 7–8 **Transcription can be visualized in the electron microscope.** The micrograph shows many molecules of RNA polymerase simultaneously transcribing two adjacent genes. Molecules of RNA polymerase are visible as a series of dots along the DNA with the transcripts (fine threads) attached to them. The RNA molecules (called rRNAs) transcribed from the genes shown in this example are not translated into protein but are instead used directly as components of ribosomes, the machines on which translation takes place. The particles at the 5′ end (the free end) of each rRNA transcript are believed to be ribosomal proteins that have assembled on the rRNA. (Courtesy of Ulrich Scheer.)

The final product of other genes, however, is the RNA itself (Table 7–1). As we shall see in later sections of this chapter, these nonmessenger RNAs, like proteins, serve as regulatory, structural, and enzymatic components of cells, and they play key parts in translating the genetic message into protein. *Ribosomal RNA* (*rRNA*) forms the core of the ribosomes, on which mRNA is translated into protein, and *transfer RNA* (*tRNA*) forms the adaptors that select amino acids and hold them in place on a ribosome for their incorporation into protein. Other small RNAs, called *microRNAs* (*miRNAs*), serve as key regulators of eucaryotic gene expression, as we discuss in Chapter 8.

In the broadest sense, the term **gene expression** refers to the process by which the information encoded in a DNA sequence is translated into a product that has some effect on a cell or organism. In cases where the final product of the gene is a protein, gene expression includes both transcription and translation. When an RNA molecule is the gene's final product, however, gene expression does not require translation.

QUESTION 7–2

In the electron micrograph in Figure 7–8, are the RNA polymerase molecules moving from right to left or from left to right? Why are the RNA transcripts so much shorter than the DNA that encodes them?

Signals in DNA Tell RNA Polymerase Where to Start and Finish

The initiation of transcription is an especially critical process because it is the main point at which the cell can select which proteins or RNAs are to be produced and at what rate. To begin transcription, RNA polymerase must be able to recognize the start of a gene and bind firmly to the DNA at this site. The way in which RNA polymerases recognize the transcription start site differs somewhat between bacteria and eucaryotes. Because the situation in bacteria is simpler, we describe it first.

When an RNA polymerase collides randomly with a piece of DNA, it sticks weakly to the double helix and then slides rapidly along. The enzyme latches on tightly only after it has encountered a region called a **promoter**, which contains a specific sequence of nucleotides indicating the starting point for RNA synthesis. Once the RNA polymerase reaches the promoter and binds tightly to the DNA, it opens up the double helix immediately in front of it to expose the nucleotides on each strand of a short stretch of DNA (Figure 7–9). One of the two exposed DNA strands

TABLE 7–1 TYPES OF RNA PRODUCED IN CELLS

TYPE OF RNA	FUNCTION
mRNAs	code for proteins
rRNAs	form the core of the ribosome and catalyze protein synthesis
miRNAs	regulate gene expression
tRNAs	serve as adaptors between mRNA and amino acids during protein synthesis
Other small RNAs	used in RNA splicing, telomere maintenance, and many other processes

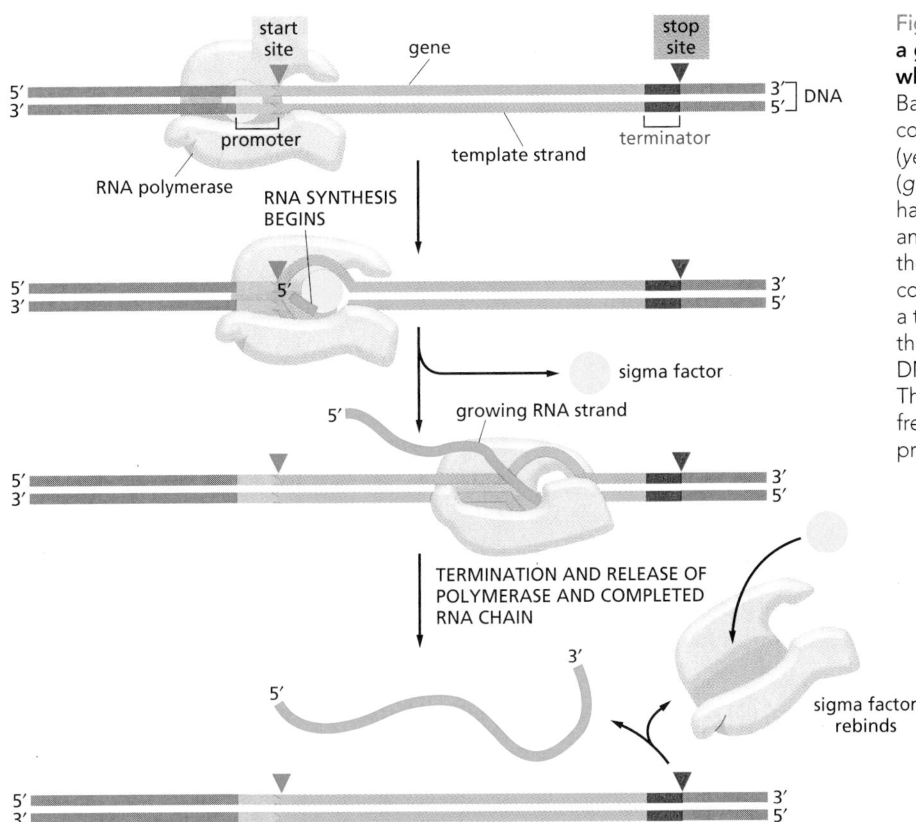

Figure 7–9 **Signals in the sequence of a gene tell bacterial RNA polymerase where to start and stop transcription.** Bacterial RNA polymerase (*light blue*) contains a subunit called the sigma factor (*yellow*) that recognizes the promoter (*green*) on the DNA. Once transcription has begun, the sigma factor is released and the polymerase continues synthesizing the RNA without it. Chain elongation continues until the polymerase encounters a termination signal (*red*) in the DNA. There the enzyme halts and releases both the DNA template and the newly made mRNA. The polymerase then reassociates with a free sigma factor and searches for another promoter to begin the process again.

then acts as a template for complementary base-pairing with incoming ribonucleotides, two of which are joined together by the polymerase to begin the RNA chain. Chain elongation then continues until the enzyme encounters a second signal in the DNA, the *terminator* (or stop site), where the polymerase halts and releases both the DNA template and the newly made RNA chain (see Figure 7–9).

A subunit of the bacterial polymerase, called *sigma* (σ) *factor,* is primarily responsible for recognizing the promoter sequence on DNA. Once the polymerase has latched onto the DNA at this site and has synthesized approximately 10 nucleotides of RNA, the sigma factor disengages, enabling the polymerase to move forward and continue transcribing without it. After the polymerase has been released at a terminator, it reassociates with a free sigma factor and searches for a promoter, where it can begin the process of transcription again.

The polymerase protein can recognize the promoter, even though the DNA is in its double-helical form, by making specific contacts with the portions of the bases that are exposed on the outside of the helix. The nucleotide sequences of a typical promoter—and a typical terminator—are presented in Figure 7–10.

Because DNA is double-stranded a promoter could, in principle, direct the synthesis of two different RNA transcripts, one transcribed left-to-right and the other right-to-left. However, the promoter is asymmetrical and binds the polymerase in only one orientation; thus, once properly positioned on a promoter, the RNA polymerase has no option but to transcribe the appropriate DNA strand, since transcription can proceed only in the 5′-to-3′ direction. Overall, the direction of transcription with respect to the chromosome as a whole varies from gene to gene (Figure 7–11). Because tight binding is required for RNA polymerase to begin transcription, a segment of DNA will be transcribed only if it is preceded by a

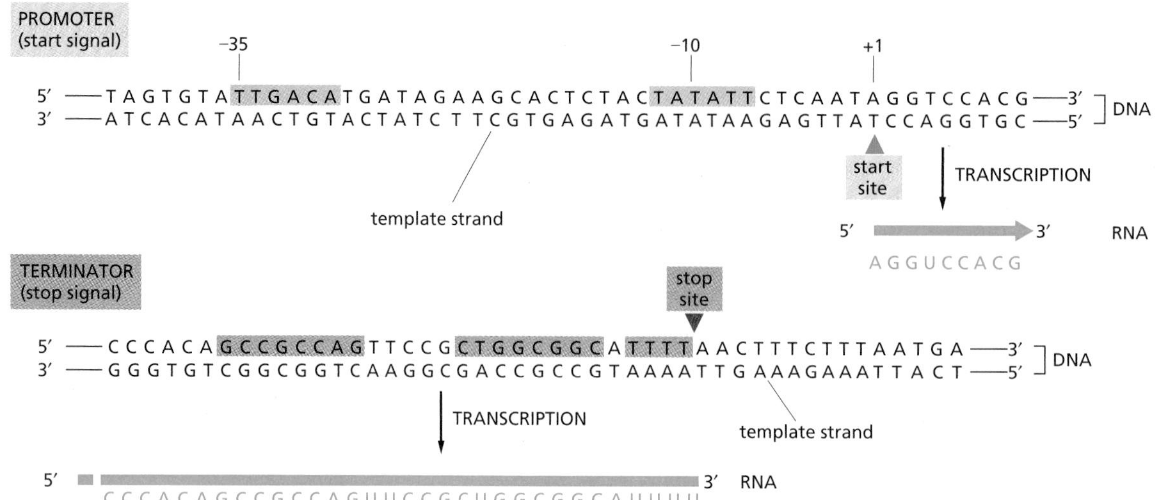

Figure 7–10 **Bacterial promoters and terminators have specific nucleotide sequences that are recognized by RNA polymerase.** The *green*-shaded regions in the top diagram represent the DNA sequences that specify a promoter. The numbers represent the positions of nucleotides counting from the first nucleotide transcribed, which is designated +1. The asymmetry of the promoter—for example, the fact that the conserved sequence at –35 lies upstream of the –10 sequence—orients the polymerase and determines the direction of transcription. All bacterial promoters contain DNA sequences at –10 and –35 that closely resemble those shown here. The *red*-shaded regions in the bottom diagram represent sequences that signal the RNA polymerase to terminate transcription.

promoter sequence. This ensures that only those parts of a DNA molecule that contain a gene will be transcribed into RNA.

Initiation of Eucaryotic Gene Transcription Is a Complex Process

Many of the principles outlined above for bacterial transcription also apply to eucaryotes. However, transcription initiation in eucaryotes differs in several important ways from that in bacteria:

- The first difference lies in the RNA polymerases themselves. While bacteria contain a single type of RNA polymerase, eucaryotic cells have three—*RNA polymerase I*, *RNA polymerase II*, and *RNA polymerase III*. These polymerases are responsible for transcribing different types of genes. RNA polymerases I and III transcribe the genes encoding transfer RNA, ribosomal RNA, and small RNAs that play structural and catalytic roles in the cell (Table 7–2). RNA polymerase II transcribes the vast majority of eucaryotic genes, including all those that encode proteins (Movie 7.3). Our subsequent discussion will therefore focus on this enzyme.
- A second difference is that whereas the bacterial RNA polymerase (along with its sigma subunit) is able to initiate transcription on its own, eucaryotic RNA polymerases require the assistance of a large set of accessory proteins. Principal among these are the *general transcription factors*, which must assemble at each promoter along with the polymerase before the polymerase can begin transcription.
- A third distinctive feature of transcription in eucaryotes is that the mechanisms that control its initiation are much more elaborate than those in prokaryotes—a point we discuss in detail in Chapter 8. In bacteria, genes tend to lie very close to one another in the DNA, with only very short lengths of nontranscribed DNA between them. But in plants and animals, including humans, individual genes are spread out along the DNA, with stretches of up

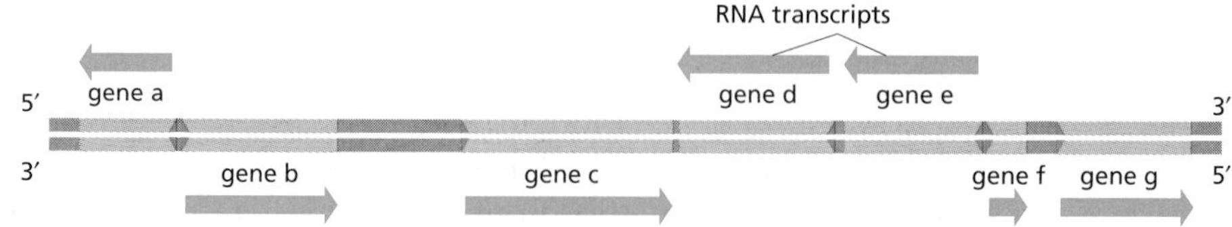

Figure 7–11 **Some genes are transcribed using one DNA strand as a template, whereas others are transcribed using the other DNA strand.** The direction of transcription is determined by the orientation of the promoter at the beginning of each gene (*green* arrowheads). The genes transcribed from left to right use the bottom DNA strand as the template; those transcribed from right to left use the top strand as the template (see Figure 7–10). Approximately 0.2% (10,000 nucleotide pairs) of the *E. coli* chromosome is depicted here.

TABLE 7–2 THE THREE RNA POLYMERASES IN EUCARYOTIC CELLS

TYPE OF POLYMERASE	GENES TRANSCRIBED
RNA polymerase I	most rRNA genes
RNA polymerase II	protein-coding genes, miRNA genes, plus genes for some small RNAs (e.g., those in spliceosomes)
RNA polymerase III	tRNA genes 5S rRNA gene genes for many other small RNAs

to 100,000 nucleotide pairs between one gene and the next. This architecture allows a single gene to be controlled by an almost unlimited number of regulatory sequences scattered along the DNA, and enables eucaryotes to engage in more complex forms of transcriptional regulation than bacteria do.

- Last but not least, eucaryotic transcription initiation must take into account the packing of DNA into nucleosomes and more compact forms of chromatin structure, as we describe in Chapter 8.

We now turn our attention to the general transcription factors and discuss how they help eucaryotic RNA polymerase II to initiate transcription.

QUESTION 7–3

Could the RNA polymerase used for transcription be used as the polymerase that makes the RNA primer required for replication (discussed in Chapter 6)?

Eucaryotic RNA Polymerase Requires General Transcription Factors

The initial finding that, unlike bacterial RNA polymerase, purified eucaryotic RNA polymerase II could not on its own initiate transcription *in vitro* led to the discovery and purification of the **general transcription factors**. These accessory proteins assemble on the promoter, where they position the RNA polymerase, pull apart the double helix to expose the template strand, and launch the RNA polymerase, to begin transcribing.

Figure 7–12 shows how the general transcription factors assemble at a promoter used by RNA polymerase II. The assembly process typically begins with the binding of the general transcription factor TFIID to a short double-helical DNA sequence primarily composed of T and A nucleotides; because of its composition, this sequence is known as the TATA sequence, or *TATA box*. Upon binding to DNA, TFIID causes a dramatic local distortion in the DNA (Figure 7–13), which helps to serve as a landmark for the subsequent assembly of other proteins at the promoter. The TATA box is a key component of many promoters used by RNA polymerase II, and it is typically located 25 nucleotides upstream from the transcription start site. Once the first general transcription factor has bound to this DNA site, the other factors are assembled, along with RNA polymerase II, to form a

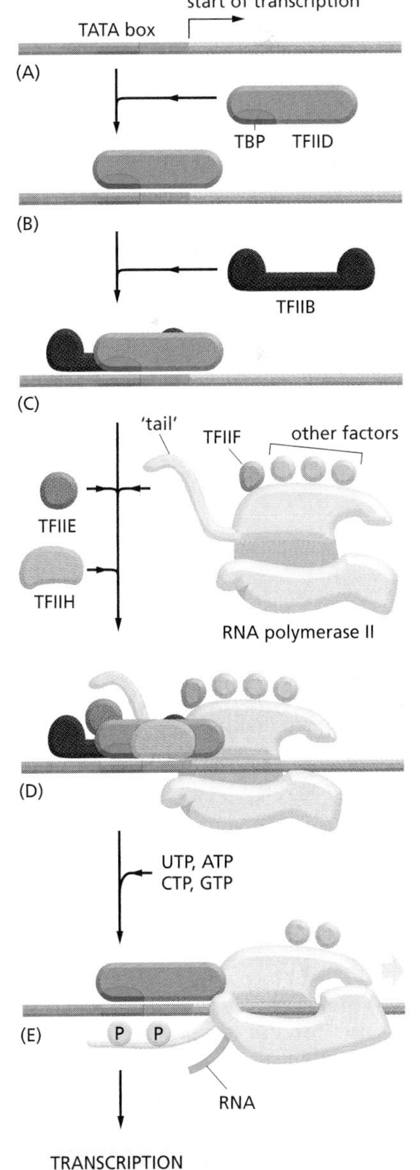

Figure 7–12 **To begin transcription, eucaryotic RNA polymerase II requires a set of general transcription factors.** These transcription factors are called TFIIA, TFIIB, and so on. (A) Many promoters contain a DNA sequence called the TATA box. (B) The TATA box is recognized and bound by transcription factor TFIID, which then enables the adjacent binding of TFIIB (C). For simplicity, the DNA distortion produced by the binding of TFIID (see Figure 7–13) is not shown. (D) The rest of the general transcription factors as well as the RNA polymerase itself assemble at the promoter. (E) TFIIH then pries apart the double helix at the transcription start point, using the energy of ATP hydrolysis, allowing the template strand to be exposed. TFIIH also phosphorylates RNA polymerase II, releasing it from the general factors so it can begin the elongation phase of transcription. The site of phosphorylation is a long polypeptide 'tail' that extends from the polymerase molecule.

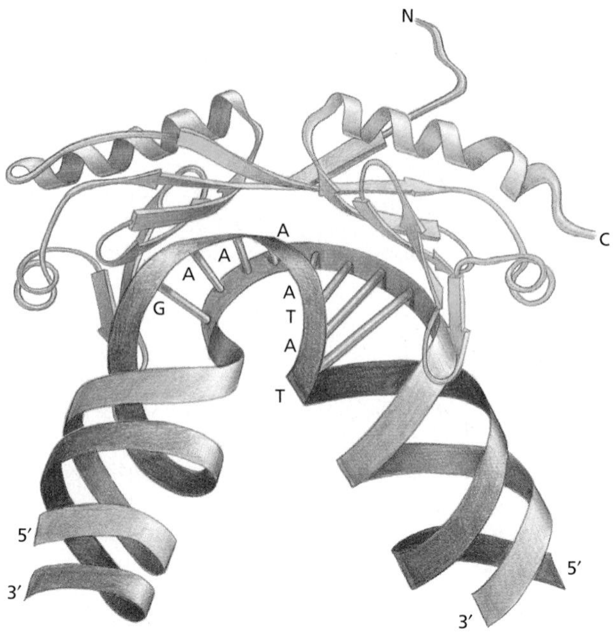

Figure 7–13 **TATA-binding protein (TBP) binds to TATA box sequences and distorts the DNA.** TBP is the subunit of the general transcription factor TFIID that is responsible for recognizing and binding to the TATA box sequence (indicated by letters). The unique DNA bending caused by TBP—two kinks in the double helix separated by partially unwound DNA—may help attract the other general transcription factors. TBP is a single polypeptide chain that is folded into two very similar domains (*blue* and *green*). Its eight β sheets sit atop the DNA helix like a saddle on a horse (Movie 7.4). (Adapted from J.L. Kim et al., *Nature* 365:520–527, 1993. With permission from Macmillan Publishers Ltd.)

complete *transcription initiation complex*. Although Figure 7–12 shows the general transcription factors piling onto the promoter in a certain order, the actual order of assembly probably differs from one promoter to the next in the cell.

After RNA polymerase II has been tethered to the promoter DNA in the transcription initiation complex, it must be released from the complex of transcription factors to begin its task of making an RNA molecule. A key step in this release is the addition of phosphate groups to the 'tail' of RNA polymerase, an act performed by the general transcription factor TFIIH, which contains a protein kinase enzyme as one of its subunits (see Figure 7–12E). This phosphorylation is thought to help the polymerase disengage from the cluster of transcription factors, allowing transcription to commence. Once transcription has begun, most of the general transcription factors are released from the DNA so that they are available to initiate another round of transcription with a new RNA polymerase molecule. When RNA polymerase II finishes transcribing, it is released from the DNA, the phosphates on its tail are stripped off by phosphatases, and it can reinitiate transcription. Only the dephosphorylated form of RNA polymerase II can initiate RNA synthesis at a promoter.

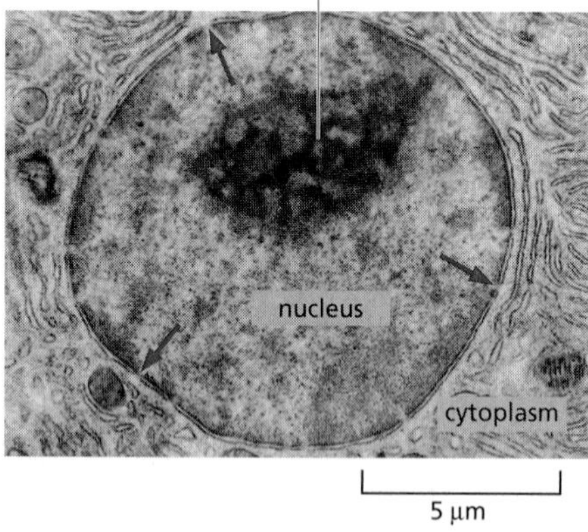

Figure 7–14 **Before they can be translated, mRNA molecules made in the nucleus move out into the cytoplasm via pores in the nuclear envelope (*arrows*).** Shown here is a section of a liver cell nucleus. (From D.W. Fawcett, A Textbook of Histology, 11th ed. Philadelphia: Saunders, 1986. With permission from Elsevier.)

Eucaryotic RNAs Are Transcribed and Processed Simultaneously in the Nucleus

Although the templating principle by which DNA is transcribed into RNA is the same in all organisms, the way in which the RNA transcripts are handled before they can be used by the cell differs greatly between bacteria and eucaryotes. Bacterial DNA lies directly exposed to the cytoplasm, which contains the *ribosomes* on which protein synthesis takes place. As mRNA molecules in bacteria are transcribed, ribosomes immediately attach to the free 5′ end of the RNA transcript and protein synthesis starts.

In eucaryotic cells, by contrast, DNA is enclosed within the *nucleus*. Transcription takes place in the nucleus, but protein synthesis takes place on ribosomes in the cytoplasm. So, before a eucaryotic mRNA can be translated, it must be transported out of the nucleus through small pores in the nuclear envelope (Figure 7–14). Before a eucaryotic RNA exits the nucleus, however, it must go through several different **RNA processing** steps. These reactions take place as the RNA is being transcribed. The enzymes responsible for RNA processing ride on the 'tail' of the eucaryotic RNA polymerase as it transcribes an RNA, processing the transcript as it emerges from the RNA polymerase (Figure 7–15).

Depending on which type of RNA is being produced—an mRNA or some other type—the transcripts are processed in various ways before leaving the nucleus. Two processing steps that occur only on transcripts destined to become mRNA molecules are *capping* and *polyadenylation* (Figure 7–16):

1. RNA capping involves a modification of the 5′ end of the mRNA transcript, the end that is synthesized first during transcription. The RNA is capped by the addition of an atypical nucleotide—a guanine (G) nucleotide with a methyl group attached. This capping occurs after the RNA polymerase has produced about 25 nucleotides of RNA, long before it has completed transcribing the whole gene.
2. Polyadenylation provides newly transcribed mRNAs with a special structure at their 3′ ends. In contrast with bacteria, where the 3′ end of an mRNA is simply the end of the chain synthesized by the RNA polymerase, the 3′ ends of eucaryotic RNAs are first trimmed by an enzyme that cuts the RNA chain at a particular sequence of nucleotides and are then finished off by a second enzyme that adds a series of repeated adenine (A) nucleotides to the cut end. This *poly-A tail* is generally a few hundred nucleotides long.

These two modifications—capping and polyadenylation—are thought to increase the stability of the eucaryotic mRNA molecule, to aid its export from the nucleus to the cytoplasm, and generally to identify the RNA molecule as an mRNA. They are also used by the protein-synthesis machinery to make sure that both ends of the mRNA are present and that the message is therefore complete before protein synthesis begins.

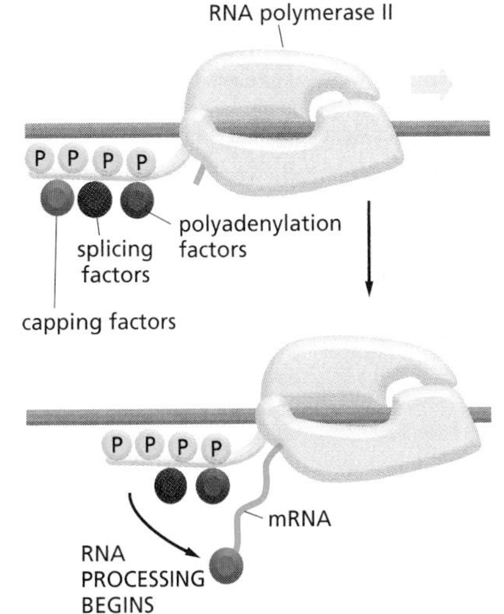

Figure 7–15 **Phosphorylation of RNA polymerase II allows RNA-processing proteins to assemble on its tail.** Not only does the polymerase transcribe DNA into RNA, it also carries RNA-processing proteins that act on the newly made RNA. Some RNA-processing proteins bind to the RNA polymerase tail when it is phosphorylated late in the process of transcription initiation (see Figure 7–12). Capping, polyadenylation, and splicing (described later in this chapter) are modifications made to RNA during processing.

Eucaryotic Genes Are Interrupted by Noncoding Sequences

Most eucaryotic RNAs have to undergo an additional processing step before they are functional. This step involves a far more radical modification of the primary RNA transcript than capping or polyadenylation, and

Figure 7–16 **Eucaryotic mRNA molecules are modified by capping and polyadenylation.** (A) The ends of a eucaryotic mRNA are modified by the addition of a cap at the 5′ end and by cleavage of the primary transcript and the addition of a poly-A tail at the 3′ end. (B) The structure of the cap at the 5′ end of eucaryotic mRNA molecules. Note the unusual 5′-to-5′ linkage of the 7-methyl G to the remainder of the RNA. Many eucaryotic mRNA caps carry an additional modification: the 2′-hydroxyl group on the second ribose sugar in the mRNA is methylated (not shown).

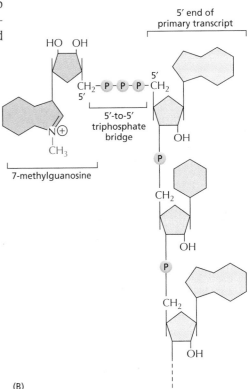

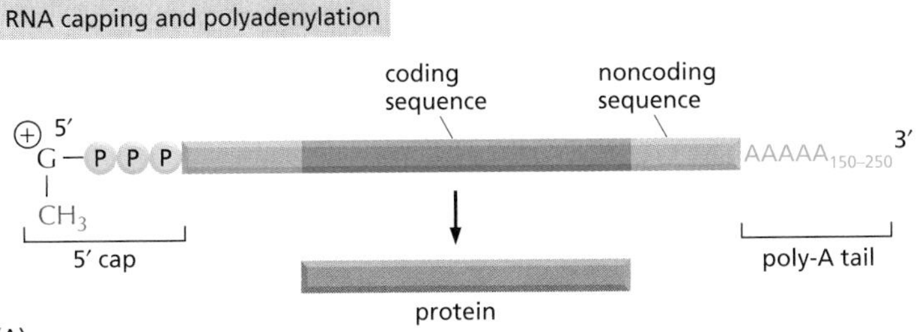

5′ end of primary transcript
HO OH
5′
CH2 – P – P – P – CH2
5′
5′-to-5′ triphosphate bridge
N⊕
CH3
7-methylguanosine
OH
P
CH2
OH
P
CH2
OH
(B)

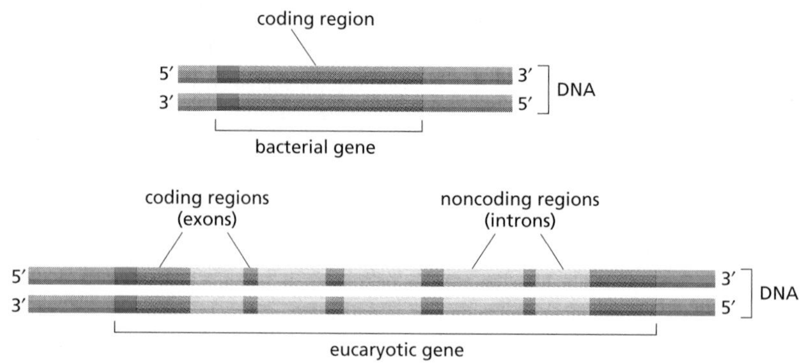

Figure 7–17 **Eucaryotic and bacterial genes are organized differently.** A bacterial gene consists of a single stretch of uninterrupted nucleotide sequence that encodes the amino acid sequence of a protein. In contrast, the coding sequences of most eucaryotic genes (*exons*) are interrupted by noncoding sequences (*introns*). Promoters for transcription are indicated in *green*.

it is the consequence of a surprising feature of eucaryotic gene arrangement. In bacteria, most proteins are encoded by an uninterrupted stretch of DNA sequence that is transcribed into an RNA that, without any further processing, can serve as an mRNA. Most eucaryotic genes, in contrast, have their coding sequences interrupted by long, noncoding *intervening sequences*, called **introns** (Figure 7–17). The scattered pieces of coding sequences, or *expressed sequences*, called **exons**, are usually shorter than the introns, and the coding portion of a eucaryotic gene is often only a small fraction of the total length of the gene. Introns range in length from a single nucleotide to more than 10,000 nucleotides; some eucaryotic genes lack introns altogether, some have only a few, but most have many (Figure 7–18).

Introns Are Removed by RNA Splicing

To produce an mRNA in a eucaryotic cell, the entire length of the gene, introns as well as exons, is transcribed into RNA. After capping, and as the RNA polymerase continues to transcribe the gene, the process of **RNA splicing** begins, in which the intron sequences are removed from the newly synthesized RNA and the exons are stitched together. Each transcript ultimately receives a poly-A tail; in some cases, this happens after splicing, and in other cases it occurs before the final splicing reactions have been completed. Once a transcript has been spliced and its 5′ and 3′ ends have been modified, the RNA is a functional mRNA molecule that can now leave the nucleus and be translated into protein.

How does the cell determine which parts of the RNA transcript to remove during splicing? Unlike the coding sequence of an exon, most of the nucleotide sequence of an intron is unimportant. Although there is little overall resemblance between the nucleotide sequences of different introns, each intron contains a few short nucleotide sequences that act as cues for its removal. These sequences are found at or near each end of the intron and are the same or very similar in all introns (Figure 7–19). Guided by these sequences, an elaborate splicing machine cuts out the intron sequence in the form of a 'lariat' structure (Figure 7–20), formed by the reaction of the 'A' highlighted in red in Figures 7–19 and 7–20.

Although we will not describe the splicing machinery in detail, we note that, unlike the other steps of mRNA production we have discussed, RNA splicing is carried out largely by RNA molecules instead of proteins. RNA

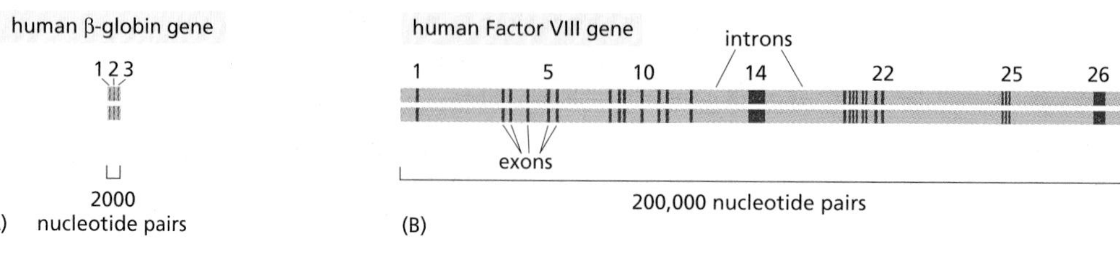

Figure 7–18 **Most human genes are broken into exons and introns.** (A) The β-globin gene, which encodes one of the subunits of the oxygen-carrying protein hemoglobin, contains 3 exons. (B) The Factor VIII gene, which codes for a protein (Factor VIII) that functions in the blood-clotting pathway, contains 26 exons. Mutations in this large gene are responsible for the most prevalent form of hemophilia.

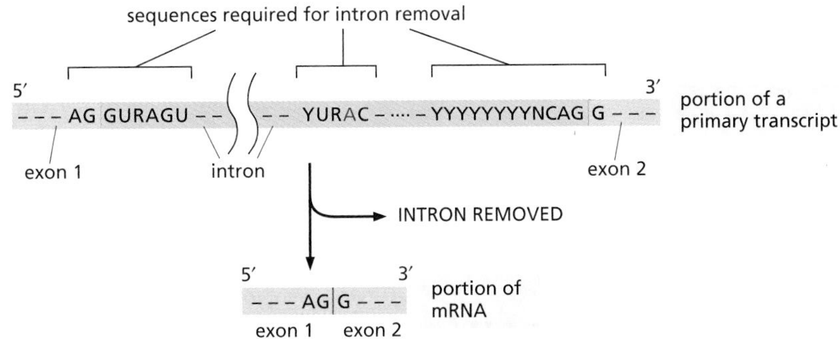

Figure 7–19 **Special nucleotide sequences signal the beginning and the end of an intron.** Only the nucleotide sequences shown are required to remove an intron. The other positions in an intron can be occupied by any nucleotide. The special sequences are recognized by small nuclear ribonucleoproteins (snRNPs), which cleave the RNA at the intron–exon borders and covalently link the exon sequence together. Here R stands for either A or G; Y stands for either C or U; N stands for any nucleotide. The A highlighted in *red* forms the branch point of the lariat produced in the splicing reaction (see Figure 7–20). The distances along the RNA between the three splicing sequences are highly variable; however, the distance between the branch point and the 5′ splice junction is typically much longer than that between the 3′ splice junction and the branch point. The splicing sequences shown are from humans; similar sequences direct RNA splicing in other eucaryotes.

molecules recognize intron–exon boundaries (through complementary base-pairing) and participate intimately in the chemistry of splicing. These RNA molecules, called **small nuclear RNAs** (**snRNAs**), are packaged with additional proteins to form *small nuclear ribonucleoprotein particles* (*snRNPs*, pronounced "snurps"). The snRNPs form the core of the **spliceosome**, the large assembly of RNA and protein molecules that carries out RNA splicing in the cell. To watch the spliceosome in action, see Movie 7.5.

The intron–exon type of gene arrangement in eucaryotes at first seems wasteful, but it does have positive consequences. First, the transcripts of many eucaryotic genes can be spliced in different ways, each of which can produce a distinct protein. Such **alternative splicing** thereby allows many different proteins to be produced from the same gene (Figure 7–21). An estimated 60% of human genes probably undergo alternative splicing. Thus RNA splicing enables eucaryotes to increase the already enormous coding potential of their genomes.

RNA splicing also provides another advantage to eucaryotes, one that is likely to have been profoundly important in the early evolutionary history of genes. As we discuss in detail in Chapter 9, the intron–exon structure of genes is thought to have speeded up the emergence of new and useful proteins. The inclusion of long introns makes genetic recombination between exons of different genes much more likely. This means that genes for new proteins could have evolved quite rapidly by the combination of parts of preexisting genes, a mechanism resembling the assembly of a new type of machine from a kit of preexisting functional components. Indeed, many proteins in present-day cells resemble patchworks composed from a common set of protein pieces, called protein *domains* (see Figure 4–16).

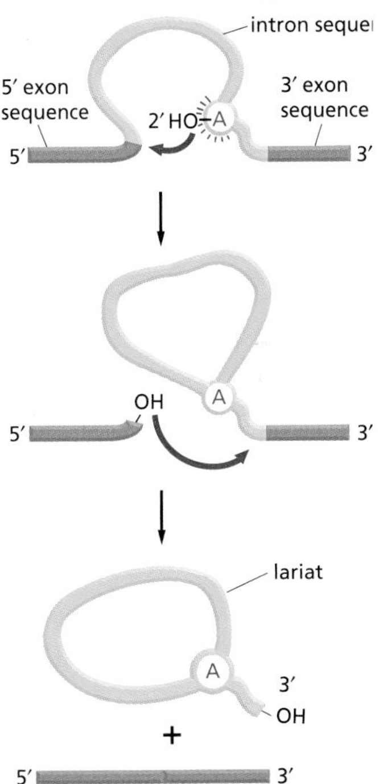

Figure 7–20 **An intron forms a branched structure during splicing.** In the first step, the branch point adenine (*red* A) in the intron sequence attacks the 5′ splice site and cuts the sugar–phosphate backbone of the RNA at this point (this is the same A highlighted in red in Figure 7–19). The cut 5′ end of the intron becomes covalently linked to the 2′-OH group of the ribose of the A to form a branched structure. The free 3′-OH end of the exon sequence then reacts with the start of the next exon sequence, joining the two exons together into a continuous coding sequence and releasing the intron in the form of a lariat structure, which is eventually degraded.

Mature Eucaryotic mRNAs Are Selectively Exported from the Nucleus

We have seen how eucaryotic mRNA synthesis and processing takes place in an orderly fashion within the cell nucleus. However, these events create a special problem for eucaryotic cells: of the total mRNA that is synthesized, only a small fraction—the mature mRNA—is useful to the cell. The remaining RNA fragments—excised introns, broken RNAs, and aberrantly spliced transcripts—are not only useless but could be dangerous to the cell if not destroyed. How, then, does the cell distinguish between the relatively rare mature mRNA molecules it needs to keep and the overwhelming amount of debris generated by RNA processing?

The answer is that the transport of mRNA from the nucleus to the cytoplasm, where it is translated into protein, is highly selective: only correctly processed RNAs are allowed to pass. This reliance on proper

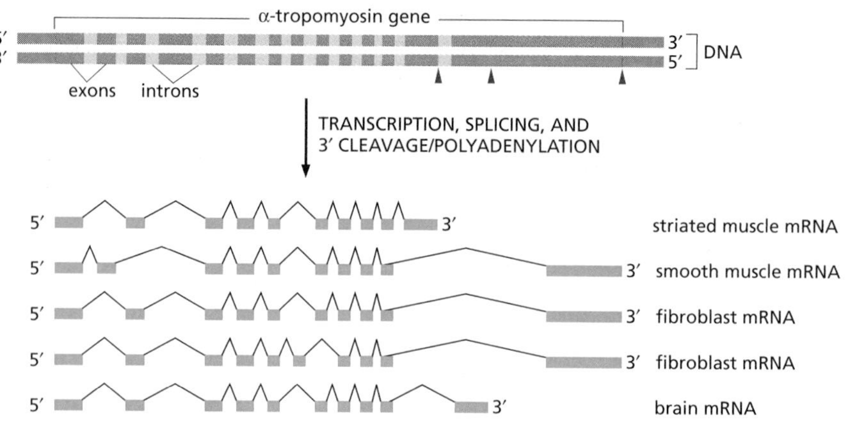

Figure 7–21 **The α-tropomyosin gene can be spliced in different ways.** α-Tropomyosin is a coiled-coil protein (see Figure 4–13) that regulates contraction in muscle cells. Its transcript can be spliced in different ways, as indicated in the figure, to produce distinct mRNAs that then give rise to variant proteins. Some of the splicing patterns are specific for certain types of cells. For example, the α-tropomyosin made in striated muscle is different from that made in smooth muscle, even though both proteins come from the same gene. The *red* arrowheads in the top part of the figure represent sites where poly-A addition can occur.

processing for RNA transport is mediated by the *nuclear pore complex*, which recognizes and exports only completed mRNAs. These aqueous pores connect the nucleoplasm with the cytosol, and as we discuss in Chapter 15, they act as gates that control which macromolecules can enter or leave the nucleus. To be 'export ready,' an mRNA molecule must be bound to an appropriate set of proteins, each of which signals that the mRNA has been correctly processed. These proteins include poly-A–binding proteins, a cap-binding complex, and proteins that mark completed RNA splices (Figure 7–22). It is presumably the entire set of bound proteins, rather than any single protein, that ultimately determines whether an RNA molecule will leave the nucleus. The 'waste RNAs' that remain behind in the nucleus are degraded, and the building blocks are reused for transcription.

mRNA Molecules Are Eventually Degraded by the Cell

Because a single mRNA molecule can be translated many times (see Figure 7–2), the length of time that a mature mRNA molecule persists in the cell affects the amount of protein it produces. Each mRNA molecule is eventually degraded into nucleotides by RNases, but the lifetimes of mRNA molecules differ considerably—depending on the nucleotide sequence of the mRNA and the type of cell in which the mRNA is produced. Most mRNAs produced in bacteria are degraded rapidly, having a

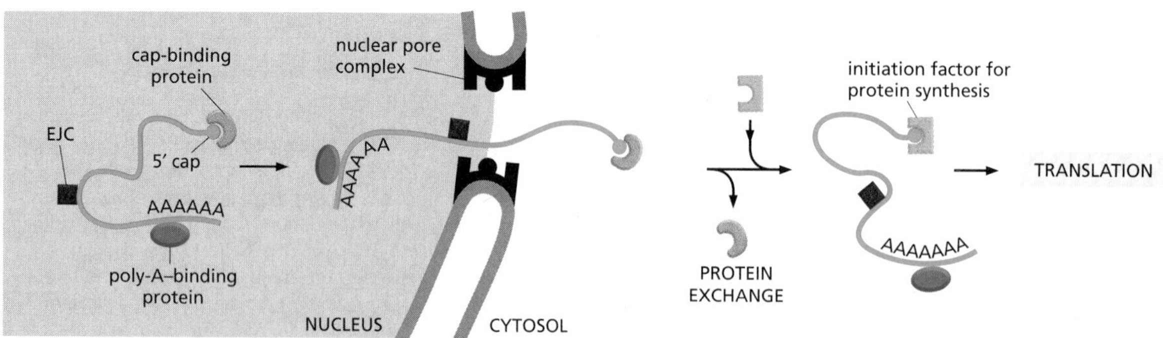

Figure 7–22 **A specialized set of RNA-binding proteins signal that a mature mRNA is ready for export to the cytoplasm.** As indicated on the left, the cap and poly-A tail of a mature mRNA molecule are 'marked' by proteins that recognize these modifications. In addition, a group of proteins called the *exon junction complex (EJC)* is deposited on the mRNA after successful RNA splicing has occurred. Once the mRNA is deemed 'export ready,' a nuclear transport receptor (discussed in Chapter 15) associates with it and guides it through the nuclear pore. Once in the cytosol, the mRNA can shed previously bound proteins and acquire new ones.

typical lifetime of about 3 minutes. The mRNAs in eucaryotic cells usually persist for longer. Some transcripts, such as that encoding β-globin, have lifetimes of more than 10 hours, whereas other eucaryotic mRNAs have lifetimes of less than 30 minutes.

These different lifetimes are in part controlled by nucleotide sequences that are present in the mRNA itself, most often in the portion of RNA called the 3′ untranslated region, that lies between the 3′ end of the coding sequence and the poly-A tail. The different lifetimes of mRNA help the cell specify the amount of each protein that it synthesizes. In general, proteins made at high levels, such as β-globin, are translated from mRNAs that have long lifetimes, whereas those proteins present at low levels, or those whose levels must change rapidly in response to signals, are typically synthesized from short-lived mRNAs. These different lifetimes are the outcome of evolutionary fine-tuning in which the stability of each mRNA is tied to the needs of the cell.

The Earliest Cells May Have Had Introns in Their Genes

The process of transcription is universal: all cells use RNA polymerase, coupled with complementary base-pairing, to synthesize RNA from DNA. Indeed, bacterial and eucaryotic RNA polymerases are almost identical in overall structure and clearly evolved from a shared ancestral polymerase. It may therefore seem puzzling that the resulting transcript is handled very differently in eucaryotes and in procaryotes (Figure 7–23). In particular, RNA splicing seems to mark a fundamental difference between those two types of cells. But how did this dramatic difference arise?

As we have seen, RNA splicing provides eucaryotes with the ability to produce a variety of proteins from a single gene, and grants them a certain degree of evolutionary flexibility. However, these advantages come with a cost: the cell has to maintain a larger genome and has to throw out a large fraction of the RNA it synthesizes. According to one school of

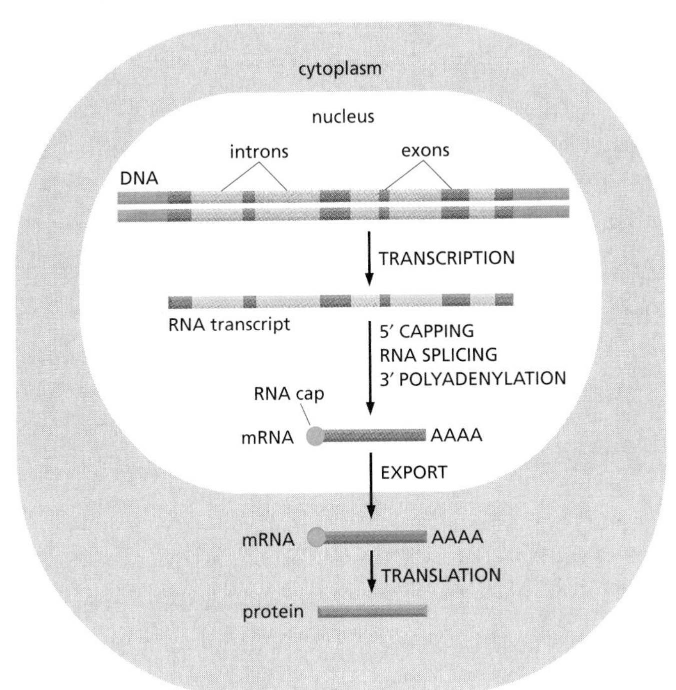

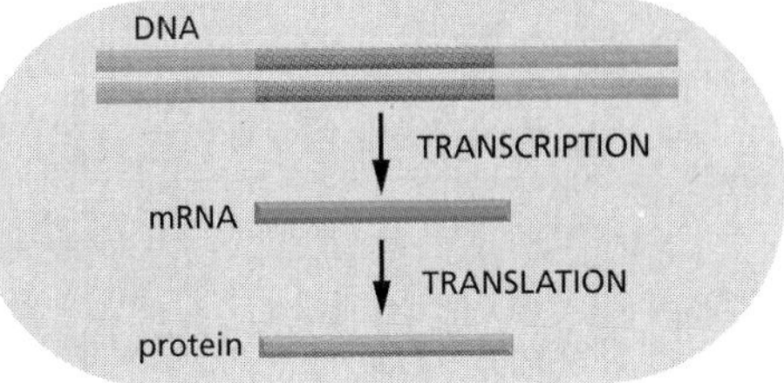

Figure 7–23 **Procaryotes and eucaryotes handle their RNA transcripts differently.** (A) In eucaryotic cells, the initial RNA molecule produced by transcription contains both intron and exon sequences. Its two ends are modified, and the introns are removed by an enzymatically catalyzed RNA splicing reaction. The resulting mRNA is then transported from the nucleus to the cytoplasm, where it is translated into protein. Although these steps are depicted as occurring one at a time, in a sequence, in reality they occur simultaneously. For example, the RNA cap is usually added and splicing often begins before the transcript has been completed. Because of this coupling, transcripts of the entire gene (including all introns and exons) do not typically exist in the cell. (B) In procaryotes, the production of mRNA molecules is simpler. The 5′ end of an mRNA molecule is produced by the initiation of transcription by RNA polymerase, and the 3′ end is produced by the termination of transcription. Because procaryotic cells lack a nucleus, transcription and translation take place in a common compartment. Translation of a bacterial mRNA can therefore begin before its synthesis has been completed. The amount of protein in a cell depends on the efficiency of each of these steps and on the rates of degradation of the RNA and protein molecules.

thought, early cells—the common ancestors of procaryotes and eucaryotes—contained introns that were lost in procaryotes during subsequent evolution. By shedding their introns and adopting a smaller, more streamlined genome, procaryotes would have been able to reproduce more rapidly and efficiently. Consistent with this idea, simple eucaryotes that reproduce rapidly (some yeasts, for example) have relatively few introns, and these introns are usually much shorter than those found in higher eucaryotes.

On the other hand, some argue that introns were originally parasitic mobile genetic elements (discussed in Chapter 6) that happened to invade an early eucaryotic ancestor, colonizing its genome. These host cells then unwittingly replicated the selfish nucleotide snippets along with their own DNA, and modern eucaryotes have never bothered to sweep away the genetic clutter left from that ancient infection. The issue, however, is far from settled; whether introns evolved early—and were lost in procaryotes—or evolved later in eucaryotes is still a topic of scientific debate, and we return to it in Chapter 9.

FROM RNA TO PROTEIN

By the end of the 1950s, biologists had demonstrated that the information encoded in DNA is copied first into RNA and then into protein. The debate then shifted to the 'coding problem': how is the information in a linear sequence of nucleotides in RNA translated into the linear sequence of a chemically quite different set of subunits—the amino acids in proteins? This fascinating question stimulated much excitement among scientists at the time. Here was a cryptogram set up by nature that, after more than 3 billion years of evolution, could finally be solved by one of the products of evolution—human beings! Indeed, scientists have not only cracked the code but have revealed, in atomic detail, the precise workings of the machinery by which cells read this code.

An mRNA Sequence Is Decoded in Sets of Three Nucleotides

Transcription as a means of information transfer is simple to understand, since DNA and RNA are chemically and structurally similar, and DNA can act as a direct template for the synthesis of RNA through complementary base-pairing. As the term transcription signifies, it is as if a message written out by hand were being converted, say, into a typewritten text. The language itself and the form of the message do not change, and the symbols used are closely related.

In contrast, the conversion of the information in RNA into protein represents a **translation** of the information into another language that uses quite different symbols. Because there are only 4 different nucleotides in mRNA but 20 different types of amino acids in a protein, this translation cannot be accounted for by a direct one-to-one correspondence between a nucleotide in RNA and an amino acid in protein. The rules by which the nucleotide sequence of a gene, through the medium of mRNA, is translated into the amino acid sequence of a protein are known as the **genetic code**.

The sequence of nucleotides in the mRNA molecule is read consecutively in groups of three. Because RNA is a linear polymer made of four different nucleotides, there are thus $4 \times 4 \times 4 = 64$ possible combinations of three nucleotides: AAA, AUA, AUG, and so on. However, only 20 different amino acids are commonly found in proteins. Either some nucleotide triplets are never used, or the code is redundant and some amino acids are

GCA GCC GCG GCU	AGA AGG CGA CGC CGG CGU	GAC GAU	AAC AAU	UGC UGU	GAA GAG	CAA CAG	GGA GGC GGG GGU	CAC CAU	AUA AUC AUU	UUA UUG CUA CUC CUG CUU	AAA AAG	AUG	UUC UUU	CCA CCC CCG CCU	AGC AGU UCA UCC UCG UCU	ACA ACC ACG ACU	UGG	UAC UAU	GUA GUC GUG GUU	UAA UAG UGA
Ala	Arg	Asp	Asn	Cys	Glu	Gln	Gly	His	Ile	Leu	Lys	Met	Phe	Pro	Ser	Thr	Trp	Tyr	Val	stop
A	R	D	N	C	E	Q	G	H	I	L	K	M	F	P	S	T	W	Y	V	

Figure 7–24 **The nucleotide sequence of an mRNA is translated into the amino acid sequence of a protein via the genetic code.** All the three-nucleotide codons that specify a given amino acid are listed above that amino acid, which is given in both its three-letter and one-letter abbreviations (see Panel 2–5, pp. 72–73, for the full name of each amino acid and its structure). By convention, codons are always written with the 5′-terminal nucleotide to the left. Note that most amino acids are represented by more than one codon, and that there are some regularities in the set of codons that specify each amino acid. Codons for the same amino acid tend to contain the same nucleotides at the first and second positions and to vary at the third position. There are three codons that do not specify any amino acid but act as termination sites (stop codons), signaling the end of the protein-coding sequence. One codon—AUG—acts both as an initiation codon, signaling the start of a protein-coding message, and as the codon that specifies methionine.

specified by more than one triplet. The second possibility is, in fact, correct, as shown by the completely deciphered genetic code in Figure 7–24. Each group of three consecutive nucleotides in RNA is called a **codon**, and each specifies one amino acid. The strategy by which this code was cracked is described in How We Know, pp. 248–249.

The genetic code shown in Figure 7–24 is used in nearly all present-day organisms. Although a few slight differences in the code have been found, these occur chiefly in the DNA of mitochondria and of some fungi and protozoa. Mitochondria have their own transcription and protein synthesis machinery that operate quite independently from those of the rest of the cell (discussed in Chapter 14), and they have been able to accommodate minor changes to the otherwise universal code. Even in fungi and protozoa, the similarities in the code far outweigh the differences.

In principle, an RNA sequence can be translated in any one of three different **reading frames**, depending on where the decoding process begins (Figure 7–25). However, only one of the three possible reading frames in an mRNA specifies the correct protein. We shall see in a later section how a special punctuation signal at the beginning of each RNA message sets the correct reading frame.

5′ 3′
1 CUC AGC GUU ACC AU
—Leu—Ser—Val—Thr—

2 C UCA GCG UUA CCA U
—Ser—Ala—Leu—Pro—

3 CU CAG CGU UAC CAU
—Gln—Arg—Tyr—His—

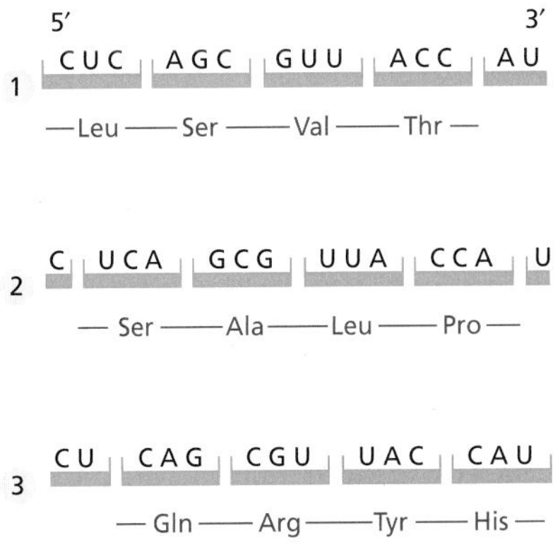

Figure 7–25 **An RNA molecule can be translated in three possible reading frames.** In the process of translating a nucleotide sequence (*blue*) into an amino acid sequence (*red*), the sequence of nucleotides in an mRNA molecule is read from the 5′ to the 3′ end in sequential sets of three nucleotides. In principle, therefore, the same RNA sequence can specify three completely different amino acid sequences, depending on the reading frame. In reality, however, only one of these reading frames encodes the actual message.

tRNA Molecules Match Amino Acids to Codons in mRNA

The codons in an mRNA molecule do not directly recognize the amino acids they specify: the group of three nucleotides does not, for example, bind directly to the amino acid. Rather, the translation of mRNA into protein depends on adaptor molecules that can recognize and bind to a codon at one site on their surface and to an amino acid at another site. These adaptors consist of a set of small RNA molecules known as **transfer RNAs (tRNAs)**, each about 80 nucleotides in length.

We saw earlier that an RNA molecule will generally fold into a three-dimensional structure by forming base pairs between different regions of the molecule. If the base-paired regions are sufficiently extensive, they will form a double-helical structure, like that of double-stranded DNA. The tRNA molecule provides a striking example of this. Four short segments of the folded tRNA are double-helical, producing a molecule that

HOW WE KNOW: CRACKING THE GENETIC CODE

By the beginning of the 1960s, the *central dogma* had been accepted as the pathway along which information flows from gene to protein. It was clear that genes encode proteins, that genes are made of DNA, and that mRNA serves as an intermediary, carrying the information from the nucleus—where DNA is stored—to the cytoplasm, where translation into protein takes place.

Even the general format of the genetic code had been worked out: each of the 20 amino acids found in proteins is represented by a triplet codon in an mRNA molecule. But an even greater challenge remained: biologists, chemists, and even physicists set their sights on breaking the genetic code—attempting to figure out which amino acid each of the 64 possible nucleotide triplets designates. The most straightforward path to the solution would have been to compare the sequence of a segment of DNA or mRNA with its corresponding polypeptide product. Techniques for sequencing nucleic acids, however, would not be devised for another 10 years.

So researchers decided that, to crack the genetic code, they would have to synthesize their own simple messages. If they could feed these molecules to ribosomes—the machines that make proteins—and then analyze the resulting polypeptide product, they would be on their way to deciphering which triplets encode which amino acids.

Losing the cells

Before researchers began preparing their synthetic mRNAs, they wanted to perfect a cell-free system for protein synthesis. This would allow them to translate their messages into polypeptides in a test tube. (Generally speaking, when working in the laboratory, the simpler the system, the easier it is to interpret the results.) To isolate the molecular machinery they needed for such a cell-free translation system, researchers broke open *E. coli* cells and loaded their contents into a centrifuge. Spinning these samples at high speed caused the membranes and other large chunks of cellular debris to be dragged to the bottom of the tube; the lighter cellular components required for protein synthesis—including mRNA, the tRNA adaptors, ribosomes, enzymes, and other small molecules—were left floating in the supernatant. Researchers found that simply adding radioactive amino acids to this cell "soup" would trigger the production of radiolabeled proteins. By centrifuging this supernatant again, at a somewhat higher speed, the researchers could force the ribosomes and any newly synthesized peptides to the bottom of the tube; the labeled polypeptides could then be detected by measuring the radioactivity in the sediment remaining in the tube after the top layer had been discarded.

The trouble with this particular system was that it produced proteins encoded by cellular mRNAs already present in the extract. But researchers wanted to use their own synthetic messages to direct protein synthesis. This problem was solved when Marshall Nirenberg discovered that he could destroy the cellular mRNA in the extract by adding a small amount of ribonuclease—an enzyme that degrades RNA—to the mix. Now all he needed to do was prepare large quantities of his own message, load this synthetic mRNA into the cell-free system, and see what peptides came out.

Faking the message

Producing a synthetic polynucleotide with a defined sequence was not as simple as it sounds. Again, it would be years before chemists developed techniques that could be used to synthesize any given string of nucleic acids. Nirenberg decided to use polynucleotide phosphorylase, an enzyme that would join ribonucleotides together in the absence of a template. The sequence of the resulting RNA would then depend entirely on which nucleotides were presented to the enzyme. A mixture of nucleotides would be sewn into a random sequence; but a single type of nucleotide would yield a homogeneous polymer string containing only that one nucleotide. Thus Nirenberg, working with his collaborator Heinrich Matthaei, first produced an mRNA made entirely of uracil—poly U.

Together the researchers fed this poly U to their cell-free translation system. They then added a single type of radioactively labeled amino acid to the mix. After testing each amino acid—one at a time, in 20 different experiments—they determined that poly U directs the synthesis of a peptide containing only phenylalanine (Figure 7–26). Because UUU is the only triplet codon in this message, UUU, they reasoned, encodes phenylalanine. The first word in the genetic code had been deciphered.

Nirenberg and Matthaei then repeated the experiment with poly A and poly C and determined that AAA codes for lysine and CCC for proline. The meaning of poly G could not be ascertained by this method because this polynucleotide forms an odd triple-stranded helix and would not cooperate as a template in the cell-free system.

Feeding ribosomes with synthetic mRNA seemed a fruitful technique. But with the single-nucleotide possibilities exhausted, researchers had three codons down, and 61 still to go. The other codon combinations, however, were harder to construct, and a new synthetic approach was needed. In the 1950s, the organic chemist Gobind Khorana had been developing methods for preparing mixed polynucleotides of defined sequence—but his techniques worked only for DNA. When he learned

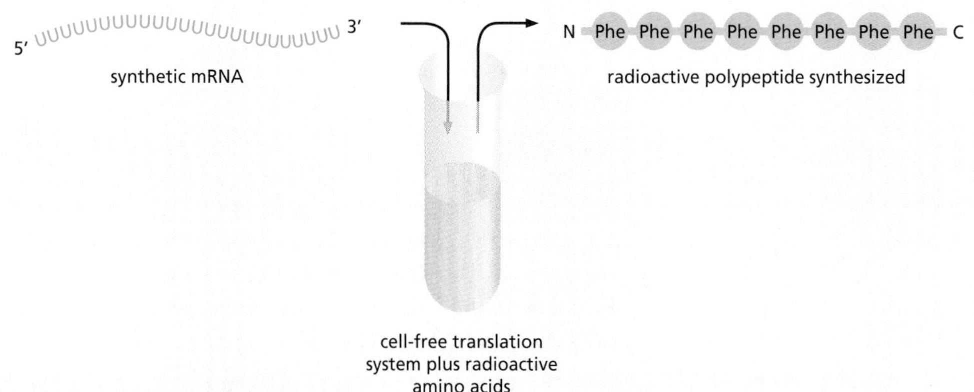

Figure 7–26 **UUU codes for phenylalanine.** Synthetic mRNAs are fed into a cell-free translation system containing ribosomes, tRNAs, enzymes, and other small molecules. Radioactive amino acids are added to this mix and the resulting polypeptides analyzed. In this case, poly U is shown to encode a polypeptide containing only phenylalanine.

of Nirenberg's work with synthetic messages, Khorana directed his energies and skills to producing RNAs. He found that if he made DNAs of a defined sequence, he could then use RNA polymerase to produce RNAs from those. In this way, Khorana prepared a collection of different mRNAs of defined repeating sequence. He generated sequences of repeating dinucleotides (such as poly UC), trinucleotides (such as poly UUC), or tetranucleotides (such as poly UAUC).

These mixed polynucleotides, however, yielded results that were much more difficult to decipher than the mononucleotide messages that Nirenberg had used. Take poly UG, for example. When this repeating dinucleotide is added to the translation system, researchers discovered that it codes for a polypeptide of alternating cysteine and valine residues. This RNA, of course, contains two different alternating codons: UGU and GUG. So researchers could say that UGU and GUG code for cysteine and valine, although they could not tell which went with which. Thus, these mixed messages provided useful information, but did not definitively reveal which codons specified which amino acids (Figure 7–27).

Trapping the triplets

These final ambiguities in the code were resolved when Nirenberg and a young medical graduate named Phil Leder discovered that RNA fragments that were only three nucleotides in length—the size of a single codon—could bind to a ribosome and attract the appropriate amino-acid-containing tRNA molecule to the protein-making machinery. These complexes—containing one ribosome, one mRNA codon, and one radiolabeled aminoacyl-tRNA—could then be captured on a piece of filter paper and the attached amino acid identified.

Their trial run with UUU—the first word—worked splendidly. Leder and Nirenberg primed the usual cell-free translation system with snippets of UUU. These trinucleotides bound to the ribosomes, and Phe-tRNAs bound to the UUU. The new system was up and running, and the researchers had confirmed that UUU codes for phenylalanine.

All that remained was for researchers to produce all 64 possible codons—a task that was quickly accomplished in both Nirenberg's and Khorana's laboratories. Because these small trinucleotides were much simpler to synthesize chemically, and the triplet-trapping tests were easier to perform and analyze than the previous decoding experiments, the researchers were able to work out the complete genetic code within the next year.

MESSAGE	PEPTIDES PRODUCED	CODON ASSIGNMENTS	
poly UG	...Cys–Val–Cys–Val...	UGU GUG	Cys, Val*
poly AG	...Arg–Glu–Arg–Glu...	AGA GAG	Arg, Glu
poly UUC	...Phe–Phe–Phe... + ...Ser–Ser–Ser... + ...Leu–Leu–Leu...	UUC UCU CUU	Phe, Ser, Leu
poly UAUC	...Tyr–Leu–Ser–Ile...	UAU CUA UCU AUC	Tyr, Leu, Ser, Ile

* One codon specifies Cys, the other Val. The same ambiguity exists for the other codon assignments shown here.

Figure 7–27 **Messages of mixed repeating sequences further narrowed the coding possibilities.** Although these mixed messages reveal the composition of the encoded peptides, they did not permit the unambiguous assignment of a single codon to a specific amino acid. For example, the results of the poly-UG experiment cannot distinguish whether UGU or GUG encodes cysteine. As indicated, the same type of ambiguity exists for all the experiments using di-, tri-, and tetranucleotides.

looks like a cloverleaf when drawn schematically (Figure 7–28A). For example, a 5′-GCUC-3′ sequence in one part of a polynucleotide chain can form a relatively strong association with a 5′-GAGC-3′ sequence in another region of the same molecule. The cloverleaf undergoes further folding to form a compact L-shaped structure that is held together by additional hydrogen bonds between different regions of the molecule (Figure 7–28B and C).

Two regions of unpaired nucleotides situated at either end of the L-shaped molecule are crucial to the function of tRNA in protein synthesis. One of these regions forms the **anticodon**, a set of three consecutive nucleotides that through base-pairing bind the complementary codon in an mRNA molecule. The other is a short single-stranded region at the 3′ end of the molecule; this is the site where the amino acid that matches the codon is attached to the tRNA.

We saw in the previous section that the genetic code is redundant; that is, several different codons can specify a single amino acid (see Figure 7–24). This redundancy implies either that there is more than one tRNA for many of the amino acids or that some tRNA molecules can base-pair with more than one codon. In fact, both situations occur. Some amino acids have more than one tRNA and some tRNAs are constructed so that they require accurate base-pairing only at the first two positions of the codon and can tolerate a mismatch (or *wobble*) at the third position. This wobble base-pairing explains why so many of the alternative codons for an amino acid differ only in their third nucleotide (see Figure 7–24). Wobble base-pairings make it possible to fit the 20 amino acids to their 61 codons with as few as 31 kinds of tRNA molecules. The exact number of different kinds of tRNAs, however, differs from one species to the next. For example, humans have nearly 500 different tRNA genes, but only 48 anticodons are represented among them.

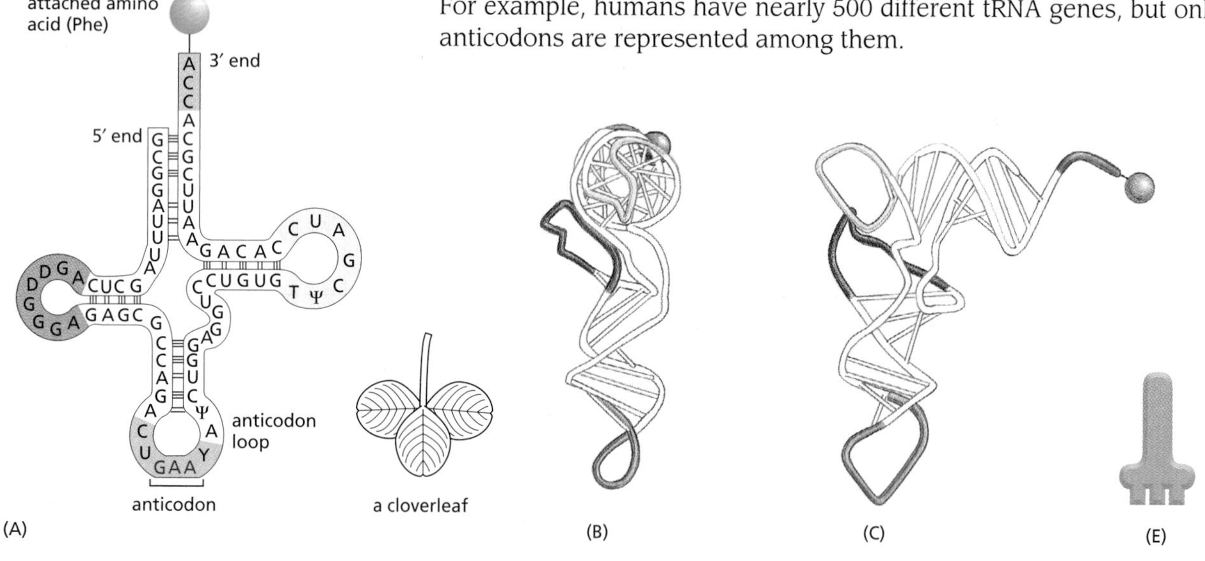

5′ GCGGAUUUAGCUCAGDDGGGAGAGCGCCAGACUGAAYAΨCUGGAGGUCCUGUGTΨCGAUCCACAGAAUUCGCACCA 3′

(D) anticodon

Figure 7–28 **tRNA molecules are molecular adaptors, linking amino acids to codons.** In this series of diagrams, the same tRNA molecule—in this case a tRNA specific for the amino acid phenylalanine (Phe)—is depicted in various ways. (A) The cloverleaf structure, a convention used to show the complementary base-pairing (*red lines*) that creates the double-helical regions of the molecule. The anticodon (*red*) is the sequence of three nucleotides that base-pairs with a codon in mRNA. The amino acid matching the codon/anticodon pair is attached at the 3′ end of the tRNA. tRNAs contain some unusual bases, which are produced by chemical modification after the tRNA has been synthesized. The bases denoted ψ (for pseudouridine) and D (for dihydrouridine) are derived from uracil. (B and C) Views of the actual L-shaped molecule, based on X-ray diffraction analysis. These images are rotated 90° with respect to one another. (D) The linear nucleotide sequence of the molecule, color-coded to match A, B, and C. (E) Schematic representation of tRNA, emphasizing the anticodon, which is used in subsequent figures.

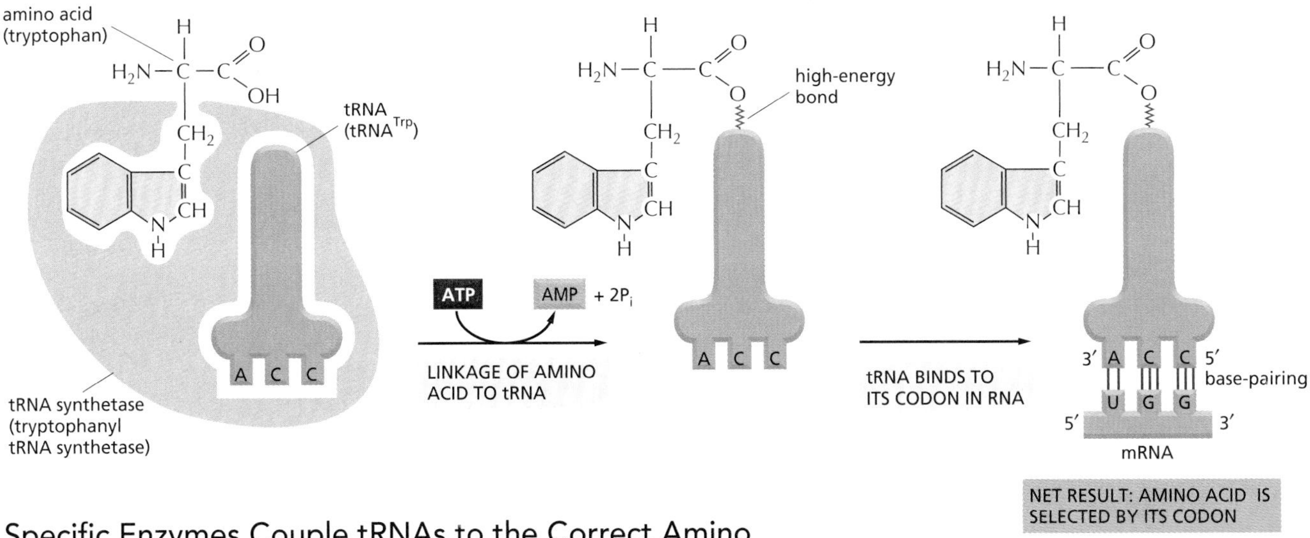

amino acid (tryptophan)
tRNA ($tRNA^{Trp}$)
high-energy bond
ATP AMP + $2P_i$
LINKAGE OF AMINO ACID TO tRNA
tRNA BINDS TO ITS CODON IN RNA
base-pairing
mRNA
tRNA synthetase (tryptophanyl tRNA synthetase)
NET RESULT: AMINO ACID IS SELECTED BY ITS CODON

Figure 7–29 **The genetic code is translated by means of two adaptors that act one after another.** The first adaptor is the aminoacyl-tRNA synthetase, which couples a particular amino acid to its corresponding tRNA; this coupling process is called charging. The second adaptor is the tRNA molecule itself, whose anticodon forms base pairs with the appropriate codon on the mRNA. A tRNA coupled with its amino acid is also called a charged tRNA. An error in either step—the charging or the binding of the charged tRNA to its codon—will cause the wrong amino acid to be incorporated into a protein chain. In the sequence of events shown, the amino acid tryptophan (Trp) is selected by the codon UGG on the mRNA.

Specific Enzymes Couple tRNAs to the Correct Amino Acid

We have seen that in order to read the genetic code in DNA, cells make many different tRNAs. We now must consider how each tRNA molecule becomes *charged*—linked to the one amino acid in 20 that is its right partner. Recognition and attachment of the correct amino acid depend on enzymes called **aminoacyl-tRNA synthetases**, which covalently couple each amino acid to its appropriate set of tRNA molecules. In most organisms, there is a different synthetase enzyme for each amino acid (that is, there are 20 synthetases in all); one attaches glycine to all tRNAs that recognize codons for glycine, another attaches phenylalanine to all tRNAs that recognize codons for phenylalanine, and so on. Specific nucleotides in both the anticodon and the amino-acid–accepting arm allow the correct tRNA to be recognized by the synthetase enzyme (Movie 7.6). The synthetases are equal in importance to the tRNAs in the decoding process, because it is the combined action of synthetases and tRNAs that allows each codon in the mRNA molecule to associate with its proper amino acid (Figure 7–29).

The synthetase-catalyzed reaction that attaches the amino acid to the 3′ end of the tRNA is one of many reactions coupled to the energy-releasing hydrolysis of ATP (see Figure 3–30), and it produces a high-energy bond between the charged tRNA and the amino acid. The energy of this bond is used at a later stage in protein synthesis to link the amino acid covalently to the growing polypeptide chain.

QUESTION 7–4

In a clever experiment performed in 1962, a cysteine already attached to its tRNA was chemically converted to an alanine. These "hybrid" tRNA molecules were then added to a cell-free translation system from which the normal cysteine-tRNAs had been removed. When the resulting protein was analyzed, it was found that alanine had been inserted at every point in the protein chain where cysteine was supposed to be. Discuss what this experiment tells you about the role of aminoacyl-tRNA synthetases during the normal translation of the genetic code.

The RNA Message Is Decoded on Ribosomes

The recognition of a codon by the anticodon on a tRNA molecule depends on the same type of complementary base-pairing used in DNA replication and transcription. However, accurate and rapid translation of mRNA into protein requires a large molecular machine that moves along the mRNA, captures complementary tRNA molecules, holds them in position, and covalently links the amino acids that they carry so as to form a protein chain. This protein-manufacturing machine is the **ribosome**—a large complex made from more than 50 different proteins (the *ribosomal proteins*) and several RNA molecules called **ribosomal RNAs (rRNAs)**. A typical living cell contains millions of ribosomes in its cytoplasm (Figure 7–30).

Eucaryotic and procaryotic ribosomes are very similar in structure and in function. Both are composed of one large and one small subunit that fit together to form a complete ribosome with a mass of several mil-

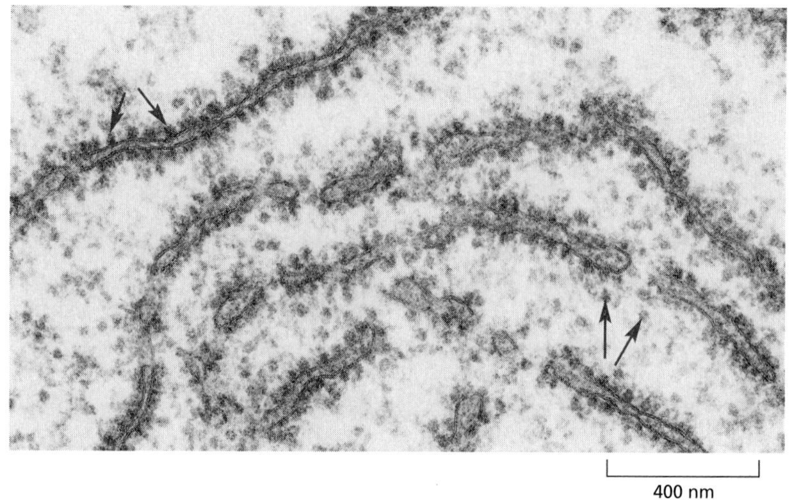

Figure 7–30 **Ribosomes are found in the cytoplasm of a eucaryotic cell.** This electron micrograph shows a thin section of a small region of cytoplasm. The ribosomes appear as black dots (*red* arrows). Some are free in the cytosol; others are attached to membranes of the endoplasmic reticulum. (Courtesy of George Palade.)

lion daltons (Figure 7–31); for comparison, an average-sized protein has a mass of 30,000 daltons. The small subunit matches the tRNAs to the codons of the mRNA, while the large subunit catalyzes the formation of the peptide bonds that covalently link the amino acids together into a polypeptide chain. The two subunits come together on an mRNA molecule, usually near its beginning (5′ end), to begin the synthesis of a protein. The mRNA is then pulled through the ribosome like a long piece of tape. As the mRNA moves through it, the ribosome translates the nucleotide sequence into an amino acid sequence one codon at a time, using the tRNAs as adaptors. Thus, each amino acid is added in the correct sequence to the end of the growing polypeptide chain (Movie 7.7). Finally, the two subunits of the ribosome separate when synthesis of

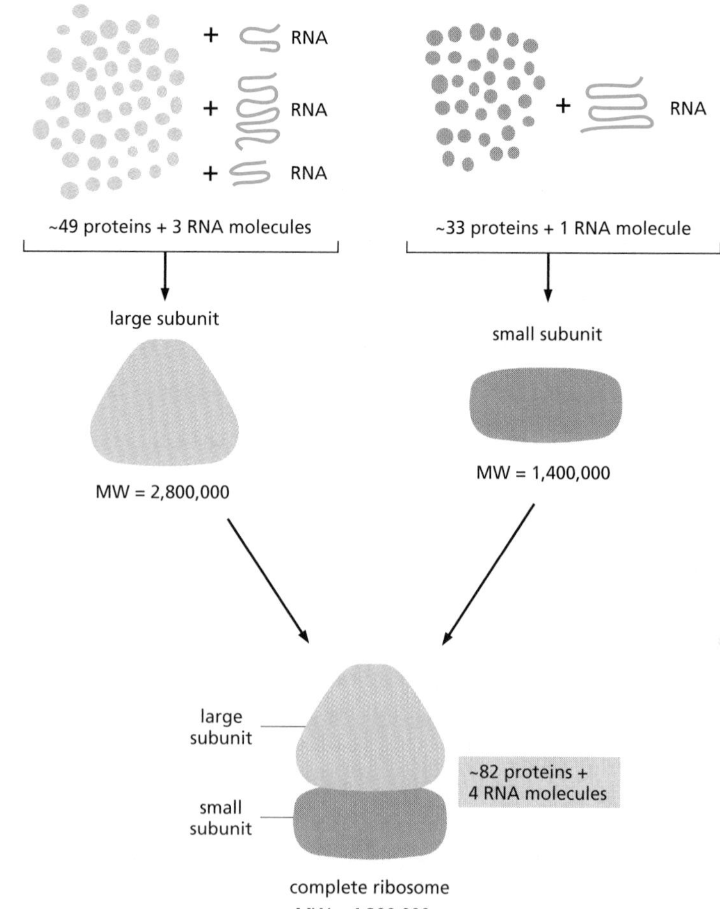

Figure 7–31 **A ribosome is a large complex of four RNAs and more than 80 proteins.** Shown here are the components of eucaryotic ribosomes. Procaryotic ribosomes are very similar. Although ribosomal proteins greatly outnumber ribosomal RNAs, the RNAs account for more than half the mass of the ribosome.

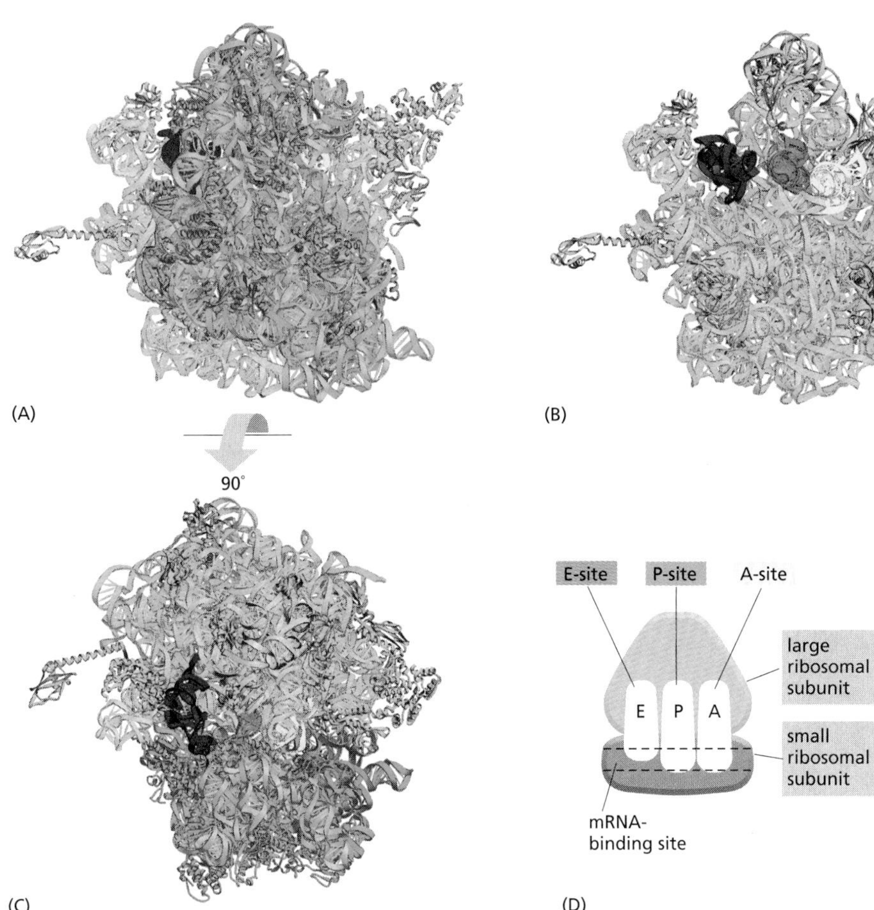

Figure 7–32 **Each ribosome has a binding site for mRNA and three binding sites for tRNA.** The tRNA sites are designated the A-, P-, and E-sites (short for aminoacyl-tRNA, peptidyl-tRNA, and exit, respectively). (A) Three-dimensional structure of a bacterial ribosome, as determined by X-ray crystallography with the small subunit in *dark green* and the large subunit in *light green*. Both the rRNAs and the ribosomal proteins are shown. tRNAs are shown bound in the E-site (*red*), the P-site (*orange*), and the A-site (*yellow*). Although all three tRNA sites are shown occupied here, during the process of protein synthesis only two of these sites are occupied at any one time (see Figure 7–33). (B) Structure of the large (*left*) and small (*right*) ribosomal subunits arranged as though the ribosome in (A) were opened like a book. (C) Structure of the ribosome in (A) viewed from the top. (D) Highly schematized representation of a ribosome (in the same orientation as C), which will be used in subsequent figures. (A, B, and C, adapted from M.M. Yusupov et al., *Science* 292:883–896, 2001, with permission from AAAS. Courtesy of Albion Bausom and Harry Noller.)

the protein is finished. Ribosomes operate with remarkable efficiency; in one second a ribosome in a eucaryotic cell adds about 2 amino acids to a polypeptide chain; bacterial ribosomes operate even faster, at a rate of about 20 amino acids per second.

How does the ribosome choreograph all the movements required for translation? Each ribosome contains a binding site for an mRNA molecule and three binding sites for tRNA molecules, called the A-site, the P-site, and the E-site (Figure 7–32). To add an amino acid to a growing peptide chain, the appropriate charged tRNA enters the A-site by base-pairing with the complementary codon on the mRNA molecule. Its amino acid is then linked to the peptide chain held by the tRNA in the neighboring P-site. Next, the ribosome shifts, and the spent tRNA is moved to the E-site before being ejected (Figure 7–33). This cycle of reactions is repeated each time an amino acid is added to the polypeptide chain, with the chain growing from its amino to its carboxyl end until a stop codon is encountered.

The Ribosome Is a Ribozyme

The ribosome is one of the largest and most complex structures in the cell, composed of two-thirds RNA and one-third protein. The determination of the entire three-dimensional structure of its large and small subunits in 2000 was a major triumph of modern biology. The structure strongly confirms earlier evidence that the rRNAs—not the proteins—are responsible for the ribosome's overall structure and its ability to choreograph protein synthesis.

The rRNAs are folded into highly compact, precise three-dimensional structures that form the core of the ribosome (Figure 7–34). In marked contrast to the central positioning of the rRNA, the ribosomal proteins

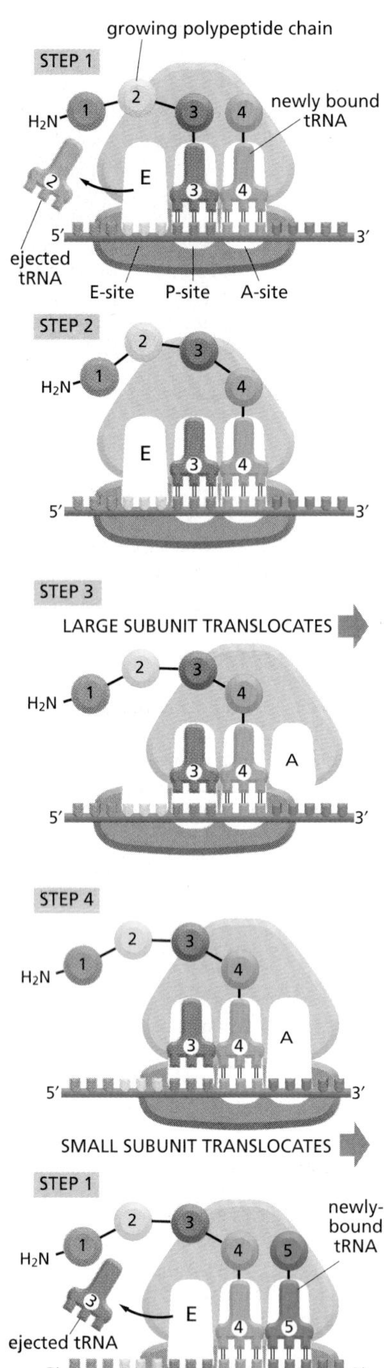

Figure 7–33 **Translation takes place in a four-step cycle.** This cycle is repeated over and over during the synthesis of a protein. In step 1, a tRNA carrying the next amino acid in the chain binds to the vacant A-site on the ribosome by forming base pairs with the codon that is exposed there. Because only one of the many types of tRNA molecules in a cell can base-pair with each codon, this codon determines the specific amino acid to be added to the growing polypeptide chain. The A- and P-sites are sufficiently close together that their two tRNA molecules are forced to form base pairs with codons that are contiguous, with no stray bases in between. This positioning of the tRNAs ensures that the correct reading frame will be preserved throughout the synthesis of the protein. In step 2, the carboxyl end of the polypeptide chain is uncoupled from the tRNA at the P-site and joined by a peptide bond to the free amino group of the amino acid linked to the tRNA at the A-site. This reaction is catalyzed by an enzymatic site in the large subunit. In step 3, a shift of the large subunit relative to the small subunit moves the two tRNAs into the E- and P-sites of the large subunit. In step 4, the small subunit moves exactly three nucleotides along the mRNA molecule, bringing it back to its original position relative to the large subunit. This movement resets the ribosome with an empty A-site so that the next aminoacyl-tRNA molecule can bind (Movie 7.8). As indicated, the mRNA is translated in the 5′-to-3′ direction, and the N-terminal end of a protein is made first, with each cycle adding one amino acid to the C-terminus of the polypeptide chain. To watch the translation cycle in atomic detail, see Movie 7.9.

are generally located on the surface, where they fill the gaps and crevices of the folded RNA. The main role of the ribosomal proteins seems to be to fold and stabilize the RNA core, while permitting the changes in rRNA conformation that are necessary for this RNA to catalyze efficient protein synthesis.

Not only are the three tRNA-binding sites (the A-, P-, and E-sites) on the ribosome formed primarily by the rRNAs, but the catalytic site for peptide bond formation is formed by the 23S RNA of the large subunit; the nearest amino acid is located too far away to make contact with the incoming aminoacyl-tRNA or with the growing polypeptide chain. The catalytic site in this RNA-based peptidyl transferase is similar in many respects to that found in some protein enzymes: it is a highly structured pocket that precisely orients the two reactants, the elongating peptide and the charged tRNA, thereby greatly increasing the probability of a productive reaction.

RNA molecules that possess catalytic activity are called **ribozymes**. In the final section of this chapter, we consider other ribozymes and discuss what the existence of RNA-based catalysis might mean for the early evolution of life on Earth. Here we need only note that there is good reason to suspect that RNA rather than protein molecules served as the first catalysts for living cells. If so, the ribosome, with its RNA core, could be viewed as a relic of an earlier time in life's history, when cells were run almost entirely by ribozymes.

QUESTION 7–5

The following sequence of nucleotides in a DNA strand was used as a template to synthesize an mRNA that was then translated into protein: 5′-TTAACGGCTTTTTTC-3′. Predict the C-terminal amino acid and the N-terminal amino acid of the resulting polypeptide. Assume that the mRNA is translated without the need for a start codon.

Codons in mRNA Signal Where to Start and to Stop Protein Synthesis

Although in the test tube, ribosomes can be forced to translate any RNA (see How We Know, pp. 248–249), in the cell a specific start signal is required. The site at which protein synthesis begins on an mRNA is crucial, because it sets the reading frame for the whole length of the message. An error of one nucleotide either way at this stage will cause every subsequent codon in the message to be misread, resulting in a nonfunctional protein with a garbled sequence of amino acids (see Figure 7–27).

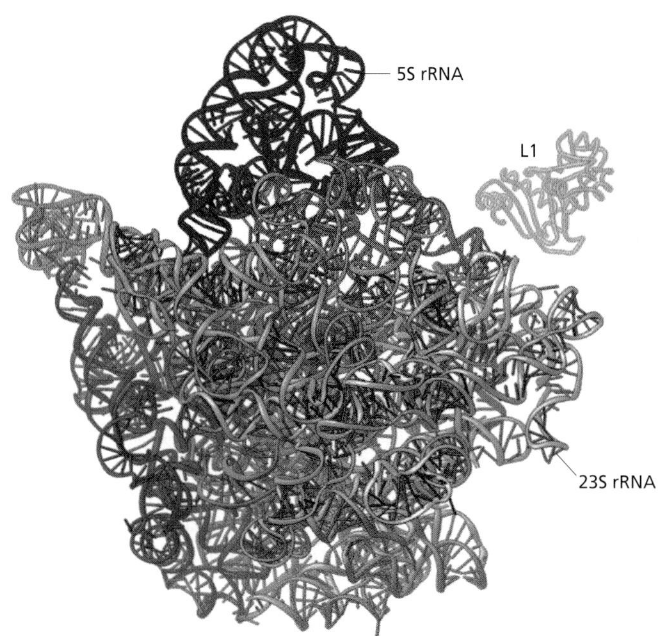

Figure 7–34 **Ribosomal RNAs give the ribosome its overall shape.** Shown here are the detailed structures of the two rRNAs—the 23S rRNA (*blue*) and the 5S rRNA (*purple*)—that form the core of the large subunit of a bacterial ribosome. One of the protein subunits of the ribosome (L1) is included as a reference point, as this protein forms a characteristic protrusion on the ribosome surface. Ribosomal components are commonly designated by their "S values," which refer to their rate of sedimentation in an ultracentrifuge. (Adapted from N. Ban et al., *Science* 289:905–920, 2000. With permission from AAAS.)

The initiation step is also of great importance in another respect, because it is the last point at which the cell can decide whether the mRNA is to be translated; the rate of initiation thus determines the rate at which the protein is synthesized from the RNA.

The translation of an mRNA begins with the codon AUG, and a special tRNA is required to initiate translation. This **initiator tRNA** always carries the amino acid methionine (or a modified form of methionine, formylmethionine, in bacteria) so that newly made proteins all have methionine as the first amino acid at their N-terminal end, the end of a protein that is synthesized first. This methionine is usually removed later by a specific protease. The initiator tRNA is distinct from the tRNA that normally carries methionine.

In eucaryotes, the initiator tRNA, coupled to methionine, is first loaded into a small ribosomal subunit, along with additional proteins called **translation initiation factors** (Figure 7–35). Of all the charged tRNAs in the cell, only the charged initiator tRNA is capable of binding tightly to the P-site of the small ribosomal subunit. Next, the loaded ribosomal subunit binds to the 5′ end of an mRNA molecule, which is signaled by the cap that is present on eucaryotic mRNA (see Figure 7–16). The small ribosomal subunit then moves forward (5′ to 3′) along the mRNA searching for the first AUG. When this AUG is encountered, several initiation factors dissociate from the small ribosomal subunit to make way for the large ribosomal subunit to assemble and complete the ribosome. Because the initiator tRNA is bound to the P-site, protein synthesis is ready to begin with the addition of the next charged tRNA to the A-site (see Figure 7–35).

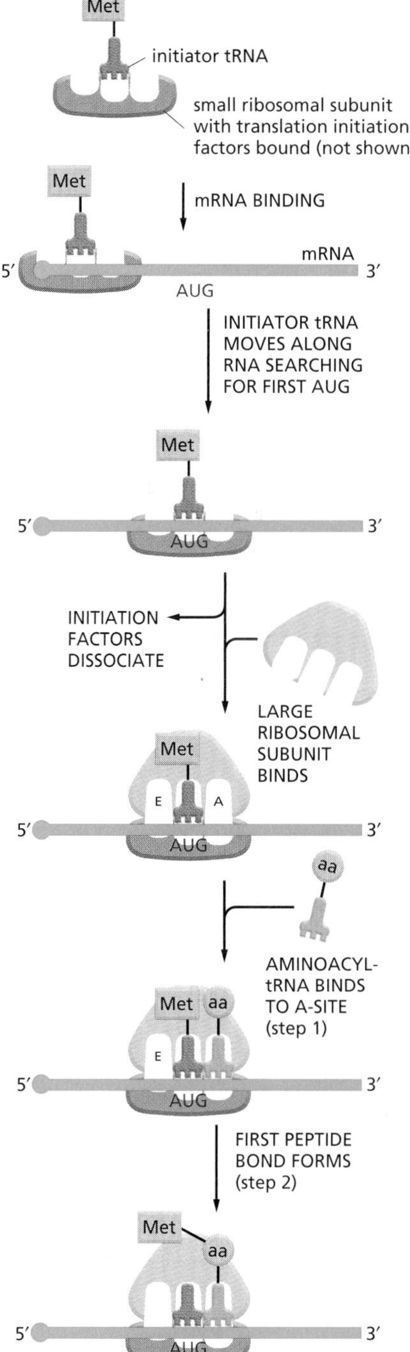

Figure 7–35 **Initiation of protein synthesis in eucaryotes requires initiation factors and a special initiator tRNA.** Although not shown here, efficient translation initiation also requires additional proteins (shown in Figure 7–22) that are bound at the 5′ cap and poly-A tail of the mRNA. In this way, the translation apparatus can ascertain that both ends of the mRNA are intact before initiating translation. Following initiation, the protein is elongated by the reactions outlined in Figure 7–33.

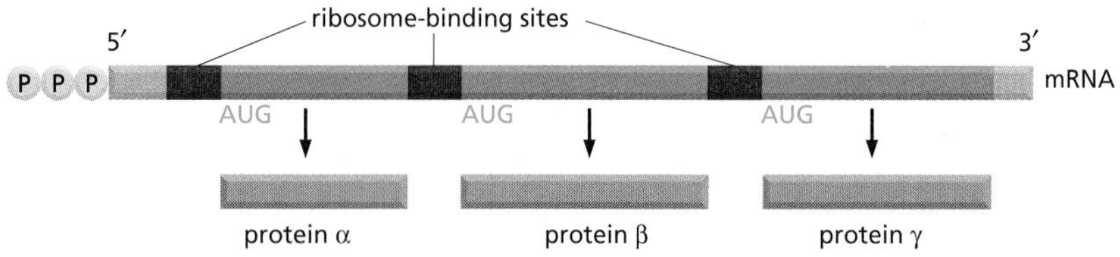

ribosome-binding sites
5′ 3′
P P P mRNA
AUG AUG AUG
protein α protein β protein γ

Figure 7–36 **A single procaryotic mRNA molecule can encode several different proteins.** In procaryotes, genes directing the different steps in a process are often organized into clusters (operons) that are transcribed together into a single mRNA. Unlike eucaryotic ribosomes, which recognize a 5′ cap, procaryotic ribosomes initiate transcription at ribosome-binding sites *(red)*, which can be located in the interior of an mRNA molecule. This feature enables procaryotes to synthesize several separate proteins from a single mRNA molecule.

The mechanism for selecting a start codon is different in bacteria. Bacterial mRNAs have no 5′ caps to tell the ribosome where to begin searching for the start of translation. Instead, they contain specific ribosome-binding sequences, up to six nucleotides long, that are located a few nucleotides upstream of the AUGs at which translation is to begin. Unlike a eucaryotic ribosome, a procaryotic ribosome can readily bind directly to a start codon that lies in the interior of an mRNA, as long as a ribosome-binding site precedes it by several nucleotides. Such ribosome-binding sequences are necessary in bacteria, as procaryotic mRNAs are often *polycistronic*—that is, they encode several different proteins, each of which is translated from the same mRNA molecule (Figure 7–36). In contrast, a eucaryotic mRNA usually carries the information for a single protein.

The end of the protein-coding message in both procaryotes and eucaryotes is signaled by the presence of one of several codons called *stop codons* (see Figure 7–26). These special codons—UAA, UAG, and UGA—are not recognized by a tRNA and do not specify an amino acid, but instead signal to the ribosome to stop translation. Proteins known as *release factors* bind to any stop codon that reaches the A-site on the ribosome, and this binding alters the activity of the peptidyl transferase in the ribosome, causing it to catalyze the addition of a water molecule instead of an amino acid to the peptidyl-tRNA (Figure 7–37). This reaction frees the carboxyl end of the polypeptide chain from its attachment to a tRNA molecule, and because this is the only attachment that holds the growing polypeptide to the ribosome, the completed protein chain is immediately released into the cytosol. The ribosome releases the mRNA and dissociates into its two separate subunits, which can then assemble on another mRNA molecule to begin a new round of protein synthesis.

We saw in Chapter 4 that many proteins can fold into their three-dimensional shape spontaneously, and some do so as they are spun out of the ribosome. Most proteins, however, require *molecular chaperones* to help them fold correctly in the cell. The best-studied chaperones use successive rounds of ATP hydrolysis to continually bind and release newly synthesized proteins until they are properly folded. This handling 'steers' the proteins along productive folding pathways by preventing them from forming aggregates or other misfolded structures. Newly synthesized proteins are typically met by their chaperones as they emerge from the ribosome.

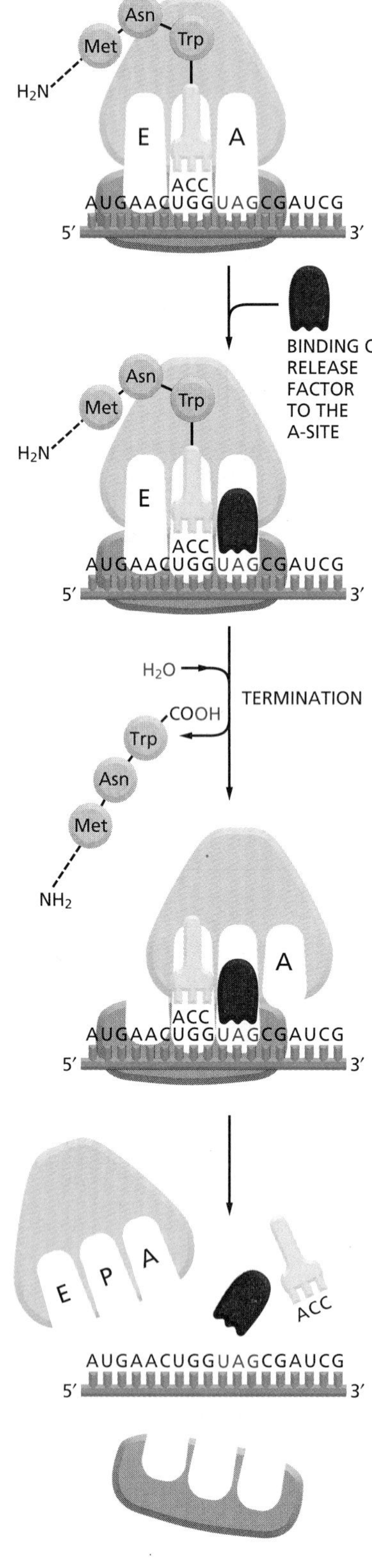

Figure 7–37 **In the final phase of protein synthesis, the binding of release factor to an A-site bearing a stop codon terminates translation.** The completed polypeptide is released, and the ribosome dissociates into its two separate subunits.

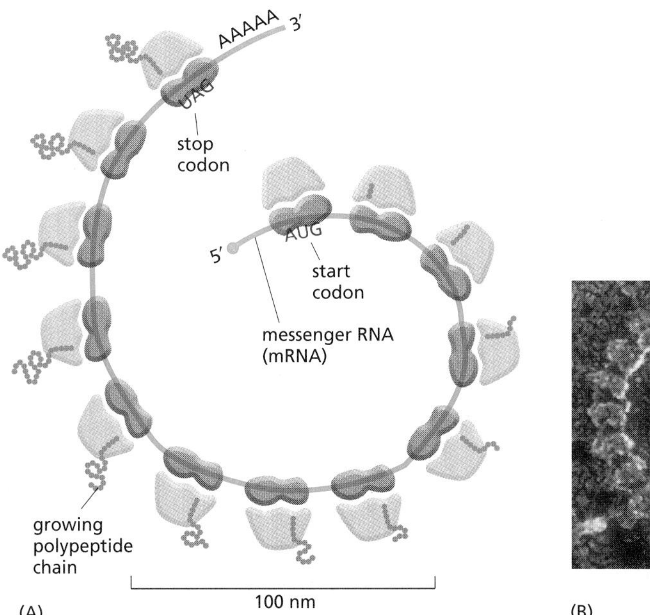

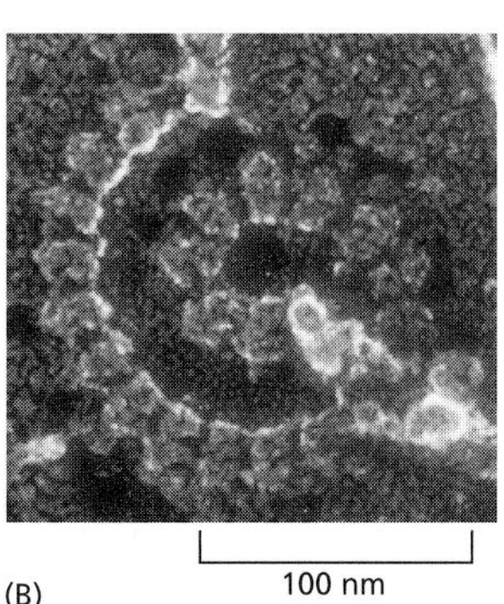

Figure 7–38 **Proteins are translated by polyribosomes.** (A) Schematic drawing showing how a series of ribosomes can simultaneously translate the same mRNA molecule (Movie 7.10) (B) Electron micrograph of a polyribosome from a eucaryotic cell. (B, courtesy of John Heuser.)

Proteins Are Made on Polyribosomes

The synthesis of most protein molecules takes between 20 seconds and several minutes. But even during this very short period, multiple initiations usually take place on each mRNA molecule being translated. If the mRNA is being translated efficiently, a new ribosome hops onto the 5′ end of the mRNA molecule almost as soon as the preceding ribosome has translated enough of the nucleotide sequence to move out of the way. The mRNA molecules being translated are therefore usually found in the form of *polyribosomes* (also known as *polysomes*), large cytoplasmic assemblies made up of several ribosomes spaced as close as 80 nucleotides apart along a single mRNA molecule (Figure 7–38). These multiple initiations mean that many more protein molecules can be made in a given time than would be possible if each had to be completed before the next could start.

Both bacteria and eucaryotes make use of polysomes, but bacteria can speed up the rate of protein synthesis even further. Because bacterial mRNA does not need to be processed and is also physically accessible to ribosomes while it is being made, ribosomes will typically attach to the free end of a bacterial mRNA molecule and start translating it even before the transcription of that RNA is complete; these ribosomes follow closely behind the RNA polymerase as it moves along DNA.

Inhibitors of Procaryotic Protein Synthesis Are Used as Antibiotics

The ability to translate mRNAs accurately into proteins is a fundamental feature of all life on Earth. Although the ribosome and other molecules that carry out this enormous task are very similar among organisms, there are some subtle differences, as we have seen, in the way that bacteria and eucaryotes synthesize proteins. Through a quirk of evolution, these differences form the basis of one of the most important advances in modern medicine.

Many of our most effective antibiotics are compounds that act by inhibiting bacterial, but not eucaryotic, protein synthesis. Some of these drugs exploit the small structural and functional differences between bacterial and eucaryotic ribosomes, so that they interfere preferentially with bacterial protein synthesis. These compounds can thus be taken in high

TABLE 7–3 ANTIBIOTICS THAT INHIBIT PROTEIN OR RNA SYNTHESIS

ANTIBIOTIC	SPECIFIC EFFECT
Tetracycline	blocks binding of aminoacyl-tRNA to A-site of ribosome (step 1 in Figure 7–33)
Streptomycin	prevents the transition from initiation complex to chain-elongating ribosome (see Figure 7–35); also causes miscoding
Chloramphenicol	blocks the peptidyl transferase reaction on ribosomes (step 2 in Figure 7–33)
Cycloheximide	blocks the translocation reaction on ribosomes (step 3 in Figure 7–33)
Rifamycin	blocks initiation of RNA chains by binding to RNA polymerase

doses without being toxic to humans. Because different antibiotics bind to different regions of the bacterial ribosome, these drugs often inhibit different steps in protein synthesis. A few of the antibiotics of this kind are listed in Table 7–3.

Many common antibiotics were first isolated from fungi. Fungi and bacteria often occupy the same ecological niches; to gain a competitive edge, fungi have evolved, over time, potent toxins that kill bacteria but are harmless to themselves. Because fungi and humans are both eucaryotes, and are thus more closely related to one another than either is to bacteria (see Figure 1–29), we have been able to borrow these compounds to combat our own bacterial enemies.

Carefully Controlled Protein Breakdown Helps Regulate the Amount of Each Protein in a Cell

After a protein is released from the ribosome, it becomes subject to a number of controls by the cell. The number of copies of a protein in a cell depends, like the human population, not only on how quickly new individuals are made but also on how long they survive. So the breakdown of proteins into their constituent amino acids by cells is a way of regulating the amount of a particular protein present at a given time. Proteins vary enormously in their life span. Structural proteins that become part of a fairly permanent tissue such as bone or muscle may last for months or even years, whereas other proteins, such as metabolic enzymes and those that regulate the cycle of cell growth, mitosis, and division (discussed in Chapter 18), last only for days, hours, or even seconds. How does the cell control these lifetimes?

Cells possess specialized pathways to enzymatically break proteins down into their constituent amino acids (a process termed *proteolysis*). The enzymes that degrade proteins, first to short peptides and finally to individual amino acids, are known collectively as **proteases**. Proteases act by cutting (hydrolyzing) the peptide bonds between amino acids (see Panel 2–5, pp. 72–73). One function of proteolytic pathways is to rapidly degrade those proteins whose lifetimes must be short. Another is to recognize and eliminate proteins that are damaged or misfolded. Eliminating improperly folded proteins is critical for an organism, as neurodegenerative disorders such as Huntington's, Alzheimer's, and Creutzfeldt–Jacob diseases are caused by the aggregation of misfolded proteins. These protein aggregates can severely damage cells and tissues and even trigger cell death.

Although most damaged proteins are broken down in the cytosol, important protein degradation pathways also operate in other compart-

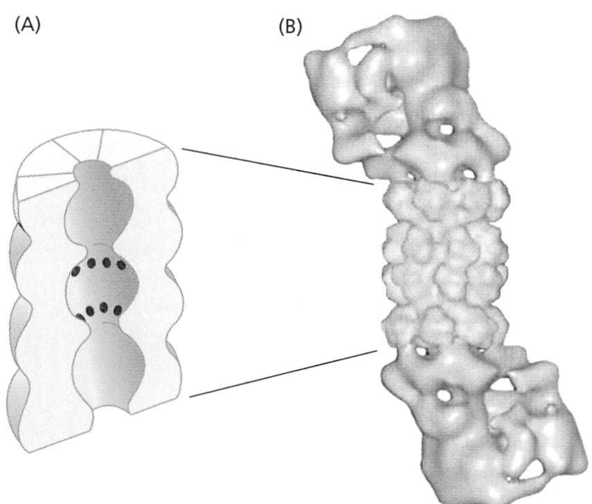

Figure 7–39 **The proteasome degrades short-lived and misfolded proteins.** (A) A cutaway view of the structure of the central cylinder of the proteasome, determined by X-ray crystallography, with the active sites of the proteases indicated by *red dots*. (B) The structure of the entire proteasome, in which the central cylinder (*yellow*) is topped by a stopper (*purple*) at each end. (B, from W. Baumeister et al., *Cell* 92:367–380, 1998. With permission from Elsevier.)

ments in eucaryotic cells, such as lysosomes (discussed in Chapter 15). In eucaryotic cells, most proteins are broken down by machines called **proteasomes**. A proteasome contains a central cylinder formed from proteases whose active sites face into an inner chamber. Each end of the cylinder is stoppered by a large protein complex formed from at least 10 different types of protein subunits (Figure 7–39). These protein stoppers bind the proteins destined for digestion and then—using ATP hydrolysis to fuel their activity—feed the doomed molecules into the inner chamber of the cylinder; there the proteases chop the proteins into short peptides, which are then released from the end of the proteasome. Housing proteases inside this molecular chamber of destruction makes sense, as it prevents them from running rampant throughout the cell.

How does the proteasome select which proteins in the cell should enter the cylinder and be degraded? Proteasomes act primarily on proteins that have been marked for destruction by the covalent attachment of a small protein called *ubiquitin*. Specialized enzymes tag selected proteins with one or more ubiquitin molecules; these ubiquitylated proteins are then recognized and sucked into the proteasome by proteins in the stopper.

Proteins that are meant to be short-lived often sport a short amino acid sequence that identifies the protein as one to be ubiquitylated and ultimately degraded by the proteasome. Denatured or misfolded proteins, as well as proteins containing oxidized or otherwise abnormal amino acids, are also recognized and degraded by this ubiquitin-dependent proteolytic system. The enzymes that add ubiquitin to such proteins recognize signals that become exposed on these proteins as a result of the misfolding or chemical damage—for example, amino acid sequences or conformational motifs that are normally buried and inaccessible.

There Are Many Steps Between DNA and Protein

We have seen so far in this chapter that many different types of chemical reactions are required to produce a protein from the information contained in a gene (Figure 7–40). The final concentration of a protein in a cell therefore depends on the efficiency with which each of the many steps is carried out. In addition, many proteins—once they leave the ribosome—require further attention before they are useful to the cell. For example, covalent modification (such as phosphorylation), the binding of small-molecule cofactors, or association with other protein subunits are often needed to activate a newly synthesized protein (Figure 7–41).

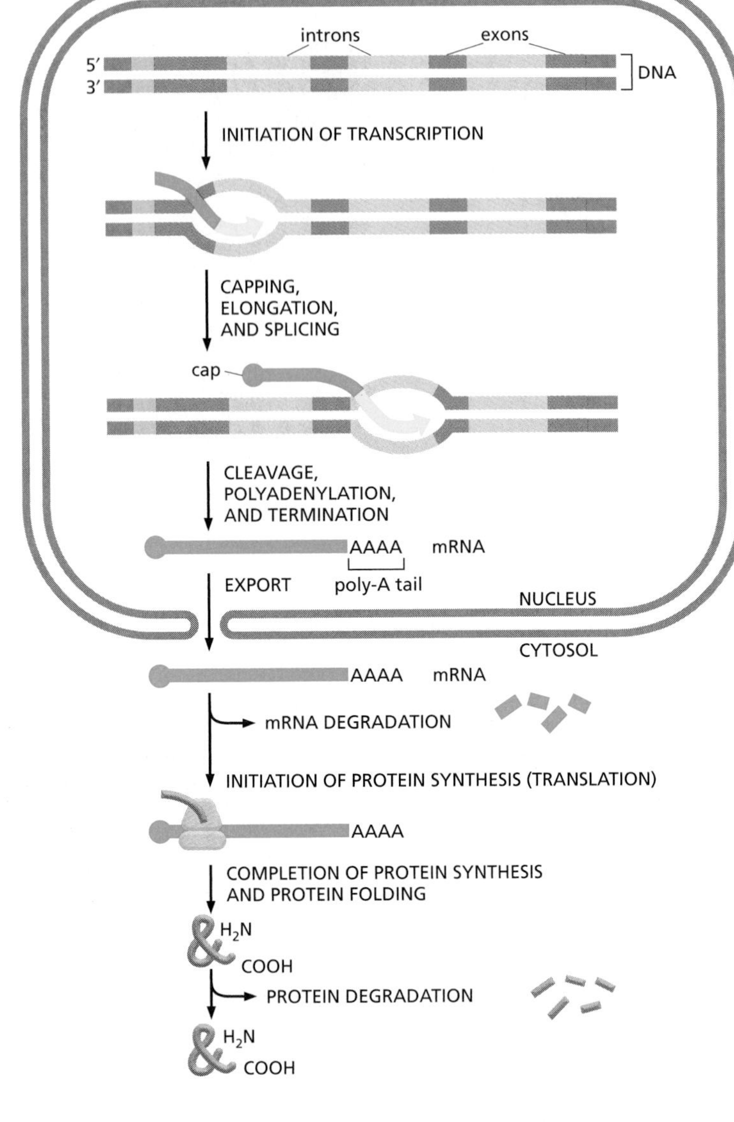

Figure 7–40 **Protein production in a eucaryotic cell requires many steps.** The final concentration of each protein depends on the efficiency of each step depicted. Even after a protein has been translated, its concentration can be regulated by degradation, and its activity regulated by the binding of small molecules and other post-translational modifications (see Figure 7–41).

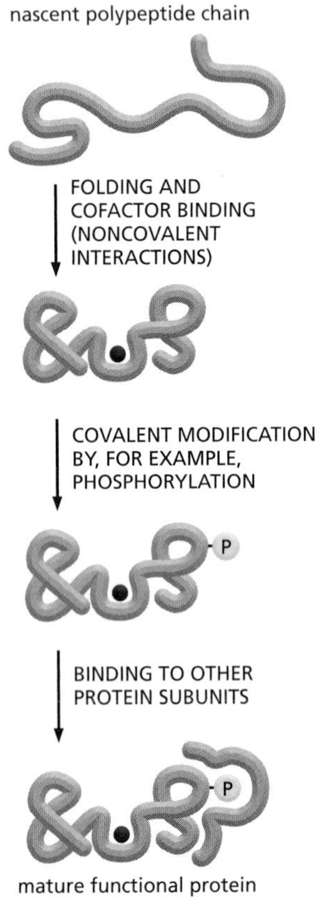

nascent polypeptide chain

FOLDING AND COFACTOR BINDING (NONCOVALENT INTERACTIONS)

COVALENT MODIFICATION BY, FOR EXAMPLE, PHOSPHORYLATION

P

BINDING TO OTHER PROTEIN SUBUNITS

P

mature functional protein

We will see in the next chapter that cells have the ability to change the concentrations of most of their proteins according to their needs. In principle, all of the steps in Figure 7–40 can be regulated by the cell. However, as we shall see in the next chapter, the initiation of transcription is the most common point for a cell to regulate the expression of each of its genes.

Transcription and translation are universal processes that lie at the heart of life. However, when scientists came to consider how the flow of information from DNA to protein might have originated, they came to some unexpected conclusions.

Figure 7–41 **Many proteins require additional modification to become fully functional.** To be useful to the cell, a completed polypeptide must fold correctly into its three-dimensional conformation, bind any cofactors required, and assemble with any protein partners. Noncovalent bond formation drives these changes. Many proteins also require covalent modification to become active. Although phosphorylation and glycosylation are the most common, more than 100 different types of covalent modifications of proteins are known.

RNA AND THE ORIGINS OF LIFE

To understand fully the processes occurring in present-day cells, we need to consider how cells evolved. But one key process—that of gene expression—presents a particular challenge: if nucleic acids are required to direct the synthesis of proteins, and proteins are required to synthesize nucleic acids, how could this system of interdependent components have arisen? One view is that an *RNA world* existed on Earth before modern cells appeared (Figure 7–42). According to this hypothesis, RNA—which today serves as an intermediate between genes and proteins—both stored genetic information and catalyzed chemical reactions in primitive cells. Only later in evolutionary time did DNA take over as the genetic material and proteins become the major catalysts and structural components of cells. If this idea is correct, then the transition out of the RNA world was never completed; as we have seen in this chapter, RNA still catalyzes several fundamental reactions in modern cells. These RNA catalysts, including the ribosome and RNA-splicing machinery, can thus be viewed as molecular fossils of an earlier world.

Life Requires Autocatalysis

The origin of life requires molecules that possess, if only to a small extent, one crucial property: the ability to catalyze reactions that lead—directly or indirectly—to the production of more molecules like themselves. Catalysts with this special self-promoting property, once they had arisen by chance, would reproduce themselves and would therefore divert raw materials from the production of other substances. In this way one can envisage the gradual development of an increasingly complex chemical system of organic monomers and polymers that function together to generate more molecules of the same types, fueled by a supply of simple raw materials in the environment. Such an *autocatalytic* system would have many of the properties we think of as characteristic of living matter: the system would contain a far from random selection of interacting molecules; it would tend to reproduce itself; it would compete with other systems dependent on the same raw materials; and, if deprived of its raw materials or maintained at a temperature that upset the balance of reaction rates, it would decay toward chemical equilibrium and 'die.'

But what molecules could have had such autocatalytic properties? In present-day living cells the most versatile catalysts are proteins, which are able to adopt diverse three-dimensional forms that bristle with chemically reactive sites. However, there is no known way in which a protein can reproduce itself directly. Nucleic acids, however, can do both of these things.

RNA Can Both Store Information and Catalyze Chemical Reactions

We have seen that complementary base-pairing enables one nucleic acid to act as a template for the formation of another. Thus, a single strand

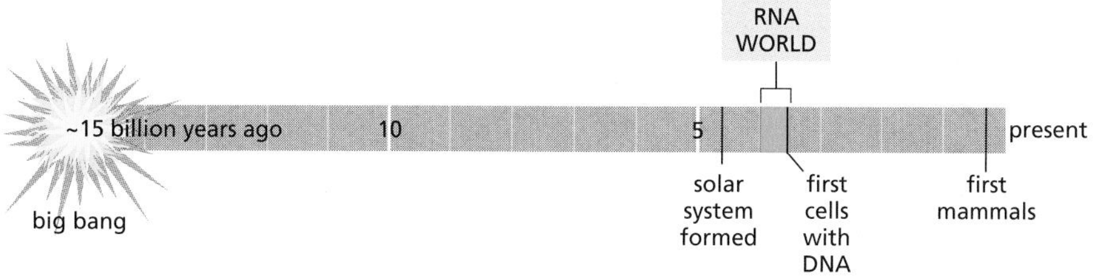

Figure 7–42 **An RNA world may have existed before modern cells arose.**

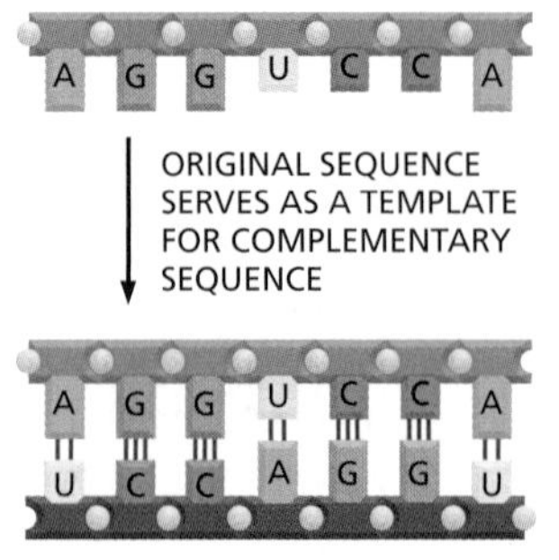

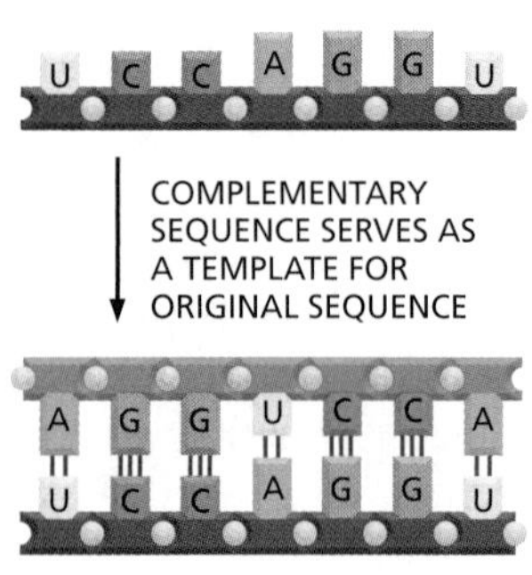

Figure 7–43 **An RNA molecule can in principle guide the formation of an exact copy of itself.** In the first step the original RNA molecule acts as a template to form an RNA molecule of complementary sequence. In the second step this complementary RNA molecule itself acts as a template, forming RNA molecules of the original sequence. Since each templating molecule can produce many copies of the complementary strand, these reactions can result in the "multiplication" of the original sequence.

of RNA or DNA can specify the sequence of a complementary polynucleotide, which, in turn, can specify the sequence of the original molecule, allowing the original nucleic acid to be replicated (Figure 7–43). Such complementary templating mechanisms lie at the heart of DNA replication and transcription in modern-day cells.

But the efficient synthesis of polynucleotides by such complementary templating mechanisms also requires catalysts to promote the polymerization reaction: without catalysts, polymer formation is slow, error-prone, and inefficient. Today, nucleotide polymerization is rapidly catalyzed by protein enzymes—such as the DNA and RNA polymerases. But how could this reaction be catalyzed before proteins with the appropriate catalytic ability existed? The beginnings of an answer to this question were obtained in 1982, when it was discovered that RNA molecules themselves can act as catalysts. We have already seen in this chapter, for example, that a molecule of RNA catalyses the peptidyl transferase reaction that takes place on the ribosome. The unique potential of RNA molecules to act both as information carriers and as catalysts is thought to have enabled them to play the central role in the origin of life.

In present-day cells, RNA is synthesized as a single-stranded molecule, and we have seen that complementary base-pairing can occur between nucleotides in the same chain (see Figure 7–5). This base-pairing, along with "nonconventional" hydrogen bonds, can cause each RNA molecule to fold up in a unique way that is determined by its nucleotide sequence. Such associations produce complex three-dimensional patterns of folding, where the molecule as a whole adopts a unique shape.

As we saw in Chapter 4, protein enzymes are able to catalyze biochemical reactions because they have surfaces with unique contours and chemical properties on which a given substrate can react. In the same way, RNA molecules, with their unique folded shapes, can serve as enzymes (Figure 7–44), although the fact that they are constructed of only four different subunits limits their catalytic efficiency and their range of chemical reactions compared with proteins. Nonetheless, ribozymes can catalyze many types of chemical reactions. Most of the ribozymes that have been studied were constructed in the laboratory (Table 7–4); relatively few catalytic RNAs exist in present-day cells. But the processes in which catalytic RNAs still seem to have major roles include some of the most fundamen-

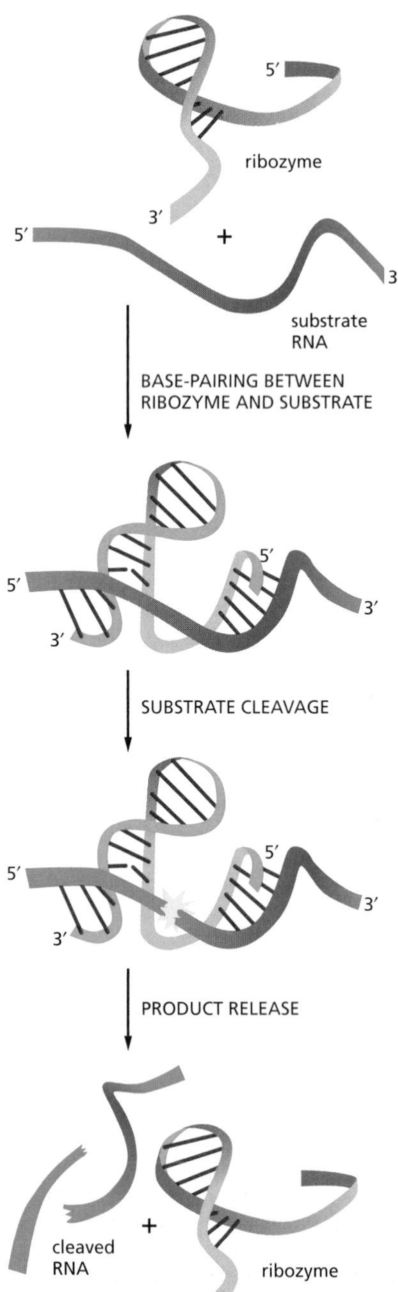

Figure 7–44 **A ribozyme is an RNA molecule that possesses catalytic activity.** The simple RNA molecule shown catalyzes the cleavage of a second RNA at a specific site. This ribozyme is found embedded in larger RNA genomes—called viroids—that infect plants. The cleavage, which occurs in nature at a distant location on the same RNA molecule that contains the ribozyme, is a step in the replication of the RNA genome. This reaction also requires a magnesium ion (not shown), which is brought next to the site of cleavage on the substrate. (Adapted from T.R. Cech and O.C. Uhlenbeck, *Nature* 372:39–40, 1994. With permission from Macmillan Publishers Ltd.)

TABLE 7–4 BIOCHEMICAL REACTIONS THAT CAN BE CATALYZED BY RIBOZYMES

ACTIVITY	RIBOZYMES
RNA cleavage, RNA ligation	self-splicing RNAs
DNA cleavage	self-splicing RNAs
Peptide bond formation in protein synthesis	ribosomal RNA
DNA ligation	*in vitro* selected RNA
RNA splicing	self-splicing RNAs, RNAs of the spliceosome (?)
RNA polymerization	*in vitro* selected RNA
RNA phosphorylation	*in vitro* selected RNA
RNA aminoacylation	*in vitro* selected RNA
RNA alkylation	*in vitro* selected RNA
C–C bond rotation (isomerization)	*in vitro* selected RNA

tal steps in the expression of genetic information—especially those steps where RNA molecules themselves are spliced or translated into protein.

RNA, therefore, has all the properties required of a molecule that could catalyze its own synthesis (Figure 7–45). Although self-replicating systems of RNA molecules have not been found in nature, scientists appear to be well on the way to constructing them in the laboratory. Although this demonstration would not prove that self-replicating RNA molecules were essential in the origin of life on Earth, it would demonstrate that such a scenario is possible.

RNA Is Thought to Predate DNA in Evolution

The first cells on Earth would presumably have been much less complex and less efficient in reproducing themselves than even the simplest present-day cells. They would have consisted of little more than a simple membrane enclosing a set of self-replicating molecules and a few other components required to provide the materials and energy for their replication. If the evolutionary speculations about RNA outlined above are correct, these earliest cells would also have differed fundamentally from the cells we know today in having their hereditary information stored in RNA rather than DNA.

Evidence that RNA arose before DNA in evolution can be found in the chemical differences between them. Ribose (see Figure 7–3A), like glucose and other simple carbohydrates, is readily formed from formaldehyde (HCHO), which is one of the principal products of experiments simulating conditions on the primitive Earth. The sugar deoxyribose is harder to make, and in present-day cells it is produced from ribose in a reaction catalyzed by a protein enzyme, suggesting that ribose predates deoxy-

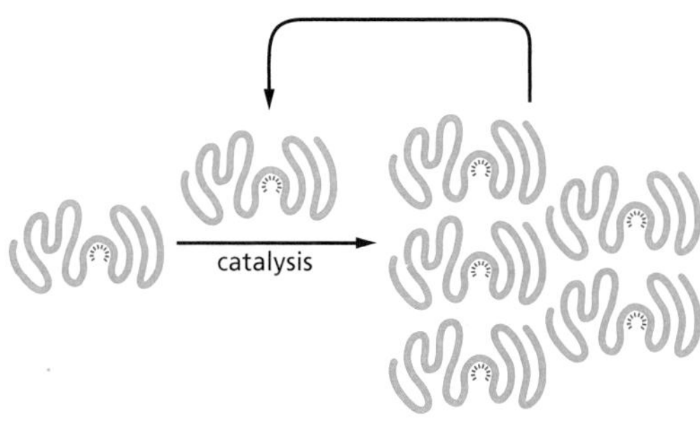

Figure 7–45 **Could an RNA molecule catalyze its own synthesis?** This hypothetical process would require the catalysis of both steps shown in Figure 7–43. The *red* rays represent the active site of this RNA enzyme.

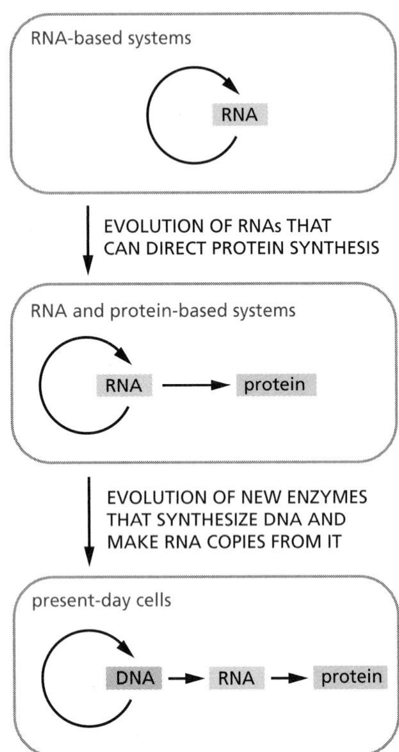

Figure 7–46 **RNA may have preceded DNA and proteins in evolution.** According to this idea, RNA molecules provided genetic, structural, and catalytic functions in the earliest cells. DNA is now the repository of genetic information, and proteins carry out almost all catalysis in cells. RNA now functions mainly as a go-between in protein synthesis, while remaining a catalyst for a few crucial reactions.

ribose in cells. Presumably, DNA appeared on the scene later, and then proved more suited than RNA as a permanent repository of genetic information. In particular, the deoxyribose in its sugar–phosphate backbone makes chains of DNA chemically much more stable than chains of RNA, so that greater lengths of DNA can be maintained without breakage.

The other differences between RNA and DNA—the double-helical structure of DNA and the use of thymine rather than uracil—further enhance DNA stability by making the molecule easier to repair. We saw in Chapter 6 that a damaged nucleotide on one strand of the double helix can be repaired by using the other strand as a template. Furthermore, deamination, one of the most common unwanted chemical changes occurring in polynucleotides, is easier to detect and repair in DNA than in RNA (see Figure 6–23). This is because the product of the deamination of cytosine is, by chance, uracil, which already exists in RNA, so that such damage would be impossible for repair enzymes to detect in an RNA molecule. However, in DNA, which has thymine rather than uracil, any uracil produced by the accidental deamination of cytosine is easily detected and repaired.

QUESTION 7–6

Discuss the following: "During the evolution of life on Earth, RNA lost its glorious position as the first self-replicating catalyst. Its role now is as a mere messenger in the information flow from DNA to protein."

Taken together, the evidence we have discussed points to the idea that RNA, having both genetic and catalytic properties, preceded DNA in evolution. As cells more closely resembling present-day cells appeared, it is believed that many of the functions originally performed by RNA were taken over by molecules more specifically fitted to the tasks required. Eventually DNA took over the primary genetic function, and proteins became the major catalysts, while RNA remained primarily as the intermediary connecting the two (Figure 7–46). With the advent of DNA, cells were able to become more complex, for they could then carry and transmit more genetic information than could be stably maintained in an RNA molecule. Because of the greater chemical complexity of proteins and the variety of chemical reactions they can catalyze, the shift (albeit incomplete) from RNA to proteins also provided a much richer source of structural components and enzymes. This enabled cells to evolve the great diversity of structure and function that we see in life today.

ESSENTIAL CONCEPTS

- **The flow of genetic information in all living cells is DNA → RNA → protein. The conversion of the genetic instructions in DNA into RNAs and proteins is termed gene expression.**
- **To express the genetic information carried in DNA, the nucleotide sequence of a gene is first transcribed into RNA. Transcription is catalyzed by the enzyme RNA polymerase. Nucleotide sequences in the DNA molecule indicate to the RNA polymerase where to start and stop transcribing.**
- **RNA differs in several respects from DNA. It contains the sugar ribose instead of deoxyribose and the base uracil (U) instead of thymine (T). RNAs in cells are synthesized as single-stranded molecules, which often fold up into precise three-dimensional shapes.**
- **Cells make several different functional types of RNAs, including messenger RNA (mRNA), which carries the instructions for making**

proteins; ribosomal RNA (rRNA), which is a component of ribosomes; and transfer RNA (tRNA), which acts as an adaptor molecule in protein synthesis.

- Transcription begins at DNA sites called promoters. To initiate transcription, eucaryotic RNA polymerases require the assembly of a complex of general transcription factors at the promoter, whereas bacterial RNA polymerase requires only an additional subunit, called sigma factor.
- In eucaryotic DNA, most genes are composed of a number of smaller coding regions (exons) interspersed with noncoding regions (introns). When a eucaryotic gene is transcribed from DNA into RNA, both the exons and introns are copied.
- Introns are removed from the RNA transcripts in the nucleus by the process of RNA splicing. In a reaction catalyzed by small ribonucleoprotein complexes known as snRNPs, the introns are excised from the RNA and the exons are joined together.
- Eucaryotic mRNAs go through several additional RNA processing steps before they leave the nucleus, including RNA capping and polyadenylation. These reactions, along with splicing, take place as the RNA is being transcribed. The mature mRNA is then transported to the cytoplasm.
- Translation of the nucleotide sequence of mRNA into a protein takes place in the cytoplasm on large ribonucleoprotein assemblies called ribosomes. As the mRNA is threaded through a ribosome, its message is translated into protein.
- The nucleotide sequence in mRNA is read in sets of three nucleotides (codons), each codon corresponding to one amino acid.
- The correspondence between amino acids and codons is specified by the genetic code. The possible combinations of the 4 different nucleotides in RNA give 64 different codons in the genetic code. Most amino acids are specified by more than one codon.
- tRNA acts as an adaptor molecule in protein synthesis. Enzymes called aminoacyl-tRNA synthetases link amino acids to their appropriate tRNAs. Each tRNA contains a sequence of three nucleotides, the anticodon, which matches a codon in mRNA by complementary base-pairing between codon and anticodon.
- Protein synthesis begins when a ribosome assembles at an initiation codon (AUG) in mRNA, a process that is regulated by proteins called translation initiation factors. The completed protein chain is released from the ribosome when a stop codon (UAA, UAG, or UGA) is reached.
- The stepwise linking of amino acids into a polypeptide chain is catalyzed by an rRNA molecule in the large ribosomal subunit. Thus, the ribosome is an example of a ribozyme, an RNA molecule that can catalyze a chemical reaction.
- The degradation of proteins in the cell is carefully controlled. Some proteins are degraded in the cytosol by large protein complexes called proteasomes.
- From our knowledge of present-day organisms and the molecules they contain, it seems likely that living systems began with the evolution of RNA molecules that could catalyze their own replication.
- It has been proposed that, as cells evolved, the DNA double helix replaced RNA as a more stable molecule for storing genetic information, and proteins replaced RNAs as major catalytic and structural components. However, important reactions such as peptide bond formation are still catalyzed by RNA; these are thought to provide a glimpse into an ancient, RNA-based world.

KEY TERMS

alternative splicing
aminoacyl-tRNA synthetase
anticodon
codon
exon
gene expression
genetic code
general transcription factors
initiator tRNA
intron
messenger RNA (mRNA)
promoter
protease
proteasome
reading frame
ribosomal RNA (rRNA)
ribosome
ribozyme
RNA
RNA polymerase
RNA processing
RNA splicing
small nuclear RNA (snRNA)
spliceosome
transcription
transfer RNA (tRNA)
translation
translation initiation factor

QUESTIONS

QUESTION 7–7

Which of the following statements are correct? Explain your answers.

A. An individual ribosome can make only one type of protein.

B. All mRNAs fold into particular three-dimensional structures that are required for their translation.

C. The large and small subunits of an individual ribosome always stay together and never exchange partners.

D. Ribosomes are cytoplasmic organelles that are encapsulated by a single membrane.

E. Because the two strands of DNA are complementary, the mRNA of a given gene can be synthesized using either strand as a template.

F. An mRNA may contain the sequence ATTGACCCCGGTCAA.

G. The amount of a protein present in the cell depends on its rate of synthesis, its catalytic activity, and its rate of degradation.

QUESTION 7–8

The Lacheinmal protein is a hypothetical protein that causes people to smile more often. It is inactive in many chronically unhappy people. The mRNA isolated from a number of different unhappy persons in the same family was found to lack an internal stretch of 173 nucleotides that is present in the Lacheinmal mRNA isolated from a control group of happy people. The DNA sequences of the *Lacheinmal* genes from the happy and unhappy families were determined and compared. They differed by a single nucleotide substitution, which lay in an intron. What can you say about the molecular basis of unhappiness in this family?

(Hints: [1] Can you hypothesize a molecular mechanism by which a single nucleotide substitution in a gene could cause the observed deletion in the mRNA? Note that the deletion is *internal* to the mRNA. [2] Assuming the 173 base-pair deletion removes coding sequences from the Lacheinmal mRNA, how would the Lacheinmal protein differ between the happy and unhappy people?)

QUESTION 7–9

Use the genetic code shown in Figure 7–24 to identify which of the following nucleotide sequences would code for the polypeptide sequence arginine-glycine-aspartate:

1. 5′-AGA-GGA-GAU-3′
2. 5′-ACA-CCC-ACU-3′
3. 5′-GGG-AAA-UUU-3′
4. 5′-CGG-GGU-GAC-3′

QUESTION 7–10

"The bonds that form between the anticodon of a tRNA molecule and the three nucleotides of a codon in mRNA are _____." Complete this sentence with each of the following options and explain why each of the resulting statements is correct or incorrect.

A. Covalent bonds formed by GTP hydrolysis

B. Hydrogen bonds that form when the tRNA is at the A-site

C. Broken by the translocation of the ribosome along the mRNA

QUESTION 7–11

List the ordinary, dictionary definitions of the terms *replication, transcription,* and *translation.* By their side, list the special meaning each term has when applied to the living cell.

QUESTION 7–12

In an alien world, the genetic code is written in pairs of nucleotides. How many amino acids could such a code specify? In a different world a triplet code is used but the sequence of nucleotides is not important. It only matters which nucleotides are present. How many amino acids could this code specify? Would you expect to encounter any problems translating these codes?

QUESTION 7–13

One remarkable feature of the genetic code is that amino acids with similar chemical properties often have similar codons. Thus, codons with U or C as the second nucleotide tend to specify hydrophobic amino acids. Can you suggest a possible explanation for this phenomenon in terms of the early evolution of the protein synthesis machinery?

QUESTION 7–14

A mutation in DNA generates a UGA stop codon in the middle of the RNA coding for a particular protein. A second mutation in the cell leads to a single nucleotide change in a tRNA that allows the correct translation of the protein; that is, the second mutation "suppresses" the defect caused by the first. The altered tRNA translates the UGA as tryptophan. What nucleotide change has probably occurred in the mutant tRNA molecule? What consequences would the presence of such a mutant tRNA have for the translation of the normal genes in this cell?

QUESTION 7–15

The charging of a tRNA with an amino acid can be represented by the following equation:

amino acid + tRNA + ATP → aminoacyl-tRNA + AMP + PP_i

where PP_i is pyrophosphate (see Figure 3–40). In the aminoacyl-tRNA, the amino acid and tRNA are linked with a high-energy bond; a large portion of the energy derived from the hydrolysis of ATP is thus stored in this bond and is available to drive peptide-bond formation at the later stages of protein synthesis. The free-energy change of the charging reaction shown in the equation is close to zero and therefore would not be expected to favor attachment of the amino acid to tRNA. Can you suggest a further step that could drive the reaction to completion?

QUESTION 7–16

A. The average molecular weight of a protein in the cell is about 30,000 daltons. A few proteins, however, are much larger. The largest known polypeptide chain made by any cell is a protein called titin (made by mammalian muscle cells), and it has a molecular weight of 3,000,000 daltons. Estimate how long it will take a muscle cell to translate an mRNA coding for titin (assume the average molecular weight of an amino acid to be 120, and a translation rate of two amino acids per second for eucaryotic cells).

B. Protein synthesis is very accurate: for every 10,000 amino acids joined together, only one mistake is made. What is the fraction of average-sized protein molecules and of titin molecules that are synthesized without any errors? (Hint: the probability P of obtaining an error-free protein is given by $P = (1 - E)^n$, where E is the error frequency and n the number of amino acids.)

C. The molecular weight of all eucaryotic ribosomal proteins combined is about 2.5×10^6 daltons. Would it be advantageous to synthesize them as a single protein?

D. Transcription occurs at a rate of about 30 nucleotides per second. Is it possible to calculate the time required to synthesize a titin mRNA from the information given here?

QUESTION 7–17

Which of the following types of mutations would be predicted to harm an organism? Explain your answers.

A. Insertion of a single nucleotide near the end of the coding sequence

B. Removal of a single nucleotide near the beginning of the coding sequence

C. Deletion of three consecutive nucleotides in the middle of the coding sequence

D. Deletion of four consecutive nucleotides in the middle of the coding sequence

E. Substitution of one nucleotide for another in the middle of the coding sequence

CHAPTER **EIGHT** 8

Control of Gene Expression

AN OVERVIEW OF GENE EXPRESSION

HOW TRANSCRIPTIONAL SWITCHES WORK

THE MOLECULAR MECHANISMS THAT CREATE SPECIALIZED CELL TYPES

POST-TRANSCRIPTIONAL CONTROLS

An organism's DNA encodes all of the RNA and protein molecules that are needed to make its cells. Yet a complete description of the DNA sequence of an organism—be it the few million nucleotides of a bacterium or the few billion nucleotides in each human cell—does not enable us to reconstruct the organism any more than a list of all the English words in a dictionary enables us to reconstruct a play by Shakespeare. We need to know how the elements in the DNA sequence or the words on a list work together to make the masterpiece.

In cell biology, the question comes down to **gene expression**. Even the simplest single-celled bacterium can use its genes selectively—for example, switching genes on and off to make the enzymes needed to digest whatever food sources are available. And, in multicellular plants and animals, gene expression is under even more elaborate control. Over the course of embryonic development, a fertilized egg cell gives rise to many cell types that differ dramatically in both structure and function. The differences between a mammalian neuron and a lymphocyte, for example, are so extreme that it is difficult to imagine that the two cells contain the same DNA (Figure 8–1). For this reason, and because cells in an adult organism rarely lose their distinctive characteristics, biologists originally suspected that genes might be selectively lost when a cell becomes specialized. We now know, however, that nearly all the cells of a multicellular organism contain the same genome. Cell *differentiation* is instead achieved by changes in gene expression.

Hundreds of different cell types carry out a range of specialized functions that depend upon genes that are only switched on in that cell type: for

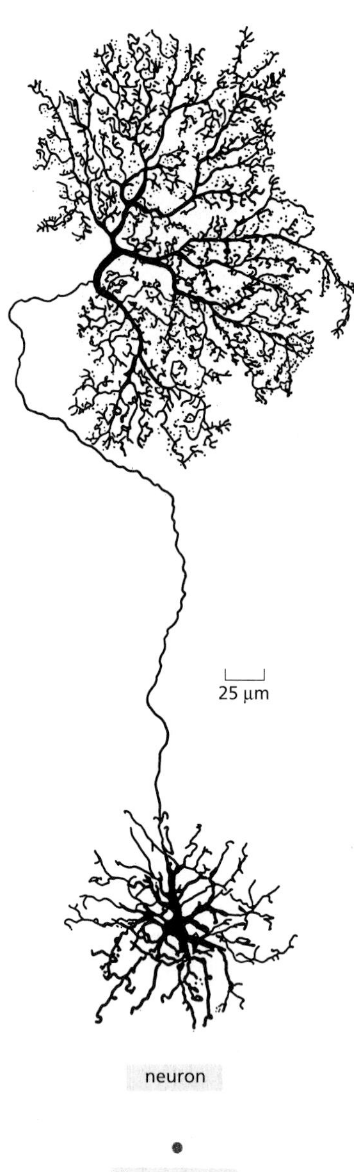

Figure 8–1 **A neuron and a lymphocyte share the same genome.** The long branches of this neuron from the retina enable it to receive electrical signals from many cells and carry them to many neighboring cells. The lymphocyte, a white blood cell involved in the immune response to infection (drawn to scale), moves freely through the body. Both of these mammalian cells contain the same genome, but they express different RNAs and proteins. (Neuron from B.B. Boycott in Essays on the Nervous System [R. Bellairs and E.G. Gray, eds.]. Oxford, U.K.: Clarendon Press, 1974. With permission from Oxford University Press.)

example, the β cells of the pancreas make the protein hormone insulin, while the α cells of the pancreas make the hormone glucagon; the lymphocytes of the immune system are the only cells in the body to make antibodies, while developing red blood cells are the only cells that make the oxygen-transport protein hemoglobin. The differences between a neuron, a lymphocyte, a liver cell, and a red blood cell depend upon the precise control of gene expression. In each case the cell is using only some of the genes in its total repertoire.

In this chapter, we shall discuss the main ways in which gene expression is controlled in bacterial and eucaryotic cells. Although some mechanisms of control apply to both sorts of cells, eucaryotic cells, through their more complex chromosomal structure, have ways of controlling gene expression that are not available to bacteria.

AN OVERVIEW OF GENE EXPRESSION

How does an individual cell specify which of its many thousands of genes to express? This decision is an especially important problem for multicellular organisms because, as the animal develops, cell types such as muscle, nerve, and blood cells become different from one another, eventually leading to the wide variety of cell types seen in the adult. Such **differentiation** arises because cells make and accumulate different sets of RNA and protein molecules: that is, they express different genes.

The Different Cell Types of a Multicellular Organism Contain the Same DNA

As discussed above, cells have the ability to change which genes they express without altering the nucleotide sequence of their DNA. But how do we know this? If DNA were altered irreversibly during development, the chromosomes of a differentiated cell would be incapable of guiding the development of the whole organism. To test this idea, a nucleus from a skin cell of an adult frog was injected into a frog egg whose own nucleus had been removed. In at least some cases the egg developed normally into a tadpole, indicating that the transplanted skin cell nucleus cannot have lost any critical DNA sequences (Figure 8–2). Such nuclear transplantation experiments have also been carried out successfully using differentiated cells taken from adult mammals, including sheep, cows, pigs, goats, and mice. And in plants, individual cells removed from a carrot, for example, can be shown to regenerate an entire adult carrot plant. These experiments all show that the DNA in specialized cell types still contains the entire set of instructions needed to form a whole organism. The cells of an organism therefore differ not because they contain different genes, but because they express them differently.

Different Cell Types Produce Different Sets of Proteins

The extent of the differences in gene expression between different cell types may be roughly gauged by comparing the protein composition of cells in liver, heart, brain, and so on using the technique of two-dimensional gel electrophoresis (see Panel 4–6, p. 167). Experiments of this kind reveal that many proteins are common to all the cells of a multicellular organism. These *housekeeping* proteins include the structural proteins of chromosomes, RNA polymerases, DNA repair enzymes, ribosomal proteins, enzymes involved in glycolysis and other basic metabolic processes, and many of the proteins that form the cytoskeleton. Each different cell type also produces specialized proteins that are responsible for the cell's distinctive properties. In mammals, for example, hemoglobin is

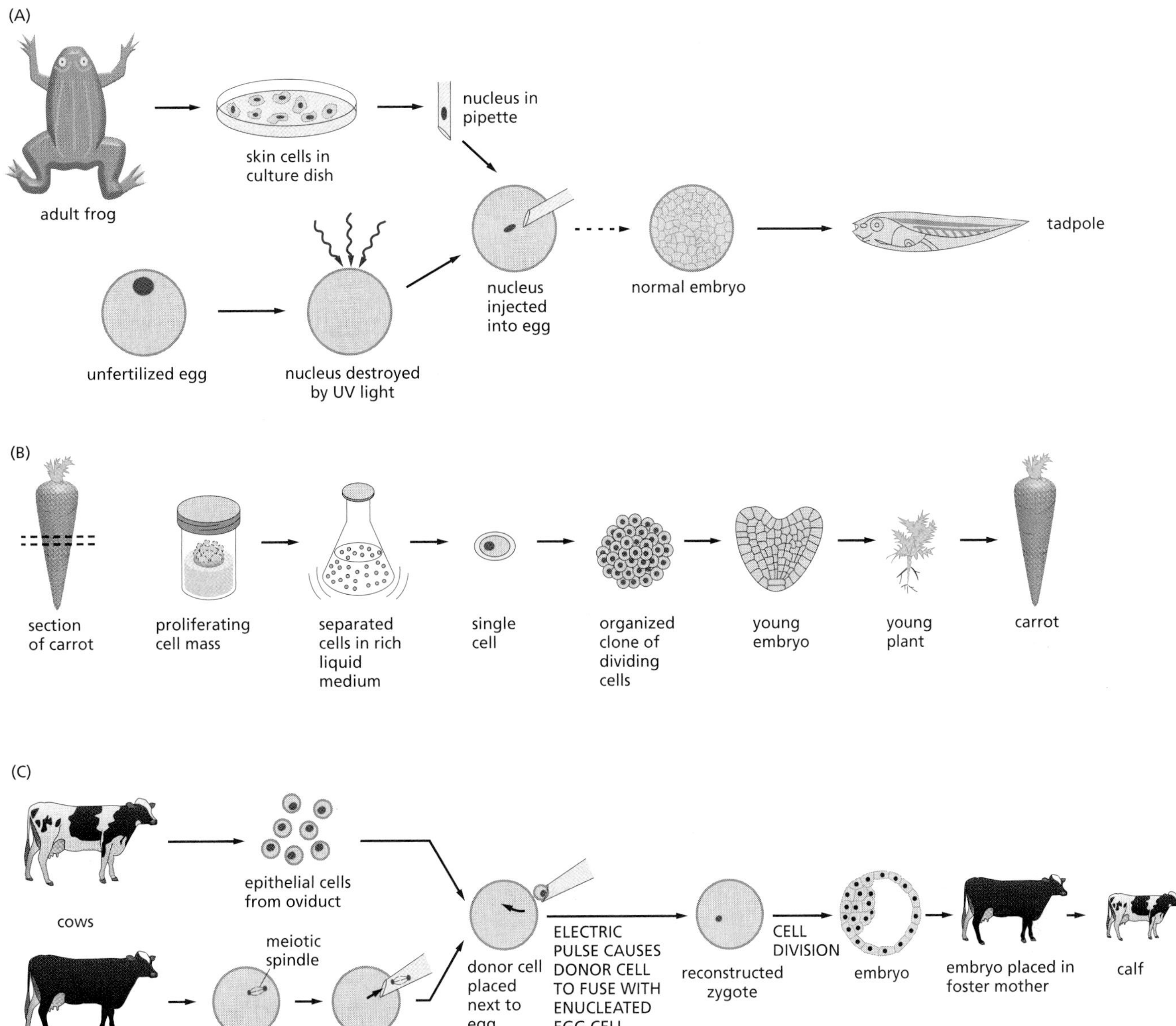

Figure 8–2 **Differentiated cells contain all the genetic instructions necessary to direct the formation of a complete organism.** (A) The nucleus of a skin cell from an adult frog transplanted into an egg whose nucleus has been removed can give rise to an entire tadpole. The broken arrow indicates that to give the transplanted genome time to adjust to an embryonic environment, a further transfer step is required in which one of the nuclei is taken from the early embryo that begins to develop and is put back into a second enucleated egg. (B) In many types of plants, differentiated cells retain the ability to "dedifferentiate," so that a single cell can form a clone of progeny cells that later give rise to an entire plant. (C) A differentiated cell from an adult cow introduced into an enucleated egg from a different cow can give rise to a calf. Different calves produced from the same differentiated cell donor are genetically identical and are therefore clones of one another. (A, modified from J.B. Gurdon, *Sci. Am.* 219(6):24–35, 1968. With permission from the Estate of Bunji Tagawa.)

made in reticulocytes, the cells that develop into red blood cells, but it cannot be detected in any other cell type.

Many proteins in a cell are produced in such small numbers that they cannot be detected by the technique of gel electrophoresis. A more sensitive technique, called mass spectrometry (see Figure 4–45) can be used to detect even rare proteins, and can also provide information about whether the proteins are covalently modified (for example by phosphorylation). Gene expression can also be studied by monitoring the mRNAs that encode proteins, rather than the proteins themselves. Estimates of the number of different mRNA sequences in human cells suggest that, at any one time, a typical differentiated human cell expresses perhaps 5000–15,000 genes from a repertoire of about 25,000. It is the expression of a different collection of genes in each cell type that causes the large variations seen in the size, shape, behavior, and function of differentiated cells.

A Cell Can Change the Expression of Its Genes in Response to External Signals

Most of the specialized cells in a multicellular organism are capable of altering their patterns of gene expression in response to extracellular cues. For example, if a liver cell is exposed to a glucocorticoid hormone (a type of steroid), the production of several specific proteins is dramatically increased. Released in the body during periods of starvation or intense exercise, glucocorticoids signal the liver to increase the production of glucose from amino acids and other small molecules. The set of proteins whose production is induced by glucocorticoids includes enzymes such as tyrosine aminotransferase, which helps to convert tyrosine to glucose. When the hormone is no longer present, the production of these proteins drops to its normal level.

Other cell types respond to glucocorticoids differently. In fat cells, for example, the production of tyrosine aminotransferase is reduced, while some other cell types do not respond to glucocorticoids at all. These examples illustrate a general feature of cell specialization: different cell types often respond in different ways to the same extracellular signal. Underlying such adjustments are features of the gene expression pattern that do not change and give each cell type its permanently distinctive character.

Gene Expression Can Be Regulated at Many of the Steps in the Pathway from DNA to RNA to Protein

If differences among the various cell types of an organism depend on the particular genes that the cells express, at what level is the control of gene expression exercised? As we saw in the last chapter, there are many steps in the pathway leading from DNA to protein, and all of them can in principle be regulated. Thus a cell can control the proteins it makes by (1) controlling when and how often a given gene is transcribed, (2) controlling how an RNA transcript is spliced or otherwise processed, (3) selecting which mRNAs are exported from the nucleus to the cytosol, (4) selectively degrading certain mRNA molecules, (5) selecting which mRNAs are translated by ribosomes, or (6) selectively activating or inactivating proteins after they have been made (Figure 8–3).

Gene expression can be regulated at each of these steps, and throughout this chapter we will describe some of the key control points along their pathway from DNA to protein. For most genes, however, the control of transcription (step number 1 in Figure 8–3) is paramount. This makes sense because only transcriptional control can ensure that no unnecessary intermediates are synthesized. So it is the regulation of transcription—and the DNA and protein components that determine which genes a cell transcribes into RNA—that we address first.

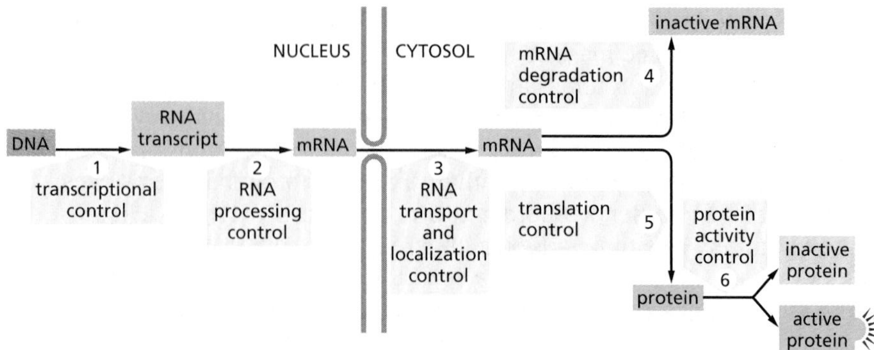

Figure 8–3 **Eucaryotic gene expression can be controlled at several different steps.** Examples of regulation at each of the steps are known, although for most genes the main site of control is step 1: transcription of a DNA sequence into RNA.

HOW TRANSCRIPTIONAL SWITCHES WORK

Only 50 years ago the idea that genes could be switched on and off was revolutionary. This concept was a major advance, and it came originally from the study of how *E. coli* bacteria adapt to changes in the composition of their growth medium. Many of the same principles apply to eucaryotic cells. However, the enormous complexity of gene regulation in higher organisms, combined with the packaging of their DNA into chromatin, creates special challenges and some novel opportunities for control—as we shall see. We begin with a discussion of the *transcription regulators,* proteins that control gene expression at the level of transcription.

Transcription Is Controlled by Proteins Binding to Regulatory DNA Sequences

Control of transcription is usually exerted at the step at which the process is initiated. In Chapter 7, we saw that the promoter region of a gene attracts the enzyme RNA polymerase and correctly orients the enzyme to begin its task of making an RNA copy of the gene. The promoters of both bacterial and eucaryotic genes include an *initiation site*, where transcription actually begins, and a sequence of approximately 50 nucleotides that extends upstream from the initiation site (if one likens the direction of transcription to the flow of a river). This region contains sites that are required for the RNA polymerase to bind to the promoter. In addition to the promoter, nearly all genes, whether bacterial or eucaryotic, have **regulatory DNA sequences** that are used to switch the gene on or off.

Some regulatory DNA sequences are as short as 10 nucleotide pairs and act as simple gene switches that respond to a single signal. Such simple switches predominate in bacteria. Other regulatory DNA sequences, especially those in eucaryotes, are very long (sometimes more than 10,000 nucleotide pairs) and act as molecular microprocessors, integrating information from a variety of signals into a command that dictates how often transcription is initiated.

Regulatory DNA sequences do not work by themselves. To have any effect, these sequences must be recognized by proteins called **transcription regulators**,which bind to the DNA. It is the combination of a DNA sequence and its associated protein molecules that acts as the switch to control transcription. The simplest bacterium codes for several hundred transcription regulators, each of which recognizes a different DNA sequence and thereby regulates a distinct set of genes. Humans make many more—several thousand—signifying the importance and complexity of this form of gene regulation in producing a complex organism.

Proteins that recognize a specific DNA sequence do so because the surface of the protein fits tightly against the special surface features of the double helix in that region. These features will vary depending on the nucleotide sequence, and thus different proteins will recognize different nucleotide sequences. In most cases, the protein inserts into the major groove of the DNA helix (see Figure 5–7) and makes a series of molecular contacts with the base pairs. The protein forms hydrogen bonds, ionic bonds, and hydrophobic interactions with the edges of the bases, usually without disrupting the hydrogen bonds that hold the base pairs together (Figure 8–4). Although each individual contact is weak, the 20 or so contacts that are typically formed at the protein–DNA interface combine to ensure that the interaction is both highly specific and very strong; indeed, protein–DNA interactions are among the tightest and most specific molecular interactions known in biology.

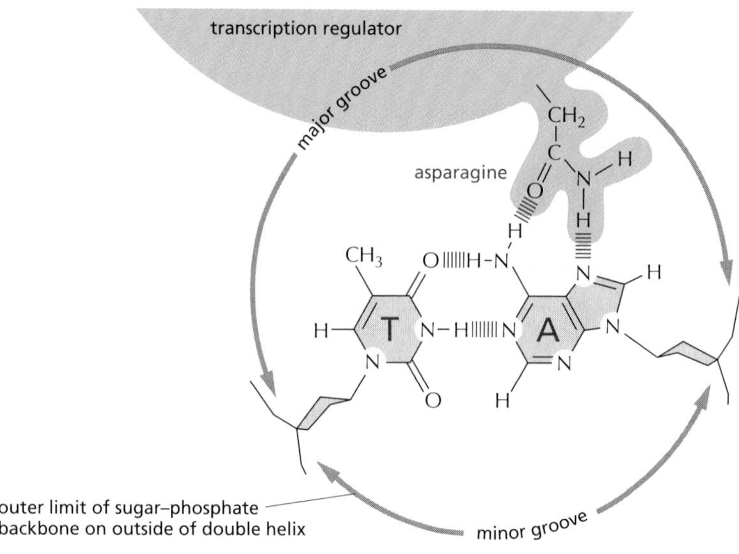

Figure 8–4 **A transcription regulator binds to the major groove of a DNA helix.** Only a single contact between the protein and one base pair in DNA is shown. Typically, the protein–DNA interface would consist of 10–20 such contacts, each involving a different amino acid and each contributing to the strength of the protein–DNA interaction.

Although each example of protein–DNA recognition is unique in detail, many of the proteins responsible for gene regulation recognize DNA through one of several structural motifs. These fit into the major groove of the DNA double helix and form tight associations with a short stretch of DNA base pairs. The *DNA-binding motifs* shown in Figure 8–5—the homeodomain, the zinc finger, and the leucine zipper—are found in transcription regulators that control the expression of thousands of different genes in virtually all eucaryotic organisms. Frequently, DNA-binding proteins bind in pairs (dimers) to the DNA helix. Dimerization roughly doubles the area of contact with the DNA, thereby greatly increasing the strength and specificity of the protein–DNA interaction. Because two different proteins can pair in different combinations, dimerization also makes it possible for many different DNA sequences to be recognized by a limited number of proteins.

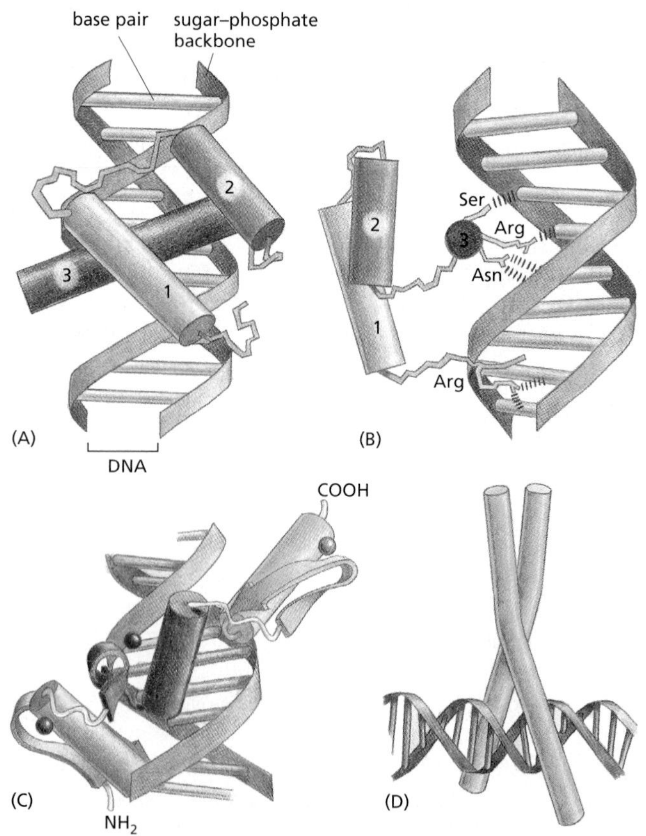

Figure 8–5 **Transcription regulators contain a variety of DNA-binding motifs.** (A and B) Front and side views of the *homeodomain*—a structural motif found in many eucaryotic DNA-binding proteins (Movie 8.1). It consists of three consecutive α helices, which are shown as *cylinders* in this figure. Most of the contacts with the DNA bases are made by helix 3 (which is seen end-on in B). The asparagine (Asn) in this helix contacts an adenine in the manner shown in Figure 8–4. (C) The *zinc finger* motif is built from an α helix and a β sheet (the latter shown as a *twisted arrow*) held together by a molecule of zinc (indicated by the *colored spheres*). Zinc fingers are often found in clusters covalently joined together to allow the α helix of each finger to contact the DNA bases in the major groove (Movie 8.2). The illustration here shows a cluster of three zinc fingers. (D) A *leucine zipper* motif. This DNA-binding motif is formed by two α helices, each contributed by a different protein molecule. Leucine zipper proteins thus bind to DNA as dimers, gripping the double helix like a clothespin on a clothesline (Movie 8.3).

Each of these motifs makes many contacts with DNA. For simplicity, only the hydrogen-bond contacts are shown in (B), and none of the individual protein–DNA contacts are shown in (C) and (D).

Transcription Switches Allow Cells to Respond to Changes in the Environment

The simplest and most completely understood examples of gene regulation occur in bacteria and in the viruses that infect them. The genome of the bacterium *E. coli* consists of a single circular DNA molecule of about 4.6×10^6 nucleotide pairs. This DNA encodes approximately 4300 proteins, although only a fraction of these are made at any one time. Bacteria regulate the expression of many of their genes according to the food sources that are available in the environment. For example, in *E. coli*, five genes code for enzymes that manufacture the amino acid tryptophan. These genes are arranged in a cluster on the chromosome and are transcribed from a single promoter as one long mRNA molecule from which the five proteins are translated (Figure 8–6). When tryptophan is present in the surroundings and enters the bacterial cell, these enzymes are no longer needed and their production is shut off. This situation arises, for example, when the bacterium is in the gut of a mammal that has just eaten a meal rich in protein. These five coordinately expressed genes are part of an *operon*—a set of genes that are transcribed into a single mRNA. Operons are common in bacteria but are not found in eucaryotes, where genes are transcribed and regulated individually (see Figure 7–36).

We now understand in considerable detail how the tryptophan operon functions. Within the promoter is a short DNA sequence (15 nucleotides in length) that is recognized by a transcription regulator. When this protein binds to this nucleotide sequence, termed the *operator*, it blocks access of RNA polymerase to the promoter; this prevents transcription of the operon and production of the tryptophan-producing enzymes. The transcription regulator is known as the *tryptophan repressor*, and it is controlled in an ingenious way: the repressor can bind to DNA only if it has also bound several molecules of the amino acid tryptophan (Figure 8–7).

The tryptophan repressor is an allosteric protein (see Figure 4–37): the binding of tryptophan causes a subtle change in its three-dimensional structure so that the protein can bind to the operator sequence. When the concentration of free tryptophan in the cell drops, the repressor no longer binds tryptophan and thus no longer binds to DNA, and the tryptophan operon is transcribed. The repressor is thus a simple device that switches production of a set of biosynthetic enzymes on and off according to the availability of the end product of the pathway that the enzymes catalyze.

The bacterium can respond very rapidly to the rise in tryptophan concentration because the tryptophan repressor protein itself is always present in the cell. The gene that encodes it is continuously transcribed at a low level, so that a small amount of the repressor protein is always being made. Such unregulated gene expression is known as *constitutive* gene expression.

QUESTION 8–1

Bacterial cells can take up the amino acid tryptophan (Trp) from their surroundings, or if there is an insufficient external supply they can synthesize tryptophan from other small molecules. The Trp repressor is a transcription regulator that shuts off the transcription of genes that code for the enzymes required for the synthesis of tryptophan (see Figure 8–7).

A. What would happen to the regulation of the tryptophan operon in cells that express a mutant form of the tryptophan repressor that (1) cannot bind to DNA, (2) cannot bind tryptophan, or (3) binds to DNA even in the absence of tryptophan?

B. What would happen in scenarios (1), (2), and (3) if the cells, in addition, produced normal tryptophan repressor protein from a second, normal gene?

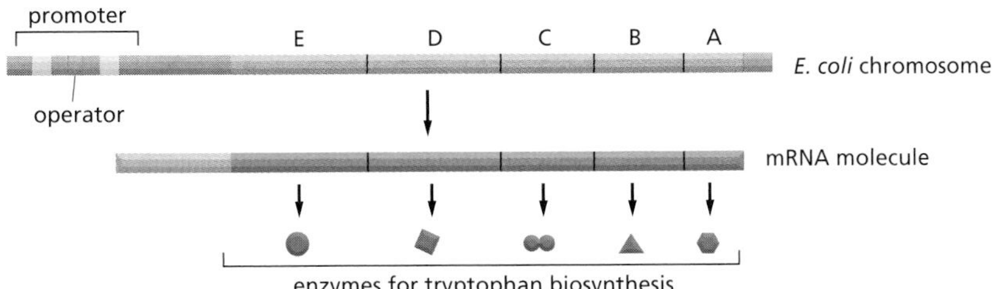

Figure 8–6 **A cluster of bacterial genes can be transcribed from a single promoter.** Each of these five genes encodes a different enzyme; all of the enzymes are needed to synthesize the amino acid tryptophan. The genes are transcribed as a single mRNA molecule, a feature that allows their expression to be coordinated. Clusters of genes transcribed as a single mRNA molecule are common in bacteria. Each such cluster is called an *operon*; expression of the tryptophan operon shown here is controlled by a regulatory DNA sequence called the *operator*, situated within the promoter.

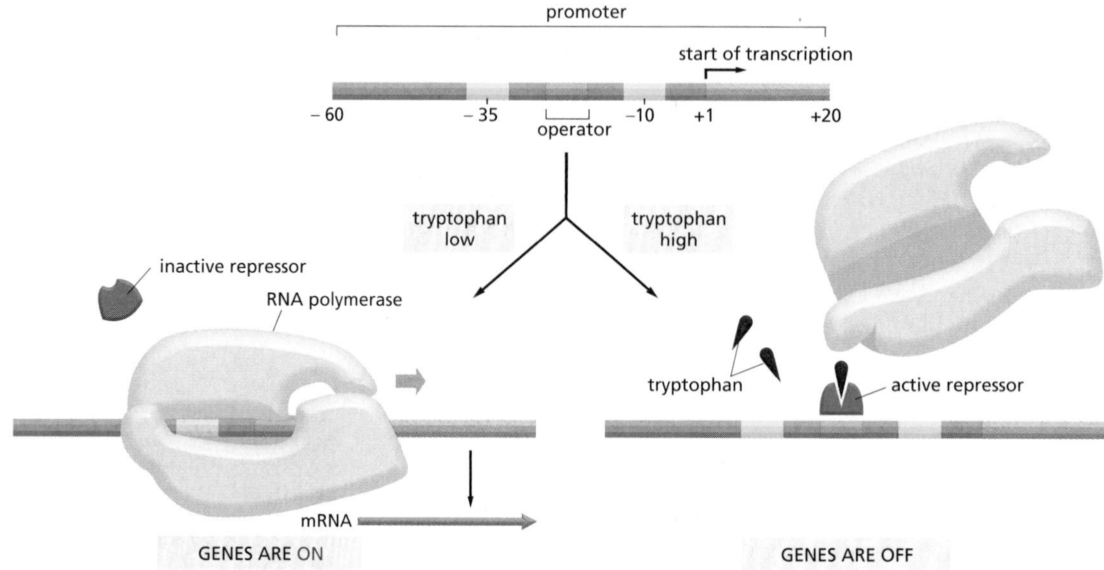

Figure 8–7 **Genes can be switched on and off with repressor proteins.** If the concentration of tryptophan inside the cell is low, RNA polymerase (*blue*) binds to the promoter and transcribes the five genes of the tryptophan operon (*left*). If the concentration of tryptophan is high, however, the repressor protein (*dark green*) becomes active and binds to the operator (*light green*), where it blocks the binding of RNA polymerase to the promoter (*right*). Whenever the concentration of intracellular tryptophan drops, the repressor releases its tryptophan and falls off the DNA, allowing the polymerase to again transcribe the operon. The promoter is marked by two key blocks of DNA sequence information, the –35 and –10 regions, highlighted in *yellow* (see Figure 7–10). The complete operon is shown in Figure 8–6.

Repressors Turn Genes Off, Activators Turn Them On

The tryptophan repressor, as its name suggests, is a **repressor** protein: in its active form, it switches genes off, or *represses* them. Some bacterial transcription regulators do the opposite: they switch genes on, or *activate* them. These **activator** proteins work on promoters that—in contrast to the promoter for the tryptophan operator—are, on their own, only marginally able to bind and position RNA polymerase; they may, for example, be recognized only poorly by the polymerase. However, these poorly functioning promoters can be made fully functional by activator proteins that bind to a nearby site on the DNA and contact the RNA polymerase to help it initiate transcription (Figure 8–8). In some cases, a bacterial transcription regulator can repress transcription at one promoter and activate transcription at another; whether the regulatory protein acts as an activator or repressor depends, in large part, on exactly where the regulatory sequences to which it binds are located with respect to the promoter.

Like the tryptophan repressor, activator proteins often have to interact with a second molecule to be able to bind DNA. For example, the bacterial activator protein CAP has to bind cyclic AMP (cAMP) before it can

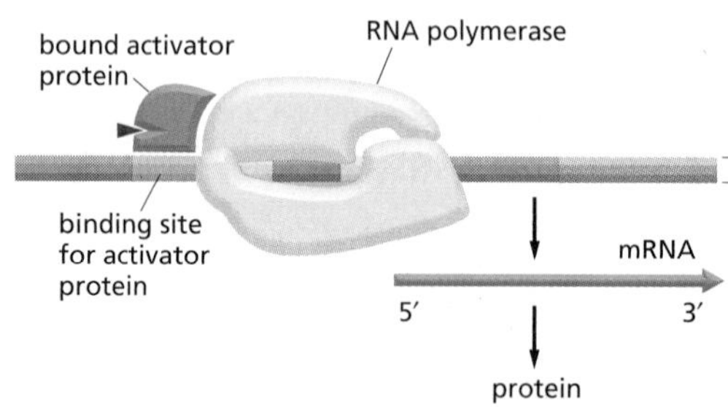

Figure 8–8 **Gene expression can also be controlled with activator proteins.** An activator protein binds to a regulatory sequence on the DNA and then interacts with the RNA polymerase to help it initiate transcription. Without the activator, the promoter fails to initiate transcription efficiently. In bacteria, the binding of the activator to DNA is often controlled by the interaction of a metabolite or other small molecule (*red* triangle) with the activator protein. For example, the bacterial catabolite activator protein (CAP) must bind cyclic AMP (cAMP) before it can bind to DNA; thus CAP allows genes to be switched on in response to increases in intracellular cAMP concentration.

bind to DNA. Genes activated by CAP are switched on in response to an increase in intracellular cAMP concentration, which signals to the bacterium that glucose, its preferred carbon source, is no longer available; as a result, CAP drives the production of enzymes capable of degrading other sugars.

An Activator and a Repressor Control the *Lac* Operon

In many instances, the activity of a single promoter is controlled by two different transcription regulators. The *Lac* operon in *E. coli*, for example, is controlled by both the Lac repressor and the activator protein CAP. The *Lac* operon encodes proteins required to import and digest the disaccharide lactose. In the absence of glucose, CAP switches on genes that allow the cell to utilize alternative sources of carbon—including lactose. It would be wasteful, however, for CAP to induce expression of the *Lac* operon when lactose is not present. Thus the Lac repressor shuts off the operon in the absence of lactose. This arrangement enables the control region of the *Lac* operon to integrate two different signals, so that the operon is highly expressed only when two conditions are met: lactose must be present and glucose must be absent (Figure 8–9). This genetic circuit thus behaves like a switch that carries out a logic operation in a computer. When lactose is present AND glucose is absent, the cell executes the appropriate program: in this case, transcription of the genes that permit the uptake and utilization of lactose.

The elegant logic of the *Lac* operon first attracted the attention of biologists more than 50 years ago. The molecular basis of the switch was uncovered by a combination of genetics and biochemistry, providing the first insight into how gene expression is controlled. In a eucaryotic cell, similar gene regulatory devices are combined to generate increasingly complex circuits. Indeed, the developmental program that takes a fertilized egg to adulthood can be viewed as an exceedingly complex circuit composed of simple components like those that control the *Lac* and tryptophan operons.

QUESTION 8–2

Explain how DNA-binding proteins can make sequence-specific contacts to a double-stranded DNA molecule without breaking the hydrogen bonds that hold the bases together. Indicate how, through such contacts, a protein can distinguish a T-A from a C-G pair. Give your answer in a form similar to Figure 8–4, and indicate what sorts of noncovalent bonds—hydrogen bonds, electrostatic attractions, or hydrophobic interactions (see Panel 2–7, pp. 76–77)—would be made. There is no need to specify any particular amino acid on the protein. The structures of all the base pairs in DNA are given in Figure 5–6.

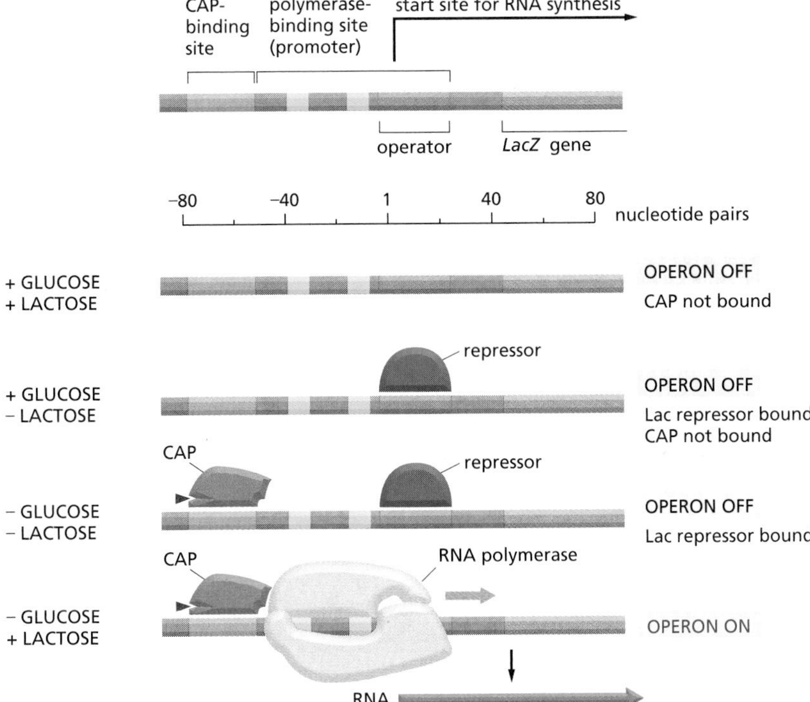

Figure 8–9 **The *Lac* operon is controlled by two signals.** Glucose and lactose concentrations control the initiation of transcription of the *Lac* operon through their effects on the Lac repressor protein and CAP. When lactose is absent, the *Lac* repressor binds the *Lac* operator and shuts off expression of the operon. Addition of lactose increases the intracellular concentration of a related compound, allolactose. Allolactose binds to the Lac repressor, causing it to undergo a conformational change that releases its grip on the operator DNA (not shown). When glucose is absent, cyclic AMP (*red* triangle) is produced by the cell and CAP binds to DNA. *LacZ*, the first gene of the operon, encodes the enzyme β-galactosidase, which breaks down lactose to galactose and glucose.

Eucaryotic Transcription Regulators Control Gene Expression from a Distance

Eucaryotes, too, use transcription regulators—both activators and repressors—to regulate the expression of their genes. The DNA sites to which eucaryotic gene activators bound were originally termed *enhancers*, because their presence dramatically enhanced, or increased, the rate of transcription. It was surprising to biologists when, in 1979, it was discovered that these activator proteins could enhance transcription even when they are bound thousands of nucleotide pairs away from a gene's promoter. They also work when bound either upstream or downstream from the gene. These observations raised several questions. How do enhancer sequences and the proteins bound to them function over such long distances? How do they communicate with the promoter?

Many models for this 'action at a distance' have been proposed, but the simplest of these seems to apply in most cases. The DNA between the enhancer and the promoter loops out to allow eucaryotic activator proteins to directly influence events that take place at the promoter (Figure 8–10). The DNA thus acts as a tether, causing a protein bound to an enhancer even thousands of nucleotide pairs away to interact with the proteins in the vicinity of the promoter—including RNA polymerase II and the general transcription factors (see Figure 7–12). Often, additional proteins serve to link the distantly bound transcription regulators to these proteins at the promoter; the most important is a large complex of proteins known as *Mediator* (see Figure 8–10). One of the ways in which eucaryotic activator proteins function is by aiding in the assembly of the general transcription factors and RNA polymerase at the promoter. Eucaryotic repressor proteins do the opposite: they decrease transcription by preventing or sabotaging the assembly of the same protein complex.

In addition to promoting—or repressing—the assembly of a transcription initiation complex, eucaryotic transcription regulators have an additional mechanism of action: they attract proteins that modulate chromatin structure and thereby affect the accessibility of the promoter to the general transcription factors and RNA polymerase, as we discuss next.

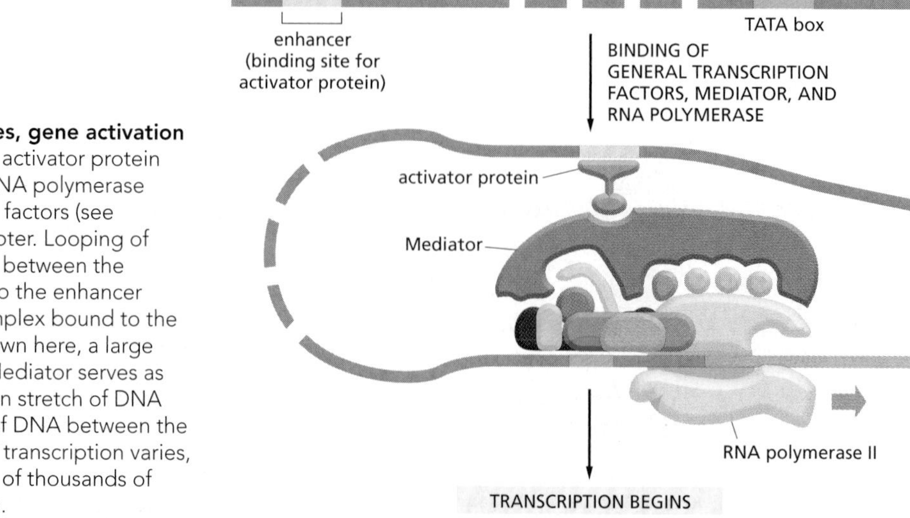

Figure 8–10 **In eucaryotes, gene activation occurs at a distance.** An activator protein bound to DNA attracts RNA polymerase and general transcription factors (see Figure 7–12) to the promoter. Looping of the DNA permits contact between the activator protein bound to the enhancer and the transcription complex bound to the promoter. In the case shown here, a large protein complex called Mediator serves as a go-between. The broken stretch of DNA signifies that the length of DNA between the enhancer and the start of transcription varies, sometimes reaching tens of thousands of nucleotide pairs in length.

Packing of Promoter DNA into Nucleosomes Affects Initiation of Transcription

Initiation of transcription in eucaryotic cells must also take into account the packaging of DNA into chromosomes. As we saw in Chapter 5, the genetic material in eucaryotic cells is packed into nucleosomes, which, in turn, are folded into higher-order structures. How do transcription regulators, general transcription factors, and RNA polymerase gain access to such DNA? Nucleosomes can inhibit the initiation of transcription if they are positioned over a promoter, probably because they physically block the assembly of the general transcription factors or RNA polymerase on the promoter. In fact, such chromatin packaging may have evolved in part to prevent leaky gene expression—initiation of transcription in the absence of the proper activator proteins.

In eucaryotic cells, activator and repressor proteins exploit chromatin structure to help turn genes on and off. As we saw in Chapter 5, chromatin structure can be altered by chromatin-remodeling complexes and by enzymes that covalently modify the histone proteins that form the core of the nucleosome (see Figures 5–27 and 5–28). Many gene activators take advantage of these mechanisms by recruiting these proteins to promoters (Figure 8–11). For example, many transcription activators attract *histone acetylases*, which attach an acetyl group to selected lysines in the tail of histone proteins. This modification alters chromatin structure, probably allowing greater accessibility to the underlying DNA; moreover, the acetyl groups themselves are recognized by proteins that promote transcription, including some of the general transcription factors.

Likewise, gene repressor proteins can modify chromatin in ways that reduce the efficiency of transcription initiation. For example, many repressors attract *histone deacetylases*—enzymes that remove the acetyl groups from histone tails, thereby reversing the positive effects that acetylation has on transcription initiation. Although some eucaryotic repressor proteins work on a gene-by-gene basis, others can orchestrate the formation of large swaths of transcriptionally inactive chromatin containing many

QUESTION 8–3

Some transcription regulators bind to DNA and cause the double helix to bend at a sharp angle. Such "bending proteins" can stimulate the initiation of transcription without contacting either the RNA polymerase, any of the general transcription factors, or any other transcription regulators. Can you devise a plausible explanation for how these proteins might work to modulate transcription? Draw a diagram that illustrates your explanation.

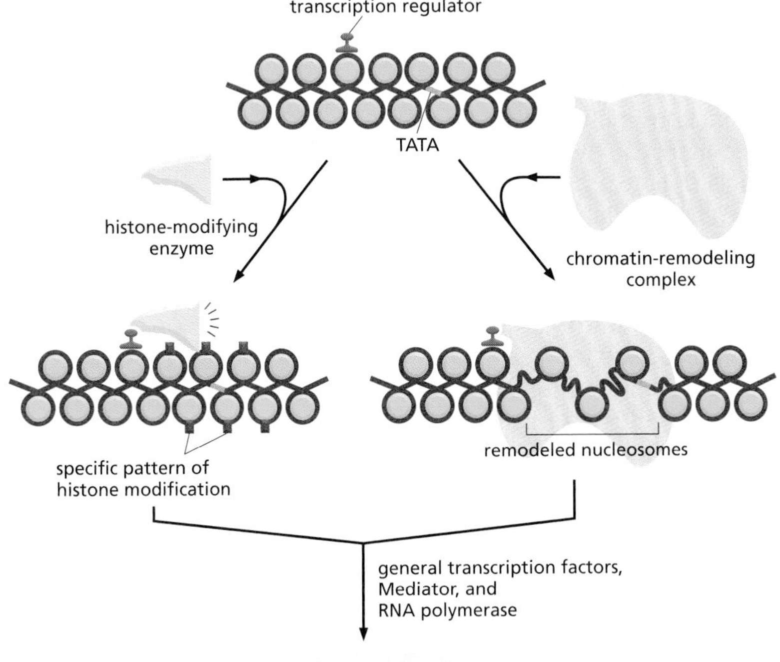

Figure 8–11 **Eucaryotic gene activator proteins can direct local alterations in chromatin structure.** Activator proteins can recruit histone-modifying enzymes and chromatin-remodeling complexes to the promoter region of a gene. The action of these proteins renders the DNA packaged in chromatin more accessible to other proteins in the cell, including those required for transcription initiation. In addition, the covalent histone modifications can serve as binding sites for proteins that stimulate transcription initiation.

genes. As discussed in Chapter 5, these transcription-resistant regions of DNA include the heterochromatin found in interphase chromosomes and the entire X chromosome in female mammals.

THE MOLECULAR MECHANISMS THAT CREATE SPECIALIZED CELL TYPES

All cells must be able to switch genes on and off in response to signals in their environments. But the cells of multicellular organisms have evolved this capacity to an extreme degree and in highly specialized ways to form an organized array of differentiated cell types. In particular, once a cell in a multicellular organism becomes committed to differentiate into a specific cell type, the choice of fate is generally maintained through many subsequent cell generations. This means that the changes in gene expression, which are often triggered by a transient signal, must be remembered. This phenomenon of *cell memory* is a prerequisite for the creation of organized tissues and for the maintenance of stably differentiated cell types. In contrast, the simplest changes in gene expression in both eucaryotes and bacteria are often only transient; the tryptophan repressor, for example, switches off the tryptophan genes in bacteria only in the presence of tryptophan; as soon as the amino acid is removed from the medium, the genes are switched back on, and the descendants of the cell will have no memory that their ancestors had been exposed to tryptophan.

In this section, we discuss some of the special features of transcriptional regulation that are found in multicellular organisms. Our focus will be on how these mechanisms create and maintain the specialized cell types that give a worm, a fly, or a human its distinctive characteristics.

Eucaryotic Genes Are Regulated by Combinations of Proteins

Because eucaryotic transcription regulators can control transcription initiation when bound to DNA many base pairs away from the promoter, the nucleotide sequences that control the expression of a gene can be spread over long stretches of DNA. In animals and plants it is not unusual to find the regulatory DNA sequences of a gene dotted over tens of thousands of nucleotide pairs, although much of this DNA serves as "spacer" sequence and is not directly recognized by the transcription regulators.

So far in this chapter we have treated transcription regulators as though each functions individually to turn a gene on or off. While this idea holds true for many simple bacterial activators and repressors, most eucaryotic transcription regulators work as part of a "committee" of regulatory proteins, all of which are necessary to express the gene in the right cell, in response to the right conditions, at the right time, and in the required amount.

The term **combinatorial control** refers to the way that groups of regulatory proteins work together to determine the expression of a single gene. We saw a simple example of such regulation by multiple signals when we discussed the bacterial *Lac* operon (see Figure 8–9). In eucaryotes, the regulatory inputs have been amplified, and a typical gene is controlled by dozens of transcription regulators (Figure 8–12). Often, some of these regulatory proteins are repressors and some are activators; the molecular mechanisms by which the effects of all of these proteins are added up to determine the final level of expression for a gene are only now beginning to be understood. An example of such a complex regulatory system—one that participates in the development of a fruit fly from a fertilized egg—is described in How We Know, pp. 282–284.

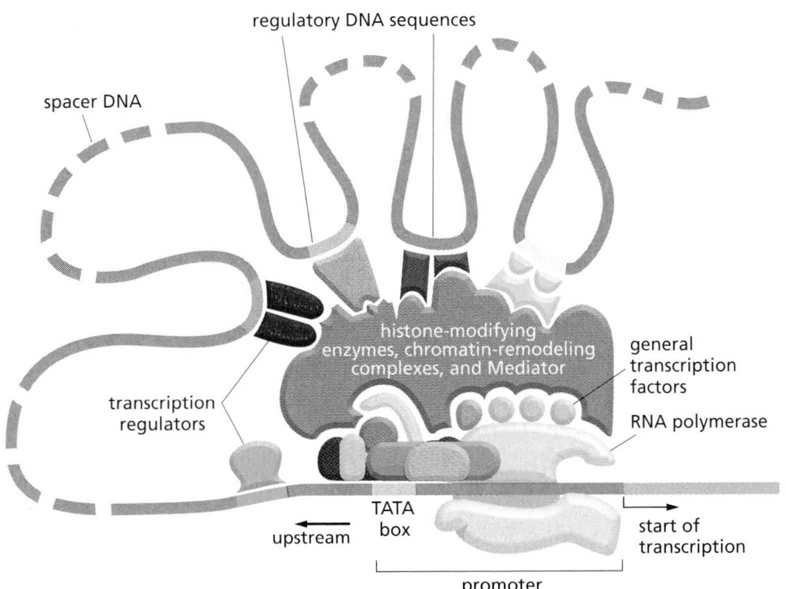

Figure 8–12 **Transcription regulators work together as a "committee" to control the expression of a eucaryotic gene.** Whereas the general transcription factors that assemble at the promoter are the same for all genes transcribed by polymerase II, the transcription regulators and the locations of their binding sites relative to the promoters are different for different genes. The effects of multiple transcription regulators combine to determine the final rate of transcription initiation. It is not yet understood in detail how these multiple inputs are integrated.

The Expression of Different Genes Can Be Coordinated by a Single Protein

In addition to being able to switch individual genes on and off, all organisms—whether procaryote or eucaryote—need to coordinate the expression of different genes. When a eucaryotic cell receives a signal to divide, for example, a number of hitherto unexpressed genes are turned on together to set in motion the events that lead eventually to cell division (discussed in Chapter 18). One way in which bacteria coordinate the expression of a set of genes is by having them clustered together in an operon under the control of a single promoter (see Figure 8–6). This is not the case in eucaryotes, in which each gene is transcribed and regulated individually. So how do eucaryotes coordinate gene expression? In particular, given that a eucaryotic cell uses a committee of transcription regulators to control each of its genes, how can it rapidly and decisively switch whole groups of genes on or off? The answer is that even though control of gene expression is combinatorial, the effect of a single transcription regulator can still be decisive in switching any particular gene on or off, simply by completing the combination needed to activate or repress that gene. This is like dialing in the final number of a combination lock: the lock will spring open if the other numbers have been previously entered. Just as the same number can complete the combination for different locks, the same protein can complete the combination for several different genes. As long as different genes contain DNA sequences recognized by the same transcription regulator, they can be switched on or off together, as a unit.

An example of this style of regulation in humans is seen with the *glucocorticoid receptor protein*. In order to bind to regulatory sites in DNA this transcription regulator must first form a complex with a molecule of a glucocorticoid hormone (for example, cortisol; see Table 16–1, p. 535). In response to glucocorticoids, liver cells increase the expression of many different genes, one of which encodes the enzyme tyrosine aminotransferase, as discussed earlier. These genes are all regulated by the binding of the glucocorticoid hormone–receptor complex to a regulatory sequence in the DNA of each gene. When the body has recovered and the

GENE REGULATION—THE STORY OF *EVE*

The ability to regulate gene expression is crucial to the proper development of a multicellular organism from a fertilized egg to a fertile adult. Beginning at the earliest moments in development, a succession of programs controls the differential expression of genes that allows an animal to form a proper body plan—helping to distinguish its back from its belly, and its head from its tail. These cues ultimately direct the correct placement of a wing or a leg, a mouth or an anus, a neuron or a sex cell.

A central problem in development, then, is understanding how an organism generates these patterns of gene expression, which are laid down within hours of fertilization. A large part of the story rests on the action of transcription regulators. By interacting with different regulatory DNA sequences, these proteins instruct every cell in the embryo to switch on the genes that are appropriate for that cell at each time point during development. How can a protein binding to a piece of DNA help direct the development of a complex multicellular organism? To see how we can address that large question, we now review the story of *Eve*.

In the Big Egg

Even-skipped—*Eve*, for short—is a gene whose expression plays an important role in the development of the *Drosophila* embryo. If this gene is inactivated by mutation, many parts of the embryo fail to form and the fly larva dies early in development. At the stage of development when *Eve* is first switched on, the embryo developing within the egg is still a single, giant cell containing multiple nuclei afloat in a common cytoplasm. This embryo, which is some 400 μm long and 160 μm in diameter, is formed from the fertilized egg through a series of rapid nuclear divisions that occur without cell division. Eventually each nucleus will be enclosed in a plasma membrane and become a cell; however, the events that concern us happen before this cellularization.

The cytoplasm of this giant egg is far from uniform: the anterior (head) end of the embryo contains different proteins from those in the posterior (tail) end. The presence of these asymmetries in the fertilized egg and the early embryo was originally demonstrated by experiments in which *Drosophila* eggs were made to leak. If the front end of an egg is punctured carefully and a small amount of the anterior cytoplasm is allowed to ooze out, the embryo will fail to develop head segments. Further, if cytoplasm taken from the posterior end of another egg is then injected into this somewhat depleted anterior area, the animal will develop a second set of abdominal segments where its head parts should have been (Figure 8–13).

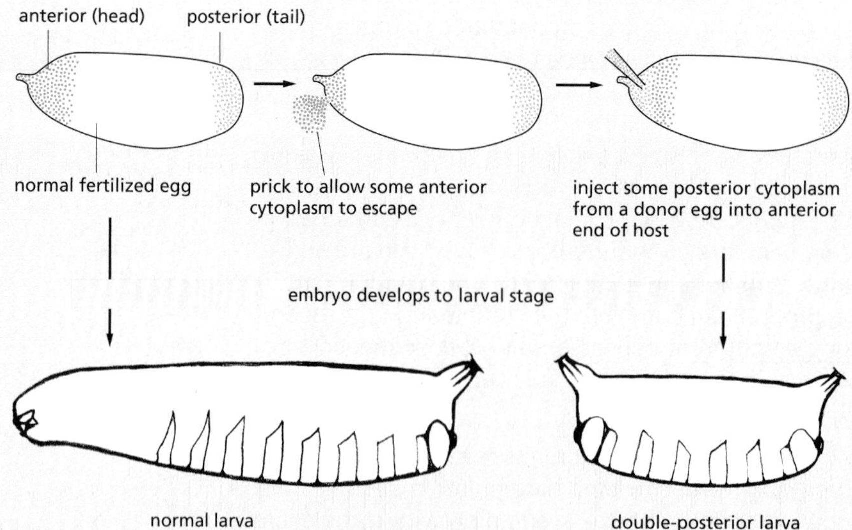

Figure 8–13 **Molecules localized at the ends of the *Drosophila* egg control its anterior–posterior polarity.** A small amount of cytoplasm is allowed to leak out of the anterior end of the egg and is replaced by an injection of posterior cytoplasm. The resulting double-tailed embryo (*right*) shows a duplication of the last three abdominal segments. A normal embryo (*left*) is shown for comparison. (Adapted from C. Nüsslein-Volhard, H.G. Frohnhöfer, and R. Lehmann, *Science* 238:1675–1681, 1987. With permission from AAAS.)

Finding the Proteins

This egg-draining experiment shows that the normal head-to-tail pattern of development is controlled by substances located at each end of the embryo. And researchers were betting that these substances were proteins. To identify them, investigators subjected eggs to a treatment that would inactivate genes at random. They then searched for embryos whose head-to-tail body plan looked abnormal. In these mutant animals, the genes that were disrupted must encode proteins that are important for establishing proper anterior–posterior polarity.

Using this approach, researchers discovered many genes required for setting up anterior–posterior polarity, including genes encoding four key transcription regulators: Bicoid, Hunchback, Krüppel, and Giant. (*Drosophila* genes are often given colorful names that reflect the appearance of flies in which the gene is inactivated by mutation.) Once these proteins had been identified, researchers could prepare antibodies that would recognize each. These antibodies, coupled to fluorescent markers, were then used to determine where in the early embryo each protein is localized (see Panel 1–1, pp. 8–9).

The results of these antibody-staining experiments are striking. The cytoplasm of the early embryo, it turns out, contains a mixture of these transcription regulators, each distributed in a unique pattern along the length of the embryo (Figure 8–14). As a result, the nuclei inside this giant, multinucleate cell begin to express different genes depending on which transcription regulators they are exposed to, which in turn depends on the location of each nucleus along the embryo. Nuclei near the anterior end of the embryo, for example, encounter a set of transcription regulators that is distinct from the set that bathe nuclei at the posterior end. Thus the differing amounts of these proteins provide the many nuclei in the developing embryo with positional information along the anterior–posterior axis of the embryo.

This is where *Eve* comes in. The regulatory DNA sequences of the *Eve* gene can read the concentrations of the transcription regulators at each position along the length of the embryo. Based on this information, *Eve* is expressed in seven stripes, each at a precise location along the anterior–posterior axis of the embryo. To find out how these regulatory proteins control the expression of *Eve* with such precision, researchers next set their sights on the regulatory region of the *Eve* gene.

Dissecting the DNA

As we have seen in this chapter, regulatory DNA sequences control which cells in an organism will express a particular gene, and at what point that gene will be turned on. One way to learn when and where a regulatory DNA sequence is active is to hook the sequence up to a **reporter gene**—a gene encoding a protein whose activity is easy to monitor experimentally. The regulatory DNA sequences will now drive the expression of the reporter gene. This artificial DNA construct is then reintroduced into a cell or an organism, and the activity of the reporter protein is measured.

By coupling various portions of the regulatory sequence of *Eve* to a reporter gene, researchers discovered that the *Eve* gene contains a series of seven regulatory modules, each of which is responsible for specifying a single stripe of *Eve* expression along the embryo. So, for example, researchers could remove the regulatory module that specifies stripe 2 from its normal setting upstream of *Eve*, place it in front of a reporter gene, and reintroduce this engineered DNA sequence into the *Drosophila* genome (Figure 8–15A). When embryos carrying this genetic construct are examined, the reporter gene is found to be expressed in precisely the position

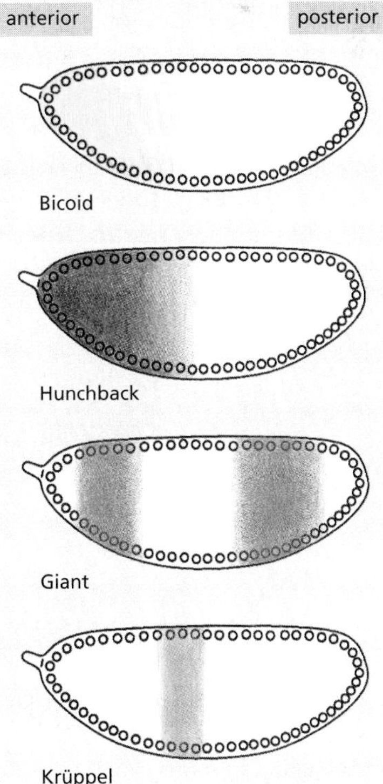

Figure 8–14 **The early *Drosophila* embryo shows a nonuniform distribution of four transcription regulators.**

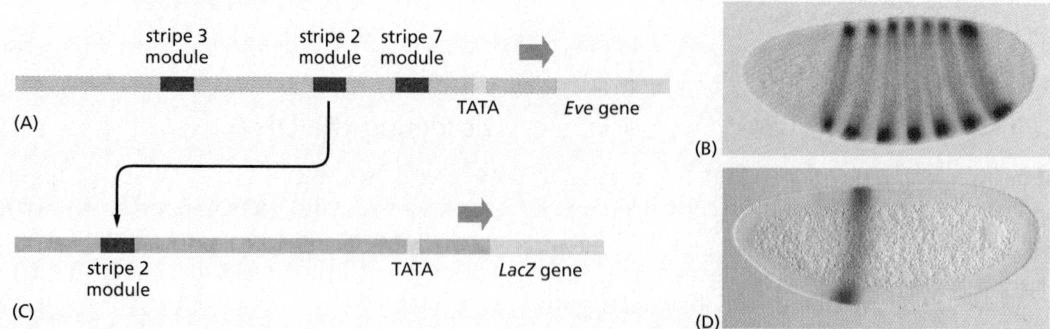

Figure 8–15 **A reporter gene reveals the modular construction of the *Eve* gene regulatory region.** (A) The *Eve* gene contains regulatory sequences that direct the production of Eve protein in stripes along the embryo. (B) Embryos stained with antibodies to the Eve protein show the seven characteristic Eve stripes. (C) In this experiment, a 480-nucleotide piece of the *Eve* regulatory region (the stripe 2 module from A) is removed and inserted upstream of the *E. coli LacZ* gene, which encodes the enzyme β-galactosidase (see Figure 8–9). (D) When the engineered DNA containing a single regulatory region is reintroduced into the genome of a *Drosophila* embryo, the resulting embryo expresses β-galactosidase precisely in the position of the second of the seven *Eve* stripes. Enzyme activity is assayed by the addition of X-gal, a modified sugar that when cleaved by β-galactosidase generates an insoluble blue product. (B and D, courtesy of Stephen Small and Michael Levine.)

of stripe 2 (Figure 8–15B). Similar experiments revealed the existence of other regulatory modules, one for each of the other six stripes.

The question then becomes: how does each of these modules direct the formation of a single stripe in a specific position? The answer, researchers found, is that each module contains a unique combination of regulatory sequences that bind different combinations of the four transcription regulators that are present in gradients in the early embryo. The stripe 2 unit, for example, contains recognition sequences for all four regulators—Bicoid and Hunchback activate *Eve* transcription, while Krüppel and Giant repress it (Figure 8–16). The concentrations of these four proteins vary across the embryo (see Figure 8–14), and these patterns determine which of the proteins are bound to the *Eve* stripe 2 module at each position along the embryo. The combination of bound proteins then 'tells' the appropriate nuclei to express *Eve,* and stripe 2 is formed.

The other stripe regulatory modules are thought to function along similar lines; each module reads positional information provided by some unique combination of transcription regulators and expresses *Eve* on the basis of this information. The entire gene control region of *Eve* is strung out over 20,000 nucleotide pairs of DNA and binds more than 20 transcription regulators, including the four discussed here. A large and complex control region is thereby formed from a series of smaller modules, each of which consists of a unique arrangement of short DNA sequences recognized by specific transcription regulators. In this way, a single gene can respond to an enormous number of combinatorial inputs. Eve itself is a transcription regulator and it—in combination with many other regulatory proteins—controls key events later in the development of the fly. This organization begins to explain how the development of a complex organism can be orchestrated by repeated applications of a few basic principles.

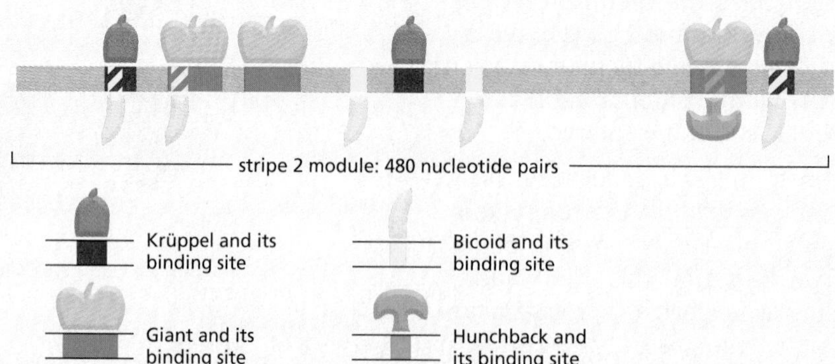

Figure 8–16 **The regulatory module for *Eve* stripe 2 contains binding sites for four different transcription regulators.** All four regulators are responsible for the proper expression of *Eve* in stripe 2. Flies that are deficient in the two activators, Bicoid and Hunchback, fail to form stripe 2 efficiently; in flies deficient in either of the two repressors, Giant or Krüppel, stripe 2 expands and covers an abnormally broad region of the embryo. As indicated in the top diagram, in some cases the binding sites for the transcription regulators overlap and the proteins compete for binding to the DNA. For example, the binding of Bicoid and Krüppel to the site at the far right is thought to be mutually exclusive.

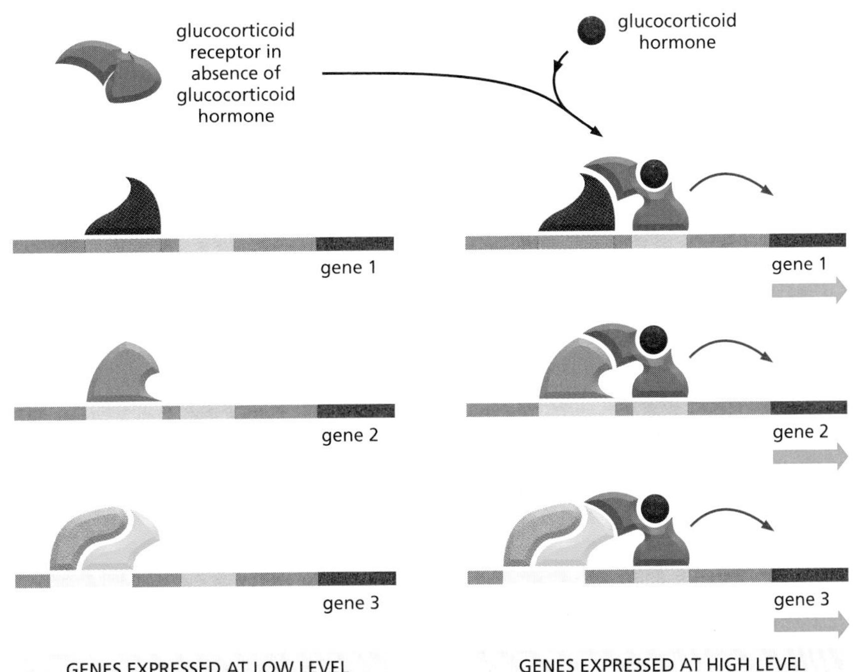

Figure 8–17 **A single transcription regulator can coordinate the expression of many different genes.** The action of the glucocorticoid receptor is illustrated. On the left is shown a series of genes, each of which has various gene activator proteins bound to its regulatory region. However, these bound proteins are not sufficient on their own to activate transcription efficiently. On the right is shown the effect of adding an additional transcription regulator—the glucocorticoid receptor in a complex with glucocorticoid hormone—that can bind to the regulatory sequences of each gene. The glucocorticoid receptor completes the combination of transcription regulators required for efficient initiation of transcription, and the genes are now switched on as a set.

hormone is no longer present, the expression of all of these genes drops to its normal level. In this way a single transcription regulator can control the expression of many different genes (Figure 8–17).

Combinatorial Control Can Create Different Cell Types

The ability to switch many different genes on or off using just one protein is not only useful in the day-to-day regulation of cell function. It is also one of the means by which eucaryotic cells differentiate into particular types of cells during embryonic development.

A striking example of the effect of a single transcription regulator on differentiation comes from studying the development of muscle cells. A mammalian skeletal muscle cell is a highly distinctive cell type. It is typically an extremely large cell that is formed by the fusion of many muscle precursor cells called *myoblasts*. The mature muscle cell is distinguished from other cells by the production of a large number of characteristic proteins, such as the actin and myosin that make up the contractile apparatus (discussed in Chapter 17) as well as the receptor proteins and ion channel proteins in the cell membranes that make the muscle cell sensitive to nerve stimulation. Genes encoding these muscle-specific proteins are all switched on coordinately as the myoblasts begin to fuse. Studies of muscle cells differentiating in culture have identified key transcription regulators, expressed only in potential muscle cells, that coordinate gene expression and thus are crucial for muscle-cell differentiation. These regulators activate the transcription of the genes that code for the muscle-specific proteins by binding to specific DNA sequences present in their regulatory regions.

These key transcription regulators can convert nonmuscle cells to myoblasts by activating the changes in gene expression typical of differentiating muscle cells. For example, when one of these regulators, MyoD, is artificially expressed in fibroblasts cultured from skin connective tissue, the fibroblasts start to behave like myoblasts and fuse to form musclelike cells. The dramatic effect of expressing the *MyoD* gene in fibroblasts is shown in Figure 8–18. It appears that the fibroblasts, which are derived from the same broad class of embryonic cells as muscle cells,

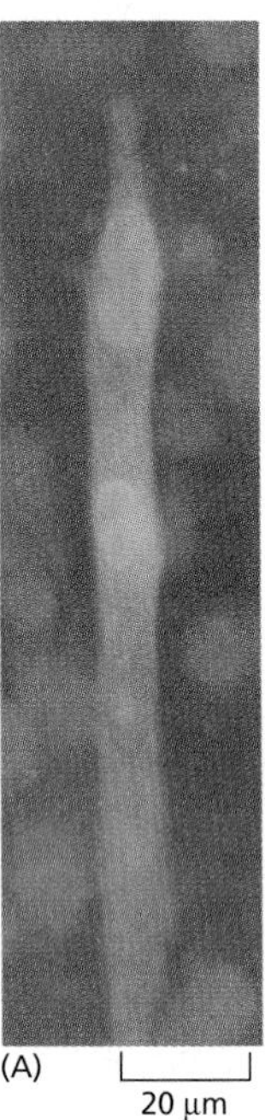

Figure 8–18 **Fibroblasts can be converted to muscle cells by a single transcription regulator.** As shown in this immunofluorescence micrograph, fibroblasts from the skin of a chick embryo have been converted to muscle cells by the experimentally induced expression of the *MyoD* gene. The fibroblasts that expressed the *MyoD* gene have fused to form elongated multinucleate musclelike cells, which are stained *green* with an antibody that detects a muscle-specific protein. Fibroblasts that do not express *MyoD* are barely visible in the background. (Courtesy of Stephen Tapscott and Harold Weintraub.)

have already accumulated many of the other necessary transcription regulators required for the combinatorial control of the muscle-specific genes, and that addition of MyoD completes the unique combination that directs the cells to become muscle. Some other cell types fail to be converted to muscle by the addition of MyoD; these cells presumably have not accumulated the other required transcription regulators during their developmental history.

How the accumulation of different transcription regulators can lead to the generation of different cell types is illustrated schematically in Figure 8–19. This figure also illustrates how, thanks to the possibilities of combinatorial control and shared regulatory DNA sequences, a limited set of transcription regulators can control the expression of a much larger number of genes.

The conversion of one cell type (fibroblast) to another (muscle) by a single transcription regulator emphasizes one of the most important principles discussed in this chapter: the dramatic differences between cell types—such as size, shape, and function—are produced by differences in gene expression.

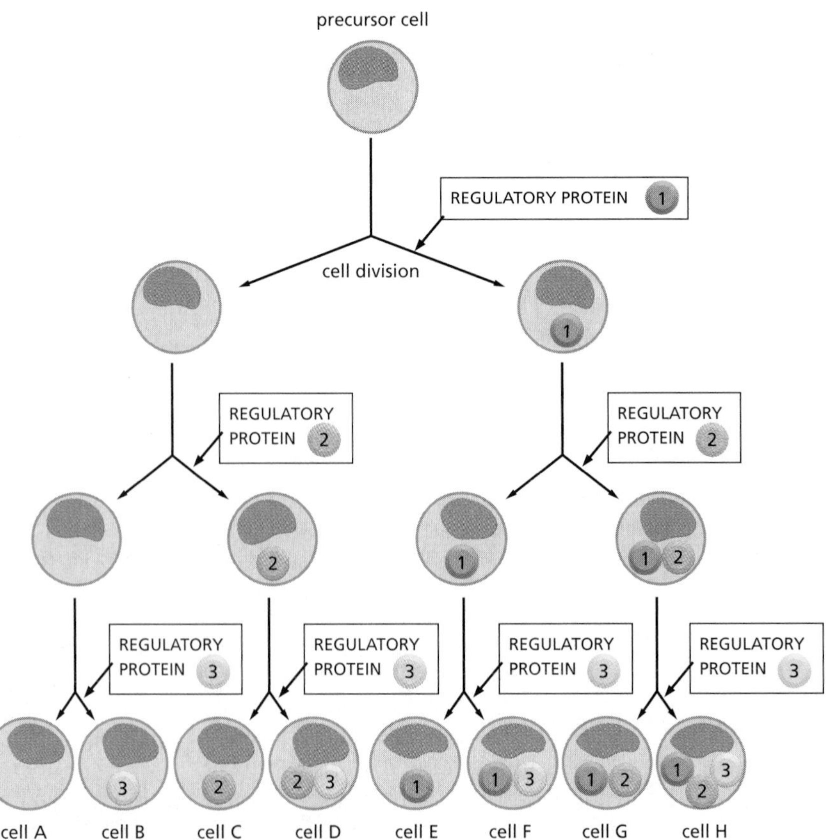

Figure 8–19 **Combinations of a few transcription regulators can generate many different cell types during development.** In this simple scheme a "decision" to make a new regulator (shown as a *numbered* circle) is made after each cell division. Repetition of this simple rule enables eight cell types (A through H) to be created using only three different transcription regulators. Each of these hypothetical cell types would then express different genes, as dictated by the combination of transcription regulators that are present within it.

Stable Patterns of Gene Expression Can Be Transmitted to Daughter Cells

As discussed earlier in this chapter, once a cell in a multicellular organism has become differentiated into a particular cell type, it will generally remain differentiated, and all its progeny cells will remain that same cell type. Some highly specialized cells never divide again once they have differentiated; for example, skeletal muscle cells and neurons. But many other differentiated cells, such as fibroblasts, smooth muscle cells, and liver cells (hepatocytes), will divide many times in the life of an individual. All of these cell types give rise only to cells like themselves when they divide: smooth muscle does not give rise to liver cells, nor liver cells to fibroblasts.

This preservation of cellular identity means that the changes in gene expression that give rise to a differentiated cell must be remembered and passed on to its daughter cells through all subsequent cell divisions. For example, in the cells illustrated in Figure 8–19, the production of each transcription regulator, once begun, has to be perpetuated in the daughter cells of each cell division. How might this be accomplished?

Cells have several ways of ensuring that their daughters "remember" what kind of cells they are supposed to be. One of the simplest is through a **positive feedback loop**, where a key transcription regulator activates transcription of its own gene in addition to that of other cell-type–specific genes (Figure 8–20). The MyoD protein discussed earlier functions in such a positive feedback loop. Another way of maintaining cell type is through the faithful propagation of a condensed chromatin structure from parent to daughter cell. We saw an example of this in Figure 5–30, where the same X chromosome is inactive through many cell generations.

A third way in which cells can transmit information about gene expression to their progeny is through **DNA methylation**. In vertebrate cells, DNA methylation occurs exclusively on cytosine bases (Figure 8–21). This covalent modification of cytosines generally turns off genes by attracting proteins that block gene expression. DNA methylation patterns are passed on to progeny cells by the action of an enzyme that copies the methylation pattern on the parent DNA strand to the daughter DNA strand immediately after replication (Figure 8–22). Because each of these mechanisms—positive feedback loops, certain forms of condensed chromatin,

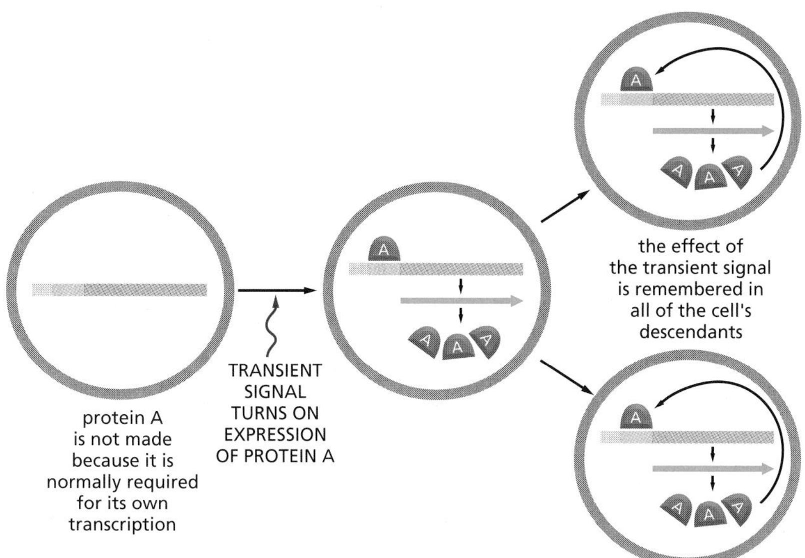
A
A
A
A
protein A
is not made
because it is
normally required
for its own
transcription
TRANSIENT
SIGNAL
TURNS ON
EXPRESSION
OF PROTEIN A
the effect of
the transient signal
is remembered in
all of the cell's
descendants

Figure 8–20 **A positive feedback loop can create cell memory.** Protein A is a transcription regulator that activates its own transcription. All of the descendants of the original cell will therefore "remember" that the progenitor cell had experienced a transient signal that initiated the production of the protein.

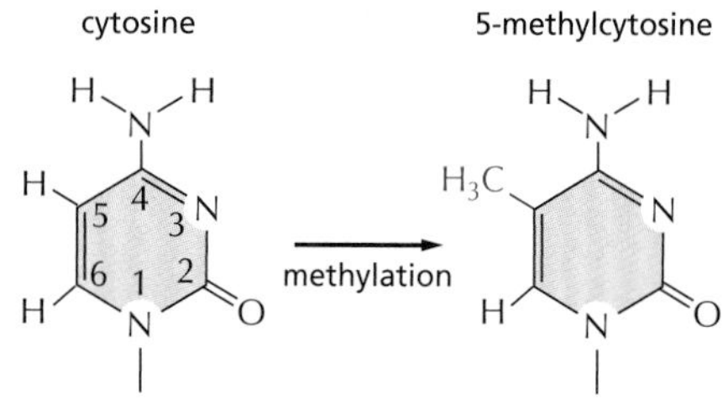

Figure 8–21 **Formation of 5-methylcytosine occurs by methylation of a cytosine base in the DNA double helix.** In vertebrates this event is confined to selected cytosine (C) nucleotides that fall next to a guanine (G).

and DNA methylation—transmits information from parent to daughter cell without altering the actual nucleotide sequence of the DNA, they are considered forms of *epigenetic inheritance* (see p. 192).

The Formation of an Entire Organ Can Be Triggered by a Single Transcription Regulator

We have seen that even though combinatorial control is the norm for eucaryotic genes, a single transcription regulator can be decisive in switching a whole set of genes on or off, and can convert one cell type into another. A dramatic extension of this principle comes from studies of eye development in *Drosophila*, mice, and humans. Here, a transcription regulator, called Ey in flies and Pax-6 in vertebrates, is crucial for eye development. When expressed in the proper type of cell, Ey can trigger the formation of not just a single cell type but a whole organ—the eye—composed of different types of cells all properly organized in three-dimensional space.

The best evidence for the action of Ey comes from experiments in fruit flies in which the *Ey* gene is artificially expressed early in development in cells that normally go on to form legs. This abnormal gene expression causes eyes to develop in the middle of the legs (Figure 8–23). The *Drosophila* eye is composed of thousands of cells, and how the Ey protein coordinates the specification of each cell in the eye is an actively studied topic in developmental biology. Here, we shall simply note that Ey directly controls the expression of many genes by binding to DNA sequences in their regulatory regions. Some of the genes controlled by Ey encode additional transcription regulators that, in turn, control the expression of other genes. Moreover, some of these regulators act back on *Ey* itself to create a positive feedback loop that ensures the continued production of the Ey protein. So the action of just one transcription regulator can produce a cascade of regulators whose combined actions lead to the formation of an organized group of many different types of cells. One can begin to imagine how, by repeated applications of this principle, a complex organism is built piece by piece.

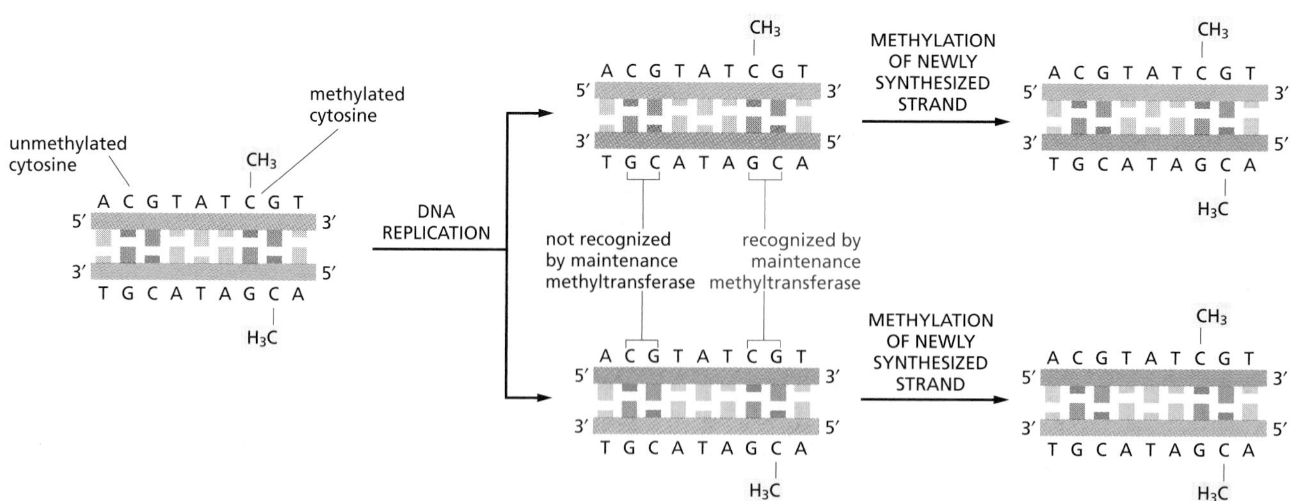

Figure 8–22 **DNA methylation patterns can be faithfully inherited.** An enzyme called a maintenance methyltransferase guarantees that once a pattern of DNA methylation has been established, it is inherited by progeny DNA. Immediately after replication, each daughter helix will contain one methylated DNA strand—inherited from the parent helix—and one unmethylated, newly synthesized strand. The maintenance methyltransferase interacts with these hybrid helices, where it methylates only those CG sequences that are base-paired with a CG sequence that is already methylated. In vertebrate DNA, a large portion of the cytosines in CG sequences are methylated.

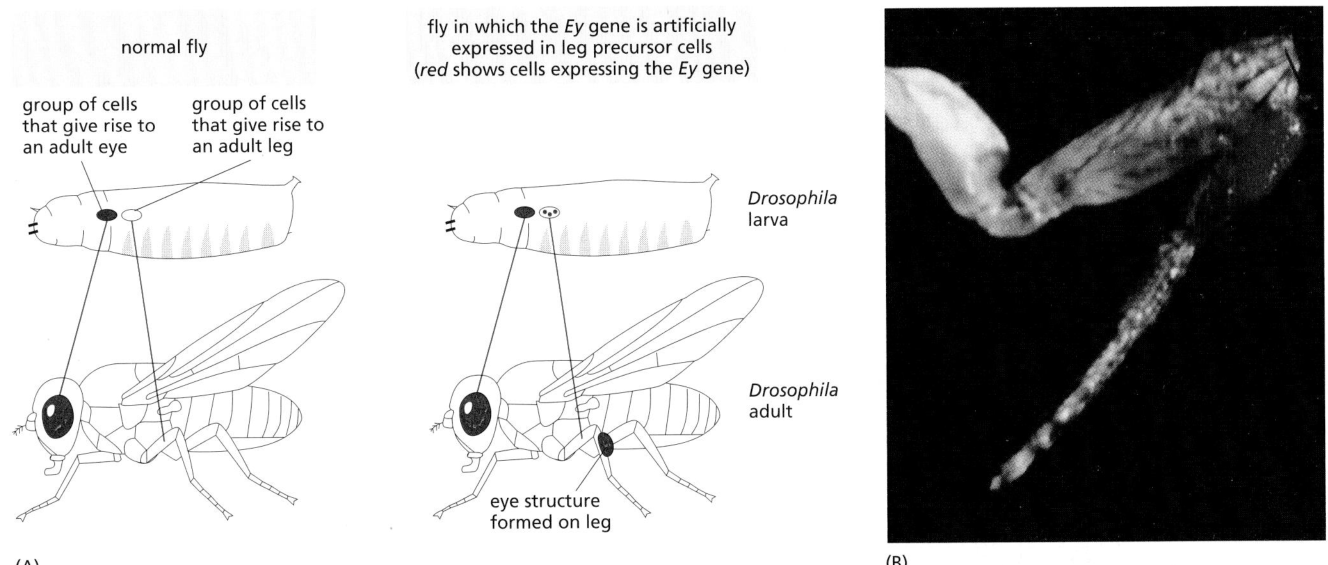

Figure 8–23 **Expression of the *Drosophila Ey* gene in the precursor cells of the leg triggers the development of an eye on the leg.** (A) Simplified diagrams showing the result when a fruit fly larva contains either the normally expressed *Ey* gene (*left*) or an *Ey* gene that is additionally expressed artificially in cells that will give rise to legs (*right*). (B) Photograph of an abnormal leg that contains a misplaced eye. (B, courtesy of Walter Gehring.)

POST-TRANSCRIPTIONAL CONTROLS

We have seen that transcription regulators control gene expression by switching on or off transcription initiation. The vast majority of genes in all organisms are regulated in this way. But additional points of control can come into play later in the pathway from DNA to protein, giving cells a further opportunity to manage the amount of gene product that is made. These **post-transcriptional controls**, which operate after RNA polymerase has bound to a gene's promoter and started to synthesize RNA, are crucial for the regulation of many genes.

In Chapter 7, we described one type of post-transcriptional control: alternative splicing, which allows different forms of a protein to be made in different tissues (Figure 7–21). Here we discuss a few more examples of the many ways in which cells can manipulate gene expression after transcription has begun.

Riboswitches Provide An Economical Solution to Gene Regulation

The mechanisms for controlling gene expression we have described thus far all involve the participation of a regulatory protein. But scientists have recently discovered a number of mRNAs that can regulate their own transcription and translation. These self-regulating mRNAs contain **riboswitches**: short sequences of RNA that change their conformation when bound to small molecules such as metabolites. Many riboswitches have been discovered, and each recognizes a specific small molecule. The conformational change that is driven by the binding of that molecule can regulate gene expression (Figure 8–24). This mode of gene regulation is particularly common in bacteria, where riboswitches sense key small metabolites in the cell and adjust gene expression accordingly.

Riboswitches are perhaps the most economical examples of gene control devices, because they bypass the need for regulatory proteins altogether. The fact that short sequences of RNA can form such highly efficient gene

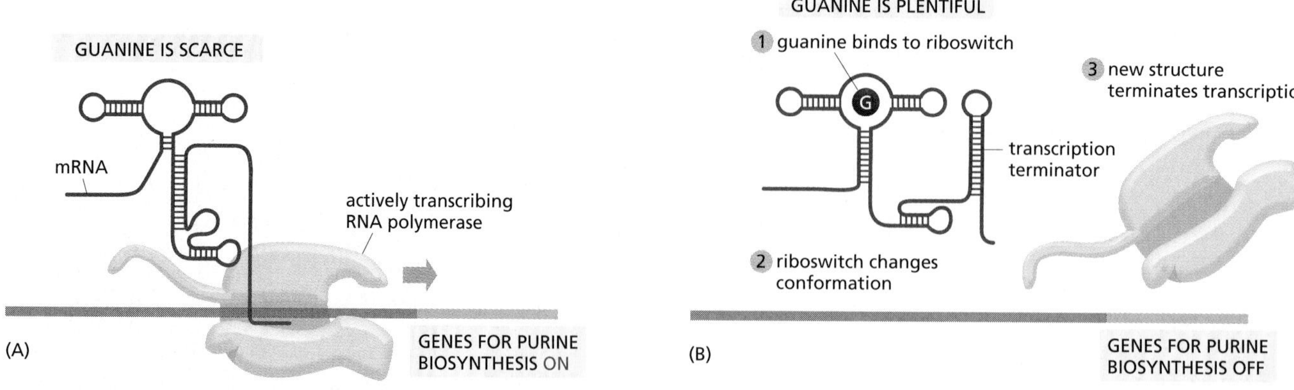

Figure 8–24 **A riboswitch controls purine biosynthesis genes in bacteria.** (A) When guanine is scarce, the riboswitch adopts a structure that allows the elongating RNA polymerase, which has already initiated transcription, to continue transcribing into the purine biosynthetic genes. The enzymes needed for guanine synthesis are thereby expressed. (B) When guanine is abundant, it binds to the riboswitch, causing it to undergo a conformational change. The new conformation includes a double-stranded structure (*red*) that forces the polymerase to terminate transcription before it reaches the purine biosynthetic genes. In the absence of guanine, formation of this double-stranded structure is blocked because one of the RNA strands that forms it pairs with a different region of the riboswitch (A). In this example, the riboswitch blocks completion of an mRNA. Other riboswitches control translation of mRNA molecules once they have been synthesized. (Adapted from M. Mandal and R.R. Breaker, *Nat. Rev. Mol. Cell Biol.* 5:451–63, 2004. With permission from Macmillan Publishers Ltd.)

control devices offers further evidence that, before modern cells arose, a world run by RNAs may have reached a high level of sophistication (see pp. 261–264).

The Untranslated Regions of mRNAs Can Control Their Translation

Once an mRNA has been synthesized, one of the most common ways of regulating how much of its protein product is made is to control the initiation of translation. Although the details of translation initiation differ between eucaryotes and bacteria, both use the same basic strategies for regulating gene expression at this step.

Bacterial mRNAs contain a short ribosome-binding sequence located a few nucleotides upstream of the AUG codon where translation begins. This recognition sequence forms base pairs with the RNA in the small ribosomal subunit, correctly positioning the initiating AUG codon within the ribosome. Because this interaction is needed for efficient translation initiation, it provides an ideal target for translational control. By blocking—or exposing—the ribosome recognition sequence, the bacterium can either inhibit—or promote—the translation of an mRNA (Figure 8–25).

Eucaryotic mRNAs possess a 5′ cap that helps guide the ribosome to the first AUG, the codon where translation will start (see Figure 7–35). In eucaryotic cells, repressors can inhibit translation initiation by binding to specific RNA sequences in the 5′ untranslated region of the mRNA and keeping the ribosome from finding the first AUG. When conditions change, the cell can inactivate the repressor and thereby increase translation of the mRNA.

Small Regulatory RNAs Control the Expression of Thousands of Animal and Plant Genes

As we saw in Chapter 7, RNAs perform many critical tasks in cells. In addition to acting as intermediate carriers of genetic information, they play key structural and catalytic roles, particularly in protein synthesis (see pp. 253–254). But a recent series of striking discoveries has revealed that noncoding RNAs—those that do not direct the production of a protein product—are far more prevalent than previously imagined and play unanticipated, widespread roles in regulating gene expression.

One particularly important type of noncoding RNA, found in plants and animals, is called **microRNA** (**miRNA**). Humans, for example, produce more than 400 different miRNAs, which seem to regulate at least one-third of all protein-coding genes. These short, regulatory RNAs control gene expression by base-pairing with specific mRNAs and controlling their stability and their translation.

Like other noncoding RNAs, such as tRNA and rRNA, the precursor miRNA transcript undergoes a special type of processing to yield the mature miRNA. This miRNA is then assembled with specialized proteins to form an *RNA-induced silencing complex* (*RISC*). The RISC patrols the cytoplasm, searching for mRNAs that are complementary to the miRNA it carries (Figure 8–26). Once a target mRNA forms base pairs with an miRNA, it is destroyed immediately by a nuclease present within the RISC or else its translation is blocked and it is delivered to a region of the cytoplasm where other nucleases will eventually degrade it. Once the RISC has taken care of an mRNA molecule, it is released and is free to seek out additional mRNA molecules. Thus a single miRNA—as part of a RISC—can eliminate one mRNA molecule after another, thereby efficiently blocking production of the protein that the mRNA encodes.

Two features of miRNAs make them especially useful regulators of gene expression. First, a single miRNA can regulate a whole set of different mRNAs so long as the mRNAs carry a common sequence; these sequences are often located in their 5′ and 3′ untranslated regions. In humans, some individual miRNAs control hundreds of different mRNAs in this manner. Second, a gene that encodes an miRNA occupies relatively little space in the genome compared with one that encodes a transcription regulator. Indeed, their small size is one reason that miRNAs were discovered only recently. Although we are only beginning to understand the full impact of miRNAs, it is clear that they represent a critical part of the cell's equipment for regulating the expression of its genes.

RNA Interference Destroys Double-Stranded Foreign RNAs

Some of the proteins that process and package miRNAs also serve as a cell defense mechanism: they orchestrate the destruction of 'foreign' RNA molecules, specifically those that are double-stranded. Many viruses—and transposable genetic elements—produce double-stranded RNA

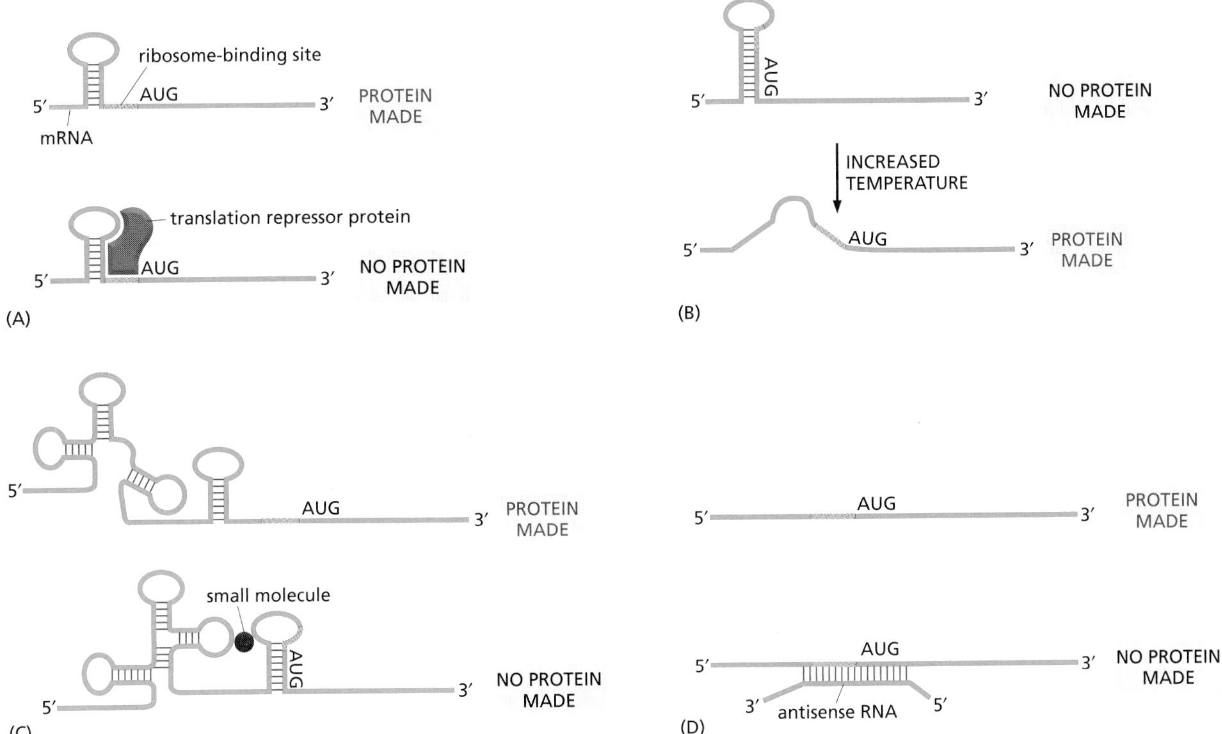

Figure 8–25 **Gene expression can be controlled by regulating translation initiation.** (A) Sequence-specific RNA-binding proteins can repress the translation of specific mRNAs by keeping the ribosome from binding to the ribosome-recognition sequence (*orange*) found at the start of a bacterial protein-coding gene. Some ribosomal proteins use this mechanism to inhibit the translation of their own mRNA. (B) An mRNA from the pathogen *Listeria monocytogenes* contains a 'thermosensor' RNA sequence that controls the translation of a set of virulence genes. At the warmer temperature that the bacterium encounters inside its human host, the thermosensor sequence is denatured and the virulence genes are expressed. (C) Binding of a small molecule to a riboswitch causes a structural rearrangement of the RNA, sequestering the ribosome-recognition sequence and blocking translation initiation. (D) A complementary, 'antisense' RNA produced by another gene base-pairs with a specific mRNA and blocks its translation. Although these examples of translational control are from bacteria, many of the same principles operate in eucaryotes.

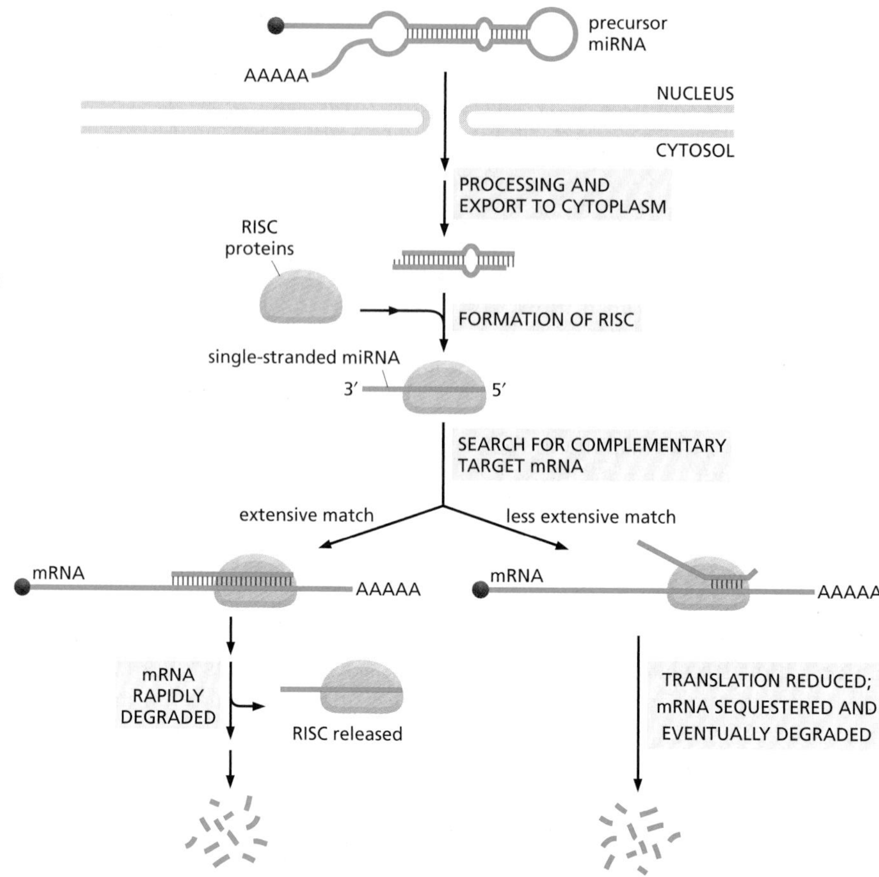

Figure 8–26 **An miRNA targets a complementary mRNA transcript for destruction.** The precursor miRNA is processed to form a mature miRNA. It then assembles with a set of proteins into a complex called RISC. The miRNA then guides the RISC to mRNAs that have a complementary nucleotide sequence. Depending on how extensive the region of complementarity is, the target mRNA is either rapidly degraded by a nuclease within the RISC or transferred to an area of the cytoplasm where other cellular nucleases will destroy it.

some time in their life cycles. This targeted RNA degradation mechanism, called **RNA interference** (**RNAi**), helps to keep these potentially dangerous invaders in check.

The presence of foreign, double-stranded RNA in the cell triggers RNAi by first attracting a protein complex containing a nuclease called Dicer. Dicer cleaves the double-stranded RNA into short fragments (approximately 23 nucleotide pairs in length) called **small interfering RNAs** (**siRNAs**). These short, double-stranded RNAs are then incorporated into RISCs, the same complexes that can carry miRNAs. The RISC discards one strand of the siRNA duplex and uses the remaining single-stranded RNA to locate a complementary foreign RNA molecule (Figure 8–27). This target RNA molecule is then rapidly degraded, leaving the RISC free to search out more of the same foreign RNA molecules.

RNAi is found in a wide variety of organisms, including single-celled fungi, plants, and worms, indicating that it is evolutionarily ancient. In some organisms, including plants, the RNAi activity can spread from tissue to tissue by the movement of RNA between cells. This RNA transfer allows the entire plant to become resistant to a virus after only a few of its cells have been infected. In a broad sense, the RNAi response resembles certain aspects of the human immune system. In both cases, an infectious organism elicits the production of 'attack' molecules (either siRNAs or antibodies) that are custom designed to inactivate the invader and thereby protect the host.

Scientists Can Use RNA Interference to Turn Off Genes

The discovery of miRNAs, siRNAs, and the mechanism of RNAi has been greeted with great enthusiasm. In a practical sense, RNAi has become

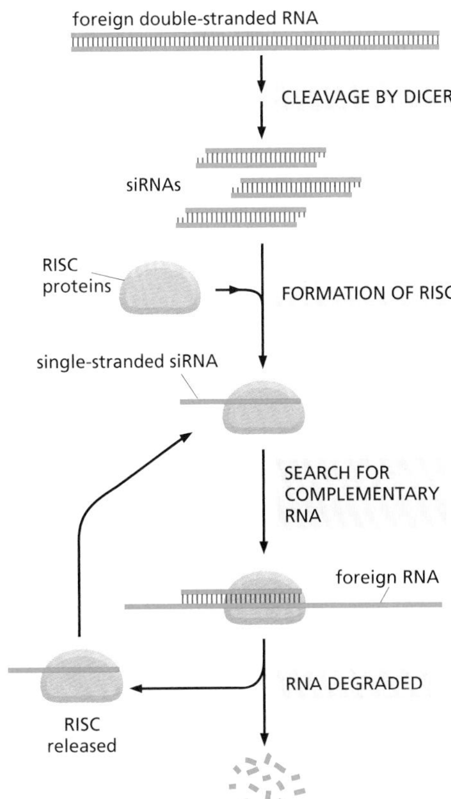

Figure 8–27 **siRNAs destroy foreign RNAs.** Double-stranded RNAs from a virus or transposable genetic element are first cleaved by a nuclease called Dicer. The resulting double-stranded fragments are incorporated into RISCs, which discard one strand of the duplex and use the other strand to locate and destroy complementary RNAs. This mechanism forms the basis for RNA interference (RNAi).

a powerful experimental tool that allows scientists to inactivate almost any gene in cultured cells or, in some cases, a whole plant or animal. We discuss how this method is being used to help determine the function of individual genes in Chapter 10.

At the same time, RNAi shows real potential as a powerful new approach for treating human disease. Because many human disorders result from the inappropriate expression of genes, the ability to turn these genes off by introducing complementary siRNA molecules holds great medical promise.

Finally, the discovery that RNAs play such a key role in controlling gene expression expands our understanding of the types of regulatory networks that cells have at their command. One of the great challenges of biology in this century will be to determine how these networks cooperate to specify the development of complex organisms, including ourselves.

ESSENTIAL CONCEPTS

- A typical eucaryotic cell expresses only a fraction of its genes, and the distinct types of cells in multicellular organisms arise because different sets of genes are expressed as cells differentiate.
- Although all of the steps involved in expressing a gene can in principle be regulated, for most genes the initiation of transcription is the most important point of control.
- The transcription of individual genes is switched on and off in cells by transcription regulators. These act by binding to short stretches of DNA called regulatory DNA sequences.
- Although each transcription regulator has unique features, most bind to DNA using one of a small number of structural motifs. The precise amino acid sequence that is folded into the DNA-binding motif determines the particular DNA sequence that is recognized.
- In bacteria, transcription regulators usually bind to regulatory DNA sequences close to where RNA polymerase binds. They can either activate or repress transcription of the gene. In eucaryotes, regulatory DNA sequences are often separated from the promoter by many thousands of nucleotide pairs.
- Eucaryotic transcription regulators act in two fundamental ways: (1) they can directly affect the assembly process of RNA polymerase and the general transcription factors at the promoter, and (2) they can locally modify the chromatin structure of promoter regions.
- In eucaryotes, the expression of a gene is generally controlled by a combination of transcription regulators.
- In multicellular plants and animals, the production of different transcription regulators in different cell types ensures the expression of only those genes appropriate to the particular type of cell.
- Cells in multicellular organisms have mechanisms that enable their progeny to 'remember' what type of cell they should be.
- A single transcription regulator, if expressed in the appropriate precursor cell, can trigger the formation of a specialized cell type or even an entire organ.

- Cells can also regulate gene expression by controlling events that occur after transcription has begun. Many of these mechanisms rely on RNA molecules that can influence their own transcription or translation.
- MicroRNAs (miRNAs) control gene expression by base-pairing with specific mRNAs and regulating their stability and their translation.
- Cells have a defense mechanism for destroying 'foreign' double-stranded RNAs, many of which are produced by viruses. Scientists can take advantage of this mechanism, called RNA interference, to inactivate genes of interest by simply injecting cells with double-stranded RNAs that target the mRNAs produced by those genes.

KEY TERMS

activator
combinatorial control
differentiation
DNA methylation
epigenetic inheritance
gene expression
microRNA (miRNA)
positive feedback loop
post-transcriptional control
regulatory DNA sequence
reporter gene
repressor
riboswitch
RNA interference (RNAi)
small interfering RNA (siRNA)
transcription regulator

QUESTIONS

QUESTION 8–4

A virus that grows in bacteria (bacterial viruses are called bacteriophages) can replicate in one of two ways. In the prophage state, the viral DNA is inserted into the bacterial chromosome and is copied along with the bacterial genome each time the cell divides. In the lytic state, the viral DNA is released from the bacterial chromosome and replicates many times in the cell. This viral DNA then produces viral coat proteins that together with the replicated viral DNA form many new virus particles that burst out of the bacterial cell. These two forms of growth are controlled by two transcription regulators, called cI ("c one") and Cro, that are encoded by the virus. In the prophage state, cI is expressed; in the lytic state, Cro is expressed. In addition to regulating the expression of other genes, cI represses the *Cro* gene, and Cro represses the *cI* gene (Figure Q8–4). When bacteria containing a phage in the prophage state are briefly irradiated with UV light, cI protein is degraded.

A. What will happen next?

B. Will the change in (A) be reversed when the UV light is switched off?

C. Why might this response to UV light have evolved?

QUESTION 8–5

(True/False) When the nucleus of a fully differentiated carrot cell is injected into a frog egg whose nucleus has been removed, the injected donor nucleus is capable of programming the recipient egg to produce a normal carrot. Explain your answer.

QUESTION 8–6

Which of the following statements are correct? Explain your answers.

A. In bacteria, but not in eucaryotes, most mRNAs encode more than one protein.

B. Most DNA-binding proteins bind to the major groove of the double helix.

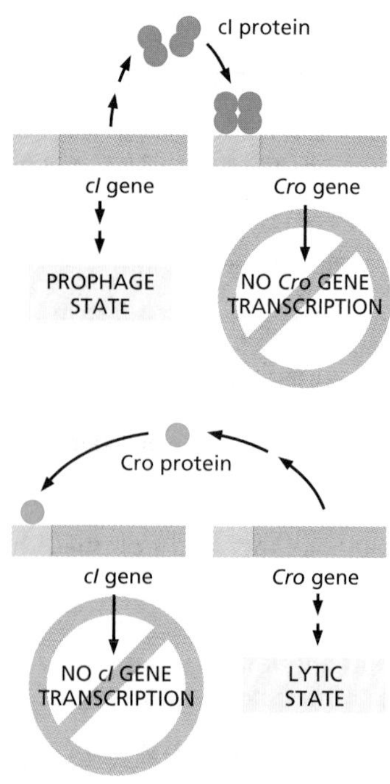

Figure Q8–4

C. Of the major control points in gene expression (transcription, RNA processing, RNA transport, translation, and control of a protein's activity), transcription initiation is one of the most common.

D. The zinc atoms in DNA-binding proteins that contain zinc finger domains contribute to the binding specificity through sequence-specific interactions that they form with the bases.

QUESTION 8–7

Your task in the laboratory of Professor Quasimodo is to determine how far an enhancer (a binding site for an activator protein) could be moved from the promoter of the *straightspine* gene and still activate transcription. You systematically vary the number of nucleotide pairs between these two sites and then determine the amount of transcription by measuring the production of Straightspine mRNA. At first glance, your data look confusing (Figure Q8–7). What would you have expected for the results of this experiment? Can you save your reputation and explain these results to Professor Quasimodo?

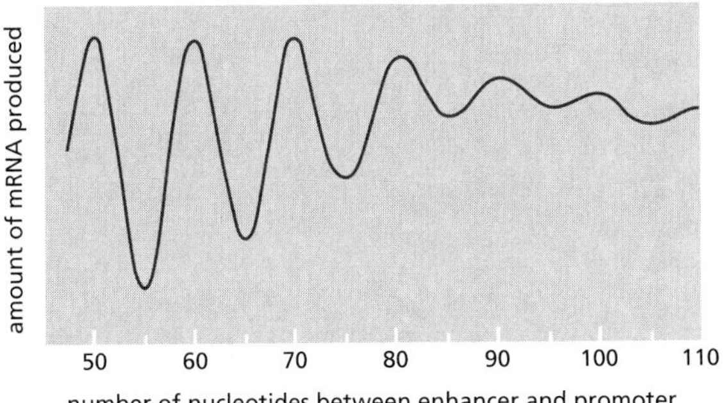

Figure Q8–7

QUESTION 8–8

Many transcription regulators form dimers of identical subunits. Why is this advantageous? Describe three structural motifs that are often used to contact DNA. What are the particular features that suit them for this purpose?

QUESTION 8–9

The λ repressor binds as a dimer to critical sites on the λ genome to repress the virus's lytic genes. This is necessary to maintain the prophage (integrated) state. Each molecule of the repressor consists of an N-terminal DNA-binding domain and a C-terminal dimerization domain (Figure Q8–9). Upon induction (for example, by irradiation with UV light), the genes for lytic growth are expressed, λ progeny are produced, and the bacterial cell is lysed (see Question 8–4). Induction is initiated by cleavage of the λ repressor at a site between the DNA-binding domain and the dimerization domain, which causes the repressor to dissociate from the DNA. In the absence of bound repressor, RNA polymerase binds and initiates lytic growth. Given that the number (concentration) of DNA-binding domains is unchanged by cleavage of the repressor, why do you suppose its cleavage results in its dissociation from the DNA?

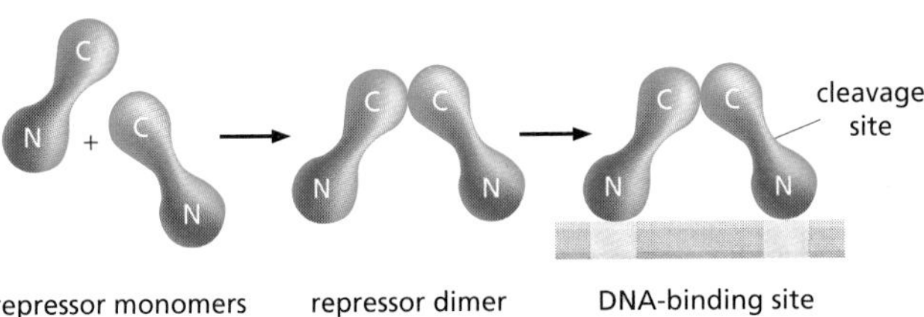

Figure Q8–9

QUESTION 8–10

The enzymes for arginine biosynthesis are located at several positions around the genome of *E. coli*, and they are regulated coordinately by a transcription regulator encoded by the *ArgR* gene. The activity of ArgR is modulated by arginine. Upon binding arginine, ArgR alters its conformation, dramatically changing its affinity for the DNA sequences in the promoters of the genes for the arginine biosynthetic enzymes. Given that ArgR is a repressor protein, would you expect that ArgR would bind more tightly or less tightly to the DNA sequences when arginine is abundant? If ArgR functioned instead as an activator protein, would you expect the binding of arginine to increase or to decrease its affinity for its regulatory DNA sequences? Explain your answers.

QUESTION 8–11

When enhancers were initially found to influence transcription many thousands of nucleotide pairs from the promoters they control, two principal models were invoked to explain this action at a distance. In the 'DNA looping' model, direct interactions between proteins bound at enhancers and promoters were proposed to stimulate transcription initiation. In the 'scanning' or 'entry-site' model, RNA polymerase (or another component of the transcription machinery) was proposed to bind at the enhancer and then scan along the DNA until it reached the promoter. These two models were tested using an enhancer on one piece of DNA and a β-globin gene and promoter on a separate piece of DNA (Figure Q8–11). The β-globin gene was not expressed from the mixture of pieces. However, when the two segments of DNA were joined via a protein linker, the β-globin gene was expressed.

Does this experiment distinguish between the DNA looping model and the scanning model? Explain your answer.

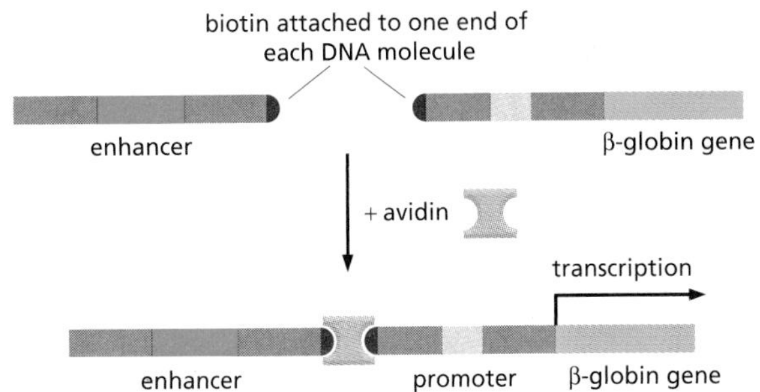

Figure Q8–11

QUESTION 8–12

Differentiated cells of an organism contain the same genes. (Among the few exceptions to this rule are the cells of the mammalian immune system, in which the formation of specialized cells is based on limited rearrangements of the genome.) Describe an experiment that substantiates the first sentence of this question, and explain why it does.

QUESTION 8–13

Figure 8–19 shows a simple scheme by which three transcription regulators are used during development to create eight different cell types. How many cell types could you create, using the same rules, with four different transcription regulators? As described in the text, MyoD is a transcription regulator that by itself is sufficient to induce muscle-specific gene expression in fibroblasts. How does this observation fit the scheme in Figure 8–19?

QUESTION 8–14

Imagine the two situations shown in Figure Q8–14. In cell I, a transient signal induces the synthesis of protein A, which is a gene activator that turns on many genes including its own. In cell II, a transient signal induces the synthesis of protein R, which is a gene repressor that turns off many genes including its own. In which, if either, of these situations will the descendants of the original cell "remember" that the progenitor cell had experienced the transient signal? Explain your reasoning.

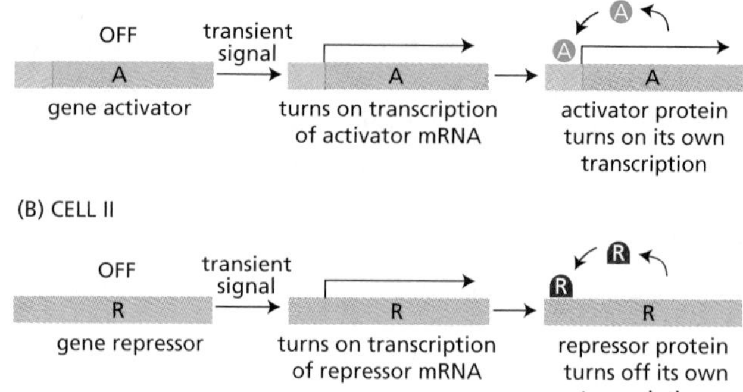

Figure Q8–14

QUESTION 8–15

Discuss the following argument: "If the expression of every gene depends on a set of transcription regulators, then the expression of these regulators must also depend on the expression of other regulators, and their expression must depend on the expression of still other regulators, and so on. Cells would therefore need an infinite number of genes, most of which would code for transcription regulators." How does the cell get by without having to achieve the impossible?

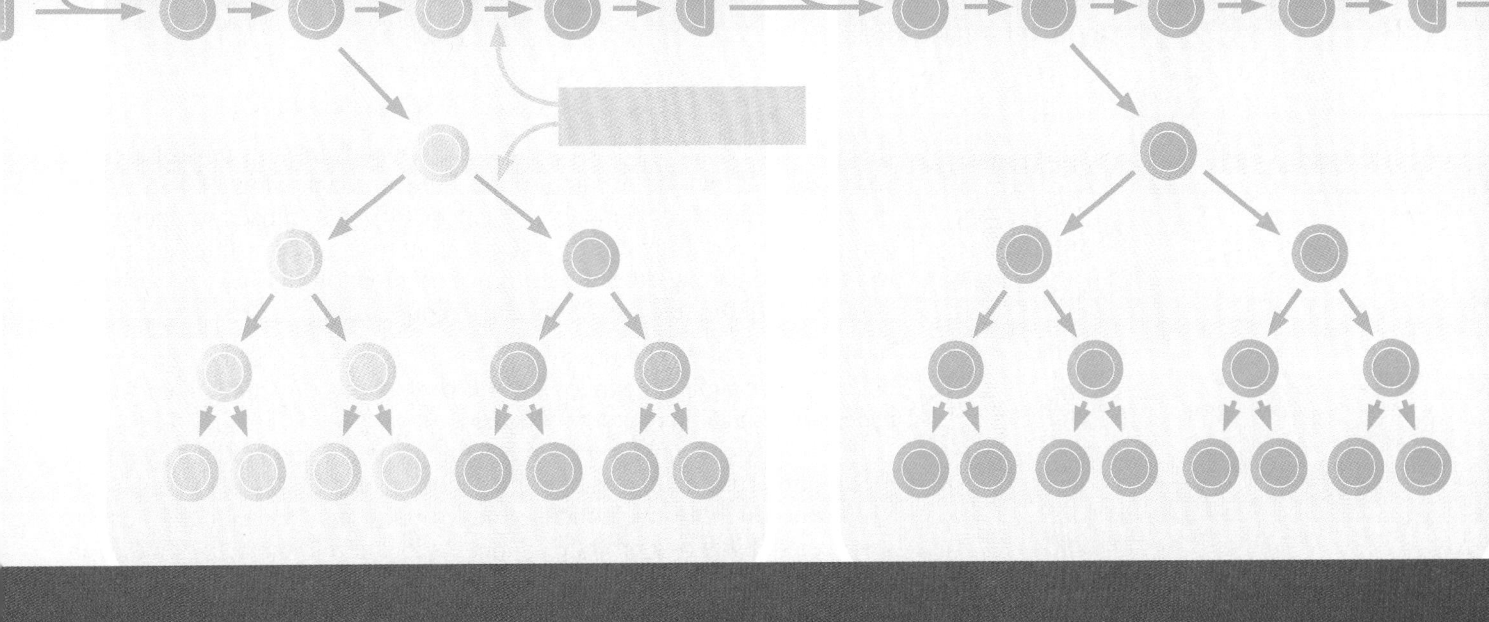

CHAPTER **NINE** 9

How Genes and Genomes Evolve

GENERATING GENETIC VARIATION

RECONSTRUCTING LIFE'S FAMILY TREE

EXAMINING THE HUMAN GENOME

No two people are exactly alike. Look at any crowd in a classroom or on a bus: each individual differs in a host of heritable characteristics—in hair color, eye color, skin color, height, build, and so on (Figure 9–1). Although we are all members of the same species, plainly our genomes do not contain exactly the same information.

Such differences in the nucleotide sequences between one organism and the next provide the raw material upon which evolution works. Sculpted by selective pressures over billions of cell generations since the beginning of life on Earth, these variations have given rise to the spectacular menagerie of present-day life forms, from bacteria to whales. The diversity of species thus depends on a delicate balance between the conservative accuracy of DNA replication that enables progeny to inherit the virtues of their parents, and the errors of genome replication and maintenance that allow organisms to 'try out' novel features and evolve new capabilities. If this balance had been struck differently, the whole history of life on Earth would have been different.

In this chapter, we discuss how genes and genomes change over time. We examine the molecular mechanisms by which genetic changes occur, and we consider how the information in present-day genomes can be deciphered to yield a historical record of the evolutionary processes that have shaped them. We end the chapter by taking a closer look at the human genome to see what our own DNA sequences tell us about who we are and where we come from.

Figure 9–1 **A group of English schoolchildren reflects the concept of variation.** Small differences in nucleotide sequence account for differences in appearance between one individual and the next. (Courtesy of Fiona Pragoff, Wellcome Images.)

GENERATING GENETIC VARIATION

Evolution works on the DNA sequences that each organism inherits from its ancestors: there is no natural mechanism for making long stretches of entirely novel nucleotide sequences. In this sense, no gene—or genome—is ever truly new. Evolution is more a tinkerer than an inventor: the astonishing diversity in form and function that we see in living systems is all the result of variations on preexisting themes.

As these variations pile up, over millions of generations, the cumulative effect can be radical change. Several basic types of genetic change are especially crucial in evolution (Figure 9–2):

- *Mutation within a gene:* an existing gene can be modified by mutations that change single nucleotides or that delete or duplicate one or more nucleotides. These mutations can alter the activity or stability of a protein, change its location in the cell, or affect its interactions with other proteins.
- *Mutation within the regulatory DNA of a gene:* when and where a gene is expressed can be affected by mutations in the stretches of DNA sequence that regulate its activity (described in Chapter 8). For example, humans and fish have a surprisingly large number of genes in common, but changes in the regulation of those shared genes underlie many of the most dramatic differences between those species.
- *Gene duplication:* an existing gene, a larger segment of DNA, or even a whole genome can be duplicated, creating a set of closely

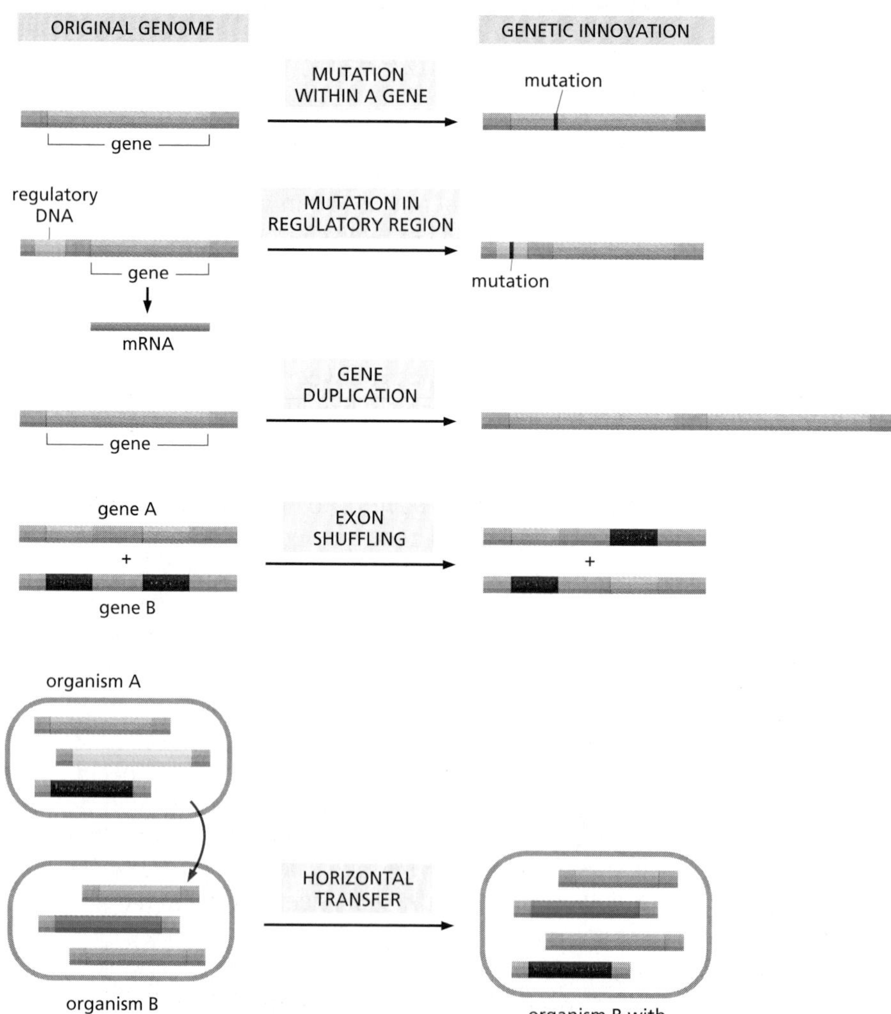

Figure 9–2 **Genes and genomes can be altered by several different mechanisms.** Small mutations, duplications, deletions, rearrangements, and even the infusion of fresh genetic material all contribute to genome evolution.

related genes within a single cell. As this cell and its progeny divide, the original gene and its duplicate can acquire additional mutations and assume functions and patterns of expression distinct from each other and from the ancestral gene.

- *Exon shuffling:* two or more existing genes can be broken and rejoined to make a hybrid gene containing DNA segments that originally belonged to separate genes. In eucaryotes, the breaking and rejoining often occurs within the long intron sequences. Because these sequences are eventually removed by RNA splicing, the breaking and joining do not have to be precise to result in a functional gene.
- *Horizontal gene transfer:* a piece of DNA can be transferred from the genome of one cell to that of another—even to that of another species. This process, rare among eucaryotes but common among procaryotes, differs from the usual "vertical" transfer of genetic information from parent to progeny.

Each of these forms of genetic variation—from the simple mutations that occur within a gene to the more extensive duplications, deletions, rearrangements, and additions that occur within a genome—has played an important part in the evolution of modern organisms. And they still play that part today, as organisms continue to evolve. In this section, we discuss these basic mechanisms of genetic change, and we consider their consequences for genome evolution. But first, we pause to consider the complications of sex—the mechanism that many organisms use to pass genetic information on to future generations.

QUESTION 9–1

In this chapter it is argued that genetic variability is beneficial for a species because it enhances that species' ability to adapt to changing conditions. Why, then, does a cell go to great lengths to ensure the fidelity of DNA replication?

In Sexually Reproducing Organisms, Only Changes to the Germ Line Are Passed Along To Progeny

For bacteria and other unicellular organisms that reproduce asexually, the inheritance of genetic information is fairly straightforward. Each individual copies its genome and then splits in two, giving rise to two daughters. Such organisms have a family tree that is simply a branching diagram of cell divisions directly linking each individual to its progeny and to its ancestors.

For a multicellular organism that reproduces sexually, the family connections are considerably more complex. For a start, sex mixes and matches the genomes of two individuals to generate offspring that are genetically distinct from either parent. It also makes use of genetic ambassadors—cells designed specifically to carry a copy of the individual's genome to the next generation. These specialized reproductive cells—the **germ cells** or *gametes*—come together during fertilization to give rise to a new individual (as we discuss in Chapter 19). All the other cells of the body—the **somatic cells**—are doomed to die without leaving descendants of their own (Figure 9–3). The cell lineage that gives rise to the germ cells is called the **germ line**, and it is through a series of germ-line cell divisions that every organism traces its descent back to its ancestors and, ultimately, back to the ancestors of us all—the first cells that existed, at the origin of life more than 3.5 billion years ago. In this sense, somatic cells exist only to help the germ line survive and propagate.

All this means that a mutation is passed on to the next generation only if it occurs in a germ-line cell. A mutation that occurs in a somatic cell—although it might have unfortunate consequences for the individual in which it occurs (causing cancer, for example)—is not transmitted to the organism's offspring (Figure 9–4). Thus, in tracking the genetic changes that accumulate during evolution, we must concentrate on events that occur in the germ line.

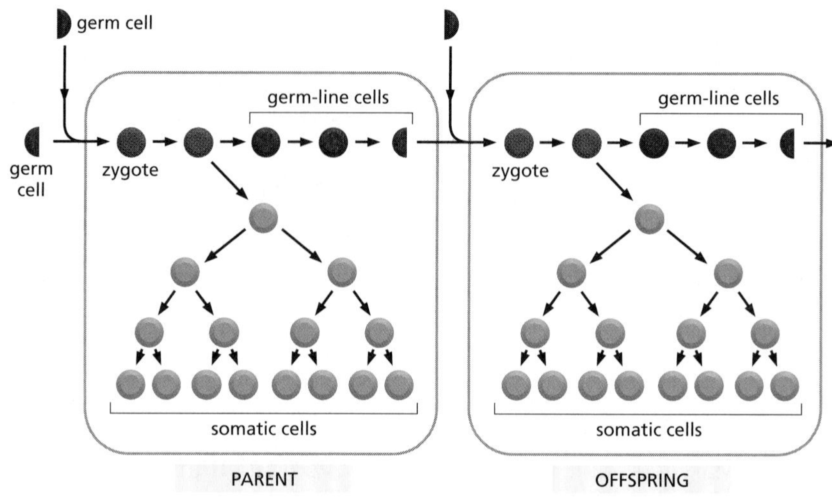

Figure 9–3 **Germ-line cells and somatic cells have fundamentally different functions.** In sexually reproducing organisms, genetic information is propagated into the next generation by specialized cells called germ-line cells (*red*). These give rise to the reproductive cells—the germ cells (*red* half circles). Germ cells are shown as half circles because they contain only half the genetic complement of the other cells in the body (full circles). When two germ cells come together during fertilization, they form a fertilized egg or zygote (*purple*) that once again contains a full set of chromosomes. This gives rise to more germ-line cells and to somatic cells (*blue*). Somatic cells form the body of the organism but do not contribute their DNA to the next generation.

Although sexual reproduction involves the mixing of genomes—which influences how genetic variations are propagated—many of the basic mechanisms that generate genetic change are the same for both sexually and asexually reproducing organisms, as we now discuss.

Point Mutations Are Caused by Failures of the Normal Mechanisms for Copying and Maintaining DNA

Despite the elaborate mechanisms that exist to faithfully copy and maintain DNA sequences, each nucleotide pair in an organism's genome runs a certain small risk of changing each time a cell divides. Changes that affect a single nucleotide pair are called **point mutations**. These typically arise from rare errors in DNA replication or repair (discussed in Chapter 6, pp. 211–217).

The nucleotide mutation rate has been determined directly in experiments with bacteria such as *E. coli.* Under laboratory conditions, *E. coli* divides about once every 20–25 minutes; in less than a day, a single *E. coli* can produce more descendants than there are humans on Earth—enough to provide a good chance for almost any conceivable point mutation to occur. A culture containing 10^9 *E. coli* cells thus harbors millions of mutant cells whose genomes differ subtly from the ancestor cell. Some of these mutations may confer a selective advantage on individual cells: resistance to a poison, for example, or the ability to survive when deprived of a standard nutrient. By exposing the culture to a selective

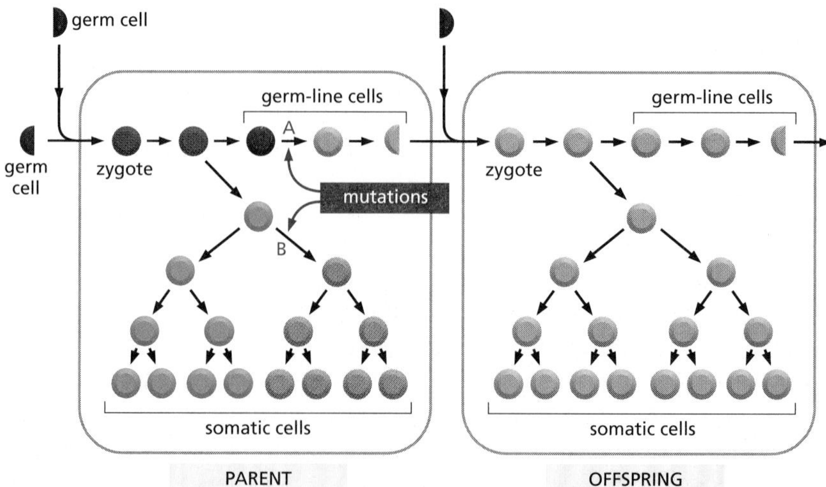

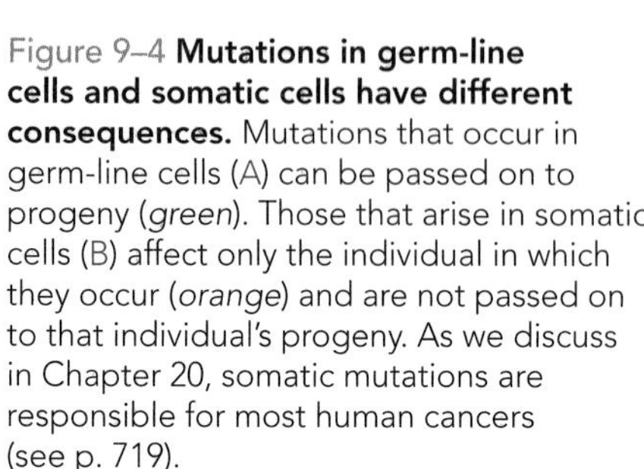
Figure 9–4 **Mutations in germ-line cells and somatic cells have different consequences.** Mutations that occur in germ-line cells (A) can be passed on to progeny (*green*). Those that arise in somatic cells (B) affect only the individual in which they occur (*orange*) and are not passed on to that individual's progeny. As we discuss in Chapter 20, somatic mutations are responsible for most human cancers (see p. 719).

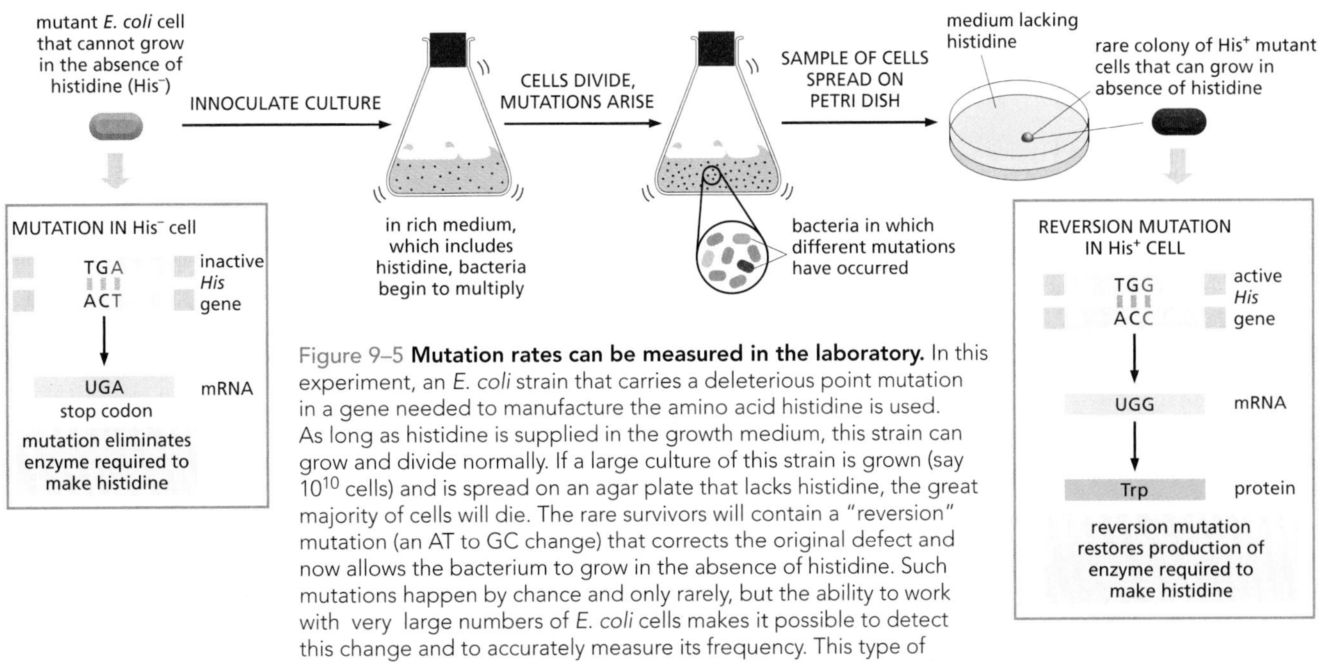

Figure 9–5 **Mutation rates can be measured in the laboratory.** In this experiment, an *E. coli* strain that carries a deleterious point mutation in a gene needed to manufacture the amino acid histidine is used. As long as histidine is supplied in the growth medium, this strain can grow and divide normally. If a large culture of this strain is grown (say 10^{10} cells) and is spread on an agar plate that lacks histidine, the great majority of cells will die. The rare survivors will contain a "reversion" mutation (an AT to GC change) that corrects the original defect and now allows the bacterium to grow in the absence of histidine. Such mutations happen by chance and only rarely, but the ability to work with very large numbers of *E. coli* cells makes it possible to detect this change and to accurately measure its frequency. This type of experiment revealed that the spontaneous mutation rate in *E. coli* was approximately 1 mistake in every 10^9 nucleotides copied.

condition—adding an antibiotic or removing an essential nutrient—one can find these needles in the haystack: cells that have undergone a specific mutation enabling them to survive in conditions where the original cells cannot (Figure 9–5). In this way, one can determine how frequently the specific mutation occurs, and from this one can estimate the overall mutation frequency in *E. coli*: it is about 1 nucleotide change per 10^9 nucleotide pairs per cell generation. Humans have a mutation rate of 0.1 nucleotide change for every 10^9 nucleotide pairs copied—which is one-tenth that of *E. coli*.

Point mutations provide a way of fine tuning a gene's function by making small adjustments to its sequence. They can also destroy a gene. Very often, however, they do neither of these things. At many sites in the genome, a point mutation has absolutely no effect on the appearance or viability of the organism; such *neutral mutations* might fall in regions whose sequence is unimportant, such as introns. They can also change the third position of a codon such that the amino acid specified is the same, or can direct the replacement of one amino acid with a related one at a position in the protein that can tolerate either.

Point Mutations Can Change the Regulation of a Gene

Mutations in the coding sequences of genes are fairly easy to spot because they change the amino acid sequence of the resulting protein in predictable ways. But mutations in regulatory DNA are more difficult to recognize, because they look like minor variations within long stretches of DNA whose actual nucleotide sequence does not seem to matter. And because regulatory sequences can be located some distance from the coding sequence of the gene, it can be particularly challenging to identify these elements and the changes that affect them.

Despite these difficulties, many examples have been discovered where point mutations in regulatory DNA have profound consequences for the organism. For example, a small number of people are resistant to malaria because of a point mutation that affects the expression of a cell-surface receptor to which the malaria parasite *Plasmodium vivax* binds.

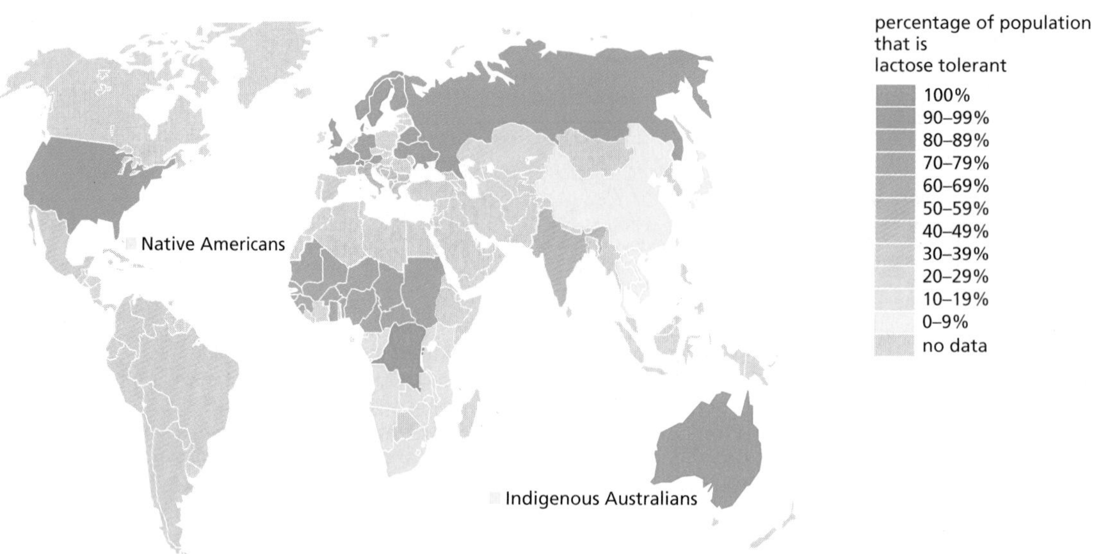

Figure 9–6 **The ability of adult humans to digest milk follows the domestication of cattle.** Approximately 10,000 years ago, humans in northern Europe and central Africa domesticated cattle, and the subsequent availability of milk—particularly during periods of starvation—gave a selective advantage to those humans able to digest lactose as adults. Two independent point mutations that allow the expression of lactase in adults arose in human populations—one in northern Europe and another in central Africa. These mutations have since spread through different regions of the world. For example, the migration of Northern Europeans to North America and Australia explains why most people living on these continents can digest lactose as adults although the native populations remain lactose intolerant.

The mutation prevents the receptor from being produced in red blood cells, rendering the individuals who carry this mutation immune to infection. Point mutations in regulatory DNA also play a role in our ability to digest the sugar lactose. Our earliest ancestors were lactose intolerant, because the enzyme that breaks down lactose—called lactase—was made only during infancy. Adults, who were no longer exposed to breast milk did not need the enzyme. When humans began to get milk from domestic animals some 10,000 years ago, variant genes appeared in the human population, enabling those who carried them to continue to express lactase as adults. We now know that people who can digest milk as adults contain a point mutation in the regulatory DNA of the lactase gene, allowing it to be efficiently transcribed throughout their lives. In a sense, adults who can digest lactose are "mutants" with respect to this trait. It is remarkable how quickly this new characteristic spread through the human population, especially in societies that depended heavily on milk for nutrition (Figure 9–6).

These evolutionary changes in the regulatory sequence of the lactase gene occurred relatively recently (10,000 years ago), well after humans became a distinct species. However, much more ancient changes in regulatory sequences have occurred in other genes; some of these are thought to underlie many of the profound differences among species (Figure 9–7).

DNA Duplications Give Rise to Families of Related Genes

Point mutations can influence the activity of an existing gene, but how do new genes come into being? Gene duplication is perhaps the most important mechanism for generating new genes from old ones. Once a gene has been duplicated, one of the two gene copies can mutate and become specialized to perform a different function, while the other continues to

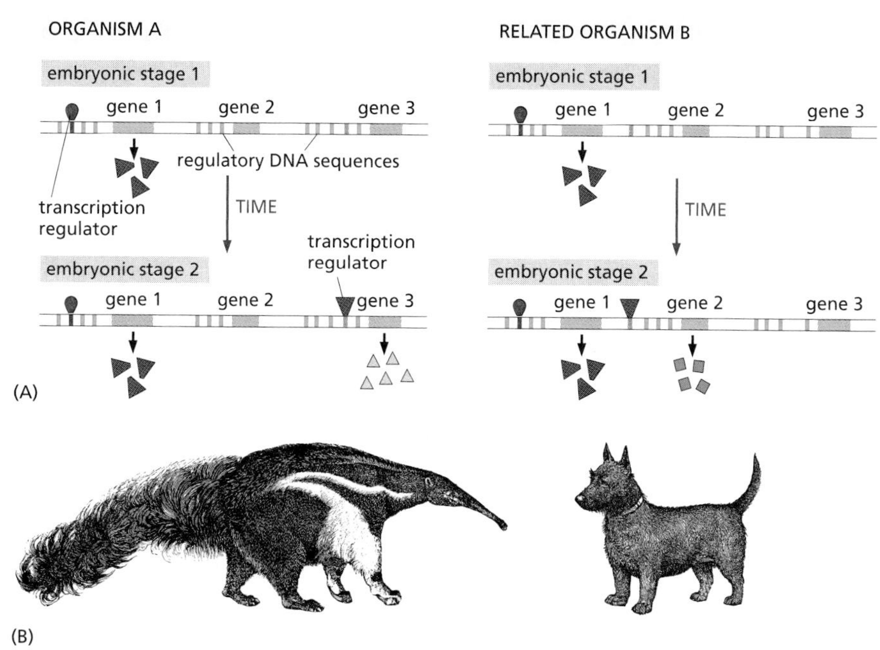

Figure 9–7 **Changes in regulatory DNA can have dramatic consequences for the development of an organism.** (A) In this hypothetical example, the genomes of organisms A and B code for the same set of regulatory proteins, but the regulatory DNA controlling expression of the proteins is different. Both organisms express the same proteins at stage 1, but because of the differences in their regulatory DNA they follow different pathways in stage 2. (B) Differences in gene regulation during development can have profound effects on the appearance of the adult organisms.

carry out the original function. Or both copies may evolve so that the function of the ancestral gene is divided between them. This specialization of duplicated genes occurs gradually, as mutations accumulate in the descendants of the original cell in which gene duplication occurred. Repeated rounds of this process of **gene duplication and divergence** over many millions of years can allow one gene to give rise to a whole family of genes, each with a specialized function, within a single genome. Analysis of genome sequences reveals many examples of such gene families: in *Bacillus subtilis*, for example, nearly half of the genes have one or more obvious relatives elsewhere in the genome (Figure 9–8). And in vertebrates, the globin gene family clearly arose from a single primordial gene, as we shall see shortly. But how does gene duplication occur in the first place?

It is thought that many gene duplications are generated by homologous recombination. Homologous recombination normally takes place only when two long stretches of nearly identical DNA become paired, most often the same region of DNA on two homologous chromosomes (discussed in Chapter 6, pp. 218–221). But on rare occasions, a recombination event will instead occur between two short repeated DNA sequences that fall on either side of a gene. If a crossover occurs, one chromosome will end up with an extra copy of the gene and the other will lose it (Figure 9–9). Once a gene has been duplicated in this way, subsequent crossovers can readily add extra copies to the duplicated set by the same mechanism. As a result, entire sets of closely related genes, arranged in series, are commonly found in genomes.

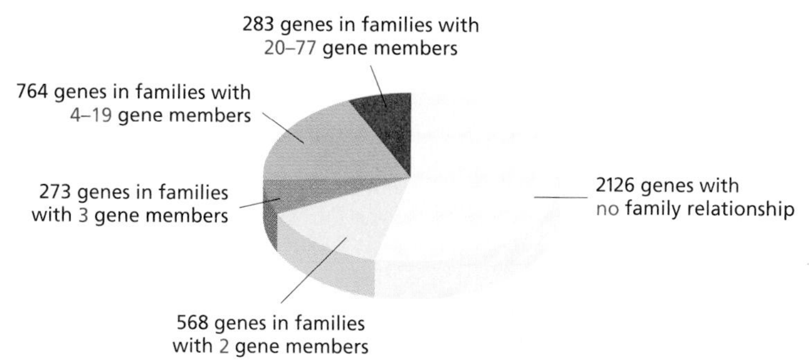

Figure 9–8 **The *Bacillus subtilis* genome contains many families of evolutionarily related genes.** The largest gene family in the *B. subtilis* genome contains 77 genes coding for a variety of transporters, proteins that move materials across the cell membrane (discussed in Chapter 12). (After F. Kunst et al., *Nature* 390:249–256, 1997. With permission from Macmillan Publishers Ltd.)

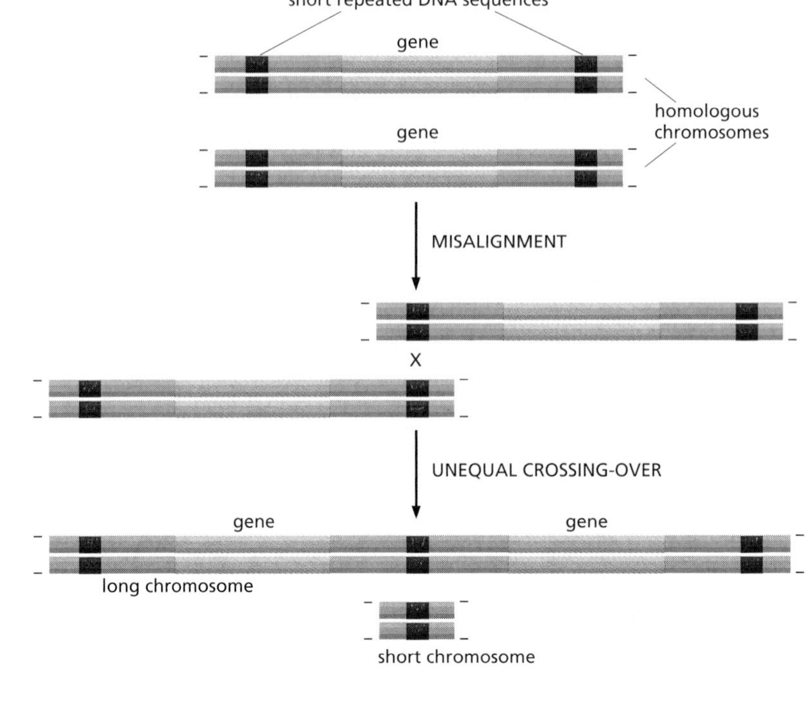

Figure 9–9 **Gene duplication can occur by crossovers between short repeated sequences on DNA.** A pair of homologous chromosomes undergoes recombination at a short sequence (*red*) that brackets a gene (*orange*). In some cases, such repeated sequences are remnants of mobile genetic elements, which are present in many copies in the human genome (see Figure 6–35). When crossing-over occurs as shown, the long chromosome will get two copies of the intervening gene, while the short chromosome will have the gene deleted. The type of crossing-over that produces gene duplications is called *unequal crossing-over* because the resulting products are unequal in size. If this process occurs in the germ line, some progeny will inherit the long chromosome, while others will inherit the short one.

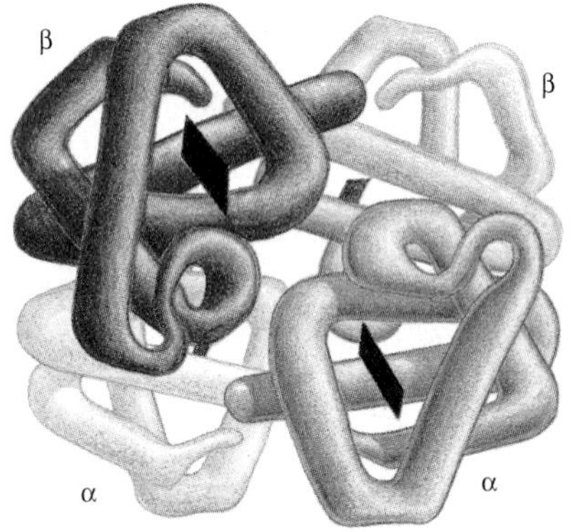

Figure 9–10 **A single-chain globin molecule gave rise to the four-chain hemoglobin used by humans and other mammals.** The mammalian hemoglobin molecule is a complex of two α- and two β-globin chains. The single-chain globin found in some primitive vertebrates forms a dimer that dissociates when it binds oxygen, representing an intermediate stage in the evolution of the four-chain molecule.

The Evolution of the Globin Gene Family Shows How Gene Duplication and Divergence Can Give Rise to Proteins Tailored to an Organism and Its Development

The evolutionary history of the globin gene family provides a striking example of how gene duplication and divergence has generated new proteins. The unmistakable similarities in amino acid sequence and structure among the present-day globins indicate that they all must derive from a common ancestral gene.

The simplest and most evolutionarily ancient globin molecule is a polypeptide chain of about 150 amino acids, which is found in many marine worms, insects, and primitive fish. Like our hemoglobin, this protein transports oxygen molecules throughout the animal's body. The oxygen-carrying molecule in the blood of adult mammals and most other vertebrates, however, is more complex; this protein is composed of four globin chains of two distinct types—α globin and β globin (Figure 9–10). The four oxygen-binding sites in the $\alpha_2\beta_2$ molecule interact, allowing an allosteric change in the molecule as it binds and releases oxygen. This structural shift enables the four-chain hemoglobin molecule to efficiently take up and release four oxygen molecules in an all-or-none fashion, a feat not possible for the single-chain version. This efficiency is particularly important for large multicellular animals, which cannot rely on the simple diffusion of oxygen through the body to oxygenate their tissues adequately.

The α and β globin genes are the result of gene duplications that occurred early in vertebrate evolution. About 500 million years ago, gene duplications followed by mutation gave rise to two slightly different globin genes, one encoding α globin, the other encoding β globin (Figure 9–11). Still later, as the different mammals began diverging from their common ancestor, the β-globin gene apparently underwent its own duplication and divergence to give rise to a second β-like globin gene that is expressed specifically in the fetus (see Figure 9–11). The resulting hemoglobin molecule has a higher affinity for oxygen compared with adult hemoglobin, a property that helps transfer oxygen from mother to fetus.

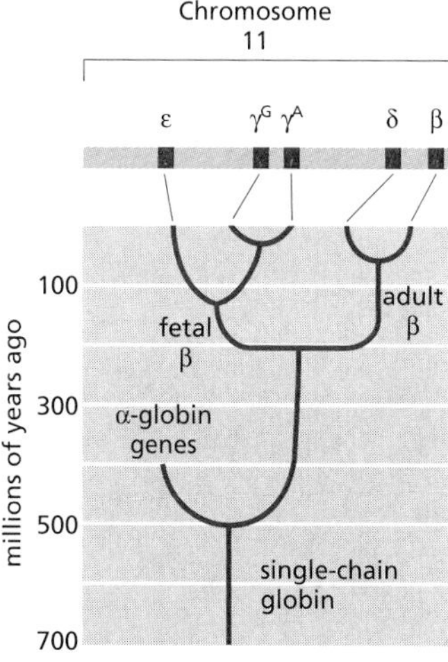

Figure 9–11 **Repeated rounds of duplication and mutation generated the globin gene family in humans.** About 500 million years ago, an ancestral globin gene duplicated and gave rise to the β-globin gene family and to the related α-globin gene family. In vertebrates, a molecule of hemoglobin is formed from two α chains and two β chains, as shown in Figure 9–10. The evolutionary scheme shown was worked out by comparing globin genes from many different organisms. For example, the nucleotide sequences of the γ^G and γ^A genes, which produce globin chains that form fetal hemoglobin, are much more similar to each other than either of them is to the adult β gene. In humans, the β-globin genes are located in a cluster on Chromosome 11. A chromosome breakage event that occurred about 300 million years ago separated the α- and β-globin genes; the α-globin genes now reside on human Chromosome 16 (not shown).

Subsequent rounds of duplication in both the α- and β-globin genes gave rise to additional members of these families. Each of these duplicated genes has been modified by point mutations that affect the properties of the final hemoglobin molecule, and by changes in regulatory DNA that determine when—and how strongly—each gene is expressed. As a result, each globin differs slightly in its ability to bind and release oxygen and is well suited to the stage in development at which it is expressed.

In addition to these specialized globin genes, there are several duplicated DNA sequences in the α- and β-globin gene clusters that are not functional genes. They are similar in DNA sequence to the functional globin genes, but they have been disabled by the accumulation of many mutations that destroy the encoded proteins and prevent their expression. The existence of such *pseudogenes* makes it clear that, as might be expected, not every DNA duplication leads to a new functional gene. Although we have focused here on the role of gene duplication and divergence as it relates to the evolution of the globin genes, similar histories apply to the many other gene families present in the human genome.

Whole Genome Duplications Have Shaped the Evolutionary History of Many Species

Almost every gene in the genomes of vertebrates exists in multiple versions, suggesting that rather than individual genes being duplicated in a piecemeal fashion, the whole vertebrate genome was duplicated in one fell swoop. Early on in vertebrate evolution, it appears that the entire genome actually underwent duplication twice in succession, giving rise to four copies of every gene. In some groups of vertebrates, such as the salmon and carp families (including the zebrafish; see Figure 1–39), there may have been yet another duplication, creating an eightfold multiplicity of genes.

The precise history of whole genome duplications in vertebrate evolution is difficult to chart because many other changes have occurred since these ancient evolutionary events. In some organisms, however, full genome duplications are especially obvious as they have occurred relatively recently—evolutionarily speaking. The frog genus *Xenopus*, for example, comprises a set of closely similar species related to one another by repeated duplications or triplications of the whole genome (Figure 9–12). Such large-scale duplications can happen if cell division fails to occur following a round of genome replication in the germ line of a particular individual. Once an accidental doubling of the genome occurs in the germ line, it will be faithfully passed on to other germ cells in the individual and, ultimately, to its progeny.

Figure 9–12 **Different species of the frog *Xenopus* have different DNA contents.** *X. tropicalis* (*above*) has an ordinary diploid genome with two sets of chromosomes in every somatic cell; the tetraploid *X. laevis* (*below*) has a duplicated genome containing twice as much DNA per cell. (Courtesy of Enrique Amaya.)

New Genes Can Be Generated by Repeating the Same Exon

The role of DNA duplication in evolution is not confined to the expansion of gene families. It can also act on a smaller scale to modify single genes by creating internal duplications. As we discussed in Chapter 4, many proteins are composed of a set of smaller protein *domains*. Some proteins—such as antibodies (see Figure 4–29) or fibrous proteins such as collagen—are built from multiple copies of the same domain linked together in series. These proteins are encoded by genes that have evolved by repeated duplications of a single DNA segment.

In eucaryotes, each individual protein domain in such genes is usually encoded by a separate exon. Duplicating these domains within a gene can then occur by breaking and rejoining the DNA anywhere in the long introns on either side of the exon that codes for the protein domain (Figure 9–13). Without introns there would be very few sites in the original gene at which a recombinational crossover between homologous chromosomes could duplicate the domain without damaging it. The evolution of new proteins is therefore thought to have been greatly facilitated by the organization of eucaryotic DNA coding sequences as a series of relatively short exons separated by long, noncoding introns (see Figures 7–17 and 7–18).

Novel Genes Can Also Be Created by Exon Shuffling

The type of recombination that allows exons to be duplicated within a gene can also occur between two different genes, joining together two initially separate exons that code for quite different protein domains—an important process called **exon shuffling**. The lack of precision that can be tolerated in a recombination event that breaks and rejoins two introns greatly increases the probability that a chance recombination between introns in different genes will generate a hybrid gene that joins exons. The presumed results of such recombinations are seen in many present-day proteins, which are a patchwork of many different protein domains (Figure 9–14).

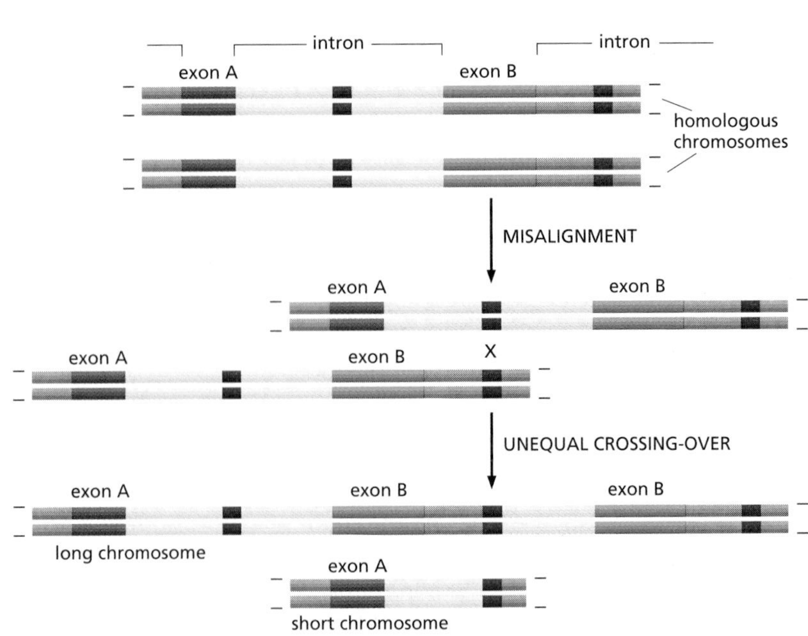

Figure 9–13 **An exon can be duplicated by a recombination event.** The general scheme is the same as that of Figure 9–9, with a short repeated sequence indicated in *red*; however, here an exon within a gene (*blue*), rather than an entire gene, is duplicated. An mRNA transcribed from the original gene will contain two exons, A and B, whereas the long chromosome will produce an mRNA with three exons (A, B, and a second copy of B). Because the duplicated exon is joined by an intron (*gray/yellow*) with its splicing sequences intact, the modified nucleotide sequence can be readily spliced after transcription to produce a functional mRNA.

It has been proposed that all the proteins encoded by the human genome (approximately 24,000) arose from the duplication and shuffling of a few thousand distinct exons, each encoding a protein domain of approximately 30–50 amino acids. This remarkable idea suggests that the great diversity of protein structures is generated from a quite small universal "list of parts" pieced together in different combinations.

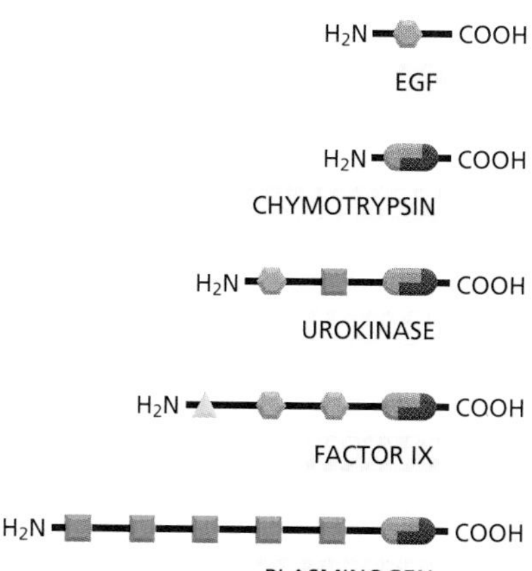

Figure 9–14 **Exon shuffling can generate proteins with new combinations of protein domains.** Each type of symbol represents a different type of protein domain. During evolution, these have been joined together end-to-end to create the modern-day human proteins shown here.

The Evolution of Genomes Has Been Accelerated by the Movement of Mobile Genetic Elements

The mobile genetic elements described in Chapter 6 are another important source of genomic change and have profoundly affected the structure of modern genomes. These parasitic DNA sequences can colonize a genome and can spread within it. During this process, they often disrupt the function or alter the regulation of existing genes; sometimes they even create novel genes through fusions between mobile sequences and segments of existing genes.

The insertion of a mobile genetic element into the coding sequence of a gene or into its regulatory region is a frequent cause of the "spontaneous" mutations that are observed in many organisms. Mobile genetic elements can severely disrupt a gene's activity if they land directly within its coding sequence. Such an insertion mutation destroys the gene's capacity to encode a useful protein. For example, a number of the mutations in the Factor VIII gene that cause hemophilia in humans result from the insertion of mobile genetic elements into the gene.

The activity of mobile genetic elements can also change the way in which existing genes are regulated. For example, an insertion of an element into the regulatory region of a gene will often have a striking effect on where and when that gene is expressed (Figure 9–15). Many mobile genetic elements carry DNA sequences that are recognized by specific transcription regulators; if these elements insert themselves near a gene, that gene can be brought under the control of these new transcription regulators, thereby changing the gene's expression pattern. Thus, mobile genetic elements can be a significant source of developmental changes, and they are thought to have been particularly important in the evolution of the body plans of multicellular plants and animals.

Finally, mobile genetic elements provide opportunities for genome rearrangements by serving as targets of homologous recombination. For example, the duplications that gave rise to the β-globin gene cluster are thought to have occurred by crossovers between *Alu*-like sequences that are sprinkled throughout the genome (see Figures 9–9 and 6–35). However, mobile genetic elements also have more direct roles in the evolution of genomes. In addition to relocating themselves, these para-

(A)

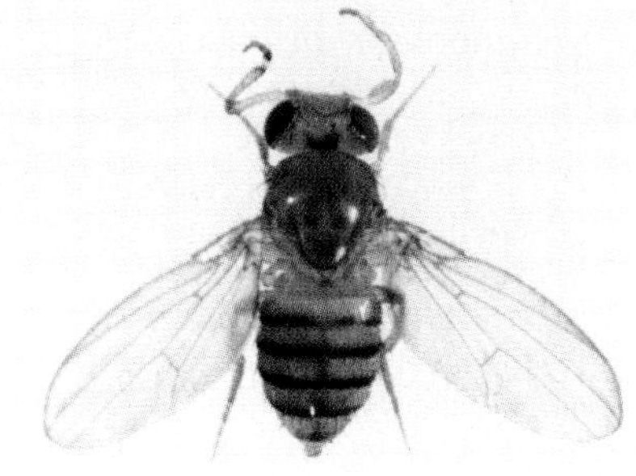

(B)

Figure 9–15 **Mutation due to a mobile genetic element can induce dramatic alterations in the body plan of an organism.** (A) A normal fruit fly (*Drosophila melanogaster*). (B) The fly's antennae have been transformed into legs because of a mutation in regulatory DNA sequences that cause genes for leg formation to be activated in the positions normally reserved for antennae. Although this particular change is not advantageous to the fly, it does illustrate how the movement of a transposable element can produce a major change in the appearance of an organism. (A, courtesy of E.B. Lewis; B, courtesy of Matthew Scott.)

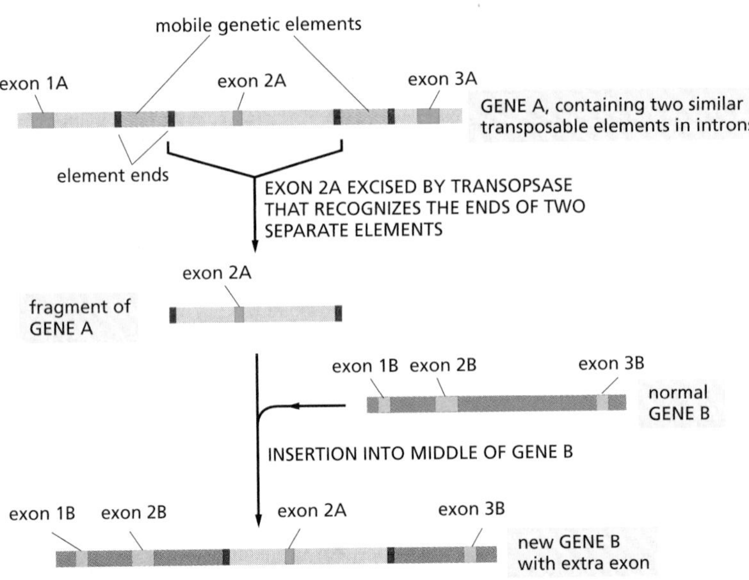

Figure 9–16 **Mobile genetic elements can create new exon arrangements.** When two mobile genetic elements of the same type (*red*) happen to insert near each other in a chromosome, the transposition mechanism occasionally uses the ends of two different elements (instead of the two ends of the same element) and thereby moves the chromosomal DNA between them to a new site. Because introns are very large relative to exons, the illustrated insertion of a new exon into a preexisting intron is a frequent outcome of such a transposition event.

sitic elements occasionally rearrange neighboring DNA sequences of the host genome. When two mobile genetic elements that are recognized by the same transposase integrate into neighboring chromosomal sites, the DNA between them can itself be transposed. In eucaryotic genomes, this provides a pathway for the movement of exons, generating new genes by creating novel arrangements of existing exons (Figure 9–16).

QUESTION 9–2

Why do you suppose that horizontal gene transfer is more prevalent in single-celled organisms than in multicellular organisms?

Genes Can Be Exchanged Between Organisms by Horizontal Gene Transfer

So far we have considered genetic changes that take place within the genome of an individual organism. However, genes and other portions of genomes can also be exchanged between individuals of different species. This mechanism, known as **horizontal gene transfer**, is rare among eucaryotes but common among bacteria (Figure 9–17).

E. coli, for example, has acquired about one-fifth of its genome from other species within the past 100 million years. And such genetic exchanges are currently responsible for the rise of new and potentially dangerous strains of drug-resistant bacteria. For example, genes that confer resistance to antibiotics can be transferred from species to species. This DNA exchange provides the recipient bacterium with an enormous selective advantage in evading the antimicrobial compounds that constitute modern medicine's frontline attack against bacterial infection. As a result, many antibiotics are no longer effective against the common bacterial infections for which they were originally used. For example, most strains of *Neisseria gonorrhoeae*, the bacterium that causes gonorrhea, are now resistant to penicillin; this antibiotic is therefore no longer the primary treatment for this disease.

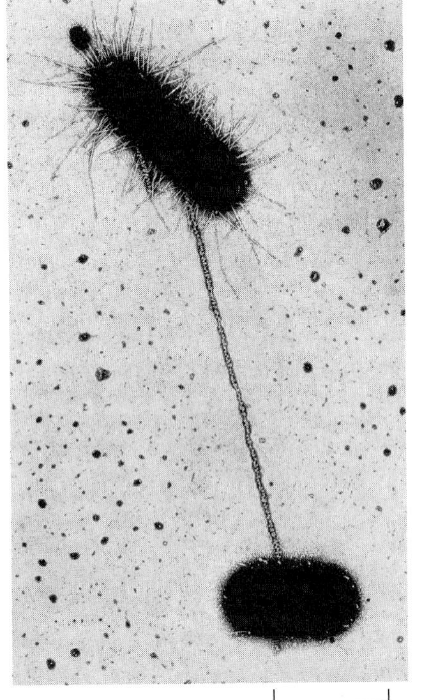

Figure 9–17 **Bacterial cells can share DNA by a process called conjugation.** Conjugation begins when a donor cell (*top*) attaches to a recipient cell (*bottom*) by a fine appendage, called a sex pilus. DNA from the donor cell then moves through the pilus into the recipient cell. In this electron micrograph, the sex pilus has been labeled along its length by viruses that specifically bind to it to make the structure more visible. Conjugation is one of several ways in which bacteria carry out horizontal gene transfer. (Courtesy of Charles C. Brinton Jr. and Judith Carnahan.)

RECONSTRUCTING LIFE'S FAMILY TREE

Given an understanding of the basic molecular mechanisms by which genomes change, we can begin to decipher clues to our evolutionary history by comparing and analyzing genome sequences. Perhaps the most astonishing revelation of such genome analyses has been that **homologous genes**—genes that are similar in their nucleotide sequence because of common ancestry—can often be recognized across vast evolutionary distances. Unmistakable homologs of many human genes are easy to detect in such organisms as worms, fruit flies, yeasts, and even bacteria. The nematode worm *Caenorhabditis elegans,* the fly *Drosophila melanogaster,* and the vertebrate *Homo sapiens*—the first three animals for which a complete genome sequence was obtained—are very distant relatives: the lineage leading to vertebrates is thought to have diverged from that leading to nematodes and insects more than 600 million years ago. Yet when the 19,000 genes of *C. elegans*, the 14,000 genes of *Drosophila*, and the 25,000 or so genes of *Homo sapiens* are systematically compared, we find that about 50% of the genes in each of these species have clear homologs in one or both of the other two species. In other words, clearly recognizable versions of at least half of all human genes were already present in the common ancestor of worms, flies, and humans.

By tracing such relationships among genes, we can begin to define the evolutionary relationships among different species, placing each bacterium, animal, plant, or fungus in a single vast family tree of life. In this section, we discuss how these relationships are determined and what they tell us about our genetic heritage.

Genetic Changes That Provide a Selective Advantage Are Likely to Be Preserved

Evolution is commonly thought of as progressive, but at the molecular level the process is random. Consider the fate of a point mutation. As we discussed earlier, on rare occasions such a mutation may represent a change for the better; very often, it will cause no significant difference in the organism's prospects; and sometimes it will cause serious damage—by disrupting the coding sequence for a key protein, for example. Changes due to mutations of the first type will tend to be perpetuated, because the organism that inherits them will have an increased likelihood of reproducing itself. Changes due to mutations of the second type—*selectively neutral* changes—may or may not be passed on, because it is a matter of chance whether the mutant organism or its cousins will secure the available resources and breed successfully. In contrast, mutations that cause serious damage lead nowhere: the organism that inherits them dies, leaving no progeny. Through endless repetition of this cycle of error and trial—of mutation and natural selection—organisms gradually evolve. Their genetic specifications change and they develop new ways to exploit the environment, to survive in competition with others, and to reproduce successfully.

Clearly, some parts of the genome can accumulate mutations more easily than others in the course of evolution. A segment of DNA that does not code for protein or RNA and has no significant regulatory role is free to change at a rate limited only by the frequency of random mutation. In contrast, alterations in a gene that codes for a highly optimized essential protein or RNA molecule cannot be accommodated so easily: when mutations occur, the faulty organism is almost always eliminated. Genes of this latter sort are therefore *highly conserved*; that is, the proteins they encode are very similar from organism to organism. Throughout 3.5 billion years or more of evolutionary history, the most highly conserved genes remain

QUESTION 9–3

Highly conserved genes such as those for ribosomal RNA are present as clearly recognizable relatives in all organisms on Earth; thus, they have evolved very slowly over time. Were such genes "born" perfect?

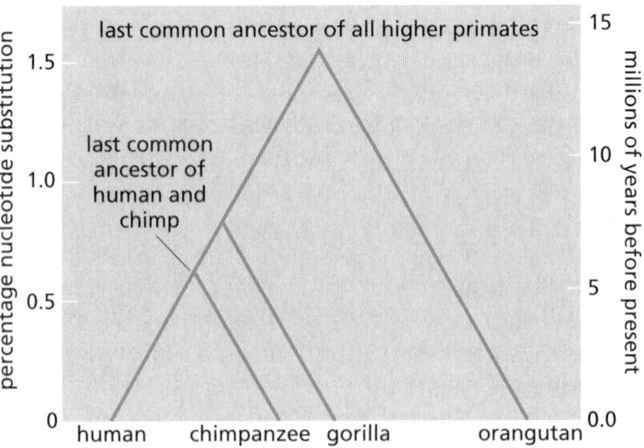

Figure 9–18 **Phylogenetic trees display the relationships among modern life forms.** In this family tree of higher primates, humans fall closer to chimpanzees than to gorillas or orangutans, as there are fewer differences between human and chimp nucleotide sequences than there are between those of humans and gorillas or humans and orangutans. As indicated, the genome sequences of all four species are estimated to differ from the sequence of their last common ancestor by about 1.5%. Because changes occur independently in each lineage, the divergence between any two species will be twice as much as the amount of change that takes place between each of the species and their last common ancestor. For example, humans and orangutans typically show sequence divergences of slightly more than 3%, while human and chimp genomes differ by about 1.2%. Although this phylogenetic tree is based solely on nucleotide sequences, the estimated times of divergence, shown on the *right* side of the graph, derive from data obtained from the fossil record. (Modified from F.C. Chen and W.H. Li, *Am. J. Hum. Genet.* 68:444–456, 2001. With permission from Elsevier.)

perfectly recognizable in all living species. These latter genes (which encode crucial proteins such as DNA and RNA polymerases) are the ones that must be examined if we wish to trace family relationships among the most distantly related organisms in the tree of life.

For species that are more closely related, however, it is often more informative to focus on the selectively neutral changes. These changes accumulate steadily at a rate that is unconstrained by selection pressures. They therefore provide us with a simple and easily readable evolutionary clock that can be used to estimate the amount of time that has passed since two species diverged from a common ancestor. For example, comparisons of such nucleotide changes have allowed us to construct a highly accurate **phylogenetic tree** that shows the evolutionary relationships among higher primates (Figure 9–18).

Human and Chimpanzee Genomes Are Similar in Organization As Well As in Detailed Sequence

Humans and chimpanzees are so closely related that it is possible to reconstruct gene sequences of the extinct common ancestor of the two species (Figure 9–19). Not only do humans and chimpanzees seem to have essentially the same set of 25,000 genes, but these genes are arranged in nearly the same way along the chromosomes of the two species. The only substantial exception is human Chromosome 2, which arose from a fusion of two chromosomes that remain separate in the chimpanzee, the gorilla, and the orangutan.

Even the massive resculpting of genomes by mobile genetic elements (discussed earlier in this chapter) has produced only minor differences between the human and chimp genomes. More than 99% of the million copies of the *Alu* retrotransposon sequences that are present in both genomes are in corresponding positions, indicating that most of the *Alu* sequences in our genome underwent transposition before humans and chimpanzees diverged. However, as described earlier, members of the *Alu* family are still capable of transposing, as is evident from a small number of observed cases in which new *Alu* insertions have caused human genetic disease. These cases involve transposition of this DNA into sites that were unoccupied in the genomes of the patients' parents.

Functionally Important Regions Show Up As Islands of Conserved DNA Sequence

As we delve back further into our evolutionary history and compare our genomes with those of more distant relatives, the picture changes. The lineages of humans and mice, for example, diverged about 75 million

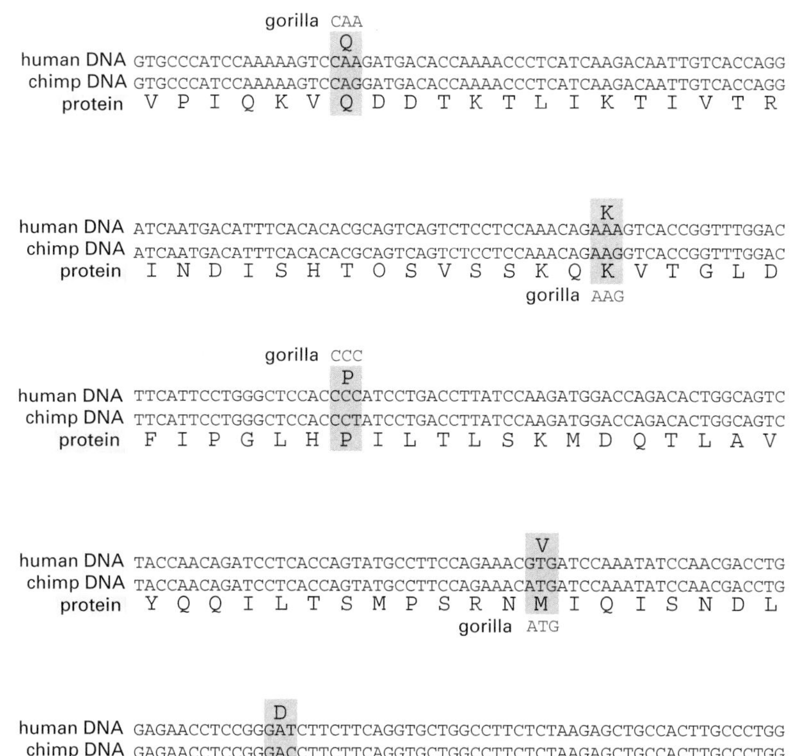

Figure 9–19 **Ancestral gene sequences can be reconstructed by comparing closely related present-day species.** Shown here in five contiguous sections is a sequence from the coding region of the leptin gene from humans and chimpanzees. Leptin is a hormone that regulates food intake and energy utilization. As indicated by the codons boxed in *green*, only 5 (of 441 nucleotides total) differ between the chimp and human sequences. Only one of these changes results in a change in the amino acid sequence. The sequence of the last common ancestor was probably the same as the human and chimp sequences where they agree; in the few places where they disagree, the gorilla sequence (*pink*) can be used as a 'tiebreaker.' This strategy is based on the relationship shown in Figure 9–18: differences between humans and chimpanzees reflect relatively recent events in evolutionary history, and the gorilla sequence reveals the most likely precursor sequence. For convenience, only the first 300 nucleotides of the leptin coding sequences are shown. The last 141 nucleotides are identical between humans and chimpanzees.

years ago. We both have genomes of the same size containing practically the same genes. Both genomes are also peppered with mobile genetic elements. However, the mobile genetic elements in the mouse and human genomes, although similar in sequence, are distributed differently. This observation implies that the elements have been independently proliferating and moving around the genome in each lineage since the two species diverged (Figure 9–20). In addition to the movement of mobile genetic elements, the large-scale organization of the genomes has become scrambled by many episodes of chromosome breakage and recombination—it is estimated that there has been a total of about 180 "break-and-join" events. As a result, the overall structures of the chromosomes have changed dramatically. For example, in humans most centromeres lie near the middle of the chromosome, whereas those of mouse are located at the chromosome ends.

Nevertheless, in spite of the genetic shuffling, one can still recognize many blocks of **conserved synteny**, regions where corresponding genes are strung together in the same order in both species. These genes were neighbors in the ancestral species and, despite all the chromosomal upheavals, they remain neighbors in the two present-day species. More

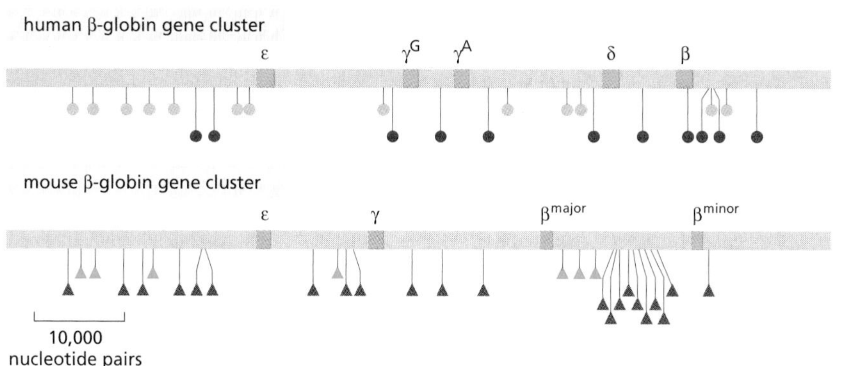

Figure 9–20 **The positions of mobile genetic elements in the human and mouse genomes reflect the long evolutionary time separating the two species.** This stretch of human Chromosome 11 (introduced in Figure 9–11) contains five functional β-globin–like genes (*orange*); the comparable region from the mouse genome contains only four. In the human β-globin gene cluster, the positions of human *Alu* sequences (*green* circles) and human *L1* sequences (*red* circles) are shown. In the mouse genome, the positions of *B1* elements (relatives of the human *Alu* elements; *blue* triangles) and mouse *L1* elements (relatives of the human *L1* elements; *brown* triangles) are shown. The absence of mobile genetic elements within the globin structural genes can be attributed to natural selection, which would have eliminated any insertion that compromised gene function. The *Alu* and *L1* mobile elements are described in more detail in Chapter 6 (pp. 222–223). (Courtesy of Ross Hardison and Webb Miller.)

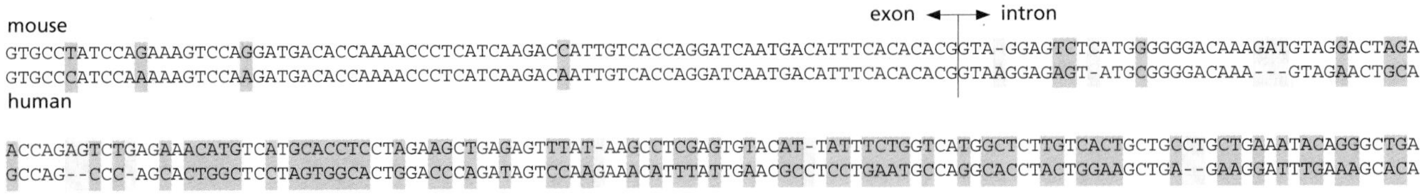

```
                                                                               exon ◄─┬─► intron
mouse
GTGCCTATCCAGAAAGTCCAGGATGACACCAAAACCCTCATCAAGACCATTGTCACCAGGATCAATGACATTTCACACACGGTA-GGAGTCTCATGGGGGACAAAGATGTAGGACTAGA
GTGCCCATCCAAAAAGTCCAAGATGACACCAAAACCCTCATCAAGACAATTGTCACCAGGATCAATGACATTTCACACACGGTAAGGAGAGT-ATGCGGGGACAAA---GTAGAACTGCA
human

ACCAGAGTCTGAGAAACATGTCATGCACCTCCTAGAAGCTGAGAGTTTAT-AAGCCTCGAGTGTACAT-TATTTCTGGTCATGGCTCTTGTCACTGCTGCCTGCTGAAATACAGGGCTGA
GCCAG--CCC-AGCACTGGCTCCTAGTGGCACTGGACCCAGATAGTCCAAGAAACATTTATTGAACGCCTCCTGAATGCCAGGCACCTACTGGAAGCTGA--GAAGGATTTGAAAGCACA
```

Figure 9–21 **Accumulated mutations have resulted in considerable divergence in the nucleotide sequences of the human and the mouse genomes.** Shown here in two contiguous sections are portions of the human and mouse leptin gene sequences. Positions where the sequences differ by a single nucleotide substitution are boxed in *green*, and positions where they differ by the addition or deletion of nucleotides are boxed in *yellow*. Note that the coding sequence of the exon is much more conserved than the adjacent intron sequence.

than 90% of the mouse and human genomes can be partitioned into such corresponding regions of conserved synteny. Within these regions, we can align the DNA of mouse with that of humans and compare the nucleotide sequences in detail. Such genome-wide sequence comparisons reveal that in the roughly 80 million years since humans and mice diverged from their common ancestor, about 50% of the nucleotides have changed. Against this background of dissimilarity, however, one can now begin to see very clearly the regions where changes are not tolerated and the human and mouse sequences have remained nearly the same (Figure 9–21). Here, the sequences have been conserved by **purifying selection**—that is, the elimination of individuals carrying mutations that interfere with important functions.

The power of *comparative genomics* can be increased by stacking our genome up against the genomes of additional animals, including the rat, chicken, and dog. Such comparisons take advantage of the results of the 'natural experiment' that has lasted for hundreds of millions of years, and they highlight some of the most interesting regions of these genomes. These comparisons reveal that roughly 5% of the human genome consists of DNA sequences that are highly conserved in many other mammals (Figure 9–22). Surprisingly, only about one-third of these sequences code for proteins. Some of the conserved noncoding sequences correspond to regulatory DNA, whereas others produce RNA molecules that are not translated into protein. But the function of the majority of these sequences remains unknown. This unexpected discovery has led scientists to conclude that we understand much less about the cell biology of vertebrates than we had previously imagined. With enormous opportunities for new discoveries, we should expect many surprises ahead.

Genome Comparisons Show That Vertebrate Genomes Gain and Lose DNA Rapidly

Going back further in evolution, we can compare our genome with those of fishes. The fish and mammalian lineages diverged about 400 million years ago. This is long enough for random sequence changes and differing selection pressures to have obliterated almost every trace of similarity in nucleotide sequence except where purifying selection has operated to prevent change. Regions of the genome conserved between humans and fishes thus stand out even more strikingly. In fishes, one can still recognize most of the same genes as in humans and even many of the same segments of regulatory DNA. On the other hand, the extent of duplication of any given gene is often different, resulting in different numbers of members of gene families in the two species.

Another thing that stands out is that although all vertebrate genomes contain roughly the same number of genes, their overall size varies considerably. Whereas human, dog, and mouse are all in the same size range

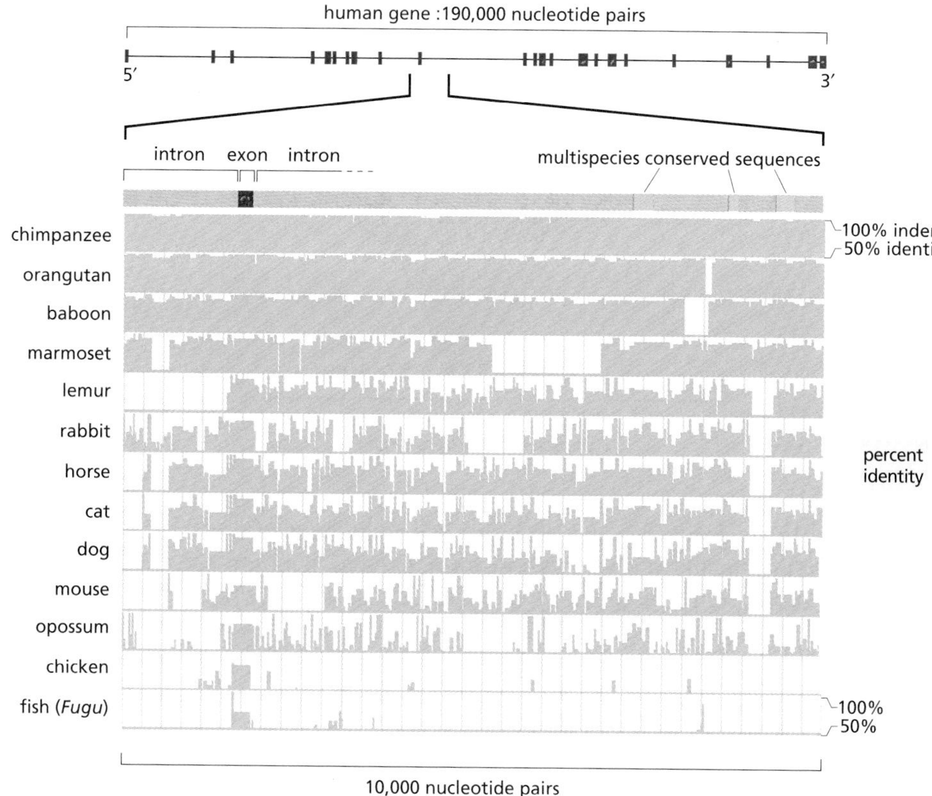

Figure 9–22 **Comparison of nucleotide sequences from many different vertebrates reveals regions of high conservation.** The nucleotide sequence examined in this diagram is of a small segment of the human gene for a plasma membrane transporter protein. Exons in the complete gene (*top*) are indicated in *magenta*. In the expanded region of the gene, the position of the exon is indicated in *red*. In the lower part of the figure, the human DNA sequence is aligned with that of different vertebrates; the percent identity with the human gene for successive stretches of 100 nucleotide pairs is plotted in *green*, with only identities above 50% shown. The sequence of the exon is conserved in all the species, including the chicken and fish. Three blocks of intron sequence that are conserved in the mammals, but not in the chicken or fish, are shown in *blue*. The functions of most conserved intron sequences in the human genome (including these three) are not known. (Courtesy of Eric D. Green.)

(3×10^9 nucleotide pairs), the chicken genome is only one-third this size. An extreme example of genome compression is the puffer fish, *Fugu rubripes* (Figure 9–23), whose tiny genome is one-tenth the size of mammalian genomes, largely because of the small size of its introns. *Fugu* introns, as well as other noncoding segments in the animal's genome, lack the repetitive DNA that makes up a large portion of most mammalian genomes. Nonetheless, the positions of most *Fugu* introns are perfectly conserved when compared with their positions in mammalian genomes (Figure 9–24). Clearly, the intron structure of most vertebrate genes was already in place in the common ancestor of fish and mammals.

What factors could be responsible for the size differences among modern vertebrate genomes? Detailed comparisons of many genomes have led to the unexpected finding that small blocks of sequence are being lost from and added to genomes at a surprisingly rapid rate. It seems likely that the *Fugu* genome is so tiny because it lost DNA sequences faster than it gained them. Over long periods, this imbalance would result in a major "cleansing" of those DNA sequences whose disappearance could be tolerated. In retrospect, this process has been enormously helpful to biologists: by "trimming the fat" from the *Fugu* genome, evolution has generously provided a slimmed-down version of a vertebrate genome in which the only DNA sequences that remain are those that are very likely to have important functions.

Sequence Conservation Allows Us to Trace Even the Most Distant Evolutionary Relationships

As we go back further still to the genomes of even more distant relatives—beyond apes, mice, fish, flies, worms, plants, and yeasts, all the way to bacteria—we find fewer and fewer resemblances to our own genome. Yet even across this enormous evolutionary divide, purifying selection has maintained a few hundred fundamentally important genes in all domains

Figure 9–23 **The puffer fish, *Fugu rubripes*, has a remarkably compact genome.** At 400 million nucleotide pairs, the *Fugu* genome is only one-quarter the size of the zebrafish genome, even though the two species have nearly the same genes. (From a woodcut by Hiroshige, courtesy of Arts and Designs of Japan.)

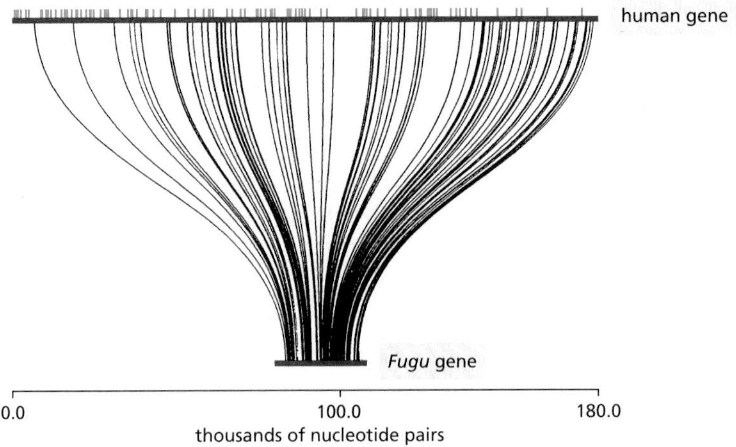

Figure 9–24 **The positions of introns are conserved between *Fugu* and humans.** Comparison of the nucleotide sequences of the human and *Fugu* genes encoding the huntingtin protein. Both genes (*red*) contain 67 short exons that align in 1:1 correspondence with one another; these exons are connected by the curved black lines. The human gene is 7.5 times larger than the *Fugu* gene (180,000 versus 24,000 nucleotide pairs), due entirely to the larger introns in the human sequence. The larger size of the human introns is due in part to mobile genetic elements, whose positions are represented by the *blue* vertical bars. In humans, mutation of the huntingtin gene causes Huntington's disease, an inherited neurodegenerative disorder. (Adapted from S. Baxendale et al., *Nat. Genet.* 10:67–76, 1995. With permission from Macmillan Publishers Ltd.)

of the living world. By comparing the sequences of these genes in different organisms and seeing how far they have diverged, we can attempt to construct a phylogenetic tree that goes all the way back to the ultimate ancestors—the cells at the origins of life from which we all derive.

To construct such a tree, biologists have focused on one particular gene that is conserved in all living species: the gene that codes for one of the ribosomal RNAs (rRNAs) of the small ribosomal subunit (see Figure 7–31). Because the process of translation is fundamental to all living cells, this component of the ribosome has been well conserved since early in the history of life on Earth (Figure 9–25).

By applying the same principles used to construct the primate family tree (see Figure 9–18), the small subunit rRNA nucleotide sequences have been used to create a single all-encompassing tree of life (Figure 9–26). Although many aspects of this phylogenetic tree were anticipated by classical taxonomy (which is based on the outward appearance of organisms), there were also many surprises. Perhaps the most important was the realization that some of the organisms that were traditionally classed as "bacteria" are as widely divergent in their evolutionary origins as is any procaryote from any eucaryote. As discussed in Chapter 1, it is now apparent that the procaryotes comprise two distinct groups—the *bacteria* and the *archaea*—that diverged early in the history of life on Earth. The living world therefore has three major divisions or *domains*: bacteria, archaea, and eucaryotes (see Figure 9–26).

Although we humans have been classifying the visible world since antiquity, we now realize that most of life's genetic diversity lies in the world of invisible microbes. These organisms have tended to go unnoticed, unless they cause disease or rot the timbers of our houses. Yet microorganisms make up most of the total mass of living matter on our planet.

```
GTTCCGGGGGGAGTATGGTTGCAAAGCTGAAACTTAAAGGAATTGACGGAAGGGCACCACCAGGAGTGGAGCCTGCGGCTTAATTTGACTCAACACGGGAAACCTCACCC  human
GCCGCCTGGGGAGTACGGTCGCAAGACTGAAACTTAAAGGAATTGGCGGGGGAGCACTACAACGGGTGGAGCCTGCGGTTTAATTGGATTCAACGCCGGGCATCTTACCA  Methanococcus
ACCGCCTGGGGAGTACGGCCGCAAGGTTAAAACTCAAATGAATTGACGGGGGCCCGC.ACAAGCGGTGGAGCATGTGGTTTAATTCGATGCAACGCGAAGAACCTTACCT  E. coli
GTTCCGGGGGGAGTATGGTTGCAAAGCTGAAACTTAAAGGAATTGACGGAAGGGCACCACCAGGAGTGGAGCCTGCGGCTTAATTTGACTCAACACGGGAAACCTCACCC  human
```

Figure 9–25 **Some genetic information has been conserved since the beginnings of life.** A part of the gene for the small subunit of rRNA (see Figure 7–31) is shown. Corresponding segments of nucleotide sequence from three distantly related species (*Methanococcus jannaschii*, an archaeon; *Escherichia coli*, a bacterium; and *Homo sapiens*, a eucaryote) are aligned in parallel. Sites where the nucleotides are identical between species are indicated by *green* shading; the human sequence is repeated at the bottom of the alignment so that all three two-way comparisons can be seen. The dot halfway along the *E. coli* sequence denotes a site where a nucleotide has been either deleted from the bacterial lineage in the course of evolution or inserted in the other two lineages. Note that the sequences from these three distantly related organisms have all diverged from one another to a roughly similar extent, while still retaining unmistakable similarities.

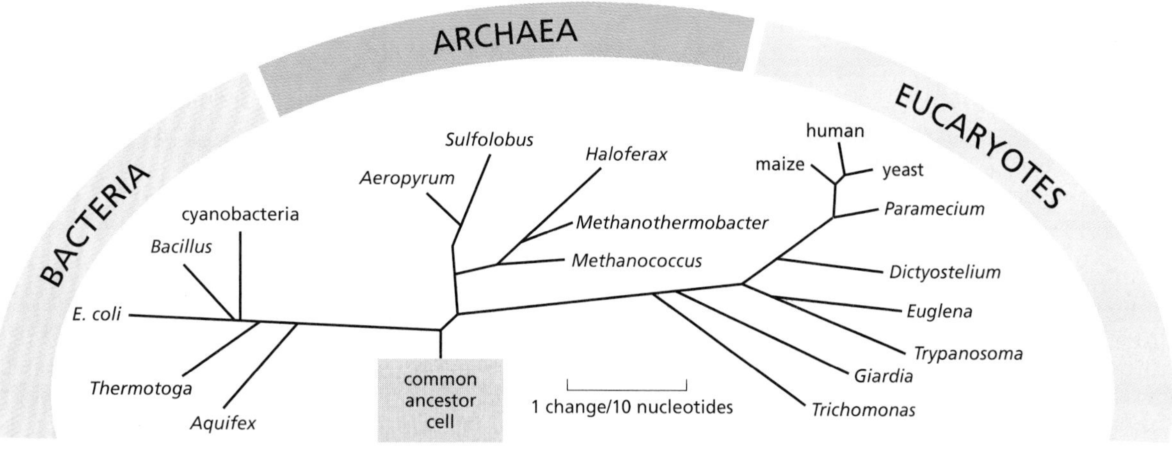

Figure 9–26 **The tree of life has three major divisions.** Each branch on the tree is labeled with the name of a representative member of that group, and the length of each branch corresponds to the degree of difference in their small subunit rRNA sequences (see Figure 9–25). Note that all the organisms we can see with the unaided eye—animals, plants, and fungi (highlighted in *yellow*)—represent only a small subset of the diversity of life.

Only now—through the analysis of DNA sequences—are we beginning to glimpse a picture of life on Earth that is not grossly distorted by our biased perspective as large animals living on dry land.

EXAMINING THE HUMAN GENOME

We have seen how genomes change gradually over time, and how comparing the genomes of different species can highlight crucial events in their evolutionary histories. We shall now turn our attention to our own genome, which is full of clues about where we came from and who we are.

The human genome—all 3.2×10^9 nucleotide pairs—is distributed over 22 autosomes and 2 sex chromosomes. The *human genome sequence* refers to the complete nucleotide sequence of the DNA contained in these 24 chromosomes. A wide variety of humans contributed DNA for the genome-sequencing project, and because individual humans differ from one another by an average of 1 nucleotide in 1000, the published human genome sequence is a composite of many individual sequences. It represents at once both our unity and our diversity as a species.

At its peak, the Human Genome Project generated raw nucleotide sequences at a rate of 1000 nucleotides per second, around the clock. The sheer quantity of information provided by this effort is staggering (Figure 9–27). Although it will be many decades before the data are fully analyzed, information gleaned from the human sequence has already affected the content of every chapter in this book. In this section, we describe some of our genome's most striking features.

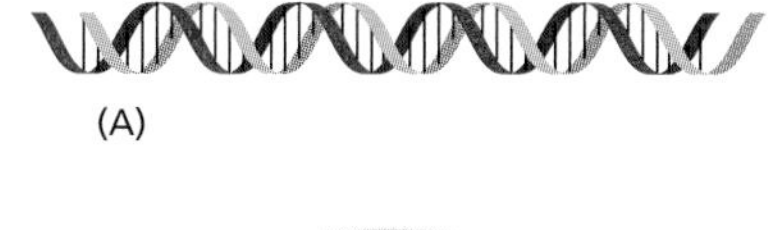

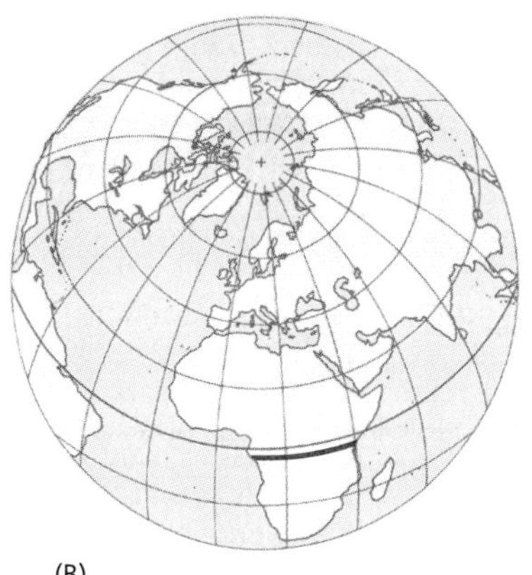

Figure 9–27 **The human genome stretches back to our origins.** If each nucleotide pair is drawn to span 1 mm, the scale in (A), then the human genome would extend 3200 km (approximately 2000 miles)—far enough to stretch across the center of Africa, the site of our human origins (*red* line in B). At this scale, there would be, on average, a protein-coding gene every 130 m. An average gene would extend for 30 m, but the coding sequences in this gene would add up to only just over a meter.

The Nucleotide Sequence of the Human Genome Shows How Our Genes Are Arranged

When the DNA sequence of human Chromosome 22, one of the smallest human chromosomes, was completed in 1999, it became possible for the first time to see exactly how genes are arranged along an entire vertebrate chromosome (Figure 9–28). With the publication of the "first draft" of the entire human genome in 2001, and the finished draft sequence in 2004, we now have a panoramic view of the genetic landscape of all human chromosomes (Table 9–1).

The first striking feature of the human genome is how little of it—only a few percent—codes for proteins, structural RNAs, and catalytic RNAs (Figure 9–29). Almost half of the remaining DNA is made up of mobile genetic elements that have colonized our genome over evolutionary time. Because they have accumulated mutations, most of these elements can no longer move; rather, they are relics from an earlier evolutionary era when mobile genetic elements ran rampant through our genome.

A second notable feature of the human genome is the very large average gene size of 27,000 nucleotide pairs. Only about 1300 nucleotide pairs are required to encode a protein of average size (about 430 amino acids in humans). Most of the remaining DNA in a gene consists of long stretches of noncoding DNA in the introns between the relatively short protein-coding exons (see Figure 9–28D). In addition to the introns and exons, each gene is associated with regulatory DNA sequences that ensure that the gene is expressed at the proper level and time, and in the proper type of cell. In humans, these regulatory DNA sequences are typically spread out over tens of thousands of nucleotide pairs, much of which seems to be 'spacer' DNA.

QUESTION 9–4

Mobile genetic elements, such as the *Alu* sequences, are found in many copies in human DNA. In what ways could the presence of an *Alu* sequence affect a nearby gene?

Exons and regulatory sequences comprise less than 2% of the human genome. Yet from comparative genomic studies, we estimate that 5% of the human genome is highly conserved with other mammalian genomes

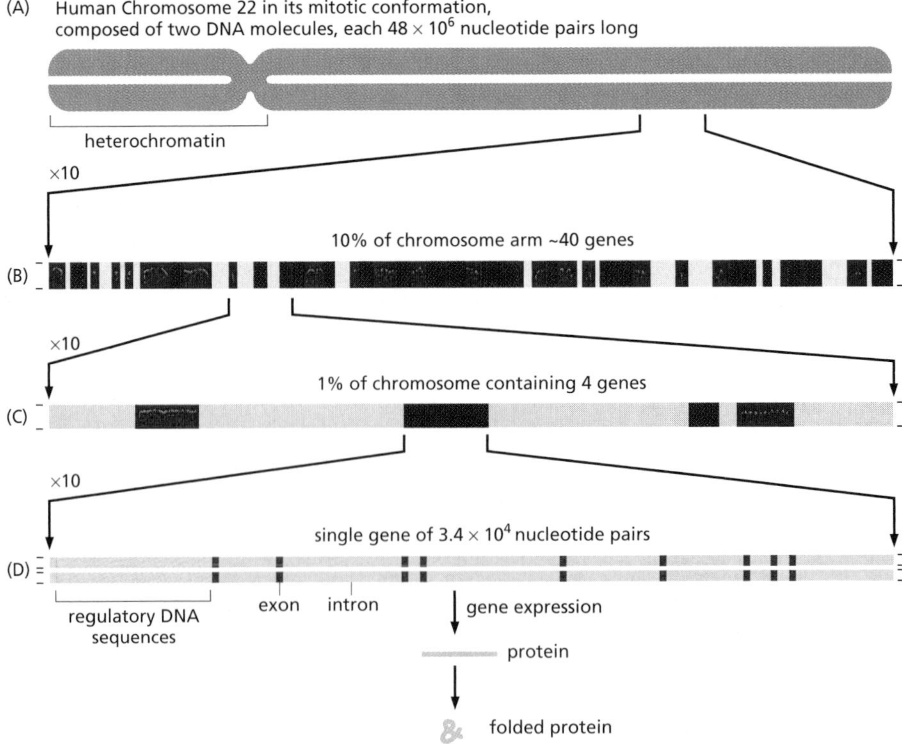

Figure 9–28 **The sequence of Chromosome 22 shows how human chromosomes are organized.** (A) Chromosome 22, one of the smallest human chromosomes, contains 48×10^6 nucleotide pairs and makes up approximately 1.5% of the entire human genome. Most of the left arm of Chromosome 22 consists of short repeated sequences of DNA that are packaged in a particularly compact form of chromatin (heterochromatin), as discussed in Chapter 5. (B) A tenfold expansion of a portion of Chromosome 22 shows about 40 genes. Those in *dark brown* are known genes and those in *red* are predicted genes. (C) An expanded portion of (B) shows the entire length of several genes. (D) The intron–exon arrangement of a typical gene is shown after a further tenfold expansion. Each exon (*red*) codes for a portion of the protein, while the DNA sequence of the introns (*gray*) is relatively unimportant. (Adapted from The International Human Genome Sequencing Consortium, *Nature* 409:860–921, 2001. With permission from Macmillan Publishers Ltd.)

TABLE 9–1 SOME VITAL STATISTICS FOR THE HUMAN GENOME	
DNA length	3.2×10^9 nucleotide pairs*
Number of genes	approximately 25,000
Largest gene	2.4×10^6 nucleotide pairs
Mean gene size	27,000 nucleotide pairs
Smallest number of exons per gene	1
Largest number of exons per gene	178
Mean number of exons per gene	10.4
Largest exon size	17,106 nucleotide pairs
Mean exon size	145 nucleotide pairs
Number of pseudogenes**	more than 20,000
Percentage of DNA sequence in exons (protein coding sequences)	1.5%
Percentage of DNA in other highly conserved sequences***	3.5%
Percentage of DNA in high-copy repetitive elements	approximately 50%

*The sequence of 2.85 billion nucleotides is known precisely (error rate of only about one in 100,000 nucleotides). The remaining DNA consists primarily of short highly repeated sequences that are tandemly repeated, with repeat numbers differing from one individual to the next.
**A pseudogene is a nucleotide sequence of DNA closely resembling that of a functional gene but containing numerous mutations that prevent its proper expression. Most pseudogenes arise from the duplication of a functional gene followed by the accumulation of damaging mutations in one copy.
***These include DNA encoding 5′ and 3′ UTRs (untranslated regions of mRNAs), genes for structural and catalytic RNAs, regulatory DNA, and conserved regions of unknown function.

and is therefore likely to be functionally important (see Figure 9–22). We simply do not know the function of this other DNA.

Another surprising feature of the human genome is its relatively small number of genes. Earlier estimates had been in the neighborhood of 100,000 (see How We Know, pp. 318–319). Although the exact number is still not certain, revised estimates place the number of human genes at about 25,000, bringing us much closer to the gene numbers for simpler multicellular animals, such as *Drosophila* (14,000) and *C. elegans* (19,000).

Finally, the nucleotide sequence of the human genome has revealed that the critical information it carries seems to be in an alarming state of disarray. As one commentator described our genome: "In some ways it may resemble your garage/bedroom/refrigerator/life: highly individualistic, but unkempt; little evidence of organization; much accumulated clutter (referred to by the uninitiated as 'junk'); and the few patently valuable items indiscriminately, apparently carelessly, scattered throughout."

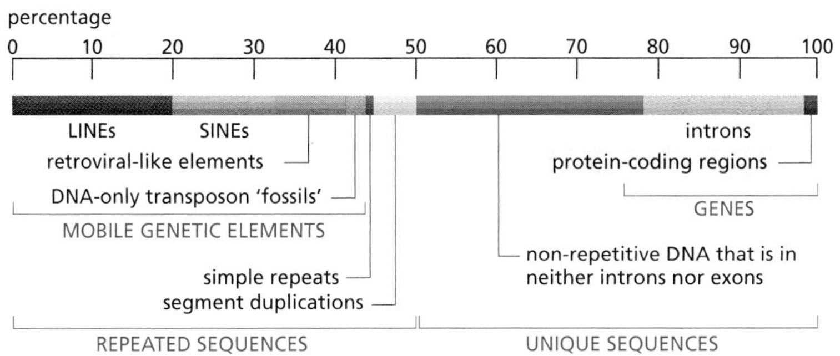

Figure 9–29 **The bulk of the human genome is made of noncoding and repetitive nucleotide sequences.** The LINEs, SINEs, retroviral-like transposons, and DNA-only transposons are mobile genetic elements that have multiplied in our genome by replicating themselves and inserting the new copies in different positions. These mobile genetic elements are discussed in Chapter 6 (pp. 222–223). Simple repeats are short nucleotide sequences (less than 14 nucleotide pairs) that are repeated again and again for long stretches. Segment duplications are large blocks of the genome (1000–200,000 nucleotide pairs) that are present at two or more locations in the genome. The unique sequences that are not part of any introns or exons (*dark green*) include gene regulatory elements, as well as sequences whose functions are not known. The most highly repeated blocks of DNA in heterochromatin have not yet been completely sequenced; therefore about 10% of human DNA sequences are not represented in this diagram. (Data courtesy of E.H. Margulies.)

COUNTING GENES

How many genes does it take to make a human? It seems a natural thing to wonder. If 6000 genes can produce a yeast, and 14,000 a fly, how many are needed to encode a human being—a complex creature, curious and clever enough to study its own genome? Until researchers completed the first draft of the human genome sequence, the most frequently cited estimate was 100,000. But where did that figure come from? And how was the revised estimate of only 25,000 genes derived?

Walter Gilbert, a physicist-turned-biologist who won a Nobel Prize for developing techniques for sequencing DNA, was one of the first to throw out a ballpark estimate of the number of human genes. In the mid-1980s, Gilbert suggested that humans could have 100,000 genes, an estimate based on the average size of the few human genes known at the time (about 3×10^4 nucleotide pairs) and the size of our genome (3×10^9 nucleotide pairs). This back-of-the-envelope calculation yielded a number with such a pleasing roundness that it wound up being quoted widely in articles and textbooks.

The calculation provides an estimate of the number of genes a human could have in principle, but it does not address the question of how many genes we actually do have. As it turns out, that question is not so easy to answer, even with the complete human genome sequence in hand. The problem is, how does one identify a gene? We saw in Chapter 5 that genes are defined as regions of DNA that determine the characteristics of a cell or organism, and that these DNA segments usually encode a protein or a functional RNA. We now know that such coding segments comprise only a few percent of the human genome. So, looking at a given piece of raw DNA sequence—an apparently random string of As, Ts, Gs, and Cs—how can one tell which parts are genes and which parts are 'junk'? Being able to accurately and reliably distinguish the coding sequences from the non-coding sequences in a genome is necessary before one can hope to locate and count its genes.

Signals and chunks

As always, the situation is simplest in procaryotes and simple eucaryotes such as yeasts. To identify genes in such a genome, one essentially searches through the entire DNA sequence looking for **open reading frames** (**ORFs**). These are long sequences—say, 100 codons or more—that lack stop codons. A random sequence of nucleotides will by chance encode a stop signal for protein synthesis about once every 20 codons (as there are three stop codons in the set of 64 possible codons). So finding an ORF—a continuous nucleotide sequence that encodes more than 100 amino acids—is the first step in identifying a good candidate for a gene. Today computer programs are used to search for such ORFs, which begin with an initiation codon, usually ATG, and end with a termination codon, TAA, TAG, or TGA (Figure 9–30).

In animals and plants, the process of identifying ORFs is complicated by the presence of large intron sequences that interrupt the coding portions of genes. As we have seen, these sequences are generally much larger than the exons themselves, which might represent only a few percent of the gene. In human DNA, exons sometimes contain as few as 50 codons (150 nucleotide pairs), while introns may exceed 10,000 nucleotide pairs in length. Fifty codons is too short to generate a statistically significant 'ORF signal,' as it is not all that unusual for 50 random codons to lack a stop signal. Moreover, introns are so long that they are likely to contain by chance quite a bit of 'ORF noise,' numerous stretches of sequence lacking stop signals. Finding the true ORFs in this sea of information in which the noise often outweighs the signal is quite difficult. Thus, to identify genes in eucaryotic DNA, it is also necessary to search for other distinctive features that mark the presence of a gene. These include the splicing sequences that signal an intron–exon boundary (see Figure 7–19) or distinctive DNA regulatory sequences that lie upstream of a gene.

But the most powerful approach to identifying genes is through homology with genes from other organisms. For example, even a very short ORF is likely to be an exon if the amino acid sequence it encodes matches up with a known protein from another organism. In addition, if a presumptive ORF is highly conserved in several different genomes, it is very likely to code for a protein even though the gene that contains it may not yet have been identified or studied (see Figure 9–22). Through such comparisons (for example, human versus mouse versus zebrafish) it is possible to identify short ORFs of unknown function and, with more work, to piece them together into whole genes.

In 1992, researchers used a computer program to predict protein-coding regions in preliminary human sequence data. They found two genes in a 58,000-nucleotide-pair segment of Chromosome 4, and five genes in a 106,000-nucleotide-pair segment of Chromosome 19. That works out to an average of 1 gene every 23,000 nucleotide pairs. Extrapolating from that density to the whole genome would give humans nearly 130,000 genes. It turned out, however, that the chromosomes the researchers analyzed had been chosen for sequencing precisely because they appeared to be gene-rich. When the estimate was adjusted to take into account the gene-poor regions of the human genome—guessing that half of the human genome had maybe one-tenth of that gene-rich density—the estimated number dropped to 71,000.

Matching tags

Of course, these estimates are based on what we think genes look like; however, we are still learning how to recognize genes. A different but complementary approach to counting the coding regions in a genome involves determining experimentally how many genes are actually expressed.

To determine which protein-coding genes are expressed in a particular cell type or tissue, mRNAs are isolated and converted into complementary DNAs, or cDNAs (discussed in Chapter 10). Because mRNAs are produced by transcription and splicing from protein-coding genes, this collection of cDNAs represents the sequences of all genes that were being expressed in the cells from which the mRNA was prepared. The cDNAs are prepared from a variety of tissues because the goal of this approach is to identify every gene, and different tissues express different genes. An additional benefit of working with cDNAs stems from the fact that mRNAs lack the introns and the stretches of DNA that lie between genes; thus, cDNA sequences correspond directly to the coding sequences in the genome.

Short fragments of these cDNAs, called *expressed sequence tags*, or ESTs, are then sequenced; the resulting EST sequences are compared with the nucleotide sequence of the entire genome to locate the gene that corresponds to each EST. By carefully analyzing how ESTs map onto the human genome, researchers arrived at an estimate of about 24,000 genes that code for proteins.

Human gene countdown

Now the most accurate approach to predicting genes combines different types of data, including (1) EST analyses, (2) computerized searches of the genome for ORFs and for sequences that signal the splice sites at the ends of every exon, and (3) comparisons with genome sequences from other organisms, especially other mammals. This last approach is particularly powerful because, as we have seen in this chapter, mammalian genomes are sufficiently divergent that only the most crucial portions of their genomes—the exons and regulatory sequences, for example—are similar (see Figure 9–22).

Although the estimates are all converging around 24,000 to 25,000, it could be many years before we have the final answer to how many genes it takes to make a human. For example, genes that code for short RNAs such as microRNAs are particularly difficult to identify. As computational methods are refined, and as we collect more data on the human genome and on the genome sequences of other organisms, our ability to predict where genes reside within a particular DNA sequence can only improve. In the end, however, knowing the exact number of genes is not nearly as important as understanding the functions of each gene and how it interacts with other genes to build a human being. These are central questions that are likely to occupy biologists for at least the next century.

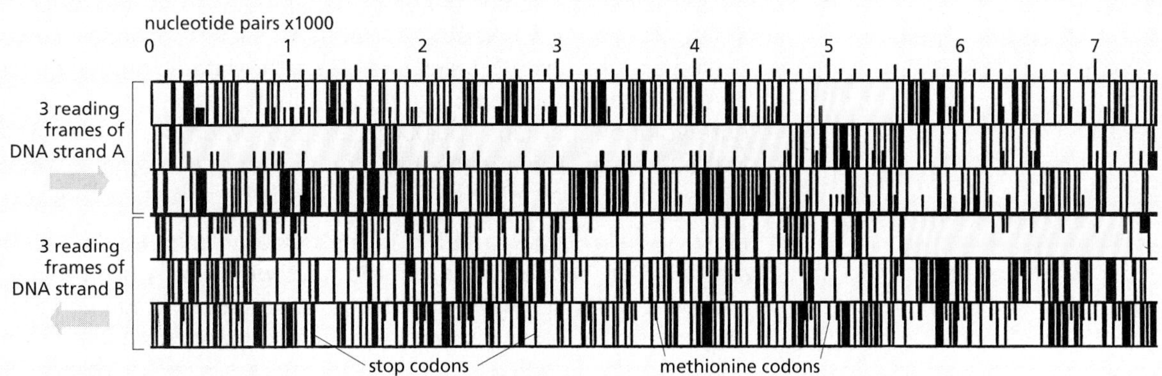

Figure 9–30 **Computer programs are used to find genes.** In this example, a DNA sequence of 7500 nucleotide pairs from the pathogenic yeast *Candida albicans* was fed into a computer, which then translated the entire sequence in all six possible open reading frames (ORFs), three from each strand (see Figure 7–25). The output displays each reading frame as a horizontal column, with the stop codons (TGA, TAA, and TAG) marked by the longer vertical lines and methionine codons (ATG) marked by the shorter lines. Four ORFs (*yellow*) can be clearly identified by the statistically significant absence of stop codons. For each ORF, the presumptive initiation codon (ATG) is indicated in *red*. The additional ATG codons in the ORFs code for methionine.

Accelerated Changes in Conserved Genome Sequences Help Reveal What Makes Us Human

As soon as both the human and the chimpanzee genome sequences became available, scientists began searching for DNA sequence changes that might account for the striking differences between us and them. With 3 billion nucleotide pairs to compare in the two species, this might seem an impossible task. But the job was made much easier by confining the search to the sequences that are highly conserved across multiple mammalian species (see Figure 9–22), representing parts of the genome that are most likely to be functionally important. Although these sequences are conserved, they are not identical: when the version in one species is compared with that in another, they are typically found to have drifted apart by a small amount corresponding simply to the time elapsed since the last common ancestor. In a small proportion of cases, however, one sees signs of a sudden evolutionary spurt. For example, some DNA sequences that have been highly conserved in other mammalian species are found to have changed exceptionally fast during the six million years of human evolution since we diverged from the chimpanzees. Such *human accelerated regions* are thought to reflect functions that have been especially important in making us different in some useful way.

In one study, about 50 such sites were identified—one-quarter of which were located near genes associated with neural development. The sequence exhibiting the most rapid change (18 changes between human and chimp, compared with only two changes between chimp and chicken) was examined further and found to encode a short noncoding RNA molecule that is produced in the human cerebral cortex at a critical time during brain development. Although the function of this RNA is not yet known, this exciting finding is stimulating further studies that will hopefully shed light on features of the human brain that distinguish us from closely related species.

Genetic Variation Within the Human Genome Contributes to Our Individuality

With the exception of identical twins, no two people have exactly the same genome. When the same region of the genome from two different humans is compared, the nucleotide sequences typically differ by about 0.1%. That might seem an insignificant degree of variation, but considering the size of the human genome, that amounts to some 3 million genetic differences in each maternal or paternal chromosome set between one person and the next. Detailed analysis of the data on human genetic variation suggests that the bulk of this variation was already present early in our evolution, perhaps 100,000 years ago, when the human population was still small. This means that a great deal of the genetic variation we possess today was inherited from our early human ancestors.

Most of the genetic variation in the human genome takes the form of single base changes called **single-nucleotide polymorphisms** (**SNPs**, pronounced snips). These polymorphisms are simply points in the genome that differ in nucleotide sequence between one portion of the population and another—positions where one large fraction of the population has a G-C nucleotide pair, for example, while another has an A-T (Figure 9–31). Two human genomes chosen at random from the world's population will differ by approximately 2.5×10^6 SNPs that are scattered throughout the genome. Because SNPs are present at such a high density, they provide useful markers in genetic analyses in which one attempts to link a specific trait (such as disease susceptibility) with a particular pattern of SNPs (as we discuss in Chapter 19). This type of analysis should

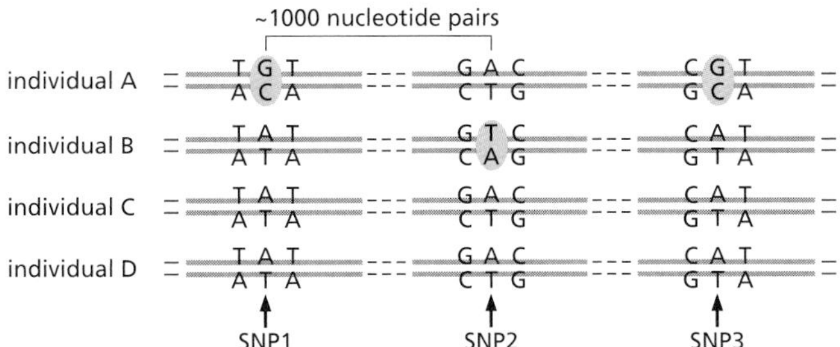

Figure 9–31 **Single-nucleotide polymorphisms (SNPs) are sequences in the genome that differ by a single nucleotide pair between one portion of the population and another.** Most, but not all, such changes in the human genome occur in regions where they do not affect the function of a gene.

lead to improvements in health care by allowing doctors to determine whether an individual is susceptible to a disease, such as heart disease, long before he or she shows any symptoms. The person can then change his or her behavior to help prevent the disease before it arises.

Another important source of variation inherited from our ancestors is the duplication and deletion of large blocks of DNA. When the genome of any person is compared with the standard reference genome, one observes roughly 100 differences involving long blocks of sequence. Some of these differences are very common, whereas others are present in only a small minority of people. From an initial sampling, nearly half of these sequences contain known genes. In retrospect, this type of variation is not surprising, given the extensive history of DNA addition and DNA loss in vertebrate genomes discussed earlier. Exactly how it contributes to our individuality, however, remains a mystery.

In addition to the SNPs and the duplications and deletions that we inherited from our ancestors, humans also possess repetitive nucleotide sequences that are particularly prone to new mutations. CA repeats, for example, are ubiquitous in the human genome. Nucleotide sequences containing large numbers of CA repeats are often replicated inaccurately (imagine trying to copy a word that is nothing more than a string of CACACACACACACACA...); hence the precise number of repeats can vary widely from one individual to the next. Because they show such exceptional variability, and because this variability has arisen so recently, CA repeats, and others like them, make ideal markers for differences between individual humans. Differences in the numbers of CA and other types of repeats at different positions in the genome are used to identify individuals by *DNA fingerprinting* in crime investigations, paternity suits, and other forensic applications (see Figure 10–19).

Most of the variations in the human genome sequence are genetically silent, as they fall within DNA sequences in noncritical regions of the genome. Such variations have no effect on how we look or how our cells function. This means that only a small subset of the variation we observe in our DNA is responsible for the heritable differences from one human to the next. As we discuss in Chapter 19, a major challenge in human genetics is to learn to recognize those relatively few variations that are functionally important against the large background of neutral variation.

QUESTION 9–5

You are interested in finding out the function of a particular gene in the mouse genome. You have determined the nucleotide sequence of the gene, defined the portion that codes for its protein product, and searched the relevant database for similar sequences; however, neither the gene nor the encoded protein resembles anything previously described. What types of additional information about the gene and the encoded protein would you like to know in order to narrow down its function, and why? Focus on the information you would want, rather than on the techniques you might use to get that information.

The Human Genome Contains Copious Information Yet to Be Deciphered

Even with the human genome sequence in hand, many questions will continue to challenge cell biologists throughout the next century. Perhaps the most perplexing one is this: given that a human, a chimp, and a mouse contain essentially the same genes, and therefore the same proteins, what makes these creatures so different from each other?

The answer, it seems, will come in large part from studies of gene regulation. The proteins encoded in the genome are like the components of a construction kit. By assembling the components in different combina-

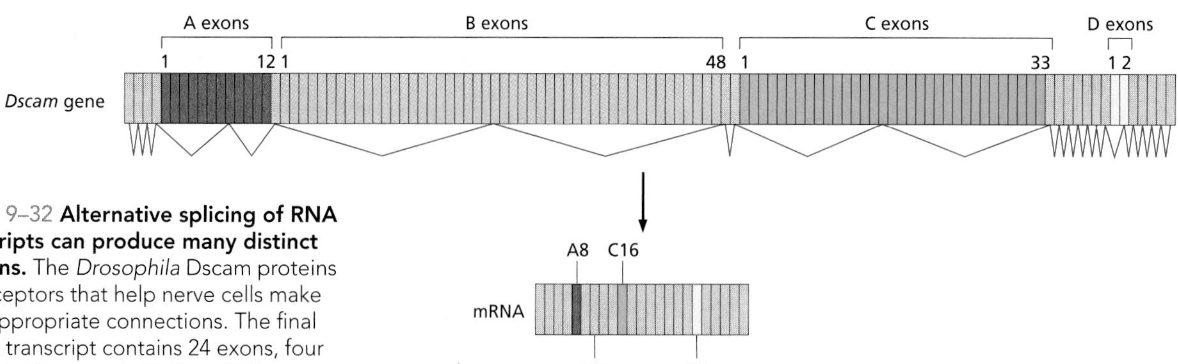

Figure 9–32 **Alternative splicing of RNA transcripts can produce many distinct proteins.** The *Drosophila* Dscam proteins are receptors that help nerve cells make their appropriate connections. The final mRNA transcript contains 24 exons, four of which (denoted A, B, C, and D) are present in the *Dscam* gene as arrays of alternative exons. Each mature mRNA contains 1 of 12 alternatives for exon A (*red*), 1 of 48 alternatives for exon B (*green*), 1 of 33 alternatives for exon C (*blue*), 1 of 2 alternatives for exon D (*yellow*), and all of the 19 invariant exons (*gray*). If all possible splicing combinations were used, 38,016 different proteins could in principle be produced from the *Dscam* gene. Only one of the many possible splicing patterns (indicated by the *magenta line*) and the mature mRNA it produces is shown. (Adapted from D.L. Black, *Cell* 103: 367–370, 2000. With permission from Elsevier.)

tions and in different orders, many different things can be built with the same kit. In the end, however, the overall shape of the final object is determined by the instructions that prescribe how the components are to be put together.

To a large extent, the instructions needed to produce a multicellular animal are contained in the noncoding regulatory DNA that is associated with each gene. As discussed in Chapter 8, this DNA contains, scattered within it, dozens of separate regulatory elements—short DNA segments that serve as binding sites for specific transcription regulators. This regulatory DNA can be said to define the sequential program of development: the rules that cells follow as they proliferate, assess their positions in the embryo, and switch on new sets of genes accordingly.

Although comparisons among many different species are a powerful way to pinpoint key regulatory sequences in a vast excess of unimportant DNA, we still do not know how to read these sequences accurately. For example, different transcription regulators can bind to the same short stretch of DNA, so that knowing the DNA sequence is not enough to specify which protein or proteins might actually regulate the gene. In addition, the fact that control of gene expression occurs in complex and combinatorial ways (see Figure 8–12) complicates our attempts to decipher when in development and in which type of cell each gene is expressed.

Another challenge in interpreting the information encoded in the human genome comes from the prevalence of alternative splicing. We know that most human genes undergo alternative splicing, allowing cells to produce a range of related but distinct proteins from a single gene (see Figure 7–21). Often this splicing is regulated, so that one form of the protein is produced in one type of tissue, while other forms are produced preferentially in other tissues. In one extreme case, from *Drosophila*, a single gene might produce as many as 38,000 different protein variants through alternative splicing (Figure 9–32). Thus an organism can produce far more protein products than it has genes. We do not yet know enough about the biology of alternative splicing to predict exactly which human genes are subject to this process, and when and where during development such regulation might occur. However, it does seem that alternative splicing is especially prevalent in the developing brain.

Another uncertainty in interpreting the human genome concerns the precise roles of microRNAs (see pp. 290–291). Discovered relatively recently, these short RNAs regulate as many as one-third of all human genes, yet few of them have been studied in any detail. Finally, although an estimated 1.5% of the human genome codes for protein, an additional 3.5% is highly conserved when compared with that of other mammalian genomes—and is therefore presumed to be important (see p. 312). Some of this conserved DNA produces RNA molecules of known function and some is regulatory DNA; the rest remains mysterious.

The human body is formed as the result of complex sequences of decisions that cells make as they proliferate and specialize during our development: which RNA molecules and which proteins are to be made where, and exactly when and how much of each is to be produced. The information for all these decisions is contained within the human genome sequence, but we are only beginning to learn how to read this type of code. Deciphering this information is one of the great challenges for the next generation of cell biologists.

ESSENTIAL CONCEPTS

- By comparing the nucleotide or amino acid sequences of contemporary organisms we are beginning to reconstruct how genomes have evolved in the billions of years that have elapsed since the appearance of the first cells.
- Genetic variation—the raw material for evolutionary change—occurs by a variety of mechanisms that alter the nucleotide sequences of genomes. These changes range from simple point mutations to larger-scale duplications and rearrangements.
- Genetic changes that offer an organism a selective advantage or those that are selectively neutral are the most likely to be perpetuated. Changes that seriously compromise an organism's fitness are eliminated through natural selection.
- Gene duplication is one of the most important sources of genetic diversity. Once a gene has been duplicated, the two copies can accumulate different mutations and thereby diversify to perform different functions. Over evolutionary time, repeated rounds of gene duplication and divergence can produce large gene families.
- The evolution of new proteins is thought to have been greatly facilitated by the swapping of exons between genes to create hybrid proteins with new functions.
- The human genome contains 3.2×10^9 nucleotide pairs divided among 22 autosomes and 2 sex chromosomes. Only a few percent of that DNA codes for proteins and for structural, regulatory, and catalytic RNAs.
- Individual humans differ from one another by an average of 1 nucleotide pair in every 1000; this variation underlies our individuality and provides the basis for identifying individuals by DNA analysis.
- Comparing genome sequences of different species provides a powerful way to identify genes and to highlight other functionally important parts of the genome.
- Because related species (such as human and mouse) share many genes in common, evolutionary changes that affect how these genes are regulated are especially important in understanding the differences between species.

KEY TERMS

conserved synteny
divergence
exon shuffling
germ cell
germ line
gene duplication and divergence
homologous gene
horizontal gene transfer
open reading frame (ORF)
phylogenetic tree
point mutation
purifying selection
single-nucleotide polymorphism (SNP)
somatic cell

QUESTIONS

QUESTION 9–6

Discuss the following statement: "Mobile genetic elements are parasites. They are harmful to the host organism and therefore place it at an evolutionary disadvantage."

QUESTION 9–7

Human Chromosome 22 (48 Mb in length) has about 700 protein-coding genes, which average 19,000 nucleotide pairs in length and contain an average of 5.4 exons, each of which averages 266 nucleotide pairs. What fraction of the average protein-coding gene is converted into mRNA? What fraction of the chromosome do these genes occupy?

QUESTION 9–8

(True/False) The majority of human DNA is unimportant junk. Explain your answer.

QUESTION 9–9

Mobile genetic elements make up nearly half of the human genome and are inserted more or less randomly throughout it. However, in some spots these elements are rare, as illustrated for a cluster of genes called HoxD, which lies on Chromosome 2 (Figure Q9–9). This cluster is about 100 kb in length and contains nine genes whose differential expression along the anteroposterior axis of the developing embryo helps establish the basic body plan for humans (and for other animals). Why do you suppose that mobile genetic elements are so rare in this cluster? In Figure Q9–9, lines that project *upward* indicate exons of known genes. Lines that project *downward* indicate mobile genetic elements; they are so numerous they merge into nearly a solid block outside the HoxD cluster. For comparison, an equivalent region of Chromosome 22 is shown.

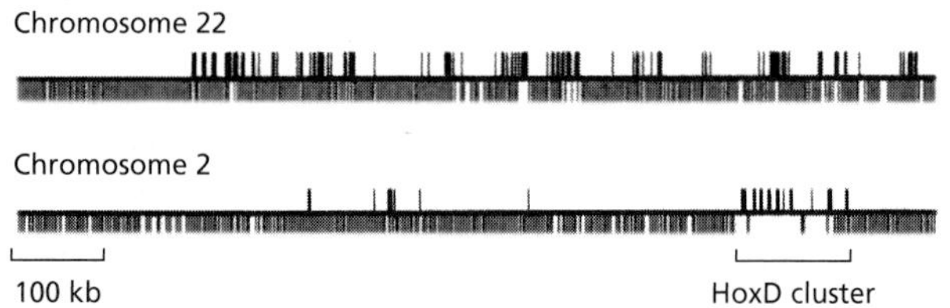

Figure Q9–9

QUESTION 9–10

An early graphical method for comparing nucleotide sequences—the so-called diagon plot—still yields one of the best visual comparisons of sequence relatedness. An example is illustrated in Figure Q9–10, in which the human β-globin gene is compared with the human cDNA for β globin (which contains only the coding portion of the gene; Figure Q9–10A) and to the mouse β-globin gene (Figure Q9–10B). Diagon plots are generated by comparing blocks of sequence, in this case blocks of 11 nucleotides at a time. If 9 or more of the nucleotides match, a dot is placed on the diagram at the coordinates corresponding to the blocks being compared. A comparison of all possible blocks generates diagrams such as the ones shown in Figure Q9–10, in which sequence similarities show up as diagonal lines.

A. From the comparison of the human β-globin gene with the human β-globin cDNA (Figure Q9–10A), can you deduce the positions of exons and introns in the β-globin gene?

B. Are the exons of the human β-globin gene (indicated by shading in Figure Q9–10B) similar to those of the mouse β-globin gene? Identify and explain any key differences.

C. Is there any sequence similarity between the human and mouse β-globin genes that lies outside the exons? If so, identify its location and offer an explanation for its preservation during evolution.

D. Did the mouse or human gene undergo a change of intron length during their evolutionary divergence? How can you tell?

QUESTION 9–11

Your advisor, a brilliant bioinformatician, has high regard for your intellect and industry. She suggests that you write a computer program that will identify the exons of protein-coding genes directly from the sequence of the human genome. In preparation for that task, you decide to write down a list of the features that might distinguish protein-coding sequences from intronic DNA and from other

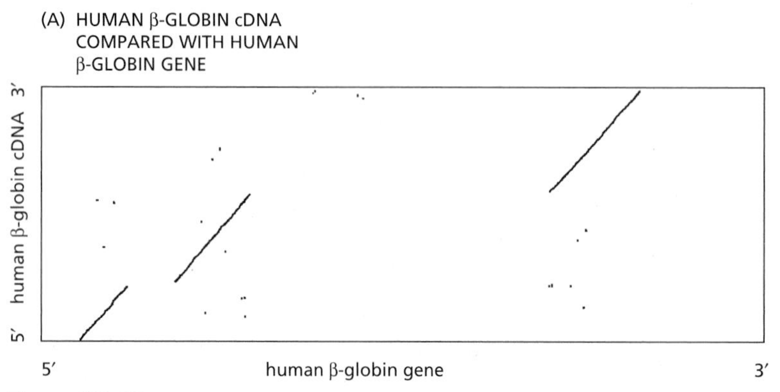

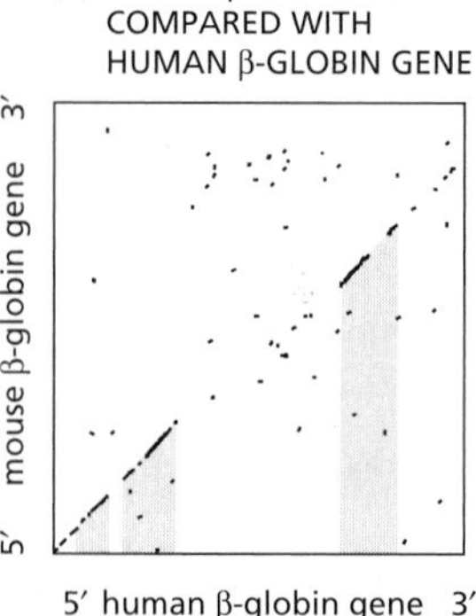

Figure Q9–10

sequences in the genome. What features would you list? (You may wish to review basic aspects of gene expression in Chapter 7.)

QUESTION 9–12

Why do you expect to encounter a stop codon about every 20 codons or so in a random sequence of DNA?

QUESTION 9–13

The genetic code (see Figure 7–24) relates the nucleotide sequence of mRNA to the amino acid sequence of encoded proteins. Ever since the code was deciphered, some have claimed it must be a frozen accident—that is, the system randomly fell into place in some ancestral organism and was then perpetuated unchanged throughout evolution; others have argued that the code has been shaped by natural selection.

A striking feature of the genetic code is its inherent resistance to the effects of mutation. For example, a change in the third position of a codon often specifies the same amino acid or one with similar chemical properties. But is the natural code more resistant to mutation than other possible versions? The answer is an emphatic "Yes," as illustrated in Figure Q9–13. Only one in a million computer-generated "random" codes is more error-resistant than the natural genetic code.

Does the resistance to mutation of the actual genetic code argue in favor of its origin as a frozen accident or as a result of natural selection? Explain your reasoning.

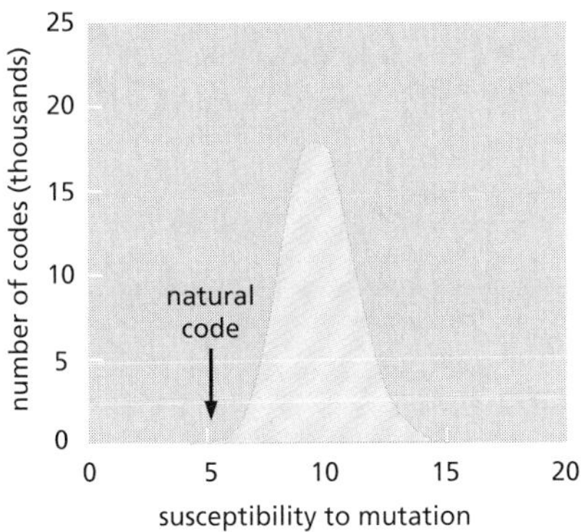

Figure Q9–13

QUESTION 9–14

Which one of the processes listed below is NOT thought to contribute significantly to the evolution of new protein-coding genes?

A. Duplication of genes to create extra copies that can acquire new functions.

B. Formation of new genes *de novo* from noncoding DNA in the genome.

C. Horizontal transfer of DNA between cells of different species.

D. Mutation of existing genes to create new functions.

E. Shuffling of protein domains by gene rearrangement.

Gene	Amino Acids	Rates of Change	
		Nonsynonymous	Synonymous
Histone H3	135	0.0	4.5
Hemoglobin α	141	0.6	4.4
Interferon γ	136	3.1	5.5

Rates were determined by comparing rat and human sequences and are expressed as nucleotide changes per site per 10^9 years. The average rate of nonsynonymous changes for several dozen rat and human genes is about 0.8.

QUESTION 9–15

Some genes evolve more rapidly than others. But how can this be demonstrated? One approach is to compare several genes from the same two species, as shown for rat and human in the table above. Two measures of rates of nucleotide substitution are indicated in the table. Nonsynonymous changes refer to single-nucleotide changes in the DNA sequence that alter the encoded amino acid (ATC → TTC, which gives isoleucine → phenylalanine, for example). Synonymous changes refer to those that do not alter the encoded amino acid (ATC → ATT, which gives isoleucine → isoleucine, for example). (As is apparent in the genetic code, Figure 7–24, there are many cases where several codons correspond to the same amino acid.)

A. Why are there such large differences between the synonymous and nonsynonymous rates of nucleotide substitution?

B. Considering that the rates of synonymous changes are about the same for all three genes, how is it possible for the histone H3 gene to resist so effectively those nucleotide changes that alter its amino acid sequence?

C. In principle, a protein might be highly conserved because its gene exists in a 'privileged' site in the genome that is subject to very low mutation rates. What feature of the data in the table argues against this possibility for the histone H3 protein?

QUESTION 9–16

Plant hemoglobins were found initially in legumes, where they function in root nodules to lower the oxygen concentration, allowing the resident bacteria to fix nitrogen. These hemoglobins impart a characteristic pink color to the root nodules.The discovery of hemoglobin in plants was initially surprising because scientists regarded hemoglobin as a distinctive feature of animal blood. It was hypothesized that the plant hemoglobin gene was acquired by horizontal transfer from an animal. Many more hemoglobin genes have now been sequenced from a variety of organisms, and a phylogenetic tree of hemoglobins is shown in Figure Q9–16.

A. Does the evidence in the tree support or refute the hypothesis that the plant hemoglobins arose by horizontal gene transfer?

B. Supposing that the plant hemoglobin genes were originally derived by horizontal transfer (from a parasitic nematode, for example); what would you expect the phylogenetic tree to look like?

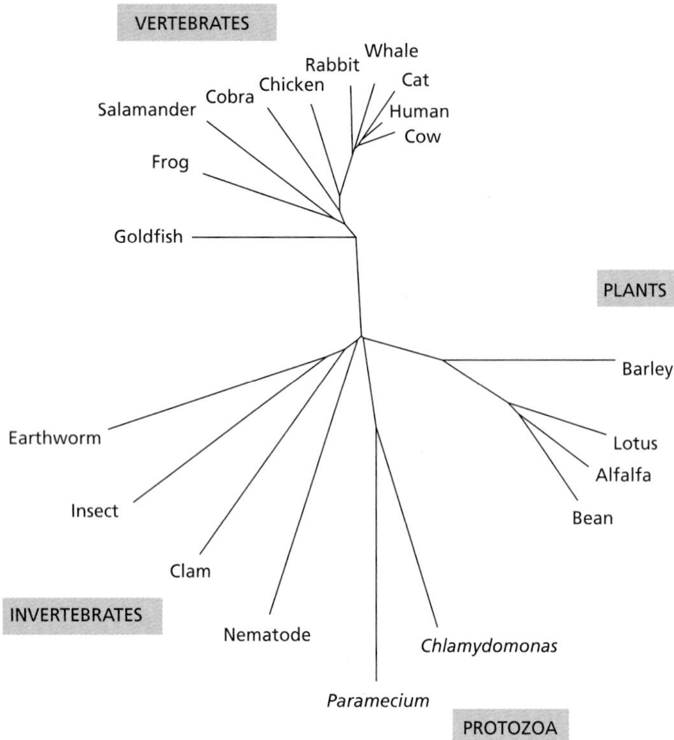

Figure Q9–16

QUESTION 9–17

The accuracy of DNA replication in the human germ cell line is such that on average only about 0.6 out of the 6 billion nucleotides is altered at each cell division. Because most of our DNA is not subject to any precise constraint on its sequence, most of these changes are selectively neutral. Any two modern humans picked at random will show about 1 difference of nucleotide sequence in every 1000 nucleotides. Suppose we are all descended from a single pair of ancestors—Adam and Eve—who were genetically identical and homozygous (each chromosome was identical to its homolog). Assuming that all germ-line mutations that arise are preserved in descendants, how many cell generations must have elapsed since the days of Adam and Eve for 1 difference per 1000 nucleotides to have accumulated in modern humans? Assuming that each human generation corresponds on average to 200 cell-division cycles in the germ-cell lineage and allowing 30 years per human generation, how many years ago would this ancestral couple have lived?

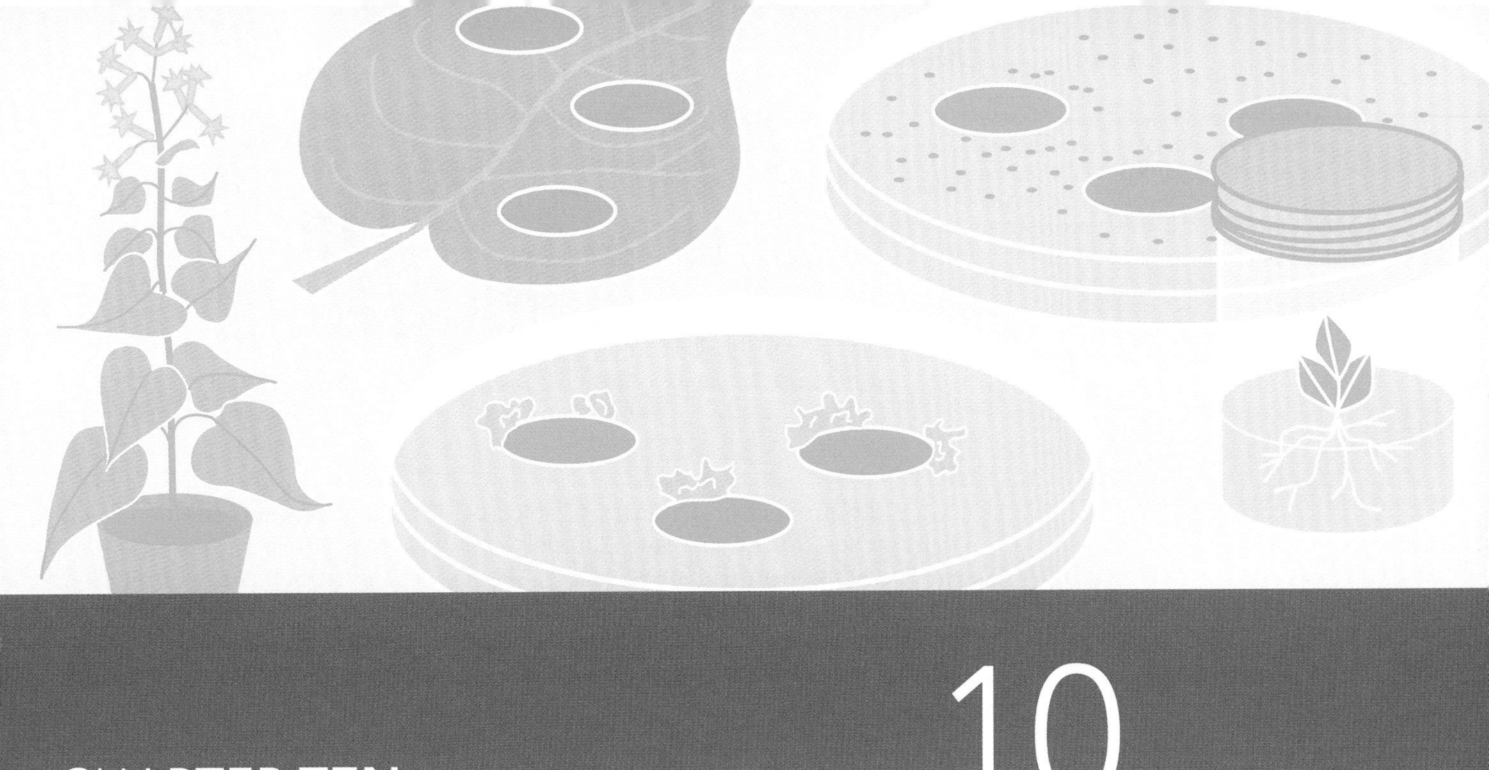

CHAPTER **TEN** 10

Analyzing Genes and Genomes

MANIPULATING AND ANALYZING DNA MOLECULES

DNA CLONING

DECIPHERING AND EXPLOITING GENETIC INFORMATION

The twenty-first century promises to be an exciting time for cell biology. New methods for analyzing and manipulating DNA, RNA, and proteins are fueling an information explosion and allowing us to study genes and cells in previously unimagined ways. We now have access to the sequences of many billions of nucleotides, providing the molecular blueprints for dozens of organisms—from microbes and plants to birds, insects, humans, and other mammals. And powerful new techniques are helping us to decipher this information, allowing us not only to compile huge, detailed catalogs of genes and proteins but to begin to unravel how these components work together to form functional cells, tissues, and organisms. The goal is nothing short of obtaining a complete understanding of the process that takes place, moment to moment, inside all living things.

This technological revolution has been powered, in large part, by the development of methods that have dramatically increased our ability to handle DNA. In the early 1970s, it became possible, for the first time, to isolate a given piece of DNA out of the many millions of nucleotide pairs in a typical chromosome. This in turn made it possible to generate new DNA molecules in the test tube and to introduce this custom-made genetic material back into living organisms. These developments—dubbed **recombinant DNA technology** or genetic engineering—make it possible to create chromosomes with combinations of genes that could never have formed naturally, or combinations that could conceivably occur in nature but might take thousands of years of chance events to come together.

Of course, humans have been experimenting with DNA, albeit without realizing it, for millennia. Modern garden-rose varieties, for example, are

(A)

(B)

Figure 10–1 **Humans have been experimenting with DNA for millennia.** (A) The oldest known depiction of a rose in Western art, from the palace of Knossos in Crete, around 2000 BC. Modern roses are the result of centuries of breeding between such wild roses. (B) A pug and a poodle illustrate the range of dog breeds. All dogs, regardless of breed, belong to a single species. (B, courtesy of Heather Angel.)

the product of centuries of selective breeding between strains of wild roses from China and Europe (Figure 10–1A). Similarly, the enormous variation in the sizes, colors, shapes, and even behaviors of different breeds of dogs is the result of deliberate genetic experiments—with selection for desired traits—carried out since the gray wolf, the ancestor of the modern dog, was first domesticated some 10,000–15,000 years ago (Figure 10–1B).

Modern genetic engineering techniques, however, allow us to alter DNA with much greater precision and speed. With the latest equipment and techniques, even a beginning student can now collect an organism's genetic material, isolate a region of DNA containing a specific gene, produce a virtually unlimited number of exact copies of this DNA, and determine its nucleotide sequence with relative ease. Using variations of these engineering techniques, the isolated gene can be redesigned in the laboratory and then transferred back into cells in culture to elucidate its function. With slightly more sophisticated methods, the engineered genes can be inserted into animals and plants, so that they become a functional and heritable part of the organism's genome.

These technical breakthroughs have had a dramatic impact on all aspects of cell biology. They have made possible our present knowledge of the organization and evolutionary history of the complex genomes of eucaryotes (as discussed in Chapter 9) and have led to the discovery of whole new classes of genes, RNAs, and proteins. They provide new ways of determining the functions of proteins and of individual domains within proteins, revealing a host of unexpected relationships among them. And they give biologists an important set of tools for unraveling the mechanisms by which a whole animal or plant can develop from a single cell.

Recombinant DNA technology has also had a profound influence on many aspects of human life outside scientific research: it is used to detect the mutations in DNA that are responsible for inherited disorders and to diagnose an individual's predisposition to genetic diseases such as cancer; it is used in forensic science to identify or acquit suspects in a crime; it is used to produce an increasing number of human pharmaceuticals, including insulin for diabetics and the blood-clotting protein Factor VIII for hemophiliacs. Even our laundry detergents, which contain heat-stable proteases that digest food spills and blood spots, make use of the products of DNA technology. Of all the discoveries described in this book, those that led to the development of recombinant DNA technology are likely to have the greatest impact on our everyday lives.

QUESTION 10–1

DNA sequencing of your own two β-globin genes (one from each of your two Chromosome 11s) reveals a mutation in one of the genes. Given this information alone, should you worry about being a carrier of an inherited disease that could be passed on to your children? What other information would you like to have to assess your risk?

In this chapter we describe the most common methods for manipulating genes and determining their function. We first survey the principal methods of the revolutionary field of recombinant DNA technology, beginning with a discussion of the basic techniques of DNA analysis. Next we describe how DNA sequences can be isolated and produced in large numbers by the techniques of *DNA cloning* and the *polymerase chain reaction (PCR)*, and we explain how these sequences can be used to produce and study proteins. In the final section of the chapter we examine how DNA technology is used to discover the roles of individual genes and proteins in cells and in organisms. Together, these techniques have made possible our current understanding of biology, including much of the material presented between the covers of this book.

MANIPULATING AND ANALYZING DNA MOLECULES

Until the development of recombinant DNA techniques, crucial clues for understanding how cells work remained locked in the genome. Once scientists realized that genetic information was encoded in the sequence of nucleotides in DNA, they needed to get their hands on individual genes to see what they look like and determine how they function. Before the revolution in DNA technology that took place in the 1970s, this task was almost impossible. Part of the problem was that isolating a single gene was not a trivial matter. Unlike a protein, a gene does not exist as a discrete entity in cells, but rather as a part of a much larger DNA molecule. Even bacterial genomes, which are much less complex than the chromosomes of eucaryotes, are enormously long. The *E. coli* genome, for example, contains 4.6 million nucleotides.

Large pieces of DNA can be broken into small pieces by mechanical shear; however, the fragment containing a particular gene will still be only one among a hundred thousand or more DNA fragments that would be obtained, for example, from a mammalian genome by these means. And in a sample containing many identical copies of the same large DNA molecule, each molecule would be broken up differently by shear, producing a confusing set of random fragments. How, then, can a gene be isolated and identified?

The solution to this problem emerged, in large part, with the discovery of a class of bacterial enzymes known as *restriction nucleases*. A nuclease catalyzes the hydrolysis of a phosphodiester bond in a nucleic acid. But these enzymes have a property that is distinct from other nucleases: they cut double-stranded DNA only at particular sites, determined by a short sequence of nucleotide pairs. Restriction nucleases can therefore be used to produce a reproducible set of specific DNA fragments from any genome. We begin this section by describing how these enzymes work and how the DNA fragments produced by them can be separated from each other. We then discuss how these fragments can be probed to identify the ones that contain the DNA of interest.

Restriction Nucleases Cut DNA Molecules at Specific Sites

Like most of the tools of DNA technology, restriction nucleases were discovered by researchers studying a specialized biological problem that had gripped their interest. Certain bacteria, it was noticed, always degraded 'foreign' DNA that was introduced into them experimentally. A search for the cause of this targeted destruction revealed a novel class of nucleases, present inside the host bacterium, that cleaves DNA only at certain nucleotide sequences. The bacterium's own DNA is protected from cleavage by chemical modification of these same sequences. Because these enzymes restricted the transfer of DNA between certain strains of bacteria, the name **restriction nuclease** was given to them. This solution to a seemingly arcane problem in bacterial biology forever changed the way in which cell and molecular biologists study life.

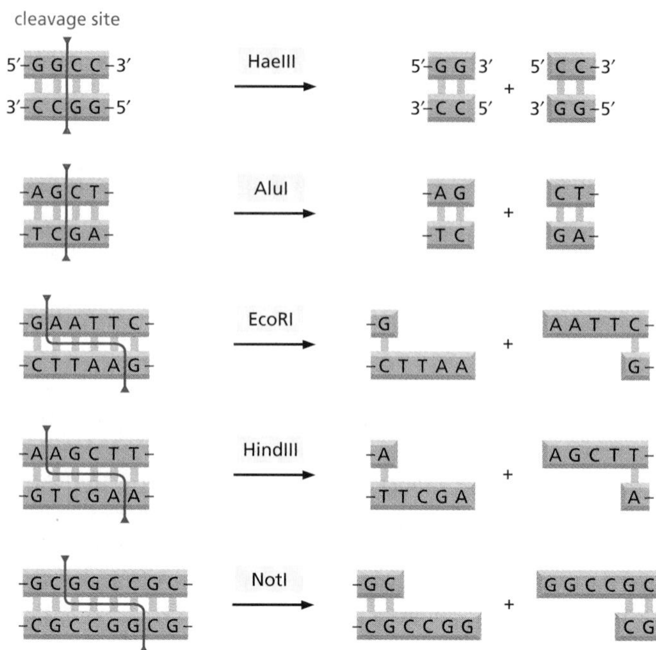

Figure 10–2 **Restriction nucleases cleave DNA at specific nucleotide sequences.** Target sequences are often palindromic (that is, the nucleotide sequence is symmetrical around a central point). In these examples, both strands of DNA are cut at specific points within the target sequence. Some enzymes, such as HaeIII and AluI, cut straight across the DNA double helix and leave two blunt-ended DNA molecules; with others, such as EcoRI, HindIII, and NotI, the cuts on each strand are staggered. These staggered cuts generate "sticky ends," short, single-stranded overhangs that help the cut DNA molecules join back together through complementary base-pairing. This rejoining of DNA molecules becomes important for DNA cloning, as we discuss later in the chapter. Restriction nucleases are usually obtained from bacteria, and their names reflect their origins: for example, the enzyme EcoRI comes from *Escherichia coli*.

Different bacterial species contain different restriction nucleases, each cutting at a different, specific sequence of nucleotides (Figure 10–2). Because these target sequences are short—generally four to eight nucleotide pairs—sites of cleavage will occur, purely by chance, in any long DNA molecule. Thus restriction nucleases can be used on DNA from any source. The main reason they are so useful is that a given enzyme will always cut a given DNA molecule at the same sites. Thus, for a sample of DNA from a human, treatment with a given restriction nuclease will always produce the same set of DNA fragments. Restriction nucleases are now sold by many companies and are typically ordered through the mail; one online catalog lists hundreds of such enzymes, each able to cut a different DNA sequence.

The target sequences of restriction nucleases vary in the frequency with which they will occur in DNA. As shown in Figure 10–2, the enzyme HaeIII cuts at a particular sequence of four nucleotide pairs; this sequence would be expected to occur purely by chance approximately once every 256 nucleotide pairs (1 in 4^4). By similar reasoning, the enzyme NotI—which has a target sequence that is eight nucleotides long—would be expected to cleave DNA on average once every 65,536 nucleotide pairs (1 in 4^8). The average sizes of the DNA fragments produced by different restriction nucleases can thus be very different. This feature makes it possible to cleave a long DNA molecule into the fragment sizes that are most suitable for a given application.

Gel Electrophoresis Separates DNA Fragments of Different Sizes

After a large DNA molecule is cleaved into smaller pieces with a restriction nuclease, it is often desirable to separate the DNA fragments from one another. This is usually accomplished using gel electrophoresis, which separates the fragments on the basis of their length. The mixture of DNA fragments is loaded at one end of a slab of agarose or polyacrylamide gel, which contains a microscopic network of pores. A voltage is then applied across the gel slab. Because DNA is negatively charged, the fragments migrate toward the positive electrode; the larger fragments migrate more slowly because their progress is more impeded by the agarose matrix. Over several hours, the DNA fragments become spread out

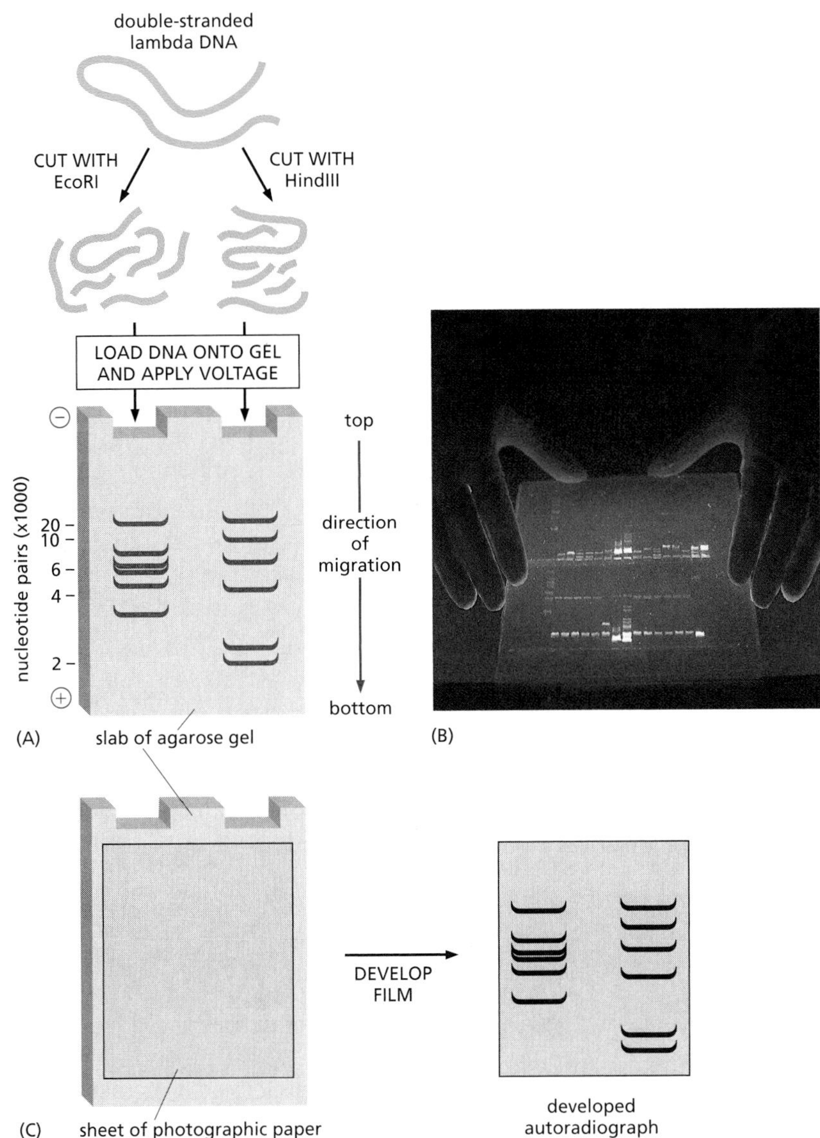

Figure 10–3 **DNA molecules can be separated by size using gel electrophoresis.** (A) This schematic illustration compares the results of cutting the same DNA molecule (in this case the genome of a bacteria-infecting virus called lambda) with two different restriction nucleases: EcoRI *(left)* and HindIII *(right)*. The fragments are then separated by gel electrophoresis. Because larger fragments migrate more slowly than smaller ones, the lowermost bands on the gel contain the smallest DNA fragments. (B) To visualize the DNA bands, the gel is soaked in a dye—such as ethidium bromide—that binds to DNA and fluoresces brightly under ultraviolet light. (C) An alternative method for visualizing the DNA bands is autoradiography. Prior to cleavage with restriction enzymes, the DNA can be labeled with the radioisotope ^{32}P by substituting ^{32}P for some of the nonradioactive phosphorus atoms. This could be done, for example, by replicating the lambda virus in the presence of ^{32}P. Because the β particles emitted from ^{32}P will expose photographic film, a sheet of film placed flat on top of the agarose gel in a darkroom will, when developed, show the position of all the DNA bands. (B, courtesy of Science Photo Library.)

across the gel according to size, forming a ladder of discrete bands, each composed of a collection of DNA molecules of identical length (Figure 10–3A). Physically isolating a particular DNA fragment from this agarose slab is fairly simple: a small section of the gel containing the band can be cut out using a scalpel or a razor blade and the DNA can then be extracted.

DNA bands on agarose or polyacrylamide gels are invisible unless the DNA is labeled or stained in some way. One sensitive method of staining DNA is to expose it to a dye that fluoresces under ultraviolet light when it is bound to DNA (Figure 10–3B). An even more sensitive detection method involves incorporating a radioisotope into the DNA molecules before electrophoresis; ^{32}P is often used, as it can be incorporated into the phosphates of DNA and emits an energetic β particle that is easily detected by the technique of autoradiography (Figure 10–3C).

QUESTION 10–2

Which products result when the piece of double-stranded DNA on the *left* is digested with (A) EcoRI, (B) AluI, (C) NotI, or (D) all three of these enzymes together? (See Figure 10–2 for the target sequences of these enzymes.)

5′-AAGAATTGCGGAATTCGAGCTTAAGGGCCGCGCCGAAGCTTTAAA-3′
3′-TTCTTAACGCCTTAAGCTCGAATTCCCGGCGCGGCTTCGAAATTT-5′

Hybridization Provides a Sensitive Way to Detect Specific Nucleotide Sequences

Exposing a gel to a fluorescent dye that binds to DNA—or labeling the DNA with ^{32}P—will allow every band on the gel to be seen. But it does not reveal which of those bands contains a sought-after segment of DNA. To identify a fragment that has the nucleotide sequence of interest, one can take advantage of the fact that any single strand of DNA will form Watson–Crick base pairs with a second strand of complementary nucleotide sequence.

Under normal conditions, the two strands of a DNA double helix are held together by hydrogen bonds, which can be broken by heating the DNA to around 90°C or by subjecting it to extremes of pH. Such treatments release the two strands from each other but do not break the covalent bonds that link nucleotides together within each strand. If this process is slowly reversed (that is, by slowly lowering the temperature to normal body temperature or by bringing the pH back to neutral), the complementary strands will readily reform double helices. This process is called **hybridization** or *renaturation,* and it results from a restoration of the complementary hydrogen bonds (Figure 10–4).

A similar hybridization reaction will occur between any two single-stranded nucleic acid chains (DNA/DNA, RNA/RNA, or RNA/DNA), provided they have complementary nucleotide sequences. The fundamental capacity of a single-stranded nucleic acid molecule to form a double helix only with a molecule complementary to it provides a powerful technique to detect specific nucleotide sequences in both DNA and RNA.

Hybridization Is Carried Out Using DNA Probes Designed to Recognize a Desired Nucleotide Sequence

To search for a nucleotide sequence by hybridization, however, one first needs a piece of nucleic acid with which to search. This *DNA probe* is a single-stranded DNA molecule, typically 10–1000 nucleotides long, that is used in hybridization reactions to detect nucleic acid molecules containing a complementary sequence. Today, DNA probes of any desired sequence can be synthesized nonenzymatically. Machines the size of a microwave oven can be programmed to string nucleotides together by chemical synthesis to produce single-stranded DNA chains of any sequence up to several hundred nucleotides in length.

Of course, to know what sequence to synthesize, one needs to know the nucleotide sequence of the DNA fragment of interest. Say, for example, one wishes to identify a DNA fragment that contains a portion of the β-globin gene. Because the sequence of this gene is known, producing a probe that will recognize it is a straightforward exercise in looking up the sequence in a database and then programming the nucleic acid synthesizer. Nowadays, such probes are usually ordered from companies that synthesize custom-made pieces of DNA. Simply send the company an email containing the desired sequence, and the DNA molecules can be delivered the next day.

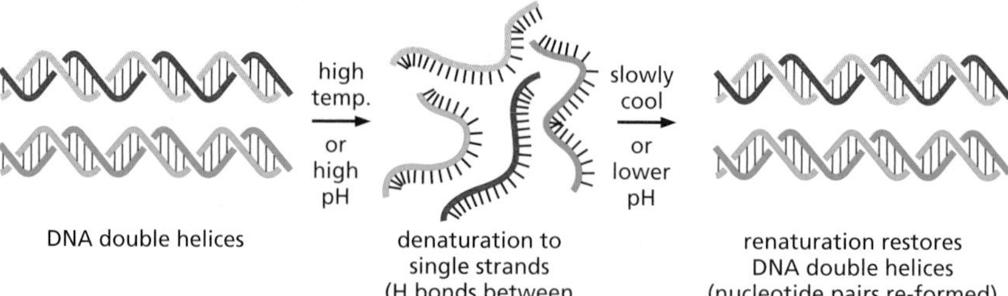

Figure 10–4 **A molecule of DNA can undergo denaturation and renaturation (hybridization).** For hybridization to occur, the two single strands must have complementary nucleotide sequences that allow base-pairing. In this example, the *red* and *orange* strands are complementary to each other, and the *blue* and *green* strands are complementary to each other.

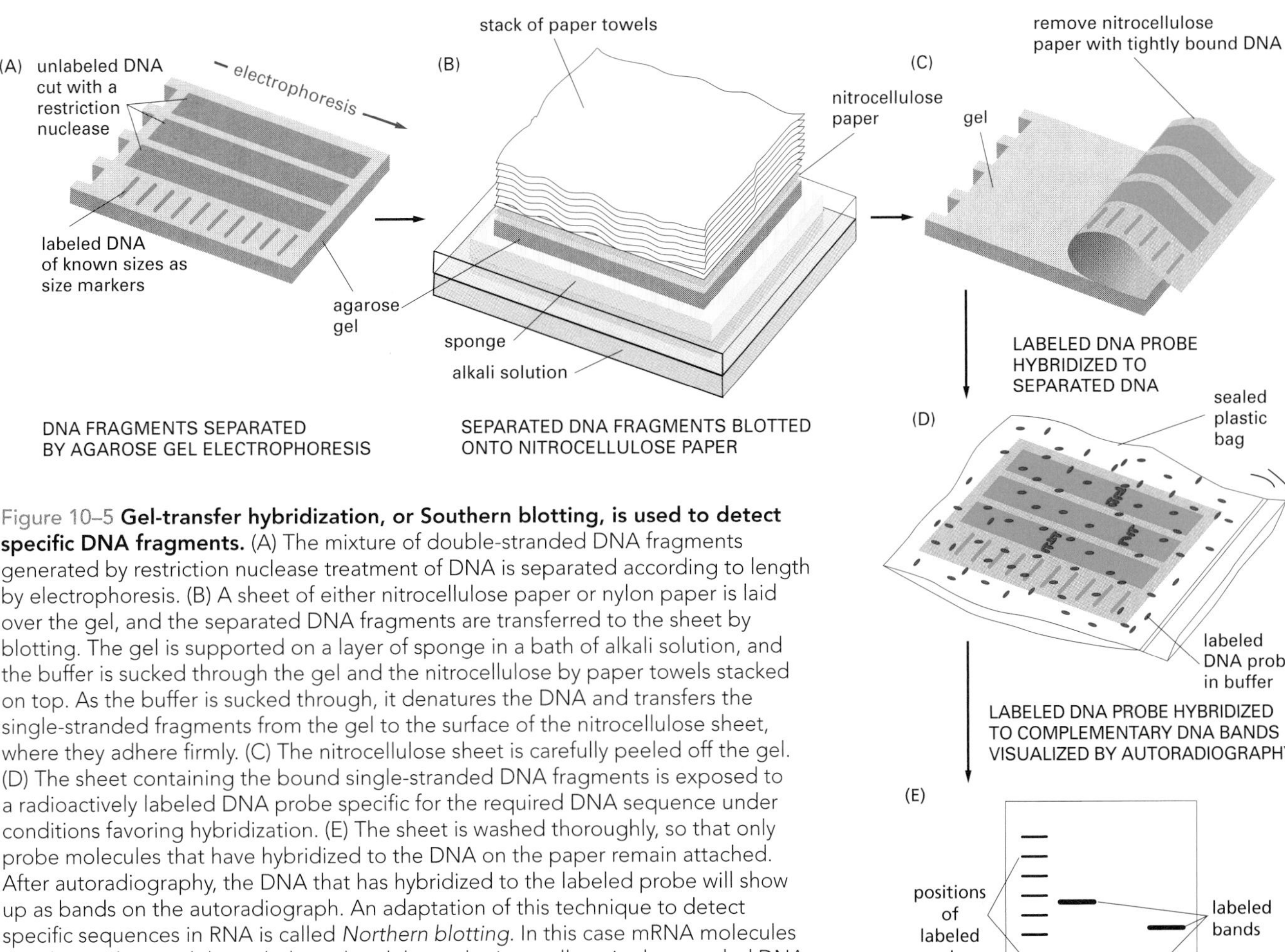

Figure 10–5 **Gel-transfer hybridization, or Southern blotting, is used to detect specific DNA fragments.** (A) The mixture of double-stranded DNA fragments generated by restriction nuclease treatment of DNA is separated according to length by electrophoresis. (B) A sheet of either nitrocellulose paper or nylon paper is laid over the gel, and the separated DNA fragments are transferred to the sheet by blotting. The gel is supported on a layer of sponge in a bath of alkali solution, and the buffer is sucked through the gel and the nitrocellulose by paper towels stacked on top. As the buffer is sucked through, it denatures the DNA and transfers the single-stranded fragments from the gel to the surface of the nitrocellulose sheet, where they adhere firmly. (C) The nitrocellulose sheet is carefully peeled off the gel. (D) The sheet containing the bound single-stranded DNA fragments is exposed to a radioactively labeled DNA probe specific for the required DNA sequence under conditions favoring hybridization. (E) The sheet is washed thoroughly, so that only probe molecules that have hybridized to the DNA on the paper remain attached. After autoradiography, the DNA that has hybridized to the labeled probe will show up as bands on the autoradiograph. An adaptation of this technique to detect specific sequences in RNA is called *Northern blotting*. In this case mRNA molecules are electrophoresed through the gel and the probe is usually a single-stranded DNA molecule.

Once the probe is in hand, it can be used to search for nucleic acids with a complementary sequence—including the fragment of interest among a collection of DNA fragments that have been separated by size on an agarose gel. The resulting hybridization with bands on an agarose gel can be visualized by *Southern blotting,* a common laboratory procedure named after the scientist who invented it (Figure 10–5).

DNA probes are widely used in cell biology. Later in the chapter we will see how hybridization with specific probes can be used to determine in which tissues and at what stages of development a gene is transcribed. But first, we discuss how hybridization facilitates the process of DNA cloning.

DNA CLONING

We have seen that DNA molecules can be cleaved into shorter fragments by restriction nucleases and that these fragments can be separated by gel electrophoresis. We have also seen that hybridization can be used to pick out a match with a DNA probe of known sequence. In this section of the chapter, we shall see how these procedures are combined to obtain a physical piece of DNA from a genome. In other words, we will discuss how a particular piece of any genome can be *cloned*. The term **DNA cloning** literally refers to the act of making many identical copies of a DNA molecule. It is this amplification that makes it possible to separate a particular stretch of DNA (often a gene of interest) physically from the rest of a cell's DNA—which is present in vanishingly small quantities compared with the final mass of the amplified sequence. Producing many copies of a defined segment of DNA from a genome is one of the most important feats of recombinant DNA technology, as it is the starting point for understanding the function of each stretch of DNA within the genome.

DNA Ligase Joins DNA Fragments Together to Produce a Recombinant DNA Molecule

Modern DNA technology depends both on the ability to break long DNA molecules into conveniently sized fragments and on the ability to join these fragments back together in new combinations. Any DNA molecule so constructed in the laboratory is called **recombinant DNA**. The cell itself has provided the tools with which to perform these molecular manipulations. As discussed in Chapter 6, the enzyme **DNA ligase** reseals the nicks in the DNA backbone that arise during DNA replication and DNA repair (see Figures 6–16 and 6–26). This enzyme has become one of the most important tools of recombinant DNA technology, as it allows scientists to join together any two DNA fragments (Figure 10–6). Because DNA has the same chemical structure in all organisms, this simple maneuver allows DNAs from any source to be united. In this way, isolated DNA fragments can be recombined in the test tube to produce DNA molecules not found in nature. Once two DNA molecules have been joined by ligase, the cell cannot tell that the two DNAs were originally separate and will treat the resulting DNA as a single molecule. If such a foreign stretch of DNA is appropriately introduced into the DNA of a host cell, it will be replicated and transcribed as if it were a normal part of the cell's own DNA.

Recombinant DNA Can Be Copied Inside Bacterial Cells

Obtaining many identical copies of a defined stretch of DNA, often a gene, can be accomplished in several ways. One way is to introduce the DNA to be copied into a rapidly dividing bacterium; each time the bacterium replicates its own DNA it also copies the introduced DNA.

(A) JOINING TWO COMPLEMENTARY STAGGERED ENDS

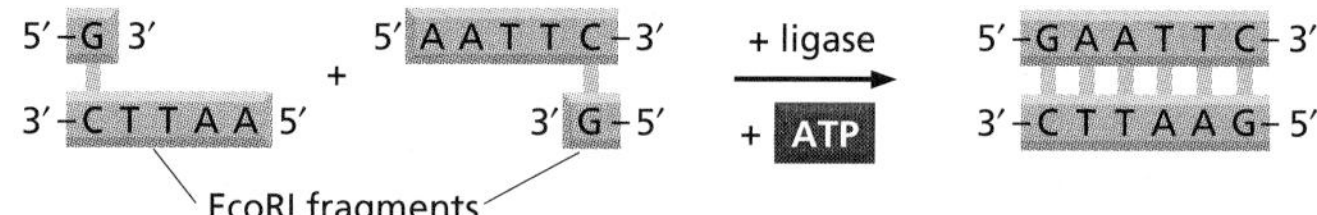

(B) JOINING TWO BLUNT ENDS

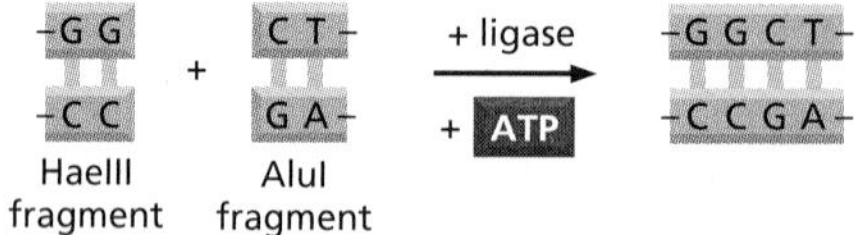

(C) JOINING A BLUNT END WITH A STAGGERED END

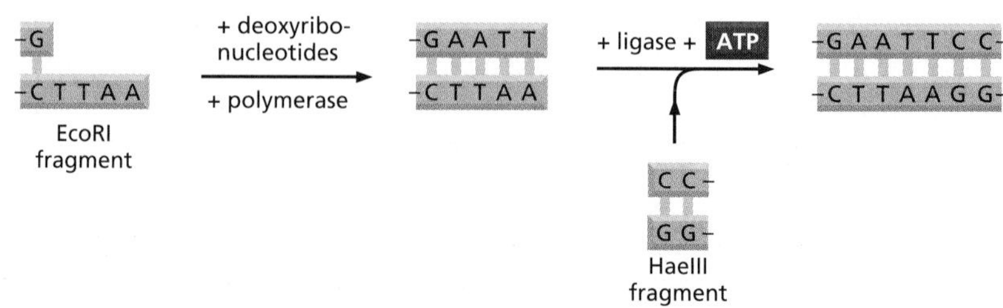

Figure 10–6 **Recombinant DNA molecules can be formed *in vitro*.** The enzyme DNA ligase can join any two DNA fragments together. ATP provides the energy necessary for the ligase to reseal the sugar–phosphate backbone of DNA. (A) The joining of two DNA fragments produced by the restriction nuclease EcoRI. Note that the staggered ends produced by this enzyme enable the ends of the two fragments to base-pair correctly with each other, greatly facilitating their rejoining. This ligation reaction also reconstructs the original restriction-nuclease cutting site. (B) The joining of a DNA fragment produced by the restriction nuclease HaeIII to one produced by AluI. (C) The joining of DNA fragments produced by EcoRI and HaeIII, respectively, using DNA polymerase to fill in the staggered cut produced by EcoRI. Each DNA fragment shown in the figure is oriented so that its 5′ ends are the left end of the upper strand and the right end of the lower strand, as indicated in (A).

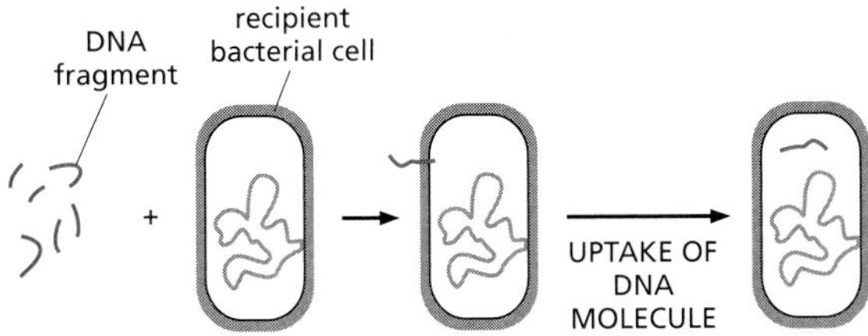

Figure 10–7 **Some bacteria can efficiently take up foreign DNA from their surroundings, a phenomenon called transformation.** In the laboratory, the DNA for transformation can come from any source, even from humans. Once inside the recipient cell, the donor DNA can become a part of the recipient genome (through the process of homologous recombination; see Chapter 6) or—in special cases—it can be maintained as a piece of DNA independent of the bacterial chromosome (as we discuss next).

DNA can be introduced into bacteria by a mechanism called **transformation**. Some bacteria naturally take up DNA molecules present in their surroundings by pulling the DNA through their cell membrane to the inside of the cell. The incoming DNA is often then incorporated into the genome by homologous recombination. The term 'transformation' originated from early observations of this phenomenon in which it appeared that one bacterial strain had become transformed into another. The deliberate transformation of one strain of bacterium with purified DNA derived from another strain provided one of the first experimental proofs that DNA is indeed the genetic material (see How We Know, pp. 174–176).

In a natural bacterial population, a source of DNA for transformation is provided by bacteria that have died and released their contents, including DNA, into the environment. In a test tube, however, bacteria such as *E. coli* can be coaxed to take up recombinant DNA that has been created in the laboratory (Figure 10–7). A great advantage to the experimenter is that naked DNA from any source, not just the DNA of the same bacterial species, can be introduced by this route. Bacterial transformation thus allows DNA from complex organisms, such as humans, to be studied easily in the laboratory.

Specialized Plasmid Vectors Are Used to Clone DNA

As mentioned previously, DNA introduced into bacteria under natural circumstances often becomes part of the bacterial genome. Investigators interested in cloning, however, find it easier to manipulate, copy, and purify their recombinant DNA when it is maintained as an independent molecule, separate from the bacterial chromosome. To maintain foreign DNA in a bacterial cell, a bacterial **plasmid** is used as a carrier, or *vector*. The plasmids typically used for gene cloning are relatively small circular DNA molecules of several thousand nucleotide pairs that can replicate inside a bacterium (Figure 10–8). A plasmid vector contains a replication origin, which enables it to replicate in the bacterial cell independently of the bacterial chromosome. It also has cleavage sites for restriction nucleases, so that the plasmid can be conveniently opened and a foreign DNA fragment can be inserted. Plasmids also usually contain a gene for some selectable property, such as antibiotic resistance, which enables bacteria that take up the recombinant DNA to be identified easily.

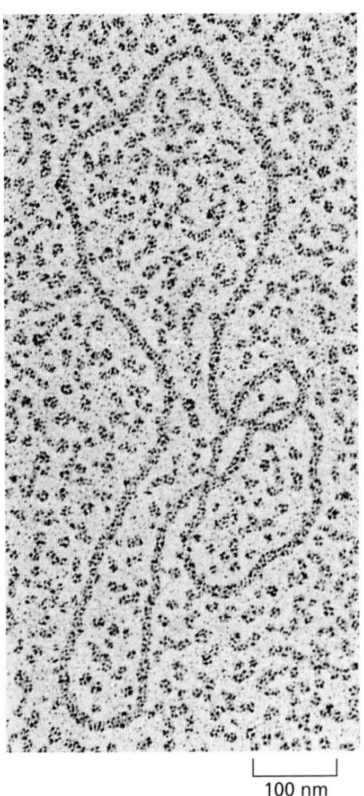

Figure 10–8 **Bacterial plasmids are commonly used as cloning vectors.** This circular, double-stranded DNA molecule consists of several thousand nucleotide pairs. The staining required to make the DNA visible in this electron micrograph makes the DNA appear much thicker than it actually is. (Courtesy of Brian Wells.)

Plasmids occur naturally in many bacteria. They were first recognized by physicians and scientists because they often carry genes that render their host bacteria resistant to one or more antibiotics. Indeed, historically potent antibiotics—penicillin, for example—are no longer effective against many of today's bacterial infections because plasmids that confer resistance to this class of antibiotic have spread among bacterial species by horizontal transfer (see Figure 9–17). The plasmids used for recombinant DNA research are basically streamlined versions of these naturally occurring plasmids.

To insert a piece of DNA into a plasmid, the purified plasmid DNA is exposed to a restriction nuclease that cleaves it in just one place, and the

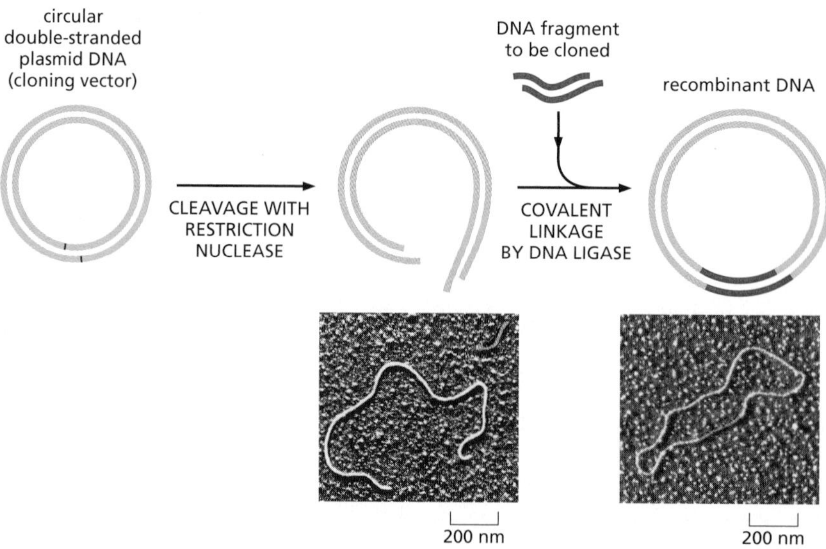

Figure 10–9 **A DNA fragment is inserted into a bacterial plasmid by using the enzyme DNA ligase.** The plasmid is cut open with a restriction nuclease (in this case one that produces staggered ends) and is mixed with the DNA fragment to be cloned (which has been prepared with the same restriction nuclease). DNA ligase and ATP are also added to the mix. The staggered ends base-pair, and the nicks in the DNA backbone are sealed by the DNA ligase to produce a complete recombinant DNA molecule. In the accompanying micrographs, we have colored the DNA fragment *red* to make it easier to see. (Micrographs courtesy of Huntington Potter and David Dressler.)

DNA fragment to be cloned is inserted into it by using DNA ligase (Figure 10–9). This recombinant DNA molecule is then introduced into a bacterium (usually *E. coli*) by transformation, and the bacterium is allowed to grow in nutrient broth, where it doubles in number every 30 minutes. Each time it doubles, the number of copies of the recombinant DNA molecule also doubles, and after just a day, hundreds of millions of copies of the plasmid will have been produced. The bacteria are then lysed, and the plasmid DNA is purified (by virtue of its small size) from the rest of the cell contents, including the large bacterial chromosome. The purified preparation of plasmid DNA will contain millions of copies of the original DNA fragment (Figure 10–10). This DNA fragment can be recovered by cutting it cleanly out of the plasmid DNA with the appropriate restriction enzyme and separating it from the plasmid DNA by gel electrophoresis (see Figure 10–3). These steps effectively allow the purification of a given stretch of DNA from the genome of any organism.

Genes Can Be Isolated from a DNA Library

We have seen how any DNA fragment can be produced in large numbers by inserting it into a bacterium. But how are these DNA fragments identified and chosen in the first place? More specifically, without knowing each gene's sequence in advance, how were individual human genes first isolated? As an example, we will describe the cloning of the gene for the human blood-clotting protein Factor VIII. Although the methods by which human genes are isolated today differ from the case described here, the example illustrates many of the general features of DNA cloning.

Defects in the gene for Factor VIII are the cause of the most common type of hemophilia—hemophilia A. This genetically determined disease has been recognized for over a thousand years and affects approximately

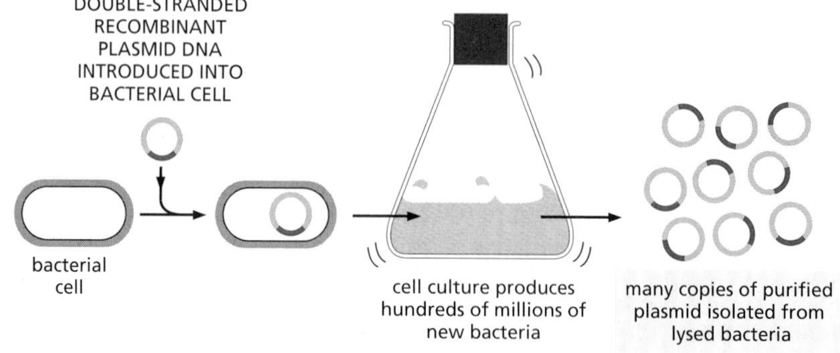

Figure 10–10 **A DNA fragment can be replicated inside a bacterial cell.** To clone a particular fragment of DNA, it is first inserted into a plasmid vector, as shown in Figure 10–9. The resulting recombinant plasmid DNA is then introduced into a bacterium via transformation; it can then be replicated many millions of times as the bacterium multiplies. For simplicity, the genome of the bacterial cell is not shown.

one in 10,000 males. People with hemophilia A fail to produce fully active Factor VIII, and thus have repeated episodes of uncontrolled bleeding. Until recently, the standard treatment for this disease was the injection of concentrated Factor VIII protein, pooled from many blood samples. Tragically, before blood was routinely screened for HIV—the virus that causes AIDS—this treatment exposed many hemophiliacs to HIV infection. The commercial production of pure Factor VIII using recombinant DNA technology thus offered a significant improvement in the treatment of hemophilia. This achievement required the cloning of the normal human gene that codes for Factor VIII and the piecing together of its coding sequence. This DNA sequence was then used to produce large amounts of the purified protein, as we shall describe.

Dealing with the 3×10^9 nucleotide pairs of the complete human genome is a daunting task, and the first step in cloning any human gene is to break up the total genomic DNA into smaller, more manageable pieces. One general procedure for doing this is outlined in Figure 10–11. Human DNA is first extracted from a tissue sample or cell culture and cleaved with a restriction nuclease, which produces millions of different DNA fragments. The mixture of DNA fragments is then ligated into plasmid vectors under conditions that favor the insertion of a single DNA fragment into each plasmid molecule. These recombinant plasmids are mixed with a culture of *E. coli* at a concentration that ensures that no more than one plasmid molecule is taken up by each bacterium. The collection of cloned DNA fragments in the resulting bacterial culture is known as a **DNA library**. In this case it is called a *genomic library,* as the DNA fragments are derived directly from the chromosomal DNA. If colonies derived from a single bacterium are isolated on Petri dishes, each bacterial colony will represent a clone of one particular stretch of human DNA. A collection of several million colonies in this library should thus represent all of the human genome.

To find a particular gene, we now face a problem analogous to that of entering a real library with millions of books and realizing that there is no card catalog or computerized listing to help us find the volume we want. How do we find a particular stretch of DNA (in our case the Factor VIII gene) in the vast human DNA library? The key is to exploit the hybridization properties of nucleic acids discussed earlier in this chapter. If we had a DNA probe for the Factor VIII gene, we could use it to find the matching clone in the library. But where did such a probe come from before the gene itself has been identified and cloned?

In the case of Factor VIII, a small amount of the protein was purified from human blood donors, using blood clotting as the biochemical assay. The partial amino acid sequence of the protein was determined (today this would be done using mass spectrometry; see Figure 4–45). Applying the genetic code in reverse, this amino acid sequence information was then used to deduce the partial nucleotide sequence of the gene. This nucleotide sequence was then chemically synthesized to create a DNA probe. Using this probe, the rare bacterial clones in the DNA library containing the complementary Factor VIII fragment were identified by hybridization (Figure 10–12).

When such a Factor VIII probe was first used on a human genomic library, it picked out a single complementary clone. The nucleotide sequence of this cloned DNA showed that it contained only a small portion of the Factor VIII gene, and the entire gene had to be laboriously pieced together. We now know that the Factor VIII gene is 180,000 nucleotide pairs long and contains many introns (see Figure 7–18B), so it is hardly surprising that no single genomic clone contained the entire gene.

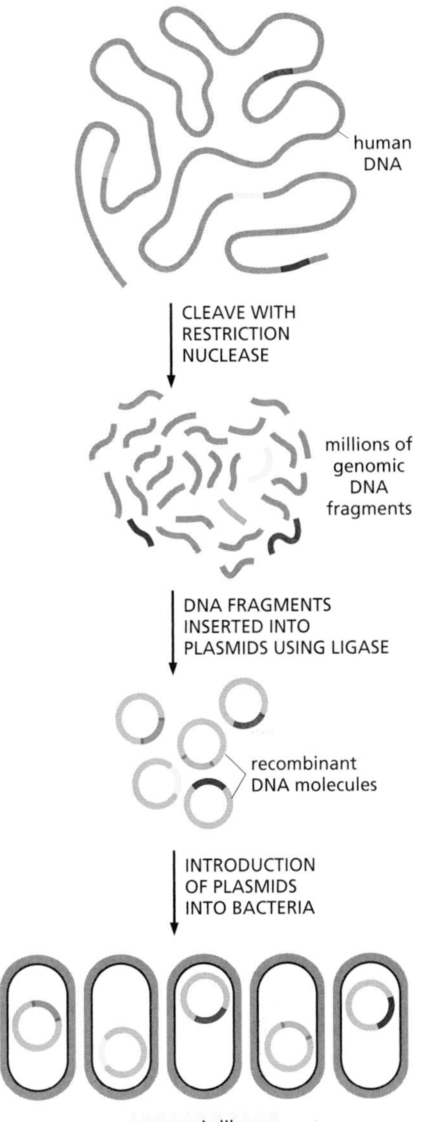

Figure 10–11 **Libraries of human genomic DNA fragments can be constructed using restriction nucleases and ligase.** A genomic library comprises a set of bacteria, each carrying a different small fragment of human DNA. For simplicity, only the *colored* DNA fragments are shown in the library; in reality, all of the *gray* fragments will also be represented.

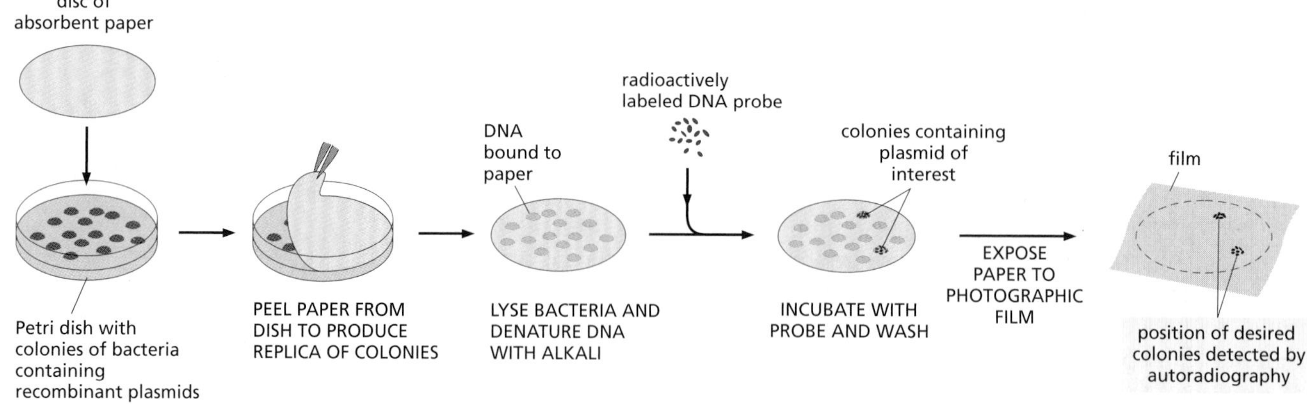

Figure 10–12 **A bacterial colony carrying a particular DNA clone can be identified by hybridization.** A replica of the arrangement of colonies on the Petri dish is made by pressing a piece of absorbent paper against the surface of the dish. This replica is treated with alkali (to lyse the cells and separate the plasmid DNA into single strands), and the paper is then hybridized to a highly radioactive DNA probe. Those bacterial colonies that have bound the probe are identified by autoradiography. Living cells containing the plasmid can then be isolated from the original dish.

Many human genes were originally identified and cloned using variations on the procedure described for Factor VIII. Now that the complete human genome sequence is known, however, cloning a particular gene is often much easier. For example, once the partial amino acid sequence of a protein of interest is known, it can, with the aid of computers, be used to directly search the human genome sequence for the gene that encodes it. Once the gene has been found in the electronic database, one can then design probes to pull the gene from a DNA library. It is even possible, as we describe shortly, to clone the gene directly from a sample of human DNA, bypassing the use of a library entirely.

cDNA Libraries Represent the mRNA Produced by a Particular Tissue

For many applications, it is advantageous to obtain a gene in a form that contains only the coding sequence; that is, a form that lacks the intron DNA. For example, in the case of the Factor VIII gene the complete genomic clone—introns and exons—is so large and unwieldy that it is necessary to work with the gene in pieces. Also, if one wanted to deduce the complete amino acid sequence of the Factor VIII protein solely from the nucleotide sequence of its gene, it would be difficult to figure out where each exon begins and ends; after all, the great majority of the gene sequence is intron sequences (see Figure 7–18B). Fortunately, it is relatively simple to isolate a gene free of all its introns. For this purpose, a different type of DNA library, called a *cDNA library,* is used.

A human cDNA library is similar to a human genomic library in that it also contains numerous clones containing many different human DNA sequences. But it differs in one important respect. The DNA that goes into a cDNA library is not genomic DNA, but instead is DNA copied from the mRNAs present in a particular tissue or cell culture. To prepare a **cDNA** library, the total mRNA is extracted from the cells, and DNA copies of the mRNA molecules are produced by the enzyme reverse transcriptase (Figure 10–13). These complementary DNA—or cDNA—molecules are then cloned, just like the genomic DNA fragments described earlier, to produce the cDNA library. For example, using such a cDNA library prepared from liver, the organ that normally produces Factor VIII, it was possible to isolate the complete coding sequence of the Factor VIII gene, devoid of introns, and present on one piece of DNA. The Factor VIII cDNA was isolated from a cDNA library, using a portion of the genomic Factor VIII DNA as a probe, by the procedure shown in Figure 10–12. We will see in the final part of this chapter how the coding sequence of the gene was used to produce purified human Factor VIII protein on a commercial scale.

There are several important differences between genomic DNA clones and cDNA clones, as illustrated in Figure 10–14. Genomic clones represent a random sample of all of the DNA sequences found in an organism's genome and, with very rare exceptions, will contain the same sequences regardless of the cell type from which the DNA came. Also, genomic clones from eucaryotes contain large amounts of repetitive DNA sequences, introns, regulatory DNA, and spacer DNA; sequences that code for proteins will make up only a few percent of the library (see Figure 9–29). By contrast, cDNA clones contain predominantly coding sequences, and only those for genes that have been transcribed into mRNA in the tissue from which the RNA came. As the cells of different tissues produce distinct sets of mRNA molecules, a different cDNA library will be obtained for each type of tissue. Patterns of gene expression change during development, so cDNA libraries will also reflect the genes expressed by cells at different stages in their development.

By far the most important advantage of cDNA clones is that they contain the uninterrupted coding sequence of the gene. Thus, if the aim of cloning the gene is either to deduce the amino acid sequence of the protein from the DNA or to produce the protein in bulk by expressing the cloned gene in a bacterial or yeast cell (neither of which can remove introns from mammalian RNA transcripts), it is essential to start with cDNA.

The main advantage of genomic clones, on the other hand, is that they contain introns as well as exons, and they include the regulatory sequences that determine when and where genes are expressed. For this reason, genomic clones are used to determine the complete nucleotide sequences of genomes, as we see later in the chapter.

QUESTION 10–3

Discuss the following statement: "From the nucleotide sequence of a cDNA clone, the complete amino acid sequence of a protein can be deduced by applying the genetic code. Thus, protein biochemistry has become superfluous because there is nothing more that can be learned by studying the protein."

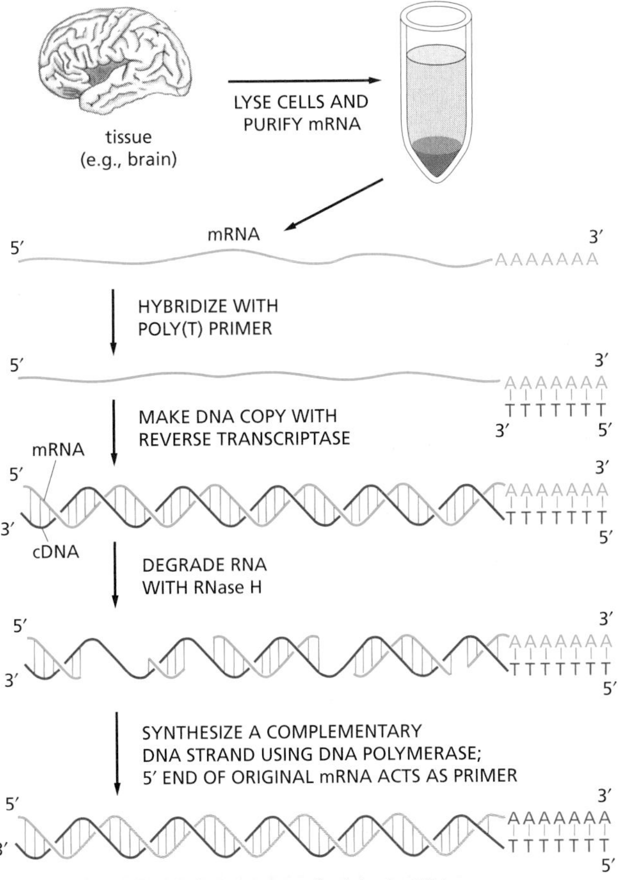

Figure 10–13 **Complementary DNA (cDNA) can be prepared from mRNA.** Total mRNA is extracted from a particular tissue, and DNA copies (cDNA) of the mRNA molecules are produced by the enzyme reverse transcriptase (see Figure 6–38). For simplicity, the copying of just one of these mRNAs into cDNA is illustrated here.

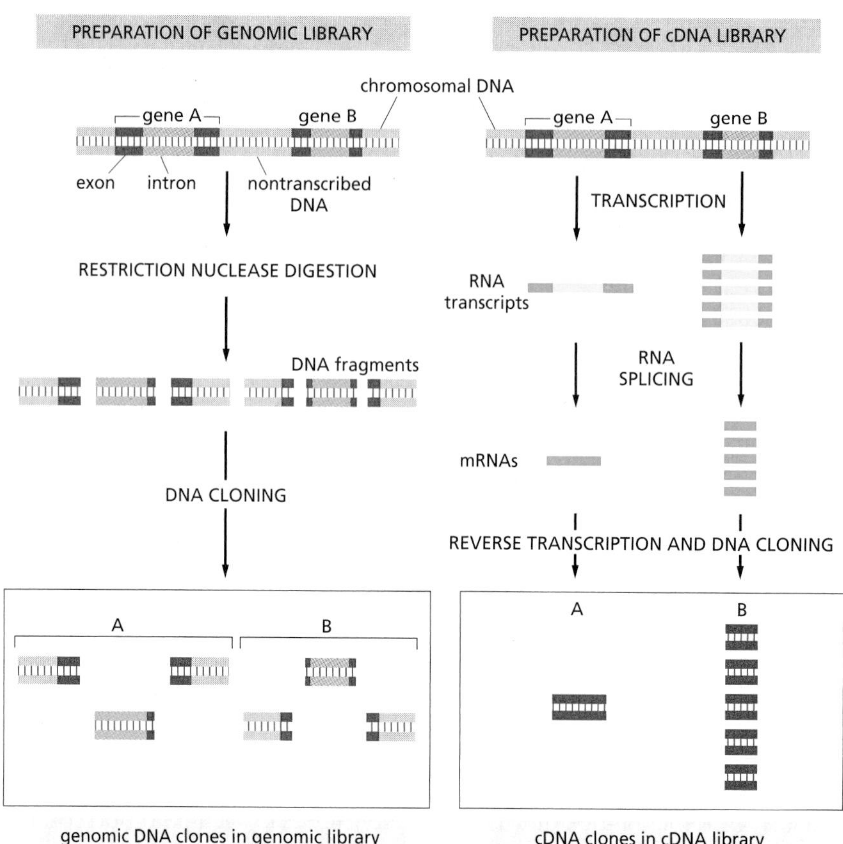

Figure 10–14 **Genomic DNA clones and cDNA clones derived from the same region of DNA are different.** In this example, gene A is infrequently transcribed, whereas gene B is frequently transcribed, and both genes contain introns *(orange)*. In the genomic DNA library, both introns and the nontranscribed DNA *(gray)* are included in the clones, and most clones will contain only part of the coding sequence of a gene *(red)*. In the cDNA clones the intron sequences have been removed by RNA splicing during the formation of the mRNA *(blue)*, and a continuous coding sequence is therefore present in each clone. Because gene B is transcribed more frequently than gene A in the cells from which the cDNA library was made, it will be represented much more often than A in the cDNA library. In contrast, A and B should be represented equally in the genomic library.

The Polymerase Chain Reaction Amplifies Selected DNA Sequences

Cloning via DNA libraries was once the only route to gene isolation, and it is still used in sequencing whole genomes and in dealing with very large genes. However, a method known as the **polymerase chain reaction (PCR)** provides a much quicker and simpler alternative for many cloning applications, particularly for those organisms whose complete genome sequence is known. Today the majority of cloning is done using PCR.

Invented in the 1980s, PCR can be carried out entirely *in vitro* without the use of cells. Using this technique, a given nucleotide sequence can be selectively and rapidly replicated in large amounts from any DNA sample that contains it. For example, the polymerase chain reaction is now widely used to provide large amounts of any gene from a small sample of human DNA. It also has many other applications, including amplifying DNA for use in diagnostic tests for disease genes and in forensic medicine, as we discuss shortly.

PCR is based on the use of DNA polymerase to copy a DNA template in repeated rounds of replication. The polymerase is guided to the sequence to be copied by short DNA primers that are added to the reaction and hybridize to the DNA template at the beginning and end of the desired DNA sequence. These primers provide the 3′ ends for the DNA polymerase to start replication on each DNA strand. Primers must be designed by the experimenter and synthesized to order, so PCR can be used only to clone a DNA whose beginning and end sequences are known. During each round of replication, the two strands of the double-stranded DNA template are separated and copied independently. Figure 10–15 shows the separate steps in the first round of replication. After many such rounds of replication, many copies of the sequence—typically billions—will have been made (Figure 10–16). PCR is extremely sensitive; it can detect a

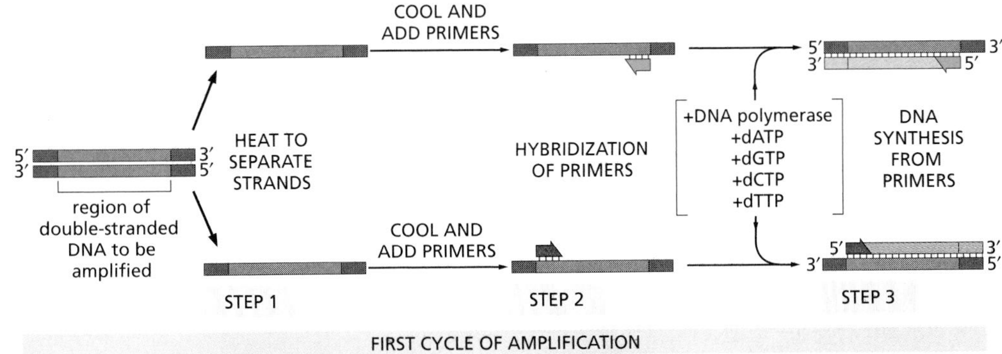

Figure 10–15 **PCR primers direct the amplification of the desired piece of DNA.** Knowledge of the DNA sequence to be amplified is used to design two short, synthetic DNA molecules, each complementary to the sequence on one strand of the DNA double helix at opposite ends of the region to be amplified. These DNA molecules serve as primers for *in vitro* DNA synthesis, which is carried out by a DNA polymerase, and they determine the segment of the DNA that is amplified. Each cycle of PCR includes three steps. First, the double-stranded DNA is heated briefly to separate the two strands (step 1). After strand separation, cooling of the DNA in the presence of a large excess of the two primers allows the primers to hybridize to complementary sequences in the two DNA strands (step 2). This mixture is then incubated with DNA polymerase and the four deoxyribonucleoside triphosphates so that DNA is synthesized, starting from the two primers (step 3). The cycle is then begun again by a heat treatment to separate the newly synthesized DNA strands. The technique depends on the use of a special DNA polymerase isolated from a thermophilic bacterium; this polymerase is stable at much higher temperatures than eucaryotic DNA polymerases, so it is not denatured by the heat treatment shown in step 1. It therefore does not have to be added again after each cycle of PCR.

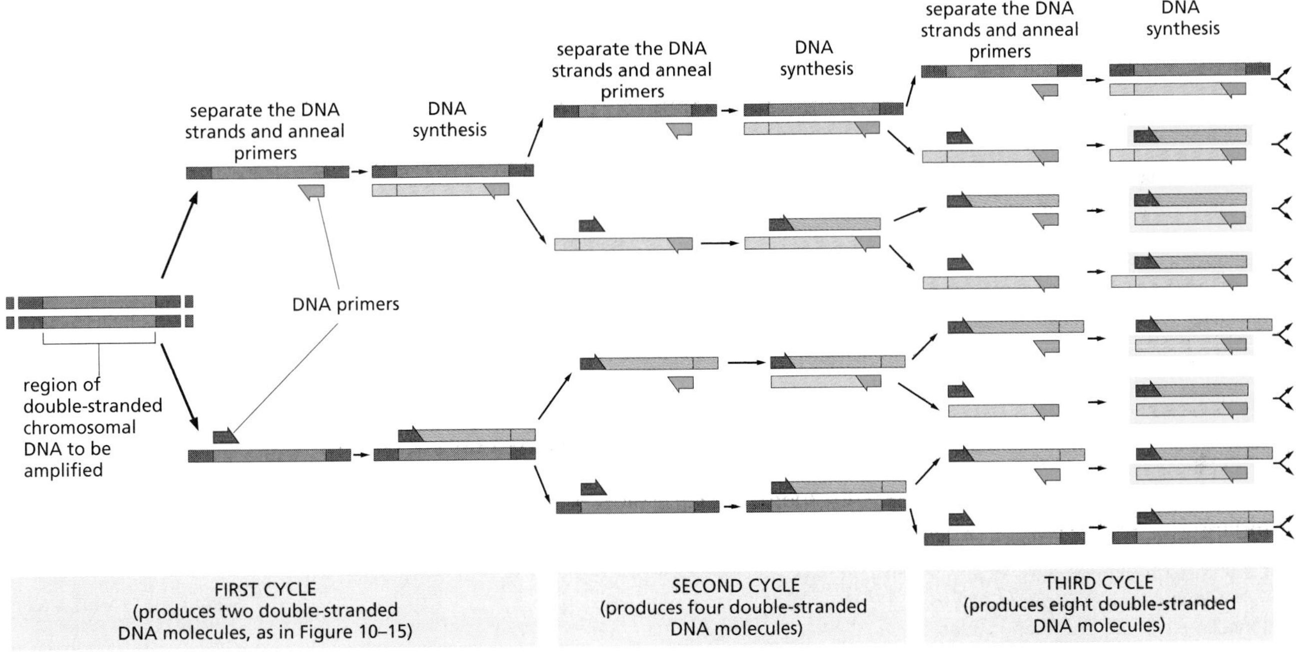

Figure 10–16 **PCR uses repeated rounds of strand separation, hybridization, and synthesis to amplify DNA.** As the procedure outlined in Figure 10–15 is carried out over and over again, the newly synthesized fragments serve as templates in their turn. Each cycle doubles the amount of DNA synthesized in the previous cycle, and within a few cycles the predominant DNA is identical to the sequence bracketed by and including the two primers in the original template. In the example illustrated here, three cycles of reaction produce 16 DNA chains, 8 of which (boxed in *yellow*) are the same length as and correspond exactly to one or the other strand of the original bracketed sequence shown at the far left; the other strands contain extra DNA downstream of the original sequence, which is replicated in the first few cycles. After four more cycles, 240 of the 256 DNA chains will correspond exactly to the original sequence, and after several more cycles, essentially all of the DNA strands will have this unique length. Although all the DNA present at the beginning of the PCR still remains, it is so vastly outnumbered that its presence is insignificant. In practice, 20–30 cycles are required for useful DNA amplification. Each cycle takes only about 5 minutes, and automation of the whole procedure now enables cell-free cloning of a DNA fragment in a few hours, compared with the several days required for standard cloning methodologies. The whole procedure is shown in Movie 10.1.

single copy of a DNA sequence in a sample by amplifying it so much that it becomes detectable by, for example, staining after separation by gel electrophoresis (see Figure 10–3).

There are several especially useful applications of PCR. First, PCR is now the method of choice for cloning relatively short DNA fragments (say, under 10,000 nucleotide pairs) from a cell. The original template for the reaction can be either DNA or RNA, so PCR can be used to obtain either a full genomic copy (complete with introns and exons) or a cDNA copy of the gene (Figure 10–17). The beauty of this method is that genes can be cloned directly from any piece of DNA or RNA without the time and effort needed to first construct a DNA library.

Another use for PCR, which relies on its extraordinary sensitivity, is the detection of infections by pathogens at very early stages. In this case, short sequences complementary to the pathogen's genome are used as primers, and following many cycles of amplification, the presence or absence of even a few copies of an invading genome in a sample of blood can be ascertained (Figure 10–18). For many infections, PCR is the most sensitive method of detection; already it is replacing the use of antibodies against surface proteins to detect the presence of pathogens in human samples.

Finally, PCR has great potential in forensic medicine. Its extreme sensitivity makes it possible to work with a very small sample—minute traces of blood and tissue that may contain the remnants of only a single cell—

QUESTION 10–4

A. If the PCR reaction shown in Figure 10–16 is carried through an additional two rounds of amplification, how many of the DNA fragments labeled in gray, green, or red or outlined in yellow are produced? If many additional cycles are carried out, which fragments will predominate?

B. Assume you start with one double-stranded DNA molecule and amplify a 500-nucleotide-pair sequence contained within it. Approximately how many cycles of PCR amplification will you need to produce 100 ng of this DNA? 100 ng is an amount that can be easily detected after staining with a fluorescent dye. (Hint: for this calculation, you need to know that each nucleotide has an average molecular weight of 330 g/mole.)

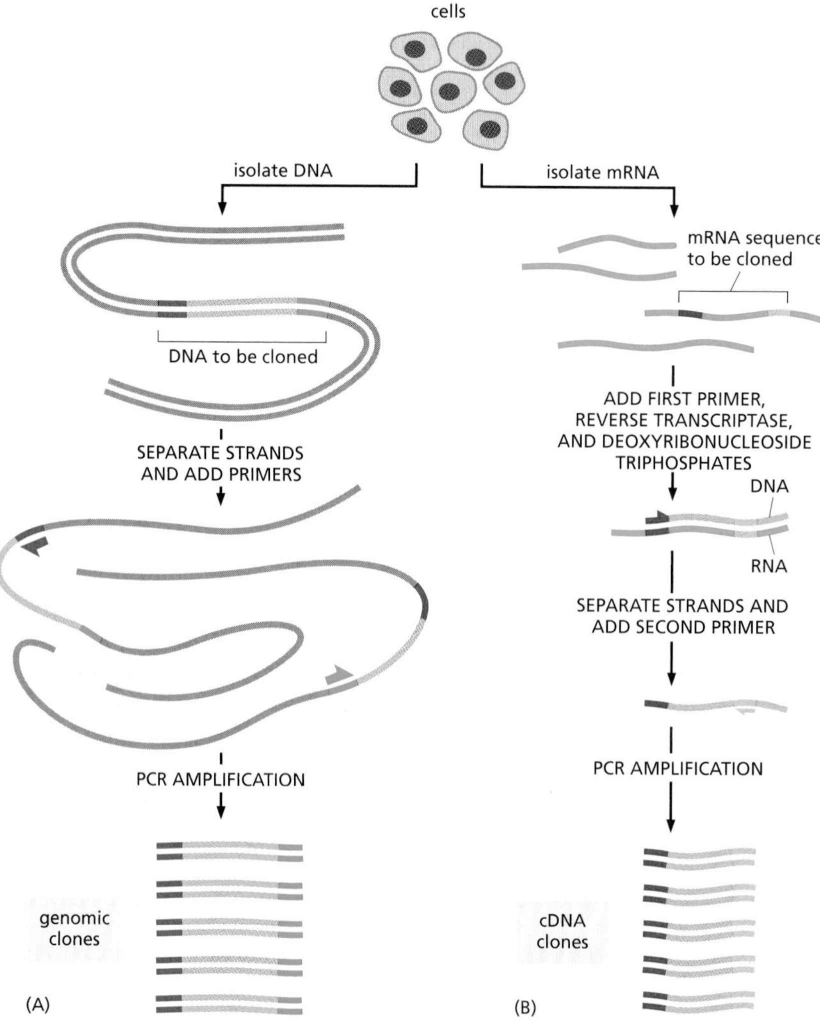

Figure 10–17 **PCR can be used to obtain genomic or cDNA clones.** (A) To obtain a genomic clone using PCR, chromosomal DNA is first purified from cells. PCR primers that flank the stretch of DNA to be cloned are added, and many cycles of the PCR reaction are completed (see Figure 10–16). Because only the DNA between (and including) the primers is amplified, PCR provides a way to obtain selectively a short stretch of chromosomal DNA in an effectively pure form. (B) To use PCR to obtain a cDNA clone of a gene, mRNA is first purified from cells. The first primer is then added to the population of mRNAs, and reverse transcriptase is used to make a complementary DNA strand. The second primer is then added, and the single-stranded DNA molecule is amplified through many cycles of PCR.

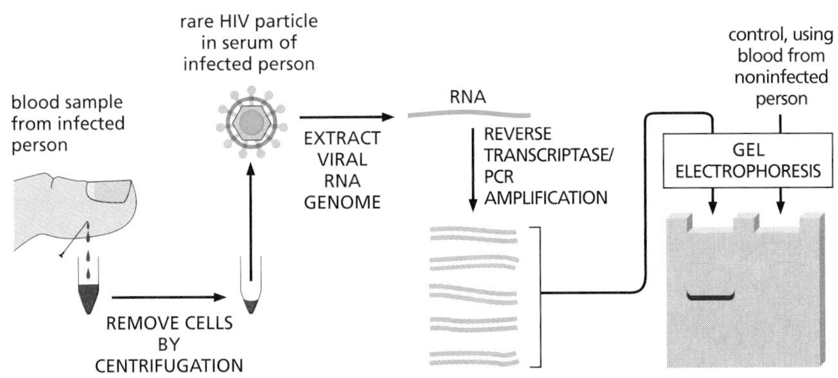

Figure 10–18 **PCR can be used to detect the presence of a viral genome in a sample of blood.** Because of its ability to enormously amplify the signal from every single molecule of nucleic acid, PCR is an extraordinarily sensitive method for detecting trace amounts of virus in a sample of blood or tissue without the need to purify the virus. For HIV, the virus that causes AIDS, the genome is a single-stranded molecule of RNA. In addition to HIV, many viruses that infect humans are now monitored in this way.

and still obtain a *DNA fingerprint* of the person from whom it came. The genome of each human (with the exception of identical twins) differs in DNA sequence from the genome of every other human; DNA amplified by PCR using a particular primer pair is therefore quite likely to differ in sequence from one individual to another. Using a carefully selected set of primer pairs that cover known highly variable parts of the human genome, PCR can generate a distinctive DNA fingerprint for each individual (Figure 10–19).

DECIPHERING AND EXPLOITING GENETIC INFORMATION

So far in this chapter we have assumed that the function of a gene we may wish to study is understood—at least in rough outline. But suppose that you have discovered a gene that codes for a protein of unknown function. How do you determine what the protein does? Now that genome-sequencing projects are rapidly identifying new genes from the DNA sequence alone, this has become a common question in cell biology. For example, of the 25,000 human genes that have been identified (see How We Know, pp. 318–319) more than 10,000 have functions that remain unknown.

The procedures described so far enable biologists to obtain large amounts of DNA in a form that is easy to work with in the laboratory. Whether it is present as fragments stored in a DNA library or as a collection of PCR products nestled in the bottom of a test tube, this DNA provides the raw material for perhaps the most exciting experiments: those designed to unravel how individual genes—and the RNA molecules and proteins they encode—function inside living organisms.

At this point in the process biologists can be especially creative, because there are as many ways to approach the problem of gene function as there are scientists who wish to tackle it. The techniques that investigators use to explore the activity of a gene often depend on their background and training: geneticists might choose to engineer mutant organisms in which the activity of the gene has been disrupted; biochemists might use the gene to produce large amounts of protein to determine its enzymatic activities and three-dimensional structure; and those who are more computationally inclined might begin by searching computer databases for genes with similar sequences or expression patterns. Although these approaches differ in terms of the expertise and equipment needed to execute them, all are designed to yield clues about what a gene does inside a cell or organism.

In this section we present a few of the methods that are used to determine the function of a gene. Because all of these approaches begin with

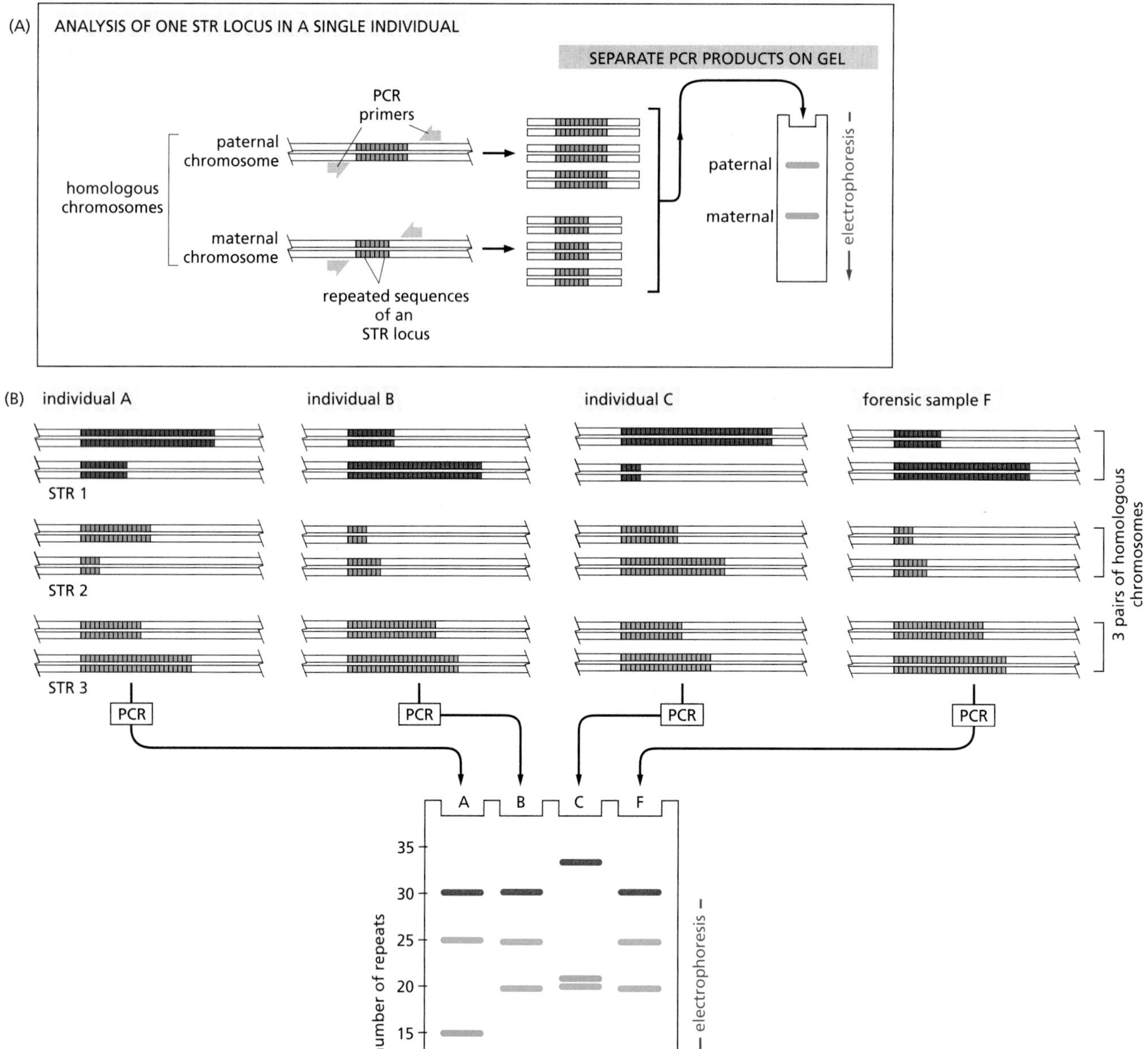

Figure 10–19 **PCR is used in forensic science to distinguish one individual from another.** (A) The DNA sequences used in this analysis are short tandem repeats (STRs) composed of sequences such as CACACA... or GTGTGT..., which are found in various positions (loci) in the human genome. The number of repeats in each STR is highly variable in the population, ranging from 4 to 40 in different individuals. Because of the variability in these sequences, individuals will usually inherit a different variant of each STR locus from their mother and from their father; two unrelated individuals therefore do not usually contain the same pair of sequences. A PCR reaction using primers that bracket the locus produces a pair of bands of amplified DNA from each individual, one band representing the maternal STR variant and the other representing the paternal STR variant. The length of the amplified DNA, and thus its position after electrophoresis, will depend on the exact number of repeats at the locus. (B) In the schematic example shown here, the same three STR loci are analyzed from three suspects (individuals A, B, and C), producing six bands for each person after polyacrylamide gel electrophoresis. Although different people can have several bands in common, the overall pattern is quite distinctive for each person. The band pattern can therefore serve as a *fingerprint* to identify an individual nearly uniquely. The fourth lane (F) contains the products of the same PCR reactions carried out on a forensic DNA sample. DNA can be obtained from a single hair or a tiny spot of blood left at the crime scene. The more loci that are examined, the more confident we can be about the results. When examining the variability at 5–10 different STR loci, the odds that two random individuals would share the same fingerprint by chance are approximately one in 10 billion. In the case shown here, individuals A and C can be eliminated from inquiries, while B remains a clear suspect. A similar approach is now used routinely for paternity testing.

the nucleotide sequence of a gene, we begin by describing the techniques used to sequence a stretch of DNA. We then discuss how scientists can begin to interpret the information encoded in this nucleotide sequence. Next, we review how, once a gene is in hand, recombinant DNA technology can be used to produce enough of its RNA or protein product to conduct structural and functional studies. Finally, we describe several techniques for investigating what a gene does inside a cell, tissue, or even a whole plant or animal. These methods have revolutionized all aspects of cell biology by providing new ways to study the functions of genes, RNA molecules, and proteins.

DNA Can Be Rapidly Sequenced

In the late 1970s, researchers developed methods that allow the nucleotide sequence of any purified DNA fragment to be determined simply and quickly. These techniques have made it possible to determine the complete nucleotide sequences of hundreds of thousands of genes and the complete genome sequences of many organisms, including the budding yeast *Saccharomyces cerevisiae*, the nematode worm *Caenorhabditis elegans*, the fruit fly *Drosophila melanogaster*, the model plant *Arabidopsis thaliana*, and the dog, rat, chimpanzee, gorilla, and human.

Several schemes for sequencing DNA have been developed, but the most widely used is **dideoxy DNA sequencing**, a method based on DNA synthesis carried out *in vitro* in the presence of special, chain-terminating dideoxyribonucleoside triphosphates. In this technique, DNA polymerase is used to make partial copies of the DNA fragment to be sequenced. These DNA replication reactions are performed under conditions that ensure that the new DNA strands terminate when a given nucleotide (A, G, C, or T) is reached (Figure 10–20). These reactions ultimately produce a collection of different DNA copies that terminate at every position in the original DNA, and thus differ in length by a single nucleotide. These DNA copies can be separated on the basis of their length by gel electrophoresis, and the nucleotide sequence of the original DNA can be determined from the order of these DNA fragments in the gel (Figure 10–21).

Although the same basic method is still used today, many improvements have been made. DNA sequencing is now completely automated: robotic

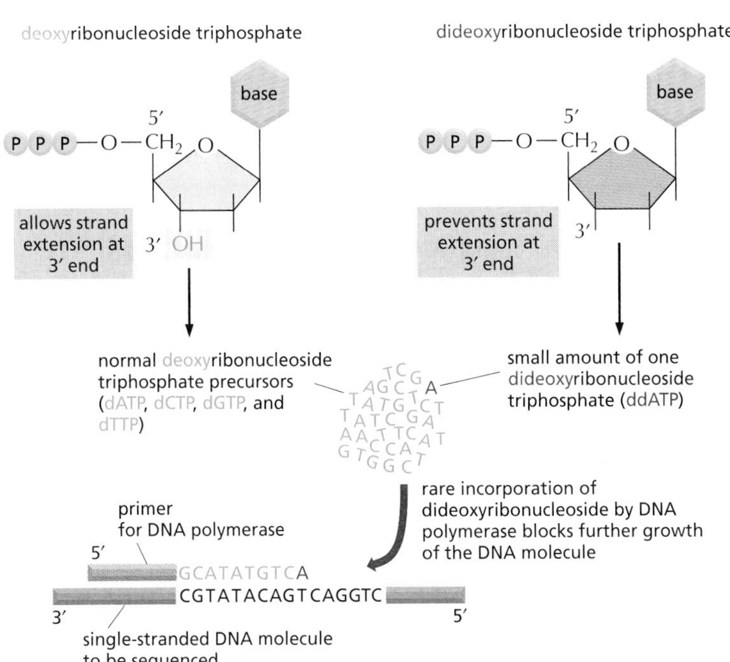

Figure 10–20 **The dideoxy method of sequencing DNA relies on chain-terminating dideoxynucleoside triphosphates.** These dideoxyribonucleoside triphosphates, derivatives of the normal deoxyribonucleoside triphosphates, lack the 3′ hydroxyl group. Purified DNA is synthesized *in vitro* in a mixture that contains single-stranded molecules of the DNA to be sequenced *(gray)*, the enzyme DNA polymerase, a short primer DNA *(orange)* to enable the polymerase to start replication, and the four deoxyribonucleoside triphosphates (dATP, dCTP, dGTP, dTTP: *blue* A, C, G, and T). If a dideoxyribonucleotide analog of one of these nucleotides (here, a *red* ddATP) is also present in the nucleotide mixture, it can become incorporated into a growing DNA chain. The chain then lacks a 3′-OH group, the addition of the next nucleotide is blocked, and the DNA chain terminates at that point. The dideoxy ATP (*red* A) competes with an excess of the normal deoxyATP (*blue* As), so that ddATP is occasionally incorporated, at random, into a growing DNA strand. This reaction mixture will eventually produce a set of DNAs of different lengths complementary to the template DNA that is being sequenced and terminating at each of the different As (see Figure 10–21). Only one of the many potential products is shown here.

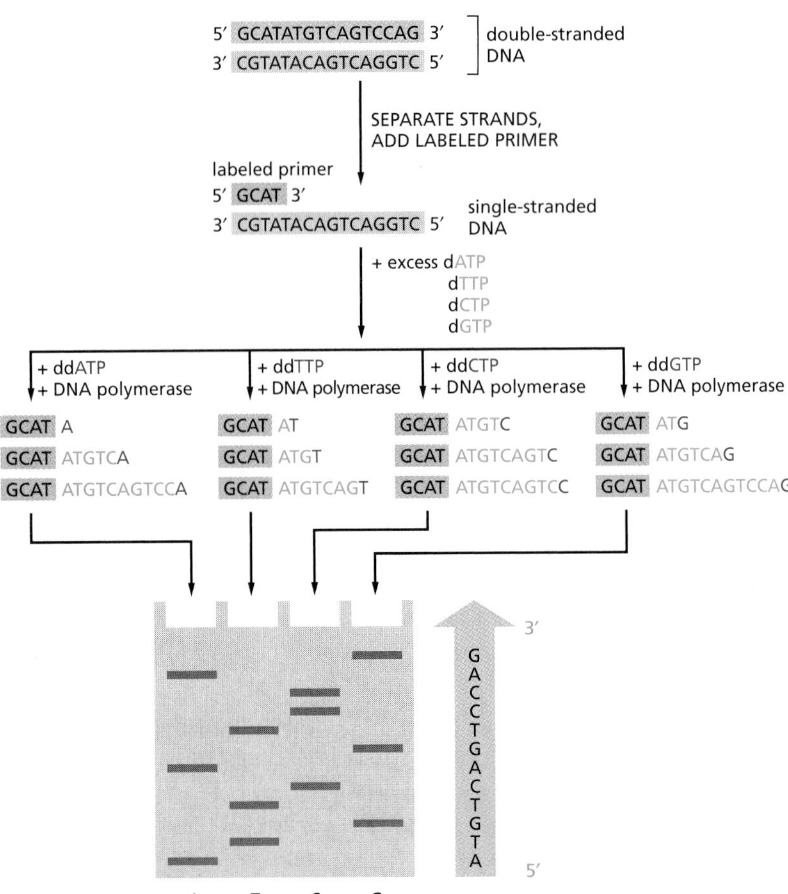

Figure 10–21 **The dideoxy method produces a collection of DNA molecules that differ by a single nucleotide.** To determine the complete sequence of a DNA fragment, the double-stranded DNA is first separated into its single strands and one of the strands is used as the template for sequencing. The four different chain-terminating dideoxyribonucleoside triphosphates (ddATP, ddCTP, ddGTP, ddTTP, again shown in *red)* are used in four separate DNA synthesis reactions on copies of the same single-stranded DNA template *(gray).* Each reaction produces a set of DNA copies that terminate at different points in the sequence. The products of these four reactions are separated by electrophoresis in four parallel lanes of a polyacrylamide gel (labeled here A, T, C, and G). The newly synthesized fragments are detected by a label (either radioactive or fluorescent) that has been incorporated either into the primer or into one of the deoxyribonucleoside triphosphates used to extend the DNA chain. In each lane, the bands represent fragments that have terminated at a given nucleotide (e.g., A in the leftmost lane) but at different positions in the DNA. By reading off the bands in order, starting at the bottom of the gel and working across all lanes, the DNA sequence of the newly synthesized strand can be determined. The sequence, which is given in the *green arrow* to the right of the gel, is identical to that of the 5′ → 3′ strand *(green)* of the original double-stranded DNA.

devices mix the reagents and then load, run, and read the order of the nucleotide bases from the gel. This process is facilitated by the use of chain-terminating nucleotides that are each tagged with a different-colored fluorescent dye; all four synthesis reactions can thus be performed in the same tube, and the products can be separated in a single lane on a gel. A detector positioned near the bottom of the gel reads and records the color of the fluorescent label on each band as it moves past, and a computer stores the sequence for subsequent analysis (Figure 10–22). How researchers can piece together a complete genome sequence from this information is described in How We Know, pp. 348–349.

Of course, obtaining the sequence of a gene or an entire genome is only a beginning. As we discussed in the previous chapter, it is a great challenge

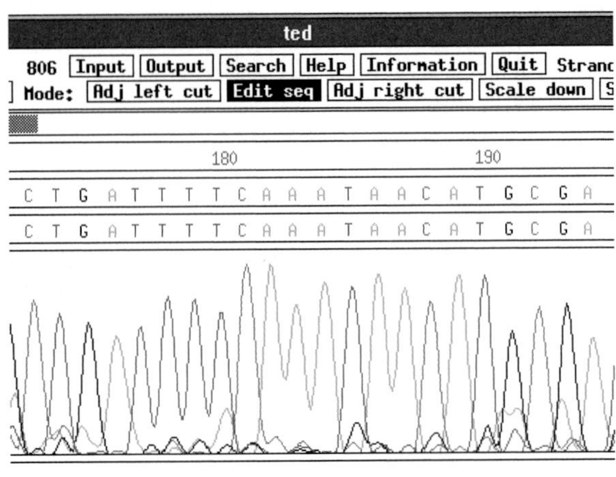

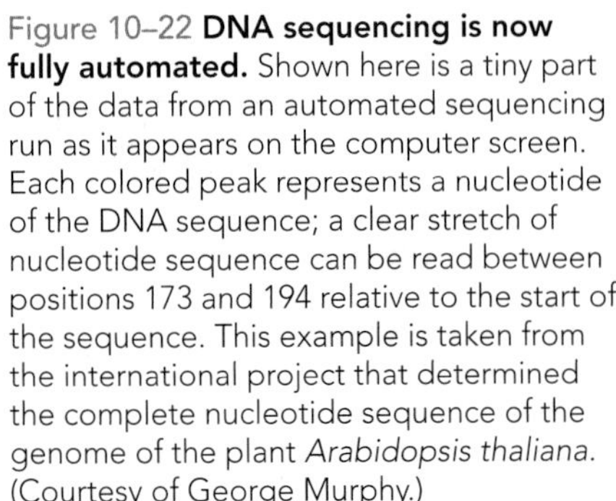

Figure 10–22 **DNA sequencing is now fully automated.** Shown here is a tiny part of the data from an automated sequencing run as it appears on the computer screen. Each colored peak represents a nucleotide of the DNA sequence; a clear stretch of nucleotide sequence can be read between positions 173 and 194 relative to the start of the sequence. This example is taken from the international project that determined the complete nucleotide sequence of the genome of the plant *Arabidopsis thaliana.* (Courtesy of George Murphy.)

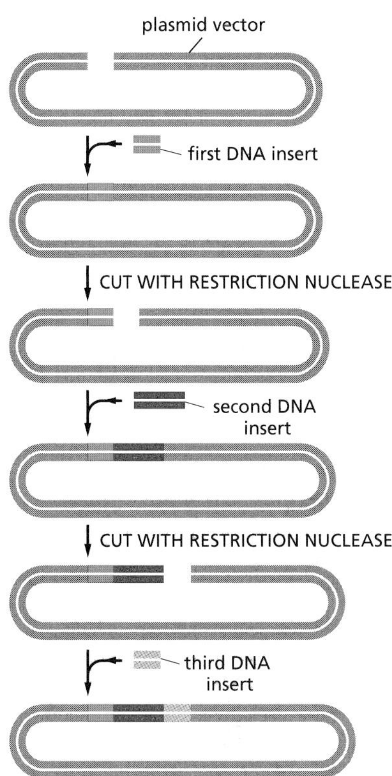

Figure 10–23 **Serial DNA cloning can be used to splice together a set of DNA fragments derived from different sources.** After each DNA insertion step, the recombinant plasmid is cloned to purify and amplify the new DNA (see Figure 10–10). The recombinant molecule is then cut once with a restriction nuclease, as indicated, and used as a cloning vector for the next DNA fragment.

to look at a nucleotide sequence and determine, for example, where a gene begins and ends, and which bits are important for regulating its activity. Even with this information, it is not necessarily apparent what role the gene plays in the physiology of an organism. To do that, biologists need to manipulate each gene more directly.

Completely Novel DNA Molecules Can Be Constructed

Studying gene function usually involves the construction of recombinant DNA molecules. As we have discussed, such recombinant molecules are generally made by using DNA ligase to join together two pieces of DNA—including DNA that comes from different organisms or DNA that has been chemically synthesized (see Figure 10–6). By repeated DNA cloning, DNAs from any source can be isolated and joined together in various combinations to produce DNA molecules of any desired sequence (Figure 10–23).

The ability to generate custom-designed DNA molecules allows scientists to manipulate genes in a variety of useful ways. They can introduce mutations that alter the activity of a gene, an approach that could reveal the normal function of the gene. They can also engineer hybrid proteins that contain, for example, a fluorescent marker protein attached to the protein of interest. Such protein fusions allow scientists to visualize where a protein is located within a cell or organism. Finally, in one of the most immediately practical uses of recombinant technology, scientists can direct the high-level production of what are normally rare cellular proteins, as we shall see next.

Rare Proteins Can Be Made in Large Amounts Using Cloned DNA

Most of the thousands of different proteins in a eucaryotic cell, including many with crucially important functions, are present in very small amounts. For these proteins, it used to be extremely difficult, if not impossible, to obtain enough of them in pure form to be able to study them. One of the most important contributions of DNA cloning and genetic engineering to cell biology is that they make it possible to produce any protein, including the rare ones, in nearly unlimited amounts.

This high-level production is usually accomplished by using specially designed vectors known as *expression vectors*. Unlike the cloning vectors discussed earlier, these vectors also include appropriate transcription and translation signals so that an inserted gene is expressed at very high levels (Figure 10–24). Different expression vectors are designed for use in

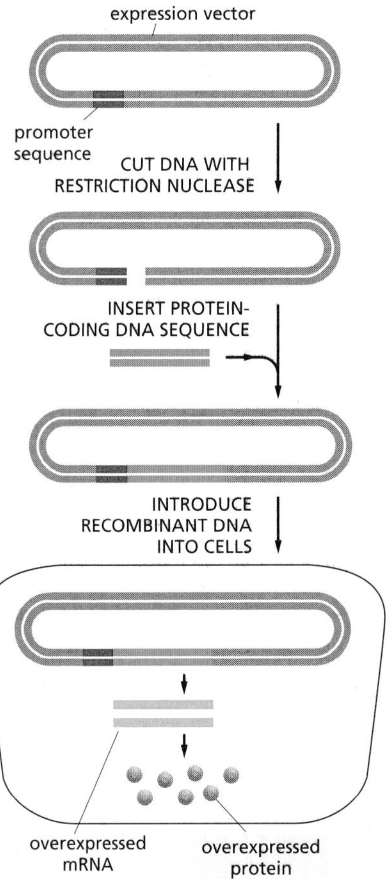

Figure 10–24 **Large amounts of a protein can be produced from a protein-coding DNA sequence cloned into an expression vector and introduced into cells.** A plasmid vector has been engineered to contain a highly active promoter, which causes unusually large amounts of mRNA to be produced from an adjacent protein-coding gene inserted into the plasmid vector. Depending on the characteristics of the cloning vector, the plasmid is introduced into bacterial, yeast, insect, or mammalian cells, where the inserted gene is efficiently transcribed and translated into protein.

SEQUENCING THE HUMAN GENOME

When DNA sequencing techniques became fully automated, determining the order of the nucleotides in a piece of DNA went from being an elaborate Ph.D. thesis project to a routine laboratory chore. Feed DNA into the sequencing machine, add the necessary reagents, and out comes the sought-after result: the order of As, Ts, Gs, and Cs. Nothing could be simpler.

So why was sequencing the human genome such a formidable task? Largely because of its size. Today's DNA sequencing methods are limited by the physical size of the gel that is used to separate the labeled fragments (see Figure 10–21). At most, only a few hundred nucleotides can be read from a single gel. How, then, do you handle a genome that contains billions of nucleotides?

The solution is to break the genome into fragments and sequence those smaller pieces. The main challenge then comes in piecing the short fragments together in the correct order to yield a comprehensive sequence of a whole chromosome, and ultimately a whole genome. To accomplish this breakage and reassembly, researchers have generally adopted two different strategies for sequencing genomes: the shotgun method and the clone-by-clone approach.

Shotgun sequencing

The most straightforward approach to sequencing a genome is to break it into random fragments, sequence each of the fragments, and then use a powerful computer to order these pieces using sequence overlaps to guide the assembly (Figure 10–25). This approach is called the shotgun sequencing strategy. As an analogy, imagine shredding several copies of *Essential Cell Biology (ECB)* into small pieces, mixing the pieces up, and then trying to put one whole copy of the book back together again by matching up the words or phrases or sentences that appear on the different slips of paper. (Several copies would be needed to generate the overlap necessary for reassembly.) It could be done, but it would be much easier if the book were, say, only two pages long.

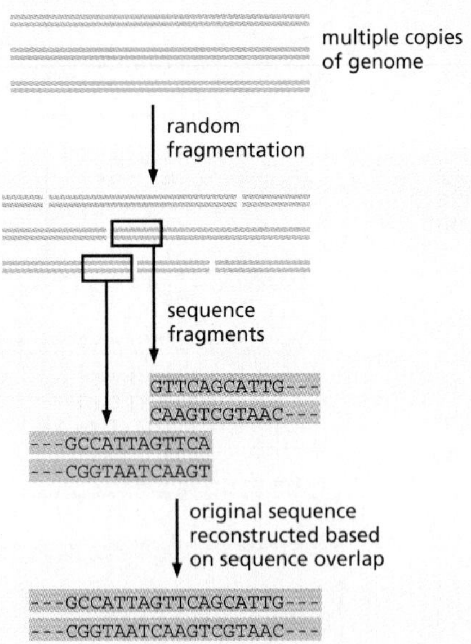

Figure 10–25 **Shotgun sequencing is the method of choice for small genomes.** The genome is first broken into much smaller, overlapping fragments. Each fragment is then sequenced, and the genome is assembled based on overlapping sequences.

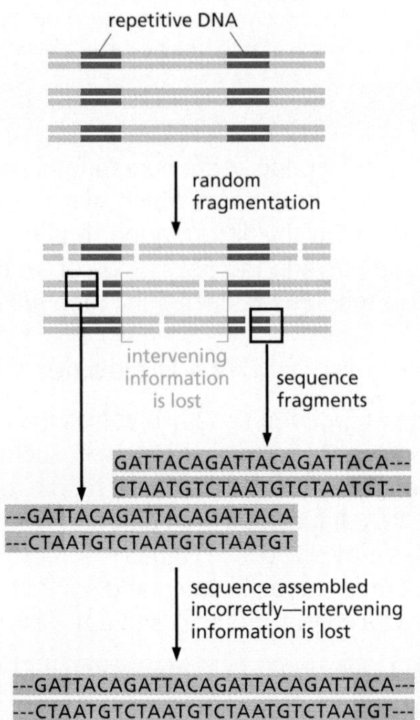

Figure 10–26 **Repetitive sequences make correct assembly difficult.** In this example, the DNA contains two segments of repetitive DNA, each made of many copies of the sequence GATTACA. When the resulting sequences are examined, two fragments from different parts of the DNA appear to overlap. Assembling these sequences incorrectly would result in a loss of the information (in brackets) that lies between the original repeats.

For this reason, a straight-out shotgun approach is the strategy of choice only for sequencing small genomes. The method proved its worth in 1995, when it was used to sequence the genome of the infectious bacterium *Haemophilus influenzae,* the first organism to have its complete genome sequence determined. The trouble with shotgun sequencing is that the reassembly process can be derailed by repetitive nucleotide sequences (Figure 10–26). Although rare in bacteria, these sequences make up a large fraction of vertebrate genomes (see Figure 9–29). Highly repetitive DNA segments make it difficult to piece DNA sequences back together accurately. Returning to the *ECB* analogy, this chapter alone contains more than a dozen instances of the phrase "the human genome." Imagine that one slip of paper from the shredded *ECB*s contains the information: "So why was sequencing the human genome" (which appears at the start of this section); another contains the information: "the human genome sequence consortium combined shotgun sequencing with a clone-by-clone approach" (which appears in the next paragraph). You might be tempted to join these two segments together based on the overlapping phrase "the human genome." But you would wind up with the nonsensical statement: "So why was sequencing the human genome sequence consortium combined shotgun sequencing with a clone-by-clone approach." You would also lose the several paragraphs of important text that originally appeared between these two instances of "the human genome."

And that's just in this section. The phrase "the human genome" appears in nearly every chapter of this book. Such repetition compounds the problem of placing each fragment in its correct context. To circumvent these assembly problems, researchers in the human genome sequence consortium combined shotgun sequencing with a clone-by-clone approach.

Clone-by-clone

In this approach, researchers started by preparing a genomic library. They broke the human genome into overlapping fragments, 100–200 kilobase pairs in size. They then plugged these segments into BACs (bacterial artificial chromosomes) and inserted them into *E. coli.* (BACs are similar to the bacterial plasmids discussed earlier in this chapter, except they can carry much larger pieces of DNA.) As the bacteria divided, they copied the BACs, thus producing a collection of overlapping cloned fragments (see Figure 10–11).

The researchers then mapped each of these DNA fragments to its correct position in the genome. To do this, investigators used restriction enzymes to generate a 'signature' of each clone (Figure 10–27). The locations of the restriction enzyme sites in each fragment allowed researchers to map each BAC clone onto a previously generated restriction map of the entire human genome.

Knowing the relative positions of the cloned fragments, the researchers then selected some 30,000 BACs, sheared each into smaller fragments, and determined the nucleotide sequence of each BAC separately using the shotgun method. They could then assemble the whole genome sequence by stitching together the sequences of thousands of individual BACs that span the length of the genome.

The beauty of this approach is that it is relatively easy to accurately determine where the BAC fragments belong in the genome. This mapping step reduces the likelihood that regions containing repetitive sequences will be assembled incorrectly, and it virtually eliminates the possibility that sequences from different chromosomes will be mistakenly joined together. Returning to the textbook analogy, the BAC-based approach is akin to first separating your copies of *ECB* into individual pages and then shredding each page into its own separate pile. It should be much easier to put the book back together when one pile of fragments contains words from page 1, a second pile from page 2, and so on. And there's virtually no chance of mistakenly sticking a sentence from page 40 into the middle of a paragraph on page 412.

All together now

The clone-by-clone approach produced the first draft of the human genome sequence in 2000 and the completed sequence in 2004. As the set of instructions that specify all of the RNA and protein molecules needed to build a human being, this string of genetic bits holds the secrets to human development, physiology, and medicine. But the human sequence will also be of great value to researchers interested in comparative genomics or in the physiology of other organisms: it has eased the assembly of nucleotide sequences from other mammalian genomes—mice, rats, dogs, and other primates. It also made it possible to determine the nucleotide sequences of the genomes of individual humans by providing a framework in which the new sequences can be simply superimposed. Thus the human sequence is likely to be the only mammalian genome to be completed in this detailed and methodical way. Thanks to the human genome project, we are beginning to glimpse the high degree of molecular diversity present in our own species.

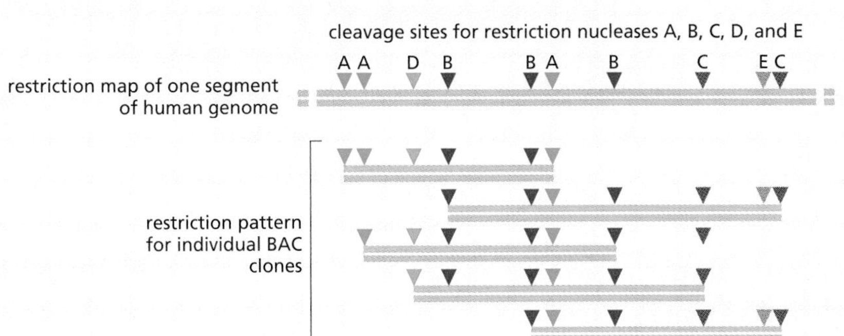

Figure 10–27 **Individual BAC clones are positioned on the physical map of the human genome sequence on the basis of their restriction digest "signatures."** Clones are digested with restriction endonucleases, and the sites at which the different enzymes cut each clone are recorded. The distinctive pattern of restriction sites allows investigators to order the fragments and place them on a previously generated restriction map of the human genome.

bacterial, yeast, insect, or mammalian cells, each containing the appropriate regulatory sequences for transcription and translation in these cells. The expression vector is replicated at each round of cell division, giving rise to a cell culture able to synthesize very large amounts of the protein of interest. Because the protein encoded by the expression vector is typically produced inside the cell, it must be purified from the host cell proteins after cell lysis by chromatography; but because it is so plentiful in the cell lysate (often comprising 1–10% of the total cell protein), purification is usually easy to accomplish in only a few steps.

This technology is now used to make large amounts of many medically useful proteins. The Factor VIII protein, for example, is now made commercially from cultures of genetically engineered mammalian cells and is thus free of viral contamination. Many other useful proteins, including hormones (such as insulin), growth factors, anticancer agents, and viral coat proteins for use in vaccines, are also produced in this fashion.

The use of expression vectors also allows scientists to produce many proteins of biological interest in large enough amounts for the detailed structural and functional studies that were once impossible. With large quantities of a protein, scientists can analyze its biological or biochemical activity and perhaps even grow crystals that would be suitable for determining its three-dimensional structure by X-ray crystallography (see How We Know, pp. 158–160). Recombinant DNA techniques allow scientists to begin with a protein of unknown function, isolate the gene that encodes it, and churn out enough of the protein to be able to investigate its structure and activity (Figure 10–28).

Reporter Genes and *In Situ* Hybridization Can Reveal When and Where a Gene Is Expressed

Neither the complete nucleotide sequence of a gene nor even the three-dimensional structure of the protein is sufficient to deduce a protein's function unless it is closely related to a protein whose function is already known. Many proteins—such as those that have a structural role in the cell or normally form part of a large multienzyme complex—have no obvi-

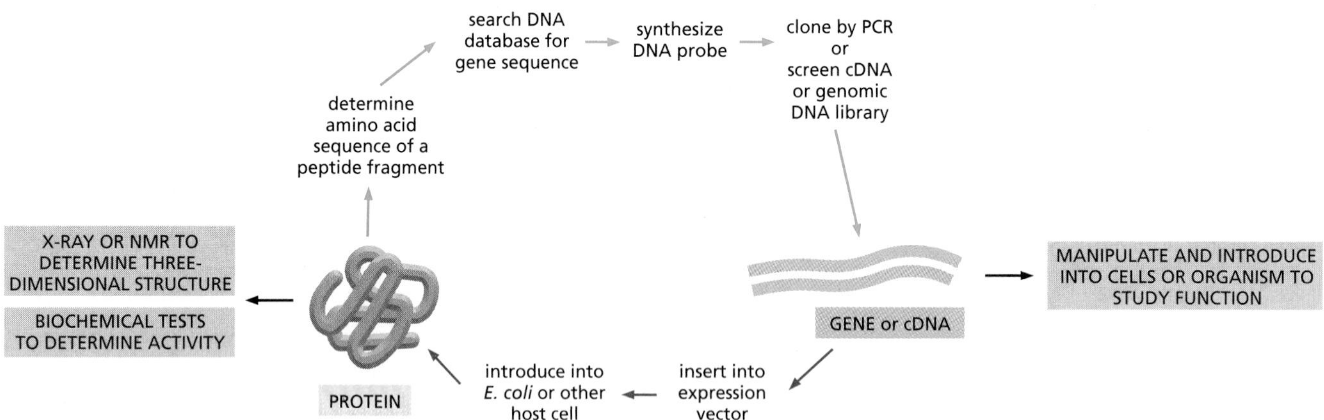

Figure 10–28 **Recombinant DNA techniques make it possible to move experimentally from gene to protein and from protein to gene.** A small quantity of a purified protein is used to obtain a partial amino acid sequence. This provides sequence information that enables the corresponding gene to be picked out of a DNA library by DNA hybridization (see Figure 10–12) or cloned and amplified by PCR from a sequenced genome (see Figure 10–17). Once the gene has been isolated and sequenced, its protein-coding sequence can be used to design a DNA molecule that can then be used to produce large quantities of the protein from genetically engineered cells (see Figure 10–24). This protein can then be studied biochemically or structurally. In addition to producing protein, the gene or DNA can also be manipulated and introduced into cells or organisms to study its function.

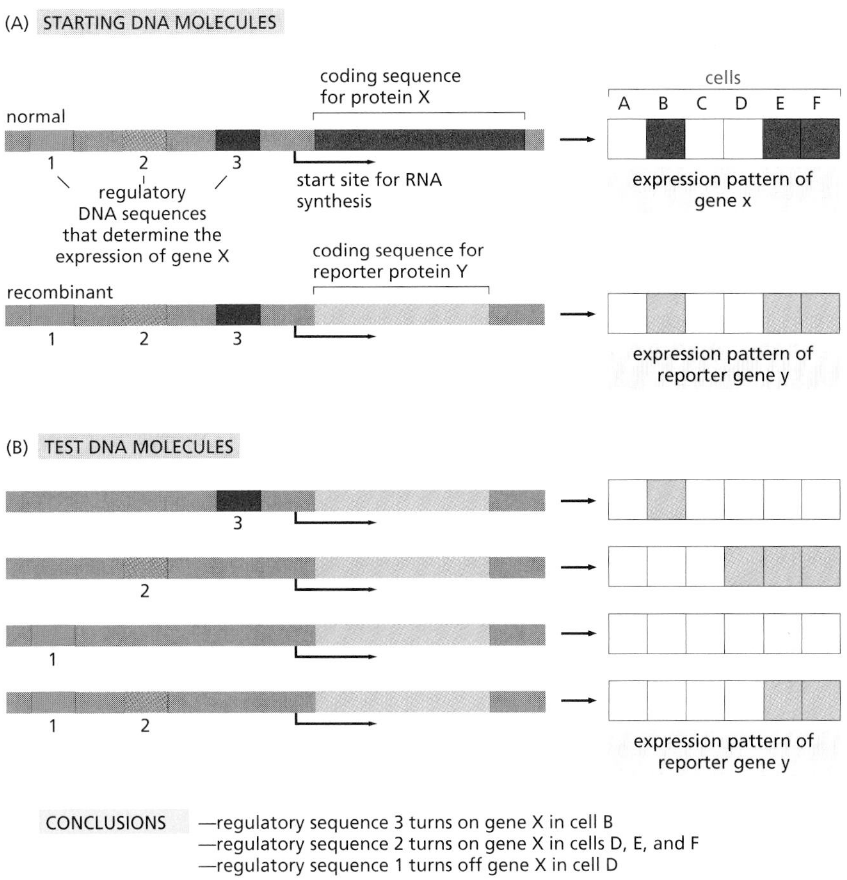

Figure 10–29 **Reporter genes can be used to determine the pattern of a gene's expression.** (A) Let us suppose we want to find out which cell types express protein X, but it is difficult to detect the protein directly. The coding sequence for protein X is replaced by the coding sequence for reporter protein Y, which can be easily monitored visually; for example, protein Y might be fluorescent. Protein Y will now be expressed under the control of the regulatory regions of gene X (here labeled 1, 2 and 3). (B) To test which regulatory regions control expression in particular cell types, various combinations of coding region Y and the regulatory regions are made. These recombinant DNA molecules are then tested for expression after their introduction into the different cell types. For experiments in eucaryotic cells, two commonly used reporter proteins are the enzymes β-galactosidase (see Figure 8–15C) and green fluorescent protein (GFP) (see Figure 10–30).

ous biochemical activity on their own. Even those that do have known activities (protein kinases, for example) could in principle participate in any number of different pathways in the cell; in other words, it is not always clear from their biochemical activities how the proteins are actually used by the cell.

In many cases, clues to a protein's function can be obtained by examining when and where its gene is expressed in the cell, or in the organism as a whole. Determining the pattern and timing of a gene's expression can be accomplished by joining the regulatory region of the gene under study to a **reporter gene**—one whose activity can be easily monitored. As we discussed in detail in Chapter 8, gene expression is controlled by regulatory DNA sequences, usually located upstream of the coding region, that are not transcribed themselves. These regulatory sequences, which control which cells will express a gene and under what conditions, can also be made to control the expression of a reporter gene. The level, timing, and cell specificity of reporter protein production will reflect the function of the original gene as well as the action of the regulatory sequences that belong to it (Figure 10–29). In most cases, the expression of the reporter gene is monitored by tracking the fluorescence or enzymatic activity of its protein product.

One of the most popular reporter proteins used today is **green fluorescent protein** (**GFP**), the molecule that gives luminescent jellyfish their greenish glow. In many cases, the GFP gene can simply be attached to one end of the gene that encodes a protein of interest. Often, the resulting *GFP fusion protein* behaves in the same way as the original protein, and its distribution in the cell or in the organism can be monitored simply by following its fluorescence by microscopy (Figure 10–30). The GFP fusion protein strategy has become a standard way to determine the distribution and to track the movement of any protein of interest in a living organism. From this information, many insights about a protein's function in the organism can be obtained.

It is also possible to directly observe the time and place that the mRNA product of a gene is expressed. In most cases, this strategy provides the same overall information as the reporter gene approach. In some situations, however, monitoring RNA is the method of choice: for example, if the gene's final product is RNA rather than protein. The technique, which relies on the principles of nucleic acid hybridization described earlier, is called ***in situ* hybridization** (from the Latin *in situ*, "in place"), because it allows specific nucleic acid sequences to be located while they are still in place within cells or within chromosomes.

In situ hybridization uses nucleic acid probes, labeled with either fluorescent dyes or radioactive isotopes, to detect the presence of RNA or DNA of a particular sequence within a cell or tissue (Figure 10–31). This technique can reveal gene expression patterns, and has led to great advances in our understanding of embryonic development, by making easily visible the many changes in gene expression that occur in different cells of the developing embryo. *In situ* hybridization also allows the visualization of specific DNA sequences within chromosomes. In this case the probes are hybridized to whole chromosomes that have been exposed briefly to a very high pH to separate the two DNA strands. The chromosomal regions that bind the labeled probe can then be seen (Figure 10–32). This technique can be used medically to identify, early in a pregnancy, fetuses that carry abnormal chromosomes.

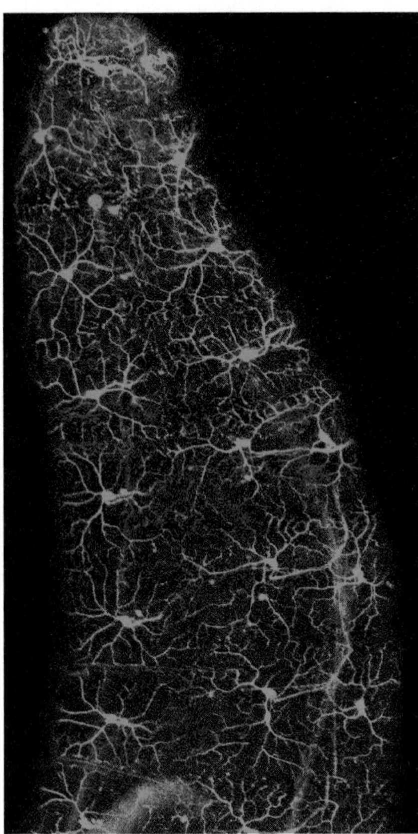

Figure 10–30 **Green fluorescent protein (GFP) can be used to identify specific cells in a living animal.** For this experiment, carried out in the fruit fly, the GFP gene was joined (using recombinant DNA techniques) to a fly promoter that is active only in a specialized set of neurons. This image of a live fly embryo was captured by a fluorescence microscope and shows approximately 20 neurons, each with long projections (axons and dendrites) that communicate with other (nonfluorescent) cells. These neurons, located just under the embryo's surface, allow the organism to sense its immediate environment. (From W.B. Grueber et al., *Curr. Biol.* 13:618–626, 2003. With permisison from Elsevier.)

Hybridization on DNA Microarrays Monitors the Expression of Thousands of Genes at Once

We saw in Chapter 8 that a cell expresses only a subset of the genes available in its genome. One of the most important uses of nucleic acid hybridization is to determine, for a population of cells, exactly which genes are actively being transcribed into mRNA and which are transcriptionally silent. *In situ* hybridization methods allow scientists to monitor the expression of one gene—or relatively few genes—at a time. In the 1990s, however, investigators developed a new tool, called a **DNA microarray**, that allows the RNA products of tens of thousands of genes to be monitored at the same time. By examining the expression of so many genes simultaneously, we can begin to identify and study the complex gene expression patterns that underlie cellular physiology, visualizing which genes are switched on (or off) as cells grow, divide, or respond to hormones, toxins, or infection.

DNA microarrays are little more than glass microscope slides studded with a large number of DNA fragments, each containing a nucleotide sequence that serves as a probe for a specific gene. The most dense arrays contain hundreds of thousands of these fragments in an area smaller than a postage stamp, allowing the expression patterns of entire genomes to

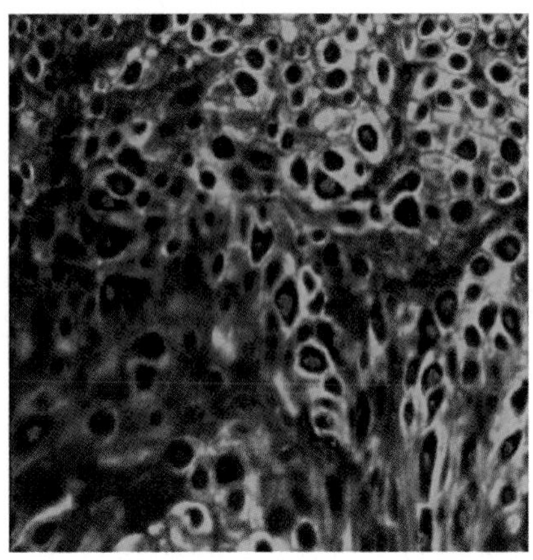

Figure 10–31 ***In situ* hybridization can be used to detect the presence of a virus in cells.** In this micrograph, the nuclei of epithelial cells infected with the human papillomavirus (HPV) are stained *pink* by a probe that recognizes a viral DNA sequence. The nuclei of all cells are stained *blue*, although this is masked by the *pink* stain in the infected cells. The cytoplasm of all cells is stained *green*. (Courtesy of Hogne Røed Nilsen.)

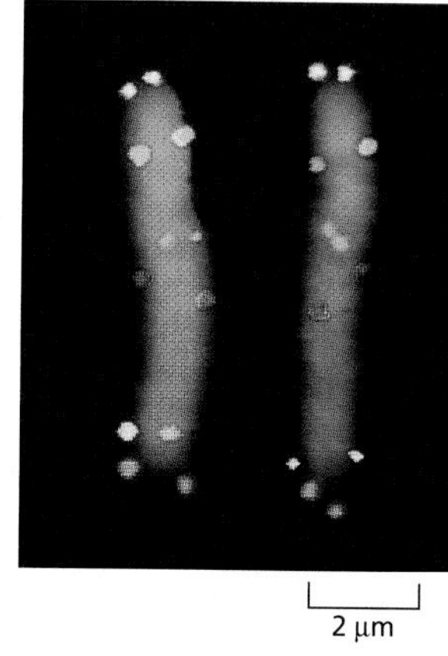

Figure 10–32 ***In situ* hybridization is used to locate genes on chromosomes.** Here, six different DNA probes have been used to mark the locations of their respective nucleotide sequences on human Chromosome 5 isolated in the metaphase stage of mitosis (see Figure 5–16 and Panel 18–1, pp. 626–627). The DNA probes have been chemically labeled and are detected using fluorescent antibodies specific for the chemical label. Both the maternal and paternal copies of Chromosome 5 are shown, aligned side by side. Each probe produces two dots on each chromosome because chromosomes undergoing mitosis have already replicated their DNA and therefore each chromosome contains two identical DNA helices. The technique employed here is nicknamed FISH, for *fluorescence* in situ *hybridization*. (Courtesy of David C. Ward.)

be monitored in a single experiment. Some types of microarrays carry DNA fragments corresponding to entire genes that are spotted onto the slides by a robot. Other types contain short, single-stranded DNA molecules that are synthesized on the surface of the wafer with techniques similar to those that are used to etch circuits onto computer chips. In either case, the exact sequence—and position—of every DNA probe on the chip is known.

To use a DNA microarray to monitor the expression of every gene in a cell simultaneously, mRNA from the cells being studied is extracted and converted to cDNA (see Figure 10–13). The cDNA is then labeled with a fluorescent probe. The microarray is incubated with the labeled cDNA sample, and hybridization is allowed to occur (Figure 10–33). The array is then washed to remove unbound molecules, and the positions to which fluorescently labeled DNA fragments have hybridized, through complementary base-pairing, are identified by an automated microscope. The array positions are then matched to the particular genes whose DNA was originally spotted at each location.

DNA microarrays have been used to examine everything from the changes in gene expression that make strawberries ripen to the genetic 'signatures' of different types of human cancer cells. Comparisons of the gene expression profiles of human cancers, for example, can be used to readily distinguish one type of cancer cell from another. By relating these expression patterns to clinical data gathered for each cancer—including how rapidly it progresses and whether it responds to treatment—it may be possible to predict whether a particular patient will respond to a specific therapy. Thus microarray-based 'profiles' of cancer cells are likely to lead to much more precise and effective treatments for this often fatal disease.

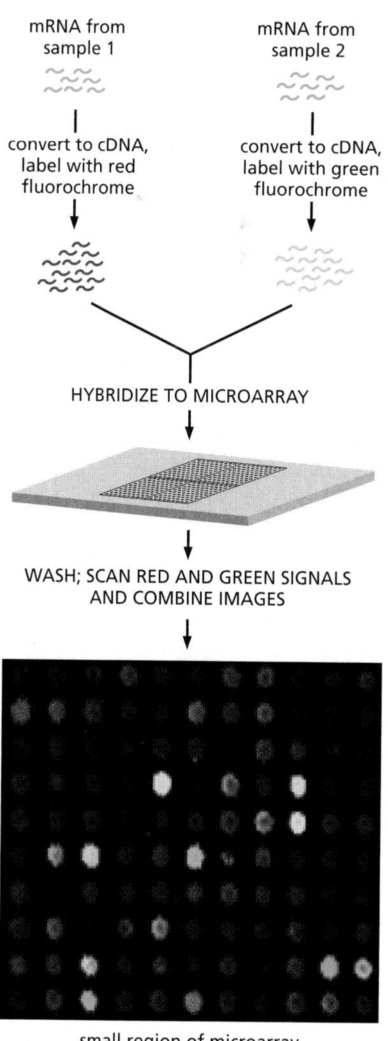

Figure 10–33 **DNA microarrays are used to monitor the expression of many thousands of genes simultaneously.** In this example, mRNA is collected from two different cell samples for direct comparison of their relative levels of gene expression—for example, cells treated with a hormone and untreated cells of the same type. The samples are converted to cDNA and labeled, one with a red fluorescent dye, the other with a green fluorescent dye. The labeled samples are mixed and then allowed to hybridize to the microarray. After incubation, the array is washed and the fluorescence scanned. Only a small proportion of the microarray, representing 110 genes, is shown. *Red* spots indicate that the gene in sample 1 is expressed at a higher level than the corresponding gene in sample 2, and *green* spots indicate that expression of the gene is more vigorous in sample 2 than in sample 1. *Yellow* spots reveal genes that are expressed at equal levels in both cell samples. *Dark* spots indicate little or no expression of the gene whose fragment is located at that position in the array.

Genetic Approaches Can Reveal the Function of a Gene

Ultimately, cell biologists wish to determine how genes, and the RNAs and proteins they encode, function in an intact organism. Before the advent of gene cloning, the functions of most genes were discovered by identifying and studying mutant organisms. In this 'classical' genetic approach, one begins by isolating mutants that have an interesting or unusual appearance: fruit flies with white eyes or curly wings, for example. Working backward from this **phenotype**—the appearance or behavior of the individual—one then determines the organism's **genotype**, the form of the gene responsible for that characteristic. The classical genetic approach is most easily applicable to organisms that reproduce rapidly and can be easily mutated in the laboratory—such as bacteria, yeasts, nematode worms, and fruit flies.

Recombinant DNA technology has made possible a different type of genetic approach. Instead of beginning with a randomly generated mutant and using it to identify a gene and its protein, one can start with a cloned gene and proceed to make mutations in it *in vitro*. Then, by reintroducing the altered gene back into the organism from which it originally came, one can produce a mutant organism in which the gene's function may be revealed. Using techniques to be discussed shortly, the coding sequence of a cloned gene can be altered to change the functional properties of the protein product or even to eliminate it altogether. Alternatively, the regulatory region of the gene can be altered so that the amount of protein made is altered, or so that the gene is expressed in a different cell type from normal or at a different time during development.

The ability to manipulate DNA *in vitro* makes it possible to introduce precise mutations, and genes can thus be altered in very subtle ways. It is often desirable, for example, to change the protein that the gene encodes by just one or a few amino acids. The use of the technique of **site-directed mutagenesis** to achieve this is outlined in Figure 10–34. By changing selected amino acids in this way, one can determine which parts of the polypeptide chain are crucial to such fundamental processes as protein folding, protein–ligand interactions, and enzymatic catalysis. Moreover, site-directed mutagenesis allows one to determine the biological roles of each part of a given protein.

QUESTION 10–5

To engineer a mutation *in vitro* as shown in Figure 10–34, a mismatch is created in the DNA. Would you expect this mismatch to be recognized and repaired by DNA mismatch repair enzymes (see Figure 6–22) when the plasmid that contains the mismatch is introduced into cells? Would this cause a problem in the site-directed mutagenesis procedure? Explain your answers.

Animals Can be Genetically Altered

To study the function of a gene that has been mutated *in vitro,* ideally one would like to generate an organism in which the normal gene has been replaced by the altered one. In this way, the function of the mutant protein can be analyzed in the absence of the normal protein. In many organisms, such **gene replacement** can be accomplished quite easily by homologous recombination between the introduced mutant DNA and the chromosomal DNA (Figure 10–35A). With the same strategy, the gene can be deleted entirely, creating a **gene knockout** (Figure 10–35B). A third possibility is that the mutant gene can be added to the genome without any alteration being made to the normal gene in the process (Figure 10–35C). Organisms into which a new gene has been introduced, or those whose genomes have been altered in other ways using recombinant DNA techniques, are known as **transgenic organisms** or *genetically modified organisms* (GMOs).

In sexually reproducing organisms, these alterations are usually made in the germ line; that is, in the reproductive cells. Such transgenic animals will then be able to pass the altered gene on to at least some of their progeny as a permanent part of their genome (Figure 10–36). Technically, even the human germ line could now be altered in this way, although this

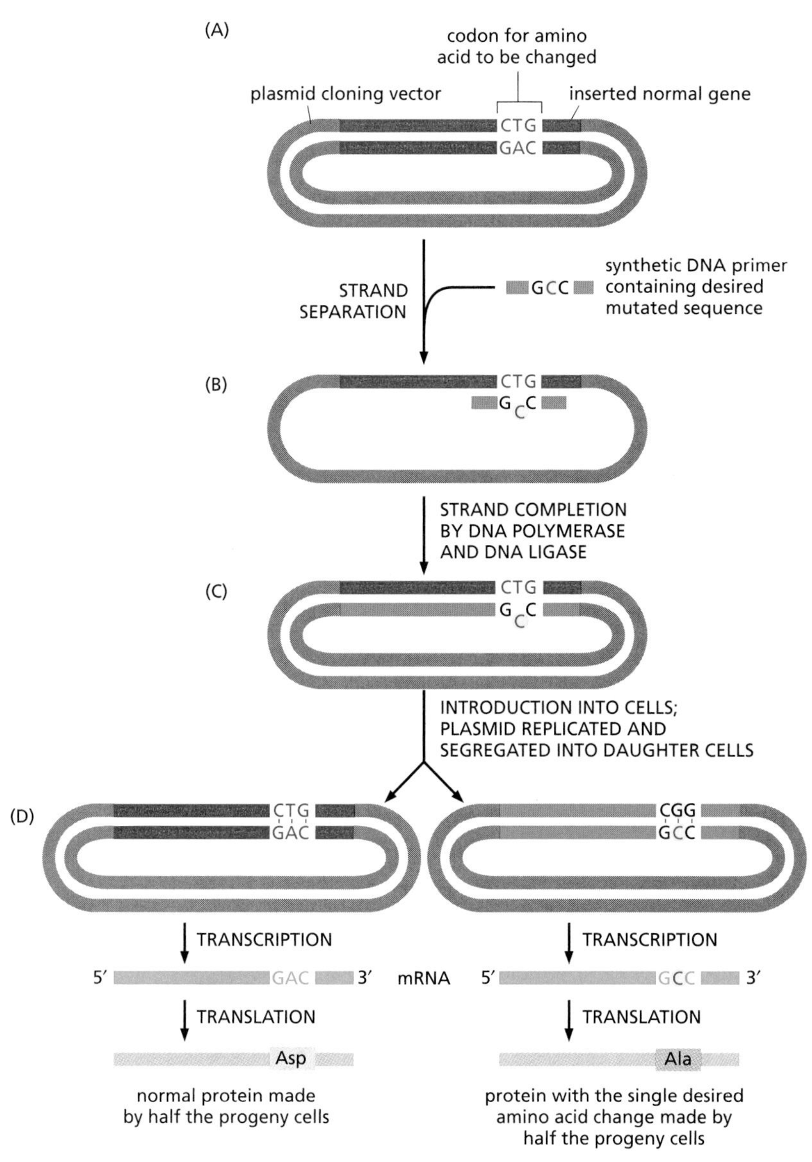

Figure 10–34 **Synthetic DNA molecules are used to modify the protein-coding region of a gene by site-directed mutagenesis.** (A) A recombinant plasmid containing a gene insert is separated into its two DNA strands. A synthetic DNA primer that contains a single altered nucleotide at a predetermined point is added to the single-stranded DNA under conditions that permit less than perfect DNA hybridization. (B) The primer hybridizes to the DNA, forming a single mismatched nucleotide pair. (C) The recombinant plasmid is made double-stranded by *in vitro* DNA synthesis starting from the primer and sealing by DNA ligase. (D) The double-stranded DNA is introduced into a cell, where it is replicated. Replication of one strand produces a normal DNA molecule, but replication of the other (the strand that contains the primer) produces a DNA molecule carrying the desired mutation. Only half of the progeny cells will end up with a plasmid that contains the desired mutant gene; however, a progeny cell that contains the mutated gene can be identified, separated from other cells, and cultured to produce a pure population of cells, all of which carry the mutated gene. With synthetic DNA molecules of the appropriate sequence, more than one amino acid substitution can be made at a time, or one or more amino acids can be added or deleted.

is unlawful for a variety of ethical reasons. Similar techniques are, however, being explored to correct genetic defects in human somatic cells. Somatic cells, such as the cells that form the liver, pancreas, bone, or skin, reproduce within an individual human being but are not passed on to progeny (see Figure 9–3). Some genetic diseases could be alleviated or even cured by the introduction of genetically corrected somatic cells into the tissue most affected by the disease; however, the alterations would not be passed on to descendants.

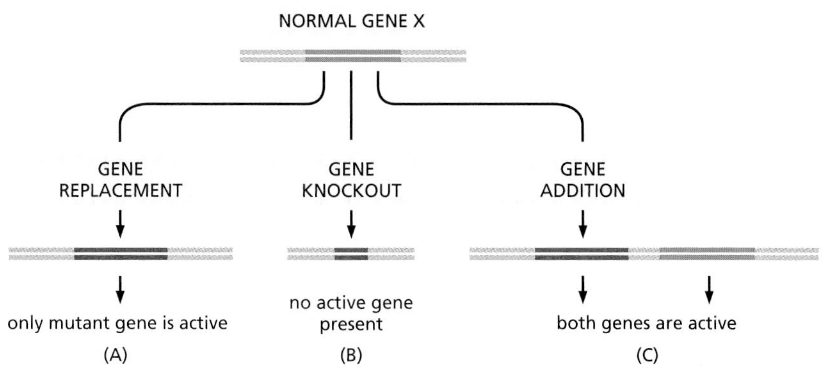

Figure 10–35 **Several types of gene alterations can be made in genetically engineered organisms.** (A) The normal gene can be completely replaced by a mutant copy of the gene, a process called gene replacement. This will provide information on the activity of the mutant gene without interference from the normal gene, and thus the effects of small and subtle mutations can be determined. (B) The normal gene can be inactivated completely, for example, by making a large deletion in it; in this case, the gene is said to have been knocked out. This type of alteration is very widely used to obtain information on the function of the normal gene in the whole animal. (C) A mutant gene can simply be added to the genome. This alteration can provide useful information when the introduced mutant gene overrides the function of the normal gene.

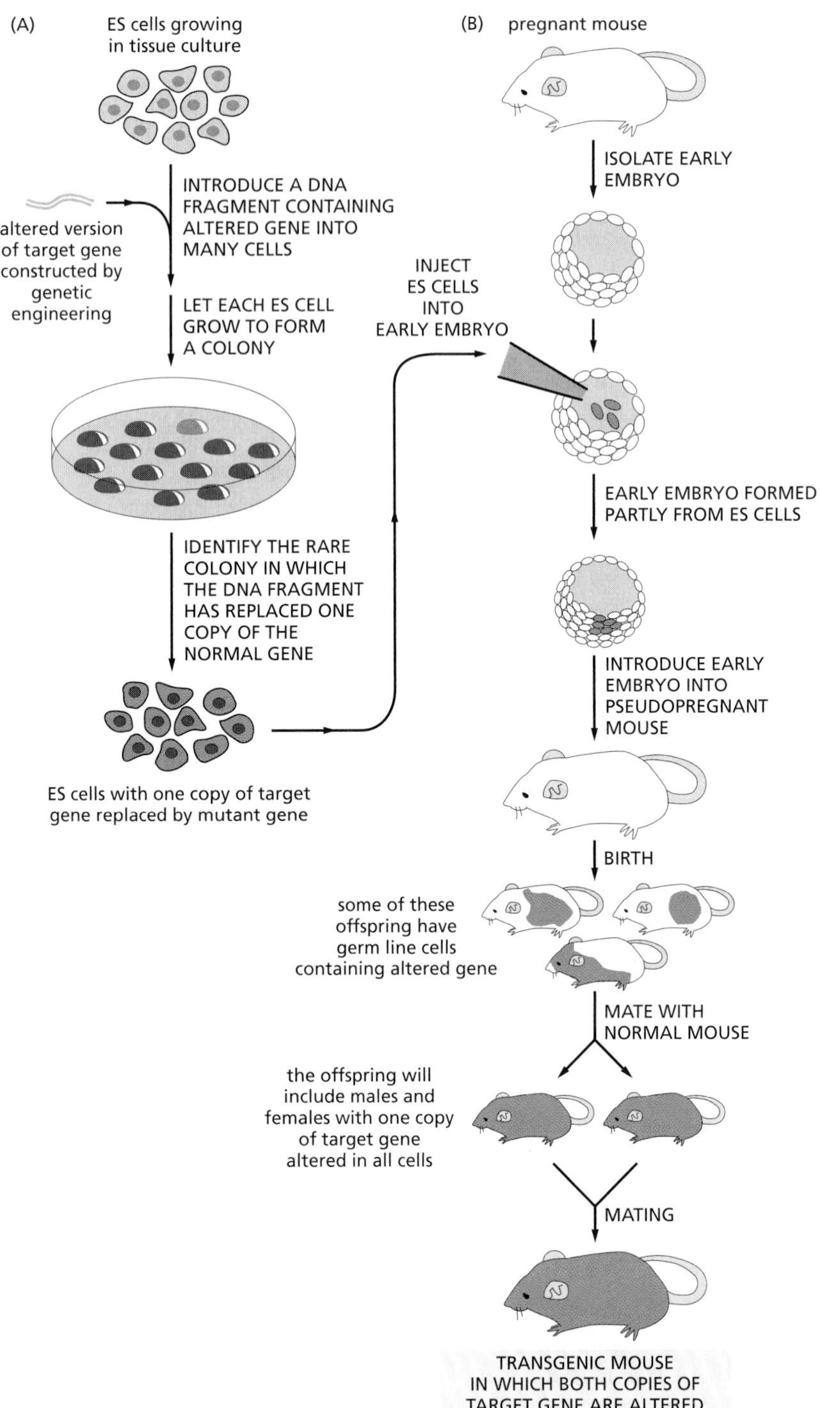

Figure 10–36 **Gene replacement in mice utilizes embryonic stem (ES) cells.** In the first step (A), an altered version of the gene is introduced into cultured ES cells. In a few rare ES cells, the corresponding normal genes will be replaced by the altered gene through homologous recombination. Although the procedure is often laborious, these rare cells can be identified and cultured to produce many descendants, each of which carries an altered gene in place of one of its two normal corresponding genes. In the next step of the procedure (B), these altered ES cells are injected into a very early mouse embryo; the cells are incorporated into the growing embryo, and a mouse produced by such an embryo will contain some somatic cells (colored *orange*) that carry the altered gene. Some of these mice may contain germ-line cells that contain the altered gene. When bred with a normal mouse, some of the progeny of these mice will contain the altered gene in all of their cells. If two such mice are in turn bred, some of the progeny will contain two altered genes (one on each chromosome) in all of their cells. If the original gene alteration completely inactivates the function of the gene, these mice are known as knockout mice. It is often the case that mice missing genes that function during development die long before they reach adulthood.

Transgenic techniques make it possible to produce complex organisms that contain altered genes or that lack certain genes entirely. For example, many *knockout mice*—strains of mice that have a particular gene permanently inactivated—have now been produced. By observing the effect of eliminating or mutating a given gene, it is often possible to infer its normal function (Figure 10–37).

RNA Interference Provides a Simple Way to Test Gene Function

Knocking out a gene in an organism and studying the consequences is perhaps the most powerful approach for understanding the function of a gene, but there is a much faster and easier way to inactivate genes

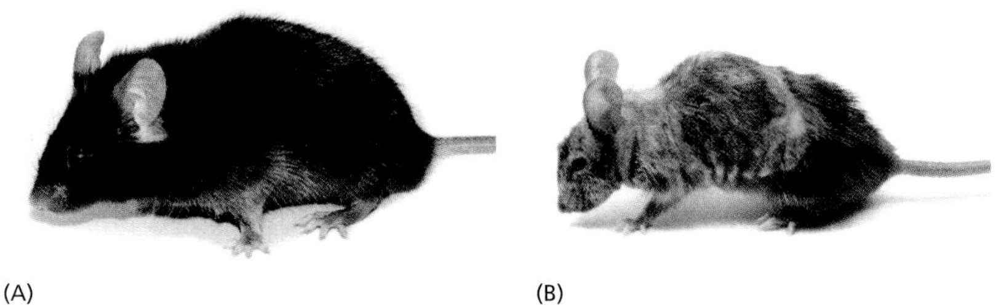

Figure 10–37 **Transgenic mice with a mutant DNA helicase show premature aging.** The helicase, encoded by the *Xpd* gene, is involved in both transcription and DNA repair. Compared with a wild-type mouse (A), a transgenic mouse that expresses a defective version of *Xpd* (B) exhibits many of the symptoms of premature aging, including osteoporosis, emaciation, early graying, infertility, and reduced life-span. The mutation in *Xpd* used here impairs the activity of the helicase and mimics a mutation that in humans causes trichothiodystrophy, a disorder characterized by brittle hair, skeletal abnormalities, and a greatly reduced life expectancy. These results support the hypothesis that an accumulation of DNA damage contributes to the aging process in both humans and mice. (From J. de Boer et al., *Science* 296:1276–1279, 2002. With permission from the AAAS.)

in cells and organisms. Called **RNA interference** (**RNAi**), this method exploits a natural mechanism used in a wide variety of plants and animals to regulate selected genes and to destroy 'foreign' RNA molecules (see Figures 8–26 and 8–27). The technique relies on introducing into a cell or organism a double-stranded RNA molecule whose nucleotide sequence matches that of the gene to be inactivated. This double-stranded RNA is recognized as being foreign, and the cell is tricked into degrading not only it but also the mRNA whose sequence it matches. Small fragments of these degraded RNAs are subsequently used by the cell to produce more double-stranded RNA that directs the continued elimination of the target mRNA. Because these short RNA fragments can be passed on to progeny cells, RNAi can cause heritable changes in gene expression.

RNAi is frequently used to inactivate genes in mammalian cell culture lines. It has also been widely used to study gene function in the nematode *C. elegans*. When working with worms, introducing the double-stranded RNA is simple: the RNA can be injected directly into the intestine of the animal, or the worm can be fed with *E. coli* engineered to produce the RNA (Figure 10–38). The RNA is distributed throughout the body of the worm, where it inhibits expression of the target gene in different tissue types. Because the entire genome of *C. elegans* has been sequenced, RNAi is being used to help in assigning functions to the entire complement of worm genes.

Transgenic Plants Are Important for Both Cell Biology and Agriculture

Although we tend to think of recombinant DNA research in terms of animal biology, these techniques have also had a profound impact on our study of plants. In fact, certain features of plants make them especially amenable to recombinant DNA methods.

When a piece of plant tissue is cultured in a sterile medium containing nutrients and appropriate growth regulators, some of the cells are stimulated to proliferate indefinitely in a disorganized manner, producing a mass of relatively undifferentiated cells called a *callus*. If the nutrients and

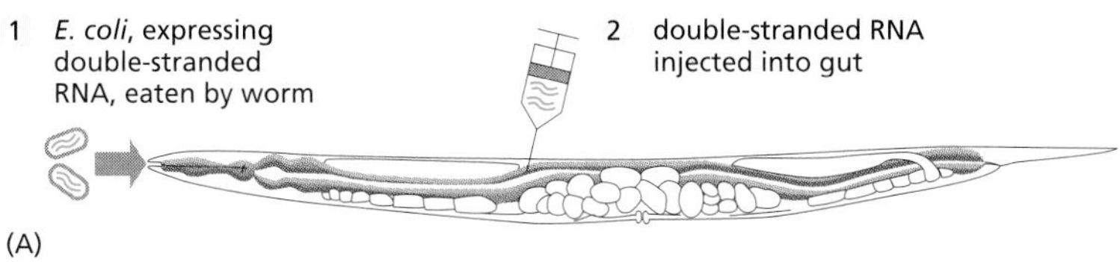

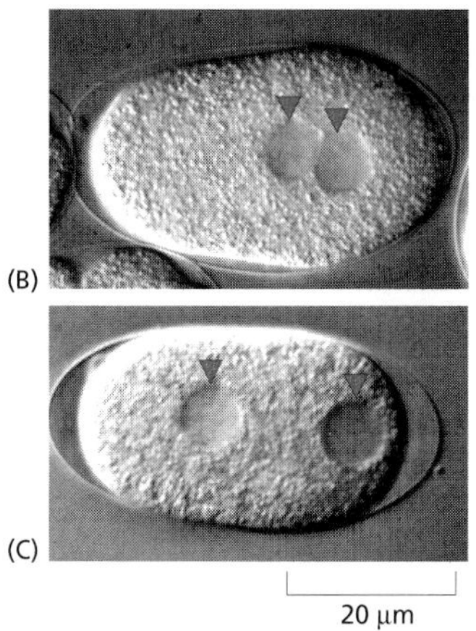

Figure 10–38 **Gene function can be tested by RNA interference.** (A) Double-stranded RNA (dsRNA) can be introduced into *C. elegans* by (1) feeding the worms *E. coli* that express the dsRNA or (2) injecting the dsRNA directly into the animal's gut. (B) In a wild-type worm embryo, the egg and sperm pronuclei (*red* arrowheads) come together in the posterior half of the embryo shortly after fertilization. (C) In an embryo in which a particular gene has been inactivated by RNAi, the pronuclei fail to migrate. This experiment revealed an important but previously unknown function of this gene in embryonic development. (B and C, from P. Gönczy et al., *Nature* 408:331–336, 2000. With permission from Macmillan Publishers Ltd.)

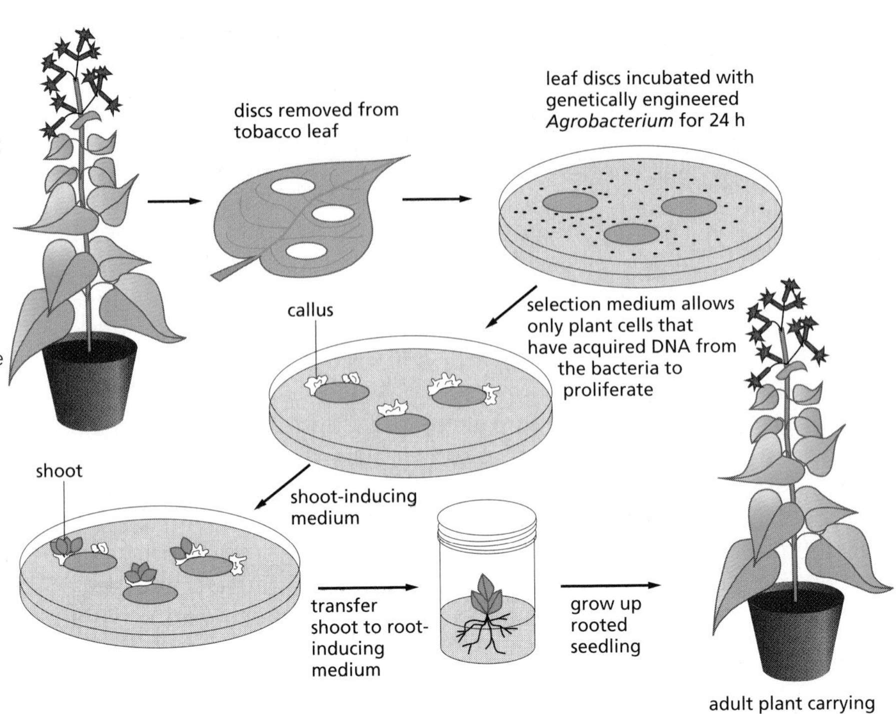
discs removed from tobacco leaf
leaf discs incubated with genetically engineered *Agrobacterium* for 24 h
callus
selection medium allows only plant cells that have acquired DNA from the bacteria to proliferate
shoot
shoot-inducing medium
transfer shoot to root-inducing medium
grow up rooted seedling
adult plant carrying transgene that was originally present in the bacteria

Figure 10–39 **Transgenic plants can be made using recombinant DNA techniques.** A disc is cut out of a leaf and incubated in an *Agrobacterium* culture in which the bacterial cells carry a recombinant plasmid with both a selectable marker and a desired engineered gene. The wounded cells at the edge of the disc release substances that attract the bacteria and cause them to inject DNA into these cells. Only those plant cells that take up the appropriate DNA and express the selectable marker gene survive to proliferate and form a callus. The manipulation of growth factors supplied to the callus induces it to form shoots that subsequently root and grow into adult plants carrying the engineered gene.

growth regulators are carefully manipulated, one can induce the formation of a shoot within the callus, and in many species a whole new plant can be regenerated from such shoots. In a number of plants—including tobacco, petunia, carrot, potato, and *Arabidopsis*—a single cell from such a callus can be grown into a small clump of cells from which a whole plant can be regenerated (see Figure 8–2B). Just as mutant mice can be derived by the genetic manipulation of embryonic stem cells in culture, so transgenic plants can be created from plant cells transfected with DNA in culture (Figure 10–39).

The ability to produce transgenic plants has greatly accelerated progress in many areas of plant cell biology. It has played an important part, for example, in isolating receptors for growth regulators and in analyzing the mechanisms of morphogenesis and of gene expression in plants. These techniques have also opened up many new possibilities in agriculture that could benefit both the farmer and the consumer. They have made it possible, for example, to modify the ratio of lipid, starch, and protein in seeds, to impart pest and virus resistance to plants, and to create modified plants that tolerate extreme habitats such as salt marshes or water-stressed soil. One variety of rice has been genetically engineered to produce β-carotene, the precursor of vitamin A. If it replaced conventional rice, this 'golden rice'—so called because of its faint yellow color—could help to alleviate severe vitamin A deficiency, which causes blindness in hundreds of thousands of children in the developing world each year.

ESSENTIAL CONCEPTS

- **Recombinant DNA technology has revolutionized the study of the cell, making it possible for researchers to pick out any gene at will from the thousands of genes in a cell and to determine the exact molecular structure of the gene.**

- A crucial element in this technology is the ability to cut a large DNA molecule into a specific and reproducible set of DNA fragments by using restriction nucleases, each of which cuts the DNA double helix only at a particular nucleotide sequence.
- DNA fragments can be separated from one another on the basis of size by gel electrophoresis.
- Nucleic acid hybridization can detect any given DNA or RNA sequence in a mixture of nucleic acid fragments. This technique relies on the fact that a single strand of DNA or RNA will form a double helix only with another nucleic acid strand of the complementary nucleotide sequence.
- Single-stranded DNAs of known sequence and labeled with fluorescent dyes or radioisotopes are used as probes in hybridization reactions.
- Short DNA strands of any sequence can be made by chemical synthesis in the laboratory.
- DNA cloning techniques enable a DNA sequence to be selected from millions of other sequences and produced in unlimited amounts in pure form.
- DNA fragments can be joined together *in vitro* by using DNA ligase to form recombinant DNA molecules that are not found in nature.
- DNA fragments can be maintained and amplified by inserting them into a DNA molecule capable of replication, such as a plasmid. This recombinant DNA molecule is then introduced into a rapidly dividing host cell, usually a bacterium, so that the DNA is replicated at each cell division.
- A collection of cloned fragments of chromosomal DNA representing the complete genome of an organism is known as a genomic library. The library is often maintained as millions of clones of bacteria, each clone carrying a different DNA fragment.
- cDNA libraries contain cloned DNA copies of the total mRNA of a particular cell type or tissue. Unlike genomic DNA clones, cloned cDNAs contain predominantly protein-coding sequences; they lack introns, regulatory DNA sequences, and promoters. They are thus most suitable for use when the cloned gene is to be expressed to make a protein.
- The polymerase chain reaction (PCR) is a powerful form of DNA amplification that is carried out *in vitro* using a purified DNA polymerase. PCR requires prior knowledge of the sequence to be amplified, because two synthetic oligonucleotide primers must be synthesized that bracket the portion of DNA to be replicated.
- Historically, genes were cloned using hybridization techniques to identify the plasmid carrying the desired sequence from a DNA library. Today, most genes are cloned using PCR to greatly amplify them and thereby obtain a specific sequence from a sample of DNA or mRNA.
- Techniques are now available for rapidly determining the nucleotide sequence of any piece of DNA.
- The complete nucleotide sequences of the genomes of hundreds of different organisms have been determined. These include bacteria, archaea, yeasts, insects, fish, plants, and mammals.
- Bacteria, yeasts, and mammalian cells can be engineered to synthesize large quantities of any protein from any organism, thus making it possible to study proteins that are otherwise rare or difficult to isolate.
- Using recombinant DNA techniques, a protein can be joined to a molecular tag, such as the green fluorescent protein (GFP), which allows the tracking of its movement inside the cell. In the case of GFP, the protein can be monitored over time in living organisms.

- *In situ* nucleic acid hybridization can be used to detect the precise location of genes in chromosomes, or RNAs in cells and tissues.
- By presenting a platform for performing a large number of simultaneous hybridization reactions, DNA microarrays can be used to monitor the expression of tens of thousands of genes at once.
- Cloned genes can be permanently inserted into the genome of a cell or an organism by using recombinant DNA technology. Cloned DNA can be altered *in vitro* to create mutant genes that can then be reinserted into a cell or an organism to study gene function.
- A straightforward strategy for studying the function of a gene is to delete it from the organism's genome and then to study the effect of this knockout on the behavior or appearance of the organism.
- The expression of particular genes can be inhibited in cells or organisms by the technique of RNA interference (RNAi), which prevents an mRNA from being translated into protein.

KEY TERMS

cDNA
dideoxy DNA sequencing
DNA cloning
DNA library
DNA ligase
DNA microarray
gene knockout
gene replacement
genotype
green fluorescent protein (GFP)
hybridization
in situ hybridization
phenotype
plasmid
polymerase chain reaction (PCR)
recombinant DNA
recombinant DNA technology
reporter gene
restriction nuclease
RNA interference (RNAi)
site-directed mutagenesis
transformation
transgenic organism

QUESTIONS

QUESTION 10–6

What are the consequences for a DNA sequencing reaction if the ratio of dideoxyribonucleoside triphosphates to deoxyribonucleoside triphosphates is increased? What happens if this ratio is decreased?

QUESTION 10–7

Almost all the cells in an individual animal contain identical genomes. In an experiment, a tissue composed of several different cell types is fixed and subjected to *in situ* hybridization with a DNA probe to a particular gene. To your surprise, the hybridization signal is much stronger in some cells than in others. How might you explain this result?

QUESTION 10–8

After decades of work, Dr. Ricky M. isolated a small amount of attractase—an enzyme that produces a powerful human pheromone—from hair samples of Hollywood celebrities. To take advantage of attractase for his personal use, he obtained a complete genomic clone of the attractase gene, connected it to a strong bacterial promoter on an expression plasmid, and introduced the plasmid into *E. coli* cells. He was devastated to find that no attractase was produced in the cells. What is a likely explanation for his failure?

QUESTION 10–9

Which of the following statements are correct? Explain your answers.

A. Restriction nucleases cut DNA at specific sites that are always located between genes.

B. DNA migrates toward the positive electrode during electrophoresis.

C. Clones isolated from cDNA libraries contain promoter sequences.

D. PCR utilizes a heat-stable DNA polymerase because for each amplification step, double-stranded DNA must be heat-denatured.

E. Digestion of genomic DNA with AluI, a restriction enzyme that recognizes a four-nucleotide sequence, produces fragments that are all exactly 256 nucleotides in length.

F. To make a cDNA library, both a DNA polymerase and a reverse transcriptase must be used.

G. DNA fingerprinting by PCR relies on the fact that different individuals have different numbers of repeats in STR regions in their genome.

H. It is possible for a coding region of a gene to be present in a genomic library prepared from a particular tissue but to be absent from a cDNA library prepared from the same tissue.

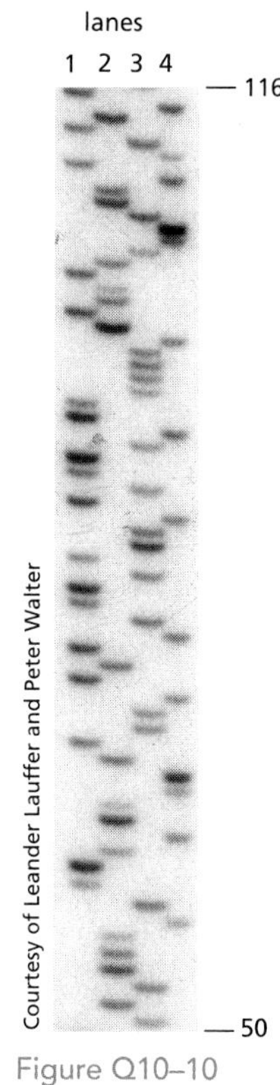

Figure Q10–10

QUESTION 10–10

A. What is the sequence of the DNA that was used in the sequencing reaction shown in Figure Q10–10? The four lanes show the products of sequencing reactions that contained ddG (lane 1), ddA (lane 2), ddT (lane 3), and ddC (lane 4). The numbers to the right of the autoradiograph represent the positions of marker DNA fragments of 50 and 116 nucleotides.

B. This DNA was derived from the middle of a cDNA clone of a mammalian protein. Using the genetic code table (see Figure 7–24), can you determine the amino acid sequence of this portion of the protein?

QUESTION 10–11

A. How many different DNA fragments would you expect to obtain if you cleaved human genomic DNA with HaeIII? (Recall that there are 3×10^9 nucleotide pairs per haploid genome.) How many fragments would you expect with EcoRI? With NotI?

B. Human genomic libraries used for DNA sequencing are often made from fragments obtained by cleaving human DNA with HaeIII in such a way that the DNA is only partially digested; that is, not all the possible HaeIII sites have been cleaved. What is a possible reason for doing this?

QUESTION 10–12

A molecule of double-stranded DNA was cleaved with three different restriction nucleases, and the resulting products were separated by gel electrophoresis (Figure Q10–12).

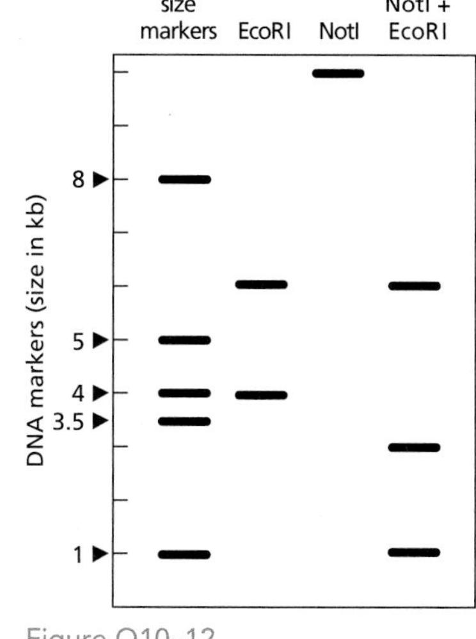

Figure Q10–12

DNA fragments of known sizes were electrophoresed on the same gel for use as size markers (*left* lane). The size of the DNA markers is given in kilobase pairs (kb), where 1 kb = 1000 nucleotide pairs. Using the size markers as a guide, estimate the length of each restriction fragment obtained. From this information, construct a map of the original DNA molecule indicating the relative positions of all the restriction enzyme cleavage sites.

QUESTION 10–13

You have isolated a small amount of a rare protein. You cleaved the protein into fragments using proteases, separated some of the fragments by chromatography, and determined their amino acid sequence. Unfortunately, as is often the case when only small amounts of protein are available, you obtained only three short stretches of amino acid sequence from the protein:

1. Trp-Met-His-His-Lys
2. Leu-Ser-Arg-Leu-Arg
3. Tyr-Phe-Gly-Met-Gln

A. Using the genetic code (see Figure 7–24), design a collection of DNA probes specific for each peptide that could be used to detect the gene in a cDNA library by hybridization. Which of the three collections of oligonucleotide probes would it be preferable to use first? Explain your answer. (Hint: the genetic code is redundant, so each peptide has multiple potential coding sequences.)

B. You have also been able to determine that the Gln of your peptide #3 is the C-terminal (i.e., the final) amino acid of your protein. How would you go about designing oligonucleotide primers that could be used to amplify a portion of the gene from a cDNA library using PCR?

C. Suppose the PCR amplification in (B) yields a DNA that is precisely 300 nucleotides long. Upon determining the nucleotide sequence of this DNA, you find the sequence CTATCACGCTTTAGG approximately in its middle. What would you conclude from these observations?

QUESTION 10–14

Assume that a DNA sequencing reaction is carried out as shown in Figure 10–21, except that the four different dideoxyribonucleoside triphosphates are modified so that each contains a covalently attached dye of a different color (which does not interfere with its incorporation into the DNA chain). What would the products be if you added a mixture of all four of these labeled dideoxyribonucleoside triphosphates along with the four unlabeled deoxyribonucleoside triphosphates into a single sequencing reaction? What would the results look like if you electrophoresed these products in a single lane of a gel?

QUESTION 10–15

Genomic DNA clones are often used to "walk" along a chromosome. In this approach, one cloned DNA is used to isolate other clones that contain overlapping DNA sequences (Figure Q10–15). Using this method, it is possible to build up a long stretch of DNA and thus identify new genes in near proximity to a previously cloned gene.

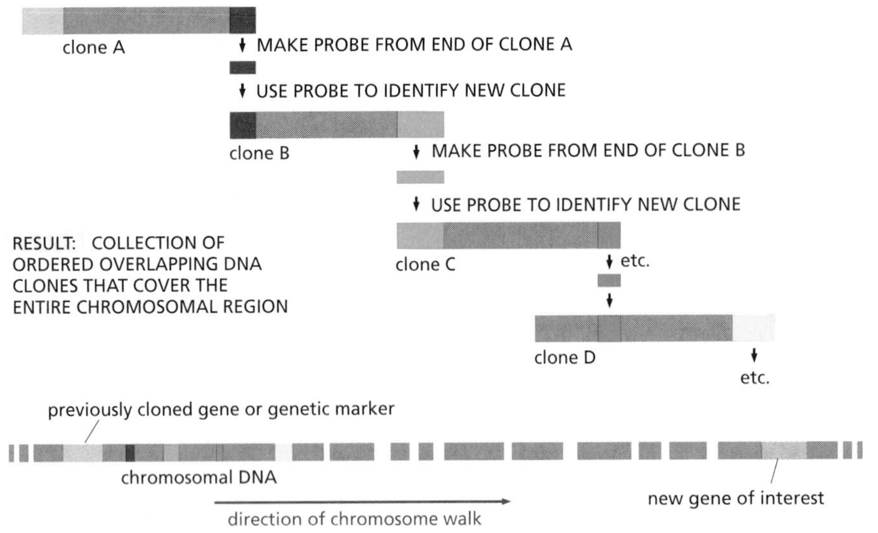

Figure Q10–15

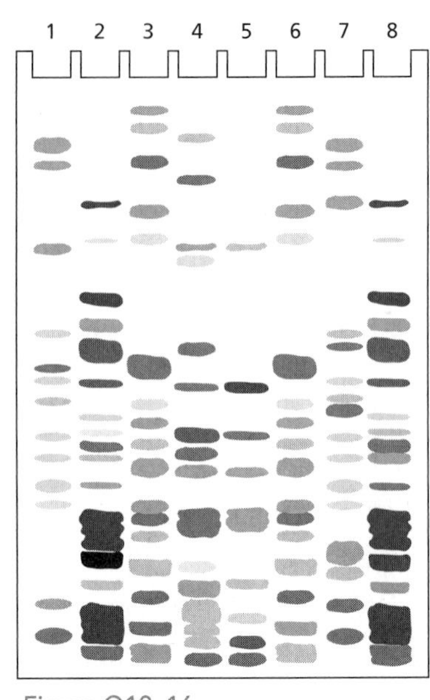

Figure Q10–16

A. Would it be faster to use cDNA clones in this method, because they do not contain any intron sequences?

B. What would happen if you encountered a repetitive DNA sequence, like the L1 transposon (see Figure 6–35), which is found in many copies and in many different places in the genome?

QUESTION 10–16

There has been a colossal snafu in the maternity ward of your local hospital. Four sets of male twins, born within an hour of each other, were inadvertently shuffled in the excitement occasioned by that unlikely event. You have been called in to set things straight. As a first step, you would like to match each baby with his twin. (Many newborns look alike so you don't want to rely on appearance alone.) To that end you analyze a small blood sample from each infant using a hybridization probe that detects short tandem repeats (STRs) located in widely scattered regions of the genome. The results are shown in Figure Q10–16.

A. Which infants are twins? Which are identical twins?

B. How could you match a pair of twins to the correct parents?

QUESTION 10–17

One of the first organisms that was genetically modified using recombinant DNA technology was a bacterium that normally lives on the surface of strawberry plants. This bacterium makes a protein, called ice-protein, that causes the efficient formation of ice crystals around it when the temperature drops to just below freezing. Thus, strawberries harboring this bacterium are particularly susceptible to frost damage because their cells are destroyed by the ice crystals. Consequently, strawberry farmers have a considerable interest in preventing ice crystallization.

A genetically engineered version of this bacterium was constructed in which the ice-protein gene was knocked out. The mutant bacteria were then introduced in large numbers into strawberry fields, where they displaced the normal bacteria by competition for their ecological niche. This approach has been successful: strawberries bearing the mutant bacteria show a much reduced susceptibility to frost damage.

At the time they were first carried out, the initial open-field trials triggered an intense debate because they represented the first release into the environment of an organism that had been genetically engineered using recombinant DNA technology. Indeed, all preliminary experiments were carried out with extreme caution and in strict containment (Figure Q10–17).

Do you think that bacteria lacking the ice-protein could be isolated without the use of modern DNA technology? Is it likely that such mutations have already occurred in nature? Would the use of a mutant bacterial strain isolated from nature be of lesser concern? Should we be concerned about the risks posed by the application of recombinant DNA techniques in agriculture and medicine? Explain your answers.

(Courtesy of John Bedbrook and DNA Plant Technology Corporation.)

Figure Q10–17

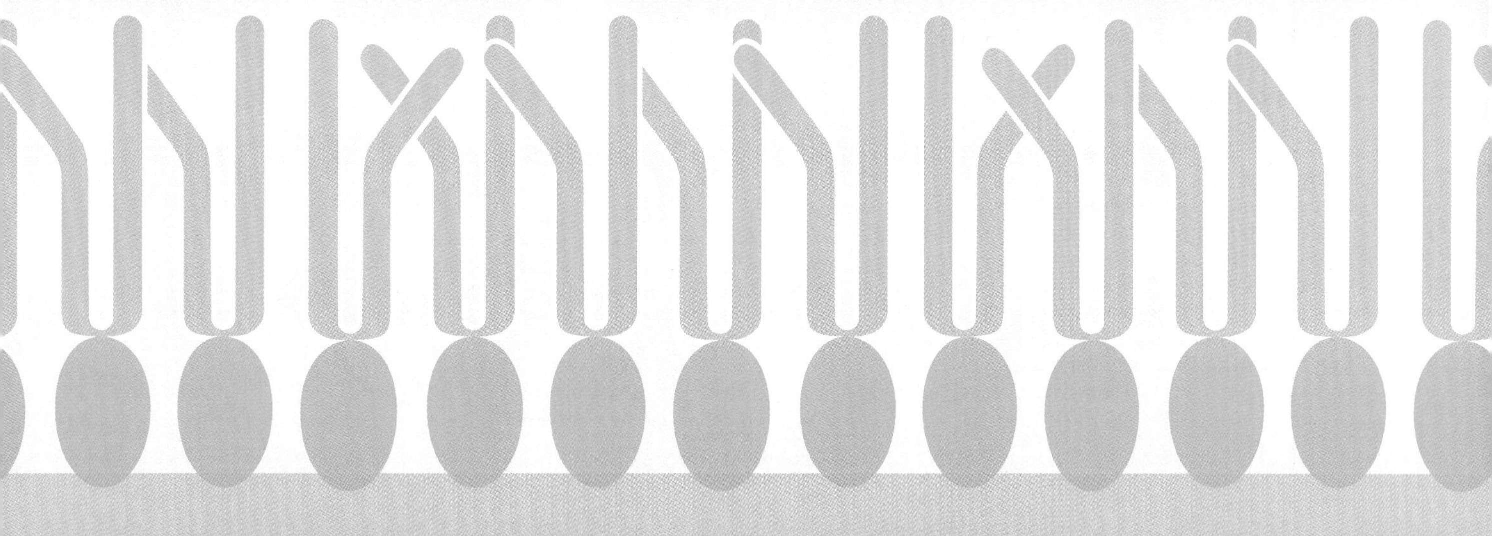

CHAPTER ELEVEN

11

Membrane Structure

THE LIPID BILAYER

MEMBRANE PROTEINS

A living cell is a self-reproducing system of molecules held inside a container. That container is the **plasma membrane**—a fatty film so thin and transparent that it cannot be seen directly in the light microscope. Every cell on Earth uses a membrane to separate and protect its chemical components from the outside environment. Without membranes there would be no cells, and thus no life.

The plasma membrane is simple in form: its structure is based on a two-ply sheet of lipid molecules about 5 nm—or 50 atoms—thick. Its properties, however, are unlike those of any sheet of material we are familiar with in the everyday world. Although the plasma membrane serves as a barrier to prevent the contents of the cell from escaping and mixing with the surrounding medium (Figure 11–1A), it does much more than that. If a cell is to survive and grow, nutrients must pass inward, across the plasma membrane, and waste products must pass out. To facilitate this exchange, the membrane is penetrated by highly selective channels and pumps—protein molecules that allow specific substances to be imported

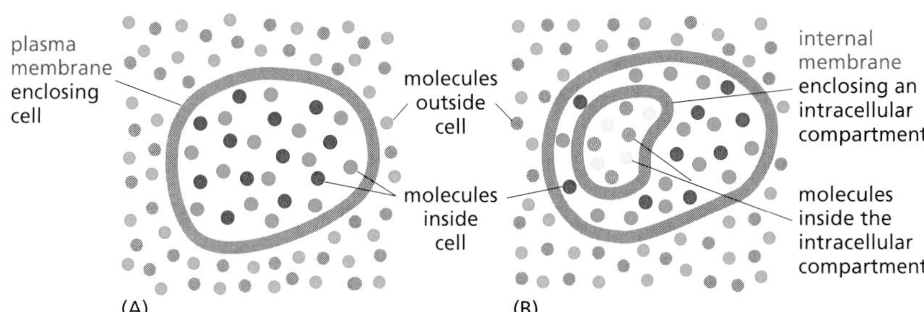

Figure 11–1 **Cell membranes act as selective barriers.** (A) The plasma membrane separates a cell from the outside and is the only membrane in most bacterial cells. It enables the molecular composition of a cell to differ from that of the cell's environment. (B) In eucaryotic cells, additional internal membranes enclose individual organelles. In both cases, the membrane prevents molecules on one side from mixing with those on the other.

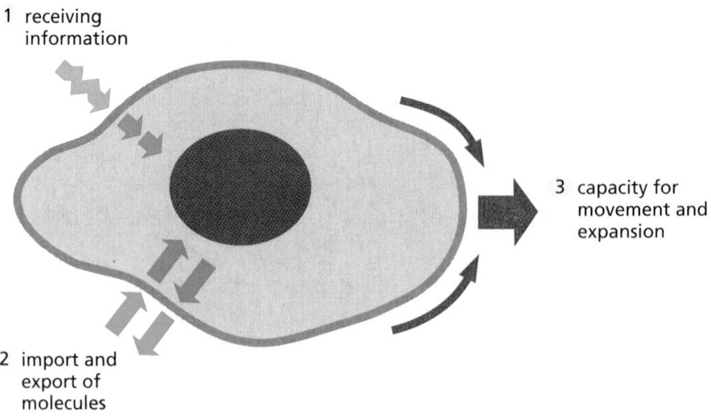

Figure 11–2 **The plasma membrane is involved in cell communication, import and export of molecules, and cell growth and motility.** (1) Receptor proteins in the plasma membrane enable the cell to receive signals from the environment; (2) transport proteins in the membrane enable the import and export of small molecules; (3) the flexibility of the membrane and its capacity for expansion allow cell growth and cell movement.

and others to be exported. Other proteins in the membrane act as sensors that enable the cell to receive information about changes in its environment and respond to them (Figure 11–2). The mechanical properties of the membrane are equally remarkable. When a cell grows or changes shape, so does its membrane: it enlarges in area by adding new membrane without ever losing its continuity, and it can deform without tearing. If the membrane is pierced, it neither collapses like a balloon nor remains torn; instead, it quickly reseals.

The simplest bacteria have only a single membrane—the plasma membrane. Eucaryotic cells, however, also contain an abundance of *internal membranes* that enclose intracellular compartments to form the various organelles, including the endoplasmic reticulum, Golgi apparatus, and mitochondria (Figure 11–3). These internal membranes are constructed on the same principles as the plasma membrane, and they, too, serve as highly selective barriers between spaces containing distinct collections of molecules (see Figure 11–1B). Subtle differences in the composition of these membranes, especially in their resident proteins, give each organelle its distinctive character.

Regardless of their location, all cell membranes are composed of lipids and proteins and share a common general structure (Figure 11–4). The lipids are arranged in two closely apposed sheets, forming a *lipid bilayer* (see Figure 11–4B and C). This lipid bilayer gives the membrane its basic structure and serves as a permeability barrier to most water-soluble molecules. The proteins carry out most of the other functions of the membrane and give different membranes their individual characteristics.

In this chapter we consider the structure of biological membranes and the organization of their two main constituents: lipids and proteins. Although we focus mainly on the plasma membrane, most of the concepts we discuss also apply to internal membranes. The functions of cell membranes, including their role in cell communication, the transport of small molecules, and energy generation, are considered in later chapters.

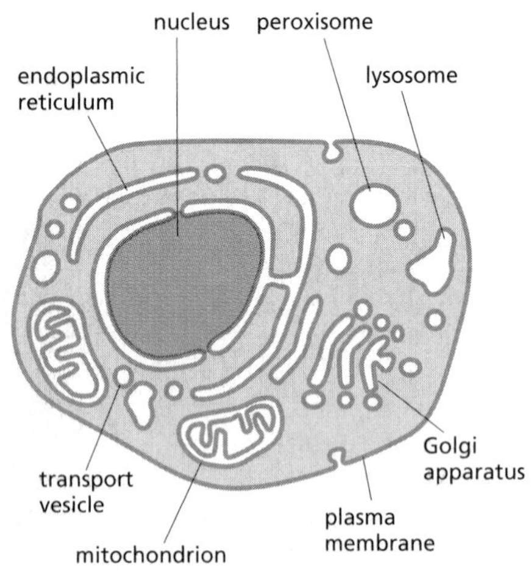

Figure 11–3 **Membranes form the many different compartments in a eucaryotic cell.** The membrane-enclosed organelles in a typical animal cell are shown here. Note that the nucleus and mitochondria are each enclosed by two membranes.

THE LIPID BILAYER

The **lipid bilayer** has been firmly established as the universal basis of membrane structure, and its properties are responsible for the general properties of all cell membranes. Because cells are filled with—and surrounded by—solutions of molecules in water, we begin this section by considering how the structure of cell membranes is a consequence of the way membrane lipids behave in a watery (aqueous) environment.

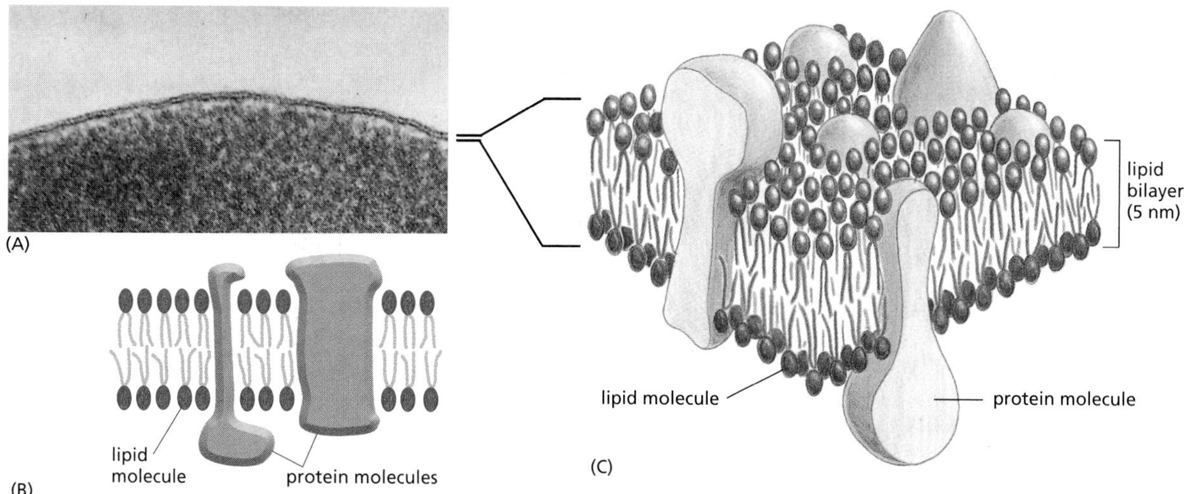

Figure 11–4 **A cell membrane can be viewed in a number of ways.** (A) An electron micrograph of a plasma membrane of a human red blood cell seen in cross section. (B and C) Schematic drawings showing two-dimensional and three-dimensional views of a cell membrane. (A, courtesy of Daniel S. Friend.)

Membrane Lipids Form Bilayers in Water

The lipids in cell membranes combine two very different properties in a single molecule: each lipid has a hydrophilic ("water-loving") *head* and one or two hydrophobic ("water-fearing") *hydrocarbon tails* (Figure 11–5). The most abundant lipids in cell membranes are the **phospholipids**, molecules in which the hydrophilic head is linked to the rest of the lipid through a phosphate group. The most common type of phospholipid in most cell membranes is **phosphatidylcholine**, which has the small molecule choline attached to a phosphate as its hydrophilic head and two long hydrocarbon chains as its hydrophobic tails (Figure 11–6).

Molecules with both hydrophilic and hydrophobic properties are termed **amphipathic**. This chemical property is also shared by other types of membrane lipids, including the *sterols* (such as the cholesterol found in animal cell membranes) and the *glycolipids*, which have sugars as part of their hydrophilic head (Figure 11–7). Having both hydrophilic and hydrophobic parts plays a crucial part in driving these lipid molecules to assemble into bilayers in an aqueous environment.

As discussed in Chapter 2, hydrophilic molecules dissolve readily in water because they contain charged atoms or polar groups, that is, chemical groups with an uneven distribution of positive and negative charges; these charged atoms can form electrostatic attractions or hydrogen bonds with water molecules, which are themselves polar (Figure 11–8). Hydrophobic molecules, by contrast, are insoluble in water because all—or almost all—of their atoms are uncharged and nonpolar; they therefore cannot form favorable interactions with water molecules. Instead, these nonpolar atoms force adjacent water molecules to reorganize into a cagelike structure around the hydrophobic molecule (Figure 11–9). Because the cagelike structure is more highly ordered than the surrounding water, its formation requires energy. The energy cost is minimized, however, if the hydrophobic molecules cluster together, limiting their contact with water to the smallest possible number of water molecules. Thus, purely hydrophobic molecules, like the fats found in animal fat cells and the oils found in plant seeds (Figure 11–10A), coalesce into a single large drop when dispersed in water.

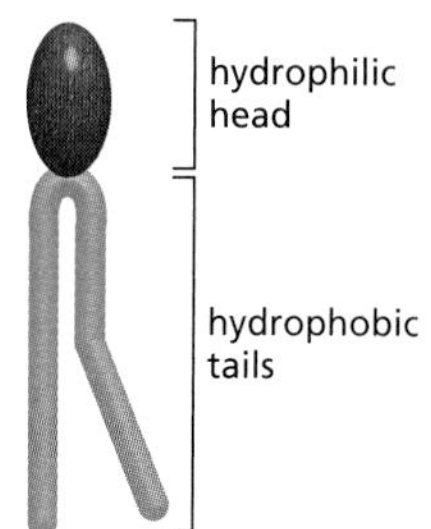

Figure 11–5 **A typical membrane lipid molecule has a hydrophilic head and hydrophobic tails.**

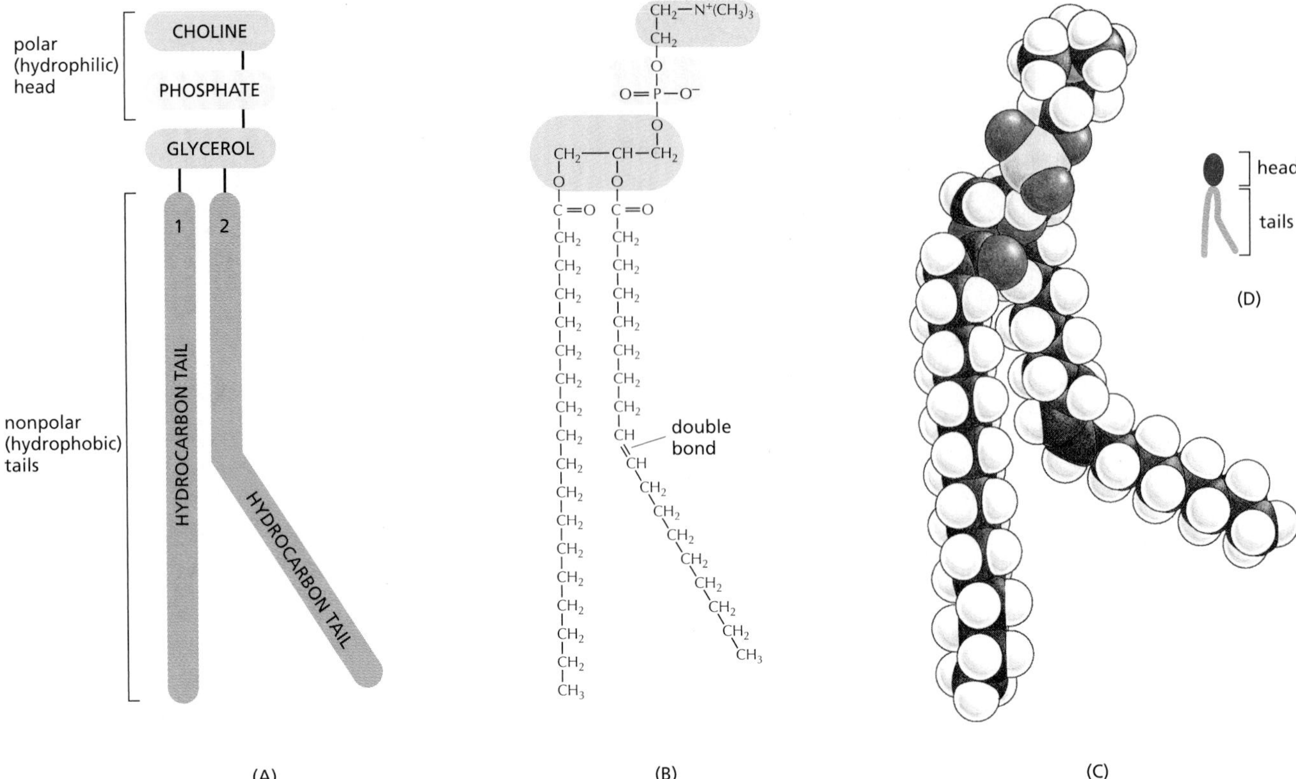

Figure 11–6 **Phosphatidylcholine is the most common phospholipid in cell membranes.** It is represented (A) schematically, (B) in formula, (C) as a space-filling model, and (D) as a symbol. This particular phospholipid is built from five parts: the hydrophilic head, *choline*, is linked via a *phosphate* to *glycerol*, which in turn is linked to two *hydrocarbon chains*, forming the hydrophobic tail. The two hydrocarbon chains originate as *fatty acids*—that is, hydrocarbon chains with a –COOH group at one end—which become attached to glycerol via their –COOH groups. A kink in one of the hydrocarbon chains occurs where there is a double bond between two carbon atoms; it is exaggerated in these drawings for emphasis. The 'phosphatidyl' part of the name of phospholipids refers to the phosphate–glycerol–fatty acid portion of the molecule.

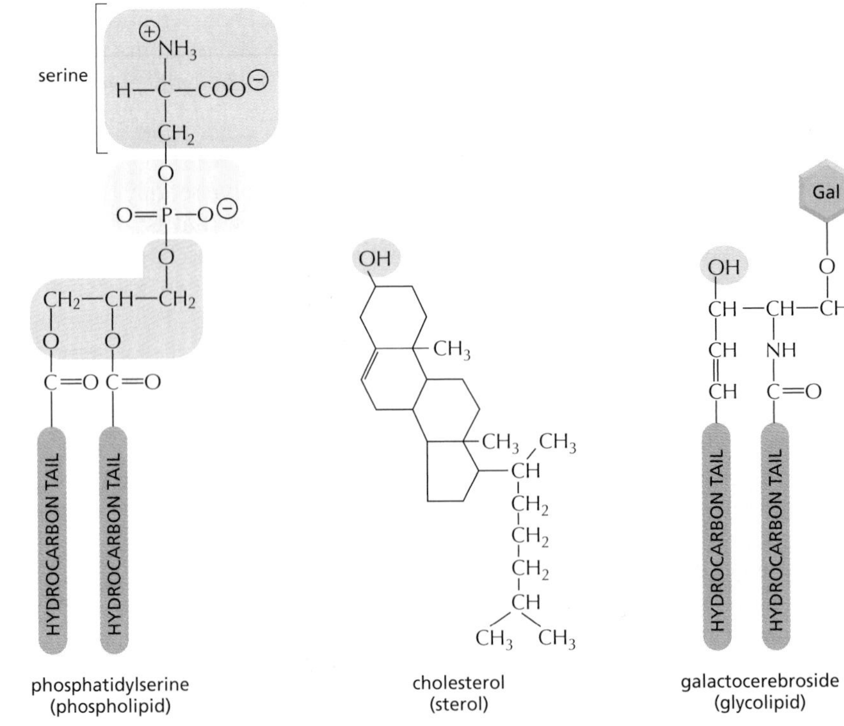

Figure 11–7 **Different types of membrane lipids are all amphipathic.** Each of the three types shown here has a hydrophilic head and one or two hydrophobic tails. The hydrophilic head (shaded *blue* and *yellow*) is serine phosphate in phosphatidylserine, an –OH group in cholesterol, and a sugar (galactose) and an –OH group in galactocerebroside. See also Panel 2–4, pp. 70–71.

Figure 11–8 **A hydrophilic molecule attracts water molecules.** Because acetone is polar, it can form favorable interactions with water molecules, which are also polar. Thus acetone readily dissolves in water. δ^- indicates a partial negative charge, and δ^+ indicates a partial positive charge. Polar atoms are shown in color (*red* and *blue*); nonpolar groups are shown in *gray*.

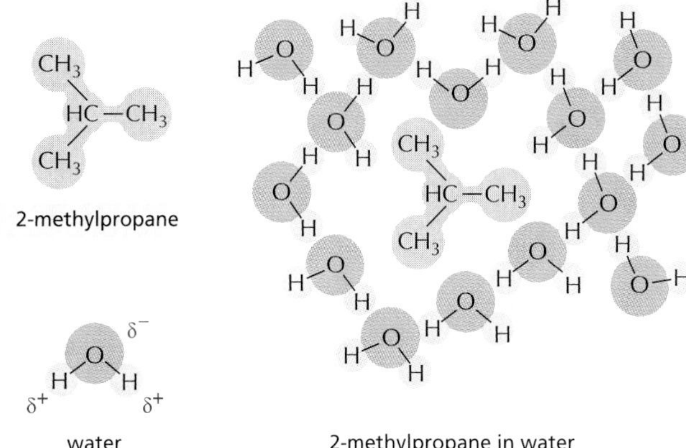

Figure 11–9 **A hydrophobic molecule tends to avoid water.** Because the 2-methylpropane molecule is entirely hydrophobic, it cannot form favorable interactions with water, and forces adjacent water molecules to reorganize into a cagelike structure around it.

In contrast, amphipathic molecules, such as phospholipids (Figure 11–10B), are subject to two conflicting forces: the hydrophilic head is attracted to water, while the hydrophobic tail shuns water and seeks to aggregate with other hydrophobic molecules. This conflict is beautifully resolved by the formation of a lipid bilayer—an arrangement that satisfies all parties and is energetically most favorable. The hydrophilic heads face the water from both surfaces of the bilayer sheet; the hydrophobic tails are all shielded from the water as they lie next to one another in the interior, like the filling in a sandwich (Figure 11–11).

The same forces that drive the amphipathic molecules to form a bilayer make the bilayer self-sealing. Any tear in the sheet will create a free edge that is exposed to water. Because this situation is energetically unfavorable,

QUESTION 11–1

Water molecules are said "to reorganize into a cagelike structure" around hydrophobic compounds (e.g., see Figure 11–9). This seems paradoxical because water molecules do not interact with the hydrophobic compound. So how could they 'know' about its presence and change their behavior to interact differently with one another? Discuss this argument and in doing so develop a clear concept of what is meant by a 'cagelike' structure. How does it compare to ice? Why would this cagelike structure be energetically unfavorable?

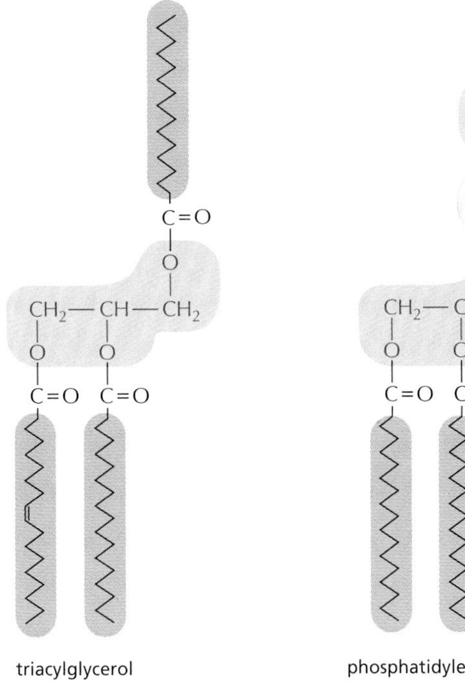

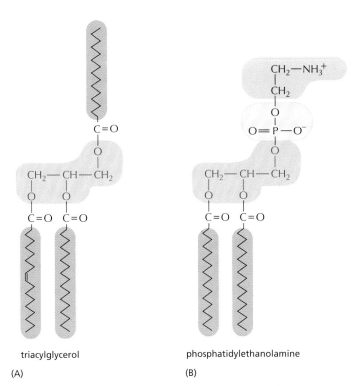

Figure 11–10 **Fat molecules are hydrophobic, whereas phospholipids are amphipathic.** (A) Triacylglycerols, which are the main constituents of animal fats and plant oils, are entirely hydrophobic. (B) Phospholipids such as phosphatidylethanolamine are amphipathic, containing both hydrophobic and hydrophilic portions. The hydrophobic parts are shaded *red*, and the hydrophilic parts are shaded *blue* and *yellow*. (The third hydrophobic tail of the triacylglycerol molecule is drawn here facing upward for comparison with the phospholipid, although normally it is depicted facing down.)

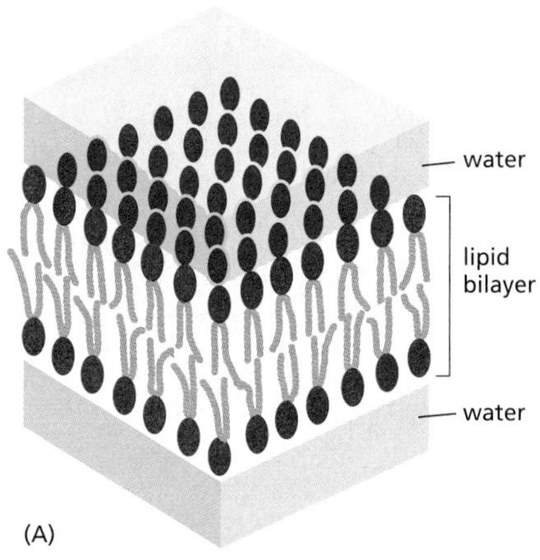

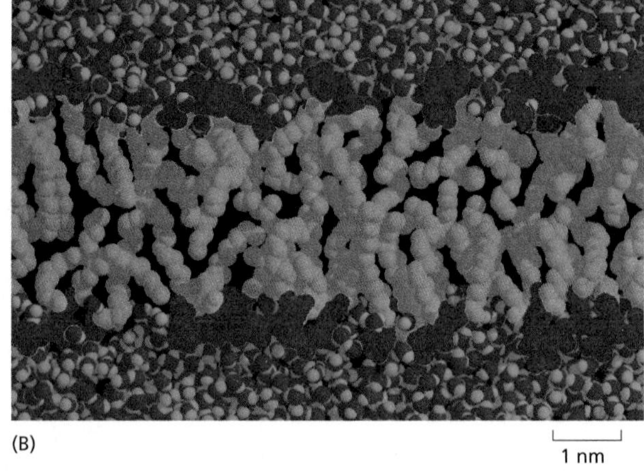

Figure 11–11 **Amphipathic phospholipids form a bilayer in water.** (A) Schematic drawing of a phospholipid bilayer in water. (B) Computer simulation showing the phospholipid molecules (*red* heads and *orange* tails) and the surrounding water molecules (*blue*) in a cross section of a lipid bilayer. (B, adapted from *Science* 262:223–228, 1993, with permission from the AAAS; courtesy of R. Venable and R. Pastor.)

the molecules of the bilayer will spontaneously rearrange to eliminate the free edge. If the tear is small, this spontaneous rearrangement will exclude the water molecules and lead to repair of the bilayer, restoring a single continuous sheet. If the tear is large, the sheet may begin to fold in on itself and break up into separate closed vesicles. In either case, the overriding principle is that free edges are quickly eliminated.

The prohibition on free edges has a profound consequence: the only way a finite sheet can avoid having free edges is to bend and seal, forming a boundary around a closed space (Figure 11–12). Therefore, amphipathic molecules such as phospholipids necessarily assemble into self-sealing containers that define closed compartments. This remarkable behavior, fundamental to the creation of a living cell, is in essence simply a result of the property that each molecule is hydrophilic at one end and hydrophobic at the other.

The Lipid Bilayer Is a Two-dimensional Fluid

The aqueous environment inside and outside a cell prevents membrane lipids from escaping from the bilayer, but nothing stops these molecules from moving about and changing places with one another within the plane of the bilayer. The membrane therefore behaves as a two-dimensional fluid, which is crucial for membrane function and integrity (Movie 11.1). This property is distinct from *flexibility*, which is the ability of the membrane to bend. Membrane flexibility is also important, and it sets a lower limit of about 25 nm to the size of vesicle that cell membranes can form.

The fluidity of lipid bilayers can be studied using synthetic lipid bilayers, which are easily produced by the spontaneous aggregation of amphipathic lipid molecules in water. Two types of synthetic lipid bilayers are commonly used in experiments. Closed spherical vesicles, called *liposomes,* form if pure phospholipids are added to water; they vary in size from about 25 nm to 1 mm in diameter (Figure 11–13). Alternatively, flat phospholipid bilayers can be formed across a hole in a partition between two aqueous compartments (Figure 11–14).

These simple artificial bilayers allow measurements of the movements of the lipid molecules, revealing that some types of movement are rare while others are frequent and rapid. Thus, in synthetic lipid bilayers, phospholipid molecules very rarely tumble from one half of the bilayer, or monolayer, to the other. Without proteins to facilitate the process and under conditions similar to those in a cell, it is estimated that this event, called 'flip-flop,' occurs less than once a month for any individual lipid molecule. On the other hand, as the result of thermal motions, lipid molecules within a monolayer continuously exchange places with their

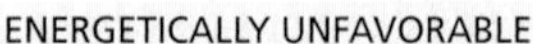

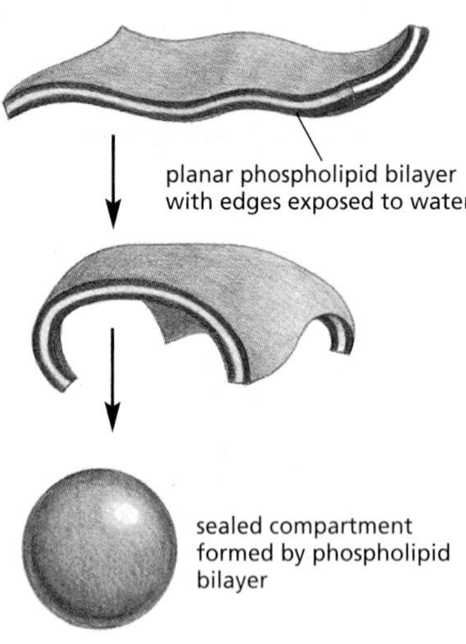

Figure 11–12 **Phospholipid bilayers spontaneously close in on themselves to form sealed compartments.** The closed structure is stable because it avoids the exposure of the hydrophobic hydrocarbon tails to water, which would be energetically unfavorable.

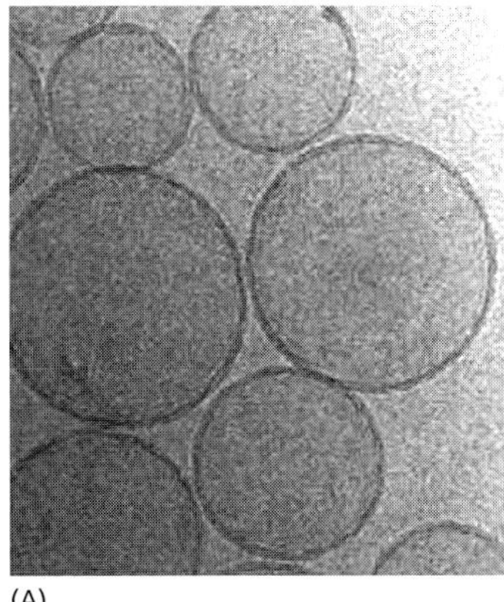
(A)

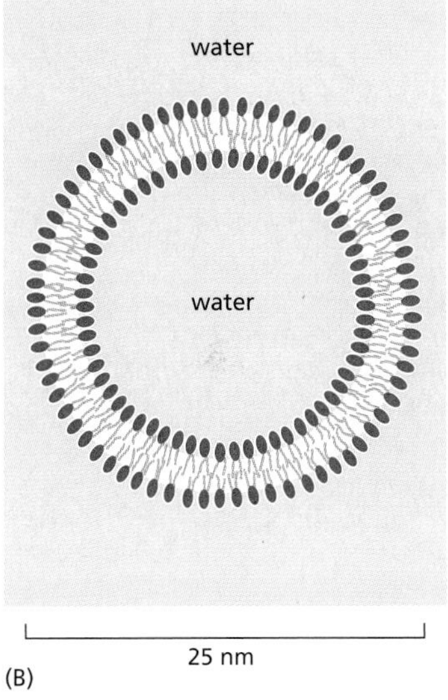

(B)

Figure 11–13 **Pure phospholipids can form closed, spherical liposomes.** (A) An electron micrograph of phospholipid vesicles (liposomes) showing the bilayer structure of the membrane. (B) A drawing of a small spherical liposome seen in cross section. (A, courtesy of Jean Lepault.)

neighbors (Figure 11–15). This exchange leads to rapid diffusion of lipid molecules in the plane of the membrane so that, for example, a lipid in an artificial bilayer may diffuse a length equal to that of a large bacterial cell (~2 μm) in about one second. If the temperature is decreased, the drop in thermal energy decreases the rate of lipid movement, making the bilayer less fluid.

Similar results are obtained when one examines isolated cell membranes and whole cells, indicating that the lipid bilayer of a cell membrane also behaves as a two-dimensional fluid in which the constituent lipid molecules are free to move within their own layer in any direction in the plane of the membrane. These studies also show that lipid hydrocarbon chains are flexible and that individual lipid molecules within a monolayer rotate very rapidly about their long axis—some reaching speeds of 30,000 revolutions per minute (see Figure 11–15). In cells, as in synthetic bilayers, individual phospholipid molecules are normally confined to their own monolayer and do not flip-flop spontaneously.

The Fluidity of a Lipid Bilayer Depends on Its Composition

The fluidity of a cell membrane—the ease with which its lipid molecules move within the plane of the bilayer—is important for membrane function and has to be maintained within certain limits. Just how fluid a lipid bilayer is at a given temperature depends on its phospholipid composition and, in particular, on the nature of the hydrocarbon tails: the closer and more regular the packing of the tails, the more viscous and less fluid the bilayer will be. Two major properties of hydrocarbon tails affect how tightly they pack in the bilayer: their length and the number of double bonds they contain.

A shorter chain length reduces the tendency of the hydrocarbon tails to interact with one another and therefore increases the fluidity of the bilayer. The hydrocarbon tails of membrane phospholipids vary in length between 14 and 24 carbon atoms, with 18–20 atoms being most usual. Most phospholipids contain one hydrocarbon tail that has one or more double bonds between adjacent carbon atoms, and a second tail with single bonds only (see Figure 11–6). The chain that harbors a double bond does not contain the maximum number of hydrogen atoms that could, in principle, be attached to its carbon backbone; it is thus said to be **unsaturated** with respect to hydrogen. The fatty acid tail with no double bonds has a full complement of hydrogen atoms; it is said to be **saturated**. Each double bond in an unsaturated tail creates a small kink in the hydro-

QUESTION 11–2

Five students in your class always sit together in the front row. This could be because (A) they really like each other or (B) nobody else in your class wants to sit next to them. Which explanation holds for the assembly of a lipid bilayer? Explain. Suppose instead that the other explanation held for lipid molecules. How would the properties of the lipid bilayer be different?

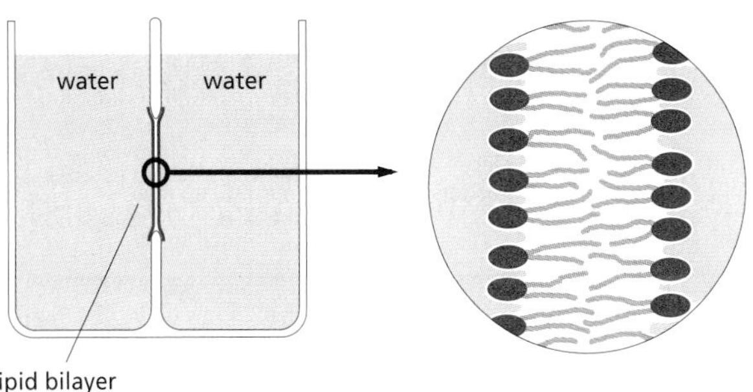

Figure 11–14 **A synthetic phospholipid bilayer can be formed across a small hole (about 1 mm in diameter) in a partition that separates two aqueous compartments.** To make the planar bilayer, the partition is submerged in an aqueous solution and a phospholipid solution (in a nonaqueous solvent) is painted across the hole with a paintbrush.

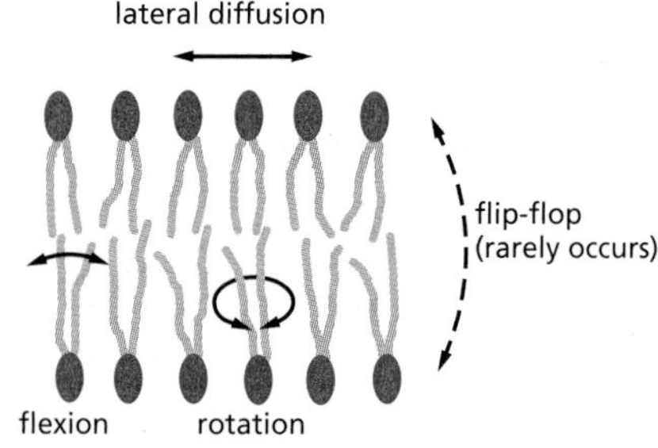

Figure 11–15 **Phospholipids can move within the plane of the membrane.** The drawing shows the types of movement possible for phospholipid molecules in a lipid bilayer.

carbon tail (see Figure 11–6), which makes it more difficult for the tails to pack against one another. For this reason, lipid bilayers that contain a large proportion of unsaturated hydrocarbon tails are more fluid than those with lower proportions.

In bacterial and yeast cells, which have to adapt to varying temperatures, both the lengths and the unsaturation of the hydrocarbon tails in the bilayer are constantly adjusted to maintain the membrane at a relatively constant fluidity: at higher temperatures, for example, the cell makes membrane lipids with tails that are longer and that contain fewer double bonds. A similar trick is used in the manufacture of margarine from vegetable oils. The fats produced by plants are generally unsaturated and therefore liquid at room temperatures, unlike animal fats such as butter or lard, which are generally saturated and therefore solid at room temperature. Margarine is made of hydrogenated vegetable oils; their double bonds have been removed by the addition of hydrogen so that they are more solid and butterlike at room temperature.

In animal cells, membrane fluidity is modulated by the inclusion of the sterol **cholesterol** (Figure 11–16A). This molecule is present in especially large amounts in the plasma membrane, where it constitutes approximately 20% of the lipids in the membrane by weight. Because cholesterol molecules are short and rigid, they fill the spaces between neighboring phospholipid molecules left by the kinks in their unsaturated hydrocarbon tails (Figure 11–16B). In this way, cholesterol tends to stiffen the bilayer, making it more rigid and less permeable. The chemical properties of membrane lipids—and how they affect membrane fluidity—are reviewed in Movie 11.2.

For all cells, membrane fluidity is important for many reasons. It enables membrane proteins to diffuse rapidly in the plane of the bilayer and to interact with one another, as is crucial, for example, in cell signaling (discussed in Chapter 16). It permits membrane lipids and proteins to diffuse from sites where they are inserted into the bilayer after their synthesis to other regions of the cell. It allows membranes to fuse with one another and mix their molecules, and it ensures that membrane molecules are distributed evenly between daughter cells when a cell divides. If biological membranes were not fluid, it is hard to imagine how cells could live, grow, and reproduce.

The Lipid Bilayer Is Asymmetrical

Cell membranes are generally asymmetrical: they present a very different face to the interior of the cell or organelle than they show to the exterior. The two halves of the bilayer often include strikingly different sets of

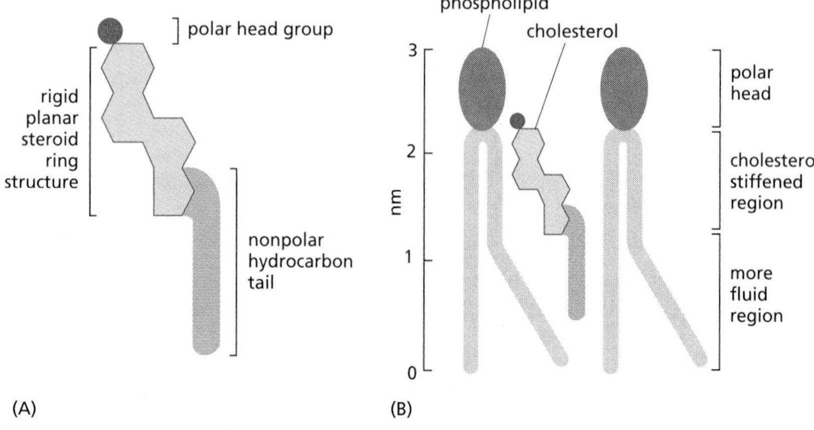

Figure 11–16 **Cholesterol tends to stiffen cell membranes.** (A) The structure of cholesterol. (B) How cholesterol fits into the gaps between phospholipid molecules in a lipid bilayer. The chemical formula of cholesterol is shown in Figure 11–7.

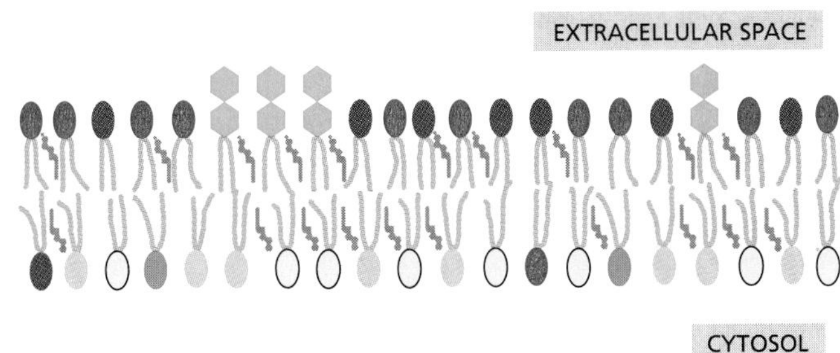
EXTRACELLULAR SPACE
CYTOSOL

Figure 11–17 **Phospholipids and glycolipids are distributed asymmetrically in the plasma membrane lipid bilayer.** Five types of phospholipid molecules are shown in different colors: phosphatidylcholine (*red*), sphingomyelin (*brown*), phosphatidylserine (*light green*), phosphatidylinositol (*dark green*), and phosphatidylethanolamine (*yellow*). The glycolipids are drawn with *blue* hexagonal head groups to represent sugars. All of the glycolipid molecules are in the external monolayer of the membrane, while cholesterol (*gray*) is distributed almost equally in both monolayers. Phosphatidylinositol (not shown) is a minor lipid always found in the cytosolic monolayer of the plasma membrane, where it is used in cell signaling. Since its head group is an inositol sugar it is therefore an exception, being localized differently from all other glycolipids.

phospholipids and glycolipids (Figure 11–17). Moreover, membrane proteins are embedded in the bilayer with a specific orientation, which is crucial for their function.

The lipid asymmetry is established and maintained as the membrane grows. In eucaryotic cells, new phospholipids are manufactured by enzymes bound to the part of the endoplasmic reticulum membrane that faces the cytosol. These enzymes, which use free fatty acids as substrates (see Panel 2–4, pp. 70–71), deposit all newly made phospholipids into the cytosolic half of the bilayer. To enable the membrane as a whole to grow evenly, half of the new phospholipid molecules then have to be transferred to the opposite monolayer. This transfer is catalyzed by enzymes called *flippases* (Figure 11–18). In the plasma membrane, flippases transfer specific phospholipids selectively, so that different types become concentrated in each monolayer.

Using selective flippases is not the only way to produce asymmetry in lipid bilayers, however. In particular, a different mechanism operates for glycolipids—the lipids that show the most striking and consistent asymmetric distribution in animal cells (see Figure 11–17). To explain their distribution, it is necessary to take a more detailed look at how new membrane is produced in eucaryotic cells.

Lipid Asymmetry Is Preserved During Membrane Transport

Nearly all new membrane synthesis in eucaryotic cells occurs in the membrane of one intracellular compartment—the *endoplasmic reticulum* (ER; see Figure 11–3). The new membrane assembled there is exported to the other membranes of the cell through a cycle of membrane budding and fusion: bits of the bilayer pinch off from the ER to form small spheres called *vesicles*, which then become incorporated into another membrane, such as the plasma membrane, by fusing with it (Figure 11–19). The orientation of the bilayer relative to the cytosol is preserved during vesicle formation and fusion. This preservation of orientation means that all cell membranes, whether the external plasma membrane or an intracellular membrane around an organelle, have distinct 'inside' and 'outside' faces that are established at the time of membrane synthesis: the *cytosolic* face is always adjacent to the cytosol, while the *noncytosolic* face is exposed to either the cell exterior or the interior space of an organelle (see Figure 11–19).

Glycolipids are located mainly in the plasma membrane, and they are found only in the noncytosolic half of the bilayer. Their sugar groups are therefore exposed to the exterior of the cell (see Figure 11–17), where they form part of a continuous protective coat of carbohydrate that surrounds most animal cells. The glycolipid molecules acquire their sugar groups in the Golgi apparatus, the organelle to which proteins and membranes made in the ER often go next (discussed in Chapter 15). The enzymes that add the sugar groups are confined to the inside of the Golgi apparatus, so that the sugars are added only to lipid molecules in the noncytosolic half

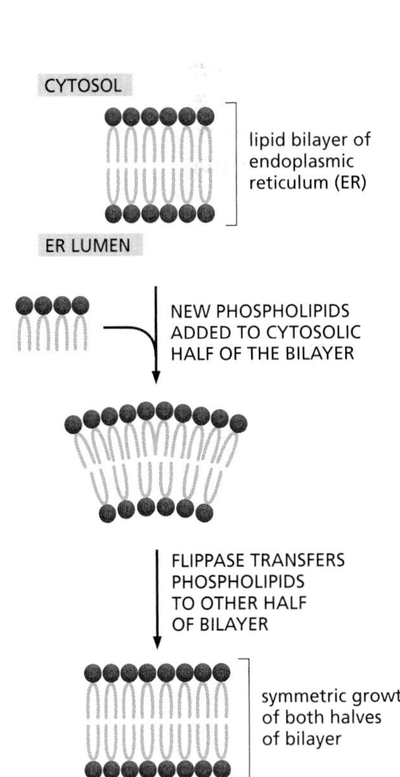
CYTOSOL
lipid bilayer of endoplasmic reticulum (ER)
ER LUMEN
NEW PHOSPHOLIPIDS ADDED TO CYTOSOLIC HALF OF THE BILAYER
FLIPPASE TRANSFERS PHOSPHOLIPIDS TO OTHER HALF OF BILAYER
symmetric growth of both halves of bilayer

Figure 11–18 **Flippases play a role in synthesizing the lipid bilayer.** Newly synthesized phospholipid molecules are all added to the cytosolic side of the ER membrane. Flippases then transfer some of these molecules to the opposite monolayer, so that the entire bilayer expands.

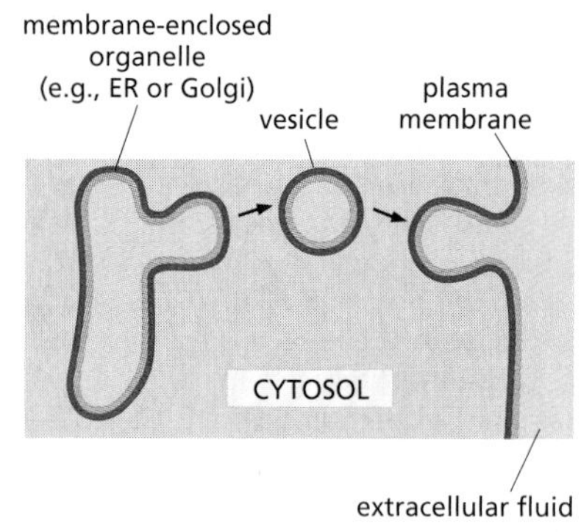

Figure 11–19 **Membranes retain their orientation even after transfer between cell compartments.** Membranes are transported by a process of vesicle budding and fusing. Here, a vesicle is shown budding from a membrane-enclosed organelle (such as the endoplasmic reticulum or Golgi) and fusing with the plasma membrane. Note that the orientation of the membrane is preserved during the process: the original cytosolic surface (*red*) remains facing the cytosol, and the noncytosolic surface (*orange*) continues to face away from the cytosol, toward the lumen of the organelle or vesicle or toward the extracellular fluid.

of the lipid bilayer. Once a glycolipid molecule has been created in this way, it remains trapped in this monolayer, as there are no flippases that transfer glycolipids to the cytosolic monolayer. Thus, when a glycolipid molecule is finally delivered to the plasma membrane, it faces away from the cytosol and displays its sugar on the exterior of the cell (see Figure 11–19).

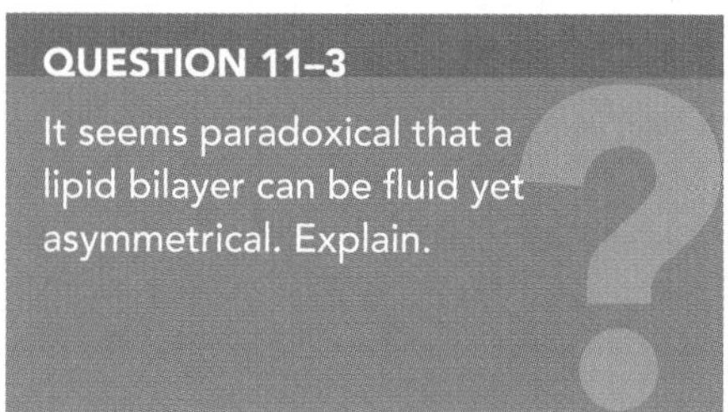

QUESTION 11–3

It seems paradoxical that a lipid bilayer can be fluid yet asymmetrical. Explain.

Other lipid molecules show different types of asymmetric distributions, related to other functions. The *inositol phospholipids*, for example, are minor components of the plasma membrane, but they play a special role in relaying signals from the cell surface to the intracellular components that respond to those signals (discussed in Chapter 16). They act only after the signal has been transmitted across the plasma membrane; thus they are concentrated in the cytosolic half of this lipid bilayer (see Figure 11–17).

MEMBRANE PROTEINS

Although the lipid bilayer provides the basic structure of all cell membranes and serves as a permeability barrier to the molecules on either side of it, most membrane functions are carried out by **membrane proteins**. In animals, proteins constitute about 50% of the mass of most plasma membranes, the remainder being lipid plus the relatively small amounts of carbohydrate found on glycolipids and glycosylated proteins. Because lipid molecules are much smaller than proteins, however, a cell membrane typically contains about 50 times more lipid molecules than protein molecules (see Figure 11–4).

Membrane proteins not only transport particular nutrients, metabolites, and ions across the lipid bilayer; they serve many other functions. Some anchor the membrane to macromolecules on either side. Others function as receptors that detect chemical signals in the cell's environment and relay them to the cell interior, and still others work as enzymes to catalyze specific reactions (Figure 11–20 and Table 11–1). Each type of cell membrane contains a different set of proteins, reflecting the specialized functions of the particular membrane. In this section, we discuss the structure of membrane proteins and illustrate the different ways that they can be associated with the lipid bilayer.

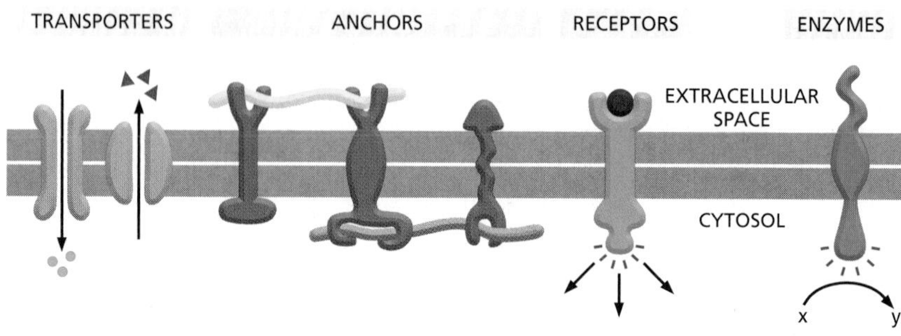

Figure 11–20 **Plasma membrane proteins have a variety of functions.**

TABLE 11–1 SOME EXAMPLES OF PLASMA MEMBRANE PROTEINS AND THEIR FUNCTIONS

FUNCTIONAL CLASS	PROTEIN EXAMPLE	SPECIFIC FUNCTION
Transporters	Na^+ pump	actively pumps Na^+ out of cells and K^+ in (as described in Chapter 12)
Anchors	integrins	link intracellular actin filaments to extracellular matrix proteins (as discussed in Chapter 20)
Receptors	platelet-derived growth factor (PDGF) receptor	binds extracellular PDGF and, as a consequence, generates intracellular signals that cause the cell to grow and divide (as discussed in Chapter 18)
Enzymes	adenylyl cyclase	catalyzes the production of intracellular signaling molecule cyclic AMP in response to extracellular signals (as detailed in Chapter 16)

Membrane Proteins Associate with the Lipid Bilayer in Various Ways

Proteins can be associated with the lipid bilayer of a cell membrane in several ways (Figure 11–21).

1. Many membrane proteins extend through the bilayer, with part of their mass on either side (Figure 11–21A). Like their lipid neighbors, these *transmembrane proteins* have both hydrophobic and hydrophilic regions. Their hydrophobic regions lie in the interior of the bilayer, nestled against the hydrophobic tails of the lipid molecules. Their hydrophilic regions are exposed to the aqueous environment on either side of the membrane.
2. Other membrane proteins are located entirely in the cytosol, associated with the inner leaflet of the lipid bilayer by an amphipathic α helix exposed on the surface of the protein (Figure 11–21B).
3. Some proteins lie entirely outside the bilayer, on one side or the other, attached to the membrane only by one or more covalently attached lipid groups (Figure 11–21C).
4. Yet other proteins are bound indirectly to one or the other face of the membrane, held in place only by their interactions with other membrane proteins (Figure 11–21D).

Proteins that are directly attached to a lipid bilayer—whether they are transmembrane, monolayer-associated, or lipid-linked—can be removed only by disrupting the bilayer with detergents, as discussed shortly. Such proteins are known as *integral membrane proteins*. The remaining membrane proteins are known as *peripheral membrane proteins*; they can be released from the membrane by more gentle extraction procedures that interfere with protein–protein interactions but leave the lipid bilayer intact.

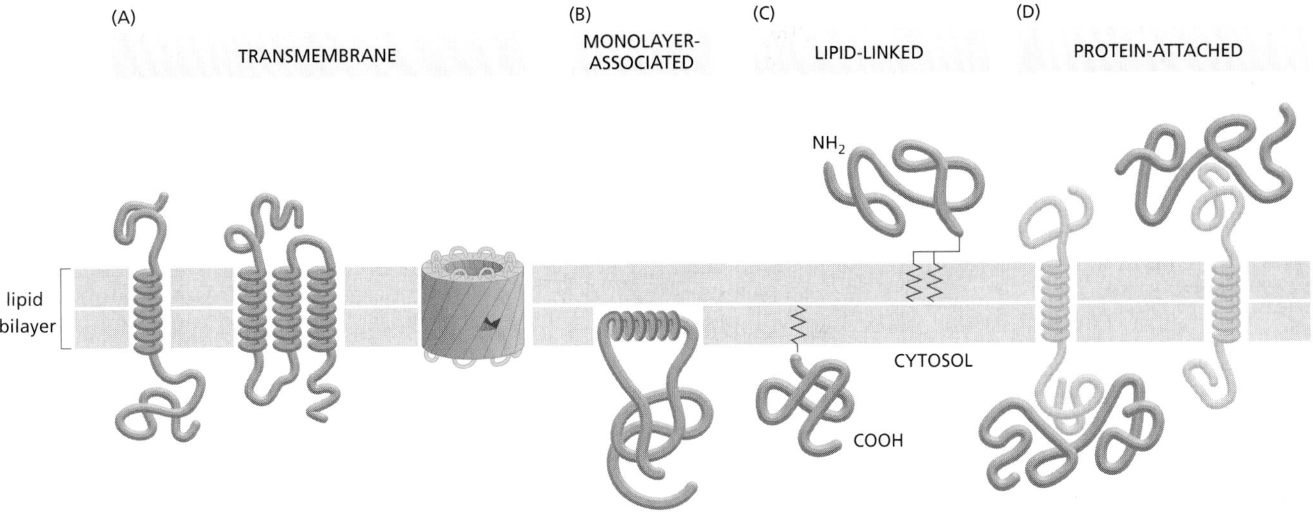

Figure 11–21 **Membrane proteins can associate with the lipid bilayer in several different ways.** (A) Transmembrane proteins can extend across the bilayer as a single α helix, as multiple α helices, or as a rolled-up β sheet (called a β barrel). (B) Some membrane proteins are anchored to the cytosolic surface by an amphipathic α helix. (C) Others are attached to either side of the bilayer solely by a covalent attachment to a lipid molecule (*red* zigzag lines). (D) Finally, many proteins are attached to the membrane only by relatively weak, noncovalent interactions with other membrane proteins.

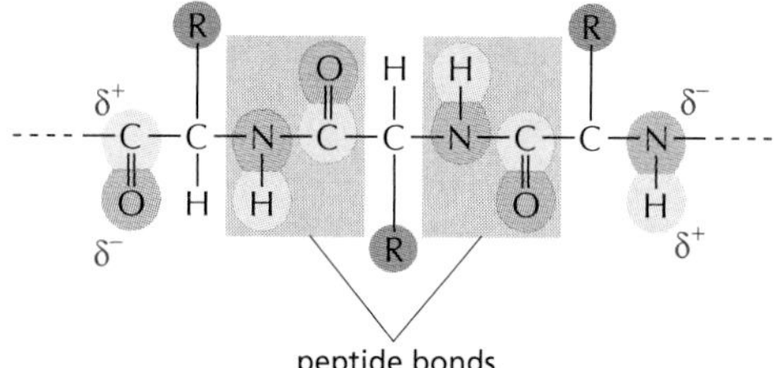

Figure 11–22 **The peptide bonds (shaded in *gray*) that join adjacent amino acids together in a polypeptide chain are polar and therefore hydrophilic.** The partial charges (δ^- indicates a partial negative charge and δ^+ indicates a partial positive charge) allow these atoms to hydrogen bond with one another when the polypeptide folds into an α helix that spans the membrane (see Figure 11–23).

A Polypeptide Chain Usually Crosses the Bilayer as an α Helix

All membrane proteins have a unique orientation in the lipid bilayer, which is essential for their function. For a transmembrane receptor protein, for example, the part of the protein that receives a signal from the environment must be on the outside of the cell, and the part that passes along the signal must be in the cytosol (see Figure 11–20). This orientation is a consequence of the way in which membrane proteins are synthesized (as discussed in Chapter 15). The portions of a transmembrane protein located on either side of the lipid bilayer are connected by specialized membrane-spanning segments of the polypeptide chain (see Figure 11–21A). These segments, which run through the hydrophobic environment of the interior of the lipid bilayer, are composed largely of amino acids with hydrophobic side chains. Because these side chains cannot form favorable interactions with water molecules, they prefer the lipid environment, where no water is present.

In contrast to the hydrophobic side chains, however, the peptide bonds that join the successive amino acids in a protein are normally polar, making the polypeptide backbone hydrophilic (Figure 11–22). Because water is absent from the bilayer, atoms forming the backbone are driven to form hydrogen bonds with one another. Hydrogen-bonding is maximized if the polypeptide chain forms a regular α helix, and so the great majority of the membrane-spanning segments of polypeptide chains traverse the bilayer as α helices (see Figure 4–10). In these membrane-spanning α helices, the hydrophobic side chains are exposed on the outside of the helix, where they contact the hydrophobic lipid tails, while atoms in the polypeptide backbone form hydrogen bonds with one another on the inside of the helix (Figure 11–23).

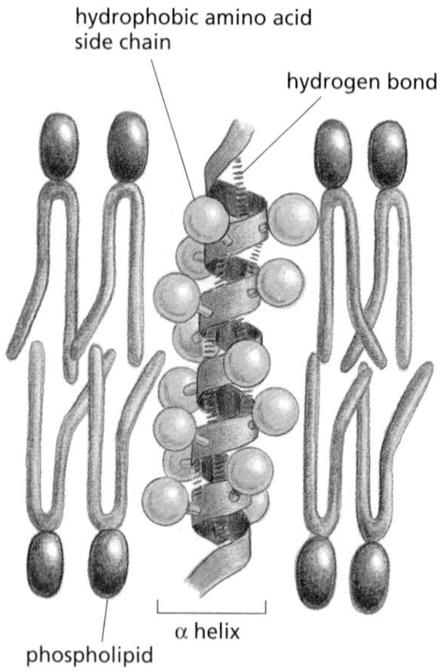

hydrophobic amino acid side chain
hydrogen bond
phospholipid
α helix

Figure 11–23 **A segment of α helix crosses a lipid bilayer.** The hydrophobic side chains of the amino acids forming the α helix contact the hydrophobic hydrocarbon tails of the phospholipid molecules, while the hydrophilic parts of the polypeptide backbone form hydrogen bonds with one another in the interior of the helix. An α helix containing about 20 amino acids is required to completely traverse a membrane.

In many transmembrane proteins the polypeptide chain crosses the membrane only once (see Figure 11–21A). Many of these proteins are receptors for extracellular signals. Other transmembrane proteins form aqueous pores that allow water-soluble molecules to cross the membrane. Such pores cannot be formed by proteins with a single, uniformly hydrophobic, transmembrane α helix. The proteins that form pores are more complicated, usually possessing a series of α helices that cross the bilayer a number of times (see Figure 11–21A). In many of these proteins, one or more of the transmembrane regions are formed from α helices that contain both hydrophobic and hydrophilic amino acid side chains. These amino acids tend to be arranged so that the hydrophobic side chains fall on one side of the helix, while the hydrophilic side chains are concentrated on the other side. In the hydrophobic environment of the lipid bilayer, α-helices of this sort pack side by side in a ring, with the hydrophobic side chains exposed to the lipids of the membrane and the hydrophilic side chains forming the lining of a hydrophilic pore through the lipid bilayer (Figure 11–24). How such pores function in the selective transport of small water-soluble molecules across membranes is discussed in Chapter 12.

Although the α helix is by far the most common form in which a polypeptide chain crosses a lipid bilayer, the polypeptide chain of some transmembrane proteins crosses the lipid bilayer as a β sheet that is curved into a cylinder, forming an open-ended keglike structure called a *β barrel*. As expected, the amino acid side chains that face the inside of the barrel, and therefore line the aqueous channel, are mostly hydrophilic, while those on the outside of the barrel, which contact the hydrophobic core of the lipid bilayer, are exclusively hydrophobic. The most striking example of a β barrel structure is found in the *porin* proteins, which form large, water-filled pores in mitochondrial and bacterial membranes (Figure 11–25). Mitochondria and some bacteria are surrounded by a double membrane, and porins allow the passage of nutrients and small ions across their outer membranes while preventing the entry of larger molecules such as antibiotics and toxins. Unlike α helices, β barrels can form only wide channels, because there is a limit to how tightly the β sheet can be curved to form the barrel. In this respect, a β barrel is less versatile than a collection of α helices.

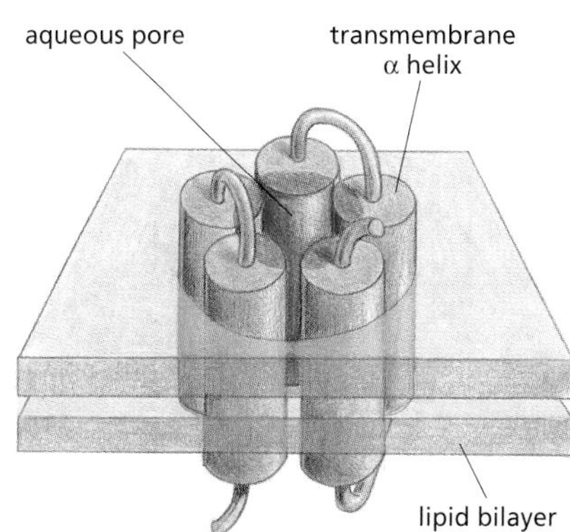

Figure 11–24 **A transmembrane hydrophilic pore can be formed by multiple α helices.** In this example, five transmembrane α helices form a water-filled channel across the lipid bilayer. The hydrophobic amino acid side chains (*green*) on one side of each helix contact the hydrophobic hydrocarbon tails, while the hydrophilic side chains (*red*) on the opposite side of the helices form a water-filled pore.

Membrane Proteins Can Be Solubilized in Detergents and Purified

To understand a protein fully one needs to know its structure in detail, and for membrane proteins this presents special problems. Most biochemical procedures are designed for studying molecules dissolved in water or a simple solvent; membrane proteins, however, are built to operate in an environment that is partly aqueous and partly fatty, and taking them out of this environment and purifying them while preserving their essential structure is no easy task.

Before an individual protein can be studied in detail, it must be separated from all the other cell proteins. For most membrane proteins, the first step in this separation process involves solubilizing the membrane with agents that destroy the lipid bilayer by disrupting hydrophobic associations. The most widely used disruptive agents are **detergents** (Movie 11.3). These are small, amphipathic, lipidlike molecules that have both a hydrophilic and a hydrophobic region (Figure 11–26). Detergents differ from membrane phospholipids in that they have only a single hydrophobic tail and, consequently, behave in a significantly different way. Because they have one tail, detergent molecules are shaped like cones; in water, they tend to aggregate into small clusters called *micelles*, rather than forming a bilayer as do the phospholipids, which are more cylindrical in shape.

QUESTION 11–4

Explain why the polypeptide chain of most transmembrane proteins crosses the lipid bilayer as an α helix or a β barrel.

When mixed in great excess with membranes, the hydrophobic ends of detergent molecules bind to the membrane-spanning hydrophobic region of the transmembrane proteins, as well as to the hydrophobic tails of the phospholipid molecules, thereby disrupting the lipid bilayer and separating the proteins from most of the phospholipids. Because the other end of the detergent molecule is hydrophilic, this association brings the membrane proteins into solution as protein–detergent complexes (Figure 11–27). At the same time, the detergent solubilizes the phospholipids. The protein–detergent complexes can then be separated from one another and from the lipid–detergent complexes by a technique such as polyacrylamide-gel electrophoresis (discussed in Chapter 4).

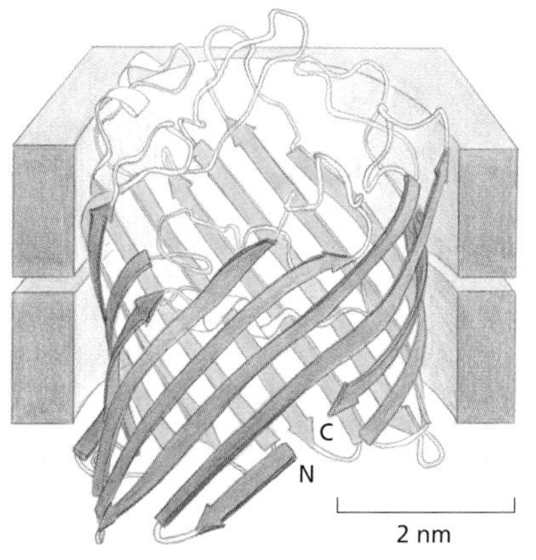

Figure 11–25 **Porin proteins form water-filled channels in the outer membrane of a bacterium (*Rhodobacter capsulatus*).** The protein consists of a 16-stranded β sheet curved around on itself to form a transmembrane water-filled channel, as shown in this three-dimensional structure, determined by X-ray crystallography. Although not shown in the drawing, three porin proteins associate to form a trimer, which has three separate channels. (From S.W. Cowan, *Curr. Opin. Struct. Biol.* 3:501–507, 1993. With permission from Elsevier.)

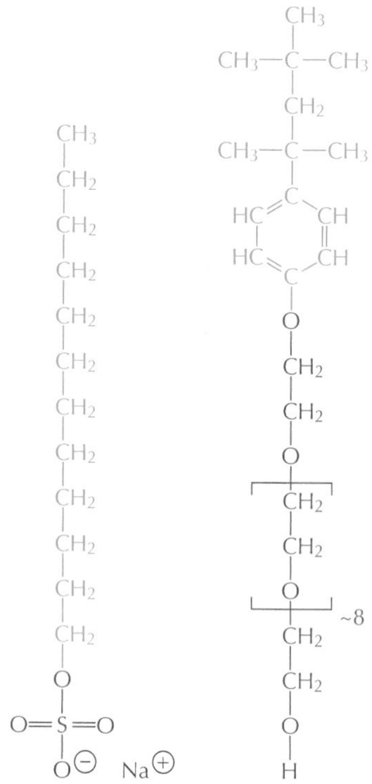

sodium dodecyl sulfate (SDS)

Triton X-100

Figure 11–26 **SDS and Triton X-100 are two commonly used detergents.** Sodium dodecyl sulfate (SDS) is a strong ionic detergent (that is, it has an ionized group at its hydrophilic end), and Triton X-100 is a mild nonionic detergent (that is, it has a nonionized but polar structure at its hydrophilic end). The hydrophobic portion of each detergent is shown in *blue*, and the hydrophilic portion in *red*. The bracketed portion of Triton X-100 is repeated about eight times. Strong ionic detergents like SDS not only displace lipid molecules from proteins but also unfold the proteins (see Panel 4–5, p. 166).

The Complete Structure Is Known for Relatively Few Membrane Proteins

For many years, much of what we knew about the structure of membrane proteins was learned by indirect means. The standard method for determining protein structure directly is X-ray crystallography (see Figure 4–48), but this requires ordered crystalline arrays of the molecule, and membrane proteins proved harder to crystallize than the soluble proteins that inhabit the cell cytosol or extracellular fluids. Nevertheless, with recent advances in crystallography, the X-ray structures of many membrane proteins have now been determined to high resolution, including *bacteriorhodopsin* and a *photosynthetic reaction center*—membrane proteins that have important roles in the capture and use of energy from sunlight, an ability we review in Chapter 14. The structure of bacteriorhodopsin first revealed exactly how α helices cross the lipid bilayer, and the structure of the photosynthetic reaction center showed in detail how a set of different protein molecules can assemble to form a functional complex in a membrane.

Bacteriorhodopsin is a small protein (about 250 amino acids) found in large amounts in the plasma membrane of an archaean, called *Halobacterium halobium*, that lives in salt marshes. Bacteriorhodopsin acts as a membrane transport protein that pumps H^+ (protons) out of the bacterium. Pumping requires energy, and bacteriorhodopsin gets its energy directly from sunlight. Each bacteriorhodopsin molecule contains a single light-absorbing nonprotein molecule—called *retinal*—that gives the protein (and the bacterium) a deep purple color. This small hydrophobic molecule is covalently attached to one of bacteriorhodopsin's seven transmembrane α helices and lies in the plane of the lipid bilayer, entirely

QUESTION 11–5

For the two detergents shown in Figure 11–26, explain why the red portions of the molecules are hydrophilic and the blue portions hydrophobic. Draw a short stretch of a polypeptide chain made up of three amino acids with hydrophobic side chains (see Panel 2–5, pp. 72–73) and apply a similar color scheme.

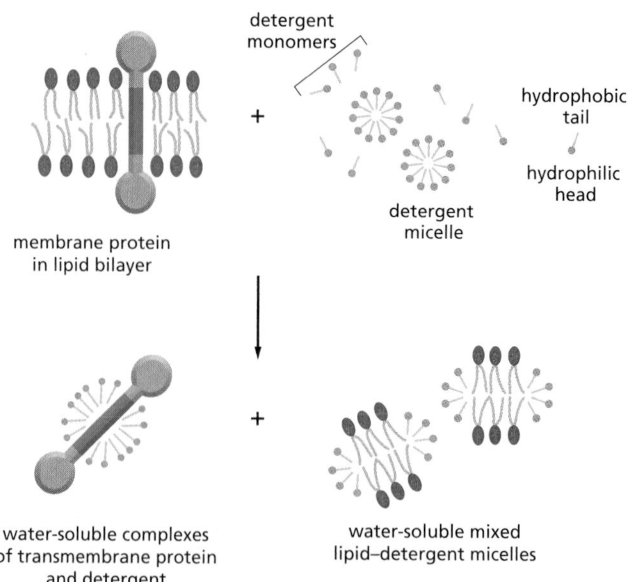

Figure 11–27 **Membrane proteins can be solubilized by a mild detergent such as Triton X-100.** The detergent molecules (*gold*) are shown as both monomers and micelles, the form in which detergent molecules tend to aggregate in water. The hydrophilic head of the detergent is the end with the circle. The detergent disrupts the lipid bilayer and brings the proteins into solution as protein–detergent complexes. The phospholipids in the membrane are also solubilized by the detergents.

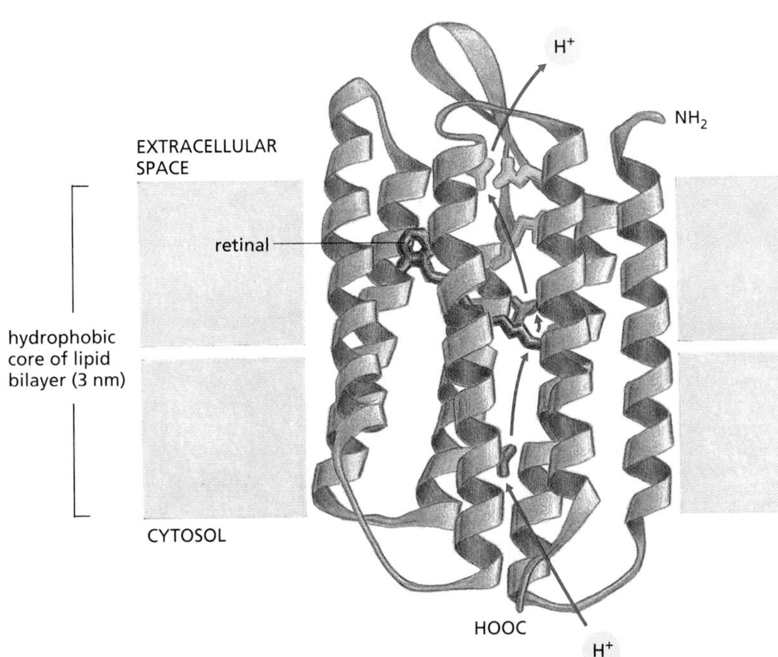

Figure 11–28 **Bacteriorhodopsin acts as a proton pump.** The polypeptide chain crosses the lipid bilayer as seven α helices. The location of the retinal (*purple*) and the probable pathway taken by protons during the light-activated pumping cycle are shown; polar amino acid side chains involved in the H^+ transfer process are shown in *red, yellow,* and *blue*. Note that the pathway taken by the protons (*red* arrows) enables them to avoid contact with the lipid bilayer. The proton transfer steps are shown in Movie 11.4. Retinal is also used to detect light in our own eyes, where it is attached to a protein with a structure very similar to bacteriorhodopsin. (Adapted from H. Luecke et al., *Science* 286:255–260, 1999. With permission from the AAAS.)

surrounded by the seven α helices (Figure 11–28). When retinal absorbs a photon of light, it changes its shape, and in doing so it causes the protein embedded in the lipid bilayer to undergo a series of small conformational changes. These changes result in the transfer of one H^+ from the retinal to the outside of the bacterium: the H^+ moves across the bilayer along a pathway of strategically placed polar amino acid side chains (see Figure 11–28). The retinal is then regenerated by taking up a H^+ from the cytosol, returning the protein to its original conformation so that it can repeat the cycle. The overall outcome is the movement of one H^+ from inside to outside the cell. In the presence of sunlight, thousands of bacteriorhodopsin molecules pump H^+ out of the cell, generating a concentration gradient of H^+ across the plasma membrane. The cell uses this proton gradient to store energy and then convert it into ATP, as we discuss in detail in Chapter 14. Bacteriorhodopsin is a type of *transporter protein*, a class of transmembrane proteins that move molecules and ions into and out of cells (see Figure 11–20). We will meet other transporters in Chapter 12.

The structure of a bacterial photosynthetic reaction center is shown in Figure 11–29. It is a large complex composed of four protein subunits. Three are transmembrane proteins; two of these (M and L) have multiple α helices passing through the lipid bilayer, while the other (H) has only one. The fourth protein (cytochrome) is associated with the outer surface of the membrane, bound to the transmembrane proteins. The entire protein complex serves as a molecular machine, taking in light energy absorbed by chlorophyll and producing high-energy electrons required for photosynthetic reactions (discussed in Chapter 14). Many membrane proteins are arranged in large complexes, and the structure of the photosynthetic reaction center reveals many design principles that apply to thousands of other membrane proteins whose structures are not yet known.

The Plasma Membrane Is Reinforced by the Cell Cortex

A cell membrane by itself is extremely thin and fragile. It would require nearly 10,000 cell membranes laid on top of one another to achieve the thickness of this paper. Most cell membranes are therefore strengthened and supported by a framework of proteins, attached to the membrane via transmembrane proteins. In particular, the shape of a cell and the

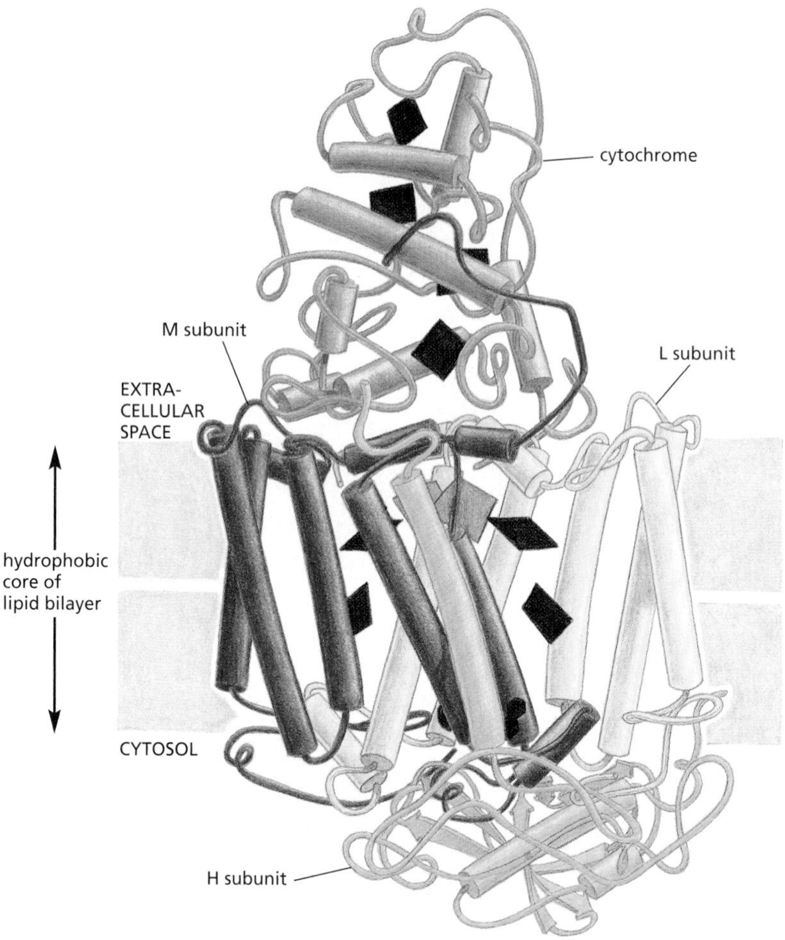

Figure 11–29 **The photosynthetic reaction center of the bacterium *Rhodopseudomonas viridis* captures energy from sunlight.** The three-dimensional structure was determined by X-ray diffraction analysis of crystals of this transmembrane protein complex. The complex consists of four subunits—L, M, H, and a cytochrome. The L and M subunits form the core of the reaction center, and each contains five α helices that span the lipid bilayer. All the α helices are shown as cylinders. The locations of the various electron carrier groups, which are covalently bound to the protein subunits, are shown in *black*, except for the special pair of chlorophyll molecules that are excited by light, which are shown in *dark green*. Note that the cytochrome is bound to the outer surface of the membrane solely by its attachment to the transmembrane subunits (see Figure 11–21D). (Adapted from a drawing by J. Richardson based on data from J. Deisenhofer et al., *Nature* 318:618–624, 1985. With permission from Macmillan Publishers Ltd.)

QUESTION 11–6

Look at the structure of the photosynthetic reaction center in Figure 11–29. As you would expect, many α helices span the membrane. At the lower right-hand corner, however, there is a stretch of the polypeptide chain of the L subunit that forms a disordered loop within the hydrophobic core of the lipid bilayer. Does this invalidate the general rule that transmembrane proteins span the lipid bilayer as α helices or β sheets?

mechanical properties of its plasma membrane are determined by a meshwork of fibrous proteins, called the *cell cortex*, that is attached to the cytosolic surface of the membrane.

The cortex of human red blood cells is a relatively simple and regular structure and is by far the best understood cell cortex. Red blood cells are small and have a distinctive flattened shape (Figure 11–30). The main component of their cortex is the protein *spectrin*, a long, thin, flexible rod about 100 nm in length. It forms a meshwork that provides support for the plasma membrane and maintains the cell's shape. The spectrin meshwork is connected to the membrane through intracellular attachment proteins that link the spectrin to specific transmembrane proteins (Figure 11–31). The importance of this meshwork is seen in mice and

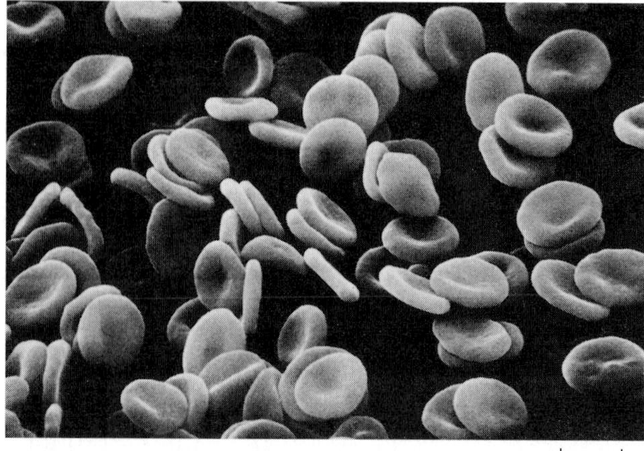

Figure 11–30 **Human red blood cells have a distinctive flattened shape, as seen in this scanning electron micrograph.** These cells lack a nucleus and other intracellular organelles. (Courtesy of Bernadette Chailley.)

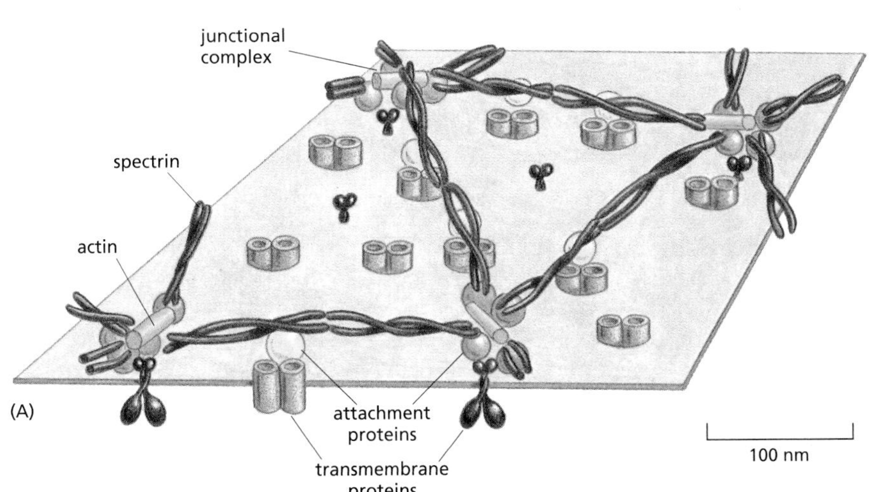

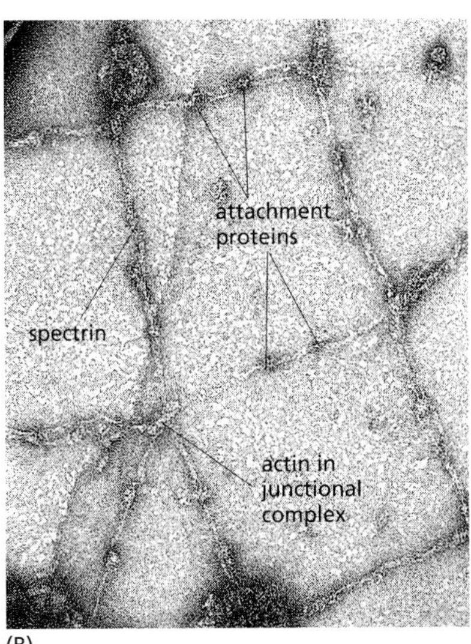

Figure 11–31 **A spectrin meshwork forms the cell cortex in human red blood cells.** (A) Spectrin dimers, together with a smaller number of actin molecules, are linked together into a netlike meshwork that is attached to the plasma membrane by the binding of at least two types of attachment proteins (shown here in *yellow* and *blue*) to two kinds of transmembrane proteins (shown here in *green* and *brown*). (B) Electron micrograph showing the spectrin meshwork on the cytoplasmic side of a red blood cell membrane. The meshwork has been stretched out to show the details of its structure; in the normal cell the meshwork shown would be much more crowded and would occupy only about one-tenth of this area. (B, courtesy of T. Byers and D. Branton, *Proc. Natl. Acad. Sci. U.S.A.* 82:6153–6157, 1985. With permission from National Academy of Sciences.)

humans that have genetic abnormalities in spectrin structure. These individuals are anemic: they have fewer red blood cells than normal, and the red cells they do have are spherical instead of flattened and are abnormally fragile.

Proteins similar to spectrin and to its associated attachment proteins are present in the cortex of most animal cells, but the cortex in these cells is much more complex than that of red blood cells. While red blood cells need their cortex mainly to provide mechanical strength as they are pumped through blood vessels, other cells also need their cortex to allow them to change their shape actively and to move, as we discuss in Chapter 17. In addition, many cells use their cortical network to restrain the diffusion of proteins within the membrane, as we see next.

Cells Can Restrict the Movement of Membrane Proteins

Because a membrane is a two-dimensional fluid, many of its proteins, like its lipids, can move freely within the plane of the lipid bilayer. This diffusion can be neatly demonstrated by fusing a mouse cell with a human cell to form a double-sized hybrid cell and then monitoring the distribution of mouse and human plasma membrane proteins. At first the mouse and human proteins are confined to their own halves of the newly formed hybrid cell, but within half an hour or so the two sets of proteins become evenly mixed over the entire cell surface (Figure 11–32).

The picture of a membrane as a sea of lipid in which all proteins float freely is too simple, however. Cells have ways of confining particular

QUESTION 11–7

Look carefully at the transmembrane proteins shown in Figure 11–31. What can you say about their mobility in the membrane?

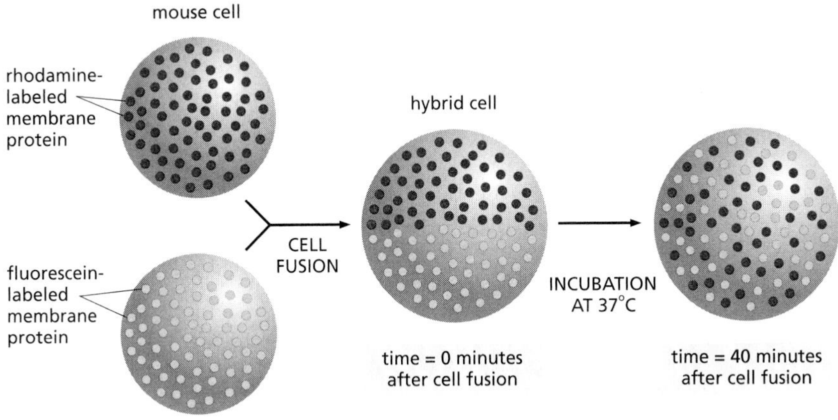

Figure 11–32 **Formation of mouse–human hybrid cells shows that plasma membrane proteins can move laterally in the lipid bilayer.** The mouse and human proteins are initially confined to their own halves of the newly formed hybrid-cell plasma membrane, but they intermix within a short time. To reveal the proteins, two antibodies that bind to human and mouse proteins, respectively, are labeled with different fluorescent tags (rhodamine or fluorescein) and added to the cells. The two fluorescent antibodies can be distinguished in a fluorescence microscope because fluorescein is *green*, whereas rhodamine is *red*. (Based on observations of L.D. Frye and M. Edidin, *J. Cell Sci.* 7:319–335, 1970. With permission from The Company of Biologists Ltd.)

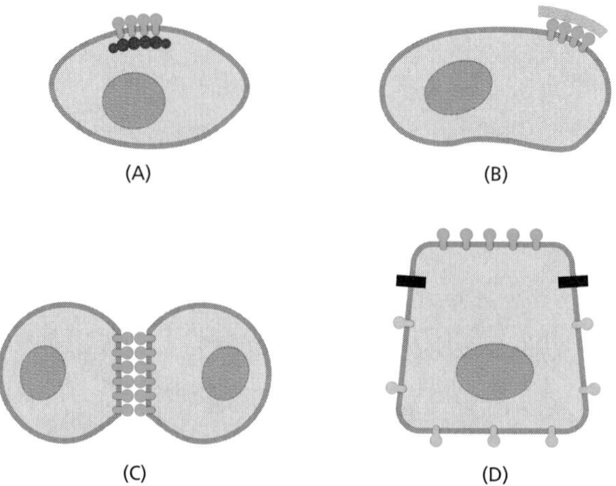

Figure 11–33 **The lateral mobility of plasma membrane proteins can be restricted in several ways.** Proteins can be tethered to the cell cortex inside the cell (A), to extracellular matrix molecules outside the cell (B), or to proteins on the surface of another cell (C). Diffusion barriers (shown as *black* bars) can restrict proteins to a particular membrane domain (D).

plasma membrane proteins to localized areas within the bilayer, thereby creating functionally specialized regions, or **membrane domains**, on the cell or organelle surface. We describe some techniques for studying the movement of membrane proteins in How We Know, pp. 382–383.

As illustrated in Figure 11–33, plasma membrane proteins can be linked to fixed structures outside the cell—for example, to molecules in the extracellular matrix (discussed in Chapter 20)—or to relatively immobile structures inside the cell, especially to the cell cortex (see Figure 11–31). Finally, cells can create barriers that restrict particular membrane components to one membrane domain. In epithelial cells that line the gut, for example, it is important that transport proteins involved in the uptake of nutrients from the gut be confined to the *apical* surface of the cells (the surface that faces the gut contents) and that other proteins involved in the transport of solutes out of the epithelial cell into the tissues and bloodstream be confined to the *basal* and *lateral* surfaces (Figure 11–34). This asymmetric distribution of membrane proteins is maintained by a barrier formed along the line where the cell is sealed to adjacent epithelial cells by a so-called *tight junction*. At this site, specialized junctional proteins form a continuous belt around the cell where it contacts its neighbors, creating a seal between adjacent cell membranes (see Figure 20–23). Membrane proteins cannot diffuse past the junction.

The Cell Surface Is Coated with Carbohydrate

We saw earlier that many of the lipids in the outer layer of the plasma membrane have sugars covalently attached to them. The same is true for most of the proteins in the plasma membrane. The great majority of

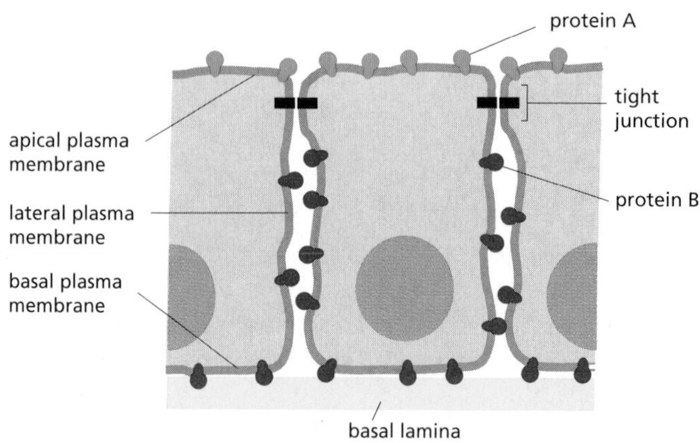

Figure 11–34 **A membrane protein is restricted to a particular domain of the plasma membrane of an epithelial cell in the gut.** Protein A (in the apical membrane) and protein B (in the basal and lateral membranes) can diffuse laterally in their own membrane domains but are prevented from entering the other domain by a specialized cell junction called a tight junction.

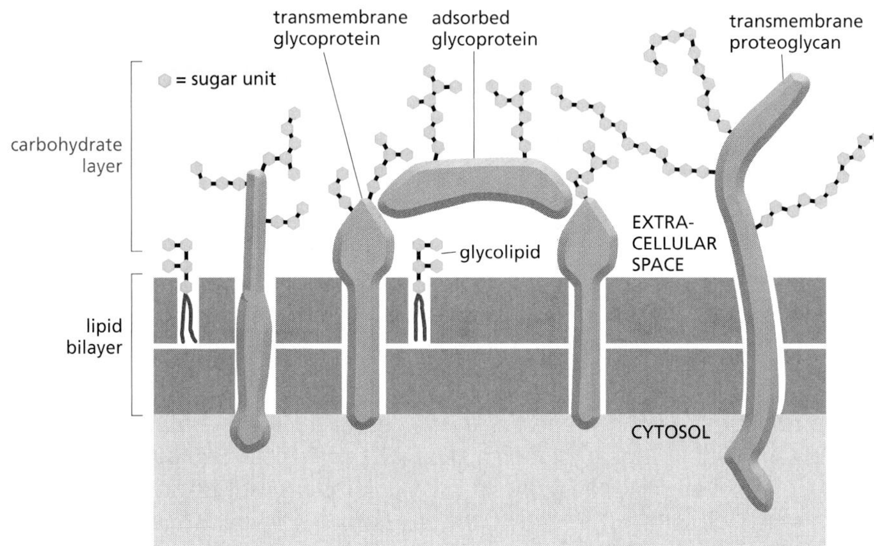

Figure 11–35 **Eucaryotic cells are coated with sugars.** The carbohydrate layer is made of the oligosaccharide side chains attached to membrane glycolipids and glycoproteins, and of the polysaccharide chains on membrane proteoglycans. Glycoproteins and proteoglycans that have been secreted by the cell and then adsorbed back onto its surface can also contribute to the carbohydrate layer. Note that all the carbohydrate is on the external (noncytosolic) surface of the plasma membrane.

these proteins have short chains of sugars, called *oligosaccharides*, linked to them; they are called *glycoproteins*. Other membrane proteins have one or more long polysaccharide chains attached to them; they are called *proteoglycans*. All of the carbohydrate on the glycoproteins, proteoglycans, and glycolipids is located on one side of the membrane, the noncytosolic side, where it forms a sugar coating called the **carbohydrate layer** (Figure 11–35).

By forming a layer of material covering the lipid bilayer, the carbohydrate layer helps to protect the cell surface from mechanical and chemical damage. As the oligosaccharides and polysaccharides in the carbohydrate layer adsorb water, they give the cell a slimy surface. This coating helps motile cells such as white blood cells to squeeze through narrow spaces, and it prevents blood cells from sticking to one another or to the walls of blood vessels.

Cell-surface carbohydrates do more than just protect and lubricate the cell, however. They have an important role in cell–cell recognition and adhesion. Just as many proteins will recognize and bind to a particular site on another protein, some proteins (called *lectins*) are specialized to recognize particular oligosaccharide side chains and bind to them. The oligosaccharide side chains of glycoproteins and glycolipids, although short (typically fewer than 15 sugar units), are enormously diverse. Unlike polypeptide (protein) chains, in which the amino acids are all joined together linearly by identical peptide bonds (see Figure 11–22), sugars can be joined together in many different arrangements, often forming branched chains (see Panel 2–3, pp. 68–69). Even three sugar groups can be put together in enough different combinations of covalent linkages that they can form hundreds of different trisaccharides.

In a multicellular organism, the carbohydrate layer thus serves as a kind of distinctive clothing, like a police officer's uniform, that is characteristic of cells specialized for a particular function and that is recognized by other cells with which each must interact. Specific oligosaccharides in the carbohydrate layer are involved, for example, in the recognition of an egg by a sperm (discussed in Chapter 19). They are also involved in our responses to infection. In the early stages of a bacterial infection, for instance, the carbohydrate on the surface of white blood cells called *neutrophils* is recognized by a lectin on the cells lining the blood vessels at the site of infection. This recognition causes the neutrophils to

HOW WE KNOW: MEASURING MEMBRANE FLOW

An essential feature of the lipid bilayer is its fluidity. This vital molecular flow is crucial for cell membrane integrity and function. It allows the resident proteins to float about the bilayer, coupling and uncoupling, engaging in the molecular interactions on which cells depend. The dynamic nature of cell membranes is so central to their proper function that our working model of membrane structure is commonly called the *fluid-mosaic model.*

Given its importance for membrane structure and function, how do we measure and study the fluidity of cell membranes? The most common methods are visual: simply label some of the molecules native to the membrane and then watch them move. Such an approach first demonstrated the diffusion of membrane proteins that had been tagged with labeled antibodies (see Figure 11–32). This experiment, however, left researchers with the impression that membrane proteins drift freely, without restriction, in an open sea of lipids. We now know that this image is not entirely accurate. To probe membrane dynamics more thoroughly, researchers had to invent more precise methods for tracking the movement of proteins within a membrane such as the plasma membrane of a living cell.

The FRAP attack

One such technique, called *fluorescence recovery after photobleaching (FRAP),* involves uniformly labeling proteins across the cell surface, bleaching the label from a small region in this fluorescent sea, and then seeing how quickly the surrounding labeled proteins seep into this bleached patch of membrane. To start, the membrane protein of interest is tagged with a specific fluorescent group. This labeling can be done either with a fluorescent antibody or by fusing the membrane protein with a fluorescent protein such as green fluorescent protein (GFP) using recombinant DNA techniques (discussed in Chapter 10).

Once the cell has been labeled, it is placed under a microscope and a small patch of its membrane is irradiated with an intense pulse from a sharply focused laser beam. This treatment irreversibly bleaches the fluorescent groups in a spot, typically 1 μm square, on the cell surface (Figure 11–36). The time it takes for fluorescent proteins to migrate from the adjacent areas into the bleached region of the membrane can then be measured. The rate of this 'fluorescence recovery' is a direct measure of the rate at which the surrounding protein molecules can diffuse within the membrane (Movie 11.5). Such experiments reveal that, generally speaking, the cell membrane is about as viscous as olive oil.

One-by-one

One drawback to the FRAP approach is that the technique monitors the movement of fairly large populations of proteins—hundreds or thousands—across relatively large areas of the membrane. With this technique it is

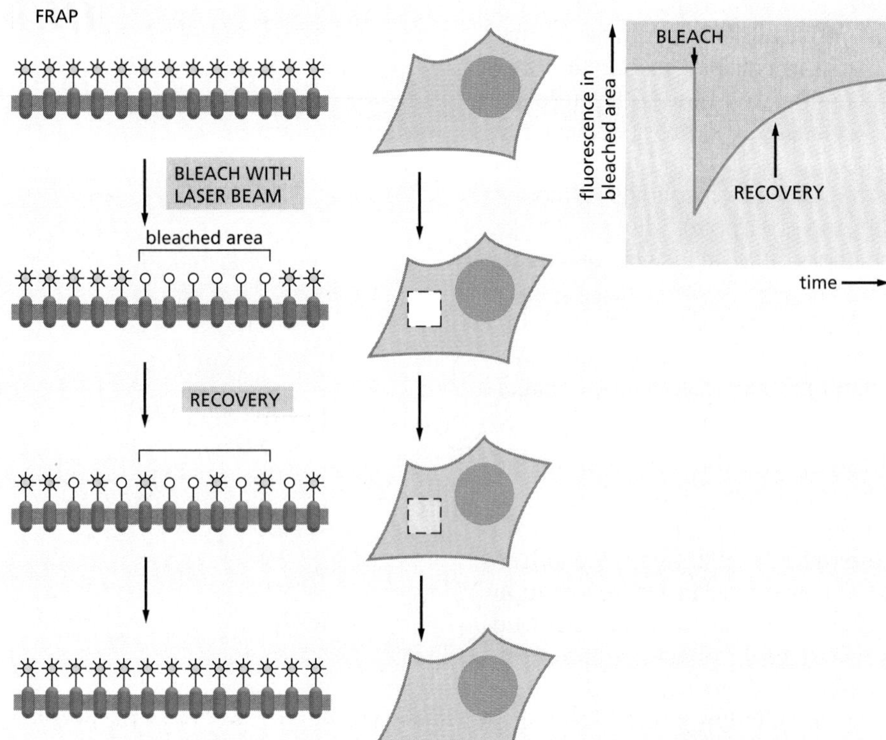

Figure 11–36 **Photobleaching techniques can be used to measure the rate of lateral diffusion of a membrane protein.** A specific protein of interest can be labeled with a fluorescent antibody (as shown here) or can be expressed as a fusion protein with green fluorescent protein (GFP), which is intrinsically fluorescent. In the FRAP technique, fluorescent molecules are bleached in a small area using a laser beam. The fluorescence intensity recovers as the bleached molecules diffuse away and unbleached molecules diffuse into the irradiated area (shown here in side and top views). The diffusion coefficient is calculated from a graph of the rate of recovery: the greater the diffusion coefficient of the membrane protein, the faster the recovery.

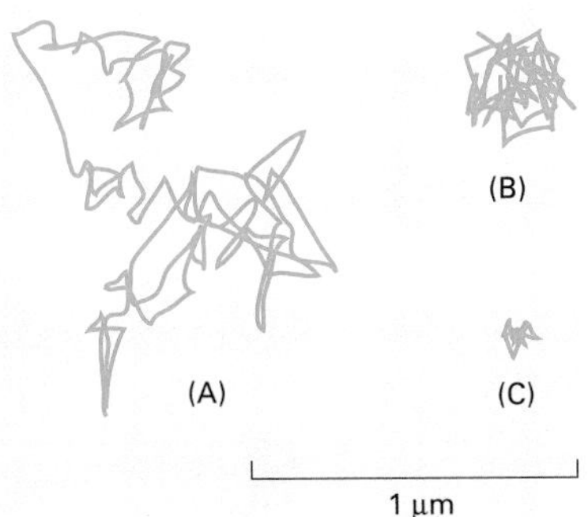

Figure 11–37 **Proteins show different patterns of motion.** Single-particle tracking studies reveal some of the pathways that real proteins follow on the surface of a living cell. Shown here are some trajectories representative of different kinds of proteins in the plasma membrane. (A) Tracks made by a protein that is free to diffuse randomly around the plasma membrane. (B) Tracks made by a protein that is corralled within a small membrane domain by other proteins. (C) Tracks made by a protein that is tethered to the cytoskeleton and hence is essentially immobile. The movement of the proteins is monitored over a period of seconds.

impossible to see what individual molecules are doing. If a protein fails to migrate into the bleached zone over the course of a FRAP study, for example, is it because the molecule is immobile, essentially anchored in one place in the membrane? Or is it restricted to movement within a very small region—fenced in by cytoskeletal proteins—and thus it only appears motionless?

To get around this problem, researchers have developed methods for labeling and tracking the movement of individual molecules or small clusters of molecules. One such technique, dubbed *single-particle tracking (SPT) microscopy,* relies on labeling protein molecules with antibodies coated with gold particles. The gold spheres appear as tiny black dots under the microscope, and their movement, and thus the movement of individually tagged proteins, can be tracked using video microscopy.

From the studies carried out to date, it appears that membrane proteins can display a variety of patterns of movement, from random diffusion to complete immobility (Figure 11–37). Some proteins switch between these different kinds of motion.

Freed from cells

In the end, researchers often wish to study the behavior of a particular protein in isolation, in the absence of molecules that might restrain its movement or activity. For such studies, membrane proteins can be removed from cells and reconstituted in artificial phospholipid vesicles (Figure 11–38). The lipids allow the isolated proteins to maintain their proper structure, and the activity and behavior of these purified proteins can then be analyzed in detail.

It is apparent from these studies that membrane proteins diffuse more freely and rapidly in artificial lipid bilayers than in cell membranes. The fact that proteins show restricted mobility in the cell membrane makes sense, as these membranes are crowded with proteins and contain a greater variety of lipids than an artificial lipid bilayer. Furthermore, many membrane proteins are tethered to proteins in the extracellular matrix, anchored to cytoskeletal elements tucked just under the plasma membrane, or both (as illustrated in Figure 11–33).

Taken together, such studies of the movement of membrane molecules reveal information about the architecture and organization of cell membranes, giving us a more accurate portrait of the membrane as a dynamic fluid mosaic.

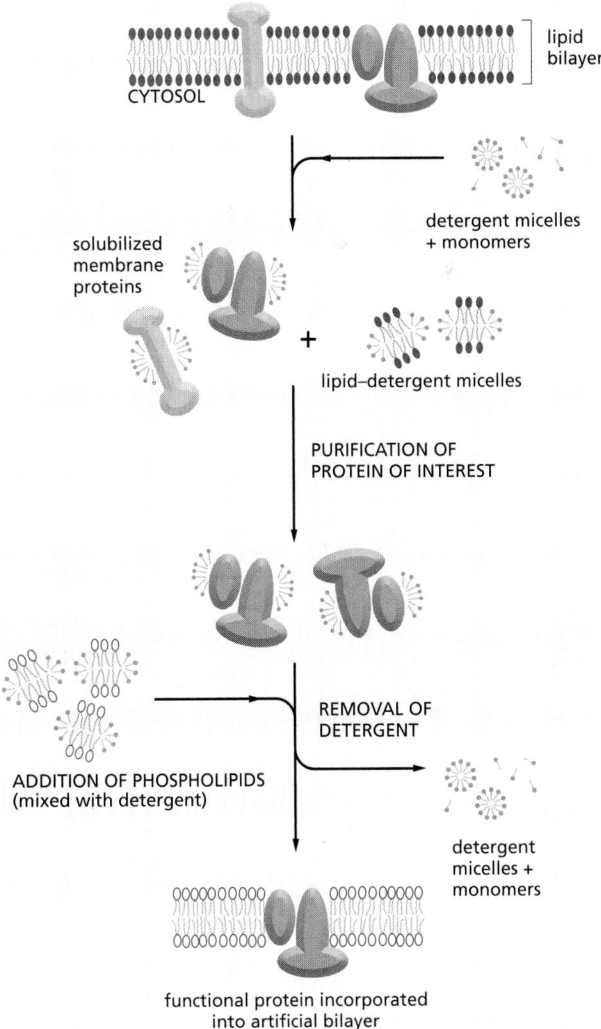

Figure 11–38 **Mild detergents can be used to solubilize and reconstitute functional membrane proteins.**

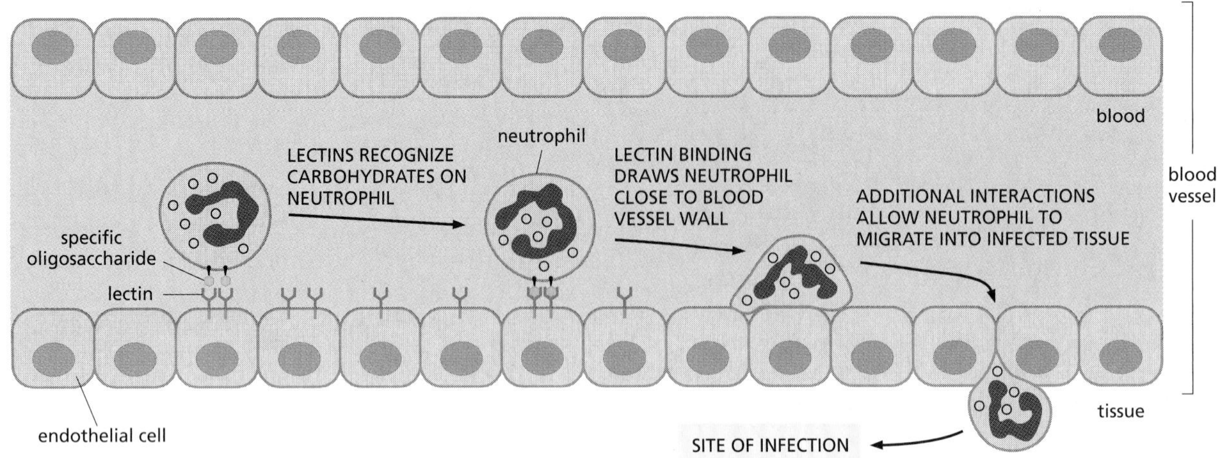

Figure 11–39 **The recognition of the cell-surface carbohydrate on neutrophils is the first stage of their migration out of the blood at sites of infection.** Specialized transmembrane proteins (called lectins) are made by the cells lining the blood vessel (called endothelial cells) in response to chemical signals emanating from the site of infection. These proteins recognize particular groups of sugars carried by glycolipids and glycoproteins on the surface of neutrophils circulating in the blood. The neutrophils consequently stick to the blood vessel wall. This association is not very strong, but it leads to another, much stronger protein–protein interaction (not shown) that helps the neutrophil migrate out of the bloodstream between the endothelial cells into the tissue at the site of infection (Movie 11.6).

adhere to the blood vessels and then migrate from the bloodstream into the infected tissue, where they help remove the invading bacteria (Figure 11–39).

Glycoproteins are important members in the family of membrane proteins. In the next chapter, we further examine the sophisticated functions of transmembrane proteins that transport molecules into and out of the cell.

ESSENTIAL CONCEPTS

- **Cell membranes enable a cell to create barriers that confine particular molecules to specific compartments.**
- **Cell membranes consist of a continuous double layer—a bilayer—of lipid molecules in which proteins are embedded.**
- **The lipid bilayer provides the basic structure and barrier function of all cell membranes.**
- **Membrane lipid molecules have both hydrophobic and hydrophilic regions. They assemble spontaneously into bilayers when placed in water, forming closed compartments that reseal if torn.**
- **There are three major classes of membrane lipid molecules: phospholipids, sterols, and glycolipids.**
- **The lipid bilayer is fluid, and individual lipid molecules are able to diffuse within their own monolayer; they do not, however, spontaneously flip from one monolayer to the other.**
- **The two layers of the plasma membrane have different lipid compositions, reflecting the different functions of the two faces of a cell membrane.**
- **Some cells adjust their membrane fluidity by modifying the lipid composition of their membranes.**

- Membrane proteins are responsible for most of the functions of a membrane, such as the transport of small water-soluble molecules across the lipid bilayer.
- Transmembrane proteins extend across the lipid bilayer, usually as one or more α helices but sometimes as a β sheet curved into the form of a barrel.
- Other membrane proteins do not extend across the lipid bilayer but are attached to one or the other side of the membrane, either by non-covalent association with other membrane proteins or by covalent attachment to lipids.
- Most cell membranes are supported by an attached framework of proteins. An example is the meshwork of fibrous proteins forming the cell cortex underneath the plasma membrane.
- Although many membrane proteins can diffuse rapidly in the plane of the membrane, cells have ways of confining proteins to specific membrane domains and of immobilizing particular proteins by attaching them to intracellular or extracellular macromolecules.
- Many of the proteins and some of the lipids exposed on the surface of cells have attached sugars, which help protect and lubricate the cell surface and are involved in cell–cell recognition.

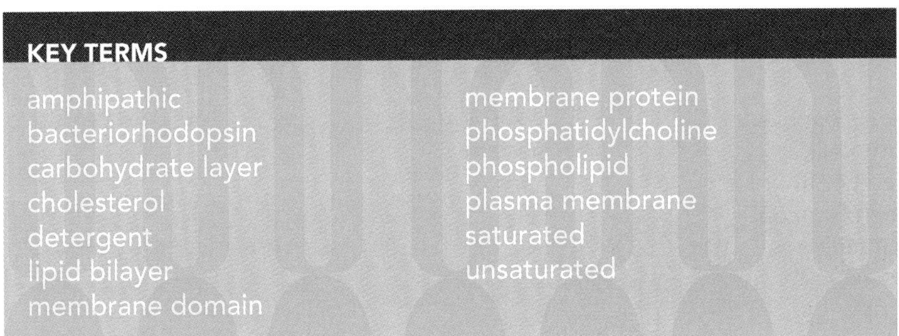

KEY TERMS
amphipathic
bacteriorhodopsin
carbohydrate layer
cholesterol
detergent
lipid bilayer
membrane domain
membrane protein
phosphatidylcholine
phospholipid
plasma membrane
saturated
unsaturated

QUESTIONS

QUESTION 11–8

Describe the different methods that cells use to restrict proteins to specific regions of the plasma membrane. Is a membrane with many anchored proteins still fluid?

QUESTION 11–9

Which of the following statements are correct? Explain your answers.

A. Lipids in a lipid bilayer spin rapidly around their long axis.

B. Lipids in a lipid bilayer rapidly exchange positions with one another in the plane of the membrane.

C. Lipids in a lipid bilayer do not flip-flop readily from one lipid monolayer to the other.

D. Hydrogen bonds that form between lipid head groups and water molecules are continually broken and re-formed.

E. Glycolipids move between different membrane-enclosed compartments during their synthesis but remain restricted to one side of the lipid bilayer.

F. Margarine contains more saturated lipids than the vegetable oil from which it is made.

G. Some membrane proteins are enzymes.

H. The sugar coat that surrounds all cells makes cells more slippery.

QUESTION 11–10

What is meant by the term "two-dimensional fluid"?

QUESTION 11–11

The structure of a lipid bilayer is determined by the particular properties of its lipid molecules. What would happen if

A. Phospholipids had only one hydrocarbon chain instead of two?

B. The hydrocarbon chains were shorter than normal, say, about 10 carbon atoms long?

C. All of the hydrocarbon chains were saturated?

D. All of the hydrocarbon chains were unsaturated?

E. The bilayer contained a mixture of two kinds of lipid molecules, one with two saturated hydrocarbon tails and the other with two unsaturated hydrocarbon tails?

F. Each lipid molecule were covalently linked through the end carbon atom of one of its hydrocarbon chains to a lipid molecule in the opposite monolayer?

QUESTION 11–12

What are the differences between a lipid molecule and a detergent molecule? How would the structure of a lipid molecule need to change to make it a detergent?

QUESTION 11–13

A. Lipid molecules exchange places with their lipid neighbors every 10^{-7} second. A lipid molecule diffuses from one end of a 2-μm-long bacterial cell to the other in about 1 second. Are these two numbers in agreement (assume that the diameter of a lipid head group is about 0.5 nm)? If not, can you think of a reason for the difference?

B. To get an appreciation for the great speed of molecular motions, assume that a lipid head group is about the size of a ping-pong ball (4 cm in diameter) and that the floor of your living room (6 m × 6 m) is covered wall-to-wall with these balls. If two neighboring balls exchanged positions once every 10^{-7} second, what would their speed be in kilometers per hour? How long would it take for a ball to move from one end of the room to the other?

QUESTION 11–14

Why does a red blood cell membrane need proteins?

QUESTION 11–15

Consider a transmembrane protein that forms a hydrophilic pore across the plasma membrane of a eucaryotic cell, allowing Na^+ to enter the cell when it is activated upon binding a specific ligand on its extracellular side. It is made of five similar transmembrane subunits, each containing a membrane-spanning α helix with hydrophilic amino acid side chains on one surface of the helix and hydrophobic amino acid side chains on the opposite surface. Considering the function of the protein as a channel for Na^+ ions to enter the cell, propose a possible arrangement of the five membrane-spanning α helices in the membrane.

QUESTION 11–16

In the membrane of a human red blood cell the ratio of the mass of protein (average molecular weight 50,000) to phospholipid (molecular weight 800) to cholesterol (molecular weight 386) is about 2:1:1. How many lipid molecules are there for every protein molecule?

QUESTION 11–17

Draw a schematic diagram that shows a close-up view of two plasma membranes as they come together during cell fusion, as shown in Figure 11–32. Show membrane proteins in both cells that were labeled from the outside by the binding of differently colored fluorescent antibody molecules. Indicate in your drawing the fates of these color tags as the cells fuse. Will they still be only on the outside of the hybrid cell (A) after cell fusion and (B) after the mixing of membrane proteins that occurs during the incubation at 37°C? How would the experimental outcome be different if the incubation were done at 0°C?

QUESTION 11–18

Compare the hydrophobic forces that hold a membrane protein in the lipid bilayer with those that help proteins fold into a unique three-dimensional structure.

QUESTION 11–19

Predict which of the following organisms will have the highest percentage of unsaturated phospholipids in their membranes. Explain your answer.

A. Antarctic fish

B. Desert snake

C. Human being

D. Polar bear

E. Thermophilic bacterium that lives in hot springs at 100°C.

QUESTION 11–20

Which of the three 20-amino acid sequences listed below in the single-letter amino acid code is the most likely candidate to form a transmembrane region (α-helix) of a transmembrane protein? Explain your answer.

A. I T L I Y F G N M S S V T Q T I L L I S

B. L L L I F F G V M A L V I V V I L L I A

C. L L K K F F R D M A A V H E T I L E E S

CHAPTER **TWELVE** 12

Membrane Transport

PRINCIPLES OF MEMBRANE TRANSPORT

TRANSPORTERS AND THEIR FUNCTIONS

ION CHANNELS AND THE MEMBRANE POTENTIAL

ION CHANNELS AND SIGNALING IN NERVE CELLS

Cells live and grow by exchanging molecules with their environment. The plasma membrane acts as a barrier that controls the transit of molecules into and out of the cell. Because the interior of the lipid bilayer is hydrophobic, as we saw in Chapter 11, the plasma membrane tends to block the passage of almost all water-soluble molecules. But various water-soluble molecules must be able to cross the plasma membrane: cells must import nutrients such as sugars and amino acids, eliminate metabolic waste products such as CO_2, and regulate the intracellular concentrations of a variety of inorganic ions. A few of these solutes, CO_2 and O_2 for example, can simply diffuse across the lipid bilayer, but the vast majority cannot. Instead, their transfer depends on specialized **membrane transport proteins** that span the lipid bilayer, providing private passageways across the membrane for select substances (Figure 12–1).

In this chapter, we consider how membranes control the traffic of small molecules into and out of cells. Cells can also selectively transfer macromolecules such as proteins across their membranes, but this transport requires more elaborate machinery and is discussed in Chapter 15. Here, we begin by outlining some of the general principles that guide the passage of small, water-soluble molecules through cell membranes. We then examine, in turn, the two main classes of membrane proteins that mediate this transfer. A *transporter*, which has moving parts, can shift small molecules from one side of the membrane to the other by changing its shape. Solutes transported in this way can be either small organic molecules or inorganic ions. *Channels*, in contrast, form tiny hydrophilic pores in the membrane through which solutes can pass by diffusion. Most channels let through inorganic ions only and are therefore called *ion channels*.

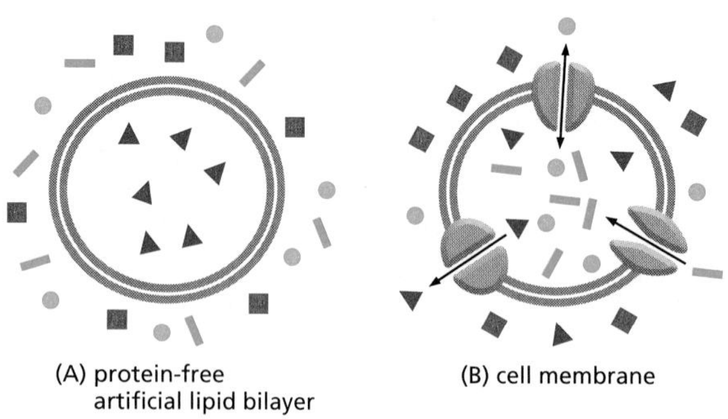

Figure 12–1 **Specialized membrane transport proteins are responsible for transferring small water-soluble molecules across cell membranes.** Whereas protein-free artificial lipid bilayers are impermeable to most water-soluble molecules (A), cell membranes are not (B). Note that each type of transport protein in a cell membrane transfers a particular type of molecule, causing a selective set of solutes to end up inside the membrane-enclosed compartment.

Because these ions are electrically charged, their movements can create powerful electric forces across the membrane. In the final part of the chapter, we discuss how these forces enable nerve cells to communicate, ultimately carrying out the astonishing range of behaviors of which the human brain is capable.

PRINCIPLES OF MEMBRANE TRANSPORT

To provide a foundation for discussing membrane transport, we first consider the differences in ion composition between a cell's interior and its environment. This will help make it clear why the transport of ions by both transporters and ion channels is of such fundamental importance to cells.

The Ion Concentrations Inside a Cell Are Very Different from Those Outside

Living cells maintain an internal ion composition that is very different from the ion composition in the fluid around them, and these differences are crucial for a cell's survival and function. Inorganic ions such as Na^+, K^+, Ca^{2+}, Cl^-, and H^+ (protons) are the most plentiful of all the solutes in a cell's environment, and their movements across cell membranes play an essential part in many biological processes, including the activity of nerve cells, as we discuss later in this chapter, and the production of ATP by all cells, as we discuss in Chapter 14.

Na^+ is the most plentiful positively charged ion (cation) outside the cell, while K^+ is the most plentiful inside (Table 12–1). For a cell to avoid being

TABLE 12–1 A COMPARISON OF ION CONCENTRATIONS INSIDE AND OUTSIDE A TYPICAL MAMMALIAN CELL

COMPONENT	INTRACELLULAR CONCENTRATION (mM)	EXTRACELLULAR CONCENTRATION (mM)
Cations		
Na^+	5–15	145
K^+	140	5
Mg^{2+}	0.5	1–2
Ca^{2+}	10^{-4}	1–2
H^+	7×10^{-5} ($10^{-7.2}$ M or pH 7.2)	4×10^{-5} ($10^{-7.4}$ M or pH 7.4)
Anions*		
Cl^-	5–15	110

* The cell must contain equal quantities of positive and negative charges (that is, be electrically neutral). Thus, in addition to Cl^-, the cell contains many other anions not listed in this table; in fact, most cellular constituents are negatively charged (HCO_3^-, PO_4^{3-}, proteins, nucleic acids, metabolites carrying phosphate and carboxyl groups, etc.). The concentrations of Ca^{2+} and Mg^{2+} given are for the free ions. There is a total of about 20 mM Mg^{2+} and 1–2 mM Ca^{2+} in cells, but this is mostly bound to proteins and other substances and, for Ca^{2+}, stored within various organelles.

torn apart by electrical forces, the quantity of positive charge inside the cell must be balanced by an almost exactly equal quantity of negative charge there, and the same is true for the charge in the surrounding fluid. However, tiny excesses of positive or negative charge, concentrated in the neighborhood of the plasma membrane, do occur, and they have important electrical effects, as we discuss later. The high concentration of Na^+ outside the cell is balanced chiefly by extracellular Cl^-. The high concentration of K^+ inside is balanced by a variety of negatively charged intracellular ions (anions).

This differential distribution of ions inside and outside the cell is controlled in part by the activity of membrane transport proteins and in part by the permeability characteristics of the lipid bilayer itself.

Lipid Bilayers Are Impermeable to Solutes and Ions

The hydrophobic interior of the lipid bilayer creates a barrier to the passage of most hydrophilic molecules, including ions. They are as reluctant to enter a fatty environment as hydrophobic molecules are reluctant to enter water. But given enough time, virtually any molecule will diffuse across a lipid bilayer. The *rate* at which it diffuses, however, varies enormously depending on the size of the molecule and its solubility properties. In general, the smaller the molecule and the more soluble it is in oil (that is, the more hydrophobic, or nonpolar, it is), the more rapidly it will diffuse across. Thus:

1. *Small nonpolar molecules,* such as molecular oxygen (O_2, molecular mass 32 daltons) and carbon dioxide (44 daltons), readily dissolve in lipid bilayers and therefore rapidly diffuse across them; indeed, cells require this permeability to gases for the cell respiration processes discussed in Chapter 14.
2. *Uncharged polar molecules* (molecules with an uneven distribution of electric charge) also diffuse rapidly across a bilayer, if they are small enough. Water (18 daltons) and ethanol (46 daltons), for example, cross fairly rapidly; glycerol (92 daltons) crosses less rapidly; and glucose (180 daltons) crosses hardly at all (Figure 12–2).
3. In contrast, lipid bilayers are highly impermeable to all *ions and charged molecules,* no matter how small. The molecules' charge and their strong electrical attraction to water molecules inhibit them from entering the hydrocarbon phase of the bilayer. Thus, synthetic bilayers are a billion (10^9) times more permeable to water than they are to even such small ions as Na^+ or K^+.

Cell membranes allow water and small nonpolar molecules to permeate by simple diffusion. But for cells to take up nutrients and release wastes, membranes must also allow the passage of many other molecules, such as ions, sugars, amino acids, nucleotides, and many cell metabolites. These molecules cross lipid bilayers far too slowly by simple diffusion; thus, specialized membrane transport proteins are required to transfer them efficiently across cell membranes.

Membrane Transport Proteins Fall into Two Classes: Transporters and Channels

Membrane transport proteins occur in many forms and in all types of biological membranes. Each protein provides a private passageway across the membrane for a particular class of molecule—ions, sugars, or amino acids, for example. Most of these protein portals are even more exclusive, allowing entrance of only select members of a particular molecular class: some, for example, are open to Na^+ but not K^+, others to K^+ but not Na^+. The set of membrane transport proteins present in the plasma

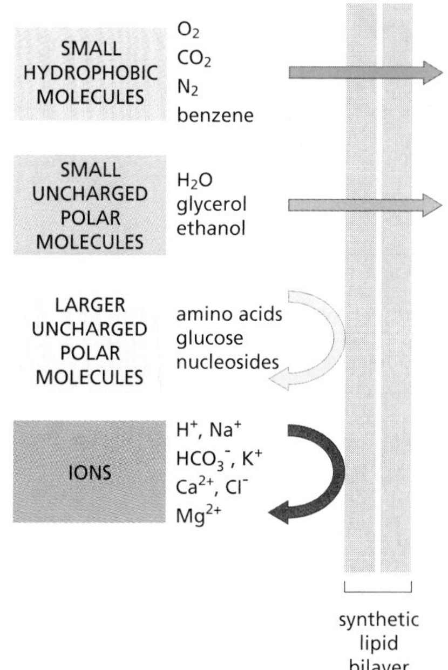

Figure 12–2 **The rate at which a molecule diffuses across a synthetic lipid bilayer depends on its size and solubility.** The smaller the molecule and, more importantly, the fewer its favorable interactions with water (that is, the less polar it is), the more rapidly the molecule diffuses across the bilayer. Note that many of the molecules that the cell uses as nutrients are too large and polar to pass through a pure lipid bilayer.

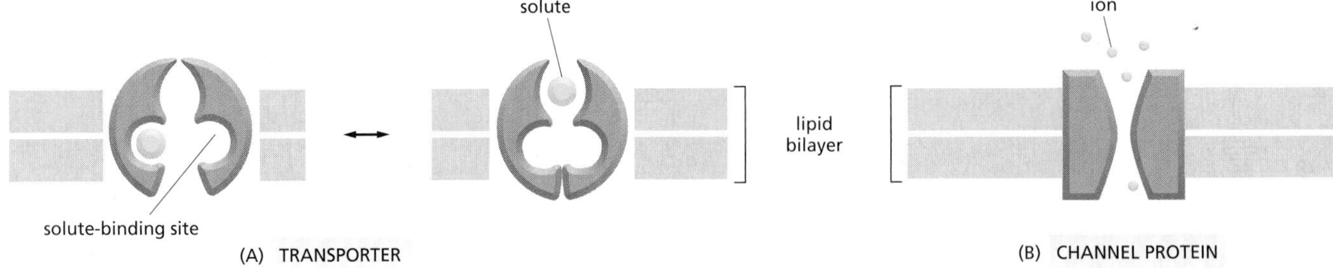

Figure 12–3 **Small molecules and ions can enter the cell through a transporter or a channel.** (A) A transporter undergoes a series of conformational changes to transfer small water-soluble molecules across the lipid bilayer. (B) A channel, in contrast, forms a hydrophilic pore across the bilayer through which specific inorganic ions or in some cases other small molecules can diffuse. As would be expected, channels transfer molecules at a much greater rate than transporters. Ion channels can exist in either an open or a closed conformation, and they transport only in the open conformation, which is shown here. Channel opening and closing is usually controlled by an external stimulus or by conditions within the cell.

membrane or in the membrane of an intracellular organelle determines exactly which solutes can pass into and out of that cell or organelle. Each type of membrane therefore has its own characteristic set of transport proteins.

As discussed in Chapter 11, the membrane transport proteins that have been studied in detail have polypeptide chains that traverse the lipid bilayer multiple times—that is, they are multipass transmembrane proteins (see Figure 11–24). By crisscrossing back and forth across the bilayer, the polypeptide chain forms a continuous protein-lined pathway that allows selected small hydrophilic molecules to cross the membrane without coming into direct contact with the hydrophobic interior of the lipid bilayer.

Membrane transport proteins can be divided into two main classes: transporters and channels. The basic difference between transporters and channels is the way they discriminate between solutes, transporting some solutes but not others (Figure 12–3). **Channels** discriminate mainly on the basis of size and electric charge: if a channel is open, an ion or a molecule that is small enough and carries the appropriate charge can slip through, as through a narrow trapdoor. A **transporter**, on the other hand, allows passage only to those molecules or ions that fit into a binding site on the protein; it then transfers these molecules across the membrane one at a time by changing its own conformation, acting more like a turnstile than an open door. Transporters bind their solutes with great specificity in the same way that an enzyme binds its substrate, and it is this requirement for specific binding that makes the transport selective.

Solutes Cross Membranes by Passive or Active Transport

Transporters and channels allow small molecules to cross the cell membrane, but what controls whether these solutes move into the cell or out of it? In many cases, the direction of transport depends on the relative concentrations of the solute. Molecules will spontaneously flow 'downhill' from a region of high concentration to a region of low concentration, provided a pathway exists. Such movements are called *passive*, because they need no other driving force. If, for example, a solute is present at a higher concentration outside the cell than inside and an appropriate channel or transporter is present in the plasma membrane, the solute will move spontaneously across the membrane down its *concentration gradient* into the cell by **passive transport** (sometimes called *facilitated diffusion*), without expenditure of energy by its membrane transport protein. All channels and many transporters act as conduits for such passive transport.

To move a solute against its concentration gradient, however, a membrane transport protein must do work: it has to drive the flow 'uphill' by coupling it to some other process that provides energy (as discussed in Chapter 3 for enzyme reactions). Transmembrane solute movement

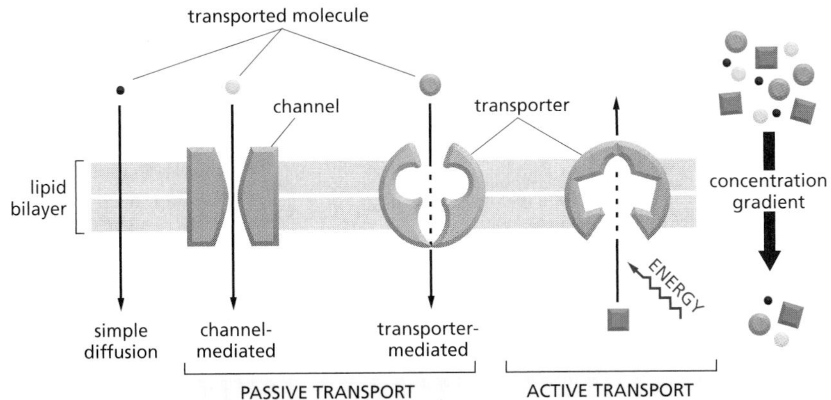

Figure 12–4 **Most solutes cross cell membranes by passive or active transport.** Some small uncharged molecules can move down their concentration gradient across the lipid bilayer by simple diffusion. But most solutes require the assistance of a channel or transporter. As indicated, movement of molecules in the same direction as their concentration gradient—passive transport—occurs spontaneously, whereas transport against a concentration gradient—active transport—requires an input of energy. Only transporters can carry out active transport, but both transporters and channels can carry out passive transport.

driven in this way is termed **active transport**, and it is carried out only by special types of transporters that can harness some energy source to the transport process (Figure 12–4). Because they drive the transport of solutes against their concentration gradient, many of these transporters are called *pumps*. We now examine a variety of transporters, both active and passive, and see how they function to move molecules across cell membranes.

TRANSPORTERS AND THEIR FUNCTIONS

Transporters are required for the movement of almost all small organic molecules across cell membranes, with the exception of fat-soluble molecules and small uncharged molecules that can pass directly through the lipid bilayer by simple diffusion (see Figure 12–2). Each transporter is highly selective, often transferring just one type of molecule. To guide and propel the complex traffic of small molecules into and out of the cell, and between the cytosol and the different membrane-enclosed organelles, each cellular membrane contains a set of different transporters appropriate to that particular membrane. For example, the plasma membrane contains transporters that import nutrients such as sugars, amino acids, and nucleotides; the lysosome membrane contains an H^+ transporter that acidifies the lysosome interior; and the inner membrane of mitochondria contains transporters for importing the pyruvate that mitochondria use as fuel for generating ATP and for exporting ATP once it is synthesized (Figure 12–5).

Although the detailed molecular mechanisms that underlie the movement of solutes are known for only a few transporters, the general principles that govern the function of these proteins are well understood.

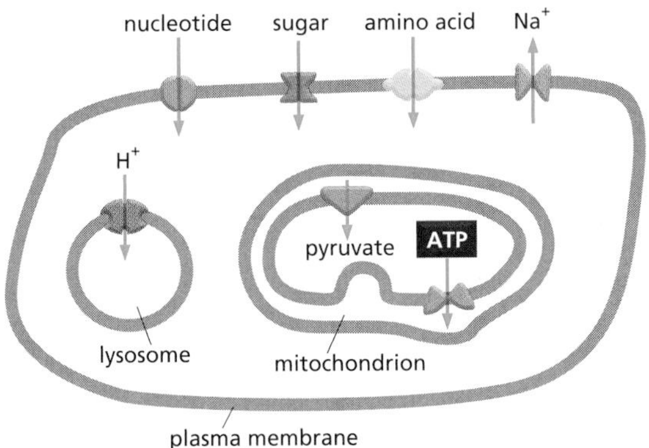

Figure 12–5 **Each cell membrane has its own characteristic set of transporters.**

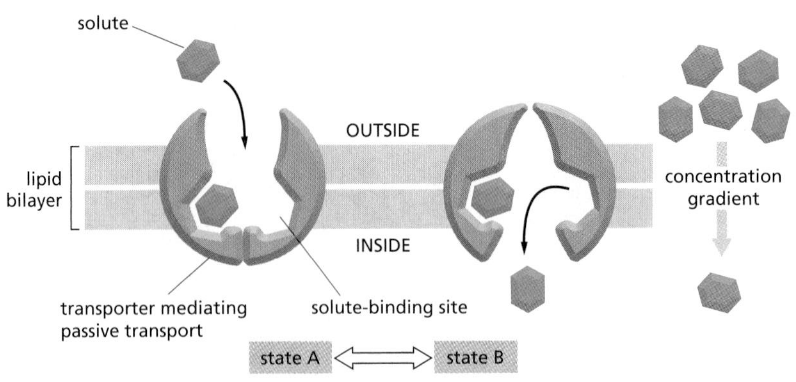

Figure 12–6 **A conformational change in a transporter could mediate the passive transport of a solute such as glucose.** In this model, the transporter can exist in two conformational states: in state A the binding sites for the solute are exposed on the outside of the membrane; in state B the same sites are exposed on the other side of the membrane. The transition between the two states is proposed to occur randomly and independently of whether the solute is bound and to be completely reversible. If the concentration of the solute is higher on the outside of the membrane, it will be more often caught up in A → B transitions that carry it into the cell than in B → A transitions that carry it out. There will therefore be a net transport of the solute down its concentration gradient.

Concentration Gradients and Electrical Forces Drive Passive Transport

Solutes can cross the membrane by passive or active transport—and transporters are capable of facilitating both types of movement (see Figure 12–4). A simple example of a transporter that mediates passive transport is the *glucose transporter* found in the plasma membrane of mammalian liver cells and many other cell types. The protein consists of a polypeptide chain that crosses the membrane at least 12 times. It is thought that the transporter can adopt at least two conformations and switches reversibly and randomly between them. In one conformation, the transporter exposes binding sites for glucose to the exterior of the cell; in the other, it exposes these sites to the interior of the cell (Figure 12–6).

When sugar is plentiful outside a liver cell, as it is after a meal, glucose molecules bind to the transporter's externally displayed binding sites; when the protein switches conformation, it carries these molecules inward and releases them into the cytosol, where the glucose concentration is low. Conversely, when blood sugar levels are low—when you are hungry—the hormone glucagon stimulates the liver cell to produce large amounts of glucose by the breakdown of glycogen. As a result, the glucose concentration is higher inside the cell than outside, and glucose binds to any internally displayed binding sites on the transporter; when the protein switches conformation in the opposite direction, the glucose is transported out of the cell. The flow of glucose can thus go either way, according to the direction of the glucose concentration gradient across the membrane: inward if glucose is more concentrated outside the cell than inside, and outward if the opposite is true. Transporters of this type, which permit a flux of solute but play no part in determining its direction, carry out passive transport. Although passive, the transport is highly selective: the binding sites in the glucose transporter bind only D-glucose and not, for example, its mirror image L-glucose, which the cell cannot use for glycolysis.

For glucose, which is an uncharged molecule, the direction of passive transport is determined solely by its concentration gradient. For electrically charged molecules, either small organic ions or inorganic ions, an additional force comes into play. For reasons we explain later, most cell membranes have a voltage across them, a difference in the electrical potential on each side of the membrane, which is referred to as the *membrane potential*. This difference in potential exerts a force on any molecule that carries an electric charge. The cytoplasmic side of the plasma membrane is usually at a negative potential relative to the outside, and this tends to pull positively charged solutes into the cell and drive negatively charged ones out. At the same time, a charged solute will also tend to move down its concentration gradient.

QUESTION 12–1

A simple enzyme reaction can be described by the equation $E + S \rightleftharpoons ES \rightarrow E + P$, where E is the enzyme, S the substrate, P the product, and ES the enzyme–substrate complex.
A. Write a corresponding equation describing the workings of a transporter (T) that mediates the transport of a solute (S) down its concentration gradient.
B. What does this equation tell you about the function of a transporter?
C. Why would this equation be an inappropriate description of channel function?

The net force driving a charged solute across the membrane is therefore a composite of two forces, one due to the concentration gradient and the other due to the voltage across the membrane. This net driving force is called the **electrochemical gradient** for the given solute. This gradient determines the direction of passive transport across the membrane. For some ions, the voltage and concentration gradient work in the same direction, creating a relatively steep electrochemical gradient (Figure 12–7A). This is the case for Na^+, which is positively charged and at a higher concentration outside cells than inside. Na^+ therefore tends to enter cells if given an opportunity. If, however, the voltage and concentration gradients have opposing effects, the resulting electrochemical gradient can be small (Figure 12–7B). This is the case for K^+, a positively charged ion that is present at a much higher concentration inside cells than outside. Because of these opposing effects, K^+ has a small electrochemical gradient across the membrane, despite its large concentration gradient, and therefore there is little net movement of K^+ across the membrane.

(A) (B)
OUTSIDE
INSIDE
electrochemical gradient when voltage and concentration gradients work in the same direction
electrochemical gradient when voltage and concentration gradients work in opposite directions

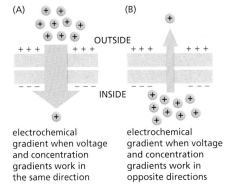

Figure 12–7 **An electrochemical gradient has two components.** The net driving force (the electrochemical gradient) tending to move a charged solute (ion) across a membrane is the sum of the concentration gradient of the solute and the voltage across a membrane (the membrane potential, which is represented here by the + and – signs at the membrane). The width of the *green* arrow represents the magnitude of the electrochemical gradient for a positively charged solute in two different situations. In (A), the concentration gradient is supplemented by a membrane potential that increases the driving force. In (B), the membrane potential acts against the concentration gradient, decreasing the driving force for movement of the solute.

Active Transport Moves Solutes Against Their Electrochemical Gradients

Of course, cells cannot rely solely on passive transport. Active transport of solutes against their electrochemical gradient is essential to maintain the intracellular ionic composition of cells and to import solutes that are at a lower concentration outside the cell than inside. Cells carry out active transport in three main ways (Figure 12–8): (i) *Coupled transporters* couple the uphill transport of one solute across the membrane to the downhill transport of another. (ii) *ATP-driven pumps* couple uphill transport to the hydrolysis of ATP. (iii) *Light-driven pumps*, which are found mainly in bacterial cells, couple uphill transport to an input of energy from light, as discussed for bacteriorhodopsin (see Figure 11–28).

Because a substance has to be carried uphill before it can flow downhill, the different forms of active transport are necessarily linked. Thus, in the plasma membrane of an animal cell, an ATP-driven pump transports Na^+ out of the cell against its electrochemical gradient, and this Na^+ can then flow back in, down its electrochemical gradient. Because the ion flows through Na^+-coupled transporters, the influx of Na^+ provides an energy source that drives the active movement of many other substances into the cell against their electrochemical gradients. If the Na^+ pump ceased operating, the Na^+ gradient would soon run down, and transport through Na^+-coupled transporters would come to a halt. The ATP-driven Na^+ pump, therefore, has a central role in membrane transport in animal cells. In plant cells, fungi, and many bacteria, a similar role is played by ATP-driven H^+ pumps that create an electrochemical gradient of H^+ ions by pumping H^+ out of the cell, as we discuss later.

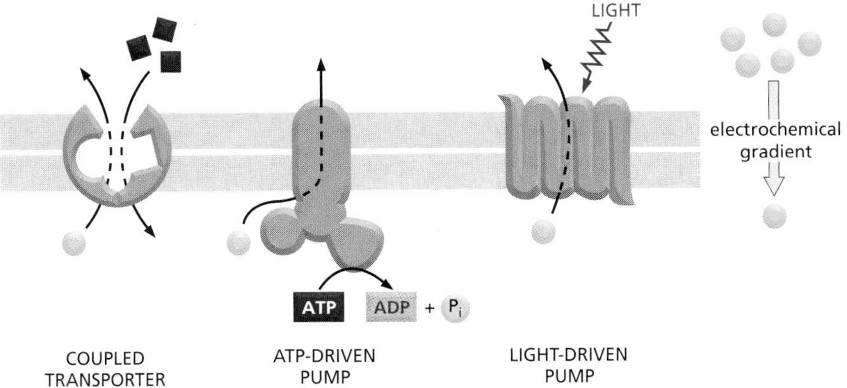

Figure 12–8 **Cells drive active transport in three main ways.** The actively transported molecule is shown in *yellow*, and the energy source is shown in *red*.

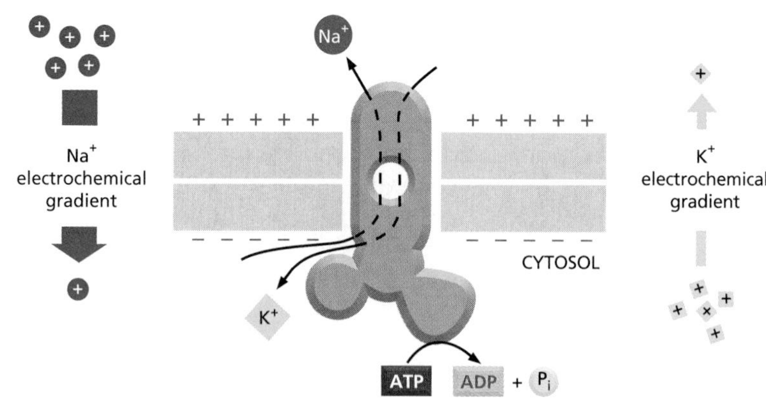

Figure 12–9 **The Na^+-K^+ pump plays a central role in membrane transport in animal cells.** This transporter uses the energy of ATP hydrolysis to pump Na^+ out of the cell and K^+ in, both against their electrochemical gradients, although the electrochemical gradient for K^+ is close to zero.

Animal Cells Use the Energy of ATP Hydrolysis to Pump Out Na^+

The ATP-driven Na^+ pump in animal cells hydrolyzes ATP to ADP to transport Na^+ out of the cell; this pump is therefore not only a transporter, but also an enzyme—an ATPase. At the same time, the protein couples the outward transport of Na^+ to an inward transport of K^+. The pump is therefore commonly known as the *Na^+-K^+ ATPase*, or the **Na^+-K^+ pump** (Figure 12–9).

This transporter plays a central part in the energy economy of animal cells, typically accounting for 30% or more of their total ATP consumption. Like a bilge pump in a leaky ship, it operates ceaselessly to expel the Na^+ that is constantly entering through other transporters and ion channels. In this way, the pump keeps the Na^+ concentration in the cytosol about 10–30 times lower than in the extracellular fluid and the K^+ concentration about 10–30 times higher (see Table 12–1, p. 388). Under normal conditions, the interior of most cells is at a negative electric potential compared with the exterior, so that positive ions tend to be pulled into the cell. This means that the inward electrochemical driving force for Na^+ is large, as it includes the driving force due to the concentration gradient and a driving force in the same direction due to the voltage gradient (see Figure 12–7A).

The Na^+ outside the cell, on the uphill side of its electrochemical gradient, is like a large volume of water behind a high dam: it represents a very large store of energy (Figure 12–10). Even if one artificially halts the operation of the Na^+-K^+ pump with a toxin such as the plant glycoside *ouabain*, the energy in this store is sufficient to sustain for many minutes the other transport processes that are driven by the downhill flow of Na^+.

Figure 12–10 **Na^+ outside the cell is like water behind a high dam.** The water in the dam has potential energy, which can be used to drive energy-requiring processes. In the same way, an ion gradient across a membrane can be used to drive active processes in a cell, including the active transport of other molecules. Shown here is the Blyde River dam in South Africa. (Courtesy of Paul Franklin. With permission from Oxford Scientific Films.)

For K^+ the situation is different. The electric force is the same as for Na^+, because it depends only on the charge carried by the ion. The concentration gradient, however, is in the opposite direction. The result, under normal conditions, is that the net driving force for movement of K^+ across the membrane is close to zero: the electric force pulling K^+ into the cell is almost exactly balanced by the concentration gradient tending to drive it out.

The Na^+-K^+ Pump Is Driven by the Transient Addition of a Phosphate Group

The Na^+-K^+ pump provides a beautiful illustration of how a protein couples one reaction to another, following the principles discussed in Chapter 3. The pump works in a cycle, as illustrated schematically in Figure 12–11. Na^+ binds to the pump at sites exposed inside the cell (stage

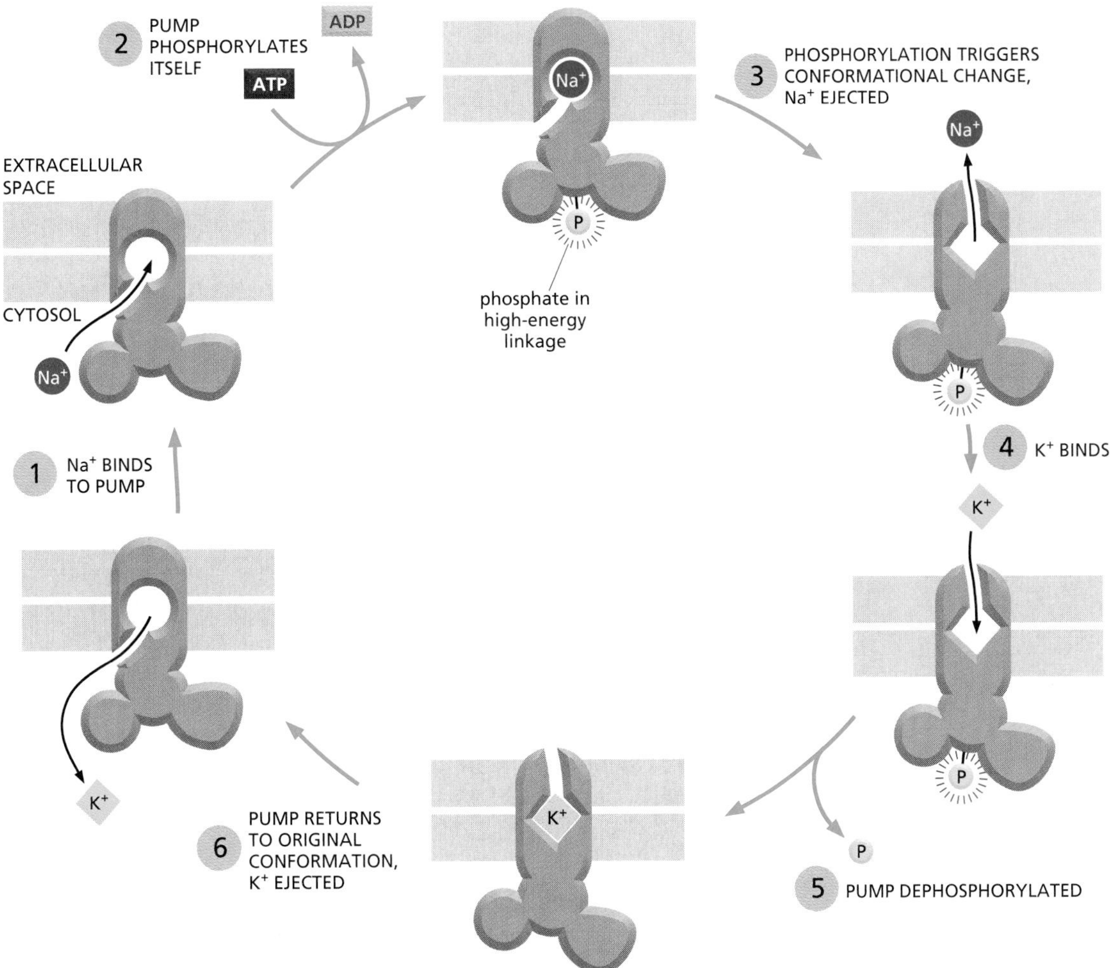
2 PUMP PHOSPHORYLATES ITSELF
ATP
ADP
3 PHOSPHORYLATION TRIGGERS CONFORMATIONAL CHANGE, Na+ EJECTED
Na+
EXTRACELLULAR SPACE
CYTOSOL
Na+
P
phosphate in high-energy linkage
1 Na+ BINDS TO PUMP
4 K+ BINDS
K+
K+
6 PUMP RETURNS TO ORIGINAL CONFORMATION, K+ EJECTED
K+
P
5 PUMP DEPHOSPHORYLATED

Figure 12–11 **The Na^+-K^+ pump transports ions in a cyclic manner.** The binding of cytosolic Na^+ (1) and the subsequent phosphorylation by ATP of the cytosolic face of the pump (2) induce the protein to undergo a conformational change that transfers the Na^+ across the membrane and releases it on the outside (3). The high-energy linkage of the phosphate to the protein provides the energy to drive the conformational change. The binding of K^+ on the extracellular surface (4) and the subsequent dephosphorylation (5) return the protein to its original conformation, which transfers the K^+ across the membrane and releases it into the cytosol (6). The cycle is shown in Movie 12.1. The changes in conformation are analogous to the A ⇌ B transitions shown in Figure 12–6, except that here the Na^+-dependent phosphorylation and the K^+-dependent dephosphorylation of the protein cause the conformational transitions to occur in an orderly manner, enabling the protein to do useful work. For simplicity, only one Na^+- and one K^+-binding site are shown. In the real pump in mammalian cells, there are thought to be three Na^+- and two K^+-binding sites. The net result of one cycle of the pump is therefore to transport three Na^+ out of the cell and two K^+ in. Ouabain inhibits the pump by preventing K^+ binding.

1), activating the ATPase activity. ATP is split, with the release of ADP and the transfer of a phosphate group into a high-energy linkage to the pump itself—that is, the pump phosphorylates itself (stage 2). Phosphorylation causes the pump to switch its conformation so as to release Na^+ at the exterior surface of the cell and, at the same time, to expose a binding site for K^+ at the same surface (stage 3). The binding of extracellular K^+ (stage 4) triggers the removal of the phosphate group (stage 5), causing the pump to switch back to its original conformation, discharging the K^+ into the cell interior (stage 6). The whole cycle, which takes about 10 milliseconds, can then be repeated. Each step in the cycle depends on the one before, so that if any of the individual steps is prevented from occurring, all the functions of the pump are halted. This tight coupling ensures that the pump operates only when the appropriate ions are available to be transported, thereby avoiding useless ATP hydrolysis.

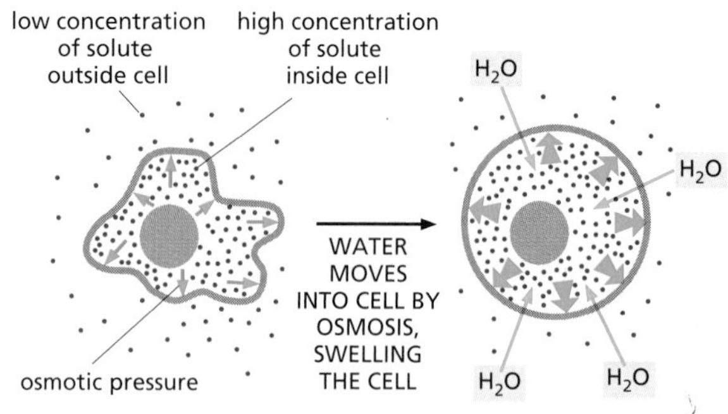

Figure 12–12 **The diffusion of water is known as osmosis.** If the concentration of solutes inside a cell is higher than that outside, water will move in by osmosis, causing the cell to swell. If a cell is placed in a high-salt solution, however, water will rush out (Movie 12.2).

The Na^+-K^+ Pump Helps Maintain the Osmotic Balance of Animal Cells

The plasma membrane is permeable to water (see Figure 12–2), and if the total concentration of solutes is low on one side of the membrane and high on the other, water will tend to move across it until the solute concentrations are equal. The movement of water from a region of low solute concentration (high water concentration) to a region of high solute concentration (low water concentration) is called **osmosis**. Cells contain specialized water channels (called *aquaporins*) in their plasma membrane that facilitate this flow. The driving force for the water movement is equivalent to a difference in water pressure and is called the **osmotic pressure**. In the absence of any counteracting pressure, the osmotic movement of water into a cell will cause it to swell (Figure 12–12).

In the tissues of the animal body, cells are bathed by a fluid that is rich in solutes, especially Na^+ and Cl^-. This balances the concentration of organic and inorganic solutes confined inside the cell. But the osmotic balance is always in danger of being upset, as the external solutes are constantly leaking into the cell down their individual electrochemical gradients. Animal cells thus have to do continuous work, pumping out unwanted solutes to maintain the osmotic equilibrium (Figure 12–13A). This function is performed mainly by the Na^+-K^+ pump, which pumps out the Na^+ that leaks in. At the same time, the Na^+-K^+ pump helps to maintain a membrane potential (as we explain later). This membrane potential tends to prevent the entry of Cl^-, which is negatively charged and would need to move against the electrical gradient generated by the pump to enter the cell.

Different cells cope with osmotic challenges in different ways. Plant cells are prevented from swelling by their tough cell walls and so can tolerate a large osmotic difference across their plasma membrane (Figure 12–13B). The cell wall exerts a counteracting pressure that tends to balance the osmotic pressure created by the solutes in the cell and thereby limits the movement of water into the cell. Osmosis, together with the active trans-

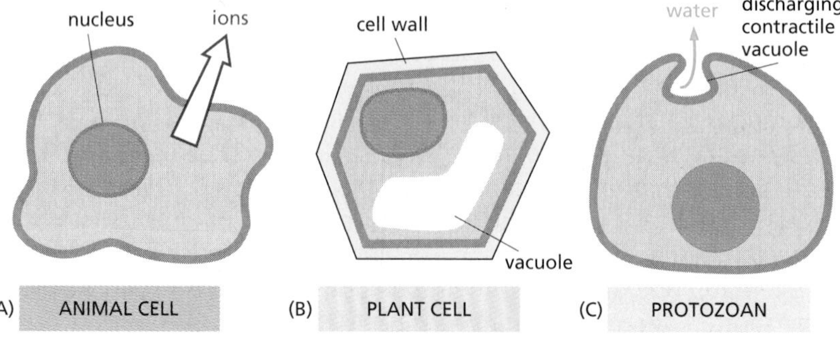

Figure 12–13 **Cells use different tactics to avoid osmotic swelling.** The animal cell keeps the intracellular solute concentration low by pumping out ions (A). The plant cell's tough wall prevents swelling (B). The protozoan avoids swelling by periodically ejecting the water that moves into the cell (C).

port of ions into the cell, results in a *turgor pressure* that keeps plant cells distended with water, with their cell wall tense. Thus, plant cells are like footballs, in which a leather outer case is held taut by the pressure in the pumped-up rubber bladder inside; the cell wall acts like the leather outer case, and the plasma membrane acts like the rubber bladder. The turgor pressure serves various functions. It holds plant stems rigid and leaves extended. It also plays a part in regulating gas exchange through the stomata—the microscopic 'mouths' in the surface of a leaf; these pores are opened and closed by the guard cells that surround them (Figure 12–14). Guard cells control their own turgor pressure by regulating the movement of K^+ across their plasma membranes.

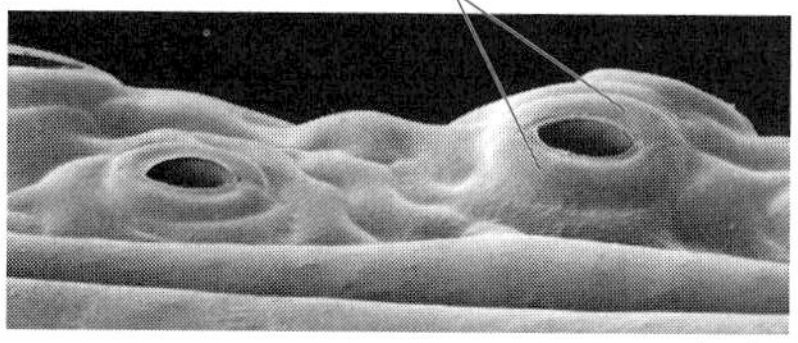

Figure 12–14 **Stomata open on the underside of a leaf.** The opening and closing of these pores is controlled by the sausage-shaped guard cells that surround them. (Courtesy of Kim Findlay.)

In some protozoans living in fresh water, such as amoebae, the excess water that continually flows into the cell by osmosis is collected in contractile vacuoles that periodically discharge their contents to the exterior (Figure 12–13C). The cell first allows the vacuole to fill with a solution rich in solutes, which causes water to follow by osmosis. The cell then retrieves the solutes by actively pumping them back into the cytosol before emptying the vacuole to the exterior.

Intracellular Ca^{2+} Concentrations Are Kept Low by Ca^{2+} Pumps

Ca^{2+}, like Na^+, is also kept at a low concentration in the cytosol compared with its concentration in the extracellular fluid, but it is much less plentiful than Na^+, both inside and outside cells. The movement of Ca^{2+} across cell membranes, however, is crucially important because Ca^{2+} can bind tightly to a variety of proteins in the cell, altering their activities. An influx of Ca^{2+} into the cytosol through Ca^{2+} channels, for example, is often used as a signal to trigger other intracellular events, such as the secretion of signal molecules and the contraction of muscle cells.

The lower the background concentration of free Ca^{2+} in the cytosol, the more sensitive the cell is to an increase in cytosolic Ca^{2+}. Thus, eucaryotic cells in general maintain very low concentrations of free Ca^{2+} in their cytosol (about 10^{-4} mM) in the face of very much higher extracellular Ca^{2+} concentrations (typically 1–2 mM). This huge concentration difference is achieved mainly by means of ATP-driven Ca^{2+} pumps in both the plasma membrane and the endoplasmic reticulum membrane, which actively pump Ca^{2+} out of the cytosol.

Like the Na^+-K^+ pump, the Ca^{2+} pump is an ATPase that is phosphorylated and dephosphorylated during its pumping cycle (Figure 12–15). It is

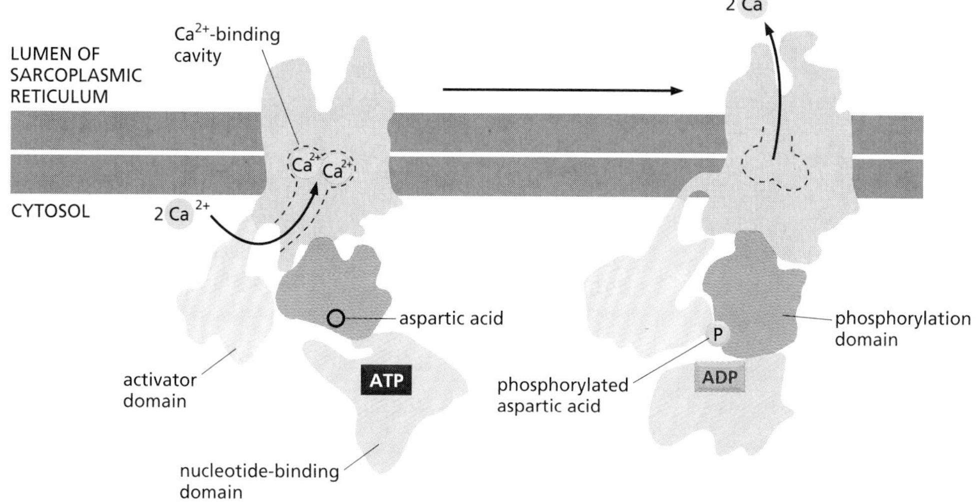

Figure 12–15 **A Ca^{2+} pump returns Ca^{2+} to the sarcoplasmic reticulum in a skeletal muscle cell.** The three-dimensional structure of this membrane transport protein has been determined by X-ray crystallography and electron microscopy. The calcium pump is a single protein composed of four discrete domains with different functions. When a muscle cell is stimulated, Ca^{2+} floods from the sarcoplasmic reticulum—a specialized form of endoplasmic reticulum—into the cytosol, allowing the cell to contract; to recover from the contraction, Ca^{2+} is returned to the sarcoplasmic reticulum by this Ca^{2+} pump. The polypeptide chain of the protein crosses the membrane as 10 α helices. ATP binding and the consequent phosphorylation of an aspartic acid in the transporter trigger conformational changes that bring the nucleotide-binding and activator domains into close proximity. This movement in turn leads to a rearrangement of the transmembrane helices, which eliminates the Ca^{2+}-binding sites and releases Ca^{2+} ions into the lumen of the sarcoplasmic reticulum. Note that the pathway Ca^{2+} ions take through the protein allows the ions to avoid contact with the lipid bilayer. (Adapted from C. Toyoshima et al., *Nature* 405:647–655, 2000. With permission from Macmillan Publishers Ltd.)

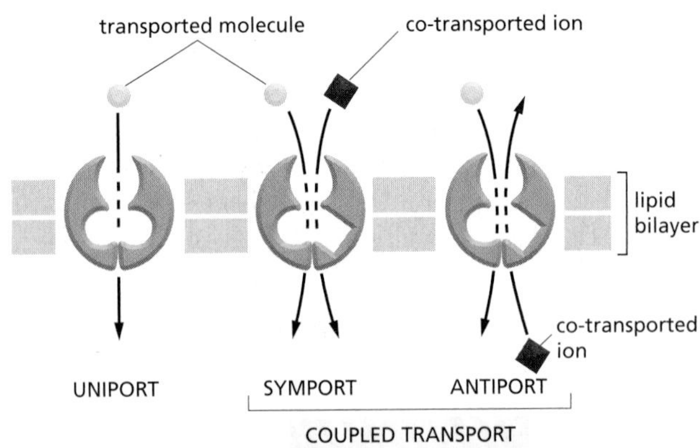

Figure 12–16 **Some transporters carry a single solute across the membrane (uniports); others couple the uphill transport of one solute across to the downhill transport of another.** In coupled transport, the solutes can be transferred either in the same direction, by symports, or in the opposite direction, by antiports (Movie 12.3). Uniports, symports, and antiports are used for both passive and active transport.

thought to work in much the same way as depicted for the Na^+-K^+ pump in Figure 12–11 except that it returns to its original conformation without binding and transporting a second ion. These two ATP-driven pumps have similar amino acid sequences and structures, suggesting that they share a common evolutionary origin.

Coupled Transporters Exploit Gradients to Take Up Nutrients Actively

A gradient of any solute across a membrane, like the Na^+ gradient generated by the Na^+-K^+ pump, can be used to fuel the active transport of a second molecule. The downhill movement of the first solute down its gradient provides the energy to drive the uphill transport of the second. The transporters that do this are called **coupled transporters** (see Figure 12–8). They may couple the movement of one inorganic ion to that of another, the movement of an inorganic ion to that of an organic molecule, or the movement of one organic molecule to that of another. If the transporter moves both solutes in the same direction across the membrane, it is called a *symport* (Figure 12–16). If it moves them in opposite directions, it is called an *antiport*. A transporter that ferries only one type of solute across the membrane (and is therefore not a coupled transporter) is called a *uniport*. The passive glucose transporter described earlier (see Figure 12–6) is a uniport.

In animal cells, an especially important role is played by symports that use the inward flow of Na^+ down its steep electrochemical gradient to drive the import of other solutes into the cell. The epithelial cells that line the gut, for example, transfer glucose from the gut across the gut epithelium. If these cells had only the passive, uniport glucose transporters, they would release glucose into the gut after a sugar-free meal as freely as they take it up from the gut after a sugar-rich meal. But these epithelial cells also possess a *glucose–Na^+ symport,* which they can use to take up glucose from the gut lumen by active transport, even when the concentration of glucose is higher inside the cell than in the gut. Because the electrochemical gradient for Na^+ is steep, when Na^+ moves into the cell down its gradient, the sugar is, in a sense, "dragged" into the cell with it (Figure 12–17). Because the binding of Na^+ and glucose is cooperative—the binding of one enhances the binding of the other—both molecules must be present for coupled transport to occur.

If the gut epithelial cells had *only* this symport, however, they could never release glucose for use by the other cells of the body. These cells, therefore, have two types of glucose transporters. In the apical domain of the plasma membrane, which faces the lumen of the gut, they have the glucose–Na^+ symports. These take up glucose actively, creating a high glucose con-

QUESTION 12–2

A rise in the intracellular Ca^{2+} concentration causes muscle cells to contract. In addition to an ATP-driven Ca^{2+} pump, muscle cells that contract quickly and regularly, such as those of the heart, have an antiport that exchanges Ca^{2+} for extracellular Na^+ across the plasma membrane. The majority of the Ca^{2+} ions that have entered the cell during contraction are rapidly pumped back out of the cell by this antiport, thus allowing the cell to relax. Ouabain and digitalis are used for treating patients with heart disease because they make the heart muscle contract more strongly. Both drugs function by partially inhibiting the Na^+-K^+ pump in the membrane of the heart muscle cell. Can you propose an explanation for the effects of the drugs in the patients? What will happen if too much of either drug is taken?

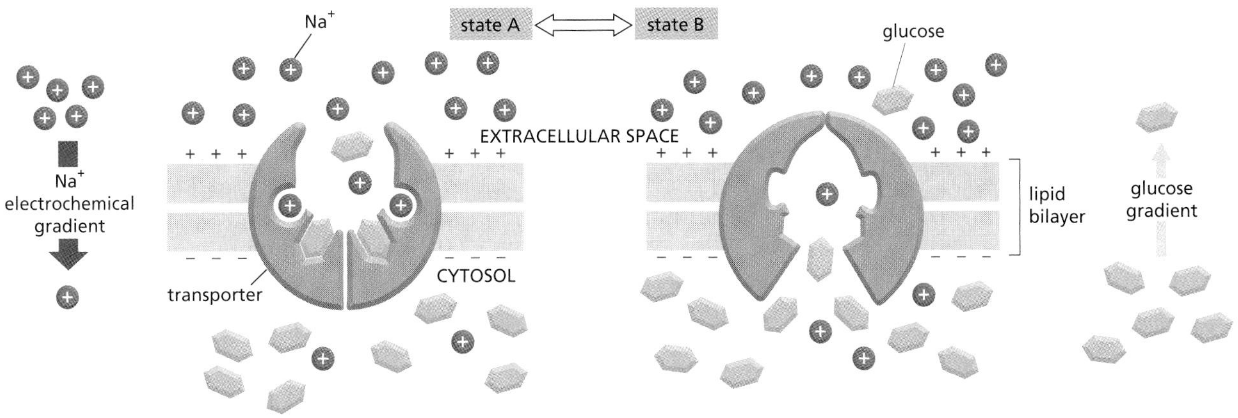

Figure 12–17 **The glucose–Na^+ symport protein uses the electrochemical Na^+ gradient to drive the import of glucose.** Glucose can be moved across epithelial cell membranes using both active and passive transporters. Shown here is one way in which the glucose–Na^+ symport protein could actively pump glucose across the membrane using the influx of Na^+ down its electrochemical gradient to drive glucose transport. The pump oscillates randomly between two alternate states, A and B. In the A state, the protein is open to the extracellular space; in the B state, it is open to the cytosol. Although Na^+ and glucose each bind to the protein in either state, they bind effectively only if both are present together: the binding of Na^+ induces a conformational change in the protein that greatly increases the protein's affinity for glucose and vice versa. Because the Na^+ concentration is much higher in the extracellular space than in the cytosol, glucose is more likely to bind to the pump in the A state; therefore, both Na^+ and glucose enter the cell (via an A → B transition) much more often than they leave it (via a B → A transition). The overall result is the net transport of both glucose and Na^+ into the cell. Note that, because the binding is cooperative, if one of the two solutes is missing, the other will fail to bind to the pump, and it will not be transported. An alternative way in which coupled transport may work is considered in Question 12–3.

centration in the cytosol. In the basal and lateral domains of the plasma membrane, they have the passive glucose uniports, which release the glucose down its concentration gradient for use by other tissues (Figure 12–18). The two types of glucose transporters are kept segregated in their proper domains of the plasma membrane by a diffusion barrier formed by a tight junction around the apex of the cell, which prevents mixing

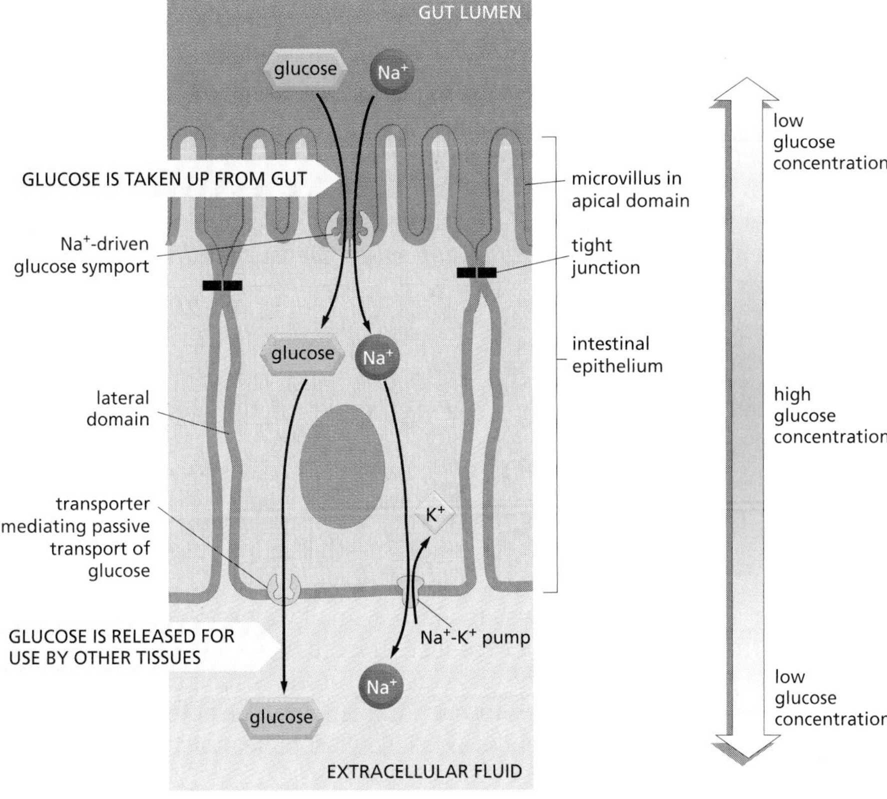

Figure 12–18 **Two types of glucose transporters enable gut epithelial cells to transfer glucose across the gut lining.** Glucose is actively transported into the cell by Na^+-driven glucose symports at the apical surface, and it is released from the cell down its concentration gradient by passive glucose uniports at the basal and lateral surfaces (Movie 12.4). The two types of glucose transporters are kept segregated in the plasma membrane by the tight junction. (See Figure 11–34; we discuss tight junctions further in Chapter 20.) To keep the concentration of Na^+ in the cytosol low, Na^+ that enters the cell via the Na^+-driven glucose symport is pumped out by Na^+-K^+ pumps. There is ample Na^+ in the gut lumen, provided by the diet.

of membrane components between the apical and the basal and lateral domains, as discussed in Chapter 11 (see Figure 11–34).

QUESTION 12–3

A transmembrane protein has the following properties: it has two binding sites, one for solute A and one for solute B. The protein can undergo a conformational change to switch between two states: either both binding sites are exposed exclusively on one side of the membrane or both binding sites are exposed exclusively on the other side of the membrane. The protein can switch between the two conformational states only if both binding sites are occupied or if both binding sites are empty, but cannot switch if only one binding site is occupied.

A. What kind of protein do these properties define?

B. Do you need to specify any additional properties to turn this protein into a symport that couples the movement of solute A up its concentration gradient to the movement of solute B down its electrochemical gradient?

C. Write a set of rules that defines an antiport.

Cells in the lining of the gut and in many other organs, such as the kidney, contain a variety of symports in their plasma membrane that are similarly driven by the electrochemical gradient of Na^+; each of these transporters specifically imports a small group of related sugars or amino acids into the cell. But Na^+-driven antiports are also important for cell function. For example, the *Na^+–H^+ exchanger* in the plasma membranes of many animal cells uses the downhill influx of Na^+ to pump H^+ out of the cell and is one of the main devices that animal cells use to control the pH in their cytosol.

H^+ Gradients Are Used to Drive Membrane Transport in Plants, Fungi, and Bacteria

Plant cells, fungi (including yeasts), and bacteria do not have Na^+-K^+ pumps in their plasma membrane. Instead of an electrochemical gradient of Na^+, they rely mainly on an electrochemical gradient of H^+ to drive the transport of solutes into the cell. The gradient is created by H^+ pumps in the plasma membrane, which pump H^+ out of the cell, thus setting up an electrochemical proton gradient, with H^+ higher outside than inside; in the process, the H^+ pump also creates an acid pH in the medium surrounding the cell. The uptake of many sugars and amino acids into bacterial cells, then, is driven by H^+ symports, which use the electrochemical gradient of H^+ across the plasma membrane in much the same way that animal cells use the electrochemical gradient of Na^+.

In some photosynthetic bacteria, the H^+ gradient is created by the activity of light-driven H^+ pumps such as bacteriorhodopsin (see Figure 11–28). In other bacteria, the gradient is created by the activities of plasma membrane proteins that carry out the final stages of cell respiration that lead to ATP synthesis, as discussed in Chapter 14. But plants, fungi, and many other bacteria set up their H^+ gradient by means of ATPases in their plasma membranes that use the energy of ATP hydrolysis to pump H^+ out of the cell; these ATPases resemble the Na^+-K^+ pumps and Ca^{2+} pumps in mammalian cells discussed earlier.

A different type of H^+ ATPase is found in the membranes of some intracellular organelles, such as the lysosomes of animal cells and the central vacuole of plant and fungal cells. Their function is to pump H^+ out of the cytosol into the organelle, thereby helping to keep the pH of the cytosol neutral and the pH of the interior of the organelle acidic. The acid environment in many organelles is crucial to their function, as we discuss in Chapter 15.

Some of the transporters considered in this chapter are shown in Figure 12–19 and are listed in Table 12–2.

We now turn to the transport of ions through channels and discuss how this ion flow can generate a membrane potential.

ION CHANNELS AND THE MEMBRANE POTENTIAL

In principle, the simplest way to allow a small water-soluble molecule to cross from one side of a membrane to the other is to create a hydrophilic channel through which the molecule can pass. Channels perform this function in cell membranes, forming transmembrane aqueous pores that allow the passive movement of small water-soluble molecules into or out of the cell or organelle.

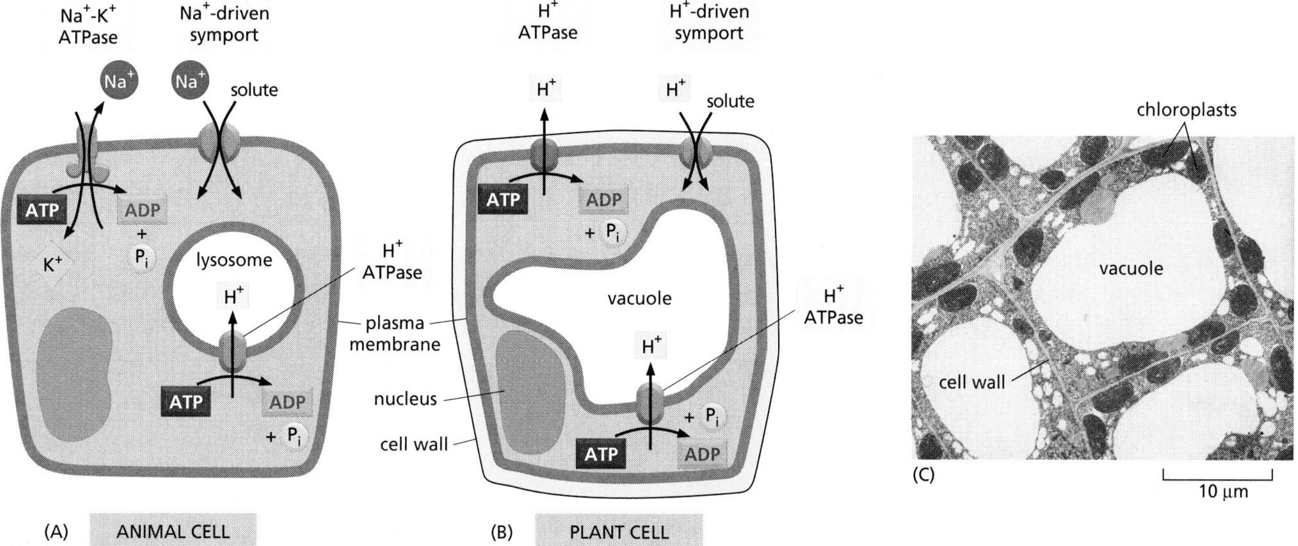

Figure 12–19 **There are similarities and differences in transporter-mediated solute movement in animal and plant cells.** In animal cells, an electrochemical gradient of Na^+, generated by the Na^+-K^+ pump (Na^+-K^+ ATPase), is often used to drive the active transport of solutes across the plasma membrane (A). An electrochemical gradient of H^+, usually set up by an H^+ ATPase, is often used for this purpose in plant cells (B), as well as in bacteria and fungi (not shown). The lysosomes in animal cells and the vacuoles in plant and fungal cells contain an H^+ ATPase in their membrane that pumps in H^+, helping to keep the internal environment of these organelles acidic. (C) An electron micrograph shows the vacuole in plant cells in a young tobacco leaf. (C, courtesy of J. Burgess.)

A few channels form relatively large pores: examples are the proteins that form *gap junctions* between two adjacent cells (see Figure 21–28) and the *porins* that form channels in the outer membrane of mitochondria and some bacteria (see Figure 11–25). But such large, permissive channels would lead to disastrous leaks if they directly connected the cytosol of a cell to the extracellular space. Thus, most of the channels in the plasma membrane of animal and plant cells have narrow, highly selective pores. One specialized channel, called *aquaporin*, facilitates the flow of water across the plasma membrane. The structure of this protein allows the rapid passage of uncharged water molecules, while prohibiting the movement of ions, including H^+. But the bulk of the cell's channels enable the transport of inorganic ions, mainly Na^+, K^+, Cl^-, and Ca^{2+}. It is these **ion channels** that we discuss next.

Ion Channels Are Ion-selective and Gated

Two important properties distinguish ion channels from simple holes in the membrane. First, they show *ion selectivity*, permitting some inorganic ions to pass but not others. Ion selectivity depends on the diameter and shape of the ion channel and on the distribution of the charged amino

TABLE 12–2 SOME EXAMPLES OF TRANSPORTERS

TRANSPORTER	LOCATION	ENERGY SOURCE	FUNCTION
Glucose transporter	plasma membrane of most animal cells	none	passive import of glucose
Na^+-driven glucose pump	apical plasma membrane of kidney and intestinal cells	Na^+ gradient	active import of glucose
Na^+-H^+ exchanger	plasma membrane of animal cells	Na^+ gradient	active export of H^+ ions, pH regulation
Na^+-K^+ pump (Na^+-K^+ ATPase)	plasma membrane of most animal cells	ATP hydrolysis	active export of Na^+ and import of K^+
Ca^{2+} pump (Ca^{2+} ATPase)	plasma membrane of eucaryotic cells	ATP hydrolysis	active export of Ca^{2+}
H^+ pump (H^+ ATPase)	plasma membrane of plant cells, fungi, and some bacteria	ATP hydrolysis	active export of H^+
H^+ pump (H^+ ATPase)	membranes of lysosomes in animal cells and of vacuoles in plant and fungal cells	ATP hydrolysis	active export of H^+ from cytosol into vacuole
Bacteriorhodopsin	plasma membrane of some bacteria	light	active export of H^+

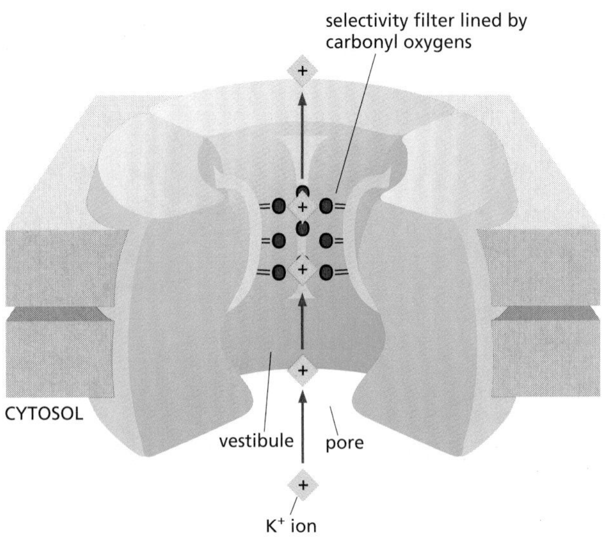

Figure 12–20 **A K^+ channel possesses a selectivity filter that controls which ion it will transport across the membrane.** Shown here is a portion of a bacterial K^+ channel. One of the four protein subunits has been omitted from the drawing to expose the interior structure of the pore. From the cytosolic side, the pore opens into a vestibule that sits in the middle of the membrane. K^+ ions in the vestibule are still cloaked in their associated water molecules (not shown). The narrow selectivity filter, which links the vestibule with the outside of the cell, is lined with carbonyl oxygen atoms (*red*) that bear a partial negative charge and form transient binding sites for the K^+ ions that have shed their watery shell. (Adapted from D.A. Doyle et al., *Science* 280:69–77, 1998. With permission from the AAAS.)

acids that line it. An ion channel is narrow enough in places to force ions into contact with the channel wall so that only those of appropriate size and charge are able to pass (Figure 12–20). Narrow channels, for example, will not pass large ions, and channels with a negatively charged lining will deter negative ions from entering because of the mutual electrostatic repulsion between like charges. In this way, channels have evolved that are selective for just one type of ion, such as Cl^- or K^+ (Movie 12.5). Each ion in aqueous solution is surrounded by a small shell of water molecules, and the ions have to shed most of their associated water molecules in order to pass, in single file, through the selectivity filter in the narrowest part of the channel. There, ions make important but very transient contacts with atoms in the amino acids that line the walls of the selectivity filter (see Figure 12–20). These precisely positioned atoms allow the channel to discriminate between ions that differ only minutely in size. This step in the transport process also limits the maximum rate of passage of ions through the channel. Thus, as ion concentrations are increased, the flow of ions through a channel at first increases proportionally but then levels off (saturates) at a maximum rate.

The second important distinction between simple pores and ion channels is that ion channels are not continuously open. Ion transport would be of no value to the cell if there were no means of controlling the flow and if all of the many thousands of ion channels in a cell membrane were open all of the time. Instead, ion channels open briefly and then close again (Figure 12–21). As we discuss later, most ion channels are *gated*: a specific stimulus triggers them to switch between a closed and an open state by a change in their conformation.

Because an open ion channel does not need to undergo conformational changes with each ion it passes, ion channels have a large advantage over transporters with respect to their maximum rate of transport. More than a million ions can pass through an open channel each second, which

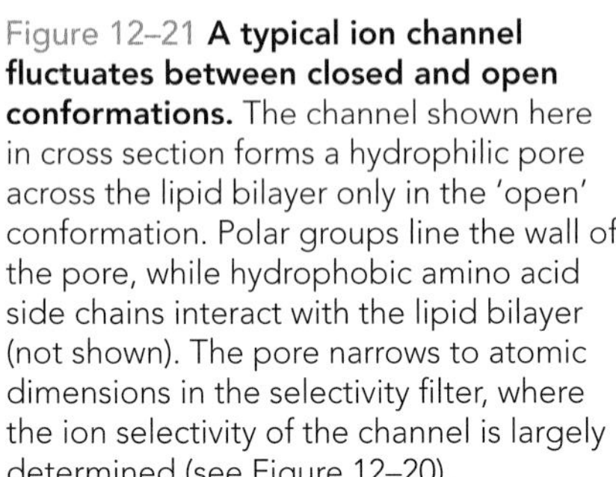

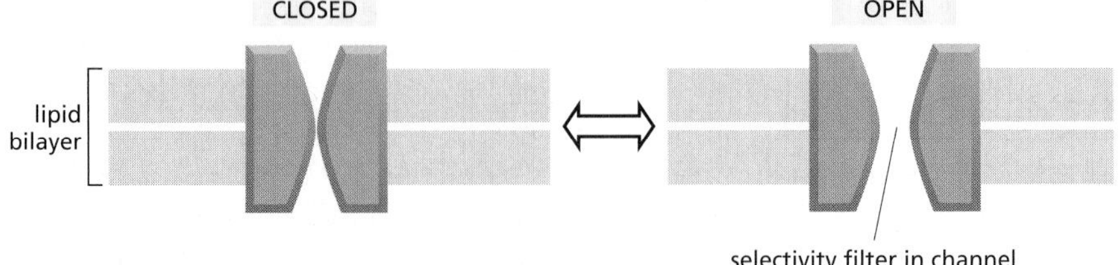

Figure 12–21 **A typical ion channel fluctuates between closed and open conformations.** The channel shown here in cross section forms a hydrophilic pore across the lipid bilayer only in the 'open' conformation. Polar groups line the wall of the pore, while hydrophobic amino acid side chains interact with the lipid bilayer (not shown). The pore narrows to atomic dimensions in the selectivity filter, where the ion selectivity of the channel is largely determined (see Figure 12–20).

is a rate 1000 times greater than the fastest rate of transfer known for any transporter. On the other hand, channels cannot couple the ion flow to an energy source to carry out active transport. The function of most ion channels is simply to make the membrane transiently permeable to selected inorganic ions, mainly Na^+, K^+, Ca^{2+}, or Cl^-, allowing these to diffuse rapidly down their electrochemical gradients across the membrane when the channel gates are open.

Thanks to active transport by pumps and other transporters, most ion concentrations are far from equilibrium across the membrane. When a channel opens, therefore, ions rush through it. The rush of ions amounts to a pulse of electric charge delivered either into the cell (as ions flow in) or out of the cell (as ions flow out). The ion flow changes the voltage across the membrane—the membrane potential—thus altering the electrochemical driving forces for transmembrane movements of all the other ions. It also forces other ion channels, which are specifically sensitive to changes in the membrane potential, to open or close in a matter of milliseconds. The resulting flurry of electrical activity can spread rapidly from one region of the cell membrane to another, conveying an electrical signal, as we discuss later in the context of nerve cells. This type of electrical signaling is not restricted to animals; it also occurs in protozoans and plants. Carnivorous plants such as the Venus flytrap use electrical signaling to sense and trap insects (Figure 12–22).

Figure 12–22 **A Venus flytrap uses electrical signaling to capture its prey.** The leaves snap shut in less than half a second when an insect moves on them. The response is triggered by touching any two of the three trigger hairs in succession in the center of each leaf. This mechanical stimulation opens ion channels and thereby sets off an electrical signal, which, by an unknown mechanism, leads to a rapid change in turgor pressure that closes the leaf. (Courtesy of J.S. Sira, Garden Picture Library.)

The membrane potential is the basis of all electrical activity in cells, whether they are plant cells, animal cells, or protozoans. Before we discuss how the membrane potential is generated, however, we look at how ion-channel activity is measured.

Ion Channels Randomly Snap Between Open and Closed States

Measuring changes in electrical current is the main method used to study ion movements and ion channels in living cells. Amazingly, electrical recording techniques have been refined to the point where it is now possible to detect and measure the electric current flowing through a single channel molecule. The procedure for doing this is known as **patch-clamp recording**, and it provides a direct and surprising picture of how individual ion channels behave.

In patch-clamp recording, a fine glass tube is used as a *microelectrode* to isolate and make electrical contact with a small area of the membrane at the surface of the cell (Figure 12–23). The technique makes it possible to record the activity of ion channels in all sorts of cell types—not only in large nerve cells, which are famous for their electrical activities, but also in cells such as yeasts that are too small for the electrical events in them to be detected by other means. By varying the concentrations of ions in the medium on either side of the membrane patch, one can test which ions will go through its resident channels. With the appropriate electronic circuitry, the voltage across the membrane patch—that is, the membrane potential—can also be set and held clamped at any chosen value (hence the term "patch-clamp"). The ability to expose the membrane to different voltages makes it possible to see how changes in membrane potential affect the opening and closing of the channels in the membrane.

With a sufficiently small area of membrane in the patch, sometimes only a single ion channel will be present. Modern electrical instruments are sensitive enough to reveal the ion flow through a single channel, detected as a minute electric current (of the order of 10^{-12} ampere). These currents typically behave in a surprising way: even when conditions are held constant, the currents switch abruptly on and abruptly off again, as though

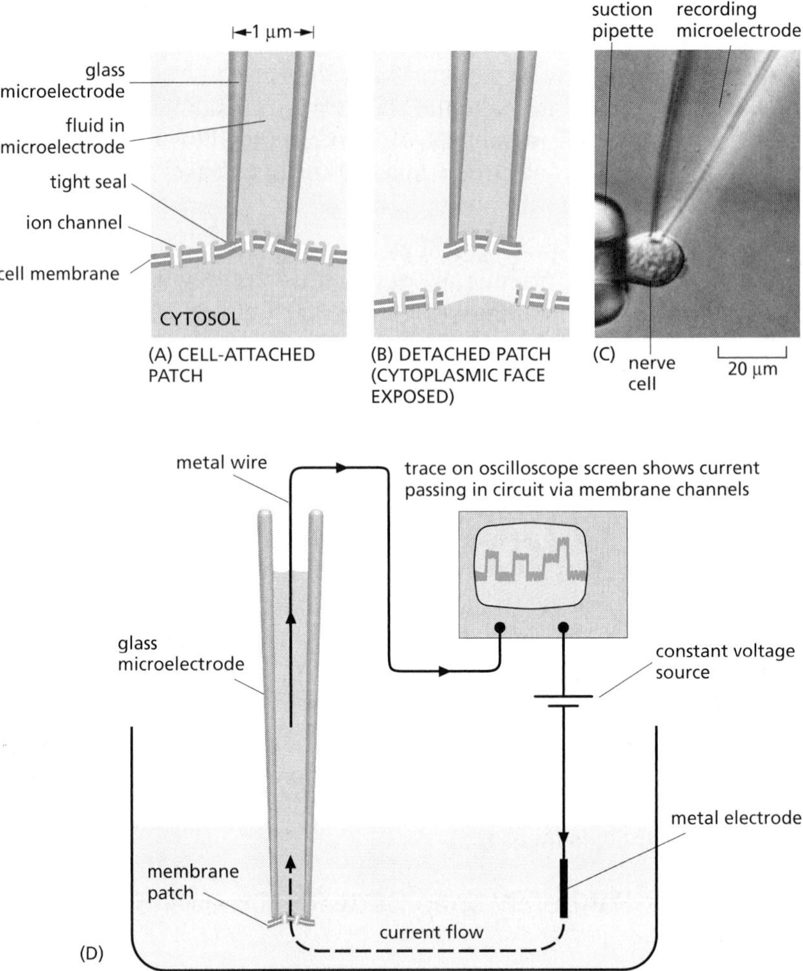

Figure 12–23 **Patch-clamp recording is used to monitor ion channel activity.** First, a microelectrode is made by heating a glass tube and pulling it to create an extremely fine tip with a diameter of no more than a few micrometers; the tube is then filled with an aqueous conducting solution, and the tip is pressed against the cell surface. With gentle suction, a tight electrical seal is formed where the cell membrane contacts the mouth of the microelectrode (A). Because of the extremely tight seal between the mouth of the microelectrode and the membrane, current can enter or leave the microelectrode only by passing through the channels in the patch of membrane covering its tip. (B) To expose the cytosolic face of the membrane, the patch of membrane held in the microelectrode can be gently detached from the cell. The advantage of the detached patch is that it is easy to alter the composition of the solution on either side of the membrane to test the effect of various solutes on channel behavior. (C) A micrograph shows a nerve cell from the eye held in a suction pipette (the tip of which is shown on the left), while a microelectrode is being used for patch-clamp recording. (D) The circuitry for patch-clamp recording. At the open end of the microelectrode, a metal wire is inserted. Current that enters the microelectrode through ion channels in the small patch of membrane covering its tip passes via the wire, through measuring instruments, back into the bath of medium in which the cell or the detached patch is sitting. (C, from T.D. Lamb, H.R. Mathews, and V. Torre, *J. Physiol.* 37:315–349, 1986. With permission from Blackwell Publishing.)

an on/off switch were being jiggled randomly (Figure 12–24). This behavior implies that the channel has moving parts and is snapping back and forth between open and closed conformations (see Figure 12–21). As this behavior is seen even when conditions are constant, it presumably indicates that the channel is being knocked from one conformation to the other by the random thermal movements of the molecules in its environment. Single-channel recording is one of a very few techniques that can be used to monitor the conformational changes of a single protein molecule. The picture it paints, of a jerky piece of machinery subjected to a constant, violent buffeting, is certain to apply also to other proteins with moving parts.

If ion channels randomly snap between open and closed conformations even when conditions on each side of the membrane are held constant, how can their state be regulated by conditions inside or outside the cell?

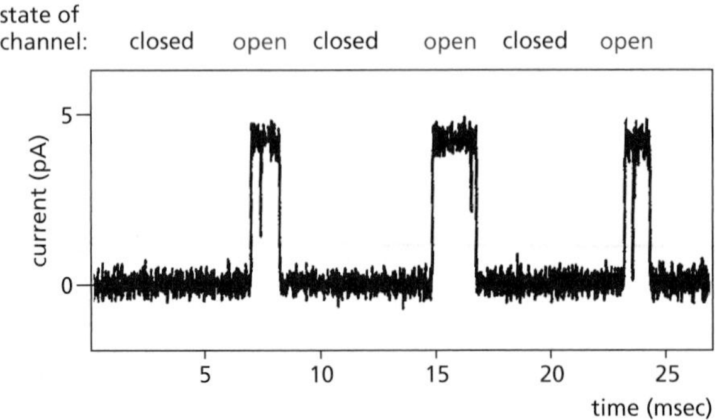

Figure 12–24 **The behavior of a single ion channel can be observed using the patch-clamp technique.** The voltage (the membrane potential) across the isolated patch of membrane is held constant during the recording. In this example, the membrane is from a muscle cell and contains a single channel protein that is responsive to the neurotransmitter acetylcholine. This ion channel opens to allow passage of positive ions when acetylcholine binds to the exterior face of the channel, as is the case here. Even when acetylcholine is bound to the channel, as is the case during the three bursts of channel opening shown here, the channel does not remain open all the time. Instead, the channel flickers between open and closed states, as seen in the brief transient drops in current in each of the three bursts. If acetylcholine were not bound to the channel, the channel would rarely open. (Courtesy of David Colquhoun.)

The answer is that when the appropriate conditions change, the random behavior continues but with a greatly changed probability: if the altered conditions tend to open the channel, for example, the channel will now spend a much greater proportion of its time in the open conformation, although it will not remain open continuously (see Figure 12–24). When an ion channel is open, it is fully open, and when it is closed, it is fully closed.

Different Types of Stimuli Influence the Opening and Closing of Ion Channels

There are more than a hundred types of ion channels, and even simple organisms can possess many different channels. The nematode worm *C. elegans*, for example, has genes that encode 68 different but related K^+ channels alone. Ion channels differ from one another primarily with respect to their *ion selectivity*—the type of ions they allow to pass; and *gating*—the conditions that influence their opening and closing. For a **voltage-gated channel**, the probability of being open is controlled by the membrane potential (Figure 12–25A). For a **ligand-gated channel**, it is controlled by the binding of some molecule (the ligand) to the channel (Figure 12–25B and C). For a **stress-gated channel**, opening is controlled by a mechanical force applied to the channel (Figure 12–25D). The *auditory hair cells* in the ear are an important example of cells that depend on stress-gated channels. Sound vibrations pull the channels open, causing ions to flow into the hair cells; this sets up an electrical signal that is transmitted from the hair cell to the auditory nerve, which conveys the signal to the brain (Figure 12–26).

Voltage-gated Ion Channels Respond to the Membrane Potential

Voltage-gated ion channels play the major role in propagating electrical signals in nerve cells. They are present in many other cells, too, including muscle cells, egg cells, protozoans, and even plant cells, where they enable electrical signals to travel from one part of the plant to another, as in the leaf-closing response of the mimosa tree (Figure 12–27). Voltage-gated ion channels have specialized charged protein domains called *voltage sensors* that are extremely sensitive to changes in the membrane potential: changes above a certain threshold value exert sufficient electri-

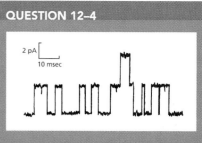

QUESTION 12–4

2 pA
10 msec

Figure Q12–4 (above) shows a recording from a patch-clamp experiment in which the electrical current passing across a patch of membrane is measured as a function of time. The membrane patch was plucked from the plasma membrane of a muscle cell by the technique shown in Figure 12–23 and contains molecules of the acetylcholine receptor, which is a ligand-gated cation channel that is opened by the binding of acetylcholine to the extracellular face of the channel. To obtain a recording, acetylcholine was added to the solution in the microelectrode. Describe what you can learn about the channels from this recording. How would the recording differ if acetylcholine were (i) omitted or (ii) added to the solution outside the microelectrode only?

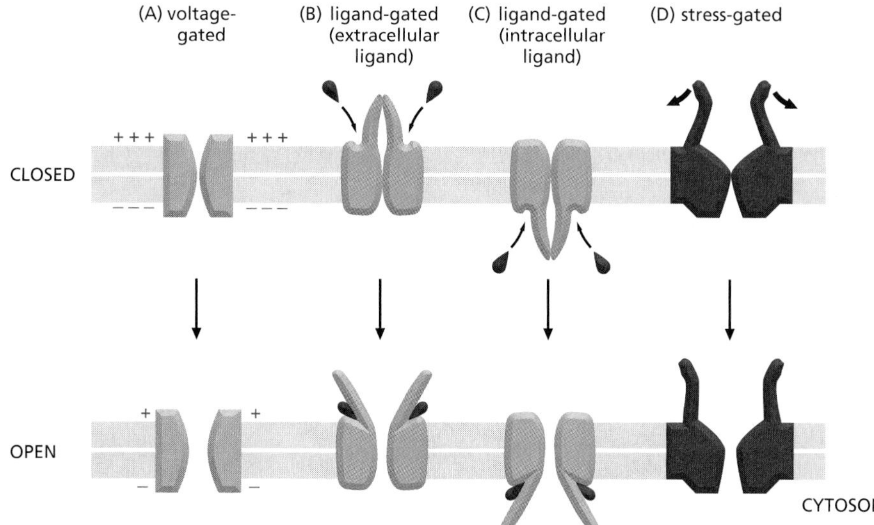

Figure 12–25 **Gated ion channels respond to different types of stimuli.** Depending on the type of ion channel, the gates open in response to a change in the voltage difference across the membrane (A), to the binding of a chemical ligand to the extracellular face (B) or the intracellular face (C) of a channel, or to mechanical stress (D).

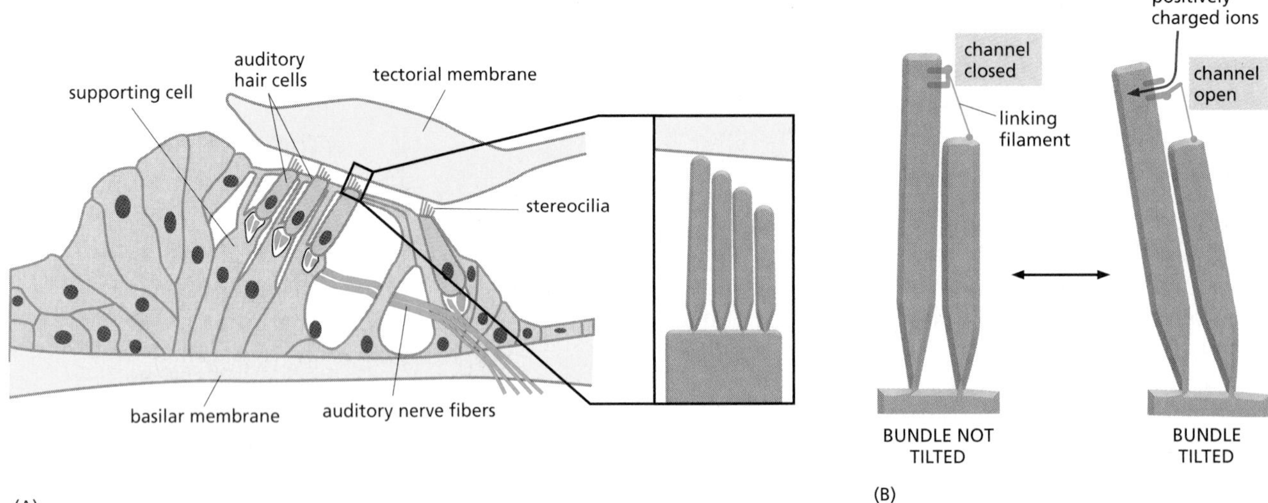

Figure 12–26 **Stress-gated ion channels allow us to hear.** (A) A section through the organ of Corti, which runs the length of the cochlea, the auditory portion of the inner ear. Each auditory hair cell has a tuft of spiky extensions called stereocilia projecting from its upper surface. The hair cells are embedded in a sheet of supporting cells, which is sandwiched between the *basilar membrane* below and the *tectorial membrane* above. (These are not lipid bilayer membranes but sheets of extracellular matrix.) (B) Sound vibrations cause the basilar membrane to vibrate up and down, causing the stereocilia to tilt. Each stereocilium in the staggered array on each hair cell is attached to the next shorter stereocilium by a fine filament. The tilting stretches the filaments, which pull open stress-gated ion channels in the stereocilium membrane, allowing positively charged ions to enter from the surrounding fluid (Movie 12.6). The influx of ions activates the hair cells, which stimulate underlying endings of the auditory nerve fibers that convey the auditory signal to the brain. The hair-cell mechanism is astonishingly sensitive: the force required to open a single channel is estimated to be about 2×10^{-13} newton, and the faintest sounds we can hear have been estimated to stretch the filaments by an average of about 0.04 nm, which is less than the diameter of a hydrogen ion. Movie 12.7 shows the remarkable flexibility of these structures.

cal force on these domains to encourage the channel to switch from its closed to its open conformation, or vice versa. A change in the membrane potential does not affect how wide the channel is open but alters the probability that it will be found in its open conformation. Thus, in a large patch of membrane, containing many molecules of the channel protein, one might find that on average 10% of them are open at any instant when the membrane is held at one potential, while 90% are open when it is held at another potential.

To appreciate the function of voltage-gated ion channels in a living cell, we have to consider what controls the membrane potential. The simple answer is that ion channels themselves control it, and the opening and closing of these channels is what makes it change. This control loop, from ion channels → membrane potential → ion channels, is fundamental to all electrical signaling in cells. Having seen how the membrane potential can regulate ion channels, we now discuss how ion channels can control the membrane potential. In the last part of the chapter, we consider how this control loop works to propagate signals in nerve cells.

Figure 12–27 **Voltage-gated ion channels underlie the leaf-closing response in mimosa.** (A) Resting leaf. (B and C) Successive responses to touch. A few seconds after the leaf is touched, the leaflets snap shut. The response involves the opening of voltage-gated ion channels, generating an electric impulse. When the impulse reaches specialized hinge cells at the base of each leaflet, a rapid loss of water by these cells occurs, causing the leaflets to fold closed suddenly and progressively down the leaf stalk. (Courtesy of G.I. Bernard, with permission from Oxford Scientific Films.)

Membrane Potential Is Governed by Membrane Permeability to Specific Ions

All cells have an electrical potential difference, or **membrane potential**, across their plasma membrane. To understand how this potential arises, it is helpful to recall some basic principles of electricity. While electricity in metals is carried by electrons, electricity in aqueous solutions is carried by ions, which are either positively (cations) or negatively (anions) charged. An ion flow across a cell membrane is detectable as an electric current, and an accumulation of ions, if not exactly balanced by an accumulation of oppositely charged ions, is detectable as an accumulation of electric charge, or a membrane potential (Figure 12–28).

To see how the membrane potential is generated and maintained, consider the ion movements into and out of a typical animal cell in an unstimulated 'resting' state. The negative charges on the organic molecules confined within the cell are largely balanced by K^+, the predominant positive ion inside the cell. The high intracellular concentration of K^+ is in part generated by the Na^+-K^+ pump, which actively pumps K^+ into the cell. This leads to a large concentration difference for K^+ across the plasma membrane, with the concentration of K^+ being much higher inside the cell than outside. The plasma membrane, however, also contains a set of K^+ channels known as *K^+ leak channels*. These channels randomly flicker between open and closed states no matter what the conditions are inside or outside the cell, and when they are open, they allow K^+ to move freely. In a resting cell, these are the main ion channels open in the plasma membrane, thus making the resting plasma membrane much more permeable to K^+ than to other ions.

K^+ therefore has a tendency to flow out of the cell through these channels down its steep concentration gradient. But any transfer of positive charge to the exterior leaves behind unbalanced negative charge within the cell, thereby creating an electrical field, or membrane potential, which will oppose any further movement of K^+ out of the cell. Within a millisecond or so, an equilibrium condition is established in which the membrane potential is just strong enough to counterbalance the tendency of K^+ to move out down its concentration gradient—that is, in which the electrochemical gradient for K^+ is zero, even though there is still a much higher concentration of K^+ inside the cell than out (Figure 12–29).

The *resting membrane potential* is the membrane potential in such steady-state conditions, in which the flow of positive and negative ions across the plasma membrane is precisely balanced, so that no further difference in charge accumulates across the membrane. The membrane potential is measured as a voltage difference across the membrane. In animal cells,

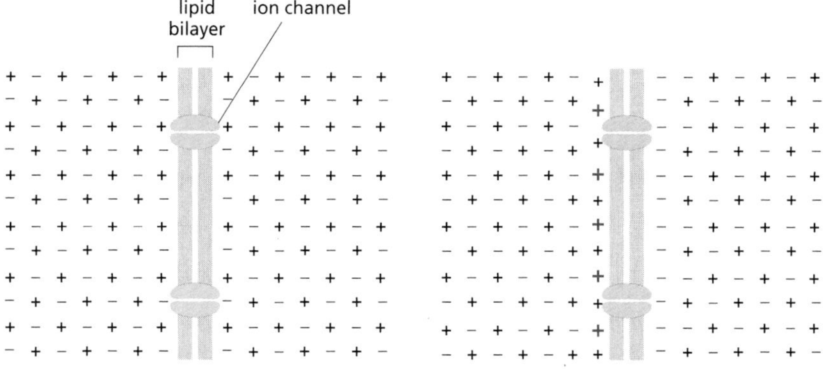
lipid bilayer
ion channel
(A) exact balance of charges on each side of the membrane
membrane potential = 0
(B) a few positive ions (*red*) cross the membrane from right to left, setting up a nonzero membrane potential

Figure 12–28 **The distribution of ions on either side of the lipid bilayer gives rise to the membrane potential.** The membrane potential results from a thin (<1 nm) layer of ions close to the membrane, held in place by their electrical attraction to oppositely charged counterparts on the other side of the membrane. (A) When there is an exact balance of charges on either side of the membrane, there is no membrane potential. (B) When ions of one type cross the membrane, they set up a charge difference between the two sides of the membrane that creates a membrane potential. The number of ions that must move across the membrane to set up a membrane potential is a tiny fraction of those present. (6000 K^+ ions crossing 1 μm^2 of cell membrane are enough to shift the membrane potential by about 100 mV; the number of K^+ ions in 1 μm^3 of bulk cytoplasm is 70,000 times larger than this.)

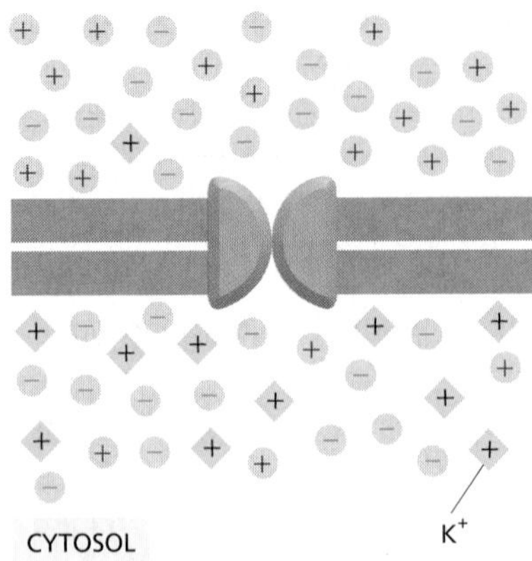

(A) K^+ channel closed, membrane potential = 0; more K^+ inside the cell than outside, but zero *net* charge on each side (positive and negative charges balanced exactly)

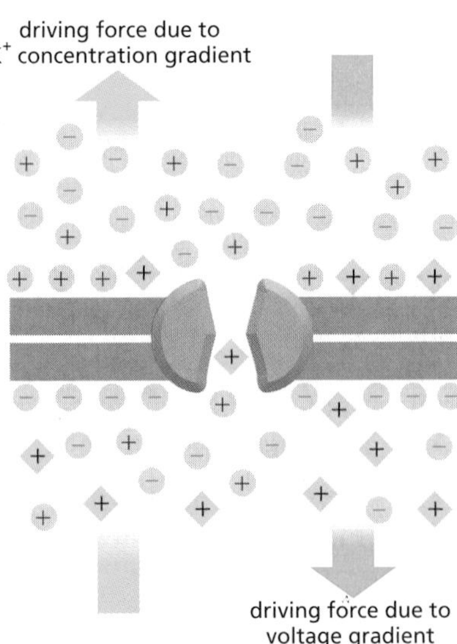

(B) K^+ channel open; K^+ moves out, leaving negative ions behind, and this charge distribution creates a membrane potential that balances the tendency of K^+ to move out

Figure 12–29 **K^+ leak channels play a major role in generating the membrane potential across the plasma membrane.** (A) A hypothetical situation where the K^+ leak channel is closed and the membrane potential is zero. (B) As soon as the channel opens, K^+ will tend to leave the cell, moving down its concentration gradient. Assuming the membrane contains no open channels permeable to other ions, K^+ ions will cross the membrane, but negative ions will be unable to follow them. The result will be an excess of positive charge on the outside of the membrane and of negative charge on the inside. This gives rise to a membrane potential that tends to drive K^+ back in. At equilibrium, the effect of the K^+ concentration gradient is exactly balanced by the effect of the membrane potential, and there is no net movement of K^+.

the resting membrane potential varies between –20 and –200 millivolts (mV), depending on the organism and cell type. It is expressed as a negative value because the interior of the cell is negative with respect to the exterior, as the negative charges inside the cell are in slight excess over positive charges. The actual value of the resting membrane potential in animal cells is chiefly a reflection of the K^+ concentration gradient across the plasma membrane, because, at rest, this membrane is chiefly permeable to K^+, and K^+ is the main positive ion inside the cell. A simple formula called the **Nernst equation** (Figure 12–30) expresses the equilibrium quantitatively and makes it possible to calculate the theoretical resting membrane potential if the ratio of internal to external ion concentrations is known.

Suppose now that other channels permeable to some other ion—say, Na^+—are suddenly opened in the plasma membrane. Because Na^+ is at a higher concentration outside the cell than inside, Na^+ will move into the cell through these channels and the membrane potential will become less negative, maybe even reversing sign to become positive (so that the interior of the cell is positive with respect to the exterior). The membrane potential will shift toward a new value that is a compromise between the negative value that would correspond to equilibrium for K^+ and the positive value that would correspond to equilibrium for Na^+. Any change in the membrane's permeability to specific ions—that is, any change in the numbers of ion channels of different sorts that are open—thus causes a change in the membrane potential. The membrane potential, therefore, is determined by both the state of the ion channels in the membrane and the ion concentrations in the cytosol and extracellular medium. Because the electrical processes at the plasma membrane occur very quickly compared with changes in the bulk ion concentrations, over the short term—milliseconds as compared with seconds or minutes—it is the ion channels that are most important in controlling the membrane potential.

To see how the interplay between the membrane potential and ion channels is used for electrical signaling, we now turn from the behavior of ions and ion channels to the behavior of entire cells. We take nerve cells as our prime example, for they, more than any other cell type, have made a profession of electrical signaling and employ ion channels in the most sophisticated ways.

The force tending to drive an ion across a membrane is made up of two components: one due to the electrical membrane potential and one due to the concentration gradient of the ion. At equilibrium, the two forces are balanced and satisfy a simple mathematical relationship given by the

Nernst equation

$$V = 62 \log_{10} (C_o/C_i)$$

where V is the membrane potential in millivolts, and C_o and C_i are the outside and inside concentrations of the ion, respectively. This form of the equation assumes that the ion carries a single positive charge and that the temperature is 37°C.

Figure 12–30 **The Nernst equation can be used to calculate the resting potential of a membrane.**

ION CHANNELS AND SIGNALING IN NERVE CELLS

The fundamental task of a nerve cell, or **neuron**, is to receive, conduct, and transmit signals. Neurons carry signals inward from sense organs to the central nervous system—the brain and spinal cord. In the central nervous system, neurons signal from one to another through networks of enormous complexity, allowing the brain and spinal cord to analyze, interpret, and respond to the signals coming in from the sense organs. From the central nervous system, neurons extend processes outward to convey signals for action to muscles and glands. To perform these functions, neurons are often extremely elongated: the motor neurons in a human that carry signals from the spinal cord to a muscle in the foot, for example, may be a meter long.

Every neuron consists of a *cell body* (containing the nucleus) that has a number of long, thin extensions radiating outward from it. Usually, a neuron has one long **axon**, which conducts signals away from the cell body toward distant target cells; it also usually has several shorter, branching *dendrites*, which extend from the cell body like antennae and provide an enlarged surface area to receive signals from the axons of other neurons (Figure 12–31). The axon commonly divides at its far end into many branches, each of which ends in a **nerve terminal**, so that the neuron's message can be passed simultaneously to many target cells—either other neurons or muscle or gland cells. Likewise, the branching of the dendrites can be extensive, in some cases sufficient to receive as many as 100,000 inputs on a single neuron.

No matter what the meaning of the signal a neuron carries—whether it is visual information from the eye, a motor command to a muscle, or one step in a complex network of neural processing in the brain—the form of the signal is always the same: it consists of changes in the electrical potential across the neuron's plasma membrane.

QUESTION 12–5

From the concentrations given in Table 12–1 (p. 388), calculate the equilibrium membrane potential of K^+ and Na^+ (assume that the concentration of intracellular Na^+ is 10 mM). What membrane potential would you predict in a resting cell? Explain your answer. What would happen if a large number of Na^+ channels suddenly opened, making the membrane much more permeable to Na^+ than to K^+? (Note that because few ions need to move across the membrane to change the charge distribution across the membrane drastically, you can safely assume that the ion concentrations on either side of the membrane do not change significantly.) What would you predict would happen next if the Na^+ channels closed again?

Action Potentials Provide for Rapid Long-Distance Communication

A neuron is stimulated by a signal—typically from another neuron—delivered to a localized site on its surface. This signal initiates a change in the membrane potential at that site. To transmit the signal onward, however, the change in membrane potential has to spread from this point, which is usually on a dendrite or the cell body, to the axon terminals, which relay the signal to the next cells in the pathway. Although a local change in membrane potential will spread passively along an axon or a dendrite to adjacent regions of the plasma membrane, it rapidly becomes weaker with increasing distance from the source. Over short distances, this weakening is unimportant. But for long-distance communication,

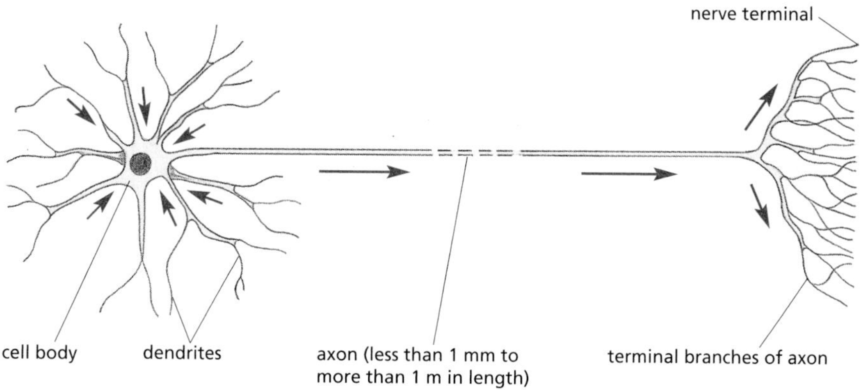

Figure 12–31 **A typical neuron has a cell body, a single axon, and multiple dendrites.** The axon conducts signals away from the cell body toward its target cells, while the multiple dendrites receive signals from the axons of other neurons. The *red* arrows indicate the direction in which signals are conveyed.

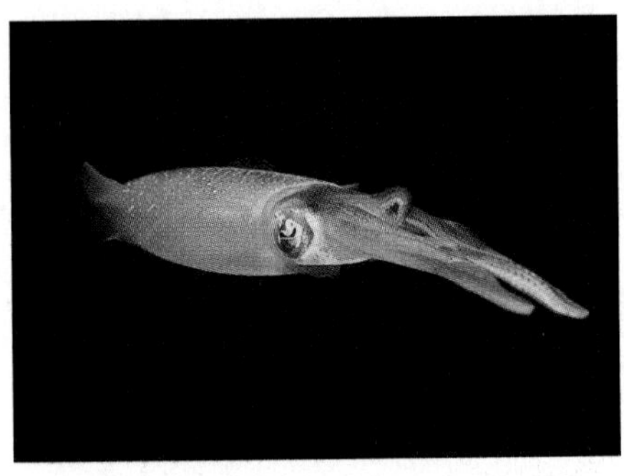

Figure 12–32 **The squid *Loligo* has a nervous system that is set up to allow the animal to respond rapidly to threats in its environment.** Among the nerve cells that make up this escape system is one that possesses a 'giant axon,' which is very large in diameter. Long before patch clamping allowed recordings from single channels in small cells, scientists were able to record action potentials in the squid giant axon and deduce the existence of ion channels in its membrane. (Courtesy of Howard Hall, with permission from Oxford Scientific Films.)

such *passive spread* is inadequate. In the same way, a telephone signal can be transmitted without amplification the short distances through the wires in your home town, but for transmission across an ocean by an undersea cable, the strength of the signal has to be boosted at intervals.

Neurons solve this long-distance communication problem by employing an active signaling mechanism: a local electrical stimulus of sufficient strength triggers an explosion of electrical activity in the plasma membrane that is propagated rapidly along the membrane of the axon and sustained by automatic renewal all along the way. This traveling wave of electrical excitation, known as an **action potential**, or a *nerve impulse*, can carry a message, without the signal weakening, from one end of a neuron to the other at speeds of up to 100 meters per second.

All of the early research that established the mechanism of electrical signaling along nerve axons was done on the giant axon of the squid (Figure 12–32). This axon has such a large diameter that it is possible to record its electrical activity from an electrode inserted directly into it (How We Know, pp. 412–413). From such studies, it was deduced that action potentials are the direct consequence of the properties of voltage-gated ion channels (see Figure 12–25A) in the nerve cell membrane, as we now explain.

Action Potentials Are Usually Mediated by Voltage-gated Na^+ Channels

An action potential in a neuron is typically triggered by a sudden local *depolarization* of the plasma membrane—that is, by a shift in the membrane potential to a less negative value (that is, a shift towards zero). We discuss later how such a depolarization is caused by the action of signal molecules—*neurotransmitters*—released by another neuron. A stimulus that causes a sufficiently large depolarization to pass a certain threshold value, promptly causes **voltage-gated Na^+ channels** to open temporarily at that site, allowing a small amount of Na^+ to enter the cell down its electrochemical gradient. The influx of positive charge depolarizes the membrane further (that is, it makes the membrane potential even less negative), thereby opening more voltage-gated Na^+ channels, which admit more Na^+ ions and cause still further depolarization. This process continues in a self-amplifying fashion until, within about a millisecond, the membrane potential in the local region of membrane has shifted from its resting value of about –60 mV to about +40 mV (Figure 12–33). This voltage is close to the membrane potential at which the electrochemical driving force for movement of Na^+ across the membrane is zero—that is, at which the effects of the membrane potential and the concentration gradient for Na^+ are equal and opposite and Na^+ has no further tendency

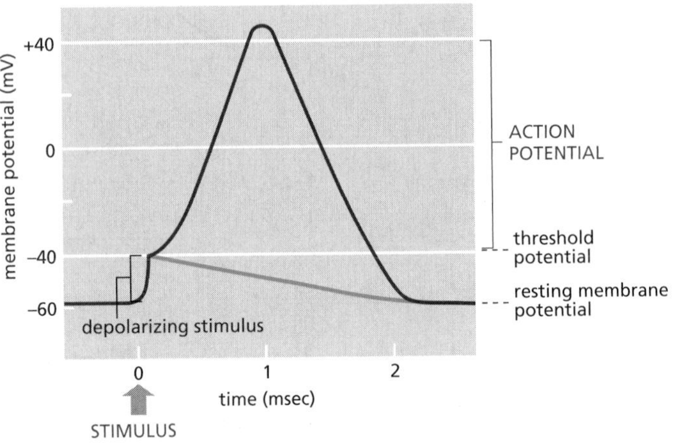

Figure 12–33 **An action potential is triggered by a rapid change in membrane potential.** The resting membrane potential in this neuron is –60 mV. The action potential is triggered when a stimulus depolarizes the plasma membrane by about 20 mV, making the membrane potential –40 mV, which is the threshold value in this cell for initiating an action potential. Once an action potential is triggered, the membrane rapidly depolarizes further: the membrane potential (*red* curve) swings past zero and reaches +40 mV before it returns to its resting negative value, as the action potential terminates. The *green* curve shows how the membrane potential would simply have relaxed back to the resting value after the initial depolarizing stimulus if there had been no voltage-gated ion channels in the plasma membrane.

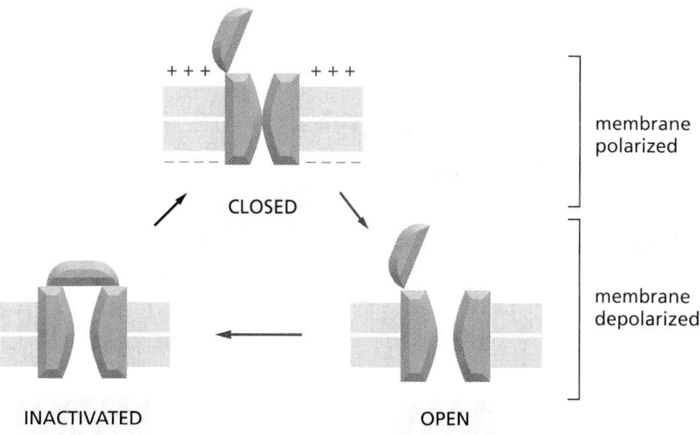

Figure 12–34 **A voltage-gated Na^+ channel can adopt at least three conformations.** The channel can flip from one conformation to another, depending on the membrane potential. When the membrane is at rest (highly polarized), the closed conformation is the most stable. When the membrane is depolarized, however, the open conformation is more stable, and so the channel has a high probability of opening; but in the depolarized membrane the inactivated conformation is more stable still, and so, after a brief period spent in the open conformation, the channel becomes inactivated and cannot open. The *red* arrows indicate the sequence that follows a sudden depolarization, and the *black* arrow indicates the return to the original conformation after the membrane is repolarized.

to enter or leave the cell. If the channels continued to respond indefinitely in this way to the altered membrane potential, the cell would get stuck at this point with all of its voltage-gated Na^+ channels predominantly open.

The cell is saved from this fate because the Na^+ channels have an automatic inactivating mechanism, which causes them to rapidly adopt (within a millisecond or so) a special inactive conformation, where the channel is unable to open again. Even though the membrane is still depolarized, the Na^+ channels will remain in this *inactivated state* until a few milliseconds after the membrane potential returns to its initial negative value. A schematic illustration of these three distinct states of the voltage-gated Na^+ channel—*closed, open,* and *inactivated*—is shown in Figure 12–34. How they contribute to the rise and fall of the action potential is shown in Figure 12–35.

The membrane is further helped to return to its resting value by the opening of *voltage-gated K^+ channels.* These also open in response to depolarization of the membrane, but not as promptly as the Na^+ channels, and they then stay open as long as the membrane remains depolarized. As the action potential reaches its peak, K^+ ions (carrying positive charge) therefore start to flow out of the cell through these K^+ channels down their electrochemical gradient, temporarily unhindered by the negative membrane potential that restrains them in the resting cell. The rapid outflow of K^+ through the voltage-gated K^+ channels brings the membrane back to its resting state more quickly than could be achieved by K^+ outflow through the K^+ leak channels alone.

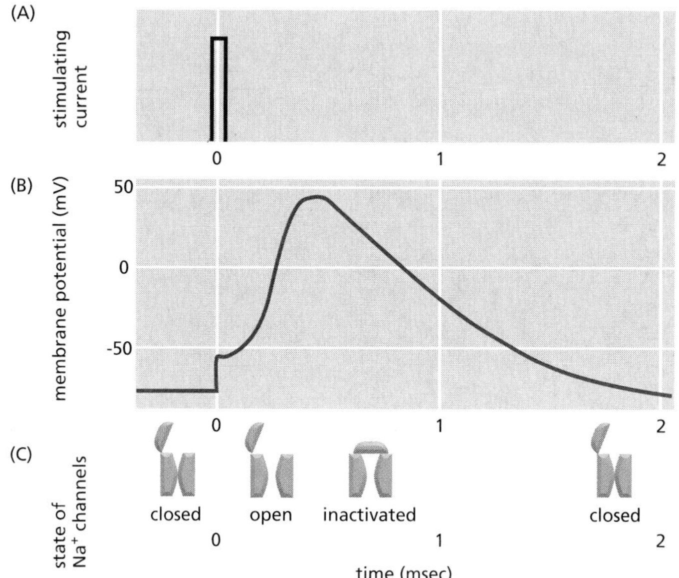

Figure 12–35 **Ion flows dictate the rise and fall of an action potential.** In this example, the action potential is triggered by a brief pulse of electric current (A), which partially depolarizes the membrane, as shown in the plot of membrane potential versus time (B). (B) shows the course of the action potential (*red* curve) that is caused by the opening and subsequent inactivation of voltage-gated Na^+ channels, whose state is shown in (C). Even if restimulated, the membrane cannot produce a second action potential until the Na^+ channels have returned from the inactivated to the simply closed conformation (see Figure 12–34). Until then, the membrane is resistant, or refractory, to stimulation.

SQUID REVEAL SECRETS OF MEMBRANE EXCITABILITY

Each spring, *Loligo pealei* migrate to the shallow waters off Cape Cod on the eastern coast of the United States. There they spawn, launching the next generation of squid. But more than just meeting and breeding, these animals provide neuroscientists summering at the Marine Biological Laboratory in Woods Hole, Massachusetts, with an excellent system for studying the mechanism of electrical signaling along nerve axons.

Like most animals, squid survive by catching prey and escaping predators. Fast reflexes and an ability to accelerate rapidly and make sudden changes in swimming direction help them avoid danger while chasing down a decent meal. Squid derive their speed and agility from a specialized biological jet propulsion system: they draw water into their mantle cavity and then contract their muscular body wall to rapidly expel the collected water through a tubular siphon, thus propelling themselves through the water.

Controlling such quick and coordinated muscle contraction requires a nervous system that can convey signals with great speed down the length of the animal's body. Indeed, *Loligo pealei* possesses some of the largest nerve fibers found in nature. Squid giant axons can reach 10 cm in length and are over 100 times the diameter of a mammalian axon—about the width of a pencil lead. Generally speaking, the larger the diameter of an axon, the more rapidly signals can travel along its length.

In the 1930s, scientists first started to take advantage of the squid giant axon for studying the electrophysiology of the nerve cell. Because of its relatively large size, researchers can insert electrodes into the axon to measure its electrical activity and monitor its action potentials. This experimental system allowed researchers to address a variety of questions about membrane conductance in neurons, 'including which ions are important for initiating and propagating an action potential, how membrane permeability changes as an action potential sweeps by, and how these changes in membrane potential control the opening and closing of ion channels.

Setup for action

Because the squid axon is so long and wide, an electrode made from a glass capillary tube containing a conducting solution can be thrust down the axis of the axon so that its tip lies deep in the cytoplasm (Figure 12–36A). This setup then allows one to measure the voltage difference between the inside and the outside of the axon—that is, the membrane potential—as an action potential sweeps past the tip of the electrode (Figure 12–36B). The action potential itself is triggered by applying a brief electrical stimulus to one end of the axon. It does not really matter which end is stimulated, as the excitation can travel in either direction; it also does not matter how big the stimulus is, as long as it exceeds a certain threshold: an action potential is all or nothing.

Once researchers could reliably generate and measure an action potential, they could use the squid axon system to answer other questions about membrane excitability. For example, which ions are critical for an action potential? The three most plentiful ions, both inside and outside the axon, are Na^+, K^+, and Cl^-. Do they have

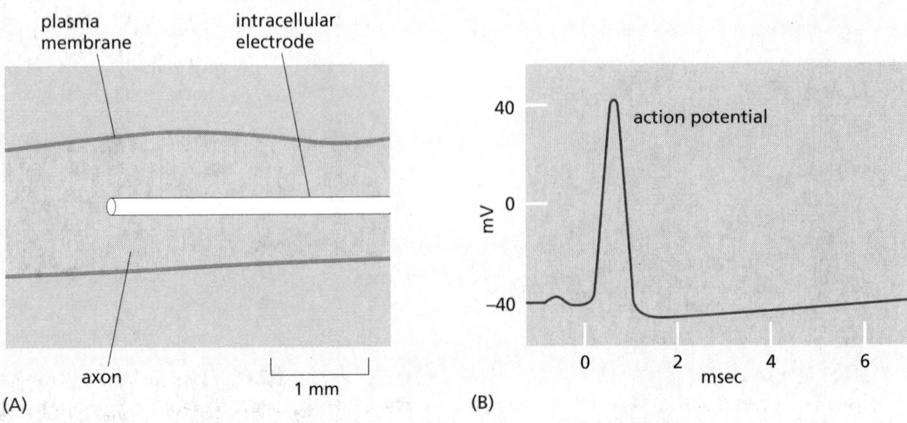

Figure 12–36 **An electrode can be inserted into the squid giant axon (A) to measure action potentials (B).**

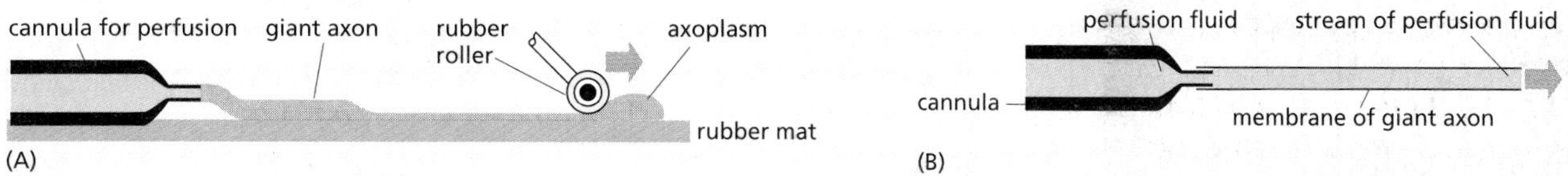

Figure 12–37 **The cytoplasm in an axon can be removed and replaced with an artificial solution of pure ions.** (A) The axon cytoplasm is extruded using a rubber roller. (B) A perfusion fluid containing the desired concentration of ions is pumped gently through the axon.

equal importance when it comes to the action potential? Because the squid axon is so large and robust, it is possible to extrude the cytoplasm from the axon like toothpaste from a tube (Figure 12–37A). The axon can then be perfused internally with a pure solution of Na^+, K^+, Cl^-, or SO_4^{2-} (Figure 12–37B); the ions in the bath solution can be varied independently. Remarkably, researchers performing this experiment discovered that the axon will generate a normal action potential if and only if the concentrations of Na^+ and K^+ approximate the natural concentrations found inside and outside the cell. Thus, the cellular components crucial to the action potential are the plasma membrane, Na^+ and K^+ ions, and the energy provided by the concentration gradients of these ions across the membrane; all other components, including other sources of metabolic energy, were presumably removed by the perfusion.

Channel traffic

Once Na^+ and K^+ had been singled out as critical to an action potential, the question then became: What does each of these ions contribute to the action potential? How permeable is the membrane to each, and how does the membrane permeability change as an action potential sweeps by? Again, the squid giant axon provided some answers. The concentrations of Na^+ and K^+ outside the membrane could be altered, and the effects of these changes on membrane potential could be measured directly. From such studies it was determined that, at rest, the membrane potential of an axon is close to the equilibrium potential for K^+. When the external concentration of K^+ is varied, the resting potential of the axon changes roughly in accordance with the Nernst equation (see Figure 12–30). At rest, therefore, the membrane is chiefly permeable to K^+; as we now know, K^+ leak channels provide the main pathway these ions can take through the cell membrane.

The situation for Na^+ is very different. When the external concentration of Na^+ is varied, there is no effect on the resting potential of the axon. However, the height of the peak of the action potential varies with the concentration of Na^+ outside the membrane (Figure 12–38). During the action potential, therefore, the membrane appears to be chiefly permeable to Na^+, as the result of the opening of Na^+ channels. In the aftermath of the action potential, the sodium channels close and the membrane potential reverts to a negative value that depends on the external concentration of K^+. As the membrane loses its permeability to Na^+, it becomes even more permeable to K^+ than before. The opening of additional K^+ channels helps speed this resetting of the membrane potential to the resting state. This readies the membrane for the next action potential.

These studies on the squid giant axon made an enormous contribution to our understanding of neuronal excitability, and the researchers who set up the system and made the discoveries reviewed here were rewarded with a Nobel Prize in 1963.

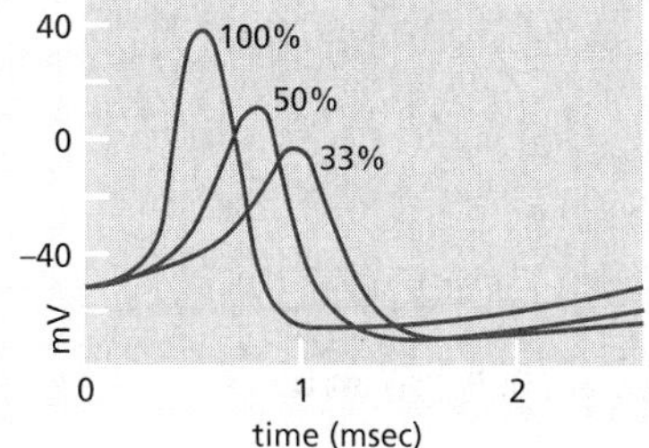

Figure 12–38 **The shape of the action potential depends on the concentration of Na^+ outside the membrane.** Shown here are action potentials recorded when the external medium contains 100%, 50%, or 33% of the normal concentration of Na^+.

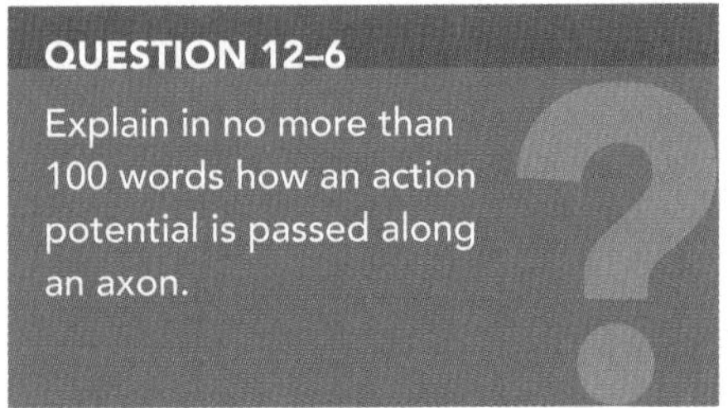

QUESTION 12–6

Explain in no more than 100 words how an action potential is passed along an axon.

This description of an action potential concerns only a small patch of plasma membrane. The self-amplifying depolarization of the patch, however, is sufficient to depolarize neighboring regions of membrane, which then go through the same self-amplifying cycle. In this way, the action potential spreads outward as a traveling wave from the initial site of depolarization, eventually reaching the extremities of the axon (Figure 12–39). Faced with the consequences of the Na^+ and K^+ fluxes caused by a passing action potential, local Na^+/K^+ ATPase molecules labor continuously to restore the ion gradients across the plasma membrane of the axon.

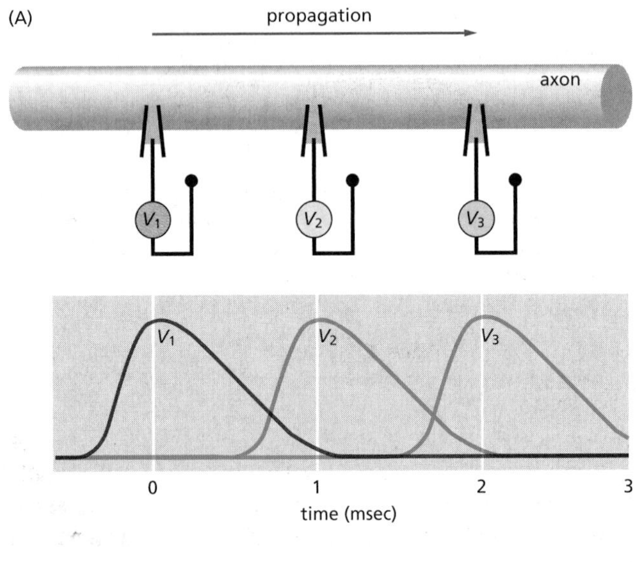

(B)

Figure 12–39 **An action potential can be propagated along the length of an axon.** (A) This schematic figure shows the voltages (V_1, V_2, and V_3) that would be recorded from a set of intracellular electrodes placed at intervals along the axon, whose width is greatly exaggerated here. Note that the action potential does not weaken as it travels. The activating stimulus is administered at time $t = 0$, and the direction in which the action potential is traveling is indicated by the *red* arrow. (B) The changes in the Na^+ channels and the consequent flows of electric current across the membrane (*orange* arrows) disturbs the membrane potential and gives rise to the traveling action potential, as shown here and in Movie 12.8. The region of the axon with a depolarized membrane is shaded in *blue*. Note that an action potential can only travel forward from the site of depolarization. This is because Na^+-channel inactivation prevents the depolarization from spreading backward (see also Figure 12–35).

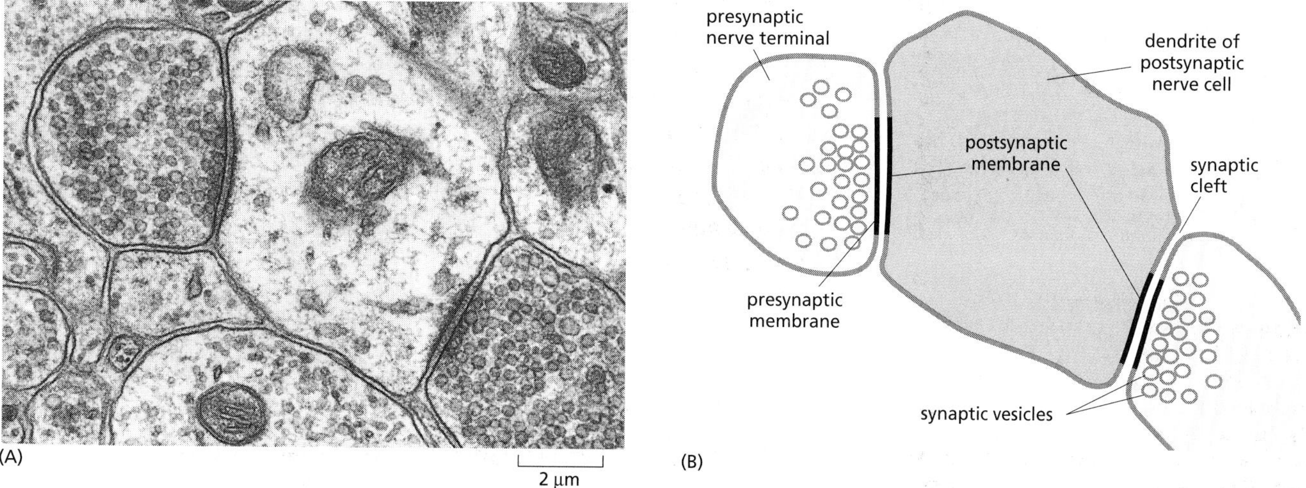

Figure 12–40 **Neurons transmit chemical signals across synapses.** An electron micrograph (A) and drawing (B) of a cross section of two nerve terminals (*yellow*) forming synapses on a single nerve cell dendrite (*blue*) in the mammalian brain. Note that both the presynaptic and postsynaptic membranes are thickened at the synapse. (A, courtesy of Cedric Raine.)

Voltage-gated Ca^{2+} Channels Convert Electrical Signals into Chemical Signals at Nerve Terminals

When an action potential reaches the *nerve terminals* at the end of an axon, the signal must somehow be relayed to the *target cells* that the nerve terminals contact: usually neurons or muscle cells. The signal is transmitted to the target cells at specialized junctions known as **synapses**. At most synapses, the plasma membranes of the transmitting and receiving cells—the *presynaptic* and the *postsynaptic* cells, respectively—are separated from each other by a narrow *synaptic cleft* (typically 20 nm across), which the electrical signal cannot cross (Figure 12–40). For the message to be transmitted from one neuron to another, the electrical signal is converted into a chemical signal, in the form of a small signal molecule known as a **neurotransmitter**.

Neurotransmitters are stored ready-made in the nerve terminals, packaged in membrane-enclosed **synaptic vesicles** (see Figure 12–40). When the action potential reaches the terminal, the neurotransmitters are released from the nerve ending by exocytosis (discussed in Chapter 15). This link between the action potential and secretion involves the activation of yet another type of voltage-gated cation channel. The depolarization of the nerve-terminal plasma membrane caused by the arrival of the action potential transiently opens *voltage-gated Ca^{2+} channels*, which are concentrated in the plasma membrane of the presynaptic nerve terminal. Because the Ca^{2+} concentration outside the cell is more than 1000 times greater than the free Ca^{2+} concentration in the cytosol, Ca^{2+} rushes into the nerve terminal through the open channels. The resulting increase in Ca^{2+} concentration in the cytosol of the presynaptic cell triggers the fusion of some of the synaptic vesicles with the plasma membrane, releasing the neurotransmitter into the synaptic cleft. Thanks to the voltage-gated Ca^{2+} channels, the electrical signal has now been converted into a chemical signal (Figure 12–41).

Transmitter-gated Channels in Target Cells Convert Chemical Signals Back into Electrical Signals

The released neurotransmitter rapidly diffuses across the synaptic cleft and binds to *neurotransmitter receptors* concentrated in the postsynap-

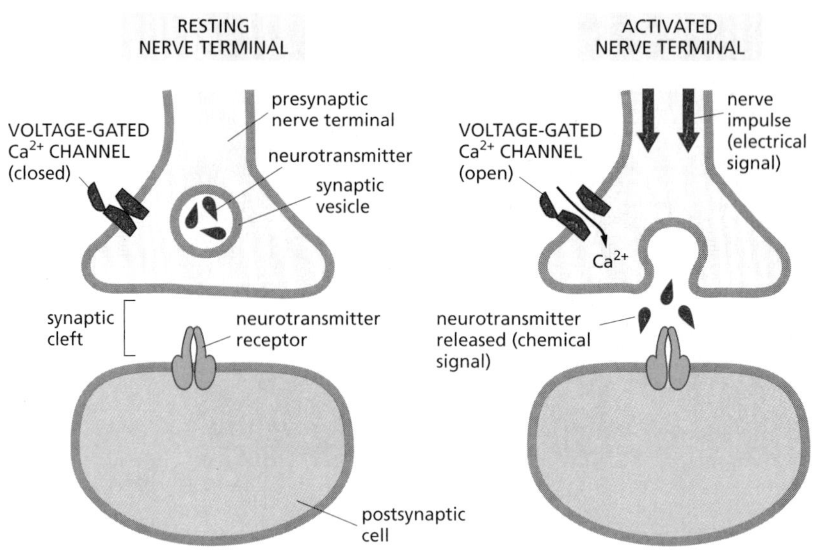

Figure 12–41 **An electrical signal is converted into a chemical signal at a nerve terminal.** When an action potential reaches a nerve terminal, it opens voltage-gated Ca^{2+} channels in the plasma membrane, allowing Ca^{2+} to flow into the terminal. The increased Ca^{2+} in the nerve terminal stimulates the synaptic vesicles to fuse with the plasma membrane, releasing their neurotransmitter into the synaptic cleft.

tic membrane of the target cell. The binding of neurotransmitter to its receptors causes a change in the membrane potential of the target cell, which can trigger the cell to fire an action potential. The neurotransmitter is then quickly removed from the synaptic cleft—either by enzymes that destroy it, or by reuptake into the nerve terminals that released it or into neighboring cells. This rapid removal of the neurotransmitter limits the signal and ensures that, when the presynaptic cell falls quiet, the postsynaptic cell will fall quiet as well.

Neurotransmitter receptors can be of various types; some mediate relatively slow effects in the target cell, whereas others trigger more rapid responses. Rapid responses—on a time scale of milliseconds—depend on receptors that are *transmitter-gated ion channels* (also called ion-channel-coupled receptors). These constitute a subclass of ligand-gated ion channels (see Figure 12–25B), and their function is to convert the chemical signal carried by a neurotransmitter back into an electrical signal. The channels open transiently in response to the binding of the neurotransmitter, thus changing the ion permeability of the postsynaptic membrane. This in turn causes a change in the membrane potential (Figure 12–42); if the change is big enough, it can trigger an action potential in the postsynaptic cell. A well-studied example of a transmitter-gated ion channel is found at the *neuromuscular junction*—the specialized type of synapse formed between a neuron and a muscle cell. In vertebrates, the neuro-

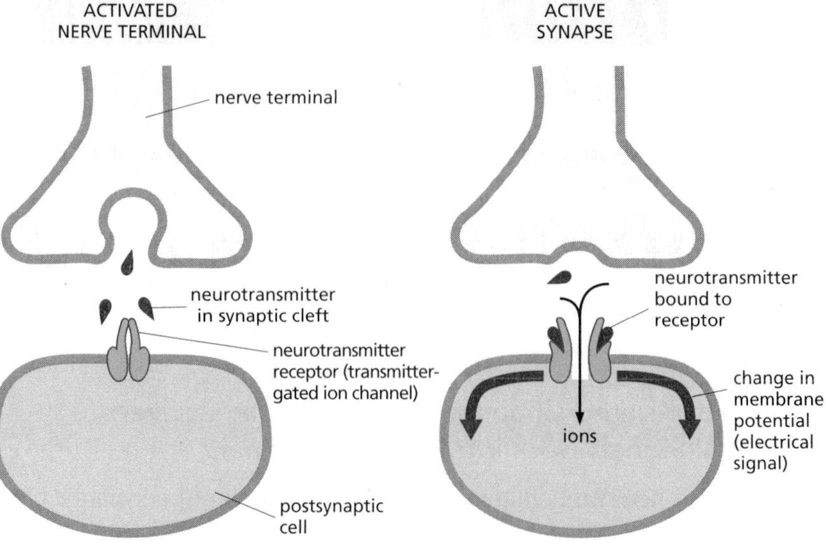

Figure 12–42 **A chemical signal is converted into an electrical signal by transmitter-gated ion channels at a synapse.** The released neurotransmitter binds to and opens the transmitter-gated ion channels in the plasma membrane of the postsynaptic cell. The resulting ion flows alter the membrane potential of the postsynaptic cell, thereby converting the chemical signal back into an electrical one (Movie 12.9).

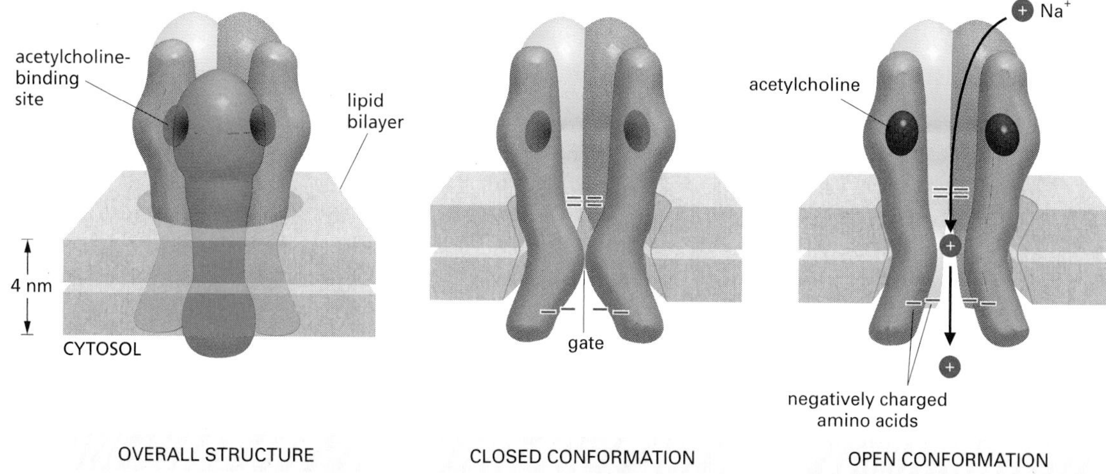

Figure 12–43 **The acetylcholine receptor, present in the plasma membrane of muscle cells, opens when it binds to the neurotransmitter acetylcholine released by a nerve.** (A) This transmitter-gated ion channel is composed of five transmembrane protein subunits that combine to form an aqueous pore across the lipid bilayer. The pore is lined by five transmembrane α helices, one contributed by each subunit. (B) Negatively charged amino acid side chains at either end of the pore (indicated here by *red* minus signs) ensure that only positively charged ions, mainly Na^+ and K^+, can pass. The blue subunit has been removed here to show the interior of the pore. When the channel is in its closed conformation, the pore is occluded (blocked) by hydrophobic amino acid side chains in the region called the gate. (C) When acetylcholine binds, the channel undergoes a conformational change in which these side chains move apart and the gate opens, allowing Na^+ to flow across the membrane down its electrochemical gradient. K^+ (not shown here) will flow in the opposite direction down its electrochemical gradient. Even with acetylcholine bound, the channel flickers randomly between the open and closed states (see Figure 12–24); without acetylcholine bound, however, it rarely opens.

transmitter here is *acetylcholine*, and the transmitter-gated ion channel is the *acetylcholine receptor* (Figure 12–43).

Neurons Receive Both Excitatory and Inhibitory Inputs

The response produced by a neurotransmitter at a synapse can be either excitatory or inhibitory. Excitatory neurotransmitters (delivered by axon terminals of *excitatory neurons*) stimulate the postsynaptic cell, encouraging it to fire an action potential. Inhibitory neurotransmitters (delivered by axon terminals of *inhibitory neurons*) do the opposite, discouraging the postsynaptic cell from firing. The drug curare, which South American Indians used to make poison arrows and surgeons use to relax muscles during an operation, causes paralysis by blocking the delivery of excitatory signals at neuromuscular junctions. By contrast, strychnine, a common ingredient in rat poisons, causes muscle spasms, convulsions, and death by blocking the delivery of inhibitory signals.

Excitatory and inhibitory neurotransmitters bind to different receptors, and it is the character of the receptor that makes the difference between excitation and inhibition. The chief receptors for excitatory neurotransmitters, mainly *acetylcholine* and *glutamate*, are ligand-gated cation channels. When the neurotransmitter binds, the channels open to allow an influx of cations, which depolarizes the plasma membrane toward the threshold potential required for triggering an action potential. Stimulation of these receptors thus tends to activate the postsynaptic cell. The receptors for inhibitory neurotransmitters, mainly *γ-aminobutyric acid (GABA)* and *glycine*, by contrast, are ligand-gated Cl^- channels. When the neurotransmitter binds, the channels open, but very little Cl^- enters the cell at this point because the driving force for movement of Cl^- across the membrane is close to zero at the resting membrane potential. If an excitatory neurotransmitter opens Na^+ channels at the same time, however, the resulting depolarization caused by the Na^+ influx will cause Cl^- to

QUESTION 12–7

In the disease myasthenia gravis, the human body makes—by mistake—antibodies to its own acetylcholine receptor molecules. These antibodies bind to and inactivate acetylcholine receptors on the plasma membrane of muscle cells. The disease leads to a devastating progressive weakening of the people affected. Early on, they may have difficulty opening their eyelids, for example, and, in an animal model of the disease, rabbits have difficulty holding their ears up. As the disease progresses, most muscles weaken, and people with myasthenia gravis have difficulty speaking and swallowing. Eventually, impaired breathing can cause death. Explain which step of muscle function is affected.

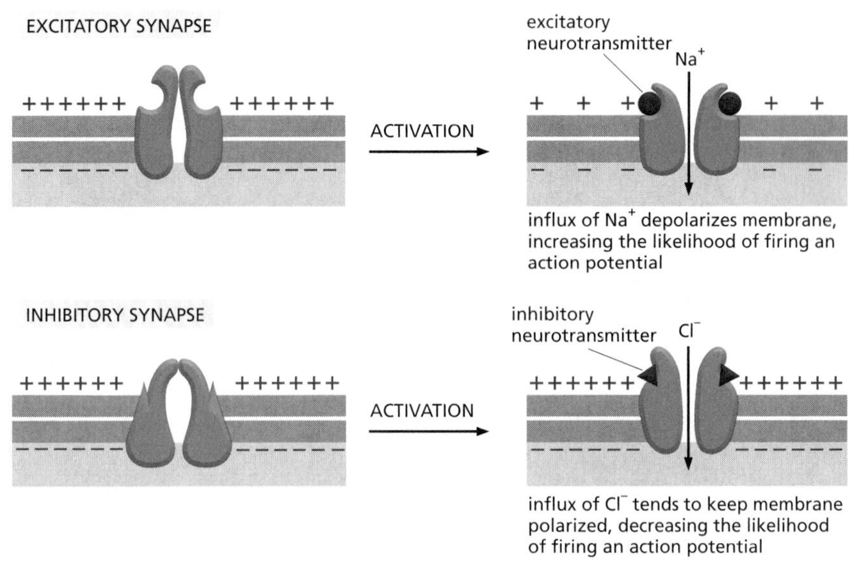
EXCITATORY SYNAPSE
++++++ ++++++
ACTIVATION
excitatory neurotransmitter
Na+
influx of Na+ depolarizes membrane, increasing the likelihood of firing an action potential
INHIBITORY SYNAPSE
++++++ ++++++
ACTIVATION
inhibitory neurotransmitter
Cl−
influx of Cl− tends to keep membrane polarized, decreasing the likelihood of firing an action potential

Figure 12–44 **Synapses can be excitatory or inhibitory.** Excitatory neurotransmitters activate ion channels that allow the passage of Na^+ and Ca^{2+}, whereas inhibitory neurotransmitters activate ion channels that allow the passage of Cl^-.

move into the cell through the open Cl^- channels, neutralizing the effect of the Na^+ influx (Figure 12–44). In this way, inhibitory neurotransmitters suppress the production of an action potential by making the target cell membrane much harder to depolarize.

The locations and functions of these ion channels, and of some of the other ion channels discussed in this chapter, are summarized in Table 12–3.

Transmitter-gated Ion Channels Are Major Targets for Psychoactive Drugs

Most drugs used in the treatment of insomnia, anxiety, depression, and schizophrenia exert their effects at synapses in the brain, and many of them act by binding to transmitter-gated ion channels. The barbiturates and tranquilizers such as Valium, Ambien, and temazepam, for example, bind to GABA-gated Cl^- channels. Their binding makes the channels easier to open by GABA, thus making the cell more sensitive to GABA's inhibitory action. By contrast, the antidepressant Prozac blocks the reuptake of an excitatory neurotransmitter, *serotonin*, increasing the amount of serotonin available at those synapses that use this transmitter. Why this should relieve depression is still a mystery.

TABLE 12–3 SOME EXAMPLES OF ION CHANNELS

ION CHANNEL	TYPICAL LOCATION	FUNCTION
K^+ leak channel	plasma membrane of most animal cells	maintenance of resting membrane potential
Voltage-gated Na^+ channel	plasma membrane of nerve cell axon	generation of action potentials
Voltage-gated K^+ channel	plasma membrane of nerve cell axon	return of membrane to resting potential after initiation of an action potential
Voltage-gated Ca^{2+} channel	plasma membrane of nerve terminal	stimulation of neurotransmitter release
Acetylcholine receptor (acetylcholine-gated Na^+ and Ca^{2+} channel)	plasma membrane of muscle cell (at neuromuscular junction)	excitatory synaptic signaling
Glutamate receptors (glutamate-gated Na^+ and Ca^{2+} channels)	plasma membrane of many neurons (at synapses)	excitatory synaptic signaling
GABA receptor (GABA-gated Cl^- channel)	plasma membrane of many neurons (at synapses)	inhibitory synaptic signaling
Glycine receptor (glycine-gated Cl^- channel	plasma membrane of many neurons (at synapses)	inhibitory synaptic signaling
Stress-activated cation channel	auditory hair cell in inner ear	detection of sound vibrations

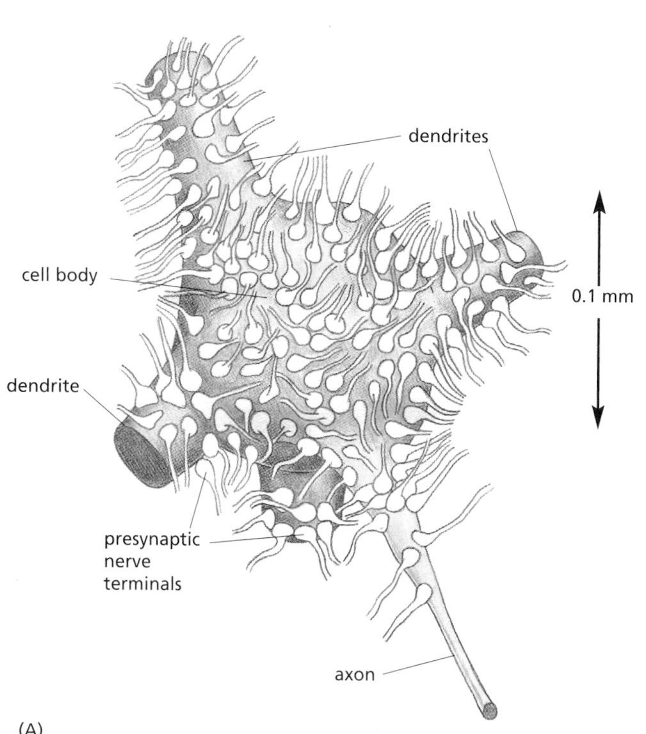

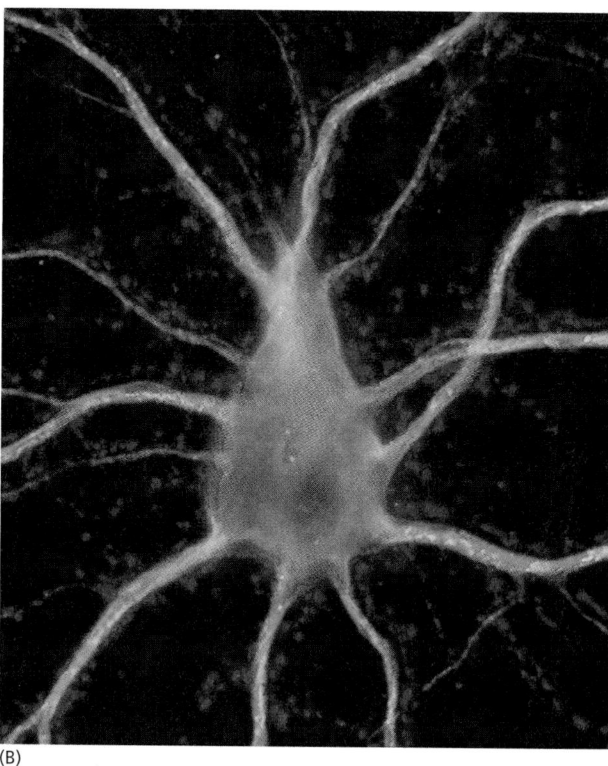

Figure 12–45 **Thousands of synapses form on the cell body and dendrites of a motor neuron in the spinal cord.** (A) Many thousands of nerve terminals synapse on the neuron, delivering signals from other parts of the animal to control the firing of action potentials along the neuron's axon. (B) A rat nerve cell in culture. Its cell body and dendrites (*green*) are stained with a fluorescent antibody that recognizes a cytoskeletal protein. Thousands of axon terminals (*red*) from other nerve cells (not visible) make synapses on the cell's surface; they are stained with a fluorescent antibody that recognizes a protein in synaptic vesicles. Electrical signals are sent out along axons, relayed across synapses, and passed in along dendrites toward the nerve cell body. The signaling depends on movements of ions across the plasma membranes of the nerve cells. (B, courtesy of Olaf Mundigl and Pietro de Camilli.)

The number of distinct types of neurotransmitter receptors is very large, although they fall into a small number of families. There are, for example, many subtypes of acetylcholine, glutamate, GABA, glycine, and serotonin receptors; they are usually located in different neurons and often differ only subtly in their properties. With such a large variety of receptors, it may be possible to design a new generation of psychoactive drugs that will act more selectively on specific sets of neurons to alleviate the mental illnesses that devastate so many people's lives. One percent of the human population, for example, have schizophrenia, another 1% have bipolar disorder, and many more suffer from anxiety or depression.

Synaptic Connections Enable You to Think, Act, and Remember

At a synapse, the nerve terminal of the presynaptic cell converts an electrical signal into a chemical one, and the postsynaptic cell converts the chemical signal back into an electrical one. Interference with these processes, for good or ill, is of enormous practical importance to us. But why has evolution favored such an apparently inefficient way to pass on an electrical signal? It would seem more efficient to have a direct electrical connection between the pre- and postsynaptic cells, or to do away with the synapse altogether and use a single continuous cell.

The value of synapses that can handle chemical signals becomes clear when we consider how they function in the context of the nervous system—a huge network of neurons, interconnected by many branching pathways, performing complex computations, storing memories, and generating plans for action. To carry out these functions, neurons have to do more than merely generate and relay signals: they must also combine them, interpret them, and record them. Chemical synapses make these activities possible. A motor neuron in the spinal cord, for example, receives inputs from hundreds or thousands of other neurons that make synapses on it (Figure 12–45). Some of these signals tend to stimulate the neuron, while others tend to inhibit it. The motor neuron has to combine all of the information it receives and react either by firing action potentials along its axon to stimulate a muscle or by remaining quiet.

This task of computing an appropriate output from a cacophonous babble of inputs is achieved by a complicated interplay between different types of ion channels in the neuron's plasma membrane. Each of the hundreds of types of neurons in your brain has its own characteristic set of receptors and ion channels that enables the cell to respond in a particular way to a certain set of inputs and thus to perform its specialized task. Furthermore, the ion channels and other components at a synapse can also undergo lasting modifications according to the usage they have experienced, thereby preserving traces of past events. In this way, memories are stored. Ion channels, therefore, are at the heart of the machinery that enables you to act, think, feel, speak, and—perhaps most important of all—to remember everything you read in books like this one.

ESSENTIAL CONCEPTS

- **The lipid bilayer of cell membranes is permeable to small nonpolar molecules such as oxygen and carbon dioxide and to very small polar molecules such as water. It is highly impermeable to most large, water-soluble molecules and all ions. Transfer of nutrients, metabolites, and ions across the plasma membrane and internal cell membranes is carried out by membrane transport proteins.**
- **Cell membranes contain a variety of transport proteins, each of which is responsible for transferring a particular type of solute across the membrane. There are two classes of membrane transport proteins: transporters and channels.**
- **The electrochemical gradient represents the net driving force on an ion due to its concentration gradient and the electric field.**
- **In passive transport, an uncharged solute moves spontaneously down its concentration gradient, a charged solute (an ion) moves spontaneously down its electrochemical gradient, and water moves down its osmotic gradient. In active transport, an uncharged solute or an ion is transported against its concentration or electrochemical gradient in a process that requires energy.**
- **Transporters bind specific solutes (inorganic ions, small organic molecules, or both) and transfer them across the lipid bilayer by undergoing conformational changes that expose the solute-binding site first on one side of the membrane and then on the other.**
- **Transporters can act as pumps to move a solute uphill against its electrochemical gradient, using energy provided by ATP hydrolysis, by a downhill flow of Na^+ or H^+ ions, or by light.**
- **The Na^+-K^+ pump in the plasma membrane of animal cells is an ATPase that actively transports Na^+ out of the cell and K^+ in, maintaining the steep Na^+ gradient across the plasma membrane that is used to drive other active transport processes and to convey electrical signals.**
- **Channels form aqueous pores across the lipid bilayer through which solutes can diffuse. Whereas solute transfer carried out by transporters can be active or passive, transport by channels is always passive.**
- **Most channels are selective ion channels, which allow inorganic ions of appropriate size and charge to cross the membrane down their electrochemical gradients. Transport through ion channels is at least 1000 times faster than movement through any known transporter. Other channels conduct water or other small metabolites.**
- **Most ion channels are gated; they open transiently in response to a specific stimulus, such as a change in membrane potential (voltage-gated channels) or the binding of a ligand (ligand-gated channels).**

- Even when opened by their specific stimulus, ion channels do not remain continuously open: they flicker randomly between open and closed conformations. An activating stimulus increases the proportion of time that the channel spends in the open state.
- The membrane potential is determined by the unequal distribution of electric charge on the two sides of the plasma membrane and is altered when ions flow through open channels. In most animal cells, K^+-selective leak channels hold the resting membrane potential at a negative value, close to the value at which the driving force for movement of K^+ across the membrane is almost zero.
- Neurons propagate signals in the form of action potentials, which can travel long distances along an axon without weakening. Action potentials are usually mediated by voltage-gated Na^+ channels that open in response to depolarization of the plasma membrane.
- Voltage-gated Ca^{2+} channels in nerve terminals couple electrical signals to transmitter release at synapses. Transmitter-gated ion channels convert these chemical signals back into electrical signals in the postsynaptic target cell.
- Excitatory neurotransmitters open transmitter-gated channels that are permeable to Na^+ and thereby depolarize the postsynaptic cell membrane toward the threshold potential for firing an action potential. Inhibitory neurotransmitters open transmitter-gated Cl^- channels and thereby suppress firing by keeping the postsynaptic cell membrane polarized.

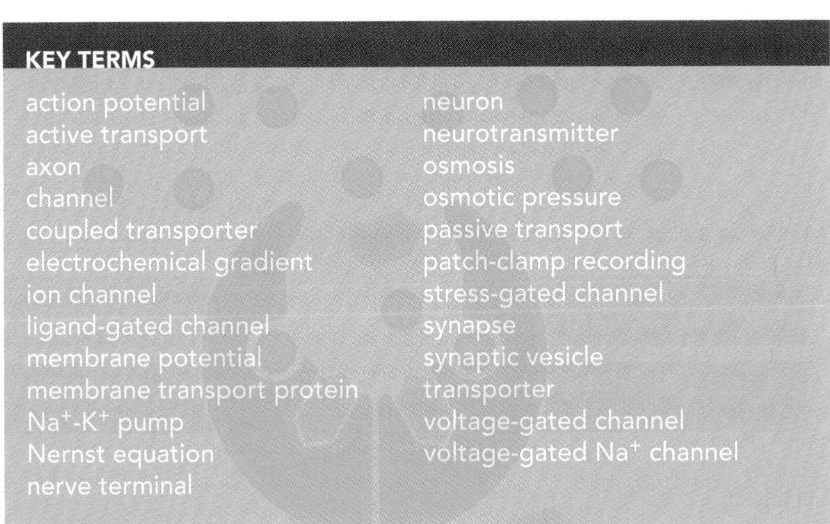

KEY TERMS

action potential
active transport
axon
channel
coupled transporter
electrochemical gradient
ion channel
ligand-gated channel
membrane potential
membrane transport protein
Na^+-K^+ pump
Nernst equation
nerve terminal
neuron
neurotransmitter
osmosis
osmotic pressure
passive transport
patch-clamp recording
stress-gated channel
synapse
synaptic vesicle
transporter
voltage-gated channel
voltage-gated Na^+ channel

QUESTIONS

QUESTION 12–8

The diagram in Figure 12–6 shows a passive transporter that mediates the transfer of a solute down its concentration gradient across the membrane. How would you need to change the diagram to convert the transporter into a pump that moves the solute up its electrochemical gradient by hydrolyzing ATP? Explain the need for each of the steps in your new illustration.

QUESTION 12–9

Which of the following statements are correct? Explain your answers.

A. The plasma membrane is highly impermeable to all charged molecules.

B. Channels must first bind to solute molecules before they can select those that they allow to pass.

C. Transporters allow solutes to cross a membrane at much faster rates than do channels.

D. Certain H^+ pumps are fueled by light energy.

E. The plasma membrane of many animal cells contains open K^+ channels, yet the K^+ concentration in the cytosol is much higher than outside the cell.

F. A symport would function as an antiport if its orientation in the membrane were reversed (i.e., if the portion of the molecule normally exposed to the cytosol faced the outside of the cell instead).

G. The membrane potential of an axon temporarily becomes more negative when an action potential excites it.

QUESTION 12–10

List the following compounds in order of increasing lipid bilayer permeability: RNA, Ca^{2+}, glucose, ethanol, N_2, water.

QUESTION 12–11

Name at least one similarity and at least one difference between the following (it may help to review the definitions of the terms using the Glossary):

A. Symport and antiport

B. Active transport and passive transport

C. Membrane potential and electrochemical gradient

D. Pump and transporter

E. Axon and telephone wire

F. Solute and ion

QUESTION 12–12

Discuss the following statement: "The differences between a channel and a transporter are like the differences between a bridge and a ferry."

QUESTION 12–13

The neurotransmitter acetylcholine is made in the cytosol and then transported into synaptic vesicles, where its concentration is more than 100-fold higher than in the cytosol. When synaptic vesicles are isolated from neurons, they can take up additional acetylcholine added to the solution in which they are suspended, but only when ATP is present. Na^+ ions are not required for acetylcholine uptake, but, curiously, raising the pH of the solution in which the synaptic vesicles are suspended increases the rate of acetylcholine uptake. Furthermore, transport is inhibited when drugs are added that make the membrane permeable to H^+ ions. Suggest a mechanism that is consistent with all of these observations.

QUESTION 12–14

The resting membrane potential of a cell is about –70 mV, and the thickness of a lipid bilayer is about 4.5 nm. What is the strength of the electric field across the membrane in V/cm? What do you suppose would happen if you applied this field strength to two metal electrodes separated by a 1-cm air gap?

QUESTION 12–15

Phospholipid bilayers form sealed spherical vesicles in water (discussed in Chapter 11). Assume you have constructed lipid vesicles that contain Na^+-K^+ pumps as the sole membrane protein, and assume for the sake of simplicity that each pump transports one Na^+ one way and one K^+ the other way in each pumping cycle. All the Na^+-K^+ pumps have the portion of the molecule that normally faces the cytosol oriented toward the outside of the vesicles. With the help of Figure 12–11, determine what would happen if:

A. Your vesicles were suspended in a solution containing both Na^+ and K^+ ions and had a solution with the same ionic composition inside them.

B. You add ATP to the suspension described in (A).

C. You add ATP, but the solution—outside as well as inside the vesicles—contains only Na^+ ions and no K^+ ions.

D. Half of the pump molecules embedded in the membrane of each vesicle were oriented the other way around so that the normally cytosolic portions of these molecules faced the inside of the vesicles. You then add ATP to the suspension.

E. You add ATP to the suspension described in (A), but in addition to Na^+-K^+ pumps, the membrane of your vesicles also contains K^+ leak channels.

QUESTION 12–16

Name the three ways in which an ion channel can be gated.

QUESTION 12–17

One thousand Ca^{2+} channels open in the plasma membrane of a cell that is 1000 μm^3 in size and has a cytosolic Ca^{2+} concentration of 100 nM. For how long would the channels need to stay open in order for the cytosolic Ca^{2+} concentration to rise to 5 μM? There is virtually unlimited Ca^{2+} available in the outside medium (the extracellular Ca^{2+} concentration in which most animal cells live is a few millimolar), and each channel passes 10^6 Ca^{2+} ions per second.

QUESTION 12–18

Amino acids are taken up by animal cells using a symport in the plasma membrane. What is the most likely ion whose electrochemical gradient drives the import? Is ATP consumed in the process? If so, how?

QUESTION 12–19

We shall see in Chapter 15 that an acidic pH inside endosomes, which are membrane-enclosed intracellular organelles need their insides to have an acidic pH in order to function. Acidification is achieved by an H^+ pump in the endosomal membrane. The endosomal membrane also contains Cl^- channels. If the channels do not function properly (e.g., because of a mutation in the genes encoding the channel proteins), acidification is also impaired.

A. Can you explain how Cl^- channels might help acidification?

B. According to your explanation, would the Cl^- channels be absolutely required to lower the pH inside the endosome?

QUESTION 12–20

Some bacterial cells can grow on either ethanol (CH_3CH_2OH) or acetate (CH_3COO^-) as their only carbon source. Dr. Schwips measured the rate at which the two compounds traverse the bacterial plasma membrane but, due to excessive inhalation of one of the compounds (which one?), failed to label his data accurately.

CONCENTRATION OF CARBON SOURCE (mM)	RATE OF TRANSPORT (μmol/min)	
	COMPOUND A	COMPOUND B
0.1	2.0	18
0.3	6.0	46
1.0	20	100
3.0	60	150
10.0	200	182

A. Plot the data from the table above.

B. Determine from your graph whether the data describing compound A correspond to the uptake of ethanol or acetate.

C. Determine the rates of transport for compounds A and B at 0.5 mM and 100 mM. (This part of the question requires that you be familiar with the principles of enzyme kinetics discussed in Chapter 3.)

Explain your answers.

QUESTION 12–21

Acetylcholine-gated cation channels do not discriminate among Na^+, K^+, and Ca^{2+} ions, allowing all to pass through them freely. So why is it that when acetylcholine binds to this protein complex in muscle cells, the channel opens and there is a large net influx of primarily Na^+ ions?

QUESTION 12–22

The ion channels that are regulated by binding of neurotransmitters, such as acetylcholine, glutamate, GABA, or glycine, have a similar overall structure. Yet, each class of these channels consists of a very diverse set of subtypes with different ligand affinities, different channel conductances, and different rates of opening and closing. Do you suppose that such extreme diversity is a good or a bad thing from the standpoint of the pharmaceutical industry?

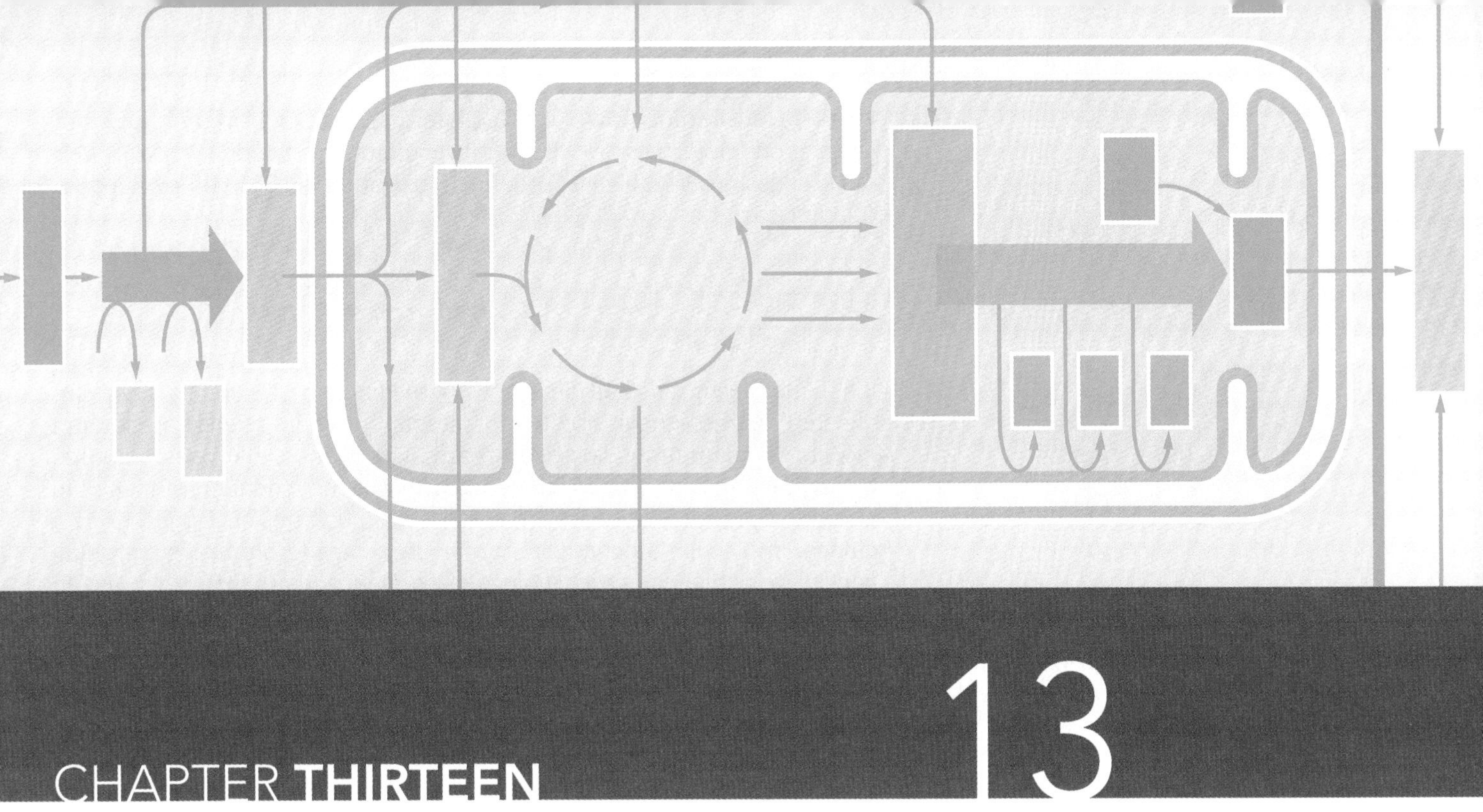

CHAPTER **THIRTEEN** 13

How Cells Obtain Energy from Food

THE BREAKDOWN AND UTILIZATION OF SUGARS AND FATS

REGULATION OF METABOLISM

As we discussed in Chapter 3, cells require a constant supply of energy to generate and maintain the biological order that keeps them alive. This energy comes from the chemical bond energy in food molecules, which thereby serve as fuel for cells.

Perhaps the most important fuel molecules are the sugars. Plants make their own sugars from CO_2 by photosynthesis. Animals obtain sugars—and other molecules, such as starch, that are easily broken down into sugars—by eating other organisms. Nevertheless, the process whereby these sugars are oxidized to generate energy is very similar in both animals and plants. In both cases, the cells that form the organism harvest useful energy from the chemical bond energy locked in sugars as the sugar molecule is broken down and oxidized to carbon dioxide (CO_2) and water (H_2O). This energy is stored as "high-energy" chemical bonds—covalent bonds that release large amounts of energy when hydrolyzed—in activated carrier molecules, such as ATP and NADPH. These carrier molecules in turn serve as portable sources of the chemical groups and electrons needed for biosynthesis (discussed in Chapter 3).

In this chapter we trace the major steps in the breakdown—or *catabolism*—of sugars and show how this oxidation produces ATP, NADH, and other activated carrier molecules in cells. We concentrate on the breakdown of glucose because these reactions dominate energy production in most animal cells. A very similar pathway operates in plants, fungi, and many bacteria. Other molecules, such as fatty acids and proteins, can also serve as energy sources if they are funneled through appropriate enzymatic pathways. We also see how many of the molecules generated from the breakdown of sugars and fats can be used to build the macromolecules in cells.

Finally, we examine how cells regulate their metabolism and how they store food molecules for their future metabolic needs. We will save our discussion of how cells produce most of the ATP that they need for Chapter 14.

THE BREAKDOWN AND UTILIZATION OF SUGARS AND FATS

If a fuel molecule such as glucose were oxidized to CO_2 and H_2O in a single step (as happens in nonliving systems), it would release an amount of energy many times larger than any carrier molecule could capture. Instead, living cells use enzymes to carry out the oxidation of sugars in a tightly controlled series of reactions. As illustrated in Figure 13–1, a glucose molecule is degraded step by step, paying out energy in small packets to activated carrier molecules by means of coupled reactions. In this way, much of the energy released by oxidizing glucose is saved in the high-energy bonds of ATP and other activated carrier molecules and made available to do useful work for the cell.

Animal cells make ATP in two ways. First, certain steps in a series of enzyme-catalyzed reactions are directly coupled to the energetically unfavorable reaction **ADP** + P_i → **ATP**. The oxidation of food molecules provides the energy that allows this unfavorable reaction to proceed. Most ATP synthesis, however, takes place in mitochondria and uses the energy from activated carrier molecules to drive ATP production; this process involves the mitochondrial membrane and will be described in detail in Chapter 14. In this chapter we focus on the first sequence of reactions by which food molecules are oxidized—both in the cytosol and inside the mitochondria. These reactions produce both ATP and the activated carrier molecules that will subsequently drive the production of much larger amounts of ATP in the mitochondrial membrane.

Food Molecules Are Broken Down in Three Stages

The proteins, lipids, and polysaccharides that make up most of the food we eat must be broken down into smaller molecules before our cells can use them—either as a source of energy or as building blocks for other molecules. This breakdown process—which uses enzymes to degrade complex molecules into simpler ones—is dubbed **catabolism**. Catabolism must act on food taken in from outside, but not on the macromolecules inside our own cells. Therefore stage 1 in the enzymatic breakdown of

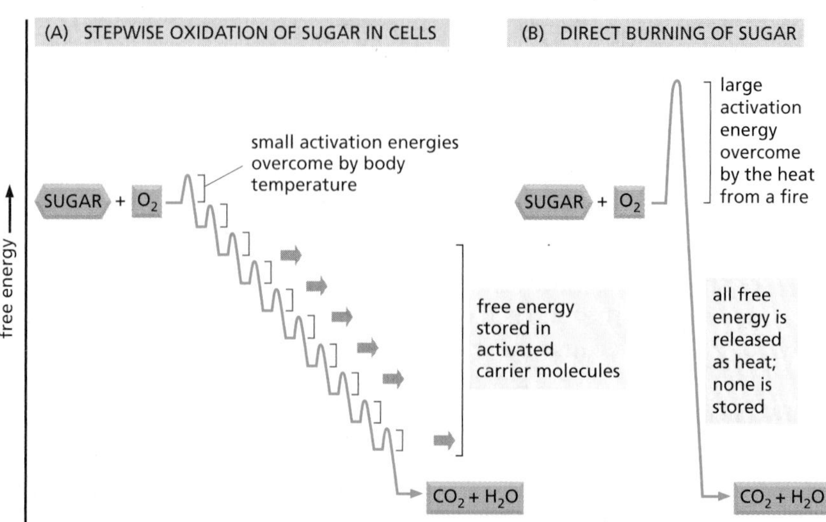

Figure 13–1 **The controlled stepwise oxidation of sugar that occurs in the cell preserves useful energy, unlike the simple burning of the same fuel molecule.** In the cell, enzymes catalyze oxidations via a series of small steps in which free energy is transferred in conveniently sized packets to carrier molecules—most often ATP and NADH. At each step, an enzyme controls the reaction by reducing the activation energy barrier that must be surmounted by the random collision of molecules at the temperature of cells (body temperature), so as to allow the specific reaction to occur. The total free energy released by the oxidation of glucose—686 kcal/mole (2880 kJ/mole)—is exactly the same in (A) and (B).

food molecules—*digestion*—occurs either outside cells (in our intestine) or in a specialized organelle within cells called the lysosome. A membrane that surrounds the lysosome keeps its digestive enzymes separated from the cytosol (discussed in Chapter 15).

Digestive enzymes reduce the large polymeric molecules in food into their monomeric subunits: proteins into amino acids, polysaccharides into sugars, and fats into fatty acids and glycerol. After digestion, the small organic molecules derived from food enter the cytosol of a cell, where their gradual oxidation begins. As illustrated in Figure 13–2, this oxidation occurs in two further stages: stage 2 starts in the cytosol and ends in mitochondria, while stage 3 is confined to the mitochondria.

In stage 2 of cellular catabolism, a chain of reactions called *glycolysis* converts each molecule of *glucose* into two smaller molecules of *pyruvate*. Sugars other than glucose can also be used after being first converted to one of the sugar intermediates in this glycolytic pathway. During the formation of pyruvate, two types of activated carrier molecules are produced: ATP and NADH. The pyruvate is then transported from the cytosol into the mitochondrion's large, internal compartment or *matrix*. There a giant enzyme complex converts each pyruvate molecule into CO_2 plus **acetyl CoA**, another of the activated carrier molecules discussed in Chapter 3 (see Figure 3–36). Large amounts of acetyl CoA are also produced by the stepwise breakdown and oxidation of fatty acids derived from fats.

Stage 3 of the oxidative breakdown of food molecules takes place entirely in mitochondria. The acetyl group in acetyl CoA is transferred to a molecule called oxaloacetate to form citrate, which enters a series of reactions called the *citric acid cycle*. The transferred acetyl group is oxidized to CO_2 in these reactions, and large amounts of the high-energy electron carrier NADH are generated. Finally, the high-energy electrons from NADH are passed along a series of enzymes within the mitochondrial inner membrane called an *electron-transport chain*, where the energy released by their transfer is used to drive a process that produces ATP and consumes molecular oxygen (O_2 gas). It is in these final steps that most of the energy released by oxidation is harnessed to produce most of the cell's ATP.

Through the production of ATP, the energy derived from the breakdown of sugars and fats is redistributed as packets of chemical energy in a form convenient for use in the cell. Roughly 10^9 molecules of ATP are in solution in a typical cell at any instant, and in many cells all of this ATP is turned over (that is, used up and replaced) every 1–2 minutes. An average person at rest will hydrolyze his or her weight in ATP molecules every 24 hours.

In total, nearly half of the energy that could, in theory, be derived from the oxidation of glucose or fatty acids to H_2O and CO_2 is captured and used to drive the energetically unfavorable reaction $P_i + ADP \rightarrow ATP$. By contrast, a modern combustion engine, such as a car engine, can convert no more than 20% of the available energy in its fuel into useful work. In both cases, the remaining energy is released as heat, which in animals helps to make our bodies warm.

Glycolysis Is a Central ATP-producing Pathway

The central process in stage 2 of the breakdown of food molecules is the degradation of **glucose** in the sequence of reactions known as **glycolysis**—from the Greek *glykys*, "sweet," and *lysis*, "splitting." Glycolysis produces ATP without the involvement of O_2. It occurs in the cytosol of most cells, including many anaerobic microorganisms. Glycolysis probably evolved early in the history of life, before photosynthetic organisms introduced oxygen into the atmosphere.

During glycolysis, a glucose molecule, which has six carbon atoms, is cleaved into two molecules of **pyruvate**, each of which contains three carbon atoms. For each molecule of glucose, two molecules of ATP are consumed to provide energy to drive the early steps, but four molecules of ATP are produced in the later steps. Thus, at the end of glycolysis, there is a net gain of two molecules of ATP for each glucose molecule broken down.

Glycolysis involves a sequence of 10 separate reactions, each producing a different sugar intermediate and each catalyzed by a different enzyme. These reactions are presented in outline in Figure 13–3 and in detail in

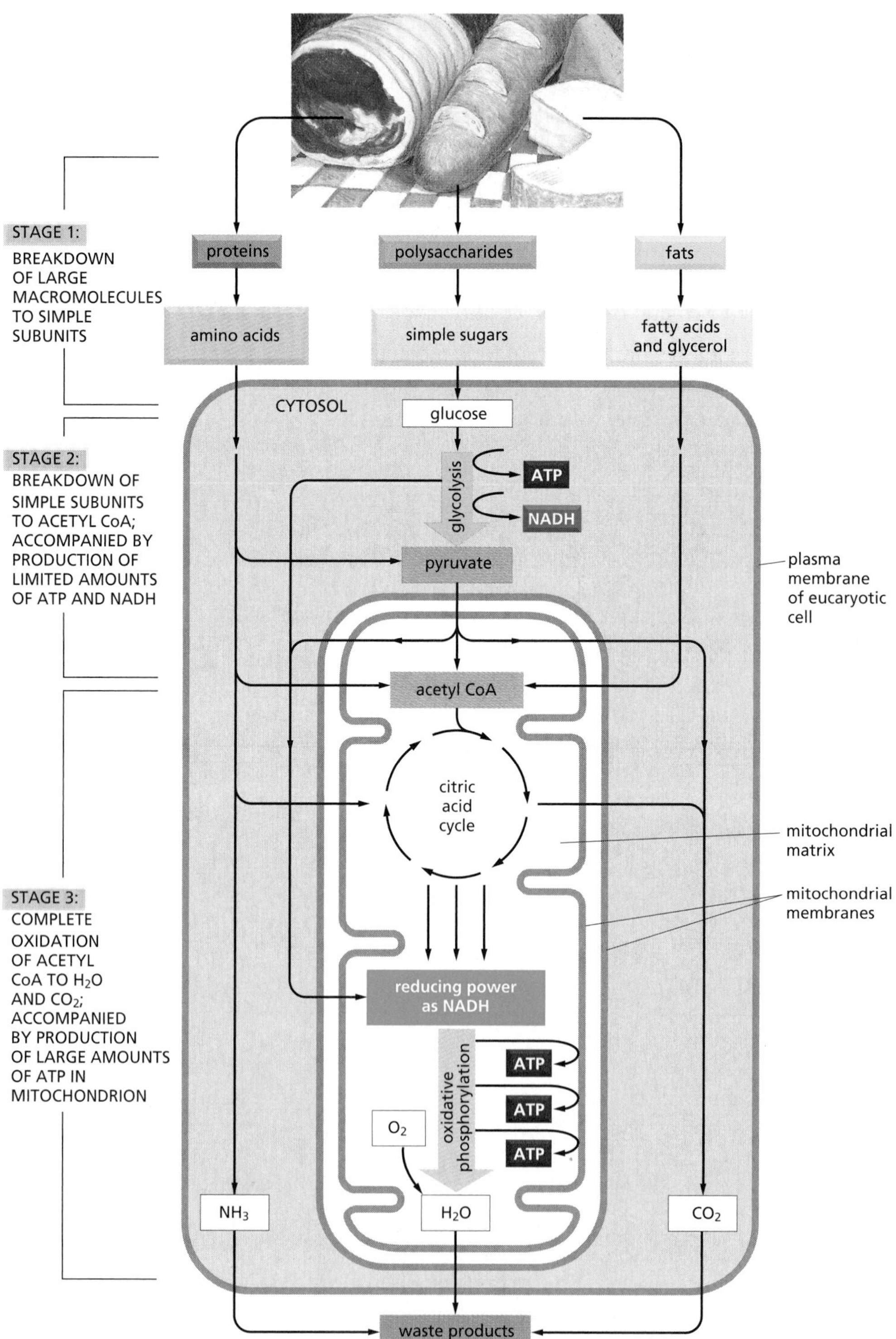

Figure 13–2 **Three stages of cellular metabolism lead from food to waste products in animal cells.** This series of reactions produces ATP, which is then used to drive biosynthetic reactions and other energy-requiring processes in the cell. Stage 1 mostly occurs outside cells—although special organelles called lysosomes can digest large molecules in the cell interior. Stage 2 occurs mainly in the cytosol, except for the final step of conversion of pyruvate to acetyl groups on acetyl CoA, which occurs in mitochondria. Stage 3 occurs entirely in mitochondria.

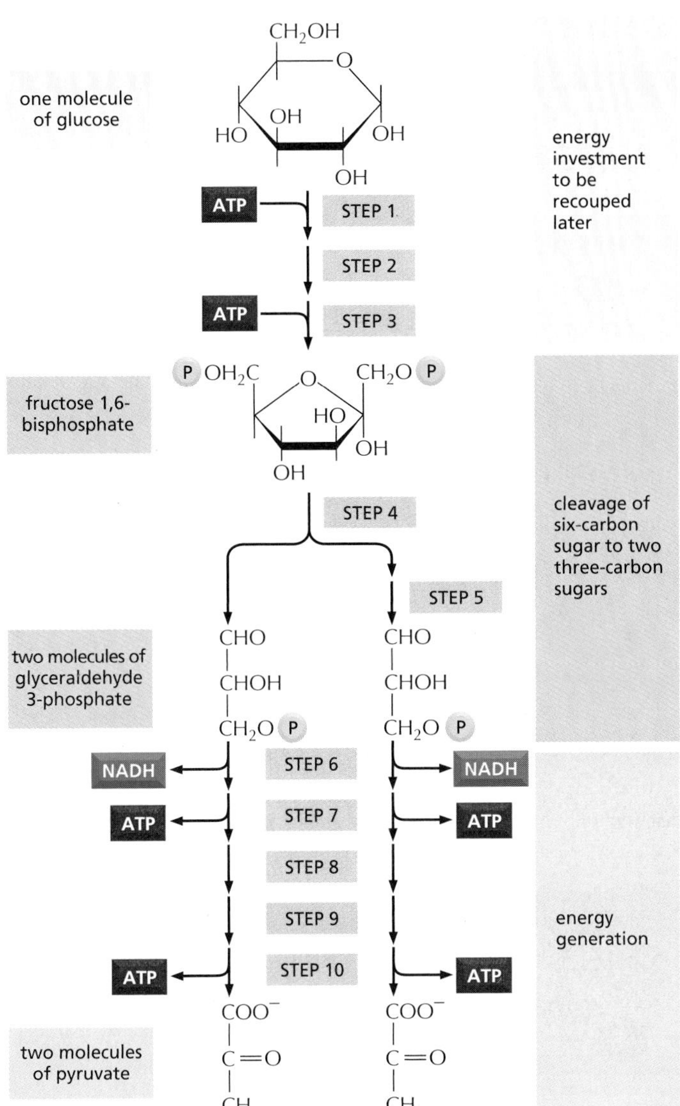

Figure 13–3 **The stepwise oxidation of sugars begins with glycolysis.** Each of the 10 steps of glycolysis is catalyzed by a different enzyme. Note that step 4 cleaves a six-carbon sugar into two three-carbon sugars, so that the number of molecules at every stage after this doubles. As indicated, step 6 begins the energy-generation phase of glycolysis, which results in the net synthesis of ATP and NADH (see also Panel 13–1). Glycolysis is also sometimes referred to as the Embden–Meyerhof pathway, named for the chemists who first described it. All the steps of glycolysis are reviewed in Movie 13.1.

Panel 13–1 (pp. 430–431). Like most enzymes (discussed in Chapter 4), the enzymes that catalyze glycolysis all have names ending in *-ase*—like *isomerase* and *dehydrogenase*—which specify the type of reaction they catalyze.

Although no molecular oxygen is involved in glycolysis, oxidation occurs: electrons are removed from some of the carbons derived from glucose by **NAD^+**, producing **NADH**. The stepwise nature of the process allows the energy of oxidation to be released in small packets, so that much of it can be stored in carrier molecules rather than all of it being released as heat (see Figure 13–1). Some of the energy released by this oxidation drives the synthesis of ATP molecules from ADP and P_i. The synthesis of ATP in glycolysis is known as *substrate-level phosphorylation* because it occurs by the transfer of a phosphate group directly from a substrate molecule—a sugar intermediate—to ADP. The remainder of the energy harnessed during glycolysis is stored in the electrons in NADH.

Two molecules of NADH are formed per molecule of glucose in the course of glycolysis. In aerobic organisms these NADH molecules donate their electrons to the electron-transport chain, as described in detail in Chapter 14. The electrons are passed along this chain to O_2, forming water, and the NAD^+ formed from the NADH is used again for glycolysis

PANEL 13–1 Details of the 10 steps of glycolysis

For each step, the part of the molecule that undergoes a change is shadowed in blue, and the name of the enzyme that catalyzes the reaction is in a yellow box. To watch a video of the reactions of glycolysis, see Movie 13.1.

Step 1 Glucose is phosphorylated by ATP to form a sugar phosphate. The negative charge of the phosphate prevents passage of the sugar phosphate through the plasma membrane, trapping glucose inside the cell.

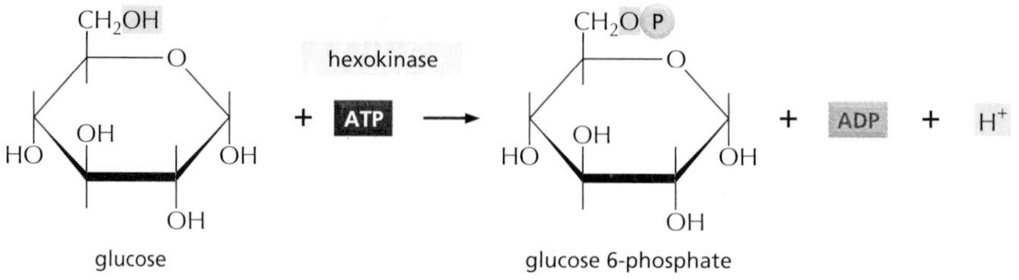

glucose + ATP → (hexokinase) glucose 6-phosphate + ADP + H^+

Step 2 A readily reversible rearrangement of the chemical structure (isomerization) moves the carbonyl oxygen from carbon 1 to carbon 2, forming a ketose from an aldose sugar. (See Panel 2–3, pp. 68–69.)

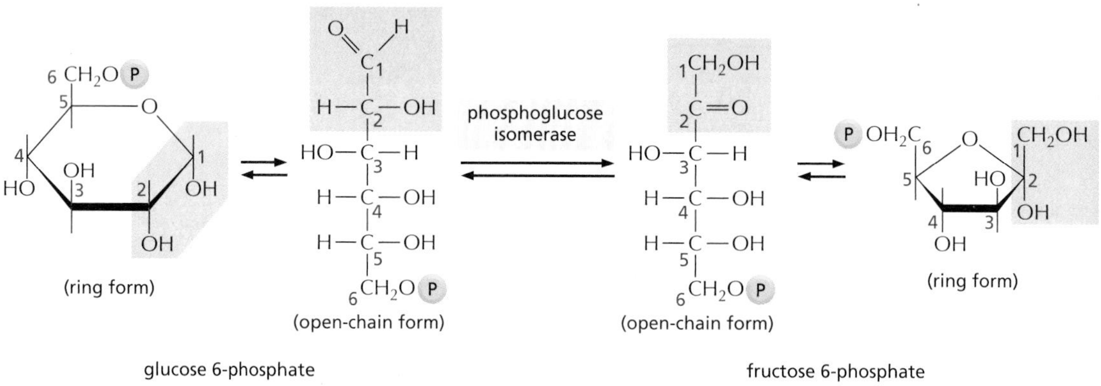

glucose 6-phosphate (ring form) ⇌ (open-chain form) ⇌ (phosphoglucose isomerase) fructose 6-phosphate (open-chain form) ⇌ (ring form)

Step 3 The new hydroxyl group on carbon 1 is phosphorylated by ATP, in preparation for the formation of two three-carbon sugar phosphates. The entry of sugars into glycolysis is controlled at this step, through regulation of the enzyme *phosphofructokinase*.

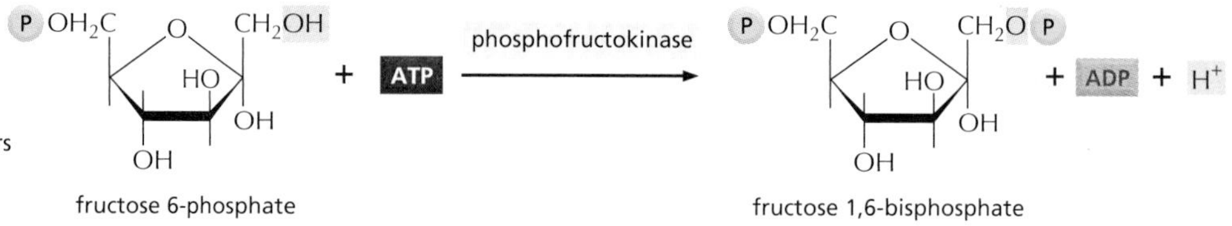

fructose 6-phosphate + ATP → (phosphofructokinase) fructose 1,6-bisphosphate + ADP + H^+

Step 4 The six-carbon sugar is cleaved to produce two three-carbon molecules. Only the glyceraldehyde 3-phosphate can proceed immediately through glycolysis.

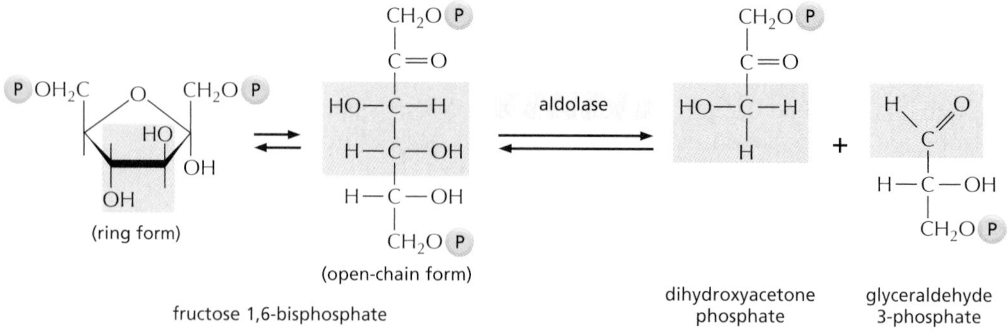

fructose 1,6-bisphosphate (ring form) ⇌ (open-chain form) ⇌ (aldolase) dihydroxyacetone phosphate + glyceraldehyde 3-phosphate

Step 5 The other product of step 4, dihydroxyacetone phosphate, is isomerized to form glyceraldehyde 3-phosphate.

dihydroxyacetone phosphate ⇌ (triose phosphate isomerase) glyceraldehyde 3-phosphate

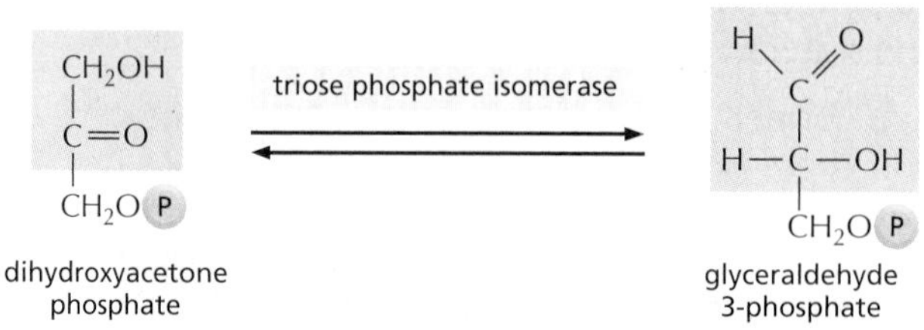

Step 6 The two molecules of glyceraldehyde 3-phosphate are oxidized. The energy-generation phase of glycolysis begins, as NADH and a new high-energy anhydride linkage to phosphate are formed (see Figure 13–5).

glyceraldehyde 3-phosphate + NAD^+ + P_i ⇌ (glyceraldehyde 3-phosphate dehydrogenase) 1,3-bisphosphoglycerate + NADH + H^+

Step 7 The transfer to ADP of the high-energy phosphate group that was generated in step 6 forms ATP.

1,3-bisphosphoglycerate + ADP ⇌ (phosphoglycerate kinase) 3-phosphoglycerate + ATP

Step 8 The remaining phosphate ester linkage in 3-phosphoglycerate, which has a relatively low free energy of hydrolysis, is moved from carbon 3 to carbon 2 to form 2-phosphoglycerate.

3-phosphoglycerate ⇌ (phosphoglycerate mutase) 2-phosphoglycerate

Step 9 The removal of water from 2-phosphoglycerate creates a high-energy enol phosphate linkage.

2-phosphoglycerate ⇌ (enolase) phosphoenolpyruvate + H_2O

Step 10 The transfer to ADP of the high-energy phosphate group that was generated in step 9 forms ATP, completing glycolysis.

phosphoenolpyruvate + ADP + H^+ → (pyruvate kinase) pyruvate + ATP

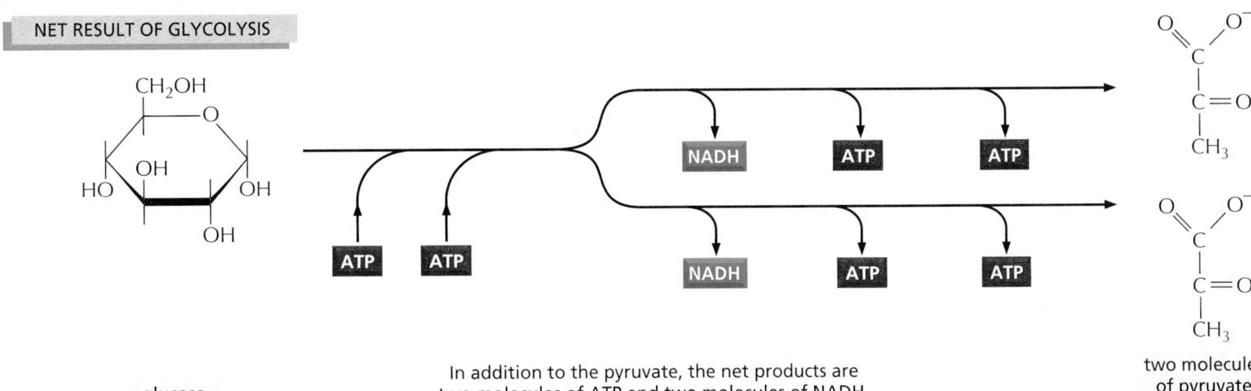

In addition to the pyruvate, the net products are two molecules of ATP and two molecules of NADH.

(see step 6 in Panel 13–1, p. 431).The piecing together of the complete glycolytic pathway in the 1930s was a major triumph of biochemistry, and it was quickly followed by the recognition of the central role of ATP in cellular processes.

Fermentations Allow ATP to Be Produced in the Absence of Oxygen

For most animal and plant cells, glycolysis is only a prelude to the third and final stage of the breakdown of food molecules, in which large amounts of ATP are generated by oxidative phosphorylation in mitochondria with the consumption of oxygen. However, for many anaerobic microorganisms, which do not use O_2 and can grow and divide in its absence, glycolysis is the principal source of ATP. The same is true in certain animal tissues, such as skeletal muscle, that can continue to function at low levels of O_2. In these anaerobic conditions, the pyruvate and the NADH stay in the cytosol. The pyruvate is converted into products that are excreted from the cell: for example, lactate in muscle or ethanol and CO_2 in the yeasts used in brewing and breadmaking. In the process, the NADH gives up its electrons and is converted back into NAD^+. This regeneration of NAD^+ is required to maintain the reactions of glycolysis (Figure 13–4). Anaerobic energy-yielding pathways like these are called **fermentations**. Studies of

QUESTION 13–1

At first glance, the final steps in fermentation appear to be unnecessary: the generation of lactate or ethanol does not produce any additional energy for the cell. Explain why cells growing in the absence of oxygen could not simply discard pyruvate as a waste product. Which products derived from glucose would accumulate in cells unable to generate either lactate or ethanol by fermentation?

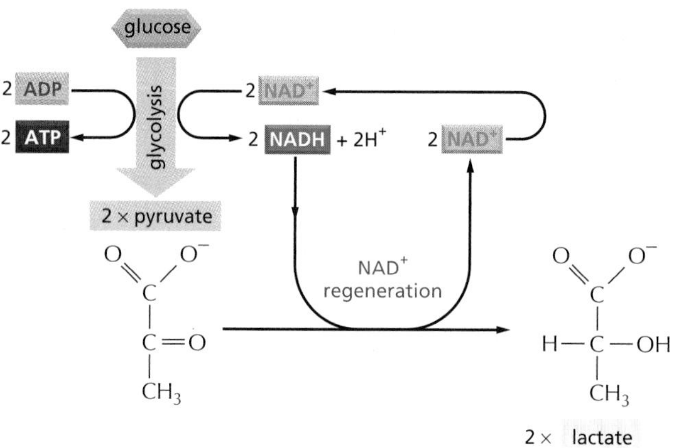

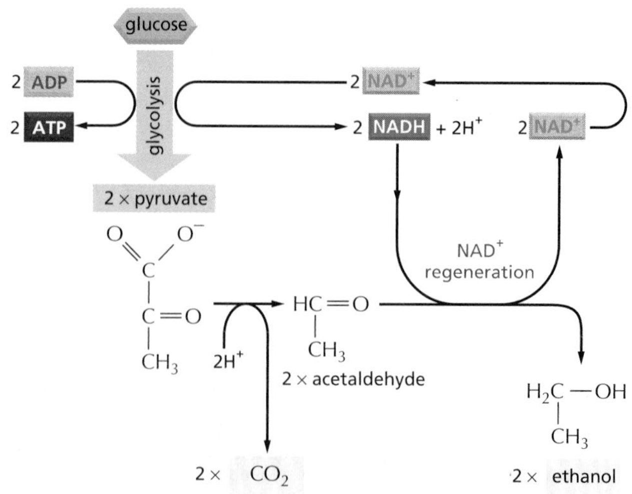

Figure 13–4 **Pyruvate can be broken down by fermentation in the absence of oxygen.** (A) When inadequate oxygen is present, for example, in a muscle cell undergoing vigorous contraction, the pyruvate produced by glycolysis is converted to lactate as shown. This reaction restores the NAD^+ consumed in step 6 of glycolysis, but the whole pathway yields much less energy overall than complete oxidation. (B) In some organisms that can grow anaerobically, such as yeasts, pyruvate is converted via acetaldehyde into carbon dioxide and ethanol. Again, this pathway regenerates NAD^+ from NADH, as required to enable glycolysis to continue. Both (A) and (B) are examples of fermentations.

the commercially important fermentations carried out by yeasts inspired much of early biochemistry.

Many bacteria and archaea can also generate ATP in the absence of oxygen by *anaerobic respiration*, a process that uses a molecule other than oxygen as a final electron acceptor. Anaerobic respiration differs from fermentation in that it involves an electron-transport chain embedded in a membrane.

Glycolysis Illustrates How Enzymes Couple Oxidation to Energy Storage

The "paddle wheel" analogy in Chapter 3 explained how cells harvest useful energy from the oxidation of organic molecules by coupling an energetically unfavorable reaction to an energetically favorable one (see Figure 3–30). To illustrate exactly how enzymes—the paddle wheel in our analogy—allow coupled reactions to occur, we take a closer look at a key pair of glycolytic reactions.

These reactions—steps 6 and 7 in Panel 13–1—convert the three-carbon sugar intermediate glyceraldehyde 3-phosphate (an aldehyde) into 3-phosphoglycerate (a carboxylic acid). This conversion entails the oxidation of an aldehyde group to a carboxylic acid group, which occurs in two steps. The overall reaction releases enough free energy to convert a molecule of ADP to ATP and to transfer two electrons from the aldehyde to NAD^+ to form NADH, while still releasing enough heat to the environment to make the overall reaction energetically favorable: the $\Delta G°$ for the overall reaction is –3.0 kcal/mole (–12.5 kJ/mole).

This remarkable feat of energy harvesting is detailed in Figure 13–5. The chemical reactions are precisely guided by two enzymes to which the sugar intermediates are tightly bound. In fact, as shown in Figure 13–5, the first enzyme (glyceraldehyde 3-phosphate dehydrogenase) forms a short-lived covalent bond to the aldehyde through a reactive –SH group on the enzyme, and catalyzes its oxidation in this attached state. The reactive enzyme–substrate bond is then displaced by an inorganic phosphate ion to produce a high-energy phosphate intermediate, which is released from the enzyme. The intermediate, 1,3-bisphosphoglycerate, binds to the second enzyme (phosphoglycerate kinase). This enzyme then catalyzes the energetically favorable transfer of the intermediate's high-energy phosphate to ADP, forming ATP and completing the process of oxidizing an aldehyde to a carboxylic acid.

We have discussed this particular oxidation process in some detail because it provides a clear example of enzyme-mediated energy storage through coupled reactions (Figure 13–6). These reactions (steps 6 and 7) are the only ones in glycolysis that create a high-energy phosphate linkage directly from inorganic phosphate. As such, they account for the net yield of two ATP molecules and two NADH molecules per molecule of glucose. As we have mentioned, this NADH must be reoxidized to the NAD^+ required for these coupled reactions. If NAD^+ is not available, glycolysis will stop (see Figure 13–4).

As we have just seen, ATP can be readily generated from ADP when reaction intermediates are formed with phosphate bonds of higher energy than those in ATP. The energy of phosphate bonds can be ordered by determining the standard free-energy change ($\Delta G°$) for the breakage of each bond by hydrolysis; Figure 13–7 compares the high-energy phosphoanhydride bonds in ATP with some other phosphate bonds that are generated during glycolysis. As explained in Panel 3–1 (p. 95), we describe bonds as "high energy" only in the sense that their hydrolysis is particularly energetically favorable.

QUESTION 13–2

Arsenate (AsO_4^{3-}) is chemically very similar to phosphate (PO_4^{3-}) and is used as an alternative substrate by many phosphate-requiring enzymes. In contrast to phosphate, however, an anhydride bond between arsenate and carbon is very quickly hydrolyzed in water. Knowing this, suggest why arsenate is a compound of choice for murderers but not for cells. Formulate your explanation in the context of Figure 13–6.

Figure 13–5 **Energy is harvested in steps 6 and 7 of glycolysis.** In these steps the oxidation of an aldehyde to a carboxylic acid is coupled to the formation of ATP and NADH. (A) In step 6, the enzyme glyceraldehyde 3-phosphate dehydrogenase couples the energetically favorable oxidation of an aldehyde to the energetically unfavorable formation of a high-energy phosphate bond. At the same time, it enables energy to be stored in NADH. The formation of the high-energy phosphate bond is driven by the oxidation reaction, and the enzyme thereby acts like the "paddle wheel" coupler in Figure 3–30B. In reaction step 7, the newly formed high-energy phosphate bond in 1,3-bisphosphoglycerate is transferred to ADP, forming a molecule of ATP and leaving a free carboxylic acid group on the oxidized sugar. This reaction is catalyzed by the enzyme phosphoglycerate kinase. Note that the shaded portion of the glyceraldehyde 3-phosphate molecule (*gray*) remains unchanged throughout all these reactions. (B) Summary of the overall chemical change produced by reactions 6 and 7.

(A) STEPS 6 AND 7 OF GYCOLYSIS

glyceraldehyde 3-phosphate

STEP 6 — glyceraldehyde 3-phosphate dehydrogenase

ENZYME —SH

NAD⁺

ENZYME —S

NADH + H⁺

high-energy bond (thioester)

phosphate

1,3-bisphosphoglycerate

STEP 7 — phosphoglycerate kinase

ADENINE

RIBOSE

ADP

ATP

3-phosphoglycerate

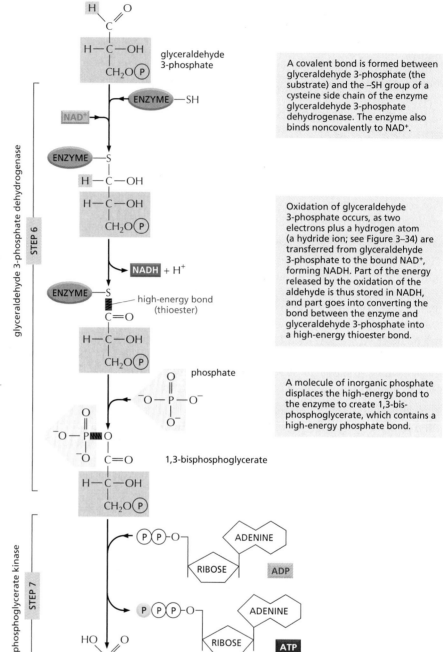

A covalent bond is formed between glyceraldehyde 3-phosphate (the substrate) and the –SH group of a cysteine side chain of the enzyme glyceraldehyde 3-phosphate dehydrogenase. The enzyme also binds noncovalently to NAD⁺.

Oxidation of glyceraldehyde 3-phosphate occurs, as two electrons plus a hydrogen atom (a hydride ion; see Figure 3–34) are transferred from glyceraldehyde 3-phosphate to the bound NAD⁺, forming NADH. Part of the energy released by the oxidation of the aldehyde is thus stored in NADH, and part goes into converting the bond between the enzyme and glyceraldehyde 3-phosphate into a high-energy thioester bond.

A molecule of inorganic phosphate displaces the high-energy bond to the enzyme to create 1,3-bis-phosphoglycerate, which contains a high-energy phosphate bond.

The high-energy bond to phosphate is transferred to ADP to form ATP.

(B)

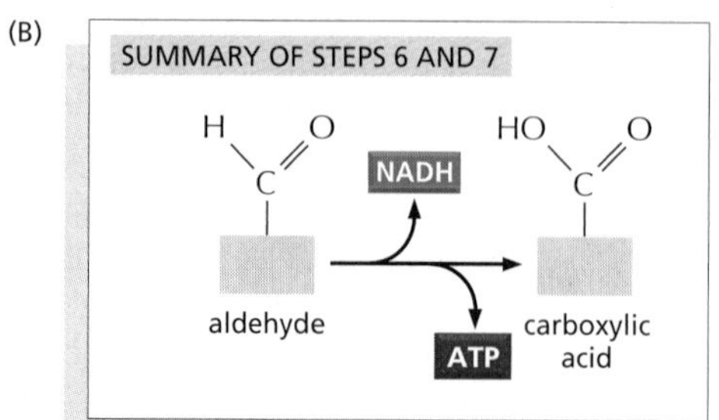

Much of the energy of oxidation has been stored in the activated carriers ATP and NADH.

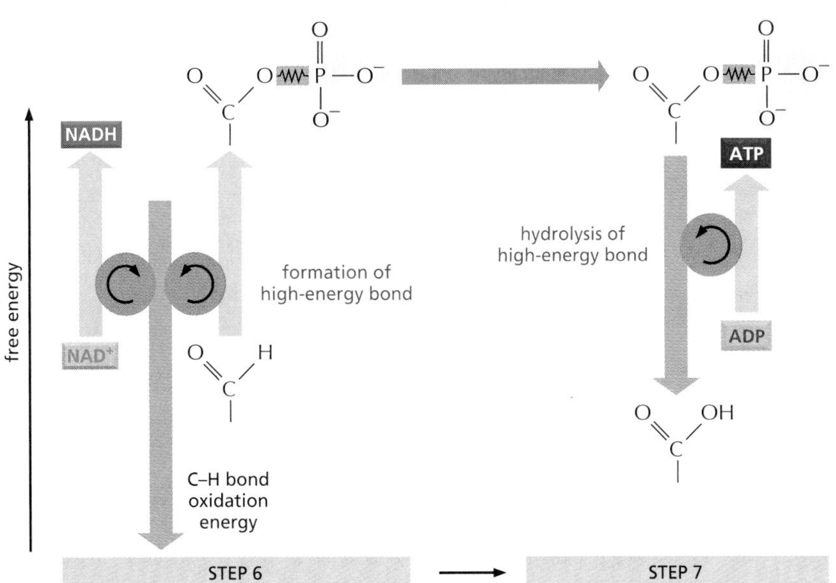

Figure 13–6 **Coupling of the reactions in steps 6 and 7 of glycolysis allows the energetically unfavorable formation of a high-energy phosphate bond.** In step 6, the C–H bond oxidation energy drives the formation of both NADH and a high-energy phosphate bond. The breakage of the high-energy bond in step 7 then drives ATP formation.

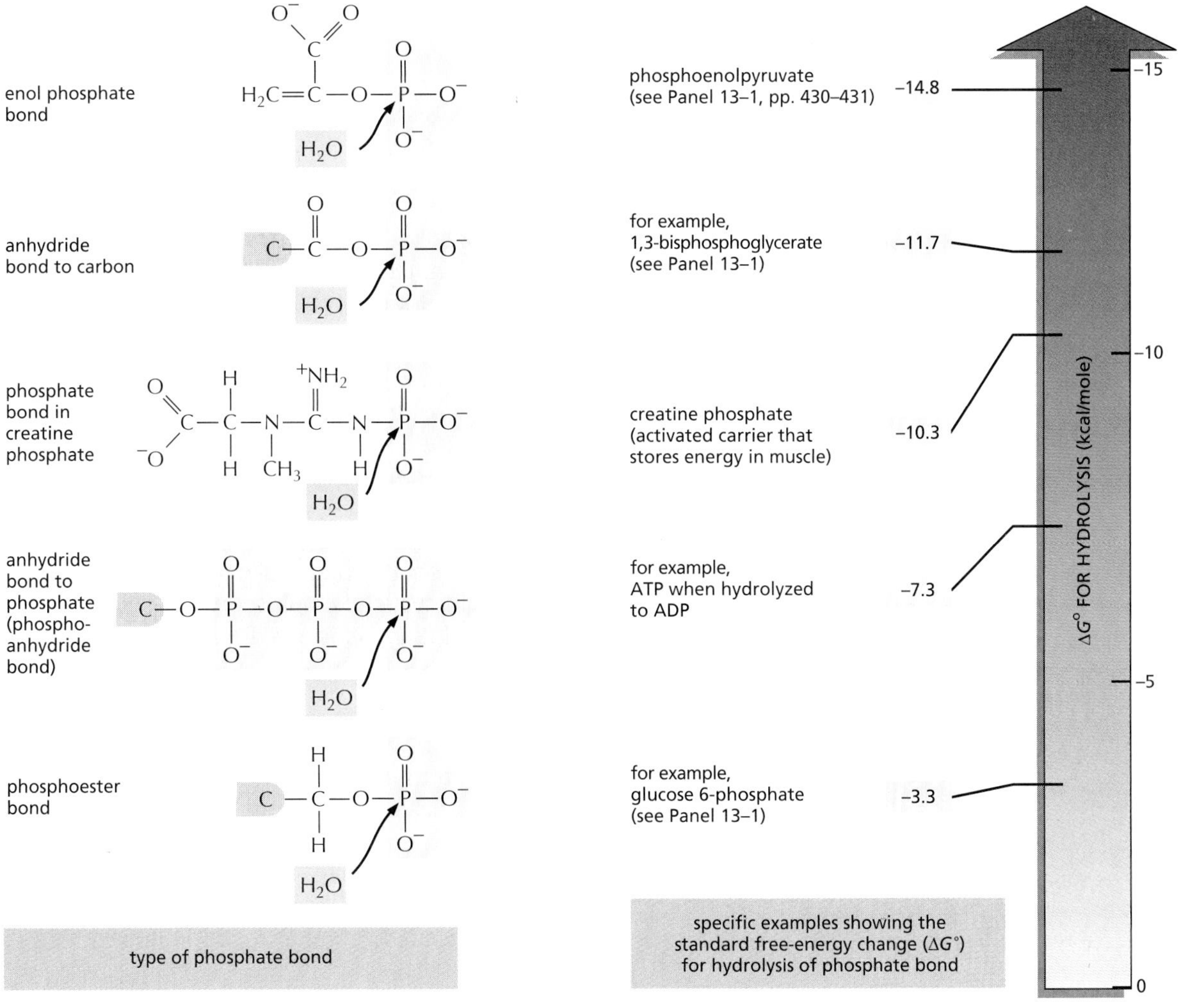

type of phosphate bond

enol phosphate bond

anhydride bond to carbon

phosphate bond in creatine phosphate

anhydride bond to phosphate (phospho-anhydride bond)

phosphoester bond

specific examples showing the standard free-energy change ($\Delta G°$) for hydrolysis of phosphate bond

phosphoenolpyruvate (see Panel 13–1, pp. 430–431) –14.8

for example, 1,3-bisphosphoglycerate (see Panel 13–1) –11.7

creatine phosphate (activated carrier that stores energy in muscle) –10.3

for example, ATP when hydrolyzed to ADP –7.3

for example, glucose 6-phosphate (see Panel 13–1) –3.3

$\Delta G°$ FOR HYDROLYSIS (kcal/mole)

–15 –10 –5 0

Figure 13–7 **Differences in the energies of different phosphate bonds allow the formation of ATP by substrate-level phosphorylation.** Examples of the different types of phosphate bonds and the sites of hydrolysis are shown in the molecules depicted on the left. Those starting with a *gray* carbon atom show only part of a molecule. Examples of molecules containing such bonds are given on the right, with the free-energy change for hydrolysis in kcal/mole (1 kcal = 4.184 kJ). The transfer of a phosphate group from one molecule to another is energetically favorable if the standard free-energy change ($\Delta G°$) for hydrolysis of the phosphate bond of the first molecule is more negative than that for hydrolysis of the phosphate bond (once present) in the second. Thus, a phosphate group is readily transferred from 1,3-bisphosphoglycerate to ADP to form ATP. The hydrolysis reaction can be viewed as the transfer of the phosphate group to water.

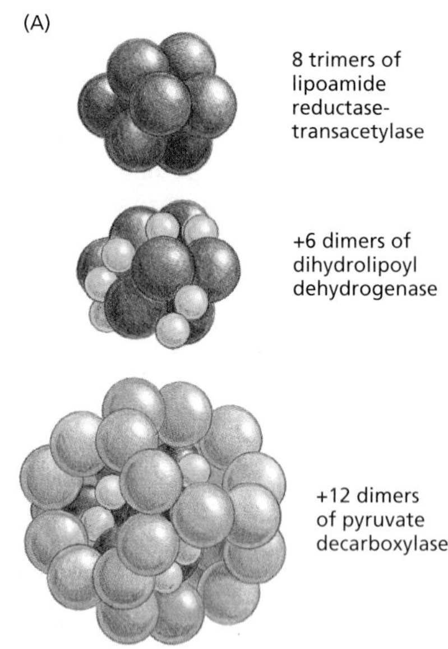

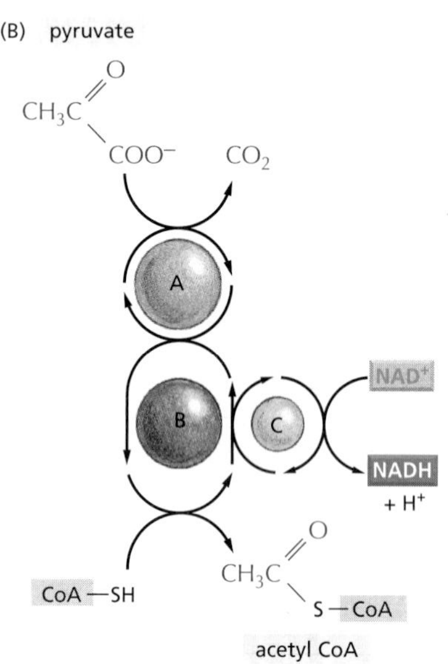

Figure 13–8 **Pyruvate is oxidized to acetyl CoA and CO_2 by pyruvate dehydrogenase.** (A) General structure of the pyruvate dehydrogenase complex, which contains three different enzymes and some 60 polypeptide chains. In this large multienzyme complex, reaction intermediates are passed directly from one enzyme to another. (B) The reactions carried out by the pyruvate dehydrogenase complex. The complex converts pyruvate to acetyl CoA in the mitochondrial matrix; NADH is also produced in this reaction. A, B, and C are the three enzymes *pyruvate decarboxylase, lipoamide reductase-transacetylase*, and *dihydrolipoyl dehydrogenase*, respectively, as illustrated in (A); their activities are linked as shown. Pyruvate and its products are shown in *red*.

Sugars and Fats Are Both Degraded to Acetyl CoA in Mitochondria

In aerobic metabolism in eucaryotic cells, the pyruvate produced by glycolysis is actively pumped into the mitochondrial matrix, the major internal compartment of this organelle (see Figure 14–12). There it is rapidly decarboxylated by a giant complex of three enzymes, called the *pyruvate dehydrogenase complex.* The products of pyruvate decarboxylation are a molecule of CO_2 (a waste product), a molecule of NADH, and a molecule of acetyl CoA. The structure and function of pyruvate dehydrogenase are outlined in Figure 13–8.

Fatty acids, derived from **fat**, are an alternative fuel to sugars for energy generation. Like the pyruvate derived from glycolysis, fatty acids are converted into acetyl CoA in mitochondria. Each long molecule of fatty acid (in the form of the activated molecule, fatty acyl CoA) is broken down completely by a cycle of reactions that trims two carbons at a time from its carboxyl end, generating one molecule of acetyl CoA for each turn of the cycle. A molecule of NADH and a molecule of another electron carrier, *$FADH_2$*, are also produced in this process (Figure 13–9).

Sugars and fats provide the major energy sources for most nonphotosynthetic organisms, including humans. In the course of their processing to acetyl CoA, only a small part of the useful energy stored in these foodstuffs is extracted and converted into ATP or NADH. Most of the energy is still locked up in acetyl CoA. The next stage in respiration, in which the acetyl group in acetyl CoA is oxidized to CO_2 and H_2O in the citric acid cycle, is therefore central to the energy metabolism of aerobic organisms. In eucaryotes the citric acid cycle takes place in mitochondria, the organelles to which pyruvate and fatty acids are directed for acetyl CoA production (Figure 13–10).

In addition to pyruvate and fatty acids, some amino acids are transported from the cytosol into mitochondria, where they are also converted into acetyl CoA or one of the other intermediates of the citric acid cycle (see Figure 13–2). Thus, in the eucaryotic cell, the mitochondrion is the center toward which all energy-yielding processes lead, whether they begin with sugars, fats, or proteins. In aerobic bacteria, which have no mitochondria, all of these reactions—glycolysis, acetyl CoA production, and the citric acid cycle—take place in the single compartment of the cytosol.

QUESTION 13–3

Many catabolic and anabolic reactions are based on reactions that are similar but work in opposite directions, such as the hydrolysis and condensation reactions described in Figure 3–38. This is true for fatty acid breakdown and fatty acid synthesis. From what you know about the mechanism of fatty acid breakdown outlined in Figure 13–9, would you expect the fatty acids found in cells to most commonly have an even or an odd number of carbon atoms?

The Citric Acid Cycle Generates NADH by Oxidizing Acetyl Groups to CO_2

The third and final stage in the oxidative breakdown of food molecules to generate energy requires abundant O_2. Although living organisms have inhabited Earth for the past 3.5 billion years, the planet is thought to have developed an atmosphere containing O_2 gas only some 1 to 2 bil-

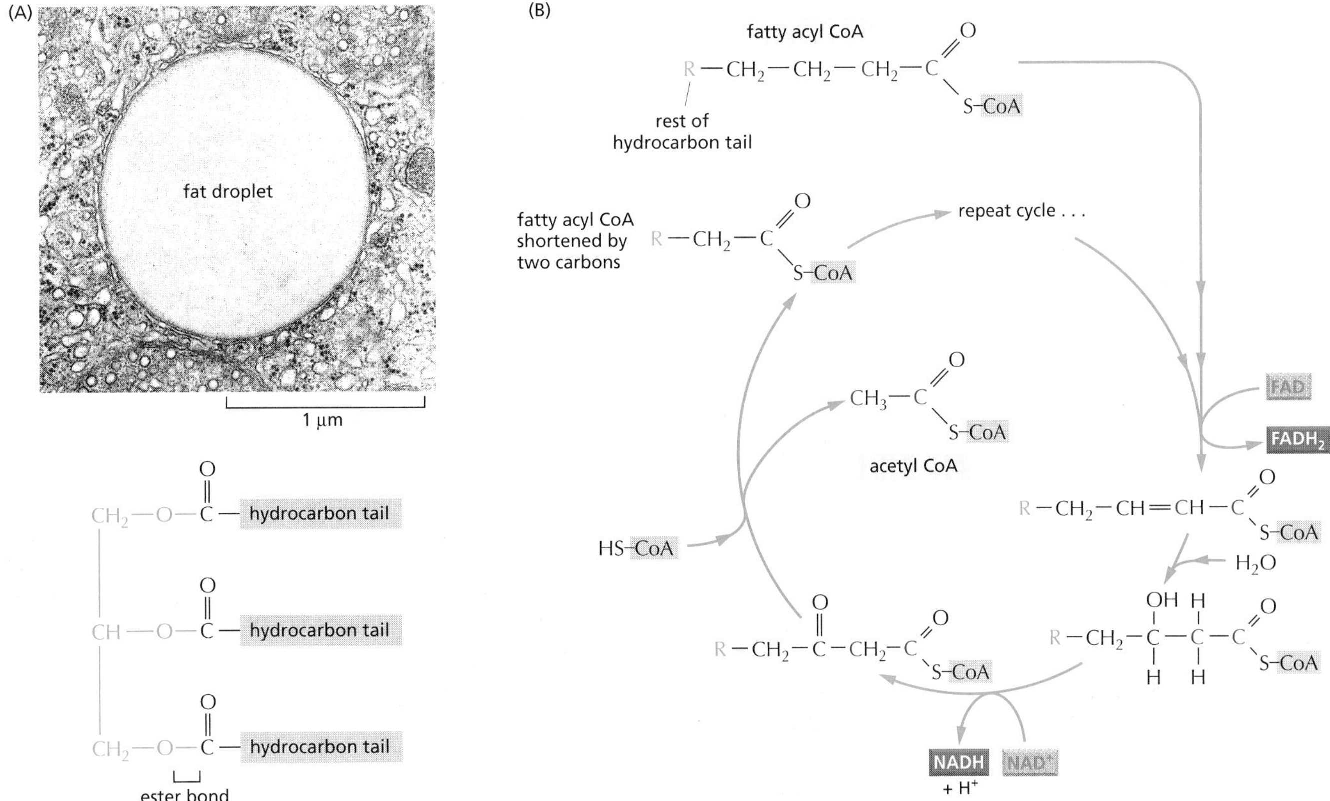

Figure 13–9 **Fatty acids are also oxidized to acetyl CoA.** (A) Electron micrograph of a lipid droplet in the cytoplasm (*top*), and the structure of fats (*bottom*). Fats are triacylglycerols. The glycerol portion, to which three fatty acids are linked through ester bonds, is shown here in *blue*. Fats are insoluble in water and form large lipid droplets in the specialized fat cells (called adipocytes) in which they are stored. (B) The fatty acid oxidation cycle. The cycle is catalyzed by a series of four enzymes in the mitochondrion. Each turn of the cycle shortens the fatty acid chain by two carbons (shown in *red*) and generates one molecule of acetyl CoA and one molecule each of NADH and $FADH_2$. (A, courtesy of Daniel S. Friend.)

lion years ago (see Figure 14–43). The oxygen-consuming reactions we discuss next are therefore likely to be of relatively recent origin. In contrast, the mechanism used to produce ATP in Figure 13–5 does not require oxygen, and relatives of that elegant pair of coupled reactions could have arisen very early in the history of life on Earth.

In the nineteenth century, biologists noticed that in the absence of air (anaerobic conditions) cells produce lactic acid (for example, in muscle) or ethanol (for example, in yeast), while in the presence of air (aerobic conditions) these cells consume O_2 and produce CO_2 and H_2O. Intensive

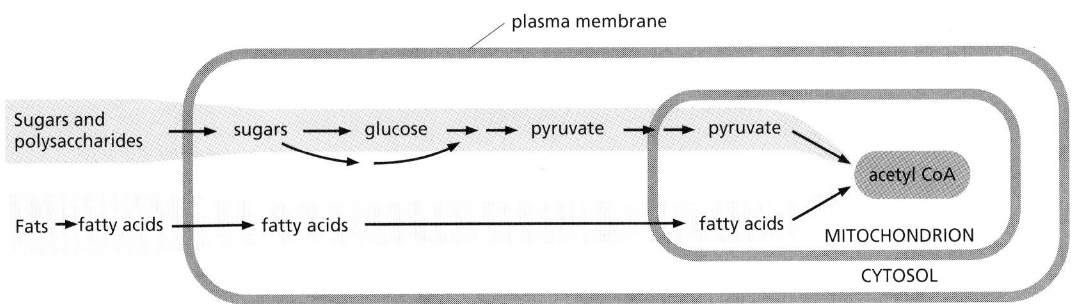

Figure 13–10 **In eucaryotic cells, acetyl CoA is produced in the mitochondria from molecules derived from sugars and fats.** Most of the cell's oxidation reactions occur in these organelles, and most of its ATP is made here.

efforts to define the pathways of aerobic metabolism eventually focused on the oxidation of pyruvate and led in 1937 to the discovery of the **citric acid cycle**, also known as the *tricarboxylic acid cycle* or the *Krebs cycle* (see How We Know, pp. 440–441). The citric acid cycle accounts for about two-thirds of the total oxidation of carbon compounds in most cells, and its major end products are CO_2 and high-energy electrons in the form of NADH. The CO_2 is released as a waste product, while the high-energy electrons from NADH are passed to a series of membrane-bound enzymes known collectively as the *electron-transport chain*. At the end of the chain, these electrons combine with O_2 to produce H_2O. The citric acid cycle itself does not use O_2. However, it requires O_2 to proceed because the electron-transport chain allows NADH to get rid of its electrons and thus regenerate the NAD^+ that is needed to keep the cycle going.

The citric acid cycle, which takes place in the mitochondrial matrix, catalyzes the complete oxidation of the carbon atoms of the acetyl groups in acetyl CoA, converting them into CO_2. The acetyl group is not oxidized directly, however. Instead, it is transferred from acetyl CoA to a larger four-carbon molecule, *oxaloacetate*, to form the six-carbon tricarboxylic acid *citric acid*, for which the subsequent cycle of reactions is named. The citric acid molecule is then gradually oxidized, and the energy of this oxidation is harnessed to produce energy-rich carrier molecules, in much the same manner as we described for glycolysis. The chain of eight reactions forms a cycle, because the oxaloacetate that began the process is regenerated at the end, as shown in outline in Figure 13–11.

We have so far discussed only one of the three types of activated carrier molecules that are produced by the citric acid cycle—NADH. In addition to three molecules of NADH, each turn of the cycle also produces one molecule of **$FADH_2$** (reduced flavin adenine dinucleotide) from FAD and one molecule of the ribonucleotide **GTP** (guanosine triphosphate) from GDP (see Figure 13–11). The structures of these two activated carrier molecules are illustrated in Figure 13–12. GTP is a close relative of ATP, and the transfer of its terminal phosphate group to ADP produces one ATP

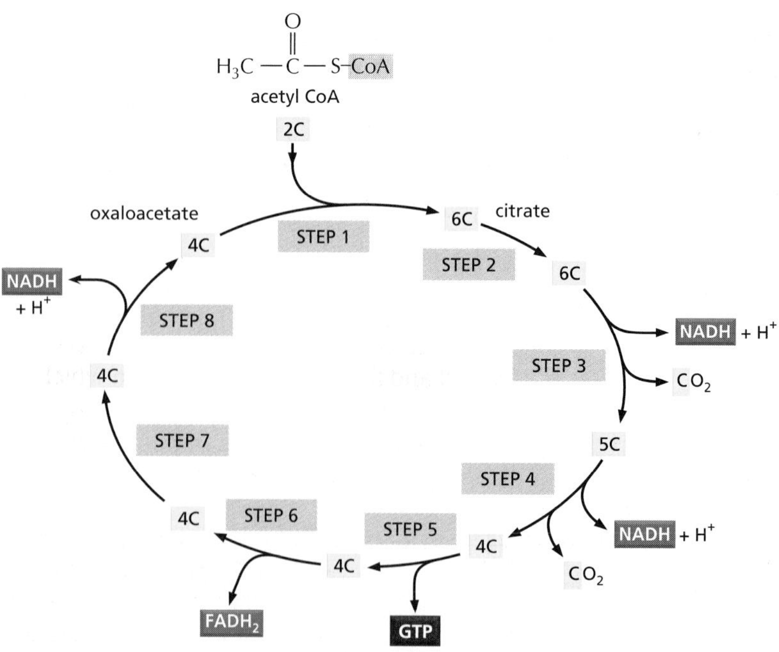

Figure 13–11 **The citric acid cycle catalyzes the complete oxidation of the carbon atoms in acetyl CoA.** The reaction of acetyl CoA with oxaloacetate starts the cycle by producing citrate (citric acid). Each turn of the cycle produces two molecules of CO_2 (as waste products), three molecules of NADH, one molecule of GTP, and one molecule of $FADH_2$. The number of carbon atoms in each intermediate is shown in a *yellow box*. (See also Panel 13–2, pp. 442–443.) All the steps of the citric acid cycle are reviewed in Movie 13.2.

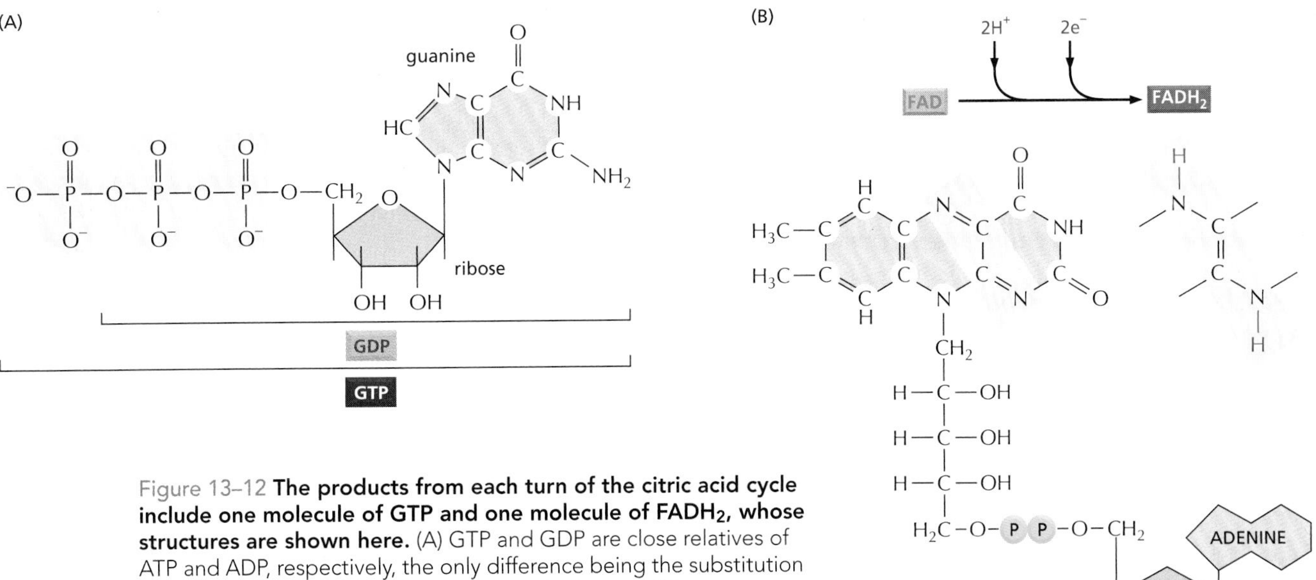

Figure 13–12 **The products from each turn of the citric acid cycle include one molecule of GTP and one molecule of $FADH_2$, whose structures are shown here.** (A) GTP and GDP are close relatives of ATP and ADP, respectively, the only difference being the substitution of the base guanine for adenine. (B) Despite its very different structure, $FADH_2$, like NADH and NADPH, is a carrier of hydrogens and high-energy electrons. It is shown here in its oxidized form (FAD) with the hydrogen-carrying atoms highlighted in *yellow*. These same atoms are shown reduced in the excerpt to the right.

molecule in each cycle. Like NADH, $FADH_2$ is a carrier of high-energy electrons and hydrogen. As we discuss shortly, the energy that is stored in the readily transferred high-energy electrons of NADH and $FADH_2$ are subsequently used to produce ATP through the process of *oxidative phosphorylation*, which occurs in the mitochondrial membrane. Oxidative phosphorylation is the only step in the oxidative catabolism of foodstuffs that directly requires O_2 from the atmosphere.

The complete citric acid cycle is presented in Panel 13–2 (pp. 442–443). Note that the oxygen atoms required to make CO_2 from the acetyl groups entering the citric acid cycle are supplied not by O_2 but by water. As illustrated in the panel, three molecules of water are split in each cycle, and the oxygen atoms of some of them are ultimately used to make CO_2. A common misconception about aerobic respiration is that the atmospheric O_2 required for the process is converted into the CO_2 that is released as a waste product. In fact, as we shall see, the molecules of O_2 gas are reduced to water, not incorporated directly into CO_2.

Many Biosynthetic Pathways Begin with Glycolysis or the Citric Acid Cycle

Catabolic reactions, such as those of glycolysis and the citric acid cycle, produce both energy for the cell and the building blocks from which many other molecules are made (see Figure 3–2). So far we have emphasized energy production rather than the provision of starting materials for biosynthesis. But many of the intermediates formed in glycolysis and the citric acid cycle are siphoned off by biosynthetic, or *anabolic*, pathways, where they are converted by series of enzyme-catalyzed reactions into amino acids, nucleotides, lipids, and other small organic molecules that the cell needs. Oxaloacetate and α-ketoglutarate from the citric acid cycle, for example, are transferred from the mitochondrion back to the cytosol, where they serve as precursors for many essential molecules, such as the amino acids aspartate and glutamate, respectively. An idea of

HOW WE KNOW: UNRAVELING THE CITRIC ACID CYCLE

"I have often been asked how the work on the citric acid cycle arose and developed," stated the biological chemist Hans Krebs in a lecture and review article in which he described his Nobel Prize-winning discovery of the cycle of reactions that lies at the center of cell metabolism. Did the concept stem from a sudden inspiration, a revelatory vision? "It was nothing of the kind," answered Krebs. Instead his realization that these reactions occur in a cycle—rather than a set of linear pathways—arose from a "very slow evolutionary process" that occurred over a five-year period, during which Krebs coupled insight and reasoning to careful experimentation to discover one of the central pathways that underlies energy metabolism in cells.

Minced tissues, curious catalysis

By the early 1930s, Krebs and other investigators had discovered that a select set of molecules are oxidized extraordinarily rapidly in various types of tissue preparations—slices of kidney or liver, or suspensions of minced pigeon muscle. Because these reactions depend on the presence of oxygen, the researchers surmised that this set of compounds might include intermediates that are important in cellular respiration—the consumption of O_2 and production of CO_2 that accompanies the metabolism of foodstuffs.

Using the minced-tissue preparations, Krebs and others made the following observations. First, that in the presence of oxygen certain organic acids—citrate, succinate, fumarate, and malate—are readily oxidized to carbon dioxide. These reactions depend on a continuous supply of oxygen.

Second, that the oxidation of these compounds falls into a pair of linear, sequential pathways:

citrate → α-ketoglutarate → succinate

and

succinate → fumarate → malate → oxaloacetate

Third, that small amounts of several of these compounds, when added to minced-muscle suspensions, stimulated an unusually large uptake of oxygen—far greater than that needed to oxidize only the added molecules. To explain this surprising observation, Albert Szent-Györgyi (the Nobel laureate who worked out the second pathway above) suggested that a single molecule of each compound must somehow act catalytically to stimulate the oxidation of many molecules of some endogenous substance in the muscle.

At this point, most of the reactions central to the citric acid cycle were known. What was not yet clear—and caused great confusion, even to future Nobel laureates—was how these apparently linear reactions could drive such a catalytic consumption of oxygen, where each molecule of metabolite fuels the oxidation of many more molecules. To simplify the discussion of how Krebs ultimately solved this puzzle—by linking these linear reactions together into a circle—we will now refer to the molecules involved by a sequence of letters, A through H (Figure 13–13).

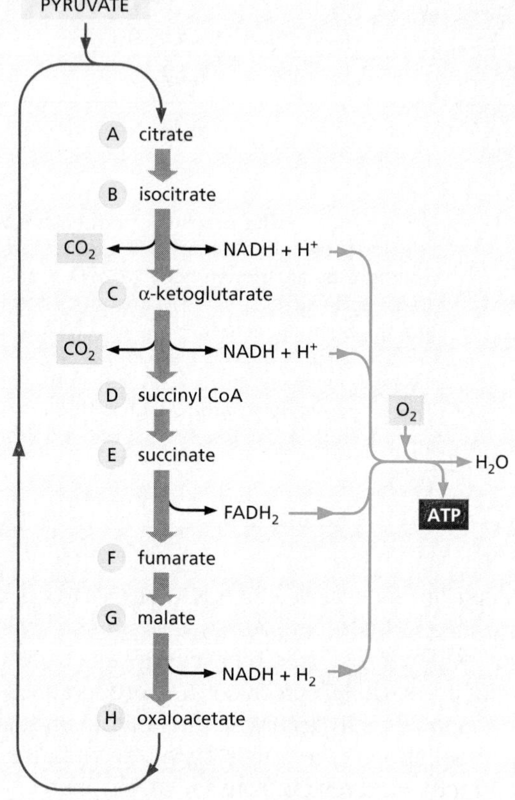

Figure 13–13 **In this simplified representation of the citric acid cycle, O_2 is consumed and CO_2 is liberated as the molecular intermediates become oxidized.** Krebs and others did not initially realize that these oxidation reactions occur in a cycle, as shown here.

A poison suggests a cycle

Many of the clues that Krebs used to formulate the citric acid cycle came from experiments using malonate—a poisonous compound that specifically inhibits the enzyme succinate dehydrogenase, which converts E to F. Malonate closely resembles succinate (E) in its structure (Figure 13–14), and it serves as a competitive inhibitor of the enzyme. Because the addition of malonate poisons respiration in tissues, Krebs concluded that succinate dehydrogenase (and the entire pathway linked to it) must play a critical role in cellular respiration.

malonate	succinate
COO^-	COO^-
CH_2	CH_2
COO^-	CH_2
	COO^-

Figure 13–14 **The structure of malonate closely resembles that of succinate.**

Krebs then discovered that when A, B, or C is added to malonate-poisoned tissue suspensions, E accumulates (Figure 13–15A). This observation reinforces the importance of succinate dehydrogenase for successful respiration. However, he found that E also accumulates when F, G, or H is added to malonate-poisoned muscle (Figure 13–15B). The latter result suggests that an additional set of reactions must exist that can convert F, G, and H molecules into E, since E was previously shown to be a precursor for F, G, and H, rather than a product of their reactions.

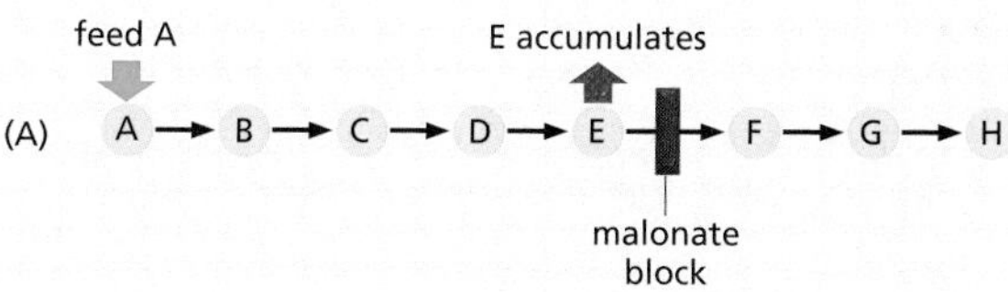

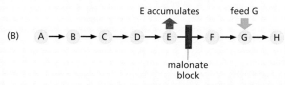

Figure 13–15 **Poisoning muscle preparations with malonate provided clues to the cyclic nature of these oxidative reactions.** (A) Adding A, B, or C to malonate-poisoned muscle causes an accumulation of E. (B) Addition of F, G, or H to a malonate-poisoned preparation also causes an accumulation of E, suggesting that enzymatic reactions can convert these molecules into E. The discovery that citrate (A) can be formed from oxaloacetate (H) and pyruvate allowed Krebs to join these two reaction pathways into a complete circle.

At about this time Krebs also determined that when muscle suspensions are incubated with pyruvate and oxaloacetate, citrate forms:

$$\text{pyruvate} + \text{oxaloacetate} \rightarrow \text{citrate}$$

$$\text{or, pyruvate} + \text{H} \rightarrow \text{A}$$

This observation led Krebs to postulate that when oxygen is present, pyruvate and H condense to form A, converting the previously delineated string of linear reactions into a cyclic sequence (see Figure 13–13). The discovery that acetyl CoA acts as an intermediary between pyruvate and oxaloacetate in this reaction, however, did not come for another decade.

Explaining the mysterious stimulatory effects

The cycle of reactions that was proposed by Krebs clearly explained how the addition of small amounts of any of the intermediates A through H could cause the large increase in the uptake of O_2 that had been observed. Pyruvate is abundant in minced tissues, being readily produced by glycolysis using glucose derived from stored glycogen (see Figure 13–3). Its oxidation requires a functioning citric acid cycle, in which each turn of the cycle results in the oxidation of one molecule of pyruvate. If the intermediates A through H are in small enough supply, the rate at which the entire cycle turns will be restricted. Adding a supply of any one of these intermediates will then have a dramatic effect on the rate at which the entire citric acid cycle operates (Figure 13–16). Thus it is easy to see how a large number of pyruvate molecules can be oxidized, and a great deal of oxygen consumed, for every molecule that is added of a citric acid cycle intermediate.

Krebs went on to demonstrate that all of the individual enzymatic reactions in his postulated cycle take place in tissue preparations. Furthermore, they occur at rates high enough to account for the rate of pyruvate and oxygen consumption in these tissues. Krebs therefore concluded that this series of reactions is the major, if not the sole, pathway for the oxidation of pyruvate—at least in muscle. By fitting together pieces of information like a jigsaw puzzle, and by searching for missing links, Krebs arrived at a coherent picture of the metabolic processes that underlie the oxidation of foodstuffs. Remarkably, he worked out this intricate metabolic pathway without the aid of reagents and techniques considered essential by modern biochemists: radioactive markers that allow one to trace labeled compounds through these reaction pathways, or mass spectrometry, a powerful method for rapidly identifying the various chemical intermediates that occur along the way.

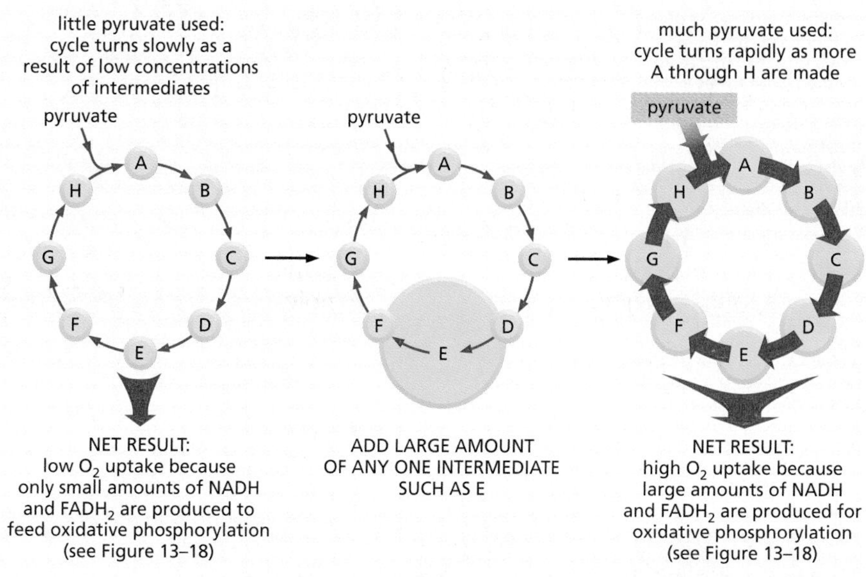

Figure 13–16 **Replenishing the supply of any single intermediate has a dramatic effect on the rate at which the entire citric acid cycle operates.**

PANEL 13–2 The complete citric acid cycle

The complete citric acid cycle. The two carbons from acetyl CoA that enter this turn of the cycle (shadowed in red) will be converted to CO_2 in subsequent turns of the cycle: it is the two carbons shadowed in blue that are converted to CO_2 in this cycle.

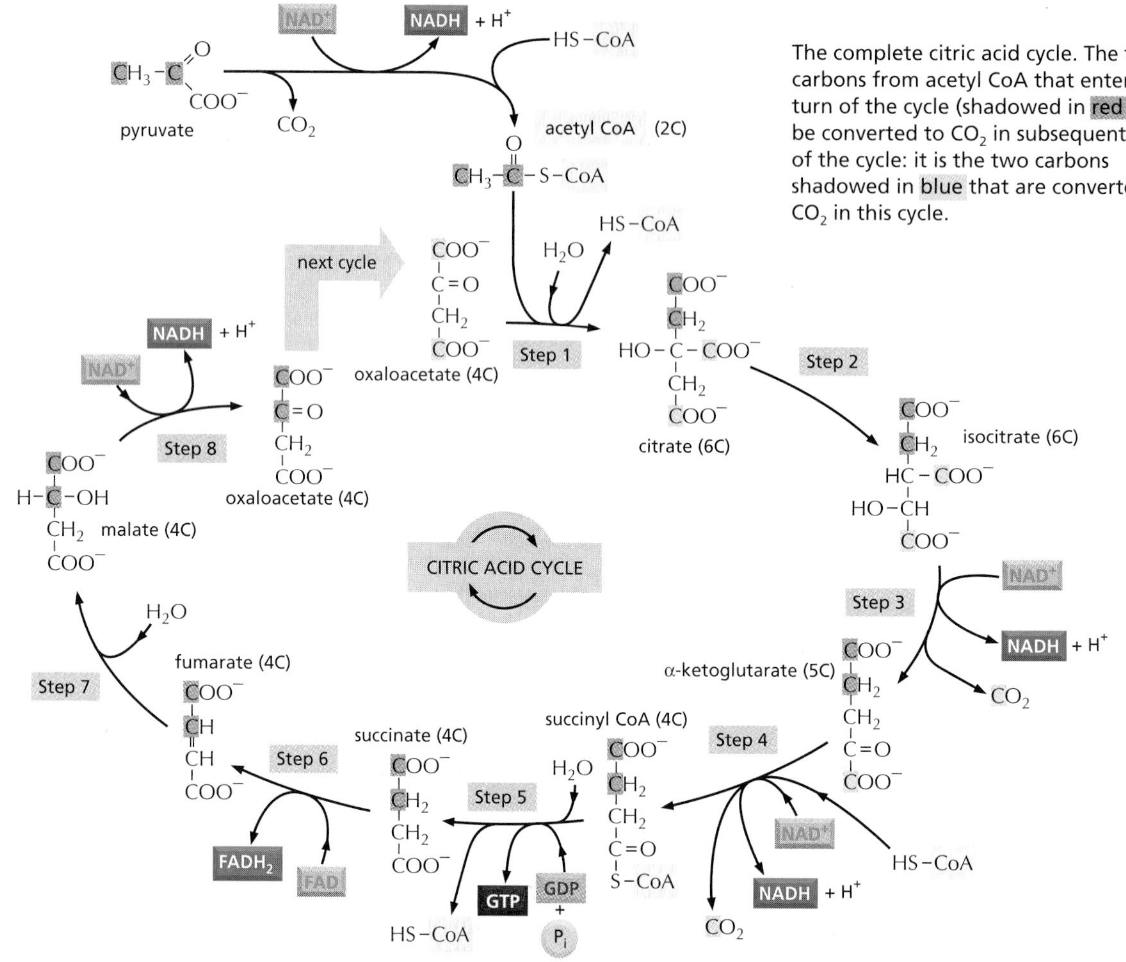

Details of these eight steps are shown below. In this part of the panel, for each step, the part of the molecule that undergoes a change is shadowed in blue, and the name of the enzyme that catalyzes the reaction is in a yellow box.
To watch a video of the reactions of the citric acid cycle, see **Movie 13.2**.

Step 1 After the enzyme removes a proton from the CH_3 group on acetyl CoA, the negatively charged CH_2^- forms a bond to a carbonyl carbon of oxaloacetate. The subsequent loss by hydrolysis of the coenzyme A (CoA) drives the reaction strongly forward.

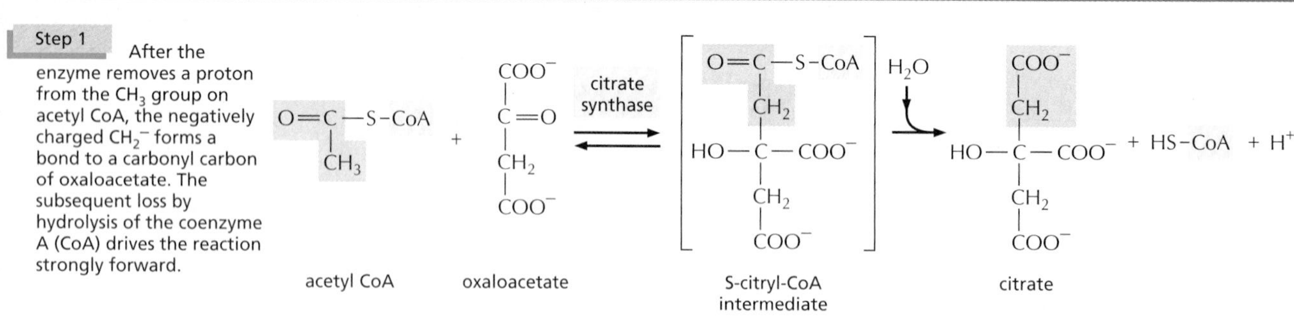

Step 2 An isomerization reaction, in which water is first removed and then added back, moves the hydroxyl group from one carbon atom to its neighbor.

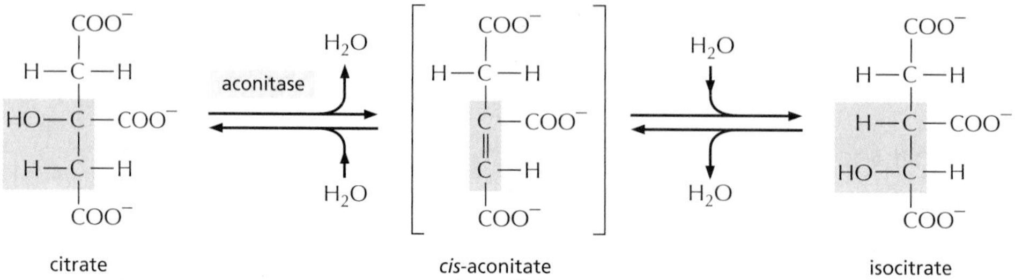

citrate — aconitase — *cis*-aconitate intermediate — isocitrate

Step 3 In the first of four oxidation steps in the cycle, the carbon carrying the hydroxyl group is converted to a carbonyl group. The immediate product is unstable, losing CO_2 while still bound to the enzyme.

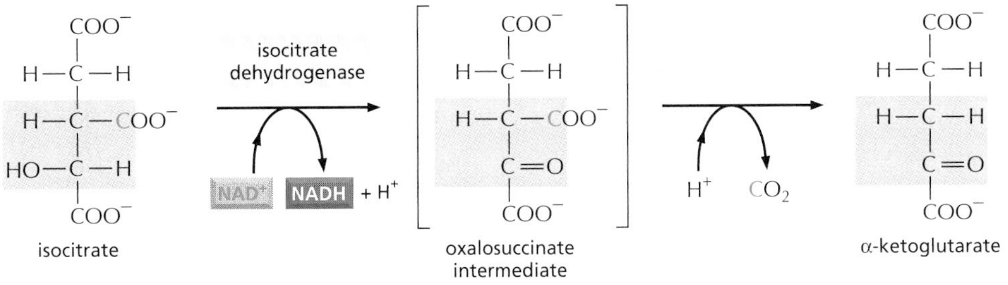

isocitrate → (isocitrate dehydrogenase; NAD^+ → NADH + H^+) → oxalosuccinate intermediate → (H^+; CO_2) → α-ketoglutarate

Step 4 The *α-ketoglutarate dehydrogenase complex* closely resembles the large enzyme complex that converts pyruvate to acetyl CoA (*pyruvate dehydrogenase*). It likewise catalyzes an oxidation that produces NADH, CO_2, and a high-energy thioester bond to coenzyme A (CoA).

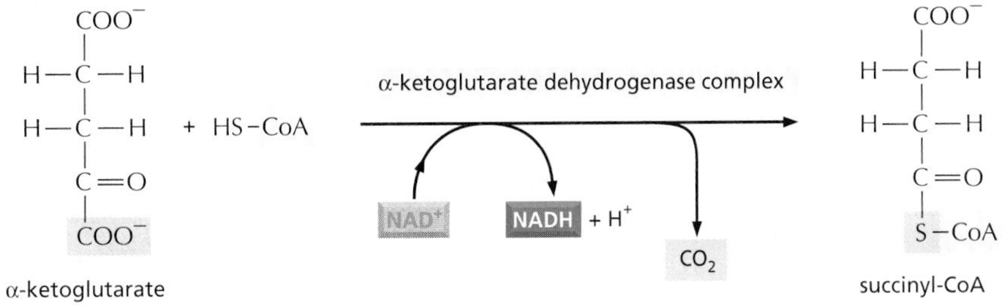

α-ketoglutarate + HS–CoA → (α-ketoglutarate dehydrogenase complex; NAD^+ → NADH + H^+; CO_2) → succinyl-CoA

Step 5 A phosphate molecule from solution displaces the CoA, forming a high-energy phosphate linkage to succinate. This phosphate is then passed to GDP to form GTP. (In bacteria and plants, ATP is formed instead.)

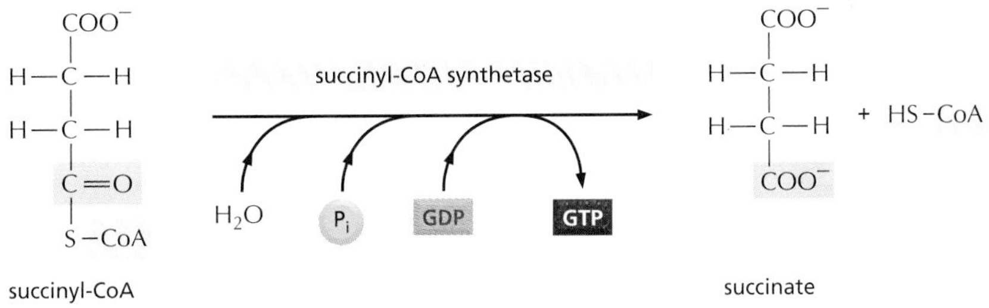

succinyl-CoA → (succinyl-CoA synthetase; H_2O, P_i, GDP → GTP) → succinate + HS–CoA

Step 6 In the third oxidation step in the cycle, FAD removes two hydrogen atoms from succinate.

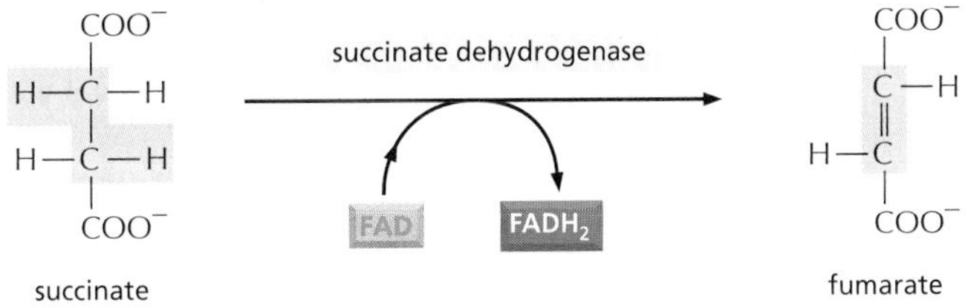

succinate → (succinate dehydrogenase; FAD → $FADH_2$) → fumarate

Step 7 The addition of water to fumarate places a hydroxyl group next to a carbonyl carbon.

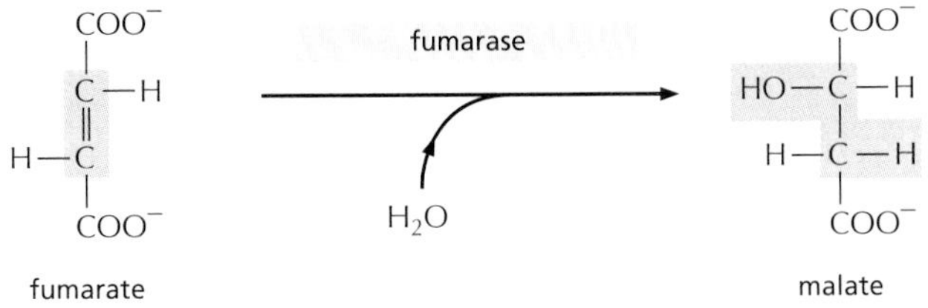

fumarate → (fumarase; H_2O) → malate

Step 8 In the last of four oxidation steps in the cycle, the carbon carrying the hydroxyl group is converted to a carbonyl group, regenerating the oxaloacetate needed for step 1.

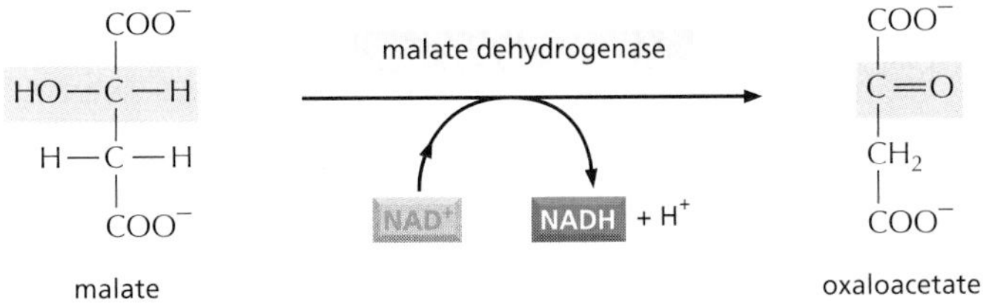

malate → (malate dehydrogenase; NAD^+ → NADH + H^+) → oxaloacetate

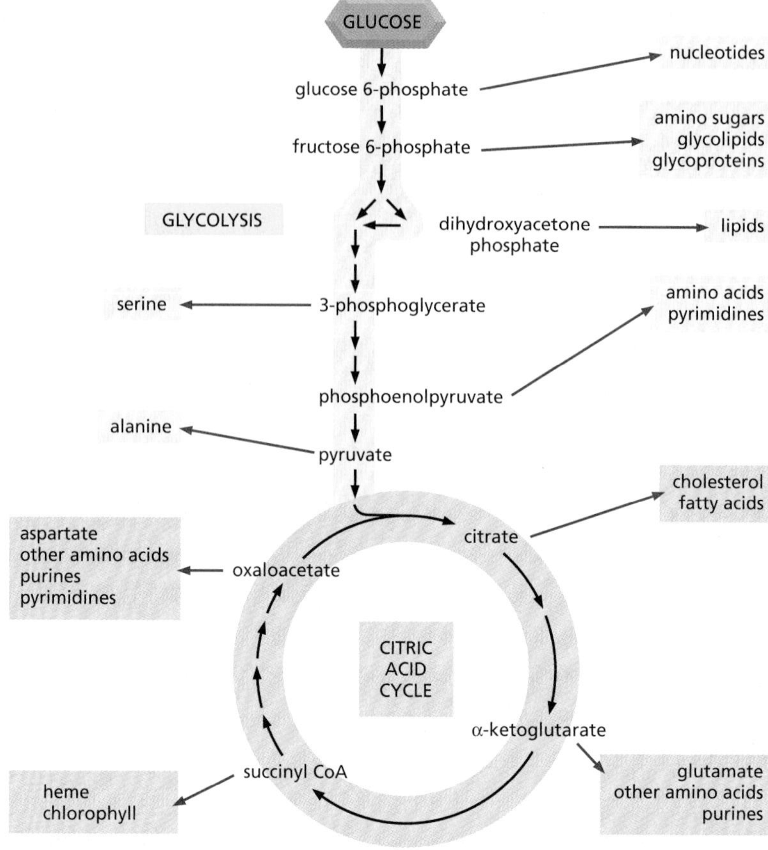

Figure 13–17 **Glycolysis and the citric acid cycle provide the precursors needed to synthesize many important biological molecules.** The amino acids, nucleotides, lipids, sugars, and other molecules—shown here as products—in turn serve as the precursors for the many macromolecules of the cell. Each *black* arrow in this diagram denotes a single enzyme-catalyzed reaction; the *red* arrows generally represent pathways with many steps that are required to produce the indicated products.

the complexity of this process can be gathered from Figure 13–17, which illustrates some of the branches leading from the central catabolic reactions to biosyntheses.

Electron Transport Drives the Synthesis of the Majority of the ATP in Most Cells

We now return to the last stage in the oxidation of a food molecule—the stage in which most of its chemical energy is released. In this final process, the electron carriers NADH and $FADH_2$ transfer the electrons they have gained by oxidizing other molecules to the **electron-transport chain**. This specialized chain of electron carriers is embedded in the inner membrane of the mitochondrion in eucaryotic cells (in the plasma membrane of bacteria). As the electrons pass through the series of electron acceptor and donor molecules that form the chain, they fall to successively lower energy states. The energy released is used to drive H^+ ions (protons) across the membrane, from the inner mitochondrial compartment to the outside. This generates a transmembrane gradient of H^+ ions that serves as a source of energy (like a battery) that can be tapped to drive a variety of energy-requiring reactions (see Chapter 12). In mitochondria, the most prominent of these reactions is the phosphorylation of ADP to generate ATP.

QUESTION 13–4

Looking at the chemistry detailed in Panel 13–2 (pp. 442–443), why do you suppose it is useful to link the acetyl group first to another carbon skeleton, oxaloacetate, before completely oxidizing both carbons to CO_2?

At the end of the transport chain, the electrons are added to molecules of O_2 that have diffused into the mitochondrion; the resulting reduced O_2 molecules simultaneously combine with protons (H^+) from the surrounding solution to produce water. The electrons have now reached their lowest energy level, and all the available energy has been extracted from the food molecule being oxidized. The oxygen-requiring generation of ATP is termed **oxidative phosphorylation** (Figure 13–18). Oxidative phosphorylation occurs in the mitochondria of eucaryotic cells and in the

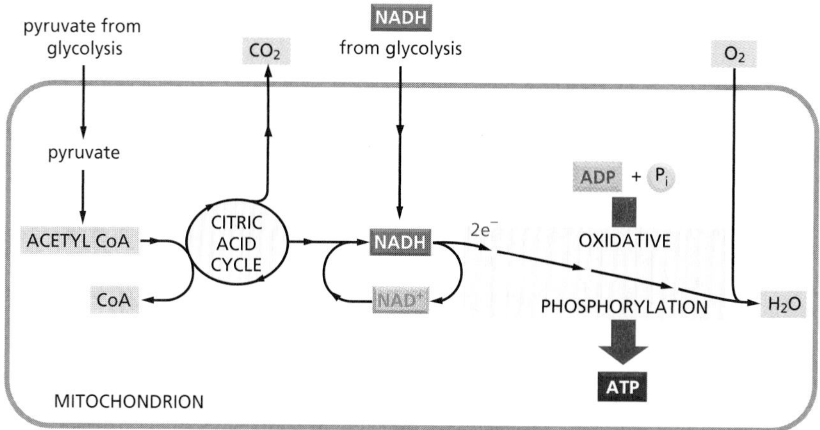

Figure 13–18 **In the final stages of oxidation of food molecules, NADH (and $FADH_2$, not shown) produced by the citric acid cycle donate high-energy electrons that eventually reduce O_2 to water.** A major portion of the energy released during an elaborate series of electron transfers in the mitochondrial inner membrane (or in the plasma membrane of bacteria) is harnessed to drive the synthesis of ATP through the process of oxidative phosphorylation.

plasma membrane of aerobic bacteria, and it is one of the most remarkable achievements of cellular evolution. As such, it will be a central topic of Chapter 14.

In total, the complete oxidation of a molecule of glucose to H_2O and CO_2 produces about 30 molecules of ATP. (For an explanation of exactly where these ATPs come from, see Table 14–1, p. 465.) In contrast, only two molecules of ATP are produced per molecule of glucose by glycolysis alone.

QUESTION 13–5

What, if anything, is wrong with the following statement: "The oxygen consumed during the oxidation of glucose in animal cells is returned as part of CO_2 to the atmosphere." How could you support your answer experimentally?

REGULATION OF METABOLISM

A cell is an intricate chemical machine, and our discussion of metabolism—with a focus on glycolysis and the citric acid cycle—has addressed only a tiny fraction of the many enzymatic reactions that can take place in a cell at any time (Figure 13–19). For all these pathways to work together smoothly, as is required to allow the cell to survive and to respond to its environment, the choice of which pathway each metabolite will follow must be carefully regulated at every branch point.

Many sets of reactions need to be carefully controlled. For example, to maintain order within their cells, all organisms need to constantly replenish their ATP pools through sugar or fat oxidation (see Figure 13–10). Yet animals have only periodic access to food, and plants need to survive overnight without sunlight, when they are unable to produce sugar through photosynthesis. Plants and animals have evolved several ways to get around this problem. One is to synthesize food reserves in times of plenty that can be later consumed when other energy sources are scarce. Thus, a cell must control whether key metabolites will be routed into anabolic or catabolic pathways—in other words, whether they will be commissioned to build other molecules or burned to provide immediate energy. In this section, we discuss how cells regulate the intricate interconnected pathways that form the basis of metabolism.

Catabolic and Anabolic Reactions Are Organized and Regulated

All the reactions shown in Figure 13–19 occur in a cell that is less than 0.1 mm in diameter, and each step requires a different enzyme. To add to the complications, the same molecule is often a part of many different pathways. Pyruvate, for example, is a substrate for half a dozen or more different enzymes, each of which modifies it chemically in a different way. As we have already seen, pyruvate dehydrogenase converts pyruvate to acetyl CoA, and lactate dehydrogenase converts it to lactate; a third enzyme changes pyruvate to oxaloacetate, a fourth to the amino

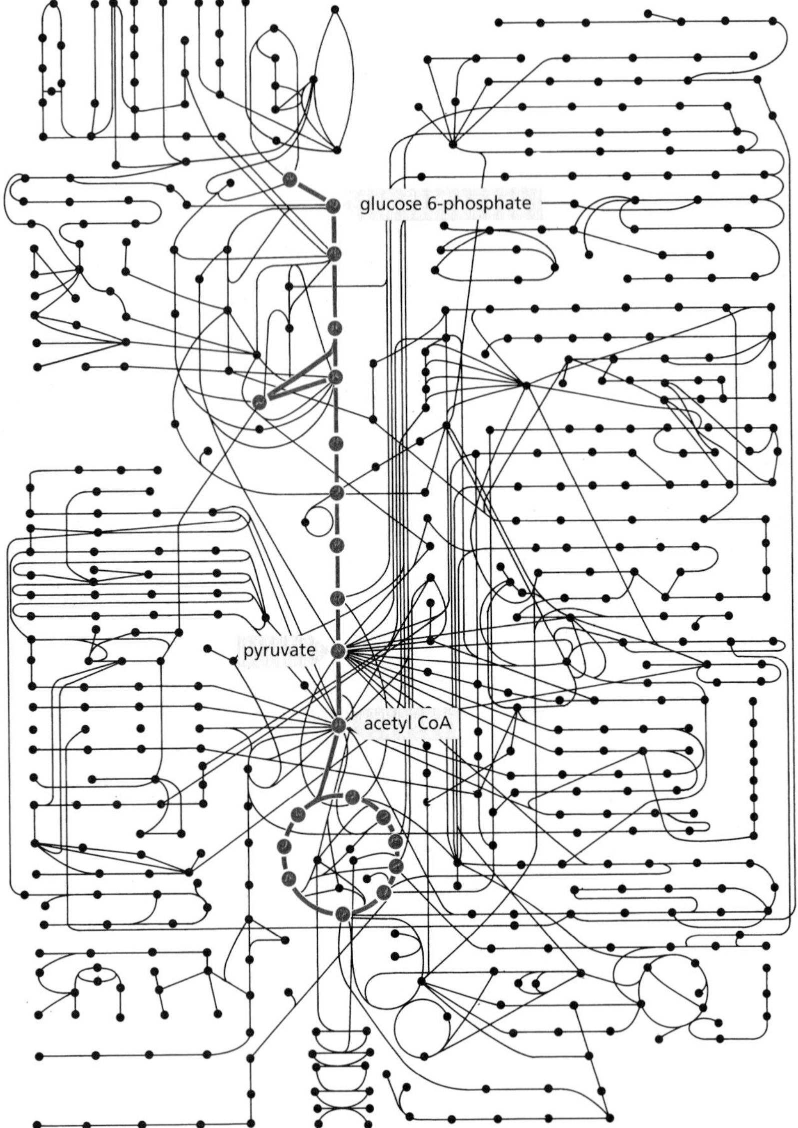

Figure 13–19 **Glycolysis and the citric acid cycle lie at the center of metabolism.** Some 500 metabolic reactions of a typical cell are shown schematically with the reactions of glycolysis and the citric acid cycle in *red*. Other reactions either lead into these two central pathways—delivering small molecules to be oxidized for energy—or they lead outward and thereby supply carbon compounds for biosynthesis. The filled circles represent molecules in various metabolic pathways and the lines that connect them represent the enzymatic reactions that convert one metabolite to another.

acid alanine, and so on. All these different pathways compete for the same pyruvate molecule, and similar competitions for thousands of other small molecules go on at the same time. One might think that the whole system would be so finely balanced that any minor upset, such as a temporary change in food intake, would be disastrous.

In fact, the metabolic balance of a cell is amazingly stable. Whenever the balance is perturbed, the cell reacts so as to restore the initial state: cells can adapt and continue to function during starvation or disease. This resilience is made possible by an elaborate network of *control mechanisms* that act on enzymes to regulate and coordinate the rates of the many metabolic reactions in a cell.

As we saw in Chapter 4, the activity of enzymes can be controlled in many different ways. Many proteins are switched on—or off—by covalent modification, such as the addition or removal of a phosphate group (see Figure 4–38). Alternatively, their activities can be controlled by the binding of a small regulatory molecule—often a metabolite—to an allosteric enzyme (see pp. 151–152). Such regulation can be positive, enhancing the activity of the enzyme, or negative, inhibiting it. As we see next, both types of regulation—positive and negative—control the activity of key enzymes involved in glycolysis.

QUESTION 13–6

A cyclic reaction pathway requires that the starting material be regenerated and available at the end of each cycle. If compounds of the citric acid cycle are siphoned off as building blocks used in a variety of metabolic reactions, why does the citric acid cycle not quickly grind to a halt?

Feedback Regulation Allows Cells to Switch from Glucose Degradation to Glucose Biosynthesis

The body needs a continuous supply of glucose to meet its metabolic needs. For example, brain cells depend almost completely on glucose for respiration. During periods without food and during hard physical exercise, the glucose in the blood is used up faster than it is being replaced from food. One way to replenish blood glucose is to synthesize it from small non-carbohydrate organic molecules such as lactate, pyruvate, or amino acids in a process called **gluconeogenesis**. An intricate pattern of feedback regulation enables cells to switch from breaking down glucose through glycolysis to synthesizing it through gluconeogenesis.

Most of the reactions involved in the breakdown of glucose to pyruvate are readily reversible. However, three of the reactions—steps 1, 3, and 10 in Panel 13–1—are effectively irreversible. In fact, it is the large negative free-energy change that occurs in these reactions that normally drives the breakdown of glucose. For the pathway to go in the opposite direction—to make glucose from pyruvate—these three reactions must be bypassed. This detour is achieved by substituting a set of alternative, enzyme-catalyzed "bypass reactions" that require an input of chemical energy (reactions A, B, C, and D in Figure 13–20). The reactions that synthesize a molecule of glucose in gluconeogenesis thus require the hydrolysis of four ATP and two GTP molecules, compared with the overall generation of two molecules of ATP for each molecule of glucose consumed during glycolysis.

In humans and other mammals, gluconeogenesis occurs mainly in liver cells, which can keep the blood supplied with glucose by using many different molecules as the starting point. One common input is lactate: this molecule, produced by overworked muscle cells, is taken up by the liver where it gets converted back into glucose to replenish depleted muscles. The balance between glycolysis and gluconeogenesis must be highly regulated, so that glucose is broken down rapidly when energy reserves run low, but is synthesized and exported to other tissues when the liver cell has sufficient energy reserves in the form of pyruvate, citrate, or ATP. If both the forward and reverse reactions in Figure 13–20 were allowed to proceed without restraint, they would shuttle metabolites backward and forward in futile cycles that would consume large amounts of energy and generate heat for no purpose.

One of the key control points in the breakdown of glucose lies in step 3 of glycolysis, the production of fructose 1,6-bisphosphate by the enzyme *phosphofructokinase*. This is one of the reactions that must be bypassed in gluconeogenesis (see step 3 in Figure 13–20 or in Panel 13–1, p. 430). Phosphofructokinase is allosterically activated by AMP, ADP, and inorganic phosphate—the byproducts of ATP hydrolysis; it is allosterically inhibited by ATP, citrate, and alternative fuels for respiration, such as fatty acids, which can be liberated from stored fat when glucose is not

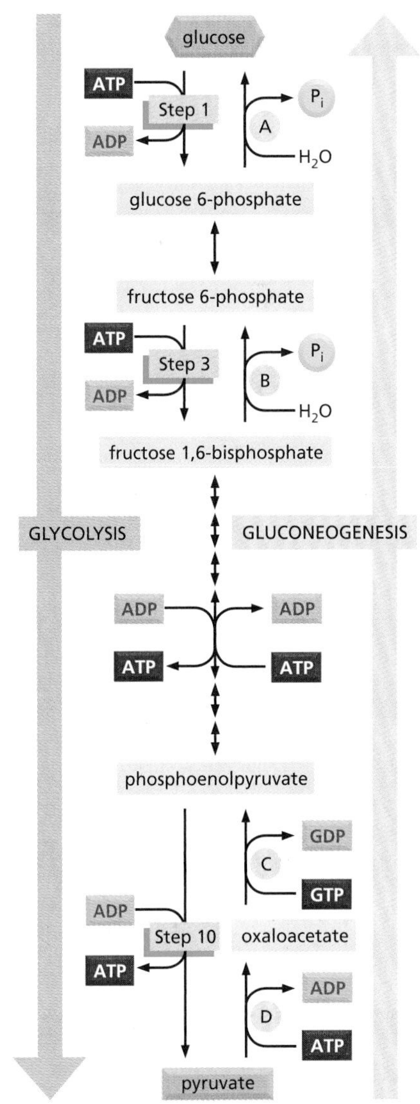

Figure 13–20 **Gluconeogenesis effectively "reverses" reactions that occur during glycolysis.** A set of four bypass reactions (labeled A through D) is needed to get around steps 1, 3, and 10 in glycolysis, which are essentially irreversible. As can be seen, synthetic reactions carried out in gluconeogenesis require an input of energy, whereas glycolysis as a whole is an energetically favorable set of reactions. To keep track of the energy produced or consumed in these processes, recall that during glycolysis, fructose 1, 6-bisphosphate gets cleaved to form two 3-carbon sugars (not shown). Thus, all the reactions that follow, whether they are part of glycolysis or gluconeogenesis, involve two sugars—and twice the number of energy carriers—for each molecule of glucose that is consumed or produced.

available. Thus, when energy reserves are low and the products of ATP hydrolysis accumulate, phosphofructokinase is activated and glycolysis proceeds. On the other hand, when ATP or fuel sources—represented by citrate and fatty acids—are abundant, phosphofructokinase is turned off, favoring gluconeogenesis and, ultimately, the storage of food molecules. To add an additional level of control, the enzyme that catalyzes the reverse reaction (fructose 1, 6–bisphosphatase; see bypass reaction B in Figure 13–20), is regulated by the same molecules—but in the opposite direction. Thus, this enzyme is activated when phosphofructokinase is turned off. Such coordinated regulatory mechanisms allow a cell to respond rapidly to changing environmental conditions and to adjust its metabolism accordingly.

Cells Store Food Molecules in Special Reservoirs to Prepare for Periods of Need

As we have seen, gluconeogenesis is a costly process, requiring substantial amounts of energy from the hydrolysis of ATP and GTP. Thus it cannot operate indefinitely. To compensate for long periods when food is unavailable, animals store food reserves within their cells. Glucose is stored as the subunits of the large, branched polysaccharide **glycogen**, which is present as small granules in the cytoplasm of many cells, mainly liver and muscle (Figure 13–21; see also Panel 2–3, pp. 68–69). The synthesis and degradation of glycogen occur by quite separate metabolic pathways, which can be rapidly and coordinately regulated according to need. When more ATP is needed than can be generated from food molecules taken in from the bloodstream, cells break down glycogen in a reaction that produces *glucose 1-phosphate*, which is then converted to the glucose 6-phosphate that feeds into the glycolytic pathway.

The glycogen synthetic and degradative pathways are coordinated by enzymes in each pathway that are allosterically regulated by glucose 6-phosphate, but in opposite directions: *glycogen synthase* in the synthetic pathway is activated by glucose 6-phosphate, whereas the *glycogen phosphorylase* that catalyzes the breakdown of glycogen is inhibited by both glucose 6-phosphate and ATP. This regulation helps to prevent the breakdown of glycogen when ATP is plentiful and favors its synthesis when glucose 6-phosphate concentration is high. The balance between

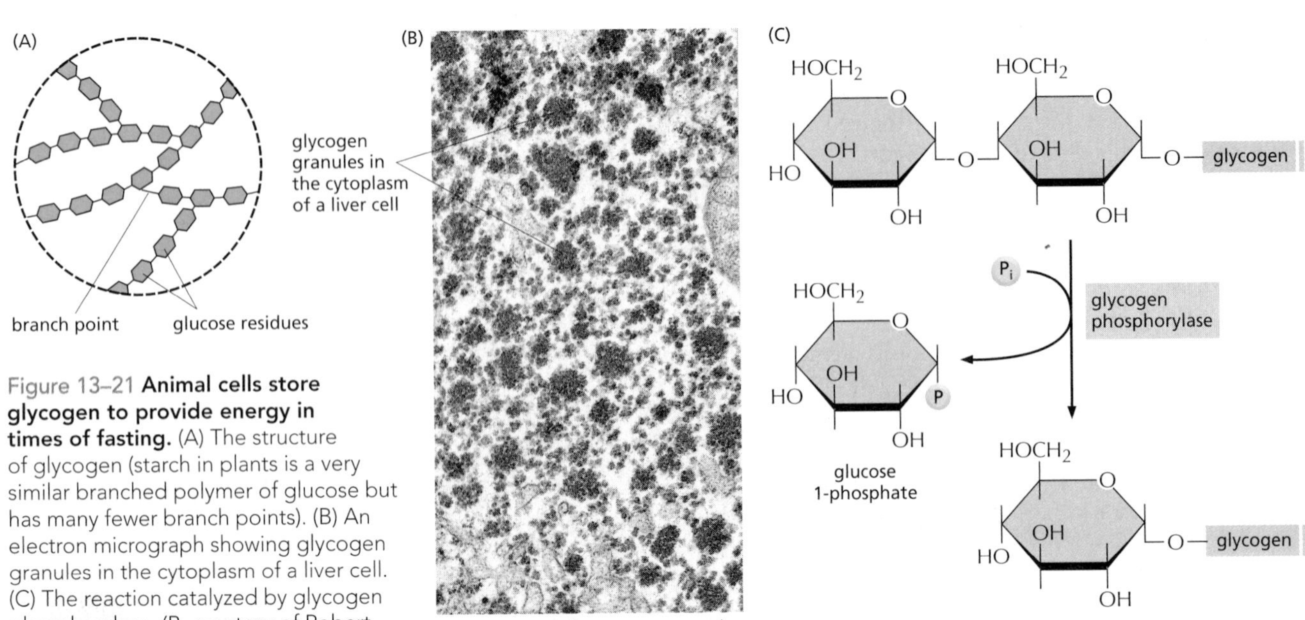

Figure 13–21 **Animal cells store glycogen to provide energy in times of fasting.** (A) The structure of glycogen (starch in plants is a very similar branched polymer of glucose but has many fewer branch points). (B) An electron micrograph showing glycogen granules in the cytoplasm of a liver cell. (C) The reaction catalyzed by glycogen phosphorylase. (B, courtesy of Robert Fletterick and Daniel S. Friend.)

glycogen synthesis and breakdown is also regulated by intracellular signaling pathways that are controlled by the hormones insulin, adrenaline, and glucagon (see Table 16–1, p. 535 and Figure 16–23, p. 550).

Quantitatively, fat is a far more important storage material than glycogen, in part because the oxidation of a gram of fat releases about twice as much energy as the oxidation of a gram of glycogen. Moreover, glycogen binds a great deal of water, producing a sixfold difference in the actual mass of glycogen required to store the same amount of energy as fat. An average adult human stores enough glycogen for only about a day of normal activity, but enough fat to last nearly a month. If our main fuel reserves had to be carried as glycogen instead of fat, body weight would need to be increased by an average of about 60 pounds (nearly 30 kilograms).

Most of our fat is stored as droplets of water-insoluble triacylglycerols in specialized adipose tissue (Figure 13–22 and see Panel 2–4, pp. 70–71). In response to hormonal signals, fatty acids can be released from these depots into the bloodstream for other cells to use as required. Such a need arises after a period of not eating; even a normal overnight fast results in the mobilization of fat. In the morning, most of the acetyl CoA that enters the citric acid cycle is derived from fatty acids rather than from glucose. After a meal, however, most of the acetyl CoA entering the citric acid cycle comes from glucose derived from food, and any excess glucose is used to replenish depleted glycogen stores or to synthesize fats. (Although animal cells can readily convert sugars to fats, they cannot convert fatty acids to sugars.)

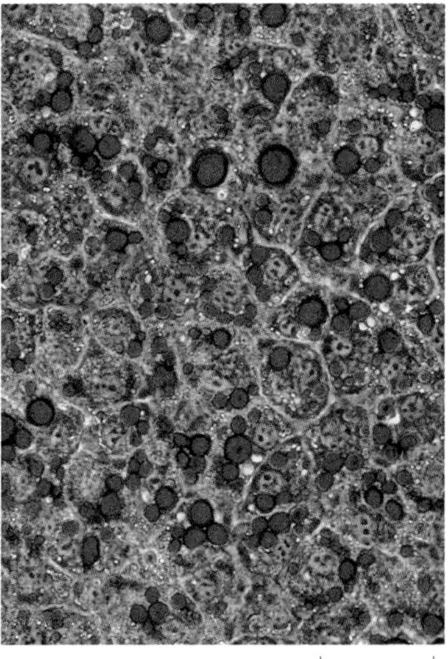

Figure 13–22 **Fats are stored in the form of fat droplets in animal cells.** Fat droplets (stained *red*) beginning to accumulate in developing adipose cells. (Courtesy of Peter Tontonoz and Ronald M. Evans.)

The food reserves in both animals and plants form a vital part of the human diet. Plants convert some of the sugars that they make through photosynthesis during daylight into fats and into **starch**, a branched polymer of glucose very similar to the glycogen of animals. The fats in plants are triacylglycerols, just like the fats in animals, and they differ only in the types of fatty acids that predominate.

The embryos inside plant seeds must live on stored food reserves for a long time, until they germinate to produce leaves that can harvest the energy in sunlight. The embryo uses these stores as sources of energy and of small molecules to build cell walls and synthesize many other biological molecules as it develops. For this reason, the seeds of plants often contain especially large amounts of fats and starch—which make them a major food source for animals, including ourselves (Figure 13–23). Germinating seeds convert the stored fat and starch into glucose as needed.

QUESTION 13–7

After looking at the structures of sugars and fatty acids (discussed in Chapter 2), give an intuitive explanation as to why oxidation of a sugar yields only about half as much energy as the oxidation of an equivalent dry weight of a fatty acid.

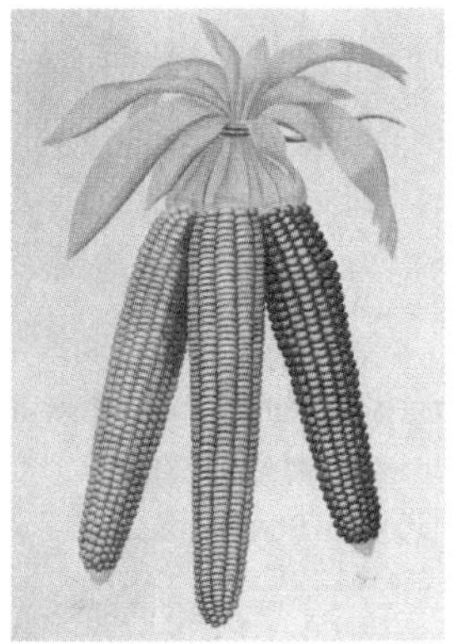

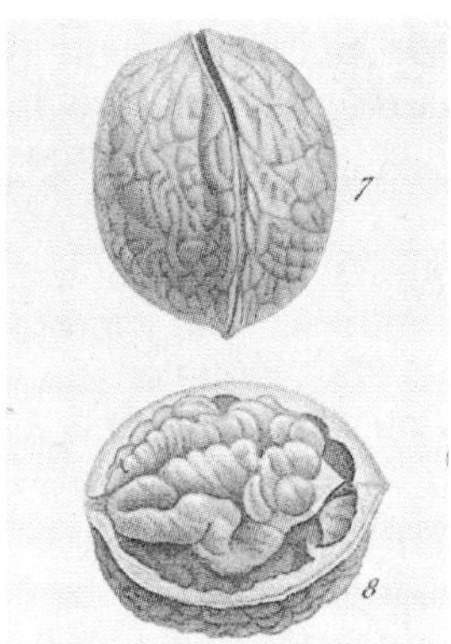

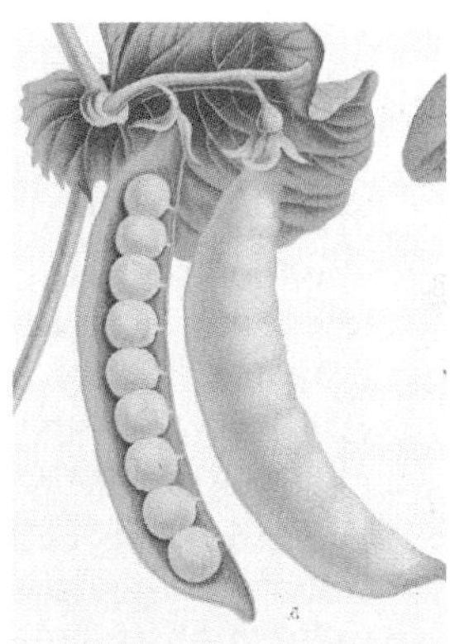

Figure 13–23 **Some plant seeds serve as important foods for humans.** Corn, nuts, and peas all contain rich stores of starch and fats that provide the young plant embryo in the seed with energy and building blocks for biosynthesis. (Courtesy of the John Innes Foundation.)

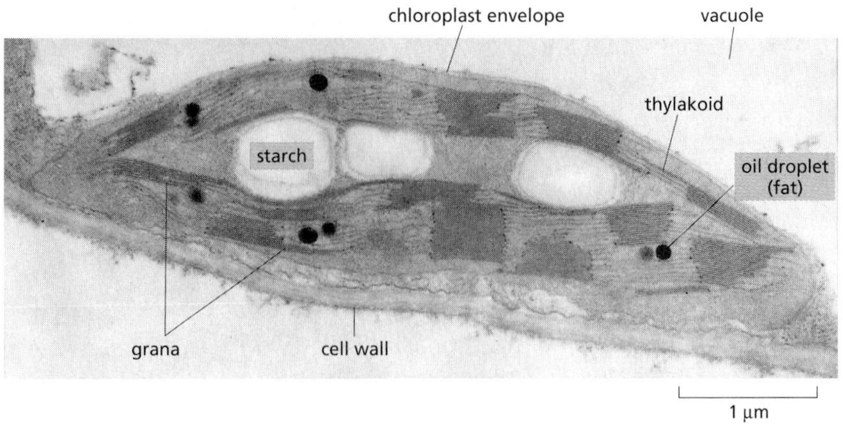

Figure 13–24 **Plant cells store both starch and fats in their chloroplasts.** A thin section of a single chloroplast from a plant cell, showing the starch granules and lipid droplets that have accumulated as a result of the biosyntheses occurring there. (Courtesy of K. Plaskitt.)

In plants, fats and starch are both stored in the chloroplast—a specialized organelle that carries out photosynthesis in plant cells (Figure 13–24). There they serve as food reservoirs that are mobilized to produce ATP during periods of darkness.

ESSENTIAL CONCEPTS

- Glucose and other food molecules are broken down by controlled stepwise oxidation to provide useful chemical energy in the form of the activated carriers ATP and NADH.
- Sugars derived from food are broken down by distinct sets of reactions: glycolysis (which occurs in the cytosol), the citric acid cycle (in the mitochondrial matrix), and oxidative phosphorylation (in the inner mitochondrial membrane).
- The reactions of glycolysis degrade the six-carbon sugar glucose to two molecules of the three-carbon sugar pyruvate, producing a relatively small amount of ATP and NADH.
- In the presence of oxygen, pyruvate is converted to acetyl CoA plus CO_2. The citric acid cycle then converts the acetyl group in acetyl CoA to CO_2 and H_2O. Much of the energy released in these oxidation reactions is stored as high-energy electrons in the activated carriers NADH and $FADH_2$. In eucaryotic cells, all these reactions occur in mitochondria.
- The other major energy source in foods is fat. The fatty acids produced from the digestion of fats are imported into mitochondria and converted to acetyl CoA molecules. These acetyl CoA molecules are then further oxidized through the citric acid cycle, producing NADH and $FADH_2$, just like the acetyl CoA derived from pyruvate.
- NADH and $FADH_2$ pass their high-energy electrons to an electron-transport chain in the inner mitochondrial membrane, where a series of electron transfers is used to drive the formation of ATP. Most of the energy captured during the breakdown of food molecules is harvested during this process of oxidative phosphorylation (described in detail in Chapter 14).
- The food we eat is not only a source of metabolic energy but also of raw materials for biosynthesis. Many intermediates of glycolysis and the citric acid cycle are starting points for pathways that lead to the synthesis of proteins, nucleic acids, and the many other specialized molecules of the cell.
- The thousands of different reactions carried out simultaneously by a cell are closely coordinated, enabling the cell to adapt and continue to function under a wide range of external conditions.

- During periods when food is scarce, regulation of the activities of a few key enzymes allows the cell to switch from glucose breakdown to glucose biosynthesis (gluconeogenesis).
- Cells store food molecules in special reserves. Glucose subunits are stored as glycogen in animals and as starch in plants; both animals and plants store fatty acids as fats. The food reserves stored by plants are major sources of food for animals, including humans.

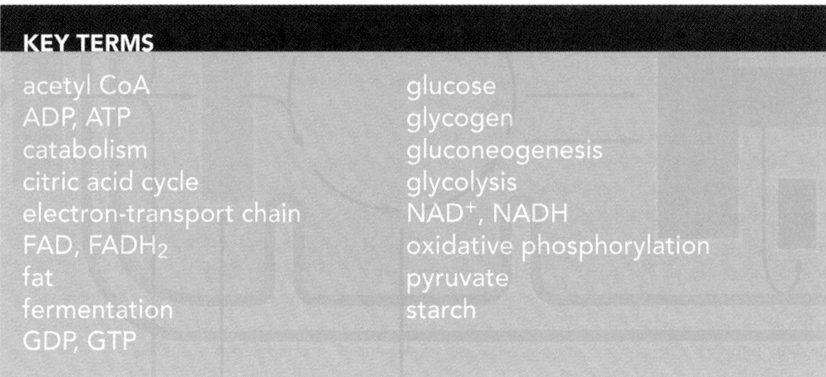

KEY TERMS
acetyl CoA
ADP, ATP
catabolism
citric acid cycle
electron-transport chain
FAD, $FADH_2$
fat
fermentation
GDP, GTP
glucose
glycogen
gluconeogenesis
glycolysis
NAD^+, NADH
oxidative phosphorylation
pyruvate
starch

QUESTIONS

QUESTION 13–8

The oxidation of sugar molecules by the cell takes place according to the general reaction $C_6H_{12}O_6$ (glucose) + $6O_2 \rightarrow 6CO_2 + 6H_2O$ + energy. Which of the following statements are correct? Explain your answers.

A. All of the energy produced is in the form of heat.

B. None of the produced energy is in the form of heat.

C. The energy is produced by a process that involves the oxidation of carbon atoms.

D. The reaction supplies the cell with essential water.

E. In cells the reaction takes place in more than one step.

F. Many steps in the oxidation of sugar molecules involve reaction with oxygen gas.

G. Some organisms carry out the reverse reaction.

H. Some cells that grow in the absence of O_2 produce CO_2.

QUESTION 13–9

An exceedingly sensitive instrument (yet to be devised) shows that one of the carbon atoms in Charles Darwin's last breath is resident in your bloodstream, where it forms part of a hemoglobin molecule. Suggest how this carbon atom might have traveled from Darwin to you, and list some of the molecules it could have entered en route.

QUESTION 13–10

Yeast cells can grow both in the presence of O_2 (aerobically) and in its absence (anaerobically). Under which of the two conditions could you expect the cells to grow better? Explain your answer.

QUESTION 13–11

During movement, muscle cells require large amounts of ATP to fuel their contractile apparatus. These cells contain high levels of creatine phosphate (shown in Figure 13–7). Why is this a useful compound to store energy? Justify your answer with the information shown in Figure 13–7.

QUESTION 13–12

Identical pathways that make up the complicated sequence of reactions of glycolysis, shown in Panel 13–1 (pp. 430–431), are found in most living cells, from bacteria to humans. One could envision, however, countless alternative chemical reaction mechanisms that would allow the oxidation of sugar molecules and that could, in principle, have evolved to take the place of glycolysis. Discuss this fact in the context of evolution.

QUESTION 13–13

Assume that an animal cell is a cube that has a side length of 10 μm. The cell contains 10^9 ATP molecules that it uses up every minute. ATP is regenerated by oxidizing glucose molecules. After what amount of time will the cell have used up an amount of oxygen gas that is equal to its own volume? (Recall that one mole contains 6×10^{23} molecules. One mole of a gas has a volume of 22.4 liters.)

QUESTION 13–14

Under the conditions existing in the cell, the free energies of the first few reactions in glycolysis (in Panel 13–1, pp. 430–431) are:

step 1 ΔG = –8.0 kcal/mole (–33.4 kJ/mole)

step 2 ΔG = –0.6 kcal/mole (–2.5 kJ/mole)

step 3 ΔG = –5.3 kcal/mole (–22.2 kJ/mole)

step 4 ΔG = –0.3 kcal/mole (–1.3 kJ/mole)

Are these reactions energetically favorable? Using these values, draw to scale an energy diagram (A) for the overall reaction and (B) for the pathway composed of the four individual reactions.

QUESTION 13–15

The chemistry of most metabolic reactions was deciphered by synthesizing metabolites containing atoms that are different isotopes from those occurring naturally. The products of reactions starting with isotopically labeled metabolites can be analyzed to determine precisely which atoms in the products are derived from which atoms in the starting material. The methods of detection exploit, for example, the fact that different isotopes have different masses that can be distinguished using biophysical techniques such as mass spectrometry. Moreover, some isotopes are radioactive and can therefore be readily recognized with electronic counters or photographic film that becomes exposed by radiation.

A. Assume that pyruvate containing radioactive ^{14}C in its carboxyl group is added to a cell extract that can support oxidative phosphorylation. Which of the molecules produced should contain the vast majority of the ^{14}C that was added?

B. Assume that oxaloacetate containing radioactive ^{14}C in its keto group (refer to Panel 13–2, pp. 442–443) is added to the extract. Where should the ^{14}C atom be located after precisely one turn of the cycle?

QUESTION 13–16

In cells that can grow both aerobically and anaerobically, fermentation is inhibited in the presence of O_2. Suggest a reason for this observation.

CHAPTER **FOURTEEN** 14

Energy Generation in Mitochondria and Chloroplasts

MITOCHONDRIA AND OXIDATIVE PHOSPHORYLATION

MOLECULAR MECHANISMS OF ELECTRON TRANSPORT AND PROTON PUMPING

CHLOROPLASTS AND PHOTOSYNTHESIS

THE ORIGINS OF CHLOROPLASTS AND MITOCHONDRIA

The fundamental need to generate energy efficiently has had a profound influence on the history of life on Earth. Much of the structure, function, and evolution of cells and organisms can be related to their need for energy. The earliest cells may have produced ATP by breaking down organic molecules, left by earlier geochemical processes, using some form of fermentation. Fermentation reactions occur in the cytosol of present-day cells. As discussed in Chapter 13, these reactions use the energy derived from the partial oxidation of energy-rich food molecules to form ATP, the chemical energy currency of cells.

But very early in the history of life, a much more efficient method for generating energy and synthesizing ATP appeared. This process is based on the transport of electrons along membranes. Billions of years later, it is so central to the survival of life on Earth that we devote this entire chapter to it. As we shall see, this membrane-based mechanism is used by cells to acquire energy from a wide variety of sources: for example, it is central to the conversion of light energy into chemical-bond energy in photosynthesis, and to the aerobic respiration that enables us to use oxygen to produce large amounts of ATP from food molecules. The mechanism we will describe first appeared in bacteria more than 3 billion years ago. The descendants of these pioneering cells crowd every corner and crevice of the land and the oceans with a wild menagerie of living forms, and they survive within eucaryotic cells in the form of chloroplasts and mitochondria.

Where we come from and how we are related to other living things are puzzles that have fascinated humans since the beginning of recorded time. The story that we can tell now, worked out through a long chain of scientific investigation, is one of the most dramatic and exciting histories ever told. And we are not yet done. Each year, further discoveries in cell biology enable us to add more details through molecular detective work of dramatically increasing power.

Absolutely central to life's progression was the ability to provide an abundant source of energy for cells. In this chapter, we discuss the remarkable mechanism that made this possible.

Cells Obtain Most of Their Energy by a Membrane-based Mechanism

The main chemical energy currency in cells is ATP (see Figure 3–32). In eucaryotic cells, small amounts of ATP are generated during glycolysis in the cytosol, but most ATP is produced by oxidative phosphorylation in mitochondria (as outlined in Chapter 13). The mechanism by which the bulk of ATP is generated in the mitochondria differs from the way in which ATP is produced by glycolysis in that it involves a membrane: oxidative phosphorylation depends on electron transport within the mitochondrial membrane and the transport of ions across it. The same type of ATP-generating process occurs in the plasma membrane of bacteria. The membrane-based mechanism for making ATP arose very early in life's history and was so successful that its essential features have been retained in the long evolutionary journey from early procaryotes to modern cells. In photosynthetic bacteria, plants, and algae, a related membrane-based process produces ATP during photosynthesis.

The membrane-based process for making ATP consists of two linked stages; both are carried out by protein complexes in the membrane.

Stage 1. Electrons derived from the oxidation of food molecules (as discussed in Chapter 13) or from other sources (discussed later) are transferred along a series of electron carriers—called an **electron-transport chain**—embedded in the membrane. These electron transfers release energy that is used to pump protons (H^+), derived from the water that is ubiquitous in cells, across the membrane and thus generate an electrochemical proton gradient (Figure 14–1A). An ion gradient across a membrane is a form of stored energy that can be harnessed to do useful work when the ions are allowed to flow back across the membrane down their gradient (as discussed in Chapter 12).

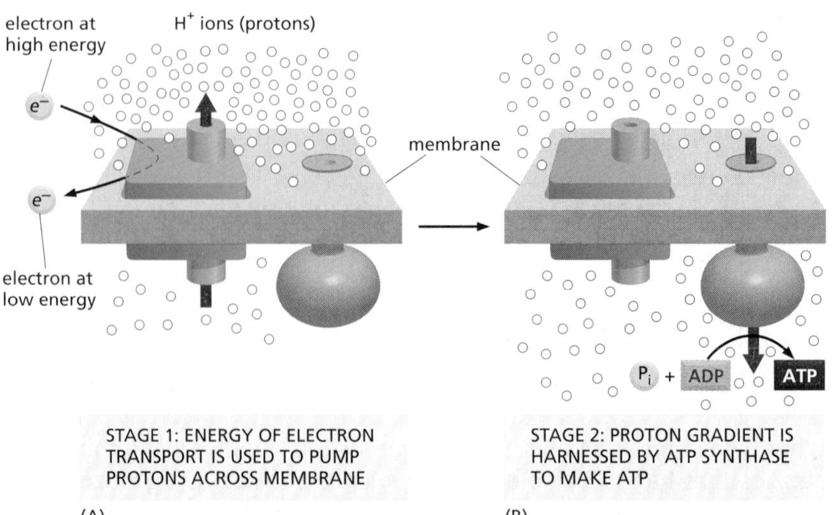

Figure 14–1 **Cells have evolved systems for harnessing the energy required for life.** (A) The essential requirements for this process are a membrane, in which are embedded a pump protein and an ATP synthase, and sources of high-energy electrons (e^-) and of protons (H^+). The pump harnesses the energy of electron transfer (details not shown here) to pump protons derived from water, creating a proton gradient across the membrane. The high-energy electrons can be provided by organic or inorganic molecules, or they can be produced by the action of light on special molecules such as chlorophyll. (B) The gradient produced in (A) serves as a versatile energy store. It is used to drive a variety of energy-requiring reactions in mitochondria, chloroplasts, and bacteria—including the synthesis of ATP by the ATP synthase. The *red arrow* shows the direction of proton movement at each stage.

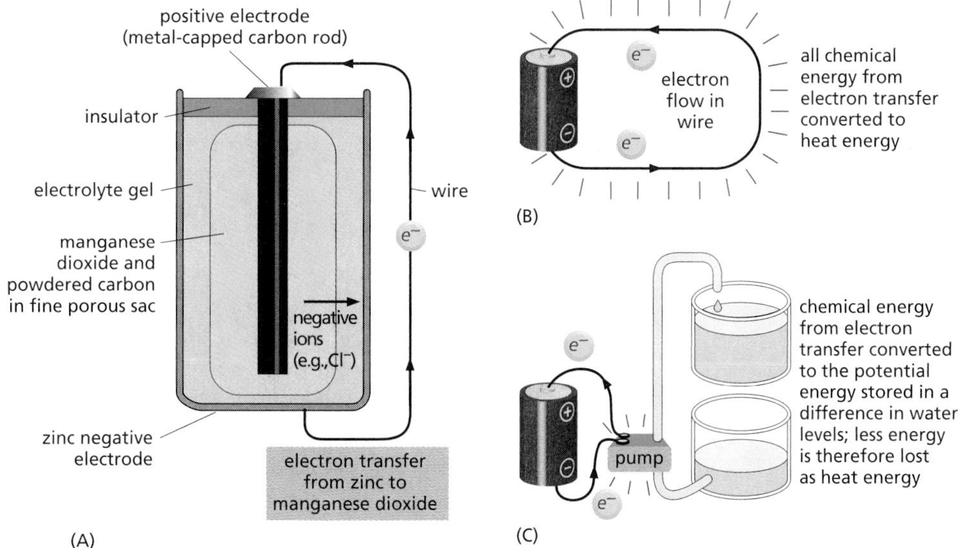

Figure 14–2 **Batteries are powered by chemical reactions based on electron transfers.** (A) When a standard flashlight battery is connected into a circuit, electrons flow from the metal container, which is made of zinc (Zn), to the manganese atom in manganese dioxide (MnO_2). Zn^{2+} and manganous oxide (MnO) are formed as products. (The carbon in the battery simply serves to conduct electrons.) (B) If the battery terminals are directly connected to each other, the energy released by electron transfer is all converted into heat. (C) If the battery is connected to a pump, much of the energy released by electron transfers can be harnessed to do work instead (in this case, to pump water). Cells can similarly harness the energy of electron transfer to a pumping mechanism, as illustrated in Figure 14–1.

Stage 2. H^+ flows back down its electrochemical gradient through a protein complex called *ATP synthase*, which catalyzes the energy-requiring synthesis of ATP from ADP and inorganic phosphate (P_i). This ubiquitous enzyme serves the role of a turbine, permitting the proton gradient to drive the production of ATP (Figure 14–1B).

The linkage of electron transport, proton pumping, and ATP synthesis was called the *chemiosmotic hypothesis* when it was first proposed in the 1960s, because of the link between the chemical bond-forming reactions that synthesize ATP ("chemi-") and the membrane transport processes ("osmotic," from the Greek *osmos*, "to push"). It is now known as **chemiosmotic coupling**. Chemiosmotic mechanisms allow cells to harness the energy of electron transfers in much the same way that the energy stored in a battery can be harnessed to do useful work (Figure 14–2).

Chemiosmotic coupling first evolved in bacteria. Aerobic eucaryotic cells appear to have adopted the bacterial chemiosmotic mechanisms intact, first by engulfing aerobic bacteria to form mitochondria, and somewhat later—in the lineages leading to algae and plants—by engulfing cyanobacteria to form chloroplasts, as described in Chapter 1 (see Figures 1–19 and 1–21).

In this chapter we shall consider energy generation in both mitochondria and chloroplasts, emphasizing the common principles by which proton gradients are created and used in these organelles and in the plasma membranes of bacteria. We start by describing the structure and function of mitochondria, looking in detail at the events that occur in the mitochondrial membrane to create the proton gradient and generate ATP. We next consider photosynthesis in the chloroplasts of plant cells. Finally, we trace the evolutionary pathways that gave rise to these mechanisms of energy generation. By examining the lifestyles of a variety of single-celled organisms—including those that might resemble our early ancestors—we can begin to see the role that chemiosmotic coupling has played in the rise of complex eucaryotes and in the development of all life on Earth.

QUESTION 14–1

Dinitrophenol (DNP) is a small molecule that renders membranes permeable to protons. In the 1940s, small amounts of this highly toxic compound were given to patients to induce weight loss. DNP was effective in melting away the pounds, especially promoting the loss of fat reserves. Can you explain how it might cause such loss? As an unpleasant side reaction, however, patients had an elevated temperature and sweated profusely during the treatment. Provide an explanation for these symptoms.

MITOCHONDRIA AND OXIDATIVE PHOSPHORYLATION

Mitochondria are present in nearly all eucaryotic cells—in plants, animals, and most eucaryotic microorganisms—and most of a cell's ATP is produced in these organelles. Without them, present-day eucaryotes would be dependent on the relatively inefficient process of glycolysis for all of their ATP production, and it seems unlikely that complex multicellular organisms could have been supported in this way. When glucose is converted to pyruvate by glycolysis, only two molecules of ATP are produced per glucose molecule (less than 10% of the total free energy potentially available). By contrast, in mitochondria, the metabolism of sugars is completed, and the energy released is harnessed so efficiently that about 30 molecules of ATP are produced for each molecule of glucose oxidized.

Defects in mitochondrial function can have serious repercussions for an organism. Consider, for example, an inherited disorder called *myoclonic epilepsy and ragged red fiber disease (MERRF)*. This disease, caused by a mutation in one of the mitochondrial transfer RNA (tRNA) genes, is characterized by a decrease in synthesis of the mitochondrial proteins required for electron transport and ATP production. As a result, patients with this disorder typically experience muscle weakness or heart problems (from effects on cardiac muscle) and epilepsy or dementia (from effects on nerve cells). Muscle and nerve cells suffer most when mitochondria are defective, because they need especially large amounts of ATP to function well.

The same metabolic reactions that occur in mitochondria also take place in aerobic bacteria, which do not possess these organelles; in these organisms the plasma membrane carries out the chemiosmotic coupling. Of course, a bacterial cell also has to carry out many other functions; the mitochondrion, by contrast, has become highly specialized for energy generation.

A Mitochondrion Contains an Outer Membrane, an Inner Membrane, and Two Internal Compartments

Mitochondria are generally similar in size and shape to bacteria, although these attributes can vary depending on the cell type. They contain their own DNA and RNA, and a complete transcription and translation system including ribosomes, which allows them to synthesize some of their own proteins. Time-lapse movies of living cells reveal mitochondria as remarkably mobile organelles, constantly changing shape and position. Present in large numbers—1000 to 2000 in a liver cell, for example—these organelles can form long, moving chains in association with the microtubules of the cytoskeleton (discussed in Chapter 17). In other cells, they remain fixed in one location to target ATP directly to a site of unusually high ATP consumption. In a heart muscle cell, for example, mitochondria are located close to the contractile apparatus, whereas in a sperm they are wrapped tightly around the motile flagellum (Figure 14–3). The number of mitochondria present in different cell types varies dramatically, and can change with the energy needs of the cell. In skeletal muscle cells, for example, the number of mitochondria may increase five- to tenfold, due to mitochondrial growth and division that occur if the muscle has been repeatedly stimulated to contract.

An individual mitochondrion is bounded by two highly specialized membranes—one surrounding the other—that play a crucial part in its activities. The outer and inner mitochondrial membranes create two mito-

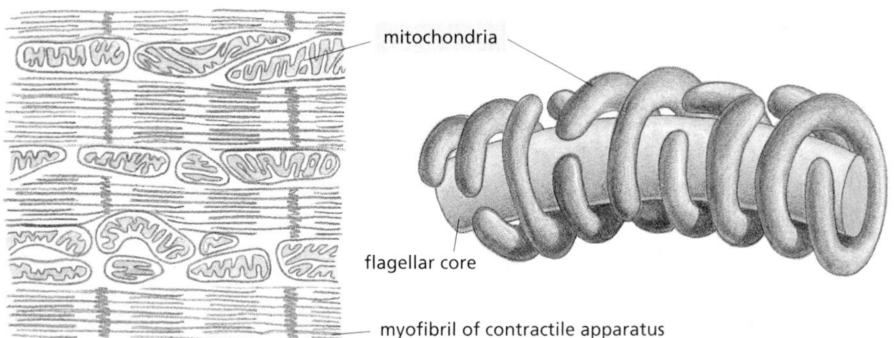

Figure 14–3 **Mitochondria are located near sites of high ATP utilization.** (A) In a cardiac muscle cell, mitochondria are located close to the contractile apparatus, in which ATP hydrolysis provides the energy for contraction. (B) In a sperm, mitochondria are located in the tail, wrapped around a portion of the motile flagellum that requires ATP for its movement.

chondrial compartments: a large internal space called the **matrix** and the much narrower *intermembrane space* (Figure 14–4). If purified mitochondria are gently processed and fractionated into separate components by differential centrifugation (see Panel 4–4, pp. 164–165), the biochemical composition of each of the two membranes and of the spaces enclosed by them can be determined. Each contains a unique collection of proteins.

The *outer membrane* contains many molecules of a transport protein called porin, which, as described in Chapter 11, forms wide aqueous channels through the lipid bilayer. As a result, the outer membrane is like a sieve that is permeable to all molecules of 5000 daltons or less, including small proteins. This makes the intermembrane space chemically equivalent to the cytosol with respect to the small molecules it contains. In contrast, the *inner membrane,* like other membranes in the cell, is impermeable to the passage of ions and most small molecules, except where a path is provided by membrane transport proteins. The mitochondrial matrix therefore contains only molecules that can be selectively transported into the matrix across the inner membrane, and its contents are highly specialized.

The inner mitochondrial membrane is the site of electron transport and proton pumping, and it contains the ATP synthase. Most of the proteins embedded in the inner mitochondrial membrane are components of the electron-transport chains required for oxidative phosphorylation. This

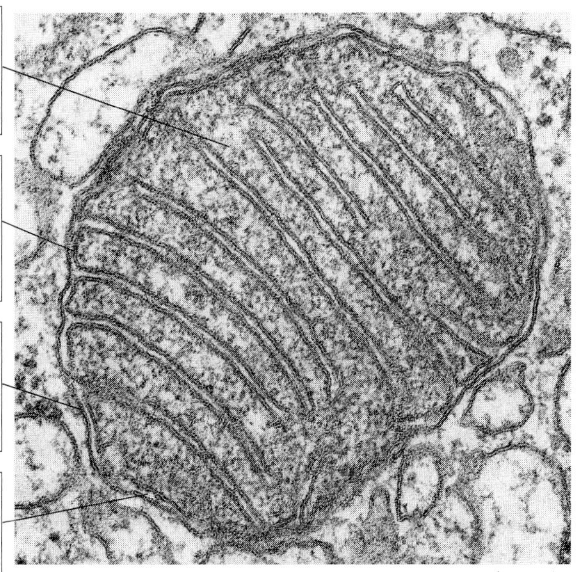

Matrix. This space contains a highly concentrated mixture of hundreds of enzymes, including those required for the oxidation of pyruvate and fatty acids and for the citric acid cycle.
Inner membrane. Folded into numerous cristae, the inner membrane contains proteins that carry out the oxidation reactions of the electron-transport chain and the ATP synthase that makes ATP in the matrix.
Outer membrane. Because it contains a large channel-forming protein (called porin), the outer membrane is permeable to all molecules of 5000 daltons or less.
Intermembrane space. This space contains several enzymes that use the ATP passing out of the matrix to phosphorylate other nucleotides.
100 nm

Figure 14–4 **A mitochondrion is organized into four separate compartments.** Each compartment contains a unique set of proteins enabling it to perform its distinct functions. In liver mitochondria, an estimated 67% of the total mitochondrial protein is located in the matrix, 21% in the inner membrane, 6% in the outer membrane, and 6% in the intermembrane space. (Courtesy of Daniel S. Friend.)

membrane also contains a variety of transport proteins that allow the entry of selected small molecules, such as pyruvate and fatty acids, into the matrix.

The inner membrane is usually highly convoluted, forming a series of infoldings, known as *cristae*, that project into the matrix space to greatly increase the surface area of the inner membrane (see Figure 14–4). These folds provide a large surface on which ATP synthesis can take place; in a liver cell, for example, the inner mitochondrial membranes of all the mitochondria constitute about one-third of the total membranes of the cell. And the number of cristae is three times greater in a mitochondrion of a cardiac muscle cell than in a mitochondrion of a liver cell.

QUESTION 14–2

Electron micrographs show that mitochondria in heart muscle have a much higher density of cristae than mitochondria in skin cells. Suggest an explanation for this observation.

The Citric Acid Cycle Generates High-Energy Electrons

Mitochondria use both pyruvate and fatty acids as fuel, the pyruvate coming mainly from glucose and other sugars, and the fatty acids from fats. These fuel molecules are transported across the inner mitochondrial membrane and then converted to the crucial metabolic intermediate acetyl CoA by enzymes located in the mitochondrial matrix (see Figure 13–10). The acetyl groups in acetyl CoA are then oxidized in the matrix via the citric acid cycle (see Panel 13–2, pp. 442–443). The cycle converts the carbon atoms in acetyl CoA to CO_2, which is released from the cell as a waste product. In addition, the cycle generates high-energy electrons, carried by the activated carrier molecules NADH and $FADH_2$ (Figure 14–5).

Although the citric acid cycle is considered to be part of aerobic metabolism, it does not itself use molecular oxygen (O_2). Oxygen is directly consumed only in the final catabolic reactions that take place on the inner mitochondrial membrane, as we see next.

A Chemiosmotic Process Converts the Energy From Activated Carrier Molecules into ATP

Nearly all the energy available from burning carbohydrates, fats, and other foodstuffs in the earlier stages of their oxidation is initially saved in the form of the activated carrier molecules generated during glycolysis and the citric acid cycle—NADH and $FADH_2$. These carrier molecules donate their high-energy electrons to the electron-transport chain in the mitochondrial membrane, and thus become oxidized to NAD^+ and FAD. The electrons are quickly passed along the chain to molecular oxygen (O_2) to form water (H_2O). The passage of the high-energy electrons along the electron-transport chain releases energy that is harnessed to pump protons across the inner mitochondrial membrane (Figure 14–6).

Figure 14–5 **NADH donates its electrons to the electron-transport chain.** In this drawing, the high-energy electrons are shown as two *red dots* on a *yellow* hydrogen atom. A hydride ion (a hydrogen atom with an extra electron) is removed from NADH and is converted into a proton and two high-energy electrons. Only the ring that carries the electrons in a high-energy linkage is shown; for the complete structure and the conversion of NAD^+ back to NADH, see the structure of the closely related NADPH in Figure 3–34. Electrons are also carried in a similar way by $FADH_2$, whose structure is shown in Figure 13–12B.

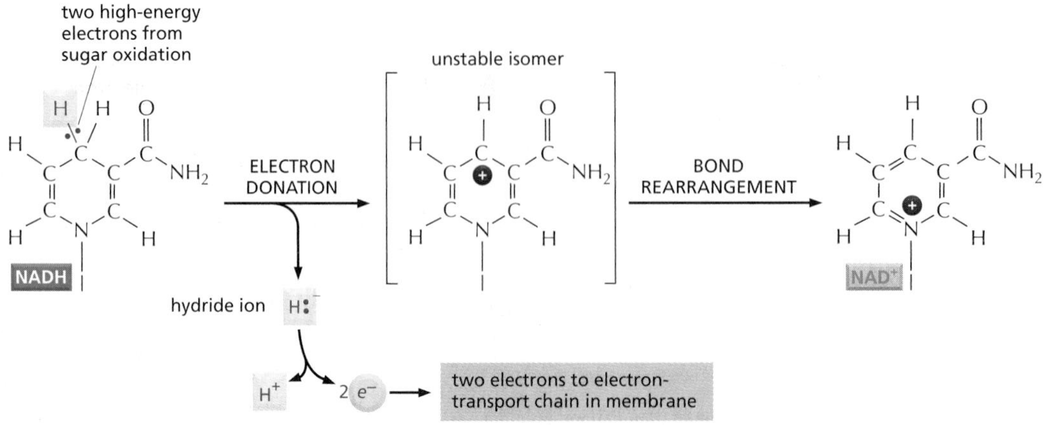

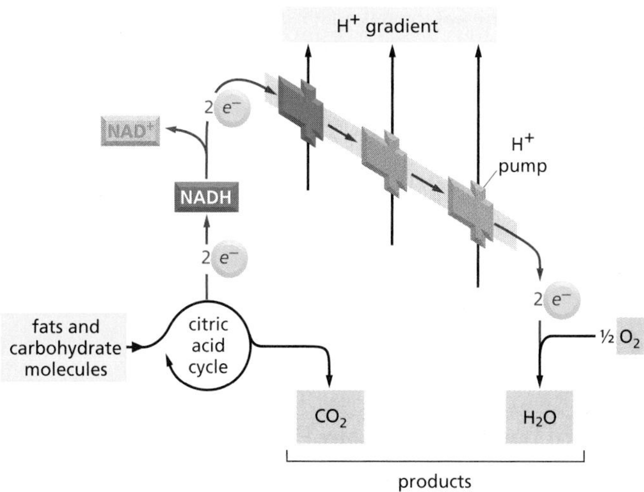

Figure 14–6 **Protons are pumped across the inner mitochondrial membrane.** Only stage 1 of chemiosmotic coupling is shown (see Figure 14–1). The path of electron flow is indicated by *red arrows*.

The resulting proton gradient in turn drives the synthesis of ATP. The full sequence of reactions is shown in Figure 14–7. The inner mitochondrial membrane thus serves as a device that converts the energy contained in the high-energy electrons of NADH into the high-energy phosphate bond of ATP (Figure 14–8). This chemiosmotic mechanism of ATP synthesis is called **oxidative phosphorylation**, because it involves both the consumption of O_2 and the addition of a phosphate group to ADP to form ATP.

Although chemiosmotic coupling escaped detection for many years, the vast majority of living organisms use this mechanism to generate ATP. The source of the electrons that power the proton pumping differs widely between different organisms and different processes. In aerobic respiration in mitochondria and aerobic bacteria, the electrons are ultimately derived from glucose or fatty acids. In photosynthesis, the required

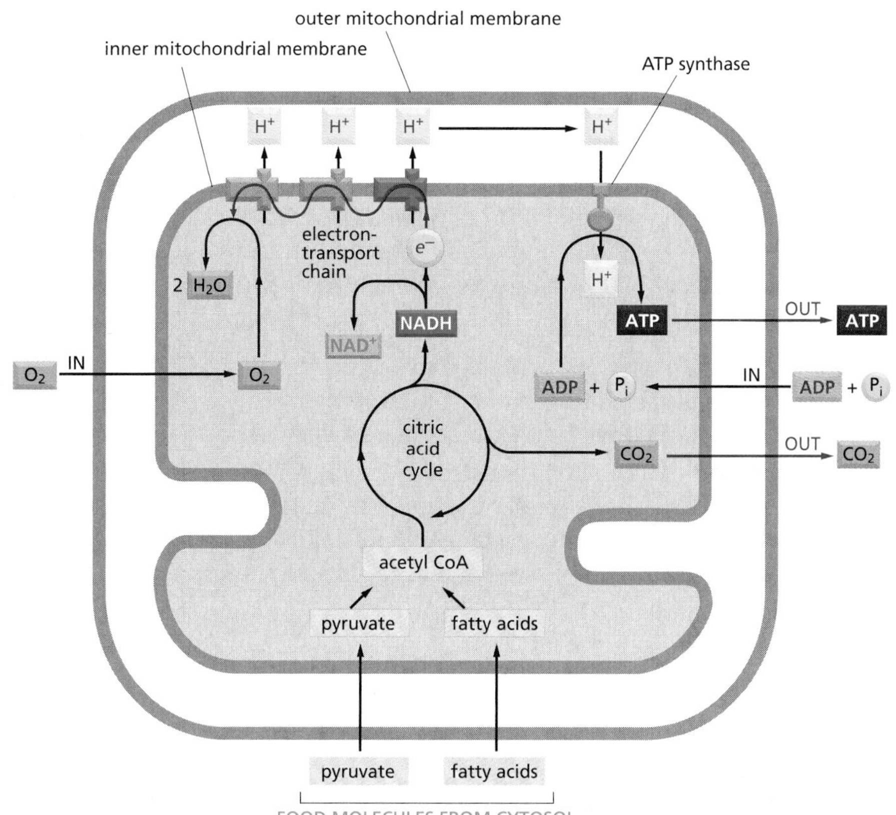

Figure 14–7 **High-energy electrons, generated during the citric acid cycle, power the production of ATP.** Pyruvate and fatty acids enter the mitochondrion (*bottom*), are converted to acetyl CoA, and are then metabolized by the citric acid cycle, which reduces NAD^+ to NADH (and FAD to $FADH_2$, not shown). In the process of oxidative phosphorylation, high-energy electrons from NADH (and $FADH_2$) are then passed along the electron-transport chain in the inner membrane to oxygen (O_2). This electron transport generates a proton gradient across the inner membrane, which is used to drive the production of ATP by ATP synthase. In this diagram, the exact ratios of "reactants" and "products" have been omitted. For example, we will see shortly that it requires four electrons from four NADH molecules to convert O_2 to two H_2O molecules.

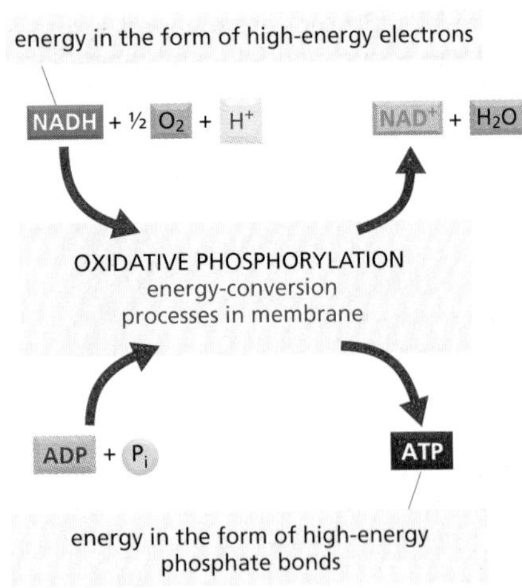

Figure 14–8 **Mitochondria catalyze a major conversion of energy.** In oxidative phosphorylation the energy released by the oxidation of NADH to NAD^+ is harnessed—through energy-conversion processes in the membrane—to drive the energy-requiring phosphorylation of ADP to form ATP. The net equation for this process, in which two electrons pass from NADH to oxygen, is $NADH + ½O_2 + H^+ \rightarrow NAD^+ + H_2O$.

electrons are derived from the action of light on the green pigment *chlorophyll.* And many bacteria use inorganic substances such as hydrogen, iron, and sulfur as the source of the high-energy electrons that they need to make ATP.

The Electron-Transport Chain Pumps Protons Across the Inner Mitochondrial Membrane

The electron-transport chain—or *respiratory chain*—that carries out oxidative phosphorylation is present in many copies in the inner mitochondrial membrane. Each chain contains over 40 proteins, most of which are embedded in the lipid bilayer and function only in the intact membrane, making them difficult to study. However, the components of the electron-transport chain, like other membrane proteins, can be solubilized using nonionic detergents (see Figure 11–27), purified, and then reconstituted in operational form in small membrane vesicles. Such studies reveal that most of the proteins involved in the mitochondrial electron-transport chain are grouped into three large *respiratory enzyme complexes,* each containing multiple individual proteins. Each complex includes transmembrane proteins that hold the entire complex firmly in the inner mitochondrial membrane.

The three respiratory enzyme complexes, in the order in which they receive electrons, are: (1) the *NADH dehydrogenase complex,* (2) the *cytochrome b-*c_1 *complex,* and (3) the *cytochrome oxidase complex.* Each contains metal ions and other chemical groups that form a pathway for the pass-age of electrons through the complex. The respiratory complexes are the sites of proton pumping, and each can be thought of as a protein machine that pumps protons across the membrane as electrons are transferred through it.

Electron transport begins when a hydride ion (H^-) is removed from NADH and is converted into a proton and two high-energy electrons: $H^- \rightarrow H^+ + 2e^-$, as shown previously in Figure 14–5. This reaction is catalyzed by the first of the respiratory enzyme complexes, the NADH dehydrogenase, which accepts electrons from NADH (Figure 14–9). The electrons are then passed along the chain to each of the other enzyme complexes in turn, using mobile electron carriers to ferry electrons between complexes. The transfer of electrons along the chain is energetically favorable: the electrons start out at very high energy and lose energy at each transfer step, eventually entering cytochrome oxidase where they combine with a molecule of O_2 to form water. This is the oxygen-requiring step of cellular respiration, and it consumes nearly all of the oxygen that we breathe.

Proton Pumping Creates a Steep Electrochemical Proton Gradient Across the Inner Mitochondrial Membrane

Without a mechanism for harnessing the energy released by electron transfers, this energy would simply be liberated as heat. But cells utilize much of the energy of electron transfer by having the transfers take place within proteins that are capable of pumping protons. In this way, the energetically favorable flow of electrons along the electron-transport

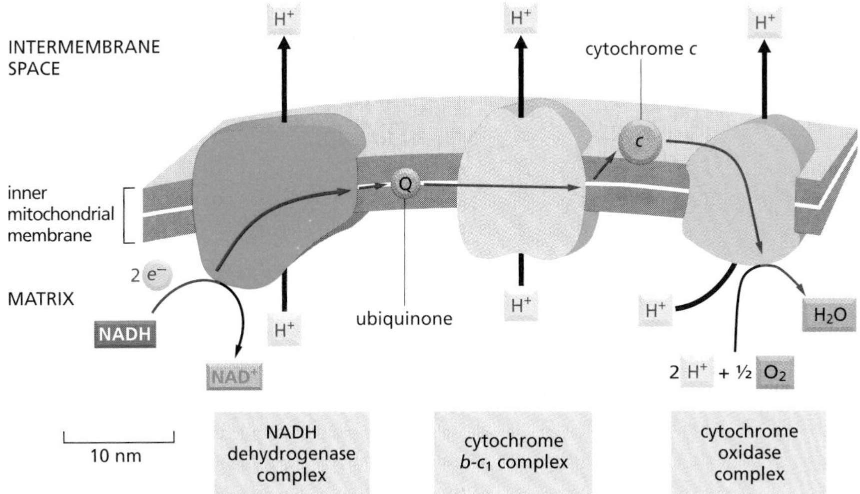

Figure 14–9 **Electrons are transferred through three respiratory enzyme complexes in the inner mitochondrial membrane.** The relative size and shape of each complex are indicated. During the transfer of electrons from NADH to oxygen (*red lines*), protons derived from water are pumped across the membrane from the matrix into the intermembrane space by each of the respiratory enzyme complexes (Movie 14.2). Ubiquinone (Q) and cytochrome *c* (*c*) serve as mobile carriers that ferry electrons from one complex to the next.

chain results in the pumping of protons across the membrane out of the mitochondrial matrix and into the space between the inner and outer mitochondrial membranes (see Figure 14–9).

Later in the chapter we review the detailed molecular mechanisms that couple electron transport to the movement of protons. For now, we focus on the consequences of this nifty biological maneuver. First, the active pumping of protons generates a gradient of H^+ concentration—a pH gradient—across the inner mitochondrial membrane, where the pH is about 0.5 unit higher in the matrix (around pH 7.5) than in the intermembrane space (which is close to 7, the same pH as the cytosol). Second, proton pumping generates a membrane potential across the inner mitochondrial membrane, with the inside (the matrix side) negative and the outside positive as a result of the net outflow of H^+.

> **QUESTION 14–3**
>
> When the drug dinitrophenol (DNP) is added to mitochondria, the inner membrane becomes permeable to protons (H^+). In contrast, when the drug nigericin is added to mitochondria, the inner membrane becomes permeable to K^+. (A) How does the electrochemical proton gradient change in response to DNP? (B) How does it change in response to nigericin?

As discussed in Chapter 12, the force driving the passive flow of an ion across a membrane is proportional to the electrochemical gradient for the ion across the membrane. This in turn depends on the voltage across the membrane, which is measured as the membrane potential, and on the concentration gradient of the ion (see Figure 12–7). Because protons are positively charged, they will move more readily across a membrane if the membrane has an excess of negative electrical charges on the other side. In the case of the inner mitochondrial membrane, the pH gradient and membrane potential work together to create a steep electrochemical proton gradient that makes it energetically very favorable for H^+ to flow back into the mitochondrial matrix. In the energy-producing membranes we discuss in this chapter, the membrane potential adds to the driving force pulling H^+ back across the membrane, which is called the *proton-motive force;* hence the membrane potential increases the amount of energy stored in the proton gradient (Figure 14–10).

The Electrochemical Proton Gradient Drives ATP Synthesis

As explained previously, the electrochemical proton gradient across the inner mitochondrial membrane is used to drive ATP synthesis. The device that makes this possible is a large enzyme called **ATP synthase**, which is also embedded in the inner mitochondrial membrane. ATP synthase creates a hydrophilic pathway across the inner mitochondrial membrane that allows protons to flow back across the membrane down their electrochemical gradient (Figure 14–11). As these ions thread their way through the enzyme, they are used to drive the energetically unfavora-

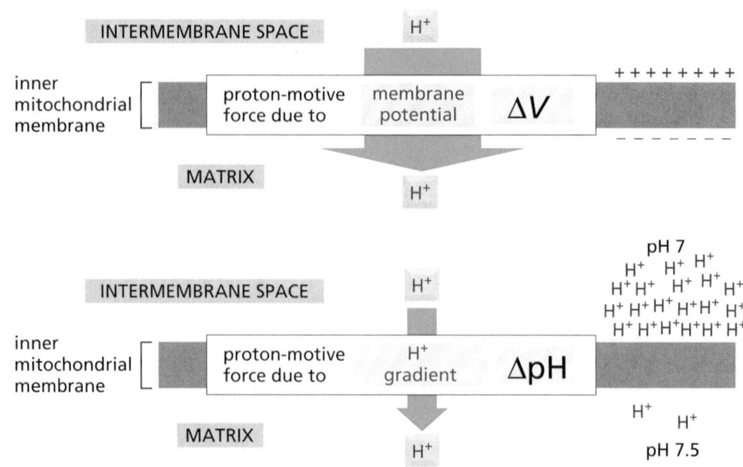

Figure 14–10 **The total electrochemical gradient of H^+ across the inner mitochondrial membrane consists of a large force due to the membrane potential (ΔV) and a smaller force due to the H^+ concentration gradient (ΔpH).** Both forces combine to produce the total proton-motive force that drives H^+ into the matrix space. The relationship between these forces is expressed by the Nernst equation (see Figure 12–30).

ble reaction between ADP and P_i that makes ATP (see Figure 2–24). The ATP synthase is of ancient origin; the same enzyme occurs in the mitochondria of animal cells, the chloroplasts of plants and algae, and in the plasma membrane of bacteria.

ATP synthase is a large, multisubunit protein (Figure 14–12). A large enzymatic portion, shaped like a lollipop head, projects into the matrix and carries out the phosphorylation reaction. This enzymatic structure is attached through a thinner multisubunit "stalk" to a transmembrane proton carrier. As protons pass through a narrow channel within the carrier, their movement causes the stalk to spin rapidly within the head, inducing the head to make ATP (see Figure 14–12A). The synthase essentially acts as an energy-generating molecular motor, converting the energy of proton flow down a gradient into the mechanical energy of two sets of proteins rubbing against one another—rotating stalk proteins pushing against stationary head proteins. The movement of the stalk changes the conformation of subunits within the head. This mechanical deformation gets converted into chemical bond energy as the subunits produce ATP. This marvelous device can produce more than 100 molecules of ATP per second, and about three protons need to pass through the synthase to make each molecule of ATP.

The ATP synthase is a reversible coupling device. It can either harness the flow of protons down their electrochemical gradient to make ATP (its normal role in mitochondria and the plasma membrane of bacteria growing aerobically) or use the energy of ATP hydrolysis to pump protons across a membrane (Figure 14–13). In the latter mode, ATP synthase functions like the H^+ pumps described in Chapter 12. Whether the ATP synthase primarily makes or consumes ATP depends on the magnitude of the electrochemical proton gradient across the membrane in which it

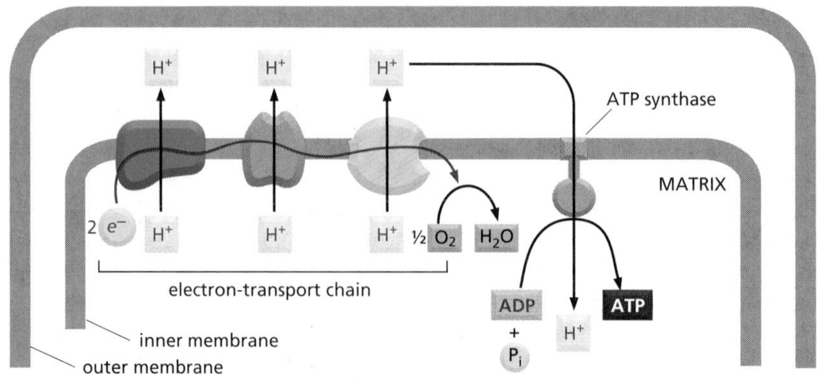

Figure 14–11 **The electrochemical proton gradient across the inner mitochondrial membrane allows ATP synthase to generate ATP.** As a high-energy electron is passed along the electron-transport chain, much of the energy released is used to drive the three respiratory enzyme complexes that pump H^+ out of the matrix space. The resulting electrochemical proton gradient across the inner membrane drives H^+ back through the ATP synthase, a transmembrane protein complex that uses the energy of the H^+ flow to synthesize ATP from ADP and P_i in the matrix.

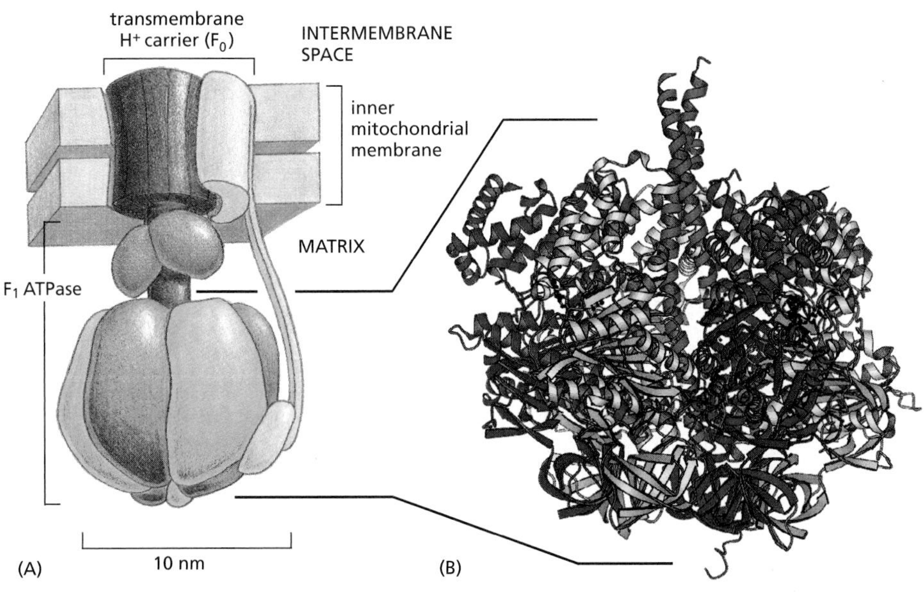

Figure 14–12 **ATP synthase is shaped like a lollipop.** (A) The enzyme is composed of a head portion, called the F_1 ATPase, and a transmembrane H^+ carrier, called F_0. Both F_1 and F_0 are formed from multiple subunits, as indicated. The F_0 portion, consisting of a rotor plus a stalk (*red*), rotates in the membrane driven by the proton gradient. The stator (*light green*) is formed from transmembrane subunits, tied to other subunits that create an elongated arm. This arm fixes the stator to a ring of subunits that forms the stationary head of the ATPase. Its β subunits (*dark green*) generate the ATP. (B) The three-dimensional structure of the F_1 ATPase, as determined by X-ray crystallography. This part of the ATP synthase derives its name from its ability to carry out the reverse of the ATP synthesis reaction, namely, the hydrolysis of ATP to ADP and P_i, when detached from the transmembrane portion. (B, courtesy of John Walker, from J.P. Abrahams et al., *Nature* 370:621–628, 1994. With permission from Macmillan Publishers Ltd.)

sits. In many bacteria that can grow either aerobically or anaerobically, the direction in which the ATP synthase works is routinely reversed when the bacterium runs out of O_2. At this point, the ATP synthase uses some of the ATP generated inside the cell by glycolysis to pump protons out of the cell, creating the proton gradient that the bacterial cell needs to import its essential nutrients by coupled transport, as we see next.

Coupled Transport Across the Inner Mitochondrial Membrane Is Also Driven by the Electrochemical Proton Gradient

The synthesis of ATP is not the only process driven by the electrochemical proton gradient. In mitochondria, many charged molecules, such as pyruvate, ADP, and P_i, are pumped into the matrix from the cytosol, while others, such as ATP, must be moved in the opposite direction. Carrier proteins that bind these molecules can couple their transport to the energetically favorable flow of H^+ into the mitochondrial matrix. Pyruvate and inorganic phosphate (P_i), for example, are individually co-transported inward with H^+ as the latter moves down its electrochemical gradient, into the matrix.

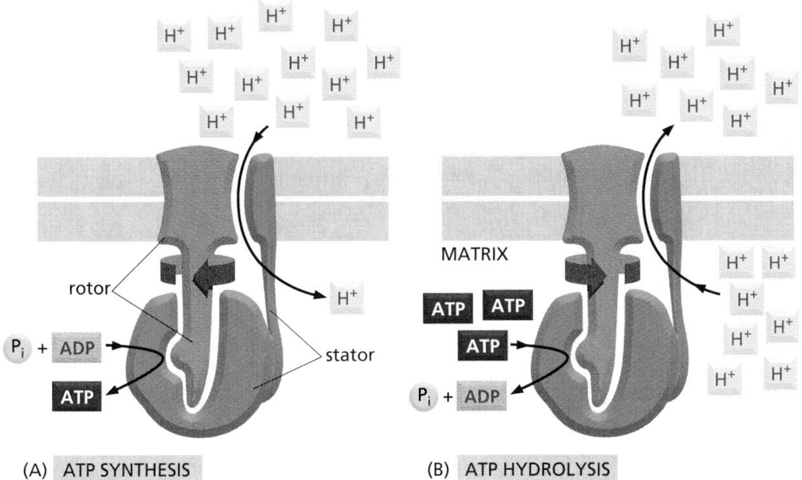

Figure 14–13 **ATP synthase is a reversible coupling device that can convert the energy of the electrochemical proton gradient into chemical-bond energy or vice versa.** The ATP synthase can either synthesize ATP by harnessing the H^+ gradient (A) or pump protons against their electrochemical gradient by hydrolyzing ATP (B). The direction of operation at any given instant depends on the net free-energy change (Δ*G*, discussed in Chapter 3) for the coupled processes of H^+ translocation across the membrane and the synthesis of ATP from ADP and P_i. For example, if the electrochemical proton gradient falls below a certain level, the Δ*G* for H^+ transport into the matrix will no longer be large enough to drive ATP production. Instead, ATP will be hydrolyzed by the ATP synthase to rebuild the gradient. The action of ATP synthase is shown in Movie 14.3 and Movie 14.4.

QUESTION 14–4

The remarkable properties that allow ATP synthase to run in either direction allow the interconversion of energy stored in the H^+ gradient and energy stored in ATP in either direction. (A) If ATP synthase making ATP can be likened to a water-driven turbine producing electricity, what would be an appropriate analogy when it works in the opposite direction? (B) Under what conditions would one expect the ATP synthase to stall, running neither forward nor backward? (C) What determines the direction in which the ATP synthase operates?

Other transporters take advantage of the fact that the electrochemical proton gradient generates a membrane potential, such that the matrix side of the inner mitochondrial membrane is more negatively charged than the intermembrane space on the other side. An antiport carrier protein exploits this voltage gradient to expel ATP from—and import ADP to—the mitochondrial matrix. Because an ATP molecule has one more negative charge than ADP, swapping these nucleotides results in the movement of one negative charge out of the mitochondrion. This nucleotide exchange—which sends ATP to the cytosol—is thus driven by the charge difference across the inner mitochondrial membrane (Figure 14–14).

In eucaryotic cells, therefore, the electrochemical proton gradient is used to drive both the formation of ATP and the transport of certain metabolites across the inner mitochondrial membrane. In bacteria, the proton gradient across the bacterial plasma membrane serves all of these functions. But, in bacteria, this gradient is also an important source of directly usable energy: in motile bacteria, a flow of protons into the cell drives the rapid rotation of the bacterial flagellum, which propels the bacterium along (Movie 14.5).

Oxidative Phosphorylation Produces Most of the Cell's ATP

As we mentioned earlier, glycolysis on its own produces a net yield of two molecules of ATP for every molecule of glucose, whereas the complete oxidation of glucose—which includes glycolysis and oxidative phosphorylation—generates about 30 ATPs. In glycolysis, it is obvious where those ATP molecules come from: two molecules of ATP are consumed early in the process and four molecules of ATP are produced toward the end (see Figure 13–3). But for oxidative phosphorylation, the accounting is less straightforward, because the ATPs are not produced directly, as they are in glycolysis. Instead, they are produced from the energy carried by NADH and $FADH_2$, which are generated during glycolysis and the citric acid cycle. These activated carrier molecules donate their electrons to the electron transport chain that lies in the inner mitochondrial membrane. These movement of these electrons along the respiratory chain fuels the formation of the proton gradient, which in turn powers the production of ATP.

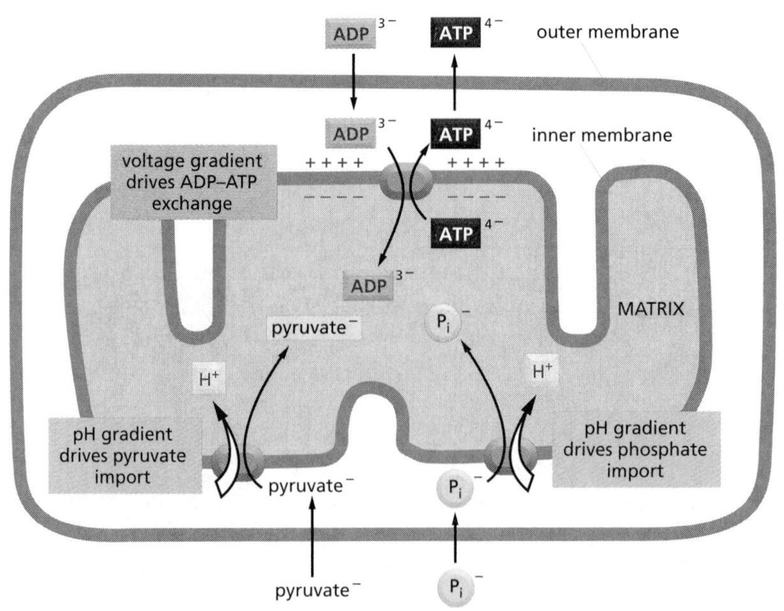

Figure 14–14 **The electrochemical proton gradient across the inner mitochondrial membrane is used to drive some coupled transport processes.** Pyruvate and inorganic phosphate (P_i) are moved into the matrix along with H^+ ions as they move down their electrochemical gradient. ADP is pumped in and ATP is pumped out by an antiport process (ADP–ATP exchange) that depends on a voltage gradient across the membrane (the membrane potential). The charge on each of the transported molecules is indicated for comparison with the membrane potential, which is negative inside as shown. The outer membrane is freely permeable to all of these compounds. The active transport of molecules across membranes by carrier proteins and the generation of a membrane potential are discussed in Chapter 12.

TABLE 14–1 PRODUCT YIELDS FROM GLUCOSE OXIDATION

PROCESS	DIRECT PRODUCT	FINAL ATP YIELD PER MOLECULE OF GLUCOSE
Glycolysis	2 NADH (cytosolic)	3*
	2 ATP	2
Pyruvate oxidation to acetyl CoA (two per glucose)	2 NADH (mitochondrial matrix)	5
Complete acetyl CoA oxidation (two per glucose)	6 NADH (mitochondrial matrix)	15
	2 $FADH_2$	3
	2 GTP	2
	TOTAL	30

*NADH produced in the cytosol yields fewer ATP molecules than NADH produced in the mitochondrial matrix because the mitochondrial inner membrane is impermeable to NADH. Transporting NADH into the mitochondrial matrix—where it encounters NADH hydrogenase—thus requires energy.

How much ATP each carrier molecule ultimately produces depends on several factors, including where its electrons enter the respiratory chain. The NADH molecules produced during the citric acid cycle, which takes place inside the mitochondria, pass their electrons to NADH dehydrogenase—the first respiratory enzyme complex in the chain. These electrons then pass from one enzyme complex to the next, promoting the pumping of protons across the inner mitochondrial membrane at each step along the way. These NADH molecules provide energy for the net formation of about 2.5 molecules of ATP (see Question 14–5 and its answer).

The $FADH_2$ generated during the citric acid cycle, on the other hand, produces a net of only 1.5 molecules of ATP. This is because $FADH_2$ molecules bypass the NADH dehydrogenase complex and pass their electrons to the membrane-embedded mobile carrier ubiquinone (see Figure 14–9). These electrons enter further down the respiratory chain, and they therefore promote the pumping of fewer protons and generate less ATP. Table 14–1 provides a complete accounting of the ATP produced by the oxidation of each molecule of glucose.

The oxidation of fatty acids also produces large amounts of NADH and $FADH_2$, which in turn produce large amounts of ATP via oxidative phosphorylation (see Figures 13–9 and 13–10). Thus, the vast majority of the ATP produced in an animal cell is produced by chemiosmotic mechanisms on the mitochondrial membrane.

QUESTION 14–5

Calculate the number of ATP molecules produced per pair of electrons transferred from NADH to oxygen, if (i) five protons are pumped across the inner mitochondrial membrane for each electron passed through the three respiratory enzyme complexes, (ii) three protons must pass through the ATP synthase for each ATP molecule that it produces from ADP and inorganic phosphate inside the mitochondrion, and (iii) one proton is used to produce the voltage gradient needed to transport each ATP molecule out of the mitochondrion to the cytosol where it is used.

The Rapid Conversion of ADP to ATP in Mitochondria Maintains a High ATP/ADP Ratio in Cells

As a result of the co-transport process discussed earlier, ADP molecules produced by ATP hydrolysis in the cytosol are rapidly drawn back into mitochondria for recharging, and the bulk of the ATP molecules formed in the mitochondrial matrix by oxidative phosphorylation are pumped into the cytosol where they are needed. A small amount of ATP is used within the mitochondrion itself to power the replication of its DNA, protein synthesis, and other energy-consuming reactions. All in all, a typical ATP molecule in the human body shuttles out of a mitochondrion and back into it (as ADP) for recharging more than once a minute, keeping the concentration of ATP in the cell about 10 times higher than that of ADP.

As discussed in Chapter 3, biosynthetic enzymes often drive energetically unfavorable reactions by coupling them to the energetically favorable hydrolysis of ATP (see Figure 3–33A). The ATP pool is thus used to drive cellular processes in much the same way that a battery can drive an electric engine. If the activity of the mitochondria were halted, ATP levels would fall and the cell's battery would run down; eventually, energetically unfavorable reactions could no longer take place and the cell would die. The poison cyanide, which blocks electron transport in the inner mitochondrial membrane, causes death in exactly this way.

MOLECULAR MECHANISMS OF ELECTRON TRANSPORT AND PROTON PUMPING

We have already considered in general terms how a mitochondrion couples electron transport to the generation of ATP. We now examine in more detail the molecular mechanisms that underlie its membrane-based energy-conversion processes. In doing so, we will also be accomplishing a larger purpose. As emphasized at the beginning of this chapter, very similar energy-conversion devices are used by mitochondria, chloroplasts, and bacteria, and the basic principles that we shall discuss next therefore underlie the function of nearly all living things.

For many years, the reason that electron-transport chains were embedded in membranes eluded the biochemists who were struggling to understand them. The process of chemiosmotic coupling entails an interplay between chemical and electrical forces that is not easy to decipher at a molecular level. The puzzle was essentially solved as soon as the fundamental role of transmembrane proton gradients in energy generation was proposed in the early 1960s. However, the idea was so novel that it was not widely accepted until many years later, after additional supporting evidence had accumulated from experiments designed to test rigorously the chemiosmotic hypothesis (see How We Know, pp. 468–469).

Although investigators today are still unraveling the details of chemiosmotic coupling at the atomic level, the fundamentals are now clear. In this part of the chapter we shall look at some of the principles that underlie the electron-transport process and explain in detail how it can generate a proton gradient.

Protons Are Readily Moved by the Transfer of Electrons

Although protons resemble other positive ions such as Na^+ and K^+ in the way they move across membranes, in some respects they are unique. Hydrogen atoms are by far the most abundant type of atom in living organisms and are plentiful not only in all carbon-containing biological molecules but also in the water molecules that surround them. The protons in water are highly mobile, flickering through the hydrogen-bonded network of water molecules by rapidly dissociating from one water molecule in order to associate with its neighbor. Thus, water, which is everywhere in cells, serves as a ready reservoir for donating and accepting protons.

Whenever a molecule is reduced by acquiring an electron, the electron (e^-) brings with it a negative charge. In many cases, this charge is rapidly neutralized by the addition of a proton from water, so that the net effect of the reduction is to transfer an entire hydrogen atom, $H^+ + e^-$ (Figure 14–15). Similarly, when a molecule is oxidized, the hydrogen atom can be readily dissociated into its constituent electron and proton, allowing the electron to be transferred separately to a molecule that accepts electrons, while the proton is passed to the water. Therefore, in a membrane

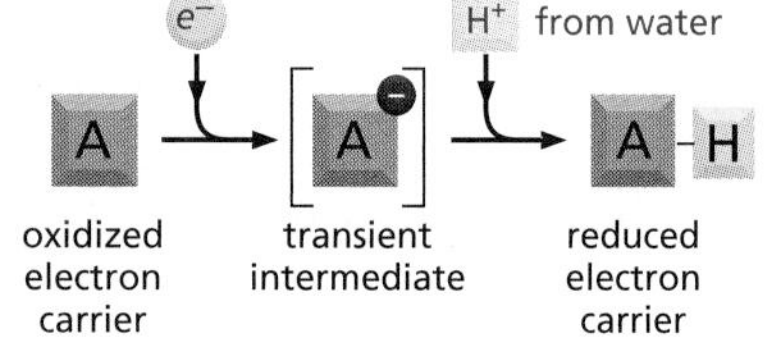

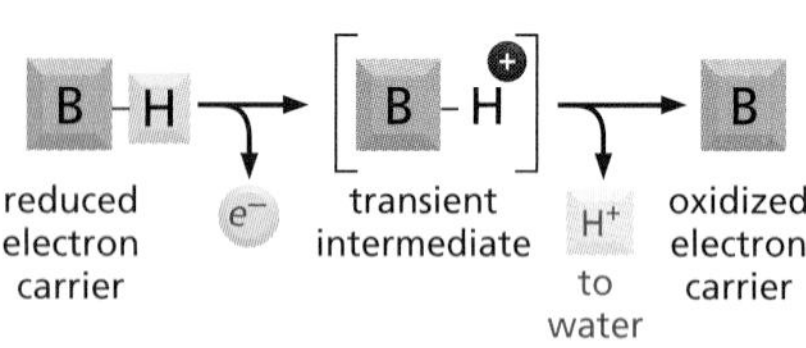

Figure 14–15 **The protons in water are highly mobile.** Electron transfer can cause the transfer of entire hydrogen atoms, because protons are readily accepted from or donated to water. In this example, A picks up an electron plus a proton when it is reduced, and B loses an electron plus a proton when it is oxidized.

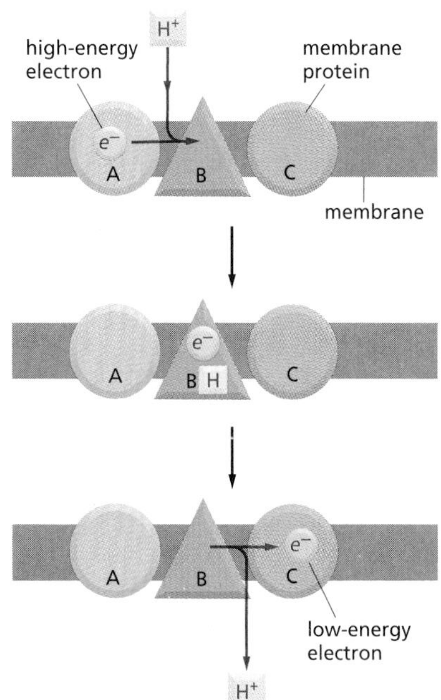

Figure 14–16 **The orientation of the electron carrier allows electron transfer to drive proton pumping.** As an electron passes along an electron-transport chain, it can bind and release a proton at each step. In this schematic diagram, electron carrier B picks up a proton (H^+) from one side of the membrane when it accepts an electron (e^-) from carrier A; it releases the proton to the other side of the membrane when it donates its electron to carrier C.

in which electrons are being passed along an electron-transport chain, it is a relatively simple matter, in principle, to pump protons from one side of the membrane to another. All that is required is that the electron carrier be arranged in the membrane in a way that causes it to pick up a proton from one side of the membrane when it accepts an electron, while releasing the proton on the other side of the membrane as the electron is passed on to the next carrier molecule in the chain (Figure 14–16).

The Redox Potential Is a Measure of Electron Affinities

The proteins of the respiratory chain guide the electrons so that they move sequentially from one enzyme complex to another—with no short circuits that skip a complex. Each electron transfer is an oxidation–reduction reaction: as described in Chapter 3, the molecule or atom donating the electron becomes oxidized, while the receiving molecule or atom becomes reduced (see pp. 87–88). Electrons will pass spontaneously from molecules that have a relatively low affinity for their available electrons, and thus lose them easily, to molecules with a higher electron affinity. For example, NADH with its high-energy electrons has a low electron affinity, so that its electrons are readily passed to the NADH dehydrogenase. The electrical batteries of our common experience are based on similar electron transfers between two chemical substances with different electron affinities.

In biochemical reactions, any electrons removed from one molecule are always passed to another, so that whenever one molecule is oxidized, another is reduced. Like any other chemical reaction, the tendency of such oxidation–reduction reactions, or **redox reactions**, to proceed spontaneously depends on the free-energy change (ΔG) for the electron transfer, which in turn depends on the relative affinities of the two molecules for electrons. (The role of free energy in chemical reactions is discussed in Chapter 3, pp. 91–92.)

Because electron transfers provide most of the energy for living things, it is worth spending a little time to understand them. Many readers are already familiar with acids and bases, which donate and accept protons (see Panel 2–2, pp. 66–67). Acids and bases exist in conjugate acid–base pairs, where the acid is readily converted into the base by the loss of a proton. For example, acetic acid (CH_3COOH) is converted into its conjugate base (CH_3COO^-) in the reaction

$$CH_3COOH \rightleftharpoons CH_3COO^- + H^+$$

In exactly the same way, pairs of compounds such as NADH and NAD^+ are called **redox pairs**, because NADH is converted to NAD^+ by the loss of electrons in the reaction

$$NADH \rightleftharpoons NAD^+ + H^+ + 2e^-$$

NADH is a strong electron donor: because its electrons are held in a high-energy linkage, the ΔG for passing its electrons to many other molecules is favorable. Conversely, it is difficult to form the high-energy linkage in NADH, so its partner, NAD^+, is of necessity a weak electron acceptor.

HOW CHEMIOSMOTIC COUPLING DRIVES ATP SYNTHESIS

In 1861, Louis Pasteur discovered that yeast cells grow and divide more vigorously when air is present, the first demonstration that aerobic metabolism is more efficient than anaerobic metabolism. His observations make sense now that we know that oxidative phosphorylation is a much more efficient means of generating ATP than is glycolysis: electron-transport systems produce about 30 molecules of ATP for each molecule of glucose oxidized, compared with the two molecules of ATP generated by glycolysis alone. But it took another hundred years for researchers to determine that it is the process of chemiosmotic coupling—using proton pumping to power ATP synthesis—that allows cells to generate energy with such efficiency.

Imaginary intermediates

In the 1950s, many researchers believed that the oxidative phosphorylation that takes place in mitochondria generates ATP via a mechanism similar to that used in glycolysis. During glycolysis, ATP is produced when a molecule of ADP receives a phosphate group directly from a high-energy intermediate. Such substrate-level phosphorylation occurs in steps 7 and 10 of glycolysis, where the high-energy phosphate groups from 1,3-bisphosphoglycerate and phosphoenolpyruvate, respectively, are transferred to ADP to form ATP (see Panel 13–1, pp. 430–431). It was assumed that the electron-transport chain in mitochondria would similarly generate some high-energy intermediate that could then donate its phosphate group directly to ADP. This model inspired a frustrating search for this mysterious intermediate that lasted for years. Investigators occasionally claimed to have discovered the missing intermediate, but the compounds turned out to be either unrelated to electron transport or, as one researcher put it in a review of the history of bioenergetics, "products of high-energy imagination."

Harnessing the force

It wasn't until 1961 that Peter Mitchell suggested that the "high-energy intermediate" his colleagues were seeking was, in fact, the electrochemical proton gradient generated by the electron-transport system. His proposal, dubbed the chemiosmotic hypothesis, stated that the energy of a H^+ gradient formed during the transfer of electrons through the transport chain could be tapped to drive ATP synthesis.

Several lines of evidence offered support for such chemiosmotic coupling. First, mitochondria do generate a proton gradient across their inner membrane. But what does this gradient do? If the H^+ electrochemical gradient (also called the proton-motive force) is required to drive ATP synthesis, as the chemiosmotic hypothesis posits, then destruction of that gradient—or of the membrane itself—should inhibit energy generation. In fact, researchers found this to be true. Physical disruption of the inner mitochondrial membrane halts ATP synthesis. Similarly, dissipation of the proton gradient by chemical "uncoupling" agents such as 2,4-dinitrophenol (DNP) also prevents ATP from being made. These gradient-busting chemicals carry H^+ across the inner mitochondrial membrane, forming a shuttle system for the movement of H^+ that bypasses the ATP synthase (Figure 14–17). In this way they uncouple electron transport from ATP synthesis. As a result of this short-circuiting, the proton-motive force is dissipated completely and ATP can no longer be made.

Such uncoupling occurs naturally in some specialized fat cells. In these cells, called *brown fat cells*, most of the energy from oxidation is dissipated as heat rather than converted into ATP. The inner membranes of the large mitochondria in these cells contain a special transport protein that allows protons to move down their electrochemical gradient, circumventing ATP synthase.

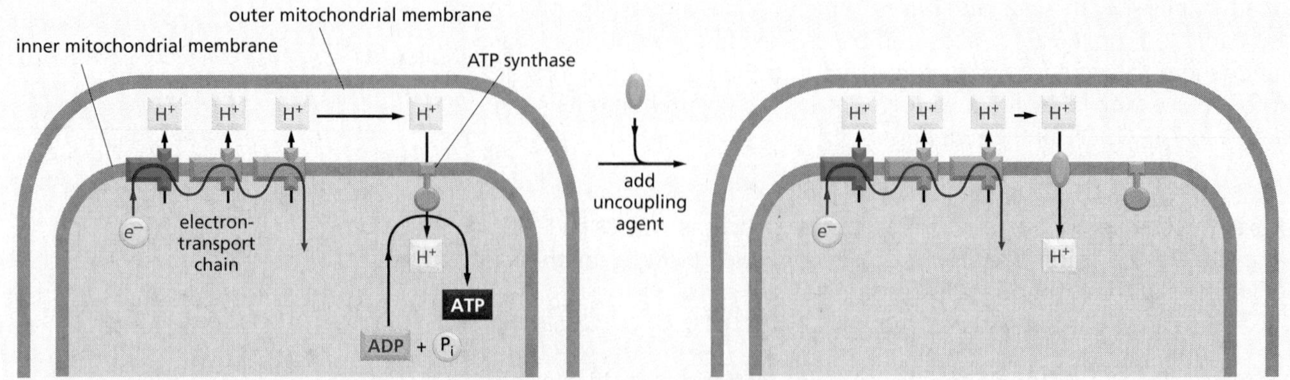

Figure 14–17 **Uncoupling agents are H^+ carriers that can insert into the mitochondrial inner membrane.** They render the membrane permeable to protons, allowing H^+ to flow into the mitochondrion without passing through ATP synthase. This short circuit effectively uncouples electron transport from ATP synthesis.

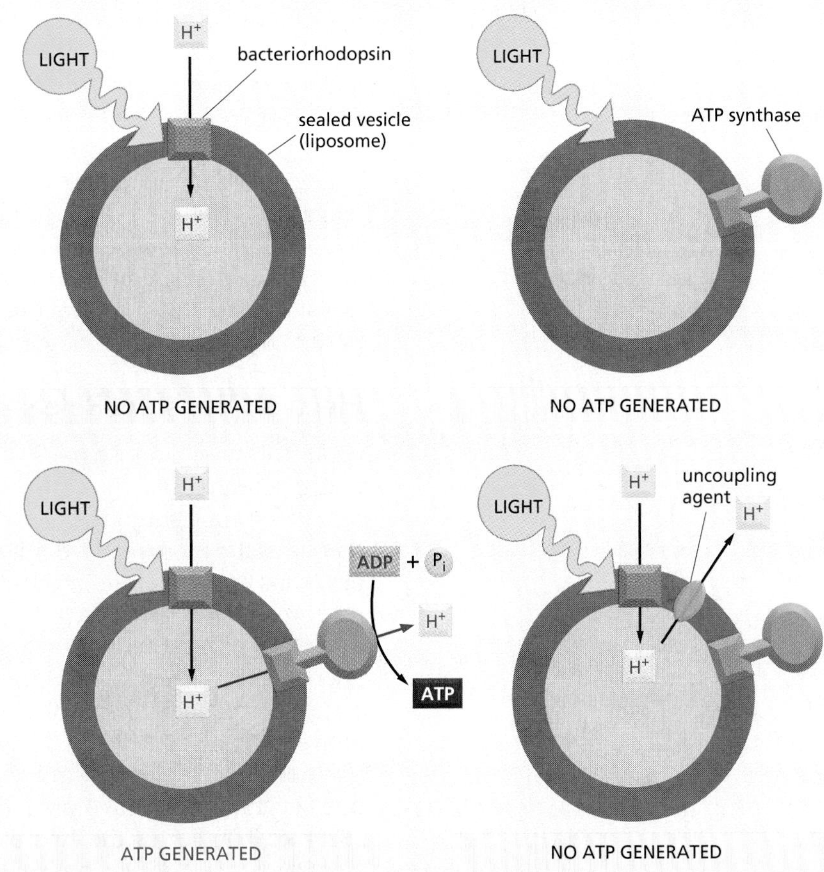

Figure 14–18 **Experiments with bacteriorhodopsin and an ATP synthase from cow-heart mitochondria provided strong evidence that proton gradients can power ATP production.** When bacteriorhodopsin is added to artificial vesicles, the protein generates a proton gradient in response to light. In artificial vesicles containing both bacteriorhodopsin and an ATP synthase, this proton gradient drives the formation of ATP. Uncoupling agents that abolish the gradient eliminate the ATP synthesis.

As a result, the cells oxidize their fat stores at a rapid rate and produce more heat than ATP. Tissues containing brown fat serve as biological heating pads, helping to revive hibernating animals and to protect sensitive areas of newborn human babies (such as the backs of their necks) from the cold.

Artificial ATP generation

If disrupting the proton gradient across the mitochondrial membrane terminates ATP synthesis, then, conversely, generating an artificial proton gradient should stimulate the production of ATP. Again, this is exactly what happens. When a H^+ gradient is imposed artificially by lowering the pH on the cytoplasmic side of the mitochondrial membrane, ATP is synthesized, even in the absence of an oxidizable substrate.

How does this proton gradient drive ATP production? This is where the ATP synthase comes in. In 1974 Efraim Racker and Walther Stoeckenius demonstrated elegantly that the combination of an ATP synthase plus a proton gradient will produce ATP. They found that they could reconstitute a complete artificial energy-generating system by combining an ATPase from cow-heart mitochondria with a protein from the purple membrane of the procaryote *Halobacterium halobium*. As discussed in Chapter 11, the plasma membrane of this archaean is packed with bacteriorhodopsin, a protein that pumps H^+ out of the cell in response to sunlight (see Figure 11–28). Thus, this membrane protein generates a proton gradient when exposed to light.

When bacteriorhodopsin is reconstituted into artificial lipid vesicles, Racker and Stoeckenius showed that in the presence of light it pumps H^+ into the vesicles, generating a proton gradient. (For some reason the orientation of the protein is reversed in these membranes, so that H^+ ions are transported into the vesicles; in the bacterium, protons are pumped out.) And when an ATPase purified from mitochondria is incorporated into these vesicles, the system catalyzes ATP synthesis in response to light. This ATP formation requires the H^+ gradient, as the researchers found that eliminating bacteriorhodopsin from the system or adding uncoupling agents abolished ATP synthesis (Figure 14–18).

Thus, although Mitchell's hypothesis initially met with considerable resistance—biochemists had hoped to discover a high-energy intermediate rather than having to settle for an elusive electrochemical force—the experimental evidence that eventually accumulated to support the importance of chemiosmotic coupling in cellular energy generation could not be ignored. Mitchell was awarded a Nobel Prize in 1978.

The tendency to transfer electrons from any redox pair can be measured experimentally. All that is required is to form an electrical circuit that links a 1:1 (equimolar) mixture of the redox pair to a second redox pair that has been arbitrarily selected as a reference standard, so that the voltage difference between them can be measured (Panel 14–1, p. 471). This voltage difference is defined as the **redox potential**; as defined, electrons will move spontaneously from a redox pair such as NADH/NAD^+ with a low redox potential (a low affinity for electrons) to a redox pair such as O_2/H_2O with a high redox potential (a high affinity for electrons). Thus, NADH is a good molecule to donate electrons to the respiratory chain, while O_2 is well suited to act as the "sink" for electrons at the end of the pathway. As explained in Panel 14–1, the difference in redox potential, $\Delta E_0'$, is a direct measure of the standard free-energy change ($\Delta G°$) for the transfer of an electron from one molecule to another. In fact, $\Delta E_0'$ is simply equal to $\Delta G°$ times a negative number that is a constant.

Electron Transfers Release Large Amounts of Energy

As just discussed, those pairs of compounds that have the most negative redox potential (E_0') have the weakest affinity for electrons and therefore the strongest tendency to donate electrons. Conversely, those pairs that have the most positive redox potential have the strongest affinity for electrons and therefore the strongest tendency to accept electrons. A 1:1 mixture of NADH and NAD^+ has a redox potential of –320 mV, indicating that NADH has a strong tendency to donate electrons; a 1:1 mixture of H_2O and ½O_2 has a redox potential of +820 mV, indicating that O_2 has a strong tendency to accept electrons. The difference in redox potential between these two pairs is 1.14 volts (1140 mV), which means that the transfer of each electron from NADH to O_2 under these standard conditions is enormously favorable: the $\Delta G° = -26.2$ kcal/mole per electron or –52.4 kcal/mole for the two electrons transferred per NADH molecule (see Panel 14–1). If we compare this free-energy change with that for the formation of the phosphoanhydride bonds in ATP ($\Delta G° = +7.3$ kcal/mole; see Figure 13–7), we see that more than enough energy is released by the oxidization of one NADH molecule to synthesize several molecules of ATP from ADP and P_i.

Living systems could certainly have evolved enzymes that would allow NADH to donate electrons directly to O_2 to make water in the reaction

$$2H^+ + 2e^- + \tfrac{1}{2}O_2 \rightarrow H_2O$$

But because of the huge free-energy drop, this reaction would proceed with almost explosive force and nearly all of the energy would be released as heat. Instead, the energetically favorable reaction $2H^+ + 2e^- + \tfrac{1}{2}O_2 \rightarrow H_2O$ is made to occur in many small steps, enabling nearly half of the released energy to be stored rather than being lost to the environment as heat.

Metals Tightly Bound to Proteins Form Versatile Electron Carriers

Within each of the three respiratory enzyme complexes, electrons move mainly between metal atoms that are tightly bound to the proteins, traveling by skipping from one metal ion to another one with a greater affinity for electrons. In contrast, electrons are carried between the different respiratory complexes by molecules that diffuse along the lipid bilayer, picking up electrons from one complex and delivering them to another in an orderly sequence. In both the respiratory and photosynthetic electron-transport chains, one of these carriers is a **quinone**, a small hydrophobic molecule that dissolves in the lipid bilayer; in the mitochondrial respi-

PANEL 14–1 Redox potentials

HOW REDOX POTENTIALS ARE MEASURED

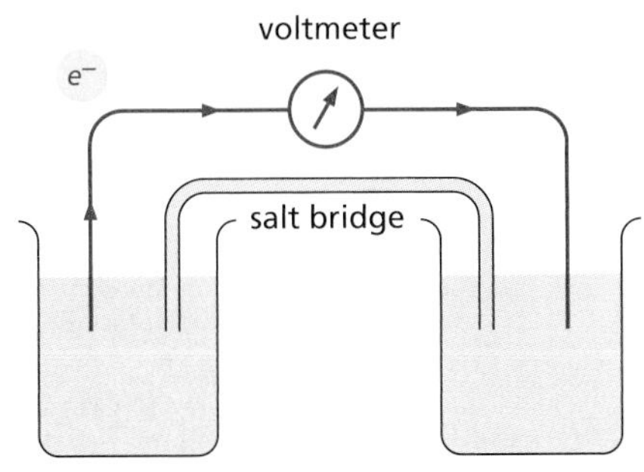

$A_{reduced}$ and $A_{oxidized}$ in equimolar amounts

1 M H^+ and 1 atmosphere H_2 gas

One beaker *(left)* contains substance A with an equimolar mixture of the reduced ($A_{reduced}$) and oxidized ($A_{oxidized}$) members of its redox pair. The other beaker contains the hydrogen reference standard ($2H^+ + 2e^- \rightleftharpoons H_2$), whose redox potential is arbitrarily assigned as zero by international agreement. (A salt bridge formed from a concentrated KCl solution allows K^+ and Cl^- to move between the beakers and neutralize the charges when electrons flow between the beakers.) The metal wire *(red)* provides a resistance-free path for electrons, and a voltmeter then measures the redox potential of substance A. If electrons flow from $A_{reduced}$ to H^+, as indicated here, the redox pair formed by substance A is said to have a negative redox potential. If they instead flow from H_2 to $A_{oxidized}$, this redox pair is said to have a postive redox potential.

By convention, the redox potential for a redox pair is designated as *E*. Since biological reactions occur at pH 7, biologists define the standard state as $A_{reduced} = A_{oxidized}$ and $H^+ = 10^{-7}$ M and use it to determine the standard redox potential E_0'.

examples of redox reactions	redox potential E_0'
$NADH \rightleftharpoons NAD^+ + H^+ + 2e^-$	–320 mV
reduced ubiquinone $\rightleftharpoons$ oxidized ubiquinone $+ 2H^+ + 2e^-$	+30 mV
reduced cytochrome *c* $\rightleftharpoons$ oxidized cytochrome *c* $+ e^-$	+230 mV
$H_2O \rightleftharpoons \frac{1}{2}O_2 + 2H^+ + 2e^-$	+820 mV

CALCULATION OF $\Delta G°$ FROM REDOX POTENTIALS

$$\Delta E_0' = E_0'(\text{acceptor}) - E_0'(\text{donor})$$

$$\Delta E_0' = +30 - (-320) = +350$$

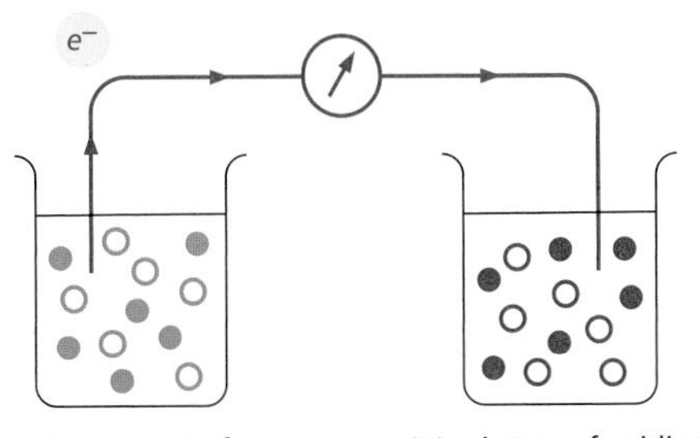

1:1 mixture of NADH and NAD^+

1:1 mixture of oxidized and reduced ubiquinone

$\Delta G° = -n(0.023)\,\Delta E_0'$, where *n* is the number of electrons transferred across a redox potential change of $\Delta E_0'$ millivolts (mV)

Example: The transfer of one electron from NADH to ubiquinone has a favorable $\Delta G°$ of –8.0 kcal/mole, calculated as follows:

$$\Delta G° = -n(0.023)\Delta E_0' = -1(0.023)(350) = -8.0 \text{ kcal/mole}$$

The same calculation reveals that the transfer of one electron from ubiquinone to oxygen has an even more favorable $\Delta G°$ of –18.2 kcal/mole. The $\Delta G°$ value for the transfer of one electron from NADH to oxygen is the sum of these two values, –26.2 kcal/mole.

EFFECT OF CONCENTRATION CHANGES

As explained in Chapter 3 (see p. 92), the actual free-energy change for a reaction, ΔG, depends on the concentration of the reactants and generally will be different from the standard free-energy change, $\Delta G°$. The standard redox potentials are for a 1:1 mixture of the redox pair. For example, the standard redox potential of –320 mV is for a 1:1 mixture of NADH and NAD^+. But when there is an excess of NADH over NAD^+, electron transfer from NADH to an electron acceptor becomes more favorable. This is reflected by a more negative redox potential and a more negative ΔG for electron transfer.

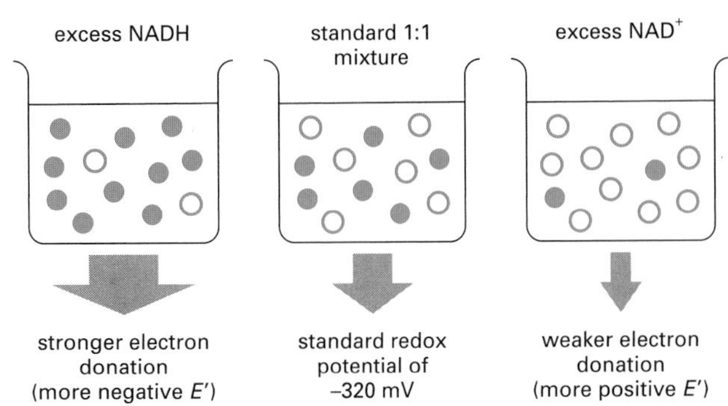

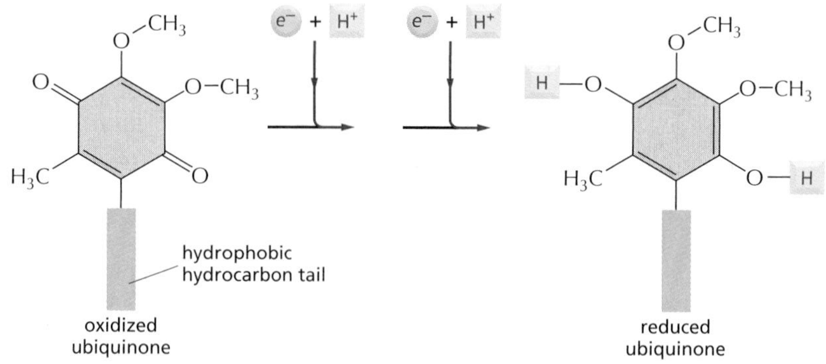

Figure 14–19 **Quinones carry electrons within the lipid bilayer.** The quinone in the mitochondrial electron-transport chain is called ubiquinone. It picks up one H^+ from the aqueous environment for every electron it accepts, and it can carry two electrons as part of a hydrogen atom (*yellow*). When reduced ubiquinone donates its electrons to the next carrier in the chain, the protons are released. The long hydrophobic tail confines ubiquinone to the membrane and consists of 6–10 five-carbon isoprene units, the number depending on the organism.

ratory chain, the quinone is called *ubiquinone*. Quinones are the only electron carriers in electron-transport chains that can function without being tightly bound to a protein.

Ubiquinone picks up electrons from the NADH dehydrogenase complex and delivers them to the cytochrome b-c_1 complex (see Figure 14–9). Ubiquinone can pick up or donate either one or two electrons, and it picks up one H^+ from the surroundings with each electron that it carries (Figure 14–19). Its redox potential of +30 mV places ubiquinone about one-quarter of the way down the chain from NADH in terms of energy loss (Figure 14–20). Ubiquinone can also receive electrons directly from the $FADH_2$ generated by the citric acid cycle or by fatty acid oxidation. Because these electrons bypass NADH hydrogenase—which is one of the proton pumps in the electron transport chain—they cause less proton pumping than do the two electrons transported from NADH.

The rest of the electron carriers in the electron-transport chain are either small molecules or metal-containing groups that are all tightly bound to proteins. To get from NADH to ubiquinone, for example, the electrons are passed inside the NADH dehydrogenase complex between a flavin group (see Figure 13–12 for its structure) bound to one of the proteins and a set of **iron–sulfur centers** of increasing redox potentials. The final iron–sulfur center in the dehydrogenase donates its electrons to ubiquinone.

Iron–sulfur centers have relatively low affinities for electrons and thus are prominent in the early part of the electron-transport chain. Later, in

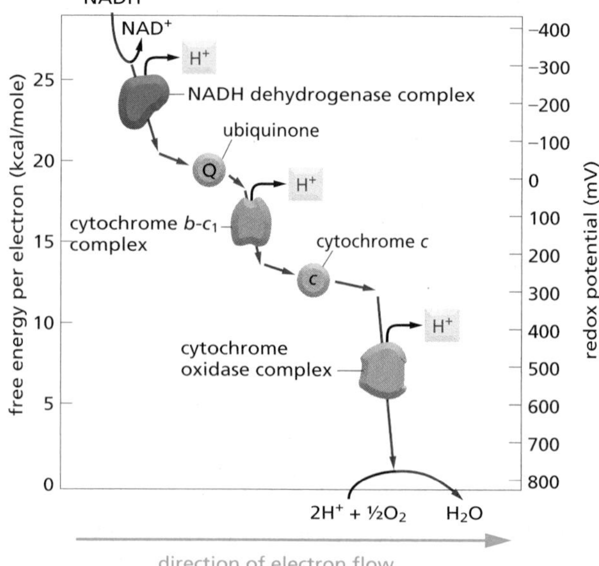

Figure 14–20 **Redox potential increases along the mitochondrial electron-transport chain.** The big increases in redox potential occur across each of the three respiratory enzyme complexes, as required for each of them to pump protons. To convert free energy values to kJ/mole, recall that 1 kilocalorie is equal to about 4.2 kilojoules.

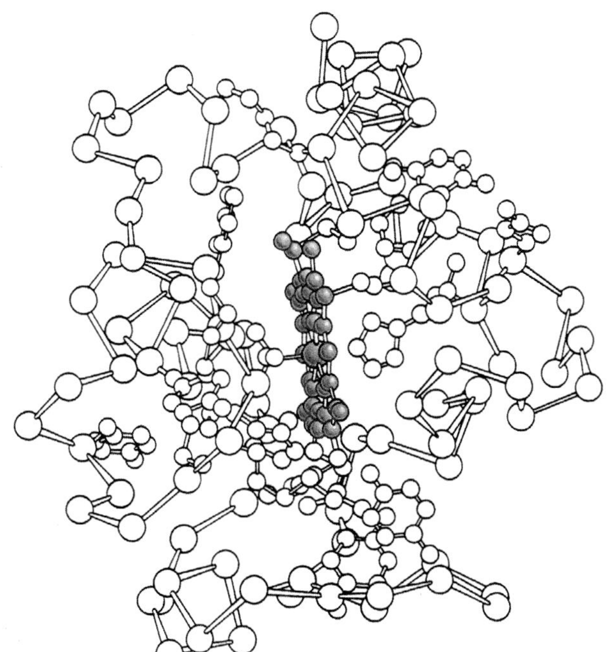

Figure 14–21 **Cytochrome *c* is an electron carrier in the electron-transport chain.** This small protein contains just over 100 amino acids and is held loosely on the outer face of the inner membrane by ionic interactions (see Figure 14–9). The iron atom (*orange*) in the bound heme (*blue*) can carry a single electron. The structure of the heme group in hemoglobin, which reversibly binds O_2 rather than an electron, was shown in Figure 4–33.

the pathway from ubiquinone to O_2, iron atoms in heme groups that are tightly bound to cytochrome proteins are commonly used as electron carriers, as in the cytochrome b-c_1 and cytochrome oxidase complexes. The **cytochromes** constitute a family of colored proteins (hence their name, from the Greek *chroma*, "color"); each contains one or more heme groups whose iron atom changes from the ferric (Fe^{3+}) to the ferrous (Fe^{2+}) state whenever it accepts an electron. As one would expect, the various cytochromes increase in redox potential as one progresses down the mitochondrial electron-transport chain towards O_2. The structure of *cytochrome* c, a small protein that shuttles electrons between the cytochrome b-c_1 complex and the cytochrome oxidase complex, is shown in Figure 14–21: its redox potential is +230 mV.

At the very end of the respiratory chain, just before O_2, the electron carriers are those in the cytochrome oxidase complex. The carriers here are either iron atoms in heme groups or copper atoms that are tightly bound to the complex in specific ways that give them a high redox potential.

QUESTION 14–6

At many steps in the electron-transport chain, Fe ions are used as part of heme or FeS clusters to bind the electrons in transit. Why do these functional groups that carry out the chemistry of electron transfer need to be bound to proteins? Provide several different reasons why this is necessary.

Cytochrome Oxidase Catalyzes the Reduction of Molecular Oxygen

Cytochrome oxidase is a protein complex that receives electrons from cytochrome *c*, thus oxidizing it (hence the name cytochrome oxidase). It then donates these electrons to O_2. In brief, four electrons from cytochrome *c* and four protons from the aqueous environment are added to each O_2 molecule in the reaction $4e^- + 4H^+ + O_2 \rightarrow 2H_2O$. In addition to the protons that couple with O_2, four other protons are pumped across the membrane during electron transfer, further increasing the electrochemical proton gradient.

Of course, for proton pumping to occur, it must be coupled in some way to energetically favorable reactions. In the case of cytochrome oxidase, the energy comes from the transfer of a series of four electrons to an O_2 molecule that is bound tightly to the protein; these electron transfers drive allosteric changes in the conformation of the protein that move protons out of the mitochondrial matrix. At its active site, where O_2 is bound, cytochrome oxidase contains a complex of a heme iron atom juxtaposed with a tightly bound copper atom (Figure 14–22). It is here that nearly all of the oxygen we breathe is used, serving as the final repository for the electrons that NADH donated at the start of the electron-transport chain.

Oxygen is useful as an electron sink because of its very high affinity for electrons. However, once O_2 picks up one electron, it forms the superoxide radical O_2^-; this radical is dangerously reactive and will avidly take up another three electrons wherever it can find them, a tendency that can cause serious damage to nearby DNA, proteins, and lipid membranes. One of the roles of cytochrome oxidase is to hold on tightly to an oxygen molecule until all four electrons needed to convert it to two H_2O molecules are in hand, thereby preventing a random attack on cellular macromolecules by superoxide radicals—damage that has been postulated to be a cause of human aging.

The evolution of cytochrome oxidase was crucial to the formation of cells that could use O_2 as an electron acceptor, and today this protein is estimated to account for 90% of the total uptake of O_2 in humans. This protein complex is therefore essential for all aerobic life. The poisons cyanide and azide are extremely toxic because they bind tightly to the cell's cytochrome oxidase complexes to stop electron transport, thereby greatly reducing ATP production.

The Mechanism of H^+ Pumping Can Be Studied in Atomic Detail

The detailed mechanism by which electron transport is coupled to H^+ pumping is different in each of the three different respiratory enzyme complexes. For example, ubiquinone has a central role in proton pumping through the cytochrome b-c_1 complex, as the quinone picks up an H^+ from the aqueous medium along with each electron it carries and liberates it when it releases the electron (see Figure 14–19). Because ubi-

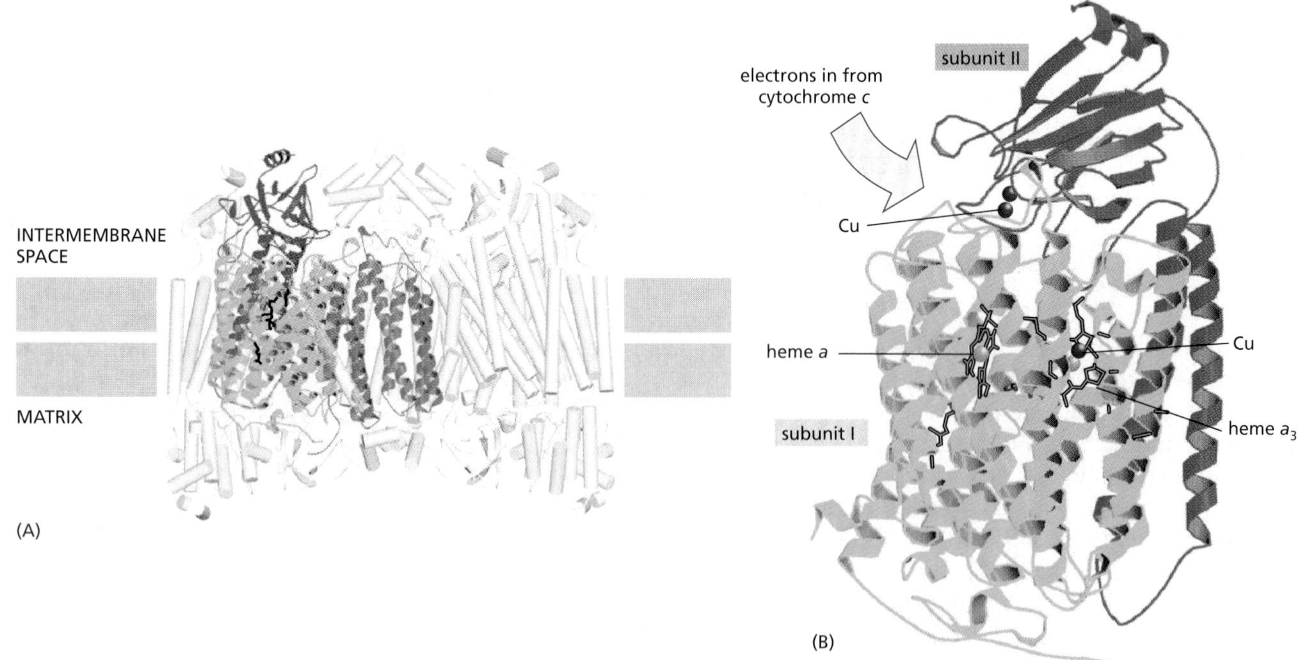

Figure 14–22 **Cytochrome oxidase is a finely tuned protein machine.** The protein is a dimer formed from a monomer with 13 different protein subunits. The three colored subunits that form the functional core of the complex are encoded by the mitochondrial genome. As electrons pass through this protein on the way to its bound O_2 molecule, they cause the protein to pump protons across the membrane. (A) The entire protein is shown, positioned in the inner mitochondrial membrane. (B) The electron carriers are located in subunits I and II, as indicated.

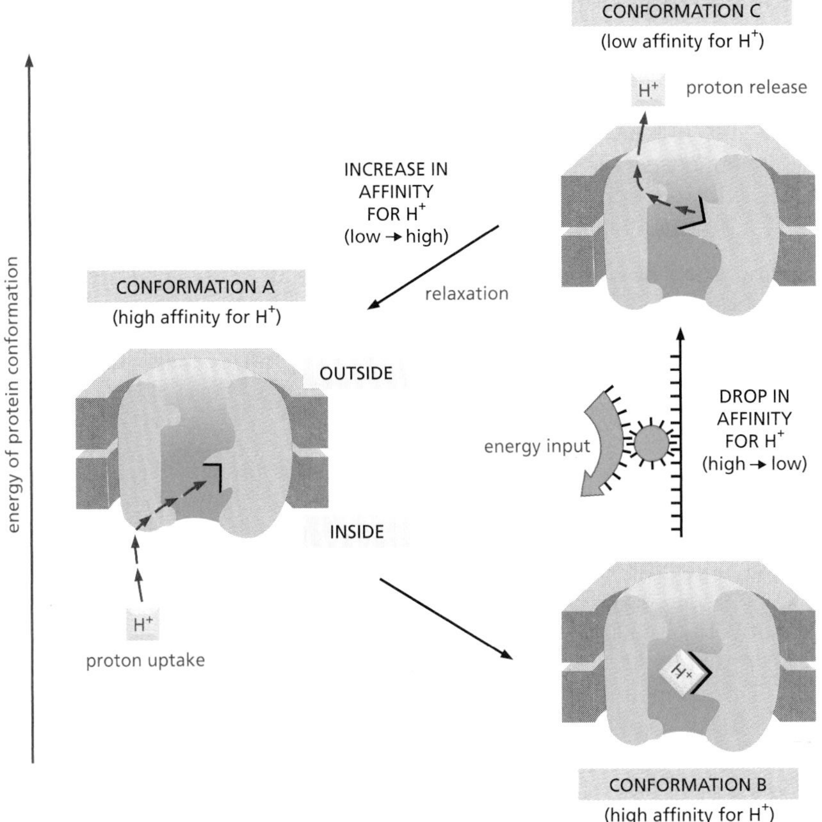

Figure 14–23 **H^+ pumping can be caused by a conformational change in a protein pump driven by an energetically favorable reaction.** This mechanism for H^+ pumping by a transmembrane protein is thought to be used by NADH dehydrogenase and cytochrome oxidase, and by many other proton pumps. The protein is driven through a cycle of three conformations: A, B, and C. As indicated by their vertical spacing, these protein conformations have different energies. In conformations A and B, the protein has a high affinity for H^+, causing it to pick up an H^+ on the inside of the membrane. In conformation C, the protein has a low affinity for H^+, causing it to release an H^+ on the outside of the membrane. The transition from conformation B to conformation C that releases the H^+ is energetically unfavorable, and it occurs only because it is driven by being allosterically coupled to an energetically favorable reaction occurring elsewhere on the protein (*blue* arrow). For cytochrome oxidase and NADH dehydrogenase, this reaction is electron transport. The other two conformational changes, A → B and C → A, lead to states of lower energy, and they proceed spontaneously. Because the overall cycle A → B → C → A releases free energy, H^+ is pumped from inside the mitochondrial matrix out to the intermembrane space. For other proton pumps, such as the Ca^{2+} pump in muscle cells, the energy required for the conformational change B → C is provided by ATP hydrolysis (see Figure 12–15). For bacteriorhodopsin, this energy comes from sunlight (see Figure 11–28).

quinone is freely mobile in the lipid bilayer, it can accept electrons near the inside surface of the membrane and donate them to the cytochrome b-c_1 complex near the outside surface. Thus ubiquinone transfers one H^+ across the bilayer for every electron it transports. However, two protons are pumped per electron in the cytochrome b-c_1 complex, and there is good evidence for a so-called *Q-cycle,* in which ubiquinone is recycled through the protein complex in a complicated, but ordered way that makes this two-for-one transfer possible.

Allosteric changes in protein conformations can also pump H^+. For both the NADH dehydrogenase complex and the cytochrome oxidase complex, electron transport drives sequential allosteric changes in the protein that cause it to pump H^+ across the mitochondrial inner membrane. A general mechanism for this type of H^+ pumping is presented in Figure 14–23.

Respiration Is Amazingly Efficient

The free-energy changes for burning fats and carbohydrates directly to CO_2 and H_2O can be compared with the total amount of energy generated and stored in the phosphate bonds of ATP during the corresponding biological oxidations. When this is done, one finds that the efficiency with which oxidation energy is converted into ATP bond energy is often greater than 40%. This is considerably better than the efficiency of most nonbiological energy-conversion devices. If cells worked only with the efficiency of an electric motor or a gasoline engine (10–20%), an organism would have to eat voraciously to maintain itself. Moreover, because wasted energy is liberated as heat, large organisms (including ourselves) would need more efficient mechanisms than they presently have for giving up heat to the environment.

QUESTION 14–7

Two different diffusible electron carriers, ubiquinone and cytochrome c, shuttle electrons between the three protein complexes of the electron-transport chain. Could the same diffusible carrier, in principle, be used for both steps? Explain your answer.

Students sometimes wonder why the chemical interconversions in cells follow such complex pathways. The oxidation of sugars to CO_2 plus H_2O could certainly be accomplished more directly, eliminating the citric acid cycle and many of the steps in the respiratory chain. This would make respiration easier for students to learn, but it would be a disaster for the cell. Oxidation produces huge amounts of free energy, which can be utilized efficiently only in small bits. Biological oxidative pathways involve many intermediates, each differing only slightly from its predecessor. The energy released is thereby parceled out into small packets that can be efficiently converted to high-energy bonds in useful molecules, such as ATP and NADH, by means of coupled reactions (see Figure 13–1).

Having seen how chemiosmotic coupling is used to generate ATP in mitochondria, we now look at how it harnesses light energy for the generation of ATP in chloroplasts.

CHLOROPLASTS AND PHOTOSYNTHESIS

Virtually all of the organic material required by present-day living cells is produced by **photosynthesis**—the series of light-driven reactions that creates organic molecules from atmospheric carbon dioxide (CO_2). Plants, algae, and the most advanced photosynthetic bacteria, such as the cyanobacteria, use electrons from water and the energy of sunlight to convert atmospheric CO_2 into organic compounds. In the course of splitting water they liberate into the atmosphere vast quantities of O_2 gas. This oxygen is in turn required for cellular respiration—not only in animals but also in plants and many bacteria. Thus, the activity of early photosynthetic bacteria, which filled the atmosphere with oxygen, enabled the evolution of life forms that use aerobic metabolism to make their ATP (Figure 14–24).

In plants, photosynthesis is carried out in a specialized intracellular organelle—the **chloroplast**, which contains light-capturing pigments such as the green pigment chlorophyll. All green parts of a plant contain chloroplasts, but for most plants the leaves are the major sites of photosynthesis. Chloroplasts perform photosynthesis during the daylight hours. The process produces ATP and NADPH, which in turn are used to convert CO_2 into sugar inside the chloroplast. Thus we begin our discussion of photosynthesis by describing the structure of this specialized organelle.

(A)

(B)

(C)

Figure 14–24 **Microorganisms that carry out oxygen-producing photosynthesis changed Earth's atmosphere.** (A) Living stromatolites from a lagoon in western Australia. These structures are formed in specialized environments by large colonies of oxygen-producing photosynthetic cyanobacteria, which lay down successive layers of material. (B) Cross section of a modern stromatolite, showing its layered structure. (C) Cross section through a fossil stromatolite in a rock 3.5 billion years old. Note the layered structure similar to that in (B). Fossil stromatolites are thought to have been formed by photosynthetic bacteria very similar to modern cyanobacteria. The activities of bacteria like these, which liberate O_2 gas as a waste product of photosynthesis, would have slowly changed Earth's atmosphere. (A, courtesy of Sally Birch, with permission from Oxford Scientific Films; B and C, courtesy of S.M. Awramik, University of California/Biological Photo Service.)

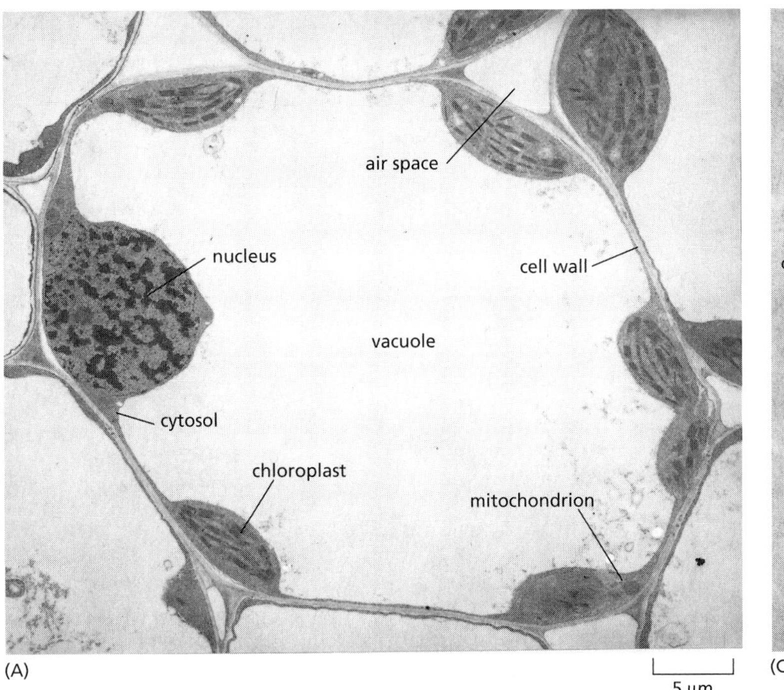

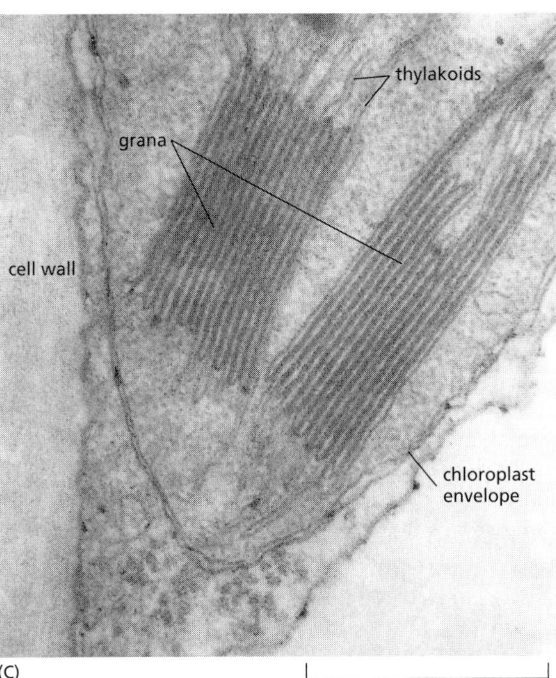

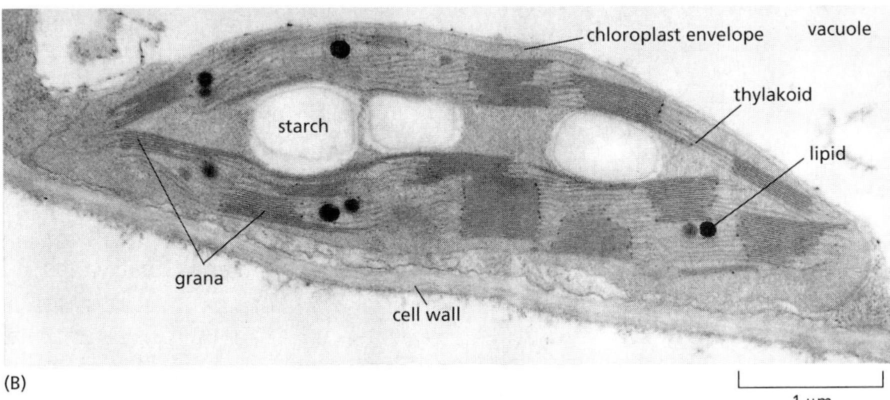

Figure 14–25 **Photosynthesis takes place in chloroplasts.** Electron micrographs show structures of chloroplasts. (A) A wheat leaf cell in which a thin rim of cytoplasm containing nucleus, chloroplasts, and mitochondria surrounds a large vacuole. (B) A thin section of a single chloroplast, showing the chloroplast envelope, starch granules, and lipid (fat) droplets that have accumulated in the stroma as a result of the biosyntheses occurring there. (C) A high-magnification view of two *grana*; a *granum* is the name given to a stack of thylakoids. (Courtesy of K. Plaskitt.)

Chloroplasts Resemble Mitochondria but Have an Extra Compartment

Chloroplasts carry out their energy interconversions by means of proton gradients in much the same way that mitochondria do. Although chloroplasts are larger (Figure 14–25A), they are organized on the same principles as mitochondria. Chloroplasts have a highly permeable outer membrane and a much less permeable inner membrane, in which membrane transport proteins are embedded. Together these membranes—and the narrow, intermembrane space that separates them—form the chloroplast envelope (Figure 14–25B). The inner membrane surrounds a large space called the **stroma**, which is analogous to the mitochondrial matrix and contains many metabolic enzymes.

There is, however, an important difference between the organization of mitochondria and that of chloroplasts. The inner membrane of the chloroplast does not contain the electron-transport chains. Instead, the light-capturing systems, the electron-transport chains, and ATP synthase are all contained in the *thylakoid membrane*, a third membrane that forms a set of flattened disclike sacs, called the *thylakoids* (Figure 14–25C). These are arranged in stacks, and the space inside each thylakoid is thought to be connected with that of other thylakoids, thereby defining a continuous third internal compartment that is separated from the stroma by the

QUESTION 14–8

Chloroplasts have a third internal compartment, the thylakoid space, bounded by the thylakoid membrane. This membrane contains the photosystems, reaction centers, electron-transport chain, and ATP synthase. In contrast, mitochondria use their inner membrane for electron transport and ATP synthesis. In both organelles, protons are pumped out of the largest internal compartment (the matrix in mitochondria and the stroma in chloroplasts). The thylakoid space is completely sealed off from the rest of the cell. Why does this arrangement allow a larger H^+ gradient in chloroplasts than can be achieved for mitochondria?

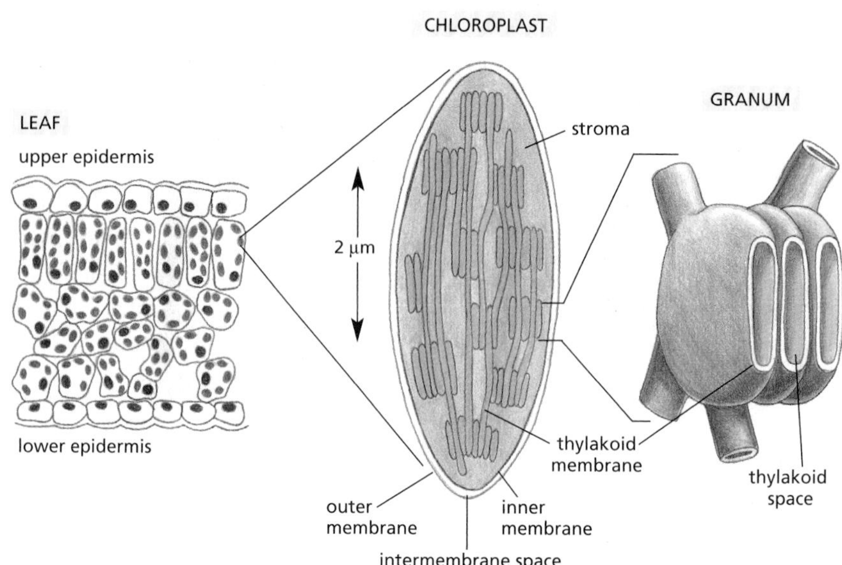

Figure 14–26 **A chloroplast contains a third internal compartment.** This photosynthetic organelle contains three distinct membranes (the outer membrane, the inner membrane, and the thylakoid membrane) that define three separate internal compartments (the intermembrane space, the stroma, and the thylakoid space). The thylakoid membrane contains all of the energy-generating systems of the chloroplast, including its light-capturing chlorophyll. In electron micrographs this membrane appears to be broken up into separate units that enclose individual flattened vesicles (see Figure 14–25C), but these are probably joined into a single, highly folded membrane in each chloroplast. As indicated, the individual thylakoids are interconnected, and they tend to stack to form grana.

thylakoid membrane (Figure 14–26). The structural similarities and differences between mitochondria and chloroplasts are illustrated in Figure 14–27.

Chloroplasts Capture Energy from Sunlight and Use It to Fix Carbon

As we saw in Chapter 3, the overall equation that summarizes the net result of photosynthesis can be written as follows:

$$\text{light energy} + CO_2 + H_2O \rightarrow \text{sugars} + O_2 + \text{heat energy}$$

Although the equation is quite simple, the reactions that allow the process to occur are fairly elaborate. Generally speaking, however, the many reactions that make up photosynthesis in plants take place in two stages (see Figure 3–8):

1. In the first stage, which is dependent on light, energy from sunlight is captured and stored transiently in the high-energy bonds of ATP and the activated carrier molecule NADPH. These energy-producing *photosynthetic electron-transfer reactions*, also called the 'light reactions,' occur entirely within the thylakoid membrane of the chloroplast. In this series of reactions, energy derived from sunlight energizes an electron in the green organic pigment **chlorophyll**, enabling the electron to move along an electron-transport chain in the thylakoid membrane in much the same way that an electron moves along the respiratory chain in mitochondria. The

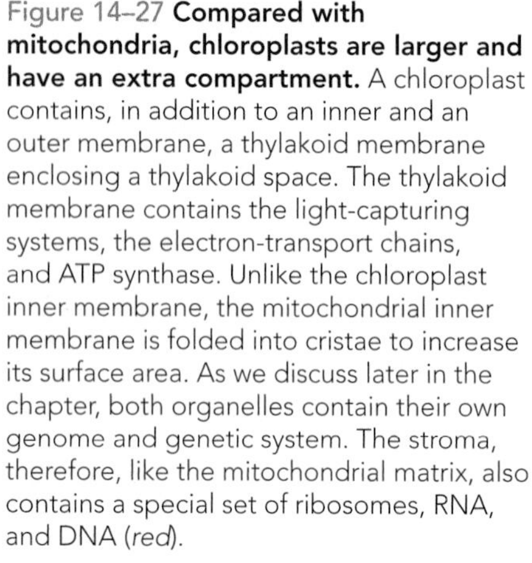
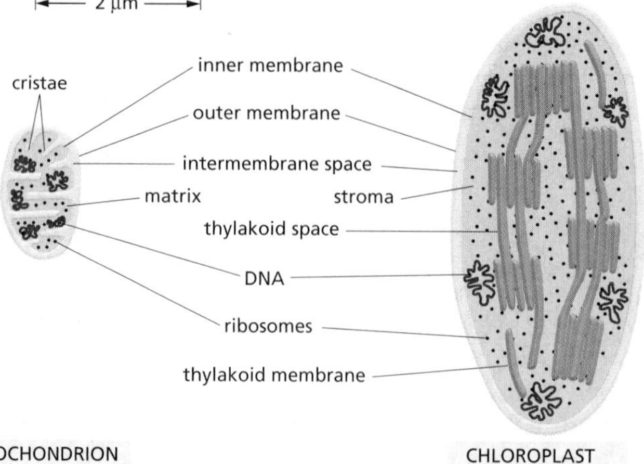

Figure 14–27 **Compared with mitochondria, chloroplasts are larger and have an extra compartment.** A chloroplast contains, in addition to an inner and an outer membrane, a thylakoid membrane enclosing a thylakoid space. The thylakoid membrane contains the light-capturing systems, the electron-transport chains, and ATP synthase. Unlike the chloroplast inner membrane, the mitochondrial inner membrane is folded into cristae to increase its surface area. As we discuss later in the chapter, both organelles contain their own genome and genetic system. The stroma, therefore, like the mitochondrial matrix, also contains a special set of ribosomes, RNA, and DNA (*red*).

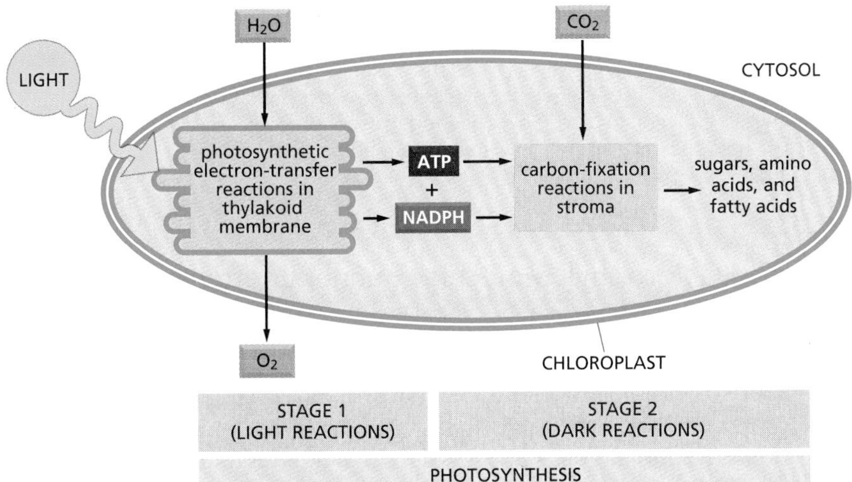

Figure 14–28 **Both stages of photosynthesis depend on the chloroplast.** Water is oxidized and oxygen is released in the photosynthetic electron-transfer reactions that produce ATP and NADPH (stage 1), and carbon dioxide is assimilated (fixed) to produce sugars and a variety of other organic molecules in the carbon-fixation reactions (stage 2). Stage 2 begins in the chloroplast stroma (as shown) and continues in the cytosol.

electron that chlorophyll donates to the electron-transport chain is ultimately replaced by an electron extracted from water. This electron shuffle splits a molecule of water (H_2O), producing O_2 as a by-product. During the electron-transport process, H^+ is pumped across the thylakoid membrane, and the resulting electrochemical proton gradient drives the synthesis of ATP in the stroma. As the final step in this series of reactions, high-energy electrons are loaded (together with H^+) onto $NADP^+$, converting it to NADPH (Figure 14–28).

2. In the second, light-independent, stage of photosynthesis, the ATP and the NADPH produced by the photosynthetic electron-transfer reactions serve as the source of energy and reducing power, respectively, to drive the manufacture of sugars from CO_2 (see Figure 14–28. These carbon-fixation reactions, also called the 'dark reactions,' begin in the chloroplast stroma and continue in the plant cell cytosol. They produce sucrose and many other organic molecules in the leaves of the plant. The sucrose is exported to other tissues as a source of both organic molecules and energy for growth.

Thus, the formation of ATP, NADPH, and O_2 (which requires light energy directly) and the conversion of CO_2 to carbohydrate (which requires light energy only indirectly) are separate processes, although elaborate feedback mechanisms interconnect the two sets of reactions. Several of the chloroplast enzymes required for carbon fixation, for example, are inactivated in the dark and reactivated by light-stimulated electron-transport processes.

Sunlight is Absorbed by Chlorophyll Molecules

Visible light is a form of electromagnetic radiation composed of many different wavelengths, ranging from violet (wavelength 400 nm) to deep red (700 nm). But when we consider events at the level of a single molecule—such as the absorption of light by a molecule of chlorophyll—we have to picture light as being composed of discrete packets of energy called *photons*. Light of different colors is distinguished by photons of different energy, with longer wavelengths corresponding to lower energies. Thus photons of red light have a lower energy than photons of green light.

When sunlight is absorbed by a molecule of chlorophyll, electrons in the molecule interact with photons of light and are raised to a higher energy level. The electrons in the extensive network of alternating single and double bonds in the chlorophyll molecule (Figure 14–29) absorb red light most strongly, which is why chlorophyll looks green to us.

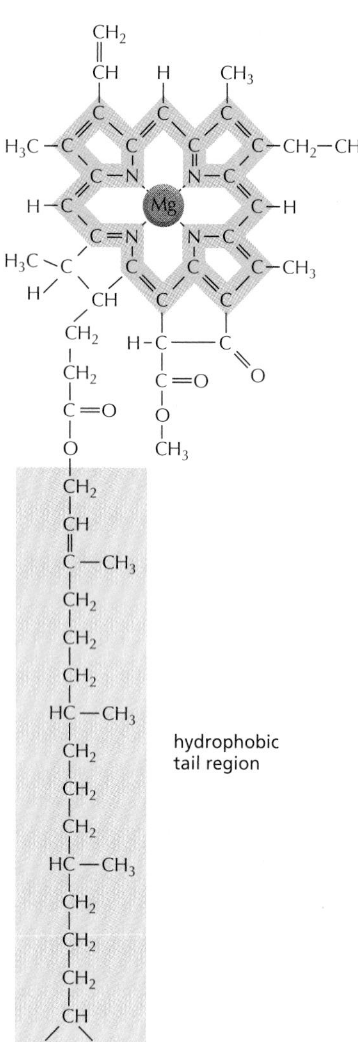

Figure 14–29 **Chlorophyll is a green pigment that absorbs energy from photons of light.** A magnesium atom (*orange*) is held in the center of a porphyrin ring, which is structurally similar to the porphyrin ring that binds iron in heme. Light is absorbed by electrons within the bond network shown in *blue*, while the long hydrophobic tail (*green*) helps to hold the chlorophyll in the thylakoid membrane.

Excited Chlorophyll Molecules Funnel Energy into a Reaction Center

An isolated molecule of chlorophyll is incapable of converting the light it absorbs to a form of energy useful to living systems. It can accomplish this feat only when it is associated with the appropriate proteins and embedded in a membrane. In plant thylakoid membranes and in the membranes of photosynthetic bacteria, the light-absorbing chlorophylls are held in large multiprotein complexes called **photosystems**. Each photosystem consists of an **antenna complex** that captures light energy and a **reaction center** that enables this light energy to be converted into chemical energy. The antenna portion of a photosystem consists of hundreds of chlorophyll molecules that capture light energy in the form of excited (high-energy) electrons. These chlorophylls are arranged so that the energy of an excited electron can be passed from one molecule to another, until finally the energy is funneled into two chlorophyll molecules called the *special pair* (Figure 14–30). These two chlorophyll molecules are located in the reaction center, a protein complex that sits adjacent to the antenna complex in the membrane. There the energy is trapped and used to energize one electron in the special pair of chlorophyll molecules.

The reaction center is a transmembrane complex of proteins and organic pigments that lies at the heart of photosynthesis. It is thought to have first evolved more than 3 billion years ago in primitive photosynthetic bacteria. Detailed structural and functional studies have revealed how it functions at an atomic level of detail (Movie 14.6). The reaction center acts as an irreversible trap for an excited electron, because the special pair of chlorophylls are poised to pass the high-energy electron to a precisely positioned neighboring molecule in the same protein complex. By moving the energized electron rapidly away from the chlorophylls, a process known as *charge separation*, the reaction center transfers this high-energy electron to an environment where it is much more stable.

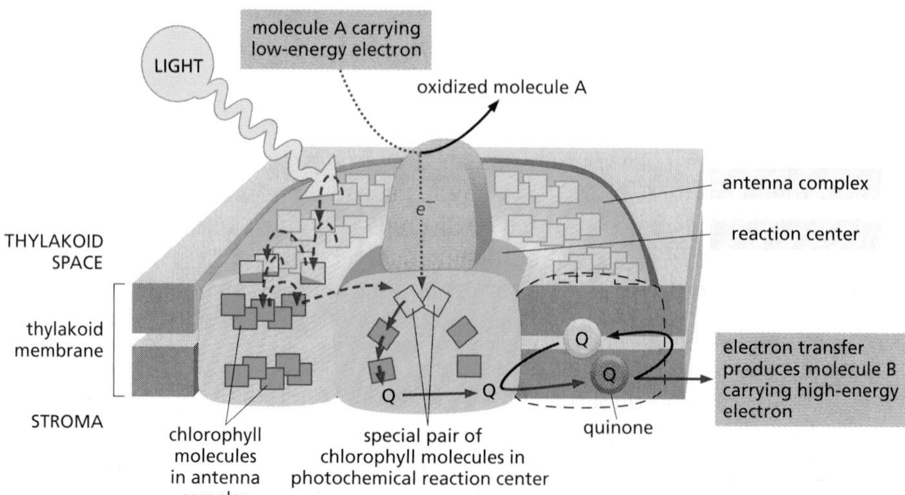

Figure 14–30 **A photosystem contains a reaction center and an antenna.** The antenna collects the energy of electrons that have been excited by light and funnels this energy (energy transfers shown by *red* dashed arrows) to a special pair of chlorophyll molecules in the reaction center. This produces a high-energy electron in the special pair that can be passed rapidly (electron transfers shown as *red* solid arrows) to the electron-transport chain in the thylakoid membrane, via the quinone (Q). In addition to chlorophyll, the antenna contains additional accessory pigments (not shown) that help to capture light energy of different wavelengths. A protein in the reaction center (*orange*) collects the low-energy electrons needed to return the system to its original unexcited state (electron transfers shown as *red* dotted arrows), as we shall see in Figure 14–31.

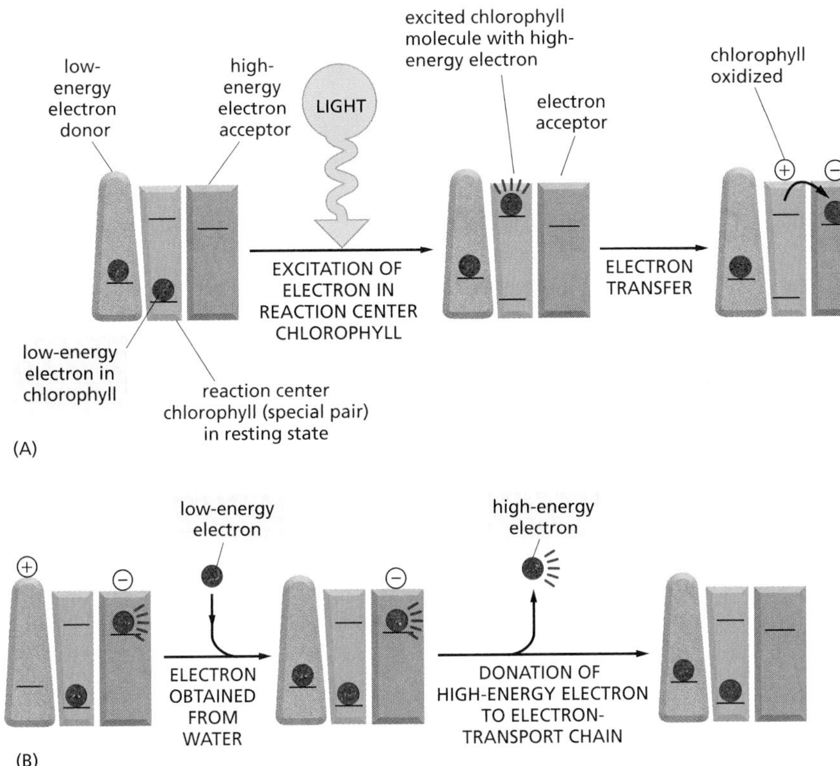

Figure 14–31 **Light energy is harvested by a reaction center chlorophyll molecule.** (A) The two chlorophyll molecules in the special pair (*blue*) are tightly held in a pigment–protein complex, positioned so that both a protein that can collect low-energy electrons (*orange*) and a high-energy electron acceptor (*green*) are available (see Figure 14–30). When light energizes an electron in the special pair, the excited electron (*red*) is rapidly passed to an electron acceptor in the reaction center, stabilizing it as a high-energy electron. The positively charged, oxidized chlorophyll then quickly attracts a low-energy electron from the electron donor, thereby returning to a resting state. These reactions, which create a charge separation between the electron donor and the electron acceptor, require less than 10^{-6} second to complete. (B) In the final stage of the process, which follows the steps in (A), the entire reaction center is restored to its original state by both extracting a new, low-energy electron (from water in this case) and passing its high-energy electron on to the electron-transport chain. Thus, in the overall reaction, low-energy electrons obtained from water are being consumed and high-energy electrons are being produced in the plant thylakoid membrane.

When the chlorophyll molecule in the reaction center loses an electron, it becomes positively charged; it then rapidly regains an electron from an adjacent electron donor to return to its unexcited, uncharged state (Figure 14–31A). Then, in slower reactions, the electron donor has its missing electron replaced with an electron removed from water. The high-energy electron that was generated by the excited chlorophyll is then transferred to the electron-transport chain. This transfer leaves the reaction center ready to receive the next high-energy electron excited by sunlight (Figure 14–31B).

Light Energy Drives the Synthesis of Both ATP and NADPH

In mitochondria, the electron-transport chain functions solely to generate ATP. But in the chloroplast, and in free-living photosynthetic organisms such as cyanobacteria, electron transport has an additional role: it also produces the activated carrier molecule NADPH (see Figure 3–34). NADPH is needed because photosynthesis is ultimately a biosynthetic process. To build organic molecules from CO_2, a cell requires a huge input of energy, in the form of ATP, and a very large amount of reducing power, in the form of NADPH. To produce this NADPH from $NADP^+$, the cell uses energy captured from sunlight to convert the low-energy electrons in water to the high-energy electrons in NADPH.

To produce both ATP and NADPH, plant cells and cyanobacteria use two photons of light: ATP is made after the first photon is absorbed, NADPH after the second. These photons are absorbed by two different photosystems that operate in series. Working together, these photosystems impart to an electron a high enough energy to produce NADPH. Along the way, a proton gradient is generated, allowing ATP to be made.

In outline, the process works as follows: the first photon of light is absorbed by one photosystem (which is paradoxically called photosystem II for historical reasons). As we have seen, that photon is used to produce a high-energy electron that is handed off to an electron-transport chain (see Figure 14–31). While traveling down the electron-transport chain, the electron drives an H⁺ pump in the thylakoid membrane and creates a proton gradient in the manner described previously for oxidative phosphorylation. An ATP synthase in the thylakoid membrane then uses this proton gradient to drive the synthesis of ATP on the stromal side of the membrane (Figure 14–32).

In the meantime, the electron-transport chain delivers the electron generated by photosystem II to the second photosystem in the pathway (called photosystem I). There the electron fills the positively charged 'hole' that was left in the reaction center of photosystem I when it absorbed the second photon of light. Because photosystem I starts at a higher energy level than photosystem II, it is able to boost electrons to the very high energy level needed to make NADPH from $NADP^+$ (see Figure 14–32). The redox potentials of the components along this electron-transport chain are shown in Figure 14–33.

In the overall process described thus far, we have seen that an electron removed from a chlorophyll molecule at the reaction center of photosystem II travels all the way through the electron-transport chain in the thylakoid membrane until it winds up being donated to NADPH. This initial electron must be replaced to return the system to its unexcited state. The replacement electron comes from a low-energy electron donor, which, in plants and many photosynthetic bacteria, is water (see Figure 14–31B). The reaction center of photosystem II includes a water-splitting enzyme that holds the oxygen atoms of two water molecules bound to a cluster of manganese atoms in the protein (Figure 14–34; see also Figure 14–32). This enzyme removes electrons one at a time from the water to fill the holes created by light in the chlorophyll molecules of the reaction center. When four electrons have been removed from two water molecules (which requires four photons of light), O_2 is released. It is this critical process, occurring over billions of years, that has generated all of the O_2 in the Earth's atmosphere.

QUESTION 14–9

Both NADPH and the related carrier molecule NADH are strong electron donors. Why might plant cells have evolved to rely on NADPH, rather than NADH, to provide the reducing power for photosynthesis?

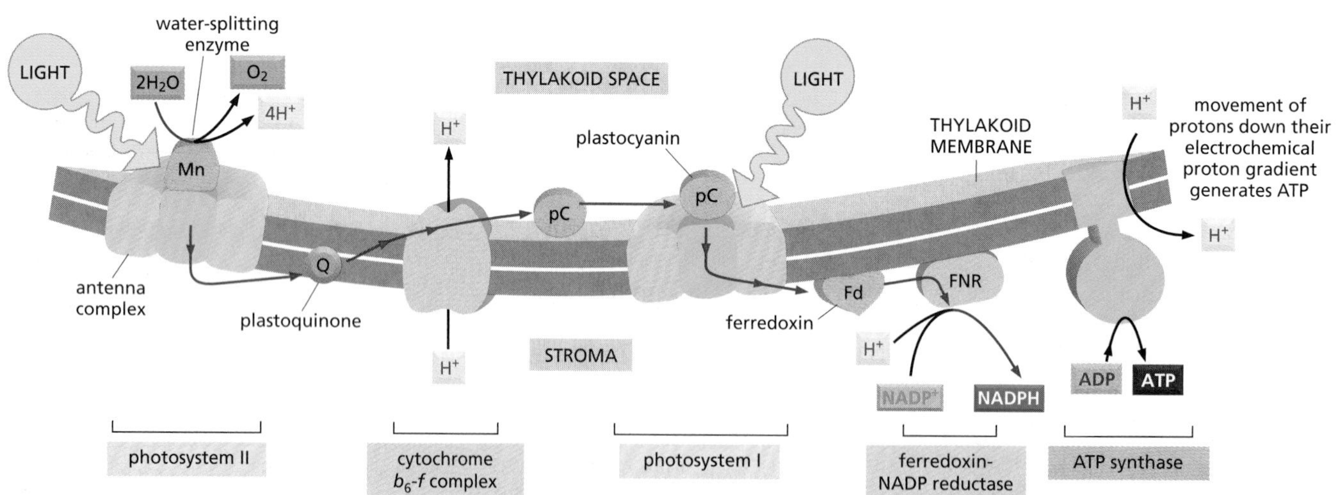

Figure 14–32 **During photosynthesis, electrons travel down an electron transport chain in the thylakoid membrane.** Light energy is harvested by the antenna complexes in each of the two membrane-embedded photosystems and funneled to a special pair of chlorophyll molecules in the reaction center. As shown in Figure 14–30, an excited, high-energy electron is thereby created in the special pair, which is transferred through a series of electron acceptors inside the reaction center before being passed to the electron-transport chain in the chloroplast membrane. The mobile electron carriers in the chloroplast electron-transport chain are plastoquinone (Q, which closely resembles the ubiquinone of mitochondria), plastocyanin (pC, a small, copper-containing protein), and ferredoxin (Fd, a small protein containing an iron–sulfur center). The cytochrome b_6-f complex resembles the cytochrome b-c_1 complex of mitochondria, and it is the sole site of active H⁺ pumping in the chloroplast electron-transport chain. The H⁺ released by the splitting of water by photosystem II and the H⁺ taken up during NADPH formation by the protein ferredoxin-NADP reductase (FNR), the last protein in the electron transport chain, also contribute to generating the electrochemical proton gradient. As indicated, the proton gradient drives an ATP synthase located in the same membrane to generate ATP. An overview of the photosynthetic reactions is shown in Movie 14.7.

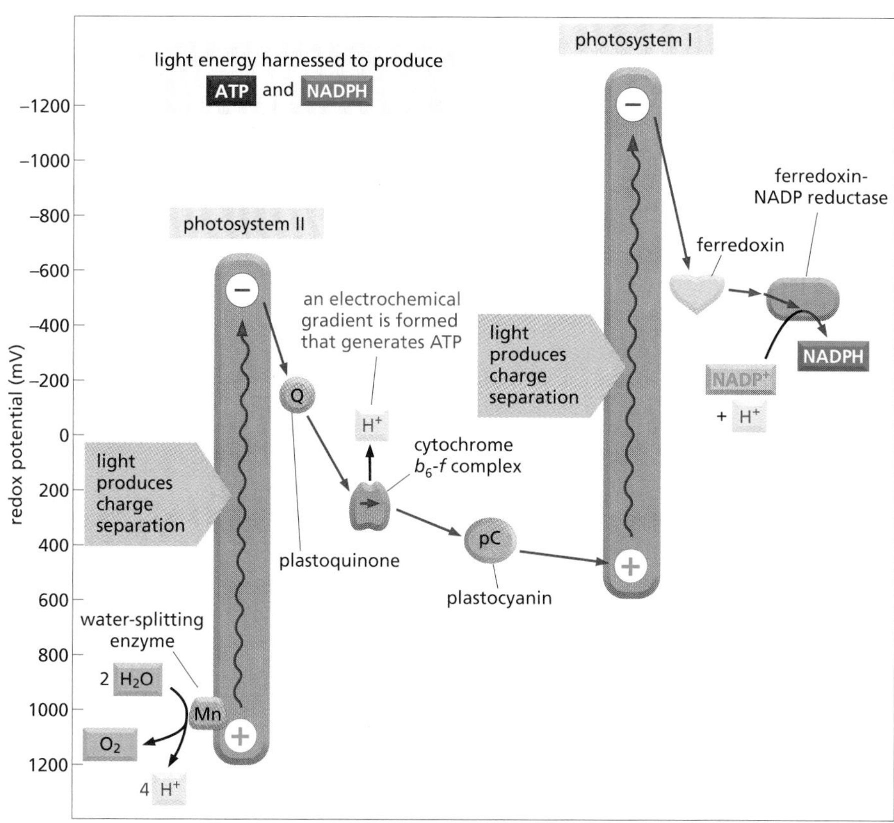

Figure 14–33 **The coupling of photosystems I and II boosts electrons to the energy level needed to produce NADPH.** The redox potential for each molecule is indicated by its position on the vertical axis. Photosystem II passes electrons from its excited chlorophyll special pair through an electron-transport chain in the thylakoid membrane that leads to photosystem I. The net electron flow through the two photosystems linked in series is from water to $NADP^+$, and it produces NADPH as well as ATP. The ATP is synthesized by an ATP synthase (not shown here) that harnesses the electrochemical proton gradient produced by electron transport.

Chloroplasts Can Adjust their ATP Production

In addition to carrying out the photosynthetic process outlined so far, chloroplasts can also generate ATP without making NADPH. To produce this extra ATP, chloroplasts can switch photosystem I into a cyclic mode so that it produces ATP instead of NADPH. In this process, called **cyclic photophosphorylation**, the high-energy electrons produced by light activation of photosystem I are transferred back to the cytochrome b_6-f complex rather than being passed on to $NADP^+$. From the b_6-f complex, the electrons are handed back to photosystem I at low energy (Figure 14–35). The net result, aside from the conversion of some light energy to heat, is that H^+ is pumped across the thylakoid membrane by the b_6-f complex as electrons pass through it. This cycle increases the electrochemical proton gradient that drives the synthesis of ATP. Cells adjust the relative amounts of cyclic photophosphorylation (which involves only photosystem I) and the standard, noncyclic form of phosphorylation

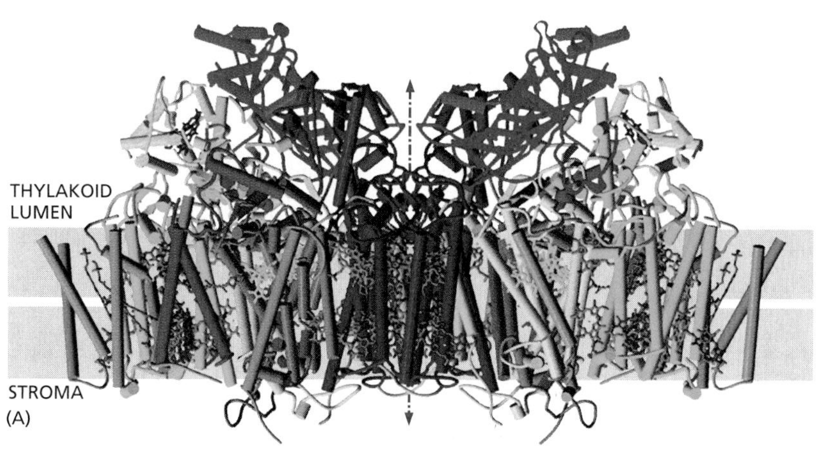

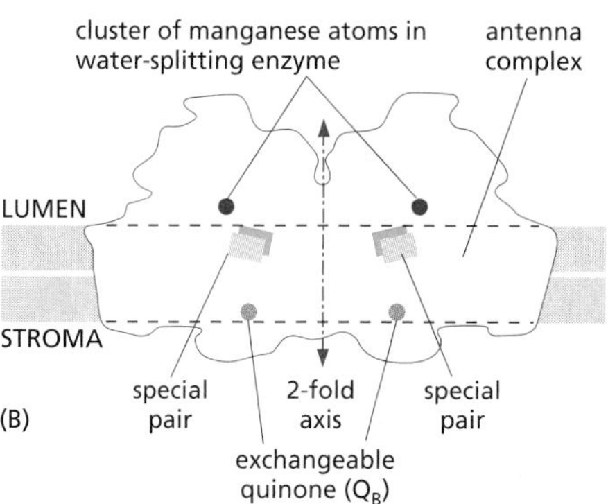

Figure 14–34 **The complete three-dimensional structures of photosystems I and II are known.** This diagram shows the structure of photosystem II, which is a dimeric complex that contains more than 20 proteins and several dozen chlorophylls. (Adapted from K. N. Ferreira et al., *Science* 303:1831–1838, 2004. With permission from AAAS.)

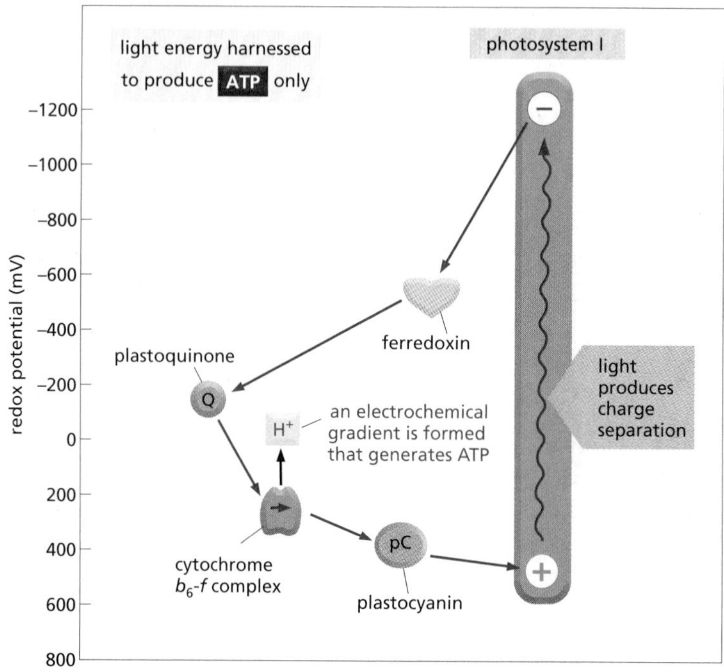

Figure 14–35 **Chloroplasts can also produce ATP by cyclic photophosphorylation.** This pathway allows ATP to be made without producing either NADPH or O_2. When NADPH accumulates, the cell tends to favor this cyclic scheme.

(which involves both photosystems I and II) depending on their relative need for reducing power (in the form of NADPH) and high-energy phosphate bonds (in the form of ATP).

Carbon Fixation Uses ATP and NADPH to Convert CO_2 into Sugars

The light reactions of photosynthesis generate ATP and NADPH in the chloroplast stroma. But the inner membrane of the chloroplast is impermeable to both of these compounds, which means that they cannot be exported directly to the cytosol. To provide reducing power and energy for the rest of the cell, ATP and NADPH are instead used within the chloroplast stroma to produce sugars that can then be directly exported. This sugar production, which occurs during the dark reactions of photosynthesis, is called **carbon fixation**.

The central reaction of photosynthetic carbon fixation, in which an atom of inorganic carbon (as CO_2) is converted to organic carbon, is illustrated in Figure 14–36. CO_2 from the atmosphere combines with the five-carbon sugar derivative *ribulose 1,5-bisphosphate* plus water to give two molecules of the three-carbon compound 3-phosphoglycerate. This carbon-fixing reaction, which was discovered in 1948, is catalyzed in the chloroplast stroma by a large enzyme called *ribulose bisphosphate carboxylase* (also called *ribulose bisphosphate carboxylase/oxygenase* or *Rubisco*). Because this enzyme works extremely sluggishly compared with most other enzymes (processing about three molecules of substrate per second

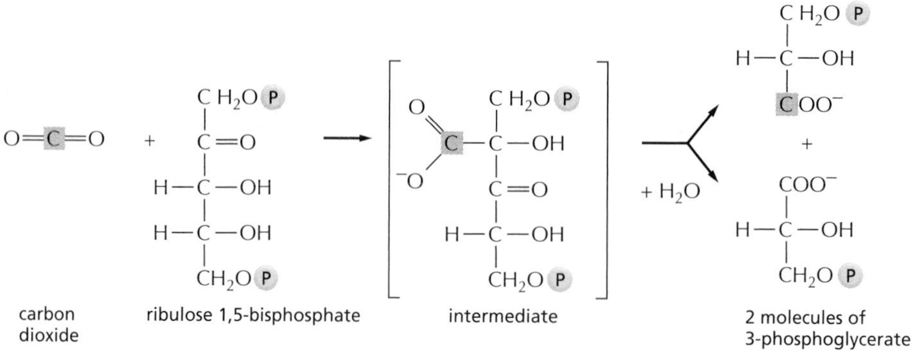

carbon dioxide + ribulose 1,5-bisphosphate → intermediate + H_2O → 2 molecules of 3-phosphoglycerate

Figure 14–36 **Carbon fixation involves the formation of a covalent bond that attaches carbon dioxide to ribulose 1,5-bisphosphate.** The reaction is catalyzed in the chloroplast stroma by the abundant enzyme ribulose bisphosphate carboxylase. As shown, the product is two molecules of 3-phosphoglycerate.

compared with 1000 molecules per second for a typical enzyme), very large numbers of enzyme molecules are needed by the plant. Ribulose bisphosphate carboxylase often represents more than 50% of the total chloroplast protein, and it is widely claimed to be the most abundant protein on Earth.

When carbohydrates are broken down and oxidized to CO_2 and H_2O by cells, a large amount of free energy is released. Clearly, the reverse overall reaction—in which CO_2 and H_2O combine to make carbohydrate during photosynthesis—must be energetically very unfavorable. For this process to occur, it must be coupled to an energetically favorable reaction that drives it. The reaction in which CO_2 is fixed by Rubisco is in fact energetically favorable, but only because it receives a continuous supply of the energy-rich compound ribulose 1,5-bisphosphate, to which each molecule of CO_2 is added (see Figure 14–36). The energy and reducing power required for the elaborate metabolic pathway by which this compound is regenerated are provided by the ATP and NADPH produced by the photosynthetic light reactions.

The series of reactions that allows cells to incorporate CO_2 into sugars forms a cycle that begins and ends with ribulose 1,5-bisphosphate (Figure 14–37). For every three molecules of CO_2 that enter the cycle, one new molecule of *glyceraldehyde 3-phosphate* is produced—the three-carbon sugar that is the net product—and three molecules of ATP and two molecules of NADPH are consumed. Glyceraldehyde 3-phosphate then provides the starting material for the synthesis of many other sugars and organic molecules. The *carbon-fixation cycle* (or Calvin cycle) was worked out in the 1940s and 1950s in one of the first successful applications of radioisotopes as tracers in biochemistry.

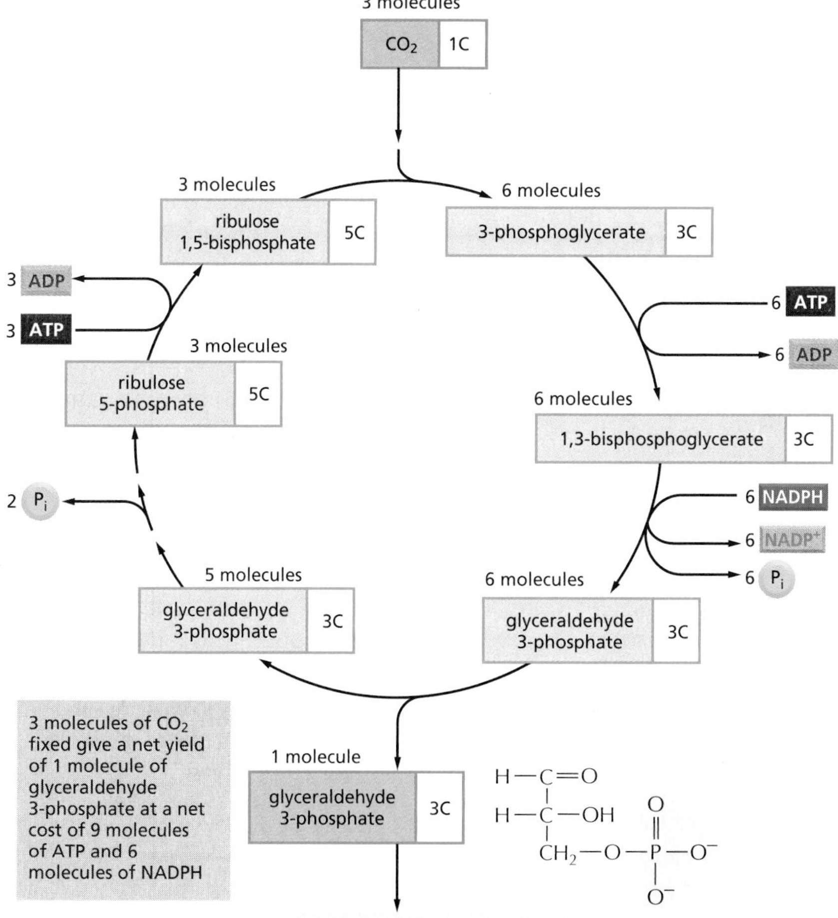

Figure 14–37 **The carbon-fixation cycle forms organic molecules from CO_2 and H_2O.** The cycle begins with the reaction shown in Figure 14–36, and it produces glyceraldehyde 3-phosphate. The number of carbon atoms in each type of molecule is indicated in the *white box*. There are many intermediates between glyceraldehyde 3-phosphate and ribulose 5-phosphate, but they have been omitted here for clarity. The entry of water into the cycle is also not shown.

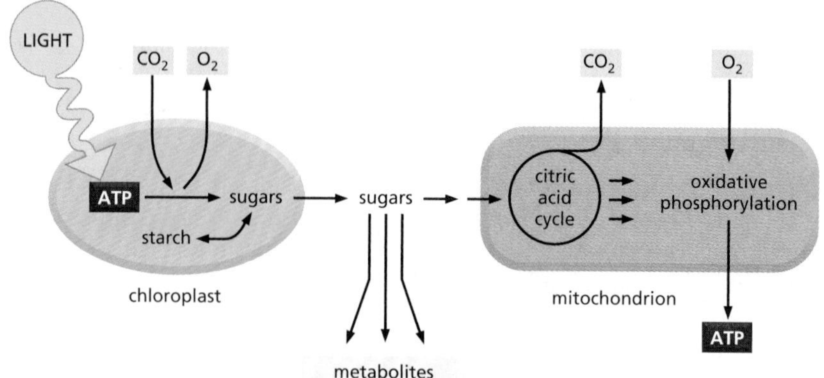

Figure 14–38 **In plants, the chloroplasts and mitochondria collaborate to supply cells with metabolites and ATP.** The chloroplast's inner membrane is impermeable to the ATP and NADPH that are produced during the light reactions of photosynthesis. These molecules are therefore funneled into the carbon-fixation cycle, where they are used to make sugars. The resulting sugars are either stored within the chloroplast—in the form of starch—or exported to the rest of the plant cell. There, they can enter the energy-generating pathway that ends in ATP synthesis in the mitochondria. Mitochondrial membranes are permeable to ATP, as indicated.

Sugars Generated by Carbon Fixation Can Be Stored As Starch or Consumed to Produce ATP

The glyceraldehyde 3-phosphate generated by carbon fixation in the chloroplast can be used in a number of different ways, depending on the needs of the plant. During periods of excess photosynthetic activity, glyceraldehyde 3-phosphate is retained in the chloroplast, where it is mainly converted to *starch* in the stroma (Figure 14–38). Like glycogen in animal cells, starch is a large polymer of glucose that serves as a carbohydrate reserve. Starch is stored as large grains in the chloroplast stroma (see Figure 14–25B), and at night it is broken down to sugars to help support the metabolic needs of the plant. Starch forms an important part of the diet of all animals that eat plants.

But the chloroplast is not merely a storage depot. Much of the glyceraldehyde 3-phosphate produced in chloroplasts is moved out of the chloroplast into the cytosol. Some of it enters the glycolytic pathway (see Figure 13–5), where it is converted to pyruvate; this pyruvate then enters the citric acid cycle in the plant cell mitochondria and leads to the production of ATP by oxidative phosphorylation (see Figure 14–38). This is the ATP the plant uses for its general metabolism, and it is synthesized in mitochondria in the same way as in animal cells and other nonphotosynthetic organisms.

The glyceraldehyde 3-phosphate exported from chloroplasts can also be converted into many other metabolites, including the disaccharide *sucrose*. Sucrose is the major form in which sugar is transported between plant cells: just as glucose is transported in the blood of animals, sucrose is exported from the leaves via the vascular bundle to provide carbohydrate to the rest of the plant.

QUESTION 14–10

A. How do cells in plant roots survive, since they contain no chloroplasts and are not exposed to light?

B. Unlike mitochondria, chloroplasts do not have a transporter that allows them to export ATP to the cytosol. How, then, do plant cells obtain the ATP that they need to carry out energy-requiring metabolic reactions in the cytosol?

THE ORIGINS OF CHLOROPLASTS AND MITOCHONDRIA

It is now widely accepted that chloroplasts and mitochondria most likely evolved from bacteria that were engulfed by ancestral eucaryotic cells more than a billion years ago (see Figures 1–19 and 1–21). As a relic of this evolutionary past, both types of organelles contain their own genomes, as well as their own biosynthetic machinery for making RNA and organelle proteins. The way in which mitochondria and chloroplasts reproduce—through the growth and division of preexisting organelles—provides additional evidence of their bacterial ancestry (Figure 14–39).

The growth and proliferation of mitochondria and chloroplasts is complicated by the fact that their component proteins are encoded by two separate genetic systems—one in the organelle and one in the cell

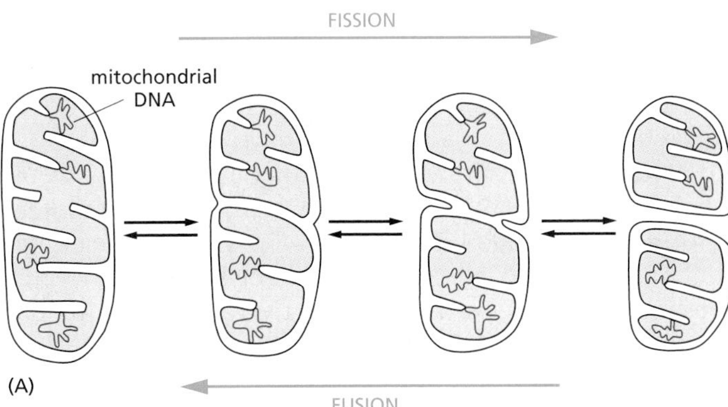

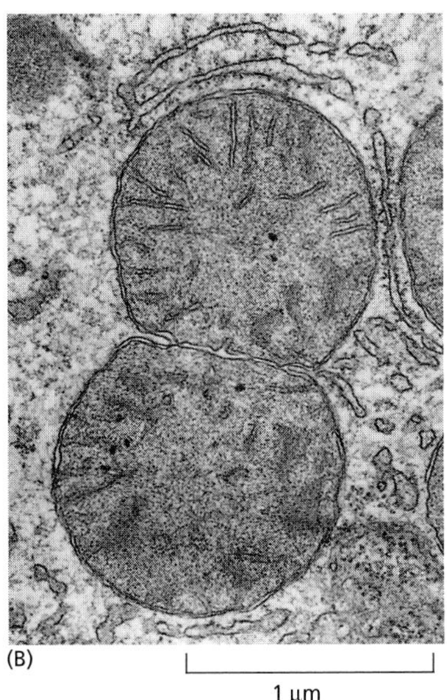

Figure 14–39 **A mitochondrion divides like a bacterium.** (A) Both mitochondrial fission and mitochondrial fusion are observed to occur. The fission process is conceptually similar to bacterial division processes. (B) An electron micrograph of a dividing mitochondrion in a liver cell. (B, Courtesy of Daniel S. Friend.)

nucleus. In the mitochondrion, most of the original bacterial genes have become transposed to the cell nucleus, leaving only relatively few genes inside the organelle itself. Animal mitochondria in fact contain a uniquely simple genetic system: the human mitochondrial genome, for example, contains only 16,569 nucleotide pairs of DNA encoding 37 genes. The vast majority of mitochondrial proteins—including those needed to make the mitochondrion's RNA polymerase and ribosomal proteins, and all of the enzymes of its citric acid cycle—are instead produced from nuclear genes, and these proteins must therefore be imported into the mitochondria from the cytosol, where they are made (discussed in Chapter 15).

Like the mitochondrion, the chloroplast contains many of its own genes, as well as a complete transcription and translation system for producing proteins from these genes. Chloroplast genomes are considerably larger than mitochondrial genomes: in higher plants, for example, the chloroplast genome contains about 120 genes in 120,000 nucleotide pairs. These genes are strikingly similar to those of cyanobacteria, the photosynthetic bacteria from which chloroplasts are thought to have been derived. Even so, many chloroplast proteins are now encoded by nuclear genes and must be imported from the cytosol.

The same techniques that have allowed us to analyze the genomes of mitochondria and chloroplasts have also permitted us to identify and explore the molecular biology of many microorganisms on the Earth. Some of these organisms thrive in the most inhospitable habitats on the planet. These include sulfurous hot springs or hydrothermal vents deep on the ocean floor. In these seemingly odd modern microbes, we can readily find clues to life's history—in the form of the many molecules from which they are made. Like the fingerprints left at the scene of a crime, these molecules provide powerful evidence that allows us to trace the history of ancient events, permitting speculations on the origin of the ATP-generating systems that are found in today's mitochondria and chloroplasts. We therefore end this chapter with a discussion of the evolution of the energy-harvesting systems that we have discussed in detail previously in this chapter.

Oxidative Phosphorylation Might Have Given Ancient Bacteria an Evolutionary Advantage

As we mentioned earlier, the first living cells on Earth—both procaryotes and primitive eucaryotes—may have consumed geochemically produced organic molecules and generated ATP by fermentation. Because oxygen

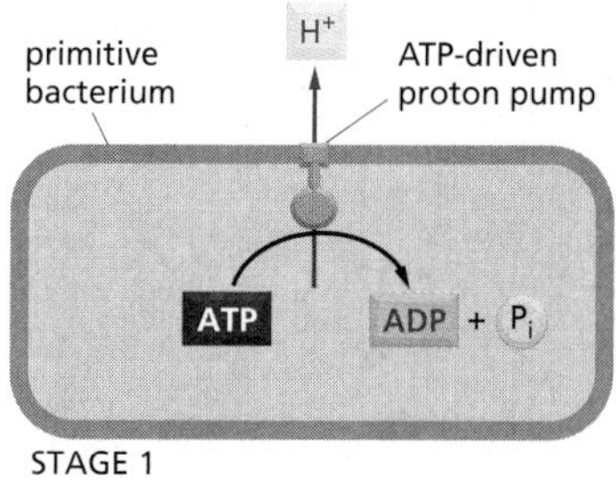

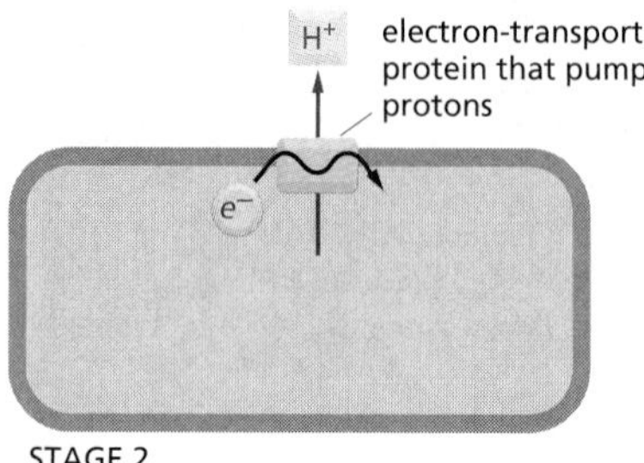

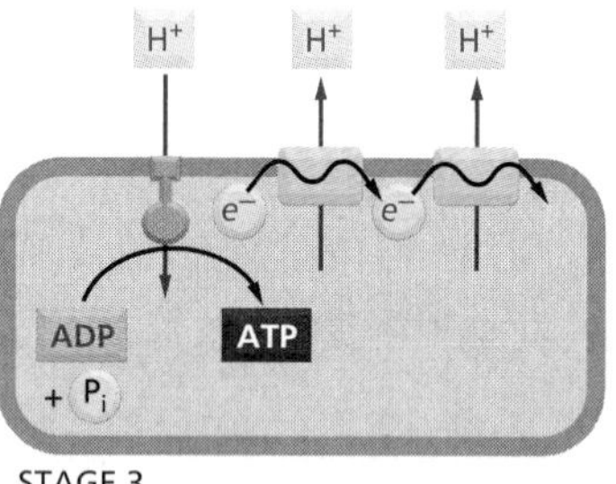

Figure 14–40 **Oxidative phosphorylation might have evolved in stages.** The first stage could have involved evolution of an ATPase that pumped protons out of the cell using the energy of ATP hydrolysis; stage 2 could have involved the evolution of a different proton pump driven by an electron-transport chain; stage 3 would have been the linking of these two systems together to generate an ATP synthase that uses the protons pumped by the electron-transport chain to synthesize ATP. A bacterium with this final system would have had a selective advantage over bacteria with only one of the systems or none.

was not yet present in the atmosphere, such anaerobic fermentation reactions would have dumped organic acids—such as lactic or formic acids, for example—into the environment (see Figure 13–4A).

Such an excretion of organic acids would have lowered the pH of the environment, favoring the survival of cells that evolved transmembrane proteins that could pump H^+ out of the cytosol, keeping the cell from becoming too acidified (stage 1 in Figure 14–40). One of these pumps may have used the energy available from ATP hydrolysis to eject H^+ from the cell; such a protein pump could have been the ancestor of the present-day ATP synthase.

As the Earth's supply of geochemically produced nutrients began to dwindle, organisms that could find a way to pump H^+ without consuming ATP would have been at an advantage: they could save the small amounts of ATP they derived from the fermentation of increasingly scarce foodstuffs to fuel other important activities. Selective pressures such as the scarcity of nutrients might therefore have led to the evolution of electron-transport proteins; these proteins allowed cells to use the movement of electrons between molecules of different redox potentials as an energy source for transporting H^+ across the plasma membrane (stage 2 in Figure 14–40). Some of these cells might have used the nonfermentable organic acids that neighboring cells had excreted as waste to provide the electrons needed to feed the system. Some present-day bacteria grow on formic acid, for example, using the small amount of redox energy derived from the transfer of electrons from formic acid to fumarate to pump H^+.

Eventually some bacteria would have developed H^+-pumping electron-transport systems that were so efficient that they could harvest more redox energy than they needed to maintain their internal pH. These cells most likely generated large electrochemical proton gradients, which they could then use to produce ATP. Protons could leak back into the cell through the ATP-driven H^+ pumps, essentially running them in reverse so that they synthesized ATP (stage 3 in Figure 14–40). Because such cells required much less of the dwindling supply of fermentable nutrients, they would have proliferated at the expense of their neighbors.

Photosynthetic Bacteria Made Even Fewer Demands on Their Environment

The major evolutionary breakthrough in energy metabolism, however, was almost certainly the formation of photochemical reaction centers that could use the energy of sunlight to produce molecules such as NADH. It is thought that this development occurred early in the process of evolution—more than 3 billion years ago, in the ancestors of the green sulfur bacteria. Present-day green sulfur bacteria use light energy to transfer hydrogen atoms (as an electron plus a proton) from H_2S to NADPH, thereby creating the strong reducing power required for carbon fixation (Figure 14–41).

The next step, which is thought to have occurred with the rise of cyanobacteria (see Figure 14–24), was the evolution of organisms capable of using water as the electron source for photosynthesis. This entailed the evolution of a water-splitting enzyme and the addition of a second photosystem, acting in tandem with the first, to bridge the enormous gap in redox potential between H_2O and NADPH (see Figure 14–33). The biological consequences of this evolutionary step were far-reaching. For the first time, there were organisms that made only very minimal chemical demands on their environment. These cells could spread and evolve in ways denied to the earlier photosynthetic bacteria, which needed H_2S or organic acids as a source of electrons. Consequently, large amounts

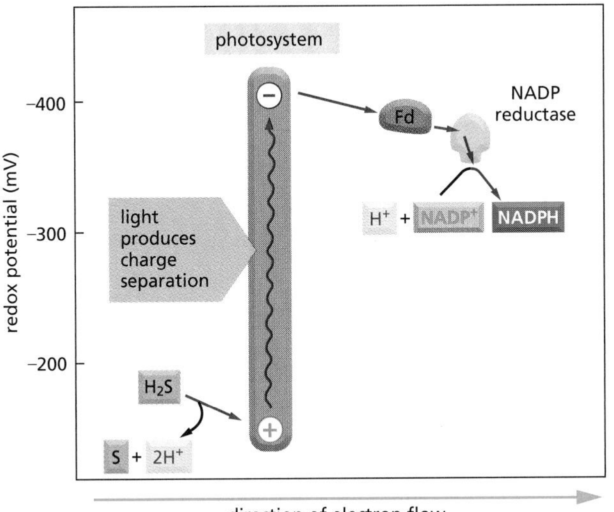

Figure 14–41 **Photosynthesis in green sulfur bacteria uses hydrogen sulfide (H_2S) as an electron donor rather than water.** Electrons are easier to extract from H_2S than from H_2O, because H_2S has a much higher redox potential (see Figure 14–33). Therefore, only one photosystem is needed to produce NADPH, and elemental sulfur is formed as a byproduct instead of O_2. The photosystem in green sulfur bacteria resembles photosystem I in plants and cyanobacteria in that they all use a series of iron–sulfur centers as the electron acceptors that eventually donate their high-energy electrons to ferredoxin (Fd). A bacterium of this type is *Chlorobium tepidum*, which can thrive at high temperatures and low light intensities in hot springs.

of biologically synthesized, fermentable organic materials accumulated. Moreover, O_2 entered the atmosphere in large amounts (Figure 14–42).

The availability of O_2 made possible the development of bacteria that relied on aerobic metabolism to make their ATP. As explained previously, these organisms could harness the large amount of energy released by breaking down carbohydrates and other reduced organic molecules all the way to CO_2 and H_2O.

As organic materials accumulated as a by-product of photosynthesis, some photosynthetic bacteria—including the ancestors of *E. coli*—lost their ability to survive on light energy alone and came to rely entirely on cellular respiration. Mitochondria probably arose when a primitive eucaryotic cell engulfed such a respiration-dependent bacterium. And plants arose somewhat later, when a descendant of this early aerobic eucaryote captured a photosynthetic bacterium, which became the precursor of the chloroplast. Once eucaryotes had acquired the bacterial

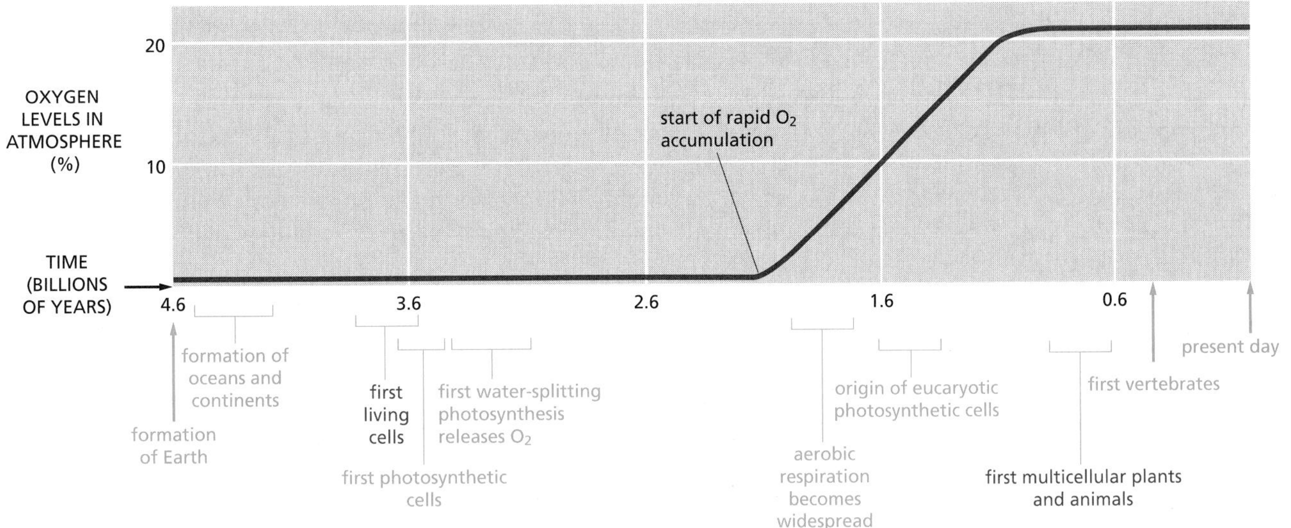

Figure 14–42 **Life on Earth has evolved over billions of years.** With the evolution of the membrane-based process of photosynthesis more than 3 billion years ago, organisms were no longer dependent on preformed organic chemicals. They could now make their own organic molecules from CO_2 gas. The delay of more than a billion years between the appearance of bacteria that split water and released O_2 during photosynthesis and the accumulation of high levels of O_2 in the atmosphere is thought to have been due to the initial reaction of the oxygen with abundant ferrous iron (Fe^{2+}) dissolved in the early oceans. Only when the iron was used up would oxygen have started to accumulate in the atmosphere. In response to the rising amount of oxygen in the atmosphere, nonphotosynthetic oxygen-using organisms appeared, and the concentration of oxygen in the atmosphere eventually leveled out.

symbionts that became mitochondria and chloroplasts, they could then embark on the amazing pathway of evolution that eventually led to complex multicellular organisms.

The Lifestyle of *Methanococcus* Suggests That Chemiosmotic Coupling Is an Ancient Process

The conditions today that most resemble those under which cells are thought to have lived 3.5–3.8 billion years ago may be those near deep-ocean hydrothermal vents. These vents represent places where the Earth's molten mantle is breaking through the crust, expanding the width of the ocean floor. Indeed, the modern organisms that appear to be most closely related to the hypothetical cells from which all life evolved live at high temperatures (75°C to 95°C, close to the temperature of boiling water). This ability to thrive at such extreme temperatures suggests that life's common ancestor—the cell that gave rise to bacteria, archaea, and eucaryotes—lived under very hot, anaerobic conditions.

One of the archaea that live in this environment today is *Methanococcus jannaschii.* Originally isolated from a hydrothermal vent more than a mile beneath the ocean surface, the organism grows entirely on inorganic nutrients in the complete absence of light and gaseous oxygen, utilizing as nutrients hydrogen gas (H_2), CO_2, and nitrogen gas (N_2) that bubble up from the vent. Its mode of existence gives us a hint of how early cells might have used electron transport to derive their energy and their carbon molecules from inorganic materials that were freely available on the hot early Earth.

Methanococcus relies on N_2 gas as its source of nitrogen for organic molecules such as amino acids. The organism reduces N_2 to ammonia (NH_3) by the addition of hydrogen, a process called **nitrogen fixation**. Nitrogen fixation requires a large amount of energy, as does the carbon-fixation process that the bacterium needs to convert CO_2 into sugars. Much of the energy required for both processes is derived from the transfer of electrons from H_2 to CO_2, with the release of large amounts of methane (CH_4) as a waste product (thus producing natural gas and giving the organism its name; Figure 14–43). Part of this electron transfer occurs in the membrane and results in the pumping of protons (H^+) across it. The resulting electrochemical proton gradient drives an ATP synthase in the same membrane to make ATP.

The fact that such chemiosmotic coupling exists in an organism as primitive as *Methanococcus* suggests that the storage of energy derived from electron transport in an H^+ gradient is an extremely ancient process. Thus, chemiosmotic coupling is likely to have fueled the evolution of nearly all life forms on Earth.

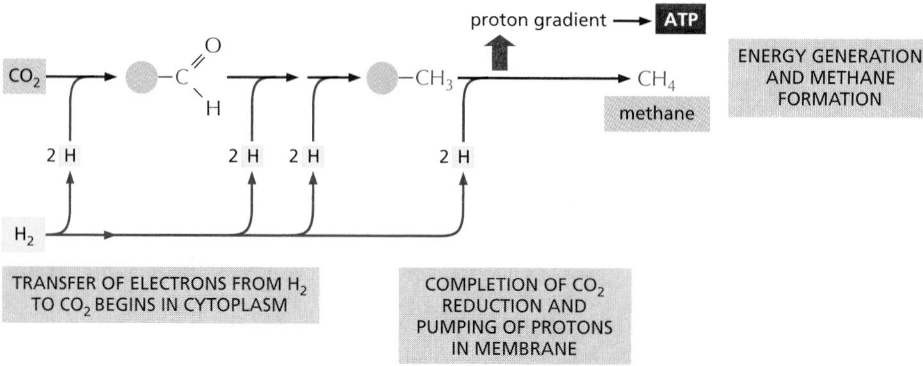

Figure 14–43 ***Methanococcus* uses chemiosmotic coupling to generate energy.** This deep-sea archaean uses hydrogen gas (H_2) as the source of reducing power for energy generation. The initial reduction steps take place via enzyme-catalyzed reactions in the cytoplasm. In contrast, the final reduction step involves a membrane-based electron transfer that generates a proton gradient driving ATP synthesis, while producing methane as a waste product. The *green* circles represent special coenzymes to which the metabolic intermediates are bound.

ESSENTIAL CONCEPTS

- Mitochondria, chloroplasts, and many bacteria produce ATP by a membrane-based mechanism known as chemiosmotic coupling.
- Mitochondria produce most of an animal cell's ATP, using energy derived from oxidation of sugars and fatty acids.
- Mitochondria have an inner and an outer membrane. The inner membrane encloses the mitochondrial matrix, a compartment that contains many enzymes, including those of the citric acid cycle. These enzymes produce large amounts of NADH and $FADH_2$ from the oxidation of acetyl CoA.
- In the inner mitochondrial membrane, high-energy electrons donated by NADH and $FADH_2$ pass along an electron-transport chain—the respiratory chain—eventually combining with molecular oxygen (O_2) in an energetically favorable reaction.
- Much of the energy released by electron transfers along the respiratory chain is harnessed to pump H^+ out of the matrix, thereby creating a transmembrane proton (H^+) gradient. The proton pumping is carried out by three large respiratory enzyme complexes embedded in the membrane.
- The resulting electrochemical proton gradient across the inner mitochondrial membrane is harnessed to make ATP when H^+ ions flow back into the matrix through ATP synthase, an enzyme located in the inner mitochondrial membrane.
- The proton gradient also drives the active transport of metabolites into and out of the mitochondrion.
- In photosynthesis in chloroplasts and photosynthetic bacteria, high-energy electrons are generated when sunlight is absorbed by chlorophyll; this energy is captured by protein complexes known as photosystems, which in plant cells are located in the thylakoid membranes of chloroplasts.
- Electron-transport chains associated with photosystems transfer electrons from water to $NADP^+$ to form NADPH. O_2 is generated as a by-product.
- The electron-transport chains in chloroplasts also generate a proton gradient across the thylakoid membrane. As in mitochondria, this electrochemical proton gradient is used by an ATP synthase embedded in the membrane to generate ATP.
- The ATP and the NADPH made by photosynthesis are used within the chloroplast to drive the carbon-fixation cycle in the chloroplast stroma, thereby producing carbohydrate from CO_2.
- Carbohydrate is exported to the cell cytosol, where it is metabolized to provide organic carbon, ATP (mostly via mitochondria), and reducing power for the rest of the cell.
- Both mitochondria and chloroplasts are thought to have evolved from bacteria that were endocytosed by primitive eucaryotic cells. Each retains its own genome and divides by processes that resemble a bacterial cell division.
- Chemiosmotic coupling mechanisms are widespread and of ancient origin. Modern microorganisms that live in environments similar to those thought to have been present on the early Earth also use chemiosmotic coupling to produce ATP.

KEY TERMS

antenna complex
ATP synthase
carbon fixation
chemiosmotic coupling
chlorophyll
chloroplast
cyclic photophosphorylation
cytochrome
cytochrome oxidase
electron-transport chain
iron–sulfur center
matrix
mitochondrion
nitrogen fixation
oxidative phosphorylation
photosynthesis
photosystem
quinone
reaction center
redox pair
redox potential
redox reaction
stroma

QUESTIONS

QUESTION 14–11

Which of the following statements are correct? Explain your answers.

A. After an electron has been removed by light, the affinity for electrons of the positively charged chlorophyll in the reaction center of the first photosystem (photosystem II) is even greater than the electron affinity of O_2.

B. Photosynthesis is the light-driven transfer of an electron from chlorophyll to a second molecule with a much lower affinity for electrons.

C. Because it requires the absorption of four photons to release one O_2 molecule from two H_2O molecules, the water-splitting enzyme has to keep the reaction intermediates tightly bound so as to prevent partly reduced, and therefore hazardous, superoxide radicals from escaping.

QUESTION 14–12

Which of the following statements are correct? Explain your answers.

A. Many, but not all, electron-transfer reactions involve metal ions.

B. The electron-transport chain generates an electrical potential across the membrane because it moves electrons from the intermembrane space into the matrix.

C. The electrochemical proton gradient consists of two components: a pH difference and an electrical potential.

D. Ubiquinone and cytochrome *c* are both diffusible electron carriers.

E. Plants have chloroplasts and therefore can live without mitochondria.

F. Both chlorophyll and heme contain an extensive system of double bonds that allows them to absorb visible light.

G. The role of chlorophyll in photosynthesis is equivalent to that of heme in mitochondrial electron transport.

H. Most of the dry weight of a tree comes from the minerals that are taken up by the roots.

QUESTION 14–13

A single proton moving down its electrochemical gradient into the mitochondrial matrix space liberates 4.6 kcal/mole of free energy. How many protons have to flow across the inner mitochondrial membrane to synthesize one molecule of ATP if the ΔG for ATP synthesis under intracellular conditions is between 11 and 13 kcal/mole? (ΔG is discussed in Chapter 3, pp. 91–98.) Why is a range given for this latter value, and not a precise number? Under which conditions would the lower value apply?

QUESTION 14–14

In the following statement, choose the correct one of the alternatives in italics and justify your answer. "If no O_2 is available, all components of the mitochondrial electron-transport chain will accumulate in their *reduced/oxidized* form. If O_2 is suddenly added again, the electron carriers in cytochrome oxidase will become *reduced/oxidized before/after* those in NADH dehydrogenase."

QUESTION 14–15

Assume that the conversion of oxidized ubiquinone to reduced ubiquinone by NADH dehydrogenase occurs on the matrix side of the inner mitochondrial membrane and that its oxidation by cytochrome *b*-c_1 occurs on the intermembrane space side of the membrane (see Figures 14–9 and 14–19). What are the consequences of this arrangement for the generation of the H^+ gradient across the membrane?

QUESTION 14–16

If a voltage is applied to two platinum wires (electrodes) immersed in water, then water molecules become split into H_2 and O_2 gas. At the negative electrode, electrons are donated and H_2 gas is released; at the positive electrode, electrons are accepted and O_2 gas is produced. When photosynthetic bacteria and plant cells split water, they produce O_2, but no H_2. Why?

QUESTION 14–17

In an insightful experiment performed in the 1960s, chloroplasts were first soaked in an acidic solution at pH 4, so that the stroma and thylakoid space became acidified

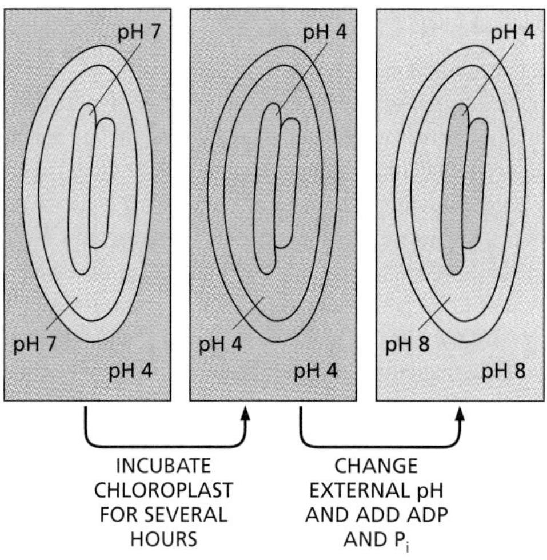

Figure Q14–17

(Figure Q14–17). They were then transferred to a basic solution (pH 8). This quickly increased the pH of the stroma to 8, while the thylakoid space temporarily remained at pH 4. A burst of ATP synthesis was observed, and the pH difference between the thylakoid and the stroma then disappeared.

A. Explain why these conditions lead to ATP synthesis.

B. Is light needed for the experiment to work?

C. What would happen if the solutions were switched so that the first incubation is in the pH 8 solution and the second one in the pH 4 solution?

D. Does the experiment support or question the chemiosmotic model?

Explain your answers.

QUESTION 14–18

As your first experiment in the laboratory, your adviser asks you to reconstitute purified bacteriorhodopsin, a light-driven H^+ pump from the plasma membrane of photosynthetic bacteria, and purified ATP synthase from ox-heart mitochondria together into the same membrane vesicles—as shown in Figure Q14–18. You are then asked to add ADP and P_i to the external medium and shine light into the suspension of vesicles.

A. What do you observe?

B. What do you observe if not all the detergent is removed and the vesicle membrane therefore remains leaky to ions?

C. You tell a friend over dinner about your new experiments, and he questions the validity of an approach that utilizes components from so widely divergent, unrelated organisms: "Why would anybody want to mix vanilla pudding with brake fluid?" Defend your approach against his critique.

QUESTION 14–19

$FADH_2$ is produced in the citric acid cycle by a membrane-embedded enzyme complex, called succinate dehydrogenase, that contains bound FAD and carries out the reactions

$$\text{succinate} + \text{FAD} \rightarrow \text{fumarate} + \text{FADH}_2$$

and

$$\text{FADH}_2 \rightarrow \text{FAD} + 2\text{H}^+ + 2e^-$$

The redox potential of $FADH_2$, however, is only –220 mV. Referring to Panel 14–1 (p. 471) and Figure 14–20, suggest a plausible mechanism by which its electrons could be fed into the electron-transport chain. Draw a diagram to illustrate your proposed mechanism.

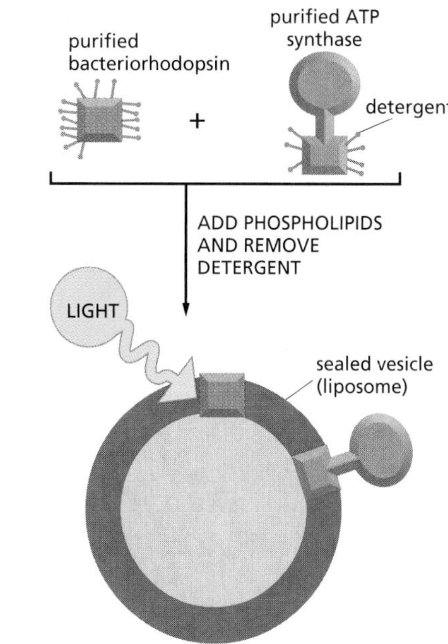

Figure Q14–18

QUESTION 14–20

Some bacteria have become specialized to live in an environment of high pH (pH ~10). Do you suppose that these bacteria use a proton gradient across their plasma membrane to produce their ATP? (Hint: all cells must maintain their cytoplasm at a pH close to neutrality.)

QUESTION 14–21

Figure Q14–21 summarizes the circuitry used by mitochondria and chloroplasts to interconvert different forms of energy. Is it accurate to say

A. that the products of chloroplasts are the substrates for mitochondria?

B. that the activation of electrons by the photosystems enables chloroplasts to drive electron transfer from H_2O to carbohydrate, which is the opposite direction of electron transfer in the mitochondrion?

C. that the citric acid cycle is the reverse of the normal carbon-fixation cycle?

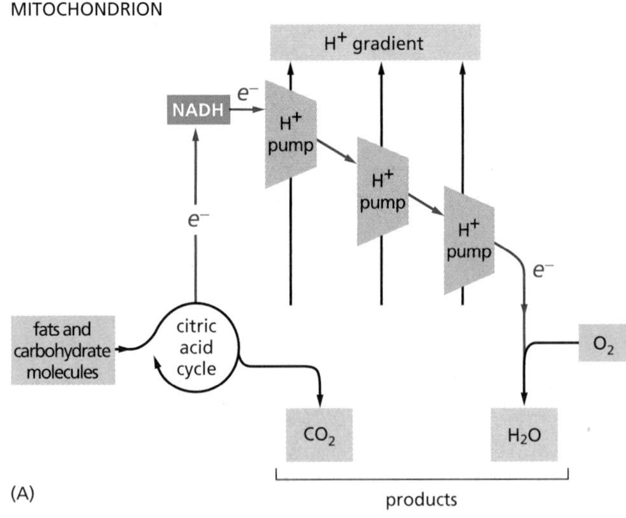

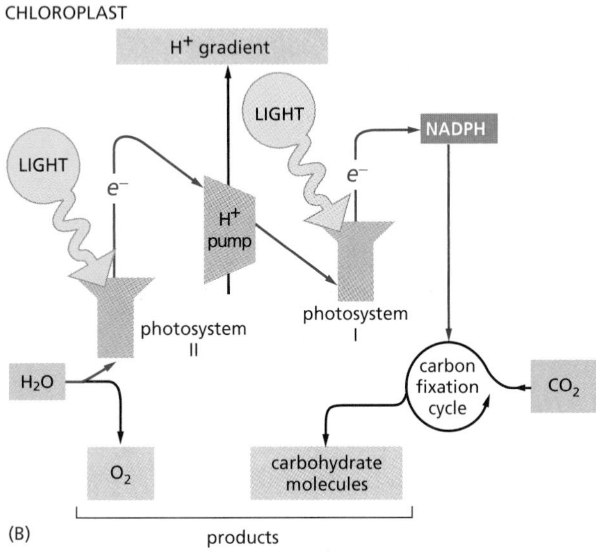

Figure Q14–21

QUESTION 14–22

A manuscript has been submitted for publication to a prestigious scientific journal. In the paper the authors describe an experiment in which they have succeeded in trapping an individual ATP synthase molecule and then mechanically rotating its head by applying a force to it. The authors show that upon rotating the head of the ATP synthase, ATP is produced, in the absence of an H^+ gradient. What might this mean about the mechanism whereby ATP synthase functions? Should this manuscript be considered for publication in one of the best journals?

QUESTION 14–23

You mix the following components in a solution. Assuming that the electrons must follow the path specified in Figure 14–9, in which experiments would you expect a net transfer of electrons to cytochrome *c*? Discuss why electron transfer does not occur in the other experiments.

A. reduced ubiquinone and oxidized cytochrome *c*

B. oxidized ubiquinone and oxidized cytochrome *c*

C. reduced ubiquinone and reduced cytochrome *c*

D. oxidized ubiquinone and reduced cytochrome *c*

E. reduced ubiquinone, oxidized cytochrome *c*, and cytochrome *b*-c_1 complex

F. oxidized ubiquinone, oxidized cytochrome *c*, and cytochrome *b*-c_1 complex

G. reduced ubiquinone, reduced cytochrome *c*, and cytochrome *b*-c_1 complex

H. oxidized ubiquinone, reduced cytochrome *c*, and cytochrome *b*-c_1 complex

CHAPTER **FIFTEEN** 15

Intracellular Compartments and Transport

MEMBRANE-ENCLOSED ORGANELLES

PROTEIN SORTING

VESICULAR TRANSPORT

SECRETORY PATHWAYS

ENDOCYTIC PATHWAYS

At any one time, a typical eucaryotic cell carries out thousands of different chemical reactions, many of which are mutually incompatible. One series of reactions makes glucose, for example, while another breaks down glucose; some enzymes synthesize peptide bonds, whereas others hydrolyze them, and so on. Indeed, if the cells of an organ such as the liver are broken apart and their contents mixed together in a test tube, chemical chaos results, and the cells' enzymes and other proteins are quickly degraded by their own proteolytic enzymes. For a cell to operate effectively, the different intracellular processes that occur simultaneously must somehow be segregated.

Cells have evolved several strategies for isolating and organizing their chemical reactions. One strategy used by both procaryotic and eucaryotic cells is to aggregate the different enzymes required to catalyze a particular sequence of reactions into large, multicomponent complexes. Such complexes are used, for example, in the synthesis of DNA, RNA, and proteins. A second strategy, which is most highly developed in eucaryotic cells, is to confine different metabolic processes, and the proteins required to perform them, within different membrane-enclosed compartments. As discussed in Chapters 11 and 12, cell membranes provide selectively permeable barriers through which the transport of most molecules can be controlled. In this chapter, we consider the strategy of compartmentalization and some of its consequences.

In the first section, we describe the principal membrane-enclosed compartments, or *membrane-enclosed organelles*, of eucaryotic cells and briefly consider their main functions. In the second section, we discuss how the protein composition of the different compartments is set up and

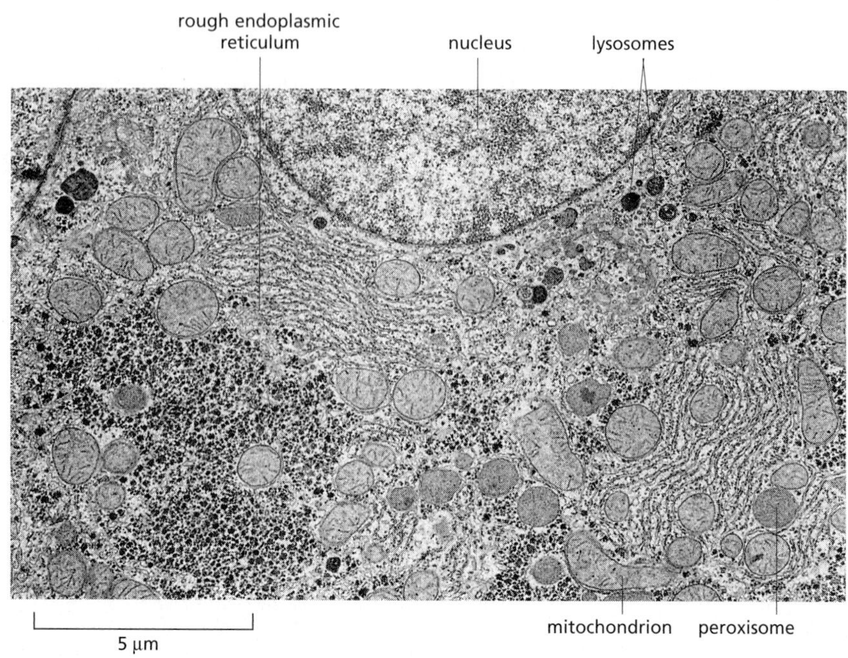

Figure 15–1 **In eucaryotic cells, internal membranes create enclosed compartments and organelles in which different metabolic processes are segregated.** Examples of many of the major membrane-enclosed organelles can be identified in this electron micrograph of part of a liver cell, seen in cross section. The small black granules between the membrane-enclosed compartments are aggregates of glycogen and the enzymes that control its synthesis and breakdown. (Courtesy of Daniel S. Friend.)

maintained. Each compartment contains a unique set of proteins that have to be transferred selectively from the cytosol, where they are made, to the compartment in which they are used. This transfer process, called *protein sorting*, depends on signals built into the amino acid sequence of the proteins. In the third section, we describe how certain membrane-enclosed compartments in a eucaryotic cell communicate with one another by forming small membrane-enclosed sacs, or *vesicles*. These pinch off from one compartment, move through the cytosol, and fuse with another compartment in a process called *vesicular transport*. In the last two sections, we discuss how this constant vesicular traffic also provides the main routes for releasing proteins from the cell by the process of *exocytosis* and for importing them by the process of *endocytosis*.

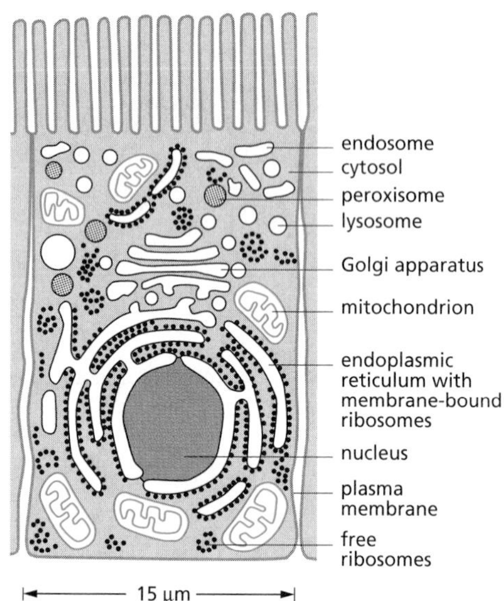

Figure 15–2 **A cell from the lining of the intestine contains the basic set of organelles found in most animal cells.** The nucleus, endoplasmic reticulum (ER), Golgi apparatus, lysosomes, endosomes, mitochondria, and peroxisomes are distinct compartments separated from the cytosol (*gray*) by at least one selectively permeable membrane. Ribosomes are also shown, even though they are not enclosed by a membrane and are too small to be seen in a light microscope and therefore do not fit the original definition of an organelle. Some ribosomes are found free in the cytosol, while others are bound to the cytosolic surface of the ER.

MEMBRANE-ENCLOSED ORGANELLES

Whereas a procaryotic cell usually consists of a single compartment, the **cytosol**, enclosed by the plasma membrane, eucaryotic cells are elaborately subdivided by internal membranes. These membranes create enclosed compartments in which sets of enzymes can operate without interference from reactions occurring in other compartments. When a cross section through a plant or an animal cell is examined in the electron microscope, numerous small, membrane-enclosed sacs, tubes, spheres, and irregularly shaped structures can be seen, often arranged without much apparent order (Figure 15–1). These structures are all distinct membrane-enclosed organelles, or parts of such organelles, each of which contains a unique set of large and small molecules and carries out a specialized function. In this section, we review these functions and discuss how different membrane-enclosed organelles may have evolved.

Eucaryotic Cells Contain a Basic Set of Membrane-enclosed Organelles

The major **membrane-enclosed organelles** of an animal cell are illustrated in Figure 15–2, and their functions are summarized in Table 15–1. These organelles are surrounded by the cytosol, which is enclosed by the plasma membrane. The *nucleus* is generally the most prominent organelle

TABLE 15–1 THE MAIN FUNCTIONS OF THE MEMBRANE-ENCLOSED COMPARTMENTS OF A EUCARYOTIC CELL	
COMPARTMENT	MAIN FUNCTION
Cytosol	contains many metabolic pathways (Chapters 3 and 13); protein synthesis (Chapter 7)
Nucleus	contains main genome (Chapter 5); DNA and RNA synthesis (Chapters 6 and 7)
Endoplasmic reticulum (ER)	synthesis of most lipids (Chapter 11); synthesis of proteins for distribution to many organelles and to the plasma membrane (this chapter)
Golgi apparatus	modification, sorting, and packaging of proteins and lipids for either secretion or delivery to another organelle (this chapter)
Lysosomes	intracellular degradation (this chapter)
Endosomes	sorting of endocytosed material (this chapter)
Mitochondria	ATP synthesis by oxidative phosphorylation (Chapter 14)
Chloroplasts (in plant cells)	ATP synthesis and carbon fixation by photosynthesis (Chapter 14)
Peroxisomes	oxidation of toxic molecules

in eucaryotic cells. It is surrounded by a double membrane, known as the *nuclear envelope*, and communicates with the cytosol via *nuclear pores* that perforate the envelope. The outer nuclear membrane is continuous with the membrane of the *endoplasmic reticulum (ER)*, a system of interconnected sacs and tubes of membrane that often extends throughout most of the cell. The ER is the major site of synthesis of new membranes in the cell. Large areas of the ER have ribosomes attached to the cytosolic surface and are designated *rough endoplasmic reticulum (rough ER)*. The ribosomes are actively synthesizing proteins that are delivered into the ER lumen or ER membrane. The *smooth endoplasmic reticulum (smooth ER)* lacks ribosomes. It is scanty in most cells but is highly developed for performing particular functions in others: for example, it is the site of steroid hormone synthesis in cells of the adrenal gland and the site where a variety of organic molecules, including alcohol, are detoxified in liver cells. In many eucaryotic cells, the smooth ER also sequesters Ca^{2+} from the cytosol; the release and reuptake of Ca^{2+} from the ER is involved in the rapid response to many extracellular signals, as discussed in Chapters 12 and 16.

The *Golgi apparatus*, which is usually situated near the nucleus, receives proteins and lipids from the ER, modifies them, and then dispatches them to other destinations in the cell. Small sacs of digestive enzymes called *lysosomes* degrade worn-out organelles, as well as macromolecules and particles taken into the cell by endocytosis. On their way to lysosomes, endocytosed materials must first pass through a series of compartments called *endosomes*, which sort the ingested molecules and recycle some of them back to the plasma membrane. **Peroxisomes** are small organelles enclosed by a single membrane. They contain enzymes used in a variety of oxidative reactions that break down lipids and destroy toxic molecules. Mitochondria and (in plant cells) chloroplasts are each surrounded by a double membrane and are the sites of oxidative phosphorylation and photosynthesis, respectively (discussed in Chapter 14); both contain membranes that are highly specialized for the production of ATP.

Many of the membrane-enclosed organelles, including the ER, Golgi apparatus, mitochondria, and chloroplasts, are held in their relative locations in the cell by attachment to the cytoskeleton, especially to microtubules. Cytoskeletal filaments provide tracks for moving the organelles around and for directing the traffic of vesicles between them. These movements are driven by motor proteins that use the energy of ATP hydrolysis to propel the organelles and vesicles along the filaments, as discussed in Chapter 17.

TABLE 15–2 THE RELATIVE VOLUMES OCCUPIED BY THE MAJOR MEMBRANE-ENCLOSED ORGANELLES IN A LIVER CELL (HEPATOCYTE)

INTRACELLULAR COMPARTMENT	PERCENTAGE OF TOTAL CELL VOLUME	APPROXIMATE NUMBER PER CELL
Cytosol	54	1
Mitochondria	22	1700
Endoplasmic reticulum	12	1
Nucleus	6	1
Golgi apparatus	3	1
Peroxisomes	1	400
Lysosomes	1	300
Endosomes	1	200

On average, the membrane-enclosed organelles together occupy nearly half the volume of a eucaryotic cell (Table 15–2), and the total amount of membrane associated with them is enormous. In a typical mammalian cell, for example, the area of the endoplasmic reticulum membrane is 20–30 times greater than that of the plasma membrane. In terms of its area and mass, the plasma membrane is only a minor membrane in most eucaryotic cells.

Much can be learned about the composition and function of an organelle once it has been isolated from other cell structures. For the most part, organelles are far too small to be isolated by hand, but it is possible to separate one type of organelle from another by differential centrifugation (described in Panel 4–4, pp. 164–165). Once a purified sample of one type of organelle has been obtained, the organelle's proteins can be identified. In many cases, the organelle itself can be incubated in a test tube under conditions that allow its functions to be studied. Isolated mitochondria, for example, can produce ATP from the oxidation of pyruvate to CO_2 and water, provided they are adequately supplied with ADP and O_2.

Membrane-enclosed Organelles Evolved in Different Ways

In trying to understand the relationships between the different compartments of a modern eucaryotic cell, it is helpful to consider how they might have evolved. The compartments probably evolved in stages. The precursors of the first eucaryotic cells are thought to have been simple microorganisms, resembling bacteria, which had a plasma membrane but no internal membranes. The plasma membrane in such cells would have provided all membrane-dependent functions, including ATP synthesis and lipid synthesis, as does the plasma membrane in most modern bacteria. Bacteria can get by with this arrangement because of their small size and thus their high surface-to-volume ratio: their plasma membrane area is sufficient to sustain all the vital functions for which membranes are required. Present-day eucaryotic cells, however, have volumes 1000 to 10,000 times greater than that of a typical bacterium such as *E. coli*. Such a large cell has a small surface-to-volume ratio and presumably could not survive with a plasma membrane as its only membrane. Thus, the increase in size typical of eucaryotic cells probably could not have occurred without the development of internal membranes.

Membrane-enclosed organelles are thought to have arisen in evolution in at least two ways. The nuclear membranes and the membranes of the ER, Golgi apparatus, endosomes, and lysosomes are believed to have originated by invagination of the plasma membrane (Figure 15–3). These membranes, and the organelles they enclose, are all part of what is collec-

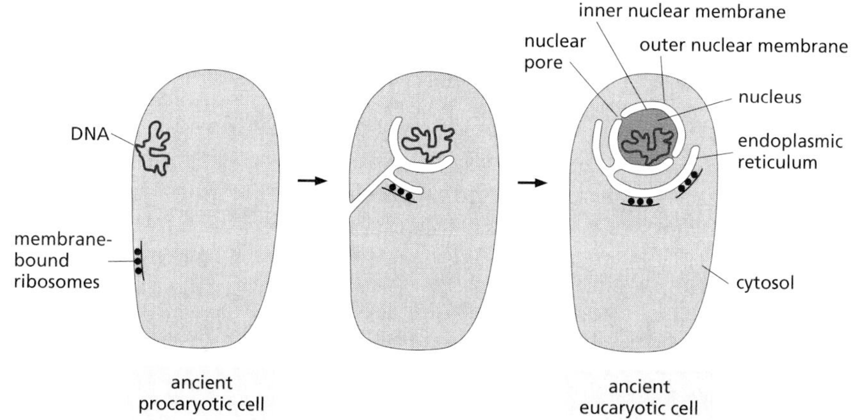

Figure 15–3 **Nuclear membranes and the ER may have evolved through invagination of the plasma membrane.** In bacteria, the single DNA molecule is typically attached to the plasma membrane. It is possible that in a very ancient procaryotic cell, the plasma membrane, with its attached DNA, could have invaginated and eventually formed a two-layered envelope of membrane completely surrounding the DNA. This envelope is presumed to have eventually pinched off completely from the plasma membrane, producing a nuclear compartment surrounded by a double membrane. This nuclear envelope is penetrated by channels called nuclear pores, which enable it to communicate directly with the cytosol. Other portions of the same membrane formed the ER, to which some of the ribosomes became attached. This hypothetical scheme would explain why the space between the inner and outer nuclear membranes is continuous with the lumen of the ER.

tively called the *endomembrane system*. As we discuss later, the interiors of these organelles (with the exception of the nucleus) communicate extensively with one another and with the outside of the cell by means of small vesicles that bud off from one of these organelles and fuse with another. Consistent with this proposed evolutionary origin, the interiors of these organelles are treated by the cell in many ways as 'extracellular,' as we shall see. The hypothetical scheme shown in Figure 15–3 would also explain why the nucleus is surrounded by two membranes. Although membrane invagination is rare in present-day bacteria, it does occur in some photosynthetic bacteria in which the regions of the plasma membrane containing the photosynthetic apparatus are internalized, forming intracellular vesicles.

Mitochondria and chloroplasts are thought to have originated in a different way. They differ from all other organelles in that they possess their own small genomes and can make some of their own proteins, as discussed in Chapter 14. The similarity of these genomes to those of bacteria and the close resemblance of some of their proteins to bacterial proteins strongly suggest that mitochondria and chloroplasts evolved from bacteria that were engulfed by primitive eucaryotic cells with which they initially lived in symbiosis (Figure 15–4). As might be expected from their origins, mitochondria and chloroplasts remain isolated from the extensive vesicular traffic that connects the interiors of most of the other membrane-enclosed organelles to one another and to the outside of the cell.

Having briefly reviewed the main membrane-enclosed organelles of the eucaryotic cell, we turn now to the question of how each organelle acquires its unique set of proteins.

QUESTION 15–1

As shown in the drawings in Figure 15–3, the lipid bilayer of the inner and outer nuclear membranes forms a continuous sheet, joined around the nuclear pores. As membranes are two-dimensional fluids, this would imply that membrane proteins can diffuse freely between the two nuclear membranes. Yet each of these two nuclear membranes has a different protein composition, reflecting different functions. How could you reconcile this apparent paradox?

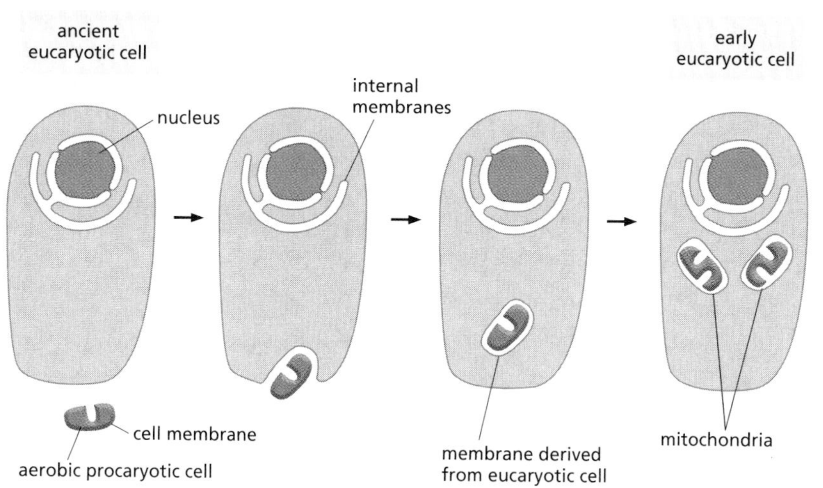

Figure 15–4 **Mitochondria are thought to have originated when a procaryote was engulfed by a larger eucaryotic cell.** Chloroplasts are thought to have originated later in a similar way, when a eucaryotic cell engulfed a photosynthetic procaryote. This theory would explain why these organelles have two membranes and why they do not participate in the vesicular traffic that connects many other intracellular compartments. It would also explain why they have their own genomes.

PROTEIN SORTING

Before a eucaryotic cell reproduces by dividing in two, it has to duplicate its membrane-enclosed organelles. A cell cannot make these organelles from scratch: it requires information and materials contained in the organelle itself. Thus, most of the organelles are formed from preexisting organelles, which grow and then divide. As cells grow, membrane-enclosed organelles enlarge by incorporation of new molecules; the organelles then divide and, during cell division, are distributed between the two daughter cells. Organelle growth requires a supply of new lipids to make more membrane and a supply of the appropriate proteins—both membrane proteins and the soluble proteins that will occupy the interior of the organelle. Even in cells that are not dividing, proteins are being produced continually. These newly synthesized proteins must be accurately delivered to their appropriate organelle—some for eventual secretion from the cell and some to replace organelle proteins that have been degraded. Directing newly made proteins to their correct organelle is therefore necessary for a cell to be able to grow, divide, and function properly.

For some organelles, including the mitochondria, chloroplasts, and the interior of the nucleus, proteins are delivered directly from the cytosol. For others, including the Golgi apparatus, lysosomes, endosomes, and the nuclear membranes, proteins and lipids are delivered indirectly via the ER, which is itself a major site of lipid and protein synthesis. Proteins enter the ER directly from the cytosol: some are retained there, but most are transported by vesicles to the Golgi apparatus and then onward to other organelles or the plasma membrane.

In this section, we discuss the mechanisms by which proteins directly enter membrane-enclosed organelles from the cytosol. Proteins made in the cytosol are dispatched to different locations in the cell according to specific address labels that they contain in their amino acid sequence. Once at the correct address, the protein enters the organelle.

Proteins Are Imported into Organelles by Three Mechanisms

The synthesis of virtually all proteins in the cell begins on ribosomes in the cytosol. The exceptions are the few mitochondrial and chloroplast proteins that are synthesized on ribosomes inside these organelles; most mitochondrial and chloroplast proteins, however, are made in the cytosol and subsequently imported. The fate of any protein molecule synthesized in the cytosol depends on its amino acid sequence, which can contain a *sorting signal* that directs the protein to the organelle in which it is required. Proteins that lack such signals remain as permanent residents in the cytosol; those that possess a sorting signal move from the cytosol to the appropriate organelle. Different sorting signals direct proteins into the nucleus, mitochondria, chloroplasts (in plants), peroxisomes, and the ER.

When a membrane-enclosed organelle imports proteins from the cytosol or from another organelle, it faces a problem: how can it draw the protein across membranes that are normally impermeable to hydrophilic macromolecules? This task is accomplished in different ways for different organelles.

1. Proteins moving from the cytosol into the nucleus are transported through the nuclear pores that penetrate the inner and outer nuclear membranes. The pores function as selective gates that actively transport specific macromolecules but also allow free diffusion of smaller molecules (mechanism 1 in Figure 15–5).

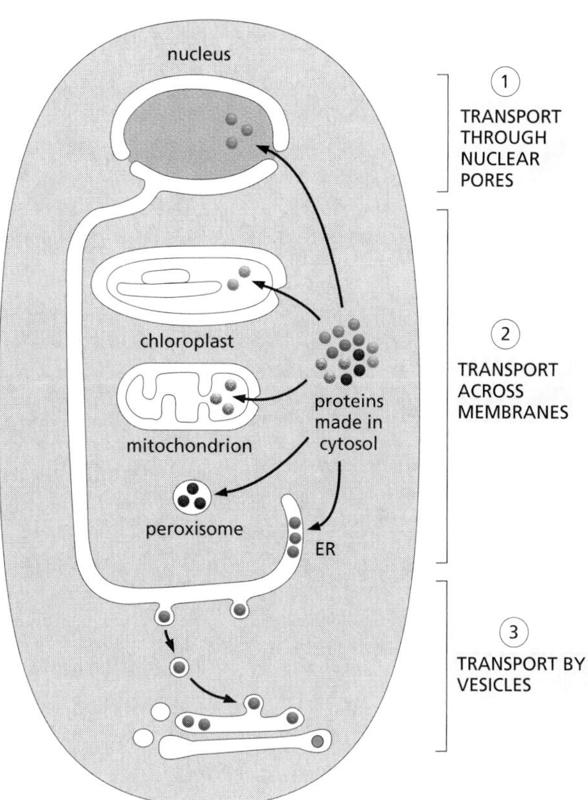

Figure 15–5 **Membrane-enclosed organelles import proteins by one of three mechanisms.** All of these processes require energy. The protein remains folded during the transport steps in mechanisms 1 and 3 but usually has to be unfolded in mechanism 2.

2. Proteins moving from the cytosol into the ER, mitochondria, or chloroplasts are transported across the organelle membrane by *protein translocators* located in the membrane. Unlike transport through nuclear pores, the transported protein molecule must usually unfold in order to snake through the membrane (mechanism 2 in Figure 15–5). Bacteria have similar protein translocators in their plasma membrane, which they use to export proteins from their cytosol.
3. Proteins moving from the ER onward and from one compartment of the endomembrane system to another are transported by a mechanism that is fundamentally different from the other two. These proteins are ferried by *transport vesicles*, which become loaded with a cargo of proteins from the interior space, or *lumen*, of one compartment, as they pinch off from its membrane. The vesicles subsequently discharge their cargo into a second compartment by fusing with its membrane (mechanism 3 in Figure 15–5). In the process, membrane lipids and membrane proteins are also delivered from the first compartment to the second.

Signal Sequences Direct Proteins to the Correct Compartment

The typical sorting signal on proteins is a continuous stretch of amino acid sequence, typically 15–60 amino acids long. This **signal sequence** is often (but not always) removed from the finished protein once it has been sorted. Some of the signal sequences used to specify different destinations in the cell are shown in Table 15–3.

Signal sequences are both necessary and sufficient to direct a protein to a particular organelle. This has been shown by experiments in which the sequence is either deleted or transferred from one protein to another by genetic engineering techniques (discussed in Chapter 10). Deleting

TABLE 15–3 SOME TYPICAL SIGNAL SEQUENCES

FUNCTION OF SIGNAL	EXAMPLE OF SIGNAL SEQUENCE
Import into ER	$^{+}H_3N$-Met-Met-Ser-Phe-Val-Ser-Leu-Leu-Leu-Val-Gly-Ile-Leu-Phe-Trp-Ala-Thr-Glu-Ala-Glu-Gln-Leu-Thr-Lys-Cys-Glu-Val-Phe-Gln-
Retention in lumen of ER	-Lys-Asp-Glu-Leu-COO^-
Import into mitochondria	$^{+}H_3N$-Met-Leu-Ser-Leu-Arg-Gln-Ser-Ile-Arg-Phe-Phe-Lys-Pro-Ala-Thr-Arg-Thr-Leu-Cys-Ser-Ser-Arg-Tyr-Leu-Leu-
Import into nucleus	-Pro-Pro-Lys-Lys-Lys-Arg-Lys-Val-
Import into peroxisomes	-Ser-Lys-Leu-

Positively charged amino acids are shown in *red*, and negatively charged amino acids in *blue*. An extended block of hydrophobic amino acids is shown in *green*. $^{+}H_3N$ indicates the N-terminus of a protein; COO^- indicates the C-terminus. The ER retention signal is commonly referred to by its single-letter amino acid abbreviation, KDEL.

a signal sequence from an ER protein, for example, converts it into a cytosolic protein, while placing an ER signal sequence at the beginning of a cytosolic protein redirects the protein to the ER (Figure 15–6). The signal sequences specifying the same destination can vary greatly even though they have the same function: physical properties such as hydrophobicity or the placement of charged amino acids often appear to be more important for the function of these signals than the exact amino acid sequence.

Proteins Enter the Nucleus Through Nuclear Pores

The **nuclear envelope** encloses the nuclear DNA and defines the nuclear compartment. It is formed from two concentric membranes. The inner nuclear membrane contains proteins that act as binding sites for the chromosomes (discussed in Chapter 5) and provide anchorage for the *nuclear lamina*, a finely woven meshwork of protein filaments that lines the inner face of this membrane and provides a structural support for the nuclear envelope (discussed in Chapter 17). The composition of the outer nuclear membrane closely resembles the membrane of the ER, with which it is continuous (Figure 15–7).

The nuclear envelope in all eucaryotic cells is perforated by **nuclear pores** that form the gates through which all molecules enter or leave the

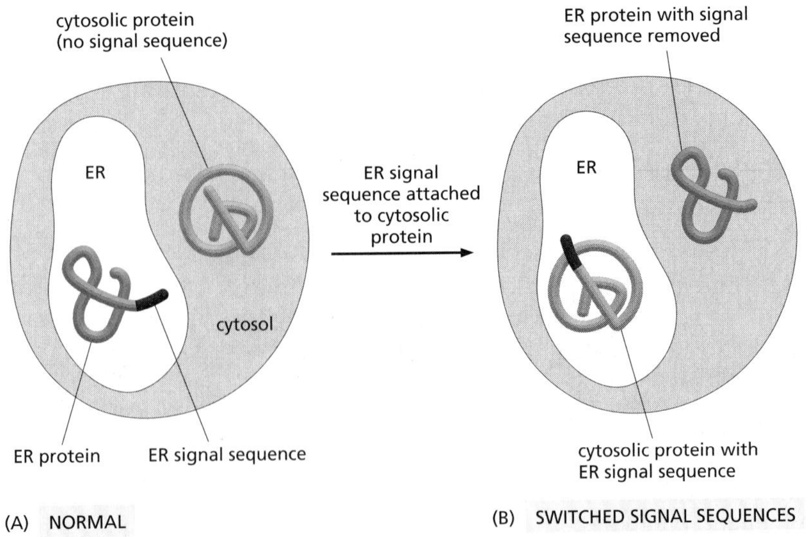

cytosolic protein (no signal sequence)
ER
cytosol
ER protein
ER signal sequence
(A) NORMAL
ER signal sequence attached to cytosolic protein
ER protein with signal sequence removed
ER
cytosolic protein with ER signal sequence
(B) SWITCHED SIGNAL SEQUENCES

Figure 15–6 **Signal sequences direct proteins to the correct organelle.** (A) Proteins destined for the ER possess an N-terminal signal sequence that directs them to that organelle, whereas those destined to remain in the cytosol lack this sequence. (B) Recombinant DNA techniques can be used to change the location of the two proteins: if the signal sequence is removed from the ER protein and attached to the cytosolic protein, the proteins end up in an abnormal location in the cell. Such experiments indicate that the ER signal sequence is both necessary and sufficient to direct a protein to the ER.

nucleus. Traffic occurs in both directions through the pores: newly made proteins destined for the nucleus enter from the cytosol (Movie 15.1); RNA molecules, which are synthesized in the nucleus, and ribosomal subunits, which are assembled in the nucleus, are exported. Messenger RNA molecules that are incompletely spliced are not exported from the nucleus, suggesting that nuclear transport serves as a final quality-control step in mRNA synthesis and processing (discussed in Chapter 7).

A nuclear pore is a large, elaborate structure composed of about 30 different proteins (Figure 15–8). Each pore contains water-filled passages through which small water-soluble molecules can pass freely and nonselectively between the nucleus and the cytosol. Many of the proteins that line the nuclear pore contain extensive, unstructured regions, which are thought to form a disordered tangle—much like a kelp bed in the ocean. This jumbled meshwork fills the center of the channel, preventing the passage of large molecules but allowing smaller molecules to slip through.

Larger molecules (such as RNAs and proteins) and macromolecular complexes must display an appropriate sorting signal to pass through the nuclear pore. The signal sequence that directs a protein from the cytosol into the nucleus, called a *nuclear localization signal*, typically consists of one or two short sequences containing several positively charged lysines or arginines (see Table 15–3).

Cytosolic proteins called *nuclear transport receptors* bind to the nuclear localization signal on newly synthesized proteins destined for the nucleus. These receptors help direct the new protein to a nuclear pore by interacting with the tentacle-like fibrils that extend from the rim of the

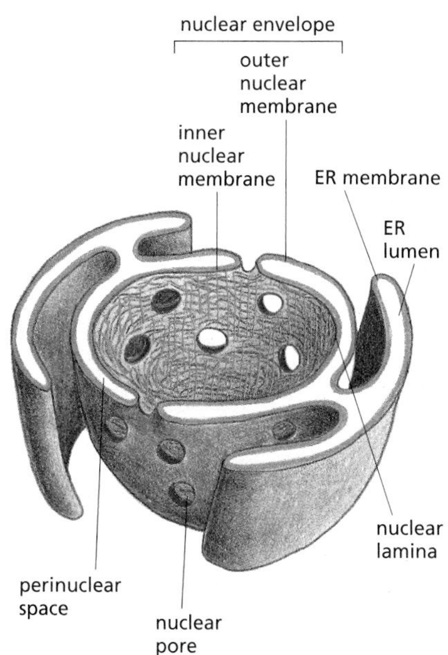

Figure 15–7 **The outer nuclear membrane is continuous with the ER.** The double membrane of the nuclear envelope is penetrated by nuclear pores. The ribosomes that are normally bound to the cytosolic surface of the ER membrane and outer nuclear membrane are not shown.

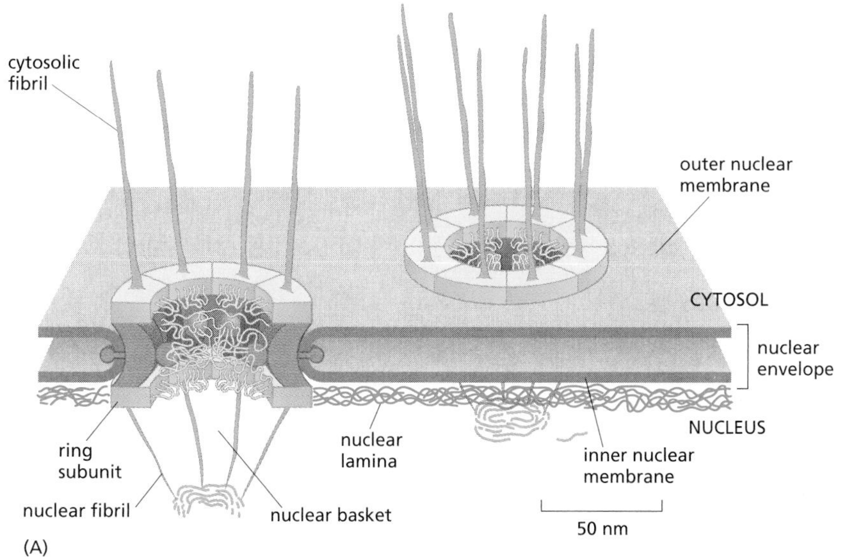

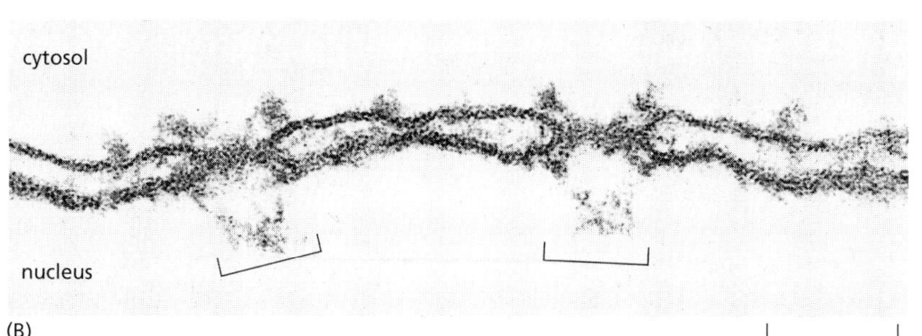

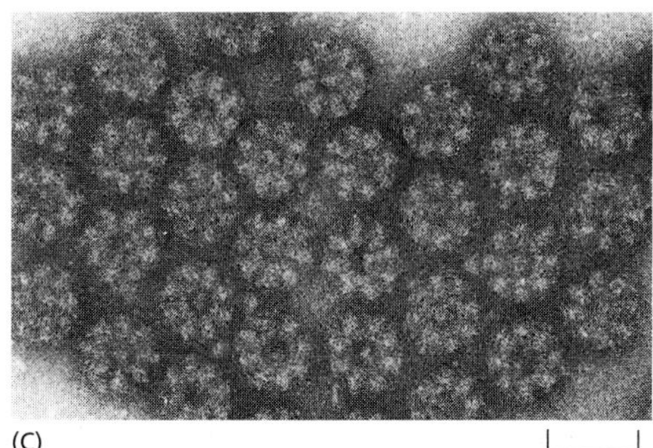

Figure 15–8 **The nuclear pore complex forms a gate through which molecules enter or exit from the nucleus.** (A) Drawing of a small region of the nuclear envelope showing two pore complexes. Each complex is composed of a large number of distinct protein subunits. Protein fibrils protrude from both sides of the complex; on the nuclear side, they converge to form a basketlike structure. The spacing between the fibrils is wide enough that the fibrils do not obstruct access to the pores. (B) Electron micrograph of a region of nuclear envelope showing a side view of two nuclear pore complexes (brackets). (C) Electron micrograph showing a face-on view of nuclear pore complexes; the membranes have been extracted with detergent. (B, courtesy of Werner W. Franke; C, courtesy of Ron Milligan.)

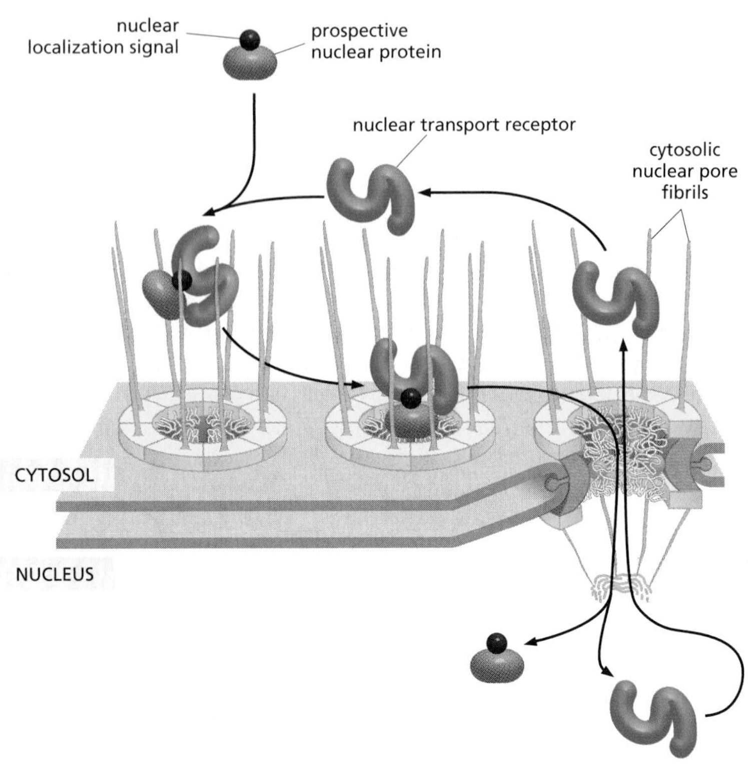

Figure 15–9 **Proteins bound for the nucleus are actively transported through nuclear pores.** The nuclear protein, with its bound receptor, is actively transported into the nucleus. For clarity, the basketlike structure of fibrils that extends into the nucleus (see Figure 15–8) is not shown. A similar type of transport receptor, operating in the reverse direction, exports mRNAs from the nucleus (see Figure 7–20); both groups of receptors have a similar basic structure. The cutaway view of one nuclear pore complex shows how the unstructured regions of the proteins that line the central pore form a tangled meshwork that blocks the passive diffusion of large macromolecules into and out of the nucleus.

pore (Figure 15–9). During transport, the nuclear transport receptors grab onto repeated amino acid sequences within the tangle of nuclear pore proteins, pulling themselves from one to the next, to carry their cargo protein into the nucleus. Once the protein has been delivered, the nuclear transport receptor is returned to the cytosol via the nuclear pore for reuse (see Figure 15–9). Like any process that creates order, importing proteins into the nucleus requires energy. In this case, the energy is provided by GTP hydrolysis, which drives nuclear transport in the appropriate direction (Figure 15–10). Nuclear pore proteins operate this molecular gate at an amazing speed, pumping macromolecules in both directions through each pore.

Nuclear pores transport proteins in their fully folded conformation and transfer ribosomal components as assembled particles. This feature distinguishes the nuclear transport mechanism from the mechanisms that

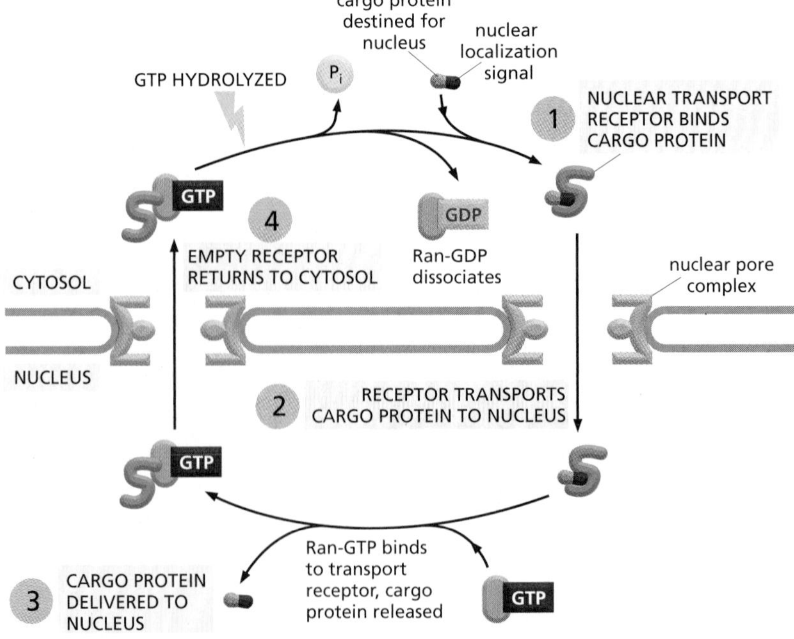

Figure 15–10 **The energy supplied by GTP hydrolysis drives nuclear transport.** A nuclear transport receptor picks up its cargo protein in the cytosol and enters the nucleus. There it encounters a small protein called Ran, which carries a molecule of GTP. This Ran-GTP binds to the nuclear transport receptor, causing it to release its cargo. Having discharged its cargo in the nucleus, the nuclear receptor—still carrying the Ran-GTP—is transported back through the pore to the cytosol. There, an accessory protein (not shown) triggers Ran to hydrolyze its bound GTP. Ran-GDP falls off the transport receptor, which is then free to bind another protein destined for the nucleus. A similar cycle operates to export mRNAs and other large molecules from the nucleus into the cytosol.

transport proteins into other organelles. Proteins have to unfold during their transport across the membranes of other organelles such as mitochondria, chloroplasts, and the ER, as we discuss next.

Proteins Unfold to Enter Mitochondria and Chloroplasts

Both mitochondria and chloroplasts are surrounded by inner and outer membranes, and both organelles specialize in the synthesis of ATP. Chloroplasts also contain a third membrane system, the thylakoid membrane (discussed in Chapter 14). Although both organelles contain their own genomes and make some of their own proteins, most mitochondrial and chloroplast proteins are encoded by genes in the nucleus and are imported from the cytosol. These proteins usually have a signal sequence at their N-terminus that allows them to enter their specific organelle. Proteins destined for either organelle are translocated simultaneously across both the inner and outer membranes at specialized sites where the two membranes are in contact with each other. Each protein is unfolded as it is transported, and its signal sequence is removed after translocation is complete (Figure 15–11). Chaperone proteins (discussed in Chapter 4) inside the organelles help to pull the protein across the two membranes and to refold the protein once it is inside. Subsequent transport to a particular site within the organelle, such as the inner or outer membrane or the thylakoid membrane in chloroplasts, usually requires further sorting signals in the protein, which are often only exposed after the first signal sequence has been removed. The insertion of transmembrane proteins into the inner membrane, for example, is guided by signal sequences in the protein that start and stop the transfer process across the membrane, as we describe later for the insertion of transmembrane proteins in the ER membrane.

The growth and maintenance of mitochondria and chloroplasts require not only the import of new proteins but also the incorporation of new lipids into their membranes. Most of their membrane phospholipids are thought to be imported from the ER, which is the main site of lipid synthesis in the cell. Phospholipids are transported individually to these organelles by water-soluble lipid-carrying proteins that extract a phospholipid molecule from one membrane and deliver it into another. Thanks to these proteins, the different cellular membranes are able to retain their characteristic lipid composition.

Proteins Enter the Endoplasmic Reticulum While Being Synthesized

The **endoplasmic reticulum** (**ER**) is the most extensive membrane system in a eucaryotic cell (Figure 15–12A). Unlike the organelles discussed so far, it serves as an entry point for proteins destined for other

QUESTION 15–2

Why do eucaryotic cells require a nucleus as a separate compartment when procaryotic cells can manage perfectly well without?

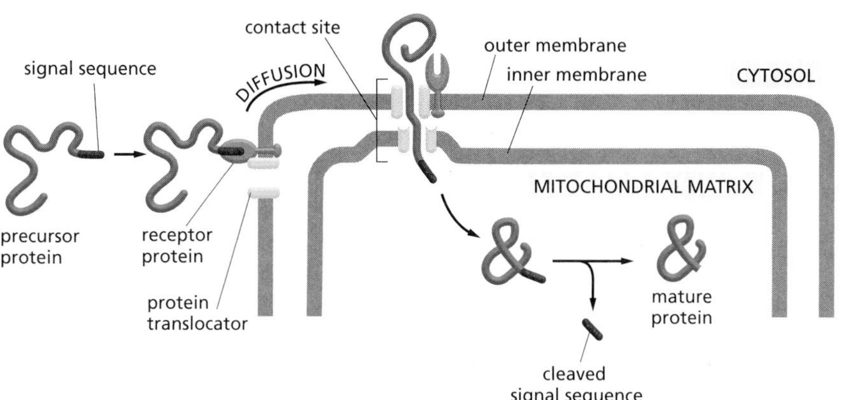

Figure 15–11 **Proteins are imported into mitochondria in an unfolded form.** The mitochondrial signal sequence is recognized by a receptor in the outer mitochondrial membrane. The complex of receptor and attached protein diffuses laterally in the membrane to a contact site, where the protein is translocated across both the outer and inner membranes by a protein translocator (Movie 15.2). The signal sequence is cleaved off by a signal peptidase inside the organelle. Proteins are imported into chloroplasts by a similar mechanism. The chaperone proteins that help to pull the protein across the membranes and help it to refold are not shown.

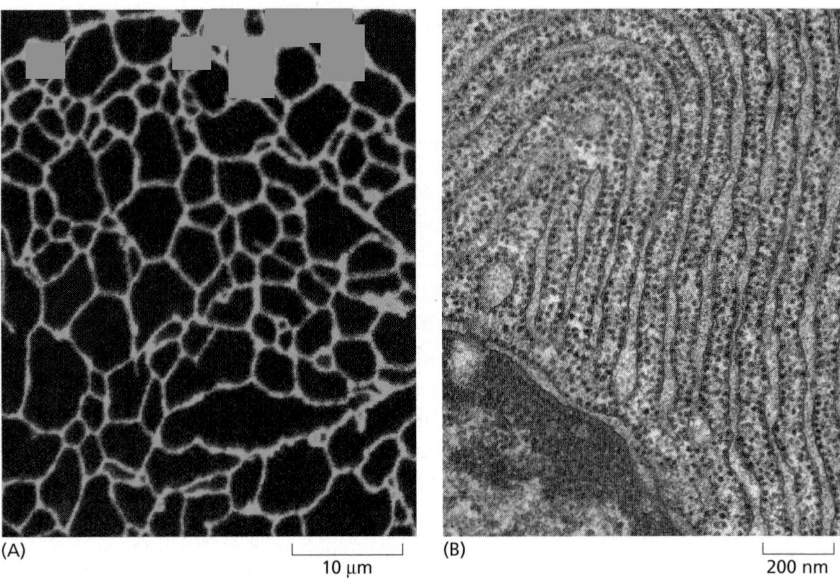

Figure 15–12 **The endoplasmic reticulum is the most extensive membrane network in eucaryotic cells.** (A) Fluorescence micrograph of a living plant cell showing the ER as a complex network of tubes. The cell shown here has been genetically engineered so that it contains a fluorescent protein in its ER. Only part of the ER network in the cell is shown. (B) An electron micrograph showing the rough ER in a cell from a dog's pancreas, which makes and secretes large amounts of digestive enzymes. The cytosol is filled with closely packed sheets of ER, studded with ribosomes. At the bottom left is a portion of the nucleus and its nuclear envelope; note that the outer nuclear membrane, which is continuous with the ER, is also studded with ribosomes. For a dynamic view of the ER network, watch Movie 15.3. (A, courtesy of Jim Haseloff; B, courtesy of Lelio Orci.)

organelles, as well as for the ER itself. Proteins destined for the Golgi apparatus, endosomes, and lysosomes, as well as proteins destined for the cell surface, all first enter the ER from the cytosol. Once inside the ER or embedded in the ER membrane, individual proteins will not reenter the cytosol during their onward journey. They will be ferried by transport vesicles from organelle to organelle and, in some cases, from organelle to the plasma membrane or to the exterior of the cell.

Two kinds of proteins are transferred from the cytosol to the ER: (1) water-soluble proteins are completely translocated across the ER membrane and are released into the ER lumen; (2) prospective transmembrane proteins are only partly translocated across the ER membrane and become embedded in it. The water-soluble proteins are destined either for secretion (by release at the cell surface) or for the lumen of an organelle; the transmembrane proteins are destined to reside in either the ER membrane, the membrane of another organelle, or the plasma membrane. All of these proteins are initially directed to the ER by an *ER signal sequence*, a segment of eight or more hydrophobic amino acids (see Table 15–3, p. 502) that is also involved in the process of translocation across the membrane.

Unlike the proteins that enter the nucleus, mitochondria, chloroplasts, and peroxisomes, most of the proteins that enter the ER begin to be threaded across the ER membrane before the polypeptide chain has been completely synthesized. This requires that the ribosome synthesizing the protein be attached to the ER membrane. These membrane-bound ribosomes coat the surface of the ER, creating regions termed **rough endoplasmic reticulum** because of the characteristic beaded appearance when viewed in an electron microscope (Figure 15–12B).

There are, therefore, two separate populations of ribosomes in the cytosol. *Membrane-bound ribosomes* are attached to the cytosolic side of the ER membrane (and outer nuclear membrane) and are making proteins that are being translocated into the ER. *Free ribosomes* are unattached to any membrane and are making all of the other proteins encoded by the nuclear DNA. Membrane-bound ribosomes and free ribosomes are structurally and functionally identical; they differ only in the proteins they are making at any given time. When a ribosome happens to be making a protein with an ER signal sequence, the signal sequence directs the ribosome to the ER membrane. As an mRNA molecule is translated, many ribosomes bind to it, forming a *polyribosome* (discussed in Chapter

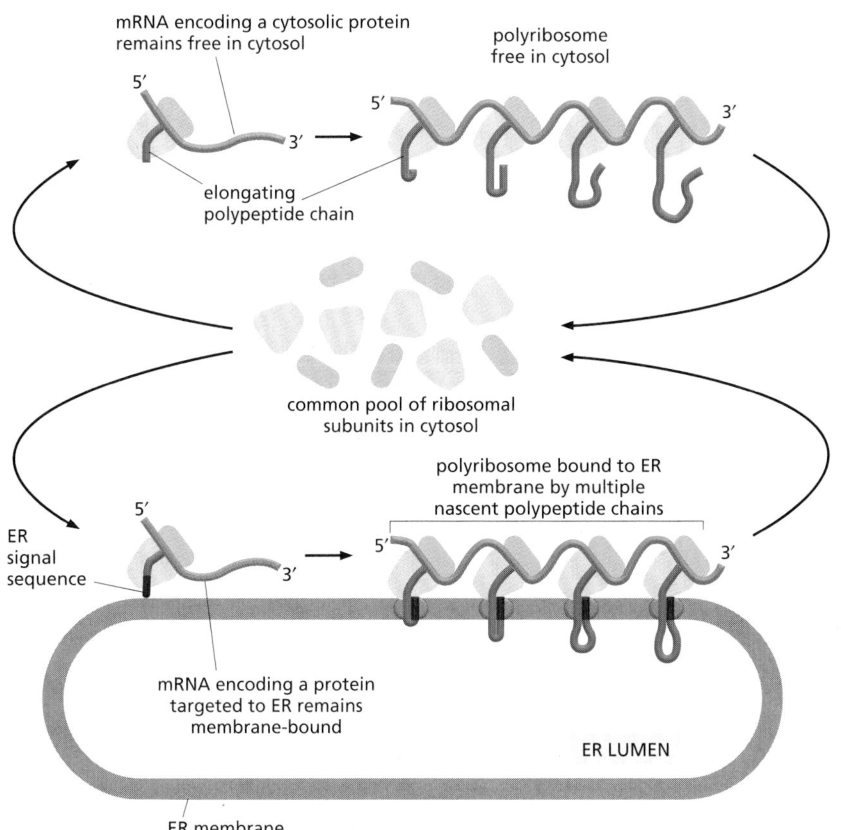

Figure 15–13 **A common pool of ribosomes is used to synthesize both the proteins that stay in the cytosol and those that are transported into membrane-enclosed organelles, including the ER.** Ribosomes that are translating cytosolic proteins remain free in the cytosol. For proteins that will be sent to the ER, a signal sequence (*red*) on the growing polypeptide chain directs the ribosome to the ER membrane. Many ribosomes bind to each mRNA molecule, forming a polyribosome. At the end of each round of protein synthesis, the ribosomal subunits are released and rejoin the common pool in the cytosol.

7). In the case of an mRNA molecule encoding a protein with an ER signal sequence, the polyribosome becomes riveted to the ER membrane by the growing polypeptide chains, which have become inserted into the membrane (Figure 15–13).

QUESTION 15–3

Explain how an mRNA molecule can remain attached to the ER membrane while individual ribosomes translating it are released and rejoin the cytosolic pool of ribosomes after each round of translation.

Soluble Proteins Are Released into the ER Lumen

The ER signal sequence is guided to the ER membrane with the aid of at least two protein components: (1) a *signal-recognition particle (SRP)*, present in the cytosol, which binds to the ER signal sequence when it is exposed on the ribosome, and (2) an *SRP receptor*, embedded in the membrane of the ER, which recognizes the SRP. Binding of an SRP to a signal sequence causes protein synthesis by the ribosome to slow down, until the ribosome and its bound SRP locate an SRP receptor on the ER. After binding to its receptor, the SRP is released and protein synthesis recommences, with the polypeptide now being threaded into the lumen of the ER through a *translocation channel* in the ER membrane (Figure 15–14). Thus, the SRP and SRP receptor function as molecular matchmakers, connecting ribosomes that are synthesizing proteins containing ER signal sequences to available ER translocation channels.

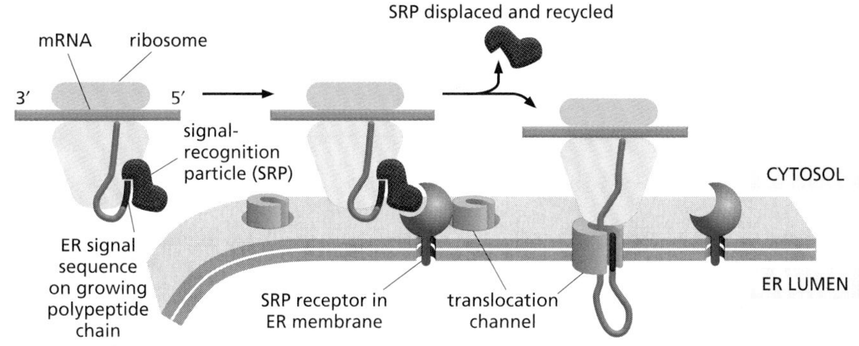

Figure 15–14 **An ER signal sequence and an SRP direct a ribosome to the ER membrane.** The SRP binds to the exposed ER signal sequence and to the ribosome, thereby slowing protein synthesis by the ribosome. The SRP–ribosome complex then binds to an SRP receptor in the ER membrane. The SRP is released, passing the ribosome to a translocation channel in the ER membrane. Finally, the translocation channel inserts the polypeptide chain into the membrane and starts to transfer it across the lipid bilayer.

In addition to directing proteins to the ER, the signal sequence—which for soluble proteins is almost always at the N-terminus—functions to open the translocation channel. The signal peptide remains bound to the channel while the rest of the protein chain is threaded through the membrane as a large loop. At some stage during translocation, the signal sequence is cleaved off by a signal peptidase located on the lumenal side of the ER membrane; the signal peptide is then released from the translocation channel and rapidly degraded. Once the C-terminus of the protein has passed through the membrane, the protein is released into the ER lumen (Figure 15–15).

Start and Stop Signals Determine the Arrangement of a Transmembrane Protein in the Lipid Bilayer

Not all proteins that enter the ER are released into the ER lumen. Some remain embedded in the ER membrane as transmembrane proteins. The translocation process for such proteins is more complicated than it is for soluble proteins, as some parts of the polypeptide chain must be translocated clear across the lipid bilayer while others remain fixed in the membrane.

In the simplest case, that of a transmembrane protein with a single membrane-spanning segment, the N-terminal signal sequence initiates translocation, just as for a soluble protein. But the transfer process is halted by an additional sequence of hydrophobic amino acids, a *stop-transfer sequence*, further into the polypeptide chain (Figure 15–16). This second sequence is released from the translocation channel and drifts into the plane of the lipid bilayer, where it forms an α-helical membrane-spanning segment that anchors the protein in the membrane. Simultaneously, the N-terminal signal sequence is also released from the channel into the lipid bilayer and is cleaved off. As a result, the translocated protein ends up as a transmembrane protein inserted in the membrane with a defined orientation—the N-terminus on the lumenal side of the lipid bilayer and the C-terminus on the cytosolic side (see Figure 15–16). As discussed in Chapter 11, once inserted into the membrane, a transmembrane protein does not change its orientation, which is retained throughout any subsequent vesicle budding and fusion events.

In some transmembrane proteins, an internal, rather than an N-terminal, signal sequence is used to start the protein transfer; this internal signal sequence, called a *start-transfer sequence,* is never removed from the polypeptide. This arrangement occurs in some transmembrane proteins in which the polypeptide chain passes back and forth across the

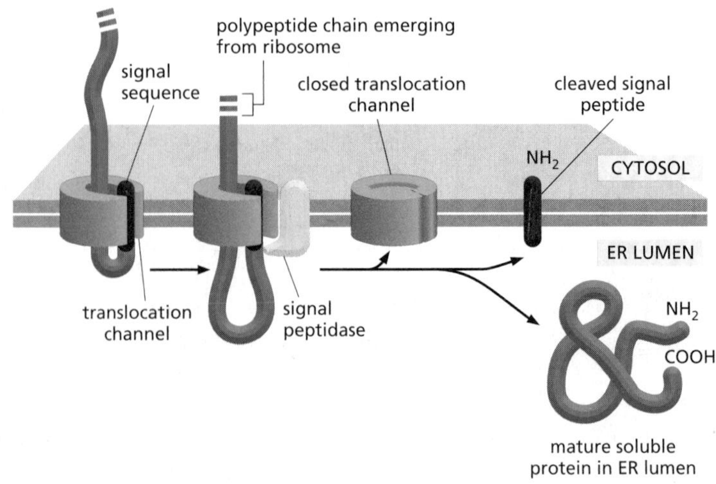

Figure 15–15 **A soluble protein crosses the ER membrane and enters the lumen.** A translocation channel binds the signal sequence and actively transfers the rest of the polypeptide across the lipid bilayer as a loop. At some point during the translocation process, the signal peptide is cleaved from the growing protein by a signal peptidase. This cleaved signal sequence is ejected into the bilayer, where it is degraded, and the translocated polypeptide is released as a soluble protein into the ER lumen. Once the protein has been released, the pore of the translocation channel closes. The membrane-bound ribosome is omitted from this and the following two figures for clarity.

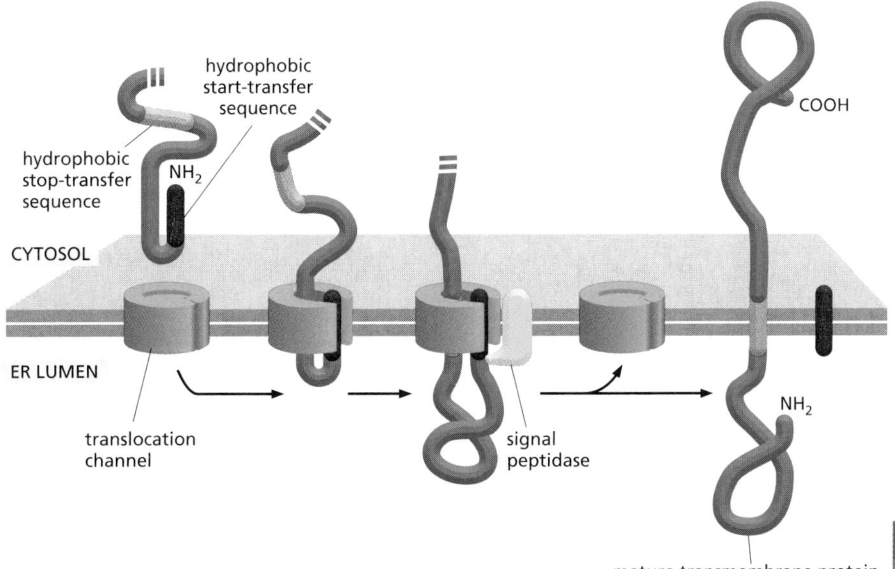

Figure 15–16 **A single-pass transmembrane protein is integrated into the ER membrane.** An N-terminal ER signal sequence (*red*) initiates transfer as in Figure 15–15. In addition, the protein also contains a second hydrophobic sequence, a stop-transfer sequence (*orange*). When this sequence enters the translocation channel, the channel discharges the protein sideways into the lipid bilayer. The N-terminal signal sequence is cleaved off, leaving the transmembrane protein anchored in the membrane (Movie 15.4). Protein synthesis on the cytosolic side continues to completion.

lipid bilayer. In these cases, hydrophobic signal sequences are thought to work in pairs: an internal start-transfer sequence serves to initiate translocation, which continues until a stop-transfer sequence is reached; the two hydrophobic sequences are then released into the bilayer, where they remain as membrane-spanning α helices (Figure 15–17). In complex multipass proteins, in which many hydrophobic α helices span the bilayer, additional pairs of stop and start sequences come into play: one sequence reinitiates translocation further down the polypeptide chain, and the other stops translocation and causes polypeptide release, and so on for subsequent starts and stops. Thus, multipass membrane proteins are stitched into the lipid bilayer as they are being synthesized, by a mechanism resembling the workings of a sewing machine.

Having considered how proteins enter the ER lumen or become embedded in the ER membrane, we now discuss how they are carried onward by vesicular transport.

QUESTION 15–4

A. Predict the membrane orientation of a protein that is synthesized with an uncleaved, internal signal sequence (shown as the *red* start-transfer sequence in Figure 15–17) but does not contain a stop-transfer peptide.

B. Similarly, predict the membrane orientation of a protein that is synthesized with an N-terminal cleaved signal sequence followed by a stop-transfer sequence, followed by a start-transfer sequence.

C. What arrangement of signal sequences would enable the insertion of a multipass protein with an odd number of transmembrane segments?

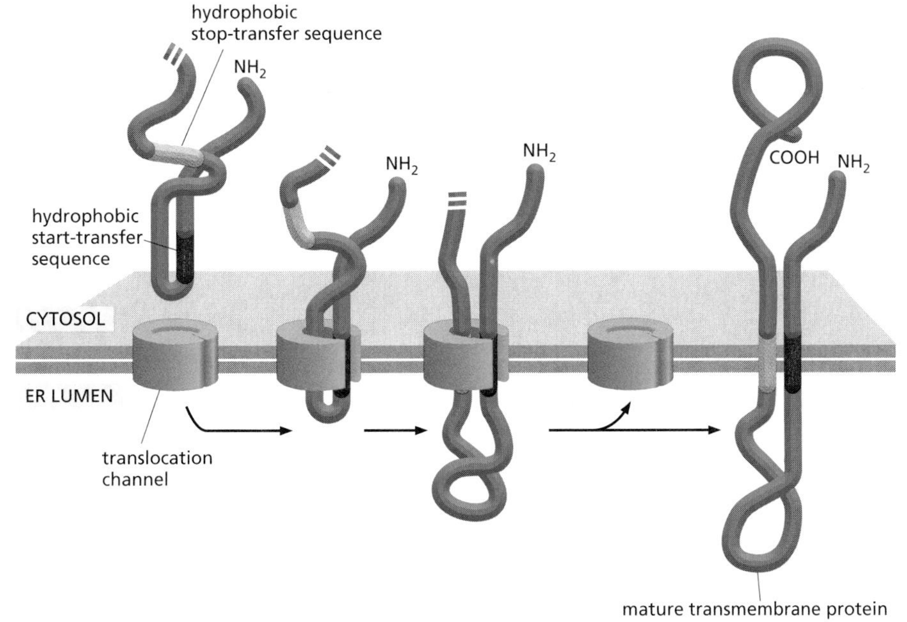

Figure 15–17 **A double-pass transmembrane protein uses an internal start-transfer sequence to integrate into the ER membrane.** An internal ER signal sequence (*red*) acts as a start-transfer signal and initiates the transfer of the polypeptide chain. Like the N-terminal ER signal sequence, the internal start-transfer signal is recognized by an SRP that brings the ribosome to the ER membrane (not shown). When a stop-transfer sequence (*orange*) enters the translocation channel, the channel discharges both sequences into the plane of the lipid bilayer. Neither the start-transfer nor the stop-transfer sequence is cleaved off, and the entire polypeptide chain remains anchored in the membrane as a double-pass transmembrane protein. Proteins that span the membrane more times contain further pairs of stop and start sequences, and the same process is repeated for each pair.

VESICULAR TRANSPORT

Entry into the ER is usually only the first step on a pathway to another destination. That destination, initially at least, is the Golgi apparatus. Transport from the ER to the Golgi apparatus and from the Golgi apparatus to other compartments of the endomembrane system is carried out by the continual budding and fusion of **transport vesicles**. The transport pathways mediated by these vesicles extend outward from the ER to the plasma membrane, and inward from the plasma membrane to lysosomes, and thus provide routes of communication between the interior of the cell and its surroundings. As proteins and lipids are transported outward along these pathways, many of them undergo various types of chemical modification, such as the addition of carbohydrate side chains (to both proteins and lipids) and the formation of disulfide bonds (in polypeptides) that stabilize protein structure.

In this section, we discuss how vesicles shuttle proteins and membranes between intracellular compartments, allowing cells to eat and secrete. We also consider how these transport vesicles are directed to their proper destination, be it the ER, Golgi apparatus, plasma membrane, or some other membrane-enclosed compartment.

Transport Vesicles Carry Soluble Proteins and Membrane Between Compartments

Vesicular transport between membrane-enclosed compartments of the endomembrane system is highly organized. A major outward *secretory pathway* starts with the synthesis of proteins on the ER membrane and their entry into the ER, and it leads through the Golgi apparatus to the cell surface; at the Golgi apparatus, a side branch leads off through endosomes to lysosomes (Figure 15–18). A major inward *endocytic pathway,* which is responsible for the ingestion and degradation of extracellular molecules, moves materials from the plasma membrane, through endosomes, to lysosomes.

To function correctly, each transport vesicle that buds off from a compartment must take with it only the proteins appropriate to its destination and must fuse only with the appropriate target membrane. A vesicle carrying

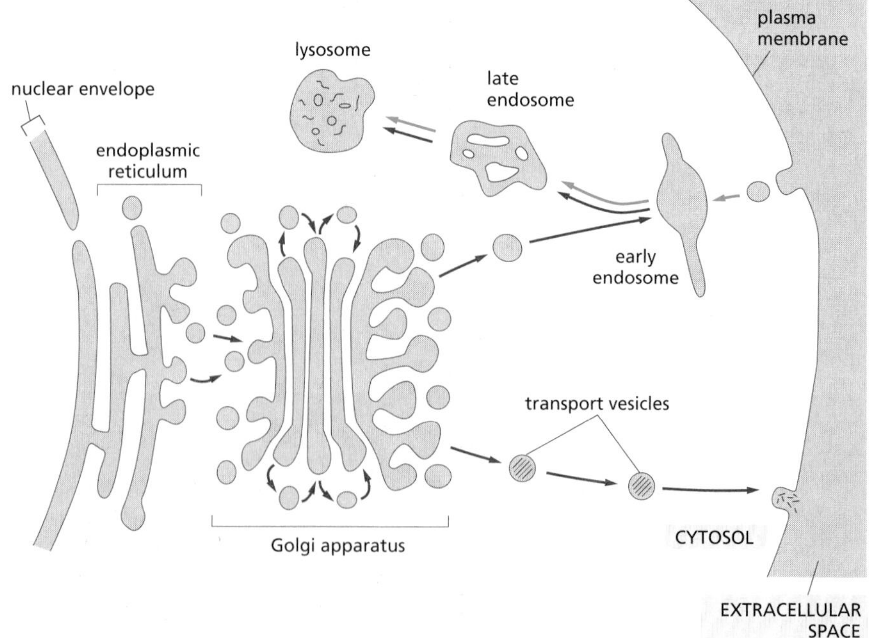

Figure 15–18 **Vesicles bud from one membrane and fuse with another, carrying membrane components and soluble proteins between cell compartments.** Each compartment encloses a space, or lumen, that is topologically equivalent to the outside of the cell (see Figure 11–19). The extracellular space and each of the membrane-enclosed compartments (shaded *gray*) communicate with one another by means of transport vesicles, as shown. In the outward secretory pathway (*red* arrows), protein molecules are transported from the ER, through the Golgi apparatus, to the plasma membrane or (via early and late endosomes) to lysosomes. In the inward endocytic pathway (*green* arrows), extracellular molecules are ingested in vesicles derived from the plasma membrane and are delivered to early endosomes and then (via late endosomes) to lysosomes.

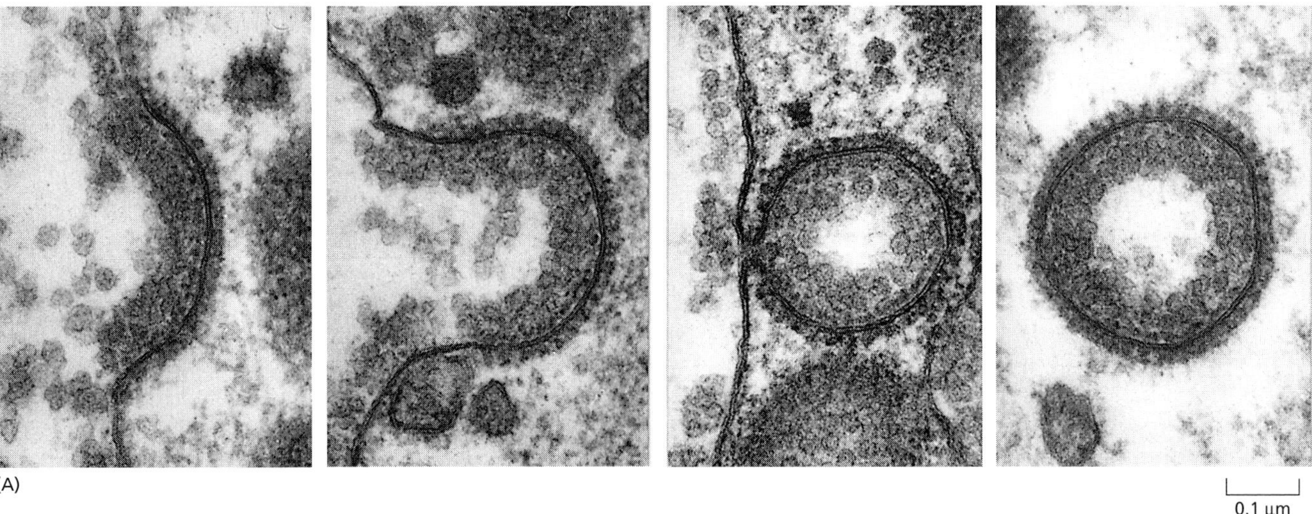

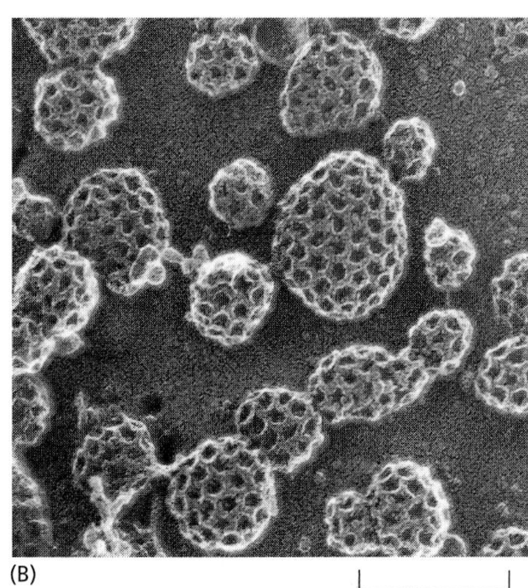

Figure 15–19 **Clathrin molecules form basketlike cages that help shape membranes into vesicles.** (A) Electron micrographs showing the sequence of events in the formation of a clathrin-coated vesicle from a clathrin-coated pit. The clathrin-coated pits and vesicles shown here are unusually large and are being formed at the plasma membrane of a hen oocyte. They are involved in taking up particles made of lipid and protein into the oocyte to form yolk. (B) Electron micrograph showing numerous clathrin-coated pits and vesicles budding from the inner surface of the plasma membrane of cultured skin cells. (A, courtesy of M.M. Perry and A.B. Gilbert, *J. Cell Sci.* 39:257–272, 1979. With permission from The Company of Biologists Ltd; B, from J. Heuser, *J. Cell Biol.* 84:560–583, 1980. With permission from Rockefeller University Press.)

cargo from the Golgi apparatus to the plasma membrane, for example, must exclude proteins that are to stay in the Golgi apparatus, and it must fuse only with the plasma membrane and not with any other organelle. While participating in this constant flow of membrane components, each organelle must maintain its own distinct identity, that is, its own distinctive protein and lipid composition. All of these recognition events depend on proteins associated with the transport vesicle membrane. As we will see, different types of transport vesicles shuttle between the various organelles, each carrying a distinct set of molecules.

Vesicle Budding Is Driven by the Assembly of a Protein Coat

Vesicles that bud from membranes usually have a distinctive protein coat on their cytosolic surface and are therefore called **coated vesicles**. After budding from its parent organelle, the vesicle sheds its coat, allowing its membrane to interact directly with the membrane to which it will fuse. Cells produce several kinds of coated vesicles, each with a distinctive protein coat. The coat serves at least two functions: it shapes the membrane into a bud, and it helps to capture molecules for onward transport.

The best-studied vesicles are those that have coats made largely of the protein **clathrin**. These *clathrin-coated vesicles* bud from the Golgi apparatus on the outward secretory pathway and from the plasma membrane on the inward endocytic pathway. At the plasma membrane, for example, each vesicle starts off as a *clathrin-coated pit*. Clathrin molecules assemble into a basketlike network on the cytosolic surface of the membrane, and it is this assembly process that starts shaping the membrane into a vesicle (Figure 15–19). A small GTP-binding protein called *dynamin* assembles as a ring around the neck of each deeply invaginated coated pit. Together with other proteins recruited to the neck of the vesicle, the dynamin causes the ring to constrict, thereby pinching off the vesicle from the membrane. Other kinds of transport vesicles, with different coat proteins, are also involved in vesicular transport. They form in a similar way and carry their own characteristic sets of molecules between the endoplasmic reticulum, the Golgi apparatus, and the plasma membrane. But how does a transport vesicle select its particular cargo? The mechanism is best understood for clathrin-coated vesicles.

Clathrin itself plays no part in capturing specific molecules for transport. This is the function of a second class of coat proteins called *adaptins,* which both secure the clathrin coat to the vesicle membrane and help select cargo molecules for transport. Molecules for onward transport carry specific *transport signals* that are recognized by *cargo receptors* in the compartment membrane. Adaptins help capture specific cargo molecules by trapping the cargo receptors that bind them. In this way, a selected set of cargo molecules, bound to their specific receptors, is incorporated into the lumen of each newly formed clathrin-coated vesicle (Figure 15–20). There are different types of adaptins: the adaptins that bind cargo receptors in the plasma membrane are not the same as those that bind cargo receptors in the Golgi apparatus, reflecting the differences in the cargo molecules from each of these sources.

Another class of coated vesicles, called *COP-coated vesicles* (COP is shorthand for 'coat protein'), is involved in transporting molecules between the ER and the Golgi apparatus and from one part of the Golgi apparatus to another (Table 15–4).

Vesicle Docking Depends on Tethers and SNAREs

After a transport vesicle buds from a membrane, it must find its way to its correct destination to deliver its contents. In most cases, the vesicle is actively transported by motor proteins that move along cytoskeletal fibers, as discussed in Chapter 17.

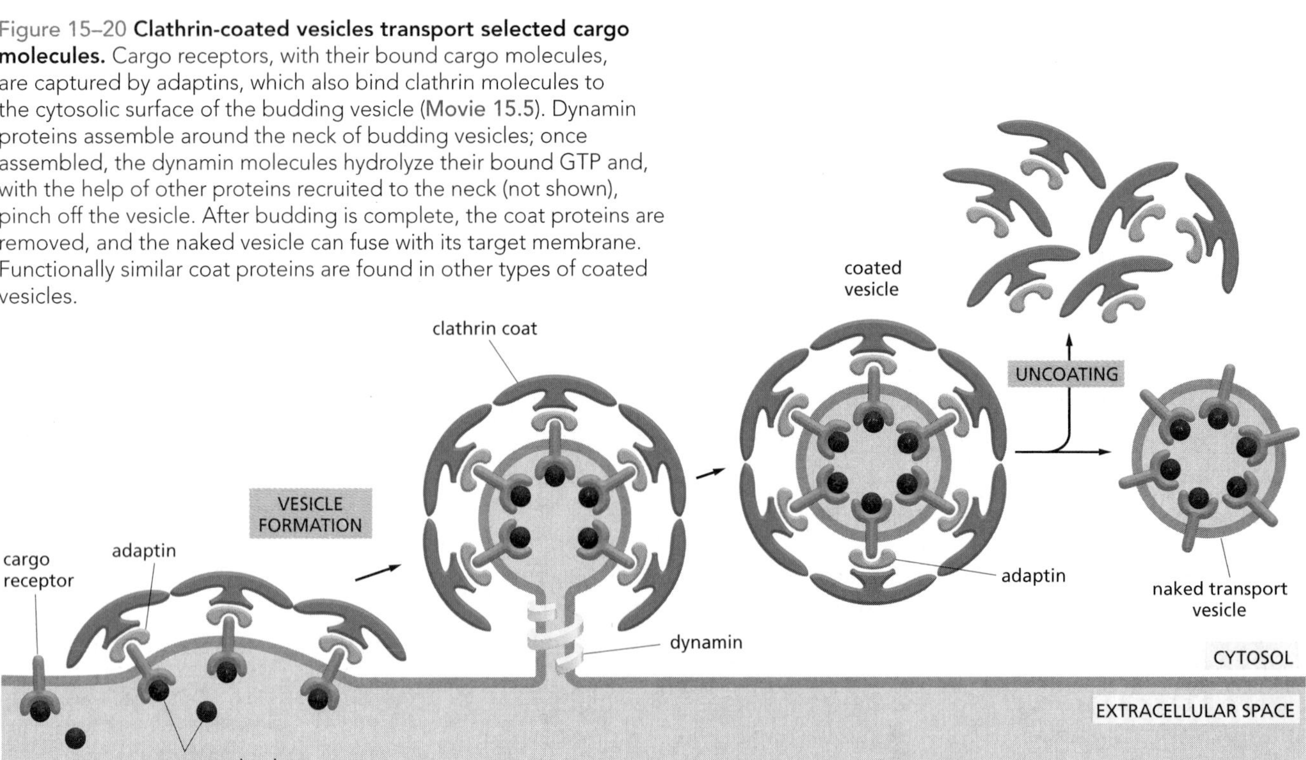

Figure 15–20 **Clathrin-coated vesicles transport selected cargo molecules.** Cargo receptors, with their bound cargo molecules, are captured by adaptins, which also bind clathrin molecules to the cytosolic surface of the budding vesicle (**Movie 15.5**). Dynamin proteins assemble around the neck of budding vesicles; once assembled, the dynamin molecules hydrolyze their bound GTP and, with the help of other proteins recruited to the neck (not shown), pinch off the vesicle. After budding is complete, the coat proteins are removed, and the naked vesicle can fuse with its target membrane. Functionally similar coat proteins are found in other types of coated vesicles.

TABLE 15–4 SOME TYPES OF COATED VESICLES

TYPE OF COATED VESICLE	COAT PROTEINS	ORIGIN	DESTINATION
Clathrin-coated	clathrin + adaptin 1	Golgi apparatus	lysosome (via endosomes)
Clathrin-coated	clathrin + adaptin 2	plasma membrane	endosomes
COP-coated	COP proteins	ER Golgi cisterna Golgi apparatus	Golgi apparatus Golgi cisterna ER

Once a transport vesicle has reached its target, it must recognize and dock with the organelle. Only then can the vesicle membrane fuse with the target membrane and unload the vesicle's cargo. The impressive specificity of vesicular transport suggests that each type of transport vesicle in the cell displays molecular markers on its surface that identify the vesicle according to its origin and cargo. These markers must be recognized by complementary receptors on the appropriate target membrane, including the plasma membrane. This identification process depends on a family of proteins called **Rab proteins**. Rab proteins on the surface of the vesicle are recognized by *tethering proteins* on the cytosolic surface of the target membrane. Each organelle and each type of transport vesicle carries a unique combination of Rab proteins, which serve as molecular markers identifying each membrane type. This coding system of Rab and tethering proteins helps to ensure that transport vesicles fuse only with the correct membrane. Additional recognition is provided by a family of related transmembrane proteins called **SNAREs**. Once the tethering protein has captured a vesicle by grabbing hold of its Rab protein, SNAREs on the vesicle (called v-SNAREs) interact with complementary SNAREs on the target membrane (called t-SNAREs), docking the vesicle in place (Figure 15–21).

Once a transport vesicle has recognized its target membrane and docked there, the vesicle has to fuse with the membrane to deliver its cargo. Fusion not only delivers the contents of the vesicle into the interior of the target organelle, but it also adds the vesicle membrane to the membrane of the organelle. Membrane fusion does not always follow immediately after docking, however; it sometimes awaits a specific molecular signal. Whereas docking requires only that the two membranes come close enough for proteins protruding from the two lipid bilayers to interact, fusion requires a much closer approach: the two bilayers must come

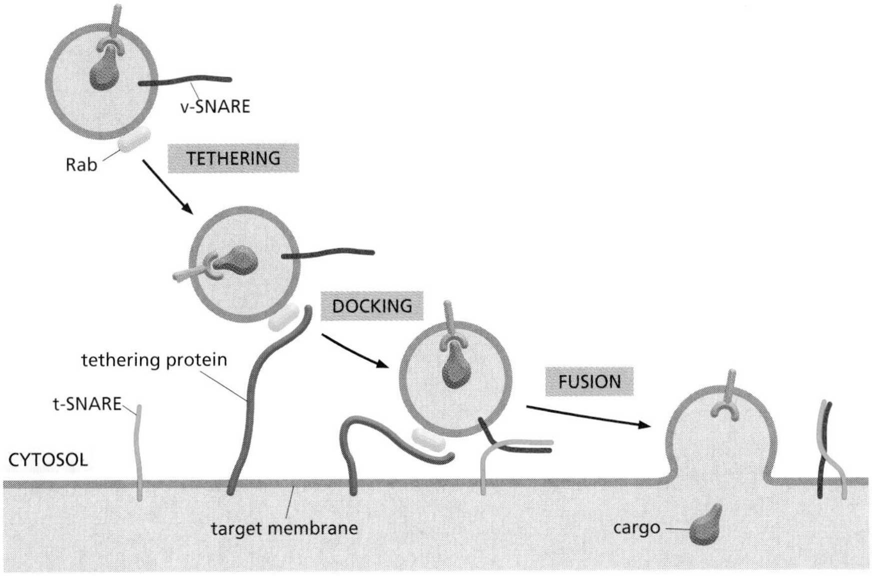

QUESTION 15–5

The budding of clathrin-coated vesicles from eucaryotic plasma membrane fragments can be observed when adaptins, clathrin, and dynamin-GTP are added to the membrane preparation. What would you observe if you omitted (A) adaptins, (B) clathrin, or (C) dynamin? (D) What would you observe if the plasma membrane fragments were from a procaryotic cell?

Figure 15–21 **Rab proteins and SNAREs help direct transport vesicles to their target membranes.** A filamentous tethering protein on a membrane binds to a Rab protein on the surface of a vesicle. This interaction allows the vesicle to dock on its target membrane. A v-SNARE on the vesicle then binds to a complementary t-SNARE on the target membrane. Whereas Rab and tethering proteins provide the initial recognition between a vesicle and its target membrane, the pairing of complementary SNAREs also helps ensure that transport vesicles reach their appropriate target membranes.

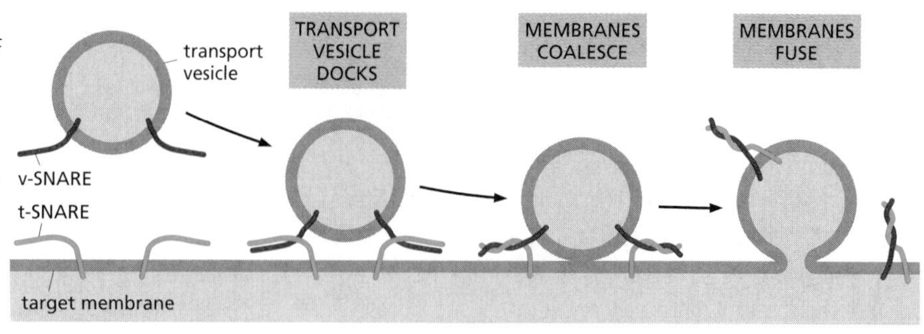

Figure 15–22 **SNARE proteins play a central role in membrane fusion.** Pairing of v-SNAREs and t-SNAREs draw the two lipid bilayers into close apposition. The force of the SNAREs winding together squeezes out any water molecules that remain trapped between the two membranes, allowing their lipids to flow together to form a continuous bilayer. In a cell, other proteins recruited to the fusion site presumably cooperate with SNAREs to initiate fusion. Additional proteins help to pry the SNAREs apart so that they can be used again.

within 1.5 nm of each other so that their lipids can intermix. For this close approach, water must be displaced from the hydrophilic surface of the membrane—a process that is energetically highly unfavorable and thus prevents membranes from fusing randomly. All membrane fusions in cells must therefore be catalyzed by specialized proteins that assemble at the fusion site to form a fusion complex, which provides the means to cross this energy barrier. The SNARE proteins themselves play a central role in the fusion process: after pairing, v-SNAREs and t-SNAREs wrap around each other, thereby acting like a winch that pulls the two membranes into close proximity (Figure 15–22).

SECRETORY PATHWAYS

Vesicular traffic is not confined to the interior of the cell. It extends to and from the plasma membrane. Newly made proteins, lipids, and carbohydrates are delivered from the ER, via the Golgi apparatus, to the cell surface by transport vesicles that fuse with the plasma membrane in a process called **exocytosis**. Each molecule that travels along this route passes through a fixed sequence of membrane-enclosed compartments and is often chemically modified en route.

In this section, we follow the outward path of proteins as they travel from the ER, where they are made and modified, through the Golgi apparatus, where they are further modified and sorted, to the plasma membrane. As a protein passes from one compartment to another, it is monitored to check that it has folded properly and assembled with its appropriate partners, so that only correctly built proteins are released at the cell surface, while all of the others, which are often the majority, are degraded in the cell.

Most Proteins Are Covalently Modified in the ER

Most proteins that enter the ER are chemically modified there. *Disulfide bonds* are formed by the oxidation of pairs of cysteine side chains (see Figure 4–26), a reaction catalyzed by an enzyme that resides in the ER lumen. The disulfide bonds help to stabilize the structure of those proteins that may encounter changes in pH and degradative enzymes outside the cell—either after they are secreted or after they are incorporated into the plasma membrane. Because of the reducing environment in the cytosol, disulfide bonds do not form there.

Many of the proteins that enter the ER lumen or ER membrane are converted to glycoproteins in the ER by the covalent attachment of short oligosaccharide side chains. This process of *glycosylation* is carried out by glycosylating enzymes present in the ER but not in the cytosol. Very few proteins in the cytosol are glycosylated, and those that are have only a single sugar attached to them. The oligosaccharides on proteins serve various functions, depending on the protein. They can protect the protein from degradation, hold it in the ER until it is properly folded, or help guide it to the appropriate organelle by serving as a transport signal for packaging the protein into appropriate transport vesicles. When displayed on the cell surface, oligosaccharides form part of the cell's carbohydrate layer (see Figure 11–35) and can function in the recognition of one cell by another.

In the ER, individual sugars are not added one by one to the protein to create the oligosaccharide side chain. Instead, a preformed, branched oligosaccharide containing a total of 14 sugars is attached en bloc to all proteins that carry the appropriate site for glycosylation. The oligosaccharide is originally attached to a specialized lipid, called *dolichol*, in the ER membrane; it is then transferred to the amino (NH_2) group of an asparagine side chain on the protein, immediately after the target asparagine emerges in the ER lumen during protein translocation (Figure 15–23). The addition takes place in a single enzymatic step that is catalyzed by a membrane-bound enzyme (an oligosaccharide protein transferase) that has its active site exposed on the lumenal side of the ER membrane—which explains why cytosolic proteins are not glycosylated in this way. A simple sequence of three amino acids, of which the asparagine is one, defines which asparagines in a protein receive the oligosaccharide. Oligosaccharide side chains linked to an asparagine NH_2 group in a protein are said to be *N-linked* and are by far the most common type of linkage found on glycoproteins.

The addition of the 14-sugar oligosaccharide in the ER is only the first step in a series of further modifications before the mature glycoprotein emerges at the other end of the outward pathway. Despite their initial similarity, the *N*-linked oligosaccharides on mature glycoproteins are remarkably diverse. All of the diversity results from extensive modification of the original precursor structure shown in Figure 15–23. This *oligosaccharide processing* begins in the ER and continues in the Golgi apparatus.

QUESTION 15–6

Why might it be advantageous to add a preassembled block of 14 sugar residues to a protein in the ER, rather than building the sugar chains step-by-step on the surface of the protein by the sequential addition of sugars by individual enzymes?

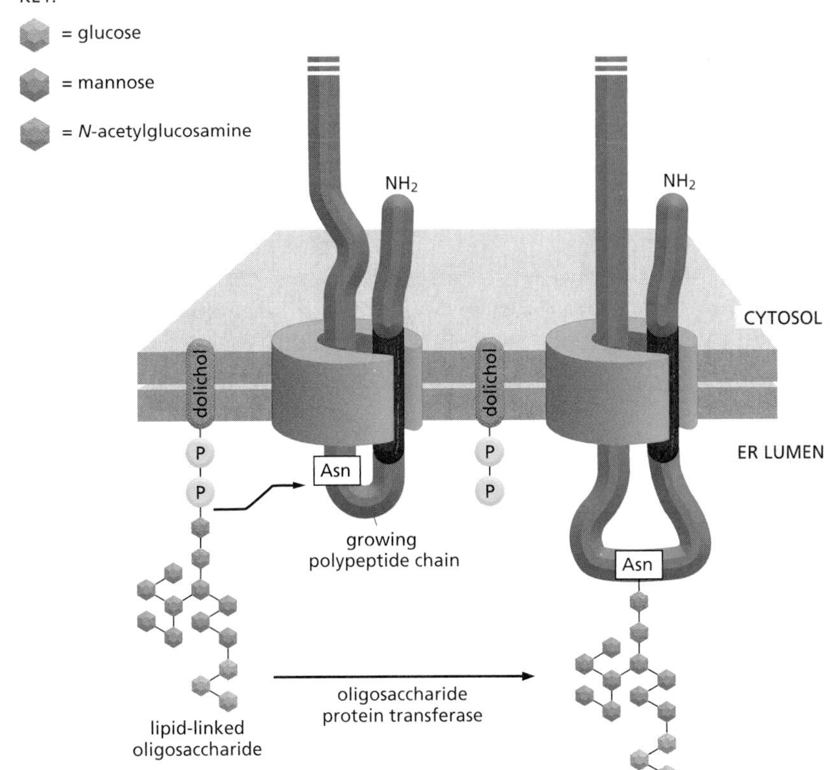

Figure 15–23 **Many proteins are glycosylated in the ER.** Almost as soon as the polypeptide chain enters the ER lumen, it is glycosylated by addition of oligosaccharide side chains to particular asparagines in the polypeptide. Each oligosaccharide chain is transferred as an intact unit to the asparagine from a lipid called dolichol, catalyzed by the enzyme oligosaccharyl transferase (not shown). Asparagines that are glycosylated are always present in the tripeptide sequences asparagine-X-serine or asparagine-X-threonine, where X can be any amino acid.

Exit from the ER Is Controlled to Ensure Protein Quality

Some proteins made in the ER are destined to function there. They are retained in the ER (and are returned to the ER when they escape to the Golgi apparatus) by a C-terminal sequence of four amino acids called an *ER retention signal* (see Table 15–3, p. 502). This retention signal is recognized by a membrane-bound receptor protein in the ER and Golgi apparatus. Most proteins that enter the ER, however, are destined for other locations; they are packaged into transport vesicles that bud from the ER and fuse with the Golgi apparatus. Exit from the ER is highly selective. Proteins that fold incorrectly, and dimeric or multimeric proteins that fail to assemble properly, are actively retained in the ER by binding to *chaperone proteins* that reside there. Interaction with chaperones holds the proteins in the ER until proper folding occurs; if this does not happen, the proteins are eventually degraded (Figure 15–24). Antibody molecules, for example, are composed of four polypeptide chains (see Figure 4–29) that assemble into the complete antibody molecule in the ER. Partially assembled antibodies are retained in the ER until all four polypeptide chains have assembled; any antibody molecule that fails to assemble properly is ultimately degraded. In this way, the ER controls the quality of the proteins that it exports to the Golgi apparatus.

Sometimes, however, this quality-control mechanism can be detrimental to the organism. For example, the predominant mutation that causes the common genetic disease *cystic fibrosis*, which causes severe degeneration of the lung, produces a plasma-membrane transport protein that is slightly misfolded; even though the mutant protein could function normally as a chloride channel if it reached the plasma membrane, it is retained in the ER, with dire consequences. The devastating disease results not because the mutation inactivates an important protein but because the active protein is discarded by the cells before it is given an opportunity to function.

The Size of the ER Is Controlled by the Amount of Protein that Flows Through It

Although chaperones help proteins in the ER fold properly and retain those that do not, when protein synthesis is vigorous, the system can become overwhelmed. When a cell's protein production exceeds the carrying—and folding—capacity of its ER, misfolded proteins begin to accumulate. These aberrant proteins actually serve as a signal to direct the cell to make more ER. They do so by activating a special set of receptors that reside in the ER membrane, which in turn activate a vast transcriptional program called the **unfolded protein response** (**UPR**). The UPR program prompts the cell to produce more ER, including all of the molecular machinery required to restore proper protein folding and processing (Figure 15–25).

The UPR program allows cells to adjust the size of the ER according to need, so that the load of proteins entering the secretory pathway will be

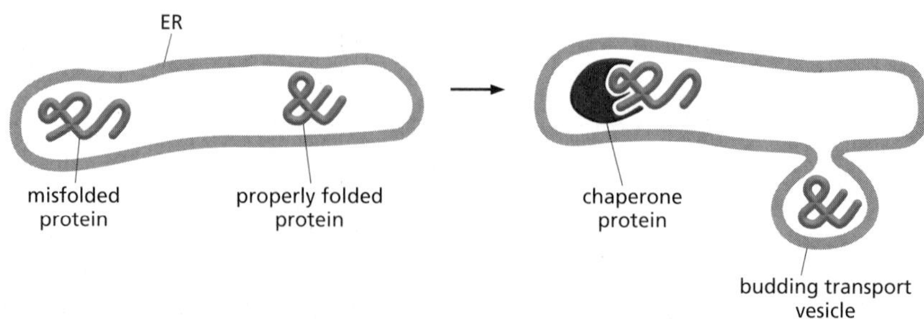

Figure 15–24 **Chaperones prevent misfolded or partially assembled proteins from leaving the ER.** Misfolded proteins bind to chaperone proteins in the ER lumen and are thereby retained, whereas normally folded proteins are transported in transport vesicles to the Golgi apparatus. If the misfolded proteins fail to refold normally, they are transported back into the cytosol, where they are degraded.

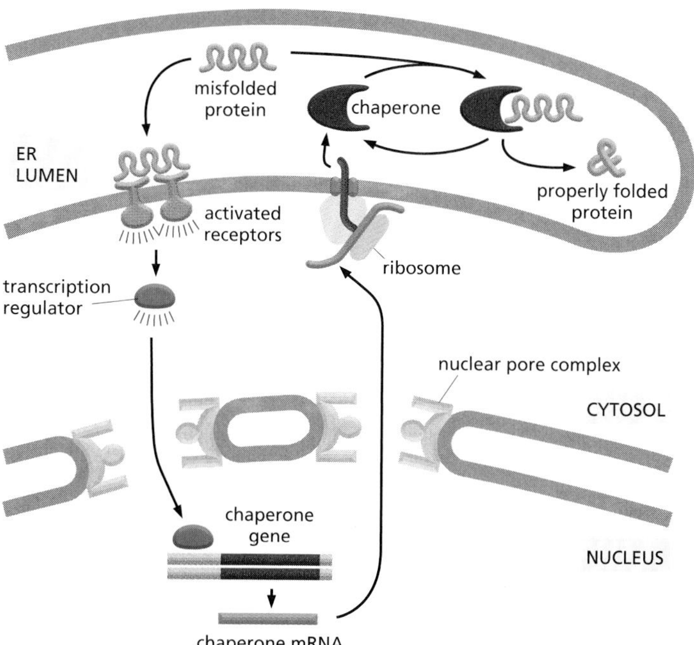

Figure 15–25 **Misfolded proteins in the ER lumen trigger the production of chaperones and the expansion of the ER.** Misfolded proteins bind to receptors that stimulate the production of a transcription regulator. This protein translocates to the nucleus where it activates genes that encode chaperones and other ER components, thus promoting the proper folding and processing of proteins. The system is known as the unfolded protein response.

folded efficiently and properly. In some cases, however, even an expanded ER can become overloaded. If a proper balance can not be reestablished—and misfolded proteins continue to accumulate—the UPR program can direct the cell to self-destruct by undergoing apoptosis. Such a situation may arise in adult-onset diabetes, where the tissues of the body gradually become resistant to the effects of insulin. As the insulin-secreting cells in the pancreas are called upon to produce more and more insulin, their ER may reach a maximum capacity, at which point further expansion becomes physiologically impossible. The UPR program may then trigger cell death. Unfortunately, as more insulin-secreting cells are eliminated, the demand to produce additional insulin will fall to the surviving cells, taxing their ERs and making them more likely to die as well. This escalating loss of insulin-producing cells exacerbates the disease.

Proteins Are Further Modified and Sorted in the Golgi Apparatus

The **Golgi apparatus** is usually located near the cell nucleus, and in animal cells it is often close to the centrosome, a small structure near the cell center. The Golgi apparatus consists of a collection of flattened, membrane-enclosed sacs (*cisternae*), which are piled like stacks of plates. Each stack contains 3–20 cisternae (Figure 15–26). The number of Golgi stacks per cell varies greatly depending on the cell type: some cells contain one large stack, while others contain hundreds of very small ones.

Each Golgi stack has two distinct faces: an entry, or *cis*, face and an exit, or *trans,* face. The *cis* face is adjacent to the ER, while the *trans* face points toward the plasma membrane. The outermost cisterna at each face is connected to a network of interconnected membranous tubes and vesicles (see Figure 15–26A). Soluble proteins and membrane enter the *cis Golgi network* via transport vesicles derived from the ER. The proteins travel through the cisternae in sequence by means of transport vesicles that bud from one cisterna and fuse with the next. Proteins exit from the *trans Golgi network* in transport vesicles destined for either the cell surface or another compartment (see Figure 15–18). Both the *cis* and *trans* Golgi networks are thought to be important for protein sorting: proteins entering the *cis* Golgi network can either move onward through the Golgi

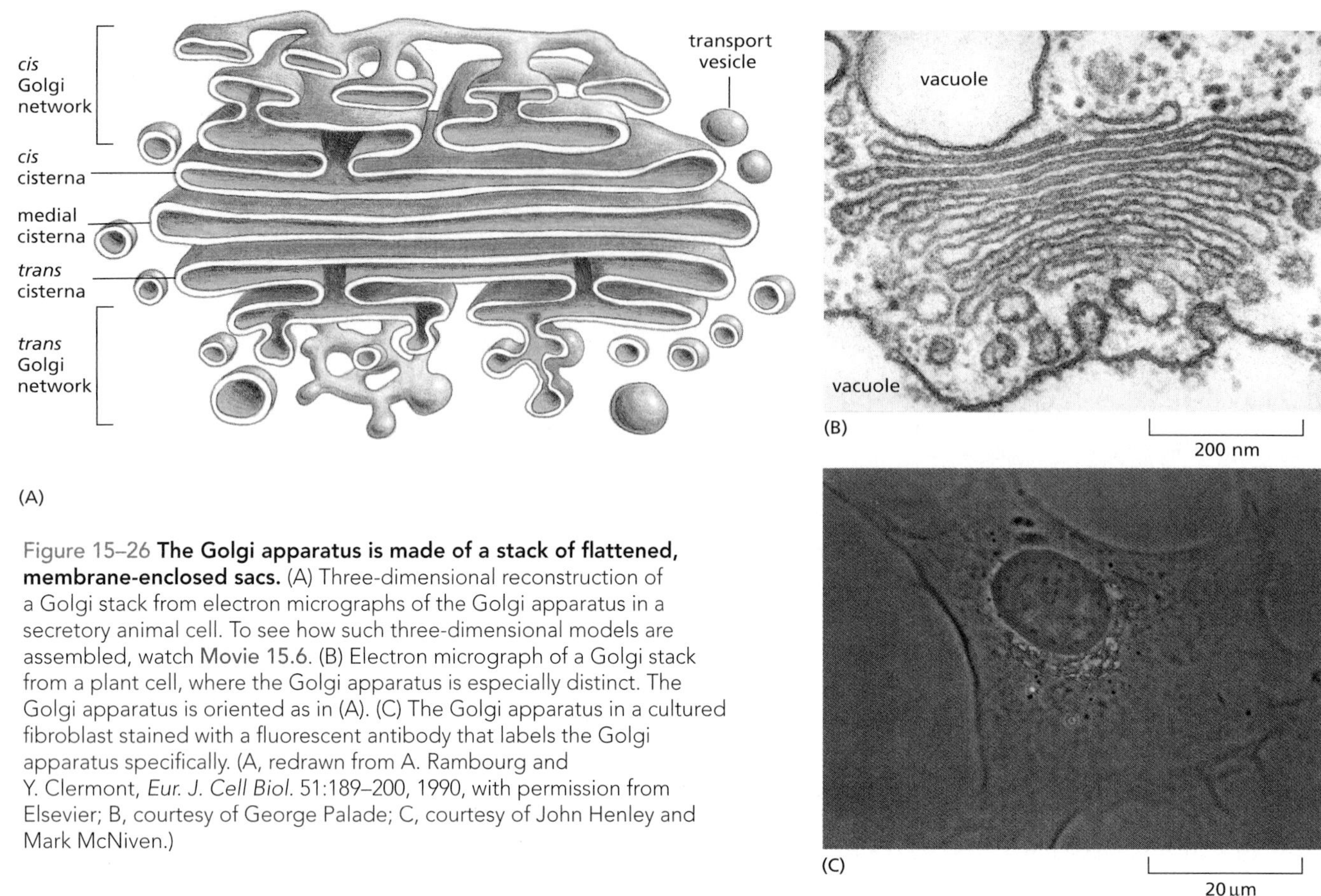

Figure 15–26 **The Golgi apparatus is made of a stack of flattened, membrane-enclosed sacs.** (A) Three-dimensional reconstruction of a Golgi stack from electron micrographs of the Golgi apparatus in a secretory animal cell. To see how such three-dimensional models are assembled, watch Movie 15.6. (B) Electron micrograph of a Golgi stack from a plant cell, where the Golgi apparatus is especially distinct. The Golgi apparatus is oriented as in (A). (C) The Golgi apparatus in a cultured fibroblast stained with a fluorescent antibody that labels the Golgi apparatus specifically. (A, redrawn from A. Rambourg and Y. Clermont, *Eur. J. Cell Biol.* 51:189–200, 1990, with permission from Elsevier; B, courtesy of George Palade; C, courtesy of John Henley and Mark McNiven.)

stack or, if they contain an ER retention signal, be returned to the ER; proteins exiting from the *trans* Golgi network are sorted according to whether they are destined for lysosomes or for the cell surface. We discuss some examples of sorting by the *trans* Golgi network later, and we present some of the methods for tracking proteins through the secretory pathways of the cell in How We Know, pp. 520–521.

Many of the oligosaccharide groups that are added to proteins in the ER undergo further modifications in the Golgi apparatus. On some proteins, for example, complex oligosaccharide chains are created by a highly ordered process in which sugars are added and removed by a series of enzymes that act in a rigidly determined sequence as the protein passes through the Golgi stack. There is a clear correlation between the position of an enzyme in the chain of processing events and its localization in the Golgi stack: enzymes that act early are found in cisternae close to the *cis* face, while enzymes that act late are found in cisternae near the *trans* face.

Secretory Proteins Are Released from the Cell by Exocytosis

In all eucaryotic cells, a steady stream of vesicles buds from the *trans* Golgi network and fuses with the plasma membrane. This *constitutive exocytosis pathway* operates continually and supplies newly made lipids and proteins to the plasma membrane (Movie 15.7); it is the pathway for plasma membrane growth when cells enlarge before dividing. The constitutive pathway also carries proteins to the cell surface to be released to the outside, a process called **secretion**. Some of the released proteins

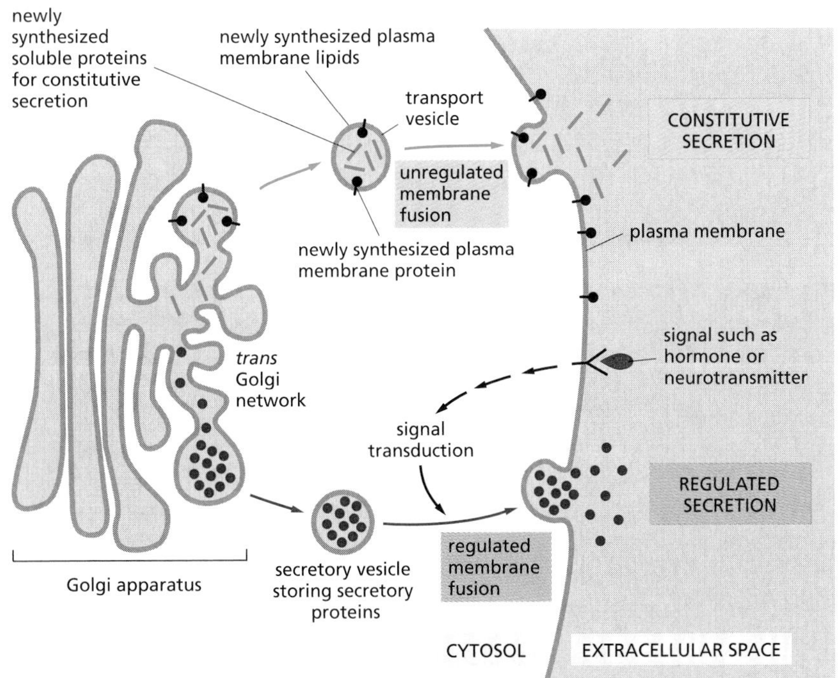

Figure 15–27 **In secretory cells, the regulated and constitutive pathways of exocytosis diverge in the *trans* Golgi network.** Many soluble proteins are continually secreted from the cell by the constitutive secretory pathway, which operates in all cells (Movie 15.8). This pathway also continually supplies the plasma membrane with newly synthesized lipids and proteins. Specialized secretory cells have, in addition, a regulated exocytosis pathway, by which selected proteins in the *trans* Golgi network are diverted into secretory vesicles, where the proteins are concentrated and stored until an extracellular signal stimulates their secretion. It is unclear how aggregates of secretory proteins are segregated into secretory vesicles. Secretory vesicles have unique proteins in their membranes; perhaps some of these proteins act as receptors for secretory protein aggregates in the *trans* Golgi network.

adhere to the cell surface, where they become peripheral proteins of the plasma membrane; some are incorporated into the extracellular matrix; still others diffuse into the extracellular fluid to nourish or to signal other cells. Because entry into this nonselective pathway does not require a particular signal sequence (like those that direct proteins to lysosomes or back to the ER), it is sometimes referred to as the *default pathway*.

In addition to the constitutive exocytosis pathway, which operates continually in all eucaryotic cells, there is a *regulated exocytosis pathway*, which operates only in cells that are specialized for secretion. Specialized *secretory cells* produce large quantities of particular products, such as hormones, mucus, or digestive enzymes, which are stored in **secretory vesicles** for later release. These vesicles bud off from the *trans* Golgi network and accumulate near the plasma membrane. There they wait for the extracellular signal that will stimulate them to fuse with the plasma membrane and release their contents to the cell exterior (Figure 15–27). An increase in blood glucose, for example, signals cells in the pancreas to secrete the hormone insulin (Figure 15–28).

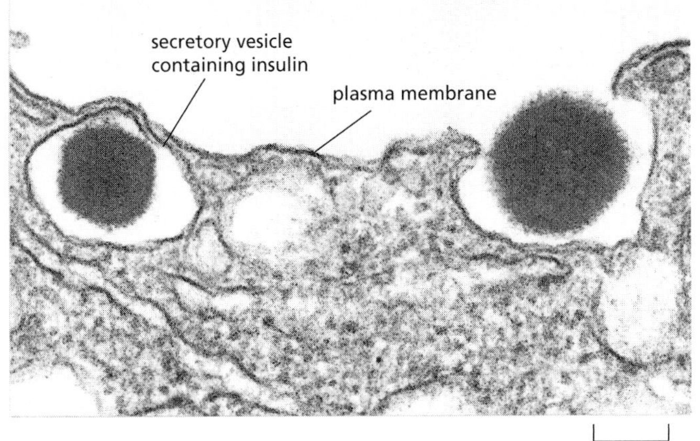

Figure 15–28 **Secretory vesicles package and discharge concentrated aggregates of protein.** The electron micrograph shows the release of insulin into the extracellular space from a secretory vesicle of a pancreatic β cell. The insulin is stored in a highly concentrated form in each secretory vesicle and is released only when the cell is signaled to secrete by an increase in glucose levels in the blood. (Courtesy of Lelio Orci, from L. Orci, J.D. Vassali, and A. Perrelet, *Sci. Am.* 259:85–94, 1988. With permission from Scientific American.)

HOW WE KNOW: TRACKING PROTEIN AND VESICLE TRANSPORT

Over the years, biologists have taken advantage of a variety of techniques to untangle the pathways and mechanisms by which proteins are sorted and transported into and out of the cell and its resident organelles. As we saw earlier, transferring an ER signal sequence to a cytosolic protein allowed researchers to confirm that such signal peptides serve to target proteins to specific intracellular compartments—in that example, the ER (see Figure 15–6). But such signal-swapping experiments are not the only way to track a protein's progress through the cell. Biochemical, genetic, and molecular biological and microscopic techniques also provide a means for studying how proteins shuttle from one cellular compartment to another. In some cases, these methods can be used to track the migration of proteins and transport vesicles in real time inside living cells.

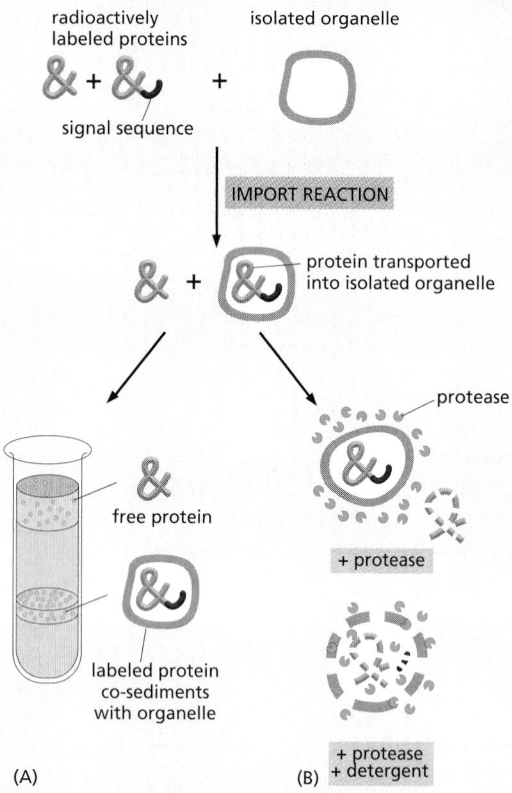

In a tube

A protein bearing a signal sequence can be introduced to a preparation of isolated organelles in a test tube. This mixture can then be tested to see whether the protein will be taken up by the organelle being examined. The protein is usually produced *in vitro* by cell-free translation of a purified mRNA encoding the polypeptide; in the process, radioactive amino acids can be used to label the protein so that it will be easy to isolate and to follow. The labeled protein is incubated with a selected organelle and its translocation is monitored by one of several different methods (Figure 15–29).

radioactively labeled proteins
isolated organelle
signal sequence
IMPORT REACTION
protein transported into isolated organelle
protease
free protein
+ protease
labeled protein co-sediments with organelle
(A)
(B)
+ protease + detergent

Figure 15–29 **Several methods can be used to determine whether a protein bearing a particular signal sequence is transported into a preparation of isolated organelles.** (A) The labeled protein with or without a signal sequence is incubated with the organelles, and the preparation is centrifuged. Only those labeled proteins that contained a signal sequence will be transported and therefore should co-fractionate with the organelle. (B) The labeled proteins are incubated with the organelle, and a protease is added to the preparation. A transported protein will be selectively protected from digestion by the protease; adding a detergent that disrupts the organelle membrane will eliminate that protection, and the transported protein will also be degraded.

Ask a yeast

Movement of proteins between different cell compartments via transport vesicles has been studied extensively using genetic techniques. Studies of mutant yeast cells that are defective for secretion at high temperatures have identified more than 25 genes that are involved in exocytosis. Many of these mutant genes encode temperature-sensitive proteins that are involved in transport and secretion. These mutant proteins may function normally at 25°C, but, when the yeast cells are shifted to 35°C, they are inactivated. As a result, when researchers raise the temperature, proteins destined for secretion instead accumulate inappropriately in the ER, the Golgi apparatus, or transport vesicles (Figure 15–30).

At the movies

Perhaps the most dramatic method for tracking a protein as it moves throughout the cell involves tagging the polypeptide with green fluorescent protein (GFP). Using the genetic engineering techniques discussed in Chapter 10, this small protein can be fused to other cell proteins. Fortunately, for most proteins studied, the addition of GFP does not perturb the molecule's normal function or transport. The movement of a GFP-tagged protein can then be monitored in a living cell with a fluorescent microscope.

GFP fusion proteins are widely used to study the location and movement of proteins in cells (Figure 15–31). GFP

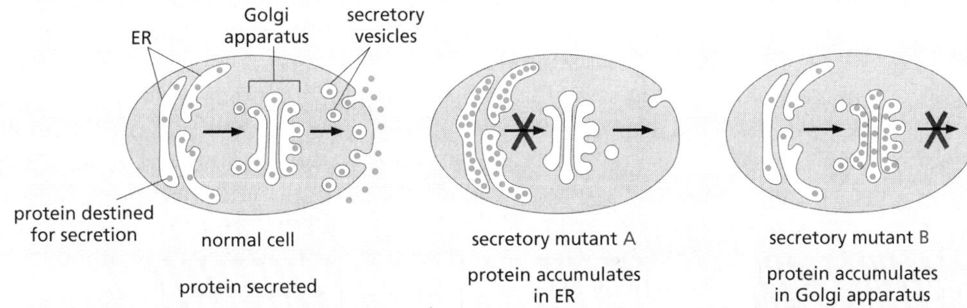

Figure 15–30 **Temperature-sensitive mutants have been used to dissect the protein secretory pathway in yeast.** Mutations in genes involved at different stages of the transport process result in the accumulation of proteins in the ER, the Golgi apparatus, or transport vesicles. For example, a mutation A that blocks transport from the ER to the Golgi apparatus will cause a buildup of proteins in the ER. A mutation B that blocks exit of proteins from the Golgi apparatus will cause proteins to accumulate within that organelle.

fused to proteins that shuttle in and out of the nucleus, for example, can be used to study nuclear transport events. GFP fused to plasma membrane proteins can be used to measure the kinetics of their movement through the secretory pathway. Movies demonstrating the power and beauty of this technique are included on the DVD that accompanies this book (Movie 15.1, Movie 15.7, Movie 15.8, and Movie 15.11).

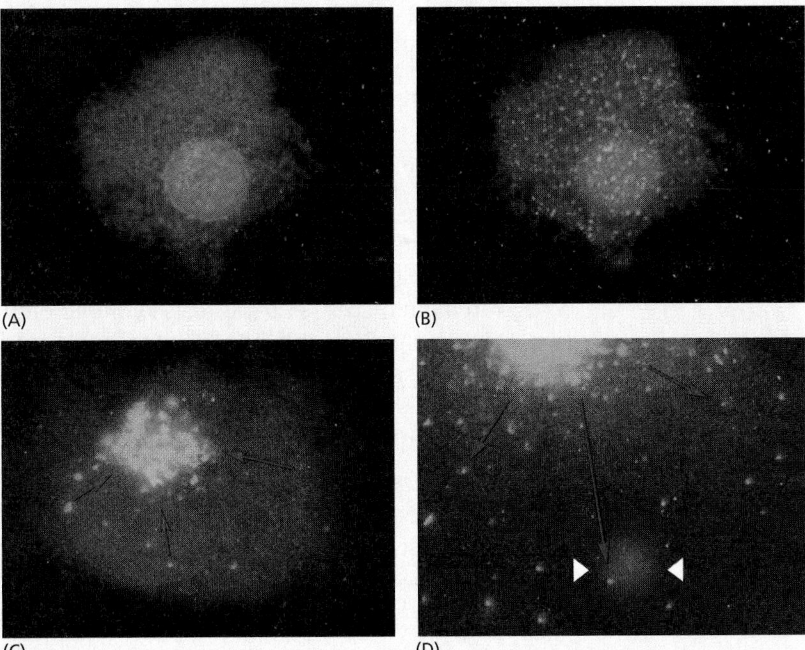

Figure 15–31 **GFP fusion allows proteins to be tracked throughout the cell.** In this experiment, GFP is fused to a viral coat protein and expressed in cultured cells. In an infected cell, the viral protein will move through the secretory pathway from the ER to the cell surface, where virus particles would be assembled. The viral coat protein used in this experiment contains a mutation that allows export from the ER only at a low temperature. (A) At high temperatures, the fusion protein labels the ER. (B) As the temperature is lowered, the GFP fusion protein rapidly accumulates at ER exit sites. (C) The fusion protein then moves to the Golgi apparatus. (D) Finally, the fusion protein is delivered to the plasma membrane, shown here in a more close-up view. The halo between the two arrowheads marks the spot where a single vesicle has fused, expelling the viral coat protein into the plasma membrane. (A–D, courtesy of Jennifer Lippincott-Schwartz.)

Proteins destined for regulated secretion are sorted and packaged in the *trans* Golgi network. Proteins that travel by this pathway have special surface properties that cause them to aggregate with one another under the ionic conditions (acidic pH and high Ca^{2+}) that prevail in the *trans* Golgi network. The aggregated proteins are packaged into secretory vesicles, which pinch off from the network and await a signal instructing them to fuse with the plasma membrane. Proteins secreted by the constitutive pathway, on the other hand, do not aggregate and are therefore carried automatically to the plasma membrane by the transport vesicles of the constitutive pathway. Selective aggregation has another function: it allows secretory proteins to be packaged into secretory vesicles at concentrations much higher than the concentration of the unaggregated protein in the Golgi lumen. This increase in concentration can reach 200-fold, enabling secretory cells to release large amounts of the protein promptly when triggered to do so (see Figure 15–28).

QUESTION 15–7

What would you expect to happen in cells that secrete large amounts of protein through the regulated secretory pathway if the ionic conditions in the ER lumen could be changed to resemble those in the lumen of the *trans* Golgi network?

When a secretory vesicle or transport vesicle fuses with the plasma membrane and discharges its contents by exocytosis, its membrane becomes part of the plasma membrane. Although this should greatly increase the surface area of the plasma membrane, it does so only transiently because membrane components are removed from other regions of the surface by endocytosis almost as fast as they are added by exocytosis. This removal returns both the lipids and the proteins of the vesicle membrane to the Golgi network, where they can be used again.

ENDOCYTIC PATHWAYS

Eucaryotic cells are continually taking up fluid, as well as large and small molecules, by the process of endocytosis. Specialized cells are also able to internalize large particles and even other cells. The material to be ingested is progressively enclosed by a small portion of the plasma membrane, which first buds inward and then pinches off to form an intracellular *endocytic vesicle*. The ingested material is ultimately delivered to lysosomes, where it is digested. The metabolites generated by digestion are transferred directly out of the lysosome into the cytosol, where they can be used by the cell.

Two main types of **endocytosis** are distinguished on the basis of the size of the endocytic vesicles formed. *Pinocytosis* ('cellular drinking') involves the ingestion of fluid and molecules via small vesicles (<150 nm in diameter). *Phagocytosis* ('cellular eating') involves the ingestion of large particles, such as microorganisms and cell debris, via large vesicles called *phagosomes* (generally >250 nm in diameter). Whereas all eucaryotic cells are continually ingesting fluid and molecules by pinocytosis, large particles are ingested mainly by specialized *phagocytic cells*.

In this final section of the chapter, we trace the endocytic pathway from the plasma membrane to lysosomes. We start by considering the uptake of large particles by phagocytosis.

Specialized Phagocytic Cells Ingest Large Particles

The most dramatic form of endocytosis, **phagocytosis**, was first observed more than a hundred years ago. In protozoa, phagocytosis is a form of feeding: these unicellular eucaryotes ingest large particles such as bacteria by taking them up into phagosomes (Movie 15.9). The phagosomes then fuse with lysosomes, where the food particles are digested. Few cells in multicellular organisms are able to ingest large particles efficiently. In the animal gut, for example, large particles of food have to be broken down to individual molecules by extracellular enzymes before they can be taken up by the absorptive cells lining the gut.

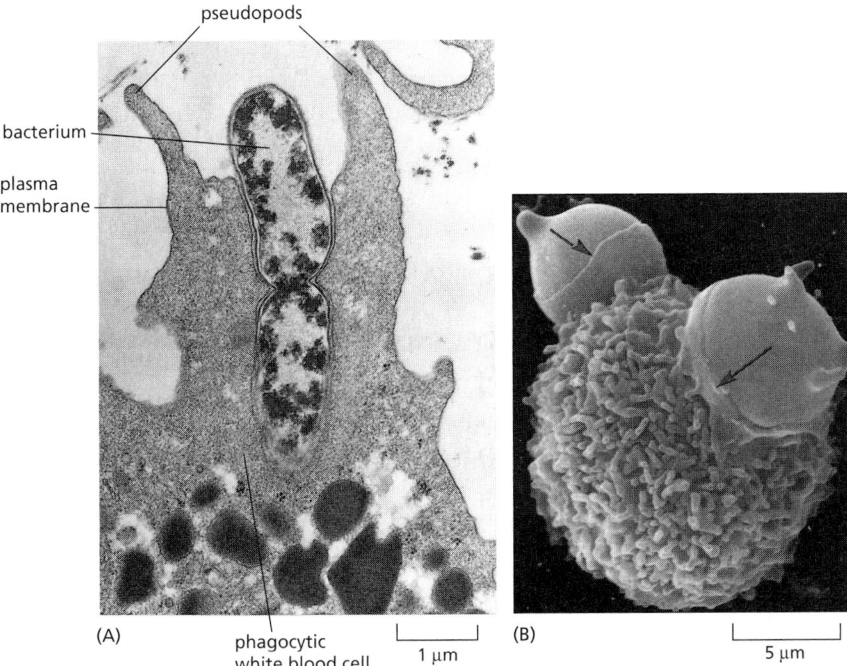

Figure 15–32 **Phagocytic cells ingest other cells.** (A) Electron micrograph of a phagocytic white blood cell (a neutrophil) ingesting a bacterium, which is in the process of dividing. (B) Scanning electron micrograph showing a macrophage engulfing a pair of red blood cells. The red blood cells are misshapen as they are being squeezed by the macrophage. The *red* arrows point to the edges of the fine sheets of membrane—called pseudopods—that the phagocytic cells are extending like collars to envelop their prey. (A, courtesy of Dorothy F. Bainton. B, courtesy of Jean Paul Revel.)

Nevertheless, phagocytosis is important in most animals for purposes other than nutrition. **Phagocytic cells**—including *macrophages*, which are widely distributed in tissues, and some other white blood cells—defend us against infection by ingesting invading microorganisms. To be taken up by a macrophage or other white blood cell, particles must first bind to the phagocytic cell surface and activate one of a variety of surface receptors. Some of these receptors recognize antibodies, the proteins that protect us against infection by binding to the surface of microorganisms. Binding of antibody-coated bacteria to these receptors induces the phagocytic cell to extend sheetlike projections of the plasma membrane, called *pseudopods,* that engulf the bacterium (Figure 15–32A) and fuse at their tips to form a phagosome. The phagosome then fuses with a lysosome, and the microbe is digested. Some pathogenic bacteria have evolved tricks for subverting the system: for example, *Mycobacterium tuberculosis*, the agent responsible for tuberculosis, can inhibit the membrane fusion that unites the phagosome with a lysosome. Instead of being destroyed, the engulfed organism survives and multiplies within the macrophage. How the bacterium accomplishes this task is still unknown.

Phagocytic cells also play an important part in scavenging dead and damaged cells and cell debris. Macrophages, for example, ingest more than 10^{11} of our worn-out red blood cells each day (Figure 15–32B).

Fluid and Macromolecules Are Taken Up by Pinocytosis

Eucaryotic cells continually ingest bits of their plasma membrane, along with small amounts of extracellular fluid, in the form of small pinocytic vesicles that are later returned to the cell surface. The rate at which plasma membrane is internalized by **pinocytosis** varies from cell type to cell type, but it is usually surprisingly large. A macrophage, for example, swallows 25% of its own volume of fluid each hour. This means that it removes 3% of its plasma membrane each minute, or 100% in about half an hour. Fibroblasts endocytose at a somewhat lower rate, whereas some phagocytic amoebae ingest their plasma membrane even more rapidly. Because a cell's total surface area and volume remain unchanged during this process, as much membrane is being added to the cell surface by vesicle fusion (exocytosis) as is being removed by endocytosis.

Pinocytosis is carried out mainly by the clathrin-coated pits and vesicles that we discussed earlier (see Figures 15–19 and 15–20). After they pinch off from the plasma membrane, clathrin-coated vesicles rapidly shed their coat and fuse with an endosome. Extracellular fluid is trapped in the coated pit as it invaginates to form a coated vesicle, and so substances dissolved in the extracellular fluid are internalized and delivered to endosomes. This fluid intake is generally balanced by fluid loss during exocytosis.

Receptor-mediated Endocytosis Provides a Specific Route into Animal Cells

Pinocytosis, as just described, is indiscriminate. The endocytic vesicles simply trap any molecules that happen to be present in the extracellular fluid and carry them into the cell. In most animal cells, however, pinocytosis via clathrin-coated vesicles also provides an efficient pathway for taking up specific macromolecules from the extracellular fluid. The macromolecules bind to complementary receptors on the cell surface and enter the cell as receptor–macromolecule complexes in clathrin-coated vesicles. This process, called **receptor-mediated endocytosis**, provides a selective concentrating mechanism that increases the efficiency of internalization of particular macromolecules more than 1000-fold compared with ordinary pinocytosis, so that even minor components of the extracellular fluid can be taken up in large amounts without taking in a correspondingly large volume of extracellular fluid. An important example of receptor-mediated endocytosis is the ability of animal cells to take up the cholesterol they need to make new membrane.

Cholesterol is extremely insoluble and is transported in the bloodstream bound to protein in the form of particles called *low-density lipoproteins*, or *LDL*. The LDL binds to receptors located on cell surfaces, and the receptor–LDL complexes are ingested by receptor-mediated endocytosis and delivered to endosomes. The interior of endosomes is more acidic than the surrounding cytosol or the extracellular fluid, and in this acidic environment the LDL dissociates from its receptor: the receptors are returned in transport vesicles to the plasma membrane for reuse, while the LDL is delivered to lysosomes. In the lysosomes the LDL is broken down by hydrolytic enzymes. The cholesterol is released and escapes into the cytosol, where it is available for new membrane synthesis. The LDL receptors on the cell surface are continually internalized and recycled, whether they are occupied by LDL or not (Figure 15–33).

This pathway for cholesterol uptake is disrupted in individuals who inherit a defective gene encoding the LDL receptor protein. In some cases, the receptors are missing; in others, they are present but nonfunctional. In either case, because the cells are deficient in taking up LDL, cholesterol accumulates in the blood and predisposes the individuals to develop atherosclerosis. Unless they take drugs (statins) to reduce their blood cholesterol, they will likely die at an early age of heart attacks resulting from cholesterol clogging the arteries that supply the heart.

Receptor-mediated endocytosis is also used to take up many other essential metabolites, such as vitamin B_{12} and iron, that cells cannot take up by the processes of membrane transport discussed in Chapter 12. Vitamin B_{12} and iron are both required, for example, for the synthesis of hemoglobin, which is the major protein in red blood cells; they enter immature red blood cells as a complex with protein. Many cell-surface receptors that bind extracellular signal molecules are also ingested by this pathway: some are recycled to the plasma membrane for reuse, whereas others are degraded in lysosomes. Unfortunately, receptor-mediated endocytosis can also be exploited by viruses: the influenza virus and HIV, which causes AIDS, gain entry into cells in this way.

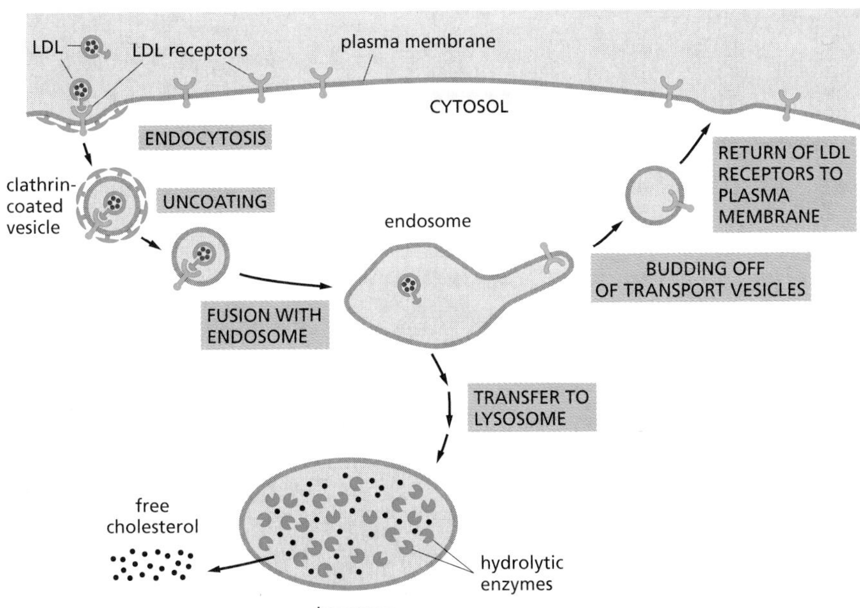

Figure 15–33 **LDL enters cells via receptor-mediated endocytosis.** LDL binds to receptors on the cell surface and is internalized in clathrin-coated vesicles. The vesicles lose their coat and then fuse with endosomes. In the acidic environment of the endosome, LDL dissociates from its receptors. Whereas the LDL ends up in lysosomes, where it is degraded to release free cholesterol, the LDL receptors are returned to the plasma membrane via transport vesicles to be used again (Movie 15.10). For simplicity, only one LDL receptor is shown entering the cell and returning to the plasma membrane. Whether it is occupied or not, an LDL receptor typically makes one round trip into the cell and back every 10 minutes, making a total of several hundred trips in its 20-hour life-span.

Endocytosed Macromolecules Are Sorted in Endosomes

Because extracellular material taken up by pinocytosis is rapidly transferred to **endosomes**, it is possible to visualize the endosomal compartment by incubating living cells in fluid containing an electron-dense marker that will show up when viewed in an electron microscope. When examined in this way, the endosomal compartment reveals itself to be a complex set of connected membrane tubes and larger vesicles. Two sets of endosomes can be distinguished in such loading experiments: the marker molecules appear first in *early endosomes*, just beneath the plasma membrane; 5–15 minutes later, they show up in *late endosomes,* closer to the nucleus (see Figure 15–18). Early endosomes mature gradually into late endosomes as they fuse with each other or with a preexisting late endosome (Movie 15.11). The interior of the endosome compartment is kept acidic (pH 5–6) by an ATP-driven H^+ (proton) pump in the endosomal membrane that pumps H^+ into the endosome lumen from the cytosol.

The endosomal compartment acts as the main sorting station in the inward endocytic pathway, just as the *trans* Golgi network serves this function in the outward secretory pathway. The acidic environment of the endosome plays a crucial part in the sorting process by causing many receptors to release their bound cargo. The routes taken by receptors once they have entered an endosome differ according to the type of receptor: (1) most are returned to the same plasma membrane domain from which they came, as is the case for the LDL receptor discussed earlier; (2) some travel to lysosomes, where they are degraded; and (3) some proceed to a different domain of the plasma membrane, thereby transferring their bound cargo molecules across the cell from one extracellular space to another, a process called *transcytosis* (Figure 15–34).

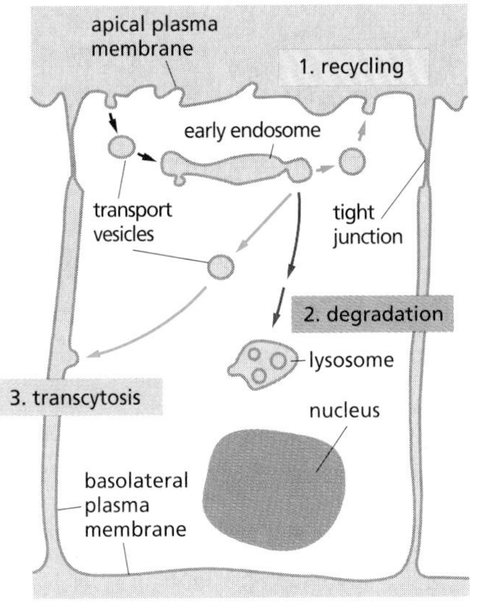

Figure 15–34 **The fate of the receptor proteins involved in endocytosis depends on the type of receptor.** Three pathways from the endosomal compartment in an epithelial cell are shown. Receptors that are not specifically retrieved from early endosomes follow the pathway from the endosomal compartment to lysosomes, where they are degraded. Retrieved receptors are returned either to the same plasma membrane domain from which they came (*recycling*) or to a different domain of the plasma membrane (*transcytosis*). If the ligand that is endocytosed with its receptor stays bound to the receptor in the acidic environment of the endosome, it will follow the same pathway as the receptor; otherwise it will be delivered to lysosomes for degradation.

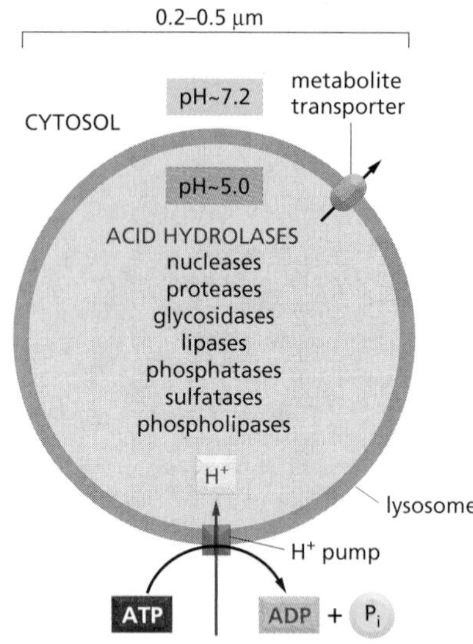

Figure 15–35 **A lysosome contains hydrolytic enzymes and an H^+ pump.** The acid hydrolases are hydrolytic enzymes that are active under acidic conditions. The lumen of the lysosome is maintained at an acidic pH by an H^+ ATPase in the membrane that pumps H^+ into the lumen.

Cargo proteins that remain bound to their receptors share the fate of their receptors. Those that dissociate from their receptors in the endosome are doomed to destruction in lysosomes, along with most of the contents of the endosome lumen. Late endosomes contain some lysosomal enzymes, so digestion of cargo proteins and other macromolecules begins in the endosome and continues as the endosome gradually matures into a lysosome.

Lysosomes Are the Principal Sites of Intracellular Digestion

Many extracellular particles and molecules ingested by cells end up in **lysosomes**, which are membranous sacs of hydrolytic enzymes that carry out the controlled intracellular digestion of both extracellular materials and worn-out organelles. They contain about 40 types of hydrolytic enzymes, including those that degrade proteins, nucleic acids, oligosaccharides, and phospholipids. All of these enzymes are optimally active in the acidic conditions (pH ~5) maintained within lysosomes. The membrane of the lysosome normally keeps these destructive enzymes out of the cytosol (whose pH is about 7.2), but the enzymes' acid dependence protects the contents of the cytosol against damage even if some of them should escape.

Like all other intracellular organelles, the lysosome not only contains a unique collection of enzymes but also has a unique surrounding membrane. The lysosomal membrane contains transporters that allow the final products of the digestion of macromolecules, such as amino acids, sugars, and nucleotides, to be transferred to the cytosol; from there, they can be either excreted or utilized by the cell. The membrane also contains an ATP-driven H^+ pump, which, like the ATPase in the endosome membrane, pumps H^+ into the lysosome, thereby maintaining its contents at an acidic pH (Figure 15–35). Most of the lysosomal membrane proteins are unusually highly glycosylated; the sugars, which cover much of the protein surfaces facing the lumen, protect the proteins from digestion by the lysosomal proteases.

The specialized digestive enzymes and membrane proteins of the lysosome are synthesized in the ER and transported through the Golgi apparatus to the *trans* Golgi network. While in the ER and the *cis* Golgi network, the enzymes are tagged with a specific phosphorylated sugar group (mannose 6-phosphate), so that when they arrive in the *trans* Golgi network they can be recognized by an appropriate receptor, the mannose 6-phosphate receptor. This tagging permits the enzymes to be sorted and packaged into transport vesicles, which bud off and deliver their contents to lysosomes via late endosomes (see Figure 15–18).

Depending on their source, materials follow different paths to lysosomes. We have seen that extracellular particles are taken up into phagosomes, which fuse with lysosomes, and that extracellular fluid and macromolecules are taken up into smaller endocytic vesicles, which deliver their contents to lysosomes via endosomes. But cells have an additional pathway for supplying materials to lysosomes; this pathway, called *autophagy,* is used for degrading obsolete parts of the cell itself. In electron micrographs of liver cells, for example, one often sees lysosomes digesting mitochondria, as well as other organelles. The process begins with the enclosure of the organelle by a double membrane, creating an *autophagosome*, which then fuses with lysosomes (Figure 15–36). It is not known what marks an organelle for such destruction.

QUESTION 15–8

Iron (Fe) is an essential trace metal that is needed by all cells. It is required, for example, for the synthesis of heme groups that are part of the active site of many enzymes involved in electron-transfer reactions; it is also required in hemoglobin, the main protein in red blood cells. Iron is taken up by cells by receptor-mediated endocytosis. The iron-uptake system has two components, a soluble protein called transferrin, which circulates in the bloodstream; and a transferrin receptor—a transmembrane protein that, like the LDL receptor in Figure 15–33, is continually endocytosed and recycled to the plasma membrane. Fe ions bind to transferrin at neutral pH but not at acidic pH. Transferrin binds to the transferrin receptor at neutral pH only when it has an Fe ion bound, but it binds to the receptor at acidic pH even in the absence of bound iron. From these properties, describe how iron is taken up, and discuss the advantages of this elaborate scheme.

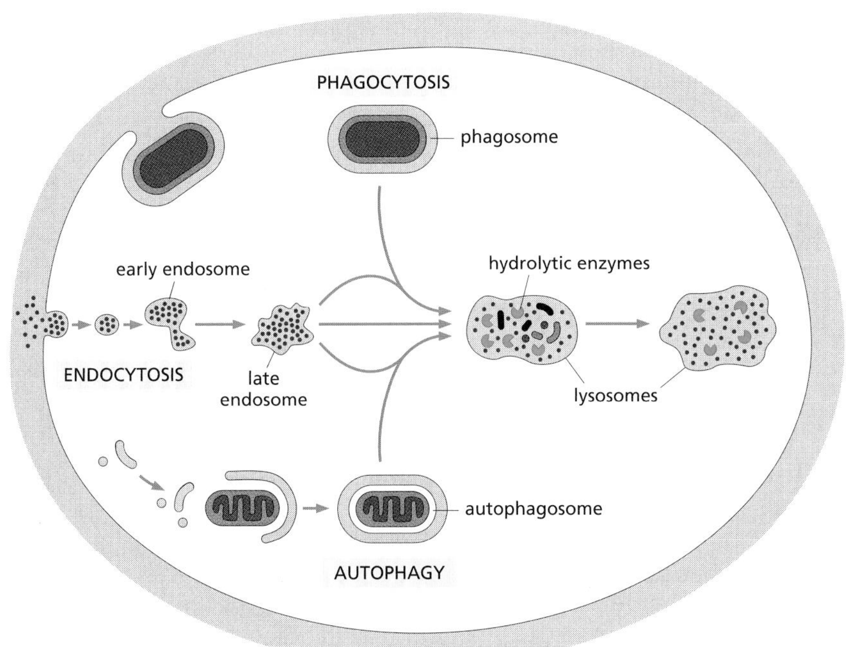

Figure 15–36 **Materials destined for degradation follow different pathways to the lysosome.** Each pathway leads to the intracellular digestion of materials derived from a different source. These organelles can fuse with lysosomes (or late endosomes), which contain hydrolytic enzymes that digest their contents. They differ from 'classical' lysosomes, however, only because of the different materials they are digesting.

ESSENTIAL CONCEPTS

- Eucaryotic cells contain many membrane-enclosed organelles, including a nucleus, an endoplasmic reticulum (ER), a Golgi apparatus, lysosomes, endosomes, mitochondria, chloroplasts (in plant cells), and peroxisomes.
- Most organelle proteins are made in the cytosol and transported into the organelle where they function. Sorting signals in the amino acid sequence guide the proteins to the correct organelle; proteins that function in the cytosol have no such signals and remain where they are made.
- Nuclear proteins contain nuclear localization signals that help direct their active transport from the cytosol into the nucleus through nuclear pores, which penetrate the double-membrane nuclear envelope. Proteins enter the nucleus without being unfolded.
- Most mitochondrial and chloroplast proteins are made in the cytosol and are then actively transported into the organelles by protein translocators in their membranes. Proteins must be unfolded to allow them to snake through the translocators in the chloroplast or mitochondrial membrane.
- The ER is the membrane factory of the cell; it makes most of the cell's lipids and many of its proteins. The proteins are made by ribosomes bound to the surface of the rough ER.
- Ribosomes in the cytosol are directed to the ER if the protein they are making has an ER signal sequence, which is recognized by a signal-recognition particle (SRP) in the cytosol; the binding of the ribosome–SRP complex to a receptor on the ER membrane initiates the translocation process that threads the growing polypeptide across the ER membrane through a translocation channel.
- Soluble proteins destined for secretion or for the lumen of an organelle pass completely into the ER lumen, while transmembrane proteins destined for the ER membrane or for other cell membranes remain anchored in the lipid bilayer by one or more membrane-spanning α helices.
- In the ER lumen, proteins fold up, assemble with other proteins, form disulfide bonds, and become decorated with oligosaccharide chains.

- Exit from the ER is an important quality-control step; proteins that either fail to fold properly or fail to assemble with their normal partners are retained in the ER by chaperone proteins and are eventually degraded.
- An accumulation of misfolded proteins triggers a response that expands the size of the ER, thus increasing its capacity to fold new proteins properly.
- Protein transport from the ER to the Golgi apparatus and from the Golgi apparatus to other destinations is mediated by transport vesicles that continually bud off from one membrane and fuse with another, a process called vesicular transport.
- Budding transport vesicles have distinctive coat proteins on their cytosolic surface; the assembly of the coat drives the budding process, and the coat proteins help incorporate receptors with their bound cargo molecules into the forming vesicle.
- Coated vesicles lose their protein coat soon after pinching off, enabling them to dock and then fuse with a particular target membrane; docking and fusion are mediated by proteins on the vesicle and on the target membranes, including Rab proteins and SNAREs.
- The Golgi apparatus receives newly made proteins from the ER; it modifies their oligosaccharides, sorts the proteins, and dispatches them from the *trans* Golgi network to the plasma membrane, lysosomes, or secretory vesicles.
- In all eucaryotic cells, transport vesicles continually bud from the *trans* Golgi network and fuse with the plasma membrane, a process called constitutive exocytosis; the process delivers plasma membrane lipids and proteins to the cell surface and also releases molecules from the cell in the process of secretion.
- Specialized secretory cells also have a regulated exocytosis pathway, where molecules stored in secretory vesicles are released from the cell by exocytosis when the cell is signaled to secrete.
- Cells ingest fluid, molecules, and sometimes even particles, by endocytosis, in which regions of plasma membrane invaginate and pinch off to form endocytic vesicles.
- Much of the material that is endocytosed is delivered to endosomes and then to lysosomes, where it is degraded by hydrolytic enzymes; most of the components of the endocytic vesicle membrane, however, are recycled in transport vesicles back to the plasma membrane for reuse.

KEY TERMS

clathrin
coated vesicle
cytosol
endocytosis
endoplasmic reticulum (ER)
endosome
exocytosis
Golgi apparatus
lysosome
membrane-enclosed organelle
nuclear envelope
nuclear pore
peroxisome
phagocytic cell
phagocytosis
pinocytosis
Rab protein
receptor-mediated endocytosis
rough endoplasmic reticulum
secretion
secretory vesicle
signal sequence
SNARE
transport vesicle
unfolded protein response (UPR)
vesicular transport

QUESTIONS

QUESTION 15–9

Which of the following statements are correct? Explain your answers.

A. Ribosomes are cytoplasmic structures that, during protein synthesis, become linked by an mRNA molecule to form polyribosomes.

B. The amino acid sequence Leu-His-Arg-Leu-Asp-Ala-Gln-Ser-Lys-Leu-Ser-Ser is a signal sequence that directs proteins to the ER.

C. All transport vesicles in the cell must have a v-SNARE protein in their membrane.

D. Transport vesicles deliver proteins and lipids to the cell surface.

E. If the delivery of prospective lysosomal proteins from the *trans* Golgi network to the late endosomes were blocked, lysosomal proteins would be secreted by the constitutive secretion pathways shown in Figure 15–27.

F. Lysosomes digest only substances that have been taken up by cells by endocytosis.

G. *N*-linked sugar chains are found on glycoproteins that face the cell surface, as well as on glycoproteins that face the lumen of the ER, *trans* Golgi network, and mitochondria.

QUESTION 15–10

Some proteins shuttle back and forth between the nucleus and the cytosol. They need a nuclear export signal to get out of the nucleus. How do you suppose they get into the nucleus?

QUESTION 15–11

Influenza viruses are surrounded by a membrane that contains a fusion protein, which is activated by acidic pH. Upon activation, the protein causes the viral membrane to fuse with cell membranes. An old folk remedy against flu recommends that one should spend a night in a horse's stable. Odd as it may sound, there is a rational explanation for this advice. Air in stables contains ammonia (NH_3) generated by bacteria in the horse's urine. Sketch a diagram showing the pathway (in detail) by which flu virus enters cells, and speculate how NH_3 may protect cells from virus infection. (Hint: NH_3 can neutralize acidic solutions by the reaction $NH_3 + H^+ \rightarrow NH_4^+$.)

QUESTION 15–12

Consider the v-SNAREs that direct transport vesicles from the *trans* Golgi network to the plasma membrane. They, like all other v-SNAREs, are membrane proteins that are integrated into the membrane of the ER during their biosynthesis and are then transported by transport vesicles to their destination. Thus, transport vesicles budding from the ER contain at least two kinds of v-SNAREs—those that target the vesicles to the *cis* Golgi cisternae, and those that are in transit to the *trans* Golgi network to be packaged in different transport vesicles destined for the plasma membrane. (A) Why might this be a problem? (B) Suggest possible ways in which the cell might solve it.

QUESTION 15–13

A particular type of *Drosophila* mutant becomes paralyzed when the temperature is raised. The mutation affects the structure of dynamin, causing it to be inactivated at the higher temperature. Indeed, the function of dynamin was discovered by analyzing the defect in these mutant fruit flies. The complete paralysis at the elevated temperature suggests that synaptic transmission between nerve and muscle cells (discussed in Chapter 12) is blocked. Suggest why signal transmission at a synapse might require dynamin. On the basis of your hypothesis, what would you expect to see in electron micrographs of synapses of flies that were exposed to the elevated temperature?

QUESTION 15–14

Edit the following statements, if required, to make them true: "Because nuclear localization sequences are not cleaved off by proteases following protein import into the nucleus, they can be reused to import nuclear proteins after mitosis, when cytosolic and nuclear proteins have become intermixed. This is in contrast to ER signal sequences, which are cleaved off by a signal peptidase once they reach the lumen of the ER. ER signal sequences cannot therefore be reused to import ER proteins after mitosis, when cytosolic and ER proteins have become intermixed; these ER proteins must therefore be degraded and resynthesized."

QUESTION 15–15

Consider a protein that contains an ER signal sequence at its N-terminus and a nuclear localization sequence in its middle. What do you think the fate of this protein would be? Explain your answer.

QUESTION 15–16

Compare and contrast protein import into the ER and into the nucleus. List at least two major differences in the mechanisms, and speculate why the ER mechanism might not work for nuclear import and vice versa.

QUESTION 15–17

During mitosis, the nuclear envelope breaks down and intranuclear proteins completely intermix with cytosolic proteins. Is this consistent with the evolutionary scheme proposed in Figure 15–3?

QUESTION 15–18

A protein that inhibits certain proteolytic enzymes (proteases) is normally secreted into the bloodstream by liver cells. This inhibitor protein, antitrypsin, is absent from the bloodstream of patients who carry a mutation that results in a single amino acid change in the protein. Antitrypsin deficiency causes a variety of severe problems, particularly in lung tissue, because of the uncontrolled activity of proteases. Surprisingly, when the mutant antitrypsin is synthesized in the laboratory, it is as active as the normal antitrypsin at inhibiting proteases. Why, then, does the mutation cause the disease? Think of more than one possibility, and suggest ways in which you could distinguish between them.

QUESTION 15–19

Dr. Outonalimb's claim to fame is her discovery of forgettin, a protein predominantly made by the pineal gland in human teenagers. The protein causes selective short-term unresponsiveness and memory loss when the auditory system receives statements like "Please take out the garbage!" Her hypothesis is that forgettin has a hydrophobic ER signal sequence at its C-terminus that is recognized by an SRP and causes it to be translocated across the ER membrane by the mechanism shown in Figure 15–14. She predicts that the protein is secreted from pineal cells into the bloodstream, from where it exerts its devastating systemic effects. You are a member of the committee deciding whether she should receive a grant for further work on her hypothesis. Critique her proposal, and remember that grant reviews should be polite and constructive.

QUESTION 15–20

Taking the evolutionary scheme in Figure 15–3 one step further, suggest how the Golgi apparatus could have evolved. Sketch a simple diagram to illustrate your ideas. For the Golgi apparatus to be functional, what else would have to have evolved?

QUESTION 15–21

If membrane proteins are integrated into the ER membrane by means of the ER protein translocation channel (which is itself composed of membrane proteins), how do the first protein translocation channels become incorporated into the ER membrane?

QUESTION 15–22

The sketch in Figure Q15–22 is a schematic drawing of the electron micrograph shown in the third panel of Figure 15–19A. Name the structures that are labeled in the sketch.

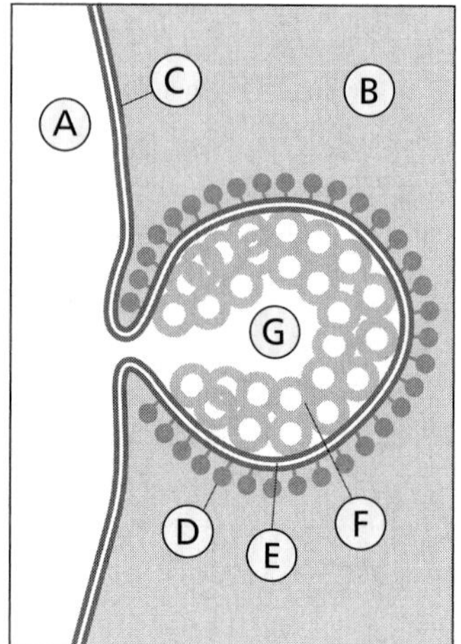

Figure Q15–22

CHAPTER **SIXTEEN** 16

Cell Communication

GENERAL PRINCIPLES OF CELL SIGNALING

G-PROTEIN-COUPLED RECEPTORS

ENZYME-COUPLED RECEPTORS

Individual cells, like multicellular organisms, need to sense and respond to their environment. A free-living cell—even a humble bacterium—must be able to track down nutrients, tell the difference between light and dark, and avoid poisons and predators. And if such a cell is to have any kind of 'social life,' it must be able to communicate with other cells. When a yeast cell is ready to mate, for example, it secretes a small protein called a mating factor. Yeast cells of the opposite 'sex' detect this chemical mating call and respond by halting their progress through the cell division cycle and reaching out toward the cell that emitted the signal (Figure 16–1).

In a multicellular organism, things are much more complicated. Cells must interpret the multitude of signals they receive from other cells to help coordinate their behaviors. During animal development, for example, cells in the embryo exchange signals to determine which specialized role each cell will adopt, what position it will occupy in the animal, and whether it will survive, divide, or die. Later in life, a large variety of signals coordinate the animal's growth and its day-to-day physiology and behavior. In plants as well, cells are in constant communication with one another. These cell–cell interactions allow a plant to respond to the conditions of light, dark, and temperature that guide the cycles of growth, flowering, and fruiting; they also allow the plant to coordinate what happens in its roots, stems, and leaves.

In this chapter, we examine some of the most important methods by which cells communicate, and we discuss how cells send signals and interpret the signals they receive. Although we concentrate on the mechanisms of signal reception and interpretation in animal cells, we also present a brief review of what is known about cell-to-cell signaling in plants. We begin

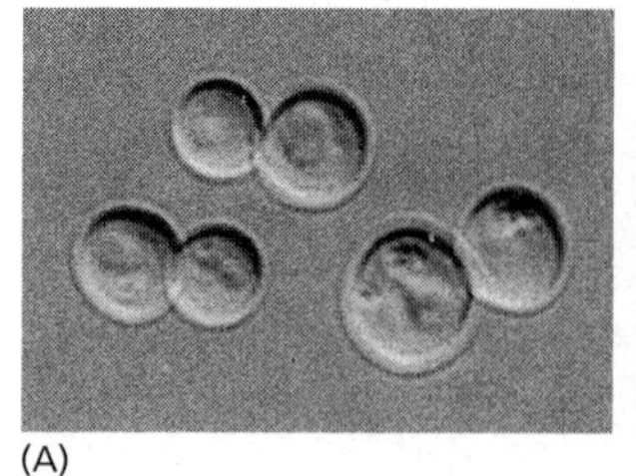

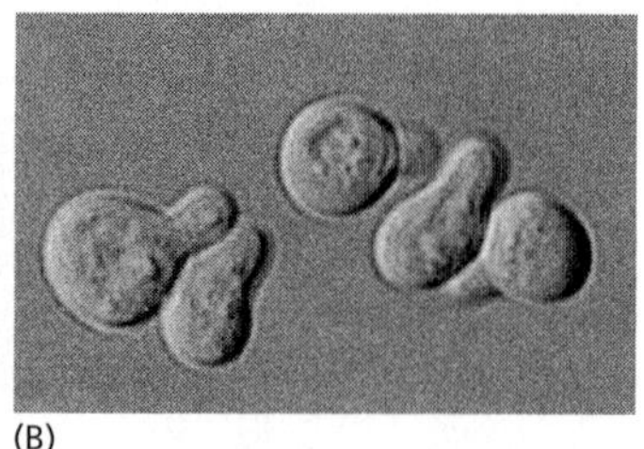

Figure 16–1 **Yeast cells respond to mating factor.** Budding yeast (*Saccharomyces cerevisiae*) cells are normally spherical (A), but when exposed to an appropriate mating factor produced by neighboring yeast cells they extend a protrusion toward the source of the factor (B). Cells that adopt this shape in response to the mating signal are called "shmoos," after a classic 1940s cartoon character created by Al Capp. (Courtesy of Michael Snyder.)

our discussion with an overview of the general principles of cell signaling and then consider two of the main systems animal cells use to receive and interpret signals.

GENERAL PRINCIPLES OF CELL SIGNALING

Information can come in a variety of forms, and communication frequently involves converting the signals that carry that information from one form to another. When you receive a call from a friend on your mobile phone, for instance, the phone converts the radio signals, which travel through the air, into sound waves, which you hear. This process of conversion is called **signal transduction** (Figure 16–2).

The signals that pass between living cells are simpler than the sorts of messages that humans ordinarily exchange. In a typical communication between cells, the *signaling cell* produces a particular type of *signal molecule* that is detected by the *target cell*. As in human conversation, most animal cells both send and receive signals, and they can therefore act as both signaling cells and target cells. Target cells possess *receptor proteins* that recognize and respond specifically to the signal molecule. Signal transduction begins when the receptor protein on a target cell receives an incoming extracellular signal and converts it to the intracellular signals that alter cell behavior. Most of this chapter is concerned with signal reception and transduction—the events that cell biologists have in mind when they refer to **cell signaling**. First, however, we look briefly at the different types of signals that cells send to one another.

Signals Can Act over a Long or Short Range

Cells in multicellular organisms use hundreds of kinds of extracellular molecules to send signals to one another. The signal molecules can be proteins, peptides, amino acids, nucleotides, steroids, fatty acid derivatives, or even dissolved gases—but they rely on only a handful of basic styles of communication for getting the message across.

In multicellular organisms, the most 'public' style of communication involves broadcasting the signal throughout the whole body by secreting it into the bloodstream (in an animal) or the sap (in a plant). Signal molecules used in this way are called **hormones**, and, in animals, the cells that produce hormones are called *endocrine* cells (Figure 16–3A). Part of

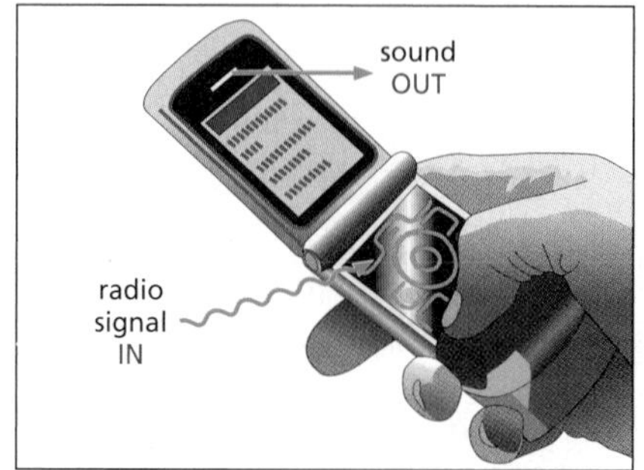

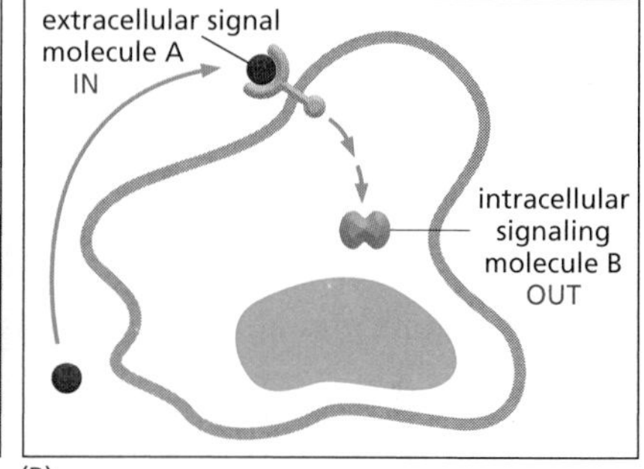

Figure 16–2 **Signal transduction is the process whereby one type of signal is converted to another.** (A) A mobile telephone converts a radio signal into a sound signal when receiving (and vice versa, when transmitting). (B) A target cell converts an extracellular signal (molecule A) into an intracellular signal (molecule B).

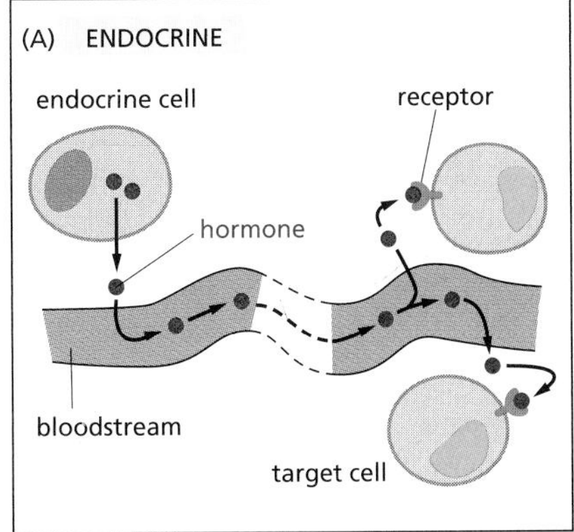

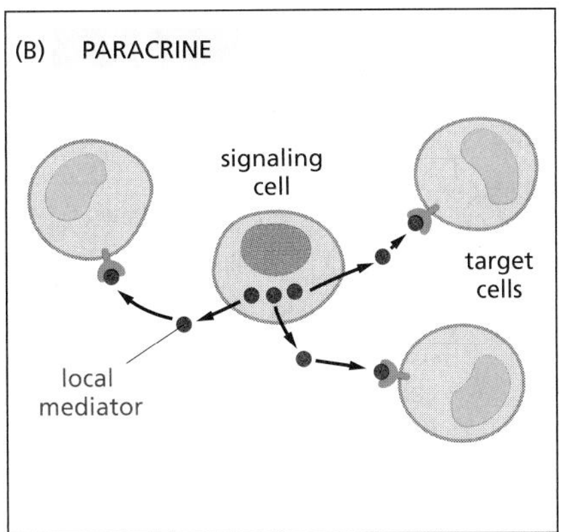

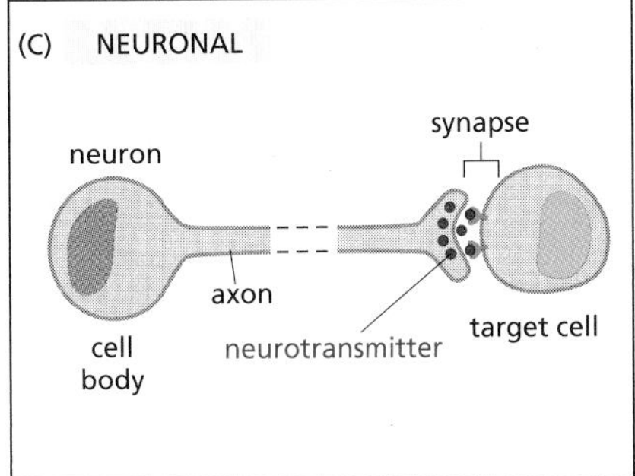

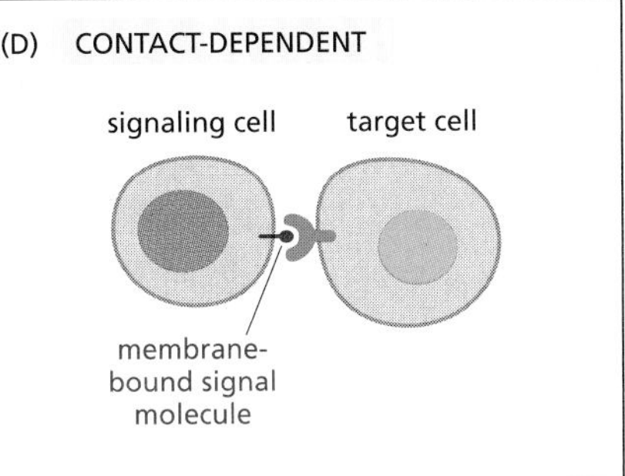

Figure 16–3 **Animal cells can signal to one another in various ways.** (A) Hormones produced in endocrine glands are secreted into the bloodstream and are distributed widely throughout the body. (B) Paracrine signals are released by cells into the extracellular fluid in their neighborhood and act locally. (C) Neuronal signals are transmitted along axons to remote target cells. (D) In contact-dependent signaling, a cell-surface-bound signal molecule binds to a receptor protein on an adjacent cell. Many of the same types of signal molecules are used for endocrine, paracrine, and neuronal signaling. The crucial differences lie in the speed and selectivity with which the signals are delivered to their targets.

the pancreas, for example, is an endocrine gland that produces the hormone insulin, which regulates glucose uptake in cells all over the body.

Somewhat less public is the process known as *paracrine signaling*. In this case, rather than entering the bloodstream, the signal molecules diffuse locally through the extracellular fluid, remaining in the neighborhood of the cell that secretes them. Thus, they act as **local mediators** on nearby cells (Figure 16–3B). Many of the signal molecules that regulate inflammation at the site of an infection or that control cell proliferation in a healing wound function in this way. In some cases, cells can respond to the local mediators that they themselves produce, a form of paracrine communication called *autocrine signaling*; cancer cells sometimes promote their own survival or proliferation in this way.

Neuronal signaling is a third form of cell communication. Like endocrine cells, nerve cells (neurons) can deliver messages over long distances. In the case of neuronal signaling, however, a message is not broadcast widely but is instead delivered quickly and specifically to individual target cells through private lines. As described in Chapter 12, the axon of a neuron terminates at specialized junctions (*synapses*) on target cells that can lie far from the neuronal cell body (Figure 16–3C). The axons that extend from the spinal cord to the big toe, for example, can be more than 1 m in length. When activated by signals from the environment or from other nerve cells, a neuron sends electrical impulses racing along its axon at speeds of up to 100 m/sec. On reaching the axon terminal, these electrical signals are converted into a chemical form: each electrical impulse stimulates the nerve terminal to release a pulse of an *extracellular signal molecule* called a **neurotransmitter**. The neurotransmitter then diffuses across the narrow (< 100 nm) gap between the axon-terminal membrane and the membrane of the target cell, reaching the target cell receptors in less than 1 msec.

A fourth style of signal-mediated cell–cell communication—the most intimate and short-range of all—does not require the release of a secreted molecule. Instead, the cells make direct physical contact through signal molecules lodged in the plasma membrane of the signaling cell and

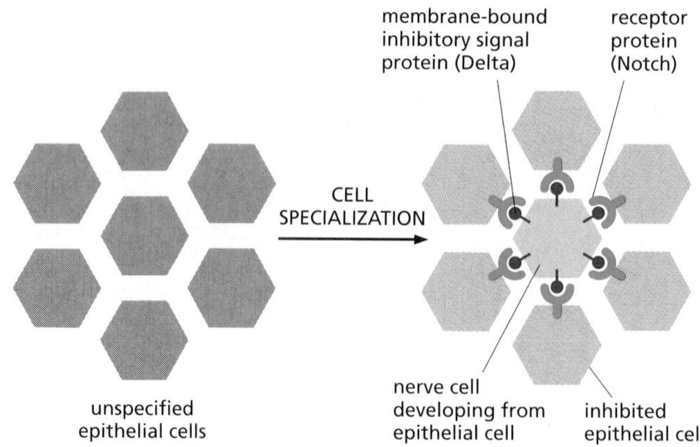

Figure 16–4 **Contact-dependent signaling controls nerve-cell production in the fruitfly *Drosophila*.** The fly nervous system originates in the embryo from a sheet of epithelial cells. Isolated cells in this sheet begin to specialize as neurons, while their neighbors remain non-neuronal and maintain the epithelial structure of the sheet. The signals that control this process are transmitted via direct cell–cell contacts: each future neuron delivers an inhibitory signal to the cells next to it, deterring them from specializing as neurons too. Both the signal molecule (in this case, Delta) and the receptor molecule (called Notch) are transmembrane proteins. The same mechanism, mediated by essentially the same molecules, controls the detailed pattern of specialized cell types in various other tissues, in both vertebrates and invertebrates. In mutant flies in which this signaling mechanism fails, some cell types (such as neurons) are produced in great excess at the expense of other cell types.

receptor proteins embedded in the plasma membrane of the target cell (Figure 16–3D). During embryonic development, for example, such *contact-dependent signaling* allows adjacent cells that are initially similar to become specialized to form different cell types (Figure 16–4).

To relate these different signaling styles, imagine trying to advertise a potentially stimulating lecture—or a concert or football game. An endocrine signal would be akin to broadcasting the information over a radio station. A localized paracrine signal would be the equivalent of posting a flyer on selected notice boards. Neuronal signals—long-distance but personal—would be similar to a phone call or an e-mail, and contact-dependent signaling would be like a good old-fashioned face-to-face conversation. In autocrine signaling, you might write a note to remind yourself to attend.

Table 16–1 lists some examples of hormones, local mediators, neurotransmitters, and contact-dependent signal molecules. The action of several of these is discussed in more detail later in the chapter.

Each Cell Responds to a Limited Set of Signals, Depending on Its History and Its Current State

A typical cell in a multicellular organism is exposed to hundreds of different signal molecules in its environment. These may be free in the extracellular fluid, embedded in the extracellular matrix in which most cells reside, or bound to the surface of neighboring cells. Each cell must respond selectively to this mixture of signals, disregarding some and reacting to others, according to the cell's specialized function.

QUESTION 16–1

To remain a local stimulus, paracrine signal molecules must be prevented from straying too far from their points of origin. Suggest different ways by which this could be accomplished. Explain your answers.

Whether a cell responds to a signal molecule depends first of all on whether it possesses a **receptor protein**, or **receptor**, for that signal. Each receptor is usually activated by only one type of signal. Without the appropriate receptor, a cell will be deaf to the signal and will not respond to it. By producing only a limited set of receptors out of the thousands that are possible, a cell restricts the types of signals that can affect it.

Of course, a small number of extracellular signal molecules can change the behavior of a target cell in a large variety of ways. They can alter the cell's shape, movement, metabolism, gene expression, or some combination of these. As we will see, the signal from a cell-surface receptor is generally conveyed into the target cell interior via a set of *intracellular signaling molecules*, which act in sequence and ultimately alter the activity of *effector proteins*, which then affect the behavior of the cell. This intracellular relay system and the intracellular effector proteins on which it acts vary from one type of specialized cell to another, so that different types of cells respond to the same signal in different ways. For example, when a heart muscle cell is exposed to the neurotransmitter

TABLE 16–1 SOME EXAMPLES OF SIGNAL MOLECULES

SIGNAL MOLECULE	SITE OF ORIGIN	CHEMICAL NATURE	SOME ACTIONS
Hormones			
Adrenaline (epinephrine)	adrenal gland	derivative of the amino acid tyrosine	increases blood pressure, heart rate, and metabolism
Cortisol	adrenal gland	steroid (derivative of cholesterol)	affects metabolism of proteins, carbohydrates, and lipids in most tissues
Estradiol	ovary	steroid (derivative of cholesterol)	induces and maintains secondary female sexual characteristics
Glucagon	α cells of pancreas	peptide	stimulates glucose synthesis, glycogen breakdown, and lipid breakdown, e.g., in liver and fat cells
Insulin	β cells of pancreas	protein	stimulates glucose uptake, protein synthesis, and lipid synthesis, e.g., in liver cells
Testosterone	testis	steroid (derivative of cholesterol)	induces and maintains secondary male sexual characteristics
Thyroid hormone (thyroxine)	thyroid gland	derivative of the amino acid tyrosine	stimulates metabolism of many cell types
Local Mediators			
Epidermal growth factor (EGF)	various cells	protein	stimulates epidermal and many other cell types to proliferate
Platelet-derived growth factor (PDGF)	various cells, including blood platelets	protein	stimulates many cell types to proliferate
Nerve growth factor (NGF)	various innervated tissues	protein	promotes survival of certain classes of neurons; promotes growth of their axons
Transforming growth factor-β (TGF-β)	many cell types	protein	inhibits cell proliferation; stimulates extracellular matrix production
Histamine	mast cells	derivative of the amino acid histidine	causes blood vessels to dilate and become leaky, helping to cause inflammation
Nitric oxide (NO)	nerve cells; endothelial cells lining blood vessels	dissolved gas	causes smooth muscle cells to relax; regulates nerve cell activity
Neurotransmitters			
Acetylcholine	nerve terminals	derivative of choline	excitatory neurotransmitter at many nerve–muscle synapses and in central nervous system
γ-Aminobutyric acid (GABA)	nerve terminals	derivative of the amino acid glutamic acid	inhibitory neurotransmitter in central nervous system
Contact-dependent Signal Molecules			
Delta	prospective neurons; various other developing cell types	transmembrane protein	inhibits neighboring cells from becoming specialized in same way as the signaling cell

acetylcholine, the rate and force of its contractions decrease. When a salivary gland is exposed to the same signal, it secretes components of saliva, even though the receptors are the same on both cell types. In skeletal muscle, acetylcholine causes the cells to contract by binding to a different receptor protein (Figure 16–5). Thus, the extracellular signal molecule alone is not the message: the information conveyed by the signal depends on how the target cell receives and interprets the signal.

A typical cell possesses many sorts of receptors—each present in tens to hundreds of thousands of copies. Such variety makes the cell simultaneously sensitive to many different extracellular signals and allows a relatively small number of signal molecules, used in different combinations, to exert subtle and complex control over cell behavior. Such combinations of signals can evoke responses that are different from the sum of the effects that each signal would trigger on its own. This is partly because the intracellular relay systems activated by the different

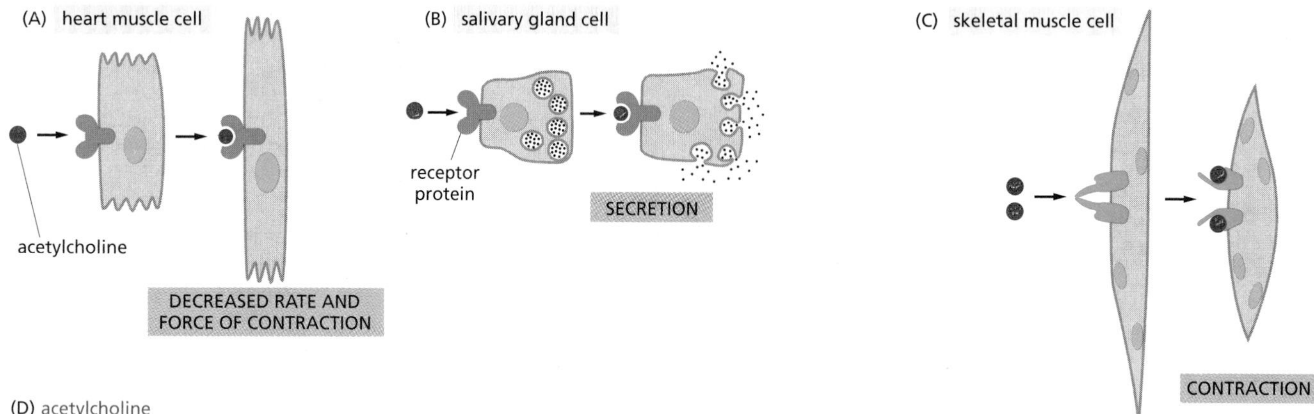

(D) acetylcholine

$$H_3C-\overset{\overset{\displaystyle O}{\|}}{C}-O-CH_2-CH_2-\underset{\underset{\displaystyle CH_3}{|}}{\overset{\overset{\displaystyle CH_3}{|}}{N^+}}-CH_3$$

Figure 16–5 **The same signal molecule can induce different responses in different target cells.** Different cell types are configured to respond to the neurotransmitter acetylcholine in different ways. Acetylcholine binds to similar receptor proteins on heart muscle cells (A) and salivary gland cells (B), but it evokes different responses in each cell type. Skeletal muscle cells (C) produce a different type of receptor protein for the same signal; this receptor generates intracellular signals that differ from those generated by the receptor type on heart muscle cells. (D) For such a versatile molecule, acetylcholine has a fairly simple chemical structure.

signals can interact, so that the presence of one signal can modify the responses to another. One combination of signals might enable a cell to survive; another might drive it to differentiate in some specialized way; and another might cause it to divide. In the absence of any signals, most animal cells are programmed to kill themselves (Figure 16–6).

A Cell's Response to a Signal Can Be Fast or Slow

The length of time a cell takes to respond to an extracellular signal can vary greatly, depending on what needs to happen once the message has been received. Some extracellular signals act swiftly: acetylcholine can stimulate skeletal muscle contraction within milliseconds and salivary gland secretion within a minute or so. Such rapid response is possible because in these cases the signal affects the activity of proteins and other molecules that are already present inside the target cell, awaiting their marching orders.

Other responses take more time. Cell growth and cell division, when triggered by the appropriate signal molecules, can take hours to execute. This is because the response to these extracellular signals requires changes in gene expression and the production of new proteins (Figure 16–7). We

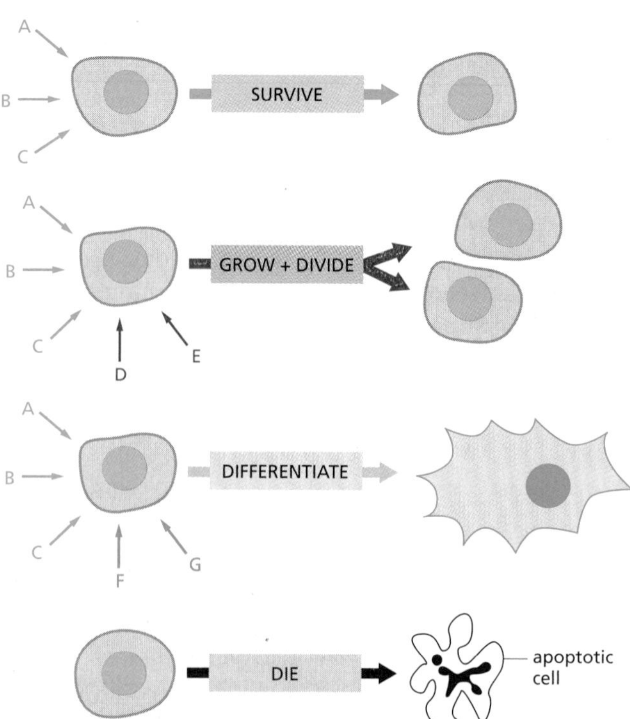

Figure 16–6 **An animal cell depends on multiple extracellular signals.** Every cell type displays a set of receptor proteins that enables it to respond to a specific set of extracellular signal molecules produced by other cells. These signal molecules work in combinations to regulate the behavior of the cell. As shown here, cells may require multiple signals (*blue* arrows) to survive, additional signals (*red* arrows) to grow and divide, and still other signals (*green* arrows) to differentiate. If deprived of survival signals, most cells undergo a form of cell suicide known as apoptosis (discussed in Chapter 18).

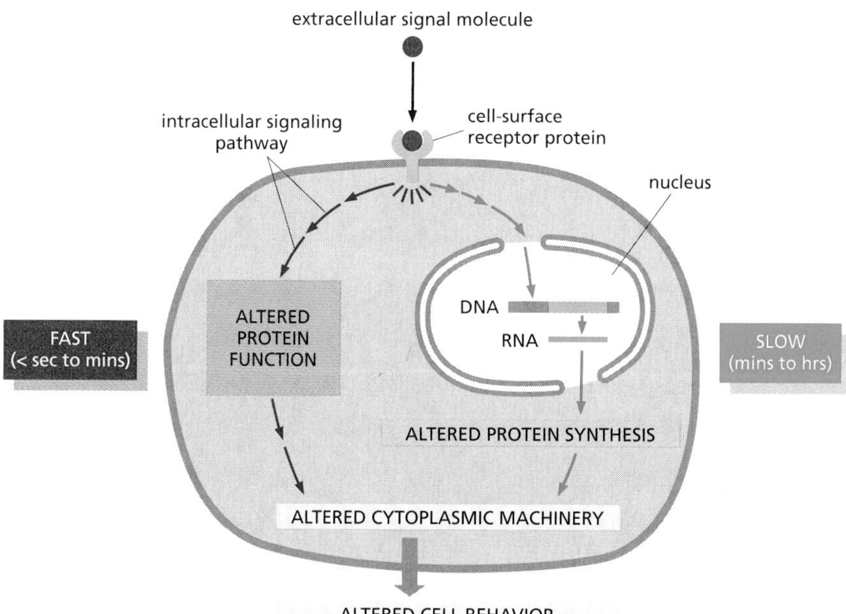

Figure 16–7 **Extracellular signals can act slowly or rapidly.** Certain types of cell responses—such as increased cell growth and division—involve changes in gene expression and the synthesis of new proteins; they therefore occur relatively slowly. Other responses—such as changes in cell movement, secretion, or metabolism—need not involve changes in gene expression and therefore occur more quickly (see Figure 16–5).

will encounter additional examples of both fast and slow responses—and the signal molecules that stimulate them—later in the chapter.

Some Hormones Cross the Plasma Membrane and Bind to Intracellular Receptors

Extracellular signal molecules generally fall into two classes. The first and largest class consists of molecules that are too large or too hydrophilic to cross the plasma membrane of the target cell. They rely on receptors on the surface of the target cell to relay their message across the membrane (Figure 16–8A). The second, and smaller, class of signals consists of molecules that are small enough or hydrophobic enough to slip easily across the plasma membrane. Once inside, these signal molecules usually activate intracellular enzymes or bind to intracellular receptor proteins that regulate gene expression (Figure 16–8B).

One important class of signal molecules that rely on intracellular receptor proteins is the **steroid hormones**—including *cortisol, estradiol,* and *testosterone*—and the *thyroid hormones* such as *thyroxine* (Figure 16–9). All of these hydrophobic molecules pass through the plasma membrane of the target cell and bind to receptor proteins located in either the cytosol or the nucleus. Both the cytosolic and nuclear receptors are referred to as *nuclear receptors,* because, when activated by hormone binding, they act as transcription regulators in the nucleus (discussed in Chapter 8). In unstimulated cells, nuclear receptors are typically present in an inactive form. When a hormone binds, the receptor undergoes a large conformational change that activates the protein, allowing it to promote or inhibit the transcription of specific target genes (Figure 16–10). Each hormone binds to a different receptor protein, and each receptor acts at a different set of regulatory sites in DNA (discussed in Chapter 8). Moreover, a given hormone usually regulates different sets of genes in different cell types, thereby evoking different physiological responses in different types of target cells.

Nuclear receptors and the hormones that activate them play an essential role in human physiology (see Table 16–1, p. 535). Loss of these signaling systems can have dramatic consequences, as illustrated by what happens in individuals who lack the receptor for the male sex hormone testosterone. Testosterone in humans shapes the formation of the external

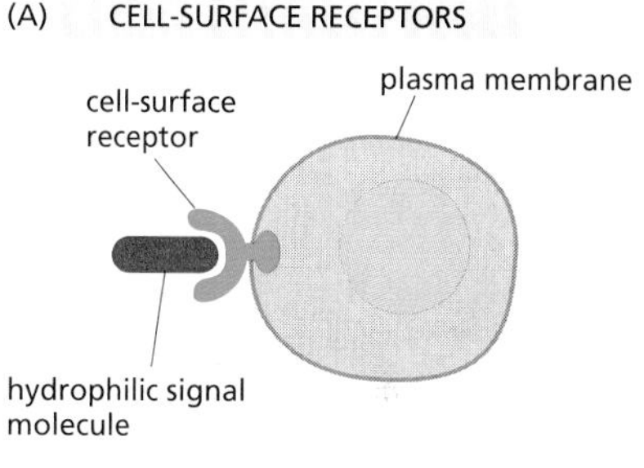

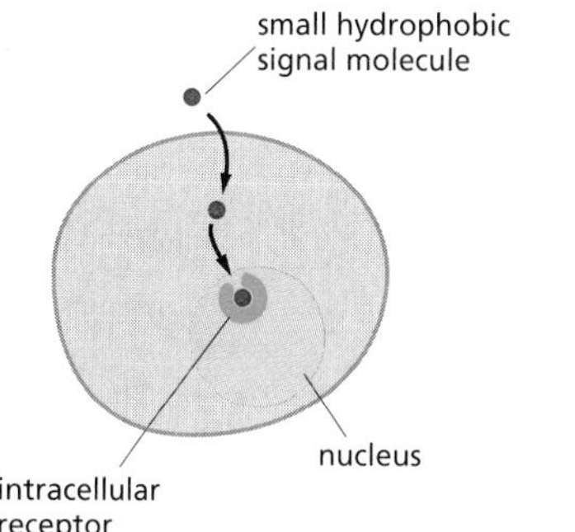

Figure 16–8 **Extracellular signal molecules bind either to cell-surface receptors or to intracellular enzymes or receptors.** (A) Most extracellular signal molecules are large and hydrophilic and are therefore unable to cross the plasma membrane directly; instead, they bind to cell-surface receptors, which in turn generate one or more signaling molecules inside the target cell. (B) Some small, hydrophobic, extracellular signal molecules, by contrast, diffuse across the target cell's plasma membrane and activate enzymes directly or bind to intracellular receptors in either the cytosol or the nucleus (as shown here).

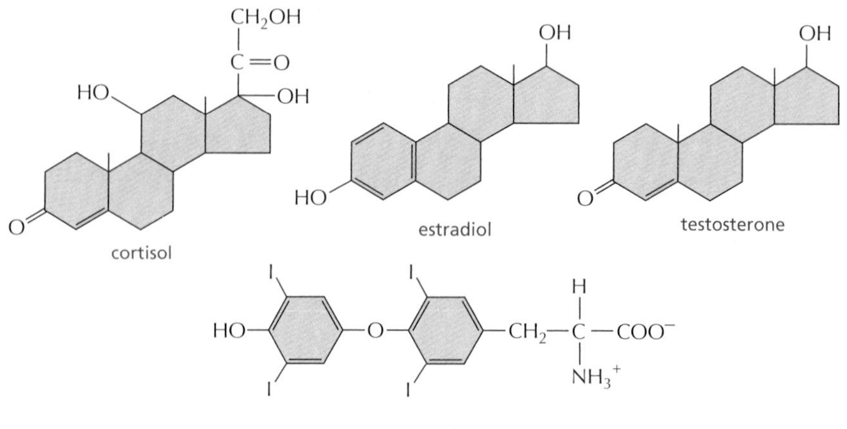

Figure 16–9 **Some small hydrophobic hormones bind to intracellular receptors that act as transcription regulators.** Although these signal molecules differ in their chemical structures and functions, they all act by binding to intracellular receptor proteins. Their receptors are not identical, but they are evolutionarily related, belonging to the *nuclear receptor superfamily* of transcription regulators. The sites of origin and functions of these hormones are given in Table 16–1 (p. 535).

cortisol

estradiol

testosterone

thyroxine

QUESTION 16–2

Consider the structure of cholesterol, a small hydrophobic molecule with a sterol backbone similar to that of three of the hormones shown in Figure 16–9, but possessing fewer polar groups such as –OH, =O, and –COO⁻. If cholesterol were not normally found in cell membranes, could it be used effectively as a hormone if an appropriate intracellular receptor evolved?

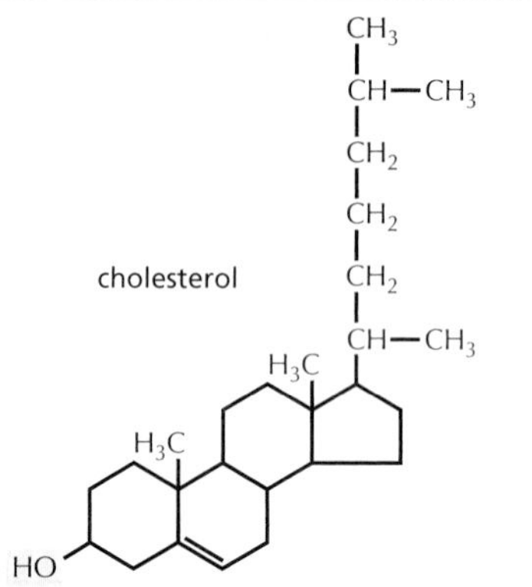

cholesterol

genitalia and influences brain development in the fetus; at puberty, the hormone triggers the development of male secondary sexual characteristics. Some very rare individuals are genetically male—that is, they have both an X and a Y chromosome—but lack the testosterone receptor as a result of a mutation in the corresponding gene; thus, they make testosterone, but their cells cannot respond to it. As a result, these individuals develop as females, which is the path that sexual and brain development would take if no male or female hormones were produced. Such a sex reversal demonstrates the crucial role of the testosterone receptor in sexual development, and it also shows that the receptor is required not just in one cell type to mediate one effect of testosterone, but in many cell types to help produce the whole range of features that distinguish men from women.

Some Dissolved Gases Cross the Plasma Membrane and Activate Intracellular Enzymes Directly

Steroid hormones and thyroid hormones are not the only extracellular signal molecules that can pass through the plasma membrane. Some dissolved gases can slip across the membrane to the cell interior and directly regulate the activity of specific intracellular proteins. This direct approach allows such signals to alter a cell within a few seconds or

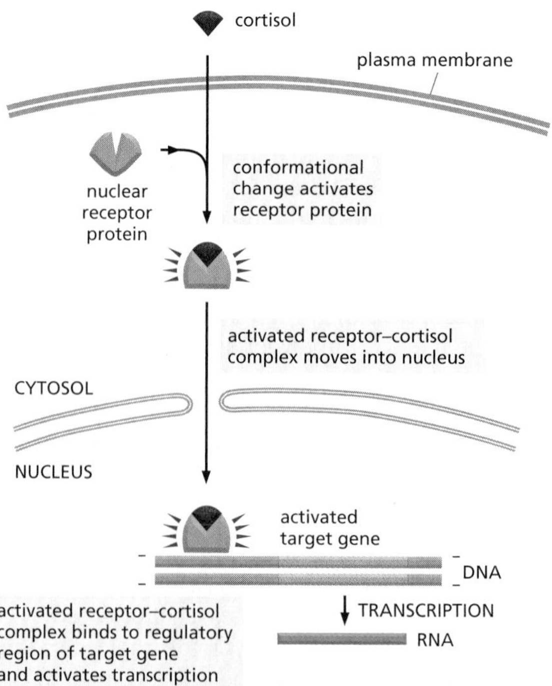

Figure 16–10 **The steroid hormone cortisol acts by activating a transcription regulator.** Cortisol is one of the hormones produced by the adrenal glands in response to stress. It diffuses directly across the plasma membrane and binds to its receptor protein, which is located in the cytosol. The hormone–receptor complex is then transported into the nucleus via the nuclear pores. Cortisol binding activates the receptor protein, which is then able to bind to specific regulatory sequences in DNA and activate (or repress, not shown) the transcription of specific target genes. Whereas the receptors for cortisol and some other steroid hormones are located in the cytosol, those for other signal molecules of this family are already bound to DNA in the nucleus even in the absence of hormone.

Figure 16–11 **Nitric oxide (NO) triggers smooth muscle relaxation in a blood-vessel wall.** (A) The drawing shows a nerve contacting a blood vessel. (B) Sequence of events leading to dilation of the blood vessel. Acetylcholine is released by nerve terminals in the blood-vessel wall. It then diffuses past the smooth muscle cells and through the basal lamina (not shown) to reach acetylcholine receptors on the surface of the endothelial cells lining the blood vessel. There it stimulates the endothelial cells to make and release NO. NO diffuses out of the endothelial cells and into adjacent smooth muscle cells, where it regulates the activity of specific proteins, causing muscle cells to relax. (C) One target protein that can be activated by NO is guanylyl cyclase. The activated cyclase catalyzes the production of cGMP from GTP. Note that NO gas is highly toxic when inhaled and should not be confused with nitrous oxide (N_2O), also known as laughing gas.

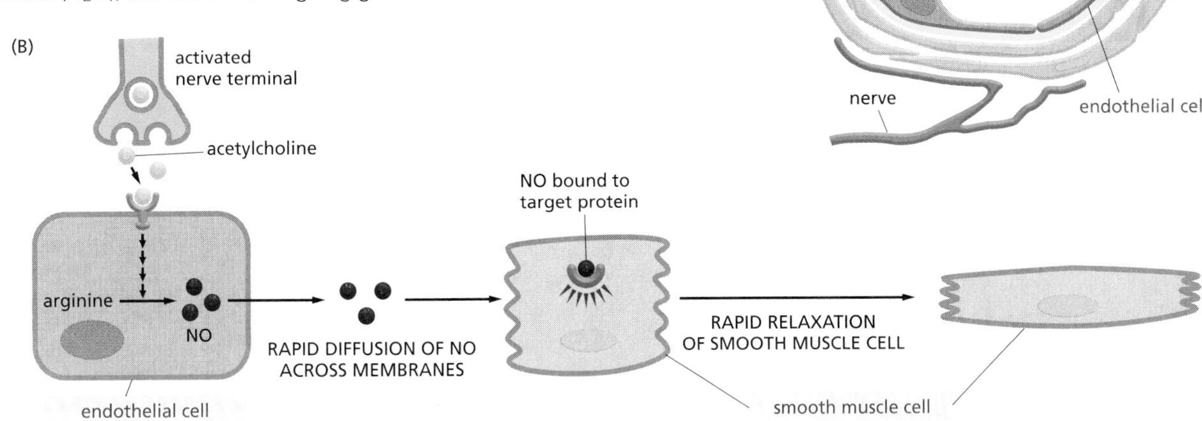

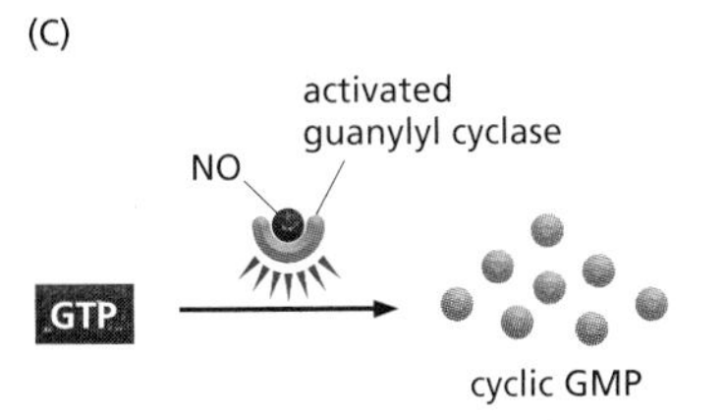

minutes. **Nitric oxide** (**NO**) acts in this way. This gas diffuses readily out of the cell that generates it and enters neighboring cells. NO is synthesized from the amino acid arginine and operates as a local mediator in many tissues. The gas acts only locally because it is quickly converted to nitrates and nitrites (with a half-life of about 5–10 seconds) by reaction with oxygen and water outside cells.

Endothelial cells—the flattened cells that line every blood vessel—release NO in response to stimulation by nerve endings. This NO signal causes smooth muscle cells in the vessel wall to relax, allowing the vessel to dilate, so that blood flows through it more freely (Figure 16–11). The effect of NO on blood vessels accounts for the action of nitroglycerine, which has been used for almost 100 years to treat patients with angina—pain caused by inadequate blood flow to the heart muscle. In the body, nitroglycerine is converted to NO, which rapidly relaxes blood vessels, thereby reducing the workload on the heart and decreasing the muscle's need for oxygen-rich blood. Many nerve cells also use NO to signal neighboring cells: NO released by nerve terminals in the penis, for instance, triggers the local blood-vessel dilation that is responsible for penile erection.

Inside many target cells, NO binds to and activates the enzyme *guanylyl cyclase*, stimulating the formation of *cyclic GMP* from the nucleotide GTP (see Figure 16–11C). Cyclic GMP is itself a small intracellular signaling molecule that forms the next link in the NO signaling chain that leads to the cell's ultimate response. The impotence drug Viagra enhances penile erection by blocking the enzyme that degrades cyclic GMP, prolonging the NO signal. Cyclic GMP is very similar in its structure and mechanism of action to *cyclic AMP*, a much more commonly used intracellular messenger molecule that we discuss later.

Cell-Surface Receptors Relay Extracellular Signals via Intracellular Signaling Pathways

In contrast to NO and the steroid and thyroid hormones, the vast majority of signal molecules are too large or hydrophilic to cross the plasma membrane of the target cell. These proteins, peptides, and small, highly

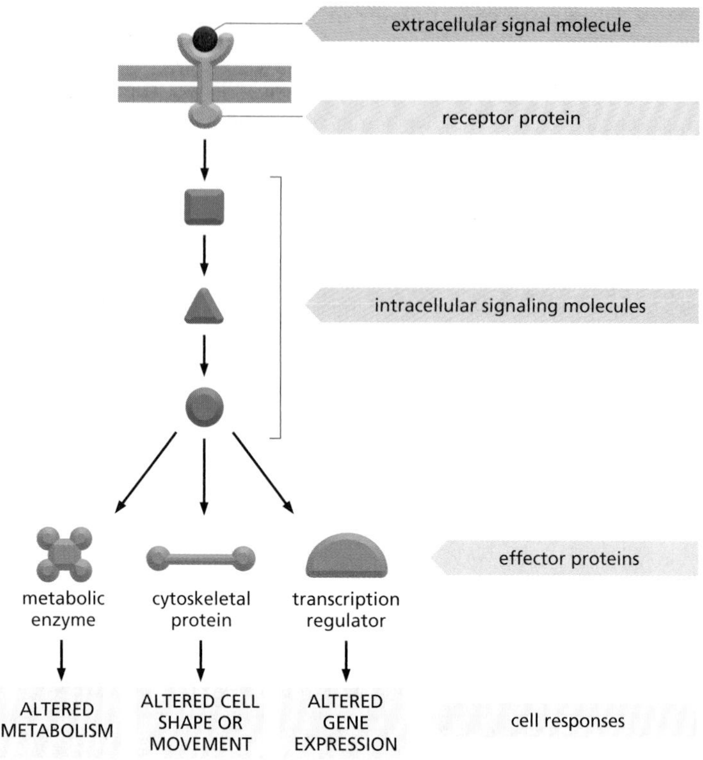

Figure 16–12 **Many extracellular signals act via cell-surface receptors to change the behavior of the cell.** The receptor protein activates one or more intracellular signaling pathways, each mediated by a series of intracellular signaling molecules, which can be proteins or small messenger molecules; only one pathway is shown. Some of these signaling molecules interact with specific effector proteins, altering them to change the behavior of the cell in various ways.

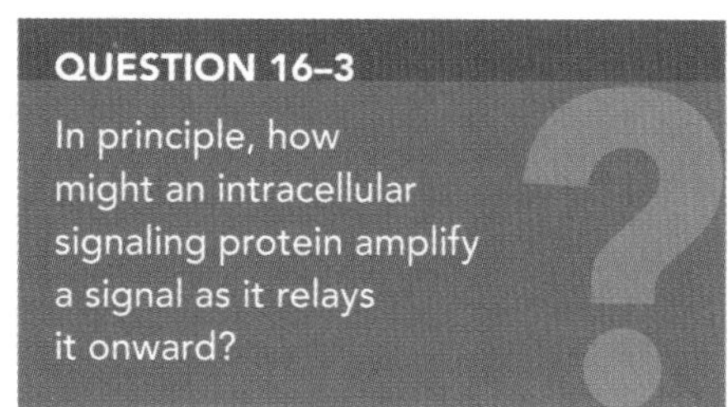

water-soluble molecules bind to cell-surface receptor proteins that span the plasma membrane (see Figure 16–8A). These transmembrane receptors detect a signal on the outside and relay the message, in a new form, across the membrane into the interior of the cell.

The receptor protein performs the primary signal transduction step: it binds to the extracellular signal and generates new intracellular signals in response (see Figure 16–2B). The resulting intracellular signaling process usually works like a molecular relay race in which the message is passed 'downstream' from one **intracellular signaling molecule** to another, each activating or generating the next signaling molecule in the pathway, until a metabolic enzyme is kicked into action, the cytoskeleton is tweaked into a new configuration, or a gene is switched on or off. This final outcome is called the *response* of the cell (Figure 16–12).

The components of these **intracellular signaling pathways** perform one or more crucial functions (Figure 16–13):

1. They can simply *relay* the signal onward and thereby help spread it through the cell.
2. They can *amplify* the signal received, making it stronger, so that a few extracellular signal molecules are enough to evoke a large intracellular response.
3. They can receive signals from more than one intracellular signaling pathway and *integrate* them before relaying a signal onward
4. They can *distribute* the signal to more than one signaling pathway or effector protein, creating branches in the information flow diagram and evoking a complex response.

As part of the integration function, many steps in a signaling pathway are open to *modulation* by other factors, including both intracellular and extracellular factors, so that the effects of each signal are tailored to the conditions prevailing inside and outside the cell.

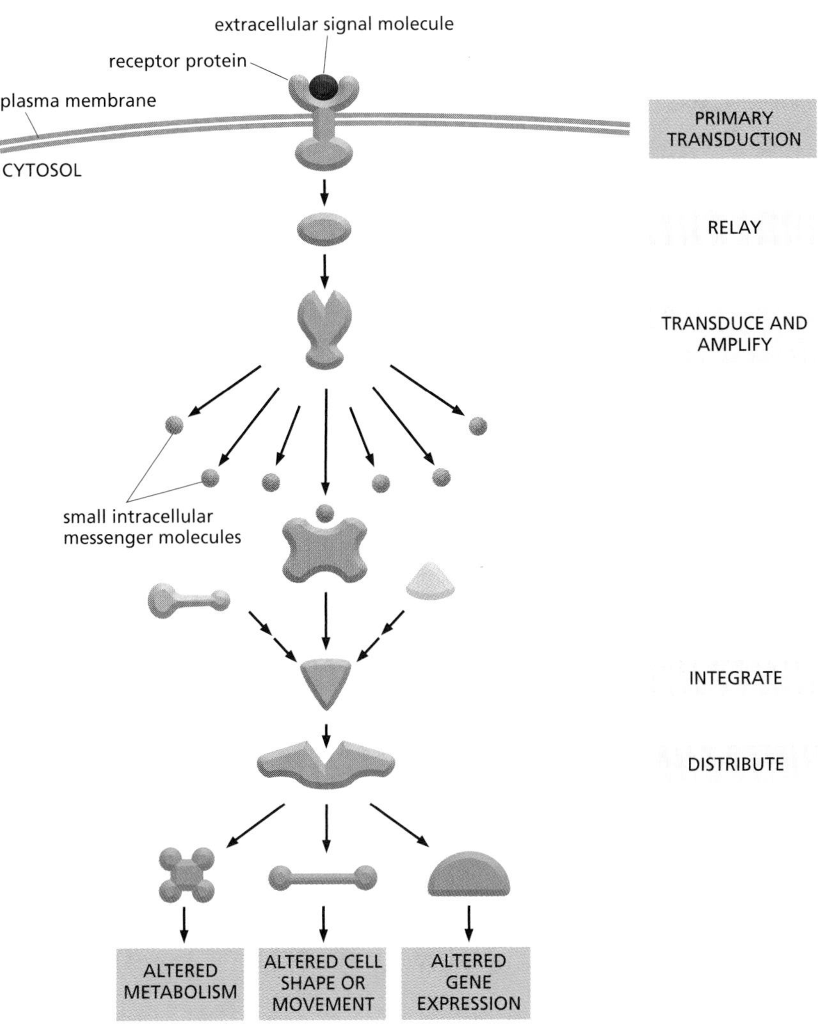

Figure 16–13 **Intracellular signaling proteins can relay, amplify, integrate and distribute the incoming signal.** A receptor protein located on the cell surface transduces an extracellular signal into an intracellular signal, which initiates one or more signaling pathways that relay the signal into the cell interior. Each pathway includes intracellular signaling proteins that can function in one of the various ways shown; some, for example, integrate signals from other signaling pathways, as illustrated. Many of the steps in the process can be modulated by other molecules or events in the cell (not shown). We discuss the production and function of small intracellular messenger molecules later in the chapter.

Some Intracellular Signaling Proteins Act as Molecular Switches

Many of the key intracellular signaling proteins behave as **molecular switches**: receipt of a signal causes them to toggle from an inactive to an active state. Once activated, these proteins can turn on other proteins in the signaling pathway. They then persist in an active state until some other process switches them off again. The importance of the switching-off process is often underappreciated. If a signaling pathway is to recover after transmitting a signal and make itself ready to transmit another, every activated protein in the pathway must be reset to its original, unstimulated state. Thus, for every activation step along the pathway, there has to be an inactivation mechanism. The two are equally important for the signaling process.

Proteins that act as molecular switches fall mostly into one of two classes. The first and by far the largest class consists of proteins that are activated or inactivated by phosphorylation, a chemical modification discussed in Chapter 4 (see Figure 4–38). For these molecules, the switch is thrown in one direction by a **protein kinase**, which tacks a phosphate group onto the switch protein, and in the other direction by a **protein phosphatase**, which plucks the phosphate off again (Figure 16–14A). The activity of any protein that is regulated by phosphorylation depends—moment by moment—on the balance between the activities of the kinases that phosphorylate it and the phosphatases that dephosphorylate it.

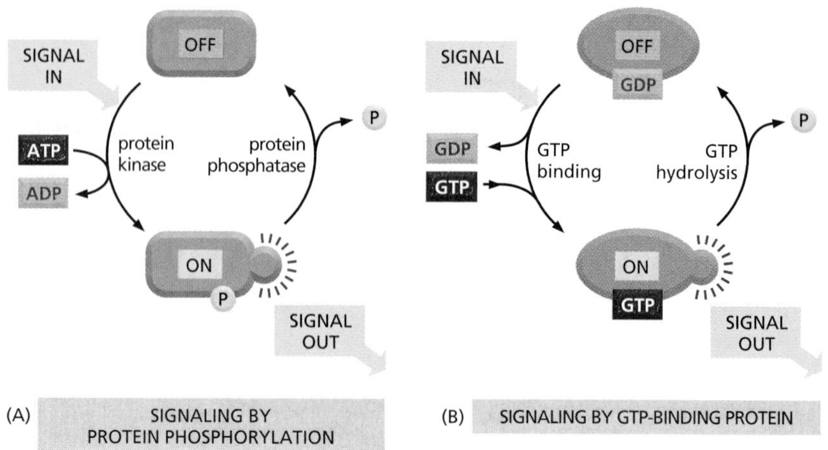

Figure 16–14 **Many intracellular signaling proteins act as molecular switches.** (A) Some intracellular signaling proteins are activated by the addition of a phosphate group and inactivated by the removal of the phosphate. In some cases, the phosphate is added covalently to the protein by a protein kinase that transfers the terminal phosphate group from ATP to the signaling protein; the phosphate is then removed by a protein phosphatase. (B) In other cases, a GTP-binding signaling protein is induced to exchange its bound GDP for GTP (which, in a sense, adds a phosphate to the protein). This activates the protein; hydrolysis of the bound GTP to GDP then switches the protein off.

Many of the switch proteins controlled by phosphorylation are themselves protein kinases, and these are often organized into *phosphorylation cascades*: one protein kinase, activated by phosphorylation, phosphorylates the next protein kinase in the sequence, and so on, transmitting the signal onward and, in the process, amplifying, distributing, and modulating it. Two main types of protein kinases operate in intracellular signaling pathways: the most common are **serine/threonine kinases**, which—as the name implies—phosphorylate proteins on serines or threonines; others are **tyrosine kinases**, which phosphorylate proteins on tyrosines.

The other main class of switch proteins involved in intracellular signaling pathways is the **GTP-binding proteins**. These switch between an active and an inactive state depending on whether they have GTP or GDP bound to them, respectively (Figure 16–14B). Once activated by GTP binding, these proteins have intrinsic GTP-hydrolyzing (*GTPase*) activity, and they shut themselves off by hydrolyzing their bound GTP to GDP. One class of GTP-activated switch proteins contains the large, trimeric GTP-binding proteins (also called *G proteins*) that relay messages from *G-protein–coupled receptors*, as we discuss in detail shortly.

Cell-Surface Receptors Fall into Three Main Classes

All cell-surface receptor proteins bind to an extracellular signal molecule and transduce its message into one or more intracellular signaling molecules that alter the cell's behavior. These receptors, however, are divided into three large families that differ in the transduction mechanism they use. (1) *Ion-channel–coupled receptors* allow a flow of ions across the plasma membrane, which changes the membrane potential and produces an electrical current (Figure 16–15A). (2) *G-protein–coupled receptors* activate membrane-bound, trimeric GTP-binding proteins (G proteins), which then activate either an enzyme or an ion channel in the plasma membrane, initiating a cascade of other effects (Figure 16–15B). (3) *Enzyme-coupled receptors* either act as enzymes or associate with enzymes inside the cell (Figure 16–15C); when stimulated, the enzymes activate a variety of intracellular signaling pathways.

The number of different types of receptors in each of these three classes is even greater than the number of extracellular signals that act on them, because for many extracellular signal molecules there is more than one type of receptor. Moreover, some signal molecules bind to receptors in more than one class. The neurotransmitter acetylcholine, for example, acts on skeletal muscle cells via an ion-channel–coupled receptor, whereas in heart muscle cells it acts through a G-protein–coupled receptor. These two types of receptors generate different intracellular signals and thus enable the two types of muscle cells to react to acetylcholine in

(A) ION-CHANNEL–COUPLED RECEPTORS

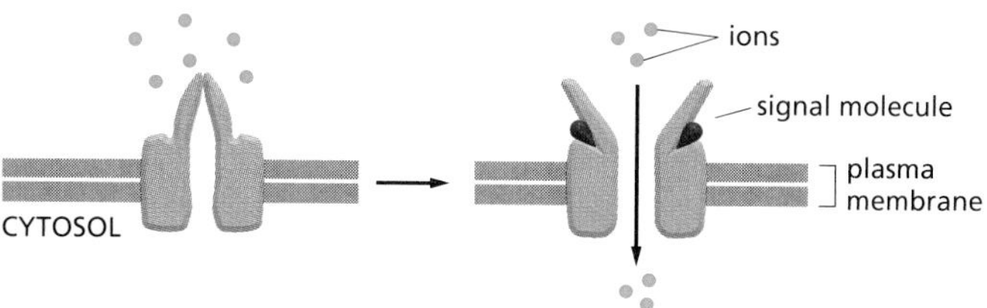

(B) G-PROTEIN–COUPLED RECEPTORS

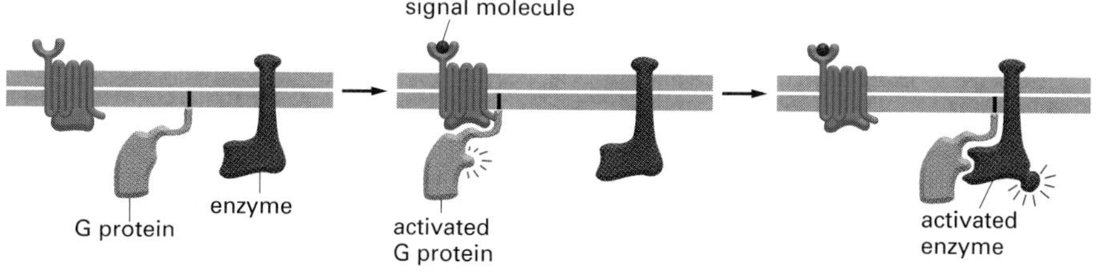

(C) ENZYME-COUPLED RECEPTORS

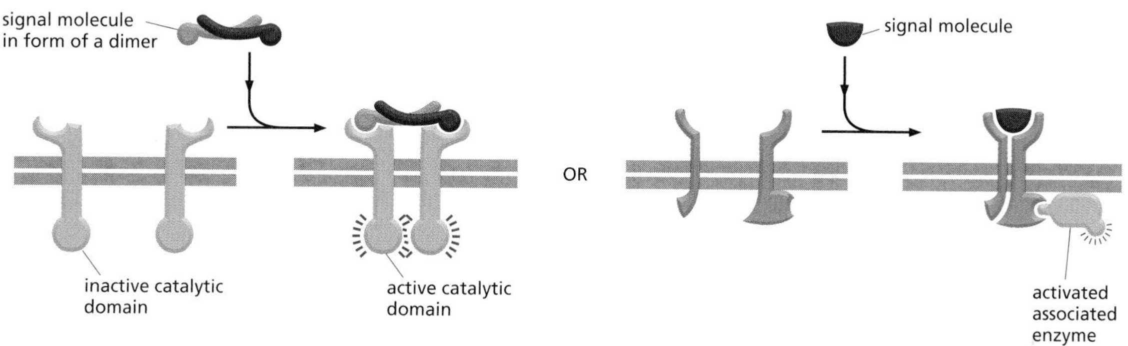

Figure 16–15 **Cell-surface receptors fall into three basic classes.** (A) An ion-channel–coupled receptor opens (or closes; not shown) in response to binding of its extracellular signal molecule. These channels are also called transmitter-gated ion channels. (B) When a G-protein–coupled receptor binds its extracellular signal molecule, the activated receptor signals to a G protein on the opposite side of the plasma membrane, which then turns on (or off) an enzyme (or an ion channel; not shown) in the same membrane. For simplicity, the G protein is shown here as a single molecule; as we see later, it is in fact a complex of three protein subunits. (C) When an enzyme-coupled receptor binds its extracellular signal molecule, an enzyme activity is switched on at the other end of the receptor, inside the cell. Many enzyme-coupled receptors have their own enzyme activity (*left*), while others rely on an enzyme that becomes associated with the activated receptor (*right*).

different ways, increasing contraction in skeletal muscle and decreasing the rate and force of contractions in heart (see Figure 16–5A and C).

The multitude of different cell-surface receptors that the body requires for signaling purposes also serve as targets for many foreign substances that interfere with our physiology and sensations, from heroin and nicotine to tranquilizers and chili peppers. These substances either mimic the natural ligand for a receptor, occupying the normal ligand-binding site, or bind to the receptor at some other site, either blocking or overstimulating the receptor's natural activity. Many drugs and poisons act in this way (Table 16–2), and a large part of the pharmaceutical industry is devoted to the search for substances that will exert a precisely defined effect by binding to a specific type of cell-surface receptor.

TABLE 16–2 SOME FOREIGN SUBSTANCES THAT ACT ON CELL-SURFACE RECEPTORS

SUBSTANCE	SIGNAL MOLECULE	RECEPTOR ACTION	EFFECT
Valium and barbiturates	γ-aminobutyric acid (GABA)	stimulate GABA-activated ion-channel–coupled receptors	relief of anxiety; sedation
Nicotine	acetylcholine	stimulates acetylcholine-activated ion-channel–coupled receptors	constriction of blood vessels; elevation of blood pressure
Morphine and heroin	endorphins and enkephalins	stimulate G-protein–coupled opiate receptors	analgesia (relief of pain); euphoria
Curare	acetylcholine	blocks acetylcholine-activated ion-channel–coupled receptors	blockage of neuromuscular transmission, resulting in paralysis
Strychnine	glycine	blocks glycine-activated ion-channel–coupled receptors	blockage of inhibitory synapses in spinal cord and brain, resulting in seizures and muscle spasm

Ion-channel–coupled Receptors Convert Chemical Signals into Electrical Ones

Of all the types of cell-surface receptors, **ion-channel–coupled receptors** (also known as transmitter-gated ion channels) function in the simplest and most direct way. These receptors are responsible for the rapid transmission of signals across synapses in the nervous system. They transduce a chemical signal, in the form of a pulse of neurotransmitter delivered to the outside of the target cell, directly into an electrical signal, in the form of a change in voltage across the target cell's plasma membrane (see Figure 12–42). When the neurotransmitter binds, this type of receptor alters its conformation so as to open or close an ion channel in the plasma membrane, allowing the flow of specific types of ions, such as Na^+, K^+, Ca^{2+}, or Cl^- (see Figure 16–15A and Movie 16.1). Driven by their electrochemical gradients, the ions rush into or out of the cell, creating a change in the membrane potential within a millisecond or so. This change in potential may trigger a nerve impulse or make it easier (or harder) for other neurotransmitters to do so. As we discuss later, the opening of Ca^{2+} channels has additional important effects, as changes in the intracellular Ca^{2+} concentration can profoundly alter the activities of many Ca^{2+}-responsive proteins in the cell. The function of ion-channel–coupled receptors is discussed in greater detail in Chapter 12.

QUESTION 16–4

The signaling mechanisms used by a steroid-hormone-type nuclear receptor and by an ion-channel–coupled receptor are relatively simple as they have few components. Can they lead to an amplification of the initial signal, and, if so, how?

Whereas ion-channel–coupled receptors are a specialty of the nervous system and of other electrically excitable cells such as muscle cells, G-protein–coupled receptors and enzyme-coupled receptors are used by practically every cell type in the body. Most of the remainder of this chapter deals with these two receptor families and with the signal transduction processes that they use.

G-PROTEIN–COUPLED RECEPTORS

G-protein–coupled receptors (**GPCRs**) form the largest family of cell-surface receptors. There are more than 700 GPCRs in humans, and mice have about 1000 concerned with the sense of smell alone. These receptors mediate responses to an enormous diversity of extracellular signal molecules, including hormones, local mediators, and neurotransmitters. The signal molecules are as varied in structure as they are in function: they can be proteins, small peptides, or derivatives of amino acids or fatty acids, and for each one of them there is a different receptor or set of receptors. Because GPCRs are involved in such a large variety of cellular processes, they are an attractive target for the development of drugs to treat a variety of disorders. About half of all known drugs work through GPCRs.

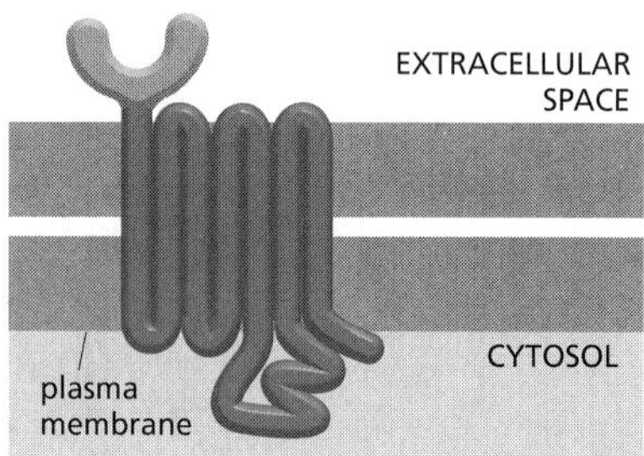

Figure 16–16 **All GPCRs possess a similar structure.** The cytoplasmic portions of the receptor are responsible for binding to the G protein inside the cell. Receptors that bind to signal molecules that are proteins usually have a large extracellular domain (*light green*). This domain, together with some of the transmembrane segments, binds the protein ligand. Receptors that recognize small signal molecules such as adrenaline or acetylcholine, however, have a small extracellular domain, and the ligand usually binds deep within the plane of the membrane to a site that is formed by amino acids from several transmembrane segments (not shown).

Despite the diversity of the signal molecules that bind to them, all GPCRs that have been analyzed have a similar structure: each is made of a single polypeptide chain that threads back and forth across the lipid bilayer seven times (Figure 16–16). This superfamily of *seven-pass transmembrane receptor proteins* includes rhodopsin (the light-activated photoreceptor protein in the vertebrate eye), the olfactory (smell) receptors in the vertebrate nose, and the receptors that participate in the mating rituals of single-celled yeasts (see Figure 16–1). Evolutionarily speaking, GPCRs are ancient: even bacteria possess structurally similar membrane proteins—such as the bacteriorhodopsin that functions as a light-driven H^+ pump (see Figure 11–28). Although they resemble eucaryotic GPCRs, these bacterial receptors do not act through G proteins; instead, they are coupled to other signal transduction systems.

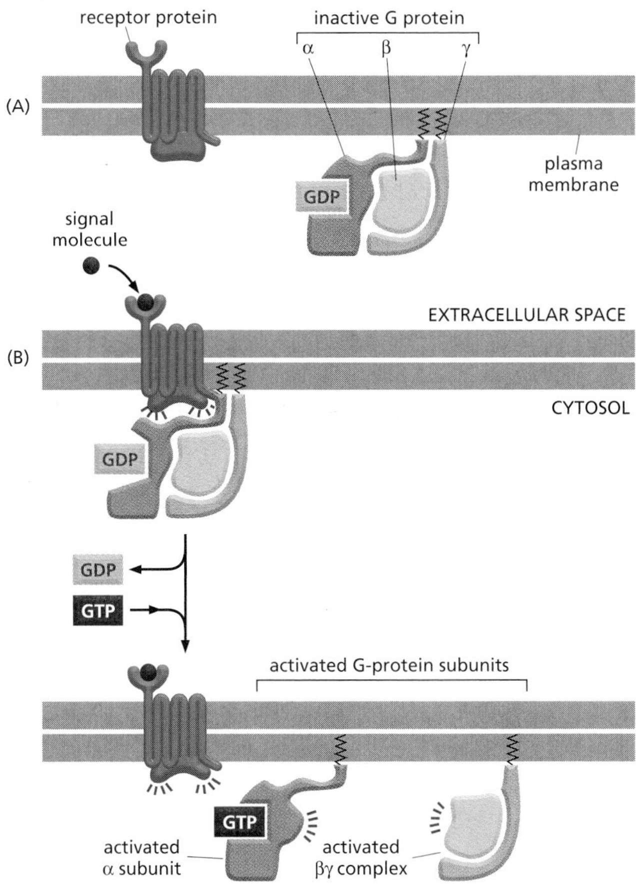

Figure 16–17 **An activated GPCR activates G proteins by encouraging the α subunit to expel its GDP and pick up GTP.** (A) In the unstimulated state, the receptor and the G protein are both inactive. Although they are shown here as separate entities in the plasma membrane, in some cases at least they are associated in a preformed complex. (B) Binding of an extracellular signal to the receptor changes the conformation of the receptor, which in turn alters the conformation of the bound G protein. The alteration of the α subunit of the G protein allows it to exchange its GDP for GTP. This exchange triggers a conformational change that activates both the α subunit and a βγ complex, which can now interact with their preferred target proteins in the plasma membrane (**Movie 16.2**). The receptor stays active while the external signal molecule is bound to it, and it can therefore catalyze the activation of many molecules of G protein. Although it was originally thought that activation of a G protein always causes the α subunit and the βγ complex to dissociate physically (as shown here), in some cases the trimer might simply open up, allowing the activated α subunit and the βγ complex to each interact with a target protein. Note that both the α and γ subunits of the G protein have covalently attached lipid molecules (*black*) that help anchor the subunits to the plasma membrane.

Stimulation of GPCRs Activates G-Protein Subunits

When an extracellular signal molecule binds to a GPCR, the receptor protein undergoes a conformational change that enables it to activate a **G protein** located on the underside of the plasma membrane. To explain how this activation leads to the transmission of a signal, we must first consider how G proteins are constructed and how they function.

There are several varieties of G proteins. Each is specific for a particular set of receptors and a particular set of target enzymes or ion channels in the plasma membrane. All of these G proteins, however, have a similar general structure and operate in a similar way. They are composed of three protein subunits—α, β, and γ—two of which are tethered to the plasma membrane by short lipid tails. In the unstimulated state, the α subunit has GDP bound to it, and the G protein is idle (Figure 16–17A). When an extracellular ligand binds to its receptor, the altered receptor activates a G protein by causing the α subunit to decrease its affinity for GDP, which is then exchanged for a molecule of GTP. In some cases, this activation is thought to break up the G-protein subunits, so that the activated α subunit, clutching its GTP, detaches from the βγ complex, which is also activated (Figure 16–17B). Regardless of whether they dissociate, the two activated parts of a G protein—the α subunit and the βγ complex—can both interact directly with target proteins in the plasma membrane, which in turn may relay the signal to yet other destinations in the cell. The longer these target proteins have an α or a βγ subunit bound to them, the stronger and more prolonged the relayed signal will be.

The amount of time that the α and βγ subunits remain 'switched on'—and hence available to relay signals—is limited by the behavior of the α subunit. The α subunit has an intrinsic GTPase activity, and it eventually hydrolyzes its bound GTP back to GDP, returning the whole G protein to its original, inactive conformation (Figure 16–18). GTP hydrolysis and inactivation occur within seconds after the G protein has been activated. The inactive G protein is now ready to be reactivated by another activated receptor.

The G-protein switch demonstrates a general principle of cell signaling mentioned earlier: the mechanisms that shut a signal off are as important as the mechanisms that turn it on (see Figure 16–14B). The shut-off mechanisms also offer as many opportunities for control, and as many dangers for mishap. Take cholera, for example. The disease is caused by a bacterium that multiplies in the intestine, where it produces a protein called *cholera toxin*. This protein enters the cells that line the intestine and modifies the α subunit of a particular G protein (called G_S, because it *stimulates* the enzyme adenylyl cyclase, which we discuss shortly) in such a way that it can no longer hydrolyze its bound GTP. The altered α subunit thus remains in the active state indefinitely, continuously transmitting a signal to its target proteins. In intestinal cells, this causes a prolonged and excessive outflow of Cl^- and water into the gut, resulting in catastrophic diarrhea and dehydration. The condition often leads to death unless urgent steps are taken to replace the lost water and ions.

A similar situation occurs in whooping cough (pertussis), a common respiratory infection against which infants are now routinely vaccinated. In this case, the disease-causing bacterium colonizes the lung, where it produces a protein called *pertussis toxin*. This protein alters the α subunit of a different type of G protein (called G_i, because it *inhibits* adenylyl cyclase). In this case, however, modification by the toxin disables the G protein by locking it into its inactive GDP-bound state. Inhibiting G_i, like activa-

QUESTION 16–5

GPCRs activate G proteins by reducing the strength of GDP binding to the G protein. This results in rapid dissociation of bound GDP, which is then replaced by GTP, because GTP is present in the cytosol in much higher concentrations than GDP. What consequences would result from a mutation in the α subunit of a G protein that caused its affinity for GDP to be reduced without significantly changing its affinity for GTP? Compare the effects of this mutation with the effects of cholera toxin.

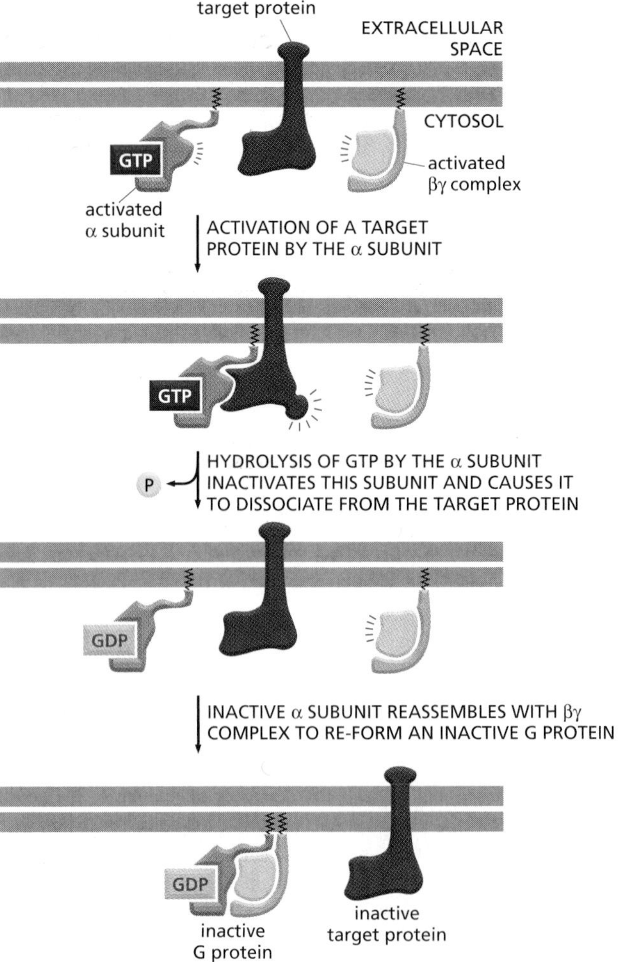

Figure 16–18 **The G-protein α subunit switches itself off by hydrolyzing its bound GTP.** When an activated α subunit binds its target protein, it activates the protein (or in some cases inactivates it; not shown) for as long as the two remain in contact. Within seconds, the α subunit hydrolyzes its bound GTP to GDP. This loss of GTP inactivates the α subunit, which dissociates from its target protein and—if the α subunit had separated from the βγ complex (as shown)—it now reassociates with a βγ complex to re-form an inactive G protein. The G protein is now ready to couple to another activated receptor, as in Figure 16–17B. Both the activated α subunit and the activated βγ complex can interact with target proteins in the plasma membrane. See also Movie 16.2.

ting G_S, results in the generation of a prolonged, inappropriate signal that in this case stimulates coughing. Both the diarrhea-producing effects of cholera toxin and the cough-provoking effects of pertussis toxin help the disease-causing bacteria move from host to host.

Some G Proteins Directly Regulate Ion Channels

The target proteins recognized by G-protein subunits are either enzymes or ion channels in the plasma membrane. There are about 20 types of mammalian G proteins, each activated by a particular set of cell-surface receptors and dedicated to activating a particular set of target proteins. In this way, binding of an extracellular signal molecule to a GPCR leads to changes in the activities of a specific subset of the possible target proteins in the plasma membrane, leading to a response that is appropriate for that signal and that type of cell.

We look first at an example of direct G-protein regulation of ion channels. The heartbeat in animals is controlled by two sets of nerves: one speeds the heart up, the other slows it down. The nerves that signal a slowdown in heartbeat do so by releasing acetylcholine, which binds to a GPCR on the surface of the heart muscle cells. This GPCR activates the G protein, G_i. In this case, the $\beta\gamma$ complex is the active signaling component: it binds to the intracellular face of a K^+ channel in the plasma membrane of the heart muscle cell, forcing the ion channel into an open conformation (Figure 16–19A). This allows K^+ to flow out of the cell, thereby inhibiting the cell's electrical excitability (Figure 16–19B). The signal is shut off—and the K^+ channel recloses—when the α subunit inactivates itself by hydrolyzing its bound GTP, returning the G protein to its inactive state (Figure 16–19C).

Some G Proteins Activate Membrane-bound Enzymes

When G proteins interact with ion channels, they cause an immediate change in the state and behavior of the cell. Their interactions with enzymes have more complex consequences, leading to the production

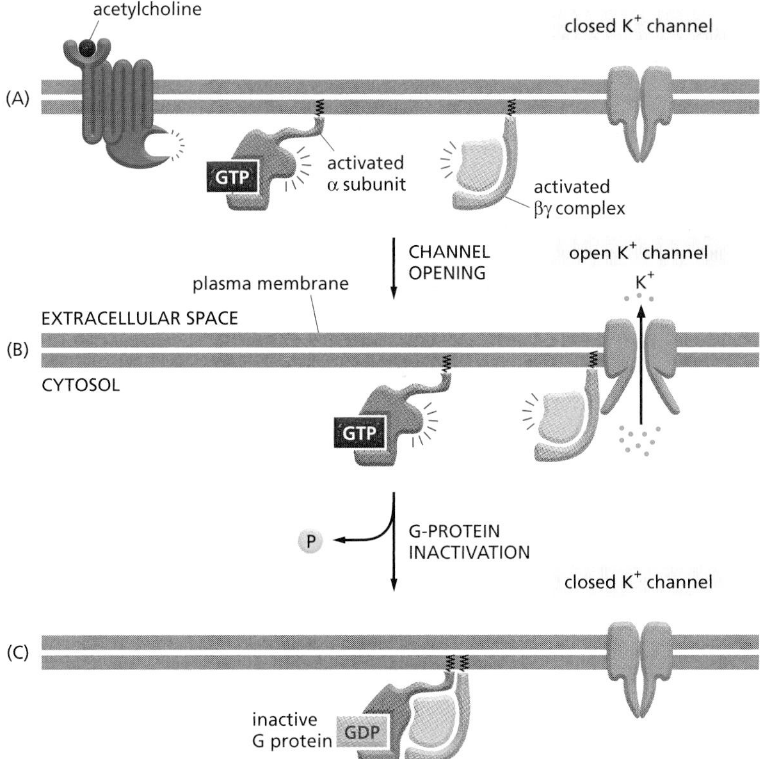

Figure 16–19 **A G_i protein directly couples receptor activation to the opening of K^+ channels in the plasma membrane of heart muscle cells.** (A) Binding of the neurotransmitter acetylcholine to its GPCR on heart muscle cells results in the activation of the G protein, G_i. (B) The activated $\beta\gamma$ complex directly opens a K^+ channel in the heart cell plasma membrane, allowing K^+ to leave the cell, decreasing the cell's excitability. (C) Inactivation of the α subunit by hydrolysis of its bound GTP returns the G protein to its inactive state, allowing the K^+ channel to close.

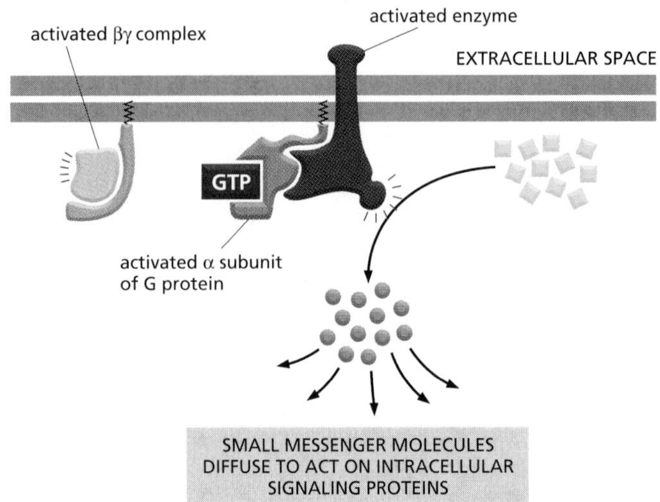

Figure 16–20 **Enzymes activated by G proteins catalyze the production of small intracellular signaling molecules.** Because each activated enzyme generates many molecules of these small messengers, the signal is greatly amplified at this step in the pathway. The signal is relayed onward by the small messenger molecules, which bind to specific signaling proteins in the cell and influence their activity.

of additional intracellular signaling molecules. The two most frequent target enzymes for G proteins are *adenylyl cyclase*, the enzyme responsible for production of the small intracellular signaling molecule *cyclic AMP*, and *phospholipase C*, the enzyme responsible for production of the small intracellular signaling molecules *inositol trisphosphate* and *diacylglycerol*. The two enzymes are activated by different types of G proteins, so that cells are able to couple the production of the small intracellular signaling molecules to different extracellular signals. As we saw earlier, the coupling may be either stimulatory or inhibitory. We concentrate here on G proteins that stimulate enzyme activity. The small intracellular signaling molecules generated in these cascades are often called **small messengers**, or **second messengers** (the "first messengers" being the extracellular signals); they are produced in large numbers when a membrane-bound enzyme—such as adenylyl cyclase or phospholipase C—is activated, and they rapidly diffuse away from their source, spreading the signal (Figure 16–20).

Different small messenger molecules, of course, produce different responses. We first examine the consequences of an increase in the intracellular concentration of cyclic AMP. This will take us along one of the main types of signaling pathways that lead from the activation of GPCRs. We then discuss the actions of inositol trisphosphate and diacylglycerol, small messenger molecules that will lead us along a different signaling route.

The Cyclic AMP Pathway Can Activate Enzymes and Turn On Genes

Many extracellular signals acting via GPCRs affect the activity of the enzyme **adenylyl cyclase** and thus alter the concentration of the small messenger molecule **cyclic AMP** inside the cell. Most commonly, the activated G-protein α subunit switches on the adenylyl cyclase, causing a dramatic and sudden increase in the synthesis of cyclic AMP from ATP (which is always present in the cell). Because it stimulates the cyclase, this G protein is called G_S. To help terminate the signal, a second enzyme, called *cyclic AMP phosphodiesterase*, rapidly converts cyclic AMP to ordinary AMP (Figure 16–21). One way that caffeine acts as a stimulant is by

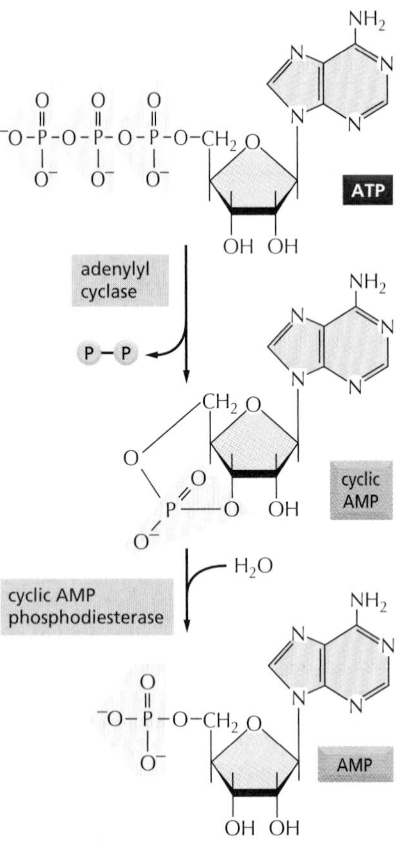

Figure 16–21 **Cyclic AMP is synthesized by adenylyl cyclase and degraded by cyclic AMP phosphodiesterase.** Cyclic AMP (also called cAMP) is formed from ATP by a cyclization reaction that removes two phosphate groups from ATP and joins the "free" end of the remaining phosphate group to the sugar part of the ATP molecule. The degradation reaction breaks this second bond, forming AMP.

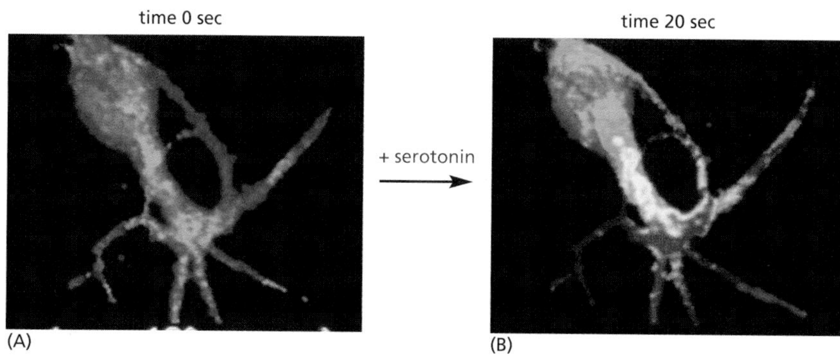

Figure 16–22 **Cyclic AMP concentration rises rapidly in response to an extracellular signal.** A nerve cell in culture responds to the binding of the neurotransmitter serotonin to a GPCR by synthesizing cyclic AMP. The concentration of intracellular cyclic AMP was monitored by injecting into the cell a fluorescent protein whose fluorescence changes when it binds cyclic AMP. *Blue* indicates a low level of cyclic AMP, *yellow* an intermediate level, and *red* a high level. (A) In the resting cell, the cyclic AMP concentration is about 5×10^{-8} M. (B) Twenty seconds after adding serotonin to the culture medium, the intracellular concentration of cyclic AMP has risen more than twentyfold (to $> 10^{-6}$ M) in the parts of the cell where the serotonin receptors are concentrated. (Courtesy of Roger Tsien.)

inhibiting this phosphodiesterase in the nervous system, blocking cyclic AMP degradation and thereby keeping the concentration of this small messenger high.

Cyclic AMP phosphodiesterase is continuously active inside the cell. Because it breaks cyclic AMP down so quickly, the concentration of this small messenger can change rapidly in response to extracellular signals, rising or falling tenfold in a matter of seconds (Figure 16–22). Cyclic AMP is a water-soluble molecule, so it can, in some cases, carry the signal throughout the cell, traveling from the site on the membrane where it is synthesized to interact with proteins located in the cytosol, in the nucleus, or on other organelles.

Cyclic AMP exerts most of its effects by activating the enzyme **cyclic-AMP–dependent protein kinase** (**PKA**). This enzyme is normally held inactive in a complex with another protein. The binding of cyclic AMP forces a conformational change that unleashes the active kinase. Activated PKA then catalyzes the phosphorylation of particular serines or threonines on certain intracellular proteins, thus altering the activity of the proteins. In different cell types, different sets of proteins are available to be phosphorylated, which largely explains why the effects of cyclic AMP vary with the type of target cell.

Many types of cell responses are mediated by cyclic AMP; a few are listed in Table 16–3. As the table shows, different target cells respond very differently to extracellular signals that change intracellular cyclic AMP concentrations. When we are frightened or excited, for example, the adrenal gland releases the hormone *adrenaline*, which circulates in the bloodstream and binds to a class of GPCRs called adrenergic receptors that are present on many types of cells. The consequences vary from one cell type to another, but all of the cell responses help prepare the body for sudden action. In skeletal muscle, for instance, adrenaline triggers a rise in the intracellular concentration of cyclic AMP, which causes the breakdown of glycogen—the polymerized storage form of glucose. It does so by activating PKA, which leads to both the activation of an enzyme that promotes glycogen breakdown (Figure 16–23) and the inhibition of an

TABLE 16–3 SOME CELL RESPONSES MEDIATED BY CYCLIC AMP

EXTRACELLULAR SIGNAL MOLECULE*	TARGET TISSUE	MAJOR RESPONSE
Adrenaline	heart	increase in heart rate and force of contraction
Adrenaline	skeletal muscle	glycogen breakdown
Adrenaline, ACTH, glucagon	fat	fat breakdown
ACTH	adrenal gland	cortisol secretion

*Although all of the signal molecules listed here are hormones, some responses to local mediators and to neurotransmitters are also mediated by cyclic AMP.

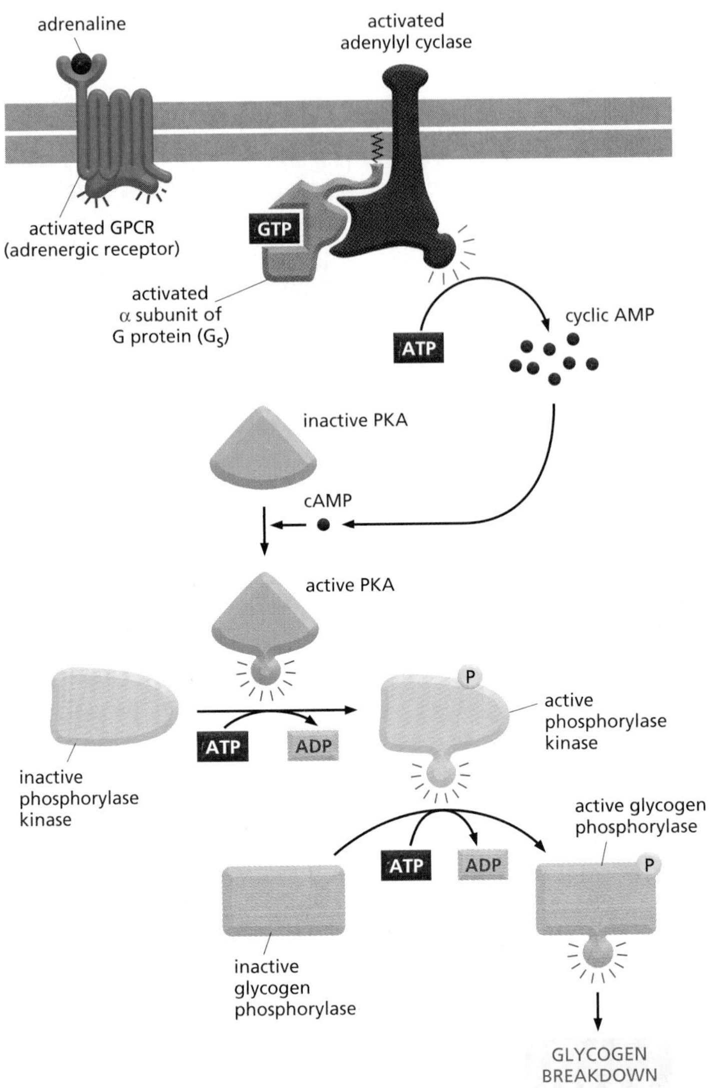

Figure 16–23 **Adrenaline stimulates glycogen breakdown in skeletal muscle cells.** The hormone activates a GPCR that turns on a G protein (G_s), which activates adenylyl cyclase, boosting the production of cyclic AMP. Cyclic AMP, in turn, activates PKA, which phosphorylates and activates an enzyme called phosphorylase kinase. This kinase activates glycogen phosphorylase, the enzyme that breaks down glycogen. Because these reactions do not involve changes in gene transcription or new protein synthesis, they occur rapidly.

QUESTION 16–6

Explain why cyclic AMP must be broken down rapidly in a cell to allow rapid signaling.

enzyme that drives glycogen synthesis. By stimulating glycogen breakdown and inhibiting its synthesis, the increase in cyclic AMP maximizes the amount of glucose available as fuel for anticipated muscular activity. Adrenaline also acts on fat cells, stimulating the breakdown of triacylglycerols (the storage form of fat) to fatty acids—an immediately usable form of fuel for ATP production (discussed in Chapter 13). These fatty acids can then be exported for use by other cells in need of energy.

In some cases, the effects of activating a cyclic AMP cascade are rapid; in skeletal muscle, for example, glycogen breakdown occurs within seconds of adrenaline binding to its receptor (see Figure 16–23). In other cases, cyclic AMP responses involve changes in gene expression that take minutes or hours to develop (see Figure 16–7). In these slow responses, PKA typically phosphorylates transcription regulators that then activate the transcription of selected genes. Thus, in endocrine cells in the hypothalamus, a rise in the amount of intracellular cyclic AMP stimulates the production of somatostatin, a peptide hormone that then suppresses the release of various hormones by other cells. Similarly, an increase in cyclic AMP concentration in some neurons in the brain controls the production of proteins involved in some forms of learning. Figure 16–24 illustrates a typical cyclic-AMP-mediated pathway from the plasma membrane to the nucleus.

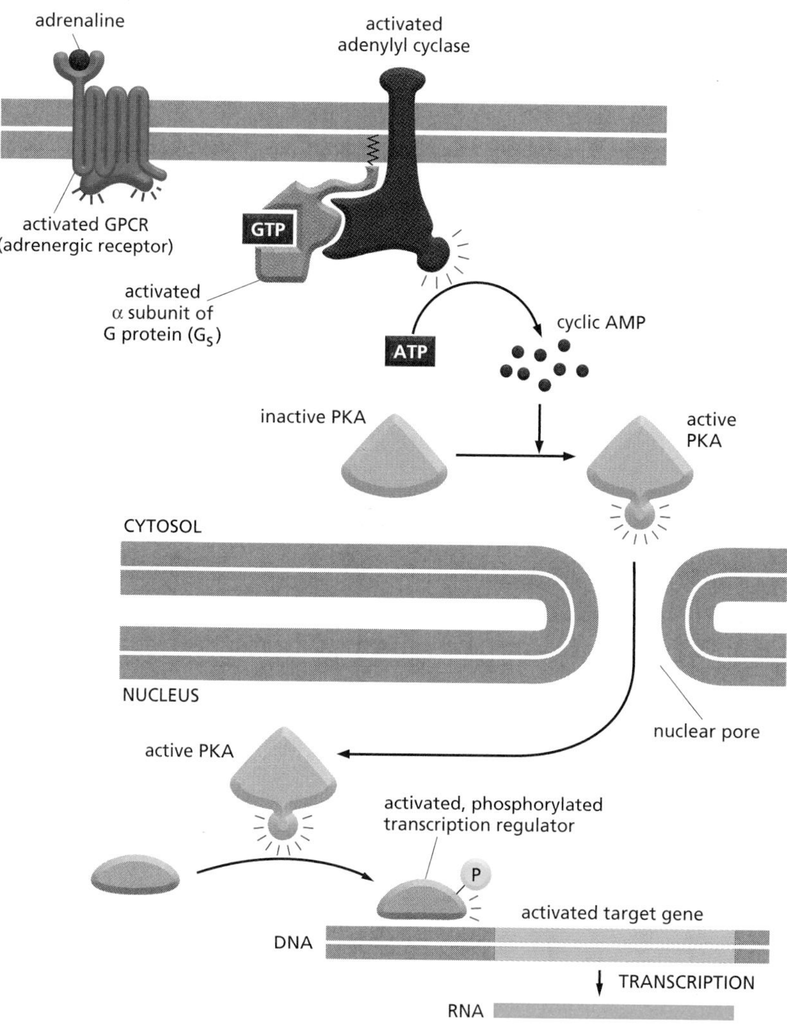

Figure 16–24 **A rise in intracellular cyclic AMP can activate gene transcription.** Binding of a signal molecule to its GPCR can lead to the activation of adenylyl cyclase and a rise in the concentration of intracellular cyclic AMP. In the cytosol, cyclic AMP activates PKA, which then moves into the nucleus and phosphorylates specific transcription regulators. Once phosphorylated, these proteins stimulate the transcription of a whole set of target genes (Movie 16.3). This type of signaling pathway controls many processes in cells, ranging from hormone synthesis in endocrine cells to the production of proteins involved in long-term memory in the brain. Activated PKA can also phosphorylate and thereby regulate other proteins and enzymes in the cytosol (as shown in Figure 16–23).

We now turn to the other enzyme-mediated signaling pathway that leads from GPCRs—the pathway that begins with the activation of the membrane-bound enzyme *phospholipase C* and leads to the generation of the second messengers *inositol trisphosphate* and *diacylglycerol*.

The Inositol Phospholipid Pathway Triggers a Rise in Intracellular Ca^{2+}

Some GPCRs exert their effects via G proteins that activate the membrane-bound enzyme **phospholipase C** instead of adenylyl cyclase. Several examples are given in Table 16–4.

Once activated, phospholipase C propagates the signal by cleaving a lipid molecule that is a component of the plasma membrane. The molecule is an **inositol phospholipid** (a phospholipid with the sugar inositol attached to its head) that is present in small quantities in the cytosolic half of the membrane lipid bilayer (see Figure 11–17). Because of the involvement of

TABLE 16–4 SOME CELL RESPONSES MEDIATED BY PHOSPHOLIPASE C ACTIVATION

SIGNAL MOLECULE	TARGET TISSUE	MAJOR RESPONSE
Vasopressin (a peptide hormone)	liver	glycogen breakdown
Acetylcholine	pancreas	secretion of amylase (a digestive enzyme)
Acetylcholine	smooth muscle	contraction
Thrombin (a proteolytic enzyme)	blood platelets	aggregation

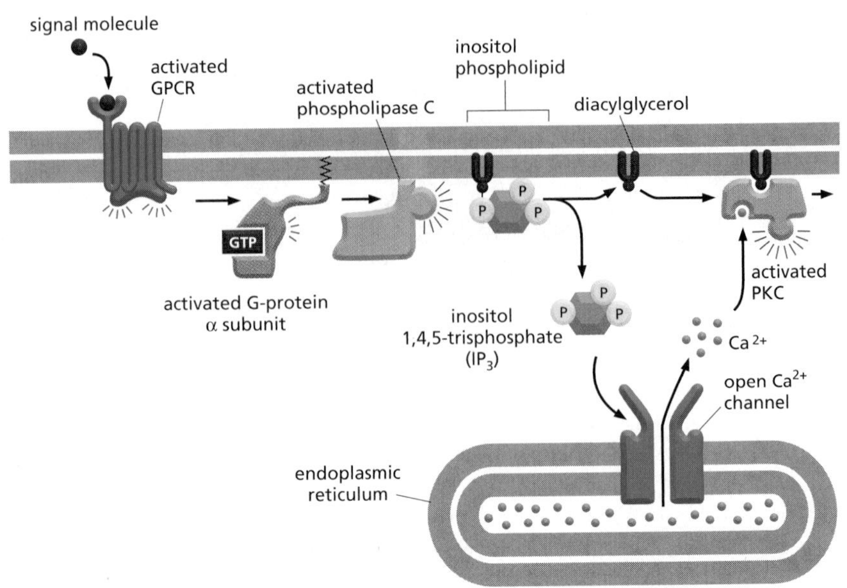

Figure 16–25 **Phospholipase C activates two signaling pathways.** Two small messenger molecules are produced when a membrane inositol phospholipid is hydrolyzed by activated phospholipase C. Inositol 1,4,5-trisphosphate (IP_3) diffuses through the cytosol and triggers the release of Ca^{2+} from the endoplasmic reticulum by binding to and opening special Ca^{2+} channels in the endoplasmic reticulum membrane. The large electrochemical gradient for Ca^{2+} across this membrane causes Ca^{2+} to rush out into the cytosol. Diacylglycerol remains in the plasma membrane and, together with Ca^{2+}, helps to activate the enzyme protein kinase C (PKC), which is recruited from the cytosol to the cytosolic face of the plasma membrane. PKC then phosphorylates its own set of intracellular proteins, further propagating the signal.

this phospholipid, the signaling pathway that begins with the activation of phospholipase C is often referred to as the *inositol phospholipid pathway*. It operates in almost all eucaryotic cells and can regulate a host of different effector proteins.

The pathway works in the following way. When phospholipase C chops the sugar-phosphate head off the inositol phospholipid, it generates two small signaling molecules: **inositol 1,4,5-trisphosphate** (IP_3) and **diacylglycerol** (**DAG**). IP_3, a water-soluble sugar phosphate, diffuses into the cytosol, while the lipid diacylglycerol remains embedded in the plasma membrane. Both molecules play a crucial part in relaying the signal, and we will consider them in turn.

The IP_3 released into the cytosol rapidly encounters the endoplasmic reticulum; there it binds to and opens Ca^{2+} channels that are embedded in the endoplasmic reticulum membrane. Ca^{2+} stored inside the endoplasmic reticulum rushes out into the cytosol through these open channels (Figure 16–25), causing a sharp rise in the cytosolic concentration of free Ca^{2+}, which is normally kept very low. This Ca^{2+} in turn signals to other proteins, as discussed below.

The diacylglycerol that is generated along with IP_3 helps recruit and activate a protein kinase, which translocates from the cytosol to the plasma membrane. This enzyme is called **protein kinase C** (**PKC**) because it also needs to bind Ca^{2+} to become active (see Figure 16–25). Once activated, PKC phosphorylates a set of intracellular proteins that varies depending on the cell type. PKC operates on the same principle as PKA, although most of the proteins it phosphorylates are different.

A Ca^{2+} Signal Triggers Many Biological Processes

Ca^{2+} has such an important and widespread role as a small intracellular messenger that we must digress to consider its functions more generally. A surge in the cytosolic concentration of free Ca^{2+} is triggered by many kinds of stimuli, not only those that act through GPCRs. When a sperm fertilizes an egg cell, for example, Ca^{2+} channels open, and the resulting rise in cytosolic Ca^{2+} triggers the egg to start development (Figure 16–26); for skeletal muscle cells, a signal from a nerve triggers a rise in cytosolic Ca^{2+} that initiates muscle contraction; and in many secretory cells, including nerve cells, Ca^{2+} triggers secretion. Ca^{2+} stimulates all these responses by binding to and influencing the activity of various Ca^{2+}-responsive proteins.

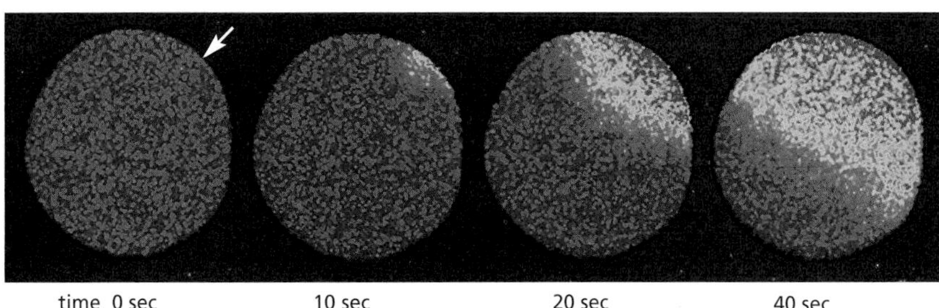

Figure 16–26 **Fertilization of an egg by a sperm triggers an increase in cytosolic Ca^{2+} in the egg.** This starfish egg was injected with a Ca^{2+}-sensitive fluorescent dye before it was fertilized. When a sperm penetrates the egg, a wave of cytosolic Ca^{2+} (*red*)—released from the endoplasmic reticulum—sweeps across the egg from the site of sperm entry (*arrow*). This Ca^{2+} wave provokes a change in the egg surface, preventing entry of other sperm, and it also initiates embryonic development. To catch this Ca^{2+} wave, go to Movie 16.4. (Courtesy of Stephen A. Stricker.)

The concentration of free Ca^{2+} in the cytosol of an unstimulated cell is extremely low (10^{-7} M) compared with its concentration in the extracellular fluid and in the endoplasmic reticulum. These differences are maintained by membrane-embedded pumps that actively pump Ca^{2+} out of the cytosol—either into the endoplasmic reticulum or across the plasma membrane and out of the cell. As a result, a steep electrochemical gradient of Ca^{2+} exists across both the endoplasmic reticulum membrane and the plasma membrane (discussed in Chapter 12). When a signal transiently opens Ca^{2+} channels in either of these membranes, Ca^{2+} rushes into the cytosol down its electrochemical gradient, triggering changes in Ca^{2+}-responsive proteins in the cytosol. The same pumps that normally operate to keep cytosolic Ca^{2+} concentrations low also help to terminate the Ca^{2+} signal.

The effects of Ca^{2+} in the cytosol are largely indirect, in that they are mediated through the interaction of Ca^{2+} with various kinds of *Ca^{2+}-responsive proteins*. The most widespread and common of these is **calmodulin**, which is present in the cytosol of all eucaryotic cells that have been examined, including those of plants, fungi, and protozoa. When Ca^{2+} binds to calmodulin, the protein undergoes a conformational change that enables it to wrap around a wide range of target proteins in the cell, altering their activities (Figure 16–27). One particularly important class of targets for calmodulin is the **Ca^{2+}/calmodulin-dependent protein kinases (CaM-kinases)**. When these kinases are activated by binding to calmodulin complexed with Ca^{2+}, they influence other processes in the cell by phosphorylating selected proteins. In the mammalian brain, for example, a neuron-specific CaM-kinase is abundant at synapses, where it is thought to play a part in some forms of learning and memory. This CaM-kinase is activated by the pulses of Ca^{2+} signals that occur during neural activity, and mutant mice that lack the kinase show a marked inability to remember where things are.

QUESTION 16–7

Why do you suppose cells have evolved intracellular Ca^{2+} stores for signaling even though there is abundant extracellular Ca^{2+}?

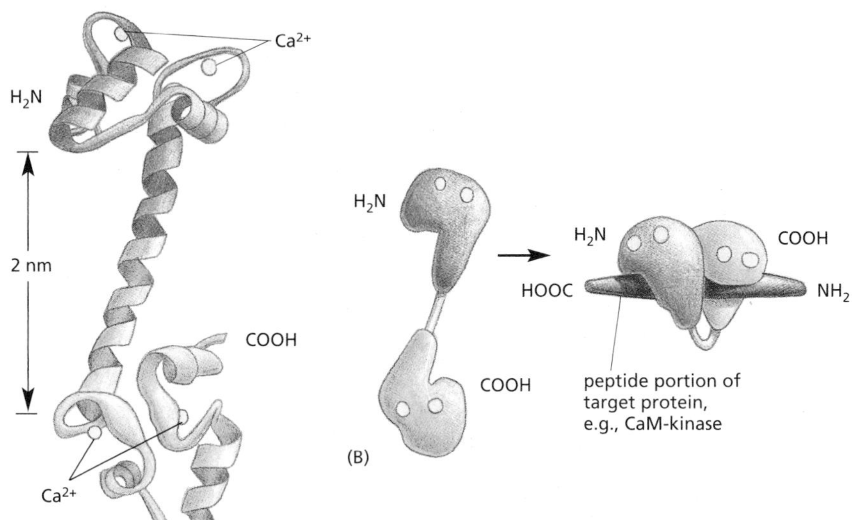

Figure 16–27 **Calcium binding changes the shape of the calmodulin protein.** (A) Calmodulin has a dumbbell shape, with two globular ends connected by a long, flexible α helix. Each end has two Ca^{2+}-binding domains. (B) Simplified representation of the structure, showing the conformational changes in Ca^{2+}/calmodulin that occur when it binds to a target protein. In this conformation, the α helix jackknifes to surround the target protein (Movie 16.5). (A, from Y.S. Babu et al., *Nature* 315:37–40, 1985. With permission from Macmillan Publishers Ltd; B, from W.E. Meador, A.R. Means, and F.A. Quiocho, *Science* 257:1251–1255, 1992, and M. Ikura et al., *Science* 256:632–638, 1992. With permission from the AAAS.)

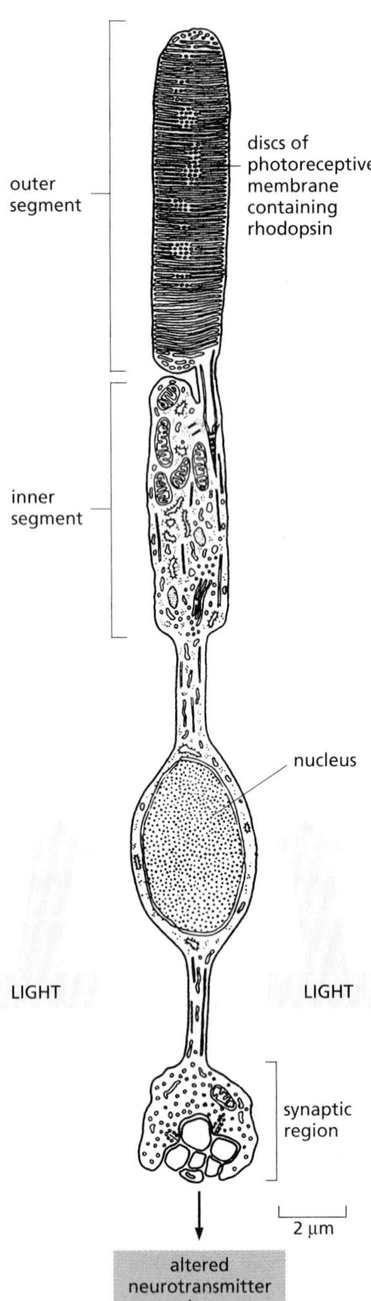

Figure 16–28 **A rod photoreceptor cell from the retina is exquisitely sensitive to light.** Drawing of a rod photoreceptor. The light-absorbing molecules of rhodopsin are embedded in many pancake-shaped vesicles (*discs*) of membrane inside the outer segment of the cell. Neurotransmitter is released from the opposite end of the cell to control firing of the retinal nerve cells that pass on the signal to the nerve cells that connect to the brain. When the rod cell is stimulated by light, a signal is relayed from the rhodopsin molecules in the discs, through the cytosol of the outer segment, to channels that allow positive ions to flow through the plasma membrane of the outer segment. These cation channels close in response to the signal, producing a change in the membrane potential of the rod cell. By mechanisms similar to those that control neurotransmitter release in ordinary nerve cells, the change in membrane potential alters the rate of neurotransmitter release from the synaptic region of the cell. (Adapted from T.L. Lentz, Cell Fine Structure. Philadelphia: Saunders, 1971. With permission from Elsevier.)

Intracellular Signaling Cascades Can Achieve Astonishing Speed, Sensitivity, and Adaptability

The steps in the *signaling cascades* associated with GPCRs take a long time to describe, but they often take only seconds to execute. Consider how quickly a thrill can make your heart beat faster (when adrenaline stimulates the GPCRs in your heart muscle cells, accelerating your heartbeat), or how fast the smell of food can make your mouth water (through the GPCRs for odors in your nose and the GPCRs for acetylcholine in salivary cells, which stimulate secretion). Among the fastest of all responses mediated by a GPCR, however, is the response of the eye to bright light: it takes only 20 msec for the most quickly responding photoreceptor cells of the retina (the cone photoreceptors, which are responsible for color vision in bright light) to produce their electrical response to a sudden flash of light.

This speed is achieved in spite of the necessity to relay the signal over several steps of an intracellular signaling cascade. But photoreceptors also provide a beautiful illustration of the positive advantages of signaling cascades: in particular, such cascades allow spectacular amplification of the incoming signal and also allow cells to adapt so as to be able to detect signals of widely varying intensity. The quantitative details have been most thoroughly analyzed for the rod photoreceptor cells in the eye, which are responsible for non-color vision in dim light (Figure 16–28). In this cell, light is sensed by *rhodopsin*, a G-protein–coupled light receptor. Light-activated rhodopsin activates a G protein called *transducin*. The activated α subunit of transducin then activates an intracellular signaling cascade that causes cation channels to close in the plasma membrane of the photoreceptor cell. This produces a change in the voltage across the cell membrane, which alters neurotransmitter release and ultimately leads to a nerve impulse being sent to the brain.

The signal is repeatedly amplified as it is relayed along this intracellular signaling pathway (Figure 16–29). When lighting conditions are dim, as on a moonless night, the amplification is enormous: as few as a dozen photons absorbed in the entire retina will cause a perceptible signal to be delivered to the brain. In bright sunlight, when photons flood through each photoreceptor cell at a rate of billions per second, the signaling cascade *adapts*, stepping down the amplification more than 10,000-fold so that the photoreceptor cells are not overwhelmed and can still register increases and decreases in the strong light. The adaptation depends on negative feedback: an intense response in the photoreceptor cell generates an intracellular signal (a change in Ca^{2+} concentration) that inhibits the enzymes responsible for signal amplification.

Adaptation frequently occurs in signaling pathways that respond to chemical signals; again, it allows cells to remain sensitive to changes of signal intensity over a wide range of background levels of stimulation. Adaptation, in other words, allows a cell to respond to both messages that are whispered and those that are shouted.

Taste and smell also depend on GPCRs. It seems likely that this mechanism of signal reception, invented early in the evolution of the eucaryotes, has its origins in the basic and universal need of cells to sense and respond to their environment. Of course, GPCRs are not the only receptors that activate intracellular signaling cascades. We now turn to another major class of cell-surface receptors—enzyme-coupled receptors—which play a key part in controlling cell numbers, cell differentiation, and cell movement in multicellular animals.

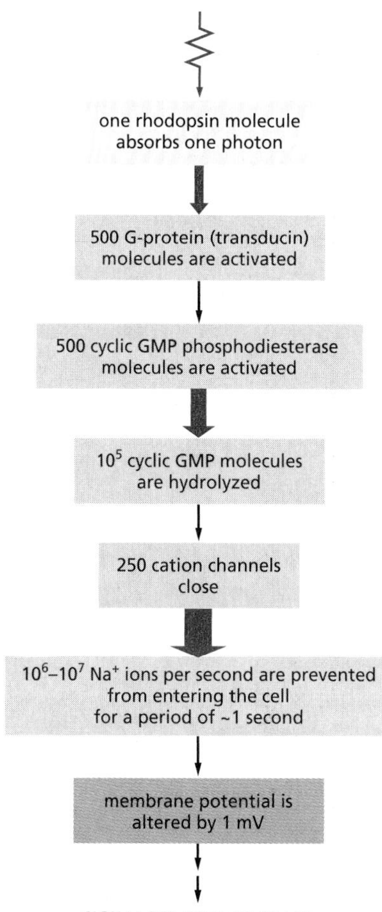
one rhodopsin molecule absorbs one photon
500 G-protein (transducin) molecules are activated
500 cyclic GMP phosphodiesterase molecules are activated
10^5 cyclic GMP molecules are hydrolyzed
250 cation channels close
10^6–10^7 Na^+ ions per second are prevented from entering the cell for a period of ~1 second
membrane potential is altered by 1 mV
SIGNAL RELAYED TO BRAIN

Figure 16–29 **The light-induced signaling cascade in rod photoreceptor cells greatly amplifies the light signal.** When rod photoreceptors are adapted for dim light, signal amplification is enormous. The intracellular signaling pathway from the G protein transducin uses components that differ from the ones previously described. The cascade functions as follows. In the absence of a light signal, the small messenger molecule cyclic GMP is continuously produced by guanylyl cyclase (see Figure 16–11C) in the photoreceptor cell. The cyclic GMP then binds to cation channels in the photoreceptor cell plasma membrane, keeping them open. Activation of rhodopsin by light results in the activation of transducin α subunits. These turn on an enzyme called cyclic GMP phosphodiesterase, which breaks down cyclic GMP to GMP (much as cyclic AMP phosphodiestease breaks down cyclic AMP; see Figure 16–21). The sharp fall in the intracellular concentration of cyclic GMP causes the bound cyclic GMP to dissociate from the cation channels, which therefore close. The *red* arrows indicate the steps at which amplification occurs, with the thickness of the arrow roughly indicating the magnitude of the amplification.

ENZYME-COUPLED RECEPTORS

Like GPCRs, **enzyme-coupled receptors** are transmembrane proteins that display their ligand-binding domains on the outer surface of the plasma membrane. Instead of associating with a G protein, however, the cytoplasmic domain of the receptor either acts as an enzyme itself or forms a complex with another protein that acts as an enzyme. Enzyme-coupled receptors (see Figure 16–15C) were discovered through their role in responses to extracellular signal proteins ('growth factors') that regulate the growth, proliferation, differentiation, and survival of cells in animal tissues (see Table 16–1, p. 535, for examples). Most of these signal proteins function as local mediators and can act at very low concentrations (about 10^{-9} to 10^{-11} M). Responses to them are typically slow (on the order of hours), and they require many intracellular transduction steps that eventually lead to changes in gene expression.

Enzyme-coupled receptors, however, can also mediate direct, rapid reconfigurations of the cytoskeleton, controlling the way a cell changes its shape and moves. The extracellular signals for these architectural alterations are often not diffusible signal proteins, but proteins attached to the surfaces over which a cell is crawling. Disorders of cell growth, proliferation, differentiation, survival, and migration are fundamental to cancer, and abnormalities in signaling via enzyme-coupled receptors have a major role in the development of this class of diseases.

The largest class of enzyme-coupled receptors is made up of those with a cytoplasmic domain that functions as a tyrosine protein kinase, phosphorylating specific tyrosines on selected intracellular proteins. Such receptors are called **receptor tyrosine kinases** (**RTKs**), and we will focus on these receptors here. Note that all of the other protein kinases we have discussed so far—including PKA, PKC, and CaM-kinases—are serine/threonine kinases.

Activated RTKs Recruit a Complex of Intracellular Signaling Proteins

To do its job as a signal transducer, an enzyme-coupled receptor has to switch on the enzyme activity of its intracellular domain (or of an associated enzyme) when an external signal molecule binds to its extracellular domain. Unlike the seven-pass GPCRs, enzyme-coupled receptor proteins usually have only one transmembrane segment, which is thought to span the lipid bilayer as a single α helix. Because a single α helix is poorly suited to transmit a conformational change across the bilayer, enzyme-coupled receptors have a different strategy for transducing the extracellular signal. In many cases, the binding of a signal molecule

QUESTION 16–8

One important feature of any intracellular signaling pathway is its ability to be turned off. Consider the pathway shown in Figure 16–29. Where would off switches be required? Which ones do you suppose would be the most important?

causes two receptor molecules to come together in the membrane, forming a dimer. Contact between the two adjacent intracellular receptor tails activates their kinase function, with the result that each receptor phosphorylates the other. In the case of RTKs, the phosphorylations occur on specific tyrosines located on the cytosolic tail of the receptors.

Tyrosine phosphorylation then triggers the assembly of an elaborate intracellular signaling complex on the receptor tails. The newly phosphorylated tyrosines serve as binding sites for a whole zoo of intracellular signaling proteins—perhaps as many as 10 or 20 different molecules (Figure 16–30). Some of these proteins become phosphorylated and activated on binding to the receptor, and they then propagate the signal; others function solely as *adaptors,* which couple the receptor to other signaling proteins, thereby helping to build the active signaling complex. All of these intracellular signaling proteins possess a specialized interaction domain, capable of recognizing specific phoshorylated tyrosines. Similar protein domains allow intracellular signaling proteins to recognize the phosphorylated lipids that are produced in a membrane in response to certain signals, as we discuss later.

While they last, the protein complexes assembled on the cytosolic tails of the RTKs can transmit the signal along several routes simultaneously to many destinations inside the cell, thus activating and coordinating the numerous biochemical changes that are required to trigger a complex response, such as cell proliferation. To help terminate the response, the tyrosine phosphorylations are reversed by *protein tyrosine phosphatases*, which remove the phosphates that were added to the tyrosines of the RTKs and other signaling proteins in response to the extracellular signal. In some cases, activated RTKs (and GPCRs) are inactivated in a more brutal way: they are dragged into the interior of the cell by endocytosis and then destroyed by digestion in lysosomes.

Different RTKs recruit different collections of intracellular signaling proteins, producing different effects; however, certain components seem to be used by most RTKs. These include, for example, a phospholipase C that functions in the same way as the phospholipase C activated by GPCRs to activate the inositol phospholipid signaling pathway discussed earlier (see Figure 16–25). Another intracellular signaling protein that is activated by almost all RTKs is a small GTP-binding protein called Ras.

Most RTKs Activate the Monomeric GTPase Ras

As we have seen, activated RTKs recruit many kinds of intracellular signaling proteins and form large signaling complexes. One of the key players in these signaling complexes is **Ras**—a small GTP-binding protein that is bound by a lipid tail to the cytoplasmic face of the plasma membrane. Virtually all RTKs activate Ras, including platelet-derived growth factor

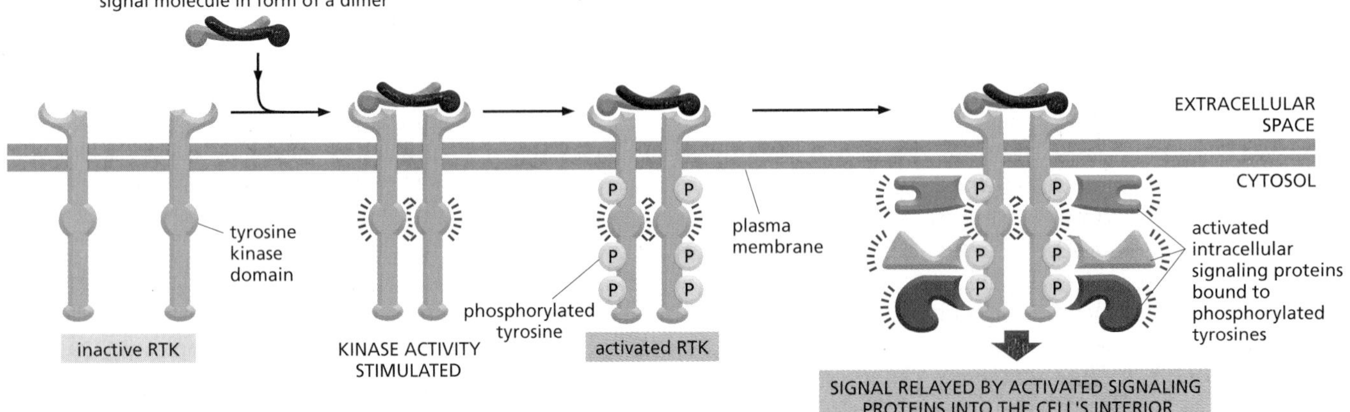

Figure 16–30 **Activation of an RTK stimulates the assembly of an intracellular signaling complex.** Typically, the binding of a signal molecule to the extracellular domain of an RTK causes two receptor molecules to associate into a dimer. The signal molecule shown here is itself a dimer and thus can physically cross-link two receptor molecules. In other cases, binding of the signal molecule changes the conformation of the receptor molecules in such a way that they dimerize. Dimer formation brings the kinase domains of each intracellular receptor tail into contact with the other; this activates the kinases and enables them to phosphorylate the adjacent tail on several tyrosines. Each phosphorylated tyrosine serves as a specific binding site for a different intracellular signaling protein, which then helps relay the signal to the cell's interior; these proteins possess a specialized interaction domain—in this case, a module called an SH2 domain—that recognizes and binds to specific phosphorylated tyrosines on an activated receptor or on another intracellular signaling protein.

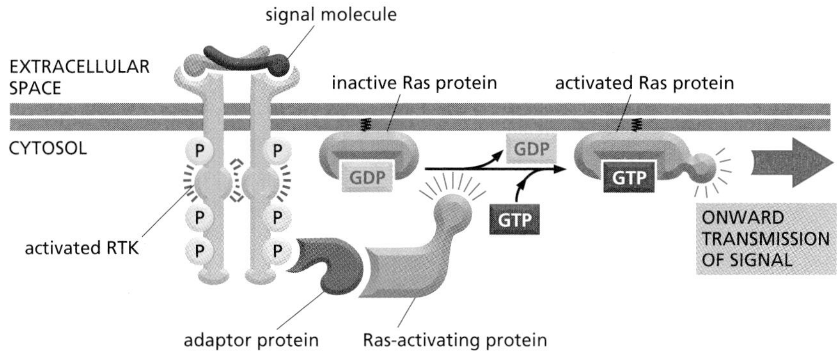

Figure 16–31 **RTKs activate Ras.** An adaptor protein docks on a particular phosphotyrosine on the activated receptor (the other signaling proteins that are shown bound to the receptor in Figure 16–30 are omitted for simplicity). The adaptor recruits a Ras-activating protein that stimulates Ras to exchange its bound GDP for GTP. The activated Ras protein can now stimulate several downstream signaling pathways, one of which is shown in Figure 16–32. Note that the Ras protein contains a covalently attached lipid group (*black*) that helps anchor the protein to the plasma membrane.

(PDGF) receptors, which mediate cell proliferation in wound healing, and nerve growth factor (NGF) receptors, which prevent certain neurons from dying in the developing nervous system.

The Ras protein is a member of a large family of small GTP-binding proteins, often called **monomeric GTPases** to distinguish them from the trimeric G proteins that we encountered earlier. Ras resembles the α subunit of a G protein and functions as a molecular switch in much the same way. It cycles between two distinct conformational states—active when GTP is bound and inactive when GDP is bound (see Figure 16–14B). Interaction with an activated signaling protein encourages Ras to exchange its GDP for GTP, thus switching Ras to its activated state (Figure 16–31). After a delay, Ras switches itself off again by hydrolyzing its bound GTP to GDP (Movie 16.6).

In its active state, Ras promotes the activation of a phosphorylation cascade in which a series of serine/threonine protein kinases phosphorylate and activate one another in sequence, like an intracellular game of dominoes (Figure 16–32). This relay system, which carries the signal from the plasma membrane to the nucleus, includes a three-kinase protein module called the **MAP-kinase signaling module**, in honor of the final kinase in the chain, the mitogen-activated protein kinase or **MAP kinase**. (*Mitogens* are extracellular signal molecules that stimulate cell proliferation.) In this pathway, MAP kinase is phosphorylated and activated by an enzyme called, logically enough, *MAP kinase kinase*. And this protein is itself switched on by a *MAP kinase kinase kinase* (which is activated by

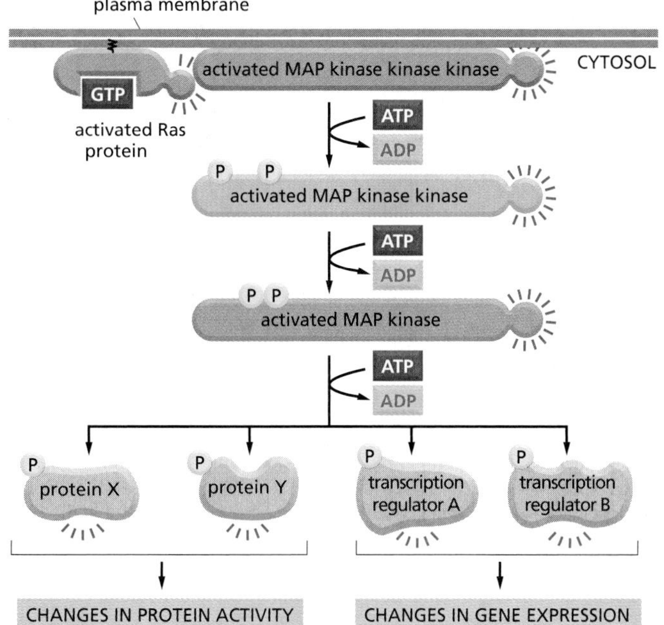

Figure 16–32 **Ras activates a MAP-kinase signaling module.** A Ras protein activated by the process shown in Figure 16–31 activates a three-kinase signaling module, which relays the signal. The final kinase in the module, MAP kinase, phosphorylates various downstream signaling or effector proteins. These proteins can include other protein kinases and, most importantly, transcription regulators that control gene expression. The resulting changes in gene expression and protein activity lead to complex changes in cell behaviors such as proliferation and differentiation.

Ras). At the end of the MAP-kinase cascade, MAP kinase phosphorylates various effector proteins, including certain transcription regulators, altering their ability to control gene transcription. This change in the pattern of gene expression may stimulate cell proliferation, promote cell survival, or induce cell differentiation: the precise outcome will depend on which other genes are active in the cell and what other signals the cell receives. How biologists unravel such complex signaling pathways is discussed in How We Know, pp. 560–561.

The importance of Ras has been demonstrated in various ways. If Ras is inhibited by an intracellular injection of Ras-inactivating antibodies, for example, a cell may no longer respond to some of the extracellular signals that it would normally respond to. Conversely, if Ras activity is permanently switched on in some cell types, the cells act as if they are being bombarded continuously by proliferation-stimulating extracellular mitogens (discussed in Chapter 18). Before Ras was discovered in normal cells, a mutant form of it was found in human cancer cells; the mutation inactivated the GTPase activity of Ras, so that the protein could not shut itself off, promoting uncontrolled cell proliferation and the development of cancer. About 30% of human cancers contain such activating mutations in *Ras* genes, and many of the cancers that do not produce mutant Ras proteins have mutations in genes whose products lie in the same signaling pathway as Ras. Many of the genes that encode these intracellular signaling proteins were initially identified in the hunt for cancer-promoting *oncogenes*, discussed in Chapter 20.

QUESTION 16–9

Would you expect to activate GPCRs and RTKs by exposing cells to antibodies that bind to the respective proteins? (Hint: review Panel 4–3, on pp. 144–145, regarding the properties of antibody molecules.)

RTKs Activate PI 3-Kinase to Produce Lipid Docking Sites in the Plasma Membrane

Many of the extracellular signal proteins that stimulate animal cells to survive, grow, and proliferate act through RTKs. These include signal proteins belonging to the insulin-like growth factor (IGF) family. One crucially important signaling pathway that RTKs activate to promote cell growth and survival relies on the enzyme **phosphoinositide 3-kinase (PI 3-kinase)**, which phosphorylates inositol phospholipids in the plasma membrane. These phosphorylated lipids become docking sites for specific intracellular signaling proteins, which relocate from the cytosol to the plasma membrane, where they can activate one another.

One of the most important of these relocated signaling proteins is the serine/threonine protein kinase *Akt*, which is also called *protein kinase B*, or *PKB* (Figure 16–33). Akt promotes the growth and survival of many cell types, often by inactivating the signaling proteins it phosphorylates. For example, Akt phosphorylates and inactivates a cytosolic protein called

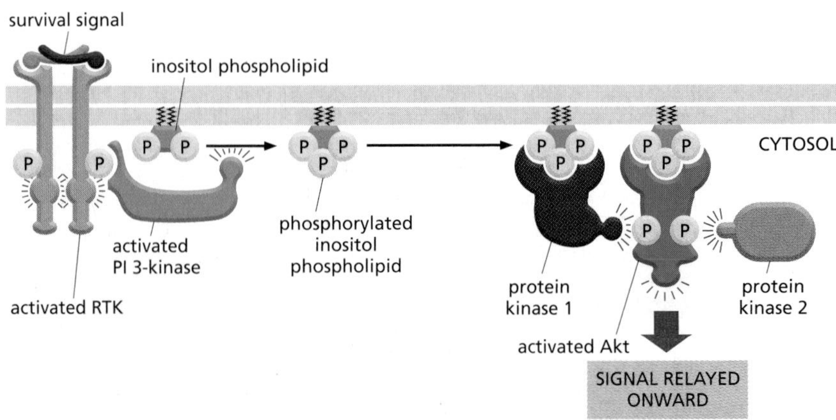

Figure 16–33 **RTKs can activate the PI-3-kinase–Akt signaling pathway.** An extracellular survival signal, such as IGF, activates an RTK, which recruits and activates PI 3-kinase. PI 3-kinase then phosphorylates a membrane-associated inositol phospholipid. The resulting phosphorylated inositol phospholipid then attracts intracellular signaling proteins that have a special domain that recognizes it. One of these signaling proteins, Akt, is a protein kinase that is activated at the membrane by phosphorylation mediated by two other protein kinases (here called protein kinases 1 and 2); protein kinase 1 is also recruited by the phosphorylated lipid docking sites. Once activated, Akt is released from the plasma membrane and phosphorylates various downstream proteins on specific serines and threonines (not shown).

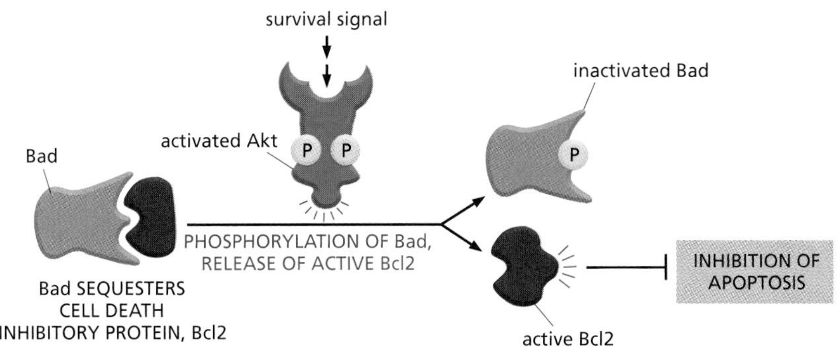

Figure 16–34 **Activated Akt promotes cell survival.** One way it does so is by phosphorylating and inactivating a protein called Bad. In its unphosphorylated state, Bad promotes apoptosis (a form of cell death) by binding to and inhibiting a protein, called Bcl2, which otherwise suppresses apoptosis. When Bad is phosphorylated by Akt, Bad releases Bcl2, which now blocks apoptosis, thereby promoting cell survival.

Bad. In its active state, Bad encourages the cell to kill itself by indirectly activating a cell-suicide program called apoptosis (discussed in Chapter 18). Phosphorylation by Akt thus promotes cell survival by inactivating a protein that otherwise promotes cell death (Figure 16–34).

In addition to promoting cell survival, the *PI-3-kinase–Akt signaling pathway* also stimulates cells to grow in size. It does so by indirectly activating a large serine/threonine kinase called *Tor*. Tor stimulates cells to grow both by enhancing protein synthesis and by inhibiting protein degradation (Figure 16–35). The anticancer drug rapamycin works by inactivating Tor—reinforcing the importance of the PI-3-kinase–Akt signaling pathway in regulating cell growth and survival.

Some Receptors Activate a Fast Track to the Nucleus

Not all enzyme-coupled receptors trigger complex signaling cascades that require the cooperation of a sequence of protein kinases to carry a message to the nucleus. Some receptors use a more direct route to control gene expression.

A few hormones and many local mediators called **cytokines** bind to receptors that can activate transcription regulators that are held in a latent, inactive state near the plasma membrane. Once turned on, these regulatory proteins—called STATs (for *s*ignal *t*ransducers and *a*ctivators of *t*ranscription)—head straight for the nucleus, where they stimulate the transcription of specific genes. This direct signaling pathway is used, for example, by *interferons*, which are cytokines produced by infected cells that instruct other cells to produce proteins that make them more resistant to viral infection.

Unlike the RTKs that stimulate elaborate signaling cascades, the cytokine and hormone receptors that rely on STATs have no intrinsic enzyme activity. Instead, they associate with cytoplasmic tyrosine kinases called *JAKs*, which are activated when a cytokine or hormone binds to the receptor. Once activated, the JAKs phosphorylate and activate STATs, which then migrate to the nucleus, where they stimulate the transcription of specific target genes. For example, the hormone prolactin, which stimulates breast cells to make milk, acts by binding to a receptor that is associ-

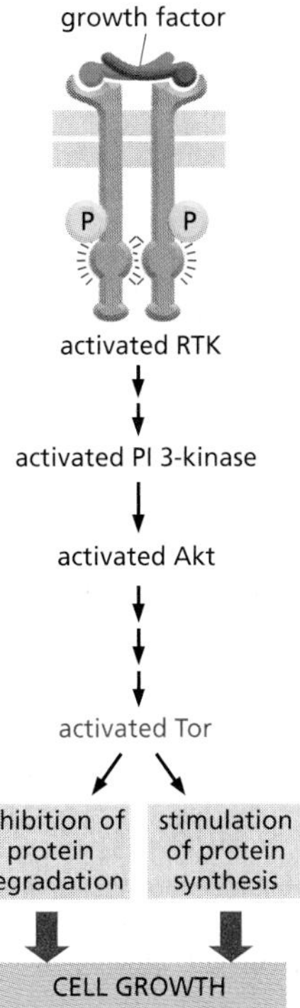

Figure 16–35 **Akt stimulates cells to grow in size by activating Tor.** The binding of a growth factor to an RTK activates the PI-3-kinase–Akt signaling pathway (as shown in Figure 16–33). Akt then indirectly activates Tor (by phosphorylating and inhibiting a protein that helps to keep Tor shut down; not shown). Tor, itself a serine/threonine kinase, stimulates protein synthesis and inhibits protein degradation (by phosphorylating key proteins in these processes; not shown). The anticancer drug rapamycin slows cell growth and proliferation by inhibiting Tor. In fact, the Tor protein derives its name from the fact that it is a *t*arget *o*f *r*apamycin.

UNTANGLING CELL SIGNALING PATHWAYS

Intracellular signaling pathways are never mapped out in a single experiment. Instead, investigators figure out, piece by piece, how all the links in the chain fit together—and how each contributes to the cell's response to an extracellular signal such as the hormone insulin. The process involves breaking down the broad questions about how a cell responds to the signal into smaller, more manageable questions. Which protein is the insulin receptor? Which intracellular proteins become activated when insulin is present? With which proteins do these activated proteins interact? How does one protein activate another? Here, we discuss the kinds of experiments that provide answers to such riddles.

Stimulated phosphorylation

When cells are exposed to an extracellular signal molecule, one result is that a number of proteins become phosphorylated. Some of these will be the intracellular signaling proteins responsible for propagating the message onward; others will be effector proteins responsible for the cell's response. To determine which molecules have been activated by phosphorylation, researchers break open the cells, separate the proteins by size on a gel (discussed in Chapter 4, Panels 4–4 to 4–6, pp. 164–167), and then use antibodies to detect phosphorylated proteins.

Another common way to visualize newly phosphorylated proteins involves supplying cells with a radioactive version of ATP when they are exposed to an extracellular signal molecule. Protein kinases activated by the signal will transfer radioactive phosphate from the labeled ATP to their protein substrates. Again, the cell proteins are separated on a gel, but the radiolabeled proteins can now be detected by exposing the gel to an X-ray film.

Close encounters

Once the activated proteins have been identified, one can determine which proteins interact with them. To identify interacting proteins, scientists often make use of *co-immunoprecipitation*. In this technique, antibodies are used to latch onto a specific protein, dragging it out of solution and down to the bottom of a test tube (discussed in Chapter 4, Panel 4–3, pp. 144–145). If the captured protein happens to be bound to other proteins, these will be dragged down as well. In this way, researchers can identify which proteins interact when an extracellular signal molecule stimulates cells.

Once two proteins are known to bind to each other, the experimenter can proceed to pinpoint which parts of the proteins are required for the interaction. This often involves using recombinant DNA technology to construct a set of mutant proteins, each of which differs slightly from the normal one. To determine which phosphorylated tyrosine on a receptor tyrosine kinase (RTK) a certain intracellular signaling protein binds to, for example, a series of mutant receptors is used, each missing a different tyrosine from its cytoplasmic domain (Figure 16–36). In this way, the specific tyrosines required for binding can be determined. Similarly, one can determine whether this tyrosine docking site is required for the receptor to transmit a signal to the cell.

Jamming the pathway

Ultimately, one wants to assess how important a particular protein is for a signaling process. A first test involves using recombinant DNA technology to introduce into cells a gene encoding a constantly active form of the protein, to see if this mimics the effect of the extracellular signal. Take Ras, for example. The form of Ras involved in human cancers is constantly active because it has lost its ability to hydrolyze the GTP that keeps it switched on. This continuously active form of Ras can stimulate some cells to proliferate even in the absence of cell-division-inducing mitogens and thereby promote the development of cancer (Figure 16–37).

The ultimate test of the importance of an intracellular protein in a signaling pathway is to inactivate the protein or its gene and see whether the signaling pathway is affected. In the case of Ras, for example, one can introduce into cells a 'dominant-negative' mutant form of Ras. This disabled form of Ras clings too tightly to GDP, and therefore cannot be activated. Because it can still bind to other signaling partners in the pathway, it jams the pathway, preventing normal Ras molecules from doing their job (which is why it is called 'dominant'). Such stalled cells do not proliferate in response to mitogens, indicating the importance of normal Ras signaling in the proliferative response. Another way to inactivate a protein in a cell is to use a small interfering RNA (siRNA) to degrade the mRNA that encodes the protein or to prevent the translation of the mRNA into protein (see Figure 8–27).

Ordering the pathway

Most signaling pathways take decades to untangle. Although insulin was first isolated from dog pancreas in the early 1920s, the molecular chain of events that links the binding of insulin to its receptor with the activation of the transporter proteins that take up glucose is still not completely understood.

One powerful strategy that scientists use to identify proteins that participate in cell signaling involves screening a large number of animals. This usually means treating tens of thousands of fruit flies or nematode worms with a mutagen and then looking for mutants in which a sig-

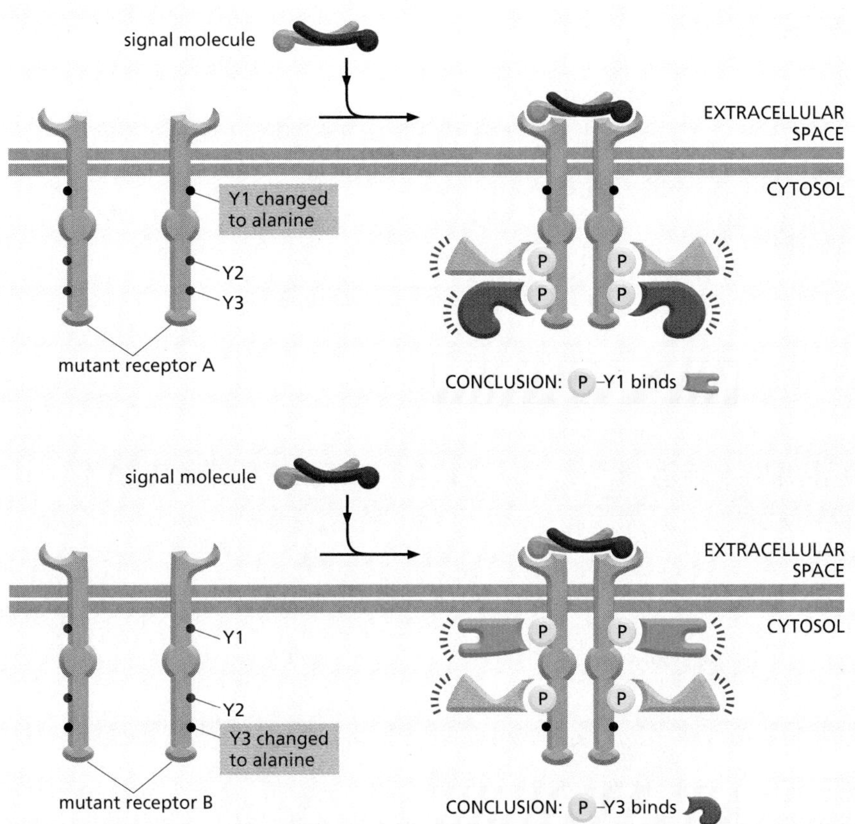

Figure 16–36 **Mutant proteins can help to determine exactly where an intracellular signaling molecule binds.** As shown in Figure 16–30, on binding their extracellular signal molecule, a pair of RTKs come together and phosphorylate specific tyrosines on each other's cytoplasmic tails. These phosphorylated tyrosines attract different intracellular signaling proteins, which then become activated and pass on the signal. To determine which tyrosine binds to a specific intracellular signaling protein, a series of mutant receptors are constructed. In the mutants shown, single tyrosines (Y1 or Y3) have been replaced by an alanine. As a result, the mutant receptors no longer bind to one of the intracellular signaling proteins shown in Figure 16–30. The effect on the cell's response to the signal can then be determined. It is important that the mutant receptor be tested in a cell that does not have its own normal receptors for the signal molecule.

naling pathway is not functioning properly. Flies and worms are useful because they reproduce rapidly and can be maintained in vast numbers in the laboratory. By examining enough mutant animals, many of the genes that encode the proteins involved in a signaling pathway can be identified—including receptors, protein kinases, transcription regulators, and so on.

Such genetic screens can also help determine the order in which intracellular signaling proteins act in a pathway. Suppose that a genetic screen reveals two new proteins, X and Y, in the Ras signaling pathway. As a working hypothesis, imagine that the receptor activates protein X, which then activates Ras, which activates protein Y (Figure 16–38A). To test that hypothesis, one could introduce an inactive, mutant form of X or Y into the cell, and then ask whether the mutant can be "rescued" by the addition of a continuously active form of Ras. If the constantly active Ras overcomes the blockage created by the mutant protein, Ras must operate downstream of that protein in the pathway (Figure 16–38B). If Ras operates upstream of the protein in the pathway, a constantly active Ras would be unable to transmit a signal past the obstruction caused by the disabled protein (Figure 16–38C).

Used together, such biochemical and genetic techniques allow even the most complex intracellular signaling pathways to be dissected.

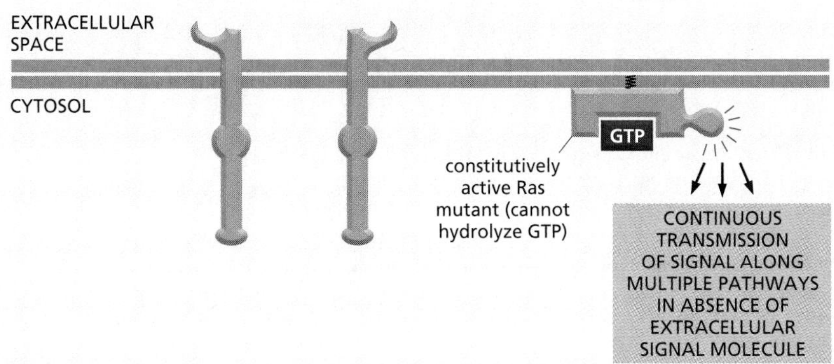

Figure 16–37 **A constitutively active form of Ras transmits a signal even in the absence of an extracellular signal molecule.** As shown in Figure 16–31, the normal Ras protein is activated in response to certain extracellular signals. The overactive mutant form of Ras shown here has lost the ability to hydrolyze GTP. Thus, it cannot shut down its activity and, as a result, is constantly (constitutively) active.

(A) NORMAL SIGNALING REQUIRES Ras AND TWO PROTEINS X AND Y

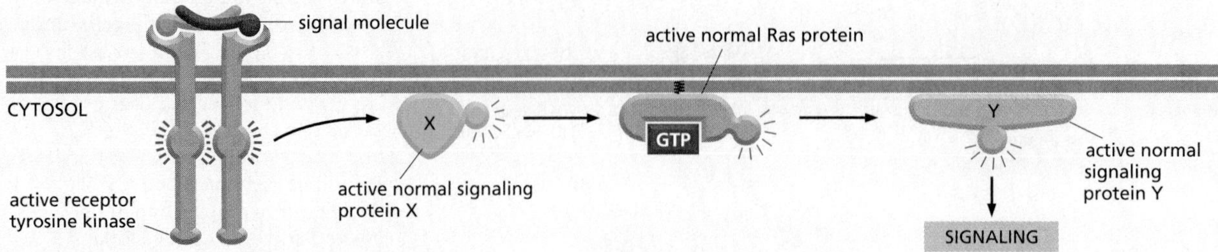

(B) MUTATION IN PROTEIN X BLOCKS SIGNALING UPSTREAM OF RAS

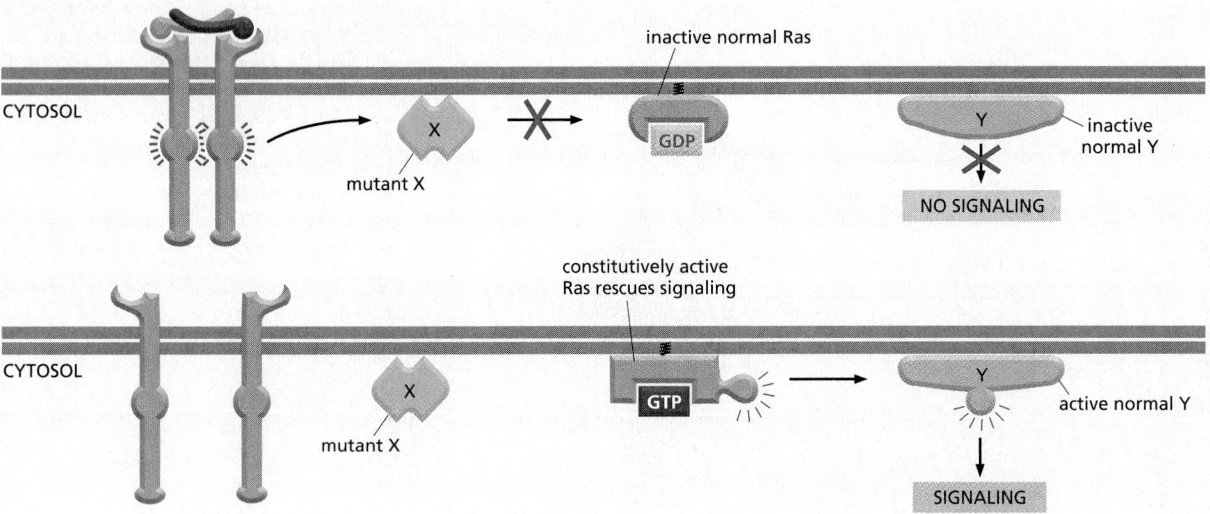

(C) MUTATION IN PROTEIN Y BLOCKS SIGNALING DOWNSTREAM OF RAS

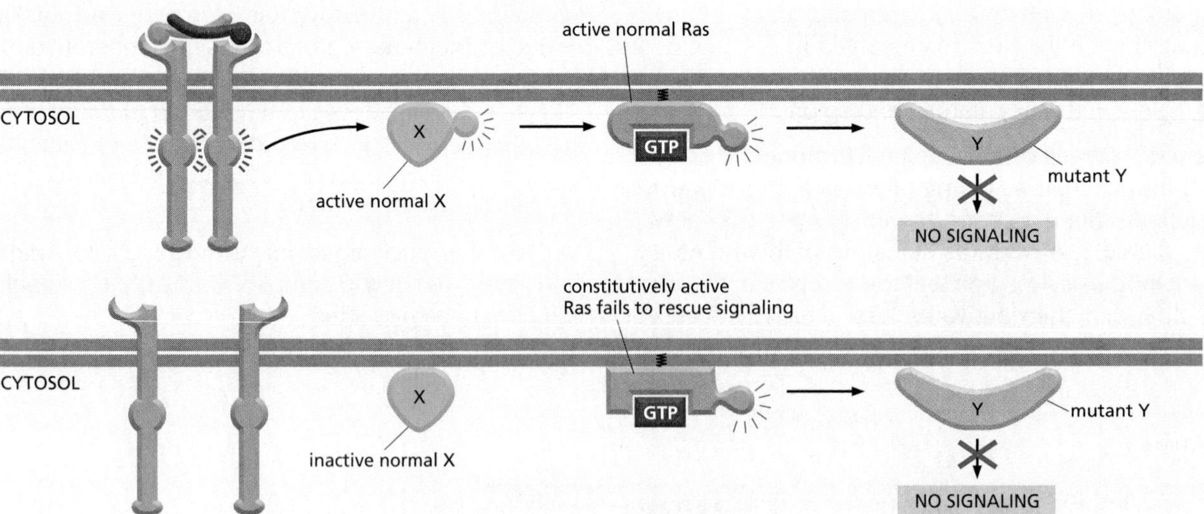

Figure 16–38 **Genetic analysis reveals the order in which intracellular signaling proteins act in a pathway.** A signaling pathway can be inactivated by mutations in any one of its components. Here, we show how a hypothetical Ras signaling pathway (A) can be shut down by a mutation in either protein X (B) or protein Y (C). Addition of a constitutively active form of Ras to these cells can help to unravel where in the pathway the mutant proteins lie. Adding a continuously active Ras to cells with a mutation in X restores activity to the pathway, allowing the signal to be transmitted even in the absence of an extracellular signal molecule (B). An overactive Ras can rescue these cells because Ras lies downstream of the mutant protein X that is jamming the pathway. By contrast, adding a continuously active Ras to cells with a mutation in protein Y fails to rescue, as Ras lies upstream of the blockage (C).

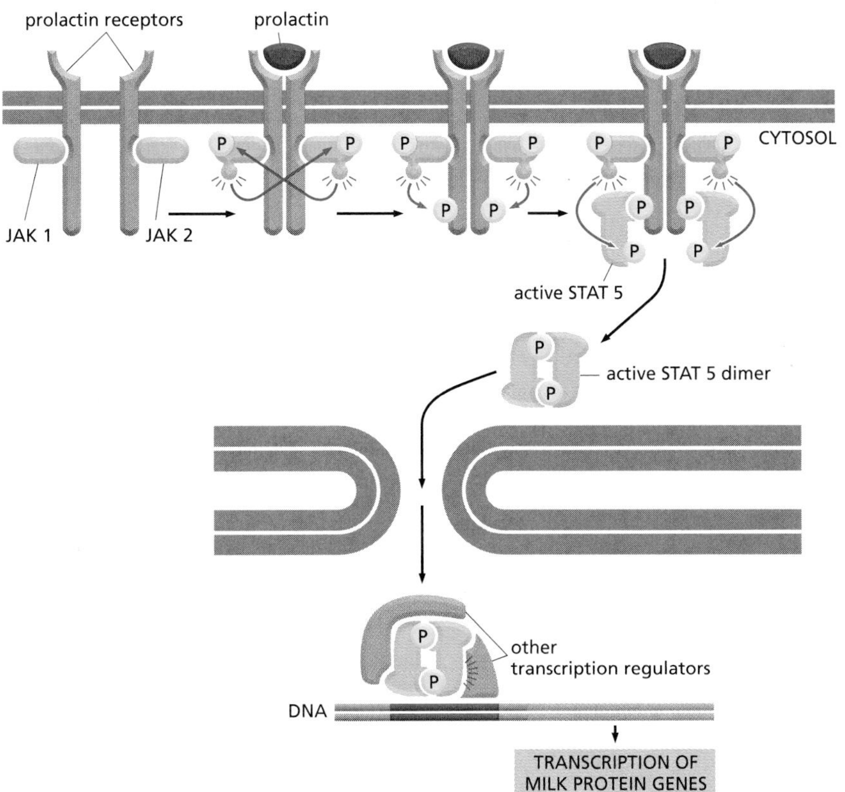

Figure 16–39 **The hormone prolactin stimulates milk production by activating a JAK–STAT signaling pathway.** Binding of prolactin to its enzyme-coupled receptors causes the associated tyrosine kinases (in this case, JAK1 and JAK2) to phosphorylate and activate each other. The activated JAKs then phosphorylate the receptor proteins. Transcription regulators called STATs (in this case STAT5), present in the cytosol, then bind to the phosphotyrosines on the receptor, and the JAKs phosphorylate and activate these proteins too. The activated STATs then dissociate from the receptor proteins, dimerize, and migrate to the nucleus. There, with the help of other transcription regulators, they activate the transcription of genes that encode milk proteins.

ated with a specific pair of JAKs. These JAKs activate a particular STAT that then turns on the transcription of the genes encoding milk proteins (Figure 16–39).

Different cytokine and hormone receptors evoke different cell responses by activating different STATs. Like any pathway that is turned on by phosphorylation, these signals are shut off by protein phosphatases that remove the phosphate groups from the activated signaling proteins.

An even more direct signaling pathway is used by the receptor protein Notch, which controls, among other things, the development of neural cells in *Drosophila* (see Figure 16–4). In this pathway, the receptor itself acts as a transcription regulator. When activated by the binding of Delta, which is attached to a neighboring cell, the Notch receptor is cleaved. This cleavage releases the cytosolic tail of the receptor, which heads to the nucleus where it helps to activate the appropriate set of Notch-responsive genes (Figure 16–40). This signaling pathway utilizes

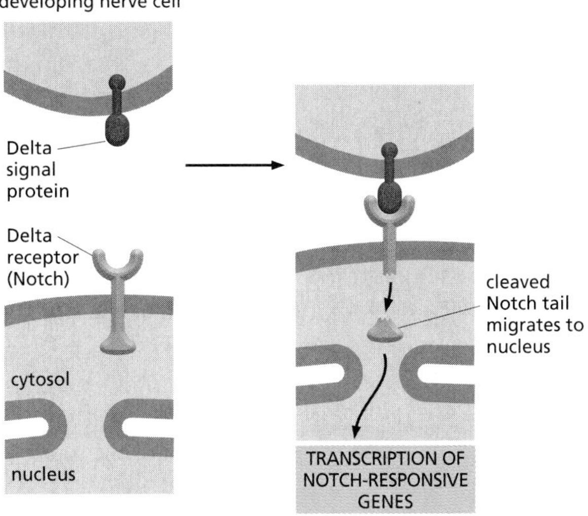

Figure 16–40 **The Notch receptor itself is a transcription regulator.** When the membrane-bound signal protein Delta binds to its receptor, Notch, on a neighboring cell, the receptor is cleaved. The released part of the cytosolic tail of Notch migrates to the nucleus, where it activates Notch-responsive genes. One consequence of this signaling process is shown in Figure 16–4).

the simplest and most direct way known to transmit a signal from a cell-surface receptor to the nucleus.

Multicellularity and Cell Communication Evolved Independently in Plants and Animals

Plants and animals have been evolving independently for more than a billion years, the last common ancestor being a single-celled eucaryote that most likely lived on its own. Because these kingdoms diverged so long ago—when it was still "every cell for itself"—each has evolved its own molecular solutions to the complex problem of becoming multicellular. Thus, the mechanisms for cell–cell communication in plants and animals evolved separately and might be expected to be quite different. At the same time, however, plants and animals started with a common set of eucaryotic genes—including some used by single-celled organisms to communicate among themselves—and so their signaling systems should show some similarities.

Like animals, plants make extensive use of transmembrane cell-surface receptors—especially enzyme-coupled receptors. The spindly weed *Arabidopsis thaliana* (see Figure 1–33) has hundreds of genes encoding receptor serine/threonine kinases. These are, however, structurally distinct from the receptor serine/threonine kinases found in animal cells (which we do not discuss in this chapter). The plant receptors are thought to play an important part in a large variety of cell signaling processes, including those governing plant growth, development, and disease resistance. In contrast to animal cells, plant cells seem not to use RTKs, steroid-hormone-type nuclear receptors, or cyclic AMP, and they seem to use few GPCRs.

One of the best-studied signaling systems in plants mediates the response of cells to ethylene—a gaseous hormone that regulates a diverse array of developmental processes, including seed germination and fruit ripening. Tomato growers use ethylene to ripen their fruit, even after it has been picked. The ethylene receptors do not belong to any of the classes of receptor proteins that we have discussed so far. They are dimeric transmembrane proteins, and, surprisingly, it is the empty receptor that is active. In the absence of ethylene, the empty receptor activates a protein kinase that ultimately shuts off the ethylene-responsive genes in the nucleus. When ethylene is present, the receptor and kinase are inactive, and the ethylene-responsive genes are transcribed (Figure 16–41). This strategy, whereby signals act to relieve transcriptional inhibition, is commonly used in plants.

Protein Kinase Networks Integrate Information to Control Complex Cell Behaviors

In this chapter, we have outlined several pathways for conveying a signal from the cell surface to the cell interior. Figure 16–42 compares five of these pathways: the routes from GPCRs via adenylyl cyclase and via phospholipase C, and the routes from RTKs via phospholipase C, via Ras, and via PI 3-kinase. Each pathway differs from the others, yet they use some common components to transmit their signals. Because all these pathways eventually activate protein kinases, it seems that each is capable in principle of regulating practically any process in the cell.

In fact, the complexity of cell signaling is much greater than we have described. First, we have not discussed many of the intracellular signaling pathways available to cells, even though many of these are critical for proper development and are deranged in cancer cells (see Figure 20–49).

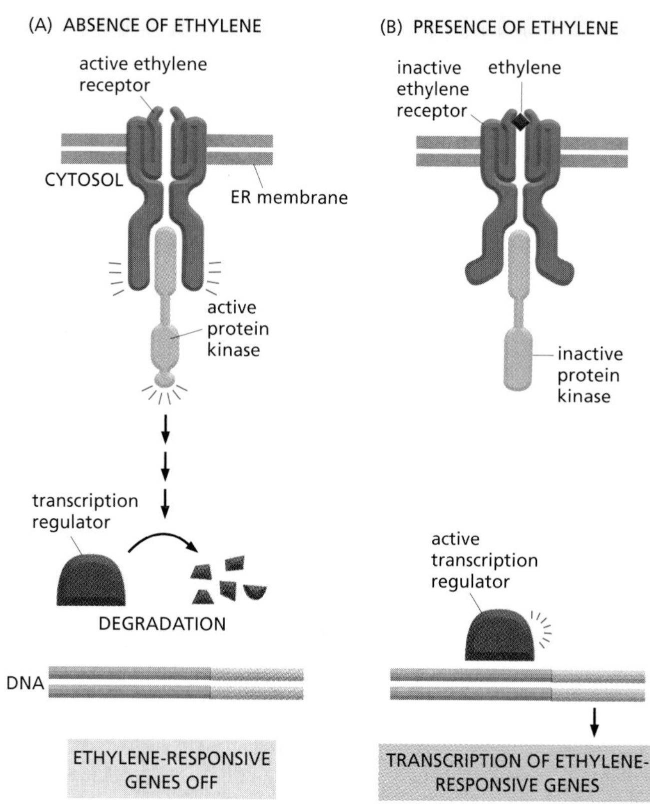

Figure 16–41 **The ethylene signaling pathway turns on genes by relieving inhibition.** (A) In the absence of ethylene, the receptor directly activates a protein kinase, which then promotes the destruction of the transcription regulator that switches on ethylene-responsive genes. As a result, the genes remain turned off. (B) In the presence of ethylene, the receptor and kinase are both inactive, and the transcription regulator remains intact and stimulates the transcription of the ethylene-responsive genes. The kinase that ethylene receptors interact with is a serine/threonine kinase that is closely related to the MAP kinase kinase kinase found in animal cells (see Figure 16–32).

Perhaps more importantly, the major signaling pathways we have discussed interact in ways that we have not described. They are connected by interactions of many sorts, but the most extensive links are those mediated by the protein kinases present in each of the pathways. These kinases often phosphorylate, and hence regulate, components in other signaling pathways, in addition to components in the pathway to which

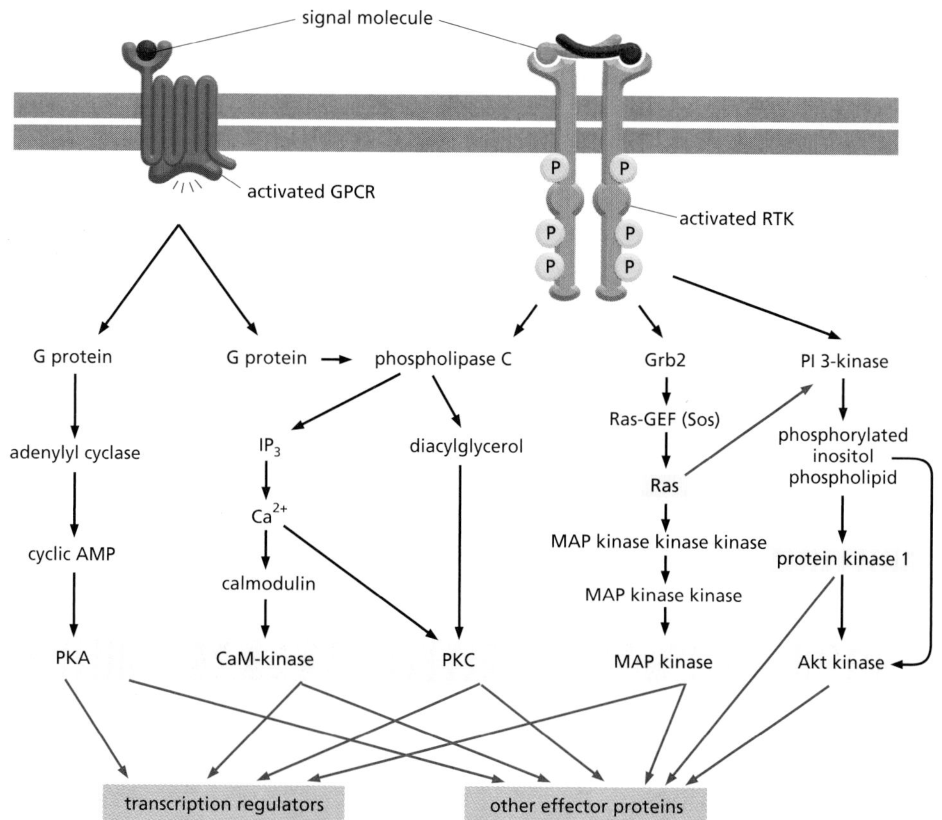

Figure 16–42 **Signaling pathways can be highly interconnected.** The diagram sketches two pathways from GPCRs—via adenylyl cyclase and via phospholipase C—and three pathways from RTKs—via phospholipase C, via Ras, and via PI 3-kinase. The protein kinases in these pathways phosphorylate many proteins, including proteins belonging to the other pathways. The resulting dense network of regulatory interconnections is symbolized by the *red* arrows radiating from each protein shaded in *yellow*; some kinases phosphorylate some of the same effector proteins. Small intracellular messengers, such as Ca^{2+}, can also influence the activity of multiple pathways.

they themselves primarily belong. Thus, a certain amount of cross-talk occurs between the different pathways (see Figure 16–42) and, indeed, between virtually all of the control systems of the cell. To give an idea of the scale of the complexity, genome sequencing studies suggest that about 2% of our ~20,000 protein-coding genes code for protein kinases; moreover, hundreds of distinct types of protein kinases are thought to be present in a single mammalian cell. How can we make sense of this tangled web of interacting signaling pathways, and what is the function of such complexity?

A cell receives messages from many sources, and it must integrate this information to generate an appropriate response: to live or die, to divide or differentiate, to change shape, to move, to send out a chemical message of its own, and so on (Movies 16.7, 16.8, and 16.9). Through the cross-talk between signaling pathways, the cell is able to put together multiple bits of information and react to the combination. Thus, some intracellular signaling proteins act as integrating devices, usually by having several potential phosphorylation sites, each of which can be phosphorylated by a different protein kinase. Information received from different sources can converge on such proteins, which then convert the input to a single outgoing signal (Figure 16–43, and see Figure 16–13). The integrating proteins in turn can deliver a signal to many downstream targets. In this way, the intracellular signaling system may act like a network of nerve cells in the brain—or like a collection of microprocessors in a computer—interpreting complex information and generating complex responses.

Our exploration of the pathways that cells use to process signals from their environment has led us from receptors on the cell surface to the proteins that form the elaborate control systems that operate deep within the cell's interior. We have examined a large array of signaling networks that allow cells to combine and process inputs from different sources, store information, and respond in an appropriate manner that benefits the organism. But our understanding of these intricate networks is still evolving: we are still discovering new links in the chains, new signaling partners, new connections, and even new pathways.

Unraveling these cell signaling pathways—in both animals and plants—is one of the most active areas of research, and new discoveries are being made every day. Genome sequencing projects continue to provide long lists of components involved in signal transduction in a large variety of organisms. Even when we have identified all the pieces, however, it will

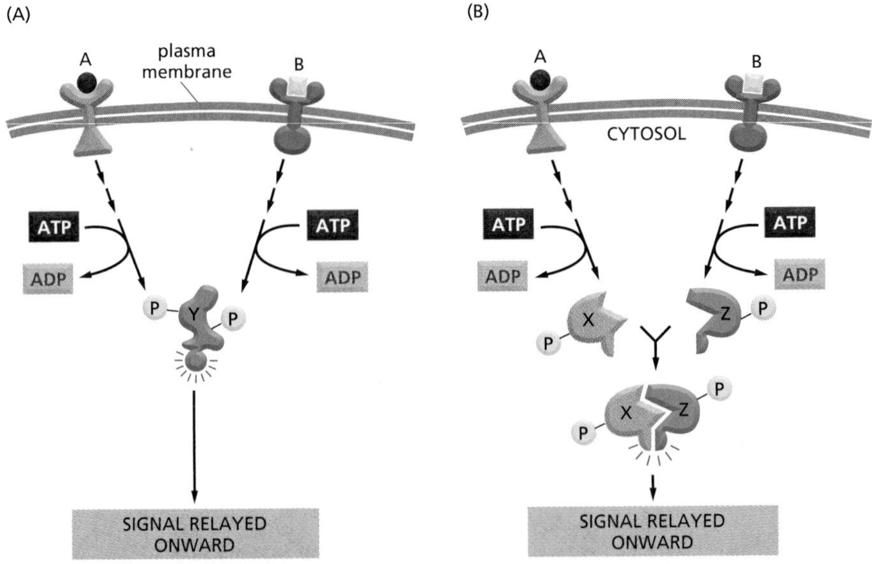

Figure 16–43 **Some intracellular signaling proteins serve to integrate incoming signals.** (A) Signals A and B may activate different cascades of protein phosphorylations, each of which leads to the phosphorylation of protein Y but at different sites on the protein. Protein Y is activated only when both of these sites are phosphorylated, and therefore it is active only when signals A and B are simultaneously present. (B) Alternatively, signals A and B could lead to the phosphorylation of two proteins, X and Z, which then bind to each other to create the active protein XZ.

remain a major challenge to figure out exactly how they fit together to allow cells to integrate the diverse array of signals in their environment and respond in the appropriate way. In a way, learning how cells 'think' is a problem akin to learning how we, as humans, think. Although we know, for example, how neurotransmitters activate certain neurons and how one neuron communicates with another, we are nowhere near having a fundamental understanding of how all these components work together to enable us to reason, converse, laugh, love, and attempt to unravel the fundamental nature of the universe and life on Earth.

ESSENTIAL CONCEPTS

- Cells in multicellular organisms communicate through a large variety of extracellular chemical signals.
- In animals, hormones are carried in the blood to distant target cells, but most other extracellular signal molecules act over only a short range. Neighboring cells often communicate through direct cell–cell contact.
- Extracellular signal molecules stimulate a target cell when they bind to and activate receptor proteins. Each receptor protein recognizes a particular signal molecule.
- Small hydrophobic extracellular signal molecules, such as steroid hormones and nitric oxide, can diffuse directly across the plasma membrane; they activate intracellular proteins, which are usually either transcription regulators or enzymes.
- Most extracellular signal molecules cannot pass through the plasma membrane; they bind to cell-surface receptor proteins that convert (transduce) the extracellular signal into different intracellular signals.
- There are three main classes of cell-surface receptors: (1) ion-channel–coupled receptors, (2) G-protein–coupled receptors (GPCRs), and (3) enzyme-coupled receptors.
- GPCRs and enzyme-coupled receptors respond to extracellular signals by activating one or more intracellular signaling pathways that alter the behavior of the cell.
- Turning off signaling pathways is as important as turning them on. Each activated component in a signaling pathway must be subsequently inactivated or removed for the pathway to function again.
- GPCRs activate a class of trimeric GTP-binding proteins called G proteins; these act as molecular switches, transmitting the signal onward for a short period and then switching themselves off by hydrolyzing their bound GTP to GDP.
- Some G proteins directly regulate ion channels in the plasma membrane. Others directly activate (or inactivate) the enzyme adenylyl cyclase, increasing (or decreasing) the intracellular concentration of the small messenger molecule cyclic AMP. Still other G proteins directly activate the enzyme phospholipase C, which generates the small messenger molecules inositol trisphosphate (IP_3) and diacylglycerol.
- IP_3 opens Ca^{2+} channels in the membrane of the endoplasmic reticulum, releasing a flood of free Ca^{2+} ions into the cytosol. Ca^{2+} itself acts as a small intracellular messenger, altering the activity of a wide range of Ca^{2+}-responsive proteins, including calmodulin, which activates various target proteins such as such as Ca^{2+}/calmodulin-dependent protein kinases (CaM-kinases) .
- A rise in cyclic AMP activates protein kinase A (PKA), while Ca^{2+} and diacylglycerol in combination activate protein kinase C (PKC).

- PKA, PKC, and CaM-kinases phosphorylate selected target proteins on serines and threonines, thereby altering protein activity. Different cell types contain different sets of signaling and effector target proteins and are therefore affected in different ways.
- Many enzyme-coupled receptors have intracellular protein domains that function as enzymes; many are receptor tyrosine kinases (RTKs), which phosphorylate themselves and selected intracellular signaling proteins on tyrosines.
- The phosphotyrosines on RTKs serve as docking sites for various intracellular signaling proteins, usually including the small GTP-binding protein Ras. Ras activates a three-protein MAP-kinase signaling module that helps relay the signal from the plasma membrane to the nucleus.
- Mutations that stimulate cell proliferation by making Ras constantly active are a common feature of many human cancers.
- Some RTKs stimulate cell growth and cell survival by activating PI 3-kinase, which phosphorylates specific inositol phospholipids to produce lipid docking sites in the plasma membrane that allow certain signaling proteins to congregate and activate one another.
- Some receptors, including Notch and cytokine receptors, activate a direct pathway to the nucleus. Instead of activating signaling cascades, they turn on transcription regulators at the plasma membrane, which then migrate to the nucleus where they activate specific genes.
- Plants, like animals, use enzyme-coupled cell-surface receptors to control their growth and development.
- Extracellular signals in plants often act by relieving the transcriptional repression of signal-responsive genes.
- Different intracellular signaling pathways interact, enabling cells to produce an appropriate response to a complex combination of signals. Some combinations of signals enable a cell to survive; other combinations signal a cell to proliferate; and, in the absence of any signals, most animal cells will kill themselves by undergoing apoptosis.

KEY TERMS

adaptation
adenylyl cyclase
Ca^{2+}/calmodulin-dependent protein kinase (CaM-kinase)
calmodulin
cell signaling
cyclic AMP
cyclic-AMP-dependent protein kinase (PKA)
cytokine
diacylglycerol (DAG)
enzyme-coupled receptor
extracellular signal molecule
G protein
G-protein–coupled receptor (GPCR)
GTP-binding protein
hormone
inositol phospholipid
inositol 1,4,5-trisphosphate (IP_3)
ion-channel–coupled receptor
intracellular signaling molecule
intracellular signaling pathway
local mediator
MAP kinase
MAP-kinase signaling module
molecular switch
monomeric GTPase
neurotransmitter
nitric oxide (NO)
nuclear receptor
phosphoinositide 3-kinase (PI 3-kinase)
phospholipase C
protein kinase
protein kinase C (PKC)
protein phosphatase
Ras
receptor, receptor protein
receptor serine/threonine kinase
receptor tyrosine kinase (RTK)
second messenger
serine/threonine kinase
signal transduction
small messenger
steroid hormone
tyrosine kinase

QUESTIONS

QUESTION 16–10

If some cell-surface receptors, including cytokine receptors and Notch, can rapidly signal to the nucleus by activating latent transcription regulators at the plasma membrane, why do most cell-surface receptors use long, indirect signaling cascades to influence gene transcription in the nucleus?

QUESTION 16–11

Which of the following statements are correct? Explain your answers.

A. The extracellular signal molecule acetylcholine has different effects on different cell types in an animal and often binds to different cell-surface receptor molecules on different cell types.

B. After acetylcholine is secreted from cells, it is long-lived, because it has to reach target cells all over the body.

C. Both the GTP-bound α subunits and nucleotide-free $\beta\gamma$ complexes—but not GDP-bound, fully assembled G proteins—can activate other molecules downstream of GPCRs.

D. IP_3 is produced directly by cleavage of an inositol phospholipid without incorporation of an additional phosphate group.

E. Calmodulin regulates the intracellular Ca^{2+} concentration.

F. Different signals originating from the plasma membrane can be integrated by cross-talk between different signaling pathways inside the cell.

G. Tyrosine phosphorylation serves to build binding sites for other proteins to bind to RTKs.

QUESTION 16–12

The Ras protein functions as a molecular switch that is set to its 'on' state by other proteins that cause it to expel its bound GDP and bind GTP. A GTPase-activating protein helps reset the switch to the 'off' state by inducing Ras to hydrolyze its bound GTP to GDP much more rapidly than it would without this encouragement. Thus, Ras works like a light switch that one person turns on and another turns off. You are given a mutant cell that lacks the GTPase-activating protein. What abnormalities would you expect to find in the way in which Ras activity responds to extracellular signals?

QUESTION 16–13

A. Compare and contrast signaling by neurons, which secrete neurotransmitters at synapses, with signaling carried out by endocrine cells, which secrete hormones into the blood.

B. Discuss the relative advantages of the two mechanisms.

QUESTION 16–14

Two intracellular molecules, X and Y, are both normally synthesized at a constant rate of 1000 molecules per second per cell. Molecule X is broken down slowly: each molecule of X survives on average for 100 seconds. Molecule Y is broken down 10 times faster: each molecule of Y survives on average for 10 seconds.

A. Calculate how many molecules of X and Y the cell contains at any time.

B. If the rates of synthesis of both X and Y are suddenly increased tenfold to 10,000 molecules per second per cell—without any change in their degradation rates—how many molecules of X and Y will there be after one second?

C. Which molecule would be preferred for rapid signaling?

QUESTION 16–15

"One of the great kings of the past ruled an enormous kingdom that was more beautiful than anywhere else in the world. Every plant glistened as brilliantly as polished jade, and the softly rolling hills were as sleek as the waves of the summer sea. The wisdom of all of his decisions relied on a constant flow of information brought to him daily by messengers who told him about every detail of his kingdom so that he could take quick, appropriate actions when needed. Despite the beauty and efficiency, his people felt doomed living under his rule, for he had an adviser who had studied cell signal transduction and accordingly administered the king's Department of Information. The adviser had implemented the policy that all messengers will be immediately beheaded whenever spotted by the Royal Guard, because for rapid signaling the lifetime of messengers ought to be short. Their plea 'Don't hurt me, I'm only the messenger!' was to no avail, and the people of the kingdom suffered terribly because of the rapid loss of their sons and daughters." Why is the analogy on which the king's adviser based his policies inappropriate? Briefly discuss the features that set cell signaling pathways apart from the human communication pathway described in the story.

QUESTION 16–16

In a series of experiments, genes that code for mutant forms of an RTK are introduced into cells. The cells also express their own normal form of the receptor from their normal gene, although the mutant genes are constructed so that the mutant RTK is expressed at considerably higher concentration than the normal RTK. What would be the consequences of introducing a mutant gene that codes for an RTK (A) lacking its extracellular domain, or (B) lacking its intracellular domain?

QUESTION 16–17

Discuss the following statement: "Membrane proteins that span the membrane many times can undergo a conformational change upon ligand binding that can be sensed on the other side of the membrane. Thus, individual protein molecules can transmit a signal across a membrane. In contrast, individual single-span membrane proteins cannot transmit a conformational change across the membrane but require oligomerization."

QUESTION 16–18

What are the similarities and differences between the reactions that lead to the activation of G proteins and the reactions that lead to the activation of Ras?

QUESTION 16–19

Why do you suppose cells use Ca^{2+} (which is kept by Ca^{2+} pumps at a cytosolic concentration of 10^{-7} M) for intracellular signaling and not another ion such as Na^+ (which is kept by the Na^+ pump at a cytosolic concentration of 10^{-3} M)?

QUESTION 16–20

It seems counterintuitive that a cell, having a perfectly abundant supply of nutrients available, would commit suicide if not constantly stimulated by signals from other cells (see Figure 16–6). What do you suppose might be the advantages of such regulation?

QUESTION 16–21

The contraction of the myosin–actin system in muscle cells is triggered by a rise in intracellular Ca^{2+}. Muscle cells have specialized Ca^{2+} channels—called ryanodine receptors because of their sensitivity to the drug ryanodine—that are embedded in the membrane of the sarcoplasmic reticulum, a specialized form of the endoplasmic reticulum. In contrast to the IP_3-gated Ca^{2+} channels in the endoplasmic reticulum shown in Figure 16–25, the signaling molecule that opens ryanodine receptors is Ca^{2+} itself. Discuss the consequences of ryanodine channels for muscle cell contraction.

QUESTION 16–22

Two protein kinases, K1 and K2, function sequentially in an intracellular signaling pathway. If either kinase contains a mutation that permanently inactivates its function, no response is seen in cells when an extracellular signal is received. A different mutation in K1 makes it permanently active, so that in cells containing that mutation a response is observed even in the absence of an extracellular signal. You characterize a double mutant cell that contains K2 with the inactivating mutation and K1 with the activating mutation. You observe that the response is seen even in the absence of an extracellular signal. In the normal signaling pathway, does K1 activate K2 or does K2 activate K1? Explain your answer.

QUESTION 16–23

A. Trace the steps of a long and indirect signaling pathway from a cell-surface receptor to a change in gene expression in the nucleus.

B. Compare this pathway with two short and direct pathways from the cell surface to the nucleus.

QUESTION 16–24

How does PI 3-kinase activate the Akt kinase after activation of RTK?

QUESTION 16–25

Animal cells and plant cells have some very different intracellular signaling mechanisms but also share some common mechanisms. Why do you think this is so?

CHAPTER **SEVENTEEN** 17

Cytoskeleton

INTERMEDIATE FILAMENTS

MICROTUBULES

ACTIN FILAMENTS

MUSCLE CONTRACTION

The ability of eucaryotic cells to adopt a variety of shapes, organize the many components in their interior, interact mechanically with the environment, and carry out coordinated movements depends on the **cytoskeleton**—an intricate network of protein filaments that extends throughout the cytoplasm (Figure 17–1). This filamentous architecture helps to support the large volume of cytoplasm in a eucaryotic cell, a function that is particularly important in animal cells, which have no cell walls. Although some cytoskeletal components are present in bacteria, the cytoskeleton is most prominent in the large and structurally complex eucaryotic cell.

Unlike our own bony skeleton, however, the cytoskeleton is a highly dynamic structure that is continuously reorganized as a cell changes shape, divides, and responds to its environment. The cytoskeleton is not only the "bones" of a cell but its "muscles" too, and it is directly responsible for large-scale movements such as the crawling of cells along a surface, contraction of muscle cells, and the changes in cell shape that take place as an embryo develops. Without the cytoskeleton, wounds would never heal, muscles would be useless, and sperm would never reach the egg.

The eucaryotic cell, like any factory making a complex product, has a highly organized interior in which specialized machines are concentrated in different areas but linked by transport systems (discussed in Chapter 15). The cytoskeleton controls the location of the organelles that conduct these specialized functions, in addition to providing the machinery for the transport between them. It is also responsible for the segregation of chromosomes into daughter cells and the pinching apart of cells at cell division, as we discuss in Chapter 18.

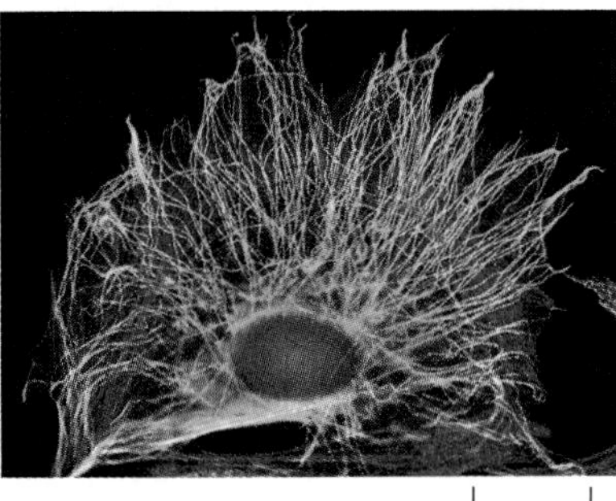

Figure 17–1 **The cytoskeleton gives a cell its shape and allows the cell to organize its internal components.** An animal cell in culture has been labeled to show two of its major cytoskeletal systems, the microtubules *(green)* and the actin filaments *(red)*. The DNA in the nucleus is labeled in *blue*. (Courtesy of Albert Tousson.)

The cytoskeleton is built on a framework of three types of protein filaments: *intermediate filaments, microtubules,* and *actin filaments.* As shown in Panel 17–1 (p. 573), each type of filament has distinct mechanical properties and is formed from a different protein subunit. A family of fibrous proteins form the intermediate filaments; *tubulin* is the subunit in microtubules; and *actin* is the subunit in actin filaments. In each case, thousands of subunits assemble into a fine thread of protein that sometimes extends across the entire cell.

In this chapter we consider the structure and function of the three types of protein filament networks in turn. We begin with the intermediate filaments that provide cells with mechanical strength. We then see how cell appendages built from microtubules propel motile cells like protozoa and sperm, and how the actin cytoskeleton provides the motive force for a crawling fibroblast. Finally, we discuss how the cytoskeleton drives one of the most obvious and best-studied forms of cell movement, the contraction of muscle.

INTERMEDIATE FILAMENTS

Intermediate filaments have great tensile strength, and their main function is to enable cells to withstand the mechanical stress that occurs when cells are stretched. The filaments are called "intermediate" because, in the smooth muscle cells where they were first discovered, their diameter (about 10 nm) is between that of the thin actin-containing filaments and the thicker myosin filaments. Intermediate filaments are the toughest and most durable of the three types of cytoskeletal filaments: when cells are treated with concentrated salt solutions and nonionic detergents, the intermediate filaments survive while most of the rest of the cytoskeleton is destroyed.

Intermediate filaments are found in the cytoplasm of most animal cells. They typically form a network throughout the cytoplasm, surrounding the nucleus and extending out to the cell periphery. There they are often anchored to the plasma membrane at cell–cell junctions such as desmosomes (discussed in Chapter 20), where the external face of the membrane is connected to that of another cell (Figure 17–2). Intermediate filaments

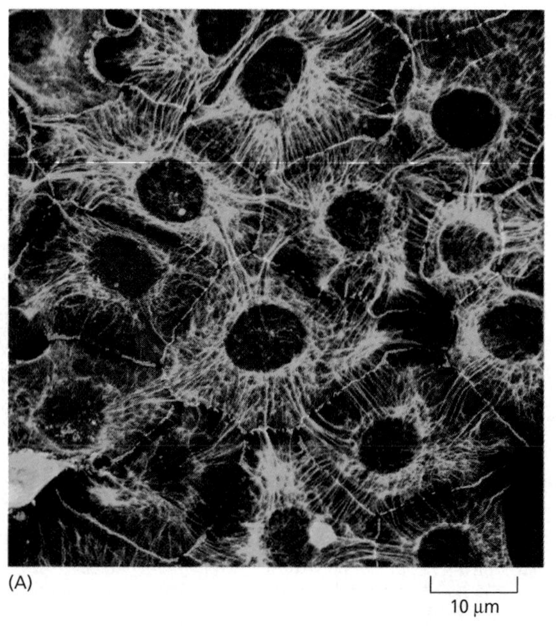

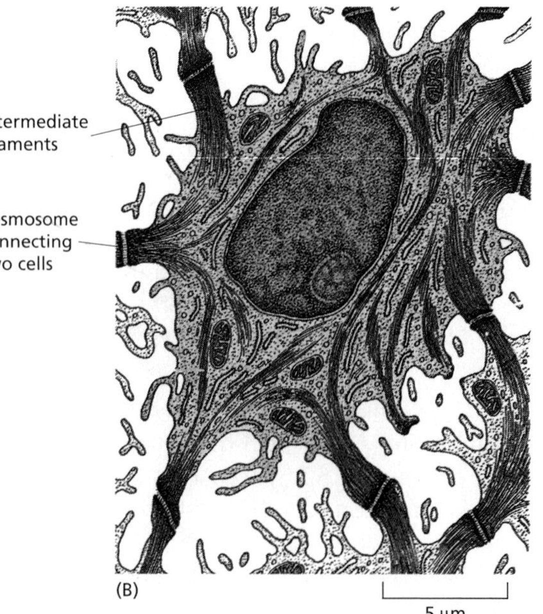

Figure 17–2 **Intermediate filaments form a strong, durable network in the cytoplasm of the cell.** (A) Immunofluorescence micrograph of a sheet of epithelial cells in culture stained to show the lacelike network of intermediate keratin filaments *(green)* that surround the nuclei and extend through the cytoplasm of the cells. The filaments in each cell are indirectly connected to those of neighboring cells through the desmosomes (discussed in Chapter 20), establishing a continuous mechanical link from cell to cell throughout the tissue. This mechanical link strengthens the epithelium, allowing its cells to form a continuous sheet lining the tissue cavity. A second protein *(blue)* has been stained to show the locations of the cell boundaries. (B) Drawing from an electron micrograph of a section of epidermis showing the bundles of intermediate filaments that traverse the cytoplasm and are inserted at desmosomes. (A, courtesy of Kathleen Green and Evangeline Amargo; B, from R.V. Krstić, Ultrastructure of the Mammalian Cell: An Atlas. Berlin: Springer, 1979. With permission from Springer-Verlag.)

PANEL 17–1 The three major types of protein filaments

INTERMEDIATE FILAMENTS

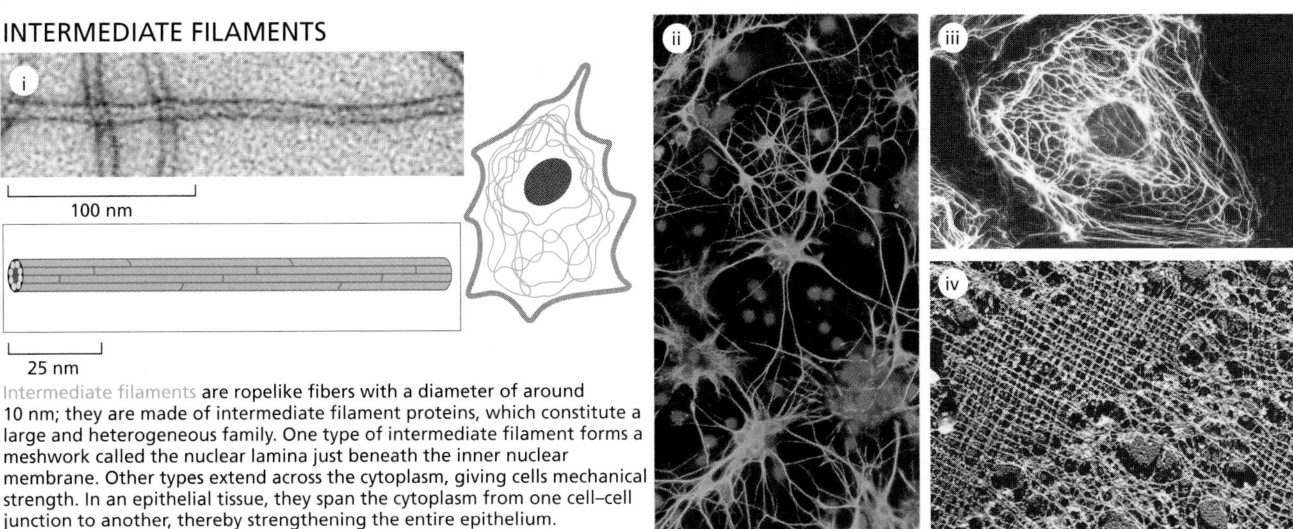

Intermediate filaments are ropelike fibers with a diameter of around 10 nm; they are made of intermediate filament proteins, which constitute a large and heterogeneous family. One type of intermediate filament forms a meshwork called the nuclear lamina just beneath the inner nuclear membrane. Other types extend across the cytoplasm, giving cells mechanical strength. In an epithelial tissue, they span the cytoplasm from one cell–cell junction to another, thereby strengthening the entire epithelium.

Micrographs courtesy of Roy Quinlan (i); Nancy L. Kedersha (ii); Mary Osborn (iii); Ueli Aebi (iv).

MICROTUBULES

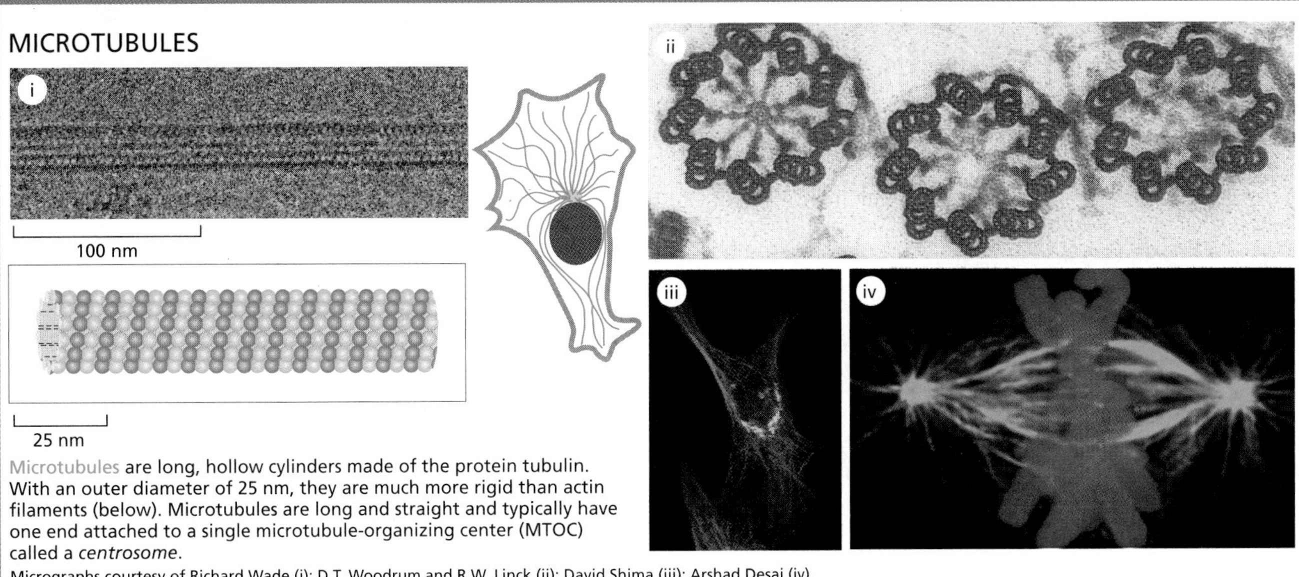

Microtubules are long, hollow cylinders made of the protein tubulin. With an outer diameter of 25 nm, they are much more rigid than actin filaments (below). Microtubules are long and straight and typically have one end attached to a single microtubule-organizing center (MTOC) called a *centrosome*.

Micrographs courtesy of Richard Wade (i); D.T. Woodrum and R.W. Linck (ii); David Shima (iii); Arshad Desai (iv).

ACTIN FILAMENTS

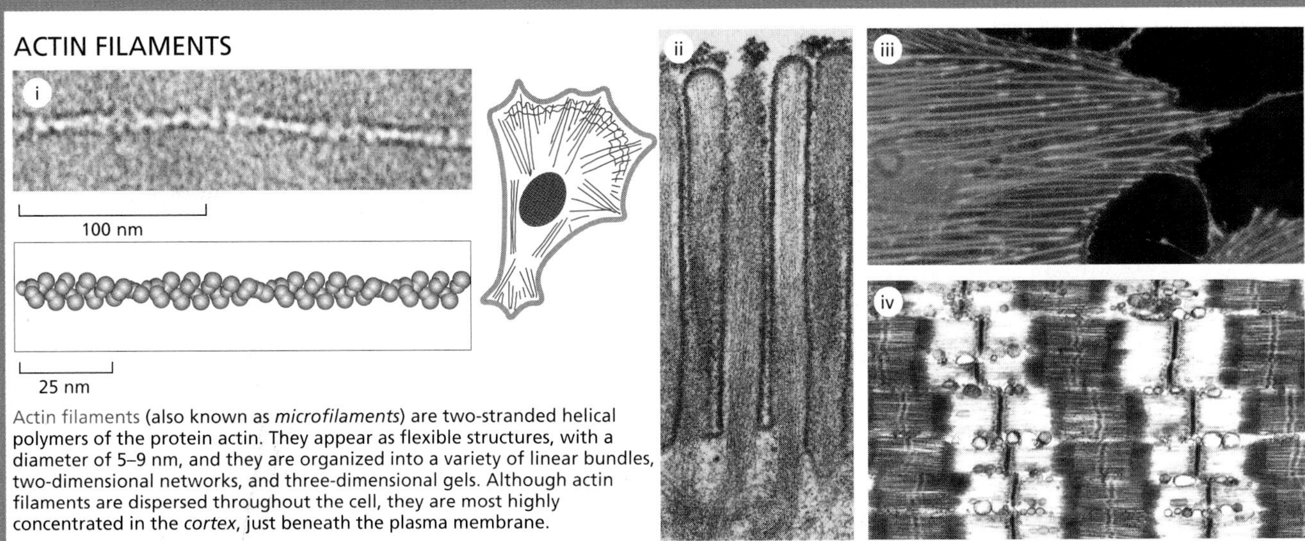

Actin filaments (also known as *microfilaments*) are two-stranded helical polymers of the protein actin. They appear as flexible structures, with a diameter of 5–9 nm, and they are organized into a variety of linear bundles, two-dimensional networks, and three-dimensional gels. Although actin filaments are dispersed throughout the cell, they are most highly concentrated in the *cortex*, just beneath the plasma membrane.

Micrographs courtesy of Roger Craig (i and iv); P.T. Matsudaira and D.R. Burgess, *Cold Spring Harb. Symp. Quant. Biol.* 46:845–854, 1982. With permission from Cold Spring Harbor Laboratory Press (ii); Keith Burridge (iii).

are also found within the nucleus; a mesh of intermediate filaments, the *nuclear lamina,* underlies and strengthens the nuclear envelope in all eucaryotic cells. In this section, we see how the structure and assembly of intermediate filaments makes them particularly suited to strengthening cells and protecting them from mechanical stress.

Intermediate Filaments Are Strong and Ropelike

Intermediate filaments are like ropes with many long strands twisted together to provide tensile strength (Movie 17.1). The strands of this rope—the subunits of intermediate filaments—are elongated fibrous proteins, each composed of an N-terminal globular head, a C-terminal globular tail, and a central elongated rod domain (Figure 17–3A). The rod domain consists of an extended α-helical region that enables pairs of intermediate filament proteins to form stable dimers by wrapping around each other in a coiled-coil configuration (Figure 17–3B), as described in Chapter 4. Two of these coiled-coil dimers then associate by noncovalent bonding to form a tetramer (Figure 17–3C), and the tetramers then bind to one another end-to-end and side-by-side, and also by noncovalent bonding, to generate the final ropelike intermediate filament (Figure 17–3D–F).

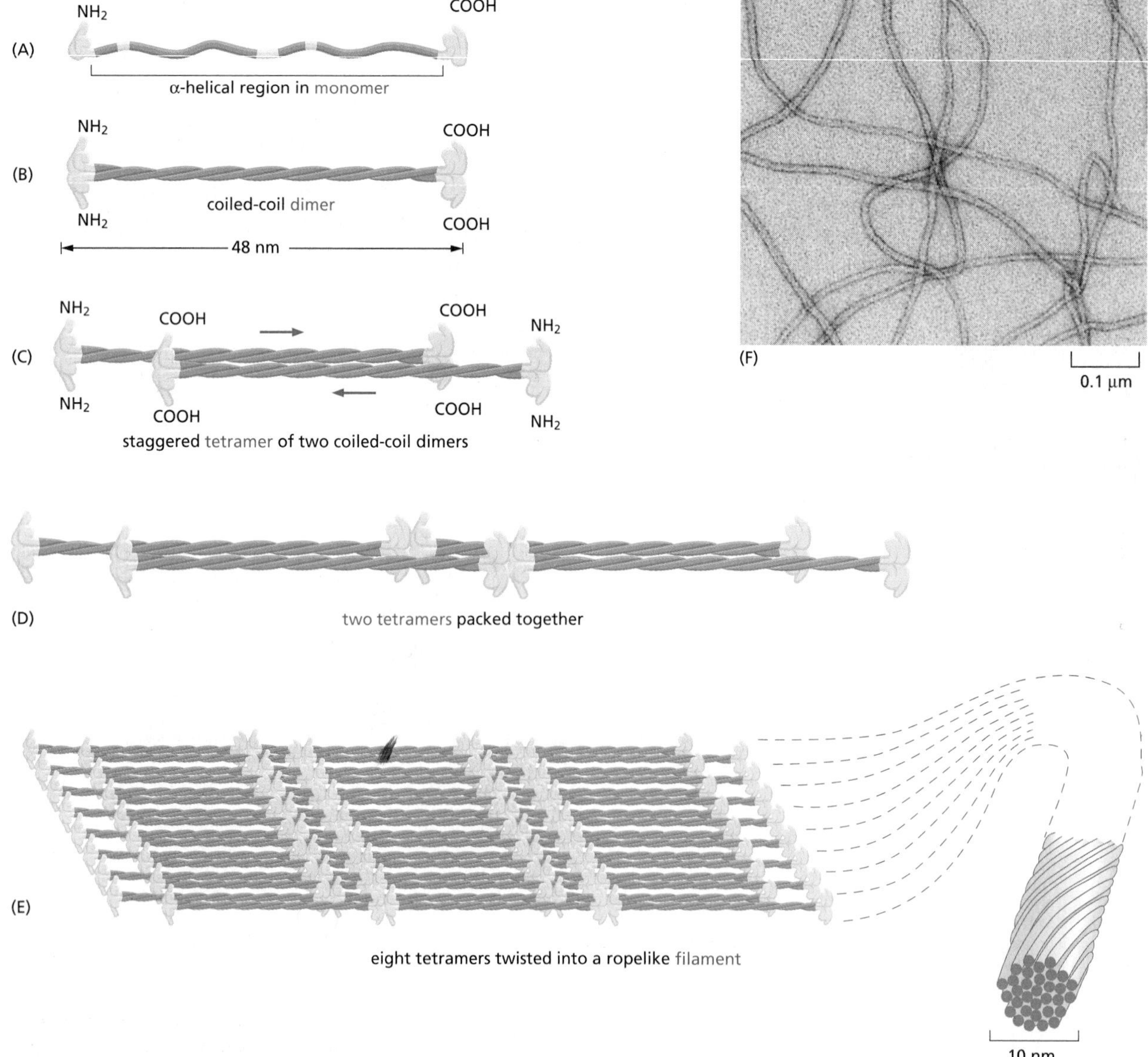

Figure 17–3 **Intermediate filaments are like ropes made of long, twisted strands of protein.** The intermediate filament protein monomer shown in (A) consists of a central rod domain with globular regions at either end. Pairs of monomers associate to form a dimer (B), and two dimers then line up to form a staggered tetramer (C). Tetramers can pack together end-to-end as shown in (D) and assemble into a helical array containing eight strands of tetramers (shown here spread out flat for clarity) that twist together to form the final ropelike intermediate filament (E). (F) Electron micrograph of the final 10-nm filament. (F, courtesy of Roy Quinlan.)

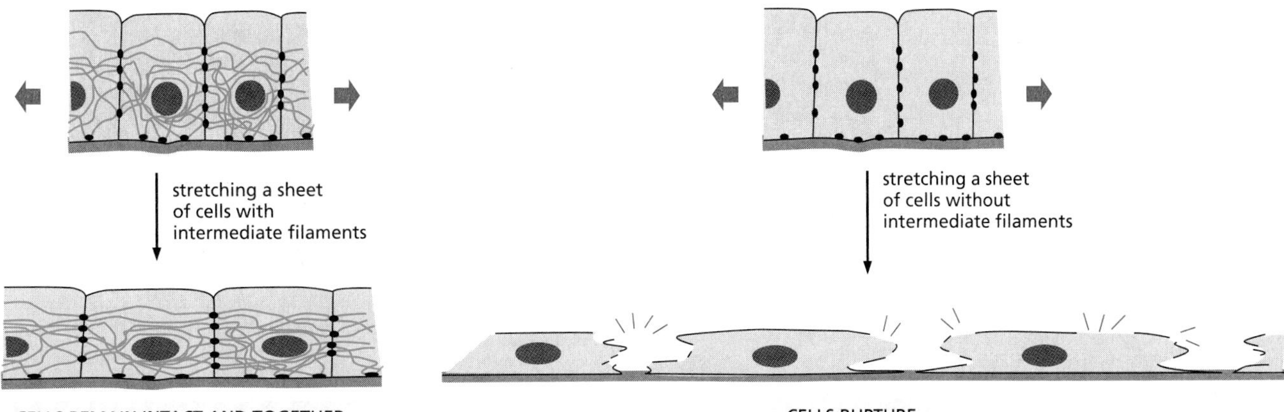

Figure 17–4 **Intermediate filaments strengthen animal cells.** If a sheet of epithelial cells is stretched by external forces (due to the growth or movements of the surrounding tissues, for example), then the network of intermediate filaments and desmosomal junctions that extends through the sheet develops tension and limits the extent of stretching. If the junctions alone were present, then the same forces would cause a major deformation of the cells, even to the extent of causing their plasma membranes to rupture.

The central rod domains of different intermediate filament proteins are all similar in size and amino acid sequence, so that when they pack together they always form filaments of similar diameter and internal structure. By contrast, the globular head and tail regions, which are exposed on the surface of the filament, allow it to interact with other components of the cytoplasm. The globular domains vary greatly in both size and amino acid sequence from one intermediate filament protein to another.

Intermediate Filaments Strengthen Cells Against Mechanical Stress

Intermediate filaments are particularly prominent in the cytoplasm of cells that are subject to mechanical stress. They are present in large numbers, for example, along the length of nerve cell axons, providing essential internal reinforcement to these extremely long and fine cell extensions. They are also abundant in muscle cells and in epithelial cells such as those of the skin. In all these cells, intermediate filaments, by stretching and distributing the effect of locally applied forces, keep cells and their membranes from breaking in response to mechanical shear (Figure 17–4). A similar principle is used to manufacture composite materials such as fiberglass or reinforced concrete, in which tension-bearing linear elements such as carbon fibers (in fiberglass) or steel bars (in concrete) are embedded in a space-filling matrix to give the material strength.

Intermediate filaments can be grouped into four classes: (1) *keratin filaments* in epithelial cells; (2) *vimentin* and *vimentin-related filaments* in connective-tissue cells, muscle cells, and supporting cells of the nervous system (glial cells); (3) *neurofilaments* in nerve cells; and (4) *nuclear lamins,* which strengthen the nuclear membrane of all animal cells (Figure 17–5). The first three filament types are found in the cytoplasm, the fourth

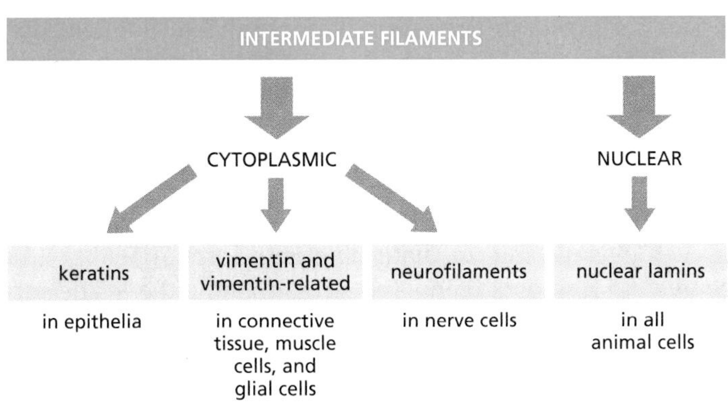

Figure 17–5 **Intermediate filaments can be divided into several different categories.**

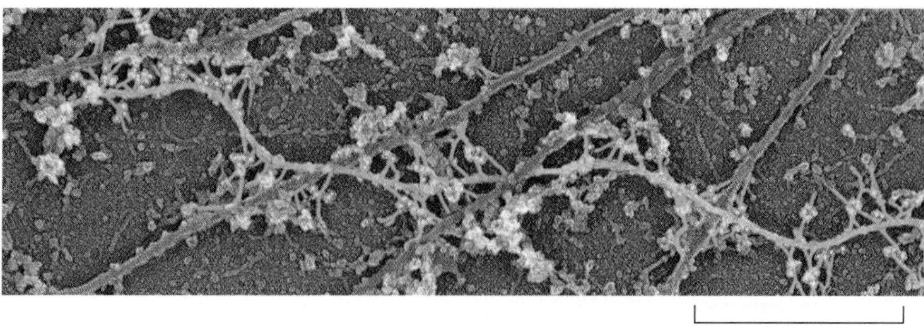

Figure 17–6 **Plectin aids in the bundling of intermediate filaments and links these filaments to other cytoskeletal protein networks.** Plectin *(green)* links intermediate filaments *(blue)* to other intermediate filaments, to microtubules *(red)*, and to actin filaments (not shown). In this electron micrograph, the *yellow dots* are gold particles linked to antibodies that recognize plectin. The actin filament network has been removed to reveal these protein linkages. (From T.M. Svitkina and G.G. Borisy, *J. Cell Biol.* 135:991–1007, 1996. With permission from The Rockefeller University Press.)

in the cell nucleus. Filaments of each class are formed by polymerization of their corresponding protein subunits.

The keratins are the most diverse class of intermediate filament. Every kind of epithelium in the vertebrate body—whether in the tongue, the cornea, or the lining of the gut—has its own distinctive mixture of keratin proteins. Specialized keratins also occur in hair, feathers, and claws. In each case, the keratin filaments are formed from a mixture of different keratin subunits. Keratin filaments typically span the interiors of epithelial cells from one side of the cell to the other, and filaments in adjacent epithelial cells are indirectly connected through cell–cell junctions called *desmosomes* (see Panel 17–1, p. 573). The ends of the keratin filaments are anchored to the desmosomes, and they associate laterally with other cell components through their globular head and tail domains, which project from the surface of the assembled filament. This cabling of high tensile strength, formed by the filaments throughout the epithelial sheet, distributes the stress that occurs when the skin is stretched. The importance of this function is illustrated by the rare human genetic disease *epidermolysis bullosa simplex,* in which mutations in the keratin genes interfere with the formation of keratin filaments in the epidermis. As a result, the skin is highly vulnerable to mechanical injury, and even a gentle pressure can rupture its cells, causing the skin to blister.

Many of the intermediate filaments are further stabilized and reinforced by accessory proteins, such as plectin, that cross-link the filament bundles into strong arrays. In addition to holding together bundles of intermediate filaments (particularly vimentin), these proteins link intermediate filaments to microtubules, to actin filaments, and to adhesive structures in the desmosomes (Figure 17–6). Mutations in the gene for plectin cause a devastating human disease that combines features of epidermolysis bullosa simplex (caused by disruption of skin keratin), muscular dystrophy (caused by disruption of intermediate filaments in muscle), and neurodegeneration (caused by disruption of neurofilaments). Mice lacking a functional plectin gene die within a few days of birth, with blistered skin and abnormal skeletal and heart muscles. Thus, although plectin may not be necessary for the initial formation of intermediate filaments, its cross-linking action is required to provide cells with the strength they need to withstand the mechanical stresses inherent to vertebrate life.

QUESTION 17–1

Which of the following types of cells would you expect to contain a high density of intermediate filaments in their cytoplasm? Explain your answers.

A. *Amoeba proteus* (a free-living amoeba)
B. Skin epithelial cell
C. Smooth muscle cell in the digestive tract
D. *Escherichia coli*
E. Nerve cell in the spinal cord
F. Sperm cell
G. Plant cell

The Nuclear Envelope Is Supported by a Meshwork of Intermediate Filaments

Whereas cytoplasmic intermediate filaments form ropelike structures, the intermediate filaments lining and strengthening the inside surface of the inner nuclear membrane are organized as a two-dimensional mesh (Figure 17–7). The intermediate filaments within this tough **nuclear lamina** are constructed from a class of intermediate filament proteins called

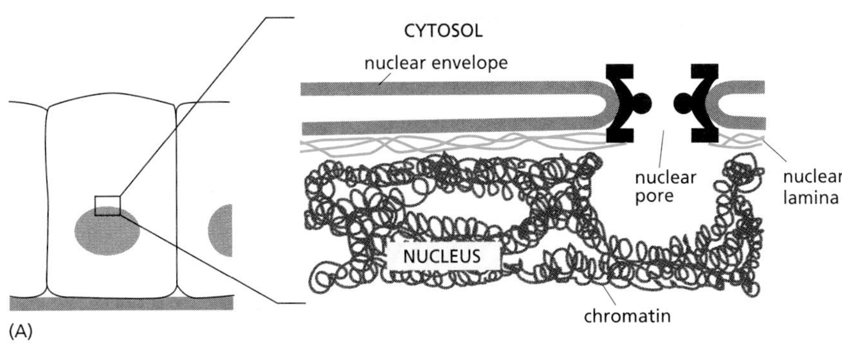

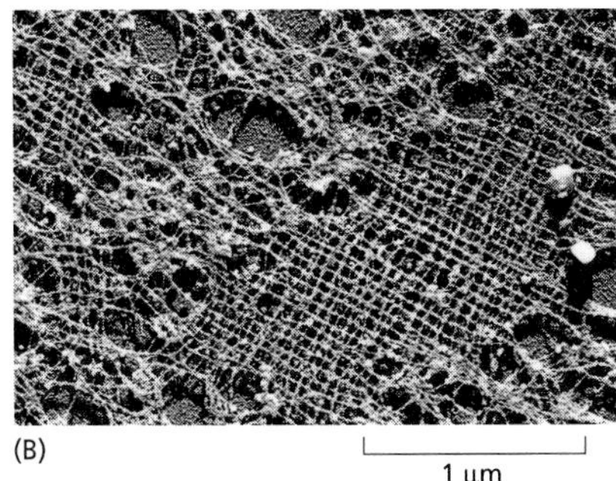

Figure 17–7 **Intermediate filaments support and strengthen the nuclear envelope.** (A) Schematic cross section through the nuclear envelope. The intermediate filaments of the nuclear lamina line the inner face of the nuclear envelope and are thought to provide attachment sites for the DNA-containing chromatin. (B) Electron micrograph of a portion of the nuclear lamina from a frog egg, or oocyte. The lamina is formed from a square lattice of intermediate filaments composed of lamins. (Nuclear laminae from other cell types are not always as regularly organized as the one shown here.) (B, courtesy of Ueli Aebi.)

lamins (not to be confused with *laminin,* which is an extracellular matrix protein). In contrast to the very stable cytoplasmic intermediate filaments found in many cells, the intermediate filaments of the nuclear lamina disassemble and re-form at each cell division, when the nuclear envelope breaks down during mitosis and then re-forms in each daughter cell (discussed in Chapter 18).

Disassembly and reassembly of the nuclear lamina are controlled by the phosphorylation and dephosphorylation (discussed in Chapter 4) of the lamins by protein kinases. When the lamins are phosphorylated, the consequent conformational change weakens the binding between the tetramers and causes the filament to fall apart. Dephosphorylation at the end of mitosis causes the lamins to reassemble (see Figure 18–31).

Defects in a particular nuclear lamin are associated with certain types of *progeria*—rare disorders that cause affected individuals to appear to age prematurely. Children with progeria have wrinkled skin, lose their teeth and hair, and often develop severe cardiovascular disease by the time they reach their teens. Although researchers do not yet know how loss of the nuclear lamins leads to these symptoms, some have suggested that the resulting nuclear instability could lead to impaired cell division or a diminished capacity for tissue repair.

MICROTUBULES

Microtubules have a crucial organizing role in all eucaryotic cells. They are long and relatively stiff hollow tubes of protein that can rapidly disassemble in one location and reassemble in another. In a typical animal cell, microtubules grow out from a small structure near the center of the cell called the *centrosome* (Figure 17–8A). Extending out toward the cell periphery, they create a system of tracks within the cell, along which vesicles, organelles, and other cell components are moved. These and other systems of cytoplasmic microtubules are the part of the cytoskeleton mainly responsible for anchoring membrane-enclosed organelles within the cell and for guiding intracellular transport.

When a cell enters mitosis, the cytoplasmic microtubules disassemble and then reassemble into an intricate structure called the *mitotic spindle.* As described in Chapter 18, the mitotic spindle provides the machinery that will segregate the chromosomes equally into the two daughter cells just before a cell divides (Figure 17–8B). Microtubules can also form permanent structures, as exemplified by the rhythmically beating hairlike structures called *cilia* and *flagella* (Figure 17–8C). These extend from the

(A) INTERPHASE CELL

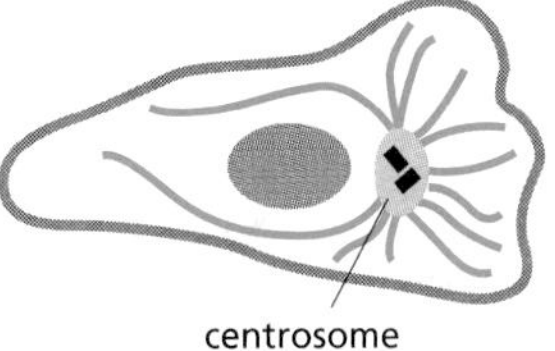

(B) DIVIDING CELL

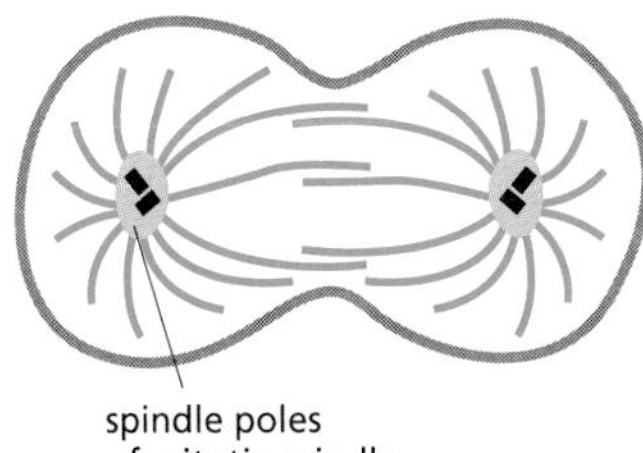

(C) CILIATED CELL

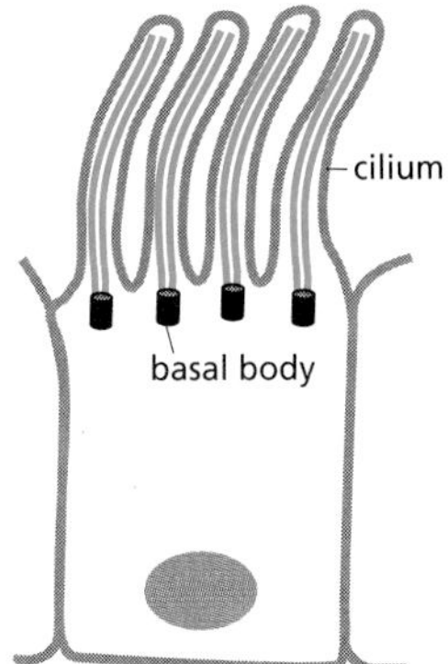

Figure 17–8 **Microtubules usually grow out of an organizing structure.** Unlike intermediate filaments, microtubules *(dark green)* extend from an organizing center such as (A) a centrosome, (B) a spindle pole, or (C) the basal body of a cilium.

surface of many eucaryotic cells, which use them either as a means of propulsion or to sweep fluid over the cell surface. The core of a eucaryotic cilium or flagellum consists of a highly organized and stable bundle of microtubules. (Bacterial flagella have an entirely different structure and act as propulsive structures by a different mechanism.)

In this section, we first look at the structure and assembly of microtubules and then discuss their role in organizing the cytoplasm. Their organizing function depends on the association of microtubules with accessory proteins, especially the *motor proteins* that propel organelles along cytoskeletal tracks. Finally, we discuss the structure and function of cilia and flagella, in which microtubules are permanently associated with motor proteins that power ciliary beating.

Microtubules Are Hollow Tubes with Structurally Distinct Ends

Microtubules are built from subunits—molecules of **tubulin**—each of which is itself a dimer composed of two very similar globular proteins called *α-tubulin* and *β-tubulin,* bound tightly together by noncovalent bonding. The tubulin dimers stack together, again by noncovalent bonding, to form the wall of the hollow cylindrical microtubule. This tubelike structure is made of 13 parallel *protofilaments,* each a linear chain of tubulin dimers with α- and β-tubulin alternating along its length (Figure 17–9). Each protofilament has a structural polarity, with α-tubulin exposed at one end and β-tubulin at the other, and this **polarity**—the directional arrow embodied in the structure—is the same for all the protofilaments, giving a structural polarity to the microtubule as a whole. One end of the microtubule, thought to be the β-tubulin end, is called its *plus end,* and the other, the α-tubulin end, its *minus end.*

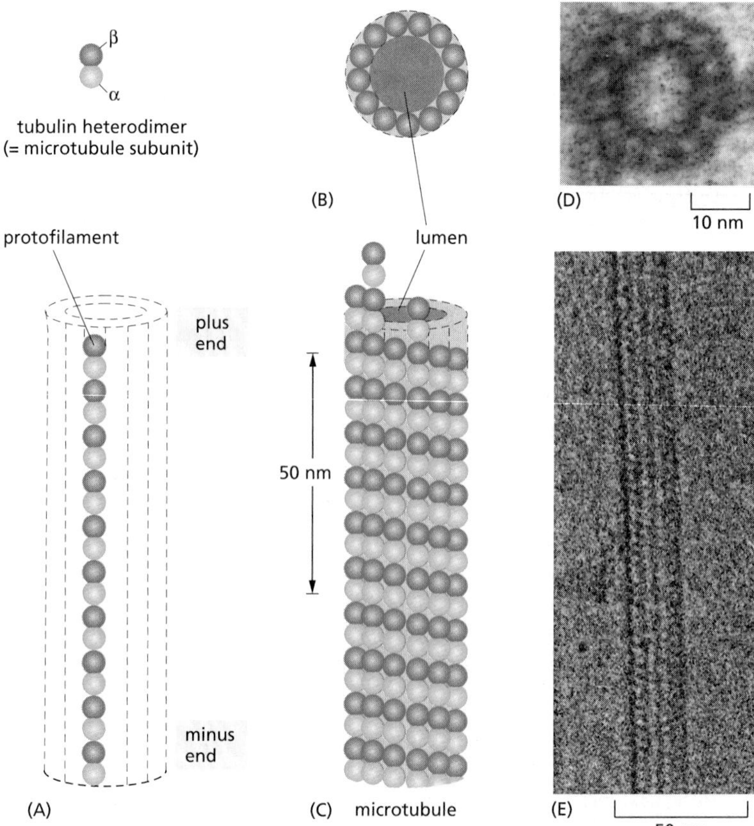

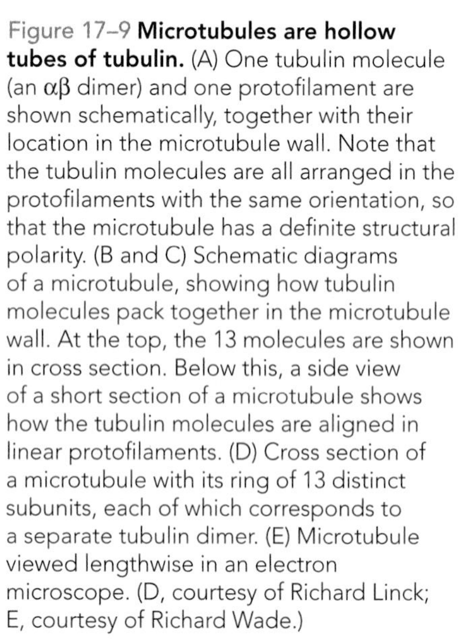

Figure 17–9 **Microtubules are hollow tubes of tubulin.** (A) One tubulin molecule (an αβ dimer) and one protofilament are shown schematically, together with their location in the microtubule wall. Note that the tubulin molecules are all arranged in the protofilaments with the same orientation, so that the microtubule has a definite structural polarity. (B and C) Schematic diagrams of a microtubule, showing how tubulin molecules pack together in the microtubule wall. At the top, the 13 molecules are shown in cross section. Below this, a side view of a short section of a microtubule shows how the tubulin molecules are aligned in linear protofilaments. (D) Cross section of a microtubule with its ring of 13 distinct subunits, each of which corresponds to a separate tubulin dimer. (E) Microtubule viewed lengthwise in an electron microscope. (D, courtesy of Richard Linck; E, courtesy of Richard Wade.)

In vitro, in a concentrated solution of pure tubulin, tubulin dimers will add to either end of a growing microtubule, although they add more rapidly to the plus end than the minus end (which is why the ends were originally named this way). The polarity of the microtubule—the fact that its structure has a definite direction, with the two ends being chemically different and behaving differently—is crucial, both for the assembly of microtubules and for their role once they are formed. If they had no polarity, they could not serve their function in defining a direction for intracellular transport, for example.

The Centrosome Is the Major Microtubule-organizing Center in Animal Cells

Microtubules in cells are formed by outgrowth from specialized organizing centers, which control the number of microtubules formed, their location and their orientation in the cytoplasm. In animal cells, for example, the **centrosome**, which is typically close to the cell nucleus when the cell is not in mitosis, organizes the array of microtubules that radiates outward from it through the cytoplasm (see Figure 17–8A). Centrosomes contain hundreds of ring-shaped structures formed from another type of tubulin, γ-tubulin, and each γ-tubulin ring serves as the starting point, or *nucleation site*, for the growth of one microtubule (Figure 17–10A). The αβ-tubulin dimers add to the γ-tubulin ring in a specific orientation, with the result that the minus end of each microtubule is embedded in the centrosome and growth occurs only at the plus end—that is, at the outward-facing end (Figure 17–10B).

In addition to its γ-tubulin rings, the centrosome in most animal cells also contains a pair of **centrioles**, curious structures each made of a cylindrical array of short microtubules. The centrioles have no role in the nucleation of microtubules in the centrosome (the γ-tubulin rings alone are sufficient), and their function remains something of a mystery, especially as plant cells lack them. Centrioles are, however, similar, if not identical, to the *basal bodies* that form the organizing centers for the microtubules in cilia and flagella (see Figure 17–8C), as discussed later in this chapter.

Microtubules need nucleating sites such as those provided by the γ-tubulin rings in the centrosome because it is much harder to start a new microtubule from scratch, by first assembling a ring of αβ-tubulin dimers, than to add such dimers to a preexisting microtubule structure. Purified free αβ-tubulin can polymerize spontaneously *in vitro* when at a high concentration, but in the living cell, the concentration of free αβ-tubulin

QUESTION 17–2

Why do you suppose it is much easier to add tubulin to existing microtubules than to start a new microtubule from scratch? Explain how γ-tubulin in the centrosome helps to overcome this hurdle.

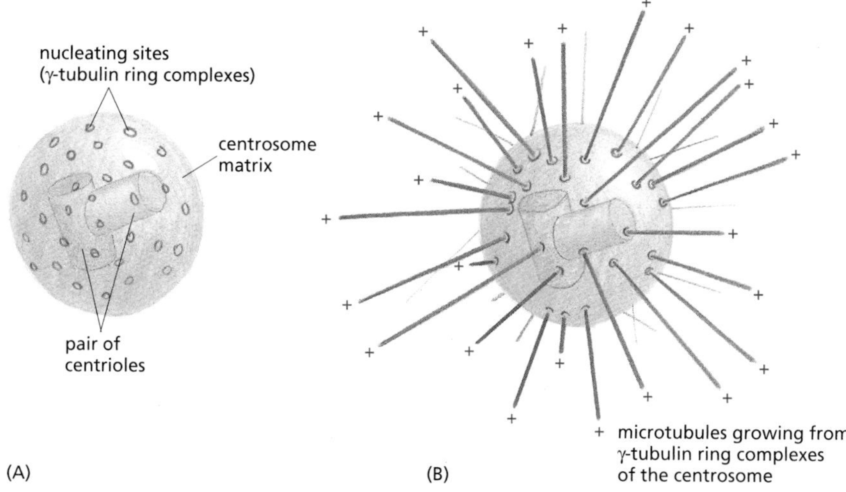

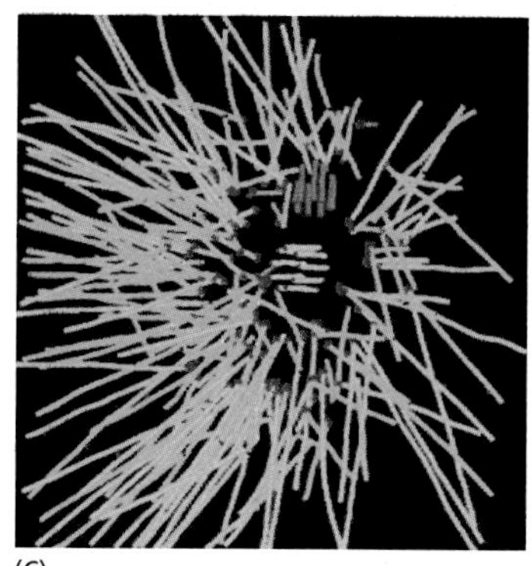

Figure 17–10 **Tubulin polymerizes from nucleation sites on a centrosome.** (A) Schematic drawing showing that a centrosome consists of an amorphous matrix of protein containing the γ-tubulin rings that nucleate microtubule growth. In animal cells, the centrosome contains a pair of centrioles, each made up of a cylindrical array of short microtubules. (B) A centrosome with attached microtubules. The minus end of each microtubule is embedded in the centrosome, having grown from a nucleating ring, whereas the plus end of each microtubule is free in the cytoplasm. (C) A reconstructed image shows a dense thicket of microtubules emanating from the centrosome of a *C. elegans* cell. (C, from E.T. O'Toole et al., *J. Cell Biol.* 163:451–456, 2003. With permission from The Rockefeller University Press.)

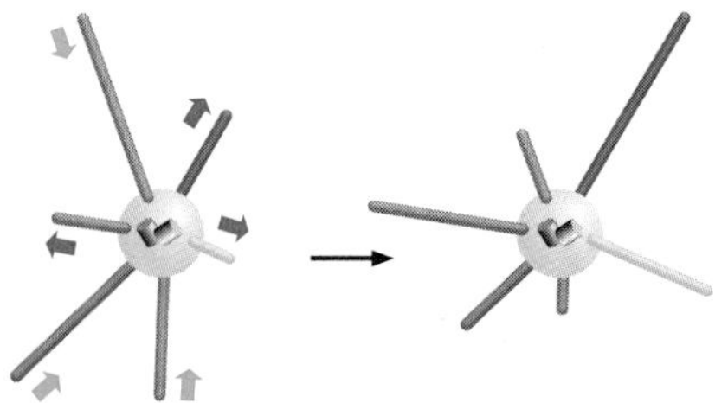

Figure 17–11 **Each microtubule filament grows and shrinks independently of its neighbors.** The array of microtubules anchored in a centrosome is continually changing as new microtubules grow *(red arrows)* and old microtubules shrink *(blue arrows)*.

is too low to drive the difficult first step of assembling the initial ring of a new microtubule. By providing organizing centers containing nucleation sites, and keeping the concentration of free αβ-tubulin dimers low, cells can thus control where microtubules form.

Growing Microtubules Show Dynamic Instability

Once a microtubule has been nucleated, its plus end typically grows outward from the organizing center by the addition of αβ-tubulin subunits for many minutes. Then, without warning, the microtubule suddenly undergoes a transition that causes it to shrink rapidly inward by losing subunits from its free end (Movie 17.2). It may shrink partially and then, no less suddenly, start growing again, or it may disappear completely, to be replaced by a new microtubule from the same γ-tubulin ring (Figure 17–11).

This remarkable behavior, known as **dynamic instability**, stems from the intrinsic capacity of tubulin molecules to hydrolyze GTP. Each free tubulin dimer contains one tightly bound GTP molecule that is hydrolyzed to GDP (still tightly bound) shortly after the subunit is added to a growing microtubule. The GTP-associated tubulin molecules pack efficiently together in the wall of the microtubule, whereas tubulin molecules carrying GDP have a different conformation and bind less strongly to each other.

When polymerization is proceeding rapidly, tubulin molecules add to the end of the microtubule faster than the GTP they carry is hydrolyzed. The end of a growing microtubule is therefore composed entirely of GTP-tubulin subunits, forming what is known as a *GTP cap*. In this situation, the growing microtubule will continue to grow (Figure 17–12A). Because of the randomness of chemical processes, however, it will occasionally happen that tubulin at the free end of the microtubule hydrolyzes its GTP before the next tubulin has been added, so that the free ends of protofilaments are now composed of GDP-tubulin subunits. This change tips the balance in favor of disassembly (Figure 17–12B). Because the rest of the microtubule is composed of GDP-tubulin, once depolymerization has started, it will tend to continue, often at a catastrophic rate; the microtubule starts to shrink rapidly and continuously, and may even disappear.

The GDP-containing tubulin molecules that are freed as the microtubule depolymerizes join the unpolymerized tubulin molecules already in the cytosol. In a typical fibroblast, for example, at any one time about half of the tubulin in the cell is in microtubules, while the remainder is free in the cytosol, forming a pool of subunits available for microtubule growth. This situation is quite unlike the arrangement with the more stable intermediate filaments, where the subunits are typically almost completely in the fully assembled form. The tubulin molecules joining the pool then exchange their bound GDP for GTP, thereby becoming competent again to add to another microtubule that is in a growth phase.

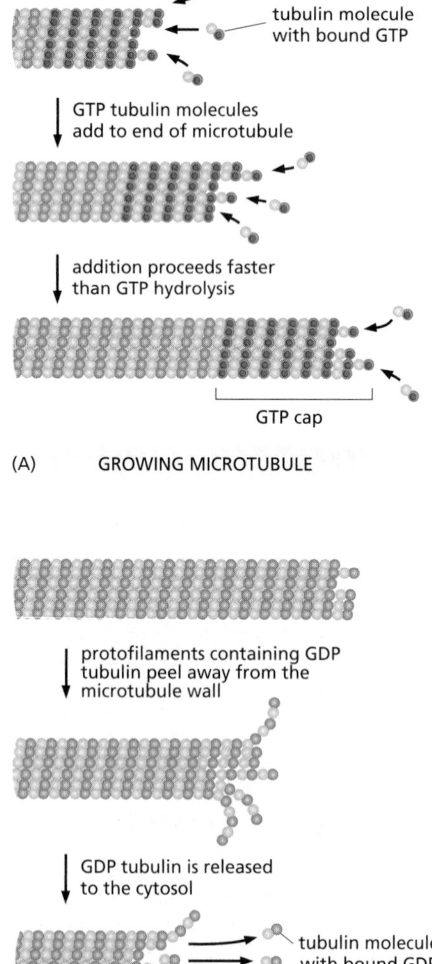
tubulin molecule with bound GTP
GTP tubulin molecules add to end of microtubule
addition proceeds faster than GTP hydrolysis
GTP cap
(A)
GROWING MICROTUBULE
protofilaments containing GDP tubulin peel away from the microtubule wall
GDP tubulin is released to the cytosol
tubulin molecule with bound GDP
(B)
SHRINKING MICROTUBULE

Figure 17–12 **GTP hydrolysis controls the growth of microtubules.** (A) Tubulin dimers carrying GTP *(red)* bind more tightly to one another than do tubulin dimers carrying GDP *(dark green)*. Therefore, microtubules that have freshly added tubulin dimers at their end with GTP bound tend to keep growing. (B) From time to time, however, especially when microtubule growth is slow, the subunits in this GTP cap will hydrolyze their GTP to GDP before fresh subunits loaded with GTP have time to bind. The GTP cap is thereby lost; the GDP-carrying subunits are less tightly bound in the polymer and are readily released from the free end, so that the microtubule begins to shrink continuously (Movie 17.3 and Movie 17.4).

Microtubules Are Maintained by a Balance of Assembly and Disassembly

The relative instability of microtubules allows them to undergo rapid remodeling, and this is crucial for microtubule function. In a normal cell, the centrosome (or other organizing center) is continually shooting out new microtubules in an exploratory fashion in different directions and retracting them. A microtubule growing out from the centrosome can, however, be prevented from disassembling if its plus end is somehow permanently stabilized by attachment to another molecule or cell structure so as to prevent tubulin depolymerization. If stabilized by attachment to a structure in a more distant region of the cell, the microtubule will establish a relatively stable link between that structure and the centrosome (Figure 17–13). The centrosome can be compared to a fisherman casting a line: if there is no bite at the end of the line, the line is quickly withdrawn and a new cast is made; but if a fish bites, the line remains in place, tethering the fish to the fisherman. This simple strategy of random exploration and selective stabilization enables the centrosome and other nucleating centers to set up a highly organized system of microtubules linking selected parts of the cell. It is also used to position organelles relative to one another.

Drugs that prevent the polymerization or depolymerization of tubulin can have a rapid and profound effect on the organization of the cytoskeleton—and the behavior of the cell. Consider the mitotic spindle, the microtubule framework that guides the chromosomes during mitosis (see Figure 17–8B). If a cell in mitosis is exposed to the drug *colchicine*, which binds tightly to free tubulin and prevents its polymerization into microtubules, the mitotic spindle rapidly disappears and the cell stalls in the middle of mitosis, unable to partition its chromosomes into two groups. This shows that the mitotic spindle is normally maintained by a continuous balanced addition and loss of tubulin subunits: when tubulin addition is blocked by colchicine, tubulin loss continues until the spindle disappears.

The drug *taxol* has the opposite action at the molecular level. It binds tightly to microtubules and prevents them from losing subunits. Because new subunits can still be added, the microtubules can grow but cannot shrink. However, despite the differences in molecular detail, taxol has the same overall effect on the cell as colchicine: it also arrests dividing cells in mitosis. We learn from this that for the spindle to function, microtubules must be able not only to assemble but also to disassemble. The behavior of the spindle is discussed in more detail in Chapter 18, when we consider mitosis.

The inactivation or destruction of the mitotic spindle eventually kills dividing cells. Cancer cells, which are dividing with less control than most other cells of the body, can sometimes be killed preferentially by microtubule-stabilizing and microtubule-destabilizing *antimitotic drugs*. Thus, drugs that interfere with microtubule polymerization or depolym-

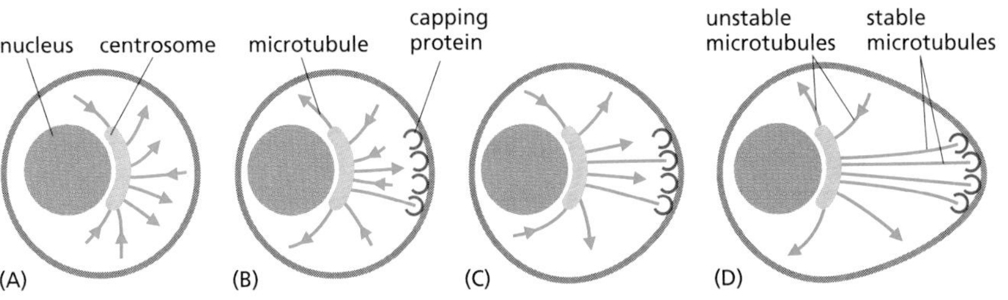

Figure 17–13 **The selective stabilization of microtubules can polarize a cell.** A newly formed microtubule will persist only if both its ends are protected from depolymerization. In cells, the minus ends of microtubules are generally protected by the organizing centers from which the filaments grow. The plus ends are initially free but can be stabilized by other proteins. Here, for example, a nonpolarized cell is depicted in (A) with new microtubules growing from and shrinking back to a centrosome in many directions randomly. Some of these microtubules happen by chance to encounter proteins (capping proteins) in a specific region of the cell cortex that can bind to and stabilize the free plus ends of microtubules (B). This selective stabilization will lead to a rapid reorientation of the microtubule arrays (C) and convert the cell to a strongly polarized form (D).

TABLE 17–1 DRUGS THAT AFFECT FILAMENTS AND MICROTUBULES	
	Action
Microtubule-specific drugs	
Taxol	binds and stabilizes microtubules
Colchicine, colcemid	binds subunits and prevents polymerization
Vinblastine, vincristine	binds subunits and prevents polymerization
Actin-specific drugs	
Phalloidin	binds and stabilizes filaments
Cytochalasin	caps filament plus ends
Latrunculin	binds subunits and prevents polymerization

erization, including colchicine, taxol, vincristine, and vinblastine, are used in the clinical treatment of cancer. As we discuss shortly, there also exist compounds that stabilize and destabilize actin filaments. Together, these drugs, listed in Table 17–1, allow biologists to study the function of the cytoskeleton.

Microtubules Organize the Interior of the Cell

Cells are able to modify the dynamic instability of their microtubules for particular purposes. As cells enter mitosis, for example, microtubules become initially more dynamic, switching between growing and shrinking much more frequently than cytoplasmic microtubules normally do. This enables them to disassemble rapidly and then reassemble into the mitotic spindle. On the other hand, when a cell has differentiated into a specialized cell type and taken on a definite fixed structure, the dynamic instability of its microtubules is often suppressed by proteins that bind to the ends of microtubules or along their length and stabilize them against disassembly. The stabilized microtubules then serve to maintain the organization of the cell.

Most differentiated animal cells are *polarized*; that is, one end of the cell is structurally or functionally different from the other. Nerve cells, for example, put out an axon from one end of the cell and dendrites from the other; cells specialized for secretion have their Golgi apparatus positioned toward the site of secretion, and so on. The cell's polarity is a reflection of the polarized systems of microtubules in its interior, which help to position organelles in their required location within the cell and to guide the streams of traffic moving between one part of the cell and another. In the nerve cell, for example, all the microtubules in the axon point in the same direction, with their plus ends toward the axon terminal (Figure 17–14). Along these oriented tracks the cell is able to send cargoes of materials, such as membrane vesicles and proteins for secretion, that are made in the cell body but required far away at the end of the axon.

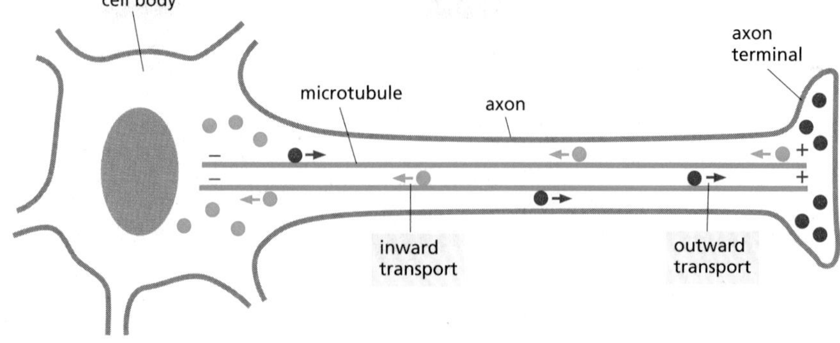

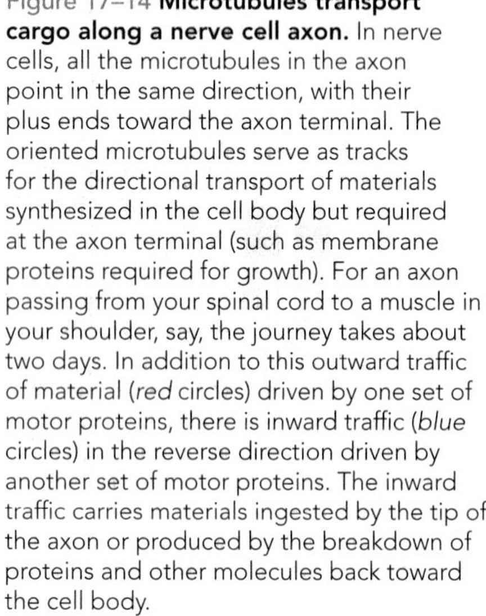

Figure 17–14 **Microtubules transport cargo along a nerve cell axon.** In nerve cells, all the microtubules in the axon point in the same direction, with their plus ends toward the axon terminal. The oriented microtubules serve as tracks for the directional transport of materials synthesized in the cell body but required at the axon terminal (such as membrane proteins required for growth). For an axon passing from your spinal cord to a muscle in your shoulder, say, the journey takes about two days. In addition to this outward traffic of material (*red* circles) driven by one set of motor proteins, there is inward traffic (*blue* circles) in the reverse direction driven by another set of motor proteins. The inward traffic carries materials ingested by the tip of the axon or produced by the breakdown of proteins and other molecules back toward the cell body.

Some of these materials move at speeds in excess of 10 cm a day, which means that they may still take a week or more to travel to the end of a long axon in larger animals. But movement along microtubules is immeasurably faster and more efficient than free diffusion. A protein molecule traveling by free diffusion would take years to reach the end of a long axon—if it arrived at all (see Question 17–12).

But the microtubules in living cells do not act alone. Their activity, like those of other cytoskeletal filaments, depends on a large variety of accessory proteins that bind to them. Some microtubule-associated proteins stabilize microtubules against disassembly, for example, while others link microtubules to other cell components, including the other types of cytoskeletal filaments. Yet other microtubule-associated proteins are motor proteins that carry organelles, vesicles and other cellular materials along the microtubules. Because the components of the cytoskeleton can interact with each other, their functions can be coordinated.

QUESTION 17–3

Dynamic instability causes microtubules either to grow or to shrink rapidly. Consider an individual microtubule that is in its shrinking phase.

A. What must happen at the end of the microtubule in order for it to stop shrinking and to start growing?

B. How would a change in the tubulin concentration affect this switch?

C. What would happen if only GDP, but no GTP, were present in the solution?

D. What would happen if the solution contained an analog of GTP that cannot be hydrolyzed?

Motor Proteins Drive Intracellular Transport

If a living cell is observed in a light microscope, its cytoplasm is seen to be in continual motion (Figure 17–15). Mitochondria and the smaller membrane-enclosed organelles and vesicles move in small, jerky steps—that is, they move for a short period, stop, and then start again. This *saltatory movement* is much more sustained and directional than the continual, small Brownian movements caused by random thermal motions. Both microtubules and actin filaments are involved in saltatory and other directed intracellular movements in eucaryotic cells. In both cases the movements are generated by **motor proteins**, which use the energy derived from repeated cycles of ATP hydrolysis to travel steadily along the actin filament or the microtubule in a single direction (see Figure 4–42). At the same time, these motor proteins also attach to other cell components and thus transport this cargo along the filaments. Dozens of motor proteins have been identified. They differ in the type of filament they bind to, the direction in which they move along the filament, and the cargo they carry.

The motor proteins that move along cytoplasmic microtubules, such as those in the axon of a nerve cell, belong to two families: the **kinesins** generally move toward the plus end of a microtubule (away from the centrosome; outward from the cell body in Figure 17–14), while the **dyneins** move toward the minus end (toward the centrosome; inward in Figure 17–14). These kinesins and dyneins are both dimers with two globular ATP-binding heads and a single tail (Figure 17–16A). The heads interact

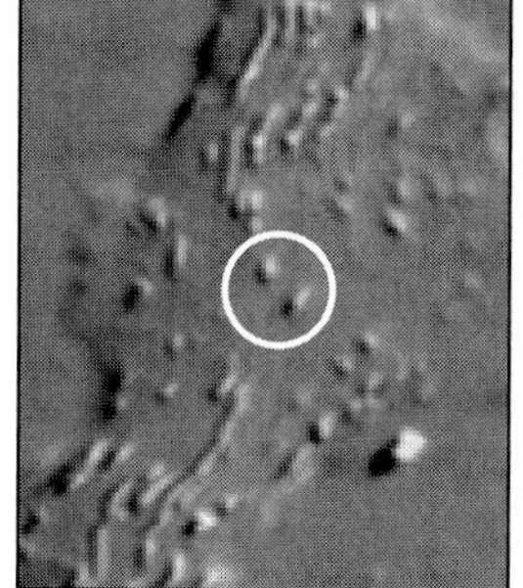
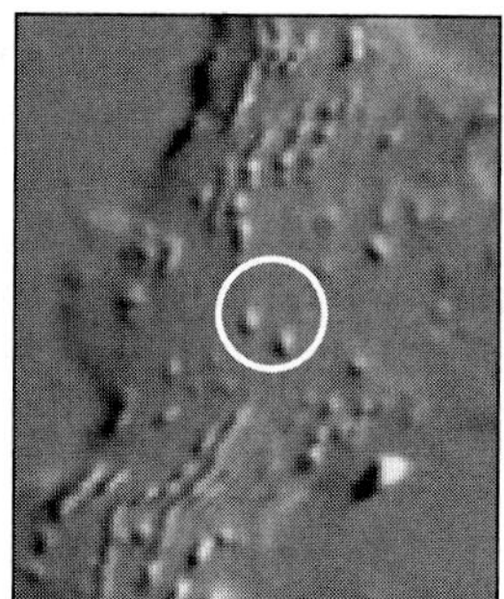
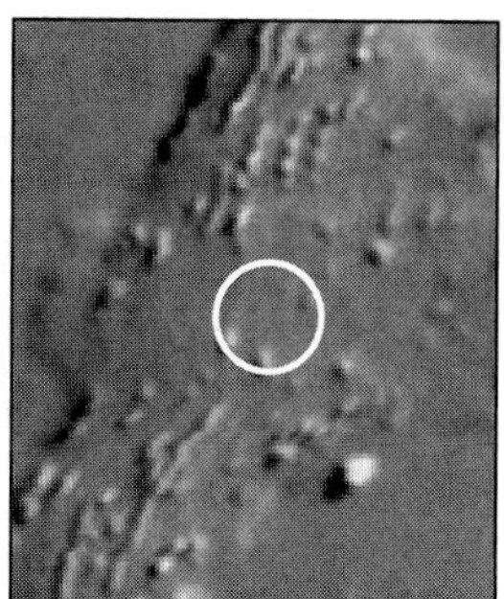
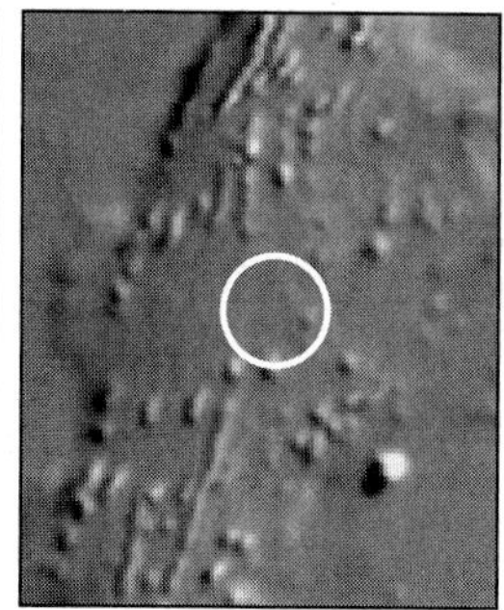
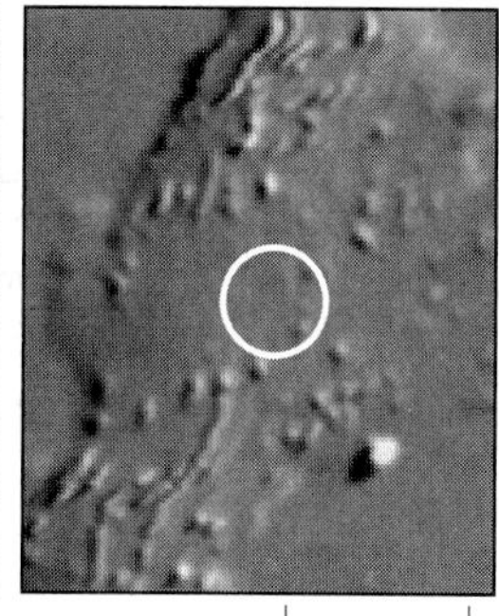

Figure 17–15 **Organelles move along microtubules at different speeds.** In this series of video-enhanced images of a flattened area of an invertebrate nerve cell, numerous membrane vesicles and mitochondria are present, many of which can be seen to move. The *white circle* provides a fixed frame of reference. These images were recorded at intervals of 400 milliseconds. (Courtesy of P. Forscher.)

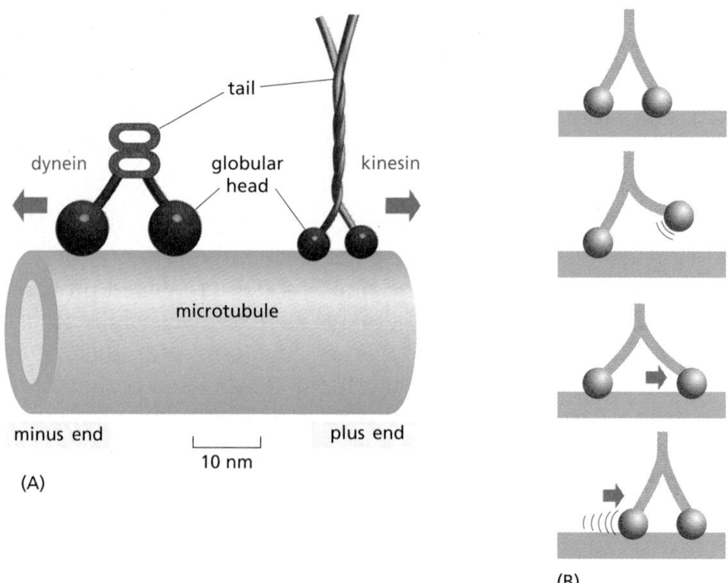

Figure 17–16 **Motor proteins move along microtubules using their globular heads.** (A) Kinesins and cytoplasmic dyneins are microtubule motor proteins that generally move in opposite directions along a microtubule. Each of these proteins (drawn here to scale) is a dimer composed of two identical molecules. Each protein has two globular heads that interact with microtubules at one end and a single tail at the other. (B) Schematic diagram of a motor protein showing ATP-dependent "walking" along a filament (see also Figure 4–42).

with microtubules in a stereospecific manner, so that the motor protein will attach to a microtubule in only one direction. The tail of a motor protein generally binds stably to some cell component, such as a vesicle or an organelle, and thereby determines the type of cargo that the motor protein can transport (Figure 17–17). The globular heads of kinesin and dynein are enzymes with ATP-hydrolyzing (ATPase) activity. This reaction provides the energy for a cycle of conformational changes in the head that enable it to move along the microtubule by a cycle of binding, release, and rebinding to the microtubule (see Figure 17–16B and Figure 4–42). For a discussion of the discovery and study of motor proteins, see How We Know, pp. 586–588.

Organelles Move Along Microtubules

Microtubules and motor proteins play an important part in positioning membrane-enclosed organelles within a eucaryotic cell. In most animal cells, for example, the tubules of the endoplasmic reticulum reach almost to the edge of the cell (Movie 17.5). The Golgi apparatus, in contrast, is located in the interior of the cell near the centrosome (Figure 17–18). Both the endoplasmic reticulum and the Golgi apparatus depend on microtubules for their alignment and positioning. The membranes of the endoplasmic reticulum extend out from their points of connection with the nuclear envelope (see Figure 1–22), aligning with microtubules that extend from the centrosome out to the plasma membrane. As the cell develops and the endoplasmic reticulum grows, kinesins attached to the outside of the endoplasmic reticulum membrane (via receptor proteins) pull it outward along microtubules, stretching it like a net. Dyneins,

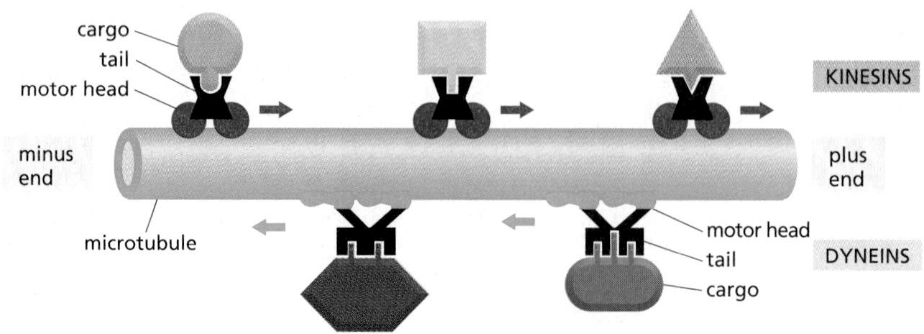

Figure 17–17 **Different motor proteins transport cargo along microtubules.** Most kinesins move toward the plus end of a microtubule, whereas dyneins move toward the minus end (Movie 17.6). Both types of microtubule motor proteins exist in many forms, each of which is thought to transport a different cargo. The tail of the motor protein determines what cargo the protein transports.

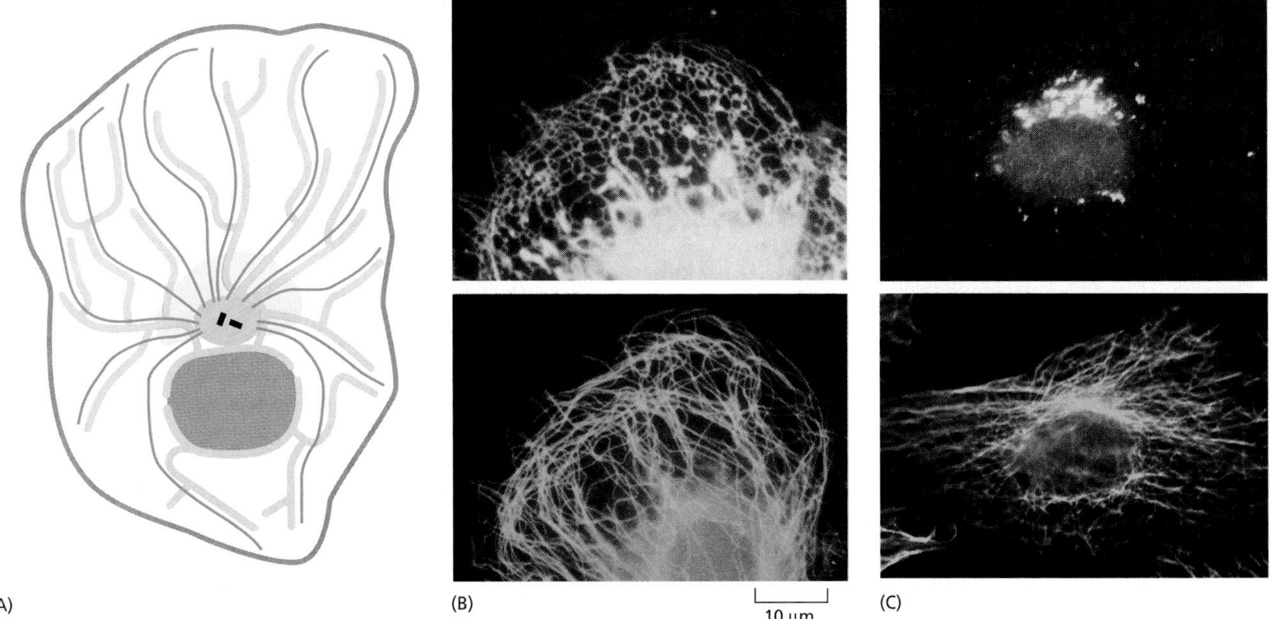

Figure 17–18 **Microtubules help to arrange the organelles in a eucaryotic cell.** (A) Schematic diagram of a cell showing the typical arrangement of microtubules *(dark green)*, endoplasmic reticulum *(blue)*, and Golgi apparatus *(yellow)*. The nucleus is shown in *brown*, and the centrosome in *light green*. (B) Cell stained with antibodies to endoplasmic reticulum *(upper panel)* and to microtubules *(lower panel)*. Motor proteins pull the endoplasmic reticulum out along microtubules. (C) Cell stained with antibodies to the Golgi apparatus *(upper panel)* and to microtubules *(lower panel)*. In this case, motor proteins move the Golgi apparatus inward to its position near the centrosome. (B, courtesy of Mark Terasaki, Lan Bo Chen, and Keigi Fujiwara; C, courtesy of Viki Allan and Thomas Kreis.)

similarly attached to the Golgi membranes, pull the Golgi apparatus the other way along microtubules, inward toward the cell center. In this way the regional differences in internal membranes, on which the successful function of the cell depends, are created and maintained.

When cells are treated with a drug such as colchicine that causes microtubules to disassemble, both of these organelles change their location dramatically. The endoplasmic reticulum, which has connections to the nuclear envelope, collapses to the center of the cell, while the Golgi apparatus, which is not attached to any other organelle, fragments into small vesicles, which disperse throughout the cytoplasm. When the drug is removed, the organelles return to their original positions, dragged by motor proteins moving along the re-formed microtubules.

Cilia and Flagella Contain Stable Microtubules Moved by Dynein

Earlier in this chapter we mentioned that many microtubules in cells are stabilized through their association with other proteins, and therefore no longer show dynamic instability. Stable microtubules are employed by cells as stiff supports on which to construct a variety of polarized structures, including the remarkable cilia and flagella that allow eucaryotic cells to move water over their surface. **Cilia** are hairlike structures about 0.25 μm in diameter, covered by plasma membrane, that extend from the surface of many kinds of eucaryotic cells (see Figure 17–8C). A single cilium contains a core of stable microtubules, arranged in a bundle, that grow from a *basal body* in the cytoplasm; the basal body serves as the organizing center for the cilium.

PURSUING MOTOR PROTEINS

The movement of organelles throughout the cell cytoplasm has been observed, measured, and speculated about since the middle of the nineteenth century. But it was not until the mid-1980s that biologists identified the molecules that drive this movement of organelles and vesicles from one part of the cell to another.

Why the lag between observation and understanding? The problem was in the proteins—or, more precisely, in the difficulty of studying them in isolation outside the cell. To investigate the activity of an enzyme, for example, biochemists first purify the polypeptide: they break open cells or tissues and separate the protein of interest from other molecular components (see Panels 4–4 and 4–5, pp. 164–167). They can then study the protein on its own, *in vitro,* controlling its exposure to substrates, inhibitors, ATP, and so on. Unfortunately, this approach did not seem to work for studies of the motile machinery that underlies intracellular transport. It is not possible to break open a cell and pull out a fully active transport system, free of extraneous material, that continues to carry mitochondria and vesicles from place to place.

The techniques needed to move the research forward came from two different sources. First, advances in microscopy allowed biologists to see that an operational transport system (with extraneous material still attached) could be squeezed from the right kind of living cell. At the same time, biochemists realized that they could assemble a working transport system from scratch—using purified cables, motors, and cargo—outside the cell. The breakthrough started with a squid.

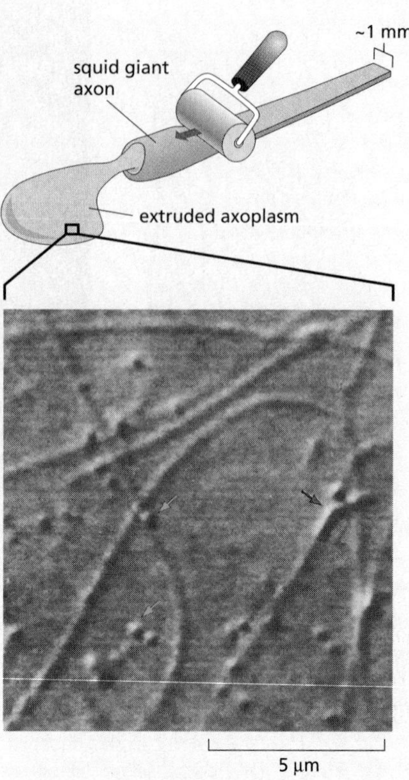

Figure 17–19 **Video-enhanced microscopy of cytoplasm squeezed from a squid giant axon reveals the motion of organelles.** In this micrograph numerous cytoskeletal filaments are visible, along with transported particles including a mitochondrion *(red arrow)* and smaller vesicles *(blue arrows).* (From R.D. Vale et al., *Cell* 40:449–454, 1985. With permission from Elsevier.)

Teeming cytoplasm

As we saw in Chapter 12, neuroscientists interested in the electrical properties of nerve cell membranes have long studied the giant axon from squid (see How We Know, pp. 412–413). Because of its large size, researchers found that they could squeeze the cytoplasm from the axon like toothpaste, and then study how ions move back and forth through various channels in the empty, tubelike membrane. The physiologists simply discarded the cytoplasmic jelly, as it appeared to be inert (and thus uninteresting) when examined under a standard light microscope.

Then along came video-enhanced microscopy. This type of microscopy, developed by Shinya Inoué, Robert Allen, and others, allows one to detect structures that are smaller than the resolution power of standard light microscopes, about 0.2 μm, or 200 nm (see Panel 1–1, pp. 8–9). Sample images are captured by a video camera and then enhanced by computer processing to reduce the background and heighten contrast. When researchers in the early 1980s applied this new technique to preparations of squid axon cytoplasm (axoplasm), they observed, for the first time, the motion of vesicles and membrane-enclosed organelles along cytoskeletal filaments.

Under the video microscope, extruded axoplasm is seen to be teeming with tiny particles—from vesicles 30–50 nm in diameter to mitochondria some 5000 nm long, all moving to and fro along cytoskeletal filaments at speeds of up to 5 μm per second. If the axoplasm is spread thinly enough, individual filaments can be seen (Figure 17–19).

The movement continues for hours, allowing researchers to manipulate the preparation and study the effects. Ray Lasek and Scott Brady discovered, for example, that organelle movement requires ATP. Substitution of ATP analogs, such as AMP-PNP, which bind to the enzyme active site but cannot be hydrolyzed (and thus provide no energy), inhibit the translocation.

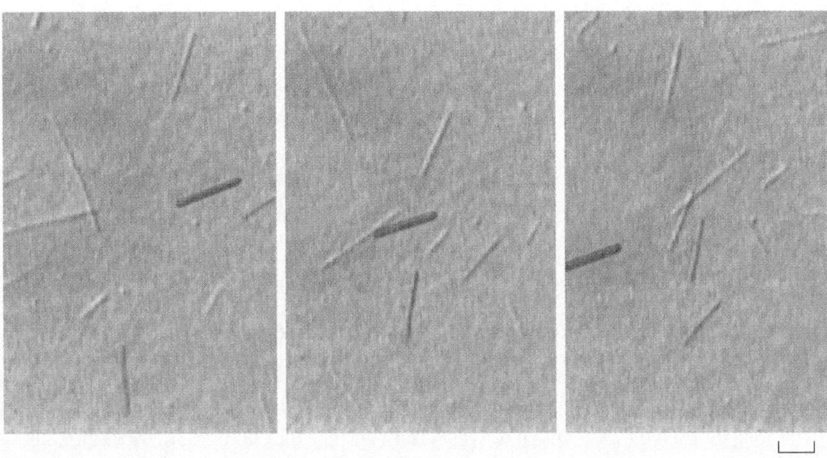

Figure 17–20 **A motor protein causes microtubule gliding.** In an active *in vitro* motility assay, purified kinesin is mixed with microtubules in a buffer containing ATP. When a drop of the mixture is placed on a glass slide and examined by video-enhanced microscopy, individual microtubules can be seen gliding over the slide driven by kinesin molecules. Images recorded at 1 second intervals. (Courtesy of Nick Carter and Rob Cross.)

Snaking tubes

More work was needed to identify the individual components that drive the transport system in squid axons. What are the filaments made of? What are the molecular machines that shuttle the vesicles and organelles along these filaments? Identifying the cables was relatively easy. Studies using antibodies to α-tubulin revealed that the filaments are microtubules. But what about the motor proteins? To find these, Ron Vale, Thomas Reese, and Michael Sheetz set up a system in which they could fish for proteins that power organelle movement.

Their strategy was simple yet elegant: add together purified cables and purified cargo and then look for molecules that induce motion. They took purified microtubules from squid optic lobe, added organelles isolated from squid axons, and showed that movement could be triggered by the addition of an extract from squid axon cytoplasm. In this preparation, the researchers could watch organelles travel along the microtubules, and watch microtubules glide snakelike over the surface of a glass coverslip (see Question 17–18). Their challenge was to isolate the protein responsible for movement in this reconstituted system.

To do that, Vale and his colleagues took advantage of Lasek and Brady's earlier work with the ATP analog AMP-PNP. Although this analog inhibits the movement of vesicles along microtubules, it still allows these components to attach to the microtubule filaments. So the researchers incubated the cytoplasm extract with microtubules in the presence of AMP-PNP; they then pulled out the microtubules with what they hoped were the motor proteins still attached. Vale and his team then added ATP to release the attached proteins, and they found a 110-kilodalton polypeptide that could bind to, and initiate movement of, microtubules *in vitro* (Figure 17–20). They dubbed the molecule kinesin (from the Greek *kinein,* "to move").

Such *in vitro* motility assays have been instrumental in the study of motor proteins and their activities. Subsequent studies showed that kinesin moves along microtubules from the minus end to the plus end, and also identified many other kinesin-related motor proteins.

Lights, camera, action

Combining these *in vitro* assays with ever more refined microscopic techniques, researchers can now monitor the movement of individual motor proteins along single microtubules, even in living cells. In an assay developed by Steven Block and his colleagues in 1990, microscopic silica beads coated with low concentrations of kinesin (so that only one molecule of kinesin is present on each bead) can be monitored as they make their way down a microtubule (Figure 17–21). Other observations of single kinesin molecules are made possible by coupling the motor protein with a fluorescent marker protein such as GFP.

Such single-molecule studies have revealed that kinesin moves along microtubules processively—that is, each molecule takes 100 or so "steps" along the filament before falling off (Figure 17–22). The length of each step is 8 nm, which corresponds to the spacing of tubulin dimers along the microtubule. Combining these observations with assays of ATP hydrolysis, researchers have confirmed that one molecule of ATP is consumed per step. Kinesin can move in a processive man-

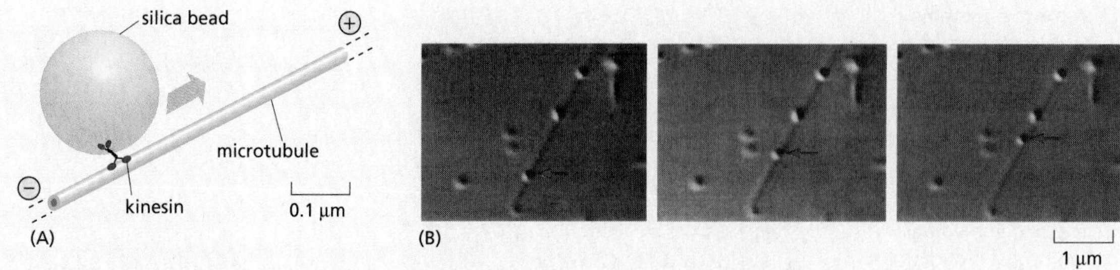

Figure 17–21 **Video microscopy can be used to track the movement of a single kinesin molecule.** (A) In this assay, silica beads are coated with kinesin molecules at a concentration such that each bead, on average, will have only one kinesin molecule attached to it. Kinesin is then allowed to walk along a microtubule, and its movement is monitored by tracking the movement of the bead. (B) In this series of images, the bead is captured by a laser-based optical tweezer, placed on a microtubule filament, and then allowed to move. Thirty seconds elapses between each frame. (From S. Block et al., *Nature* 348:348–352, 1990. With permission from Macmillan Publishers Ltd.)

ner because it has two heads (see Figure 17–22). The motor is thought to walk its way toward the plus end of the microtubule in a hand-over-hand fashion, each head binding and releasing the filament in turn. Further studies are required to refine this model, and researchers are now working to improve their methods so that they can watch not only single molecules of kinesin, but also each individual head as it moves, in relation to its partner, along the microtubule. The results will yield additional insights into the molecular movements that underlie the organization and activity of eucaryotic cells.

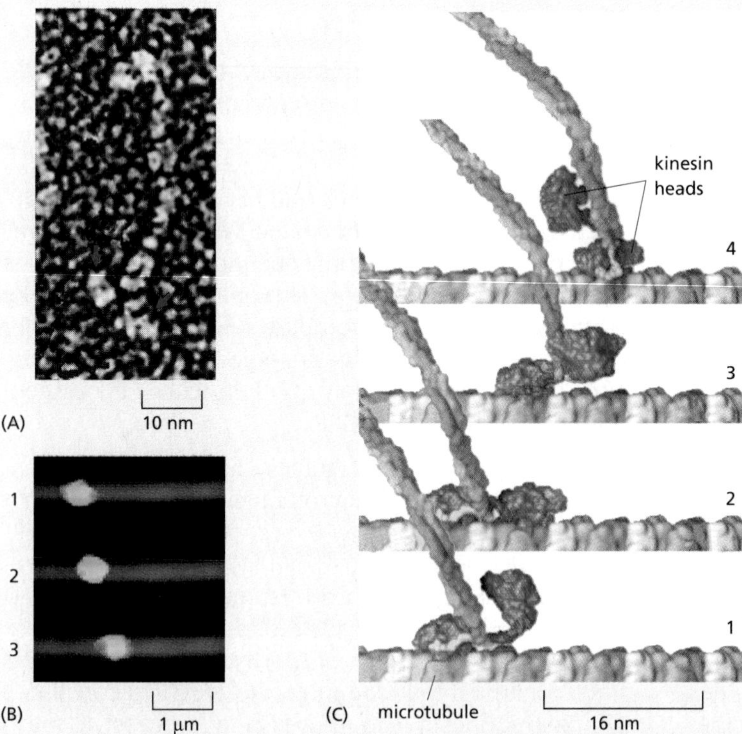

Figure 17–22 **A single molecule of kinesin moves along a microtubule.** (A) Electron micrograph of a single kinesin molecule showing the two head domains (*red* arrows). (B) Three frames, separated by intervals of 1 second, record the movement of an individual kinesin-GFP molecule (*green*) along a microtubule (*red*) at a speed of 0.3 μm/sec. (C) Series of molecular models of the two heads of a kinesin molecule, showing how they are thought to processively walk their way along a microtubule in a series of 8-nm steps (Movie 17.7). (A, courtesy of John Heuser; B and C, courtesy of Ron Vale.)

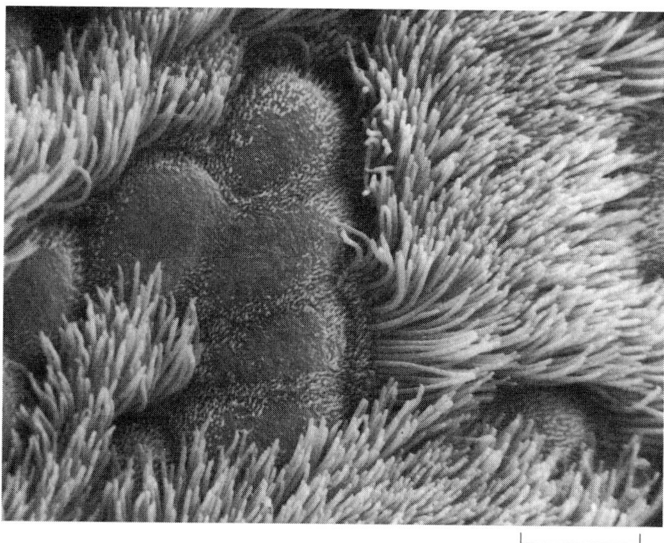

Figure 17–23 **Hairlike cilia coat the surface of many eucaryotic cells.** Scanning electron micrograph of the ciliated epithelium on the surface of the human respiratory tract. The thick tufts of cilia on the ciliated cells are interspersed with the dome-shaped surfaces of nonciliated epithelial cells. (Reproduced from R.G. Kessel and R.H. Karden, Tissues and Organs. San Francisco: W.H. Freeman & Co., 1979. With permission from W.H. Freeman & Co.)

Cilia move fluid over the surface of a cell or propel single cells through a fluid. Some protozoa, for example, use cilia to collect food particles, and others use them for locomotion. On the epithelial cells lining the human respiratory tract (Figure 17–23), huge numbers of cilia (more than a billion per square centimeter) sweep layers of mucus containing trapped dust particles and dead cells up toward the throat, to be swallowed and eventually eliminated from the body. Similarly, cilia on the cells of the oviduct wall create a current that helps to move eggs along the oviduct. Each cilium acts as a small oar, moving in a repeated cycle that generates the current that washes over the cell surface (Figure 17–24).

The **flagella** (singular **flagellum**) that propel sperm and many protozoa are much like cilia in their internal structure but are usually very much longer. They are designed to move the entire cell, and instead of generating a current, they propagate regular waves along their length that drive the cell through liquid (Figure 17–25).

The microtubules in cilia and flagella are slightly different from the cytoplasmic microtubules; they are arranged in a curious and distinctive pattern that was one of the most striking revelations of early electron microscopy. A cross section through a cilium shows nine doublet microtubules arranged in a ring around a pair of single microtubules (Figure 17–26A). This "9 + 2" array is characteristic of almost all forms of eucaryotic cilia and flagella, from those of protozoa to those found in humans.

The movement of a cilium or a flagellum is produced by the bending of its core as the microtubules slide against each other. The microtubules are associated with numerous proteins (Figure 17–26B), which project at regular positions along the length of the microtubule bundle. Some serve as cross-links to hold the bundle of microtubules together; others generate the force that causes the cilium to bend.

The most important of these accessory proteins is the motor protein *ciliary dynein*, which generates the bending motion of the core. It closely resembles cytoplasmic dynein and functions in much the same way. Ciliary dynein is attached by its tail to one microtubule, while its heads interact with an adjacent microtubule to generate a sliding force between the two filaments. Because of the multiple links that hold the adjacent microtubule doublets together, what would be a simple parallel sliding movement between free microtubules is converted to a bending motion

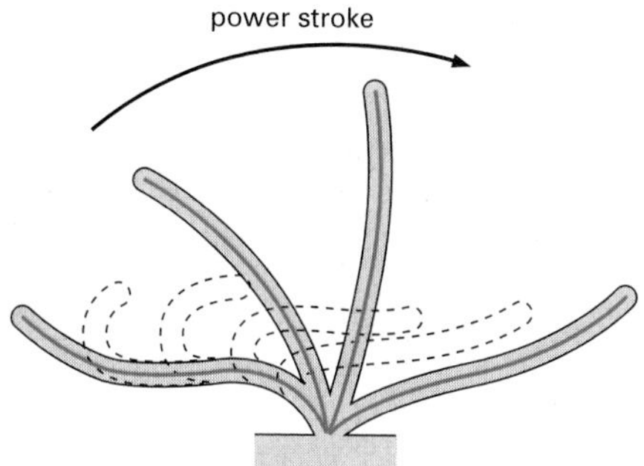

Figure 17–24 **A cilium beats by performing a repetitive cycle of movements consisting of a power stroke followed by a recovery stroke.** In the fast power stroke, the cilium is fully extended and fluid is driven over the surface of the cell; in the slower recovery stroke, the cilium curls back into position with minimal disturbance to the surrounding fluid. Each cycle typically requires 0.1–0.2 second and generates a force perpendicular to the axis of the cilium.

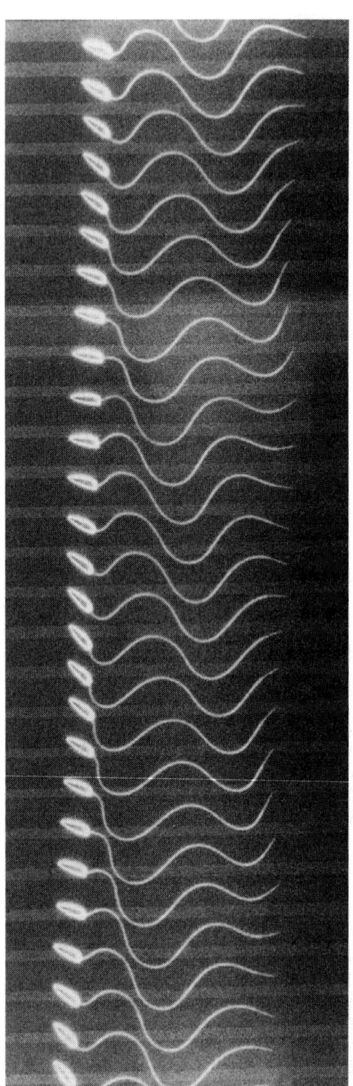

Figures 17–25 **Flagella propel a cell using a repetitive wavelike motion.** The wavelike motion of a single flagellum on a tunicate sperm is seen in a series of images captured by stroboscopic illumination at 400 flashes per second. (Courtesy of Charles J. Brokaw.)

in the cilium (Figure 17–27). In humans, hereditary defects in ciliary dynein cause Kartagener's syndrome. Men with this disorder are infertile because their sperm are non-motile, and all those affected have an increased susceptibility to bronchial infections because the cilia that line their respiratory tract are paralyzed and thus unable to clear bacteria and debris from the lungs.

ACTIN FILAMENTS

Actin filaments are found in all eucaryotic cells and are essential for many of their movements, especially those involving the cell surface. Without actin filaments, for example, an animal cell could not crawl along a surface, engulf a large particle by phagocytosis, or divide in two. Like microtubules, many actin filaments are unstable, but by associating with other proteins they can also form stable structures in cells, such as the contractile apparatus of muscle. Actin filaments interact with a large number of *actin-binding proteins* that enable the filaments to serve a variety of functions in cells. Depending on their association with different proteins, actin filaments can form stiff and relatively permanent structures, such as the *microvilli* on the brush-border cells lining the intestine (Figure 17–28A) or small *contractile bundles* in the cytoplasm that can contract and act like the "muscles" of a cell (Figure 17–28B); they can also form temporary structures, such as the dynamic protrusions formed at the leading edge of a crawling fibroblast (Figure 17–28C) or the *contractile ring* that pinches the cytoplasm in two when an animal cell divides (Figure 17–28D). In this section, we see how the arrangements of actin filaments in a cell depend on the types of actin-binding proteins present.

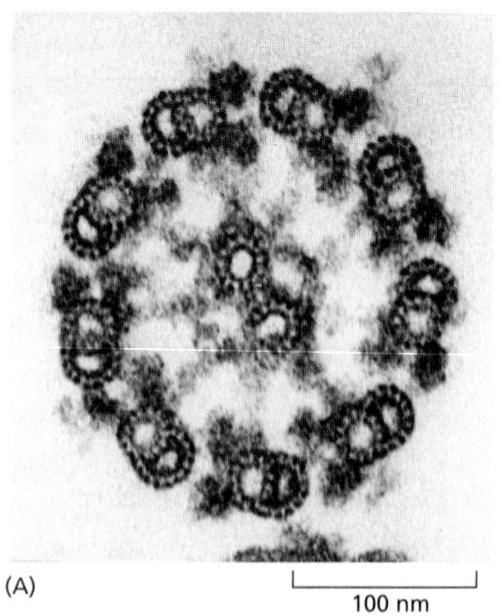

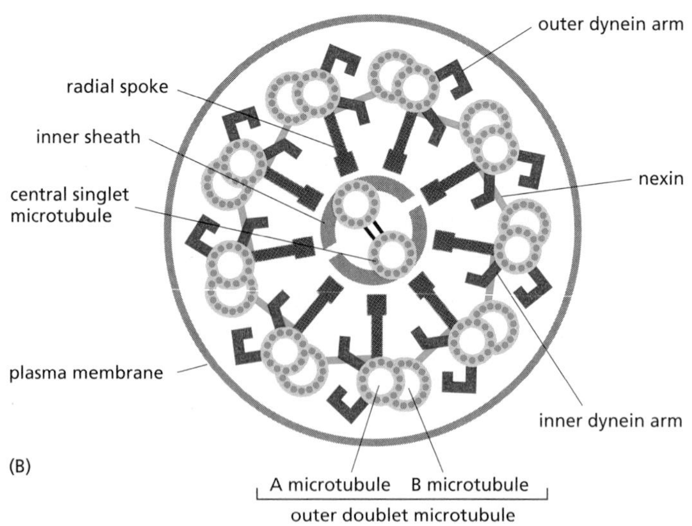

Figure 17–26 **Microtubules in a cilium or flagellum are arranged in a "9 + 2" array.** (A) Electron micrograph of a flagellum of *Chlamydomonas* shown in cross section, illustrating the distinctive 9 + 2 arrangement of microtubules. (B) Diagram of the flagellum in cross section. The nine outer microtubules (each a special paired structure) carry two rows of dynein molecules. The heads of these dyneins appear in this view like pairs of arms reaching toward the adjacent microtubule. In a living cilium, these dynein heads periodically make contact with the adjacent microtubule and move along it, thereby producing the force for ciliary beating. Various other links and projections shown are proteins that serve to hold the bundle of microtubules together and to convert the sliding motion produced by dyneins into bending, as illustrated in Figure 17–27. (A, courtesy of Lewis Tilney.)

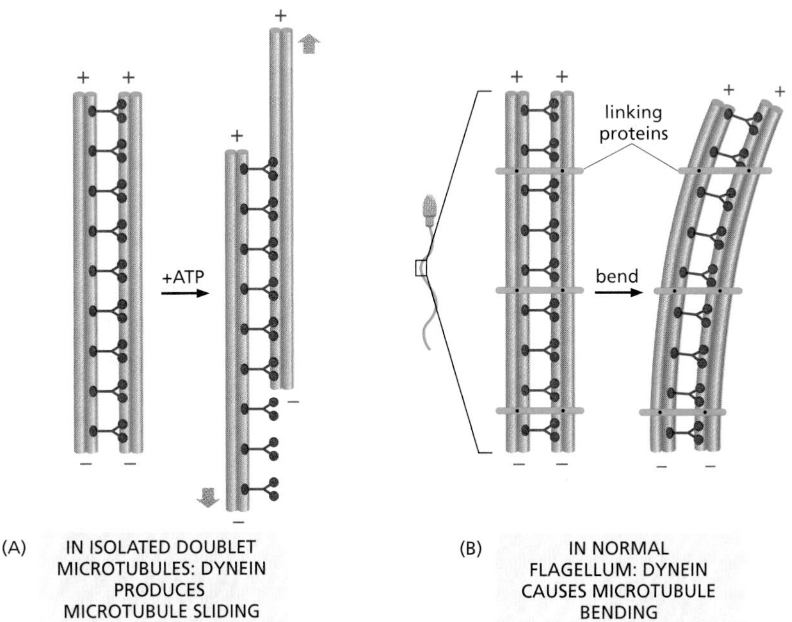

Figure 17–27 **The movement of dynein causes the flagellum to bend.** (A) If the outer doublet microtubules and their associated dynein molecules are freed from other components of a sperm flagellum and then exposed to ATP, the doublets slide against each other, telescope-fashion, due to the repetitive action of their associated dyneins. (B) In an intact flagellum, however, the doublets are tied to each other by flexible protein links so that the action of the system produces bending rather than sliding.

QUESTION 17–4

Dynein arms in a cilium are arranged so that, when activated, the heads push their neighboring outer doublet outward toward the tip of the cilium. Consider a cross section of a cilium (see Figure 17–26). Why would no bending motion of the cilium result if all dynein molecules were active at the same time? What pattern of dynein activity can account for the bending of a cilium in one direction?

Even though actin filaments and microtubules are formed from unrelated types of proteins, we shall see that the principles according to which they assemble and disassemble, control cell structure, and bring about movement are strikingly similar.

Actin Filaments Are Thin and Flexible

Actin filaments appear in electron micrographs as threads about 7 nm in diameter. Each filament is a twisted chain of identical globular actin molecules, all of which "point" in the same direction along the axis of the chain. Like a microtubule, therefore, an actin filament has a structural polarity, with a plus end and a minus end (Figure 17–29).

Actin filaments are thinner, more flexible, and usually shorter than microtubules. There are, however, many more of them, so that the total length of all the actin filaments in a cell is generally many times greater than the total length of all of the microtubules. Actin filaments rarely occur in isolation in the cell; they are generally found in cross-linked bundles and networks, which are much stronger than the individual filaments.

Actin and Tubulin Polymerize by Similar Mechanisms

Actin filaments can grow by the addition of actin monomers at either end, but the rate of growth is faster at the plus end than at the minus end. A naked actin filament, like a microtubule without associated proteins, is inherently unstable, and it can disassemble from both ends. Each free actin monomer carries a tightly bound nucleoside triphosphate, in this case ATP, which is hydrolyzed to ADP soon after the incorporation of

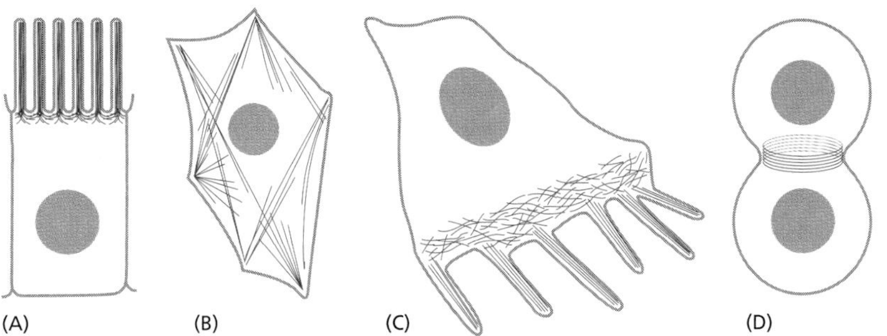

Figure 17–28 **Actin filaments allow eucaryotic cells to adopt a variety of shapes and perform a variety of functions.** Various actin-containing structures are shown here in *red:* (A) microvilli; (B) contractile bundles in the cytoplasm; (C) sheetlike *(lamellipodia)* and fingerlike *(filopodia)* protrusions from the leading edge of a moving cell; (D) contractile ring during cell division.

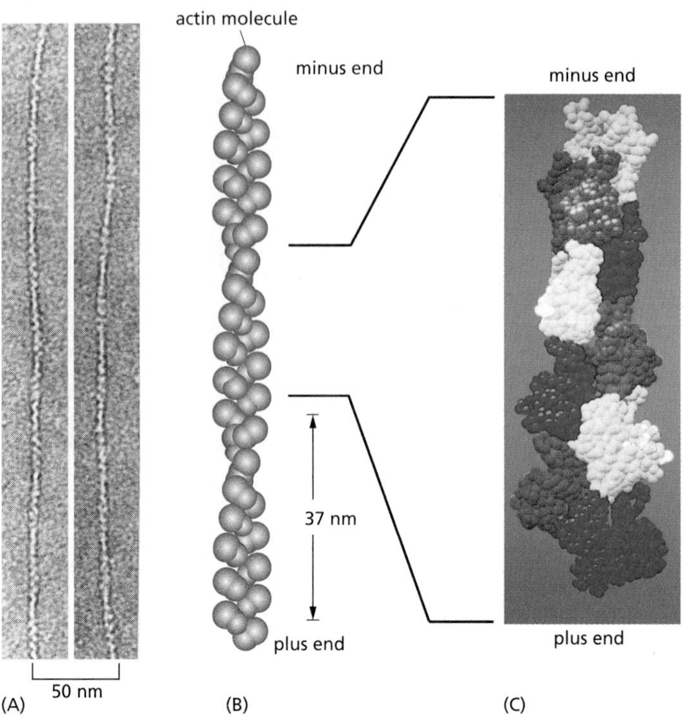

Figure 17–29 **Actin filaments are thin, flexible protein threads.** (A) Electron micrographs of negatively stained actin filaments. (B) Arrangement of actin molecules in an actin filament. Each filament may be thought of as a two-stranded helix with a twist repeating every 37 nm. Strong interactions between the two strands prevent the strands from separating. (C) The identical subunits of an actin filament are depicted in different colors to emphasize the close interaction between each actin molecule and its four nearest neighbors. (A, courtesy of Roger Craig; C, from K.C. Holmes et al., *Nature* 347:44–49, 1990. With permission from Macmillan Publishers Ltd.)

the actin monomer into the filament. As with the GTP bound to tubulin, hydrolysis of ATP to ADP in an actin filament reduces the strength of binding between monomers and decreases the stability of the polymer. Nucleotide hydrolysis thereby promotes depolymerization, helping the cell to disassemble filaments after they have formed (Figure 17–30).

As with microtubules, the ability to assemble and disassemble is required for many of the functions performed by actin filaments, such as their role in cell locomotion. Actin filament function can be perturbed experimentally by certain toxins produced by fungi or marine sea sponges. Some, such as the *cytochalasins,* prevent actin polymerization; others, such as *phalloidin,* stabilize actin filaments against depolymerization (see Table 17–1, p. 582). Addition of these toxins to the medium bathing cells or tissues, even in low concentrations, instantaneously freezes cell movements such as the crawling motion of a fibroblast. Thus, the function of actin filaments depends on a dynamic equilibrium between the actin filaments and the pool of actin monomers. Filaments often persist for only a few minutes after they are formed.

QUESTION 17–5

The formation of actin filaments in the cytosol is controlled by actin-binding proteins. Some actin-binding proteins significantly increase the rate at which the formation of an actin filament is initiated. Suggest a mechanism by which they might do this.

Many Proteins Bind to Actin and Modify Its Properties

About 5% of the total protein in a typical animal cell is actin; about half of this actin is assembled into filaments, and the other half remains as actin monomers in the cytosol. The concentration of monomer is there-

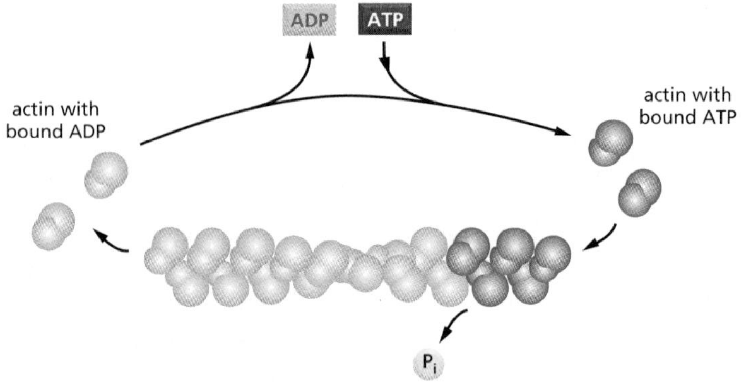

Figure 17–30 **ATP hydrolysis decreases the stability of the actin polymer.** Actin monomers in the cytosol carry ATP, which is hydrolyzed to ADP soon after assembly into a growing filament. The ADP molecules remain trapped within the actin filament, unable to exchange with ATP until the actin monomer that carries them dissociates from the filament.

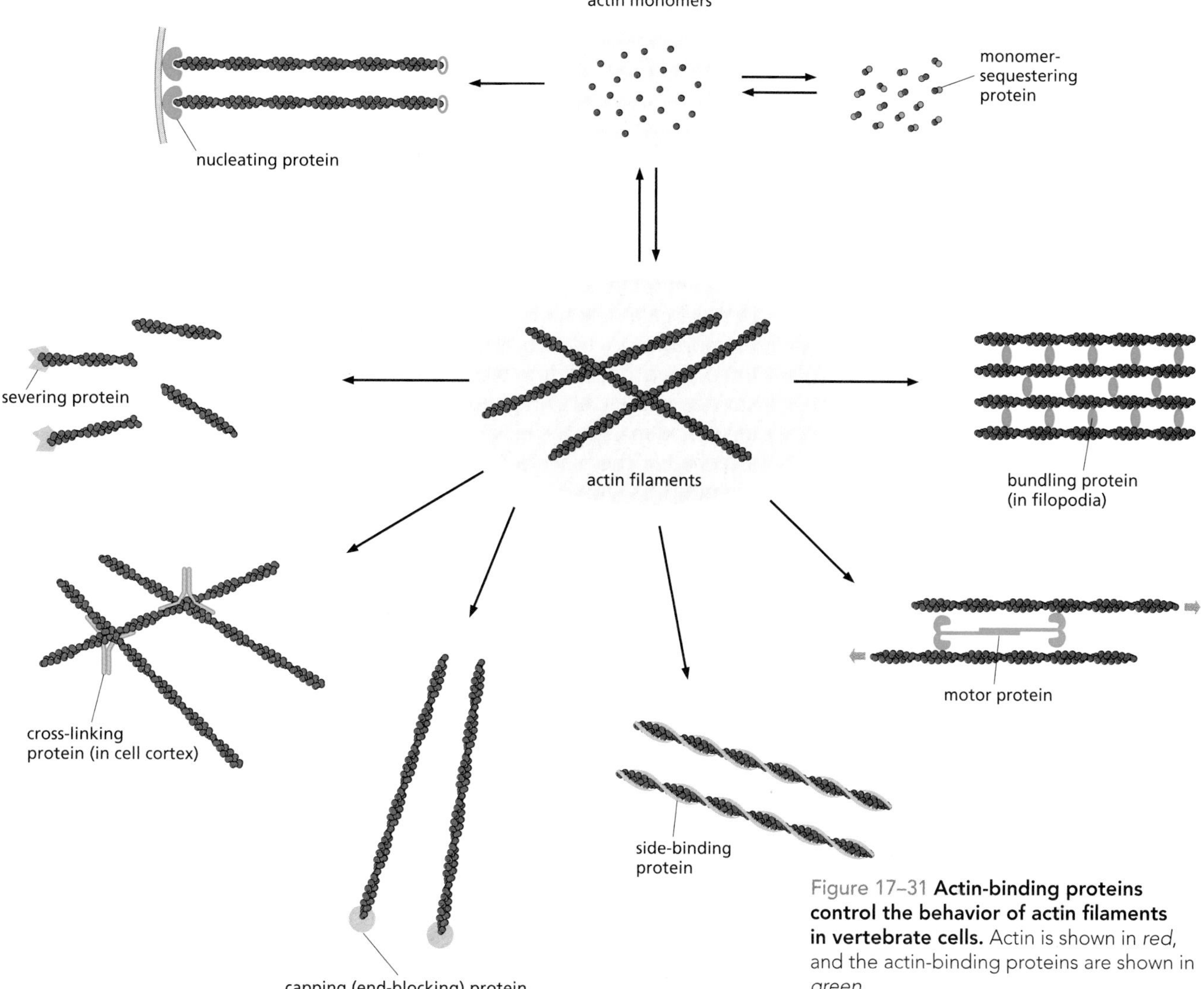

Figure 17–31 **Actin-binding proteins control the behavior of actin filaments in vertebrate cells.** Actin is shown in *red*, and the actin-binding proteins are shown in *green*.

fore high—much higher than the concentration required for purified actin monomers to polymerize *in vitro*. What, then, keeps the actin monomers in cells from polymerizing totally into filaments? The answer is that cells contain small proteins, such as *thymosin* and *profilin,* that bind to actin monomers in the cytosol, preventing them from adding to the ends of actin filaments. By keeping actin monomers in reserve until they are required, these proteins play a crucial role in regulating actin polymerization. When actin filaments are needed, other actin-binding proteins promote their assembly. Proteins called *formins* and *actin-related proteins (ARPs)* both control actin assembly at the advancing front of a migrating cell.

There are a great many actin-binding proteins in cells. Most of these bind to assembled actin filaments rather than to actin monomers and control the behavior of the intact filaments (Figure 17–31). Actin-bundling proteins, for example, hold actin filaments together in parallel bundles in microvilli; other cross-linking proteins hold actin filaments together in a gel-like meshwork within the *cell cortex*—the layer of cytoplasm just beneath the plasma membrane; filament-severing proteins, such as *gelsolin,* fragment actin filaments into shorter lengths and thus can convert an actin gel to a more fluid state. Actin filaments can also associate with motor proteins to form contractile bundles, as in muscle cells. And they often form tracks along which motor proteins transport organelles, a function that is especially conspicuous in plant cells.

In the remainder of this chapter, we consider some characteristic structures that actin filaments can form, and discuss how different types of actin-binding proteins are involved in their formation. We begin with the actin-rich cell cortex and its role in cell locomotion, and in the final section we consider the contractile apparatus of muscle cells as an example of a stable structure based on actin filaments.

An Actin-rich Cortex Underlies the Plasma Membrane of Most Eucaryotic Cells

Although actin is found throughout the cytoplasm of a eucaryotic cell, in most cells it is highly concentrated in a layer just beneath the plasma membrane. In this region, called the **cell cortex**, actin filaments are linked by actin-binding proteins into a meshwork that supports the outer surface of the cell and gives it mechanical strength. In red blood cells, as described in Chapter 11, a simple and regular network of fibrous proteins attached to the plasma membrane provides it with support necessary to maintain its simple discoid shape (see Figure 11–31). The cell cortex of other animal cells, however, is thicker and more complex and supports a far richer repertoire of shapes and movements. Like the cortex in a red cell, it contains spectrin and ankyrin; however, it also includes a dense network of actin filaments that project into the cytoplasm, where they become cross-linked into a three-dimensional meshwork. This cortical actin mesh governs the shape and mechanical properties of the plasma membrane and the cell surface: the rearrangement of actin filaments within the cortex provides the molecular basis for changes in cell shape and cell locomotion.

Cell Crawling Depends on Actin

Many cells move by crawling over surfaces, rather than by swimming by means of cilia or flagella. Carnivorous amoebae crawl continually, in search of food. The advancing tip of a developing axon migrates in response to growth factors, following a path of substrate-bound and diffusible chemicals to its eventual synaptic target. White blood cells known as *neutrophils* migrate through tissues when they 'smell' small diffusing molecules released by bacteria, which the neutrophils seek out and destroy. For these immune hunters, chemotactic molecules binding to receptors on the cell surface trigger changes in actin filament assembly that drive the cells toward their prey.

The molecular mechanisms of these and other forms of cell crawling entail coordinated changes of many molecules in different regions of the cell, and no single, easily identifiable locomotory organelle, such as a flagellum, is responsible. In broad terms, however, three interrelated processes are known to be essential: (1) the cell pushes out protrusions at its "front," or leading edge; (2) these protrusions adhere to the surface over which the cell is crawling; and (3) the rest of the cell drags itself forward by traction on these anchorage points (Figure 17–32).

All three processes involve actin, but in different ways. The first step, the pushing forward of the cell surface, is driven by actin polymerization. The leading edge of a crawling fibroblast in culture regularly extends thin, sheetlike **lamellipodia**, which contain a dense meshwork of actin filaments, oriented so that most of the filaments have their plus ends close to the plasma membrane (Figure 17–33). Many cells also extend thin, stiff protrusions called **filopodia**, both at the leading edge and elsewhere on their surface. These are about 0.1 μm wide and 5–10 μm long, and each contains a loose bundle of 10–20 actin filaments, again oriented with their plus ends pointing outward. The advancing tip (growth cone) of a

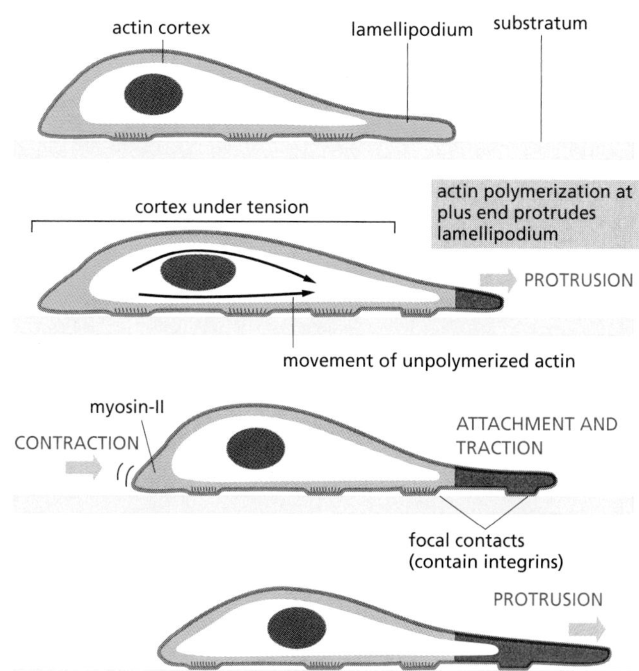

Figure 17–32 **Forces generated in the actin-rich cortex move a cell forward.** In this proposed mechanism for cell movement, actin polymerization at the leading edge of the cell *pushes* the plasma membrane forward (protrusion) and forms new regions of actin cortex, shown here in *red*. New points of anchorage are made between the actin filaments and the surface on which the cell is crawling (attachment). Contraction at the rear of the cell then draws the body of the cell forward (traction). New anchorage points are established at the front, and old ones are released at the back as the cell crawls forward. The same cycle is repeated over and over again, moving the cell forward in a stepwise fashion.

developing nerve cell axon extends even longer filopodia, up to 50 μm long, which help it to probe its environment and find the correct path to its target. Both lamellipodia and filopodia are exploratory, motile structures that form and retract with great speed, moving at around 1 μm per second. Both are thought to be generated by the rapid local growth of actin filaments, which assemble close to the plasma membrane and elongate by the addition of actin monomers at their plus ends. In this way the filaments push out the membrane without tearing it.

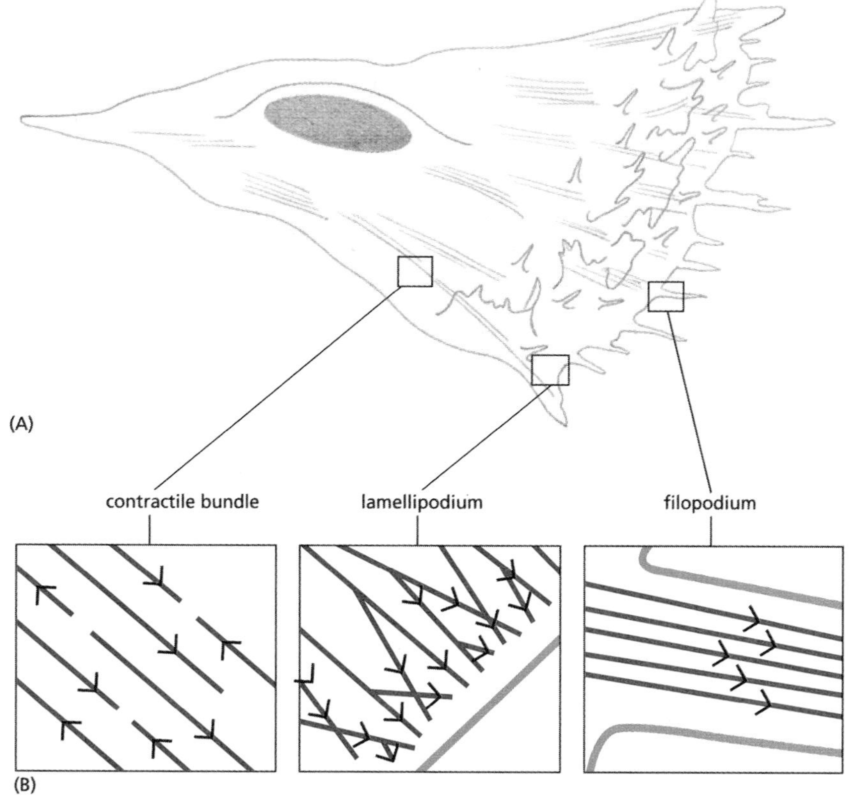

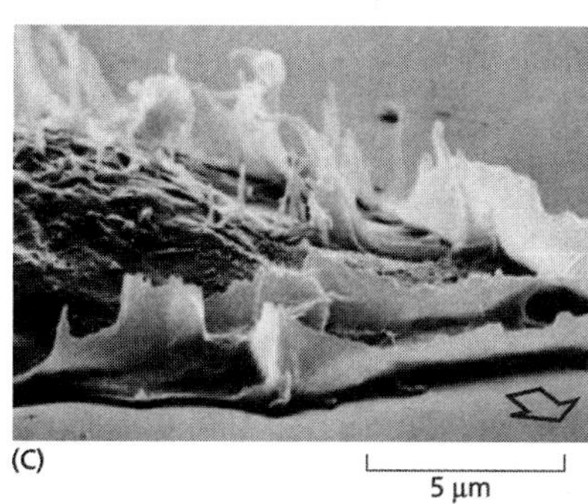

Figure 17–33 **Actin filaments allow animal cells to migrate.** (A) Schematic drawing of a fibroblast showing flattened lamellipodia and fine filopodia projecting from its surface, especially in the regions of the leading edge. (B) Details of the arrangement of actin filaments in three regions of the fibroblast are shown, with arrowheads pointing toward the plus end of the filaments. (C) Scanning electron micrograph showing lamellipodia and filopodia at the leading edge of a human fibroblast migrating in culture. An arrow shows the direction of cell movement. (C, courtesy of Julian Heath.)

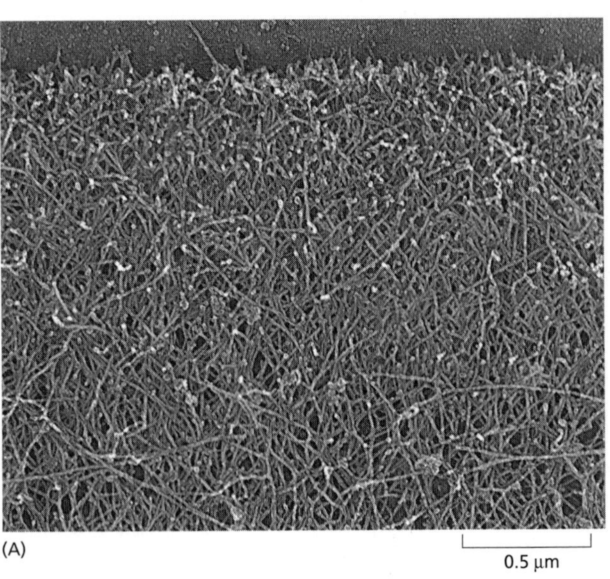

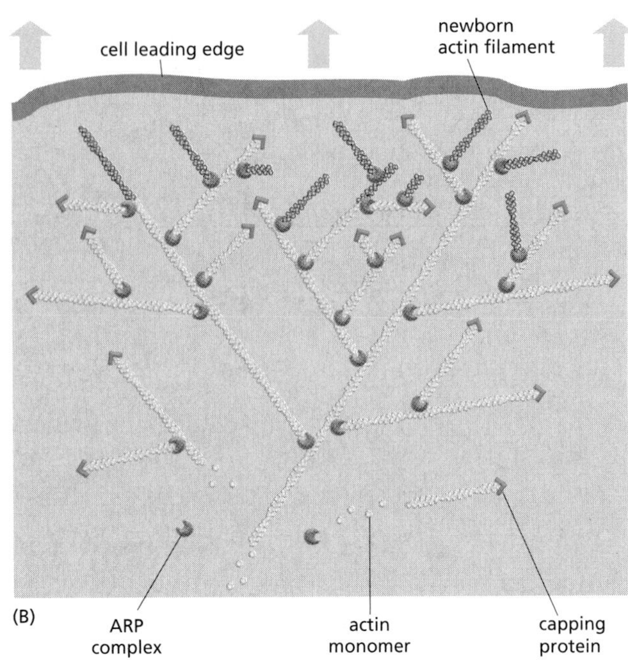

Figure 17–34 **A web of actin filaments pushes the leading edge of a lamellipodium forward.** (A) Highly motile keratocytes from frog skin were fixed, dried, and shadowed with platinum, and examined in an electron microscope. Actin filaments form a dense network, with the fast-growing ends of the filaments terminating at the leading margin of the lamellipodium (top of figure). (B) Nucleation of new actin filaments (*red*) is mediated by ARP complexes (*orange*) attached to the sides of preexisting filaments. The resulting branching structure pushes the plasma membrane forward. The plus ends of the actin filaments become protected by capping proteins (*blue*), while the minus ends of actin filaments nearer the center of the cell continually disassemble through the action of depolymerizing proteins (not shown). The web of actin as a whole thereby undergoes a continual rearward movement due to the assembly of filaments at the front and their disassembly at the rear. (A, courtesy of Tatyana Svitkina and Gary Borisy.)

The formation and growth of actin filaments at the leading edge of a cell are assisted by various actin-binding accessory proteins. One set of proteins—the actin-related proteins, or ARPs—promotes the formation of a web of branched actin filaments in lamellipodia. These proteins form complexes that bind to existing actin filaments and nucleate the formation of new filaments, which grow out at an angle to produce side branches (Figure 17–34). With the aid of additional actin-binding proteins, this web undergoes continual assembly at the leading edge and disassembly further back, pushing the lamellipodia forward. The other kind of cell protrusion, the filopodium, depends on formins, which attach to the growing ends of actin filaments and promote the addition of new monomers to form straight unbranched filaments (Figure 17–35). Formins are also used elsewhere to assemble unbranched filaments, as in the cleavage furrow of a dividing animal cell.

When the lamellipodia and filopodia touch down on a favorable patch of surface, they stick: transmembrane proteins in their plasma membrane, known as *integrins*, adhere to molecules in the extracellular matrix that surrounds cells or on the surface of a neighboring cell over which the moving cell is crawling. Meanwhile, on the intracellular face of the crawling cell's membrane, integrins capture actin filaments, thereby creating a robust anchorage for the system of actin filaments inside the crawling cell (see Figure 20–14C). To use this anchorage to drag its body forward, the cell now makes use of internal contractions to exert a pulling force (see Figure 17–32). These too depend on actin, but in a different way—through the interaction of actin filaments with motor proteins known as *myosins*.

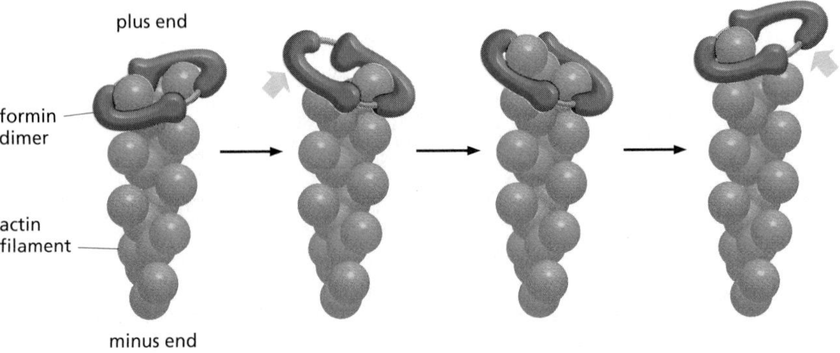

Figure 17–35 **Formins help drive the elongation of actin filaments.** Formin dimers (*green*) attach to the growing end of an actin filament (*red*). Each formin subunit binds to one actin monomer. The formin dimer promotes filament growth by holding onto one of the two actin subunits exposed at the plus end and pulling in a new actin monomer.

It is still not certain how this pulling force is produced: contraction of bundles of actin filaments in the cytoplasm or contraction of the actin meshwork in the cell cortex, or both, may be responsible. The general principles of how myosin motor proteins interact with actin filaments to cause movement is clear, however, as we now discuss.

Actin Associates with Myosin to Form Contractile Structures

All actin-dependent motor proteins belong to the **myosin** family. They bind to and hydrolyze ATP, which provides the energy for their movement along actin filaments from the minus end of the filament toward the plus end. Myosin, along with actin, was first discovered in skeletal muscle, and much of what we know about the interaction of these two proteins was learned from studies of muscle. There are several different types of myosins in cells, of which the *myosin-I* and *myosin-II* subfamilies are most abundant. Myosin-II is the major myosin found in muscle. Myosin-I is found in all types of cells, and because it is simpler in structure and mechanism of action we shall discuss it first.

Myosin-I molecules have only one head domain and a tail (Figure 17–36A). The head domain interacts with actin filaments and has an ATP-hydrolyzing motor activity that enables it to move along the filament in a cycle of binding, detachment, and rebinding (Movie 17.9). The tail varies among the different types of myosin-I, and it determines what cell components will be dragged along by the motor. For example, the tail may bind to a particular type of membrane vesicle and propel it through the cell along actin filament tracks (Figure 17–36B), or it may bind to the plasma membrane and move it relative to cortical actin filaments, thus pulling the membrane into a different shape (Figure 17–36C).

QUESTION 17–6

Suppose that the actin molecules in a cultured skin cell have been randomly labeled in such a way that 1 in 10,000 molecules carries a fluorescent marker. What would you expect to see if you examined the lamellipodium (leading edge) of this cell through a fluorescence microscope? Assume that your microscope is sensitive enough to detect single fluorescent molecules.

Extracellular Signals Control the Arrangement of Actin Filaments

We have seen that myosin and other actin-binding proteins can regulate the location, organization, and behavior of actin filaments. But the activity of these accessory proteins is, in turn, controlled by extracellular signals, allowing the cell to rearrange its cytoskeleton in response to the environment.

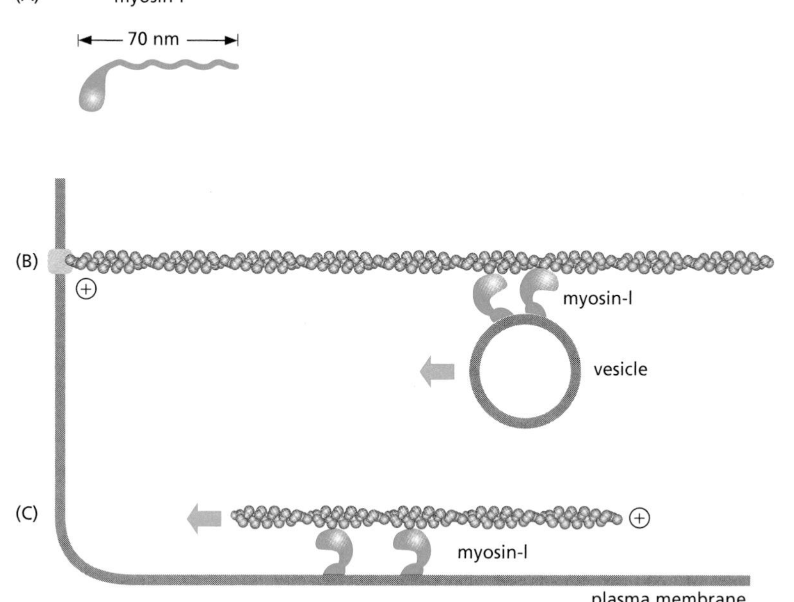

Figure 17–36 **The short tail of a myosin-I molecule contains sites that bind to various components of the cell, including membranes.** (A) Myosin-I has a single globular head and a tail that attaches to another molecule or organelle in the cell. This arrangement allows the head domain to move a vesicle relative to an actin filament (B), or an actin filament and the plasma membrane relative to each other (C). Note that the head group of the myosin always walks toward the plus end of the actin filament it contacts.

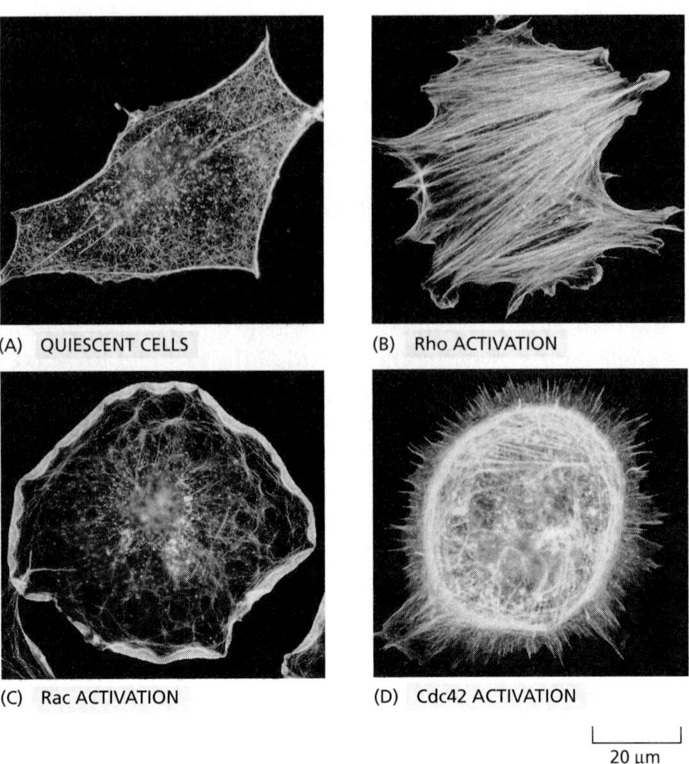

Figure 17–37 **Activation of GTP-binding proteins has a dramatic effect on the organization of actin filaments in fibroblasts.** In these micrographs, actin has been stained with fluorescently labeled phalloidin, a molecule that binds to actin filaments (see Table 17–1, p. 582). (A) Unstimulated fibroblasts have actin filaments primarily in the cortex. (B) Microinjection of an activated form of Rho promotes the rapid assembly of long, unbranched contractile bundles. (C) Microinjection of an activated form of Rac, a GTP-binding protein similar to Rho, causes the formation of an enormous lamellipodium that extends from the entire circumference of the cell. (D) Microinjection of an activated form of Cdc42, another Rho family member, stimulates the protrusion of many long filopodia at the cell periphery. (From A. Hall, *Science* 279:509–514, 1998. With permission from AAAS.)

For the actin cytoskeleton, such structural rearrangements are triggered by activation of a variety of receptor proteins embedded in the plasma membrane. All of these signals then seem to converge inside the cell on a group of closely related GTP-binding proteins called the **Rho protein family**. As we saw in Chapter 16, proteins of this kind behave as molecular switches that control cellular processes by cycling between an active, GTP-bound state and an inactive, GDP-bound state (see Figure 16–14B). In the case of the cytoskeleton, activation of different members of the Rho family affects the organization of actin filaments in different ways. For example, activation of one Rho family member triggers actin polymerization and bundling to form filopodia; activation of another promotes the formation of sheetlike lamellipodia and membrane ruffles; and activation of Rho itself drives the bundling of actin filaments with myosin II and the clustering of integrins that promotes cell crawling (Figure 17–37).

These dramatic and complex structural changes occur because the GTP-binding proteins, together with the protein kinases and accessory proteins with which they interact, act like a computational network to control actin organization and dynamics. This network receives external signals from nutrients, growth factors, and contacts with neighboring cells, along with 'inside information' regarding the cell's nutritional state, size, and readiness for division. The Rho network then processes these inputs and produces signals that shape the actin cytoskeleton—for example, by activating the formin proteins that promote the formation of filopodia (see Figure 17–35) or by enhancing the actin-nucleating activities of ARP complexes at the leading edge of the cell to generate large lamellipodia.

One of the most tightly regulated rearrangements of cytoskeletal elements occurs when actin associates with myosin in muscle fibers in response to signals from the nervous system. We now discuss how this molecular interaction generates the rapid, repetitive, forceful movements characteristic of the contraction of vertebrate muscles.

QUESTION 17–7

At the leading edge of a crawling cell, the plus ends of actin filaments are located close to the plasma membrane, and actin monomers are added at these ends, pushing the membrane outward to form lamellipodia or filopodia. What do you suppose holds the filaments at their other ends to prevent them from just being pushed into the cell's interior?

MUSCLE CONTRACTION

Muscle contraction is the most familiar and best understood of animal cell movements. In vertebrates, running, walking, swimming, and flying all depend on the ability of *skeletal muscle* to contract strongly and move various bones. Involuntary movements such as heart pumping and gut peristalsis depend on *cardiac muscle* and *smooth muscle,* respectively, which are formed from muscle cells that differ in structure from skeletal muscle but use actin and myosin in a similar way to contract. Although muscle cells are highly specialized, many cell movements—from the locomotion of whole cells down to the motion of components inside cells—depend on the interaction of actin and myosin. Much of our understanding of the mechanisms of cell movement originated from studies of muscle cell contraction. In this section, we discuss how actin and myosin interact to create coherent movement.

Muscle Contraction Depends on Bundles of Actin and Myosin

Muscle myosin belongs to the myosin-II subfamily of myosins, all of which have two ATPase heads and a long, rodlike tail (Figure 17–38A). Each myosin-II molecule is a dimer composed of a pair of identical myosin molecules held together by their tails; it has two globular ATPase heads at one end and a single coiled-coil tail at the other. Clusters of myosin-II molecules bind to each other through their coiled-coil tails, forming a bipolar *myosin filament* in which the heads project from the sides (Figure 17–38B).

The myosin filament is like a double-headed arrow, with the two sets of heads pointing in opposite directions away from the middle. One set of heads binds to actin filaments in one orientation and moves them one way; the other set of heads binds to other actin filaments in the opposite orientation and moves them in the opposite direction (Figure 17–39). The overall effect is to slide sets of oppositely oriented actin filaments past one another. We can see how, therefore, if actin filaments and myosin filaments are organized together in a bundle, the bundle can generate a contractile force. This is seen most clearly in muscle contraction, but it also occurs in the *contractile bundles* of actin filaments and myosin-II filaments (see Figure 17–28B) that assemble transiently in nonmuscle cells, and in the *contractile ring* that pinches a dividing cell in two by contracting and pulling inward on the plasma membrane (discussed in Chapter 19).

QUESTION 17–8

If both the thick and thin filaments of muscle are made up of subunits held together by weak noncovalent bonds, how is it possible for a human being to lift heavy objects?

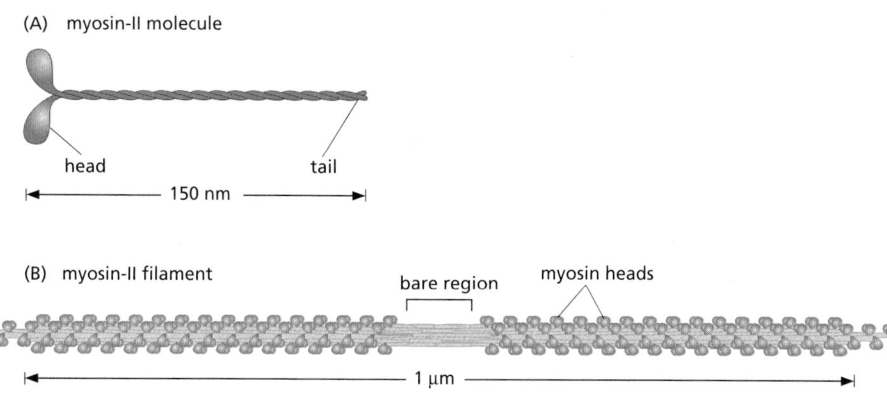

Figure 17–38 **Myosin-II molecules can associate with one another to form myosin filaments.** (A) A molecule of myosin-II has two globular heads and a coiled-coil tail. (B) The tails of myosin-II associate with one another to form a bipolar myosin filament in which the heads project outward from the middle in opposite directions. The bare region in the middle of the filament consists of tails only.

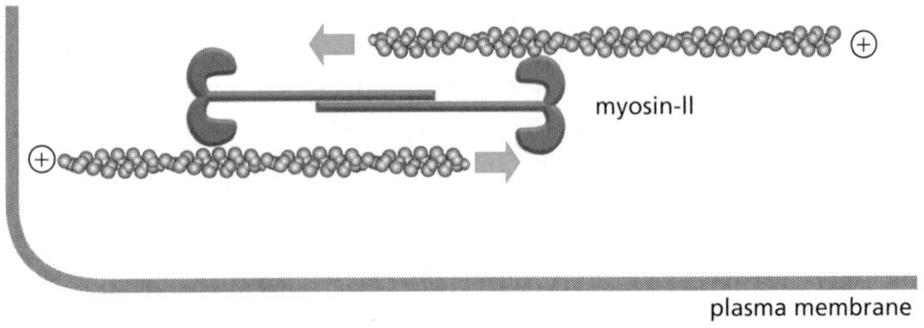

Figure 17–39 **Even small bipolar filaments composed of myosin-II molecules can slide actin filaments over each other, thus mediating local *shortening* of an actin filament bundle.** As with myosin-I, the head group of myosin-II walks toward the plus end of the actin filament it contacts.

During Muscle Contraction Actin Filaments Slide Against Myosin Filaments

The long fibers of skeletal muscle are huge single cells formed by the fusion of many separate smaller cells. The individual nuclei of the contributing cells are retained in the muscle fiber and lie just beneath the plasma membrane. The bulk of the cytoplasm is made up of **myofibrils**, the contractile elements of the muscle cell. These cylindrical structures are 1–2 μm in diameter and may be as long as the muscle cell itself (Figure 17–40A).

A myofibril consists of a chain of identical tiny contractile units, or **sarcomeres**. Each sarcomere is about 2.5 μm long, and the repeating pattern of sarcomeres gives the vertebrate myofibril a striped, or striated, appearance (Figure 17–40B). Sarcomeres are highly organized assemblies of two types of filaments—actin filaments and filaments of muscle-specific myosin-II. Myosin filaments (the *thick filaments*) are centrally positioned in each sarcomere, whereas the more slender actin filaments (the *thin filaments*) extend inward from each end of the sarcomere (where they are anchored by their plus ends to a structure known as the *Z disc*) and overlap the ends of the myosin filaments (Figure 17–41).

The contraction of a muscle cell is caused by a simultaneous shortening of all the sarcomeres, which in turn is caused by the actin filaments sliding past the myosin filaments, with no change in the length of either type of filament (Figure 17–42). The sliding motion is generated by myosin heads that project from the sides of the myosin filament and interact with adjacent actin filaments. When a muscle is stimulated to contract, the myosin heads start to walk along the actin filament in repeated cycles

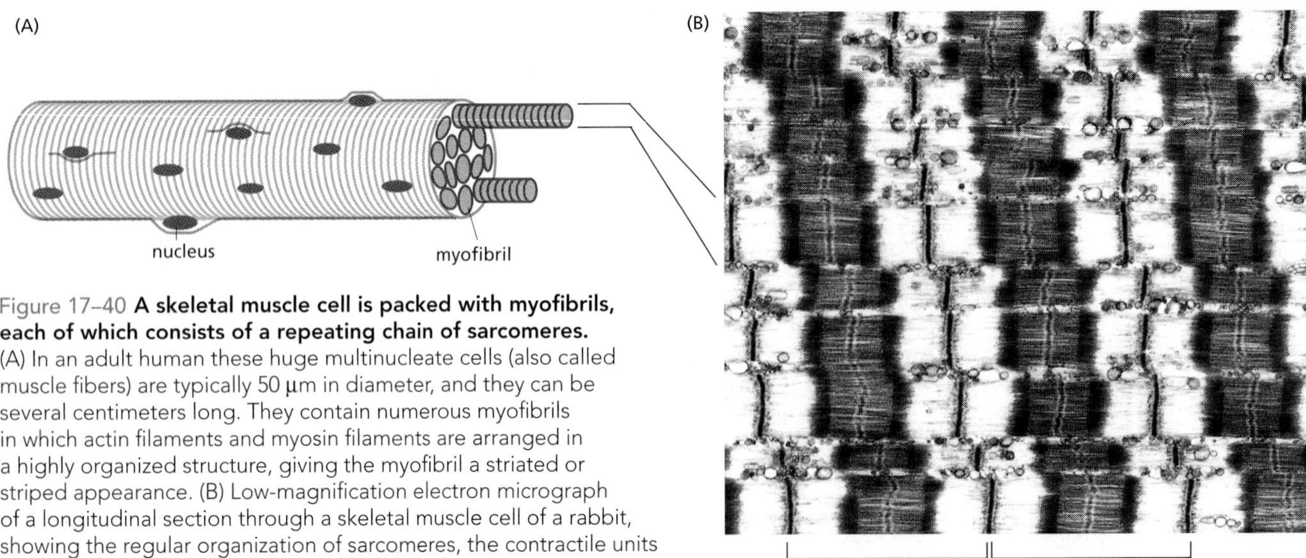

Figure 17–40 **A skeletal muscle cell is packed with myofibrils, each of which consists of a repeating chain of sarcomeres.** (A) In an adult human these huge multinucleate cells (also called muscle fibers) are typically 50 μm in diameter, and they can be several centimeters long. They contain numerous myofibrils in which actin filaments and myosin filaments are arranged in a highly organized structure, giving the myofibril a striated or striped appearance. (B) Low-magnification electron micrograph of a longitudinal section through a skeletal muscle cell of a rabbit, showing the regular organization of sarcomeres, the contractile units of the myofibrils. (B, courtesy of Roger Craig.)

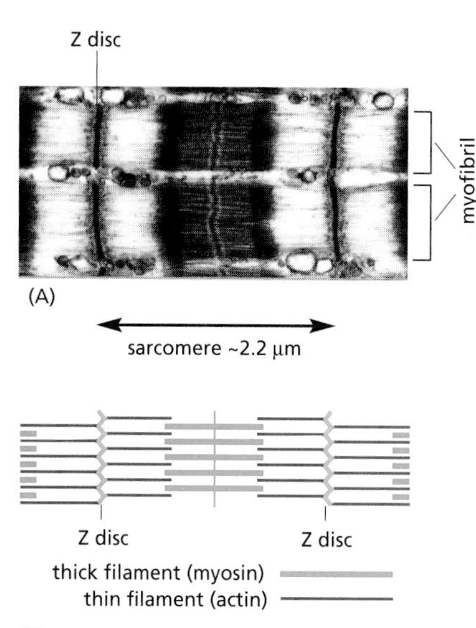

Figure 17–41 **Sarcomeres are the contractile units of muscle.** (A) Detail of the skeletal muscle cell shown in Figure 17–40 showing two myofibrils, with the extent of one sarcomere marked. (B) Schematic diagram of a single sarcomere showing the origin of the light and dark bands seen in the microscope. Z discs at either end of the sarcomere are attachment points for actin filaments; the centrally located thick filaments *(green)* are each composed of many myosin-II molecules. (A, courtesy of Roger Craig.)

of attachment and detachment. During each cycle, a myosin head binds and hydrolyzes one molecule of ATP. This causes a series of conformational changes in the myosin molecule that move the tip of the head by about 5 nm along the actin filament toward the plus end. This movement, repeated with each round of ATP hydrolysis, propels the myosin molecule unidirectionally along the actin filament (Figure 17–43). In so doing, the myosin heads pull against the actin filament, causing it to slide against the myosin filament. The concerted action of many myosin heads pulling the actin and myosin filaments past each other causes the sarcomere to contract. After a contraction is completed, the myosin heads lose contact with the actin filaments completely, and the muscle relaxes.

A myosin filament has about 300 myosin heads. Each myosin head can attach and detach from actin about five times per second, allowing the myosin and actin filaments to slide past one another at speeds of up to 15 μm per second. This speed is sufficient to take a sarcomere from a fully extended state (3 μm) to a fully contracted state (2 μm) in less than one-tenth of a second. All of the sarcomeres of a muscle are coupled together and are triggered almost instantaneously by the system of signals we shall describe next. Therefore, the entire muscle contracts extremely rapidly, usually within one-tenth of a second.

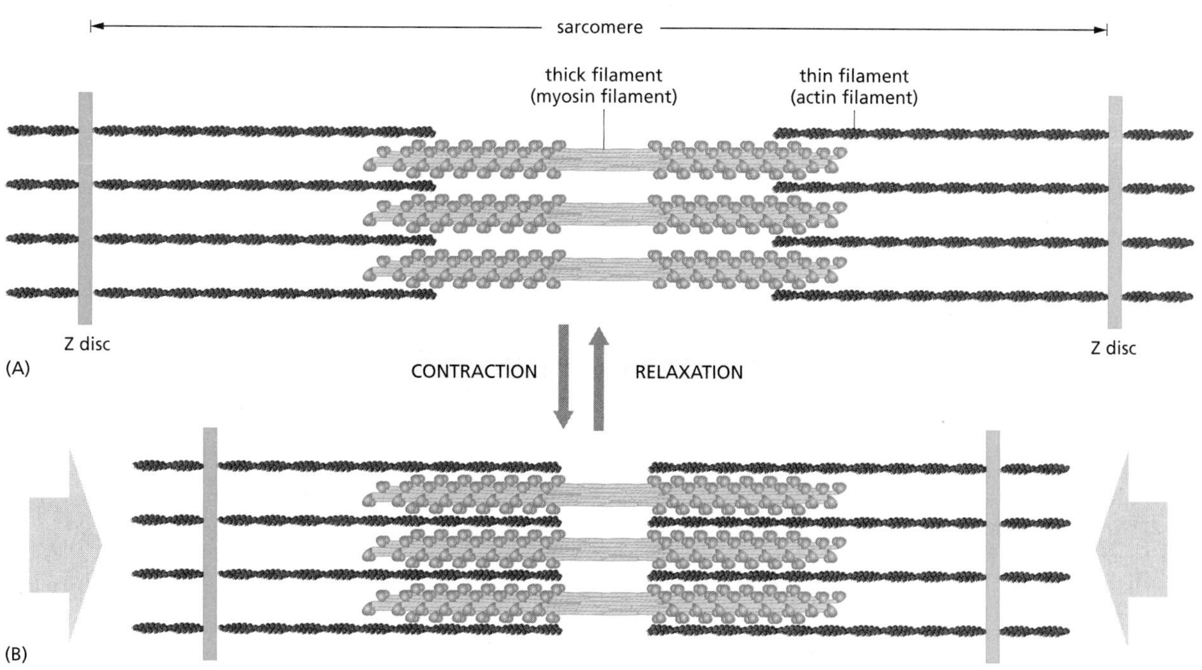

Figure 17–42 **Muscles contract by a sliding-filament mechanism.** (A) The myosin and actin filaments of a sarcomere overlap with the same relative polarity on either side of the midline. Recall that actin filaments are anchored by their plus ends to the Z disc and that myosin filaments are bipolar. (B) During contraction, the actin and myosin filaments slide past each other without shortening. The sliding motion is driven by the myosin heads walking toward the plus end of the adjacent actin filament (Movie 17.8).

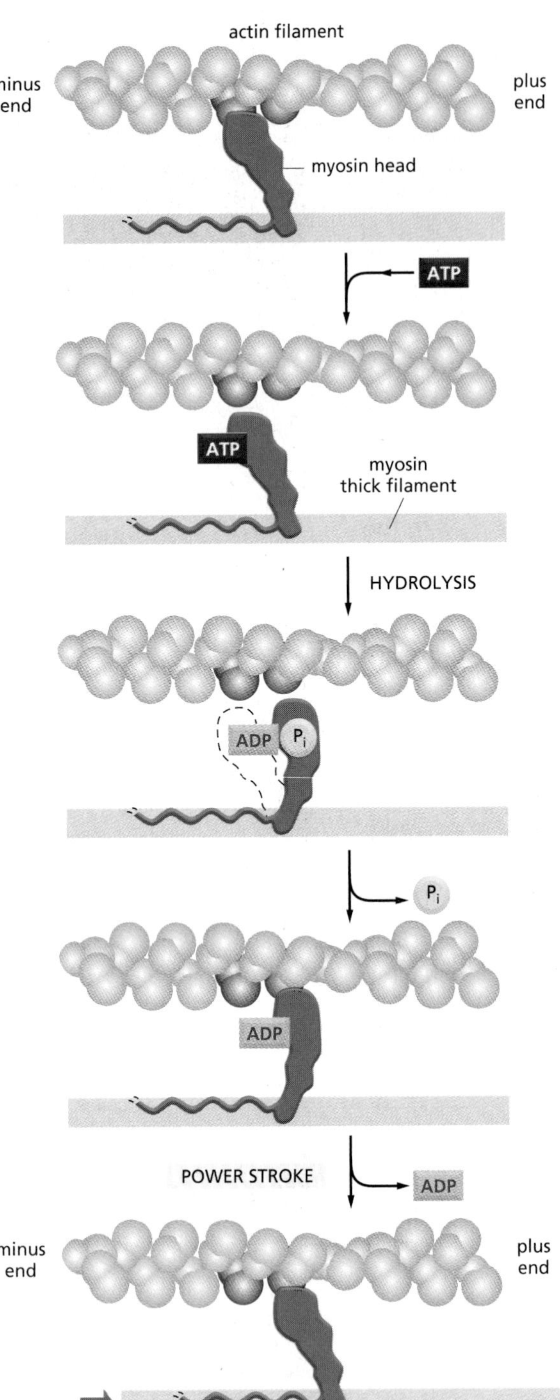

ATTACHED At the start of the cycle shown in this figure, a myosin head lacking a bound nucleotide is locked tightly onto an actin filament in a *rigor* configuration (so named because it is responsible for *rigor mortis*, the rigidity of death). In an actively contracting muscle, this state is very short-lived, being rapidly terminated by the binding of a molecule of ATP.

RELEASED A molecule of ATP binds to the large cleft on the "back" of the head (that is, on the side furthest from the actin filament) and immediately causes a slight change in the conformation of the domains that make up the actin-binding site. This reduces the affinity of the head for actin and allows it to move along the filament. (The space drawn here between the head and actin emphasizes this change, although in reality the head probably remains very close to the actin.)

COCKED The cleft closes like a clam shell around the ATP molecule, triggering a large shape change that causes the head to be displaced along the filament by a distance of about 5 nm. Hydrolysis of ATP occurs, but the ADP and inorganic phosphate (P_i) produced remain tightly bound to the protein.

FORCE-GENERATING A weak binding of the myosin head to a new site on the actin filament causes release of the inorganic phosphate produced by ATP hydrolysis, concomitantly with the tight binding of the head to actin. This release triggers the power stroke—the force-generating change in shape during which the head regains its original conformation. In the course of the power stroke, the head loses its bound ADP, thereby returning to the start of a new cycle.

ATTACHED At the end of the cycle, the myosin head is again locked tightly to the actin filament in a rigor configuration. Note that the head has moved to a new position on the actin filament.

Figure 17–43 **A myosin molecule walks along an actin filament through a cycle of structural changes.** To see actin and myosin in action, watch Movie 17.10. (Based on I. Rayment et al., *Science* 261:50–58, 1993. With permission from AAAS.)

Muscle Contraction Is Triggered by a Sudden Rise in Ca^{2+}

The force-generating molecular interaction between myosin and actin filaments takes place only when the skeletal muscle receives a signal from the nervous system. The signal from a nerve terminal triggers an action potential (discussed in Chapter 12) in the muscle cell plasma membrane. This electrical excitation spreads in a matter of milliseconds into a series of membranous tubes, called *transverse* (or *T*) *tubules*, that extend inward from the plasma membrane around each myofibril. The electrical signal is then relayed to the *sarcoplasmic reticulum*, an adjacent sheath of interconnected flattened vesicles that surrounds each myofibril like a net stocking (Figure 17–44).

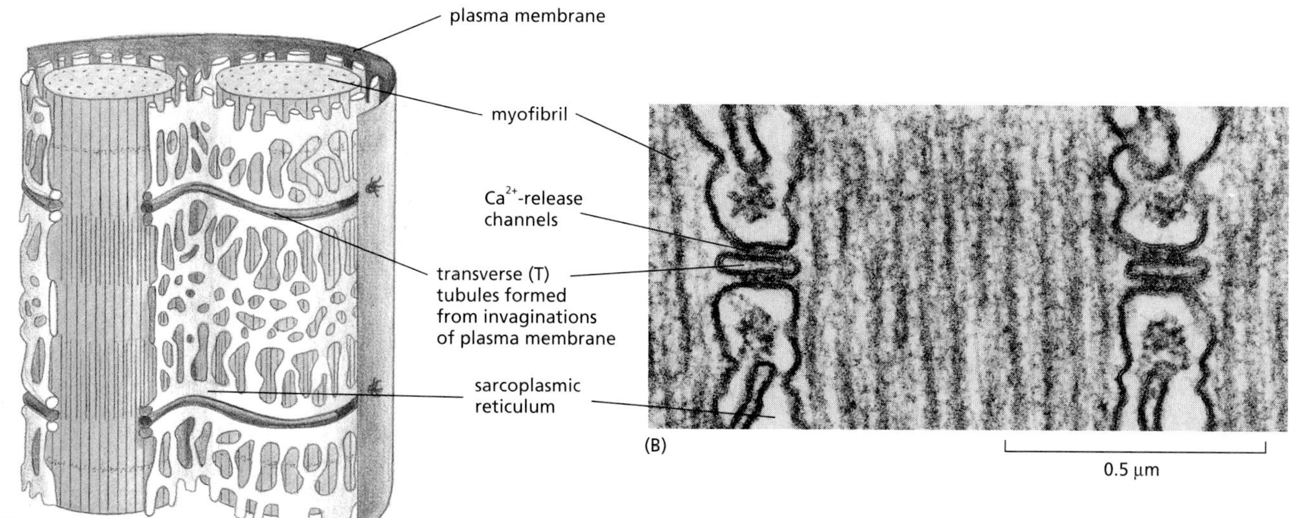

Figure 17–44 **T tubules and sarcoplasmic reticulum surround the myofibrils.** (A) Drawing of the two membrane systems that relay the signal to contract from the muscle cell plasma membrane to all of the myofibrils in the cell. (B) Electron micrograph showing a cross section of two T tubules and their adjacent sarcoplasmic reticulum compartments. (B, courtesy of Clara Franzini-Armstrong.)

The sarcoplasmic reticulum is a specialized region of the endoplasmic reticulum in muscle cells. It contains a very high concentration of Ca^{2+}, and in response to the incoming electrical excitation, much of this Ca^{2+} is released into the cytosol through ion channels that open in the sarcoplasmic reticulum membrane in response to the change in voltage across the plasma membrane (Figure 17–45). As discussed in Chapter 16, Ca^{2+} is widely used as an intracellular signal to relay a message from the exterior to the internal machinery of the cell. In muscle, the Ca^{2+} interacts with a molecular switch made of specialized accessory proteins closely associated with the actin filaments (Figure 17–46A). One of these proteins is *tropomyosin,* a rigid, rod-shaped molecule that binds in the groove of the actin helix, overlapping seven actin monomers, and prevents the myosin heads from associating with the actin filament. The other is *troponin,* a protein complex that includes a Ca^{2+}-sensitive protein associated with the end of a tropomyosin molecule. When the level of Ca^{2+} rises in the cytosol, Ca^{2+} binds to troponin and induces a change in its shape. This in turn causes the tropomyosin molecules to shift their position slightly, allowing myosin heads to bind to the actin filament and initiating contraction (Figure 17–46B).

Because the signal from the plasma membrane is passed within milliseconds (via the transverse tubules and sarcoplasmic reticulum) to every sarcomere in the cell, all the myofibrils in the cell contract at the same

QUESTION 17–9

Compare the structure of intermediate filaments with that of the myosin-II filaments in skeletal muscle cells. What are the major similarities? What are the major differences? How do the differences in structure relate to their function?

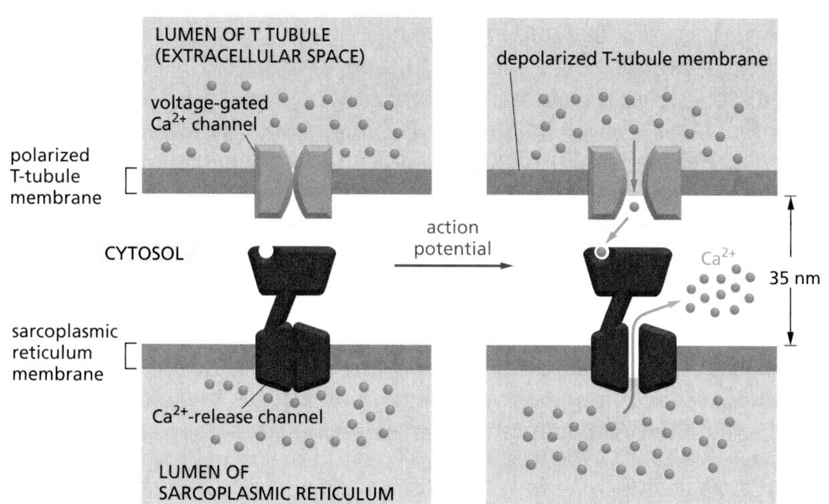

Figure 17–45 **In skeletal muscle, contraction involves Ca^{2+} signaling.** This schematic diagram shows how a Ca^{2+}-release channel in the sarcoplasmic reticulum membrane is thought to be opened by activation of a voltage-gated Ca^{2+} channel in the T-tubule membrane.

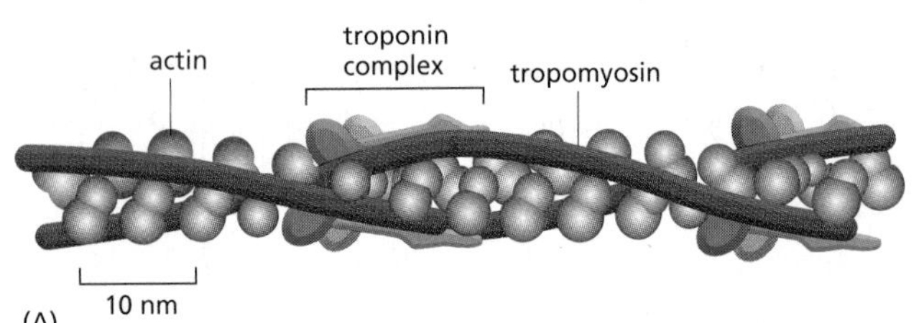

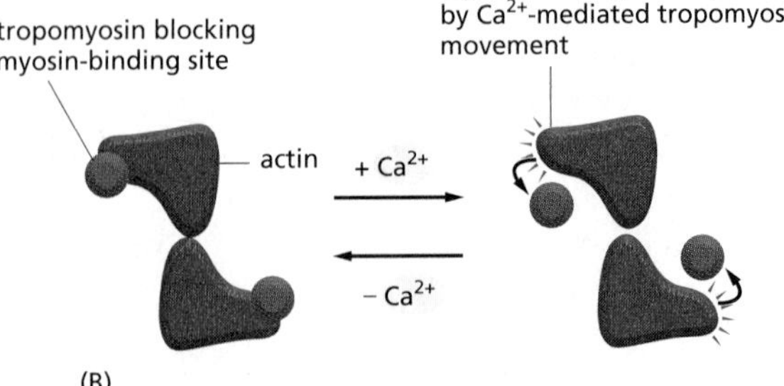

Figure 17–46 **Skeletal muscle contraction is controlled by troponin.** (A) A muscle thin filament showing the positions of tropomyosin and troponin along the actin filament. Every tropomyosin molecule has seven evenly spaced regions of homologous sequence, each of which is thought to bind to an actin subunit in the filament. (B) When Ca^{2+} binds to troponin, the troponin moves the tropomyosin that otherwise blocks the interaction of actin with the myosin heads. In this diagram, the thin filament is shown end-on.

time. The increase in Ca^{2+} in the cytosol ceases as soon as the nerve signal stops, because the Ca^{2+} is rapidly pumped back into the sarcoplasmic reticulum by abundant Ca^{2+} pumps in its membrane (discussed in Chapter 12). As soon as Ca^{2+} concentrations have returned to their resting level, troponin and tropomyosin molecules move back to their original positions, where they block myosin binding and thus end contraction.

Muscle Cells Perform Highly Specialized Functions in the Body

The highly specialized contractile machinery in muscle cells is thought to have evolved from the simpler contractile bundles of myosin and actin filaments found in all eucaryotic cells. The myosin-II in nonmuscle cells is also activated by a rise in cytosolic Ca^{2+}, but the mechanism of activation is quite different. An increase in Ca^{2+} leads to the phosphorylation of myosin-II, which alters the myosin conformation and enables it to interact with actin. A similar activation mechanism operates in *smooth muscle,* which lies in the walls of the stomach, intestine, uterus, and arteries, and in many other structures in which slow and sustained contractions are needed. Contractions produced by this second mode are slower because time is needed for enzyme molecules to diffuse to the myosin heads and carry out the phosphorylation or dephosphorylation. However, this mechanism has the advantage that it is less specialized and can be driven by a variety of incoming signals: thus smooth muscle, for example, is triggered to contract by adrenaline, serotonin, prostaglandins, and several other extracellular signals.

QUESTION 17–10

A. Note that in Figure 17–46 troponin molecules are evenly spaced along an actin filament, with one troponin found every seventh actin molecule. How do you suppose troponin molecules can be positioned this regularly? What does this tell you about the binding of troponin to actin filaments?
B. What do you suppose would happen if you mixed actin filaments with (i) troponin alone, (ii) tropomyosin alone, or (iii) troponin plus tropomyosin, and then added myosin? Would the effects be dependent on Ca^{2+}?

In addition to skeletal and smooth muscle, other forms of muscle each perform a specific mechanical function in the body. Perhaps the most familiar is the heart, or *cardiac,* muscle that drives the circulation of blood. This remarkable organ contracts autonomously for the lifetime of the organism—some 3 billion (3×10^9) times in a human (Movie 17.11). Even subtle changes in the actin and myosin of heart muscle can lead to serious heart disease. For example, mutations in cardiac myosin and other contractile proteins in the sarcomere cause familial hypertrophic cardiomyopathy, a hereditary disorder responsible for sudden death in young athletes,

The contraction of muscle cells represents a highly specialized use of the basic components of the eucaryotic cytoskeleton. In the following chapters, we see how the cytoskeleton participates in perhaps the most fundamental cell movement of all, the formation of two daughter cells during the process of cell division.

ESSENTIAL CONCEPTS

- The cytoplasm of a eucaryotic cell is supported and spatially organized by a cytoskeleton of intermediate filaments, microtubules, and actin filaments.
- Intermediate filaments are stable, ropelike polymers of fibrous proteins that give cells mechanical strength. Some types underlie the nuclear membrane to form the nuclear lamina; others are distributed throughout the cytoplasm.
- Microtubules are stiff, hollow tubes formed by the polymerization of tubulin dimer subunits. They are polarized structures with a slow-growing "minus" end and a fast-growing "plus" end.
- Microtubules are nucleated in, and grow out from, organizing centers such as the centrosome. The minus ends of the microtubules are embedded in the organizing center.
- Many of the microtubules in a cell are in a labile, dynamic state in which they alternate between a growing state and a shrinking state. These transitions, known as dynamic instability, are controlled by the hydrolysis of GTP bound to tubulin dimers.
- Each tubulin dimer has a tightly bound GTP molecule that is hydrolyzed to GDP after the tubulin has assembled into a microtubule. GTP hydrolysis reduces the affinity of the subunit for its neighbors and decreases the stability of the polymer, causing it to disassemble.
- Microtubules can be stabilized by proteins that capture the plus end—a process that influences the position of microtubule arrays in a cell. Cells contain many microtubule-associated proteins that stabilize microtubules, bind them to other cell components, and harness them for specific functions.
- Kinesins and dyneins are motor proteins that use the energy of ATP hydrolysis to move unidirectionally along microtubules. They carry specific membrane vesicles and other cargoes and in this way help to maintain the spatial organization of the cytoplasm.
- Eucaryotic cilia and flagella contain a bundle of stable microtubules. Their beating is caused by bending of the microtubules, driven by a motor protein called ciliary dynein.
- Actin filaments are helical polymers of actin molecules. They are more flexible than microtubules and are generally found in bundles or networks.
- Actin filaments are polarized structures with a fast- and a slow-growing end, and their assembly and disassembly are controlled by the hydrolysis of ATP tightly bound to each actin monomer.
- The varied forms and functions of actin filaments in cells depend on multiple actin-binding proteins. These control the polymerization of actin filaments, cross-link the filaments into loose networks or stiff bundles, attach them to membranes, or move them relative to one another.
- A concentrated network of actin filaments underneath the plasma membrane forms the cell cortex and is responsible for the shape and movement of the cell surface, including the movements involved when a cell crawls along a surface.
- Myosins are motor proteins that use the energy of ATP hydrolysis to move along actin filaments: they can carry organelles along actin-filament tracks or cause adjacent actin filaments to slide past each other in contractile bundles.
- In muscle, huge regular arrays of overlapping actin filaments and myosin filaments generate contractions by sliding over one another.

- **Muscle contraction is initiated by a sudden rise in cytosolic Ca^{2+}, which delivers a signal to the contractile apparatus via Ca^{2+}-binding proteins.**

KEY TERMS

actin filament	kinesin
cell cortex	lamellipodium
centriole	microtubule
centrosome	motor protein
cilium	myofibril
cytoskeleton	myosin
dynamic instability	nuclear lamina
dynein	polarity
filopodium	Rho protein family
flagellum	sarcomere
intermediate filament	tubulin

QUESTIONS

QUESTION 17–11

Which of the following statements are correct? Explain your answers.

A. Kinesin moves endoplasmic reticulum membranes along microtubules so that the network of ER tubules becomes stretched throughout the cell.

B. Without actin, cells can form a functional mitotic spindle and pull their chromosomes apart but cannot divide.

C. Lamellipodia and filopodia are "feelers" that a cell extends to find anchor points on the substratum that it will then crawl over.

D. GTP is hydrolyzed by tubulin to cause the bending of flagella.

E. Cells having an intermediate-filament network that cannot be depolymerized would die.

F. The plus ends of microtubules grow faster because they have a larger GTP cap.

G. The transverse tubules in muscle cells are an extension of the plasma membrane, with which they are continuous; similarly, the sarcoplasmic reticulum is an extension of the endoplasmic reticulum.

H. Activation of myosin movement on actin filaments is triggered by the phosphorylation of troponin in some situations and by Ca^{2+} binding to troponin in others.

QUESTION 17–12

The average time taken for a molecule or an organelle to diffuse a distance of x cm is given by the formula

$$t = x^2/2D$$

where t is the time in seconds and D is a constant called the diffusion coefficient for the molecule or particle. Using the above formula, calculate the time it would take for a small molecule, a protein, and a membrane vesicle to diffuse from one side to another of a cell 10 μm across. Typical diffusion coefficients in units of cm^2/sec are: small molecule, 5×10^{-6}; protein molecule, 5×10^{-7}; vesicle, 5×10^{-8}. How long would a membrane vesicle take to reach the end of an axon 10 cm long by free diffusion?

QUESTION 17–13

Why do eucaryotic cells, and especially animal cells, have such large and complex cytoskeletons? List the differences between animal cells and bacteria that depend on the eucaryotic cytoskeleton.

QUESTION 17–14

Examine the structure of an intermediate filament shown in Figure 17–3. Does the filament have a unique polarity—that is, could you distinguish one end from the other by chemical or other means? Explain your answer.

QUESTION 17–15

There are no known motor proteins that move on intermediate filaments. Suggest an explanation for this.

QUESTION 17–16

When cells enter mitosis, their existing array of cytoplasmic microtubules has to be rapidly broken down and replaced with the mitotic spindle that forms to pull the chromosomes into the daughter cells. The enzyme katanin, named after Japanese samurai swords, is activated during the onset of mitosis, and chops microtubules into short pieces. What do you suppose is the fate of the microtubule fragments created by katanin? Explain your answer.

QUESTION 17–17

The drug taxol, extracted from the bark of yew trees, has an opposite effect to the drug colchicine, an alkaloid from autumn crocus. Taxol binds tightly to microtubules and stabilizes them; when added to cells, it causes much of the free tubulin to assemble into microtubules. In contrast, colchicine prevents microtubule formation. Taxol is just as pernicious to dividing cells as colchicine, and both are used as anticancer drugs. Based on your knowledge of microtubule dynamics, suggest why both drugs are toxic to dividing cells despite their opposite actions.

QUESTION 17–18

A useful technique for studying microtubule motors is to attach them by their tails to a glass coverslip (which can be accomplished quite easily because the tails stick avidly to a clean glass surface) and then allow them to settle. The microtubules may then be viewed in a light microscope as they are propelled over the surface of the coverslip by the heads of the motor proteins. Because the motor proteins attach at random orientations to the coverslip, however, how can they generate coordinated movement of individual microtubules rather than engaging in a tug-of-war? In which direction will microtubules crawl on a 'bed' of kinesin molecules (i.e., will they move plus end first, or minus end first)?

QUESTION 17–19

A typical time course of polymerization of purified tubulin to form microtubules is shown in Figure Q17–19.

A. Explain the different parts of the curve (labeled A, B, and C). Draw a diagram that shows the behavior of tubulin molecules in each of the three phases.

B. How would the curve in the figure change if centrosomes were added at the outset?

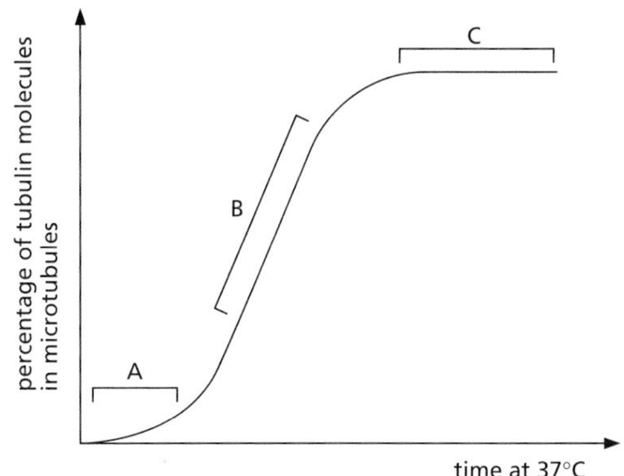

Figure Q17–19

QUESTION 17–20

The electron micrographs shown in Figure Q17–20A were obtained from a population of microtubules that were growing rapidly. Figure Q17–20B was obtained from microtubules undergoing "catastrophic" shrinking. Comment on any differences between A and B, and suggest likely explanations for the differences that you observe.

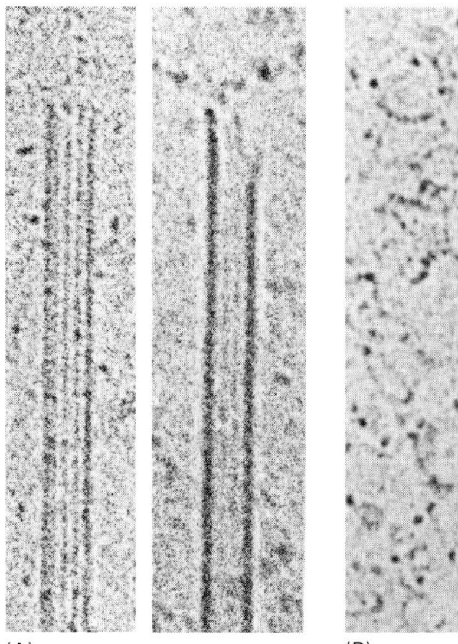
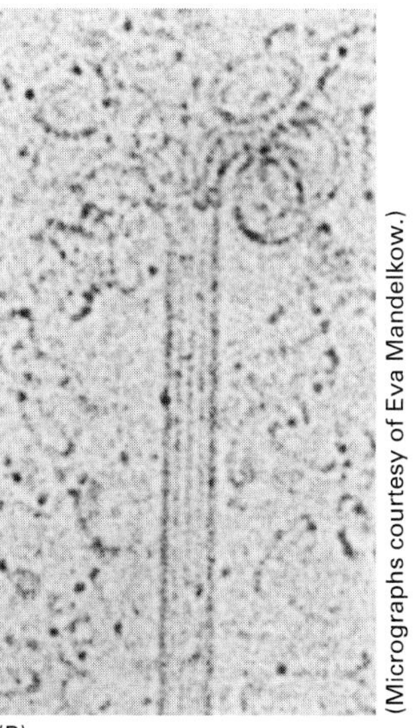
(A) (B)

(Micrographs courtesy of Eva Mandelkow.)

Figure Q17–20

QUESTION 17–21

The locomotion of fibroblasts in culture is immediately halted by the drug cytochalasin, whereas colchicine causes fibroblasts to cease to move directionally and to begin extending lamellipodia in seemingly random directions. Injection of fibroblasts with antibodies to vimentin has no discernible effect on their migration. What do these observations suggest to you about the involvement of the three different cytoskeletal filaments in fibroblast locomotion?

QUESTION 17–22

Complete the following sentence accurately, explaining your reason for accepting or rejecting each of the four phrases (more than one can be correct). The role of calcium in muscle contraction is:

A. To detach myosin heads from actin.

B. To spread the action potential from the plasma membrane to the contractile machinery.

C. To bind to troponin, cause it to move tropomyosin, and thereby expose actin filaments to myosin heads.

D. To maintain the structure of the myosin filament.

QUESTION 17–23

Which of the following changes takes place when a skeletal muscle contracts?

A. Z discs move farther apart.

B. Actin filaments contract.

C. Myosin filaments contract.

D. Sarcomeres become shorter.

CHAPTER **EIGHTEEN** 18

The Cell Division Cycle

OVERVIEW OF THE CELL CYCLE

THE CELL-CYCLE CONTROL SYSTEM

S PHASE

M PHASE

MITOSIS

CYTOKINESIS

CONTROL OF CELL NUMBER AND CELL SIZE

"Where a cell arises, there must be a previous cell, just as animals can only arise from animals and plants from plants." This *cell doctrine,* proposed by the German pathologist Rudolf Virchow in 1858, carried with it a profound message for the continuity of life. Cells are generated from cells, and the only way to make more cells is by the division of those that already exist. All living organisms, from a unicellular bacterium to a multicellular mammal, are thought to be products of repeated rounds of cell growth and division extending back in time to the beginnings of life more than 3 billion years ago.

A cell reproduces by carrying out an orderly sequence of events in which it duplicates its contents and then divides in two. This cycle of duplication and division, known as the **cell cycle**, is the essential mechanism by which all living things reproduce. The details of the cell cycle vary from organism to organism and at different times in an individual organism's life. In unicellular organisms, such as bacteria and yeasts, each cell division produces a complete new organism, whereas many rounds of cell division are required to make a new multicellular organism from a fertilized egg. Certain features of the cell cycle, however, are universal, as they allow every cell to perform its most fundamental task—to copy and pass on its genetic information to the next generation of cells. To produce two genetically identical daughter cells, the DNA in each chromosome must be faithfully replicated, and the replicated chromosomes must then be accurately distributed, or *segregated*, into the two daughter cells, so that each cell receives a complete copy of the entire genome (Figure 18–1). Most cells also duplicate their other macromolecules and organelles, and they double in size before they divide; otherwise, each time they divided

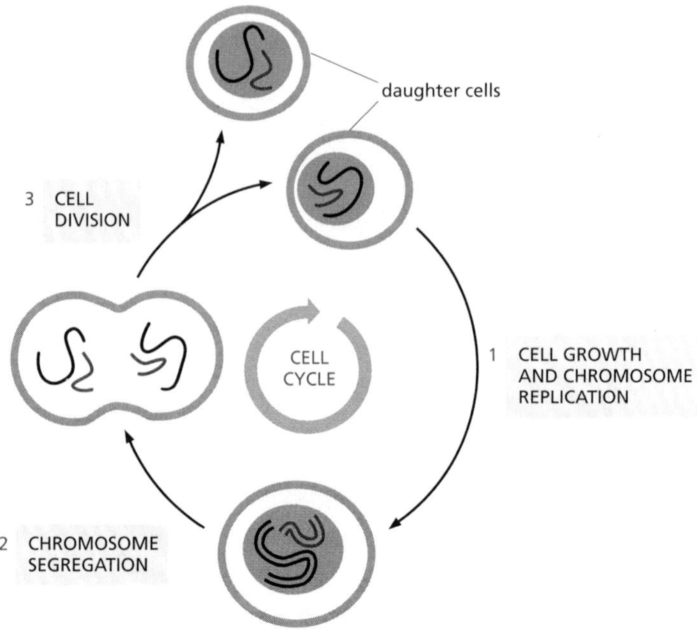

Figure 18–1 **Cells reproduce by duplicating their contents and dividing in two, a process called the cell division cycle, or cell cycle.** The division of a hypothetical eucaryotic cell with two chromosomes is shown to illustrate how each cell cycle produces two genetically identical daughter cells. Each daughter cell can divide again by going through another cell cycle.

QUESTION 18–1

Consider the following statement: "All present-day cells have arisen by an uninterrupted series of cell divisions extending back in time to the first cell division." Is this strictly true?

they would get smaller and smaller. Thus, to maintain their size, dividing cells must coordinate their growth with their division.

To explain how cells reproduce, we therefore have to consider three major questions: (1) How do cells duplicate their contents? (2) How do they partition the duplicated contents and split in two? (3) How do they coordinate all the machinery that is required for these two processes? The first question is considered elsewhere in this book: in Chapter 6, we discuss how DNA is replicated, and in Chapters 7, 11, 15, and 17, we describe how the eucaryotic cell manufactures other components, such as proteins, membranes, organelles, and cytoskeletal filaments. In this chapter, we tackle the second and third questions: how a eucaryotic cell segregates its duplicated contents to produce two daughter cells and how it coordinates the various steps of this reproductive cycle.

We begin with an overview of the events that take place during the cell cycle. We then describe the complex system of regulatory proteins called the *cell-cycle control system*, which orders and coordinates these events to ensure that they occur in the correct sequence. We next discuss in detail the major stages of the cell cycle, in which the chromosomes are duplicated and then segregated into the two daughter cells. At the end of the chapter, we consider how an animal regulates the size and number of its cells and thereby the size of the organism and its organs: we describe how animals eliminate unwanted cells by a form of programmed cell death called *apoptosis,* and we then discuss how they use extracellular signals to control cell survival, cell growth, and cell division.

OVERVIEW OF THE CELL CYCLE

The most basic function of the cell cycle is to duplicate accurately the vast amount of DNA in the chromosomes and then distribute the copies into genetically identical daughter cells. The duration of the cell cycle varies greatly from one cell type to another. A single-celled yeast can divide every two hours or so in ideal conditions, whereas a mammalian liver cell divides on average less than once a year (Table 18–1). We briefly describe here the sequence of events that occur in a fairly rapidly dividing

TABLE 18–1 SOME EUCARYOTIC CELL-CYCLE TIMES	
CELL TYPE	CELL-CYCLE TIMES
Early frog embryo cells	30 minutes
Yeast cells	1.5–3 hours
Mammalian intestinal epithelial cells	~12 hours
Mammalian fibroblasts in culture	~20 hours
Human liver cells	~1 year

(proliferating) mammalian cells. We then introduce the cell-cycle control system that ensures that the various events of the cycle take place in the correct sequence and at the correct time.

The Eucaryotic Cell Cycle Is Divided into Four Phases

Seen under a microscope, the two most dramatic events in the cycle are when the nucleus divides, a process called *mitosis,* and when the cell later splits in two, a process called *cytokinesis.* These two processes together constitute the **M phase** of the cell cycle. In a typical mammalian cell, the whole of M phase takes about an hour, which is only a small fraction of the total cell-cycle time.

The period between one M phase and the next is called **interphase**. Under the microscope, it appears, deceptively, as an uneventful interlude during which the cell simply increases in size. Interphase, however, is a very busy time for the cell, and it encompasses the remaining three phases of the cell cycle. During **S phase** (S = synthesis), the cell replicates its nuclear DNA, an essential prerequisite for cell division. S phase is flanked by two phases in which the cell continues to grow. The **G_1 phase** (G = gap) is the interval between the completion of M phase and the beginning of S phase. The **G_2 phase** is the interval between the end of S phase and the beginning of M phase (Figure 18–2). During these gap phases, the cell monitors the internal and external environments to ensure that conditions are suitable and its preparations are complete before it commits itself to the major upheavals of S phase and mitosis. At particular points in G_1 and G_2, the cell decides whether to proceed to the next phase or pause to allow more time to prepare.

During all of interphase, a cell generally continues to transcribe genes, synthesize proteins, and grow in mass. Together, G_1 and G_2 phases provide additional time for the cell to grow and duplicate its cytoplasmic

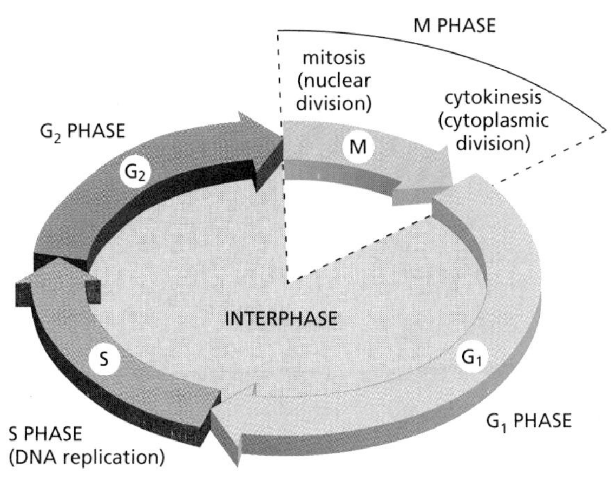

Figure 18–2 **The cell cycle is divided into four phases.** The cell grows continuously in interphase, which consists of three phases: G_1, S, and G_2. DNA replication is confined to S phase. G_1 is the gap between M phase and S phase, and G_2 is the gap between S phase and M phase. During M phase, the nucleus divides first, in a process called mitosis; then the cytoplasm divides, in a process called cytokinesis.

QUESTION 18–2

A population of proliferating cells is stained with a dye that becomes fluorescent when it binds to DNA, so that the amount of fluorescence is directly proportional to the amount of DNA in each cell. To measure the amount of DNA in each cell, the cells are then passed through a flow cytometer, an instrument that measures the amount of fluorescence in individual cells. The number of cells with a given DNA content is plotted on the graph below.

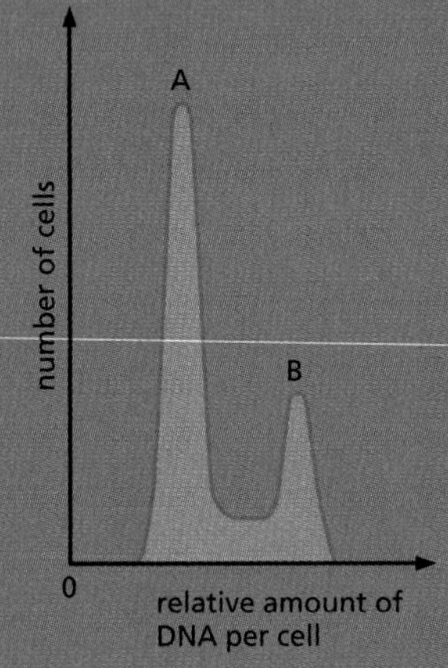

Indicate on the graph where you would expect to find cells that are in the following stages: G_1, S, G_2, and mitosis. Which is the longest phase of the cell cycle in this population of cells?

organelles: if interphase lasted only long enough for DNA replication, the cell would not have time to double its mass before it divided and would consequently shrink with each division. Indeed, in some special circumstances that is just what happens. In some animal embryos, for example, the first cell divisions after fertilization (called *cleavage divisions)* serve to subdivide a giant egg cell into many smaller cells as quickly as possible. In these embryonic cell cycles, the G_1 and G_2 phases are drastically shortened, and the cells do not grow before they divide.

Following DNA replication in S phase, the two copies of each chromosome remain tightly bound together. The first visible sign that a cell is about to enter M phase is the progressive *condensation* of its chromosomes. As condensation proceeds, the replicated chromosomes first become visible in the light microscope as long threads, which gradually get shorter and thicker. This condensation makes the chromosomes less likely to get entangled, so that they are easier to segregate to the two forming daughter cells during mitosis.

A Cell-Cycle Control System Triggers the Major Processes of the Cell Cycle

To ensure that they replicate all their DNA and organelles, and divide in an orderly manner, eucaryotic cells possess a complex network of regulatory proteins known as the *cell-cycle control system*. This system guarantees that the events of the cell cycle—DNA replication, mitosis, and so on—occur in a set sequence and that each process has been completed before the next one begins. To accomplish this, the control system is itself regulated at certain critical points of the cycle by feedback from the process being performed. Without such feedback, an interruption or a delay in any of the processes could be disastrous. All of the nuclear DNA, for example, must be replicated before the nucleus begins to divide, which means that a complete S phase must precede M phase. If DNA synthesis is slowed down or stalled, mitosis and cell division must also be delayed. Similarly, if DNA is damaged, the cycle must arrest in G_1, S, or G_2 so that the cell can repair the damage, either before DNA replication is started or completed or before the cell enters M phase. The cell-cycle control system achieves all of this by means of molecular brakes that can stop the cycle at various **checkpoints**. In this way, the control system does not trigger the next step in the cycle unless the cell is properly prepared.

Three checkpoints that control progression through the cell cycle are illustrated in Figure 18–3. One checkpoint operates in G_1 and allows the cell to confirm that the environment is favorable for cell proliferation before committing to S phase. Cell proliferation in animals requires both sufficient nutrients and specific signal molecules in the extracellular environment; if extracellular conditions are unfavorable, cells can delay progress through G_1 and may even enter a specialized resting state known as *G_0* (G zero). Many cells, including nerve cells and skeletal muscle cells, remain in G_0 for the lifetime of the organism. Another checkpoint operates in G_2 and ensures that cells do not enter mitosis until damaged DNA has been repaired and DNA replication is complete. A third checkpoint operates during mitosis and ensures that the replicated chromosomes are properly attached to a cytoskeletal machine, called the *mitotic spindle*, before the spindle pulls the chromosomes apart and distributes them into the two daughter cells.

The checkpoint in G_1 is especially important as a point in the cell cycle where the control system can be regulated by signals from other cells. In a multicellular animal, the control system is highly responsive to signals

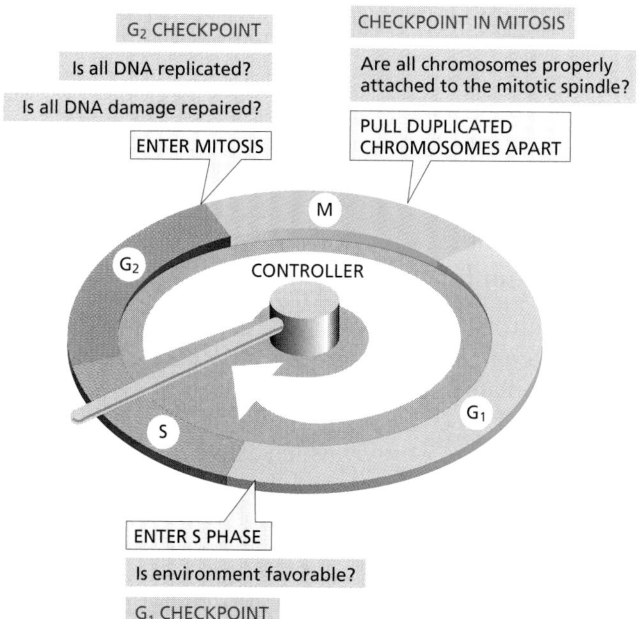
G$_2$ CHECKPOINT
Is all DNA replicated?
Is all DNA damage repaired?
ENTER MITOSIS
CHECKPOINT IN MITOSIS
Are all chromosomes properly attached to the mitotic spindle?
PULL DUPLICATED CHROMOSOMES APART
M
G$_2$
CONTROLLER
S
G$_1$
ENTER S PHASE
Is environment favorable?
G$_1$ CHECKPOINT

Figure 18–3 **Checkpoints in the cell-cycle control system ensure that key processes in the cycle occur in the proper sequence.** The cell-cycle control system is shown as a controller arm that rotates clockwise, triggering essential processes when it reaches particular points on the outer dial. These processes include DNA replication in S phase and the segregation of duplicated chromosomes in mitosis. Feedback from the intracellular events of the cell cycle, as well as signals from the cell's environment, determine whether the cycle will progress beyond certain checkpoints. Three prominent checkpoints are highlighted: the checkpoint in G_1 determines whether the cell proceeds to S phase; the one in G_2 determines whether the cell proceeds to mitosis; and the one in M phase determines whether the cell is ready to pull the duplicated chromosomes apart and segregate them into two new daughter cells.

from other cells that stimulate cell division when more cells are needed and block it when they are not. The control system therefore plays a central part in the regulation of cell numbers in the tissues of the body; if the system malfunctions such that cell division is excessive, cancer can result. We will see later how extracellular signals influence the decisions made at this checkpoint.

Cell-Cycle Control is Similar in All Eucaryotes

Some features of the cell cycle, including the time required to complete certain events, vary greatly from one cell type to another, even within the same organism. The basic organization of the cycle, however, is essentially the same in all eucaryotic cells, and all eucaryotes appear to use similar machinery and control mechanisms to drive and regulate cell-cycle events. The proteins of the cell-cycle control system first appeared more than a billion years ago, and they have been so well conserved over the course of evolution that many of them function perfectly when transferred from a human cell to a yeast (How We Know, pp. 15–16).

Because of this similarity, biologists can study the cell cycle and its regulation in a variety of organisms and use the findings from all of them to assemble a unified picture of how eucaryotic cells divide. Many discoveries about the cell cycle have come from the systematic search for mutations that inactivate essential components of the cell-cycle control system in yeasts. Studies of cultured mammalian cells and animal embryos have also been useful for examining the molecular mechanisms governing the control of cell proliferation in multicellular organisms.

THE CELL-CYCLE CONTROL SYSTEM

Two types of machinery are involved in cell division: one manufactures the new components of the growing cell, and another hauls the components into their correct places and partitions them appropriately when the cell divides in two. The **cell-cycle control system** switches all this machinery on and off at the correct times and thereby coordinates the

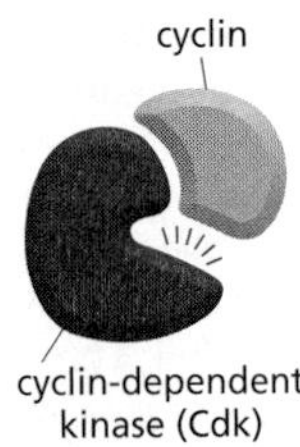

Figure 18–4 **Progression through the cell cycle depends on cyclin-dependent protein kinases (Cdks).** A Cdk must bind a regulatory protein called a cyclin before it can become enzymatically active. The active cyclin–Cdk complex phosphorylates key proteins in the cell that are required to initiate particular steps in the cell cycle. The cyclin also helps direct the Cdk to the target proteins that the Cdk phosphorylates.

various steps of the cycle. The core of the cell-cycle control system is a series of biochemical switches that operate in a defined sequence and orchestrate the main events of the cycle, including DNA replication and segregation of the duplicated chromosomes. In this section, we review the protein components of the control system and discuss how they work together to trigger the different phases of the cycle.

The Cell-Cycle Control System Depends on Cyclically Activated Protein Kinases called Cdks

The cell-cycle control system governs the cell-cycle machinery by cyclically activating and then inactivating the key proteins and protein complexes that initiate or regulate DNA replication, mitosis, and cytokinesis. As discussed in Chapter 4, phosphorylation followed by dephosphorylation is one of the most common ways by which cells switch the activity of a protein on and off (see Figure 4–38), and the cell-cycle control system uses this mechanism repeatedly. The phosphorylation reactions that control the cell cycle are carried out by a specific set of protein kinases, while dephosphorylation is performed by a set of protein phosphatases.

The protein kinases at the core of the cell-cycle control system are present in proliferating cells throughout the cell cycle. They are activated, however, only at appropriate times in the cycle, after which they quickly become deactivated again. Thus, the activity of each of these kinases rises and falls in a cyclical fashion. Some of these protein kinases, for example, become active toward the end of G_1 phase and are responsible for driving the cell into S phase; another kinase becomes active just before M phase and is responsible for driving the cell into mitosis.

Switching these kinases on and off at the appropriate times is partly the responsibility of another set of proteins in the control system—the **cyclins**. Cyclins have no enzymatic activity themselves, but they have to bind to the cell-cycle kinases before the kinases can become enzymatically active. The kinases of the cell-cycle control system are therefore known as **cyclin-dependent protein kinases**, or **Cdks** (Figure 18–4). Cyclins are so-named because, unlike the Cdks, their concentrations vary in a cyclical fashion during the cell cycle. The cyclical changes in cyclin concentrations help drive the cyclic assembly and activation of the cyclin–Cdk complexes; activation of these complexes in turn triggers various cell-cycle events, such as entry into S phase or M phase (Figure 18–5). We discuss how the Cdks and cyclins were discovered in the How We Know, pp. 615–616.

The Activity of Cdks Is Also Regulated by Phosphorylation and Dephosphorylation

The rise and fall of cyclin levels plays an important part in regulating Cdk activity during the cell cycle, but there is more to the story. Cyclin concen-

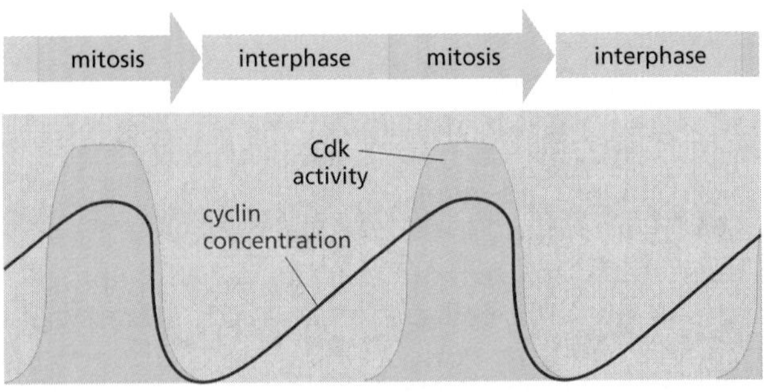

Figure 18–5 **The accumulation of cyclins regulates the activity of Cdks.** The formation of active cyclin–Cdk complexes drives various cell-cycle events, including entry into S phase or M phase. The figure shows the changes in cyclin concentration and Cdk activity responsible for controlling entry into M phase. The increase in the cyclin concentration helps form the active cyclin–Cdk complex that drives entry into M phase. Although the enzymatic activity of the cyclin–Cdk complex rises and falls during the course of the cell cycle, the concentration of the Cdk component does not (not shown).

For many years, cell biologists watched the 'puppet show' of DNA synthesis, mitosis, and cytokinesis but had no idea what was behind the curtain, controlling these events. The cell-cycle control system was simply a 'black box' inside the cell. It was not even clear whether there was a separate control system, or whether the cell-cycle machinery somehow controlled itself. A breakthrough came with the identification of the key proteins of the control system and the realization that they are distinct from the components of the cell-cycle machinery—the enzymes and other proteins that perform the essential processes of DNA replication, chromosome segregation, and so on.

The first components of the cell-cycle control system to be discovered were the cyclins and cyclin-dependent kinases (Cdks) that drive cells into M phase. They were found in studies of cell division conducted on animal eggs.

Back to the egg

The fertilized eggs of many animals are especially suitable for biochemical studies of the cell cycle because they are exceptionally large and divide rapidly. An egg of the frog *Xenopus,* for example, is just over 1 mm in diameter (Figure 18–6). After fertilization, it divides rapidly to partition the egg into many smaller cells. These rapid cell cycles consist mainly of repeated S and M phases, with very short or no G_1 or G_2 phases between them. There is no new gene transcription: all of the mRNAs, as well as most of the proteins, required for this early stage of embryonic development are already packed into the very large egg during its development as an oocycte in the ovary of the mother. In these early division cycles *(cleavage divisions),* no cell growth occurs, and all the cells of the embryo divide synchronously.

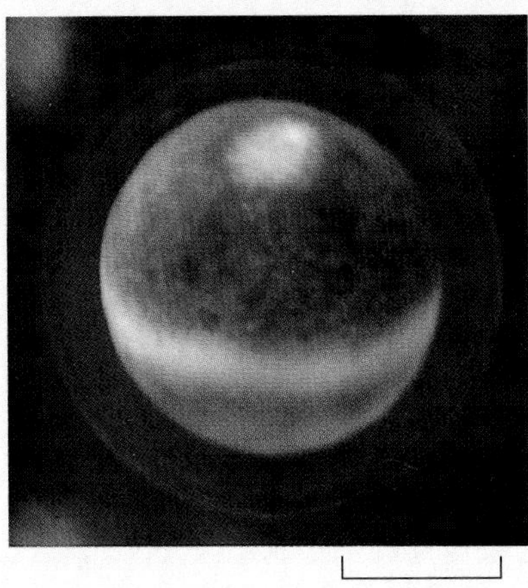

Figure 18–6 **A mature *Xenopus* egg provides a convenient system for studying cell division.** (Courtesy of Tony Mills.)

Because of the synchrony, it is possible to prepare an extract from frog eggs that is representative of the cell-cycle stage at which it is made. The biological activity of such an extract can then be tested by injecting it into a *Xenopus* oocyte (the immature precursor of the unfertilized egg) and observing, microscopically, its effects on cell-cycle behavior. The *Xenopus* oocyte is an especially convenient test system for detecting an activity that drives cells into M phase, because of its large size, and because it has completed DNA replication and is arrested at a stage in the meiotic cell cycle (discussed in Chapter 19) that is equivalent to the G_2 phase of a mitotic cell cycle.

Give us an M

In such experiments, researchers found that an extract from an M-phase egg instantly drives the oocyte into M phase, whereas cytoplasm from a cleaving egg at other phases of the cycle does not. When first discovered, the biochemical identity and mechanism of action of the factor responsible for this activity were unknown, and the activity was simply called *maturation promoting factor,* or MPF (Figure 18–7). By testing cytoplasm from different stages of the cell cycle, MPF activity was found to oscillate dramatically during the course of each cell cycle: it increased rapidly just before the start of mitosis and fell rapidly to zero toward the end of mitosis (Figure 18–8). This oscillation made MPF a strong candidate for a component involved in cell-cycle control.

When MPF was finally purified, it was found to contain a protein kinase that was required for its activity. But the kinase portion of MPF did not act alone. It had to have a specific protein (now known to be M cyclin) bound to it in order to function. M cyclin was discovered in a different type of experiment, involving clam eggs.

Fishing in clams

M cyclin was initially identified as a protein whose concentration rose gradually during interphase and then fell rapidly to zero as cleaving clam eggs went through M phase (see Figure 18–5). The protein repeated this performance in each cell cycle. Its role in cell-cycle control, however, was initially obscure. The breakthrough occurred when cyclin was found to be a component of MPF and to be required for MPF activity. Thus, MPF, which we now call M-Cdk, is a protein complex containing two subunits—a regulatory subunit, M cyclin,

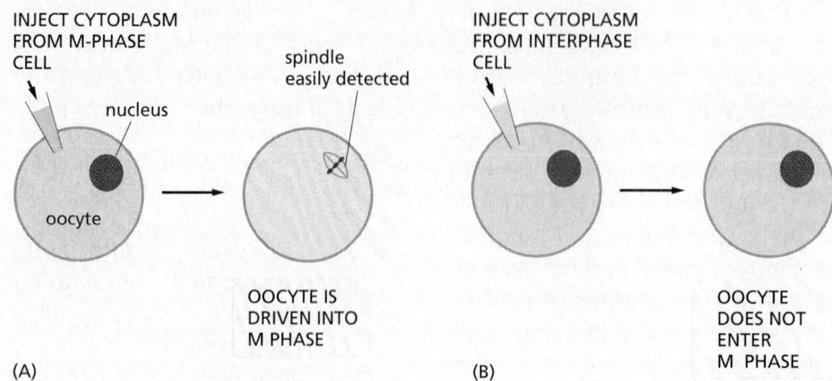

Figure 18–7 **MPF activity was discovered by injecting *Xenopus* egg cytoplasm into *Xenopus* oocytes.** (A) A *Xenopus* oocyte is injected with cytoplasm taken from a *Xenopus* egg in M phase. The cell extract drives the oocyte into M phase of the first meiotic division, causing the large nucleus to break down and a spindle to form. (B) When the cytoplasm is taken from a cleaving egg in interphase, it does not cause the oocyte to enter M phase. Thus, the extract in (A) must contain some activity—a maturation promoting factor (MPF)—that triggers entry into M phase.

and a catalytic subunit, the mitotic Cdk. After the components of M-Cdk were identified, other types of cyclins and Cdks were isolated, whose concentrations or activities, respectively, rose and fell at other stages in the cell cycle.

All in the family

While biochemists were identifying the proteins that regulate the cell cycles of frog and clam embryos, yeast geneticists were taking a different approach to dissecting the cell-cycle control system. By studying mutants that get stuck or misbehave at specific points in the cell cycle, these researchers were able to identify many genes responsible for cell-cycle control. Some of these genes turned out to encode cyclin or Cdk proteins, which were unmistakably similar—in both amino acid sequence and function—to their counterparts in frogs and clams. Similar genes were soon identified in human cells.

Many of the cell-cycle control genes have changed so little during evolution that the human version of the gene will function perfectly well in a yeast cell. For example, a yeast with a defective copy of the gene encoding its only Cdk fails to divide; the mutant will divide normally, however, if a copy of the appropriate human gene is artificially introduced into the defective cell. Surely, even Darwin would have been astonished at such clear evidence that humans and yeasts are cousins. Despite a billion years of divergent evolution, all eucaryotic cells—whether yeast, animal, or plant—use essentially the same molecules to control the events of their cell cycle.

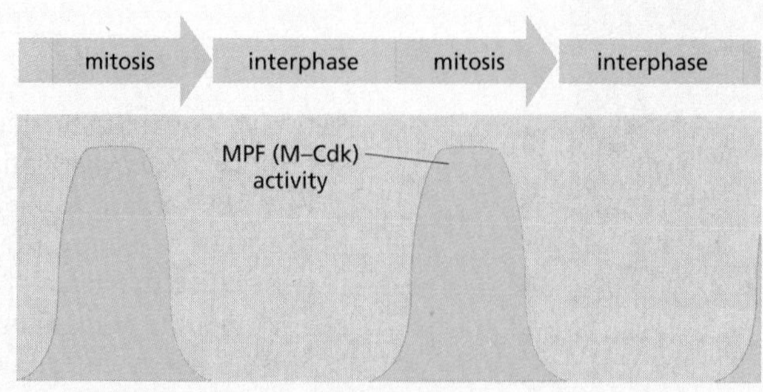

Figure 18–8 **The activity of MPF oscillates during the cell cycle in *Xenopus* embryos.** The activity assayed using the test outlined in Figure 18–7 rises rapidly just before the start of mitosis and falls rapidly to zero toward the end of mitosis.

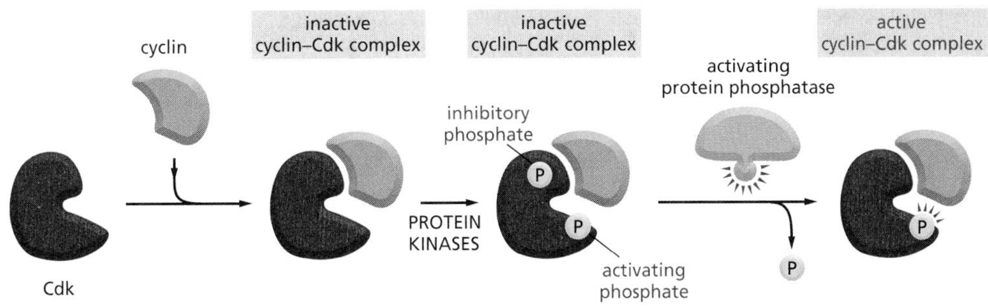

Figure 18–9 **For a Cdk to be active, it must be phosphorylated at one site and dephosphorylated at two other sites.** When it first forms, the cyclin–Cdk complex is not phosphorylated and is inactive. Subsequently, the Cdk is phosphorylated at a site that is required for its activity and at two other (overriding) sites that inhibit its activity. This phosphorylated complex remains inactive until it is finally activated by a protein phosphatase that removes the two inhibitory phosphate groups. For simplicity, only one inhibitory phosphate group is shown here.

trations increase gradually, but the activity of the associated cyclin–Cdk complexes tends to switch on abruptly at the appropriate time in the cell cycle (see Figure 18–5). So what triggers the abrupt activation of these complexes? For a cyclin–Cdk to be maximally active, the Cdk has to be phosphorylated at one site by a specific protein kinase and dephosphorylated at other sites by a specific protein phosphatase (Figure 18–9). We discuss later how these kinases and phosphatases regulate the activity of specific cyclin–Cdks and thus control progression through the cell cycle.

Different Cyclin–Cdk Complexes Trigger Different Steps in the Cell Cycle

There are several types of cyclins and, in most eucaryotes, several types of Cdks involved in cell-cycle control. Different cyclin–Cdk complexes trigger different steps of the cell cycle. The cyclin that acts in G_2 to trigger entry into M phase is called **M cyclin**, and the active complex it forms with its Cdk is called **M-Cdk**. Distinct cyclins, called **S cyclins** and **G_1/S cyclins**, bind to a distinct Cdk protein late in G_1 to form **S-Cdk** and **G_1/S-Cdk**, respectively, and trigger S phase. The action of S-Cdk and M-Cdk are shown in Figure 18–10. Other cyclins, called **G_1 cyclins**, act earlier in G_1 and bind to other Cdk proteins to form **G_1-Cdks**, which help drive the cell through G_1 toward S phase. We will see later that the formation of these G_1-Cdks in animal cells usually depends on extracellular signal molecules that stimulate cells to divide. The names of the individual cyclins and their Cdks are listed in Table 18–2.

As previously explained, the different Cdks also have to be phosphorylated and dephosphorylated in order to act (see Figure 18–9). Each of these activated cyclin–Cdk complexes in turn phosphorylates a different set of target proteins in the cell. As a result, each type of complex triggers a different transition step in the cycle. M-Cdk, for example, phosphory-

TABLE 18–2 THE MAJOR CYCLINS AND CDKS OF VERTEBRATES

CYCLIN–CDK COMPLEX	CYCLIN	CDK PARTNER
G_1-Cdk	cyclin D*	Cdk4, Cdk6
G_1/S-Cdk	cyclin E	Cdk2
S-Cdk	cyclin A	Cdk2
M-Cdk	cyclin B	Cdk1**

*There are three D cyclins in mammals (cyclins D1, D2, and D3).

**The original name of Cdk1 was Cdc2 in vertebrates.

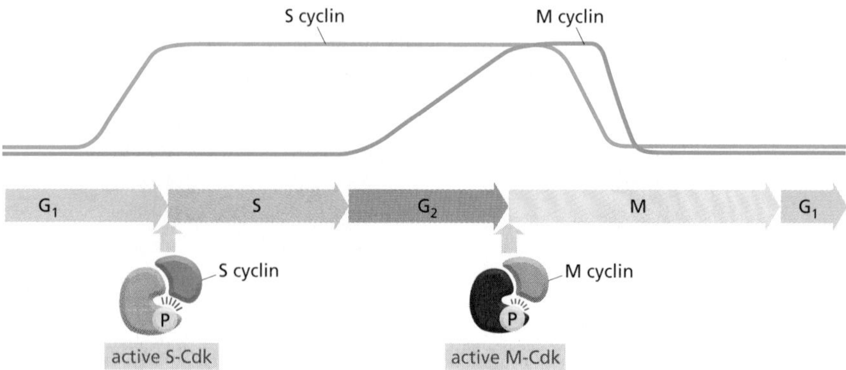

Figure 18–10 **Distinct Cdks associate with different cyclins to trigger the different events of the cell cycle.** For simplicity, only two types of cyclin–Cdk complexes are shown—one that triggers S phase and one that triggers M phase. In both cases, the activation of the Cdk requires phosphorylation and dephosphorylation as well as cyclin binding.

lates key proteins that cause the chromosomes to condense, the nuclear envelope to break down, and the microtubules of the cytoskeleton to reorganize to form the mitotic spindle. These events herald the entry into mitosis, as we discuss later.

The Cell-Cycle Control System Also Depends on Cyclical Proteolysis

The concentration of each type of cyclin rises gradually but then falls sharply at a specific time in the cell cycle (see Figure 18–10). This abrupt fall results from the targeted degradation of the cyclin protein. Specific enzyme complexes add ubiquitin chains to the appropriate cyclin, which is then directed to the proteasome for destruction (Figure 18–11). This rapid elimination of the cyclin returns the Cdk to its inactive state.

Although the activation of Cdks triggers some of the transitions from one part of the cell cycle to the next, their inactivation triggers others. For example, inactivation of M-Cdk—which is triggered by the destruction of M cyclin—leads to the molecular events that take the cell out of mitosis.

Proteins that Inhibit Cdks Can Arrest the Cell Cycle at Specific Checkpoints

We have seen that the cell-cycle control system triggers the events of the cycle in a specific order. It triggers mitosis, for example, only after all the DNA has been replicated, and it permits the cell to divide into two only after mitosis has been completed. If one of the steps is delayed, the control system delays the activation of the next step so that the normal sequence is maintained. This self-regulating property of the control system ensures, for example, that if DNA synthesis is halted for some reason during S phase, the cell will not proceed into M phase with its DNA only partly replicated. As mentioned earlier, the control system accomplishes this feat largely through the action of molecular brakes that can stop the cell cycle at specific checkpoints, allowing the cell to monitor its internal state and its environment before continuing through the cycle (see Figure 18–3).

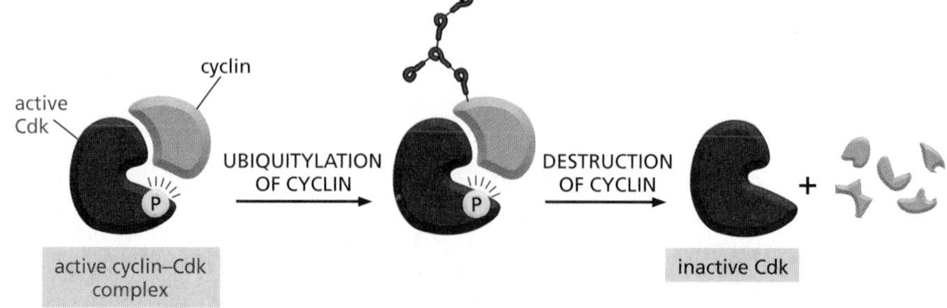

Figure 18–11 **The activity of Cdks is regulated by cyclin degradation.** Ubiquitylation of a cyclin marks the protein for destruction in proteasomes (as discussed in Chapter 7). The loss of cyclin renders its Cdk partner inactive.

Figure 18–12 **A checkpoint in G_1 offers the cell a crossroad.** The cell can commit to completing another cell cycle, pause temporarily until conditions are right, or withdraw from the cell cycle altogether and enter G_0. In some cases, cells in G_0 can re-enter the cell cycle when conditions improve, but many cell types permanently withdraw from the cell cycle when they differentiate, persisting in G_0 for the lifetime of the animal.

Some of these molecular brakes rely on **Cdk inhibitor proteins** that block the assembly or activity of one or more cyclin–Cdk complexes. Certain Cdk inhibitor proteins, for example, help maintain Cdks in an inactive state during the G_1 phase of the cycle, thus delaying progression into S phase. Pausing at this checkpoint gives the cell more time to grow, or allows it to wait until extracellular conditions are favorable for division. As a general rule, mammalian cells will multiply only if they are stimulated to do so by extracellular signals called *mitogens* produced by other cells. If deprived of such signals, the cell cycle arrests at a *G_1 checkpoint* (see Figure 18–3); and, if the cell is deprived for long enough, it will withdraw from the cell cycle and enter the non-proliferating state G_0, in which the cell can remain for days or weeks or even for the lifetime of the organism (Figure 18–12).

Most of the diversity in cell-division rates in the adult body lies in the variation in the time that cells spend in G_0 or in G_1. Some cell types, such as liver cells, normally divide only once every year or two, whereas certain epithelial cells in the gut divide more than twice a day to renew the lining of the gut continually. Many of our cells fall somewhere in between: they can divide if the need arises but normally do so infrequently. Escape from the G_1 checkpoint or from G_0 requires the accumulation of G_1 cyclins, and mitogens function by stimulating this accumulation.

Once past the G_1 checkpoint, a cell usually proceeds all the way through the rest of the cell cycle quickly—typically within 12–24 hours in mammals. The G_1 checkpoint is therefore sometimes called *Start,* because passing it represents a commitment to complete a full division cycle, although a better name might be Stop (see Figure 18–12). Some of the main checkpoints in the cell cycle are summarized in Figure 18–13.

The most radical decision that the cell-cycle control system can make is to withdraw the cell from the cell cycle permanently. This is different from withdrawing from the cell cycle temporarily, to wait for more favorable conditions, and it has a special importance in multicellular organisms. In the human body, for example, nerve cells and skeletal muscle cells permanently stop dividing when they differentiate. They enter an irreversible

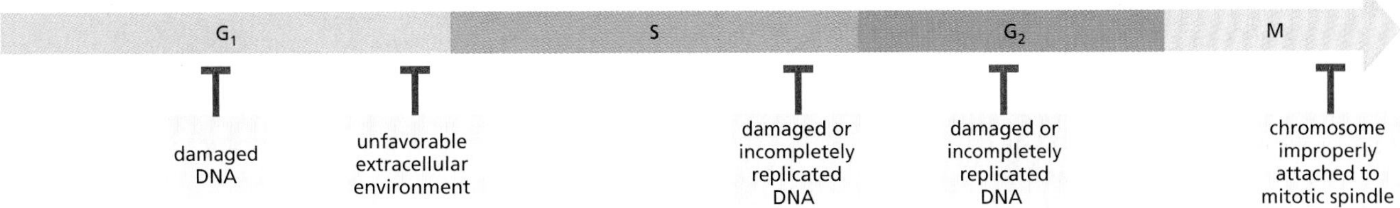

Figure 18–13 **The cell-cycle control system can arrest the cycle at various checkpoints.** The *red* "T"s represent points in the cycle where the control system can apply molecular brakes (such as Cdk inhibitor proteins) to stop progression in response to DNA damage, intracellular processes that have not been completed, or an unfavorable extracellular environment. The checkpoint indicated in M phase ensures that all of the chromosomes are appropriately attached to the mitotic spindle before the duplicated chromosomes are pulled apart.

QUESTION 18–3

Why do you suppose cells have evolved a special G_0 state to exit from the cell cycle, rather than just stopping in a G_1 state at a G_1 checkpoint?

G_0 state, in which the cell-cycle control system is largely dismantled: many of the Cdks and cyclins disappear, and the cyclin–Cdk complexes that are still present are inhibited by Cdk inhibitor proteins.

We now turn to S phase, in which cells replicate their DNA and begin to prepare their chromosomes for segregation.

S PHASE

Before a cell divides, it must duplicate its DNA. As we discuss in Chapter 6, this replication must occur with extreme accuracy to minimize the risk of mutations in the next cell generation. Of equal importance, every nucleotide in the genome must be copied once—and only once—to prevent the damaging effects of gene amplification. In this section, we consider the elegant mechanisms by which the cell-cycle control system initiates the replication process and, at the same time, prevents replication from happening more than once per cell cycle.

S-Cdk Initiates DNA Replication and Helps Block Re-Replication

As we discuss in Chapter 6, DNA replication begins at *origins of replication*, nucleotide sequences that are scattered along each chromosome. These sequences recruit specific proteins that control the initiation and completion of DNA replication. One multiprotein complex, the **origin recognition complex (ORC)**, remains bound to origins of replication throughout the cell cycle, where it serves as a sort of landing pad for additional regulatory proteins that bind before the start of S phase.

One of these regulatory proteins, called Cdc6, is present at low levels during most of the cell cycle, but its concentration increases transiently in early G_1. When Cdc6 binds to ORCs in G_1, it promotes the binding of additional proteins to form a *pre-replicative complex*. Once the pre-replicative complex has been assembled, the replication origin is ready to 'fire.' The activation of S-Cdk in late G_1 then 'pulls the trigger,' initiating DNA replication.

As shown in Figure 18–14, S-Cdk does not only initiate origin firing; it also helps prevent re-replication of the DNA. Activated S-Cdk helps phosphorylate Cdc6, causing it and the other proteins in the pre-replicative complex to dissociate from the ORC after an origin has fired. This disassembly prevents replication from occurring again at the same origin. In addition to promoting dissociation, phosphorylation of Cdc6 by S-Cdk (and by M-Cdk, which becomes active at the start of M phase) marks it

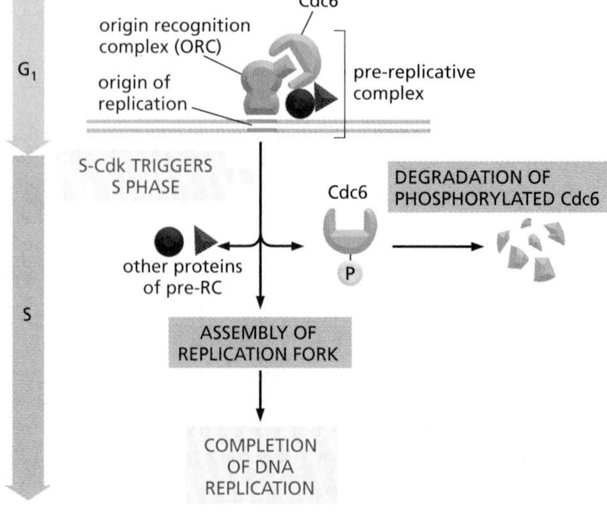

Figure 18–14 **S-Cdk triggers DNA replication and ensures that DNA replication is initiated only once per cell cycle.** ORCs remain associated with origins of replication throughout the cell cycle. In early G_1, the regulatory protein Cdc6 associates with the ORC. Aided by Cdc6, additional proteins bind to the adjacent DNA, resulting in the formation of a pre-replicative complex, which includes the proteins and the DNA to which they are bound. S-Cdk then triggers origin firing by causing the assembly of the protein complexes that initiate DNA synthesis (discussed in Chapter 6). S-Cdk also helps block re-replication by helping to phosphorylate Cdc6, which dissociates from the origin and is degraded.

for degradation, ensuring that DNA replication is not reinitiated later in the same cell cycle.

Cohesins Help Hold the Sister Chromatids of Each Replicated Chromosome Together

After the chromosomes have been duplicated in S phase, the two copies of each replicated chromosome remain tightly bound together as identical **sister chromatids**. The sister chromatids are held together by protein complexes called **cohesins**, which assemble along the length of each sister chromatid as the DNA is replicated in S phase. The cohesins form protein rings that surround the two sister chromatids, keeping them united (Figure 18–15). This cohesion between sister chromatids is crucial for proper chromosome segregation, and it is broken completely only in late mitosis to allow the sister chromatids to be pulled apart by the mitotic spindle. Defects in sister-chromatid cohesion—in yeast mutants, for example—lead to major errors in chromosome segregation.

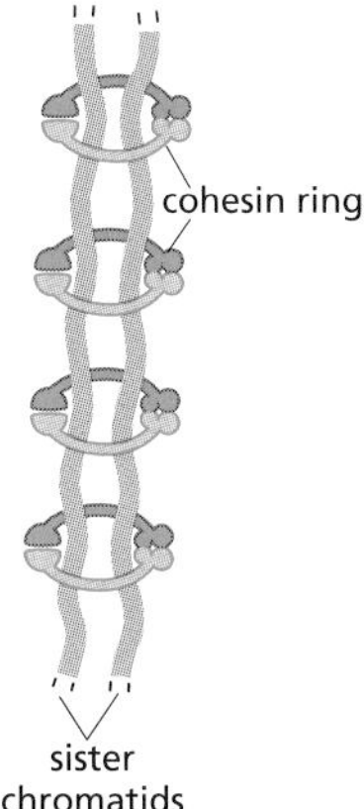

Figure 18–15 **Cohesins tie together the two adjacent sister chromatids in each replicated chromosome.** They form large protein rings that surround the sister chromatids, preventing them from coming apart, until the rings are broken late in mitosis.

DNA Damage Checkpoints Help Prevent the Replication of Damaged DNA

The cell-cycle control system uses several distinct checkpoint mechanisms to halt progress through the cell cycle if DNA is damaged. *DNA damage checkpoints* in G_1 and S phase prevent the cell from starting or completing S phase and replicating damaged DNA. Another checkpoint operates in G_2 to prevent the cell from entering M phase with damaged or incompletely replicated DNA (see Figure 18–13).

The G_1 checkpoint mechanism is especially well understood. DNA damage causes an increase in both the concentration and activity of a protein called **p53**, which is a transcription regulator that activates the transcription of a gene encoding a Cdk inhibitor protein called p21. The p21 protein binds to G_1/S-Cdk and S-Cdk, preventing them from driving the cell into S phase (Figure 18–16). The arrest of the cell cycle in G_1 gives the cell time to repair the damaged DNA before replicating it. If the DNA damage is too severe to be repaired, p53 can induce the cell to kill itself by undergoing apoptosis. If p53 is missing or defective, the unrestrained replication of damaged DNA leads to a high rate of mutation and the production of cells that tend to become cancerous. In fact, mutations in the *p53* gene are found in about half of all human cancers.

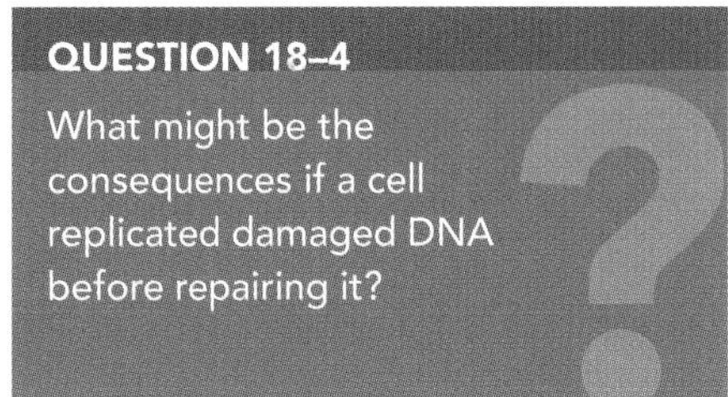

QUESTION 18–4
What might be the consequences if a cell replicated damaged DNA before repairing it?

Once DNA replication has begun, another type of checkpoint mechanism operates to prevent a cell entering M phase with damaged or incompletely replicated DNA. As we saw in Figure 18–9, the activity of cyclin–Cdk complexes is inhibited by phosphorylation at particular sites. For the cell to progress into mitosis, M-Cdk has to be activated by the removal of these inhibitory phosphates by a specific protein phosphatase. When DNA is damaged (or incompletely replicated), this activating protein phosphatase is itself inhibited, so the inhibitory phosphates are not removed from M-Cdk. As a result, M-Cdk remains inactive and M phase cannot be initiated until DNA replication is complete and any DNA damage is repaired.

Once a cell has passed through these checkpoints and has successfully replicated its DNA in S phase and progressed through G_2, it is ready to enter M phase, in which it divides its nucleus (the process of mitosis) and then its cytoplasm (the process of cytokinesis) (see Figure 18–2). In the next three sections, we focus on M phase. We first present a brief overview of M phase as a whole and then discuss in more detail the events that occur during mitosis and then those that occur during cytokinesis. Our focus will be mainly on animal cells.

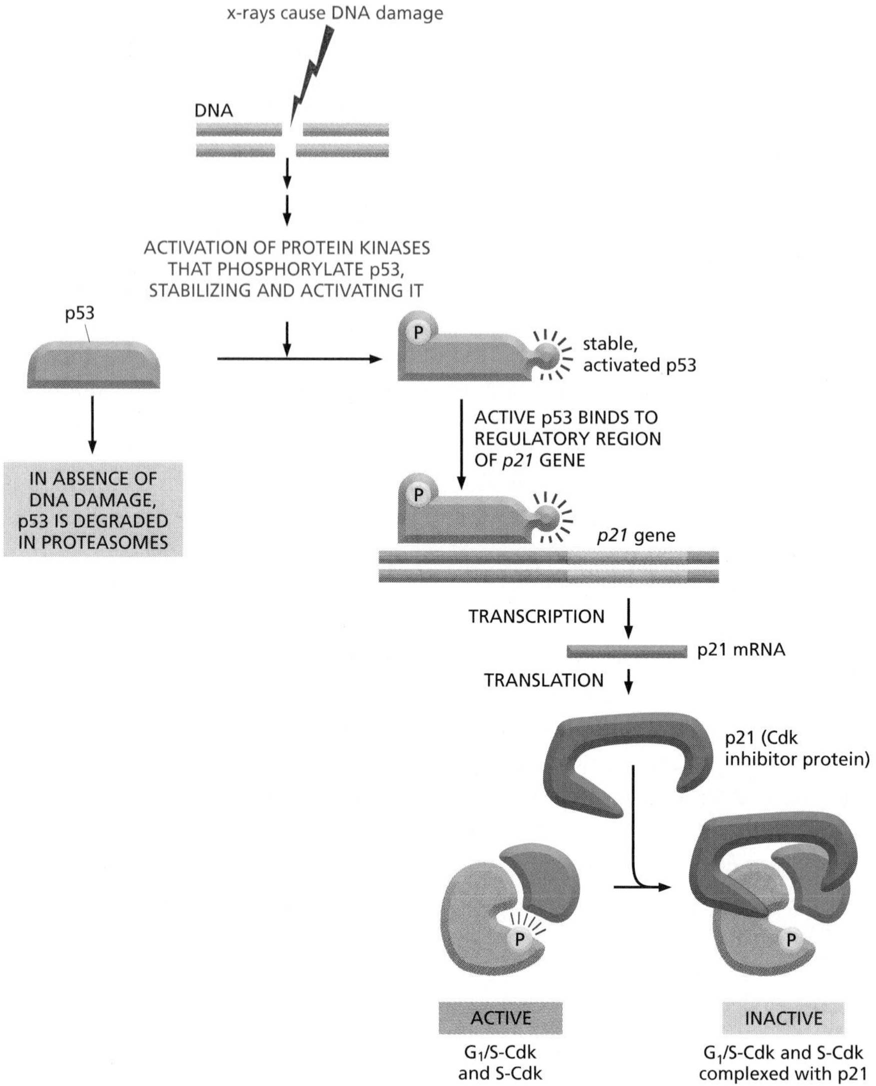

Figure 18–16 **DNA damage can arrest the cell cycle at a checkpoint in G_1.** When DNA is damaged, specific protein kinases respond by activating the p53 protein and halting its normal rapid degradation. Activated p53 protein then accumulates and binds to DNA (Movie 18.1). There it stimulates the transcription of the gene that encodes the Cdk inhibitor protein, p21. The p21 protein binds to G_1/S-Cdk and S-Cdk and inactivates them, so that the cell cycle arrests in G_1.

M PHASE

Although M phase (mitosis plus cytokinesis) occurs over a relatively short amount of time—about one hour in a mammalian cell that divides once a day, or even once a year—it is by far the most dramatic phase of the cell cycle. During this brief period, the cell reorganizes virtually all of its components and distributes them equally into the two daughter cells. The earlier phases of the cell cycle, in effect, serve to set the stage for the drama of M phase.

The central problem for a cell in M phase is to accurately segregate its chromosomes, which were replicated in the preceding S phase, so that each new daughter cell receives an identical copy of the genome. With minor variations, all eucaryotes solve this problem in a similar way: they assemble two specialized cytoskeletal machines, one that pulls the duplicated chromosome sets apart (during mitosis) and another that divides the cytoplasm into two halves (cytokinesis). We begin our discussion of M phase with an overview in which we consider how the cell sets the processes of M phase in motion. We then address mitosis and cytokinesis in turn.

M-Cdk Drives Entry Into M Phase and Mitosis

One of the most remarkable features of cell-cycle control is that a single protein complex, M-Cdk, brings about all of the diverse and intricate

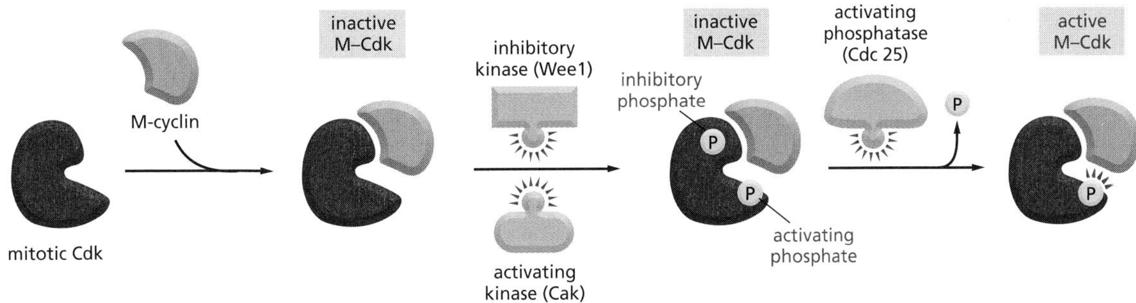

Figure 18–17 **For M-Cdk to be active, it must be phosphorylated at one site and dephosphorylated at other sites.** The M cyclin–Cdk complex is enzymatically inactive when first formed. Subsequently, the Cdk is phosphorylated at one site that is required for its activity by an enzyme called Cdk-activating kinase, Cak (Movie 18.2). It is also phosphorylated at two other, overriding sites that inhibit its activity (by an enzyme called Wee1); for simplicity, only one inhibitory phosphate group is shown. It is still not clear how the timing of this complex activation process is controlled.

rearrangements that occur in the early stages of mitosis. M-Cdk triggers the condensation of the replicated chromosomes into compact, rod-like structures, readying them for segregation, and it also induces the assembly of the mitotic spindle that will separate the condensed chromosomes and segregate them into the two daughter cells.

As discussed earlier, M-Cdk activation begins with the accumulation of M cyclin (see Figure 18–10). Synthesis of M cyclin starts immediately after S phase; its concentration then rises gradually and helps time the onset of M phase. The increase in M cyclin protein leads to a corresponding accumulation of M-Cdk complexes. But those complexes, when they first form, are inactive. The sudden activation of the M-Cdk stockpile at the end of G_2 is triggered by the activation of a protein phosphatase (Cdc25) that removes the inhibitory phosphates holding M-Cdk activity in check (Figure 18–17).

Once activated, each M-Cdk complex can indirectly activate more M-Cdk, by phosphorylating and activating more Cdc25, as illustrated in Figure 18–18. In addition, activated M-Cdk also inhibits the inhibitory kinase Wee1 (see Figure 18–17), further promoting the activation of M-Cdk. The overall consequence is that, once the activation of M-Cdk begins, there is an explosive increase in M-Cdk activity that drives the cell abruptly from G_2 into M phase.

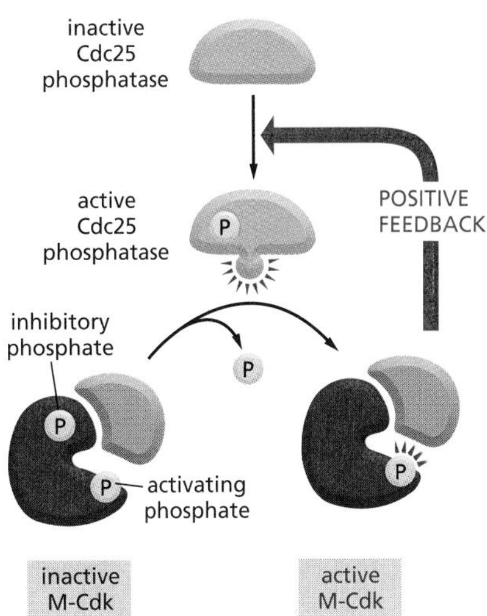

Figure 18–18 **Activated M-Cdk indirectly activates more M-Cdk, creating a positive feedback loop.** Once activated, M-Cdk phosphorylates, and thereby activates, more Cdk-activating phosphatase (Cdc25). The phosphatase can now activate more M-Cdk by removing the inhibitory phosphate groups from the Cdk subunit.

Condensins Help Configure Duplicated Chromosomes for Separation

When the cell is about to enter M phase, the replicated chromosomes condense, becoming visible as threadlike structures. Protein complexes, called **condensins**, help carry out this **chromosome condensation**. The M-Cdk that initiates entry into M phase triggers the assembly of condensin complexes onto DNA by phosphorylating some of the condensin subunits. Condensation makes the mitotic chromosomes more compact, reducing them to small physical packets that can be more easily segregated within the crowded confines of the dividing cell.

Condensins are structurally related to cohesins—the proteins that hold sister chromatids together (see Figure 18–15). Both cohesins and condensins form ring structures, and, together, the two types of protein rings help to configure the replicated chromosomes for mitosis. Cohesins assemble on the DNA as it replicates in S phase and tie together two parallel DNA molecules—the identical sister chromatids. Condensins, by

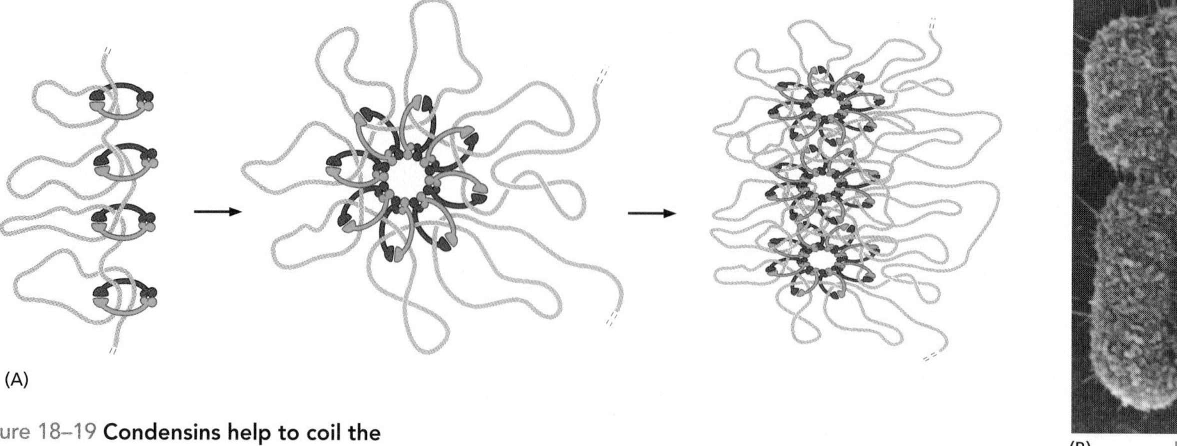

Figure 18–19 **Condensins help to coil the mitotic chromatids into smaller, more compact structures that can be more easily segregated during mitosis.** (A) A model for the way in which condensin proteins might compact a single chromatid by coiling up long loops of DNA. (B) A scanning electron micrograph of a condensed human mitotic chromosome, consisting of two sister chromatids joined along their length. The constricted region *(arrow)* is the centromere, where each chromatid will attach to the mitotic spindle, which pulls the sister chromatids apart toward the end of mitosis. (B, courtesy of Terry D. Allen.)

contrast, assemble on each individual chromatid at the start of M phase and coil up the DNA to help each chromatid condense (Figure 18–19).

The Cytoskeleton Carries Out Both Mitosis and Cytokinesis

After the replicated chromosomes have condensed, two complex cytoskeletal machines assemble in sequence to carry out the two mechanical processes that occur in M phase. The *mitotic spindle* carries out nuclear division (mitosis), and, in animal cells and many unicellular eucaryotes, the *contractile ring* carries out cytoplasmic division (cytokinesis) (Figure 18–20). Both structures rapidly disassemble after they have performed their tasks.

The mitotic spindle is composed of microtubules and the various proteins that interact with them, including microtubule-associated motor proteins (discussed in Chapter 17). In all eucaryotic cells, the mitotic spindle is responsible for separating the replicated chromosomes and allocating one copy of each chromosome to each daughter cell.

The *contractile ring* consists mainly of actin filaments and myosin filaments arranged in a ring around the equator of the cell (introduced in Chapter 17). It starts to assemble just beneath the plasma membrane toward the end of mitosis. As the ring contracts, it pulls the membrane inward, thereby dividing the cell in two (see Figure 18–20). We discuss later how plant cells, which have a cell wall to contend with, divide their cytoplasm by a very different mechanism.

M Phase Is Conventionally Divided into Six Stages

Although M phase proceeds as a continuous sequence of events, it is traditionally divided into six stages. The first five stages of M phase—prophase, prometaphase, metaphase, anaphase, and telophase—constitute

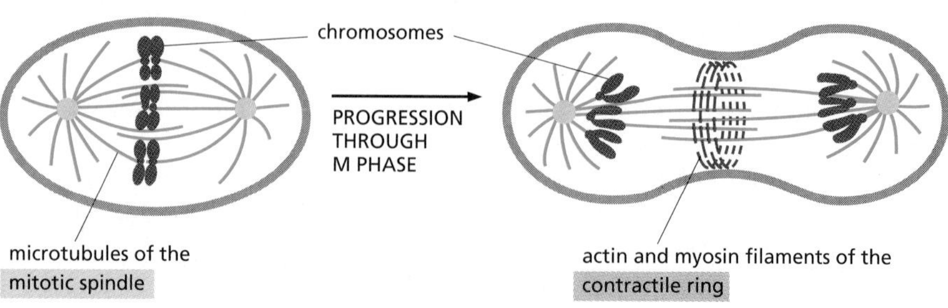

Figure 18–20 **Two transient cytoskeletal structures mediate M phase in animal cells.** The mitotic spindle assembles first to separate the replicated chromosomes. Then, the contractile ring assembles to divide the cell in two. Whereas the mitotic spindle is based on microtubules, the contractile ring is based on actin and myosin filaments. Plant cells use a very different mechanism to divide the cytoplasm, as we discuss later.

mitosis, which was originally defined as the period in which the chromosomes are visible (because they have become condensed). *Cytokinesis* constitutes the sixth stage, and it overlaps in time with the end of mitosis. The six stages of M phase are summarized in Panel 18–1 (pp. 626–627). Together, they form a dynamic sequence in which many independent cycles—involving the chromosomes, cytoskeleton, and centrosomes—are coordinated to produce two genetically identical daughter cells.

The five stages of mitosis occur in strict sequential order, whereas cytokinesis begins in anaphase and continues through telophase. During *prophase*, the replicated chromosomes condense and the mitotic spindle begins to assemble outside the nucleus. During *prometaphase*, the nuclear envelope breaks down, allowing the spindle microtubules to bind to the chromosomes. During *metaphase*, the mitotic spindle gathers all of the chromosomes to the center (equator) of the spindle. During *anaphase*, the two sister chromatids in each replicated chromosome synchronously split apart, and the spindle draws them to opposite poles of the cell. During *telophase*, a nuclear envelope reassembles around each of the two sets of separated chromosomes to form two nuclei (Movie 18.3 and Movie 18.4). Cytokinesis is complete by the end of telophase, when the nucleus and cytoplasm of each of the daughter cells return to interphase, signaling the end of M phase.

QUESTION 18–5

A small amount of cytoplasm isolated from a mitotic cell is injected into an unfertilized frog oocycte, causing the oocyte to enter M phase (see Figure 18–7A). A sample of the injected oocyte's cytoplasm is then taken and injected into a second oocyte, causing this cell also to enter M phase. The process is repeated many times until, essentially, none of the original protein sample remains, and yet, cytoplasm taken from the last in the series of injected oocytes is still able to trigger entry into M phase with undiminished efficiency. Explain this remarkable observation.

MITOSIS

Before nuclear division, or mitosis, begins, each chromosome has been replicated and consists of two identical sister chromatids, held together along their length by cohesin proteins (see Figure 18–15). During mitosis, the cohesin proteins are cleaved, the sister chromatids split apart, and the resulting daughter chromosomes are pulled to opposite poles of the cell by the mitotic spindle (Figure 18–21). In this section, we examine how the mitotic spindle assembles and functions. We discuss how the dynamic instability of microtubules and the activity of microtubule-associated motor proteins contribute to both the assembly of the spindle and its ability to segregate the sister chromatids. Finally, we review the checkpoint mechanism that operates in mitosis to ensure the synchronous separation of the sister chromatids, the proper segregation of the two chromosome sets to the two daughter cells, and the orderly and timely exit from mitosis.

Centrosomes Duplicate To Help Form the Two Poles of the Mitotic Spindle

Before M phase begins, two critical events must be completed: DNA must be fully replicated, and, in animal cells, the centrosome must be duplicated. The **centrosome** is the principal *microtubule-organizing center* in animal cells. It duplicates so that it can help form the two poles of the mitotic spindle and so that each daughter cell can receive its own centrosome.

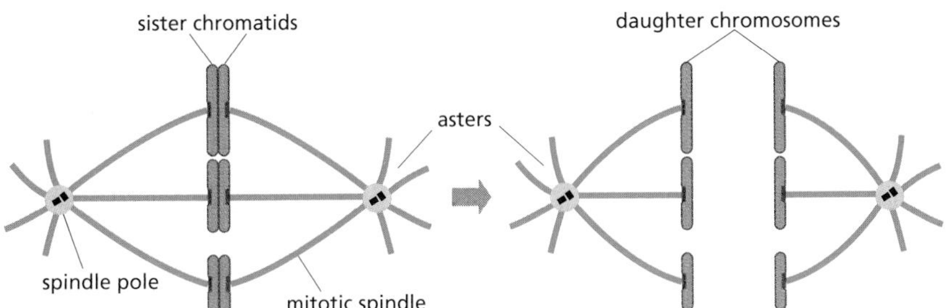

Figure 18–21 **At the beginning of anaphase, each pair of sister chromatids separates.** The resulting daughter chromosomes are then pulled to opposite poles of the cell by the mitotic spindle.

PANEL 18–1 The principal stages of M phase in an animal cell

CELL DIVISION AND THE CELL CYCLE

INTERPHASE

S

G_1

G_2

6 CYTOKINESIS

1 PROPHASE

CELL CYCLE

5 TELOPHASE

2 PROMETAPHASE

4 ANAPHASE

3 METAPHASE

M PHASE

The division of a cell into two daughters occurs in the M phase of the cell cycle. M phase consists of nuclear division, or mitosis, and cytoplasmic division, or cytokinesis. In this figure, M phase has been expanded for clarity. Mitosis is itself divided into five stages, and these, together with cytokinesis, are described in this panel.

INTERPHASE

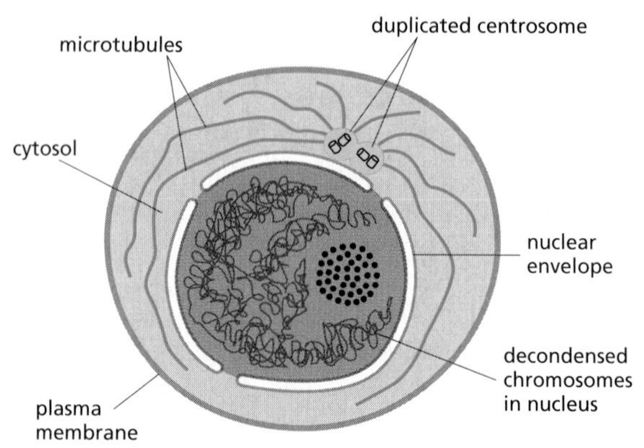

During interphase, the cell increases in size. The DNA of the chromosomes is replicated, and the centrosome is duplicated.

In the light micrographs of dividing animal cells shown in this panel, chromosomes are stained *orange* and microtubules are *green*. (Micrographs courtesy of Julie Canman and Ted Salmon; "Metaphase" from cover of *J. Cell. Sci.* 115(9), 2002, with permission from The Company of Biologists Ltd; "Telophase" from J.C. Canman et al., *Nature* 424:1074–1078, 2003, with permission from Macmillan Publishers Ltd.)

1 PROPHASE

centrosome

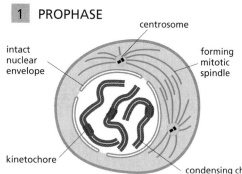

intact nuclear envelope

forming mitotic spindle

kinetochore

condensing chromosome with two sister chromatids held together along their length

At prophase, the replicated chromosomes, each consisting of two closely associated sister chromatids, condense. Outside the nucleus, the mitotic spindle assembles between the two centrosomes, which have begun to move apart. For simplicity, only three chromosomes are drawn.

time = 0 min

2 PROMETAPHASE

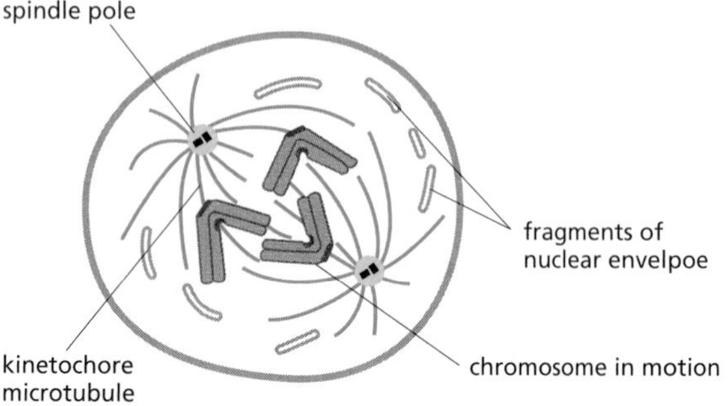

Prometaphase starts abruptly with the breakdown of the nuclear envelope. Chromosomes can now attach to spindle microtubules via their kinetochores and undergo active movement.

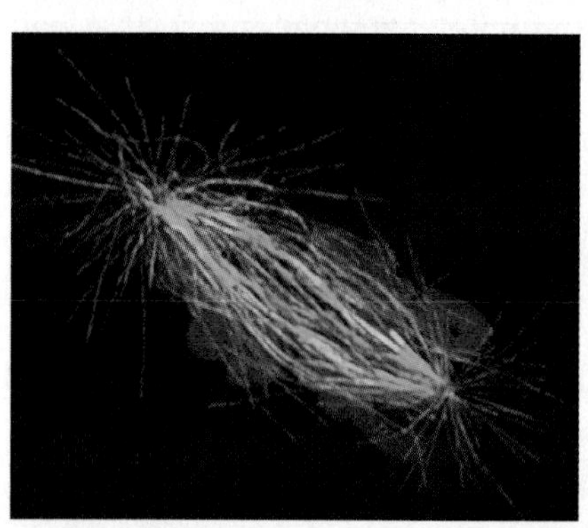

time = 79 min

3 METAPHASE

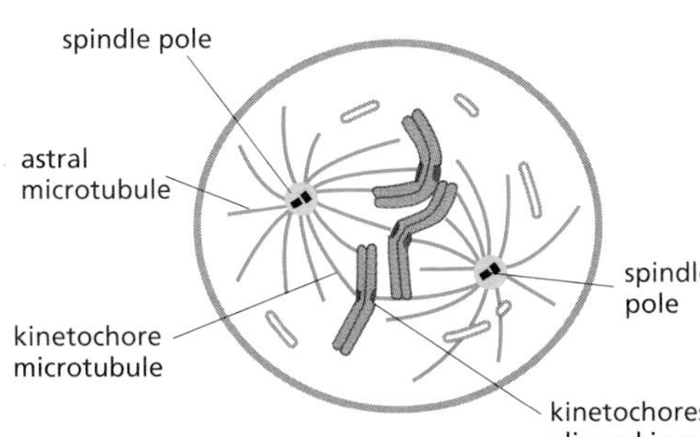

At metaphase, the chromosomes are aligned at the equator of the spindle, midway between the spindle poles. The paired kinetochore microtubules on each chromosome attach to opposite poles of the spindle.

time = 250 min

4 ANAPHASE

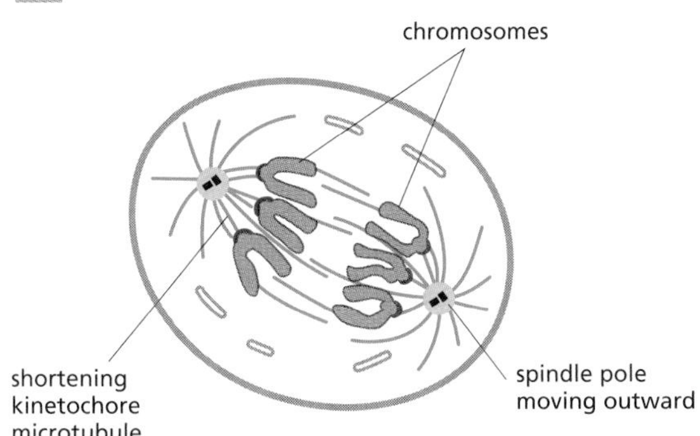

At anaphase, the sister chromatids synchronously separate, and each is pulled slowly toward the spindle pole it is attached to. The kinetochore microtubules get shorter, and the spindle poles also move apart, both contributing to chromosome segregation.

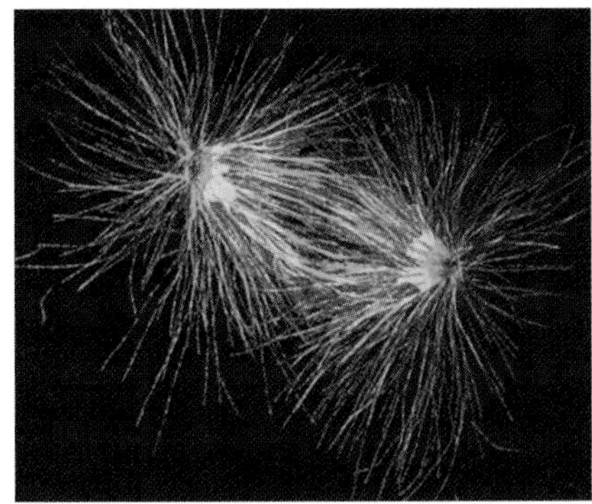

time = 279 min

5 TELOPHASE

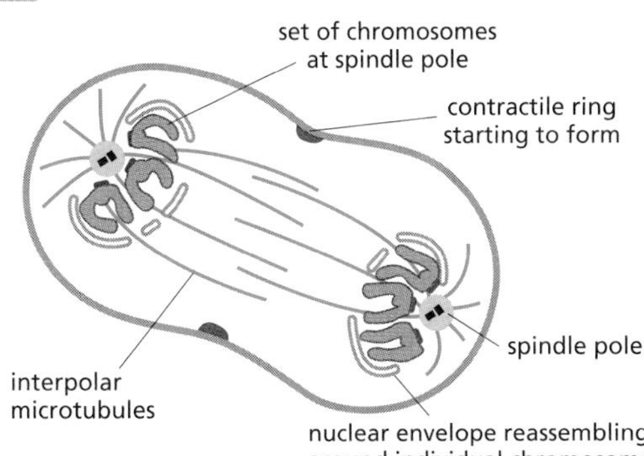

During telophase, the two sets of chromosomes arrive at the poles of the spindle. A new nuclear envelope reassembles around each set, completing the formation of two nuclei and marking the end of mitosis. The division of the cytoplasm begins with the assembly of the contractile ring.

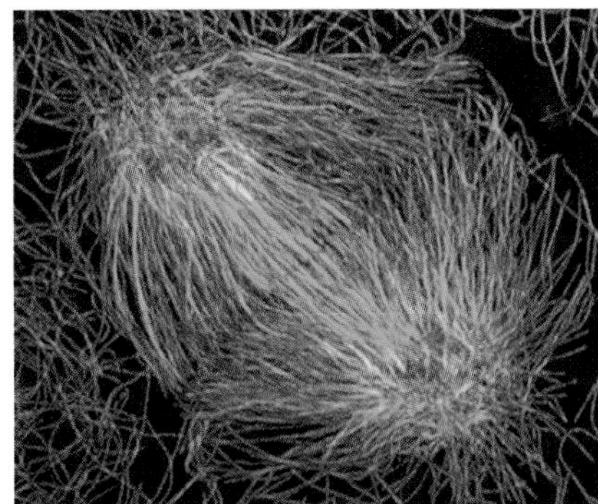

time = 315 min

6 CYTOKINESIS

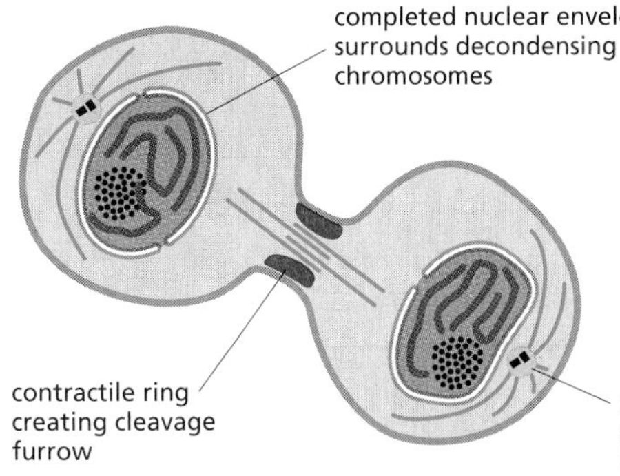

During cytokinesis of an animal cell, the cytoplasm is divided in two by a contractile ring of actin and myosin filaments, which pinches in the cell to create two daughters, each with one nucleus.

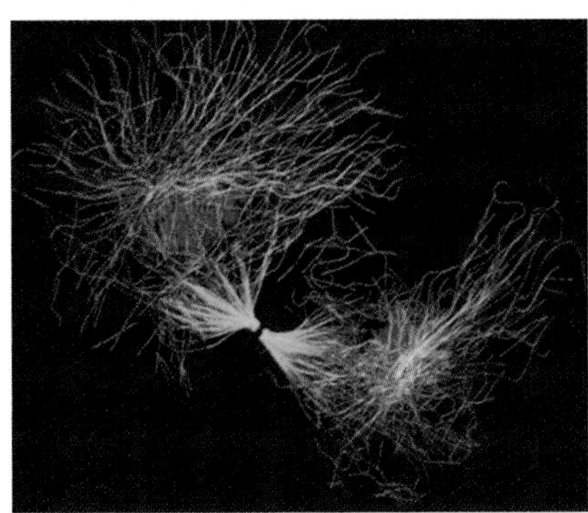

time = 362 min

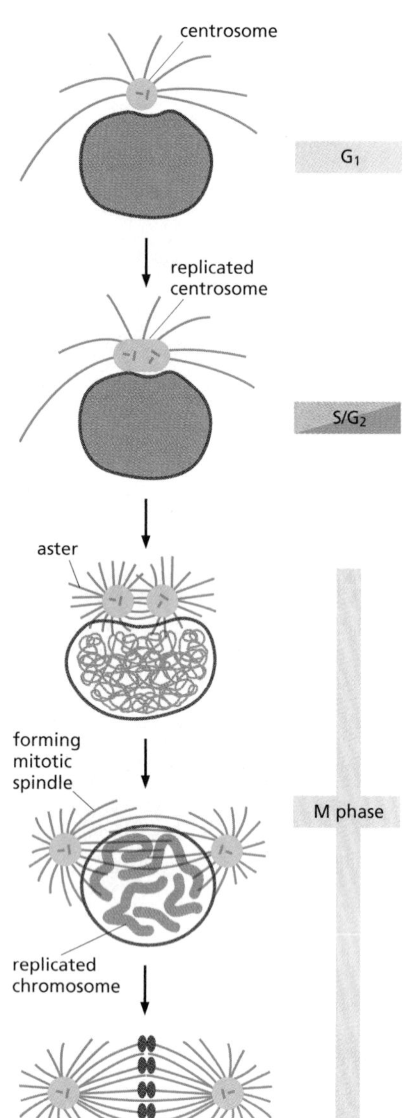

Figure 18–22 **The centrosome in an interphase cell duplicates to form the two poles of a mitotic spindle.** In most animal cells in interphase (G_1, S, and G_2), a centriole pair (shown here as a pair of *dark green bars*) is associated with the centrosome matrix *(light green)* that nucleates microtubule outgrowth. (The volume of the centrosome matrix is exaggerated in this diagram for clarity.) Centrosome duplication begins at the start of S phase and is complete by the end of G_2. Initially, the two centrosomes remain together, but, in early M phase, they separate into two, each nucleating its own aster. The two asters then move apart, and the microtubules that interact between the two asters elongate preferentially to form a bipolar mitotic spindle, with an aster at each pole. When the nuclear envelope breaks down, the spindle microtubules are able to interact with the chromosomes.

Centrosome duplication begins at the start of S phase and is triggered by the same Cdks (G_1/S-Cdk and S-Cdk) that trigger DNA replication. Initially, when the centrosome duplicates, both copies remain together as a single complex on one side of the nucleus. As mitosis begins, however, the two centrosomes separate, and each nucleates a radial array of microtubules called an **aster**. The two asters move to opposite sides of the nucleus to form the two poles of the mitotic spindle (Figure 18–22). The process of centrosome duplication and separation is known as the **centrosome cycle**.

The Mitotic Spindle Starts to Assemble in Prophase

The mitotic spindle begins to form in **prophase**. This assembly of the highly dynamic spindle depends on the remarkable properties of microtubules. As discussed in Chapter 17, microtubules continuously polymerize and depolymerize by the addition and loss of their tubulin subunits, and individual filaments alternate between growing and shrinking—a process called *dynamic instability* (see Figure 17–11). At the start of mitosis, the dynamic instability of microtubules increases, in part because M-Cdk phosphorylates microtubule-associated proteins that influence the stability of microtubule filaments. As a result, during prophase, rapidly growing and shrinking microtubules extend in all directions from the two centrosomes, exploring the interior of the cell. Some of the microtubules growing from one centrosome interact with the microtubules from the other centrosome. This interaction stabilizes the microtubules, preventing them from depolymerizing, and it joins the two sets of microtubules together to form the basic framework of the **mitotic spindle**, with its characteristic bipolar shape (Movie 18.5). The two centrosomes that give rise to these microtubules are now called **spindle poles**, and the interacting microtubules are called *interpolar microtubules* (Figure 18–23). The assembly of the spindle is driven, in part, by motor proteins associated with the interpolar microtubules that help to cross-link the two sets of microtubules.

In the next stage of mitosis, the replicated chromosomes attach to the spindle in such a way that, when the sister chromatids separate, they will be drawn to opposite poles of the cell.

Chromosomes Attach to the Mitotic Spindle at Prometaphase

Prometaphase starts abruptly with the disassembly of the nuclear envelope, which breaks up into small membrane vesicles. This process is triggered by the phosphorylation and consequent disassembly of nuclear pore proteins and the intermediate filament proteins of the nuclear lamina,

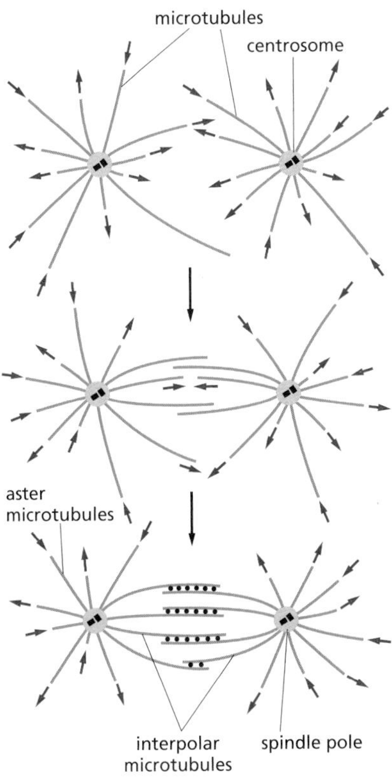

Figure 18–23 **A bipolar mitotic spindle is formed by the selective stabilization of interacting microtubules.** New microtubules grow out in random directions from the two centrosomes. The two ends of a microtubule (by convention, called the plus and the minus ends) have different properties, and it is the minus end that is anchored in the centrosome (discussed in Chapter 17). The free plus ends are dynamically unstable and switch suddenly from uniform growth (outward-pointing *red* arrows) to rapid shrinkage (inward-pointing *red* arrows). When two microtubules from opposite centrosomes interact in an overlap zone, motor proteins and other microtubule-associated proteins cross-link the microtubules together (*black* dots) in a way that stabilizes the plus ends by decreasing the probability of their depolymerization.

the network of fibrous proteins that underlies and stabilizes the nuclear envelope (see Figure 17–7). The spindle microtubules, which have been lying in wait outside the nucleus, now gain access to the replicated chromosomes and capture them (see Panel 18–1, p. 626).

The spindle microtubules end up attached to the chromosomes through specialized protein complexes called **kinetochores**, which assemble on the condensed chromosomes during late prophase (Figure 18–24). As discussed earlier, each replicated chromosome consists of two sister chromatids joined along their length, and each chromatid is constricted at a region of specialized DNA sequence called the *centromere* (see Figure 18–19B). Just before prometaphase, kinetochore proteins assemble into a large complex on each centromere. Each duplicated chromosome therefore has two kinetochores (one on each sister chromatid), which face in opposite directions. Kinetochore assembly depends on the presence of the centromere DNA sequence: in the absence of this sequence, kinetochores fail to assemble and, consequently, the chromosomes fail to segregate properly during mitosis.

Once the nuclear envelope has broken down, a randomly probing microtubule encountering a chromosome will bind to it, thereby capturing the chromosome. The microtubule eventually attaches to the kinetochore, and this *kinetochore microtubule* links the chromosome to a spindle pole (see Figure 18–24 and Panel 18–1, p. 626). Because kinetochores on sister chromatids face in opposite directions, they tend to attach to microtubules from opposite poles of the spindle, so that each replicated chromosome becomes linked to both spindle poles. The attachment to opposite poles, called **bi-orientation**, generates tension on the kineto-

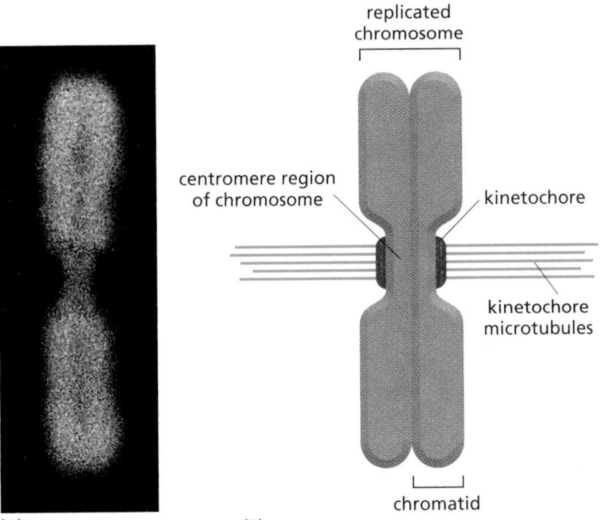

Figure 18–24 **Kinetochores attach chromosomes to the mitotic spindle.** (A) A fluorescence micrograph of a replicated mitotic chromosome. The DNA is stained with a fluorescent dye, and the kinetochores are stained *red* with fluorescent antibodies that recognize kinetochore proteins. These antibodies come from patients with scleroderma (a disease that causes progressive overproduction of connective tissue in skin and other organs), who, for unknown reasons, produce antibodies against their own kinetochore proteins. (B) Schematic drawing of a mitotic chromosome showing its two sister chromatids attached to kinetochore microtubules, which bind by their plus ends. Each kinetochore forms a plaque on the surface of the centromere. (A, courtesy of B.R. Brinkley.)

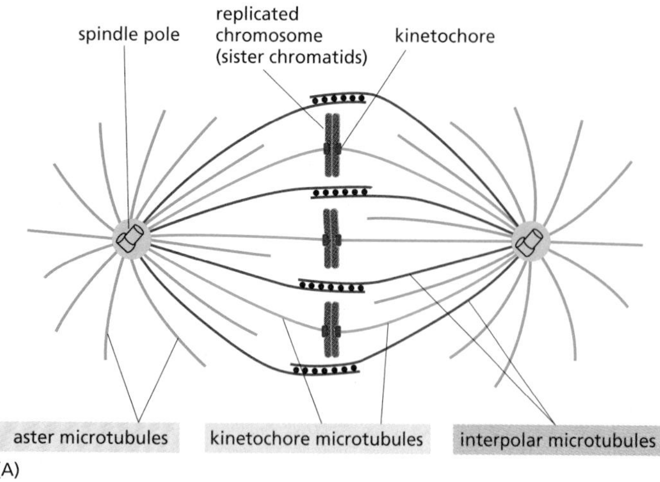

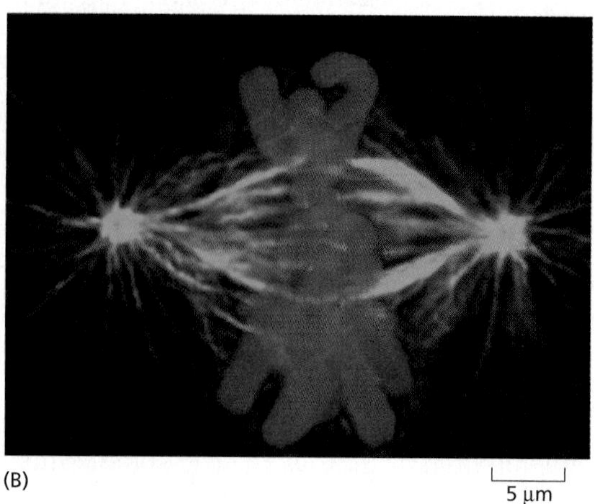

Figure 18–25 **Three classes of microtubules make up the mitotic spindle.** (A) Schematic drawing of a spindle with chromosomes attached, showing the three types of spindle microtubules: astral microtubules, kinetochore microtubules, and interpolar microtubules. In reality, the chromosomes are much larger than shown, and usually multiple microtubules are attached to each kinetochore. (B) Fluorescence micrograph of chromosomes at the metaphase plate of a real mitotic spindle. In this image, kinetochores are labeled in *red*, microtubules in *green*, and chromosomes in *blue*. (B, from A. Desai, *Curr. Biol.* 10:R508, 2000. With permission from Elsevier.)

chores, which are being pulled in opposite directions. This tension signals to the sister kinetochores that they are attached correctly, and are ready to be separated. The cell-cycle control system monitors this tension to ensure correct chromosome attachment, constituting another important cell-cycle checkpoint (see Figures 18–3 and 18–13).

The number of microtubules attached to each kinetochore varies among species: each human kinetochore binds 20–40 microtubules, for example, whereas a yeast kinetochore binds just one. The three classes of microtubules that form the mitotic spindle are shown in Figure 18–25.

Chromosomes Aid in the Assembly of the Mitotic Spindle

Chromosomes are more than passive passengers in the process of spindle assembly: they can stabilize and organize microtubules into functional mitotic spindles. In cells without centrosomes—including all plant cells and some animal cell types—the chromosomes themselves nucleate microtubule assembly, and motor proteins then move and arrange the microtubules and chromosomes into a bipolar spindle. Even in animal cells that normally have centrosomes, a bipolar spindle can still be formed by these means if the centrosomes are removed (Figure 18–26). In cells with centrosomes, the chromosomes, motor proteins, and centrosomes work together to form the mitotic spindle.

Chromosomes Line Up at the Spindle Equator at Metaphase

During prometaphase, the chromosomes, now attached to the mitotic spindle, begin to move about, as if jerked first this way and then that. Eventually, they align at the equator of the spindle, halfway between the

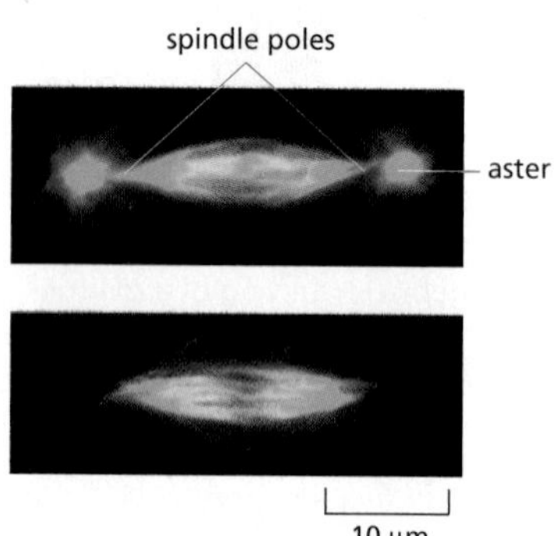

Figure 18–26 **Motor proteins and chromosomes can direct the assembly of a functional bipolar spindle in the absence of centrosomes.** In these fluorescence micrographs of embryos of the insect *Sciara*, the microtubules are stained *green* and the chromosomes *red*. The top micrograph shows a normal spindle formed with centrosomes in a normally fertilized embryo. The bottom micrograph shows a spindle formed without centrosomes in an embryo that initiated development without fertilization and thus lacks the centrosome normally provided by the sperm when it fertilizes the egg. Note that the spindle with centrosomes has an aster at each pole, whereas the spindle formed without centrosomes does not. Both types of spindles are able to segregate the daughter chromosomes. (From B. de Saint Phalle and W. Sullivan, *J. Cell Biol.* 141:1383–1391, 1998. With permission from The Rockefeller University Press.)

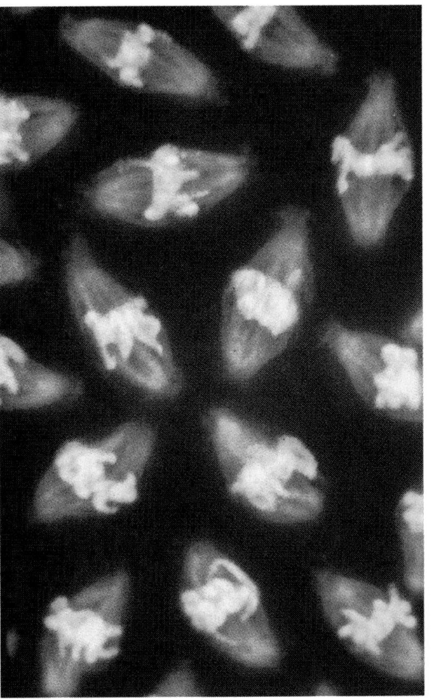

Figure 18–27 **During metaphase, chromosomes gather halfway between the two spindle poles.** This fluorescence micrograph shows multiple mitotic spindles at metaphase in a fruit fly *(Drosophila)* embryo. The microtubules are stained *red,* and the chromosomes are stained *green.* At this stage of *Drosophila* development, there are multiple nuclei in one large cytoplasmic compartment, and all of the nuclei divide synchronously, which is why all of the nuclei shown here are at the same stage of the cell cycle: metaphase (Movie 18.6). Metaphase spindles are usually pictured in two dimensions, as they are here; when viewed in three dimensions, however, the chromosomes are seen to be gathered at a platelike region at the equator of the spindle—the so-called metaphase plate. (Courtesy of William Sullivan.)

two spindle poles, thereby forming the *metaphase plate.* This defines the beginning of **metaphase** (Figure 18–27). Although the forces that act to bring the chromosomes to the equator are not well understood, both the continual growth and shrinkage of the microtubules and the action of microtubule motor proteins are thought to be involved. A continuous balanced addition and loss of tubulin subunits is also required to maintain the metaphase spindle: when tubulin addition to the ends of microtubules is blocked by the drug colchicine, tubulin loss continues until the spindle disappears.

The chromosomes gathered at the equator of the metaphase spindle oscillate back and forth, continually adjusting their positions, indicating that the tug-of-war between the microtubules attached to opposite poles of the spindle continues to operate after the chromosomes are all aligned. If one of the pair of kinetochore attachments is artificially severed with a laser beam during metaphase, the entire chromosome immediately moves toward the pole to which it remains attached. Similarly, if the attachment between sister chromatids is cut, the two daughter chromosomes separate and move toward opposite poles. These experiments show that the replicated chromosomes at the metaphase plate are held there under tension. Evidently, the forces that will ultimately pull the sister chromatids apart begin operating as soon as microtubules attach to the kinetochores.

QUESTION 18–6

If fine glass needles are used to manipulate a chromosome inside a living cell during early M phase, it is possible to trick the kinetochores on the two sister chromatids into attaching to the same spindle pole. This arrangement is normally unstable, but the attachments can be stabilized if the needle is used to gently pull the chromosome so that the microtubules attached to both kinetochores (and the same spindle pole) are under tension. What does this suggest to you about the mechanism by which kinetochores normally become attached and stay attached to microtubules from opposite spindle poles? Is the finding consistent with the possibility that a kinetochore is programmed to attach to microtubules from a particular spindle pole? Explain your answers.

Proteolysis Triggers Sister-Chromatid Separation and the Completion of Mitosis

Anaphase begins abruptly with the release of the cohesin linkage that holds the sister chromatids together (see Figure 18–15). This allows each chromatid to be pulled to the spindle pole to which it is attached (Figure 18–28). This movement segregates the two identical sets of chromosomes to opposite ends of the spindle (see Panel 18–1, p. 627).

The cohesin linkage is destroyed by a protease called *separase,* which up to the beginning of anaphase is held in an inactive state by binding to an inhibitory protein called *securin.* At the beginning of anaphase, securin is targeted for destruction by a protein complex called the **anaphase-promoting complex** (**APC**). Once securin has been removed, separase is then free to break the cohesin linkages (Figure 18–29).

The APC not only triggers the degradation of cohesins, but also targets M cyclin for destruction, thus rendering the M-Cdk complex inactive. This rapid inactivation of M-Cdk helps to initiate the exit from mitosis.

Chromosomes Segregate During Anaphase

Once the sister chromatids separate, they are pulled to the spindle pole to which they are attached. They all move at the same speed, which is typi-

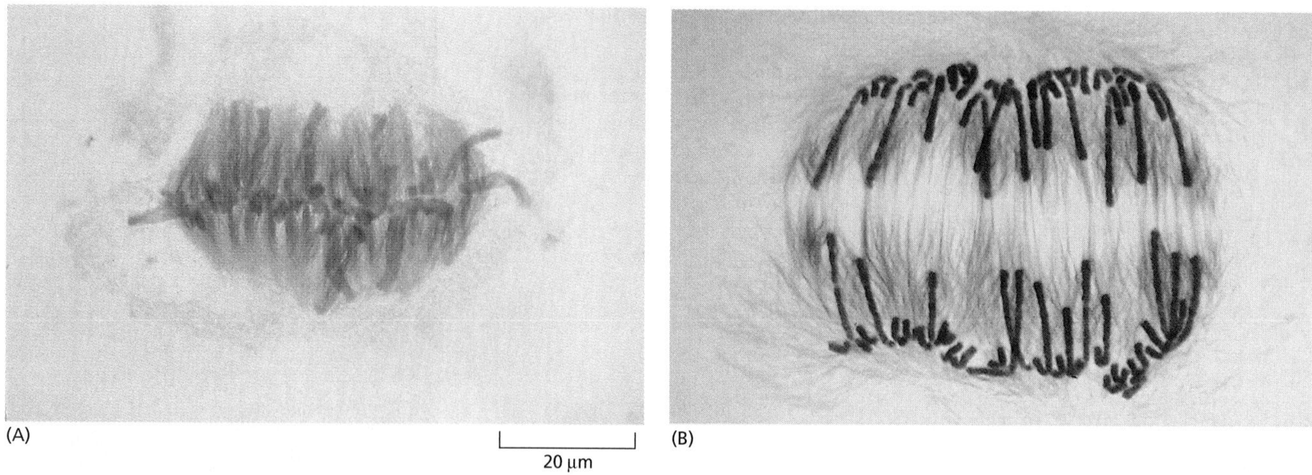

Figure 18–28 **Sister chromatids separate at anaphase.** In the transition from metaphase (A) to anaphase (B), sister chromatids (stained *blue*) suddenly separate and move toward opposite poles, as seen in these plant cells stained with gold-labeled antibodies to label the microtubules (*red*). Plant cells generally do not have centrosomes and therefore have less sharply defined spindle poles than animal cells (see Figure 18–35D); nonetheless, spindle poles are present here at the top and bottom of each micrograph, although they cannot be seen. (Courtesy of Andrew Bajer.)

cally about 1 μm per minute. The movement is the consequence of two independent processes that involve different parts of the mitotic spindle. The two processes are called *anaphase A* and *anaphase B,* and they occur more or less simultaneously. In anaphase A, the kinetochore microtubules shorten by depolymerization, and the attached chromosomes move poleward. In anaphase B, the spindle poles themselves move apart, further contributing to the segregation of the two sets of chromosomes (Figure 18–30).

The driving force for the movements of anaphase A is thought to be provided mainly by the microtubule-associated motor proteins operating at the kinetochore, aided by the shortening of kinetochore microtubules. The loss of tubulin subunits from the kinetochore microtubules depends on a motor-like protein that is bound to both the microtubule and the

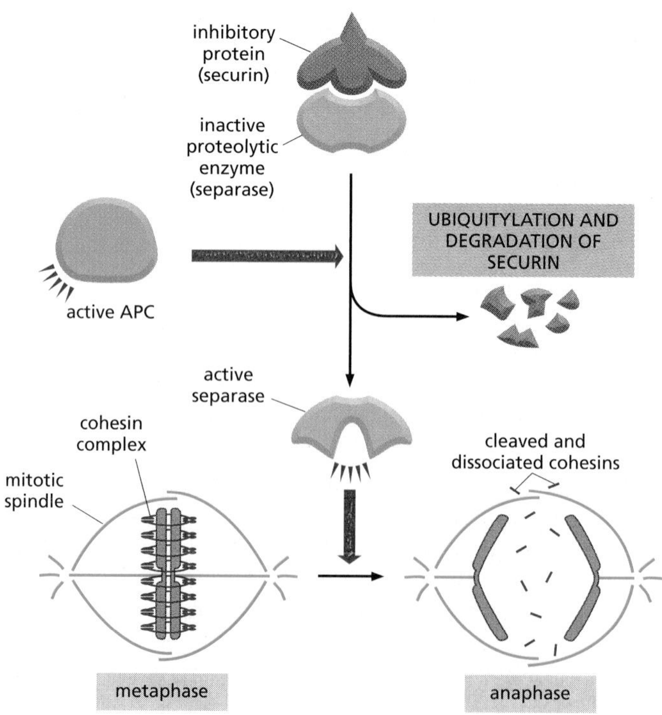

Figure 18–29 **The APC triggers the separation of sister chromatids by promoting the destruction of cohesins.** Activated APC indirectly triggers the cleavage of the cohesins that hold sister chromatids together. It catalyzes the ubiquitylation and destruction of an inhibitory protein called securin. Securin inhibits the activity of a proteolytic enzyme called separase; when freed from securin, separase cleaves the cohesin complexes, allowing the mitotic spindle to pull the sister chromatids apart.

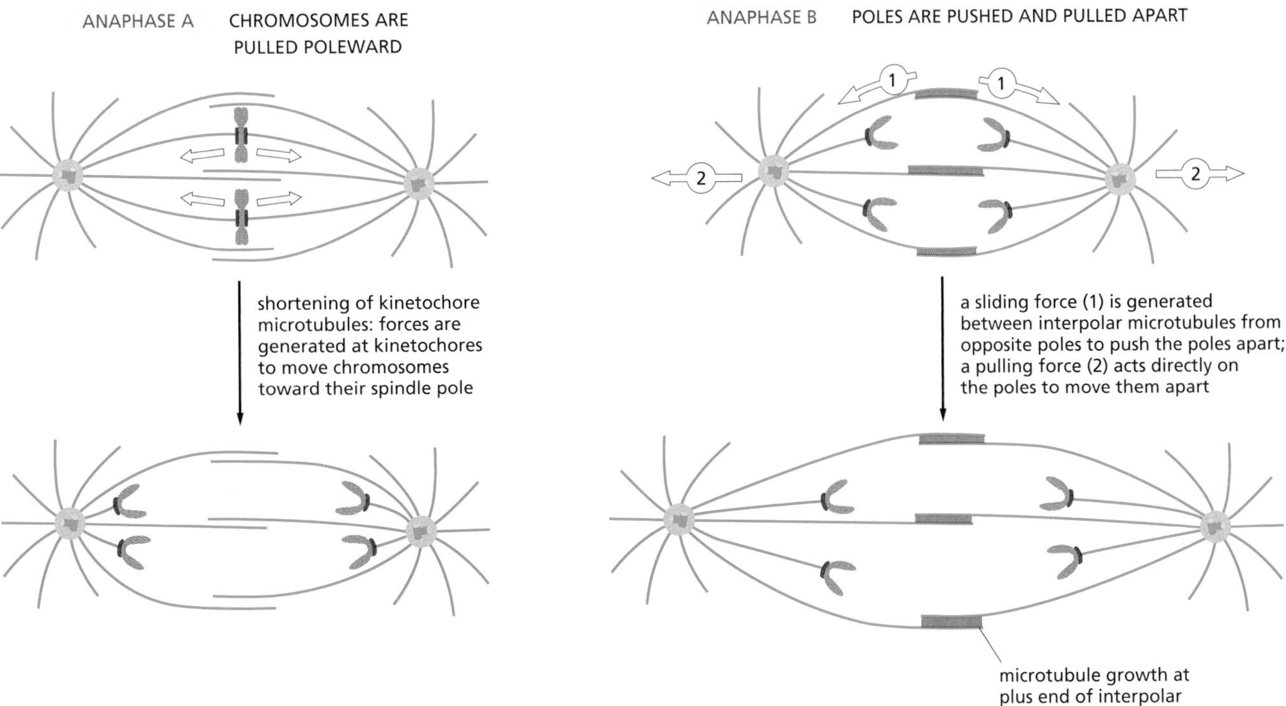
ANAPHASE A
CHROMOSOMES ARE PULLED POLEWARD
shortening of kinetochore microtubules: forces are generated at kinetochores to move chromosomes toward their spindle pole
ANAPHASE B
POLES ARE PUSHED AND PULLED APART
1
1
2
2
a sliding force (1) is generated between interpolar microtubules from opposite poles to push the poles apart; a pulling force (2) acts directly on the poles to move them apart
microtubule growth at plus end of interpolar microtubules

Figure 18–30 **Two processes segregate daughter chromosomes at anaphase.** In *anaphase A*, the daughter chromosomes are *pulled* toward opposite poles as the kinetochore microtubules depolymerize at the kinetochore. The force driving this movement is generated mainly at the kinetochore. In *anaphase B*, the two spindle poles move apart as the result of two separate forces: (1) the elongation and sliding of the interpolar microtubules past one another *pushes* the two poles apart, and (2) forces exerted by outward-pointing astral microtubules at each spindle pole *pull* the poles away from each other, toward the cell cortex. All of these forces are thought to depend on the action of motor proteins associated with the microtubules.

kinetochore and uses the energy of ATP hydrolysis to remove tubulin subunits from the microtubule.

In anaphase B, the spindle poles and the two sets of chromosomes move farther apart. The driving forces for this movement are thought to be provided by two sets of motor proteins—members of the kinesin and dynein families (see Figure 17–20)—operating on different types of spindle microtubules. One set of motor proteins acts on the long, overlapping interpolar microtubules that form the spindle itself; these motor proteins slide the interpolar microtubules from opposite poles past one another at the equator of the spindle, pushing the spindle poles apart. The other set operates on the astral microtubules that extend from the spindle poles and point away from the spindle equator and toward the cell periphery. These motor proteins are thought to be associated with the cell cortex, which underlies the plasma membrane, and they pull each pole toward the adjacent cortex and away from the other pole (see Figure 18–30).

Unattached Chromosomes Block Sister-Chromatid Separation

If a dividing cell were to begin to segregate its chromosomes before all the chromosomes were properly attached to the spindle, one daughter would receive an incomplete set of chromosomes, while the other daughter would receive a surplus. Both situations could be lethal for the cell. Thus, a dividing cell must ensure that every last chromosome is attached properly to the spindle before it completes mitosis. To monitor chromosome attachment, the cell makes use of a negative signal: unattached chromosomes send a 'stop' signal to the cell-cycle control system. Although the exact nature of the signal remains elusive, we know that it inhibits further progress through mitosis by blocking the activation of the APC. Without active APC, the sister chromatids remain glued together. Thus, none of the duplicated chromosomes can be pulled apart until every chromosome has been positioned correctly on the mitotic spindle. This so-called *spindle assembly checkpoint* controls exit from mitosis (see Figures 18–3 and 18–13).

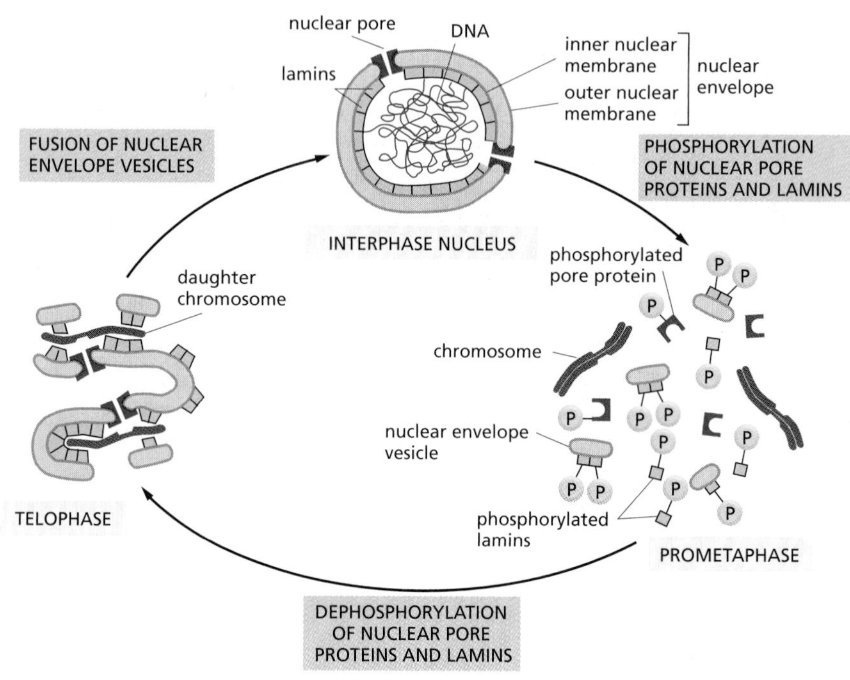

Figure 18–31 **The nuclear envelope breaks down and re-forms during mitosis.** The phosphorylation of nuclear pore proteins and lamins helps trigger the disassembly of the nuclear envelope at prometaphase. Dephosphorylation of pore proteins and lamins at telophase helps reverse the process.

The Nuclear Envelope Re-forms at Telophase

By the end of anaphase, the daughter chromosomes have separated into two equal groups, one at each pole of the spindle. During **telophase**, the final stage of mitosis, the mitotic spindle disassembles, and a nuclear envelope reassembles around each group of chromosomes to form the two daughter nuclei. Vesicles of nuclear membrane first cluster around individual chromosomes and then fuse to re-form the nuclear envelope (see Panel 18–1, p. 627). During this process, the nuclear pore proteins and nuclear lamins that were phosphorylated during prometaphase are now dephosphorylated, which allows them to re-assemble and form the nuclear envelope and nuclear lamina, respectively (Figure 18–31). Once the nuclear envelope has re-formed, the pores pump in nuclear proteins, the nucleus expands, and the condensed mitotic chromosomes decondense into their interphase state. As a consequence of decondensation, gene transcription is able to resume. A new nucleus has been created, and mitosis is complete. All that remains is for the cell to complete its division into two separate daughter cells.

QUESTION 18–7

Consider the events that lead to the formation of the new nucleus at telophase. How do nuclear and cytosolic proteins become properly re-sorted so that the new nucleus contains nuclear proteins but not cytosolic proteins?

CYTOKINESIS

Cytokinesis, the process by which the cytoplasm is cleaved in two, completes M phase. It usually begins in anaphase but is not completed until the two daughter nuclei have formed in telophase. Whereas mitosis depends on a transient microtubule-based structure, the mitotic spindle, cytokinesis in animal cells depends on a transient structure based on actin and myosin filaments, the *contractile ring* (see Figure 18–20). Both the plane of cleavage and the timing of cytokinesis, however, are determined by the mitotic spindle.

The Mitotic Spindle Determines the Plane of Cytoplasmic Cleavage

The first visible sign of cytokinesis in animal cells is a puckering and furrowing of the plasma membrane that occurs during anaphase (Figure 18–32). The furrowing invariably occurs in a plane that runs perpendicu-

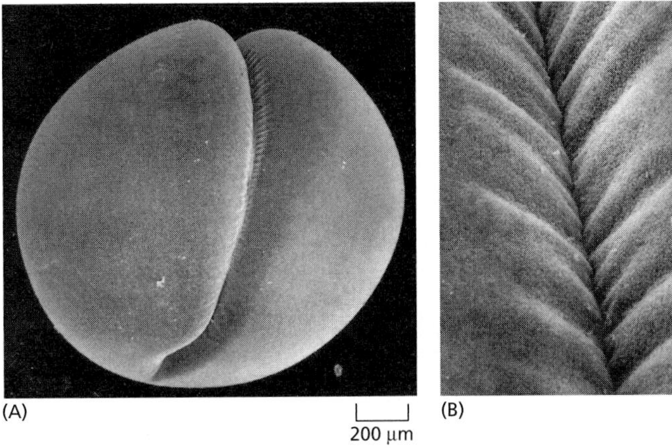

Figure 18–32 **The cleavage furrow is formed by the action of the contractile ring underneath the plasma membrane.** In these scanning electron micrographs of a dividing fertilized frog egg, the cleavage furrow is unusually well defined. (A) Low-magnification view of the egg surface. (B) A higher-magnification view of the cleavage furrow. (From H.W. Beams and R.G. Kessel, *Am. Sci.* 64:279–290, 1976. With permission of Sigma Xi.)

lar to the long axis of the mitotic spindle. This positioning ensures that the *cleavage furrow* cuts between the two groups of segregated chromosomes so that each daughter cell receives an identical and complete set of chromosomes. If the mitotic spindle is deliberately displaced (using a fine glass needle) as soon as the furrow appears, the furrow disappears and a new one develops at a site corresponding to the new spindle location and orientation. Once the furrowing process is well under way, however, cleavage proceeds even if the mitotic spindle is artificially sucked out of the cell or depolymerized using the drug colchicine. How the mitotic spindle dictates the position of the cleavage furrow is still uncertain, but it seems that, during anaphase, both the astral microtubules and the interpolar microtubules (and their associated proteins) signal to the cell cortex to initiate the assembly of the contractile ring at a position midway between the spindle poles. Because these signals originate in the anaphase spindle, this mechanism also contributes to the timing of cytokinesis in late mitosis.

When the mitotic spindle is located centrally in the cell—the usual situation in most dividing cells—the two daughter cells produced will be of equal size. During embryonic development, however, there are some instances in which the dividing cell moves its mitotic spindle to an asymmetrical position, and, consequently, the furrow creates two daughter cells that differ in size. In most cases, the daughters also differ in the molecules they inherit, and they usually develop into different cell types. Special mechanisms are required to position the mitotic spindle eccentrically in such *asymmetric divisions*.

The Contractile Ring of Animal Cells Is Made of Actin and Myosin

The contractile ring is composed mainly of an overlapping array of actin filaments and myosin filaments (Figure 18–33). It assembles at anaphase and is attached to membrane-associated proteins on the cytoplasmic face of the plasma membrane. Once assembled, the contractile ring is capable of exerting a force strong enough to bend a fine glass needle inserted into the cell before cytokinesis. The sliding of the actin filaments against the myosin filaments generates the force (see Figure 17–39), much as it does during muscle contraction. Unlike the contractile apparatus in muscle, however, the contractile ring is a transient structure: it assembles to carry out cytokinesis, gradually becomes smaller as cytokinesis progresses, and disassembles completely once the cell has been cleaved in two.

Cell division in many animal cells is accompanied by large changes in cell shape and a decrease in the adherence of the cell to its neighbors, to the extracellular matrix, or to both. These changes result, in part, from

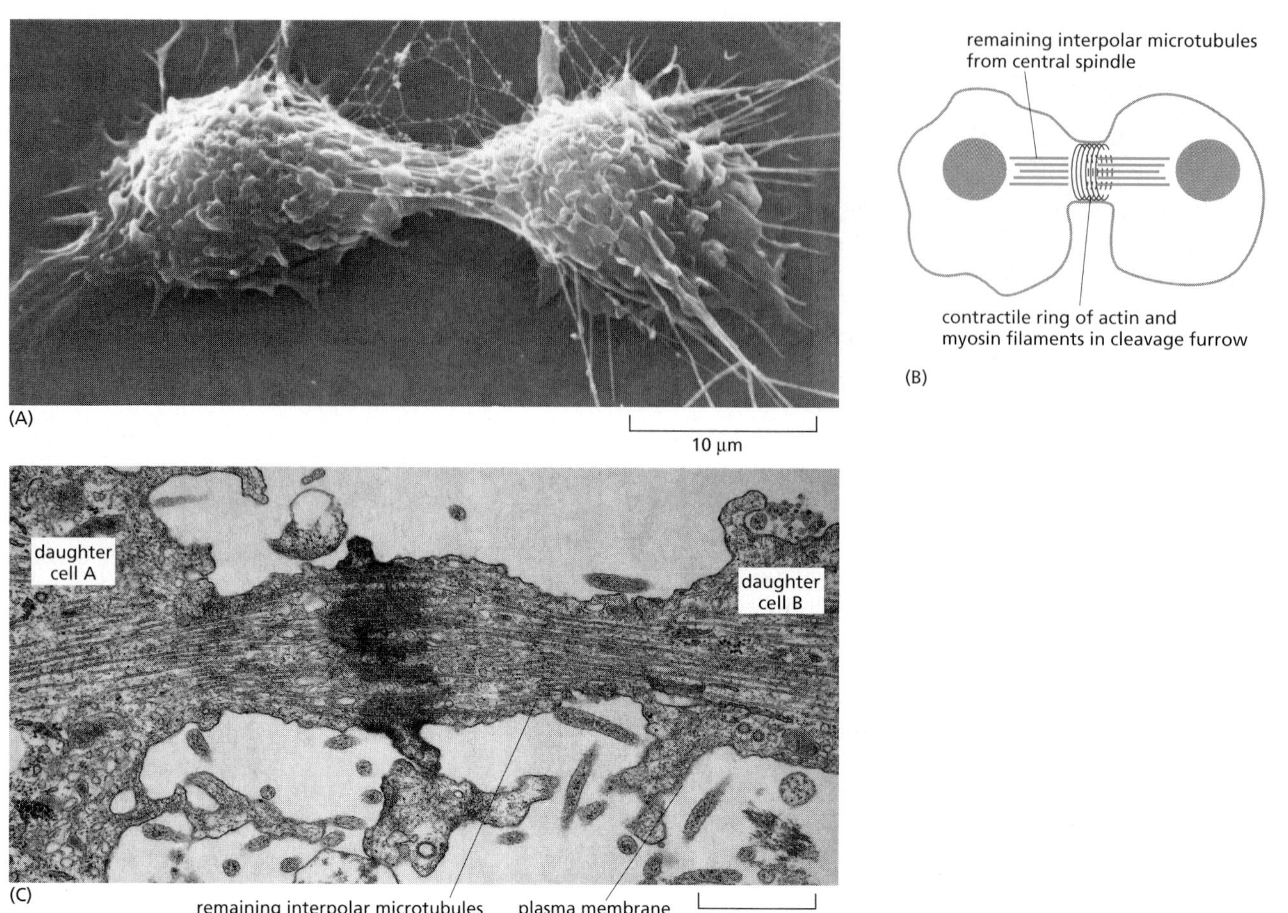

Figure 18–33 **The contractile ring divides the cell in two.** (A) Scanning electron micrograph of an animal cell in culture in the last stages of cytokinesis. (B) Schematic diagram of the midregion of a similar cell showing the contractile ring beneath the plasma membrane and the remains of the two sets of interpolar microtubules. (C) A conventional electron micrograph of a dividing animal cell. Cleavage is almost complete, but the daughter cells remain attached by a thin strand of cytoplasm containing the remains of the overlapping interpolar microtubules of the central mitotic spindle. (A, courtesy of Guenter Albrecht-Buehler; C, courtesy of J.M. Mullins.)

the reorganization of actin and myosin filaments in the cell cortex, only one aspect of which is the assembly of the contractile ring. Mammalian fibroblasts in culture, for example, spread out flat during interphase, as a result of the strong adhesive contacts they make with the surface they are growing on—called the substratum. As the cells enter M phase, however, they round up. The cells change shape in part because some of the plasma membrane proteins responsible for attaching the cells to the substratum—the *integrins* (discussed in Chapter 20)—become phosphorylated and thus weaken their grip. Once cytokinesis is complete, the daughter cells reestablish their strong contacts with the substratum and flatten out again (Figure 18–34). When cells divide in an animal tissue, this cycle of attachment and detachment presumably allows the cells to rearrange their contacts with neighboring cells and with the extracellular matrix, so that the new cells produced by cell division can be accommodated within the tissue.

Cytokinesis in Plant Cells Involves the Formation of a New Cell Wall

The mechanism of cytokinesis in higher plants is entirely different from that in animal cells, presumably because plant cells are surrounded by a tough cell wall (discussed in Chapter 20). The two daughter cells are separated not by the action of a contractile ring at the cell surface but instead by the construction of a new wall that forms inside the dividing cell. The positioning of this new wall precisely determines the position of the two daughter cells relative to neighboring cells. Thus, the planes of cell division, together with cell enlargement, determine the final form of the plant.

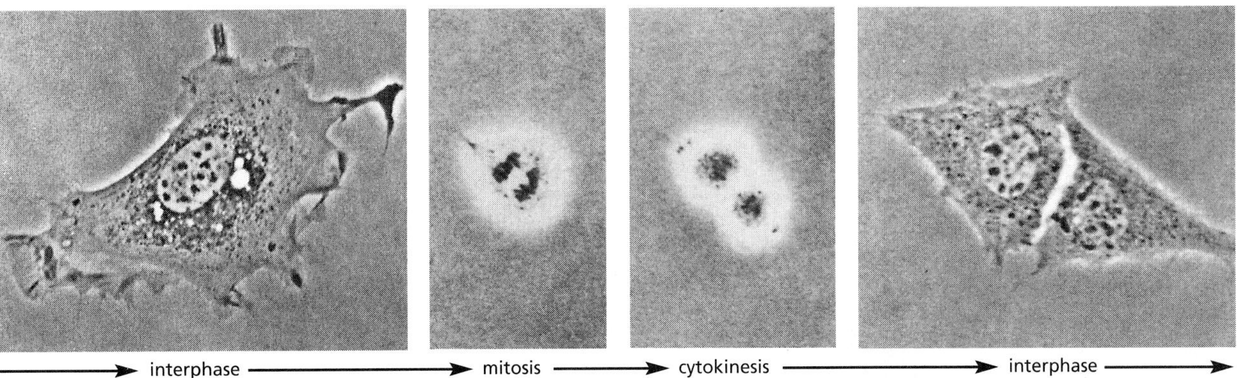

Figure 18–34 **Animal cells change shape during M phase.** In these micrographs of a mouse fibroblast dividing in culture, the same cell was photographed at successive times. Note how the cell rounds up as it enters mitosis; the two daughter cells then flatten out again after cytokinesis is complete. (Courtesy of Guenter Albrecht-Buehler.)

The new cell wall starts to assemble in the cytoplasm between the two sets of segregated chromosomes at the start of telophase. The assembly process is guided by a structure called the **phragmoplast**, which is formed by the remains of the interpolar microtubules at the equator of the old mitotic spindle. Small membrane-enclosed vesicles, largely derived from the Golgi apparatus and filled with polysaccharides and glycoproteins required for the cell-wall matrix, are transported along the microtubules to the phragmoplast. Here, they fuse to form a disclike, membrane-enclosed structure, which expands outward by further vesicle fusion until it reaches the plasma membrane and original cell wall and divides the cell in two (Figure 18–35). Later, cellulose microfibrils are laid down within the matrix to complete the construction of the new cell wall.

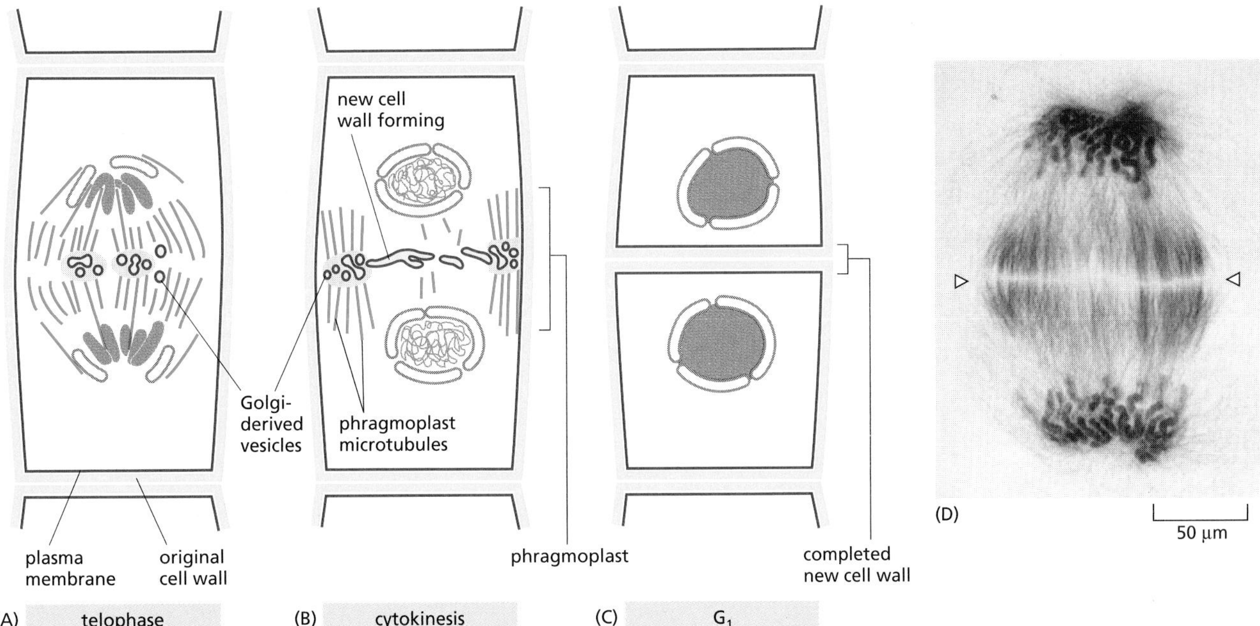

Figure 18–35 **Cytokinesis in a plant cell is guided by a specialized microtubule-based structure called the phragmoplast.** At the beginning of telophase, after the daughter chromosomes have segregated, a new cell wall starts to assemble inside the cell at the equator of the old spindle (A). The interpolar microtubules of the mitotic spindle remaining at telophase form the *phragmoplast* and guide the vesicles toward the equator of the spindle. Here, membrane-enclosed vesicles, derived from the Golgi apparatus and filled with cell-wall material, fuse to form the growing new cell wall (B), which grows outward to reach the plasma membrane and original cell wall. The plasma membrane and the membrane surrounding the new cell wall (both shown in *red*) fuse, completely separating the two daughter cells (C). A light micrograph of a plant cell in telophase is shown in (D) at a stage corresponding to (A). The cell has been stained to show both the microtubules and the two sets of daughter chromosomes segregated at the two poles of the spindle. The location of the growing new cell wall is indicated by the arrowheads. (D, courtesy of Andrew Bajer.)

QUESTION 18–8

Draw a detailed view of the formation of the new cell wall that separates the two daughter cells when a plant cell divides (see Figure 18–35). In particular, show where the membrane proteins of the Golgi-derived vesicles end up, indicating what happens to the part of a protein in the Golgi vesicle membrane that is exposed to the interior of the Golgi vesicle. (Refer to Chapter 11 if you need a reminder of membrane structure.)

Membrane-Enclosed Organelles Must Be Distributed to Daughter Cells When a Cell Divides

Organelles such as mitochondria and chloroplasts cannot assemble spontaneously from their individual components; they arise only from the growth and division of the preexisting organelles. Likewise, cells cannot make a new endoplasmic reticulum (ER) or Golgi apparatus unless some part of it is already present, which can then be enlarged. How, then, are these various membrane-enclosed organelles segregated when the cell divides so that each daughter gets some?

Organelles such as mitochondria and chloroplasts are usually present in large numbers and will be safely inherited if, on average, their numbers simply double once each cell cycle. The ER in interphase cells is continuous with the nuclear membrane and is organized by the microtubule cytoskeleton (see Figure 17–18A). Upon entry into M phase, the reorganization of the microtubules releases the ER; in most cells, the released ER remains intact during mitosis and is cut in two during cytokinesis. The Golgi apparatus fragments during mitosis; the fragments associate with the spindle microtubules via motor proteins, thereby hitching a ride into the daughter cells as the spindle elongates in anaphase. Other components of the cell, including all of the soluble proteins, are inherited randomly when the cell divides.

Having discussed how cells divide, we now turn to the general problem of how the size of an animal or an organ is determined, which leads us to consider how cell number and cell size are controlled.

CONTROL OF CELL NUMBER AND CELL SIZE

A fertilized mouse egg and a fertilized human egg are similar in size, and yet an adult mouse is much smaller than an adult human. What are the differences between the control of cell behavior in humans and mice that generate such differences in size? The same fundamental question can be asked about each organ and tissue in an individual's body. What adjustment of cell behavior explains the length of an elephant's trunk or the size of its brain or its liver? These questions are largely unanswered, but it is at least possible to say what the ingredients of an answer must be. Three fundamental processes largely determine organ and body size: cell growth, cell division, and cell death. Each of these processes, in turn, depends on programs intrinsic to the individual cell and is regulated by signals from other cells in the body.

In this section, we first consider how organisms eliminate unwanted cells by a form of programmed cell death called *apoptosis*. We then discuss how extracellular signals stimulate cell survival, cell growth, and cell division and thereby help control the size of an animal and its organs. We conclude the section with a brief discussion of the inhibitory extracellular signals that help keep these three processes in check.

QUESTION 18–9

The Golgi apparatus is thought to be partitioned into the daughter cells at cell division by a random distribution of fragments that are created at mitosis. Explain why random partitioning of chromosomes would not work.

Apoptosis Helps Regulate Animal Cell Numbers

The cells of a multicellular organism are members of a highly organized community. The number of cells in this community is tightly regulated—not simply by controlling the rate of cell division, but also by controlling the rate of cell death. If cells are no longer needed, they can commit suicide by activating an intracellular death program—a process called **programmed cell death**. In animals, by far the most common form of programmed cell death is called **apoptosis** (from a Greek word meaning 'falling off,' as leaves fall from a tree).

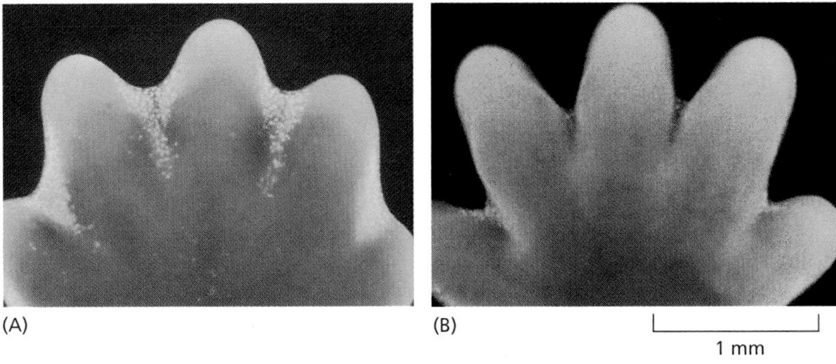

Figure 18–36 **Apoptosis in the developing mouse paw sculpts the digits.** (A) The paw in this mouse embryo has been stained with a dye that specifically labels cells that have undergone apoptosis. The apoptotic cells appear as *bright green dots* between the developing digits. (B) This cell death eliminates the tissue between the developing digits, as seen in the paw shown one day later. Here, few, if any, apoptotic cells can be seen. (From W. Wood et al., *Development* 127:5245–5252, 2000. With permission from The Company of Biologists Ltd.)

The amount of apoptosis that occurs in both developing and adult animal tissues can be astonishing. In the developing vertebrate nervous system, for example, more than half of the nerve cells produced normally die soon after they are formed. In a healthy adult human, billions of cells die in the bone marrow and intestine every hour. It seems remarkably wasteful for so many cells to die, especially as the vast majority are perfectly healthy at the time they kill themselves. What purposes does this massive cell suicide serve?

In some cases, the answers are clear. Mouse paws—and our own hands and feet—are sculpted by apoptosis during embryonic development: they start out as spadelike structures, and the individual fingers and toes separate because the cells between them die (Figure 18–36). In other cases, cells die when the structure they form is no longer needed. When a tadpole changes into a frog at metamorphosis, the cells in the tail die, and the tail, which is not needed in the frog, disappears (Figure 18–37). In these cases, the unneeded cells die by apoptosis.

In adult tissues, cell death usually exactly balances cell division, unless the tissue is growing or shrinking. If part of the liver is removed in an adult rat, for example, liver cells proliferate to make up the loss. Conversely, if a rat is treated with the drug phenobarbital, which stimulates liver cell division, the liver enlarges. However, when the phenobarbital treatment is stopped, apoptosis in the liver greatly increases until the organ has returned to its original size, usually within a week or so. Thus, the liver is kept at a constant size through regulation of both the cell death rate and the cell birth rate.

Apoptosis Is Mediated by an Intracellular Proteolytic Cascade

Cells that die as a result of acute injury typically swell and burst, spilling their contents all over their neighbors, a process called *cell necrosis* (Figure 18–38A). This eruption triggers a potentially damaging inflamma-

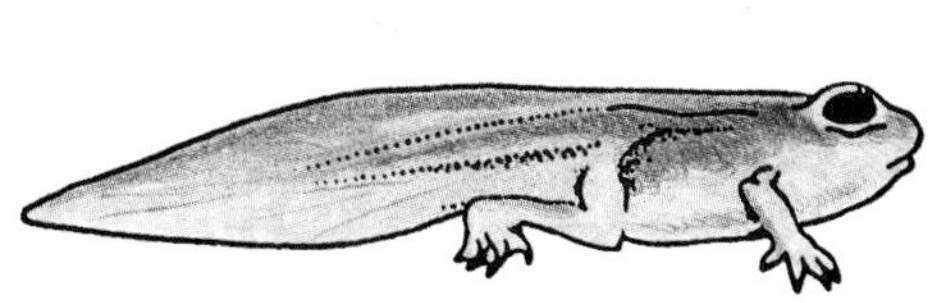

Figure 18–37 **As a tadpole changes into a frog, the cells in its tail are induced to undergo apoptosis.** All of the changes that occur during metamorphosis, including the induction of apoptosis in the tadpole tail, are stimulated by an increase in thyroid hormone in the blood.

tory response. By contrast, a cell that undergoes apoptosis dies neatly, without damaging its neighbors. A cell in the throes of apoptosis shrinks and condenses (Figure 18–38B). The cytoskeleton collapses, the nuclear envelope disassembles, and the nuclear DNA breaks up into fragments (Movie 18.7). Most importantly, the cell surface is altered in such a manner that it immediately attracts phagocytic cells, usually specialized phagocytic cells called macrophages (see Figure 15–32B). These cells engulf the apoptotic cell before it spills its contents (Figure 18–38C). This rapid removal of the dying cell avoids the damaging consequences of cell necrosis, and also allows the organic components of the apoptotic cell to be recycled by the cell that ingests it.

QUESTION 18–10

Why do you think apoptosis occurs by a different mechanism from the cell death that occurs in cell necrosis? What might be the consequences if apoptosis were not achieved in so neat and orderly a fashion, whereby the cell destroys itself from within and avoids leakage of its contents into the extracellular space?

The machinery that is responsible for apoptosis seems to be similar in all animal cells. It involves the **caspase** family of proteases, the members of which are made as inactive precursors called *procaspases*. Procaspases are typically activated by proteolytic cleavage in response to signals that induce apoptosis. The activated caspases cleave, and thereby activate, other members of the procaspase family, resulting in an amplifying proteolytic cascade (Figure 18–39). They also cleave other key proteins in the cell. One of the caspases, for example, cleaves the lamin proteins, which form the nuclear lamina underlying the nuclear envelope; this cleavage causes the irreversible breakdown of the nuclear lamina (see Figure 18–31). In this way, the cell dismantles itself quickly and cleanly, and its corpse is rapidly taken up and digested by another cell.

Activation of the apoptotic program, like entry into a new stage of the cell cycle, is usually triggered in an all-or-none fashion. The proteolytic cascade is not only destructive and self-amplifying but also irreversible; once a cell reaches a critical point along the path to destruction, it cannot turn back. Thus, it is important that the decision to die is tightly controlled.

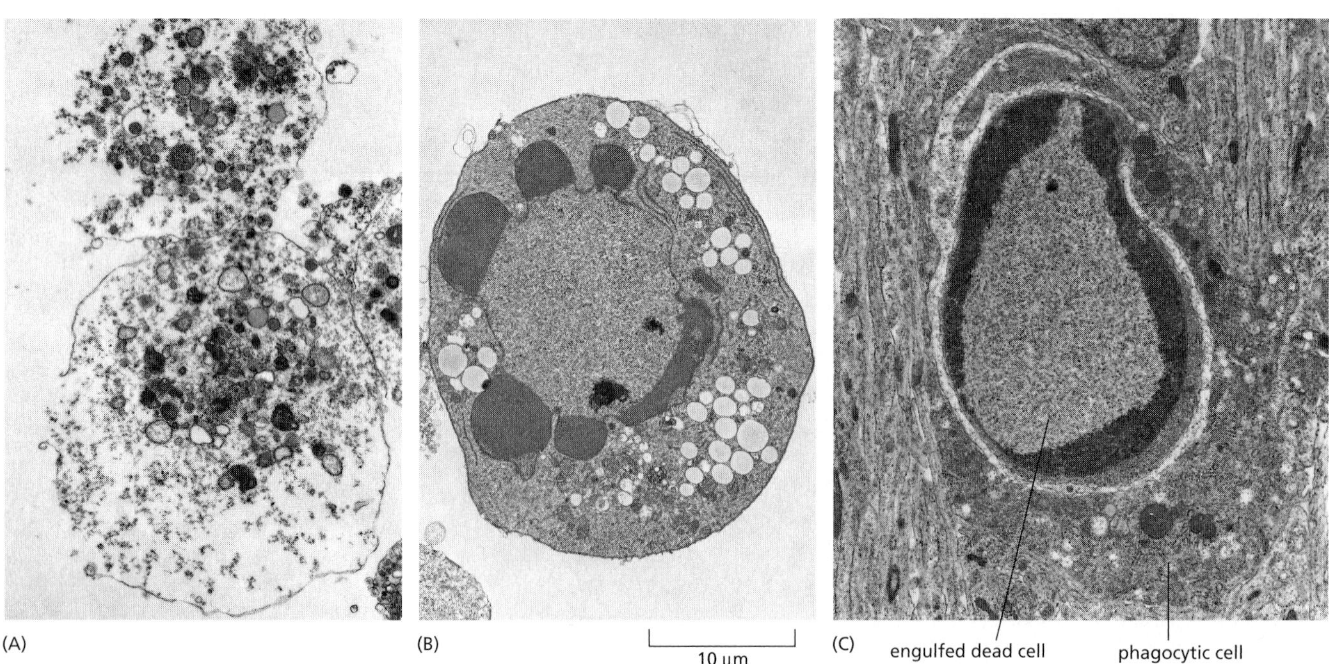

Figure 18–38 **Cells undergoing apoptosis die quickly and cleanly.** Electron micrographs showing cells that have died by necrosis (A) or by apoptosis (B and C). The cells in (A) and (B) died in a culture dish, whereas the cell in (C) died in a developing tissue and has been engulfed by a phagocytic cell. Note that the cell in (A) seems to have exploded, whereas those in (B) and (C) have condensed but seem relatively intact. The large vacuoles seen in the cytoplasm of the cell in (B) are a variable feature of apoptosis. (Courtesy of Julia Burne.)

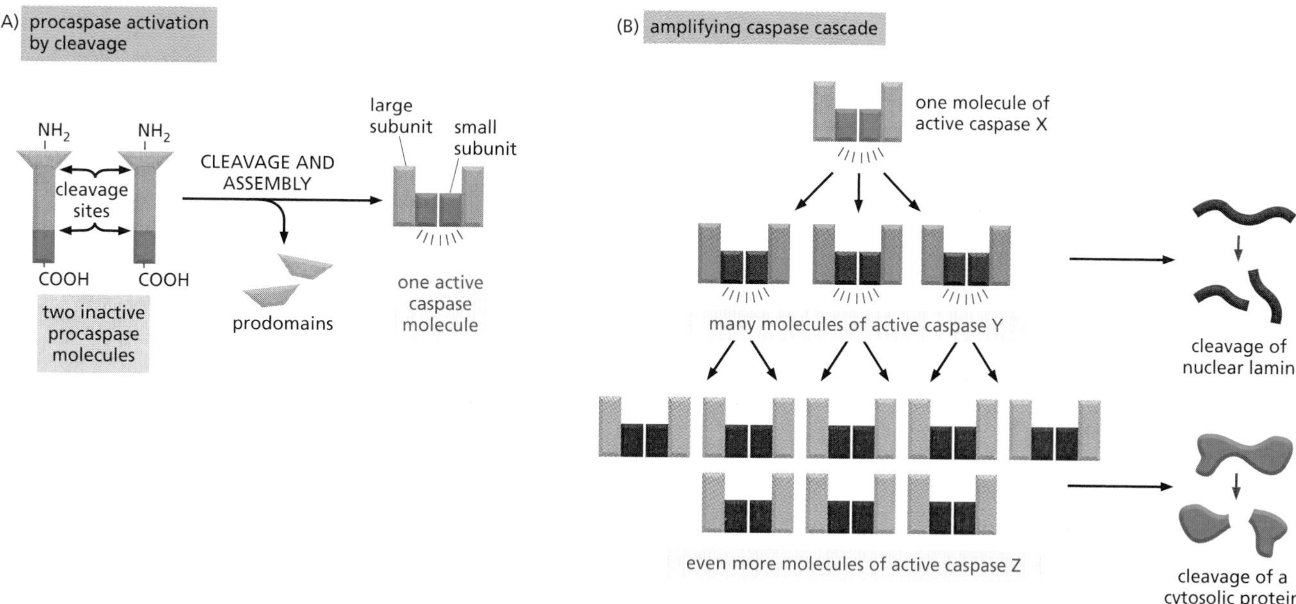

Figure 18–39 **Apoptosis is mediated by an intracellular proteolytic cascade.** (A) Each suicide protease (caspase) is made as an inactive proenzyme, a procaspase, which is itself often activated by proteolytic cleavage by another member of the same protease family: two cleaved fragments from each of two procaspase molecules associate to form an active caspase, which is formed from two small and two large subunits; the two prodomains are usually discarded. (B) Each activated caspase molecule can then cleave many procaspase molecules, thereby activating them, and these can activate even more procaspase molecules. In this way, an initial activation of a small number of protease molecules can lead, via an amplifying chain reaction (a cascade), to the explosive activation of a large number of protease molecules. Some of the activated caspases then break down a number of key proteins in the cell, such as nuclear lamins, leading to the controlled death of the cell.

The Death Program Is Regulated by the Bcl2 Family of Intracellular Proteins

All nucleated animal cells contain the seeds of their own destruction: in these cells, inactive procaspases lie waiting for a signal to destroy the cell. It is therefore not surprising that caspase activity is tightly regulated inside the cell to ensure that the death program is held in check until it is needed.

The main proteins that regulate the activation of procaspases are members of the **Bcl2 family** of intracellular proteins. Some members of this protein family promote procaspase activation and cell death, whereas others inhibit these processes. Two of the most important death-promoting family members are proteins called *Bax* and *Bak*. These proteins activate procaspases indirectly, by inducing the release of cytochrome *c* from mitochondria into the cytosol. Cytochrome *c* promotes the assembly of a large, seven-armed pinwheel-like structure that recruits specific procaspase molecules, forming a protein complex called an *apoptosome*. The procaspase molecules become activated within the apoptosome, triggering a caspase cascade that leads to apoptosis (Figure 18–40). Bax and Bak proteins are themselves activated by other death-promoting members of the Bcl2 family, which are produced or activated by various insults to the cell, such as DNA damage.

Other members of the Bcl2 family, including Bcl2 itself, act to inhibit, rather than promote, procaspase activation and apoptosis. One way in which they do so is by blocking the ability of Bax and Bak to release cytochrome *c* from mitochondria. Some of the Bcl2 family members that promote apoptosis, including a protein called Bad, do so by binding to and blocking the activity of Bcl2 and other death-suppressing members of the Bcl2 family (see Figure 16–34). The balance between the activities of pro-apoptotic and anti-apoptotic members of the Bcl2 family largely determines whether a mammalian cell lives or dies by apoptosis.

The intracellular death program is also regulated by signals from other cells, which can either activate or suppress the program. Indeed, cell survival, cell division, and cell growth are all regulated by extracellular signals, which together help multicellular organisms control cell number and cell size, as we now discuss.

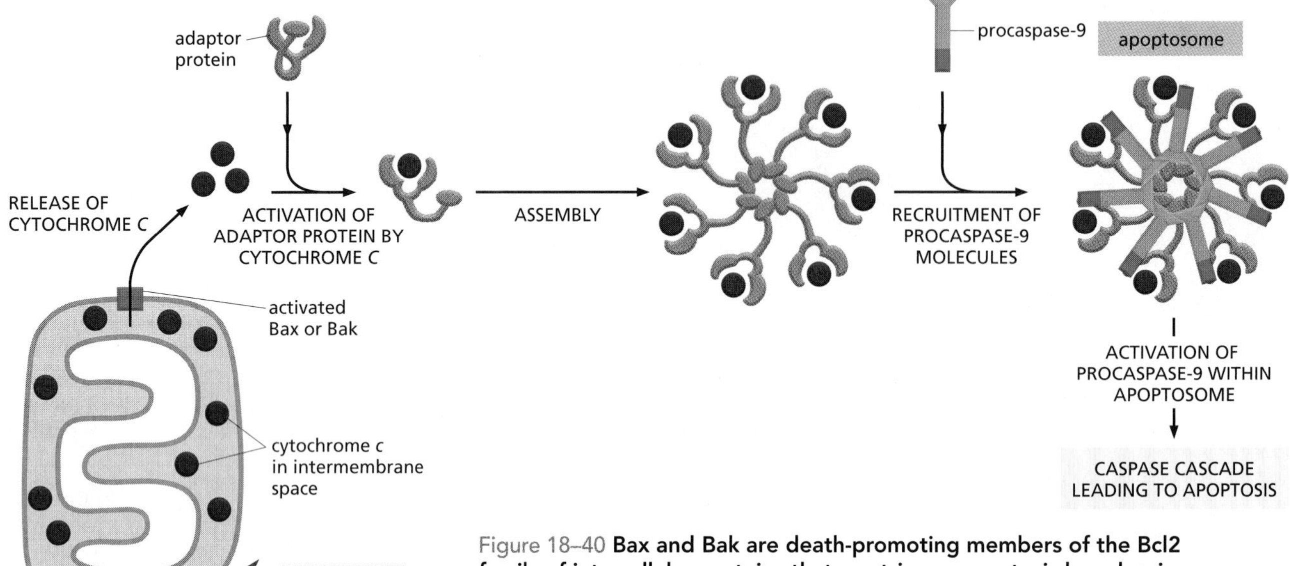

Figure 18–40 **Bax and Bak are death-promoting members of the Bcl2 family of intracellular proteins that can trigger apoptosis by releasing cytochrome c from mitochondria.** When Bak or Bax is activated by an apoptotic stimulus, it aggregates in the outer mitochondrial membrane, leading to the release of cytochrome *c* by an unknown mechanism. The cytochrome *c* is released into the cytosol from the mitochondrial intermembrane space (along with other proteins in this space—not shown). Cytochrome *c* then binds to an adaptor protein, causing it to assemble into a seven-armed complex. This complex then recruits seven molecules of a specific procaspase (called procaspase-9) to form a structure called an apoptosome. The procaspase-9 proteins become activated within the apoptosome and now activate different procaspases in the cytosol, leading to a caspase cascade and apoptosis.

Animal Cells Require Extracellular Signals to Survive, Grow, and Divide

Unicellular organisms such as bacteria and yeasts tend to grow and divide as fast as they can, and their rate of proliferation depends largely on the availability of nutrients in the environment. The cells in a multicellular organism, by contrast, are controlled so that an individual cell survives only when it is needed and divides only when another cell is required, either to allow tissue growth or to replace cell loss. For either tissue growth or cell replacement, cells must grow before they divide. Thus, for an animal cell to survive, grow, or divide, nutrients are not enough. It must also receive chemical signals from other cells, usually its neighbors.

Most of the extracellular signal molecules that influence cell survival, cell growth, and cell division are either soluble proteins secreted by other cells or proteins bound to the surface of other cells or the extracellular matrix. Although most act positively to stimulate one or more of these cell processes, some act negatively to inhibit a particular process. The positively acting signal proteins can be classified, on the basis of their function, into three major categories:

1. **Survival factors** promote cell survival, largely by suppressing apoptosis.
2. **Mitogens** stimulate cell division, primarily by overcoming the intracellular braking mechanisms that tend to block progression through the cell cycle.
3. **Growth factors** stimulate cell growth (an increase in cell size and mass) by promoting the synthesis and inhibiting the degradation of proteins and other macromolecules.

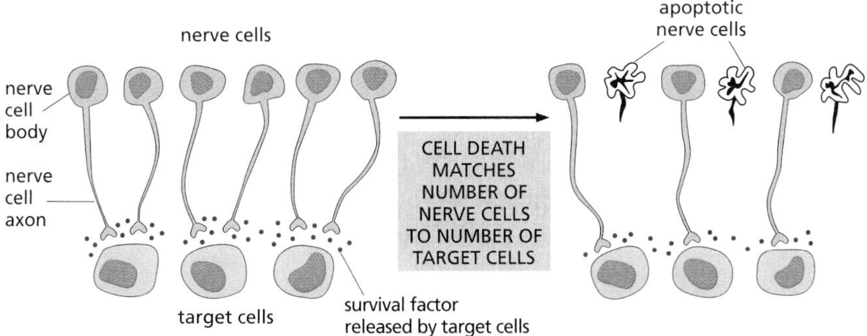

Figure 18–41 **Cell death helps adjust the number of developing nerve cells to the number of target cells they contact.** More nerve cells are produced than can be supported by the limited amount of survival factor released by the target cells. Therefore, some cells receive insufficient amounts of survival factor to keep their suicide program suppressed and, as a consequence, undergo apoptosis. This strategy of overproduction followed by culling ensures that all target cells are contacted by nerve cells and that the 'extra' nerve cells are automatically eliminated.

These categories are not mutually exclusive, as many signal molecules have more than one of these functions. Unfortunately, the term 'growth factor' is often used as a catch-all phrase to describe a protein with any of these functions. Indeed, the phrase 'cell growth' is often used incorrectly to mean an increase in cell number, which is more correctly termed 'cell proliferation.'

In the following sections, we examine each of these types of signal molecules in turn.

Animal Cells Require Survival Factors to Avoid Apoptosis

Animal cells need signals from other cells to survive. If deprived of such survival factors, cells activate their intracellular suicide program and die by apoptosis. This requirement for signals from other cells helps to ensure that cells survive only when and where they are needed. Nerve cells, for example, are produced in excess in the developing nervous system and then compete for limited amounts of survival factors that are secreted by the target cells they contact. Nerve cells that receive enough survival factor live, while the others die by apoptosis. In this way, the number of surviving nerve cells is automatically adjusted so that it is appropriate for the number of cells with which they connect (Figure 18–41). A similar dependence on survival signals from neighboring cells is thought to control cell numbers in other tissues, both during development and in adulthood.

Survival factors usually act by binding to cell-surface receptors. These activated receptors then turn on intracellular signaling pathways that keep the death program suppressed, usually by regulating members of the Bcl2 family of proteins. Some survival factors, for example, increase the production of Bcl2, a protein that suppresses apoptosis (Figure 18–42).

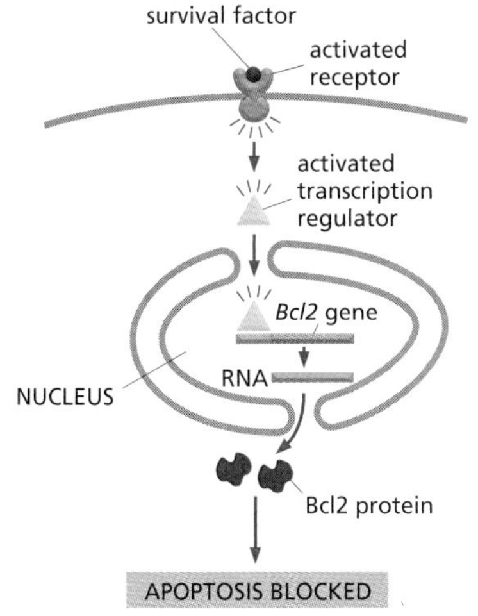

Figure 18–42 **Survival factors often suppress apoptosis by regulating Bcl2 family members.** In this case, the activated receptor activates a transcription regulator in the cytosol. This protein moves to the nucleus, where it activates the gene encoding Bcl2, a protein that inhibits apoptosis.

Mitogens Stimulate Cell Division

Most mitogens are secreted signal proteins that bind to cell-surface receptors. When activated by mitogen binding, these receptors initiate various intracellular signaling pathways (discussed in Chapter 16) that stimulate cell division. These signaling pathways act mainly by releasing the molecular brakes that block the transition from the G_1 phase of the cell cycle into S phase.

An important example of such a molecular brake is the *Retinoblastoma (Rb) protein,* first identified through studies of a rare childhood eye tumor called retinoblastoma, in which the Rb protein is missing or defective. The Rb protein is abundant in the nucleus of all vertebrate cells. It binds to particular transcription regulators, preventing them from stimulating the transcription of genes required for cell proliferation. Mitogens release the Rb brake in the following way. They activate intracellular signaling pathways that lead to the activation of the G_1-Cdk and G_1/S-Cdk complexes discussed earlier. These complexes phosphorylate the Rb protein, altering its conformation so that it releases its bound transcription regulators, which are then free to activate the genes required for cell proliferation (Figure 18–43).

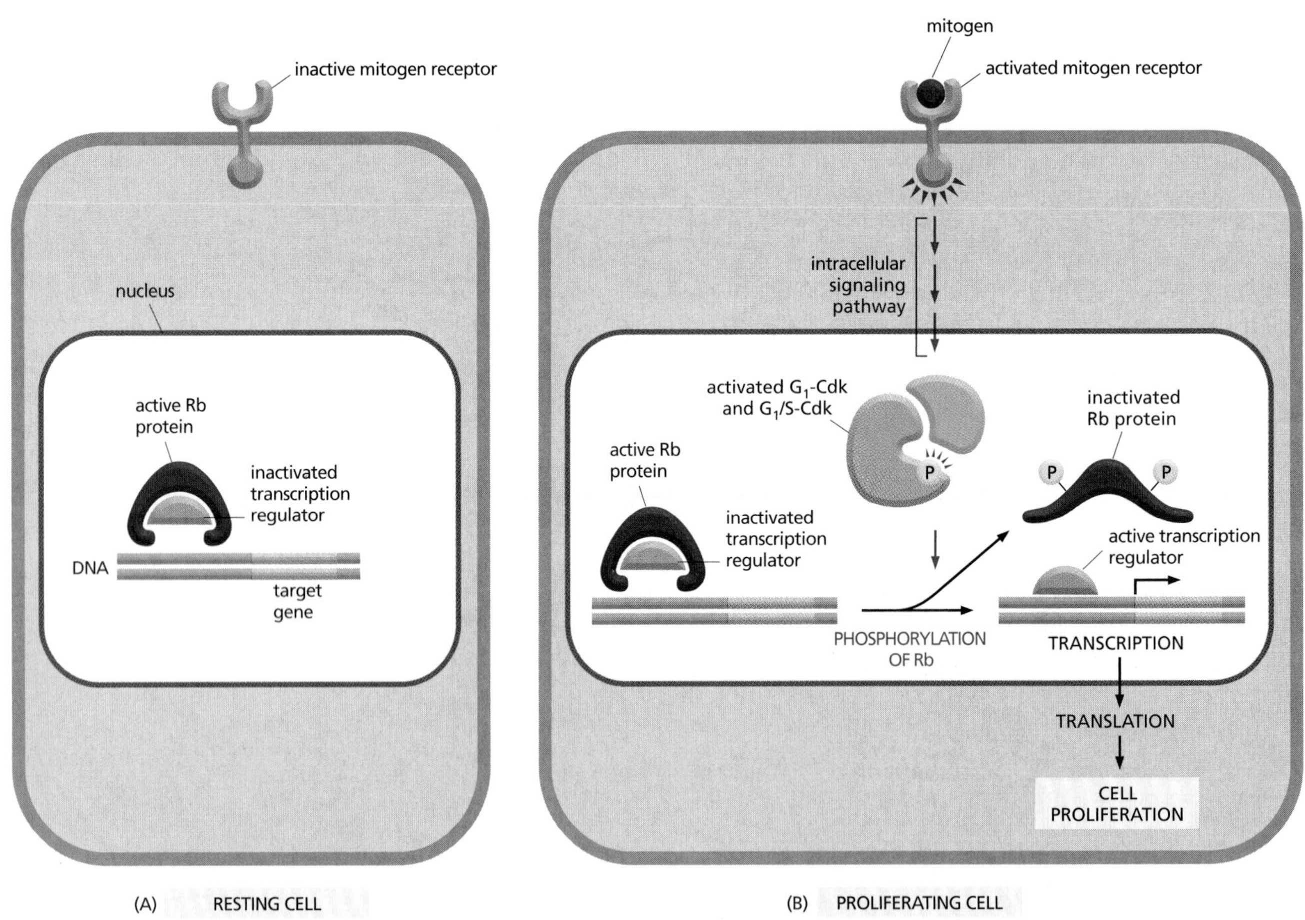

Figure 18–43 **One way in which mitogens stimulate cell proliferation is by inhibiting the Rb protein.** (A) In the absence of mitogens, dephosphorylated Rb protein holds specific transcription regulators in an inactive state; these transcription regulators are required to stimulate the transcription of target genes that encode proteins needed for cell proliferation. (B) Mitogens bind to cell-surface receptors and activate intracellular signaling pathways that lead to the formation and activation of the G_1-Cdk and G_1/S-Cdk complexes. These complexes phosphorylate, and thereby inactivate, the Rb protein. The transcription regulators are now free to activate the transcription of their target genes, leading to cell proliferation.

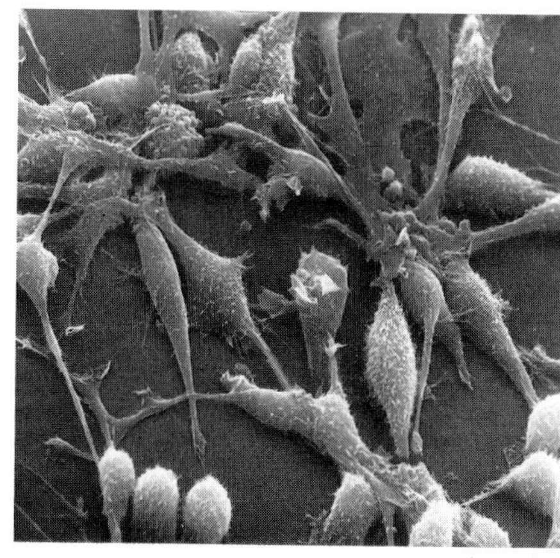

Figure 18–44 **Rat fibroblasts proliferate in response to growth factors and mitogens.** The cells in this scanning electron micrograph are cultured in the presence of calf serum, which contains growth factors and mitogens that stimulate the cells to grow and proliferate. The spherical cells at the bottom of the micrograph have rounded up in preparation for cell division (see Figure 18–34). (Courtesy of Guenter Albrecht-Buehler.)

Most mitogens have been identified and characterized by their effects on cells in culture (Figure 18–44). One of the first mitogens identified in this way was *platelet-derived growth factor,* or *PDGF,* the effects of which are typical of many others discovered since. When blood clots form (in a wound, for example), blood platelets incorporated in the clots are triggered to release PDGF. PDGF then binds to receptor tyrosine kinases (discussed in Chapter 16) in surviving cells at the wound site, stimulating them to proliferate and help heal the wound. Similarly, if part of the liver is lost through surgery or acute injury, cells in the liver and elsewhere produce a protein called *hepatocyte growth factor,* which helps stimulate the surviving liver cells to proliferate.

Growth Factors Stimulate Cells to Grow

The growth of an organism or organ depends on cell growth as much as on cell division. If cells divided without growing, they would get progressively smaller, and there would be no increase in total cell mass. In single-celled organisms such as yeasts, cell growth (like cell division) requires only nutrients. In animals, by contrast, both cell growth and cell division depend on signals from other cells. Cell growth, however, unlike cell division, does not depend on the cell-cycle control system, in either yeasts or animal cells. Indeed, many animal cells, including nerve cells and most muscle cells, do most of their growing after they have become specialized and permanently stopped dividing.

Like most survival factors and mitogens, most extracellular growth factors bind to cell-surface receptors, which activate various intracellular signaling pathways. These pathways lead to the accumulation of proteins and other macromolecules, and they do so by both increasing the rate of synthesis of these molecules, and decreasing their rate of degradation (Figure 18–45). Some extracellular signal proteins, including PDGF, can act as both growth factors and mitogens, stimulating both cell growth and progression through the cell cycle. Such proteins help ensure that cells maintain their appropriate size as they proliferate.

Compared to cell division, there has been surprisingly little study of how cell size is controlled in animals. As a result, it remains a mystery how different cell types in the same animal grow to be so different in size (Figure 18–46).

Some Extracellular Signal Proteins Inhibit Cell Survival, Division, or Growth

The extracellular signal proteins that we have discussed so far—survival factors, mitogens, and growth factors—act positively to increase the size of organs and organisms. Some extracellular signal proteins, however,

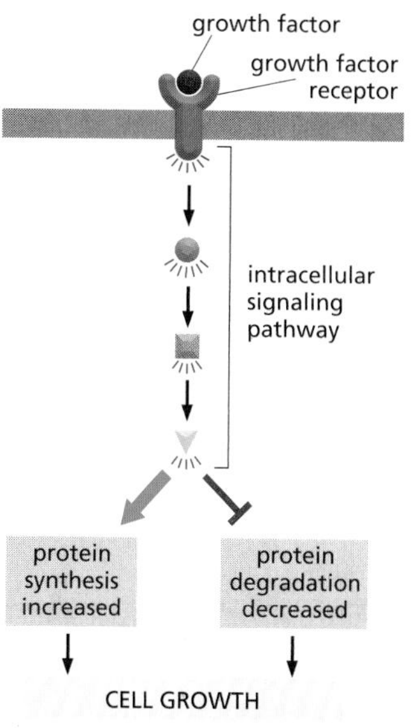

Figure 18–45 **Extracellular growth factors increase the synthesis and decrease the degradation of macromolecules.** This leads to a net increase in macromolecules and thereby cell growth (see also Figure 16–35).

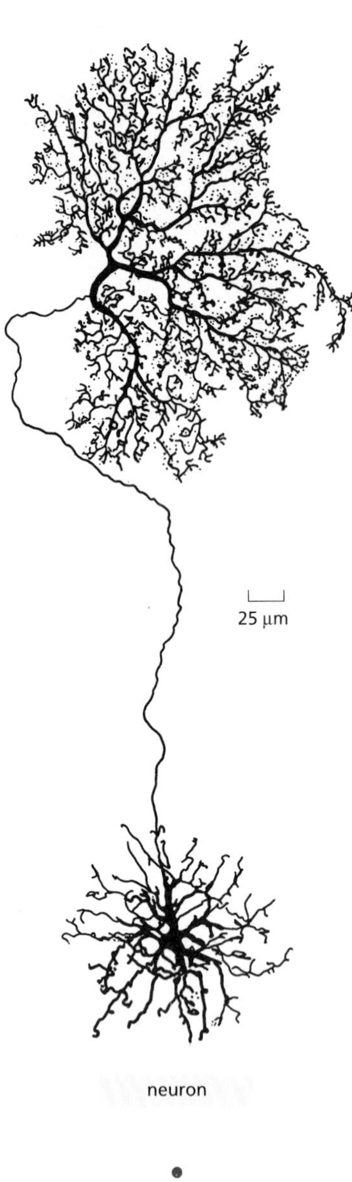

Figure 18–46 **A nerve cell and a lymphocyte are very different in size.** These two cell types, which are drawn at the same scale, both come from the same species of monkey and contain the same amount of DNA. A neuron grows progressively larger after it has permanently stopped dividing. During this time, the ratio of cytoplasm to DNA increases enormously—by a factor of more than 10^5 for some neurons. (Neuron from B.B. Boycott in Essays on the Nervous System [R. Bellairs and E.G. Gray, eds.]. Oxford, U.K.: Clarendon Press, 1974. With permission from Oxford University Press.)

act to oppose these positive regulators and thereby inhibit tissue growth. *Myostatin,* for example, is a secreted signal protein that normally inhibits the growth and proliferation of the myoblasts that fuse to form skeletal muscle cells during mammalian development. When the gene that encodes myostatin is deleted in mice, their muscles grow to be several times larger than normal, because both the number and the size of muscle cells is increased. Remarkably, two breeds of cattle that were bred for large muscles turned out to have mutations in the gene encoding myostatin (Figure 18–47).

As we discuss in the final chapter, cancers are similarly the products of mutations that set cells free from the normal 'social' controls operating on cell survival, growth, and proliferation. Because cancer cells are generally less dependent than normal cells on signals from other cells, they can out-survive, outgrow, and out-divide their normal neighbors, producing tumors that can kill their host.

In this chapter, when we have discussed cell division, we have always been referring to those ordinary divisions that produce two daughter cells, each with a full and identical complement of the parent cell's genetic material. There is, however, a different and highly specialized type of cell division called *meiosis,* which is required for sexual reproduction in eucaryotes. In the next chapter, we describe the special features of meiosis and how they underlie the genetic principles that define the laws of inheritance.

Figure 18–47 **Mutation of the *myostatin* gene leads to a dramatic increase in muscle mass.** This Belgian Blue was produced by cattle breeders and was only later found to have a mutation in the *Myostatin* gene. Mice purposely made deficient in the same gene also have remarkably big muscles. (From HL Sweeney, *Sci. Am.* 291:62–69, 2004. With permission from Scientific American.)

ESSENTIAL CONCEPTS

- The eucaryotic cell cycle consists of several distinct phases. These include S phase, during which the nuclear DNA is replicated, and M phase, during which the nucleus divides (mitosis) and then the cytoplasm divides (cytokinesis).
- In most cells, there is one gap phase (G_1) after M phase and before S phase, and another (G_2) after S phase and before M phase. These gaps give the cell more time to grow and to prepare for the events of S phase and M phase.
- The cell-cycle control system coordinates the events of the cell cycle by sequentially and cyclically switching on the appropriate parts of the cell-cycle machinery and then switching them off.
- The control system depends on a set of protein kinases, each composed of a regulatory subunit called a cyclin and a catalytic subunit called a cyclin-dependent protein kinase (Cdk).
- Cyclin concentrations rise and fall at specific times in the cell cycle, helping to trigger events of the cycle. The Cdks are cyclically activated by both cyclin binding and the phosphorylation of some amino acids and the dephosphorylation of others; when activated, Cdks phosphorylate key proteins in the cell.
- Different cyclin–Cdk complexes trigger different steps of the cell cycle: M-Cdk drives the cell into mitosis; G_1-Cdk drives it through G_1; G_1/S-Cdk and S-Cdk drive it into S phase.
- The control system also uses protein complexes that trigger the proteolysis of specific cell-cycle regulators at particular stages of the cycle.
- The cell-cycle control system can halt the cycle at specific checkpoints to ensure that intracellular and extracellular conditions are favorable and that the next step in the cycle does not begin before the previous one has finished. Some of these checkpoints rely on Cdk inhibitors that block the activity of one or more cyclin–Cdk complexes.
- S-Cdk initiates DNA replication during S phase and helps ensure that the genome is copied only once. Checkpoints in G_1, S phase, and G_2 prevent cells from replicating damaged DNA.
- M-Cdk drives the cell into mitosis with the assembly of the microtubule-based mitotic spindle, which will move daughter chromosomes to opposite poles of the cell.
- Microtubules grow out from the duplicated centrosomes, and some interact with microtubules growing from the opposite pole, thereby becoming the interpolar microtubules that form the spindle.
- Centrosomes, microtubule-associated motor proteins, and the replicated chromosomes themselves work together to assemble the spindle.
- When the nuclear envelope breaks down, the spindle microtubules invade the nuclear area and capture the replicated chromosomes. The microtubules bind to protein complexes, called kinetochores, associated with the centromere of each sister chromatid.
- Microtubules from opposite poles pull in opposite directions on each replicated chromosome, bringing the chromosomes to the equator of the mitotic spindle.
- The sudden separation of sister chromatids allows the resulting daughter chromosomes to be pulled to opposite poles by the spindle. The two poles also move apart, further separating the two sets of chromosomes.
- The movement of chromosomes by the spindle is driven both by microtubule motor proteins and by microtubule polymerization and depolymerization.

- A nuclear envelope re-forms around the two sets of segregated chromosomes to form two new nuclei, thereby completing mitosis.
- The Golgi apparatus breaks into many smaller fragments during M phase, ensuring an even distribution between the daughter cells.
- In animal cells, cytoplasmic division is mediated by a contractile ring of actin filaments and myosin filaments, which assembles midway between the spindle poles and contracts to divide the cytoplasm in two; in plant cells, by contrast, cell division occurs by the formation of a new cell wall inside the parent cell, which divides the cytoplasm in two.
- Animal cell numbers are regulated by a combination of intracellular programs and extracellular signals that control cell survival, cell growth, and cell proliferation.
- Many normal cells die by apoptosis during the lifetime of an animal; they do so by activating an internal suicide program and killing themselves.
- Apoptosis depends on a family of proteolytic enzymes called caspases, which are made as inactive precursors (procaspases). The procaspases are themselves often activated by proteolytic cleavage mediated by caspases.
- Most animal cells require continuous signaling from other cells to avoid apoptosis; this may be a mechanism to ensure that cells survive only when and where they are needed.
- Animal cells proliferate only if stimulated by extracellular mitogens produced by other cells, ensuring that a cell divides only when another cell is needed; the mitogens activate intracellular signaling pathways to override the normal brakes that otherwise block cell-cycle progression.
- For an organism or an organ to grow, cells must grow as well as divide. Animal cell growth depends on extracellular growth factors, which stimulate protein synthesis and inhibit protein degradation.
- Cell and tissue size can also be influenced by inhibitory extracellular signal proteins that oppose the positive regulators of cell survival, cell growth, and cell division.
- Cancer cells fail to obey these normal 'social' controls on cell behavior and therefore outgrow, out-divide, and out-survive their normal neighbors.

KEY TERMS

anaphase
anaphase-promoting complex (APC)
apoptosis
aster
Bcl2 family
bi-orientation
caspase
Cdk (cyclin-dependent protein kinase)
Cdk inhibitor protein
cell cycle
cell-cycle control system
centrosome
centrosome cycle
checkpoint
chromosome condensation
cohesin
condensin
cyclin
cytokinesis
G_1-Cdk
G_1 cyclin
G_1 phase
G_2 phase
G_1/S-Cdk
G_1/S cyclin
growth factor
interphase
kinetochore
M-Cdk
M cyclin
M phase
metaphase
mitogen
mitosis
mitotic spindle
origin recognition complex (ORC)
p53
phragmoplast
programmed cell death
prometaphase
prophase
S-Cdk
S cyclin
S phase
sister chromatid
spindle pole
survival factor
telophase

QUESTIONS

QUESTION 18–11

Roughly, how long would it take a single fertilized human egg to make a cluster of cells weighing 70 kg by repeated divisions, if each cell weighs 1 nanogram just after cell division and each cell cycle takes 24 hours? Why does it take very much longer than this to make a 70-kg adult human?

QUESTION 18–12

The shortest eucaryotic cell cycles of all—shorter even than those of many bacteria—occur in many early animal embryos. These so-called cleavage divisions take place without any significant increase in the weight of the embryo. How can this be? Which phase of the cell cycle would you expect to be most reduced?

QUESTION 18–13

One important biological effect of a large dose of ionizing radiation is to halt cell division.

A. How does this occur?

B. What happens if a cell has a mutation that prevents it from halting cell division after being irradiated?

C. What might be the effects of such a mutation if the cell is not irradiated?

D. An adult human who has reached maturity will die within a few days of receiving a radiation dose large enough to stop cell division. What does that tell you (other than that one should avoid large doses of radiation)?

QUESTION 18–14

If cells are grown in a culture medium containing radioactive thymidine, the thymidine will be covalently incorporated into the cell's DNA during S phase. The radioactive DNA can be detected in the nuclei of individual cells by autoradiography (i.e., by placing a photographic emulsion over the cells, radioactive cells will activate the emulsion and be labeled by black dots when looked at under a microscope). Consider a simple experiment in which cells are radioactively labeled by this method for only a short period (about 30 minutes). The radioactive thymidine medium is then replaced with one containing unlabeled thymidine, and the cells are grown for some additional time. At different time points after replacement of the medium, cells are examined in a microscope. The fraction of cells in mitosis (which can be easily recognized because the cells have rounded up and their chromosomes are condensed) that have radioactive DNA in their nuclei is then determined and plotted as a function of time after the labeling with radioactive thymidine (Figure Q18–14).

A. Would all cells (including cells at all phases of the cell cycle) be expected to contain radioactive DNA after the labeling procedure?

B. Initially there are no mitotic cells that contain radioactive DNA (see Figure Q18–14). Why is this?

C. Explain the rise and fall and then rise again of the curve.

D. Estimate the length of the G_2 phase from this graph.

QUESTION 18–15

One of the functions of M-Cdk is to cause a precipitous drop in M-cyclin concentration halfway through M phase. Describe the consequences of this sudden decrease and suggest possible mechanisms by which it might occur.

QUESTION 18–16

Figure 18–5 shows the rise of cyclin concentration and the rise of M-Cdk activity in cells as they progress through the cell cycle. It is remarkable that the cyclin concentration rises slowly and steadily, whereas M-Cdk activity increases suddenly. How do you think this difference arises?

QUESTION 18–17

What is the order in which the following events occur during cell division:

A. anaphase

B. metaphase

C. prometaphase

D. telophase

E. lunar phase

F. mitosis

G. prophase

Where does cytokinesis fit in?

QUESTION 18–18

The lifetime of a microtubule in a mammalian cell, between its formation by polymerization and its spontaneous disappearance by depolymerization, varies with the stage of the cell cycle. For an actively proliferating cell, the average lifetime is 5 minutes in interphase and 15 seconds in mitosis. If the average length of a microtubule in interphase is 20 μm, how long will it be during mitosis, assuming that the rates of microtubule elongation due to the addition of tubulin subunits in the two phases are the same?

QUESTION 18–19

The balance between plus-end-directed and minus-end-directed motor proteins that bind to interpolar microtubules in the overlap region of the mitotic spindle is thought to help determine the length of the spindle.

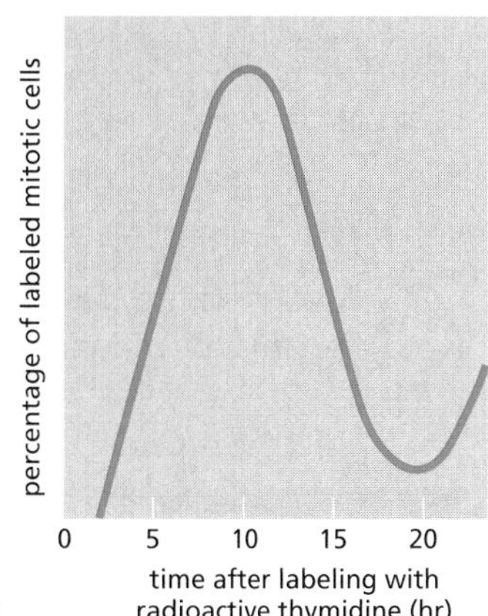

Figure Q18–14

How might each type of motor protein contribute to the determination of spindle length?

QUESTION 18–20

Sketch the principal stages of mitosis, using Panel 18–1 (pp. 626–627) as a guide. Color one sister chromatid and follow it through mitosis and cytokinesis. What event commits this chromatid to a particular daughter cell? Once initially committed, can its fate be reversed? What may influence this commitment?

QUESTION 18–21

The polar movement of chromosomes during anaphase A is associated with microtubule shortening. In particular, microtubules depolymerize at the ends at which they are attached to the kinetochores. Sketch a model that explains how a microtubule can shorten and generate force yet remain firmly attached to the chromosome.

QUESTION 18–22

Rarely, both sister chromatids of a replicated chromosome end up in one daughter cell. How might this happen? What could be the consequences of such a mitotic error?

QUESTION 18–23

Which of the following statements are correct? Explain your answers.

A. Centrosomes are replicated before M phase begins.

B. Two sister chromatids arise by replication of the DNA of the same chromosome and remain paired as they line up on the metaphase plate.

C. Interpolar microtubules attach end-to-end and are therefore continuous from one spindle pole to the other.

D. Microtubule polymerization and depolymerization and microtubule motor proteins are all required for DNA replication.

E. Microtubules nucleate at the centromeres and then connect to the kinetochores, which are structures at the centrosome regions of chromosomes.

QUESTION 18–24

An antibody that binds to myosin prevents the movement of myosin molecules along actin filaments (the interaction between actin and myosin is described in Chapter 17). How do you suppose the antibody exerts this effect? What might be the result of injecting this antibody into cells (A) on the movement of chromosomes at anaphase or (B) on cytokinesis? Explain your answers.

QUESTION 18–25

Look carefully at the electron micrographs in Figure 18–38. Describe the differences between the cell that died by necrosis and those that died by apoptosis. How do the pictures confirm the differences between the two processes? Explain your answer.

QUESTION 18–26

Which of the following statements are correct? Explain your answers.

A. Cells do not pass from G_1 into M phase of the cell cycle unless there are sufficient nutrients to complete an entire cell cycle.

B. Apoptosis is mediated by special intracellular proteases, one of which cleaves nuclear lamins.

C. Developing neurons compete for limited amounts of survival factors.

D. Some vertebrate cell-cycle control proteins function when expressed in yeast cells.

E. It is possible to study yeast mutants that are defective in cell-cycle control proteins, despite the fact that these proteins are essential for the cells to live.

F. The enzymatic activity of a Cdk protein is determined both by the presence of a bound cyclin and by the phosphorylation state of the Cdk.

QUESTION 18–27

Compare the rules of cell behavior in an animal with the rules that govern human behavior in society. What would happen to an animal if its cells behaved as people normally behave in our society? Could the rules that govern cell behavior be applied to human societies?

QUESTION 18–28

In his highly classified research laboratory Dr. Lawrence M. is charged with the task of developing a strain of dog-sized rats to be deployed behind enemy lines. In your opinion, which of the following strategies should Dr. M. pursue to increase the size of rats?

A. Block all apoptosis.

B. Block p53 function.

C. Overproduce growth factors, mitogens, or survival factors.

D. Obtain a taxi driver's license and switch careers.

Explain the likely consequences of each option.

QUESTION 18–29

PDGF is encoded by a gene that can cause cancer when expressed inappropriately. Why do cancers not arise at wounds in which PDGF is released from platelets?

QUESTION 18–30

What do you suppose happens in mutant cells that

A. cannot degrade M-cyclin?

B. always express high levels of p21?

C. cannot phosphorylate Rb?

QUESTION 18–31

Liver cells proliferate excessively both in patients with chronic alcoholism and in patients with liver cancer. What are the differences in the mechanisms by which cell proliferation is induced in these diseases?

CHAPTER **NINETEEN** 19

Sex and Genetics

Individual cells reproduce by duplicating their DNA and dividing in two. This basic process occurs in all living species—in the cells of multicellular organisms as well as in free-living cells such as bacteria and yeasts—and it allows each cell to pass along its genetic information to future generations.

Yet reproduction in a multicellular organism—in a fish or a fly, a person or a plant—is a much more complicated affair. Multicellular organisms go through elaborate developmental cycles, in which all of the cells and tissues and organs of the individual must be generated afresh from a single cell. And this initial cell is no ordinary cell. It has a very peculiar origin: for most animal and plant species, it is produced by the fusion of two cells, derived from two separate individuals—a mother and a father. As a result of this fusion—a central event in *sexual reproduction*—two genomes merge to form the genome of the new individual. The mechanisms that govern genetic inheritance in sexually reproducing organisms are therefore different, and more complex, than those that operate in organisms that pass on their genetic information simply via cell division.

In this chapter, we explore the cell biology of sexual reproduction. We begin by discussing why organisms bother with sex, and we describe how they do it. Generation of the special cells that carry genetic information from each parent involves a specialized process of cell division called *meiosis*, and we review the mechanics of that process. We then discuss how Gregor Mendel, an Austrian monk preoccupied with peas, deduced the basic logic of these genetic mechanisms. Finally, we describe how scientists can exploit the genetics of sexual reproduction to gain insights into human biology, human origins, and the molecular underpinnings of human disease.

THE BENEFITS OF SEX

MEIOSIS AND FERTILIZATION

MENDEL AND THE LAWS OF INHERITANCE

GENETICS AS AN EXPERIMENTAL TOOL

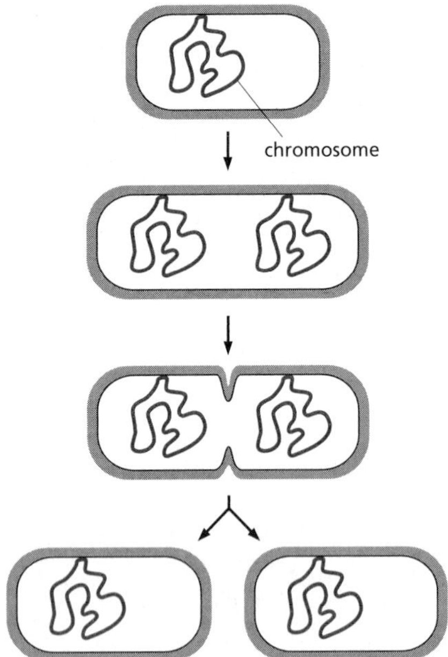

Figure 19–1 **Bacteria reproduce by simple cell division.** The division of one bacterium into two takes 20–25 minutes under ideal growth conditions.

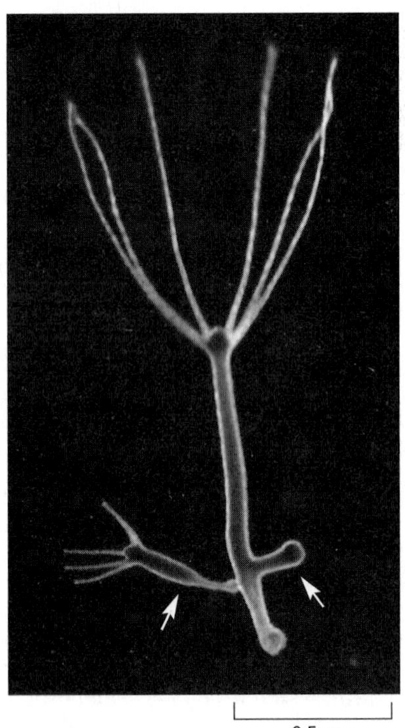

Figure 19–2 **A hydra reproduces asexually.** This relatively simple multicellular animal reproduces by forming buds (*arrows*) that are genetically identical to the parent. Eventually these buds will detach from their parent and live independently. (Courtesy of Amata Hornbruch.)

THE BENEFITS OF SEX

Most of the creatures we see around us reproduce sexually. However, many organisms, especially those invisible to the naked eye, can reproduce without sex. Bacteria and other single-celled organisms most often reproduce by simple cell division (Figure 19–1). Many plants also reproduce asexually, forming multicellular offshoots that later detach from the parent to make new independent plants. Even in the animal kingdom there are species that can reproduce by budding (Figure 19–2), and some worms can be split into two halves, each of which regenerates its missing half. And the females of some species of insects, lizards, and even birds, can produce eggs that develop *parthenogenetically*—that is, without any need for males, sperm, or fertilization—into healthy daughters that can then reproduce in the same way.

But while such **asexual reproduction** is simple and direct, it gives rise to offspring that are genetically identical to the parent organism. **Sexual reproduction**, on the other hand, involves the mixing of genomes from two individuals to produce offspring that are genetically distinct from one another and from both their parents. This mode of reproduction apparently has great advantages, as the vast majority of plants and animals have adopted it.

Sexual Reproduction Involves Both Diploid and Haploid Cells

Organisms that reproduce sexually are generally *diploid*: each cell contains two sets of chromosomes—a *maternal* chromosome set and a *paternal* chromosome set—one inherited from each parent. The two parents, as members of the same species, have chromosome sets that are similar except for the **sex chromosomes**—specialized chromosomes, present in some organisms, that distinguish males from females. Each diploid cell, therefore, carries two copies of each gene, with the exception of genes carried on the sex chromosomes, which may be present in only one copy.

What makes individuals within a species different from each other is that genes occur in variant versions, or **alleles**. Many different alleles of each gene are generally present in the collective gene pool of a species. This means that the two copies of any given gene in a particular individual are often somewhat different from each other and from those carried by other individuals. Thanks to sexual reproduction, each new individual represents a new combination of alleles.

Unlike the other cells in a diploid organism, the specialized cells that perform the central process in sexual reproduction—the **germ cells**, or **gametes**—are *haploid*; that is, they each contain only one set of chromosomes. Typically, two types of gametes are produced. In animals one is large and nonmotile and is referred to as the egg; the other is small and motile and is referred to as the sperm (Figure 19–3). These haploid germ cells are generated when a diploid cell undergoes meiosis. During meiosis the chromosomes of the double chromosome set are partitioned out, in fresh combinations, into single chromosome sets. The two different haploid gametes then fuse to make a diploid cell (the fertilized egg, or **zygote**) with a new combination of chromosomes (Figure 19–4). The zygote thus produced develops into a new individual with a diploid set of chromosomes that is distinct from that of either parent.

For almost all multicellular animals, including vertebrates, practically the whole life cycle is spent in the diploid state. The haploid cells exist only briefly, do not divide at all, and are highly specialized for their function as

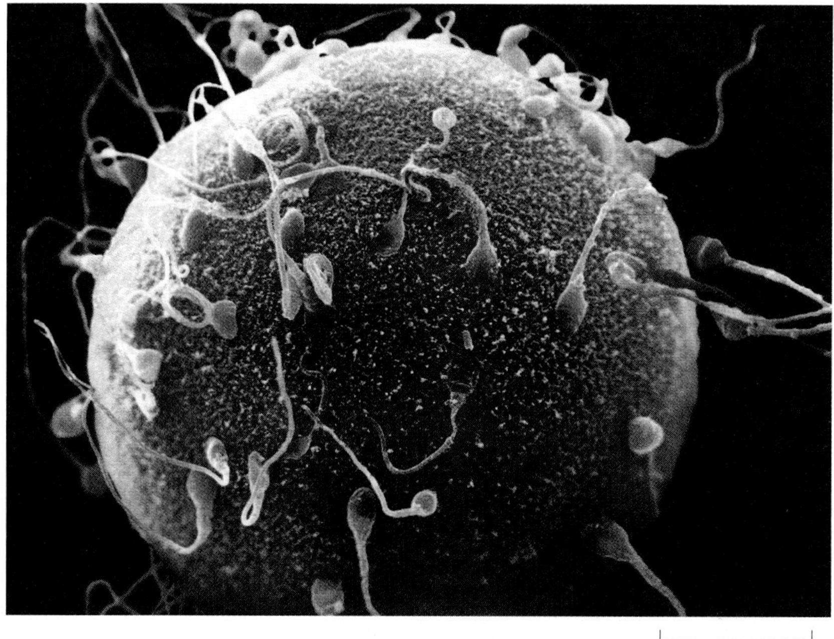

Figure 19–3 **Despite their tremendous difference in size, sperm and egg contribute equally to the genetic character of the zygote.** This difference in size (eggs contain a large quantity of cytoplasm, whereas sperm contain almost none) is consistent with our knowledge that the cytoplasm is not the basis of inheritance. If it were, the female's contribution to the make-up of the offspring would be much greater than the male's. Shown here is a scanning electron micrograph of an egg with human sperm bound to its surface. Although many sperm are bound to the egg, only one will fertilize it. (Courtesy of David M. Phillips/Photo Researchers, Inc.)

gametes; they have to be freshly generated from diploid precursor cells within the organism. This precursor cell lineage, dedicated to the production of germ cells, is called the **germ line**. The cells forming the rest

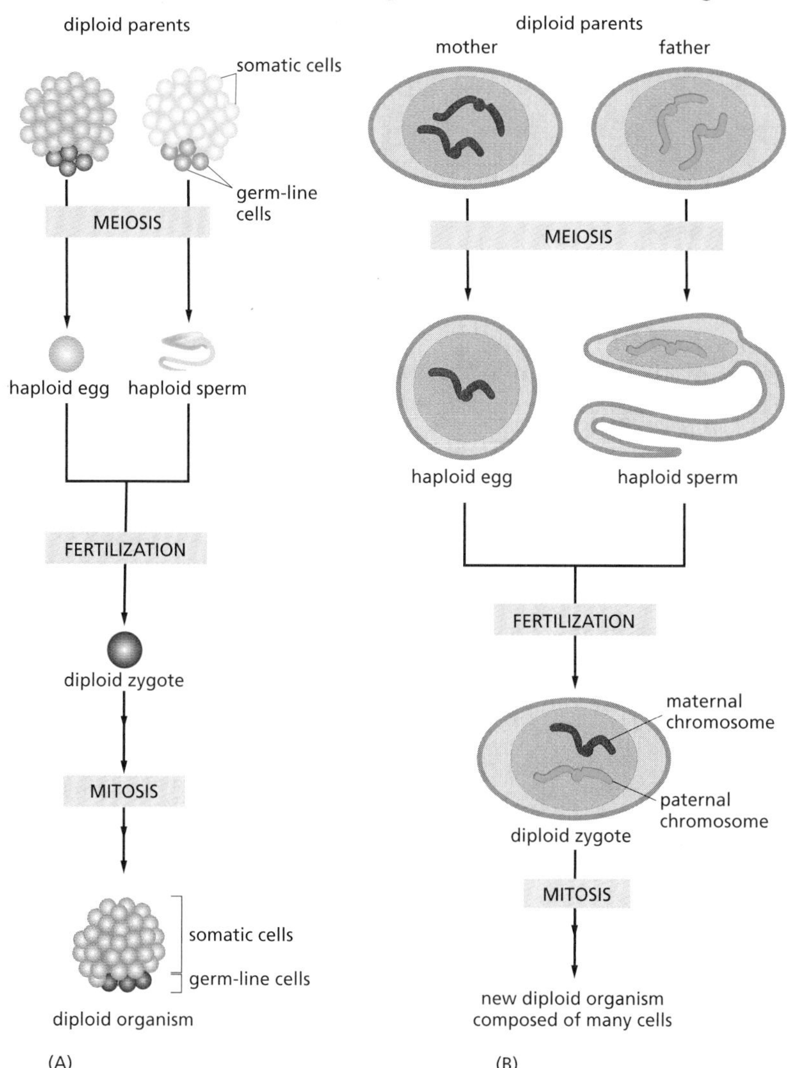

Figure 19–4 **Sexual reproduction involves both haploid and diploid cells.** (A) Cells in higher eucaryotic organisms proliferate in the diploid phase to form a multicellular organism; haploid gametes—the egg and the sperm—are formed by meiosis. These gametes reunite at fertilization to generate a diploid zygote, which will develop into a diploid organism that is genetically distinct from either parent. In the multicellular organism, the germ-line cells (*dark color*) are the precursor cells that give rise to gametes; the somatic cells (*pale color*) are the other cells of the body. (B) The same process is diagrammed with a view of the chromosomes involved. For simplicity, only one chromosome is shown for each gamete, and the sperm cell has been greatly enlarged (see Figure 19–3 for actual sizes).

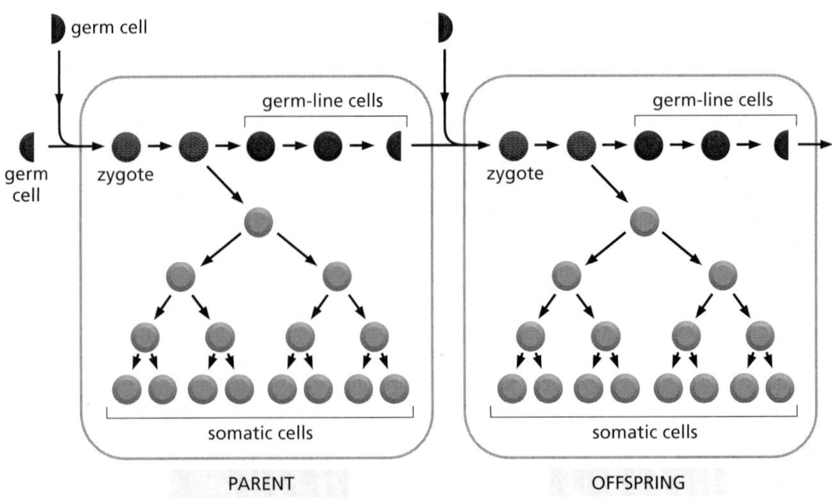

Figure 19–5 **Germ-line cells and somatic cells carry out fundamentally different functions.** In sexually reproducing animals, the germ-line cells *(red)* are specified early in development. These cells propagate genetic information into the next generation. Somatic cells *(blue)*, which form the body of the organism and are therefore necessary to support sexual reproduction, leave no progeny.

of the animal's body—the **somatic cells**—ultimately leave no progeny (Figure 19–5). They exist, in effect, only to help the cells of the germ line to survive and propagate.

The sexual reproductive cycle thus involves an alternation of haploid cells, each carrying a single set of chromosomes, with generations of diploid cells, each carrying two sets of chromosomes. The mixing of genomes that characterizes sexual reproduction is achieved by fusion of two haploid cells to form a diploid cell. In this way, through cycles of diploidy, meiosis, haploidy, and cell fusion, old combinations of genes are broken up and new combinations are created.

Sexual Reproduction Gives Organisms a Competitive Advantage

The new combinations of alleles produced in each round of sexual reproduction are generated by a random process, and they are at least as likely to represent a change for the worse as a change for the better. Why, then, should an ability to try out new genetic combinations give organisms that reproduce sexually an evolutionary advantage over those that breed true through an asexual process? This problem continues to perplex evolutionary geneticists, but one advantage seems to be that reshuffling the genes through sexual reproduction can help a species survive in an unpredictably variable environment. If two parents produce many offspring with a wide variety of gene combinations, the chance that at least one of their progeny will have the combination of features necessary for survival is increased. This may explain why even unicellular organisms, such as yeasts, intermittently indulge in a simple form of sexual reproduction. Typically, they switch on this behavior as an alternative to ordinary cell division when times are hard and starvation looms. Yeasts with a genetic defect that makes them unable to reproduce sexually show a reduced ability to evolve and adapt when they are placed in harsh conditions.

Sexual reproduction may also be advantageous for another reason. In any population, new mutations continually occur, giving rise to new alleles, and many of these new mutations will be harmful. Sexual reproduction can speed the elimination of these deleterious alleles and help to prevent them from accumulating in the population. By mating with only the fittest males, females select for good combinations of alleles and allow bad combinations to be lost from the population more efficiently than they would otherwise be. According to this theory, which has been supported by some careful calculations of cost and benefit, sexual reproduction is favored because males can serve as a genetic filtering device: the males who succeed in mating allow the best, and only the best, collections of genes to be passed on, while the males who fail to mate serve as a genetic 'trash can'—a way of discarding bad collections of alleles from the population. Of course, evolution has taken many twists and turns since sexual reproduction first arose, and, in social organisms especially, it has to be conceded that males may sometimes make themselves useful in other ways.

Whatever its advantages, sexual reproduction has clearly been favored by evolution. In the following section, we review its central features, beginning with meiosis, the process by which the sex cells are formed.

MEIOSIS AND FERTILIZATION

Our modern understanding of the fundamental cycle of events involved in sexual reproduction grew out of discoveries reported in 1888, when Theodor Boveri observed that the fertilized egg of a parasitic roundworm contains four chromosomes, whereas the worm's gametes (sperm and egg) contain only two. This observation revealed that gametes are **haploid**—they carry only a single set of chromosomes. All of the other cells of the body, including the cells that give rise to the gametes, are **diploid**—they carry two sets of chromosomes, one derived from the mother and the other from the father. Therefore, sperm and eggs must be formed by a special kind of cell division in which the number of chromosomes is precisely halved. The term **meiosis** was coined to describe this form of cell division; it comes from a Greek word meaning "diminution," or "lessening."

From Boveri's experiments on worms and other species, it became clear that the behavior of the chromosomes, which at that time were simply stainable microscopic bodies of unknown function, matched the pattern of inheritance, in which the two parents make equal contributions to determining the character of the progeny despite the enormous difference in size between egg and sperm (see Figure 19–3). This was the clue that led to the first realization that the chromosomes contain the material of heredity. The study of sexual reproduction and meiosis therefore has a central place in the history of cell biology.

In this section, we describe the underlying cell biology from a modern point of view, focusing on the elaborate dance of chromosomes that occurs when a cell undertakes meiosis. We begin with an overview of how meiosis distributes chromosomes to the gametes. We then take a closer look at how chromosomes pair, recombine, and are segregated during the process, thereby shuffling the maternal and paternal genes into novel combinations. We also discuss what happens when meiosis goes awry. Finally, we consider briefly the process of fertilization, through which gametes come together to form a new and genetically distinct individual.

Haploid Germ Cells Are Produced From Diploid Cells Through Meiosis

When diploid cells divide by mitosis in the ordinary way, they begin by duplicating their two sets of chromosomes precisely, so as to allow identical sets of chromosomes—a complete maternal set plus a complete paternal set—to be transmitted to each daughter cell (discussed in Chapter 18). Meiosis is different, because only a single set of chromosomes—consisting of part of the maternal and part of the paternal chromosome sets—is finally parceled out to each gamete from the diploid starting cell. Yet, perverse as it might seem, given that the final outcome is a reduction in the number of chromosomes, meiosis begins with a single round of DNA replication that duplicates all of the chromosomes. The ultimate reduction in chromosome number occurs because this single round of replication is followed by two successive cell divisions. One might imagine that meiosis could occur by a simple modification of a normal mitotic cell division, in which DNA replication (S phase) was omitted. In principle, a single round of cell division would then produce two haploid cells directly. But, for reasons that are still unclear, this is not the way it is done.

Meiosis in many species is a relatively long-drawn-out procedure, taking very much longer than any mitosis: in a human male, for example, the process requires 24 days; in a human female, it can go on for decades. Meiosis begins in specialized diploid germ-line cells in the ovaries or testes. Each of these cells contains two copies of each chromosome, one inherited from the organism's father (the *paternal homolog*) and one from its mother (the *maternal homolog*). In the first step of meiosis, the chromosomes of this diploid cell are duplicated: as in any cell preparing to divide, these duplicated chromosomes remain attached to one another, like a set of conjoined twins. The next phase of the process is unique to meiosis. Each duplicated paternal chromosome finds and pairs with the duplicated maternal homolog. This specialized pairing ensures that the homologs will segregate properly during the subsequent cell divisions, so that each of the final gametes receives a complete haploid set of chromosomes.

Two successive cell divisions, called meiotic division I and meiotic division II, now parcel out a complete set of chromosomes to each of the four haploid cells produced. Because the assignment of the homologs to each cell is random, the original maternal and paternal chromosomes are reshuffled into different combinations in the gametes that will eventually form from these haploid cells. During fertilization, two gametes will unite, generating a diploid zygote that is genetically distinct from each of its parents (see Figure 19–4B). The zygote then develops into a multicellular organism through repeated rounds of cell division and cell specialization.

Thus, meiosis produces four cells that are genetically dissimilar and that contain exactly half as many chromosomes as the original parent cell. In contrast, mitosis produces two genetically identical daughter cells. We now discuss the molecular events of the meiotic cycle in more detail, beginning with the pairing of maternal and paternal chromosomes, a process that is central to this specialized form of cell division.

Meiosis Involves a Special Process of Chromosome Pairing

Before the cell divides—by either meiosis or mitosis—it first duplicates all of its chromosomes, as we have just noted. The twin copies of each fully replicated chromosome, called **sister chromatids**, at first remain tightly linked along their length. The way these replicated chromosomes are handled, however, differs in meiosis and mitosis. In mitosis, as we saw in Chapter 18, the replicated chromosomes line up in random order at the metaphase plate; as mitosis continues, the two previously joined sister chromatids then separate from each other to become individual chromosomes, and the two daughter cells produced by cytokinesis inherit one copy of each paternal chromosome and one copy of each maternal chromosome. Thus, both sets of genetic information are transmitted intact to both daughter cells, which are, therefore, diploid and genetically identical.

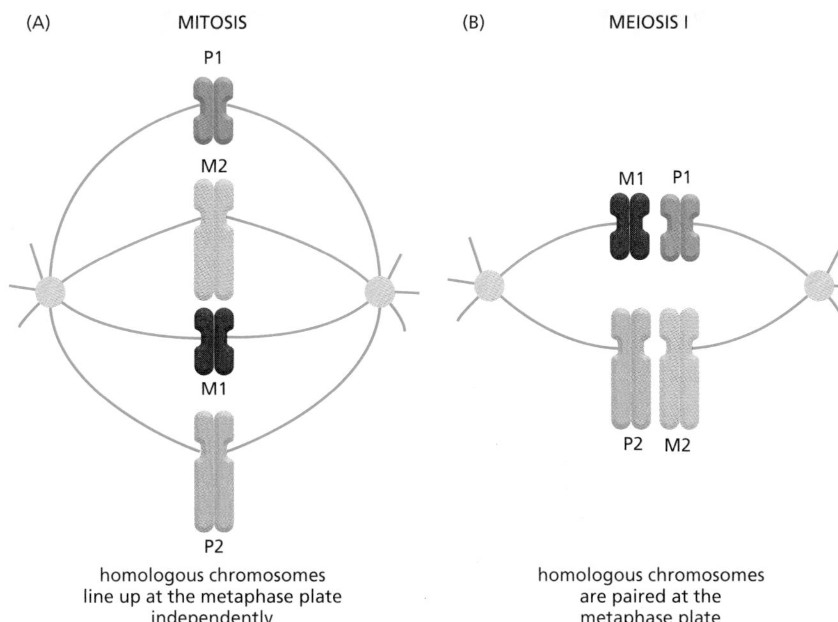

Figure 19–6 **During meiosis, homologous chromosomes pair before lining up on the spindle.** In mitosis (A), the individual maternal (M) and paternal (P) chromosomes line up independently at the metaphase plate; each of them consists of a pair of sister chromatids, which subsequently splits apart. By contrast, in division I of meiosis (B), homologous maternal and paternal chromosomes have paired before lining up at the metaphase plate; during the first meiotic division, each daughter cell receives, at random, either a maternal homolog or a paternal homolog. In both mitosis and meiosis, the chromosomes have been replicated before they become aligned. The spindle is shown in *green*.

The events that occur in the first meiotic cell division mirror the sequence of stages that a cell goes through in mitosis: in prophase, the replicated chromosomes condense; in metaphase, they align at the equator of the meiotic spindle; and in anaphase, they are segregated to the poles. For a review of these stages, see Panel 18–1 (pp. 626–627). The need to halve the number of chromosomes during meiosis, however, makes an extra demand on the cell-division machinery and leads to the first main difference between meiosis and mitosis. In division I of meiosis, the replicated paternal and maternal chromosomes (including the two replicated sex chromosomes) pair up alongside each other before they line up on the spindle (Figure 19–6). This physical pairing of matched sets of chromosomes—called **homologous chromosomes** or **homologs**—is crucial because it enables the paternal and maternal chromosomes to be segregated to different daughter cells at this first division. For each chromosome, allocation of the maternal or paternal homolog to one daughter cell or the other is random. In this way, the original maternal and paternal chromosomes—with their different sets of alleles—are reshuffled into new combinations in each daughter cell of the first meiotic division.

How the homologs (and the two sex chromosomes) recognize each other is still not fully understood. In many organisms, the initial association—the process of **pairing**—seems to be mediated by an interaction between matching maternal and paternal DNA sequences at numerous sites that are widely dispersed along the chromosomes. The structure formed when the duplicated chromosomes pair is called a **bivalent**, and it contains four chromatids (Figure 19–7). The bivalent forms and is maintained during the long meiotic prophase, a stage that can last for years in some organisms.

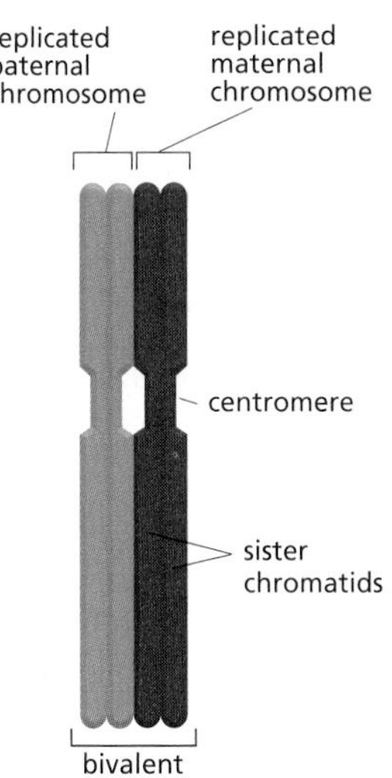

Figure 19–7 **Replicated chromosomes come together during meiosis to form a bivalent.** This structure, which contains four sister chromatids, forms during meiotic prophase.

Crossing-Over Can Occur Between Maternal and Paternal Chromosomes

The picture of meiotic division I that we have just painted is severely simplified, in that it leaves out one crucial feature. In almost all sexually reproducing organisms, the pairing of the maternal and paternal chromosomes is accompanied by **recombination,** a process in which an exchange of DNA occurs between two identical or very similar nucleotide sequences (see Figure 6–31). In a diploid germ-line cell, recombination

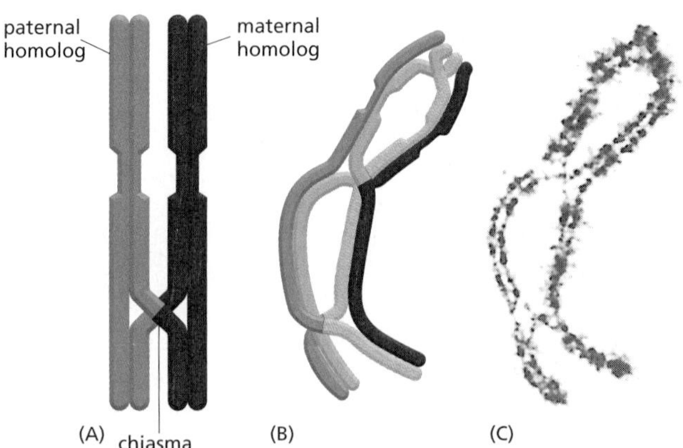

Figure 19–8 **Crossover events create chiasmata between non-sister chromatids in the bivalent.** (A) In this set of paired homologs, a single crossover event has occurred during prophase, creating a single chiasma. (B) A little later in meiosis, the maternal and paternal homologs will start to separate, remaining attached only at sites where recombination has occurred. (C) Photograph of a grasshopper bivalent at this stage, with three chiasmata. (C, courtesy of Bernard John.)

occurs during the long prophase of the first meiotic division. This recombination typically results in a physical swapping of homologous segments from the maternal and paternal chromosomes, an event known as **crossing-over** (see Figure 6–30).

Crossing-over in meiosis is a complex process catalyzed by an intricate protein machine, and it depends on the formation of a *synaptonemal complex*. As the duplicated homologs pair, this elaborate complex holds them together and aligns them so that genetic recombination can readily occur between the non-sister chromatids. The structure also serves to space out the crossover events along each chromosome.

Each of the two chromatids of a duplicated chromosome can cross over with either of the two chromatids of the other chromosome in the bivalent. By the time that prophase ends, the synaptonemal complex has disassembled, allowing the homologs to separate along most of their length. But each pair of duplicated homologs is now held together by at least one **chiasma** (plural **chiasmata**), the connection that corresponds to a crossover between two non-sister chromatids (Figure 19–8). (The structure is named after the Greek letter chi, χ, which is shaped like a cross.) Many bivalents contain more than one chiasma, indicating that multiple crossovers can occur between homologous chromosomes (see Figure 19–8B and C and Figure 19–9). On average, between two and three crossover events occur between each pair of human chromosomes during meiosis I.

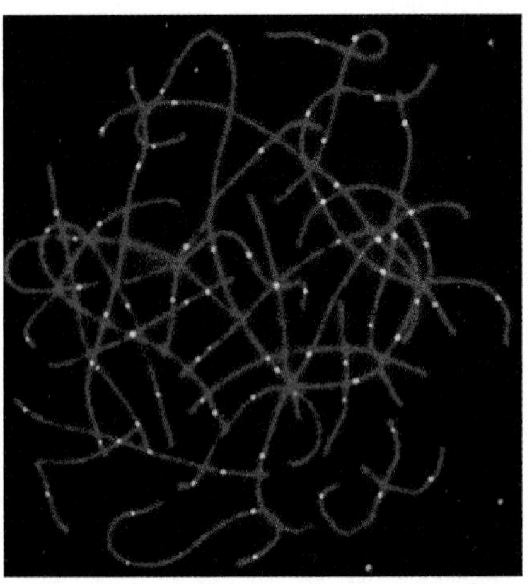

Figure 19–9 **Multiple crossovers can occur between homologous chromosomes.** Shown is a light micrograph of a spread of the chromosomes of a human oocyte (egg-cell precursor) at the stage where all four chromatids—maternal and paternal—are still tightly associated: each single long thread (stained *red*) is a bivalent containing four DNA double helices. Sites of recombination are marked by the presence of a protein (stained *green*) that is a key component of the recombination machinery. (From C. Tease et al., *Am. J. Hum. Genet.* 70:1469–1479, 2002. With permission from Elsevier.)

Crossovers that occur during meiosis are a major source of genetic variation in sexually reproducing species. By scrambling the genetic constitution of each of the chromosomes in the gamete, crossing-over helps to produce individuals with novel assortments of genes. Crossing-over not only generates new combinations of maternal and paternal genes on individual chromosomes, but also plays a second important role in meiosis. By holding homologous chromosomes together during prophase I, crossing-over ensures that the maternal and paternal homologs will segregate from one another correctly at the first meiotic division, as we discuss next.

Chromosome Pairing and Recombination Ensure the Proper Segregation of Homologs

In most organisms, recombination during meiosis is required for the correct segregation of the two duplicated homologs into separate daughter nuclei. The chiasmata created by crossover events have a crucial role in tying together the maternal and paternal homologs until the spindle separates them at anaphase I. Before anaphase I, the two poles of the spindle pull on the duplicated homologs in opposite directions, and the chiasmata resist this pulling (Figure 19–10A). In so doing, the chiasmata

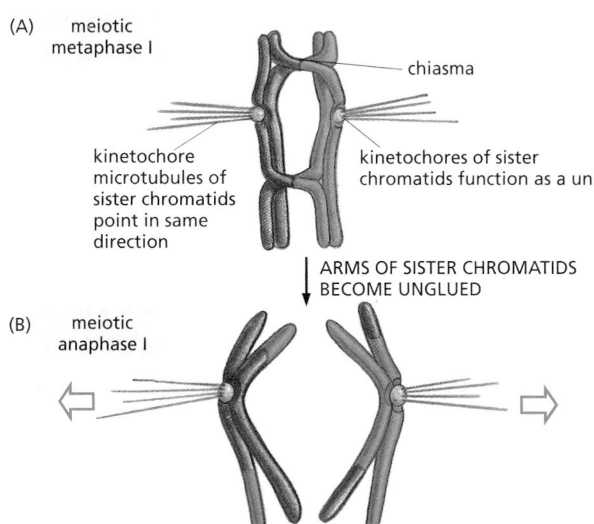

Figure 19–10 **Chiasmata ensure proper segregation of chromosomes in meiosis.** (A) In metaphase I, chiasmata created by crossing-over hold together the maternal and paternal homologs. At this stage, cohesin proteins (not shown) keep the sister chromatids glued together along their entire length. The kinetochores of sister chromatids function as a single unit in meiosis I, and microtubules that attach to them point in a single direction. (B) At anaphase I, the cohesins holding the arms of the sister chromatids together are degraded, but cohesins in the centromere persist and continue to hold sister chromatids together there. This arrangement allows the sister chromatids to remain together when the newly recombined homologs separate and are pulled to opposite poles of the spindle. In contrast, at anaphase in mitosis, both the arms and the centromeres come apart at the same time.

help to position and stabilize bivalents on the metaphase plate. In addition to the chiasmata, which hold the maternal and paternal homologs together, cohesin proteins keep the arms of the sister chromatids glued together along their length (see Figures 19–8 and 18–15). The sister chromatids become unglued when the cohesin on the arms is degraded at the start of anaphase I; this allows the newly recombined homologs to be pulled apart (Figure 19–10B).

The Second Meiotic Division Produces Haploid Daughter Cells

The first meiotic cell division does not produce cells with a haploid amount of DNA. To achieve this goal, each cell now proceeds through a second round of division, meiosis II, which occurs without further DNA replication and without any significant interphase period. A spindle forms, the chromosomes align at its equator, and the sister chromatids now separate to produce daughter cells with a haploid DNA content. In division II of meiosis, the kinetochores on each pair of sister chromatids attach to kinetochore microtubules pointing in opposite directions, as in an ordinary mitotic division. This configuration allows the individual chromatids to be drawn into different daughter cells at anaphase II (Figure 19–11). When the meiosis-specific cohesins that hold the sister chromatids together at the centromere are suddenly degraded, the chromatids separate.

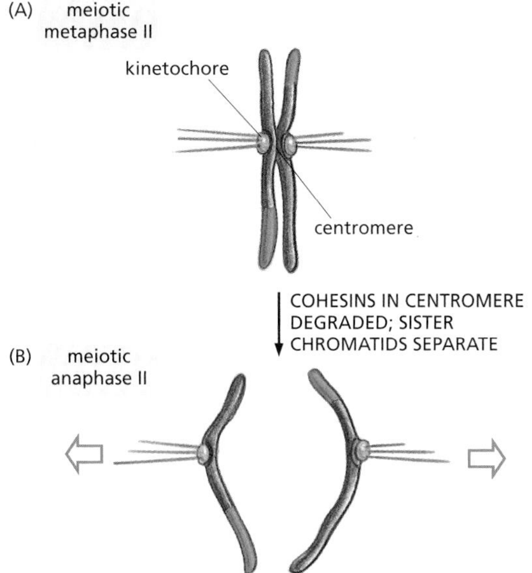

Figure 19–11 **In meiosis II, as in mitosis, the kinetochores on each sister chromatid function independently, allowing the two sister chromatids to be pulled to opposite poles.** (A) In metaphase II, the kinetochores of the sister chromatids point in opposite directions. (B) The cohesins holding the sister chromatids together at the centromere are now degraded, allowing kinetochore microtubules to pull the individual chromatids to opposite poles.

To summarize, meiosis consists of a single round of DNA replication followed by two cell divisions, so that four nonidentical haploid cells are produced from each diploid cell that enters meiosis (Movie 19.1). In contrast, the ordinary mitotic cell cycle begins with a diploid cell and produces two identical diploid cells. The two processes are compared in Figure 19–12.

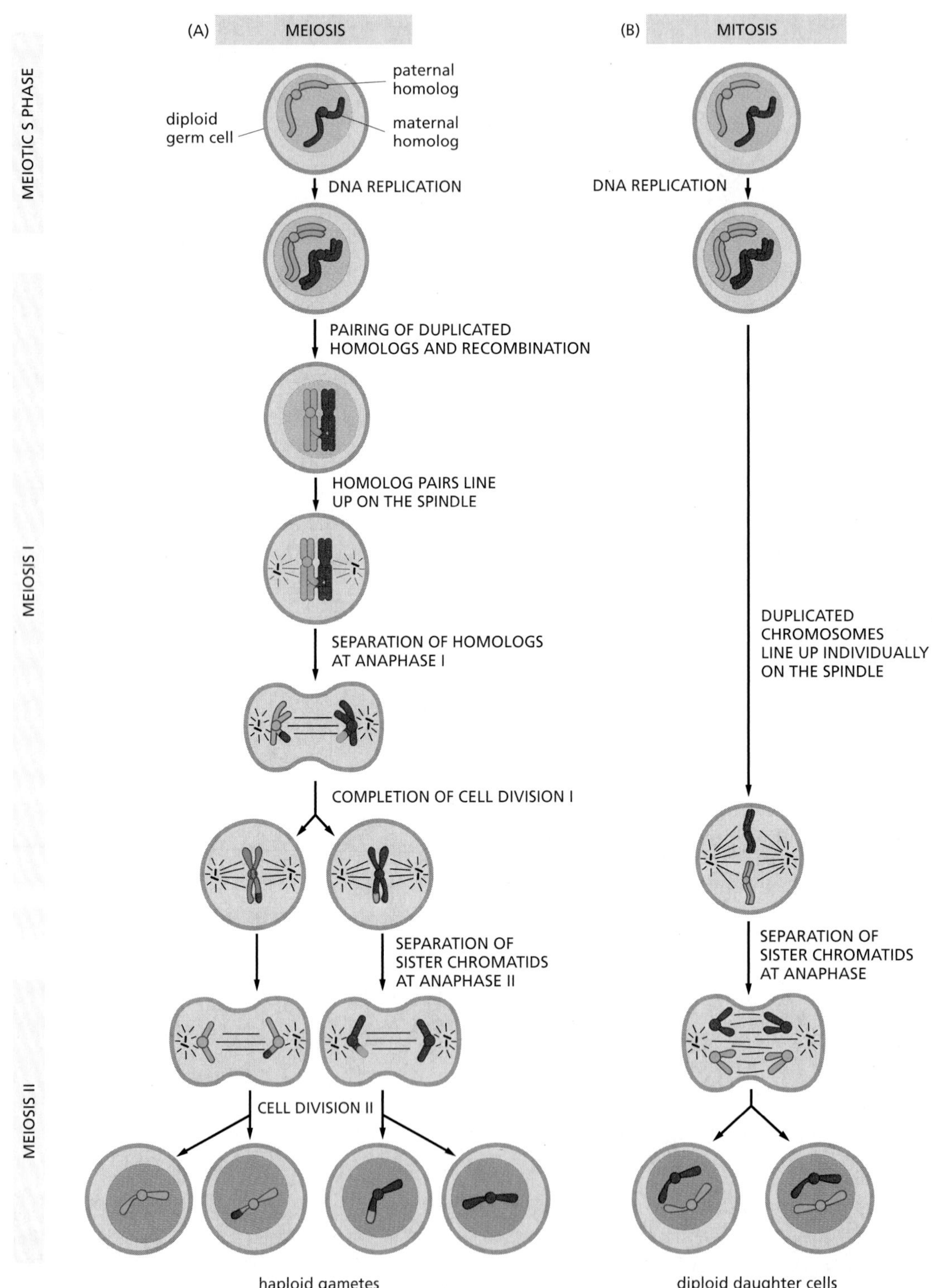

Figure 19–12 **Meiosis generates four nonidentical haploid cells, whereas mitosis produces two identical diploid cells.** As in Figure 19–4B, only one pair of homologous chromosomes is shown. In meiosis, two cell divisions are required after DNA replication to produce the haploid gametes. Each diploid cell that enters meiosis therefore produces four haploid cells, whereas each diploid cell that divides by mitosis produces two diploid cells. Although mitosis and meiosis II are usually accomplished within hours, meiosis I can last days, months, or even years, because of the long time spent in prophase.

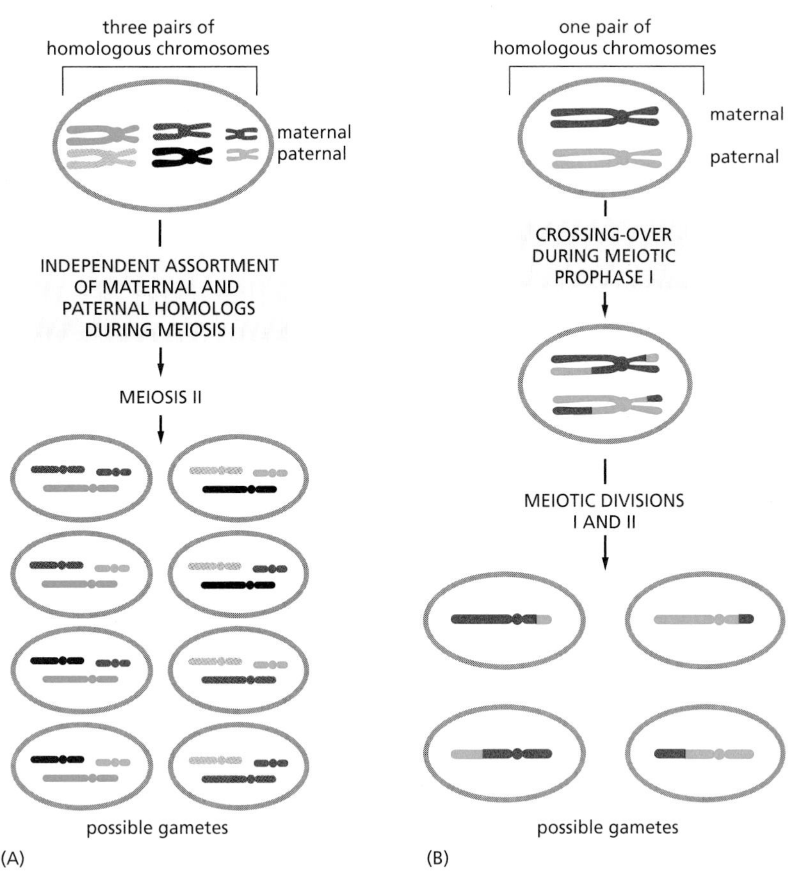

Figure 19–13 **Two kinds of reassortment generate new chromosome combinations during meiosis.** (A) The independent assortment of the maternal and paternal homologs during meiosis produces 2^n different haploid gametes for an organism with n chromosomes. Here $n = 3$, and there are 2^3, or 8, different possible gametes. For simplicity, chromosome crossing-over is not shown here. (B) Crossing-over during meiotic prophase I exchanges segments of homologous chromosomes and thereby reassorts genes on individual chromosomes. For simplicity, only a single pair of homologous chromosomes is shown. Both independent assortment and crossing-over occur during every meiosis.

Haploid Cells Contain Reassorted Genetic Information

Identical twins, which develop from a single zygote, are genetically identical; otherwise no two siblings are genetically the same. This is true because, even before fertilization takes place, meiosis has produced two kinds of randomizing genetic reassortments.

First, as we have seen, the maternal and paternal chromosomes are shuffled and dealt out among the gametes during meiosis. Although the chromosomes are carefully distributed so that each gamete receives one and only one copy of each chromosome, the choice between the two copies is made at random. Thus, each gamete contains maternal versions of some chromosomes and paternal versions of others (Figure 19–13A). The assortment is decided entirely by the way each bivalent is positioned when it lines up on the spindle during metaphase I. Whether the maternal or paternal homolog is captured by the spindles from one pole or the other depends on which way the bivalent is facing when the microtubules connect with its kinetochore (see Figure 19–10). Because the orientation of each bivalent at the moment it is captured is completely random, the assortment of maternal and paternal chromosomes is random as well.

Thanks to this type of reassortment alone, an individual could in principle produce 2^n genetically different gametes, where n is the haploid number of chromosomes. Each human, for example, can in theory produce $2^{23} = 8.4 \times 10^6$ different gametes simply from the random assortment of paternal and maternal homologs that takes place in meiosis. The actual number of different gametes each person could produce, however, is much greater than 2^{23}. This is because the crossing-over that takes place during meiosis provides a second source of randomized genetic reassortment. Between two and three crossovers occur on average on each pair of human chromosomes per meiosis. This process puts maternal and

QUESTION 19–1

Why would it not be desirable for an organism to use the first steps of meiosis (up to and including meiotic cell division I) not only for meiosis, but also for the ordinary mitotic division of somatic cells?

paternal genes that are initially on separate chromosomes onto the same chromosome, as illustrated in Figure 19–13B. Because crossing-over occurs at more or less random sites along the length of a chromosome, each meiosis will produce four sets of entirely novel chromosomes.

The reassortment of chromosomes in meiosis, together with the recombination of genes that arises from crossing-over, provides a nearly limitless source of genetic variation in the gametes produced by a single individual. Considering that each human is formed from the fusion of two such gametes, one from the father and one from the mother, the richness of human variation that we see around us, even within a single family, is not at all surprising.

QUESTION 19–2

Ignoring the effects of chromosome crossovers, an individual human can in principle produce $2^{23} = 8.4 \times 10^6$ genetically different gametes. How many of these possibilities can be realized in the average life of (A) a female? (B) a male?

Meiosis Is Not Flawless

The sorting of chromosomes that takes place during meiosis is a remarkable feat of cellular bookkeeping: in humans, each meiosis requires that the starting cell keep track of 92 chromosomes (23 pairs, each of which has duplicated), handing out one complete set to each gamete. Not surprisingly, mistakes do occur in the distribution of chromosomes during this elaborate process.

Occasionally, homologs fail to separate properly—a phenomenon known as *nondisjunction*. As a result, some of the haploid cells that are produced lack a particular chromosome, while others have more than one copy of it. Upon fertilization, such gametes form abnormal embryos, most of which die. Some survive, however. *Down syndrome*, for example, a human disease involving mental retardation and characteristic physical abnormalities, is caused by an extra copy of Chromosome 21. This error results from nondisjunction of a Chromosome 21 pair during meiosis, giving rise to a gamete that contains two copies of Chromosome 21 instead of one copy (Figure 19–14). When this abnormal gamete fuses with a normal gamete at fertilization, the resulting embryo contains three copies of Chromosome 21 instead of two. This chromosome imbalance produces an extra dose of the proteins encoded on Chromosome 21 and thereby interferes with the proper development of the embryo.

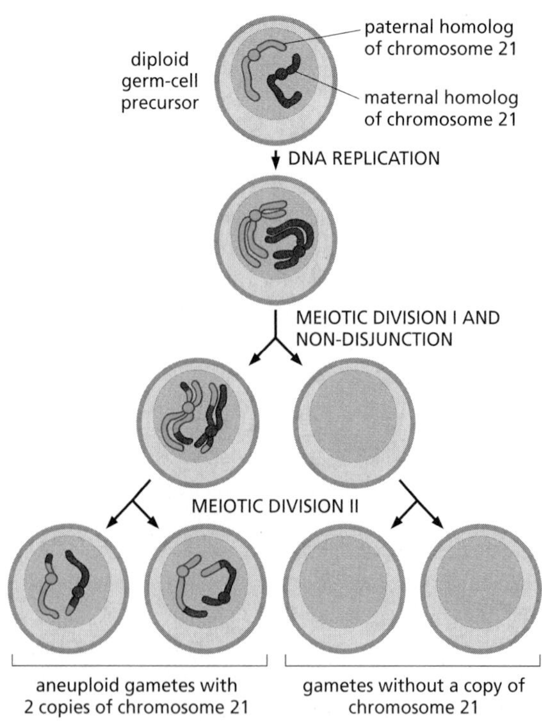

Figure 19–14 **Errors in chromosome segregation during meiosis can result in gametes with incorrect numbers of chromosomes.** In this example, the duplicated maternal and paternal copies of Chromosome 21 fail to separate normally during the first meiotic division. As a result, two of the gametes receive no copy of the chromosome, while the other two gametes receive two copies instead of the proper single copy. Gametes that receive an incorrect number of chromosomes are called *aneuploid* gametes. If one of them participates in the fertilization process, the resulting zygote will also have an abnormal number of chromosomes. If a gamete bearing two copies of Chromosome 21 fuses with a normal gamete, the resulting child will have Down syndrome.

The frequency of mis-segregation in human gametes is remarkably high, particularly in the female: nondisjunction occurs in about 10% of the meioses in human oocytes, giving rise to eggs that contain the wrong number of chromosomes (a condition called *aneuploidy*). Aneuploidy occurs less often in human sperm, perhaps because sperm development is subjected to more stringent quality control than egg development. If meiosis goes wrong in male cells, a cell-cycle checkpoint mechanism is activated, arresting meiosis and leading to cell death by apoptosis. Regardless of whether the segregation error occurs in the sperm or the egg, nondisjunction is thought to be one reason for the high rate of miscarriages (spontaneous abortions) in early pregnancy in humans.

Fertilization Reconstitutes a Complete Diploid Genome

Having seen how chromosomes are parceled out during meiosis, we now discuss briefly how they come back together during the process of **fertilization** to form a new zygote with a complete set of chromosomes.

Of the 300 million human sperm ejaculated during coitus, only about 200 reach the site of fertilization in the oviduct. There is evidence that chemical signals released by cells that surround the ovulated egg attract the sperm to the egg, but the nature of the chemoattractant molecules is unknown. Once it finds an egg, the sperm must migrate through a protective layer of cells and then bind to, and tunnel through, the egg coat, called the *zona pellucida*. Finally, the sperm must bind to and fuse with the underlying egg plasma membrane (Figure 19–15). Although fertilization normally occurs by this process of sperm–egg fusion, it can also be achieved artificially by injecting the sperm directly into the egg cytoplasm; this is sometimes done in infertility clinics when there is a problem with natural sperm–egg fusion.

Although many sperm may bind to an egg, only one normally fuses with the egg plasma membrane and introduces its DNA into the egg cytoplasm. The control of this step is especially important because it ensures that a fertilized egg will contain two, and only two, sets of chromosomes. Several mechanisms prevent multiple sperm from entering an egg. In one mechanism, the first successful sperm triggers the release of a wave of Ca^{2+} ions in the egg cytoplasm. The Ca^{2+} in turn then triggers the secretion of enzymes that cause a 'hardening' of the zona pellucida. This prevents 'runner up' sperm from penetrating the zona pellucida, thereby ensuring that in the race to fertilize the egg there is a single winner. To watch a fertilization-induced calcium wave, see Movies 16.4 and 19.2.

Once fertilized, the egg is called a **zygote**. The process is not complete, however, until the two haploid nuclei (called *pronuclei*) come together and combine their chromosomes into a single diploid nucleus. Fertilization marks the beginning of one of the most remarkable phenomena in all of biology—the process of embryogenesis, in which the zygote divides to produce large numbers of diploid cells that develop into a new individual.

Figure 19–15 **A sperm binds to the plasma membrane of an egg.** Shown here is a scanning electron micrograph of a human sperm touching a hamster egg. The egg has been stripped of its zona pellucida, exposing the plasma membrane, which is covered in finger-like microvilli. Such uncoated hamster eggs are sometimes used in infertility clinics to assess whether a man's sperm are capable of penetrating an egg. The zygotes resulting from this test do not develop. (Courtesy of David M. Phillips.)

MENDEL AND THE LAWS OF INHERITANCE

In organisms that reproduce without sex, the genetic material of the parent is transmitted faithfully to its progeny. Thus offspring are genetically identical to a single parent. Before Mendel started working with peas, some biologists suspected that inheritance might work that way in people (Figure 19–16).

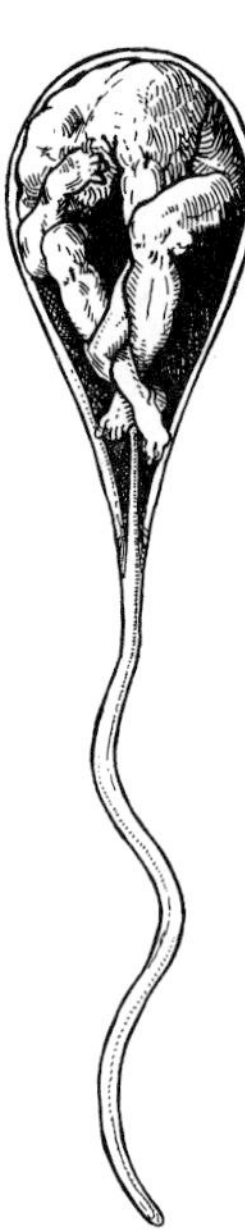

Figure 19–16 **One disproven theory of inheritance suggested that genetic traits are passed down solely from the father.** In support of this particular theory of uniparental inheritance, some early microscopists fancied that they could detect a small, perfectly formed human crouched inside the head of each sperm.

But though children resemble their parents, they are not carbon copies of either the mother or the father. For organisms that reproduce sexually, offspring tend to exhibit a mixture of traits derived from both parents. This fact helped early geneticists (scientists who study heredity) in their efforts to unravel the laws of inheritance. To understand the underlying principles of heredity, one needs to be able to track particular characteristics as they pass (or fail to pass) from parent to child. And those characteristics have to show some variation. If every blue-eyed parent gave birth only to blue-eyed children, generation after generation, it would be impossible, on the basis of studying those families, to go beyond the statement that blue eye color is inherited.

Thanks to the mechanisms of meiosis just described, sex breaks up existing collections of genetic information, shuffles alleles into new combinations, and produces individuals that have different characters as a result. It is therefore not surprising that the study of inheritance via sexual reproduction provided the first crucial insights into the mechanisms of heredity.

Humans, like other species, exhibit simple traits that can be followed from one generation to the next—such as whether an individual's earlobes are attached or pendulous, or whether a person can detect certain odors or flavors (Figure 19–17). But our family sizes are small, and human development is so slow that it takes 40 years or more just to get a couple of generations of progeny to analyze. Humans therefore don't make the best experimental material for geneticists.

The laws of heredity were instead unraveled in species that are easy to breed and that produce large numbers of offspring. Gregor Mendel, the father of genetics, chose to study peas, but similar experiments can be performed in fruit flies, worms, dogs, cats, or any other plant or animal that possesses characteristics of interest. We know now that the same basic laws of inheritance apply to all sexually reproducing organisms, from microscopic yeasts to peas to people.

In this section, we describe the logic of genetic inheritance in sexually reproducing organisms. We shall see how the behavior of chromosomes during meiosis—their segregation into gametes that then unite at random to form genetically unique offspring—explains the experimentally

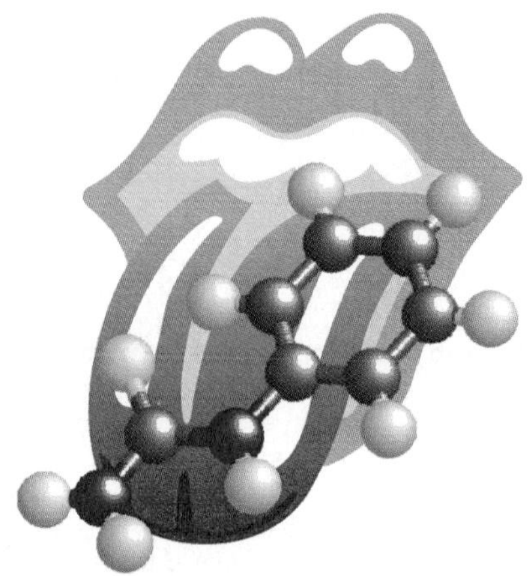

Figure 19–17 **Some people taste it, some people don't.** The ability to taste the chemical phenylthiocarbamide (PTC) is governed by a single gene. Although geneticists have known since the 1930s that the inability to taste PTC is inherited in a Mendelian fashion, researchers did not identify the gene responsible—one that encodes a bitter taste receptor—until 2003. Nontasters produce a PTC receptor protein that contains amino acid substitutions that are thought to reduce the activity of the protein.

derived laws of inheritance. But to begin we discuss how Mendel, breeding peas in his monastery garden, discovered these laws in the nineteenth century.

Mendel Chose to Study Traits That Are Inherited in a Discrete Fashion

When designing an experiment to address a scientific question, selecting the right organism can be critical. Mendel chose to carry out his studies in pea plants. Peas are easy to cultivate quickly and large numbers can be raised in a small space—such as an abbey garden. Furthermore, Mendel could control which plants mated with which. Each flower on a pea plant contains both male and female structures, and left to their own devices the plants normally self-fertilize. But Mendel found he could cross-pollinate the peas by removing the immature male parts from one flower and then fertilizing that emasculated plant with sperm (pollen) from another plant. Thus, Mendel could be certain of the parentage of every pea plant he examined.

Equally critical for Mendel's purposes, pea plants were available in many varieties. For example, one variety has purple flowers, another has white. One variety produces peas with smooth skins, another produces peas that are wrinkled. Mendel chose to follow sets of traits—such as flower color and seed shape—that were distinct, easily observable, and, most importantly, inherited in a discrete fashion (Figure 19–18). In other words, the plants have either purple flowers or white flowers—nothing in between.

Mendel Could Disprove the Alternative Theories of Inheritance

The breeding experiments that Mendel performed were straightforward. He started with stocks of genetically pure, true-breeding plants. When true-breeding plants self-pollinate, all of their offspring are of the same variety. If he were following seed color, for example, he would use plants

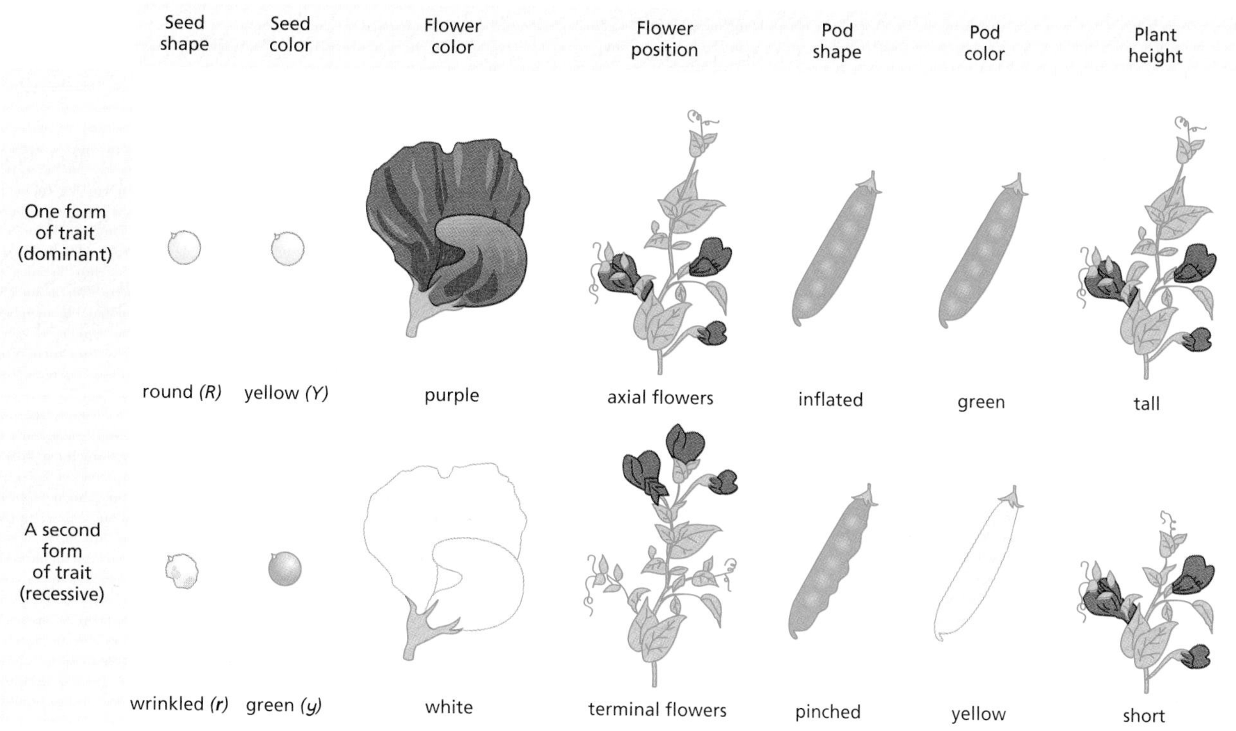

Figure 19–18 **Mendel studied seven different traits that are inherited in a discrete fashion.** For each trait, the plants display either one variation or the other. In other words, an individual pea plant produces either yellow peas or green peas—nothing in between. As we shall see shortly, one form of each trait is dominant, whereas the other is recessive.

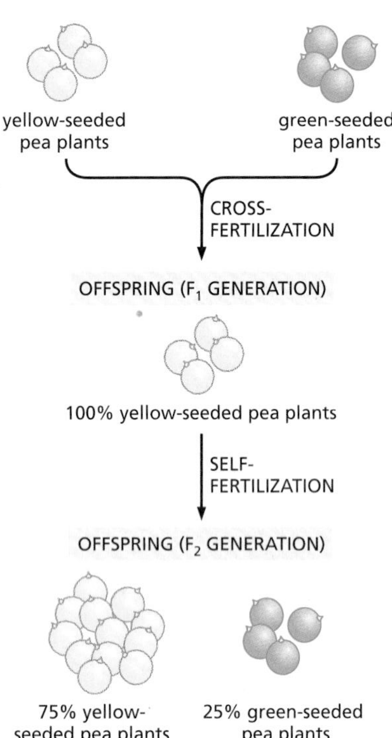

Figure 19–19 **A simple set of experiments revealed the discrete nature of heredity.** True-breeding green-pea plants, crossed with true-breeding yellow-pea plants, always produce offspring with yellow peas. But as illustrated, when these offspring (the F_1 plants) are bred with each other, or allowed to self-fertilize, 25% of the progeny produce green peas.

with yellow peas that always produced offspring with yellow peas, and plants with green peas that always produced offspring with green peas.

Mendel took the unique approach of studying each trait one at a time. His predecessors had focused on whole organisms that varied in many traits, and they often wound up trying to characterize offspring whose appearance varied in such a complex way that the progeny could not easily be compared with their parents. In a typical experiment, Mendel would take two of his true-breeding varieties and cross-pollinate them. Then he would record the inheritance of the chosen trait in the next generation. For example, Mendel crossed plants producing yellow peas with plants producing green peas. In this case, he discovered that the resulting hybrid offspring, called the first filial or *F_1* generation, all had yellow peas (Figure 19–19). A similar finding held true for every trait he followed: the F_1 hybrids all resembled only one of their two parents.

From his experiment with green- and yellow-seeded pea plants, Mendel could conclude that seed color is not passed down by blending. If it were, the peas in the F_1 generation would have all been yellowish-green. On the surface, though, the results seemed to support a theory of uniparental inheritance, which posits that the appearance of the offspring will match one parent or the other. Had Mendel stopped there—observing only the F_1 generation—he might have developed some mistaken ideas about the nature of heredity. Fortunately, he took his breeding experiments to the next step: mating the F_1 plants to one another—or allowing them to self-fertilize—and examining the results.

Mendel's Experiments Were the First to Reveal the Discrete Nature of Heredity

Mendel's experiments with the hybrid F_1 plants were designed to address an obvious question: what had happened to the traits, such as green pea color, that disappeared in the F_1 generation (see Figure 19–19)? Did the plants bearing green peas fail to make a genetic contribution to their offspring? To find out, Mendel allowed the F_1 plants to self-fertilize. If the trait for green peas, for example, had been lost, then the F_1 plants would produce only plants with yellow peas in the next, F_2, generation. He used a large sample size and kept an accurate count of the results. And he found that the 'disappearing trait' returned: although three-quarters of the offspring in the F_2 generation had yellow peas, one-quarter had green peas (see Figure 19–19).

This result definitively laid to rest the theory of blended inheritance. There is simply no way that blending could explain how a cross between one yellow-pea plant and another yellow-pea plant could yield plants with green peas. But the data also gave Mendel a clue as to what was going on. Although the green pea trait disappeared temporarily in the F_1 generation, it reappeared in F_2. This means that at least some of the F_1 plants must have still harbored a factor that specifies green peas: it was just hidden somehow. Mendel saw the same type of behavior for each of the other six traits he examined.

To account for these observations, Mendel proposed that the inheritance of traits is governed by hereditary factors (which we now call genes), and that these factors act like discrete particles that remain separate instead of blending. Furthermore, he suggested that genes come in alternative versions that account for the variations seen in inherited characteristics. The gene dictating seed color, for example, exists in two 'flavors': one that directs the production of yellow peas and one that directs production of green peas. Such alternative versions of a gene are what we now call **alleles**, and the whole collection of alleles possessed by an individual—its genetic makeup—is called its **genotype**.

Mendel reasoned that for each characteristic, a plant must inherit two copies, or alleles, of each gene—one from its mother, one from its father. The true-breeding parental strains, he theorized, each possessed a pair of identical alleles—the yellow-pea plants possessed two alleles for yellow peas, the green-pea plant two alleles for green peas. An individual that possesses two identical alleles is said to be **homozygous** for that trait. The F_1 hybrid plants, on the other hand, had received two dissimilar alleles—one specifying yellow peas and the other green. These plants were **heterozygous** for the trait of interest.

The appearance, or **phenotype**, of the plant depends on which versions of each allele it gets. To explain the disappearance of one of the traits in the F_1 generation, Mendel supposed that for any pair of alleles, one allele is *dominant* and the other is *recessive* or hidden. The dominant allele, whenever it is present, would dictate the plant's phenotype. In the case of seed color, the allele that specifies yellow peas is dominant; the green-pea allele is recessive.

One important consequence of heterozygosity, and of dominance and recessiveness, is that not all of the alleles that an individual carries can be detected in its phenotype. Humans have about 25,000 genes, and each of us is heterozygous for a very large number of these. Thus, we all carry a great deal of genetic information that remains hidden in our personal phenotype, but that can turn up in future generations.

Each Gamete Carries a Single Allele for Each Character

Mendel's theory—that for every gene, an individual inherits one copy from its mother and one copy from its father—raised some logistical issues. If an organism has two copies of every gene, how does it pass only one copy of each to its progeny? And how do these gene sets come together again in the resulting offspring?

Mendel postulated that when sperm and eggs are formed, the two copies of each gene present in the parent separate so that each gamete receives only one allele for each trait. Thus, each egg (ovum) and each sperm (pollen) receives only one allele for seed color (either yellow or green), one allele for seed shape (smooth or wrinkled), one allele for flower color (purple or white), and so on. During fertilization, sperm with one or the other allele unites with an egg carrying one or other allele, restoring the two copies of the gene for each trait in the fertilized egg or zygote. Which type of sperm unites with which type of egg at fertilization is entirely a matter of chance.

This principle of heredity is laid out in Mendel's first law, the **law of segregation**, which states that the two alleles for each trait separate (or segregate) during gamete formation, and that they then unite at random—one from each parent—at fertilization. According to the law of segregation, the F_1 hybrid plants with yellow peas will produce two classes of gametes: half the gametes will get a yellow-pea allele and half will get a green-pea allele. When the hybrid plants self-pollinate, these two classes of gametes will unite at random. Thus, an ovum with a green-pea allele has an equal chance of being fertilized by a pollen grain carrying a green-pea allele or a yellow-pea allele. The same is true for an ovum carrying a yellow-pea allele. Thus four different combinations of alleles can come together in the F_2 offspring (Figure 19–20). One-quarter of the F_2 plants will receive two alleles specifying green peas; these plants will obviously produce green peas. One-quarter of the plants will receive two yellow-pea alleles and will produce yellow peas. But one-half of the plants will inherit one allele for yellow peas and one allele for green. Because the yellow allele is dominant, these plants—like their heterozy-

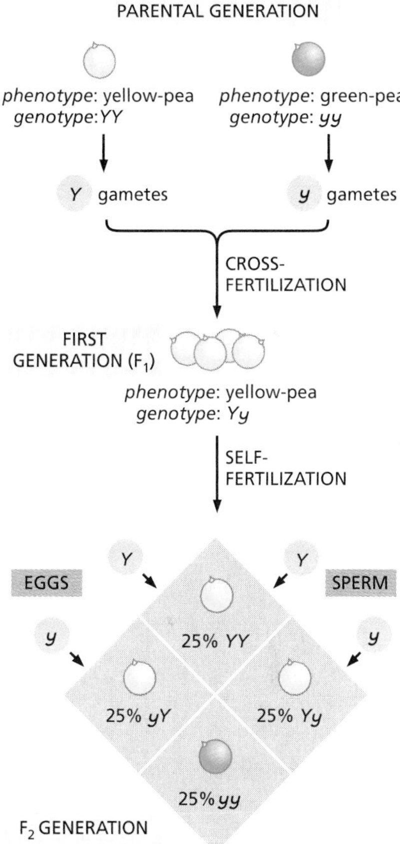

Figure 19–20 **Parent plants produce gametes that each contain one allele for each trait; the phenotype of the offspring depends on which combination of alleles it receives.** Here we see both the genotype and phenotype of the pea plants that were bred in Figure 19–19. The true-breeding yellow-pea plants produce only *Y*-bearing gametes; the true-breeding green plants produce only *y*. Their offspring, which produce yellow peas, have the genotype *Yy*. When these hybrid plants are bred with each other, 75% of the offspring have yellow peas, 25% have green. The gray box at the bottom, called a Punnett square after a British mathematician who was a follower of Mendel, allows one to track the separation of alleles during gamete formation and to predict the combinations that occur upon fertilization. According to the system invented by Mendel, capital letters symbolize a dominant allele and lower-case letters the recessive allele.

gous F_1 parents—will all produce yellow peas. All told, three-quarters of the offspring will have yellow peas and one-quarter will have green peas. Thus Mendel's law of segregation explains the 3:1 ratio that he observed in the F_2 generation.

Mendel's Law of Segregation Applies to All Sexually Reproducing Organisms

Mendel's law of segregation explained the data for all of the traits he examined in peas. He was also able to replicate his basic findings with corn and beans. But his rules governing inheritance are not limited to reproduction in plants. Mendel's concept of the gene and his law of segregation can be generalized to all sexually reproducing organisms, including humans.

Let's consider a human phenotype that reflects the action of a single gene. The major form of albinism—Type II albinism—is a rare condition that is inherited in a recessive manner in many animals, including humans. In other words, like the pea plants that produce green seeds, albinos are homozygous for the recessive allele (*a*) of a particular gene. Their genotype is *aa*. The dominant allele (*A*) of the gene codes for an enzyme involved in making melanin, the pigment that is responsible for most of the brown and black color present in hair, skin, and the retina of the eye. The recessive allele codes for a version of this enzyme that is less active or completely inactive. Without this enzyme, albinos have white hair, white skin, and pupils that look pink because the lack of coloration reveals the red color of the hemoglobin present in blood vessels in the retina.

If a Type II albino man (genotype *aa*) has children with a Type II albino woman (whose genotype is also *aa*), all of their children will be albino (*aa*). However, imagine that a nonalbino man marries and has children with an albino woman. (We will assume that the man is homozygous for the dominant *A* allele; albinism is quite rare, and if no one in his family has ever had the condition, this genotype is probably valid.) Their children should all be normally pigmented—that is, none would be albino. This result mirrors Mendel's crosses between true-breeding pea plants. The father would contribute a dominant *A* allele to each gamete, and the mother would contribute a recessive *a*. The offspring, with genotype *Aa*, would all have the dominant phenotype (Figure 19–21). If one of these children some day met and started a family with an individual of a similar genotype (a man or woman who also has an albino parent), we would expect that their children would follow the pattern seen in Mendel's F_2 generation: for every three normally pigmented children, there would be on average one albino child. Or, from the point of view of an individual child, each would have a 25% chance of receiving two recessive alleles.

Of course, humans generally don't have families large enough to guarantee accurate Mendelian ratios. (Mendel arrived at his ratios by breeding and counting thousands of peas for most of his crosses.) Geneticists interested in following the inheritance of specific traits in humans get around this problem by working with large numbers of families, or with several generations of a few large families. To keep track of this type of information and to help draw out the pattern of inheritance, geneticists prepare a **pedigree** that shows the phenotype of each family member for the relevant trait. Figure 19–22 gives an example, illustrating an important practical consequence of Mendel's laws: it shows how first-cousin marriages create a greatly increased risk of producing children that are homozygous for a recessive deleterious mutation and so display a mutant phenotype.

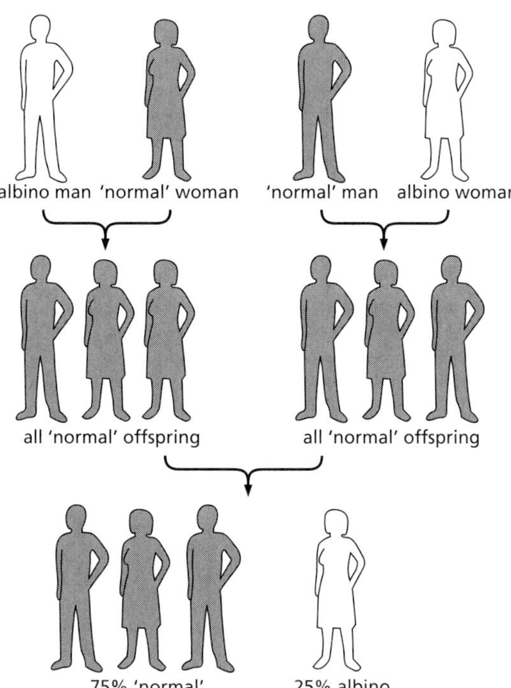

Figure 19–21 **Mendel's law of segregation applies to all sexually reproducing organisms.** Here we trace the inheritance of Type II albinism, a recessive trait that is associated with a single gene in humans.

Alleles for Different Traits Segregate Independently

Mendel deliberately simplified the problem of heredity by performing monohybrid crosses—breeding experiments that focused on the inheritance of one trait at a time. He then continued his studies, next examining the simultaneous inheritance of two or more apparently unrelated traits.

In the simplest situation, a *dihybrid cross*, Mendel followed the inheritance of two traits at once—seed color and seed shape, for example. For seed color, we have already seen that yellow is dominant over green. In the case of seed shape, round is dominant over wrinkled. What happens, Mendel wondered, when plants that differ in both of these characters are crossed? Again, he started with true-breeding parental strains: the dominant strain produced yellow round peas (its genotype is *YYRR*), the recessive strain produced green wrinkled peas (*yyrr*). One possibility is that the two characters, seed color and seed shape, would be transmitted from parents to offspring as a linked package. In other words, plants would always produce either yellow round peas or green wrinkled ones. The other possibility is that seed color and seed shape would be inherited independently of one another, which means that at some point plants that produce a novel mix of traits—yellow wrinkled peas or green round peas—would arise.

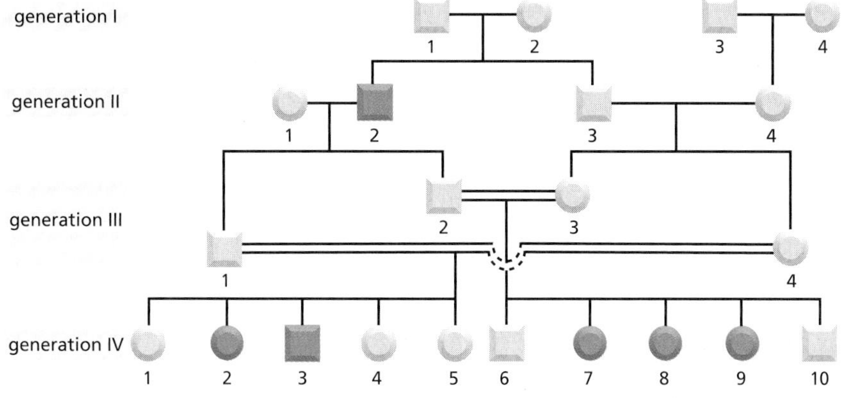

Figure 19–22 **A pedigree shows the risks of first-cousin marriages.** Shown here is an actual pedigree for a family that harbors a rare recessive mutation causing deafness. According to convention, *squares* represent males, *circles* are females. Family members that show the deaf phenotype are indicated by a filled symbol *(blue);* those that do not are represented by a hollow symbol *(white)*. A single horizontal line connecting a male and female represents a mating between unrelated individuals, and a double horizontal line represents a mating between blood relatives. The offspring of each mating are shown underneath, in order of their birth from left to right.

Individuals within each generation are labeled sequentially from left to right for purposes of identification. In the third generation in this pedigree, for example, individual 2, a man who is not deaf, marries his first cousin, individual 3, who is also not deaf. Three out of their five children (individuals 7, 8, and 9 in the fourth generation) are deaf. Meanwhile, individual 1, the brother of 2, also marries a first cousin, individual 4, the sister of 3. Two out of their five children are deaf. (Adapted from Z.M. Ahmed et al., *BMC Med. Genet.* 5:24, 2004. With permission from BMC Medical Genetics.)

Mendel bred the plants and kept a careful record of his results. As expected, the F_1 generation showed a single phenotype: all plants produced peas that were yellow and round. But this would occur regardless of whether the parental alleles were linked. When these F_1 plants were allowed to self-fertilize, the results clearly showed that each character is independently inherited; that is, the two alleles for seed color segregate independently of the two alleles for seed shape, producing four different pea phenotypes: yellow-round, yellow-wrinkled, green-round, and green-wrinkled (Figure 19–23). Mendel tried his seven pea characters in various pairwise combinations and always observed a characteristic 9:3:3:1 phenotypic ratio in the F_2 generation. The independent segregation of each pair of alleles during gamete formation is known as Mendel's second law—the **law of independent assortment**.

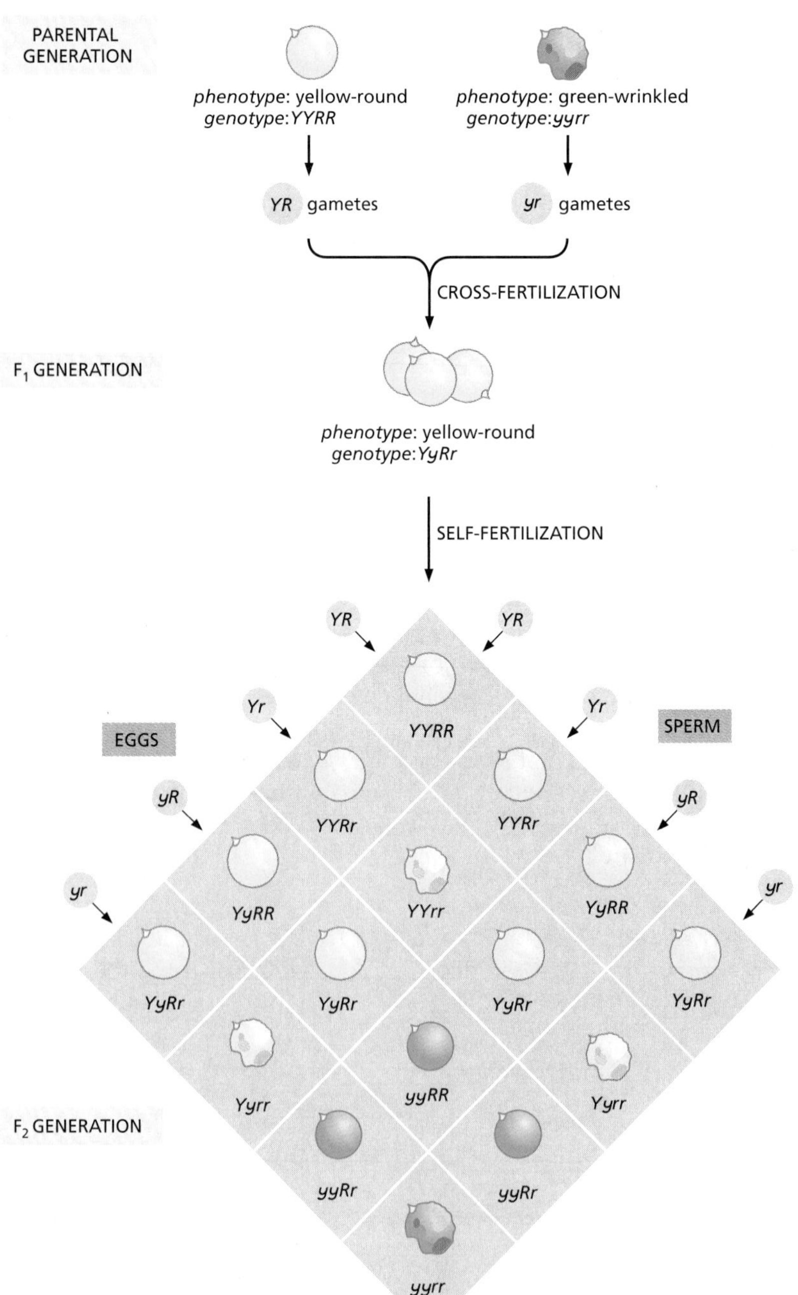

Figure 19–23 **A dihybrid cross demonstrates that alleles for different traits can segregate independently of one another.** When alleles segregate independently of each other, they will be packaged into gametes in all possible combinations. So the *Y* allele is equally likely to be packaged with the *R* or *r* allele during gamete formation; and the same holds true for the *y* allele. Thus four classes of gametes would be produced in roughly equal quantities: *YR*, *Yr*, *yR*, and *yr*. When these gametes are allowed to combine at random to produce the F_2 generation, the resulting pea phenotypes are yellow-round, yellow-wrinkled, green-round, and green-wrinkled in a ratio of 9:3:3:1.

The Behavior of Chromosomes During Meiosis Underlies Mendel's Laws of Inheritance

So far we have talked about alleles and genes as disembodied entities. As biologists, however, we are interested in heredity as more than a collection of mathematical ratios and probabilities—the likelihood that a pea plant will have purple flowers or that a child will be born an albino. We wish to understand how heredity works inside the sperm, the egg, and the resulting zygote. Mendel had assumed that genes are located in cells, but he didn't know what they were made of or where they could be found. We now know that Mendel's "factors"—which we call genes—are carried on chromosomes that are parceled out during the formation of gametes and then brought together in novel combinations in the zygote at fertilization. Chromosomes therefore provide the physical basis for Mendel's laws. As we shall now see, their behavior during meiosis and fertilization explains Mendel's laws perfectly.

During meiosis, as we discussed earlier, the maternal and paternal homologs—and the genes that lie on them—pair and then separate from one another on their way to being parceled out into gametes. These homologous chromosomes will possess different variants—or alleles—of many of the genes they carry. Take, for example, a pea plant that is heterozygous for the yellow-pea gene (*Yy*). During meiosis, the chromosomes bearing the *Y* and *y* alleles will be separated, producing two types of haploid gametes—ones that contain a *Y* allele, and others that contain a *y*. Upon self-fertilization, these haploid gametes recombine at random to produce the diploid individuals of the next generation—which may be *YY*, *Yy*, or *yy*. The meiotic mechanisms that drive the separation of alleles into gametes and the random recombination of gametes at fertilization give rise to exactly the results that Mendel's genetic laws describe.

During meiosis, each set of paired homologs attaches to the spindle independently. This random arrangement of chromosomes on the metaphase spindle is reflected in Mendel's law of independent assortment, since genes on different chromosomes will be inherited independently. Although each gamete receives one, and only one, copy of each chromosome, it winds up with a random mixture of paternal and maternal homologs (see Figure 19–13A).

Figure 19–24 diagrams this process for a pea plant that is heterozygous for both seed color (*Yy*) and seed shape (*Rr*). The chromosome pair carrying the color alleles will attach to the meiotic spindle with a certain orientation. Whether the *Y*-bearing homolog or the *y*-bearing homolog is captured by the microtubules from one pole or the other depends on which way the bivalent happens to be facing at the moment of attachment (see Figure 19–24). The same is true for the chromosome pair carrying the alleles for seed shape. Thus, whether the final gamete receives the *YR*, *Yr*, *yR*, or *yr* combination of alleles depends entirely on which way the two chromosome pairs were facing when they were captured by the meiotic spindle, which has the same degree of randomness as the tossing of a coin.

Chromosome Crossovers Can Be Used to Determine the Order of Genes

Mendel studied seven traits that were carried by seven genes, each of which segregated independently of the others. It turns out that most of these genes are located on different chromosomes, which readily explains the random assortment he observed. But Mendel's observation that different traits assort independently does not necessarily require that the genes lie on different chromosomes. Genes that are far enough away

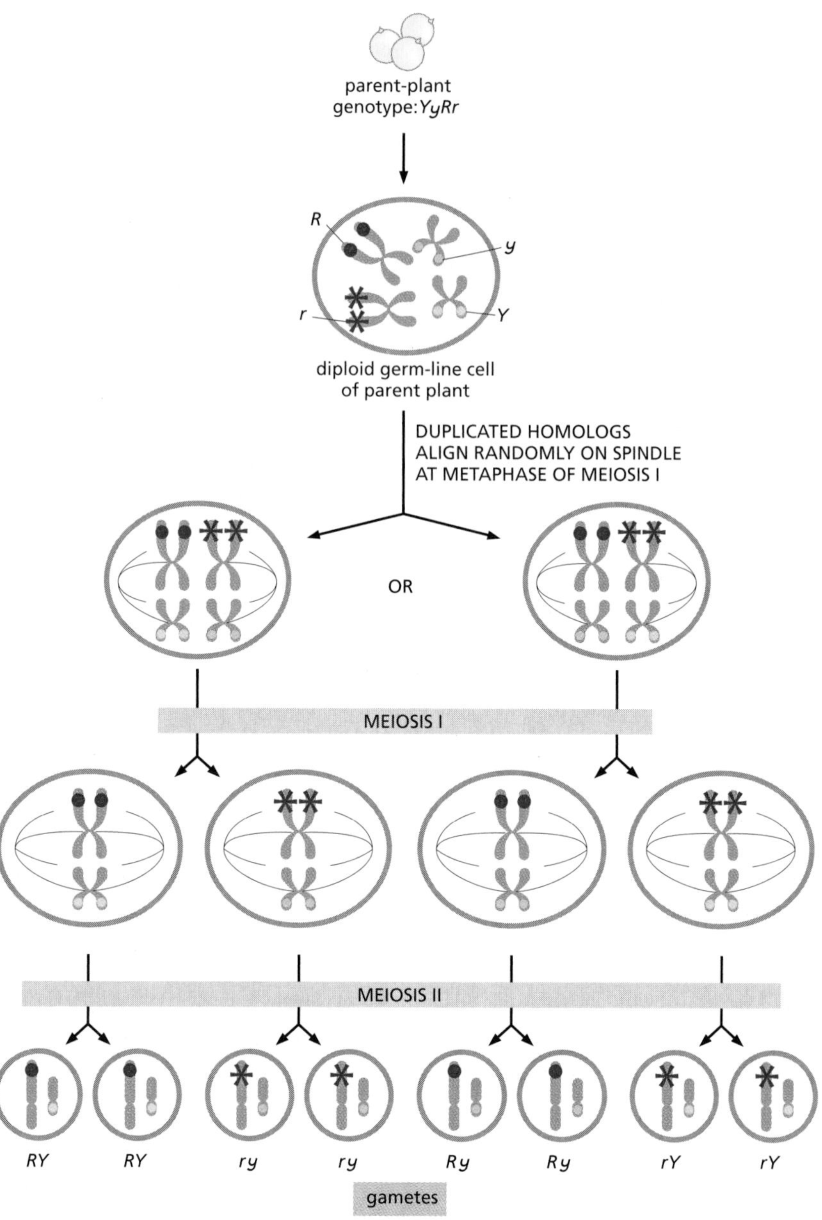

Figure 19–24 **The separation of chromosomes during meiosis explains Mendel's laws of segregation and independent assortment.** Here we show independent assortment of the alleles for seed color, yellow *(Y)* and green *(y)*, and seed shape, round *(R)* and wrinkled *(r)*, corresponding to two genes on different chromosomes. For simplicity, the figure is drawn for the case where there is no recombination between maternal and paternal homologs; the outcome—independent assortment—is the same (given that the seed-color and seed-shape genes lie on different chromosomes) if recombination occurs.

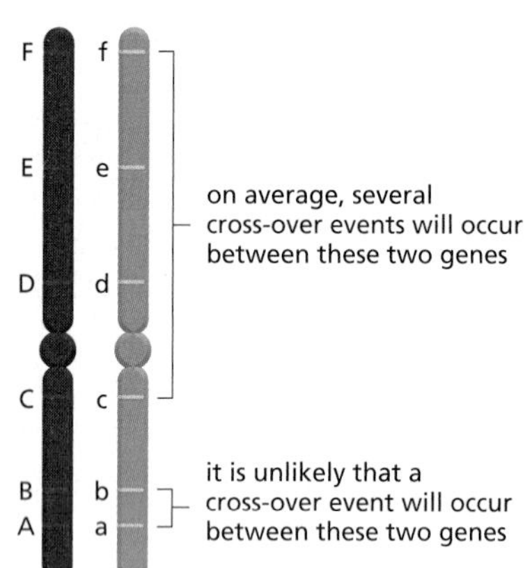

Figure 19–25 **Genes that lie far enough apart on the same chromosome can sort independently.** Because several cross-over events occur randomly along each chromosome during prophase of meiosis I, two genes on the same chromosome will obey Mendel's law of independent assortment if they are far enough apart. Thus, for example, there is a high probability of crossovers occurring in the long region between *C/c* and *F/f*, making it just as likely that a gamete carrying the *F* allele will wind up with the *c* allele as it will the *C*. In contrast, the *A/a* and *B/b* genes are close together, so there is only a small chance of crossing-over between them: thus the *A* allele is likely to be co-inherited with the *B* allele, and the *a* allele with the *b* allele. From the frequency of recombination, one can estimate the distances between the genes.

from one another on the same chromosome will also sort independently because of the crossing-over that occurs during meiosis. As we discussed earlier, when the duplicated chromosomes form bivalents and line up on the metaphase spindle, maternal and paternal homologs typically undergo several recombination events and thereby exchange genetic material. Such crossing-over events can separate alleles that were formerly together on the same chromosome, causing them to segregate into different gametes (Figure 19–25). We now know, for example, that Mendel's genes for pea shape and pod color are located on the same chromosome, but because they are far apart they segregate independently.

Of course, not all genes are inherited independently as per Mendel's second law. If genes lie close together on a chromosome, they are likely to be inherited as a unit. For example, genes associated with red–green colorblindness and hemophilia in humans are typically inherited together for this reason. By measuring how frequently genes are co-inherited, researchers can determine whether they reside on the same chromosome and, if so, how far apart they lie. These measurements of *genetic linkage* have been used to map the relative positions of the genes on each

Figure 19–26 **Gene mutations can affect the protein product in a variety of ways.** In this example, the wild-type protein has a specific cellular function denoted by the *red rays*. Mutations that eliminate this function, make it hyperactive, or render it sensitive to higher temperatures are shown. A *temperature-sensitive mutation* is a type of *conditional mutation:* the allele produces active protein under one condition (25°C, here) but inactive protein under another. Here, the protein carries an amino acid substitution *(red)* that prevents its proper folding at 37°C but permits proper folding at 25°C.

chromosome of many organisms. Such *genetic maps* have been crucial for isolating and characterizing genes whose disruption by mutation is responsible for human genetic diseases such as cystic fibrosis.

Mutations in Genes Can Cause a Loss of Function or a Gain of Function

Mutations can be classified into a number of different basic categories (Figure 19–26). Those changes that reduce or eliminate the activity of a gene are called **loss-of-function mutations**. An organism that receives two copies of a loss-of-function allele will generally show a mutant phenotype—one that differs from the 'normal,' most commonly occurring, phenotype. The heterozygote, possessing one mutant allele and one wild-type allele, generally makes enough active gene product to function normally and retain the wild-type phenotype (Panel 19–1, p. 674). Thus, loss-of-function mutations are usually recessive, because—for most genes—a halving of the normal amount of gene product still leaves enough for the cell to function normally.

As an example, in the case of Mendel's wrinkled peas, the gene that dictates seed shape codes for an enzyme that helps to convert sugars into branched starch molecules. The wild-type, dominant allele, *R*, produces the active enzyme; the recessive, mutant allele, *r*, does not. Because they lack this enzyme, plants that are homozygous for the *rr* allele contain more sugar—and produce less starch—than plants that possess the dominant *R* allele, which gives their peas a wrinkled appearance. (The frozen peas available in the supermarket are wrinkled mutants, although the alleles they carry may not be the same one that Mendel used.)

Another class of mutant alleles produces proteins that are overactive, or are active in inappropriate circumstances. Such **gain-of-function mutations** are usually dominant. For example, as we saw in Chapter 16, certain mutations in the gene encoding Ras, a protein involved in controlling cell growth and proliferation, generate a form of the protein that is always active and therefore overshadows the wild-type allele. The mutant Ras protein can stimulate cells to divide inappropriately, even in the absence of any growth signal. About 30% of all human cancers contain such dominant activating mutations in the *Ras* gene.

QUESTION 19–3

Imagine that each chromosome undergoes one and only one crossover event on each chromatid during each meiosis. How would the co-inheritance of traits that are determined by genes at opposite ends of the same chromosome compare with the co-inheritance observed for genes on two different chromosomes? How does this compare with the actual situation?

Each of Us Carries Many Potentially Harmful Recessive Mutant Alleles

As we saw in Chapter 9, mutations provide the fodder for evolution. Mutations can alter the fitness of an organism, making it either less or more likely to survive and leave progeny. Natural selection determines whether these mutations are preserved. Those changes that confer a selective advantage on an organism tend to be perpetuated, while those that compromise an organism's fitness tend to be lost.

The great majority of chance mutations are either neutral, with no effect on phenotype, or deleterious. A deleterious mutation that is dominant—one that exerts its negative effects when present even in a single copy—will be eliminated almost as soon as it arises: if a mutant organism

PANEL 19–1 Some essentials of classical genetics

GENES AND PHENOTYPES

Gene: a functional unit of inheritance, usually corresponding to the segment of DNA coding for a single protein.

Genome: all of an organism's DNA sequences.

locus: the site of the gene in the genome

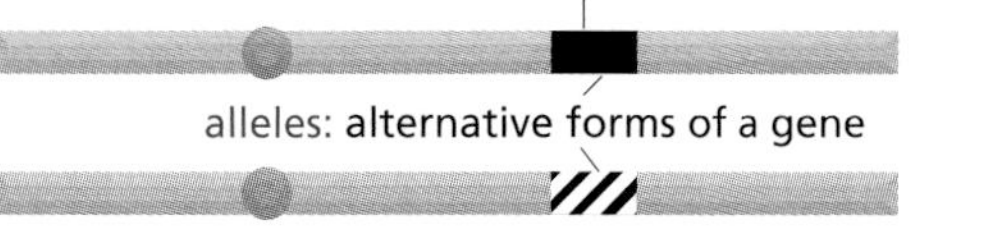

alleles: alternative forms of a gene

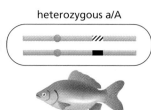

Wild type: the normal, naturally occurring type

Mutant: differing from the wild type because of a genetic change (a mutation)

GENOTYPE: the specific set of alleles forming the genome of an individual

PHENOTYPE: the visible character of the individual

homozygous A/A

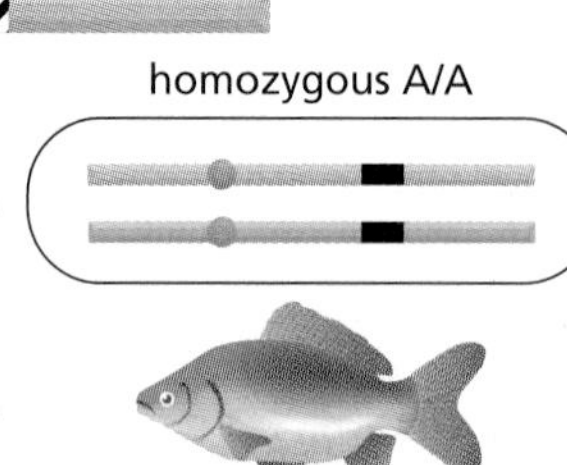

heterozygous a/A

homozygous a/a

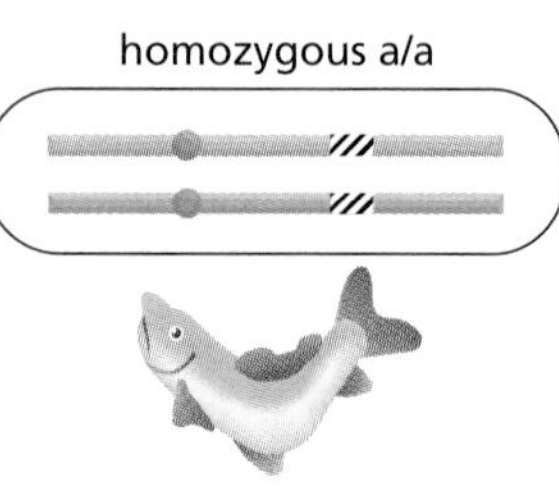

allele A is dominant (relative to a); allele a is recessive (relative to A)

In the example above, the phenotype of the heterozygote is the same as that of one of the homozygotes; in cases where it is different from both, the two alleles are said to be co-dominant.

TWO GENES OR ONE?

Given two mutations that produce the same phenotype, how can we tell whether they are mutations in the same gene? If the mutations are recessive (as they most often are), the answer can be found by a complementation test.

In the simplest type of complementation test, an individual who is homozygous for one mutation is mated with an individual who is homozygous for the other. The phenotype of the offspring gives the answer to the question.

COMPLEMENTATION: MUTATIONS IN TWO DIFFERENT GENES

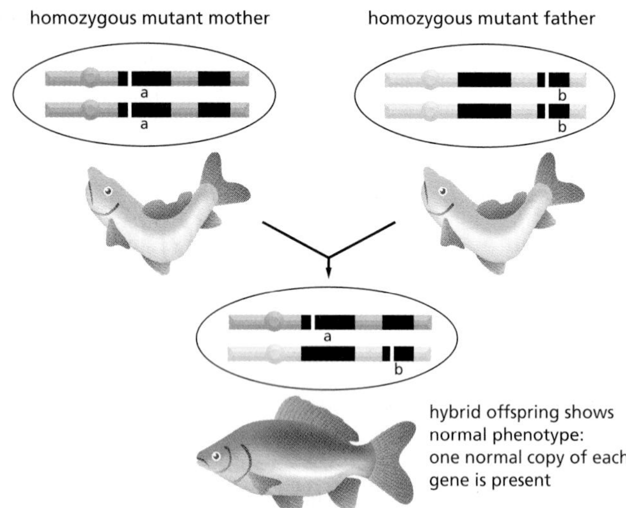

NONCOMPLEMENTATION: TWO INDEPENDENT MUTATIONS IN THE SAME GENE

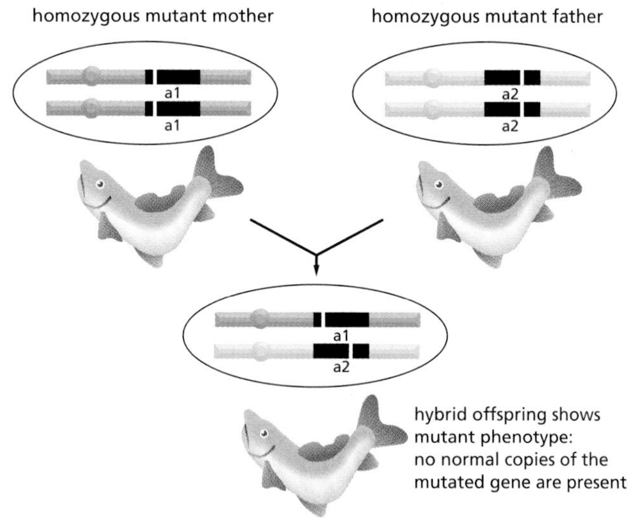

MEIOSIS AND GENETIC RECOMBINATION

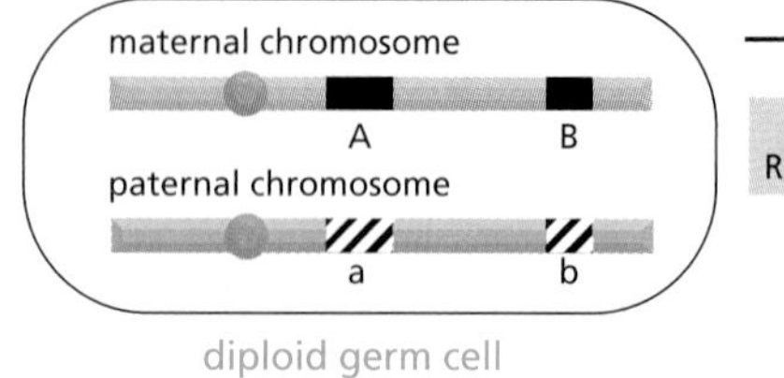

diploid germ cell

genotype $\frac{AB}{ab}$

MEIOSIS AND RECOMBINATION

genotype Ab

A b

site of crossing-over

genotype aB

a B

haploid gametes (eggs or sperm)

The greater the distance between two loci on a single chromosome, the greater is the chance that they will be separated by crossing-over occurring at a site between them. If two genes are thus reassorted in x% of gametes, they are said to be separated on a chromosome by a genetic map distance of x map units (or x centimorgans).

is unable to reproduce, the mutation that causes that failure will be lost from the population as soon as the mutant individual dies. For deleterious mutations that are recessive, things are a little more complicated. When such a mutation first arises (through some chemical accident in a DNA molecule), it will generally be present in only a single copy, and the carrier of the mutation will produce just as many progeny as other individuals. Many of these progeny will inherit a single copy of the mutation. They too will appear fit and healthy. But as they and their descendants begin to mate with one another, some individuals will inherit two copies of the mutant allele and show the harmful phenotype.

If such a homozygous individual fails to reproduce, two copies of the mutant allele will be lost from the population. Eventually, an equilibrium is reached, where the rate of creation of new copies of the allele by newly occurring mutations balances the rate of loss through matings that yield homozygous individuals. As a consequence, many deleterious recessive mutations are present at a surprisingly high frequency in a population, even though individuals showing the deleterious phenotype are rare. Thus, children with the most common form of hereditary deafness (due to mutation in a gap junction protein; see Figure 20–29) represent about one in 4000 births; but about one in 30 of us are carriers of a loss-of-function mutant allele of the gene.

GENETICS AS AN EXPERIMENTAL TOOL

The realization that chromosomes are the structures responsible for shuttling our genes from one generation to the next did more than demystify the basis of inheritance. It united the science of genetics with other disciplines: cell biology, biochemistry, physiology, and medicine. It also led to the discovery that genes are made of DNA. In this way, genetics has provided a path to new discoveries. By examining and manipulating DNA, we can begin to learn how our genes function together to create our phenotype, and how differences in genes underlie the differences between individuals. Increasingly, knowledge of genetics enables us to diagnose and treat human diseases more accurately and helps us decipher our similarities and differences, in relation both to other people and to other species.

In this section, we outline the classical genetic approach to identifying genes and determining how they influence the phenotype of an organism. The process involves intentionally generating large numbers of mutant laboratory organisms and screening them to pick out those rare individuals that show a phenotype of interest. This *genetic screen* identifies bearers of mutations in the genes that govern that phenotype. By analyzing these mutant individuals and their progeny, we can track down and characterize the genes themselves—and, ultimately, work out the chain of cause and effect that leads from the genes to the phenotype.

Modern technology has opened up additional ways to track down genes with significant functions. In the final part of this section, we discuss how we can examine DNA collected from human populations all over the world for clues to the genetics of complex traits and genetic diseases—those that are not simply governed by a single gene—as well as for hints about the evolution of our species.

The Classical Approach Begins with Random Mutagenesis

Before the advent of recombinant DNA technology (discussed in Chapter 10), most genes were identified by observing the processes disrupted when the gene was mutated. The analysis begins with the isolation of mutants that have an interesting or unusual appearance: fruit flies with

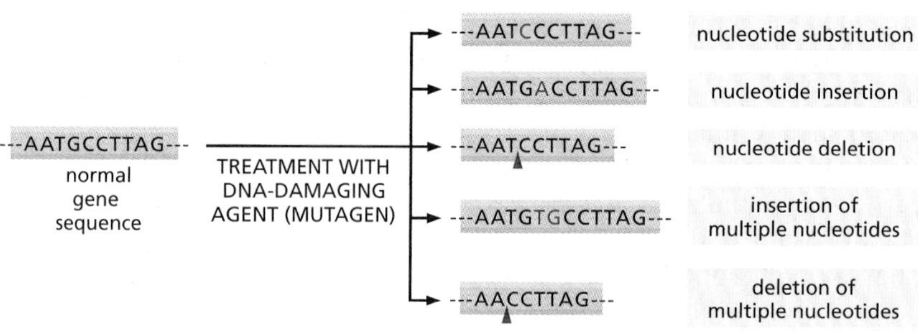

Figure 19–27 **Mutations can be caused by a variety of alterations in DNA.** Some common categories are shown here. Different mutagens tend to produce different types of changes.

white eyes or curly wings, for example. Working backward from this phenotype, one then determines the organism's genotype. This classical genetic approach—searching for mutant phenotypes, and then isolating the genes responsible for them—is most easily performed in organisms that reproduce rapidly and are amenable to genetic manipulation, such as bacteria, yeasts, worms, and fruit flies.

Although spontaneous mutants can be found by examining extremely large populations—thousands or millions of individual organisms—the process of identifying interesting mutants can be made much more efficient by generating mutations artificially with agents that damage DNA, called *mutagens*. Different mutagens can generate different types of DNA alterations (Figure 19–27). By treating organisms with mutagens, very large numbers of mutants can be created quickly and then screened for a particular defect of interest, as we will see shortly.

Mutagenesis is a powerful tool for identifying genes and relating them to phenotypes in worms and flies; but how can we identify and study genes in humans? Unlike the organisms we have been discussing, humans do not reproduce rapidly, and intentional mutagenesis of humans is out of the question. Moreover, any human with a serious defect in an essential process, such as DNA replication, will die long before birth.

There are two main answers to the question of how we study human genes. First, because genes and gene functions have been so highly conserved in the course of evolution, we can discover a great deal about human genes by investigating their counterparts in other organisms that are more suitable for experimentation. The corresponding human genes can then be studied further in cultured human cells. Second, many mutations that are not necessarily lethal, such as the mutation causing deafness discussed above, have arisen spontaneously in the human population—and, indeed, have arisen many times over, because the human population is so large. Analyses of the phenotypes of the affected individuals, together with studies of their cultured cells, have provided many unique insights into human gene functions. Although such mutants are rare, they are very efficiently discovered because of a unique human property: the mutant individuals call attention to themselves by virtue of their peculiarities and by seeking special medical care.

Genetic Screens Identify Mutants Deficient in Specific Cellular Processes

A **genetic screen** typically involves examining many thousands of mutagenized individuals to find those few who show a specific altered phenotype of interest. For example, to search for genes involved in cell metabolism, one might screen mutagenized cells to pick out those that have lost the ability to grow in the absence of a particular amino acid or other nutrient.

A problem arises, however, if we wish to study genes that are required for fundamental cell processes—RNA synthesis or cell cycle control, for instance. Defects in such genes are usually lethal, which means that special strategies are needed to propagate a stock of individuals that carry the interesting mutations. If you cannot breed the mutants, you cannot use them to study the gene.

If your organism is diploid—a mouse or a pea plant, say—and the mutant phenotype is recessive, there is a simple solution. You breed heterozygous individuals, carrying one good copy of the gene and one defective copy. They will have a normal phenotype, but when they are mated with one another, 25% of their progeny will be homozygous mutants and will show the lethal mutant phenotype, while another 50% of progeny will be heterozygous carriers of the mutation like their parents, allowing you to keep on breeding.

But what if your organism is haploid? One method hinges on the use of *temperature-sensitive* mutants. In these mutants, the protein product of the gene is only conditionally defective: it functions normally within a certain range of temperatures (the *permissive* temperature range) but can be inactivated by a small shift in temperature that takes it out of this range (see Figure 19–26). Thus the abnormality can be switched on and off experimentally simply by changing the temperature. A cell containing a temperature-sensitive mutation in an essential gene can be kept alive by growing it at the permissive temperature; it can then be driven by a temperature shift to display its mutant phenotype (Figure 19–28).

Many temperature-sensitive mutants were isolated to identify the genes that encode the bacterial proteins required for DNA replication. Here large populations of mutagen-treated bacteria were screened for cells that stop making DNA when they are warmed from 30°C to 42°C. Temperature-sensitive mutants have also been used to identify many of the proteins involved in regulating the cell cycle (discussed in Chapter 18) or in moving proteins through the secretory pathway in yeast (discussed in Chapter 15).

Genes involved in complex phenotypes, such as changes in learning or behavior, can also be identified by genetic screens in model organisms. As an example, researchers were able to isolate a gene that affects social behavior in worms by screening for animals that feed alone (Figure 19–29).

A Complementation Test Reveals Whether Two Mutations Are in the Same Gene

A large-scale genetic screen can turn up many different mutants that have the same phenotype. These defects might lie in different genes that function in the same process, or they might represent different mutations in the same gene. How can we distinguish between the two possibilities? If the mutations are recessive—if, for example, they represent a loss of function of a particular gene—a **complementation test** can reveal whether the mutations fall in the same gene or in different genes.

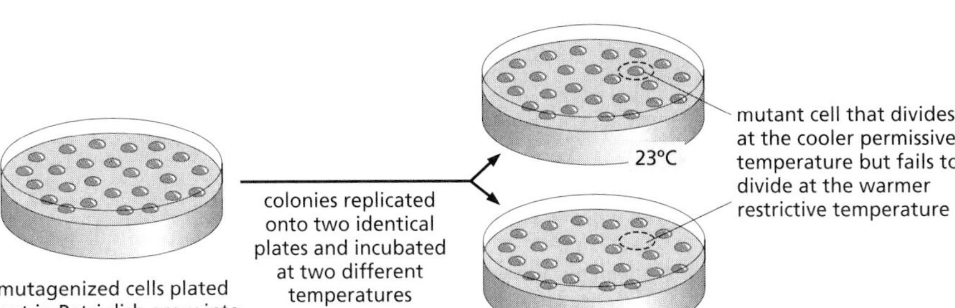
mutagenized cells plated out in Petri dish grow into colonies at 23°C
colonies replicated onto two identical plates and incubated at two different temperatures
23°C
36°C
mutant cell that divides at the cooler permissive temperature but fails to divide at the warmer restrictive temperature

Figure 19–28 **Temperature-sensitive mutants are valuable for identifying the genes and proteins involved in essential cell processes.** Here yeast cells are incubated with a chemical that generates mutations in their DNA. They are then spread on a culture plate at a relatively cool temperature and allowed to grow, forming colonies. The colonies are then transferred to two identical Petri dishes by using a technique called replica plating. One of these plates is incubated at the cool temperature, the other at a warmer temperature. Those cells that contain a temperature-sensitive mutation in a gene essential for proliferation can be readily identified, because they divide at the cooler temperature—the *permissive* temperature—but fail to divide at the warmer temperature—the *non-permissive*, or *restrictive*, temperature.

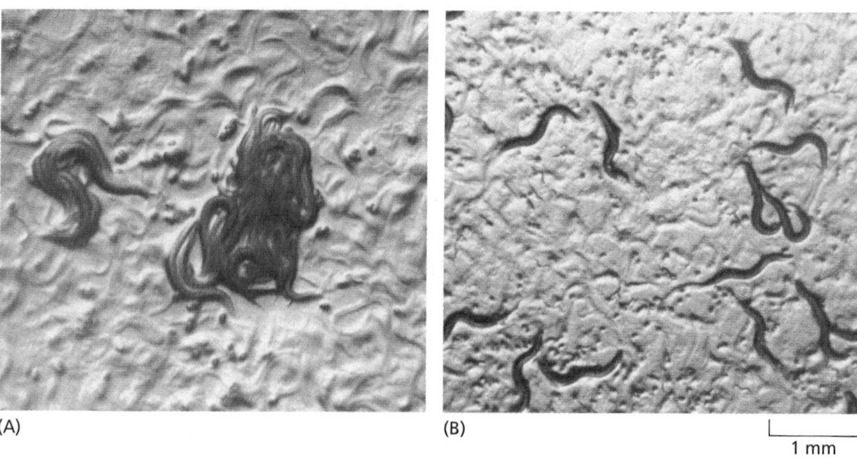

Figure 19–29 **Genetic screens can be used to identify mutations that affect an animal's behavior.** (A) Wild-type *C. elegans* engage in social feeding. The worms swim around until they encounter their neighbors and only then settle down to feed. (B) Mutant worms dine alone. (Courtesy of Cornelia Bargmann, cover of *Cell* 94, 1998. With permission from Elsevier.)

QUESTION 19–4

When two individuals from different isolated inbred subpopulations of a species come together and mate, their offspring often show "hybrid vigor": that is, they appear more robust, healthy, and fertile than either parent. Can you suggest an explanation for this phenomenon?

In the simplest type of complementation test, an individual that is homozygous for one recessive mutation—possessing two identical copies of the mutant allele—is mated with an individual that is homozygous for the other mutation. If the two mutations are in the same gene, the offspring will show the mutant phenotype, because they carry only defective copies of the gene in question (see Panel 19–1, p. 674). If, in contrast, the mutations fall in different genes, the resulting offspring will show the normal, wild-type phenotype because they will have one normal copy (and one mutant copy) of each gene.

When the normal phenotype is restored in such a test, the sets of alleles inherited from the two parents are said to *complement* one another (Figure 19–30). Complementation testing of mutants identified during genetic screens has revealed, for example, that 5 genes are required for yeast to digest the sugar galactose; that 20 genes are needed for *E. coli* to build a functional flagellum; and that many hundreds of genes are needed to guide the development of an adult nematode worm from a fertilized egg.

Single-Nucleotide Polymorphisms (SNPs) Serve as Landmarks for Genetic Mapping

With the recent determination of the complete human genome sequence, we can now study human genetics in a way that was impossible only a few years ago. Using the human genome sequence as a starting point, we can begin to identify directly, through gene sequencing, those DNA differences that distinguish one individual from another.

No two humans (with the exception of identical twins) have the same genome. Each of us carries in our genome a set of variations in nucleotide sequence that make us unique. At some sites in the genome, these variations turn out to be common and relatively harmless, so that there is a high probability that any two people chosen at random will differ at that position. In these cases, where two or more sequence variants

Figure 19–30 **A complementation test reveals that two different genes can be responsible for the same phenotype.** In this cross, a hen of the Recessive White strain (*top left*) is mated with a cock of the Silky breed (*not shown*), which is also white. The resulting offspring (shown at the *bottom*) have normal coloration, implying that the two white breeds are white because of mutations in different genes. A hen of the Silky breed is shown at *upper right*. (From W. Bateson, *Mendel's Principles of Heredity*, 1st ed. Cambridge, UK: Cambridge University Press, 1913. With permission from Cambridge University Press.)

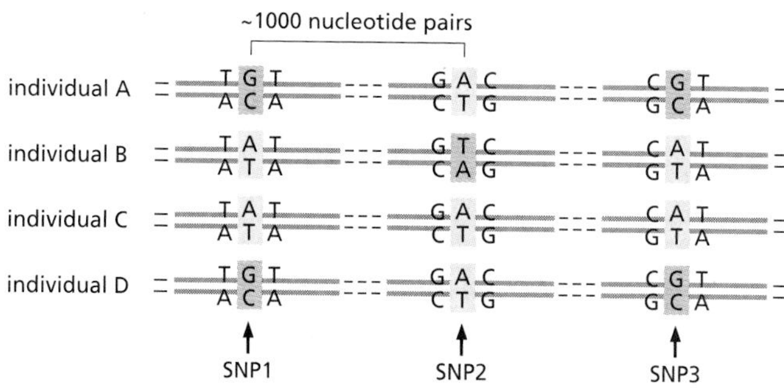

Figure 19–31 **Single-nucleotide polymorphisms (SNPs) are sites in the genome where two or more alternative choices of a nucleotide are common the population.** Most such variations in the human genome occur at locations where they do not significantly affect a gene's function.

coexist in the population and are both common, the variants are called **polymorphisms**. Some of these polymorphisms correspond to deletions or insertions of large chunks of DNA sequence, but the commonest type of variation is due to the substitution of a single nucleotide.

Typically, any two humans differ by about 0.1% in their nucleotide sequences (approximately one nucleotide difference for every 1000 nucleotides). This translates to about 3 million single-nucleotide differences between one person and another. These **single-nucleotide polymorphisms**, or **SNPs** (Figure 19–31), can be used as markers for building genetic maps or for conducting searches for mutations that correlate with specific diseases or predispositions to disease. The "How We Know" panel on pp. 680–681 explains how SNPs are used in a variety of ways for this latter purpose. Mutations that give rise, in a regular, reproducible way, to rare but clearly defined abnormalities, such as albinism or congenital deafness, can often be pinpointed through family studies. Common diseases with more complex causes, such as diabetes or arthritis, call for a different approach. For these conditions, there is no single gene that is all-important; instead, there are many genes, as well as environmental factors, that affect the risk. But with the help of SNP analysis, it is becoming possible to track these genes down.

For this purpose, researchers collect DNA from a large number of people who have the disease, and compare those samples with DNA from a parallel group of people who do not have the disease. Suppose, for example, that a particular allele of a gene creates a heightened risk of heart attack. People who inherit that allele are likely also to inherit a set of SNPs that are genetically linked to that allele. Thus, SNPs that are exceptionally common among heart-attack patients serve as flags indicating that a risky allele of a gene lies nearby in the genome (Figure 19–32).

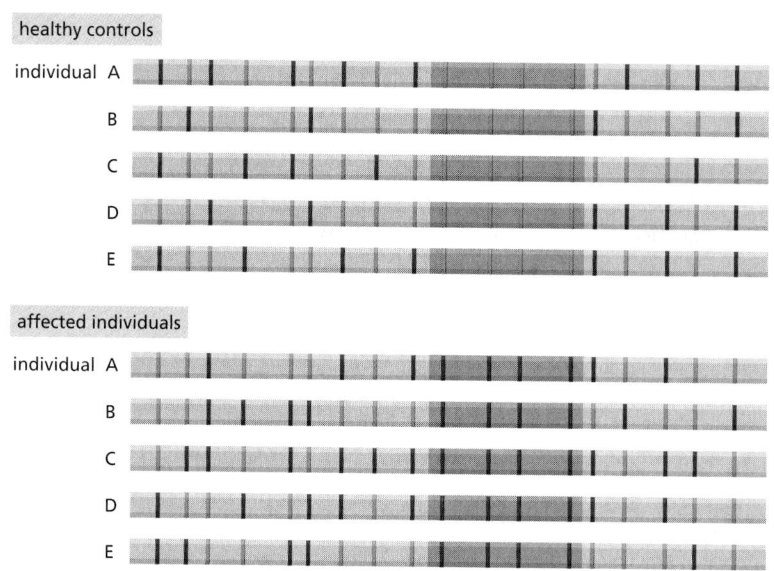

Figure 19–32 **Genes affecting propensity to common diseases can be tracked down through linkage to SNPs.** Here, the patterns of SNPs are compared between two sets of individuals—a set of healthy controls, and a set affected by a particular common disease. A segment of a typical chromosome is shown. For most polymorphic sites in this region, it is a random matter whether an individual has one SNP variant (*red*) or another (*blue*); and the same randomness is seen both for the control group and for the affected individuals. However, in the part of the chromosome that is shaded *darker gray*, a bias is seen, such that most normal individuals have the *blue* SNP variants, whereas most affected individuals have the *red* SNP variants. This strongly suggests that this region contains, or is close to, a gene that is genetically linked to these red SNP variants and that predisposes to the disease. Using carefully selected controls and large numbers (thousands) of individuals, this technique can be used to track down disease-related genes even when their common alleles (SNP variants) confer only a slightly different risk for contracting a given disease.

USING SNPS TO GET A HANDLE ON HUMAN DISEASE

For diseases that have their roots in genetics, finding the gene responsible can be the first step toward improved diagnosis and even a cure. The task is not simple, but having access to SNPs can help. In 1999, an international group of scientists set out to collect and catalogue 300,000 SNPs—the single-nucleotide polymorphisms that are common in the human population (see Figure 19–31). Today, the database has grown to include more than 10 million different SNPs. These SNPs not only help to define the differences between one individual and another. But for geneticists, they also serve as signposts that can point the way toward the genes that are involved in common human disorders, such as diabetes, obesity, asthma, arthritis, and even gallstones and restless leg syndrome.

Making a Map

One way that SNPs have facilitated the search for disease genes is by providing the physical markers needed to construct detailed genetic linkage maps. A genetic linkage map displays the relative locations of a set of genes. Such genetic linkage maps are based on the frequency with which two alleles are co-inherited—something we can discover by seeing how often the phenotypic traits associated with those alleles show up together in an individual. Genes that lie close to one another on the same chromosome will be inherited together much more frequently than those that lie farther apart. By determining how often recombination separates two genes, the relative distance between them can be calculated (see Panel 19–1, p. 674).

The same sort of analysis can be used to discover linkage between a SNP and a gene. One simply looks for co-inheritance of a SNP variant (detected by DNA sequence analysis) and a gene allele (detected by the phenotypic trait that it gives rise to). Finding linkage indicates that the gene must lie in close proximity to the SNP (Figure 19–33). And because we know the exact location of every SNP we examine, the linkage can tell us the neighborhood in which the gene physically resides. A more detailed analysis of that region—to look for deletions, insertions, or other functionally significant abnormalities in the DNA sequence of affected individuals—can then lead to a precise identification of the critical gene.

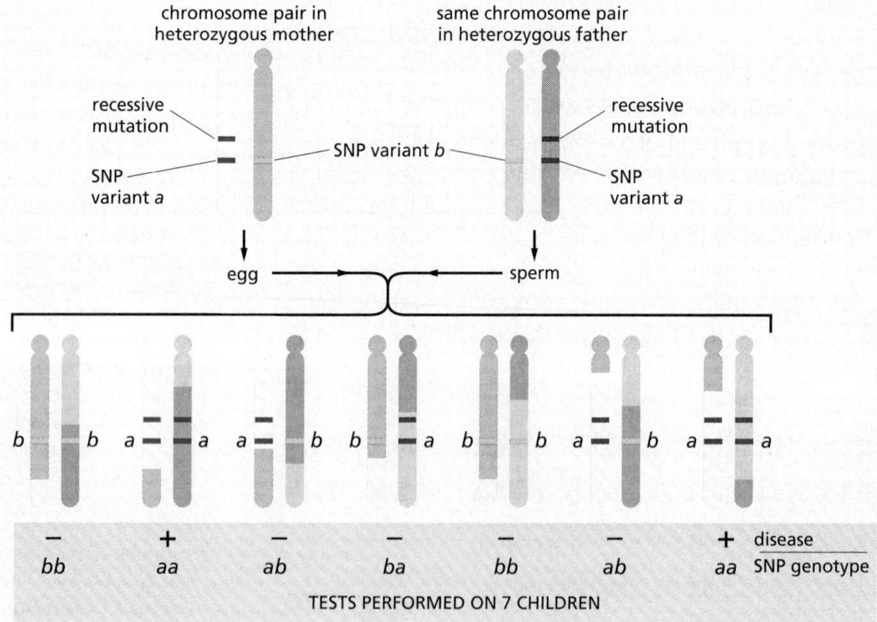

Figure 19–33 **SNP analysis can pin down the location of a mutation that causes a genetic disease.** In this procedure one studies the co-inheritance of a specific human phenotype (here a genetic disease) with a particular set of SNP variants. The figure shows the logic for the common case of a family in which both parents are carriers of a recessive mutation. If individuals who show the disease, and only such individuals, are nearly always homozygous for a particular SNP variant, then the SNP and the recessive mutation causing the disease are likely to be close together on the same chromosome, as shown here. To prove that an apparent linkage is statistically significant, a few dozen individuals from such families may need to be examined. With more individuals and using more SNPs, it is possible to locate the mutation more precisely.

Such linkage analyses are usually carried out in families that are particularly prone to a disorder—the larger the family, the better. And the method works best where there is a simple cause-and-effect relationship, such that a particular mutation directly and reliably determines whether or not a given family member will have the disorder. But most common disorders are not like this: many factors affect the disease risk—some genetic, some environmental, some just a matter of chance. For such conditions, a different approach is needed.

Making Associations

Genetic association studies allow us to discover common genetic variants that affect the risk for a common disease, even if each variant alters susceptibility only slightly. Because mutations that destroy the activity of a key gene are likely to have a disastrous effect on the fitness of the mutant individual, they tend to be eliminated from the population by natural selection and so are rarely seen. Genetic variants that make for slight differences in a gene's function, on the other hand, are much more common. By using genetic association studies to chase down these small but common changes, we can sniff out genes that play key parts in the biology of common diseases.

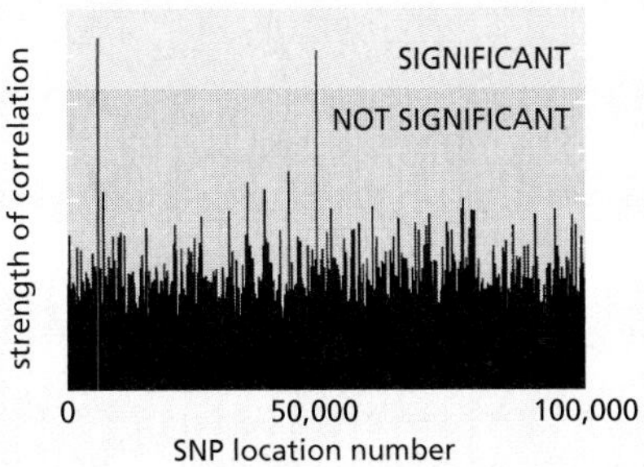

Figure 19–34 **Genetic association studies identify DNA variations that are significantly more frequent in people with a given disease.** In this study, scientists examined more than 100,000 different SNPs in each of 146 people. The x-axis of the graph shows the relative position of each SNP in the genome, starting at the left with the SNPs on chromosome 1. The y-axis shows the strength of each SNP's observed correlation with AMD. The *blue* region indicates a cutoff level for significance corresponding to a probability of less than 5% of finding that strength of correlation by pure chance anywhere among the whole set of 100,000 tested SNPs. The SNP marked in *red* is the one that led the way to the *Cfh* gene. The association of the other prominent SNP with the disease disappeared when additional sequencing at that site was performed, revealing probable errors in the original genotyping. (Adapted from R.J. Klein et al., *Science* 308:385–389, 2005. With permission from AAAS.)

Genetic association studies directly compare the DNA sequences of two populations: individuals who have a particular disease and those who do not. Association studies then look for genetic markers, such as SNP variants, that are present in the people who have the disease more often than would be expected by chance. Those SNP variants may themselves cause the increased likelihood of the disease. Or they may be linked to another polymorphism or mutation that causes the effect.

Take, for example, the case of age-related macular degeneration (AMD), a degenerative disorder that is a leading cause of blindness in the elderly. To search for genetic variations that are associated with AMD, researchers looked at a panel of just over 100,000 SNPs that spanned the entire genome. They determined the sequence at each of these SNPs in 96 people who had AMD, and 50 who did not. Among the 100,000 SNPs, they discovered one particular variant that was present significantly more often in the individuals who had the disease (Figure 19–34).

The SNP is located in a gene called *Cfh* (*complement factor H*). But it falls within one of the gene's introns and appears unlikely to have any effect on the protein product. This SNP itself, therefore, did not seem likely to be the cause of the change in disease susceptibility. But it focused the researchers' attention on the *Cfh* gene. So they resequenced the region to look for additional polymorphisms that might also be inherited more often by people with AMD, along with the SNP that they had already identified. They discovered three variants that would result in a change in the amino acid sequence of the Cfh protein. One of those alleles, which would substitute a histidine for a tyrosine at one particular place in the protein, was strongly associated with the disease (and almost always coupled with the original SNP variant that had put the researchers on the track of the *Cfh* gene). Individuals who carried two copies of this risky allele were five to seven times more likely to develop AMD than those who harbored a different allele at this site.

Several other groups of researchers, using a similar genetic association approach, have also pointed to *Cfh* variants as increasing the likelihood of developing AMD, making it almost certain that this gene has something to do with the biology of the disease. Cfh is part of the complement system, which helps mediate immunity and inflammation. The normal protein serves to rein in the complement system, preventing it from becoming overactive. Interestingly, the environmental risk factors associated with the disease—smoking, obesity, and age—all affect the activity of the complement system. Thus, whatever the detailed mechanism by which the *Cfh* gene influences the risk of AMD, the finding that complement is critical could lead to new tests for the early diagnosis of the disorder, as well as potential new avenues for treatment.

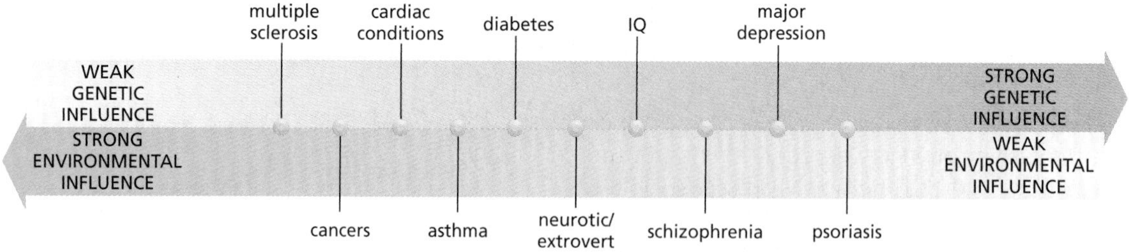

Figure 19–35 **Some human traits are strongly influenced by the environment, others less so.** Studies of identical and fraternal twins have indicated the relative genetic and environmental contributions to different human traits.

This approach has been used to search for genes affecting propensity to common human diseases such as diabetes, coronary artery disease, rheumatoid arthritis, bipolar disease (manic depressive illness), and several others. For all of these conditions, environmental as well as genetic factors play an important part (Figure 19–35). Moreover, for all of them the genetic factors themselves are complex: not one but many genes contribute to the incidence of each disease. By tracking down the genes involved—and the polymorphisms associated with the disease—it becomes possible not only to identify and help those people who run heightened risks, but also to discover more about the molecular mechanisms of the disorders.

Linked Groups of SNPs Define Haplotype Blocks

Of course, the more SNPs one has to analyze to track down the location of a gene, the more costly and laborious is the enterprise. Fortunately, for the purposes of genetic mapping, it is not necessary to examine all 3 million SNPs in the human genome. This is because SNPs are not all independent and uncorrelated: they are linked to their nearest neighbors in blocks—called **haplotype blocks**—that tend to be inherited as a unit. Thus, individuals who carry a particular variant at one polymorphic site are highly likely to carry correlated variants at all of the neighboring polymorphic sites within that haplotype block (Figure 19–36). It is enough, therefore, to examine one or two representative SNPs for each haplotype block.

To understand why haplotype blocks exist, we need to consider our evolutionary history. It is thought that modern humans expanded from a relatively small population that existed in Africa about 60,000 years ago. Among the small group of ancestors from whom we are all descended, some individuals will have carried one set of genetic variants, others a different set. The chromosomes of a present-day human represent a shuffled combination of fragments of the chromosomes of different members of this small ancestral group of people. Because only about two thousand generations separate us from the ancestral population, large segments of these ancestral human chromosomes have passed from parent to child unbroken by the cross-over events that occur during meiosis.

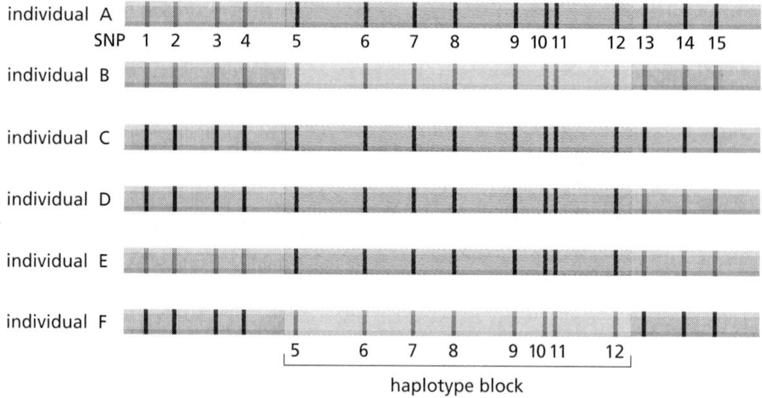

Figure 19–36 **SNPs are commonly inherited in large blocks.** Highlighted in color are 15 nucleotides at which there is polymorphism. For example, at the site labeled SNP1, part of the population might have an A (*blue*), another portion might have a G (*red*). Different people inherit different sets of variants, but SNPs 5 to 12 are usually inherited as a block. That is, if a chromosome contains the *red* variant form of any of the SNPs in this block, it will contain the *red* variant form of all of them, and if it contains the *blue* variant of one it will contain the *blue* variant of all.

(Remember, only a few crossovers occur in each human chromosome per meiosis.) As a result, certain sets of alleles—and DNA markers such as SNPs—are inherited in linked groups. These ancestral chromosome segments—sets of alleles and markers that have been inherited in clusters with little genetic rearrangement across the generations—are the haplotype blocks. Like genes and genetic markers—which exist in different allelic forms—haplotype blocks also come in a limited number of variants that are common in the human population, each representing an allele combination passed down from a particular ancestor long ago.

Haplotype Blocks Give Clues to our Evolutionary History

A detailed examination of haplotype blocks can provide some remarkable insights into the history of human populations. New alleles are continually being generated by mutation; many of these variants will be selectively neutral—that is, they will not affect the reproductive success of the individual—and so they have a chance of becoming common in the population. The more time that has elapsed since the origin of an allele, the smaller should be the haplotype block that surrounds it, because over the course of many generations, crossover events will have had many chances to separate the allele from the SNPs nearby. In fact, by comparing the sizes of haplotype blocks from different human populations, it is possible to estimate how many generations have elapsed since the origin of a specific mutation. In this way, combining genetic comparisons with archeology, we can trace our history from that small set of African ancestors and even draw deductions about the most probable routes our ancestors took when they left the continent (Figure 19–37).

Selectively neutral alleles take a long time to become established in a population, but an allele that is strongly favored by natural selection will sweep through the population and build to a high frequency rapidly. For example, a mutation or variation that makes an organism more resistant to an infection will be favored by selection because organisms with this variation will be more likely to survive and pass the mutation on to their offspring. Again, haplotype analysis can be used to estimate when such historical genetic events occurred. If a favorable mutation cropped up in the population relatively recently, there will have been fewer opportunities for recombination to break up the area around that mutation, so the surrounding haplotype block will be large. Such is the case for two different genes that confer resistance to malaria. The alleles that confer resistance are widespread in African populations, where malaria is rife. And they are embedded in unusually large haplotype blocks, suggesting that these protective variants rose to prominence recently in the African gene pool—probably about 2500 years ago for one of them and about 6500 years ago for the other. In this way, analysis of modern human

QUESTION 19–5

In a recent automated analysis, thousands of SNPs across the genome were analyzed in pooled DNA samples from humans who had been sorted into groups according to their age. For the vast majority of these sites, there was no change in the relative frequencies of different variants as these humans aged. Rarely, however, a particular variant at one position was found to decrease in frequency progressively for people over 50 years old. Which of the possible explanations seems most likely?

A. The nucleotide in that SNP at that position is unstable, and mutates with age.

B. Those people born more than 50 years ago came from a population that tended to lack the disappearing SNP variant.

C. The SNP variant alters an important gene product in a way that shortens the human life span, or is linked to a neighboring allele that has this effect.

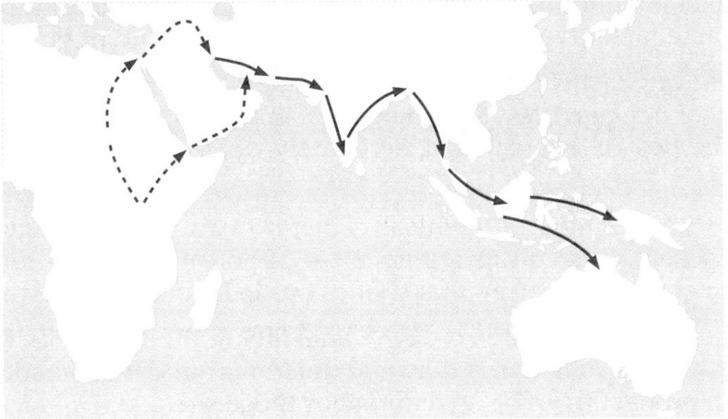

Figure 19–37 **The human populations that are now dispersed around the world originated in Africa about 60,000 to 80,000 years ago.** The map shows the routes the earliest successful migrations are thought to have followed. Dotted lines indicate two alternative routes that may have been taken out of Africa. Studies of the size of haplotype blocks indicate that modern Europeans are descended from a small group of ancestors (a 'population bottleneck') about 30,000 to 50,000 years ago. Haplotype blocks in a Nigerian population are significantly smaller, indicating that the Nigerian population was established a longer time ago. These findings agree with archeological findings, which suggest that the ancestors of modern native Australians (*solid red* arrows)—and of modern European and Middle Eastern populations (migration routes not shown)—reached their destinations about 45,000 years ago. (Modified from P. Forster and S. Matsumura, *Science* 308:965–966, 2005. With permission from AAAS.)

genomes can reveal the ancient history of human exposure to specific infections and of our development of resistance to them.

In revealing the paths along which humans evolved, the map of human haplotypes and SNPs provides a new window into our past; in helping us discover the genes that make us susceptible or resistant to disease, the map may also lead us toward new ways of coping with our present medical problems.

ESSENTIAL CONCEPTS

- Sexual reproduction involves the cyclic alternation of diploid and haploid states: diploid cells divide by meiosis to form haploid gametes, and the haploid gametes from two individuals fuse at fertilization to form a new diploid cell.
- During meiosis, the maternal and paternal chromosomes of a diploid cell are parceled out to gametes so that each gamete receives one copy of each chromosome. Because the assortment of the two members of each chromosome pair occurs at random, many genetically different gametes can be produced from a single individual.
- Crossing-over ensures the proper segregation of homologous chromosomes and enhances the genetic reassortment that occurs during meiosis by exchanging genes between maternal and paternal homologs.
- Although most of the mechanical features of meiosis are similar to those of mitosis, the behavior of the chromosomes is different: meiosis produces four genetically dissimilar haploid cells by two consecutive cell divisions, whereas mitosis produces two genetically identical diploid cells by a single cell division.
- Mendel unraveled the laws of heredity by studying the inheritance of a handful of discrete traits in garden peas.
- Mendel's law of segregation states that the maternal and paternal alleles for each trait separate from one another during gamete formation and then reunite at random during fertilization.
- Mendel's law of independent assortment states that during gamete formation, different alleles segregate independently of each other.
- The behavior of chromosomes during meiosis explains Mendel's laws.
- If two genes are close to each other on a chromosome, they tend to be inherited as a unit. The frequency of recombination between them can be used to construct a genetic map that shows the order of genes on a chromosome.
- Mutant alleles can be either dominant or recessive. If the heterozygous organism has a mutant phenotype, the mutant allele is dominant; if it has a normal phenotype, the mutant allele is recessive.
- Complementation tests reveal whether two mutations that produce the same phenotype lie in the same gene or in different genes.
- Mutant organisms can be generated by treating animals with chemicals that damage DNA. Such mutants can then be screened to identify phenotypes of interest and, ultimately, to isolate the responsible genes.
- With the exception of identical twins, no two human genomes are alike. Each of us carries a unique set of polymorphisms—variations in nucleotide sequence—that shapes our individual phenotypes.
- Polymorphisms are sites of high variability in the genome sequence, such that two individuals drawn at random from the population have a high probability of being different at those sites.

- Single-nucleotide polymorphisms (SNPs) are polymorphisms where the variability is in the choice of a single nucleotide, with two or more variants both being common in the population. SNPs provide useful markers for genetic mapping.
- Human SNPs tend to be inherited in large haplotype blocks—segments of the genome that have been passed down intact from our distant ancestors and in most individuals have not yet been broken up by meiotic recombination. The sizes of haplotype blocks give clues to our evolutionary history.

KEY TERMS

allele
asexual reproduction
bivalent
chiasma (plural chiasmata)
complementation test
complex trait
crossing-over
diploid
fertilization
gain-of-function mutation
gamete
genetic map
genetic screen
genetics
genotype
germ cell
germ line
haploid
haplotype block
heterozygous
homolog
homologous chromosome
homozygous
law of independent assortment
law of segregation
loss-of-function mutation
meiosis
pairing
pedigree
phenotype
polymorphism
recombination
sex chromosome
sexual reproduction
sister chromatid
SNP (single-nucleotide polymorphism)
somatic cell
zygote

QUESTIONS

QUESTION 19–6

It is easy to see how deleterious mutations in bacteria, which have a single copy of each gene, are eliminated by natural selection: the affected bacteria die and the mutation is thereby lost from the population. Eucaryotes, however, have two copies of most genes—that is, they are diploid. Often an individual with two normal copies of the gene (homozygous normal) is indistinguishable in phenotype from an individual with one normal copy and one defective copy of the gene (heterozygous). In such cases, natural selection can operate only against an individual with two copies of the defective gene (homozygous defective). Consider the situation in which a defective form of the gene is lethal when homozygous, but without effect when heterozygous. Can such a mutation ever be eliminated from the population by natural selection? Why or why not?

QUESTION 19–7

Which of the following statements are correct? Explain your answers.

A. The egg and sperm cells of animals contain haploid genomes.

B. During meiosis, chromosomes are allocated so that each germ cell obtains one and only one copy of each of the different chromosomes.

C. Mutations that arise during meiosis are not transmitted to the next generation.

QUESTION 19–8

What might cause chromosome disjunction, where two copies of the same chromosome end up in the same daughter cell? What could be the consequences of this event occurring (a) in mitosis and (b) in meiosis?

QUESTION 19–9

Why do sister chromatids have to remain paired in division I of meiosis? Does the answer suggest a strategy for washing your socks?

QUESTION 19–10

Distinguish between the following genetic terms:

A. Gene and allele.

B. Homozygous and heterozygous.

C. Genotype and phenotype.

D. Dominant and recessive.

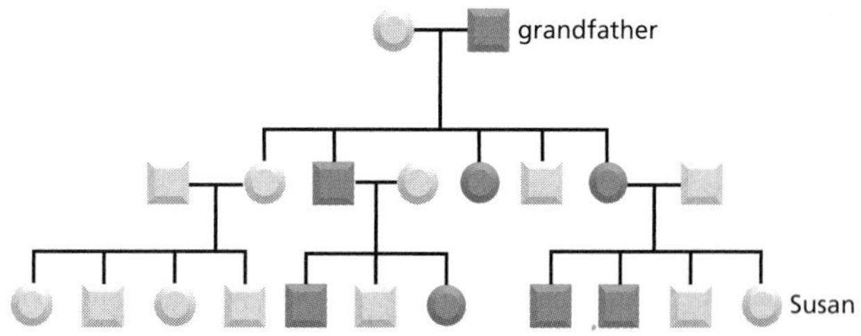

Figure Q19–12

QUESTION 19–11

You have been given three wrinkled peas, which we shall call A, B, and C, each of which you plant to produce a mature pea plant. Each of these three plants, once self-pollinated, produces only wrinkled peas.

A. Given that you know that the wrinkled-pea phenotype is recessive, as a result of a loss-of-function mutation, what can you say about the genotype of each plant?

B. Can you safely conclude that each of the three plants carries a mutation in the same gene?

C. If not, how could you rule out the possibility that each plant carries a mutation in a different gene, each of which gives the wrinkled-pea phenotype?

QUESTION 19–12

Susan's grandfather was deaf, and passed down a hereditary form of deafness within Susan's family as shown in Figure Q19–12.

A. Is this mutation most likely to be dominant or recessive?

B. Is it carried on an autosome or a sex chromosome? Why?

C. A complete SNP analysis has been done for all of the 11 grandchildren (4 affected, and 7 disease-free). In comparing these 11 SNP results, how long a haplotype block would you expect to find around the critical gene? How would you detect it?

QUESTION 19–13

Given that the mutation causing deafness in the family shown in Figure 19–22 is very rare, what is the most probable genotype of each of the four children in generation II?

QUESTION 19–14

In the pedigree shown in Figure Q19–14, the first born in each of three generations is the only person affected by a dominant genetically inherited disease, D. Your friend concludes that the first child born has a greater chance of inheriting the mutant *D* allele than do later children.

children

grandchildren

great grandchildren

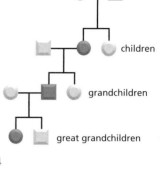

Figure Q19–14

A. According to Mendel's laws, is this conclusion plausible?

B. What is the probability of obtaining this result by chance?

C. What kind of additional data would be needed to test your friend's idea?

D. Is there any way in which your friend's hypothesis might turn out to be right?

QUESTION 19–15

Suppose one person in 100 is a carrier of a nasty recessive mutation, such that babies homozygous for the mutation die soon after birth. In a population where there are 1,000,000 births per year, how many babies per year will be born with the fatal homozygous condition?

QUESTION 19–16

Certain mutations are called *dominant-negative mutations.* What do you think this means and how do you suppose these mutations act? Explain the difference between a dominant-negative mutation and a gain-of-function mutation.

QUESTION 19–17

Discuss the following statement: "We would have no idea today of the importance of insulin as a regulatory hormone if its absence were not associated with the devastating human disease diabetes. It is the dramatic consequences of its absence that focused early efforts on the identification of insulin and the study of its normal role in physiology."

QUESTION 19–18

Early genetic studies in *Drosophila* laid the foundation for our current understanding of genes. *Drosophila* geneticists were able to generate mutant flies with a variety of easily observable phenotypic changes. Alterations from the fly's normal brick-red eye color have a venerable history because the very first mutant found by Thomas Hunt Morgan was a white-eyed fly (Figure Q19–18). Since that

brick-red flies with other eye colors white Figure Q19–18

TABLE Q19–18 COMPLEMENTATION ANALYSIS OF *DROSOPHILA* EYE-COLOR MUTATIONS

MUTATION	white	garnet	ruby	vermilion	cherry	coral	apricot	buff	carnation
white	–	+	+	+	–	–	–	–	+
garnet		–	+	+	+	+	+	+	+
ruby			–	+	+	+	+	+	+
vermilion				–	+	+	+	+	+
cherry					–	–	–	–	+
coral						–	–	–	+
apricot							–	–	+
buff								–	+
carnation									–

+ indicates that progeny of a cross between individuals showing the indicated eye colors are phenotypically normal; – indicates that the eye color of the progeny is abnormal.

time, a large number of mutant flies with intermediate eye colors have been isolated and given names that challenge your color sense: garnet, ruby, vermilion, cherry, coral, apricot, buff, and carnation. The mutations responsible for these eye-color phenotypes are all recessive. To determine whether the mutations affected the same or different genes, homozygous flies for each mutation were bred to one another in pairs and the eye colors of their progeny were noted. In Table Q19–18, a + or a – indicates the phenotype of the progeny flies produced by mating the fly listed at the top of the column with the fly listed to the left of the row; brick-red wild-type eyes are shown as (+) and other colors are indicated as (–).

A. How is it that flies with two different eye colors—ruby and white, for example—can give rise to progeny that all have brick-red eyes?

B. Which mutations are alleles of the same gene and which affect different genes?

C. How can different alleles of the same gene give different eye colors?

QUESTION 19–19

What are single-nucleotide polymorphisms (SNPs), and how can they be used to locate a mutant gene by linkage analysis?

QUESTION 19–20

Tim and John have had their entire genomes sequenced. This allows a list to be made of 3 million sites where their DNA sequences differ by a single nucleotide. Why will many of these sites of variation not prove to be useful DNA markers for genetic mapping experiments? What experiment could you do to distinguish between useful and not useful sites of variation?

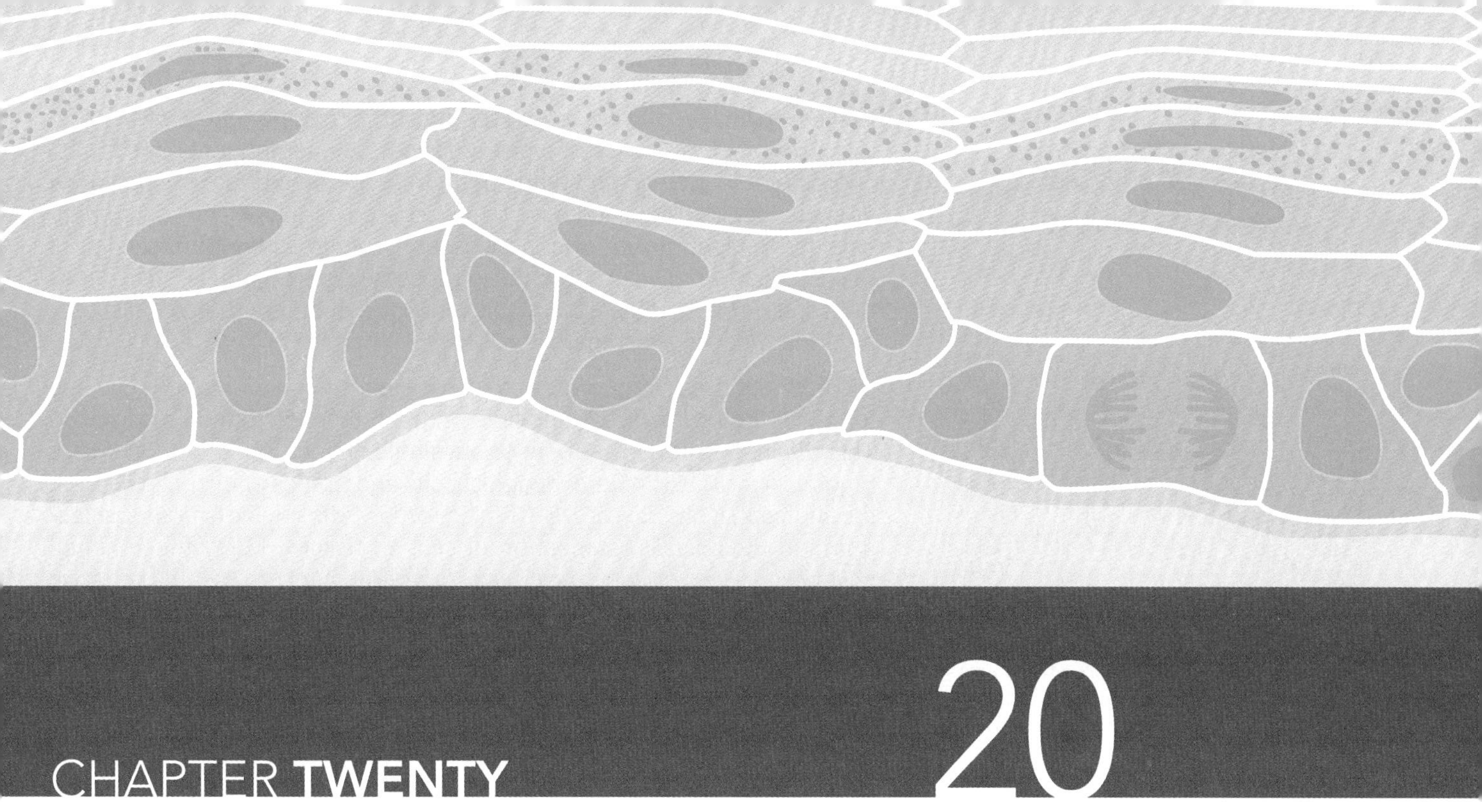

CHAPTER **TWENTY** 20

Cellular Communities: Tissues, Stem Cells, and Cancer

EXTRACELLULAR MATRIX AND CONNECTIVE TISSUES

EPITHELIAL SHEETS AND CELL JUNCTIONS

TISSUE MAINTENANCE AND RENEWAL

CANCER

Cells are the building blocks of multicellular organisms. This seems a simple statement, but it raises deep problems. Cells are not like bricks: they are small and squishy. How can they be used to construct a giraffe or a giant redwood tree? Each cell is enclosed in a flimsy membrane less than a hundred-thousandth of a millimeter thick, and it depends on the integrity of this membrane for its survival. How, then, can cells be joined together robustly to form a muscle that will lift an elephant's weight? Most mysterious of all, if cells are the building blocks, where is the builder and where are the architect's plans? How are all the different cell types in a plant or an animal produced, with each in its proper place in an elaborate pattern (Figure 20–1)?

Most of the cells in multicellular organisms are organized into cooperative assemblies called **tissues**, such as the nervous, muscle, epithelial, and connective tissues found in vertebrates (Figure 20–2). In this chapter, we begin by discussing the architecture of tissues from a mechanical point of view. We will see that tissues are composed not only of cells, with their internal framework of cytoskeletal filaments (discussed in Chapter 17), but also of **extracellular matrix**, which cells secrete around themselves; it is this matrix that gives supportive tissues such as bone or wood their strength. The matrix provides one way to bind cells together, but cells can also attach to one another directly. Thus, we also discuss the *cell junctions* that link cells together in the flexible, mobile tissues of animals, transmitting forces from the cytoskeleton of one cell to that of the next, or from the cytoskeleton of a cell to the extracellular matrix.

But there is more to the organization of tissues than mechanics. Just as a building needs plumbing, telephone lines, and other fittings, so an animal

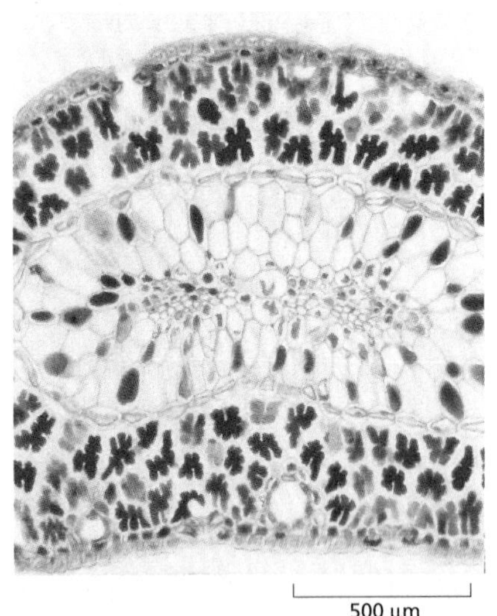

Figure 20–1 **Multicellular organisms are built from organized collections of cells.** Stained cross section through a leaf of a pine tree (a pine needle), showing the precisely organized pattern of different cell types.

tissue requires blood vessels, nerves, and other components formed from a variety of specialized cell types. All the tissue components have to be coordinated correctly, and many of them require continual maintenance and renewal. Cells die and have to be replaced with new cells of the right type, in the right places, and in the right numbers. In the third section of this chapter, we discuss how these processes are organized, as well as the crucial role that *stem cells*, self-renewing undifferentiated cells, play in tissue renewal and repair.

Disorders of tissue renewal are a major medical concern, and those due to the misbehavior of mutant cells underlie the development of *cancer*. This disorder will be the topic of our final section. The study of cancer requires a synthesis of knowledge of cells and tissues at every level, from the molecular biology of DNA repair to the principles of natural selection and the social organization of cells in tissues. Many fundamental advances in cell biology have been driven by cancer research, and we will see that, in return, the basic science has borne fruit in a deepened understanding of the disease and a new optimism about its treatment.

EXTRACELLULAR MATRIX AND CONNECTIVE TISSUES

Plants and animals have evolved their multicellular organization independently, and their tissues are constructed on different principles. Animals prey on other living things—and often are preyed on by other animals—and for this they must be strong and agile: they must possess tissues capable of rapid movement, and the cells that form those tissues must be able to generate and transmit forces and to change shape quickly. Plants, by contrast, are sedentary, their tissues are more or less rigid, and their cells are weak and fragile if isolated from their supporting tissue framework.

The strength of a plant tissue comes from the **cell walls**, formed like boxes, that enclose, protect, and constrain the shape of each of its cells (Figure 20–3). The cell wall is a type of extracellular matrix that the plant cell secretes around itself. The cell controls the composition of this material: it can be thick and hard, as in wood, or thin and flexible, as in a leaf. But the principle of tissue construction is the same in either case: many tiny boxes are cemented together, with a delicate cell living inside each one. Indeed, as we noted in Chapter 1, it was this close-packed mass of microscopic chambers, seen in a slice of cork by Robert Hooke three centuries ago, that originally gave rise to the term "cell."

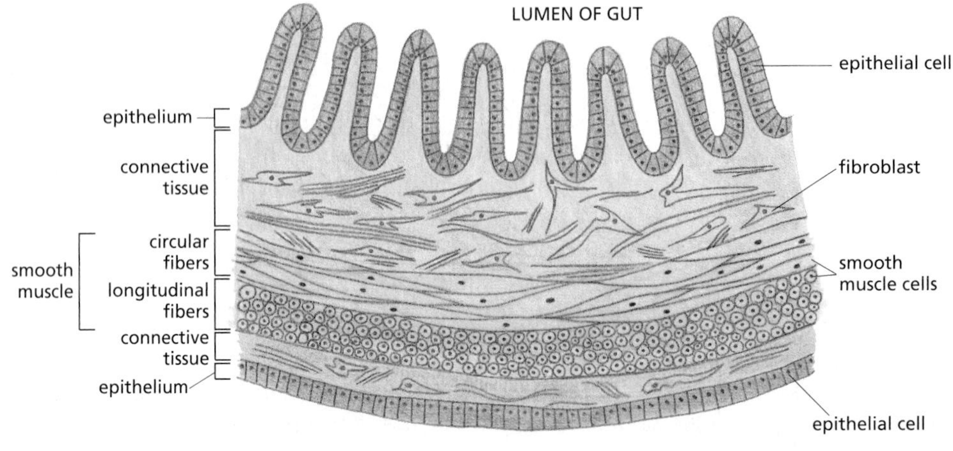

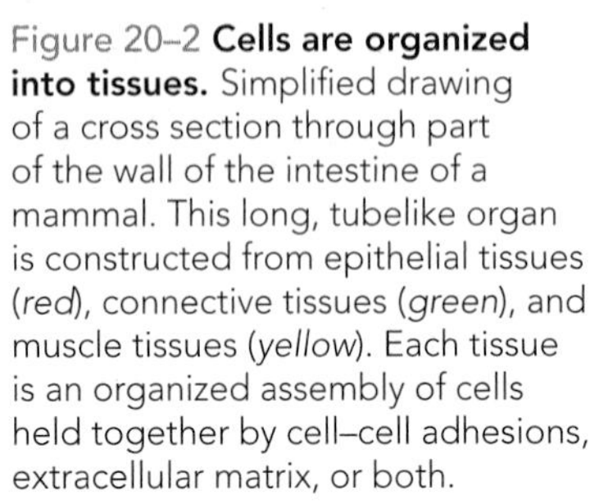
Figure 20–2 **Cells are organized into tissues.** Simplified drawing of a cross section through part of the wall of the intestine of a mammal. This long, tubelike organ is constructed from epithelial tissues (*red*), connective tissues (*green*), and muscle tissues (*yellow*). Each tissue is an organized assembly of cells held together by cell–cell adhesions, extracellular matrix, or both.

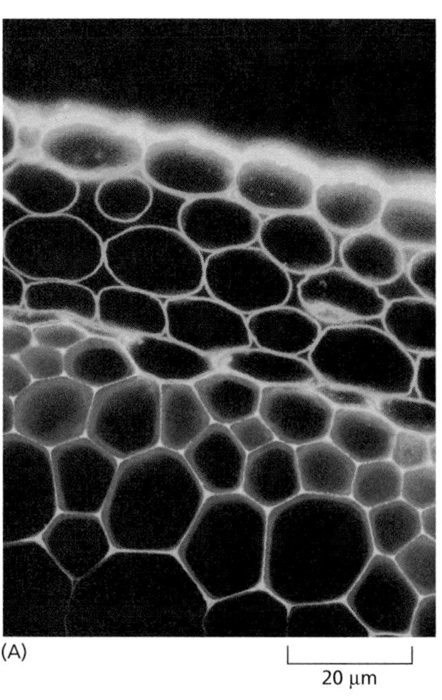

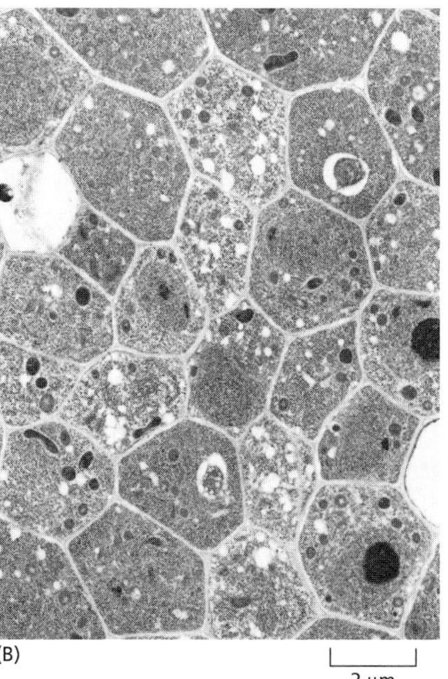

Figure 20–3 **Plant tissues are strengthened by the plant cell wall.** (A) A cross section of part of the stem of the flowering plant *Arabidopsis* is shown, stained with fluorescent dyes that label two different cell wall components—cellulose in *blue*, and another polysaccharide (pectin) in *green*. The cells themselves are unstained and invisible in this preparation. Regions rich in both cellulose and pectin appear white. Pectin predominates in the outer layers of cells, which have only primary cell walls (deposited while the cell is still growing). Cellulose is more plentiful in the inner layers, which have thicker, more rigid secondary cell walls (deposited after cell growth has ceased). (B) The cells and their walls are clearly seen in this electron micrograph of young cells in the root of the same plant. (Courtesy of Paul Linstead.)

Animal tissues are more diverse. Like plant tissues, they consist of both cells and extracellular matrix, but these components are organized in many different ways. In some tissues, such as bone or tendon, extracellular matrix is plentiful and mechanically all-important; in others, such as muscle or epidermis, extracellular matrix is scanty, and the cytoskeletons of the cells themselves carry the mechanical load. We begin with a brief discussion of plant cells and tissues, before moving on to those of animals.

Plant Cells Have Tough External Walls

A naked plant cell, artificially stripped of its wall, is a delicate and vulnerable thing. With care, it can be kept alive in culture; but it is easily ruptured, and even a small maladjustment of the osmotic strength of the culture medium can cause it to swell and burst. Its cytoskeleton lacks the tension-bearing intermediate filaments found in animal cells, and it has virtually no tensile strength. An external wall, therefore, is essential.

The plant cell wall has to be tough, but it does not necessarily have to be rigid. Osmotic swelling of the cell, limited by the resistance of the cell wall, can keep the chamber distended, and a mass of such swollen chambers cemented together forms a semirigid tissue. Such is the state of a crisp lettuce leaf (Figure 20–4). If water is lacking so that the cells shrink, the leaf wilts.

Most newly formed cells in a multicellular plant initially make relatively thin *primary cell walls* that can slowly expand to accommodate cell growth. The driving force for growth is the same as that keeping the lettuce leaf crisp—a swelling pressure, called the turgor pressure, that develops as the result of an osmotic imbalance between the interior of the cell and its surroundings (discussed in Chapter 12). Once growth stops and the wall no longer needs to expand, a more rigid *secondary cell wall* is often produced, either by thickening of the primary wall or by deposition of new layers with a different composition underneath the old ones. When plant cells become specialized, they generally produce specially adapted types of walls: waxy, waterproof walls for the surface epidermal cells of a leaf; hard, thick, woody walls for the xylem cells of the stem; and so on.

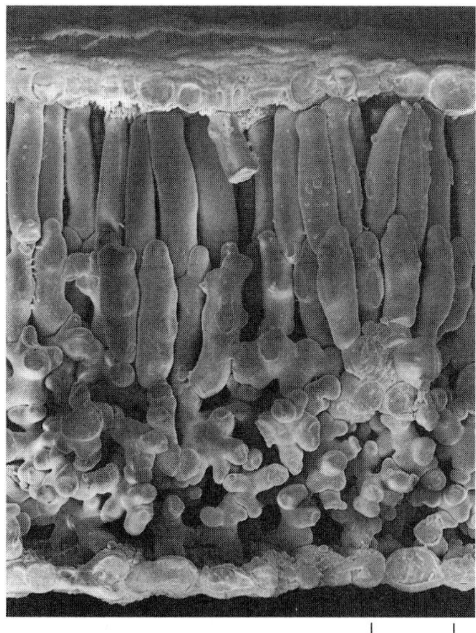

Figure 20–4 **A scanning electron micrograph shows the cells in a crisp lettuce leaf.** The cells, swollen by osmotic forces, are stuck together via their walls. (Courtesy of Kim Findlay.)

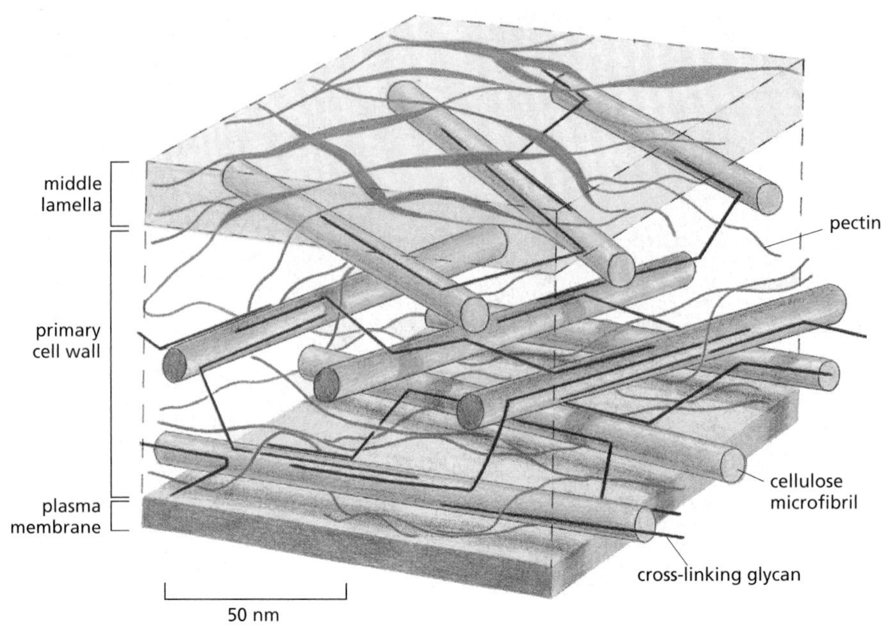

Figure 20–5 **Scale model shows a portion of a primary plant cell wall.** The *green* bars represent cellulose microfibrils, providing tensile strength. Other polysaccharides cross-link the cellulose microfibrils (*red* strands), while the polysaccharide pectin (*blue* strands) fills the spaces between them, providing resistance to compression. The middle lamella *(yellow)* is rich in pectin and is the layer that cements one cell wall to another.

Cellulose Microfibrils Give the Plant Cell Wall Its Tensile Strength

Like all extracellular matrices, plant cell walls derive their tensile strength from long fibers oriented along the lines of stress. In higher plants, the long fibers are generally made from the polysaccharide *cellulose*, the most abundant organic macromolecule on Earth. These **cellulose microfibrils** are interwoven with other polysaccharides and some structural proteins, all bonded together to form a complex structure that resists compression as well as tension (Figure 20–5). In woody tissue, a highly cross-linked network of *lignin* (which is not a polysaccharide or a protein but a different kind of polymer) is deposited within this matrix to make it more rigid and waterproof.

For a plant cell to grow or change its shape, the cell wall has to stretch or deform. Because the cellulose microfibrils resist stretching, their orientation governs the direction in which the growing cell enlarges: if, for example, they are arranged circumferentially as a corset, the cell will grow more readily in length than in girth (Figure 20–6). By controlling the way that it lays down its wall, the plant cell consequently controls its own shape and thus the direction of growth of the tissue to which it belongs.

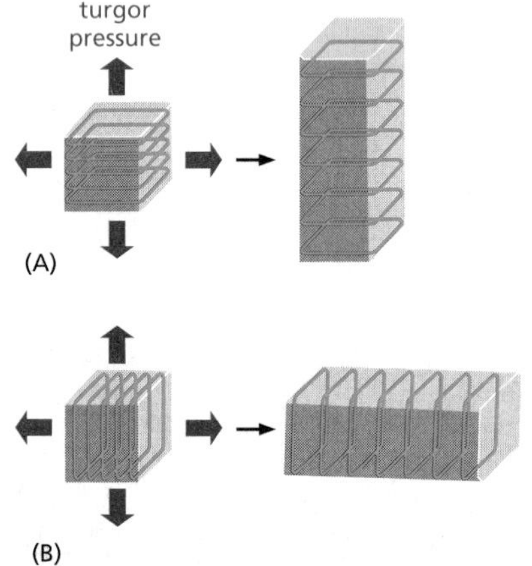

Figure 20–6 **The orientation of cellulose microfibrils within the plant cell wall influences the direction in which the cell elongates.** The cells in (A) and (B) start off with identical shapes (shown here as *cubes*) but with different orientations of cellulose microfibrils in their walls. Although turgor pressure is uniform in all directions, each cell tends to elongate in a direction perpendicular to the orientation of the microfibrils, which have great tensile strength. The final shape of an organ, such as a shoot, is determined by the direction in which its cells expand.

Cellulose is produced in a radically different way from most other extracellular macromolecules. Instead of being made inside the cell and then exported by exocytosis (discussed in Chapter 15), it is synthesized on the outer surface of the cell by enzyme complexes embedded in the plasma membrane. These complexes transport sugar monomers across the membrane and incorporate them into a set of growing polymer chains at their points of membrane attachment. Each set of chains forms a cellulose microfibril. The enzyme complexes move in the membrane, spinning out new polymers and laying down a trail of oriented cellulose microfibrils behind them.

The paths followed by the enzyme complexes dictate the orientation in which cellulose is deposited in the cell wall; but what guides the enzyme complexes? Just underneath the plasma membrane, microtubules are aligned exactly with the cellulose microfibrils outside the cell (Figure 20–7A, B). These microtubules are thought to serve as tracks to guide

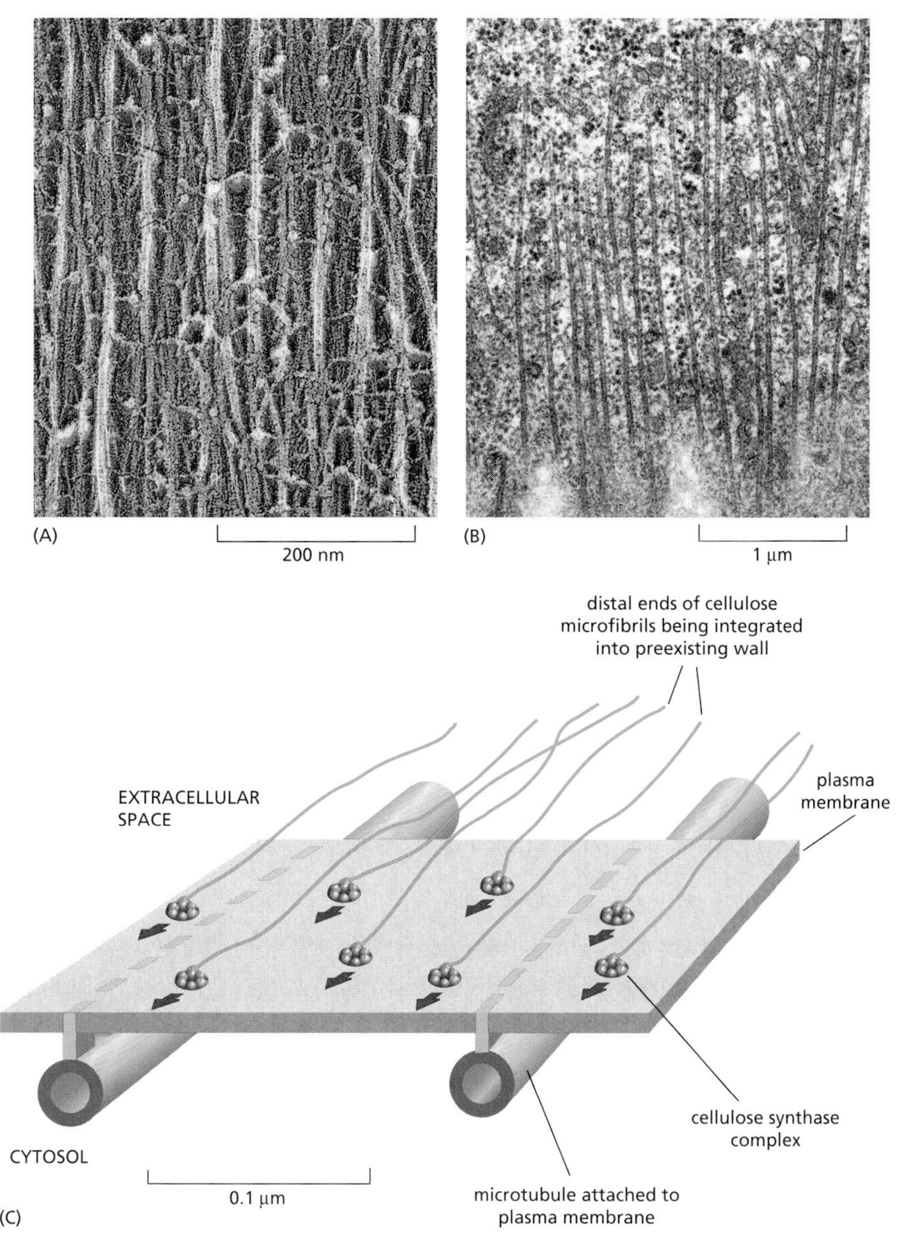

Figure 20–7 **Microtubules direct the deposition of cellulose in the plant cell wall.** (A) Oriented cellulose microfibrils in a plant cell wall, shown by electron microscopy. (B) Oriented microtubules just beneath a plant cell's plasma membrane. (C) One model of how the orientation of the newly deposited extracellular cellulose microfibrils might be determined by the orientation of the intracellular microtubules. The large *cellulose synthase* enzyme complexes are integral membrane proteins that continuously synthesize cellulose microfibrils on the outer face of the plasma membrane. The outer ends of the stiff microfibrils become integrated into the texture of the wall, and their elongation at the other end pushes the synthase complex along in the plane of the membrane. Because the cortical array of microtubules is attached to the plasma membrane in a way that confines the enzyme complex to defined membrane tracks, the microtubule orientation determines the direction in which the microfibrils are laid down. (A, courtesy of Brian Wells and Keith Roberts; B, courtesy of Brian Gunning.)

the movement of the enzyme complexes (Figure 20–7C). In this curiously indirect way, the cytoskeleton controls the shape of the plant cell and the modeling of the plant tissues. We shall see that animal cells use their cytoskeleton to control tissue architecture in a much more direct way.

QUESTION 20–1

Cells in the stem of a seedling that is grown in the dark orient their microtubules horizontally. How would you expect this to affect the growth of the plant?

Animal Connective Tissues Consist Largely of Extracellular Matrix

It is traditional to distinguish four major types of tissues in animals: connective, epithelial, nervous, and muscular. But the basic architectural distinction is between connective tissues and the rest. In **connective tissues**, extracellular matrix is plentiful and carries the mechanical load. In other tissues, such as epithelia, extracellular matrix is scanty, and the cells are directly joined to one another and carry the mechanical load themselves. We discuss connective tissues first.

Animal connective tissues are enormously varied. They can be tough and flexible like tendons or the dermis of the skin; hard and dense like bone; resilient and shock-absorbing like cartilage; or soft and transparent like

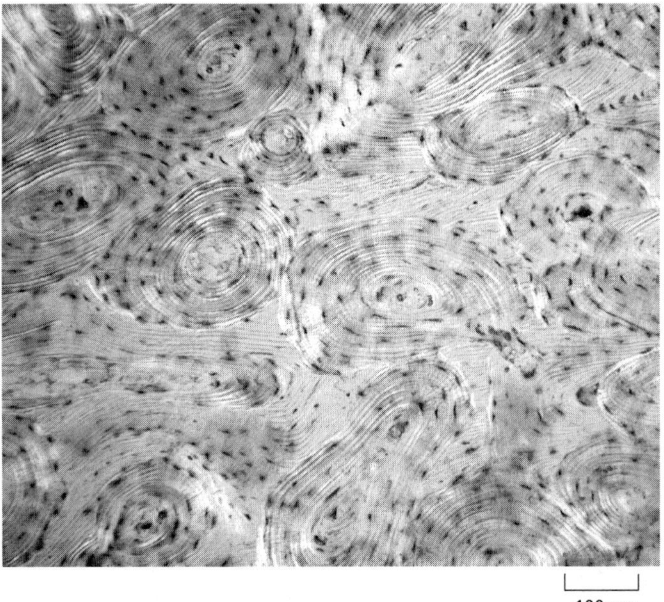

Figure 20–8 **Extracellular matrix is plentiful in connective tissue such as bone.** Cells in this cross section of bone appear as small, dark, ant-like objects embedded in the bone matrix, which occupies most of the volume of the tissue and provides all its mechanical strength. The alternating light and dark bands are layers of matrix containing oriented collagen (made visible with the help of polarized light). Calcium phosphate crystals filling the interstices between the collagen fibrils make bone matrix resistant to compression as well as tension, like reinforced concrete.

the jelly that fills the interior of the eye. In all these examples, the bulk of the tissue is occupied by extracellular matrix, and the cells that produce the matrix are scattered within it like raisins in a pudding (Figure 20–8). In all of these tissues, the tensile strength—whether great or small—is chiefly provided not by a polysaccharide, as in plants, but by a fibrous protein: collagen. The various types of connective tissues owe their specific characters to the type of collagen that they contain, to its quantity, and, most importantly, to the other molecules that are interwoven with it in varying proportions. These include the rubbery protein elastin, which gives the walls of arteries their resilience as blood pulses through them, as well as a host of specialized polysaccharide molecules, which we discuss shortly.

Collagen Provides Tensile Strength in Animal Connective Tissues

Collagen is found in all multicellular animals, and it comes in many varieties. Mammals have about 20 different collagen genes, coding for the variant forms of collagen required in different tissues. Collagens are the chief proteins in bone, tendon, and skin (leather is pickled collagen), and they constitute 25% of the total protein mass in a mammal—more than any other type of protein.

The characteristic feature of a typical collagen molecule is its long, stiff, triple-stranded helical structure, in which three collagen polypeptide chains are wound around one another in a ropelike superhelix (see Figure 4–25A). These molecules in turn assemble into ordered polymers called collagen fibrils, which are thin cables 10–300 nm in diameter and many micrometers long; these can pack together into still thicker collagen fibers (Figure 20–9). Other collagen molecules decorate the surface of collagen fibrils and link the fibrils to one another and to other components in the extracellular matrix.

The connective-tissue cells that manufacture and inhabit the matrix go by various names according to the tissue: in skin, tendon, and many other connective tissues they are called **fibroblasts** (Figure 20–10); in bone they are called *osteoblasts*. They make both the collagen and the other organic components of the matrix. Almost all of these molecules are synthesized intracellularly and then secreted in the standard way, by exocytosis (dis-

QUESTION 20–2

Mutations in the genes encoding collagens often have detrimental consequences, resulting in severely crippling diseases. Particularly devastating are mutations that change glycines, which are required at every third position in the collagen polypeptide chain so that it can assemble into the characteristic triple-helical rod (see Figure 20–9).

A. Would you expect collagen mutations to be detrimental if only one of the two copies of a collagen gene is defective?

B. A puzzling observation is that the change of a glycine residue into another amino acid is most detrimental if it occurs toward the amino terminus of the rod-forming domain. Suggest an explanation for this.

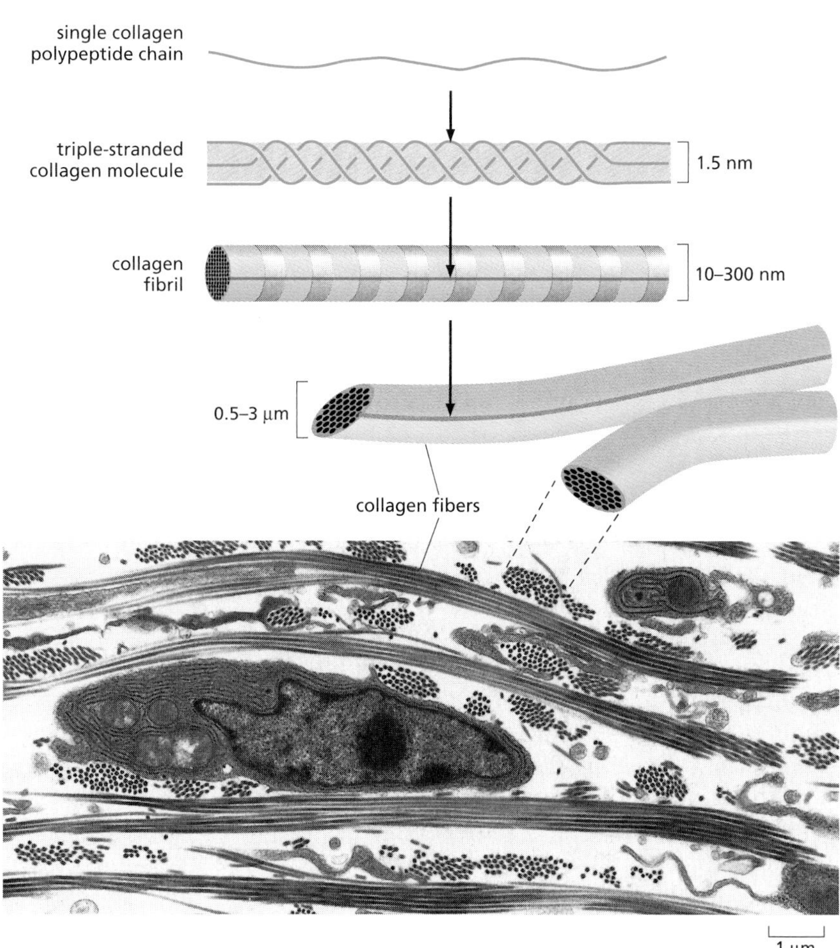

Figure 20–9 **Collagen fibrils are organized into bundles.** The drawings show the steps of collagen assembly, from individual polypeptide chains to triple-stranded collagen molecules, then to fibrils and, finally, fibers. The electron micrograph shows fully assembled collagen in the connective tissue of embryonic chick skin. The fibrils are organized into bundles (fibers), some running in the plane of the section, others approximately at right angles to it. The cell in the photograph is a fibroblast, which secretes the collagen as well as other extracellular matrix components. (Photograph from C. Ploetz et al., *J. Struct. Biol.* 106:73–81, 1991. With permission from Elsevier.)

cussed in Chapter 15). Outside the cell, they assemble into huge, cohesive aggregates. If assembly were to occur prematurely, before secretion, the cell would become choked with its own products. In the case of collagen, the cells avoid this catastrophe by secreting collagen molecules in a precursor form, called procollagen, with additional peptides at each end that obstruct assembly into collagen fibrils. Extracellular enzymes—called procollagen proteinases—cut off these terminal domains to allow assembly only after the molecules have emerged into the extracellular space.

Some people have a genetic defect in one of these proteinases, or in procollagen itself, so that their collagen fibrils do not assemble correctly. As a result, the skin and various other connective tissues have a lower tensile strength and are extraordinarily stretchable (Figure 20–11).

Cells in tissues have to be able to degrade matrix as well as make it. This ability is essential for tissue growth, repair, and renewal; it is also important where migratory cells, such as macrophages, need to burrow through the thicket of collagen and other extracellular matrix polymers. Matrix proteases that cleave extracellular proteins play a part in many disease processes, ranging from arthritis, where they contribute to the breakdown of cartilage in affected joints, to cancer, where they help the cancer cells to invade normal tissue.

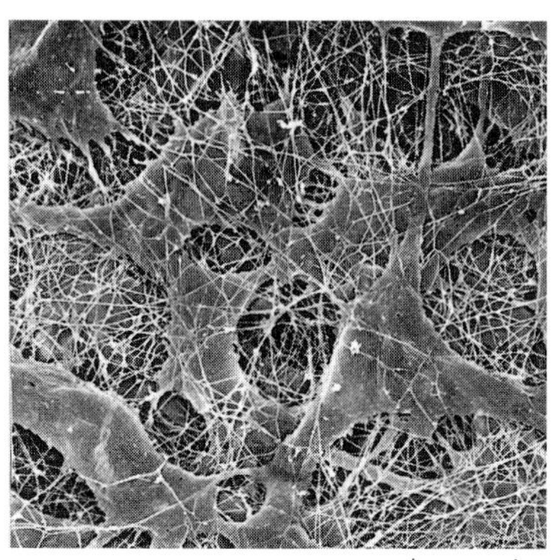

Figure 20–10 **Fibroblasts produce the extracellular matrix of connective tissue.** A scanning electron micrograph showing fibroblasts and collagen in connective tissue from the cornea of a rat. Other components that normally form a hydrated gel filling the spaces between the collagen fibrils have been removed by enzyme and acid treatment. (From T. Nishida et al., *Invest. Ophthalmol. Vis. Sci.* 29:1887–1890, 1988. With permission from ARVO.)

Figure 20–11 **Incorrect collagen assembly can cause hyperextensible skin.** James Morris, "the elastic skin man," from a photograph taken in about 1890. Abnormally stretchable skin is part of a genetic syndrome that results from a defect in the assembly or cross-linking of collagen. In some individuals, this arises from a lack of an enzyme that converts procollagen to collagen.

Cells Organize the Collagen That They Secrete

To do their job, collagen fibrils must be correctly aligned. In skin, for example, they are woven in a wickerwork pattern, or in alternating layers with different orientations so as to resist tensile stress in multiple directions (Figure 20–12). In tendons, which attach muscles to bone, they are aligned in parallel bundles along the major axis of tension.

The connective-tissue cells control this orientation, first by depositing the collagen in an oriented fashion and then by rearranging it. During development of the tissue, fibroblasts work on the collagen they have secreted, crawling over it and pulling on it—helping to compact it into sheets and draw it out into cables. This mechanical role of fibroblasts in shaping collagen matrices has been demonstrated dramatically in cell culture. When fibroblasts are mixed with a meshwork of randomly oriented collagen fibrils that form a gel in a culture dish, the fibroblasts tug on the meshwork, drawing in collagen from their surroundings and compacting it. If two small pieces of embryonic tissue containing fibroblasts are placed far apart on a collagen gel, the intervening collagen becomes organized into a dense band of aligned fibers that connect the two explants (Figure 20–13). The fibroblasts migrate out from the explants along the aligned collagen fibers. Thus, the fibroblasts influence the alignment of the collagen fibers, and the collagen fibers in turn affect the distribution of the fibroblasts. Fibroblasts presumably play a similar role in generating long-range order in the extracellular matrix inside the body—in helping to create tendons, for example, and the tough, dense layers of connective tissue that ensheathe and bind together most organs. Fibroblast migration is also important for healing wounds (Movie 20.1).

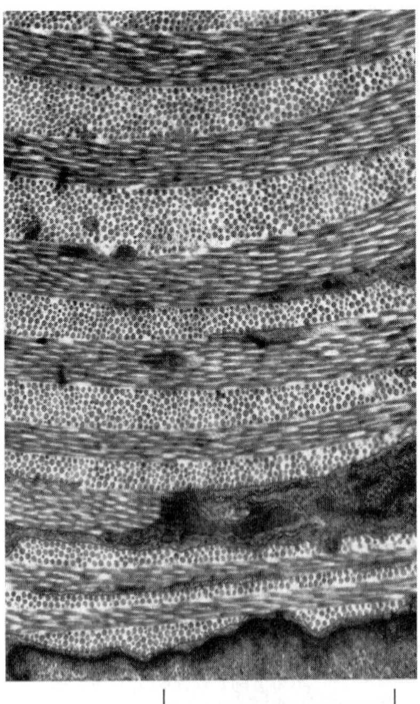

Figure 20–12 **Collagen fibrils in skin are arranged in a plywoodlike pattern.** The electron micrograph shows a cross section of tadpole skin. Successive layers of fibrils are laid down nearly at right angles to each other (see also Figure 20–9). This arrangement is also found in mature bone and in the cornea. (Courtesy of Jerome Gross.)

Integrins Couple the Matrix Outside a Cell to the Cytoskeleton Inside It

If cells are to pull on the matrix and crawl over it, they must be able to attach to it. Cells do not attach well to bare collagen. Another extracellular matrix protein, **fibronectin**, provides a linkage: one part of the fibronectin molecule binds to collagen, while another part forms an attachment site for a cell (Figure 20–14A).

A cell attaches itself to this specific site in fibronectin by means of a receptor protein, called an **integrin**, which spans the cell's plasma membrane. While the extracellular domain of the integrin binds to fibronectin, the intracellular domain binds (through a set of adaptor molecules) to actin filaments inside the cell (Figure 20–14C). Thanks to this anchorage, instead of being ripped out of the flimsy lipid bilayer membrane when there is tension between the cell and the matrix, the integrin molecule transmits that stress to the sturdier cytoskeleton.

Integrins do more than passively transmit stress: they also react to stress—and to chemical signals from inside and outside the cell that direct them to maintain their attachment to other molecules or to let go. Integrins form and break attachments, for example, as a cell crawls through a tissue, grabbing hold of the matrix at its front end and releasing its grip at the rear (see Figure 17–32).

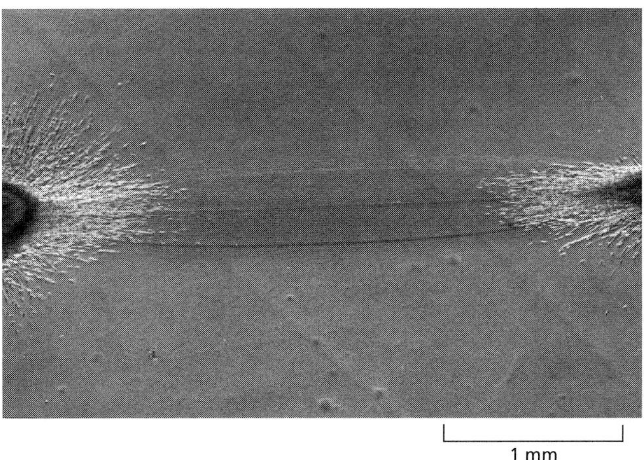

Figure 20–13 **Fibroblasts influence the alignment of collagen fibers.** This micrograph shows a region between two pieces of embryonic chick heart (rich in fibroblasts as well as heart muscle cells) that have grown in culture on a collagen gel for four days. A dense tract of aligned collagen fibers has formed between the explants, presumably as a result of the fibroblasts in the explants tugging on the collagen. Elsewhere in the culture dish the collagen remains disorganized and unaligned, so that it appears uniformly gray. (From D. Stopak and A.K. Harris, *Dev. Biol.* 90:383–398, 1982. With permission from Elsevier.)

Integrins perform these functions by undergoing remarkable conformational changes. Binding to a molecule on one side of the membrane causes the integrin molecule to stretch out into an extended, activated state so that it can then latch onto another molecule on the opposite side—an effect that operates in both directions across the membrane (Figure 20–15).

These conformational changes in integrins are used to transmit chemical as well as mechanical signals across the cell membrane. An intracellular signaling molecule can activate the integrin from inside the cell, causing it to reach out and grab hold of an extracellular structure. And binding to an external structure can activate intracellular signaling cascades via protein kinases that associate with the intracellular end of the integrin molecule. In this way, the external attachments that a cell makes help

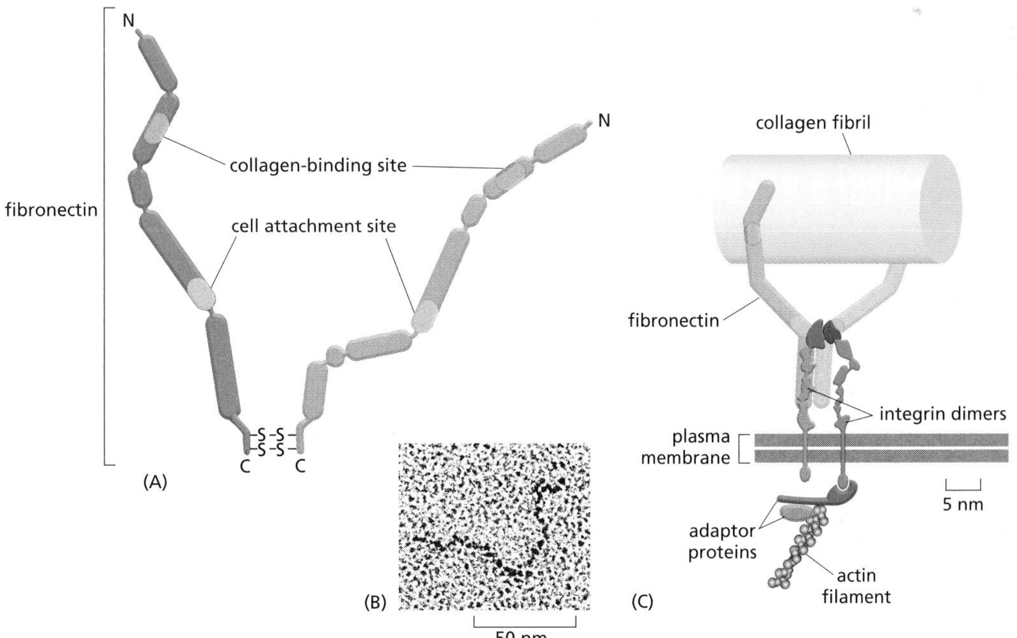

Figure 20–14 **Fibronectin and integrin molecules attach a cell to the extracellular matrix.** Fibronectin molecules outside the cell bind to collagen fibrils. Integrins in the cell membrane bind to the fibronectin and tether it to the cytoskeleton inside the cell. (A) Diagram and (B) electron micrograph of a molecule of fibronectin. (C) The transmembrane linkage mediated by an integrin molecule (*blue* and *green* dimer). The integrin molecule transmits tension across the plasma membrane: it is anchored inside the cell to the cytoskeleton and externally via fibronectin to other extracellular matrix proteins. The plasma membrane itself does not have to be strong. The integrin shown here links fibronectin to an actin filament inside the cell, but other integrins connect different extracellular proteins to the cytoskeleton (usually to actin filaments, but sometimes to intermediate filaments). (B, from J. Engel et al., *J. Mol. Biol.* 150:97–120, 1981. With permission from Elsevier.)

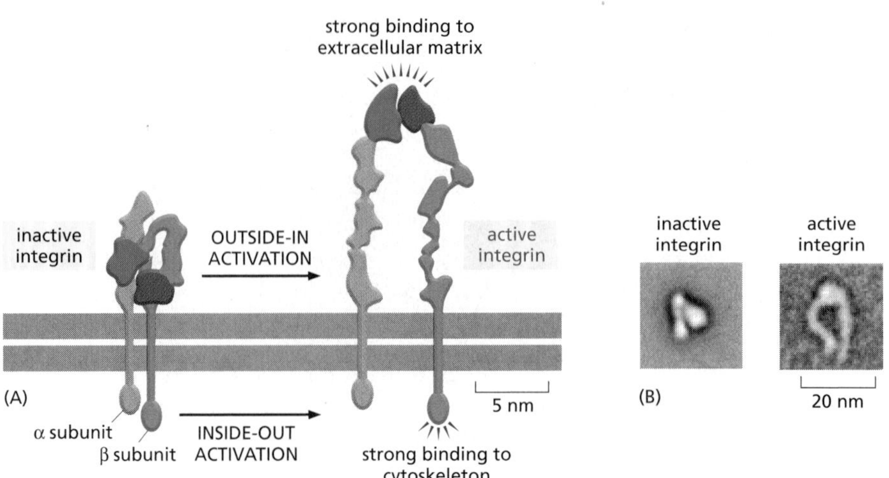

Figure 20–15 **An integrin molecule switches to an active conformation when it binds to molecules at either of its ends.** (A) The integrin molecule consists of two different subunits, α and β, and it can switch between a folded, inactive form and an extended, active form. The switch to the activated state can be triggered by binding to an extracellular matrix molecule (such as fibronectin) or to intracellular proteins that then link it to the cytoskeleton. In either case, the conformational change alters the molecule so that its opposite end rapidly forms a counterbalancing attachment to the appropriate structure. In this way, the integrin creates a mechanical linkage that spans the membrane. (B) Electron micrographs of single integrin molecules, showing how their shape changes in response to a small peptide that mimics an extracellular ligand to which integrin binds (the peptide itself is too small to be visible). (A, based on T. Xiao at al., *Nature* 432:59–67, 2004. With permission from Macmillan Publishers Ltd.; B, from J. Takagi et al., *Cell* 110:599–611, 2002. With permission from Elsevier.)

to regulate whether it lives or dies, and—if it does survive—whether it grows, divides, or differentiates.

Humans make at least 24 different kinds of integrins, which recognize different extracellular structures and have distinct functions depending on which type of cell they reside in. For example, the integrins on white blood cells help those cells to crawl out of blood vessels at sites of infection so as to deal with the marauding microbes. People who lack this type of integrin develop a disease called *leucocyte adhesion deficiency* and suffer from repeated bacterial infections. A different form of integrin is found in blood platelets. Individuals who lack this integrin bleed excessively because their platelets cannot bind to the necessary clotting factor in the extracellular matrix.

Gels of Polysaccharide and Protein Fill Spaces and Resist Compression

While collagen provides tensile strength to resist stretching, a completely different group of macromolecules in the extracellular matrix of animal tissues provides the complementary function, resisting compression and serving as space-fillers. These are the **proteoglycans**, extracellular proteins linked to a special class of complex negatively charged polysaccharides, the **glycosaminoglycans** (**GAGs**) (Figure 20–16). Proteoglycans are extremely diverse in size, shape, and chemistry. Typically, many GAG chains are attached to a single core protein, which may in turn be linked at one end to another GAG, creating an enormous macromolecule resembling a bottlebrush, with a molecular weight in the millions of daltons (Figure 20–17).

In dense, compact connective tissues such as tendon and bone, the proportion of GAGs is small, and the matrix consists almost entirely of

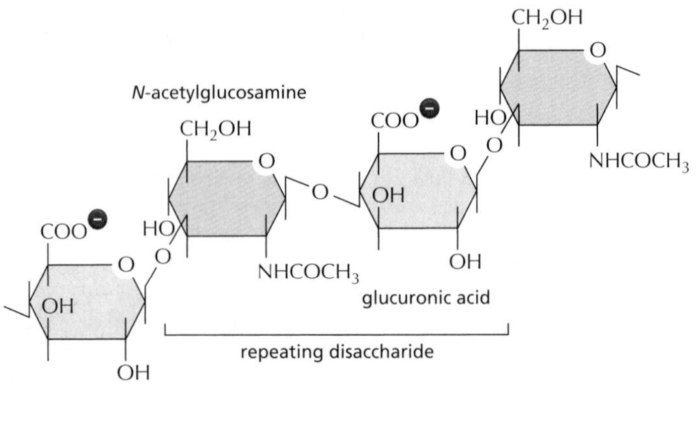

Figure 20–16 **Glycosaminoglycans (GAGs) help to fill space in the extracellular matrix of connective tissues.** Hyaluronan, a relatively simple GAG, is shown here. It consists of a single long chain of up to 25,000 repeated disaccharide units, each carrying a negative charge. As in other GAGs, one of the sugar monomers in each disaccharide unit is an amino-sugar. Many GAGs have additional negatively charged side groups, especially sulfate groups.

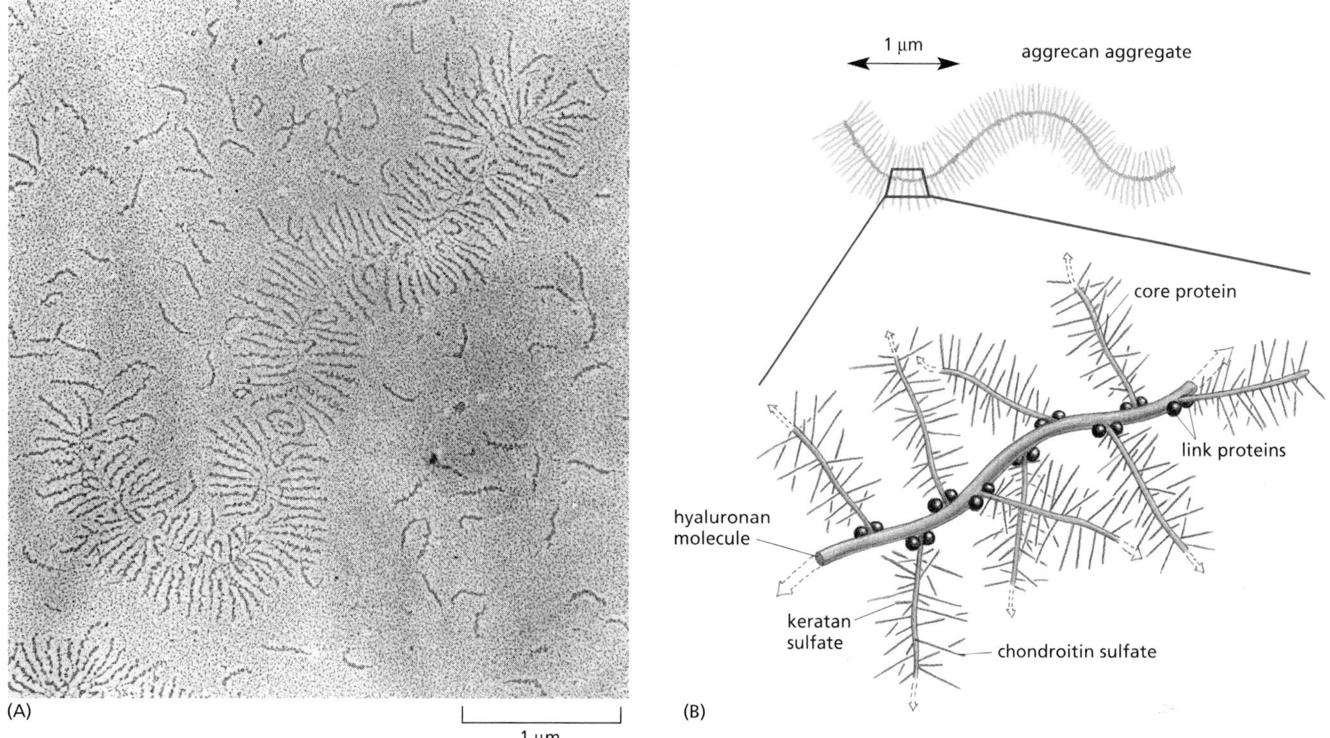

Figure 20–17 **Proteoglycans and GAGs can form large aggregates.** (A) Electron micrograph of an aggregate from cartilage spread out on a flat surface. Many free subunits—themselves large proteoglycan molecules—can also be seen. (B) Schematic drawing of the giant aggregate illustrated in (A), showing how it is built up from GAGs (*red* and *blue*) and proteins (*green* and *black*). The molecular weight of such a complex can be 10^8 daltons or more, and it occupies a volume equivalent to that of a bacterium, which is about 2×10^{-12} cm^3. (A, courtesy of Lawrence Rosenberg.)

collagen (or, in the case of bone, of collagen plus calcium phosphate crystals). At the other extreme, the jellylike substance in the interior of the eye consists almost entirely of one particular type of GAG, plus water, with only a small amount of collagen. In general, GAGs are strongly hydrophilic and tend to adopt highly extended conformations, which occupy a huge volume relative to their mass (see Figure 20–17). They form gels even at very low concentrations, their multiple negative charges attracting a cloud of cations, such as Na^+, that are osmotically active, causing large amounts of water to be sucked into the matrix. This creates a swelling pressure that is balanced by tension in the collagen fibers interwoven with the proteoglycans. When the matrix is rich in collagen and large quantities of GAGs are trapped in its meshes, both the swelling pressure and the counterbalancing tension are enormous. Such a matrix is tough, resilient, and resistant to compression. The cartilage matrix that lines the knee joint, for example, has this character: it can support pressures of hundreds of kilograms per square centimeter.

Proteoglycans perform many sophisticated functions in addition to providing hydrated space around cells. They can form gels of varying pore size and charge density that act as filters to regulate the passage of molecules through the extracellular medium. They can bind secreted growth factors and other proteins that serve as signals for cells. They can block, encourage, or guide cell migration through the matrix. In all these ways, the matrix components influence the behavior of cells, often the same cells that make the matrix—a reciprocal interaction that has important effects on cell differentiation. Much remains to be learned about how cells weave the tapestry of matrix molecules and how the chemical messages they leave in its fabric are organized and act.

QUESTION 20–3

Proteoglycans are characterized by the abundance of negative charges on their sugar chains. How would the properties of these molecules differ if the negative charges were not as abundant?

EPITHELIAL SHEETS AND CELL JUNCTIONS

There are more than 200 visibly different cell types in the body of a vertebrate. The majority of these are organized into **epithelia** (singular **epithelium**) in which the cells are joined together, side to side, to form multicellular sheets. In some cases, the sheet is many cells thick, or *stratified*, as in the epidermal covering of the skin; in other cases, it is a *simple epithelium,* only one cell thick, as in the lining of the gut. Epithelial cells can take many forms. They can be tall and columnar, or cuboidal, or squat and squamous (Figure 20–18). Within a given sheet, they may be all alike or a mixture of different types. Some epithelia, like the skin, act mainly just as a protective barrier; others have complex biochemical functions. Some secrete specialized products such as hormones, milk, or tears; others, such as the epithelium lining the gut, absorb nutrients; yet others detect signals, such as light, sensed by the layer of photoreceptors in the retina of the eye, or sound, sensed by the epithelium containing the auditory hair cells in the ear. Despite these and many other variations, one can recognize a standard set of structural features that virtually all animal epithelia share. The arrangement of cells into epithelia is so commonplace that one easily takes it for granted; yet it requires a collection of specialized devices, as we shall see, and these are common to a wide variety of different cell types.

Epithelia cover the external surface of the body and line all its internal cavities, and they must have been an early feature in the evolution of multicellular animals. Their importance is obvious. Cells joined together into an epithelial sheet create a barrier that has the same significance for the multicellular organism that the plasma membrane has for a single cell. It keeps some molecules in, and others out; it takes up nutrients and exports wastes; it contains receptors for environmental signals; and it protects the interior of the organism from invading microorganisms and fluid loss.

Epithelial Sheets Are Polarized and Rest on a Basal Lamina

An epithelial sheet has two faces: the **apical** surface is free and exposed to the air or to a watery fluid; the **basal** surface rests on some other tissue—usually a connective tissue—to which it is attached (Figure 20–19). Supporting the basal surface of the epithelium is a thin tough sheet of extracellular matrix, called the **basal lamina** (Figure 20–20), composed of a specialized type of collagen (Type IV collagen) and various other macromolecules. These include a protein called *laminin*, which provides

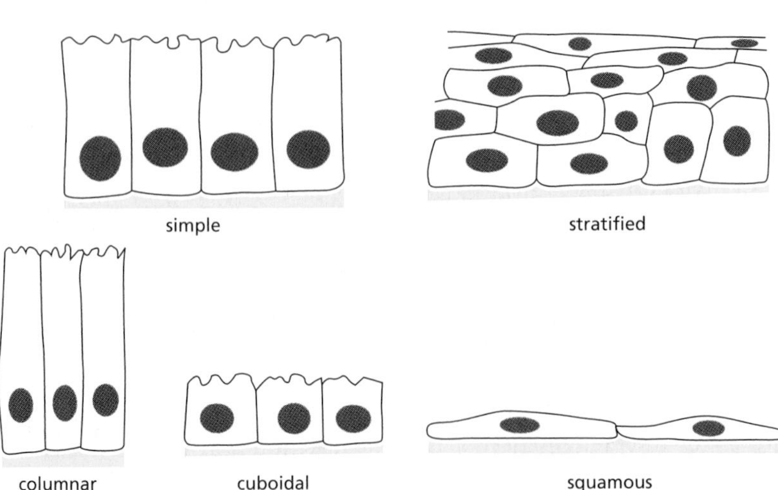

Figure 20–18 **Cells can be packed together in different ways to form an epithelial sheet.** Five basic types of epithelia are shown.

adhesive sites for integrin molecules in the plasma membranes of the epithelial cells, and thus serves a linking role like that of fibronectin in connective tissues.

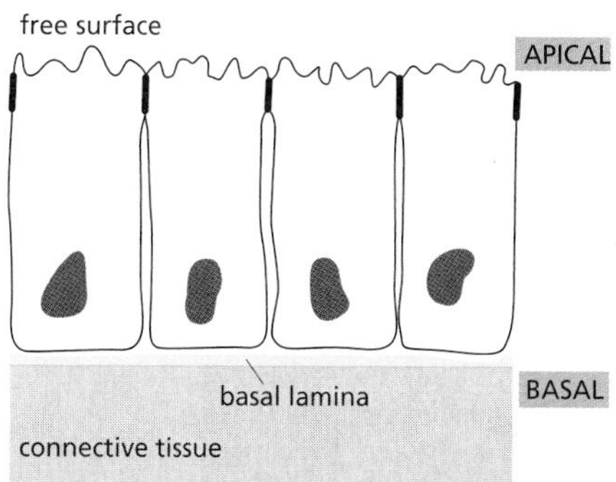

Figure 20–19 **A sheet of epithelial cells has an apical and a basal surface.** The basal surface sits on a specialized sheet of extracellular matrix called the basal lamina, while the apical surface is free.

The apical and basal faces of an epithelium are chemically different, reflecting a polarized internal organization of the individual epithelial cells: each has a top and a bottom, with different properties. This polarized organization is crucial for epithelial function. Consider, for example, the simple columnar epithelium that lines the small intestine of a mammal. It mainly consists of two intermingled cell types: absorptive cells, which take up nutrients, and goblet cells (so called because of their shape), which secrete the mucus that protects and lubricates the gut lining (Figure 20–21). Both cell types are polarized. The absorptive cells import food molecules from the gut lumen through their apical surface and export these molecules from their basal surface into the underlying tissues. To do this, absorptive cells require different sets of membrane transport proteins in their apical and basal plasma membranes (see Figure 12–17). The goblet cells also have to be polarized, but in a different way: they have to synthesize mucus and then discharge it from their apical ends only (see Figure 20–21); the Golgi apparatus, secretory vesicles, and cytoskeleton are all asymmetrically organized so as to bring this about. This organization depends on the junctions that the epithelial cells form with one another and with the basal lamina, which in turn control the arrangement of an elaborate system of membrane-associated intracellular proteins that govern the polarized organization of the cytoplasm.

Tight Junctions Make an Epithelium Leak-proof and Separate Its Apical and Basal Surfaces

Epithelial **cell junctions** can be classified according to their function. Some provide a tight seal to prevent the leakage of molecules across the epithelium through the gaps between its cells; some provide strong mechanical attachments; and some provide for a special type of intimate chemical communication. In most epithelia, all these types of junctions are present (Figure 20–22). Each type of junction is characterized by its own class of membrane proteins that hold the cells together.

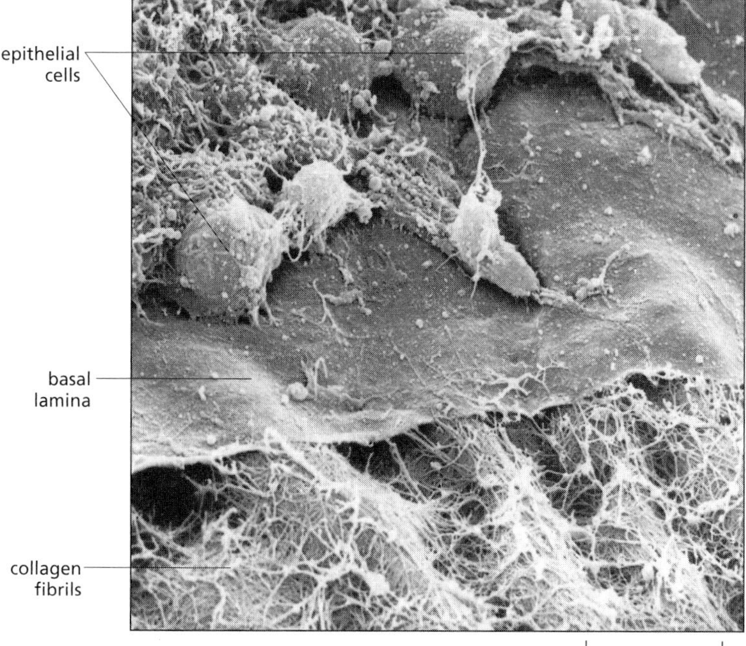

Figure 20–20 **The basal lamina supports a sheet of epithelial cells.** Scanning electron micrograph of a basal lamina in the cornea of a chick embryo. Some of the epithelial cells have been removed to expose the upper surface of the matlike basal lamina woven from type IV collagen and laminin proteins. A network of other collagen fibrils in the underlying connective tissue interacts with the lower face of the lamina. (Courtesy of Robert Trelstad.)

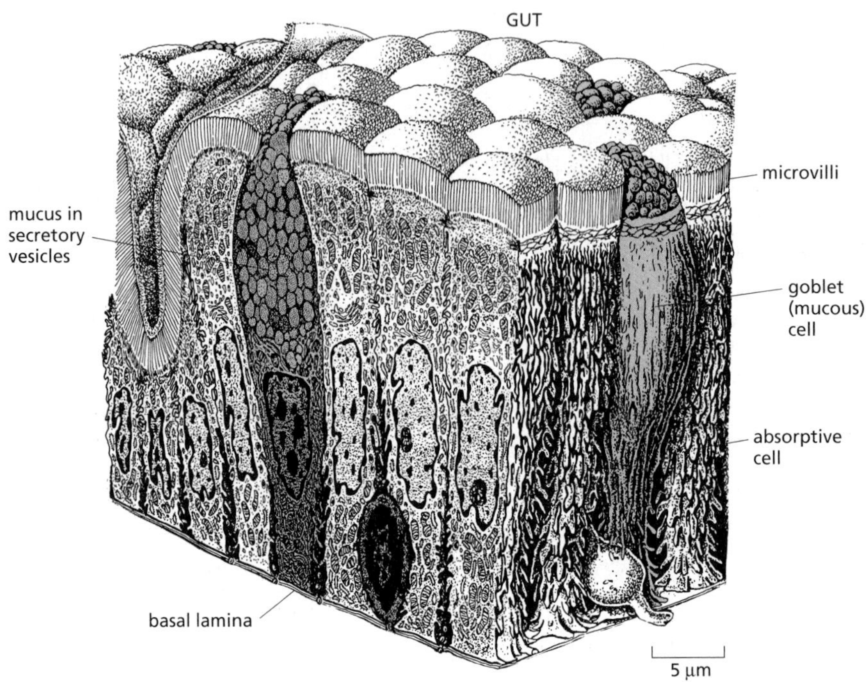

Figure 20–21 **Functionally polarized cell types line the gut.** Absorptive cells, which take up nutrients from the gut, are mingled in the gut lining with goblet cells *(brown)*, which secrete mucus into the gut. The absorptive cells are often called *brush-border cells*, because of the brushlike mass of microvilli on their apical surface, serving to increase the area of membrane for transport of small molecules into the cell. The goblet cells owe their gobletlike shape to the mass of secretory vesicles that distends their apical region. (Adapted from R. Krstić, Human Microscopic Anatomy. Berlin: Springer, 1991. With permission from Springer-Verlag.)

The sealing function is served (in vertebrates) by **tight junctions**. These junctions seal neighboring cells together so that water-soluble molecules cannot easily leak between them. If a tracer molecule is added to one side of an epithelial cell sheet, it will usually not pass beyond the tight junction (Figure 20–23). The tight junction is formed from proteins called *claudins* and *occludins*, which are arranged in strands along the lines of junction to create the seal. Without tight junctions to prevent leakage, the pumping activities of absorptive cells such as those in the gut would be futile, and the composition of the extracellular medium would become the same on both sides of the epithelium. As discussed in Chapter 11, tight junctions also play a key part in maintaining the polarity of the individual epithelial cells in two ways. First, the tight junction around the apical rim of each cell prevents diffusion of membrane proteins within the plasma membrane and so keeps the apical domain of the plasma membrane different from the basal (or baso-lateral) domain (see Figure

Figure 20–22 **Several types of cell–cell junctions are found in epithelia in animals.** Tight junctions are peculiar to epithelia; the other types also occur, in modified forms, in various nonepithelial tissues.

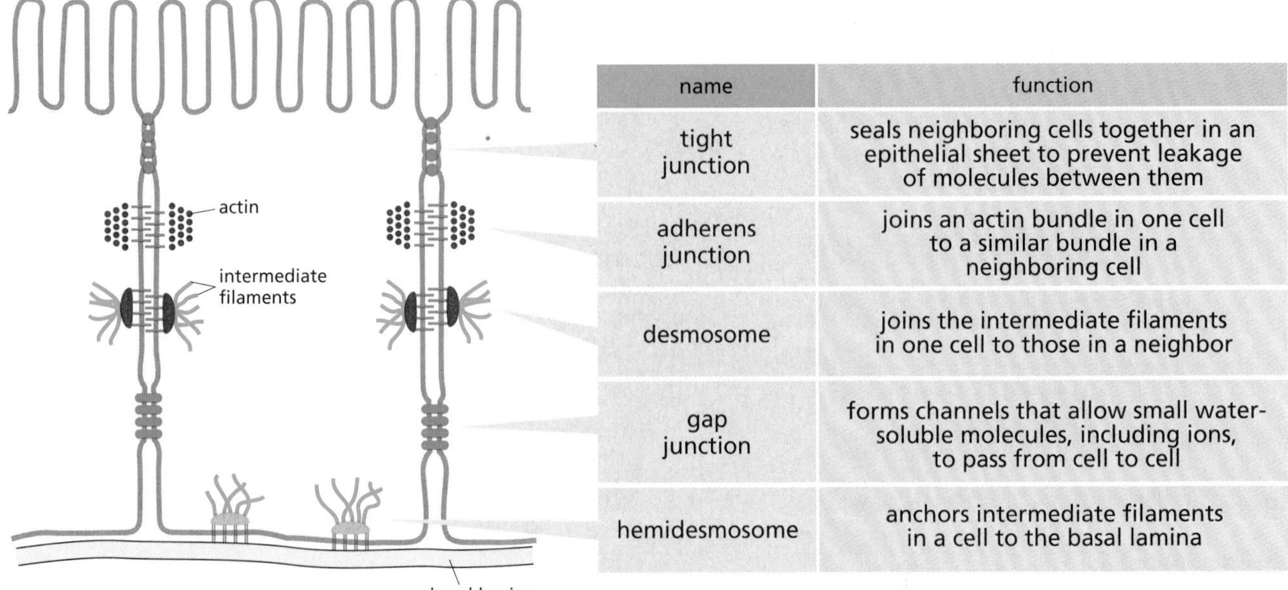

name	function
tight junction	seals neighboring cells together in an epithelial sheet to prevent leakage of molecules between them
adherens junction	joins an actin bundle in one cell to a similar bundle in a neighboring cell
desmosome	joins the intermediate filaments in one cell to those in a neighbor
gap junction	forms channels that allow small water-soluble molecules, including ions, to pass from cell to cell
hemidesmosome	anchors intermediate filaments in a cell to the basal lamina

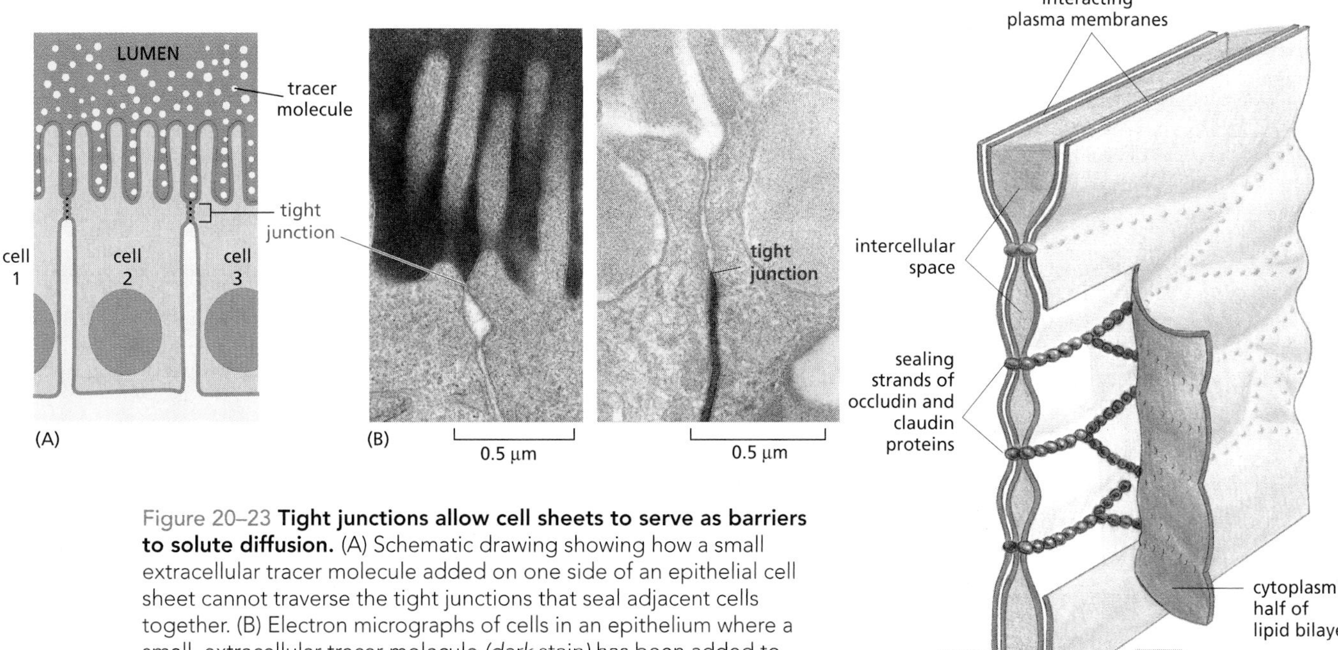

Figure 20–23 **Tight junctions allow cell sheets to serve as barriers to solute diffusion.** (A) Schematic drawing showing how a small extracellular tracer molecule added on one side of an epithelial cell sheet cannot traverse the tight junctions that seal adjacent cells together. (B) Electron micrographs of cells in an epithelium where a small, extracellular tracer molecule *(dark* stain) has been added to either the apical side (on the *left*) or the basolateral side (on the *right*); in both cases the tracer is stopped by the tight junction. (C) A model of the structure of a tight junction, showing how the cells are sealed together by proteins, called claudins and occludins *(green),* in the outer leaflet of the plasma membrane bilayer. (B, courtesy of Daniel Friend.)

11–34). Second, in many epithelia, the tight junctions are sites of assembly for the complexes of intracellular proteins that govern apico-basal polarity in the interior of the cell.

Cytoskeleton-linked Junctions Bind Epithelial Cells Robustly to One Another and to the Basal Lamina

The junctions that hold an epithelium together by forming mechanical attachments are of three main types. *Adherens junctions* and *desmosomes* bind one epithelial cell to another, while *hemidesmosomes* bind epithelial cells to the basal lamina. All of these junctions provide mechanical strength by the same strategy that we have already encountered in connective tissue (see Figure 20–14C): the molecule that forms the external adhesion spans the membrane and is linked inside the cell to strong cytoskeletal filaments. In this way, the cytoskeletal filaments are tied into a network that extends from cell to cell across the whole expanse of the epithelial sheet.

Adherens junctions and desmosomes are both built around transmembrane proteins that belong to the **cadherin** family: a cadherin molecule in the plasma membrane of one cell binds directly to an identical cadherin molecule in the plasma membrane of its neighbor (Figure 20–24). Such binding of like to like is called *homophilic* binding. In the case of cadherins, binding also requires that Ca^{2+} be present in the extracellular medium—hence its name.

At an **adherens junction**, each cadherin molecule is tethered inside its cell, via several linker proteins, to actin filaments. Often, the adherens junctions form a continuous adhesion belt around each of the interacting epithelial cells; this belt is located near the apical end of the cell, just below the tight junctions (Figure 20–25). Actin bundles are thus connected

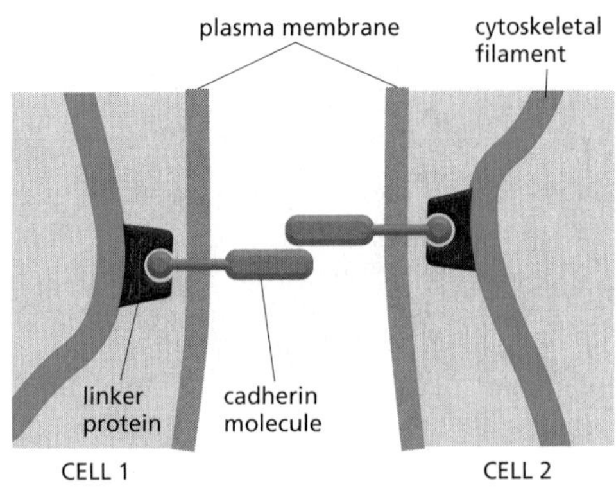

Figure 20–24 **Cadherin molecules mediate mechanical attachment of one cell to another.** Two similar cadherin molecules in the plasma membranes of adjacent cells bind to one another extracellularly; intracellularly they are attached, via linker proteins, to cytoskeletal filaments—either actin or keratin. As cells touch one another, their cadherins become concentrated at the point of attachment (Movie 20.2).

from cell to cell across the epithelium. This actin network is potentially contractile, and it gives the epithelial sheet the capacity to develop tension and to change its shape in remarkable ways. By shrinking its apical surface along one axis, the sheet can roll itself up into a tube (Figure 20–26A). Alternatively, by shrinking its apical surface locally along both axes at once, the sheet can develop a cup-shaped concavity and eventually create a vesicle that may pinch off from the rest of the epithelium. Epithelial movements such as these are important in embryonic development, where they create structures such as the neural tube, which gives rise to the central nervous system (Figure 20–26B), and the lens vesicle, which develops into the lens of the eye (Figure 20–26C).

At a **desmosome**, by contrast, a different set of cadherin molecules are anchored inside each cell. These cadherins connect to intermediate filaments—specifically, to keratins, which are the type of intermediate filament found in epithelia (Figure 20–27). Thick bundles of ropelike keratin filaments criss-cross the cytoplasm and are 'spot-welded' via desmosome junctions to the bundles of keratin filaments in adjacent cells. This arrangement confers great tensile strength on the epithelial sheet and is characteristic of tough, exposed epithelia such as the epidermis.

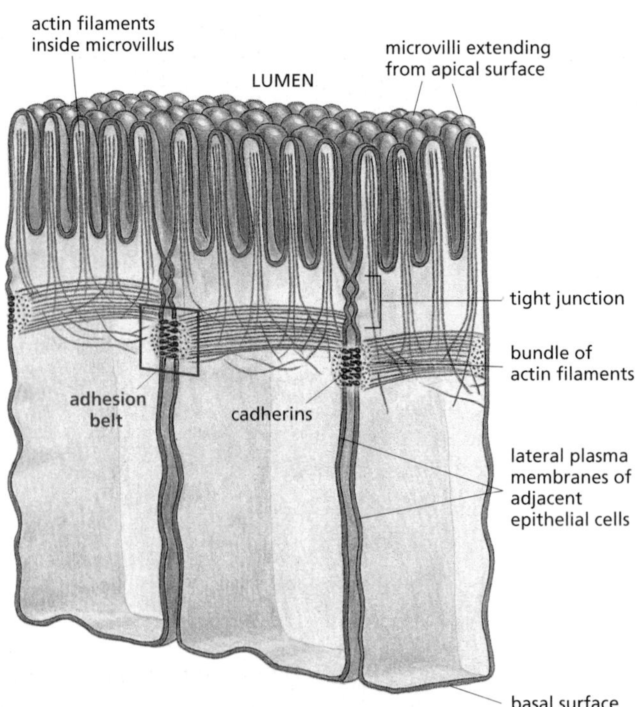

Figure 20–25 **Adherens junctions form adhesion belts around epithelial cells in the small intestine.** A contractile bundle of actin filaments runs along the cytoplasmic surface of the plasma membrane near the apex of each cell, and these bundles of actin filaments in adjacent cells are linked to one another via cadherin molecules that span the cell membranes (see Figure 20–24).

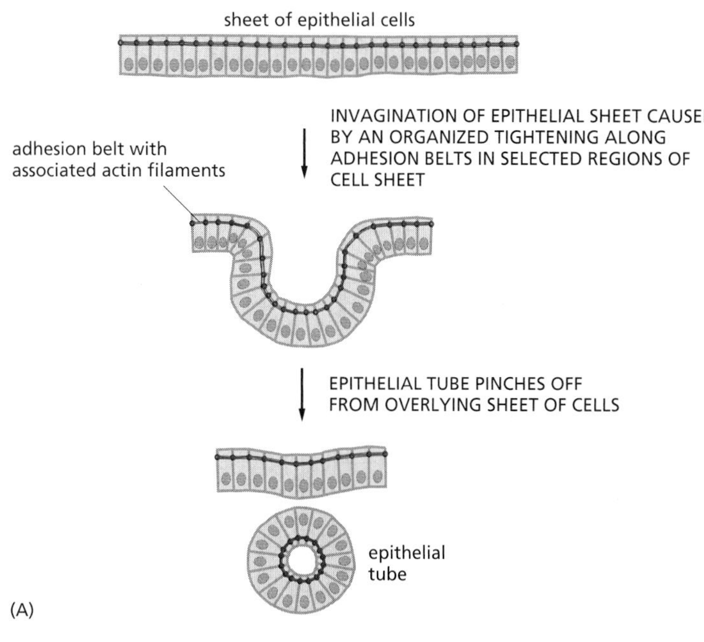

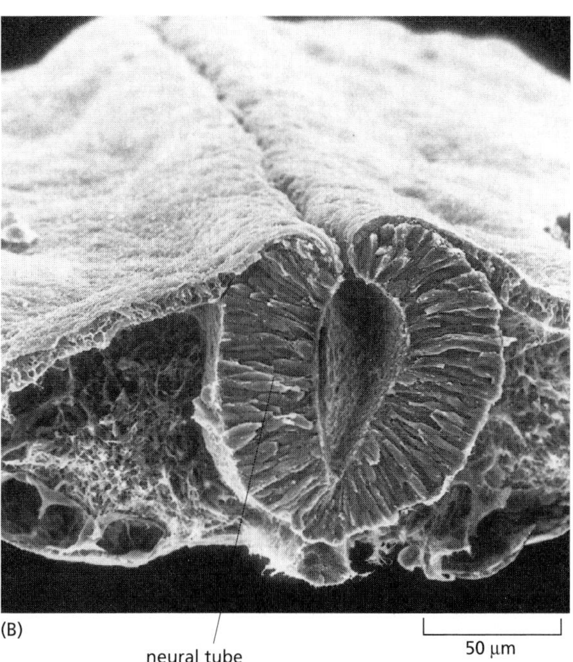

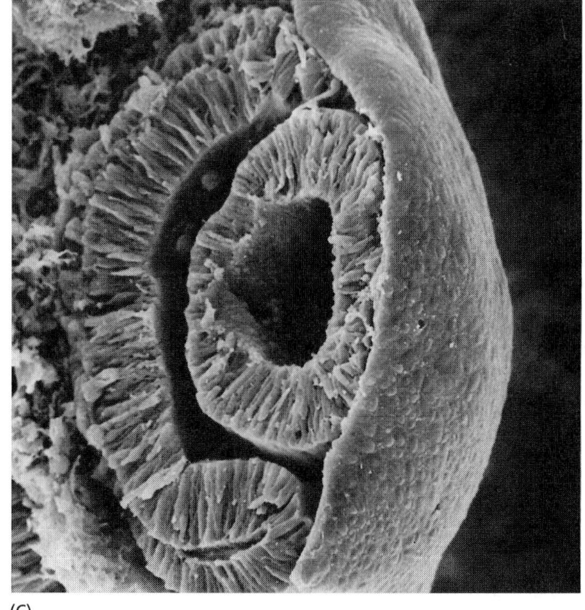

Figure 20–26 **Epithelial sheets can bend to form a tube or a vesicle.** (A) Diagram showing how contraction of apical bundles of actin filaments linked from cell to cell via adherens junctions causes the epithelial cells to narrow at their apex. According to whether the contraction is oriented along one axis or is equal in all directions, the epithelium may be caused to roll up into a tube or to invaginate to form a vesicle. (B) Formation of the neural tube; the scanning electron micrograph shows a cross section through the trunk of a two-day chick embryo. Part of the epithelial sheet that covers the surface of the embryo has thickened, has rolled up into a tube by apical contraction, and is about to pinch off to become a separate internal structure. (C) Formation of the lens; a patch of surface epithelium overlying the embryonic rudiment of the retina of the eye has become concave and has pinched off as separate vesicle—the lens vesicle—within the eye cup. (B, courtesy of Jean-Paul Revel; C, courtesy of K.W. Tosney.)

Blisters are a painful reminder that it is not enough for epithelial cells to be firmly attached to one another: they must also be anchored to the underlying tissue. As we noted earlier, the anchorage is mediated by integrins in the basal plasma membrane of the epithelial cells. Externally, these integrins bind to the extracellular matrix protein laminin in the basal lamina; inside the cell, they are linked to keratin filaments, creating a structure that looks superficially like half a desmosome. These attachments of epithelial cells to the extracellular matrix beneath them are therefore called **hemidesmosomes** (Figure 20–28).

Gap Junctions Allow Ions and Small Molecules to Pass from Cell to Cell

The final type of epithelial cell junction, found in virtually all epithelia and in many other types of tissues, serves a totally different purpose. It is called a **gap junction**. In the electron microscope, it appears as a region

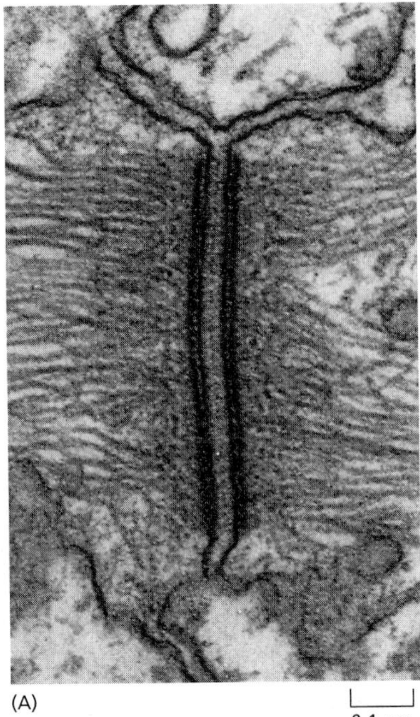

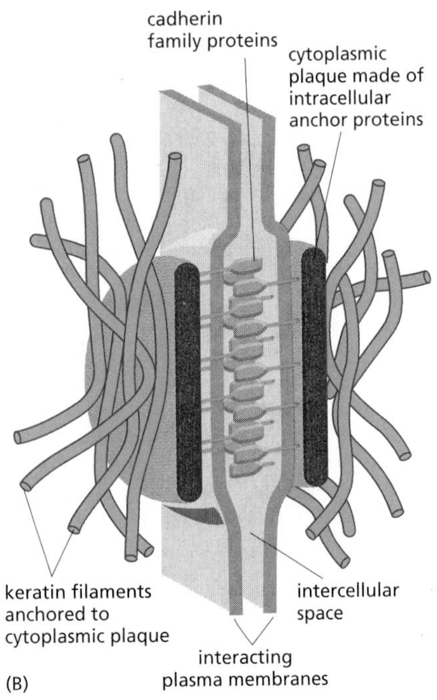

Figure 20–27 **Desmosomes link the keratin filaments of one cell to those of another.** (A) An electron micrograph of a desmosome joining two cells in the epidermis of a newt, showing the attachment of keratin filaments. (B) Schematic drawing of a desmosome. On the cytoplasmic surface of each interacting plasma membrane is a dense plaque composed of a mixture of intracellular anchor proteins. A bundle of keratin filaments is attached to the surface of each plaque. Transmembrane adhesion proteins of the cadherin family bind to the outer face of each plaque and interact through their extracellular domains to hold the adjacent cells together. (A, from D.E. Kelly, *J. Cell Biol.* 28:51–72, 1966. With permission from The Rockefeller University Press.)

where the membranes of two cells lie close together and exactly parallel, with a very narrow gap of 2–4 nm between them (Figure 20–29A). The gap is not empty but is spanned by the protruding ends of many identical protein complexes that lie in the plasma membranes of the two apposed cells. These complexes, called *connexons*, form channels across the two plasma membranes and are aligned end-to-end so as to create narrow passageways that allow inorganic ions and small water-soluble molecules (up to a molecular mass of about 1000 daltons) to move directly from the cytosol of one cell to the cytosol of another (Figure 20–29B). This creates an electrical and a metabolic coupling between the cells. Gap junctions between heart muscle cells, for example, provide the electrical coupling that allows electrical waves of excitation to spread through the tissue. These waves of excitation trigger the coordinated contraction of the cells, thus producing a regular heart beat.

Gap junctions in many tissues can be opened or closed as needed in response to extracellular signals. The neurotransmitter dopamine, for example, reduces gap-junction communication within a class of neurons in the retina in response to an increase in light intensity (Figure 20–30). This reduction in gap-junction permeability changes the pattern of electrical signaling and helps the retina switch from using rod photoreceptors, which are good detectors of low light, to cone photoreceptors, which detect color and fine detail in bright light.

QUESTION 20–4

Analogues of hemidesmosomes are the focal contact sites described in Chapter 17, which are also sites where the cell attaches to the extracellular matrix. These junctions are prevalent in fibroblasts but largely absent in epithelial cells. On the other hand, hemidesmosomes are prevalent in epithelial cells but absent in fibroblasts. In focal contact sites, intracellular connections are made to actin filaments, whereas in hemidesmosomes connections are made to intermediate filaments. Why do you suppose these two different cell types attach differently to the extracellular matrix?

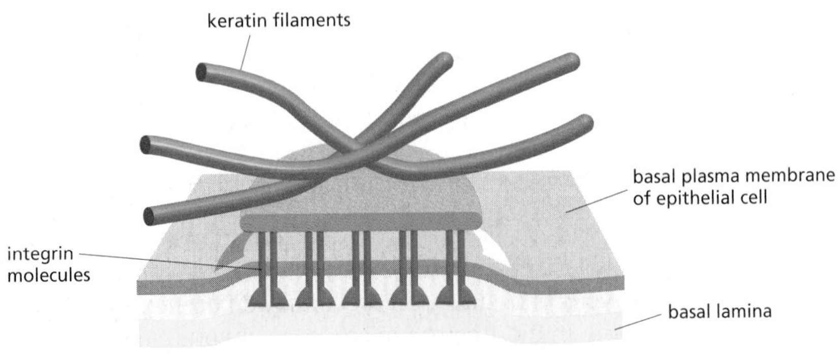

Figure 20–28 **Hemidesmosomes anchor the keratin filaments in an epithelial cell to the basal lamina.** The linkage is mediated by integrins.

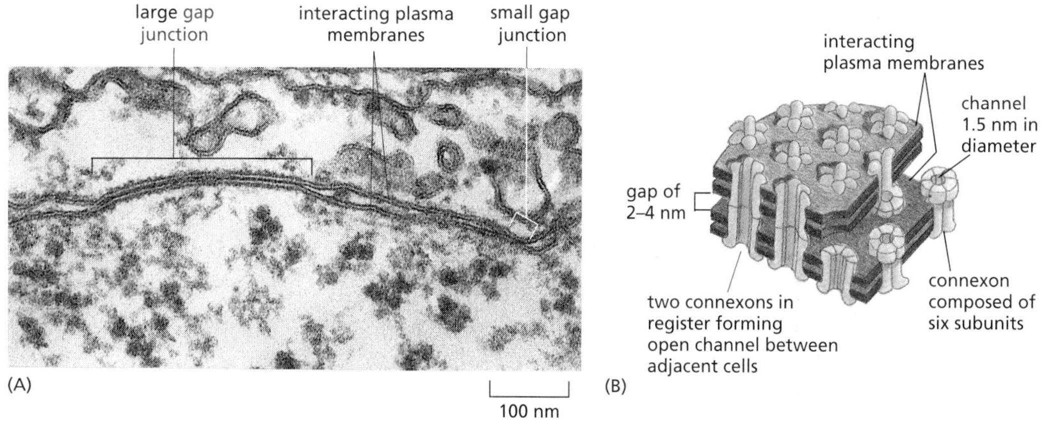

Figure 20–29 **Gap junctions provide neighboring cells with a direct channel of communication.** (A) Thin-section electron micrograph of two gap junctions between two cells in culture. (B) A model of a gap junction. The drawing shows the interacting plasma membranes of two adjacent cells. The apposed lipid bilayers *(red)* are penetrated by protein assemblies called *connexons (green)*, each of which is thought to be formed from six identical protein subunits. Two connexons join across the intercellular gap to form an aqueous channel connecting the two cells. (A, from N.B. Gilula, in Cell Communication [R.P. Cox, ed.], pp. 1–29. New York: Wiley, 1974. With permission from John Wiley & Sons, Inc.)

Curiously, plant tissues, although they lack all the other types of cell junctions we have described earlier, have a functional counterpart of the gap junction. The cytoplasms of adjacent plant cells are connected via minute communicating channels called **plasmodesmata**, which span the intervening cell walls (Figure 20–31). In contrast to gap-junctional channels, plasmodesmata are cytoplasmic channels lined with plasma membrane, and thus in plants the cytoplasm is, in principle, continuous from one cell to the next. Ions, small molecules, and even macromolecules such as some proteins and microRNAs can pass through plasmodesmata, and the regulated traffic of transcription regulators from one cell to another is important in plant development.

QUESTION 20–5

Gap junctions are dynamic structures that, like conventional ion channels, are gated: they can close by a reversible conformational change in response to changes in the cell. The permeability of gap junctions decreases within seconds, for example, when the intracellular Ca^{2+} is raised. Speculate why this form of regulation might be important for the health of a tissue.

TISSUE MAINTENANCE AND RENEWAL

One cannot contemplate the organization of tissues without wondering how these astonishingly patterned structures come into being. This raises one of the most ancient and fundamental questions in all of biology: how is a whole multicellular organism generated from a single fertilized egg?

In the process of development, the egg cell divides repeatedly to give a clone of cells—about 10,000,000,000,000 of them for a human being—all containing the same genome but specialized in different ways. This clone has a structure. It may take the form of a daisy or an oak tree, a sea

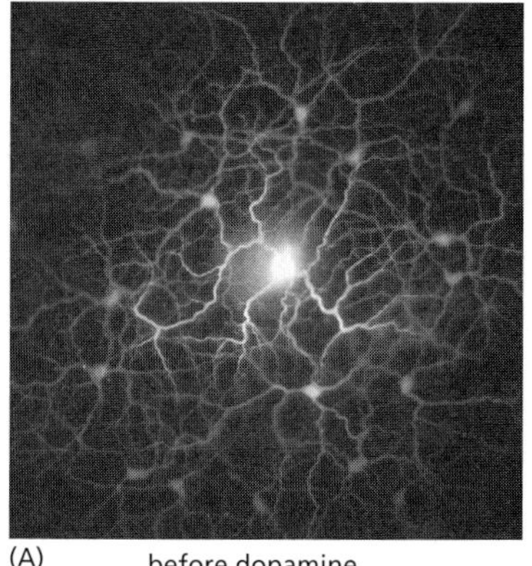

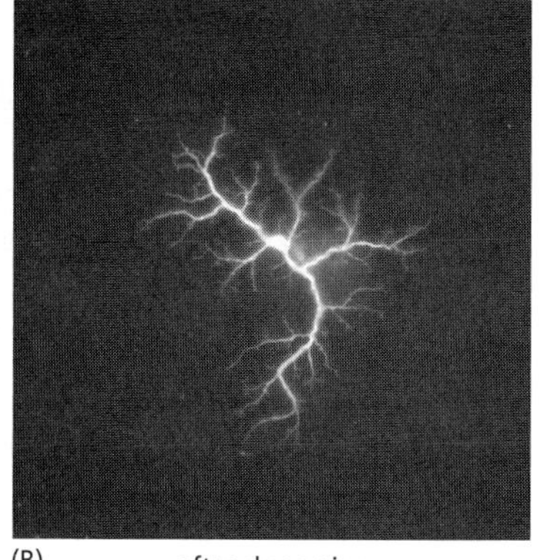

Figure 20–30 **Extracellular signals can regulate the permeability of gap junctions.** (A) A neuron in a rabbit retina was injected with the dye Lucifer yellow, which passes readily through gap junctions. The dye thus labels neurons of the same type that are connected by gap junctions. (B) The retina was treated with the neurotransmitter dopamine before the neuron was injected with Lucifer yellow. As can be seen, the dopamine treatment greatly decreased the permeability of the gap junctions. (Courtesy of David Vaney.)

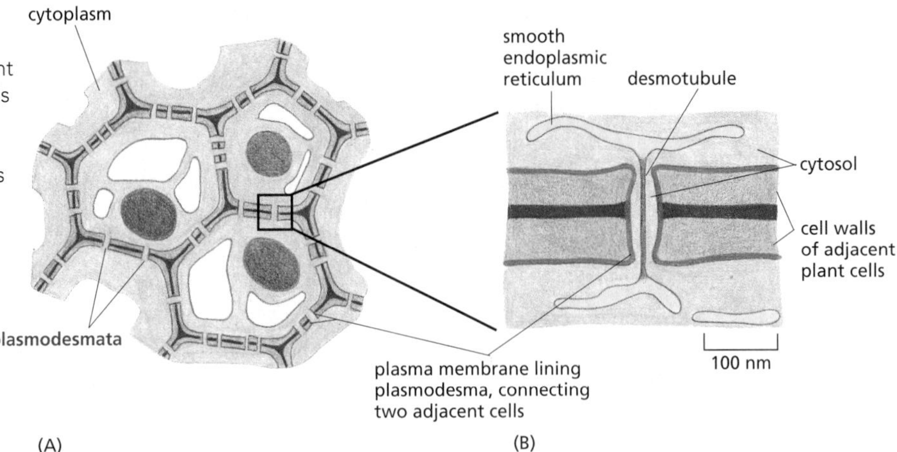

Figure 20–31 **Plant cells are connected via plasmodesmata.** (A) The cytoplasmic channels of plasmodesmata pierce the plant cell wall and connect the interiors of all cells in a plant. (B) Each plasmodesma is lined with plasma membrane common to the two connected cells. It usually also contains a fine tubular structure, the desmotubule, derived from smooth endoplasmic reticulum.

urchin, a whale, or a mouse (Figure 20–32). The structure is determined by the genome that the egg contains. The linear sequence of A, G, C, and T nucleotides in the DNA directs the production of a host of distinct cell types, each expressing different sets of genes and arranged in a precise, intricate, three-dimensional pattern.

Although the final structure of an animal's body may be enormously complex, it is generated by a limited repertoire of cell activities. Examples of all these activities have been discussed in earlier pages of this book. Cells grow, divide, and die. They form mechanical attachments and generate forces for movement. They differentiate by switching on or off the production of specific sets of proteins. They produce molecular signals to influence neighboring cells, and they respond to signals that neighboring cells deliver to them. They remember the effects of previous signals they have received, and so progressively become more and more specialized in the characteristics they adopt. The genome, identical in every cell,

Figure 20–32 **The genome of the egg determines the structure of the clone of cells that will develop from it.** (A and B) A sea-urchin egg gives rise to a sea urchin; (C and D) a mouse egg gives rise to a mouse. (A, courtesy of David McClay; B, courtesy of M. Gibbs, with permission from Oxford Scientific Films; C, courtesy of Patricia Calarco, from G. Martin, *Science* 209:768–776, 1980, with permission from AAAS; D, courtesy of O. Newman, with permission from Oxford Scientific Films.)

defines the rules according to which these various possible cell activities are called into play. Through its operation in each cell individually, the genome guides the whole intricate process by which a multicellular organism is generated from a fertilized egg. Movies 1.1, 20.3, and 20.4 offer some visual examples of how development unfurls for the embryos of a frog, a fruit fly, and a zebrafish, respectively.

For developmental biologists, the challenge is to explain in these terms the entire sequence of interlocking events that lead from the egg to the adult organism. We shall not attempt to set out an answer to this problem here: we do not have space to do it justice, even though a great deal of the process is now understood. But the same basic activities that combine to create the organism during development continue even in the adult body, where fresh cells are continually generated in precisely controlled patterns. It is this more limited topic that we discuss in this section, focusing on the organization and maintenance of the tissues of vertebrates.

Tissues Are Organized Mixtures of Many Cell Types

Although the specialized tissues in our body differ in many ways, they all have certain basic requirements, usually provided for by a mixture of cell types, as illustrated for the skin in Figure 20–33. As discussed earlier in this chapter, all tissues need mechanical strength, which is often supplied by a supporting bed or framework of connective tissue inhabited by fibroblasts. In this connective tissue, blood vessels lined with endothelial cells satisfy the need for oxygen, nutrients, and waste disposal. Likewise, most tissues are innervated by nerve-cell axons, which are ensheathed by Schwann cells that provide electrical insulation. Macrophages dispose of dying cells and other unwanted debris, and lymphocytes and other white blood cells combat infection. Most of these cell types originate out-

Figure 20–33 **Mammalian skin is made of a mixture of cell types.** (A) Schematic diagrams showing the cellular architecture of thick skin. (B) Light micrograph of a cross section through the sole of a human foot, stained with hematoxylin and eosin. The skin can be viewed as a large organ composed of two main tissues: epithelial tissue (the *epidermis*), which lies outermost, and connective tissue, which consists of the tough *dermis* (from which leather is made) and the underlying fatty *hypodermis*. Each tissue is composed of a variety of cell types. The dermis and hypodermis are richly supplied with blood vessels and nerves. Some nerve fibers extend also into the epidermis.

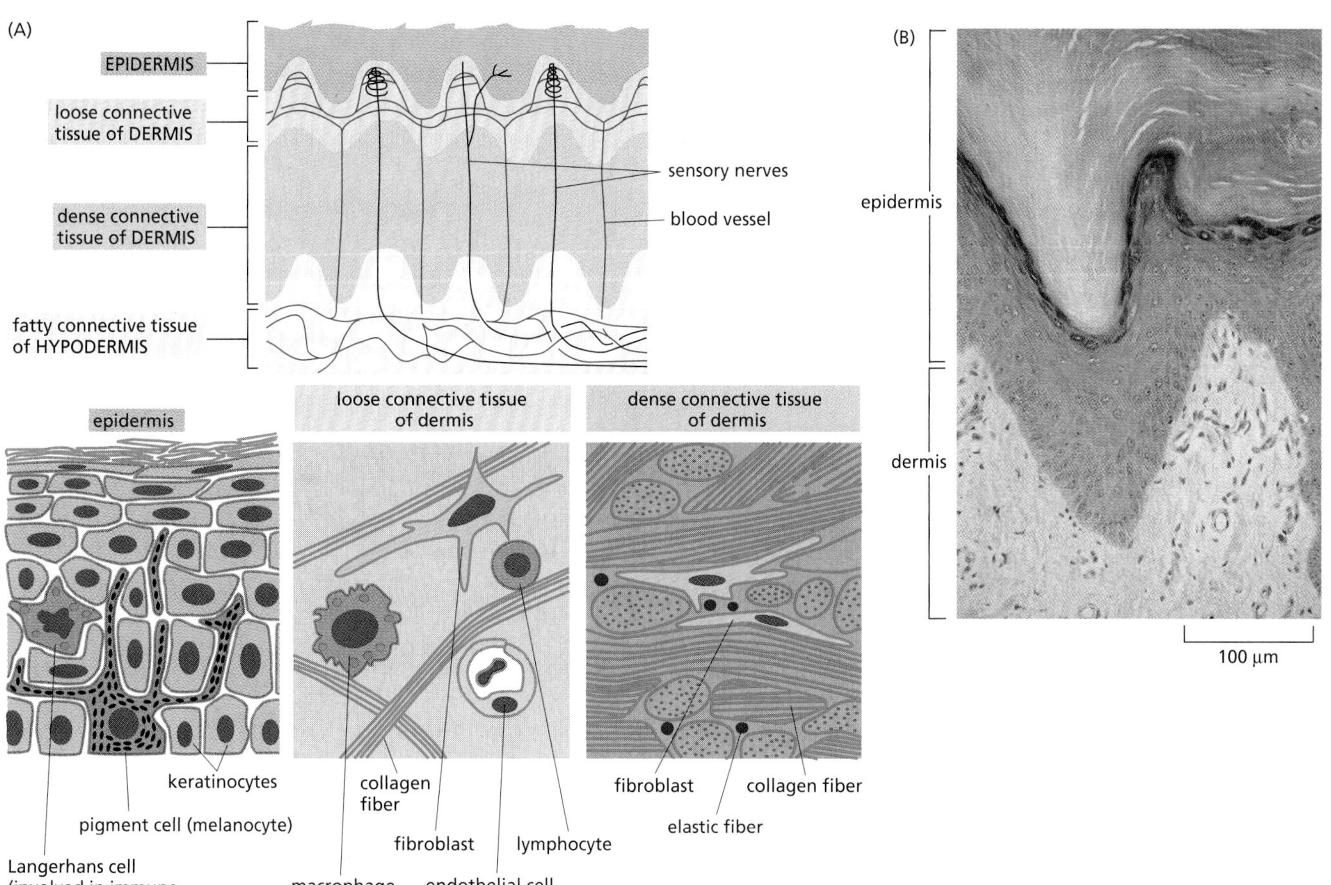

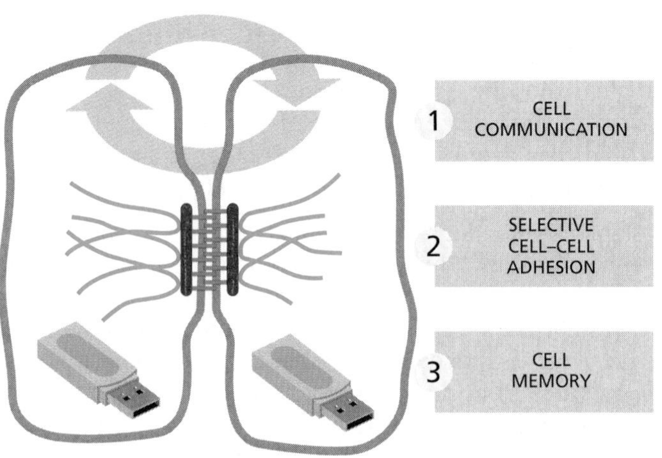

Figure 20–34 **Three key factors maintain the cellular organization of tissues.**

side the tissue and invade it, either early in the course of its development (endothelial cells, nerve-cell axons, and Schwann cells) or continually during life (macrophages and other white blood cells). This complex supporting apparatus is required to maintain the principal specialized cells of the tissue: the contractile cells of muscle, the secretory cells of glands, or the blood-forming cells of bone marrow, for example.

Almost every tissue is therefore an intricate mixture of many cell types that must remain different from one another while coexisting in the same environment. Moreover, in almost all adult tissues, cells are continually dying and being replaced; throughout this hurly-burly of cell replacement and tissue renewal, the organization of the tissue must be preserved. Three main factors contribute to this stability (Figure 20–34).

1. Cell communication: each type of specialized cell continually monitors its environment for signals from other cells and adjusts its behavior accordingly; in fact, the very survival of most cells depends on such social signals (discussed in Chapter 16). These communications ensure that new cells are produced and survive only when and where they are required.

2. Selective cell–cell adhesion: because different cell types have different cadherins and other adhesion molecules in their plasma membrane, they tend to stick selectively, by homophilic binding, to other cells of the same type. They may also form selective attachments to certain other cell types or to specific extracellular matrix components. The selectivity of adhesion prevents the different cell types in a tissue from becoming chaotically mixed.

3. Cell memory: as discussed in Chapter 8, specialized patterns of gene expression, evoked by signals that acted during embryonic development, are afterward stably maintained, so that cells autonomously preserve their distinctive character and pass it on to their progeny. A fibroblast divides to produce more fibroblasts, an endothelial cell divides to produce more endothelial cells, and so on. This principle, with elaborations that we explain later, preserves the diversity of cell types in the tissue.

Different Tissues Are Renewed at Different Rates

Cells in tissues vary enormously in their rate and pattern of turnover. At one extreme are nerve cells, most of which last a lifetime without replacement. At the other extreme are the cells that line the intestine, which are replaced every few days. Between these extremes there is a spectrum of different rates and styles of cell replacement and tissue renewal. Bone, for example, has a turnover time of about ten years in humans, involving renewal of the matrix as well as of cells: old bone matrix is slowly eaten away by a set of cells called *osteoclasts*, akin to macrophages, while new matrix is deposited by another set of cells, *osteoblasts*, akin to fibroblasts. New red blood cells in humans are continually generated in the bone marrow (from yet another class of cells) and released into the circulation, from which they are removed and destroyed after 120 days. In the skin, the outer layers of the epidermis are continually flaking off and being replaced from below, so that the epidermis is renewed with a turnover time of about two months. And so on.

Our life depends on these renewal processes. A large dose of ionizing radiation blocks cell division and thus halts renewal: within a few days, the lining of the intestine, for example, becomes denuded of cells, leading to the devastating diarrhea and water loss characteristic of acute radiation sickness.

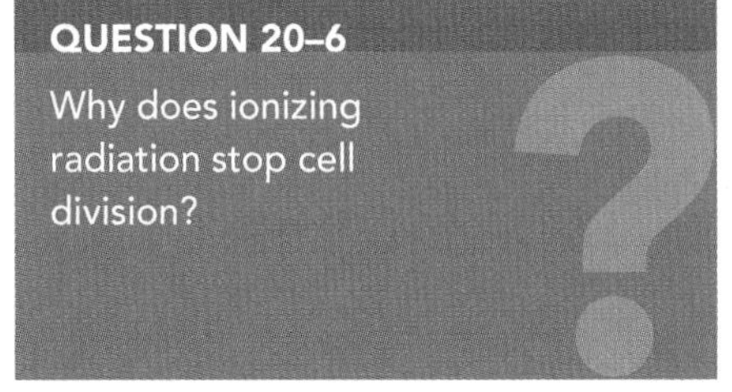

QUESTION 20–6
Why does ionizing radiation stop cell division?

Clearly, there have to be control mechanisms to keep cell production and cell loss in balance in the normal, healthy adult body. Cancers originate through violation of these controls, allowing cells in the self-renewing tissues to proliferate to excess. To understand cancer, therefore, it is important to understand the normal processes of tissue renewal that cancer perverts.

Stem Cells Generate a Continuous Supply of Terminally Differentiated Cells

Many of the differentiated cells that need continual replacement are themselves unable to divide. Red blood cells, surface epidermal cells, and the absorptive and goblet cells of the gut lining are all of this type. Such cells are referred to as *terminally differentiated*: they lie at the dead end of their developmental pathway.

Replacements for terminally differentiated cells are generated from a stock of *proliferating precursor cells*, which themselves usually derive from small numbers of dividing **stem cells**. Both stem cells and proliferating precursor cells are retained in the corresponding tissues along with the differentiated cells. Stem cells are not terminally differentiated and can divide without limit (or at least for the lifetime of the animal). When a stem cell divides, though, each daughter has a choice: either it can remain a stem cell, or it can embark on a course leading irreversibly to terminal differentiation, usually via a series of precursor cell divisions (Figure 20–35). The job of the stem cell and precursor cells, therefore, is not to carry out the specialized function of the differentiated cell, but rather to produce cells that will. Stem cells are usually present in small numbers and often have a nondescript appearance, making them difficult to identify. Although they are not terminally differentiated, stem cells of adult tissues are nonetheless specialized: under normal conditions, they stably express sets of transcription regulators that ensure that their differentiated progeny will be of the appropriate types.

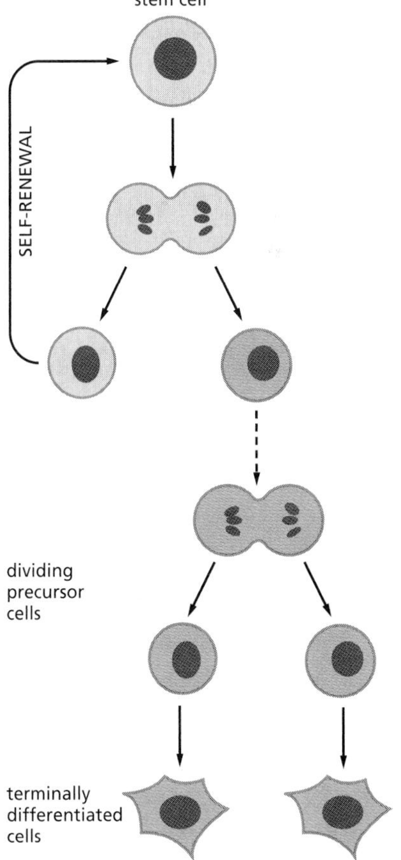

Figure 20–35 **When a stem cell divides, each daughter can either remain a stem cell or go on to become terminally differentiated.** The terminally differentiated cells usually develop from precursor cells that divide a limited number of times before they differentiate.

The pattern of cell replacement varies from one stem-cell-based tissue to another. In the lining of the small intestine, for example, the absorptive and secretory cells (mucus-producing goblet cells, as well as some other secretory cell types) together are arranged as a single-layered, simple epithelium covering the surfaces of the fingerlike villi that project into the gut lumen. This epithelium is continuous with the epithelium lining the crypts that descend into the underlying connective tissue; the stem cells lie near the bottom of the crypts. Newborn absorptive and secretory cells generated from stem cells begin to differentiate in the crypts. Most of these differentiating cells are carried upward by a sliding movement

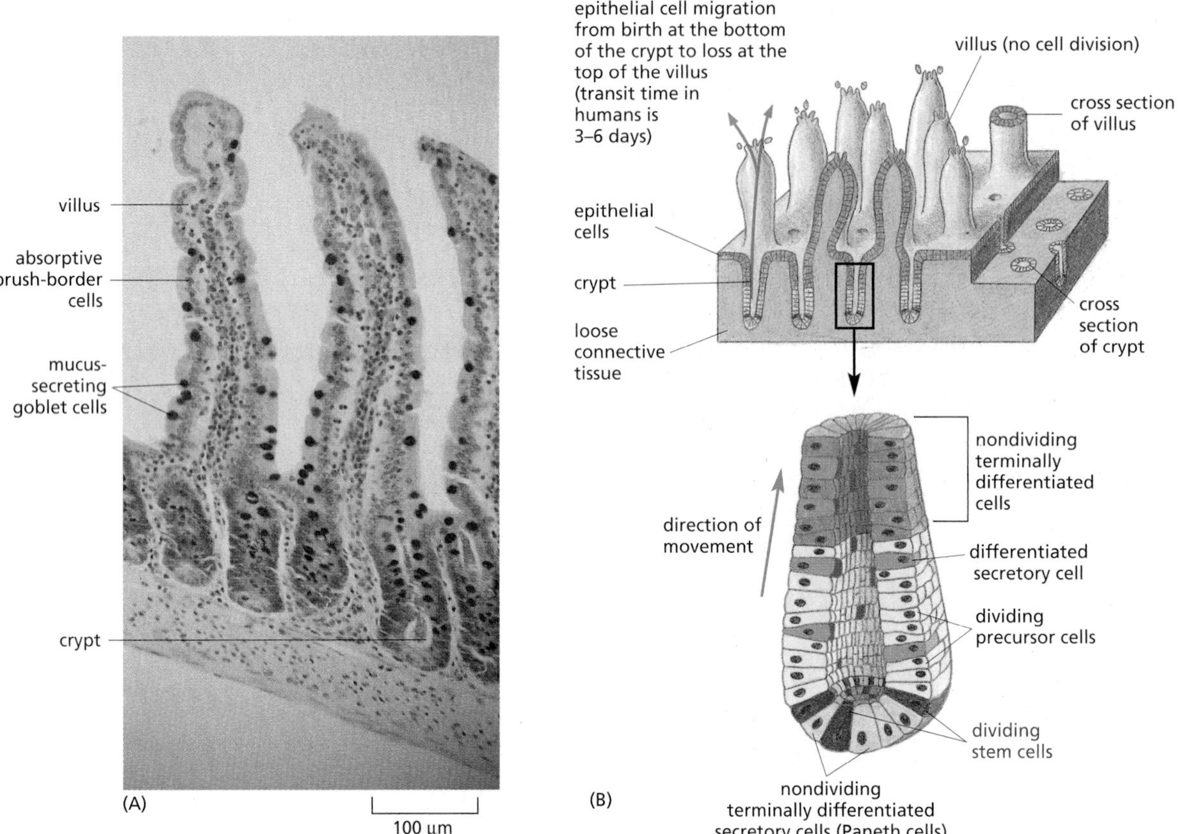

Figure 20–36 **Renewal occurs continuously in the lining of the adult intestine.** (A) Photograph of a section of part of the lining of the small intestine of a mammal, showing the villi and crypts. Mucus-secreting goblet cells (stained *purple*) are interspersed among the absorptive brush-border cells in the epithelium of the villi. Smaller numbers of two other secretory cell types—enteroendocrine cells (not visible here), which secrete gut hormones, and Paneth cells, which secrete anti-bacterial proteins—are also present and derive from the same stem cells. (B) Cartoon showing the pattern of cell turnover and the proliferation of stem cells in the epithelium that forms the lining of the mammalian small intestine. All the differentiated cell types have a finite lifetime, terminated by programmed cell death, and are continually replaced by progeny of the stem cells. The crypt-villus organization is thought to be maintained by signals from the crypt environment that keep the crypt cells in a proliferative state, as we explain later.

in the plane of the epithelial sheet until they reach the exposed surfaces of the villi; at the tips of the villi the cells die and are shed into the gut (Figure 20–36).

A contrasting example is found in the epidermis. The epidermis is a stratified epithelium, with stem cells and precursor cells in the basal layer, adhering to the basal lamina; the differentiating cells travel outward from their site of origin in a direction perpendicular to the plane of the cell sheet (Figure 20–37).

QUESTION 20–7

Why do you suppose epithelial cells lining the gut are renewed frequently, whereas most neurons last for the lifetime of the organism?

Often, a single type of stem cell gives rise to several types of differentiated progeny: the stem cells of the intestine, for example, produce absorptive cells, goblet cells, and several other secretory cell types. The process of blood-cell formation, or hemopoiesis, provides an extreme example of this phenomenon. All of the different cell types in the blood—both the red blood cells that carry oxygen and the many types of white blood cells that fight infection (Figure 20–38)—ultimately derive from a shared hemopoietic stem cell found in the bone marrow (Figure 20–39).

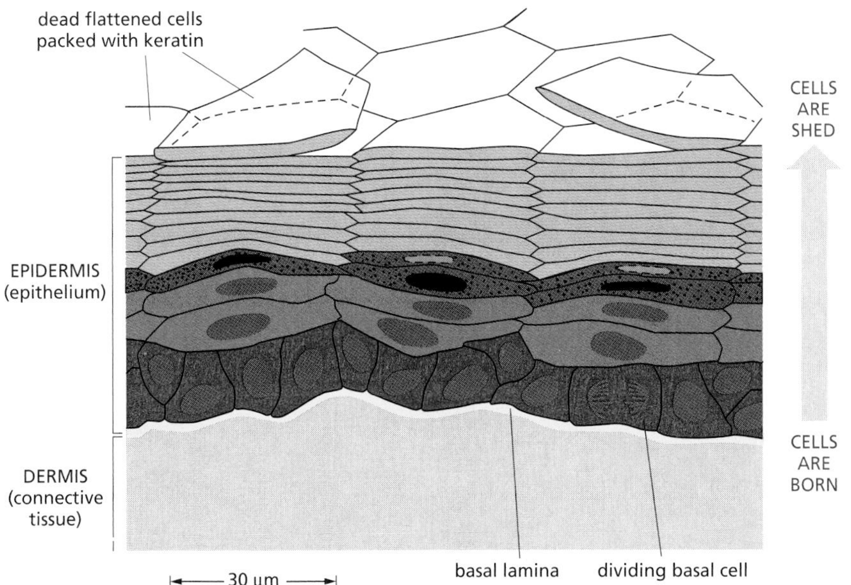

Figure 20–37 **The epidermis is renewed from stem cells in its basal layer.** The basal layer contains a mixture of stem cells and dividing precursor cells that are produced from the stem cells. On emerging from the basal layer, the cells stop dividing and move outward, differentiating as they go. Eventually, the cells undergo a special form of cell death: the nucleus disintegrates and the cell shrinks to the form of a flattened scale packed with keratin. The scale is ultimately shed from the surface of the body.

Specific Signals Maintain the Stem-Cell Populations

Every stem-cell system requires control mechanisms to ensure that new cells are generated in the appropriate places and in the right quantities. The controls depend on molecular signals exchanged between the stem cells, their progeny, and the surrounding tissues. These signals and the biochemical pathways through which they act fall into a surprisingly small number of families, corresponding to half-a-dozen basic signaling mechanisms, some of which we have discussed at length in Chapter 16. These few mechanisms are used again and again, in the embryo and in the adult, in different combinations, and evoking different responses in different contexts.

Almost all these families of signaling mechanisms contribute to the task of maintaining the complex organization of a stem-cell system such as that of the intestine. Thus, a class of signal molecules known as the **Wnt proteins** serve to keep the stem cells and precursor cells at the base of

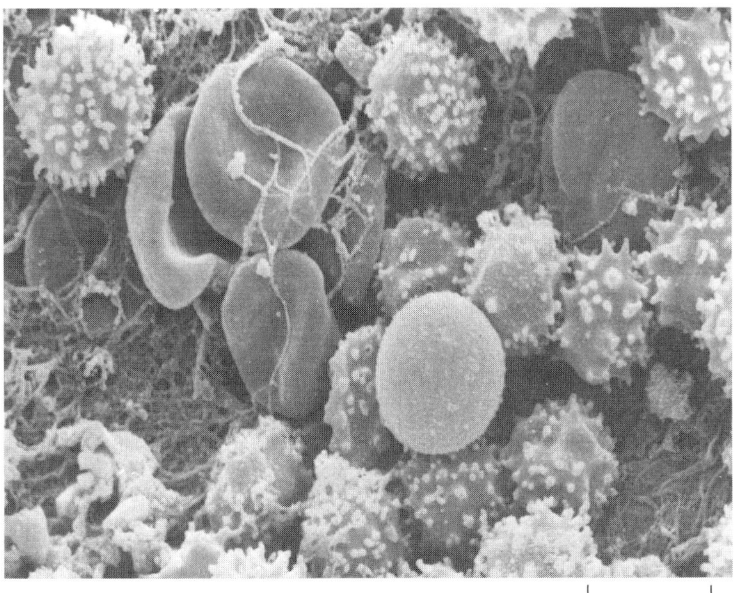

Figure 20–38 **Blood contains many circulating cell types, all derived from a single type of stem cell.** In this scanning electron micrograph, the larger, more spherical cells with a rough surface are white blood cells; the smaller, smoother, flattened cells are red blood cells. (From R.G. Kessel and R.H. Kardon, Tissues and Organs: A Text-Atlas of Scanning Electron Microscopy. San Francisco: Freeman, 1979. With permission from W.H. Freeman and Company.)

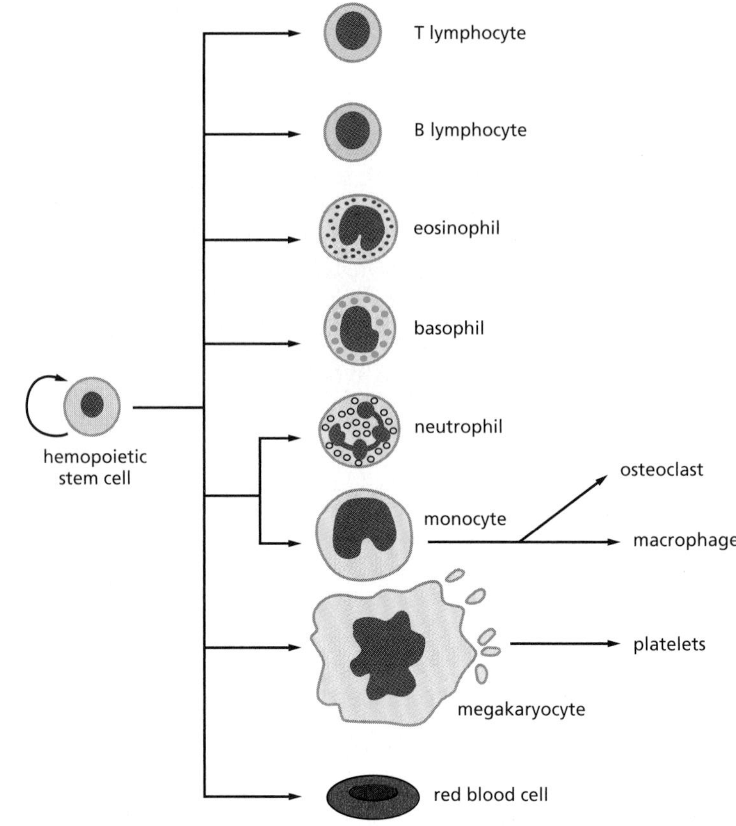

Figure 20–39 **A hemopoietic stem cell divides to generate more stem cells as well as precursor cells (not shown) that proliferate and differentiate into the mature blood cell types found in the circulation.** The macrophages found in many tissues of the body and the osteoclasts that eat away bone matrix originate from the same source, as do a few other types of tissue cells not shown in this scheme. Megakaryocytes give rise to blood platelets (Movie 20.5). A large number of different signal molecules are known to act at various points in this cell lineage to control the production of each cell type and to maintain appropriate numbers of stem cells.

each intestinal crypt in a proliferative state: the cells in these regions both secrete Wnt proteins and express the receptors for these proteins; and, apparently through positive feedback, they stimulate themselves to continue dividing (Figure 20–40). At the same time, these cells produce other signals, which act at longer range to prevent activation of the Wnt pathway outside the crypts. The cells within the crypt exchange yet other signals with one another to control their diversification, so that some differentiate into secretory cells while others become absorptive cells.

Disorders of these signaling mechanisms disrupt the structure of the gut lining. In particular, as we see later, defects in the regulation of Wnt signaling underlie the commonest forms of human intestinal cancer.

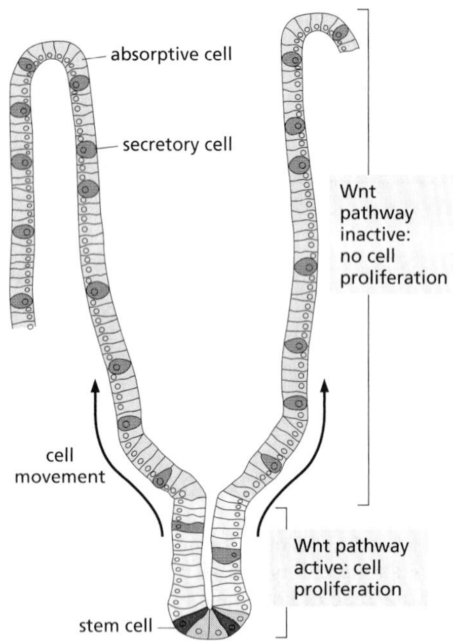

Figure 20–40 **The Wnt signaling pathway helps to control the production of differentiated cells from stem cells in the intestine.** Wnt signaling maintains proliferation in the crypt, where the stem cells reside and their progeny become committed to diverse fates.

Stem Cells Can Be Used to Repair Damaged Tissues

Because stem cells proliferate indefinitely and produce differentiated progeny, they provide for both continual renewal of normal tissue and repair of tissue lost through injury. For example, by transfusing a few hemopoietic stem cells into a mouse whose own blood stem cells have been destroyed by irradiation, it is possible to fully repopulate the animal with new blood cells and rescue it from death by anemia, infection, or both. A similar approach is used in the treatment of human leukemia with irradiation (or cytotoxic drugs) followed by bone marrow transplantation.

Stem cells taken directly from adult tissues hold promise for use in tissue repair, but another type of stem cell, first identified through experiments in mice, may have even greater potential. It is possible, through cell culture, to derive from early mouse embryos an extraordinary class of stem cells called **embryonic stem cells**, or **ES cells**. Under appropriate conditions, these cells can be kept proliferating indefinitely in culture and yet

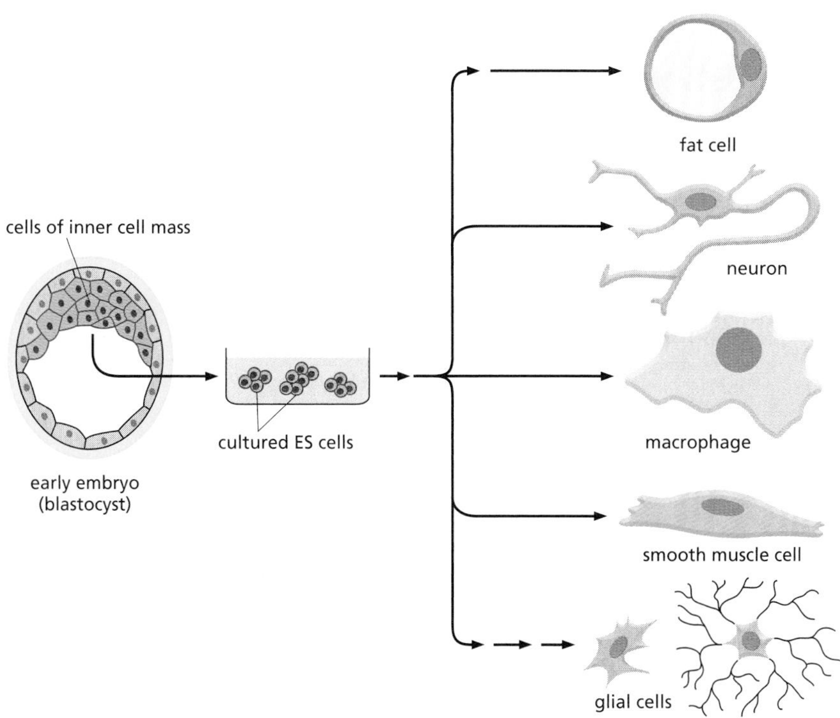

Figure 20–41 **ES cells derived from an embryo can give rise to all of the tissues and cell types of the body.** ES cells are harvested from the inner cell mass of an early embryo and can be maintained indefinitely as stem cells in culture. If they are put back into an embryo, they will integrate perfectly and differentiate to suit whatever environment they are placed in. The cells can also be kept in culture and supplied with different hormones or growth factors to encourage them to differentiate into specific cell types (**Movie 20.6**). (Based on data from E. Fuchs and J.A. Segré, *Cell* 100:143–155, 2000. With permission from Elsevier.)

retain unrestricted developmental potential and are thus said to be *pluripotent*. If the cells from the culture dish are put back into an early embryo, they can give rise to all the tissues and cell types in the body, including germ cells (Figure 20–41). Their descendants in the embryo are able to integrate perfectly into whatever site they come to occupy, adopting the character and behavior that normal cells would show at that site.

Cells with properties similar to those of mouse ES cells can now be derived from early human embryos, creating a potentially inexhaustible supply of cells that might be used for the replacement and repair of mature human tissues that are damaged. For example, experiments in mice suggest that it should be possible to use ES cells to replace the skeletal muscle fibers that degenerate in victims of muscular dystrophy, the nerve cells that die in patients with Parkinson's disease, the insulin-secreting cells that are destroyed in type I diabetics, and the cardiac muscle cells that die during a heart attack. Perhaps one day it might even become possible to grow entire organs from ES cells by a recapitulation of embryonic development.

There are, however, many hurdles to be cleared before such dreams can become reality. One major problem concerns immune rejection: if the transplanted cells are genetically different from the cells of the patient into whom they are grafted, they are likely to be rejected and destroyed by the immune system. A possible way around this problem is to use a strategy known colloquially as "therapeutic cloning," as we explain next.

Therapeutic Cloning Could Provide a Way to Generate Personalized ES Cells

The term 'cloning' has been used in confusing ways as a shorthand term for several quite distinct types of procedure, particularly in public debates about the ethics of stem-cell research. It is important to understand the distinctions.

As biologists define the term, a clone is simply a set of individuals that are genetically identical by virtue of their descent from a single ancestor. The simplest type of cloning is the cloning of cells. Thus, one can take

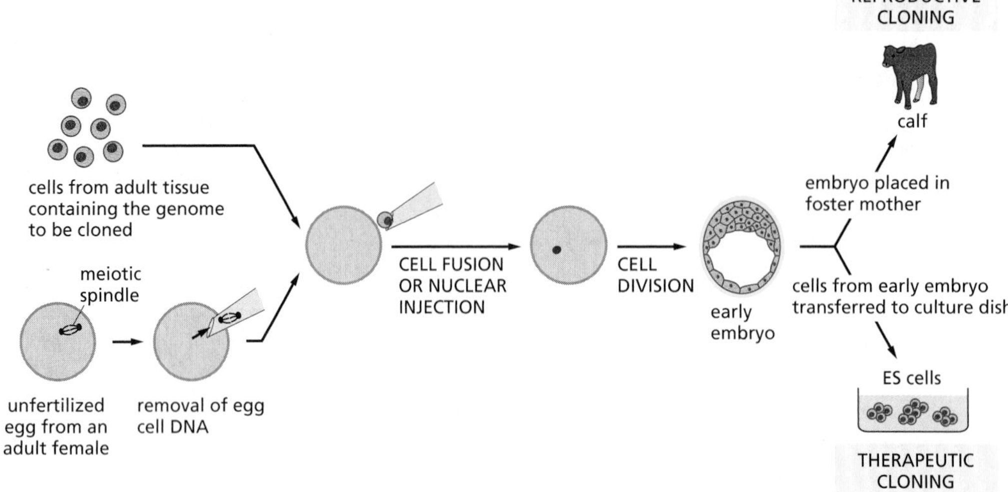

Figure 20–42 **Cells from adult tissue can be used for 'cloning' in two quite different senses of the word.** In reproductive cloning, a whole new multicellular individual is generated; in therapeutic cloning, only cells are produced. In the approach shown here, both procedures begin with nuclear transplantation: a nucleus taken from an adult cell is transferred into the cytoplasm of an egg cell, so as to create a cell that has an embryonic character but carries the genes of the adult cell.

a single epidermal stem cell from the skin and let it grow and divide in culture to obtain a large clone of genetically identical epidermal cells, which can, for example, be used to help reconstruct the skin of a badly burned patient. This kind of cloning is no more than an extension by artificial means of the processes of cell proliferation and repair that occur in a normal human body.

The cloning of entire multicellular animals, called **reproductive cloning**, is a very different enterprise, involving a far more radical departure from the ordinary course of nature. As we discuss in Chapter 19, each individual animal normally has both a mother and a father and is not genetically identical to either of them. In reproductive cloning, the need for two parents and sexual union is bypassed. For mammals, this difficult feat has been achieved in mice and sheep and some other domestic animals by a technique called *nuclear transplantation*. The procedure begins with an unfertilized egg cell. The nucleus of this haploid gamete cell is sucked out, and in its place a nucleus from a regular diploid cell is introduced. The diploid donor cell can, for example, be taken from a tissue of an adult individual. The hybrid cell, consisting of a diploid donor nucleus in a host egg cytoplasm, is allowed to develop for a short while in culture. In a small proportion of cases, this can give rise to an early embryo, which is then put into the uterus of a foster mother (Figure 20–42). If the experimenter is lucky, development continues as it would in a normal embryo, eventually giving rise to a whole new animal. An individual produced in this way, by reproductive cloning, should be genetically identical to the adult individual who donated the diploid cell (except for the small amount of genetic information in mitochondria, which are inherited with the egg cytoplasm).

Another procedure, different again from the ones just outlined, uses the technique of nuclear transplantation to produce cultured ES cells (see Figure 20–42). In this case, the cell that has received the transplanted nucleus is allowed to undergo the earliest steps of development, giving rise to a very early embryo, consisting of about 200 cells. But this embryo is not transferred to the uterus of a foster mother. Instead, it is used as a source from which ES cells are derived in culture, with the aim of generating various cell types that can be used for tissue repair. This so-called **therapeutic cloning** is an elaborate technique for generating personalized ES cells, rather than whole cloned animals. Because the cells obtained by this route are genetically identical to the original donor cell, they can be grafted back into the adult from whom the donor tissue was taken, without fear of immunological rejection. Nuclear transplan-

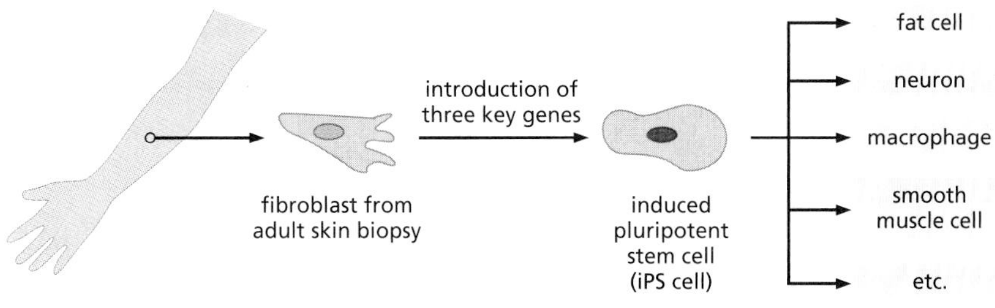

Figure 20–43 **Induced pluripotent stem cells (iPS cells) can be generated by direct transformation of cells from adult tissues.** This procedure begins by taking cells from an adult tissue—in this example, fibroblasts from a skin biopsy—and placing them in culture. Using genetically manipulated viruses as vectors, researchers then introduce into these cells artificial versions of a specific set of genes coding for transcription regulators that are normally expressed in ES cells. These gene constructs are expressed in the host fibroblasts, driving production of the corresponding proteins. After a few weeks in culture, some of the fibroblasts have transformed into cells that look and behave like ES cells and have the same ability as ES cells to differentiate into any of the cell types in the body.

tation is technically very difficult, and it has not yet been successful in human egg cells. The procedure requires a supply of human egg cells, which would have to be obtained from women donors, and it raises serious ethical problems. Indeed, nuclear transplantation into human egg cells is outlawed in some countries.

These ethical problems can be bypassed by a recent alternative approach, in which cells are taken from an adult tissue, grown in culture, and reprogrammed into an ES-like state by artificially introducing a specific set of genes, using genetically manipulated viruses as vectors. Investigators have found that expression of a set of just three genes (called *Oct3/4*, *Sox2*, and *Klf4*) is sufficient to convert fibroblasts into cells with practically all the properties of ES cells, including the ability to differentiate in diverse ways and to contribute to any tissue (Figure 20–43). These ES-like cells are called **induced pluripotent stem cells** (**iPS cells**). The conversion rate is low, however—only a small proportion of the fibroblasts make the switch—and there are serious worries about the safety of implanting derivatives of such virus-infected cells into patients. Much work remains to be done before this approach can be used to treat human diseases.

Meanwhile, however, human ES cells and especially human iPS cells promise to be immediately valuable in other ways. They can be used to generate large homogeneous populations of differentiated cells of a specific type in culture; these can serve for testing the effects of large numbers of chemical compounds in the search for new drugs with useful actions on a given human cell type. Moreover, it is possible to create iPS cells containing the genomes of patients who suffer from a given genetic disease, and to use these patient-specific stem cells for the discovery of drugs useful in the treatment of that disease. Such cells will be valuable also for analysis of the disease mechanism. And at a basic level, manipulations of ES and iPS cells in culture should help us fathom some of the many unsolved mysteries of stem-cell biology.

CANCER

We pay a price for having bodies that can renew and repair themselves. The delicately adjusted mechanisms that control these processes can go wrong, leading to catastrophic disruption of the body's structure. Foremost among the diseases of tissue renewal is **cancer**, which stands alongside infectious illness, malnutrition, war, and heart disease as a major cause of death among humans. In Europe and North America, for example, one in four of us will die of cancer.

Cancers arise from violations of the basic rules of social cell behavior. To make sense of the origins and progress of the disease, and to devise treatments, we have to draw upon almost every part of our knowledge of how cells work and interact in tissues. Conversely, much of what we know about cell and tissue biology has been discovered as a byproduct of cancer research. In this section, we examine the causes and mechanisms of cancer, the types of cell misbehavior that contribute to its progress, and the ways in which we hope to use our understanding to defeat these misbehaving cells and, hence, the disease.

Cancer Cells Proliferate, Invade, and Metastasize

If order is to be maintained as the tissues of the body grow and renew themselves, the individual cell must adjust its behavior according to the needs of the organism as a whole. The cell must divide when new cells of its specific type are needed, and refrain from dividing when they are not; it must live as long as it is required to live, and kill itself when it is required to die; it must maintain its appropriate specialized character; and it must occupy its proper place and not stray into inappropriate territories.

Of course, in a large organism, no significant harm is done if an occasional single cell misbehaves. But a potentially devastating breakdown of control occurs when a single cell suffers a genetic alteration that allows it to survive and divide when it should not, producing daughter cells that behave in the same asocial way. The organization of the tissue, and eventually that of the body as a whole, may then become disrupted by a relentlessly expanding clone of abnormal cells. It is this catastrophe that happens in cancer.

Cancer cells are defined by two heritable properties: they and their progeny (1) proliferate in defiance of the normal constraints and (2) invade and colonize territories normally reserved for other cells (Movie 20.7). It is the combination of these features that creates the lethal danger. Cells that have the first property but not the second proliferate excessively but remain clustered together in a single mass, forming a tumor, but the tumor in this case is said to be *benign*, and it can usually be removed cleanly and completely by surgery. A tumor is cancerous only if its cells have the ability to invade surrounding tissue, in which case the tumor is said to be *malignant*. Malignant tumor cells with this invasive property can break loose from the primary tumor, enter the bloodstream or lymphatic vessels, and form secondary tumors (*metastases*) at other sites in the body (Figure 20–44). The more widely the cancer spreads, the harder it becomes to eradicate.

Epidemiology Identifies Preventable Causes of Cancer

Prevention is always better than a cure, but to prevent cancer we need to know what causes it. Do factors in our environment or features of our way of life trigger the disease and help it to progress? If so, what are they? Answers to these questions come mainly from epidemiology—the statistical analysis of human populations that is used to look for factors that correlate with disease incidence. This approach has provided strong evidence that the environment plays a part in the causation of most cases of cancer. The types of cancers that are common, for example, vary from country to country, and studies of migrants show that it is usually where people live, rather than where they were born, that governs their cancer risk. Although it is still hard to discover which specific factors in the environment or lifestyle are significant, and many remain unknown, some have been identified quite precisely. Thus, it was noted long ago that cervical cancer, arising in the epithelium lining the cervix (neck) of the uterus, was much commoner in married women than in single women. This pointed to a cause related to sexual activity. We now know, through modern epidemiological studies, that most cases of cervical cancer involve infection of the cervical epithelium with certain subtypes of a common virus, called human papilloma virus. This is transmitted through sexual intercourse and can sometimes, if one is unlucky, provoke uncontrolled proliferation of the infected cells. Knowing this, we can attempt to prevent the cancer by preventing the infection—for example, by vaccina-

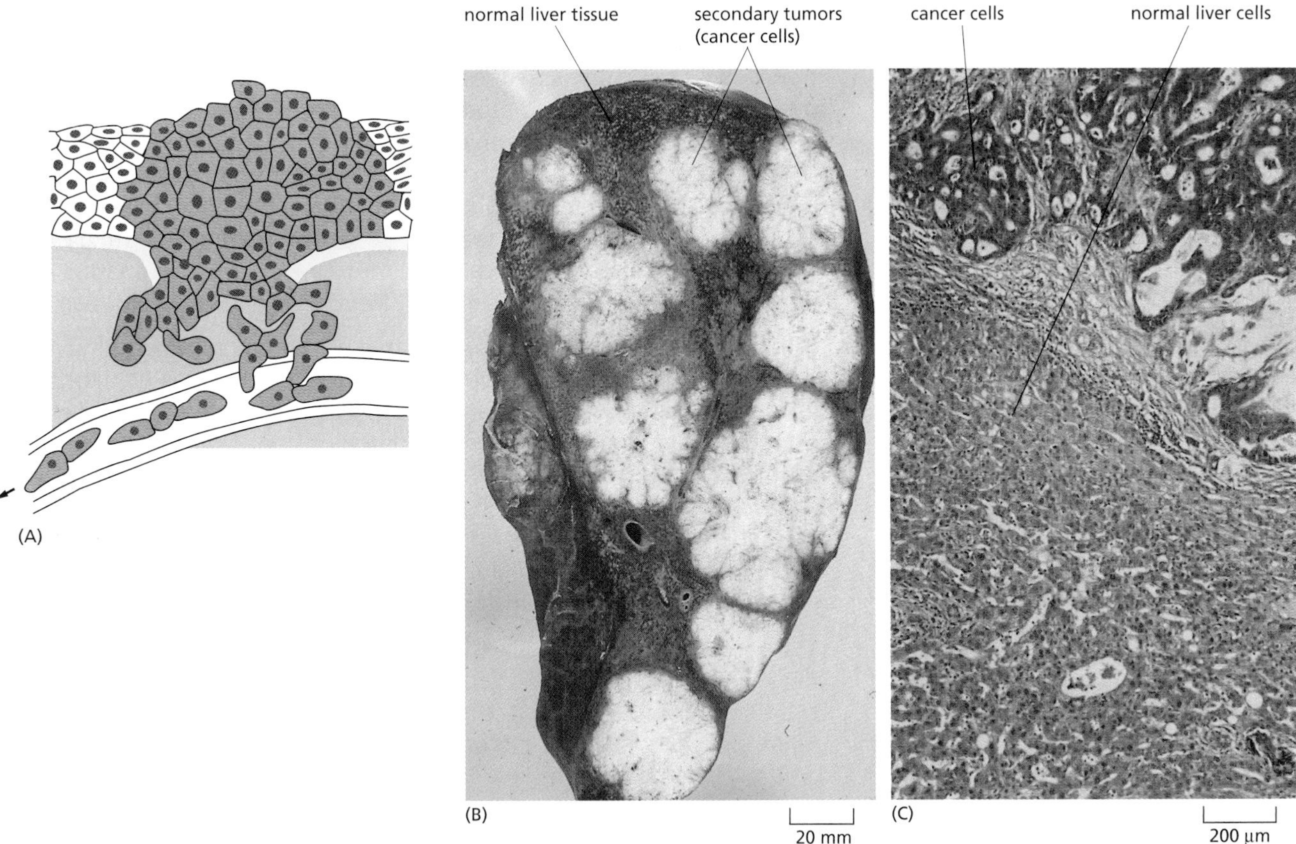

Figure 20–44 **Cancers spread to invade surrounding tissues.** (A) To give rise to a colony in a new site, the cells of a primary tumor in an epithelium must typically cross the basal lamina, migrate through connective tissue, and get into the blood or lymphatic vessels. They then have to exit from the bloodstream or lymph and settle and survive in the new location. (B) Secondary tumors in a human liver, originating from a primary tumor in the colon. (C) Higher-magnification view of one of the secondary tumors, stained differently to show the contrast between the normal liver cells and the cancer cells. (B and C, courtesy of Peter Isaacson.)

tion against papilloma virus. Such a vaccine is now available, conferring a high level of protection if given to girls when they are young, before they become sexually active.

In the great majority of human cancers, however, viruses do not appear to play a part: cancer is not an infectious disease. But epidemiology reveals other factors. Obesity, for example, is correlated with an increased cancer risk, and the relationship is suspected to be causal. By far the most important environmental cause of cancer in the modern world, however, is tobacco-smoking, which is not only responsible for almost all cases of lung cancer but also raises the incidence of several other cancers, such as those of the bladder. If we could stop the use of tobacco, it is estimated that we could prevent about 30% of all cancer deaths. No other single policy or treatment is known that would have such an impact on the cancer death rate.

As we explain below, although environmental factors affect the incidence of cancer and are critical for some forms of the disease, it would be wrong to conclude that they are the fundamental cause of cancers in general. No matter how hard we try to prevent cancer by healthy living, we will never be able to eradicate it entirely; we will always be confronted with cases that demand treatment. To devise treatments that will succeed, we need to understand the biology of cancer cells and the mechanisms that underlie the growth and spread of tumors.

Cancers Develop by an Accumulation of Mutations

Cancer is fundamentally a genetic disease: it arises as a consequence of pathological changes in the information carried by DNA. It differs from other genetic diseases in that the mutations underlying cancer are mainly somatic mutations—those that occur in individual cells of the mature body—as opposed to germ-line mutations, which are handed down via the germ cells from which the entire multicellular organism develops.

Most of the identified agents known to contribute to the causation of cancer, including ionizing radiation and most chemical carcinogens, are mutagens: they cause changes in the nucleotide sequence of DNA. But even in an environment that is free of tobacco smoke, radioactivity, and all the other external mutagens that worry us, mutations will occur spontaneously as a result of fundamental limitations on the accuracy of DNA replication and DNA repair (discussed in Chapter 6). In fact, environmental carcinogens other than tobacco probably account for only a small fraction of the mutations responsible for cancer, and elimination of all these external risk factors would still leave us prone to the disease.

Although DNA is replicated and repaired with great accuracy, an average of one mistake slips by for every 10^9 or 10^{10} nucleotides copied, as we discuss in Chapter 6. This means that spontaneous mutations occur at an estimated rate of about 10^{-6} or 10^{-7} mutations per gene per cell division, even without encouragement by external mutagens. About 10^{16} cell divisions take place in a human body in the course of a lifetime; thus, every single gene is likely to have undergone mutation on more than 10^9 separate occasions in any individual. From this point of view, the problem of cancer seems to be not why it occurs, but why it occurs so infrequently.

The explanation is that it takes more than a single mutation to turn a normal cell into a cancer cell. Precisely how many mutations are required is still a matter of debate, but it is certainly more than two or three. These mutations do not all occur at once, but sequentially, usually over a period of many years.

Cancer, therefore, is most often a disease of old age, because it takes a long time for an individual line of cells to accumulate a large number of mutations (see Figure 6–20). In fact, most human cancer cells not only contain many mutations but also are genetically unstable. This **genetic instability** results from mutations that interfere with the accurate replication and maintenance of the genome and thereby increase the mutation rate itself. Sometimes, the increased mutation rate may result from a defect in one of the many proteins needed to repair damaged DNA and to correct errors in DNA replication. Sometimes, there may be a defect in the checkpoint mechanisms that normally prevent a cell with damaged DNA from attempting to divide before it has completed a repair (discussed in Chapter 18). Sometimes, there may be a fault in the machinery of mitosis. The consequences of these defects in the way the cancer cell handles its DNA are often manifested in chromosome breaks and rearrangements, resulting in a grossly abnormal and unstable karyotype (Figure 20–45).

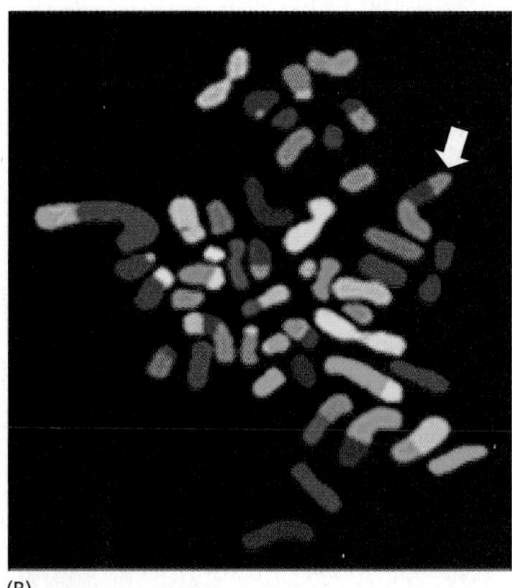

Figure 20–45 **Cancer cells often have highly abnormal chromosomes, reflecting genetic instability.** In the example shown here, chromosomes were prepared from a breast cancer cell in metaphase, spread on a glass slide, and stained with (A) a general DNA stain or (B) a combination of fluorescent stains that give a different color for each human chromosome. The staining (displayed in false color) shows multiple translocations, including a doubly translocated chromosome (*white* arrow) made up of two pieces of chromosome 8 (*green*) and a piece of chromosome 17 (*purple*). The karyotype also contains 48 chromosomes, instead of the normal 46. (Courtesy of Joanne Davidson and Paul Edwards.)

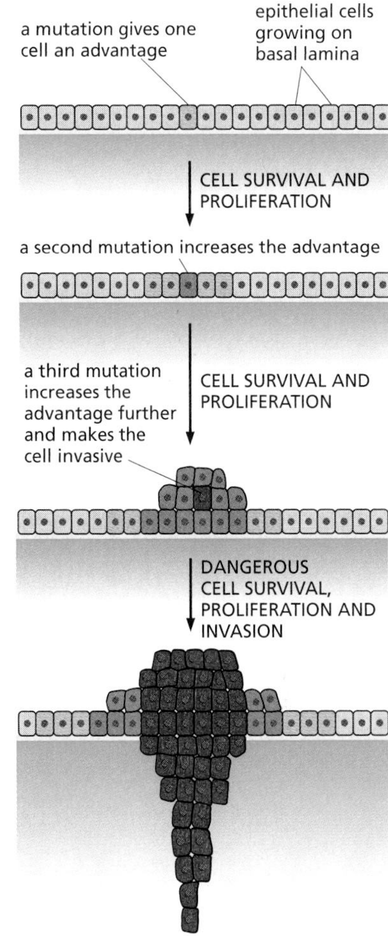

Figure 20–46 **Tumors evolve by repeated rounds of mutation and proliferation.** The final outcome is a clone of fully malignant cancer cells. At each step, a single cell undergoes a mutation that enhances its ability to proliferate, or survive, or both, so that its progeny become the dominant clone in the tumor. Proliferation of this clone then hastens occurrence of the next step of tumor progression by increasing the size of the cell population at risk of undergoing an additional mutation.

Cancer Cells Evolve Properties that Give Them a Competitive Advantage

The mutations that lead to cancer do not cripple the mutant cells. On the contrary, they give these cells a competitive advantage over their neighbors. It is this advantage enjoyed by the mutant cells that leads to disaster for the organism as a whole. Natural selection favors cells carrying mutations that enhance cell proliferation and cell survival, regardless of the effects on neighbors, and this process culminates in the genesis of cancer cells that run riot within the population of cells that form the body, upsetting its regular structure. As an initial population of mutant cells grows, it slowly evolves: new chance mutations occur in the member cells, and some of these mutations are favored by natural selection (Figure 20–46). Non-mutagenic environmental or lifestyle factors such as obesity may favor the development of cancer by altering the selection pressures that operate in the tissues of the body. Circulating nutrients or hormones, for example, may help cells with dangerous mutations to survive and proliferate. Eventually, cells emerge that have all the abnormalities required for full-blown cancer.

To be successful, a cancer cell must acquire a whole range of abnormal properties—a collection of subversive new skills. An epithelial stem cell in the lining of the gut, for example, must undergo changes that permit it to carry on dividing when it should stop. That cell and its progeny must also be able to displace their normal neighbors and to attract a blood supply to nourish continued tumor growth. For these cells to then become invasive, they must acquire the ability to digest their way through the basal lamina of the epithelium into the underlying tissue. To spread to other tissues, an ability known as *metastasis*, they must be able to get in and out of the blood or lymph circulation and settle and survive in new sites (see Figure 20–44).

Different cancers require different combinations of properties. Nevertheless, we can draw up a general list of key behaviors of cancer cells that distinguish them from normal cells.

1. They have a reduced dependence on signals from other cells for their growth, survival, and division. Often, this is because they contain mutations in components of the cell signaling pathways through which cells respond to such social cues. A mutation in a *Ras* gene (discussed in Chapter 16), for example, can cause an intracellular signal for proliferation to be produced even in the absence of the extracellular signal that would normally be needed to trigger it, like a faulty doorbell that rings even when nobody is pressing the button.

2. Cancer cells are less prone than normal cells to kill themselves by *apoptosis*. This aversion to suicide is often caused by mutations in genes that regulate the intracellular death program responsible for apoptosis (discussed in Chapter 18). For example, about 50% of all human cancers have lost or suffered a mutation in the *p53* gene. The p53 protein normally acts as part of a checkpoint mechanism that causes cells either to cease dividing (see Figure 18–13) or to die

QUESTION 20–8

About 10^{16} cell divisions take place in a human body during a lifetime, yet an adult human body consists of only about 10^{13} cells. Why are these two numbers so different?

by apoptosis when their DNA is damaged. Chromosome breakage, for example, if not repaired, will generally cause a cell to commit suicide; but if the cell is defective in p53, it may survive and divide, creating highly abnormal daughter cells that can become more malignant.

3. Unlike most normal human cells, cancer cells can often proliferate indefinitely. Most normal human somatic cells will only divide a limited number of times in culture, after which they permanently stop, apparently because these cells do not produce the enzyme telomerase, so the telomeres on the ends of their chromosomes become too short (see page 210). Cancer cells typically break through this barrier by reactivating production of the telomerase enzyme that maintains telomere length.
4. Most cancer cells, as noted above, are genetically unstable, with a greatly increased mutation rate.
5. Cancer cells are abnormally invasive, and this is often in part because they lack specific cell-adhesion molecules, such as cadherins, that hold normal cells in their proper place.
6. Cancer cells can often survive and proliferate in foreign tissues to form secondary tumors (metastases), whereas most normal cells die when misplaced. We still do not understand precisely what molecular changes are needed to confer this ability.

To understand the molecular biology of cancer, we have to be able to identify the mutations that give rise to these abnormal forms of behavior.

Many Diverse Types of Genes Are Critical for Cancer

A great variety of approaches have been used to track down the genes and mutations that are critical for cancer. Though many of the most important of these genes have been identified, the hunt for others continues.

In some cases, the dangerous mutations are ones that make the affected gene product hyperactive. These mutations have a dominant effect: only one gene copy needs to be mutated to cause trouble; the resulting mutant gene is called an **oncogene** (Figure 20–47A); the corresponding normal

(A) dominant mutation (gain of function)
normal cell
single mutation event in proto-oncogene
creates oncogene
activating mutation enables oncogene to stimulate cell survival and proliferation
excessive cell survival and proliferation
(B) recessive mutation (loss of function)
normal cell
mutation event
inactivates tumor suppressor gene
no effect of mutation in one gene copy
second mutation event
inactivates second gene copy
two inactivating mutations functionally eliminate the tumor suppressor gene, stimulating cell survival and proliferation
excessive cell survival and proliferation

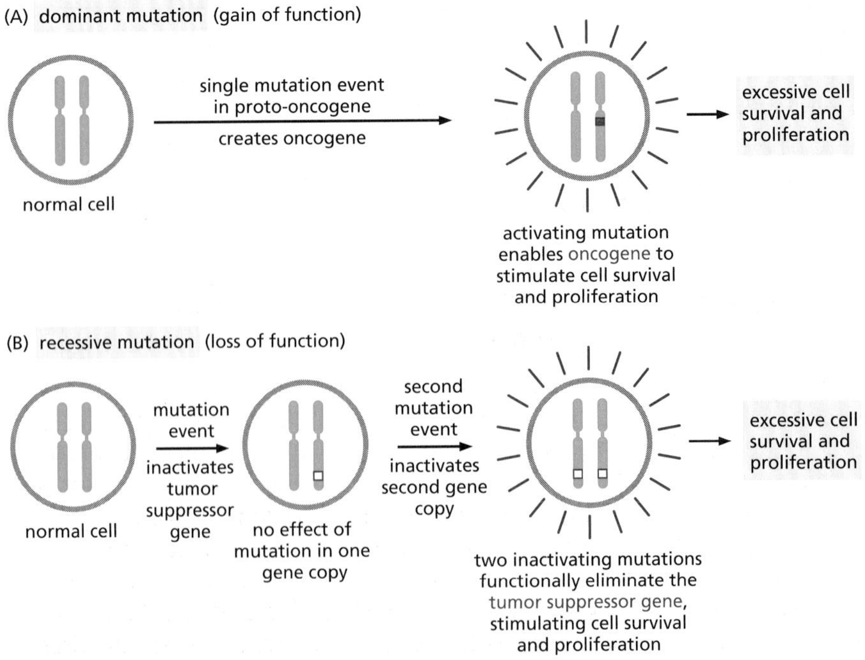

Figure 20–47 **Genes that are critical for cancer are classified as proto-oncogenes or tumor suppressor genes, according to whether the dangerous mutations are dominant or recessive.** Oncogenes act in a dominant manner: a gain-of-function mutation in a single copy of the proto-oncogene can drive a cell toward cancer. Mutations in tumor suppressor genes, on the other hand, generally act in a recessive manner: the function of both alleles of the gene must be lost to drive a cell toward cancer. In this diagram, activating mutations are represented by *solid red* boxes, inactivating mutations by *hollow red* boxes.

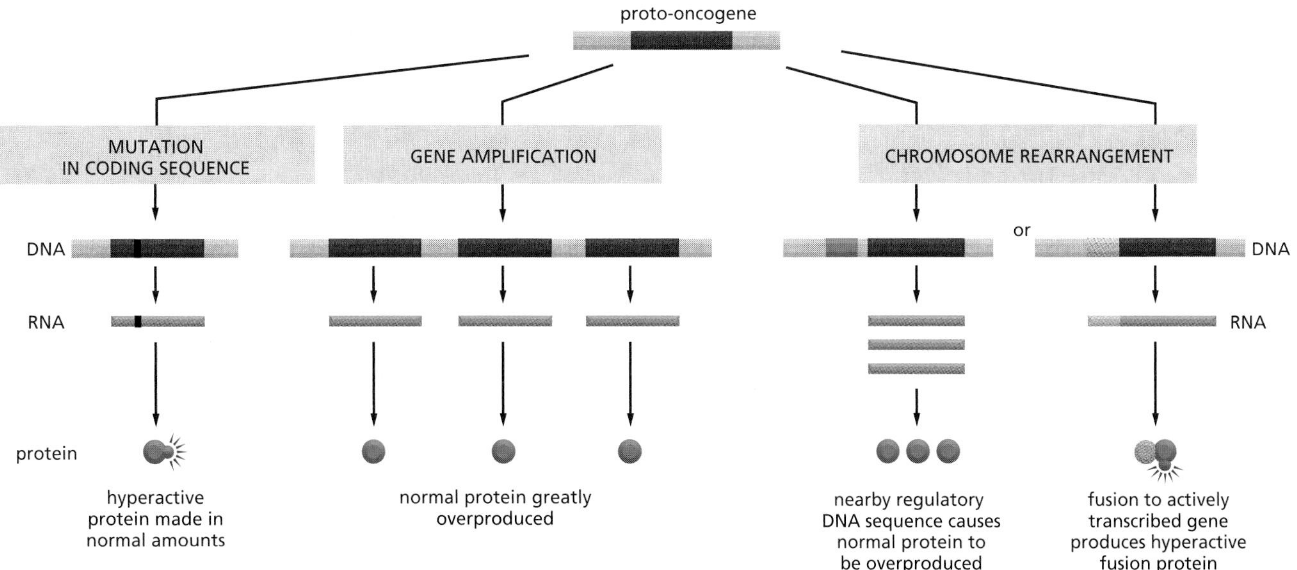

Figure 20–48 **Several kinds of genetic change can convert a proto-oncogene into an oncogene.** In each case, the change leads to an increase in the gene's function.

form of the gene is called a **proto-oncogene**. Figure 20–48 shows a variety of ways in which the conversion from proto-oncogene to oncogene can occur.

For other genes, the danger lies in mutations that destroy gene function. These mutations are generally recessive: both gene copies must be lost or inactivated before an effect is seen; the affected gene is called a **tumor suppressor gene** (see Figure 20–47B). Tumor suppressor genes were first identified by studies of human genetics. Occasionally, individuals are encountered who have inherited a mutation in a tumor suppressor gene; although one gene copy is enough for normal cell behavior, the cells of these individuals are only one mutational step away from total loss of the gene's function (as compared to two steps away for a normal person). Because the number of additional mutations required for cancer is smaller, the disease occurs with higher frequency and on average at an earlier age, sometimes in childhood. The families that carry such mutations are therefore unusually prone to cancer.

Proto-oncogenes and tumor suppressor genes are of many sorts, corresponding to the many different kinds of misbehavior that cancer cells display. Some of these genes code for growth factors, for receptors, or—like *Ras*—for components of the intracellular signaling pathways that growth factors activate. Others code for DNA repair proteins, for mediators of the DNA damage response such as p53, or for regulators of the cell cycle or of apoptosis. Still others, as we have mentioned, code for cell adhesion molecules such as cadherins. Figure 20–49 conveys some idea of this diversity.

Colorectal Cancer Illustrates How Loss of a Gene Can Lead to Growth of a Tumor

Colorectal cancer provides a well-studied example to show, first, how a tumor suppressor gene can be identified and, second, how its identification leads on to an understanding of a basic molecular mechanism underlying the growth of a common type of tumor. Colorectal cancer arises from the epithelium lining the colon and rectum; most cases are seen in old people and do not have any discernible hereditary cause. A small proportion of cases, however, occur in families that are exceptionally prone to the disease and show an unusually early onset. In one set of families, the predisposition to cancer has been traced to an inherited

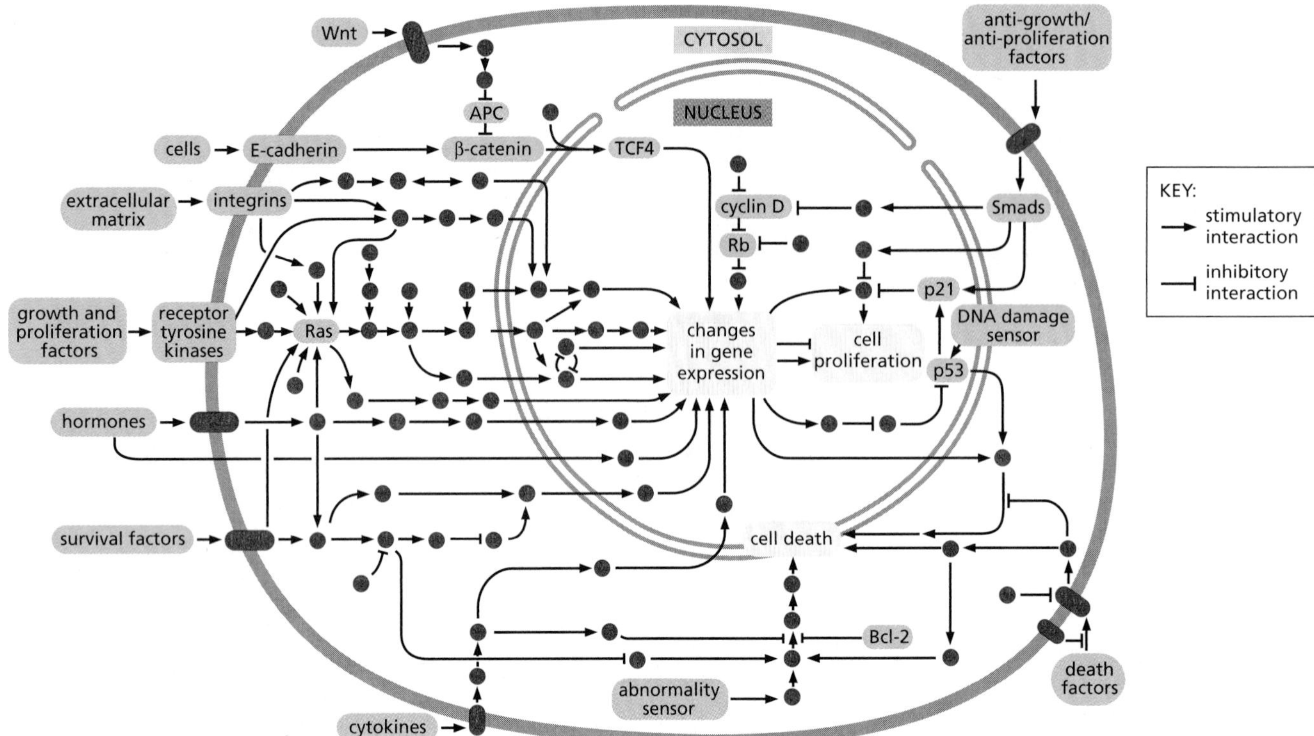

Figure 20–49 **Many types of genes are critical for cancer.** The cartoon shows the major signaling pathways relevant to cancer in human cells, indicating the intracellular locations of some of the proteins modified by mutation in cancers. Products of both oncogenes and tumor suppressor genes often occur within the same pathways. Individual signaling proteins are indicated by *solid red* circles, with the cancer-critical components and control mechanisms discussed in this book in *green.* Stimulatory and inhibitory interactions between components are indicated by arrows and bars, respectively, as shown in the key. (Adapted from D. Hanahan and R.A.Weinberg, *Cell* 100:57–70, 2000. With permission from Elsevier.)

mutation in a DNA repair enzyme, as discussed in Chapter 6. In another class of hereditary colorectal cancer patients, a different mutation is present, leading to a highly distinctive phenotype. The affected individuals develop colorectal cancer in early adult life, and the onset of their disease is foreshadowed by the development of hundreds or thousands of little tumorous growths, called polyps, in the lining of the colon and rectum. Through family studies, the abnormality can be traced to deletion or inactivation of a gene called the *Adenomatous Polyposis Coli* (*APC*) gene. Affected individuals inherit one mutant copy of the gene and one normal copy; their cancers arise from cells that can be shown to have undergone a somatic mutation that inactivates the remaining good copy. But what about the great majority of colorectal cancer patients, who have inherited two good copies of *APC* and do not have the hereditary condition or any significant family history of cancer? When their tumors are analyzed, it turns out that in more than 60% of cases, although both copies of *APC* are present in the adjacent normal tissue, the tumor cells themselves have lost both copies of this gene, presumably through two independent somatic mutations.

All this clearly identifies *APC* as a tumor suppressor gene, and knowing its sequence and mutant phenotype one can begin to decipher how its loss helps to initiate the development of cancer. As explained in How We Know (pp. 725–726), the *APC* gene was found to encode an inhibitory protein that normally restricts the activation of the Wnt signaling pathway, which is involved in stimulating cell proliferation in the crypts of the gut lining as described earlier. When *APC* is lost, the pathway is hyperactive and

HOW WE KNOW:

MAKING SENSE OF THE GENES THAT ARE CRITICAL FOR CANCER

The search for genes that are critical for the development of cancer sometimes begins with a family that shows a hereditary predisposition to a particular form of the disease. *APC*—a tumor suppressor gene that is frequently deleted or inactivated in people with colorectal cancer—was tracked down by searching for genetic defects in families prone to the disease. But identifying the gene is only half the battle. The next step is determining what the gene does in a normal cell—and why alterations in that gene promote cancer.

Guilt by association

Determining what a gene—or its encoded protein product—does inside a cell is not a simple task. Imagine isolating an uncharacterized protein and being told that it acts as a protein kinase. That information does not reveal how the protein functions in the context of a living cell. What are its protein targets? In which tissues is it active? What role does it have in the growth or development of the organism? Additional information is required to understand the context in which the biochemical activity is used.

Most proteins do not function in isolation: they interact with other proteins inside the cell. Thus one way to begin to decipher a protein's biological role is to identify its binding partners. If an uncharacterized protein interacts with a protein whose role in the cell is understood, its function is likely to be in some way related. Perhaps the simplest method for identifying proteins that bind to one another tightly is co-immunoprecipitation (see Panel 4–3, pp. 144–145). In this technique, an antibody is used to capture and precipitate a specific target protein from an extract prepared by breaking open cells; if this target protein is associated tightly with another protein, the partner protein will precipitate as well. This is the approach that was taken to characterize APC.

Two groups of researchers used antibodies against APC to isolate the protein from extracts prepared from cultured human cells. The antibodies captured APC along with a second protein. When the researchers examined the amino acid sequence of this partner, they recognized the protein as β-catenin.

The discovery that APC interacts with β-catenin initially led to wrong guesses about the role of APC in colorectal cancer. In mammals, β-catenin was known primarily for its role at adherens junctions between cells, where it serves as a linker to connect the membrane-spanning cadherin molecules to the intracellular actin cytoskeleton (see, for example, Figure 20–24). Thus, for some time scientists thought that APC might be involved in cell adhesion. But within a few years it emerged that β-catenin also has another completely different function and that APC's interaction with it is important in cancer for a quite different reason.

Wingless flies

Not long before the discovery that APC binds to β-catenin, developmental biologists working on the fruit fly *Drosophila* had noticed that the human β-catenin protein is highly similar in sequence to a *Drosophila* protein called Armadillo. Armadillo was known to be a key protein in a signaling pathway that plays an important role in normal development in fruit flies. The pathway is activated by a family of signal molecules called Wnt proteins; the founding member of the *Wnt* gene family was called *wingless*, after its mutant phenotype. Wnt proteins bind to receptors on the surface of a cell, switching on an intracellular signaling cascade that ultimately leads to the activation of a set of genes that influence cell growth, division, and differentiation. Mutations in any of the proteins in this pathway lead to developmental errors that disrupt the basic body plan of the fly. The least devastating mutations cause flies to develop without wings; most mutations, however, result in the death of the embryo. But in any case, the damage is done, it seems, through effects on gene expression. This strongly suggested that Armadillo, and hence its vertebrate homolog β-catenin, were not just nuts and bolts in the apparatus of cell adhesion, but somehow mediated the control of gene expression.

The Wnt pathway was discovered and studied intensively in fruit flies, but it turns out that a similar set of proteins controls many aspects of development in vertebrates, including mice and humans. Indeed, some of the proteins in the Wnt pathway function almost interchangeably in *Drosophila* and vertebrates. The direct link between β-catenin and gene expression became clear from work in mammalian cells. Just as APC could be used as a 'bait' to catch its partner β-catenin by immunoprecipitiation, so β-catenin could be used as a bait to catch the next protein in the chain of cause and effect. This was found to be a transcription regulator called LEF-1/TCF, or TCF for short. It too was found to have a *Drosophila* counterpart in the Wnt pathway, and a combination of *Drosophila* genetics and mammalian cell biology revealed how the gene control mechanism works.

Wnt transmits its signal by promoting the accumulation of 'free' β-catenin (or, in flies, Armadillo)—that is, of β-catenin that is not locked up in adherens junctions. This free protein migrates from the cytoplasm into the nucleus. There it binds to the TCF transcription regulator, creating a complex that activates transcription of various Wnt-responsive genes, including genes whose products stimulate cell proliferation (Figure 20–50).

APC regulates the activity of this pathway by facilitating degradation of β-catenin and thereby preventing it from activating TCF in cells where no Wnt signal has been received (see Figure 20–50). Loss of APC allows the concentration of β-catenin to rise, so that TCF is activated and Wnt-responsive genes are turned on even in the absence of Wnt. But how does this cause colorectal cancer? To find out, researchers turned to mice that lack TCF4, a member of the TCF gene family that is specifically expressed in the gut lining.

Tales from the crypt

Although it may sound counterintuitive, one of the most direct ways of finding out what a gene does is to see what happens to the organism when that gene is missing. If one can pinpoint the cellular processes that are disrupted or compromised, one can begin to decipher the gene's function.

With this in mind, researchers generated 'knockout' mice in which the gene encoding TCF4 was disrupted. The mutation is lethal: mice lacking TCF4 die shortly after birth. But the animals showed an interesting abnormality in their intestines. The intestinal crypts, with their populations of stem cells for renewal of the gut lining (see Figure 20–36), completely failed to develop. The researchers concluded that TCF4 is normally responsible for maintaining the pool of proliferating gut stem cells.

When APC is missing, we see the other side of the coin: without APC to promote its degradation, β-catenin accumulates in excessive quantities, binds to the TCF4 transcription regulator, and thereby overactivates the TCF4-responsive genes. This drives the formation of polyps by promoting the inappropriate proliferation of gut stem cells. Differentiated progeny cells continue to be produced and discarded into the gut lumen, but the crypt cell population grows too fast for this disposal mechanism to keep pace. The result is crypt enlargement and a steady increase in the number of crypts. The growing mass of tissue bulges out into the gut lumen as a polyp (see Figure 20–51). Further mutations are needed to convert this primary tumor into an invasive cancer.

More than 60% of human colorectal tumors harbor mutations in the *APC* gene. Interestingly, among the minority class of tumors that retain functional APC, about a quarter have activating mutations in β-catenin instead. These mutations tend to make the β-catenin protein more resistant to degradation and thus produce the same effect as loss of APC. In fact, mutations that enhance the activity of β-catenin have been found in a wide variety of other tumor types, including melanomas, stomach cancers, and liver cancers. Thus, the Wnt signaling pathway provides multiple targets for mutations that can spur the development of cancer.

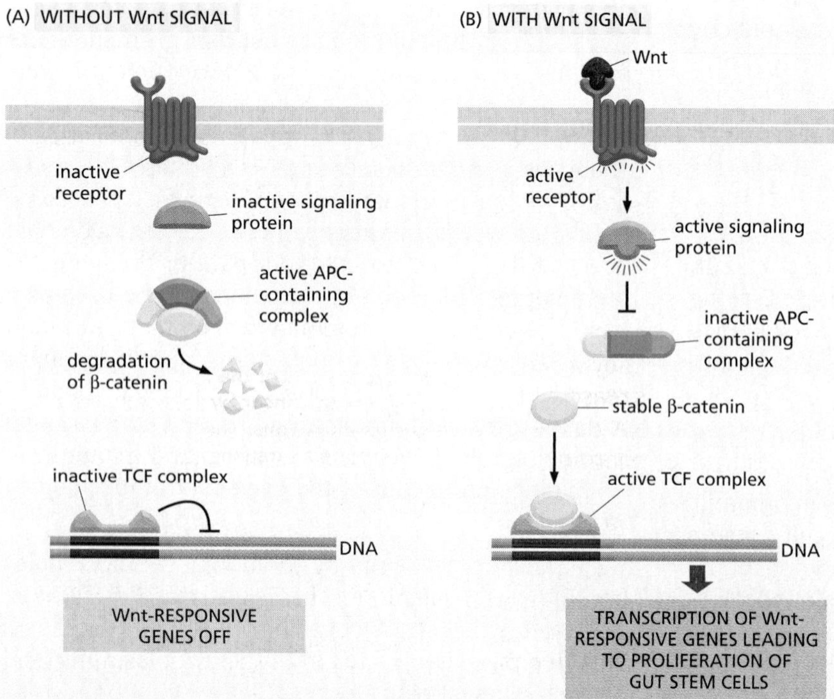

Figure 20–50 **The APC protein keeps the Wnt signaling pathway inactive when the cell is not exposed to Wnt protein.** It does this by promoting degradation of the signaling molecule β-catenin. In the presence of Wnt, or in the absence of active APC, free β-catenin becomes plentiful and combines with the transcription regulator TCF to drive transcription of Wnt target genes and ultimately, the proliferation of stem cells in the intestinal crypt. In the colon, mutations that inactivate APC initiate tumors by causing excessive activation of the Wnt pathway,

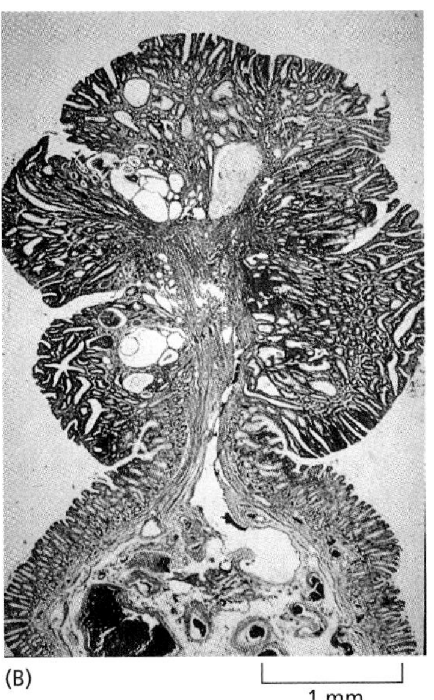

(A) (B)

Figure 20–51 **Colorectal cancer often begins with loss of the tumor suppressor gene *APC*, leading to growth of a polyp.** (A) Thousands of small polyps, and a few much larger ones, are seen in the lining of the colon of a patient with an inherited *APC* mutation (one or two polyps might be seen in a genetically normal person). Through further mutations, some of these polyps will progress to become malignant cancers, if the tissue is not removed surgically. (B) Cross section of one such polyp; note the excessive quantities of deeply infolded epithelium, corresponding to crypts full of abnormal, proliferating cells. (A, courtesy of John Northover and Cancer Research UK; B, courtesy of Anne Campbell.)

epithelial cells proliferate to excess, generating a polyp (Figure 20–51). Within this growing mass of tissue, further mutations may occur, resulting in invasive cancer (Figure 20–52).

An Understanding of Cancer Cell Biology Opens the Way to New Treatments

The better we understand the tricks that cancer cells use to survive, proliferate, and spread, the better are our chances of finding ways to defeat them. The task is hard, because cancer cells are mutable and, like weeds or parasites, rapidly evolve resistance to treatments used to exterminate them. Moreover, because mutations arise randomly, each case of each variety of cancer is liable to have its own unique combination of genes mutated. Thus, no single treatment is likely to work in every patient. Moreover, cancers generally are not detected until the primary tumor has reached a diameter of 1 cm or more, by which time it consists of hundreds of millions of cells that are already genetically diverse and often have already begun to metastasize (Figure 20–53).

In spite of the difficulties, many cancers can be treated effectively, and the future prospects for more and better treatments are bright. Surgery remains the most effective tactic, and surgical techniques are continually improving: in many cases, if a cancer has not spread far, it can often be cured by simply cutting it out. Where surgery fails, therapies based on the intrinsic peculiarities of cancer cells can be used. Lack of normal checkpoint mechanisms, for example, may help to make cancer cells particularly vulnerable to DNA damage: whereas a normal cell will halt its proliferation until damage is repaired, a cancer cell may charge ahead regardless, producing daughter cells that may die because they inherit a broken, incomplete set of chromosomes. Presumably for this reason, cancer cells can often be killed by doses of radiotherapy or DNA-damaging chemotherapy that leave adjacent normal cells relatively unharmed.

These treatments are long established, but many novel approaches are also being enthusiastically pursued. In some cases, as with loss of checkpoints, the very feature that helps to make the cancer cell dangerous also makes it vulnerable, enabling us to kill it with a properly targeted treatment. Some cancers of the breast and ovary, for example, owe their genetic instability to the lack of a protein (Brca1 or Brca2) needed for accurate repair of double-strand breaks in DNA (discussed in Chapter 6);

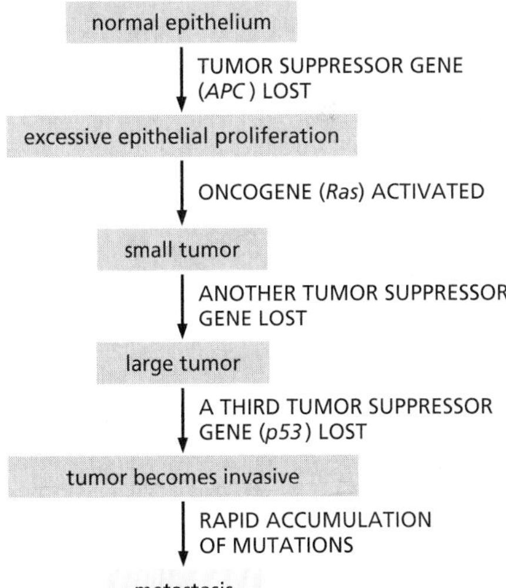

Figure 20–52 **A polyp in the gut lining, caused by loss of the *APC* gene, can progress to cancer by accumulation of further mutations.** The diagram shows a sequence of mutations that might underlie a typical case of colorectal cancer. A sequence of events such as that shown here would usually be spread over 10 to 20 years or more. Though most colorectal cancers are thought to begin with loss of the *APC* tumor suppressor gene, the subsequent sequence of mutations is quite variable; indeed, many polyps never progress to cancer.

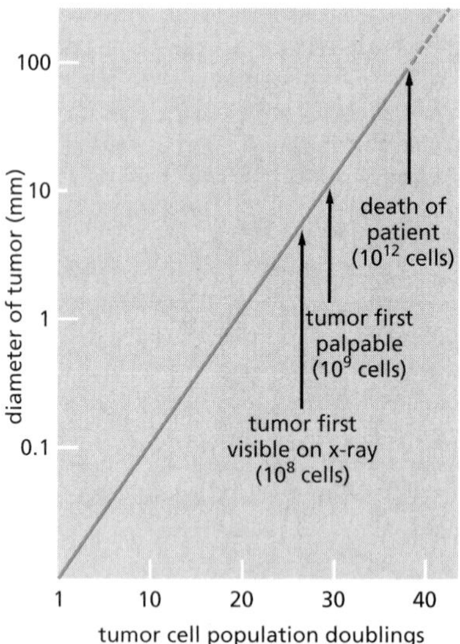

Figure 20–53 **A tumor is generally not diagnosed until it has grown to contain hundreds of millions of cells.** Here, the growth of a typical tumor is plotted on a logarithmic scale. Years may elapse before the tumor becomes noticeable. The doubling time of a typical breast tumor, for example, is about 100 days.

the cancer cells survive by relying on other machinery that provides alternative types of DNA repair. A drug that inhibits one of these alternative DNA repair pathways kills the cancer cells by raising their genetic instability to such a level that the cells die from chromosome fragmentation when they attempt to divide. Normal cells, which have intact double-strand break repair machinery, are relatively unaffected, and the drug seems to have few side-effects.

Another promising strategy is to block formation of the new blood vessels that normally invade a growing tumor (Movie 20.8). This approach should choke tumor growth by depriving the cells of their blood supply. Yet another set of strategies aim in various ways to use the immune system to kill the tumor cells, taking advantage of tumor-specific cell-surface molecules to target the attack. Vaccination with tumor-specific molecules can stimulate the patient's own immune system to turn against the tumor, or antibodies against these tumor molecules can be produced *in vitro* and injected into the patient to mark the tumor cells for destruction.

In some cancers, it is becoming possible to target the products of specific oncogenes directly so as to block their harmful action. In chronic myeloid leukemia (CML), the misbehavior of the cancerous cells is known to depend on a mutant signaling protein (a tyrosine protein kinase) that causes the cells to proliferate when they should not. A small drug molecule, called Gleevec, has been designed to block the activity of this kinase (Figure 20–54). The results have been a dramatic success: in many patients, the abnormal proliferation and survival of the leukemic cells is strongly inhibited and a prolonged remission of symptoms is achieved. The same drug is effective in some other cancers containing similar oncogenes.

With these examples before us, we can hope that soon, equipped with our modern understanding of the molecular biology of cancer, it will be possible to devise effective rational methods of treatment for a still wider

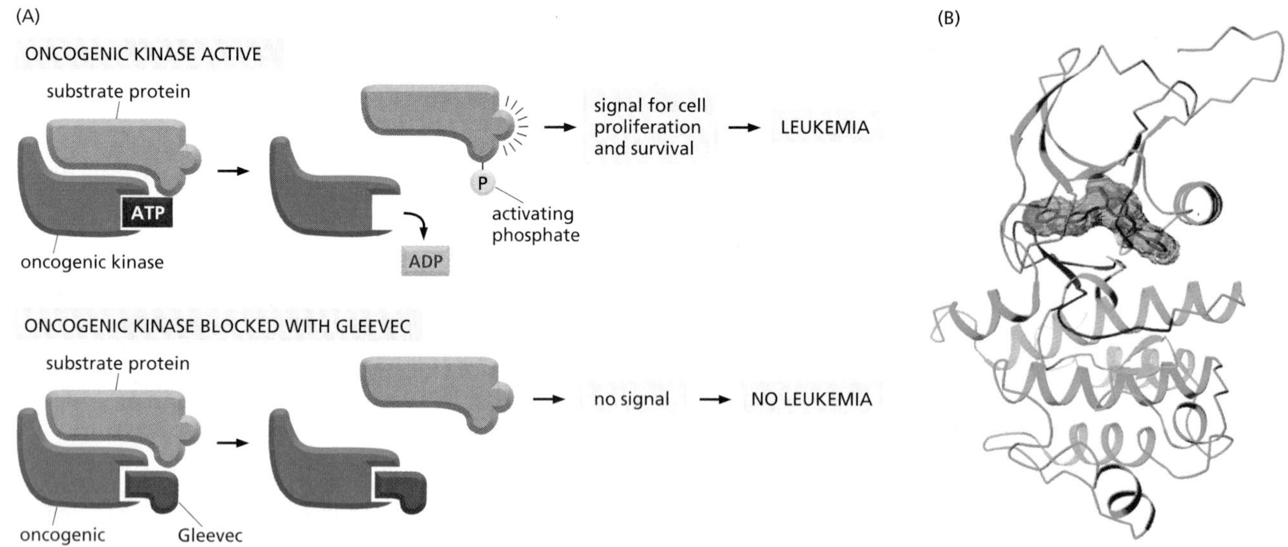

Figure 20–54 **The drug Gleevec blocks the activity of an oncogenic protein and halts certain cancers.** (A) In chronic myeloid leukemia, a cancer in which white blood cells are overproduced, the pathological cell behavior is almost always a result of a specific mutation (a chromosome translocation) affecting a gene called *Abl,* coding for a protein tyrosine kinase. The mutation creates an oncogenic form of *Abl,* whose protein product is hyperactive: it phosphorylates other proteins when it should not, thereby generating an intracellular signal that provokes excessive production of white blood cells. Gleevec sits in the ATP-binding pocket of the hyperactive kinase and prevents it from transferring a phosphate group from ATP onto a tyrosine in its target proteins. This inhibition blocks onward transmission of a signal for cell proliferation and survival. Gleevec is also effective against some other cancers involving oncogenes coding for protein tyrosine kinases similar to Abl. (B) The structure of a complex of Gleevec (solid *blue* object) with the tyrosine kinase domain of the Abl protein (ribbon diagram), as determined by X-ray crystallography. (B, from T. Schindler et al., *Science* 289:1938–1942, 2000. With permission from AAAS.)

range of forms of the disease. Conversely, the focus on cancer has taught us many important lessons about basic molecular biology. The applications of that knowledge go far beyond the treatment of cancer, for the knowledge gives us insight into the way the whole living world works.

ESSENTIAL CONCEPTS

- Tissues are composed of cells plus extracellular matrix.
- In plants, each cell surrounds itself with extracellular matrix in the form of a cell wall made chiefly of cellulose and other polysaccharides.
- Naked plant cells are fragile but can exert an osmotic swelling pressure on the enveloping wall to keep the tissue to which they belong turgid.
- Cellulose fibers in the plant cell wall confer tensile strength; other cell wall components give resistance to compression.
- The orientation in which cellulose is deposited controls the orientation of plant growth.
- Animal connective tissues provide mechanical support and consist of extracellular matrix with sparsely scattered cells.
- The protein and polysaccharide components of the matrix are made by the connective-tissue cells embedded in it; in most connective tissues, these cells are called fibroblasts.
- In the extracellular matrix of animals, tensile strength is provided by the fibrous protein collagen.
- Transmembrane integrin proteins link extracellular matrix proteins such as collagen and fibronectin to the intracellular cytoskeleton.
- Glycosaminoglycans (GAGs), covalently linked to proteins to form proteoglycans, act as space-fillers in the extracellular matrix and provide resistance to compression.
- Cells joined together in epithelial sheets line all external and internal surfaces of the animal body.
- In epithelial sheets, in contrast to connective tissues, tension is transmitted directly from cell to cell via cell–cell junctions.
- Proteins of the cadherin family span the epithelial cell membrane and bind to similar cadherins in the adjacent epithelial cell.
- At an adherens junction, the cadherins are linked intracellularly to actin filaments; at a desmosome junction, they are linked to keratin filaments.
- Actin bundles connected from cell to cell across an epithelium can contract, causing the epithelium to bend.
- Hemidesmosomes attach the basal face of an epithelial cell to the basal lamina, a specialized sheet of extracellular matrix.
- Tight junctions seal one epithelial cell to the next, barring the diffusion of water-soluble molecules across the epithelium.
- Gap junctions form channels that allow the passage of small molecules and ions from cell to cell; plasmodesmata in plants have the same function but a different structure.
- Most tissues in vertebrates are complex mixtures of cell types that are subject to continual turnover.
- The structure of the adult organism is maintained and renewed by the same basic processes that generated it in the embryo: cell proliferation, cell movement, and cell differentiation. As in the embryo, these processes are controlled by cell communication, selective cell–cell adhesion, and cell memory.

- New terminally differentiated cells are generated from stem cells, usually via the production of proliferating precursor cells.
- Embryonic stem cells (ES cells) can be maintained indefinitely in culture and remain capable of differentiating into any cell type in the body.
- Induced pluripotent stem cells (iPS cells), which resemble ES cells, can be generated from cells of the adult human body by artificially driving the expression of a small set of key genes.
- Cancer cells fail to obey the social constraints that normally maintain tissue organization: they proliferate when they should not, survive where they should not, and invade regions where they do not belong.
- Tobacco smoke causes more cancers than any other environmental mutagen.
- Cancers arise from the accumulation of many mutations in a single somatic cell lineage.
- Cancer cells are genetically unstable, having increased rates of mutation; many show gross chromosomal abnormalities.
- Cancer cells typically express telomerase, enabling them to continue dividing when normal human cells would stop.
- Most human cancer cells harbor mutations in the *p53* gene, allowing them to survive and divide even when their DNA is damaged.
- The mutations that promote cancer can do so by converting proto-oncogenes into oncogenes, which are hyperactive, or by inactivating tumor suppressor genes.
- Tumor suppressor genes can sometimes be identified through studies of rare cancer-prone families in which a mutation of one gene copy is inherited.
- Knowing the molecular abnormalities in the cells of specific cancers, we can now begin to design effective targeted treatments.

KEY TERMS

adherens junction
apical
basal
basal lamina
cadherin
cancer
cell junction
cell wall
cellulose microfibril
collagen
connective tissue
desmosome
extracellular matrix
embryonic stem (ES) cell
epithelium (plural epithelia)
fibroblast
fibronectin
gap junction
genetic instability
glycosaminoglycan (GAG)
hemidesmosome
induced pluripotent stem (iPS) cell
integrin
metastasis
oncogene
plasmodesma (plural plasmodesmata)
proteoglycan
reproductive cloning
proto-oncogene
stem cell
therapeutic cloning
tight junction
tissue
tumor suppressor gene
Wnt protein
Wnt signaling

QUESTIONS

QUESTION 20–9

Which of the following statements are correct? Explain your answers.

A. Gap junctions connect the cytoskeleton of one cell to that of a neighboring cell or to the extracellular matrix.

B. A wilted plant leaf can be likened to a deflated bicycle tire.

C. Because of their rigid structure, proteoglycans can withstand a large amount of compressive force.

D. The basal lamina is a specialized layer of extracellular matrix to which sheets of epithelial cells are attached.

E. Skin cells are continually shed and are renewed every few weeks; for a permanent tattoo, it is therefore necessary to deposit pigment below the epidermis.

F. Although stem cells are not differentiated, they are specialized and therefore give rise only to specific cell types.

QUESTION 20–10

Which of the following substances would you expect to spread from one cell to the next through (a) gap junctions and (b) plasmodesmata: glutamic acid, mRNA, cyclic AMP, Ca^{2+}, G proteins, and plasma membrane phospholipids?

QUESTION 20–11

Discuss the following statement: "If plant cells contained intermediate filaments to provide the cells with tensile strength, their cell walls would be dispensable."

QUESTION 20–12

Through the exchange of small metabolites and ions, gap junctions provide metabolic and electrical coupling between cells. Why, then, do you suppose that neurons communicate primarily through synapses rather than through gap junctions?

QUESTION 20–13

Gelatin is primarily composed of collagen, which is responsible for the remarkable tensile strength of connective tissue. It is the basic ingredient of jello; yet, as you probably experienced many times yourself while consuming the strawberry-flavored variety, jello has virtually no tensile strength. Why?

QUESTION 20–14

"The structure of an organism is determined by the genome that the egg contains." What is the evidence on which this statement is based? Indeed, a friend challenges you and suggests that you replace the DNA of a stork's egg with human DNA to see if a human baby results. How would you answer him?

QUESTION 20–15

Leukemias—that is, cancers arising through mutations that cause excessive production of white blood cells—have an earlier average age of onset than other cancers. Propose an explanation for why this might be the case.

QUESTION 20–16

Carefully consider the graph in Figure Q20–16, showing the number of cases of colon cancer diagnosed per 100,000 women per year as a function of age. Why is this graph so steep and curved, if mutations occur with a similar frequency throughout a person's life-span?

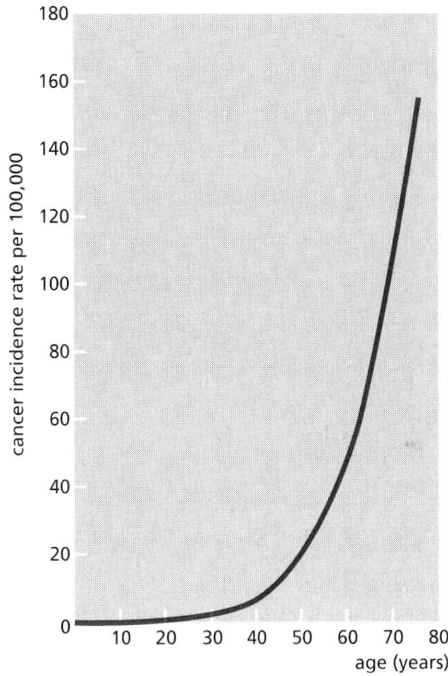

Figure Q20–16

QUESTION 20–17

Heavy smokers or industrial workers exposed for a limited time to a chemical carcinogen that induces mutations in DNA do not usually begin to develop cancers characteristic of their habit or occupation until 10, 20, or even more years after the exposure. Suggest an explanation for this long delay.

QUESTION 20–18

High levels of the female sex hormone estrogen increase the incidence of some forms of cancer. Thus, some early types of contraceptive pills containing high concentrations of estrogen were eventually withdrawn from use because this was found to increase the risk of cancer of the lining of the uterus. Male transsexuals who use estrogen preparations to give themselves a female appearance have an increased risk of breast cancer. High levels of androgens (male sex hormones) increase the risk of some other forms of cancer, such as cancer of the prostate. Can one infer that estrogens and androgens are mutagenic?

QUESTION 20–19

Is cancer hereditary?

Answers

Chapter 1

ANSWER 1–1 Trying to define life in terms of properties is an elusive business, as suggested by this scoring exercise (Table A1–1). Vacuum cleaners are highly organized objects, take matter and energy from the environment and transform the energy into motion, responding to stimuli from the operator as they do so. On the other hand, they cannot reproduce themselves, or grow and develop—but then neither can old animals. Potatoes are not particularly responsive to stimuli, and so on. It is curious that standard definitions of life usually do not mention that living organisms on Earth are largely made of organic molecules, that life is carbon based. As we now know, the key types of "informational macromolecules"—DNA, RNA, and protein—are the same in every living species.

TABLE A1–1 PLAUSIBLE "LIFE" SCORES FOR A VACUUM CLEANER, A POTATO, AND A HUMAN

CHARACTERISTIC	VACUUM CLEANER	POTATO	HUMAN
1. Organization	Yes	Yes	Yes
2. Homeostasis	Yes	Yes	Yes
3. Reproduction	No	Yes	Yes
4. Development	No	Yes	Yes
5. Energy	Yes	Yes	Yes
6. Responsiveness	Yes	No	Yes
7. Adaptation	No	Yes	Yes

ANSWER 1–2 Most random changes to the shoe design would result in objectionable defects: shoes with multiple heels, with no soles, or with awkward sizes would obviously not sell and would therefore be selected against by market forces. Other changes would be neutral, such as minor variations in color or in size. A minority of changes, however, might result in more desirable shoes: deep scratches in a previously flat sole, for example, might create shoes that would perform better in wet conditions; the loss of high heels might produce shoes that are more comfortable. The example illustrates that random changes can lead to significant improvements if the number of trials is large enough and selective pressures are imposed.

ANSWER 1–3 It is extremely unlikely that you created a new organism in this experiment. Far more probably, a spore from the air landed in your broth, germinated, and gave rise to the cells you observed. In the middle of the nineteenth century, Louis Pasteur invented a clever apparatus to disprove the then widely accepted belief that life could arise spontaneously. He showed that sealed flasks never grew anything if properly heat-sterilized first. He overcame the objections of those who pointed out the lack of oxygen or who suggested that his heat sterilization killed the life-generating principle, by using a special flask with a slender "swan's neck," which was designed to prevent spores carried in the air from contaminating the culture (Figure A1–3). The cultures in these flasks never showed any signs of life; however, they were capable of supporting life, as could be demonstrated by washing some of the "dust" from the neck into the culture.

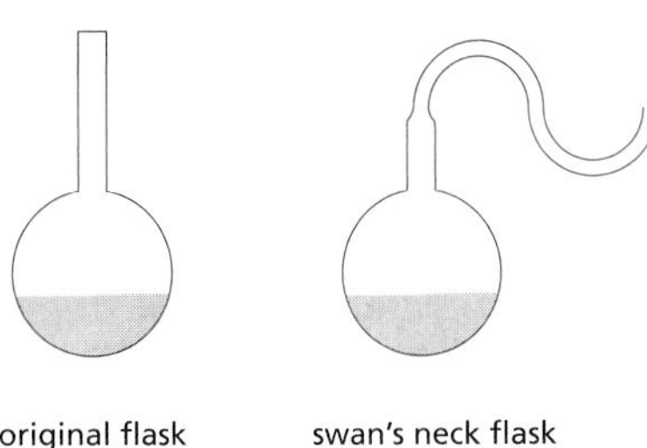

Figure A1–3

ANSWER 1–4 6×10^{39} (= 6×10^{27} g/10^{-12} g) bacteria would have the same mass as the Earth. And $6 \times 10^{39} = 2^{t/20}$, according to the equation describing exponential growth. Solving this equation for t results in t = 2642 minutes (or 44 hours). This represents only 132 generation times(!), whereas 5×10^{14} bacterial generation times have passed during the last 3.5 billion years. Obviously, the total mass of bacteria on this planet is nowhere close to the mass of the Earth. This illustrates that exponential growth can occur only for very few generations, i.e., for minuscule periods of time compared with evolution. In any realistic scenario, food supplies very quickly become limiting.

This simple calculation shows us that the ability to grow and divide quickly when food is ample is only one factor in the survival of a species. Food is generally scarce, and individuals of the same species have to compete with one another for the limited resources. Natural selection favors mutants that win the competition, or that find ways to exploit food sources that their neighbors are unable to use.

ANSWER 1–5 See Figure A1–5.

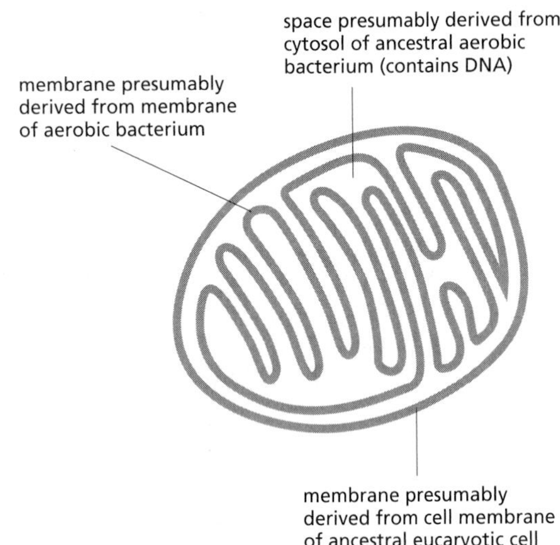

Figure A1–5

ANSWER 1–6 By engulfing substances, such as food particles, eucaryotic cells can sequester them and feed on them efficiently. Bacteria, in contrast, have no way of capturing lumps of food; they can export substances that help break down food substances in the environment, but the products of this labor must then be shared with other cells in the same neighborhood.

ANSWER 1–7 Light microscopy is much easier to use and requires much simpler instruments. Objects that are 1 μm in size can easily be resolved; the lower limit of resolution is 0.2 μm, which is a theoretical limit imposed by the wavelength of visible light. Visible light is nondestructive and passes readily through water, making it possible to observe living cells. Electron microscopy, on the other hand, is much more complicated, both in the preparation of the sample (which needs to be extremely thinly sliced, stained with electron-dense heavy metal, and completely dehydrated) and in the nature of the instrument. Living cells cannot be observed in an electron microscope. The resolution of electron microscopy is much higher, however, and objects as small as 10 nm can easily be resolved. To see any structural detail, microtubules, mitochondria, and bacteria would need to be analyzed by electron microscopy. It is possible, however, to stain them with specific dyes and then determine their location by light microscopy.

ANSWER 1–8 Because the basic workings of cells are so similar, a great deal has been learned from studying model systems. Brewer's yeast is a good model system because yeast cells are much simpler than human cancer cells. We can grow yeast inexpensively and in vast quantities, and we can manipulate yeast cells genetically and biochemically much more easily than human cells. This allows us to use yeast to decipher the ground rules governing how cells divide and grow. Cancer cells divide when they should not (and therefore give rise to tumors), and a basic understanding of how cell division is controlled is therefore directly relevant to the cancer problem. Indeed, the National Cancer Institute, the American Cancer Society, and many other institutions that are devoted to finding a cure for cancer strongly support basic research on various aspects of cell division in different model systems, such as yeast.

ANSWER 1–9 Check your answers using the Glossary and Panel 1–2 (p. 25).

ANSWER 1–10

A. False. The hereditary information is encoded in the cell's DNA, which in turn specifies its proteins.
B. True. Baceria do not have a nucleus.
C. False. Plants are composed of eucaryotic cells that contain chloroplasts as cytoplasmic organelles. The chloroplasts are thought to be evolutionarily derived from procaryotic cells.
D. True. The number of chromosomes varies from one organism to another, but is constant in all cells of the same organism.
E. False. The cytosol is the cytoplasm excluding all organelles.
F. True. The nuclear envelope is a double membrane, and mitochondria are surrounded by both an inner and an outer membrane.
G. False. Protozoans are single-cell organisms and therefore do not have different tissues. They have a complex structure, however, that has highly specialized parts.
H. Somewhat true. Peroxisomes and lysosomes contain enzymes that catalyze the breakdown of substances produced in the cytosol or taken up by the cell. One can argue, however, that many of these substances are degraded to generate food molecules, and as such are certainly not "unwanted."

ANSWER 1–11 One average brain cell weighs 10^{-9} g (= 1000 g/10^{12}). Because 1 g of water occupies 1 ml = 1 cm^3 (= 10^{-6} m^3), the volume of one cell is 10^{-15} m^3 (= 10^{-9} g × 10^{-6} m^3/g). Taking the cube root yields a side length of 10^{-5} m, or 10 μm (10^6 μm = 1 m) for each cell. The page of the book has a surface of 0.057 m^2 (= 21 cm × 27.5 cm), and each cell has a footprint of 10^{-10} m^2 (10^{-5} m × 10^{-5} m). Therefore, 57 × 10^7 (= 0.057 m^2/10^{-10} m^2) cells fit on this page when spread out as a single layer. Thus, 10^{12} cells would occupy 1750 pages (= 10^{12}/[57 × 10^7]).

ANSWER 1–12 In this plant cell, A is the nucleus, B is a vacuole, C is the cell wall, and D is a chloroplast. The scale bar is about 10 μm, the width of the nucleus.

ANSWER 1–13 The three major filaments are actin filaments, intermediate filaments, and microtubules. Actin filaments are involved in rapid cell movement, such as contraction of a muscle cell; intermediate filaments provide mechanical stability such as in epidermal cells of the skin; and microtubules function as "railroad tracks" for intracellular movements, and are responsible for the separation of chromosomes during cell division. Other functions of all these filaments are discussed in Chapter 17.

ANSWER 1–14 It takes only 20 hours, i.e., less than a day, before mutant cells become more abundant in the culture. Using the equation provided in the question, we see that the number of the original ("wild-type") bacterial cells at time t minutes after the mutation occurred is $10^6 \times 2^{t/20}$. The number of mutant cells at time t is $1 \times 2^{t/15}$. To find out when the mutant cells "overtake" the wild-type cells, we simply have to make these two numbers equal to each other (i.e., $10^6 \times 2^{t/20} = 2^{t/15}$). Taking the logarithm to base 10 of both sides of this equation and solving it for t results in $t = 1200$ minutes (or 20 hours). At this time, the culture contains 2×10^{24} cells ($10^6 \times 2^{60} + 1 \times 2^{80}$). Incidentally, 2×10^{24} bacterial cells, each weighing 10^{-12} g, would weigh 2×10^{12} g (= 2×10^9 kg, or 2 million tons!). This can only have been a thought experiment.

ANSWER 1–15 Bacteria continually acquire mutations in their DNA. In the population of cells exposed to the poison, one or a few cells may acquire a mutation that makes them resistant to the action of the drug. Antibiotics that are poisonous to bacteria because they bind to certain bacterial proteins, for example, would not work if the proteins have a slightly changed surface so that binding occurs more weakly or not at all. These mutant bacteria would continue dividing rapidly while their cousins are slowed down. The antibiotic-resistant bacteria would soon become the predominant species in the culture.

ANSWER 1–16 $10^{13} = 2^{(t/1)}$. Therefore, it would take only 43 days [$t = 13/\log(2)$]. This explains why some cancers can progress extremely rapidly. Many cancer cells divide much more slowly, however, or die because of their internal abnormalities or because they do not have sufficient blood supply, and the actual progression of cancer is therefore usually slower.

ANSWER 1–17 Living cells evolved from nonliving matter, but grow and replicate. Like the material they originated from, they are governed by the laws of physics, thermodynamics, and chemistry. Thus, for example, they cannot create energy *de novo* or build ordered structures without the expenditure of free energy. We can understand virtually all cellular events, such as metabolism, catalysis, membrane assembly, and DNA replication, as complicated chemical reactions that can be experimentally reproduced, manipulated, and studied in test tubes.

Despite this fundamental reducibility, a living cell is more than the sum of its parts. We cannot randomly mix proteins, nucleic acids, and other chemicals together in a test tube, for example, and make a cell. The cell functions by virtue of its organized structure, and this is a product of its evolutionary history. Cells always come from preexisting cells, and the division of a mother cell passes both chemical constituents and structures to its daughters. The plasma membrane, for example, never has to form *de novo*, but grows by expansion of a preexisting membrane; there will always be a ribosome, in part made up of proteins whose function it is to make more proteins including those that build more ribosomes.

ANSWER 1–18 In a multicellular organism, different cells take on specialized functions and cooperate with one another. In this way, multicellular organisms are able to exploit food sources that are inaccessible to single-cell organisms. A plant, for example, can reach the soil with its roots to take up water and nutrients and at the same time harvest light energy and CO_2 from the air through its leaves. By protecting its reproductive cells with other specialized cells, the multicellular organism can develop new ways to survive in harsh environments or to fight off predators. When food runs out, it may be able to preserve its reproductive cells by allowing them to draw upon resources stored by their companions—or even to cannibalize relatives (a common process, in fact).

ANSWER 1–19 The volume and the surface area are 5.24×10^{-19} m^3 and 3.14×10^{-12} m^2 for the bacterial cell, and 1.77×10^{-15} m^3 and 7.07×10^{-10} m^2 for the animal cell, respectively. From these numbers, the surface-to-volume ratios are 6×10^6 m^{-1} and 4×10^5 m^{-1}, respectively. In other words, although the animal cell has a 3375-fold larger volume, its membrane surface is increased only 225-fold. If internal membranes are included in the calculation, however, the surface-to-volume ratios of both cells are about equal. Thus, because of their internal membranes, eucaryotic cells can grow bigger and still maintain a sufficiently large membrane area, which—as we shall discuss in more detail in later chapters—is required for many essential functions.

ANSWER 1–20 There are many lines of evidence for a common ancestor. Analyses of modern-day living cells show an amazing degree of similarity in the basic components that make up the inner workings of otherwise vastly different cells. Many metabolic pathways, for example, are conserved from one cell to another, and the compounds that make up nucleic acids and proteins are the same in all living cells, even though it is easy to imagine that a different choice of compounds (e.g., amino acids with different side chains) would have worked just as well. Similarly, it is not uncommon to find that important proteins have closely similar detailed structures in procaryotic and eucaryotic cells. Theoretically, there would be many different ways to build proteins that could perform the same functions. The evidence overwhelmingly shows that most important processes were "invented" only once and then became fine-tuned during evolution to suit the particular needs of specialized cells.

It seems highly unlikely, however, that the first cell survived to become the primordial founder cell of today's living world. As evolution is not a directed process with a purposeful progression, it is more likely that there were a vast number of unsuccessful trial cells that replicated for a while and then became extinct because they could not adapt to changes in the environment or could not survive in competition with other types of cells. We can therefore speculate that the primordial ancestor cell was a "lucky" cell that ended up in a relatively stable environment in which it had a chance to replicate and evolve.

ANSWER 1–21 See Figure A1–21.

ANSWER 1–22 A quick inspection might reveal beating cilia on the cell surface; their presence would tell you that the cell was eucaryotic. If you don't see them—and you are quite likely not to—you will have to look for other distinguishing features. If you are lucky, you might see the cell divide. Watch it then with the right optics, and you might be able to see condensed mitotic

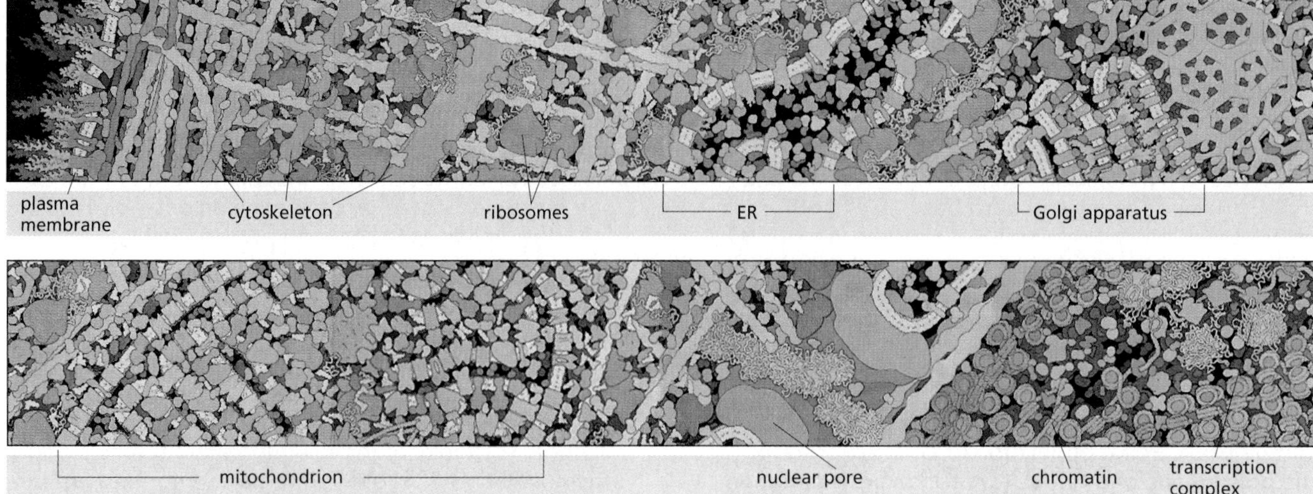

Figure A1–21 Courtesy of D. Goodsell.

chromosomes, which again would tell you that it was a eucaryote. Fix the cell and stain it with a dye for DNA: if this is contained in well defined nucleus, the cell is a eucaryote; if you cannot see a well defined nucleus, the cell may be a procaryote. Alternatively, stain it with a dye that binds actin or tubulin (proteins that are highly conserved in eucaryotes but absent in bacteria). Embed, it section it, and look with an electron microscope: can you see organelles such as mitochondria inside your cell? Try staining it with the Gram stain, which is specific for molecules in the cell wall of some classes of bacteria. But all these tests might fail, leaving you still uncertain. For a definitive answer, you could attempt to analyze the sequences of the DNA and RNA molecules that it contains, using the sophisticated methods described later in this book. The sequences of highly conserved molecules, such as those that form the core components of the ribosome, provide a molecular signature that can tell you whether your cell is a eucaryote, a bacterium, or an archaeon. If you can't detect any RNA, you are probably looking not at a cell but at a piece of dirt.

Chapter 2

ANSWER 2–1 The chances are excellent because of the enormous size of Avogadro's number. The original cup contained one mole of water, or 6×10^{23} molecules, and the volume of the world's oceans, converted to cubic centimeters, is 1.5×10^{24} cm^3. After mixing, there should be on average 0.4 of a "Greek" water molecule per cm^3
($6 \times 10^{23}/1.5 \times 10^{24}$), or 7.2 molecules in 18 g of Pacific Ocean.

ANSWER 2–2
A. The atomic number is 6; the atomic weight is 12 (= 6 protons + 6 neutrons).
B. The number of electrons is six (= the number of protons).
C. The first shell can accommodate two and the second shell eight electrons. Carbon therefore needs four additional electrons (or would have to give up four electrons) to obtain a full outermost shell. Carbon is most stable when it shares four additional electrons with other atoms (including other carbon atoms) by forming four covalent bonds.
D. Carbon-14 has two additional neutrons in its nucleus. Because the chemical properties of an atom are determined by its electrons, carbon-14 is chemically identical to carbon-12.

ANSWER 2–3 The statement is correct. Both ionic and covalent bonds are based on the same principles: electrons can be shared equally between two interacting atoms, forming a nonpolar covalent bond; electrons can be shared unequally between two interacting atoms, forming a polar covalent bond; or electrons can be completely lost from one atom and gained by the other, forming an ionic bond. There are bonds of every conceivable intermediate state, and for borderline cases it becomes arbitrary whether a bond is described as a very polar covalent bond or an ionic bond.

ANSWER 2–4 The statement is correct. The hydrogen–oxygen bond in water molecules is polar, so that the oxygen atom carries a more negative charge than the hydrogen atoms. These partial negative charges are attracted to the positively charged sodium ions, but are repelled from the negatively charged chloride ions.

ANSWER 2–5
A. Hydronium (H_3O^+) ions result from water dissociating into protons and hydroxyl ions, each proton binding to a water molecule to form a hydronium ion ($2H_2O \rightarrow H_2O + H^+ + OH^- \rightarrow H_3O^+ + OH^-$). At neutral pH, i.e., in the absence of an acid providing more H_3O^+ ions or a base providing more OH^- ions, the concentrations of H_3O^+ ions and OH^- ions are equal. We know that at neutrality the pH = 7.0, and therefore, the H^+ concentration is 10^{-7} M. The H^+ concentration equals the H_3O^+ concentration.
B. To calculate the ratio of H_3O^+ ions to H_2O molecules, we need to know the concentration of water molecules. The molecular weight of water is 18 (i.e., 18 g/mole), and 1 liter of water weighs 1 kg. Therefore, the concentration of water is 55.6 M (= 1000 [g/l]/[18 g/mole]), and the ratio of H_3O^+ ions to H_2O molecules is 1.8×10^{-9} (= $10^{-7}/55.6$); i.e., only two water molecules in a billion are dissociated at neutral pH.

ANSWER 2–6 No mistake. Note that the small hydrogen atoms are those that are linked to an oxygen atom, whereas the ones linked to a carbon atom are larger. This reflects the polarity of the respective bonds: the H–C bond is nonpolar, whereas the H–O bond is polar. Oxygen more strongly draws the shared electrons away from the hydrogen, resulting in a smaller radius of the electron cloud of the hydrogen atom.

ANSWER 2–7 The synthesis of a macromolecule with a unique structure requires that in each position only one stereoisomer is used. Changing one amino acid from its L- to its D-form would result in a different protein. Thus, if for each amino acid a random mixture of the D- and L-forms were used to build a protein, its amino acid sequence could not specify a single structure, but many different structures (2^N different structures would be formed, where N is the number of amino acids in the protein).

Why L-amino acids were selected in evolution as the exclusive building blocks of proteins is a mystery; we could easily imagine a cell in which certain (or even all) amino acids were used in the D-forms to build proteins, as long as these particular stereoisomers were used exclusively.

ANSWER 2–8 The term "polarity" can be used in two different ways. In one meaning it refers to directional asymmetry—for example, in linear polymers such as polypeptides, which have an N-terminus and a C-terminus; or nucleic acids, which have a 3′ and a 5′ end. Because bonds form only between the amino and the carboxyl groups of the amino acids in a polypeptide, and between the 3′ and the 5′ ends of nucleotides, nucleic acids and polypeptides always have two different ends, which give the chain a defined chemical polarity.

In the other meaning, polarity refers to a separation of electric charge in a bond or molecule. This kind of polarity promotes hydrogen-bonding to water molecules, and because the water solubility, or hydrophilicity, of a molecule depends upon its being polar in this sense, the term "polar" also indicates water solubility.

ANSWER 2–9 A major advantage of condensation reactions is that they are readily reversible by hydrolysis (and water is readily available in the cell). This allows cells to break down their macromolecules (or macromolecules of other organisms that were ingested as food) and to recover the subunits intact so that they can be "recycled," i.e., used to build new macromolecules.

ANSWER 2–10 Many of the functions that macromolecules perform rely on their ability to associate and dissociate readily. This allows cells, for example, to remodel their interior when they move or divide, and to transport components from one organelle to another. Covalent bonds would be too strong and too permanent for such a purpose.

ANSWER 2–11

A. True. All nuclei are made of positively charged protons and uncharged neutrons; the only exception is the hydrogen nucleus, which consists of only one proton.
B. False. Atoms are electrically neutral. The number of positively charged protons is always balanced by an equal number of negatively charged electrons.
C. True—but only for the cell nucleus (see Chapter 1), and not for the atomic nucleus discussed in this chapter.
D. False. Elements can have different isotopes, which differ only in their number of neutrons.
E. True. In certain isotopes the large number of neutrons destabilizes the nucleus, which decomposes in a process called radioactive decay.
F. True. Examples include granules of glycogen, a polymer of glucose, found in liver cells; and fat droplets, made of aggregated triacylglycerols, found in fat cells.
G. True. Individually, these bonds are weak and readily broken by thermal motion, but because interactions between two macromolecules involve a large number of such bonds, the overall binding can be quite strong, and because hydrogen bonds form only between correctly positioned groups on the interacting macromolecules, they are very specific.

ANSWER 2–12

A. One cellulose molecule has a molecular weight of $n \times (12[C] + 2 \times 1[H] + 16[O])$. We do not know n, but we can determine the ratio with which the individual elements contribute to the weight of cellulose. The contribution of carbon atoms is 40% [= 12/(12 + 2 + 16) × 100%]. Therefore, 2 g (40% of 5 g) of carbon atoms are contained in the cellulose that makes up this page. The atomic weight of carbon is 12 g/mole, and there are 6×10^{23} atoms or molecules in a mole. Therefore,
10^{23} carbon atoms [= (2 g/12 [g/mole]) × 6 × 10^{23} (molecules/mole)] make up this page.
B. The volume of the page is 4×10^{-6} m^3 (= 21.2 cm × 27.6 cm × 0.07 mm), which is the same as the volume of a cube with a side length of 1.6 cm (= $\sqrt[3]{4 \times 10^{-6}\text{ m}^3}$). Because we know from part A that the page contains 10^{23} carbon atoms, geometry tells us that there could be about 4.6×10^7 carbon atoms (= $\sqrt[3]{10^{23}}$) lined up along each side of this cube. Therefore, in cellulose, about 200,000 carbon atoms (= $4.6 \times 10^7 \times 0.07 \times$
10^{-3} m/1.6×10^{-2} m) span the thickness of the page.
C. If tightly stacked, 350,000 carbon atoms with a 0.2-nm diameter would span the 0.07-mm thickness of the page.
D. The 1.7-fold difference in the two calculations reflects (1) that carbon is not the only atom in cellulose and
(2) that paper is not an atomic lattice of precisely arranged molecules (as a diamond would be for precisely arranged carbon atoms), but a random meshwork of fibers containing many voids.

ANSWER 2–13

A. The occupancies of the three innermost electron levels are 2, 8, 8.
B.

helium	already has full level
oxygen	gain 2
carbon	gain 4 or lose 4
sodium	lose 1
chlorine	gain 1

C. Helium with its fully occupied electron level is chemically unreactive. Sodium and chlorine, on the other hand, are extremely reactive and readily form stable Na^+ and Cl^- ions that form ionic bonds, as in table salt.

ANSWER 2–14 A sulfur atom is much larger than an oxygen atom, and because of its larger size, the outermost electrons are not as strongly attracted to the nucleus of the sulfur atom as they are in an oxygen atom. Consequently, the hydrogen–sulfur bond is much less polar than the hydrogen–oxygen bond. Because of the reduced polarity, the sulfur in a H_2S molecule is not strongly attracted to the hydrogen atoms in an adjacent H_2S molecule, and the hydrogen bonds that are so predominant in water do not form.

ANSWER 2–15 The reactions are diagrammed in Figure A2–15, where R_1 and R_2 are amino acid side chains.

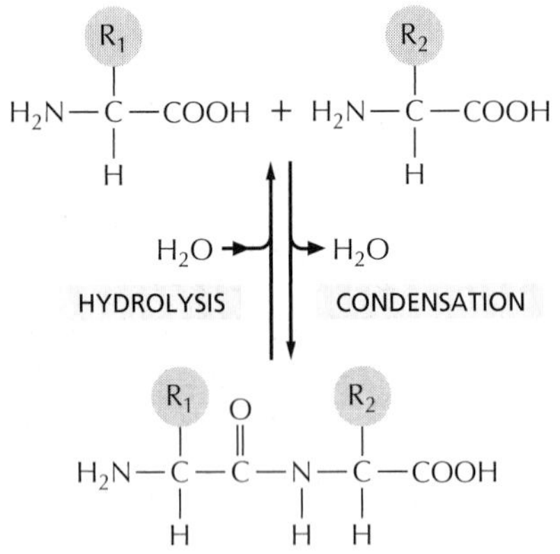

Figure A2–15

ANSWER 2–16

A. False. The properties of a protein depend on both the amino acids it contains and the order in which they are linked together. The diversity of proteins is due to the almost unlimited number of ways in which 20 different amino acids can be combined in a linear sequence.
B. False. Lipids assemble into bilayers by noncovalent forces. A membrane is therefore not a macromolecule.
C. True. The backbone of nucleic acids is made up of alternating ribose (or deoxyribose in DNA) and phosphate groups. Ribose and deoxyribose are sugars.
D. True. About half of the 20 naturally occurring amino acids have hydrophobic side chains. In folded proteins, many of these side chains face toward the inside of the folded-up proteins, because they are repelled from water.
E. True. Hydrophobic hydrocarbon tails contain only nonpolar bonds. Thus, they cannot participate in hydrogen-bonding and are repelled from water. We consider the underlying principles in more detail in Chapter 11.
F. False. RNA contains the four listed bases, but DNA contains T instead of U. T and U are very much alike, however, and differ only by a single methyl group.

ANSWER 2–17

A. (a) 400 (= 20^2); (b) 8000 (= 20^3); (c) 160,000 (= 20^4).
B. A protein with a molecular weight of 4800 daltons is made of about 40 amino acids; thus there are 1.1×10^{52} (= 20^{40}) different ways to make such a protein. Each individual protein molecule weighs 8×10^{-21} g (= $4800/6 \times 10^{23}$); thus a mixture of one molecule each weighs 9×10^{31} g (= 8×10^{-21} g $\times 1.1 \times 10^{52}$), which is 15,000 times the total weight of the planet Earth, weighing 6×10^{24} kg. You would need a quite large container, indeed.
C. Given that most cellular proteins are even larger than the one used in this example, it is clear that only a minuscule fraction of the total possible amino acid sequences are used in living cells.

ANSWER 2–18 Because all living cells are made up of chemicals and because all chemical reactions (whether in living cells or in test tubes) follow the same rules, an understanding of basic chemical principles is fundamentally important to the understanding of biology. In the course of this book, we will frequently refer back to these principles, on which all of the more complicated pathways and reactions that occur in cells are based.

ANSWER 2–19

A. Hydrogen bonds require specific groups to interact; one is always a hydrogen atom linked in a polar bond to an oxygen or a nitrogen, and the other is usually a nitrogen or an oxygen atom. Van der Waals attractions are weaker and occur between any two atoms that are in close enough proximity. Both hydrogen bonds and van der Waals attractions are short-range interactions that come into play only when two molecules are already close. Both types of bonds can therefore be thought of as means of "fine-tuning" an interaction, i.e., helping to position two molecules correctly with respect to each other once they have been brought together by diffusion.
B. Van der Waals attractions would form in all three examples. Hydrogen bonds would form in (c) only.

ANSWER 2–20 Noncovalent interactions form between the subunits of macromolecules—e.g., the side chains of amino acids in a polypeptide chain—and cause the polypeptide chain to assume a unique shape. These interactions include hydrogen bonds, ionic interactions, van der Waals interactions, and hydrophobic interactions. Because these interactions are weak, they can be broken with relative ease; thus, most macromolecules can be unfolded by heating, which increases thermal motion.

ANSWER 2–21 Amphipathic molecules have both a hydrophilic and a hydrophobic end. Their hydrophilic ends can hydrogen-bond to water, but their hydrophobic ends are repelled from water because they interfere with the water structure. Consequently, the hydrophobic ends of amphipathic molecules tend to be exposed to air at air–water interfaces, or will always cluster together to minimize their contact with water molecules—both there and in the interior of an aqueous solution. (See Figure A2–21.)

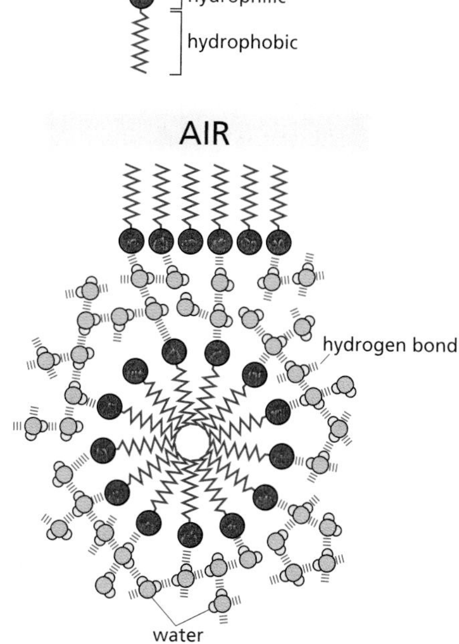

Figure A2–21

ANSWER 2–22

A,B. (A) and (B) are both correct formulas of the amino acid phenylalanine. In formula (B) phenylalanine is shown in the ionized form that exists in water solution, where the basic amino group is protonated and the acidic carboxylic group is deprotonated.

C. Incorrect. This structure of a peptide bond is missing a hydrogen atom bound to the nitrogen.

D. Incorrect. This formula of an adenine base features one double bond too many, creating a five-valent carbon atom and a four-valent nitrogen atom.

E. Incorrect. In this formula of a nucleoside triphosphate there should be two additional oxygen atoms, one between each of the phosphorus atoms.

F. This is the correct formula of ethanol.

G. Incorrect. Water does not hydrogen-bond to hydrogens bonded to carbon. The lack of the capacity to hydrogen-bond makes hydrocarbon chains hydrophobic, i.e., water-hating.

H. Incorrect. Na and Cl form an ionic bond, Na^+Cl^-, but a covalent bond is drawn.

I. Incorrect. The oxygen atom attracts electrons more than the carbon atom; the polarity of the two bonds should therefore be reversed.

J. This structure of glucose is correct.

K. Almost correct. It is more accurate to show that only one hydrogen is lost from the $-NH_2$ group and the –OH group is lost from the –COOH group.

Chapter 3

ANSWER 3–1 The equation represents the "bottom line" of a large set of individual reactions that are catalyzed by many individual enzymes. Because sugars are more complicated molecules than CO_2 and H_2O, the reaction generates a more ordered state inside the cell. As demanded by the second law of thermodynamics, heat is generated at many steps on the long pathway leading to the products summarized in this equation.

ANSWER 3–2 Oxidation is defined as removal of electrons. Therefore, (A) is an oxidation, and (B) is a reduction. The *black* carbon atom in (C) remains largely unchanged; the neighboring carbon atom, however, loses a hydrogen atom (i.e., an electron and a proton) and hence becomes oxidized. The *black* carbon atom in (D) becomes oxidized because it loses a hydrogen atom, whereas the *black* carbon atom in (E) becomes reduced because it gains a hydrogen atom.

ANSWER 3–3

A. Both states of the coin, H and T, have an equal probability. There is therefore no driving force, i.e., no energy difference, that would favor H turning to T or vice versa. Therefore, $\Delta G° = 0$ for this reaction. However, a reaction proceeds if H and T coins are not present in the box in equal numbers. In this case, the concentration difference between H and T creates a driving force, a $\Delta G \neq 0$, for the reaction until it reaches equilibrium, i.e., until there are equal numbers of H and T.

B. The amount of shaking corresponds to the temperature, as it results in the "thermal" motion of the coins. The activation energy of the reaction is the energy that needs to be expended to flip the coin, i.e., to stand it on its rim, from where it can fall back facing either side up. Jigglase would speed up the flipping by lowering the energy required for this; it could, for example, be a magnet that is suspended above the box and helps lift the coins. Jigglase would not affect where the equilibrium lies (at an equal number of H and T), but would speed up the process of reaching the equilibrium, because in the presence of jigglase more coins would flip back and forth.

ANSWER 3–4 See Figure A3–4. Note that $\Delta G°_{X \rightarrow Y}$ is positive, whereas $\Delta G°_{Y \rightarrow Z}$ and $\Delta G°_{X \rightarrow Z}$ are negative. The graph also shows that $\Delta G°_{X \rightarrow Z} = \Delta G°_{X \rightarrow Y} + \Delta G°_{Y \rightarrow Z}$. We do not know from the information given in Figure 3–12 how high the activation energy barriers are; they are therefore drawn to an arbitrary height (solid lines). They would be lowered by enzymes that catalyze these reactions, thereby speeding up the reaction rates (dotted lines).

ANSWER 3–5 The reaction rates might be limited by (1) the concentration of the substrate, i.e., how often a molecule of CO_2 collides with the active site on the enzyme; (2) how many of these collisions are energetic enough to lead to a reaction; and (3) how fast the enzyme can release the products of the reaction and therefore be free to bind the next substrate. The diagram in Figure A3–5 shows that by lowering the activation energy barrier, more molecules have sufficient energy to undergo the reaction. The area under the curve from point A to infinite energy or from point B to infinite energy indicates the total number of molecules that will react without or with the enzyme, respectively. Although not drawn to scale, the ratio of these two areas should be 10^7.

ANSWER 3–6 All reactions are reversible. If the compound AB can dissociate to produce A and B, then

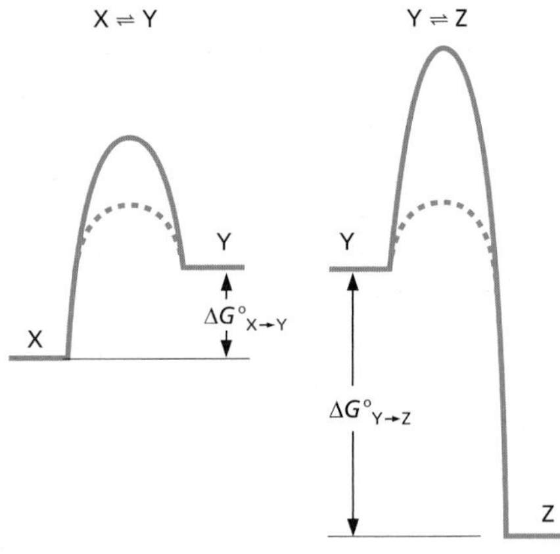

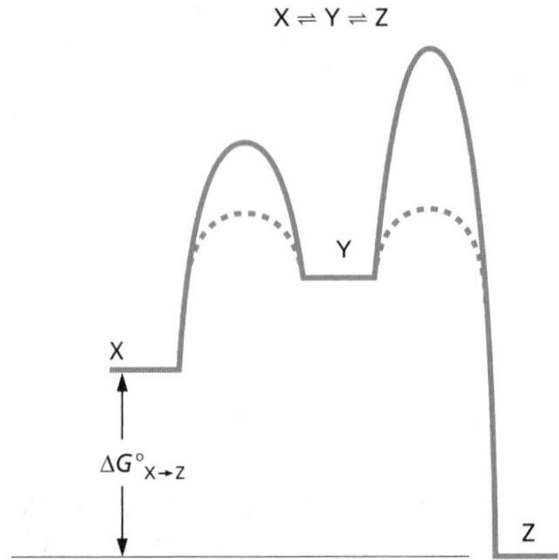

Figure A3–4

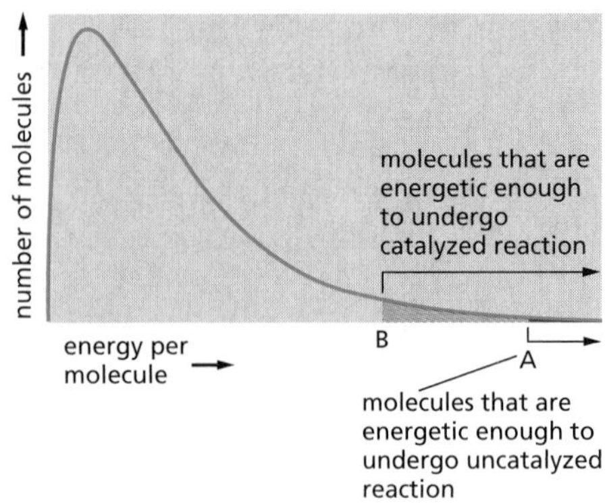

Figure A3–5

it must also be possible for A and B to associate to form AB. Which of the two reactions predominates depends on the equilibrium constant and the concentration of A, B, and AB (as discussed in Figure 3–19). Presumably, when this enzyme was isolated its activity was detected by supplying A and B in relatively large amounts and measuring the amount of AB generated. We can suppose, however, that in the cell there is a large concentration of AB so that under normal conditions the enzyme actually catalyzes AB → A + B. (This question is based on an actual example in which an enzyme was isolated and named according to the reaction in one direction, but was later shown to catalyze the reverse reaction in living cells.)

ANSWER 3–7

A. The rocks in Figure 3–30B provide the energy to lift the bucket of water. ATP is driving the reaction; thus ATP corresponds to the rocks on top of the cliff. The broken debris corresponds to ADP and P_i, the products of ATP after it has released its energy and performed its work. In the reaction, ATP hydrolysis is coupled to the conversion of X to Y. X, therefore, is the starting material, the bucket on the ground, which is converted to Y, the bucket at its highest point.

B. (i) The rock hitting the ground would be the futile hydrolysis of ATP. In the absence of an enzyme that uses the energy of ATP hydrolysis to drive an otherwise unfavorable reaction, the energy stored in the phosphoanhydride bond would be lost as heat. (ii) The energy stored in Y could be used to drive another reaction. If Y represented the activated form of amino acid X, for example, it could undergo a condensation reaction to form a peptide bond during protein synthesis.

ANSWER 3–8 The free energy ΔG derived from ATP hydrolysis depends on both $\Delta G°$ and the concentrations of substrate and products. For example, for a particular set of concentrations, one might have

$$\Delta G = -12 \text{ kcal/mole} = -7.3 \text{ kcal/mole} + 0.616 \ln \frac{[\text{ADP}] \times [\text{P}]}{[\text{ATP}]}$$

ΔG is smaller than $\Delta G°$, largely because the ATP concentration in cells is high (in the millimolar range) and the ADP concentration is low (in the 10 μM range). The concentration term of this equation is therefore smaller than 1 and its logarithm is a negative number.

$\Delta G°$ is a constant for the reaction that will not vary with reaction conditions. ΔG, in contrast, depends on the concentrations of ATP, ADP, and phosphate, which can be somewhat different between cells.

ANSWER 3–9 Reactions B, C, D, and E all require coupling to other, energetically favorable reactions. In each case, higher-order structures are formed that are more complicated and have higher-energy bonds than the starting materials. In contrast, reaction A is a catabolic reaction that leads to compounds in a lower energy state and will occur spontaneously.

ANSWER 3–10

A. Nearly true, but strictly speaking, false. Because enzymes enhance the rate but do not change the equilibrium point of a reaction, a reaction will always occur in the absence of the enzyme, though often at a minuscule rate. Moreover, competing reactions may use up the substrate more quickly, thus further impeding the desired reaction. Thus, in practical terms, without an enzyme, some reactions may never occur to an appreciable extent.

B. False. High-energy electrons are more easily transferred, i.e., more loosely bound to the donor molecule. This does not mean that they move any faster.

C. True. Hydrolysis of an ATP molecule to form AMP also produces a pyrophosphate (PP_i) molecule, which in turn is hydrolyzed into two phosphate molecules.

This second reaction releases almost the same amount of energy as the initial hydrolysis of ATP, thereby approximately doubling the energy yield.

D. True. Oxidation is the removal of electrons, which reduces the diameter of the carbon atom.
E. True. ATP, for example, can donate both chemical bond energy and a phosphate group.
F. False. Living cells have a particular kind of chemistry in which most oxidations are energy-releasing events; under different conditions, however, such as in a hydrogen-containing atmosphere, reductions would be energy-releasing events.
G. False. All cells, including those of cold- and warm-blooded animals, radiate comparable amounts of heat as a consequence of their metabolic reactions. For bacterial cells, for example, this becomes apparent when a compost pile heats up.
H. False. The equilibrium constant of the reaction $X \rightleftharpoons Y$ remains unchanged. If Y is removed by a second reaction, more X is converted to Y so that the ratio of X to Y remains constant.

ANSWER 3–11 The free-energy difference ($\Delta G°$) between Y and X due to three hydrogen bonds is –3 kcal/mole. (Note that the free energy of Y is lower than that of X, because energy would need to be expended to break the bonds to convert Y to X. The value for $\Delta G°$ for the transition $X \rightarrow Y$ is therefore negative.) The equilibrium constant for the reaction is therefore about 100 (from Table 3–1, p. 96); i.e., there are 100 times more molecules of Y than of X at equilibrium. An additional three hydrogen bonds would increase $\Delta G°$ to –6 kcal/mole and increase the equilibrium constant another 100-fold to 10^4. Thus, relatively small differences in energy can have a major effect on equilibria.

ANSWER 3–12

A. The equilibrium constant is defined as $K = [AB]/([A] \times [B])$. The square brackets indicate the concentration. Thus, if A, B, and AB are each 1 μM (10^{-6} M), K will be 10^6 M^{-1} [= $10^{-6}/(10^{-6} \times 10^{-6})$].
B. Similarly, if A, B, and AB are each 1 μM (10^{-9} M), then K will be 10^9 M^{-1}.
C. This example illustrates that interacting proteins that are present in cells in lower concentrations need to bind to each other with high affinities so that a significant fraction of the molecules are bound at equilibrium. In this particular case, lowering the concentration by three orders of magnitude (from μM to nM) requires a change in the equilibrium constant by three orders of magnitude to maintain the AB protein complex (corresponding to –4.3 kcal of free energy; see Table 3–1). This corresponds to about four or five extra hydrogen bonds.

ANSWER 3–13 The statement is correct. The criterion for whether a reaction proceeds spontaneously is ΔG, not $\Delta G°$, and takes the concentrations of the reacting components into account. A reaction with a negative $\Delta G°$, for example, would not proceed spontaneously under conditions where there is a large enough excess of products, i.e., more than at equilibrium. Conversely, a reaction with a positive $\Delta G°$ might spontaneously go forward under conditions where there is a huge excess of substrates.

ANSWER 3–14

A. A maximum of 57 ATP molecules (= 686/12) corresponds to the total energy released by the complete oxidation of glucose to CO_2 and H_2O.
B. The overall efficiency of ATP production would be about 53%, calculated as the ratio of actually produced ATP molecules (30) divided by the number of ATP molecules that could be obtained if all the energy stored in a glucose molecule could be harvested as chemical energy in ATP (57).
C. During the oxidation of 1 mole of glucose, 322 kcal (the remaining 47% of the available 686 kcal in one mole of glucose that is not stored as chemical energy in ATP) would be released as heat. This amount of energy would heat your body by 4.3°C (= 322 kcal/75 kg). This is a significant amount of heat, considering that 4°C of elevated temperature would be a quite incapacitating fever and that 1 mole (180 g) of glucose is no more than two cups of sugar.
D. If the energy yield were only 20%, then instead of 47% in the example above, 80% of the available energy would be released as heat and would need to be dissipated by your body. The heat production would be more than 1.7-fold higher than normal, and your body would certainly overheat.
E. The chemical formula of ATP is $C_{10}H_{12}O_{13}N_5P_3$, and its molecular weight is therefore 503 g/mole. Your resting body therefore hydrolyzes about 80 moles (= 40 kg/0.503 kg/mole) of ATP in 24 hours (this corresponds to about 1000 kcal of liberated chemical energy). Because every mole of glucose yields 30 moles of ATP, this amount of energy could be produced by oxidation of 480 g glucose (= 180 g/mole $\times$ 80 moles/30).

ANSWER 3–15 This scientist is definitely a fake. The 57 ATP molecules would store 684 kcal (= 57 $\times$ 12 kcal) of chemical energy, which implies that the efficiency of ATP production from glucose would have been greater than 99%. This impossible degree of efficiency would leave virtually no energy to be released as heat, and this release is required according to the laws of thermodynamics.

ANSWER 3–16

A. From Table 3–1 (p. 96) we know that a free-energy difference of 4.3 kcal/mole corresponds to an equilibrium constant of 10^{-3}, i.e., $[A^*]/[A] = 10^{-3}$. The concentration of A* is therefore 1000-fold lower than that of A.
B. The ratio of A to A* would be unchanged. Lowering the activation energy barrier would accelerate the rate of the reaction, i.e., it would allow more molecules in a given time period to convert from $A \rightarrow A^*$ and from $A^* \rightarrow A$, but would not affect the equilibrium point.

ANSWER 3–17

A. The mutant mushroom would probably be safe to eat. ATP hydrolysis can provide approximately –12 kcal/mole of energy. This amount of energy shifts the equilibrium point of a reaction by an enormous factor: about 10^8-fold (from Table 3–1, p. 96, we see that –5.7 kcal/mole corresponds to an equilibrium constant of 10^4; thus, –12 kcal/mole corresponds to about 10^8. Note that for coupled reactions energies are additive, whereas equilibrium constants are multiplied). Therefore, if the energy of ATP hydrolysis cannot be utilized by the enzyme, 10^8-fold less poison is made. This example illustrates that coupling a reaction to the hydrolysis of an activated carrier molecule can shift the equilibrium point drastically.

B. It would be risky to consume this mutant mushroom. Slowing down the reaction rate would not affect its equilibrium point, and if the reaction were allowed to proceed for a long enough time, the mushroom would likely be loaded with poison. Perhaps the equilibrium of the reaction would not be reached, but it would not be advisable to take a chance.

ANSWER 3–18 Enzyme A is beneficial. It allows the interconversion of two energy-carrier molecules, both of which are required as the triphosphate form for many metabolic reactions. Any ADP that is formed is quickly converted to ATP, and thus the cell maintains a high ATP/ADP ratio. Because of enzyme A, called nucleotide phosphokinase, some of the ATP is used to keep the GTP/GDP ratio similarly high.

Enzyme B would be highly detrimental to the cell. Cells use NAD^+ as an electron acceptor in catabolic reactions and must maintain high concentrations of this form of the carrier so as to break down glucose to make ATP. In contrast, NADPH is used as an electron donor in biosynthetic reactions and is kept at high concentration in the cells so as to allow the synthesis of nucleotides, fatty acids, and other essential molecules. Since enzyme B would deplete the cell's reserves of both NAD^+ and NADPH, it would decrease the rates of both catabolic and biosynthetic reactions.

ANSWER 3–19 Because enzymes are catalysts, enzyme reactions have to be thermodynamically feasible; the enzyme only lowers the activation energy barrier that otherwise slows the rate with which the reaction occurs. Heat confers more kinetic energy to the substrates so that a higher fraction of them can surmount the normal activation energy barrier. Many substrates, however, have many different ways in which they could react, and all of these potential pathways will be enhanced by heat, whereas an enzyme confers selectivity and facilitates only one particular pathway that, in evolution, was selected to be useful for the cell. Heat, therefore, cannot substitute for enzyme function, and chicken soup must exert its beneficial effects by other mechanisms that remain to be discovered.

ANSWER 3–20

A. When [S] << K_M, the term ([S] + K_M) approaches K_M. Therefore, the equation is simplified to rate = V_{max}/K_M [S], that is, the rate is proportional to [S].

B. When [S] = K_M, the term [S]/([S] + K_M) equals ½. Therefore, the reaction rate is half of the maximal rate V_{max}.

C. If [S] >> K_M, the term ([S] + K_M) approaches [S]. Therefore, [S]/([S] + K_M) equals 1 and the reaction occurs at its maximal rate V_{max}.

ANSWER 3–21 The substrate concentration is 1 mM. This value can be obtained by substituting values in the equation, but it is simpler to note that the desired rate (50 μmole/sec) is exactly half of the maximum rate, V_{max}, where the substrate concentration is typically equal to

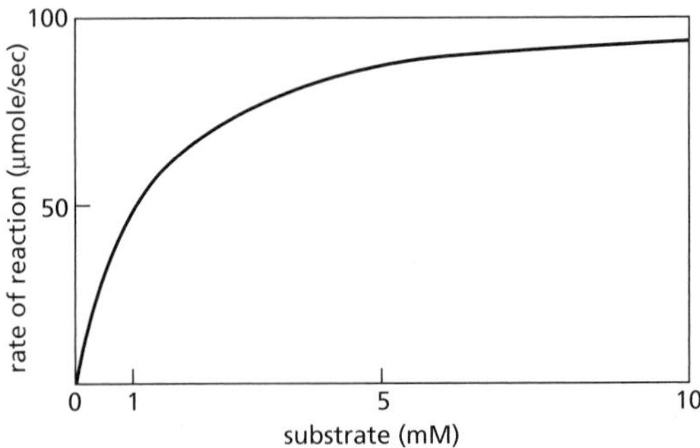

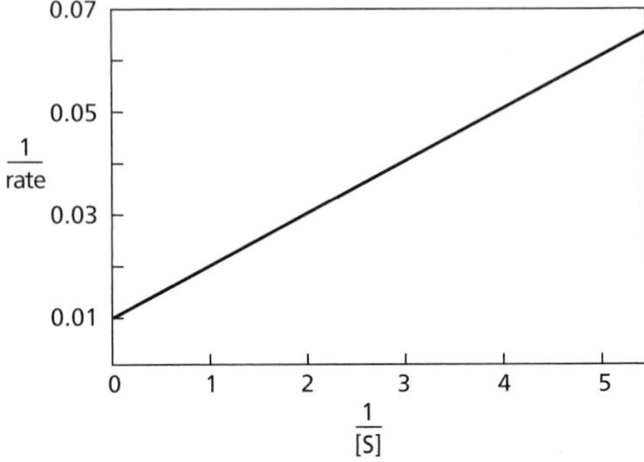

Figure A3–21

the K_M. The two plots requested are shown in Figure A3–21. A plot of 1/rate versus 1/[S] is a straight line because rearranging the standard equation yields the equation listed in Question 3–23B.

ANSWER 3–22 If [S] is very much smaller than K_M, the active site of the enzyme is mostly unoccupied. If [S] is very much greater than K_M, the reaction rate is limited by the enzyme concentration (because most of the catalytic sites are fully occupied).

ANSWER 3–23

A,B. The data in the boxes have been used to plot the red curve and red line in Figure A3–23. From the plotted data, the K_M is 1 μM and the V_{max} is 2 μmole/min. Note that the data are much easier to interpret in the linear plot, because the curve in (A) approaches, but never reaches, V_{max}.

C. It is important that only a small quantity of product is made, because otherwise the rate of reaction would decrease as the substrate was depleted and product accumulated. Thus the measured rates would be lower than they should be.

D. If the K_M increases, then the concentration of substrate needed to give a half-maximal rate is increased. As more substrate is needed to produce the same rate, the enzyme-catalyzed reaction has been inhibited by the phosphorylation. The expected data plots for the phosphorylated enzyme are the green curve and the green line in Figure A3–23.

Chapter 4

ANSWER 4–1 Urea is a very small molecule that functions both as an efficient hydrogen-bond donor (through its $-NH_2$ groups) and as an efficient hydrogen-bond acceptor (through its –C=O group). As such, it can

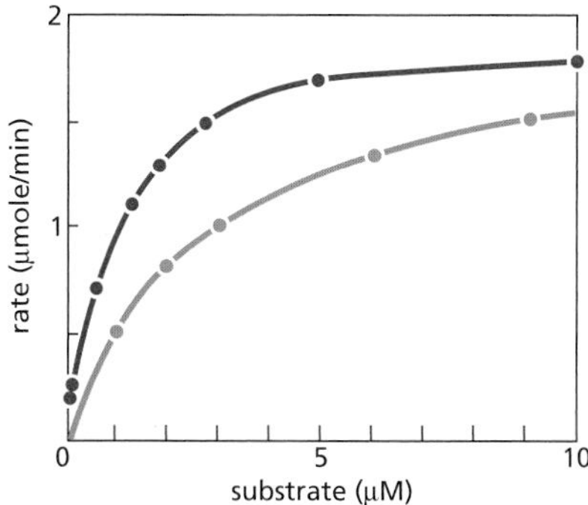

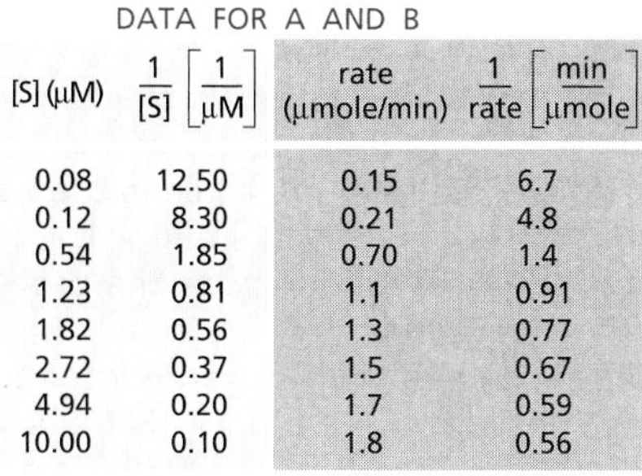

DATA FOR A AND B

[S] (μM)	$\frac{1}{[S]}\left[\frac{1}{\mu M}\right]$	rate (μmole/min)	$\frac{1}{rate}\left[\frac{min}{\mu mole}\right]$
0.08	12.50	0.15	6.7
0.12	8.30	0.21	4.8
0.54	1.85	0.70	1.4
1.23	0.81	1.1	0.91
1.82	0.56	1.3	0.77
2.72	0.37	1.5	0.67
4.94	0.20	1.7	0.59
10.00	0.10	1.8	0.56

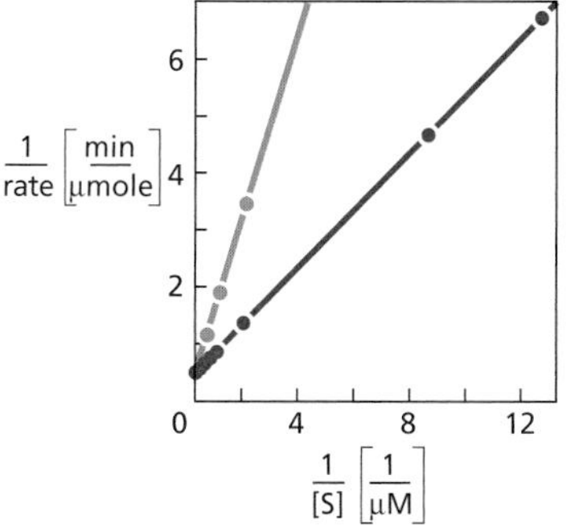

Figure A3–23

squeeze between hydrogen bonds that stabilize protein molecules and thus destabilize protein structures. In addition, nonpolar side chains are held together in the interior of folded proteins because they disrupt the structure of water if they are exposed. At high concentrations of urea, the hydrogen-bonded network of water molecules becomes disrupted so that these hydrophobic forces are significantly diminished. Proteins unfold in urea as a consequence of its effect on these two forces.

ANSWER 4–2 There are two α helices, and both are right-handed. The three chains that form the largest region of β sheet (*green*) are antiparallel. There are no knots in the polypeptide chain, presumably because a knot would interfere with the folding of the protein into its three-dimensional conformation after protein synthesis.

ANSWER 4–3 The amino acid sequence consists of alternating nonpolar and charged or polar amino acids. The resulting strand in a β sheet would therefore be polar on one side and hydrophobic on the other. Such a strand would probably be surrounded on either side by similar strands that together form a β sheet with a hydrophobic and a polar face. In a protein, such a β sheet (called "amphipathic," from the Greek *amphi*, "of both kinds," and *pathos*, "passion," because of its two surfaces with such different properties) would be positioned so that the hydrophobic side would face the protein's interior and the polar side would be on its surface, exposed to the water outside.

ANSWER 4–4 Mutations that are beneficial to an organism are selected in evolution because they confer a growth or survival advantage to the organism. Examples might be the better utilization of a food source; enhanced resistance to environmental insults, such as heat or concentrated salt; or an enhanced ability to attract a mate for sexual reproduction. In contrast, the accumulation of useless proteins is detrimental to organisms. Useless mutant proteins waste the metabolic energy required to make them. If such mutant proteins were made in excess, the synthesis of normal proteins would suffer because the capacity of the cell is limited. In more severe cases, mutant proteins interfere with the normal workings of the cell; a mutant enzyme that still binds an activated carrier molecule but does not catalyze a reaction, for example, may compete for a limited amount of this carrier and therefore inhibit normal processes. Natural selection therefore provides a strong driving force that leads to the loss of both useless and harmful proteins.

ANSWER 4–5 Strong reducing agents that break all of the S–S bonds would cause all of the keratin filaments to separate. Hair would therefore disintegrate. Indeed, strong reducing agents are used commercially in hair removal creams sold by your local pharmacist. However, mild reducing agents are used in treatments that either straighten or curl hair, the latter requiring hair curlers. (See Figure A4–5.)

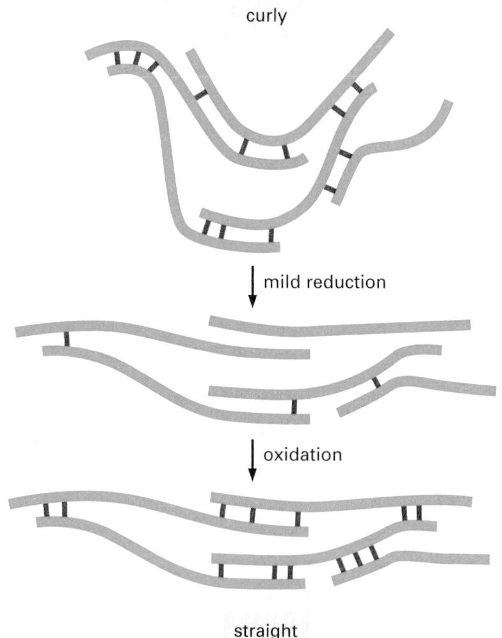

Figure A4–5

ANSWER 4–6 See Figure A4–6.

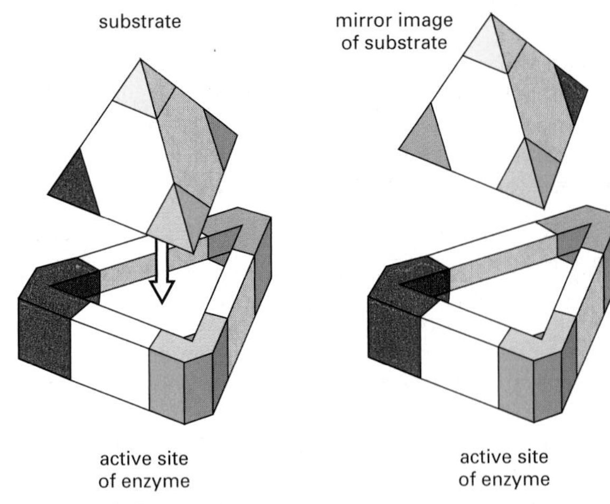

Figure A4–6

ANSWER 4–7

A. Feedback inhibition from Z that affects the reaction B → C would increase the flow through the B → X → Y → Z pathway, because the conversion of B to C is inhibited. Thus, the more Z there is, the more production of Z would be stimulated. This is likely to result in an uncontrolled "runaway" amplification of this pathway.
B. Feedback inhibition from Z affecting Y → Z controls the production of Z. In this scheme, however, X and Y are still made at normal rates, even though both of these intermediates are no longer needed at this level. This pathway is therefore less efficient than the one shown in Figure 4–34.
C. If Z is a positive regulator of the step B → X, then the more Z there is, the more B will be converted to X and therefore shunted into the pathway producing more Z. This would result in a runaway amplification similar to that described for (A).
D. If Z is a positive regulator of the step B → C, then accumulation of Z leads to a redirection of the pathway to make more C. This is a second possible way, in addition to that shown in the figure, to balance the distribution of compounds into the two branches of the pathway.

ANSWER 4–8 Both nucleotide binding and phosphorylation can induce allosteric changes in proteins. These can have a multitude of consequences, such as altered enzyme activity, drastic shape changes, and changes in affinity for other proteins or small molecules. Both mechanisms are quite versatile. An advantage of nucleotide binding is the fast rate with which a small nucleotide can diffuse to the protein; the shape changes that accompany the function of motor proteins, for example, require quick nucleotide replenishment. If the different conformational states of a motor protein were controlled by phosphorylation, for example, a protein kinase would either need to diffuse into position at each step, a much slower process, or be associated permanently with each motor. One advantage of phosphorylation is that it requires only a single amino acid residue on the protein's surface, rather than a specific binding site. Phosphates can therefore be added to many different side chains on the same protein (as long as protein kinases with the proper specificities exist), thereby vastly increasing the complexity of regulation that can be achieved for a single protein.

ANSWER 4–9 In working as a complex, all three proteins contribute to the specificity (by binding the safe and key directly), help position one another correctly, and provide the mechanical bracing that allows them to perform a task that they could not perform individually (the key is grasped by two subunits, for example). Moreover, their functions are generally coordinated in time (for example, the binding of ATP to one subunit is likely to require that ATP has already been hydrolyzed to ADP by another).

ANSWER 4–10

A. True. Only a few amino acid side chains contribute to the active site. The rest of the protein is required to maintain the polypeptide chain in the correct position, provide additional binding sites for regulatory purposes, and localize the protein in the cell.
B. True. Some enzymes form covalent intermediates with their substrates (see Figure 13–5); however, in all cases the enzyme is restored to its original structure after the reaction.
C. False. β sheets can, in principle, contain any number of strands because the two strands that form the rims of the sheet are available for hydrogen-bonding to other strands. (β sheets in known proteins contain from 2 to 16 strands.)
D. False. It is true that the specificity of an antibody molecule is exclusively contained in loops on its surface; however, these loops are contributed by both the folded light- and heavy-chain domains (see Figure 4–29).
E. False. The possible linear arrangements of amino acids that lead to a stably folded protein domain are so few that most new proteins evolve by alteration of old ones.
F. True. Allosteric enzymes generally bind one or more molecules that function as regulators at sites that are distinct from the active site.
G. False. Noncovalent bonds are a major contributor to the three-dimensional structure of macromolecules.
H. False. Affinity chromatography separates specific macromolecules because of their interactions with specific ligands, not because of their charge.
I. False. The larger an organelle is, the more centrifugal force it experiences and the faster it sediments, despite an increased frictional resistance from the fluid through which it moves.

ANSWER 4–11 In an α helix and in the central strands of a β sheet, all of the N–H and C=O groups in the polypeptide backbone are engaged in hydrogen bonds. This gives considerable stability to these secondary structure elements, and it allows them to form from many different amino acid sequences.

ANSWER 4–12 No. It would not have the same or even a similar structure, because the peptide bond has a polarity. Looking at two sequential amino acids in a polypeptide chain, the amino acid that is closer to the N-terminal end contributes the carboxyl group and the other amino acid contributes the amino group to the peptide bond that links the two. Changing their order would put the side chains into a different position with respect to the peptide backbone and therefore change their chemical environment.

ANSWER 4–13 As it takes 3.6 amino acid residues to complete a turn of an α helix, this sequence of 14 amino acids would make close to 4 full turns. It is remarkable because its polar and hydrophobic amino acids are spaced so that all polar residues are on one side of an α helix and all the hydrophobic residues are on the other. It is therefore likely that such an amphipathic α helix is exposed on the protein surface with its hydrophobic side facing the protein's interior. In addition, two such helices might wrap around each other as shown in Figure 4–13.

ANSWER 4–14

A. ES represents the enzyme–substrate complex.
B. Enzyme and substrate are in equilibrium between their free and bound states; once bound to the enzyme, a substrate molecule may either dissociate again (hence the bidirectional arrows) or be converted to product. As substrate is converted to product (with the concomitant release of free energy), however, a reaction often proceeds strongly in the forward direction, as indicated by the unidirectional arrow.
C. The enzyme is a catalyst and is therefore liberated in an unchanged form after the reaction; thus, E appears at both ends of the equation.
D. Often the products of a reaction resemble the substrates sufficiently that they can also bind to the enzyme. Any enzyme molecules that are bound to product (i.e., are part of the EP complex) are unavailable for catalysis; excess P therefore inhibits the reaction by lowering the concentration of free E.
E. Compound X is an inhibitor of the reaction and works similarly by forming an EX complex. However, since P has to be made before it can inhibit the reaction, it takes longer to act than X, which is present from the beginning of the reaction.

ANSWER 4–15 The polar amino acids Ser, Ser-P, Lys, Gln, His, and Glu are more likely to be found on a protein's surface, and the hydrophobic amino acids Leu, Phe, Val, Ile, and Met are more likely to be found in its interior. The oxidation of two cysteine residues to form a disulfide bond eliminates their potential to form hydrogen bonds and therefore makes them even more hydrophobic. Disulfide bonds are usually found in the interior of proteins. Irrespective of the nature of their side chains, the most N-terminal amino acid and the most C-terminal amino acid each contain a charged group (the amino and carboxyl groups that mark the ends of the polypeptide chain) and hence are usually found on the protein's surface.

ANSWER 4–16 Many secondary structure elements are not stable in isolation but require the presence of additional parts of the polypeptide chain. Hydrophobic regions that would normally be hidden in the inside of a folded domain would be exposed on the outside, and because such regions are energetically disfavored in water solution, the fragments tend to aggregate nonspecifically. Such fragments therefore would not have a defined structure, and they would be inactive for ligand binding even if they contained all of the amino acids that would normally contribute to the ligand-binding site. A protein domain, in contrast, is considered a folding unit, and fragments of a polypeptide chain that correspond to intact domains are often able to fold correctly. Thus, separated protein domains often retain their activities, such as ligand binding, if the binding site is contained entirely within this domain. Thus the most likely place in which the polypeptide chain of the protein in Figure 4–16 could be severed to give rise to stable fragments is at the boundary between the two domains (i.e., at the loop between the two α helices at the bottom right of the structure shown).

ANSWER 4–17 The heat inactivation of the enzyme suggests that the mutation causes the enzyme to have a less stable structure. For example, a hydrogen bond that is normally formed between two amino acid side chains might no longer be formed because the mutation replaces one of these amino acids with a different one that cannot participate in the bond. Lacking such a bond that normally helps to keep the polypeptide chain folded properly, the protein unfolds at a temperature at which it would normally be stable. Polypeptide chains that are denatured when the temperature is raised often aggregate, and they rarely refold into active proteins when the temperature is decreased.

ANSWER 4–18 The motor protein in the illustration can move just as easily to the left as to the right and so will not move steadily in one direction. However, if just one of the steps is coupled to ATP hydrolysis (for example, by making detachment of one foot dependent on binding of ATP and coupling the reattachment to hydrolysis of the bound ATP), then the protein will show unidirectional movement that requires the continued consumption of ATP. Note that, in principle, it does not matter which step is coupled to ATP hydrolysis (Figure A4–18).

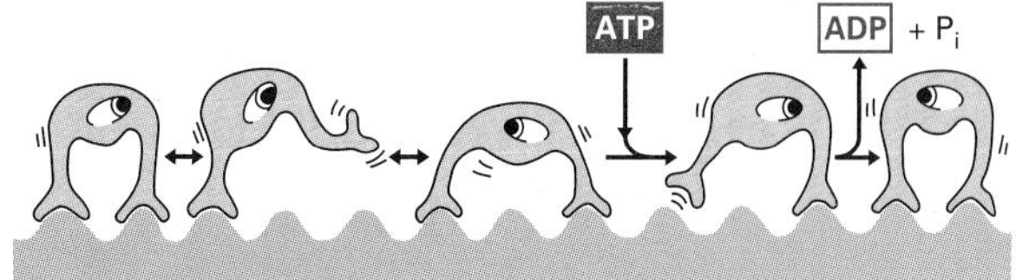

Figure A4–18

ANSWER 4–19 The slower migration of small molecules through a gel-filtration column occurs because smaller molecules have access to many more spaces in the porous beads that are packed into the column than do larger molecules. However, it is important to give the smaller molecules sufficient time to diffuse into the spaces inside the beads. At very rapid flow rates, all molecules will move rapidly around the beads, so that large and small molecules will now tend to exit together from the column.

ANSWER 4–20 The α helix is right-handed and the coiled-coil is left-handed. The reversal occurs because of the staggered positions of hydrophobic side chains in the α helix.

ANSWER 4–21 The atoms at the binding sites of proteins must be precisely located to fit the molecules that they bind. Their location in turn requires the precise positioning of many of the amino acids and their side chains in the core of the protein, distant from the binding site itself. Thus, even a small change in this core can disrupt protein function by altering the conformation at a binding site far away.

Chapter 5

ANSWER 5–1

A. False. The polarity of a DNA strand commonly refers to the orientation of its sugar–phosphate backbone.
B. True. G-C base pairs are held together by three hydrogen bonds, whereas A-T base pairs are held together by only two.

ANSWER 5–2 The scale bar in Figure 5–11 is in millions of nucleotide pairs. Using this to estimate the amount of DNA packaged into Chromosome 1 we obtain approximately 256 million nucleotide pairs. This would give a total length for the DNA of 8.7 cm ($256 \times 10^6 \times 0.34$ nm; 1 nm = $1/10^9$ m) and a compaction of 8.7 cm/10 μm = 8700-fold.

ANSWER 5–3 Histone octamers occupy about 9% of the volume of the nucleus. The volume of the nucleus is

$$V = 4/3 \times 3.14 \times (3 \times 10^3 \text{ nm})^3$$

$$V = 1.13 \times 10^{11} \text{ nm}^3$$

The volume of the histone octamers is

$$V = 3.14 \times (4.5 \text{ nm})^2 \times (5 \text{ nm}) \times (32 \times 10^6)$$

$$V = 1.02 \times 10^{10} \text{ nm}^3$$

The ratio of the volume of histone octamers to the nuclear volume is 0.09; thus, histone octamers occupy about 9% of the nuclear volume. Because the DNA also occupies about 9% of the nuclear volume, together they occupy about 18% of the volume of the nucleus.

ANSWER 5–4 In contrast to most proteins, which accumulate amino acid changes over evolutionary time, the functions of histone proteins must involve nearly all of their amino acids, so that a change in any position is deleterious to the cell. Histone proteins are exquisitely refined for their function.

ANSWER 5–5 Men have only one copy of the X chromosome; a defective gene carried on it therefore has no backup copy. Women, on the other hand, have two copies of the X chromosome, one inherited from each parent, so a defective copy of the gene on one X chromosome can generally be compensated for by a normal copy on the other chromosome. This is the case with regard to the gene that causes color blindness. However, during female development, transcription from one X chromosome is shut down because it is compacted into heterochromatin (Figure 5–30). This occurs at random in each cell to one or the other of the two X chromosomes, and therefore some cells of the woman will express the defective mutant copy of the gene, whereas others will express the normal copy. This results in a retina in which on average only every other cone cell has functional color photoreceptors, and women carrying the mutant gene on one X chromosome therefore see color images with reduced resolution.

A woman who is color-blind must have two defective copies of this gene, one inherited from each parent. Her father must therefore carry the mutation on his X chromosome; because this is his only copy of the gene, he would be color-blind. Her mother could carry the defective gene on either or both of her X chromosomes. Her mother could therefore either be color-blind (defective genes on both X chromosomes) or have color vision but reduced resolution as described above. Several different types of inherited color blindness are found in the human population; this question applies to only one type.

ANSWER 5–6

A. The complementary strand reads 5′-`TGATTGTGGACAAAAATCC`-3′. Paired DNA strands have opposite polarity, and the convention is to write a single-stranded DNA sequence in the 5′-to-3′ direction.
B. The DNA is made of four nucleotides (100% = 13% A + x% T + y% G + z% C). Because A pairs with T, the two nucleotides are represented in equimolar proportions in DNA. Therefore, the bacterial DNA in question contains 13% thymidine. This leaves 74% [= 100% – (13% + 13%)] for G and C, which also form base pairs and hence are equimolar. Thus $y = z = 74/2 = 37$.
C. A single-stranded DNA molecule that is N nucleotides long can have any one of 4^N possible sequences, but the number of possible double-stranded DNA molecules is more difficult to calculate. Many of the 4^N single-stranded sequences will be the complement of another possible sequence in the list; for example, 5′-AGTCC-3′ and 5′-GGACT-3′ form the same double-stranded DNA molecule and therefore count as a single double-stranded possibility. If N is an odd number, then every single-stranded sequence will complement another sequence in the list so that the number of double-stranded sequences will be 0.5×4^N. If N is an even number, then there will be slightly more than this, since some sequences will be self-complementary (such as 5′-ACTAGT-3′) and the actual value can be calculated to be $0.5 \times 4^N + 0.5 \times 4^{N/2}$.
D. To specify a unique sequence which is N nucleotides long, 4^N has to be larger than 3×10^6. Thus, 4^N / greater than 3×10^6, solved for N, gives N /greater than $\ln(3 \times 10^6)/\ln(4) = 10.7$. Thus, on average a sequence of only 11 nucleotides in length is unique in the genome. Performing the same calculation for the genome size of an animal cell yields a minimal stretch of 16 nucleotides. This shows that a relatively short sequence can mark a unique position in the genome and is sufficient, for example, to serve as an identity tag for one specific gene.

ANSWER 5–7 If the wrong bases were frequently incorporated during DNA replication, genetic information could not be inherited accurately. Life, as we know it, could not exist. Although the bases can form hydrogen-bonded pairs as indicated, these do not fit into the structure of the double helix. Thus, the angle with which the A residue is attached to the sugar–phosphate backbone is vastly different in the A-C pair, and the spacing between the two sugar–phosphate strands is considerably increased in the A-G pair, where two large purine rings interact. Consequently, it is energetically unfavorable to incorporate the wrong bases into the DNA chain, and such errors occur only very rarely.

ANSWER 5–8

A. The bases V, W, X, and Y can form a DNA-like double-helical molecule with virtually identical properties to those of bona fide DNA. V would always pair with X, and W with Y. Therefore, the macromolecules could be derived from a living organism using the same principles of replication of its genome. In principle, different bases, such as V, W, X, and Y, could have been selected during evolution as building blocks of DNA on Earth. (Similarly, there are many more conceivable amino acid side chains than the set of 20 selected in evolution that make up all proteins.)

B. None of the bases V, W, X, or Y can replace A, T, G, or C. To preserve the distance between the two sugar– phosphate strands in the double helix, a pyrimidine always has to pair with a purine (see, for example, Figure 5–6). Thus, the eight possible combinations are V-A, V-G, W-A, W-G, X-C, X-T, Y-C, and Y-T. Because of the positions of hydrogen-bond acceptors and hydrogen-bond donor groups, however, no stable base pairs would form in any of these combinations, as shown for the pairing of V-A in Figure A5–8, where only a single hydrogen bond could form.

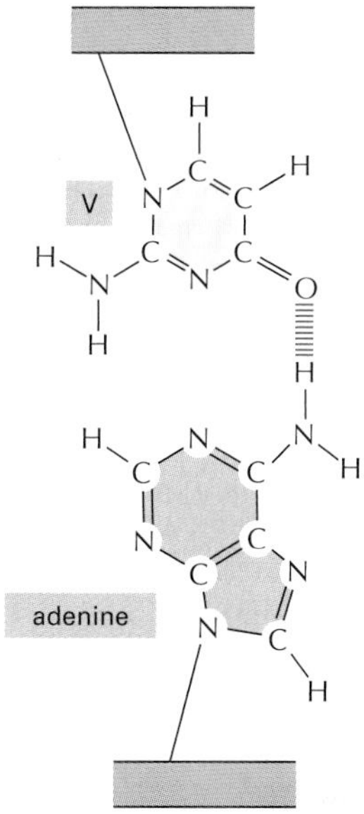

Figure A5–8

ANSWER 5–9 As the strands are held together by hydrogen bonds between the bases, the stability of the helix is largely dependent on the number of hydrogen bonds that can be formed. Thus two parameters determine the stability: the number of nucleotide pairs and the number of hydrogen bonds that each nucleotide pair contributes. As shown in Figure 5–6, an A-T pair contributes two hydrogen bonds, whereas a G-C pair contributes three hydrogen bonds. Therefore, helix C (containing a total of 34 hydrogen bonds) melts at the lowest temperature, helix B (containing a total of 65 hydrogen bonds) melts next, and helix A (containing a total of 78 hydrogen bonds) is the most stable, largely owing to its high GC content. Indeed, the DNA of organisms that grow in extreme temperature environments, such as certain bacteria that grow in geothermal vents, has an unusually high GC content.

ANSWER 5–10 The DNA would be enlarged by a factor of 2.5×10^6 (= $5 \times 10^{-3}/2 \times 10^{-9}$ m). Thus the extension cord would be 2500 km long. This is approximately the distance from London to Istanbul, San Francisco to Kansas City, Tokyo to the southern tip of Taiwan, and Melbourne to Cairns. Adjacent nucleotides would be about 0.85 mm apart (which is only about the thickness of a stack of 12 pages of this book). A gene that is 1000 nucleotide pairs long would be about 85 cm in length.

ANSWER 5–11

A. It takes two bits to specify each nucleotide pair (for example, 00, 01, 10, and 11 would be the binary codes for the four different nucleotides, each paired with its appropriate partner).

B. The entire human genome (3×10^9 nucleotide pairs) could be stored on two CDs ($3 \times 10^9 \times 2$ bits/4.8×10^9 bits).

ANSWER 5–12

A. True.

B. False. Nucleosome core proteins are approximately 11 nm in diameter. A model for the way they are packed to form a 30-nm-diameter filament is shown in Figure 5–25.

ANSWER 5–13 The definitions of the terms can be found in the Glossary. DNA assembles with specialized proteins to form *chromatin.* At a first level of packing, *histones* form the core of *nucleosomes.* In a nucleosome, the DNA is wrapped twice around this core. Between nuclear divisions, that is, in interphase, the *chromatin* of the *interphase chromosomes* is in a relatively extended form and is dispersed in the nucleus, although some regions of it, the *heterochromatin,* remain densely packed and are transcriptionally inactive. During nuclear division—that is, in mitosis—replicated chromosomes become condensed into *mitotic chromosomes,* which are transcriptionally inactive and are designed to be distributed between the daughter cells.

ANSWER 5–14 Colonies are clumps of cells that originate from a single founder cell and grow outward as the cells divide again and again. In the lower colony of Figure Q5–14, the *Ade2* gene is inactivated when placed near a telomere, but apparently it can become spontaneously activated in a few cells, which then turn white. Once spontaneously activated in a cell, the *Ade2* gene continues to be active in the descendants of that cell, resulting in clumps of white cells (the white sectors) in the colony. This result shows both that the inactivation of a gene positioned close to a telomere can be reversed and that this change is passed on to further generations. This change in *Ade2* expression probably results from a spontaneous decondensation of the chromatin structure around the gene.

ANSWER 5–15 In the electron micrographs, one can detect chromatin regions of two different densities; the densely stained regions correspond to heterochromatin, while less condensed chromatin is more lightly stained. The chromatin in nucleus A is mostly in the form of condensed, transcriptionally inactive heterochromatin, whereas most of the chromatin in nucleus B is decondensed and therefore potentially active for transcription. Nucleus A is from a reticulocyte, a red blood cell precursor, which is largely devoted to making a single protein, hemoglobin. Nucleus B is from a lymphocyte, which is active in transcribing many different genes.

ANSWER 5–16 Helix A is right-handed. Helix C is left-handed. Helix B has one right-handed strand and one left-handed strand. There are several ways to tell the handedness of a helix. For a vertically oriented helix, like the ones in Figure Q5–14, if the strands in front point up to the right, the helix is right handed; if they point up to the left, the helix is left handed. Once you are comfortable identifying the handedness of a helix, you will be amused to note that nearly 50% of the 'DNA' helices in advertisements are left handed, as are a surprisingly high number of the ones in books. Amazingly, a version of Helix B was used in advertisements for a prominent international conference, celebrating the 30-year anniversary of the discovery of the DNA helix.

ANSWER 5–17 The packing ratio within a nucleosome core is 4.5 [(146 bp × 0.34 nm/bp)/(11 nm) = 4.5]. If there is an additional 54 bp of linker DNA, then the packing ratio for 'beads-on-a-string' DNA is 2.3 [(200 bp × 0.34 nm/bp)/(11 nm + {54 bp × 0.34 nm/bp}) = 2.3]. This first level of packing represents only 0.023% (2.3/10,000) of the total condensation that occurs at mitosis.

ANSWER 5–18 All of the mechanisms that are known to alter chromatin structure appear to be readily reversible. Thus, in response to specific signals, any of the covalent modifications to the lysine, arginine, and serine side chains in histones can be removed. This reversibility retains an important flexibility in developmental pathways, which would be much more difficult to achieve for the hypothetical DNA-based scheme.

Chapter 6

ANSWER 6–1

A. The distance between replication forks 4 and 5 is about 280 nm, corresponding to 824 nucleotides (= 280/0.34). These two replication forks would collide in about 8 seconds. Forks 7 and 8 move away from each other and would therefore never collide.
B. The total length of DNA shown in the electron micrograph is about 1.5 μm, corresponding to 4400 nucleotides. This is only about 0.002% [= (4400/1.8 × 10^8) × 100%] of the total DNA in a fly cell.

ANSWER 6–2 Although the process may seem wasteful, it is not possible to proofread during the initial stages of primer synthesis. To start a new primer on a piece of single-stranded DNA, one nucleotide needs to be put in place and then linked to a second and then to a third, and so on. Even if these first nucleotides were perfectly matched to the template strand, they would bind with very low affinity, and it would consequently be difficult to distinguish the correct from incorrect bases by a hypothetical primase with proofreading activity; the enzyme would therefore stall. The task of the primase is to "just polymerize nucleotides that bind reasonably well to the template without worrying too much about accuracy." Later these sequences are removed and replaced by DNA polymerase, which uses newly synthesized (and therefore proofread) DNA as its primer.

ANSWER 6–3

A. Without DNA polymerase, no replication can take place at all. RNA primers will be laid down at the origin of replication.
B. DNA ligase links the DNA fragments that are produced on the lagging strand. In the absence of ligase, the newly replicated DNA strands will remain as fragments, but no nucleotides will be missing.
C. Without the sliding clamp, the DNA polymerase will frequently fall off the DNA template. In principle, it can rebind and continue, but the continual falling off and rebinding will be time-consuming and will greatly slow down DNA replication.
D. In the absence of RNA excision enzymes, the RNA fragments will remain covalently attached to the newly replicated DNA fragments. No ligation will take place, because the ligase will not link DNA to RNA. The lagging strand will therefore consist of fragments composed of both RNA and DNA.
E. Without DNA helicase, the DNA polymerase will stall because it cannot separate the strands of the template DNA ahead of it. Little or no new DNA will be synthesized.
F. In the absence of primase, RNA primers cannot begin on either the leading or the lagging strand. DNA replication therefore cannot begin.

ANSWER 6–4 DNA damage by deamination and depurination reactions occurs spontaneously. This type of damage is not the result of replication errors and is therefore equally likely to occur on either strand. If DNA repair enzymes recognized such damage only on newly synthesized DNA strands, half of the defects would go uncorrected. The statement is therefore incorrect.

ANSWER 6–5 The AIDS virus (the human immunodeficiency virus, HIV) is a retrovirus, and thus synthesizes DNA from an RNA template using reverse transcriptase. This leads to frequent mutation of the viral genome. In fact, AIDS patients often carry many different genetic variants of HIV that are distinct from the original virus that infected them. This poses great problems in treating the infection: drugs that block essential viral enzymes work only temporarily, because new strains of the virus resistant to these drugs arise rapidly by mutation.

RNA replicases (enzymes that synthesize RNA using RNA as a template) do not proofread either. Thus, RNA viruses that replicate their RNA genomes directly (that is, without using DNA as an intermediate) also mutate frequently. In such a virus, this tends to produce changes in the coat proteins that cause the mutated virus to appear "new" to our immune systems; the virus is therefore not suppressed by immunity that has arisen to the previous version. This is part of the explanation for the new strains of the influenza (flu) virus and the common cold virus that regularly appear.

ANSWER 6–6 If the old strand were "repaired" using the new strand that contains a replication error as the template, then the error would become a permanent mutation in the genome. The old information would be erased in the process. Therefore, if repair enzymes did not distinguish between the two strands, there would be only a 50% chance that any given replication error would be corrected.

ANSWER 6–7 The argument is severely flawed. You cannot transform one species into another simply by introducing 1% random changes into the DNA. It is exceedingly unlikely that the 5000 mutations that would accumulate every day in the absence of DNA repair would be in the very positions where human and ape DNA sequences are different. It is also very likely that at such a high mutation frequency many essential genes would be inactivated, leading to cell death. Furthermore, your body is made up of about 10^{13} cells. For you to turn into an ape, not just one but many of these cells would need to be changed. And even then, many of these changes would have to occur during development to effect changes in your body plan (making your arm longer than your legs, for example).

ANSWER 6–8

A. False. Identical DNA polymerase molecules catalyze DNA synthesis of the leading and lagging strands of a bacterial replication fork. The replication fork is asymmetrical because the lagging strand is synthesized in pieces that are then stitched together.
B. False. Only the RNA primers are removed by a RNA nuclease; the Okazaki fragments are the pieces of newly synthesized DNA that are eventually joined to form the new lagging strand.
C. True. DNA polymerase has an error rate of one mistake in 10^7 nucleotides polymerized. 99% of its errors are corrected by DNA mismatch repair enzymes, bringing the final error rate to one in 10^9.
D. True. Mutations would accumulate rapidly, destroying the genes.
E. True. If a damaged nucleotide also occurred naturally in DNA, the repair enzyme would have no way of identifying the damage. It would therefore have only a 50% chance of fixing the right strand.
F. True. Usually, multiple mutations of specific types need to accumulate before a cell turns into a cancer cell. A mutation in a gene that codes for a DNA repair enzyme can make a cell more liable to accumulate further mutations, thereby accelerating the onset of cancer.

ANSWER 6–9 With a single origin of replication that launches two DNA polymerases in opposite directions on the DNA, each moving at 100 nucleotides per second, the number of nucleotides replicated in 24 hours will be
1.73×10^7 (= $2 \times 100 \times 24 \times 60 \times 60$). To replicate all the 6×10^9 nucleotides of DNA in the cell in this time, therefore, will require at least 348 (= $6 \times 10^9/1.73 \times 10^7$) origins of replication. The estimated 10,000 origins of replication in the human genome are therefore more than enough to satisfy this minimum requirement.

ANSWER 6–10

A. Compound A is dideoxycytosine triphosphate (ddCTP), identical to dCTP except that it lacks the 3′-hydroxyl group on the sugar ring. ddCTP is recognized by DNA polymerase as dCTP and becomes incorporated into DNA; because it lacks the crucial 3′-hydroxyl group, however, its addition to a growing DNA strand creates a dead end to which no further nucleotides can be added. Thus, if ddCTP is added in large excess, strands will be synthesized until the first G (the nucleotide complementary to C) is encountered on the template strand. ddCTP will then be incorporated instead of C, and the extension of this strand will be terminated.
B. If ddCTP is added at about 10% of the concentration of the available dCTP, there is a 1 in 10 chance of its being incorporated whenever a G is encountered on the template strand. Thus a population of DNA fragments will be synthesized, and from their lengths one can deduce where the G residues are located on the template strand. This strategy forms the basis of methods used to determine the sequence of nucleotides in a stretch of DNA (discussed in Chapter 10).
   The same chemical phenomenon is exploited by a drug, 3′-azido-3′-deoxythymidine (AZT), that is now commonly used in HIV-infected patients to treat AIDS. AZT is converted in cells to the triphosphate form and is incorporated into the growing viral DNA. Because the drug lacks a 3′-OH group, it blocks DNA synthesis and replication of the virus. AZT inhibits viral replication preferentially because reverse transcriptase has a higher affinity for the drug than for thymidine triphosphate; human cellular DNA polymerases do not show this preference.
C. Compound B is dideoxycytosine monophosphate (ddCMP), which lacks the 5′-triphosphate group as well as the 3′-hydroxyl group of the sugar ring. It therefore cannot provide the energy that drives the polymerization reaction of nucleotides into DNA and therefore will not be incorporated into the replicating DNA. Addition of this compound should not affect DNA replication.

ANSWER 6–11 To use the energy of hydrolysis of the 3′-triphosphate group for polymerization, strand growth would need to occur in the opposite, that is, the 3′-to-5′, direction. Proofreading, in principle, could then occur by a 5′-to-3′ nuclease activity. This scenario would be the same as that shown in the left side of Figure 6–15, except that triphosphate groups would be on the right side of the DNA and of the incoming nucleotide triphosphate.

ANSWER 6–12 See Figure A6–12.

ANSWER 6–13 Both strands of the bacterial chromosome contain 6×10^6 nucleotides. During the polymerization of nucleoside triphosphates into DNA, two phosphoanhydride bonds are broken for each nucleotide added: the nucleoside triphosphate is hydrolyzed to produce the nucleoside monophosphate added to the growing DNA strand, and the released pyrophosphate is hydrolyzed to phosphate. Therefore, 1.2×10^7 high-energy bonds are hydrolyzed during each

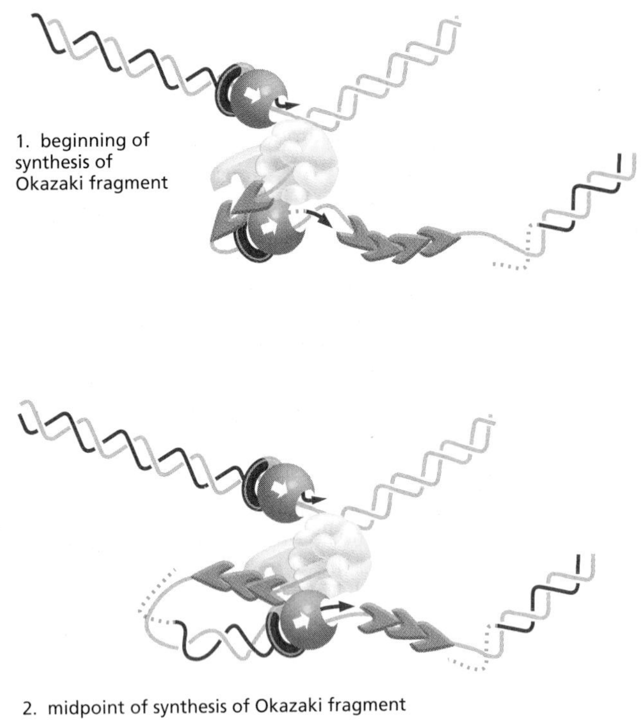

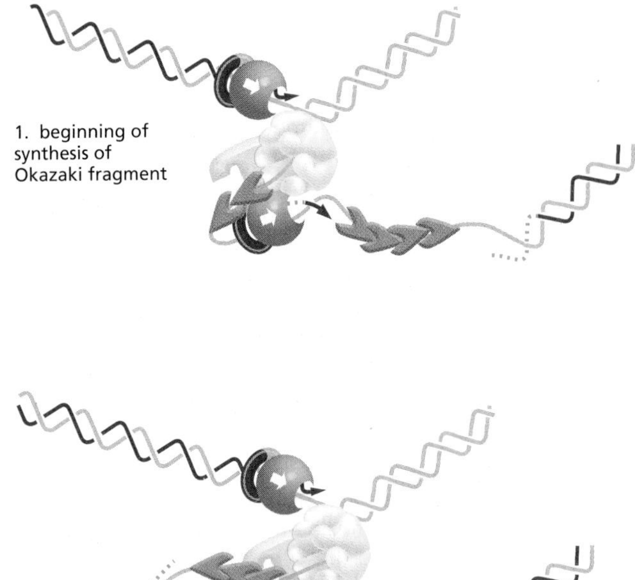

2. midpoint of synthesis of Okazaki fragment

Figure A6–12

round of bacterial DNA replication. This requires 4×10^5 (= $1.2 \times 10^7/30$) glucose molecules, which weigh 1.2×10^{-16} g (= 4×10^5 molecules × 180 g/mole/6×10^{23} molecules/mole), which is 0.01% of the total weight of the cell.

ANSWER 6–14 The statement is correct. If the DNA in somatic cells is not sufficiently stable (that is, if it accumulates mutations too rapidly), the organism dies (of cancer, for example), and because this may often happen before the organism can reproduce, the species will die out. If the DNA in reproductive cells is not sufficiently stable, many mutations will accumulate and be passed on to future generations, so that the species will not be maintained.

ANSWER 6–15 As shown in Figure A6–15, thymine and uracil lack amino groups and therefore cannot be deaminated. Deamination of adenine and guanine produces purine rings that are not found in conventional nucleic acids. In contrast, deamination of cytosine produces uracil. Therefore, if uracil were a naturally occurring base in DNA, repair enzymes could not distinguish whether a uracil is the appropriate base or whether it arose through spontaneous deamination of cytosine. This dilemma is not encountered, however, because thymine is used in DNA. Therefore, if a uracil base is found in DNA, it can be automatically recognized as a damaged base and then excised and replaced by cytosine.

ANSWER 6–16

A. Because DNA polymerase reqires a 3′-OH to synthesize DNA, without telomeres and telomerase, chromosome ends would shrink during each round of replication (Figure A6–16). For bacterial chromosomes, which have no ends, the problem does not arise; there will always be a 3′-OH group available to prime the DNA polymerase that replaces the RNA primer with DNA. Telomeres and telomerase prevent the shrinking of chromosomes because the telomeres extend the 3′ end of a DNA strand (see Figure 6–18). This extension of the lagging strand template provides the 'space' to begin the final Okazaki fragments.

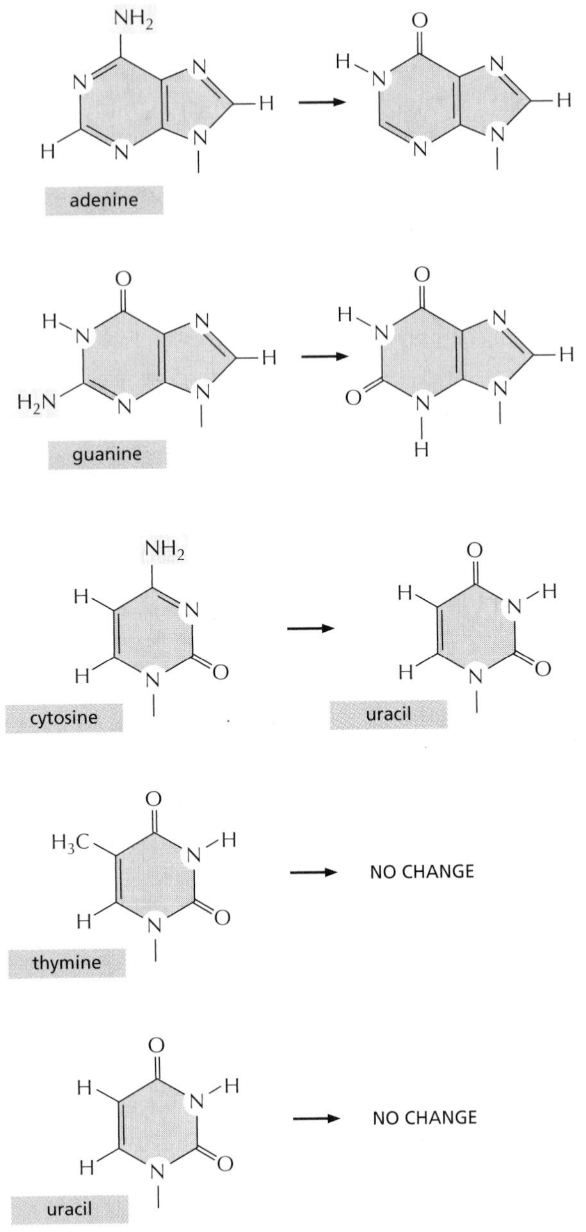

B. As shown in Figure A6–16, telomeres and telomerase are still needed even if the last fragment of the lagging strand were initiated by primase at the very 3′ end of chromosomal DNA, inasmuch as the RNA primer must be removed.

adenine → ; guanine → ; cytosine → uracil; thymine → NO CHANGE; uracil → NO CHANGE

Figure A6–15

ANSWER 6–17 Viruses cannot exist as free-living organisms: they have no metabolism, do not communicate with other viruses, and cannot reproduce themselves. They thus have none of the attributes that

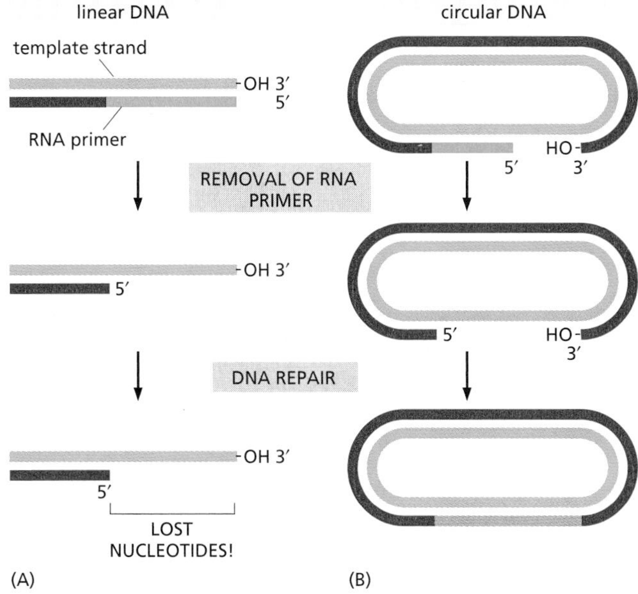

Figure A6–16

one normally associates with life. Indeed, they can even be crystallized. Only inside cells can they redirect normal cellular biosynthetic activities to the task of making more copies of themselves. Thus, the only aspect of "living" that viruses display is their capacity to direct their own reproduction once inside a cell.

ANSWER 6–18 Each time another copy of a transposon is inserted into a chromosome, the change can be either neutral, beneficial, or detrimental for the organism. Because individuals that accumulate detrimental insertions would be selected against, the proliferation of transposons is controlled by natural selection. If a transposon arose that proliferated uncontrollably, it is unlikely that a viable host organism could be maintained. For this reason, most transposons have evolved to transpose only rarely. Many transposons, for example, synthesize only infrequent bursts of very small amounts of the transposase that is required for their movement.

ANSWER 6–19

A. If the single origin of replication were located exactly in the center of the chromosome, it would take more than 8 days to replicate the DNA [= 75 × 10^6 nucleotides/(100 nucleotides/sec)]. The rate of replication would therefore severely limit the rate of cell division. If the origin were located at one end, the time required to replicate the chromosome would be approximately double this.
B. A chromosome end that is not "capped" with a telomere would lose nucleotides during each round of DNA replication and would gradually shrink. Eventually, essential genes would be lost, leading to cell death.
C. Without centromeres, which attach them to the mitotic spindle, the two new chromosomes that result from replication cannot be partitioned accurately between the two daughter cells. Therefore many daughter cells would die, because they would not receive a full set of chromosomes.

Chapter 7

ANSWER 7–1 Perhaps the best answer was given by Francis Crick himself, who coined the term in the mid-1950s: "I called this idea the central dogma for two reasons, I suspect. I had already used the obvious word hypothesis in the sequence hypothesis, which proposes that genetic information is encoded in the sequence of the DNA bases, and in addition I wanted to suggest that this new assumption was more central and more powerful.... As it turned out, the use of the word dogma caused more trouble than it was worth. Many years later Jacques Monod pointed out to me that I did not appear to understand the correct use of the word dogma, which is a belief that cannot be doubted. I did appreciate this in a vague sort of way but since I thought that all religious beliefs were without serious foundation, I used the word in the way I myself thought about it, not as the world does, and simply applied it to a grand hypothesis that, however plausible, had little direct experimental support at the time." (Francis Crick, *What Mad Pursuit: A Personal View of Scientific Discovery.* Basic Books, 1988.)

ANSWER 7–2 Actually, the RNA polymerases are not moving at all, because they have been fixed and coated with metal to prepare the sample for viewing in the electron microscope. However, before they were fixed, they were moving from left to right, as indicated by the gradual lengthening of the RNA transcripts.

The RNA transcripts are shorter because they begin to fold up (i.e., to acquire a three-dimensional structure) as they are synthesized (see, for example, Figure 7–5), whereas the DNA is an extended double helix.

ANSWER 7–3 At first glance, the catalytic activities of an RNA polymerase used for transcription could replace the primase adequately. Upon further reflection, however, there are some serious problems. (1) The RNA polymerase used to make primers would need to initiate every few hundred bases, which is much more often than promoters are spaced on the DNA. Initiation would therefore need to occur in a promoter-independent fashion or many more promoters would have to be present in the DNA, both of which would be problematic for the control of transcription. (2) Similarly, the RNA primers used in replication are much shorter than mRNAs. The RNA polymerase would therefore need to terminate much more frequently than during transcription. Termination would need to occur spontaneously, i.e., without requiring a terminator sequence in the DNA, or many more terminators would need to be present. Again, both of these scenarios would be problematic for the control of transcription.

Although it might be possible to overcome this problem if special control proteins became attached to RNA polymerase during replication, the problem has been solved during evolution by using separate enzymes with specialized properties. Some small DNA viruses, however, do utilize the host RNA polymerase to make primers for their replication.

ANSWER 7–4 This experiment demonstrates that the ribosome does not check the amino acid that is attached to a tRNA. Once an amino acid has been coupled to a tRNA, the ribosome will "blindly" incorporate that amino acid into the position according to the match between the codon and anticodon. We can therefore conclude that a significant part of the correct reading of the genetic code, i.e., the matching of a codon with the correct amino acid, is performed by the synthetase enzymes that correctly match tRNAs and amino acids.

ANSWER 7–5 The mRNA will have a 5′-to-3′ polarity opposite to that of the DNA strand that serves as a template. Thus, the mRNA sequence will read 5′-GAAAAAAGCCGUUAA-3′. The N-terminal amino acid coded for by GAA is glutamic acid. UAA specifies a stop codon, so the C-terminal amino acid is coded for by CGU and is an arginine. Note that the convention in describing the sequence of a gene is to give the sequence of the DNA strand that is *not* used as a template for RNA synthesis; this sequence is the same as that of the RNA transcript, with T written in place of U.

ANSWER 7–6 The first statement is factually correct: RNA is thought to have been the first self-replicating catalyst and in modern cells is no longer self-replicating. We can debate, however, whether this represents a 'loss.' RNA now serves many roles in the cell: as messengers, as adaptors for protein synthesis, as primers for DNA replication, and as catalysts for some of the most fundamental reactions.

ANSWER 7–7

A. False. Ribosomes can make any protein that is specified by the particular mRNA that they are translating. After translation, ribosomes are released from the mRNA and can then start translating a different mRNA.
B. False. mRNAs are translated as linear polymers; there is no requirement that they have any particular folded structure. In fact, such structures that are formed by mRNA can inhibit translation because the ribosome has to unfold the mRNA in order to read the message it contains.
C. False. Ribosomal subunits exchange partners after each round of translation. After a ribosome is released from an mRNA, its two subunits dissociate and enter a pool of free small and large subunits from which new ribosomes assemble around a new mRNA.
D. False. Ribosomes are cytoplasmic organelles, but they are not individually enclosed in a membrane.
E. False. The position of the promoter determines the direction in which transcription proceeds and which DNA strand is used as the template. Transcription in the opposite direction would produce an mRNA with a completely different (and probably meaningless) sequence.
F. False. RNA contains uracil but not thymine.
G. False. The level of a protein depends on its rate of synthesis and degradation but not on its catalytic activity.

ANSWER 7–8 Because the deletion in the Lacheinmal mRNA is internal, it is likely that the deletion arises from a splicing defect. The simplest interpretation is that the *Lacheinmal* gene contains a 173-nucleotide-long exon (labeled "E2" in Figure A7–8), and that this exon is lost during the processing of the mutant precursor mRNA. This could occur, for example, if the mutation changed the 3′ splice site in the preceding intron ("I1") so that

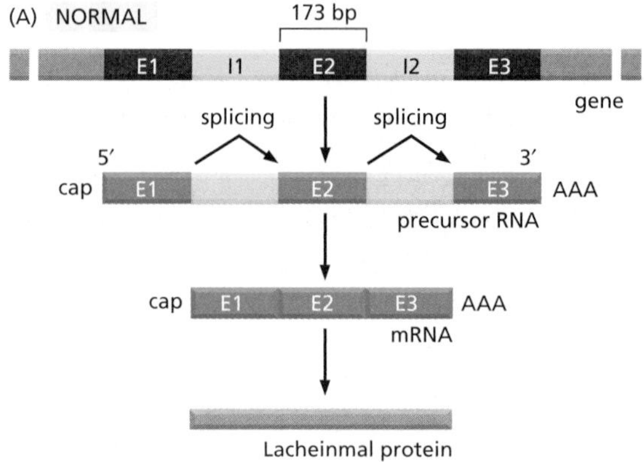

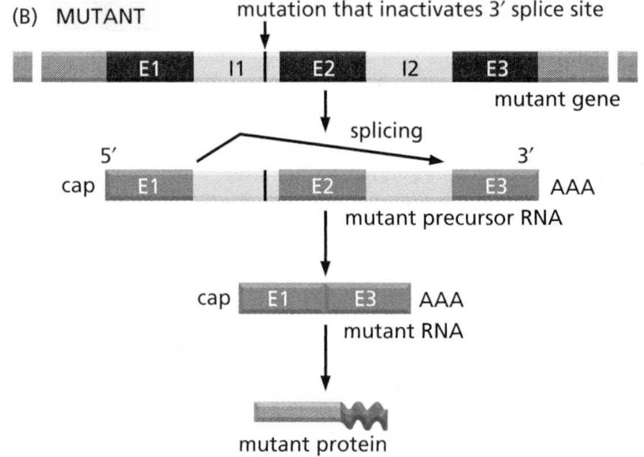

Figure A7–8

it was no longer recognized by the splicing machinery (a change in the CAG sequence shown in Figure 7–19 could do this). The snRNP would search for the next available 3′ splice site, which is found at the 3′ end of the next intron ("I2"), and the splicing reaction would therefore remove E2 together with I1 and I2, resulting in a shortened mRNA. The mRNA is then translated into a defective protein, resulting in the Lacheinmal deficiency.

Because 173 nucleotides do not amount to an integral number of codons, the lack of this exon in the mRNA will shift the reading frame at the splice junction. Therefore, the Lacheinmal protein would be made correctly only through exon E1. As the ribosome begins translating sequences in exon E3, it will be in a different reading frame and therefore will produce a protein sequence that is unrelated to the Lacheinmal sequence normally encoded by exon E3. Most likely, the ribosome will soon encounter a stop codon, which in RNA sequences that do not code for protein would be expected to occur on average about once in every 21 codons (there are 3 stop codons in the 64 codons of the genetic code).

ANSWER 7–9 Sequence 1 and sequence 4 both code for the peptide Arg-Gly-Asp. Because the genetic code is redundant, different nucleotide sequences can encode the same amino acid sequence.

ANSWER 7–10

A. Incorrect. The bonds are not covalent, and their formation does not require input of energy.

B. Correct. The aminoacyl-tRNA enters the ribosome at the A-site.
C. Correct. As the ribosome moves along the mRNA, the tRNAs that have donated their amino acid to the growing polypeptide chain are ejected from the ribosome and the mRNA. The ejection takes place two cycles after the tRNA first enters the ribosome (see Figure 7–33).

ANSWER 7–11 *Replication.* Dictionary definition: the creation of an exact copy; molecular biology definition: the act of duplicating DNA. *Transcription.* Dictionary definition: the act of writing out a copy, especially from one physical form to another; molecular biology definition: the act of copying the information stored in DNA into RNA. *Translation.* Dictionary definition: the act of putting words into a different language; molecular biology definition: the act of polymerizing amino acids into a defined linear sequence using the information provided by the linear sequence of nucleotides in mRNA. (Note that "translation" is also used in a quite different sense, both in ordinary language and in scientific contexts, to mean a movement from one place to another.)

ANSWER 7–12 A code of two nucleotides could specify 16 different amino acids (= 4^2), and a triplet code in which the position of the nucleotides is not important could specify 20 different amino acids (= 4 possibilities of 3 of the same bases + 12 possibilities of 2 bases the same and one different + 4 possibilities of 3 different bases). In both cases, these maximal amino acid numbers would need to be reduced by at least 1, because of the need to specify translation stop codons. It is relatively easy to envision how a doublet code could be translated by a mechanism similar to that used in our world by providing tRNAs with only two relevant bases in the anticodon loop. It is more difficult to envision how the nucleotide composition of a stretch of three nucleotides could be translated without regard to their order, because base-pairing can then no longer be used: AUG, for example, will not base-pair with the same anticodon as UGA.

ANSWER 7–13 It is likely that in early cells the matching between codons and amino acids was less accurate than it is in present-day cells. The feature of the genetic code described in the question may have allowed early cells to tolerate this inaccuracy by allowing a blurred relationship between sets of roughly similar codons and roughly similar amino acids. One can easily imagine how the matching between codons and amino acids could have become more accurate, step by step, as the translation machinery evolved into that found in modern cells.

ANSWER 7–14 The codon for Trp is 5′-UGG-3′. Thus, a normal Trp-tRNA contains the sequence 5′-CCA-3′ as its anticodon. If this tRNA contains a mutation so that its anticodon is changed to <u>U</u>CA, it will recognize a UG<u>A</u> codon and lead to the incorporation of a tryptophan residue instead of causing translation to stop. Many other protein-encoding sequences, however, contain UGA codons as their natural stop sites, and these stops would also be affected by the mutant tRNA. Depending on the competition between the altered tRNA and the normal translation release factors (Figure 7–37), some of these proteins would be made with additional amino acids at their C-terminal end. The additional lengths would depend on the number of codons before the ribosomes encounter a non-UGA stop codon in the mRNA in the reading frame in which the protein is translated.

ANSWER 7–15 One effective way of driving a reaction to completion is to remove one of the products, so that the reverse reaction cannot occur. ATP contains *two* high-energy bonds that link the three phosphate groups. In the reaction shown, PP_i is released, consisting of two phosphate groups linked by one of these high-energy bonds. Thus, PP_i can be hydrolyzed with a considerable gain of free energy, and thereby be efficiently removed. This happens rapidly in cells, and reactions that produce and further hydrolyze PP_i are therefore virtually irreversible (discussed in Chapter 3).

ANSWER 7–16
A. A titin molecule is made of 25,000 amino acids. It therefore takes about 3.5 hours to synthesize a single molecule of titin in muscle cells.
B. Because of its large size, the probability of making a titin molecule without any mistakes is only 0.08 [= $(1 - 10^{-4})^{25,000}$]; i.e., only 8 in 100 titin molecules synthesized are free of mistakes. In contrast, over 97% of newly synthesized proteins of average size are made correctly.
C. The error rate limits the sizes of proteins that can be synthesized accurately. Similarly, if a eucaryotic ribosomal protein were synthesized as a single molecule, a large portion (87%) of this hypothetical giant ribosomal protein would be expected to contain at least one mistake. It is more advantageous to make ribosomal proteins individually, because in this way only a small proportion of each type of protein will be defective, and these few bad molecules can be individually eliminated by proteolysis to ensure that there are no defects in the ribosome as a whole.
D. To calculate the time it takes to transcribe a titin mRNA, you would need to know the size of its gene, which is likely to contain many introns. Transcription of the exons alone requires about 42 minutes. Because introns can be quite large, the time required to transcribe the entire gene is likely to be considerably longer.

ANSWER 7–17 Mutations of the type described in (B) and (D) are often the most harmful. In both cases, the reading frame would be changed, and because this frameshift occurs near the beginning or in the middle of the coding sequence, much of the protein will contain a nonsensical and/or truncated sequence of amino acids. In contrast, a reading-frame shift that occurs toward the end of the coding sequence, as described in scenario (A), will result in a largely correct protein that may be functional. Deletion of three consecutive nucleotides, scenario (C), leads to the deletion of an amino acid but does not alter the reading frame. The deleted amino acid may or may not be important for the folding or activity of the protein; in many cases such mutations are silent, i.e., have no or only minor consequences for the organism. Substitution of one nucleotide for another, as in (E), is often completely harmless. In some cases, it will not change the amino acid sequence of the protein; in other cases it will change a single amino acid; at worst, it may create a new stop codon, giving rise to a truncated protein.

Chapter 8

ANSWER 8–1

A. Transcription of the tryptophan operon would no longer be regulated by the absence or presence of tryptophan; the enzymes would be permanently on in scenarios (1) and (2) and permanently off in scenario (3).

B. In scenarios (1) and (2), the normal tryptophan repressor molecules would completely restore the regulation of the tryptophan biosynthesis enzymes. In contrast, expression of the normal protein would have no effect in scenario 3, because the tryptophan operator would remain permanently occupied by the mutant protein.

ANSWER 8–2 Contacts can form between the protein and the edges of the base pairs that are exposed in the major groove of the DNA (Figure A8–2). The bonds responsible for sequence-specific contacts are hydrogen bonds and hydrophobic interactions with the methyl group on thymine. Note that the arrangement of hydrogen-bond donors and hydrogen-bond acceptors of a T-A pair is different from that of a C-G pair. Similarly, the arrangement of hydrogen-bond donors and hydrogen-bond acceptors of A-T and G-C pairs would be different from one another and from the two pairs shown in the figure. In addition to the contacts shown in the figure, electrostatic attractions between the positively charged amino acid side chains of the protein and the negatively charged phosphate groups in the DNA backbone usually stabilize DNA–protein interactions.

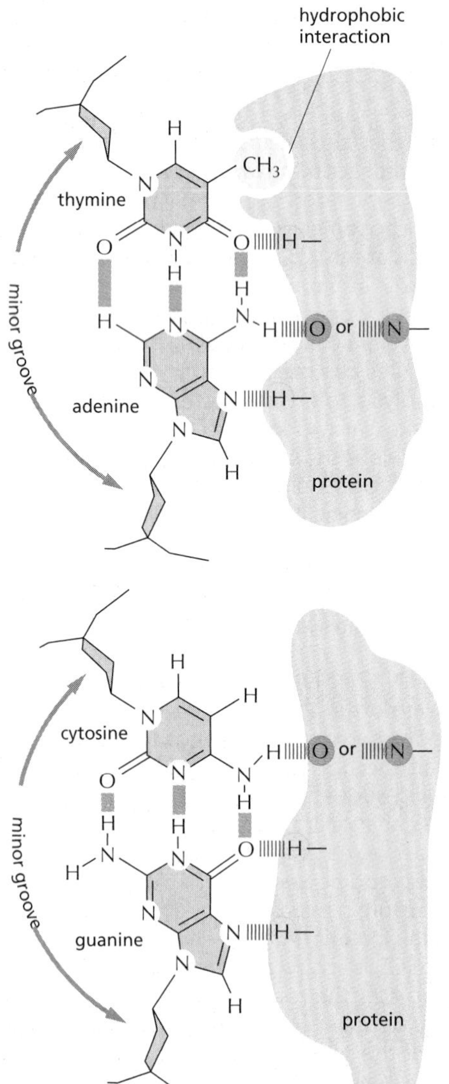

Figure A8–2

ANSWER 8–3 Bending proteins can help to bring distant DNA regions together that normally would contact each other only inefficiently (Figure A8–3). Such proteins are found in both procaryotes and eucaryotes and are involved in many examples of transcriptional regulation.

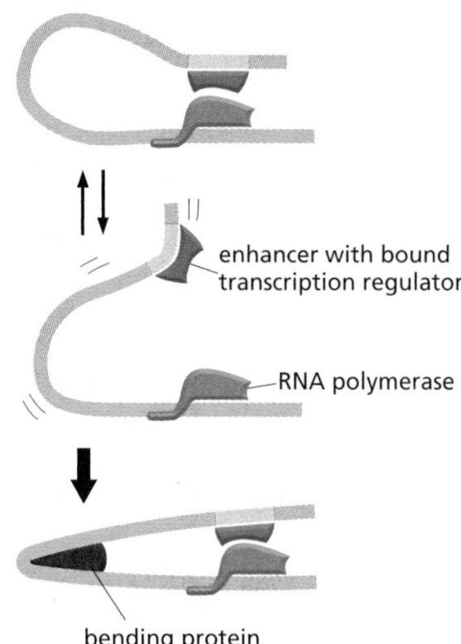

Figure A8–3

ANSWER 8–4

A. UV light throws the switch from the prophage to the lytic state: when cI protein is destroyed, Cro is made and turns off the further production of cI. The virus produces coat proteins, and new virus particles are made.

B. When the UV light is switched off, the virus remains in the lytic state. Thus, cI and Cro form a gene regulatory switch that 'memorizes' its previous setting.

C. This switch makes sense in the viral life cycle: UV light tends to damage the bacterial DNA (see Figure 6–24), thereby rendering the bacterium an unreliable host for the virus. A prophage will therefore switch to the lytic state and leave the 'sinking ship' in search for new host cells to infect.

ANSWER 8–5 False. Carrots can be grown from single carrot cells and tadpoles can be produced by injecting differentiated frog nuclei into frog eggs. But carrots cannot be produced from frog eggs, no matter what.

ANSWER 8–6

A. True. Procaryotic mRNAs are often transcripts of entire operons. Ribosomes can initiate translation at internal AUG start sites of these "polycistronic" mRNAs (see Figures 7–36 and 8–6).

B. True. The major groove of double-stranded DNA is sufficiently wide to allow a protein surface, such as one face of an α helix, access to the base pairs.

C. True. It is advantageous to exert control at the earliest possible point in a pathway. This conserves metabolic energy because unnecessary products are not made in the first place.

D. False. The zinc atoms in zinc finger domains are required for the correct folding of the protein domain; they are internal to these domains and do not contact the DNA.

ANSWER 8–7 From our knowledge of enhancers, one would expect their function to be relatively independent of their distance from the promoter—possibly weakening as this distance increases. The surprising feature of the data (which have been adapted from an actual experiment) is the periodicity: the enhancer is maximally active at certain distances from the promoter (50, 60, or 70 nucleotides), but almost inactive at intermediate distances (55 or 65 nucleotides). The periodicity of 10 suggests that the mystery can be explained by considering the structure of double-helical DNA, which has 10 base pairs per turn. Thus, placing an enhancer on the side of the DNA opposite to that of the promoter (Figure A8–7) would make it more difficult for the activator that binds to it to interact with the proteins bound at the promoter. At longer distances, there is more DNA to absorb the twist, and the effect is diminished.

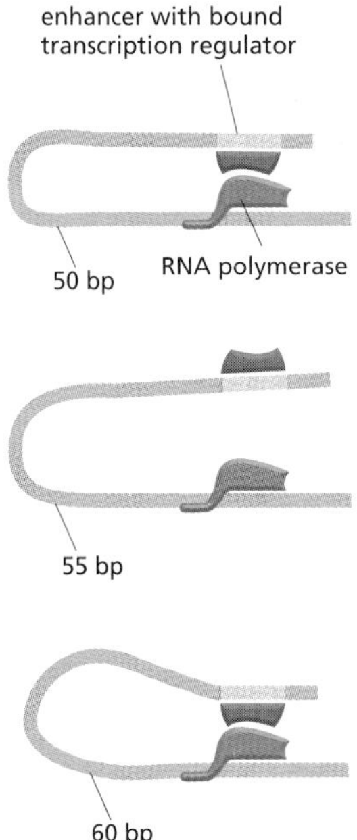

Figure A8–7

ANSWER 8–8 Two advantages of dimeric DNA-binding proteins are (1) that dimer formation greatly increases the specificity and the strength of a protein–DNA interaction by doubling the number of protein–DNA contacts, and (2) that different subunits can be combined in different combinations to increase the number of DNA-binding specificities available to the cell. Three of the most common protein domains that bind DNA are leucine zippers, homeodomains, and zinc fingers. Each provides a particularly stable fold in the polypeptide chain that positions an α helix so it can be inserted into the major groove of the DNA helix and contact the sides of the base pairs (see Figure 8–5).

ANSWER 8–9 The affinity of the dimeric λ repressor for its binding site is the sum of the interactions made by each of the two DNA-binding domains. A single DNA-binding domain can make only half the contacts and provide just half the binding energy compared with the dimer. Thus, although the concentration of binding domains is unchanged, they are no longer coupled, and their individual affinities for DNA are sufficiently weak that they cannot remain bound. As a result, the genes for lytic growth are turned on.

ANSWER 8–10 The function of these *Arg* genes is to synthesize arginine. When arginine is abundant, expression of the biosynthetic genes should be turned off. If ArgR acts as a gene repressor (which it does in reality), then binding of arginine should increase its affinity for its regulatory sites, allowing it to bind and shut off gene expression. If ArgR acted as a gene activator instead, then the binding of arginine would be predicted to reduce its affinity for its regulatory DNA, preventing its binding and thereby shutting off expression of the *Arg* genes.

ANSWER 8–11 The results of this experiment favor DNA looping, which should not be affected by the protein bridge (so long as it allowed the DNA to bend, which it does). The scanning or entry site model, however, is predicted to be affected by the nature of the linkage between the enhancer and the promoter. If the proteins enter at the enhancer and scan to the promoter, they would have to traverse the protein linkage. If such proteins are geared to scan on DNA, they would likely have difficulty scanning across such a barrier.

ANSWER 8–12 The most definitive result is one showing that a single differentiated cell taken from a specialized tissue can re-create a whole organism. This proves that the cell must contain all the information required to produce a whole organism, including all of its specialized cell types. Experiments of this type are summarized in Figure 8–2.

ANSWER 8–13 You could create 16 different cell types with 4 different transcription regulators (all the 8 cell types shown in Figure 8–19, plus another 8 created by adding an additional transcription regulator). MyoD by itself is sufficient to induce muscle-specific gene expression only in certain cell types, such as some kinds of fibroblasts. The action of MyoD is therefore consistent with the model shown in Figure 8–19: if muscle cells were specified, for example, by the combination of transcription regulators 1, 3, and MyoD, then the addition of MyoD would convert only two of the cell types of Figure 8–19 (cells F and H) to muscle.

ANSWER 8–14 The induction of a gene activator protein that stimulates its own synthesis can create a positive feedback loop that can produce cell memory. The continued self-stimulated synthesis of activator A can in principle last for many cell generations, serving as a constant reminder of an event that took place in the past. By contrast, the induction of a gene repressor that inhibits its own synthesis creates a negative feedback loop that guarantees only a transient response to the transient stimulus. Because repressor R shuts off its own synthesis, the cell will quickly return to the state that existed before the signal.

ANSWER 8–15 Many transcription regulators are continually made in the cell; that is, their expression is constitutive and the activity of the protein is controlled by signals from inside or outside the cell (e.g., the availability of nutrients, as for the tryptophan repressor, or by hormones, as for the glucocorticoid receptor), thereby adjusting the transcriptional program to the physiological needs of the cell. Moreover, a given transcription regulator usually controls the expression of many different genes. Transcription regulators are often used in various combinations and can affect each other's activity, thereby further increasing the possibilities for regulation with a limited set of proteins. Nevertheless, the cell devotes a large fraction of its genome to the control of transcription: an estimated 10% of all genes in eucaryotic cells code for transcription regulators.

Chapter 9

ANSWER 9–1 The answer lies in the need for the cell to maintain a balance between stability and change. If the mutation rate were too high, a species would eventually die out because all its individuals would accumulate too many mutations in genes essential for survival. For a species to be successful—in evolutionary terms—it is important for individual members to have good genetic memory; that is, fidelity in DNA replication. At the same time, occasional changes are needed if the species is to adapt to changing conditions. If the change leads to an improvement, it will persist by selection; if it proves disastrous, the individual organism that was the unfortunate subject of nature's experiment will die, but the species will survive.

ANSWER 9–2 In single-celled organisms the genome is the germ line and any modification is passed on to the next generation. By contrast, in multicellular organisms most of the cells are somatic cells and make no contribution to the next generation; thus, modification of those cells by horizontal gene transfer would have no consequence for the next generation. The germ-line cells are usually sequestered in the interior of multicellular organisms, minimizing their contact with foreign cells, viruses, and DNA, thus insulating the species from the effects of horizontal gene transfer. Nevertheless, horizontal gene transfer is possible for multicellular organisms. For example, the genomes of some insect species contain DNA that was horizontally transferred from bacteria that infect them.

ANSWER 9–3 It is unlikely that any gene came into existence perfectly optimized for its function. Indeed, the environment in which an organism finds itself is changeable, so no gene could be perfect indefinitely. Because ribosomal RNAs (and the products of other highly conserved genes) participate in fundamental processes, there is less leeway for change. Nonetheless, there are significant differences in ribosomal RNAs among species.

ANSWER 9–4 Mobile genetic elements could provide opportunities for homologous recombination events, thereby causing genomic rearrangements. They could insert into genes, possibly obliterating splicing signals and thereby changing the protein produced by the gene. They could also insert into the regulatory region of a gene, where insertion between an enhancer and a transcription start site could block the function of the enhancer and therefore reduce the level of expression of a gene. In addition, the mobile genetic element could itself contain an enhancer and thereby change the time and place in the organism where the gene is expressed.

ANSWER 9–5 It is not a simple matter to determine the function of a gene from scratch, nor is there a universal recipe for how to do it. Nevertheless, there are a variety of standard questions whose answers help to narrow down the possibilities. Below we list some of these questions.

In what tissues is the gene expressed? If the gene is expressed in all tissues, it is likely to have a general function. If it is expressed in one or a few tissues, its function is likely to be more specialized, perhaps related to the specialized functions of the tissues. If the gene is expressed in the embryo but not the adult, it probably functions in development.

In what compartment of the cell is the protein found? Knowing the subcellular localization of the protein—nucleus, plasma membrane, mitochondria, etc.—can also help to suggest categories of potential function. For example, a protein that is localized to the plasma membrane is likely to be a transporter, a receptor or other component of a signaling pathway, a cell-adhesion molecule, etc.

What are the effects of mutations in the gene? Mutations that eliminate or modify the function of the gene product can also provide clues to function. For example, if the gene product is critical at a certain time during development, the embryo will often die at that stage or develop obvious abnormalities. Unless the abnormality is highly specific, it is usually difficult to deduce its function. And often the links are indirect, becoming apparent only after the gene's function is known.

With what other proteins does the encoded protein interact? In carrying out their function, proteins often interact with other proteins involved in the same or closely related processes. If an interacting protein can be identified, and if its function is already known (through previous research or through the searching of databases), the range of possible functions can be narrowed dramatically.

Can mutations in other genes alter effects of mutation in the unknown gene? Searching for such mutations can be a very powerful approach to investigating gene function, especially in organisms such as bacteria and yeast, which have simple genetic systems. Although much more difficult to perform in the mouse, this type of approach can nonetheless be used. The rationale for this strategy is analogous to that of looking for interacting proteins: genes that interact genetically are often involved in the same process or in closely related processes. Identification of such an interacting gene (and knowledge of its function) would provide an important clue to the function of the unknown gene.

Addressing each of these questions requires specialized experimental expertise and a substantial time commitment from the investigator. It is no wonder that progress is made much more rapidly when a clue to a gene's function can be found simply by identifying a similar gene of known function in the database. As more and more genes are studied, this strategy will become increasingly successful.

ANSWER 9–6 With their ability to facilitate genetic recombination, mobile genetic elements have almost certainly played an important part in the evolution of modern-day organisms. They can facilitate gene duplication and the creation of new genes via exon shuffling, and they can change the way in which existing genes are expressed. Although the transposition of a mobile genetic element can be harmful for an individual organism—if, for example, it disrupts the activity of a critical gene—these agents of genetic change may well be beneficial to the species as a whole.

ANSWER 9–7 About 7.6% of each gene is converted to mRNA [(5.4 exons/gene × 266 nucleotide pairs/exon)/(19,000 nucleotide pairs/gene) = 7.6%]. Protein-coding genes occupy about 28% of Chromosome 22 [(700 genes × 19,000 nucleotide pairs/gene)/(48 × 10^6 nucleotide pairs) = 27.7%]. However, over 90% of this DNA is made of introns.

ANSWER 9–8 This statement is probably true. For example, nearly half our DNA is composed of defunct mobile genetic elements. However, it is possible that future research will uncover a function for this seemingly unimportant DNA.

ANSWER 9–9 The HoxD cluster is packed with complex and extensive regulatory sequences that direct each of its genes to be expressed at the correct time and place during development. Insertion of mobile genetic elements into the HoxD cluster is thought to be selected against because it would disrupt proper regulation of its resident genes.

ANSWER 9–10

A. The exons in the human β-globin gene correspond to the positions of sequence similarity (in this case identity) with the cDNA, which is a direct copy of the mRNA and thus contains no introns. The introns correspond to the regions between the exons. The positions of the introns and exons in the human β-globin gene are indicated in Figure A9–10A. Also shown (in open bars) are sequences present in the mature β-globin mRNA (and in the gene) that are not translated into protein.

B. From the positions of the exons, as defined in Figure A9–10A, it is clear that the first two exons of the human β-globin gene have homologous counterparts in the mouse β-globin gene (Figure A9–10B). However, only the first half of the third exon of the human β-globin gene is similar to the mouse β-globin gene. The similar portion of the third exon contains sequences that encode protein, whereas the portion that is different represents the 3′ untranslated region of the gene. Because this portion of the gene does not encode protein (nor does it contain extensive regulatory sequences), its sequence is not constrained.

C. The human and mouse β-globin genes are also homologous at their 5′ ends, as indicated by the cluster of points along the same diagonal as the first exon (Figure A9–10B). These sequences correspond to the regulatory regions upstream of the start sites for transcription. Functional sequences, which are under selective pressure, diverge much more slowly than sequences without function.

D. The diagon plot shows that the first intron is nearly the same length in the human and mouse genes, but the length of the second intron is noticeably different (Figure A9–10B). If the introns were the same length, the line segments that represent sequence similarity would fall on the same diagonal. The easiest way to test for the colinearity of the line segments is to tilt the page and sight along the diagonal. It is impossible to tell from this comparison if the change in length is due to a shortening of the mouse intron or to a lengthening of the human intron, or some combination of those possibilities.

ANSWER 9–11 Computer algorithms that search for exons are complex affairs, as you might imagine. To identify unknown genes, these programs combine statistical information derived from known genes, such as:

1. An exon that encodes protein will have an open reading frame. If the amino acid sequence specified by this open reading frame matches a protein sequence in any database, there is a high likelihood that it is an authentic exon.

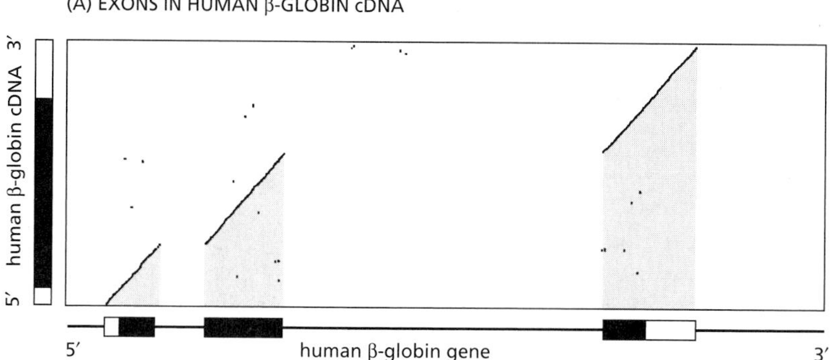

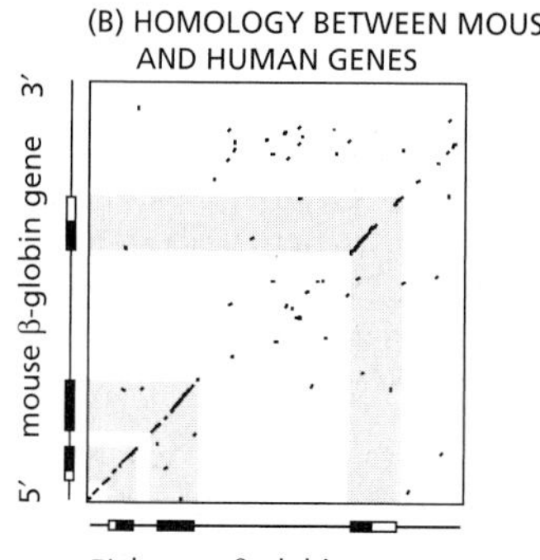

Figure A9–10

2. The reading frames of adjacent exons in the same gene will match up when the intron sequences are omitted.
3. Internal exons (excluding the first and the last) will have splicing signals at each end; most of the time (98.1%) these will be AG at the 5′ ends of the exons and GT at the 3′ ends.
4. The multiple codons for most individual amino acids are not used with equal frequency. This so-called coding bias can be factored in to aid in the recognition of true exons.
5. Exons and introns have characteristic length distributions. The median length of exons in human genes is about 120 nucleotide pairs. Introns tend to be much larger: a median length of about 2 kb in genomic regions of 30–40% GC content, and a median length of about 500 nucleotide pairs in regions above 50% GC.
6. The initiation codon for protein synthesis (nearly always an ATG) has a statistical association with adjacent nucleotides that seem to enhance its recognition by translation factors.
7. The terminal exon will have a signal (most commonly AATAAA) for cleavage and polyadenylation close to its 3′ end.

The statistical nature of these features, coupled with the low frequency of coding information in the genome (2–3%) and the frequency of alternative splicing (an estimated 60% of human genes), makes it especially impressive that current algorithms can identify more than 70% of individual exons in the human genome.

ANSWER 9–12 In a very long, random sequence of DNA, each of the 64 different codons will occur with equal frequency. Because 3 of the 64 are stop codons, they will be expected to occur on average every 21 codons (64/3 = 21.3).

ANSWER 9–13 On the surface, its resistance to mutation suggests that the genetic code was shaped by forces of natural selection. An underlying assumption, which seems reasonable, is that resistance to mutation is a valuable feature of a genetic code. This reasoning suggests that it would have been a lucky accident indeed—roughly a one-in-a-million chance—to stumble on a code as error-proof as our own.

But all is not so simple. If resistance to mutation is an essential feature of any code that can support complex organisms such as ourselves, then the only codes we *could* observe are ones that are error resistant. A less favorable frozen accident, giving rise to a more error-prone code, might have limited the complexity of life to organisms too simple to contemplate their own genetic code. This is akin to the anthropic principle of cosmology: many universes may be possible, but few are compatible with life that can ponder the nature of the universe.

Beyond these considerations, there is ample evidence that the code is not static, and thus could respond to the forces of natural selection. Altered versions of the standard genetic code have been identified in the mitochondrial and nuclear genomes of several organisms. In each case one or a few codons have taken on a new meaning.

ANSWER 9–14

B. Formation of protein-coding genes *de novo* from the vast amount of unused, noncoding DNA typical of eucaryotic genomes is not thought to be a significant contributor to gene evolution.

ANSWER 9–15

A. Because synonymous changes do not alter the amino acid sequence of the protein, they usually do not affect the overall fitness of the organism and are therefore not selected against. By contrast, nonsynonymous changes, which substitute a new amino acid in place of the original one, can alter the function of the encoded protein and change the fitness of the organism. Since most amino acid substitutions probably harm the protein, they tend to be selected against.
B. The histone H3 protein must be so exquisitely tuned to its function that virtually all amino acid substitutions are deleterious and are therefore selected against. The extreme conservation of histone H3 argues that its function is very tightly constrained, probably because of extensive interactions with other proteins and with DNA.
C. Histone H3 is clearly not in a 'privileged' site in the genome because it undergoes synonymous nucleotide changes at about the same rate as other genes.

ANSWER 9–16

A. The data embodied in the phylogenetic tree (Figure Q9–16) refutes the hypothesis that plant hemoglobin genes were acquired by horizontal transfer. Looking at the more familiar parts of the tree, we see that the hemoglobins of vertebrates (fish to human) have approximately the same phylogenetic relationships as do the species themselves. Plant hemoglobins also form a distinct group that displays accepted evolutionary relationships, with barley, a monocot, diverging before bean, alfalfa, and lotus, which are all dicots (and legumes). The basic hemoglobin gene, therefore, was in place long ago in evolution. The phylogenetic tree of Figure Q9–16 indicates that the hemoglobin genes in modern plant and animal species were inherited from a common ancestor.
B. Had the plant hemoglobin genes arisen by horizontal transfer from a parasitic nematode, then the plant sequences would have clustered with the nematode sequences in the phylogenetic tree in Figure Q9–16.

ANSWER 9–17 In each human lineage, new mutations will be introduced at a rate of 10^{-10} alterations per nucleotide per cell generation, and the differences between two human lineages will accumulate at twice this rate. To accumulate 10^{-3} differences per nucleotide will thus take $10^{-3}/(2 \times 10^{-10})$ cell generations, corresponding to $(1/200) \times 10^{-3}/(2 \times 10^{-10}) = 25{,}000$ human generations, or 750,000 years. In reality, we are not descended from one pair of genetically identical ancestral humans; rather, it is likely that we are descended from a relatively small founder population of humans who were already genetically diverse. More sophisticated analysis suggests that this founder population existed about 150,000 years ago.

Chapter 10

ANSWER 10–1 The presence of a mutation in a gene does not necessarily mean that the protein expressed from it is defective. For example, the mutation could change one codon into another that still specifies the same amino acid, and so does not change the amino acid sequence of the protein. Or, the mutation may cause a change from one amino acid to another in the protein, but in a position that is not important for the folding or function of the protein. In assessing the likelihood that such a mutation might cause a defective protein, information on the known β-globin mutations that are found in humans is essential. You would therefore want to know the precise nucleotide change in your mutant gene, and whether this change has any known or predictable consequences for the function of the encoded protein. If your mate has two normal copies of the globin gene, 50% of your children would be carriers of your mutant gene.

ANSWER 10–2

A. Digestion with EcoRI produces two products:

5′-AAGAATTGCGG AATTCGAGCTTAAGGGCCGCGCCGAAGCTTTAAA-3′
3′-TTCTTAACGCCTTAA GCTCGAATTCCCGGCGCGGCTTCGAAATTT-5′

B. Digestion with AluI produces three products:

5′-AAGAATTGCGGAATTCGAG CTTAAGGGCCGCGCCGAAG CTTTAAA-3′
3′-TTCTTAACGCCTTAAGCTC GAATTCCCGGCGCGGCTTC GAAATTT-5′

C. The sequence lacks a NotI cleavage site.

D. Digestion with all three enzymes therefore produces:

5′-AAGAATTGCGG AATTCGAG CTTAAGGGCCGCGCCGAAG CTTTAAA-3′
3′-TTCTTAACGCCTTAA GCTC GAATTCCCGGCGCGGCTTC GAAATTT-5′

ANSWER 10–3 Protein biochemistry is still very important because it provides the link between the amino acid sequence (which can be deduced from DNA sequences) and the functional properties of the protein. We are still not able to infallibly predict the folding of a polypeptide chain from its amino acid sequence, so that in many cases information regarding the function of the protein, such as its catalytic activity, cannot be deduced from the gene sequence alone. Instead, such information must be obtained experimentally by analyzing the properties of proteins biochemically. Furthermore, the structural information that can be deduced from DNA sequences is necessarily incomplete. We cannot, for example, accurately predict covalent modifications of the protein, proteolytic processing, the presence of tightly bound small molecules, or the association of the protein with other subunits. Moreover, we cannot accurately predict the effects these modifications might have on the activity of the protein.

ANSWER 10–4

A. After an additional round of amplification there will be 2 gray, 4 green, 4 red, and 22 yellow-outlined fragments; after a second additional round there will be 2 gray, 5 green, 5 red, and 52 yellow-outlined fragments. Thus the DNA fragments outlined in yellow increase exponentially and will eventually overrun the other reaction products. Their length is determined by the DNA sequence that spans the distance between the two primers plus the length of the primers.

B. The mass of one DNA molecule 500 nucleotide pairs long is 5.5×10^{-19} g [$= 2 \times 500 \times 330$ (g/mole)/6×10^{23} (molecules/mole)]. Ignoring the complexities of the first few steps of the amplification reaction (which produce longer products that eventually make an insignificant contribution to the total DNA amplified), this amount of product approximately doubles for every amplification step. Therefore, 100×10^{-9} g $= 2^N \times 5.5 \times 10^{-19}$ g, where N is the number of amplification steps of the reaction. Solving this equation for $N = \log(1.81 \times 10^{11})/\log(2)$ gives $N = 37.4$. Thus, only about 40 cycles of PCR amplification are sufficient to amplify DNA from a single molecule to a quantity that can be readily handled and analyzed biochemically. This whole procedure is automated and takes only a few hours in the laboratory.

ANSWER 10–5 If the repair enzymes act on the plasmid before it is replicated, the plasmid will indeed be repaired. However, the repair enzymes cannot distinguish which strand of the DNA contains the mutation and which contains the normal nucleotide. Therefore, in half of the cells that have been transformed with the mismatched plasmid, the normal gene would be restored, whereas in the other half of the cells the normal strand would be converted to match the mutated strand. Cells containing the plasmid with the desired mutation can be identified by hybridization with a single-stranded DNA probe that distinguishes between the normal and mutant genes.

ANSWER 10–6 If the ratio of dideoxyribonucleoside triphosphates to deoxyribonucleoside triphosphates is increased, DNA polymerization will be terminated more frequently and thus shorter DNA strands will be produced. Such conditions are favorable for determining nucleotide sequences that are close to the DNA primer used in the reaction. In contrast, decreasing the ratio of dideoxyribonucleoside triphosphates to deoxyribonucleoside triphosphates will produce longer DNA fragments, thus allowing one to determine nucleotide sequences more distant from the primer.

ANSWER 10–7 Although several explanations are possible, the simplest is that the DNA probe has hybridized predominantly with its corresponding mRNA, which is typically present in many more copies per cell than is the gene. The different extents of hybridization probably reflect different levels of gene expression. Perhaps each of the different cell types that make up the tissue expresses the gene at a different level.

ANSWER 10–8 Like the vast majority of mammalian genes, the attractase gene likely contains introns. Bacteria do not have the splicing machinery required to remove introns, and therefore the correct protein would not be expressed from the gene. For expression of most mammalian genes in bacterial cells, a cDNA version of the gene must be used.

ANSWER 10–9

A. False. Restriction sites are found at random throughout the genome, within as well as between, genes.

B. True. DNA bears a negative charge at each phosphate, giving DNA an overall negative charge.

C. False. Clones isolated from cDNA libraries do not contain promoter sequences. These sequences are not transcribed and are therefore not part of the mRNAs that are used as the templates to make cDNAs.

D. True. Each polymerization reaction produces double-stranded DNA that must, at each cycle, be denatured to allow new primers to hybridize so that the DNA strand can be copied again.

E. False. Digestion of genomic DNA with restriction nucleases that recognize four-nucleotide sequences produces fragments that are *on average* 256 nucleotides long. However, the actual lengths of the fragments produced will vary considerably on both sides of the average.

F. True. Reverse transcriptase is first needed to copy the mRNA into single-stranded DNA, and DNA polymerase is then required to make the second DNA strand.

G. True. Using a sufficient number of STRs, individuals can be uniquely 'fingerprinted' (see Figure 10–19).

H. True. If cells of the tissue do not transcribe the gene of interest, it will not be represented in a cDNA library prepared from this tissue. However, it will be represented in a genomic library prepared from the same tissue.

ANSWER 10–10

A. The DNA sequence, from its 5′ end to its 3′ end, is read starting from the bottom of the gel, where the smallest DNA fragments migrate. Each band results from the incorporation of the appropriate dideoxyribonucleoside triphosphate, and as expected there are no two bands that have the same mobility. This allows one to determine the DNA sequence by reading off the bands in strict order, proceeding upward from the bottom of the gel, and assigning the correct nucleotide according to which lane the band is in.

The nucleotide sequence of the top strand (Figure A10–10A) was obtained directly from the data of Figure Q10–10, and the bottom strand was deduced from the complementary base-pairing rules.

B. The DNA sequence can then be translated into an amino acid sequence using the genetic code. However, there are two strands of DNA that could be transcribed into RNA and three possible reading frames for each strand. Thus there are six amino acid sequences that can in principle be encoded by this stretch of DNA. Of the three reading frames possible from the top strand, only one is not interrupted by a stop codon (*yellow* blocks in Figure A10–10B).

From the bottom strand, two of the three reading frames also have stop codons (not shown). The third frame gives the following sequence:

SerAlaLeuGlySerSerGluAsnArgProArgThrProAlaArgThrGlyCysProValIle

It is not possible from the information given to tell which of the two open reading frames corresponds to the actual protein encoded by this stretch of DNA. What additional experiment could distinguish between these two possibilities?

ANSWER 10–11

A. Cleavage of human genomic DNA with HaeIII would generate about 11×10^6 different fragments [$= 3 \times 10^9/4^4$], with EcoRI about 730,000 different fragments [$= 3 \times 10^9/4^6$], and with Not I about 46,000 different fragments [$= 3 \times 10^9/4^8$]. There will also be some additional fragments generated because the maternal and paternal chromosomes are very similar but not identical in DNA sequence.

B. A set of overlapping DNA fragments will be generated. Libraries constructed from sets of overlapping fragments are valuable because they can be used to order cloned sequences in relation to their original order in the genome and thus obtain the DNA sequence of a long stretch of DNA (see Figure 10–27).

ANSWER 10–12 By comparison with the positions of the size markers, we find that EcoRI treatment gives two fragments of 4 kb and 6 kb; NotI treatment gives one fragment of 10 kb; and treatment with EcoRI + NotI gives three fragments of 6 kb, 3 kb, and 1 kb. This gives a total length of 10 kb calculated as the sum of the fragments in each lane. Thus the original DNA molecule must be 10 kb (10,000 nucleotide pairs) long. Because treatment with NotI gives a fragment 10 kb long it could be that the original DNA is a linear molecule with no cutting site for NotI. But we can rule that out by the results of the EcoRI + NotI digestion. We know that EcoRI cleavage alone produces two fragments of 6 kb and 4 kb, and in the double digest this 4-kb fragment is further cleaved by NotI into a 3-kb and a 1-kb fragment. The DNA therefore contains a single NotI cleavage site, and thus it must be circular, as a single fragment of 10 kb is produced when it is cut with NotI alone. Arranging the cutting sites on a circular DNA to give the appropriate sizes of fragments produces the map illustrated in Figure A10–12.

ANSWER 10–13

A. The genetic code is degenerate, and there is more than one possible codon for each amino acid, with the exception of tryptophan and methionine. Therefore, to detect the nucleotide sequence that codes for the amino acid sequence of the protein, many DNA molecules must be made and pooled to ensure that the mixture will contain the one that exactly matches the DNA sequence of the gene. For the three peptide sequences given in this question, the following probes need to be made (alternative bases at the same position are given in parentheses):

Peptide 1:
5′-TGGATGCA(C,T)CA(C,T)AA(A,G)-3′

(A) 5′-TATAAACTGGACAACCAGTTCGAGCTGGTGTTCGTGGTCGGTTTTCAGAAGATCCTAACGCTGACG-3′
3′-ATATTTGACCTGTTGGTCAAGCTCGACCACAAGCACCAGCCAAAAGTCTTCTAGGATTGCGACTGC-5′

(B) 5′ top strand of DNA 3′
TATAAACTGGACAACCAGTTCGAGCTGGTGTTCGTGGTCGGTTTTCAGAAGATCCTAACGCTGACG
1 LeuLysLeuGluAsnGlnPheGlnLeuValPheValValGlyPheGlnLysIleLeuThrLeuThr
2 IleAsnTrpThrThrSerSerSerTrpCysSerTrpSerValPheArgArgSer Arg
3 ThrGlyGlnProValArgAlaGlyValArgGlyArgPheSerGluAspProAsnAlaAsp

Figure A10–10

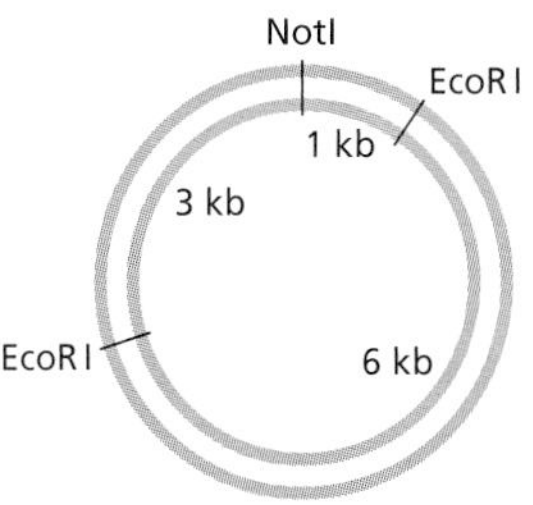

Figure A10–12

Because of the three twofold degeneracies, you would need eight (= 2^3) different DNA sequences in the mixture.

Peptide 2:
5′(T,C)T(G,A,T,C)(A,T)(G,C)(G,A,T,C)(A,C)G(G,A,T,C)(T,C)T(G,A,T,C)(A,C)G(G,A,T,C)-3′

The mixture representing peptide sequence #2 is much more complicated. Leu, Ser, and Arg are each encoded by six different codons; you would therefore need to synthesize a mixture of 7776 (= 6^5) different DNA molecules. This could not be done, however, simply by using more than one different nucleotide in any one position because the different bases in each codon are not independent. (Ser, for example, has A or T as the first base of the codon, G or C as the second base, and G, A, T, or C as the third base; when the first base is A, however, the second base is always G and the third base can be only T or C.)

Peptide 3:
5′-TA(C,T)TT(C,T)GG(G,A,T,C)ATGCA(A,G)3′

Because of three twofold and one fourfold degeneracies, you would need 32 (= $2^3 \times 4$) different sequences in the mixture.

You would probably first use probe #1 to screen your library by hybridization. Because there are only eight possible DNA sequences, the ratio of the one correct sequence to the incorrect ones is highest, giving you the best chance of finding a matching clone. Probe #2 is practically useless, because only 1/7776 of the DNA in the mixture would perfectly hybridize to your gene of interest. You could use probe #3 to verify that the clone you obtained is correct. Any library clones that hybridize to probes #1 and #3 are very likely to contain the gene of interest.

B. Knowing that peptide sequence #3 contains the last amino acid of the protein is valuable information because it tells you that the other two peptide sequences must precede it, that is, they must be located farther toward the N-terminal end of the protein. Knowing this order is important, because DNA primers can be extended by DNA polymerases only from their 3′ ends; thus, the 3′ ends of two primers need to "face" each other during a PCR amplification reaction (see Figure 10–16). A PCR primer based on peptide sequence #3 must therefore be the complementary sequence of probe #3 (so that its 3′ end corresponds to the first nucleotide of the sequence complementary to the Trp codon):
5′-(TC)TGCAT(G,A,T,C)CC(G,A)AA(G,A)TA-3′

As before, this 'primer' would contain 32 different DNA sequences, only one of which will perfectly match the gene. Probe #1 could be your choice for the second primer. Probe #2, again because of its high degeneracy, would be a much less suitable choice.

C. The ends of the final amplification product are derived from the primers, which are each 15 nucleotides long. Therefore, a 270-nucleotide segment of the cDNA of the gene has been amplified. This will encode 90 amino acids; adding the amino acids encoded by the primers gives you a protein-coding sequence of 100 amino acids. This is unlikely to represent the whole gene. To your satisfaction, however, you note that CTATCACGCTTTAGG encodes peptide sequence #2. This information therefore confirms that your PCR product indeed encodes a fragment of the protein you originally isolated.

ANSWER 10–14 The products will comprise a large number of different single-stranded DNA molecules, one for each nucleotide in the sequence. However, each DNA molecule will be one of four colors, depending on which of the four dideoxyribonucleotides terminated the polymerization reaction of that chain. Separation by gel electrophoresis will generate a ladder of bands, each one nucleotide apart, and the sequence can be read from the order of colors (Figure A10–14). The method described here forms the basis for the DNA sequencing strategy used in most automated DNA sequencing machines (see Figure 10–22).

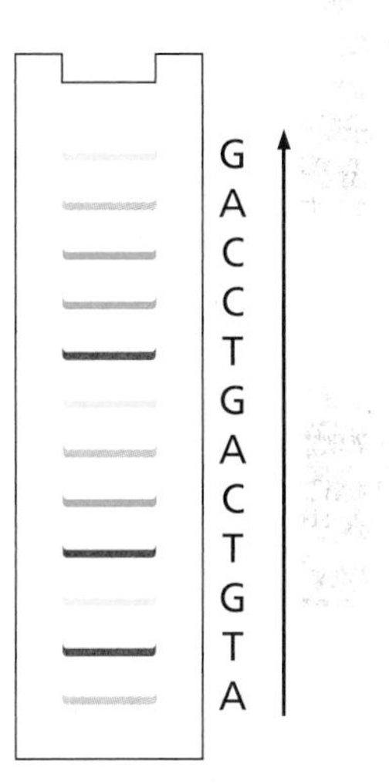

Figure A10–14

ANSWER 10–15

A. cDNA clones could not be used because there is no overlap between cDNA clones from adjacent genes.

B. Such repetitive DNA sequences can confuse chromosome walks, because the walk would appear to branch off in many different directions at once. The general strategy for avoiding these problems is to use genomic clones that are sufficiently long to span beyond the repetitive DNA sequences.

ANSWER 10–16

A. Infants 2 and 8 have identical STR patterns and therefore must be identical twins. Infants 3 and 6 also have identical STR patterns and must also be identical twins. The other two sets of twins must be fraternal twins because their STR patterns are not identical. Fraternal twins, like any pair of siblings born to the same parents, will have roughly half their genome in common. Thus, roughly half the STR polymorphisms in fraternal twins will be identical. Using this criterion, you can identify infants 1 and 7 as fraternal twins and infants 4 and 5 as fraternal twins.

B. You can match infants to their parents by using the same sort of analysis of STR polymorphisms. Every band present in the analysis of an infant should have a matching band in one or the other of the parents, and, on average, each infant will share half of its polymorphisms with each parent. Thus, the degree of match between each child and each parent will be approximately the same as that between fraternal twins.

ANSWER 10–17 Mutant bacteria that do not produce ice-protein have probably arisen many times in nature. However, bacteria that produce ice-protein have a slight growth advantage over bacteria that do not, so it would be difficult to find such mutants in the wild. Recombinant DNA technology makes these mutants much easier to obtain. In this case, the consequences, both advantageous and disadvantageous, of using a genetically modified organism are therefore nearly indistinguishable from those of a natural mutant. Indeed, bacterial and yeast strains have been selected for centuries for desirable genetic traits that make them suitable for industrial-scale applications such as cheese and wine production. The possibilities of recombinant DNA technology are endless, however, and as with any technology, there is a finite risk of unforeseen consequences. Recombinant DNA experimentation, therefore, is regulated, and the risks of individual projects are carefully assessed by review panels before permissions are granted. The state of our knowledge is sufficiently advanced that the consequences of some changes, such as the disruption of a bacterial gene in the example above, can be predicted with reasonable certainty. Other applications, such as germ-line gene therapy to correct human disease, may have far more complex outcomes, and it will take many more years of research and ethical debate to determine whether such treatments will eventually be used.

Chapter 11

ANSWER 11–1 Water is a liquid, and thus hydrogen bonds between water molecules are not static; they are continually formed and broken again by thermal motion. When a water molecule happens to be next to a hydrophobic molecule, it is more restricted in motion and has fewer neighbors with which it can interact, because it cannot form any hydrogen bonds in the direction of the hydrophobic molecule. It will therefore form hydrogen bonds to the more limited number of water molecules in its proximity. Bonding to fewer partners results in a more ordered water structure, which represents the cagelike structure in Figure 11–9. This structure has been likened to ice, although it is a more transient, less organized, and less extensive network than even a tiny ice crystal. The formation of any ordered structure decreases the entropy of the system (see Chapter 3) and is thus energetically unfavorable.

ANSWER 11–2 (B) is the correct analogy for lipid bilayer assembly because exclusion from water rather than attractive forces between the lipid molecules is involved. If the lipid molecules formed bonds with one another, the bilayer would be less fluid, and might even become rigid, depending on the strength of the interaction.

ANSWER 11–3 The fluidity of the bilayer is strictly confined to one plane: lipid molecules can diffuse laterally in their own monolayer but do not readily flip from one monolayer to the other. Specific types of lipid molecules inserted into one monolayer therefore remain in it unless they are actively transferred by an enzyme—called a flippase.

ANSWER 11–4 In both an α helix and a β barrel the polar peptide bonds of the polypeptide backbone can be completely shielded from the hydrophobic environment of the lipid bilayer by the hydrophobic amino acid side chains. Internal hydrogen bonds between the peptide bonds stabilize the α helix and β barrel.

ANSWER 11–5 The sulfate group in SDS is charged and therefore hydrophilic. The OH group and the C–O–C groups in Triton X-100 are polar; they can form hydrogen bonds with water and are therefore hydrophilic. In contrast, the blue portions of the molecules are either hydrocarbon chains or aromatic rings, neither of which has polar groups that could form hydrogen bonds with water molecules; they are therefore hydrophobic. (See Figure A11–5.)

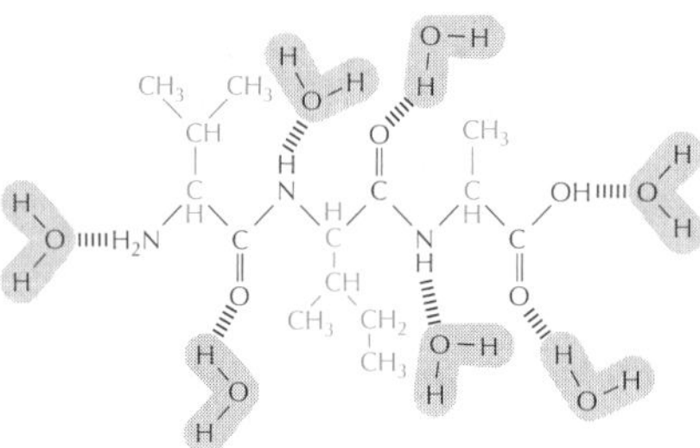

Figure A11–5

ANSWER 11–6 Alpha helices in proteins are often used to span lipid bilayers. These structures are well suited for this purpose because they expose hydrophobic amino acid side chains to the hydrophobic interior of the lipid bilayer but sequester the polar peptide bonds of the polypeptide backbone away from the hydrophobic phase (see Figures 11–22 through 11–25). There are, however, other, less regular ways to fold up a polypeptide chain to achieve the same result, as seen in the small loop in the photosynthetic reaction center. This illustrates the importance of determining three-dimensional structures, which to date are known for only a small number of membrane proteins.

ANSWER 11–7 Some of the molecules of the two different transmembrane proteins are anchored to the spectrin filaments of the cell cortex. These molecules are not free to rotate or diffuse within the plane of the membrane. There is an excess of transmembrane proteins over the available attachment sites in the cortex, however, so that some of the transmembrane protein molecules are not anchored and are free to rotate and diffuse within the plane of the membrane. Indeed, measurements of protein mobility show that there are two populations of each transmembrane protein, corresponding to those proteins that are anchored and those that are not.

ANSWER 11–8 The different ways in which membrane proteins can be restricted to different regions of the membrane are summarized in Figure 11–33. The mobility of the membrane proteins is drastically reduced if they are bound to other proteins such as those of the cytoskeleton or the extracellular matrix. Some membrane proteins are confined to membrane domains by barriers, such as tight junctions. The fluidity of the lipid bilayer is not significantly affected by the anchoring of membrane proteins; the sea of lipid molecules flows around anchored membrane proteins like water around the posts of a pier.

ANSWER 11–9 All of the statements are correct.

A, B, C, D. The lipid bilayer is fluid because the lipid molecules in the bilayer can undergo these motions.

E. Glycolipids are mostly restricted to the monolayer of membranes that faces away from the cytosol. Some special glycolipids, such as phosphatidylinositol (discussed in Chapter 16), are found specifically in the cytosolic monolayer.

F. The reduction of double bonds (by hydrogenation) allows lipid molecules to pack more tightly against one another and therefore increases the viscosity—that is, it turns oil into margarine.

G. Examples include enzymes involved in signaling (discussed in Chapter 16).

H. Polysaccharides are the main constituents of mucus and slime; the carbohydrate coat of a cell, which is made up of polysaccharides and oligosaccharides, is a very important lubricant—for example, for cells that line blood vessels or circulate in the bloodstream.

ANSWER 11–10 In a two-dimensional fluid the molecules are free to move only in one plane; the molecules in a normal fluid, in contrast, can move in three dimensions.

ANSWER 11–11

A. You would have a detergent. The diameter of the lipid head would be much larger than that of the hydrocarbon tail, so that the shape of the molecule would be a cone rather than a cylinder and the molecules would aggregate to form micelles rather than bilayers.

B. Lipid bilayers formed would be much more fluid. The bilayers would also be less stable, as the shorter hydrocarbon tails would be less hydrophobic, so the forces that drive the formation of the bilayer would be reduced.

C. The lipid bilayers formed would be much less fluid. Whereas a normal lipid bilayer has the viscosity of olive oil, a bilayer made of the same lipids but with saturated hydrocarbon tails would have the consistency of bacon fat.

D. The lipid bilayers formed would be much more fluid. Also, because the lipids would pack together less well, there would be more gaps and the bilayer would be more permeable to small water-soluble molecules.

E. If we assume that the lipid molecules are completely intermixed, the fluidity of the membrane would be unchanged. In such bilayers, however, the saturated lipid molecules would tend to aggregate with one another because they can pack so much more tightly and would therefore form patches of much-reduced fluidity. The bilayer would not, therefore, have uniform properties over its surface. Because, normally, one saturated and one unsaturated hydrocarbon tail are linked to the same hydrophilic head in membrane lipid molecules, such segregation does not occur in cell membranes.

F. The lipid bilayers formed would have virtually unchanged properties. Each lipid molecule would now span the entire membrane, with one of its two head groups exposed at each surface. Such lipid molecules are found in the membranes of thermophilic bacteria, which can live at temperatures approaching boiling water. Their bilayers do not come apart at elevated temperatures, as usual bilayers do, because the original two monolayers are now covalently linked into a single membrane.

ANSWER 11–12 Lipid molecules are approximately cylindrical in shape. Detergent molecules, by contrast, are conical or wedge-shaped. A lipid molecule with only one hydrocarbon tail, for example, would be a detergent. To make a lipid molecule into a detergent, you would have to make its hydrophilic head larger or remove one of its tails so that it could form a micelle. Detergent molecules also usually have shorter hydrocarbon tails than lipid molecules. This makes them slightly water-soluble, so that detergent molecules leave and reenter micelles frequently in aqueous solution. Because of this, some monomeric detergent molecules are always present in aqueous solution and therefore can enter lipid bilayers to solubilize membrane proteins (see Figure 11–27).

ANSWER 11–13 When lined up, there are about 4000 lipid molecules (each 0.5 nm wide) between a lipid molecule at one end of the bacterial cell and one at the other end. Thus, if one of these molecules started to move toward the other, exchanging places with a neighboring molecule every 10^{-7} sec, it would take only 4×10^{-4} sec (= 4000×10^{-7} sec) to reach the other end. In reality, however, the lipid molecule would move in a random path rather than in a defined direction, so it would take considerably longer (1 sec) to reach the other end. If a 4-cm ping-pong ball exchanged places with a neighbor every 10^{-7} sec, it would travel at a speed of 1,440,000 km/hr (= 4 cm/10^{-7} sec). If its movement were only in one direction, it would reach the other wall in 1.5×10^{-5} sec and, if it kept going, it would circle the Earth in approximately 2 minutes. In a random walk it would take considerably longer to reach the other side of the room (~2 msec).

ANSWER 11–14 Membrane proteins anchor the lipid bilayer to the cytoskeleton, strengthening the plasma membrane so that it can withstand the forces on it when the red blood cell is pumped through small blood vessels. Membrane proteins also transport nutrients and ions across the plasma membrane.

ANSWER 11–15 The hydrophilic faces of the five membrane-spanning α helices, each contributed by a different subunit, are thought to come together to form a pore across the lipid bilayer that is lined with the hydrophilic amino acid side chains (Figure A11–15). Ions can pass through this hydrophilic pore without coming

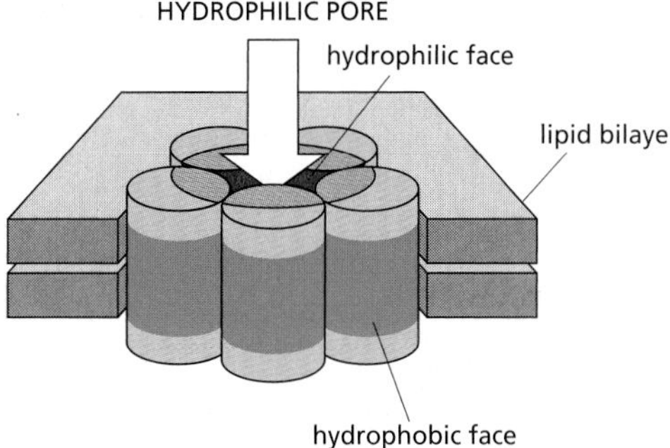

Figure A11–15

into contact with the lipid tails of the bilayer. The hydrophobic side chains interact with the hydrophobic lipid tails.

ANSWER 11–16 There are about 100 lipid molecules (i.e., phospholipid + cholesterol) for every protein molecule in the membrane [= (2/50,000)/(1/800 + 1/256)]. A similar protein/lipid ratio is seen in many cell membranes.

ANSWER 11–17 Membrane fusion does not alter the orientation of the membrane proteins with their attached color tags: the portion of each transmembrane protein that is exposed to the cytosol always remains exposed to cytosol, and the portion exposed to the outside always remains exposed to the outside (Figure A11–17). At 0°C, the fluidity of the membrane is reduced, and the mixing of the membrane proteins is significantly slowed.

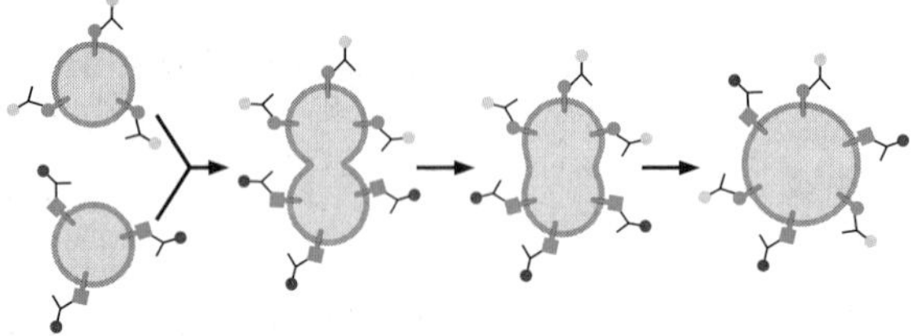

Figure A11–17

ANSWER 11–18 The exposure of hydrophobic amino acid side chains to water is energetically unfavorable. There are two ways that such side chains can be sequestered from water to achieve an energetically more favorable state. First, they can form transmembrane segments that span a lipid bilayer. This requires about 20 of them to be located sequentially in a polypeptide chain. Second, the hydrophobic amino acid side chains can be sequestered in the interior of the folded polypeptide chain. This is one of the major forces that lock the polypeptide chain into a unique three-dimensional structure. In either case, the hydrophobic forces in the lipid bilayer or in the interior of a protein are based on the same principles.

ANSWER 11–19 (A) Antarctic fish live at sub-zero temperatures and are cold-blooded. To keep their membranes fluid at these temperatures, they have a high percentage of unsaturated phospholipids.

ANSWER 11–20 Sequence B is most likely to form a transmembrane helix. It is composed primarily of hydrophobic amino acids, and therefore can be stably integrated into a lipid bilayer. In contrast, sequence A contains many polar amino acids (S, T, N, Q), and sequence C contains many charged amino acids (K, R, H, E, D), which would be energetically disfavored in the hydrophobic interior of the lipid bilayer.

Chapter 12

ANSWER 12–1

A. The movement of molecules mediated by a transporter can be described by a strictly analogous equation:

$$(1)\ T + S \rightleftharpoons TS \rightarrow T + S^*$$

where S is the solute, S* is the solute on the other side of the membrane (i.e., although it is still the same molecule, it is now located in a different environment), and T is the transporter.

B. This equation is useful because it describes a binding step, followed by a delivery step. The mathematical treatment of this equation would be very similar to that described for enzymes (see Figure 3–24); thus, transporters are characterized by a K_M value that describes their affinity for a solute and a V_{max} value that describes their maximal rate of transfer. To be more accurate, one could include the conformational change of the transporter in the reaction scheme

$$(2a)\ T + S \rightleftharpoons TS \rightleftharpoons T^*S^* \rightarrow T^* + S^*$$

$$(2b)\ T \rightleftharpoons T^*$$

where T* is the transporter after the conformational change that exposes its solute-binding site on the other side of the membrane. This account requires a second equation (2b) that allows the transporter to return to its starting conformation.

C. The equations do not describe the behavior of channels because solutes passing through channels do not bind to them in the way that a substrate binds to an enzyme.

ANSWER 12–2 If the Na^+-K^+ pump is not working at full capacity because it is partially inhibited by ouabain or digitalis, it generates an electrochemical gradient of Na^+ that is less steep than that in untreated cells. Consequently, the Ca^{2+}-Na^+ antiport works less efficiently, and Ca^{2+} is removed from the cell more slowly. When the next cycle of muscle contraction begins, there is still an elevated level of Ca^{2+} left in the cytosol. The entry of the same number of Ca^{2+} ions into the cell therefore leads to a higher Ca^{2+} concentration than in untreated cells, which in turn leads to a stronger and longer-lasting contraction. Because the Na^+-K^+ pump fulfills essential functions in all animal cells, both to maintain osmotic balance and to generate the Na^+ gradient used to power many transporters, the drugs are deadly poisons at higher concentrations.

ANSWER 12–3

A. The properties define a symport.

B. No additional properties need to be specified. The important feature that provides the coupling of the two solutes is that the protein cannot switch its conformation if only one of the two solutes is bound. Solute B, which is driving the transport of solute A, is in excess on the side of the membrane from which transport initiates and occupies its binding site most of the time. In this state, the transporter, prevented from switching its conformation, waits until a solute A molecule binds on occasion. With both binding sites occupied, the transporter switches conformation. Now exposed to the other side of the membrane, the binding site for solute B is mostly empty because there is little of it in the solution on this side of the membrane. Although the binding site for A is now more frequently occupied, the transporter can switch back only after solute A is unloaded as well.

C. An antiport could be similarly constructed as a transmembrane protein with the following properties. It has two binding sites, one for solute A and one for solute B. The protein can undergo a conformational change to switch between two states: either both binding sites are exposed exclusively on one side of the membrane or both binding sites are exposed exclusively on the other side of the membrane. The protein can switch between the two conformational states only if one binding site is occupied, but cannot switch if both binding sites are occupied or if both binding sites are empty.

   Note that these rules provide an alternative model to that shown in Figure 12–17. Thus there are two possible ways to couple the transport of two solutes: (1) provide cooperative solute-binding sites and allow the pump to switch between the two states randomly as shown in Figure 12–16 or (2) allow independent binding of both solutes and make the switch between the two states conditional on the occupancy of the binding sites. As the structure of a coupled transporter has not yet been determined, we do not know which of the two mechanisms such pumps use.

ANSWER 12–4 Each of the rectangular peaks corresponds to the opening of a single channel that allows a small current to pass. You note from the recording that the channels present in the patch of membrane open and close frequently. Each channel remains open for a very short, somewhat variable time, averaging about 10 milliseconds. When open, the channels allow a small current with a unique amplitude (4 pA; one picoampere = 10^{-12} A) to pass. In one instance, the current doubles, indicating that two channels in the same membrane patch opened simultaneously.

If acetylcholine is omitted or is added to the solution outside the pipette, you would measure only the baseline current. Acetylcholine must bind to the extracellular portion of the acetylcholine receptor molecules to allow the channel to open, and in the membrane patch shown in Figure 12–23, the cytoplasmic side of the membrane is exposed to the solution outside the microelectrode.

ANSWER 12–5 The equilibrium potential of K^+ is –90 mV [= 62 mV $\log_{10}$ (5 mM/140 mM)], and that of Na^+ is +72 mV [= 62 mV $\log_{10}$ (145 mM/10 mM)]. The K^+ leak channels in the plasma membrane of a resting cell allow K^+ to come to equilibrium; the membrane potential of the cell is therefore close to –90 mV. When Na^+ channels open, Na^+ rushes in, and, as a result, the membrane potential reverses its polarity to a value nearer to +72 mV, the equilibrium value for Na^+. Upon closure of the Na^+ channels, the K^+ leak channels allow K^+, now no longer at equilibrium, to exit from the cell until the membrane potential is restored to the equilibrium value for K^+, about –90 mV.

ANSWER 12–6 When the resting membrane potential of an axon drops below a threshold value, voltage-gated Na^+ channels in the immediate neighborhood open and allow an influx of Na^+. This depolarizes the membrane further, causing more distant voltage-gated Na^+ channels to open as well. This creates a wave of depolarization that spreads rapidly along the axon, called the action potential. Because Na^+ channels become inactivated soon after they open, the flow of K^+ through voltage-gated K^+ channels and K^+ leak channels is able to restore the original resting membrane potential rapidly after the action potential has passed. (96 words)

ANSWER 12–7 If the number of functional acetylcholine receptors is reduced by the antibodies, the neurotransmitter (acetylcholine) that is released from the nerve terminals cannot (or can only weakly) stimulate the muscle to contract.

ANSWER 12–8 By analogy to the Na^+-K^+ pump shown in Figure 12–11, ATP might be hydrolyzed and donate a phosphate group to the transporter when—and only when—it has the solute bound on the "inside" face of the membrane (step 1 → 2). The attachment of the phosphate would trigger an immediate conformational change (step 2 → 3), thereby capturing the solute and exposing it to the "outside." The phosphate would be removed from the protein when—and only when—the solute had dissociated, and the now empty, nonphosphorylated transporter would switch back to the starting position (step 3 → 4) (Figure A12–8).

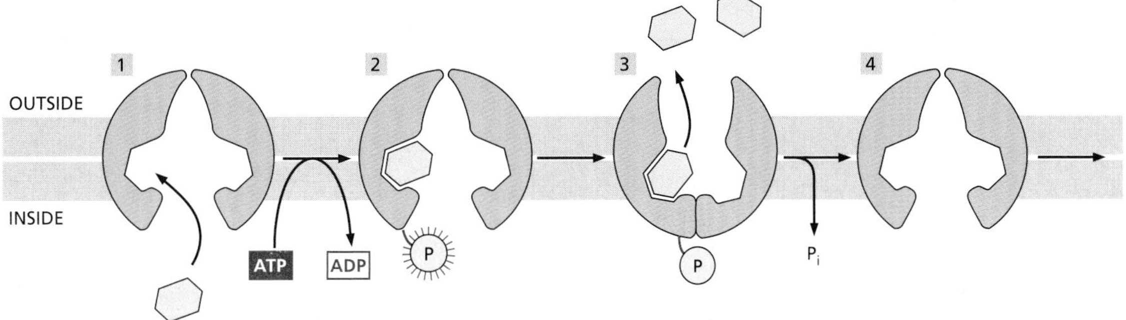

Figure A12–8

ANSWER 12–9

A. False. The plasma membrane contains proteins that confer selective permeability to many charged molecules. In contrast, a pure lipid bilayer lacking proteins is highly impermeable to all charged molecules.
B. False. Channels do not bind the solute that passes through them. Selectivity of a channel is achieved by the size of the internal pore and by charged regions at the entrance of the pore that attract or repel ions of the appropriate charge.
C. False. Transporters are slower. They have enzymelike properties, i.e., they bind solutes and need to undergo conformational changes during their functional cycle. This limits the maximal rate of transport to about 1000 solute molecules per second, whereas channels can pass up to 1,000,000 solute molecules per second.
D. True. The bacteriorhodopsin of some photosynthetic bacteria moves H^+, using energy captured from visible light.
E. True. Most animal cells contain K^+ leak channels in their plasma membrane that are predominantly open. The K^+ concentration inside the cell still remains higher than outside, because the membrane potential is negative and therefore inhibits the positively charged K^+ from leaking out. K^+ is also continually pumped into the cell by the Na^+-K^+ pump.
F. False. A symport binds two different solutes on the same side of the membrane. Turning it around would not change it into an antiport, which must also bind to different solutes, but on opposing sides of the membrane.
G. False. The peak of an action potential corresponds to a transient shift of the membrane potential from a negative to a positive value. The influx of Na^+ causes the membrane potential first to move toward zero and then to reverse, rendering the cell positively charged on its inside. Eventually, the resting potential is restored by an efflux of K^+ through voltage-gated K^+ channels and K^+ leak channels.

ANSWER 12–10 The permeabilities are N_2 (small and nonpolar) > ethanol (small and slightly polar) > water (small and polar) > glucose (large and polar) > Ca^{2+} (small and charged) > RNA (very large and charged).

ANSWER 12–11

A. Both couple the movement of two different solutes across the membrane. Symports transport both solutes in the same direction, whereas antiports transport the solutes in opposite directions.
B. Both are mediated by membrane transport proteins. Passive transport of a solute occurs downhill, in the direction of its concentration or electrochemical gradient, whereas active transport occurs uphill and therefore needs an energy source. Active transport can be mediated by transporters but not by channels, whereas passive transport can be mediated by either.
C. Both terms describe energy changes involved in moving an ion from one side of a membrane to the other. The membrane potential refers to the electrical energy change; the electrochemical gradient is a composite of this electrical energy change and the chemical energy change associated with moving between a region of high concentration and a region of low concentration. The membrane potential is defined independently of the choice of ion, whereas an electrochemical gradient depends on the concentration gradient of the particular ionic solute and is therefore a solute-specific parameter.
D. A pump is a specialized transporter that uses energy to transport a solute uphill against an electrochemical gradient.
E. Both transmit signals by electrical activity. Wires are made of copper, axons are not. The signal passing down an axon does not diminish in strength, because it is self-amplifying, whereas the signal in a wire decreases over distance (by leakage of current across the insulating sheath).
F. Both affect the osmotic pressure of the cell. An ion is a solute that bears a charge.

ANSWER 12–12 A bridge allows vehicles to pass over a river in a steady stream; the entrance can be designed to exclude, for example, oversized trucks, and it can be intermittently closed to traffic by a gate. By analogy, channels allow ions to flow in a gated stream across the membrane, imposing size and charge restrictions.

A ferry, in contrast, loads vehicles on one riverbank and then, after movement of the ferry itself, unloads on the other side of the river. This process is slower. During loading, particular vehicles could be selected from the waiting line because they fit particularly well on the car deck. By analogy, transporters bind solutes on one side of the membrane and then, after a conformational movement, release them on the other side. Specific binding leads to the selection of the molecules to be transported. As in the case of coupled transport, sometimes you have to wait until the ferry is full before you can go.

ANSWER 12–13 Acetylcholine is being transported into the vesicles by an H^+–acetylcholine antiport in the vesicle membrane. The H^+ gradient that drives the uptake is generated by an ATP-driven H^+ pump in the vesicle membrane, which pumps H^+ into the vesicle (hence the dependence of the reaction on ATP). Raising the pH of the solution surrounding the vesicles increases the H^+ gradient: at an elevated pH there are fewer H^+ ions in the solution outside the vesicles while the number inside remains the same. This explains the observed enhanced rate of uptake.

ANSWER 12–14 The voltage gradient across the membrane is about 150,000 V/cm. This extremely powerful electric field is close to the limit at which insulating materials—such as the lipid bilayer—break down and cease to act as insulators. The large field corresponds to the large amount of energy that can be stored in electrical gradients across the membrane, as well as to the extreme electrical forces that proteins can experience in a membrane. A voltage of 150,000 V would instantly discharge in an arc across a 1-cm-wide gap (that is, air would be an insufficient insulator for this strength of field).

ANSWER 12–15

A. Nothing. You require ATP to drive the Na^+-K^+ pump.
B. The ATP becomes hydrolyzed, and Na^+ is pumped into the vesicles, generating a concentration

gradient of Na^+ across the membrane. At the same time, K^+ is pumped out of the vesicles, generating a concentration gradient of K^+ of opposite polarity. When all the K^+ had been pumped out of the vesicle or the ATP ran out, the pump would stop.

C. The Na^+-K^+ pump would go through states 1, 2, and 3 in Figure 12–11. Because all reaction steps must occur strictly sequentially, however, dephosphorylation and the conformation switch cannot occur in the absence of K^+. The Na^+-K^+ pump will therefore become stuck in the phosphorylated state, waiting indefinitely for a potassium ion. The number of sodium ions transported would be minuscule, because each pump molecule would have functioned only a single time.

Similar experiments, leaving out individual ions and analyzing the consequences, were used to determine the sequence of steps by which the Na^+-K^+ pump works.

D. ATP would become hydrolyzed and Na^+ and K^+ would be pumped across the membrane as described in scenario (B). However, the pump molecules that sit in the membrane in the reverse orientation would be completely inactive (i.e., they would not—as one might have erroneously assumed—pump ions in the opposite direction), because ATP would not have access to the site on these molecules where phosphorylation occurs. This site is normally exposed to the cytosol. ATP is highly charged and cannot cross membranes without the help of specific transporters.

E. ATP becomes hydrolyzed and Na^+ and K^+ are pumped across the membrane, as described in scenario (B). K^+, however, immediately flows back into the vesicles through the K^+ leak channels. K^+ moves down the K^+ concentration gradient formed by the action of the Na^+-K^+ pump. With each K^+ that moves into the vesicle through the leak channel, a positive charge is moved across the membrane, building a membrane potential that is positive on the inside of the vesicles. Eventually, K^+ will stop flowing through the leak channels when the membrane potential balances the concentration gradient. The scenario described here is a slight oversimplification: the Na^+-K^+ pump in mammalian cells actually moves three sodium ions out of cells for each two potassium ions that it pumps into the cell, thereby driving an electric current across the membrane and making a small additional contribution to the resting membrane potential (which therefore corresponds only approximately to a state of equilibrium for K^+ moving via K^+ leak channels).

ANSWER 12–16 Ion channels can be ligand-gated, voltage-gated, or mechanically gated.

ANSWER 12–17 The cell has a volume of 10^{-12} liters (= 10^{-15} m^3) and thus contains 6×10^4 calcium ions (= 6×10^{23} molecules/mole $\times$ 100 $\times$ 10^{-9} moles/liter $\times$ 10^{-12} liters). Therefore, to raise the intracellular Ca^{2+} concentration fiftyfold, another 2,940,000 calcium ions have to enter the cell (note that at 5 μM concentration there are 3×10^6 ions in the cell, of which 60,000 are already present before the channels are opened). Because each of the 1000 channels allows 10^6 ions to pass per second, each channel has to stay open for only 3 milliseconds.

ANSWER 12–18 Animal cells drive most transport processes across the plasma membrane with the electrochemical gradient of Na^+. ATP is needed to fuel the Na^+-K^+ pump to maintain the Na^+ gradient.

ANSWER 12–19

A. If H^+ is pumped across the membrane into the endosomes, an electrochemical gradient of H^+ results—composed of both an electrical potential and a concentration gradient, with the interior of the vesicle positive. Both of these components add to the energy that is stored in the gradient and that must be supplied to generate it. The electrochemical gradient will therefore limit the transfer of more H^+. If, however, the membrane also contains Cl^- channels, the negatively charged Cl^- will flow into the endosomes and diminish the electrical potential. It therefore becomes energetically less expensive to pump more H^+ across the membrane, and the interior of the endosomes can become more acidic.

B. Yes. As explained in (A), some acidification would still occur in their absence.

ANSWER 12–20

A. See Figure A12–20A.

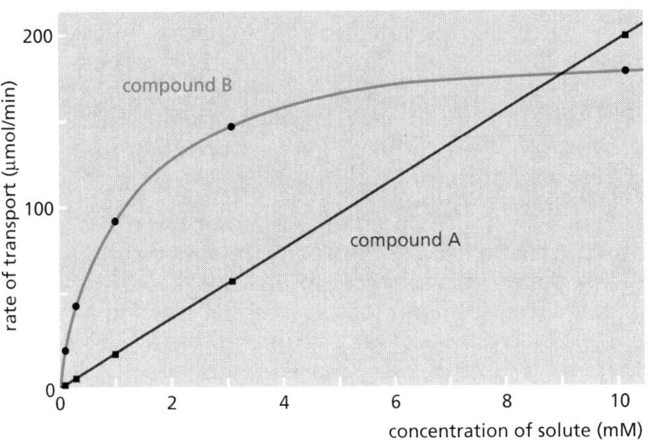

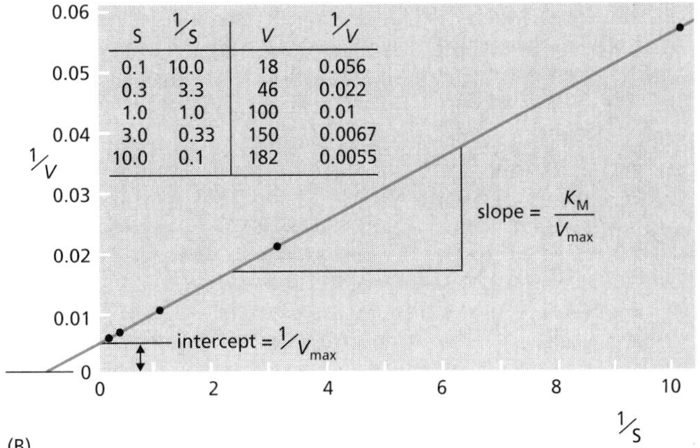

S	1/S	V	1/V
0.1	10.0	18	0.056
0.3	3.3	46	0.022
1.0	1.0	100	0.01
3.0	0.33	150	0.0067
10.0	0.1	182	0.0055

Figure A12–20

B. The transport rates of compound A are proportional to its concentration, indicating that compound A can diffuse through membranes on its own. Compound A is likely to be ethanol, because it is a small and relatively nonpolar molecule that can diffuse readily through the lipid bilayer.

In contrast, the transport rates of compound B saturate at high concentrations, indicating that compound B is transported across the membrane by some sort of membrane transport protein. Transport rates cannot increase beyond a maximal rate at which this protein can function. Compound B is likely to be acetate, because it is a charged molecule that could not cross the membrane without the help of a membrane transport protein.

C. For ethanol, we measured a linear relationship between concentration and transport rate. Thus, at 0.5 mM the transport rate would be 10 μmol/min, and at 100 mM the transport rate would be 2 μmol/min.

For the transporter–mediated movement of acetate, the relationship between concentration, S, and transport rate can be described by the Michaelis–Menten equation, which describes simple enzyme reactions:

(1) transport rate = $V_{max} \times S/[K_M + S]$

Recall from Chapter 3 (see Question 3–20, p. 116) that to determine the V_{max} and K_M, a trick is used in which the Michaelis–Menten equation is transformed so that it is possible to plot the data as a straight line. A simple transformation yields

(2) **1/rate** = (K_M/V_{max}) **(1/S)** + $1/V_{max}$

(i.e., an equation of the form $y = ax + b$)

Calculation of 1/rate and 1/S for the given data and plotting them in a new graph as in Figure A12–20B gives a straight line. The K_M (= 1.0 mM) and V_{max} (= 200 μmol/min) are determined from the intercept of the line with the y axis ($1/V_{max}$) and from its slope (K_M/V_{max}). Knowing the values for K_M and V_{max} allows you to calculate the transport rates for 0.5 mM and 100 mM acetate using equation (1). The results are 67 μmol/min and 198 μmol/min, respectively.

ANSWER 12–21 The membrane potential and the high extracellular Na^+ concentration provide a large electrochemical driving force and a large reservoir of Na^+ ions, so that mostly Na^+ ions enter the cell as acetylcholine receptors open. Ca^{2+} ions will also enter the cell, but their influx is much more limited because of their lower extracellular concentration. (Most of the Ca^{2+} that enters the cytosol upon muscle activation is released from intracellular stores, as we discuss in Chapter 17). Because of their high intracellular concentration and the opposing direction of membrane potential, K^+ ions are already close to equilibrium across the membrane. For this reason, there will be little if any movement of K^+ ions upon opening of a cation channel.

ANSWER 12–22 The diversity of neurotransmitter-gated ion channels is a good thing for the industry as it raises the possibility of developing new drugs specific for each channel type. Each of the diverse subtypes of these channels is expressed in a narrow set of neurons. This narrow range of expression should make it possible, in principle, to discover or design drugs that affect particular receptor subtypes present in a selected set of neurons, thus targeting particular brain functions with greater specificity.

Chapter 13

ANSWER 13–1 To keep glycolysis going, cells need to regenerate NAD^+ from NADH. There is no efficient way to do this without fermentation. In the absence of regenerated NAD^+, step 6 of glycolysis (the oxidation of glyceraldehyde 3-phosphate to 1,3-bisphosphoglycerate (Panel 13–1, pp. 430–431) could not occur and the product glyceraldehyde 3-phosphate would accumulate. The same thing would happen in cells unable to make either pyruvate or ethanol: neither would be able to regenerate NAD^+, and so glycolysis would be blocked at the same step.

ANSWER 13–2 Arsenate instead of phosphate becomes attached in step 6 of glycolysis to form 1-arseno-3-phosphoglycerate (Figure A13–2). Because of its sensitivity to hydrolysis in water, the high-energy bond is destroyed before the molecule that contains it can diffuse to reach the next enzyme. The product of the hydrolysis, 3-phosphoglycerate, is the same product normally formed in step 7 by the action of phosphoglycerate kinase. But because hydrolysis occurs nonenzymatically, the energy liberated by breaking the high-energy bond cannot be captured to generate ATP. In Figure 13–6, therefore, the reaction corresponding to the downward-pointing arrow would still occur, but the wheel that provides the coupling to ATP synthesis is missing. Arsenate wastes metabolic energy by uncoupling many phosphotransfer reactions by the same mechanism, which is why it is so poisonous.

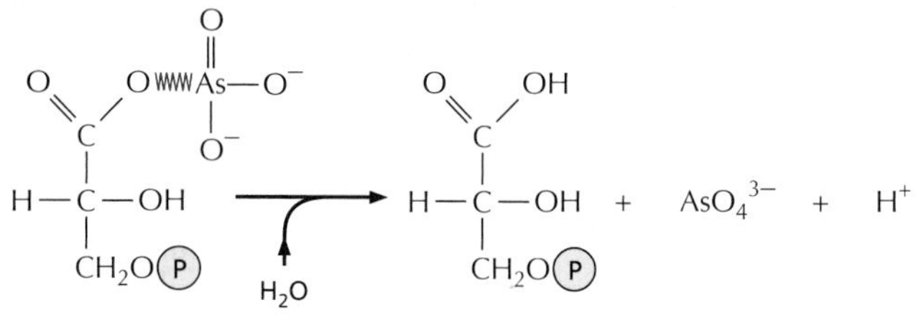

Figure A13–2

ANSWER 13–3 The oxidation of fatty acids breaks the carbon chain into two-carbon units (acetyl groups) that become attached to CoA. Conversely, during biogenesis fatty acids are constructed by linking together acetyl groups. Most fatty acids therefore have an even number of carbon atoms.

ANSWER 13–4 Because the function of the citric acid cycle is to harvest the energy released during the oxidation, it is advantageous to break the overall reaction into as many steps as possible (see Figure 13–1). Using a two-carbon compound, the available chemistry would be much more limited, and it would be impossible to generate as many intermediates.

ANSWER 13–5 It is true that oxygen atoms are returned as part of CO_2 to the atmosphere. The CO_2

released from the cells, however, does not contain those specific oxygen atoms that were consumed as part of the oxidative phosphorylation reaction and converted into water. One can show this directly by incubating living cells in an atmosphere that includes molecular oxygen containing a different isotope, ^{18}O, instead of the naturally abundant isotope, ^{16}O. In such an experiment one finds that all the CO_2 released from cells contains only ^{16}O. Therefore, the oxygen atoms in the released CO_2 molecules do not come directly from the atmosphere but from organic molecules that the cell has first made and then oxidized as fuel (see Panel 13–2, pp. 442–443).

ANSWER 13–6 The cycle continues because intermediates are replenished as necessary by reactions leading to the citric acid cycle (instead of away from it). One of the most important reactions of this kind is the conversion of pyruvate to oxaloacetate by the enzyme pyruvate carboxylase:

$$\text{pyruvate} + CO_2 + \text{ATP} + H_2O \rightarrow \text{oxaloacetate} + \text{ADP} + P_i + 2H^+$$

This is one of the many examples of how metabolic pathways are elegantly balanced and work together to maintain appropriate concentrations of all metabolites required by the cell (see Figure A13–6).

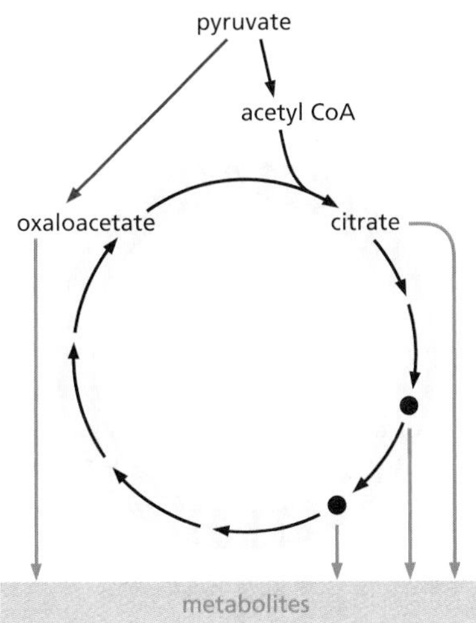

Figure A13–6

ANSWER 13–7 The carbon atoms in sugar molecules are already partially oxidized, in contrast to all but the very first carbon atoms in the acyl chains of fatty acids. Thus, two carbon atoms from glucose are lost as CO_2 during the conversion of pyruvate to acetyl CoA, and only four of the six carbon atoms of the sugar molecule are recovered and can enter the citric acid cycle, where most of the energy is captured. In contrast, all carbon atoms of a fatty acid are converted into acetyl CoA.

ANSWER 13–8

A. False. If this were the case, then the reaction would be useless for the cell. No chemical energy would be harvested in a useful form (e.g., ATP) to be used for metabolic processes. (Cells would be nice and warm, though!)

B. False. No energy conversion process can be 100% efficient. Recall that entropy in the universe always has to increase, and for most reactions this is accomplished by releasing heat.

C. True. The carbon atoms in glucose are in a reduced state compared with those in CO_2, in which they are fully oxidized.

D. False. The reaction does indeed produce some water, but water is so abundant in the biosphere that this is no more than "a drop in the bucket."

E. True. If it had occurred in only one step, then all the energy would be released at once and it would be impossible to harness it efficiently to drive other reactions, such as the synthesis of ATP.

F. False. Molecular oxygen (O_2) is used only in the very last step of the reaction.

G. True. Plants convert CO_2 into sugars by harvesting the energy of light in photosynthesis. O_2 is produced in the process and released by plant cells.

H. True. Anaerobically growing cells use glycolysis to oxidize sugars to pyruvate. Animal cells convert pyruvate to lactate, and no CO_2 is produced; yeast cells, however, convert pyruvate to ethanol and CO_2. It is this CO_2 gas, released from yeast cells during fermentation, that makes bread dough rise and that carbonates beer and champagne.

ANSWER 13–9 Darwin exhaled the carbon atom, which therefore must be the carbon atom of a CO_2 molecule. After spending some time in the atmosphere, the CO_2 molecule must have entered a plant cell, where it became "fixed" by photosynthesis and converted into part of a sugar molecule. While it is certain that these early steps must have happened this way, there are many different paths from there that the carbon atom could have taken. The sugar could have been broken down by the plant cell into pyruvate or acetyl CoA, for example, which then could have entered biosynthetic reactions to build an amino acid. The amino acid might have been incorporated into a plant protein, maybe an enzyme or a protein that builds the cell wall. You might have eaten the delicious leaves of the plant in your salad, and digested the protein in your gut to produce amino acids again. After circulating in your bloodstream, the amino acid might have been taken up by a developing red blood cell to make its own protein, such as the hemoglobin in question. If we wish, of course, we can make our food chain scenario more complicated. The plant, for example, might have been eaten by an animal that in turn was consumed by you during lunch break. Moreover, because Darwin died more than 100 years ago, the carbon atom could have traveled such a route many times. In each round, however, it would have started again as fully oxidized CO_2 gas and entered the living world following its reduction during photosynthesis.

ANSWER 13–10 Yeast cells grow much better aerobically. Under anaerobic conditions they cannot perform oxidative phosphorylation and therefore have to produce all their ATP by glycolysis, which is less efficient. Whereas one glucose molecule yields a net gain of two ATP molecules by glycolysis, the additional use of the citric acid cycle and oxidative phosphorylation boosts the energy yield up to about 30 ATP molecules.

ANSWER 13–11 The amount of free energy stored in the phosphate bond in creatine phosphate is larger than that of of the anhydride bonds in ATP. Hydrolysis of creatine phosphate can therefore be directly coupled to the production of ATP.

creatine phosphate + ADP → creatine + ATP

The $\Delta G°$ for this reaction is –3 kcal/mole, indicating that it proceeds rapidly to the right, as written.

ANSWER 13–12 The extreme conservation of glycolysis is evidence that all present cells are derived from a single founder cell as discussed in Chapter 1. The elegant reactions of glycolysis would therefore have evolved only once, and then they would have been inherited as cells evolved. The later invention of oxidative phosphorylation allowed follow-up reactions to capture 15 times more energy than is possible by glycolysis alone. This remarkable efficiency is close to the theoretical limit and hence virtually eliminates the opportunity for further improvements. Thus, the generation of alternative pathways would result in no obvious growth advantage that would have been selected in evolution.

ANSWER 13–13 As discussed in the text, 30 ATP molecules are produced from each glucose molecule that is oxidized according to the reaction $C_6H_{12}O_6$ (glucose) + $6O_2 \rightarrow 6CO_2 + 6H_2O$ + energy. Thus, one O_2 molecule is consumed for every five ATP molecules produced. The cell therefore consumes 2×10^8 O_2 molecules/min, which corresponds to the consumption of 3.3×10^{-16} moles (= $[2 \times 10^8]/[6 \times 10^{23}]$) or 7.4×10^{-15} liter (= $3.3 \times 10^{-16} \times 22.4$) each minute. The volume of the cell is 10^{-15} m^3 [= $(10^{-5})^3$], which is 10^{-12} liter. The cell therefore consumes about 0.7% (= $100 \times 7 \times 10^{-15}/10^{-12}$) of its volume of O_2 gas every minute, or its own volume of O_2 gas in 2 hours and 15 minutes.

ANSWER 13–14 The reactions each have negative ΔG values and are therefore energetically favorable (see Figure A13–14 for energy diagrams).

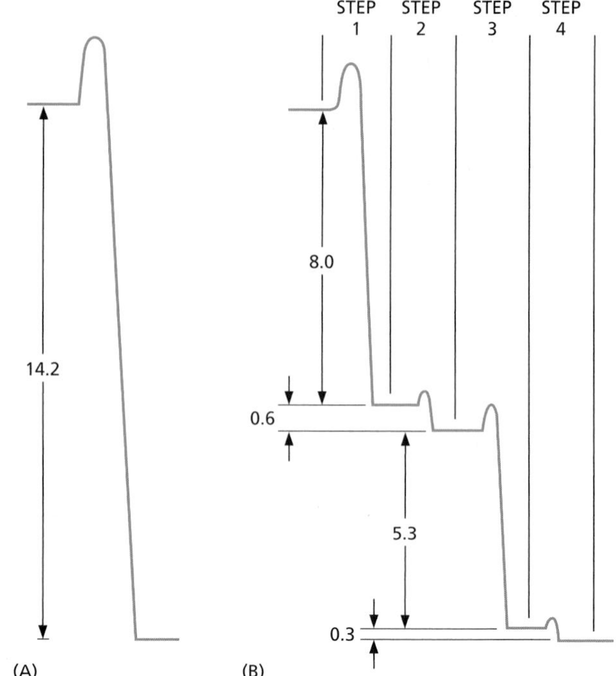

Figure A13–14

ANSWER 13–15

A. Pyruvate is converted to acetyl CoA, and the labeled ^{14}C atom is released as $^{14}CO_2$ gas (see Figure 13–8B).

B. By following the ^{14}C-labeled atom through every reaction in the cycle, shown in Panel 13–2 (pp. 442–443), you find that the added ^{14}C label would be quantitatively recovered in oxaloacetate. The analysis also reveals, however, that it is no longer in the keto group but in the methylene group of oxaloacetate (Figure A13–15).

ANSWER 13–16 In the presence of molecular oxygen, oxidative phosphorylation converts most of the cellular NADH to NAD^+. Since fermentation requires NADH, it is severely inhibited by the availability of oxygen gas.

Chapter 14

ANSWER 14–1 By making membranes permeable to protons, DNP collapses—or at very small concentrations diminishes—the proton gradient across the inner mitochondrial membrane. Cells continue to oxidize food molecules to feed high-energy electrons into the electron-transport chain, but H^+ ions pumped across the membrane flow back into the mitochondria in a futile cycle. As a result, the energy of the electrons cannot be tapped to drive ATP synthesis, and instead is released as heat. Patients who have been given small doses of DNP lose weight because their fat reserves are used more rapidly to feed the electron-transport chain, and the whole process simply "wastes" energy.

A similar mechanism of heat production is used by a specialized tissue composed of brown fat cells, which is abundant in newborn humans and in hibernating animals. These cells are packed with mitochondria that leak part of their H^+ gradient futilely back across the membrane for the sole purpose of warming up the organism. These cells are brown because they are packed with mitochondria, which contain high concentrations of pigmented proteins, such as cytochromes.

ANSWER 14–2 The inner mitochondrial membrane is the site of oxidative phosphorylation, and it produces most of the cell's ATP. Cristae are portions of the mitochondrial inner membrane that are folded inward. Mitochondria that have a higher density of cristae have

radioactive oxaloacetate added to the extract	radioactive oxaloacetate isolated after one turn of citric acid cycle
COO^- – $^{14}C{=}O$ – CH_2 – COO^-	COO^- – $C{=}O$ – $^{14}CH_2$ – COO^-

Figure A13–15

a larger area of inner membrane and therefore a greater capacity to carry out oxidative phosphorylation. Heart muscle expends a lot of energy during its continuous contractions, whereas skin cells have a smaller energy demand. An increased density of cristae therefore increases the ATP-production capacity of the heart muscle cell. This is a remarkable example of how cells adjust the abundance of their individual components according to need.

ANSWER 14–3

A. The DNP collapses the electrochemical proton gradient completely. H^+ ions that are pumped to one side of the membrane flow back freely, and therefore no energy can be stored across the membrane.

B. An electrochemical gradient is made up of two components: a concentration gradient and an electrical potential. If the membrane is made permeable to K^+ with nigericin, K^+ will be driven into the matrix by the electrical potential of the inner membrane (negative inside, positive outside). The influx of positively charged K^+ will abolish the membrane's electrical potential. In contrast, the concentration component of the H^+ gradient (the pH difference) is unaffected by nigericin. Therefore, only part of the driving force that makes it energetically favorable for H^+ ions to flow back into the matrix is lost.

ANSWER 14–4

A. Such a turbine running in reverse is an electrically driven water pump, which is analogous to what the ATP synthase becomes when it uses the energy of ATP hydrolysis to pump protons against their electrochemical gradient across the inner mitochondrial membrane.

B. The ATP synthase should stall when the energy that it can draw from the proton gradient is just equal to the ΔG required to make ATP; at this equilibrium point there will be neither net ATP synthesis nor net ATP consumption.

C. As the cell uses up ATP, the ATP/ADP ratio in the matrix falls below the equilibrium point just described, and ATP synthase uses the energy stored in the proton gradient to synthesize ATP in order to restore the original ATP/ADP ratio. Conversely, when the electrochemical proton gradient drops below that at the equilibrium point, ATP synthase uses ATP in the matrix to restore this gradient.

ANSWER 14–5 An electron pair causes 10 H^+ to be pumped across the membrane when passing from NADH to O_2 through the three respiratory complexes. Four H^+ are needed to make each ATP: three for synthesis from ADP and one for ATP export to the cytosol. Therefore, 2.5 ATP molecules are synthesized from each NADH molecule.

ANSWER 14–6 One can describe four essential roles for the proteins in the process. First, the chemical environment provided by a protein's amino acid side chains sets the redox potential of each Fe ion such that electrons can be passed in a defined order from one component to the next, giving up their energy in small steps and becoming more firmly bound as they proceed. Second, the proteins position the Fe ions so that the electrons can move efficiently between them. Third, the proteins prevent electrons from skipping an intermediate step; thus, as we have learned for other enzymes (discussed in Chapter 4), they channel the electron flow along a defined path. Fourth, the proteins couple the movement of the electrons down their energy ladder to the pumping of protons across the membrane, thereby harnessing the energy that is released and storing it in a proton gradient that is then used for ATP production.

ANSWER 14–7 It would not be productive to use the same carrier in two steps. If ubiquinone, for example, could transfer electrons directly to the cytochrome oxidase, the cytochrome b-c_1 complex would often be skipped when electrons are collected from NADH dehydrogenase. Given the large difference in redox potential between ubiquinone and cytochrome oxidase, a large amount of energy would be released as heat and thus be wasted. Electron transfer directly between NADH dehydrogenase and cytochrome *c* would similarly allow the cytochrome b-c_1 complex to be bypassed.

ANSWER 14–8 Protons pumped across the inner mitochondrial membrane into the intermembrane space equilibrate with the cytosol, which functions as a huge H^+ sink. Both the mitochondrial matrix and the cytosol support many metabolic reactions that require a pH around neutrality. The H^+ concentration difference, ΔpH, that can be achieved between mitochondrial matrix and cytosol is therefore relatively small (less than one pH unit). Much of the energy stored in the mitochondrial electrochemical proton gradient is instead due to the electrical potential of the membrane (see Figure 14–10).

In contrast, chloroplasts have a smaller, dedicated compartment into which H^+ ions are pumped. Much higher concentration differences can be achieved (up to a thousandfold, or 3 pH units), and much of the energy stored in the thylakoid H^+ gradient is due to the H^+ concentration difference between the thylakoid space and the stroma.

ANSWER 14–9 NADH and NADPH differ by the presence of a single phosphate group. That phosphate gives NADPH a slightly different shape from NADH, which allows these molecules to be recognized by different enzymes, and thus to deliver their electrons to different destinations. Such a division of labor is useful because NADPH tends to be involved in biosynthetic reactions, where high-energy electrons are used to produce energy-rich biological molecules. NADH, on the other hand, is involved in reactions that oxidize energy-rich food molecules to produce ATP. Inside the cell the ratio of NAD^+ to NADH is kept high, whereas the ratio of $NADP^+$ to NADPH is kept low. This provides plenty of NAD^+ to act as an oxidizing agent and plenty of NADPH to act as a reducing agent—as required for their special roles in catabolism and anabolism, respectively.

ANSWER 14–10

A. Photosynthesis produces sugars, most importantly sucrose, that are transported from the photosynthetic cells through the sap to root cells. There, the sugars are oxidized by glycolysis in the root cell cytoplasm and by oxidative phosphorylation in the root cell mitochondria to produce ATP, as well as being used as the building blocks for many other metabolites.

B. Mitochondria are required even during daylight hours in chloroplast-containing cells to supply the cell with ATP derived by oxidative phosphorylation. Glyceraldehyde 3-phosphate made by photosynthesis in chloroplasts moves to the cytosol and is eventually used as a source of energy to drive ATP production in mitochondria.

ANSWER 14–11 All statements are correct.

A. This is a necessary condition. If it were not true, electrons could not be removed from water and the reaction that splits water molecules ($H_2O \rightarrow 2H^+ + \frac{1}{2}O_2 + 2e^-$) would not occur.
B. This transfer allows the energy of the photon to be harnessed as energy that can be utilized in chemical conversions.
C. It can be argued that this is one of the most important obstacles that had to be overcome during the evolution of photosynthesis: partially reduced oxygen molecules, such as the superoxide radical O_2^-, are dangerously reactive and will attack and destroy almost any biologically active molecule. These intermediates therefore have to remain tightly bound to the metals in the active site of the enzyme until all four electrons have been removed from two water molecules. This requires the sequential capture of four photons by the same reaction center.

ANSWER 14–12

A. True. NAD^+ and quinones are examples of compounds that do not have metal ions but can participate in electron transfer.
B. False. The potential is due to protons (H^+) that are pumped across the membrane from the matrix to the intermembrane space. Electrons remain bound to electron carriers in the inner mitochondrial membrane.
C. True. Both components add to the driving force that makes it energetically favorable for H^+ to flow back into the matrix.
D. True. Both move rapidly in the plane of the membrane.
E. False. Not only do plants need mitochondria to make ATP in cells that do not have chloroplasts, such as root cells, but mitochondria make most of the cytosolic ATP in all plant cells.
F. True. Chlorophyll's physiological function requires it to absorb light; heme just happens to be a colored compound from which blood derives its red color.
G. False. Chlorophyll absorbs light and transfers energy in the form of an energized electron, whereas the iron in heme is a simple electron carrier.
H. False. Most of the dry weight of a tree comes from carbon derived from the CO_2 that has been fixed during photosynthesis.

ANSWER 14–13 It takes three protons. The precise value of the ΔG for ATP synthesis depends on the concentrations of ATP, ADP, and P_i (as described in Chapter 3). The higher the ratio of the concentration of ATP to ADP, the more energy it takes to make additional ATP. The lower value of 11 kcal/mole therefore applies to conditions where cells have expended a lot of energy and have therefore decreased the normal ATP/ADP ratio.

ANSWER 14–14 If no O_2 is available, all components of the mitochondrial electron-transport chain will accumulate in their *reduced* form. This is the case because electrons derived from NADH enter the chain but cannot be transferred to O_2. The electron-transport chain therefore stalls with all of its components in the reduced form. If O_2 is suddenly added again, the electron carriers in cytochrome oxidase will become *oxidized before* those in NADH dehydrogenase. This is true because, after O_2 addition, cytochrome oxidase will donate its electrons directly to O_2, thereby becoming oxidized. A wave of increasing oxidation then passes backward with time from cytochrome oxidase through the components of the electron-transport chain, as each component regains the opportunity to pass on its electrons to downstream components.

ANSWER 14–15 As oxidized ubiquinone becomes reduced, it picks up two electrons but also two protons from water (Figure 14–19). Upon oxidation, these protons are released. If reduction occurs on one side of the membrane and oxidation at the other side, a proton is pumped across the membrane for each electron transported. Electron transport by ubiquinone thereby contributes directly to the generation of the H^+ gradient.

ANSWER 14–16 Photosynthetic bacteria and plant cells use the electrons derived in the reaction $2H_2O \rightarrow 4e^- + 4H^+ + O_2$ to reduce $NADP^+$ to NADPH, which is then used to produce useful metabolites. If the electrons were used instead to produce H_2 in addition to O_2, the cells would lose any benefit they derive from carrying out the reaction, because the electrons could not take part in metabolically useful reactions.

ANSWER 14–17

A. The switch in solutions creates a pH gradient across the thylakoid membrane. The flow of H^+ ions down its electrochemical potential drives ATP synthase, which converts ADP to ATP.
B. No light is needed, because the H^+ gradient is established artificially without a need for the light-driven electron-transport chain.
C. Nothing. The H^+ gradient would be in the wrong direction; ATP synthase would not work.
D. The experiment provided early supporting evidence for the chemiosmotic model by showing that an H^+ gradient alone is sufficient to drive ATP synthesis.

ANSWER 14–18

A. When the vesicles are exposed to light, H^+ ions (derived from H_2O) pumped into the vesicles by the bacteriorhodopsin flow back out through the ATP synthase, causing ATP to be made in the solution surrounding the vesicles in response to light.
B. If the vesicles are leaky, no H^+ gradient can form and thus ATP synthase cannot work.
C. Using components from widely divergent organisms can be a very powerful experimental tool. Because the two proteins come from such different sources, it is very unlikely that they form a direct functional interaction. The experiment therefore strongly suggests that electron transport and ATP synthesis are separate events. This approach is therefore a valid one.

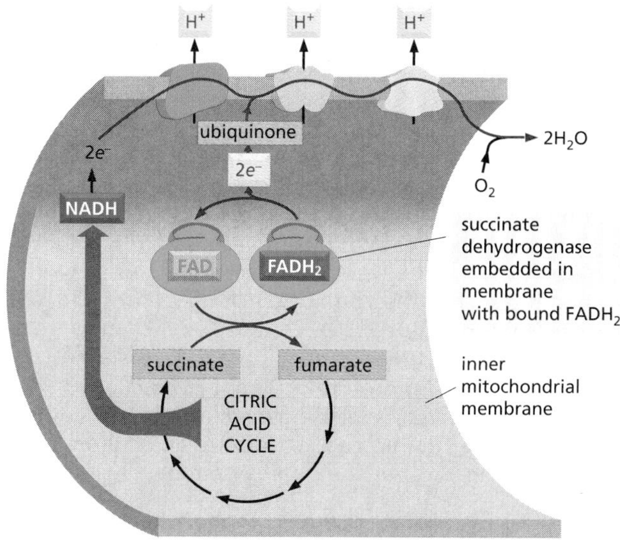

Figure A14–19

ANSWER 14–19 The redox potential of $FADH_2$ is too low to transfer electrons to the NADH dehydrogenase complex, but high enough to transfer electrons to ubiquinone (Figure 14–20). Therefore, electrons from $FADH_2$ can enter the electron-transport chain only at this step (Figure A14–19). Because the NADH dehydrogenase complex is bypassed, fewer H^+ ions are pumped across the membrane and less ATP is made. This example shows the versatility of the electron-transport chain. The ability to use vastly different sources of electrons from the environment to feed electron transport is thought to have been an essential feature in the early evolution of life.

ANSWER 14–20 If these bacteria used a proton gradient to make their ATP in a fashion analogous to that in other bacteria (that is, fewer protons inside than outside), they would need to raise their cytoplasmic pH even higher than that of their environment (pH 10). Cells with a cytoplasmic pH greater than 10 would not be viable. These bacteria must therefore use gradients of ions other than H^+, such as Na^+ gradients, in the chemiosmotic coupling between electron transport and an ATP synthase.

ANSWER 14–21 Statements A and B are accurate. Statement C is incorrect, because the chemical reactions that are carried out in each cycle are completely different, even though the net effect is the same as that expected for simple reversal.

ANSWER 14–22 This experiment would suggest a two-step model for ATP synthase function. According to this model, the flow of protons through the base of the synthase drives rotation of the head, which in turn causes ATP synthesis. In their experiment, the authors have succeeded in uncoupling these two steps. If rotating the head mechanically is sufficient to produce ATP in the absence of any applied proton gradient, the ATP synthase is a protein machine that indeed functions like a "molecular turbine." This would be a very exciting experiment, indeed, because it would directly demonstrate the relationship between mechanical movement and enzymatic activity. There is no doubt that it should be published and that it would become a "classic."

ANSWER 14–23 Only under condition (E) is electron transfer observed, with cytochrome *c* becoming reduced. A portion of the electron-transport chain has been reconstituted in this mixture, so that electrons can flow in the energetically favored direction from reduced ubiquinone to the cytochrome b-c_1 complex to cytochrome *c*. Although energetically favorable, the transfer in (A) cannot occur spontaneously in the absence of the cytochrome b-c_1 complex to catalyze this reaction. No electron flow occurs in the other experiments, whether the cytochrome b-c_1 complex is present or not: in experiments (B) and (F) both ubiquinone and cytochrome *c* are oxidized; in experiments (C) and (G) both are reduced; and in experiments (D) and (H) electron flow is energetically disfavored because reduced cytochrome c has a lower free energy than oxidized ubiquinone.

Chapter 15

ANSWER 15–1 Although the nuclear envelope forms one continuous membrane, it has specialized regions that contain special proteins and have a characteristic appearance. One such specialized region is the inner nuclear membrane. Membrane proteins can indeed diffuse between the inner and outer nuclear membranes, at the connections formed around the nuclear pores. Those proteins with particular functions in the inner membrane, however, are usually anchored there by their interaction with other components such as chromosomes and the nuclear lamina (a protein meshwork underlying the inner nuclear membrane that helps give structural integrity to the nuclear envelope).

ANSWER 15–2 Eucaryotic gene expression is more complicated than procaryotic gene expression. In particular, procaryotic cells do not have introns that interrupt the coding sequences of their genes, so that an mRNA can be translated immediately after it is transcribed, without a need for further processing (discussed in Chapter 7). In fact, in procaryotic cells, ribosomes start translating most mRNAs before transcription is finished. This would have disastrous consequences in eucaryotic cells, because most RNA transcripts have to be spliced before they can be translated. The nuclear envelope separates the transcription and translation processes in space and time: a primary RNA transcript is held in the nucleus until it is properly processed to form an mRNA, and only then is it allowed to leave the nucleus so that ribosomes can translate it.

ANSWER 15–3 An mRNA molecule is attached to the ER membrane by the ribosomes translating it. This ribosome population, however, is not static; the mRNA is continuously moved through the ribosome. Those ribosomes that have finished translation dissociate from the 3′ end of the mRNA and from the ER membrane, but the mRNA itself remains bound by other ribosomes, newly recruited from the cytosolic pool, that have attached to the 5′ end of the mRNA and are still translating the mRNA. Depending on its length, there are about 10–20 ribosomes attached to each membrane-bound mRNA molecule.

ANSWER 15–4

A. The internal signal sequence functions as a membrane anchor, as shown in Figure 15–17. Because there is no stop-transfer sequence, however, the C-terminal end of the protein continues to be translocated into the ER lumen. The resulting protein therefore has its N-terminal domain in the cytosol, followed by a single transmembrane segment, and a C-terminal domain in the ER lumen (Figure A15–4A).
B. The N-terminal signal sequence initiates translocation of the N-terminal domain of the protein until translocation is stopped by the stop-transfer sequence. A cytosolic domain is synthesized until the start-transfer sequence initiates translocation again. The situation now resembles that described in (A), and the C-terminal domain of the protein is translocated into the lumen of the ER. The resulting protein therefore spans the membrane twice. Both its N-terminal and C-terminal domains are in the ER lumen, and a loop domain between the two transmembrane regions is exposed in the cytosol (Figure A15–4B).
C. It would need a cleaved signal sequence, followed by an internal stop-transfer sequence, followed by pairs of start- and stop-transfer sequences (Figure A15–4C).

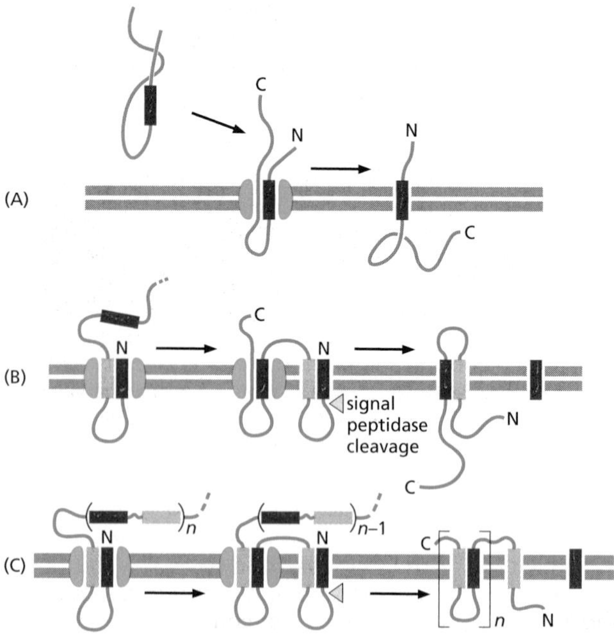

Figure A15–4

These examples demonstrate that complex protein topologies can be achieved by simple variations and combinations of the two basic mechanisms shown in Figures 15–16 and 15–17.

ANSWER 15–5

A. Clathrin coats cannot assemble in the absence of adaptins that link the clathrin to the membrane. At high clathrin concentrations and under the appropriate ionic conditions, clathrin cages assemble in solution, but they are empty shells, lacking other proteins, and they contain no membrane. This shows that the information to form clathrin baskets is contained in the clathrin molecules themselves, which are therefore able to self-assemble.
B. Without clathrin, adaptins still bind to receptors in the membrane, but no clathrin coat can form and thus no clathrin-coated pits or vesicles are produced.
C. Deeply invaginated clathrin-coated pits form on the membrane, but they do not pinch off to form closed vesicles (see Figure A15–13).
D. Procaryotic cells do not perform endocytosis. A procaryotic cell therefore does not contain any receptors with appropriate cytosolic tails that could mediate adaptin binding. Therefore, no clathrin can bind and no clathrin coats can assemble.

ANSWER 15–6 The preassembled sugar chain allows better quality control. The assembled oligosaccharide chains can be checked for accuracy before they are added to the protein; if a mistake were made in adding sugars individually to the protein, the whole protein would have to be discarded. Because far more energy is used in building a protein than in building a short oligosaccharide chain, this is a much more economical strategy. Also, once a sugar tree is added to a protein, it is more difficult for enzymes to modify its branches, compared with modifying them on the free sugar tree. This difficulty becomes apparent as the protein moves to the cell surface: although sugar chains are continually modified by enzymes in various compartments of the secretory pathway, these modifications are often incomplete and result in considerable heterogeneity of the glycoproteins that leave the cell. This heterogeneity is largely due to the restricted access that the enzymes have to the sugar trees attached to the surface of proteins. The heterogeneity also explains why glycoproteins are more difficult to study and purify than nonglycosylated proteins.

ANSWER 15–7 Aggregates of the secretory proteins would form in the ER, just as they do in the *trans* Golgi network. As the aggregation is specific for secretory proteins, ER proteins would be excluded from the aggregates. The aggregates would eventually be degraded.

ANSWER 15–8 Transferrin without Fe bound does not interact with its receptor and circulates in the bloodstream until it catches an Fe ion. Once iron is bound, the iron–transferrin complex can bind to the transferrin receptor on the surface of a cell and be endocytosed. Under the acidic conditions of the endosome, the transferrin releases its iron, but the transferrin remains bound to the transferrin receptor, which is recycled back to the cell surface, where it encounters the neutral pH environment of the blood. The neutral pH causes the receptor to release the transferrin into the circulation, where it can pick up another Fe ion to repeat the cycle. The iron released in the endosome, like the LDL in Figure 15–33, moves on to lysosomes, from where it is transported into the cytosol.

The system allows cells to take up iron efficiently even though the concentration of iron in the blood is extremely low. The iron bound to transferrin is concentrated at the cell surface by binding to transferrin receptors; it becomes further concentrated in clathrin-coated pits, which collect the transferrin receptors. In this way, transferrin cycles between the blood and endosomes, delivering the iron that cells need to grow.

ANSWER 15–9

A. True.
B. False. The signal sequences that direct proteins to the ER contain a core of eight or more hydrophobic amino acids. The sequence shown here contains many hydrophilic amino acid side chains, including the charged amino acids His, Arg, Asp, and Lys, and the uncharged hydrophilic amino acids Gln and Ser.
C. True. Otherwise they could not dock at the correct target membrane or recruit a fusion complex to a docking site.
D. True.
E. True. Lysosomal proteins are selected in the *trans* Golgi network and packaged into transport vesicles that deliver them to the late endosome. If not selected, they would enter by default into transport vesicles that move constitutively to the cell surface.
F. False. Lysosomes also digest internal organelles by autophagy.
G. False. Mitochondria do not participate in vesicular transport, and therefore *N*-linked glycoproteins, which are exclusively assembled in the ER, cannot be transported to mitochondria.

ANSWER 15–10 They must contain a nuclear localization signal as well. Proteins with nuclear export signals shuttle between the nucleus and the cytosol. An example is the A1 protein, which binds to mRNAs in the nucleus and guides them through the nuclear pores. Once in the cytosol, a nuclear localization signal ensures that the A1 protein is reimported so that it can participate in the export of further mRNAs.

ANSWER 15–11 Influenza virus enters cells by endocytosis and is delivered to endosomes, where it encounters an acidic pH that activates its fusion protein. The viral membrane then fuses with the membrane of the endosome, releasing the viral genome into the cytosol (Figure A15–11). NH_3 is a small molecule that readily penetrates membranes. Thus, it can enter all intracellular compartments, including endosomes, by diffusion. Once in a compartment that has an acidic pH, NH_3 binds H^+ to form NH_4^+, which is a charged ion and therefore cannot cross the membrane by diffusion. NH_4^+ ions therefore accumulate in acidic compartments, raising their pH. When the pH of the endosome is raised, viruses are still endocytosed, but because the viral fusion protein cannot be activated, the virus cannot enter the cytosol. Remember this the next time you have the flu and have access to a stable.

EXTRACELLULAR SPACE
plasma membrane
CYTOSOL
endosomal membrane
H^+
H^+
endocytosis
activation of viral fusion protein
fusion of viral and endosomal membranes
release of viral genome into cell

Figure A15–11

ANSWER 15–12

A. The problem is that vesicles having two different kinds of v-SNAREs in their membrane could dock on either of two different membranes.
B. The answer to this puzzle is currently not known, but we can predict that cells must have ways of turning the docking ability of SNAREs on and off. This may be achieved through other proteins that are, for example, co-packaged in the ER with SNAREs into transport vesicles and facilitate the interactions of the correct v-SNARE with the t-SNARE in the *cis* Golgi network.

ANSWER 15–13 Synaptic transmission involves the release of neurotransmitters by exocytosis. During this event, the membrane of the synaptic vesicle fuses with the plasma membrane of the nerve terminals. To make new synaptic vesicles, membrane must be retrieved from the plasma membrane by endocytosis. This endocytosis step is blocked if dynamin is defective, as the protein is required to pinch off the clathrin-coated endocytic vesicles. The first clue to deciphering the role of dynamin came from electron micrographs of synapses of the mutant flies (Figure A15–13). Note that there are many flasklike invaginations of the plasma membrane, representing deeply invaginated clathrin-coated pits that cannot pinch off. The collars visible around the necks of these invaginations are made of mutant dynamin.

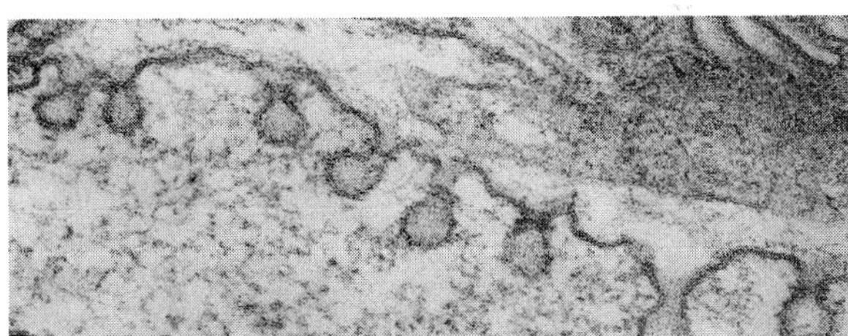

From J.H. Koenig and K. Ikeda, *J. Neurosci.* 9:3844–3860, 1989. With permission from The Society for Neuroscience.

Figure A15–13

ANSWER 15–14 The first two sentences are correct. The third is not. It should read: "Because the contents of the lumen of the ER or any other compartment in the secretory or endocytic pathways never mix with the cytosol, proteins that enter these pathways will never need to be imported again." When the nuclear envelope breaks down in mitosis, i.e., retracts from the chromatin, its membrane proteins intermix with ER membrane proteins, but its contents remain separated from the cytosol by an intact membrane.

ANSWER 15–15 The protein is translocated into the ER. Its ER signal sequence is recognized as soon as it emerges from the ribosome. The ribosome then becomes bound to the ER membrane, and the growing polypeptide chain is transferred through the ER translocation channel. The nuclear localization sequence is therefore never exposed to the cytosol. It will never encounter nuclear import receptors, and the protein will not enter the nucleus.

ANSWER 15–16 (1) Proteins are imported into the nucleus after they have been synthesized, folded, and, if appropriate, assembled into complexes. In contrast, unfolded polypeptide chains are translocated into the ER as they are being made by the ribosomes. Ribosomes are assembled in the nucleus yet function in the cytosol, and the enzyme complexes that catalyze RNA transcription and splicing are assembled in the cytosol yet function in the nucleus. Thus, both ribosomes and these enzyme complexes need to be transported through the nuclear pores intact. (2) Nuclear pores are gates, which are always open to small molecules; in contrast, translocation channels in the ER membrane are normally closed (as indicated by the "plug" in Figure 15–15), and open only after the ribosome has attached to the membrane and the translocating polypeptide chain has sealed the channel from the cytosol. It is important that the ER membrane remain impermeable to small molecules during the translocation process, as the ER is a major store for Ca^{2+} in the cell, and Ca^{2+} release into the cytosol must be tightly controlled (discussed in Chapter 16). (3) Nuclear localization signals are not cleaved off after protein import into the nucleus; in contrast, ER signal peptides are usually cleaved off. Nuclear localization signals are needed to repeatedly reimport nuclear proteins after they have been released into the cytosol during mitosis, when the nuclear envelope breaks down.

ANSWER 15–17 The transient intermixing of nuclear and cytosolic contents during mitosis supports the idea that the nuclear interior and the cytosol are indeed evolutionarily related. In fact, one can consider the nucleus as a subcompartment of the cytosol that has become surrounded by the nuclear envelope, with access only through the nuclear pores.

ANSWER 15–18 The actual explanation is that the single amino acid change causes the protein to misfold slightly so that, although it is still active as a protease inhibitor, it is prevented by chaperone proteins in the ER from exiting this organelle. It therefore accumulates in the ER lumen and is eventually degraded. Alternative interpretations might have been that (1) the mutation affects the stability of the protein in the bloodstream so that it is degraded much faster in the blood than the normal protein, or (2) the mutation inactivates the ER signal sequence and prevents the protein from entering the ER. (3) Another explanation could have been that the mutation altered the sequence to create an ER retention signal, which would have retained the mutant protein in the ER. One could distinguish between these possibilities by using fluorescently tagged antibodies against the protein or by expressing the protein as a fusion with GFP to follow its transport in the cells (see Panel 4–6, p. 167, and How We Know, pp. 520–521).

ANSWER 15–19 Critique: "Dr. Outonalimb proposes to study the biosynthesis of forgettin, a protein of significant interest. The main hypothesis on which this proposal is based, however, requires further support. In particular, it is questionable whether forgettin is indeed a secreted protein, as proposed. ER signal sequences are normally found at the N-terminus. C-terminal hydrophobic sequences will be exposed outside the ribosome only after protein synthesis has already terminated and can therefore not be recognized by an SRP during translation. It is therefore unlikely that forgettin will be translocated by an SRP-dependent mechanism; it is more likely that it will remain in the cytosol. Dr. Outonalimb should take these considerations into account when submitting a revised application."

ANSWER 15–20 The Golgi apparatus may have evolved from specialized patches of ER membrane. These regions of the ER might have pinched off, forming a new compartment (Figure A15–20), which still communicates with the ER by vesicular transport. For the newly evolved Golgi compartment to be useful, transport vesicles would also have to have evolved.

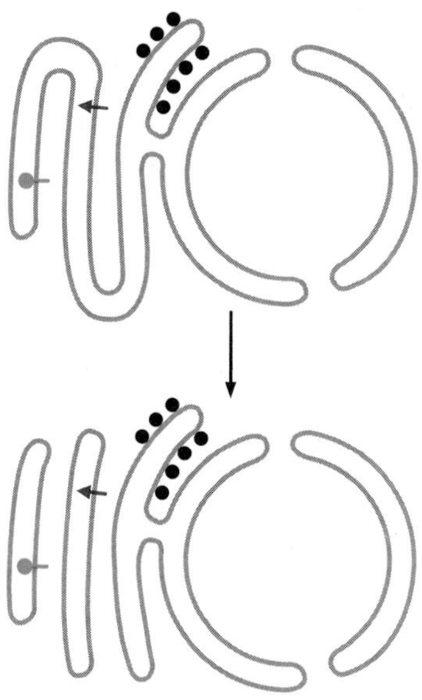

Figure A15–20

ANSWER 15–21 This is a chicken-and-egg question. In fact, the situation never arises in present-day cells, although it must have posed a considerable problem for the first cells that evolved. New cell membranes are made by expansion of existing membranes, and the ER is never made *de novo*. There will always be an existing piece of ER with translocation channels to integrate new translocation channels. Inheritance is therefore not limited to the propagation of the genome; a cell's organelles must also be passed from generation to generation. In fact, the ER translocation channels can be traced back to structurally related translocation channels in the procaryotic plasma membrane.

ANSWER 15–22

A. Extracellular space
B. Cytosol
C. Plasma membrane
D. Clathrin coat
E. Membrane of deeply invaginated clathrin-coated pit
F. Captured cargo particles
G. Lumen of deeply invaginated clathrin-coated pit

Chapter 16

ANSWER 16–1 Most paracrine signaling molecules are very short-lived after they are released from a signaling cell: they are either degraded by extracellular enzymes or are rapidly taken up by neighboring target cells. In addition, some become attached to the extracellular matrix and are thus prevented from diffusing too far.

ANSWER 16–2 Polar groups are hydrophilic, so cholesterol, with only one polar –OH group, would be too hydrophobic to be an effective hormone. Because it is virtually insoluble in water, it could not move readily as a messenger from one cell to another via the extracellular fluid.

ANSWER 16–3 The protein could be an enzyme that produces a large number of small intracellular signaling molecules such as cyclic AMP or cyclic GMP. Or, it could be an enzyme that modifies a large number of intracellular target proteins—for example, by phosphorylation.

ANSWER 16–4 In the case of the steroid hormone receptor, a one-to-one complex of steroid and receptor binds to DNA to activate or inactivate gene transcription; there is thus no amplification between ligand binding and transcriptional regulation. Amplification occurs later, because transcription of a gene gives rise to many mRNAs, each of which is translated to give many copies of the protein it encodes (discussed in Chapter 7). For the ion-channel–linked receptors, a single ion channel will let through thousands of ions in the time it remains open; this serves as the amplification step in this type of signaling system.

ANSWER 16–5 The mutant G protein would be almost continuously activated, because GDP would dissociate spontaneously, allowing GTP to bind even in the absence of an activated GPCR. The consequences for the cell would therefore be similar to those caused by cholera toxin, which modifies the α subunit of G_s so that it cannot hydrolyze GTP to shut itself off. In contrast to the cholera toxin case, however, the mutant G protein would not stay permanently activated: it would switch itself off normally, but then it would instantly become activated again as the GDP dissociated and GTP re-bound.

ANSWER 16–6 Rapid breakdown keeps the intracellular cyclic AMP concentrations low. The lower the cAMP levels are, the larger and faster the increase achieved upon activation of adenylyl cyclase, which makes new cyclic AMP. If you have $100 in the bank and you deposit another $100, you have doubled your wealth; if you have only $10 to start with and you deposit $100, you have increased your wealth tenfold, a much larger proportional increase resulting from the same deposit.

ANSWER 16–7 Recall that the plasma membrane constitutes a rather small area compared with the total membrane surfaces in a cell (discussed in Chapter 15). The endoplasmic reticulum is especially abundant and spans the entire volume of the cell as a vast network of membrane tubes and sheets. The Ca^{2+} stored in the endoplasmic reticulum can therefore be released throughout the cytosol. This is important because the rapid clearing of Ca^{2+} ions from the cytosol by Ca^{2+} pumps prevents Ca^{2+} from diffusing any significant distance in the cytosol.

ANSWER 16–8 Each reaction involved in the amplification scheme must be turned off to reset the signaling pathway to a resting level. Each of these off switches is equally important.

ANSWER 16–9 Because each antibody has two antigen-binding sites, it can cross-link the receptors and cause them to cluster on the cell surface. This clustering is likely to activate RTKs, which are usually activated by dimerization. For RTKs, clustering allows the individual kinase domains of the receptors to phosphorylate adjacent receptors in the cluster. The activation of GPCRs is more complicated, because the ligand has to induce a particular conformational change; only very special antibodies mimic receptor ligands sufficiently well to induce the conformational change that activates a GPCR.

ANSWER 16–10 The more steps there are in an intracellular signaling pathway, the more places the cell has to regulate the pathway, amplify the signal, integrate signals from different pathways, and spread the signal along divergent paths (see Figure 16–13).

ANSWER 16–11

A. True. Acetylcholine, for example, decreases the beating of heart muscle cells by binding to a GPCR and stimulates the contraction of skeletal muscle cells by binding to a different acetylcholine receptor, which is an ion-channel-coupled receptor.
B. False. Acetylcholine is short-lived and exerts its effects locally. Indeed, the consequences of prolonging its lifetime can be disastrous. Compounds that inhibit the enzyme acetylcholinesterase, which normally breaks down acetylcholine at a nerve–muscle synapse, are extremely toxic: for example, the nerve gas sarin, used in chemical warfare, is an acetylcholinesterase inhibitor.
C. True. Nucleotide-free $\beta\gamma$ complexes can activate ion channels, and GTP-bound α subunits can activate enzymes. The GDP-bound form of trimeric G proteins is the inactive state.
D. True. The inositol phospholipid that is cleaved to produce IP_3 contains three phosphate groups, one of which links the sugar to the diacylglycerol lipid. IP_3 is generated by a simple hydrolysis reaction (see Figure 16–25).
E. False. Calmodulin senses but does not regulate intracellular Ca^{2+} levels.
F. True. See Figure 16–39.
G. True. See Figure 16–30.

ANSWER 16–12

1. You would expect a high background level of Ras activity, because Ras cannot be turned off efficiently.
2. Because some Ras molecules are already GTP-bound, Ras activity in response to an extracellular signal would be greater than normal, but this activity would be liable to saturate when all Ras molecules are converted to the GTP-bound form.
3. The response to a signal would be much less rapid, because the signal-dependent increase in GTP-

bound Ras would occur over an elevated background of preexisting GTP-bound Ras (see Question 16–6).
4. The increase in Ras activity in response to a signal would also be prolonged compared to the response in normal cells.

ANSWER 16–13
A. Both types of signaling can occur over a long range: neurons can send action potentials along very long axons (think of the axons in the neck of a giraffe, for example), and hormones are carried via the bloodstream throughout the organism. Because neurons secrete large amounts of neurotransmitters at a synapse, a small, well-defined space between two cells, the concentrations of these signal molecules are high; neurotransmitter receptors, therefore, need to bind to neurotransmitters with only low affinity. Hormones, in contrast, are vastly diluted in the bloodstream, where they circulate at often minuscule concentrations; hormone receptors therefore generally bind their hormone with extremely high affinity.
B. Whereas neuronal signaling is a private affair, with one neuron talking to a select group of target cells through specific synaptic connections, endocrine signaling is a public announcement, with any target cell with appropriate receptors able to respond to the hormone in the blood. Neuronal signaling is very fast, limited only by the speed of propagation of the action potential and the workings of the synapse, whereas endocrine signaling is slower, limited by blood flow and diffusion over larger distances.

ANSWER 16–14
A. There are 100,000 molecules of X and 10,000 molecules of Y in the cell (= rate of synthesis × average lifetime).
B. After one second, the concentration of X will have increased by 10,000 molecules. The concentration of X, therefore, one second after its synthesis is increased, is about 110,000 molecules per cell—which is a 10% increase over the concentration of X before the boost of its synthesis. The concentration of Y will also increase by 10,000 molecules, which, in contrast to X, represents a full twofold increase in its concentration (for simplicity, we can neglect the breakdown in this estimation because X and Y are relatively stable during the one-second stimulation).
C. Because of its larger proportional increase, Y is the preferred signaling molecule. This calculation illustrates the surprising but important principle that the time it takes to switch a signal on is determined by the lifetime of the signaling molecule.

ANSWER 16–15 The information transmitted by a cell-signaling pathway is contained in the *concentration* of the messenger, be it a small molecule or a phosphorylated protein. Therefore, to allow the detection of a change in concentration, the original messenger has to be rapidly destroyed. The shorter the lifetime of the messenger, the faster the system can respond to changes. Human communication relies on messages that are delivered only once and that are generally not interpreted by their abundance but by their *content*. So it is a mistake to kill the human messengers; they can be used more than once.

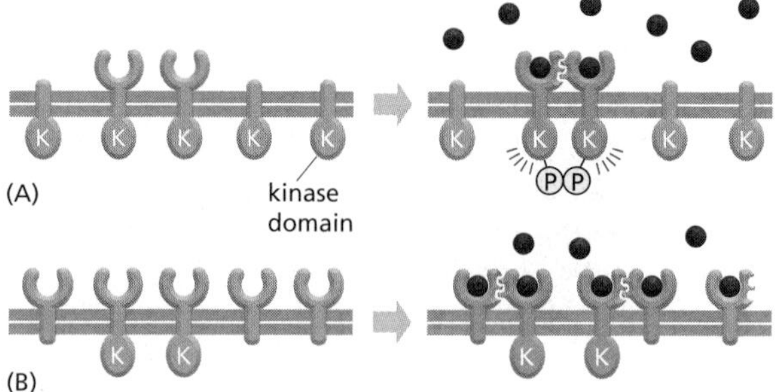

Figure A16–16

ANSWER 16–16
A. The mutant RTK lacking its extracellular ligand-binding domain is inactive. It cannot bind extracellular signals, and its presence has no consequences for the function of the normal RTK (Figure A16–16A).
B. The mutant RTK lacking its intracellular domain is also inactive, but its presence will block signaling by the normal receptors. When a signal molecule binds to either receptor, it will induce their dimerization. Two normal receptors have to come together to activate each other by phosphorylation. In the presence of an excess of mutant receptor, however, normal receptors will usually form mixed dimers, in which their intracellular domain cannot be activated because their partner is a mutant and lacks a kinase domain (Figure A16–16B).

ANSWER 16–17 The statement is correct. Upon ligand binding, transmembrane helices of multispanning receptors, like the GPCRs, shift and rearrange with respect to one another (Figure A16–17A). This conformational change is sensed on the cytosolic side of the membrane because of a change in the arrangement of the cytoplasmic loops. A single transmembrane segment is not sufficient to transmit a signal across the membrane directly; no rearrangements in the membrane are possible upon ligand binding. Upon ligand binding, single-span receptors such as most RTKs tend to dimerize, thereby bringing their intracellular kinase domains into proximity so that they can cross-phosphorylate and activate each other (Figure A16–17B).

ANSWER 16–18 Activation in both cases depends on proteins that catalyze GDP–GTP exchange on the G protein or Ras protein. Whereas activated GPCRs perform this function directly for G proteins, enzyme-linked receptors assemble multiple signaling proteins into a signaling complex when the receptors are activated by phosphorylation; one of these is an adaptor protein that recruits a Ras-activating protein that fulfills this function for Ras.

ANSWER 16–19 Because the cytosolic concentration of Ca^{2+} is so low, an influx of relatively few Ca^{2+} ions leads to large changes in its cytosolic concentration. Thus, a tenfold increase in cytosolic Ca^{2+} can be achieved by raising its concentration into the micromolar range, which would require far fewer ions than would be required to change significantly the cytosolic

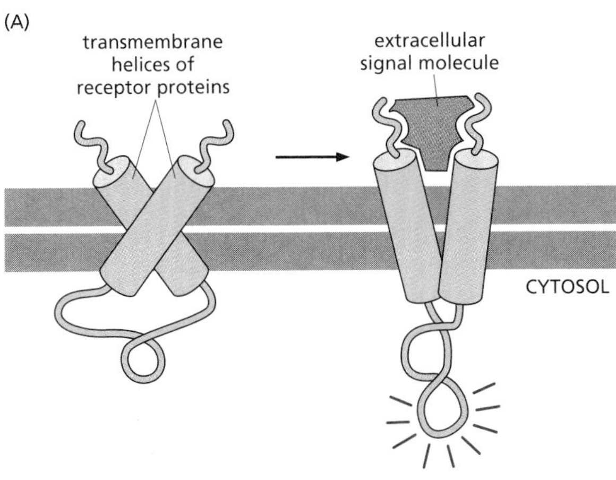

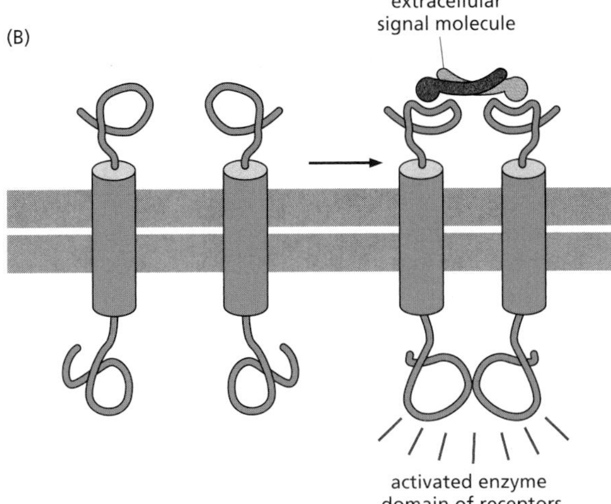

Figure A16–17

concentration of a more abundant ion such as Na^+. In muscle, a greater than tenfold change in cytosolic Ca^{2+} concentration can be achieved in microseconds by releasing Ca^{2+} from the sarcoplasmic reticulum, a task that would be difficult to accomplish if changes in the millimolar range were required.

ANSWER 16–20 In a multicellular organism such as an animal, it is important that cells survive only when and where they are needed. Having cells depend on signals from other cells may be a simple way of ensuring this. A misplaced cell, for example, would probably fail to get the survival signals it needs (as its neighbors would be inappropriate) and would therefore kill itself. This strategy can also help regulate cell numbers: if cell type A depends on a survival signal from cell type B, the number of B cells could control the number of A cells by making a limited amount of the survival signal, so that only a certain number of A cells could survive. There is indeed evidence that such a mechanism does operate to help regulate cell numbers—in both developing and adult tissues (see Figure 18–41).

ANSWER 16–21 Ca^{2+}-activated Ca^{2+} channels create a positive feedback loop: the more Ca^{2+} that is released, the more Ca^{2+} channels open. The Ca^{2+} signal in the cytosol is therefore propagated explosively throughout the entire muscle cell, thereby ensuring that all myosin–actin filaments contract almost synchronously.

ANSWER 16–22 K2 activates K1. If K1 is permanently activated, a response is observed regardless of the status of K2. If the order were reversed, K1 would need to activate K2, which cannot occur because in our example K2 contains an inactivating mutation.

ANSWER 16–23

A. Three examples of extended signaling pathways to the nucleus are (1) extracellular signal → RTK → adaptor protein → Ras-activating protein → MAP kinase kinase kinase → MAP kinase kinase → MAP kinase → transcription regulator; (2) extracellular signal → GPCR → G protein → phospholipase C → IP_3 → Ca^{2+} → calmodulin → CaM-kinase → transcription regulator; (3) extracellular signal → GPCR → G protein → adenylyl cyclase → cyclic AMP → PKA → transcription regulator.
B. Two examples of direct signaling pathways to the nucleus are (1) cytokine → cytokine receptor → JAK kinase → STAT; (2) Delta → Notch → cleaved Notch tail → transcription.

ANSWER 16–24 When PI 3-kinase is activated by an activated RTK, it phosphorylates a specific inositol phospholipid in the plasma membrane. The resulting phosphorylated inositol phospholipid then recruits to the plasma membrane both Akt and another protein kinase that helps phosphorylate and activate Akt. A third kinase that is permanently associated with the membrane also helps activate Akt (see Figure 16–33).

ANSWER 16–25 Animals and plants are thought to have evolved multicellularity independently and therefore will be expected to have evolved some distinct signaling mechanisms for their cells to communicate with one another. On the other hand, animal and plant cells are thought to have evolved from a common eucaryotic ancestor cell, and so plants and animals would be expected to share some intracellular signaling mechanisms that the common ancestor cell used to respond to its environment.

Chapter 17

ANSWER 17–1 Cells that migrate rapidly from one place to another, such as amoebae (A) and sperm cells (F), do not in general need intermediate filaments in their cytoplasm, since they do not develop or sustain large tensile forces. Plant cells (G) are pushed and pulled by the forces of wind and water, but they resist these forces by means of their rigid cell walls rather than by their cytoskeleton. Epithelial cells (B), smooth muscle cells (C), and the long axons of nerve cells (E) are all rich in cytoplasmic intermediate filaments, which prevent them from rupturing as they are stretched and compressed by the movements of their surrounding tissues.

All of the above eucaryotic cells possess at least intermediate filaments in their nuclear lamina. Bacteria, such as *Escherichia coli* (D), have none whatsoever.

ANSWER 17–2 Two tubulin dimers have a lower affinity for each other (because of a more limited number of interaction sites) than a tubulin dimer has for the end of a microtubule (where there are multiple possible interaction sites, both end-to-end of tubulin dimers adding to a protofilament and side-to-side of the tubulin dimers interacting with tubulin subunits in adjacent protofilaments forming the ringlike cross section). Thus, to initiate a microtubule from scratch, enough tubulin dimers have to come together and remain bound to one another for long enough for other tubulin molecules to add to them. Only when a number of tubulin dimers have already assembled will the binding of the next subunit be favored. The formation of these initial "nucleating sites" is therefore rare and will not occur spontaneously at cellular concentrations of tubulin.

Centrosomes contain preassembled rings of γ-tubulin (in which the γ-tubulin subunits are held together in much tighter side-to-side interactions than αβ-tubulin can form) to which αβ-tubulin dimers can bind. The binding conditions of αβ-tubulin dimers resemble those of adding to the end of an assembled microtubule. The γ-tubulin rings in the centrosome can therefore be thought of as permanently preassembled nucleation sites.

ANSWER 17–3

A. The microtubule is shrinking because it has lost its GTP cap, i.e., the tubulin subunits at its end are all in their GDP-bound form. GTP-loaded tubulin subunits from solution will still add to this end, but they will be short-lived—either because they hydrolyze their GTP or because they fall off as the microtubule rim around them disassembles. If, however, enough GTP-loaded subunits are added quickly enough to cover up the GDP-containing tubulin subunits at the microtubule end, a new GTP cap can form and regrowth is favored.
B. The rate of addition of GTP-tubulin will be greater at higher tubulin concentrations. The frequency with which shrinking microtubules switch to the growing mode will therefore increase with increasing tubulin concentration. The consequence of this regulation is that the system is self-balancing: the more microtubules shrink (resulting in a higher concentration of free tubulin), the more frequently microtubules will start to grow again. Conversely, the more microtubules grow, the lower the concentration of free tubulin will become and the rate of GTP-tubulin addition will slow down; at some point GTP hydrolysis will catch up with new GTP-tubulin addition, the GTP cap will be destroyed, and the microtubule will switch to the shrinking mode.
C. If only GDP were present, microtubules would continue to shrink and eventually disappear, because tubulin dimers with GDP have very low affinity for each other and will not add stably to microtubules.
D. If GTP is present but cannot be hydrolyzed, microtubules will continue to grow until all free tubulin subunits have been used up.

ANSWER 17–4 If all the dynein arms were equally active, there could be no significant relative motion of one microtubule to the other as required for bending (think of a circle of nine weight lifters, each trying to lift his neighbor off the ground: if they all succeeded, the group would levitate!). Thus, a few ciliary dynein molecules must be activated selectively on one side of the cilium. As they move their neighboring microtubules toward the tip of the cilium, the cilium bends away from the side containing the activated dyneins.

ANSWER 17–5 Any actin-binding protein that stabilizes complexes of two or more actin monomers without blocking the ends required for filament growth will facilitate the initiation of a new filament (nucleation).

ANSWER 17–6 Only fluorescent actin molecules assembled into filaments are visible, because unpolymerized actin molecules diffuse so rapidly they produce a dim uniform background. Since, in your experiment, so few actin molecules are labeled (1:10,000), there should be at most one labeled actin monomer per filament (see Figure 17–29). The lamellipodium as a whole has many actin filaments, some of which overlap and therefore show a random speckled pattern of actin molecules, each marking a different filament.

This technique (called "speckle fluorescence") can be used to follow the movement of polymerized actin in a migrating cell. If you watch this pattern with time, you will see that individual fluorescent spots move steadily back from the leading edge toward the interior of the cell, a movement that occurs whether or not the cell is actually migrating. Rearward movement takes place because actin monomers are added to filaments at the plus end and are lost from the minus end (where they are depolymerized) (see Figure 17–34B). In effect actin monomers "move through" the actin filaments, a phenomenon termed "treadmilling." Treadmilling has been demonstrated to occur in isolated actin filaments in solution and also in dynamic microtubules, such as those within a mitotic spindle.

ANSWER 17–7 Cells contain actin-binding proteins that bundle and cross-link actin filaments (see Figure 17–31). The filaments extending from lamellipodia and filopodia become firmly connected to the filamentous meshwork of the cell cortex, thus providing the mechanical anchorage required for the growing rodlike filaments to deform the cell membrane.

ANSWER 17–8 Although the subunits are indeed held together by noncovalent bonds that are individually weak, there are a very large number of them, distributed among a very large number of filaments. As a result, the stress a human being exerts by lifting a heavy object is dispersed over so many subunits that their interaction strength is not exceeded. By analogy, a single thread of silk is not nearly strong enough to hold a human, but a rope woven of such fibers is.

ANSWER 17–9 Both filaments are composed of subunits in the form of protein dimers that are held together by coiled-coil interactions. Moreover, in both cases, the dimers polymerize through their coiled-coil domains into filaments. Whereas intermediate filament dimers assemble head-to-head, however, and thereby create a filament that has no polarity, all myosin molecules in the same half of the myosin filament are oriented with their heads pointing in the same direction. This polarity is necessary for them to be able to develop a contractile force in muscle.

ANSWER 17–10

A. Successive actin molecules in an actin filament are identical in position and conformation. After a first protein (such as troponin) had bound to the actin filament, there would be no way in which a second protein could recognize every seventh monomer in a naked actin filament. Tropomyosin, however, binds along the length of an actin filament, spanning precisely seven monomers, and thus provides a molecular "ruler" that measures the length of seven actin monomers. Troponin becomes localized by binding to the evenly spaced ends of tropomyosin molecules.

B. Calcium ions influence force generation in the actin–myosin system only if both troponin (to bind the calcium ions) and tropomyosin (to transmit the information that troponin has bound calcium to the actin filament) are present. (i) Troponin cannot bind to actin without tropomyosin. The actin filament would be permanently exposed to the myosin, and the system would be continuously active, independently of whether calcium ions were present or not (a muscle cell would therefore be continuously contracted with no possibility of regulation). (ii) Tropomyosin would bind to actin and block binding of myosin completely; the system would be permanently inactive, no matter whether calcium ions were present, because tropomyosin is not affected by calcium. (iii) The system will contract in response to calcium ions.

ANSWER 17–11

A. True. A continual outward movement of ER is required; in the absence of microtubules, the ER collapses toward the center of the cell.
B. True. Actin is needed to make the contractile ring that causes the physical cleavage between the two daughter cells, whereas the mitotic spindle that partitions the chromosomes is composed of microtubules.
C. True. Both extensions are associated with transmembrane proteins that protrude from the plasma membrane and enable the cell to form new anchor points on the substratum.
D. False. To cause bending, ATP is hydrolyzed by the dynein motor proteins that are attached to the outer microtubules in the flagellum.
E. False. Cells could not divide without rearranging their intermediate filaments, but many terminally differentiated and long-lived cells, such as nerve cells, have stable intermediate filaments that are not known to depolymerize.
F. False. The rate of growth is independent of the size of the GTP cap. The plus and minus ends have different growth rates because they have physically distinct binding sites for the incoming tubulin subunits; the rate of addition of tubulin subunits differs at the two ends.
G. True. Both are nice examples of how the same membrane can have regions that are highly specialized for a particular function.
H. False. Myosin movement is activated by the phosphorylation of myosin, or by calcium binding to troponin.

ANSWER 17–12 The average time taken for a small molecule (such as ATP) to diffuse a distance of 10 μm is given by the calculation

$$(10^{-3})^2 / (2 \times 5 \times 10^{-6}) = 0.1 \text{ seconds}$$

Similarly, a protein takes 1 second and a vesicle 10 seconds on average to travel 10 μm. A vesicle would require on average 10^9 seconds, or more than 30 years, to diffuse to the end of a 10-cm axon. This calculation makes it clear why kinesin and other motor proteins evolved to carry molecules and organelles along microtubules.

ANSWER 17–13 (1) Animal cells are much larger and more diversely shaped, and do not have a cell wall. Cytoskeletal elements are required to provide mechanical strength and shape in the absence of a cell wall. (2) Animal cells, and all other eucaryotic cells, have a nucleus that is shaped and held in place in the cell by intermediate filaments; the nuclear lamins attached to the inner nuclear membrane support and shape the nuclear membrane, and a meshwork of intermediate filaments surrounds the nucleus and spans the cytosol. (3) Animal cells can move by a process that requires a change in cell shape. Actin filaments and myosin motor proteins are required for these activities. (4) Animal cells have a much larger genome than bacteria; this genome is fragmented into many chromosomes. For cell division, chromosomes need to be accurately distributed to the daughter cells, requiring the function of the microtubules that form the mitotic spindle. (5) Animal cells have internal organelles. Their localization in the cell is dependent on motor proteins that move them along microtubules. A remarkable example is the long-distance travel of membrane-enclosed vesicles (organelles) along microtubules in an axon that can be up to 1 m (≈3 ft) long in the case of the nerve cells that extend from your spinal cord to your feet.

ANSWER 17–14 The ends of an intermediate filament are indistinguishable from each other, because the filaments are built by the assembly of symmetrical tetramers made from two coiled-coil dimers. In contrast to microtubules and actin filaments, intermediate filaments therefore have no polarity.

ANSWER 17–15 Intermediate filaments have no polarity; their ends are chemically indistinguishable. It would therefore be difficult to envision how a hypothetical motor protein that bound to the middle of the filament could sense a defined direction. Such a motor protein would be equally likely to attach to the filament facing one end or the other.

ANSWER 17–16 Katanin breaks microtubules along their length, and at positions remote from their GTP caps. The fragments that form therefore contain GDP-tubulin at their exposed ends and rapidly depolymerize. Katanin thus provides a very quick means of destroying existing microtubules.

ANSWER 17–17 Cell division depends on the ability of microtubules both to polymerize and to depolymerize. This is most obvious when one considers that the formation of the mitotic spindle requires the prior depolymerization of other cellular microtubules to free up the tubulin required to build the spindle. This rearrangement is not possible in taxol-treated cells, whereas in colchicine-treated cells, division is blocked because a spindle cannot be assembled. On a more subtle but no less important level, both drugs block the

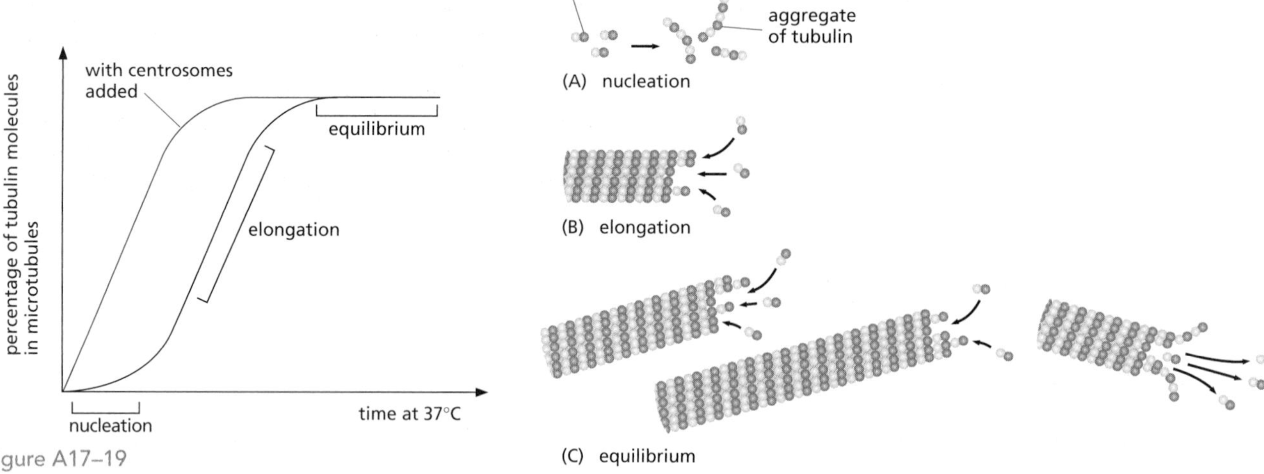

Figure A17–19

dynamic instability of microtubules and would therefore interfere with the workings of the mitotic spindle, even if one could be properly assembled.

ANSWER 17–18 Motor proteins are unidirectional in their action; kinesin always moves toward the plus end of a microtubule and dynein toward the minus end. Thus if kinesin molecules are attached to glass, only those individual motors that have the correct orientation in relation to the microtubule that settles on them can attach to the microtubule and exert force on it to propel it forward. Since kinesin moves toward the plus end of the microtubule, the microtubule will always crawl minus end first over the coverslip.

ANSWER 17–19

A. Phase A corresponds to a lag phase, during which tubulin molecules assemble to form nucleation centers (Figure A17–19A). Nucleation is followed by a rapid rise (phase B) to a plateau value as tubulin dimers add to the ends of the elongating microtubules (Figure A17–19B). At phase C, equilibrium is reached with some microtubules in the population growing while others are rapidly shrinking (Figure A17–19C). The concentration of free tubulin is constant at this point, because polymerization and depolymerization are balanced (see also Question 17–3, p. 583).
B. The addition of centrosomes introduces nucleation sites that eliminate the lag phase A as shown by the red curve in Figure A17–19. The rate of microtubule growth (i.e., the slope of the curve in the elongation phase B) and the equilibrium level of free tubulin remain unchanged, because the presence of centrosomes does not affect the rates of polymerization and depolymerization.

ANSWER 17–20 The ends of the shrinking microtubule are visibly frayed, and the individual protofilaments appear to come apart and curl as the end depolymerizes. This micrograph therefore suggests that the GTP cap (which is lost from shrinking microtubules) holds the protofilaments properly aligned with each other, perhaps by strengthening the side-to-side interactions between αβ-tubulin subunits when they are in their GTP-bound form.

ANSWER 17–21 Cytochalasin interferes with actin filament formation, and its effect on the cell demonstrates the importance of actin to cell locomotion. The experiment with colchicine shows that microtubules are required to give a cell a polarity that then determines which end becomes the leading edge (see Figure 17–13). In the absence of microtubules, cells still go through the motions normally associated with cell movement, such as the extension of lamellipodia, but in the absence of cell polarity these are futile exercises because they happen indiscriminately in all directions.

Antibodies bind tightly to the antigen (in this case vimentin) to which they were raised (see Panel 4–3, pp. 144–145). When bound, an antibody can interfere with the function of the antigen by preventing it from interacting properly with other cell components. The antibody injection experiment therefore suggests that intermediate filaments are not required for the maintenance of cell polarity or for the motile machinery.

ANSWER 17–22 Either (B) or (C) would complete the sentence correctly. The direct result of the action potential in the plasma membrane is the release of Ca^{2+} into the cytosol from the sarcoplasmic reticulum; muscle cells are triggered to contract by this rapid rise in cytosolic Ca^{2+}. Calcium ions at high concentrations bind to troponin, which in turn causes tropomyosin to move to expose myosin-binding sites on the actin filaments. (A) and (D) would be wrong because Ca^{2+} has no effect on the detachment of the myosin head from actin, which is the result of ATP hydrolysis. Nor does it have any role in maintaining the structure of the myosin filament.

ANSWER 17–23 Only (D) is correct. Upon contraction, the Z discs move closer together, and neither actin nor myosin filaments contract (see Figures 17–41 and 17–42).

Chapter 18

ANSWER 18–1 Because all cells arise by division of another cell, this statement is correct, assuming that "first cell division" refers to the division of the successful founder cell from which all life as we know it has derived. There were probably many other unsuccessful attempts to start the chain of life.

ANSWER 18–2 Cells in peak B contain twice as much DNA as those in peak A, indicating that they contain replicated DNA, whereas the cells in peak A contain unreplicated DNA. Peak A therefore contains cells that are in G_1, and peak B contains cells that are in G_2 and mitosis. Cells in S phase have begun but not finished DNA synthesis; they therefore have various intermediate amounts of DNA and are found in the region between the two peaks. Most cells are in G_1, indicating that it is the longest phase of the cell cycle (see Figure 18–2).

ANSWER 18–3 For multicellular organisms, the control of cell division is extremely important. Individual cells must not proliferate unless it is to the benefit of the whole organism. The G_0 state offers protection from aberrant activation of cell division, because the cell-cycle control system is largely dismantled. If, on the other hand, a cell just paused in G_1, it would still contain all of the cell-cycle control system and could readily be induced to divide. The cell would also have to remake the 'decision' not to divide almost continuously. To reenter the cell cycle from G_0, a cell has to resynthesize all of the components that have disappeared.

ANSWER 18–4 The cell would replicate its damaged DNA and therefore would introduce mutations to the two daughter cells when the cell divides. Such mutations could increase the chances that the progeny of the affected daughter cells would eventually become cancer cells.

ANSWER 18–5 Before injection, the frog oocytes must contain inactive M-Cdk. Upon injection of the M-phase cytoplasm, the small amount of the active M-Cdk in the injected cytoplasm activates the inactive M-Cdk by switching on the activating phosphatase (Cdc 25), which removes the inhibitory phosphate groups from the inactive M-Cdk (see Figure 18–17). An extract of the second oocyte, now in M phase itself, will therefore contain as much active M-Cdk as the original cytoplasmic extract, and so on.

ANSWER 18–6 The experiment shows that kinetochores are not preassigned to one or other spindle pole; microtubules attach to the kinetochores that they are able to reach. For the chromosomes to remain attached to a microtubule, however, tension has to be exerted. Tension is normally achieved by the opposing pulling forces from opposite spindle poles. The requirement for such tension ensures that if two sister kinetochores ever become attached to the same spindle pole, so that tension is not generated, one or both of the connections would be lost, and microtubules from the opposing spindle pole would have another chance to attach properly.

ANSWER 18–7 Recall from Figure 18–31 that the new nuclear envelope reassembles on the surface of the chromosomes. The close apposition of the envelope to the chromosomes prevents cytosolic proteins from being trapped between the chromosomes and the envelope. Nuclear proteins are then selectively imported through the nuclear pores, causing the nucleus to expand while maintaining its characteristic protein composition.

ANSWER 18–8 The membranes of the Golgi vesicles fuse to form part of the plasma membranes of the two daughter cells. The interiors of the vesicles, which are filled with cell-wall material, become the new cell-wall matrix separating the two daughter cells. Proteins in the membranes of the Golgi vesicles thus become plasma membrane proteins. Those parts of the proteins that were exposed to the lumen of the Golgi vesicle will end up exposed to the new cell wall (Figure A18–8).

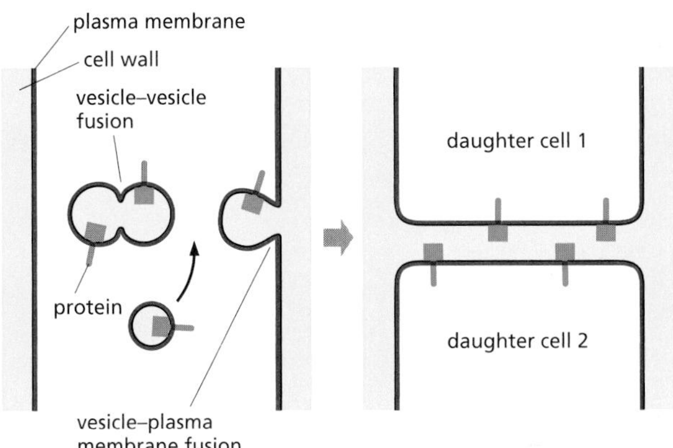

Figure A18–8

ANSWER 18–9 In a eucaryotic organism, the genetic information that the organism needs to survive and reproduce is distributed between multiple chromosomes. It is therefore crucial that each daughter cell receives a copy of each chromosome when a cell divides; if a daughter cell receives too few or too many chromosomes, the effects are usually deleterious or even lethal. Only two copies of each chromosome are produced by chromosome replication in mitosis. If the cell were to randomly distribute the chromosomes when it divided, it would be very unlikely that each daughter cell would receive precisely one copy of each chromosome. In contrast, the Golgi apparatus fragments into tiny vesicles that are all alike, and by random distribution it is very likely that each daughter cell will receive an approximately equal number of them.

ANSWER 18–10 As apoptosis occurs on a large scale in both developing and adult tissues, it must not trigger alarm reactions that are normally associated with cell injury. Tissue injury, for example, leads to the release of signal molecules that stimulate the proliferation of surrounding cells so that the wound heals. It also causes the release of signals that can cause a destructive inflammatory reaction. Moreover, the release of intracellular contents could elicit an immune response against molecules that are normally not encountered by the immune system. Such reactions would be self-defeating if they occurred in response to the massive cell death that occurs in normal development.

ANSWER 18–11 Because the cell population is increasing exponentially, doubling its weight at every cell division, the weight of the cell cluster after N cell divisions is $2^N \times 10^{-9}$ g. Therefore, 70 kg (70×10^3 g) = $2^N \times 10^{-9}$ g, or $2^N = 7 \times 10^{13}$. Taking the logarithm of both sides allows you to solve the equation for N. Therefore, $N = \ln (7 \times 10^{13}) / \ln 2 = 46$; i.e., it would take only 46 days if cells proliferated exponentially. Cell division in animals is tightly controlled, however, and most cells in the human body stop dividing when they become highly specialized. The example demonstrates that exponential cell proliferation occurs only for very brief periods, even during embryonic development.

ANSWER 18–12 The egg cells of many animals are big and contain stores of enough cell components to last for many cell divisions. The daughter cells that form during the first cell divisions after fertilization are progressively smaller in size and thus can be formed without a need for new protein or RNA synthesis. Whereas normally dividing cells would grow continuously in G_1, G_2, and S phases, until their size doubled, there is no cell growth in these early cleavage divisions, and both G_1 and G_2 are virtually absent. As G_1 is usually longer than G_2, G_1 is the most drastically reduced in these divisions.

ANSWER 18–13

A. Radiation leads to DNA damage, which activates a checkpoint mechanism (mediated by p53 and p21; see Figure 18–16) that arrests the cell cycle until the DNA has been repaired.
B. The cell will replicate damaged DNA and thereby introduce mutations in the daughter cells when the cell divides.
C. The cell will be able to divide normally, but it will be prone to mutations, because some DNA damage always occurs as the result of natural irradiation caused, for example, by cosmic rays. The checkpoint mechanism mediated by p53 is mainly required as a safeguard against the devastating effects of accumulating DNA damage, but not for the natural progression of the cell cycle in undamaged cells.
D. Cell division in humans is an ongoing process that does not cease upon reaching maturity, and it is required for survival. Blood cells and epithelial cells in the skin or lining the gut, for example, are being constantly produced by cell division to meet the body's needs; each day, your body produces about 10^{11} new red blood cells alone.

ANSWER 18–14

A. Only the cells that were in the S phase of their cell cycle (i.e., those cells making DNA) during the 30-minute labeling period contain any radioactive DNA.
B. Initially, mitotic cells contain no radioactive DNA because these cells were not engaged in DNA synthesis during the labeling period. Indeed, it takes about two hours before the first labeled mitotic cells appear.
C. The initial rise of the curve corresponds to cells that were just finishing DNA replication when the radioactive thymidine was added. The curve rises as more labeled cells enter mitosis; the peak corresponds to those cells that had just started S phase when the radioactive thymidine was added. The labeled cells then exit from mitosis, and are replaced by unlabeled mitotic cells, which were not yet in S phase during the labeling period. After 20 hours the curve starts rising again, because the labeled cells enter their second round of mitosis.
D. The intial two-hour lag before any labeled mitotic cells appear corresponds to the G_2 phase, which is the time between the end of S phase and the beginning of mitosis. The first labeled cells seen in mitosis were those that were just finishing S phase (DNA synthesis) when the radioactive thymidine was added.

ANSWER 18–15 Loss of M-cyclin leads to inactivation of M-Cdk. As a result, the M-Cdk target proteins become dephosphorylated by phosphatases, and the cells exit from mitosis: they disassemble the mitotic spindle, reassemble the nuclear envelope, decondense their chromosomes, and so on. The M-cyclin is degraded by ubiquitin-dependent destruction in proteosomes, and the activation of M-Cdk leads to the activation of the APC, which ubiquitylates the cyclin, but with a substantial delay. As discussed in Chapter 7, ubiquitylation tags proteins for degradation in proteasomes.

ANSWER 18–16 M-cyclin accumulates gradually as it is steadily synthesized. As it accumulates, it will tend to form complexes with the mitotic Cdk molecules that are present. After a certain threshold level has been reached, a sufficient amount of M-Cdk has been formed so that it is activated by the appropriate kinases and phosphatases that phosphorylate and dephosphorylate it. Once activated, M-Cdk acts to enhance the activity of the activating phosphatase; this positive feedback leads to the explosive activation of M-Cdk (see Figure 18–18). Thus, M-cyclin accumulation acts like a slow-burning fuse, which eventually helps trigger the explosive self-activation of M-Cdk. The precipitous destruction of M-cyclin terminates M-Cdk activity, and a new round of M-cyclin accumulation begins.

ANSWER 18–17 The order is G, C, B, A, D. Together, these five steps are referred to as mitosis (F). No step in mitosis is influenced by the phases of the moon (E). Cytokinesis is the last step in M phase, which overlaps with anaphase and telophase. Mitosis and cytokinesis are both part of M phase.

ANSWER 18–18 If the growth rate of microtubules is the same in mitotic and in interphase cells, their length is proportional to their lifetime. Thus, the average length of microtubules in mitosis is 1 μm (= 20 μm × 15 s/300 s).

ANSWER 18–19 As shown in Figure A18–19, the overlapping interpolar microtubules from opposite poles of the spindle have their plus ends pointing in opposite directions. Plus-end-directed motor proteins cross-link adjacent antiparallel microtubules together and tend to move the microtubules in the direction that will push the two poles of the spindle apart, as shown in the figure. Minus-end-directed motor proteins also cross-link adjacent antiparallel microtubules together but move in the opposite direction, tending to pull the spindle poles together (not shown).

ANSWER 18–20 The sister chromatid becomes committed when a microtubule from one of the spindle poles attaches to the kinetochore of the chromatid. Microtubule attachment is still reversible until a second microtubule from the other spindle pole attaches to the kinetochore of its partner sister chromatid so that the duplicated chromosome is now put under mechanical tension by pulling forces from both poles. The tension ensures that both microtubules remain attached to the chromosome. The position of a chromatid in the cell

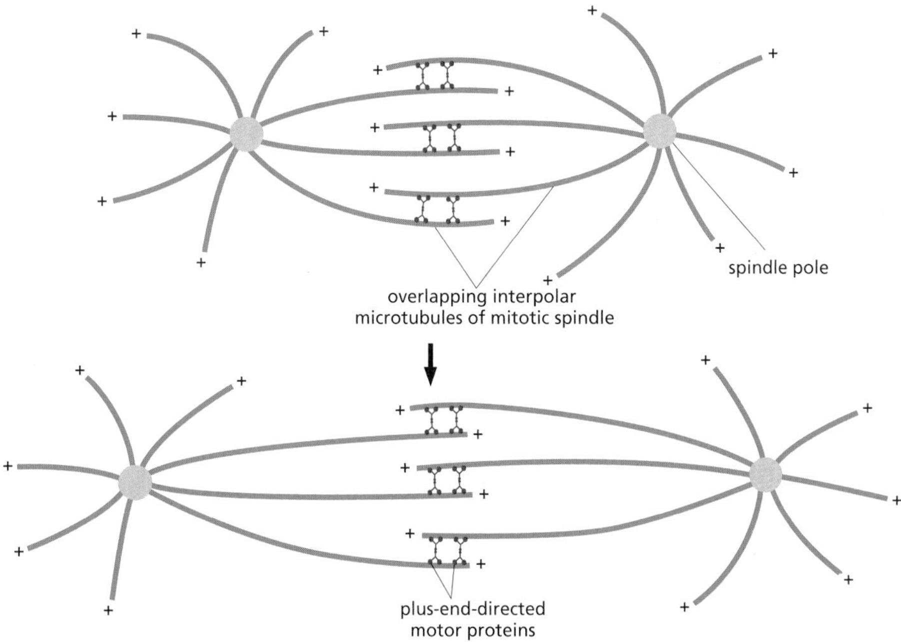

Figure A18–19

at the time that the nuclear envelope breaks down will influence which spindle pole it will be pulled to, as its kinetochore is most likely to become attached to the spindle pole toward which it is facing.

ANSWER 18–21 It is still not certain how this works. In principle, two possible models could explain how the kinetochore may generate a poleward force on its chromosome during anaphase A, as illustrated in Figure A18–21. In (A), microtubule motor proteins are part of the kinetochore and use the energy of ATP hydrolysis to pull the chromosome along its bound microtubules. The depolymerization of the microtubule at its kinetochore end would occur as a consequence of this movement. In (B), chromosome movement is driven by microtubule disassembly catalyzed by an enzyme that uses the energy of ATP hydrolysis to remove tubulin subunits from the attached end of the microtubule. As tubulin subunits dissociate, the kinetochore is obliged to slide poleward to maintain its binding to the walls of the microtubule. It is possible that both mechanisms are used, but current evidence favors the model shown in (B).

ANSWER 18–22 Both sister chromatids could end up in the same daughter cell for any of a number of reasons. (1) If the microtubules or their connections with a kinetochore were to break during anaphase, both sister chromatids could be drawn to the same pole, and hence into the the same daughter cell. (2) If microtubules from the same spindle pole attached to both kinetochores, the chromosome would be pulled to the same pole. (3) If the cohesins that link sister chromatids were not degraded, the pair of chromatids might be pulled to the same pole. (4) If a duplicated chromosome never engaged microtubules and was left out of the spindle, it would also end up in one daughter cell.

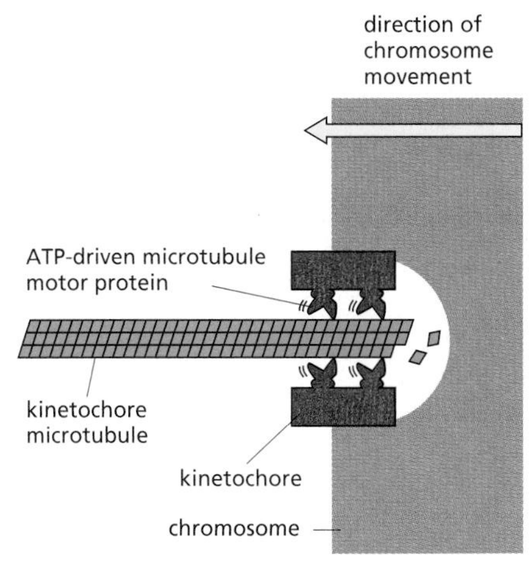

(A) ATP-driven chromosome movement drives microtubule disassembly

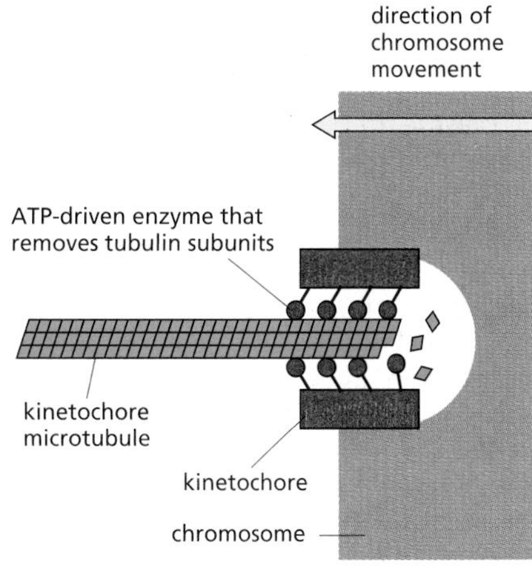

(B) microtubule disassembly drives chromosome movement

Figure A18–21

Some of these errors in the mitotic process would be expected to activate a checkpoint mechanism that delays the onset of anaphase until all chromosomes are attached properly to both poles of the spindle. This 'spindle-assembly checkpoint' mechanism should allow most chromosome attachment errors to be corrected, which is one reason why such errors are rare.

The consequences of both sister chromatids ending up in one daughter cell are usually dire. One daughter cell would contain only one copy of all the genes carried on that chromosome and the other daughter cell would contain three copies. The altered gene dosage, leading to correspondingly changed amounts of the mRNAs and proteins produced, is often detrimental to the cell. In addition, there is the possibility that the single copy of the chromosome may contain a defective gene with a critical function, which would normally be taken care of by the good copy of the gene on the other chromosome that is now missing.

ANSWER 18–23

A. True. Centrosomes replicate during interphase, before M phase begins.
B. True. Sister chromatids separate completely only at the start of anaphase.
C. False. The ends of interpolar microtubules overlap and attach to one another via proteins (including motor proteins) that bridge between the microtubules.
D. False. Microtubules and their motor proteins play no role in DNA replication.
E. False. To be a correct statement, the terms "centromere" and "centrosome" must be switched.

ANSWER 18–24 Antibodies bind tightly to the antigen (in this case myosin) to which they were raised. When bound, an antibody can interfere with the function of the antigen by preventing it from interacting properly with other cell components. (A) The movement of chromosomes at anaphase depends on microtubules and their motor proteins and does not depend on actin or myosin. Injection of an anti-myosin antibody into a cell will therefore have no effect on chromosome movement during anaphase. (B) Cytokinesis, on the other hand, depends on the assembly and contraction of a ring of actin and myosin filaments, which forms the cleavage furrow that splits the cell in two. Injection of anti-myosin antibody will therefore block cytokinesis.

ANSWER 18–25 The plasma membrane of the cell that died by necrosis in Figure 18–38A is ruptured; a clear break is visible, for example, at a position corresponding to the 12 o'clock mark on a watch. The cell's contents, mostly membranous and cytoskeletal debris, are seen spilling into the surroundings through these breaks. The cytosol stains lightly, because most soluble cell components were lost before the cell was fixed. In contrast, the cell that underwent apoptosis in Figure 18–38B is surrounded by an intact membrane, and its cytosol is densely stained, indicating a normal concentration of cell components. The cell's interior is remarkably different from a normal cell, however. Particularly characteristic are the large "blobs" that extrude from the nucleus, probably as the result of the breakdown of the nuclear lamina. The cytosol also contains many large, round, membrane-enclosed vesicles of unknown origin, which are not normally seen in healthy cells. The pictures visually confirm the notion that necrosis involves cell lysis, whereas cells undergoing apoptosis remain relatively intact until they are phagocytosed and digested by another cell.

ANSWER 18–26

A. False. There is no G_1 to M phase transition. The statement is correct, however, for the G_1 to S phase transition, in which cells commit themselves to a division cycle.
B. True. Apoptosis is an active process carried out by special proteases (caspases).
C. True. This mechanism is thought to adjust the number of neurons to the number of specific target cells to which the neurons connect.
D. True. An amazing evolutionary conservation!
E. True. Such studies employ so-called conditional mutations, which lead to the production of proteins that usually are stable and functional at one temperature, but unstable or inactive at another temperature. Cells can be grown at the temperature at which the mutant protein functions normally, and then they can be shifted to a temperature at which the protein's function is lost.
F. True. Association of a Cdk protein with a cyclin is required for its activity (hence its name cyclin-dependent kinase). Furthermore, phosphorylation at a specific site and dephosphorylation at other sites on the Cdk protein are required for the cyclin–Cdk complex to be active.

ANSWER 18–27 Cells in an animal must behave for the good of the organism as a whole—to a much greater extent than people generally act for the good of society as a whole. In the context of an organism, unsocial behavior would lead to a loss of organization and to cancer. Many of the rules that cells have to obey would be unacceptable in a human society. Most people, for example, would be reluctant to kill themselves for the good of society, yet our cells do it all the time.

ANSWER 18–28 The most likely approach to success (if that is what the goal should be called) is plan C, which should result in an increase in cell numbers. The problem is, of course, that cell numbers of each tissue must be increased similarly to maintain balanced proportions in the organism, yet different cells respond to different growth factors. As shown in Figure A18–28,

Figure A18–28 Courtesy of Ralph Brinster

however, the approach has indeed met with limited success. A mouse producing very large quantities of growth hormone *(left)*—which acts to stimulate the production of a secreted protein that acts as a survival factor, growth factor, or mitogen, depending on the cell type—grows to almost twice the size of a normal mouse *(right)*. To achieve this twofold change in size, however, growth hormone was massively overproduced (about fiftyfold). And note that the mouse did not even attain the size of a rat, let alone a dog.

The other approaches have conceptual problems:

A. Blocking all apoptosis would lead to defects in development, as rat development requires the selective death of many cells. It is unlikely that a viable animal would be obtained.
B. Blocking p53 function would eliminate an important checkpoint of the cell cycle that detects DNA damage and stops the cycle so that the cell can repair the damage; removing p53 would increase mutation rates and lead to cancer. Indeed, mice without p53 usually develop normally but die of cancer at a young age.
D. Given the circumstances, switching careers might not be a bad option.

ANSWER 18–29 The on-demand, limited release of PDGF at a wound site triggers cell division of neighboring cells for a limited amount of time, until the PDGF is degraded. This is different from the continuous release of PDGF from mutant cells, where PDGF is made in an uncontrolled way at high levels. Moreover, the mutant cells that make PDGF often express their own PDGF receptor inappropriately, so that they can stimulate their own proliferation, thereby promoting the development of cancer.

ANSWER 18–30 All three types of mutant cells would be unable to divide. The cells

A. would enter mitosis but would not be able to exit mitosis.
B. would arrest permanently in G_1 because the cyclin–Cdk complexes that act in G_1 would be inactivated.
C. would not be able to activate the transcription of genes required for cell division because the required transcription regulators would be constantly inhibited by unphosphorylated Rb.

ANSWER 18–31 In alcoholism, liver cells proliferate because the organ is overburdened and becomes damaged by the large amounts of alcohol that have to be metabolized. This need for more liver cells activates the control mechanisms that normally regulate cell proliferation. Unless badly damaged and full of scar tissue, the liver will usually shrink back to a normal size after the patient stops drinking excessively. In liver cancer, in contrast, mutations abolish normal cell proliferation control and, as a result, cells divide and keep on dividing in an uncontrolled manner, which is usually fatal.

Chapter 19

ANSWER 19–1 Although each daughter cell ends up with a diploid amount of DNA after the first meiotic division, each cell has effectively only a haploid set of chromosomes (albeit in two copies), representing only one or other homolog of each type of chromosome (although some mixing will have occurred during crossing-over). Because the maternal and paternal chromosomes of a pair will carry different versions of many of the genes, these daughter cells will not be genetically identical; each one will, however, have lost either the paternal or the maternal version of each chromosome. In contrast, somatic cells dividing by mitosis inherit a diploid set of chromosomes, and all daughter cells are genetically identical and inherit both maternal and paternal gene copies. The role of gametes produced by meiosis is to mix and reassort gene pools during sexual reproduction, and thus it is a definite advantage for each of them to have a slightly different genetic constitution. The role of somatic cells on the other hand is to build an organism that contains the same genes in all its cells and retains in each cell both maternal and paternal genetic information.

ANSWER 19–2 A typical human female produces fewer than 1000 mature eggs in her lifetime (12 per year over about 40 years); this is less than one-tenth of a percent of the possible gametes, excluding the effects of meiotic crossing-over. A typical human male produces billions of sperm during a lifetime, so in principle, every possible chromosome combination is sampled many times.

ANSWER 19–3 For simplicity, consider the situation where a father carries genes for two dominant traits, M and N, on one of his two copies of human chromosome 1. If these two genes were located at opposite ends of this chromosome, and there were one and only one crossover event per chromosome as postulated in the question, half of his children would express trait M and the other half would express trait N—with no child resembling the father in carrying both traits. This is very different from the actual situation, where there are multiple crossover events per chromosome, causing the traits M and N to be inherited as if they were on separate chromosomes. By constructing a Punnett square like that in Figure 19–23, one can see that in this latter, more realistic case, we would actually expect one-fourth of the children of this father to inherit both traits, one-fourth to inherit trait M only, one-fourth to inherit trait N only, and one-fourth to inherit neither trait.

ANSWER 19–4 Inbreeding tends to give rise to individuals who are homozygous for most genes. To see why, consider the extreme case where inbreeding takes the form of brother–sister matings (as among the Pharaohs of ancient Egypt): because the parents are closely related, there is a high probability that the maternal and paternal alleles inherited by the offspring will be the same. Inbreeding continued over many generations gives rise to individuals who are all alike and homozygous for almost every gene. Because of the randomness of the mechanism of inheritance, there is a large chance that some deleterious alleles will become prevalent in the population in this way, giving all individuals a reduced fitness. In another, separate inbred population, the same thing will happen, but the chances are that a different set of deleterious alleles will become prevalent. When individuals from the two separate inbred populations mate, their offspring will inherit deleterious alleles of genes *A*, *B*, and *C* for example, from the mother, but good alleles of those genes from the father; conversely, they will inherit deleterious alleles of genes *D*, *E*, and *F* from the father, but good alleles of those genes from the mother. Most deleterious mutations are recessive. The offspring, because they are heterozygous for all these genes, will thus escape the deleterious effects seen in the parents.

ANSWER 19–5 Although any one of the three explanations could in principle account for the observed result, A and B can be ruled out as being completely implausible.

A. There is no precedent for any instability in DNA so great as to be detectable in such a SNP analysis; in any case, the hypothesis would predict a steady decrease in the frequency of the SNP with age, not a drop in frequency that begins only at age 50.
B. Human populations change their genes only very slowly over time (unless a massive population migration brings an influx of individuals who are genetically different). People born 50 years ago will be, on average, virtually the same genetically as those being born today.
C. This hypothesis is correct. A SNP with these properties has been used to discover a gene that appears to cause a substantial increase in the probability of death from cardiac problems.

ANSWER 19–6 Natural selection alone is not sufficient to eliminate recessive lethal genes from the population. Consider the following line of reasoning. Homozygous defective individuals can arise only as the offspring of a mating between two heterozygous individuals. By the rules of Mendelian genetics, offspring of such a mating will be in the ratio of 1 homozygous normal: 2 heterozygous: 1 homozygous defective. Thus, statistically, heterozygous individuals should always be more numerous than the homozygous, defective individuals. And although natural selection effectively eliminates the defective genes in homozygous individuals through death, it can't touch the defective genes in heterozygous individuals because they do not affect the phenotype. Natural selection will keep the frequency of the defective gene low in the population, but in the absence of any other effect there will always be a reservoir of defective genes in the heterozygous individuals.

At low frequencies of the defective gene another important factor—chance—comes into play. Chance variation can increase or decrease the frequency of heterozygous individuals (and thereby the frequency of the defective gene). By chance, the offspring of a mating between heterozygotes could all be normal, which would eliminate the defective gene from that lineage. Increases in the frequency of a deleterious gene are opposed by natural selection; however, decreases are unopposed and can, by chance, lead to elimination of the defective gene from the population. On the other hand, new mutations are continually occurring, albeit at a low rate, creating fresh copies of the deleterious recessive allele. In a large population, a balance will be struck between the creation of new copies of the allele in this way, and their elimination through the death of homozygotes.

ANSWER 19–7

A. True.
B. True.
C. False. Mutations that occur during meiosis can be propagated, unless they give rise to nonviable gametes.

ANSWER 19–8 Two copies of the same chromosome can end up in the same daughter cell if one of the microtubule connections breaks before sister chromatids are separated. Alternatively, microtubules from the same spindle pole could attach to both kinetochores of the chromosome. As the consequence of this severe and rare error, one daughter cell would contain only one copy of all the genes carried on that chromosome, and the other daughter cell would contain three copies. The changed gene dosage, leading to correspondingly changed amounts of the mRNAs and proteins produced, is in many cases detrimental to the cell. If the mistake happens during meiosis, in the process of gamete formation, it will be propagated in all cells of the organism. A form of mental retardation called Down syndrome, for example, is due to the presence of three copies of Chromosome 21 in all of the nucleated cells in the body.

ANSWER 19–9 Meiosis begins with DNA replication, producing a tetraploid cell containing four copies of each chromosome. These four copies have to be distributed equally during the two sequential meiotic divisions into four haploid cells. Sister chromatids remain paired so that (1) the cells resulting from the first division receive two complete sets of chromosomes and (2) the chromosomes can be evenly distributed again in the second meiotic division. If the sister chromatids did not remain paired, it would not be possible in the second division to distinguish which chromatids belong together, and it would therefore be difficult to ensure that precisely one copy of each chromatid is pulled into each daughter cell. Keeping two sister chromatids paired in the first meiotic division is therefore an easy way to keep track of which chromatids belong together.

This biological principle suggests that you might consider clamping your socks together in matching pairs before putting them into the laundry. In this way, the cumbersome process of sorting them out afterward—and the seemingly inevitable mistakes that occur during that process—could be avoided.

ANSWER 19–10

A. A gene is a stretch of DNA that codes for a protein or functional RNA. An allele is an alternative form of a gene. Within the population, there are often several "normal" alleles, whose functions are indistinguishable. In addition, there may be many rare alleles that are defective to varying degrees. An individual, however, normally has a maximum of two alleles of a gene.
B. An individual is said to be homozygous if the two alleles of a gene are the same. An individual is said to be heterozygous if the two alleles of a gene are different.
C. The genotype is the specific set of alleles forming the genome of an individual; it is an enumeration of all the particular forms of each gene in the genome. In practice, for organisms studied in a laboratory,

the genotype is usually specified as a list of the known differences between the individual and the wild type, which is the standard, naturally occurring type. The phenotype is a description of the visible characteristics of the individual. In practice, the phenotype is usually a list of the differences in visible characteristics between the individual and the wild type.

D. An allele *A* is dominant (relative to a second allele *a*) if the presence of even a single copy of *A* is enough to affect the phenotype; that is, if heterozygotes (with genotype *Aa*) appear different from *aa* homozygotes. An allele *a* is recessive (relative to a second allele *A*) if the presence of a single copy makes no difference to the phenotype, so that *Aa* individuals look just like *AA* individuals. If the phenotype of the heterozygous individual differs from the phenotypes of individuals that are homozygous for either allele, the alleles are said to be co-dominant.

ANSWER 19–11

A. Since the pea plant is diploid, any true-breeding plant must carry two mutant copies of the same gene—both of which have lost their function.

B. No, the same phenotype will often be produced by several different genotypes.

C. If each plant carries a mutation in a different gene, this will be revealed by complementation tests (see Panel 19–1, p. 674). When plant A is crossed with plant B, all of the F_1 plants will produce only round peas. And the same result will be obtained when plant B is crossed with plant C, or when plant A is crossed with plant C. In contrast, a cross between any two true-breeding plants that carry loss-of-function mutations in the same gene (even if these mutations are different) should produce only plants with wrinkled peas.

ANSWER 19–12

A. The mutation is likely to be dominant, because roughly half of the progeny born to an affected parent—in each of three marriages to hearing partners—are deaf, and it is unlikely that all these hearing partners were heterozygous carriers of the mutation.

B. The mutation is present on an autosome. If it were instead carried on a sex chromosome, either only the female progeny should be affected (expected if the mutation arose in a gene on the grandfather's X chromosome), or only the male progeny should be affected (expected if the mutation arose in a gene on the grandfather's Y chromosome). In fact, the pedigree reveals that both some males and some females have inherited the mutant form of the gene.

C. Suppose that the mutation was present on one of the two copies of the grandfather's chromosome 12. Each of these copies of chromosome 12 would be expected to carry a different pattern of SNPs, since one of them was inherited from his father and the other was inherited from his mother. Each of the copies of chromosome 12 that was passed to his grandchildren will have gone through two meioses—one meiosis per generation.

Because two or three crossover events occur per chromosome during a meiosis, each chromosome inherited by a grandchild will have been subjected to about five crossovers since it left the grandfather, dividing it into six segments. An identical pattern of SNPs should surround whatever gene causes the deafness in each of the four affected grandchildren; moreover, this SNP pattern should be clearly different from that surrounding the same gene in each of the seven grandchildren who are normal. These SNPs would form an unusually long haplotype block—one that extends for about one-sixth of the length of chromosome 12. (One-quarter of the DNA of each grandchild will have been inherited from the grandfather, in roughly 70 segments of this length scattered among the grandchild's 46 chromosomes.)

ANSWER 19–13 Individual 1 might be either heterozygous (+/–) or homozygous for the normal allele (+/+). (Both her parents must have been heterozygous, since they had a homozygous mutant child). Individual 2 must be homozygous for the recessive deafness allele (–/–). Individual 3 is almost certainly heterozygous (+/–) and responsible for transmitting the mutant allele to his children and grandchildren. Given that the mutant allele is rare, individual 4 is most probably homozygous for the normal allele (+/+).

ANSWER 19–14 Your friend is wrong. (A) Mendel's laws, and the clear understanding that we now have concerning the mechanisms that produce them, rule out many false ideas concerning human heredity. One of them is that a first-born child has a different chance of inheriting particular traits from its parents than its siblings. (B) The probability of this type of pedigree arising by chance is one-fourth for each generation, or one in 64 for the three generations shown. (C) Data from an enlarged sampling of family members, or from more generations, would quickly reveal that the regular pattern observed in this particular pedigree arose by chance. (D) An opposite result, if it had strong statistical significance, would suggest that some process of selection was involved: for example, parents who had had a first child that was affected might regularly opt for screening of subsequent pregnancies and selectively terminate those pregnancies in which the fetus was found to be affected. Fewer second children would then be born with the abnormality.

ANSWER 19–15 Each carrier is a heterozygote, and 50% of his sperm or her eggs will carry the bad allele. When two carriers marry, there is therefore a 25% chance that any baby will inherit the bad allele from both parents and so will show the fatal phenotype. Because one person in 100 is a carrier, one partnership in 10,000 (100 x 100) will be a partnership of carriers (assuming that people choose their partners at random). Other things being equal, one baby in 40,000 will then be born with the defect, or 25 babies per year out of a total of a million babies born.

ANSWER 19–16 A dominant–negative mutation gives rise to a mutant gene product that interferes with the function of the normal gene product, causing a loss-of-function phenotype even in the presence of a normal copy of the gene. This ability of a single defective allele to determine the phenotype is the reason why such an allele is dominant. A gain-of-function mutation increases the activity of the gene or makes it active in inappropriate circumstances. The change in activity often has a phenotypic consequence, which is why such mutations are usually dominant.

ANSWER 19–17 This statement is largely true. Diabetes is one of the oldest diseases described by humans, dating at least back to the time of the ancient Greeks. Diabetes itself comes from the Greek word for siphon, which was used to describe the main symptoms: "The disease was called diabetes, as though it were a siphon, because it converts the human body into a pipe for the transflux of liquid humors" (in other words, untreated patients have constant thirst, balanced by high output of urine). If there were no human disease, the role of insulin would not have come to our attention in so demanding a way. We would have ultimately understood its role—and by now may have. Yet it is difficult to overstate the case for the role of disease in focusing our efforts toward a molecular understanding. Even today, the quest to understand and alleviate human disease is a principal driving force in biomedical research.

ANSWER 19–18

A. As outlined in Figure A19–18, if flies that are defective in different genes mate, their progeny will have one normal gene at each locus. In the case of a mating between a ruby-eyed fly and a white-eyed fly, every progeny fly will inherit one functional copy of the white gene from one parent and one functional copy of the ruby gene from the other parent. Because each of the mutant alleles is recessive to the corresponding wild-type allele, the progeny will have the wild-type phenotype—brick-red eyes.

B. Garnet, ruby, vermilion, and carnation complement one another and the various alleles of the white gene (that is, when these mutant flies are mated with each other, they produce flies with a normal eye color); thus each of these mutants defines a separate gene. In contrast, white, cherry, coral, apricot, and buff do not complement each other; thus, they must be alleles of the same gene, which has been named the white gene. Thus, these nine different eye-color mutants define five different genes.

C. Different alleles of the same gene, like the five alleles of the white gene, often have different phenotypes. Different mutations compromise the function of the gene product to different extents, depending on the location of the mutation. Alleles that do not produce any functional product (null alleles), even if they result from different DNA sequence changes, do have the same phenotype.

ANSWER 19–19 SNPs are single-nucleotide differences between individuals for which two or more variants are each found at high frequency in the population. In the human population, SNPs occur roughly once per 1000 nucleotides of sequence. Many have been identified and mapped in various organisms, including several million in the human genome. SNPs, which can be detected by oligonucleotide hybridization, serve as physical markers whose genomic locations are known. By tracking a mutant gene through different matings, and correlating the presence of the gene with the co-inheritance of particular SNP variants, one can narrow down the potential location of a gene to a chromosomal region that may contain only a few genes. These candidate genes can then be tested for the presence of a mutation that could account for the original mutant phenotype (see Figure 19–32).

ANSWER 19–20 What you immediately know is all of the nucleotide sequence differences between Tim and John. But variants that are rare in the human population are not useful for most genetic mapping analyses. Testing each variant for its frequency in a large population of humans will reveal which of them are found in at least

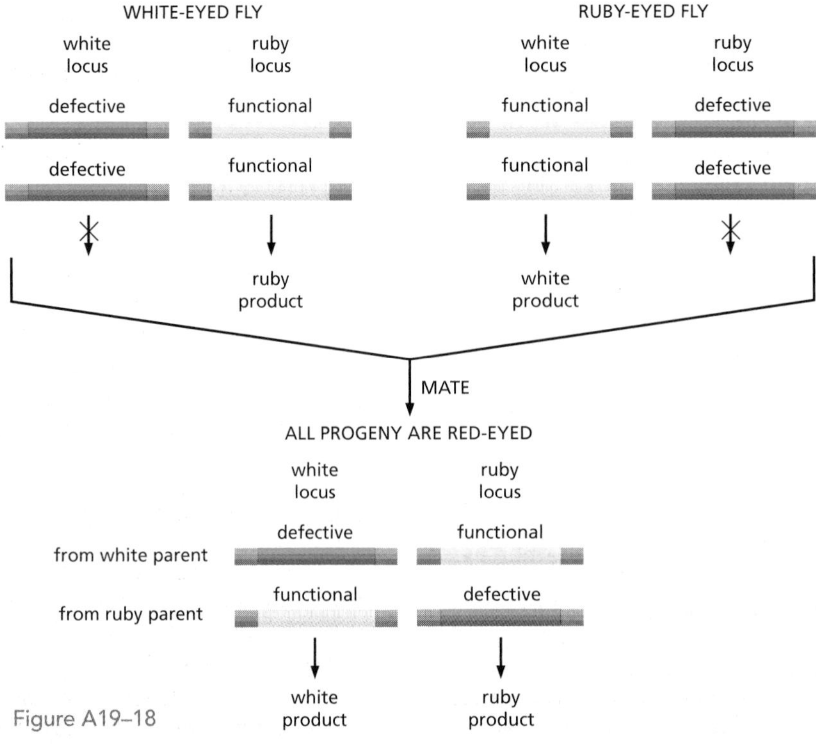

Figure A19–18

10 percent of the people in that population. These are the SNPs that will provide useful markers for future mapping analyses.

Chapter 20

ANSWER 20–1 The horizontal orientation of the microtubules will be associated with a horizontal orientation of cellulose microfibrils deposited in the cell walls. The growth of the cells will therefore be in a vertical direction, expanding the distance between the cellulose microfibrils without stretching them. In this way, the stem will rapidly elongate; in a typical natural environment, this will hasten emergence from darkness into light.

ANSWER 20–2

A. As three collagen chains have to come together to form the triple helix, a defective molecule will impair assembly, even if normal collagen chains are present at the same time. Collagen mutations are therefore dominant; that is, they have a deleterious effect even in the presence of a normal copy of the gene.
B. The different severity of the mutations results from a polarity in the assembly process. Collagen monomers assemble into the triple-helical rod starting from their amino-terminal ends. A mutation in an 'early' glycine therefore allows only short rods to form, whereas a mutation farther downstream allows for longer, more normal rods.

ANSWER 20–3 The remarkable ability to swell and thus occupy a large volume of space depends on the negative charges. These attract a cloud of positive ions, chiefly Na^+, which by osmosis draw in large amounts of water, thus giving proteoglycans their unique properties. Uncharged polysaccharides such as cellulose, starch, and glycogen, by contrast, are easily compacted into fibers or granules.

ANSWER 20–4 Focal contact sites are common in connective tissue, where fibroblasts exert traction forces on the extracellular matrix, and in cell culture, where cell crawling is observed. The forces for pulling on matrix or for driving crawling movement are generated by the actin cytoskeleton. In mature epithelium, focal contact sites are presumably less prominent because the cells are largely fixed in place and have no need to crawl over the basal lamina or actively pull on it.

ANSWER 20–5 Suppose a cell is damaged so that its plasma membrane becomes leaky. Ions present in high concentration in the extracellular fluid, such as Na^+ and Ca^{2+}, then rush into the cell, and valuable metabolites leak out. If the cell were to remain connected to its healthy neighbors, these too would suffer from the damage. But the influx of Ca^{2+} into the sick cell causes its gap junctions to close immediately, effectively isolating the cell and preventing damage from spreading in this way.

ANSWER 20–6 Ionizing (high-energy) radiation tears through matter, knocking electrons out of their orbits and breaking chemical bonds. In particular, it creates breaks and other damage in DNA, and thus causes cells to arrest in the cell cycle (discussed in Chapter 18). If the damage is so severe that it cannot be repaired, cells become permanently arrested and undergo apoptosis; that is, they activate a suicide program.

ANSWER 20–7 Cells in the gut epithelium are exposed to a quite hostile environment, containing digestive enzymes and many other substances that vary drastically from day to day depending on the food intake of the organism. The epithelial cells also form a first line of defense against potentially hazardous compounds and mutagens that are ubiquitous in our environment. The rapid turnover protects the organism from harmful consequences, as wounded and sick cells are discarded. If an epithelial cell started to divide inappropriately as the result of a mutation, for example, it and its unwanted progeny would most often simply be discarded by natural disposal from the tip of a villus: even though such mutations must occur often, they rarely give rise to a cancer.

A neuron, on the other hand, lives in a very protected environment, insulated from the outside world. Its function depends on a complex system of connections with other neurons—a system that is created during development and is not easy to reconstruct if the neuron subsequently dies.

ANSWER 20–8 Every cell division generates one additional cell; so if the cells were never lost or discarded from the body, the number of cells in the body should equal the number of divisions plus one. The number of divisions is 1000-fold greater than the number of cells because, in the course of a lifetime, 1000 cells are discarded and replaced for every cell that is retained in the body.

ANSWER 20–9

A. False. Gap junctions are not connected to the cytoskeleton; their role is to provide cell–cell communication by allowing small molecules to pass from one cell to another.
B. True. Upon wilting, the turgor pressure in the plant cell is reduced, and consequently the cell walls, having tensile but little compressive strength, like a rubber tire, no longer provide rigidity.
C. False. Proteoglycans can withstand a large amount of compressive force but do not have a rigid structure. Their space-filling properties result from their tendency to absorb large amounts of water.
D. True.
E. True.
F. True. Stem cells stably express control genes that ensure that their daughter cells will be of the appropriate differentiated cell types.

ANSWER 20–10 Small cytosolic molecules, such as glutamic acid, cyclic AMP, and Ca^{2+} ions, pass readily through both gap junctions and plasmodesmata, whereas large cytosolic macromolecules, such as mRNA and G proteins, are excluded. Plasma membrane phospholipids diffuse in the plane of the membrane through plasmodesmata because the plasma membranes of adjacent cells are continuous through these junctions. This traffic is not possible through gap junctions, because the membranes of the connected cells remain separate.

ANSWER 20–11 Plants are exposed to extreme changes in the environment, which often are accompanied by huge fluctuations in the osmotic properties of their surroundings. An intermediate filament network as we know it from animal cells would not be able to provide full osmotic support for cells: the sparse rivetlike attachment points would not be able to prevent the membrane from bursting in response to a huge osmotic pressure applied from the inside of the cell.

ANSWER 20–12 Action potentials can, in fact, be passed from cell to cell through gap junctions. Indeed, heart muscle cells are connected in this way, which ensures that they contract synchronously when stimulated. This mechanism of passing the signal from cell to cell is rather limited, however. As we discuss in Chapter 12, synapses are far more sophisticated and allow signals to be modulated and to be integrated with other signals received by the cell. Thus, gap junctions are like simple soldered joints between electrical components, while synapses are like complex relay devices, enabling systems of neurons to perform computations.

ANSWER 20–13 To make jello, gelatin is boiled in water, which denatures the collagen fibers. Upon cooling, the disordered fibers form a tangled mess that solidifies into a gel. This gel actually resembles the collagen as it is initially secreted by fibroblasts; that is, before the fibers become aligned and cross-linked.

ANSWER 20–14 The evidence that DNA is the blueprint that specifies all the structural characteristics of an organism is based on observations that small changes in the DNA by mutation result in changes in the organism. Although DNA provides the plans that specify structure, these plans need to be executed during development. This requires a suitable environment (a human baby would not fit into a stork's egg shell), suitable nourishment, suitable tools (such as the appropriate transcription regulators required for early development), suitable spatial organization (such as the asymmetries in the egg cell required to allow for appropriate cell differentiation during the early cell divisions), and so on. Thus inheritance is not restricted to the passing on of the organism's DNA, because development requires appropriate conditions to be set up by the parent. Nevertheless, when all these conditions are met, the plans that are archived in the genome will determine the structure of the organism to be built.

ANSWER 20–15 White blood cells circulate in the bloodstream and migrate into and out of tissues in performance of their normal function of defending the body against infection: they are naturally invasive. Once mutations have occurred to upset the normal controls on production of these cells, there is no need for additional mutations to enable the cells to spread through the body. Thus, the number of mutations that have to be accumulated in order to give rise to leukemia is smaller than for other types of cancer.

ANSWER 20–16 The shape of the curve reflects the need for multiple mutations to accumulate in a cell before a cancer results. If a single mutation were sufficient, the graph would be a straight horizontal line: the likelihood of occurrence of a particular mutation, and therefore of cancer, would be the same at any age. If two specific mutations were required, the graph would be a straight line sloping upward from the origin: the second mutation has an equal chance of occurring at any time, but will tip the cell into cancerous behavior only if the first mutation has already occurred in the same cell lineage; and the likelihood that the first mutation has already occurred will be proportional to the age of the individual. The steeply curved graph shown in the figure goes up approximately as the fifth power of the age, and this indicates that far more than two mutations have to be accumulated before cancer sets in. It is not easy to say precisely how many, because of the complex ways in which cancers develop. Successive mutations can alter cell numbers and cell behavior, and thereby change both the probability of subsequent mutations and the selection pressures that drive the evolution of cancer.

ANSWER 20–17 During exposure to the carcinogen, mutations are induced, but the number of relevant mutations in any one cell is usually not enough to convert it directly into a cancer cell. Over the years, the cells that have become predisposed to cancer through the induced mutations accumulate progressively more mutations. Eventually, one of them will turn into a cancer cell. The long delay between exposure and cancer has made it extremely difficult to hold cigarette manufacturers or producers of industrial carcinogens legally responsible for the damage that is caused by their products.

ANSWER 20–18 By definition, a carcinogen is any substance that promotes the occurrence of one or more types of cancer. The sex hormones can therefore be classified as naturally occurring carcinogens. Although most carcinogens act by directly causing mutations, carcinogenic effects are also often exerted in other ways. The sex hormones increase both the rate of cell division and the numbers of cells in hormone-sensitive organs such as breast, uterus, and prostate. The first effect increases the mutation rate per cell, because mutations, regardless of environmental factors, are spontaneously generated in the course of DNA replication and chromosome segregation; the second effect increases the number of cells at risk. In these and possibly other ways, the hormones can favor the development of cancer, even though they do not directly cause mutations.

ANSWER 20–19 The short answer is no—cancer in general is not a hereditary disease. It arises from new mutations occurring in our own somatic cells, rather than mutations we inherit from our parents. In some rare types of cancer, however, there is a strong heritable risk factor, so that parents and their children both show the same predisposition to a specific form of the disease. This occurs, for example, in families carrying a mutation that knocks out one of the two copies of the tumor suppressor gene *APC;* the children then inherit a propensity to colorectal cancer. Much weaker heritable tendencies are also seen in several other cancers, including breast cancer, but the genes responsible for these effects are still mostly unknown.

Glossary

acetyl CoA (acetyl coenzyme A)
Small water-soluble molecule that carries acetyl groups in cells. Contains an acetyl group linked to coenzyme A (CoA) by an easily hydrolyzable thioester bond.

acetyl group
Chemical group derived from acetic acid.

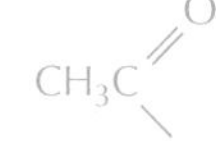

acid
In the context of cell biology, an organic molecule that dissociates in water to generate hydronium (H_3O^+) ions (thereby producing a low pH).

actin filament
Protein filament, about 7-nm wide, formed from a chain of globular actin molecules. A major constituent of the cytoskeleton of all eucaryotic cells and especially abundant in muscle cells.

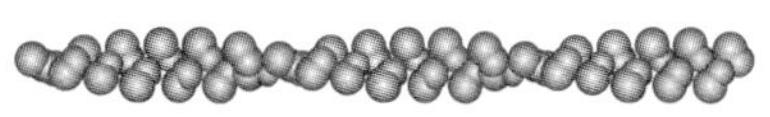

action potential
Rapid, transient, self-propagating electrical signal in the plasma membrane of a cell such as a neuron or muscle. A nerve impulse.

activated carrier
A small molecule used to carry energy or chemical groups in many different metabolic reactions. Examples include ATP, acetyl CoA and NADH.

activation energy
The extra energy that must be acquired by a molecule to undergo a particular chemical reaction.

activator
A protein that binds to a specific regulatory region of DNA to permit transcription of an adjacent gene.

active site
Specialized region of an enzyme surface to which a substrate molecule binds before it undergoes a catalyzed reaction.

active transport
Movement of a molecule across a membrane driven by ATP hydrolysis or another form of metabolic energy.

acyl group
Functional group derived from a carboxylic acid. (R represents an alkyl group, such as methyl.)

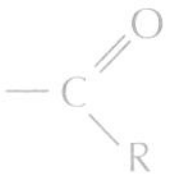

adaptation
Adjustment of sensitivity of a cell or organism following repeated stimulation. Can allow a response even when there is a high background level of stimulation.

adenylyl cyclase
Enzyme that catalyzes the formation of cyclic AMP from ATP. An important component of some intracellular signaling pathways.

adherens junction
Cell junction in which the cytoplasmic face is attached to actin filaments.

ADP (adenosine 5′-diphosphate)
Nucleotide that is produced by hydrolysis of the terminal phosphate of ATP. (*See* Figure 3–31.)

alcohol
Organic compound containing a hydroxyl group (–OH) bound to a saturated carbon atom—for example ethyl alcohol. (*See* Panel 2–1, pp. 64–65.)

aldehyde
Reactive organic compound that contains the HC=O group, for example glyceraldehyde. (*See* Panel 2–1, pp. 64–65.)

alkyl group
General term for a group of covalently linked carbon and hydrogen atoms such as methyl ($-CH_3$) or ethyl ($-CH_2CH_3$) groups. Sometimes abbreviated as R.

allele
One of a set of alternative forms of a gene. In a diploid cell each gene will have two alleles, each occupying the same position (locus) on homologous chromosomes.

allosteric
Describes a protein that can exist in two or more conformations depending on the binding of a molecule (a ligand) at a site other than the catalytic site. Allosteric proteins composed of multiple subunits often display a cooperative response to ligand binding, because the binding of a ligand to one subunit facilitates the binding of ligands to the other subunits.

alpha helix (α helix)
Common structural motif of proteins in which a linear sequence of amino acids folds into a right-handed helix stabilized by internal hydrogen bonding between backbone atoms.

alternative splicing
Splicing of RNA transcripts from the same gene in different ways, each of which produces a distinct protein.

amide
Molecule containing a carbonyl group linked to an amine. (*See* Panel 2–1, pp. 64–65.)

amine
Molecule containing an amino group ($-NH_2$). (*See* Panel 2–1, pp. 64–65.)

aminoacyl-tRNA synthetase
Enzyme that attaches the correct amino acid to a tRNA molecule to form an aminoacyl-tRNA.

amino acid
Organic molecule containing both an amino group and a carboxyl group. α-Amino acids (those in which the amino and carboxyl groups are linked to the same carbon atom) serve as the building blocks of proteins. (*See* Panel 2–5, pp. 72–73.)

amino acid sequence
The order of amino acid residues in a protein chain. Sometimes called the primary structure of a protein.

amino group
Weakly basic functional group ($—NH_2$) derived from ammonia. In aqueous solution an amino group can accept a proton and carry a positive charge. (*See* Panel 2–1, pp. 64–65.)

amino terminus—*see* **N-terminus**

AMP (adenosine 5′ monophosphate)
One of the four nucleotides in an RNA molecule. AMP is produced by the energetically favorable hydrolysis of ATP. (*See* Figure 3–40.)

amphipathic
Having both hydrophobic and hydrophilic regions, as in a phospholipid or a detergent molecule.

anabolism
Reaction pathways by which large molecules are made from smaller ones. Biosynthesis.

anaerobic
Describes a cell, organism, or metabolic process that functions in the absence of air or, more precisely, in the absence of molecular oxygen.

anaphase
Stage of mitosis during which the two sets of chromosomes separate and move away from each other. Composed of anaphase A (chromosomes move toward the two spindle poles) and anaphase B (spindle poles move apart).

anaphase-promoting complex (APC)
A protein complex that promotes the destruction of specific proteins, by catalyzing their ubiquitylation. It is a crucial component of the cell-cycle control system.

anion
Negatively charged ion, such as Cl^- or CH_3COO^-.

antibody
Protein produced by B lymphocytes in response to a foreign molecule or invading organism. Binds to the foreign molecule or cell extremely tightly, thereby inactivating it or marking it for destruction.

anticodon
Sequence of three nucleotides in a transfer RNA molecule that is complementary to the three-nucleotide codon on a messenger RNA molecule; each anticodon is matched to a specific amino acid covalently attached elsewhere on the transfer RNA molecule.

antigen
Molecule that provokes the production of specific neutralizing antibodies in an immune response.

antiparallel
Describes two similar structures arranged in opposite orientations, such as the two strands of a DNA double helix.

antiport
Type of coupled transporter that transports two different ions or small molecules across a membrane in opposite directions, either simultaneously or in sequence.

APC—*see* **anaphase-promoting complex**

apical
Describes the tip of a cell, structure, or organ. The apical surface of an epithelial cell is the exposed free surface (opposite to the basal surface).

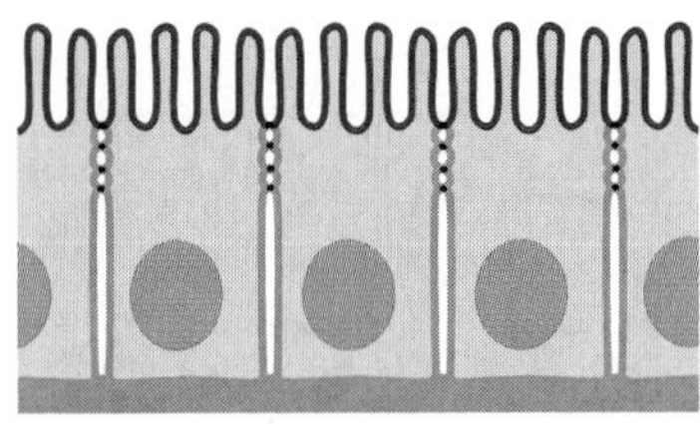

apoptosis
Normal, benign type of programmed cell death in which a cell shrinks, fragments its DNA, and alters its surface so as to activate the cell's phagocytosis by macrophages.

archaea
One of the two divisions of procaryotes, often found in hostile environments such as hot springs or concentrated brine. (*See also* bacteria.)

asexual reproduction
Any type of reproduction (such as budding in *Hydra,* binary fission in bacteria, or mitotic division in eucaryotic microorganisms) that does not involve gamete formation and fusion. It produces an individual genetically identical to the parent.

aster
Star-shaped system of microtubules emanating from a centrosome or from a pole of a mitotic spindle.

atom
The smallest particle of an element that still retains its distinctive chemical properties.

atomic weight
Mass of an atom expressed in daltons.

ATP (adenosine 5′-triphosphate)
Nucleoside triphosphate composed of adenine, ribose, and three phosphate groups that is the principal carrier of chemical energy in cells. The terminal phosphate groups are highly reactive in the sense that their hydrolysis, or transfer to another molecule, is accompanied by the release of a large amount of free energy. (*See* Figures 2–23, 3–31.)

ATP synthase
Membrane-associated enzyme complex that catalyzes the formation of ATP during oxidative phosphorylation and photosynthesis. Found in mitochondria, chloroplasts, and bacteria.

Avogadro's number
The number of molecules in a quantity of substance equal to its molecular weight in grams. Approximately 6×10^{23}.

axon
Long thin nerve cell process capable of rapidly conducting nerve impulses over long distances so as to deliver signals to other cells.

bacteriorhodopsin
Pigmented protein found in the plasma membrane of a salt-loving bacterium, *Halobacterium halobium*; it pumps protons out of the cell in response to light.

bacteria (singular bacterium)
Commonly used name for any procaryotic organism, but more precisely refers to the eubacteria, the "true bacteria," one of the three major domains of life. Most are single-celled organisms. Some species of bacteria cause disease. (*See also* archaea.)

basal
Situated near the base. The basal surface of a cell is opposite the apical surface.

basal body—*see* **centriole**

basal lamina
Thin mat of extracellular matrix that separates epithelial sheets, and many types of cells such as muscle cells or fat cells, from connective tissue. Sometimes called a basement membrane.

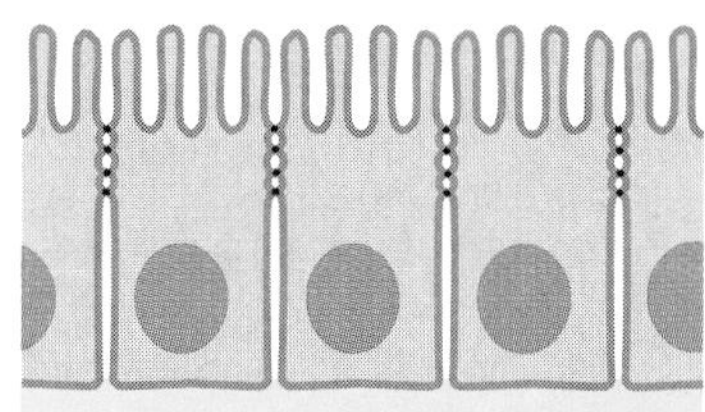

base
Molecule that accepts a proton in solution. Also used to refer to the purine or pyrimidines in DNA and RNA, which are organic bases.

base pair
Two nucleotides in an RNA or a DNA molecule that are specifically paired by hydrogen bonds—for example, G with C, and A with T or U.

Bcl2 family
Family of intracellular proteins that either promote or inhibit apoptosis by regulating the activation of caspases.

beta sheet (β sheet)
Folding pattern found in many proteins in which neighboring regions of the polypeptide chain associate side by side with each other through hydrogen bonds to give a rigid, flattened structure.

binding site
Region on the surface of a protein, typically a cavity or groove, that is complementary in shape to, and forms multiple noncovalent bonds with, a second molecule (the ligand).

bi-orientation
The symmetrical alignment of sister chromatid pairs on the mitotic spindle, such that one chromatid is attached to one spindle pole and the other chromatid to the opposite pole.

biosynthesis
The formation of complex molecules from simple substances by living cells.

bivalent
A duplicated chromosome paired with its homologous duplicated chromosome at the beginning of meiosis.

bond—*see* **chemical bond**

bond length
The distance between two atoms in a molecule, usually those linked by a covalent bond.

bond energy
The strength of the chemical linkage between two atoms, measured by the energy in kilocalories needed to break it.

buffer
Any weak acid or base that can release or take up protons, and thereby serve to maintain the pH under a variety of conditions.

C-terminus (carboxyl terminus)
The end of a polypeptide chain that carries an unattached carboxylic acid group.

cadherin
A member of a family of proteins that mediates Ca^{2+}-dependent cell–cell adhesion in animal tissues.

calmodulin (CaM)
Small Ca^{2+}-binding protein that modifies the activity of many target enzymes and membrane transport proteins in response to changes in Ca^{2+} concentration.

Ca^{2+}/calmodulin-dependent protein kinase (CaM kinase)
Enzyme that phosphorylates target proteins in response to an increase in Ca^{2+} ion concentration, through its interaction with the Ca^{2+}-binding protein calmodulin.

calorie
Unit of heat. One calorie (small "c") is the amount of heat needed to raise the temperature of 1 gram of water by 1°C.

cancer
Disease caused by abnormal and uncontrolled cell division resulting in localized growths, or tumors, which may spread throughout the body.

CaM—*see* **calmodulin**

carbohydrate
General term for sugars and related compounds with the general formula $(CH_2O)_n$. (*See* Panel 2–3, pp. 68–69.)

carbohydrate layer
A layer of sugar residues, including the polysaccharide portions of proteoglycans and oligosaccharides attached to protein or lipid molecules, on the outer surface of a cell.

carbon fixation
Process by which green plants incorporate carbon atoms from atmospheric carbon dioxide into sugars. The second stage of photosynthesis.

carbonyl group
Pair of atoms consisting of a carbon atom linked to an oxygen atom by a double bond. (*See* Panel 2–1, pp. 64–65.)

carboxyl group
Carbon atom linked both to an oxygen atom by a double bond and to a hydroxyl group. Molecules containing a carboxyl group are weak (carboxylic) acids. (*See* Panel 2–1, pp. 64–65.)

carboxyl terminus—*see* **C-terminus**

cascade—*see* **signaling cascade**

caspase
A family of proteases. Members of the family are activated as part of the pathway leading to apoptosis.

catabolism
General term for the enzyme-catalyzed reactions in a cell by which complex molecules are degraded to

simpler ones with release of energy. Intermediates in these catabolic reactions are sometimes called catabolites.

catalysis
The acceleration of a chemical reaction due to the presence of a substance (the catalyst) that itself remains unchanged after the reaction. In cells, virtually all biochemical reactions are catalyzed (by enzymes) to enable them to occur at the temperature of living matter and within the timescale required.

catalyst
Substance that accelerates a chemical reaction without itself undergoing a change. Enzymes are protein catalysts.

cation
Positively charged ion, such as Na^+ or $CH_3NH_3^+$. (Pronounced "cat-ion".)

Cdk—*see* **cyclin-dependent kinase**

Cdk inhibitor protein
Protein that inhibits cyclin-Cdk complexes, primarily to inhibit progress through the G_1 and S phases of the cell cycle.

cDNA—*see* **complementary DNA**

cell
The basic unit from which living organisms are made, consisting of an aqueous solution of organic molecules enclosed by a membrane. All cells arise from existing cells, usually by a process of division.

cell cortex
Specialized layer of cytoplasm on the inner face of the plasma membrane. In animal cells it is an actin-rich layer responsible for cell-surface movements.

cell cycle
Reproductive cycle of the cell: the orderly sequence of events by which a cell duplicates its contents and divides into two.

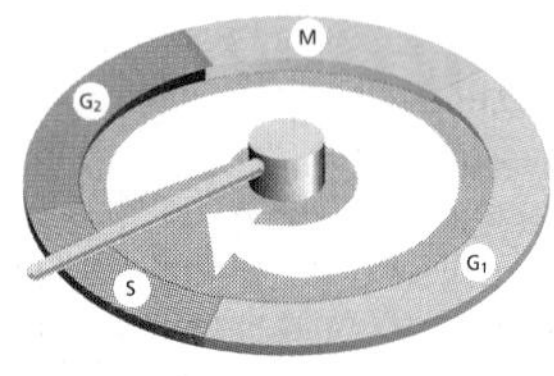

cell-cycle control system
Network of regulatory proteins that governs progression of a eucaryotic cell through the cell cycle.

cell division
Separation of a cell into two daughter cells. In eucaryotic cells it entails division of the nucleus (mitosis) closely followed by division of the cytoplasm (cytokinesis).

cell junction
Specialized region of connection between two cells or between a cell and the extracellular matrix.

cell line
Population of cells of plant or animal origin capable of dividing indefinitely in culture.

cell locomotion
Active movement of a cell from one location to another.

cell memory
The ability of cells and their descendants, without undergoing any change of DNA sequence, to retain a trace of the effects of past influences, displaying the consequences in persistently altered patterns of gene expression.

cell signaling
The molecular mechanisms by which cells detect and respond to external stimuli and send messages to other cells.

cellulose
Structural polysaccharide consisting of long chains of covalently linked glucose units. It provides tensile strength in plant cell walls.

cell wall
Mechanically strong fibrous layer deposited by a cell outside its plasma membrane. Prominent in most plants, bacteria, algae, and fungi, but not present in most animal cells.

central dogma
The principle that genetic information flows from DNA to RNA to protein.

centriole
Short cylindrical array of microtubules, usually found in pairs at the center of a centrosome in animal cells. Also found at the base of cilia and flagella (and called basal bodies).

centromere
Constricted region of a mitotic chromosome that holds sister chromatids together; also the site on the DNA where the kinetochore forms and then captures microtubules from the mitotic spindle.

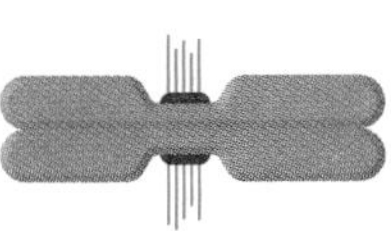

centrosome (cell center)
Centrally located organelle of animal cells that is the primary microtubule-organizing center and separates to form the two spindle poles during mitosis. In most animal cells it contains a pair of centrioles.

centrosome cycle
Duplication of the centrosome (during interphase) and separation of the two new centrosomes (at the beginning of mitosis), to form the poles of the mitotic spindle.

channel
An aqueous pore in a lipid membrane, with walls made of protein, through which selected ions or molecules can pass.

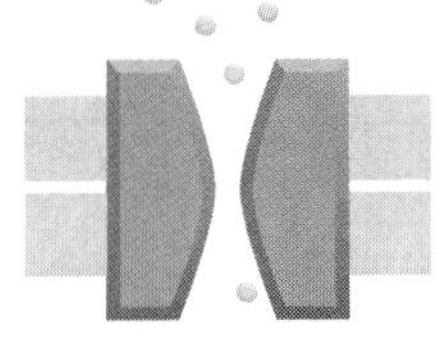

checkpoint
Point in the eucaryotic cell-division cycle where progress through the cycle can be halted until conditions are suitable for the cell to proceed to the next stage.

chemical bond
Chemical affinity between two atoms that holds them together. Types found in living cells include ionic bonds, covalent bonds, and hydrogen bonds.

chemical group
Set of covalently linked atoms, such as a hydroxyl group (–OH) or an amino group ($-NH_2$) that occurs in many different molecules and the chemical behavior of which is well characterized.

chemiosmotic coupling
Mechanism in which a gradient of hydrogen ions (a pH gradient) across a membrane is used to drive an energy-requiring process, such as ATP production or the transport of a molecule across a membrane.

chiasma (plural chiasmata)
X-shaped connection visible between paired homologous chromosomes in division I of meiosis, and which represents a site of crossing-over.

chlorophyll
Light-absorbing pigment that plays a central part in photosynthesis.

chloroplast
Specialized organelle in algae and plants that contains chlorophyll and in which photosynthesis takes place.

cholesterol
Lipid molecule with a characteristic four-ringed steroid structure that is an important component of the plasma membranes of animal cells. (See Figure 11–7.)

chromatid—*see* **sister chromatid**

chromatin
Complex of DNA, histones, and nonhistone proteins found in the nucleus of a eucaryotic cell. The material of which chromosomes are made.

chromatin-remodeling complex
Enzyme (typically multisubunit) that uses the energy of ATP hydrolysis to alter histone–DNA interactions in eucaryotic chromosomes; the resulting alteration changes the accessibility of the underlying DNA to other proteins, including those involved in transcription.

chromatography
General term for a type of technique used to separate molecules in a mixture on the basis of their size, charge, or their ability to bind to a particular chemical group. In a common form of the technique, the mixture is run through a column filled with a particulate matrix that is designed to bind (or let through) the desired molecule.

chromosome
Long threadlike structure composed of DNA and associated proteins that carries the genetic information of an organism. Especially visible when plant and animal cells undergo mitosis or meiosis.

chromosome condensation
Process by which a chromosome becomes packed into a more compact structure prior to M phase of the cell cycle.

citric acid cycle (TCA, or tricarboxylic acid cycle; Krebs cycle)
Central metabolic pathway in all aerobic organisms that oxidizes acetyl groups derived from food molecules to CO_2. In eucaryotic cells these reactions are located in the mitochondrial matrix.

cilium (plural cilia)
Hairlike extension on the surface of a cell with a core bundle of microtubules and capable of performing repeated beating movements. Cilia, in large numbers, drive the movement of fluid over epithelial sheets, as in the lungs.

cis
On the same side; for example, the *cis* Golgi network is that part closest to the endoplasmic reticulum.

clathrin
Protein that makes up the coat of one type of transport vesicle. Clathrin-coated vesicles bud from the Golgi apparatus on the outward secretory pathway and bud from the plasma membrane on the inward endocytic pathway.

cloning
Making many identical copies of a cell or a DNA molecule or an organism.

coated vesicle
Small membrane-enclosed organelle with a cage of proteins (the coat) on its cytosolic surface. It is formed by the pinching off of a protein-coated region of membrane.

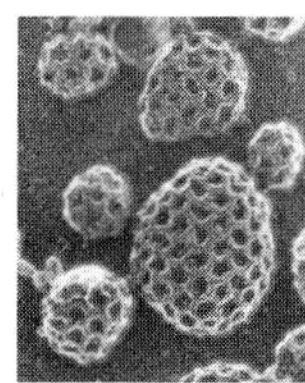

codon
Sequence of three nucleotides in a DNA or messenger RNA molecule that represents the instruction for incorporation of a specific amino acid into a growing polypeptide chain.

coenzyme A (CoA)
Small molecule used in the enzymatic transfer of acyl groups in the cell. (*See also* acetyl CoA and Figure 3–36.)

cohesin
Protein complex that forms a ring that holds sister chromatids together after DNA has been replicated in the cell cycle.

coiled-coil
Especially stable rod-like protein structure formed by two or more α helices coiled around each other.

collagen
Fibrous protein rich in glycine and proline that is a major component of the extracellular matrix and connective tissues. Exists in many forms: type I, the most common, is found in skin, tendon, and bone; type II is found in cartilage; type IV is present in basal laminae; and so on.

combinatorial control
Describes the way in which groups of proteins work together in combination to control the expression of a single gene.

complementary
Describes two molecular surfaces that fit together closely and form noncovalent bonds with each other. Examples include complementary base pairs, such A and T, and the two complementary strands of a DNA molecule.

complementary DNA (cDNA)
DNA molecule made as a copy of mRNA and therefore lacking the introns that are present in genomic DNA. Used to determine the amino acid sequence of a protein by DNA sequencing or to make the protein in large quantities by cloning followed by expression.

complex
A collection of macromolecules that are bound to each other by noncovalent bonds to form a large macromolecular structure. Associations of proteins are called protein complexes; associations of protein and nucleic acids are called nucleoprotein complexes.

complex trait
A heritable characteristic whose transmission to progeny does not obey Mendel's laws. Such traits, for example height, usually result from the interaction of multiple genes, each of which is inherited in Mendelian fashion.

condensation—*see* **chromosome condensation**

condensation reaction
Type of chemical reaction in which two organic molecules become linked to each other by a covalent bond with concomitant removal of a molecule of water.

condensin
Protein complexes with a ring-like structure that help carry out chromosome condensation.

conformation
Spatial location of the atoms of a molecule relative to each other. The precise shape of a protein or other macromolecule in three dimensions.

connective tissue
Tissues such as bone, tendons, and the dermis of the skin, in which extracellular matrix is plentiful and carries the mechanical load.

conserved synteny
Regions of the genome where corresponding genes are in the same order in the species being compared.

coupled reaction
One of a linked pair of chemical reactions in which free energy released by one reaction serves to drive the other reaction.

coupled transporter
Membrane transport protein that carries out transport in which the transfer of one molecule depends on the simultaneous or sequential transfer of a second molecule.

covalent bond
Stable chemical link between two atoms produced by sharing one or more pairs of electrons.

crossing-over
Process whereby two homologous chromosomes break at corresponding sites and rejoin to produce two recombined chromosomes, resulting in a physical exchange of DNA segments.

cyclic AMP (cAMP)
Nucleotide generated from ATP in response to hormonal stimulation of cell-surface receptors. cAMP acts as a signaling molecule by activating protein kinase A; it is hydrolyzed to AMP by a phosphodiesterase.

cyclic-AMP-dependent protein kinase (protein kinase A, PKA)
Enzyme that phosphorylates target proteins in response to a rise in intracellular cyclic AMP concentration.

cyclic photophosphorylation
Photosynthetic process involving photosystem I only, by which chloroplasts can generate ATP without making NADPH.

cyclin
Protein that periodically rises and falls in concentration in step with the eucaryotic cell cycle. Cyclins activate specific protein kinases (*see* cyclin-dependent protein kinases) and thereby help control progression from one stage of the cell cycle to the next.

cyclin-dependent protein kinase (Cdk)
Protein kinase that has to be complexed with a cyclin protein in order to act. Different Cdk–cyclin complexes trigger different steps in the cell-division cycle by phosphorylating specific target proteins.

cytochrome
Colored, heme-containing protein that transfers electrons during cellular respiration and photosynthesis.

cytokine
Small protein made and secreted by cells that acts on neighboring cells to alter their behavior. Cytokines act via cell-surface cytokine receptors.

cytokinesis
Division of the cytoplasm of a plant or animal cell into two, as distinct from the division of its nucleus (which is mitosis).

cytoplasm
Contents of a cell that are contained within its plasma membrane but, in the case of eucaryotic cells, outside the nucleus.

cytoskeleton
System of protein filaments in the cytoplasm of a eucaryotic cell that gives the cell shape and the capacity for directed movement. Its most abundant components are actin filaments, microtubules, and intermediate filaments.

cytosol
Contents of the main compartment of the cytoplasm, excluding membrane-enclosed organelles such as endoplasmic reticulum and mitochondria. The cell fraction remaining after membranes, cytoskeletal components, and other organelles have been removed.

DAG—*see* diacylglycerol

dalton
Unit of molecular mass. Defined as one-twelfth the mass of an atom of carbon-12 (1.66×10^{-24} g); approximately equal to the mass of a hydrogen atom.

denature
To cause a dramatic change in conformation of a protein or nucleic acid by heating it or by exposing it to chemicals. Usually results in the loss of biological function.

deoxyribonucleic acid—*see* DNA

desmosome
Specialized cell–cell junction, usually formed between two epithelial cells, characterized by dense plaques of protein into which intermediate filaments in the two adjoining cells are inserted.

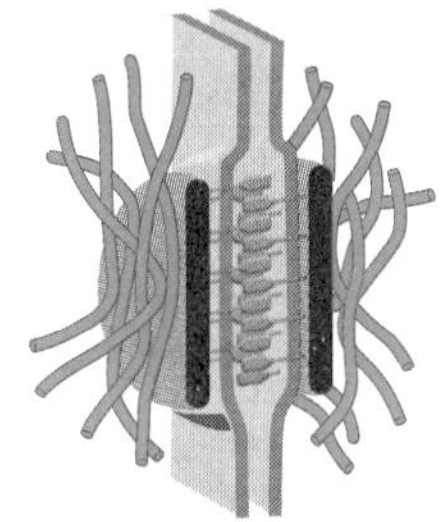

detergent
Soapy substance used by biochemists to solubilize membrane proteins.

diacylglycerol (DAG)
Lipid produced by the cleavage of membrane inositol phospholipids in response to extracellular signals. Composed of two fatty acid chains linked to glycerol, it serves as a membrane-located signaling molecule to help activate protein kinase C.

dideoxy DNA sequencing
The standard method of DNA sequencing. It utilizes DNA polymerases and chain-terminating nucleotides.

differentiation
Process by which a cell undergoes a progressive change to a more specialized and usually easily recognized cell type.

diffusion
The spread of molecules and small particles from one location to another by random, thermally driven movements.

dimer
A structure composed of two halves. A homodimer is composed of two identical subunits, a heterodimer of two different subunits.

diploid
A cell or organism containing two sets of homologous chromosomes and hence two copies of each gene or genetic locus. (*See also* haploid.)

disulfide bond (S–S bond)
Covalent linkage formed between two sulfhydryl groups on cysteines. Common way to join two proteins or to link together different parts of the same protein in the extracellular space.

divergence
The differences due to mutation that accumulate in two DNA sequences derived from a common ancestral sequence.

DNA (deoxyribonucleic acid)
Double-stranded polynucleotide formed from two separate chains of covalently linked deoxyribonucleotide units. It serves as the cell's store of genetic information that is transmitted from generation to generation.

DNA cloning—*see* cloning

DNA library
Collection of cloned DNA molecules, representing either an entire genome (genomic library) or copies of the mRNA produced by a cell (cDNA library).

DNA ligase—*see* ligase

DNA methylation
The enzymatic addition of methyl groups to cytosine bases in DNA. Methylation generally turns off genes by attracting proteins that block gene expression.

DNA microarray
A glass slide upon which a large number of short DNA molecules (typically in the tens of thousands) have been immobilized in an orderly pattern. Each of these DNA fragments acts as a probe for a specific gene, allowing the RNA products of thousands of genes to be monitored at the same time.

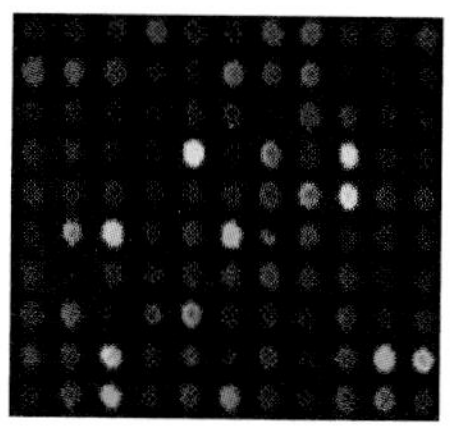

DNA polymerase—*see* polymerase

DNA repair
Collective term for the enzymatic processes that correct deleterious changes affecting the continuity or sequence of a DNA molecule.

DNA replication
The process by which a copy of a DNA molecule is made.

DNA transcription—*see* transcription

domain
Small discrete region of a structure. A protein domain is a compact and stable folded region of polypeptide. A membrane domain is a region of bilayer with a characteritistic lipid and protein composition.

double bond
A type of chemical linkage between two atoms formed by sharing four electrons.

double helix
The typical conformation of a DNA molecule in which two polynucleotide strands are wound around each other with base pairing between the strands.

dynamic instability
The property shown by microtubules of growing and shrinking repeatedly through the addition and loss of tubulin subunits from their exposed ends.

dynein
Member of a family of large motor proteins that undergo ATP-dependent movement along microtubules. Dynein is responsible for the bending of cilia.

electrochemical gradient
Driving force that causes an ion to move across a membrane. Caused by differences in ion concentration and in electrical charge on either side of the membrane.

electron
Fundamental subatomic particle with a unit negative charge (e^-).

electron acceptor
Atom or molecule that takes up electrons readily, thereby gaining an electron and becoming reduced.

electron carrier
Molecule such as cytochrome *c* that transfers an electron from a donor molecule to an acceptor molecule.

electron donor
Molecule that easily gives up an electron, becoming oxidized in the process.

electron-transport chain
A series of membrane-embedded electron carrier molecules along which electrons move from a higher to a lower energy level, as in oxidative phosphorylation and photosynthesis.

electrophoresis
Technique for separating mixture of proteins of DNA fragments by size and electric charge, by placing them on a polymer gel and subjecting them to an electric field. The molecules migrate through the gel at different speeds depending on their size and net charge.

electrostatic attraction
Attractive force that occurs between oppositely charged atoms. Examples are ionic bonds and the attractions between molecules containing polar covalent bonds.

element
Substance that cannot be broken down to any other chemical form; composed of a single type of atom.

embryonic stem cell (ES cell)
An undifferentiated cell type derived from the inner cell mass of an early mammalian embryo. Embryonic stem cells can be maintained indefinitely as a proliferating cell population (cell line) in culture, but remain capable of differentiating, when placed in an appropriate environment, to give any of the specialized cell types in the adult body.

endocytosis
Uptake of material into a cell by an invagination of the plasma membrane and its internalization in a membrane-bounded vesicle. (*See also* pinocytosis and phagocytosis.)

endoplasmic reticulum (ER)
Labyrinthine, membrane-enclosed compartment in the cytoplasm of eucaryotic cells, where lipids and secreted and membrane-bound proteins are made.

ES cell—*see* **embryonic stem cell**

endosome
Membrane-enclosed compartment of a eucaryotic cell through which endocytosed material passes on its way to lysosomes.

enhancer
Regulatory DNA sequence to which transcription regulators bind, influencing the rate of transcription of a structural gene that can be many thousands of base pairs away.

entropy
Thermodynamic quantity that measures the degree of disorder in a system; the higher the entropy, the more the disorder.

enzyme
A protein that catalyzes a specific chemical reaction.

epigenetic inheritance
Inherited that is superimposed on the information inherited in the DNA sequence itself. Often, information in the form of a particular type of chromatin structure (eg a certain pattern of histone modification or DNA methylation).

enzyme-coupled receptor
Transmembrane receptor proteins that activate an intracellular enzyme (either a separate enzyme or part of the receptor itself) in response to ligand binding to the extracellular part of the receptor.

epithelium (plural epithelia)
Sheet of cells covering an external surface or lining an internal body cavity.

equilibrium
In a chemical context, a state in which two or more reactions are proceeding at such a rate that they exactly balance each other and no net chemical change is occurring.

equilibrium constant (*K*)
A number that characterizes the steady state reached by a reversible chemical reaction. Given by the ratio of forward and reverse rate constants of a reaction. (*See* Table 3–1, p. 96.)

ER—*see* **endoplasmic reticulum**

***Escherichia coli* (*E. coli*)**
Rodlike bacterium normally found in the colon of humans and other mammals and widely used in biomedical research.

eubacteria
The proper term for the bacteria of common occurrence, used to distinguish them from archaea.

eucaryote
Living organism composed of one or more cells with a distinct nucleus and cytoplasm. Includes all forms of life except archaea and bacteria (the procaryotes) and viruses.

euchromatin
One of the two main states in which chromatin exists within an interphase cell, the other being heterochromatin. Characterized by particular histone modifications and associated proteins; genes in euchromatin are in general able to be expressed.

evolution
The gradual change in living organisms taking place over generations that results in new species being formed.

exocytosis
Process by which most molecules are secreted from a eucaryotic cell. These molecules are packaged in membrane-enclosed vesicles that fuse with the plasma membrane, releasing their contents to the outside.

exon
Segment of a eucaryotic gene that is transcribed into RNA and expressed; dictates the amino acid sequence of part of a protein.

exon shuffling
Evolutionary process by which new genes form by linking together combinations of initially separate exons encoding different protein domains.

extracellular matrix
Complex network of polysaccharides (such as glycosaminoglycans or cellulose) and proteins (such as collagen) secreted by cells. A structural component of tissues that also influences their development and physiology.

extracellular signal molecule
Any molecule present outside the cell that can elicit a response inside the cell when the molecule binds to a receptor protein. Some signal molecules, such as steroid hormones, can enter cells and act on internal receptors, whereas others, such as proteins, act at receptors embedded in the plasma membrane and exposed on the cell surface.

FAD—*see* **$FADH_2$**

$FADH_2$ (reduced flavin adenine dinucleotide)
A major electron carrier in metabolism produced by reduction of FAD during the oxidation of catabolites such as succinate.

fat
Lipids used by living cells to store metabolic energy. Mainly composed of triacylglycerols. (*See* Panel 2–4, pp. 70–71.)

fatty acid
Compound such as palmitic acid that has a carboxylic acid attached to a long hydrocarbon chain. Used as a major source of energy during metabolism and as a starting point for the synthesis of phospholipids. (*See* Panel 2–4, pp. 70–71.)

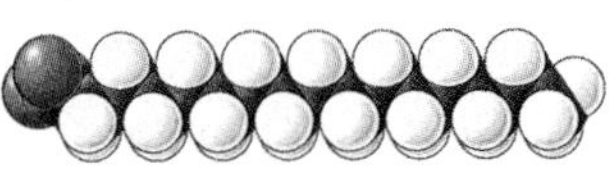

feedback inhibition
A form of metabolic control in which the end product of a chain of enzymatic reactions reduces the activity of an enzyme early in the pathway.

fermentation
The breakdown of organic molecules without the involvement of molecular oxygen. This form of oxidation is less complete than in aerobic processes and yields less energy.

fertilization
Sequence of events that starts when a sperm cell makes contact with an egg and leads to their fusion and further development.

fibroblast
Common cell type found in connective tissue that secretes an extracellular matrix rich in collagen and other extracellular matrix macromolecules. Migrates and proliferates readily in wounded tissue and in tissue culture.

fibronectin
Extracellular matrix protein that binds to integrins on cell surfaces, helping cells to adhere to the matrix.

fibrous protein
A protein with an elongated shape. Typically one such as collagen or intermediate filament protein that is able to associate into long filamentous structures.

filopodium (plural filopodia)
Long thin actin-containing extension on the surface of an animal cell. Sometimes has an exploratory function, as in a growth cone.

flagellum (plural flagella)
Long, whiplike protrusion that drives a cell through a fluid medium by its beating. Eucaryotic flagella are longer versions of cilia; bacterial flagella are completely different, being smaller and simpler in construction.

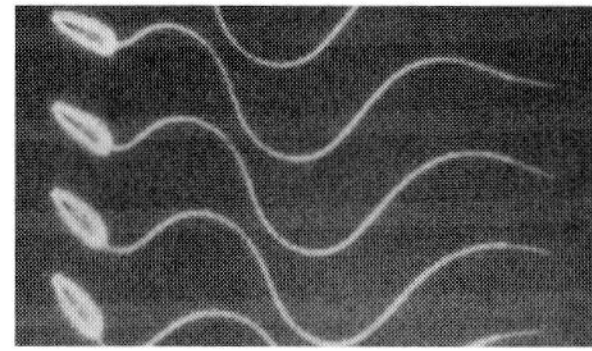

free energy (*G*)
Energy that can be extracted from a system to do useful work, such as driving a chemical reaction. The standard free energy of a substance, G^0, is its free energy measured at a defined concentration, temperature and pressure.

free-energy change (ΔG)
"Delta G": the difference in free energy between reactant and product molecules in a chemical reaction. A large negative value of ΔG indicates that the reaction has a strong tendency to occur. The standard free-energy change ($\Delta G°$) is the free-energy change measured at defined concentration, temperature and pressure.

G, ΔG, $\Delta G°$—*see* **free energy, free-energy change**

G-protein–coupled receptor
Cell-surface receptor that associates with an intracellular trimeric GTP-binding protein (G protein) after receptor activation by an extracellular ligand. These receptors are sevenpass transmembrane proteins.

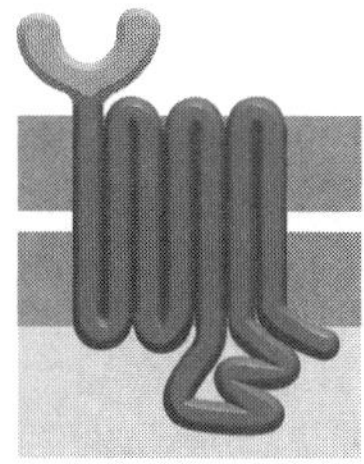

GAG—*see* **glycosaminoglycan**

gain-of-function mutation
A mutation that increases the activity of a gene, or makes it active in inappropriate circumstances; such mutations are usually dominant.

gamete—*see* **germ cell**

gap junction
Communicating cell–cell junction that allows ions and small molecules to pass from the cytoplasm of one cell to the cytoplasm of the next.

GDP (guanosine 5′-diphosphate)
Nucleotide that is produced by the hydrolysis of the terminal phosphate of GTP, a reaction that also produces inorganic phosphate. When free in solution, GDP is rapidly rephosphorylated to GTP, usually by the transfer of the terminal phosphate from ATP in the reaction ATP + GDP → ADP + GTP.

gene
Region of DNA that controls a discrete hereditary characteristic of an organism, usually responsible for specifying a single protein or RNA molecule.

gene duplication
The accidental duplication of a gene (or of a stretch of DNA containing several genes) in the genome. The resulting two copies of the gene can then diverge by the accumulation of mutations. Gene families arise through a series of gene duplication events.

gene expression
The process by which a gene makes its effect on an cell or organism by directing the synthesis of a protein or an RNA molecule with a characteristic activity.

gene replacement
The replacement of a normal gene in an organism with one that has been mutated *in vitro*; used to investigate the gene's function.

general transcription factors
Proteins that assemble on the promoters of many eucaryotic genes near the start site of transcription and load the RNA polymerase in the correct position.

genetic code
Set of rules specifying the correspondence between nucleotide triplets (codons) in DNA or RNA and amino acids in proteins.

genetic engineering—*see* **recombinant DNA technology**

genetic instability
Increased mutation rate seen, for example, in cancer cells, resulting from the presence of mutations that interfere with the accurate replication and maintenance of the genome.

genetic map
A graphic representation of the order of genes in chromosomes, spaced according to the amount of recombination that occurs between them.

genetic screen
A search through a collection of mutants for a particular phenotype.

genetics
The study of the genes of an organism based on heredity and variation.

genome
The total genetic information carried by a cell or an organism (or the DNA molecules that carry this information).

genotype
Set of genes carried by an individual cell or organism.

germ cell (gamete)
Cell type in a diploid organism that carries only one set of chromosomes and is specialized for sexual reproduction. A sperm or an egg.

germ line
The lineage of reproductive cells that contributes to the formation of a new generation of organisms, as distinct from somatic cells, which form the body and leave no descendants in the next generation.

globular protein
Any protein with an approximately rounded shape. Most enzymes are globular.

gluconeogenesis
The synthesis of glucose from small organic molecules such as lactate, pyruvate, or amino acids.

glucose
Six-carbon sugar that plays a major role in the metabolism of living cells. Stored in polymeric form as glycogen in animal cells and as starch in plant cells. (*See* Panel 2–3, pp. 68–69.)

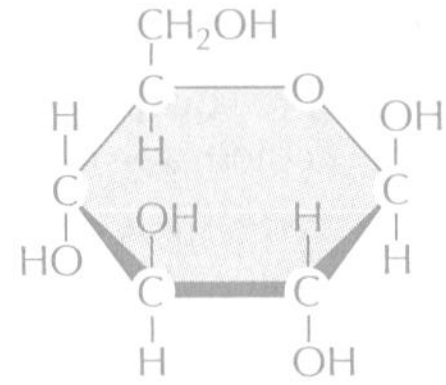

glycogen
Polysaccharide composed exclusively of glucose units used to store energy in animal cells. Large granules of glycogen are especially abundant in liver and muscle cells.

glycolipid
Membrane lipid molecule with a short carbohydrate chain attached to a hydrophobic tail.

glycolysis
Ubiquitous metabolic pathway in the cytosol in which sugars are incompletely degraded with production of ATP. (Literally, "sugar splitting.")

glycoprotein
Any protein with one or more covalently linked oligosaccharide chains. Includes most secreted proteins and most proteins exposed on the outer surface of the plasma membrane.

glycosaminoglycan (GAG)
Family of high-molecular-weight polysaccharides containing amino sugars; found as protective coats around animal cells.

Golgi apparatus
Membrane-enclosed organelle in eucaryotic cells where the proteins and lipids made in the endoplasmic reticulum are modified and sorted for transport to other sites. (Named after its discoverer, Camillo Golgi.)

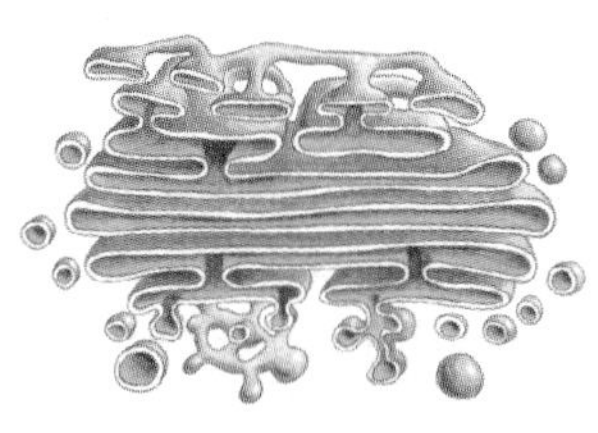

G_1 phase
Gap 1 phase of the eucaryotic cell cycle, between the end of cytokinesis and the start of DNA synthesis.

G_1-Cdk
Cyclin-dependent kinase whose activity drives the cell through G_1 phase.

G_1/S-Cdk
Cyclin-dependent kinase whose activity triggers entry into S phase of the cell cycle.

G_2 phase
Gap 2 phase of the eucaryotic cell cycle, between the end of DNA synthesis and the beginning of mitosis.

G protein
One of a large family of GTP-binding proteins composed of three different subunits (heterotrimeric GTP-binding proteins) that are important intermediaries in intracellular signaling pathways. Usually activated by the binding of a hormone or other ligand to a transmembrane receptor.

green fluorescent protein (GFP)
Fluorescent protein (from a jellyfish) that is widely used as a marker for monitoring the movement of proteins in living cells.

group—*see* chemical group

growth factor
Extracellular polypeptide signaling molecule that stimulates a cell to grow or proliferate. Examples are epidermal growth factor (EGF) and platelet-derived growth factor (PDGF).

GTP (guanosine 5′-triphosphate)
Major nucleoside triphosphate used in the synthesis of RNA and in some energy-transfer reactions. Also has a special role in microtubule assembly, protein synthesis, and cell signaling.

GTP-binding protein
An allosteric protein whose conformation and activity are determined by its association with either GTP or GDP. Includes many proteins involved in cell signaling, such as Ras and G proteins.

haploid
A cell or organism with only one set of chromosomes, as in a sperm cell or a bacterium. (*See also* diploid.)

haplotype block
A combination of alleles and other DNA markers on a chromosome that has been inherited in a large, linked block—undisturbed by genetic recombination—across many generations.

helix
An elongated structure in which a filament or thread twists in a regular fashion around a central axis.

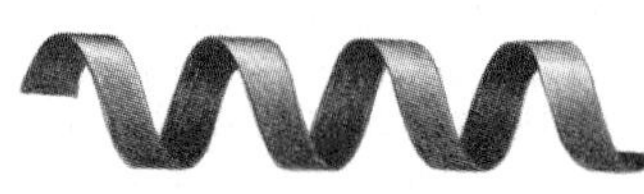

α-helix—*see* alpha helix

hemidesmosome
Specialized anchoring cell junction between an epithelial cell and the underlying basal lamina.

heredity
The transmission from one generation to another of genetic factors that determine individual characteristics. Responsible for the similarity between parents and children.

heterochromatin
Region of a chromosome that remains unusually condensed and transcriptionally inactive during interphase.

heterozygous
Describes an organism with dissimilar alleles for a given gene.

high-energy bond
Covalent bond whose hydrolysis releases an unusually large amount of free energy under the conditions existing in a cell. Examples include the phosphodiester bonds in ATP and the thioester linkage in acetyl CoA.

histone
One of a group of abundant basic proteins, rich in arginine and lysine, that are associated with DNA in chromosomes to form nucleosomes.

histone deacetylase
Enzyme that removes acetyl groups from lysines present in histones; the acetylation state of histones acts as a signal that attracts other proteins that activate or repress transcription.

homolog
(1) *See* homologous chromosome. (2) Any structure or macromolecule that has a close similarity to another as a result of common ancestry.

homologous
Describes organs or molecules that are similar because of their common evolutionary origin. Specifically it describes similarities between protein sequences or nucleic acid sequences.

homologous chromosome
One of the two copies of a particular chromosome in a diploid cell, one from the father and the other from the mother.

homologous gene—*see* **homologous**

homologous recombination
Genetic exchange between a pair of identical or very similar DNA sequences, typically located on a pair of homologous chromosomes. A similar process is used to repair double-strand breaks in DNA.

homozygous
Describes an organism having identical alleles for a given gene.

horizontal gene transfer
Process through which DNA is passed from one organism to another, permanently changing the DNA composition of the recipient. This contrasts with "vertical" gene transfer, which refers to the inheritance of genes from parent to progeny.

hormone
A chemical substance produced by one set of cells in a multicellular organism and transported via body fluids to target tissues on which it exerts a specific effect.

hybridization
Experimental process in which two complementary nucleic acid strands form a double helix; a powerful technique for detecting specific nucleotide sequences.

hydrogen bond
A weak noncovalent chemical bond between an electronegative atom such as nitrogen or oxygen and a hydrogen atom bound to another electronegative atom.

hydrogen ion
Commonly used term for a proton (H^+) in aqueous solution, the basis of acidity. Since the proton readily combines with a water molecule to form H_3O^+, it is more accurate to call it a hydronium ion. (*See* Panel 2–2, pp. 66–67.)

hydrolysis (adjective hydrolytic)
Cleavage of a covalent bond with accompanying addition of water, –H being added to one product of the cleavage and –OH to the other.

hydronium ion (H_3O^+)
The form taken by a proton (H^+) in aqueous solution. (*See* Panel 2–2, pp. 66–67.)

hydrophilic
Polar molecule or part of a molecule that forms enough hydrogen bonds to water to dissolve readily in water. (Literally, "water loving.")

hydrophobic
Nonpolar molecule or part of a molecule that cannot form favorable bonding interactions with water molecules and therefore does not dissolve in water. (Literally, "water hating.")

hydroxyl (–OH)
Chemical group consisting of a hydrogen atom linked to an oxygen, as in an alcohol. (*See* Panel 2–1, pp. 64–65.)

induced pluripotent stem cell (iPS cell)
Somatic cell reprogrammed into an embryonic stem-cell-like state by artificially introducing a particular set of genes.

initiation factor
Protein that promotes the proper association of ribosomes with mRNA and is required for the initiation of protein synthesis.

initiator tRNA
Special tRNA that initiates translation. It always carries the amino acid methionine.

inositol
Sugar molecule with six hydroxyl groups that forms the framework for inositol phospholipids.

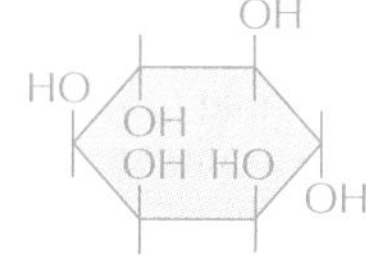

inositol 1,4,5-trisphosphate (IP_3)
Small intracellular signaling molecule produced during activation of the inositol phospholipid signaling pathway; causes Ca^{2+} release from the endoplasmic reticulum.

inositol phospholipids (phosphoinositides)
Minor lipid components of plasma membranes that contain phosphorylated inositol derivatives; important both for distinguishing different intracellular membranes and for signal transduction in eucaryotic cells.

***in situ* hybridization**
Technique in which a single-stranded RNA or DNA probe is used to locate a gene or an mRNA molecule in an entire cell or tissue.

integrin
Family of transmembrane proteins present on cell surfaces that enable cells to adhere to each other and to the extracellular matrix, being also involved in cell signaling.

intermediate filament
Fibrous protein filament (about 10 nm in diameter) that forms ropelike networks in animal cells. Often used as a structural element that resists tension applied to the cell from outside.

interphase
Long period of the cell cycle between one mitosis and the next. Includes G_1 phase, S phase, and G_2 phase.

interphase chromosome
State of a eucaryotic chromosome when the cell is between divisions; these chromosomes are active in transcription and much more extended than mitotic chromosomes.

intron
Noncoding region of a eucaryotic gene that is transcribed into an RNA molecule but is then excised by RNA splicing to produce mRNA.

intracellular signaling molecule
Molecule (usually a protein) that is part of the mechanism for transducing and transmitting signals inside a cell.

intracellular signaling pathway
The set of proteins and small-molecule second messengers that interact with each other to relay a signal from the cell membrane to its final destination in the cytoplasm or nucleus.

in vitro
Term used by biochemists to describe a process taking place in an isolated cell-free extract. Also used by cell biologists to refer to cells growing in culture (*in vitro*), as opposed to in an organism (*in vivo*). (Latin for "in glass.")

in vivo
In an intact cell or organism. (Latin for "in life.")

ion
An atom carrying an electrical charge, either positive or negative.

ion channel
Transmembrane protein or protein complex that forms a water-filled channel across the lipid bilayer through which specific inorganic ions can diffuse down their electrochemical gradients.

ion-channel-coupled receptor
Transmembrane receptor protein or protein complex that forms a gated ion channel that opens in response to the binding of a ligand to the external face of the channel.

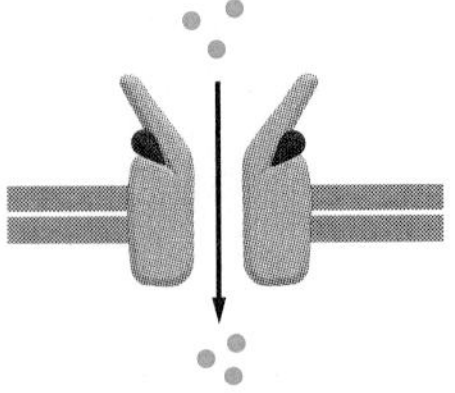

ionic bond
Attractive force that holds together two ions, one positive, the other negative.

IP_3—*see* inositol 1,4,5-trisphosphate

iron–sulfur center
One of a family of electron transporters containing iron atoms linked to sulfur atoms and cysteine side chains; found in electron-transport chains such as those in mitochondria and chloroplasts.

isomer (stereoisomer)
One of two or more substances that contain the same atoms and have the same molecular formula (such as $C_6H_{12}O_6$) but differ in the spatial arrangement of these atoms. Optical isomers differ only by being mirror images of each other.

isotopes
Two or more forms of an atom that have the same chemistry but differ in atomic weight. May be either stable or radioactive.

***K*—*see* equilibrium constant**

K^+
Potassium ion—a major ionic constituent of living cells.

K_M
The concentration of substrate at which an enzyme works at half its maximum rate. Large values of K_M usually indicate that the enzyme binds to its substrate with relatively low affinity.

karyotype
A display of the full set of chromosomes of a cell arranged with respect to size, shape, and number.

kilocalorie (kcal)
Unit of heat equal to 1000 calories. Often used to express the energy content of food or molecules: bond strengths, for example, are measured in kcal/mole. An alternative unit in wide use is the kilojoule.

kilojoule (kJ)
Standard unit of energy equal to 0.239 kilocalories.

kinase—*see* protein kinase

kinesin
A large family of motor proteins that use the energy of ATP hydrolysis to move along a microtubule.

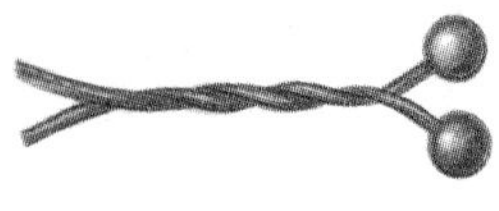

kinetochore
Complex protein-containing structure on a mitotic chromosome to which microtubules attach. The kinetochore forms on the part of the chromosome known as the centromere.

knockout mouse
A genetically engineered mouse in which a specific gene has been inactivated, for example by introducing a deletion in its DNA.

lagging strand
One of the two newly made strands of DNA found at a replication fork. The lagging strand is made in discontinuous lengths that are later joined covalently.

lamellipodium
Dynamic sheetlike extension on the surface of an animal cell, especially one migrating over a surface.

law of independent assortment
The second law of heredity, derived by Mendel, which states that during gamete formation the alleles for different traits segregate independently of one another.

law of segregation
The first law of heredity, derived by Mendel, which states that the maternal and paternal alleles for a trait separate from one another during gamete formation and then reunite during fertilization.

leading strand
One of the two newly made strands of DNA found at a replication fork. The leading strand is made by continuous synthesis in the 5′-to-3′ direction.

ligand
General term for a molecule that binds to a specific site on a protein.

ligand-gated channel
An ion channel that opens when it binds a small molecule such as a neurotransmitter.

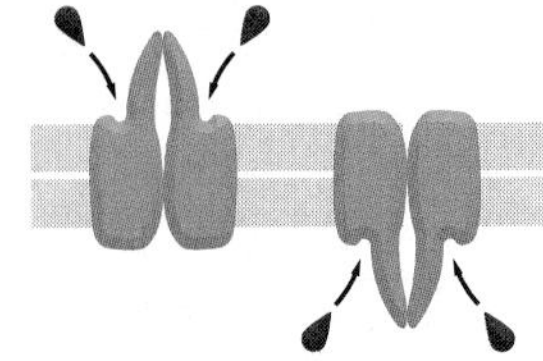

ligase
Enzyme that joins two DNA strands together end to end.

lipid
Organic molecule that is insoluble in water but dissolves readily in nonpolar organic solvents. One class, the phospholipids, forms the structural basis of biological membranes.

lipid bilayer
Thin bimolecular sheet of mainly phospholipid molecules that forms the structural basis

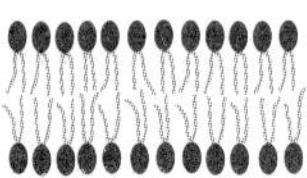

for all cell membranes. The two layers of lipid molecules are packed with their hydrophobic tails pointing inward and their hydrophilic heads outward, exposed to water.

local mediator
Secreted signal molecule that acts at a short range on adjacent cells.

loss-of-function mutation
A mutation that reduces or eliminates the activity of a gene. Such mutations are usually recessive: the organism can function normally as long as it retains at least one normal copy of the affected gene.

lumen
Cavity enclosed by an epithelial sheet (in a tissue) or by a membrane (in a cell), as in the lumen of the endoplasmic reticulum. (From Latin, *lumen*, light or opening.)

lymphocyte
White blood cell that mediates the immune response to a foreign molecule (an antigen). Lymphocytes are either of the antibody-secreting B-cell type or the T-cell type that regulates responses and also forms the cell-mediated immune response system.

lysosome
Intracellular membrane-enclosed organelle containing digestive enzymes, typically those most active at the acid pH found in these organelles.

M–Cdk
Active protein complex formed at the start of M phase of the cell cycle by an M-cyclin and the mitotic cyclin-dependent protein kinase (Cdk).

M-cyclin
Cyclin protein that binds to mitotic Cdk to form M–Cdk at the start of M phase of the cell cycle.

M phase
Period of the eucaryotic cell cycle during which the nucleus and cytoplasm divide.

macromolecule
Molecule such as a protein, nucleic acid, or polysaccharide with a molecular mass greater than a few thousand daltons. (From Greek, *makros*, large.)

macrophage
Cell found in animal tissues that is specialized for the uptake of particulate material by phagocytosis; derived from a type of white blood cell.

MAP kinase
Mitogen-activated protein kinase. Protein kinase that performs a crucial step in relaying signals from cell-surface receptors to the nucleus. It is the final kinase in a three-kinase sequence called the MAP kinase cascade.

matrix
Most generally, a space within which something is formed. In cell biology, this word often refers to the large internal compartment of the mitochondrion. The mitochondrial matrix contains a concentrated mixture of special enzymes that catalyze oxidation reactions, as well as the mitochondrial genome and the proteins needed to express mitochondrial genes. (*See* Figure 14–4.)

meiosis
Special type of cell division by which eggs and sperm cells are made. Two successive nuclear divisions with only one round of DNA replication generates four haploid daughter cells from an initial diploid cell. (From Greek, *meiosis*, diminution.)

membrane
Thin sheet of lipid molecules and associated proteins that encloses all cells and forms the boundaries of many eucaryotic organelles.

membrane-enclosed organelle
Any organelle in the eucaryotic cell that is surrounded by a lipid bilayer membrane, for example, the endoplasmic reticulum, Golgi apparatus, and lysosome.

membrane domain
Functionally specialized region in a cell membrane, characterized by the presence of particular proteins.

membrane potential
Voltage difference across a membrane due to a slight excess of positive ions on one side and of negative ions on the other. A typical membrane potential for an animal cell plasma membrane is –60 mV (inside negative), measured relative to the surrounding fluid.

membrane protein
A protein associated with a lipid bilayer; can be either integral (transmembrane) or peripheral.

membrane transport protein
Any protein embedded in a membrane that serves as a carrier of ions or small molecules from one side to the other.

messenger RNA (mRNA)
RNA molecule that specifies the amino acid sequence of a protein. Produced by RNA splicing (in eucaryotes) from a larger RNA molecule made by RNA polymerase as a complementary copy of DNA. It is translated into protein in a process catalyzed by ribosomes.

metabolic pathway
Sequence of enzymatic reactions in which the product of one reaction is the substrate of the next.

metabolism
The sum total of the chemical reactions that take place in the cells of a living organism resulting in growth, division, energy production, excretion of waste and so on.

metaphase
Stage of mitosis at which chromosomes are firmly attached to the mitotic spindle at its equator but have not yet segregated toward opposite poles.

metastasis
The spreading of cancer cells throughout the body from the initial site of the tumor.

methyl (–CH_3) group
Hydrophobic chemical group derived from methane (CH_4). (*See* Panel 2–1, pp. 64–65.)

micro
Prefix denoting 10^{-6}.

micrograph
Picture taken through a microscope. Either a light micrograph or an electron micrograph, depending upon the type of microscope used.

micrometer (μm)
Unit of measurement often applied to cells and organelles. Equal to 10^{-6} meter or 10^{-4} centimeter.

microRNA (miRNA)
Small noncoding RNAs that control gene expression by base-pairing with specific mRNAs to regulate their stability and their translation.

microscope
Instrument for viewing extremely small objects. A light microscope utilizes a focused beam of visible light and is used to examine cells and organelles. An electron microscope utilizes a beam of electrons and can be used to examine objects as small as individual molecules.

microtubule
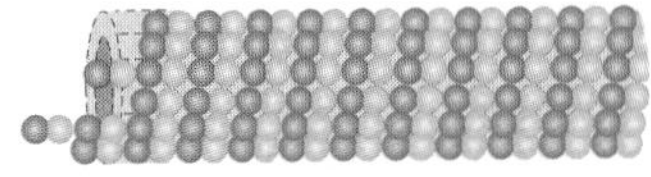
Long, stiff, cylindrical structure composed of the protein tubulin. Used by eucaryotic cells to regulate their shape and control their movements.

milli-
Prefix denoting 10^{-3}.

mismatch repair
Important error correction mechanism in DNA replication that is triggered by the misfit ("mismatch") of noncomplementary base pairs.

mitochondrion (plural mitochondria)

Membrane-enclosed organelle, about the size of a bacterium, that carries out oxidative phosphorylation and produces most of the ATP in eucaryotic cells.

mitogen
An extracellular signal molecule that stimulates cell proliferation.

mitosis
Division of the nucleus of a eucaryotic cell, which involves condensation of the DNA into visible chromosomes. (From Greek, *mitos*, a thread, referring to the threadlike appearance of the condensed chromosomes.)

mitotic chromosome
Highly condensed duplicated chromosome with the two new chromosomes (also called sister chromatids) still held together at the centromere. A chromosome during one of the stages of mitosis.

mitotic spindle
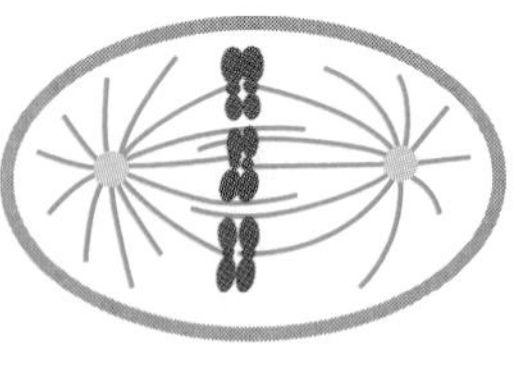
Array of microtubules and associated molecules that forms between the opposite poles of a eucaryotic cell during mitosis; during the separation of the duplicated chromosomes, the spindle serves to move the two chromosome sets apart.

mobile genetic element
Short segment of DNA that can move, sometimes through an RNA intermediate, from one location in a genome to another. They are an important source of genetic variation in most genomes.

model organism
An organism selected for intensive study as a representative of a large group of species. Examples are the mouse (representing mammals), the yeast *Saccharomyces cerevisiae* (representing a unicellular eucaryote), and *Escherichia coli* (representing bacteria).

mole
M grams of a substance, where *M* is its relative molecular mass (molecular weight); this will be 6×10^{23} molecules of the substance.

molecular switch
Protein or protein complex that operates in an intracellular signaling pathway and can reversibly switch between an active and inactive state.

molecular weight
Mass of a molecule expressed in daltons.

molecule
Group of atoms joined together by covalent bonds.

monomer
Small molecule that can be linked to others of a similar type to form a larger molecule (polymer).

monomeric GTPase
Small, single-subunit GTP-binding protein. Proteins of this family, such as Ras and Rho, are part of many different signaling pathways.

motor protein
Protein such as myosin or kinesin that uses energy derived from ATP hydrolysis to propel itself along a protein filament or polymeric molecule.

mRNA—*see* **messenger RNA**

mutation
A randomly produced, heritable change in the nucleotide sequence of a chromosome.

myofibril
Long, highly organized bundle of actin, myosin, and other proteins in the cytoplasm of muscle cells that contracts by a sliding filament mechanism.

myosin
Type of motor protein that uses ATP to drive movements along actin filaments. Myosin II is a large protein that forms the thick filaments of skeletal muscle. Smaller myosins, such as myosin I, are widely distributed and are responsible for many actin-based movements.

N-terminus (amino terminus)
The end of a polypeptide chain that carries a free α-amino group.

Na^+
Sodium ion—a major ionic constituent of living cells.

NAD^+ (nicotine adenine dinucleotide)
Activated carrier molecule that participates in an oxidation reaction by accepting a hydride ion (H^-) from a donor molecule, thereby producing NADH. Widely used in the energy-producing breakdown of sugar molecules. (*See* Figure 3–34.)

NADPH (nicotine adenine dinucleotide phosphate)
A carrier molecule closely related to NADH used as an electron donor in biosynthetic pathways. In the process it is oxidized to $NADP^+$. (*See* Figure 3–35.)

Na^+-K^+ pump (Na^+-K^+ ATPase, sodium pump)
Transmembrane carrier protein, found in the plasma membrane of most animal cells, that pumps Na^+ out of and K^+ into the cell, using the energy derived from ATP hydrolysis.

nanometer (nm)
Unit of length commonly used to measure molecules and cell organelles. 1 nm = 10^{-3} µm = 10^{-9} m.

Nernst equation
Quantitative expression that relates the equilibrium ratio of concentrations of an ion on either side of a permeable membrane to the voltage difference across the membrane.

nerve cell—*see* **neuron**

nerve terminal
The ending of an axon from which signals are sent to adjoining cells, usually at a synapse.

neuron (nerve cell)
Cell with long processes specialized to receive, conduct, and transmit signals in the nervous system.

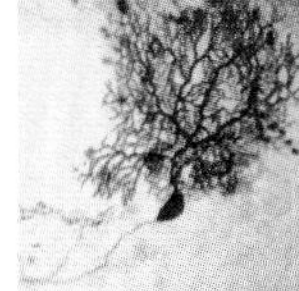

neurotransmitter
Small signaling molecule secreted by a nerve cell at a chemical synapse to signal to the postsynaptic cell. Examples include acetylcholine, glutamate, GABA, and glycine.

nitric oxide (NO)
Small highly diffusible molecule widely used as an intracellular signal.

nitrogen fixation
Conversion of nitrogen from the atmosphere into nitrogen-containing organic molecules by soil bacteria and cyanobacteria.

NO—*see* **nitric oxide**

noncovalent bond
Chemical bond in which, in contrast with a covalent bond, no electrons are shared. Noncovalent bonds are relatively weak, but they can sum together to produce strong, highly specific interactions between molecules. Examples are hydrogen bonds and van der Waals interactions.

nonhomologous end-joining
Mechanism for repairing double-strand breaks in DNA in which the two broken ends are brought together and rejoined without requiring sequence homology.

nonpolar
Describes a molecule that lacks a local accumulation of positive or negative charge. Nonpolar molecules are generally insoluble in water.

nuclear envelope
Double membrane surrounding the nucleus. Consists of outer and inner membranes perforated by nuclear pores.

nuclear lamina
Fibrous layer on the inner surface of the inner nuclear membrane formed as a network of intermediate filaments made from nuclear lamins.

nuclear magnetic resonance (NMR) spectroscopy
Technique used for determining the three-dimensional structure of a protein; it is performed in solution without requiring a protein crystal.

nuclear pore
Channel through the nuclear envelope that allows selected large molecules to move between the nucleus and the cytoplasm.

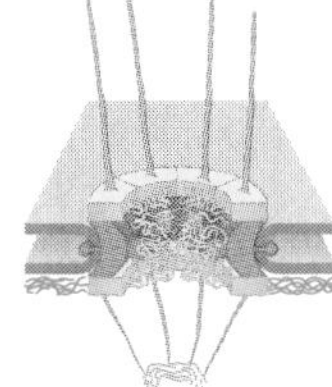

nuclear receptor
Receptor proteins present inside a eucaryotic cell that can bind to signal molecules that enter the cell, such as steroid hormones; the complex of nuclear receptor and signal molecule subsequently acts as a transcription regulator.

nucleic acid
RNA or DNA; consists of a chain of nucleotides joined together by phosphodiester bonds.

nucleolus
Large structure in the nucleus where ribosomal RNA is transcribed and ribosomal subunits are assembled.

nucleoside
Compound composed of a purine or pyrimidine base linked to either a ribose or a deoxyribose sugar. (*See* Panel 2–6, pp. 74–75.)

nucleosome
Beadlike structural unit of a eucaryotic chromosome composed of a short length of DNA wrapped around a core of histone proteins; the fundamental subunit of chromatin.

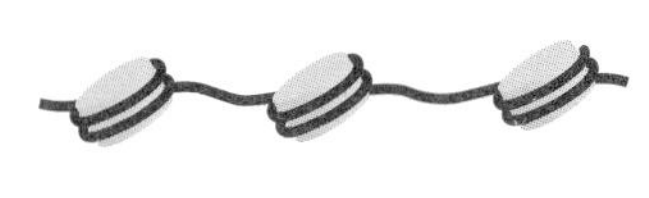

nucleotide
Nucleoside with a series of one or more phosphate groups joined by an ester linkage to the sugar moiety. DNA and RNA are polymers of nucleotides.

nucleus
The major organelle of a eucaryotic cell, which contains DNA organized into chromosomes. Also, when referring to an atom, the central mass built from neutrons and protons.

Okazaki fragment
Short length of DNA produced on the lagging strand during DNA replication. Adjacent fragments are rapidly joined together by DNA ligase to form a continuous DNA strand.

oligo-
Prefix that denotes a short polymer (oligomer). May be made of amino acids (oligopeptide), sugars (oligosaccharide), or nucleotides (oligonucleotide). (From Greek, *oligos*, few or little.)

oncogene
Any abnormally activated gene that can make a cell cancerous. Typically a mutant form of a normal gene (proto-oncogene) involved in the control of cell growth or division.

organic chemistry
The branch of chemistry concerned with compounds made of carbon. Includes essentially all of the molecules from which living cells are made, apart from water and metal ions such as Na^+.

organelle
A discrete structure or subcompartment of a eucaryotic cell (especially one that is visible in the light microscope) that is specialized to carry out a particular function. Examples include mitochondria and the Golgi apparatus.

origin recognition complex (ORC)
Large protein complex that is bound to the DNA at origins of replication in eucaryotic chromosomes throughout the cell cycle.

osmosis
Net movement of water molecules across a semipermeable membrane driven by a difference in concentration of solute on either side. The membrane must be permeable to water but not to the solute molecules.

osmotic pressure
Pressure that must be exerted on the low-solute concentration side of a semipermeable membrane to prevent the flow of water across the membrane as a result of osmosis.

oxidation
Loss of electron density from an atom, as occurs during the addition of oxygen to a carbon atom or when a hydrogen is removed from a carbon atom. The opposite of reduction. (*See* Figure 3–11.)

oxidative phosphorylation
Process in bacteria and mitochondria in which ATP formation is driven by the transfer of electrons from food molecules to molecular oxygen. Involves the intermediate generation of a pH gradient across a membrane and chemiosmotic coupling.

p53
Regulatory protein that responds to the presence of DNA damage, preventing the cell from entering S phase until the damage has been repaired.

pairing
In a genetic sense, the event early in meiosis in which two homologous chromosomes line up together to form a duplicated structure. (*See also* base pair.)

passive transport
The movement of a small molecule or ion across a membrane due to a difference in concentration or electrical charge.

patch-clamp recording
Technique in which the tip of a small glass electrode is sealed onto a patch of cell membrane, thereby making it possible to record the flow of current through individual ion channels in the patch.

PCR—*see* **polymerase chain reaction**

pedigree
The line of descent, or ancestry, of an individual animal.

peptide bond
Chemical bond between the carbonyl group of one amino acid and the amino group of a second amino acid—a special form of amide linkage. (*See* Panel 2–5, pp. 72–73.)

$$-\overset{\overset{\displaystyle O}{\|}}{C}-\underset{\underset{\displaystyle H}{|}}{N}-$$

peroxisome
Small membrane-enclosed organelle that uses molecular oxygen to oxidize organic molecules. Contains some enzymes that produce hydrogen peroxide (H_2O_2) and others that degrade it.

phagocytic cell
A cell such as a macrophage or neutrophil that is specialized to take up particles and microorganisms by phagocytosis.

phagocytosis
The process by which particulate material is engulfed ("eaten") by a cell. Prominent in predatory cells, such as *Amoeba proteus* and in cells of the vertebrate immune system such as macrophages.

phenotype
The observable character of a cell or organism.

phosphatidylcholine
Common phospholipid present in abundance in most biological membranes. (*See* Figure 11–6.)

phosphodiester bond
A covalent chemical bond in which two carbon atoms are held in ester linkage (via oxygen atoms) to the same phosphate group; phosphodiester bonds join the adjacent nucleotides in RNA or DNA. (*See* Figure 2–25.)

phosphoinositide 3-kinase (PI 3-kinase)
Enzyme that phosphorylates inositol phospholipids in the plasma membrane in response to signals received by a cell. The phosphorylated lipids become docking sites for intracellular signaling proteins,

phospholipase C
Enzyme associated with the plasma membrane that performs a crucial step in inositol phospholipid signaling pathways.

phospholipid
Type of lipid molecule used to make biological membranes. Generally composed of two fatty acids linked through glycerol phosphate to one of a variety of polar groups.

phosphorylation—*see* **protein phosphorylation**

photosynthesis
The process by which plants and some bacteria use the energy of sunlight to drive the synthesis of organic molecules from carbon dioxide and water.

photosystem
Large multiprotein complex containing chlorophyll that captures light energy.

phragmoplast
Structure made of microtubules and membrane vesicles that forms in the equatorial region of a dividing plant cell and from which the membrane that divides the daughter cells will be made.

pH scale
Scale used to measure the acidity of a solution: "p" refers to power of 10, "H" to hydrogen. Defined as the negative logarithm of the hydrogen ion concentration in moles per liter (M). Thus an acidic solution with pH 3 will contain 10^{-3} M hydrogen ions.

phylogenetic tree
Chart or "family tree" showing the evolutionary history of a group of organisms.

pinocytosis
Type of endocytosis in which soluble materials are taken up from the environment and incorporated into vesicles for digestion. (Literally, "cell drinking.")

PKA—*see* **protein kinase C**

plasma membrane
The membrane that surrounds a living cell.

plasmid
Small circular DNA molecule that replicates independently of the genome. Used extensively as a vector for DNA cloning.

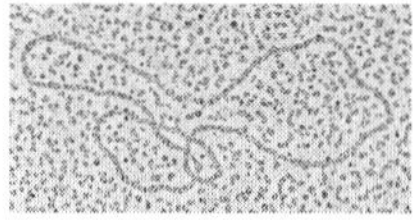

plasmodesma (plural plasmodesmata)
Cell–cell junction in plants in which a channel of cytoplasm lined by membrane connects two adjacent cells through a small pore in their cell walls.

point mutation
Change in a single nucleotide pair in a DNA sequence.

polar
Describes a molecule, or a covalent bond in a molecule, in which bonding electrons are attracted more strongly to specific atoms, thereby creating an uneven (or polarized) distribution of electric charge.

polarity
Refers to a structure such as an actin filament or a fertilized egg that has an inherent asymmetry—so that one end can be distinguished from the other.

polymer
Large and usually linear molecule made by the repetitive assembly, using covalent bonds, of multiple identical or similar subunits (monomers).

polymerase
General term for an enzyme that catalyzes addition of subunits to a polymer. DNA polymerase, for example, makes DNA, and RNA polymerase makes RNA.

polymerase chain reaction (PCR)
Technique for amplifying specific regions of DNA by multiple cycles of DNA synthesis, each followed by a brief heat treatment to separate complementary DNA strands.

polymorphism
The case where two or more variants (alleles) of a gene or variants of a DNA sequence coexist in the population and are each relatively common.

polynucleotide
A molecular chain of nucleotides chemically bonded by a series of phosphodiester linkages. A strand of RNA or DNA.

polypeptide, polypeptide chain
Linear polymer composed of multiple amino acids. Proteins are composed of one or more long polypeptide chains.

polypeptide backbone
The chain of atoms containing repeating peptide bonds that runs through a protein molecule and to which the amino acid side chains are attached.

polysaccharide
Linear or branched polymer composed of sugars. Examples are glycogen, hyaluronic acid, and cellulose.

polysome (polyribosome)
Messenger RNA molecule with multiple attached ribosomes that is engaged in protein synthesis.

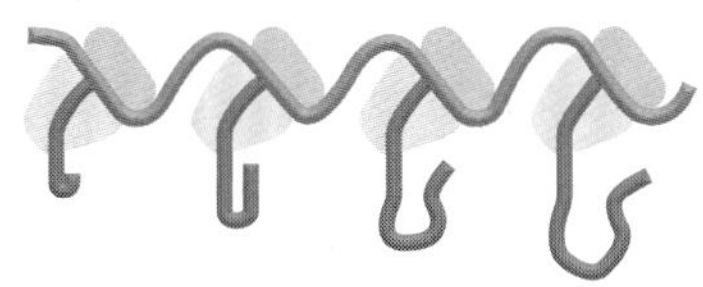

positive feedback loop
Situation in which the end product of a reaction stimulates its own production.

post-transcriptional control
Regulation of gene expression that occurs after transcription of the gene has begun; examples are regulation of RNA splicing and other RNA processing events, and regulation of translation by microRNA.

primary transcript—*see* **transcription**

primer
In DNA replication, a short length of RNA made at the beginning of a DNA synthesis event catalyzed by DNA polymerase; these RNA primers are subsequently removed and filled in with DNA.

procaryote
Major category of living cells distinguished by the absence of a nucleus. Procaryotes comprise the archaea and the eubacteria (commonly called bacteria), two of the three domains of life.

processive
Describes a protein that performs repeated rounds of catalysis or conformational changes while still attached to a polymer. A characteristic of motor proteins involved in transport, such as kinesin.

programmed cell death—*see* **apoptosis**

prometaphase
Stage of mitosis that precedes metaphase.

promoter
Nucleotide sequence in DNA to which RNA polymerase binds to begin transcription.

proofreading
The process by which DNA polymerase corrects its own errors as it moves along DNA.

prophase
First stage of mitosis during which the chromosomes are condensed but not yet attached to a mitotic spindle. Also a superficially similar stage in meiosis.

protease (proteinase, proteolytic enzyme)
Enzyme such as trypsin that degrades proteins by hydrolyzing some of their peptide bonds.

proteasome
Large protein complex in the cytosol that is responsible for degrading cytosolic proteins that have been marked for destruction by ubiquitylation or by some other means.

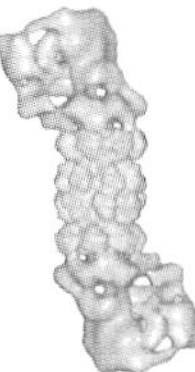

protein
The major macromolecular constituent of cells. Each protein is composed of one or more linear chains of amino acids linked together by peptide bonds in a specific sequence. The amino acid chain is folded into a three-dimensional shape that is unique to the given protein and determines its function.

protein domain—*see* **domain**

protein complex—*see* **complex**

protein family
A group of proteins in an organism with a similar amino acid sequence. The similarity is thought to reflect the evolution of the genes that encode the proteins from a common ancestor gene through a process of gene duplication followed by gene divergence. Usually, the different members of a protein family will have related but distinct functions. For example, each member of the protein kinase family carries out a similar phosphorylation reaction, but the substrates and regulation differ for each enzyme.

protein kinase
One of a very large number of enzymes that transfers the terminal phosphate group of ATP to a specific amino acid side chain on a target protein.

protein kinase A—*see* **cyclic-AMP-dependent protein kinase**

protein kinase C (PKC)
Enzyme that phosphorylates target proteins in response to a rise in diacylglycerol and Ca^{2+} ions.

protein machine
A set of protein molecules that bind to each other in specific ways, so that concerted movements within the protein complex can carry out a sequence of reactions with unusual speed and effectiveness. A large number of the central reactions of the cell are catalyzed by such protein machines, with protein synthesis and DNA replication being particularly well understood examples.

protein phosphatase (phosphoprotein phosphatase)
Enzyme that removes, by hydrolysis, a phosphate group from a protein, often with high specificity for the phosphorylated site.

protein phosphorylation
The covalent addition of a phosphate group to a side chain of a protein catalyzed by a protein kinase. Phosphorylation usually alters the activity or properties of the protein in some way.

proteoglycan
Molecule consisting of one or more glycosaminoglycan chains attached to a core protein.

proteolysis
Degradation of a protein by means of a protease.

proteomics
The large-scale study of proteins, investigating many different proteins in a cell or tissue simultaneously.

proton
Subatomic particle found in the atomic nucleus. Also exists as an independent chemical species as the positive hydrogen ion (H^+).

proto-oncogene—*see* oncogene

protozoan
A member of the protozoa—free-living, nonphotosynthetic, single-celled, motile eucaryotic organisms. Many protozoans, such as *Paramecium* or *Amoeba,* live by feeding on other organisms.

purifying selection
Process of selection during evolution in which individuals carrying mutations that interfere with important functions are eliminated.

purine
One of the two categories of nitrogen-containing ring compounds found in DNA and RNA. Examples are adenine and guanine. (*See* Panel 2–6, pp. 74–75.)

pyrimidine
One of the two categories of nitrogen-containing ring compounds found in DNA and RNA. An example is cytosine. (*See* Panel 2–6, pp. 74–75.)

pyruvate
Metabolite formed from the breakdown of glucose that provides a crucial link to the citric acid cycle and many biosynthetic pathways.

$$COO^- \\ | \\ C{=}O \\ | \\ CH_3$$

quinone
Small, lipid-soluble, mobile electron carrier molecule found in the respiratory and photosynthetic electron-transport chains. (See Figure 14–20.)

Rab protein
A family of small GTP-binding proteins present on the surfaces of transport vesicles and organelles that serve as molecular markers identifying each membrane type. Rab proteins help to ensure that transport vesicles fuse only with the correct membrane.

Ras
One of a large family of small GTP-binding proteins (also called the monomeric GTPases) that help relay signals from cell-surface receptors to the nucleus. Named for the *Ras* gene, first identified in viruses that cause rat sarcomas.

reaction center
In photosynthetic membranes, a protein complex that contains a specialized pair of chlorophyll molecules that performs photochemical reactions to convert the energy of photons (light) into high-energy electrons for transport down the photosynthetic electron-transport chain.

reading frame
The set of successive triplets in which a string of nucleotides is translated into protein. An mRNA molecule is read in one of three possible reading frames, depending on the starting point.

receptor
A cell (such as the photoreceptor cells of the eye) or cellular component (such as a receptor protein) that detects an external signal and triggers a specific cellular response.

receptor-mediated endocytosis
Mechanism of selective uptake of material by animal cells in which a macromolecule binds to a receptor in the plasma membrane and enters the cell in a clathrin-coated vesicle.

receptor protein
Protein that detects a stimulus, usually a change in concentration of a specific molecule, and then initiates a response in the cell. Cell-surface receptors, such as the acetylcholine receptor and the insulin receptor, are located in the plasma membrane, with their ligand-binding site exposed to the external medium. Intracellular receptors, such as steroid hormone receptors, bind ligands that diffuse into the cell across the plasma membrane.

receptor serine/threonine kinase
Enzyme-coupled receptor with an extracellular signal-binding domain and an intracellular kinase domain that phosphorylates signaling proteins on serine or threonine.

receptor tyrosine kinase (RTK)
Enzyme-coupled receptor in which the intracellular domain has a tyrosine kinase activity, which is activated by ligand binding to the receptor's extracellular domain.

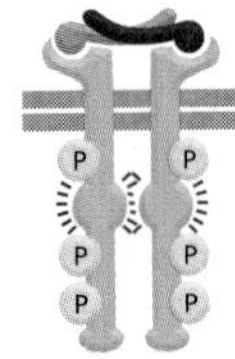

recognition—*see* molecular recognition

recombinant DNA
A DNA molecule that is composed of DNA from different sources.

recombinant DNA technology (genetic engineering)
The collection of techniques by which DNA segments from different sources are combined to make new DNA. Recombinant DNAs are widely used in the cloning of genes, in the genetic modification of organisms, and in molecular biology generally.

recombination
Process in which an exchange of genetic information occurs between two chromosomes or DNA molecules. Enzyme-mediated recombination can occur naturally

in living cells, or in a test tube using purified DNA and enzymes that break and re-ligate DNA strands.

redox pair
Pair of molecules in which one acts as an electron donor and one as an electron acceptor in an oxidation–reduction reaction; for example, NADH (electron donor) and NAD^+ (electron acceptor).

redox potential
A measure of the tendency of a given redox pair to donate electrons (act as a reducing agent) or to accept electrons (act as an oxidizing agent).

redox reaction
A reaction in which electrons are transferred from one chemical species to another. An oxidation–reduction reaction.

reduction
Addition of electron density to an atom, as occurs during the addition of hydrogen to a carbon atom or the removal of oxygen from it. The opposite of oxidation. (*See* Figure 3–11.)

regulatory DNA sequence
DNA sequence to which a transcription regulator binds to determine when, where, and in what quantities a gene is to be transcribed into RNA.

regulatory protein code
The set of covalent modifications that a protein has at any given time, which controls the behavior of the protein inside the cell

replication fork
Y-shaped region of a replicating DNA molecule at which the two daughter strands are formed and separate.

replication origin
Site on a chromosome at which DNA replication begins.

reporter gene
Introduced gene encoding a protein whose activity is easy to monitor experimentally. It is usually joined to a regulatory sequence, which will then switch on the reporter gene in the normal context in which its own gene is usually expressed.

repressor
A protein that binds to a specific regulatory region of DNA to prevent transcription of an adjacent gene.

reproductive cloning
The artificial production of genetically identical copies of an animal by, for example, the transplantation of a somatic cell nucleus into an enucleated fertilized egg.

respiration
General term for any process in a cell in which the uptake of molecular oxygen (O_2) is coupled to the production of CO_2.

restriction nuclease (restriction enzyme)
Nuclease that can cleave a DNA molecule at any site where a specific short sequence of nucleotides occurs. Different restriction nucleases cut at different sequences. Extensively used in recombinant DNA technology.

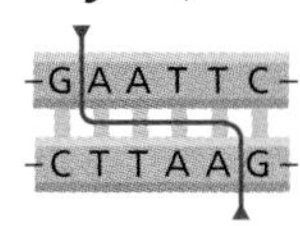

retrotransposon
Type of mobile genetic element that moves by being first transcribed into an RNA copy that is reconverted to DNA by reverse transcriptase and inserted elsewhere in the chromosomes.

retrovirus
RNA-containing virus that replicates in a cell by first making a double-stranded DNA intermediate. This DNA is inserted into the cell's chromosome, where it can be maintained for a long time and is transcribed to produce new viral genomes and mRNAs that encode viral proteins.

reverse transcriptase
Enzyme that makes a double-stranded DNA copy from a single-stranded RNA template molecule. Present in retroviruses and as part of the transposition machinery of retrotransposons.

Rho protein family
Family of small GTPases involved in signaling that causes a rearrangement of the actin cytoskeleton.

ribonucleic acid—*see* RNA

ribosomal RNA (rRNA)
Any one of a number of specific RNA molecules that form part of the structure of a ribosome and participate in the synthesis of proteins. Often distinguished by their sedimentation coefficient, such as 28S rRNA or 5S rRNA.

ribosome
Particle composed of ribosomal RNAs and ribosomal proteins that associates with messenger RNA and catalyzes the synthesis of protein.

riboswitch
Short sequences within some RNAs that change their conformation when specifically bound to small molecules such as metabolites and in this way regulate transcription or translation.

ribozyme
An RNA molecule possessing catalytic properties.

RNA (ribonucleic acid)
A (usually) single-stranded polynucleotide in the form of a chain of covalently linked ribonucleotide subunits. It is synthesized when an RNA polymerase copies the nucleotide sequence of DNA. RNA serves a variety of functions in cells. *See for example* messenger RNA, microRNA, ribosomal RNA, transfer RNA.

RNA interference (RNAi)
Cellular mechanism activated by double-stranded RNA molecules that results in the destruction of RNAs containing a similar nucleotide sequence. It is widely exploited as an experimental tool for preventing the expression of selected genes (gene silencing).

RNA polymerase
Enzyme that catalyzes the synthesis of an RNA molecule on a DNA template from nucleoside triphosphate precursors.

RNA primer—*see* primer

RNA processing
Broad term for the modifications that an RNA undergoes as it reaches its mature form. For a eucaryotic mRNA, processing typically includes capping, splicing, and polyadenylation.

RNA splicing
Process in which intron sequences are excised from RNA molecules in the nucleus during the formation of messenger RNA.

rough endoplasmic reticulum (RER)
Region of the endoplasmic reticulum associated with ribosomes and involved in the synthesis of secreted and membrane-bound proteins.

rRNA—*see* **ribosomal RNA**

RTK—*see* **receptor tyrosine kinase**

S phase
Period during a eucaryotic cell cycle in which DNA is synthesized.

sarcomere
Repeating unit of a myofibril in a muscle cell, about 2.5 μm long, composed of an array of overlapping thick (myosin) and thin (actin) filaments.

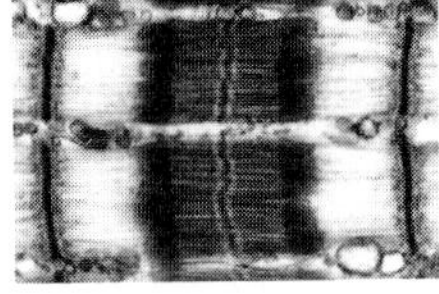

saturated
Describes an organic molecule that contains no double or triple carbon–carbon bonds. Not unsaturated.

second messenger
Small molecule formed in or released into the cytosol in response to an extracellular signal that helps to relay the signal to the interior of the cell. Examples include cAMP, IP_3, and Ca^{2+}.

secondary structure
Regular local folding pattern of a polymeric molecule. In proteins, it refers to α helices and β sheets.

secretion
Production and release of a substance from a cell.

secretory vesicle
Membrane-enclosed organelle in which molecules destined for secretion are stored prior to release. Sometimes called a secretory granule because darkly staining contents make the organelle visible as a small solid object.

sequence
The linear order of monomers in a large molecule, for example amino acids in a protein or nucleotides in DNA. In general the sequence of a macromolecule specifies its precise biological function.

SER—*see* **smooth endoplasmic reticulum**

serine/threonine kinase
Enzyme that phosphorylates specific proteins on serines or threonines.

sex chromosome
Chromosome that may be present or absent, or present in a variable number of copies, according to the sex of the individual. In mammals, the X and Y chromosomes.

sexual reproduction
Type of reproduction in which the genomes of two individuals are mixed in the formation of a new organism. Individuals produced by sexual reproduction differ from either of their parents and from each other.

β sheet—*see* **beta sheet**

side chain
Portion of an amino acid not involved in making peptide bonds; the side chain gives each amino acid its unique properties.

signaling cascade
Sequence of linked protein reactions, often including phosphorylation and dephosphorylation, that carries information within a cell, often amplifying an initial signal.

signal sequence
Amino acid sequence that directs a protein to a specific location in the cell, such as the nucleus or mitochondria.

signal transduction
Conversion of an impulse or stimulus from one physical or chemical form to another. In cell biology, the process by which a cell responds to an extracellular signal.

single nucleotide polymorphism (SNP)
Sequences in the genome that differ by a single nucleotide between one portion of the population and another.

siRNA—*see* **small interfering RNA**

sister chromatid
One copy of a chromosome (a chromatid) formed by DNA replication that is still joined at the centromere to the other copy, the pair of chromatids being known as sister chromatids.

site-directed mutagenesis
Technique by which a mutation can be made at a particular site in DNA.

site-specific recombination
Type of recombination that does not require extensive similarity in the two DNA sequences. Can occur between two different DNA molecules or within a single DNA molecule.

small interfering RNA (siRNA)
Short lengths of RNA produced from double-stranded RNA during the process of RNA interference. They base-pair with identical sequences in other RNAs, leading to the inactivation or destruction of the target RNA.

small messenger—*see* **second messenger**

small nuclear ribonucleoprotein particle (snRNP)
Structural unit of a spliceosome built of RNA and protein.

small nuclear RNA (snRNA)
RNA molecules of around 200 nucleotides involved in RNA splicing.

smooth endoplasmic reticulum (SER)
Region of the endoplasmic reticulum not associated with ribosomes; involved in the synthesis of lipids.

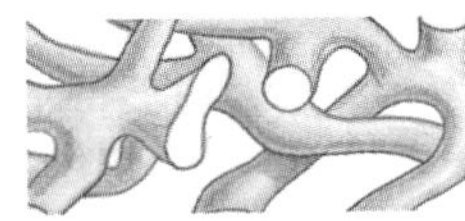

SNARE
One of a family of membrane proteins responsible for the selective fusion of vesicles with a target membrane inside the cell.

SNP—*see* **single nucleotide polymorphism**

snRNA—*see* **small nuclear RNA**

snRNP—*see* **small nuclear ribonucleoprotein**

sodium pump—*see* **Na^+-K^+ pump**

solute
Any molecule that is dissolved in a liquid. The liquid is called the solvent.

somatic cell
Any cell of a plant or animal other than a germ cell or germ-line precursor. (From Greek, *soma*, body.)

specificity
Selective affinity of one molecule for another that permits the two to bind or react, even in the presence of many unrelated molecular species.

spindle pole
One of two centrosomes in a cell undergoing mitosis. Microtubules radiating from these centrosomes form the mitotic spindle.

spliceosome
Large assembly of RNA and protein molecules that splices introns out of pre-mRNA in eucaryotic cells.

starch
Polysaccharide composed exclusively of glucose units, used as an energy store in plant cells.

stem cell
Relatively undifferentiated cell that can continue dividing indefinitely, throwing off daughter cells that undergo terminal differentiation into particular cell types.

steroid hormone
Lipophilic molecule related to cholesterol that acts as a hormone. Examples include estrogen and testosterone.

stress-gated channel
Membrane protein that allows the selective entry of specific ions into a cell and is opened by mechanical force.

stroma
(1) The connective tissue in which a glandular or other epithelium is embedded. (2) The large interior space of a chloroplast, containing enzymes that incorporate CO_2 into sugars in the carbon-fixation stage of photosynthesis.

substrate
The molecule on which an enzyme acts.

substratum
Solid surface to which a cell adheres.

subunit
A monomer that forms part of a larger molecule, such as an amino acid residue in a protein or a nucleotide residue in a nucleic acid. Can also refer to a complete molecule that forms part of a larger molecule. Many proteins, for example, are composed of multiple polypeptide chains, each of which is called a protein subunit.

sugar
A substance made of carbon, hydrogen, and oxygen with the general formula $(CH_2O)_n$. A carbohydrate or saccharide. The "sugar" of everyday usage is sucrose, a specific sweet-tasting disaccharide produced by beet or sugar cane.

sulfhydryl group (–SH, thiol)
Chemical group containing sulfur and hydrogen found in the amino acid cysteine and other molecules. Two sulfhydryls can join to produce a disulfide bond.

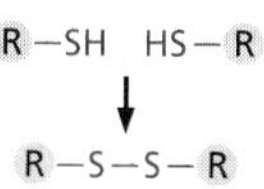

survival factor
Extracellular signaling molecule that must be present to prevent apoptosis.

symbiosis
Intimate association between two organisms of different species from which both derive a long-term selective advantage.

synapse
Specialized junction between a nerve cell and another cell (nerve cell, muscle cell, gland cell) across which the nerve impulse is transferred. In most synapses the signal is carried by a neurotransmitter, which is secreted by the nerve cell and diffuses to the target cell.

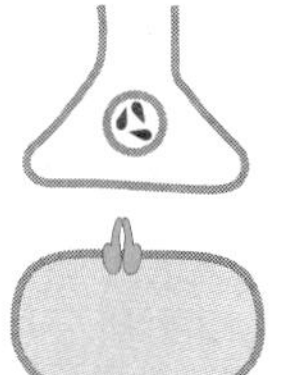

synaptic vesicle
Small membrane-enclosed sac filled with neurotransmitter that releases its contents by exocytosis at a synapse.

telomerase
Enzyme that elongates telomeres, the repetitive nucleotide sequences found at the ends of eucaryotic chromosomes.

telomere
Structure at the ends of linear chromosomes, associated with a characteristic DNA sequence that is replicated in a special way. Counteracts the tendency of the chromosome otherwise to shorten with each round of replication. (From Greek, *telos*, "end".)

telophase
Final stage of mitosis in which the two sets of separated chromosomes decondense and become enclosed by nuclear envelopes.

template
A molecular structure that serves as a pattern for the production of other molecules. Thus, a specific sequence of nucleotides in DNA can act as a template to direct the synthesis of a new strand of complementary DNA.

tight junction
Cell–cell junction that seals adjacent epithelial cells together, preventing the passage of most dissolved molecules from one side of the epithelial sheet to the other.

tissue
Organized mass of cells with a specific function that forms a distinctive part of a plant or animal.

thioester bond
High-energy bond formed by a condensation reaction between an acid (acyl) group and a thiol group (–SH); seen, for example, in acetyl CoA and in many enzyme–substrate complexes.

trans
Beyond, or on the other side.

transcription
A process that uses one strand of DNA as the template to synthesize a complementary RNA sequence, sometimes termed the primary transcript, catalyzed by the enzyme RNA polymerase.

transcription factor
Term loosely applied to any protein required to initiate or regulate transcription in eucaryotes. Includes transcription regulators as well as the general transcription factors.

transcription regulator
Protein that binds specifically to a regulatory DNA sequence and is involved in controlling whether a gene is switched on or off.

transfer RNA (tRNA)
Set of small RNA molecules used in protein synthesis as an interface (adaptor) between mRNA and amino acids.

Each type of tRNA molecule is covalently linked to a particular amino acid.

transformation
Process by which cells take up DNA molecules from their surroundings and then express genes on that DNA.

transgenic organism
A plant or animal that has stably incorporated one or more genes from another cell or organism and can pass them on to successive generations.

***trans* Golgi network (TGN)**
That part of the Golgi apparatus that is furthest from the endoplasmic reticulum and from which proteins and lipids leave for lysosomes, secretory vesicles or the cell surface.

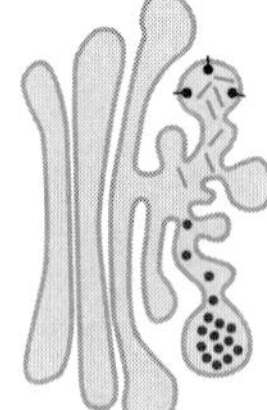

transition state
Chemical structure that forms transiently in the course of a reaction and has the highest free energy of any reaction intermediate.

translation
Process by which the sequence of nucleotides in a messenger RNA molecule directs the incorporation of amino acids into protein; occurs on a ribosome.

translation initiation factor
Protein that promotes the proper association of ribosomes with mRNA and is required for the initiation of protein synthesis.

transport vesicle
Membrane vesicles that carry proteins from one intracellular compartment to another, for example from the ER to the Golgi apparatus.

transporter
Membrane protein that transports ions or molecules across a cell membrane.

transposon
General name for short segments of DNA that can move from one location to another in the genome. Also known as mobile genetic element.

triacylglcerol
Glycerol ester of fatty acids. The main constituent of fat droplets in animal tissues (where the fatty acids are saturated) and of vegetable oil (where the fatty acids are mainly unsaturated).

tRNA—*see* **transfer RNA**

tryptophan repressor
A bacterial protein that, in the presence of tryptophan, binds to a specific region of DNA and shuts off production of the tryptophan biosynthetic enzymes.

tubulin
Protein from which microtubules are made.

γ-tubulin ring
Protein complex in centrosomes that nucleates microtubule assembly.

tumor suppressor gene
A gene that in a normal tissue cell inhibits progress through the cell cycle. Loss or inactivation of both copies of such a gene from a diploid cell can cause it to divide as a cancer cell.

turnover number
In enzyme catalysis, the number of substrate molecules processed to product per second per enzyme molecule. Although different types of enzymes can have very different turnover numbers, turnover numbers of 1000 or more are quite common—a reflection of the impressive catalytic power of enzymes.

tyrosine kinase
Enzyme that phosphorylates specific proteins on tyrosines.

unfolded protein response
Cellular response triggered by the accumulation of misfolded proteins in the endoplasmic reticulum. The cell produces more endoplasmic reticulum and more of the molecular machinery needed to restore proper protein folding and processing.

unsaturated
Describes a molecule that contains one or more double or triple carbon–carbon bonds.

V_{max}
The maximum rate of an enzymatic reaction, attained immediately after the addition of substrate at a concentration sufficient to fully occupy the active sites of all enzyme molecules present.

valence
For an atom, the number of electrons that it must either gain or lose (whether by electron sharing or by electron transfer) to achieve a filled outer shell most readily. Thus, for example, the valence of Na is one (it must lose one electron), and the valence of Cl is one (it must gain one electron). The valence of an atom is equal to the number of single bonds that the atom can form.

van der Waals force
Attractive force due to fluctuating electrical charges that comes into play between two atoms that are 0.3 to 0.4 nm apart. At a shorter distance, repulsive forces begin to operate.

vector
Genetic element, usually a bacteriophage or plasmid, that is used to carry a fragment of DNA into a recipient cell for the purpose of gene cloning.

vesicle
Small, membrane-enclosed, spherical organelle in the cytoplasm of a eucaryotic cell.

vesicular transport
Transport of material between organelles in the eucaryotic cell via membrane-enclosed vesicles.

virus
Particle consisting of nucleic acid (RNA or DNA) enclosed in a protein coat and capable of replicating within a host cell and spreading from cell to cell. Often the cause of disease.

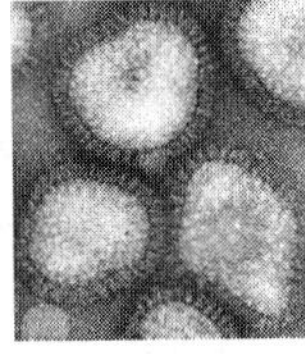

voltage-gated channel
Membrane protein that selectively allows ions such as Na^+ (carried by the voltage-gated Na^+ channel) to cross a membrane and is opened by changes in membrane potential. Found mainly in electrically excitable cells such as nerve and muscle.

wild type
Normal, nonmutant form of a species resulting from breeding under natural conditions.

Wnt protein
Member of the Wnt family of extracellular signal proteins with many roles in development, including the maintenance of stem cells in a proliferative state.

X chromosome
One of the two sex chromosomes in mammals. The cells of men possess one X and one Y chromosome.

X-ray crystallography
Technique used to determine three-dimensional protein structures by analyzing the diffraction pattern of a beam of X-rays passed through a crystal of the protein.

Y chromosome
One of the two sex chromosomes of mammals. The cells of women contain two X chromosomes.

yeast
Common term for several families of unicellular fungi (eukaryotic cells used as model organisms). Includes species used for brewing beer and making bread, as well as species that cause disease.

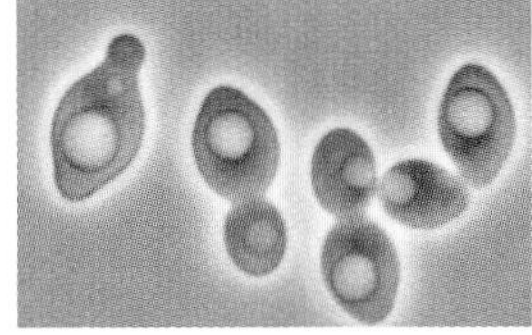

zygote
Diploid cell produced by fusion of a male and a female gamete. A fertilized egg.

Index

Page numbers in boldface refer to major discussion of a topic; page numbers followed by F refer to a figure, and with FF to consecutive figures; page numbers followed by T refer to tables.

A

- Acetyl coenzyme A (acetyl CoA)
  - citric acid cycle, 442FF
  - as energy carrier, 109–110, 110F
  - mitochondrial utilization, 458
  - production, 427, 428F, **436**, 436F, 437F
- Acetylcholine, 417, 417F, 535T
  - activation by phospholipase C, 551T
  - activation of G proteins, 547, 547F
  - blood vessel regulation, 539
  - patch-clamp recording, 404F
  - receptors, 417, 417F, 418T, 542–543
  - selectivity of cell responses, 535, 536F
  - speed of response, 536
- *N*-Acetylglucosamine, 69F
- Acid anhydride, 65F
- Acid hydrolases, lysosomal, 526, 526F
- Acids, **49–50**, 67F
  - weak, 50, 67F
- Aconitase, 442F
- ACTH, 549T
- Actin, 572
  - arrangement in filaments, 591, 592F
  - cell crawling, **594–597**, 595F, 596F
  - monomers in cytosol, 592–593
  - polymerization, **591–592**, 592F
- Actin-binding proteins, 590–591, **592–594**
  - *see also* Myosin; *other specific proteins*
  - cell cortex, 594
  - cell crawling, 596, 596F
  - extracellular signals controlling, 597–598
  - skeletal muscle, 603, 604F
- Actin filaments, 22, 22F, 572, 573F, **590–598**
  - adherens junctions, 703–704, 704F
  - assembly and disassembly, **591–592**, 592F
  - cell cortex, 573F, 593, **594**
  - cell crawling, **594–597**, 595F, 596F
  - contractile ring, 590, 591F, 635
  - drugs acting on, 582T, 592
  - extracellular signals controlling, **597–598**, 598F
  - muscle contraction, **600–601**, 601F, 602F
  - myosin association, **597**, 597F
  - myosin filament interactions, 599, 600F
  - plus and minus ends, 591, 592F
  - regulation of formation, 592–594
  - sarcomeres, 600, 601F
  - structure, 137, 137F, **591**, 592F
- Actin-related proteins (ARPs), 593, 596, 596F
- Action potentials, **409–414**
  - experimental methods, 412–413, 412F, 413F
  - generation and course, 410–411, 410F, 411F
  - neurotransmitter release, 415, 416F
  - propagation along axon, 412, 412F
- Activated carrier molecules, **104–112**, 104F
  - *see also* ATP; NADH; NADPH
  - breakdown of food producing, 427, 428F
  - chemiosmotic coupling, **458–460**, 459F
  - formation in coupled reactions, 104–105, 105F
  - types, 109–110, 109T
- Activation energy, 89–90, 89F, 146
- Activators, transcriptional *see* Transcriptional activators
- Active site, 91, 91F, 146
- Active transport, **390–391**, 391F
  - mechanisms, **393**, 393F
- Adaptation, signaling pathways, 554, 555
- Adaptins, 512, 512F
- Adaptors, receptor tyrosine kinases, 556
- Adenine (A), 57, 74F, 177
  - loss in DNA damage, 214, 214F
  - RNA, 233, 233F
- *Adenomatous Polyposis Coli (APC)* gene, 724–727, 727F
- Adenosine diphosphate *see* ADP
- Adenosine monophosphate (AMP), 112, 113F
- Adenosine triphosphate *see* ATP
- *S*-Adenosylmethionine, 109T
- Adenovirus, 225F
- Adenylyl cyclase, 548, 548F
- Adherens junctions, 702F, **703–704**, 704F
- ADP
  - ATP/ADP ratio and, 465–466
  - ATP cycle, 105, 106F
  - ATP synthesis from, 426, 427, 429
- ADP–ATP exchange, 464, 464F
- Adrenaline (epinephrine), 535T, 549–550, 549T
- Aerobic organisms, 14–15
  - chemiosmotic coupling, 455, 456
  - energy metabolism, 429, 436
- Aerobic respiration, 436–439
- Affinity chromatography, 162, 162F, 166F
- Aggrecan, 699F
- AIDS, 226
- Akt, 558–559, 558F, 559F
- Alanine, 56F, 73F
- Albinism, inheritance, 668, 669F
- Alcohol dehydrogenase, 126F
- Alcohols, 65F
- Aldehydes, 65F
- Aldolase, 430F
- Alkalinity, 50
- Alleles, 652, 674F
  - *see also* Gene(s); Mutations
  - dominant, 667–668, 667F, 674F
  - elimination of deleterious, 654–655
  - Mendel's experiments, 666–667
  - Mendel's law of independent assortment, **669–670**, 670F
  - Mendel's law of segregation, **667–668**, 667F, 669F
  - recessive, 667, 667F, 674F
  - selectively neutral, 683
- Allen, Robert, 586
- Allosteric regulation, **150–152**, 152F, 162T
- α helices, **127**, 130F, 131, 162T

coiled-coil formation, 131, 132F
membrane proteins, 131, 131F
protein domains, 133, 133F, 134F
transmembrane proteins, **374–375**, 374F
Alternative splicing, 243, 244F, 322, 322F
Alu elements, 223, 224F
evolutionary relationships and, 310, 311F
recombination involving, 307
AluI, 330F
Amides, 65F, 70F
Amines, 65F
Amino acid(s), 4, 51–52, **55–56**, 122F
acidic, 73F
basic, 72F
energy production from, 428F, 436
families, 72F
gluconeogenesis, 447
linkage together, 55–56, 56F, 72F
nonpolar, 73F, 122, 122F
nucleotides specifying *see* Genetic code
optical isomers, 56, 72F
polar, 122, 122F
side chains, 56, 72F, 73F, 122F
structure, 56F, 72FF
uncharged polar, 73F
Amino acid sequence, 59, **121–123**, 122F
see also Polypeptides; Protein structure
determination, 125, 158, 159F
evolution, 134–135, 217
signal sequence, **501–502**, 502T
specification by DNA, 231–232
Amino group, 55, 56F
Amino terminus *see* N-terminus
Aminoacyl-tRNA synthetases, 251, 251F
Amoebae, 16, 27F, 397, 594
AMP (adenosine monophosphate), 112, 113F
AMP-PNP, 586, 587
Amphipathic molecules, 54
membrane formation, 55, 55F
membrane lipids, 365, 365F, 366F
Anabaena cylindrica, 15F
Anabolism, 82, 82F
see also Biosynthesis
glycolysis/citric acid cycle and, **439–444**
NADPH role, 109F
regulation, 445–450
Anaerobic organisms, 17
energy production, 432–433, 432F
Anaerobic respiration, 433
Anaphase, 627F
A, 632–633, 633F
B, 632, 633, 633F
cleavage furrow formation, 634–635
meiosis I, 657, 658–659, 659F
meiosis II, 659, 659F
mitosis, 624–625, 625F, **631–633**
Anaphase-promoting complex (APC), 631, 632F, 633
Anchor proteins, 373T
Aneuploidy, 662F, 663
Animal(s)
asexual reproduction, 652, 652F
connective tissues, **693–699**
evolution of cell communication, **564**
model, 29–33
tissues, 689–690, 690F, 691
Animal cells
cytoskeleton, 22–23
dependence on multiple signals, 535–536, 536F
energy reserves, 448–449, 448F
membrane-enclosed organelles, 496–498, 496F
membrane potential, 407–408
membrane transporters, 401F
Na^+-K^+ pump, **394–395**, 394F
osmotic balance, **396–397**
receptor-mediated endocytosis, 524, 525F
signaling methods, 532–534, 533F
structure, 25F
survival, growth and proliferation, **642–646**
Anions, 44, 388T, 389
Ankyrin, 594
Antenna complex, 480, 480F
Antibiotic(s)
protein synthesis inhibitors, **257–258**, 258T
resistance genes, 222, 222F, 308, 335
Antibodies (immunoglobulins), 144FF
antigen binding sites, **142**, 142F
applications, 145F
assembly in ER, 516
production, 144F
variable domain, 134F
Anticodon, 250, 250F
Antidepressants, 418
Antifreeze protein, 132, 133F
Antigens, 142, 142F
Antimitotic drugs, 581–582, 582T
Antiparallel β sheets, 132, 132F
Antiparallel orientation, DNA strands, 173F, 177F, 178
Antiports, 398, 398F
APC *see* Anaphase-promoting complex
APC gene, 724–727, 727F
APC protein, 725–726, 726F
Apical surface of epithelial cells, 700, 701, 701F
membrane proteins, 380, 380F
Apoptosis, 610, **638–641**
Akt-mediated inhibition, 559, 559F
Bcl2 family proteins regulating, **641**, 642F
embryonic development, 639, 639F
inhibition in cancer cells, 721–722
necrosis compared, 639–640, 640F
signaling cascade, **640**, 641F
suppression by survival factors, **643**, 643F
Apoptosome, 641, 642F
Aquaporins, 396, 401
Arabidopsis thaliana, **28–29**, 28F
cell-surface receptors, 564
genome sequencing, 346F
Archaea, 14, 15–16
chemiosmotic coupling, 490, 490F
evolution, 314, 314F, 315F
Arginine, 72F
Armadillo, 725
Arsenate, 433
Asexual reproduction, 299, 652, 652F
Asparagine, 73F
Aspartate transcarbamoylase, 126F, 152F
Aspartic acid, 73F
Aster, 628, 628F, 630F
Ataxia, inherited, 181F
Atom(s), **40–41**, 40F
interactions between, **41–43**, 43F
Atomic number, 40, 40F
Atomic weight, 41
ATP
as energy carrier, 57, 57F, 75F, **105–107**, 106F
protein phosphorylation, 153, 153F
radiolabeled, 560
structure, 57, 57F
transport across mitochondrial membrane, 463–464, 464F
ATP/ADP ratio, **465–466**
ATP cycle, 57F, 105, 106F
ATP-driven pumps, 393, 393F, **394–398**
ATP hydrolysis, 105, 106F
actin filament disassembly, 592, 592F
alternative route, 112, 113F
biosynthetic reactions, 106–107, 107F
carbon fixation, **484–485**, 485F
carboxyl group transfer, 110, 111F
DNA replication, 209, 209F
energy transfer, 105–106, 106F, 435F
motor proteins, 154–155, 155F, 584, 586, 587–588
muscle contraction, 601, 602F
myosin, 597
Na^+-K^+ pump, 394, 394F
polymer synthesis, 111–112, 113F
ATP synthase, 454F, 455
inner mitochondrial membrane, **461–463**, 462F, 463F
historical research, 469, 469F
structure, 462, 463F
thylakoid membrane, 482, 482F
ATP synthesis, 17, 105, 106F, 426
chloroplasts, 479, **481–482**, 482F, 483F
cyclic photophosphorylation, **483–484**, 484F
electron-transport chain, **444–445**
from food molecules, 427, 428F
glycolysis, 429, 429F, 430FF, 433, 434F
membrane-based mechanisms, 454–455, 454F

mitochondria, **458–460**, 458F, 459F
citric acid cycle, 438–439
electrochemical H^+ gradients, 454F, 455, **461–463**, 462F
historical research, 469, 469F
yields of different processes, 464–465, 465T
ATPases, 143T, 394, 400
see also Calcium pumps; Na^+-K^+ pump
Auditory hair cells, 405, 406F
AUG codon, 247F, 255, 255F
Autocatalysis
living systems, **261**
RNA synthesis, 263, 263F
Autocrine signaling, 533
Autophagosome, 526, 527F
Autophagy, 526, 527F
Autoradiography, 331, 331F
Avery, Oswald, 175, 175F
Avogadro's number, 41
Axons, 409, 409F
action potential propagation, 414, 414F
electrical signaling along, 409–410
growth cones, 594–595
intermediate filaments, 575
squid giant *see* Squid giant axons
transport along, 582–583, 582F

B

B cells, 144F
B1 elements, 224F, 311F
Bacillus subtilis, related genes, 303, 303F
Bacteria (eubacteria), 15–16
see also Escherichia coli; Procaryotes
cell membranes, 363F, 364
chemical components, 51T
chromosomes, 179
discovery, 24T
DNA-only transposons, 222, 222F
evolution, 314, 314F, 315F
gene regulation, 275–277, 289–290, 290F
genome sizes, 34, 34F
H^+ gradients driving membrane transport, **400**
horizontal gene transfer, 308, 308F
membrane proteins, 375, 375F, 376–377
model organism, 27
origins of chloroplasts and mitochondria, **486–490**, 499
photosynthesis, 15, 15F, 476, 476F, **488–490**, 489F
protein import, 501
reproduction, 652, 652F
restriction nucleases, 329–330, 330F
shapes and sizes, 14, 14F
structure, **11–16**, 25F
transcription *see* Transcription, procaryotic
transformation, 174–175, 174F
translation initiation, 256, 256F
Bacterial artificial chromosomes (BAC), 349, 349F
Bacteriorhodopsin
historical research, 469, 469F
structure, 376–377, 377F
transport function, 400, 401T
Bad, 559, 641
Bak, 641, 642F
Barbiturates, 418, 543T
Basal lamina, 700–701, 701F
epithelial cell attachment, 703, 705, 706F
Basal surface of epithelial cells, 700–701, 701F
membrane proteins, 380, 380F
Base(s) (chemical), **50**, 67F
weak, 50, 67F
Base(s) (nucleotide), 56–57
DNA, 177
linking, 57–58, 58F
RNA, 233, 233F
structure, 74F
Base-pairing, complementary, 58
DNA double helix, 177–178, 177F
DNA replication, 198–199
accuracy, 205–206
RNA intramolecular, 234F
RNA primer from DNA strand, 207
RNA synthesis, 233, 234F
transfer RNAs, 250, 250F
wobble, 250
Base pairs, 177–178, 177F
Batteries, electrical, 455F, 467
Bax, 641, 642F
Bcl2, 559F
cell survival and, 643, 643F
inhibition of apoptosis, 641
Bcl2 family of proteins
regulation by survival factors, 643, 643F
regulation of apoptosis, **641**, 642F
Bdellovibrio bacteriovorus, 2, 3F
Beggiatoa, 15F
Benzene, 47, 64F
β barrels, 375, 375F
β sheets, **127**, 132, 162T
antifreeze protein, 132, 133F
antiparallel, 132, 132F
parallel, 132, 132F
protein domains, 133, 133F, 134F
Bi-orientation, kinetochores, 629–630
Bicoid, 283, 283F, 284, 284F
Binding energy, 98, 98F
Binding sites
antibody, **142**, 142F
protein, 135, 136F, 141, 141F
Biochemistry, 82
Biosphere, 87
Biosynthesis, 82, **104–112**
see also Anabolism
ATP-mediated, 106–107, 107F, 110, 111F
biological polymers, 110–112, 112F, 113F
glycolysis/citric acid cycle and, 439–444
NADPH-mediated, 108–109, 109F
regulation, 150, 150F, 151F, 445–450
Biotechnology, 163, 163F
Biotin, 149
carboxylated, 109T, 111F
1,3-Bisphosphoglycerate, 431F, 433, 434F, 435F
Bivalents, 657, 657F
chiasmata, 658, 658F
Blisters, 705
Block, Steven, 587
Body plan, evolution, 307, 307F
Body size, 638
Bone
matrix, 694, 694F, 698–699
turnover, 710–711
Boveri, Theodor, 655
Brady, Scott, 586
Brain development, human, 320
Branch points, homologous recombination, 219, 219F
Brca1/Brca2 proteins, 727–728
Breast cancer, 720F, 727–728
Breeding, selective, 328, 328F
Brewer's yeast *see Saccharomyces cerevisiae*
Brown fat cells, 468–469
Brush-border cells, absorptive, 701, 702F, 712F
Budding, reproduction by, 652, 652F
Buffers, 50

C

C-terminus, polypeptide chain, 56, 56F, 122F
CA repeats, 321
Ca^{2+} *see* Calcium
Cadherins, 703, 704F
adherens junctions, 703, 704F
desmosomes, 704, 706F
Caenorhabditis elegans, 32, 32F
genetic screens, 678F
homologous genes, 309
ion channels, 405
RNA interference, 357, 357F
Caffeine, 548–549
Calcium (Ca^{2+}), 44
intracellular concentration, 544, 552
fertilization-induced change, 552, 553F, 663
muscle contraction, **602–604**, 603F
regulation, **397–398**, 397F, 553
signaling function, 397, **552–553**, 553F
vs. extracellular, 388T, 397
stores in ER, 552
Calcium (Ca^{2+})/calmodulin-dependent protein kinases (CaM-kinases), 553
Calcium (Ca^{2+}) channels
ER, 552, 552F
sarcoplasmic reticulum, 603, 603F
voltage-gated, **415**, 416F, 418T
Calcium (Ca^{2+}) pumps, **397–398**, 398F, 401T, 553

Page numbers in boldface refer to major discussion of a topic; page numbers followed by F refer to a figure, and with FF to consecutive figures; page numbers followed by T refer to tables.

Calcium (Ca^{2+})-responsive proteins, 553
Callus, 357–358
Calmodulin, 553, 553F
  structure, 126F
Calvin cycle (carbon-fixation cycle), 485, 485F
cAMP *see* Cyclic AMP
Cancer, 212, 690, **717–729**
  see also Tumor(s)
  age-related incidence, 212F
  cells
    accumulation of mutations, **719–720**, 721, 721F
    antimitotic drugs, 581–582
    evolution, **721–722**, 721F
    genetic instability, 720, 720F, 722
    lack of normal 'social' controls, 646
    properties, 718, 719F, **721–722**, 721F
    Ras mutations, 558, 721
  critical genes in development, **722–723**, 722F, 724F
  discovery of critical genes, **725–726**, 727
  DNA microarrays, 353
  inherited predisposition, 213, **723–724**
  preventable causes, **718–719**
  treatments, **727–729**, 728F
CAP *see* Catabolite activator protein
Capping, mRNA, 241, 241F
Carbohydrates, 52
  see also Sugars
  cell surface layer, **380–384**, 381F, 384F
  photosynthetic production, 479, **484–485**, 484F, 485F
Carbon (C)
  atoms, 40, 40F, 41
  chemical bonds, 43
  compounds, **50–51**, 64FF
  covalent bonds, 44, 46F
  double bonds, 47F
  isotopes, 41
  skeletons, 64F
Carbon-14, 41
Carbon cycle, 87, 87F
Carbon dioxide (CO_2)
  assimilation *see* Carbon fixation
  diffusion across lipid bilayers, 389, 389F
  production, **436–439**, 436F, 438F
Carbon fixation, 86, 476, **478–479**, 479F
  cycle, 485, 485F
  evolutionary origins, 488–489, 489F
  reactions, **484–485**, 484F, 485F
Carbonic anhydrase, 99
Carbonyl group, 65F
Carboxyl groups, 56F, 65F, 70F, 110, 111F
Carboxyl terminus *see* C-terminus
Carboxylic acid, 54, 65F
Carboxypeptidase, 149
Carcinogens, chemical, 720
Cardiac muscle cells, 599, 604
  gap junctions, 706
  G_i protein signaling, 547, 547F
  mitochondria, 456, 457F, 458
  response to acetylcholine, 534–535, 536F, 542–543
Cardiomyopathy, familial hypertrophic, 604
Cargo receptors, 512, 512F
Carsonella ruddii, 34
Cartilage, 699, 699F
Caspases, 640, 641, 641F, 642F
Catabolism, 82, 82F, 425
  NADH role, 109F
  regulation, 445–450
  stages, 426–427, 428F
Catabolite activator protein (CAP)
  gene regulation, 276–277, 276F, 277F
  structure, 133, 133F, 136, 136F
Catalase, 126F
Catalysis, 81–82, **88–100**
  enzyme-mediated *see* Enzyme(s), catalysis
Catalysts, 90
β-Catenin, 725–726, 726F
Cations, 44, 388–389, 388T
Cdc2 gene mutations, rescue studies, 31, 31F
Cdc6, 620–621, 620F
Cdc25, 623, 623F
Cdc42, 598F
Cdk-activating kinase (Cak), 623F
Cdks (cyclin-dependent proteins), **614–618**, 614F, 617T
  see also specific Cdks
  cyclin associations *see* Cyclin–Cdk complexes
  cyclin-mediated regulation, 614–617, 617F, 618, 618F
  discovery, **615–616**, 616F
  inhibitor proteins, 619, 621, 622F
cDNA
  DNA microarrays, 353, 353F
  preparation method, 338, 339F
cDNA libraries, **338–339**, 339F
  genomic libraries compared, 339, 340F
  obtained by PCR, 342, 342F
Cell(s), **1–38**
  basic chemistry, 3–5, 4F
  chemical components, **39–80**, 51T
  communication, **531–570**, 710
  communities, **689–731**
  compartments within *see* Compartments, intracellular
  components, sizes, 10F, 13F
  differentiated *see* Differentiated cells
  discovery, 6–7, 24T
  diversity, 2–6, 3F
  eucaryotic, **16–26**
  evolution, 5
  historical landmarks, 24T
  microscopy, 6–11, 8–9FF
  molecules in, **50–58**
  organelles *see* Organelles
  procaryotic, **11–16**
  similarities, **2–6**
  specialized, 2–3, 269–270
  structure, 10–11, 11F, 12F
Cell adhesion
  extracellular matrix, **696–698**, 697F
  to other cells, 710
Cell biology, 1, 6–7
Cell cortex, **377–379**, 378F, 379F
  actin filaments, 573F, 593, **594**
  cell locomotion and, 594, 595F
Cell crawling, **594–597**, 595F, 596F
Cell culture, **157–161**, 161F
Cell cycle, **609–638**, 610F
  see also specific phases
  arrest, **618–620**, 619F
  checkpoints, 612–613, 613F
    cell cycle arrest, 618–620, 619F
    DNA damage, **621**, 622F
    G_1 *see* G_1 phase, checkpoint
    G_2, 612, 613F, 621
    mitosis, 612, 613F, 633
  chromosomes during, 182–183, 182F
  control system, 610, 612, **613–620**
    see also Cdks; Cyclin(s)
    evolutionary conservation, 613, 616
  discovery, **615–616**, 616F
  duration, 610–611, 611T
  experimental analysis, **30–31**, 31F
  overview, **610–613**
  phases, 611–612, 611F
  Start, 619
Cell death
  see also Apoptosis
  necrotic, 639–640, 640F
  programmed, 32, 638
Cell division, 7F, 10
  see also Cell proliferation
  asymmetric, 635
  balance with cell death, 639
  cell shape changes, 635–636, 637F
  cleavage, 612, 615
  contractile ring *see* Contractile ring
  cycle *see* Cell cycle
  duplication of organelles, 500
  experimental analysis, 30–31
  extracellular signals stimulating *see* Mitogens
  inhibitors, 645–646
  meiosis, 656
  reproduction by, 652, 652F
  role of cytoskeleton, 22, 23, 23F
Cell doctrine, 609
Cell fractionation, 24T, 161, 164FF
Cell growth
  cell cycle phases, 611–612, 611F
  factors stimulating, 642, **645**, 645F
  inhibitors, 645–646
  plants, 691, 691F, 692, 692F
Cell homogenates, 161, 164F
Cell junctions, 689, **700–707**, 702F
  see also specific types

Cell membranes *see* Membrane(s)
Cell memory, 280, 287–288, 287F, 710
Cell necrosis, 639–640, 640F
Cell numbers
  extracellular signals influencing, **642–646**
  regulation by apoptosis, 638–639, 643, 643F
Cell proliferation, 643
  see also Cell cycle; Cell division
  cancer cells, 722
  factors stimulating, 644–645, 644F, 645F
  Ras signaling, 558, 560
Cell responses, 540, 540F
Cell shape
  changes during M phase, 635–636, 637F
  cytoskeleton role, 571, 571F
  diversity, 2, 3F
Cell signaling, **532–544**
  see also Signal molecules
  complexity and integration, **564–567**, 565F
  contact-dependent, 533–534, 533F, 534F, 535T
  intracellular, **539–542**, 540F, 541F
  modes, **532–534**, 533F
  plants, **564**, 565F
  selectivity of response, **534–536**, 536F
  speed of response, **536–537**, 537F
Cell size, 7, 10F, 13F
  diversity, 2, 3F
  regulation, 638, 645, 646F
Cell-surface receptors, 537, 537F, **539–540**, 540F
  see also Enzyme-coupled receptors; G-protein–coupled receptors; Ion-channel–coupled receptors
  classes, **542–543**, 543F
  foreign substances acting on, 543, 543T
  growth factors, 645, 645F
  intracellular signaling pathways, 540, 541F
  mitogens, 644, 644F
  plants, 564
Cell survival
  factors stimulating, 642, **643**, 643F
  inhibitors, 645–646
  promotion by Akt, 559, 559F
Cell theory, 7, 7F, 24T
Cell walls, 2
  plant *see* Plant cell walls
  procaryotic, 14
Cellular respiration *see* Respiration
Cellulase, 160F
Cellulose, 53, 691F, **692–693**
  deposition in plant cell wall, 692–693, 693F
  microfibrils, 692, 692F
Cellulose synthase, 693F
Central dogma, 232, 248
Centrifugation, 164FF
  density gradient, 165F
  differential, 165F
  equilibrium sedimentation, 165F
  velocity sedimentation, 165F
Centrioles, 25F, 579
Centromeres, 180F, 624F
  kinetochore assembly, 629, 629F
  mitotic chromosomes, 183, 183F
Centrosome(s), 573F
  cycle, 628
  duplication, **625–628**, 628F
  microtubule nucleation from, **579–580**, 579F
  microtubule organization, 577, 577F
  mitosis in cells lacking, 630, 630F
  selective microtubule stabilization, 581
Cervical cancer, 718–719
cGMP *see* Cyclic GMP
Channels, 387–388, 390, 390F, 400–401
  see also Ion channels
Chaperones, molecular, 125
  folding of newly synthesized proteins, 256
  protein import into organelles, 505
  protein quality control in ER, 516, 516F
Charge separation, 480, 481F
Chase, Martha, 176, 176F
Checkpoints, cell cycle *see under* Cell cycle
Chemical bonds, **40–50**, 64FF
  see also Covalent bonds; Ionic bonds; *other specific bond types*
  high energy, 95F, 425
  weak, 76FF
Chemical groups, 51, 64FF
Chemical reactions
  activation energy, 89–90, 89F
  coupled, 95F, **104–105**, 105F
  energetically favorable/unfavorable, 91, 91F, 92F
  enzyme-catalyzed, 81–82
  producing disorder, 83–84, 94F
  rates, 94F
  sequential, 97–98, 98F
  spontaneous, 91, 91F, 94F
Chemiosmotic coupling, 455, 455F
  ATP synthesis in mitochondria, **458–460**, 459F
    historical research, **468–469**
  evolutionary origins, **490**, 490F
Chemiosmotic hypothesis, 455
Chiasma (chiasmata), 658–659, 658F, 659F
Chimpanzees, **310**, 310F, 311F, 320
Chitin, 53
Chloramphenicol, 258T
Chloride (Cl^-) channels, transmitter-gated, 417–418, 418F
Chloride ions (Cl^-), 44, 388T, 389
Chlorine atoms (Cl), 44, 45F
Chlorobium tepidum, 489F
Chlorophyll, 18, 19F, 478–479
  absorption of sunlight, **479**, 480F
  energy transfer to reaction center, **480–481**, 480F, 481F
  special pair of molecules, 480, 480F
Chloroplasts, **18–19**, 19F, **476–486**
  carbon fixation, **484–485**, 484F, 485F
  cyclic photophosphorylation, **483–484**, 484F
  energy storage, 450, 450F
  envelope, 477, 477F
  evolution, 15, 19, 19F, 24F, **486–490**, 499
  functions, 497, 497T
  genetic system, 486–487
  mitochondria compared, 477–478, 478F
  mitochondrial interrelationships, 486, 486F
  photosynthesis *see* Photosynthesis
  protein import, 501, 501F, **505**
  segregation at cell division, 638
  stroma, 477, 478F
  structure, **477–478**, 477F, 478F
Cholera toxin, 546
Cholesterol
  amphipathic nature, 365, 366F
  biosynthesis, 109F
  membrane fluidity and, 370, 370F
  structure, 71F, 366F, 538
  uptake into cells, 524, 525F
Chondroitin sulfate, 699F
Chromatids, 183F, 188F
  see also Sister chromatids
Chromatin, 179, 185
  see also Euchromatin; Heterochromatin; Nucleosomes
  30-nm fibers, 185, 185F, 187, 187F
  beads-on-a-string form, 185, 185F, 187, 187F
  condensed and extended forms, 190–191, 190F, 191F
  inherited changes in structure, 191–192, 192F
  levels of organization, 187–188, 188F
  modification by transcription regulators, **279–280**, 279F
  regulation, **188–191**
Chromatin-remodeling complexes, 188, 189F
Chromatography, 161–162, 162T, 166FF
  affinity, 162, 162F, 166F
  column, 166F
  gel-filtration, 166F
  ion-exchange, 166F
Chromosome(s), 172, **179–192**
  see also Sister chromatids
  abnormalities, 181, 181F
    aneuploidy, 662F, 663
    cancer cells, 720, 720F
  arrangement of genes, 181–182, 181F, 316, 316F
  bacterial, 179

Page numbers in boldface refer to major discussion of a topic; page numbers followed by F refer to a figure, and with FF to consecutive figures; page numbers followed by T refer to tables.

bacterial artificial (BAC), 349, 349F
banding, 180–181, 180F
bivalents *see* Bivalents
condensation, **184–185**, 185F, 612, **623–624**, 624F
dividing cells, 17, 17F, 172F
ends, replication, 210, 210F
fluorescence *in situ* hybridization, 353F
historical landmarks, 24T
homologous *see* Homologous chromosomes
interphase, 182–183, 182F
chromatin packing, 187–188
condensation of chromatin, 190–191, 190F
DNA condensation, 185, 185F
organization in nucleus, **184**, 184F
maternal, 652, 656
meiotic, **656–659**
see also Homologous chromosomes
crossing-over *see* Crossing-over
nondisjunction, 662–663, 662F
pairing, **657**, 657F
segregation, **658–659**, 659F
mitotic, 182F, 183, 183F
alignment at spindle equator, **630–631**, 631F
attachment to spindle, **629–630**, 629F
chromatin packing, 187–188, 188F
condensation, 184–185, 185F, **623–624**, 624F
decondensation, 634
DNA packing, 187–188, 187F
mitotic spindle assembly, **630**, 630F
monitoring spindle attachment, 633
segregation, **631–633**, 633F
numbers, 182, 182F
painting, 180, 180F
paternal, 652, 656
proteins, 185
structure, **179–188**
see also Chromatin; Nucleosomes
levels of DNA packing, 187–188, 187F
regulation, **188–192**
role of histones, 185–187, 186F, 187F
Chronic myeloid leukemia (CML), 728, 728F
Chymotrypsin, 126F, 135, 135F
Cilates, 27F
Cilia, 2, 3F, **585–590**, 589F
basal body, 585
microtubules, 577–578, 577F, 589, 590F
movement, 589–590, 589F
Citrate synthase, 442F
Citric acid (citrate), 438, 442F
Citric acid cycle, 427, **436–439**, 438F
biosynthetic role, **439–444**, 444F
central role in metabolism, 445, 446F
discovery, **440–441**, 440F, 441F
high-energy electrons generated, 458, 459F
reactions, 442FF
Clam eggs, cell cycle studies, 615–616
Clamp loader, 209
Clathrin, 511
Clathrin-coated pits, 511–512, 511F, 524
Clathrin-coated vesicles, 511–512, 511F, 513T
capture of cargo molecules, 512, 512F
pinocytosis, 524
receptor-mediated endocytosis, 524, 525F
Claudins, 702, 703F
Cleavage divisions, 612, 615
Cleavage furrow, 634–635, 635F
Clone-by-clone approach, genome sequencing, 349, 349F
Cloning, **715–717**
DNA *see* DNA cloning
reproductive, 716, 716F
therapeutic, **716–717**, 716F
Co-immunoprecipitation, 560, 725
Coated vesicles, 511–512, 512F, 513T
Codons, 246–247, 247F
start, 247F, 255–256, 255F
stop, 247F, 256, 256F
tRNA base-pairing to, 250, 250F
Coenzyme A (CoA), 75F
Cohesins, **621**, 621F
degradation, 625, 631, 632F, 659, 659F
similarity to condensins, 623–624
Coiled-coils, 131, 132F
Colcemid, 582T
Colchicine, 581, 582T, 585
Collagen, **694–696**
assembly, 695
defective, 695, 696F
cell attachment to, 696, 697F
fibers, 695F
alignment, 696, 697F
fibrils, 138, 139F, 694, 695F
alignment, **696**, 696F
gene mutations, 694
secretion, 694–695, 695F
structure, 126F, 694
type IV, basal lamina, 700–701, 701F
Colorectal cancer, **723–727**
Column chromatography, 166F
Combinatorial control
developmental genes, 285–286, 286F
gene expression, 280, 281F
Communication, cell, **531–570**, 710
Comparative genomics, 33–35, **310–315**
gene identification, 318
human accelerated regions, 320
human–chimpanzee, 310, 311F, 320
human–mouse, 310–312, 311F, 312F
tracing evolutionary relationships, 313–315, 314F
vertebrates, 312–313, 313F
Compartments, intracellular, **19–21**, **495–526**
see also Organelles, membrane-enclosed
protein sorting, **500–509**
vesicular transport between, **510–511**
Complementary DNA *see* cDNA
Complementation test, 674F, **677–678**, 678F
Concentration gradients, passive transport, 390, 391F, **392–393**, 392F
Condensation reactions, 53, 53F
ATP coupling, 105–106, 106F, 107, 107F
energetics, 110, 111F
polymer formation, 59, 59F, 110–111
Condensins, **623–624**, 624F
Confocal microscopy, 9F, 11F
Conformations, 62, 62F
see also Protein conformation
Conjugation, bacterial, 308F
Connective tissues, **690–699**
see also Extracellular matrix
animal, **693–699**
plant, 690, **691–693**
skin, 709–710, 709F
tensile strength, 694
Connexons, 706, 707F
Conserved synteny, 311–312
Constitutive gene expression, 275
Contact-dependent signaling, 533–534, 533F, 534F, 535T
Contractile bundles, cytoplasmic, 590, 591F, 599
Contractile ring, 599, 624, 624F, **635–636**
composition, 590, 591F, 635
cytokinesis, 635, 636F
COP-coated vesicles, 512, 513T
Cortisol, 535T, 537, 538F
Coupled reactions, 95F, 104–105, 105F
Coupled transporters, 393, 393F, **398–400**, 398F, 399F
Covalent bonds, 42–43, 43F, **45–47**, 64F
bond angles, 46, 46F
bond lengths, 45, 45F, 47T
bond strengths, **46**, 47T
cross-linkages, 138–140, 139F
double, 47, 47F, 64FF
formation, **45–46**, 45F, 46F
polar *see* Polar covalent bonds
single, 47, 64F
types, **47**
Creatine phosphate, 435F
Crick, Francis, 172, 173, 200
Crossing-over, **657–658**, 658F

genetic information exchange, 220, 220F, 221F
Mendelian segregation and, **671–673**, 672F
unequal, 303, 304F
Cultured cells, **157–161**, 161F
Curare, 417, 543T
Cyanobacteria, 476, 476F
evolution, 488–489
photosynthesis, 481–482
Cyclic AMP (cAMP), **548–551**
CAP interaction, 276–277, 276F
effects on target cells, 549–550, 549T, 550F, 551F
formation, 548, 549, 549F
inactivation, 548–549, 548F
structure, 75F, 548F
Cyclic AMP-dependent protein kinase (PKA), 549, 550, 551F
Cyclic AMP phosphodiesterase, 548–549, 548F
Cyclic GMP (cGMP), 539, 539F
light-induced signaling, 555F
Cyclic GMP phosphodiesterase, 555F
Cyclic photophosphorylation, **483–484**, 484F
Cyclin(s), 614, 614F, 617T
see also specific cyclins
Cdk association *see* Cyclin–Cdk complexes
discovery, **615–616**, 616F
ubiquitin-mediated degradation, 618, 618F
Cyclin–Cdk complexes, 614, 614F, **617–618**, 617T
see also specific complexes
cell-cycle control, 617–618, 618F
regulation of activity, 617, 617F
Cyclin-dependent protein kinases *see* Cdks
Cycloheximide, 258T
Cysteine, 73F
Cystic fibrosis, 516
Cytochalasin, 582T, 592
Cytochrome *b*-c_1 complex, 460, 472, 472F, 473
proton pumping, 474–475
Cytochrome b_6-*f* complex, 482F, 483–484, 484F
Cytochrome b_{562}, 134F
Cytochrome *c*
oxidation, 473–474
regulation of apoptosis, 641, 642F
structure, 126F, 473, 473F
Cytochrome oxidase, **473–474**, 474F
Cytochrome oxidase complex, 460, 472F, 473
proton pumping, 475, 475F
Cytochromes, 473
Cytokines, 559–563
Cytokinesis, 611, 624, 627F, **634–638**, 636F
see also M phase
cleavage furrow, **634–635**, 635F
contractile ring *see* Contractile ring
plant cells, **636–637**, 637F
segregation of organelles, **638**
timing within M phase, 625
Cytoplasm, 10
diffusion through, 99, 99F
dynamic nature, 23
organelles within, 19–22, 22–23F
Cytosine (C), 57, 65F, 74F, 177
deamination, 214, 214F, 215F
methylation, 287, 288F
RNA, 233, 233F
Cytosine triphosphate (CTP), 152F
Cytoskeleton, **22–23**, 22F, **571–607**
see also Actin filaments; Intermediate filaments; Microtubule(s)
dynamic nature, 23, 571
filament types, 572, 573FF
functions, 571, 571F
membrane-enclosed organelles and, 497
role in mitosis and cytokinesis, **624**, 624F
Cytosol, **21–22**, 21F, 496, 497T
diffusion through, 99
protein import by organelles from, 500–501, 501F
volume, 498T

D

D-isomers, 52, 56
DAG *see* Diacylglycerol
Daltons, 41
Dark reactions, 479, 484–485
see also Carbon fixation
Darwin, Charles, 7
Deafness, hereditary, 675
Deamination, DNA, 214, 214F, 215F
Dedifferentiation, 271F
Dehydrogenation reactions, 88
Delbrück, Max, 200, 201
Delta, 534F, 535T, 563, 563F
ΔG *see* Free-energy change
$\Delta G°$ *see* Standard free-energy change
Dendrites, 409, 409F
Density gradient centrifugation, 165F
Deoxyribonuclease, structure, 126F
Deoxyribonucleic acid *see* DNA
Deoxyribonucleotides, 56
Deoxyribose, 74F, 177, 233F
origin of life and, 263–264
Dephosphorylation, 153
Depolarization, neuronal membrane, 410, 411, 411F
Depurination, DNA, 214, 214F, 215F
Dermis, 709F
Desmosomes, 576, 702F, 703, **704**, 706F
Detergents, **375**, 376F, 383
Development, 707–709, 708F
see also Zygotes
apoptosis, 639, 639F
combinatorial control of gene expression, **285–286**, 286F
egg polarity, **282–284**, 282F, 283F, 284F
epithelial movements, 704, 705F
eye, 288, 289F
human brain, 320
Diabetes, adult-onset, 517
Diacylglycerol (DAG), 548, 552, 552F
Dicer, 292, 293F
Dideoxy DNA sequencing, 345, 345F, 346F
Didinium, 26, 26F
Differentiated cells, 2–3, 269–270
cell cycle withdrawal, 619–620
dedifferentiation, 271F
differences in gene expression, 5–6, **270–271**, 271F
mechanisms of development, **280–288**
cell memory, 287–288, 287F
combinatorial control, **285–286**, 286F
nuclear transplantation studies, 270, 271F
terminally, 711, 711F
Diffusion, **98–99**
across cell membranes, 389, 389F
facilitated, 390
Digestion, 427, 526
Dihybrid crosses, 669–670, 670F
Dihydrofolate reductase, 148
Dihydrolipoyl dehydrogenase, 436F
Dihydroxyacetone, 68F
Dihydroxyacetone phosphate, 430F
Dimers, 136, 136F
Dinitrophenol (DNP), 455, 468
Dinoflagellate, 27F
Diploid cells, **652–654**, 653F, 655
formation of haploid cells from, **655–663**, 660F
Disaccharides, 52, 53F, 69F
Diseases, human
environmental factors, 682, 682F
genetic predisposition, 679–682, 679F
inherited *see* Genetic disease
Disulfide bonds, 73F
formation in ER, 514
proteins, **139–140**, 139F
Divergence
see also Phylogenetic trees
gene duplication and, 303, 304–305, 305F
DNA, 4, 58, **171–195**
chloroplast, 19
condensation, 183, **184–185**, 185F
double-stranded, 173F
electron microscopy, 12F
evolution, **263–264**, **264F**
historical research, 171–172, 174–176
linker, 186, 186F
manipulating and analyzing, 329–333
mechanism of heredity, **178–179**, 179F, 231
mitochondrial *see* Mitochondrial DNA
noncoding *see* Noncoding DNA

Page numbers in boldface refer to major discussion of a topic; page numbers followed by F refer to a figure, and with FF to consecutive figures; page numbers followed by T refer to tables.

nuclear, 16F, 17
packing in chromosomes, 179, **185–188**, 187F
recombinant *see* Recombinant DNA
repetitive sequences, 317F, 321
DNA-binding proteins *see* Transcription regulators
DNA chains *see* DNA strand(s)
DNA cloning, 329, **333–343**
DNA libraries, 336–339, 337F, 340F
plasmid vectors, 335–336, 335F
polymerase chain reaction, 340–343
serial, recombinant DNA creation, 347, 347F
DNA damage, 197, **213–215**
see also DNA repair; Mutations
cancer treatments, 727
checkpoints, cell cycle, **621**, 622F
depurination and deamination reactions, 214, 214F
double-strand breaks, 216–217, 217F, **218–220**
induced by mutagens, 676, 676F
mutations resulting from, 215, 215F
DNA double helix, 173F, **177–178**
complementary base-pairing, 177–178, 177F
major and minor grooves, 178F
unzipping, 208–209, 209F, 234
DNA fingerprinting, 321, 343, 344F
DNA helicases, 208–209, 209F
DNA hybridization *see* Hybridization
DNA libraries, **336–338**, 337F
see also cDNA libraries; Genomic libraries
identifying specific DNA clones, 337–338, 338F
DNA ligase
DNA repair, 216, 216F
DNA replication, 208, 208F
recombinant DNA technology, **334**, 334F, 336, 336F
DNA ligation, homologous recombination, 219, 219F
DNA methylation, 287–288, 288F
DNA microarrays, **352–353**, 353F
DNA polymerase
associated proteins, 208–210, 209F
DNA synthesis, **203–204**, 204F
polymerase chain reaction, 340, 341F
proofreading, 206, 206F, 207F
repair, 215–216, 216F
homologous recombination, 219
lagging strand DNA synthesis, 208, 208F
RNA polymerase compared, 235
DNA probes, **332–333**, 333F
DNA libraries, 337, 338F
in situ hybridization, 352, 352F, 353F
DNA repair, 197, **211–217**
see also DNA damage
basic pathway, 215–216, 216F
homologous recombination, 217, **218–220**, 219F
mismatch system, **212–213**, 213F
non-homologous end-joining, 216–217, 217F
polymerase *see* DNA polymerase, repair
DNA replication, 197, **198–210**
see also DNA synthesis
accuracy, 205–206, 206F, 212T
bidirectional nature, 203, 203F
cell cycle checkpoints controlling, **621**, 622F
compared to transcription, 234
conservative model, 200, 200F
dispersive model, 200, 200F
end-replication problem, 210, 210F
hypotheses, 200–202
initiation, 199–203, 199F
machine, 199, 203, **208–210**, 209F
meiosis, 656, 660, 660F
Meselson–Stahl experiment, 201–202, 201F, 202F
mismatches, repair, 212–213, 213F
origins *see* Replication origins
rate, 203
replication forks *see* Replication forks
S phase, 611, **620–621**, 620F
semiconservative nature, 199, 199F, **200–202**, 200F
DNA sequence(s)
comparisons
see also Comparative genomics; Phylogenetic trees
closely related species, 310, 311F
more distantly related species, 312, 312F
conserved, 217, 217F
evolutionary relationships, 313–315, 314F
functional importance, **310–312**, 311F
detection of specific, **332–333**, 333F
genetic code *see* Genetic code
homology, homologous recombination, 218
regulatory *see* Regulatory DNA sequences
transcription into RNA *see* Transcription
DNA sequencing, **345–347**
see also Genome, sequencing
automation, 345–346, 346F
dideoxy method, 345, 345F, 346F
DNA strand(s), **173–178**, 173F
3′ end, 177, 177F
5′ end, 177, 177F
antiparallel orientation, 173F, 177F, 178
complementary base-pairing, 177–178, 177F
hybridization *see* Hybridization
lagging, 205, 205F, 208, 208F
leading, 205, 205F
nicks
DNA mismatch repair, 213, 213F
sealing by DNA ligase, 208F, 216, 216F
RNA primer synthesis, 207
separation, 199, 199F, 208–209, 209F
single-strand binding protein, 208–209, 209F
template *see* DNA template
DNA structure, **172–179**
see also DNA double helix; DNA strand(s)
building blocks, 173–177, 173F
compared to RNA, 233, 233F
mechanism for heredity, 178–179, 179F
Watson–Crick model, 172, 173
DNA synthesis
see also DNA replication
5′-to-3′ direction, 203, 204–205, 204F, 206, 207F
chemistry, 111, 113F
in vitro, 332, 340, 341F
lagging strand, 205, 205F, 208, 208F
leading strand, 205, 205F
at replication forks, **203–204**, 203F
replication machine (proteins), 208–210, 209F
RNA primers, **206–208**, 208F
DNA template
DNA replication, 198–199, 198F
transcription, 234, 235F
Dolichol, 515, 515F
Dolichol phosphate, 71F
Dominant alleles, 667–668, 667F, 674F
Dominant mutations, 673, 722–723, 722F
Dopamine, 706, 707F
Down syndrome, 662, 662F
Drosophila melanogaster, 29, 29F
alternative splicing, 322, 322F
contact-dependent signaling, 534F, 563–564, 563F
egg polarity genes, **282–284**, 282F, 283F, 284F
eye development, 288, 289F
homologous genes, 309
Wnt signaling pathway, 725
Drugs
as enzyme inhibitors, **148**
psychoactive, **418–419**
Dscam proteins, 322F
Dynamin, 511–512, 512F
Dyneins
ciliary, 589–590, 590F, 591F
movement along microtubules, 583–584, 584F
transport of organelles, 584–585

E

EcoRI, 330F, 331F, 334F
EF-Tu, 154, 154F
Effector proteins, intracellular, 534–535
Eggs, 652, 653F
aneuploidy, 663
cell cycle studies, 615–616, 616F
fertilization, 663, 663F

fertilized *see* Zygotes
nuclear transplantation, 716–717, 716F
Elastase, 135, 135F
Elastin, elastic fiber formation, 138, 139F
Electrical signaling, 403, 403F
neurons, **409–414**
voltage-gated ion channels, 405–406
Electrochemical gradients, 393, 393F
active transport against, **393**, 393F
proton *see* Proton gradients, electrochemical
Electron(s), 40, 40F
affinities, redox potential as measure, **467–470**
atomic interactions, **41–43**, 43F
covalent bond formation, **45–46**
high-energy
capture in photosystems, 480, 480F
donation to electron-transport chain, 458–460, 458F
mitochondrial sources, 458, 458F, 459–460
ionic bond formation, **44**, 45F
shells, 41–43, 43F
Electron carriers, **470–473**, 472F, 473F
chloroplast electron-transport chain, 482F
mitochondrial electron-transport chain, 472
NADH and NADPH as, 107–109, 110F
Electron microscopy, 6, 9FF, 10–11, 12F
development, 24T
procaryotic cells, 14, 14F, 15F
resolution, 10, 10F, 13F
Electron transfers
driving proton pumping, 466–467, 467F
energy released, **470**
oxidation and reduction, 87–88, 88F
photosynthesis, 480–481, 480F, 481F
respiratory chain, 460, 461F
Electron-transport chain
electron carriers, **470–473**, 472F, 473F
energy generation mechanism, 454–455, 454F
evolutionary origins, 488, 490
mitochondrial (respiratory chain), 427, **444–445**, **458–465**
ATP yield, 464–465
chemiosmotic coupling, **458–460**, 459F
citric acid cycle linkage, 438, 458, 459F
components, **460**
electron donation to, 458, 458F, 459–460
historical research, **468–469**
location, 457–458, 457F
proton pumping, **460–461**, 461F, 462F
redox potential along, 472, 472F
molecular mechanisms, **466–476**
photosynthesis, 478–479, 480–481, 481F
location in chloroplasts, 477–478
NADPH and ATP synthesis, **481–482**, 482F, 483F
redox potentials, 483F
redox potential, 467–470
Electrophoresis, 162, 162T, 167F
see also Gel electrophoresis
Electrostatic attractions, 44, **47–48**, 77F
see also Ionic bonds; Polar covalent bonds
aqueous solutions, 77F
within macromolecules, 62
protein folding, 123, 123F
Elements, 40
chemical reactivity, 42, 43F
periodic table, 44F
relative abundance, 41, 42F
Embden–Meyerhof pathway *see* Glycolysis
Embryonic development *see* Development
Embryonic stem (ES) cells, **714–715**, 715F
production, 716–717, 716F
repair of damaged tissue, 715
transgenic mice, 356F
Endocrine cells, 532–533, 533F
Endocytic pathways, 510, 510F, **522–526**
Endocytic vesicles, 522
Endocytosis, 21, 21F, 496, **522–524**
see also Phagocytosis; Pinocytosis
endosomal compartment, 525–526, 525F
lysosomal pathway, 526, 527F
receptor-mediated, **524**, 525F
Endomembrane system, 499
Endoplasmic reticulum (ER), 12F, 20, 20F, 497
Ca^{2+} stores, 552
evolutionary origins, 498–499, 499F
functions, 497, 497T, 505–506, 506F
luminal release of soluble proteins, 508, 508F
membrane synthesis, 371
microtubule-associated transport, 584–585, 585F
partitioning at cell division, 638
protein export via transport vesicles, 501, 510, 510F, 516
protein modification, **514–515**, 515F
protein translocation channels, 507–508, 507F, 508F
protein translocation into, 501, 501F, **505–509**
during synthesis, 506–507, 507F
transmembrane proteins, 506, **508–509**, 509F
water-soluble proteins, 506, **507–508**, 507F, 508F
retention signal, 502T, 516
rough, 497, 506
secretory pathways, **514–518**
signal sequences, 502, 502F, 502T, 506, 507–508, 507F
size control mechanism, 516–517, 517F
smooth, 497
unfolded protein response, **516–517**, 517F
volume, 498T
Endosomes, 21, 524
early, 510F, 525, 525F
functions, 497, 497T
late, 510F, 525, 526
number per cell, 498T
sorting of macromolecules, **525–526**, 525F
Endothelial cells, 709–710
nitric oxide release, 539, 539F
Energy
activated carriers *see* Activated carrier molecules
activation, 89–90, 89F, 146
binding, 97, 97F
chemical bonds, 46, 95F, 104
free *see* Free energy
heat, **82–84**, 84F, 89, 427
production
chloroplasts, **476–486**
efficiency, 475–476
electron transfers, **470**
from food, **425–445**
glucose oxidation, 426, 426F
glycolysis, 429–432, 429F, **433**, 434F
membrane-based mechanism, 453, **454–455**, 454F
mitochondria, **456–476**
oxidation reactions, **86–87**
sources, 15, 18–19
see also Fats; Food; Protein(s); Sugars
sunlight, **84–86**, 85F, 86F
storage
enzyme-mediated coupling to oxidation, **433**, 434F, 435F
intracellular, 53, 54–55, **448–450**
utilization, **82–88**
Enhancers, 278, 278F
Enol phosphate bonds, 431F, 435F
Enolase, 431F
Enteroendocrine cells, 712F
Entropy, 83, 84
Environmental change, gene expression and, **275**
Environmental factors
cancer causation, 718–719, 720
common human diseases, 682, 682F
Enzyme(s), 46, 59, 81–82, **143–148**
active site, 91, 91F, 146
attached nonproteins, 149
catalysis
energetics, **88–91**, 89F, 90F

Page numbers in boldface refer to major discussion of a topic; page numbers followed by F refer to a figure, and with FF to consecutive figures; page numbers followed by T refer to tables.

equilibrium points, 100, 100F
mechanisms, **146–148**, 148F
competitive inhibitors, 102–103, 103F
control of activity, 102–103, **150**, 446
allosteric regulation, **150–152**, 152F, 162T
phosphorylation, 152–153
design, 103
diffusion within cells, 98–99, 99F
drugs inhibiting, **148**
feedback inhibition, 150, 150F, 151F
functional classes, 143, 143T
functions, 90–91, 90F, 91F, 120F
kinetics, **99–100**, 100F
see also K_M; V_{max}
experimental analysis, **101–103**
stopped-flow apparatus, 102F
lock-and-key analogy, 162T
mechanisms of action, **143–148**
membrane, 373T
substrate binding, 99, 146
substrates *see* Substrates
turnover number, 100
Enzyme-coupled receptors, 542, 543F, **555–567**
see also Receptor tyrosine kinases
direct signaling pathways, 559–563
Enzymology, 162T
Epidemiology, 718
Epidermal growth factor (EGF), 535T
Epidermis, 709F, 711
desmosomes, 704, 706F
intermediate filaments, 572F
renewal from stem cells, 712, 713F
Epidermolysis bullosa simplex, 576
Epigenetic inheritance, 192, 288
Epinephrine (adrenaline), 535T, 549–550, 549T
Epithelia, **700–707**
barrier function, 701–703, 703F
polarization, **700–701**, 701F
tube/vesicle formation, 704, 705F
types, 700, 700F
Epithelial cells, 700, 700F
apical surface, 380, 380F, 700, 701, 701F
attachments between, 703–704, 704F
basal lamina attachment, 703, 705, 706F
basal surface, 380, 380F, 700–701, 701F
cilia, 589, 589F
communication between, 705–707, 707F
cultured, 157–161, 161F
intermediate filaments, 573F, 575F, 576
junctions between, **700–707**, 702F
plasma membrane proteins, 380, 380F
Epithelial sheets, 700–707
Equilibrium, chemical, 92–93, 93F, 95F
Equilibrium constant *(K)*, **93–97**, 95F
concentration dependence, 96
$\Delta G°$ and, 93–96, 96F
energy of binding interactions and, 96–97, 97F
ER *see* Endoplasmic reticulum
Erythrocytes *see* Red blood cells
ES cells *see* Embryonic stem cells
Escherichia coli, 14F, **27**
DNA libraries, 337, 337F
gene regulation, 275, 277
genome size, 34, 34F, 329
horizontal gene transfer, 308
human genome sequencing, 349
mutation rates, 300–301, 301F
T2 viruses, 176, 176F
transformation, 335, 335F, 336
Esters, 65F, 70F
Estradiol, 535T, 537, 538F
Ethane, 47F
Ethanol
diffusion across lipid bilayers, 389
production, 432–433, 432F
Ethene, 47F, 62
Ethidium bromide, 331F
Ethylene, 564, 565F
Ethylene glycol poisoning, 103
Eubacteria *see* Bacteria
Eucaryotes, 14, **16–26**
cell cortex, 594
cell cycle *see* Cell cycle
cell membranes, 363F, 364, 364F
chromosomes, **179–192**
evolution, **23–26**, 24F
gene regulation, **278**, 278F, **280–288**
genome sizes, 34, 34F
membrane-enclosed organelles, 496–498, 496F
model organisms, 28–35
single-celled, 16, 26
transcription *see* Transcription, eucaryotic
translation initiation, 255, 255F
Euchromatin, 191, 191F
Euglenoid, 27F
Even-skipped (eve), **282–284**, 284F
Evolution, 5, **297–326**
see also Comparative genomics; Genetic variation; Phylogenetic trees; Selection
cancer, **721–722**, 721F
chloroplasts *see under* Chloroplasts
Darwin's theory, 7
gene duplication events, 303, **304–305**, 305F
genome duplication events, 305
human populations, **683–684**, 683F
lactose tolerance, 302, 302F
membrane-enclosed organelles, **498–499**, 499F
mitochondria *see under* Mitochondria
mobile genetic elements and, **307–308**, 308F
protein domains, 243, 306–307, 307F
sexual reproduction and, 654–655
Evolutionary conservation, 217, 217F
cell-cycle control system, 613, 616
Excitatory neurons, 417
Exocytosis, 21, 21F, 496, 514, **518–522**
constitutive (default) pathway, 518–519, 519F
regulated pathway, 519–522, 519F
Exon junction complex (EJC), 244F
Exon shuffling, 298F, 299, **306–307**, 307F
Exons, 242, 242F
duplication events, **306**, 306F
splice site detection, 318
transposition-induced rearrangements, 307–308, 308F
Expressed sequence tags (ESTs), 319
Expression vectors, 347–350, 347F
Extracellular matrix, 7, 10F, 689, **690–699**
animal tissues, 691, **693–694**, 694F
basal lamina, 700–701, 701F
cell attachment, **696–698**, 697F
cell crawling on, 596
collagen *see* Collagen
degradation, 695
plant tissues, 690
production, 694–695, 695F
proteins, **138–140**, 139F
proteoglycans, 698–699
Extracellular signal molecules
controlling actin filaments, **597–598**, 598F
controlling cell survival, growth and division, **642–646**
mechanisms of action, **537–540**, 537F
regulating gap junctions, 706, 707F
Ey gene, 288, 289F
Eye
development, 288, 289F
intracellular signaling, 554, 554F, 555F
lens development, 704, 705F

F

F_1 generation, 666, 666F, 667, 667F
F_2 generation, 666, 666F, 667–668, 667F
Factor VIII gene, 242F
cloning, 336–337, 338
mutations, 307
Factor VIII protein, production, 350
FAD, 438, 439F
$FADH_2$, 109
ATP yields, 464–465, 465T
donation of electrons, 458–460, 459F
production, 436, 437F, 438–439, 438F
structure, 439F
Fats
see also Lipids
breakdown and utilization, 425, **426–445**

    - oxidation to acetyl CoA, 436, 437F
    - stages, 426–427, 428F
  - hydrophobicity, 365, 367F
  - solid and liquid, 370
  - storage in cells, 449–450, 449F, 450F
- Fatty acids, 51–52, **54–55**, 70FF
  - membrane phospholipids, 366F, 369–370
  - mitochondrial utilization, 458, 459F
  - oxidation to acetyl CoA, 436, 437F, 465
  - saturated *see* Saturated fatty acids
  - structure, 54, 54F
  - unsaturated *see* Unsaturated fatty acids
- Feedback regulation
  - glucose biosynthesis/breakdown, 447, **447–448**
  - negative (feedback inhibition), 150, 150F, 151F
    - conformational changes, 151, 152F
  - positive, 150
    - maintaining cell type, 287, 287F
- Fermentations, **432–433**, 432F, 487–488
- Ferredoxin, 482F
- Ferredoxin-NADP reductase (FNR), 482F
- Fertilization, 656, **663**
  - *see also* Zygotes
  - Ca^{2+} signaling, 552, 553F, 663
- Fibroblasts, 8F
  - collagen organization, 696, 697F
  - collagen secretion, 694–695, 695F
  - conversion to muscle cells, 285–286, 286F
  - crawling, 590, 594–595, 595F, 598F
  - in culture, 157, 161F, 645F
  - focal contact sites, 706
  - shape changes during M phase, 636, 637F
  - transformation into ES-like cells, 717, 717F
- Fibroin, 127
- Fibronectin, 696, 697F
- Fibrous proteins, 138, 139F
  - intermediate filaments, 574–575, 574F
- Filopodia, 591F, 594–595, 595F
  - extracellular signals regulating, 598, 598F
  - formins, 596, 596F
  - surface adhesion, 596
- Fischer, Emil, 60
- Flagella (flagellum), 2, 3F, 25F, **585–590**
  - mechanism of movement, 589–590, 591F
  - microtubules, 577–578, 589, 590F
  - source of energy, 464
- Flavin adenine dinucleotides *see* FAD; $FADH_2$
- Flippases, 371, 371F
- Flow cytometry, 612
- Fluid-mosaic model, 382
- Fluorescence *in situ* hybridization (FISH), 353F
- Fluorescence microscopy, 8–9FF, 10
- Fluorescence recovery after photobleaching (FRAP), 382, 382F
- Fluorescent probes, 8F
- Focal contact sites, crawling cells, 595F, 596, 706
- Food
  - digestion, 427
  - energy production from, **425–445**
  - stages of breakdown, **426–427**, 428F
  - storage in cells, **448–450**
- Forensic medicine, use of PCR, 342–343, 344F
- Formic acid, 488
- Formins, 593, 596, 596F
- Free energy (*G*), 88–100, 94FF
- Free-energy change (ΔG), **91–93**, 94FF
  - chemical equilibrium, 92–93, 93F
  - concentration of reactants and, 92
  - positive and negative, 91, 91F, 92F, 94F
  - standard *see* Standard free-energy change
- Fructose, 68F
- Fructose 1,6-bisphosphatase, 448
- Fructose 1,6-bisphosphate, 429F, 430F
- Fructose 6-phosphate, 430F
- *Fugu rubripes*
  - genome compaction, 313, 313F
  - intron conservation, 313, 314F
- Fumarase, 443F
- Fumarate, 443F
- Fungi
  - *see also* Yeasts
  - antibiotics from, 258
  - membrane transport, **400**, 401F

G

- *G see* Free energy
- G protein(s), 542, **545–547**
  - activation, 545–546, 545F
  - activation of membrane-bound enzymes, **547–548**, 548F
  - α subunit, 545–546, 545F, 546F
  - $\beta\gamma$ complex, 545, 545F, 546F
  - G_i, 546–547, 547F
  - G_s, 546, 548, 550F
  - inactivation, 546–547, 546F
  - ion channel regulation, **547**, 547F
- G-protein–coupled receptors (GPCRs), 542, 543F, **544–555**
  - activation of G protein, **545–547**, 545F
  - intracellular signaling pathways, 547–555
  - structure, 544, 544F
- G_0 state, 612, 619–620, 619F
- G_1-Cdk, 617, 617T
- G_1 cyclins, 617
- G_1 phase, 611–612, 611F
  - checkpoint, 612–613, 613F
    - cell cycle arrest, 619
    - mechanism, **621**, 622F
- G_1/S-Cdk, 617, 617T, 621, 622F
- G_1/S cyclins, 617
- G_2 phase, 611–612, 611F
  - checkpoint, 612, 613F, 621
- GABA *see* γ-aminobutyric acid
- Galactocerebroside, 366F
- Galactose, 68F
- β-Galactosidase, 277F, 284F
- Gametes, 299, 652
  - *see also* Eggs; Germ cells; Sperm
  - Mendel's law of segregation, **667–668**, 667F, 669F
  - production via meiosis, **655–663**, 660F
  - reassorted genetic information, **661–662**, 661F
- γ-aminobutyric acid (GABA), 535T
- γ-aminobutyric acid (GABA)-gated Cl^- channels, 417–418, 418T
  - drugs acting on, 418, 419
- Gap junctions, 401, 702F, **705–707**, 707F
- GDP, 438, 439F
  - inactivated G proteins, 545–546
- Gel electrophoresis, 162, 167F
  - DNA fragments, **330–331**, 331F
- Gel-filtration chromatography, 166F
- Gelsolin, 593
- Gene(s), 4, 5–6, 171, 674F
  - *see also* Alleles
  - arrangement in chromosomes, 181–182, 181F, 316, 316F
  - bacterial, 242, 242F
  - cloning *see* DNA cloning
  - eucaryotic, **241–242**, 242F
  - function, methods of studying, **343–358**
  - highly conserved, 309–310
  - historical studies of composition, 171–172, 174–176
  - homologous, 33, 309
  - horizontal transfer *see* Horizontal gene transfer
  - human, 316, 316F
  - identification methods, **318–319**, 319F, 675–677
  - jumping *see* Mobile genetic elements
  - manipulating and analyzing, **329–333**
  - numbers of, 34, 181, 317, **318–319**
  - promoters *see* Promoters
  - regulatory DNA sequences *see* Regulatory DNA sequences
  - silencing, 190–191
- Gene duplication, 298–299, 298F, 302–303
  - and divergence, 303, **304–305**, 305F
  - globin genes, **304–305**, 304F, 305F

  mechanism, 303, 304F
- Gene expression, 4F, 172, 178, 179F
  - *see also* Transcription; Translation
  - cAMP-mediated changes, 550, 551F
  - constitutive, 275
  - control *see* Gene regulation
  - different cell types, 5–6, **270–271**, 271F
  - expressed sequence tag (EST) analysis, 319
  - methods of analysis, 271, **350–353**
  - position effect, 190F
  - processes mediating, 232, 236
  - varying efficiencies, 232, 232F
- Gene families
  - gene duplication creating, **302–303**, 303F
  - globin genes, **304–305**, 305F
- Gene knockout, 354, 355F
- Gene regulation, **269–296**
  - combinatorial, 280, 281F, **285–286**, 286F
  - coordinate, **281–285**, 285F
  - environmental influences, 275
  - experimental analysis, **282–284**, 282F, 283F, 284F
  - external signals, 272
  - heterochromatin, 190–191, 190F, 191F
  - post-transcriptional, **289–293**
  - sites, 272, 272F
  - transcriptional *see* Transcription, regulation
- Gene regulatory proteins *see* Transcription regulators
- Gene replacement, 354, 355F, 356F
- General transcription factors *see* Transcription factors, general
- Genetic code, 178, 246–247, 247F
  - *see also* Translation
  - cracking the, **248–249**, 249F
  - redundancy, 247F, 250
- Genetic disease
  - chromosome abnormalities, 181F
  - gene therapy, 355
  - SNP analysis, 679
- Genetic engineering *see* Recombinant DNA technology
- Genetic instability, cancer cells, 720, 720F, 722
- Genetic linkage, 672–673
- Genetic maps, 672–673
  - common human diseases, 679–682, 679F
  - distance, 674F
- Genetic screens, 675, **676–677**, 678F
  - cell signaling molecules, 560–561
- Genetic traits, 664
  - discrete, 665, 665F
  - inheritance *see* Inheritance
- Genetic variation, 297, 297F, **298–308**
  - human genome, **320–321**, 321F
  - mechanisms of generation, **298–308**, 298F
  - meiosis, 658, **661–662**, 661F
  - sexual reproduction, 654
- Genetically modified organisms (GMOs), 354–356
- Genetics, 651
  - essential concepts, 674FF
  - as experimental tool, **675–684**
- Genome, 5, 172, 178–179
  - definition, 674F
  - determination of development, 707–709, 708F
  - duplications, **305**
  - evolution
    - highly conserved regions, 312–313, 313F
    - mobile genetic elements and, **307–308**
    - phylogenetic tree construction, 313–315, 315F
  - sequence comparisons *see* Comparative genomics
  - sequencing, 345, **348–349**
    - clone-by-clone approach, 349, 349F
    - shotgun method, 348–349, 348F
  - size, 34, 34F, 181–182
    - variability, 312–313, 313F
- Genomic libraries, 337, 337F
  - cDNA libraries compared, 339, 340F
  - obtained by PCR, 342, 343F
- Genotype, 354, 666, 674F
- Germ cells, 299, 652
  - *see also* Gametes
  - mutations, 212, 299, 300F
  - production via meiosis, **655–663**
  - transfer of genetic information, **299–300**, 300F
- Germ line, **299–300**, 300F, 653–654
- Germ-line cells, 653–654, 654F
- Giant, 283, 283F, 284, 284F
- *Giardia*, 17
- Gibbs free energy *see* Free energy
- Giemsa staining, 180F
- Gilbert, Walter, 318
- Gleevec, 148, 728, 728F
- α-Globin genes, evolution, **304–305**, 304F, 305F
- β-Globin genes
  - evolution, **304–305**, 304F, 305F
    - *Alu*-mediated recombination, 307–308
  - exons, 242F
  - heterochromatin and, 190
  - *L1* and *Alu* elements, 224F
  - mutations, 211, 211F
  - transcription, 245
- γ-Globin genes, evolution, 305F
- Globular proteins, 138
- Glucagon, 535T
- Glucocorticoid receptor protein, 281–285, 285F
- Glucocorticoids, control of gene expression, 272, 281–285, 285F
- Gluconeogenesis, 447–448, 447F
- Glucosamine, 69F
- Glucose
  - biosynthesis, 447–448, 447F
  - as energy source, 53
  - oxidation (breakdown), 427, 428F
    - ATP yields, 464–465, 465T
    - feedback regulation, 447–448, **447F**
    - glycolysis *see* Glycolysis
    - stepwise nature, 426, 426F
  - structure, 52, 52F, 68F
- Glucose 1-phosphate, 448, 448F
- Glucose 6-phosphate, 430F, 435F, 448–449
- Glucose–Na^+ symport, 398–399, 399F, 401T
- Glucose transporters, 401T
  - gut epithelial cells, 398–400, 399F
  - Na^+-driven symport, 401T
  - passive uniport, 392, 392F
- Glucuronic acid, 69F
- Glutamate receptors, 417, 418T
- Glutamic acid, 73F
- Glutamine, 73F
  - biosynthesis, 107, 107F
- Glyceraldehyde, 68F
- Glyceraldehyde 3-phosphate
  - carbon-fixation cycle, 485, 485F
  - glycolysis, 429F, 430FF, 433, 434F
  - utilization pathways in plants, 486
- Glyceraldehyde 3-phosphate dehydrogenase, 431F, 433, 434F
- Glycerol, 54–55, 54F, 70F
- Glycine, 73F
  - receptors, 417, 418T
- Glycogen, 53, 448–449, 448F
  - adrenaline-mediated breakdown, 549–550, 550F
  - structure, 69F, 448F
- Glycogen phosphorylase, 448–449, 448F
- Glycogen synthase, 448
- Glycolipids, 53–54, 55, 71F
  - amphipathic nature, 365, 366F
  - distribution in lipid bilayer, 371, 371F, 372
  - synthesis, 371–372
- Glycolysis, **427–432**, 429F
  - ATP yield, 456, 464, 465T
  - biosynthetic role, **439–444**, 444F
  - central role in metabolism, 445, 446F
  - oxidation–energy storage coupling, **433**, 435F
  - reactions, 430FF
  - regulation, **447–448**, 447F
- Glycoproteins, 53–54
  - glycosylation in ER, 514–515, 515F
  - membrane, 380–384, 381F
  - modifications in Golgi apparatus, 518
- Glycosaminoglycans (GAGs), 698–699, 698F, 699F
- Glycosidic bonds, 52
- Glycosylation, 514–515, 515F
- Goblet cells, 701, 702F, 712F
- Golgi apparatus, 20–21, 20F, **517–518**, 518F
  - *cis* network, 517–518, 518F

cisternae, 517, 518F
discovery, 24T
evolutionary origins, 498–499
functions, 497, 497T
microtubule-associated transport, 584–585, 585F
partitioning at cell division, 638
secretory pathway, **517–518**, 518F
trans network, 517–518, 518F
exocytosis pathways, 519–522, 519F
vesicular transport, 510–511, 510F
volume, 498T
Green fluorescent protein (GFP), 24T, 351–352, 352F
protein transport studies, 520–521, 521F
Green sulfur bacteria, photosynthesis, 488, 489F
Griffith, Fred, 174–175, 174F
Growth, cell *see* Cell growth
Growth factors, 642, **645**, 645F
GTP (guanosine triphosphate)
cap, microtubules, 580
production, 438–439, 438F
structure, 439F
GTP-binding proteins (GTPases), **153–154**
as molecular switches, 153–154, 153F, 542, 542F
monomeric (small), 557
regulating actin cytoskeleton, 598, 598F
trimeric *see* G protein(s)
GTP hydrolysis
G proteins, 153, 545–546, 546F
nuclear transport, 504, 504F
tubulin, 580, 580F
GTPases *see* GTP-binding proteins
Guanine (G), 57, 74F, 177
biosynthesis, control in bacteria, 290F
loss in DNA damage, 214, 214F
RNA, 233, 233F
Guanosine diphosphate *see* GDP
Guanosine triphosphate *see* GTP
Guanylyl cyclase, 539, 539F
Guard cells, 397, 397F
Gut epithelium *see* Intestinal epithelium

H

HaeIII, 330, 330F, 334F
Haemophilus influenzae, genome sequencing, 348
Hair cells, auditory, 405, 406F
Halobacterium halobium, 376, 469
Haploid cells, **652–654**, 653F, 655
see also Gametes; Germ cells
production via meiosis, **655–663**, 660F
reassorted genetic information, **661–662**, 661F
Haplotype blocks, **682–684**, 682F, 683F
Heart muscle *see* Cardiac muscle
Heat
energy, **82–84**, 84F, 89, 427
production, brown fat cells, 468–469
Helices
see also α helices
formation, **131**, 131F
protein filaments, 137, 137F
Heliozoan, 27F
Heme groups, 149, 149F, 473–474, 473F
Hemidesmosomes, 702F, 703, 705, 706F
Hemoglobin
evolution, 304–305, 304F
globin subunits, 136, 136F
heme groups, 149, 149F
historical landmarks, 162T
macromolecular structure, 60–61, 61F
shape and size, 126F
sickle-cell mutation, 211, 211F
Hemophilia, 223, 242F, 307, 336–337
see also Factor VIII gene
Hemopoiesis, 712, 713F, 714F
Hemopoietic stem cells, 712, 714F
transplantation, 714
Hepatocyte growth factor, 645
Heredity
see also Genetics; Inheritance
DNA code, **178–179**, 179F, 231
Heroin, 543T
Hershey, Alfred, 176, 176F
Heterochromatin, **190–191**, 190F, 191F
nuclear localization, 184F
X-chromosome inactivation, 190–191, 191F
Heterozygotes, 667
Hexokinase, 143, 430F
Hexoses, 68F
HindIII, 330F, 331F
Histamine, 535T
Histidine, 72F
Histone(s), **185–187**
acetylation, 189F, 279
covalent tail modifications, 188–190, 189F
inheritance, 192, 192F
evolutionary conservation, 187
H1 (linker), 187, 187F
methylation, 189F, 190
octamer (core), 186, 186F
phosphorylation, 189F
Histone acetylases, 279
Histone deacetylases, 279
HMG-CoA reductase inhibitors, 148
Holliday junctions, 221F
Holmes, Frederic Lawrence, 202
Homeodomain proteins, 120F, 274, 274F
Homogenization, 164F
Homologous chromosomes (homologs), 179, 657
crossing-over *see* Crossing-over
genetic reassortments, 661–662, 661F
maternal and paternal, 656
nondisjunction, 662–663, 662F
pairing during meiosis, **657**, 657F
segregation, **658–659**, 659F
Homologous genes, 33, 309
Homologous recombination, 198, **218–220**
DNA repair, 217, 218–220, 219F
gene duplication, 303, 304F
meiosis, 218, **220**, 221F, 657–658
mobile genetic elements as targets, 307
Homophilic binding, 703, 710
Homozygotes, 667
Hooke, Robert, 6, 24T, 690
Horizontal gene transfer, 220, 298F, 299, 308, 308F
Hormones, 532–533, 533F, 535T
see also specific hormones
intracellular receptors, **537–538**, 537F
Housekeeping proteins, 270
Human(s), 32–33
chromosome 22, arrangement of genes, 316, 316F
diseases *see* Diseases, human
evolutionary history, **683–684**, 683F
evolutionary relationships, 310F, 314–315, 315F
genes, methods of studying, 676
Mendel's law of segregation, 668, 669F
Human accelerated regions, 320
Human genome, **315–323**, 317T
comparative genomics
chimpanzee, **310**, 311F, 320
mouse, 310–312, 311F, 312F
vertebrates, 312–313, 313F
conserved synteny, 311–312
genetic variation within, **320–321**, 321F
highly conserved regions, 312, 313F
homologous genes, 309
mobile genetic elements, 222–223, 224F, 316, 317F
see also Alu elements; *L1* elements
noncoding DNA, 316–317, 317F
number of genes, 317, **318–319**
repetitive DNA, 317F, 321
sequence, 315, 315F
sequencing, **348–349**
size, 34, 34F
undeciphered information, 321–323
Human Genome Project, 315
Human immunodeficiency virus (HIV), 226
Human papilloma virus, 718–719
Hunchback, 283, 283F, 284, 284F
Hyaluronan, 699F
Hybrid vigor, 678
Hybridization (nucleic acid), **332–333**, 332F
chromosome painting, 180, 180F

Page numbers in boldface refer to major discussion of a topic; page numbers followed by F refer to a figure, and with FF to consecutive figures; page numbers followed by T refer to tables.

DNA libraries, 337, 338F
DNA microarrays, **352–353**, 353F
with DNA probes, 332–333, 333F
gel-transfer (Southern blotting), 333, 333F
in situ, 352, 352F, 353F
Hydra, 652F
Hydrocarbons, 64F
hydrophobic, 49, 66F
Hydrochloric acid, 50, 67F
Hydrogen atoms (H), 40, 40F, 41
carbon compounds, 64F
chemical bonds, 43
chemical reactivity, 42, 43F
Hydrogen bonds, **48–49**, 76F
α helices, 127, 130F, 131, 131F
β sheets, 127, 130F, 132
DNA double helix, 177–178, 177F
length and strength, 47T, 66F
within macromolecules, 62
protein folding, 123, 123F, 124F
water, 48–49, 66F, 76F
Hydrogen ions (H^+) *see* Proton(s)
Hydrogen molecules (H_2), 45, 45F
Hydrogen sulfide (H_2S), 488, 489F
Hydrogenation reactions, 88
Hydrolases, 143T
Hydrolysis, 53, 53F
energetics, 95F, 110, 111F
lysozyme-mediated, 146–148
Hydronium ions (H_3O^+), 49–50, 49F, 67F
Hydrophilic interactions
lipid bilayer, 365, 367F
membrane proteins, 373, 374–375, 374F, 375F
Hydrophilic molecules, 49, 66F
Hydrophobic interactions, 62, 77F
lipid bilayer, 365, 367F
membrane proteins, 373, 374–375, 374F, 375F
protein folding, 123, 124F
Hydrophobic molecules, 49, 66F
Hydrothermal vents, 490
Hydroxyl group, 65F
Hydroxyl (OH^-) ions, 50
Hypodermis, 709F

I

Immunoaffinity column chromatography, 145F
Immunoglobulins *see* Antibodies
Immunoprecipitation, 145F
In situ hybridization, 352, 352F, 353F
In vitro experiments, 157
In vivo experiments, 157
Influenza virus, 225F
Inheritance, **664–675**
blended, theory of, 666
chromosome cross-overs and, **671–673**, 672F
discrete (particulate) nature, **666–667**, 666F
disproven theories, 664, 664F, 666
meiotic mechanisms underlying, 671, 672F
Mendel's experiments, **665–667**, 666F
Mendel's law of independent assortment, **669–670**, 670F
Mendel's law of segregation, **667–668**, 667F, 669F
uniparental, 664, 664F, 666
Inhibitory neurons, 417
Initiator transfer RNA (tRNA), 255
Inorganic molecules, 51
Inositol 1,4,5-trisphosphate (IP_3), 548, 552, 552F
Inositol phospholipids, 372
phosphorylation by PI 3-kinase, 558, 558F
signaling pathway, **551–552**, 552F
Inoué, Shinya, 586
Insulin, 125, 126F, 162T, 535T
receptors, 120F
secretion, 519, 519F
Insulin-like growth factor (IGF) family, 558
Integrase, retroviral, 226, 226F
Integrins, **696–698**
conformational changes, 697–698, 698F
crawling cells, 596
deficiency disorders, 698
dividing cells, 636
fibronectin binding, 696, 697F
hemidesmosomes, 705, 706F
Interference-contrast microscopy, 8F, 11F
Interferons, 559
Intermediate filaments, 22, 22F, 138, **572–577**, 573F
accessory proteins, 576, 576F
cell strengthening function, **575–576**, 575F
classes, 575–576, 575F
network formation, 572–574, 572F
structure, 574–575, 574F
supporting nuclear envelope, **576–577**, 577F
Interphase, 611–612, 626FF
chromosomes *see* Chromosome(s), interphase
Intestinal epithelium
glucose transporters, 398–400, 399F
plasma membrane proteins, 380, 380F
polarized cells, 701, 702F
renewal from stem cells, 711–712, 712F
Wnt signaling, 713–714, 714F
Intracellular compartments *see* Compartments, intracellular
Intracellular signaling molecules, 534–535, 540, 540F
as integrating devices, 566, 566F
molecular switches, **541–542**, 542F
Intracellular signaling pathways *see* Signaling pathways, intracellular
Intracellular transport, **495–526**
see also Endocytic pathways; Secretory pathways; Vesicular transport
microtubule-associated, 583–585, 583F, 584F, 585F
Introns, 242, 242F
conservation, 313, 314F
evolutionary origins, **245–246**
removal by splicing, **242–243**, 243F
Ion(s), 44
extra- and intracellular concentrations, **388–389**, 388T
permeability of lipid bilayers, 389, 389F
solubility in water, 48–49, 66F
Ion-channel–coupled receptors (transmitter-gated ion channels), 416–417, 416F
excitatory and inhibitory, 417–418, 418F
intracellular signaling, 542, 543F, **544**
as psychoactive drug targets, **418–419**
Ion channels, 387–388, 390F, **400–420**, 418T
see also Ligand-gated ion channels; Stress-gated ion channels; Voltage-gated ion channels
G-protein-regulated, 547, 547F
gating, 402–403, 405, 405F
open and closed conformations, 402–403, 402F
patch-clamp recordings, 403–405, 404F
selectivity, 401–402, 402F
signaling in nerve cells and, **409–420**
Ion-exchange chromatography, 166F
Ionic bonds, 42, 43F
formation, **44**, 45F
length and strength, 47T
Iron, uptake into cells, 524
Iron–sulfur centers, 472–473
Isocitrate, 442FF
Isocitrate dehydrogenase, 443F
Isoelectric focusing, 167F
Isoelectric point, 167F
Isoleucine, 73F
Isomerases, 143T
Isomers, 52
optical, 52, 56
Isoprenes, 55, 71F
Isotopes, 40–41

J

JAKs, 559–563, 563F
"Junk" DNA *see* Noncoding DNA

K

K see Equilibrium constant
K^+ *see* Potassium ions
Kartagener's syndrome, 590
Karyotype, 181
Keratan sulfate, 699F

Keratinocytes, 596F
Keratins
α-keratin, 127, 131, 138
desmosomes, 704, 706F
hemidesmosomes, 705, 706F
network formation, 572F
structure, 138, 575–576, 575F
α-Ketoglutarate, 439–444, 443F
α-Ketoglutarate dehydrogenase complex, 443F
Ketones, 65F
Khorana, Gobind, 248–249
Kilocalories (kcal), 46
Kilojoules (kJ), 46
Kinases, 143T
Kinesins, 155
methods of studying, 587–588, 587F, 588F
movement along microtubules, 583–584, 584F
transport of organelles, 584–585
Kinetochores, 629–630, 629F, 630F
meiosis, 659, 659F
K_M, 100, 100F, 140
measurement, 101–102, 101F, 102F
Knockout mice, 356, 357F, 726
Krebs, Hans, 440–441
Krebs cycle *see* Citric acid cycle
Krüppel, 283, 283F, 284, 284F

L

L-isomers, 52, 56
L1 elements (LINE-1), 222–223, 224F, 311F
Lac operon, 277, 277F
Lactate
conversion to glucose, 447
production, 432, 432F
Lactic dehydrogenase, NAD-binding domain, 134F
Lactose, 69F
Lac repressor, 120F, 277, 277F
tolerance, evolution, 302, 302F
Lagging strand, 205, 205F, 208, 208F
Lamellipodia, 591F, 594–595, 595F
actin-related proteins, 596, 596F
extracellular signals regulating, 598, 598F
surface adhesion, 596
Laminin, 577
basal lamina, 700–701, 701F
hemidesmosomes, 705, 706F
Lamins, nuclear *see* Nuclear lamins
Lariat structure, RNA splicing, 242, 243F
Lasek, Roy, 586
Latrunculin, 582T
Leading strand, 205, 205F
Lectins, 381–384, 384F
Leder, Phil, 249
Leeuwenhoek, Antoni van, 6, 24T
LEF-1/TCF, 725–726
Leptin gene, sequence comparisons, 311F, 312F
Leucine, 73F
Leucine zipper motif, 274, 274F
Leucocyte adhesion deficiency, 698
Leucocytes *see* White blood cells
Leukemia, chronic myeloid (CML), 728, 728F
Life, 1, 3–4
ordered nature, 82, 83F
origin of *see* Origin of life
tree of, **313–315**, 315F
Ligand-gated ion channels, 405, 405F
neurotransmitter-gated *see* Ion-channel–coupled receptors
Ligands, 140
binding sites, 141, 141F
stabilizing protein conformation, 152, 152F
Light-driven pumps, 393, 393F
Light energy, **478–479**, 479F
see also Sunlight
absorption by chlorophyll, **479**, 480F
driving ATP and NADPH synthesis, **481–482**, 482F, 483F
transfer to reaction center, **480–481**, 480F, 481F
Light-induced signaling cascade, 554, 554F, 555F
Light microscopy, 6, 7–10, 8FF
history, 6–7
internal cell structure, 10, 11F
resolution, 10F
sample preparation, 8F, 10
Light reactions *see* Photosynthesis, electron-transfer reactions
Lignin, 692
LINE-1 *see L1* elements
Linkage, genetic, 672–673
Linker DNA, 186, 186F
Lipid bilayer, 55, 55F, 71F, **364–372**
see also Membrane lipids
asymmetry, 370–372, 371F
barrier function, 368, 368F
disruption by detergents, 375, 376F, 383F
ER, integration of transmembrane proteins, **508–509**, 509F
fluidity, 368–370, 382
formation in water, 365–368, 368F
membrane protein associations, **373**, 373F
permeability, 387, 388F, **389**
structure of polypeptide chains spanning, **374–375**, 374F
synthetic, 368, 369F
Lipids, 55, 70FF
see also Fats
aggregates, 71F
membrane *see* Membrane lipids
Lipoamide reductase-transacetylase, 436F
Liposomes, 368, 369F
Listeria monocytogenes, thermosensor, 291F
Liver cells, gluconeogenesis, 447
Local mediators, 533, 533F, 535T
Locomotion, cell, **594–597**, 595F, 596F
Locus, 674F
Loligo see Squid
Low-density lipoproteins (LDL), receptor-mediated endocytosis, 524, 525F
Lymphocytes, 709–710
Lysine, 72F
Lysosomes, 12F, 21, **526**
endocytic pathway, 522
evolutionary origins, 498–499
functions, 497, 497T
membrane transporters, 391, 391F
number per cell, 498T
pathways to, 526, 527F
receptor-mediated endocytosis, 524, 525F
Lysozyme, **143–148**
active site, 146
catalytic mechanism, 146–148, 146F, 147F
disulfide bonds, 139
structure, 126F

M

M-Cdk, 617–618, 617T
cyclin-mediated activation, 623, 623F
discovery, 615–616, 616F
inactivation, 631
M phase initiation, 621, **622–623**, 623F
M cyclin, 615–616, 617
degradation, 631
M-Cdk activation, 623, 623F
M phase, 611, 611F, **622–638**
see also Cytokinesis; Mitosis
cell shape changes, 635–636, 637F
chromosomes, 182F, 183, 183F
control of entry, 621, **622–623**, 623F
role of cytoskeleton, **624**, 624F
signs of entry, 612
stages, **624–625**, 626FF
MacLeod, Colin, 175, 175F
Macromolecules, 51, **58–63**
abundance in cells, 58, 58F
biosynthesis, 110–112, 112F, 113F
experimental analysis, **60–61**, 60F, 61F
formation from subunits, 59, 59F
formation of complexes, 63, 63F
functions, 59
interactions between, 63, 63F
three-dimensional shape, 59–62, 62F
Macrophages, 2, 640, 714F
phagocytosis, 523, 523F
skin, 709–710
Magnesium ions (Mg^{2+}), 44, 388T
Malaria
resistance mutation, 301–302, 683
sickle-cell anemia and, 211F
Malate, 443F
Malate dehydrogenase, 443F
Malonate, 440–441, 440F, 441F
Maltose, 69F

Page numbers in boldface refer to major discussion of a topic; page numbers followed by F refer to a figure, and with FF to consecutive figures; page numbers followed by T refer to tables.

Mammals, 32–33, 33F
Mannose, 68F
Mannose 6-phosphate receptors, 526
MAP kinase, 557–558, 557F
MAP-kinase signaling module, 557–558, 557F
Margarine, 370
Marriages, first-cousin, 668, 669F
Mass spectrometry, protein sequencing, 158, 159F
Mating factor, yeast, 531, 532F
Matthaei, Heinrich, 248
Maturation promoting factor (MPF), 615–616, 616F
McCarty, Maclyn, 175, 175F
Mechanical stress
  role of intermediate filaments, 575–576
  transmission by integrins, 696, 697F
Mechanical stress-gated ion channels, 405, 405F, 406F, 418T
Mediator complex, 278, 278F
Megakaryocytes, 714F
Meiosis, 646, 651, 655–663, 660F, 674F
  I, 656, 657–659, 657F, 659F
  II, 656, 659, 659F
  chromosome pairing, 656–657, 657F
  chromosome segregation, 658–659, 659F
  crossing over *see* Crossing-over
  duration, 656
  errors, 662–663, 662F
  function, 652, 653F
  genetic reassortment, 661–662, 661F
  homologous recombination, 218, 220, 221F, 657–658
  mitosis compared, 656–657, 657F, 660F
  role in Mendelian inheritance, 671, 672F
Membrane(s)
  see also Plasma membrane
  flexibility, 368
  fluid-mosaic model, 382
  fluidity, 368–370
    lipid composition and, 369–370, 370F
    measurement, 382–383, 382F, 383F
    types of lipid movements, 368–369, 370F
  internal, 11, 19–21, 21F, 363F, 364, 496
  lipid bilayer *see* Lipid bilayer
  lipids *see* Membrane lipids
  microscopy, 11
  permeability, 387, 388F, 389
  proteins *see* Membrane proteins
  structure, 363–386, 365F
  synthesis of new, 371–372, 372F
  vesicular transport, 371–372, 372F, 510–511, 510F
Membrane domains, 380, 380F
Membrane-enclosed organelles *see* Organelles, membrane-enclosed
Membrane lipids, 55, 55F, 365–368, 365F
  see also Cholesterol; Glycolipids; Lipid bilayer; Phospholipids
  amphipathic nature, 365, 365F, 366F
  asymmetric distribution, 371, 371F
  flip-flop, 368, 370F
  import into organelles, 505
  movements within bilayer, 368–369, 370F
  saturation of hydrocarbon tails, 369–370
  types, 365, 366F
Membrane potential, 403, 407–408, 407F, 408F
  changes, triggering action potentials, 410–411, 410F
  control of voltage-gated ion channels, 405–406
  effect on passive transport, 392–393, 393F
  proton-motive force, 461, 462F
  resting, 407–408, 408F
Membrane proteins, 372–384
  see also Transmembrane proteins
  carbohydrate linkages, 380–384, 381F, 384F
  functions, 372, 372F, 373T
  integral, 373
  lipid bilayer associations, 373, 373F
  localization within domains, 380, 380F
  mobility within membrane, 379–380, 379F
    methods of studying, 382–383, 382F
    patterns shown, 383, 383F
  peripheral, 373
  reconstitution in artificial vesicles, 383, 383F
  solubilization in detergent, 375, 376F, 383F
  structure, 376–377, 377F, 378F
    determination, 376
    membrane-spanning segments, 374–375, 374F, 375F
Membrane transport, 387–423
  passive and active, 390–391, 391F
  principles, 388–391
Membrane transport proteins, 387, 388F
  see also Channels; Ion channels; Transmembrane proteins; Transporters
  classes, 389–390, 390F
Mendel, Gregor, 651, 664
  chromosome cross-overs and, 671–673, 672F
  discrete nature of heredity, 666–667, 666F
  experimental methods, 665–666, 666F
  law of independent assortment, 669–670, 670F
  law of segregation, 667–668, 667F, 669F
  meiotic mechanisms explaining laws, 671, 672F
  pea traits studied, 665, 665F
Meselson, Max, 200–202
Meselson–Stahl experiment, 201–202, 201F, 202F
Messenger RNA (mRNA), 4, 235, 236T
  amino acid coding, 246–247, 247F
  analysis of expression, 271
  capping, 241, 241F
  cDNA preparation from, 338, 339F
  control of translation, 290, 291F
  degradation, 244–245, 292
  export from nucleus, 240, 240F, 243–244, 244F, 503
  Northern blotting, 333F
  polyadenylation, 241, 241F
  procaryotic, polycistronic nature, 256, 256F
  processing, 240–241, 241F, 245F
  reading frames, 247, 247F
  regulation by miRNAs, 290–291, 292F
  ribosome binding sites, 253, 253F
  riboswitches, 289–290, 290F
  splicing *see* Splicing
  translation *see* Translation
  untranslated regions, 290
Metabolic pathways, 82, 82F, 143
Metabolism, 82
  see also Anabolism; Catabolism
  complexity, 445, 446F
  regulation, 150, 150F, 151F, 445–450
Metamorphosis, apoptosis during, 639, 639F
Metaphase, 627F
  meiosis I, 657, 658–659, 659F
  meiosis II, 659F
  mitosis, 624–625, 630–631
Metaphase plate
  meiosis, 657, 657F
  mitosis, 631, 631F, 656, 657F
Metastases (secondary tumors), 718, 719F, 722
Metastasis, 721
Methane, 64F
Methanococcus jannaschii, 490, 490F
Methionine, 73F, 255
Methotrexate, 148
Methyl group, 64F
Mice, 32–33, 33F
  human genome comparison, 310–312, 311F, 312F
  knockout, 356, 357F, 726
  mouse–human hybrid cells, 379, 379F
  transgenic, 357F
Micelles, 71F, 375, 376F
Michaelis' constant *see* K_M
Michaelis–Menten equation, 101F
Microelectrodes, 403, 404F

Microfilaments *see* Actin filaments
Micrometers, 2
MicroRNAs (miRNAs), 236, 236T
  regulation of gene expression, **290–291**, 292F
  uncertainties about, 322F
Microscopy, 6–11, 8–9FF
  history of development, 6–7, 24T
  resolution, 10, 10F
Microtubule(s), 11F, 22, 22F, 572, 573F, **577–590**
  see also Tubulin
  actin filaments compared, 591–592
  assembly–disassembly balance, **581–582**, 581F
  astral, 628, 628F, 630F, 633
  cilia and flagella, **585–590**, 590F
  dividing cells, 22, 23F
  drugs acting on, 581–582, 582T, 585
  dynamic instability, **580**, 580F, 628
  growth, 580, 580F
  interpolar, 628, 629F, 630F, 633, 633F
  intracellular transport
    within axons, 582–583, 582F
    methods of studying, **586–588**, 587F, 588F
    motor proteins, **583–584**, 584F
    organelles, 583F, **584–585**, 585F
  kinetochore, 629–630, 629F, 630F
    shortening, 632–633, 633F
  mitotic spindle formation, **628–630**, 629F, 630F
  nucleation sites, 579–580, 579F
  organization of cell interior, **582–583**
  plant cell wall formation, 692–693, 693F
  plus and minus ends, 578, 578F
  polarity, 578, 579
  structure, **578–579**, 578F
Microtubule-associated proteins, 583
  see also Dyneins; Kinesins
  cilia and flagella, 589–590, 590F
Microtubule-organizing centers, 577, 577F, **579–580**, 579F, 625
  see also Centrosome(s)
Microvilli, 590, 591F
Mimosa, leaf-closing response, 405, 406F
Mitchell, Peter, 468–469
Mitochondria, **17**, 17F, **456–466**
  acetyl CoA production, **436**, 436F, 437F, 458
  breakdown of food molecules, 427, 428F
  chloroplast interrelationships, 486, 486F
  chloroplasts compared, 477–478, 478F
  compartments, **456–458**, 457F
  discovery, 24T
  disorders of function, 456
  electron microscopy, 12F
  evolution, 14–15, 17, 18F, 24F, **486–490**, 499, 499F
  functions, 17, 497, 497T
  genetic system, 486–487
  growth and reproduction, 486–487, 487F
  inner membrane, 17, 18F, **456–458**, 457F
    ATP synthesis, 461–463, 462F
    coupled transport across, **463–464**, 464F
    cristae, 457F, 458
    electron-transport chain *see under* Electron-transport chain
  intermembrane space, 457, 457F
  location within cells, 456, 457F
  matrix, 427, 457, 457F
  membrane transporters, 391, 391F
  outer membrane, **456–457**, 457F
    channels, 375
  oxidative phosphorylation, 444–445, 445F
  partitioning at cell division, 638
  protein import, 501, 501F, **505**, 505F
  signal sequences, 502T
  structure, 17, 18F, **456–458**, 457F
  volume/number per cell, 498T
Mitochondrial DNA, 17, 487
  genetic code, 247
Mitogen-activated protein kinase *see* MAP kinase
Mitogens, 619, 642, **644–645**, 644F, 645F
Mitosis, 611, **625–634**
  see also M phase; *individual stages*
  checkpoint, 612, 613F, 633
  chromosomes *see* Chromosome(s), mitotic
  completion, 634
  control of entry into, 622–623
  DNA condensation, 184–185
  meiosis compared, 656–657, 657F, 660F
  stages, 624–625
Mitotic chromosomes *see* Chromosome(s), mitotic
Mitotic spindle, 624, 624F
  assembly, **628–630**, 629F, 630F
    cells lacking centrosomes, 630, 630F
    checkpoint, 612, 613F, 633
    chromosome attachment, 629–630, 629F
    initiation, **628**, 628F
    role of chromosomes, 630, 630F
  chromosome alignment at equator, **630–631**, 631F
  cleavage plane determination, **634–635**
  composition, 577, 577F
  disassembly, 634
  drugs acting on, 581–582
  microtubule classes, 630, 630F
  poles *see* Spindle poles
Mobile genetic elements, 198, **221–226**
  see also Retrotransposons; Transposons; Viruses
  bacterial, 222, 222F
  evolutionary relationships and, 310, 311, 311F
  human genome, 222–223, 224F, 316, 317F
  introns as, 246
  role in evolution, **307–308**, 307F, 308F
Model organisms, **26–35**
Molar solutions, 41, 41F
Mole, 41, 41F
Molecular switches, 153–154, 153F, **541–542**, 542F
  switch-off mechanism, 541, 546–547
Molecular weight, 41
Molecules, 40, 45
  ball-and-stick models, 46F
  in cells, **50–58**
  methods of representing, 52F
Monoclonal antibodies, 145F
Monohybrid crosses, 669
Monomers, 51, 58, 59F
  see also Subunits
Monosaccharides, 52, 68FF
Monounsaturated fatty acids, 54F
Morphine, 543T
Morris, James, 696F
Motor neurons, 419–420, 419F
Motor proteins, 120F, **583–584**
  actin dependent *see* Myosin
  chromosome segregation, 632–633
  mechanism of movement, **154–155**, 154F, 155F, 588, 588F
  methods of studying, **586–588**
  mitotic spindle assembly, 630, 630F
  movement along microtubules, 583–584, 584F
  transport of organelles, 584–585, 585F
Mouse *see* Mice
mRNA *see* Messenger RNA
Multicellular organisms, 1, 2–3, 4F, 16
  see also Differentiated cells
  cellular communities, 689–690, 690F
  control of cell numbers and size, 642–646
  creation of specialized cell types, **280–288**
  evolution of cell communication, **564**
  model organisms, 28–33
  reproduction, 651
  tissues, 689–690, 690F
Muscle cells, 599, **604**
  see also Cardiac muscle cells; Skeletal muscle cells; Smooth muscle
  acetylcholine receptors, 416–417, 417F
  anaerobic metabolism, 432, 432F
Muscle contraction, 552, **599–604**
  actin and myosin bundles, **599**, 600F
  Ca^{2+} signaling, **602–604**, 603F

Page numbers in boldface refer to major discussion of a topic; page numbers followed by F refer to a figure, and with FF to consecutive figures; page numbers followed by T refer to tables.

sliding-filament mechanism, 600–601, 601F, 602F
Mutagenesis
random, 675–676
site-directed, 354, 355F
Mutagens, 676, 676F, 720
Mutant organisms, 354, 674F
generation, 675–676
screening, 560–561, 675, 676–677, 678F
Mutations, 5, 197
see also Alleles; DNA damage
beneficial, 683
cancer causation, 719–720, 721, 721F
complementation test, 674F, 677–678, 678F
conditional, 30, 673F
consequences, 211–212, 211F
deleterious, 673–675
elimination, 217, 654–655, 673–675
DNA damage, 215, 215F
dominant, 673, 722–723, 722F
gain-of-function, 673, 673F, 722–723, 722F
within gene regulatory regions, 298, 298F, 301–302, 303F
within genes, 298, 298F, 300–301
germ cell, 212, 299, 300F
insertion, 307
loss-of-function, 673, 673F, 722F, 723
point, 300–302
rates, 300–301, 301F, 720
recessive, 673, 722F, 723
selection pressures on, 309–310
selectively neutral, 301, 309, 310, 673, 683
somatic cell, 212, 299, 300F
temperature-sensitive *see* Temperature-sensitive mutations
Myasthenia gravis, 417
Mycobacterium tuberculosis, 523
Myoblasts
cultured, 161F
muscle cell development, 285
Myoclonic epilepsy and ragged red fiber disease (MERRF), 456
Myocytes *see* Muscle cells
MyoD
muscle cell development, 285–286, 286F
positive feedback loop, 287
Myofibrils, 600, 600F, 601F
Myoglobin, 126F, 134
Myosin, 131, 596
actin association, 597
generation of movement, 155
myosin-I subfamily, 597, 597F
myosin-II subfamily, 597, 599, 599F, 604
Myosin filaments, 599, 599F
actin filament association, 599, 600F
contractile ring, 635
muscle contraction, 600–601, 601F, 602F
sarcomeres, 600, 600F, 601F
Myostatin, 646, 646F

N

N-acetylglucosamine, 69F
N-terminus, polypeptide chain, 56, 56F, 122F
Na^+ *see* Sodium ions
Na^+-H^+ exchanger, 400, 401T
Na^+-K^+ pump, 394–395, 401T
control of osmotic balance, 396
mechanism of action, 394–395, 394F, 395F
NAD^+, 107–109, 110F
as electron acceptor, 467–470, 471F
glycolysis, 429–432
regeneration, 432, 432F, 433
NADH, 107–109, 110F
ATP yields, 464–465, 465T
donation of electrons, 458–460, 458F, 459F, 470, 471F
electron affinity, 467–470, 471F
synthesis, 427, 428F
acetyl CoA production, 436, 436F, 437F
citric acid cycle, 436–439, 438F
glycolysis, 429–432, 429F, 433, 434F
NADH dehydrogenase complex, 460, 472, 472F
proton pumping, 475, 475F
$NADP^+$, 107–109, 110F
NADPH, 107–109, 110F
carbon fixation utilizing, 484–485, 485F
generation by bacteria, 488, 489F
photosynthetic generation, 479, 481–482, 482F, 483F
Nanometers, 10
Natural selection, 309–310
Necrosis, cell, 639–640, 640F
Negative feedback regulation *see* Feedback regulation, negative
Negative regulation, 150
Neisseria gonorrhoeae, 308
Nernst equation, 408, 408F
Nerve cells, 2, 3F, 409–420
see also Neurons
regulation of numbers, 643, 643F
skin, 709–710
staining, 24T
Nerve growth factor (NGF), 535T
receptors, 557
Nerve impulse, 410
see also Action potentials
Nerve terminals, 409, 409F
neurotransmitter release, 415, 416F
Neural tube, 704, 705F
Neuraminidase, 136, 136F
Neurodegenerative disorders, 125
Neurofilaments, 575–576, 575F
Neuromuscular junction, 416–417
Neurons
see also Axons
action potentials, 409–414
anatomy, 409, 409F
excitatory and inhibitory inputs, 417–418, 418F
gene expression, 270F
growth, 646F
longevity, 710
polarization, 582, 582F
retinal, gap junctions, 706, 707F
signaling, 409–420, 533
synaptic signaling, 415, 415F
Neurotransmitters, 410
excitatory, 417, 418F
gated ion channels *see* Ion-channel–coupled receptors
inhibitory, 417–418, 418F
intracellular signaling, 544
receptors, 415–417, 416F
see also Ion-channel–coupled receptors
drugs acting on, 419
release at nerve terminals, 415, 416F
removal from synaptic cleft, 416
signaling function, 533, 533F, 535T
Neutrons, 40–41
Neutrophils, 2, 3F
cell-surface carbohydrate, 381–384, 384F
migration, 594
phagocytosis, 523F
Nexin, 590F
Nicks, DNA strand *see* DNA strand(s), nicks
Nicotinamide adenine dinucleotide phosphates *see* $NADP^+$; NADPH
Nicotinamide adenine dinucleotides *see* NAD^+; NADH
Nicotine, 543T
Nirenberg, Marshall, 248, 249
Nitric oxide (NO), 535T, 539, 539F
Nitrogen (N)
carbon compounds, 65F
covalent bonds, 45, 46F
fixation, 15, 490
Nitroglycerine, 539
Noncoding ("junk") DNA, 34–35, 181
see also Regulatory DNA sequences
distinction from coding DNA, 318
human genome, 316–317, 317F
Noncovalent bonds, 44, 47–49
bond strengths, 46, 47T
energy of binding interactions, 97–98, 98F
macromolecular interactions, 63, 63F
macromolecular structure, 59–62, 62F
protein binding, 48F, 140, 141F
protein folding, 123, 123F
weak, 76FF
Nondisjunction, 662–663, 662F

Nonhomologous end-joining, 216–217, 217F
Northern blotting, 333F
Notch receptor, 534F, 563–564, 563F
NotI, 330, 330F
Nuclear envelope, 16F, 17
- chromosome organization and, 184, 184F
- composition, 502–503, 503F
- disassembly in prometaphase, 628–629, 634F
- intermediate filaments supporting, **576–577**, 577F
- re-formation at telophase, **634**, 634F

Nuclear lamina, 502, 503F
- breakdown at apoptosis, 640
- intermediate filaments, 574, **576–577**, 577F
- interphase nucleus, 184

Nuclear lamins, 575, 575F, 576–577
Nuclear localization signals, 502T, 503
Nuclear magnetic resonance (NMR) spectroscopy, 158, **159**, 160F, 163
Nuclear membranes, 502, 503F
- evolutionary origins, 498–499, 499F

Nuclear pores, 502–503, 503F
- mRNA export, 244, 503
- protein import, 500, 501F, **502–505**, 504F
- structure, 503, 503F

Nuclear receptors, 537–538, 538F
Nuclear transplantation
- differentiated cells, 270, 271F
- reproductive cloning, 716–717, 716F

Nuclear transport
- mRNA export, 240, 240F, **243–244**, 244F, 503
- protein import, **502–505**, 504F
- receptors, 503–504, 504F

Nucleases, 143T
- DNA repair mechanism, 215, 216F
- homologous recombination, 218–219, 219F
- lagging strand DNA synthesis, 208, 208F
- restriction *see* Restriction nucleases

Nucleic acids, 57–58, 74F, 75F
- *see also* DNA; RNA
- biosynthesis, 111, 112F, 113F
- hybridization *see* Hybridization

Nucleolus, 184, 184F
Nucleoside triphosphates, DNA synthesis, 203–204, 204F
Nucleosides, 56, 75F
Nucleosomes, **185–187**, 185F
- *see also* Chromatin
- core particles, 185–187, 186F
- gene regulation and, 279–280, 279F
- regulation of structure, **188–190**

Nucleotide(s), 4, 51–52, **56–58**, 74FF
- *see also* Genetic code
- basic sugar linkage, 74F
- DNA strand formation, **173–177**, 173F
- functions, 75F
- nomenclature, 75F
- phosphates, 74F
- phosphodiester linkage, 57–58, 58F, 75F
- sugars, 74F

Nucleotide sequences
- *see also* DNA sequence(s)
- detection of specific, 332–333, 333F
- mechanism for heredity, 178–179
- translation into amino acid sequence *see* Translation

Nucleus (atomic), 40, 40F
Nucleus (cell), 10, 11F, 16–17, 496–497, 497T
- discovery, 24T
- evolutionary origin, 498–499, 499F
- interphase chromosomes, **184**, 184F
- structure, 16F
- volume, 498T

O

Obesity, as cause of cancer, 719
Occludins, 702, 703F
Okazaki fragments, 205, 205F, 209, 209F
Oleic acid, 54, 70F
Oligodendrocyte precursor cells, 161F
Oligosaccharides, 52–53, 69F
- addition to proteins in ER, 514–515, 515F
- complex, 69F
- membrane proteins, 381, 381F
- *N*-linked, 515
- processing, 515

Oncogenes, 558, **722–723**, 722F, 723F
Oocytes *see* Eggs
Open reading frames (ORFs), 318, 319F
Operons, 256F
- *Lac*, **277**, 277F
- tryptophan, 275, 275F, 276F

Optical isomers, 52, 56
Order, biological, 82, 83F
Organelles, **16–21**, 21F
- *see also specific organelles*
- electron microscopy, 11, 12F
- eucaryotic cells, **496–498**, 496F
- evolution, **498–499**, 499F
- functions, 497T
- growth and duplication, 500
- historical landmarks, 24T
- isolation, 498
- membrane-enclosed, 364, 364F, 495, **496–499**, 496F
  - compartment formation, 19–21
  - segregation at cell division, **638**
- motion, 23, 583, 583F, 586, 586F
- movement along microtubules, 583F, **584–585**, 585F
- protein import mechanisms, **500–501**, 501F
- protein sorting, **500–509**
- saltatory movement, 583
- vesicular transport between, **510–526**
- volume/number per cell, 498, 498T

Organic chemistry, 39
Organic molecules, 51
- large *see* Macromolecules
- small cellular, **51–58**, 51T, 52F

Origin of life, 488–490, 489F
- common ancestor, 5
- RNA and, **261–264**

Origin recognition complex (ORC), 620–621, 620F
Origins of replication *see* Replication origins
Osmosis, **396–397**, 396F
Osmotic pressure, 396
Osteoblasts, 694–695, 711
Osteoclasts, 711, 714F
Ouabain, 394, 395F
Ovarian cancer, 727–728
Oxaloacetate
- as biosynthetic precursor, 439–444
- citric acid cycle, 438, 442FF
- synthesis, 111F, 443F

Oxidation
- electron transfers, 87–88, 88F
- energy production, **86–87**

Oxidation–reduction reactions *see* Redox reactions
Oxidative phosphorylation, 439, 444–445, 445F
- ATP yield, **464–465**, 465T
- chemiosmotic coupling, 459–460, 459F, 460F
- evolution, **487–488**, 488F
- historical research, 468–469
- membrane-based mechanism, 454–455, 454F

Oxido-reductases, 143T
Oxygen (O_2)
- atmospheric, evolutionary origins, 488–489, 489F
- carbon compounds, 65F
- covalent bonds, 45, 46F, 48F
- diffusion across lipid bilayers, 389, 389F
- as electron acceptor, **473–474**
- energy-producing reactions requiring, 436–439, 444–445, 458, 460
- mitochondrial utilization, 17
- release during photosynthesis, 476, 478–479, 479F, 482

P

p21 protein, 621, 622F
p53 protein
- covalent modification, 156–157, 156F
- defects in cancer cells, 721–722
- DNA damage checkpoint, 621, 622F

Paddle wheel analogy, 104–105, 105F
Pairing, homologous chromosomes, **657**, 657F
Palmitic acid, 54, 54F, 70F
Pancreatic β cells, 519F

Page numbers in boldface refer to major discussion of a topic; page numbers followed by F refer to a figure, and with FF to consecutive figures; page numbers followed by T refer to tables.

Paneth cells, 712F
Paracrine signaling, 533, 533F
Parallel β sheets, 132, 132F
Paramecium, 2, 3F
Parthenogenesis, 652
Passive transport, **390–391**, 391F
  mechanisms, **392–393**, 392F
Pasteur, Louis, 7, 468
Patch-clamp recordings, 403–405, 404F
Pauling, Linus, 201
Pax-6, 288
PCR *see* Polymerase chain reaction
Peas, Mendel's experiments, 665–667, 665F, 666F
Pectin, 691F
Pedigrees, 668, 669F
Pentoses, 68F, 74F
Peptide bonds, 56, 56F, 72F, 121F
  ribosomal catalysis, 252, 254
Peptidyl transferase, 252–253, 254, 254F
Periodic table of elements, 44F
Peroxisomes, 12F, 21, 497, 497T
  number per cell, 498T
  signal sequences, 502T
Pertussis toxin, 546–547
pH, 49, 67F
  see also Proton(s)
Phagocytic cells, 522, **523**, 523F
Phagocytosis, **522–523**, 523F
  apoptotic cells, 640, 640F
Phagosomes, 522, 523, 526, 527F
Phalloidin, 582T, 592
Phase-contrast microscopy, 8F
Phenotype, 354, 667, 674F
Phenylalanine, 73F
Phenylthiocarbamide (PTC) taste, 664F
Phormidium laminosum, 15F
Phosphatases, 143T
Phosphate, inorganic (P_i), 65F
  ATP cycle, 105, 106F, 426
  nucleotides, 74F
  transport into mitochondria, 463–464, 464F
Phosphate bonds, 65F
  see also specific types
  formation during glycolysis, 433, 434F, 435F
  types and energy released, 435F
Phosphatidylcholine, 365, 366F, 371F
Phosphatidylethanolamine, 367F, 371F
Phosphatidylinositol, 371F
Phosphatidylserine, 366F, 371F
Phosphoanhydride bonds, 65F, 435F
  ATP, 57, 57F, 106F
  nucleotides, 75F
Phosphodiester bonds, 57–58, 58F, 75F
Phosphoenolpyruvate, 431F, 435F
Phosphoester bonds, 106F, 435F
Phosphofructokinase, 430F, 447–448
Phosphoglucose isomerase, 430F
2-Phosphoglycerate, 431F
3-Phosphoglycerate
  glycolysis, 431F, 433, 434F
  photosynthetic carbon fixation, 484–485, 484F, 485F
Phosphoglycerate kinase, 431F, 433, 434F
Phosphoglycerate mutase, 431F
Phosphoinositide 3-kinase (PI 3-kinase), **558–559**, 558F
Phospholipase C, 548, 556
  effects on cells, 551–552, 551T, 552F
Phospholipids, 55, 70F, 365, 366F
  see also Membrane lipids
  amphipathic properties, 365, 367F
  bilayer formation in water, 367–368, 368F
  biosynthesis, 371, 371F
  compartment formation, 368, 368F
  distribution in lipid bilayer, 371, 371F
  import into organelles, 505
  membrane, 55, 55F, 71F
  movements within bilayer, 368–369, 370F
  saturation of hydrocarbon tails, 369–370
  synthetic bilayers, 368, 369F, 383, 383F
Phosphorus (P), 45
Phosphoryl groups, 65F
Phosphorylation
  cascades, 542
  oxidative *see* Oxidative phosphorylation
  protein *see* Protein phosphorylation
  reactions, 105–106, 106F
  substrate-level, 429, 435F, 468
Photobleaching techniques, 382, 382F
Photons, 479, 481–482
Photophosphorylation, cyclic, **483–484**, 484F
Photoreceptor cells, rod, 554, 554F, 555F
Photosynthesis, 19, **476–486**
  bacterial, 15, 15F, 476, 476F, **488–490**, 489F
  carbon fixation *see* Carbon fixation
  cellular respiration and, 86–87, 87F
  electron-transfer reactions (light reactions), 478–479, 479F
  evolutionary origins, **488–490**, 489F
  source of electrons, 459–460
  two-stage process, 478–479, 479F
  use of solar energy, **84–86**, 478–479
Photosynthetic reaction centers, **480–481**, 480F
  bacterial, 376, 377, 378F
  light energy transfer, 480–481, 481F
Photosystem(s), 480–481, 480F
  I, 482, 482F, 483F
    cyclic photophosphorylation, 483–484, 484F
  II, 481–482, 482F
    structure, 483F
  ATP and NADPH synthesis, 481–482, 482F, 483F
  green sulfur bacteria, 489F
  structure, 483F
Phragmoplast, 637, 637F
Phylogenetic trees
  of life, **313–315**, 315F
  primates, 310, 310F
Pinocytosis, 522, **523–524**
Plant(s)
  asexual reproduction, 652
  cell communication, **564**, 565F
  chloroplasts *see* Chloroplasts
  energy sources, 15
  model, 28–29, 28F
  seeds, 449, 449F
  transgenic, **357–358**, 358F
Plant cell(s), 2, 3F
  connections between, 707, 708F
  cytokinesis, **636–637**, 637F
  cytoskeleton, 22–23
  energy reserves, 449–450, 449F, 450F
  membrane transport, **400**, 401F
  organization into tissues, 690–693, 690F
  osmotic balance, **396–397**, 396F, 691, 691F
  structure, 25F
  turgor pressure, 397, 403F, 691
Plant cell walls, 690, **691–693**, 691F
  cellulose deposition, 692–693, 693F
  formation during cell division, **636–637**, 637F
  primary, 691, 691F, 692F
  secondary, 691, 691F
  tensile strength, 692–693, 692F
Plasma membrane, 11, 363–364, 363F, 364F
  cell cortex reinforcing, 377–379, 378F, 379F
  cholesterol content, 370
  ingestion by pinocytosis, 523
  lipids *see* Membrane lipids
  origin of organelles from, 498–499, 499F
  proteins *see* Membrane proteins; Transmembrane proteins
  transport *see* Membrane transport
  transporters, 391, 391F
  underlying layer *see* Cell cortex
  vesicle fusion with, 518–519, 519F, 522
Plasmid vectors
  cloning, **335–336**, 335F, 336F, 337
  expression, 347–350, 347F
Plasmids, 335
Plasmodesmata, 707, 708F
Plasmodium vivax, cell-surface receptor mutation, 301–302
Plastocyanin, 482F
Plastoquinine, 482F
Platelet-derived growth factor (PDGF), 535T, 645
  receptors, 556–557
Platelets, blood, 645, 698, 714F

Plectin, 576, 576F
Pluripotent cells, 715
Pluripotent stem cells, induced (iPS cells), 717, 717F
Point mutations, **300–302**
Polar covalent bonds, 42–43, 48
  formation, 47, 48F
  oxidation and reduction, 88, 88F
  water, 66F
Polar molecules
  acids and bases, 49–50
  permeability of lipid bilayers, 389, 389F
  solubility in water, 66F
Polarized cells
  epithelial sheets, **700–701**, 701F
  gut epithelium, 701, 702F
  role of microtubules, 582–583, 582F
Polyadenylation, mRNA, 241, 241F
Polycistronic mRNA, 256, 256F
Polyisoprenoids, 71F
Polymerase chain reaction (PCR), 329, **340–343**
  applications, 342–343, 342F, 343F, 344F
  primers, 340, 341F
  technique, 340–342, 341F
Polymerases, 143T
Polymers (polymeric molecules), 39, 58, 59F
  see also Macromolecules
  biosynthesis, 110–112, 112F, 113F
  formation from monomers, **59**, 59F, 110, 111F
Polymorphisms, 679
  see also Single-nucleotide polymorphisms
Polypeptide backbone, 122, 122F
Polypeptides (polypeptide chains), 56, 56F, 121, 122F
  see also Amino acid sequence
  C-terminus, 56, 56F, 122F
  evolution, 134–135
  folding, 122–123, 123F
  N-terminus, 56, 56F, 122F
  as protein subunits, **135–136**, 136F
Polyps, colorectal, 724, 727, 727F
Polyribosomes (polysomes), **257**, 257F, 506–507, 507F
Polysaccharides, 52–53, 69F
  biosynthesis, 111–112, 112F
  side chains, membrane proteins, 381, 381F
Polyunsaturated fatty acids, 54, 54F
Pores, transmembrane, 374, 375, 375F
  see also Channels; Ion channels
Porins, 401
  mitochondrial outer membrane, 457
  structure, 375, 375F
Positive feedback regulation *see* Feedback regulation, positive
Post-transcriptional controls, **289–293**
Postsynaptic membrane, 415–416, 415F
  neurotransmitter signaling, 416–417, 416F
Potassium (K^+) channels
  activation by G proteins, 547, 547F
  leak, 407, 408F, 418T
  selectivity, 402, 402F
  voltage-gated, 411, 418T
Potassium ions (K^+), 44
  action potentials, 413
  electrochemical gradient, 393
  extra- and intracellular concentrations, 388–389, 388T
  membrane potential generation, 407, 407F, 408, 408F
  transport by Na^+-K^+ pump, 394–395, 394F, 395F
Potato virus X, 225F
Pre-replicative complex, 620, 620F
Precursor cells, 711, 711F
Presynaptic membrane, 415, 415F
Primase, 207, 208
Primates, phylogenetic tree, 310, 310F
Primers, polymerase chain reaction, 340, 341F
Prion diseases, 125, 125F
Procaryotes, **11–16**
  see also Bacteria
  diversity, 14–15, 15F
  domains, 15–16
  energy sources, 15, 15F
  evolution, 314, 314F, 315F
  genome sizes, 34, 34F
  transcription *see* Transcription, procaryotic
Procaspases, 640, 641, 641F, 642F
Procollagen, 695
Procollagen proteinases, 695, 696F
Profilin, 593
Progeria, 577
Programmed cell death, 32, 638
  see also Apoptosis
Prolactin, 563F
Proline, 73F
Prometaphase, 624–625, 626F, **628–630**, 634F
Promoters
  bacterial, 236–238, 237F, 238F
  eucaryotic, 239–240, 239F
  operons, 275, 275F
  transcription activators, 276, 276F
  transcription initiation site, 273
  transcription repressors, 275, 276F
Pronuclei, 663
Proofreading, 206, 206F, 207F
Propane, 46F
Prophase, 626F
  meiosis I, 657–658
  mitosis, 624–625, 628
Proteases, 143T, 258
  proteasomes, 259, 259F
  protein sequencing, 158
Proteasomes, 259, 259F
Protein(s), 4, **119–170**
  aggregates, 125
  amino acid sequence *see* Amino acid sequence
  attached small molecules, **148–149**
  complexes *see* Protein assemblies
  covalent modification, **156–157**, 156F
    Golgi apparatus, **517–518**, 518F
    processes in ER, **514–515**, 515F
  crystallization, 158
  denatured, 124, 125F
  digestion and utilization, 426–427, 428F
  filaments, 136–137, 137F, 138F
  fingerprints, 158, 162T
  function *see* Protein function
  glysosylation, 514–515, 515F
  historical landmarks, 162T
  isolation and purification, **161–162**, 162F, 164FF
  mass production, **163**, 163F, **347–350**, 347F
  membrane *see* Membrane proteins
  recombinant DNA methods of study, 350, 350F
  renaturation, 124, 125F
  structure *see* Protein structure
  vesicular transport, **510–511**, 510F
Protein assemblies (complexes), **136–137**, 137F
  see also Protein machines
  molecular organization, 137, 138F
  noncovalent bonds, 63
Protein binding, **140–141**
  see also Protein interactions
  noncovalent bonds, 48F, 140, 141F
  sites, 135, 136F, 141, 141F
  specificity, 140
Protein code, regulatory, 157
Protein conformation, 4, **124–125**, 125F
  see also Protein folding
  changes
    feedback inhibition triggering, 151–152, 152F
    GTP-binding proteins, 153–154, 153F, 154F
    motor proteins, 154–155, 154F, 155F
    phosphorylation mediating, 152
  cross-linkages stabilizing, 138–140, 139F
  determination, 127
  diversity, **125–127**, 126F
  evolution, 134–135
  ligand binding and, 152, 152F
Protein degradation, **258–259**
  digestive, 426–427, 428F
  proteasomes, 259, 259F
  ubiquitin-dependent, 259
Protein domains, 127, **133–134**, 133F, 134F
  evolution, 243, 306–307, 307F
Protein families, **135**
Protein folding, 122–123, 123F
  α helices and β sheets, 127, 130F
  chaperones, 125, 256, 516

Page numbers in boldface refer to major discussion of a topic; page numbers followed by F refer to a figure, and with FF to consecutive figures; page numbers followed by T refer to tables.

disruption (denaturation), 124, 125F
energetics, 124–125
hydrogen bonds stabilizing, 123, 124F
hydrophobic interactions, 123, 124F
incorrect (misfolding), 125, 125F
patterns, 127–132
quality control in ER, 516, 516F
unfolded protein response (UPR), 516–517, 517F
unfolding before import into organelles, **505**, 505F
Protein function, 59, 119, 120F
automated analysis, **163**
mechanisms, **140–149**
methods of studying, 157–163, 164FF, 725
regulation, **149–157**
Protein interactions, **140–149**
see also Protein binding
Protein kinase(s), 153, 153F
intracellular signaling, 541–542, 542F
networks, **564–567**, 565F
Protein kinase A (PKA) *see* Cyclic AMP-dependent protein kinase
Protein kinase B (PKB) *see* Akt
Protein kinase C (PKC), 552, 552F
Protein machines, **155–156**, 155F
see also Protein assemblies
covalent modifications controlling, 156–157, 156F
DNA replication *see* DNA replication, machine
Protein phosphatases, 153, 153F
intracellular signaling, 541–542, 542F
Protein phosphorylation, **152–153**, 153F
cAMP triggered, 549
cascades, 542
experimental analysis, 560
intracellular signaling via, 541–542, 542F
Na^+-K^+ pump, 394–395, 395F
promoting assembly of proteins, 156
Protein sorting, 496, 500–509
see also Protein translocation
endoplasmic reticulum, 505–509
Golgi apparatus, **517–518**, 518F
into lysosomes, 526
into mitochondria and chloroplasts, **505**, 505F
nucleus, **502–505**, 504F
signal sequences, 501–502, 502F, 502T
signals, 500, 501
Protein structure, **121–140**
see also Protein conformation
primary, 133
see also Amino acid sequence
secondary, 133
see also α helices; β sheets
tertiary, 133
quaternary, 133, 135–136, 136F
binding sites, 135, 136F
determination, 125–127
automated methods, **163**
historical experiments, 60–61, 61F
methods, **158–159**, 160F
diversity, **125–127**, 126F
historical landmarks, 24T, 162T
levels of organization, 133–134
models representing, 127, 128FF
Protein subunits, 135–136, 136F
Protein synthesis, **246–260**
see also Translation
antibiotic actions, **257–258**, 258T
bacterial, 256, 256F, 257
cell-free system, 248
coding problem, 246
energy input, 111, 112F
further modification after, 259, 260F
multiple steps, 259–260, 260F
ribosomal function, 252–253
translocation into ER during, **506–507**, 507F
Protein translocation
channels, ER membrane, 507–508, 507F, 508F
into mitochondria and chloroplasts, 505, 505F
soluble proteins into ER, 505–507, 507F, 508F
transmembrane proteins into ER, 508–509, 509F
Protein translocators, 501, 501F, 505F
Protein transport
see also Protein sorting; Protein translocation; Vesicular transport
experimental analysis, 520–521, 520F, 521F
mechanisms, 500–501
through nuclear pores, **502–505**, 504F
Protein tyrosine phosphatases, 556
Proteoglycans
extracellular matrix, **698–699**, 699F
membrane, 381, 381F
Proteolysis, 258–259
see also Protein degradation
Proteomics, 163
Proto-oncogenes, 722F, 723, 723F
Proton(s) (H^+)
atoms, 40, 40F
extra- and intracellular concentrations, 388T
mobility in water, 466, 466F
release from acids, **49–50**, 49F, 67F
Proton gradients, electrochemical, 444, 454–455, 454F
chloroplast membrane, 482, 482F
driving membrane transport, **400**
evolutionary origins, 488, 490
inner mitochondrial membrane, **460–461**, 462F
driving ATP synthesis, 459, 459F, **461–463**, 462F
driving coupled transport, **463–464**, 464F
historical research, 468–469, 469F
uncoupling agents, 468, 468F
Proton-motive force, 461, 462F
see also Proton gradients, electrochemical
Proton pumping
chloroplast membrane, 482, 482F
electron transfer driving, 466–467, 467F
inner mitochondrial membrane, **460–461**, 461F
atomic mechanism, **474–475**, 475F
chemiosmotic coupling, 458–459, 459F
forces driving, 461, 462F
O_2 reduction reaction, 473–474
molecular mechanisms, **466–476**
photosynthesis, 479
Proton (H^+) pumps, 400, 401T
ATP synthase function as, 462–463, 463F
bacteriorhodopsin, 376–377, 377F
endosomal membrane, 525
evolution, 488
inner mitochondrial membrane, 460
lysosomal membrane, 526, 526F
mechanisms of action, 475F
Protozoans
discovery, 24T
diversity, 26, 27F
flagella, 589
osmotic balance, 396F, 397
phagocytosis, 522
as predators, 26, 26F
Provirus, 226
Pseudogenes, 305
Pseudopods, 523, 523F
Psychoactive drugs, **418–419**
Pumps, 391
ATP-driven, 393, 393F, **394–398**
light-driven, 393, 393F
Punnett square, 667F
Purifying selection, 312, 313–314
Purines, 57, 74F
biosynthesis, control in bacteria, 290F
Pyrimidines, 57, 74F
Pyrophosphate (PP_i), DNA synthesis, 203–204, 204F
Pyruvate, 427, 445–446
fermentation, 432–433, 432F
mitochondrial utilization, 458, 459F
oxidation/decarboxylation, 436, 436F, 441
production, 428, 429F, 431F
transport into mitochondria, 463–464, 464F
Pyruvate carboxylase, 111F
Pyruvate decarboxylase, 436F
Pyruvate dehydrogenase complex, 436, 436F
Pyruvate kinase, 431F

Q

Q-cycle, 475
Quinones, 470–472, 472F

R

Rab, 513, 513F
Rac, 598F
Racker, Efraim, 469
Radiation, ionizing, 711
Ran, 504F
Random walk, 98, 98F
Rapamycin, 559, 559F
Ras, **556–558**
 activating mutations, 558, 673, 721
 activation, 556–557, 557F
 methods of studying, 560, 561, 561F, 562F
 signaling cascade, 557–558, 557F
Reaction centers, photosynthetic *see* Photosynthetic reaction centers
Reading frames, 247, 247F
Receptor(s), 120F, 532, 534–535
 see also specific receptors
 cargo, 512, 512F
 cell surface *see* Cell-surface receptors
 degradation, 525, 525F
 fate in endosomes, 525–526, 525F
 intracellular, **537–538**, 537F
 membrane, 373T
 nuclear, 537–538, 538F
 recycling, 524, 525, 525F
 transcytosis, 525, 525F
Receptor-mediated endocytosis, **524**, 525F
Receptor serine/threonine kinases, 564, 565F
Receptor tyrosine kinases (RTKs), **555–559**
 activation, 555–556, 556F
 experimental studies, 560, 561F
 intracellular signaling complex, 556, 556F
 phosphorylation, 156, 556, 556F
 PI-3-kinase–Akt signaling, **558–559**, 558F, 559F
 Ras activation by, 556–558, 557F
Recessive alleles, 667, 667F, 674F
Recessive mutations, 673, 722F, 723
Recombinant DNA
 formation *in vitro*, **334**, 334F
 introduction into bacteria, **334–335**, 335F
Recombinant DNA technology, **327–360**
 applications, 103, 328, 350
 DNA cloning, 333–343
 manipulating and analyzing DNA molecules, 329–333
 production of proteins, **163**, 163F, **347–350**, 347F
Recombination, 657–658, 674F
 homologous *see* Homologous recombination
Red blood cells, 711, 713F
 see also Hemoglobin
 cortex, 378–379, 378F, 379F, 594
 generation, 712, 714F
 vitamin B_{12} and iron uptake, 524
Redox pairs, 467–470
Redox potential, **467–470**
 chloroplast electron-transport chain, 483F
 concentration changes and, 471F
 difference ($\Delta E_0'$), 470, 471F
 measurement, 471FF
 mitochondrial electron-transport chain, 472, 472F
Redox reactions (oxidation–reduction reactions)
 electron carriers, 107–109
 electron-transport chain, 467–470
Reduction, electron transfers, 87–88, 88F
Reese, Thomas, 587
Regulatory DNA sequences, 273–274, 274F
 current uncertainties, 322
 human genome, 316, 316F
 mutations, 298, 298F
  insertion, 307
  point, **301–302**, 303F
 operator, 275, 275F
 proteins binding *see* Transcription regulators
Regulatory protein code, 157
Renaturation, DNA, 332, 332F
Repetitive DNA sequences, 317F, 321
 fingerprinting, 344F
 genome sequencing and, 348–349, 348F
Replication forks, **203–205**
 asymmetry, 204–205, 204F, 205F
 proteins forming replication machine, 208–210, 209F
Replication machine *see* DNA replication, machine
Replication origins, 183, **199–203**, 199F
 regulation of activity, 620, 620F
Reporter genes, **350–353**, 351F
 Drosophila eve gene, 283–284, 284F
Repressors, transcriptional *see* Transcriptional repressors
Reproduction, 651, 652
 asexual, 299, 652, 652F
 sexual *see* Sexual reproduction
Resonance, 64F
Respiration (cellular), 17
 aerobic, 438–439
 anaerobic, 433
 efficiency, **475–476**
 photosynthesis and, 86–87, 87F
Respiratory chain *see* Electron-transport chain, mitochondrial
Respiratory enzyme complexes, 460
Respiratory tract cells, cilia, 589, 589F
Restriction nucleases, **329–330**, 330F, 331F
 DNA libraries, 337, 337F
 plasmid vectors, 335–336, 336F
Retina, 706, 707F
Retinal, 148–149, 149F
 bacteriorhodopsin, 376–377, 377F
Retinoblastoma (Rb) protein, 644, 644F
Retrotransposons, 222–223, 223F, 224F
Retroviruses, **225–226**, 226F
Reverse transcriptase
 cDNA preparation, 338, 339F
 retrotransposition, 222–223, 223F
 retroviruses, 225–226, 226F
Rho protein family, 598
Rhodobacter capsulatus, 375F
Rhodopseudomonas viridis, 378F
Rhodopsin, 544
 association with retinal, 148–149, 149F
 intracellular signaling cascade, 554, 554F, 555F
Ribonucleic acid *see* RNA
Ribonucleoside triphosphates, 235, 235F
Ribonucleotides, 56, 233
Ribose, 68F, 74F, 233, 233F
 origin of life and, 263–264
Ribosomal proteins, 251, 253–254, 255F
Ribosomal RNA (rRNA), 236, 236T, 251
 5S, 255F
 23S, 254, 255F
 genes
  chromosomal localization, 180F
  nucleolus, 184, 184F
  phylogenetic tree construction, 314, 314F, 315F
 structure, 253–254, 255F
 transcription, 236F
Ribosomes, 22, **251–254**
 bacterial, 240
 electron microscopy, 12F, 20F, 252F
 eucaryotic, 251–252, 252F, 253
 free, 506, 507F
 as macromolecular complexes, 63, 63F
 membrane-bound, 506, 507F
 mRNA and tRNA binding sites, 253, 253F
 polyribosome formation, **257**, 257F
 procaryotic, 251–252, 253
 protein synthesis mechanism, 252–253, 254F
 protein translocation into ER and, 506–507, 507F
 release of newly synthesized proteins, 256, 256F
 subunit structure, 251–252, 252F
 three-dimensional structure, 253–254, 255F
Riboswitches, **289–290**, 290F
Ribozymes, 254, **262–263**, 262F, 263T
Ribulose, 68F

Page numbers in boldface refer to major discussion of a topic; page numbers followed by F refer to a figure, and with FF to consecutive figures; page numbers followed by T refer to tables.

Ribulose 1,5-bisphosphate, 484–485, 484F, 485F
Ribulose bisphosphate carboxylase (Rubisco), 160F, 484–485, 484F, 485F
Rice, genetically engineered, 358
Rifamycin, 258T
Rigor mortis, 602F
RISC *see* RNA-induced silencing complex
RNA, 4, 58
  see also specific types of RNA
  catalytic activity
    see also Ribozymes
    ribosomes, 252–253, 254, 254F
    RNA world hypothesis, **261–263**, 263F
  double-stranded foreign, degradation, **291–292**, 293F
  evolution, **263–264**, 264F
  folding, 233, 234F
  genetic information transfer, 4F, 232, 232F
  origin of life and, **261–264**
  processing, **240–241**, 241F, 245F
  replication (in principle), 262F
  self-replication potential, **261–263**, 263F
  short noncoding, human brain development, 320
  small regulatory, **290–291**
  splicing *see* Splicing
  structure, 233, 233F
  synthesis
    see also Transcription
    chemical reactions, 111, 112F, 113F
    from DNA template, **234–235**, 235F
  types, **235–236**, 236T
RNA-induced silencing complex (RISC)
  mRNA targeting, 291, 292F
  RNA interference, 292, 293F
RNA interference (RNAi), **291–293**
  scientific applications, 292–293, **356–357**, 357F
  transfer between cells, 292
RNA polymerase(s)
  I, 238, 239T
  II, 238, 239T
    initiation of transcription, 239–240, 239F
    phosphorylation of tail, 240, 241F
  III, 238, 239T
  bacterial
    regulation, 275, 276, 276F
    sigma (σ) factor, 237, 237F
    start and stop signals, 236–238, 237F
  eucaryotic, 238, 239T
  primase, 207
  retrovirus replication, 226, 226F
  transcription of DNA, 234–235, 235F
RNA primers, DNA synthesis, **206–208**, 208F
RNA probes, *in situ* hybridization, 352
RNA world hypothesis, **261–264**, 261F
Rod photoreceptor cells, 554, 554F, 555F
Rough endoplasmic reticulum, 497, 506
rRNA *see* Ribosomal RNA
Rubisco *see* Ribulose bisphosphate carboxylase

S

S-adenosylmethionine, 109T
S-Cdk, 617, 617T
  control of DNA replication, **620–621**, 620F
  inactivation by DNA damage, 621, 622F
S cyclins, 617
S phase, 611, 611F, **620–621**
  see also DNA replication
  DNA damage checkpoint, 621
S–S bonds *see* Disulfide bonds
Saccharomyces cerevisiae, 16F, 28, 28F
  arrangement of genes, 181F
  cell cycle analysis, 30–31, 31F
  mating factor, 532F
Salivary gland, response to acetylcholine, 535, 536F
Salmonella, 14F
Saltatory movement, 583
Salts, 44
Sarcomeres, 600–601, 600F
  structure, 600, 601F
Sarcoplasmic reticulum, 397F, 602–604, 603F
Saturated fatty acids, 54, 54F, 70F
  membrane fluidity and, 369–370
Scanning electron microscopy (SEM), 9F, 11
Schizosaccharomyces pombe, cell cycle analysis, 30–31, 31F
Schleiden, Matthias, 6–7
Schwann, Theodor, 6–7
Schwann cells, 709–710
SDS polyacrylamide-gel electrophoresis (SDS-PAGE), 167F
Second messengers, 548
Secretion, 518–519
Secretory cells, 519–522, 519F
Secretory pathways (protein), 510, 510F, **514–522**
  exit from ER and quality control, 516, 516F
  exocytosis, 518–522
  methods of studying, 520–521, 521F
  modification and sorting in Golgi apparatus, 517–518
  protein modification in ER, 514–515, 515F
  unfolded protein response, 516–517, 517F
Secretory vesicles, 519–522, 519F
Securin, 631, 632F
Seeds, plant, 449, 449F
Selection
  see also Evolution
  natural, 309–310
  purifying, 312, 313–314
Separase, 631, 632F
Sequence, polymer, 59
Serine, 73F
Serine proteases, 135, 135F
Serine/threonine kinases, 542, 555
Serotonin, 418, 549F
Seven-pass transmembrane receptor proteins, 544
Sex chromosomes, 179, 180F, 652
  see also X chromosome; Y chromosome
Sex-determination genes, evolutionary conservation, 217F
Sexual reproduction, **651–687**
  see also Fertilization; Meiosis
  advantages, **654–655**
  laws of inheritance, **664–675**
  outline, 652–654, 653F
  transfer of genetic information, 299–300, 300F
SH2 domain, 126F, 132, 556F
  models, 127, 128FF
Sheetz, Michael, 587
Short tandem repeats (STRs), 344F
Shotgun sequencing, 348–349, 348F
Sickle-cell anemia, 211, 211F
Side chains, amino acid *see* Amino acid(s), side chains
Sigma (σ) factor, 237, 237F
Signal molecules, 532
  cellular dependence on, 535–536, 536F
  extracellular, 537–540, 537F
  intracellular *see* Intracellular signaling molecules
  selectivity of cell responses, 534–535, 536F
  types, **532–534**, 533F, 535T
Signal peptidase, 508, 508F
Signal proteins, 120F
Signal-recognition particle (SRP), 507, 507F
  receptor, 507, 507F
Signal sequences, **501–502**, 502F, 502T
  experimental analysis, 520, 520F
Signal transduction, 532, 532F
Signaling cascades, intracellular, **554–555**, 555F
Signaling pathways, intracellular, 540, 541F
  enzyme-coupled receptors, 555–567
  experimental analysis, **560–561**, 561F, 562F

G-protein–coupled receptors (GPCRs), 547–553
integration, **564–567**, 565F
Single-nucleotide polymorphisms (SNPs), 320–321, 321F, **678–682**
common human diseases, 679–682, 679F
haplotype blocks, **682–684**, 682F, 683F
Single-particle tracking (SPT) microscopy, 383, 383F
Single-strand binding protein, 208–209, 209F
Sister chromatids, 625
see also Bivalents
bi-orientation, 629–630
cohesion, **621**, 621F, 623–624, 659
condensation, 623–624, 624F
meiosis and mitosis compared, 656–657, 657F
separation, 625, 625F, **631–633**, 632F
meiosis II, 659, 659F
negative regulation, 633
Site-directed mutagenesis, 354, 355F
Skeletal muscle, 599
Skeletal muscle cells
adrenaline actions, 549–550, 550F
Ca^{2+} signaling, 397F, 552, **602–604**, 603F
control of differentiation, **285–286**, 286F
mechanism of contraction, 600–601, 601F, 602F
response to acetylcholine, 535, 536F, 542–543
structure, 600, 600F
Skin
cell types, 709–710, 709F
collagen organization, 696, 696F
Sliding clamp protein, 209, 209F
Small interfering RNAs (siRNAs), 292
Small intestinal epithelium *see* Intestinal epithelium
Small messengers, 548
Small nuclear ribonucleoproteins (snRNPs), 243, 243F
Small nuclear RNAs (snRNAs), 243
Smell, sense of, 555
Smoking, tobacco, 719, 720
Smooth muscle, 599, 604
nitric oxide actions, 539, 539F
SNARE proteins, 513, 513F, 514, 514F
SNPs see Single-nucleotide polymorphisms
Sodium atoms (Na), 44, 45F
Sodium (Na^+) channels, voltage-gated *see* Voltage-gated sodium channels
Sodium chloride (NaCl), 44, 45F, 66F
Sodium dodecyl sulfate (SDS), 167F, 376F
Sodium hydroxide, 50, 67F
Sodium ions (Na^+), 44
action potentials, 410–411, 411F, 413
active transport, 393
ATP-driven pump *see* Na^+-K^+ pump
electrochemical gradient, 393
extra- and intracellular concentrations, 388–389, 388T
glucose–Na^+ symport, 398–399, 399F
Na^+-H^+ exchanger, 400
Solutes, 67F
passive and active transport, **390–391**, 391F
permeability of lipid bilayers, 389, 389F
Solutions, 67F
Solvent, 67F
Somatic cells, 299, 300F
functions, 654, 654F
mutations, 212, 299, 300F
Somatostatin, 550
Sorting signals, protein, 500
see also Protein sorting
Southern blotting, 333, 333F
Specialized cells *see* Differentiated cells
Spectrin, 378–379, 379F, 594
Spectrophotometry, 101
Sperm, 652, 653F
aneuploidy, 663
fertilization, 663, 663F
flagellum, 589, 590F, 591F
mitochondria, 456, 457F
Sphingomyelin, 371F
Spindle, mitotic *see* Mitotic spindle
Spindle poles, 628, 629F
formation, 625–628, 628F
separation, 633, 633F
Spliceosome, 243
Splicing (RNA), 232, **242–243**
alternative, 243, 244F, 322, 322F
evolutionary significance, 243, 245–246, 245F
lariat structure, 242, 243F
Squid giant axons, 410, 410F, 412
experimental methods, **412–413**, 412F, 413F
motion of organelles, 586, 586F, 587
SRP *see* Signal-recognition particle
Stahl, Frank, 200–202, 202F
Standard free-energy change ($\Delta G°$), 92, 95FF
calculation from redox potential, 470, 471F
equilibrium constant and, **93–96**, 95F, 96T
sequential reactions, 97–98, 98F
Starch, 53
storage in plant cells, 449–450, 450F
synthesis, 486, 486F
Start-transfer sequence, ER transmembrane proteins, 508–509, 509F
Statins, 148, 524
STATs, 559–563, 563F
Stearic acid, 54, 70F
Stem cells, 690, **711–715**, 711F
see also Embryonic stem cells
control mechanisms, **713–714**, 714F
induced pluripotent (iPS cells), 717, 717F
tissue renewal, 711–712, 712F, 713F, 714F
Steroid hormones, 537–538, 538F
Steroids, 55, 71F
Sterols, amphipathic nature, 365, 366F
Stoeckenius, Walther, 469
Stomata, 397, 397F
Stop codons, 247F, 256, 256F
Stop-transfer sequence, ER transmembrane proteins, 508, 509F
Storage proteins, 120F
Streptococcus, 14F
Streptococcus pneumoniae, 174–175, 174F
Streptomycin, 258T
Stress-gated ion channels, 405, 405F, 406F, 418T
Stromatolites, 476F
Structural proteins, 120F
Strychnine, 417, 543T
Substrate-level phosphorylation, 429, 435F, 468
Substrates, 90, 91F, 143
encounter by enzymes, 98–99
enzyme binding to, 99, 146
Subunits, 39, 58, **59**, 59F, 110
protein, 135–136, 136F
Succinate, 440F, 443F
Succinate dehydrogenase, 440–441, 443F
Succinyl-CoA, 443F
Succinyl-CoA synthetase, 443F
Sucrose, 52, 69F
synthesis in plants, 486
Sugars, 51, **52–54**, 68FF
see also Glucose; *other specific sugars*
α and β links, 69F
breakdown and utilization, 425, **426–445**
see also Glycolysis
acetyl CoA production, 436, 436F, 437F
stages, 426–427, 428F
derivatives, 69F
fate in plant cells, 486, 486F
isomers, 52, 68F
nucleotides, 74F
production during photosynthesis, 479, **484–485**, 484F, 485F
ring formation, 68F
stepwise oxidation, 426, 426F
Sulfur (S), 45
Sulfur bacteria, 15F
photosynthesis, 488, 489F

Page numbers in boldface refer to major discussion of a topic; page numbers followed by F refer to a figure, and with FF to consecutive figures; page numbers followed by T refer to tables.

Sunlight
see also Light energy
absorption by chlorophyll, **479**, 480F
DNA damage, 214, 214F
energy capture from, **478–479**, 479F
as energy source, **84–86**, 85F, 86F
Superoxide radical, 474
Survival factors, 642, **643**, 643F
Svedberg, Theodor, 60–61, 61F
Switches *see* Molecular switches; Riboswitches
Symbiotic relationship, ancestors of mitochondria, 17, 18F
Symports, 398, 398F
Synapses, 415, 415F, 533
biological functions, **419–420**, 419F
excitatory and inhibitory, 417–418, 418F
Synaptic cleft, 415, 415F, 416F
Synaptic vesicles, 415, 415F
Synaptonemal complex, 658
Synteny, conserved, 311–312
Synthases, 143T
Szent-Györgyi, Albert, 440

T

T tubules, 602, 603F
T2 viruses, 176, 176F
T4 virus, 225F
Taste, 555
TATA-binding protein (TBP), 240F
TATA box, 239, 239F
Taxol, 581, 582T
TCF, 725–726
Telomerase, **210**, 210F, 722
Telomeres, 183, **210**, 210F
Telophase, 624–625, 627F, **634**, 634F
Temperature-sensitive mutations, 673F
gene identification, 677, 677F
yeast, 30–31, 30F, 31F
Template, DNA *see* DNA template
Tendons, 696, 698–699
Terminally differentiated cells, 711, 711F
see also Differentiated cells
Testosterone, 535T
intracellular receptors, 537–538
structure, 71F, 538F
Tethering proteins, 513, 513F
Tetracycline, 258T
TFIIB transcription factor, 239F
TFIID transcription factor, 239, 239F
TFIIH transcription factor, 239F, 240
Thermodynamics, **82–100**, 94FF
see also Free energy
first law, 84, 85F
second law, 82–84, 84F, 94F
Threonine, 73F
Thrombin, 143, 551T
Thylakoid membrane, 477, 478F
photosystems, 480–481, 480F
Thylakoid space, 477–478, 478F
Thylakoids, 477–478, 477F, 478F
Thymine (T), 57, 74F, 177
compared to uracil, 233, 233F
dimers, 214–215, 214F, 216
DNA stability and, 264
Thymosin, 593
Thyroid hormone (thyroxine), 535T, 537, 538F
Tight junctions, **702–703**, 702F
barrier function, 702, 703F
maintaining cell polarity, 380, 380F
Tissues, 7–10, 10F, 689–690, 690F
see also Connective tissues
maintenance and renewal, 690, **707–717**
maintenance of cellular organization, 710, 710F
mixture of cell types, **709–710**, 709F
renewal, **710–712**, 712F, 713F
repair, **714–715**
Tobacco cells, cultured, 161F
Tobacco-smoking, 719, 720
Tor, 559, 559F
Tranquilizers, 418
Transcription, 4, 4F, **232–246**
see also Messenger RNA; RNA polymerase(s)
direction, 237–238, 238F
electron microscopy, 236F
error rate, 235
eucaryotic, **238–244**, 245, 245F
evolution, 245–246
general transcription factors, 239–240
initiation, 238–240
regulation, 278–280
evolutionary origins, 245–246
initiation, **236–240**
bacterial, 236–238, 237F
eucaryotic, 238–240
regulation, **273–280**
site, 273
initiation complex, 240
procaryotic (bacterial), 244–245, 245F
initiation, 236–238, 237F
regulation, 275–277
terminator (stop site), 237, 237F, 238F
regulation, **273–280**
eucaryotic, 278–280, 280–288
procaryotic (bacterial), 275–277
RNA strand formation, 234–235, 235F
Transcription factors, general, 238, **239–240**, 239F
Transcription regulators (DNA-binding proteins), **273–274**
see also Transcriptional activators; Transcriptional repressors
bacterial, 275–277
chromatin modification, 279–280, 279F
combinatorial control, 280, 281F, **285–286**, 286F
control of organ formation, 288, 289F
dimerization, 274
DNA binding, 273–274, 274F
DNA-binding motifs, 274, 274F
eucaryotic, **278–280**, 278F, 279F
experimental analysis, **282–284**, 282F, 283F, 284F
positive feedback loops, 287, 287F
Transcriptional activators
bacterial, **276–277**, 276F, 277F
eucaryotic, 278, 278F, 279, 279F
Transcriptional repressors
bacterial, 275, 276, 276F, 277, 277F
eucaryotic, 278, 279–280
Transcytosis, 525, 525F
Transducin, 554, 555F
Transfer RNAs (tRNAs), 236, 236T, **247–251**
codon base-pairing, 250, 250F
initiator, 255
mechanism of charging, 251, 251F
ribosome binding sites, 253, 253F
structure, 247–250, 250F
translation cycle, 253, 254F
wobble base-pairing, 250
Transformation, bacterial
historical studies, 174–175, 174F
by recombinant DNA, 335, 335F
Transforming growth factor-β (TGF-β), 535T
Transgenic mice, 357F
Transgenic organisms, 354–356
Transgenic plants, **357–358**, 358F
Transition state, 146
Translation, 4, 4F, **246–260**
cell-free system, 248
experimental analysis, 248–249
four-step cycle, 253, 254F
initiation, **254–256**, 255F
regulation, **290**, 291F
initiation factors, 255, 255F
release factors, 256, 256F
ribosomal location, **251–253**
termination, 256, 256F
Transmembrane proteins, 373, 373F
see also Membrane proteins; Membrane transport proteins
linkage to cell cortex, 378–379, 379F
pore formation, 374, 375F
start-transfer sequence, 508–509, 509F
stop-transfer sequence, 508, 509F
structure, **374–375**, 374F, **376–377**, 377F, 378F
translocation and integration into ER, 506, **508–509**, 509F
Transmission electron microscopy (TEM), 9F, 11, 12F
Transmitter-gated ion channels
see Ion-channel–coupled receptors
Transport *see* Intracellular transport; Membrane transport; Vesicular transport

Transport proteins, 120F
see also Membrane transport proteins
Transport signals, 512
Transport vesicles, 501, **510–511**
see also Clathrin-coated vesicles; Vesicular transport
budding, **511–512**, 511F
capture of cargo molecules, 512, 512F
docking with target membranes, **512–514**, 513F, 514F
Transporters, 373T, 387, **391–400**, 401T
bacteriorhodopsin, 377
basic principles, 390, 390F
coupled *see* Coupled transporters
lysosomal membrane, 526
passive transport, **392–393**, 392F
Transposase, 222F
Transposons
see also Mobile genetic elements; Retrotransposons
DNA-only, 222, 222F, 223F
Transverse (T) tubules, 602, 603F
Trees, phylogenetic *see* Phylogenetic trees
Treponema pallidum, 14F
Triacylglycerols, 54–55, 54F, 70F
adrenaline-mediated breakdown, 550
aggregation, 71F
hydrophobicity, 367F
Tricarboxylic acid cycle *see* Citric acid cycle
Triose phosphate isomerase, 430F
Trioses, 68F
Triton X-100, 376F
tRNAs *see* Transfer RNAs
Tropomyosin, 603, 604F
α, alternative splicing, 244F
Troponin, 603, 604F
Trypsin, 135, 158
Tryptophan
regulation of synthesis, 275, 275F, 276F
structure, 73F
Tryptophan repressor, 275, 276, 276F, 280
Tubulin, 572, **578–579**, 578F
see also Microtubule(s)
α-tubulin, 578, 579–580
β-tubulin, 578, 579–580
drugs acting on, 581–582, 582T
γ-tubulin, 579, 579F
nucleation sites, 579–580, 579F
polymerization, 579–580, 579F
protofilaments, 578, 578F
Tumor(s)
see also Cancer
benign, 718
malignant, 718
secondary, 718, 719F, 722
size at diagnosis, 727, 728F
Tumor suppressor genes, 722F, 723
Turgor pressure, 397, 403F, 691
Turnover number, 100
Twins, identical, 661
Two-dimensional polyacrylamide-gel electrophoresis, 162, 167F
Tyrosine, 73F
Tyrosine kinases, 542, 555

U

Ubiquinone, 472–473, 472F
proton pumping role, 474–475
Ubiquitin, 259
Ultracentrifugation, 60–61, 61F, 164FF
Ultraviolet radiation, DNA damage, 214, 214F
Uncoupling agents, 468, 468F
Unfolded protein response (UPR), **516–517**, 517F
Uniports, 398, 398F
Unsaturated fatty acids, 54, 54F, 70F
membrane fluidity and, 369–370
Uracil (U), 57, 74F
base-pairing, 233, 234F
DNA repair and, 264
RNA, 233, 233F
Uridine diphosphate glucose, 109T

V

Vale, Ron, 587
Valine, 73F
Valium, 543T
van der Waals attractions, 76F
bond length and strength, 47T
within macromolecules, 62
protein folding, 123, 123F
van der Waals radius, 76F
Vasopressin, 551T
Vectors, plasmid *see* Plasmid vectors
Venus flytrap, 403, 403F
Vertebrates
gene duplications, 304–305, 305F
genome comparisons, **312–313**, 313F
Vesicles, 21, 21F, 496
clathrin-coated *see* Clathrin-coated vesicles
coated, 511–512, 512F, 513T
COP-coated, 512, 513T
endocytic, 522
preserved membrane asymmetry, 371, 372F
secretory, 519–522, 519F
transport *see* Transport vesicles
Vesicular transport, 496, 501, 501F, **510–526**
endocytic pathway, 510, **522–526**
experimental methods, **520–521**, 520F, 521F
membranes, 371–372, 372F, **510–511**, 510F
pathways, **510–511**, 510F
secretory pathway, 510, **514–522**
Viagra, 539
Video-enhanced microscopy, 583F, 586, 586F
Vimentin, 575–576, 575F
Vimentin-related filaments, 575–576, 575F
Vinblastine, 582T
Vincristine, 582T
Virchow, Rudolf, 609
Viruses, 5, 222, **223–226**
assembly, 137, 138F
detection using PCR, 342, 344F
diversity, 225, 225F
genomes, 224, 224T
in situ hybridization, 352F
replication, 224, 225F
Vision, intracellular signaling, 554, 554F, 555F
Vitamin B_{12}, 524
Vitamins, 149
V_{max}, 100, 100F
competitive inhibition and, 102, 103F
measurement, 101–102, 101F
Voltage-gated ion channels, **405–406**, 405F, 406F, 418T
mediating action potentials, **410–411**, 411F
mediating neurotransmitter release, **415**, 415F
Voltage-gated sodium (Na^+) channels, 418T
closed, open and inactivated conformations, 411, 411F
mediating action potentials, **410–414**, 414F
Voltage sensors, 405–406

W

Water (H_2O)
acids and bases, 49–50
channels, 401
chemical properties, 66FF
diffusion across lipid bilayers, 389
diffusion across membranes *see* Osmosis
electrostatic attractions in, 77F
hydrogen bonds, **48–49**, 66F, 76F
lipid bilayer formation, **365–368**, 368F
mobility of protons in, 466, 466F
molecules, 46F, 66F
photosynthetic electron transfer, 478–479, 479F, 482, 482F
polar covalent bonds, 48, 48F
as solvent, 48–49
Water-splitting enzyme, 482, 482F, 483F
Watson, James, 172, 173, 200
Wee1, 623, 623F
Whales, 217, 217F
White blood cells, 713F
generation, 712, 714F
in tissues, 709–710
Whooping cough, 546–547
Wild type, 674F
Wingless, 725
Wnt proteins, 713–714, 725
Wnt signaling pathway, 714, 714F

Page numbers in boldface refer to major discussion of a topic; page numbers followed by F refer to a figure, and with FF to consecutive figures; page numbers followed by T refer to tables.

cancer cells, 724
discovery, 725–726, 726F

X

X chromosome, 179, 180F
inactivation, 190–191, 191F
X-ray crystallography, 11, 158–159, 160F, 162T
automation, 163
membrane proteins, 376
Xenopus
eggs, cell cycle studies, 615, 615F, 616F
genome duplication, 305, 306F
Xeroderma pigmentosum, 216

Y

Y chromosome, 179, 180F
Yeasts, 16, 16F
see also Saccharomyces cerevisiae; Schizosaccharomyces pombe
cell-cycle studies, 30–31, 31F, 616
mating factor, 531, 532F
protein transport studies, 520, 521F
sexual reproduction, 654
temperature-sensitive mutants, 30, 30F

Z

Z discs, 600, 601F
Zebrafish, 32, 32F
Zinc finger motif, 274, 274F
Zona pellucida, 663
Zygotes (fertilized eggs), 652, 653F, 656, 663
see also Development; Eggs
cells derived from, 707–709, 708F
cleavage divisions, 612, 615
regulation of polarity, 282–284, 282F, 283F, 284F